8TH INTERNATIONAL KIMBERLITE CONFERENCE
Selected Papers

VOLUME 2

The J. Barry Hawthorne Volume

8th international

KIMBERLITE
conference

This volume, and its companion, contain selected papers from the 8th International Kimberlite Conference (in the same format as they appeared in Lithos volumes 76 and 77). The Conference was held in Victoria, BC, Canada, 22-27 June 2003.

8TH INTERNATIONAL KIMBERLITE CONFERENCE
Selected Papers

VOLUME 2

The J. Barry Hawthorne Volume

Edited by

R.H. MITCHELL
H.S. GRÜTTER
L.M. HEAMAN
B.H. SCOTT SMITH
T. STACHEL

2004

ELSEVIER

Amsterdam – Boston – Heidelberg – London – New York – Oxford – Paris
San Diego – San Francisco – Singapore – Sydney – Tokyo

ELSEVIER B.V.   ELSEVIER Inc.   ELSEVIER Ltd   ELSEVIER Ltd
Sara Burgerhartstraat 25 525 B Street, Suite 1900 The Boulevard, Langford Lane 84Theobalds Road
P.O. Box 211, 1000 AE   San Diego, CA 92101-4495 Kidlington, Oxford OX5 1GB   London WC1X 8RR
Amsterdam, The Netherlands USA   UK   UK

First edition 2004

Library of Congress Cataloging in Publication Data
A catalog record is available from the Library of Congress.

British Library Cataloguing in Publication Data
A catalogue record is available from the British Library.

ISBN: 0 444 51776 6 (Set)
ISBN: 0 444 51775 8 (Volume 1)
ISBN: 0 444 51777 4 (Volume 2)

Volume 1 is reprinted from *Lithos,* volume 76/1-4, and Volume 2 from *Lithos,* 77/1-4

∞ The paper used in this publication meets the requirements of ANSI/NISO Z39.48-1992 (Permanence of Paper).
Printed in The Netherlands.

Special Issue

**Selected Papers from the 8th International Kimberlite Conference,
Victoria, BC, Canada, 22–27 June 2003
Volume 2: The J. Barry Hawthorne Volume**

edited by

ROGER H. MITCHELL
HERMAN S. GRÜTTER
LARRY M. HEAMAN
BARBARA H. SCOTT SMITH
THOMAS STACHEL

Selected papers from the 8th International Kimberlite Conference, Victoria, BC, Canada, 22–27 June 2003

Available online at www.sciencedirect.com

SCIENCE DIRECT®

ELSEVIER

Lithos 77 (2004) vii–ix

LITHOS

www.elsevier.com/locate/lithos

Contents

Special Issue: Selected papers from the 8th International Kimberlite Conference, Victoria, BC, Canada, 22–27 June 2003
Volume 2: The J. Barry Hawthorne volume

Available online at www.sciencedirect.com

Lithos 77 (2004) xv–xvi

www.elsevier.com/locate/lithos

8 IKC Committees and Sponsors

Organizing Committee

Convenor—Roger Mitchell
Co-convenor—Barbara Scott Smith

Members

John Armstrong
Ken Armstrong
Dante Canil
Jon Carlson
Howard Coopersmith
Buddy Doyle
Roy Eccles
Brian Grant
Herman Grütter
Larry Heaman
Tony Irving
Bruce Kjarsgaard
Maya Kopolova
Tom McCandless
Richard Molyneux
Rory Moore
George Poling
George Read
George Simandl

Thomas Stachel
Eira Thomas
Mary Lou Willows
Bruce Wyatt

International Kimberlite Conference Advisory Committee (1997–2003)

Chairman—Roger H. Mitchell

Members

F.(Joe) R. Boyd (USA)
Gerhardt Brey (Germany)
J. Barry Dawson (UK)
Jose C. Gaspar (Brazil)
John J. Gurney (South Africa)
Stephen H. Haggerty (USA)
William F. McKechnie (South Africa)
Peter H. Nixon (UK)
Sue Y. O'Reilly (Australia)
Roberta L. Rudnick (USA)
Barbara H. Scott Smith (Canada)
Craig B. Smith (South Africa)
Nickolai V. Sobolev (Russia)

doi:10.1016/j.lithos.2004.06.003

The 8th International Organizing Committee gratefully acknowledges the financial support of the following Conference Sponsors (ordered alphabetically)

Major Sponsors

BHP Billiton
De Beers Canada
Diavik Diamond Mines
Mineral Services International
Mineralogical Society of America
Natural Resources Canada

Sponsors

Aber Diamonds
AMEC
Ashton Mining of Canada
Boart Longyear
Canabrava Diamond
Chuck Fipke
Diamondex Resources
Eira Thomas
Geoanalytical Laboratories, SRC
Grenville Thomas
Haywood Securities
Kensington Resources
Majescor Resources
Major Drilling Group International
Overburden Drilling Management

Rescan Environmental Services
Robert Gannicott
Scott-Smith Petrology
Seventh International Kimberlite Conf.
SGS Lakefield Research
Southern Era Resources
Stornoway Ventures
SRK Consulting
Vancouver Petrographics

Contributors

Apex Geoscience
Band-Ore Resources
Black Swan Resources
Braden-Burry Expediting
Bradley Brothers
Canaccord Capital
Diamonds North Resources
Fugro Airborne Surveys
GGL Diamond
Golder Associates
Great Slave Helicopters
International Samuel Exploration
Navigator Exploration
Pure Gold Minerals
Rhonda
SDS Drilling
Shear Minerals
Superior Diamonds
Tahera
Terraquest

Available online at www.sciencedirect.com

Lithos 77 (2004) xvii

www.elsevier.com/locate/lithos

Foreword

The Eighth International Kimberlite Conference was held in Victoria, British Columbia, Canada from June 22 to 27th, 2003. These two volumes record some of the presentations made at the conference and are dedicated to Roger Clement (Volume 1) and Barry Hawthorne (Volume 2); in recognition of their contributions to, influence on, and encouragement of, kimberlite and upper mantle studies over the past 35 years.

The conference was attended by 585 full delegates who listened to 86 oral presentations and perused 185 posters. Many of these presentations gave for the first time detailed information on the geology and petrology of the kimberlites discovered in Canada during the past decade, the mantle-derived xenoliths and xenocrysts they contain, together with data on the diamonds from the recently opened Ekati and Diavik mines. An especially innovative feature of the conference was the opportunity for delegates to examine some of these newly discovered kimberlites at the "Large Core Exhibit" where approximately 2 km of drillcore was on display. The conference was preceded and followed by field excursions to the Ekati and Diavik diamond mines in northern Canada. Other field excursions visited areas in which kimberlites, lamproites and alkaline rocks occur in Colorado, Wyoming, Montana, British Columbia and Ontario.

Kimberlite conferences, which typically are held every four years, are unusual in that they bring industrial and academic geoscientists together in a symbiotic forum. This is a direct consequence of both groups realizing that this mutual cooperation is a catalyst leading to improved exploration techniques for kimberlites and better evaluation methods of diamond deposits, coupled with an increased understanding of kimberlite geology, diamond genesis and upper mantle petrology. Each conference culminates with the publication of a proceedings volume. Papers presented in these volumes usually record important steps in our understanding of a particular topic. Commonly, these build upon the results of the preceding conference. This incremental approach results in the proceedings volumes having lasting scientific value. Thus, geoscientists today still quote the seminal papers published in the proceedings of the first conference held in 1973!

These two volumes present 86 papers drawn from the oral and poster presentations. The volumes are organized according to themes reflecting the character of the conference. Volume 1—the Clement Volume—details advances in kimberlite geology, mineralogy and petrogenesis. Volume 2—the Hawthorne Volume—describes studies of diamonds, eclogites, the upper mantle and cratons together with exploration methods for diamond-bearing rocks.

The editors of these proceedings consider the papers included in these volumes to be novel and substantive, and that they will stand the test of time and be widely quoted in future studies of kimberlites, diamonds and the upper mantle.

8IKC Editorial Committee

Available online at www.sciencedirect.com

LITHOS

Lithos 77 (2004) xix–xxi

www.elsevier.com/locate/lithos

Preface

J. Barry Hawthorne

Volume 2 of the Proceedings of the Eighth International Kimberlite Conference is dedicated to John Barry Hawthorne in recognition of his pivotal role and visionary contributions to kimberlite geology, diamond studies and upper mantle petrology.

Early in his long and successful career with De Beers, Barry Hawthorne provided fundamental insights to kimberlite geology. His 1975 paper proposing the composite geological model for many southern African pipes is still applicable. His earlier but less well known work on kimberlite sills, some of it undertaken with Barry Dawson, showed that kimberlites were undoubtedly magmas, a crucial but unresolved issue at the time. Although Barry Haw-

0024-4937/$ - see front matter © 2004 Elsevier B.V. All rights reserved.
doi:10.1016/j.lithos.2004.06.007

thorne was an excellent scientist, his greatest contributions to science have been as an *eminence grise* who had the foresight to recognize that the diamond industry and academic research workers could form a mutually beneficial association. Barry had vision and the desire to change things. With far-sighted leadership, he revolutionised what was happening in De Beers, especially during his time as Chief Geologist in Kimberley starting in 1970. Barry's recognition of the importance of sound science in exploration, evaluation and mining eventually led to the establishment of the De Beers Kimberlite Petrographic Unit (KPU). The KPU, while providing a training ground for a future generation of kimberlite petrologists, became a Mecca for visiting scientists from all over the world and acted as a focal point for kimberlite and upper mantle studies. For the first time, international academic scientists had nearly unlimited access to kimberlites, diamonds and mantle-derived material and the possibility of unsurpassed diverse research projects. He forged associations with many academic institutions such as the Universities of Cape Town and of the Witwatersrand. Many of the scientists who are now prominent in kimberlite and upper mantle studies visited Kimberley and benefited from the access provided to mines and samples. It was the investigation of this material and the research he fostered and fast-tracked that lead to the enormous increases in our knowledge of the upper mantle and kimberlite petrology over the last three decades. The success, and indeed survival,

of kimberlite conferences has depended directly on these study materials. Without the cooperation of De Beers, and particularly Barry's enthusiastic support, the 1st International Kimberlite Conference would have not been the phenomenal success that it was. Barry brought an enlightened attitude into this field of geology around the world. He awakened kimberlite research into a new era. He opened doors and created opportunities. He is largely responsible for the unusual relationship that exists between industry and academia in the kimberlite conference community. He also crossed political barriers and initiated contact with countries such as the Soviet Union and China. Barry had a particular interest in encouraging young scientists and there is no doubt that many senior scientists today owe their scientific reputations and ability to gather research funds for students to their earlier association with Barry and the KPU. Barry initiated endless research studies most from an economic perspective. The advances made in applied research in turn stimulated the diamond industry as a whole.

Barry is an extraordinary role model and mentor. He inspired the best in everyone, motivated, encouraged, listened, offered guidance, challenged, communicated well with everyone, covertly monitored progress and was always supportive. With his wife Margie, he is always incredibly hospitable and friendly to all in the 'kimberlite family'.

8 IKC Editorial Committee

Having been out of mainstream kimberlite research for 15 years and having attended only the first four Kimberlite Conferences, my participation in 8 IKC was both an exciting and rewarding experience. This was particularly so as many of the papers presented at the Conference represent the culmination of studies initiated between 1970 and 1990 and many of the researchers who pioneered these studies of kimberlite, its associated minerals and the Upper Mantle continue to make major contributions in this unique field of Earth Science.

Sadly, some of the stalwarts of the early Conferences have passed away or have retired from active involvement in research. Despite this, the most

encouraging aspect of the Conference was to see the number of young researchers whose contributions made a profound impact on the proceedings. Clearly, the "Old Guard" are being adequately replaced. In addition, as the number of Conference participants continues to increase at each meeting, it would appear that the future of the IKC movement is assured.

However, before lapsing into a possibly unwarranted mood of complacency on this score, the reasons for the ongoing success of the Conferences should be examined. Of paramount importance has been the discovery of new fields of kimberlite and other diamond-bearing rock types, which, in most instances, have their origins in the Upper Mantle. Along with these new discoveries, there has been a change in the perception of many exploration and mining companies regarding the value of what was previously regarded as irrelevant "academic research". This has allowed improved access to mining sites and increased levels of funding and support for research.

Another extremely important factor has been the quality, capability and willingness of the organisers to make each Conference a success in a unique manner; this along with the support of a host of both young and old dedicated volunteers who seem to relish the challenges posed by running a meeting of this nature. There has also been an impressive series of venues chosen for the Conferences as well as the organisation of outstanding field excursions, often in the face of formidable logistical obstacles.

The development of a wide variety of new techniques in the fields of exploration, geophysics and rock and mineral analysis, amongst others, has also played a significant role in stimulating research and achieving a clearer understanding of the true nature of kimberlites, their related rock types and their origins.

So what does the future hold for Upper Mantle and kimberlite research? The venue for the next Conference is India, which hosts some of the least visited kimberlites and oldest known diamond fields. The organisers have shown enthusiasm and willingness to take up the burden and they clearly intend to make 9 IKC a success. The opportunity to visit the Subcontinent and obtain rock and mineral specimens and undertake field studies should provide considerable motivation to continue research on a broad front. In the future, it is hoped that new discoveries will be made but even if this does not happen there are still many African occurrences, which have not been adequately studied. However, political and security considerations may play a role in selecting the venue for 10 IKC.

On balance, it would appear that provided funding for research does not diminish dramatically Upper Mantle and kimberlite studies are likely to continue for at least the next decade.

J. Barry Hawthorne

Available online at www.sciencedirect.com

Lithos 77 (2004) 1–19

ELSEVIER

LITHOS

www.elsevier.com/locate/lithos

The trace element composition of silicate inclusions in diamonds: a review

Thomas Stachel[a,*], Sonja Aulbach[b,c], Gerhard P. Brey[b], Jeff W. Harris[d],
Ingrid Leost[b], Ralf Tappert[a,b], K.S. (Fanus) Viljoen[e]

[a] *Department of Earth and Atmospheric Sciences, University of Alberta, Edmonton, AB, Canada T6G 2E3*
[b] *Institut für Mineralogie, Universität Frankfurt, 60054 Frankfurt, Germany*
[c] *GEMOC, Macquarie University, Sydney, NSW 2109, Australia*
[d] *Division of Earth Sciences, University of Glasgow, Glasgow G12 8QQ, UK*
[e] *De Beers GeoScience Centre, P.O. Box 82232, Southdale 2135, South Africa*

Received 27 June 2003; accepted 17 February 2004
Available online 1 June 2004

Abstract

On a global scale, peridotitic garnet inclusions in diamonds from the subcratonic lithosphere indicate an evolution from strongly sinusoidal REE_N, typical for harzburgitic garnets, to mildly sinusoidal or "normal" patterns (positive slope from $LREE_N$ to $MREE_N$, fairly flat $MREE_N$–$HREE_N$), typical for lherzolitic garnets. Using the Cr-number of garnet as a proxy for the bulk rock major element composition it becomes apparent that strong LREE enrichment in garnet is restricted to highly depleted lithologies, whereas flat or positive LREE–MREE slopes are limited to less depleted rocks. For lherzolitic garnet inclusions, there is a positive relation between equilibration temperature, enrichment in MREE, HREE and other HFSE (Ti, Zr, Y), and decreasing depletion in major elements. For harzburgitic garnets, relations are not linear, but it appears that lherzolite style enrichment in MREE–HREE only occurs at temperatures above 1150–1200 °C, whereas strong enrichment in Sr is absent at these high temperatures. These observations suggest a transition from melt metasomatism (typical for the lherzolitic sources) characterized by fairly unfractionated trace and major element compositions to metasomatism by CHO fluids carrying primarily incompatible trace elements. Melt and fluid metasomatism are viewed as a compositional continuum, with residual CHO fluids resulting from primary silicate or carbonate melts in the course of fractional crystallization and equilibration with lithospheric host rocks.

Eclogitic garnet inclusions show "normal" REE_N patterns, with LREE at about $1 \times$ and HREE at about $30 \times$ chondritic abundance. Clinopyroxenes approximately mirror the garnet patterns, being enriched in LREE and having chondritic HREE abundances. Positive and negative Eu anomalies are observed for both garnet and clinopyroxene inclusions. Such anomalies are strong evidence for crustal precursors for the eclogitic diamond sources. The trace element composition of an "average eclogitic diamond source" based on garnet and clinopyroxene inclusions is consistent with derivation from former oceanic crust that lost about 10% of a partial melt in the garnet stability field and that subsequently experienced only minor reenrichment in the most incompatible trace elements. Based on individual diamonds, this simplistic picture becomes more complex, with evidence for both strong enrichment and depletion in LREE.

Trace element data for sublithospheric inclusions in diamonds are less abundant. REE in majoritic garnets indicate source compositions that range from being similar to lithospheric eclogitic sources to strongly LREE enriched. Lower mantle sources, assessed based on CaSi–perovskite as the principal host for REE, are not primitive in composition but show moderate to strong

---

* Corresponding author.
E-mail address: tstachel@ualberta.ca (T. Stachel).

LREE enrichment. The bulk rock $LREE_N$–$HREE_N$ slope cannot be determined from CaSi–perovskites alone, as garnet may be present in these shallow lower mantle sources and then would act as an important host for HREE. Positive and negative Eu anomalies are widespread in CaSi–perovskites and negative anomalies have also been observed for a majoritic garnet and a coexisting clinopyroxene inclusion. This suggests that sublithospheric diamond sources may be linked to old oceanic slabs, possibly because only former crustal rocks can provide the redox gradients necessary for diamond precipitation in an otherwise reduced sublithospheric mantle.

Keywords: Inclusion in diamond; REE; Metasomatism; Lithosphere; Garnet; Majorite; Lower mantle; Subduction

1. Introduction

In a first study of REE patterns of inclusions in diamonds, Shimizu and Richardson (1987) analyzed two harzburgitic garnets from Finsch and Kimberley Pool Mines. Little more trace and ultra trace element measurements were published until the 6th International Kimberlite Conference in 1995. Subsequently, garnet inclusions were recognized as the most useful mineral from which to obtain REE information and a number of studies included trace element analyses obtained mainly by SIMS (ion microprobe) and, more recently, also by laser ablation ICP-MS (e.g., Kaminsky et al., 2001; Davies et al., 2004). So far, these data have been interpreted in the context of their specific diamond sources only. For the purpose of this review, we have compiled a data base of major and trace element analyses for inclusions in diamonds worldwide representing both the lithospheric mantle and sublithospheric sources (asthenosphere, transition zone and lower mantle). The data base is used to constrain the evolution of these lithospheric and sublithospheric source rocks and to examine the possible presence of fluids or melts during diamond formation.

1.1. Data base

For the peridotitic suite a data set of 145 major and trace element analyses of garnet inclusions was assembled. Trace element data on peridotitic clinopyroxene inclusions are still fairly scarce (e.g., Hutchison, 1997; Stachel and Harris, 1997a; Stachel et al., 1999, 2000a; Wang et al., 2000b; Wang and Gasparik, 2001), principally because clinopyroxenes are rare as inclusions in diamond. Furthermore, they are restricted to the lherzolitic (and wehrlitic) inclusions paragenesis and therefore cannot be used to constrain differences and similarities between harzburgitic and lherzolitic diamond sources. Thus, peridotitic clinopyroxene analyses are not included here.

Outliers are a common problem with analytical data bases: a few exotic samples with extreme compositions determine the scale of most plots, making it virtually impossible to display compositional variations affecting the bulk of the samples. We therefore filtered the garnet data base using exceptionally high and low Nd and Ho (both are turning points of sinusoidal patterns) concentrations to exclude 10 aberrant samples. The remaining 135 garnet analyses (100 harzburgitic, 35 lherzolitic) represent diamonds from the Siberian Craton (Aikhal, Mir and Udachnaya: Shimizu et al., 1997), the Sino-Korean Craton (Wang et al., 2000b), the Kalahari Craton including the Kaapvaal Block (Jwaneng: Stachel et al., 2004; Namibian placer deposits: Harris et al., 2004), the Limpopo Belt (Venetia, this study) and the Zimbabwe Block (Orapa: Stachel et al., 2004), the East African Craton (Mwadui: Stachel et al., 1999), the West African Craton (Birim deposits, Ghana: Stachel and Harris, 1997a; Kankan deposits, Guinea: Stachel et al., 2000a), the Guayana Shield, Brazil (Boa Vista: Tappert et al., 2004) and the Slave Craton (Panda: Tappert et al., 2004; DO-27: Davies et al., 2004). The data set covers the compositional space observed for garnet inclusions worldwide with the exception that no garnets with less than 4 wt.% Cr_2O_3 have been analyzed so far (Fig. 1).

For the eclogitic inclusion suite, after exclusion of four aberrant analyses (see above), an analytical data set comprising 39 garnet and 22 clinopyroxene inclusions was established. The samples are derived from the Siberian Craton (Mir and Udachnaya: Taylor et al., 1996), the Kalahari Craton (Jwaneng: Stachel et al., 2004; Namibian placer deposits: this study; Venetia: Aulbach et al., 2002), the East African Craton (Mwadui: Stachel et al., 1999), the

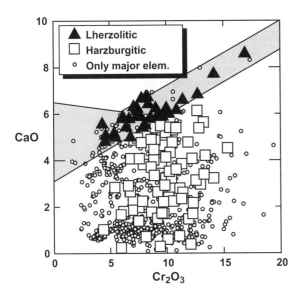

Fig. 1. CaO vs. Cr_2O_3 (wt.%) in garnet. The harzburgitic and lherzolitic garnets in the trace element data base almost completely span the compositional ranges observed for garnet inclusions worldwide (for references see Stachel and Harris, 1997b; Stachel et al., 2000a), with the exception that trace element data are absent for garnet with Cr_2O_3 <4 wt.%. The outline of the lherzolite field (shaded area) is taken from Sobolev et al. (1973).

West African Craton (Kankan deposits, Guinea: Stachel et al., 2000a), the Guayana Shield (Boa Vista: Tappert et al., 2004) and the Slave Craton (DO-27: Davies et al., 2004).

Major and trace element analyses for 15 majoritic garnet inclusions from the asthenosphere and transition zone were obtained for Monastery (Moore et al., 1991), Juina-São Luiz (Wilding, 1990; Harte, 1992; Kaminsky et al., 2001) and Kankan (Stachel et al., 2000a). Trace element analyses of 13 lower mantle CaSi–perovskites are available for Juina-São Luiz (Hutchison, 1997; Harte et al., 1999; Kaminsky et al., 2001) and Kankan (Stachel et al., 2000b).

Previously unpublished data for peridotitic inclusions in diamonds from Venetia (Table 1) and eclogitic inclusions from the placer deposits along the Namibian coast (Table 2) were obtained at the IMS-4f ion probe facility at the University of Edinburgh. Analytical procedures and precision are the same as described in Stachel and Harris (1997a) and Harte and Kirkley (1997). Major elements were determined by electron microprobe analysis (for analytical details, see Stachel et al., 2000a).

2. Peridotitic suite

Garnet inclusions of the peridotitic suite can be further subdivided into a harzburgitic and a lherzolitic paragenesis based on their Ca and Cr contents (Fig. 1, Sobolev et al., 1973; Gurney, 1984). Based on this subdivision, Fig. 2 shows that harzburgitic garnet inclusions are characterized by sinusoidal REE_N patterns ($_N$ stands for normalization to the C1-chondrite composition of McDonough and Sun, 1995), whereas lherzolitic garnets show both sinusoidal and "normal" patterns. Normal patterns—positive slope within the $LREE_N$, flat and enriched $MREE_N$–$HREE_N$—are typical for lherzolitic garnet from mantle xenoliths (Stachel et al., 1998), in particular from sheared peridotites (Shimizu, 1975) and in off-craton occurrences (Hoal et al., 1994).

Hoal et al. (1994) and Shimizu et al. (1997) explained sinusoidal REE_N patterns with disequilibrium models involving modification of preexisting garnet or precipitation from a supersaturated melt. An essential prerequisite for these disequilibrium models is that diffusion of REE in garnet decreases significantly from LREE to HREE. However, Van Orman et al. (2002) showed experimentally (at 2.8 GPa and 1200–1450 °C) that diffusion coefficients for Ce, Sm, Dy, and Yb in pyrope garnet are indistinguishable from each other within analytical uncertainty. Normalization of REE concentrations to a primitive garnet composition (Fig. 3) reveals that sinusoidal garnet patterns are less complex than apparent from normalization to C1-chondrite. The steep positive slope within chondrite normalized LREE is an artifact of rapidly increasing compatibility within the garnet structure due to decreasing ionic radius. Compared to garnet from primitive mantle, average harzburgitic and lherzolitic garnets have LREE enriched, V-shaped REE-patterns and lherzolitic garnets with flat $MREE_N$–$HREE_N$ (e.g., from the Birim deposits in Ghana) actually approach a primitive trace element composition. In addition to the recognition of constant diffusion speeds for LREE and HREE in garnet (Van Orman et al., 2002), the consistency of certain characteristics, such as a fixed turning point at Er over a large compositional range from highly depleted to almost primitive compositions (Fig. 3) clearly is not in support of disequilibrium models either.

Table 1

Major (EPMA) and trace element analyses (SIMS) of peridotitic garnet and clinopyroxene inclusions from Venetia (South Africa)

Sample Mineral Assembly	V-64a Garnet Grt, 2ol	V-87b Garnet Grt, ol	V-95 Garnet Grt	V-112a Garnet Grt, 2ol	V-149 Cpx Cpx	V-167b Garnet Grt, Cpx	V-167c Cpx Grt, cpx	V-169a Garnet Grt	V-175a Cpx 3cpx, opx	V-175b Cpx 3cpx, opx	V-175c Cpx 3cpx, opx	V-195a Garnet 2grt, ol	V-197ab Garnet Grt, ol
P_2O_5	0.01	0.08	≤0.01	0.07	≤0.01	0.06	≤0.01	0.03	≤0.01	≤0.01	≤0.01	0.03	0.02
SiO_2	41.87	40.79	41.03	39.56	55.17	41.78	54.72	40.95	55.17	54.80	55.30	41.76	41.24
TiO_2	≤0.01	0.10	0.09	0.05	0.08	0.23	0.04	0.20	0.08	0.06	0.08	0.61	0.10
Al_2O_3	17.65	11.91	17.49	10.16	1.46	19.85	1.34	14.81	1.23	1.23	1.22	17.04	15.92
Cr_2O_3	9.40	14.75	9.21	18.49	1.58	4.63	0.83	11.43	1.04	0.86	1.04	7.85	9.06
FeO	4.30	5.52	6.41	5.47	3.27	5.92	2.79	5.07	2.42	2.70	2.46	4.98	6.33
MnO	0.20	0.27	0.34	0.31	0.13	0.28	0.13	0.23	0.11	0.11	0.10	0.25	0.30
NiO	≤0.01	0.01	≤0.01	≤0.01	0.07	0.02	0.08	0.02	0.06	0.07	0.06	0.01	0.02
MgO	25.94	20.79	22.92	23.23	18.10	21.78	19.89	20.38	19.19	21.05	19.68	22.37	20.11
CaO	0.29	6.06	2.64	1.64	18.37	5.07	18.23	6.59	18.85	17.14	18.28	4.24	6.23
Na_2O	≤0.01	0.03	0.03	≤0.01	1.20	0.03	0.90	0.03	0.84	0.75	0.81	0.05	0.02
K_2O	≤0.01	≤0.01	≤0.01	≤0.01	0.07	≤0.01	0.15	≤0.01	0.05	0.04	0.06	≤0.01	≤0.01
Total	99.66	100.30	100.18	98.99	99.51	99.65	99.08	99.74	99.05	98.82	99.08	99.19	99.35
Ti	14.20	602.00	463.00	327.00	453.00	1130.00	218.00	1140.00	351.00	351.00	363.00	3300.00	547.00
Sr	9.97	3.42	0.42	31.30	177.00	0.89	168.00	0.63	39.80	40.80	41.60	8.30	0.47
Y	0.29	2.22	1.63	2.25	0.78	4.30	0.35	4.07	0.80	0.79	0.82	24.30	2.24
Zr	1.63	29.40	27.90	24.20	0.90	13.20	0.38	15.40	0.32	0.31	0.29	119.00	1.54
Nb	0.16	8.52	0.49	0.67	0.51	6.00	1.49	4.25	0.42	0.46	0.53	0.45	1.94
Ba	0.02	0.01	0.02	0.08	0.69	0.04	2.28	0.01	1.80	1.31	2.87	0.01	0.01
La	0.44	0.45	0.01	0.74	3.00	0.34	8.91	0.20	1.01	0.98	1.32	0.23	0.06
Ce	5.33	4.05	0.16	21.60	8.74	2.34	21.90	1.60	1.59	1.76	2.05	2.48	0.76
Pr	1.17	1.28	0.08	7.88	1.30	0.57	2.54	0.38	0.22	0.21	0.28	0.82	0.24
Nd	6.12	11.30	0.89	39.50	6.06	3.57	9.42	3.87	1.36	1.32	1.37	8.01	1.89
Sm	0.73	4.89	0.55	3.42	0.74	0.89	0.86	2.61	0.30	0.35	0.49	4.66	0.36
Eu	0.11	1.71	0.22	0.72	0.22	0.28	0.21	0.85	0.12	0.12	0.10	1.83	0.10
Gd	0.40	3.60	0.88	2.71	0.43	0.74	0.31	2.07	0.36	0.39	0.32	6.35	0.23
Tb	0.01	0.32	0.10	0.21	0.07	0.15	0.05	0.20	0.06	0.07	0.03	0.96	0.04
Dy	0.05	0.87	0.55	1.16	0.36	0.75	0.04	0.98	0.16	0.27	0.36	5.83	0.36
Ho	0.01	0.15	0.08	0.12	0.04	0.17	0.02	0.15	0.04	0.04	0.05	1.13	0.09
Er	0.04	0.35	0.17	0.20	0.02	0.57	0.04	0.65	0.13	0.16	0.11	2.96	0.39
Yb	0.09	0.56	0.18	0.08	n.a.	0.90	n.a.	0.80	n.a.	n.a.	n.a.	2.24	0.68
Lu	0.04	0.09	0.06	0.01	0.01	0.13	0.00	0.15	0.00	0.01	0.01	0.28	0.15
Hf	0.02	0.58	0.66	0.40	0.07	0.41	0.05	0.25	0.03	0.01	0.04	2.47	0.05

Trace element concentrations are given in wt. ppm and are rounded to the second decimal place, concentrations of 0.00 ppm therefore refer to values < 0.005 ppm. "n.a." stands for "not analyzed".

Thus, only three hypotheses for the origin of peridotitic REE patterns are further considered: (i) the patterns are an inherent characteristic of cratonic garnet peridotites, related to their primary formation; (ii) they are the result of a reenrichment event that also modified the major element composition of these rocks, i.e., melt infiltration; or (iii) they were caused by fluid metasomatism involving CHO agents enriched in incompatible trace elements, but without significant impact on major elements.

2.1. Relationship between garnet trace element and bulk rock major element composition

The first two hypotheses outlined above require that discernible correlations between major and trace element compositions exist. The Cr/Al ratio (or Cr content) of garnet is a measure of the Cr/Al ratio of the source rock which in turn is a proxy for the degree of depletion in major elements. Griffin et al. (1999a) have shown that this relationship is sufficiently strong to employ Cr in garnet to predict the major element

Table 2
Major (EPMA) and trace element analyses (SIMS) of eclogitic garnet and clinopyroxene inclusions from Namibia (alluvial mines along the Orange River close to Oranjemund, along the Namibian coast to Elisabeth Bay and from Namibian offshore deposits)

Sample Mineral	Nam-13 Garnet	Nam-34 Cpx	Nam-35 Garnet	Nam38 Garnet	Nam-38 Cpx	Nam-43 Cpx	Nam-47 Garnet	Nam-5 Garnet	Nam-59 Garnet	Nam-63 Garnet	Nam-68 Garnet	Nam74 Garnet
P_2O_5	0.05	0.01	0.14	0.03	0.01	≤0.01	0.03	0.14	0.04	0.05	0.05	0.08
SiO_2	39.43	53.55	39.10	39.67	54.71	54.52	40.79	39.78	39.37	40.30	38.73	39.41
TiO_2	0.67	0.53	1.14	0.17	0.16	0.26	0.24	1.01	0.30	0.53	0.36	0.30
Al_2O_3	22.18	7.92	20.81	22.92	6.29	4.12	22.80	21.52	21.98	21.95	22.12	22.92
Cr_2O_3	0.03	0.05	0.02	0.19	0.18	0.10	0.14	0.15	0.06	0.18	0.04	0.04
FeO	15.37	6.41	21.76	19.75	6.51	7.08	15.57	15.93	22.48	19.23	19.67	19.24
MnO	0.27	0.07	0.41	0.43	0.18	0.21	0.34	0.29	1.02	0.40	0.67	0.23
NiO	≤0.01	0.03	≤0.01	0.01	≤0.08	0.02	0.01	≤0.01	≤0.01	≤0.01	≤0.01	≤0.01
MgO	7.35	10.11	6.95	14.34	12.71	14.93	16.41	10.52	9.21	13.77	8.98	10.89
CaO	14.36	15.70	9.74	2.66	11.92	15.42	3.59	10.41	5.86	3.66	8.45	6.59
Na_2O	0.25	4.19	0.37	0.09	4.18	2.43	0.15	0.40	0.17	0.14	0.18	0.17
K_2O	≤0.01	0.26	≤0.01	≤0.01	0.49	0.13	≤0.01	≤0.01	≤0.01	≤0.01	≤0.01	≤0.01
Total	99.96	98.83	100.45	100.26	97.42	99.20	100.07	100.16	100.49	100.21	99.26	99.88
Sr	2.27	134.00	3.42	0.37	169.00	81.20	0.29	3.26	3.69	0.50	4.78	1.50
Y	42.90	1.57	47.10	33.35	8.03	6.72	11.20	39.50	21.20	37.10	20.40	68.30
Zr	49.20	20.00	63.60	2.10	7.79	7.78	7.98	176.00	9.70	25.30	17.80	69.20
Nb	0.00	0.14	0.00	0.04	0.00	0.00	0.00	0.19	0.00	0.00	0.00	0.00
Ba	0.07	0.84	0.01	3.22	19.00	5.93	0.00	0.01	0.02	0.18	0.05	0.00
La	0.02	1.96	0.10	0.06	6.88	1.16	0.00	0.10	0.05	0.03	0.16	0.28
Ce	0.38	6.02	1.05	0.15	12.00	2.50	0.03	1.98	0.43	0.16	0.85	2.78
Pr	0.15	1.17	0.43	0.04	1.40	0.54	0.02	0.91	0.13	0.07	0.26	0.77
Nd	1.55	7.08	4.17	0.47	7.45	2.94	0.14	9.30	1.50	1.17	3.15	6.80
Sm	1.62	1.62	2.80	0.64	2.77	1.52	0.23	5.22	1.76	1.19	2.92	4.31
Eu	0.94	0.45	1.33	0.44	0.91	0.48	0.10	1.48	0.92	0.47	1.47	1.34
Gd	4.95	0.98	5.25	2.04	4.95	2.44	0.50	7.32	2.62	2.52	4.13	8.56
Tb	1.22	0.09	1.16	0.64	0.50	0.28	0.14	1.31	0.55	0.70	0.72	1.91
Dy	8.49	0.44	8.50	5.27	2.57	1.47	1.56	8.10	3.62	6.03	3.98	12.80
Ho	1.93	0.09	1.90	1.28	0.40	0.37	0.45	1.67	0.86	1.38	0.85	2.72
Er	5.40	0.09	6.14	4.20	0.73	0.77	1.89	5.01	2.30	4.45	2.39	8.48
Yb	5.39	n.a.	7.44	5.11	n.a.	n.a.	2.91	4.75	2.99	5.98	2.12	10.00
Lu	0.65	0.01	1.11	0.82	0.06	0.07	0.46	0.68	0.38	0.86	0.30	1.47

Sample Mineral	Nam-78 Cpx	Nam-80 Garnet	Nam-86 Garnet	Nam-89 Cpx	Nam-89 Garnet	Nam-102 Cpx	Nam-202 Cpx	Nam-202 Garnet	Nam-203 Cpx	Nam-203 Garnet	Nam-205 Cpx	Nam-207 Cpx
P_2O_5	0.02	0.05	0.04	≤0.01	0.02	0.01	≤0.01	0.00	0.01	0.08	≤0.01	0.02
SiO_2	54.99	40.90	39.59	53.86	40.54	56.81	55.80	37.53	53.62	39.51	54.44	55.78
TiO_2	0.62	0.25	0.58	0.20	0.33	0.43	0.53	0.12	0.46	0.57	0.21	0.45
Al_2O_3	12.51	23.29	22.55	5.50	22.67	19.32	9.66	20.85	7.20	21.00	3.77	6.17
Cr_2O_3	0.03	0.24	0.02	0.10	0.18	0.08	0.09	0.05	0.08	0.10	0.30	0.20
FeO	5.36	8.35	10.53	4.75	13.50	2.79	3.41	30.92	6.54	18.09	5.99	7.22
MnO	0.06	0.13	0.20	0.08	0.31	0.03	0.05	0.60	0.08	0.37	0.10	0.13
NiO	0.01	0.01	≤0.01	0.06	0.02	0.00	0.04	≤0.01	0.03	≤0.01	0.08	0.06
MgO	7.28	14.44	7.57	13.36	14.45	3.97	10.69	1.73	10.45	9.62	15.46	14.13
CaO	11.77	11.89	18.53	17.76	7.87	5.99	13.92	8.60	15.61	9.07	15.82	11.28
Na_2O	6.92	0.12	0.19	2.74	0.10	9.59	5.71	0.02	4.23	0.16	2.12	4.55
K_2O	0.24	≤0.01	≤0.01	0.10	≤0.01	0.10	0.08	≤0.01	0.03	≤0.01	0.59	0.09
Total	99.80	99.68	99.82	98.51	99.97	99.11	99.97	100.42	98.32	98.56	98.89	100.08
Sr	272.00	1.28	3.81	76.20	0.59	44.10	183.35	0.12	79.00	0.83	166.80	200.00
Y	2.46	5.79	22.00	1.04	12.90	1.04	0.54	60.00	1.70	33.00	1.76	8.30

(continued on next page)

Table 2 (*continued*)

Sample Mineral	Nam-78 Cpx	Nam-80 Garnet	Nam-86 Garnet	Nam-89 Cpx	Nam-89 Garnet	Nam-102 Cpx	Nam-202 Cpx	Nam-202 Garnet	Nam-203 Cpx	Nam-203 Garnet	Nam-205 Cpx	Nam-207 Cpx
Zr	13.70	8.38	77.80	3.41	7.24	28.00	13.52	2.50	31.00	43.00	2.25	46.00
Nb	0.00	0.00	0.00	0.28	0.27	0.00	0.02	0.00	0.00	0.00	1.05	0.10
Ba	0.88	0.06	0.02	0.56	0.04	0.22	27.01	0.02	0.35	0.01	67.28	0.20
La	2.97	0.00	0.03	1.00	0.03	0.38	4.85	0.01	0.27	0.01	19.04	2.10
Ce	8.48	0.00	0.38	3.47	0.30	1.26	8.48	0.02	1.70	0.16	18.94	10.00
Pr	1.40	0.17	0.15	0.50	0.10	0.25	0.98	0.00	0.47	0.11	1.24	2.20
Nd	6.50	0.98	2.23	2.38	0.75	2.13	4.23	0.08	3.80	2.30	4.62	13.00
Sm	1.17	0.54	1.79	0.52	0.59	0.62	0.69	0.30	1.30	2.20	0.58	3.00
Eu	0.33	0.35	1.04	0.19	0.42	0.16	0.20	0.21	0.44	1.10	0.16	0.91
Gd	0.71	1.57	4.04	0.35	1.12	0.48	0.43	3.00	1.20	4.40	0.99	3.20
Tb	0.12	0.25	0.77	0.07	0.32	0.08	0.03	1.50	0.16	0.88	0.10	0.47
Dy	1.05	1.40	5.15	0.26	2.52	0.29	0.23	12.00	0.87	6.90	0.63	2.60
Ho	0.11	0.29	1.03	0.09	0.52	0.05	0.05	2.00	0.10	1.50	0.10	0.45
Er	0.46	0.38	2.41	0.00	1.70	0.12	0.09	4.50	0.30	4.40	0.36	0.94
Yb	n.a.	0.39	1.93	n.a.	2.09	n.a.	n.a.	n.a.	n.a.	n.a.	n.a.	n.a.
Lu	0.01	0.08	0.30	0.00	0.36	0.00	0.02	0.52	0.03	0.80	0.01	0.09

Trace element concentrations are given in wt. ppm and are rounded to the second decimal place, concentrations of 0.00 ppm therefore refer to values < 0.005 ppm. "n.a." stands for "not analyzed".

and modal composition of cratonic garnet peridotites. Thus, the molar Cr-number ($100Cr/[Cr + Al]$) of garnet inclusions will be used to examine possible correlations between major and trace elements.

The shape of garnet REE_N patterns (Fig. 2) is determined by the existence and position of a peak within the $LREE_N$ and by the slopes (i) within the $LREE_N$, (ii) from $LREE_N$ to $MREE_N$ (MREE: Sm–Ho) and (iii) from $MREE_N$ to $HREE_N$. The actual concentrations of REE are less diagnostic as they will be strongly influenced by the amount of modal garnet present. In addition, for lherzolitic garnets LREE concentrations (and consequently, ratios of LREE to MREE and HREE) will be influenced by the presence of clinopyroxene.

No significant linear correlations between Cr-number and relative and absolute REE concentrations are observed for peridotitic garnets of both parageneses. However, for the harzburgitic garnets it is noted (i) that the highest La (LREE) contents occur at Cr-numbers greater than 25 (Fig. 4) and (ii) that enrichment in MREE and Y (i.e., positive $LREE_N$–$MREE_N$ slopes, indicated by superchrondritic Y/Nd in Fig. 4), and in HREE is restricted to Cr-numbers below 30 (Fig. 4). The apparent relationship between high Cr-number and high average La content possibly reflects highly depleted rocks with low modal garnet, thus being very sensitive to metasomatic modification.

Such effects may have been enhanced by a moderate increase in garnet/liquid distribution coefficients for LREE with increasing Cr content in garnet (Wang et al., 1998; HREE are not affected).

For the lherzolitic garnets positive slopes from $LREE_N$ to $MREE_N$ (similar to harzburgitic garnets, see Fig. 4) and the highest concentrations in MREE–HREE (from Tb onwards, see Yb in Fig. 4) are restricted to Cr-numbers below 30. The highest contents in strictly incompatible elements such as Ce (LREE) and Sr are found for the three lherzolitic garnets with Cr-numbers above 40. Usually, Sr contents above 2 ppm are restricted to harzburgitic garnets, suggesting that very low modal clinopyroxene (as the principal host of Sr and LREE in lherzolite) in the source of Cr-rich lherzolitic garnets may be the cause of elevated Ce and Sr.

Despite the observation that a few harzburgitic garnets with low Cr-number show distinct MREE–HREE enrichment, the REE_N patterns for most of the harzburgitic paragenesis are independent of the bulk rock major element composition. For lherzolitic garnets, it is evident that positive slopes from $LREE_N$ to $MREE_N$–$HREE_N$ are restricted to less Cr-rich samples, but this relationship does not take the form of a linear correlation and a number of samples with low Cr-number still have sinusoidal patterns.

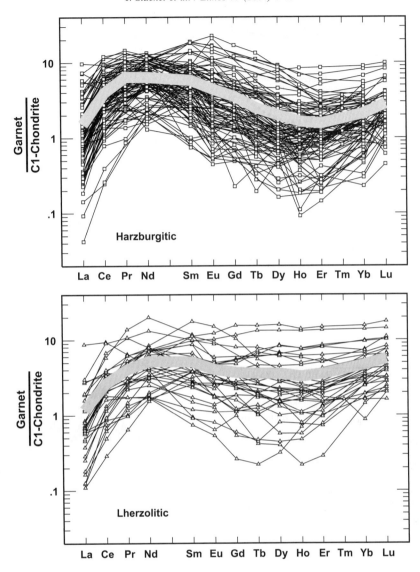

Fig. 2. Chrondrite-normalized REE patterns of harzburgitic and lherzolitic garnet inclusions from worldwide sources. Average compositions are indicated by thick shaded lines. The validity of the calculated average for harzburgitic garnets is difficult to assess from this diagram alone because of a large number of overlapping analyses. However, we have verified from the REE concentrations that we are dealing with unimodal distributions and that the median REE pattern is not significantly different from the average shown above.

As a first conclusion, the absence of linear correlations between the bulk rock major element composition (inferred from the Cr-number of garnet) and garnet REE_N excludes the primary processes (which cause the chemical depletion of the subcratonic lithospheric mantle) as the determining factor for the observed variations in trace composition. This is in agreement with a two-stage model of primary depletion and secondary reenrichment first proposed by Frey and Green (1974). However, for some lherzolitic and a very few harzburgitic garnets, it appears that there is a relationship between MREE–HREE enrichment and decreasing depletion in bulk rock major element composition. The nonlinearity of this relationship may be attributed to variations either in the style of metasomatic overprint or in the degree of primary depletion of

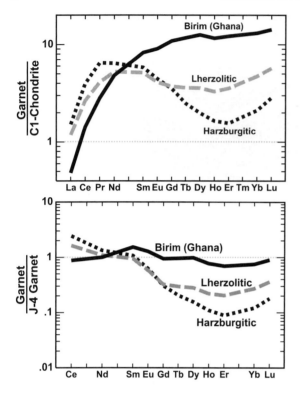

Fig. 3. Average compositions of harzburgitic and lherzolitic garnet inclusions from worldwide sources and of three lherzolitic garnets with "flat" MREE–HREE from the Birim deposits in Ghana (Stachel and Harris, 1997a). The data are normalized to C1-chondrite and to a garnet from a primitive mantle bulk rock composition (J-4 of Jagoutz and Spettel, see Stachel et al., 1998 for details).

the source rock (leading to different starting compositions for metasomatic overprint), or both.

2.2. Garnet composition and equilibration temperature

To test if there is a possible dependence of the style of metasomatic overprint on the thermal regime compositional parameters are plotted against garnet–olivine equilibration temperatures (O'Neill and Wood, 1979; O'Neill, 1980), calculated for a fixed pressure of 5 GPa. For the data base evaluated here, there is a negative correlation between equilibration temperature and Cr-number (Fig. 5) for lherzolitic garnets. This negative correlation is less well developed for harzburgitic garnets where it appears to break down at temperatures below about 1100 °C. This relationship cannot be linked to increased solution of Cr-rich

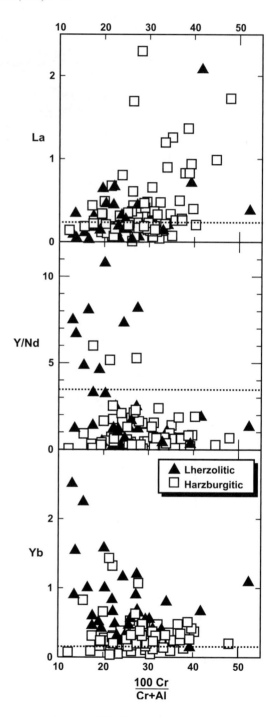

Fig. 4. Covariations of REE (in wt. ppm; Y is used as a substitute for the less abundant MREE) and molar Cr-number of garnet inclusions worldwide. Dotted lines indicate chondritic abundances or, in the case of Y/Nd, the chondritic ratio. Garnets with superchondritic Y/Nd have a positive $LREE_N–MREE_N$ slope.

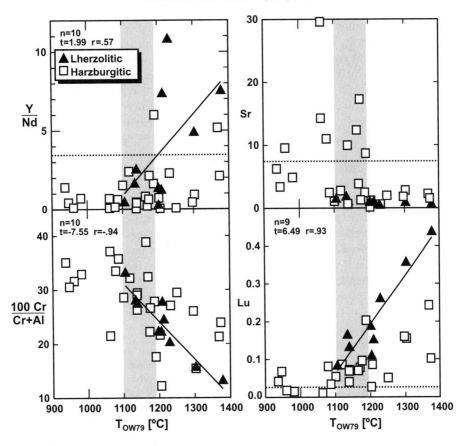

Fig. 5. Covariations between compositional parameters of garnet inclusions and equilibration temperature (calculated from garnet–olivine equilibria for a fixed pressure of 5 GPa). Regression lines are based on lherzolitic inclusions (▲) only. Dotted lines indicate chondritic abundance or ratio. The shaded area indicates the approximate temperature where a change in metasomatic regime appears to take place.

spinel into garnet with increasing pressure and temperature (Doroshev et al., 1997; Grütter and Sweeney, 2000), as this would lead to a positive correlation of temperature and Cr-number. Therefore, a decreasing degree of chemical depletion of the diamond source rocks with increasing temperature is indicated for all samples that formed above about 1100 °C.

For the harzburgitic paragenesis, the fairly crude correlation between increasing temperature and source fertility (Fig. 5) is not accompanied by linear correlations with the garnet trace element composition. However, some nonlinear relationships are observed: high Sr (>3 ppm, see Fig. 5) and Ce (>4 ppm, not shown) are restricted to equilibration temperatures below about 1150–1200 °C, whereas high HREE (Yb>50 ppb and Lu>100 ppb, see Fig. 5) occur only in some of the samples which formed above 1190 °C. A similar

relationship exists for the $LREE_N–MREE_N$ slope (represented in Fig. 5 by the Y/Nd ratio), where positive slopes for harzburgitic and also for lherzolitic garnets only occur above 1190 °C.

For the lherzolitic garnets where all 10 samples follow a linear relation between temperature and source fertility, positive correlations with equilibration temperature also exist with the MREE–HREE from Eu onwards (the HREE Lu is shown in Fig. 5) and with the other HFSE: Ti, Y and Zr.

2.3. Model

The garnet data indicate that the metasomatic processes reenriching the harzburgitic diamond sources in incompatible trace elements generally show little dependence on equilibration temperature and leave the major element composition largely unaffected. This suggests metasomatism by CHO fluids with highly fractionated trace element compositions (very high $LREE_N/HREE_N$, see Fig. 6). This interpretation is consistent with the high solidus temperature of harzburgite, which effectively prevents grain boundary percolation of *dry* silicate and carbonate melts at the PT conditions of diamond formation, as they would freeze upon equilibration with the host rock (Nielson and Wilshire, 1993; Stachel and Harris, 1997a). However, the observed nonlinear relationships, i.e., the apparent restriction of strong preferential enrichment in highly incompatible elements (increased Sr and Ce) to harzburgitic sources at temperatures below about 1150–1200 °C, indicate that such highly fractionated fluids may be absent at high temperatures. A few harzburgitic garnets, which all equilibrated at temperatures above 1190 °C, appear to be influenced by "lherzolite style" metasomatism introducing HREE and probably also refertilizing the bulk rock major element composition (all of these garnets have Cr-numbers below 30). Equilibration temperatures are still below the harzburgitic solidus,

but these samples may be derived from the vicinity of magmatic intrusions or close to the base of the lithosphere, where silicate melts may penetrate for some distance into harzburgite before they freeze. The observation that harzburgitic garnets show a (poor) linear correlation between equilibration temperature and Cr-number without accompanying trace element trends suggests the possible operation of an additional process which cannot be constrained based on this data set.

The observations for lherzolitic garnet inclusions imply that their sources were affected by metasomatism that increased in intensity with temperature and affected both major and trace elements. This coincides with the fact that diamond formation generally takes place above the "wet" lherzolite solidus (about 1100–1150 °C at 5 GPa in the presence of CHO, e.g., Wyllie, 1987) facilitating percolation of silicate melts along grain boundaries.

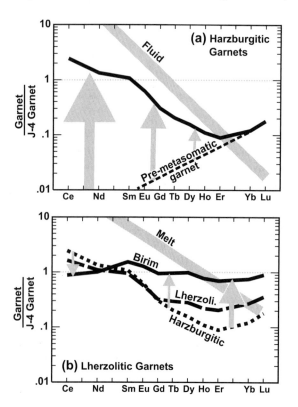

Fig. 6. Schematic illustration of the metasomatic reenrichment affecting harzburgitic and lherzolitic diamond sources. Average garnet compositions are normalized to garnet from a primitive mantle bulk rock composition (see Fig. 3). (a) Harzburgitic garnets have V-shaped patterns with the positive slope within the HREE reflecting the strongly LREE depleted composition of a protolith that experienced a major melt extraction event (c.f. Stachel et al., 1998). The premetasomatic REE pattern of garnet in equilibrium with this protolith is shown as a dashed line. The evolution of the harzburgitic diamond source thus requires interaction with an extremely fractionated metasomatic agent that introduces mainly LREE, comparatively little MREE and almost no HREE. The thick grey line indicates a possible REE pattern for such a fluid. (b) This scenario is based on an origin of lherzolitic diamond sources through metasomatic enrichment of former harzburgite (Stachel et al., 1998; Griffin et al., 1999b). The figure shows a two-stage evolution from harzburgitic garnet (dotted line) to average lherzolitic garnet (dashed line) and finally to fully refertilized lherzolitic garnet (solid line, "Birim") with primitive REE pattern. Transition from harzburgitic to average lherzolitic garnet involves enrichment in MREE and HREE. High garnet–liquid partition coefficients for HREE cause HREE enrichment in garnet even through melts with approximately chondritic HREE abundances. Conversion of harzburgite to lherzolite is accompanied by introduction of increasing modal clinopyroxene which leads to an apparent depletion in LREE relative to harzburgitic garnet. Continuous introduction of melt (approximate composition indicated as a thick grey line) finally leads to garnet that mimics the trace element composition of garnet from primitive mantle. Calculated melt compositions in equilibrium with such "primitive" garnets correspond to typical low-volume mantle melts (e.g., kimberlite, lamproite or MORB source megacryst magma, see Stachel and Harris, 1997a; Burgess and Harte, 1999).

The two styles of metasomatic enrichment identified here (Fig. 6), (i) subsolidus infiltration of strongly fractionated CHO fluids with a very high ratio of LREE to MREE, HREE and other HFSE and (ii) supersolidus percolation of melts that refertilize the diamond sources both in major and trace elements and that are characterized by a moderate enrichment of LREE over MREE, HREE and other HFSE, probably should be viewed as end-members of a compositional continuum rather than two strictly separate processes.

3. Eclogitic suite

The very light carbon isotopic composition of some eclogitic diamonds has been interpreted by numerous authors to reflect diamond formation from subducted organic matter (e.g., Frank, 1969; Kirkley et al., 1991; McCandless and Gurney, 1997). However, high equilibration temperatures of eclogitic inclusions are inconsistent with diamond formation within cold subducting slabs (Gurney, 1989; Stachel et al., 2002) and it appears possible that isotopic fraction-

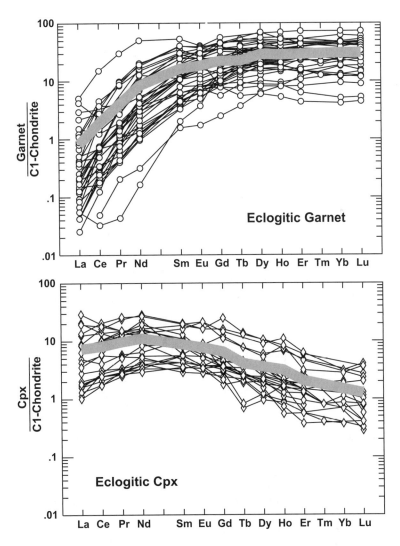

Fig. 7. Chondrite normalized REE concentrations in eclogitic garnet (top) and clinopyroxene inclusions in diamonds from worldwide sources. Average compositions are indicated by thick shaded lines.

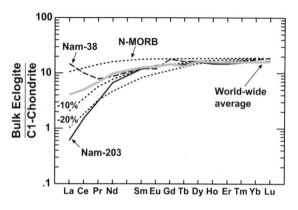

Fig. 8. Based on a garnet–clinopyroxene ratio of 1:1, whole-rock REE patterns are calculated (i) from the average compositions shown in Fig. 7 and (ii) for coexisting garnet–clinopyroxene pairs in two diamonds from Namibia. In addition, the composition of N-MORB is shown together with residues of an original N-MORB composition after 10% and 20% of partial melt were removed in the stability field of eclogite (batch melting with a garnet–cpx residue, assuming a garnet–cpx ratio of 1:1. $D^{Cpx/L}$ from Hart and Dunn, 1993; $D^{Grt/L}$ from Zack et al., 1997).

ation may be the true cause of the observed range in $\delta^{13}C$ (Cartigny et al., 1998), which is conceivable because of the poor buffering capacity of eclogite for hydrous CO_2 fluids (Luth, 1993). Oxygen (e.g., Macgregor and Manton, 1986; Jacob and Foley, 1999; Schulze et al., 2003) and sulfur isotopic analyses (Farquhar et al., 2002) nevertheless provide strong indications that eclogitic diamond sources probably have crustal protoliths. It is generally assumed that cratonic eclogites are not simply the metamorphosed equivalent of Archean seafloor, but that partial melting, probably in the eclogite stability field, lead to chemical depletion (Ireland et al., 1994), thereby explaining the absence of a free SiO_2 phase. Based on REE analyses of garnet and clinopyroxene inclusions in eclogitic diamonds, it is possible to revisit the question of possible oceanic precursors.

Eclogitic garnets (Fig. 7) show REE_N patterns that are similar in shape to the most fertile lherzolitic inclusions (see Fig. 3, Birim garnets), i.e., a steep positive slope within the $LREE_N$ and fairly flat $MREE_N$–$HREE_N$, but at higher MREE–HREE concentrations (averaging at about $30 \times$ chondritic abundance). Eclogitic clinopyroxenes (Fig. 7) have positive slopes within the $LREE_N$, peaking at Nd and then slowly decrease in $MREE_N$ and $HREE_N$ to about chondritic abundance for Lu. This is distinctive

from the majority of peridotitic clinopyroxenes which often have negative slopes within the $LREE_N$ and subchondritic Lu. It has been shown that eclogitic diamond sources, after emplacement in the cratonic lithosphere, are affected by both metasomatic overprint and partial melting (e.g., Taylor et al., 1996; Sobolev et al., 1998). However, the overall consistency of the majority of analyses shown in Fig. 7 suggests that metasomatic overprint may not have completely eradicated the primary signature of the eclogitic sources and that perturbations for the bulk of the data may be limited to the most incompatible elements. Assuming that eclogite represents approximately equal proportions of garnet and clinopyroxene, an average source composition can be calculated (Fig. 8, c.f. Ireland et al., 1994) based on the mean compositions shown in Fig. 7. This calculated average composition compares extremely well with an N-MORB precursor that has lost about 10% of a partial melt in the eclogite stability field and that subsequently experienced some reenrichment in LREE. Two eclogitic garnet–clinopyroxene inclusion pairs in diamonds from Namibia may be used to support this interpretation. Both pairs yield similar bulk rock compositions for the MREE and HREE but differ in LREE, with Nam-203 resembling a strongly (ca 20%) melt depleted oceanic protolith and Nam-38 showing the effect of metasomatic reenrichment in LREE.

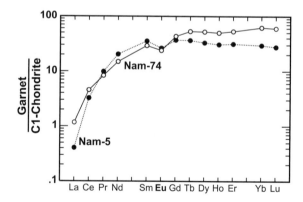

Fig. 9. Negative Eu anomalies in eclogitic garnet inclusions from Namibia. Negative and positive Eu anomalies have also been observed in eclogitic garnet and clinopyroxene inclusions from Venetia (Aulbach et al., 2002) and Kankan (Stachel et al., 2000a). Such anomalies indicate separation of Eu^{2+} from Eu^{3+} (and the other REE) and probably relate to plagioclase fractionation (or accumulation in the case of positive anomalies) during formation of crustal protoliths.

Additional support for the presence of "primary" trace element signatures comes from the observation of negative and positive Eu anomalies in eclogitic inclusions (garnet and clinopyroxene) in diamonds from Kankan (Guinea), Venetia (S.A.) and Namibia (Fig. 9). We therefore conclude that the trace element signature of eclogitic inclusions in diamonds is in support of crustal protoliths. The close spatial relationship of eclogitic diamonds to metasomatic veins (Schulze et al., 1996; Taylor et al., 2000) suggests that diamond precipitation occurred in the course of infiltration by fluids/melts, as it is the case for peridotitic sources.

4. Sublithospheric diamonds

4.1. Asthenosphere and transition zone

Compositional heterogeneities in the Earth's upper mantle are too large to employ the rather subtle compositional variations that are associated with the conversion of orthopyroxene to low-Ca clinopyroxene and of olivine to wadsleyite and ringwoodite for the recognition of deep asthenospheric and transition zone (410–660 km) inclusions (c.f. Stachel, 2001). Therefore, the evidence for diamond formation in the asthenosphere and the transition zone rests exclusively on the observation of inclusions of majorite garnet. With increasing depth pyroxene becomes soluble in the garnet structure (Ringwood, 1967) via simultaneous accommodation of four-valent and divalent cations on the octahedral garnet sites, resulting in the majorite end-member $M_6(Al_2M_1Si_1)^{[VI]}Si_6^{[IV]}O_{24}$. The majorite transition has a negative pressure–temperature slope (Fei and Bertka, 1999) and in particular the pressure dependence of the reaction is well established experimentally (e.g. Irifune, 1987). A second type of pressure-dependent substitution is $Na^+Si^{4+} = M^{2+}Al^{3+}$ (e.g., Irifune et al., 1989) which enables accommodation of the Na content of omphacitic clinopyroxene in majorite garnet.

The first find of majoritic garnet inclusions in diamonds was reported by Moore and Gurney (1985, 1989) for the Monastery Mine. Apart from scattered occurrences of single majorite inclusions, so far only three additional diamond sources with a significant proportion of majoritic garnet inclusions have been

recognized: Jagersfontein (Deines et al., 1991) and the secondary deposits at Juina-São Luiz (Wilding, 1990; Harte, 1992; Hutchison, 1997; Kaminsky et al., 2001) and Kankan (Stachel et al., 2000a). The majority of majoritic garnet inclusions in diamonds have less than 6.4 Si atoms per formula unit which implies an origin well within the upper mantle (s.s.). However, diamonds from the four main deposits also contain inclusions which are likely to be derived from below 410 km, i.e., from the transition zone (e.g. Moore and Gurney, 1985, 1989; Deines et al., 1991).

Compositionally, almost all garnets containing a significant majorite component have eclogitic chemistries and show the same large spread in Ca contents observed for "normal" (lithospheric) eclogitic inclusions. More subtle compositional differences between lithospheric and sublithospheric eclogitic garnets are discussed in Stachel (2001).

Trace element data of majoritic garnet inclusions are scarce with the most detailed study being carried out by Moore et al. (1991) for the Monastery mine. Monastery garnets have relatively low LREE (0.02–2 times chondritic, Fig. 10) and high HREE (five out of seven majorites have 20–30 times chondritic HREE abundances). HREE show significant negative correlations with Si content (ranging from 6.23 to 6.58 cations). Two majoritic garnets from Kankan (Fig. 10) show a sharp rise from La_N to Ce_N and then flat or negative $LREE_N$–$HREE_N$ patterns (HREE at 10–30 times chondritic). For Kankan, two clinopyroxenes

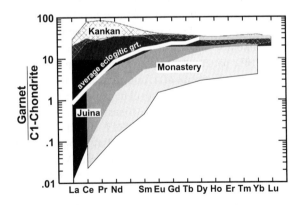

Fig. 10. Compositional fields for REE_N of majoritic (Si>6.15 cations at [O]=24) garnets from Monastery (seven diamonds), Juina-São Luiz (six diamonds) and Kankan (two diamonds). For references see Data base section in text. The average composition of lithospheric eclogitic garnet is taken from Fig. 7.

coexisting with majorite garnets have LREE contents of 100–400 times chondritic abundance (not shown), underlining the prominent LREE enrichment in this specific source. Majoritic eclogitic garnets from Juina-São Luiz are transitional in their LREE between Monastery and Kankan and enclose the average REE composition of "normal" eclogitic garnets (Fig. 10). Silicon contents for the garnets from Juina-São Luiz analyzed for trace elements (Wilding, 1990; Harte, 1992) range from 6.19 to 6.41 cations (indicating a purely asthenospheric origin) and REE from Ce onwards decrease with increasing Si.

In order to elucidate the origin of "basaltic" diamond sources in the asthenosphere and transition zone, Moore et al. (1991) applied garnet/melt partition coefficients to invert the REE composition of Monastery majorites into the composition of melts that may have been in equilibrium with such garnets. The resulting melts have negative $LREE_N/HREE_N$ slopes at high $LREE_N$, typical for low-volume mantle melts such as OIB, alkaline basalts and kimberlites. However, experimental data (Yurimoto and Ohtani, 1992; Draper et al., 2003), which only became available after the work of Moore et al. (1991), indicate that the partitioning behaviour of garnet/melt and majorite/melt is considerably different. Most notably, partition coefficients for HREE drop to values ≤ 1, with the opposite effect (a slight increase in partition coefficients) being observed for LREE. Thus, possible melts in equilibrium with Monastery garnets would have REE_N patterns which are fairly flat and less enriched compared to the results of Moore et al. (1991).

The coexistence of garnet and clinopyroxene in two diamonds from Kankan allows calculation of bulk rock REE_N patterns (assuming an approximate modal relationship of grt:cpx of 1:1) which show flat HREE at about 10–20 chondritic level but strong enrichment in LREE. The majority of Monastery garnets have similar HREE but much lower LREE and, therefore, a fairly flat REE_N pattern with $La_N/Yb_N < 1$ and $HREE_N \sim 10$ may be predicted for the Monastery diamond source which is similar to normal MORB (see Fig. 8). A genetic link between subducted oceanic crust and formation of eclogitic diamonds in the asthenosphere and transition zone is also indicated by negative Eu anomalies in majoritic garnet and clinopyroxene included together in a diamond from Kankan (Fig. 11).

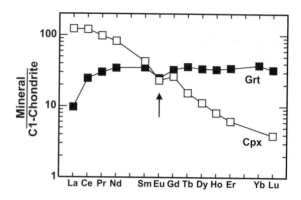

Fig. 11. Negative Eu anomalies in coexisting inclusions of majoritic garnet and clinopyroxene in diamond KK-81 from Kankan (Guinea).

4.2. Lower mantle

Harte et al. (1994) obtained the first trace element data for lower mantle inclusions discovered in diamonds from Rio São Luiz in the Juina area (Brazil). Since then, the data set for Juina-São Luiz has been expanded (Hutchison, 1997; Harte et al., 1999; Kaminsky et al., 2001) and additional REE analyses of lower mantle inclusions have become available for the Kankan deposits in Guinea (Stachel et al., 2000b).

Assuming a pyrolitic bulk composition, the mineralogy of the lower mantle will be dominated by MgSi–perovskite (>70%) and ferropericlase (almost 20%), with only minor amounts (about 8%) of CaSi–perovskite (Ringwood, 1991). For the top of the lower mantle (uppermost about 50 km), garnet is expected to be present as an additional phase, as the garnet–perovskite transition for pyrolitic bulk compositions takes place over a pressure interval, with garnet gradually dissolving in increasingly aluminous MgSi–perovskite (Irifune and Ringwood, 1987a; Wood, 2000). The role of tetragonal almandine–pyrope phase (TAPP; Harris et al., 1997) as a possible substitute for garnet in the topmost lower mantle is not entirely clear with both a possible important role of high ferric iron ratios in the source and an entirely retrograde origin being discussed (Finger and Conrad, 2000; Brenker et al., 2002).

Trace element data on inclusions in lower mantle diamonds (Harte et al., 1994, 1999; Stachel et al., 2000b) show that for lower mantle parageneses (incl.

TAPP), only CaSi–perovskite has to be considered as a significant depository of REE. This result has recently been confirmed experimentally by Wang et al. (2000a) and Corgne and Wood (2002). The presence of garnet, instead of TAPP, probably would require modification of this simplistic picture of CaSi–perovskite as the sole host of the lower mantle REE budget with respect to the HREE. Exsolution of CaSi–perovskite from majorite garnet begins in the deeper parts of the transition zone (Irifune and Ringwood, 1987b; Wood, 2000) and, therefore, garnet at the top of the lower mantle is not expected to contain significant Si^{4+} on octahedral sites, thus reducing electrostatic effects that lead to the suppressed compatibility of HREE in strongly majoritic garnets (see above).

REE_N patterns for CaSi–perovskite inclusions from Juina-São Luiz and Kankan are shown in Fig. 12. All inclusions have high $LREE_N/HREE_N$ and high LREE in common. Based on a modal proportion of 8% CaSi–perovskite in the lower mantle, Ce contents of 200–2000 times chondritic abundance (Fig. 12) indicate enrichment in the source rock to 10–100 the value of primitive mantle. Besides these common characteristics, on a more detailed level the REE_N patterns in Fig. 12 may be split into three groups, with both deposits

(Juina-São Luiz and Kankan) being represented in all groups:

(1) A group of only two samples with REE patterns (shown as dotted lines in Fig. 12) that share the most extreme enrichment in LREE and consequently steep $LREE_N–HREE_N$ slopes. No significant changes in the slope of $MREE_N$ are associated with Eu (i.e., no Eu anomalies).
(2) A group of eight samples (solid lines in Fig. 12) that have flat $LREE_N$ at around 300–400 chondritic abundance. All samples in this group have positive Eu anomalies.
(3) A group of three samples (shown as dashed lines in Fig. 12) that share a "depletion" in MREE relative to the other CaSi–perovskites. All samples in this group have pronounced negative Eu anomalies.

REE_N patterns for samples in groups 1 and 3 were determined both by SIMS ("Kankan" and "Edinburgh") and by LA-ICP-MS ("Macquarie"), indicating that analytical uncertainties cannot be invoked to explain the observed Eu anomalies. Fractionation of Eu^{2+}/Eu^{3+} between lower mantle phases also is not a likely explanation (i) as the LREE–MREE budget of the lower mantle sources seems to be quantitatively

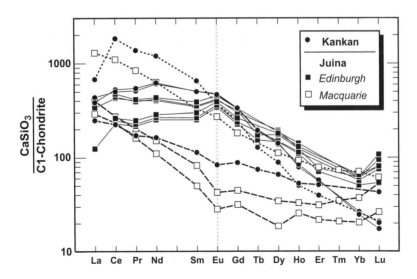

Fig. 12. REE_N patterns of $CaSiO_3$ inclusions in diamonds presumed to have originally crystallized in the lower mantle in the structure of CaSi–perovskite. For references see Data base section in text. The Kankan data and the "Edinburgh" portion of the Juina-São Luiz data set are ion probe analyses; the "Macquarie" data represent LA-ICP-MS analyses. Note that both methods are consistent with respect to Eu anomalies.

hosted in CaSi–perovskite and (ii) as other lower mantle phases do not show corresponding ("mirrored") Eu anomalies (Harte et al., 1999; Stachel et al., 2000b).

CaSi–perovskites thus reflect source compositions that are highly enriched in incompatible trace elements and that may show positive or negative Eu anomalies. Combined with indications (low Al in MgSi–perovskite) that lower mantle diamonds are probably preferentially derived from the topmost part of the lower mantle, a relationship to former oceanic slabs which accumulated at the top of the lower mantle (megalith model of Ringwood, 1991) appears likely (Harte et al., 1999; Stachel et al., 2000b).

In such a subduction scenario, negative Eu anomalies could indicate a protolith that experienced plagioclase fractionation (e.g., volcanic rocks in the upper part of oceanic crust), whereas positive anomalies would indicate feldspar accumulation (cumulate rocks in gabbroic reservoirs). In two cases (Kankan: KK-66a and 87a), positive Eu anomalies are accompanied by extreme concentrations in Sr (about 7000 ppm), which would be consistent with a cumulate model. However, lack of knowledge about possible variations in the modal composition of the lower mantle sources and in particular about the presence or absence of garnet, make a detailed evaluation of the different types of REE_N patterns in CaSi–perovskites impossible. In addition, the possibility exists that instead of diamond formation in ancient slabs, the enriched REE patterns (including Eu anomalies) were imprinted on "normal" lower mantle rocks during metasomatic alteration through slab derived fluids and melts.

In any case, it appears that diamond formation beneath the lithosphere (asthenosphere, transition zone and lower mantle) is intimately linked to subduction processes. Oxygen fugacity in the sublithospheric mantle is expected to decrease with increasing pressure for crystallochemical reasons (O'Neill et al., 1993; Wood et al., 1996), which may lead to conditions that are too reducing for the formation of macro diamonds. In such a scenario former crustal rocks may form a necessary prerequisite providing the redox gradients necessary for diamond precipitation in an otherwise reduced sublithospheric mantle.

Acknowledgements

The data set presented in this review only exists because of continuous generous support through De Beers Consolidated Mines who supplied all the diamonds from which the inclusions were analyzed by us and provided financial assistance for costly ion probe work. The ion probe facility at Edinburgh University and in particular the outstanding support of John Craven and Richard Hinton was instrumental in carrying out most of the research reviewed here. T.S. is grateful to Ben Harte (Edinburgh) for first introducing him to the interpretation of REE in mantle minerals and his continued advise (including numerous reviews). Grants by Deutsche Forschungsgemeinschaft (DFG), NSERC and the Canada Research Chairs Program are gratefully acknowledged. John Gurney and an anonymous reviewer are thanked for their constructive criticisms of the manuscript.

References

Aulbach, S., Stachel, T., Viljoen, K.S., Brey, G.P., Harris, J.W., 2002. Eclogitic and websteritic diamond sources beneath the Limpopo Belt—is slab-melting the link? Contrib. Mineral. Petrol. 143, 56–70.

Brenker, F.E., Stachel, T., Harris, J.W., 2002. Exhumation of lower mantle inclusions in diamond: ATEM investigation of retrograde phase transitions, reactions and exsolution. Earth Planet. Sci. Lett. 198, 1–9.

Burgess, S.R., Harte, B., 1999. Tracing lithospheric evolution trough the analysis of heterogeneous G9/G10 garnets in peridotite xenoliths: I. Major element chemistry. In: Gurney, J.J., Gurney, J.L., Pascoe, M.D., Richardson, S.H. (Eds.), The J.B. Dawson Volume, Proceedings of the VIIth International Kimberlite Conference. Red Roof Design, Cape Town, pp. 66–80.

Cartigny, P., Harris, J.W., Phillips, D., Girard, M., Javoy, M., 1998. Subduction-related diamonds? The evidence for a mantle-derived origin from coupled $\delta^{13}C–\delta^{15}N$ determinations. Chem. Geol. 147, 147–159.

Corgne, A., Wood, B.J., 2002. $CaSiO_3$ and $CaTiO_3$ perovskite-melt partitioning of trace elements: implications for gross mantle differentiation. Geophys. Res. Lett. 29 (art. no. 1933). doi:10.1029/2001GL014398.

Davies, R.M., Griffin, W.L., O'Reilly, S.Y., Doyle, B.J., 2004. Mineral inclusions and geochemical characteristics of microdiamonds from the DO27, A154, A21, A418, DO18, DD17 and Ranch Lake kimberlites at Lac de Gras, Central Slave Craton, Canada. Lithos, these proceedings.

Deines, P., Harris, J.W., Gurney, J.J., 1991. The carbon isotopic composition and nitrogen content of lithospheric and asthenospheric diamonds from the Jagersfontein and Koffiefontein kimberlitev, South Africa. Geochim. Cosmochim. Acta 55, 2615–2625.

Doroshev, A.M., Brey, G.P., Girnis, A.V., Turkin, A.I., Kogarko, L.N., 1997. Pyrope–knorringite garnets in the Earth's upper mantle: experiments in the $MgO–Al_2O_3–SiO_2–Cr_2O_3$ system. Russ. Geol. Geophys. 38, 559–586.

Draper, D.S., Xirouchakis, D., Agee, C.B., 2003. Trace element partitioning between garnet and chondritic melt from 5 to 9 GPa: implications for the onset of the majorite transition for the Martian mantle. Phys. Earth Planet. Inter. 139, 149–169.

Farquhar, J., et al., 2002. Mass-independent sulfur of inclusions in diamond and sulfur recycling on early Earth. Science 298 (5602), 2369–2372.

Fei, Y., Bertka, C.M., 1999. Phase transitions in the Earth's mantle and mantle mineralogy. In: Fei, Y., Bertka, C.M., Mysen, B.O. (Eds.), Mantle Petrology: Field Observations and High-Pressure Experimentation. A Tribute to Francis R. (Joe) Boyd. Special Publication. The Geochemical Society, Houston, pp. 189–207.

Finger, L.W., Conrad, P.G., 2000. The crystal structure of "tetragonal almandine–pyrope phase" (TAPP): a reexamination. Am. Mineral. 85, 1804–1807.

Frank, F.C., 1969. Diamonds and deep fluids in the upper mantle. In: Runcorn, S.K. (Ed.), The Application of Modern Physics to the Earth's and Planetary Interiors. Wiley, New York, pp. 247–250.

Frey, F.A., Green, D.H., 1974. The mineralogy, geochemistry and origin of lherzolite inclusions in Victorian basanites. Geochim. Cosmochim. Acta 38, 1023–1059.

Griffin, W.L., O'Reilly, S.Y., Ryan, C.G., 1999a. The composition and origin of subcontinental lithospheric mantle. In: Fei, Y., Bertka, C.M., Mysen, B.O. (Eds.), Mantle Petrology: Field Observations and High Pressure Experimentation: A Tribute to Francis R. (Joe) Boyd. Special Publication. The Geochemical Society, Houston, pp. 13–45.

Griffin, W.L., Shee, S.R., Ryan, C.G., Win, T.T., Wyatt, B.A., 1999b. Harzburgite to lherzolite and back again: metasomatic processes in ultramafic xenoliths from the Wesselton kimberlite Kimberley, South Africa. Contrib. Mineral. Petrol. 134 (2–3), 232–250.

Grütter, H.S., Sweeney, R.J., 2000. Tests and constraints on single-grain Cr-pyrope barometer models: some initial results. GAC-MAC GeoCanada 2000 Conference, Calgary, 2000. CD, not paginated.

Gurney, J.J., 1984. A correlation between garnets and diamonds in kimberlites. Publ.-Geol. Dept. Univ. Ext., Univ. West. Aust. 8, 143–166.

Gurney, J.J., 1989. Diamonds. In: Ross, J., et al. (Eds.), Kimberlites and Related Rocks. Spec. Publ.-Geol. Soc. Aust., vol. 14. Blackwell, Carlton, pp. 935–965.

Harris, J.W., Hutchison, M.T., Hursthouse, M., Light, M., Harte, B., 1997. A new tetragonal silicate mineral occurring as inclusions in lower-mantle diamonds. Nature 387 (6632), 486–488.

Harris, J.W., Stachel, T., Léost, I., Brey, G.P., 2004. Peridotitic diamonds from Namibia: constraints on the composition and evolution of their mantle source. Lithos 77, 209–223 (this volume).

Hart, S., Dunn, T., 1993. Experimental cpx/melt partitioning of 24 trace elements. Contrib. Mineral. Petrol. 113, 1–8.

Harte, B., 1992. Trace element characteristics of deep-seated eclo-

gite parageneses—an ion microprobe study of inclusions in diamonds. V.M. Goldschmidt Conference. The Geochemical Society, Reston, VA, p. A-48.

Harte, B., Kirkley, M.B., 1997. Partitioning of trace elements between clinopyroxene and garnet: data from mantle eclogites. Chem. Geol. 136, 1–24.

Harte, B., Hutchison, M.T., Harris, J.W., 1994. Trace element characteristics of the lower mantle: an ion probe study of inclusions in diamonds from São Luiz, Brazil. Min. Mag. 58A, 386–387.

Harte, B., Harris, J.W., Hutchison, M.T., Watt, G.R., Wilding, M.C., 1999. Lower mantle mineral associations in diamonds from Sao Luiz, Brazil. In: Fei, Y., Bertka, C.M., Mysen, B.O. (Eds.), Mantle Petrology: Field Observations and High Pressure Experimentation: A Tribute to Francis R. (Joe) Boyd. Special Publication. The Geochemical Society, Houston, pp. 125–153.

Hoal, K.E.O., Hoal, B.G., Erlank, A.J., Shimizu, N., 1994. Metasomatism of the mantle lithosphere recorded by rare-earth elements in garnets. Earth Planet. Sci. Lett. 126, 303–313.

Hutchison, M.T., 1997. Constitution of the deep transition zone and lower mantle shown by diamonds and their inclusions. Unpubl. PhD thesis, University of Edinburgh, vol. 1. 340 pp., vol 2. 306 pp.

Ireland, T.R., Rudnick, R.L., Spetsius, Z., 1994. Trace elements in diamond inclusions from eclogites reveal link to Archean granites. Earth Planet. Sci. Lett. 128, 199–213.

Irifune, T., 1987. An experimental investigation of the pyroxene–garnet transformation in a pyrolite composition and its bearing on the constitution of the mantle. Earth Planet. Sci. Lett. 45, 324–336.

Irifune, T., Ringwood, A.E., 1987a. Phase transformations in a harzburgite composition to 26 GPa: implications for dynamical behaviour of subducting slab. Earth Planet. Sci. Lett. 86, 365–376.

Irifune, T., Ringwood, A.E., 1987b. Phase transformations in primitive MORB and pyrolite compositions to 25 GPa and some geophysical implications. In: Manghnani, M., Syono, Y. (Eds.), High Pressure Research in Geophysics. AGU, Washington, pp. 231–242.

Irifune, T., Hibberson, W.O., Ringwood, A.E., 1989. Eclogite–garnetite transformation at high pressure and its bearing on the occurrence of garnet inclusions in diamond. In: Ross, J., et al. (Eds.), Kimberlites and Related Rocks. Spec. Publ.-Geol. Soc. Aust., vol. 14. Blackwell, Carlton, pp. 877–882.

Jacob, D.E., Foley, S.F., 1999. Evidence for Archean ocean crust with low high field strength element signature from diamondiferous eclogite xenoliths. Lithos 48, 317–336.

Kaminsky, F.V., et al., 2001. Superdeep diamonds from the Juina area, Mato Grosso State, Brazil. Contrib. Mineral. Petrol. 140, 734–753.

Kirkley, M.B., Gurney, J.J., Otter, M.L., Hill, S.J., Daniels, L.R., 1991. The application of C isotope measurements to the identification of the sources of C in diamonds—a review. Appl. Geochem. 6, 477–494.

Luth, R.W., 1993. Diamonds, eclogites, and the oxidation state of the Earth's mantle. Science 261 (5117), 66–68.

Macgregor, I.D., Manton, W.I., 1986. Roberts-Victor eclogites—ancient oceanic-crust. J. Geophys. Res.-Solid Earth Planets 91 (B14), 14063–14079.

McCandless, T.E., Gurney, J.J., 1997. Diamond eclogites: comparison with carbonaceous chondrites, carbonaceous shales, and microbial carbon-enriched MORB. Geol. Geofiz. 38 (2), 371–381.

McDonough, W.F., Sun, S.-S., 1995. The composition of the Earth. Chem. Geol. 120, 223–253.

Moore, R.O., Gurney, J.J., 1985. Pyroxene solid solution in garnets included in diamonds. Nature 318, 553–555.

Moore, R.O., Gurney, J.J., 1989. Mineral inclusions in diamond from Monastery kimberlite, South Africa. In: Ross, J., et al. (Eds.), Kimberlites and Related Rocks. Spec. Publ.-Geol. Soc. Aust., vol. 14. Blackwell, Carlton, pp. 1029–1041.

Moore, R.O., Gurney, J.J., Griffin, W.L., Shimizu, N., 1991. Ultra-high pressure garnet inclusions in Monastery diamonds—trace element abundance patterns and conditions of origin. Eur. J. Mineral. 3, 213–230.

Nielson, J.E., Wilshire, H.G., 1993. Magma transport and metasomatism in the mantle: a critical review of current models. Am. Mineral. 78, 1117–1134.

O'Neill, H.S.C., 1980. An experimental study of the iron–magnesium partitioning between garnet and olivine and its calibration as a geothermometer: corrections. Contrib. Mineral. Petrol. 72, 337.

O'Neill, H.S.C., Wood, B.J., 1979. An experimental study of the iron–magnesium partitioning between garnet and olivine and its calibration as a geothermometer. Contrib. Mineral. Petrol. 70, 59–70.

O'Neill, H.S.C., et al., 1993. Mössbauer spectroscopy of mantle transition zone phases and determination of minimum Fe^{3+} content. Am. Mineral. 78, 456–460.

Ringwood, A.E., 1967. The pyroxene garnet transformation in the Earth's mantle. Earth Planet. Sci. Lett. 2, 255–263.

Ringwood, A.E., 1991. Phase transformations and their bearing on the constitution and dynamics of the mantle. Geochim. Cosmochim. Acta 55 (8), 2083–2110.

Schulze, D.J., Wiese, D., Steude, J., 1996. Abundance and distribution of diamonds in eclogite revealed by volume visualization of CT X-ray scans. J. Geol. 104, 109–114.

Schulze, D.J., Harte, B., Valley, J.W., Brenan, J.M., Channer, D.M.D., 2003. Extreme crustal oxygen isotope signatures preserved in coesite in diamond. Nature 423 (6935), 68–70.

Shimizu, N., 1975. Rare earth elements in garnets and clinopyroxenes from garnet lherzolite nodules in kimberlites. Earth Planet. Sci. Lett. 25, 26–32.

Shimizu, N., Richardson, S.H., 1987. Trace element abundance patterns of garnet inclusions in peridotite-suite diamonds. Geochim. Cosmochim. Acta 51, 755–758.

Shimizu, N., Sobolev, N.V., Yefimova, E.S., 1997. Chemical heterogeneities of inclusion garnets and juvenile character of peridotitic diamonds from Siberia. Russ. Geol. Geophys. 38-2, 356–372.

Sobolev, N.V., Lavrent'ev, Y.G., Pokhilenko, N.P., Usova, L.V., 1973. Chrome-rich garnets from the kimberlites of Yakutia and their paragenesis. Contrib. Mineral. Petrol. 40, 39–52.

Sobolev, N.V., et al., 1998. Extreme chemical diversity in the mantle during eclogitic diamond formation: Evidence from 35 garnet and 5 pyroxene inclusions in a single diamond. Int. Geol. Rev. 40 (7), 567–578.

Stachel, T., 2001. Diamonds from the asthenosphere and the transition zone. Eur. J. Mineral. 13, 883–892.

Stachel, T., Harris, J.W., 1997a. Diamond precipitation and mantle metasomatism—evidence from the trace element chemistry of silicate inclusions in diamonds from Akwatia, Ghana. Contrib. Mineral. Petrol. 129 (2–3), 143–154.

Stachel, T., Harris, J.W., 1997b. Syngenetic inclusions in diamond from the Birim field (Ghana)—a deep peridotitic profile with a history of depletion and re-enrichment. Contrib. Mineral. Petrol. 127 (4), 336–352.

Stachel, T., Viljoen, K.S., Brey, G., Harris, J.W., 1998. Metasomatic processes in lherzolitic and harzburgitic domains of diamondiferous lithospheric mantle: REE in garnets from xenoliths and inclusions in diamonds. Earth Planet. Sci. Lett. 159 (1–2), 1–12.

Stachel, T., Harris, J.W., Brey, G.P., 1999. REE patterns of peridotitic and eclogitic inclusions in diamonds from Mwadui (Tanzania). In: Gurney, J.J., Gurney, J.L., Pascoe, M.D., Richardson, S.H. (Eds.), The P.H. Nixon Volume, Proceedings of the VIIth International Kimberlite Conference. Red Roof Design, Cape Town, pp. 829–835.

Stachel, T., Brey, G.P., Harris, J.W., 2000a. Kankan diamonds (Guinea) I: From the lithosphere down to the transition zone. Contrib. Mineral. Petrol. 140, 1–15.

Stachel, T., Harris, J.W., Brey, G.P., Joswig, W., 2000b. Kankan diamonds (Guinea) II: Lower mantle inclusion parageneses. Contrib. Mineral. Petrol. 140, 16–27.

Stachel, T., Harris, J.W., Aulbach, S., Deines, P., 2002. Kankan diamonds (Guinea) III: $\delta^{13}C$ and nitrogen characteristics of deep diamonds. Contrib. Mineral. Petrol. 142 (4), 465–475.

Stachel, T., Viljoen, K.S., McDade, P., Harris, J.W., 2004. Diamondiferous lithospheric roots along the western margin of the Kalahari Craton—the peridotitic inclusion suite in diamonds from Orapa and Jwaneng. Contrib. Mineral. Petrol. 147, 32–47.

Tappert, R., et al., 2004. Mineral inclusions in diamonds from the Panda kimberlite, Slave Province, Canada. Eur. J. Mineral. (submitted).

Taylor, L.A., et al., 1996. Eclogitic inclusions in diamonds: evidence of complex mantle processes over time. Earth Planet. Sci. Lett. 142 (3–4), 535–551.

Taylor, L.A., et al., 2000. Diamonds and their mineral inclusions, and what they tell us: a detailed "pull-apart" of a diamondiferous eclogite. Int. Geol. Rev. 42, 959–983.

Van Orman, J.A., Grove, T.L., Shimizu, N., Layne, G.D., 2002. Rare earth element diffusion in a natural pyrope single crystal at 2.8 GPa. Contrib. Mineral. Petrol. 142, 416–424.

Wang, W.Y., Gasparik, T., 2001. Metasomatic clinopyroxene inclusions in diamonds from the Liaoning province, China. Geochim. Cosmochim. Acta 65 (4), 611–620.

Wang, W., Sueno, S., Takahashi, E., 1998. Influence of Cr on REE partitioning between garnet and silicate melt: application to metasomatism of mineral inclusions in diamonds. Rev. High Press. Sci. Technol. 7, 92–94.

Wang, W.Y., Gasparik, T., Rapp, R.P., 2000a. Partitioning of rare earth elements between $CaSiO_3$ perovskite and coexisting phases: constraints on the formation of $CaSiO_3$ inclusions in diamonds. Earth Planet. Sci. Lett. 181, 291–300.

Wang, W.Y., Sueno, S., Takahashi, E., Yurimoto, H., Gasparik, T.,

2000b. Enrichment processes at the base of the Archean lithospheric mantle: observations from trace element characteristics of pyropic garnet inclusions in diamonds. Contrib. Mineral. Petrol. 139, 720–733.

Wilding, M.C., 1990. Untitled. Unpubl. PhD thesis, thesis, University of Edinburgh, UK. 281 pp.

Wood, B.J., 2000. Phase transformations and partitioning relations in peridotite under lower mantle conditions. Earth Planet. Sci. Lett. 174, 341–351.

Wood, B.J., Pawley, A., Frost, D.R., 1996. Water and carbon in the Earth's mantle. Philos. Trans.-Royal Soc., Math. Phys. Eng. Sci. 354, 1495–1511.

Wyllie, P.J., 1987. Metasomatism and fluid generation in mantle xenoliths. In: Nixon, P.H. (Ed.), Mantle Xenoliths. Wiley, Chichester, pp. 609–621.

Yurimoto, H., Ohtani, E., 1992. Element partitioning between majorite and liquid—a secondary ion mass-spectrometric study. Geophys. Res. Lett. 19, 17–20.

Zack, T., Foley, S.F., Jenner, G.A., 1997. A consistent partition coefficient set for clinopyroxene, amphibole and garnet from laser ablation microprobe analysis of garnet pyroxenites from Kakanui New Zealand. Neues Jahrb. Mineral. Abh. 172 (1), 23–41.

Available online at www.sciencedirect.com

SCIENCE @ DIRECT®

Lithos 77 (2004) 21–38

ELSEVIER

LITHOS

www.elsevier.com/locate/lithos

The morphological characteristics of diamonds from the Ekati property, Northwest Territories, Canada

John J. Gurney[a,b,*], Peter R. Hildebrand[a], Jon A. Carlson[c],
Yana Fedortchouk[d], Darren R. Dyck[c]

[a] Mineral Services, South Africa
[b] University of Cape Town, South Africa
[c] BHP Billiton Diamonds Inc., Canada
[d] University of Victoria, Canada

Received 27 June 2003; accepted 17 January 2004
Available online 8 July 2004

Abstract

Examination of exploration diamond parcels from 13 kimberlites on the Ekati Diamond Mine™ property revealed abundant octahedra and dodecahedra, significant numbers of cubes (fibrous and non-fibrous), minor cubo-octahedra and aggregates, and rare macles and pseudo-hemimorphic crystals. Some octahedra have a fibrous diamond coat. The diamonds are predominantly colourless or brown with minor yellow and very few other colours.

A striking feature of the diamonds from all 13 localities is their mixed character, evidenced by the variety of colours, crystal forms and surface textures and as previously documented for Siberia (Orlov, Yu L., 1977. Mineralogy of the Diamond. Translated from the Russian language and published by Wiley, New York. Original (1973) published in Russian by Izdatel'stvo, Nauka, USSR.) and by Robinson (Robinson, D.N., 1979. Surface textures and other features of diamonds. PhD thesis, University of Cape Town, South Africa.) for the Kaapvaal craton. This mixed character can be largely accounted for by variations in the proportions of components (termed building blocks [BB]) of diamonds with similar characteristics that are represented at each of the 13 localities.

These descriptions reveal three very strongly developed regional associations. The highest proportion of both colourless octahedra and total octahedra is present in the northwest part of the study area, which includes the commercially exploited kimberlites, Panda, Koala and Beartooth. In contrast, diamonds in the centrally placed kimberlites Arnie and Mark are dominated by opaque fibrous cubes. In the southeast part of the study area, the diamonds from four kimberlites including Misery are characterized by dodecahedra that are dominantly brown. The most southerly pipe, Piranha, is unique within this group of 13 kimberlites in that it has a high proportion of cubo-octahedra and of colourless cubes.
© 2004 Elsevier B.V. All rights reserved.

Keywords: Colour; Diamonds; Ekati; Kimberlite; Morphology

1. Introduction

A unique feature of the economics of diamond mining is the variation in value of the product. In

---

* Corresponding author. Fax: +27-21-531-9887.
E-mail address: john.gurney@minserv.co.za (J.J. Gurney).

extreme cases, this can exceed four orders of magnitude for diamonds of equal weight. Whilst the spread in value per carat is less when considering average values of run-of-mine production, it is still greater than two orders of magnitude. Essentially, the reasons for the variation in value come down to parameters that are important in the presentation of diamonds for incorporation in jewelry, such as colour, transparency, fractures, inclusions, interpenetrant aggregation and twinning.

Some kimberlites produce run-of-mine diamonds that are quite homogenous in appearance. For instance, diamonds from the Messina mine (Bobbejaan dyke, Barkly West area, South Africa) are almost entirely colourless octahedra, showing minimal signs of resorption. At the other end of the value scale, the diamonds from the Monastery kimberlite (Marquard, South Africa) are highly included fragments. Some of the dominant features of diamond populations are regional. Mbuji Mayi diamonds (Democratic Republic of Congo [DRC]) are characterized by fibrous, opaque cubes and coated stones with a range of colours. Diamonds from Udachnaya, Mir and several other Siberian kimberlites have high proportions of octahedra. Most of the major mines in southern Africa, in contrast, are dominated by resorbed forms, dodecahedra and tetrahexahedra (Robinson, 1979, Robinson et al., 1989).

Little is known, however, about how diamonds vary in appearance from kimberlites within a kimberlite province, a kimberlite cluster or even in adjacent diatremes. Sutton (1928) documented that the differences can be major even in adjacent pipes such as the five large mines in Kimberley (South Africa). In a sample of 50,000 carats from De Beers, Kimberley, Dutoitspan, Wesselton and Bultfontein, no 'fine, white' diamonds were recorded from De Beers, Kimberley or Dutoitspan, whilst they predominated at Wesselton, particularly at Bultfontein. On the other hand, yellow diamonds dominated in the latter three diatremes.

Sutton (1928) used now dated terminology in describing the above diamonds. Modern diamond descriptions (e.g., Harris, 1992), focus on individual economic diamond-producing localities and regional characteristics. Recent (1991–2003) exploration for kimberlites on the Ekati property (Northwest Territories, Canada) has provided the opportunity to assess the characteristics of geographically associated kimberlite diatremes in a manner not previously possible. Ap-

proximately 150 kimberlite occurrences have been discovered in the Ekati Diamond Mine™ property since 1991 of which one quarter have been tested for macro-diamonds to some extent. Parcel sizes vary from less than 50 carats to greater than 5000 carats in individual cases. Large variations in the overall appearance and value of parcels of diamonds from individual localities are common. The initial impression is that each locality has its own specific diamond population, as might be expected if diamonds were phenocrysts forming in the host kimberlite, and not accidental xenocrysts. However, closer inspection revealed that the differences between each locality's population can primarily be accounted for by variations in the proportions of sub-populations of diamonds (termed building blocks [BB] in this manuscript).

An opportunity to describe diamond populations in detail and further assess such a possibility arose during the period that the Ekati Diamond Mine™ was being brought through feasibility and permitting to production. Observational data are presented here for 13 kimberlites that occur within an area of approximately 30 × 30 km in and around the present active mining operations at Panda, Koala and Misery (see Fig. 1).

2. Diamond descriptions

The requirement for this purely descriptive diamond study was to devise a simple procedure that would usefully categorise the diamonds in a reasonably quick and reproducible manner. The scheme that evolved was partly based on the system used in the TERRAC diamond valuation programme (Terraconsult: L. Rombouts) with some additional information added to the comments column. Each stone was weighed, described in terms of shape (A, B, C), colour, colour intensity (1–4), clarity (1–5), crystal morphology and pseudo-hemimorphism.

Primary crystal morphology includes octahedral, cubic, cubo-octahedral, macle (contact twin) or aggregate. Secondary morphology ranged from modified primary forms to rounded dodecahedral, tetrahexahedroidal and anhedral, corroded and sculpted surfaces. Any resorbed diamond with remnant octahedral or cubic faces was described as a 'resorbed' primary crystal. Any other resorbed form, with no recognizable remnant primary surface was termed a dodeca-

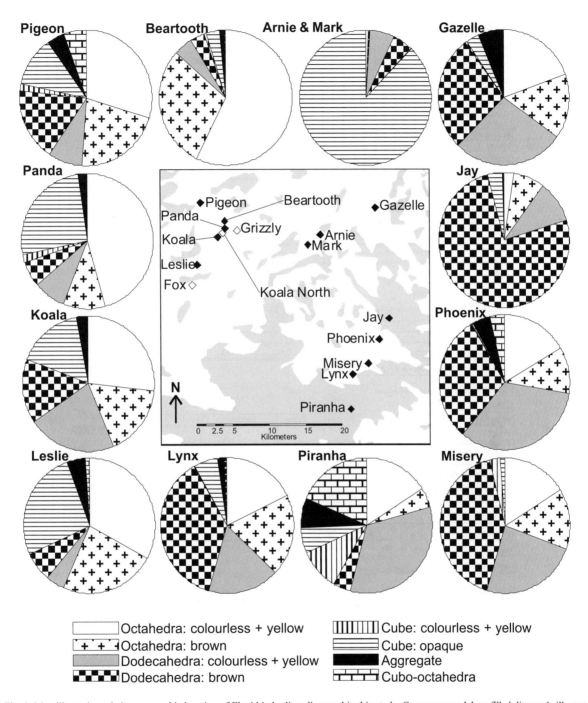

Fig. 1. Map illustrating relative geographic location of Ekati kimberlites discussed in this study. Grey areas are lakes; filled diamonds illustrate kimberlites whose data is presented in detail within this report; open diamonds illustrate kimberlites whose data is not presented in detail within this report. Surrounding the map are pie charts showing the proportions of different diamond types at each locality. The data for these charts are from Tables 1–3, excluding all the unassigned stones. Also excluded for diagrammatic simplification are the 'pseudohemimorphic' morphologies and 'other' colours (which account for < ~ 3% of the total assigned stones) from Tables 2 and 3.

Table 1

Diamond-parcel characteristics for Panda, Misery, Jay, Arnie and Mark kimberlites. In the case of Misery data, a representative split of 1000 smaller sized stones was described

Panda

Morphology	Colourless	Brown	Yellow	Opaque	Total	%
Octahedra: flat faced, sharp edged	138	6	0		144	30.32%
Octahedra: step faced	0	13	0		13	2.74%
Octahedra: others	16	27	2		45	9.47%
Octahedral twins	4	0	0		4	0.84%
Octahedra with fibrous coats	58	3	0		61	12.84%
Octahedra: total	216	49	2		267	56.21%
Dodecahedra	37	28	0		65	13.68%
Cubes	0	0	10	*122	132	27.79%
Aggregates	9	2	0		11	2.32%
Total assigned (106.67 carats)	262	79	12	122	475	100.00%
† Unassigned (17.70 carats=14.23%)					304	39.02%
Total (124.37 carats)					779	

Misery

Morphology	Colourless	Brown	Yellow	Opaque	Total	%
Octahedra: flat faced, sharp edged	22	32	16		70	1.54%
Octahedra: step faced	324	249	8		581	12.79%
Octahedra: others	200	266	21		487	10.72%
Octahedral twins	124	104	0		228	5.02%
Octahedra with fibrous coats	19	8	1		28	0.62%
Octahedra: total	689	659	46		1394	30.68%
Dodecahedra	998	1883	90		2971	65.38%
Cubes	97	0	33	*45	175	3.85%
Aggregates	4	0	0		4	0.09%
Total assigned (189.94 carats)	1788	2542	169	45	4544	100.00%
† Unassigned (18.72 carats=8.97%)					719	13.66%
Total (208.66 carats)					5263	

Jay

Morphology	Colourless	Brown	Yellow	Opaque	Total	%
Octahedra: flat faced, sharp edged	2	0	0		2	0.08%
Octahedra: step faced	0	32	0		32	1.27%
Octahedra: others	42	115	0		157	6.24%
Octahedral twins	9	60	0		69	2.74%
Octahedra with fibrous coats	10	1	0		11	0.44%
Octahedra: total	63	208	0		271	10.76%
Dodecahedra	257	1983	0		2240	88.96%
Cubes	0	0	0	*1	1	0.04%
Aggregates	5	1	0		6	0.24%
Total assigned (190.56 carats)	325	2192	0	1	2518	100.00%
† Unassigned (33.88 carats=15.10%)					910	26.55%
Total (224.44 carats)					3428	

Arnie & Mark

Morphology	Colourless	Brown	Yellow	Opaque	Total	%
Octahedra: flat faced, sharp edged	0	0	0		0	0.00%
Octahedra: step faced	0	0	0		0	0.00%
Octahedra: others	13	11	0		24	1.14%
Octahedral twins	1	0	0		1	0.05%
Octahedra with fibrous coats	0	0	0		0	0.00%
Octahedra: total	14	11	0		25	1.19%
Dodecahedra	126	102	1		229	10.89%
Cubes	0	0	1	*1848	1849	87.92%
Aggregates	0	0	0		0	0.00%
Total assigned (52.56 carats)	140	113	2	1848	2103	100.00%
† Unassigned (0.07 carats=0.13%)					1	0.05%
Total (52.63)					2104	

* = All non-white and non-yellow cubes. These are mostly brown or black and sometimes green or grey.

† = Unassigned are mostly fragments with no discernable morphology.
    Rare stones of unusual colour or shape are included.

Table 2

Diamond-parcel characteristics for the NW group of kimberlites: Beartooth, Koala, Leslie amd Pigeon

Beartooth

Morphology	Colourless	Light Brown	Dark Brown	Light Yellow	Dark Yellow	Opaque	Other	Total	%
Octahedra	1361	565	197	5	0		2	2130	77.17%
Octahedral twins	54	22	5	0	0		0	81	2.93%
Octahedra with fibrous coats	149	19	4	0	0		0	172	6.23%
Octahedra: total	1564	606	206	5	0		2	2383	86.34%
Dodecahedra	121	77	23	0	0		0	221	8.01%
Cubes	5	0	0	4	1	*95	0	105	3.80%
Cubo-octahedra	4	2	2	0	0		0	8	0.29%
Aggregates	17	6	5	0	0		0	28	1.01%
Pseudohemimorphic	11	3	1	0	0		0	15	0.54%
Total assigned (331.94 carats)	1722	694	237	9	1	95	2	2760	100.00%
† Unassigned (32.85 carats=9.00%)	121	164	75	0	1		95	456	14.18%
Total (364.79 carats)	1843	858	312	9	2	95	97	3216	

Koala

Morphology	Colourless	Light Brown	Dark Brown	Light Yellow	Dark Yellow	Opaque	Other	Total	%
Octahedra	66	31	15	4	0		0	116	34.52%
Octahedral twins	16	2	4	0	0		0	22	6.55%
Octahedra with fibrous coats	3	2	4	0	0		0	9	2.68%
Octahedra: total	85	35	23	4	0		0	147	43.75%
Dodecahedra	70	31	20	2	0		1	124	36.90%
Cubes	0	0	0	0	0	*54	0	54	16.07%
Cubo-octahedra	0	0	0	0	0		0	0	0.00%
Aggregates	5	0	5	0	0		0	10	2.98%
Pseudohemimorphic	1	0	0	0	0		0	1	0.30%
Total assigned (36.25 carats)	161	66	48	6	0	54	1	336	100.00%
† Unassigned (21.91carats=37.67%)	81	99	111	1	0		60	352	51.16%
Total (58.16 carats)	242	165	159	7	0	54	61	688	

Leslie

Morphology	Colourless	Light Brown	Dark Brown	Light Yellow	Dark Yellow	Opaque	Other	Total	%
Octahedra	105	51	20	7	0		0	183	47.78%
Octahedral twins	8	10	4	1	0		0	23	6.01%
Octahedra with fibrous coats	6	2	3	0	0		0	11	2.87%
Octahedra: total	119	63	27	8	0		0	217	56.66%
Dodecahedra	17	14	13	0	0		0	44	11.49%
Cubes	0	0	0	0	0	*101	0	101	26.37%
Cubo-octahedra	2	0	0	0	0		1	3	0.78%
Aggregates	10	3	2	0	0		3	18	4.70%
Pseudohemimorphic	0	0	0	0	0		0	0	0.00%
Total assigned (36.08 carats)	148	80	42	8	0	101	4	383	100.00%
† Unassigned (23.38 carats=39.32%)	137	159	76	0	1		111	484	55.82%
Total (59.46 carats)	285	239	118	8	1	101	115	867	

Pigeon

Morphology	Colourless	Light Brown	Dark Brown	Light Yellow	Dark Yellow	Opaque	Other	Total	%
Octahedra	360	214	49	5	0		22	650	49.54%
Octahedral twins	14	11	0	0	0		0	25	1.91%
Octahedra with fibrous coats	1	0	0	0	0		0	1	0.08%
Octahedra: total	375	225	49	5	0		22	676	51.52%
Dodecahedra	104	185	37	0	0		13	339	25.84%
Cubes	11	0	0	5	3	*153	0	172	13.11%
Cubo-octahedra	57	5	1	1	0		3	67	5.11%
Aggregates	34	16	4	0	1		3	58	4.42%
Pseudohemimorphic								0	0.00%
Total assigned (158.19 carats)	581	431	91	11	4	153	41	1312	100.00%
† Unassigned (14.67 carats=8.49%)	39	47	14	0	0		23	123	8.57%
Total (172.86 carats)	620	478	105	11	4	153	64	1435	

* = All non-white and non-yellow cubes. These are mostly brown or black and sometimes green or grey.

† = Unassigned are mostly fragments with no discernable morphology. Rare stones of unusual shape are included.

Table 3

Diamond-parcel characteristics for Gazelle, Lynx, Phoenix, and Piranha kimberlites

Gazelle

Morphology	Colourless	Light Brown	Dark Brown	Light Yellow	Dark Yellow	Opaque	Other	Total	%
Octahedra	273	207	50	33	13		5	581	29.72%
Octahedral twins	41	34	6	11	7		1	100	5.12%
Octahedra with fibrous coats	1	2	0	0	0		0	3	0.15%
Octahedra: total	315	243	56	44	20		6	684	34.99%
Dodecahedra	357	452	83	115	60		5	1072	54.83%
Cubes	2	0	0	0	5	*64	0	71	3.63%
Cubo-octahedra	0	0	0	0	0		0	0	0.00%
Aggregates	64	32	14	11	3		0	124	6.34%
Pseudohemimorphic	2	2	0	0	0		0	4	0.20%
Total assigned (105.16 carats)	740	729	153	170	88	64	11	1955	100.00%
† Unassigned (39.56 carats=27.34%)	142	182	74	36	23		116	573	22.67%
Total (144.73 carats)	882	911	227	206	111	64	127	2528	

Lynx

Morphology	Colourless	Light Brown	Dark Brown	Light Yellow	Dark Yellow	Opaque	Other	Total	%
Octahedra	210	203	135	11	0		5	564	27.13%
Octahedral twins	22	26	12	1	0		1	62	2.98%
Octahedra with fibrous coats	121	12	7	0	0		3	143	6.88%
Octahedra: total	353	241	154	12	0		9	769	36.99%
Dodecahedra	326	464	295	26	1		16	1128	54.26%
Cubes	9	0	0	0	0	*117	0	126	6.06%
Cubo-octahedra	7	5	1	1	0		0	14	0.67%
Aggregates	12	12	6	0	0		1	31	1.49%
Pseudohemimorphic	2	5	3	0	0		1	11	0.53%
Total assigned (229.30 carats)	709	727	459	39	1	117	27	2079	100.00%
† Unassigned (48.90 carats=17.58%)	286	293	221	20	0		80	900	30.21%
Total (278.20 carats)	995	1020	680	59	1	117	107	2979	

Phoenix

Morphology	Colourless	Light Brown	Dark Brown	Light Yellow	Dark Yellow	Opaque	Other	Total	%
Octahedra	232	125	40	12	5		13	427	21.97%
Octahedral twins	44	34	8	12	1		0	99	5.09%
Octahedra with fibrous coats	9	1	0	1	0		0	11	0.57%
Octahedra: total	285	160	48	25	6		13	537	27.62%
Dodecahedra	496	355	222	112	30		28	1243	63.94%
Cubes	6	0	0	0	2	*5	0	13	0.67%
Cubo-octahedra	42	11	4	6	3		1	67	3.45%
Aggregates	45	18	13	3	2		3	84	4.32%
Pseudohemimorphic	0	0	0	0	0		0	0	0.00%
Total assigned (129.60 carats)	874	544	287	146	43	5	45	1944	100.00%
† Unassigned (24.00 carats=15.63%)	237	184	119	79	14		35	668	25.57%
Total (153.60 carats)	1111	728	406	225	57	5	80	2612	

Piranha

Morphology	Colourless	Light Brown	Dark Brown	Light Yellow	Dark Yellow	Opaque	Other	Total	%
Octahedra	297	68	31	5	0		5	406	17.85%
Octahedral twins	14	2	2	4	1		1	24	1.05%
Octahedra with fibrous coats	40	4	0	0	0		0	44	1.93%
Octahedra: total	351	74	33	9	1		6	474	20.84%
Dodecahedra	728	73	23	26	3		4	857	37.67%
Cubes	218	0	0	8	2	*140	0	368	16.18%
Cubo-octahedra	384	23	2	3	2		1	415	18.24%
Aggregates	123	32	0	4	2		0	161	7.08%
Pseudohemimorphic	0	0	0	0	0		0	0	0.00%
Total assigned (169.15 carats)	1804	202	58	50	10	140	11	2275	100.00%
† Unassigned (31.01 carats=15.49%)	271	128	66	21	7		89	582	20.37%
Total (200.16 carats)	2075	330	124	71	17	140	100	2857	

* = All non-white and non-yellow cubes. These are mostly brown or black and sometimes green or grey.

† = Unassigned are mostly fragments with no discernable morphology. Rare stones of unusual shape are included.

hedron. The implication being that the diamond was a resorbed form that had lost more than 45% of its original mass and volume. Frequently, resorption is an imperfect surface process that produces complex shapes, particularly from non-ideal primary shapes. The intention in applying the label 'dodecahedron' to a diamond was to denote extensive resorption, not to accurately define the crystal shape. Information on clarity, whilst noted in the diamond descriptions, has not been utilised in this study.

This descriptive procedure has been completed for the recoveries from mini-bulk samples of the 13 kimberlites, amounting to over 30,000 individual diamond records, of which representative data sets are presented in Tables 1–3. Over the roughly 10-year period that these observations have been catalogued, the database has been adapted to include more detail. As a result, it has been necessary to present summary tables of the data in two slightly different formats. Table 1 (Panda, Misery, Jay, and Mark and Arnie combined) has no subdivisions between light and dark brown, nor between light and dark yellow diamonds. The eight remaining localities described in Tables 2 and 3 do have this information, but lack information on the occurrence of unresorbed octahedra and their step-faced or flat-faced/sharp-edged distinctions.

In describing the diamonds, variable amounts of material have remained unassigned within the above outlined descriptive procedure at each locality. Unassigned particles are all small to very small. Most are fragments of diamond without readily discernible remnant external surfaces, making it impossible to assess crystal morphology. Many refract light internally, making it difficult to assess body colour as well. Unassigned diamonds are therefore an unavoidable reality of the diamond parcels.

A particular concern is whether preferential breakage of susceptible diamonds will materially alter the make-up of a particular diamond parcel when fragments are not classified. It is reasonable to expect, for example, that the brittle, often thin and fragile fibrous coat on some octahedral diamonds will splinter off the main crystal and contribute to the fragments in a greater proportion than its overall presence would suggest. Inspection of the fragments, however, showed that they have the appearance of fragments of the general sample inclusive of fibrous fragments, and it was concluded that any such effect was not major.

The diamond samples described in this study have the additional limitation that even though more than 30,000 individual diamonds have been described, parcels from individual localities are too small to completely define the contributing populations in a fully quantifiable manner. Consideration of these two factors has led to the utilisation of the data in only a semi-quantitative manner therefore.

2.1. The diamond sample

The localities from which data are presented and discussed are shown in Fig. 1. In the NW of Fig. 1, the Koala, Panda, Beartooth, Pigeon and Leslie diamonds fall into a related group, as separately do Jay, Phoenix, Misery and Lynx. Piranha to the SE, and Gazelle fall into a category of its own, whilst diamonds from Arnie and Mark are quite distinctly different from any of the other diamond populations. The groupings will be discussed in the above order.

2.2. The NW group of kimberlites: Beartooth, Koala, Leslie, Panda and Pigeon

2.2.1. General characteristics of the diamonds

Diamond characteristics are presented in Tables 1 and 2 and these data are summarised in Fig. 1. The diamonds are dominated by relatively unresorbed octahedra in the larger size fraction (e.g., Fig. 2A), with fairly abundant cubes in the smaller sizes only (e.g., Fig. 1 and Fig. 2K). Many of the octahedra have a coat, which is usually opaque and fibrous, but occasionally brown, not fibrous and transparent. The fibrous coat varies in thickness from several millimetres on some diamonds (e.g., Fig. 2M) to a thin, barely visible coat, often over only part of the diamond core (e.g., Fig. 2I). Fibrous coat colour varies with thickness from very dark, black appearance to grey. This appears to be an effect related to translucency rather than a real colour change. The coat is brittle and fragments break off the surface readily. In addition, where coated diamonds are resorbed, the coat is thinned and eventually removed, particularly at crystal edges. As a result, it is often possible to look through a window in the coat. These windows, together with fragments of coated diamond, can be used to demonstrate that the internal cores of coated diamonds are octahedra, predominantly colourless, but occasionally brown. A coated dodecahedron has yet to be found.

Uncoated octahedra are also predominantly colourless with subordinate light browns, whilst yellow is very rare. The light brown colour of some octahedra appears to be due to a thin surface skin. Octahedral diamonds with excellent primary shapes have flat crystal surfaces and sharp edges (e.g., Fig. 2A). These octahedra frequently have one or more imperfect terminations at the crystal points, giving them a very distinctive appearance (e.g., Fig. 2B).

A second distinctive set of octahedral diamonds show a different primary growth feature: stepped faces with sharp, step edges. At Panda, where noted in this study, they were brown, but at nearby Beartooth and Koala, step-faced octahedra are also regularly colourless (e.g., Fig. 2D).

Dodecahedra are a minor component of the diamond population at Panda, and are on average smaller than the octahedra. Whilst many are white (e.g., Fig. 2N), the brown component is more strongly represented than for the octahedra, both in number and range of intensity (e.g., Fig. 1 and Fig. 2M, O and P). The cubic diamonds are predominantly opaque, have fibrous growth structure and show a range of colours from black through brown to yellow to grey, and occasionally green (e.g., Fig. 2K). They are confined to the smaller size ranges, rarely exceeding 0.2 carats in weight. Rare aggregates complete the commonly seen diamonds. Not noted in Table 1, but included in Table 2, are the rare, but important pseudo-hemimorphic crystals that show differential resorption on a single crystal. These are a small component at all localities, and they are evidence, along with plastic deformation, of a xenolithic origin for at least some diamonds (Robinson, 1979).

In summary, the NW group of kimberlites has relatively simple diamond populations. The larger diamonds are predominantly octahedral, colourless or light brown, and often coated with a fibrous diamond overgrowth. Dodecahedra are subordinate. Cubes almost always show fibrous growth structures and are opaque and small in size. Yellow diamonds are very rare, and are pale yellow, except for occasional lemon yellow cubic diamonds (e.g., Fig. 2J).

Although the northwestern group of five diatremes have broadly similar diamonds, there are nevertheless some fairly substantial differences between them. All five localities have more than 43% octahedra by abundance, but Beartooth is outstandingly high at 86.34%. Whilst fibrous cubes always are significantly present, and always of small average size, they vary in number from 3.8% to 27.8%. Yellow diamonds are very rare at all five locations. Colourless dodecahedra range from 8.0% at Beartooth to 36.9% at Koala. These differences can be inspected in more detail in Fig. 1 and Tables 1 (Panda) and 2 (the other four localities in this group).

Fox and Grizzly are two larger than average kimberlite pipes in the same area as the NW group of five kimberlites described in detail here. Fox has been extensively sampled and falls within the currently approved mine plan. A large number of Fox diamonds have been described from several samples (1896 diamonds; 790 carats). These did not cover the full size range, being focussed on the diamond sizes larger than 2 mm. The results cannot therefore be directly compared with those tabled here for other localities. However, it is clear that Fox diamonds have within the same range of abundance of colourless octahedral character ($\sim$ 65%) as the other five localities, with a small, but distinct proportion of pale yellow crystals ($\pm$ 8%). The remaining diamonds are mostly brown. Cubes, macles and dodecahedra make up approximately 8% of the overall population. Coated stones are quite common and always octahedral, usually single crystals, with occasional contact twins (macles).

Fig. 2. (A) Colourless flat-faced octahedra from Fox. Note the imperfect termination on the lower right crystal. (B) Colourless flat-faced octahedron from Fox, showing imperfect crystal terminations. (C) Light brown octahedra from Grizzly. The left diamond is brown at the rim and colourless in the core. (D) Colourless step-faced octahedron from Koala. (E) Colourless macles (contact twinned octahedra) from Fox. (F) Brown step-faced octahedron from Misery. (G) Dark brown octahedron from Koala. (H) Grey-coated flat-faced octahedron from Panda. (I) Remnant grey-coated flat-faced octahedron from Panda. The coat has been sufficiently resorbed at the edges and corners so as to reveal windows into a colourless interior. (J) Translucent lemon yellow cube from Sable. (K) Black, pale grey and dark grey (left to right) opaque cubes from Grizzly. (L) Colourless cubo-octahedra from Piranha. The diamonds have a translucent core and partly transparent rims. (M) Light brown rounded resorbed dodecahedron from Fox with relic octahedral surface features. (N) Colourless dodecahedron from Koala. (O) Light brown dodecahedron with well developed curvilinear faces from Misery. (P) Dark brown rutted and rounded dodecahedron from Misery.

Geographically adjacent Grizzly provided a diamond population that was significantly different to those from Fox and the other NW localities. Despite the fact that the number of diamonds available for study was smaller than can accurately define the full population (778 diamonds; 60 carats), certain important factors have been noted. There are fewer octahedra, a modest increase in macles, coated stones are very rare, although fibrous cubes are present, and dodecahedra are more common than octahedra in contrast to the neighbouring localities. Most critical of all, ~ 80% of the diamonds are brown (e.g., Fig. 2C); a feature most readily explained by a deformation event in the mantle (Urusovskaya and Orlov, 1964; Robinson, 1979), which is less in evidence at the immediately adjacent Panda, Koala and Beartooth localities.

2.3. The SE group of kimberlites: Jay, Lynx, Misery, Phoenix and Piranha

Diamond characteristics are presented in Tables 1 and 3 and in Fig. 1. Four of these five kimberlites, roughly aligned in a NNE–SSW direction (see Fig. 1), can also be grouped together with respect to their diamond population characteristics. All of them have >50% dodecahedra (such as those shown in Fig. 2M–P), except Piranha, which has 38% only, a figure negatively affected by the presence of a very unusual abundance (18%) of cubo-octahedra (Fig. 2L), and a high proportion of colourless transparent cubes, giving the Piranha diamond population a very different appearance. Diamond parcels from this group include colourless octahedra, a minority of which are flat-faced and step-faced octahedra.

Brown dodecahedra dominate the parcels, particularly at Jay, but also to a large extent at Misery, whilst overall dodecahedra (colourless, yellow plus brown) are highly dominant at Jay, Misery and Phoenix, dominant at Lynx and important even at Piranha. Lynx has the highest proportion of octahedra in this group of kimberlites and the highest percentage of colourless coated stones. Piranha has the highest proportion of colourless diamonds, but many cubes and cubo-octahedra have fibrous nuclei and crystal proportions that are not commercially favourable. Whilst the earlier described diamond populations from the NW kimberlite group were characterized by primary crystal shapes (octahedra and cubes), which to a greater or lesser extent had survived storage in the mantle and transport to the surface whilst retaining their primary crystal form, the SE group is dominated by dodecahedra that have lost >45% of their original mass to resorption processes (Robinson, 1979). Of equal interest is the fact that the full range of common diamond shapes, colours and surface features can be found at any one particular locality (i.e., from virtually no resorption to >45% resorption).

2.4. The Gazelle kimberlite

Diamond characteristics are presented in Table 3 and Fig. 1. East of the Panda group and north of the Misery group, the Gazelle kimberlite diamonds, by some characteristics, lie intermediate between the two main groupings in the NW and SE of the study area (see Fig. 1). Gazelle is dominated by dodecahedra, but has 35% octahedra, 3.6% cubes, 6.3% aggregates and a high proportion of colourless and yellow octahedra and dodecahedra, to which both light and dark yellow contribute more than usual at Ekati (see Table 3 and Fig. 1). As with the previously described diamond populations, examples of all the diamonds shown in Fig. 2 can be found at Gazelle.

2.5. Mark and Arnie kimberlites

Geographically central to the area of interest (see Fig. 1), the Mark and Arnie kimberlites have not been extensively sampled. Diamond characteristics are presented in Fig. 1 and Table 1. Only 34.2 carats of diamonds recovered from Mark and 18.4 carats from Arnie. However, since the average size of the diamonds was small, in total, 2104 diamonds were available for study (Table 1; Fig. 1). Of these, 1848 were small, opaque, fibrous cubes. Both the Mark and Arnie parcels were similarly dominated by fibrous opaque cubes, and for the purposes of this study, the results were added together to increase the size of the parcel of diamonds described.

Although greatly dominated by fibrous cubic diamonds, flat-faced, step-faced, colourless, yellow and brown octahedra, colourless, yellow and brown dodecahedra, and aggregates, all report in the very small non-fibrous diamond component recovered from these two localities combined. Even here, therefore, examples can be found of all the common and most of the rare diamond types shown in Fig. 2.

2.6. Kimberlite ages

Nine of the kimberlites in this study, plus Grizzly, have been dated, although only two of them have an isochron age, the rest being model ages (Creaser et al., 2004). The ages fall within the range ~ 48–56 Ma: ~ 56 Ma in the SE (two kimberlites), ~ 53 Ma in the NW (four kimberlites) and ~ 48 Ma in the central area (three kimberlites; see Fig. 3). Whilst more isochron ages are desirable, the existing data are suggestive of an approximately 10 Ma period available for changes in the cratonic mantle to take place and account for some of the variations observed in the diamond parcels. Alternatively, the cratonic mantle has not changed, but has been erratically sampled, and/or is highly heterogeneous.

The diamonds at Snap Lake (~ 140 km south of the Ekati property) are very similar to those in the NW cluster of kimberlites in this study area (Stachel et al., 2003). This similarity extends to the presence of coated diamonds at both localities, even though Snap Lake has been dated at 523 Ma and the Ekati group at 53 ± 1 Ma. The diamond coats at Snap Lake differ in colour from those seen at Ekati and, in view of the very different age of the kimberlite intrusion, and the small mantle residence time of coat, Snap Lake coat must

have formed at an earlier time than the Ekati fibrous coat. Coat apart, the similarity between diamonds sampled at 523 Ma and at 53 Ma is supportive of the long-term existence of laterally extensive multiple reservoirs of diamond within the craton root, that were available to be sampled at both Snap Lake and Ekati.

3. Discussion

The descriptions of the diamonds from the 13 localities at Ekati have illustrated three regional associations that are very strongly developed. In simple terms, these can be represented as predominantly octahedral, cubic and dodecahedral diamond populations in the western, central and eastern parts of the study area, respectively. However, underlying this dominant trend is the observation that the principal differences between the diamond populations at each and every locality are the product of variations in the proportions of a number of basic components (termed building blocks in this study).

It is well known that all diamond populations have both peridotitic and eclogitic components (Meyer, 1987), and that the peridotitic inventory includes both harzburgitic and lherzolitic diamonds, which may have formed at different times (Richardson et al., 1993). Within the eclogitic paragenesis, a calc–silicate assemblage has been identified as distinct from the more common mafic component (Sobolev, 1984). Dia-

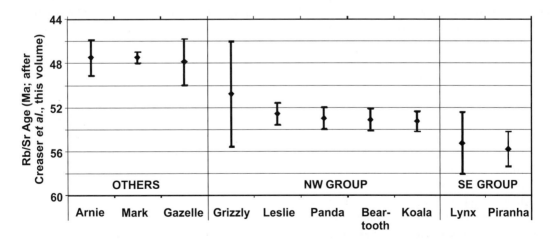

Fig. 3. Published Rb/Sr ages of kimberlites discussed in this study (after Creaser et al., 2004). Note that only the Mark and Panda ages are isocron ages, the others are all model ages.

monds may also have a websteritic association (Gurney et al., 1984), whilst cubic diamonds and overgrowths on diamonds are distinguishable from other populations not only by their fibrous growth mechanism but also by their restricted range in carbon isotopic ratio and their comparatively young age relative to the host kimberlite (Navon, 1999). Rarely diamonds have an ultra-deep origin, quite distinct from those mentioned above. There may be even more than these seven different parageneses represented in a single kimberlite occurrence. McDade and Harris (1999) have documented the existence of 10 diamond sources in four lithological units in the low grade Letseng La Terai kimberlite in Northern Lesotho.

Diamond inclusions from all these source rocks, plus diamonds with ultra-deep origins, have already been described from diamonds from localities on the Slave craton, despite the comparatively short time since their initial discovery (Davies et al., 1999; Kopylova et al., 1999; Pokhilenko et al., 2001; Stachel et al., 2003; Westerlund et al., 2003; Chinn, personal communication). Current indications are that diamonds from the Slave craton have similar origins to those demonstrated on other cratons elsewhere in the world.

If a full knowledge of diamond genesis and distribution is to be attained, it must be important to understand as fully as possible the implications of these diverse origins. Given the results of this study of diamonds from the Ekati property, it is tempting to attempt to assign the described diamonds to specific 'building blocks'. Each block corresponding to a subset of diamonds that are likely to have formed in the same or similar processes. Such a project is fraught with difficulties since it is well known that whilst many diamonds have simple growth histories, many others do not. Episodic growth, declining growth rates, increasing growth rates, sudden fluctuations of volatile species such as H, H_2O, CO_2 and N, changes in morphology, changes in nucleation rate, and incorporation of fluid and other inclusions are amongst the recorded complications relevant to diamond growth.

Furthermore, several different crystal shapes have been described from within single primary source rocks such as eclogite and peridotite (e.g., Robinson, 1979; Shee et al., 1982; Hills and Haggerty, 1989; Afanasiev et al., 2000). These include flat-faced octahedra, step-faced octahedra, cubes, fibrous coated octahedra and aggregates.

Deformation nitrogen aggregation and diamond resorption are post-formation processes that can affect size, shape and/ or colour.

Consequently, it is unlikely that purely visual criteria will successfully discriminate between all the various diamond populations that occur within a single kimberlite. However, such observations may form a useful base study from which to develop a detailed understanding of the diamond populations in adjacent ore bodies. This concept is given encouragement by the observations made in this study that the gross variations seen within the various diamond populations study are chiefly a product of variations in the proportions of a finite number of these building blocks.

3.1. Diamond formation

It is worthwhile considering what evidence there is about when and where diamonds have formed in the mantle, and how, when and where secondary processes have subsequently affected them. At present, and consistent with the general picture worldwide, the earliest formed identified diamonds in the Slave craton were peridotitic diamonds, formed in the Archaean (Westerlund et al., 2003). These are most probably predominantly harzburgitic in origin, octahedral in shape and colourless at formation. They required long-term storage in the diamond stability field, and in at least some cases are transported to the Earth's surface with minimal weight loss. This could be achieved by transport upward within a swiftly emplaced kimberlite with a suitably low redox potential. Alternatively, the well-preserved diamonds in this paragenesis could have been transported for a substantial portion of the journey from depth to the Earth's surface, protected within a harzburgite xenolith (Robinson, 1979).

Other known xenolithic sources of diamonds in the Slave craton include lherzolite, eclogite and websterite from the lithosphere, and ultra-deep diamonds (Chinn et al., 1998; Kopylova et al., 1999; Pokhilenko et al., 2001; Stachel et al., 2003).

3.2. Diamond deformation

Deformation of diamonds can occur at any time during storage within the host rock, but it is considered very likely that any brown colour due to deformation, and seen in diamonds at the Earth's surface (Orlov, 1977; Robinson, 1979; Collins, 2001), must be a remnant of a young deformation event, otherwise, the diamond would have annealed by analogy to the General Electric [GE] industrial process for enhancing the colour of certain diamonds (Anthony et al., 2001; Collins, 2001; Van Royen and Palyanov, 2002).

Grain–boundary contact together with deviatoric stress is required to generate plastic deformation in rock minerals. It is therefore not going to occur in the kimberlite magma, but it could be generated at any stage by crack propagation in lithospheric mantle associated with metasomatic activity (Hills and Haggerty, 1989), diapiric uprising of aesthenospheric melts, or strain developed in the mantle envelope surrounding the intrusion of a megacryst magma or a kimberlite precursor or the kimberlite itself. Mantle diapirs, metasomatic crack propagation and a megacryst magma might be locally focussed and affect only one diamondiferous horizon, whilst the host kimberlite intrusion could influence all the diamond parageneses that fall within the volume of physically sampled cratonic lithosphere during kimberlite emplacement. It is of interest therefore to note that colourless (relatively undeformed) and a variety of brown diamonds (variably deformed) occur at all the localities studied in the Ekati property.

The proportion of brown diamonds can also change markedly between geographically closely spaced kimberlites, as illustrated by the example of brown-dominated Grizzly (mentioned above). Whilst ~ 80% of the diamonds at Grizzly are brown, nearby Panda (Fig. 1) has only ~ 15% brown diamonds. The age of Grizzly has a 10 Ma uncertainty (Fig. 3), spanning the age for Panda and precludes putting any time constraints on the deformation event(s) that have affected the diamonds. A similarly large difference between the abundances of brown and colourless diamonds can be seen by comparing Misery with Piranha. These examples are further illustrations of the fact that fundamentally significant variations in diamond populations can occur within adjacent kimberlites.

3.3. Diamond resorption

In contrast to deformation of diamonds, resorption is a process that can affect diamonds by interaction with the kimberlite during emplacement. Diamondiferous mantle eclogites have been described where diamonds partly protruding from the xenolith surface have developed a prominent resorbed surface on the exposed portion of the crystal (Gurney, 1989). At Ekati, pseudo-hemimorphic crystals, interpreted to have formed in a like manner (Robinson, 1979) are regularly seen. The regular observation of partially resorbed fibrous coat on octahedral diamonds throughout the Ekati Diamond Mine ™ confirms that some resorption is a late-stage process, since coat is itself young based on nitrogen aggregation studies (Navon, 1999).

On the basis of a detailed study of some George Creek diamonds, which documented episodic diamond growth, interrupted by resorption, Chinn, 1995 has demonstrated that significant resorption can occur in the mantle. Given the need to preserve diamonds in the craton root for long periods of time dating back to the Archaean, it appears highly likely that diamond resorption in the mantle is, like deformation, a young event or series of events; perhaps even the same events formulated to cause plastic deformation, but excluding the kimberlite emplacement event itself. In order to allow the kimberlite to sample and transport to the Earth's surface diamonds with such a complex and long history, it is considered highly likely that separate domains exist with respect to redox conditions at different levels in the subcratonic mantle (Haggerty, 1986).

Gurney and Zweistra (1995) noted that fibrous coated diamonds at Mbuji Mayi (DRC) and Koidu (Sierra Leone) were associated with extremely reduced high-MgO, high-Cr_2O_3 ilmenites. That is also true at Ekati, where in addition, there are also very resorbed diamonds, both in the same localities as the ilmenites and widely spread over the study area. It is difficult to envisage a process whereby diamonds with young fibrous coats, along with well-preserved octahedra that grew by spiral atomic growth, and resorbed dodecahedral diamonds were all originally sampled from the same mantle source. Suggestive of the same conclusion is the fact that, in general, and overall, far fewer octahedral diamonds are brown than are dodecahedral diamonds (see Tables 1, 2 and 3). Does that mean diamonds that are more susceptible to plastic deformation are more prone to resorption, or alternatively have at least some of the dodecahedral diamonds experienced different processes through formation in a different part of the mantle stratigraphy?

The fact that dodecahedra never have fibrous coats indicates that either resorption post-dates the formation of the coat or that the dodecahedra were never present where the coat was formed. On the other hand, coated diamonds are never extensively resorbed, most are very well preserved. Is that because coated diamonds are armoured against resorption, or did they never experience the requisite redox conditions where they resided in the mantle? An attempt has been made to summarise these concepts in cartoon form as they relate to the study area in Fig. 4.

3.4. Building blocks

The previous discussion leads to the conclusion that in an initial visual assessment such as this one, the

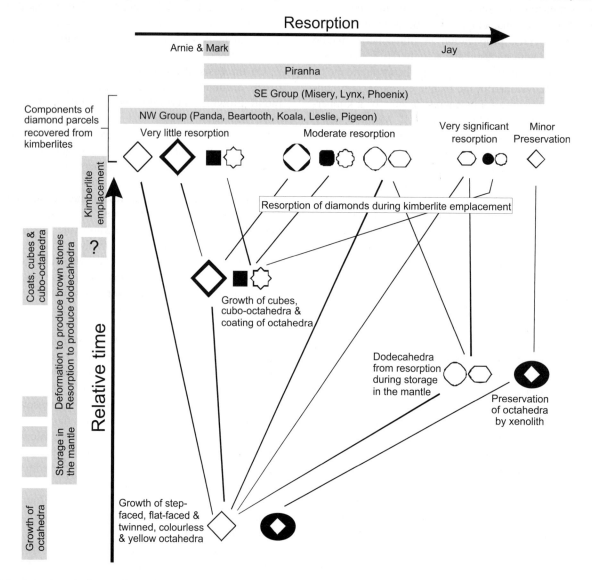

Fig. 4. A schematic diagram illustrating some of the diamond growth and evolution processes and complexities involved in time and resorption space. Along the top of the diagram are the different types of diamonds recovered from kimberlites, along with an indication of the ranges recovered from the different Ekati kimberlites considered in this study.

number of diamond populations contribution to the overall run-of-mine production may be over-estimated. This is due to the difficulty in seeing through the secondary processes affecting the diamonds post-formation. Despite the difficulties, a preliminary attempt to identify 'building blocks' can be justified on the grounds that it could represent the initial stages of a project to clearly identify the important sources of diamonds within a kimberlite or suite of kimberlites.

In the case of these 13 Ekati diamond sources, a simple starting point would be as follows.

(1) Octahedra:
  (a) Colourless, light brown, light yellow. Flat-faced (e.g., Fig. 2A and B).
  (b) Colourless, light brown, light yellow. Step-faced (e.g., Fig. 2D).
  (c) Coated (e.g., Fig. 2H and I).
  (d) Macles (i.e., contact twins; e.g., Fig. 2E).
  (e) Other (visibly resorbed). Colourless, light brown, light yellow (e.g., Fig. 2C).
  (f) Dark brown (e.g., Fig. 2F and G).
(2) Cubes:
  (a) Opaque fibrous, any colour (e.g., Fig. 2K).
  (b) Colourless, transparent, translucent.
  (c) Yellow, transparent, translucent (e.g., Fig. 2J).
(3) Cubo-octahedra:
  (a) Colourless, transparent, translucent (e.g., Fig. 2L).
(4) Dodecahedra:
  (a) Colourless, light brown, light yellow (e.g., Fig. 2N, M and O).
  (b) Dark brown, sometimes rutted (e.g., Fig. 2P).

These sub-groups account for all the assigned diamonds in Table 1. In Tables 2 and 3, they account for greater than ~ 97% of all the assigned diamonds, excluding only the 'pseudo-hemimorphic' morphologies and the 'other' colours. Many of these sub-groups could share a common origin, some more likely than others. In the octahedral category, the macles cannot justify a separate building block since both can clearly form together on the basis of yields from diamond-bearing mantle eclogites and peridotites. (Robinson, 1979; Afanasiev et al., 2000). Category 1e octahedra may well be resorbed variants of 1a, b or c. Category 1c could be a variant of 1a or b.

Octahedral class 1a has been observed to have mineral inclusions of purple garnets, chromites and colourless to pale green silicates interpreted as olivines and/or orthopyroxene. In addition, recovered sulphides have proved to be of peridotitic (high nickel) composition and Archaean age (Westerlund et al., 2003). Class 1a diamonds are very abundant in the Panda kimberlites, which carries a strong diamond peridotitic signature (Stachel et al., 2003). There is, therefore, a peridotitic association between these octahedra and garnet peridotite, probably harzburgite, which is not necessarily unique.

Two step-faced octahedra (Class 1b) have been seen to include an eclogitic garnet, demonstrating an eclogitic association for some step-faced octahedra at least. No paragenesis-related observations have been made for coated diamonds, but it has been observed that the interiors are always octahedral, usually flat-faced, usually colourless and sometimes brown. They could therefore be related to Class 1a, but that would be in contrast compared to diamonds in the Koidu kimberlite, Sierra Leone and Mbuji Mayi kimberlite, DRC, where coated diamonds are associated with eclogitic diamonds (Hills and Haggerty, 1989; Schrauder and Navon, 1994, and Navon, 1999).

It is well known that plastic deformation, which can take place in diamond-bearing rocks at depth (De Vries 1975), can and does turn colourless diamonds brown (Urusovskaya and Orlov, 1964). However, octahedral class 1f is a deeper brown than commonly associated with plastic deformation. On dodecahedra, plastic deformation can be visually recognized from the presence of lamination lines on the crystal surface (Robinson, 1979), whilst on octahedra, the same slip planes can be recognized by alignments of inverted trigons on crystal faces (Afanasiev et al., 2000). Neither feature has been associated with this very dark brown colour, suggesting that not all brown diamonds at Ekati are deformed. So no link has been established between deformation and this very dark brown colour. It is therefore suggested that as an initial set of building blocks for octahedral diamonds, Classes 1a, b, c and f should be retained.

The three cubic categories all appear to be sufficiently discrete in appearance to justify separate investigation (Classes 2a, b and c). The cubo-octahedra (Class 3) may well be closely linked to, or even be a variety of Classes 1 and 2b, where diamond crystallisation proceeds at a slower rate as the crystal grows; octahedral faces starting to develop in the later growth stages. However, the cubo-octahedral morphology is distinctive enough to justify an initial attempt at characterisation of these diamonds.

The sub-classification of the dodecahedra is the most problematic of the 'building-block' exercises, because the variations in primary characteristics seen in the other categories can all be over-written by resorption to produce dodecahedra. This leads to the suggestion that only two categories of dodecahedra should be investigated as potential 'building blocks' at this stage: colourless or light brown/light yellow dodecahedra, and dark brown dodecahedra. Both these categories are possibly resorbed versions of the appropriate octahedron. However, within the dodecahedra, there are at least two further possible building blocks: a pale brown population of diamonds with well-developed curvilinear faces (e.g., Fig. 2O), and a population of exceptionally fine colourless diamonds that are distinctive on the basis of a combination of completely developed tetra-hexahedral shape and a grading in the categories D/E colour of the GIA grading system.

In contrast to diamonds from Kimberley, Finsch, Premier, Jagersfontein, (South Africa), the marine and beach deposits of southern Africa and lamproitic diamonds from Ellendale, Ekati diamonds lack the strong yellow component that is thought to be produced by N_3 aggregation in the diamond lattice (Collins, 2001). In this, Ekati diamonds are similar to diamonds from the kimberlite dykes in the Barkly West and Theunissen and Swartruggens areas of South Africa (Gurney; personal observation). The kinetics of nitrogen aggregation are now quite well understood, and it should be possible to explain whether the N_3 colour centre is poorly developed in diamonds from this Ekati study area because of low nitrogen contents, mantle storage at low temperature or short residence time in the mantle. In unravelling the building blocks for these diamonds, it will also be necessary to clarify the rate of plastic deformation in turning diamonds brown and the rate at which plastic deformation is annealed at mantle temperatures and pressures. Is deformation responsible for all brown diamonds or only some? What role, if any, do nitrogen content and platelet development play in facilitating plastic deformation? Without answers to these and other questions,

the relationship between diamonds with a range of characteristics will not be fully evaluated.

4. Conclusions

These descriptions reveal three very strongly developed regional associations. The highest proportion of both colourless octahedra and total octahedra are present in the northwest part of the study area, which includes the commercially exploited kimberlites, Panda, Koala and Beartooth. In contrast, diamonds in the centrally placed kimberlites Arnie and Mark are dominated by opaque fibrous cubes. In the southeast part of the study area, the diamonds from four kimberlites including Misery are characterized by dodecahedra that are dominantly brown. The most southerly pipe, Piranha, is unique within this group of 13 kimberlites in that it has a high proportion of cubo octahedra and of colourless cubes.

It is suggested, on the basis of the observations made in this study, that the variations in diamond parcels recovered from kimberlites on the Ekati property are mainly the product of differential sampling of a number of diamondiferous rock types from various parts of the cratonic lithosphere. Ten components or building blocks have been visually identified as contributing to the overall variations in the study area. These are listed under Section 3.4 as categories 1a, b, c and f; 2a, b and c; 3a; and 4a and b. There may be additional building blocks within 1e and 4a, raising the postulated number of building blocks to 13.

This study has demonstrated that geographically associated, penecontemporaneously emplaced kimberlites can have diamond populations with significantly different physical appearances. However, these differences are due to variations in the proportion of a finite number of visually identifiable diamond subpopulations termed 'building blocks'. Elucidating the origins of the various identified building blocks within those populations is essential to understanding the overall picture including the size distribution of diamonds within a kimberlite occurrence. Detailed cathodoluminescence, Fourier transform infrared (FTIR), stable isotope ratio, fluid inclusion, mineral inclusion and crystallographic studies will help to identify the building blocks with greater certainty and possibly identify additional building blocks and probably provide evidence of common origins for some octahedral and dodecahedral diamond populations particularly. These additional investigations should also contribute to resolving debate over the origins of the regional variation from relatively unresorbed diamonds (NW) to resorbed (SE) documented in this study. In particular, is it a pervasive time-related variation throughout the cratonic lithosphere, or is it a localised process that affects only a specific mantle domain?

Acknowledgements

The Ekati Diamond Mine™ is gratefully acknowledged for making this study possible by allowing access to the diamonds described. The authors would like to extend their thanks to Barbara Crawford for assistance with Fig. 2, and Louise Hildebrand for figure preparation and manuscript typing. Peter Hildebrand and John Gurney gratefully acknowledge the financial assistance of BHP Billiton in carrying out this study, and thank the organising committee of the VIIth International kimberlite conference for the award of a travel grant to attend IKC VIII and present this paper.

References

Afanasiev, V.P., Yefimova, E.S., Zinchuk, N.N., Koptil, V.I., 2000. In: Sobolev, N.V. (Ed.), Atlas of Morphology of Diamonds from Russian Sources. Russian Academy of Sciences, Siberian Branch, Novosibirsk.

Anthony, T.R., Vagarali, S.S., Casey, J.K., Smith, A.C., 2001. International Patent Applications WO 01/14050 A1 (01 March 2001) WO 01/33203 A1 (10 May 2001).

Chinn, I.L., 1995. A study of unusual diamonds from the George Creek K1 Kimberlite Dyke, Colorado. PhD thesis, University of Cape Town, pp. 94.

Chinn, I.L., Gurney, J.J., Kyser, K.T., 1998. Diamonds and mineral inclusions from the N.W.T., Canada. Proceedings of the VIIth International Kimberlite Conference, University of Cape Town, South Africa, vol. 2. Red Roof Design, Cape Town. addendum–not paginated.

Collins, A.T., 2001. The colour of diamonds and how it may be changed. J. Gemmol. 27 (6), 341–359.

Creaser, R.A., Grütter, H., Carlson, J.C., Crawford, B., 2004. Macrocrystal phlogopite Rb–Sr dates for the Ekati property kimberlites, Slave Province, Canada: evidence for multiple intrusive episodes in the Paleocene and Eocene. Proceedings of the VIIIth International Kimberlite Conference, Victoria, British Columbia, Canada. This Volume.

Davies, R.M., Griffin, W.L., Pearson, N.J., Andrew, A.S., Doyle, B.J., O'Reilly, S.Y., 1999. Diamonds from the deep: pipe DO-27, Slave Craton, Canada. Proceedings of the VIIth International Kimberlite Conference, University of Cape Town, South Africa, vol. 1. Red Roof Design, Cape Town, 148–155.

Gurney, J.J., 1989. Diamonds. In: Ross, J. (Ed.), Kimberlites and Related Rocks Geol. Soc. Spec. Publ. vol. 14. Blackwell, Carlton, pp. 966–989.

Gurney, J.J., Harris, J.W., Rickard R.S., 1984. Silicate and oxide inclusions in diamonds from the Orapa Mine, Botswana. In: Kornprobst, J. (Ed.), Proceedings of the 3rd International Kimberlite Conference. Kimberlites II; The mantle and crust relationship. Elsevier, Amsterdam, pp. 3–9.

Gurney, J.J., Zweistra, P., 1995. The interpretation of major element compositions of mantle minerals in diamond exploration. J. Geochem. Explor. 53, 293–309.

Haggerty, S.E., 1986. Diamond genesis in a multiply-constrained model. Nature 320, 34–38.

Harris, J.W., 1992. Diamond geology. In: Field, J.E. (Ed.), The Properties of Natural and Synthetic Diamonds, pp. 345–393 Academic Press, London.

Hills, D.V., Haggerty, S.E., 1989. Petrochemistry of eclogites from the Koidu kimberlite complex, Sierra Leone. Contrib. Mineral. Petrol. 103, 397–422.

Kopylova, M.G., Russell, J.K., Cookenboo, H., 1999. Petrology of peridotite and pyroxenite xenoliths from the Jericho kimberlite: implications for the thermal state of the mantle beneath the Slave craton, Northern Canada. J. Petrol. 40 (1), 79–104.

McDade, P., Harris, J.W., 1999. Syngenetic inclusion bearing diamonds from Letseng-la-Terai, Lesotho. Proceedings of the VIIth International Kimberlite Conference, University of Cape Town, South Africa, vol. 2. Red Roof Design, Cape Town, 557–565.

Meyer, H.O.A., 1987. Inclusions in Diamonds. In: Nixon, P.H. (Ed.), Mantle xenoliths. Wylie, New York.

Navon, O., 1999. Diamond formation in the Earth's mantle. Proceedings of the VIIth International Kimberlite Conference, University of Cape Town, South Africa, vol. 2. Red Roof Design, Cape Town, 584–684.

Orlov, Yu L., 1977. Mineralogy of the Diamond. Translated from the Russian language and published by John Wiley & Sons, New York. Original (1973) published in Russian by Izdatel'stvo, Nauka, USSR.

Pokhilenko, et al., 2001. Crystalline inclusions in diamonds from kimberlites from the Snap Lake area (Slave Craton, Canada). New evidence for anomalous lithospheric structure. Dokl. Earth Sci. 380 (7), 806–811.

Richardson, S.H., Harris, J.W., Gurney, J.J., 1993. Three generations of diamonds from old continental mantle. Nature 366, 256–258.

Robinson, D.N., 1979. Surface textures and other features of diamonds. PhD thesis, University of Cape Town, South Africa.

Robinson, D.N., Scott, J.A., Van Niekerk, A., Anderson, V.G., 1989. The sequence of events reflected in the diamonds of some southern African kimberlites Kimberlites and related rocks: vol. 2. Geological Society of Australia Special Publication, vol. 14, pp. 990–1000.

Schrauder, M., Navon, O., 1994. Hydrous and carbonatitic fluids in fibrous diamonds from Jwaneng, Botswana. Geochem. Comochem. Acta 58, 761–771.

Shee, S.R., Gurney, J.J., Robinson, D.N., 1982. Two diamond-bearing peridotite xenoliths from the Finsch Kimberlite, South Africa. Contrib. Mineral. Petrol. 81, 79–87.

Sobolev, N.V., 1984. Crystalline inclusions in diamonds from New South Wales, Australia. Kimberlite occurrence and origin: A basis for conceptual models in exploration. University of Western Australia, Perth, pp. 143–166. Publication No. 8.

Stachel, T., Harris, J.W., Tappert, R., Brey, G.P., 2003. Peridotitic inclusions in diamonds from the Slave and the Kaapvaal cratons—similiarities and differences based on a preliminary data set. Lithos 71, 489–503.

Sutton, J.R., 1928. Kimberley diamonds: especially cleavage diamonds. Trans. Royal Soc. S. Afr. 7, 65–96.

Urusovskaya, A.A., Orlov, Yu L., 1964. Nature of Plastic Deformation of Diamond Crystals, vol. 154. Doklady Akademii, Nauk, USSR, pp. 112–115.

Van Royen, J., Palyanov, Yu N., 2002. High-pressure–high-temperature treatment of natural diamonds. J. Phys. Condens. Matter 14, 10953–10956.

Westerlund, K.J., Shirey, S.B., Richardson, S.H., Gurney, J.J., Harris, J.W., 2003. Re/Os Isotope systematics of peridotitic diamond inclusion sulphides from the Panda Kimberlite, Slave Craton. Extended Abstracts, VIIIth International Kimberlite Conference, Victoria, Canada.

Available online at www.sciencedirect.com

Lithos 77 (2004) 39–55

ELSEVIER

LITHOS

www.elsevier.com/locate/lithos

Mineral inclusions and geochemical characteristics of microdiamonds from the DO27, A154, A21, A418, DO18, DD17 and Ranch Lake kimberlites at Lac de Gras, Slave Craton, Canada[☆]

Rondi M. Davies[a,*], William L. Griffin[a,b], Suzanne Y. O'Reilly[a], Buddy J. Doyle[c]

[a] GEMOC National Key Centre, School of Earth and Planetary Sciences, Macquarie University, Sydney NSW 2109, Australia
[b] CSIRO Exploration and Mining, P.O. Box 136, North Ryde, NSW 1670, Australia
[c] Kennecott Canada Exploration Inc., 200 Granville Street, Vancouver, BC, Canada V6C 1S4

Received 27 June 2003; accepted 6 December 2003
Available online 2 June 2004

Abstract

A mineral inclusion, carbon isotope, nitrogen content, nitrogen aggregation state and morphological study of 576 microdiamonds from the DO27, A154, A21, A418, DO18, DD17 and Ranch Lake kimberlites at Lac de Gras, Slave Craton, was conducted. Mineral inclusion data show the diamonds are largely eclogitic (64%), followed by peridotitic (25%) and ultradeep (11%). The paragenetic abundances are similar to macrodiamonds from the DO27 kimberlite (Davies, R.M., Griffin, W.L., O'Reilly, S.Y., 1999. Diamonds from the deep: pipe DO27, Slave craton, Canada. In: Gurney, J.J., Gurney, J.L., Pascoe, M.D., Richardson, S.H. (Eds.), The J. B. Dawson Vol., Proc. 7th Internat. Kimberlite Conf., Red Roof Designs, Cape Town, pp. 148–155) but differ to diamonds from nearby kimberlites at Ekati (e.g., Lithos (2004); Tappert, R., Stachel, T., Harris, J.W., Brey, G.P., 2004. Mineral Inclusions in Diamonds from the Panda Kimberlite, S. P., Canada. 8th International Kimberlite Conference, extended abstracts) and Snap Lake to the south (Dokl. Earth Sci. 380 (7) (2001) 806), that are dominated by peridotitic stones.

Eclogitic diamonds with variable inclusion compositions and temperatures of formation (1040–1300 °C) crystallised at variable lithospheric depths sometimes in changing chemical environments. A large range to very ^{13}C-depleted C-isotope compositions (δ^{13}C = − 35.8‰ to − 2.2‰) and an NMORB bulk composition, calculated from trace elements in garnet and clinopyroxene inclusions, are consistent with an origin from subducted oceanic crust and sediments. Carbon isotopes in the peridotitic diamonds have mantle compositions (δ^{13}C mode − 4.0‰). Mineral inclusion compositions are largely harzburgitic. Variable temperatures of formation (garnet T_{Ni} = 800–1300 °C) suggest the peridotitic diamonds originate from the shallow ultra-depleted and deeper less depleted layers of the central Slave lithosphere. Carbon isotopes (δ^{13}C av. = − 5.1‰) and mineral inclusions in the ultradeep diamonds suggest they formed in peridotitic mantle (~ 670 km). The diamonds may have been entrained in a plume and subcreted to the base of the central Slave lithosphere.

Poorly aggregated nitrogen (IaA without platelets) in a large number of eclogitic (67%) and peridotitic (32%) diamonds, with similar nitrogen contents, indicates the diamonds were stored in the mantle at low temperatures (1060– < 1100 °C) following crystallisation in the Archean. Type IaA diamonds have largely cubo-octahedral growth forms, and Type II and Type IaAB diamonds, with higher nitrogen aggregation states, mostly have octahedral morphologies. However, no correlation between

[☆] Supplementary data associated with this article can be found, in the online version, at doi: 10.1016/j.lithos.2004.04.016.

* Corresponding author. Present Address: Department of Earth and Planetary Sciences, American Museum of Natural History, New York, NY 10024-5192, USA. Tel.: +1-212-769-5314; fax: +1-212-769-5339.

E-mail address: rdavies@amnh.org (R.M. Davies).

these groups and their mineral inclusion compositions, C-isotopes, and N-contents rules out the possibility of unique source origins and suggests eclogitic and peridotitic diamonds experienced variable mantle thermal states. Variation in mineral inclusion chemistries in single diamonds, possible overgrowths of ^{13}C-depleted eclogitic diamond on diamonds with peridotitic and ultradeep inclusions, and Type I ultradeep diamond with low N-aggregation is consistent with diamond growth over time in changing chemical environments.

Keywords: Slave Craton; Lac de Gras; Inclusions in diamond; C-isotopes; N-contents; N-aggregation states

1. Introduction

The discovery of diamondiferous kimberlites in the Slave Craton, Northwest Territories, Canada, in the last decade, has provided the opportunity to investigate the composition of the mantle beneath this craton, using xenolithic material brought to the Earth's surface in these kimberlites. Diamonds and mineral inclusions trapped in diamond at the time of growth are inert to the many processes that modify their host mantle through time. Provided the minerals included in diamond remain unaltered, they represent pristine syngenetic samples of cratonic mantle environments.

This paper presents new data on 576 microdiamonds from seven kimberlite pipes in the Lac de Gras (LDG) area of the central Slave Craton, Canada (Fig. 1). More than half (343) of the microdiamonds are from the DO27 kimberlite pipe. The study focuses on major and trace elements from mineral inclusions, carbon isotopes, nitrogen contents and nitrogen aggregation states, and diamond morphologies. Each of these data sets, when combined, allows us to define the characteristics of diamonds from the different parageneses. The objectives of the study are to determine the nature and origin of microdiamonds from the central Slave Craton, characterise mantle environments in which the diamonds have formed, compare the microdiamonds to macrodiamond populations in the area, and further constrain the tectonic history of the central Slave Craton, with applications to diamond exploration in the area.

At present, there are only a few studies of diamonds from the Slave Craton including diamonds from the central Slave (Chinn et al., 1998; Davies et al., 1999; Stachel et al., 2003; Tappert et al., 2004; Westerlund et al., 2003, 2004), and

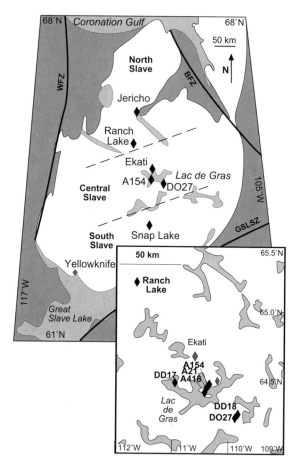

Fig. 1. Maps of the Slave Craton and central Slave Craton (inset) in the LDG area. The Slave Craton map shows major geological boundaries (GSLSZ = Great Slave Lake Shear Zone, WFZ = Wopmay Fault Zone, and BFZ = Bathurst Fault Zone), north, central and southern craton areas, major lakes (shaded) and the location of some kimberlite pipes (black diamonds). The central Slave Craton map shows the seven kimberlites from which diamonds in this study are sampled (black diamonds) and the Ekati mine property (grey diamonds).

southeastern Slave (Pokhilenko et al., 2001, 2004; Promprated et al., 2003). In macrodiamonds from the DO27 kimberlite, Davies and others identified diamonds containing inclusions of transition zone (TZ) and lower mantle (LM) origin ($\sim$ 670 km), and a high proportion of eclogitic ($\sim$ 50%) compared with peridotitic (35%) lithospheric diamonds. This led Griffin et al. (1999) to suggest that plume transport of LM material to the base of the craton was critical to the development of the craton's structure. Other studies have also identified ultra-deep diamonds in Slave kimberlites (Pokhilenko et al., 2001; Tappert et al., 2004), however without an abundance of eclogitic diamonds.

2. Samples and methods

Microdiamonds (343) from the DO27 pipe, weighing 19.1 carats in total, were selected under microscopic examination from approximately 1085 carats of diamonds obtained from bulk sampling by Kennecott Diamonds, Canada. In addition, 233 microdiamonds from the A154 (110), A21 (43), A418 (58), DO18 (6), DD17 (1) and Ranch Lake (15) kimberlites were provided for study (Table 1). The selected DO27 diamonds occur in the 1.18- to 1.7-mm sieve classes. The other diamonds have a sieve range between 0.60 and 1.7 mm; most stones are between 0.85 and 1.18 mm diameter. Although a size of 1 mm or less is the strict definition of a microdiamond, we refer to all the diamonds in this study as microdiamonds. In later sections, we compare them to macrodiamonds from the DO27 kimberlite pipe (Davies et al., 1999).

Surface and morphological features of the diamonds were described from microscopic examination. These observations were lodged in a database with other geochemical characteristics and an image of each diamond.

Syngenetic mineral inclusions (172) were extracted from 155 microdiamonds by crushing. Epigenetic inclusions were extracted from another 56 microdiamonds. Syngenetic inclusions were identified by their crystalline geometric shapes, unaltered appearance and observed absence of cracks around the inclusion whilst in the diamond. Epigenetic inclusions are secondary replacements of primary

Table 1
Syngenetic inclusion abundances and parageneses of microdiamonds from the DO27, A154, A21, A418, DO18, DD17 and Ranch Lake kimberlites (Sf = sulfide, Gt = garnet, Omph = Omphacite, Diop = diopside, fPer = ferropericlase, MgPvk = Mg, Si-perovskite, CaPvk = Ca, Si-perovskite)

	Mono-mineralic		Multi-mineralic	
Ecl.	Sf	31	Cpx, Sf $\pm$ Diop	7
	Gt	23	Gt, Sf	4
	Omph	15	Gt, Cpx $\pm$ Sf	6
	Rutile	9	Sf, SiO$_2$	2
			Sf, rutile	1
			Ilmenite, Sf	1
			Omph, diop	1
Total		78		22
Perid.	Olivine	9	Sf-SiO$_2$	1
	Cr-pyrope	5	Cr-spinel, SiO$_2$	1
	Cr-spinel	8	Gt, Olivine	1
	Sf	7	Olivine, Cr-Diop-Opx	1
	Cr-diopside	2	Olivine, Cr-spinel	1
	Opx	2		
Total		33		5
UD	fPer	9	fPer, CaPvk, MgPvk	1
	CaPvk	1	fPer, SiO$_2$	2
			fPer, silicate	2
			MgSiPvk, fPer, Mg$_2$SiO$_4$	1
			MgPvk, Diop	1
Total		10		7
Total diamonds		121		34

syngenetic inclusions, and internal crack fillings. Mineral inclusions were mounted individually in petropoxy resin 154 on glass slides. The epoxy was hardened for 20 min at 200 °C and polished by hand to expose the inclusions using fine wet and dry sandpaper and diamond paste.

Mineral inclusions were analysed for major elements by electron microprobe (EMP: CAMECA SX50 at Macquarie University) using an accelerating voltage of 15 kV for silicates, and 20 kV for sulfides, a sample current of 20 nA and a beam width of about 3 μm. Standards were a mix of natural and synthetic minerals and matrix corrections follow the Pouchou and Pichoir (1984) method. Trace elements were analysed by laser-ablation ICP-MS [Elan 6000 ICPMS with a Nd-YAG laser (266 nm) at Macquarie University; see Norman et al. 1996, 1998 for methodology, accuracy and standards]. The NIST612 and NIST610 glasses were used as external standards and CaO as the internal standard. The data were processed using the GLITTER software (van Achterbergh et al. 1999). Data were normalised to chondrite using values recommended by GERM (http://www-ep.es.llnl.gov/germ/). Electron microprobe data of mineral inclusions in diamond from occurrences worldwide have been compiled from the literature for comparison and are referred to here as the "worldwide data" (including the following sources: Bulanova, 1995; Chinn, 1995; Daniels and Gurney, 1989; Davies et al. 1999 and refs. therein, this volume; Gurney, 1979; Gurney and Hatton, 1989; Gurney et al., 1984a,b, 1985; Hall and Smith, 1984; Harris et al., 1994; Hervig et al., 1980; Jaques et al., 1989, 1994; McCandless et al., 1989; Meyer, 1987; Meyer and Boyd, 1972; Meyer and McCallum, 1986; Meyer and Svisero, 1975; Moore and Gurney, 1989; Moore et al., 1989; Otter and Gurney, 1989; Prinz et al., 1975; Rickard et al., 1989; Sobolev, 1977; Sobolev et al., 1986a,b; Stachel and Harris, 1997; Stachel et al., 2000a; Tsai et al., 1979; Wilding et al., 1994; Zhang and Meyer, 1989).

Carbon isotopic compositions of 185 microdiamonds were analysed at the Centre for Isotope Studies, CSIRO, North Ryde (see Davies et al., 1999 for methodology). FTIR spectra of 421 microdiamonds were obtained at CSIRO Division of Coal and Energy Technology, North Ryde, using methods described in detail by Davies et al. (1999). The spectra were deconvoluted to calculate N-contents using the absorption coefficients, %IaB estimates and equations of Mendelssohn and Milledge (1995). Nitrogen contents calculated from Ib to IaA spectra used the absorption coefficients of Kiflawi et al. (1994). Diamond characteristics, C-isotope, N-content and N-aggregation state data for inclusion bearing diamonds are presented in electronic supplementary Table 1.

3. Mineral inclusion chemistry

Syngenetic mineral inclusions in microdiamonds in this study occur in the following paragenetic abundances: eclogitic (64%), peridotitic (25%) and ultradeep (11%). Inclusions recovered from the eclogitic paragenesis include low-Ni sulfide, garnet, omphacitic clinopyroxene, rutile, SiO_2 and ilmenite in decreasing order of abundance (Table 1 and electronic supplementary Tables 2 and 3).

3.1. Eclogitic garnet and clinopyroxene inclusions

Thirty-nine garnet inclusions were recovered from 33 microdiamonds as single inclusions or as discrete inclusions coexisting with clinopyroxene and sulfide. Mineral compositions are highly variable with respect to Ca, Mg#, Na and Ti, with grossular (Gr: $100Ca/(Ca + Fe + Mg)$) contents ranging from 14 to 54 (av. 28), Mg# ($100Mg/(Mg + Fe)$) ranging from 38 to 72 (av. 50) (Fig. 2a), Na_2O ranging from 0.05 to 0.39 wt.% and TiO_2 ranging from 0.27 to 1.24 wt.% (Fig. 2b). One diamond (DO27290) contains four large garnets with variable Gr (14–22) and Mg# (50–72).

Eclogitic clinopyroxene inclusions were extracted from 29 microdiamonds. In 14 of the diamonds, clinopyroxene occurs with sulfide and/or garnet. Jadeite contents range from 9 to 39 mol% Jd (av. 25 mol%; Fig. 3) and K_2O contents range from below detection (< 0.04 wt.%) to 1.60 wt.% (av. 0.68 wt.%); the higher values are some of the highest recorded for clinopyroxene inclusions in diamond and may indicate a high-pressure origin (Harlow and Veblen, 1991). Mg#'s for omphacitic inclusions range between 70 and 86 (av. 77). One diamond (VR43706) contains six clinopyroxene inclusions with extremely variable compositions: 13–36 mol% Jd; 0.04 to 1.3 wt.% K_2O and Mg# 73–85 (Fig. 3). Clinopyroxene inclusions have low Na relative to Al, compared to other localities worldwide. K is negatively correlated with Na and may form a coupled substitution, with ratios of $(Na + K)/Al$ being similar to that of K/Al for clinopyroxene inclusions in the worldwide data.

Diopsidic clinopyroxenes with high Ti (0.16 to 1.27; av. 0.55 wt.% TiO_2) occur in five diamonds and in additional diamonds with eclogitic and ultradeep phases. The diopside inclusions are often large (≤ 300 μm), show exsolution textures, and are dark in appearance. These inclusions have been classified as epigenetic, however they are noteworthy because in some cases they may represent syngenetic eclogitic inclusions or decomposed transition zone or lower mantle phases (e.g., $CaSiO_3 + (Fe,Mg)SiO_3$ perovskite;

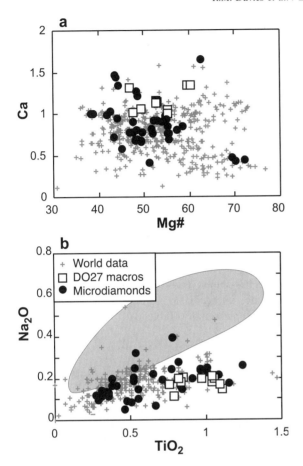

Fig. 2. (a) Ca versus Mg# (cations per formula unit) and (b) TiO_2 versus Na_2O (wt.%) in eclogitic garnet inclusions in this study compared to diamond inclusions from localities worldwide ($n = 522$). The shaded area represents data from the Argyle locality and majoritic garnet inclusions in diamond.

3.2. Geothermometry

Temperature estimates for mineral inclusion entrapment, calculated from garnet–clinopyroxene pairs in four eclogitic diamonds, at a pressure of 50 kbar, yielded values from 1040 to 1300 °C (av. 1182 °C; Ellis and Green, 1979). This assumes that the inclusions are in equilibrium and were trapped in the diamond at the same time. The diamond with the highest temperature (DO27155) has poorly aggregated nitrogen defects (Type IaA).

3.3. Rutile, sulfide and ilmenite

Nineteen rutile inclusions were recovered from nine diamonds. The rutile commonly contains Fe-rich lamellae (≤6.3 wt.% FeO). Fe-rutile in diamond has been reported previously (Prinz et al., 1975; Meyer and McCallum, 1986).

Low-Ni sulfide, the most common inclusion in the eclogitic microdiamonds in this study, typically occurs as central rosettes. Inclusions were extracted from 48 diamonds, 17 of which also contain clinopyroxene, garnet, SiO_2, ilmenite or rutile. Most sulfides are pyrrhotites, with nickel contents up to 7 wt.% (electronic supplementary Table 4). A predominance of sulfide among inclusions in diamond is also observed in studies of diamonds from Siberia

Canil, 1994, or $CaSiO_3+(Fe, Mg)_2SiO_4$ silicate spinel; Irifune et al., 1986).

Garnet–clinopyroxene inclusion pairs occur as discrete inclusions in five microdiamonds (DO27 (2), A418 (2) and A154 (1)) (Table 1). Garnets are Ca-rich (Gr 22–43; av. 32) and clinopyroxenes have low to moderate jadeite contents (7–25, av. 19 mol%). The coexisting compositions have lower Ca in garnet and Na in clinopyroxene compared to diamondiferous eclogite xenoliths from LDG (Gr>30; >50 mol% Jd; Pearson et al., 1999) but have compositions within the range of other eclogite xenoliths from LDG (Aulbach et al., 2003).

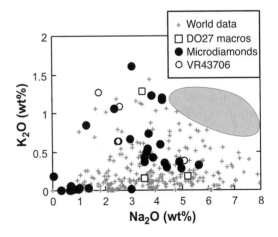

Fig. 3. K_2O versus Na_2O (wt.% oxides) in clinopyroxene inclusions in eclogitic diamonds in this study and other data from localities worldwide ($n = 361$). The shaded area represents data from the Argyle locality.

(e.g., Bulanova et al., 1998) and South Africa (Gurney et al., 1984b).

Ilmenite occurs in one diamond where it coexists with pyrrhotite. The ilmenite is Fe-rich, rather than Mg-rich, which is typical of diamond inclusions in association with the eclogitic paragenesis (Meyer and McCallum, 1986; Chinn, 1995). Fe-rich ilmenite is a common inclusion in diamonds of ultradeep origin from Juina, Brazil (Kaminsky et al., 2001).

3.4. SiO₂ and diamond

SiO_2 inclusions occur as single inclusions (1) and with eclogitic (sulfide (2)), peridotitic (Cr-diopside (1) and Ni-sulfide (1)) and ultradeep (ferropericlase (2)) phases. The SiO_2 inclusions do not comprise equilibrium assemblages with their coexisting peridotitic and ultradeep phases. It is possible that SiO_2 was trapped in diamond at different times to their other phases. Diamond inclusions with cubo-octahedral shapes and variable colour, relative to their host diamond, were observed in nine stones.

3.5. Trace element characteristics of eclogitic garnet and clinopyroxene

Eclogitic garnet inclusions from 10 DO27 diamonds are enriched in heavy rare earth elements (HREE) and depleted in light rare earth elements (LREE) (La_N/Lu_N ratios < 1; Fig. 4, electronic supplementary Table 5). All but two inclusions show a small negative Eu anomaly and one inclusion has a positive Eu anomaly. HREE, MREE and yttrium values are more elevated than most eclogitic garnet diamond inclusions in the literature (Ireland et al., 1994; Taylor et al., 1996; Sobolev et al. 1998; Stachel et al., 2000a; Aulbach et al., 2002).

Six eclogitic clinopyroxene inclusions from four DO27 and one A154 diamond show positive LREE slopes to Nd and shallow negative slopes from Nd to Lu (Fig. 5). Two of the inclusions have weak positive Eu anomalies while one has a weak negative Eu anomaly. Three inclusions have a peak at Ho, for which we have no explanation.

From the average trace element compositions of garnet and clinopyroxene inclusions, a whole rock composition for a theoretical bulk eclogite was calculated. The calculation assumes the modal composition of the eclogite represents a 1:1 mixture of garnet and clinopyroxene, and ignores minor phases in the rock such as rutile. The bulk rock eclogite composition is similar to a typical mid-ocean ridge basalt (NMORB). Eu anomalies observed in garnet and clinopyroxene inclusions are not evident in the bulk eclogite. Negative anomalies for Nb, Zr, Hf and Ti are likely to indicate the unaccounted presence of rutile, which is a common inclusion in these diamonds.

4. Peridotitic inclusions in diamond

Peridotitic inclusions in this study are olivine, sulfide, Cr-spinel, garnet, orthopyroxene and clinopyroxene, in decreasing order of abundance (electronic supplementary Table 6).

4.1. Olivine

Olivine inclusions from 12 diamonds have harzburgitic compositions as defined by their Mg# and Ni contents (Fo 91.3–93.3, av. 92.4; 0.33–0.41 wt.% NiO, av. 0.34; Fig. 6a; Meyer, 1987). Two olivine inclusions in one Ranch Lake diamond have low Ni (0.12, 0.24 wt.% NiO; VR16382) and

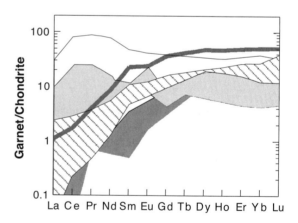

Fig. 4. REE (C1-chondrite normalised) plot of eclogitic garnet inclusions in diamond. An average value for 10 DO27 garnets (black line) is plotted with fields from Kankan (white; Stachel et al., 2000a), Venetia (grey; Aulbach et al., 2002), Udachnaya and Mir (patterned; Ireland et al., 1994; Taylor et al., 1996) and Mir (dark grey; Sobolev et al., 1998).

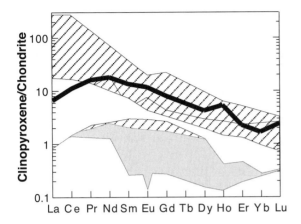

Fig. 5. Average REE compositions of eclogitic clinopyroxene inclusions in diamond compared to data from Kankan (white), Venetia (grey) and Udachnaya and Mir (patterned; see Fig. 4 for references).

and Ti, which behave as incompatible elements, have similar abundances in all garnets.

Similar trends in trace element patterns occur in harzburgitic and lherzolitic garnets from xenoliths and mineral concentrates in the central Slave kimberlites (Griffin et al., 1999; Fig. 8b). However, the harzburgitic garnet diamond inclusions are more enriched in U, Th,

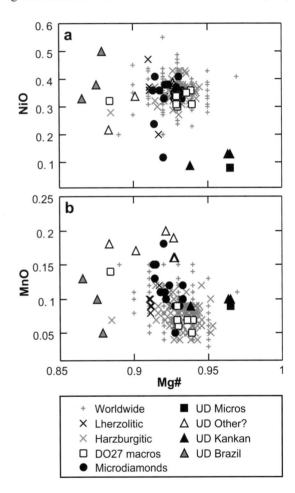

Fig. 6. MgSiO$_4$ inclusions in diamond with upper mantle (olivine) and transition zone or lower mantle origins (UD; inferred to have formed as β- or γ-spinel). Olivine grouped as worldwide, harzburgitic and lherzolitic are from the worldwide diamond inclusion data (n = 380). The harzburgitic and lherzolitic groups are defined by coexisting inclusion phases. 'Other UD' inclusions are from Guinea (Stachel and Harris, 1997), Koffiefontein (Rickard et al., 1989) and Ellendale (Jacques et al., 1994). Inclusions from the latter two localities are inferred to be UD based on composition. The DO27 macrodiamond olivine data are from Davies et al. (1999), the UD Kankan data are from Stachel et al (2000b), and the UD Brazil data are from Harte et al. (1999) and Kaminsky et al. (2001).

unusually high Mn contents (0.15, 0.18 wt.% MnO). The inclusion with lowest Ni also has high Ca (0.15 wt.% CaO). Similar high Ca and Mn contents are observed in an A21 diamond with normal Mg# and Ni (VR16303; Fig. 6b). One Mg$_2$SiO$_4$ inclusion occurring in a microdiamond with inclusion phases of ultradeep origin is also depleted in Ni but does not have elevated Ca or Mn (DO27300). High Ca and Mn in olivine may indicate a dunitic compositions or Ca and Mn-rich micro-inclusions.

4.2. Garnet

Chrome-pyrope garnet inclusions in three from four DO27 microdiamonds have high Cr contents and are strongly subcalcic, indicating depleted harzburgitic compositions (Fig. 7, $T_{Ni} = 800 - 1040°C$; Ryan et al., 1996). One inclusion with lower Cr and higher Ca is lherzolitic ($T_{Ni} = 1300$ °C; DO27216).

All of the peridotitic garnets are HREE depleted, indicative of growth in a highly depleted environment. The lherzolitic garnet (DO27216) is LREE depleted compared to the harzburgitic garnets. One harzburgitic inclusion (DO27069) has a sinuous REE pattern (Fig. 8a). The harzburgitic garnets also are more enriched in incompatible elements than the lherzolitic garnet, with higher contents of Sr, LREE and most HFSE, in particular Zr and Hf (Fig. 8b). Ga

Sr and LREE, the lherzolitic inclusion is depleted in Zr and Hf, and all inclusions are relatively depleted in HREE.

4.3. Cr-spinel, orthopyroxene and Cr-diopside

Compared to Cr-spinel in Siberian and Kaapvaal diamonds, Cr-spinel inclusions from 10 microdiamonds in this study have high Cr# ((100Cr/(Cr + Al)) 86–92) and low Mg# (51–64), similar to inclusions in diamond from other Slave Craton localities (Sobolev et al., 1997; Stachel et al., 2003). One chromite inclusion from the A154 kimberlite contains low Cr and Al (Cr# = 72; VR43369). Another, from Ranch Lake contains 2.4 wt.% TiO_2 and is Fe-rich (Mg# 50; VR16390). A spinel of similar composition was found in a diamond from the Sloan locality (Meyer and McCallum, 1986). Ti- and Fe-rich Cr-spinels occur in a suite of ultradeep diamonds from Brazil (Kaminsky et al., 2001).

Orthopyroxene (Mg# 92.8 and 93.1) occurs in two microdiamonds. One inclusion is in contact with a Cr-diopside suggesting lherzolitic compositions. Cr-diopside occurs in three microdiamonds.

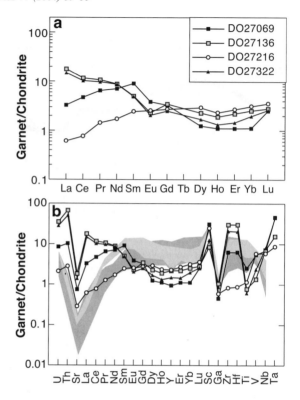

Fig. 8. (a) REE and (b) other trace element compositions of Cr-pyrope inclusions in DO27 microdiamonds. Data in (b) are compared to garnet in harzburgite (light field) and lherzolite xenoliths (dark field) from kimberlites at LDG.

4.4. Sulfide

Peridotitic sulfide inclusions with Ni-rich compositions (8.7–60 wt.% Ni) were analysed from 11 microdiamonds (electronic supplementary Table 3). Most inclusions are monosulfide solid solutions (mss) with variable contents of Ni, Fe, Co and minor Cu ($(Ni,Co,Fe)_{3-x}S_2$). Cobalt-rich inclusions (up to 14.7 wt.% Co) in diamond have not been reported previously. However, they have been found as inclusions in olivine macrocrysts from central Slave kimberlites (Aulbach et al., submitted for publication).

5. Ultradeep inclusions in diamond

Ultradeep inclusions in microdiamonds from this study are ferropericlase (fPer), $MgSiO_3$-perovskite

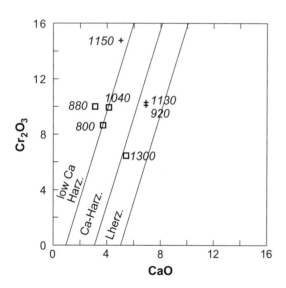

Fig. 7. Cr_2O_3 versus CaO (wt.% oxides) in Cr-pyrope garnets. The lines of Gurney et al. (1984b) separate garnet fields. Squares represent garnets from the DO27 microdiamonds and crosses represent the DO27 macrodiamonds (Davies et al., 1999). Nickel temperatures are plotted with the data points.

(MgPvk), CaSiO$_3$-perovskite (CaPvk), (Fe,Mg)$_2$SiO$_4$ silicate spinel (γ-spinel?), and SiO$_2$ (electronic supplementary Table 7). X-ray identification of the perovskite and silicate spinel structures have not been conducted and thus these primary structures are inferred.

5.1. Ferropericlase (fPer)

Eighteen fPer inclusions were extracted from 15 microdiamonds. Ferropericlase occurs with MgPvk + γ-spinel?, CaPvk + MgPvk and SiO$_2$ in four diamonds. Mg# ranges between 83.4 and 89.6 (av. 85.7), extending the range reported in the DO27 macrodiamonds to higher values (80–87; Davies et al., 1999). The fPer compositions are similar to fPer inclusions from other ultradeep diamond localities, except for Guinea (Stachel et al., 2000b) and Brazil (Harte et al., 1999), which have a broader range of Mg#.

Notable substitutions in the Slave fPer inclusions are Cr$_2$O$_3$ (0.41 to 2.56 wt.%), NiO (1.13 to 1.57 wt.%), Na$_2$O (0.04 to 1.11 wt.%; av. 0.35 wt.%) and MnO (0.14 to 0.38 wt.%; av. 0.18 wt.%). Ni correlates positively with Mg#. Na and Mn correlate negatively with Ni and positively with Cr. Ferropericlase coexisting with SiO$_2$ in two microdiamonds has an Mg# of 83.6 and 85.8. According to experimental data, these compositions are too magnesian to be in equilibrium with SiO$_2$ at 670 km (Ito and Takahashi, 1989).

The association of silicate phases (γ-spinel?, MgPvk, CaPvk and SiO$_2$) with fPer provides evidence that the microdiamonds have a transition zone or lower mantle origin rather than a possible lower pressure origin as suggested for solitary fPer inclusions (Stachel and Harris, 1997; Brey et al., 2004). Microdiamonds in this study that contain single fPer inclusions have similar morphological, C-isotope and IR-characteristics to other ultradeep diamonds (see below), which suggests that they too have an ultradeep origin.

5.2. MgSi-perovskite (MgPvk) and CaSi-perovskite (CaPvk)

Inclusions with MgSiO$_3$ compositions occur in microdiamonds containing fPer + γ-spinel?, fPer + CaPvk, and as a single inclusion in one diamond.

MgPvk differs from enstatite inclusions having lower Ni and higher Mg# (93.4–96.0). Low Al contents (0.96–1.2 wt.% Al$_2$O$_3$) suggest formation in the lowermost pressure range for MgPvk, where majorite has not significantly dissolved into the perovskite structure (Wood, 2000). Low Ni suggests equilibration with ferropericlase into which Ni is strongly partitioned (Stachel et al., 2000b).

CaPvk (CaSiO$_3$) inclusions, extracted from an A21 microdiamond and a DO27 microdiamond containing MgPvk + fPer, are essentially pure CaSiO$_3$ with traces of FeO (0.14 wt.%) and K$_2$O (0.07 wt.%).

5.3. Mg$_2$SiO$_4$ (γ-spinel?)

The Mg$_2$SiO$_4$ inclusion, occurring in a microdiamond with fPer and MgPvk (DO27-300), has a high Mg# (96.5) and low Ni (0.08 wt.% NiO; Fig. 6). Experimental data indicates these three phases can form an equilibrium assemblage in a narrow P–T window at the spinel–perovskite boundary (~ 670 km) (Ito and Takahashi, 1989). However, experiments show that the Mg# of the Mg$_2$SiO$_4$ inclusion should be lower than MgPvk (95.8) and in the order of 90.

Low Ni in the Mg$_2$SiO$_4$ inclusion suggests it could have crystallised in equilibrium with ferropericlase into which Ni is partitioned. The fPer-Mg$_2$SiO$_4$ assemblage is stable in a number of mantle environments including the spinel–perovskite transition (~ 670 km); and in the upper mantle or transition zone as the lower pressure transformation product of MgPvk + fPer.

Brey et al. (2004) suggest the Mg$_2$SiO$_4$ inclusion in this study has characteristics of olivine from the upper mantle, namely because it lacks traces of Al and Cr providing evidence that the phase was formerly MgPvk at higher pressures (Stachel et al., 2000b). To date, we have not been able to identify the high-pressure polymorphs of olivine (β- and γ-spinel) in diamond through chemical or X-ray analysis (Stachel, 2001).

6. Carbon isotopes

Carbon isotope analyses of 185 microdiamonds yield a large range of values (δ^{13}C = − 35.8‰

to $+0.2‰$). Excluding the DO27 data, values for 105 microdiamonds range from $-16.8‰$ to $-2.1‰$ with a strong peak at $-4.7‰$ (Fig. 9a). Microdiamonds from individual kimberlites show a similar isotopic range and distribution to this sample set. Not including the DO27 data, the microdiamonds eclogitic, peridotitic and ultradeep $\delta^{13}C$ range and mode are as follows. $E = -15.0‰$ to $-2.2‰$; $-6.5‰$; $P = -11.5‰$ to $-2.3‰$; $-4.7‰$; $UD = -8.8‰$ to $-2.1‰$; av. $-5.0‰$. C-isotope ranges and modes for the DO27 microdiamonds are: $E = -35.8‰$ to $-3.6‰$, with peaks at $-5‰$, $-12‰$ and $-19‰$; $P = -18.7‰$ to $+0.2‰$; $-4.2‰$. $UD = -14.3‰$ to $-2.5‰$; av. $-5.1‰$ (Fig. 9b). The DO27 microdiamonds show a greater range to more depleted values, however modal values for all groups are similar ($\sim -4.5‰$). The exception is the DO27 eclogitic diamonds with three modes. Carbon isotope analyses of 53 DO27 macrodiamonds show a similar distribution of values (Davies et al., 1999; Fig. 9b).

7. Diamond morphologies

Eclogitic microdiamonds are dominantly brown (72%), followed by colourless (19%), yellow (5%) and opaque (4%). Peridotitic microdiamonds are largely colourless (53%), followed by brown (32%), yellow (10%) and opaque (5%). Equal proportions of the ultradeep microdiamonds are colourless and brown, with minor yellow stones.

Crystal forms of eclogitic microdiamonds are cubo-octahedra (50%), dodecahedra (45%) or octahedra (4%). Dodecahedra consist of resorbed cubo-octahedral or octahedral primary growth forms. Peridotitic diamonds consist of fewer cubo-octahedra (17%) and more octahedra (33%); the rest are dodecahedra. Ultradeep diamonds have resorbed octahedral morphologies (elongate dodecahedra with low-relief glossy surfaces), except for one cubo-octahedral Type I diamond with a ferropericlase inclusion. Aggregates and fragments are common in eclogitic and peridotitic diamonds.

Plastic $\pm$ brittle deformation features are common on the surfaces of the DO27 microdiamonds (75%), similar to the DO27 macrodiamonds. However, deformation is rarely observed on the surfaces of the other Slave microdiamonds (ca. 15%).

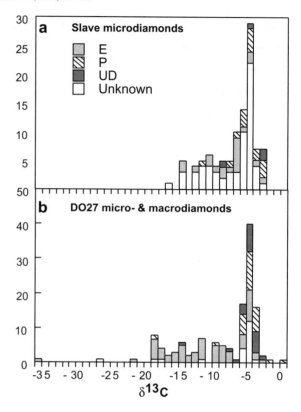

Fig. 9. (a) Carbon isotopic compositions of microdiamonds from the A154, A21, A418, DO18, DD17 and Ranch Lake kimberlites and (b) DO27 micro- and macrodiamonds, with paragenetic fields defined.

The high proportion of brown diamonds however most likely indicates these diamonds are plastically deformed (Robinson et al., 1989).

8. Nitrogen contents and nitrogen aggregation states

Nitrogen is a common impurity substituting in the lattice of diamond (Type I diamond). At high temperatures, nitrogen aggregates from single substitutional atoms (Type Ib) to pairs of N-atoms (Type IaA) and then to four N-atoms arranged around a vacancy (Type IaB). The kinetics of N-aggregation in diamond is experimentally determined (Evans and Harris, 1989 and references therein). However, using these calibrations to extrapolate an age or thermal storage history of a diamond in the mantle is problematic because factors such as deformation and thermal perturbations enhance N-aggregation states. N-contents and N-aggregation characteristics in diamond, calculated from IR-spectral measurements, however is a useful tool for estimating relative diamond ages and mantle storage temperatures, and for defining diamond population groups.

Nitrogen content and aggregation data are summarised in Table 2 and plotted in Fig. 10. In diamonds with a mix of IaA and IaB peaks (Type IaAB), the degree of N-aggregation is given in the notation %IaB where 100% IaB is equivalent to the presence of B-aggregates only. Average N-contents for eclogitic Type I diamonds are about 500 ppm. Eclogitic diamonds are N-free (Type II; 16%), Type IaA–IaB mixes (17%) and Type IaA with no platelet development (67%). Thirty percent of IaA diamonds (all from DO27) show evidence of Ib peaks. Almost half of the peridotitic microdiamonds are Type II. Type I diamonds are Type IaA (av. N = 440 ppm) followed by Type IaAB (av. N = 660 ppm; 46% IaB). The DO27 macrodiamonds consist of more Type IaAB peridotitic and eclogitic diamonds. The ultradeep microdiamonds are Type II except for two Type IaAB diamonds with low N-aggregation (av. 340 ppm; 15% IaB). Rare

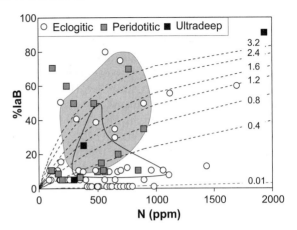

Fig. 10. Nitrogen contents versus nitrogen aggregation states for the Type I microdiamonds in this study. Circled areas show data for the DO27 macrodiamonds (shaded = peridotitic, clear = eclogitic; Davies et al., 1999). The isochron lines (Ga) are calculated for an arbitrary temperature of 1150 °C to show that diamonds with low aggregation states have short mantle residence histories and/or were stored in the mantle at low temperatures.

Type I ultradeep macrodiamonds from DO27 and other localities show advanced N-aggregation states (Davies et al., 1999; Hutchison et al., 1999; Kaminsky et al., 2001). Hydrogen peaks occur in IaA and IaAB eclogitic and peridotitic IaAB and Type II diamonds. CO_2 peaks were detected in ~ 20% of the eclogitic IaA and Type II diamonds.

9. Discussion

A study of microdiamonds from seven kimberlite pipes in the central Slave Craton has combined a number of techniques, including morphological observations, N-defect impurities, C-isotopes and compositional data from mineral inclusions, to define the characteristics of microdiamonds from each paragenesis. This information assists us in identifying the diamonds growth and mantle storage environments, and the timing of their growth relative to the formation history of the craton and kimberlite emplacement. Below we discuss the results from the previous sections, integrating the data obtained from the various techniques.

The dominance of eclogitic microdiamonds in this study (64%) suggests the diamond-bearing rock types beneath the central Slave craton at Lac de Gras are

Table 2
N-content, N-aggregation and other FTIR characteristics for microdiamonds in this study compared to DO27 macrodiamonds

Locality	Slave microdiamonds (%)	DO27 macrodiamonds (%)
FTIR group		
Eclogitic	$n = 69$	$n = 15$
Type II	6 (CO_2)	6
Type IaA	67 (490 ppm; 30% Ib–IaA; CO_2, H)	47 (540 ppm)
Type IaA–IaB	17 (650 ppm; 42% IaB; H)	47 (480 ppm; 23% IaB)
Peridotitic	$n = 31$	$n = 15$
Type II	43 (H)	40
Type IaA	32 (440 ppm)	13 (440 ppm)
Type IaA–IaB	25 (660 ppm; 46% IaB; H)	47 (600 ppm; 40% IaB)
Ultradeep	$n = 16$	$n = 5$
Type II	88	80
Type IaA	–	–
Type IaA–IaB	12 (340 ppm; 15% IaB)	20 (1950; 95% IaB)

Type II = N-free; Type IaA (av. N ppm; < 10% IaB); Type IaA–IaB (av. N ppm; %IaB).

largely eclogitic, whilst peridotitic (25%) and ultra-deep (11%) lithologies comprise a lesser proportion of the diamondiferous mantle. This observation is consistent with the study of macrodiamonds from the DO27 kimberlite (Davies et al., 1999). However, the abundance of eclogitic diamonds is apparently a localized feature. Nearby in the Ekati kimberlite field (e.g. Chinn et al., 1998, Stachel et al., 2003; Tappert et al., 2004), and south at Snap Lake (Pokhilenko et al., 2001), peridotitic diamonds are most common, eclogitic diamonds are fewer and ultradeep diamonds are rare.

9.1. Eclogitic and peridotitic diamonds

Eclogitic microdiamonds in this study show compositional variation in mineral inclusions and C-isotopes to suggest the diamonds formed in variable eclogitic source environments, $P–T$ space and depths within the lithospheric mantle. Both garnet and clinopyroxene inclusions show a wide range in composition. In garnet, Ca and Na contents encompass the compositional field of the DO27 macrodiamonds. The DO27 macrodiamonds are proposed to have formed at the base of the Slave lithosphere ($\sim$ 250 km) based on the occurrence of majoritic garnet and Ca-rich garnet with compositions that correspond to garnets in diamondiferous eclogites from nearby kimberlites with constrained $P–T$ origins (Davies et al., 1999; Pearson et al., 1999). Elevated Na in garnet and K in clinopyroxene in some inclusions suggests higher $P–T$ origins for these diamonds (McCandless and Gurney, 1989; Harlow and Veblen, 1991). Variable in inclusion entrapment temperatures (1040 to 1300 °C) suggests the eclogitic microdiamonds crystallised over a range of depths.

Six eclogitic clinopyroxenes included in one diamond cover almost the full compositional range for all clinopyroxene inclusions. The A154 diamond (VR43706) is an unresorbed cubo-octahedral fragment with a high N-content (960 ppm-N) and low N-aggregation state (IaA). Another cubo-octahedral fragment (DO27290) contains four eclogitic garnets with variable grossular and Mg#. The diamonds show no obvious evidence for growth in multiple events although their cathodoluminescence properties were not observed. The variable inclusion compositions may indicate metasomatic diamond growth (Sobolev

et al., 1998), or they could indicate diamond growth in changing chemical environments.

We favour diamond growth in changing chemical environments. We observe such evidence from other diamonds in this study. That is, cubo-octahedral aggregate growth forms with colourless to brown hues in the various attached crystals; diamonds with peridotitic and ultradeep inclusions and ^{13}C-depleted C-isotopes to suggest eclogitic diamond overgrowths; and Type I ultradeep diamonds with low N-aggregation states suggesting upper mantle diamond overgrowths on ultradeep diamond that is typically Type II or Type I with high N-aggregation states.

The protolith of the eclogitic diamonds may be subducted oceanic crust as shown from trace elements that include Eu anomalies in garnet and clinopyroxene inclusions and an NMORB trace element signature calculated for the bulk rock eclogite. Furthermore, the distribution and range of C-isotope compositions, some of which are exceptionally ^{13}C-depleted, is consistent with a subducted carbon origin (e.g. Kirkley et al. 1991).

The peridotitic microdiamonds in this study have C-isotopes with typical mantle values. Mineral inclusions are largely harzburgitic with minor lherzolitic compositions. The DO27 macrodiamonds also are dominantly harzburgitic (Davies et al., 1999). Diamonds from the Panda and Snap Lake kimberlites however show less depleted peridotitic inclusion compositions (Pokhilenko et al., 2001; Stachel et al., 2003). Lherzolitic inclusion compositions from this study overlap with those from Panda and Snap Lake indicating spatial similarity in the composition of the peridotitic mantle beneath the Slave Craton. Compositional and temperature data from Cr-pyrope garnets (T_{Ni} 800–1300 °C) suggests the peridotitic microdiamonds formed both in the upper ultra-depleted and lower less depleted layers of the central Slave Craton (Griffin et al., 1999).

Nitrogen aggregation characteristics in the eclogitic and peridotitic microdiamonds divide into three groups: Type II, Type IaA, and Type IaAB. Nitrogen contents in the Type I diamonds are largely constant ($\sim$ 500 ppm). The number of diamonds with poorly aggregated nitrogen (IaA with no platelets) is unusual, totaling 67% of eclogitic and 32% of the peridotitic microdiamonds. The IaA diamonds can be roughly distinguished from other diamonds by their cubo-octahedral growth forms. The IaAB and Type II diamonds largely have octahedral growth forms. The high proportion ($\sim 50\%$) of resorbed dodecahedral diamonds in which their primary form cannot be identified limits us from determining if this trend holds for the majority of diamonds.

The data suggests the poorly aggregated IaA diamonds with cubo-octahedral primary forms may be younger than the IaAB and Type II diamonds with octahedral growth forms that possibly formed closer to the time of craton stabilisation. However, Archean Re–Os ages from sulfide inclusions in an A154 eclogitic IaAB diamond ($T_{RD} = 3.03$ Ga, $T_{MA} = 3.35$ Ga; VR16367A; electronic supplementary Table 1; Pearson et al., 2002) and peridotitic IaA diamonds from Panda (Westerlund et al., 2003), suggests both the IaA and IaAB diamonds in this study may record Archean ages.

In addition to the similar Re–Os ages, we find these groups defined by IR-characteristics and growths forms do not hold when we compare their N-contents, mineral inclusion (including major and trace elements) and C-isotope compositions. The data suggests a common eclogitic and peridotitic origin for the respective diamond groups. Furthermore, the strong temperature dependence of N-aggregation in diamond means this is not a robust proxy for identifying unique diamond groups of characteristic features. That is, N-aggregation will be dramatically slower in a diamond stored at low mantle temperatures. For example, a diamond with 500 ppm-N and 5% IaB aggregation yields a mantle storage time of 3 Ga at 1060 °C and only 1 Ma at 1100 °C.

The IaA diamonds in this study may not represent a younger population of diamonds, but may represent diamonds stored in the mantle at low temperature; that is in the narrow low-T range noted above (1060– <1100 °C). Temperatures may be even lower if the effects of plastic deformation, common to most of the diamonds, are considered. Low mantle storage temperatures are consistent with inclusion entrapment temperatures for some, but not all, peridotitic and eclogitic diamonds (800–1300 °C). This suggests mantle storage temperatures are lower than crystallisation temperatures (Stachel et al., 2003; Tappert et al., 2004).

9.2. Formation history of ultradeep diamonds

Microdiamonds with lower mantle inclusions have resorbed octahedral forms with evidence of deformation, mostly Type II IR-characteristics, and C-isotopes with peridotitic mantle compositions. These characteristics are also observed in ultradeep macrodiamonds from DO27 and other localities (e.g. Davies et al., 1999; Hutchison et al., 1999; Stachel et al., 2000b; Kaminsky et al., 2001).

Coexisting lower mantle inclusion assemblages of fPer + MgPvk + CaPvk and fPer + MgPvk + γ-spinel?, and minimal Al-substitution in MgPvk, suggest formation in P–T conditions corresponding approximately to 670 km that defines the upper and lower mantle boundary, and marks the spinel-perovskite transition. The ultradeep microdiamonds formed in bulk rocks of fertile peridotitic composition based on the occurrence of ferropericlase in all diamonds, and the nature of coexisting silicate phases (Fei and Bertka, 1999).

The ultradeep diamonds show evidence for diamond overgrowths in the transition zone and/or upper mantle. This includes the occurrence of an Mg_2SiO_4 inclusion in a diamond containing fPer and MgPvk, which may have been included at lower pressures as olivine, rather than ringwoodite; upper mantle diamond overgrowths could account for two ultradeep Type I diamonds with low N-aggregation. One of these diamonds is a cubo-octahedron with a depleted C-isotope composition ($-14.3‰$), possibly indicating the diamond overgrowth is eclogitic. An alternative explanation for Type IaA ultradeep diamond raises the possibility of lower mantle diamond growth in a proto-kimberlite melt (Haggerty, 1994).

9.3. Synthesis, ultradeep diamonds and eclogites

Plume tectonics has been identified as a key component in the formation of the lithosphere beneath the central Slave Craton (Griffin et al., 1999) that has a layered chemical and thermal structure and an unusual representation of ultradeep diamond. This study confirms the presence of a high proportion of ultradeep diamonds in the central Slave Craton, supporting the role of a plume as a mechanism for the upward transport of diamonds from the lower mantle to the base of the cratonic lithosphere.

Here we show that the eclogitic and peridotitic microdiamonds from the central Slave Craton formed in the lithospheric mantle over a range of depths and chemical environments, most likely early in formation history of the craton. Harzburgitic diamonds may have formed at shallower depths than lherzolitic diamonds, while eclogitic diamonds formed throughout. Low N-aggregation states in a large proportion of diamonds suggest the thermal state of the diamondiferous mantle has been low since diamond formation. We propose diamond growth continued over time in order to explain variations in inclusion chemistries in single diamonds, possible overgrowths of ^{13}C-depleted eclogitic diamond on peridotitic and ultradeep diamond, and Type I ultradeep diamonds with low N-aggregation states.

This and earlier studies show a strong relationship between the occurrence of ultradeep diamond and an abundance of eclogitic diamonds and xenoliths in kimberlites (Davies et al., 1999; Pearson et al., 1999). At nearby localities, where ultradeep diamonds are less common, diamonds of the peridotitic paragenesis are more common (Chinn et al., 1998; Pokhilenko et al., 2001; Stachel et al., 2003), as is typical for most kimberlitic diamond localities worldwide, which lack evidence of having sampled ultradeep mantle. This suggests that an intrinsic relationship exists between the accumulation of basaltic material and eclogitic diamond formation in the cratonic upper mantle in regions of plume activity. Our observation is extended to other lower mantle diamond localities where ultradeep diamonds occur with a high proportion of eclogitic diamonds (e.g. Sao Luiz, Harte et al., 1999; Kaminsky et al., 2001; Sloan, Moore et al., 1986; Otter and Gurney, 1989; Koffiefontein, Rickard et al., 1989; and Kankan, Stachel et al., 2000b).

Acknowledgements

We thank Kennecott Canada Exploration for financial support and access to diamonds. Juanita Bellinger and Kevin Kivi at Kennecott Canada, Tony Vassalo for helpful assistance with FTIR, Anita Andrew and Brad McDonald at CSIRO, Norm Pearson, and Ashwini Sharma at Macquarie. The manuscript has been improved by reviews from Thomas Stachel, Steve Richardson, Ralf Tappert and Larry Taylor. This is publication number 364 form the ARC National Key Centre for the Geochemical Evolution and Metallogeny of Continents (GEMOC) http://www.es.mq.edu.au/gemoc/.

References

Aulbach, S., Stachel, T., Viljoen, K.S., Brey, G.P., Harris, J.W., 2002. Eclogitic and websteritic diamond sources beneath the Limpopo Belt—remnants of an ancient slab? Contrib. Mineral. Petrol. 143, 56–70.

Aulbach, S., Griffin, W.L., Pearson, N.J., O'Reilly, S.Y., Kivi, K., Doyle, B.J., 2003. Origins of eclogites beneath the central Slave Craton. 8th Internat. Kimberlite Conf., Victoria, Canada.

Aulbach, S., Griffin, W.L., Pearson, N.J., O'Reilly, S.Y., Kivi, K., Doyle, B.J., 2004. Constraints on mantle formation and evolution of HSE abundances and Re–Os isotope systematics of sulfide inclusions in mantle xenocrysts. Chem. Geol. (submitted for publication).

Brey, G.P., Bulatov, V., Girnis, A., Harris, J.W., Stachel, T., 2004. Ferropericlase—a lower mantle phase in the upper mantle. Lithos (this vol.).

Bulanova, G.P., 1995. The formation of diamond. In: Griffin, W.L. (Ed.), Diamond Exploration: Into the 21st Century. Spec. Vol. J. Geochem. Explor. 53, 1–23.

Bulanova, G.P., Griffin, W.L., Ryan, C.G., 1998. Nucleation environment of diamonds from Yakutian kimberlites. Mineral. Mag. 62, 409–419.

Canil, D., 1994. Stability of clinopyroxene at pressure–temperature conditions of the transition region. Phys. Earth Planet. Inter. 86, 25–34.

Chinn, I.L., 1995. A study of unusual diamonds from the George Creek K1 kimberlite dyke, Colorado, PhD thesis, Univ. Cape Town, South Africa.

Chinn, I.L., Gurney, J.J., Kyser, K.T., 1998. Diamonds and mineral inclusions from the NWT, Canada. 7th Internat. Kimberlite Conf., Cape Town. Aaddendum, not paginated.

Daniels, L.R.M., Gurney, J.J., 1989. The chemistry of the garnets, chromites and diamond inclusions of Dokolwayo kimberlite, Kingdom of Swaziland. In: Ross, J., et al. (Ed.), Kimberlites and Related Rocks v. 2. Their Mantle/Crust Setting. Spec. Publ. Geol. Soc. Aust., vol. 14, pp. 1012–1021.

Davies, R.M., Griffin, W.L., O'Reilly, S.Y., 1999. Diamonds from the deep: pipe DO27, Slave craton, Canada. In: Gurney, J.J., Gurney, J.L., Pascoe, M.D., Richardson, S.H. (Eds.), The J.B. Dawson Vol. Proc. 7th Internat. Kimberlite Conf. Red Roof Designs, Cape Town, pp. 148–155.

Ellis, D.J., Green, D.H., 1979. An experimental study of the effect of Ca upon garnet–clinopyroxene Fe–Mg exchange equilibria. Contrib. Mineral. Petrol. 71, 13–22.

Evans, T., Harris, J.W., 1989. Nitrogen aggregation, inclusion equilibration temperatures and the age of diamonds. In: Ross, J., et al. (Ed.), Kimberlites and Related Rocks v. 2. Their

Mantle/Crust SettingSpec. Publ. Geol. Soc. Aust., vol. 14, pp. 1002–1006.

Fei, Y., Bertka, C.M., 1999. Phase transitions in the Earth's mantle and mantle mineralogy. In: Fei, Y., Bertka, C.M., Mysen, B.O. (Eds.), Mantle Petrology: Field Observations and High Pressure Experimentation: A tribute to Francis R (Joe) Boyd. Spec. Publ.-Geochem. Soc., vol. 6, pp. 189–207.

Griffin, W.L., Doyle, B.J., Ryan, C.G., Pearson, N.J., O'Reilly, S.Y., Davies, R.M., Kivi, K., Van Achterbergh, E., Natapov, L.M., 1999. Layered mantle lithosphere in the Lac de Gras area, Slave Craton: composition, structure and origin. J. Petrol. 40, 705–727.

Gurney, J.J., 1979. Silicate and oxide inclusions in diamond from Finsch kimberlite pipe. In: Boyd, F.R., Meyer, H.O.A. (Eds.), Kimberlites, Diatremes and Diamonds: Their Geology, Petrology and Geochemistry. AGU, Washington, pp. 1–15.

Gurney, J.J., Hatton, C.J., 1989. Diamondiferous minerals from the Star Mine, South Africa. In: Ross, J. (Ed.), Kimberlites and Related Rocks v. 2. Their Mantle/Crust Setting. Spec. Publ. Geol. Soc. Aust. vol. 14, pp. 1022–1028.

Gurney, J.J., Harris, J.W., Rickard, R.S., 1984a. Minerals associated with diamonds from the Roberts Victor Mine. In: Kornprobst, J. (Ed.), Kimberlites II: The Mantle and Crust–Mantle Relationships. Elsevier, Amsterdam, pp. 25–32.

Gurney, J.J., Harris, R.S., Rickard, R.S., 1984b. Silicate and oxide inclusions in diamonds from the Orapa Mine, Botswana. In: Kornprobst, J. (Ed.), Kimberlites II: The Mantle and Crust–Mantle Relationships. Elsevier, Amsterdam, pp. 3–10.

Gurney, J.J., Harris, J.W., Rickard, R.S., Moore, R.O., 1985. Inclusions in Premier Mine diamonds. Trans. Proc. Geol. Soc. S. Afr. 88, 301–310.

Haggerty, S.E., 1994. Superkimberlites: a geodynamic window to the Earth's core. Earth Planet. Sci. Lett. 122, 57–69.

Hall, A.E., Smith, C.B., 1984. Lamproite diamonds—are they different?. In: Glover, J.E., Harris, P.G. (Eds.), Kimberlite Occurrence and Origin. Publ.-Univ. West Austr. Geol. Dept., vol. 8, pp. 167–212.

Harlow, G.E., Veblen, D.R., 1991. Potassium in clinopyroxene inclusions from diamond. Science 251, 652–655.

Harris, J.W., Duncan, D.J., Zhang, F., Miao, Q., Zhu, Y., 1994. The physical characterisation and syngenetic inclusion geochemistry of diamonds from Pipe 50, Liaoning Province, Peoples Republic of China. In: Meyer, H.O.A., Leonardos, O.H. (Eds.), Diamonds: Characterisation, Genesis and Exploration, Vol. 2, CPRM (Brazil), Spec. Publ. 1A, pp. 106–115.

Harte, B., Harris, J.W., Hutchison, M.T., Watt, G.R., Wilding, M.C., 1999. Lower mantle mineral associations in diamonds from Sao Luiz, Brazil. In: Fei, Y., Bertka, C.M., Mysen, B.O. (Eds.), Mantle Petrology: Field Observations and High Pressure Experimentation: A Tribute to Francis R (Joe) Boyd. Spec. Publ.-Geochem. Soc., vol. 6, pp. 125–153.

Hervig, R.L., Smith, J.V., Steele, I.M., Gurney, J.J., Meyer, O.A., Harris, J.W., 1980. Diamonds: minor elements in silicate inclusions: pressure–temperature implications. J. Geophys. Res. 85, 6919–6929.

Hutchison, M.T., Cartigny, P., Harris, J.W., 1999. Carbon and nitrogen compositions and physical characteristics of transition zone and lower mantle diamonds from Sao Luiz, Brazil. In: Gurney, J.J., Gurney, J.L., Pascoe, M.D., Richardson, S.H. (Eds.), The J.B. Dawson Vol. Proc. 7th Internat. Kimberlite Conf. Red Roof Designs, Cape Town, pp. 372–382.

Ireland, T.R., Rudnick, R.L., Spetsius, S., 1994. Trace elements in diamond inclusions from eclogites reveal link to Archaean granites. Earth. Planet. Sci. Lett. 128, 199–213.

Irifune, T., Sekine, T., Ringwood, A.E., Hibberson, W.O., 1986. The eclogite–garnetite transformation at high pressure and some geophysical implications. Earth Planet. Sci. Lett. 77, 245–256.

Ito, E., Takahashi, E., 1989. Postspinel transformations in the system Mg_2SiO_4–Fe_2SiO_4 and some geophysical implications. J. Geophys. Res. 94, 10637–10646.

Jaques, A.L., Hall, A.E., Sheraton, J.W., Smith, C.B., Sun, S.S., Drew, R.M., Foudoulis, C., Ellingsen, K., 1989. Composition of crystalline inclusions and C-isotopic composition of Argyle and Ellendale diamonds. In: Ross, J. (Ed.), Kimberlites and Related Rocks v. 2, Their Mantle/Crust Setting. Spec. Publ. Geol. Soc. Aust., vol. 14, pp. 966–989.

Jaques, A.L., Hall, A.E., Sheraton, J., Smith, C.B., Roksandic, Z., 1994. Peridotitic planar octahedral diamonds from the Ellendale lamproite pipes, Western Australia. In: Meyer, H.O.A., Leonardos, O.H. (Eds.), Diamonds: Characterisation, Genesis and Exploration, Vol. 2, CPRM (Brazil), Spec. Publ. 1A, pp. 69–77.

Kaminsky, F.V., Zakharchenko, O.D., Davies, R.M., Griffin, W.L., Khachatryan-Blinova, G.K., Shiryaev, A.A., 2001. Superdeep diamonds from the Juina Area, Mato Grosso State, Brazil. Contrib. Mineral. Petrol. 140, 734–753.

Kiflawi, I., Mayer, A.E., Spear, P.M., van Wyk, J.A., Woods, G.S., 1994. Infrared absorption by the single nitrogen and A defect centres in diamond. Phila. Mag. B69, 1141–1147.

Kirkley, M.B., Gurney, J.J., Otter, M.L., Hill, S.J., Daniels, L.R., 1991. The application of C isotope measurements to the identification of the sources of C in diamonds: a review. Appl. Geochem. 6, 477–494.

McCandless, T.E., Gurney, J.J., 1989. Sodium in garnet and potassium in clinopyroxene: Criteria for classifying mantle eclogites. Internal Rept. 10, August 1996, KRG, Univ. Cape Town, South Africa.

McCandless, T.E., Kirkley, M.B., Robinson, D.N., Gurney, J.J., Griffin, W.L., Cousens, D.R., Boyd, F.R., 1989. Some initial observations on polycrystalline diamonds mainly from Orapa. In: Boyd, F.R., Meyer, H.O.A., Sobolev, N.V. (Eds.), Workshop on Diamonds. 28th Internal. Geol. Cong., Carnegie Inst. Washington, D.C., pp. 47–51.

Mendelssohn, M.J., Milledge, H.J., 1995. Geologically significant information from routine analysis of the mid-infrared spectra of diamonds. Int. Geol. Rev. 37, 95–110.

Meyer, H.O.A., 1987. Inclusions in diamond. In: Nixon, P.H. (Ed.), Mantle Xenoliths. Wiley, New York, pp. 501–522.

Meyer, H.O.A., Boyd, F.R., 1972. Composition and origin of crystalline inclusions in natural diamonds. Geochim. Cosmochim. Acta 36, 1255–1273.

Meyer, H.O.A., McCallum, M.E., 1986. Mineral inclusions in diamonds from the Sloan Kimberlites, Colorado. J. Geol. 94, 600–612.

Meyer, H.O.A., Svisero, D.P., 1975. Mineral inclusions in Brazilian diamonds. Phys. Chem. Earth 9, 785–796.

Moore, R.O., Gurney, J.J., 1989. Mineral inclusions in diamond from the Monastery kimberlite, South Africa. In: Ross, J. (Ed.), Kimberlites and Related Rocks v. 2, Their Mantle/Crust Setting. Spec. Publ. Geol. Soc. Aust., vol. 14, pp. 1029–1041.

Moore, R.O., Otter, M.L., Rickard, R.S., Harris, J.W., Gurney, J.J., 1986. The occurrence of moissanite and ferro-periclase in inclusions in diamond. 4th Internat. Kimberlite Conf., Perth, Australia, Geol. Soc. Australia, No. 16, pp. 409–411.

Moore, R.O., Gurney, J.J., Fipke, C.E., 1989. The Development of Advanced Technology to distinguish between diamondiferous and barren diatremes. Open File Rep. Geol. Surv. Canada 2124, pp. 1183.

Norman, M.D., Pearson, N.J., Sharma, A., Griffin, W.L., 1996. Quantitative analysis of trace elements in geological materials by laser ablation ICPMS: instrumental operating conditions and calibration values of NIST glasses. Geostand. Newsl. 20, 247–261.

Norman, M.D., Griffin, W.L., Pearson, N.J., Garcia, M.O., O'Reilly, S.Y., 1998. Quantitative analysis of trace element abundances in glasses and minerals: a comparison of laser ablation ICPMS, solution ICPMS, proton microprobe, and electron microprobe data. J. Anal. At. Spectrom. 13, 477–482.

Otter, M.L., Gurney, J.J., 1989. Mineral inclusions in diamonds from Sloan diatremes, Colorado–Wyoming State Line kimberlite district, North America. In: Ross, J. (Ed.), Kimberlites and Related Rocks v. 2, Their Mantle/Crust Setting. Spec. Publ. Geol. Soc. Aust., vol. 14, pp. 1042–1053.

Pearson, N.J., Griffin, W.L., Doyle, B.J., O'Reilly, S.Y., van Achterbergh, E., Kivi, K., 1999. Xenoliths from kimberlite pipes of the Lac de Gras area, Slave Craton, Canada. In: Gurney, J.J., Gurney, J.L., Pascoe, M.D., Richardson, S.H. (Eds.), The P.H. Nixon Vol. Proc. 7th Internat. Kimberlite Conf. Red Roof Designs, Cape Town, pp. 644–658.

Pearson, N.J., Alard, O., Griffin, W.L., Jackson, S.E., O'Reilly, S.Y., 2002. In situ measurements of Re–Os isotopes in mantle sulfides by laser ablation multicollector-inductively coupled mass spectrometry: analytical methods and preliminary results. Geochim. Cosmochim. Acta 66, 1037–1050.

Pokhilenko, N.P., et al., 2001. Crystalline inclusions in diamonds from kimberlites of the Snap Lake area (Slave Craton, Canada): New evidences for the anomalous lithospheric structure. Dokl. Earth Sci. 380 (7), 806–811.

Pokhilenko, N.P., McDonald, J.A., Sobolev, N.V., Reutsky, V.N., Hall, A.E., Logvinova, A.M., Reimers, L.F., 2004. Crystalline inclusions and C-isotope composition of diamonds from the Snap Lake/King Lake Kimberlite dyke system: evidence for an ultradeep and enriched lithospheric mantle. Lithos (this vol.).

Pouchou, J.L., Pichoir, F., 1984. A new model for quantitative X-ray microanalysis: Part 1. Application to the analysis of homogeneous samples. Rech. Aérosp. 5, 13–38.

Prinz, M., Manson, D.V., Hlava, P.F., Keil, K., 1975. Inclusions in diamonds: garnet lherzolite and eclogite assemblages. Phys. Chem. Earth 9, 797–815.

Promprated, P., Taylor, L.A., Floss, C., Malkovets, V.G., Anand, M., Griffin, W.L., Pokhilenko, N.P., Sobolev, N.V., 2003. Dia-mond inclusions from Snap Lake, NWT, Canada. 8th Internat. Kimberlite Conf., Victoria, Canada (http://www.venuewest.com/8IKC/s3post.htm#3.P10).

Rickard, R.S., Harris, J.W., Gurney, J.J., Cardoso, P., 1989. Mineral inclusions in diamonds from Koffiefontein mine. In: Ross, J. (Ed.), Kimberlites and Related Rocks v. 2, Their Mantle/Crust Setting, vol. 14. Geol. Soc. of Australia Spec. Publ, pp. 1054–1062.

Robinson, D.N., Scott, J.A., Van Nierkerk, A., Anderson, V.G., 1989. The sequence of events reflected in the diamonds of some southern African kimberlites. In: Ross, J. (Ed.), Kimberlites and Related Rocks v. 2, Their Mantle/Crust Setting. Spec. Publ. Geol. Soc. Aust., vol. 14, pp. 990–1000.

Ryan, C.G., Griffin, W.L., Pearson, N.J., 1996. Garnet geotherms: a technique for derivation of P–T data from Cr-Pyrope garnets. J. Geophys. Res. 101, 5611–5625.

Sobolev, N.V., 1977. Deep-Seated Inclusions in Kimberlites and the Problems of the Composition of the Upper Mantle. AGU, Washington. 279 p.

Sobolev, N.V., Galimov, E.M., Smith, C.B., Yefinova, E.S., Maltsev, K.A., Hall, A.E., Usova, L.V., 1986a. Comparative characteristics of morphology, inclusions and carbon isotope composition of diamonds in alluvial deposits of the King George River and lamproite deposits of Argyle. Geol. Geofiz. 12 (in Russian).

Sobolev, N.V., Yefimova, E.S., Shemanina, E.I., 1986b. Crystalline inclusions in alluvial diamonds from the Urals, USSR. Abstr.-Geol. Soc. Austr. 16, 429.

Sobolev, N.V., Yefimova, E.S., Reimers, L.F., Zakharchenko, O.D., Makhin, A.I., Usova, L.V., 1997. Mineral inclusions in diamonds of the Arkhangelsk kimberlite province. Russ. Geol. Geophys. 38, 379–393.

Sobolev, N.V., Snyder, G.A., Taylor, L.A., Keller, R.A.Yefimova, E.S., Sobolev, V.N., Shimizu, N., 1998. Extreme chemical diversity in the mantle during eclogitic diamond formation: evidence form 35 garnet and 5 pyroxene inclusions in a single diamond. Int. Geol. Rev. 40, 567–578.

Stachel, T., 2001. Diamonds from the asthenosphere and transition zone. Eur. J. Mineral. 13, 883–892.

Stachel, T., Harris, J.W., 1997. Diamond precipitation and mantle metasomatism—evidence from the trace element chemistry of silicate inclusions in diamonds from Akwatia, Ghana. Contrib. Mineral. Petrol. 129, 143–154.

Stachel, T., Brey, G.P., Harris, J.W., 2000a. Kankan diamonds (Guinea): I. From lithosphere down to the transition zone. Contrib. Mineral. Petrol. 140, 1–15.

Stachel, T., Harris, J.W., Brey, G.P., Joswig, W., 2000b. Kankan diamonds (Guinea): II. Lower mantle inclusion parageneses. Contrib. Mineral. Petrol. 140, 16–27.

Stachel, T., Harris, J.W., Tappert, R., Brey, G.P., 2003. Peridotitic diamonds from the Slave and the Kaapvaal cratons—similarities and differences based on a preliminary data set. Lithos 71, 489–503.

Tappert, R., Stachel, T., Harris, J.W., Brey, G.P., 2004. Mineral inclusions in diamonds from the Panda Kimberlite, S.P., Canada. 8th International Kimberlite Conference, Extended Abstracts.

Taylor, L.A., Snyder, G.A., Crozaz, G., Sobolev, V.N., Yefimova, E.S., Sobolev, N.V., 1996. Eclogitic inclusions in diamonds:

evidence of complex mantle processes over time. Earth Planet. Sci. Lett. 142, 535–551.

Tsai, H.M., Meyer, H.O.A., Moreau, J., Milledge, H., 1979. Mineral inclusions in diamond: premier, Jagersfontein, and Finsch Kimberlites, South Africa, and Williamson Mine, Tanzania. In: Boyd, F.R., Meyer, H.O.A. (Eds.), Kimberlites, Diatremes, and Diamonds: Their Geology, Petrology, and Geochemistry-Proc. 2nd Int. Kimb. Conf., vol. 1, pp. 16–26.

van Achterbergh, E., Ryan, C.G., Griffin, W.L., 1999. GLITTER: on-line interactive data reduction for the laser ablation ICP-MS microprobe. Proc. 9th V.M. Goldschmidt Conference, Cambridge, MA, 305.

Westerlund, K.J., Hauri, E.H., Gurney, J.J., 2003. FTIR absorption and stable nitrogen and carbon isotope microanalysis of Mid-Archean diamonds from the Panda Kimberlite, Slave Craton. 8th Internat. Kimberlite Conf., Victoria, Canada..

Westerlund, K.J., Shirey, S.B., Richardson, S.H., Gurney, J.J., Harris, J.W., 2004. Re–Os Isotope Systematics of Peridotitic Diamond Inclusion Sulfides from the Panda Kimberlite, Slave Craton. Lithos (this vol.).

Wilding, M.C., Harte, B., Fallick, A.E., Harris, J.W., 1994. Inclusions chemistry, carbon isotopes and nitrogen distribution in diamonds from the Bultfontein Mine, South Africa. In: Meyer, H.O.A., Leonardos, O.H. (Eds.), Diamonds: Characterisation, Genesis and Exploration, Vol. 2, CPRM (Brazil), Spec. Publ. 1A, pp. 116–125.

Wood, B.J., 2000. Phase transformations and partitioning relations in peridotite under lower mantle conditions. Earth Planet. Sci. Lett. 174, 341–354.

Zhang, A., Meyer, H.O.A., 1989. Inclusions in diamonds from Chinese kimberlites. In: Boyd, F.R., Meyer, H.O.A., Sobolev, N.V. (Eds.), Workshop on Diamonds. 28th Internal. Geol. Conf., Carnegie Inst. Washington, pp. 1–3.

Available online at www.sciencedirect.com

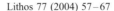

Lithos 77 (2004) 57–67

www.elsevier.com/locate/lithos

Crystalline inclusions and C isotope ratios in diamonds from the Snap Lake/King Lake kimberlite dyke system: evidence of ultradeep and enriched lithospheric mantle

N.P. Pokhilenko[a,b,*], N.V. Sobolev[a], V.N. Reutsky[a], A.E. Hall[b], L.A. Taylor[c]

[a] *United Institute of Geology, Geophysics and Mineralogy, Novosibirsk 630090, Russia*
[b] *Diamondex Resources Ltd., 1410-650 W. Georgia Street, Vancouver, BC, Canada V6B 4N8*
[c] *Planetary Geosciences Institute, Department of Earth and Planetary Sciences, University of Tennessee, Knoxville, TN 37996-1410, USA*

Received 27 June 2003; accepted 14 December 2003
Available online 19 May 2004

Abstract

U-type paragenesis inclusions predominate (94.7%) among the crystalline inclusion suite of 115 diamonds ($-4+2$ mm) obtained from the recently discovered Snap Lake/King Lake (SKL) kimberlite dyke system, Southern Slave, Canada. The most common inclusions are olivine (90) and enstatite (22). Sulfide, Cr-pyrope, chromite and Cr-diopside inclusion are less abundant (15, 10, 5 and 1, respectively). Results of the inclusion composition study demonstrate the following. (a) The relatively enriched character of the mantle parent rocks of the U-type diamonds. The average Mg# of olivine inclusions is 92.1, and of enstatite inclusions average 93.3. CaO content in Cr-pyrope inclusions is relatively high (3.73–5.75 wt.%). (b) Four of ten U-type Cr-rich pyrope inclusions contain a majoritic component up to 16.8 mol.% which requires pressures of ~ 110 kbar. Carbon isotopes compositions for 34 diamonds with U-type inclusions have a $\delta^{13}C$ range from $-3.2‰$ to $-9‰$ with a strong peak around $-3.5‰$. This is much heavier than the ratios of U-type diamonds from Siberia and South Africa (~ 4.5‰). Diamonds with olivine inclusions can be divided into two groups based on their $\delta^{13}C$ values as well as the Mg# and Ni/Fe ratio in the olivines. Most show a narrow range of $\delta^{13}C$ values from $-3.2‰$ to $-4.8‰$ (average $-3.72‰$) and have olivine inclusions with Mg# less than 92.3 and relatively high Fe/Ni ratios. A second group is characterized by a much wider variation of C isotope composition ($\delta^{13}C$ varies from $-3.8‰$ to $-9.0‰$, average $-5.97‰$), and the olivine inclusions having a higher Mg# (up to 93.6) and relatively low Fe/Ni ratios. This difference in the C isotope composition may have several explanations: (a) peculiarities of asthenosphere degassing coupled with an abnormal thickness of lithosphere; (b) the abnormal thickness and enriched character of lithospheric mantle; (c) involvement of subducted C of crustal origin in the processes of the diamond formation. The presence of subcalcic Cr-rich majorite (up to 17 mol.%) pyropes of low-Ca harzburgite paragenesis among the crystalline inclusion suite of SKL diamonds is strong evidence for the existence of diamondiferous depleted peridotite in lithospheric mantle at depth near 300 km beneath Southern Slave area and is postulated to be one of the main reasons for the much heavier C isotope composition of SKL U-type diamonds in comparison with those from Siberian and South African kimberlites.
© 2004 Elsevier B.V. All rights reserved.

Keywords: Crystalline inclusions; Carbon isotopes; Diamonds; Lithospheric mantle; Slave Craton

* Corresponding author. Institute of Mineralogy and Petrography, 3 Koptyuga Avenue, Novosibirsk 630090, Russia. Tel.: +7-3832-332-108; fax: +7-3832-332-792.
  E-mail address: chief@uiggm.nsc.ru (N.P. Pokhilenko).

0024-4937/$ - see front matter © 2004 Elsevier B.V. All rights reserved.
doi:10.1016/j.lithos.2004.04.019

1. Introduction

The kimberlite dyke system of the Snap Lake/King Lake (SKL) area, Southern Slave Craton, Canada was discovered by Winspear Resources, a junior Canadian exploration company, in 1997. In terms of kimberlite characteristics and geometry of ore-bodies it represents a new type of large primary diamond deposits (Pokhilenko et al., 1998, 2000). The age of SKL kimberlite emplacement was estimated as near 540 Ma (Agashev et al., 2001), much older compared to kimberlites of Central and Northern Slave (47–84 and − 170 Ma, respectively) and similar to neighboring Kennady Lake area kimberlites (Carlson et al., 1999). Petrochemical, geochemical and isotope features of SKL kimberlites are different from both Group 1 and Group 2 kimberlites (Agashev et al., 2001). An abnormally wide range of Cr_2O_3 (up to 17 wt.%) in garnets, and a very high proportion of high-Cr chromites (up to 27% of chromites with >62 wt.% Cr_2O_3) together with other specific geochemical and isotope characteristics of the kimberlites allowed to propose an abnormal thickness of lithosphere beneath the SKL area (Pokhilenko et al., 2000, 2001; Agashev et al., 2001; McLean et al., 2001).

A preliminary study of the SKL diamonds and their inclusions provided additional evidence for a relatively undepleted and abnormally thick predominantly peridotitic lithosphere beneath the Southern Slave (Pokhilenko et al., 2001). It showed a significant increase of ^{13}C proportion in the SKL U-type diamonds isotope composition (Reutsky et al., 2002) compared to world-wide data for diamonds of the same type of paragenesis (Van Heerden et al., 1995), and this feature can also be connected with abnormal thickness of the SKL area lithosphere. Further evidence for an increasing thickness of the lithosphere beneath the Slave Craton from north to south is based on a comparative study of peculiarities of the pyrope composition distribution in the craton kimberlites (Grütter et al., 1999), and the petrology and geochemistry of upper mantle xenoliths in these kimberlites (Kopylova and Garo, 2001; Kopylova et al., 1999).

A preliminary data suggesting about the abnormally thick and relatively undepleted lithospheric mantle beneath the Southern Slave, and especially beneath the Snap Lake area, is an important additional information on this region lithosphere structure and com-

position. A valuable data of this kind for the most deep-seated portion of the lithosphere cross-section can be obtained from a complex study of diamonds from the Southern Slave kimberlites. This paper includes results of detail study of crystalline inclusions and C isotope composition of representative collection of the SKL diamonds.

2. Methods

Major elements were obtained with a CAMEBAX electron microprobe using a wide range of natural minerals and synthetic glass standards at the Analytical Center of the United Institute of Geology, Geophysics and Mineralogy (UIGGM), Siberian Branch of Russian Academy of Sciences, Novosibirsk, Russia. Most inclusions were recovered for analysis after burning of their host diamond crystals, but more than 20 diamonds were polished to expose their inclusions. Special attention was paid to the analysis of the majorite-bearing subcalcic Cr-pyropes: 7–23 high precision analyses each were performed for the accurate determination of Si excess.

Some of inclusions were re-analyzed for a major-elements with a CAMECA SX-50 electron microprobe at the University of Tennessee. As it was shown before (Taylor et al., 1996) there are no signs of any influence of the burning of diamonds on composition of their garnet and Cpx inclusions.

C-isotopes were also analyzed at the UIGGM Analytical Center using a Finigan-MAT Delta mass spectrometer with a typical precision of $\delta^{13}C < 0.2\%o$. Methods and techniques of diamond crystal preparation were described in detail previously (Reutsky et al., 1999). Isotopic composition is given in delta notation (parts per thousand) as the deviation from the VPDB standard.

3. Distribution and composition of mineral inclusions in SKL diamonds

Mineral inclusions from 115 diamonds from the SKL dyke system have been studied. All studied diamonds are − 4 + 2 mm in size, 75% of them are colorless octahedrons, the rest are crystals of transitional habit (octahedral to dodecahedra) and spinel twins. Rare crystals have a pale smoky and greenish tint.

Inclusions of U-type paragenesis predominate (94.7%) and most common are olivines. They were found in 90 diamonds. In 65 crystals olivine is present alone with up to five separate inclusions in a single diamond, and in 25 diamonds olivine coexist with other U-type paragenesis mineral. Enstatites were found in 22 crystals, and in 10 it is present as a monomineral inclusion (up to three separate inclusions within a single diamond crystal). Sulfide, Cr-pyrope, chromite and Cr-diopside are less abundant (in 15, 10, 5 and 1 diamond, respectively). U-type mineral associations in single diamonds are: olivine + sulfide (9), olivine + enstatite (8), Cr-pyrope + olivine (4), chromite + sulfide (2), Cr-pyrope + olivine + enstatite (2), chromite + olivine + enstatite (1), olivine + chromite (1), Cr-pyrope + sulfide (1) and enstatite + Cr-diopside (1).

E-type inclusions are represented by clinopyroxene (5 diamonds), yellow-orange garnet (2) and sulfide (1) with the following associations in single diamonds: garnet + clinopyroxene + sulfide (1) and garnet + clinopyroxene (1).

3.1. Olivine inclusions

Olivine inclusions are characterized by a relatively narrow range in composition (Fig. 1; Table 1). About 80% have Mg# from 91.8 to 93.0, and one inclusion has Mg# less than 91 (SL_5-44 = 90.7). Olivine inclusions in two diamonds have Mg# higher than 93 (SL_3-20 = 93.6; SL_5-14 = 95.2). NiO varies in from 0.29 to 0.43 wt.% but more than 80% have NiO contents from 0.33 to 0.37 wt.%. There is unusual tendency of the negative correlation of NiO content and Mg# (Fig. 1).

Cr_2O_3 varies from 0.01 to 0.07 wt.%; CaO-from 0.01 to 0.05 wt.%. The maximum compositional variation of separate multiple olivine inclusions within a single diamond is 0.3 in their Mg#, and usually they are less than 0.3, and only in two samples these differences reach 0.5: SL_5-51 (five separate olivine inclusions, Mg# varies from 92.2 to 92.7); SL-00/133 (two inclusions, Mg#: 92.2–92.7).

3.2. Enstatite inclusions

Thirty separate enstatites from 20 diamonds (one to three within a single diamond) mostly have a narrow

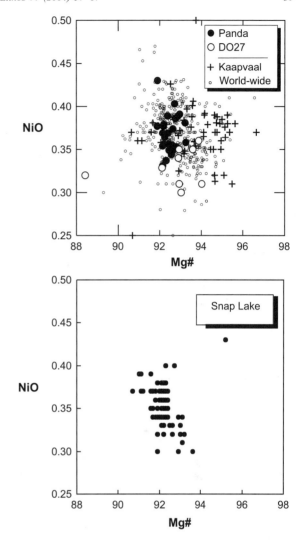

Fig. 1. Mg# vs. NiO for olivine inclusions from diamonds of Panda and DO27 (Slave Craton, Canada), the Kaapvaal Craton and World-wide (all from Stachel et al., 2003), and from the SKL diamonds.

range of composition: in 15 diamonds FeO content ranges from 4.4 to 4.8 wt.%, and only in inclusions from five diamonds these values are outside of this range (3.01, 3.97, 3.98, 4.34 and 6.29 wt.%, respectively; Figs. 2 and 3; Table 2). Al_2O_3 contents of enstatites inclusions from 18 diamonds vary from 0.4 to 0.6 wt.%. The lowest value (0.35 wt.%) was obtained for an enstatite-Cr-diopside intergrowth (SL5-100, Table 2), which also has lowest Cr_2O_3 content (0.26 wt.%) but highest-CaO (0.42 wt.%)

Table 1
Representative analyses of olivine inclusions in SKL diamonds

	Sl$_5$-14	Sl$_3$-20	Sl$_5$-64	Sl$_3$-28	Sl-00/133	Sl$_3$-31	Sl$_5$-8	Sl$_3$-12	Sl$_3$-37	Sl$_5$-44
SiO$_2$	41.7	41.0	41.7	40.9	40.6	41.7	41.3	40.9	41.6	41.0
Cr$_2$O$_3$	0.07	0.02	0.02	0.02	0.03	0.07	0.02	0.03	0.07	0.07
FeO	4.70	6.34	6.69	7.13	7.13	7.51	7.83	7.83	7.42	9.09
MnO	0.05	0.10	0.09	0.09	0.08	0.11	0.12	0.10	0.11	0.13
MgO	52.4	52.1	51.2	51.7	51.0	50.2	50.0	50.8	50.0	49.5
CaO	0.01	0.01	0.01	0.02	0.02	0.03	0.02	0.01	0.03	0.01
NiO	0.43	0.29	0.32	0.30	0.40	0.34	0.38	0.34	0.37	0.37
Total	99.4	99.9	100.0	100.2	99.3	100.0	99.7	100.0	100.1	100.2
Mg#	95.2	93.6	93.2	92.8	92.7	92.3	91.9	92.0	91.8	90.7

and FeO (6.29 wt.%). Sample SL5-96 has the highest Al$_2$O$_3$ (0.68 wt.%) and Cr$_2$O$_3$ (0.49 wt.%) content and the lowest CaO (0.16 wt.%) and FeO (3.98 wt.%). There is definite negative correlation of Al$_2$O$_3$ and FeO contents (Fig. 2).

The majority of the enstatites have Cr$_2$O$_3$ and CaO contents between 0.3 and 0.4 wt.%. Practically all the multiple (2–3) separate enstatite inclusions within a single diamond have very similar composition. Only in one diamond crystal three enstatite inclusions show a range in Mg# from 92.9 to 93.7.

3.3. Garnet inclusions

Thirteen garnet inclusions from 12 diamond crystals were analyzed (Table 3, Figs. 4 and 5), 11 of them are Cr-pyropes (from 10 diamonds), and two are E-type garnets. Cr$_2$O$_3$ content in Cr-pyrope

inclusions varies from 7.71 to 12.8 wt.%, CaO from 3.73 to 5.75 wt.%, TiO$_2$ from <0.01 to 0.19 wt.%, and Mg# from 82.9 to 86.7. One E-type garnet (SL$_5$-

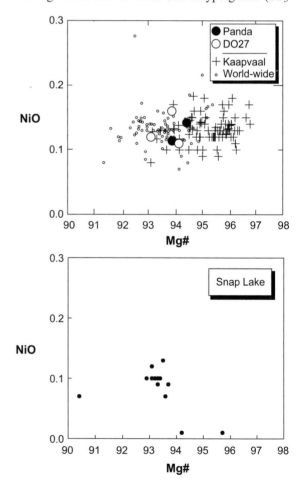

Fig. 3. Mg# vs. NiO for enstatite inclusions in diamonds from Panda and DO27 (Slave Craton, Canada), the Kaapvaal Craton and World-wide (all from Stachel et al., 2003), and from the SKL diamonds.

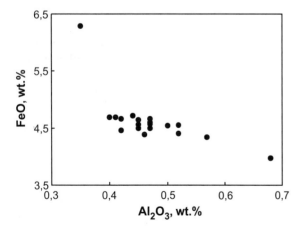

Fig. 2. Al$_2$O$_3$ vs. FeO for enstatite inclusions from the SKL diamonds: there is negative correlation of Al and Fe contents.

Table 2
Representative analyses of enstatite and clynopyroxene inclusions in SKL diamonds

	Sl₅-14	Sl₅-96	Sl₃-41	Sl₅-8	Sl₃-12	Sl₅-100		Sl₅-86	Sl₅-6	Sl₅-52
SiO_2	57.2	57.3	58.7	57.4	58.4	58.0	55.4	55.0	55.9	55.1
TiO_2	<0.01	<0.01	<0.01	<0.01	0.01	<0.01	<0.01	0.49	0.45	0.48
Al_2O_3	0.32	0.68	0.57	0.42	0.41	0.35	0.82	7.42	8.52	8.66
Cr_2O_3	0.46	0.49	0.34	0.30	0.33	0.26	1.36	0.02	0.09	0.04
FeO	3.01	3.98	4.34	4.66	4.69	6.29	2.53	8.18	6.73	5.64
MnO	0.09	0.08	0.12	0.12	0.12	0.15	0.12	0.10	0.12	0.06
MgO	38.3	36.4	35.7	36.0	36.1	33.4	16.7	9.10	11.8	9.27
CaO	0.26	0.16	0.35	0.37	0.35	0.42	21.2	15.0	9.20	14.1
Na_2O	0.04	0.02	0.04	0.06	0.06	0.03	0.42	4.16	5.16	4.53
K_2O	–	–	–	–	–	–	0.71	0.03	0.23	1.37
NiO	n.d.	n.d.	0.07	0.10	0.10	0.07	n.d.	n.d.	n.d.	
Total	99.7	99.1	100.2	99.4	100.6	99.0	99.3	99.5	98.2	99.3
Mg#	93.4	94.2	93.6	93.2	93.2	90.4	92.2	66.5	75.8	74.6

86) has very low Mg# (38.8) combined with high TiO_2 (1.44 wt.%), CaO (10.6 wt.%) Na_2O (0.38 wt.%). The second (SL₅-6) is practically twice in Mg# (65.5), has less than 1/3 TiO_2 (0.44 wt.%) and CaO (3.03 wt.%), but comparable Na_2O (0.33 wt.%). Both have Si excess (3.061 and 3.050, respectively) corresponding to a majorite component of 6.1 and 5.0 mol%, respectively.

The majorite component is highest in two Cr-rich, high-Mg, subcalcic pyrope inclusions (SL3-31 and SL3-30, Table 3). Excess Si is 3.116 and 3.168 corresponding to 11.6 and 16.8 mol% majorite, respectively). Excess of Si in the structural formulae of these Cr-pyropes perfectly corresponds to the deficit of Al+Cr (1.755 and 1.590).

3.4. Clinopyroxene inclusions

Clinopyroxene inclusions were found in 6 SKL diamonds, and in five they are E-type clinopyroxenes (Table 2). They are omphacites (jadeite component content from 24.9 to 36.2 mol%, Mg# from 66.5 to 75.8) with high TiO_2 contents (up to 0.49 wt.% and most of all high K_2O: two discrete inclusions in a single diamond (SL₅-52) contain 1.27 and 1.37 wt.% K_2O and the omphacite in diamond SL₅-92 contains 1.05 wt.% of K_2O. The highest K_2O content in the studied omphacite inclusion is more than extreme values for the clinopyroxene inclusions in diamonds of the Guaniamo area, Venezuela, and less than in diamonds from the Argyle pipe (Fig. 6).

Table 3
Representative analyses of garnet inclusions in SKL diamonds

Sample	Sl-00/133	Sl₅-8	Sl₅-5	Sl₅-2	Sl₃-12	Sl₃-5	Sl₃-3	Sl₃-31 $\bar{x}_{23}(\delta)$	Sl₃-30 $\bar{x}_7(\delta)$	Sl₅-6	Sl₅-86
Paragenesis	U-type Cr-pyropes									E-type garnets	
SiO_2	41.2	41.1	42.0	41.5	41.3	40.5	40.9	42.2 (0.28)	42.3 (0.22)	42.1	40.1
TiO_2	0.05	<0.01	0.06	0.06	0.13	0.16	0.03	0.19 (0.01)	0.06 (0.01)	0.44	1.44
Al_2O_3	16.7	16.3	17.2	16.2	15.2	13.9	14.7	12.3 (0.12)	9.46 (0.09)	21.7	19.9
Cr_2O_3	7.71	8.01	8.37	9.03	10.2	11.2	11.5	11.8 (0.19)	12.8 (0.16)	0.07	0.02
FeO	7.38	6.46	6.06	6.38	6.5	6.52	21.2	6.52 (0.09)	7.64 (0.18)	15.5	20.2
MnO	0.33	0.29	0.28	0.32	0.32	0.32	0.28	0.32 (0.02)	0.33 (0.02)	0.34	0.36
MgO	20.0	21.5	22.2	21.5	21.1	19.9	21.2	21.1 (0.17)	21.2 (0.10)	16.6	7.19
CaO	5.75	4.68	3.73	4.39	4.76	5.91	4.44	4.68 (0.10)	5.11 (0.07)	3.03	10.6
Na_2O	0.02	0.02	<0.01	0.03	0.03	0.02	0.02	0.03 (0.02)	0.01 (0.01)	0.33	0.38
Total	99.1	99.0	99.9	99.4	99.5	98.4	99.4	99.1 (0.32)	99.0 (0.21)	100.1	100.2
Mg#	82.9	85.5	86.7	85.7	85.2	84.4	85.9	85.2	83.2	65.5	38.8
MJ. mol%	2.2	–	2.3	1.0	0.9	0.8	0.4	11.6	16.8	5.0	6.1

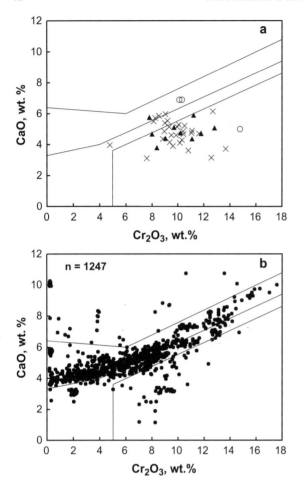

Fig. 4. Cr_2O_3 vs. CaO (plot after Sobolev et al., 1973) for Cr-pyrope inclusions from the SKL diamonds (solid triangles), Panda Pipe diamonds (×, from Stachel et al., in press), and DO27 Pipe diamonds (open circles, from Davies et al., 1999) (a), and Cr-pyropes from concentrate of the SKL kimberlite (b).

A single inclusion of U-type clinopyroxene in diamond SL$_5$-100 has a quite unusual composition: K_2O (0.71 wt.%) is significantly higher than Na_2O (0.42 wt.%). This feature makes the inclusion unique among previously studied U-type clinopyroxene inclusions in diamonds.

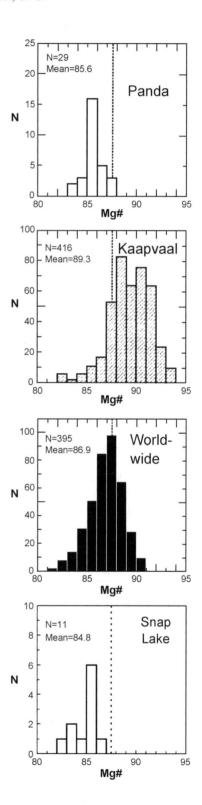

Fig. 5. Histogram showing the distribution of Mg# for Cr-pyrope inclusions from diamonds of the Panda pipe, Kaapvaal Craton, World-wide data (all from Stachel et al., 2003), and from the SKL diamonds.

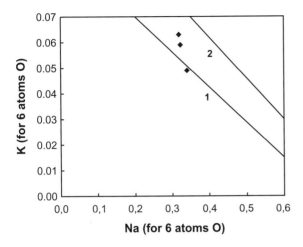

Fig. 6. Na vs. K in molecules of K-enriched clinopyroxene inclusions in the SKL diamonds (see Table 2 for analyses). Lines 1 and 2 bound estimated solubility limits of K in clinopyroxenes from Guaniamo diamonds (Venezuela), and Argyle diamonds (Western Australia), respectively (Sobolev et al., 1998).

3.5. Mg-chromite inclusions

Chromite inclusions were identified in six diamonds: in one sample as monomineral inclusion, in two samples in association with sulfides, in one in association with separate olivines, and in another as intergrowth with olivine and enstatite (SL$_5$-14). Quite unusual is the presence of relatively low-Cr inclusions in two diamonds (SL$_5$-64-55.6 wt.% Cr$_2$O$_3$ and SL$_5$-104-59.1 wt.% Cr$_2$O$_3$, Table 4). Two separate inclusions in diamond SL$_5$-64 are significantly different in composition (SL$_5$-64a and SL$_5$-64b) with 55.6 and 64.4 wt.% Cr$_2$O$_3$, 12.5 and 4.79 wt.% Al$_2$O$_3$, − 14.3 and 13.8 wt.% MgO, respectively. An abnormal difference in Cr$_2$O$_3$ and Al$_2$O$_3$ contents in chromite inclusions from a single diamond makes this situation unique worldwide diamond makes this situation unique worldwide (Sobolev and Yefimova, 1998; Bulanova, 1995) and demonstrate a significant evolution of the upper mantle matter composition along the diamond growth history. All chromite inclusions have low TiO$_2$ content (0.05–0.11 wt.%).

4. Carbon isotope composition

Carbon isotopes have been determined for 34 diamonds with U-type inclusions (olivine, enstatite), one

with E-type inclusions (garnet + cpx), and four without inclusions (Table 5). The E-type diamond has $\delta^{13}C = −$ 13.4‰. Most of U-type diamonds have a narrow range of $\delta^{13}C$ − 3.2‰ to − 5.1‰ (37 diamonds) with a strong peak at − 3.5‰ to − 4.5‰ (Fig. 7), and only one diamond has $\delta^{13}C$ of − 9.0‰. An average for the U-type SKL diamonds is − 4.02‰. This is definitely heavier compared to U-type diamonds from Siberia and South Africa (around − 4.5‰) (Fig. 7).

The predominate diamonds with the narrow range of $\delta^{13}C$ from − 3.2‰ to − 4.8‰ (n = 24, average − 3.72‰) have olivine inclusions with Mg# less than 92.3 and relatively high Fe/Ni ratios (Reutsky et al., 2002). Diamonds with higher $\delta^{13}C$ but also some from the heavier group have higher Mg# of their olivines (up to 93.6) and relatively low Fe/Ni ratios (Fig. 8).

5. Discussion

The small proportion of E-type mineral inclusions in SKL diamonds, and the dominance of depleted peridotites point to a mostly harzburgitic lithospheric mantle beneath the SKL area at depth corresponding to the diamond stability field. Also previous studies of the Slave Craton kimberlites, diamond inclusions and upper mantle xenoliths point to a significant increase in lithosphere thickness toward the south of the craton which may reach 300 km at time of kimberlite emplacement (Cambrian) (Pokhilenko et al., 1998, 2000, 2001;

Table 4
Representative analyses of chromite inclusions in SKL diamonds

	Sl$_5$-64a	Sl$_5$-64b	Sl$_5$-104	Sl$_3$-14	Sl$_5$-14
SiO$_2$	0.26	0.24	0.10	0.11	0.09
TiO$_2$	0.09	0.11	0.05	0.08	0.11
Al$_2$O$_3$	12.5	4.79	11.2	7.37	7.40
Cr$_2$O$_3$	55.6	64.4	59.1	62.6	63.3
FeO	15.5	15.3	14.5	15.7	13.6
MnO	0.12	0.19	0.14	0.20	0.17
MgO	14.3	13.8	14.9	13.8	14.4
NiO	0.13	0.13	0.13	0.11	0.11
Total	98.5	99.0	100.1	100.0	99.2
Mg#	62.2	61.6	64.7	61.0	65.4
Ca/(Ca + Al)	74.9	90.0	78.0	85.1	85.2

Table 5
Carbon isotope composition of the SKL diamonds

Sample	$\delta^{13}C$, ‰	Mineral inclusions (type of paragenesis)	Sample	$\delta^{13}C$, ‰	Mineral inclusions (type of paragenesis)
SL_5-21/00	− 3.2	Ol + Enst (U)	SL_5-40/00	− 3.8	Ol (U)
SL_5-25/00	− 3.2	Ol (U)	SL_3-13/00	− 3.9	Ol (U)
SL_5-36/00	− 3.2	Ol (U)	SL_5-37/00	− 3.9	Ol + Enst (U)
SL_5-22/00	− 3.3	Ol (U)	SL_5-49/00	− 3.9	Ol (U)
SL_5-38/00	− 3.3	Ol (U)	SL_3-7/00	− 4.0	Ol (U)
SL_5-42/00	− 3.3	Enst (U)	SL_5-31/00	− 4.0	Ol + Enst (U)
SL_5-58/00	− 3.3	Ol (U)	SL_5-13/00	− 4.2	Ol (U)
SL_5-60/00	− 3.3	Ol (U)	SL_5-45/00	− 4.2	Ol (U)
SL_5-95/00	− 3.3	−(?)	SL_5-94/00	− 4.2	Ol (U)
SL_3-27/00	− 3.4	Enst (U)	SL_5-74/00	− 4.3	Ol (U)
SL_3-3/00	− 3.5	Pyr (U)	SL_5-61/00	− 4.5	Ol + Enst (U)
SL_5-76/00	− 3.5	Ol (U)	SL-00/47	− 4.8	−(?)
SL_3-18/00	− 3.6	Ol (U)	SL_5-24/00	− 4.8	Ol (U)
SL_5-59/00	− 3.6	Ol + Enst (U)	SL_5-32/00	− 4.9	Ol (U)
SL_3-4/00	− 3.7	Ol (U)	SL_5-39/00	− 4.9	Enst (U)
SL_3-12/00	− 3.7	Ol + Enst + Pyr (U)	SL-00/10	− 5.1	−(?)
SL_3-29/00	− 3.7	Ol (U)	SL_3-20/00	− 5.1	Ol (U)
SL_5-8/00	− 3.7	Ol + Enst + Pyr (U)	SL_3-28/00	− 9.0	Ol (U)
SL_5-30/00	− 3.7	Ol (U)	SL_5-86/00	− 13.9	Gt + Cpx (E)
SL-00/12	− 3.8	−(?)			

Agashev et al., 2001). Upper mantle xenoliths studies allow to conclude that lithospheric thickness was 160–190 km in the northern part of the Slave Craton (Kopylova et al., 1999), near 200 km in the central Slave (Pearson et al., 1999), and at a minimum 230 km in the southern part (Kennady Lake area) of the Slave Craton (Kopylova and Garo, 2001). Also evaluation of the garnet compositions in Slave Craton kimberlites demonstrate a progressive increase of maximum Cr_2O_3 content in pyropes from northern to southern kimberlites of the craton (Grütter et al., 1999).

Presence of high-Cr high-Mg subcalcic pyrope inclusions in SKL diamonds is definitely related to diamondiferous depleted ultramafic rocks of the lithospheric mantle, and SKL diamonds were at least in part formed at depths exceeding 300 km. From experiments modeling natural ultramafic systems, pressures of at least 110 kbar are required to achieve dissolution of 16–17 mol% majorite component in a magnesian garnet (Irifune, 1987, Fig. 9), and such pressures correspond to depth over 300 km. Other indications that some SKL diamonds were formed at very high pressures are: (a) cpx inclusions with high K_2O in both E-type (up to

1.37 wt.%) and U-type (0.71 wt.%) paragenesis: the values which are among the highest recorded for diamond inclusions worldwide (Fig. 6); (b) both E-type garnets contain significant Na_2O, and one of them (SL_5-6) has Na content (0.046 for 12 atoms of O) significantly higher than Ti (0.024 for 12 atoms of O). Furthermore absence of P (< 0.001 for 12 atoms of O) suggests that the reaction $R^{2+}Al \leftrightarrow NaSi$ may occur accompanied by partitioning of Si into octahedrally coordinated sites (Sobolev and Lavrent'ev, 1971). Thus about 1/3 of all analyzed U-type garnets from SKL diamonds demonstrate the presence of majorite component. This is in contrast with the data for diamond inclusions from Siberian kimberlites where only few microdiamonds contain majoritic garnet inclusions (Sobolev et al., 2004).

The lithospheric mantle beneath the SKL area is depleted but not to such an extent as, e.g., beneath the Kaapvaal Craton or the Siberian Craton as shown by: (a) by relatively low Mg# of olivine, enstatite and pyrope inclusions as well as by relatively high CaO contents of Cr-pyrope inclusions (average = 4.6 wt.%); b) by the very uniform composition of olivines and enstatites (Figs. 1–3), as well as a

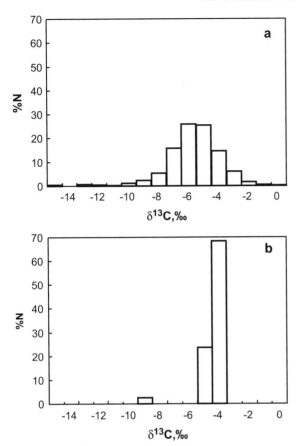

Fig. 7. Histogram shoving distribution of C isotope composition of U-paragenesis diamonds worldwide (Van Heerden et al., 1995) (a), and U-paragenesis diamonds of SKL kimberlites (b).

in CO_2 toward the surface in the lithospheric mantle. It means that diamonds of the most deep-seated origin will be enriched by ^{13}C if compare with less deep-seated ones.

The unusual distribution of $\delta^{13}C$ and the heavy nature the carbon isotope composition of one group of SKL diamonds may be due to a number of reasons: (a) peculiarities of asthenosphere degassing causing fractionation of carbon isotopes during the oxidation of CH_4 producing decrease of role of ^{13}C with decrease of depth inside the lithospheric mantle cross-section (Galimov, 1991) coupled with abnormal thickness of the SKL region lithosphere; (b) abnormal thickness and relatively enriched character of the SKL area lithospheric mantle, and the second can be related with increase of ^{13}C role in diamond C isotope composition (Van Heerden et al., 1995); (c) involve of subducted C of crustal origin in processes of diamond formation (Sobolev and Sobolev, 1980).

Formation of subcalcic Cr-pyrope inclusions with up to 17 mol% of a majorite component is firm evidence of existence of diamondiferous depleted peridotites within lithospheric mantle at depth not less than 300 km beneath the SKL area, and this feature may be the most important among the main reasons of

comparative high abundance of enstatite inclusions in SKL diamonds.

As it was shown in the Galimov's model (Galimov, 1991) the rising carbon-bearing fluid of asthenosphere origin is reduced and consists mainly of methane, and its carbon isotope composition is around $\delta^{13}C$-4.6‰. Methane oxidizes during its moving to the Earth surface with generation of H_2O and CO. A coefficient of carbon fractionation between the CO_2 and CH_4 in the diamond stability field is 1.004 (Galimov, 1991), and their carbon isotope composition will be related to proportion of these components. So, at $CO_2/CH_4 = 0.1$ $\delta^{13}C$ of CO_2 will be around $-1‰$ and CH_4 $\sim -5‰$, and at $CO_2/CH_4 = 0.5$ $\delta^{13}C$ will be $\sim -3‰$ and $-7‰$, respectively. Thus it is possible to expect a decrease of ^{13}C

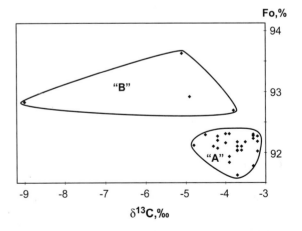

Fig. 8. Carbon isotope composition of SKL diamonds vs. Fo content of their olivine inclusions. "A"-group of diamonds ($n = 24$) with narrow variation of C-isotope composition and relatively low Fo olivine inclusions; "B"-diamonds ($n = 4$) with wide variation of C-isotope composition and relatively high-Fo olivine inclusions.

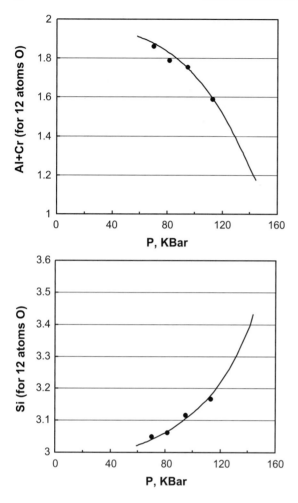

Fig. 9. Distribution of Si and Al+Cr in the structural formulae of majorite bearing garnet inclusions in the SKL diamonds. Calibration curves from (Irifune, 1987).

much heavier C isotope composition of the SKL kimberlite diamond population.

Acknowledgements

Tom Stachel is gratefully thanked for instructive comments on an early version of this manuscript. Gerhard Brey and Bill Griffin are acknowledged for a thoughtful and very constructive review. Authors are thankful to Winspear Diamonds for providing the diamond samples, and to Diamondex Resources for financial support of studies. This study was also supported by Russian Foundation for Basic Research (Grant No. 01-05-65166).

References

Agashev, A.M., Pokhilenko, N.P., McDonald, J.A., Takazawa, E., Vavilov, M.A., Sobolev, N.V., Watanabe, T., 2001. A unique kimberlite-carbonatite primary association in the Snap Lake dyke system, Slave Craton: evidence from geochemical and isotopic studies. The Slave-Kaapvaal Workshop, Ext. Abstr. Vol., Merrickville, Ontario, Canada.

Bulanova, G.P., 1995. The formation of diamond. J. Geochem. Explor. 53, 1–23.

Carlson, J.A., Kirkley, M.B., Thomas, E.M., Hillier, W.D., 1999. Recent Canadian kimberlite discoveries. In: Gurney, J.J., Gurney, J.L., Pascoe, M.D., Richardson, S.H. (Eds.), The J.B. Dawson Volume, Proc. 7th Int. Kimberlite Conf., Red Roof Design, Cape Town, pp. 81–89.

Davies, R.M., Griffin, W.L., Pearson, N.J., Andrew, A.S., Doyle, B.J., O'Reilly, S.Y., 1999. Diamonds from the Deep: Pipe DO-27, Slave Craton, Canada. In: Gurney, J.J., Gurney, J.L., Pascoe, M.D., Richardson, S.H. (Eds.), Proc. 7th Int. Kimberlite Conf., Red Roof Design, Cape Town, vol. 1, pp. 148–155.

Galimov, E.M., 1991. Isotope fractionation related to kimberlite magmatism and diamond formation. Geochim. Cosmochim. Acta 55, 1697–1708.

Grütter, H.S., Apter, D.B., Kong, J., 1999. Crust–mantle coupling: evidence from mantle derived xenocrystic garnets. In: Gurney, J.J., Gurney, J.L., Pascoe, M.D., Richardson, S.H. (Eds.), Proc. 7th Int. Kimberlite Conf., Red Roof Design, Cape Town, vol. 1, pp. 307–312.

Irifune, T., 1987. An experimental investigation of the pyroxene–garnet transformation and its bearing on the constitution of the mantle. Earth Planet. Sci. Lett. 45, 324–336.

Kopylova, M.G., Garo, G., 2001. Lithospheric terranes of the Slave Craton: contrasting North and South. The Slave-Kaapvaal Workshop, Ext. Abstr. Vol., Merrickville, Ontario, Canada.

Kopylova, M.G., Russell, J.K., Cookenboo, H., 1999. Mapping the lithosphere beneath the North Central Slave Craton. In: Gurney, J.J., Gurney, J.L., Pascoe, M.D., Richardson, S.H. (Eds.), Proc. 7th Int. Kimberlite Conf., Red Roof Design, Cape Town, vol. 1, pp. 468–479.

McLean, R.C., Pokhilenko, N.P., Hall, A.E., Luth, R., 2001. Pyropes and chromites from kimberlites of the Snap Lake area, Southeast Slave Craton: garnetization reaction of depleted peridotites at extremely deep levels of the lithospheric mantle. The Slave-Kaapvaal Workshop, Ext. Abstr. Vol., Merrickville, Ontario, Canada.

Pearson, N.J., Griffin, W.L., Doyle, B.J., O'Reilly, S.Y., van Achtenbergh, E., Kivi, K., 1999. Xenoliths from kimberlite pipes of the Lac de Gras area, Slave Craton, Canada. In: Gurney, J.J., Gurney, J.L., Pascoe, M.D., Richardson, S.H. (Eds.), Proc. 7th Int. Kimberlite Conf., Red Roof Design, Cape Town, vol. 2, pp. 307–312.

Pokhilenko, N.P., McDonald, J.A., Melnyk, W., Hall, A.E., Shimizu, N., Vavilov, M.A., Afanasiev, V.P., Reimers, L.F., Irvin,

J., Pokhilenko, L.N., Vasilenko, V.B., Kuligin, S.S., Sobolev, N.V., 1998. Kimberlites of Camsell Lake field and some features of structure and composition of lithosphere roots of southeastern part of Slave Craton, Canada. 7th Int. Kimberlite Conf., Ext. Abstr. Vol., Red Roof Design, Cape Town, pp. 699–701.

Pokhilenko, N.P., Sobolev, N.V., Cherny, S.D., Mityukhin, S.I., Yanygin, Y.T., 2000. Pyrope and chromite from kimberlites of the Nakyn Field (Yakutia) and the Snap Lake region (Slave Craton, Canada): evidence for anomalous lithospheric structure. Dokl. Earth Sci. 372, 356–360.

Pokhilenko, N.P., Sobolev, N.V., McDonald, J.A., Hall, A.E., Yefimova, E.S., Zedgenizov, D.A., Logvinova, A.M., Reimers, L.F., 2001. Crystalline inclusions in diamonds from kimberlites of the Snap Lake area (Slave Craton, Canada): new evidences for the anomalous lithospheric structure. Dokl. Earth Sci. 380, 806–811.

Reutsky, V.N., Logvinova, A.M., Sobolev, N.V., 1999. Carbon isotope composition of the polycrystslline diamond aggregates with chromite inclusions from Mir kimberlite pipe Yakutia. Geokhimiya 11, 1186–1191 (in Russian).

Reutsky, V.N., Pokhilenko, N.P., Hall, A.E., Sobolev, N.V., 2002. Polygeneity of diamonds from kimberlites of the Snap Lake area (Slave Craton, Canada): the results of olivine inclusions and C isotope composition study. Dokl. Earth Sci. 386 (7), 791–794.

Sobolev, N.V., Lavrent'ev, Y.G., 1971. Isomorphic sodium admixture in garnets formed at high pressures. Contrib. Mineral. Petrol. 31, 1–12.

Sobolev, N.V., Yefimova, E.S., 1998. Compositional variations of chromite inclusions as indicators of the zonation of diamond crystals. Dokl. Akad. Nauk 358, 649–652.

Sobolev, V.S., Sobolev, N.V., 1980. New proof on very deep subsidence of eclogitized crustal rocks. Dokl. Akad. Nauk 250, 683–685.

Sobolev, N.V., Lavrent'ev, Y.G., Pokhilenko, N.P., Usova, L.V., 1973. Chrome-rich garnets from the kimberlites of Yakutia and their parageneses. Contrib. Mineral. Petrol. 40, 39–52.

Sobolev, N.V., Yefimova, E.S., Channer, D.M. DeR., Anderson, P.F.N., Barron, K.M., 1998. Unusual upper mantle beneath Guaniamo, Guyana shield, Venezuela: evidence from diamond inclusions. Geology 26, 971–974.

Sobolev, N.V., Logvinova, A.M., Zedgenizov, D.A., Seryotkin, Y.V., Yefimova, E.S., Taylor, L.A., 2004. Mineral inclusions in microdiamonds and macrodiamonds from kimberlites of Yakutia: a comparative study. Lithos 77, 225–242 (this volume). doi:10.1016/j.lithos.2004.04.001

Stachel, T., Harris, J.W., Tappert, R., Brey, G.P., 2003. Peridotitic inclusions in diamonds from the Slave and Kaapvaal cratons-similarities and differences based on a preliminary data set. Lithos 71, 489–503.

Taylor, L.A., Snyder, G.A., Crozaz, G., Sobolev, V.N., Yefimova, E.S., Sobolev, N.V., 1996. Eclogitic inclusions in diamonds: evidence of complex mantle processes over time. Earth Planet. Sci. Lett. 142, 535–551.

Van Heerden, L.A., Gurney, J.J., Deines, P., 1995. The carbon isotopic composition of harzburgitic, lherzolitic, websteritic and eclogitic paragenesis diamonds from southern Africa: a comparison of genetic models. S. Afr. J. Geol. 98 (2), 119–125.

Available online at www.sciencedirect.com

SCIENCE DIRECT®

Lithos 77 (2004) 69–81

LITHOS

www.elsevier.com/locate/lithos

ELSEVIER

Multiple-mineral inclusions in diamonds from the Snap Lake/King Lake kimberlite dike, Slave craton, Canada: a trace-element perspective

Prinya Promprated[a], Lawrence A. Taylor[a,*], Mahesh Anand[a], Christine Floss[b], Nikolai V. Sobolev[c], Nikolai P. Pokhilenko[c]

[a]Department of Earth & Planetary Sciences, Planetary Geosciences Institute, University of Tennessee, Knoxville, TN 37996, USA
[b]Laboratory for Space Sciences, Washington University, St. Louis, MO 63130, USA
[c]Institute of Mineralogy and Petrography, Russian Academy of Sciences-Siberian Branch, Novosibirsk 630090, Russia

Received 27 June 2003; accepted 14 December 2003
Available online 9 June 2004

Abstract

Multiple inclusions of minerals in diamonds from the Snap Lake/King Lake kimberlites of the southeastern Slave craton in Canada have been analyzed for trace elements to elucidate the petrogenetic history of these inclusions, and of their host diamonds. As observed worldwide, the harzburgitic-garnet diamond inclusions (DIs) possess sinusoidal REE patterns that indicate an early depletion event, followed by metasomatism by LREE-enriched, HREE-depleted fluids. Furthermore, these fluids appear to contain appreciable concentrations of LILE and HFSE, based on the increasing abundances of these elements in the olivine inclusion that occurs at the outer portion of a diamond compared to that near the core. The compositions of these fluids are probably a mixture of hydrous-silicic melt, carbonatitic melt, and brine, similar to the compositions of micro-inclusions in diamonds reported by Navon et al. (2003). Comparison between the compositions of majoritic and normal harzburgitic garnets shows that the former are more depleted in terms of major/minor elements (higher Cr#) but significantly more enriched in the REE (up to ~ 10×). This characteristic may indicate the higher susceptibility for metasomatic enrichment of previously more depleted garnets. Garnets of eclogitic paragenesis show strong LREE-depleted patterns, whereas the coexisting omphacite inclusion has relatively flat light- and middle-REE but depleted HREE. Whole-rock reconstruction from coexisting garnet and omphacite inclusions indicates that the protolith of these inclusions was probably the extrusive section of an oceanic crust, subducted beneath the Slave craton.
© 2004 Elsevier B.V. All rights reserved.

Keywords: Snap Lake; Diamond inclusions; Trace element; REE; In situ analysis; Slave craton

1. Introduction

Mineral inclusions in diamonds are generally accepted to be pristine samples of the mantle, because they are protected from retrograde reactions and re-equilibration with the surrounding environment by the extremely strong and chemically inert host diamonds. Unraveling the petrogenetic history of diamond inclusions (DIs) is crucial to understanding the processes involved in diamond formation, as well as the evolution of the mantle beneath a craton. Understanding of the nature of the mantle beneath the Slave craton,

* Corresponding author. Tel.: +1-865-974-6013; fax: +1-865-974-6022.
E-mail address: lataylor@utk.edu (L.A. Taylor).

0024-4937/$ - see front matter © 2004 Elsevier B.V. All rights reserved.
doi:10.1016/j.lithos.2004.04.009

Northwest Territories, Canada, was limited in the past largely because of lack of direct samples from the mantle. Within the last decade, however, the Slave craton has become the "epicenter" of diamond exploration and mining, including a series of kimberlite dikes; one such dike is highly diamondiferous (Pokhilenko et al., 1998; Carlson et al., 1999). This Snap Lake/King Lake diamondiferous dike has been dated at approximately 540 Ma (Agashev et al., 2001). Limited data on diamonds and diamond inclusions from this kimberlite dike have been presented by Pokhilenko et al. (2001, 1998, 2000).

Pokhilenko et al. (in press) summarize, in a companion paper, the present state of knowledge on Snap Lake/King Lake (SKL) diamonds, their carbon isotope composition, the nature of inclusions in these diamonds (DIs), and their major- and minor-element chemistry. From this DI suite, we have selected for detailed trace-element study those inclusions that were recovered as multiple inclusions by the burning of 11 diamonds. In addition, we present new major- and trace-element data of DIs analyzed in situ on polished surfaces of selected diamonds. This investigation is an attempt to provide new insights into diamond petrogenesis and the nature of the mantle beneath the SE Slave craton.

2. Analytical methods

Mineral inclusions were liberated from diamonds both by crushing and burning, as well as by polishing host diamonds exposing inclusions. Twenty-four inclusions (10 garnet—Gt; 4 clinopyroxene—Cpx; 10 orthopyroxene—Opx) from the sample set described by Pokhilenko et al. (in press) were recovered by the "burning" technique. Fourteen additional olivine (Ol), Opx, and Cpx inclusions were examined from five diamonds using an in situ technique (Taylor et al., 2000). As shown earlier by Taylor et al. (1996) in a comparative study of pyroxene inclusions liberated both by diamond burning versus simply crushing, the pyroxene from different fragments of the same diamond showed no differences in major- and trace-element contents—i.e., the burning process did not modify the mineral composition. During the present study, inclusion grains released by burning, as well as minerals exposed on polished

surfaces of diamonds, were mounted in epoxy resin for analysis.

With the in situ technique, the diamonds were polished so as to expose the encapsulated inclusion(s) for analysis with EMP and ion microprobe. Cathodoluminescence (CL) images of the polished surfaces were initially obtained, in an attempt to correlate the inclusion chemistry with the growth features of diamonds. As observed in this study and in the companion paper Pokhilenko et al. (in press), most inclusions showed no petrographic or chemical signs of alteration—e.g., chemical zonation, as determined by electron microprobe analyses. We believe that the chemical characteristics of the DIs reported here are those at the instant of the diamond capture and encapsulation—i.e., the DIs are pristine.

Major-element compositions of DIs were determined with a fully automated CAMECA SX-50 electron microprobe at the University of Tennessee. The EMP analytical conditions employed an accelerating potential of 15 kV, 30-nA beam current, a 5-μm beam size, and 20-s counting time. All data were corrected for matrix effects using the CAMECA PAP procedure.

Trace-element analyses of the diamond inclusions were performed with the modified CAMECA IMS-3f ion microprobe at Washington University. Details of the experimental procedures are described by Zinner and Crozaz (1986a,b), Fahey et al. (1987), Alexander (1994), and Hsu (1995). Detection limits are variable, depending on the element and phase being analyzed, but may be as low as a few ppb in favorable cases. The precision of the trace-element measurements is limited primarily by the number of counts obtained. Errors (1σ) are generally less than 10%, but may range up to 50% for some of the REEs present in low concentrations.

3. Chemistry of the mineral inclusions

3.1. Garnet

Chondrite-normalized REE (rare earth element) concentrations of harzburgitic-garnet inclusions from two diamonds display sinusoidal patterns that peak at Nd and reach the minimum at Ho or Er (Fig. 1A). The

REE patterns are identical, within error, for garnets from the two individual diamonds. Some differences in the chemistry of garnets from these two diamonds should be noted. First, the garnet inclusions in dia-

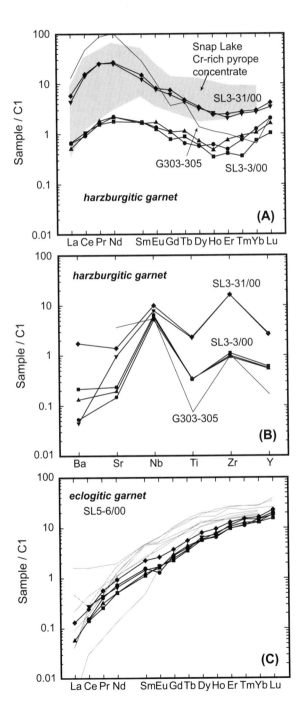

Fig. 1. Chondrite-normalized REE patterns of garnet DIs from the Snap/King Lake kimberlites. (A) The characteristic sinusoidal REE patterns of harzburgitic garnets. Notice the higher REE abundance in "majoritic" garnets (SL3-31/00), compared to the normal ones (SL3-3/00). A majoritic garnet diamond inclusion from Ghana (G303-305; Stachel and Harris, 1997) is also plotted for comparison. The shaded area represents the range of REE patterns of garnet concentrates from SKL kimberlites (Pokhilenko et al., 2001). (B) Plot of LILE and HFSE. (C) LREE-depleted patterns typical for low-Ca garnets. Lines represent DIs from Siberia (Taylor et al., 1996; Sobolev et al., 1998).

mond SL3-31/00 contain significant majoritic components as much as 3.072 pfu Si (Table 1), corresponding to the pressure of around 8 GPa (Irifune, 1987), which suggests the greater depth of origin of this diamond compared to diamond SL3-3/00, whose garnet has no discernable Si excess. Second, garnets in SL3-31/00 diamond are distinctly more enriched in the REEs (up to an order of magnitude), compared to those in SL3-3/00. Third, the REE-elevated garnets in the former are also Cr-rich, with the Cr# (100Cr/Cr + Al) around 40 compared to ~34 in garnets from the latter diamond (Table 1). Moreover, both LILE and HFSE of garnets in diamond SL3-31/00 are noticeably higher, with an exception for Ba in one inclusion (Fig. 1B).

The REE patterns of the multiple eclogitic-garnet inclusions from diamond SL5-6/00 display a strong depletion in LREEs (Fig. 1C), all with identical REE patterns, within errors. Such LREE-depleted patterns are typical for low-Ca eclogitic-garnet DIs from other world localities (e.g., Taylor et al., 1996; Sobolev et al., 1998; Stachel et al., 2000).

3.2. Clinopyroxene

Four omphacitic clinopyroxenes were recovered from three Snap Lake/King Lake diamonds, one of which coexists with five eclogitic garnets in diamond SL5-6/00. Chondrite-normalized REE patterns of the Cpx inclusions are distinctly different between the two K_2O-rich clinopyroxenes versus the two with low K_2O: (1) the K_2O-rich clinopyroxenes have a steeply LREE-enriched pattern with a hump at Pr, and (2) the K_2O-poor pyroxenes have relatively flat light-middle REEs, with negative slopes towards the HREEs (Fig. 2). Similar correlations between the REE patterns and K_2O contents have also been observed in the clino-

Table 1
Trace-element compositions (ppm) of garnet and clinopyroxene diamond inclusions

Sample	SL3-3/00			SL3-31/00		SL5-6/00					SL3-32/00	SL5-52/00		SL5-6/00
Grain no.	1	2	3	35	36	112	113	114	115	116	39	91	92	117
Mineral	Gt	Gt	Gt	Gt	Gt	Gt	Gt	Gt	Gt	Gt	Cpx	Cpx	Cpx	Cpx
Paragenesis	harz	harz	harz	harz	harz	eclo	eclo	eclo	eclo	eclo	eclo	eclo	eclo	eclo
Sc	156	157	158	148	146	48.0	46.5	47.1	45.3	45.9	14.2	11.9	12.2	14.5
Ti	144	148	150	966	1001	2510	2167	2056	2177	2077	3030	2322	2286	2429
V	297	310	314	341	387	113	118	122	122	125	324	251	247	244
Sr	1.84	1.15	1.50	7.53	10.9	0.79	4.04	0.53	1.93	1.80	177	711	681	71.5
Y	0.93	0.86	0.84	4.29	4.14	15.0	13.2	12.5	13.3	11.8	3.84	3.62	3.02	3.91
Zr	4.32	3.89	3.65	64.0	64.1	28.7	17.1	16.0	17.2	15.9	16.3	13.0	9.58	14.8
Nb	1.58	1.28	1.44	1.94	2.45	0.20	0.17	0.098	0.21	0.19	0.18	0.31	0.18	0.50
Ba	0.50	0.12	0.31	0.11	4.06	0.17	1.25	0.097	6.31	2.42	5.14	11.7	4.54	11.0
La	0.15	0.16	0.12	1.03	1.36	0.031	n.d.	0.014	n.d.	n.d.	0.60	4.27	4.21	0.81
Ce	0.64	0.56	0.59	8.66	9.36	0.15	0.17	0.089	0.093	0.086	1.82	15.0	13.1	5.81
Pr	0.14	0.14	0.16	2.20	2.22	0.051	0.039	0.028	0.037	0.023	0.39	2.45	2.25	0.44
Nd	0.81	1.01	0.97	11.3	12.0	0.42	0.30	0.23	0.33	0.23	1.85	10.9	9.66	2.00
Sm	0.26	0.25	0.24	1.81	2.19	0.33	0.21	0.18	0.23	0.17	0.62	2.01	2.03	0.57
Eu	0.069	0.075	0.095	0.43	0.45	0.15	0.095	0.089	0.07	0.098	0.26	0.58	0.59	0.16
Gd	0.16	0.22	0.21	1.22	1.46	0.73	0.48	0.57	0.55	0.44	0.93	1.32	1.45	0.50
Tb	0.032	0.024	0.041	0.16	0.17	0.20	0.13	0.16	0.15	0.13	0.19	0.18	0.14	0.10
Dy	0.14	0.14	0.17	0.77	0.83	1.95	1.58	1.40	1.38	1.38	0.89	0.77	0.88	0.66
Ho	0.019	0.034	0.026	0.15	0.13	0.53	0.41	0.44	0.36	0.38	0.17	0.14	0.15	0.17
Er	0.065	0.078	0.12	0.33	0.39	2.05	1.79	1.69	1.56	1.55	0.28	0.34	0.42	0.38
Tm	0.009	0.018	0.022	0.060	0.066	0.37	0.36	0.31	0.27	0.28	0.023	0.024	0.060	0.051
Yb	0.12	0.20	0.17	0.44	0.48	2.67	2.39	2.13	2.43	2.11	n.d.	0.19	0.29	0.34
Lu	0.025	0.050	0.039	0.084	0.10	0.56	0.44	0.45	0.49	0.38	0.011	0.028	n.d.	0.047
Si (pfu)	3.004	2.994	2.994	3.050	3.072	3.032	3.023	3.038	3.041	3.035	2.000	1.989	1.990	2.007
Cr#	34.3	34.4	34.1	39.8	39.5	0.21	0.37	0.32	0.27	0.32	0.48	0.26	0.27	0.56

Major-element compositions of these DIs are presented in Pokhilenko et al. (2003). The Si (per formula unit) and Cr# ($100Cr/Cr + Al$) obtained from the major-element compositions are reanalyzed for this study, which are in general agreement with the analysis done by Pokhilenko et al. (2003). Paragenesis: harz. = harzburgitic, eclo. = eclogitic. Gt = garnet; Cpx = clinopyroxene. n.d. = not detected.

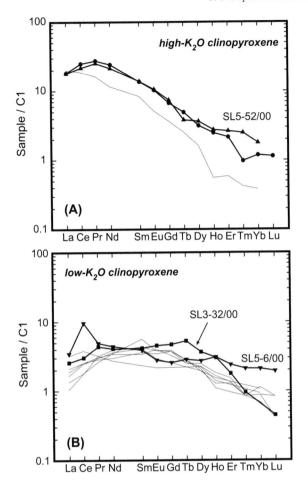

Fig. 2. REE patterns of clinopyroxene DIs from the Snap/King Lake area. (A) LREE-enriched patterns occur in high-K_2O Cpx. (B) Low-K clinopyroxenes possess flat light-middle REE patterns that become depleted towards HREE. Lines represent DI data from Siberia (Mir and Udachanaya; Taylor et al., 1996; Sobolev et al., 1998).

in polished diamonds, and their major-element compositions are shown in Table 2. The Cpxs are small (< 5 μm) and occur at the edges of the inclusions to which they are attached. These Cpxs have high Mg# (94.2 and 94.6), relatively high Cr- and low Ti-contents, and are classified as Cr-diopsides. Their K_2O contents (0.14 and 0.33 wt.%) are generally lower than those in the eclogitic counterparts, but high for peridotitic inclusions. $P–T$ estimates of the Opx–Cpx pair yield 63 kb (Finnerty and Boyd, 1987) and 1214 °C (Brey and Köhler, 1990).

3.3. Orthopyroxene

The composition of the three orthopyroxenes from diamond SL5-75 obtained from the in situ

Table 2
Major-elements compositions (wt.%) of clinopyroxene and orthopyroxene from diamond SL5-75, analyzed in situ

Mineral	Cpx[a]	Opx[a]	Cpx[b]	Opx	Opx
Grain no.	5	4	6	2	3
SiO_2	55.1	57.6	55.3	57.7	57.7
TiO_2	< 0.03	< 0.03	< 0.03	< 0.03	< 0.03
Al_2O_3	0.63	0.46	1.01	0.48	0.48
Cr_2O_3	1.03	0.32	1.37	0.32	0.33
MgO	19.3	35.6	17.6	35.8	35.7
CaO	20.6	0.38	20.9	0.37	0.37
MnO	0.08	0.11	0.09	0.11	0.11
FeO	1.98	4.65	1.94	4.63	4.68
Na_2O	0.48	0.13	0.71	0.13	0.13
K_2O	0.14	0.03	0.33	0.03	< 0.03
Total	99.39	99.31	99.24	99.60	99.52
Oxygen	6	6	6	6	6
Si	1.995	1.986	2.008	1.984	1.984
Ti	0.000	0.000	0.000	0.000	0.000
Al	0.027	0.019	0.044	0.019	0.020
Cr	0.029	0.009	0.039	0.009	0.009
Mg	1.042	1.832	0.952	1.835	1.833
Ca	0.798	0.014	0.812	0.014	0.014
Mn	0.002	0.003	0.003	0.003	0.003
Fe	0.060	0.134	0.059	0.133	0.135
Na	0.034	0.004	0.050	0.004	0.004
K	0.007	0.002	0.015	0.002	0.001
Total	3.995	4.003	3.981	4.003	4.002
Mg#	94.6	93.2	94.2	93.2	93.2
En	54.9	92.5	52.2	92.6	92.5
Wo	42.0	0.7	44.5	0.7	0.7
Fs	3.2	6.8	3.2	6.7	6.8

[a] Touching Opx–Cpx pair.
[b] In contact with olivine.

pyroxene DIs from Siberia (Taylor et al., 1996, 1998, 2000; Sobolev et al., 1998; Anand et al., 2004 this volume). The depletion of Tm for SL5-52/00 is within error of the analysis because of its low abundance approaching the detection limit. In contrast, the Ce spike for SL5-6/00 is difficult to explain but may be attributed to weathering (Floss et al., 2000), in which case, this would appear to be indication of "open-system behavior" of the diamond with respect to the clinopyroxene.

One clinopyroxene was found in contact with Opx and another with Ol during our in situ study of DIs

technique (Table 2) are essentially identical, and fall within the restricted compositional range of the Opx compositions reported by Pokhilenko et al. (in press). However, subtle differences can be seen in some trace elements (Table 3). In the Cr-Ni plot (Fig. 3), the orthopyroxenes DIs can be divided into two distinct groups: low Cr (<2000 ppm) and high Cr (>2000 ppm), each showing a positive slope. Furthermore, the three Opx inclusions analyzed in situ (SL5-75) appear to form extension to the high-Cr group. These differences may indicate compositional variations of harzburgitic source rocks, at least with respect to Ni and Cr, as commonly observed in garnet (Cr) and olivine (Ni) DIs. Significant variations of Ti, Zr, and Sr were observed (6–179, 0.18–16.9, 0.2–9.8 ppm, respectively), and their concentrations are notably higher than those in some Opx DIs from West Africa (Stachel and Harris, 1997).

3.4. Olivine

In five of the diamonds polished to expose mineral inclusions for in situ analysis, nine olivines were encountered, whose major-element chemistry (Table 4) generally lies within the range observed for olivines recovered by the burning technique. However, NiO contents mostly fall outside this range, which suggests a higher variation in trace-element abundances among olivine DIs than previously recognized (Fig. 4).

Diamond SL5-62 contains two relatively large (~ 200 μm) olivine inclusions occurring in the core and intermediate growth zones, as revealed by cathodoluminescence (CL) imaging of the diamond (Fig. 5). Well-defined "stratigraphy" of the diamond is clearly visible; however, the continuity of the growth layers is disrupted by the obscured regions ("dead zones") surrounding the inclusions, a typical occurrence for DIs (Taylor et al., 2003a). Major-element

Table 3

Trace-element compositions (ppm) of orthopyroxene diamond inclusions

Sample	SL5-21/00		SL5-39/00			SL5-42/00		
Grain no.	63	64	74	75	76	77	78	79
Ca	2710	2720	1849	2522	3105	3133	2166	4906
Sc	1.24	1.11	1.62	1.88	1.38	1.77	1.47	1.87
Ti	30.9	26.1	5.66	179	102	19.2	31.0	90.7
V	26.1	27.9	27.8	28.8	28.4	28.1	25.4	28.3
Cr	2430	2532	1908	1928	1958	2570	2488	2511
Mn	629	660	541	574	584	699	652	659
Ni	918	973	893	921	968	1033	918	967
Sr	2.90	2.55	0.41	2.29	4.33	4.62	1.07	9.82
Y	0.066	0.073	0.024	0.19	0.061	0.075	0.023	0.065
Zr	0.90	1.33	0.18	16.9	1.39	2.13	0.30	1.15
Nb	0.15	0.20	0.060	1.77	0.089	0.092	0.089	0.20
Sample	SL5-59/00		SL5-54/00		SL3-27/00	SL5-75[a]		
Grain no.	97	99	104	105	122	2	3	4
Ca	1964	4675	2892	2167	2192	3709	3136	2572
Sc	1.31	1.59	1.80	1.28	1.77	0.60	1.43	1.44
Ti	15.8	20.8	26.1	27.5	40.1	36.1	35.4	43.3
V	24.4	25.8	24.4	23.2	23.7	27.0	25.6	27.6
Cr	1768	1791	2343	2298	1926	1955	1952	1894
Mn	610	622	614	598	620	582	585	575
Ni	847	854	851	847	866	649	730	686
Sr	0.20	7.32	4.32	1.23	1.19	6.86	4.21	2.85
Y	0.022	0.10	0.077	0.050	0.039	0.062	0.056	0.075
Zr	0.19	1.42	1.69	0.76	0.55	4.74	2.98	7.33
Nb	0.032	0.066	0.13	0.11	0.042	0.053	0.089	0.063

[a] Analyzed in situ.

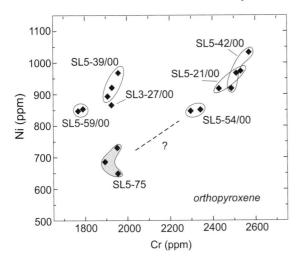

Fig. 3. Cr versus Ni in orthopyroxene inclusions from Snap/King Lake area. Two positive trends can be established from the data points. The three data points within the shaded envelope are in situ analyses of sample SL5-75. These three inclusions are probably part of the high-Cr data array (joined with a dashed line).

chemistry of the two olivine DIs in diamond SL5-62 is essentially identical, but certain trace-element abundances are different (Table 4). That is, Ol-2 that occurs in the outer portion of the diamond contains higher abundances of incompatible trace elements (Ca, Sc, Ti, Sr, Y, Zr, Nb, Ba), when compared to Ol-1, which resides in the core region (Fig. 6).

Two additional olivine inclusions were analyzed for trace elements (Table 4). With few exceptions, both HFSE and LILE vary in restricted ranges, consistent with the relatively uniform major-element chemistry. The low-Ti contents in conjunction with low- and variable-Sr contents are similar with harzburgitic-olivine DIs from Akwatia, Ghana (Stachel and Harris, 1997).

4. Discussion

4.1. Constraints from trace-element chemistry

The harzburgitic-garnet diamond inclusions from the Snap Lake/King Lake kimberlites display the sinusoidal REE patterns, typical of garnet diamond inclusions worldwide. It is generally agreed that such REE patterns are the result of metasomatism of depleted garnets by LREE-enriched, HREE-depleted

fluids; however, various models have been proposed. To explain the hump formed by the LREE and HREE, Hoal et al. (1994) and Shimizu and Richardson (1987) invoked disequilibrium processes that imply higher diffusivities of the LREE compared to the MREE. In contrast, Griffin et al. (1999) called for the progressive misfit in the garnet structure of the larger LREE to explain the hump. These two models, however, appear unlikely in light of the recent experiment by Van Orman et al. (2002) that shows no perceptible difference in diffusion rates among the REEs in the garnet structure. Stachel et al. (2003) pointed out that: (1) the apparent complexity of the sinusoidal REE patterns is merely an artifact of rapidly increasing compatibility, from LREE to MREE, within the garnet structure due to decreasing ionic radius, and (2) the consistency of the patterns over the large compositional range, particularly the inflection points on the sinusoidal curve, is difficult to explain by disequilibrium models. Ivanic et al. (2003) have also demonstrated that garnets with the humped REE profiles could be in equilibrium with extremely LREE-enriched melt. The consistency of REE profiles obtained in this study with those of the worldwide data (Stachel et al., 2003) appears to support the view of equilibrium metasomatic enrichment proposed by both Stachel and Ivanic et al. (2003).

Noteworthy in this study is that both "majoritic" and normal garnet inclusions (SL3-31/00 and SL3-3/00, respectively) possess similar REE patterns, although the REE concentrations in the former are significantly more enriched, up to ~ 10 × higher for LREEs but 2 × for HREEs. It is also important to note that the Cr# (the depletion index) is higher in majoritic garnets compared to the normal ones (~ 40 versus 34, respectively; Table 1). A number of possibilities may be used to explain this difference in the REE concentrations. As demonstrated by Stachel and Harris (1997), the partition coefficients of REEs, especially the light and middle REE, increase with increasing pressure and temperature (i.e., depth), thus resulting in the higher uptakes of the REEs in the majoritic garnets. This would appear to explain the present situation. However, it should also be pointed out that Stachel et al. (2000), based on the partitioning data of Zack et al. (1997), stated that majoritic garnets typically contain lower REE concentrations, which would seem to be the opposite

Table 4
Compositions of major (wt.%) and trace elements (ppm) in olivine diamond inclusions, analyzed in situ

Sample	SL5-62		SL5-75[a]	SL5-47			SL3-53	SL00/173	
Grain no.	1	2		1	2	3		1	2
SiO_2	40.8	41.0	40.6	40.7	40.6	40.7	41.4	40.8	40.6
Cr_2O_3	0.04	0.04	<0.03	0.05	<0.03	0.03	0.03	0.07	0.06
MgO	50.7	51.0	50.52	49.9	49.83	50.3	51.9	50.8	51.0
CaO	<0.03	0.06	<0.03	0.05	<0.03	0.03	0.05	<0.03	0.03
MnO	0.10	0.09	0.10	0.11	0.10	0.10	0.08	0.10	0.10
FeO	7.67	7.80	7.86	8.03	7.85	7.85	6.85	6.37	6.42
NiO	0.39	0.40	0.40	0.38	0.36	0.37	0.32	0.38	0.37
Total	99.7	100.37	99.49	99.20	98.78	99.36	100.58	98.58	98.65
Oxygen	4	4	4	4	4	4	4	4	4
Si	0.994	0.993	0.993	0.998	0.999	0.997	0.996	1.001	0.996
Cr	0.001	0.001	0.000	0.001	0.000	0.001	0.001	0.001	0.001
Mg	1.844	1.844	1.843	1.827	1.829	1.834	1.861	1.856	1.865
Ca	0.000	0.002	0.000	0.001	0.001	0.001	0.001	0.001	0.001
Mn	0.002	0.002	0.002	0.002	0.002	0.002	0.002	0.002	0.002
Fe	0.156	0.158	0.161	0.165	0.162	0.161	0.138	0.131	0.132
Ni	0.008	0.008	0.008	0.007	0.007	0.007	0.006	0.008	0.007
Total	3.005	3.008	3.006	3.001	2.999	3.003	3.004	2.999	3.004
Fo	92.2	92.1	92.0	91.7	91.9	91.9	93.1	93.4	93.4
Ca (ppm)	281	320	462	366					
Sc	0.44	0.64	0.55	0.54					
Ti	4.83	6.75	4.65	4.67					
V	11.0	9.84	9.92	11.3					
Cr	572	351	298	493					
Mn	559	572	558	581					
Ni	2389	2391	2416	2220					
Sr	0.46	1.16	0.66	0.78					
Y	0.025	0.034	0.016	0.081					
Zr	0.35	0.56	0.18	6.34					
Nb	0.015	0.036	0.030	0.093					
Ba	0.74	1.11	0.22	2.68					

[a] In contact with Cpx.

from that we have observed. Alternatively, the higher degree of depletion of these majoritic garnets results in their greater susceptibility to metasomatic enrichment that significantly increases their trace-element concentrations, without modifying the major-element abundances (Stachel et al., 2003). It may also be possible that the fluids interacting with majoritic garnets in this study contain higher incompatible trace-element concentrations than those interacted with normal garnets.

Nature of the fluids/melts that may have enriched harzburgitic garnets is not well understood at present (e.g., Taylor et al., 2003b). Stachel and Harris (1997) believed that the fluids must be very rich in LREE, and to a lesser degree Rb and Sr, and low or void in HFSE and HREE. These requirements seem to ex-

clude kimberlitic, lamproitic, or carbonatitic melts, and they favored methane-rich fluids as the most likely enriching media, as well as being the carbon source for diamond formation. In contrast, Griffin et al. (1999) interpreted that fluids such as carbonatite cause the depletion in HFSE but enrichment in LREE observed in some harzburgitic garnets from South African ultramafic xenoliths.

The "majoritic" harzburgitic diamond inclusions in this study (SL3-31/00) differ from the previous investigations in their enrichment in HFSE, particularly Zr (Fig. 2B), possibly reflecting the compositions of metasomatic fluids. It is unlikely that the relatively high HFSE concentrations are the inherent nature of pre-metasomatized garnets because the "normal" harzburgitic garnets (SL3-3/00), which are

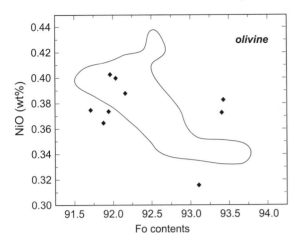

Fig. 4. Fo versus NiO plot of olivine inclusions in this study, compared with the data from Pokhilenko et al. (2003), as denoted by the envelope.

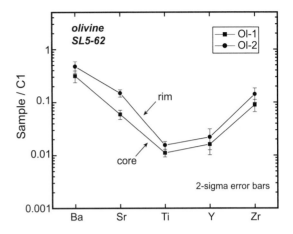

Fig. 6. Plot of LILE and HFSE in olivine inclusions from sample SL5-62 that shows the higher abundances of these elements in olivne-2 (rim), compared to those in olivine-1 (core).

less depleted (lower Cr#), contain lower abundances of these elements. The increase in HFSE in SKL diamond inclusions, perhaps, is best seen in the in situ analysis of two olivines in diamond SL5-62 (Fig. 6). Olivine-2 that occurs in the outer portion of the diamond contains consistently higher abundances of HFSE and LILE than olivine-1 that occurs in the core of the diamond, albeit overlaps in 2σ error bars of the analyses. Based on these observations, the fluids that interacted with at least some of the SKL diamond

inclusions are probably enriched in LREE, and to a lesser degree LILE and HFSE, and depleted in HREE. Perhaps, these fluids are akin to those occurring as micro-inclusions in diamonds studied by Navon et al. (2003). They suggested that the major-element compositions of these micro-inclusions correspond to three end-members, including hydrous-silicic melt, carbonatitic melt, and brine. The silicic component of this fluid may contain high abundances of HFSE similar to silicate melts, and therefore, may supply HFSE to these SKL diamond inclusions.

4.2. Remnant of oceanic crust beneath SE Slave craton

Eclogitic diamond inclusions and eclogite xenoliths are common in kimberlitic fields worldwide, and kimberlites of the Slave craton have also carried eclogitic materials from the great depths (Pearson et al., 1999a,b; Davies et al., 1999a,b; Pokhilenko et al., 2001). Based on the major-element chemistry of garnets and omphacitic clinopyroxenes (Pokhilenko et al., 2003), E-type diamond inclusions from the Snap Lake kimberlite have been classified as Group B, suggesting their derivation from an ancient oceanic crust. This may have been associated with the 1.9-Ga collision of the Slave craton with the Hottah Terrane, as described by Bowring and Grotzinger (1992).

Eclogites are essentially bimineralic rocks, consisting mainly of garnet and omphacitic clinopyroxene,

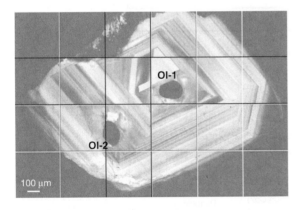

Fig. 5. Cathodoluminescence (CL) image of a plate of diamond SL5-62 polished parallel to (110) plane. Two olivine inclusions occur at core (Ol-1) and at intermediate (Ol-2) growth zones of the diamond. Well-defined growth layers of the diamond are clearly visible, except at the regions surrounding the two DIs.

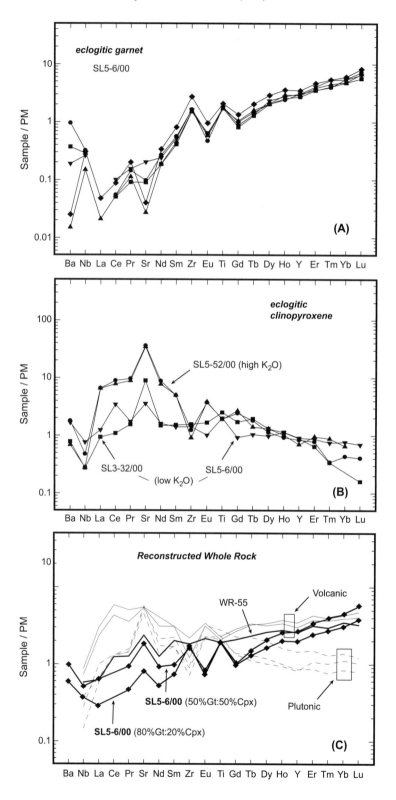

often with accessory minerals such as rutile, kyanite, coesite, etc. In diamond SL5-6/00, both eclogitic garnets and omphacite were found, thereby enabling the reconstruction of whole-rock composition. The compositions of the five garnet inclusions in this diamond were averaged for reconstruction, since they have similar major- and trace-element compositions. Assuming the modal proportions of garnet/cpx at 50:50 and 80:20 (for comparison), the trace-element pattern for the reconstructed compositions (PM normalized) of the Snap Lake eclogite, along with that of constituent minerals, is illustrated in Fig. 7, together with a similar reconstruction on diamondiferous eclogites from Udachanaya (Jacob and Foley, 1999). This reconstructed composition is generally considered as representative of an oceanic crust that has lost partial melts during subduction to produce trondhjemite–tonalite–granodiorite (TTG) magmas in the Archean (Ireland et al., 1994). In comparison, the composition of the reconstructed eclogite from Snap Lake is very similar to that of WR-55, with the exception of Zr and Ti depletions. This eclogite is believed to be derived from the volcanic section of an ancient oceanic crust (Jacob and Foley, 1999) and, analogously, this may also hold true for the protolith of the inclusions in diamond SL5-6/00.

Trace-element patterns of all eclogitic clinopyroxene diamond inclusions in this study show prominent positive Sr anomaly, which is also present in the reconstructed whole-rock composition (Fig. 7). This anomaly has often been interpreted as the consequence of carbonate depositions on the ocean floor, coupled with remobilization of the Sr during hydrothermal alterations of the eclogite protoliths (Ireland et al., 1994; Jacob and Foley, 1999), and these Sr-enrichment processes may be applicable to the eclogitic clinopyroxenes of the present study as well. Hydrothermal alteration, however, is unlikely to cause positive anomalies of Ti and Zr in eclogitic garnets from diamond SL5-6/00 (Fig. 7A), since these HFSEs are essentially immobile in the hydrothermal fluids (Jacob and Foley, 1999). Metasomatic enrichment by a silicate melt will also increase the abundances of other incompatible trace elements, such as LREE, which is not observed in the present data. Based on the experiment by Konzett (1997), a possible cause of these HFSE anomalies in garnet is the increase in solubility of these elements in garnets upon increasing $P–T$ during the subduction (Klemme et al., 2002) that eventually produced the eclogite protoliths of the DIs.

5. Conclusion

Trace-element chemistry of multiple diamond inclusions presented here provides insights into the nature of processes involved in the formation of diamonds in the Snap Lake/King Lake area. The harzburgitic-garnet inclusions possess sinusoidal REE patterns that are consistent in both concentrations and shapes (positions of peaks and troughs) with the worldwide data (Stachel et al., 2003). Such consistency most likely suggests the *equilibrium* metasomatic enrichment of pre-existing garnets that previously experienced partial-melting episode(s). The data in this study also support the interpretation that the most-depleted garnets (high Cr#) are most sensitive to metasomatism, thus capable of absorbing high concentrations of incompatible trace-element abundances (Stachel et al., 2003). Moreover, it seems that "majoritic" garnets, which were derived from the greater depths, experience more intense melting and metasomatic enrichment, compared to the normal garnets. The relatively high concentrations of HFSE in majoritic garnets and the increase from core to rim in HFSE and LILE of olivine diamond inclusions may indicate the relatively high abundances of these components in metasomatic fluids. Such fluids may be represented by the three end-member components, including hydrous-silicic melt, carbonatitic melt, and brines (Navon et al., 2003). Reconstructed whole-rock trace-element composition of eclogitic DIs suggests that the eclogite protolith was probably the volcanic layer of an oceanic crust

---

Fig. 7. Primitive Mantle (PM) normalized trace-element patterns of eclogitic DIs (A and B) and the reconstructed whole-rock compositions, using two different proportions of garnet and Cpx (C). Note that Ce is not included in the whole-rock reconstruction due to its abnormally high concentration in the omphacite. Also plotted for comparison in (C) are the reconstructed compositions of diamondiferous eclogite xenoliths from Udachanaya (WR-55), Siberia, which are separated into volcanic and plutonic suites (Jacob and Foley, 1999).

that had undergone hydrothermal alteration and partial melting during subduction.

Acknowledgements

Allan Patchen is gratefully acknowledged for his assistance in obtaining electron microprobe data. Constructive criticisms and critical reviews by Gerhard Brey, Sonja Aulbach, and Thomas Stachel assisted greatly in the improvement of this paper. This research was partially supported by funds from NSF Grant EAR 99-09430 (LAT). The SIMS analyses at Washington University were supported by NSF Grant EAR 99-80394 (G. Crozaz).

References

Alexander, C.M.O.D., 1994. Trace element distributions with ordinary chondrite chondrules: Implications for chondrule formation conditions and precursors. Geochim. Cosmochim. Acta 58, 3451–3467.

Agashev, A.M., Pokhilenko, N.P., McDonald, J.A., Takazawa, E., Vavilov, M.A., Sobolev, N.V., Watanabe, T., 2001. A unique kimberlite-carbonatite primary association in the Snap Lake dyke system, Slave Craton: evidence from geochemical and isotopic studies. The Slave Kaapvaal Workshop, Ext. Abstr. Vol. Merrickville, Canada, unpaged.

Anand, M., Taylor, L.A., Carlson, W.D., Taylor, D.-H., Sobolev, N.V., 2004. Diamond genesis revealed by X-ray tomography of diamondiferous eclogites. Lithos, Proceedings of the 8th International Kimberlite Conference. This volume.

Bowring, S.A., Grotzinger, J.P., 1992. Implications of new chronostratigraphy for tectonic evolution of Wopmay Orogen, Northwest Canadian Shield. Amer. Jour. Sci. 292, 1–20.

Brey, G.P., Köhler, T., 1990. Geothermobarometry in four-phase lherzolites II: new thermobarometers and practical assessment of existing thermobarometers. J. Petrol. 31, 1353–1378.

Carlson, R.W., Pearson, D.G., Boyd, F.R., Shirey, S.B., Irvine, G., A.H., A.H., Menzies, A.H., Gurney, J.J., 1999. Regional age variation of the southern African mantle: significance for models of lithospheric mantle formation. In: Gurney, J.J., Gurney, M.D., Pascoe, M.D., Richardson, S.H. (Eds.), Proceedings 7th International Kimberlite Conference vol. 1. Red Roof Design, Cape Town, pp. 99–108.

Davies, R., Griffin, W.L., Pearson, N.J., Andrew, A., Doyle, B.J., O'Reilly, S.Y., 1999a. Diamonds from the Deep: pipe DO-27, Slave Craton, Canada. Proc. of the 7th Int'l Kimberlite Conf., vol. 1, pp. 148–155.

Davies, R.M., O'Reilly, S.Y., Griffin, W.L., 1999b. Growth structures and nitrogen characteristics of group B alluvial diamond crystals from Bingara and Wellington, Eastern Australia. Proceedings of the 7th International Kimberlite Conference, vol. 1, pp. 156–163.

Fahey, A.J., Zinner, E.K., Crozaz, G., Kornacki, A.S., 1987. Microdistribution of Mg isotopes and REE abundances in a Type A calcium-aluminum-rich inclusion from Efremovka. Geochim. Cosmochim. Acta 51, 3215–3229.

Finnerty, A.A., Boyd, F.R., 1987. Thermobarometry for garnet peridotite xenoliths: a basis for upper mantle stratigraphy. In: Nixon, P.H. (Ed.), Mantle Xenoliths, vol. 1, 381–402.

Floss, C., Crozaz, G., Yamaguchi, A., Keil, K., 2000. Trace element constraints on the origins of highly metamorphosed Antarctic eucrites. Antarct. Meteor. Res. 13, 222–237.

Griffin, W.L., Shee, S.R., Ryan, C.G., Win, T.T., Wyatt, B.A., 1999. Harzburgite to lherzolite and back again: metasomatic processe in ultramafic xenoliths from the Wesselton kimberlite, Kimberley, South Africa. Contrib. Mineral. Petrol. 134, 232–250.

Hoal, K.E.O., Hoal, B.G., Erlank, A.J., Shimizu, N., 1994. Metasomatism of the mantle lithosphere recorded by rare earth elements in garnets. Earth. Plan. Sci. Lett. 126, 303–313.

Hsu, W., 1995. Ion Microprobe Studies of the Petrogenesis of Enstatite Chondrites and Eucrites. Washington University, St. Louis, MO, United States, p. 380.

Ireland, T.R., Rudnick, R.L., Spetsius, Z.V., 1994. Trace elements in diamond inclusions from eclogites reveal link to Archean granites. Earth Planet. Sci. Lett. 128, 199–213.

Irifune, T., 1987. An experimental investigation of the pyroxene-garnet transformation in a pyrolite composition and its bearing on the composition of the mantle. Earth Planet Sci. Lett. 45, 324–336.

Ivanic, T.J., Harte, B., Burgess, S.R., Gurney, J.J., 2003. Factors in the formation of sinuous and humped REE patterns in garnets from mantle harzburgitic assemblages (Abstract). Proc. 8th IKC, vol. 4, pp. 14.

Jacob, D.E., Foley, S.F., 1999. Evidence for Archean ocean crust with low high field strength element signature from diamondiferous eclogite xenoliths. Lithos 48, 317–336.

Klemme, S., Blundy, J.D., Wood, B.J., 2002. Experimental constraints on major and trace element partitioning during partial melting of eclogite. Geochim. Cosmochim. Acta 66, 3109–3123.

Konzett, J., 1997. Phase relations and chemistry of Ti-rich K-richterite-bearing mantle assemblages: an experimental study to 8.0 GPa in Ti-KNCMASH system. Contrib. Mineral. Petrol. 128, 385–404.

Navon, O., Izaeli, E.S., Klein-BenDavid, O., 2003. Fluid inclusions in diamods: the carbonatitic connection. Proc. 8th Int. Kimb. Conf. long abstracts # 107.

Pearson, N.J., Griffin, W.L., Doyle, B.J., O'Reilley, S.Y., Van Achterbergh, E., Kivi, K., 1999a. Xenoliths from kimberlite pipes of the Lac de Gras area, Slave craton, Canada. Proc. 7th Int'l Kimberlite Conf., Extnd. Abstr., vol. 2, pp. 644–658.

Pearson, D.G., Shirey, S.B., Bulanova, G.P., Carlson, R.W., Milledge, H.J., 1999b. Dating and paragenetic distinction of diamonds using the Re-Os isotope system: application to some Siberian diamonds. Proceedings of the 7th International Kimberlite Conference, vol. 2, pp. 637–643.

Pokhilenko, N.P., Mcdonald, J.A., Melnyk, W., Hall, A.E., Shimizu, N., Vavilov, M.A., Afanasiev, V.P., Reimers, L.F., Irvin, J., Pokhilenko, L.N., Vasilenko, V.B., Kuligin, S.S., Sobolev, N.V., 1998. Kimberlites of Camsell Lake Field and some features of structure and composition of lithosphere roots of southeastern part of Slave Craton, Canada. Proc. 7th Int. Kimberlite Conf. Ext. Abstr. Vol., Cape Town, pp. 699–701.

Pokhilenko, N.P., Sobolev, N.V., Cherny, S.D., Mityukhin, S.I., Yanygin, Y.T., 2000. Pyrope and chromite from kimberlites of the Nakyn Field (Yakutia) and the Snap Lake region (Slave Craton, Canada) evidence for anomalous lithospheric structure. Doklady Earth Sci. 372, 356–360.

Pokhilenko, N.P., Sobolev, N.V., McDonald, J.A., Hall, A.E., Yefimova, E.S., Zedgenizov, D.A., Logvinova, A.M., Reimers, L.F., 2001. Crystalline inclusions in diamonds from kimberlites of the Snap Lake area (Slave Craton, Canada): new evidences for the anomalous lithospheric structure. Dokl. Earth Sci. 380, 806–811.

Pokhilenko, N.P., Sobolev, N.V., Reutsky, V.N., Hall, A.E., Taylor, L.A., 2003. Crystalline inclusions and C-isotope ratios of diamonds from the Snap Lake/King Lake kimberlite dyke system: Evidence for an ultradeep and enriched lithosphere mantle, Lithos. Proceedings of the 8th International Kimberlite Conference. This volume.

Pokhilenko, N.P., Sobolev, N.V., Reutsky, V.N., Hall, A.E., Taylor, L.A., 2004. Crystalline inclusions and c isotope ratios in diamonds from the Snap Lake/King Llake kimberlite dyke system: evidence of ultradeep and enriched lithospheric mantle. Proc. 8th Int'l. Kimb. Conf., Lithos. in press.

Shimizu, N., Richardson, S.H., 1987. Trace element abundance patterns of garnet inclusions in peridotite suite diamonds. Geochim. Cosmochim. Acta 51, 755–758.

Sobolev, N.V., Snyder, G.A., Taylor, L.A., Keller, R.A., Yefimova, E.S., Sobolev, V.N., Shimizu, N., 1998. Extreme chemical diversity in the mantle during eclogite diamond formation: evidence from 35 garnet and 5 pyroxene inclusions in a single diamond. Int. Geol. Rev. 40, 567–578.

Stachel, T., Harris, J.W., 1997. Diamond precipitation and mantle metasomatism: evidence from the trace element chemistry of silicate inclusions in diamonds from Akwatia, Ghana. Contrib. Mineral. Petrol. 129, 143–154.

Stachel, T., Brey, G.P., Harris, J.W., 2000. Kankan diamonds (Guines): I. from the lithosphere down to the transition zone. Contrib. Mineral. Petrol. 140, 1–15.

Stachel, T., Aulbach, S., Brey, G.P., Harris, J.W., Leost, I., Tappert, R., Viljoen, K.S., 2003. Diamond formation and mantle metasomatism: a trace element perspective. 8th Int'l Kimberlite Conf. Extnd. Abstr., Victoria BC, Canada.

Taylor, L.A., Snyder, G.A., Crozaz, G., Sobolev, V.N., Yefimova, E.S., Sobolev, N.V., 1996. Eclogitic Inclusions in diamonds: Evidence of complex mantle processes over time. Earth Planet. Sci. Lett. 142, 535–551.

Taylor, L.A., Milledge, H.J., Bulanova, G.P., Snyder, G.A., Keller, R.A., 1998. Metasomatic eclogitic diamond growth: evidence from multiple diamond inclusions. Int. Geol. Rev. 40, 592–604.

Taylor, L.A., Keller, R.A., Snyder, G.A., Wang, W., Carlson, W.D., Hauri, E.H., McCandless, T., Kim, K.-R., Sobolev, N.V., Bezborodov, S.M., 2000. Diamonds and their mineral inclusions, and what they tell us: a detailed "pull-apart" of a diamondiferous eclogite. Int. Geol. Rev. 42, 959–983.

Taylor, L.A., Anand, M., Promprated, P., 2003a. Diamonds and their inclusions: are the criteria for syngenesis valid? 8th Internat. Kimb. Conf., Ext Abstr.

Taylor, L.A., Anand, M., Promprated, P., Floss, C., Sobolev, N.V., 2003b. The significance of mineral inclusions in large diamonds from Yakutia, Russia. Am. Mineral. 88, 912–920.

Van Orman, J.A., Grove, T.L., Shimizu, N., 2002. Diffussive fractionation of trace elements during production and transport of melt in Earth's upper mantle. Earth Plan. Sci. Lett. 198, 93–112.

Zack, T., Foley, S.F., Jenner, G.A., 1997. A consistent partition coeffcient set for clinopyroxene, amphibole and garnet from laser ablation microprobe analysis of garnet pyroxenites from Kakanui, New Zealand. Neues Jahrb. Mineral. Abh. 172, 23–41.

Zinner, E., Crozaz, G., 1986a. A method for the quantitative measurement of rare earth elements in the ion microprobe. Int. J. Mass Spectrom. Ion Process. 69, 17–38.

Zinner, E., Crozaz, G., 1986b. Ion probe determination of the abundances of all the rare earth elements in single mineral grains. Int. J. Mass Spectrom. Ion Process. 69, 444–446.

Available online at www.sciencedirect.com

Lithos 77 (2004) 83–97

www.elsevier.com/locate/lithos

Features of coated diamonds from the Snap Lake/King Lake kimberlite dyke, Slave craton, Canada, as revealed by optical topography

A.P. Yelisseyev[a,*], N.P. Pokhilenko[a], J.W. Steeds[b], D.A. Zedgenizov[a], V.P. Afanasiev[a]

[a] *Siberian Division, Institute of Mineralogy and Petrography, 3 Ac.Koptyug av., 630090 Novosibirsk, Russia*
[b] *H.H. Wills Physics Laboratory, University of Bristol, Tyndall Avenue, Bristol BS8 1TL, UK*

Received 27 June 2003; accepted 14 December 2003

Abstract

Confocal photoluminescence (PL) and local absorption spectroscopy were used to study the types and spatial distribution of point defects in coated diamonds, the input of which is about 30% in the Snap Lake deposit, Canada. Nitrogen concentration is on the level of several hundreds of ppm in the core, with a nitrogen-poor layer in its outer part, whereas in the coat it is usually several times higher as a result of fast growth. Nitrogen defects in the core are strongly aggregated with N3, B and B′-forms dominating, whereas A-defects are typical of the coat. The rounded shape of the coated diamonds is a result of the combined effect of partial dissolution of the octahedral core and the "abnormal" growth of the coat, which produces a fibrous structure. Analysis of PL and PL excitation spectra showed that structureless yellow-green PL of the coat is likely to be due to nickel-nitrogen complexes with their fine structure broadened in the strain fields. The presence of irradiation/annealing products such as vacancies V^0 and nitrogen-vacancy complexes NV^-, N_2V_2 shows that the diamonds studied have undergone post-growth ionizing irradiation with further low-temperature annealing in natural conditions.
© 2004 Elsevier B.V. All rights reserved.

Keywords: Coated diamonds; Absorption; Luminescence; Nickel; Nitrogen; Impurity defects

1. Introduction

Coated diamonds are specific diamonds, which consist of a clear colorless core, often a well-formed octahedron of good quality, enveloped by a yellow, green or grey coat. The coat material is also diamond, but filled with a particulate matter, the individual particles of which are generally of submicron size and are not clearly resolved with an optical microscope. Most of the recorded coated diamonds have been from the Democratic Republic of the Congo (DRC), where they form about 90% of the population (Kamiya and Lang, 1965), and Sierra Leone (up to 50% in some deposits). They are very characteristic of Mbuji Mayi in the DRC, which is the world's largest producer of this type of diamond (Boyd et al., 1987). The systematic study of coated diamonds began in 1950 with a general description by Custers (1950). Using X-ray topographyKamiya and Lang (1965) have shown that the coat has a fibrous structure

---

* Corresponding author. Tel.: +7-3832-333-843; fax: +7-3832-333-843.

E-mail address: elis@mail.nsk.ru (A.P. Yelisseyev).

interrupted by layers of fine particles. Later Lang (1974) concluded that the structural difference resulted from different growth mechanisms for the core and coat. The core grew continuously in the octahedral habit, by concentric, complete {111} growth layers, while the coat demonstrated a much rarer, "abnormal" mode whereby the growing crystal is broken up into a bundle of columns, growing outwards independently but with equal velocities. Faulkner et al. (1965) found in an Electron Spin Resonance (ESR) study that the concentration of single nitrogen atoms (P1 centers in ESR or C-defects in infrared (IR) spectroscopy) is low in the coated diamonds: only about 30 ppm in the samples cut from the coat. Angress and Smith (1965) discovered, using IR spectroscopy, that a considerable part of the nitrogen in the coat (up to 1700 ppm) is aggregated into pairs (A-defects) with a dominant 1282 cm^{-1} peak. In the core the nitrogen was found in A and/or B-form, the latter is believed to be a nitrogen-vacancy complex of the N_4V structure and is usually accompanied by platelets (B′ defects). There were some additional bands at 475, 1086 and 1105 cm^{-1} in the IR spectra for the coat. Numerous narrow lines, related to OH, HOH, CH, CO_2, CO_3^- and to silicate and other compounds in the numerous inclusions, were found in the coat related IR spectra of diamonds with strongly colored gray coat (Navon et al., 1988; Schrauder and Navon, 1994). The same authors also found that inclusions in fibrous/coated diamonds are rich in various oxides such as SiO_2, K_2O, CaO, FeO, MgO and P_2O_5. Logvinova et al. (2003) found S–Fe–Ni-rich inclusions in fibrous diamonds from Yubileinaya pipe, Yakutia. Inclusions are characterized by a wide S/(S + Ni + Fe) range and the high Ni content (up to 40 ± 7 wt.%). Broad compositional variations of S–Ni–Fe inclusions, the constant Ni/(Ni + Fe), the presence of varying amounts of alkalis and Mg and the association with carbonatitic melts suggest that the melt-bearing micro-inclusions do not contain a mineral phase, but should be regarded as sulfide melts.

The geochemical features and isotopic characteristics of the Snap Lake/King Lake (SL/KL) kimberlites combined with the results of the study of crystalline inclusions in the diamonds suggest that their mantle sources are unusual for known kimberlites (Pokhilenko et al., 2001): The lithosphere beneath the Snap Lake area is abnormally thick (at least 300 km). This means that the pressure interval of the natural diamond formation in the lithosphere was from 37 to at least 110 kbar while the normal pressure interval is from ~ 37 to ~ 65 kbar. As a result differences in the internal structure and physical characteristics of the Snap Lake diamonds, formed at abnormally high pressures, would be expected.

In the present paper the similarity between the coated diamonds from SL/KL and other deposits (DRC, etc.) is demonstrated using a microscopic study, birefringence, photoluminescence (PL) and X-ray topography. Using different methods of optical spectroscopy we show for the first time that in diamonds with fibrous structure nickel is present not only in inclusions but is also incorporated into diamond lattice. Its concentration is particularly high in the coat, and nickel–nitrogen complexes are responsible for its yellow-green luminescence. An attempt to reconstruct the growth history of coated diamonds was undertaken.

2. Experimental

A group of several hundred samples from the SL/KL was studied using optical and electron scanning microscopy. Two typical coated diamonds (#SL-00/31 and #SL-00/14) were prepared as {110} plates, about 0.5 mm thick. For comprehensive studies we used all available techniques: transmission, birefringence, PL patterns, X-ray topography and optical spectroscopy.

In the case of (110) sections the octahedron faces in the core are perpendicular to the diamond plate and a good spatial resolution is achieved when using all mentioned techniques. Transmission patterns were obtained in white and in blue light; the latter allowed us to increase the image contrast when investigating the yellow color distribution along samples. A Bruker IFS 113 FTIR spectrometer combined with an optical microscope was used to measure the IR absorption spectra with a 2 cm^{-1} resolution at room temperature for a set of points along specific lines across the sample. The typical diameter of the aperture spot was 50 μm with 100 μm distance between analysis points. In the one-phonon region, the spectra were resolved into individual components related to A- and B-centers and the nitrogen concentration was calculated based on the relations determined by Woods et al. (1990) and Boyd et al. (1995). At several most important points absorption spectra were recorded at 80 K and in the UV to near region using a metal 200×200 μm^2 diaphragm. The PL spectra were recorded using a Renishaw micro-Raman spectrometer fitted with Oxford Instruments Microstats. When measuring PL spectra, the 325 nm line of a He–Cd laser, and the 488 and 514 nm lines of an Ar^+ laser were used for excitation. The spectra were recorded for a set of points equally spaced with a 50 μm steps along the sample diameter. The laser beam was focused to 5 μm in the layer ~ 10 μm under the diamond plate surface. The PL excitation spectra (PLE) were measured using a 1 kW Xe lamp with continuous emission spectrum: the diffraction MDR2 monochromator and SDL-1 diffraction spectrometer were used for separation of excitation spectral range and for detection of the PL emission, respectively. In the case of PLE spectra the emission band/line is fixed and detected by the SDL-1, whereas the MDR2 is monitored along the Xe lamp emission region. As a result we record a spectrum, which is similar to the absorption spectrum. The difference between the absorption and PLE spectra is: in the first case total absorbed energy is recorded whereas in the second case we record only the part, which excites a certain PL emission (a certain type of luminescence centers). One of the main advantages of the PLE technique is its very high sensitivity (many orders higher than for absorption), which allows one to obtain the absorption spectrum related to some defect, even in the case where a sample is of complicated form, has a poor transparency or when its absorption is very low. For diamond, which is characterized by weak electron–phonon interaction, typical PL and PLE/absorption spectra consist of a narrow zero-phonon line (ZPL) and a broad phonon-related band, which has a red or a blue shift relative to ZPL, respectively (Zaitsev, 2001). For centers with the simplest two-level energy diagram, such as H2, H3, H4, N3 and some others, one observes the only ZPL both in PL and PLE/ absorption spectra, but the vibronic structure is different. For the overwhelming majority of optically active defects in diamond the energy level diagram is more complicated and both ZPLs and vibronic structures are different in PL and PLE: as a result the PLE technique can be used as an additional method, which gives independent information and could help when comparing different defects.

3. Results and discussion

3.1. Samples description

Examination of the sample morphology using optical and ES microscopy showed that the coated diamonds are wide spread among diamonds from SL/ KL: according to results of visual study their input is about 30%. Such crystals are somewhat rounded, but their unambiguous identification is possible only in the case when the coat is partly destroyed. Six samples of this type with sizes in the 1.2–3 mm range are shown in Fig. 1. The coat thickness varies considerably from very thin, which is hardly revealed on the crystal (Fig. 1B,C for example) to that up to ~ 1 mm thick (Fig. 1A,D).

The overwhelming majority of the examined stones demonstrate no PL at UV excitation, apparently because of high concentration of nitrogen impurity: the latter is known to quench PL. We selected two rounded luminescent stones (#SL-00/31 and #SL-00/ 14) and studied 0.5 mm thick plates from them. In both cases one can see high quality colorless crystals of octahedral habit as a core, which is surrounded with a yellow translucent coat of about 0.5 mm thick (Figs. 2 and 3). The coat was of a pronounced yellow color for the first stone (Fig. 2A) and only slightly yellowish for the second (Fig. 3A). The inclusion density and the intensity of the resulting grey color were much lower than for samples described by Navon et al. (1988), Schrauder and Navon (1994). The coloration is much stronger near the core/coat boundary (Figs. 2A and 3A) and its intensity becomes weaker inside the coat. The birefringence correlates with cracks and inclusions: it is maximum in the coat and relatively weak in the core although sometimes the stress field extends from coat into core (Fig. 2B and 3B). Other sources of birefringence are boundaries between single crystal needles, which form the fibrous structure of the coat (Fig. 3B). The latter is particularly well pronounced in the X-ray topograms (Figs. 2C and 3C) with individual needles resolved.

Analysis of Fig. 1 shows, that such coat of fibrous structure is unstable and can be easily destroyed by interaction with surrounding kimberlite melt. In Section 1C and D one can see that a thin coat layer just near a stressed core/coat boundary, which contains maximum number of inclusions,

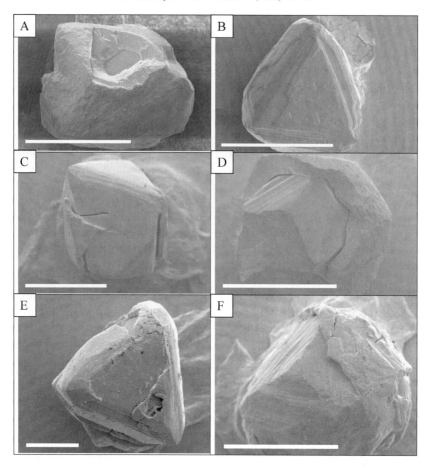

Fig. 1. The shape of six coated diamonds with partly removed coats from the Snap Lake/King Lake kimberlite dyke system. The length of a light bar is 1 mm in all sections.

undergoes maximum etching and can be removed completely; at the same time there are no traces of any etching such as reverse trigons on the core surface. Such easy destruction of the coat in kimberlite melt allows us to suppose that many diamonds (maybe all) had coats originally and lost them later: thus an abnormal growth on the last stage is likely typical of all SL/KL diamond association.

The PL patterns show that both the core and coat are not homogeneous: there is an outer nonluminescent layer in the core and several layers, sometimes with different PL color in the coat (Figs. 2D and 3D). An intense blue PL is typical of the central part of the core (area 1 in Figs. 2D and 3D), whereas the outer nonluminescent layer is black in PL patterns (area 2). The nonluminescent layer is ~ 20 μm thick for #SL-

00/31 and is much more pronounced (up to 250 μm) for #SL-00/14. The coat PL is mainly of yellow-green color with a pronounced blue component, which is more intense for #SL-00/14. For both diamonds one found an intense white PL in the near-surface layer. In Fig. 4 a fragment of the cathodoluminescence (CL) pattern for # SL-00/14 is presented. In this case, the 20 keV electrons excite emission in a thin (only several microns thick), near-surface layer of the diamond plate and much higher spatial resolution is available. As in the case of PL the internal part of the core (area 1) demonstrates an intense white-bluish emission with well-pronounced octahedral boundaries. The following layer (2) has a weak CL intensity as in the case of PL: nevertheless, there are also some thin luminescent layers and as a result one sees a

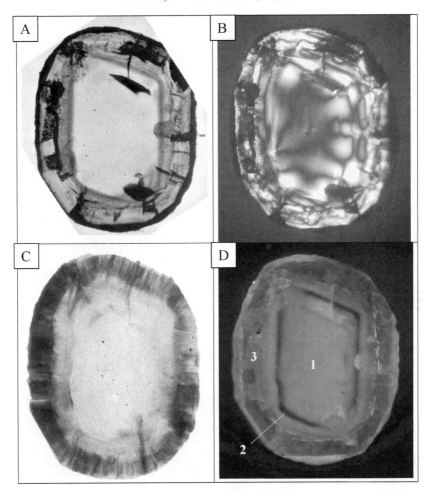

Fig. 2. Photographs of natural coated diamond #SL-00/31, obtained in the transmitted white light (A) and with crossed polarizers (B); X-ray topogram (C) and PL pattern at UV excitation (D). In the X-ray topogram (C) the needles, which form the coat, are resolved. The core demonstrates an intense blue PL (area 1 in section D) with the outer nonluminescent layer (black area 2). PL in the coat (3) is of yellow-green color with more intense white emission in a thin near-surface layer.

striation pattern. The light lines are straight in this area and repeat mainly the octahedral boundary of the internal area 1. A concave fragment, at right on the boundary of the luminescent area 1, as well as a rounded outer boundary of the striated area 2, show that there were at least two breaks in crystal growth, which were accompanied with partial crystal dissolution. Before the second break the diamond grew layer-by-layer in equilibrium conditions and the shape of the resulting stone should also be octahedral. After the following dissolution stage crystals became rounded and the coat only repeated this new shape. It is

interesting that in contrast to PL cathodoluminescence is weak in the coat body. An intense CL is typical of a thin near-surface layer whereas its monotonic increase when moving from the core/coat boundary to the surface may be due to refraction effects inside the coat.

Many of diamonds from SL/KL have green pigmentation spots on their surface and sometimes the crystal surface is homogeneously colored into blue or green. According to Orlov (1973) such coloration shows that diamonds have been irradiated with α- or β-particles.

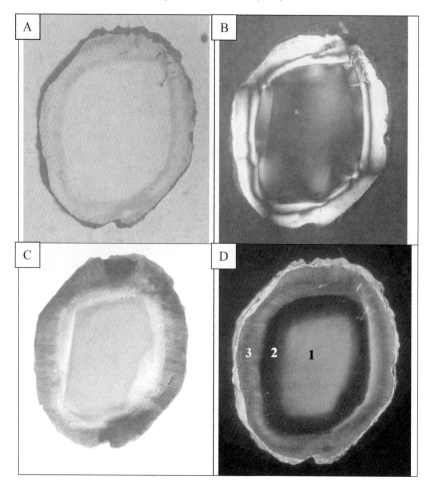

Fig. 3. Photographs of natural coated diamond #SL-00/14, obtained in the transmitted blue light (A) and with crossed polarizers (B), X-ray topogram (C) and PL pattern at UV excitation (D). The same areas as in Fig. 2D are enumerated.

3.2. Absorption spectroscopy

3.2.1. IR absorption

Typical IR absorption spectra for the core and the coat are shown in Fig. 5. For the core in the one-phonon region each spectrum is formed by two well-known components, related to A and B nitrogen defects with dominating peaks at 1282 and 1175 cm^{-1}, respectively (curve 1), which are accompanied by the 1360 cm^{-1} absorption band of platelets or B′ defects. Platelets are thought to be {100} planes of interstitial carbon and usually accompany B defects in natural diamonds (Zaitsev, 2001). In the coat the only system, which is identified reliably, is a system due to A-defects. Nevertheless, experimental spectra (curve

3) differ from those, which are given in literature for stones with the only A system (curve 2, Zaitsev, 2001): one can see that there is an additional broad band of unknown origin in the 900–1300 cm^{-1} region. For example it can be absorption caused by numerous silicate submicron inclusions similar to those, which have been described by Navon et al. (1988). Two extremely strong lines at 1405 and 3107 cm^{-1} with intensity up to 10 cm^{-1} for the latter were found only in the coat: they are related to bending and stretching vibrations of C–H bonds probably located on the interfacial surfaces (Clark et al., 1992). The lines at 3237 and 3309 cm^{-1} are also intense in the coat. The first one has been found in grey-violet hydrogen-rich diamonds and also attributed to C–H

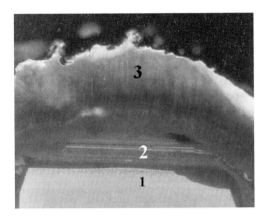

Fig. 4. A fragment of the cathodoluminescence pattern for coated diamond #SL-00/14. A light area with a white-bluish CL corresponds to the internal part of the core, whereas in its external part (2) CL has much lower intensity and well-pronounced octahedral striation. CL in the coat (3) is also relatively weak, but it increases when moving to the crystal surface. Maximum CL intensity in the coat is typical for a near-surface layer.

vibration, whereas the second is observed in some natural diamonds showing type Ib character (Zaitsev, 2001).

Spatial distribution of nitrogen in A/B form and of different spectroscopic features mentioned above along the plates is demonstrated in Fig. 6. One can see that in the coat nitrogen concentration is always high (more than 1000 ppm in A form) whereas in the core it may be higher (~ 1200 ppm in #SL-00/31) or several times lower (~ 300 ppm, #SL-00/14) with the main part of nitrogen in B-form (Fig. 6). For #SL-00/14 with a wide nonluminescent layer in the core nitrogen concentration was found to be very low, less than 20 ppm. The maximum intensity of the C–H vibrations is typical of the narrow layer near the core-to-coat boundary. Real spatial resolution of IR absorption topography was found to be lower than that in the PL images (Figs. 2D and 3D): the core/coat boundaries are smooth and we could not reveal the low nitrogen layer in #SL-00/31. The reasons are (1) a large thickness of the plate (500 μm, which is several times larger than the aperture), (2) a curvature/slope of typical growth structures and (3) a deviation of the beam form from cylinder when using the short-focus optics of the IR microscope: as a result a light beam goes through layers with

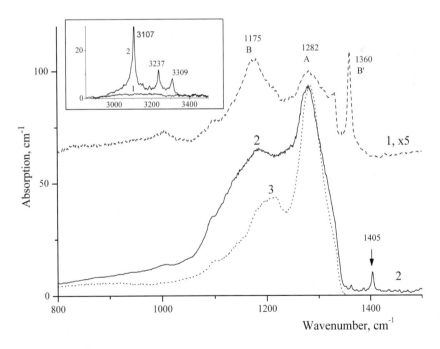

Fig. 5. Typical absorption spectra in the mid-IR region for the core (1) and the coat (2) with #SL-00/14 sample as an example, the latter is compared with spectrum (3), obtained for diamond with the only type (A) of nitrogen defects. The spectra are shifted upwards for clarity. Detail with (C–H)-related features is given in the inset.

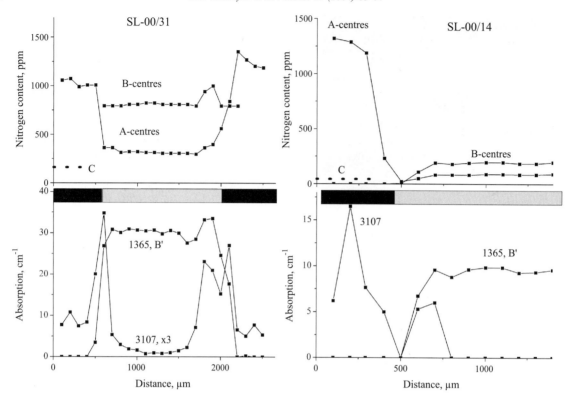

Fig. 6. The spatial distribution of the main defects across #SL-00/31 (at left) and a part of #SL-00/14 sample (at right) according to the IR absorption spectroscopy: nitrogen concentration in A and B form (above) and absorption intensity at 1365 cm^{-1} for B′-defect and 3107 cm^{-1}, C–H bonds (below). The block diagram shows a core (light grey) and a coat (grey).

different properties simultaneously and their spectra are superimposed.

3.2.2. Absorption in the UV–visible region

In the luminescent central part of the core the diamond's transparency is determined by a broad band near fundamental absorption edge with narrow lines at 306, 316 nm, which are due to A-defects: as a result crystals become transparent at $\lambda > 270$ nm. At longer wavelengths, the well-pronounced N3 vibronic system with ZPL at 415.2 nm and the N_3V structure is observed (Fig. 7). This defect is typical of natural diamonds with a high degree of nitrogen aggregation and usually accompanies B and B′ defects. In the nonluminescent periphery layer of the core the crystal is transparent from 225 nm with typical N9 features as 230, 236 nm lines near the fundamental absorption edge: this system is associated with electronic transitions in B defects. In the coat the diamond becomes transparent only from

$\lambda \sim 320$ nm with a broad structureless shoulder extending up to 600 nm. The latter is similar to the well-known absorption of donor nitrogen (C-centers) and is in agreement with previous data of Faulkner et al. (1965), who revealed these defects in coated diamonds using ESR. We estimated input of C-centres based on the absorption value at 477 nm and using the $\alpha_{477} = 1.4\alpha_{1135}$ equation, given by Sobolev et al. (1968): thus we obtained nitrogen values of ~ 190 and 62 ppm for #SL-00/31 and #SL-00/14. We propose that a relatively low concentration of C-defects, which is less than 1/5 of the total nitrogen content, is the reason we do not see the corresponding component in the IR absorption spectra. These values are shown by dotted lines for both samples in Fig. 6. In the nitrogen-poor layer of the core the concentration in form of the A-defects was estimated from the absorption intensity at 306 nm using equations given by Kaiser and Bond (1959): it is lower than 40 ppm.

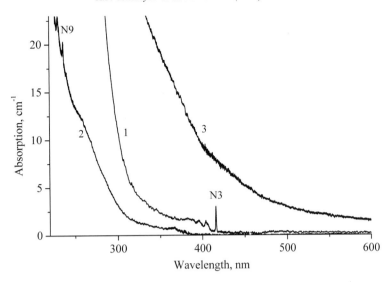

Fig. 7. The absorption spectra in the UV-visible region, recorded at 80 K for the inner (1) and outer nonluminescent (2) parts of the core and for the coat (3) of #SL-00/14 sample.

3.3. PL spectroscopy

The PL spectra, recorded at 325 and 488 nm excitation, for #SL-00/31 plate are given in Fig.

8 and in Fig. 9 intensity of different spectroscopic features (ZPLs or bands) is plotted versus distance from the surface. For the central part of the core, the PL spectra with dominating N3 system and weak

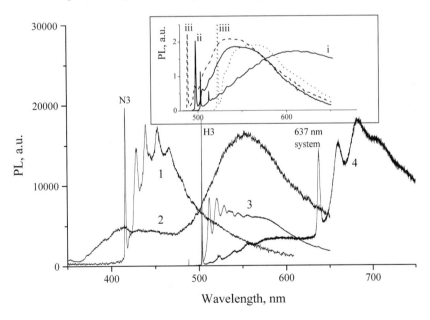

Fig. 8. The low-temperature PL spectra, recorded for #SL-00/31 at UV (spectra 1, 2) and 488 nm excitation (3, 4). Spectrum (1) is the N3 emission of the core. Spectra 2–4 demonstrate PL of the coat: broad band PL in its central part (2), H3 emission in the core/coat boundary (3) and 637 nm system in the near-surface layer of the diamond. In the inset: The individual vibronic systems with ZPLs at 503.4/510.7 nm (spectrum i, S1 system), 496.7 (ii), 488.9 (iii), 523.2 nm (iiii). These spectra were been recorded at 80 K, with a 325 nm excitation, for a natural Ib type diamond (i) and for a high-quality synthetic diamond, annealed at $T > 1700$ °C (ii, iii, iiii).

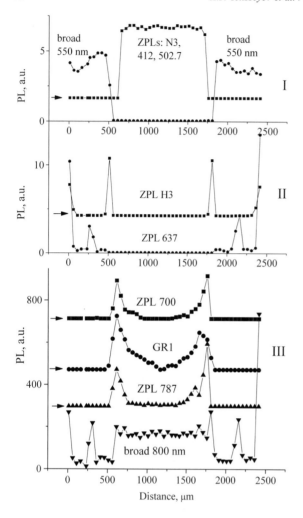

Fig. 9. The spatial distribution of the main features in PL spectra of #SL-00/31 diamond at 325 (section I), 488 (II) and 514 nm (III) excitation, at $T = 8$ K. The curves are shifted upwards for clarity. Spectra are obtained with light beam diameter of about 5 μm from the near-surface layer of the diamond plate. The spatial resolution is higher than that in Fig. 6: here boundaries between different layers are narrow and the nonluminescent outer layer of the core is well resolved.

ZPLs at 412.0 and 502.7 nm were practically identical for both #SL-00/31 and #SL-00/14 (Fig. 8, curve 1) and their intensity was approximately uniform in the central luminescent area of the core. The PL spectra for the coats show mainly structureless bands at ~ 430 and 550 nm (curve 2) although some layers also demonstrate a fine structure. A shortwave broad band centered at ~ 420 nm is assumed to be a

dislocation A-band, which is well known both for synthetic and natural diamonds (Zaitsev, 2001): it is responsible for a bluish tint in the coat emission. The other broad band centered at 550 nm has a half width of about 50 nm: it is responsible for a yellow-green PL of the coat. An additional H3 system with a 503.2 nm ZPL is superimposed on this band in a thin layer, corresponding to core-to-coat boundary (3), whereas both H3 and "637 nm" system (see spectrum 4 in Fig. 8) is emitted in the outer, near-surface layer of the stone at 488/514 nm excitation (Fig. 9). These systems are believed to be due to nitrogen-vacancy complexes VNNV and NV⁻, respectively (Zaitsev, 2001). The width of the 415.2 nm ZPL is ~ 0.5 nm in the core, whereas those for H3 and "637 nm" system in the coat are considerably broadened (1.3 and 3.5 nm, respectively). The first one is split into several components with a distance between end-members of about 3.9 meV. The strong stresses in the coat, which are also revealed by X-ray topography and birefringence (Figs. 2B,C and 3B,C) are the main reason of the ZPL broadening for H3 and 637 nm systems and the complete removal of the ZPL of the 420 and 550 nm bands. Based on ZPL split value the stresses were estimated to be of about 0.3 GPa following Davies et al. (1976). The H3 defects can be induced by strong mechanical deformations, for instance in the regions of indentations, around slip traces, or dislocation lines (Clark et al., 1992) and inside coated diamonds the H3 emission traces a layer in the core-to-coat boundary with maximum inclusions concentration and maximum stresses.

It is interesting to understand what is the origin of a yellow-green emission of the coat. According to Zaitsev's data handbook (2001) there are six optical systems, which could be responsible for yellow or yellow-greenish PL in diamonds. Three of them are due to nitrogen-vacancy defects: (1) S1 system with ZPLs at 503.4 and 510.7 nm, which is due to NV pair, (2) H3 and (3) H4 (496.2 nm, complex of B defect and vacancy). The others are associated with nickel-nitrogen-vacancy complexes: S3 with ZPL at 496.7 nm and structure N–VNiV–N plus two systems with ZPLs at 488.9 and 523.2 nm, which both are due to isomers of N–VNiV–N₂ (Nadolinny and Yelisseyev, 1993). The last three nickel–nitrogen complexes were studied recently in detail using optical spectroscopy and ESR in synthetic diamonds grown in the Fe–Ni–

C system and annealed at $T>1700$ °C to make nitrogen mobile: corresponding paramagnetic centers are NE1, NE2 and NE3, respectively. The individual PL spectra with fine structure obtained for high quality synthetic diamonds are given in the inset in Fig. 8. All three nickel–nitrogen complexes have also been found in natural diamonds from different deposits including Siberia, Canada, Australia, etc. Since their structure and formation conditions are very similar they are present together in the overwhelming majority of both synthetic and natural diamonds.

The S1 vibronic system is given in the inset in Fig. 8: one can see that the broad band maximum is considerably shifted to long waves (to ~ 600 nm) and this band is much wider than 550 nm PL band of the coat. This system is typical mainly of Ib type diamonds, whereas the coat is related to IaA type. Nevertheless, the presence of C-centres was established using UV–Vis absorption and ESR (Section 3.2) and input from the S1 system cannot be excluded. Both H3 and H4 systems have a broad band centered at shorter wavelengths (~ 520 nm) and are narrower (~ 25 nm half-widths) relative to experimentally detected 550 nm band. Moreover, as shown in Fig. 8 (spectrum 3), the H3 system saves its fine structure under stresses although the ZPL is broadened. H4 is a product of B defects after diamond irradiation and annealing but we do not identify this primary defect in the coat from the IR absorption spectra.

The above mentioned nickel-containing complexes are typical of IaA type synthetic diamonds, their phonon-related structure is weakly pronounced and nothing is known about their behaviour under stress. As one can see in Fig. 8 the position and width of the broad bands in their spectra are similar to that for the coat PL. On the other hand Lang et al. (in press) showed recently that the cuboid-shaped fragments with fibrous structure inside octahedral natural diamonds demonstrate very intense yellow-green PL of the $N-VNiV-N_2$ isomers with well-pronounced ZPLs at 488.9 and 523.2 nm. The outer layers of such diamonds, which cover the cuboid interior and are responsible for the final octahedral shape, have strong N3 luminescence and are related to IaAB type. One may suppose that the stresses typical of fibrous structure and appearing during the "abnormal" growth have been removed during further HPHT annealing of these stones in natural conditions. Thus,

broadened S3, 488.9 and 523.2 nm systems due to nickel–nitrogen-vacancy complexes and maybe S1 are likely to be responsible for intense yellow-green PL of the coat.

Additional spectroscopic features were found in the core at 514 nm excitation: a ZPL doublet (GR1) with the main component at 741 nm, two ZPLs at 700 and 787 nm and a broad band at 800 nm. GR1 system is due to a neutral single vacancy V^0, which is a primary radiation defect and appears after diamond irradiation with fast electrons, γ-rays or heavy particles (Zaitsev, 2001). Primary radiation defects V^0 are annealed at ~ 600 °C, with ~ 2.5 eV activation energy in ideal diamond lattice, whereas in polycrystalline diamond films the energy was found to be much lower (~ 1.5 eV, Zaitsev, 2001). Annealing stimulates vacancy migration and formation of their complexes with C and A-defects: resulting defects are responsible for "637 nm" and H3 systems in PL. The fact that primary radiation defects V^0 are saved in the core whereas products of radiation + annealing are concentrated in the near surface layer of coated diamonds shows that (1) after growth coated diamonds were irradiated both with fast electrons/γ-rays, which go through the whole stone, and also with heavy particles, producing damage in a thin near-surface layer, and (2) diamonds were annealed afterwards at a relatively low temperature of about 300–400 °C, which is not enough for vacancy migration in the perfect diamond lattice of the core.

Plotnikova et al. (1980) showed that 700 and 787 nm systems are typical of some natural Ia type diamonds, where they always accompany the nickel related systems with ZPLs at 496.7 (S3), 488.9 and 523.2 nm. Yelisseyev et al. (2002) found 700 and 787 nm systems in synthetic diamond grown in the Ni–Fe–C systems after further HPHT annealing. Temperature, which is necessary to produce these systems, is higher than that for systems with 496.7 (S3), 488.9 and 523.2 nm ZPLs. Thus 700 and 787 nm systems should be related to some more complicated nickel–nitrogen complexes, into which the $N-VNiV-N$ and $N-VNiV-N_2$ defects transform after capturing additional nitrogen atoms. The specific feature of coated diamonds studied is that 700 and 787 nm systems are present in the absence of PL systems with ZPLs at 496.7 (S3), 488.9 and 523.2 nm, which they usually accompany. This fact may be a result of the extreme temperatures and pressures during formation/annealing of the core part of examined diamonds and the resulting maximum degree of impurity aggregation. This assumption is in accord both with maximum aggregation degree of purely nitrogen defects (N3, B-defects) and petrological estimations. The latter gives maximum pressure of about 110 kbar and this value is supported by the presence of subcalcic chromric piropes with up to 17 mol% of majoritic component among the crystalline inclusions in the Snap Lake diamonds (Pokhilenko et al., 2001).

3.4. PLE spectroscopy

To verify the presence of PL from nickel–nitrogen complexes we recorded the photoluminescence excitation spectra (PLE) for the main PL systems: they are given in the upper half of Fig. 10. Curve 1 with a pronounced 415.2 nm ZPL system confirms that a blue emission of the core is N3. Curve 2, related to yellow-green emission in the coat demonstrates two broad bands centered at ~ 320 and 430 nm. These two bands may be considered as β and α-bands in the

absorption/PLE spectra associated with some of the NE1-NE3 nickel–nitrogen complexes (Nadolinny and Yelisseyev, 1993). The individual PLE spectra for these three emission systems with typical fine structure, which can be observed only in the high-quality stones, are given in the lower half of Fig. 10: here spectra 4, 5, 6 correspond to NE1, NE2, NE3 defects, respectively. The presence of several broad bands in the absorption/PLE spectrum indicates that there are two or more excited state levels in their energy diagrams. For the lower excited state, α, there is a single ZPL for each system and its position is 472.8 nm for NE1, 477.0 (NE2) and 478.5(NE3) as was established by Yelisseyev et al. (2003). As in the case of PL, the PLE spectra are broadened for the coat and it is difficult to determine the input of each defect. The PLE spectrum for S1 emission is considerably different: it contains the only broad band in the 450–300 nm spectral range with maximum at about 360 nm (Zaitsev, 2001).

Thus it is obvious that nickel–nitrogen complexes of NE1–NE3 type are responsible for yellow-green PL in the coat. We cannot definitely select, which of

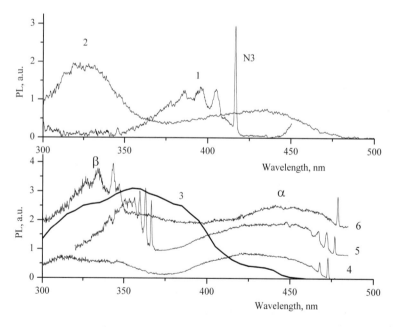

Fig. 10. Above are the PLE spectra for #SL-00/14 coated diamond: for blue PL in the core (1, N3 system, which is close to a mirror image of analogous system in PL spectra) and for yellow-green PL in the coat (2). Below are the PLE spectra for S1 emission of a natural type Ib diamond (3) and for emission in systems with ZPLs at 496.7 nm (4), 488.9 nm (5), and 523.2 nm (6) in a high quality synthetic diamond. All spectra are recorded at 80 K. Spectra 5, 6 are shifted upwards for clarity.

these three defects is responsible for 550 nm PL of the coat. Nevertheless, taking into account a long-wave shift in position of 523.2 nm PL system and the fact that 488.9 and 523.2 nm systems are due to isomers with similar formation conditions, one can suppose that in PL spectra of the coat we have a combination of all three systems or a sum of 488.9 and 523.2 nm systems. Our previous examination of synthetic diamonds showed that all three mentioned systems cannot be excited by electron beam and it is in accord with a weak CL intensity in the coat in Fig. 4.

3.5. Growth history of coated diamonds

All coated diamonds studied consist of a core of octahedral habit, which contains nitrogen aggregated in A, B forms and platelets B′ in its main central nitrogen-rich part, surrounded by a nitrogen-poor layer. In the coat nitrogen is mainly in the A form as deduced by IR spectroscopy whereas some nitrogen is present also in C-form as detected by UV–Vis absorption and ESR. Total nitrogen concentration is usually high in the coat (~ 1500 ppm in examined luminescent stones) and may be comparable or several times lower, in the central part of the core. The core diamond grew under equilibrium conditions by a layer-by-layer mechanism and there were at least two growth stages in the core history with considerably different growth conditions. In the first stage nitrogen concentration was high (hundreds to thousand ppm), in the following stage it was reduced to several tens ppm, which could be caused by low nitrogen content or presence of nitrogen getters such as Zr, Ti, etc., in the surrounding Ni-containing melt or by an extremely slow growth rate. Each growth stage was followed by a partial dissolution and as a result the originally octahedral diamonds became rounded. The core fragment of the samples underwent HPHT annealing, after which all nitrogen is highly aggregated in A, N_3V and B complexes, whereas there is some Ni in the form of most complicated nickel–nitrogen complexes responsible for 700 and 787 nm PL systems. The latter is likely to be a result of the extreme temperatures and pressures, the latter up to 110 kbar according to petrological estimations (Pokhilenko et al., 2001), in the lithosphere during diamond formation/annealing.

The last growth stage was characterized by high supersaturation and a high growth rate: at this stage diamond grew as needles from numerous seeding points (Kamiya and Lang, 1965) and captured a lot of impurities both as inclusions (visible or submicron invisible ones) and as point (structural) defects. In this area of abnormal or nonequilibrium growth, where diamond is formed at relatively low temperatures, impurities are captured mainly as individual atoms (ions) and the aggregates appear to be a result of nitrogen migration. Nickel plays an important role in the aggregation process: According to Fisher and Lawson (1998) nickel is known to accelerate considerably the nitrogen aggregation process and to lower the specific temperatures. Thus, Ni is the reason why nitrogen is present as A defects in the coat. On the other hand high Ni content in inclusions is typical of ultrabasic inclusions association in diamond. This conclusion is in accord with a prevalence of peridotite-type paragenesis for inclusions in diamonds from the SL/KL deposit in Canada (Pokhilenko et al., 2001). The very fast growth of the diamond coat is associated with the processes of metasomatism and partial melting in the lithosphere roots, which in turn may be a part of the cycle of kimberlite melt formation.

The as-grown coated diamonds were irradiated with electrons and/or γ-rays, which produced primary V^0 defects throughout the diamond volume. Particularly high concentrations of such defects were produced during irradiation with heavy particles (α) in the near-surface layer. The fact that primary radiation defects V^0 are saved in the core shows that further annealing occurred at a relatively low temperature (~ 300–400 °C), which was enough for vacancy migration with formation of different nitrogen-vacancy complexes only in the Ni enriched and stressed coat. For the perfect diamond lattice of the core a temperature of about 600 °C is necessary to stimulate vacancy migration (Zaitsev, 2001), but it was not reached. It is obvious that irradiation took place after formation of the kimberlite body. On the other hand the radiation dose is expected to be high enough for the Snap Lake diamonds because (1) Snap Lake kimberlites are located in ancient granitoids and the dykes thicknesses are small enough (~ 2 m); (2) these kimberlites are of Cambrian age (~ 540 Ma) of emplacement and the exposure time to radiation is very large.

4. Conclusions

1. Coated diamonds are a result of a multistage growth process with the stages separated in time. Examination of a large batch of diamonds from the SL/KL deposit in Canada allows us to suppose that many of these stones have or initially had a fibrous coat: thus an abnormal growth on the last stage is likely to be typical of all SL/KL diamond association.
2. Nickel in the form of point defects was found for the first time in coated diamonds: in the coat it is present in the form of nickel–nitrogen complexes with nitrogen number $n = 2$ or 3, whereas in the core nickel-containing defects are more complicated: it is likely to be a result of ultra-high temperatures and pressures during formation/annealing of the core part of the diamonds.
3. The Ni presence in coated diamonds is in accord with results of a petrologic study, which showed the absolute majority of ultrabasic type diamonds among the Snap Lake diamond population.
4. Coated diamonds as well as many other diamonds from Canada underwent high-energy ionizing irradiation after formation of the kimberlite body.

Acknowledgements

This work was supported partly by the Grants nos 98-05-65283, 02-05-65075 and 03-05-64040 of the Russian Foundation for Basic Research. A.P.Y. thanks the Royal Society for the Visiting Research Fellowship. Authors are grateful to G. Rylov (IMP SB RAS) for X-ray topography and H.J. Milledge (University College, London) for the cathodoluminescence study of the examined stones. The authors are greatly thankful to I. Chinn from De Beers Group Exploration Macrodiamond Laboratory, reviewer, for kind edition of the manuscript.

References

Angress, J.F., Smith, S.D., 1965. The observation of defect-activated one-phonon infra-red absorption in diamond coat. Philos. Mag. 12, 415–417.

Boyd, S.R., Mattey, D.P., Pillinger, C.T., Milledge, H.J., Mendelssohn, M., Seat, M., 1987. Multiple growth events during diamond genesis: an integrated study of carbon and nitrogen isotopes and nitrogen aggregation state in coated stones. Earth Planet. Sci. Lett. 86, 341–357.

Boyd, S.R., Kiflawi, I., Woods, G.S., 1995. Infrared absorption by the B nitrogen aggregate in diamond. Philos. Mag., B 72, 351–361.

Clark, C.D., Collins, A.T., Woods, G.S., 1992. Absorption and luminescence spectroscopy. In: Field, J.E. (Ed.), The Properties of Natural and Synthetic Diamonds. Academic Press, London, pp. 35–80.

Custers, J.F.H., 1950. On the nature of the opal-like layer of coated diamonds. Am. Miner. 35, 51–58.

Davies, G., Nazare, M.N., Hamer, M.F., 1976. The H3 (2.463 eV) vibronic band in diamond: uniaxial stress effects and the breakdown of mirror symmetry. Proc. R. Soc. Lond. A351, 245.

Faulkner, E.A., Whippey, P.W., Newman, R.C., 1965. Electron spin resonance in diamond coat. Philos. Mag. 12, 413–414.

Fisher, D., Lawson, S., 1998. The effect of nickel and cobalt on the aggregation of nitrogen in diamond. Diam. Rel. Mat. 7, 299–304.

Kaiser, W., Bond, W.L., 1959. Nitrogen, a major impurity in common type I diamond. Phys. Rev. 115, 857–863.

Kamiya, Y., Lang, A.R., 1965. On the structure of coated diamonds. Philos. Mag. 11, 347–356.

Lang, A.R., 1974. Glimpses into the growth history of natural diamonds. J. Cryst. Growth 24/25, 108–115.

Lang, A.R., Yelisseyev, A.P., Pokhilenko, N.P., Steeds, J.W., Wotherspoon, A., 2003. Is dispersed nickel in natural diamonds associated with cuboid growth sectors in diamonds that exhibit a history of mixed-habit growth? J. Cryst. Growth. 263, 575–589.

Logvinova, A.M., Klein BenDavid, O., Izraeli, E.S., Navov, O., Sobolev, N.V., 2003. Microinclusions in fibrous diamonds from Yubileinaya kimberlite pipe (Yakutia). 8th Kimberlite Conference, Victoria, Canada, Long Abstracts, Paper FLA-0025.

Nadolinny, V.A., Yelisseyev, A.P., 1993. New paramagnetic nickel-containing centers in diamond. Diam. Rel. Mat. 3, 17–22.

Navon, O., Hutcheon, I.D., Rossman, G.R., Wasserburg, G.J., 1988. Mantle-derived fluids in diamond micro-inclusions. Nature 335, 784–789.

Orlov, Yu.L., 1973. The mineralogy of diamond. Izv. Nauk. SSSR, 235 (translated in 1977 from Russian, Wiley, New York).

Plotnikova, S.P., Kluyev, Yu.A., Parfianovich, I.A., 1980. Long-wave photoluminescence of natural diamonds. Mineral. J. 4, 75–80.

Pokhilenko, N.P., Sobolev, N.V., McDonald, J.A., Hall, A.E., Yefimova, E.S., Zedgenizov, D.A., Logvinova, A.M., Reimers, L.F., 2001. Crystalline inclusions in diamonds from kimberlites of the Snap Lake area (Slave Craton, Canada): New evidences for the anomalous lithospheric structure. Dokl. Earth Sci. 380 (7), 806–811.

Schrauder, M., Navon, O., 1994. Hydrous and carbonatitic mantle fluids in fibrous diamonds from Jwaneng, Botswana. Geochim. Cosmochim. Acta 58 (2), 761–771.

Sobolev, E.V., Lisoivan, V.I., Samsonenko, N.D., 1968. On the state of impurity nitrogen in synthetic diamonds. Fis. Tverd. Tela 10, 2266–2268 (in Russian).

Woods, G.S., Purser, G.C., Mtimculu, A.S.S., Collins, A.T., 1990.

The nitrogen content of type Ia natural diamonds. J. Phys. Chem. Solids 51, 1191–1197.

Yelisseyev, A.P., Nadolinny, V.A., Feigelson, B.N., Babich, Yu.V., 2002. Spectroscopic features due to Ni-related defects in HPHT synthetic diamonds. Int. J. of Mod. Phys. B, Condens. Matter Phys. 16, 900–905.

Yelisseyev, A.P., Lawson, S., Sildos, I., Osvet, A., Nadolinny, V.,

Feigelson, B., Baker, J.M., Newton, M., Yuryeva, O., 2003. Effect of HPHT annealing on the photoluminescence of synthetic diamonds, grown in the Fe–Ni–C system. Diam. Rel. Mat. 12, 2147–2168.

Zaitsev, A.M., 2001. Optical Properties of Diamonds: A Data Handbook. Springer Verlag, Berlin.

Available online at www.sciencedirect.com

Lithos 77 (2004) 99–111

www.elsevier.com/locate/lithos

Inclusions in diamonds from the K14 and K10 kimberlites, Buffalo Hills, Alberta, Canada: diamond growth in a plume?

Rondi M. Davies[a,*], William L. Griffin[a,b], Suzanne Y. O'Reilly[a], Tom E. McCandless[c]

[a] GEMOC ARC National Key Centre, School of Earth and Planetary Sciences, Macquarie University, Sydney, NSW 2109, Australia
[b] CSIRO Exploration and Mining, P.O. Box 136, North Ryde, NSW 1670, Australia
[c] Ashton Mining of Canada, Unit 123-930 West 1st Street, North Vancouver, BC, Canada VTP3N4

Received 27 June 2003; accepted 31 October 2003
Available online 1 June 2004

Abstract

Analyses of mineral inclusions, carbon isotopes, nitrogen contents and nitrogen aggregation states in 29 diamonds from two Buffalo Hills kimberlites in northern Alberta, Canada were conducted. From 25 inclusion bearing diamonds, the following paragenetic abundances were found: peridotitic (48%), eclogitic (32%), eclogitic/websteritic (8%), websteritic (4%), ultradeep? (4%) and unknown (4%). Diamonds containing mineral inclusions of ferropericlase, and mixed eclogitic-asthenospheric-websteritic and eclogitic-websteritic mineral associations suggests the possibility of diamond growth over a range of depths and in a variety of mantle environments (lithosphere, asthenosphere and possibly lower mantle).

Eclogitic diamonds have a broad range of C-isotopic composition ($\delta^{13}C = -21\permil$ to $-5\permil$). Peridotitic, websteritic and ultradeep diamonds have typical mantle C-isotope values ($\delta^{13}C = -4.9\permil$ av.), except for two ^{13}C-depleted peridotitic ($\delta^{13}C = -11.8\permil$, $-14.6\permil$) and one ^{13}C-depleted websteritic diamond ($\delta^{13}C = -11.9\permil$). Infrared spectra from 29 diamonds identified two diamond groups: 75% are nitrogen-free (Type II) or have fully aggregated nitrogen defects (Type IaB) with platelet degradation and low to moderate nitrogen contents (av. 330 ppm-N); 25% have lower nitrogen aggregation states and higher nitrogen contents ($\sim 30\%$ IaB; <1600 ppm-N).

The combined evidence suggests two generations of diamond growth. Type II and Type IaB diamonds with ultradeep, peridotitic, eclogitic and websteritic inclusions crystallised from eclogitic and peridotitic rocks while moving in a dynamic environment from the asthenosphere and possibly the lower mantle to the base of the lithosphere. Mechanisms for diamond movement through the mantle could be by mantle convection, or an ascending plume. The interaction of partial melts with eclogitic and peridotitic lithologies may have produced the intermediate websteritic inclusion compositions, and can explain diamonds of mixed parageneses, and the overlap in C-isotope values between parageneses. Strong deformation and extremely high nitrogen aggregation states in some diamonds may indicate high mantle storage temperatures and strain in the diamond growth environment. A second diamond group, with Type IaA–IaB nitrogen aggregation and peridotitic inclusions, crystallised at the base of the cratonic lithosphere. All diamonds were subsequently sampled by kimberlites and transported to the Earth's surface.
Crown Copyright © 2004 Published by Elsevier B.V. All rights reserved.

Keywords: Inclusions in diamond; Alberta diamonds; Buffalo Hills; Carbon isotopes; Nitrogen contents; Nitrogen aggregation states

* Corresponding author. Present address: Department of Earth and Planetary Sciences, American Museum of Natural History, New York, NY 10024-5192, USA. Tel.: +1-212-769-5314; fax: +1-212-769-5339.
E-mail address: rdavies@amnh.org (R.M. Davies).

0024-4937/$ - see front matter. Crown Copyright © 2004 Published by Elsevier B.V. All rights reserved.
doi:10.1016/j.lithos.2004.04.008

1. Introduction

Syngenetic mineral inclusions trapped in diamond represent pristine samples of mantle material incorporated in the diamond during its growth. Such inclusions provide direct evidence of the composition and physical conditions of ancient mantle and diamond growth environments, and allow us to place constraints on the tectonic histories of the diamond-bearing mantle beneath cratons. Diamond inclusion studies can also be applied to diamond exploration by identifying minerals of characteristic compositions that can be targeted as prospective indicators.

Diamond morphology (including surface resorption and deformation features), C-isotopes, defect and impurity abundances (mainly nitrogen) measured by FTIR spectroscopy provide additional evidence that can shed light on the origins of diamond; through identifying the origin of the carbon and the conditions of diamond growth, mantle storage and emplacement.

This paper reports the results of a study of 29 diamonds, 25 of which syngenetic inclusions were extracted and analysed, from the Buffalo Hills K14 (24) and K10 (1) kimberlite pipes in Alberta, Canada. An additional 10 diamonds have been analysed for C-isotopes. The results provide evidence that the diamonds have a complex mantle evolutionary history having grown in changing chemical environments of the lower mantle, asthenospheric and lithospheric mantle.

2. Geological setting

The Buffalo Hills kimberlite field, comprising 36 kimberlites discovered to date, occurs within the Precambrian Buffalo Head Terrane of northern Alberta, Canada (Ross et al., 1991; Carlson et al., 1998; McNicoll et al., 2000; Ross, 2002; Fig. 1). The terrane has no surface exposure, and is overlain by a thick sequence of Devonian and Cretaceous sedimentary rocks through which the Cretaceous kimberlites have erupted. Sm–Nd and U–Pb ages suggest the terrane formed between 2.3 and 2.0 Ga (Ross et al., 1991; Villeneuve et al., 1993; McNicoll et al., 2000). Bounded by the structural discontinuities of the Great Slave Lake shear zone to the north, the Snowbird Tectonic Zone to the south and the Taltson plutonic

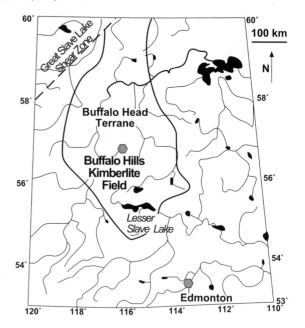

Fig. 1. Locality map of the Buffalo Head terrane with the Buffalo Hills Kimberlite Field shaded (modified from Carlson et al., 1998).

belt to the east, the Buffalo Head Terrane is thought to represent a crustal sliver that was incorporated with other Archean and Proterozoic crustal blocks through subduction, accretion and orogenesis around 2.0 Ga.

3. Samples and analytical techniques

The morphological features of 29 diamonds starting with Ash- are described (Table 1) and mineral inclusions identified prior to diamond crushing. The 10 diamonds with sample numbers starting with BH- are only considered in the C-isotope section of this paper. From 25 of the 29 diamonds, 53 crystalline inclusions were extracted (Table 2). Fifteen diamonds also contained epigenetic phases (Table 2). Crystalline inclusions were identified prior to destruction of the sample by the absence of cracks around the inclusion, by their geometric shapes, and by their unaltered appearance once extracted. Mineral inclusions are between 20 and 200 μm across. Epigenetic inclusions are typically larger (≤400 μm across).

The mineral inclusions were extracted by breaking individual diamonds in a closed cell in a custom-built stainless steel vice with a tungsten carbide piston.

Table 1
Syngenetic inclusion parageneses, morphological characteristics, carbon isotopes and nitrogen characteristics diamonds from the K14 kimberlite

Sample	Weight (g)	Col.	Shape	RC	Defm	Abr.	Para.	Inclusions	$\delta^{13}C$	N (ppm)	% IaB	Comments
Ash-101	0.0057	C-l	D F	1		–	P	Fo	– 4.7	750	100	H, no P
Ash-102A	0.0244	C-l	D	1			E	Gnt	– 18.8	450	100	H, no P
Ash-102B	0.008	Y	D	1		–	P	Fo	– 5.0	1620	20	H
Ash-104A	0.0061	B	E D	1	LL, Sh, Btl	–	P	Fo (3), cpx	– 11.8	0	0	Type II
Ash-104B	0.0138	B	D F	1	LL, Sh	–	P	Fo	– 4.7	0	0	Type II
Ash-104C	0.0184	B	D Ag	1	LL, Sh	–	–	–	– 2.9	<50	100	H, no P
Ash-105A	0.0122	C-l	D	3	LL, Sh, Btl	–	E	Majorite (2)	– 14.8	0	0	Type II
Ash-105B	0.0174	B	D Tw F	1	LL	–	P	Fo	– 3.6	0	0	Type II
Ash-105pm	0.0027	B	D F	1	Sh	–	P	Gnt, Fo	– 2.7	<50	100	H, no P
Ash-106A	0.0176	C-l	D Tw	3		–	E?	Fe-Gnt	– 6.9	720	~ 30	Type I
Ash-106B	0.0196	pY	D	1		–	–	–	– 5.6	a	a	Type I
Ash-106C	0.0061	Y	D Tw	1	LL, Sh	–	P	Fo	– 6.4	900	15	H
Ash-106D	0.0031	C-l	D Tw	1		–	P	Cr-sp (>10)	– 2.7	0	0	Type II
Ash-106pm	0.0084	B	D F	1	Sh	–	P	Fo	– 14.6	230	100	H, no P
Ash-108A	0.0576	B	D F	2	LL, Sh, Btl	bf	E	Gnt	– 4.8	65	100	H, no P
Ash-108B	0.0013	B	D Tw F	–		–	E/W	Gnt-cpx-ru, Gnt (2), E-cpx (2), W-cpx, Maj	– 4.8	0	0	Type II
Ash-109A	0.003	pB	D F	1	Sh	–	UD	fPer	– 4.6	0	0	Type II
Ash-109pm	0.0031	C-l	D	1		–	Web	Cpx	– 11.9	830	100	H, no P
Ash-110	0.0067	C-l	D Tw	1	Sh	–	P	Cpx	– 8.8	0	0	Type II
Ash-111A	0.0216	C-l	D Tw	1		–	–	–	– 5.8	1070	<10	no P, H
Ash-111B	0.0127	C-l	D F	1		–	E	Ru	– 4.7	0	0	Type II
Ash-111C	0.0082	B	F D	1	Btl, Sh	–	E	Gnt	– 14.6	0	0	Type II
Ash-111D	0.0073	C-l	F O	4	LL, Sh, Btl	bf	E	Cpx (2)	– 14.7	0	0	Type II
Ash-111E	0.0053	pB	D Tw	1	Sh	–	–	–	– 3.7	0	0	Type II
Ash-111F	0.0036	B	D F	2	LL, Sh, Btl	–	E	Cpx	– 20.6	a	a	Type I
Ash-111G	0.0023	C-l	Ir D F	2	LL, Sh, Btl	–	E	Cpx	– 13.7	400	100	H, no P
Ash-111H	0.0363	pB	D F	1	LL, Sh, Btl	bf	E/W	Cpx (>3)	– 4.6	135	100	H, no P
Ash-113	0.0051	pB	Ir D	1	LL, Sh, Btl	–	P	Fo (2)	– 5.7	560	12	P
Ash-K10-2	0.0019	C-l	O Tw	5		–	P	Gnt, Fo (8) Cpx (3),	– 3.8	0	0	Type II
BH14-1	0.0041	C-l	D Ag	1					– 7.2	–	–	
BH14-2	0.0023	C-l	O F	4		bf			– 4.9	–	–	–
BH14-3	0.0041	C-l	O F	4		bf			– 6.1	–	–	–
BH14-4	0.0012	C-l	O	5					– 3.5	–	–	–
BH14-5	0.0031	pB	C F	4		bf			– 2.6	–	–	–
BH14-6	0.0023	C-l	F	1		bf			– 4.1	–	–	–
BH14-7	0.0023	C-l	F	2		bf			– 7.1	–	–	–
BH14-8	0.0046	C-l	F Ir	1		bf			– 12.0	–	–	–
BH14-9	0.0008	C-l	F	1		bf			– 3.9	–	–	–
BH14-10	0.0012	C-l	C-O F	3		bf			– 6.4	–	–	–

Col. = colour, C-l = colourless, B = brown, Y = yellow, p = pale, D = dodecahedron, O = octahedron, E = elongate, Ag = aggregate, Tw = twinned, F = fragment, Ir = irregular, RC = resorption class (Robinson et al., 1989), LL = lamination lines, Sh = shagreen texture, Btl. = brittle displacement laminae, Defm = deformation, Abr. = abrasion, bf = broken face, Para = paragenesis, E = eclogitic, P = peridotitic, Web = websteritic, UD = ultradeep, Gnt = garnet, Cpx = clinopyroxene, Fo = olivine, Ru = rutile, Cr. sp. = Cr-spinel, Maj = majoritic garnet, fPer = ferropericlase. P = platelets, H = hydrogen.

[a] Data unresolved due to poor quality spectra.

Following crushing, the diamond rubble was tipped into a petrie dish and inclusions were identified using binocular and petrographic microscopes. Inclusions were then individually placed in petropoxy resin 154 on a glass slide. The epoxy was heat-hardened at 200 °C for 15 min. The inclusions in epoxy were polished by hand using fine wet and dry sandpaper and diamond paste.

Table 2
Inclusion abundances in diamonds from the K10 and K14 kimberlites, Alberta, Canada

Para.	Single phases	n	Para.	Multiple phases	n	Secondary phases	
E	Gnt	3	P	Gnt, Fo (not recovered)	1	Sphene + serpentine	1
E	Maj	1	P	Fo-Cr. diop-Gnt	1	Carbonate	7
E	Rutile	1	P	Fo-Cr. diop	1	Perovskite	2
E?	Fe-gnt	1	P	Cr-diop-serpentine	1	Phlogopite	1
E	Cpx	3	E-Web	Gnt-E. cpx-Ru, Maj,	2	Phlogopite-carbonate-perovskite	2
Web	Cpx	1		W. cpx, E. Cpx		Serpentine	2
P	Fo	7					
P	Cr-sp.	1					
UD	fPer	1					
Total		19			6		15

Major-element compositions of mineral inclusions were analysed using a Cameca SX 50 electron microprobe at Macquarie University. Operating conditions included an accelerating voltage of 15 kV, a sample current of 20 nA and a beam width about 3 μm, with a mixture of natural and synthetic minerals standards. Minor and trace elements in garnet inclusions were determined by laser ablation ICP-MS using a Perkin-Elmer Sciex ELAN 6000 coupled with a 266nm Nd:YAG laser at Macquarie University. The NIST 610 glass standard was used for calibration of element sensitivities and Ca determined by electron probe was used as an internal standard. Analytical techniques, accuracy and precision are discussed in detail by Norman et al. (1996, 1998).

Carbon isotope analyses for diamonds with sample numbers starting with Ash- were carried out at the Centre for Isotope Studies, CSIRO, North Ryde. Fragments of diamond weighing ~ 1 mg were placed into a baked clean quartz glass tube with 10–20 mg of conditioned CuO agent. Tubes were evacuated, sealed and reacted in a furnace at 1000 °C for 5 h. CO_2 was collected in a gas sample tube for mass spectrometric analysis (Finnigan 252 mass spectrometer using dual inlet mode). An anthracite standard ($\delta^{13}C = -23.1‰$) was run with most batches and international graphite standard NBS21 ($\delta^{13}C = -28.1‰$) was analysed at regular intervals. Carbon isotope analyses of the 10 diamonds with sample numbers beginning with BH- were contracted by Ashton Mining of Canada.

Fourier-transform-infrared spectra of diamonds were measured using a Digilab BIO-RAD FTS-60A attached to a UMA 300A IR-microscope with a liquid nitrogen reservoir at the Division of Coal and Energy Technology, North Ryde. Spectra were measured between wave numbers of 4400 and 550 cm^{-1}. The beam aperture was ≈ 90 μm. Sixty-four scans were made per analysis at a resolution of 8 cm^{-1}. Nitrogen contents in diamond were calculated from the spectra using the methods outlined in Mendelssohn and Milledge (1995). Nitrogen calculations assume pure IaA and IaB defects equal 160 and 750 at. ppm, respectively, at a fixed absorption coefficient of 1 per mm for diamond of 1 mm thickness (Woods et al., 1990; Mendelssohn and Milledge, 1995).

4. Diamond characteristics

Twelve of the twenty-nine diamonds are colourless, fourteen are brown and three are yellow. Twenty-seven diamonds are strongly resorbed octahedra (dodecahedra) and two are octahedra. Nine of the twenty-nine diamonds are twinned, one is an aggregate and eleven are fragments.

Common surface features on dodecahedral faces include large and small hillocks, polished low-relief surfaces, dissolution laminae and resorbed plates. Four diamonds have well-defined faces and edges. Eleven diamonds have freshly broken faces; fracturing may have occurred during extraction of the diamond from its host rock. Octahedral surfaces are planar and host negative trigonal and hexagonal etch pits. Etch channels (ruts) are prevalent (21%) and are commonly filled with serpentine.

Deformation is evident on the surfaces of two-thirds of the 29 diamonds. Nine of the diamonds with evidence of deformation show plastic deformation features with lamination lines and shagreen textures (micro-hillocks indicating the intersection of two or more sets of lamination lines). The other nine show both plastic and brittle deformation with displacements along slip planes expressed as steplike lineations with herringbone textures, indicating stronger deformation during their growth and/or mantle storage. Most of the nondeformed diamonds are colourless, whereas 13 of the 18 deformed diamonds are shades of brown. This observation supports other studies, which have observed that brown colouration is prevalent in deformed diamond (e.g., Robinson et al., 1989).

5. Mineral inclusions in diamonds

5.1. Eclogitic garnet

Nine low-Cr garnets were extracted and analysed from five diamonds (Table 3). Except for the inclusion in diamond 106A, compositions are within the range of eclogitic inclusions worldwide with moderate contents of Fe, Mg and Ca (Mg# ($100Mg/(Mg+Fe_{Total})$) 44 to 57; XCa ($100Ca/(Ca+Fe+Mg)$) 21 to 30). Three garnets in two eclogitic diamonds (105A, 108B) are moderately majoritic (garnet and clinopyr-

oxene in solid solution). One of these diamonds (108B) also contains two normal garnet inclusions, one of which is in contact with clinopyroxene and a rutile, and three clinopyroxenes.

The three majoritic garnets are characterised by excess Si (3.12–3.19 cations per formula unit (pfu) for 12 oxygens) at the expense of Al, which falls below the ideal value of 2 (1.56 to 1.72 cations pfu; Fig. 2a). The majoritic garnet inclusions have very high Na (0.22–0.23) and Ti (0.072–0.093; Fig. 2b) and higher Ca and lower Mg and Fe contents than other garnet inclusions. Elemental relationships may be explained by the high-pressure coupled substitution of $Si^{4+}-Na^{+}$ for $Al^{3+}-Mg^{2+}$ as shown in experiments by Irifune et al. (1986), with compositions corresponding to experimentally calibrated maximum pressures of formation of about 9 GPa (Fig. 2a). Such pressures correspond to asthenospheric depths of about 275 km.

Na and Ti contents in the Buffalo Hills eclogitic garnet diamond inclusions are high and show a bimodal distribution corresponding to the majoritic

Table 3
Representative electron microprobe analyses of eclogitic and peridotitic garnet inclusions in diamond

Sample	106A	102A	105A(1)	105A(2)	108A	108B(1)	108B(2)	108B(3)	111C	K10	105C
Mineral	Alm-gt	Gt	Maj	Maj	Gt	Maj	Gt	Gt	Gt	Gt	Gt
Para.	E	E	E	E	E	E	E	E cpx-gt	E	P	P
Assoc. Anal.	4	4	3	3	5	6	3	4	4	3	3
SiO_2	35.7	39.8	42.2	41.5	40.1	42.4	39.9	39.4	39.4	41.0	41.6
TiO_2	0.02	1.01	1.28	1.33	1.10	1.64	1.00	1.01	1.06	0.21	0.05
Al_2O_3	20.8	21.4	19.5	19.6	20.7	17.6	21.1	21.4	21.1	17.7	16.7
Cr_2O_3	<0.06	<0.06	0.07	<0.06	<0.06	0.08	<0.06	<0.06	<0.06	7.02	7.89
FeO	32.0	16.58	14.49	14.44	16.92	14.72	16.60	16.66	18.28	6.62	7.17
MnO	6.67	0.29	0.31	0.28	0.40	0.26	0.29	0.28	0.42	0.17	0.28
MgO	3.87	10.86	9.54	9.63	12.42	10.39	11.04	10.87	8.10	20.34	19.93
CaO	0.46	9.73	10.66	10.82	8.16	10.69	9.86	9.93	11.06	5.76	6.06
Na_2O	0.02	0.28	1.55	1.53	0.21	1.57	0.31	0.32	0.48	0.02	0.01
K_2O	<0.04	<0.04	<0.04	<0.04	<0.04	<0.04	<0.04	<0.04	<0.04	<0.04	<0.04
NiO	<0.06	<0.06	<0.06	<0.06	<0.06	<0.06	<0.06	<0.06	<0.06	<0.06	<0.06
Total [12O]	99.7	100.0	99.7	99.2	100.0	99.6	100.1	100.0	99.9	98.9	99.7
Si	2.916	2.984	3.155	3.122	3.000	3.188	2.989	2.964	2.999	2.998	3.032
Al	2.004	1.893	1.721	1.741	1.822	1.564	1.865	1.895	1.888	1.524	1.434
Ti	0.001	0.057	0.072	0.075	0.062	0.093	0.056	0.057	0.060	0.011	0.003
Na	0.004	0.041	0.224	0.224	0.031	0.228	0.045	0.046	0.071	0.002	0.001
Mg#	17.7	53.9	54.0	54.3	56.7	55.7	54.2	53.8	44.1	84.6	83.2
XCa	0.01	0.26	0.30	0.30	0.21	0.29	0.26	0.26	0.30	0.15	0.15
XMg	0.15	0.40	0.38	0.38	0.45	0.39	0.40	0.40	0.31	0.72	0.70
XFe	0.69	0.34	0.32	0.32	0.34	0.31	0.34	0.34	0.39	0.13	0.14
XMn	0.15										

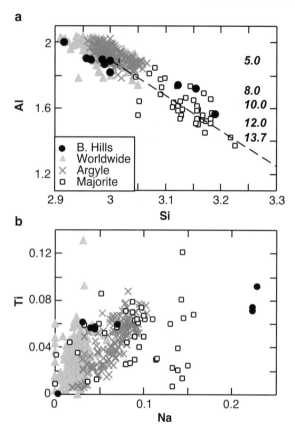

Fig. 2. Cation ratios (for 12 oxygen atoms) of eclogitic garnet inclusions in the Buffalo Hills diamonds and other diamond inclusions worldwide. The Argyle inclusion data are from Hall and Smith (1984), Sobolev et al. (1986) and Moore et al. (1989). In (a), numbers indicate pressure in GPa from which garnet with these Si/Al ratios are synthesised from pyrolite minus olivine (Irifune et al., 1986).

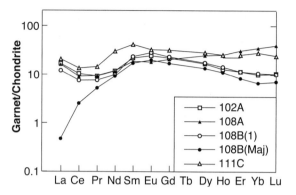

Fig. 3. Chondrite normalised rare earth element abundances of eclogitic garnet inclusions.

La, creating a sinusoidal LREE pattern. The majoritic garnet inclusion has lower REE, Zr and Hf and higher Ti than the other garnet inclusions. All inclusions show significant negative anomalies in Sr, Ga and V (Table 4).

The concave REE pattern for the majoritic garnet suggests a complex history of melt depletion (deple-

and normal garnet groups (Fig. 2b). Garnets show values similar to those in the Argyle diamonds from Western Australia that also occur in Proterozoic mantle (Jaques et al., 1989). High Na and Ti may indicate both high-pressure formation (e.g., McCandless and Gurney, 1989) and/or an Na- and Ti-rich eclogite bulk composition.

The trace element patterns of six eclogitic garnet inclusions from four diamonds are relatively flat, with heavy rare earth element (HREE) enrichment in two diamonds (108A, 111C) and HREE depletion in the others (Fig. 3). In all but one inclusion (the majoritic garnet from 108B), light rare earth elements (LREE) Ce, Pr and Nd are depleted except for

Table 4
Trace element abundances in eclogitic garnet inclusions from Buffalo Hills diamonds

Sample	102A	108A	108B(1)	108BMaj	111C	Average
U	31.5	43.2	24.7	–	38.3	34.4
Th	57.0	38.3	36.4	–	64.6	49.1
Sr	1.51	1.71	1.23	1.04	5.74	2.25
La	16.62	12.36	11.93	0.47	20.5	12.36
Ce	10.28	9.37	7.46	2.50	13.43	8.61
Pr	9.15	8.36	7.52	5.16	14.25	8.89
Nd	11.60	8.51	9.73	9.06	29.6	13.70
Sm	21.8	12.6	23.1	17.00	40.8	23.1
Eu	23.9	13.3	28.0	19.64	32.1	23.4
Gd	21.6	15.0	22.9	16.79	31.5	21.6
Dy	16.48	20.0	17.31	13.19	28.0	18.99
Ho	12.68	20.0	14.03	10.97	25.2	16.58
Y	11.92	19.55	11.92	9.36	24.1	15.37
Er	11.33	22.7	11.33	8.18	25.2	15.73
Yb	9.85	29.5	10.46	6.52	27.7	16.81
Lu	10.29	29.6	9.88	7.00	23.9	16.13
Sc	4.74	12.18	4.91	4.43	7.90	6.84
Ga	1.19	1.18	0.98	0.90	1.09	1.07
Zr	36.0	31.5	28.0	11.17	38.1	29.0
Hf	33.7	37.0	26.0	11.44	38.5	29.3
Ti	12.71	14.70	12.30	20.6	11.55	14.36
V	1.15	4.66	1.22	2.06	2.94	2.41
Nb	5.53	24.8	0.69	14.23	5.61	10.17
Ta	21.8	7.75	4.93	14.08	16.20	12.96

tion of LREE) with subsequent metasomatic enrichment of the MREE (Griffin et al., 1999 and references therein). The enrichment in La and Ce in the normal garnet inclusions is unusual and distinctive from the eclogitic majoritic garnet and implies a complex history for the eclogite protolith.

A crystalline almandine-rich garnet (32.0 wt.% FeO) with high Mn (6.67 wt.%) and no Na was found in one diamond (analysis 106A; Table 3). Although it is unlikely that the extraction methodology used could account for contamination, the inclusion had not been identified prior to the destruction of the diamond so it is considered with caution. We cannot rule out the possibility that the inclusion may have been embedded in a rut or crack on the diamond's surface. However, the diamond contained a large graphite rosette, which could be where the inclusion resided.

5.2. Peridotitic garnet

Two chrome-pyrope garnets were analysed from two diamonds (105C and K10, Table 3). Based on their low ratios of Cr_2O_3 (7.0 and 7.9 wt.%) to CaO (5.8 and 6.1 wt.%), and the coexistence of one inclusion with chrome-diopside and relatively Fe-rich olivine (Fo 90–91), the garnets are classified as lherzolitic.

5.3. Clinopyroxene

Sixteen inclusions were extracted from nine diamonds. Thirteen representative analyses are shown in Table 5. The compositions of the clinopyroxene inclusions show a continuum from Cr-diopside with peridotitic affinity (0.60–1.48 wt.% Cr_2O_3; Mg# = 93) to intermediary clinopyroxene compositions with lower Cr_2O_3 (0.20–0.48 wt.%; Mg# = 74–87) and moderate jadeite ($\leq$21 mol% Jd), to eclogitic compositions (<0.05–0.09 wt.% Cr_2O_3; 26–35 mol% Jd; Mg# = 76–89; Fig. 4a and b). The intermediary clinopyroxene compositions containing both Cr and jadeite share characteristics with peridotitic and eclogitic diamond inclusions and here are classified as websteritic. The eclogitic and websteritic clinopyroxenes both are distinguished from the peridotitic ones by higher Ti contents (Fig. 4c).

Four websteritic clinopyroxenes occur in three diamonds (108B, 109P, 111H). 108B also contains eclogitic clinopyroxene, garnet and majoritic garnet. This may suggest a compositional variation in the bulk chemistry of rocks encasing diamond through time with inclusions trapped at different stages in the diamonds growth history. Alternatively the variable clinopyroxene compositions in diamonds 108B and

Table 5
Representative electron microprobe analyses of eclogitic, websteritic and peridotitic clinopyroxene inclusions in diamond

Sample	110	104A	108B-1	108B-2	108B-3	109P	111D-1	111D-2	111F	111G	111H-1	111H-2	K10
Para.	P	P	E	E	Web	Web	E	E	E	E	Web	Web	P
Anal.	3	3	3	3	3	4	2	3	3	3	3	5	3
SiO_2	55.0	55.0	54.7	55.1	54.7	48.4	56.4	55.6	53.6	54.3	54.2	52.8	55.2
TiO_2	0.03	0.05	0.81	0.80	0.62	1.60	0.29	0.40	0.16	0.21	0.60	0.20	<0.03
Al_2O_3	1.02	0.74	8.61	8.71	4.81	3.79	6.40	5.71	7.49	7.38	4.79	0.78	1.04
Cr_2O_3	1.48	0.64	0.06	0.05	0.22	0.34	0.05	0.09	<0.05	<0.05	0.20	0.48	1.08
FeO	2.39	2.73	4.97	5.18	7.26	7.38	3.11	5.56	5.83	6.05	7.34	4.36	2.34
MnO	<0.05	0.12	<0.05	<0.05	0.07	0.12	0.08	0.10	0.06	0.08	0.08	0.06	0.06
MgO	17.28	18.51	9.83	9.66	11.98	13.99	13.62	12.76	10.36	10.46	11.91	16.69	17.49
CaO	21.1	20.9	13.97	13.79	16.57	22.3	16.27	14.88	17.55	17.46	16.51	22.4	22.1
Na_2O	1.11	0.74	6.01	6.09	3.34	0.40	3.86	3.86	3.74	3.69	3.31	0.59	0.80
K_2O	<0.04	<0.04	<0.04	<0.04	0.14	<0.04	<0.04	0.45	<0.04	<0.04	0.14	<0.04	0.19
NiO	0.09	0.08	<0.06	<0.06	<0.06	<0.06	<0.06	0.06	<0.06	<0.06	<0.06	<0.06	0.07
Total	99.6	99.6	99.0	99.5	99.7	98.4	100.1	99.5	98.8	99.7	99.1	98.4	100.4
Mg#	92.8	92.4	77.9	76.9	74.6	77.2	88.6	80.4	76.0	75.5	74.3	87.2	93.0
Jd	0.04	0.03	0.34	0.35	0.21	0.00	0.27	0.26	0.27	0.26	0.20	0.00	0.03
Di + Hd	0.78	0.79	0.51	0.51	0.62	0.84	0.62	0.56	0.66	0.65	0.63	0.87	0.82
XsNa	0.04	0.03	0.08	0.08	0.03	0.03		0.02			0.03	0.04	0.02

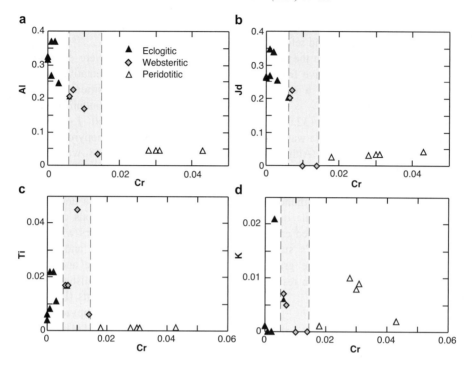

Fig. 4. Cation ratios (for six oxygen atoms) for eclogitic, websteritic (shaded area) and peridotitic clinopyroxene inclusions in diamond.

111H could be a result of metasomatic processes (e.g., Sobolev et al., 1998).

Potassium substitution in clinopyroxene is a function of increased formation pressure (Harlow and Veblen, 1991; Luth, 1997). In omphacite inclusions in eclogitic diamond, potassium contents typically range between 0.07 and 0.15 wt.% K_2O (McCandless and Gurney, 1989). The potassium content of clinopyroxenes in the Buffalo Hills diamonds is variable (<0.04 and 0.45 wt.%) and independent of composition and paragenesis (Fig. 4d). In one eclogitic diamond (111D), one omphacite inclusion is potassium-free, while another contains 0.45 wt.% K_2O. In a websteritic diamond (111H) containing both jadeite-rich and jadeite-free clinopyroxene, K_2O is present in the jadeite-rich but not in the jadeite-free clinopyroxene. In addition, a peridotitic Cr-diopside contains 0.19 wt.% K_2O.

5.4. Olivine

Sixteen olivine inclusions were analysed from eight diamonds (representative analyses in Table 6). Forsterite contents range between 90.3 and 92.9, reflecting a mixture of diamonds from the harzburgitic

and lherzolitic parageneses (Meyer, 1987). The Mg# and abundances of Ca and Mn in lherzolitic olivine, including one from a diamond with coexisting garnet and clinopyroxene, suggest fertile compositions.

5.5. Chrome spinel

Six inclusions were extracted and analysed from a single diamond. Their compositions show some variation, the largest being in Mg# (52 to 59). Compared to chromite inclusions in diamonds from South Africa and Siberia, chromites in the Buffalo Hills diamond are more Fe-rich (mean Mg# = 58.2 versus 66 worldwide (n = 156)). However, they are similar to chromites in diamonds from the Slave Craton (Mg# 56–64; Davies et al., 2004). The Cr# (100Cr/(Cr + Al) = 86) of these inclusions matches the average value of inclusions from diamond localities worldwide.

5.6. Ferropericlase

One ferropericlase inclusion was extracted from one diamond (Table 6). Ferropericlase is a phase formed from fertile peridotite at pressures and tem-

Table 6
Representative electron microprobe analyses of olivine (Fo) and a ferropericlase (fPer) inclusion

Sample	101	113	102B	104A	104B	105B	106C	K10	K10	109A
Mineral	Fo	Fo	Fo	Fo	Fo	Fo	Fo	Fo	Fo	fPer
Anal.	3	4	2	4	3	3	3	2	2	3
SiO_2	40.7	41.1	41.2	40.3	41.3	41.2	39.7	40.5	40.7	<0.02
TiO_2										0.03
Al_2O_3	0.02	<0.02	<0.02	<0.02	0.02	0.02	<0.02	<0.02	0.02	0.21
Cr_2O_3	0.08	0.05	0.12	0.07	0.07	0.06	0.08	0.09	0.05	1.00
FeO	7.23	7.88	7.06	8.30	7.45	7.76	8.10	9.55	8.50	20.0
MnO	0.08	0.09	0.06	0.15	0.12	0.10	0.10	0.10	0.15	0.22
MgO	51.3	50.7	51.6	50.3	51.1	50.5	50.9	49.7	50.1	75.8
CaO	0.06	0.05	<0.03	0.09	0.03	0.10	0.05	0.11	0.09	0.01
Na_2O										0.50
NiO	0.33	0.37	0.34	0.41	0.34	0.38	0.36	0.38	0.44	1.43
Total	99.8	100.3	100.4	99.7	100.5	100.1	99.4	100.6	100.1	99.3
Mg#	92.7	92.0	92.9	91.5	92.4	92.1	91.8	90.3	91.3	87.1

peratures corresponding to the lower mantle ($\sim$ 670 km; e.g., Ito and Takahashi, 1989), or in the upper mantle in conditions of low Si activity (Stachel et al., 1998, 2000). The Buffalo Hills ferropericlase has a composition typical of other ferropericlase inclusions in diamond of definite lower mantle origin, with respect to Mg#, Ni and Cr. In addition C-isotope ($\delta^{13}C = -4.6‰$) and IR-characteristics (Type II) of the diamond are typical of lower mantle diamonds. The occurrence of a diamond suite containing both ferropericlase and majoritic garnet inclusions draws parallel to other ultradeep (asthenosphere to lower mantle) diamond localities (e.g., Moore et al., 1991; Davies et al., 1999a; Harte et al., 1999; Hutchison et al., 1999; Kaminsky et al., 2000; Stachel et al., 2000). However, a second lower mantle phase to distinguish the paragenesis is lacking in this instance.

5.7. Rutile

Three large rutile inclusions of the eclogitic paragenesis were analysed from one diamond (111B). The inclusions contain trace amounts of FeO, Cr_2O_3, and Al_2O_3. In another diamond, a small rutile (not analysed) occurs in contact with eclogitic garnet and clinopyroxene.

5.8. Epigenetic inclusions

Epigenetic inclusions occurred as crack fillings and altered primary phases, consist dominantly of an assemblage of phlogopite, serpentine, Ca–Ti perovskite and carbonate. Analyses were carried out on polished mineral surfaces.

5.9. Geothermometry

Temperature calculations were carried out on eclogitic inclusions by the statistical approach of combining each discrete garnet inclusion ($n = 3$) with each clinopyroxene inclusion ($n = 2$) within a single eclogitic diamond (sample 108B). This procedure yielded a temperature of 1280 °C at a pressure of 5 GPa (Ellis and Green, 1979). In the same diamond, a garnet and clinopyroxene inclusion pair in mutual contact yields a slightly lower temperature for the same pressure value (1210 °C) indicating the inclusion pair re-equilibrated to lower temperatures during mantle storage. The temperature calculated from garnet and olivine in a lherzolitic diamond (sample K-10) is similar to the eclogitic diamond (1255 °C; O'Neill and Wood, 1979; assumed $P = 5$ GPa).

6. Carbon isotopes

On the basis of C-isotope values alone, and including the BH series of unknown paragenesis in Table 1, the Buffalo Hills diamonds cluster in three zones across a large range ($\delta^{13}C = -20.6‰$ to $-2.7‰$; Fig. 5). Most samples group around a peak with a typical average mantle value ($\delta^{13}C = -5$ av.), includ-

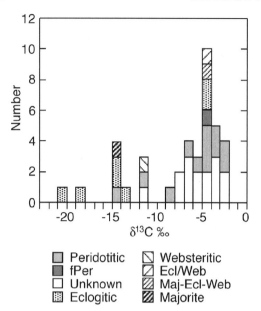

Fig. 5. Carbon isotopes and diamond parageneses.

ing diamonds containing ferropericlase, peridotitic, eclogitic and eclogitic + majoritic garnet + websteritic phases. Other diamonds with majoritic garnet, eclogitic, peridotitic and websteritic inclusions have intermediate compositions ($\delta^{13}C = -14.7$ to -11.8). Two eclogitic diamonds have the most ^{13}C-depleted values ($\sim -20‰$). From this data, we can link diamonds with eclogitic-majoritic garnet and websteritic inclusions into both the main and intermediate groups.

Two peridotitic diamonds and one websteritic diamond are ^{13}C-depleted ($-14.6‰$ and $-11.8‰$), and lie outside the $\delta^{13}C$ range of most peridotitic diamonds worldwide ($\delta^{13}C = -10‰$ to $0‰$; e.g., Kirkley et al., 1991). These compositions may indicate C-isotope heterogeneity in peridotitic mantle source. In view of the present inclusion mineralogy, they may also represent diamonds of mixed paragenesis, where the inclusions in the diamond were entrapped during growth in a peridotitic environment, while the bulk of the diamond grew at a different time in an eclogitic host rock with a ^{13}C-depleted carbon source.

The large range in ^{13}C-values in the Buffalo Hills eclogitic diamonds, which includes ^{13}C-depleted compositions, is a feature seen in other eclogitic diamonds

globally. A simple explanation for this range is a source derivation from subducted ocean crust mixed with organic sedimentary material (e.g., Kirkley et al., 1991). An alternative model explains the range of $\delta^{13}C$ values in eclogitic diamonds by Rayleigh fractionations through continuous extraction of mantle melts or fluids from the diamond source (Cartigny et al., 1998).

7. Nitrogen contents and nitrogen aggregation states

Nitrogen is the most common impurity in diamond and its substitution in kimberlitic diamonds is as high as 3000 ppm. The aggregation of nitrogen via diffusion in the diamond lattice is a time-temperature dependent process (Evans and Harris, 1989), but is influenced by the concentration of nitrogen, vacancies and by deformation. Because of the many factors that influence nitrogen aggregation in natural diamond, the thermal maturation histories cannot be accurately constrained by this method. Despite this limitation, FTIR-spectroscopy is a useful tool for characterising and comparing diamond groups of similar nitrogen defect characteristics relative to other diamonds from the same or related kimberlites.

From the 29 Buffalo Hills diamond set (Table 1), three groups are identified from their infrared spectral characteristics: (1) about 45% are nitrogen-free (Type II), (2) 33% have fully aggregated nitrogen defects (Type IaB) with platelet degradation and low to moderate nitrogen contents (65–830; av. 330 ppm-N), and (3) about 20% have low nitrogen aggregation states and moderate to high nitrogen contents (<30% IaB (aggregation from IaA to IaB); <1600 ppm-N). Type II diamonds (1) are grouped with (2) diamonds showing Type IaB characteristics because in most of the Type IaB diamonds nitrogen occurs in low amounts. The occurrence of pure IaB defects suggests that the diamonds were affected by storage in the mantle at high temperatures, for a significant residence time, and/or with significant strain. Strong deformation evident in many of these diamonds suggests the high degrees of nitrogen aggregation could reflect this process. The occurrence of such a high proportion of Type II and pure IaB diamonds with degraded platelets is rare in diamond populations

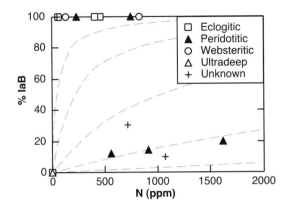

Fig. 6. Relationships between nitrogen contents, nitrogen aggregation states and inclusion parageneses in the Buffalo Hills diamonds from this study. The dashed isochron–isotherm lines define areas of similar nitrogen contents and aggregation states, inferring similar temperature/time/deformation histories (e.g., Evans and Harris, 1989).

worldwide and is in contrast to Type I diamonds from the Slave Craton with unusually low nitrogen aggregation states (Stachel et al., 2003; Davies et al., 2004). The Type IaA–IaB diamond group witnessed a different mantle storage history. These diamonds experienced shorter or cooler mantle storage conditions. Half of these diamonds also show strong deformation features.

Diamonds from all parageneses in this study are represented in the Type II and IaB IR-group (eclogitic, peridotitic, websteritic, and diamonds containing ferropericlase and majoritic garnet). The Type I diamonds with lower nitrogen aggregation states are peridotitic and have average mantle C-isotope values (Fig. 6).

8. Discussion

Mineral inclusion data from the Buffalo Hills diamonds suggest that the diamonds grew in host rocks of changing bulk chemistry and pressure (depth). These changes are recorded in the highly variable Cr, Na and K contents in clinopyroxenes, majoritic garnet, with compositions indicating formation at high pressure in the asthenosphere, and eclogitic and websteritic inclusions in single diamonds. Additional evidence for diamond growth in various and changing chemical environments comes from

variable Cr-spinel compositions in a single diamond, and a diamond containing ferropericlase indicating a possible lower mantle origin. Compositional differences between isolated minerals in single diamonds could indicate diamond growth in a metasomatic environment (Sobolev et al., 1998). However, the occurrence of majoritic garnet with eclogitic and websteritic phases in a single diamond rules out this possibility.

Carbon isotope compositions show considerable overlap in diamonds containing eclogitic, peridotitic, websteritic, asthenospheric and ferropericlase inclusions. These diamond types are spread across two of three clusters in the $\delta^{13}C$ distribution ($-20.6‰$ to $-2.7‰$). Morphological and IR-characteristics also do not conform to paragenetic groupings. These lines of evidence support the proposal that diamond growth occurred in a dynamic mantle environment (Davies et al., 1999b; Taylor et al., 1999).

8.1. A model

The combined mineral inclusion, nitrogen content and aggregation state and C-isotope data provides evidence for at least two generations of diamond growth beneath the Buffalo Head terrane. (1) Type II diamonds containing majoritic garnet and ferropericlase formed in the asthenosphere and possibly the uppermost lower mantle, respectively, and were subsequently transported to the base of the lithosphere. Here, Type II and Type I diamonds crystallised or continued to grow from fluids in peridotitic and eclogitic mantle material.

Mechanisms by which diamonds are transported from the asthenosphere and lower mantle to the lithosphere could be by either mantle convection (Stachel et al., 2003), an ascending plume (Davies et al., 1999a) or in deep-seated proto-kimberlite melts. The interaction of small volume melts with eclogitic and peridotitic rocks may have promoted diamond growth in fluids and produced the intermediate websteritic inclusion compositions, and the overlap in C-isotope values between eclogitic, peridotitic, websteritic, asthenospheric and ultradeep? diamonds (e.g., Wang, 1998). Dynamic transport of diamond at high T can explain the strong deformation and the extremely high nitrogen aggregation states. (2) A second generation of diamonds with Type IaA–IaB nitrogen aggregation, high nitrogen contents and peridotitic inclusions do not show evidence of dynamic mantle growth histories except for strong plastic deformation in some stones. These diamonds are likely to have crystallised in a cooler mantle, and possibly at a later time, at the base of the cratonic lithosphere. The two groups of diamonds were later sampled and emplaced at the Earth's surface in kimberlite magmas.

Acknowledgements

We thank Ashton Mining of Canada for samples, in particular Wayne Hillier and Shawn Carlson. Norman Pearson, Carol Lawson and Ashwini Sharma (Macquarie), and Tony Vassalo, Anita Andrew and Brad McDonald (CSIRO) for analytical assistance. Reviews by Buddy Doyle, Jeff Harris and Thomas Stachel significantly improved the manuscript. Funding for this research was provided by grants from the ARC. This is publication number 343 from the ARC National Key Centre for the Geochemical Evolution and Metallogeny of Continents (GEMOC, http://www.es.mq.edu.au/gemoc/).

References

Carlson, S.M., Hillier, W.D., Hood, C.T., Pryde, R.P., Skelton, D.H., 1998. The Buffalo Hills kimberlites: a newly-discovered diamondiferous kimberlite province in North-Central Alberta, Canada. Proc. 7th Internat. Kimberlite Conf. Red Roof Designs, Cape Town, pp. 109–116.

Cartigny, P., Harris, J.W., Javoy, M., 1998. Eclogitic diamond formation at Jwaneng: no room for a recycled component. Science 280, 1421–1424.

Davies, R.M., Griffin, W.L., O'Reilly, S.Y., 1999a. Diamonds from the deep: pipe DO27, Slave craton, Canada. Proc. 7th Internat. Kimberlite Conf. Red Roof Designs, Cape Town, pp. 148–155.

Davies, R.M., Griffin, W.L., O'Reilly, S.Y., 1999b. Growth structures and nitrogen characteristics of Group B diamonds from Wellington and Bingara, eastern Australia. Proc. 7th Internat. Kimberlite Conf. Red Roof Designs, Cape Town, pp. 156–163.

Davies, R.M., Griffin, W.L., O'Reilly, S.Y., Doyle, B.J., 2004. Mineral inclusions and geochemical characteristics of microdiamonds from the DO27, A154, A21, A418, DO18, DD17 and Ranch Lake kimberlites at Lac de Gras, Slave Craton, Canada. Lithos 77, 39–55 this volume [doi:10.1016/j.lithos.2004.016].

Ellis, D.J., Green, D.H., 1979. An experimental study of the effect of Ca upon garnet-clinopyroxene Fe–Mg exchange equilibria. Contrib. Mineral. Petrol. 71, 13–22.

Evans, T., Harris, J.W., 1989. Nitrogen aggregation, inclusion equilibration temperatures and the age of diamonds. In: Ross, J., et al., (Eds.), Kimberlites and Related Rocks v.2. Their Mantle/Crust Setting. Spec. Publ. Geol. Soc. Aust., vol. 14, pp. 1002–1006.

Griffin, W.L., Shee, S.R., Ryan, C.G., Win, T.T., Wyatt, B.A., 1999. Harzburgite to lherzolite and back again: metasomatic processes in ultramafic xenoliths from Wesselton kimberlite, Kimberly, South Africa. Contrib. Mineral. Petrol. 134, 232–250.

Hall, A.E., Smith, C.B., 1984. Lamproite diamonds—are they different? In: Glover, J.E., Harris, P.G. (Eds.), Kimberlite Occurrence and Origin. Publ. Univ.West. Aust., Geol. Dept., vol. 8, pp. 167–212.

Harlow, G.E., Veblen, D.R., 1991. Potassium in clinopyroxene inclusions from diamond. Science 251, 652–655.

Harte, B., Harris, J.W., Hutchison, M.T., Watt, G.R., Wilding, M.C., 1999. Lower mantle mineral associations in diamonds from Sao Luiz, Brazil. In: Fei, Y., Bertka, C.M., Mysen, B.O. (Eds.), Mantle Petrology: Field Observations and High Pressure Experimentation: A tribute to Francis R. (Joe) Boyd. Spec. Publ.-Geochem. Soc., vol. 6, pp. 125–153.

Hutchison, M.T., Cartigny, P., Harris, J.W., 1999. Carbon and nitrogen compositions and physical characteristics of transition zone and lower mantle diamonds from Sao Luiz, Brazil. Proc. 7th Internat. Kimberlite Conf. Red Roof Designs, Cape Town, pp. 372–382.

Irifune, T., Hibberson, W.O., Ringwood, A.E., 1986. Eclogite–garnetite transformations at high pressure and its bearing on the occurrence of garnet inclusions in diamond. In: Ross, J., et al., (Eds.), Kimberlites and Related Rocks v. 2. Their Mantle/Crust Setting. Spec. Publ. Geol. Soc. Aust., vol. 14, pp. 877–882.

Ito, E., Takahashi, E., 1989. Postspinel transformations in the system $Mg_2SiO_4 - Fe_2SiO_4$ and some geophysical implications. J. Geophys. Res. 94, 10637–10646.

Jaques, A.L., Hall, A.E., Sheraton, J.W., Smith, C.B., Sun, S.-S., Drew, R.M., Foudoulis, C., Ellingsen, K., 1989. Composition of crystalline inclusions and C-isotope composition of Argyle and Ellendale diamonds. In: Ross, J., et al., (Eds.), Kimberlites and Related Rocks v. 2. Their Mantle/Crust Setting. Spec. Publ. Geol. Soc. Aust., vol. 14, pp. 966–989.

Kaminsky, F.V., Zakharchenko, O.D., Davies, R.M., Griffin, W.L., Khachatryan-Blinova, G.K., Shiryaev, A.A., 2000. Superdeep diamonds from the Juina Area, Mato Grosso State, Brazil. Contrib. Mineral. Petrol. 140, 734–753.

Kirkley, M.B., Gurney, J.J., Otter, M.L., Hill, S.J., Daniels, L.R., 1991. The application of C isotope measurements to the identification of the sources of C in diamonds: a review. Appl. Geochem. 6, 477–494.

Luth, R.W., 1997. Experimental study of the system phlogopite–diopside from 3.5 to 17 GPa. Am. Mineral. 82, 1198–1209.

McCandless, T.E., Gurney, J.J., 1989. Sodium in garnet and potassium in clinopyroxene: criteria for classifying mantle eclogites. Internal Rept. 10, August 1986, KRG, Univ. Cape Town.

McNicoll, V.J., Theriault, R.J., McDonough, M.R., 2000. Taltson basement gneissic rocks: U–Pb and Nd isotopic constraints on the basement to the Paleoproterozoic Taltson magmatic zone, northeastern Alberta. Can. J. Earth Sci. 37, 1475–1596.

Mendelssohn, M.J., Milledge, H.J., 1995. Geologically significant information from routine analysis of the mid-infrared spectra of diamonds. Int. Geol. Rev. 37, 95–110.

Meyer, H.O.A., 1987. Inclusions in diamond. In: Nixon, P.H. (Ed.), Mantle Xenoliths. John Wiley and Sons, New York, pp. 501–522.

Moore, R.O., Gurney, J.J., Fipke, C.E., 1989. The development of advanced technology to distinguish between diamondiferous and barren diatremes. Geol. Surv. Canada Open File Rept. 2124, 4 vols.

Moore, R.O., Gurney, J.J., Griffin, W.L., Shimuzu, N., 1991. Ultra-high pressure garnet inclusions in Monastery diamonds: trace element abundance patterns and conditions of origin. Eur. J. Mineral. 3, 213–230.

Norman, M.D., Pearson, N.J., Sharma, A., Griffin, W.L., 1996. Quantitative analysis of trace elements in geological materials by laser ablation ICPMS: instrumental operating conditions and calibration values of NIST glasses. Geostand. Newsl. 20, 247–261.

Norman, M.D., Griffin, W.L., Pearson, N.J., Garcia, M.O., O'Reilly, S.Y., 1998. Quantitative analysis of trace element abundances in glasses and minerals: a comparison of laser ablation ICPMS, solution ICPMS, proton microprobe, and electron microprobe data. J. Anal. At. Spectrom. 13, 477–482.

O'Neill, H.S.C., Wood, B.J., 1979. An experimental study of Fe–Mg partitioning between garnet and olivine and its calibration as a geothermometer. Contrib. Mineral. Petrol. 70, 59–70.

Robinson, D.N., Scott, J.A., Van Niekerk, A., Anderson, V.G., 1989. The sequence of events reflected in the diamonds of some southern African kimberlites. In: Ross, J., et al., (Eds.), Kimberlites and Related Rocks v.2. Their Mantle/Crust Setting. Spec. Publ. Geol. Soc. Aust., vol. 14, pp. 990–1000.

Ross, G.M., 2002. Introduction to special issue of Canadian Journal of Earth Sciences: the Alberta Basement Transect of Lithoprobe. Can. J. Earth Sci. 39, 287–288.

Ross, G.M., Parrish, R.R., Villeneuve, M.E., Bowring, S.A., 1991. Geophysics and geochronology of the crystalline basement of the Alberta Basin, western Canada. Can. J. Earth Sci. 28, 512–522.

Sobolev, N.V., Galimov, E.M., Smith, C.B., Yefinova, E.S., Maltsev, K.A., Hall, A.E., Usova, L.V., 1986. Comparative characteristics of morphology, inclusions and carbon isotope composition of diamonds in alluvial deposits of the King George River and lamproite deposits of Argyle. Geol. Geofiz. 12 (in Russ.).

Sobolev, N.V., Snyder, G.A., Taylor, L.A., Keller, R.A., Yefimova, E.S., Sobolev, V.N., Shimizu, N., 1998. Extreme chemical diversity in the mantle during eclogitic diamond formation: evidence form 35 garnet and 5 pyroxene inclusions in a single diamond. Int. Geol. Rev. 40, 567–578.

Stachel, T., Harris, J.W., Brey, G.P., 1998. Rare and unusual mineral inclusions in diamonds from Mwadui, Tanzania. Contrib. Mineral. Petrol. 132, 34–47.

Stachel, T., Harris, J.W., Brey, G.P., Joswig, W., 2000. Kankan diamonds (Guinea): II. Lower mantle inclusion parageneses. Contrib. Mineral. Petrol. 140, 16–27.

Stachel, T., Harris, J.W., Tappert, R., Brey, G.P., 2003. Peridotitic diamonds from the Slave and the Kaapvaal cratons— similarities and differences based on a preliminary data set. Lithos 71, 489–503.

Taylor, L.A., Snyder, G.A., Camacho, A., 1999. Diamond: just another metamorphic mineral? Abst. 9th Ann. Goldschmidt Conf., LPI Contribution No. 971, Lunar and Planetary Institute, Houston, pp. 291–292.

Villeneuve, M.E., Ross, G.M., Parrish, R.R., Therault, T.J., Miles, W., Broome, J., 1993. Geophysical subdivision, U–Pb geochronology and Sm–Nd isotope geochemistry of the crystalline basement of the Western Canada sedimentary Basin, Alberta and northeastern British Columbia. Bull. Geol. Surv. Can. 447, p. 85.

Wang, W., 1998. Formation of diamond with mineral inclusions of "mixed" eclogite and peridotite paragenesis. Earth Planet. Sci. Lett. 160, 831–843.

Woods, G.S., Purser, G.C., Mtimkulu, S.S., Collins, A.T., 1990. The nitrogen content of Type Ia natural diamonds. Phys. Chem. Solids 51, 1191–1197.

Available online at www.sciencedirect.com

Lithos 77 (2004) 113–124

ELSEVIER

LITHOS

www.elsevier.com/locate/lithos

Ar–Ar age determinations of eclogitic clinopyroxene and garnet inclusions in diamonds from the Venetia and Orapa kimberlites

R. Burgess[a,*], G.B. Kiviets[b], J.W. Harris[c]

[a] Department of Earth Sciences, University of Manchester, Oxford Road, Manchester, M13 9PL, UK
[b] DeBeers Geoscience Centre, P.O. Box 82232, Southdale, 2135, South Africa
[c] Division of Earth Sciences, University of Glasgow, Glasgow, G12 8QQ, UK

Received 27 June 2003; accepted 17 February 2004
Available online 19 May 2004

Abstract

Ar–Ar age measurements are reported for selected eclogitic clinopyroxene and garnet inclusions in Orapa diamonds and clinopyroxene inclusions in Venetia diamonds. Laser drilling of encapsulated clinopyroxene inclusions within Venetia diamonds released a maximum of 3% of the total ^{40}Ar, indicating little diffusive transfer and storage of radiogenic ^{40}Ar at the diamond–inclusion boundary. Apparent ages obtained during stepped heating of three diamonds are consistent with diamond crystallisation occurring just prior to the kimberlite eruption ~ 520 Ma ago. Stepped heating of three clinopyroxene-bearing Orapa diamonds gave ages of 906–1032 Ma, significantly above the eruption age, but consistent with previously determined isotopic ages. A few higher apparent ages hint at the presence an older generation of Orapa diamonds that formed >2500 Ma ago. Orapa garnets also contain measurable K contents, and record a range of ages between 1000 and 2500 Ma. The old apparent ages and lack of significant interface ^{40}Ar released by the laser probe, suggests that pre-eruption radiogenic ^{40}Ar and mantle-derived ^{40}Ar components are trapped in microinclusions within the pyroxene and garnet inclusions.
© 2004 Elsevier B.V. All rights reserved.

Keywords: Diamond; Dating; Ar–Ar; Clinopyroxene; Garnet

1. Introduction

The Ar–Ar method was the first radiometric dating method to be applied to individual diamonds (Burgess et al., 1989; Phillips et al., 1989), but so far, it is only applicable to clinopyroxene inclusions in eclogitic diamonds which contain sufficient potassium for age determinations. In the past, there have been problems in making geological sense of the data obtained.

Apparent ages of cleaved diamonds, which exposes the inclusion, range upwards from the eruption age (Burgess et al., 1992). We have suggested that this is due to diffusion (at mantle temperatures) of ^{40}Ar to the interface between the inclusion and diamond followed by partial loss of argon when the diamond is cleaved (Burgess et al., 1992). To overcome this problem, it is necessary to analyse the Ar content of the inclusion without cleaving the diamond. Another possible problem arises from the fact that the concentration of "ambient" ^{40}Ar in the mantle is equivalent to several hundreds of million years of radiogenic in-growth of ^{40}Ar in a clinopyroxene inclusion containing a few

---

* Corresponding author. Tel.: +44-161-275-3958; fax: +44-161-275-3947.

E-mail address: Ray.Burgess@man.ac.uk (R. Burgess).

hundred ppm K. However, based on known partition coefficients of ^{40}Ar between minerals and melts, it is unlikely that much of the ambient argon component is incorporated directly into the clinopyroxene minerals, although it may be present in defect structures such as fluid inclusions. On a worst case scenario, the age based on total ^{40}Ar/^{39}Ar$_K$ is an absolute upper limit to the time at which the clinopyroxene was encased by the diamond. Recently, Phillips et al. (2004) carried out laser Ar–Ar age measurements of clinopyroxene inclusions that were removed from the host diamonds prior to analysis. The apparent ages obtained ranged between kimberlite eruption and the diamond genesis ages constrained by Re–Os or Sm–Nd dating. Phillips et al. (2004) discussed several alternatives for the high apparent ages including the partial retention or re-diffusion of pre-eruption radiogenic ^{40}Ar, but favoured an explanation involving excess (ambient) Ar derived from the mantle that is trapped in defects or micro-inclusions within the pyroxene.

In this study, we present Ar–Ar results from clinopyroxene- and garnet-bearing eclogitic diamonds from the Orapa kimberlite, Botswana, and eclogitic clinopyroxene-bearing diamonds from the Venetia kimberlite, South Africa. In order to account for any interface Ar, the diamonds were not cleaved prior to analysis. A two-stage analytical protocol was applied to Venetia diamonds in an attempt to separately extract interface and volume-related Ar components from the pyroxene inclusions. Since garnet normally has a low K content and should not contain appreciable levels of radiogenic ^{40}Ar, it was anticipated that the effects of ambient mantle ^{40}Ar, either dissolved in the silicate or held in defects such as microinclusions, could be assessed by comparing results from pyroxene and garnet inclusions in Orapa diamonds.

2. Samples

Eclogitic diamonds from the Orapa kimberlite, Botswana, consisted of six specimens containing green clinopyroxene inclusions, and four containing orange garnet inclusions. The Orapa kimberlite was erupted 93 Ma ago (Davis, 1977), however, based on Sm–Nd and Re–Os studies of silicate and sulfide inclusions in diamonds, there were at least two discrete eclogitic diamond-forming events occurring 990

and 2900 Ma ago (Richardson et al., 1990; Shirey et al., 1999). Previous Ar–Ar analyses of ten cleaved clinopyroxene-bearing diamonds from Orapa, gave a wide range in ages between 96 and 1580 Ma (Burgess et al., 1992). This age variation was attributed to partial loss of radiogenic ^{40}Ar from the inclusion–diamond interface during cleaving, therefore, the Orapa diamonds analysed in the present study were not cleaved prior to Ar–Ar analysis.

Five clinopyroxene-bearing eclogitic diamonds from the Venetia kimberlite were analysed during this study. The Venetia kimberlite field is located in northern South Africa and comprises a cluster of 12 pipes and dykes intruding the Central Zone of the Limpopo Belt (Seggie et al., 1999). Rb–Sr isochron ages of the K1 pipe are 533 ± 4 Ma (mica and whole rock) and 510 ± 16 Ma (mica only). Ar–Ar stepped heating of groundmass phlogopite grains from the western hypabyssal complex yielded an age of 519 ± 6 Ma (2σ error; Phillips et al., 1999).

3. Experimental methods

Diamonds were selected on the basis of the mineralogy and size of their inclusion content (i.e., clinopyroxene or garnet with sizes estimated to be between 20 and 100 μm). Prior to irradiation each Venetia diamond sample was cut and polished down to a distance of

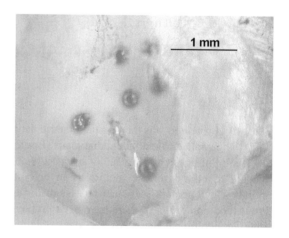

Fig. 1. Photograph of a clinopyroxene inclusion in a Venetia diamond. The diamond has been cut and polished to within approximately 0.5 mm of the inclusion. Dark circles are UV laser ablation pits used to help locate the inclusion following irradiation.

approximately 0.5 mm above the surface of the clino-pyroxene inclusion (Fig. 1). This was necessary because irradiation renders the diamond opaque making it difficult to determine the exact location of the inclusion. The ability to accurately locate the inclusion is an essential requirement for the laser drilling experiments described later. After polishing, the precise position of the inclusion was marked using laser pits on the flat surface (Fig. 1); these pits remain visible after irradiation. Visual examination of each sample was made using a binocular microscope ($\times 60$ magnification) in order to check for the presence of any fractures passing through the diamond that may intersect with the inclusions and provide pathways for Ar loss. Only minor fractures surrounding the inclusions were observed. These features most likely formed by cooling during kimberlite eruption.

The samples were placed in evacuated silica vials and Venetia diamonds were irradiated in position B2-West of the Safari-1 reactor, Pelindaba, for a period of approximately 20 h (DeBeers irradiation DB22). Orapa diamonds were irradiated for 50 h in position L67 of the Ford reactor, University of Michigan (Manchester irradiation MN4). In both irradiations, the sample canisters were rotated $180°$ half-way through the irradiation period. All samples received a similar fast neutron flux of 2×10^{18} n/cm^2 as determined from Hb3gr monitors ($t = 1072 \pm 11$ Ma: Turner et al., 1971) placed in close proximity to the diamonds. Calculated J values from monitors in both irradiations showed a 6% neutron flux variation over the vertical distance encompassing the samples within the vials. Neutron interference reactions were not monitored during the irradiation of Venetia diamond, and are assumed to be the same as reported by Phillips et al. (1999) under nearly identical irradiation conditions. Interference corrections for Orapa diamonds were monitored using CaF$_2$ and K$_2$SO$_4$ and values obtained are indistinguishable from those previously reported for the Ford reactor (McDougall and Harrison, 1999).

During initial experiments, we attempted to use an infrared continuous laser to selectively melt clinopyroxene inclusions in diamond; however, this proved difficult to accomplish due to the high thermal conductivity of the diamond, and was often accompanied by significant atmospheric argon release from the surface of the diamond and the aluminium holder. Argon was therefore extracted from the clinopyroxene inclusions using an ultraviolet (UV) laser followed by furnace stepped heating. For Orapa diamonds that were not polished prior to irradiation, the inclusions were deeply buried in the diamonds and attempts to probe the inclusions with the laser proved unsuccessful; only data for furnace stepped heating are reported for these samples. Argon was extracted from polished Venetia diamonds, using both procedures. For the first stage of Ar extraction, Venetia diamonds were placed in a UHV laser port and an UV Nd-YAG laser (266 nm) was used to drill down to the inclusion. This was done in a series of 10-min extractions with the laser operating for a period of 5–8 min during that time period. During laser extraction, Ar was purified using a hot (250 °C) Zr–Al getter. Following the 10-min extraction and gas purification step, Ar isotopic analysis was carried out on the MS1 mass spectrometer equipped with a Baur–Signer source and electron multiplier detector. Procedural blanks were measured at the start of each day and after every fifth samples analysis. Penetration to the inclusion took between two and four laser extractions. It was assumed that the laser beam had penetrated to an inclusion when there was a significant increase in the amounts of ^{39}Ar$_K$ and ^{37}Ar$_{Ca}$ above blank levels. Once the laser pierced an inclusion, the diamond was removed from the laser port for the next stage of the analysis.

During the second stage of Ar extraction, the diamonds were step-heated using a Ta-resistance furnace. Argon was extracted using six temperature increments over the interval 600–2150 °C to release volume-related Ar isotopes from the inclusions. During each step, the temperature was increased to the required temperature over a 10-min period, followed by 15 min at the step temperature, then finally 5 min of cooling. During heating, Ar was purified using a Zr–Al getter at 250 °C. Furnace stepped heating of the diamond to extract Ar was used in preference to laser heating because the former has the potential to discriminate between Ar release from pyroxene and the host diamond. Based on our experience, release of radiogenic ^{40}Ar occurs during melting of pyroxene below 1800 °C and is accompanied by the release of Ca-derived ^{37}Ar$_{Ca}$. This compares with the release of any trapped Ar from the host diamond which occurs during graphitisation >2000 °C. Furthermore, any surface adsorbed atmospheric Ar (indicated by a high ^{36}Ar content) should be released during the low temperature steps, prior to significant release of radiogenic ^{40}Ar. During stepped heating, most of the clinopyroxene and garnet-bearing diamonds showed a bimodal release of Ar with peaks at 1200–1800 and >2000 °C, as exemplified by Ar release from VC1 shown in Fig. 2.

Furnace blanks were measured at the start of each set of sample analyses and a high temperature blank (2150 °C) was measured between each sample. Argon blanks released between 600 and 1800 °C were in a narrow range of $0.7–0.9 \times 10^{-12}$ cm^3 STP, rising to 80×10^{-12} cm^3 STP at 2150 °C. All blanks had an atmospheric Ar isotopic composition and the amounts are typically <10% of the Ar released from those sample heating steps where it was possible to calculate apparent ages (Tables 1 and 2).

Once the two-stage analysis was complete, Ar released from laser and furnace experiments was combined and an age calculated from the total ^{40}Ar/^{39}Ar$_K$ ratio; this should give the time at which the inclusion was trapped in the diamond. Amounts of K, Cl and Ca were calculated, after corrections for decay and neutron interference, from the release of neutron-produced ^{39}Ar$_K$, ^{38}Ar$_{Cl}$ and ^{37}Ar$_{Ca}$, respectively. Calcium data could not be obtained for the Orapa samples because these diamonds were analysed more than a year after irradiation and the ^{37}Ar$_{Ca}$ had decayed (half-life 35 days) to levels below detection. As clinopyroxene

inclusions have high Ca/K ratios, neglecting Ca-derived isotopic interferences (^{36}Ar$_{Ca}$ and ^{39}Ar$_{Ca}$) leads to an underestimation of apparent ages. The magnitude of these isotopic interferences was assessed by assuming a K/Ca value of 3.67×10^{-2}, the average value obtained from nine clinopyroxene inclusions reported by Burgess et al. (1992), and using a previously determined value of α for the MN4 irradiation defined as: $\alpha=(K/Ca) \times (^{37}$Ar/^{39}Ar$)=0.542 \pm 0.001$. For samples OC4, OC5 and OC6 the interference ^{36}Ar$_{Ca}$ and ^{39}Ar$_{Ca}$ have a negligible effect on apparent ages (<1%) and are not considered further, however for samples OC1, OC2 and OC3 it is estimated that the apparent ages are underestimated by approximately 3% due to these interferences and are discussed further in the next section.

Analytical data are given in Tables 1 and 2, where all errors are reported at the two standard deviation level of uncertainty. Errors on apparent ages include a contribution from the uncertainty on the monitor age.

4. Results

4.1. Orapa diamonds

Results of stepped heating Orapa diamonds are given in Table 1. Most of the clinopyroxene and

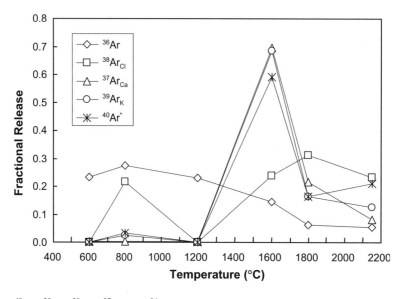

Fig. 2. Release profile of ^{40}Ar*, ^{39}Ar$_K$, ^{38}Ar$_{Cl}$, ^{37}Ar$_{Ca}$ and ^{36}Ar from Venetia clinopyroxene sample VC1. The bimodal release is interpreted to reflect separate releases from pyroxene at 1200–1800 °C and from host diamond during graphitisation at >2000 °C.

Table 1
Ar–Ar data for eclogitic clinopyroxene and garnet inclusions in Orapa diamonds

Sample/temperature step (°C)	Cl (× 10^{-12} mol)	K (× 10^{-12} mol)	$^{40}Ar^*$ (× 10^{-18} mol)	%K	Cl/K	$^{40}Ar^*$/K (× 10^{-6})	Age[a] (Ma)
Clinopyroxene							
OC1							
800	0.11 ± 0.73	2 ± 3	nd	0.3	–	–	–
1200	2.61 ± 0.79	194 ± 8	1547 ± 23	35.2	0.013 ± 0.004	8.0 ± 0.3	896 ± 30
1600	4.86 ± 0.75	263 ± 12	2211 ± 23	47.6	0.018 ± 0.003	8.4 ± 0.4	935 ± 33
1800	nd	nd	nd		–	–	–
2050	5.01 ± 1.76	34 ± 6	nd	6.1	0.148 ± 0.059	–	–
2150	13.84 ± 1.20	60 ± 14	12,159 ± 387	10.8	0.232 ± 0.057	203.4 ± 47.5	5158 ± 397
Total	26.43 ± 2.50	553 ± 21	15,916 ± 388		0.048 ± 0.005	28.8 ± 1.3	2169 ± 57
1200–1800	7.47 ± 1.08	458 ± 14	3757 ± 32	82.8	0.016 ± 0.002	8.2 ± 0.3	918 ± 23
OC2							
800	nd	10 ± 4	3377 ± 925	0.1	–	343.1 ± 178.5	6059 ± 906
1200	nd	nd			–	–	–
1600	3.14 ± 2.28	5484 ± 285	52,997 ± 589	67.6	0.0006 ± 0.0004	9.7 ± 0.5	1042 ± 42
1800	nd	1599 ± 49	12,222 ± 642	19.7	–	7.6 ± 0.5	868 ± 42
2050	nd	494 ± 17	4012 ± 142	6.1	–	8.1 ± 0.4	910 ± 35
2150	nd	527 ± 23	3823 ± 590	6.5	–	7.3 ± 1.2	833 ± 107
Total	3.14 ± 2.28	8114 ± 290	76,431 ± 1408		0.0004 ± 0.0003	9.4 ± 0.4	1021 ± 31
1200–1800	3.14 ± 2.28	7083 ± 289	65,219 ± 871	87.3	0.0004 ± 0.0003	9.2 ± 0.4	1004 ± 33
OC3							
800	nd	nd	nd		–	–	–
1200	nd	nd	nd		–	–	–
1600	1.32 ± 1.09	500 ± 16	28,358 ± 688	44.2	0.003 ± 0.002	56.7 ± 2.3	3103 ± 60
1800	0.57 ± 1.08	628 ± 25	4822 ± 138	55.5	0.001 ± 0.002	7.7 ± 0.4	871 ± 34
2050	0.66 ± 2.26	3 ± 24	nd	0.3	–	–	–
2150	nd	nd	nd		–	–	–
Total	2.55 ± 2.73	1131 ± 38	33,180 ± 702		0.002 ± 0.002	29.3 ± 2.3	2193 ± 101
1200–1800	1.90 ± 1.54	1128 ± 30	33,180 ± 702	99.7	0.002 ± 0.001	29.4 ± 1.0	2196 ± 43
OC4							
800	0.79 ± 0.36	nd	nd		–	–	–
1200	nd	14 ± 6	nd	1.0	–	–	–
1600	1.72 ± 0.65	734 ± 24	48,352 ± 417	50.8	0.002 ± 0.001	65.9 ± 2.3	3328 ± 52
1800	7.62 ± 0.82	562 ± 17	35,121 ± 440	38.9	0.014 ± 0.002	62.5 ± 2.1	3249 ± 50
2050	nd	45 ± 5	746 ± 28	3.1	–	16.6 ± 1.9	1536 ± 117
2150	3.29 ± 0.83	90 ± 9	43,015 ± 570	6.2	0.037 0.010	477.6 ± 47.5	6637 ± 175
Total	13.42 ± 1.38	1445 ± 32	127,233 ± 833		0.009 0.001	88.1 ± 2.0	3779 ± 37
1200–1800	9.33 ± 1.05	1310 ± 30	83,472 ± 606	90.7	0.007 0.001	63.7 ± 1.6	3278 ± 37
OC5							
800	nd	nd	nd		–	–	–
1200	0.55 ± 0.53	9 ± 4	480 ± 500	3.2	0.059 ± 0.063	51.7 ± 58.6	2968 ± 1650
1600	6.53 ± 0.78	102 ± 7	7522 ± 475	35.0	0.064 ± 0.009	73.7 ± 7.0	3501 ± 146
1800	0.65 ± 0.50	100 ± 5	1803 ± 514	34.3	0.007 ± 0.005	18.1 ± 5.2	1624 ± 309
2050	nd	nd	nd		–	–	–
2150	5.98 ± 1.53	80 ± 4	15,053 ± 487	27.5	0.075 ± 0.019	187.6 ± 10.6	5021 ± 96
Total	13.70 ± 1.87	291 ± 10	24,859 ± 988		0.047 ± 0.007	85.3 ± 4.5	3728 ± 83
1200–1800	7.72 ± 1.06	211 ± 9	9806 ± 860	72.5	0.037 ± 0.005	46.4 ± 4.6	2813 ± 140
OC6							
800	1.56 ± 1.77	15 ± 9	4257 ± 874	3.6	0.102 ± 0.131	279.4 ± 177.6	5702 ± 1098
1600	7.00 ± 1.40	342 ± 28	93,277 ± 787	81.2	0.020 ± 0.004	272.4 ± 22.4	5659 ± 142
1800	nd	20 ± 21	5871 ± 227	4.8	–	287.7 ± 291.7	5753 ± 1754
2050	2.92 ± 2.24	19 ± 36	3755 ± 1270	4.4	0.156 ± 0.324	200.6 ± 392.7	5135 ± 3326

(continued on next page)

Table 1 (*continued*)

Sample/temperature step (°C)	Cl ($\times 10^{-12}$ mol)	K ($\times 10^{-12}$ mol)	^{40}Ar* ($\times 10^{-18}$ mol)	%K	Cl/K	^{40}Ar*/K ($\times 10^{-6}$)	Age[a] (Ma)
Clinopyroxene							
OC6							
2150	3.16 ± 3.55	25 ± 49	9315 ± 699	5.9	0.126 ± 0.286	372.2 ± 730.9	6200 ± 3429
Total	14.64 ± 4.77	422 ± 71	$116{,}476 \pm 1880$		0.035 ± 0.013	276.1 ± 46.6	5682 ± 291
1200–1800	8.55 ± 2.26	378 ± 36	$103{,}406 \pm 1198$	89.6	0.023 ± 0.006	273.5 ± 26.2	5666 ± 166
Garnet							
OG1							
800	nd	26 ± 6	2386 ± 806	3.0	–	90.7 ± 37.5	3825 ± 657
1200	1.34 ± 0.68	10 ± 4	2222 ± 7499	1.2	0.128 ± 0.082	212.6 ± 722.4	5233 ± 5793
1600	2.05 ± 0.61	529 ± 21	9774 ± 489	60.9	0.004 ± 0.001	18.5 ± 1.2	1649 ± 69
1800	1.10 ± 0.50	183 ± 16	2501 ± 86	21.1	0.006 ± 0.003	13.7 ± 1.3	1342 ± 89
2050	2.26 ± 1.68	44 ± 12	1879 ± 1998	5.0	0.052 ± 0.041	42.9 ± 47.2	2702 ± 1539
2150	nd	75 ± 22	2223 ± 2299	8.7	–	29.6 ± 31.8	2204 ± 1369
Total	6.76 ± 1.97	867 ± 37	$20{,}985 \pm 8149$		0.008 ± 0.002	24.2 ± 9.5	1955 ± 466
1200–1800	4.50 ± 1.04	722 ± 27	$14{,}497 \pm 7516$	83.3	0.006 ± 0.001	20.1 ± 10.4	1740 ± 580
OG3							
800	0.93 ± 0.70	13 ± 2	nd	2.1	0.070 ± 0.054	–	–
1200	4.38 ± 0.98	254 ± 12	2491 ± 643	40.2	0.017 ± 0.004	9.8 ± 2.6	1053 ± 209
1600	2.94 ± 1.90	327 ± 11	2724 ± 142	51.8	0.009 ± 0.006	8.3 ± 0.5	928 ± 45
1800	0.50 ± 1.23	13 ± 6	174 ± 31	2.0	0.039 ± 0.098	13.7 ± 7.2	1343 ± 499
2050	2.41 ± 1.62	4 ± 11	515 ± 971	0.6	0.602 ± 1.770	128.8 ± 441.5	4392 ± 5644
2150	0.62 ± 1.17	21 ± 13	1785 ± 647	3.3	0.030 ± 0.060	85.9 ± 63.7	3738 ± 1169
Total	11.78 ± 3.25	632 ± 25	7689 ± 1340		0.019 ± 0.005	12.2 ± 2.2	1235 ± 160
1200–1800	7.83 ± 2.46	594 ± 17	5389 ± 659	94.0	0.013 ± 0.004	9.1 ± 1.1	992 ± 96
OG4							
800	nd	nd	nd		–	–	–
1200	0.29 ± 0.57	10 ± 4	142 ± 33	45.8	0.029 ± 0.059	14.5 ± 6.5	1402 ± 436
1600	0.71 ± 0.31	5 ± 2	88 ± 26	22.4	0.149 ± 0.087	18.4 ± 9.1	1644 ± 536
1800	nd	nd	nd		–	–	–
2050	0.87 ± 1.07	7 ± 7	3040 ± 45	31.8	0.128 ± 0.203	447.7 ± 448.8	6524 ± 1760
2150	nd	nd	nd		–	–	–
Total	1.86 ± 1.25	21 ± 8	3270 ± 62		0.087 ± 0.067	153.2 ± 57.5	4681 ± 626
1200–1800	1.00 ± 0.65	15 ± 4	230 ± 42	68.2	0.069 ± 0.049	15.8 ± 5.4	1485 ± 346
OG5							
800	nd	nd	nd		–	–	–
1200	nd	nd	nd		–	–	–
1600	1.40 ± 0.93	48 ± 4	8374 ± 5871	32.4	0.029 ± 0.020	176.2 ± 124.4	4915 ± 1190
1800	nd	nd	–		–	–	–
2050	1.63 ± 0.43	82 ± 5	3298 ± 597	55.8	0.020 ± 0.005	40.3 ± 7.7	2612 ± 264
2150	1.50 ± 0.97	17 ± 8	3336 ± 49	11.9	0.086 ± 0.069	191.6 ± 89.6	5057 ± 792
Total	4.53 ± 1.41	147 ± 10	$15{,}008 \pm 5901$		0.031 ± 0.010	102.2 ± 40.8	4016 ± 643
1200–1800	1.40 ± 0.93	48 ± 4	8374 ± 5871	32.4	0.029 ± 0.020	176.2 ± 124.4	4915 ± 1190

All errors are 2σ, nd = not determinable.

[a] Apparent ages of clinopyroxene have not been corrected for Ca-derived ^{36}Ar and ^{39}Ar formed during irradiation. Based on an assumed K/Ca ratio of 3.67×10^{-2}, it is estimated that apparent ages of OC1, OC2 and OC3 are 3–4% higher than given here, but <1% higher for OC4, OC5 and OC6.

garnet-bearing diamonds show a bimodal release of Ar with peaks at 1200–1800 °C and >2000 °C. The high temperature release accounts for 0–68% of the ^{39}Ar$_K$ release and can most simply be attributed to diamond graphitisation. The majority of the ^{39}Ar$_K$ (32–100%) is released between 1200 and 1800 °C, most likely by mechanical rupture of the host diamonds during melting of the silicate inclusions. Re-

Table 2
Ar–Ar data for eclogitic clinopyroxene inclusions in Venetia diamonds

Sample/ temperature step (°C)	Cl (× 10^{-12} mol)	Ca (× 10^{-9} mol)	K (× 10^{-12} mol)	^{40}Ar* (× 10^{-18} mol)	%K	Ca/K	Cl/K	^{40}Ar*/K (× 10^{-6})	Age (Ma)
VC1									
600	nd	0.33 ± 0.09	nd	nd	–	–	–	–	–
800	3.52 ± 3.24	0.20 ± 0.09	33 ± 17	1803 ± 672	2.4	6.1 ± 4.2	0.105 ± 0.111	53.9 ± 19.6	3041 ± 534
1200	nd	0.21 ± 0.11	nd	nd	–	–	–	–	–
1600	3.89 ± 1.59	65.99 ± 1.68	941 ± 32	31,669 ± 830	68.6	70.1 ± 3.0	0.004 ± 0.002	33.7 ± 0.7	2382 ± 28
1800	5.08 ± 2.16	20.39 ± 0.58	226 ± 27	8813 ± 644	16.4	90.4 ± 11.0	0.023 ± 0.010	39.1 ± 3.6	2583 ± 128
2150	3.80 ± 2.23	7.67 ± 0.25	172 ± 24	11,307 ± 1310	12.6	44.5 ± 6.4	0.022 ± 0.013	65.7 ± 5.3	3336 ± 122
Total	16.29 ± 4.76	94.78 ± 1.80	1372 ± 51	53,591 ± 284		69.1 ± 2.9	0.012 ± 0.003	39.1 ± 0.7	2583 ± 23
1200–1800	8.97 ± 2.68	86.59 ± 1.78	1166 ± 41	40,482 ± 1051	87.4	74.2 ± 3.0	0.008 ± 0.002	34.7 ± 1.5	2411 ± 58
VC2									
800	7.35 ± 2.66	nd	67 ± 17	nd	1.1	–	–	–	–
1200	164.78 ± 7.78	0.48 ± 0.09	3577 ± 114	13,529 ± 20	57.2	0.1 ± 0.0	0.046 ± 0.003	3.8 ± 0.1	485 ± 13
1600	7.02 ± 1.67	3.28 ± 0.16	376 ± 15	2168 ± 108	6.0	8.7 ± 0.6	0.019 ± 0.005	5.8 ± 0.4	696 ± 37
1800	2.50 ± 1.67	35.59 ± 1.05	1795 ± 49	7290 ± 83	29.0	19.8 ± 0.8	0.001 ± 0.001	4.1 ± 0.1	516 ± 13
2150	0.64 ± 2.42	6.87 ± 0.20	415 ± 33	15,958 ± 96	6.7	16.6 ± 1.4	0.002 ± 0.006	38.5 ± 3.1	2561 ± 109
Total	182.29 ± 8.88	46.23 ± 1.08	6230 ± 130	38,937 ± 168		7.4 ± 0.2	0.029 ± 0.002	6.3 ± 0.1	748 ± 14
1200–1800	174.30 ± 8.13	39.36 ± 1.07	5747 ± 125	22,988 ± 137	93.3	6.8 ± 0.2	0.030 ± 0.002	4.0 ± 0.1	505 ± 10
VC3									
600	1.22 ± 2.69	0.23 ± 0.08	nd	990 ± 22		–	–	–	–
800	3.11 ± 2.43	0.20 ± 0.09	12 ± 18	3379 ± 84	4.2	16.6 ± 26.3	0.265 ± 0.450	287.4 ± 433.5	5766 ± 2610
1200	7.09 ± 1.82	nd	44 ± 16	1665 ± 83	15.5	–	–	38.0 ± 13.8	2544 ± 497
1600	1.84 ± 1.55	0.28 ± 0.10	15 ± 11	1466 ± 34	5.3	18.7 ± 15.2	0.123 ± 0.138	98.1 ± 72.3	3963 ± 1183
1800	nd	6.97 ± 0.33	nd	nd	–	–	–	–	–
2150	4.25 ± 1.65	7.47 ± 0.33	208 ± 19	1762 ± 158	74.9	35.9 ± 3.7	0.020 ± 0.008	8.5 ± 1.1	947 ± 95
Total	17.52 ± 4.65	15.01 ± 0.51	279 ± 34	9262 ± 201		53.8 ± 6.8	0.063 ± 0.018	33.2 ± 4.1	2117 ± 154
1200–1800	8.93 ± 2.39	7.25 ± 0.34	59 ± 19	3131 ± 89	25.1	123.3 ± 40.9	0.152 ± 0.064	53.3 ± 17.5	3011 ± 482
VC5									
600	0.82 ± 2.49	0.32 ± 0.14	nd	nd		–	–	–	–
800	1.12 ± 2.88	0.22 ± 0.10	27 ± 21	3368 ± 174	12.3	84 ± 7.7	–	126.4 ± 101.8	4362 ± 1323
1200	0.81 ± 2.03	0.14 ± 0.09	nd	509 ± 70		–	–	–	–
1600	nd	0.05 ± 0.09	49 ± 12	nd	22.7	1.1 ± 1.8	–		
1800	nd	0.01 ± 0.07	14 ± 4	nd	6.7	1.0 ± 4.9	–	–	–
2150	18.16 ± 2.55	0.48 ± 0.13	126 ± 20	27,215 ± 68	58.2	3.8 ± 1.2	0.144 ± 0.031	216.0 ± 35.1	5261 ± 278
Total	20.90 ± 5.01	1.23 ± 0.26	216 ± 32	31,092 ± 200	35.1	5.7 ± 1.5	0.097 ± 0.027	143.7 ± 21.4	4574 ± 247
VC6									
V6 800C	0.70 ± 1.99	nd	20 ± 10	nd	0.5	–	–	–	–
V6 1200C	148.93 ± 8.37	0.53 ± 0.13	1987 ± 60	7623 ± 2573	46.3	0.3 + 0.1	0.075 + 0.005	3.8 + 1.3	491 + 146
V6 1600C	6.47 ± 2.29	30.51 ± 0.91	1829 ± 70	7959 ± 738	44.4	16.1 ± 0.8	0.003 ± 0.001	4.2 ± 0.4	532 ± 46
V6 1800C	nd	0.16 ± 0.08	nd	nd	–	–	–	–	–
V6 2150C	32.06 ± 6.98	8.96 ± 0.27	376 ± 53	44,3710 ± 64,595	8.8	23.8 ± 3.4	0.085 ± 0.022	75.5 ± 14.7	4134 ± 302
Total	188.17 ± 11.31	40.05 ± 0.97	4275 ± 110	45,9292 ± 64,651		9.4 ± 0.3	0.044 ± 0.003	107.4 ± 15.4	4110 ± 232
1200–1800	155.40 ± 8.68	31.20 ± 0.92	3879 ± 92	15,581 ± 2677	90.7	8.0 ± 0.3	0.040 ± 0.002	4.0 ± 0.7	507 ± 77

All errors are 2σ, nd = not determinable.

lease over a similar temperature interval has been observed in previous stepped heating experiments of irradiated diamonds containing volatile-rich microinclusions (Turner et al., 1990). Thus, relatively low temperature argon release may be characteristic of diamonds containing inclusions having sizes ranging over several orders of magnitude.

Apparent ages obtained during stepped heating of both garnet and pyroxene varied significantly, but were usually lowest over the 1200–1800 °C interval (Table 1; Fig. 3). Three of the pyroxene-bearing diamonds gave similar minimum ages (Fig. 3), defined in each sample by one to three heating steps, of 918 ± 23 Ma (OC1, 1200–1800 °C), 1004 ± 33 Ma (OC2, 1200–1800 °C) and 871 ± 34 Ma (OC3, 1800 °C step). Due to the effects of Ca-derived $^{36}Ar_{Ca}$ and $^{39}Ar_{Ca}$ interferences formed during irradiation, these ages are probably underestimated by 3–4%. Using the approximate correction for these interferences discussed earlier, the

apparent ages increase to 948 Ma (OC1), 1032 Ma (OC2) and 906 Ma (OC3). These ages compare favourably with the Sm–Nd isochron age for composites of garnet and clinopyroxene inclusions of 990 ± 50 Ma (2σ error; Richardson et al., 1990). The Re–Os systematics of some sulfide inclusions in eclogitic Orapa diamonds are consistent with the Sm–Nd age, but also indicate the presence of an older population of diamonds that formed 2900 Ma ago (Shirey et al., 1999). The three remaining pyroxene-bearing diamonds gave higher apparent ages (Table 1): Both OC4 and OC5 gave similar minimum ages of 1536 ± 117 Ma (2050 °C) and 1624 ± 309 Ma (1800 °C), respectively (Fig. 3). However, apparent ages obtained from Ar released between 1200 and 1800 °C are 3278 ± 37 Ma (OC4) and 2813 ± 140 Ma (OC5); broadly similar to the 2900 Ma Re–Os ages of sulfide inclusions in Orapa eclogitic diamonds (Shirey et al., 1999). Finally, OC6 gave anomalously old apparent ages >4500 Ma for all temperature steps and must therefore be affected by excess ^{40}Ar (Table 1).

An unexpected finding was the relatively high amount of $^{39}Ar_K$ released from most of the garnet inclusions (Table 1), with OG1 and OG3 having a higher $^{39}Ar_K$ content than three of the clinopyroxene-bearing diamonds, and only OG4 having a negligible release. Sample OG3 was the garnet containing the highest $^{39}Ar_K$ concentration, giving an apparent age of 992 ± 96 Ma from Ar released between 1200 and 1800 °C and accounting for 94% of the total $^{39}Ar_K$ released from this sample. The minimum apparent ages obtained from the remaining garnets were: OG4 = 1485 ± 346 (1200–1800 °C; 68% $^{39}Ar_K$ release) and OG5 = 2612 ± 264 Ma (2050 °C step; 56% $^{39}Ar_K$ release).

$^{39}Ar_K$ released from microscopic inclusions other than garnet is a possibility although none were observed during a visual inspection using a binocular microscope. Total Cl/K values of the garnets, determined from measured $^{38}Ar_{Cl}/^{39}Ar_K$ ratios, are in the range 0.008–0.087 (Table 1) quite similar to values obtained from phlogopite micas in kimberlites (e.g., Matson et al., 1986). However, the Cl/K ratios are also comparable to the values obtained from Orapa clinopyroxenes (Cl/K = 0.0004–0.048, Table 1), so without accompanying $^{37}Ar_{Ca}$ data, it is difficult to be certain about the identity of the K-rich phase in the garnet. We note that our previous Ar–Ar determinations of Cl/K in

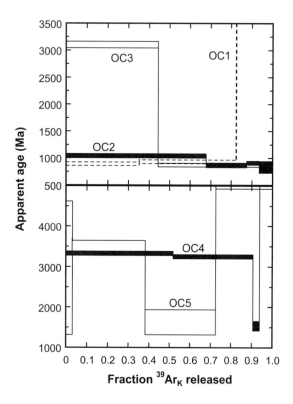

Fig. 3. Apparent age spectrum and K/Ca versus fractional $^{39}Ar_K$ release obtained by stepped heating eclogitic clinopyroxene-bearing diamonds from Orapa. Note the difference in apparent age scale between the upper and lower diagrams.

cleaved Orapa clinopyroxene inclusions were lower at between 0.0004 and 0.0093 (Burgess et al., 1992).

4.2. Venetia diamonds

Ultraviolet laser drilling of clinopyroxene in Venetia diamonds released a maximum of only 3% of the total ^{40}Ar, and yielded apparent ages of between 500 and 2500 Ma. Ar–Ar stepped heating data for Venetia clinopyroxene inclusions are given in Table 2. The Ar released during stepped heating can be broadly divided into a low temperature release at 1200–1800 °C, associated with the major release of ^{37}Ar$_{Ca}$, and attributed to the melting of pyroxene, and a high temperature release at 2150 °C during diamond graphitisation (Fig. 2).

For VC2 and VC6, the Ar released during pyroxene melting yielded apparent ages of VC2 and VC6 gave ages of 505 ± 10 Ma (Fig. 4a) and 507 ± 77 Ma (Fig. 4b), respectively. These ages are indistinguish-

able from the Ar–Ar mica age of the Venetia kimberlite of 519 ± 6 Ma (Phillips et al., 1999). Sample VC3 released only 25% of the total ^{39}Ar$_K$ over the low temperature release, with a high apparent age of 3011 ± 482 Ma (Fig. 4b). However, this was not associated with any release of ^{37}Ar$_{Ca}$ and is therefore considered unlikely to be from clinopyroxene. It is not possible to establish the identity of the phase releasing Ar at low temperature, thus at present we do not feel justified in attaching geological significance to the 3000 Ma age. A lower apparent age of 947 ± 95 Ma was obtained at 2150 °C corresponding to 75% of the total ^{39}Ar$_K$ and most of the ^{37}Ar$_{Ca}$ release and may represent Ar release from a pyroxene grain that had not been exposed by laser drilling (Fig. 4b).

The 1200–1800 °C release from VC1 diamond gave a high apparent age of 2411 ± 58 Ma (Fig. 4a). This age was associated with 87% and 91% of the ^{39}Ar$_K$ and ^{37}Ar$_{Ca}$ release respectively, making pyrox-

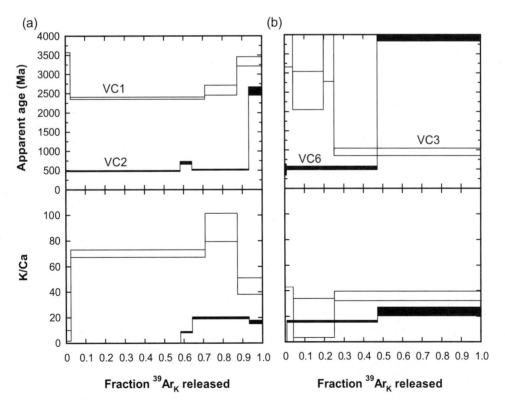

Fig. 4. Apparent age spectrum and K/Ca versus fractional ^{39}Ar$_K$ release obtained by stepped heating eclogitic clinopyroxene-bearing diamonds from Venetia: (a) VC1 and VC2; (b) VC3 and VC6.

ene the most likely source of this argon. The remaining sample (VC5) released only small amounts of Ar at all temperature steps, leading to imprecise ages. Combining gas released from all temperature steps gives an apparent age of 4574 ± 247 Ma (Table 2).

Higher temperature releases at 2150 °C from all the diamonds gave much older apparent ages (some >4500 Ma). This is attributed to the presence of excess ^{40}Ar probably trapped in microinclusions in the host diamond. The total ages given in Table 2 are obtained by combining Ar released from all the temperature steps, and should therefore be considered as upper limits to the time of diamond crystallisation.

The Ca/K ratios of Venetia diamonds, derived from the measured ^{39}Ar$_K$/^{37}Ar$_{Ca}$ ratios, are between 4 and 69 mol (Table 2; Fig. 4), within the range determined previously for pyroxene inclusions in diamonds from southern African kimberlites (Burgess et al., 1992). Release of chlorine-derived ^{38}Ar$_{Cl}$ was correlated with release of ^{39}Ar$_K$ with Cl/K values in the range 0.012– 0.097, similar to that determined for the Orapa clinopyroxene inclusions.

5. Discussion

The good agreement between the Ar–Ar ages of pyroxene inclusions VC2 and VC6 with the published mica ages of the Venetia kimberlite (Phillips et al., 1999) provides evidence for at least one population of diamonds crystallising within a period of no more than about 50 Ma prior to kimberlite emplacement. The Ar–Ar ages of clinopyroxene inclusions in some eclogitic diamonds from Orapa, are between 906 and 1032 Ma, close to the Sm–Nd isochron age of 990 Ma, giving confidence in the methods used and providing support for the diffusive loss model to explain the low and variable ages obtained previously from laser probe Ar–Ar dating of cleaved Orapa diamonds (Burgess et al., 1992). The results from one Venetia and a few Orapa diamonds hint at the presence of older populations of diamonds >2000 Ma in age. Unfortunately, laser drilling experiments were not carried out on Orapa diamonds, and none of the Venetia diamonds, that were probed with the laser released significant amounts of ^{40}Ar from the diamond–inclusion interface. Only a relatively small number of samples so far give apparent ages of >2000 Ma, making it difficult to

judge whether more ancient populations of eclogitic diamonds are present at Orapa and Venetia. However, the high apparent ages obtained from these diamonds merits further discussion of the possible reasons for the lack of interface ^{40}Ar released by the laser, and the underlying reasons for the high apparent ages obtained during stepped heating.

(1) Radiogenic ^{40}Ar formed within pyroxene did not diffuse to the diamond–inclusion interface, and the age can be interpreted as the time at which the diamond crystallised. This scenario gains some support from the recent contention of Kelley and Wartho (2000) that geologically meaningful pre-eruption Ar– Ar ages can be obtained from mantle xenoliths transported rapidly to the surface in kimberlites. Arguments in favour of this model are based on the apparent correspondence of Ar–Ar ages of xenoliths with those obtained using isotope systems more resistant to thermal or metasomatic events (Kelley and Wartho, 2000). There is uncertainty regarding the mechanism whereby ^{40}Ar is retained in minerals that are normally open system under mantle conditions, but anhydrous conditions are likely to be key factor. For diffusive loss to occur from inclusions encased in diamond, the presence of void space and/ or cracks surrounding the inclusions are necessary as storage sites for the diffusing ^{40}Ar. These are commonly observed under surface conditions, but may not have been as abundant within the diamonds under the high temperature and pressure conditions during mantle residence. Many of these storage sites may develop by differential cooling and volume changes during rapid ascent of the kimberlite during eruption. The lack of any physical space at the contact between the inclusion and diamond, may therefore prevent significant loss of radiogenic ^{40}Ar from the pyroxene.

(2) Phillips et al. (2004) discuss an alternative view whereby a proportion of the radiogenic ^{40}Ar produced during mantle residence is transferred to defects or microinclusions in clinopyroxene, either during mantle residence or at the time of kimberlite eruption. The amount of ^{40}Ar retained at the pyroxene–diamond interface is considered to depend upon the number and volume of available storage sites within the pyroxene inclusions. Some additional transfer of ^{40}Ar to the interface may subsequently occur by decrepitation of defects and microinclusions during kimberlite eruption. Whatever the detailed mechanism at work, we

may anticipate relatively low temperature release and old apparent ages from any radiogenic ^{40}Ar that is retained in defects or microinclusions in pyroxene. This idea gains some support from the finding that old apparent ages are obtained during the low temperature steps (≤ 1200 °C) from some of the pyroxene and garnet inclusions (e.g., OC2, OG1, VC3) analysed in this study.

(3) The presence of excess or ambient mantle ^{40}Ar in the inclusions. The most conclusive evidence for the presence of ambient Ar is from sample OC6, which gave anomalously old apparent ages (>4500 Ma) at all temperature steps during stepped heating. The presence of mantle ^{40}Ar is difficult to establish in other inclusions because they give reasonable apparent ages (i.e., <4500 Ma). The measured amounts of ^{36}Ar are generally low, being close to blank levels, however the mantle ^{40}Ar/^{36}Ar value is high being typically >10,000, making it a potentially important contaminant whose presence is difficult to detect. Some insight into the possible presence of excess ^{40}Ar can be gained from Fig. 5, in which apparent age is plotted against the ^{39}Ar$_K$ content of the Orapa garnet and clinopyroxene inclusions analysed in this study, together with previously published data for cleaved Orapa pyroxene inclusions (Burgess et al., 1992). If it is assumed that pyroxene grains in different diamonds contain approximately equal amounts of mantle ^{40}Ar, and also that they are approximately the same size, then a negative

correlation between apparent age and ^{39}Ar$_K$ content is expected. In spite of the limitations associated with using a small sample set and the above assumptions, there is an overall negative trend in Fig. 5 providing some support for an ambient mantle ^{40}Ar component in the pyroxene and garnet inclusions.

6. Conclusions

The combined use of laser drilling and stepped heating has been applied to eclogitic clinopyroxene inclusions in diamonds from the Venetia kimberlite, and eclogitic clinopyroxene and garnet-bearing diamonds from the Orapa kimberlite. Attempts to release any radiogenic ^{40}Ar that has diffused to the inclusion–diamond interface, by laser drilling of Venetia diamonds, has been inconclusive because most of the diamonds give ages in close agreement with the host kimberlite age of 520 Ma. Stepped heating ages obtained from most pyroxene inclusions in Orapa diamonds are in the range 906–1032 Ma, much older than the host kimberlite, but in good agreement with the Sm–Nd isochron age of 990 Ma (Richardson et al., 1990). Unfortunately, the presence of diamond–inclusion interface Ar could not be assessed in these diamonds because the inclusions were too deeply buried within the diamonds to be accessible for laser drilling. A few of the Orapa diamonds, and a single

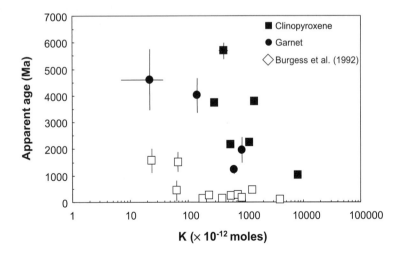

Fig. 5. Plot of apparent age versus K content for eclogitic clinopyroxene and garnet inclusions in Orapa diamonds analysed in this study and for previously published laser Ar–Ar determinations of cleaved clinopyroxene inclusions in Orapa diamonds (data from Burgess et al., 1992).

Venetia sample, gave apparent ages >2500 Ma. Due to the limited size of the data set, it is difficult to establish whether these high apparent ages record an earlier period of diamond crystallisation. The lack of inclusion–diamond interface Ar released by laser drilling of the Venetia diamonds, may be interpreted to indicate retention of either radiogenic or excess (ambient) mantle ^{40}Ar, within the pyroxene inclusion. In the absence of sufficient interface space under mantle conditions, it is possible that this Ar was retained in defect structures such as microinclusions (Phillips et al., 2004).

Acknowledgements

We thank DeBeers for enabling us to undertake this work in Manchester, both by provision of diamond samples and for GK to undertake experimental work during an extended visit. Technical support provided by B. Clementson and D. Blagburn is gratefully acknowledged. We thank D. Phillips and S.H. Richardson for helpful reviews and L. Heaman for editorial support. The Royal Society provided financial assistance for RB to attend the 8IKC.

References

Burgess, R., Turner, G., Laurenzi M., Harris, J.W., 1989. ^{40}Ar–^{39}Ar laser probe dating of individual clinopyroxene inclusions in Premier eclogitic diamonds. Earth Planet. Sci. Lett. 94, 22–28.

Burgess, R., Turner, G., Harris, J.W., 1992. ^{40}Ar–^{39}Ar laser probe studies of clinopyroxene inclusions in eclogitic diamonds. Geochim. Cosmochim. Acta 56, 389–402.

Davis, G.L., 1977. The ages and uranium contents of zircons from kimberlites and associated rocks. Carnegie Inst. Wash. Yearb. 76, 631–635.

Kelley, S.P., Wartho, J.A., 2000. Rapid kimberlite ascent and the significance of Ar–Ar ages in xenolith phlogopites. Science 289, 609–611.

Matson, D.W., Muenow, D.W., Garcia, M.O., 1986. Volatile contents of phlogopite micas from South Africa. Contrib. Mineral. Petrol. 93, 399–408.

McDougall, I., Harrison, T.M., 1999. Geochronology and Thermochronology by the ^{40}Ar/^{39}Ar Method. Oxford Univ. Press, New York. 212 pp.

Phillips, D., Onstott, T.C., Harris, J.W., 1989. ^{40}Ar/^{39}Ar laser-probe dating of diamond inclusions from the Premier kimberlite. Nature 340, 460–462.

Phillips, D., Kiviets, G.B., Barton, E.S., Smith, C.B., Viljoen, K.S., Fourie, L.F., 1999. ^{40}Ar/^{39}Ar dating of kimberlites and related rocks: problems and perspectives. In: Gurney, J.J., Gurney, J.L., Pascoe, M.D., Richardson, S.H. (Eds.), Proc 7th Kimb. Conf., vol. 2. Red Roof Designs, Cape Town, pp. 677–688.

Phillips, D., Harris, J.W., Kiviets, G.B., 2004. ^{40}Ar/^{39}Ar analyses of clinopyroxene inclusions in African diamonds: implications for source ages of detrital diamonds. Geochim. Cosmochim. Acta 68, 151–165.

Richardson, S.H., Erlank, A.J., Harris, J.W., Hart, S.R., 1990. Eclogitic diamonds of Proterozoic age from Cretaceous kimberlites. Nature 346, 54–56.

Seggie, A.G., Hannweg, G.W., Colgan, E.A., Smith, C.B., 1999. The geology and geochemistry of the Venetia kimberlite cluster, Northern Province, South Africa. In: Gurney, J.J., Gurney, J.L., Pascoe, M.D., Richardson, S.H. (Eds.), Proc 7th Kimb. Conf., vol. 27. Red Roof Designs, Cape Town, pp. 750–756.

Shirey, S.B., Harris, J.W., Carlson, R.W., 1999. Re–Os systematics of sulfide inclusions in diamonds from the Orapa kimberlite, Botswana: implications for multiple generations of diamond growth. Eos, Trans. Am. Geophys. Union. 80 (Suppl. 46), 1191.

Turner, G., Huneke, J.C., Podosek, F.A., Wasserburg, G.J., 1971. ^{40}Ar–^{39}Ar ages and cosmic ray exposure age of Apollo 14 samples. Earth Planet. Sci. Lett. 12, 19–35.

Turner, G., Burgess, R., Bannon, M., 1990. Volatile-rich mantle fluids inferred from inclusions in diamond and mantle xenoliths. Nature 344, 653–655.

Available online at www.sciencedirect.com

Lithos 77 (2004) 125–142

www.elsevier.com/locate/lithos

New insights into the occurrence of ^{13}C-depleted carbon in the mantle from two closely associated kimberlites: Letlhakane and Orapa, Botswana

P. Deines[a,*], J.W. Harris[b]

[a] *Department of Geosciences, The Pennsylvania State University, 210 Deike Building, University Park, PA 16802, USA*
[b] *Division of Earth Sciences, University of Glasgow, Glasgow G128QQ, Scotland, UK*

Received 27 June 2003; accepted 28 November 2003
Available online 28 May 2004

Abstract

Carbon isotope measurements on diamonds from the Letlhakane kimberlite, and the analyses of their inclusions, permit the examination of km-scale mantle-composition variations by comparing the results with those for the nearby Orapa kimberlite. Diamonds from Letlhakane have a wide range in carbon isotopic composition ($-3‰$ to $-21‰$); however, the relative abundance of diamonds depleted in ^{13}C is significantly lower than in the Orapa kimberlite. Most of the ^{13}C-depleted diamonds belong to the eclogictic or websteritic paragenesis. The relative abundance of inclusions in diamonds and their composition indicate that there are significant differences in petrology in the mantle below the two locations. At Letlhakane, peridotitic compositions are more prevalent than at Orapa and the protolith of P-Type inclusions in diamonds may have experienced a higher degree of partial melting at Letlhakane compared to Orapa. P/T estimates for both W- and E-Type diamonds indicate that a region of ^{13}C-depletion may exist beneath the two kimberlites. The relationships between carbon isotopic composition of the host diamond and the Al_2O_3/Cr_2O_3 ratios of their websteritic and eclogitic garnet inclusions indicate that the low $\delta^{13}C$ regions may represent a primary mantle feature, unrelated to a crustal component.
© 2004 Elsevier B.V. All rights reserved.

Keywords: Mantle; Diamond; Inclusion; Carbon isotopic composition; ^{13}C-depletion; Websterite

1. Introduction

The Letlhakane and Orapa kimberlites, separated by 40 km, are located about 250 km west of Francis-town, northeastern Botswana. They are intruded into the Proterozoic Magondi Belt, part of the mobile belts bounding the Zimbabwe craton and separating it from the Kaapvaal craton. The Orapa kimberlite has been dated at 93 Ma (Davies, 1977) and the Letlhakane diatremes have probably the same age (Stiefenhofer et al., 1997). The latter authors have recently reviewed the geology of the occurrences. While at Orapa the crater facies of the kimberlite is exposed, at Letlhakane the diatreme facies of the kimberlites, consisting of tuffisitic kimberlite, is encountered.

Mantle xenoliths found in the Letlhakane diatremes include peridotites, pyroxenites, eclogites, megacrysts, MARID and glimmerites (Stiefenhofer et al.,

---

* Corresponding author. Tel.: +1-814-865-7152; fax: +1-814-863-7823.

E-mail address: p7d@psu.edu (P. Deines).

Table 1

Carbon isotopic composition (‰ vs. PDB), shape, color, state of plastic deformation (PD) of diamond host and mineral inclusion assemblage, paragenesis (Par.) and mineral (Min.) composition (wt.%) of the Letlhakane sample suite

Sample	$\delta^{13}C$	Shape	Color	PD	Assemblage	Par.	Min.	SiO_2	TiO_2	Al_2O_3	Cr_2O_3	FeO	MnO	MgO	CaO	NiO	Na_2O	K_2O	P_2O_5	Total
lk 1a	-3.66	dodec.	coll.	no	chr.	p	chr.	0.10	0.02	7.00	64.3	14.40	0.26	13.92		0.10	0.03			100.13
lk 2a	-4.32	ag.	coll.	yes	2 chr.	p	chr.	0.13	0.46	7.53	64.01	14.73	0.25	10.82		0.10				98.03
lk 2b					2chr.	p	chr.	0.12	0.46	7.49	63.61	15.60	0.26	9.71		0.09				97.34
lk 3a	-5.6	flat. dodec.	coll.	yes	chr.	p	chr.	0.24	0.25	5.91	64.46	14.91	0.26	13.92		0.11				100.06
lk 4a	-5.04	irr.	coll.	yes	chr.	p	chr.	0.16	0.07	8.73	62.87	12.14	0.22	15.4		0.09				99.68
lk 5c	-6.27	irr. mac.	coll.	yes	chr.	p	chr.	0.12	0.08	7.80	63.79	13.35	0.24	15.21		0.10				100.69
lk 6	-1.91	oct.	br.	no			gr.													
lk 7a	-4.59	dodec.	coll.	no	chr.	p	chr.	0.14	0.05	7.67	62.81	14.32	0.24	14.06		0.11				99.40
lk 8a	-5.93	dodec.	coll.	yes	2chr.	p	chr.	0.30	0.12	5.43	64.82	15.48	0.26	13.12		0.10				99.63
lk 8b					2chr.	p	chr.	0.28	0.11	5.35	64.59	15.45	0.25	12.88		0.10				99.01
lk 9a	-3.99	irr. mac.	coll.	no	gt.	p	gt.	42.2		19.18	6.42	5.63	0.26	24.07	1.87					99.63
lk 10a	-3.96	triang.mac.	coll.	no	2 gt.,ol.	p	gt.	41.21	0.18	14.10	11.51	5.75	0.25	20.57	6.11					99.68
lk 10b					2 gt.,ol.	p	gt.	40.95	0.17	14.24	11.36	5.72	0.26	20.41	6.05	0.02				99.18
lk 10c					2 gt.,ol.	p	ol.	40.92		0.03	0.05	7.82	0.12	50.85	0.04	0.36				100.19
lk 11a	-1.36	irr.	coll.	no	gt.	p	gt.	42.04		20.11	4.73	5.77	0.26	22.92	3.39			0.02		99.24
lk 12a	-3.85	irr.	coll.	no	3chr.	p	chr.	0.11	0.14	7.99	64.55	13.45	0.25	12.71		0.11				99.31
lk 12b					3chr.	p	chr.	0.11	0.13	7.90	64.55	13.52	0.24	12.56		0.11				99.12
lk 12c					3chr.	p	chr.	0.10	0.12	8.03	64.73	13.08	0.23	12.33		0.11		0.02		98.75
lk 13a	-6.23	irr.	coll.	no	2chr.	p	chr.	0.18	0.10	7.65	65.26	13.36	0.23	11.94		0.11				98.83
lk 13b					2chr.	p	chr.	0.19	0.11	7.51	65.48	13.46	0.23	11.42		0.10				98.50
lk 14a	-3.08	irr.	br.	yes	chr.	p	chr.	0.13	0.07	8.30	63.91	11.96	0.22	15.44		0.10				100.13
lk 15a	-7.88	dodec.mac.	br.	yes	2 gt, 1chr.	p	gt.	41.68	0.04	16.07	10.89	6.26	0.34	23.16	2.22			0.02		100.68
lk 15b					2 gt, 1chr.	p	gt.	41.33	0.05	12.08	13.71	6.29	0.29	23.67	1.60	0.02				99.04
lk 15c					2 gt, 1chr.	p	chr.	0.10	0.17	5.04	68.05	12.22	0.25	12.12		0.08				98.03
lk 16	-3.99	irr.	coll.	no	gt.	p	gt.	41.05		15.36	11.11	5.91	0.30	23.74	1.50	0.02				98.99
lk 17a	-6.61	hemim.	br.	yes	gt.	p	gt.	40.39	0.05	14.39	12.54	5.82	0.28	22.12	2.87	0.02				99.10
lk 18a	-8.9	hemim.	br.	no	gt.	p	gt.									0.5	0.06			
lk 19a	-4.35	irr.mac.	br.	yes	gt.,ol.	p	ol.	40.25		0.02	0.04	6.93	0.10	50.82	0.03	0.33	0.03			98.55
lk 19b					gt.,ol.	p	gt.	41.44	0.08	16.19	9.66	5.52	0.26	22.39	3.72	0.01	0.03			99.33
lk 20a (l)	-6.35	dodec.	br.	yes	gt.–opx	p	gt.	41.48		17.31	9.49	6.08	0.30	24.38	1.01	0.02	0.03			100.07
lk 20a (s)					gt.–opx	p	opx.	57.29		0.52	0.47	3.66	0.09	36.22	0.08	0.13				98.46
lk 21a	-5.99	flat. dodec.	br.	yes	gt.	p	gt.	41.51	0.05	20.41	4.24	7.84	0.31	21.13	3.98		0.06			99.53
lk 22a	-4.76	irr.mac.	coll.	no	gt.	p	gt.	40.12	0.38	12.87	13.84	5.44	0.29	21.48	4.16	0.02	0.08			99.03
lk 23a	-8.79	dodec.mac.	br.	yes	gt.	p	gt.	40.82	0.06	11.93	13.75	6.28	0.28	23.62	1.40	0.02			0.35	98.16
lk 24a	-4.65	dodec.	br.	yes	gt.,ol.	p	gt.	42.39	0.05	18.39	6.17	5.74	0.22	22.57	4.16	0.03			0.02	99.74
lk 24b					gt.,ol.	p	ol.	39.85		0.03	0.04	8.08	0.12	49.89	0.05	0.38	0.03			98.47
lk 25a	-6.2	dodec.mac.	br.	yes	gt.	p	gt.	42.28		17.28	8.13	5.39	0.25	25.18	0.74	0.02				99.27
lk 26a	-5.65	dodec.	yel.	no	2 gt.	e	gt.	40.47	0.51	22.73	0.05	14.29	0.25	10.63	11.81	0.01			0.07	100.82
lk 26b					2 gt.	e	gt.	40.32	0.38	17.01	0.05	12.31	0.21	20.76	8.84	0.09	0.16		0.04	100.17
lk 27a	-5.03	oct.	coll.	no	gt.,ol.	p	ol.	40.89		0.02	0.03	6.80	0.10	51.83	0.03	0.34				100.01

Sample	$\delta^{13}C$	habit	colour	incl.	incl. minerals	par.	min.	SiO_2	TiO_2	Al_2O_3	Cr_2O_3	FeO	MnO	MgO	CaO	Na_2O	K_2O	NiO	Total
lk 27b					gt.,ol.	p	gt.	41.93	0.03	19.35	6.15	5.62	0.27	23.13	3.09				99.57
lk 28a	−5.28	dodec.	br.	yes	gt.	p	gt.	42.00	0.03	16.14	9.97	5.68	0.27	24.51	0.97			0.02	99.59
lk 29b		irr.	br.	yes	gt.,ol.	p	ol.	40.90		0.02	0.04	6.72	0.10	51.48	0.02			0.35	99.63
lk 29a	−5.46			no	gt.,ol.	p	gt.	41.95	0.30	20.24	4.94	5.81	0.26	22.67	3.64			0.02	99.53
lk 30a	−7.02	irr.	coll.	no	gt.	p	gt.	41.75	0.03	17.88	7.44	5.40	0.26	21.45	5.41			0.04	99.93
lk 31a	−5.73	oct.	br.	no	gt.	p	gt.	42.16		20.59	5.02	5.41	0.23	23.91	2.43				99.78
lk 32a	−8.95	oct.	br.	yes	gt.	p	gt.	40.51	0.09	14.97	11.63	7.09	0.35	19.98	4.86			0.09	99.61
lk 33a	−6.68	macled hemim.	br.	yes	ol.	p	ol.	40.29		0.02	0.03	6.54	0.10	51.03	0.03			0.40	98.44
lk 34	−6.18	dodec.	coll.	yes	?		coll.												
lk 35	−5.51	mac. oct.	coll.	no	?		coll.												
lk 36	−4.31	mac.	coll.	yes	?		coll.												
lk 37	−4.11	irr. ag.	br.	yes	?		coll.												
lk 38a	−3.35	flat. dodec.mac.	br.	yes	opx.	p	opx.	57.18		0.76	0.49	4.60	0.13	35.1	0.68	0.13		0.04	99.11
lk39	−2.79	irr.	coll.	no	?		ol.												
lk40	−5.33	irr.	coll.	no	?		sulphide												
lk 41a	−4.64	flat. dodec.mac.	coll.	yes	opx.	p	opx.	57.13		0.84	0.68	3.74	0.10	36.09	0.18	0.12		0.12	98.88
lk 42a	−9.83	triang.mac.	coll.	no	ol.	p	ol.	40.43			0.06	6.20	0.10	51.21		0.36		0.10	98.36
lk 43a	−10.56	irr.mac.	coll.	yes	ol.	p	ol.	40.81		1.6	0.06	6.16	0.09	51.56	0.86	0.35		0.35	101.49
lk 44a	−4.05	dodec.	coll.	no	opx.	p	opx.	57.75		0.82	0.53	4.55	0.34	35.63	0.65	0.12		0.03	100.21
lk 45a	−4.23	oct.	br.	no	gt., cpx.	e	gt.	41.11	0.51	22.54	0.09	14.98	0.34	12.99	8.24	0.21	0.02	0.05	101.08
lk 45c					gt., cpx.	e	cpx.	54.31	0.38	8.97	0.06	6.07	0.09	10.63	14.47	4.12	0.11	0.11	99.23
lk 46a	−3.91	oct.	br.	no	gt., 2 cpx.	e	gt.	40.85	0.93	21.60	0.73	11.84	0.23	19.00	3.63	0.13		0.11	99.07
lk 46b					gt., 2 cpx.	e	cpx.	55.39	0.48	3.37	0.28	8.12	0.14	20.38	9.55	1.76	0.06	0.06	99.64
lk 46c					gt., 2 cpx.	e	cpx.	55.32	0.47	3.48	0.30	7.95	0.14	20.35	9.51	1.78	0.06	0.04	99.51
lk 47a	−5.75	irr.mac.	coll.	no	2 gt.	e	gt.	40.06	0.47	22.78	0.05	15.38	0.27	9.91	11.54	0.18		0.07	100.71
lk 47b					2 gt.	e	gt.	39.61	0.47	22.46	0.06	15.34	0.27	9.74	11.60	0.18		0.04	99.77
lk 48a	−6.63	irr.	coll.	no	2 gt.	e	gt.	40.03	0.45	22.81	0.02	14.96	0.27	10.86	10.78	0.18		0.04	100.42
lk 48b					2 gt.	e	gt.	39.82	0.46	22.72	0.02	14.93	0.27	10.70	10.77	0.19		0.07	99.95
lk 49b	−4.01	oct.	coll.	no	gt.	e	gt.	38.74	0.50	22.07	0.05	17.42	0.37	9.79	9.64	0.21		0.05	98.84
lk 50a	−6.08	irr.mac.	coll.	no	2 gt.	e	gt.	39.40	0.47	22.17	0.06	15.66	0.27	9.63	11.45	0.19		0.06	99.36
lk 50c					2 gt.	e	gt.	40.25	0.48	22.49	0.06	15.43	0.27	9.85	11.46	0.18		0.05	100.52
lk 51a	−5.68	dodec.	yel.	no	2 gt.	e	gt.	39.51	0.57	22.38	0.03	13.46	0.24	9.42	13.39	0.20		0.07	99.27
lk 51b					2 gt.	e	gt.	39.31	0.46	22.25	0.06	15.43	0.28	9.67	11.53	0.18		0.06	99.23
lk 52a	−4.28	irr.	coll.	yes	gt.	e	gt.	40.97	0.57	22.58	0.10	14.04	0.31	14.68	6.94	0.17		0.04	100.42
lk 53a	−5.57	dodec.	br.	yes	gt.	e	gt.	40.45	0.56	22.48	0.08	13.96	0.30	14.53	6.96	0.16		0.04	99.54
lk 54a	−4.26	oct. ag.	br.	no	gt.	e	gt.	40.06	0.63	22.21	0.11	13.53	0.29	10.92	11.89	0.17		0.06	99.87
lk 55a	−7.1	irr.	yel.	no	gt.	e	gt.	39.53	0.59	22.62	0.04	14.52	0.25	9.05	13.14	0.19		0.06	99.99
lk 56a	−4.11	oct.	br.	no	2 gt.	e	gt.	39.71	0.65	22.17	0.07	17.25	0.39	13.20	5.71	0.26		0.05	99.46
lk 56b					2 gt.	e	gt.	40.52	0.65	22.49	0.06	17.4	0.40	13.39	5.68	0.27		0.05	100.93
lk 57a	−7.84	irr.	gr. c.	no	gt.	e	gt.	39.15	0.54	22.27	0.07	16.35	0.29	10.12	10.36	0.19		0.07	99.41
lk 58a	−4.42	oct.	gr. c.	no	cpx.	e	cpx.	53.88	0.38	8.97	0.08	5.69	0.09	10.51	14.78	4.05	0.15	0.02	98.63
lk 59a	−7.06	dodec.	coll.	no	4 gt.	p	gt.	41.44	0.03	16.73	9.93	5.50	0.27	24.41	0.96				99.27
lk 59b (n)					4 gt.	p	gt.	40.38	0.03	17.61	10.92	6.05	0.29	23.59	1.08	0.02			99.95
lk 59c					4 gt.	p	gt.	41.61		16.61	9.89	5.39	0.26	24.45	0.97	0.02		0.03	99.23

(continued on next page)

Table 1 (continued)

Sample	$\delta^{13}C$	Shape	Color	PD	Assemblage	Par.	Min.	SiO_2	TiO_2	Al_2O_3	Cr_2O_3	FeO	MnO	MgO	CaO	NiO	Na_2O	K_2O	P_2O_5	Total
lk 59d					4 gt.	p	gt.	41.64	0.03	16.54	9.97	5.46	0.27	24.49	0.96	0.02	0.02			99.40
lk 60a	−8.91	ag.	coll.	no	2 gt.	p	gt.	42.05		15.56	10.13	5.56	0.25	25.12	0.49	0.02				99.18
lk 60b					2 gt.	p	gt.	41.60		16.57	10.43	5.62	0.26	24.63	0.47					99.58
lk 61	−4.03	oct.	coll.	no	?	p	cpx.													
lk 62a	−4.38	oct.	coll.	yes	opx.	p	opx.	57.17		0.86	0.48	4.10	0.10	35.63	0.38	0.13	0.04		0.02	98.91
lk 63a	−11.73	oct.	coll.	no	2 gt.	e	gt.	40.51	0.34	22.82	0.06	16.56	0.37	13.85	5.87	0.2	0.14		0.03	100.70
lk 63b					2 gt.	e	gt.	40.61	0.34	23.05	0.05	16.59	0.38	13.88	5.89		0.15		0.03	100.97
lk 64a	−10.75	oct.	br.	yes	gt.	e	gt.	39.64	0.55	22.36	0.04	17.49	0.38	9.78	9.71		0.22		0.12	100.29
lk 65a	−8.55	oct.	br.	no	gt.	e	gt.	39.55	0.51	22.29	0.08	17.56	0.37	9.31	10.07		0.20		0.07	100.02
lk 66a	−16.86	flat. dodec.	coll.	no	2 cpx.	e	cpx.	55.11	0.77	16.37	0.07	4.28	0.10	5.30	7.60		8.14	0.28	0.02	98.05
lk 66b					2 cpx.	e	cpx.	55.12	0.74	16.30	0.07	4.22	0.11	5.36	7.59		8.04	0.28	0.03	97.87
lk 67a	−21.32	irr.	coll.	no	gt.	w?	gt.	40.96	0.82	20.71	1.79	11.91	0.46	18.21	4.83		0.06		0.03	99.78
lk 68a	−14.07	irr.	coll.	no	2 cpx.	e	cpx.	55.06	0.59	14.33	0.06	3.67	0.04	6.96	11.10	0.02	6.83	0.21		98.87
lk 68c					2 cpx.	e	cpx.	54.74	0.48	12.21	0.05	3.23	0.05	8.88	14.10		5.07	0.54		99.35
lk 69a	−7.33	irr.	coll.	no	gt., cpx.	e	gt.	39.85	0.55	22.48	0.04	15.92	0.31	10.16	10.56		0.19		0.06	100.12
lk 69b					gt., cpx.	e	cpx.	53.99	0.46	12.24	0.05	3.13	0.05	8.61	14.07		4.97	0.53	0.02	98.12
lk 70a	−8.84	flat. dodec.mac.	coll.	yes	gt., gt.–altered	e	gt.	39.43	0.87	21.98	0.06	17.33	0.38	9.67	9.90		0.27		0.08	99.97
lk 70b					gt., gt.–altered	e	gt.	39.62	0.76	22.01	0.04	17.14	0.40	9.84	9.97		0.26		0.06	100.10
lk 70b*					gt., gt.–altered	e	alt.?	37.79	0.03	8.72	0.14	28.04	0.53	24.42	0.17		0.07	0.05	0.04	100.00
lk 71a	−3.62	oct. ag.	br.	no	gt., cpx.	e	gt.	39.54	0.65	21.75	0.06	19.01	0.44	10.82	7.27		0.25		0.09	99.88
lk 71b					gt., cpx.	e	cpx.	53.8	0.42	9.32	0.05	8.25	0.13	9.47	12.17	0.03	4.80	0.19		98.63
lk 72a	−16.66	ag.	coll.	no	2 cpx.	e	cpx.	54.28	0.87	8.93	0.07	6.59	0.11	9.86	12.45	0.02	4.89	0.64	0.06	98.77

Sample	δ13C	Shape	Habit	Color	Norm.	Inclusions	Parag.	Mineral	SiO2	TiO2	Al2O3	Cr2O3	FeO	MnO	MgO	CaO	Na2O	K2O		Total
lk 72b						2 cpx.	e	cpx.	53.02	0.89	8.48	0.07	6.92	0.13	9.71	12.47	4.64	0.62	0.07	97.02
lk 73b	−6.41	flat.	dodec.:mac.	coll.	yes	gt.	e	gt.	40.10	0.44	22.63	0.07	11.95	0.21	9.76	14.56	0.02	0.17	0.07	99.98
lk 74a	−18.21	oct.		br.	yes	gt., 2 cpx.,alt., opx.	w	cpx.	54.77	0.54	10.62	0.08	4.72	0.07	8.92	12.94	5.59	0.30	0.05	98.60
lk 74a*						gt., 2 cpx.,alt. opx.	w	alt.?	50.27	0.17	8.49	0.15	12.03	0.06	28.05	0.42	0.19	0.06	0.12	100.01
lk 74b (n)						gt., 2 cpx., alt. opx.	w	opx.	52.45	0.21	5.55	0.18	9.65	0.03	31.60	0.15	0.10	0.06		99.98
lk 74c						gt., 2 cpx, alt., opx.	w	cpx.	54.68	0.59	11.14	0.08	4.73	0.07	9.02	13.00	5.84	0.09	0.03	99.27
lk 74d						gt., 2 cpx~alt., opx.	w	gt.	39.59	0.56	22.42	0.08	17.18	0.41	11.31	8.28	0.23	0.11	0.01	100.18
lk 75a	−10.94	irr.		coll.	yes	gt., cpx.	e	gt.	40.16	0.24	23.10	0.06	13.28	0.32	14.36	7.36	0.19	0.22		99.29
lk 75d (n)						gt., cpx.	e	cpx.	55.81	0.26	12.86	0.07	3.13	0.05	8.54	11.99	6.97	0.24	0.05	99.99
lk 76a	−3.99	oct.		br.	no	gt., 2 cpx.	e	gt.	40.81	0.43	22.53	0.15	12.62	0.29	13.94	8.96	0.14	0.06		99.93
lk 76b						gt., 2 cpx.	e	cpx.	54.19	0.28	8.15	0.11	4.60	0.08	11.77	15.66	3.49	0.11	0.02	98.46
lk 76c						gt., 2 cpx.	e	cpx.	53.48	0.27	8.03	0.10	4.45	0.07	11.77	16.15	3.38	0.12	0.03	97.85
lk 77a	−18.3	irr.		br.	yes	gt.	e	gt.	38.70	0.72	21.93	0.04	19.33	0.51	8.47	9.51	0.27	0.13	0.08	99.69
lk 78a	−4.95	oct.		br.	yes	gt., cpx., 2 rut.	e	gt.	39.56	0.66	21.85	0.07	17.87	0.35	9.56	9.85	0.20	0.03		100.00
lk 78b						3 gt., cpx., 2 rut.	e	gt.	38.63	0.51	21.99	0.06	18.49	0.36	9.16	9.38	0.15	0.02		98.75
lk 78c						3 gt., cpx., 2 rut.	e	cpx.	54.04	0.68	9.34	0.05	5.04	0.06	9.48	14.90	5.13	0.16		98.88
lk 78d						3 gt., cpx., 2 rut.	e	rut.	0.04	96.22	0.17	0.08	0.77	0.02	0.04					97.34
lk 78e						3 gt., cpx., 2 rut.	e	gt.	38.77	0.65	21.79	0.07	17.81	0.35	9.47	9.80	0.17			98.88
lk 78f						3 gt., cpx, 2 rut.	e	rut.	0.03	95.69	0.88	0.09	1.06	0.01	0.07					97.83
lk 79a	−4.22	irr.		br.	yes	gt.	e	gt.	39.23	0.83	21.69	0.08	17.12	0.38	10.73	8.74	0.27	0.05		99.12

Detection limit 0.02 for all major elements except for Na$_2$O (0.03).

Parageneses: p=peridotitic; e=eclogitic; w=websteritic. Shape: dodec.=dodec.ahedron; ag.=aggregate; flat.=flattened; irr.=irregular; oct.=octahedron; triang.=triangular; mac.=macle; hemim.=hemimorphic. Color: coll.=colorless; br.=brown; yel.=yellow; gr.c.=green coated. Minerals: chr.=chromite; gr=graphite; gt.=garnet; ol.=olivine; cpx.=clinopyroxene; opx.=orthopyroxene; rut.=rutile; alt.=altered. Sample comments: lk 70b*=normalized, unpolished; lk 74a*=normalized, unpolished rim; n=normalized; l=large; s=small.

1997), whereas at Orapa almost exclusively eclogites and megacrysts have been observed (Shee, 1978; Robinson et al., 1984; Shee and Gurney, 1979). However, the difference in the xenoliths suites between the two kimberlites may not necessarily reflect differences in the mantle petrology underneath Orapa and Letlhakane. Stiefenhofer et al. (1997) attribute the absence of peridotite xenoliths at Orapa and their presence at Letlhakane to differences in the weathering of the two kimberlites. The authors consider that in the crater lake of the Orapa diatreme the peridotitic xenoliths would have been quickly decomposed, because of the susceptibility of olivine and orthopyroxene to weathering. Because garnet and clinopyroxene are relatively more stable under these conditions, there could have been a preferential preservation of eclogitic xenoliths. At Letlhakane, on the other hand, deeper levels of the diatreme, not exposed to extensive surface weathering and therefore favoring the preservation of peridotitic xenoliths, are mined.

Based on their studies of the petrology and geochemistry of the peridotitic xenoliths from Letlhakane, Stiefenhofer et al. (1997) conclude that, in general terms, the Letlhakane xenoliths suite is very similar to other xenoliths suites from the Kaapvaal craton, although there are some exceptions. The authors also conclude that the Letlhakane, and by inference Orapa, kimberlites are underlain by Archean mantle which is related to the Zimbabwe craton.

In view of the suggested broad similarity in the mantle beneath Letlhakane and Orapa, it is of interest to compare diamonds and their inclusions recovered at the two locations. Because these mantle samples are unaffected by weathering processes, they will provide direct evidence for any differences in the petrology and geochemistry of the mantle underlying the two kimberlites.

2. Experimental

The major element compositions of the mineral inclusions were determined on a CAMECA SX50 electron microprobe, with an operating voltage of 15 kV and 20 nA beam current using silicate, oxides and metal standards. Count times were generally 30 s on peaks and 15 s for backgrounds.

The carbon isotopic compositions of 79 individual stones were determined using the techniques described by Deines et al. (1984). The sample CO_2 gases were compared with a working reference (PSU-2) whose isotopic composition with respect to the original PDB reference was measured by H. Craig (personal communications 1971) and checked, at intervals, with secondary carbon isotope reference samples. The isotopic compositions are reported in the conventional delta notation with respect to the PDB reference.

3. Results

The carbon isotope measurements are summarized in Table 1, along with information on the shape of the diamonds, their color, the presence of plastic deformation, and the inclusion mineral paragenesis.

The summary carbon isotope frequency-distribution (Fig. 1A) is skewed toward lower ^{13}C-contents. P-Type diamond $\delta^{13}C$ values range from $-1.36‰$ to $-10.56‰$, E-Type diamonds have $\delta^{13}C$ values between $-3.62‰$ and $-18.3‰$, while the two W-Type diamonds have very low ^{13}C-contents ($-18.21‰$, $-21.3‰$). As in our earlier studies no relationship of the carbon isotopic composition of the diamonds to their shape, color or state of deformation was found.

Mean values for the inclusion mineral compositions are given in Tables 2 and 3. Although for all minerals examined the means fall in the general range of compositions reported for other southern African diamond suites (Finsch, Jagersfontein, Jwaneng, Koffiefontein, Orapa, Premier, Roberts Victor and Venetia, Deines et al., 1984, 1987, 1991b, 1993, 1997, 2001), significant differences between Letlhakane diamond inclusions and those from individual kimberlites are observed. A detailed comparison of the Letlhakane composition means with those found for Orapa is carried out below.

4. Discussion

The study of diamond suites from the eight southern African kimberlites has revealed that each kimberlite has a characteristic $\delta^{13}C$-distribution and that within a particular kimberlite systematic associations

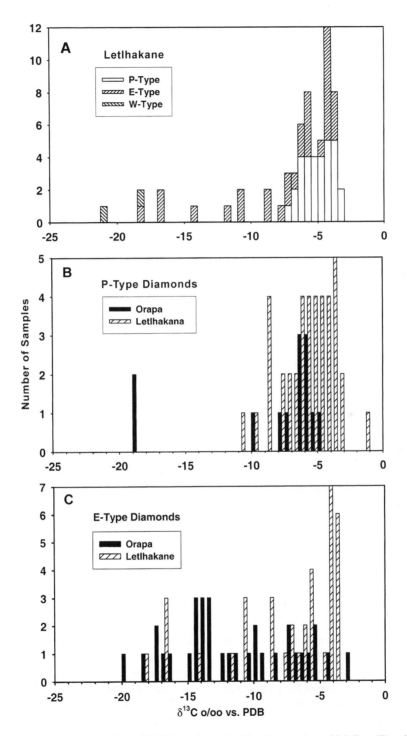

Fig. 1. Summary of the carbon isotopic composition of Letlhakane diamonds (A) and comparison with P-Type (B) and E-Type (C) diamonds from Orapa.

Table 2

Summary of the chemical composition of P-Type diamond inclusions for Letlhakane and comparison with Orapa

	Orapa			Letlhanane			t-test				
	Mean	S.D.	n	Mean	S.D.	n	Or-Let	t	DF	$t_{0.01}$	Sig.
Olivine											
$\delta^{13}C$	−9.24	5.72	4	−7.29	2.50	5	−1.95	−0.69	7	3.5	n
SiO_2	40.66	0.17	6	40.54	0.22	8	0.12	1.08	12	3.06	n
Al_2O_3	0.02	0.02	6	0.02	0.00	8	−0.01	−0.67	12	3.06	n
Cr_2O_3	0.06	0.01	6	0.04	0.02	8	0.02	1.57	12	3.06	n
FeO	7.33	0.35	6	6.91	0.30	8	0.42	2.44	12	3.06	y
MnO	0.08	0.01	6	0.10	0.00	8	−0.02	−4.32	12	3.06	y
MgO	51.45	0.46	6	51.10	0.31	8	0.34	1.68	12	3.06	n
CaO	0.08	0.04	6	0.03	0.01	8	0.05	3.41	12	3.06	y
Mg#	0.92589	0.00	6	0.9295	0.0029	8	−0.0036	−2.02	12	3.06	n
Garnet											
$\delta^{13}C$	−9.27	5.58	4	−6.20	1.96	19	−3.07	−2.01	21	2.83	n
SiO_2	41.29	0.56	4	41.48	0.61	27	−0.19	−0.59	29	2.76	n
TiO_2	0.38	0.45	4	0.06	0.09	27	0.32	3.50	29	2.76	y
Al_2O_3	19.95	0.54	4	16.60	2.40	27	3.35	2.74	29	2.76	p
Cr_2O_3	4.32	1.96	4	9.41	2.81	27	−5.09	−3.47	29	2.76	y
FeO	8.76	3.15	4	5.85	0.54	27	2.91	4.77	29	2.76	y
MnO	0.38	0.11	4	0.27	0.03	27	0.11	4.40	29	2.76	y
MgO	19.56	2.16	4	23.10	1.46	27	−3.54	−4.27	29	2.76	y
CaO	5.09	0.67	4	2.60	1.72	27	2.49	2.83	29	2.76	y
Mg#	0.7981	0.0761	4	0.8751	0.0148	27	−0.0770	−5.10	29	2.76	y
Ca	0.130	0.017	4	0.066	0.044	27	0.064	2.85	29	2.76	y
Mg	0.695	0.075	4	0.818	0.048	27	−0.122	−4.47	29	2.76	y
Fe	0.175	0.064	4	0.116	0.011	27	0.059	4.78	29	2.76	y
Si/Al	2.072	0.072	4	2.548	0.368	27	−0.476	−2.54	29	2.76	n
Al/Cr	6.673	4.306	4	2.083	1.104	27	4.590	4.94	29	2.76	y
Chromite											
$\delta^{13}C$	−6.43	1.34	7	−4.86	1.08	10	−1.57	−2.68	15	2.95	n
TiO_2	0.21	0.14	15	0.15	0.13	16	0.06	1.24	29	2.76	n
Al_2O_3	6.40	1.26	15	7.21	1.10	16	−0.81	−1.92	29	2.76	n
Cr_2O_3	64.53	1.27	15	64.50	1.17	16	0.03	0.08	29	2.76	n
FeO	13.28	0.83	15	13.80	1.16	16	−0.52	−1.43	29	2.76	n
MnO	0.77	0.06	15	0.24	0.01	16	0.53	36.63	29	2.76	y
MgO	13.93	1.07	15	13.00	1.58	16	0.93	1.91	29	2.76	n
Mg#	0.6508	0.0253	15	0.6237	0.0414	16	0.0271	2.18	29	2.76	n
Al_2O_3/Cr_2O_3	0.0994	0.0204	15	0.1120	0.0181	16	−0.0126	−1.81	29	2.76	n

Mean, standard deviation (S.D.) and number of samples (n) are given. Or-Let = Difference in means between Orapa and Letlhakane, t = computed t value, DF = degrees of freedom, $t_{0.01}$ = t value at the 1% level for the given degrees of freedom, Sig. = indicates whether the difference in means is significant at the 1% level, y = yes, n = no.

between mineral inclusion chemistry and host carbon isotopic composition can occur. The question over what scale such regularities could extend in the mantle, and hence how far the results for a particular kimberlite could be generalized for the mantle as a whole has not been answered as yet. It is therefore interesting to compare the carbon isotope record of diamonds and the geochemistry of their inclusions for Letlhakane with those from of Orapa, because the diatremes occur in the same geologic setting and the intrusion of the two kimberlites is closely associated in time (93 Ma, Davies, 1977) and space (the two kimberlites are separated by about 40 km).

When the summary $\delta^{13}C$-distribution for Letlhakane (Fig. 1A) is compared to that for Orapa (Fig. 1, Deines et al., 1993) it is immediately apparent that

Table 3
Summary of the chemical composition of E-Type diamond inclusions for Letlhakane and comparison with Orapa

| | Orapa | | | Letlhakane | | | t-test | | | | |
	Mean	S.D.	n	Mean	S.D.	n	Or-Let	t	DF	$t_{0.01}$	Sig.
Garnet											
$\delta^{13}C$	− 12.48	3.73	17	− 6.71	3.00	35	− 5.77	− 6.00	50	2.68	y
SiO_2	39.91	0.48	17	39.80	0.65	36	0.11	0.61	51	2.678	n
TiO_2	0.54	0.14	17	0.55	0.14	36	− 0.01	− 0.27	51	2.678	n
Al_2O_3	22.27	0.35	17	22.20	0.95	36	0.07	0.28	51	2.678	n
Cr_2O_3	0.08	0.04	17	0.08	0.11	36	0.00	0.15	51	2.678	n
FeO	16.56	1.55	17	15.70	1.98	36	0.86	1.57	51	2.678	n
MnO	0.37	0.11	17	0.32	0.07	36	0.05	1.93	51	2.678	n
MgO	11.79	2.10	17	11.40	2.74	36	0.39	0.52	51	2.678	n
CaO	7.90	2.40	17	9.52	2.39	36	− 1.62	− 2.30	51	2.678	n
Na_2O	0.13	0.09	17	0.19	0.04	36	− 0.06	− 3.56	51	2.678	y
Mg#	0.5557	0.0560	17	0.5594	0.0705	36	− 0.004	− 0.19	51	2.678	n
Ca	21.26	6.47	17	25.4	6.64	36	− 4.14	− 2.14	51	2.678	n
Mg	43.98	7.33	17	42	8.24	36	1.98	0.85	51	2.678	n
Fe	34.76	3.60	17	32.6	4.76	36	2.16	1.66	51	2.678	n
Al_2O_3/Cr_2O_3	292.5	129.6	17	397.7	157.2	36	− 105.21	− 2.40	51	2.678	n
CPX											
$\delta^{13}C$	− 11.35	3.95	21	− 9.15	5.53	16	− 2.20	− 1.41	35	2.72	n
SiO_2	54.93	0.51	22	54.50	0.75	16	0.43	2.09	36	2.718	n
TiO_2	0.50	0.16	22	0.53	0.20	16	− 0.03	− 0.53	36	2.718	n
Al_2O_3	9.47	3.31	22	10.08	3.68	16	− 0.61	− 0.54	36	2.718	n
Cr_2O_3	0.07	0.04	22	0.10	0.08	16	− 0.03	− 1.52	36	2.718	n
FeO	5.40	1.32	22	5.33	1.74	16	0.07	0.15	36	2.718	n
MnO	0.10	0.04	22	0.09	0.03	16	0.01	0.46	36	2.718	n
MgO	10.66	2.51	22	10.50	4.16	16	0.16	0.15	36	2.718	n
CaO	13.31	1.97	22	12.40	2.64	16	0.91	1.22	36	2.718	n
Na_2O	4.83	1.37	22	4.88	1.83	16	− 0.05	− 0.10	36	2.718	n
K_2O	0.26	0.23	22	0.27	0.19	16	− 0.01	− 0.14	36	2.718	n
Mg#	0.7763	0.0347	22	0.7709	0.0550	16	0.0054	0.37	36	2.718	n
Ca	41.55	3.75	22	41.00	7.96	16	0.55	0.28	36	2.718	n
Mg	45.42	4.14	22	45.50	7.26	16	− 0.08	− 0.04	36	2.718	n
Fe	13.03	1.94	22	13.50	3.59	16	− 0.47	− 0.52	36	2.718	n

Mean, standard deviation (S.D.) and number of samples (n) are given. Or-Let = Difference in means between Orapa and Letlhakane, t = computed t value, DF = degrees of freedom, $t_{0.01}$ = t value at the 1% level for the given degrees of freedom, Sig. = indicates whether the difference in means is significant at the 1% level, y = yes, n = no.

the two differ significantly as a result of the higher abundance of low $\delta^{13}C$ diamonds at Orapa. In making such a comparison, however, one needs to recognize that, because different categories of diamonds can have different $\delta^{13}C$-distributions, biases may occur in the absence of proper weighting of these categories. Therefore, in Fig. 1B and C, P-Type and E-Type distributions are compared separately. In Fig. 1B, a tendency of the Orapa analyses to be displaced toward lower $\delta^{13}C$ values may be noted, however, a nonparametric test (Kolmogorov–Smirnov statistic, Miller and Kahn, 1962) indicates

that there is no significant difference between the two locations. The number of P-Type samples analyzed for Orapa is just too small. However, there is a significant difference between the $\delta^{13}C$-distributions for E-Type diamonds (Fig. 1C) (Kolmogorov–Smirnov statistic, 1% level). Orapa E-Type diamonds are systematically lower in $\delta^{13}C$ compared to Letlhakane E-Type diamonds.

The relative proportions of the syngenetic inclusions within the diamonds differ between the two kimberlites. In an unpublished study, Harris and Gurney (private communication) examined the abundance of P-Type, E-Type and sulfide inclusions in one diamond size (just less than two mm) from Orapa; the W-paragenesis was not recognized then. Two substantial examinations were completed, although the numbers of diamonds involved were not recorded. From this work, 160 and 107 inclusion-bearing diamonds were selected and the percentages observed respectively were: P-type, 10.0 and 19.6%, E-type 65.4 and 69.4% and sulphides with 20.0 and 15.0%. The fraction of E-Type diamonds (87.4% and 87.2%, respectively) is greater than that recorded (70%) by Deines et al. (1993). There is the possibility, therefore, that in the 1993 study, the E-Type paragenesis is slightly under represented, which means that low $\delta^{13}C$ diamonds are also under represented in the summary $\delta^{13}C$-distribution of E- and P-Type diamonds.

At Letlhakane the inclusion abundance study, of the same diamond size as that for Orapa, resulted in 126 inclusion-bearing diamonds being selected from 8500 diamonds. Inclusion proportions were 31.7% P-type, 29.4% E-type, and 38.9% sulphides. Clearly, the fraction of E-Type diamonds from Letlhakane (E/(E + P) = 48.1%) is significantly lower than for Orapa. In the present study 43% of the diamonds investigated were E-Type, hence the observed summary $\delta^{13}C$-distribution represents an unbiased sample of the carbon isotope distribution of Letlhakane diamonds of the size range from which the specimens were selected.

We may conclude from these observations that there is a significant difference in the carbon isotope distribution in the mantle below the Orapa and Letlhakane kimberlites. This observation is not necessarily in accord with the expectations for the occurrence of low $\delta^{13}C$ values in the mantle based on the conclusions by Shirey et al. (2002).

We found that at Letlhakane the abundance of P-Type diamonds outweighs that of E-Type diamonds, while the reverse is true for Orapa. This parallels the observation made for peridotitic and eclogitic xenoliths (Stiefenhofer et al., 1997). Since the relative abundance of E- and P-Type diamonds is unaffected by weathering processes, it is unlikely that differential weathering alone would account for the difference the relative abundances of peridotitic and eclogitic xenoliths in these two kimberlites. It is probable then that, in addition to the difference in

the carbon isotope distribution, there are also differences in the relative proportions of peridotitic and eclogitic mantle beneath the two kimberlites.

Tables 2 and 3 permit to examine whether there are composition differences between the eclogitic as well as the peridotitic inclusions in diamond from the mantle beneath the two kimberlites. In these tables, *t*-tests are reported for the comparison of the various inclusion minerals. There is a significant difference in the P-Type garnet composition between the inclusion from Letlhakane and Orapa. Compared to Orapa, Letlhakane garnets are significantly lower in TiO_2, FeO, MnO, and CaO and significantly higher in Cr_2O_3, and MgO. The significantly lower FeO content of Letlhakane olivines, and the significantly lower Mn content of Letlhakane chromites are consistent with the garnet data. Due to the lack of orthopyroxene analyses for the Orapa suite no comparison could be made for this mineral. Because Mg and Cr are refractory elements while Ti, Fe Mn and Ca would enter partial melts, the difference in chemistry for the growth environment of diamonds beneath the two kimberlites can be interpreted as reflecting a higher degree of partial melting in the source of the P-Type diamond inclusions of Letlhakane compared to Orapa.

A similar comparison was made for E-Type garnets and clinopyroxenes (Table 3). No significant differences between the samples from the two locations were observed, except that the Na content of Letlhakane E-Type garnets may be slightly higher. It is interesting to note that although the chemistries of the E-Type garnets are indistinguishable, the E-Type garnet hosts at Letlhakane (− 6.71‰) are significantly enriched in ^{13}C compared to their counter parts at Orapa (− 9.15‰). The hosts of E-Type clinopyroxenes from the two kimberlites have, on average, the same $\delta^{13}C$ values, however, note the large standard deviations. This curious difference let us to test whether diamonds hosting coexisting clinopyroxene–garnet pairs ($\delta^{13}C = − 5.57‰ ± 2.68$, $n = 7$), have an isotopic composition which is distinct from that of diamonds hosting only clinopyroxene ($\delta^{13}C = − 14.2‰ ± 4.14$, $n = 7$). Using a *t*-test, the answer is an unambiguous yes. At Letlhakane diamonds hosting clinopyroxene only are significantly depleted in the heavy isotope. In the Orapa sample site such a difference is not observed. We may conclude that within the eclogitic paragenesis of Letlhakane, based on the carbon isotopic composition, a subgroup of lower $\delta^{13}C$ values can be recognized.

If two or more different mineral inclusions coexist within a single diamond, P/T conditions of equilibration may be assessed, provided the necessary experimental calibrations have been carried out. Possible P/T conditions for P-Type diamonds from the two kimberlites have been compiled in Fig. 2. With one exception, the Letlhakane samples tend to show higher P/T conditions than those from Orapa, which would be consistent with the greater degree of partial melting indicated by the P-Type garnet compositions. There also may be a trend to lower $\delta^{13}C$ values lower pressures and temperatures of equilibration. However, the data are rather limited and these tentative conclusions remain to be confirmed. The $P–T$ computations for the olivine/garnet pair of sample lk24 yield temperatures of equilibration at the extreme of temperature estimates for P-Type dia-

monds (compare Gurney, 1989). The olivine of this sample is characterized by the highest FeO and lowest MgO and SiO_2 content among the olivines from Letlhakane, yet has a $Mg/Mg + Fe$ ratio close to the mean of the set olivines from this kimberlite. The garnet from lk24 has the highest SiO_2 content of the P-Type garnets in the Letlhakane suite, the FeO, MgO and CaO concentrations do not fall at the extremes of the composition range. The question of whether the olivine and garnet are not in chemical equilibrium, or whether the diamond lk24 formed in special mantle setting cannot be settled unambiguously. There is nothing unusual about the carbon isotopic composition of lk24, however.

Possible conditions for the equilibration of W-Type diamonds are shown in Fig. 3. Diamond lk74 includes garnet, clinopyroxene and orthopyroxene. Hence, two P/T estimates are possible. While that based on the garnet–clinopyroxene geothermobarometer of Krogh (1988) is consistent with the P/T estimates for Orapa,

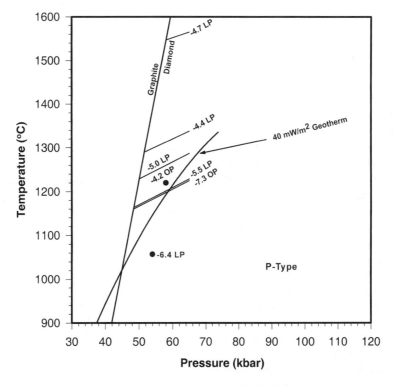

Fig. 2. Temperature and pressure conditions of equilibration for P-Type diamonds from Letlhakane and Orapa. Possible P/T conditions for garnet/olivine equilibration were computed after O'Neill and Wood (1979, 1980), the isotopic composition of the host diamond is indicated, L = Letlhakane, O = Orapa. The graphite diamond transition is from Kennedy and Kennedy (1976) and the 40 mW/m² cratonic geotherm from Pollack and Chapman (1977) are shown.

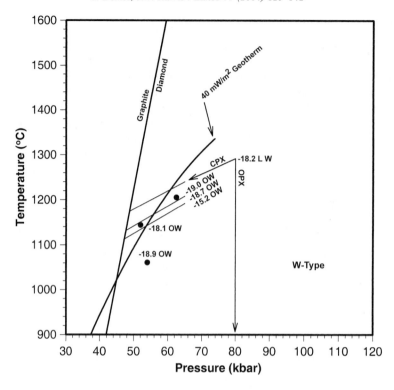

Fig. 3. Temperature and pressure conditions of equilibration for W-Type diamonds from Letlhakane and Orapa. Possible *P/T* conditions for garnet/olivine equilibration were computed after O'Neill and Wood (1979, 1980), for garnet/clinopyroxene after Krogh (1988), and for garnet/orthopyroxene after Finnerty and Boyd (1987) and Harley (1984); the isotopic composition of the host diamond is indicated, LW = Letlhakane, OW = Orapa. The graphite diamond transition is from Kennedy and Kennedy (1976) and the 40 mW/m² cratonic geotherm from Pollack and Chapman (1977) are shown.

that based on the garnet–orthopyroxene pair (Harley, 1984; Finnerty and Boyd, 1987) (7 kb, 753 °C) is not. The garnet–clinopyroxene–orthopyroxene triplet does not represent an equilibrium assemblage. Considering the Orapa and Letlhakane data jointly, W-Type diamonds tend show lower *P/T* conditions of equilibration than P-Type diamonds. Because the ^{13}C-content of these diamonds is consistently low and covers only a very limited δ^{13}C range (− 19‰ to − 15‰) the results suggest that ^{13}C-depletion is more frequently observed in a restricted *P/T* range beneath these two kimberlites.

For E-Type diamonds *P/T* conditions of equilibration have been outlined in Fig. 4 on a plot of the distribution coefficient K_D for the Fe/Mg partitioning between coexisting garnet–clinopyroxene pairs and the mole fraction of Ca in the garnet. The solid lines are drawn at the pressure corresponding to the graphite/diamond transition for the indicated temper-

ature (Kennedy and Kennedy, 1976) and are based on the calibration of Krogh (1988). These lines represent hence the highest temperatures and lowest pressures at which garnet and clinopyroxene could have equilibrated during diamond growth. The data shown include E-Type diamonds from Letlhakane and Orapa, as well as results for diamond and graphite eclogite xenoliths from Orapa. The δ^{13}C values of this data set cover a wide range of ^{13}C-depletion, however, samples with carbon isotopic compositions below about − 8‰ are restricted to a limited region in the diagram (shaded). This suggests that at Orapa and Letlhakane E-Type diamonds of low ^{13}C-content are derived from a limited *P/T* regime also. The Letlhakane E-Type diamonds containing only clinopyroxene (δ^{13}C = − 14.2‰) may have derived their carbon from this region as well. In so far as there is a greater relative abundance of low δ^{13}C E-Type diamonds at Orapa

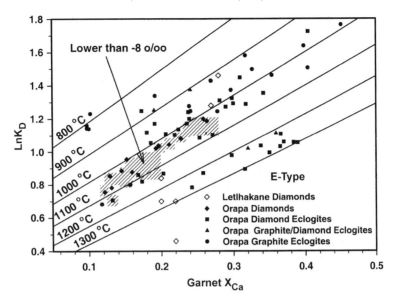

Fig. 4. The relationship between the Mg/Fe fractionation of coexisting eclogitic garnet/clinopyroxene pairs (K_D=(Fe/Mg)$_{Gt}$/(Fe/Mg)$_{CPX}$ and the mole fraction of Ca in garnet (X_{Ca}). The solid lines correspond to equilibration at the graphite/diamond transition pressure (Kennedy and Kennedy, 1976) for the indicated temperature (Krogh, 1988) and hence to the maximum temperature at which inclusions could have been in equilibrium within a diamond. The isotherms are derived for a given P and T by computing ln(K_D) at a series of mole fractions of Ca in garnet using the relationship of Krogh (1988). Diamonds and diamond eclogites with $\delta^{13}C$ values below − 8‰ are restricted to the shaded area. The Orapa data are from Deines et al. (1993) and Deines et al. (1991a) and this study.

compared to Letlhakane, one may speculate that this region of ^{13}C-depletion is less extensive beneath Letlhakane.

While E- and P-Type mineral parageneses in diamonds were recognized early (Harris, 1987; Meyer, 1987), the presence of a separate websteritic paragenesis (which is rare among diamond inclusions) was not recognized until detailed work on the chemistry of the diamonds inclusions was initiated. In order to understand the origin of the W-Type inclusion paragenesis, one draws on the observations made on the petrogenesis of websterites. Websterites are found among mantle xenoliths, as well as in orogenic peridotites, and occur as dikes, veins or layers and are thought to constitute on the order of 2% to 5% of the upper mantle (Hirschmann and Stolper, 1996). Two broad categories of geneses have been proposed, one linking them to the subduction of oceanic crust, the other to the crystallization of primary mantle melts. Their composition is transitional between peridotites and eclogites.

The intermediate nature of the W-Type garnet compositions between P- and E-Type is demonstrated

in Fig. 5. The compositional ranges have been indicated by shading and it is apparent that while there may be a compositional transition between W- and P-Type garnets there is a distinct composition gap between W- and E-Type garnets. The carbon isotopic composition of the garnet hosts has been indicated in the figure as well. While one might postulate a smooth transition in chemistry from P- to W-Type diamonds and hence a petrogenetic link, the carbon isotopic composition changes abruptly between the two parageneses and certainly indicates otherwise. The picture is quite the opposite for the relationship between W- and E-Type garnets and the host carbon isotopic composition. While there is a significant overlap between the $\delta^{13}C$ of the W- and E-Type hosts, there is a significant chemical gap between the two parageneses. Of the two samples classified original as W-Type lk67 plots in the websteritic field while sample lk74 plots squarely in the eclogitic composition field. In view of the general systematic composition difference between W- and E-Type garnets, one may question the original classification of sample lk74 as websteritic. During the microprobe analyses of this

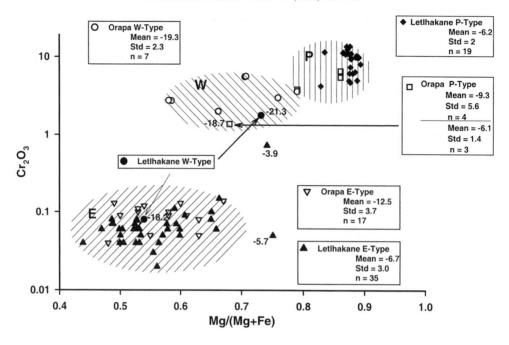

Fig. 5. Chrome content and magnesium number (Mg/(Mg + Fe)) for P-, E- and W-Type garnets from Letlhakane and Orapa. Means, standard deviations (S.D.) and number of samples (*n*) are given for each kimberlite and paragenesis. For Orapa P-Type diamonds two sets of values have been computed, the second excludes one outlying sample.

sample signs of alteration of the mineral provisionally classified as OPX were observed which make the classification of this sample uncertain and it was hence set aside in the further analyses of the data. We conclude that the combination of chemical and carbon isotopic composition of the websteritic paragenesis observed at Letlhakane and Orapa cannot be produced through a simple combination of a peridotitic and eclogitic source or a simple magmatic fractionation processes.

The observations concerning the E-Type and W-Type parageneses that can be most closely compared to those of Orapa and Letlhakane are those by Aulbach et al. (2002). These authors examined the trace-element characteristics of eclogitic and websteritic inclusions in diamonds from Venetia, South Africa. Concerning the E-Type paragenesis the authors conclude that about 10% partial melting of a broadly MORB gabbroic precursor, under eclogite facies conditions, can produce the trace-element abundance-patterns in the residual garnet and clinopyroxene, similar to those observed in the E-Type garnet and clinopyroxene inclusions of Ven-

etia diamonds. Concerning the websteritic association the authors suggest that their trace-element distribution-pattern can be best explained by assuming that a websteritic melt was formed by mixing of an eclogitic with a peridotitic component. This could occur if the eclogitic portion of a subducting slab would melt, separate, and assimilate peridotitic mantle.

In order to examine how the carbon isotope-record of the diamond hosts might relate to these conclusions, we considered that there are significant differences in the Al_2O_3/Cr_2O_3 and SiO_2/Al_2O_3 between crustal and mantle rocks. Because the exact mechanisms of the formation of inclusions in diamonds are still a matter of debate, detailed deductions about the chemistry of their growth environment based on their composition are not possible. However, there is evidence from the analyses of mantle xenoliths, as well as experimental studies that the overall composition of garnets can reflect that of the environment in which they formed.

Griffin et al. (1999) observed that there is a strong correlation between the chrome content of garnets from ultra mafic rocks and the Al_2O_3 content of their

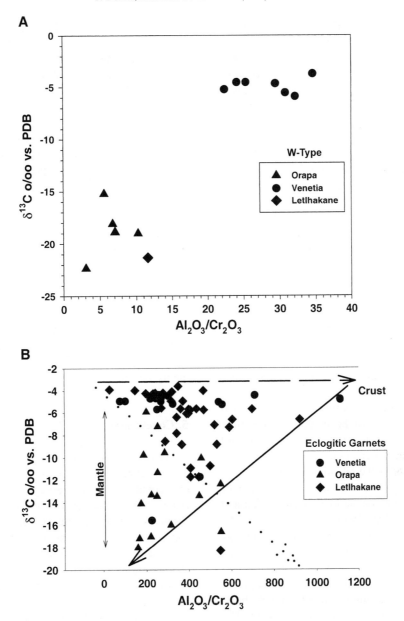

Fig. 6. (A) Relationship of carbon isotopic composition of the diamond host and the Al_2O_3/Cr_2O_3 ratio for websteritic garnet inclusions, from the Letlhakane, Orapa and Venetia kimberlites. (B) Al_2O_3/Cr_2O_3 ratio of E-Type garnets from Letlhakane, Orapa and Venetia and the $\delta^{13}C$ of their hosts. Data: Deines et al. (1993, 2001). Dashed line: low Al_2O_3/Cr_2O_3 source (mantle) mixed with high Al_2O_3/Cr_2O_3 (crust) of the same $\delta^{13}C$ (e.g. $-4‰$ to $-6‰$, average mantle, average crust). Dotted line: low Al_2O_3/Cr_2O_3 source (mantle) $\delta^{13}C = -4‰$ to $-6‰$, mixed with high Al_2O_3/Cr_2O_3 source (crust) of $\delta^{13}C = -25‰$ (subduction of organic carbon). Solid line: high Al_2O_3/Cr_2O_3 source (Crust) of $\delta^{13}C = -4‰$ to $-6‰$ (average crust) mixed with low Al_2O_3/Cr_2O_3 source (mantle) of $\delta^{13}C = -22‰$.

hosts. These authors observed furthermore that the Al_2O_3 concentration of these rocks correlates with the concentration of other major oxides in them. This permitted them to develop equations for the estimation of host rock compositions on the basis of the Cr content of their garnets.

Laboratory experiments by Klemme et al. (2002) on the partitioning of major and trace elements during partial melting of eclogite demonstrate that the TiO_2/SiO_2, SiO_2/Al_2O_3, MgO/Al_2O_3, CaO/Al_2O_3 and Na_2O/Al_2O_3 ratios of the garnets in the run products are well correlated with these ratios in the partial melts that were formed.

In experiments simulating melt–rock interaction at the slab–mantle wedge interface during subduction, Rapp et al. (1999) allowed basaltic melts to assimilate perdotites of varying composition. The results demonstrate that the Al_2O_3/Cr_2O_3, SiO_2/Al_2O_3, and MgO/Al_2O_3 rations in the garnets are reflecting those of the melts with which they are in equilibrium.

Thus both field and laboratory studies support the concept that the composition of garnets, and that of the environment in which they form are related.

In Fig. 6A, the carbon isotopic composition of the host of websteritic garnets has been plotted as a function of the Al_2O_3/Cr_2O_3 ratio for all of the W-Type diamonds we have studied to date. Low Al_2O_3/Cr_2O_3 ratios (characteristic for the mantle) are associated with low $\delta^{13}C$ values, while high Al_2O_3/Cr_2O_3 ratios (characteristic for the crust) are associated with high $\delta^{13}C$ values. The relationship can be interpreted to indicate that there must be mantle sources (low Al_2O_3/Cr_2O_3), which have $\delta^{13}C$ values around $-20‰$ and that are not eclogitic.

When one considers the relationship between $\delta^{13}C$ of E-Type diamonds and the Al_2O_3/Cr_2O_3 ratio of their garnet inclusions for the Letlhakane, Orapa and Venetia kimberlites (Fig. 6B) one comes also to the conclusion that there ought to be sources with low Al_2O_3/Cr_2O_3 ratios (mantle) that are low in ^{13}C. The compositions plotted in Fig. 6B are fairly well restricted to a triangular area, which can be interpreted as a result of mixing of multiple (at a minimum three) sources. Schematic mixing trends are indicated. The dashed line indicates compositions that would result by mixing a low (mantle) and a high (crust) Al_2O_3/Cr_2O_3 source of more or less the same $\delta^{13}C$ (about $-5‰$). The solid line represents the trend expected for mixing of a crustal source of $\delta^{13}C$ of about $-5‰$ with a mantle source of $-20‰$. Most of the data of the figure could be explained by assuming mixing of a crustal component of a high Al_2O_3/Cr_2O_3 and $\delta^{13}C$ of $-5‰$ with a range of low Al_2O_3/Cr_2O_3 (mantle) $\delta^{13}C$ components with $\delta^{13}C$ values between $-5‰$ and

approximately $-20‰$. The data of the figure do not support very well the most frequently cited hypothesis, i.e. that the low $\delta^{13}C$ values of diamonds are exclusively the result of the subduction of crustal organic carbon. The applicable mixing trend for this hypothesis has been shown as a dotted line in the figure.

The present study provides hence additional evidence that there must be several reasons for the low ^{13}C-content of diamonds. Potential causes that have been discussed previously include thermodynamic equilibrium and kinetic isotope effects, mantle degassing, subduction of surface organic-carbon, as well as isotopic heterogeneity surviving from the earliest stages of the Earth's history (Deines, 1980, 2002; Deines et al., 2001). The systematic ^{13}C-depletion in the websteritic diamonds may hold the key to further our understanding of the carbon isotopic composition variability within the mantle. An expansion of the database on W-Type diamonds and the characterization of minor- and trace-elements and radiogenic isotopes of their inclusions would be highly desirable.

5. Conclusions

(1) There are differences in petrology in the mantle below Letlhakane and Orapa. At Letlhakane peridotitic compositions are more prevalent compared to elcogitic compositions, whereas the reverse is true at Orapa. In addition, the source region of P-Type inclusions of Letlhakane diamonds experienced a higher degree of partial melting, compared to that of Orapa P-Type inclusions.

(2) There are significant differences in the carbon isotope distribution in the mantle below the Orapa and Letlhakane kimberlites. Diamonds of low $\delta^{13}C$ values are more frequently observed at Orapa than at Letlhakane.

(3) While there is no significant difference in the mean carbon isotopic composition of P-Type diamonds from Orapa and Letlhakane, garnet containing E-Type diamonds from Orapa are significantly depleted in ^{13}C compared to Letlhakane. Letlhakane diamonds containing only eclogitic clinopyroxene inclusions are systematically depleted in ^{13}C compared to diamonds containing both eclogitic pyroxenes and garnet.

(4) Estimates of P/T conditions of equilibration indicate that ^{13}C-depleted websteritic diamonds and eclogitic diamonds may be derived from a restricted mantle region.

(5) The relationships between carbon isotopic composition and the Al_2O_3/Cr_2O_3 ratio of websteritic and eclogitic garnets indicate that there are mantle regions low in $\delta^{13}C$, and that the low ^{13}C content is unrelated to the introduction of a crustal component.

Acknowledgements

The authors would like to thank DeBeers Consolidated Mines for the provision of the diamond samples. We are particularly grateful for the support and encouragement shown by J.B. Hawthorne. Financial support for this work was provided through NSF grants EAR 96 27324 and EAR 0229551 to P. Deines.

References

Aulbach, S., Stachel, T., Viljoen, K.S., Brey, G.P., Harris, J.W., 2002. Eclogitic and websteritic diamond sources beneath the Limpopo Belt; is slab-melting the link? Contrib. Mineral. Petrol. 143, 56–70.

Davies, G.L., 1977. The ages and uranium contents of zircons from kimberlites and associated rocks. Extended Abstracts 2nd International Kimberlite Conference, Santa Fe. Not Paginated.

Deines, P., 1980. The carbon isotopic composition of diamonds: relationship to diamond shape, color, occurrence and vapor composition. Geochim. Cosmochim. Acta 44, 943–961.

Deines, P., 2002. The carbon isotope geochemistry of mantle xenoliths. Earth Sci. Rev. 58, 247–278.

Deines, P., Gurney, J.J., Harris, J.W., 1984. Associated chemical and carbon isotopic composition variations in diamonds from Finsch and Premier kimberlite, South Africa. Geochim. Cosmochim. Acta 48, 325–342.

Deines, P., Harris, J.W., Gurney, J.J., 1987. Carbon isotopic composition. Nitrogen content and inclusion composition of diamonds from the Roberts Victor kimberlite, South Africa: evidence for ^{13}C depletion in the mantle. Geochim. Cosmochim. Acta 51, 1227–1243.

Deines, P., Harris, J.W., Robinson, D.N., Gurney, J.J., Shee, S.R., 1991a. Carbon and oxygen isotope variations in diamond and graphite eclogites from Orapa, Botswana and the nitrogen content of their diamonds. Geochim. Cosmochim. Acta 55, 515–524.

Deines, P., Harris, J.W., Gurney, J.J., 1991b. The carbon isotopic composition and nitrogen content of lithospheric and asthenopsheric diamonds from the Jagersfontein and Koffiefontein kimberlite, South Africa. Geochim. Cosmochim. Acta 55, 2615–2625.

Deines, P., Harris, J.W., Gurney, J.J., 1993. Depth-related carbon isotope and nitrogen concentration variability in the mantle below the Orapa kimberlite. Botswana. Geochim. Cosmochim. Acta 57, 2781–2796.

Deines, P., Harris, J.W., Gurney, J.J., 1997. Carbon isotope ratios, nitrogen content and aggregation state and inclusion chemistry of diamonds from Jwaneng. Botswana. Geochim. Cosmochim. Acta 61, 3993–4005.

Deines, P., Viljoen, F., Harris, J.W., 2001. Implications of the carbon isotope and mineral inclusion record for the formation of diamonds in the mantle underlying a mobile belt: Venetia, South Africa. Geochim. Cosmochim. Acta 65, 813–838.

Finnerty, A.A., Boyd, F.R., 1987. Thermobarometry for garnet peridotites: basis for the determination of thermal and compositional structure of the upper mantle. In: Nixon, P.H. (Ed.), Mantle Xenoliths. Wiley, Chichester, pp. 381–412.

Griffin, W.L., O'Reilly, S.Y., Ryan, C.G., 1999. The composition and origin of sub-continental lithospheric mantle. In: Fei, Y., Bertka, C.M., Mysen, B.O. (Eds.), Mantle Petrology: Field Observations and High-Pressure Experimentation, A Tribute to Francis R. (Joe) Boyd. The Geochemical Society Special Publication, vol. 6. The Geochemical Society, Houston, pp. 13–45.

Gurney, J.J., 1989. Diamonds. Proceedings of he Fourth International Kimberlite Conference, Perth 1986, V2. The Geochemical Society, Houston, pp. 935–965.

Harley, S.L., 1984. An experimental study of the partitioning of Fe and Mg between garnet and orthopyroxene. Contrib. Mineral. Petrol. 86, 359–373.

Harris, J.W., 1987. Recent physical, chemical, and isotopic research of diamond. In: Nixon, P.H. (Ed.), Mantle Xenoliths. Wiley, Chichester, UK, pp. 477–500.

Hirschmann, M.M., Stolper, E.M., 1996. A possible role for garnet pyroxenite in the origin of the "garnet signature" in MORB. Contrib. Mineral. Petrol. 124, 185–208.

Kennedy, C.S., Kennedy, G.C., 1976. The equilibrium boundary between graphite and diamond. J. Geophys. Res. 81, 2467–2470.

Klemme, S., Blundy, J.D., Wood, B.J., 2002. Experimental constraints on major and trace element partitioning during partial melting of eclogite. Geochim. Cosmochim. Acta 66, 3109–3123.

Krogh, E.J., 1988. The garnet–clinopyroxene Fe–Mg geothermometer, a reinterpretation of existing experimental data. Contrib. Mineral. Petrol. 99, 44–48.

Meyer, H.O.A., 1987. Inclusions in diamond. In: Nixon, P.H. (Ed.), Mantle Xenolith. Wiley, Chichester, UK, pp. 501–522.

Miller, R.L., Kahn, J.S., 1962. Statistical Analysis in the Geologic Sciences. Wiley, New York. 484 pp.

O'Neill, H.St.C., Wood, B.J., 1979. An experimental study of Fe–Mg partitioning between garnet and olivine and its calibration as a geothermometer. Contrib. Mineral. Petrol. 70, 59–70.

O'Neill, H.St.C., Wood, B.J., 1980. An experimental study of Fe–

Mg partitioning between garnet and olivine and its calibration as a geothermometer: corrections. Contrib. Mineral. Petrol. 72, 337.

Pollack, H.N., Chapman, D.S., 1977. On the regional variation of heat flow, geotherms and lithospheric thickness. Tectonophysics 38, 279–296.

Rapp, R.P., Shimizu, N., Norman, M.D., Applegate, G.S., 1999. Reaction between slab-derived melts and peridotite in the mantle wedge: experimental constraints at 3.8 GPa. Chem. Geol. 160, 335–356.

Robinson, D.N., Gurney, J.J, Shee, S.R., 1984. Diamond eclogite and graphite eclogite xenoliths from Orapa, Botswana. In: Kornprobst, J. (Ed.), Kimberlites II. The Mantle and Crust–Mantle Relationships. Elsevier, Amsterdam, pp. 11–24.

Shee, S.R., 1978. The mineral chemistry of xenoliths from the Orapa kimberlite pipe, Botswana. MSc Thesis, Dept. of Geochemistry, Univ Cape Town.

Shee, S.R., Gurney, J.J., 1979. The mineralogy of xenoliths from Orapa, Botswana. In: Boyd, F.R., Meyer, H.O.A. (Eds.), The Mantle Sample: Inclusions in Kimberlites and Related Rocks. American Geophysical Union, Washington, pp. 37–49.

Stiefenhofer, J., Viljoen, K.S., Marsh, J.S., 1997. Petrology and geochemistry of peridotite xenoliths from the Letlhakane kimberlites, Botswana. Contrib. Mineral. Petrol. 127, 147–158.

Shirey, S.B., Harris, J.W., Richardson, S.H., Fouch, M.J., James, D.E., Cartigny, P., Deines, P., Viljoen, F., 2002. Diamond genesis, seismic structure, and evolution of the Kaapvaal–Zimbabwe Craton. Science 297, 1683–1686.

Available online at www.sciencedirect.com

Lithos 77 (2004) 143–154

www.elsevier.com/locate/lithos

Episodic diamond genesis at Jwaneng, Botswana, and implications for Kaapvaal craton evolution

S.H. Richardson[a,*], S.B. Shirey[b], J.W. Harris[c]

[a] Department of Geological Sciences, University of Cape Town, Rondebosch 7701, South Africa
[b] Department of Terrestrial Magnetism, Carnegie Institution of Washington, Washington, DC 20015, USA
[c] Division of Earth Sciences, University of Glasgow, Glasgow G12 8QQ, UK

Received 27 June 2003; accepted 14 December 2003
Available online 18 May 2004

Abstract

Major element and Re–Os isotope analysis of single sulfide inclusions in diamonds from the 240 Ma Jwaneng kimberlite has revealed the presence of at least two generations of eclogitic diamonds at this locality, one Proterozoic (ca. 1.5 Ga) and the other late Archean (ca. 2.9 Ga). The former generation is considered to be the same as that of eclogitic garnet and clinopyroxene inclusion bearing diamonds from Jwaneng with a Sm–Nd isochron age of 1.54 Ga. The latter is coeval with the 2.89 Ga subduction-related generation of eclogitic sulfide inclusion bearing diamonds from Kimberley formed during amalgamation of the western and eastern Kaapvaal craton near the Colesberg magnetic lineament.

The Kimberley, Jwaneng, and Premier kimberlites are key localities for characterizing the relationship between episodic diamond genesis and Kaapvaal craton evolution. Kimberley has 3.2 Ga harzburgitic diamonds associated with creation of the western Kaapvaal cratonic nucleus, and 2.9 Ga eclogitic diamonds resulting from its accretion to the eastern Kaapvaal. Jwaneng has two main eclogitic diamond generations (2.9 and 1.5 Ga) reflecting both stabilization and subsequent modification of the craton. Premier has 1.9 Ga lherzolitic diamonds that postdate Bushveld–Molopo magmatism (but whose precursors have Archean Sm–Nd model ages), as well as 1.2 Ga eclogitic diamonds. Thus, Jwaneng provides the overlap between the dominantly Archean vs. Proterozoic diamond formation evident in the Kimberley and Premier diamond suites, respectively. In addition, the 1.5 Ga Jwaneng eclogitic diamond generation is represented by both sulfide and silicate inclusions, allowing for characterization of secular trends in diamond type and composition. Results for Jwaneng and Kimberley eclogitic sulfides indicate that Ni- and Os-rich end members are more common in Archean diamonds compared to Proterozoic diamonds. Similarly, published data for Kimberley and Premier peridotitic silicates show that Ca-rich (lherzolitic) end members are more likely to be found in Proterozoic diamonds than Archean diamonds. Thus, the available diamond distribution, composition, and age data support a multistage process to create, stabilize, and modify Archean craton keels on a billion-year time scale and global basis.
© 2004 Elsevier B.V. All rights reserved.

Keywords: Diamond; Eclogite; Peridotite; Sulfide; Inclusion; Isotope; Craton

* Corresponding author. Tel.: +27-21-650-2916; fax: +27-21-650-3783.
E-mail addresses: shr@geology.uct.ac.za (S.H. Richardson), shirey@dtm.ciw.edu (S.B. Shirey), jwh@earthsci.gla.ac.uk (J.W. Harris).

0024-4937/$ - see front matter © 2004 Elsevier B.V. All rights reserved.
doi:10.1016/j.lithos.2004.04.027

1. Introduction

Isotopic dating of syngenetic inclusion bearing diamonds is feasible because silicate and sulfide inclusions are the major carrier phases of radiogenic isotopes in peridotitic and eclogitic diamonds (e.g., Pearson and Shirey, 1999). For us, *syngenetic* simply means that the inclusion minerals crystallized or recrystallized at the time of diamond formation, as demonstrated by their xenohedral morphology. The corollary is that nontouching inclusions are effectively closed to diffusive exchange of radiogenic isotopes by encapsulation in diamond. However, the scale of isotopic resetting in the precursor before encapsulation and the extent to which multiple inclusions in the same and different diamonds are *cogenetic* (i.e., inherit the same initial isotope ratios) is another issue that needs to be addressed for each isotope system and host assemblage under consideration.

There are four possibilities regarding the incorporation of parent and daughter elements of radiogenic isotope systems during the (re)crystallization of inclusion minerals:

(1) The mineral includes neither parent nor daughter, thus precluding dating.
(2) The mineral includes the parent but very little of the daughter. Absolute ages are obtained for individual grains (or composites of cogenetic grains) as for U–Pb in zircon (e.g., Kinny and Meyer, 1994), K–Ar in clinopyroxene (e.g., Burgess et al., this volume), and Re–Os in eclogitic sulfides with negligible common Os (e.g., Richardson et al., 2001).
(3) The mineral includes the daughter but very little of the parent. Model ages are obtained regardless of the number of single grains (or composites of cogenetic grains) as for U–Pb in sulfide (e.g., Kramers, 1979; Rudnick et al., 1993), Rb–Sr in garnet (e.g., Richardson et al., 1984), and Re–Os in peridotitic sulfides (e.g., Pearson et al., 1999).
(4) The mineral includes both parent and daughter. In this case, a single grain (or composite of cogenetic grains) will always give a model age but a combination of two or more nontouching cogenetic inclusions (or composites of cogenetic inclusions) with a range in parent/daughter ratio from the same or different diamonds will give an isochron age. This applies to the Sm–Nd system in lherzolitic and eclogitic garnet and clinopyroxene (with the same initial Nd and/or Sr isotope ratios; e.g., Richardson et al., 1990) and the Re–Os system in eclogitic sulfides (with the same

initial Os and/or Pb isotope ratios; e.g., Richardson et al., 2001). Furthermore, comparable age results on the same paragenesis may be obtained using two or more independent chronometers, which is the approach adopted in this paper.

Isotopic dating of such cogenetic inclusions coupled with nitrogen aggregation studies of the host diamonds thus far indicates that most lithospheric diamonds of octahedral habit and peridotitic or eclogitic paragenesis are around 1–3 Ga old (e.g., Richardson and Harris, 1997; Pearson et al., 1999; Richardson et al., 2001). Yet, the interpretation of old diamond ages continues to be debated based on the idea that old lithospheric mineral grains may be captured by young (relative to kimberlite emplacement) diamond or zircon growth (e.g., Shimizu and Sobolev, 1995; Spetsius et al., 2002). Nevertheless, there is general agreement that Archean lithospheric mantle is dominated by harzburgites with unradiogenic Os and Nd, and radiogenic Sr isotope signatures as a consequence of early melt depletion and metasomatism (e.g., Carlson et al., 1999). In our view, the preservation of relatively unradiogenic Sr isotope signatures in garnet inclusions vs. highly radiogenic Sr isotope signatures in unencapsulated garnet macrocrysts from disaggregated diamond host rocks (e.g., Richardson et al., 1984; Pearson and Shirey, 1999) is compelling evidence for ancient diamond crystallization. In this case, Rb–Sr model ages represent encapsulation ages for the garnet inclusions, which became isolated from further diffusive exchange with their low Re/Os and Sm/Nd, and high Rb/Sr host rocks.

In the Shimizu and Sobolev (1995) study, unusual incompatible trace element zoning was found in peridotitic garnet inclusions and interpreted as a lack of time (< 70 kyr) for diffusive reequilibration following young diamond formation immediately before kimberlite emplacement. However, the possibility of recent metasomatism via cracks in the host diamonds could not be excluded and the requirement for multimillion- to billion-year mantle storage times to explain the typical nitrogen aggregation state of lithospheric diamonds (e.g., Richardson and Harris, 1997; Pearson et al., 1999; Navon, 1999) was not addressed.

In the Spetsius et al. (2002) study, sulfides with variable Archean Re–Os model ages (not encapsulation or isochron ages) were found in zircon mega-crysts with Paleozoic U–Pb ages comparable to the age of kimberlite emplacement. This is not unexpected given the extensive interaction between young megacryst magmas and old lithospheric mantle (e.g., Burgess and Harte, 1999) and the growth of kimberlitic zircon from such differentiated magmas. Nevertheless, this is of little relevance to diamond formation because, to our knowledge, there is no evidence of a genetic connection between kimberlitic megacrysts and lithospheric diamonds, nor any indication that macrodiamonds are phenocrysts in kimberlite. Furthermore, we note that the scatter in Re–Os isotope data for lithospheric peridotites at any given locality is far greater than that for peridotitic sulfide inclusions in diamonds (Pearson et al., 1995a, 1999; Westerlund et al., in preparation). Given the exposure of craton keels to ongoing mantle metasomatism and metamorphic reequilibration, we consider that consistent isochron relationships for independent chronometers are unlikely to be preserved in random mineral grains stored in old lithospheric mantle for extended periods before capture by young diamonds.

The only direct isotopic evidence suggesting young peridotitic diamond formation is for one possibly lherzolitic diamond from the Mesozoic Koffiefontein kimberlite that gave an internal Re–Os isochron age within error of pipe emplacement but with a chondritic initial Os isotope composition clearly atypical of cratonic mantle (Pearson et al., 1998). Conversely, young (relative to kimberlite emplacement) eclogitic diamond formation is indicated for at least one Proterozoic locality, the Premier kimberlite, where eclogitic silicate and sulfide inclusions give consistent Sm–Nd isochron, U–Pb model, and Ar–Ar closure ages of ca. 1.2 Ga, within error of pipe emplacement at 1.18 Ga (Kramers, 1979; Richardson, 1986; Burgess et al., 1989; Phillips et al., 1989).

More recent studies of sulfide inclusions in eclogitic diamonds involving the Re–Os isotope system indicate that this system is a superior tracer for the source of eclogitic materials (Shirey et al., 2001). Furthermore, the high Re and Os concentrations in sulfides make the system uniquely useful for the analysis of single sulfide inclusions in diamonds (Pearson et al., 1998; Pearson and Shirey, 1999; Richardson et al., 2001). Here we report Fe, Ni, and Cu contents and Re–Os isotope data for sulfide

inclusions in eclogitic diamonds recovered from the 240 Ma Jwaneng kimberlite and compare them with previous composition and age data for eclogitic and peridotitic diamonds from the 85 Ma De Beers Pool (Kimberley) and 1180 Ma Premier kimberlites. When combined with the results of the Southern Africa Seismic Experiment (James et al., 2001), these data support a multistage process to make the Kaapvaal–Zimbabwe craton keel (Shirey et al., 2002).

2. Jwaneng diamond geology

The Jwaneng DK2 kimberlite in southeastern Botswana, probably the most profitable mine in the world, is otherwise notable in several respects. First, the kimberlite carries two generations of kimberlitic zircon, one that gives the 240 Ma age of emplacement and the other around 2.5 Ga old (Kinny et al., 1989; Griffin et al., 2000). Second, a significant proportion of the diamonds are polycrystalline aggregates (Kirkley et al., 1994) as well as cubic or fibrous forms containing microinclusions of mantle fluids (Schrauder and Navon, 1994; Schrauder et al., 1996). Third, the majority of octahedral diamonds with sulfide and silicate inclusions are eclogitic (Gurney et al., 1995) with a wide range in carbon and nitrogen isotope composition and significantly aggregated nitrogen (Deines et al., 1997; Cartigny et al., 1998).

The first work on dating eclogitic silicate inclusion bearing diamonds from Jwaneng involved the ^{40}Ar–^{39}Ar laser probe method applied to selected clinopyroxene inclusions (Burgess et al., 1992). Ages for a set of six cleaved diamonds ranged from a minimum overall age of 240 Ma to a maximum individual age of 1890 ± 450 Ma. Although the diamonds are xenocrysts in the host kimberlite, diffusion of radiogenic argon to the inclusion–diamond interface during mantle storage results in apparent ages converging on that of kimberlite emplacement (Burgess et al., 1992). Subsequently, Richardson et al. (1999) used the Sm–Nd and Rb–Sr isotope systems to date composites of larger numbers of eclogitic garnet (47) and clinopyroxene (50) inclusions from this locality selected on the basis of MgO content as determined by electron microprobe. A single large clinopyroxene inclusion was also analyzed. The data gave a two-point garnet-clinopyroxene Sm–Nd isochron age of 1540 ± 20 Ma with an initial ratio ($\varepsilon_{Nd}=+1.0$) indicative of a mildly depleted precursor (or mixture of depleted asthenospheric and enriched lithospheric precursors).

3. Materials and methods

Some 27 diamonds (JWR1-27) with 100–400 μm sulfide inclusions were selected from run-of-mine production from the Jwaneng kimberlite. The diamonds are 3.0 mm in maximum dimension and comprise octahedra (13), dodecahedra (11), an octahedral macle, an octahedral aggregate and an irregular. They are all colorless except for one brown stone (JWR13) and free of any obvious evidence of plastic deformation. Preliminary FTIR measurements indicate that significant nitrogen is present in all the diamonds.

Sulfide inclusions can generally be distinguished from silicate and oxide inclusions by the presence of large rosette fracture systems surrounding each sulfide grain, attributable to differential decompression during kimberlite emplacement. Thin films of pyrrhotite have been analyzed from on these fracture surfaces in a previous study of Orapa specimens (Shirey et al., in preparation), consistent with sulfide extrusion along fracture planes, but systematic studies of such films to see if this composition is typical have yet to be undertaken. The diamonds were broken in a steel cracker to release the inclusions. Several specimens were also found to contain small pale green eclogitic clinopyroxene inclusions and in two cases tiny orange eclogitic garnet inclusions (Table 1; Fig. 1b). In practice, most sulfide inclusions break up when liberated from the host diamond, allowing for the characterization of interior fragments using electron

Table 1
Fe, Ni, and Cu composition of eclogitic sulfide inclusions from Jwaneng diamonds

	Weight (μg)	Fe (at.%)	Ni (at.%)	Cu (at.%)	Fe (wt.%)	Ni (wt.%)	Cu (wt.%)
JWR1	21.1	93.3	3.7	3.0	56.9	2.6	2.2
JWR2	24.6	93.0	3.9	3.1	56.7	2.8	2.3
JWR3	7.0	88.5	2.6	8.9	53.7	1.8	6.5
JWR6	19.7	83.1	12.6	4.4	50.8	8.8	3.2
JWR7[a]	41.0	94.5	2.4	3.2	57.6	1.7	2.3
JWR8	22.2	93.8	3.7	2.5	57.2	2.6	1.8
JWR9	22.5	86.7	9.0	4.4	53.0	6.2	3.2
JWR10	15.0	86.6	8.7	4.8	52.9	6.0	3.5
JWR13[a]	19.5	94.6	3.0	2.5	57.7	2.1	1.8
JWR16	11.0	88.3	7.6	4.1	54.0	5.3	3.0
JWR18	28.3	95.4	2.3	2.3	58.2	1.6	1.7
JWR19[a]	9.9	91.6	4.6	3.8	55.9	3.2	2.8
JWR21	8.3	84.0	9.8	6.2	51.3	6.8	4.5
JWR22[b]	25.6	94.6	2.6	2.8	57.7	1.8	2.0
JWR24	8.6	92.6	4.2	3.2	56.5	2.9	2.4
JWR25	9.4						
JWR26[a]	5.9						
JWR27	6.0						

Fe, Ni and Cu concentrations are normalized to a total of 100 atom % as analyzed by ICP-MS at DTM. No solutions were available for inclusions JWR25-27. To calculate the wt.% composition of sulfide, all the Ni is assigned to pentlandite (pn) component with a Fe, Ni, and S composition of 32, 35, and 33 wt.% and recalculated to give a wt.% Ni in the bulk sulfide. All the Cu is assigned to chalcopyrite (cp) component with a Fe, Cu, and S composition of 32, 33, and 35 wt.% and recalculated to give a wt.% Cu in the bulk sulfide. The Fe content of pn and cp is subtracted from the total Fe and the remainder is assigned to pyrrhotite (po) component with a Fe and S composition of 60 and 39 wt.%. Sulfur (not analyzed on the ICP-MS) is included in the calculation and partitioned according to the above compositions of po, pn, and cp. Estimated by difference, it would range from 36.3 to 38.0 wt.% with added uncertainties due to the unknown extent of Fe deficiency in pyrrhotite. The average phase compositions used in the calculations are typical of eclogitic sulfide inclusions in diamonds from southern Africa (Deines and Harris, 1995).

[a] Diamonds also containing eclogitic clinopyroxene inclusions.
[b] Diamond also containing eclogitic garnet inclusion.

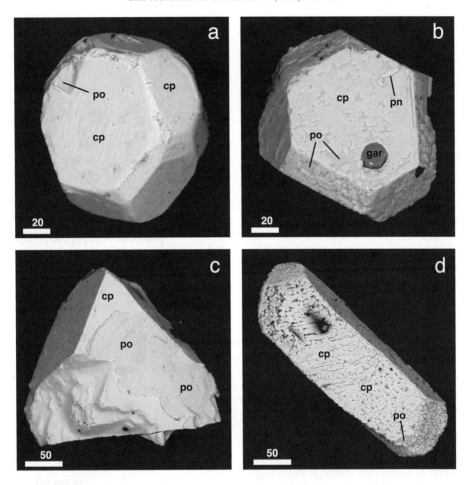

Fig. 1. Back-scattered electron images of eclogitic sulfide inclusions liberated from Jwaneng diamonds (JWR1-27) with pyrrhotite (po), chalcopyrite (cp), and pentlandite (pn) identified on the basis of energy-dispersive electron probe microanalysis (scale bars in μm). (a) Smaller of two inclusions from JWR8 showing syngenetic diamond-induced cubo-octahedral morphology and exterior surface covered by cp except for the small nick (upper left) where po is exposed (not included in Re–Os analysis of larger inclusion). (b) Fragment of inclusion from JWR4, showing cp-dominated surface with minor po and pn, and small eclogitic garnet (inclusion not analyzed for Re–Os). (c) Fragment of inclusion from JWR11 showing cp-covered surface with large missing slivers where po is exposed (inclusion not analyzed for Re–Os). (d) Intact inclusion from JWR17 with contaminant sodium chloride crystal casting a shadow on cp-dominated surface (saved for later cleaning and analysis).

microbeam techniques (see Fig. 1) before recombination for isotope analysis.

The Jwaneng sulfides are all low-Ni, pyrrhotite–pentlandite–chalcopyrite assemblages derived from monosulfide solid solution by exsolution during cooling (Craig and Kullerud, 1969). While pyrrhotite is the dominant phase, the proportions of chalcopyrite and pentlandite vary considerably between fragments, with chalcopyrite occurring mostly on exterior surfaces (Fig. 1). After inclusion characterization and recovery, 18 of the 27 specimens, representing inclusions with-

out touching silicate phases (see Fig. 1) or obvious missing fragments, were selected for Re–Os analysis. A mixed $^{185}Re/^{190}Os$ tracer solution was added and Re and Os analyzed using microchemistry and negative thermal ionization mass spectrometry (NTIMS) techniques described in Pearson et al. (1998) and Shirey et al. (in preparation). Fe, Ni, and Cu contents of the sulfide inclusions (Table 1) were measured by inductively coupled plasma mass spectrometry (ICP-MS) using aliquots of solutions washed off the anion columns before Re elution (Shirey et al., in preparation). The remaining portions of these solutions have been saved for later Pb isotope analysis.

4. Results

Recalculated bulk inclusion Ni contents fall in the range 1.6–8.8 wt.% (Table 1; Fig. 2). This range extends to significantly lower values than that seen at Kimberley (5.2–14.5 wt.%; Richardson et al., 2001) and is typical of sulfides coexisting with eclogitic garnet and/or clinopyroxene in diamonds from these and other localities in southern Africa (Deines and Harris, 1995). Furthermore, the one garnet and four clinopyroxene inclusions in specimens JWR7, 13, 19, 22, and 26 were analyzed by electron microprobe and plot within the Ca–Mg–Fe compositional fields for inclusions of each of these eclogitic minerals at Jwaneng (Richardson et al., 1999).

The range in Re (320–1090 ppb) and low 'common' Os (1–700 ppb) concentrations and high Re/Os (Table 2) are also consistent with eclogitic sulfide values (Pearson et al., 1998; Richardson et al., 2001). The Re–Os isotope data for the set of 18 Jwaneng sulfides (Table 2) span a wide range of Re/Os and Os isotope space (Fig. 3). The variation of Os with Ni concentration for the Jwaneng sulfides shows similarities to that for the 'high Os' eclogitic sulfide inclusion suite identified at Kimberley although extending to lower Ni values (Fig. 2; Richardson et al., 2001).

At least two generations of sulfides are distinguishable in the Re–Os isotope data set (Table 2; Fig. 3). Nine of the 18 Jwaneng sulfides with convecting mantle model ages ≥ 2.9 Ga approximate a 2.9 Ga isochron drawn on the basis of the Kimberley sulfide Re–Os data (Richardson et al., 2001) with a higher than chondritic initial ratio. Formal regression of the data for these nine specimens gives an age of 3.01 ± 0.33 Ga and initial $^{187}Os/^{188}Os$ of 0.9 ± 1.1 (Isoplot Model 3). A lower initial $^{187}Os/^{188}Os$ value of 0.43 at 2.9 Ga is given by the two least radiogenic specimens (JWR2, 9). This is still substantially more radiogenic than the Kimberley sulfide initial value (0.156 ± 0.011). Five of the nine Jwaneng specimens with the lowest $^{187}Re/^{188}Os$ values (6–65) have model ages $\geq$ the age of the Earth and scatter to the low Re–Os side of the 2.9 Ga reference isochron. This is primarily due to their elevated initial

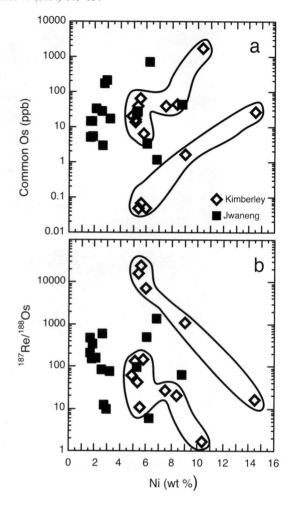

Fig. 2. Variation of (a) common Os concentration, and (b) $^{187}Re/^{188}Os$, with Ni concentration for eclogitic sulfide inclusions in Jwaneng and Kimberley diamonds.

$^{187}Os/^{188}Os$ relative to convecting mantle, particularly for the specimens with the least radiogenic Os isotope composition. Nevertheless, back-scattered electron imaging shows that chalcopyrite is concentrated on exterior surfaces of inclusions (see Fig. 1) during exsolution at the time of kimberlite emplacement. Because Re is preferentially concentrated in chalcopyrite relative to pyrrhotite and/or pentlandite (see Richardson et al., 2001; Brenan, 2002), the scatter to low Re–Os may also be due in part to analysis of some incomplete inclusions. Furthermore, the composition of the small volume of unrecovered sulfide in rosette fractures is considered to be the same as that

Table 2
Re and Os concentrations and Os isotope composition of eclogitic sulfide inclusions in diamonds from Jwaneng, Botswana

	Weight (µg)	Re (ppb)	Os (ppb)	Os_i (ppb)	$^{187}Re/^{188}Os$	$^{187}Os/^{188}Os$	T (Ga)
JWR1	21.1	491	40.5	28.7	83.0(14)	3.347(27)	2.29
JWR2	24.6	434	190	171	12.33(58)	1.037(4)	4.41
JWR3	7.0	456	25.8	14.8	149.1(23)	5.866(99)	2.27
JWR6	19.7	591	74.7	44.1	65.0(26)	5.509(61)	4.80
JWR7	41.0	627	29.2	14.7	206.9(43)	7.763(45)	2.18
JWR8	22.2	372	8.85	3.02	597.6(86)	15.12(54)	1.49
JWR9	22.5	860	755	703	5.93(41)	0.7340(11)	6.26
JWR10	15.0	333	13.6	3.30	489(13)	24.3(12)	2.90
JWR13	19.5	1091	69.7	33.4	158.4(42)	8.559(60)	3.12
JWR16	11.0	546	35.4	27.3	97.0(38)	2.464(47)	1.43
JWR18	28.3	516	13.6	5.28	473.6(60)	12.35(24)	1.53
JWR19	9.9	270	23.8	17.0	76.9(12)	3.276(59)	2.42
JWR21	8.3	323	4.26	1.15	1350(260)	21(9)	0.92
JWR22	25.6	389	19.9	5.38	350.5(85)	20.91(45)	3.46
JWR24	8.6	445	245	217	9.95(25)	1.187(5)	6.33
JWR25	9.4	836	180	137	29.7(13)	2.579(16)	4.82
JWR26	5.9	566	9.39	1.53	1790(64)	40(16)	1.32
JWR27	6.0	688	65.7	41.3	80.8(16)	4.735(60)	3.34

Re and Os concentrations are in ng g^{-1} after correction for minimum blanks: Re, 40 fg; Os, 2 fg. Common Os concentrations (Os_i) are calculated assuming chondritic $^{187}Os/^{188}Os$. Errors in $^{187}Re/^{188}Os$ are the sum of Re and Os run precision ($2\sigma_{mean}$) after relevant isotope dilution error magnification but disregarding uncertainty in blank correction. Errors in $^{187}Os/^{188}Os$ are the sum of run precision ($2\sigma_{mean}$) and uncertainty in Os blank (4 ± 2 fg) correction. Errors for specimens with $^{187}Re/^{188}Os > 1000$ are dominated by uncertainty in Os blank correction. Individual ages (T) are model ages relative to a convecting mantle with $^{187}Re/^{188}Os = 0.4353$ and $^{187}Os/^{188}Os = 0.1296$ (Meisel et al., 2001) using a ^{187}Re decay constant $= 1.666 \times 10^{-11}$ $year^{-1}$.

for the bulk inclusion, particularly if the sulfide was molten during extrusion into the fractures. This may not apply in all cases. While every effort was made to select only complete specimens, missing material is not always obvious and may have been overlooked in a few cases beyond those already excluded for Re–Os analysis (see Section 3 and Fig. 1).

Another five of the Jwaneng sulfides with model ages ≤1.5 Ga, including the two specimens with the highest $^{187}Re/^{188}Os$ values, approximate a 1.5 Ga reference isochron (Fig. 3) corresponding to the 1.54 Ga Sm–Nd age for eclogitic garnet and clinopyroxene inclusions in Jwaneng diamonds (Fig. 4; Richardson et al., 1999). The one Kimberley sulfide with a Proterozoic model age also falls on this isochron although no equivalent tectonothermal event has yet been identified in the Kaapvaal crust. It should be noted that the large (and potentially underestimated) errors for the two most radiogenic middle Proterozoic sulfide specimens (JWR21, 26) are the result of the very small osmium sample loads (≤10 fg common Os) and hence large relative blank corrections (4 ± 2 fg common Os). Nevertheless, given that blank-cor-

rection-related errors for high values of $^{187}Re/^{188}Os$ and $^{187}Os/^{188}Os$ are strongly correlated (not shown in Fig. 3), the probability that error bars for these samples might extend to overlap the 2.9 Ga reference isochron is exceedingly low.

The remaining four Jwaneng sulfides with model ages between 2.2 and 2.4 Ga and relatively low Re–Os (Fig. 3, inset) may be outliers from either of the above generations. Alternatively, they may represent an early Proterozoic generation for which more radiogenic counterparts have not yet been sampled. Regression of the data for these four specimens gives a Bushveld age of 2.10 ± 0.26 Ga and initial $^{187}Os/^{188}Os$ of 0.50 ± 0.59 (Isoplot Model 3) within error of that for the 2.9 Ga generation.

5. Discussion

5.1. Sulfide inclusion major element compositions

The recalculated bulk compositions of Jwaneng and Kimberley sulfide inclusions are comparable to

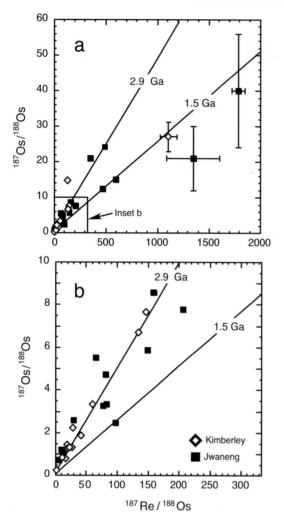

Fig. 3. Re–Os isochron diagrams (a, b) for single eclogitic sulfide inclusions in Jwaneng and Kimberley diamonds. Error bars are smaller than the size of plotted points except as indicated. The large errors in $^{187}Os/^{188}Os$ for specimens with $^{187}Re/^{188}Os > 1000$ are dominated by blank correction effects (see Table 2). The 2.9 and 1.5 Ga reference lines are drawn with convecting mantle initial ratios.

ern African localities. As discussed in Deines and Harris (1995), the low Ni content of Jwaneng and Kimberley sulfides leads to Ni/Fe ratios that are too low to be in equilibrium with mantle olivine. Thus, their composition supports the affinity to a basaltic precursor discussed below (see also Shirey et al., this volume).

5.2. Subduction-related additions to the craton keel

The basaltic composition of eclogite xenoliths and their density at high pressure have long suggested links between mantle eclogites and subduction (Mac-Gregor and Carter, 1970; Helmstaedt and Gurney, 1984). Sulfur and Pb isotope studies of sulfide inclusions in diamond have also suggested subduction-related eclogitic diamond formation (Eldridge et al., 1991; Rudnick et al., 1993; Farquhar et al., 2002). Evidence that the late Archean is a time of significant subduction-related eclogite formation comes from

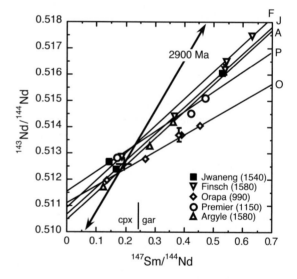

Fig. 4. Sm–Nd isochron diagram for eclogitic clinopyroxene (lowest Sm/Nd) and garnet (higher Sm/Nd) inclusions in Jwaneng, Premier, Orapa, Finsch, and Argyle diamonds (after Richardson et al., 1999). For each locality, groups of inclusions from separate diamonds were sorted on the basis of color (pale bluish green clinopyroxene; pale, medium, dark, and reddish orange garnet) and deemed to be cogenetic or otherwise on the basis of initial Sr isotope ratios. Error bars are smaller than the size of plotted points except as indicated. Isochron ages are given in parentheses (Ma). The 2900 Ma reference line is drawn through a chondritic mantle composition shown by the bold cross.

the Fe, Ni, and Cu contents of a much broader suite of sulfide inclusions from southern Africa reported in Deines and Harris (1995) although lacking in the highest Ni (i.e. peridotitic) end member. The fact that no high-Ni inclusions were found in the Jwaneng and Kimberley suites in spite of the relative abundance of peridotitic silicate inclusions, particularly at Kimberley, also matches the relative rarity of peridotitic sulfides previously noted for the wider suite of south-

reconstructed whole-rock Sm–Nd, U–Pb, and Re–Os isochron ages of 2.6–3.1 Ga for some eclogite xenoliths from the Kaapvaal, West African, and Siberian cratons (Jagoutz et al., 1984; Jacob et al., 1994; Pearson et al., 1995b; Jacob and Foley, 1999; Shirey et al., 2001; Barth et al., 2002) although regression errors are relatively large (0.1–0.4 Ga) and initial ratios poorly constrained. However, the timing of diamond crystallization in these rocks remains unconstrained. The late Archean Re–Os ages for eclogitic sulfide inclusions discussed below extend the previously established Proterozoic age range for eclogitic diamonds (e.g., Richardson, 1986; Richardson et al., 1990; Smith et al., 1991), based largely on Sm–Nd isochron ages for silicate inclusions (Fig. 4), and allow for multiple generations of subduction-related diamond formation and preservation in the cratonic keel.

The relatively well-constrained Re–Os isochron age (2.89 ± 0.06 Ga) and radiogenic initial Os isotope composition (0.156 ± 0.011; $\gamma_{Os} = +45$) for eclogitic sulfide inclusions in Kimberley diamonds (Richardson et al., 2001) and corresponding parameters for Jwaneng diamonds, provide important new constraints on subduction-related additions to the craton keel. Evidence that this was not merely a local phenomenon is now available from Kimberley and Jwaneng, as well as Koffiefontein and Orapa where comparable, although less well-constrained, late Archean ages are distinguishable (Pearson et al., 1998; Shirey et al., 2001). Late Archean subduction could also account for the mass-independent sulfur isotope fractionation observed in comparable eclogitic sulfide inclusions in Orapa diamonds (Farquhar et al., 2002). In particular, late Archean additions to the keel are considered to have occurred during suturing of the Kimberley block to the rest of the craton near the Colesberg magnetic lineament with a westward dipping slab (Richardson et al., 2001; Schmitz, 2002; Schmitz et al., in press).

The Jwaneng, Orapa, Koffiefontein, and Premier kimberlites also carry Proterozoic generations of eclogitic diamonds with ages of approximately 1–2 Ga based on both Sm–Nd (Fig. 4) and Re–Os isochron relationships (Richardson, 1986; Richardson et al., 1990; Pearson et al., 1998; Richardson et al., 1999; Shirey et al., 2001, 2002, 2003, 2004; this work). This suggests that further additions to the keel occurred

during Proterozoic modification of the Kaapvaal–Zimbabwe craton keel, although their relationship to subduction remains controversial (e.g., Cartigny et al., 1998; Navon, 1999; Cartigny et al., 2001; see also Shirey et al., 2004).

5.3. Episodic diamond genesis and Archean craton evolution

From a Kaapvaal craton perspective, the Kimberley, Jwaneng, and Premier kimberlites are key localities with respect to characterizing the relationship between episodic diamond genesis and Archean craton evolution. The occurrence of 3.2–3.3 Ga diamonds with depleted harzburgitic silicate inclusions (Sm–Nd, Rb–Sr model ages; Richardson et al., 1984) and 2.9 Ga diamonds with enriched eclogitic sulfide inclusions (Re–Os isochron age; Richardson et al., 2001) in the same Kimberley kimberlites indicates that formation of the Kaapvaal craton was at least a two-stage process (Shirey et al., 2002, 2003). Considering Re–Os, Sm–Nd, and Rb–Sr model age relationships for both inclusions and macrocryst minerals, a time gap of 300 ± 200 million years is required between the two Archean diamond formation events recorded in Kimberley diamonds (see also Shirey et al., this volume, regarding the reliability of these ages). Thus, the Kimberley kimberlites, which penetrate 3.2 Ga crust, carry both 3.2 Ga harzburgitic diamonds associated with creation of the western Kaapvaal cratonic nucleus, and 2.9 Ga eclogitic diamonds formed during subduction-related amalgamation of the western and eastern Kaapvaal along a N–S axis coincident with the Colesberg magnetic lineament (Richardson et al., 2001; Schmitz et al., in press). The Jwaneng kimberlite, on the western side of the combined Kaapvaal craton, carries both 2.9 and 1.5 Ga eclogitic diamonds reflecting stabilization and subsequent modification of the craton. The Premier kimberlite, on the edge of the massive Bushveld Complex in the north-central part of the craton, carries 1.9 Ga lherzolitic diamonds that postdate Bushveld–Molopo magmatism (but whose precursors have Archean Sm–Nd model ages; Richardson et al., 1993; Richardson and Harris, 1997), as well as 1.2 Ga eclogitic diamonds (whose age is within error of that of pipe emplacement; Kramers, 1979; Richardson, 1986; Burgess et al., 1989; Phillips et al., 1989).

The available diamond distribution, composition, and Sm–Nd and Re–Os age data support a multistage process to create, stabilize, and modify Archean cratons (Shirey et al., 2002). Mantle keels to early continental nuclei may have been created by severe middle Archean depletion events with high degrees of melting producing komatiites (e.g., Richardson et al., 1984; Wilson et al., 2003), followed by metasomatism and early harzburgitic diamond formation (3.2–3.3 Ga). Late Archean accretionary events (2.9 Ga) involving a subducted oceanic lithosphere component stabilized the craton and contributed eclogitic diamonds to the existing harzburgitic diamond population. Subsequent Proterozoic tectonothermal events (notably Bushveld–Molopo magmatism at 2.05 Ga) altered the composition of the cratonic keel by adding a basaltic component and subsequently introducing new generations of lherzolitic and eclogitic diamonds to an already extensive Archean diamond suite (see also Shirey et al., 2004).

From an intercraton perspective, there are interesting parallels although similar treatment of results from the Slave and Siberian cratons is still at an early stage. For example, there is clear evidence of a middle Archean peridotitic diamond formation event at Ekati on the Slave craton based on a sulfide inclusion Re–Os isochron age of 3.4 ± 0.3 Ga (Westerlund et al., 2003). Furthermore, published Re–Os and Sm–Nd data for sulfide and silicate inclusions from Udachnaya on the Siberian craton indicate both harzburgitic and lherzolitic diamond formation events at ca. 3.3 and 2.0 Ga, respectively (Richardson and Harris, 1997; Pearson et al., 1999). In both cases, the host diamonds display nitrogen aggregation characteristics consistent with lithospheric mantle storage at temperatures of around 1150 °C for a comparable period. As at Premier on the Kaapvaal craton, the precursors of the Udachnaya lherzolitic diamonds have Archean Sm–Nd model ages (see also Shirey et al., this volume).

6. Conclusions

The Re–Os and Sm–Nd ages and isotopic signatures obtained for sulfide and/or silicate inclusions in eclogitic diamonds from the Jwaneng, Kimberley, Orapa, Koffiefontein, Finsch, and Premier kimberlites

allow for multiple generations of subduction-related diamond formation and preservation in the Kaapvaal–Zimbabwe keel. Inclusions in diamonds of different age from the same kimberlite, as well as diamonds of the same age from different kimberlites, or even different continents, are providing key constraints on craton evolution. The evidence for a global connection between episodic diamond genesis and Archean craton evolution is now as compelling as that for the xenocrystic relationship between diamond and kimberlite.

Acknowledgements

We are grateful to De Beers Consolidated Mines for donating the diamond specimens and to the NRF, NSF and De Beers for financial assistance. We thank V. Anderson, G. Parker and E. van Blerck for diamond sorting, C. Hadidiacos and D. George for GL electron microprobe support, T. Mock and M. Horan for DTM isotope laboratory support, and R. Rudnick and S. Graham for thoughtful reviews.

References

Barth, M.G., Rudnick, R.L., Carlson, R.W., Horn, I., McDonough, W.F., 2002. Re–Os and U–Pb geochronological constraints on the eclogite–tonalite connection in the Archean Man Shield, West Africa. Precambrian Research 118, 267–283.

Brenan, J.M., 2002. Re–Os fractionation in magmatic sulfide melt by monosulfide solid solution. Earth and Planetary Science Letters 199, 257–268.

Burgess, S.R., Harte, B., 1999. Tracing lithosphere evolution through the analysis of heterogeneous G9/G10 garnets in peridotite xenoliths: I. Major element chemistry. In: Gurney, J.J., Gurney, J.L., Pascoe, M.D., Richardson, S.H. (Eds.), The J.B. Dawson Volume—Proceedings of the Seventh International Kimberlite Conference, Cape Town. Red Roof Design, Cape Town, pp. 66–80.

Burgess, R., Turner, G., Laurenzi, M., Harris, J.W., 1989. $^{40}Ar/^{39}Ar$ laser probe dating of individual clinopyroxene inclusions in Premier eclogitic diamonds. Earth and Planetary Science Letters 94, 22–28.

Burgess, R., Turner, G., Harris, J.W., 1992. $^{40}Ar/^{39}Ar$ laser probe studies of clinopyroxene inclusions in eclogitic diamonds. Geochimica et Cosmochimica Acta 56, 389–402.

Burgess, R., Kiviets, G.B., and Harris, J.W. Ar–Ar age determinations of syngenetic clinopyroxene and garnet inclusions in eclogitic diamonds from the Venetia and Orapa kimberlites. Proceedings of the Eighth International Kimberlite Conference, Victoria BC, Canada, this volume.

Carlson, R.W., Pearson, D.G., Boyd, F.R., Shirey, S.B., Irvine, G., Menzies, A.H., Gurney, J.J., 1999. Re–Os systematics of lithospheric peridotites implications for lithosphere formation and preservation. In: Gurney, J.J., Gurney, J.L., Pascoe, M.D., Richardson, S.H. (Eds.), The J.B. Dawson Volume—Proceedings of the Seventh International Kimberlite Conference, Cape Town. Red Roof Design, Cape Town, pp. 99–108.

Cartigny, P., Harris, J.W., Javoy, M., 1998. Eclogitic diamond formation at Jwaneng; no room for a recycled component. Science 280, 1421–1424.

Cartigny, P., Harris, J.W., Javoy, M., 2001. Diamond genesis, mantle fractionations and mantle nitrogen content: a study of $\delta^{13}C$–N concentrations in diamonds. Earth and Planetary Science Letters 185, 85–98.

Craig, J.R., Kullerud, G., 1969. Phase relations in the Cu–Fe–Ni–S system and their application to magmatic ore deposits. Economic Geology Monograph 4, 344–358.

Deines, P., Harris, J.W., 1995. Sulfide inclusion chemistry and carbon isotopes of African diamonds. Geochimica et Cosmochimica Acta 59, 3173–3188.

Deines, P., Harris, J.W., Gurney, J.J., 1997. Carbon isotope ratios, nitrogen content and aggregation state, and inclusion chemistry of diamonds from Jwaneng, Botswana. Geochimica et Cosmochimica Acta 61, 3993–4005.

Eldridge, C.S., Compston, W., Williams, I.S., Harris, J.W., Bristow, J.W., 1991. Isotope evidence for the involvement of recycled sediments in diamond formation. Nature 353, 649–653.

Farquhar, J., Wing, B.A., McKeegan, K.D., Harris, J.W., Cartigny, P., Thiemens, M.H., 2002. Mass-independent sulfur of inclusions in diamond and sulfur recycling on early Earth. Science 298, 2369–2372.

Griffin, W.L., Pearson, N.J., Belousova, E., Jackson, S.E., van Achterbergh, E., O'Reilly, S.Y., Shee, S.R., 2000. The Hf isotope composition of cratonic mantle: LAM-MC-ICPMS analysis of zircon megacrysts in kimberlites. Geochimica et Cosmochimica Acta 64, 133–147.

Gurney, J.J., Harris, J.W., Otter, M.L., Rickard, R.S., 1995. Jwaneng diamond inclusions. Extended Abstract, Sixth International Kimberlite Conference, Novosibirsk, Russia. United Institute of Geology, Geophysics and Mineralogy, Novosibirsk, Russia, pp. 208–210.

Helmstaedt, H.H., Gurney, J.J., 1984. Kimberlites of southern Africa—are they related to subduction processes? In: Kornprobst, J. (Ed.), Kimberlites and Related Rocks. Elsevier, Amsterdam, pp. 425–434.

Jacob, D., Foley, S.F., 1999. Evidence for Archean oceanic crust with low high field strength element signature from diamondiferous eclogite xenoliths. Lithos 48, 317–336.

Jacob, D., Jagoutz, E., Lowry, D., Mattey, D., Kudrjavtseva, G., 1994. Diamondiferous eclogite xenoliths from Siberia; remnants of Archean oceanic crust. Geochimica et Cosmochimica Acta 58, 5191–5207.

Jagoutz, E., Dawson, J.B., Hoernes, S., Spettel, B., Wanke, H., 1984. Anorthositic oceanic crust in the Archean Earth. Lunar and Planetary Science Institute, Houston, Lunar and Planetary Science XV, 395–396.

James, D.E., Fouch, M.J., VanDecar, J.C., van der Lee, S., and

the Kaapvaal Seismic Group, 2001. Tectospheric structure beneath southern Africa. Geophysical Research Letters 28, 2485–2488.

Kinny, P.D., Meyer, H.O.A., 1994. Zircon from the mantle: a new way to date old diamonds. Journal of Geology 102, 475–481.

Kinny, P.D., Compston, W., Bristow, J.W., Williams, I.S., 1989. Archaean mantle xenocrysts in a Permian kimberlite: two generations of kimberlitic zircon in Jwaneng DK2, southern Botswana. In: Ross, J., et al. (Eds.), Kimberlites and Related Rocks. Geological Society of Australia Special Publication, vol. 14, pp. 833–842.

Kirkley, M.B., Gurney, J.J., Rickard, R.S., 1994. Jwaneng framesites: carbon isotopes and intergrowth compositions. In: Meyer, H.O.A., Leonardos, O.H. (Eds.), Diamonds: Characterization, Genesis and Exploration. CPRM Special Publication, vol. 1/B, pp. 127–135.

Kramers, J., 1979. Lead, uranium, strontium, potassium and rubidium in inclusion-bearing diamonds and mantle-derived xenoliths from southern Africa. Earth and Planetary Science Letters 42, 58–70.

MacGregor, I.D., Carter, J.L., 1970. The chemistry of clinopyroxenes and garnets of eclogite and peridotite xenoliths from the Roberts Victor Mine, South Africa. Physics of Earth and Planetary Interiors 3, 391–397.

Meisel, T., Walker, R.J., Irving, A.J., Lorand, J.-P., 2001. Osmium isotopic compositions of mantle xenoliths: a global perspective. Geochimica et Cosmochimica Acta 65, 1311–1323.

Navon, O., 1999. Diamond formation in the Earth's mantle. In: Gurney, J.J., Gurney, J.L., Pascoe, M.D., Richardson, S.H. (Eds.), The P.H. Nixon Volume—Proceedings of the Seventh International Kimberlite Conference, Cape Town. Red Roof Design, Cape Town, pp. 584–604.

Pearson, D.G., Shirey, S.B., 1999. Isotopic dating of diamonds. In: Lambert, D.D., Ruiz, J. (Eds.), Application of Radiogenic Isotopes to Ore Deposit Research and Exploration. Reviews in Economic Geology. Society of Economic Geologists, Denver, pp. 143–172.

Pearson, D.G., Shirey, S.B., Carlson, R.W., Boyd, F.R., Pokhilenko, N.P., Shimizu, N., 1995a. Re–Os, Sm–Nd and Rb–Sr isotope evidence for thick Archaean lithospheric mantle beneath the Siberian craton modified by multi-stage metasomatism. Geochimica et Cosmochimica Acta 59, 959–977.

Pearson, D., Snyder, G., Shirey, S., Taylor, L., Carlson, R., Sobolev, N., 1995b. Archaean Re–Os age for Siberian eclogites and constraints on Archaean tectonics. Nature 374, 711–713.

Pearson, D.G., Shirey, S.B., Harris, J.W., Carlson, R.W., 1998. Sulfide inclusions in diamonds from the Koffiefontein kimberlite, S. Africa: constraints on diamond ages and mantle Re–Os systematics. Earth and Planetary Science Letters 160, 311–326.

Pearson, D.G., Shirey, S.B., Bulanova, G.P., Carlson, R.W., Milledge, H.J., 1999. Re–Os isotope measurements of single sulfide inclusions in a Siberian diamond and its nitrogen aggregation systematics. Geochimica et Cosmochimica Acta 63, 703–711.

Phillips, D., Onstott, T.C., Harris, J.W., 1989. $^{40}Ar/^{39}Ar$ laser-probe dating of diamond inclusions from the Premier kimberlite. Nature 340, 460–462.

Richardson, S.H., 1986. Latter-day origin of diamonds of eclogitic paragenesis. Nature 322, 623–626.

Richardson, S.H., Harris, J.W., 1997. Antiquity of peridotitic diamonds from the Siberian craton. Earth and Planetary Science Letters 151, 271–277.

Richardson, S.H., Gurney, J.J., Erlank, A.J., Harris, J.W., 1984. Origin of diamonds in old enriched mantle. Nature 310, 198–202.

Richardson, S.H., Erlank, A.J., Harris, J.W., Hart, S.R., 1990. Eclogitic diamonds of Proterozoic age from Cretaceous kimberlites. Nature 346, 54–56.

Richardson, S.H., Harris, J.W., Gurney, J.J., 1993. Three generations of diamonds from old continental mantle. Nature 366, 256–258.

Richardson, S.H., Chinn, I.L., Harris, J.W., 1999. Age and origin of eclogitic diamonds from the Jwaneng kimberlite, Botswana. In: Gurney, J.J., Gurney, J.L., Pascoe, M.D., Richardson, S.H. (Eds.), The P.H. Nixon Volume—Proceedings of the Seventh International Kimberlite Conference, Cape Town. Red Roof Design, Cape Town, pp. 734–736.

Richardson, S.H., Shirey, S.B., Harris, J.W., Carlson, R.W., 2001. Archean subduction recorded by Re–Os isotopes in eclogitic sulfide inclusions in Kimberley diamonds. Earth and Planetary Science Letters 191, 257–266.

Rudnick, R.L., Eldridge, C.S., Bulanova, G.P., 1993. Diamond growth history from in situ measurement of Pb and S isotopic compositions of sulphide inclusions. Geology 21, 13–16.

Schmitz, M.D., 2002. Geology and thermochronology of the lower crust of southern Africa. PhD Thesis. Massachusetts Institute of Technology, Cambridge. 269 pp.

Schmitz, M.D., Bowring, S.A., de Wit, M.J., Gartz, V., 2004. Subduction and terrane collision stabilize the western Kaapvaal craton tectosphere 2.9 billion years ago. Earth and Planetary Science Letters 222, 363–376.

Schrauder, M., Navon, O., 1994. Hydrous and carbonatitic mantle fluids in fibrous diamonds from Jwaneng, Botswana. Geochimica et Cosmochimica Acta 58, 761–771.

Schrauder, M., Koeberl, C., Navon, O., 1996. Trace element analyses of fluid-bearing diamonds from Jwaneng, Botswana. Geochimica et Cosmochimica Acta 60, 4711–4724.

Shimizu, N., Sobolev, N.V., 1995. Young peridotitic diamonds from the Mir kimberlite pipe. Nature 375, 394–397.

Shirey, S.B., Carlson, R.W., Richardson, S.H., Menzies, A.H., Gurney, J.J., Pearson, D.G., Harris, J.W., Wiechert, U., 2001. Archean emplacement of eclogitic components into the lithospheric mantle during formation of the Kaapvaal Craton. Geophysical Research Letters 28, 2509–2512.

Shirey, S.B., Harris, J.W., Richardson, S.H., Fouch, M.J., James, D.E., Cartigny, P., Deines, P., Viljoen, F., 2002. Diamond genesis, seismic structure, and evolution of the Kaapvaal–Zimbabwe craton. Science 297, 1683–1686.

Shirey, S.B., Harris, J.W., Richardson, S.H., Fouch, M.J., James, D.E., Cartigny, P., Deines, P., Viljoen, F., 2003. Regional patterns in the paragenesis and age of inclusions in diamond, diamond composition and the lithospheric seismic structure of southern Africa. Lithos 71, 243–258.

Shirey, S.B., Richardson, S.H., Harris, J.W., 2004. Age, paragenesis, and composition of diamonds and evolution of the Precambrian lithosphere of southern Africa. South African Journal of Geology 107, 91–106.

Shirey, S.B., Richardson, S.H., Harris, J.W., 2004. Integrated models of diamond formation and craton evolution. Proceedings of the Eighth International Kimberlite Conference, Victoria BC, Canada, 77, 923–944 (this volume).

Smith, C.B., Gurney, J.J., Harris, J.W., Otter, M.L., Robinson, D.N., Kirkley, M.B., Jagoutz, E., 1991. Neodymium and strontium isotope systematics of eclogite and websterite paragenesis inclusions from single diamonds. Geochimica et Cosmochimica Acta 55, 2579–2590.

Spetsius, Z.V., Belousova, E.A., Griffin, W.L., O'Reilly, S.Y., Pearson, N.J., 2002. Archean sulfide inclusions in Paleozoic zircon megacrysts from the Mir kimberlite, Yakutia: implications for the dating of diamonds. Earth and Planetary Science Letters 199, 111–126.

Westerlund, K.J., Shirey, S.B., Richardson, S.H., Gurney, J.J., Harris, J.W., 2003. Re–Os isotope systematics of peridotitic diamond inclusion sulfides from the Panda kimberlite, Slave Craton. Extended Abstract, Eighth International Kimberlite Conference, Victoria BC, Canada.

Wilson, A.H., Shirey, S.B., Carlson, R.W., 2003. Archean ultradepleted komatiites formed by hydrous melting of cratonic mantle. Nature 423, 858–861.

Available online at www.sciencedirect.com

Lithos 77 (2004) 155–179

www.elsevier.com/locate/lithos

Mineral chemistry and thermobarometry of inclusions from De Beers Pool diamonds, Kimberley, South Africa

D. Phillips[a,*], J.W. Harris[b], K.S. Viljoen[c]

[a] School of Earth Sciences, The University of Melbourne, Parkville, VIC, 3010, Australia
[b] Division of Earth Sciences, University of Glasgow, Glasgow G12 8QQ, Scotland, UK
[c] GeoScience Centre, De Beers Consolidated Mines Ltd., P.O. Box 82232, Southdale, 2135, South Africa

Received 27 June 2003; accepted 14 December 2003
Available online 1 July 2004

Abstract

Silicate and oxide mineral inclusions in diamonds from the geologically and historically important De Beers Pool kimberlites in Kimberley, South Africa, are characterised by harzburgitic compositions (>90%), with lesser abundances from eclogitic and websteritic parageneses. The De Beers Pool diamonds contain unusually high numbers of inclusion intergrowths, with garnet + orthopyroxene ± chromite ± olivine and chromite + olivine assemblages dominant. More unusual intergrowths include garnet + olivine + magnesite and an eclogitic assemblage comprising garnet + clinopyroxene + rutile. The mineral chemistry of the De Beers Pool inclusions overlaps that of most worldwide localities. Peridotitic garnet inclusions exhibit variable CaO (<5.8 wt.%) and Cr_2O_3 contents (3.0–15.0 wt.%), although the majority are harzburgitic with very low calcium concentrations (<2 wt.% CaO). Eclogitic garnet inclusions are characterised by a wide range in CaO (3.3–21.1 wt.%) with low Cr_2O_3 (<1 wt.%). Websteritic garnets exhibit intermediate compositions. Most chromite inclusions contain 63–67 wt.% Cr_2O_3 and <0.5 wt.% TiO_2. Olivine and orthopyroxene inclusions are magnesium-rich with Mg-numbers of 93–97. Olivine inclusions in chromite exhibit the highest Mg-numbers and also contain elevated Cr_2O_3 contents up to 1.0 wt.%. Peridotitic clinopyroxene inclusions are Cr-diopsides with up to 0.8 wt.% K_2O. Eclogitic and websteritic clinopyroxene inclusions exhibit overlapping compositions with a wide range in Mg-numbers (66–86).

Calculated temperatures for non-touching inclusion pairs from individual diamonds range from 1082 to 1320 °C (average = 1197 °C), whereas pressures vary from 4.6 to 7.7 GPa (average = 6.3 GPa). Touching inclusion assemblages are characterised by equilibration temperatures of 995 to 1182 °C (average = 1079 °C) and pressures of 4.2–6.8 GPa (average = 5.4 GPa). Provided that the non-touching inclusions represent equilibrium assemblages, it is suggested that these inclusions record the conditions at the time of diamond crystallisation (~ 1200 °C; ~ 3.0 Ga). The lower average temperatures for touching inclusions are attributed to re-equilibration in a cooling mantle (~ 1050 °C) prior to kimberlite eruption at ~ 85 Ma. Pressure estimates for touching garnet–orthopyroxene inclusions are also skewed towards lower values than most non-touching inclusions. This apparent difference may be an artefact of the Al-exchange geobarometer and/ or the result of sampling bias, due to limited numbers of non-touching garnet–orthopyroxene inclusions. Alternatively pressure differences could be caused by differential uplift in the mantle or possibly variations in thermal compressibility

---

* Corresponding author.

between diamond and silicate inclusions. However, thermodynamic modelling suggests that thermal compressibility differences would cause only minor changes in internal inclusion pressures (<0.2 GPa/100 °C).

Keywords: Mineral chemistry; Thermobarometry; De Beers Pool inclusions

1. Introduction

Diamond provides an impermeable barrier that can insulate mineral inclusions from interaction with mantle melts/fluids, the transporting magma and crustal alteration processes. Consequently, studies of syngenetic inclusions in diamonds provide important information on the mineralogy and geochemistry of the sub-continental mantle lithosphere at the time of diamond formation. Most inclusions can be sub-divided into peridotitic and eclogitic parageneses, coincident with the major rock types observed in mantle xenoliths (cf. Meyer, 1987; Harris, 1992). There are, however, important differences between the mineral chemistry of inclusions in diamonds and typical mantle xenolith/xenocryst phases, with inclusions generally exhibiting more depleted major element compositions (e.g. Gurney et al., 1979). Recent isotopic work indicates that most diamonds are significantly older than the host kimberlite/lamproite magma, with some retaining mantle signatures for ~ 3.2 Ga (e.g. Richardson et al., 1984). Therefore, geochemical differences between inclusions and xenolith phases may be related to continuing compositional evolution of the mantle surrounding diamonds (e.g. metasomatism), changing temperature/pressure conditions and/or unique environments of diamond formation (e.g. Taylor et al., 1996; Sobolev et al., 1997).

Unfortunately, mineral inclusions are restricted to a small proportion of diamonds ($<1\%$) and most occur as monomineralic entities. Multi-phase or polymineralic inclusion assemblages are far less common, but are key to determining equilibrium mantle assemblages as well as the physical conditions prevailing in the mantle during diamond growth (e.g. Wang et al., 1996; Viljoen et al., 1999; Stachel et al., 2000). In most cases, co-existing inclusions from individual diamonds occur either as separate (non-touching) inclusions or as intergrowths (touching) of two or more minerals (e.g. Otter and Gurney, 1989; Phillips and Harris, 1995; Sobolev et al., 1997). If trapped at the same time under equilibrium conditions, non-touching inclusions of appropriate mineralogy can provide estimates of the temperature, pressure and oxygen fugacity at the time of diamond growth and inclusion encapsulation (e.g. Gurney et al., 1984a,b; Griffin et al., 1992; Stachel and Harris, 1997). However, establishing that co-existing inclusions are in equilibrium is a non-trivial problem and there is increasing evidence that at least some diamonds grew over prolonged time periods and under changing physico-chemical conditions (e.g. Griffin et al., 1992; Bulanova, 1995; Stachel et al., 1998). Mineral intergrowths, which may be encapsulated as multiphase inclusions or formed by exsolution processes, are more likely to reflect equilibrium mantle assemblages (e.g. Prinz et al., 1975; Stachel et al., 2000). Importantly, however, inclusion intergrowths may re-equilibrate over time in response to changing external pressure/temperature conditions. Consequently, touching inclusion assemblages may provide information on mantle conditions subsequent to the time of diamond growth (Phillips and Harris, 1995; Stachel et al., 2000; Leost et al., 2003).

In this study, we document the chemistry of a large suite of mineral inclusions extracted from diamonds from the geologically and historically important De Beers Pool kimberlites in Kimberley, South Africa (28°46′ S, 24°48′ E). The Kimberley region is significant as the site of the first diamond discoveries in South Africa, the type locality for Group I (basaltic) kimberlites (Smith, 1983), the source of a vast array of mantle xenoliths (cf. Nixon, 1987 and references therein) and the discovery site for ancient xenolithic diamonds (Richardson et al., 1984, 2001). The De Beers Pool kimberlites were emplaced into the Archaean Kaapvaal craton at ~ 84 Ma (Allsopp et al., 1989) and overlie an unusually thick section of continental lithosphere (James et al., 2001). Peridotitic inclusions from De Beers Pool diamonds have been dated at 3.2–3.3 Ga (Richardson et al., 1984), whereas recent Re–Os isotopic studies of eclogitic inclusions suggest formation of eclogitic diamonds at ~ 2.9 Ga (Richardson et al., 2001). Although inclusions in diamond have been studied from a large number of kimberlites, lamproites and detrital localities worldwide (e.g. Stachel et al., 2000 and references therein), little information is available for the De Beers Pool kimberlites, apart from a limited investigation of Bultfontein diamonds by Wilding (1990).

The De Beers Pool diamond production is derived from four kimberlites, namely De Beers, Du Toit's Pan, Bultfontein and Wesselton, which have been mined for over a century and are reaching the end of their minable resources. Therefore, the current study represents one of the last opportunities to analyse inclusions from this historic locality. The current inclusion suite is also unusual in that it contains a high proportion of touching and non-touching inclusion assemblages (Fig. 1), that provide a rare opportunity to identify equilibrium mantle

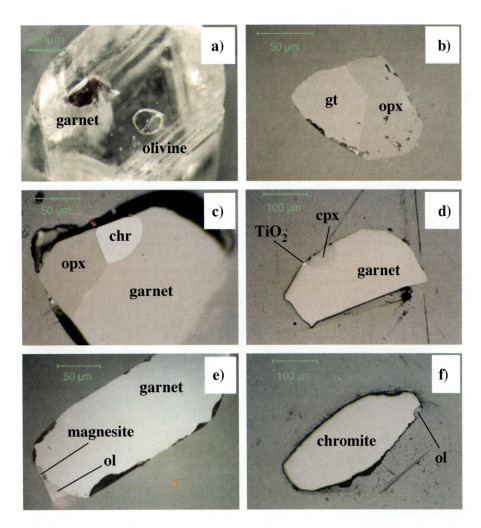

Fig. 1. Photomicrographs of inclusion assemblages in De Beers Pool diamonds; (a) non-touching peridotitic garnet and olivine inclusions; (b) peridotitic garnet intergrown with an orthopyroxene inclusion; (c) intergrowth of peridotitic garnet + chromite + orthopyroxene inclusions; (d) eclogitic garnet intergrown with clinopyroxene and rutile; (e) polymineralic intergrowth of peridotitic garnet + olivine + magnesite; (f) peridotitic chromite inclusion intergrown with a small olivine.

assemblages and evaluate potential changes in temperature and pressure conditions from diamond growth to exhumation.

2. Samples and analytical methods

Inclusion-bearing diamonds were recovered from the $-6+5$ (1.5 mm), $-7+6$ (2.2 mm), $-9+7$ (2.5 mm) and $-13+12$ (2.9 mm) sieve fractions of the pooled run-of-mine production from the De Beers, Du Toit's Pan, Bultfontein and Wesselton (De Beers Pool) mines. Diamonds from the four kimberlites exhibit very similar population characteristics and the close proximity of the pipes may indicate a common source at depth (Richardson et al., 1984). The diamonds utilised in this study were originally collected for an internal De Beers Consolidated Mines research project. As a result, the suite does not represent a complete cross-section of inclusion assemblages, as emphasis was placed on the selection of garnet, chromite and clinopyroxene inclusions, which are the minerals of prime importance to diamond exploration. However,

inclusion abundances for each of the De Beers Pool mines have been collated from previous studies and these are summarised in Table 1.

Mineral inclusions were extracted from a total of 425 De Beers Pool diamonds (Table 2). Initially, attempts were made to select 200 each of peridotitic garnets, chromites and clinopyroxenes and eclogitic garnets and clinopyroxenes. However, eclogitic inclusions and peridotitic clinopyroxenes proved to be rare and fewer numbers were ultimately recovered. Although olivine and orthopyroxene inclusions were not sought specifically, a number of these inclusions were recovered either in association with garnets or chromites or were mistaken for clinopyroxene inclusions.

The inclusions were released from their host diamonds using a stainless steel crusher equipped with a glass window to observe the process under a binocular microscope. All diamonds were first examined for evidence of fractures connecting inclusions with the surface. After the recovery of the first composite inclusion, attempts were made to carefully cleave the diamonds to remove intact inclusions and minimise the potential for loss of intergrowth phases.

Table 1
Visual inclusion abundance data from previous studies of diamonds from the Kimberley region, South Africa

	Bultfontein[a,b]	De Beers Pool		Wesselton[b]	Roberts Victor[c]	Koffiefontein[d]	Finsch[e]
		De Beers[b]	DuToit's Pan[b]				
Relative proportion of inclusions in diamonds							
Peridotitic	88.3	79.8	81.2	89.9	71.1	31.8	56.2
Eclogitic	4.8	7.8	5.0	3.2	8.7	2.5	1.1
Sulphides	6.2	12.4	8.8	4.2	20.2	45.5	40.1
Clouds	0.7	0.0	5.0	2.7	nd	20.2	2.6
	100.0	100.0	100.0	100.0	100.0	100.0	100.0
Ecl/(Ecl + Per)	5.2	8.9	5.8	3.4	7.4	10.8	0.1
Relative proportion of peridotitic silicate and oxide inclusions							
Colourless	44.7	46.5	46.1	50.1			81.0
Olivine					44.9	44.0	
Orthopyroxene					21.8	36.8	
Garnet	21.7	13.7	15.4	41.1	11.5	17.1	18.2
Chromite	32.8	39.9	36.9	7.3	20.5	0.3	0.8
Clinopyroxene	0.8	0.0	1.6	1.6	1.3	1.8	0.0
	100.0	100.0	100.0	100.0	100.0	100.0	100.0

Ecl: eclogitic; Per: peridotitic; nd=not determined.
[a] Wilding (1990).
[b] Harris, unpublished data.
[c] Gurney et al. (1984a).
[d] Rickard et al. (1989).
[e] Gurney et al. (1979).

Table 2
Number of inclusions and mineral assemblages in the De Beers Pool diamond suite

Inclusion assemblage	Monomineralic	Polymineralic touching	Polymineralic non-touching
Peridotitic			
Garnet (gt)	163		
Chromite (chr)	154		
Olivine (ol)	0		
Orthopyroxene (opx)	14		
Clinopyroxene (cpx)	2		
gt + opx		21	7
gt + ol		2	3
gt + cpx		0	1
gt + sulphide		1	
chr + ol		7	2
chr + opx		2	0
opx + ol		0	1
gt + opx + chr		3	0
gt + opx + ol		3	0
gt + ol + magnesite		1	0
Eclogitic/websteritic			
Garnet	23		
Clinopyroxene	23		
gt + cpx		1	6
gt + cpx + TiO$_2$		1	0

This proved possible for many diamonds of predominantly octahedral morphology. However, rounded dodecahedral diamonds invariably shattered, often causing fragmentation of the mineral inclusions inside. Where composite inclusions could be identified, these were mounted so as to expose multiple phases on polishing. As many composite inclusions proved to be very delicate and, with one mineral dominant, it is likely that at least some multi-phase inclusions were either overlooked or lost during sample preparation. It is perhaps noteworthy that Wilding (1990) reported no composite inclusions in his study of Bultfontein diamonds.

After extraction from the host diamond, the inclusions were embedded in epoxy resin within small brass cylinders (5-mm diameter). These mounts were ground by hand to expose the inclusion(s) and then polished using successive diamond grits to 0.25 μm. Major and minor element data were obtained using a Cameca SX-50 electron microprobe at the De Beers Geoscience Centre, following procedures outlined by

Viljoen et al. (1996). Operating conditions included an accelerating voltage of 20 kV and a beam current of 20 nA. Matrix corrections were carried out using on-line software with PAP corrections. Detection limits for low abundance elements such as Na$_2$O and K$_2$O were typically 0.02 wt.%. Counting times were 40 s for all peaks and 10 s for backgrounds. Each inclusion was analysed at least three times from core to rim. None of the inclusions exhibited detectable spatial variations in chemistry and all are considered to be unzoned. Consequently, electron microprobe compositions are reported as averaged values of at least three analyses per inclusion (Table 3).

3. Inclusion chemistry

Previous visual abundance studies have shown that diamonds from the De Beers Pool mines are characterised by a predominance of inclusions belonging to the peridotitic paragenesis (Table 1). Similar eclogitic/peridotitic proportions have also been reported in diamonds from other kimberlites in the region, such as Roberts Victor, Koffiefontein and Finsch (Table 1). With the exception of the Wesselton mine, the most common peridotitic inclusions in De Beers Pool diamonds are colourless phases (olivine + orthopyroxene) and chromite. Chromite inclusions are less common in Wesselton and Roberts Victor diamonds and rare in Koffiefontein and Finsch samples. Wesselton diamonds are unusual in hosting a relatively high proportion of peridotitic garnet inclusions. Clinopyroxene inclusions of peridotitic association (Cr-diopside) are uncommon in diamonds from all listed localities.

The majority of diamonds from the current De Beers Pool collection contained single, or occasionally multiple inclusions of the same mineral (Table 2). Eighteen diamonds, however, yielded two or more non-touching inclusions of different mineralogy. In addition, an unusually high number of diamonds (42) produced inclusion intergrowths, with the most common associations being garnet + orthopyroxene and chromite + olivine (Table 2). Representative electron microprobe analyses for selected touching and non-touching inclusion assemblages are listed in Table 3. The complete De Beers Pool inclusion dataset is available from the Lithos on-line data repository.

Table 3
Representative electron microprobe analyses of selected inclusions from De Beers Pool diamonds

Sample no.	DP014		DP042		DP049		DP070		DP168		DP182		DP185			DP209			DP211	
Mineral	Garnet	Opx	Garnet	Olivine	Garnet	Opx	Garnet	Opx	Garnet	Olivine	Garnet	Opx	Garnet	Opx	Olivine	Garnet	Opx	Chr	Garnet	Opx
Suite	Per (lhz)		Per (hz)		Per (hz)		Per (hz)		Per (hz)		Per (hz)		Per (hz)			Per (hz)			Per (hz)	
Assemblage	gt, opx		gt, ol		gt, opx		gt, opx		gt, ol		gt, opx		gt, opx, ol			gt, opx, chr			gt, opx	
Touching	n	n	n	n	y	y	y	y	y	y	y	y	y	y	y	y	y	y	y	y
SiO_2	42.56	59.01	41.16	41.32	42.50	58.82	43.61	58.86	43.23	41.89	42.52	57.91	42.92	57.42	40.79	42.09	58.63	nd	41.58	58.50
Al_2O_3	19.33	0.80	14.06	0.01	19.89	0.57	21.84	0.63	19.80	0.00	21.49	0.67	20.27	0.65	0.05	16.86	0.54	5.58	17.72	0.49
TiO_2	0.01	0.00	0.15	0.00	0.05	0.01	0.01	0.01	0.00	0.02	0.01	0.00	0.00	0.01	0.00	0.02	0.01	0.05	0.02	0.01
Cr_2O_3	6.75	0.44	12.96	0.05	5.41	0.38	3.68	0.23	5.31	0.07	3.69	0.29	4.82	0.38	0.19	10.22	0.49	66.43	8.09	0.46
FeO	4.28	3.13	5.25	6.48	5.34	3.54	3.66	2.22	5.04	5.17	5.84	3.82	4.43	2.78	4.58	5.53	3.24	12.60	5.11	3.09
MnO	0.15	0.07	0.05	0.08	0.16	0.07	0.10	0.04	0.16	0.02	0.16	0.07	0.17	0.04	0.04	0.08	0.07	0.15	0.11	0.08
NiO	0.01	0.12	0.01	0.35	0.02	0.15	0.02	0.17	0.01	0.38	0.02	0.11	0.01	0.15	0.55	0.01	0.16	0.11	0.02	0.11
MgO	25.49	36.89	21.34	50.37	22.75	37.10	26.75	37.24	25.49	50.99	23.49	36.78	25.65	38.43	54.65	23.94	36.07	13.49	26.05	38.06
CaO	0.94	0.14	4.10	0.02	3.48	0.35	0.66	0.07	1.00	0.00	3.39	0.35	1.04	0.12	0.03	1.26	0.08	na	0.70	0.06
Na_2O	0.00	0.02	0.01	0.01	0.02	0.07	0.01	0.01	0.01	0.01	0.02	0.07	0.02	0.03	0.00	0.00	0.01	na	0.02	0.05
K_2O	0.00	0.01	0.00	0.00	0.01	0.00	0.00	0.00	0.00	0.01	0.00	0.00	0.01	0.00	0.00	0.00	0.00	na	0.00	0.01
V_2O_5	0.03	0.01	0.07	0.00	0.03	0.01	0.02	0.01	0.03	0.00	0.04	0.01	0.02	0.02	0.00	0.07	0.00	0.32	0.04	0.01
ZnO	na	na	na	na	na	na	na	na	na	na	na	na	na	na	na	na	na	0.05	na	na
Total	99.55	100.64	99.16	98.69	99.66	101.07	100.36	99.48	100.08	98.55	100.67	100.08	99.36	100.03	100.88	100.08	99.30	98.78	99.46	100.93

Sample no.	DP218			DP231		DP326			DP348		DP364		DP374		DP379		DP387			DP392
Mineral	Garnet	Opx	Olivine	Garnet	Olivine	Garnet	Opx	Olivine	Garnet	Cpx	Garnet	Opx	Garnet	Cpx	Garnet	Cpx	Garnet	Olivine	Mag	Garnet
Suite	Per (lhz)			Per (hz)		Per (hz)			Ecl		Per (hz)		Per (lhz)		Ecl		Per (hz)			Per (hz)
Assemblage	gt, opx, ol			gt, ol		gt, opx, ol			gt, cpx		gt, opx, chr		gt, cpx		gt, cpx		gt, ol, mag			gt, opx
Touching	y	y	y	y	y	y	y	y	y	y	y	y	n	n	y	y	y	y	y	n
SiO_2	41.66	58.78	41.50	41.81	41.72	41.22	58.34	39.96	39.88	56.35	42.78	59.15	41.23	54.76	39.34	52.78	41.93	41.93	0.04	41.13
Al_2O_3	20.78	0.60	0.00	19.85	0.02	18.06	0.52	0.04	22.11	10.16	19.61	0.60	19.77	1.11	20.96	2.40	20.59	0.03	0.02	20.35
TiO_2	0.02	0.01	0.00	0.01	0.01	0.06	0.01	0.00	0.45	0.54	0.00	0.00	0.08	0.01	0.61	0.34	0.21	0.01	0.00	0.01
Cr_2O_3	4.31	0.30	0.10	4.97	0.17	7.67	0.43	0.20	0.03	0.06	5.46	0.33	5.53	0.89	0.38	0.17	4.74	0.07	0.05	4.92
FeO	6.69	4.53	7.45	5.88	6.27	5.56	3.55	5.84	16.53	3.98	5.36	3.48	6.26	2.37	18.58	6.90	5.82	6.00	3.12	5.71
MnO	0.20	0.07	0.08	0.18	0.08	0.19	0.08	0.05	0.29	0.02	0.15	0.06	0.15	0.10	0.42	0.10	0.16	0.04	0.09	0.15
NiO	0.01	0.13	0.39	0.01	0.51	0.02	0.14	0.51	0.02	0.07	0.01	0.16	0.02	0.07	0.01	0.06	0.01	0.47	0.10	0.01
MgO	21.30	36.19	51.45	23.39	52.46	21.77	34.99	50.66	11.98	9.80	23.79	36.68	20.59	17.46	12.75	13.40	23.01	51.89	46.07	23.57
CaO	4.68	0.49	0.04	3.39	0.04	4.47	0.39	0.06	7.93	13.96	2.79	0.27	5.45	21.46	5.82	18.85	3.53	0.05	1.00	2.55
Na_2O	0.02	0.08	0.01	0.01	0.01	0.02	0.06	0.01	0.14	6.10	0.01	0.02	0.02	0.70	0.10	1.82	0.03	0.01	0.01	0.01
K_2O	0.00	0.01	0.00	0.00	0.01	0.01	0.00	0.01	0.01	0.05	0.00	0.00	0.01	0.28	0.00	0.09	0.01	0.00	0.01	0.01
V_2O_5	0.08	0.02	0.00	0.04	0.00	0.05	0.01	0.00	0.03	0.07	0.04	0.00	0.07	0.03	0.09	0.09	0.08	0.00	0.01	0.03
ZnO	na	na	na	na	na	na	na	na	na	na	na	na	na	na	na	na	na	na	na	na
Total	99.75	101.21	101.02	99.54	101.30	99.10	98.52	97.34	99.40	101.16	100.00	100.75	99.18	99.24	99.06	96.92	100.12	100.51	50.52	98.45

Sample no.	Mineral	Suite	Assemblage	Touching	SiO$_2$	Al$_2$O$_3$	TiO$_2$	Cr$_2$O$_3$	FeO	MnO	NiO	MgO	CaO	Na$_2$O	K$_2$O	V$_2$O$_5$	ZnO	Total
DP392	Opx	Per (hz)	gt, opx	n	57.86	0.43	0.00	0.22	3.57	0.07	0.13	36.63	0.25	0.02	0.01	0.00	na	99.19
DP400	Garnet	Per (hz)	gt, opx	y	42.87	20.01	0.00	5.65	5.11	0.10	0.01	26.13	0.95	0.01	0.00	0.03	na	100.87
DP400	Opx			y	58.94	0.56	0.00	0.34	3.14	0.06	0.13	38.51	0.09	0.02	0.00	0.01	na	101.80
DP414	Garnet	Ecl	gt, cpx	n	38.71	21.68	0.48	0.11	21.72	0.51	0.02	11.74	3.65	0.13	0.01	0.05	na	98.81
DP414	Cpx			n	54.91	5.61	0.54	0.09	9.94	0.16	0.03	10.93	12.41	4.17	0.09	0.07	na	98.95
DP418	Garnet	Per (hz)	gt, opx	n	42.30	16.66	0.02	9.48	4.97	0.06	0.01	23.91	1.30	0.00	0.01	0.04	na	98.76
DP418	Opx			n	58.67	0.72	0.01	0.51	3.58	0.07	0.12	36.48	0.19	0.01	0.01	0.00	na	100.37
DP422	Garnet	Per (hz)	gt, opx	y	42.33	18.69	0.01	7.21	5.30	0.10	0.01	25.24	1.04	0.01	0.00	0.06	na	100.00
DP422	Opx			y	58.63	0.52	0.00	0.36	3.24	0.05	0.12	37.72	0.08	0.01	0.01	0.00	na	100.74
DP440	Garnet	Per (hz)	gt, opx	n	42.94	21.68	0.00	3.56	3.61	0.07	0.03	27.01	0.66	0.01	0.01	0.02	na	99.60
DP440	Opx			n	59.60	0.67	0.00	0.22	2.43	0.05	0.13	36.35	0.10	0.01	0.00	0.01	na	99.57
DP442	Garnet	Per (hz)	gt, ol	n	41.91	20.47	0.01	4.67	4.51	0.18	0.01	25.74	0.94	0.01	0.01	0.02	na	98.48
DP442	Olivine			n	41.33	0.01	0.00	0.03	5.36	0.08	0.39	53.17	0.01	0.03	0.01	0.00	na	100.42
DP444	Garnet	Ecl	gt, cpx	n	39.57	21.11	0.17	0.15	21.44	1.35	0.02	11.64	3.27	0.12	0.00	0.03	na	98.86
DP444	Cpx			n	54.62	6.63	0.15	0.11	10.16	0.49	0.07	10.98	10.42	4.23	0.31	0.06	na	98.23
DP448	Garnet	Per (lhz)	gt, opx	y	42.13	20.42	0.08	4.28	6.27	0.19	0.02	21.81	5.03	0.03	0.00	0.05	na	100.31
DP448	Opx			y	58.08	0.61	0.01	0.29	4.13	0.09	0.17	35.79	0.57	0.10	0.00	0.01	na	99.85
DP457	Garnet	Per (hz)	gt, opx, chr	y	41.61	15.64	0.06	11.52	4.96	0.03	0.01	24.26	2.21	0.00	0.00	0.04	na	100.34
DP457	Opx			y	58.68	0.57	0.01	0.63	2.93	0.05	0.14	37.98	0.14	0.01	0.00	0.02	na	101.16
DP457	Chr			y	na	4.70	0.15	68.03	11.52	0.13	0.10	13.81	na	na	na	0.22	0.02	98.68
DP460	Garnet	Per (hz)	gt, ol	n	41.64	16.27	0.06	9.90	4.62	0.19	0.01	23.86	2.61	0.02	0.00	0.05	na	99.23
DP460	Olivine			n	41.80	0.01	0.00	0.04	5.23	0.07	0.32	51.89	0.02	0.00	0.01	0.01	na	99.40
DP464	Chr	Per (hz)	chr, opx	y	na	3.84	0.36	68.26	11.14	0.18	0.08	14.26	na	na	0.25	0.02	na	98.39
DP464	Opx			y	59.30	0.36	0.03	0.54	2.65	0.06	0.10	38.35	0.14	0.02	0.01	0.00	na	101.56
DP473	Garnet	Per (hz)	gt, opx	n	42.67	19.29	0.00	6.64	4.39	0.10	0.00	25.79	0.82	0.02	0.01	0.03	na	99.76
DP473	Opx			y	59.51	0.54	0.00	0.37	2.62	0.04	0.15	37.41	0.07	0.02	0.00	0.01	na	100.74
DP481	Garnet	Per (hz)	gt, opx	n	42.98	19.20	0.01	6.64	4.24	0.09	0.02	25.64	0.83	0.01	0.00	0.03	na	99.69
DP481	Opx			y	58.59	0.65	0.00	0.33	2.96	0.06	0.13	37.78	0.16	0.01	0.01	0.01	na	100.69
DP486	Garnet	Per (hz)	gt, opx	n	42.44	17.82	0.01	8.11	5.16	0.19	0.01	24.57	1.46	0.01	0.01	0.07	na	99.86
DP486	Opx			n	58.79	0.62	0.01	0.37	2.97	0.07	0.13	36.85	0.13	0.01	0.01	0.00	na	99.96
DP488	Garnet	Per (hz)	gt, opx	y	42.12	20.20	0.01	4.91	4.74	0.11	0.03	25.33	1.01	0.01	0.01	0.03	na	98.51
DP488	Opx			y	58.70	0.75	0.01	0.48	3.40	0.07	0.13	36.91	0.21	0.01	0.01	0.01	na	100.69
DP497	Garnet	Per (hz)	gt, opx	n	42.60	18.16	0.00	7.45	5.13	0.21	0.01	25.42	0.69	0.01	0.00	0.05	na	99.73
DP497	Opx			n	58.87	0.16	0.00	0.11	2.86	0.03	0.13	37.71	0.06	0.01	0.01	0.01	na	99.96
DP539	Garnet	Web	gt, cpx	n	40.27	21.47	0.54	0.29	16.85	0.36	0.02	14.50	4.28	0.07	0.00	0.08	na	98.73
DP539	Cpx			n	54.18	3.66	0.22	0.14	7.99	0.14	0.06	13.74	16.32	1.87	0.41	0.07	na	98.80
DP542	Garnet	Ecl	gt, cpx	n	39.69	22.33	0.30	0.12	11.52	0.19	0.01	10.13	14.31	0.10	0.01	0.03	na	98.74
DP542	Cpx			n	55.72	13.84	0.14	0.10	2.51	0.01	0.03	7.28	11.99	6.48	0.01	0.03	na	98.14

Per: peridotitic; Ecl: eclogitic; Web: websteritic; hz: harzburgitic; lhz: lherzolitic; gt: garnet; ol: olivine; opx: orthopyroxene; cpx: clinopyroxene; chr: chromite; mag: magnesite; total iron is given as FeO; na: not analysed; y: yes; n: no.

3.1. Garnet

A total of 271 garnets were analysed from 236 De Beers Pool diamonds. Some 186 diamonds contained only garnet, whereas 50 diamonds contained additional phases either as intergrowths (touching inclusions) or as separate entities (non-touching assemblages) (Table 2, Fig. 1). Bi-mineralic combinations include garnet + orthopyroxene, garnet + olivine, garnet + clinopyroxene and garnet + sulphide. Tri-mineralic intergrowths include garnet + olivine + orthopyroxene, garnet + orthopyroxene + chromite, garnet + clinopyroxene + rutile(?) and one unusual assemblage of garnet + olivine + magnesite. In all, garnets from 33 diamonds (12.5%) were found to be intergrown with other phases (Table 2). This is likely to be a minimum number as many intergrowth phases are very small and some may have been overlooked or lost during sample preparation procedures.

Although individual garnet inclusions and intergrowths were found to be compositionally homogeneous, minor variations were detected between garnets from 9 of 35 diamonds (26%) hosting multiple garnet inclusions. Non-equilibrium between co-existing phases has been noted by a number of workers previously and attributed to changes in mantle geochemistry and/or temperature during prolonged diamond growth (e.g. Griffin et al., 1992; Sobolev et al., 1997).

Peridotitic garnets from the De Beers Pool diamonds exhibit typical Cr-pyrope chemistries and are dominated by harzburgitic compositions (94%) (Figs. 2 and 3). Only 12 diamonds (6%) yielded garnets that plot in the lherzolitic compositional field (Fig. 2). Two lherzolitic garnets are intergrown with orthopyroxene and one with orthopyroxene and olivine. A single lherzolitic garnet was found to co-exist with a clinopyroxene inclusion. The high proportion of Ca-poor harzburgitic garnets (78% have < 2 wt.% CaO) indicates a highly depleted mantle source and is typical of a number of localities in this region of South Africa, such as Finsch (Gurney et al., 1979), Roberts Victor (Gurney et al., 1984a) and Koffiefontein (Rickard et al., 1989). Overall, peridotitic garnets contain 3.0–15.0 wt.% Cr_2O_3 0.2–5.8 wt.% CaO, 3.5–7.2 wt.% FeO and 19.3–28.0 wt.% MgO. These compositions fall within the worldwide database compiled by Stachel et al. (2000). Two harzburgitic garnets intergrown with chromite exhibit high Cr_2O_3 contents of 10.2 and 11.5 wt.%. A third garnet, with a very small chromite inclusion (DP364), exhibits a lower Cr_2O_3 content of 5.5 wt.%. TiO_2 concentrations range from below detection limits to a maximum of 0.44 wt.%, with the higher values derived from more calcic garnets

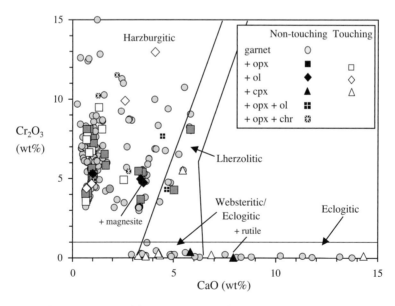

Fig. 2. Plot of Cr_2O_3 versus CaO for garnets recovered from De Beers Pool diamonds. Also shown is the lherzolite field defined by Sobolev (1977). Eclogite and websteritic fields modified after Aulbach et al. (2002).

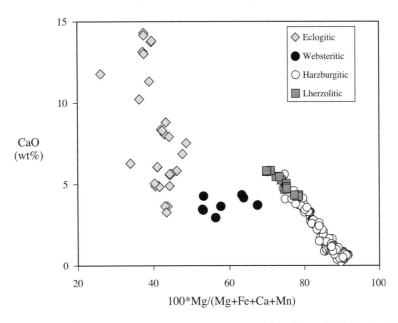

Fig. 3. Plot of pyrope content [100*Mg/(Mg + Fe + Ca + Mn)] versus the CaO concentration of garnet inclusions from De Beers Pool diamonds. This diagram effectively separates compositions belonging to the peridotitic, eclogitic and websteritic parageneses.

(lherzolitic and high-Ca harzburgitic garnets). NiO and Na_2O values are generally below detection limits, although some Ca-rich garnets contain up to 0.06 wt.% Na_2O. Si values for all garnets range from 5.84 to 6.14 (based on 24 oxygens per formula unit) and there is no clear evidence of a majoritic garnet component (Si>6.15; Stachel et al., 2000).

Pyrope–almandine garnets were recovered from 31 diamonds. These inclusions contain less than 1 wt.% Cr_2O_3 and belong to the eclogitic or websteritic paragenesis. The websteritic inclusion paragenesis (garnet + clinopyroxene + orthopyroxene) was first recognised in Orapa diamonds by Gurney et al. (1984b). As none of the garnets from the De Beers Pool samples co-exist with orthopyroxene, an assessment of websteritic affinity must be based on the compositions of individual garnets and any co-existing clinopyroxene inclusions. A distinct compositional gap is evident between the De Beers Pool lherzolitic and websteritic/eclogitic inclusions in terms of Cr_2O_3 content (Fig. 2), and the main difficulty lies in distinguishing websteritic from eclogitic garnets. Aulbach et al. (2002) defined a compositional field for websteritic garnets within the lherzolite field of Sobolev et al. (1973), with limits of <6.5 wt.% CaO and <1 wt.% Cr_2O_3. This field is not exclusive to web-

steritic garnets, however, but also includes garnets from orthopyroxene-free, low-CaO eclogites (e.g. Beard et al., 1996). The current study suggests that the pyrope content of garnets [100*Mg/(Mg + Fe^t + Ca + Mn)] may be a more useful additional discriminant between eclogitic and websteritic compositions. Aulbach et al. (2002) noted that websteritic garnets from their study exhibit distinct pyrope contents of 53–67 mol%.

On the basis of the above Cr_2O_3, CaO and pyrope constraints, garnets from eight diamonds are classed as websteritic inclusions (Fig. 3). These garnets exhibit pyrope contents of 53.0–67.5 mol% and contain 0.15–0.97 wt.% Cr_2O_3 and 2.94–4.35 wt.% CaO. Garnets from two other samples (DP379, DP415) exhibit low pyrope contents (46.2; 47.7 mol%), but elevated Cr_2O_3 levels (0.38; 0.20 wt.%) and are associated with low-Na_2O omphacite inclusions that are typical of a websteritic paragenesis. The latter two samples highlight the difficulties in assigning parageneses based on incomplete inclusion assemblages.

The assignment of eight garnet-bearing diamonds to the websteritic paragenesis leaves 23 diamonds containing garnets of nominally eclogitic affinity. One eclogitic garnet is intergrown with clinopyroxene, whereas a second touching inclusion assemblage comprises garnet, clinopyroxene and a TiO_2-phase, possibly rutile (Fig. 1d). Overall, the eclogitic garnets exhibit a wide compositional range with pyrope contents of 16.1–48.7 mol%, CaO levels of 3.3–21.1, up to 0.68 wt.% TiO_2, and <0.29 wt.% Na_2O. These values are fairly typical of eclogitic garnet inclusions from worldwide sources (cf. Stachel et al., 2000).

3.2. Chromite

Chromites were recovered from 165 De Beers Pool diamonds, 43 of which contained multiple spinel inclusions. Chromites from eight of these diamonds yielded minor differences in composition, indicating non-equilibrium growth conditions. Co-existing monomineralic chromite and olivine inclusions were recovered from two diamonds. Inclusions of olivine were identified in six chromites with orthopyroxene inclusions found in another two chromites (Fig. 1, Table 2). The exposure of silicate inclusions in chromite is fortuitous as they are not visible prior to polishing and may be far more common than current numbers suggest. One chromite grain was also located as a small inclusion in an olivine.

The majority of chromite inclusions define a narrow compositional range, analogous to chromites from localities worldwide (cf. Stachel et al., 2000), and contain 63–67 wt.% Cr_2O_3, <0.3 wt.% TiO_2, 9.2–15.3 wt.% FeO, 12.3–16.6 wt.% MgO, 0.09–0.24 wt.% MnO, 0.05–0.16 wt.% NiO (mean = 410 ppm), 0.08–0.37 wt.% V_2O_3 and up to 0.12 wt.% ZnO. Three chromites exhibit elevated Cr_2O_3 contents up to 68.3 wt.%, one of which is intergrown with orthopyroxene and one with orthopyroxene and garnet (Fig. 4). Such high Cr_2O_3 values are rare, but have been reported from Akwatia diamonds by Stachel and Harris (1997). The co-existing garnet also contains high Cr_2O_3, which suggests a depleted source and elevated Cr_2O_3 contents in the local mantle (Girnis et al., 1999). The lowest Cr_2O_3 content of 50 wt.% was recorded from a small chromite inclusion in a harzburgitic olivine. Apart from the latter example, chromites that co-exist with olivine are characterised by intermediate Cr- [$100*Cr/(Cr+Al)$] and Mg-numbers [$100*Mg/(Mg+Fe)$], and low TiO_2 (<0.2 wt.%) (Fig. 4). In contrast, chromites associated with orthopyroxene contain higher Cr_2O_3 or TiO_2, but lower Mg-numbers. The compositions of the co-existing garnet, olivine and orthopyroxene phases indicate that most

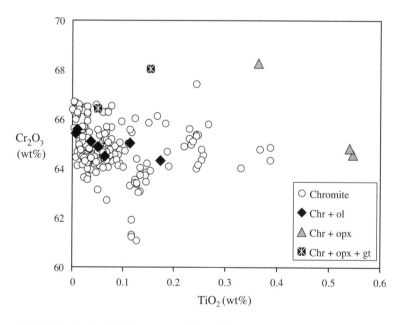

Fig. 4. Plot of TiO_2 versus Cr_2O_3 for chromite inclusions separated from De Beers Pool diamonds. Touching inclusions are shown as filled symbols.

chromites belong to the harzburgitic paragenesis. However, the common association of chromite with high-Mg olivine raises the possibility that many chromites may be derived from a dunitic paragenesis.

3.3. Olivine

Olivine-free diamonds were sought specifically for the present study. Nineteen diamonds, however, contained olivine co-existing with other inclusion mineralogies (Table 2). The forsterite content of all olivine inclusions ranges from 92.5 to 96.5 (Fig. 5). The highest values were recorded from olivine + chromite intergrowths, whereas the lowest forsterite content is associated with a lherzolitic olivine + garnet composite. All other olivine inclusions are considered to belong to the harzburgitic (co-exist with garnet and/ or enstatite), or possibly dunitic (chromite + olivine), parageneses.

NiO concentrations have a similar mean value to harzburgitic olivine inclusions from localities worldwide, but extend to higher values (up to 0.55 wt.%). MnO (0.02–0.08 wt.%) and CaO (<0.02–0.06 wt.%) contents are within the range reported for olivine inclusions previously. The single lherzolitic olivine inclusion contains 0.04 wt.% CaO, which is consistent

with diamond formation conditions of ~ 1100 °C and 4.5–6.5 GPa (Köhler and Brey, 1990).

Cr_2O_3 concentrations for olivines associated with orthopyroxene or garnet vary from 0.03 to 0.19 wt.%, which is typical of olivines from diamonds worldwide (e.g. Stachel and Harris, 1997). One olivine grain with a small chromite inclusion exhibits very low Cr_2O_3 (0.02 wt.%). In contrast, olivine grains included in chromite samples exhibit exceptionally high Cr_2O_3 values up to 1.10 wt.% (average of two inclusions). Although many of these inclusions are quite small, impingement of the electron microprobe beam on the enveloping chromite grains is unlikely, as Al_2O_3 levels are not elevated. High Cr_2O_3 concentrations, up to 0.34 wt.%, have also been reported from Akwatia diamonds by Stachel and Harris (1997), although chromite + olivine intergrowths, analysed by Sobolev et al. (1997) from two Sputnick diamonds, yielded low Cr_2O_3 concentrations (up to 0.05 wt.%). A number of authors have suggested that Cr in olivine may exist in the divalent state in the octahedral site (e.g. Hervig et al., 1980). High Cr^{2+} levels in olivine may be caused by low oxygen fugacity (fO_2) conditions (Ryabchikov et al., 1981) and/or elevated temperatures (Li et al., 1995). Experimental studies by Li et al. (1995) indicate that at typical mantle

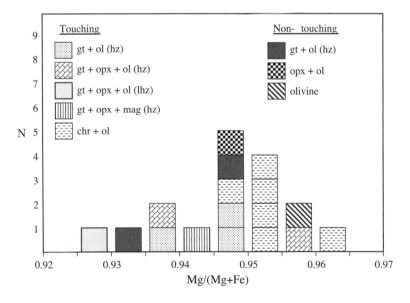

Fig. 5. Histogram showing Mg-numbers [100*Mg/(Mg + Fe)] for olivine inclusions from De Beers Pool diamonds. Olivine inclusions contained in chromite inclusions exhibit the most depleted compositions.

conditions of ~ 1500 °K and fO_2 close to the IW buffer, olivine in equilibrium with spinel hosting 30% $MgCr_2O_4$ should contain less than 0.2 wt.% Cr_2O_3. One possible explanation for the high Cr_2O_3 in the Kimberley Pool olivine inclusions is formation at relatively high mantle temperatures and exceptionally low fO_2. Alternatively, co-existence of olivine with high-Cr_2O_3 chromites in a depleted mantle environment may permit increased uptake of Cr in olivine under 'normal' mantle lithosphere conditions.

3.4. Orthopyroxene

Orthopyroxene inclusions were recovered from 50 diamonds, seven of which also contained non-touching garnet inclusions. Touching inclusion assemblages of orthopyroxene + garnet ± chromite were recovered from another 29 stones (Table 2).

With one exception, Mg-numbers [100*Mg/(Mg + Fe)] for the De Beers Pool orthopyroxene inclusions vary from 93.0 to 96.8 (Fig. 6). This range is similar to orthopyroxene inclusions from occurrences worldwide, although current values extend the field to slightly higher Mg-numbers. The lowest Mg-numbers are associated with lherzolitic garnets (93.0–93.9), with harzburgitic orthopyroxenes having values of 94.5 to 96.8. One orthopyroxene, with a Mg-number of 89.7, also yielded elevated MnO (0.13

wt.%) and CaO (0.62 wt.%) contents. The latter composition is similar to that recorded for websteritic orthopyroxene inclusions from Venetia diamonds (Viljoen et al., 1999; Aulbach et al., 2002). Therefore, this inclusion is considered to be of websteritic association and provides additional evidence for the presence of a minor websteritic paragenesis amongst De Beers Pool diamonds.

TiO_2 contents of most orthopyroxene inclusions are below detection limits (< 0.02 wt.%), although one lherzolitic specimen has an unusually high value of 0.10 wt.% TiO_2. Cr_2O_3 concentrations are typical of worldwide compositions, but extend to slightly higher values (0.11–0.86 wt.%). Although the orthopyroxene inclusions in chromite exhibit elevated Cr_2O_3 levels, these levels are significantly below those recorded by olivine inclusions in chromite. NiO contents vary from 0.08 to 0.18 wt.%, which are within the range of worldwide orthopyroxene inclusions. MnO and CaO concentrations range up to 0.13 and 0.60 wt.%, respectively, with the highest values recorded by lherzolitic inclusions.

3.5. Clinopyroxene

Compositional information was obtained on 40 clinopyroxene inclusions from 34 De Beers Pool

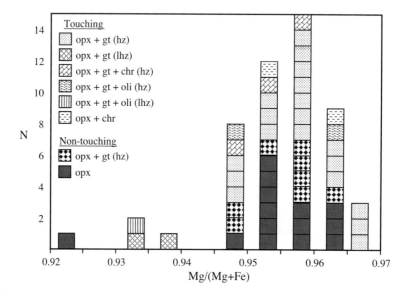

Fig. 6. Histogram of Mg-numbers [Mg/(Mg + Fe)] for orthopyroxene inclusions from De Beers Pool diamonds.

diamonds. Two diamonds contained composite inclusions of eclogitic clinopyroxene + garnet and clinopyroxene + garnet + rutile(?). Seven diamonds (6 eclogitic; 1 lherzolitic) contained non-touching pairs of clinopyroxene and garnet inclusions.

Despite a prolonged search, peridotitic clinopyroxene inclusions were recovered from only three diamonds, one of which is associated with a separate garnet inclusion. The clinopyroxene inclusions have Mg-numbers [$100*Mg/(Mg+Fe)$] of 92.5 to 93.9, Ca-numbers [$100*Ca/Ca+Mg+Fe$)] of 44.4 to 45.3, and plot along the diopside/endiopside boundary of the pyroxene quadrilateral. Due to their elevated Cr_2O_3 contents (0.89–2.69 wt.%), these inclusions are usually termed Cr-diopsides. Other characteristics of the De Beers Pool Cr-diopsides include low Al_2O_3 (1.11–1.44 wt.%), Na_2O (0.42–1.47 wt.%), and TiO_2 (<0.07 wt.%), but high K_2O (up to 0.80 wt.%) concentrations. The latter K_2O value is one of the highest recorded from Cr-diopside inclusions, with only inclusions from Koffiefontein exhibiting higher values (1.6–1.7 wt.%) (Rickard et al., 1989).

Viljoen et al. (1999) and Aulbach et al. (2002) recognised a suite of websteritic clinopyroxene inclusions from the Venetia mine, which were distinguished from lherzolitic and eclogitic compositions by having intermediate Mg-numbers, Na_2O and Cr_2O_3 levels. The absence of associated phases (garnet + orthopyroxene) in most of the De Beers Pool clinopyroxene-bearing diamonds necessitated identification of websteritic clinopyroxene inclusions on the basis of mineral chemistry. However, unlike the garnet inclusions, the De Beers Pool clinopyroxenes exhibit a continuum from distinctly peridotitic to definitely eclogitic compositions (Fig. 7). Four inclusions exhibit intermediate Na, Al and Mg values, with slightly elevated Cr_2O_3 contents (>0.3 wt.%) and have been assigned nominally to the websteritic paragenesis (Fig. 7). These compositions overlap the websteritic field identified by Aulbach et al. (2002) and Deines et al. (1993). Several other De Beers Pool inclusions exhibit Na_2O and Mg values that fall within the websterite field of Aulbach et al. (2002); however, these inclusions have low Cr_2O_3 and are here considered to belong to the eclogitic paragenesis. These arguments again show the difficulties involved in identifying websteritic inclusions solely on the basis of mineral chemistry, but also highlight the close chemical association between the eclogitic and websteritic parageneses.

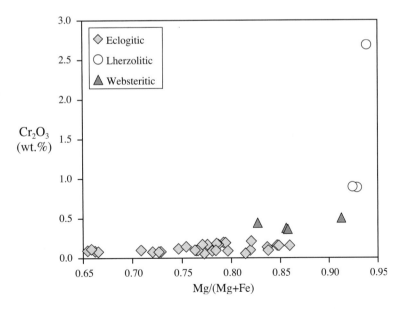

Fig. 7. Plot of Mg-number versus the Cr_2O_3 content for clinopyroxene inclusions from De Beers Pool diamonds. Four inclusions with intermediate compositions are assigned to the websteritic paragenesis.

Clinopyroxenes from the remaining diamonds are considered to be omphacitic in composition with elevated Al_2O_3 (up to 13.8 wt.%) and Na_2O (up to 6.5 wt.%) contents, but relatively low Mg- (65.5–86.0) and Ca-numbers (31.1–49.8). The correlation of Na and Al, plus sufficient Si to fill the tetrahedral site, indicates that most Al is in octahedral coordination and is accommodated largely as the jadeite component. Many of the omphacite inclusions exhibit higher Cr_2O_3 contents than usual (0.06–0.21 wt.%) and there is a correlation between the Cr_2O_3 concentration of the clinopyroxene and that of co-existing garnet inclusions.

3.6. Magnesite

An unusual composite inclusion of Cr-pyrope garnet, orthopyroxene and magnesite was recovered from diamond DP387 (Table 2; Fig. 1e). Electron microprobe analyses of the magnesite grain yielded concentrations of 46.1% MgO, 3.1% FeO, 1.0% CaO, 0.09 wt.% MnO and 0.10 wt.% NiO. Semi-quantitative analyses using a P2 crystal analyser confirmed the presence of carbon and oxygen in the correct proportions for magnesite. The co-existing garnet exhibits a harzburgitic composition (Fig. 2).

3.7. Rutile

Diamond DP379 contained an intergrowth of eclogitic garnet, omphacitic clinopyroxene and a TiO_2 phase (Fig. 1d). The majority of TiO_2 phases reported in the literature are associated with eclogitic assemblages and are usually assumed to be rutile (cf. Meyer, 1987). However, the only structural verification of rutile is recorded by Meyer and Svisero (1975) for an inclusion from a Brazilian diamond. Wang et al. (1996) reported the presence of the anatase polymorph in an unusual magnesite-bearing peridotitic assemblage from a Finsch diamond. Due to its higher density, rutile should be the more stable polymorph under mantle conditions; therefore, Wang et al. (1996) suggested that the Finsch anatase may be a metastable phase that is only preserved due to topotaxic growth on magnesite and a short mantle residence time. Consequently, it seems likely that the DP379 TiO_2 inclusion has a rutile structure.

4. Geothermobarometry

Equilibration temperatures and pressures can be estimated from co-existing De Beers Pool inclusion assemblages using published geothermobarometers. In the case of non-touching inclusions, the application of appropriate thermometers and barometers should constrain the temperature and pressure regime that prevailed at the time of diamond crystallisation—provided that the inclusions represent unchanged equilibrium assemblages. As some diamonds probably grew over protracted time periods under evolving geochemical conditions, non-touching inclusion pairs will not necessarily record equilibrium conditions, as has been highlighted in recent studies by Stachel and Harris (1997) and Stachel et al. (1998). In contrast, chemically homogeneous, touching inclusions should reflect mineral equilibration, although calculated temperatures and pressures may not represent conditions at the time of diamond formation. Because touching inclusions are able to re-equilibrate in response to changing mantle temperatures and pressures, these assemblages could potentially record decreasing temperatures related to a cooling mantle (e.g. Phillips and Harris, 1995), or even increased temperatures relating to interaction with the transporting kimberlite magma (Sobolev et al., 1997). As the De Beers Pool diamonds are believed to be ~ 3.2 Ga old (Richardson et al., 1984), geothermobarometry of both touching and non-touching inclusions affords the opportunity to evaluate the thermal and tectonic history of the diamondiferous mantle over a considerable time period.

Initial insights into this question can be gained from the mineral chemistry of co-existing inclusions. Co-existing mineral pairs from the touching and non-touching inclusion groups exhibit subtle compositional variations that likely relate to differing equilibration conditions. For example, the Mg-numbers of touching garnet–orthopyroxene mineral pairs form a narrow linear compositional array, whereas non-touching inclusions exhibit a broad range in values, with most samples plotting at higher garnet Mg-numbers for any given orthopyroxene Mg-number (Mg_{opx}) (Fig. 8). Although fewer in number, non-touching garnet–olivine pairs also plot at higher garnet Mg-numbers than touching gar-

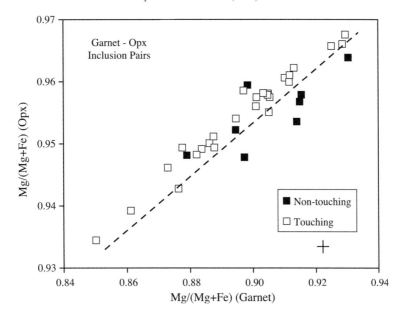

Fig. 8. Diagram showing Mg-numbers for co-existing garnet and orthopyroxene inclusions from De Beers Pool diamonds. The dashed line defines the limits for touching inclusions, which plot at lower Mg_{gt} for any given Mg_{opx} value, consistent with lower equilibration temperatures. A representative error bar is shown at the 95% confidence level.

net–olivine assemblages (at constant Mg_{opx}) (Table 3). This relationship is again observed for garnet–clinopyroxene assemblages (Fig. 9), with non-touching inclusions exhibiting lower ln(Kd) values than touching assemblages at fixed $X_{Ca}(gt)$ (Fig. 9) [Kd= $(Fe^{2+}/Mg)_{gt}/(Fe^{2+}/Mg)_{cpx}$; $X_{Ca}(gt) = Ca/(Ca + Mn +$

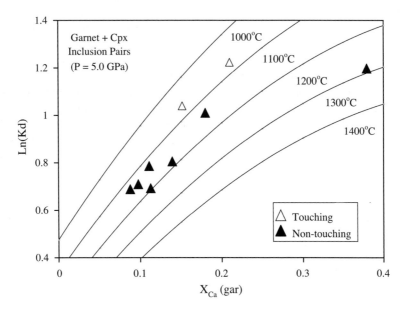

Fig. 9. Plot of ln(Kd) versus $X_{Ca}(gt)$ for touching and non-touching clinpyroxene–garnet inclusion pairs from De Beers Pool diamonds. [Kd=$(Fe^{2+}/Mg)_{gt}/(Fe^{2+}/Mg)_{cpx}$; $X_{Ca}(gt) = Ca/(Ca + Mn + Fe + Mg)$]. Solid lines are isotherms calculated from the Fe–Mg exchange geothermometer of Krogh (1988), assuming a pressure of 5.0 GPa. Error bars are smaller than symbol sizes.

Fe + Mg) in garnet; Krogh, 1988]. These compositional variations suggest that the touching inclusions experienced lower equilibration temperatures than the non-touching inclusions. This argument is supported by the observation that orthopyroxene intergrowths with garnet contain, on average, lower Al_2O_3 contents than monomineralic orthopyroxene inclusions, again indicating lower equilibration temperatures (Girnis et al., 1999). The magnitude of these differences can be estimated from geothermobarometry.

As with most diamond inclusion suites, the De Beers Pool samples encompass a limited array of mineral systems that are suitable for major element thermobarometry. As a result, temperature determinations were restricted to the commonly used Fe–Mg exchange thermometers for garnet–orthopyroxene (Harley, 1984), garnet–olivine (O'Neill and Wood, 1979; O'Neill, 1980) and garnet–clinopyroxene (Krogh, 1988) assemblages. The only applicable geobarometer involves the Al-exchange reaction between garnet and orthopyroxene; in this study, pressures were determined using the formulations of Brey and Köhler (1990) and MacGregor (1974) (Table 4). [Note: unless otherwise stated, listed pressures were calculated using the Brey and Köhler (1990) barometer]. In the case of garnet–orthopyroxene mineral pairs, both temperature and pressure can be calculated iteratively. For garnet–olivine and garnet–clinopyroxene assemblages, an equilibration pressure of 5.0 GPa was assumed, in accordance with previous inclusion and mantle xenolith studies. In terms of uncertainties, Girnis et al. (1999) have suggested that garnet–orthopyroxene thermobarometry is associated with absolute errors of 100 °C and 0.4 GPa; these estimates, however, include uncertainties in thermodynamic parameters as well as conservative electron microprobe errors. In the relative comparison of the De Beers Pool touching and non-touching inclusions, uncertainties in the model parameters have been neglected. Based on the reproducibility of electron microprobe analyses only, typical precision errors for the De Beers Pool garnet–orthopyroxene inclusion pairs are estimated at ± 25 °C and ± 0.2 GPa at 95% confidence levels (Fig. 10). Precision estimates for garnet–olivine and garnet–clinopyroxene temperatures are on the order of ± 20 and ± 7 °C, respectively, at an assumed pressure of 5.0 GPa.

Non-touching garnet–orthopyroxene inclusion pairs ($n = 8$) yielded a wide range of temperatures (1082–1320 °C; average = 1203 °C) and pressures (4.6–7.7 GPa; average = 6.3 GPa), with most values straddling the conductive geothermal gradient for a 40 mW/m² surface heat flow (Pollack and Chapman, 1977) (Table 4; Fig. 10). Calculated temperatures for non-touching garnet–olivine ($n = 3$) and garnet–clinopyroxene ($n = 7$) assemblages range from 1175 to 1276 °C (average = 1225 °C) and 1126 to 1292 °C (average = 1169 °C), respectively, for an assumed pressure of 5.0 GPa (Table 4). If the above inclusions represent equilibrium mantle assemblages, then the results indicate a broad array of diamond crystallisation conditions. Although it is not possible to rule out non-equilibrium for all non-touching inclusion assemblages, equilibrium is indicated for the majority of inclusions, because: (i) all garnet–orthopyroxene results plot within the diamond stability field; (ii) none of the non-touching inclusion pairs exhibit temperatures below those of the touching inclusion group; (iii) several diamonds contained multiple mineral pairs which exhibit analogous compositions and temperatures—for example, two garnet and two clinopyroxene inclusions from DP414, give temperatures of 1143, 1128, 1133, 1118 °C (average = 1131 °C; $P = 5.0$ GPa). Disequilibrium is, however, indicated for one garnet–orthopyroxene inclusion pair (DP497), which produced an unusually high pressure of 7.6 GPa, but a relatively low corresponding temperature of 1140 °C, thus plotting well below the 40 mW/m² reference line (Fig. 10). Two other non-

Notes to Table 4:

Minerals: garnet (gt), orthopyroxene (opx), olivine (ol), clinopyroxene (cpx), chromite (chr), magnesite (mag).

[a] Harley (1984).
[b] Brey and Köhler (1990).
[c] MacGregor (1974).
[d] O'Neill and Wood (1979) and O'Neill (1980).
[e] Assumed pressure = 5.0 GPa.
[f] Krogh (1988).

Table 4
Thermobarometry of touching and non-touching silicate inclusions from De Beers Pool diamonds

Sample no.	Paragenesis	Co-existing assemblages	Garnet–Opx			Garnet–Olivine		Garnet–Cpx	
			T (°C)[a]	P (GPa)[b]	P (GPa)[c]	T (°C)[d]	P (GPa)[e]	T (°C)[f]	P (GPa)[e]
Touching mineral assemblages									
DP023	Harzburgitic	gt + opx	1112	6.0	6.0				
DP049	Harzburgitic	gt + opx	1061	5.0	5.3				
DP068	Harzburgitic	gt + opx	1122	6.2	5.5				
DP070	Harzburgitic	gt + opx	1116	5.4	5.5				
DP168	Harzburgitic	gar + ol				1102	5.0		
DP180	Harzburgitic	gt + opx	1106	5.9	5.6				
DP182	Harzburgitic	gt + opx	1054	4.6	5.0				
DP185	Harzburgitic	gt + opx + ol	1062	5.0	5.1	1046	5.0		
DP206	Harzburgitic	gt + opx	1037	5.0	5.0				
DP209	Harzburgitic	gt + opx + chr	1039	5.5	5.2				
DP211	Harzburgitic	gt + opx	1157	6.8	6.1				
DP218	Lherzolitic	gt + opx + ol	1040	4.7	5.1	1080	5.0		
DP231	Harzburgitic	gt + ol				1080	5.0		
DP325	Harzburgitic	gt + opx	995	4.2	4.6				
DP326	Harzburgitic	gt + opx + ol	1068	5.2	5.4	1062	5.0		
DP348	Eclogitic	gt + cpx						1072	5.0
DP364	Harzburgitic	gt + opx + chr	1092	5.2	5.4				
DP370	Lherzolitic	gt + opx	997	5.1	5.0				
DP379	Eclogitic	gt + cpx						1066	5.0
DP387	Harzburgitic	gt + ol + mag				1046	5.0		
DP398	Harzburgitic	gt + opx	1111	5.5	5.5				
DP400	Harzburgitic	gt + opx	1087	5.5	5.5				
DP405	Harzburgitic	gt + opx	1142	6.2	5.8				
DP406	Harzburgitic	gt + opx	1106	6.1	5.8				
DP422	Harzburgitic	gt + opx	1087	5.7	5.6				
DP448	Lherzolitic	gt + opx	1042	4.6	5.0				
DP449	Harzburgitic	gt + opx	1093	5.5	5.1				
DP457	Harzburgitic	gt + opx + chr	1038	5.7	5.8				
DP473	Harzburgitic	gt + opx	1083	5.6	5.5				
DP477	Harzburgitic	gt + opx	1073	5.6	5.5				
DP488	Harzburgitic	gt + opx	1102	5.5	5.5				
DP503	Harzburgitic	gt + opx	1182	5.9	5.9				
Non-touching mineral assemblages									
DP014	Harzburgitic	gt + opx	1293	6.2	6.3				
DP042	Harzburgitic	gt + ol				1223	5.0		
DP374	Lherzolitic	gt + cpx						1181	5.0
DP392	Harzburgitic	gt + opx	1082	5.7	5.8				
DP414	Eclogitic	gar + cpx						1131	5.0
DP415	Eclogitic	gar + cpx						1137	5.0
DP418	Harzburgitic	gt + opx	1255	6.4	6.3				
DP440	Harzburgitic	gt + opx	1289	6.4	6.6				
DP442	Harzburgitic	gt + ol				1276	5.0		
DP444	Eclogitic	gt + cpx						1131	5.0
DP460	Harzburgitic	gt + ol				1175	5.0		
DP473	Harzburgitic	gt + opx	1174	5.7	5.8				
DP481	Harzburgitic	gt + opx	1320	7.2	6.9				
DP486	Harzburgitic	gt + opx	1082	4.7	5.0				
DP497	Harzburgitic	gt + opx	1131		7.6				
DP539	Websteritic	gt + cpx						1194	5.0
DP542	Eclogitic	gt + cpx						1290	5.0
DP544	Websteritic	gt + cpx						1122	5.0

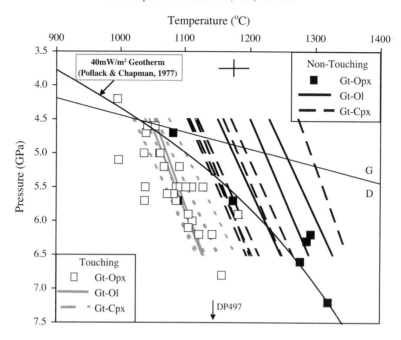

Fig. 10. Thermobarometry results obtained from co-existing inclusion assemblages from De Beers Pool diamonds. Also plotted are the diamond/ graphite univariant reaction line of Kennedy and Kennedy (1976) and the curve for a conductive geothermal gradient based on a surface heat flow of 40 mW/m² (Pollack and Chapman, 1977). Temperatures were calculated from the geothermometers of Harley (1984), O'Neill (1980) and Krogh (1988). Pressure estimates for garnet–orthopyroxene pairs are based on the Al-exchange geobarometer of Brey and Köhler (1990). Pressures for the remaining assemblages were assumed to be 5.0 GPa. The error bar represents 95% confidence level uncertainties for garnet– orthopyroxene temperature/pressure estimates.

touching inclusion pairs (DP392, DP486) plot within the field of touching inclusions (Fig. 10), raising the possibility that one or more of these inclusions may have been intergrown with other phases that were either lost or escaped detection during sample preparation. Alternatively, these samples could reflect a cooler diamond crystallisation event or non-equilibrium conditions.

Compared to the non-touching inclusion group, touching garnet–orthopyroxene assemblages ($n=27$) are characterised by a fairly restricted range in temperature (995–1182 °C; average = 1082 °C), but a similar range in pressure (4.2–6.8 GPa; average = 5.4 GPa) (Table 4, Fig. 10). Temperatures for touching garnet–olivine ($n=6$) and garnet–clinopyroxene ($n=2$) inclusions are estimated at 1046–1102 °C (average = 1069 °C) and 1066–1072 °C (average = 1069 °C), respectively ($P=5.0$ °C). In contrast to the non-touching inclusions, almost all touching inclusions plot below the 40 mW/m² reference curve, suggesting lower equilibration temper-

atures (Fig. 10). Evaluating possible pressure differences between the two inclusion groups is more difficult, due to the small number of non-touching garnet–orthopyroxene inclusion pairs. Based on the current dataset, pressure estimates for non-touching inclusions are skewed towards higher pressures than the touching inclusion group.

5. Discussion

5.1. Diamond formation and mantle geochemistry

Perhaps the most striking feature of the De Beers Pool inclusion suite is the unusually high proportion of inclusion intergrowths. Although touching inclusions have been reported from other localities, they are mostly limited to a handful of examples in each case (e.g. Otter and Gurney, 1989; Stachel et al., 1998; Viljoen et al., 1999). Sobolev et al. (1997) documented a number of inclusion intergrowths from larger Yakutian diamonds; however, these samples constitute a very small proportion of the Yakutian inclusion-bearing diamond population. In contrast, more than 12% of garnet-bearing diamonds from the De Beers Pool collection were found to contain touching inclusions. The reasons for this apparent anomaly are not entirely clear. As the constituents of such intergrowths are often quite small, one possible explanation is the non-recognition of inclusion intergrowths in previous studies. However, the existence of touching inclusion assemblages has been known since the earliest studies of diamonds (e.g. Meyer and Svisero, 1975). Furthermore, a recent dedicated search of inclusion-bearing diamonds from the Venetia mine in South Africa yielded very few intergrowths (Viljoen et al., 1999).

An alternative explanation is that the De Beers Pool inclusion intergrowths represent exsolution products of high temperature/pressure phases. Experimental studies of harzburgitic compositions indicate increased dissolution of pyroxene into garnet to form majorite garnet ($A_4^{2+}Si^{(VI)}Si^{(IV)}O_{12}$) as a function of increasing pressure (Irifune and Ringwood, 1987). Therefore, it is possible that the De Beers Pool garnet–pyroxene intergrowths represent exsolution products of high-pressure majorite garnet. If correct, this would imply an aesthenospheric origin for these samples. The transformation to majorite garnet involves the substitution of silica into octahedral sites, with the amount of excess silica required to fill available tetrahedral sites increasing towards higher pressure (Irifune and Ringwood, 1987). The De Beers Pool intergrowths generally comprise garnet/pyroxene ratios of ~ 80:20 to ~ 60:40. A recombination of garnet and pyroxene components in these proportions produces mixtures with majoritic compositions and elevated Si values of 6.3–6.6 (based on 24 oxygens p.f.u.). These values correspond to aethenospheric pressures on the order of 13–18 GPa, equivalent to depths of ~ 400–500 km (Irifune and Ringwood, 1987).

Majorite garnet inclusions have been reported from a number of localities previously, including the Monastery kimberlite in South Africa (Moore and Gurney, 1989), the Sao Luiz alluvials in Brazil (Wilding, 1990) and the Kankan region of Guinea (Stachel et al., 2000). In all these examples, majorite garnet occurs either as a metastable phase or as complex fine exsolution intergrowths of garnet and pyroxene.

If the De Beers Pool intergrowths represent exsolved majorite garnet, then transport to the lithosphere must have occurred at an early stage to allow formation of discrete garnet and pyroxene phases (Fig. 1). There are, however, a number of arguments against an exsolution origin for the De Beers Pool intergrowths: (i) several diamonds contained garnet–pyroxene intergrowths as well as monomineralic garnet and/or pyroxene inclusions (e.g. DP473, DP414)—in all of these cases, the monomineralic garnets contain no excess silica (i.e. Si < 6.1 atoms p.f.u.) suggesting formation in the mantle lithospheric; (ii) no inclusions of unequivocal deep mantle origin (e.g. majorite garnet, ferropericlase + opx; opx + cpx lamellar intergrowths; Leost et al., 2003) were recovered from any De Beers Pool diamonds; (iii) aside from small compositional variations that can be attributed to minor differences in temperature and pressure equilibration conditions, the touching and non-touching inclusion groups exhibit overlapping major and minor element compositions (Fig. 2). Although an exsolution origin cannot be totally discounted for all De Beers Pool intergrowths, it seems more likely that most were simply encapsulated as polymineralic inclusions in the mantle lithosphere.

The variety of inclusion intergrowths recovered from the De Beers Pool samples permits an evaluation of equilibrium assemblages in the diamondiferous mantle lithosphere. This is more difficult to determine from monomineralic inclusion assemblages, as these may have been incorporated into the diamond at different times in an evolving geochemical system (e.g. Stachel et al., 1998). The most common mineral associations in the De Beers Pool diamonds are garnet + orthopyroxene ± olivine ± chromite and chromite + olivine. The latter association has been noted in several Yakutian diamonds; however, most localities worldwide are characterised by garnet + olivine or garnet + clinopyroxene (usually eclogitic) intergrowths. The common association of garnet + orthopyroxene and chromite + olivine assemblages in the De Beers Pool diamond suite appears to be unique. These associations suggest that the diamond-forming environment beneath the Kimberley area may have been dominated by garnet-bearing harzburgite and Cr-rich dunite compositions.

The De Beers Pool inclusion suite also includes more unusual intergrowths of garnet + olivine + magnesite and eclogitic garnet + clinopyroxene + TiO_2 (rutile). The latter association confirms the eclogitic affinity of rutile (cf. Meyer, 1987). The magnesite-bearing intergrowth indicates that this phase forms part of the peridotitic (harzburgitic) paragenesis. In general, carbonate inclusions are rare in diamond and magnesite inclusions have only been reported from four localities previously, namely Mir (Bulanova and Pavlova, 1987), Bultfontein (Wilding et al., 1994), Finsch (Wang et al., 1996) and Namibia (Leost et al., 2003). Magnesite is believed to be a major reservoir for carbon in the mantle and its rarity in mantle xenoliths is attributed to decarbonation reactions during ascent to surface (cf. Wang et al., 1996). Thus, the occurrence of magnesite, plus dolomite (Leung, 1984; Stachel et al., 1998) and calcite (Meyer and McCallum, 1986; Leung, 1984; Leost et al., 2003), inclusions in diamonds may provide the only unequivocal evidence for mantle carbonates. At the same time, the rarity of carbonate inclusions argues against an abundance of magnesite in the diamondiferous mantle. Possible explanations are that carbonate stabilisation requires more oxidised conditions than are typical for diamond crystallisation, or that ferro-magnesite reacts to form ferro-periclase and diamond (Liu, 2002).

A further characteristic feature of the De Beers Pool diamonds is the high proportion of inclusions of harzburgitic paragenesis, a feature that is also characteristic of other diamondiferous kimberlites in the Kimberley region such as Roberts Victor, Koffiefontein and Finsch (Table 1). In contrast, other diamond sources in southern Africa exhibit much higher proportions of eclogitic (e.g. Orapa and Jwaneng; Deines et al., 1993; Richardson et al., 1999) or lherzolitic (e.g. Premier; Gurney et al., 1985) inclusions. The similarities in inclusion abundances for the Kimberley region also extends to inclusion geochemistry, with samples from these sources characterised by very depleted major element compositions (e.g. high Mg-numbers). In particular, the Kimberley region diamonds (most notably De Beers Pool and Finsch samples) feature unusually high proportions of extremely Ca-poor (< 2 wt.% CaO) garnets, some of which contain very high Cr_2O_3 levels up to 15 wt.% (Fig. 2). The overall similarities in inclusion abundances and mineral chemistry in the Kimberley region suggest a similar diamond growth environment and source, dominated by highly depleted harburgite or dunite substrates.

Aside from the more common peridotitic and eclogitic inclusions, the De Beers Pool diamonds also contain a small websteritic inclusion population. Inclusions belonging to this paragenesis are uncommon, but have been reported previously from localities such as Orapa, Monastery and Venetia (Gurney et al., 1984b; Moore and Gurney, 1989; Deines et al., 1993; Viljoen et al., 1999). It is possible that the paucity of this paragenesis in other localities is a result of the limited numbers of inclusions available for study. The current study highlights the difficulty in identifying this paragenesis solely on the basis of monomineralic inclusions. Websteritic garnets are characterised by compositions that are intermediate between typical lherzolitic and eclogitic garnets, with relatively low Cr_2O_3 (0.2–1.0 wt.%) and CaO (< 6.5 wt.%) levels, and pyrope contents of 52–68 mol%. In the absence of co-existing garnet ± orthopyroxene inclusions, the identification of websteritic clinopyroxene inclusions is even more difficult, as there is a continuum between websteritic, lherzolitic and eclogitic compositions, with the former exhibiting intermediate, but overlapping Mg-numbers, Na_2O and Cr_2O_3 contents. Models proposed for the origin of mantle websterites include subducted oceanic crust, crystallisation from mantle melts and variations on these themes (see review by Aulbach et al., 2002). Trace element analyses of websteritic diamond inclusions from Venetia show similarities to eclogitic inclusions, but with lower MREE, Sr and Zr contents (Aulbach et al., 2002). Both parageneses exhibit Eu anomalies suggesting a history of plagioclase fractionation. Aulbach et al. (2002) postulate that the major and trace element compositions of websteritic diamond inclusions may reflect subducted crust that has undergone melting at depth, but favour a model involving derivation by mixing of slab-derived magmas and mantle peridotite.

5.2. Temperatures, pressures and mantle evolution

Non-touching De Beers Pool inclusions are characterised by a wide range in calculated temperatures (1082–1320 °C); however, these are skewed towards higher values, such that the average temperature (1196 °C) exceeds the worldwide average of ~ 1050 °C (cf. Stachel et al., 1998). This discrepancy could be related, in part, to the number of De Beers Pool garnet–orthopyroxene inclusions that exhibit relatively high pressures (up to 7.2 GPa, excluding DP497). However, even if a pressure of 5.0 GPa is assumed for all inclusions, the average temperature only decreases to 1142 °C. Other localities characterised by elevated equilibration temperatures include River Ranch (~ 1200 °C), Venetia (1152 °C) and Mwadui (~ 1150 °C) (Kopylova et al., 1997; Viljoen et al., 1999; Stachel et al., 1998). A further distinctive feature of the De Beers Pool samples is the similarity in temperatures obtained from garnet–orthopyroxene/olivine and garnet–clinopyroxene inclusion pairs (P = 5.0 GPa); most other localities show temperature discrepancies of up to 200 °C between these inclusion assemblages (cf. Viljoen et al., 1999). Given the lack of pressure constraints on garnet–clinopyroxene assemblages, it is uncertain whether these discrepancies relate to different depths of origin or to localised temperature fluctuations.

The difference in mineral chemistry (e.g. Mg-, Cr-numbers) between the touching and non-touching inclusion groups must either be due to disequilibrium among the non-touching inclusions or to variations in equilibration temperature (and possibly pressure). Arguments in support of the majority of non-touching inclusions representing equilibrium assemblages have been outlined in Section 4. If the temperature estimates listed in Table 4 are taken at face value, then almost 90% of non-touching inclusions are characterised by temperatures above 1125 °C (average = 1200 °C), whereas ~ 90% of touching inclusions exhibit temperatures below 1125 °C (average = 1080 °C). Of course, this comparison may not be fully justified, because a pressure must be assumed for the garnet–olivine and garnet–clinopyroxene temperature calculations (5.0 GPa). Nonetheless, within the limits of the diamond stability field and maximum lithospheric pressures (6.5–7.0 GPa), several non-touching garnet–olivine/clinopyroxene inclusion pairs exhibit higher temperatures than touching inclusions, regardless of assumed pressure (Fig. 10). In other cases, the temperatures for non-touching assemblages only approach those of touching inclusions if the latter inclusions formed at much higher pressures. The latter scenario is at odds with the garnet–orthopyroxene results, which indicate that most touching pairs were encapsulated at similar or lower pressures than touching inclusions.

If the non-touching inclusion temperatures are representative of diamond formation conditions in the sub-continental mantle beneath Kimberley, then one explanation for the observed temperature differences is that the touching inclusions re-equilibrated in response to a cooling mantle. An alternative scenario is that the touching inclusions were encapsulated at lower temperatures than the non-touching assemblages. Although this possibility cannot be totally refuted, it is not supported by results from diamond DP473, which contained both touching and non-touching inclusions with temperatures of 1083 and 1174 °C, respectively. Consequently, the preferred interpretation of the temperature difference between the two inclusion groups involves diamond crystallisation at ~ 3 Ga, under high temperature conditions (>1125 °C), as recorded by the non-touching inclusions. In contrast, the touching inclusions were able to re-equilibrate to changing mantle conditions and likely reflect subsequent cooling of the mantle prior to kimberlite eruption at ~ 85 Ma. Mantle cooling could result from a return to ambient mantle conditions subsequent to localised heating during diamond crystallisation, slow regional cooling over 3.0 Ga and/or physical transfer of diamonds or diamond-bearing horizons to cooler regions of the mantle. The above findings are supported by results from other localities, in which touching inclusions yielded lower equilibration temperatures (e.g. Meyer and Tsai, 1976; Stachel et al., 1998; Viljoen et al., 1999). It is also noteworthy that mantle xenoliths from the Kimberley region, that plot within the diamond stability field, define similar mantle conditions to the Kimberley Pool touching inclusion suite (e.g. Finnerty and Boyd, 1987), again supporting the re-equilibration model. Other inclusion studies, however, showed no temperature differences between touching and non-touching assemblages (Hervig et al., 1980; Sobolev et al., 1997), suggesting limited mantle cooling and/or subsequent reheating by the kimberlite melt.

Although the variations in mineral chemistry for co-existing inclusions are consistent with an interpretation of mantle cooling subsequent to diamond crystallisation, the case for a difference in pressure between the two inclusion groups is less certain. It is clear that pressure estimates for non-touching garnet–opx inclusion pairs are skewed towards higher pressures

($P_{avg} = 6.3$ GPa) than touching inclusions ($P_{avg} = 5.4$ GPa). Unfortunately, the number of non-touching garnet–opx inclusions is relatively small ($n = 8$), and it is possible that the apparent pressure difference is simply due to sampling bias. Another consideration is the sensitivity of the Al-in-orthopyroxene geobarometer to temperature and the lack of clear mineral chemistry signatures to support differences in equilibration pressures. The only sample to contain both touching and non-touching inclusions (DP374) is characterised by a pressure difference of only 0.2 GPa, which is within analytical uncertainties (Table 4). Therefore, there appears to be little independent support for a pressure difference between the two inclusion pairs.

Nonetheless, if the apparent pressure difference is representative of the De Beers Pool inclusion population, one possible cause could be differential movement in the mantle and/or uplift of the Kimberley region. The Kaapvaal craton is considered to have remained relatively stable for the past ~ 3.0 Ga; however, exposed Archaean sediments are greenstones, suggesting possible uplift/erosion levels of ~ 15 km, depending of paleo-geothermal gradients (cf. de Wit et al., 1992). An alternative explanation is that the pressure difference is due to differential thermal expansion between the diamond and encapsulated inclusions. The thermal expansivity of diamond is significantly less than that of silicate inclusions such as garnet and pyroxene (e.g. Zhang, 1998). Therefore, if the diamond does not fracture or undergo viscous flow, the inclusion volume will be dictated by the compressibility of the host diamond (see Zhang, 1998, for theoretical discussion). This attribute of diamond has been used to determine mantle conditions from residual pressures in fluid and mineral inclusions in diamond (e.g. Navon, 1991). At the time of diamond crystallisation, mineral inclusions will be in equilibrium with external thermal conditions and the internal inclusion pressure will be the same as the confining pressure. However, subsequent changes in mantle temperature/pressure could produce internal pressures that differ from external conditions, leading to erroneous pressure estimates for touching inclusions in diamonds. For example, cooling of the mantle may result in internal inclusion pressures (for touching inclusions) that are lower than actual external pressures. Assuming no plastic defor-

mation in diamond, Zhang's (1998) elastic stress model suggests that cooling (or heating) of the mantle would cause a reduction (or increase) in internal inclusion pressure on the order of 0.2 GPa/100 °C. If plastic deformation is a factor (e.g. Weidner, 1998), then the calculated pressure change will be a maximum estimate. The value of 0.2 GPa/°C is also similar to geobarometer analytical uncertainties, suggesting that differential compressibility may only be a minor contributor to the apparent pressure differences. In summary and based on current results, it is not possible to determine whether the apparent pressure difference relates to sampling bias or to other factors that may include differential uplift/erosion, uncertainties in the geobarometer determinations and differential compressibility.

6. Conclusions

(1) The De Beers Pool diamond population contains an abundance of colourless (olivine and orthopyroxene) inclusions and chromite, with lesser amounts of Cr-pyrope garnet, eclogitic inclusions (garnet and clinopyroxene) and rare Cr-diopside. These abundance patterns are broadly similar to other kimberlite localities in the region, such as Finsch, Roberts Victor and Koffiefontein.

(2) Silicate and oxide inclusions extracted from De Beers Pool diamonds are dominated by harzburgitic compositions (>90%), with eclogitic and websteritic parageneses being of lesser importance.

(3) A characteristic feature of the De Beers Pool diamonds is the relatively high proportion of touching inclusion intergrowths. The most common intergrowths are garnet + orthopyroxene ± chromite ± olivine and chromite + olivine. Unusual touching assemblages include garnet + olivine + magnesite, garnet + clinopyroxene + rutile (eclogitic) and garnet + sulphide (peridotitic).

(4) The mineral chemistry of the De Beers Pool garnet, chromite, orthopyroxene, olivine and clinopyroxene inclusions is typical of that from worldwide occurrences, for both touching and non-touching inclusions.

(5) Peridotitic garnet inclusions span a broad compositional range from abundant, depleted harzburgitic inclusions (<2 wt.% CaO), to Ca-saturated lherzolitic compositions. Eclogitic garnets display a wide range in CaO (3–21 wt.%), elevated TiO_2 and low Cr_2O_3. A websteritic garnet population is clearly identified, with intermediate major and minor element contents. No majoritic garnets were identified.

(6) Chromite compositions overlap with worldwide sources and are characterised by high Cr_2O_3 (63–67 wt.%) and low TiO_2.(<0.3 wt.%). Three inclusions contain elevated Cr_2O_3 (up to 68.3 wt.%), suggesting a highly depleted Cr-rich mantle source.

(7) The Mg-numbers of olivine and orthopyroxene inclusions range from 93–97, with a mode of 96, indicating a predominance of harzburgitic compositions.

(8) Only three peridotitic clinopyroxene inclusions were recovered, with compositions typical of Cr-diopsides and elevated K_2O levels (up to 0.8 wt.%). Eclogitic clinopyroxene inclusions, with omphacitic compositions, are more common. Four clinopyroxene inclusions were assigned to the websteritic paragenesis; however, there is considerable overlap between eclogitic and websteritic compositions.

(9) Olivine and pyroxene inclusions intergrown with garnet exhibit higher Mg-numbers than non-touching inclusions, for any given garnet Mg-number or $X_{Ca}(gt)$. These variations correspond to thermobarometry estimates for non-touching inclusion pairs of 1082–1320 °C (average = 1197 °C) and 4.6–7.7 GPa (average = 6.3 GPa). Touching inclusion assemblages are skewed towards lower overall temperatures (995–1182 °C; average = 1079 °C) and pressures (4.2–6.8 GPa; average = 5.4 GPa).

(10) The difference in temperature between the two inclusion groups is attributed to re-equilibration of the touching inclusion assemblages to changing mantle conditions. Thus, it is suggested that non-touching inclusions record the conditions at the time of diamond crystallisation (~ 1200 °C, ~ 3.0 Ga), whereas the touching inclusions re-equilibrated in a cooling mantle (~ 1050 °C) until the time of kimberlite eruption (~ 85 Ma).

(11) Compared to touching garnet–orthopyroxene inclusions, pressure estimates for non-touching assemblages are skewed towards higher pressures. This apparent discrepancy may reflect sampling bias due to limited numbers of non-touching garnet–orthopyroxene inclusions, or alternatively differential uplift in the Kimberley region, inaccuracies in the geobarometry calculations and possibly differential thermal compressibility between the diamond and silicate inclusions. Thermodynamic modelling of the latter option suggests maximum changes in internal inclusion pressures of ~ 0.2 GPa/100 °C.

Acknowledgements

We thank De Beers Consolidated Mines for the supply of samples and permission to publish this manuscript. We are especially indebted to V.G. Anderson, J. Parker and E. van Blerk, of Harry Openheimer House in Kimberley, for their efforts in selecting the inclusion-bearing diamonds. H. Horsch is acknowledged for assistance with the electron microprobe analyses. R. Powell is thanked for the discussions on matters thermodynamic. Statistical calculations were undertaken using K. Ludwig's ISOPLOT software package. The manuscript also benefited from the thought-provoking reviews of R.W. Luth and F.E. Brenker.

References

Allsopp, H.L., Bristow, J.W., Smith, C.B., Brown, R., Gleadow, A.J.W., Kramers, J.D., Garvie, O.G., 1989. A summary of radiometric dating methods applicable to kimberlites and related rocks. In: Ross, J. (Ed.), Kimberlites and Related Rocks, Volume 1. GSA Spec. Publ. No. 14. Blackwell Scientific, Carlton. pp. 343–357.

Aulbach, S., Stachel, T., Viljoen, K.S., Brey, G.P., Harris, J.W., 2002. Eclogitic and websteritic diamond sources beneath the Limpopo Belt—is slab-melting the link? Contrib. Mineral. Petrol. 143. 56–70.

Beard, B.L., Fraracci, K.N., Taylor, L.A., Snyder, G.A., Clayton, R.A., Mayeda, T.K., Sobolev, N.V., 1996. Petrography and geochemistry of eclogites from the Mir kimberlite, Yakutia, Russia. Contrib. Mineral. Petrol. 125. 293–310.

Brey, G., Köhler, T., 1990. Geothermobarometry in four-phase lherzolites: II. New thermobarometers, and practical assessment of existing thermobarometers. J. Petrol. 31. 1353–1378.

Bulanova, G.P., 1995. The formation of diamond. J. Geochem. Explor. 53. 1–23.

Bulanova, G.P., Pavlova, L.P., 1987. Magnesite peridotite mineral association in a diamond from the Mir pipe. Trans. (Dokl.) USSR Acad. Sci., Earth Sci. Sect. 295. 176–179.

Deines, P., Harris, J.W., Gurney, J.J., 1993. Depth-related carbon isotope and nitrogen concentration variability in the mantle below the Orapa kimberlite, Botswana, Africa. Geochim. Cosmochim. Acta 57. 2781–2796.

de Wit, M.J., Roering, C., Hart, R.J., Armstrong, R.A., de Ronde, C.E.J., Green, R.W.E., Tredoux, M., Peberdy, E., Hart, R.A., 1992. Formation of an Archaean continent. Nature 357. 553–562.

Finnerty, A.A., Boyd, F.R., 1987. Thermobarometry for garnet peridotites: basis for the determination of thermal and compositional structure of the upper mantle. In: Nixon, P.H. (Ed.), Mantle Xenoliths. Wiley, Chichester. pp. 381–402.

Girnis, A.V., Stachel, T., Brey, G.P., Harris, J.W., Phillips, D., 1999. Internally consistent geothermobarometers for garnet harzburgites: model refinement and application. Proc. 7th Int. Kimberlite Conf., Cape Town, The J.B. Dawson Volume. Red Roof Design, Cape Town. pp. 247–254.

Griffin, W.L., Gurney, J.J., Ryan, C.G., 1992. Variations in trapping temperatures and trace elements in peridotitic-suite inclusions from African diamonds: evidence for two inclusion suites, and implications for lithosphere stratigraphy. Contrib. Mineral. Petrol. 110. 1–15.

Gurney, J.J., Harris, J.W., Rickard, R.S., 1979. Silicate and oxide inclusions in diamonds from the Finsch mine. In: Boyd, F.R., Meyer, H.O.A. (Eds.), Kimberlites, Diatremes and Diamonds: Their Geology, Petrology and Geochemistry. AGU, Washington. pp. 1–15.

Gurney, J.J., Harris, J.W., Rickard, R.S., 1984a. Minerals associated with diamonds from the Roberts Victor mine. In: Kornprobst, J.B. (Ed.), Kimberlites II: The Mantle and Crust Relationships. Elsevier, Amsterdam. pp. 25–33.

Gurney, J.J., Harris, J.W., Rickard, R.S., 1984b. Silicate and oxide inclusions in diamonds from the Orapa mine, Botswana. In: Kornprobst, J. (Ed.), Kimberlites II: The Mantle and Crust–Mantle Relationships. Elsevier, Amsterdam. pp. 3–10.

Gurney, J.J., Harris, J.W., Rickard, R.S., Moore, R.O., 1985. Inclusions in Premier mine diamonds. Trans. Proc. Geol. Soc. S. Afr. 88. 301–310.

Harley, S.L., 1984. An experimental study of the partitioning of Fe and Mg between garnet and orthopyroxene. Contrib. Mineral. Petrol. 86. 359–373.

Harris, J.W., 1992. Diamond geology. In: Field, J.E. (Ed.), The Properties of Natural and Synthetic Diamond. Academic Press, London. pp. 345–394.

Hervig, R.L., Smith, J.V., Steele, I.M., Gurney, J.J., Meyer, H.O.A., Harris, J.W., 1980. Diamonds: minor elements in silicate inclusions: pressure–temperature implications. J. Geophys. Res. 85. 6919–6940.

Irifune, T., Ringwood, A.E., 1987. Phase transformations in a harzburgite composition to 26 GPa: implications for dynamical behaviour of the subducting slab. Earth Planet. Sci. Lett. 86. 365–376.

James, D.E., Fouch, M.J., VanDecar, J.C., van der Lee, S., Group, K.S., 2001. Tectospheric structure beneath southern Africa. Geophys. Res. Lett. 28. 2485–2488.

Kennedy, C.S., Kennedy, G.C., 1976. The equilibrium boundary between graphite and diamond. J. Geophys. Res. 81. 2467–2470.

Köhler, T., Brey, G.P., 1990. Calcium exchange between olivine and clinopyroxene calibrated as a geothermobarometer for natural peridotites from 2 to 60 kb with applications. Geochim. Cosmochim. Acta 54. 2375–2388.

Kopylova, M.G., Gurney, J.J., Daniels, L.R.M., 1997. Mineral inclusions in diamonds from the River Ranch kimberlite. Contrib. Mineral. Petrol. 129. 366–384.

Krogh, E.J., 1988. The garnet–clinopyroxene Fe–Mg geothermometer—a reinterpretation of existing experimental data. Contrib. Mineral. Petrol. 99. 44–48.

Leost, I., Stachel, T., Brey, G.P., Harris, J.W., Ryabchikov, I.D., 2003. Diamond formation and source carbonation: mineral associations in diamonds from Namibia. Contrib. Mineral. Petrol. 145. 15–24.

Leung, I.S., 1984. The discovery of calcite inclusion in natural diamond, kimberlite and carbonatite. Abstr. Programs-Geol. Soc. Am. 16. 574.

Li, J.-P., O'Neill, H.StC., Seifert, F., 1995. Subsolidus phase relations in the system $MgO-SiO_2-Cr-O$ in equilibrium with metallic Cr, and their significance for the petrochemistry of chromium. J. Petrol. 36. 107–132.

Liu, L., 2002. An alternative interpretation of lower mantle mineral associations in diamonds. Contrib. Mineral. Petrol. 144. 16–21.

MacGregor, I.D., 1974. The system $MgO-Al_2O_3-SiO_2$: solubility of Al_2O_3 in enstatite for spinel and garnet peridotite compositions. Am. Mineral. 59. 110–119.

Meyer, H.O.A., 1987. Inclusions in diamond. In: Nixon, P.H. (Ed.), Mantle Xenoliths. Wiley, Chichester. pp. 501–522.

Meyer, H.O.A., McCallum, M.E., 1986. Mineral inclusions in diamonds from the Sloan kimberlites, Colorado. J. Geol. 94. 600–612.

Meyer, H.O.A., Svisero, D.P., 1975. Mineral inclusions in Brazilian diamonds. Phys. Chem. Earth 9. 785–795.

Meyer, H.O.A., Tsai, H.M., 1976. Mineral inclusions in diamond: temperature and pressure of equilibration. Science 91. 849–851.

Moore, R.O., Gurney, J.J., 1989. Mineral inclusions in diamond from Monastery kimberlite, South Africa. In: Ross, J. (Ed.), Kimberlites and Related Rocks, Volume 1. GSA Spec. Publ. No. 14, vol. 2. Blackwell, Carlton. pp. 1029–1041.

Navon, O., 1991. High internal pressures in diamond fluid inclusions determined by infrared absorption. Nature 353. 746–748.

Nixon, P.H. (Ed.), 1987. Mantle Xenoliths. Wiley, Chichester. 844 pp.

O'Neill, H.StC., 1980. An experimental study of the iron–magnesium partitioning between garnet and olivine and its calibration as a geothermometer: corrections. Contrib. Mineral. Petrol. 72. 337.

O'Neill, H.StC., Wood, B.J., 1979. An experimental study of Fe–Mg partitioning between garnet and olivine and its calibration as a geothermometer. Contrib. Mineral. Petrol. 70. 59–70.

Otter, M.L., Gurney, J.J., 1989. Mineral inclusions in diamonds from the Sloan diatremes, Colorado–Wyoming State Line district, North America. In: Ross, J. (Ed.), Kimberlites and Related Rocks. GSA Spec. Publ. 14, vol. 2. Blackwell, Carlton. pp. 1042–1053.

Phillips, D., Harris, J.W., 1995. Geothermometry of diamond inclusions from the De Beers Pool Mines, Kimberley, South Africa. Extended Abstracts. Proc. 6th Int. Kimberlite Conf., Novosibirsk, (Extd. Abstr.). UIGMM, Novosibirsk. pp. 441–443.

Pollack, H.N., Chapman, D.S., 1977. On the regional variation of heat flow, geotherms and lithospheric thickness. Tectonophysics 38. 279–296.

Prinz, M., Manson, D.V., Hlava, P.F., Keil, K., 1975. Inclusions in diamonds. Garnet lherzolite and eclogite assemblages. Phys. Chem. Earth 9. 797–815.

Richardson, S.H., Gurney, J.J., Erlank, A.J., Harris, J.W., 1984. Origin of diamonds in old enriched mantle. Nature 310. 198–202.

Richardson, S.H., Chinn, I.L., Harris, J.W., 1999. Age and origin of eclogitic diamonds from the Jwaneng kimberlite, Botswana. In: Gurney, J.J., Gurney, J.L., Pascoe, M.D., Richardson, S.H. (Eds.), Proc. 7th Int. Kimberlite Conf., The P.H. Nixon Volume. Red Roof Design, Cape Town 709–713.

Richardson, S.H., Shirey, S.B., Harris, J.W., Carlson, R.W., 2001. Archean subduction recorded by Re–Os isotopes in eclogitic sulfide inclusions in Kimberley diamonds. Earth Planet. Sci. Lett. 191. 257–266.

Rickard, R.S., Harris, J.W., Gurney, J.J., Cardoso, P., 1989. Mineral inclusions in diamonds from Koffiefontein Mine. In: Ross, J. (Ed.), Kimberlites and Related Rocks, Volume 1. GSA Spec. Publ. No. 14, vol. 2. Blackwell Scientific, Carlton. pp. 1054–1062.

Ryabchikov, I.D., Green, D.H., Wall, V., Brey, G.P., 1981. The oxidation state of carbon in the reduced velocity zone. Geochem. Int. 18. 148–158.

Smith, C.B., 1983. Pb, Sr and Nd isotopic evidence for sources of southern African Cretaceous kimberlites. Nature 304. 51–54.

Sobolev, N.V., 1977. Deep-Seated Inclusions in Kimberlites and the Problem of the Composition of the Upper Mantle. (English translation of Russian edition, 1974. Izdatel'stvo Mauka) Am. Geophys. Union, Washington. 279 pp.

Sobolev, N.V., Lavrent'ev, Yu.G., Pokilenko, N.P., Usova, L.V., 1973. Chrome-rich garnets from the kimberlites of Yakutia and their paragenesis. Contrib. Mineral. Petrol. 40. 39–52.

Sobolev, N.V., Kaminsky, F.V., Griffin, W.L., Yefimova, E.S., Win, T.T., Ryan, C.G., Botkunov, A.I., 1997. Mineral inclusions in diamonds from the Sputnik kimberlite pipe, Yakutia. Lithos 39. 135–157.

Stachel, T., Harris, J.W., 1997. Syngenetic inclusions in diamond from the Birim field (Ghana)—a deep peridotitic profile with a history of depletion and re-enrichment. Contrib. Mineral. Petrol. 127. 336–352.

Stachel, T., Harris, J.W., Brey, G.P., 1998. Rare and unusual mineral inclusions in diamonds from Mwadui, Tanzania. Contrib. Mineral. Petrol. 132. 34–47.

Stachel, T., Brey, G.P., Harris, J.W., 2000. Kankan diamonds (Guinea) I: from the lithosphere down to the transition zone. Contrib. Mineral. Petrol. 140. 1–15.

Taylor, L.W., Snyder, G.A., Grozaz, G., Sobolev, N.V., Yefimova, E.S., Sobolev, N.V., 1996. Eclogitic inclusions in diamonds: evidence of complex mantle processes over time. Earth Planet. Sci. Lett. 142. 535–551.

Viljoen, K.S., Smith, C.B., Sharp, Z.D., 1996. Stable and radiogenic isotope study of eclogite xenoliths from the Orapa kimberlite, Botswana. Chem. Geol. 131. 235–255.

Viljoen, K.S., Phillips, D., Harris, J.W., Robinson, D.N., 1999. Mineral inclusions in diamonds from the Venetia kimberlites, Northern Province, South Africa. Proc. 7th Int. Kimberlite Conf., Cape Town, The P.H. Nixon Volume. Red Roof Design, Cape Town. 888–895.

Wang, A., Pasteris, J.D., Meyer, H.O.A., Deleduboi, M.L., 1996. Magnesite-bearing inclusion assemblages in natural diamond. Earth Planet. Sci. Lett. 141. 293–306.

Weidner, D.J., 1998. Rheological studies at high pressure. In: Hemley, R.J. (Ed.), Ultrahigh-Pressure Mineralogy: Physics and Chemistry of the Earth's Deep Interior. Rev. Mineral., vol. 37 493–524.

Wilding, M.C., 1990. A study of diamonds with syngenetic inclusions. Unpubl PhD thesis. University of Edinburgh. 281 pp.

Wilding, M.C., Harte, B., Fallick, A.E., Harris, J.W., 1994. Inclusion chemistry, carbon isotopes and nitrogen distribution in diamonds from the Bultfontein mine, South Africa. In: Meyer, H.O.A., Leonardes, O.H. (Eds.), Diamonds: Characterisation, Genesis and Exploration. CPRM Spec. Publ. 1/B Jan/94. CPRM, Brasilia 116–126.

Zhang, Y., 1998. Mechanical and phase equilibria in inclusion-host systems. Earth Planet. Sci. Lett. 157. 209–222.

Available online at www.sciencedirect.com

SCIENCE DIRECT®

Lithos 77 (2004) 181–192

ELSEVIER

LITHOS

www.elsevier.com/locate/lithos

Syngenetic inclusions of yimengite in diamond from Sese kimberlite (Zimbabwe) — evidence for metasomatic conditions of growth

G.P. Bulanova[a,*], E. Muchemwa[b], D.G. Pearson[c], B.J. Griffin[d], S.P. Kelley[e], S. Klemme[f], C.B. Smith[g]

[a] Department of Earth Sciences, University of Bristol, Wills Memorial Bg., Queens Rd., Bristol BS81RG, UK
[b] Rio Tinto Zimbabwe Ltd., Kenilworth Garden, 1 Kenilworth Rd., Newlands, Harare, Zimbabwe
[c] Department of Geological Sciences, Durham University, Durham DH13LE, UK
[d] University of Western Australia, CMM, 35, Stirling Highway, Crawley, WA 6009, Australia
[e] Department of Earth Sciences, Open University, Milton Keynes, MK7 6AA, UK
[f] Institut für Mineralogie, Universität Heidelberg, Im Neuenheimer Feld 236, Heidelberg 6910, Germany
[g] Rio Tinto, 10 Upper Camden Place, Bath BA15HX, UK

Received 27 June 2003; accepted 4 December 2003
Available online 25 June 2004

Abstract

Syngenetic inclusions of yimengite K (Cr, Ti, Mg, Fe, Al)$_{12}$O$_{19}$, a potassium member of the magnetoplumbite mineral group, have been recorded in an octahedral macrodiamond from the Sese kimberlite (50 km south of Masvingo, Zimbabwe). One yimengite inclusion carries lamellae of chromite suggesting peridotitic diamond paragenesis. The diamond and inclusions were studied in situ in a plate polished parallel to (011). Cathodoluminescence (CL) imaging has shown blue colour and octahedral zonation of the diamond, lack of cracks and the location of five yimengites in different growth zones. Nitrogen (N) contents (at. ppm) in the diamond determined by Fourier transform infrared spectroscopy (FTIR) steadily decrease from 576 (core) to 146 (rim). N aggregation (%1aB) is correspondingly 40% in the core and 30% in the rim. Hydrogen (H) content is high in the core, moderate in the intermediate and very high in the rim zones. Four yimengites were dated using the laser ^{40}Ar/^{39}Ar method. Three inclusions yielded total gas ages that agree with, or are younger than, or within error of, the Sese kimberlite eruption age (538 ± 11 Ma) but may be compromised by gas loss. One inclusion, with the highest tapped interface gas yield, gave a total gas age of 892 ± 21 Ma that is a likely minimum yimengite age. Time–T °C constraints from N aggregation systematics give a range of possible ages from kimberlite eruption date back to Archean and do not resolve the variable results of the ^{40}Ar/^{39}Ar dating. Compared with the published chemistry of yimengite from kimberlites, inclusions from the Sese diamond contain higher Al, Mg, and Sr and have lower concentration of Fe^{3+}. The chondrite-normalised REE pattern of the yimengite shows enrichment in LREE and depletion in HREE, but LREE/HREE fractionations are lower than for lindsleyite–mathiasite series mantle titanates and rather similar to the

* Corresponding author. 10 Upper Camden Place, Bath BA1 5HX, UK. Tel.: +44-1225-481588.
E-mail address: galina_bulanova@hotmail.com (G.P. Bulanova).

0024-4937/$ - see front matter © 2004 Elsevier B.V. All rights reserved.
doi:10.1016/j.lithos.2004.04.002

REE concentrations in kimberlite and lamproite rocks. It is suggested that Sese yimengite formed in the lithospheric mantle from metasomatism of chrome spinel by a fluid rich in Ti, K, Ba and LREE.

Keywords: Yimengite; Diamond; $^{40}Ar/^{39}Ar$; Kimberlite; Mantle; Metasomatism; Zimbabwe

1. Introduction

Yimengite, a very rare Large Ion Lithophile Element (LILE) oxide K (Cr, Ti, Mg, Fe, Al)$_{12}$O$_{19}$, was named due to its first finding in kimberlite dykes in the Yimengshan area of Shandong, China (Dong et al., 1983). Yimengite belongs to the magnetoplumbite mineral group which has the general formula of PbFe$_{12}$O$_{19}$ (Haggerty, 1991). The main feature of the magnetoplumbite crystalline structure is the presence of a large cation site in which K, Ba and other LILE can be placed. Yimengite is the K end member of this group.

The Shandong yimengite is associated with olivine, pyrope, magnesian chromite, phlogopite, ilmenite, chromian diopside, apatite, zircon, moissanite and perovskite (Dong et al., 1983). Subsequently, yimengite together with Mg chromite was found in a heavy mineral concentrate from a kimberlite sill in the Guaniamo district, Venezuela (Nixon and Condliffe, 1989). The mineral has been identified also as a metasomatic alteration product of chromium spinel macrocrysts from the Turkey Well kimberlites, Yilgarn Craton, Australia (Kiviets et al., 1998). Dating of Turkey Well yimengite by $^{40}Ar/^{39}Ar$ laser probe analysis gave a mean age of 2188 ± 11 Ma (Kiviets et al., 1998), consistent with the Rb–Sr data for the time of emplacement of the kimberlites.

A titanate of complex composition associated with chromite, subcalcic G10 garnet and enstatite was identified by Sobolev et al. (1988) in diamond from the Sputnik kimberlite (Yakutia), classified later by Haggerty (1991) as approaching Sr-magnetoplumbite in composition. Another titanate, hawthorneite, Ba(Ti$_3$ Cr$_4$ Fe$_4$ Mg)$_{12}$O$_{19}$, found in a metasomatized vein in a harzburgite xenolith from kimberlite (Kimberley, South Africa), has also a magnetoplumbite type of structure (Peng and Lu, 1985). Grey et al. (1987) and Haggerty et al. (1989) determined that up to 20% of the hawthorneite molecule could be present in yimen-

gite compositions and concluded there is broad isomorphism between the two minerals.

Synthetic yimengite coexisting with a hawthorneite phase was grown under high pressure (4.3 to 5.0 GPa) and high temperature (1150 to 1350 °C) experiments, confirming the mantle origin of these minerals and the possibility of the association of both minerals with diamond (Foley et al., 1994). The origin of yimengite in Upper Mantle rocks has been interpreted as resulting from deep mantle metasomatism generated by K- and Ba-rich fluids (Haggerty, 1987; Nixon and Condliffe, 1989).

Although the mantle association and metasomatic genesis of yimengite has been described before, the origin and significance of this mineral in diamondiferous mantle rocks is still not clear. The occurrence of yimengite in the Sese diamond is the first finding of this mineral in a natural diamond. In addition, we report the first trace element determinations of yimengite. The aim of the present paper is to establish whether yimengite is a syngenetic inclusion in diamond and to document the coupled major and trace element geochemistry of yimengite. This will allow us to further investigate its possible link with diamond origin and general metasomatic processes in the continental lithospheric mantle.

2. Description of the diamond

Sese kimberlite (Smith et al., these proceedings) is located on the Zimbabwe Craton in South Central Zimbabwe (Fig. 1) and its emplacement has been dated at 538 ± 11 Ma (Rb–Sr mica isochron; RioTinto, unpublished). A pale-brown, slightly resorbed, step-layered, octahedral macrodiamond containing five black inclusions (40–150 μm in size) was selected for study during investigation of a small diamond collection from Sese kimberlite pipe K1. Inclusions of yimengite within the diamond were not visually

Fig. 1. Location of Sese kimberlite. CR—Zimbabwe Craton; LMB—Limpopo Mobile Belt.

distinguishable from chromites, having the same black, dark-cherry colour. The shape of the inclusions is typical for those of diamond syngenetic inclusions, which are in general "negative diamond crystals". Some of the yimengites are represented by hexagonal plates and some are composed of numerous inductive (vicinal) small faces, leading to more complicated final morphology. Only local microcracks in the immediate vicinity of some inclusions were observed under most careful optical study (transmitted light and birefringence), but no cracks leading to the diamond surface were present, supporting our interpretation that the inclusions are syngenetic.

The diamond was polished down along parallel dodecahedral crystallographic planes until two yimengite inclusions were exposed on the two opposite sides of the relatively thick diamond plate. The internal structure of the diamond and location of inclusions inside the different growth zones were identified by imaging of cathodoluminescence (CL) and anomalous birefringence (ABR) (Fig. 2a, b). The diamond has blue CL colour and simple octahedral zonation. The core growth zone, located off the geometrical centre of the diamond, contains three yimengite inclusions (1, 2 and 4) one of which was exposed by the polishing (1). Unexposed inclusions 2

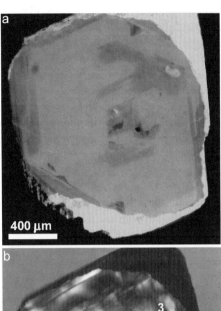

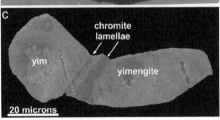

Fig. 2. Inclusions of yimengite within a central polished plate of Sese diamond. (a) CL image of the diamond showing octahedral zonation and dark-blue (dark-grey on the photo) and light-blue (light grey) colours. (b) Birefringence image of diamond displaying the location of inclusions along the different octahedral growth zones. (c) SEM image of the yimengite-5 inclusion intergrown with chromite lamellae. Visual inhomogeneities on the surface of the inclusion are the result of mechanical polishing and surface damage after SIMS analysis.

and 4 from the central zone and inclusion 3 located within the intermediate diamond area are presumed to be yimengites as well. This preliminary identification is consistent with high concentration of ^{39}Ar and hence parent K in these inclusions revealed during ^{40}Ar/^{39}Ar dating (Section 4.4). Yimengite-5 from the rim zone was exposed on the other surface of the diamond plate (Fig. 2b, c).

ABR and CL images of both sides of the diamond plate further demonstrate the absence of cracks, either fresh or rehealed, leading to the surface of the diamond and show the coherent location of inclusions along the octahedral diamond growth zones (Fig. 2b). Such orientation of the inclusions in the mineral-host is a strong indication of their syngenetic nature and synchronous growth with the diamond.

3. Methods of study

Nitrogen (N) content and aggregation state in different diamond growth zones were determined by Fourier transform infrared spectroscopy (FTIR) at the University of Western Australia, using the standard methods and data reduction procedures described by Mendelssohn and Milledge (1995).

The exposed inclusions were analysed by electron microprobe (EMP) for major elements at University of Western Australia. Simple and compound oxides were used as primary standards and cross-checked against a range of compound oxide secondary standards, e.g., barium titanate to check Ba–Ti interference. Correction for internal secondary fluorescence effects was made by the data reduction software using standard "ZAF" procedures, following Reed (1997). The consistency of the yimengite analyses supports a view that external fluorescence effects are minimal as it is highly unlikely otherwise that the effects could be consistent between randomly positioned analysis points within the yimengite. Furthermore, the only mineral surrounding the inclusions is diamond; hence, external effects should be absent.

Trace elements in one of the yimengites were determined by secondary ion mass spectrometry (SIMS) on a Cameca ion microprobe at the University of Edinburgh, using a 10 kV primary beam of ^{16}O-ions. Positive secondary ions were acceler-

ated to 4.5 keV, with an offset of 75 eV to reduce the transmission of molecular ion species. The energy window was set at 25 eV and primary beam currents were set at 2 nA, resulting in a spatial resolution of ca. 15 μm diameter at the sample surface. Cr, as determined by electron microprobe analysis, was used as an internal standard. Ion yields were calibrated on NIST standard SRM 610. Mass 130.5 was used to monitor the background: all analyses reported here have zero background counts. Replicate analyses on in-house garnet and clinopyroxene standards (Blundy and Dalton, 2000; Klemme et al., 2002) demonstrate that calibration on SRM610 yields results for a wide range of trace elements that are accurate to better than ± 20% relative. It should be noted, however, that there is very little information about the ion yield of yimengites or similar minerals, which may be a source of some additional uncertainty. Count times were adjusted so as to yield a minimum of 50 counts for most isotopes per analysis. The SIMS data for major elements such as Fe, Mg, Si and Ca roughly agree with the EMP data; however, standard deviations (errors) are larger during SIMS analysis (more than 20%).

After EMP and SIMS analyses, the diamond plate was cut by a laser into two parts. The largest part of the plate containing four inclusions was used for Ar/Ar dating of yimengites. The smallest piece of the diamond plate with rim yimengite inclusion 5 was preserved for further study. Dating of yimengites was made by the ^{40}Ar/^{39}Ar method using techniques described in Burgess et al. (1992). The largest part of the diamond plate with one exposed and three unexposed yimengites was irradiated. The unexposed inclusions were breached

Table 1
FTIR data for Sese diamond with yimengite

Location	N (at. ppm)	1aB (%)	1aA (%)	H(3107)	Hydrogen level
Core/int.	576	44	56	0.0403	high
Core/int.	311	38	62	0.0248	moderate
Interm.	304	40	60	0.0139	moderate
Interm.	266	41	59	0.0303	moderate
Rim	220	28	72	0.0617	very high
Rim	146	32	68	0.2093	very high

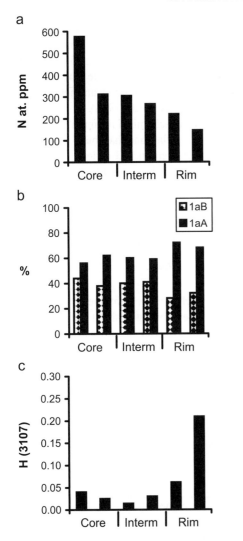

Fig. 3. FTIR data for Sese diamond in core–rim traverse. (a) Nitrogen content and aggregation. (b) Nitrogen aggregation. (c) Relative hydrogen distribution, as reflected by the 3107 peak intensity.

using a UV laser and the inclusion fused using an IR laser.

4. Results

4.1. Nitrogen content and aggregation

The abundance of N in the diamond is moderate to low. Its concentration decreases systematically from

576–311 at. ppm in the core to 304–266 at. ppm in the intermediate area, falling to 220–146 at. ppm in the rim zone (Table 1, Fig. 3a). The degree of N aggregation is moderate, reducing very slightly from the core-intermediate areas (~ 40% of 1aB) to the rim zones (30% of 1aB; Fig. 3b). In general, the observed distribution of N content and aggregation in the diamond is similar to the core–rim trends found in other diamonds worldwide and with theoretical considerations of nitrogen aggregation (Evans and Harris, 1989; Taylor et al., 1995).

Hydrogen (H) content within the diamond is high in the core, moderate in the intermediate zone and very elevated in the rim zone (Fig. 3c). The high content of H in the rim is unusual. The only significant concentration of H in diamond populations worldwide is normally restricted to the core zones. The anomalously high H contents in the rim zone may reflect specific diamond growth conditions, such as the presence of fluids saturated in H-bearing species, e.g., CH_4 and H_2O. It has been speculated that H catalyses carbon vapour deposition diamond growth and may be encapsulated by accumulating carbon atoms under rapid growth conditions (Frenklach et al., 1994). Chinn et al. (2000) have suggested this

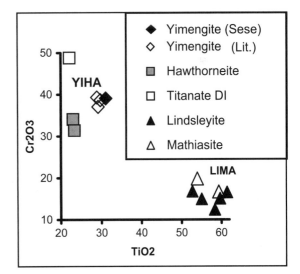

Fig. 4. Cr_2O_3 and TiO_2 variations for alkali mantle titanates of the LIMA (lindsleyite–mathiasite)–YIHA (yimengite–hawthorneite) series. Analysis for LIMA from Haggerty (1991); for YIHA—from Dong et al. (1983), Nixon and Condliffe (1989) and Sobolev et al. (1988).

process may cause rapid nucleation of natural diamonds. However, the Sese diamond shows simple octahedral zonation and layer by layer mechanism of growth, which is evidence of a slow growth rate.

Applying the most reasonable temperature range of 1050–1200 °C for formation of lithospheric mantle diamonds to time–temperature constraints from the N aggregation systematics of Sese diamond gives a wide range of age possibilities from 530 Ma emplacement time to Archean.

4.2. Major element chemistry

Minerals of the yimengite–hawthorneite series have less Ti but Cr-enriched compositions compared with other mantle alkali titanates such as the lindsleyite–mathiasite group (Fig. 4). The compositions of yimengite and hawthorneite are compared in Table 2. Yimengite is distinguished from hawthorneite by its higher concentration of Cr, Ti, Al, Mg and K and by

lower contents of Ba and Sr. Inclusions from the Sese diamond contain very little Fe^{3+} and have higher Al, Mg and Sr concentrations (Fig. 5) when compared with the chemistry of yimengite from kimberlites and Sr-magnetoplumbite found as an inclusion in Sputnik diamond (Sobolev et al., 1988).

The crystallo-chemical formula of the yimengite from the Sese diamond has been calculated on the basis of 19 oxygen and 13 cations, with iron redistribution between Fe^{3+} and Fe^{2+} by the method of Finger (1972). The result is:

$$(K_{0.723}Ba_{0.109}Sr_{0.038}Ca_{0.015})_{0.885}$$
$$(Cr_{4.5}Ti_{3.291}Al_{1.264}Si_{0.156}Mg_{1.849}Fe^{2+}_{0.991}$$
$$Fe^{3+}_{0.063})_{12.114}O_{19}.$$

The stoichiometry obtained approaches ideal proportions, exhibiting a slight deficiency of cations on the A-site (K, Ba, Sr and Ca) and a correspond-

Table 2

Chemistry of yimengite syngenetic inclusion with chromite lamellae in diamond from Sese kimberlite compared with published data

N	V8.K1-107-9				1	2	3	4
Locality	Sese, Zimbabwe				China	Venezuela	Yakutia	Bultfontein, SA
Mineral	Yimengite 5[a]	Standard deviation	Chromite lamellae	Yimengite + chromite	Yimengite	Yimengite	Sr-titanate[b]	Hawthorneite
Position	DI	± 2sigma	In yim. DI	Bulk an.[c]	Kimb. dyke	Kimberlite	DI	Harzburgite xen.
SiO$_2$	1.06	0.02	0.21	0.98	0.55	0.55	0.78	0.06
TiO$_2$	29.72	0.17	3.12	27.11	29.15	31.05	22.1	23.27
Al$_2$O$_3$	7.28	0.06	11.43	7.71	1.61	3.87	3.29	0.22
Cr$_2$O$_3$	38.66	0.44	55.52	40.43	37.06	39.08	48.8	31.41
Fe$_2$O$_3$				0.00	0	1.20		
FeO	8.55	0.16	12.25	8.94	18.36	11.21	10	24.8
MnO	0.00		0.21	0.02	0	tr.	0.1	0.24
MgO	8.42	0.07	15.42	9.14	7.89	5.88	5.45	3.5
CaO	0.10	0.01	0.01	0.09	0	tr.		
K$_2$O	3.85	0.07		3.47	3.75	4.47	2.46	0.45
BaO	1.90	0.05		1.71	1.61	1.85	0.8	13.61
SrO	0.44	0.02		0.40	0	0.31	4	
Nb$_2$O$_5$	n.a.			0.00			0.3	
V$_2$O$_3$	n.a.			0.00			0.9	
NiO	n.a.			0.00			0.07	
Total	99.99		98.16	100.00	99.98	99.47	99.05	97.56

Analysis: 1—Dong et al. (1983), 2—Nixon and Condliffe (1989); 3—Sobolev et al. (1988); 4—Haggerty (1987). DI—diamond inclusion; bulk an.—bulk analysis; xen—xenolith; n.a.—not analysed.

[a] Average of four EMP analyses for yimengite-5.

[b] In association with G10 garnet, chromite and enstatite.

[c] 90% yimengite, 10% chromite.

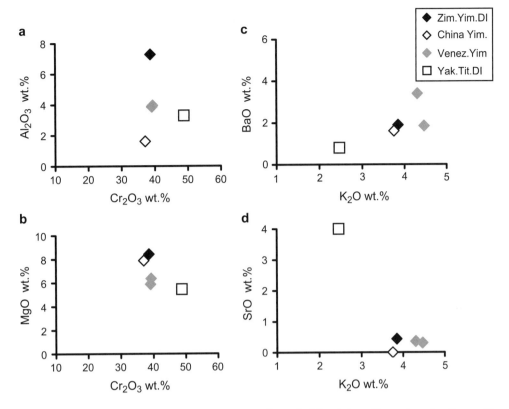

Fig. 5. Chemistry of major elements in yimengite inclusion from Sese diamond. Abbreviations: Yim.—yimengite; Venez.—Venezuela; Yak. Tit.—Yakutian titanate; DI—diamond inclusion. Source references are given in Table 2.

ingly minor excess on the B-site (Cr, Ti, Al, Mg, Fe and Si). Small amounts of Si and Ca are present in the Sese yimengite analysis, but the presence of these elements as ultrasmall inclusions cannot be excluded. The high BaO content of the Sese yimengite is similar to that found in yimengite from Venezuela (Nixon and Condliffe, 1989) and shows that the hawthorneite molecule is present in solid solution. These observations contrast with experimental results that find a high pressure solvus between pure yimengite (K-rich) and hawthorneite (Ba-rich) members of the solid solution series (Foley et al., 1994).

SEM imaging of the rim inclusion-5 revealed compositional inhomogeneity. In the central part of the yimengite inclusion, two oriented lamellae of chromium spinel were found indicating a peridotitic diamond paragenesis (Fig. 2c). The boundary between spinel and yimengite is fairly sharp; no features of reaction between the two phases were observed. In

obtaining quantitative EMP analysis of the 5-μm-wide spinel lamellae, secondary fluorescence effects may be a possible cause of their elevated Ti contents, e.g., of Ti Kα by Cr Kα on TiKα X-rays, due to close proximity of the yimengite host. However, the interaction volume for the probe spot used should not be more than $2–3$ μm^3, so we suggest that most of analysed volume was inside the spinel lamellae and hence the spinel Ti contents are more likely to be real. The analysis of a spinel lamella has shown that it contains only 55 wt.% Cr_2O_3 and very high concentration of TiO_2 (3 wt.%, Table 2). There is negligible Fe_2O_3 content, assuming normal spinel stoichiometry. These features are similar to the composition of some chrome spinel inclusions in other Sese diamonds (RioTinto, unpublished data), but are unusual for chromite diamond inclusions worldwide (Fipke et al., 1995). Such low Cr and high Ti chemistry may be considered as a transitional type between typical high chromium spinel and yimengite.

Table 3
SIMS analysis of yimengite from Sese diamond

Elements	Li	Be	B	F	Cs	Rb	Ba	Th	U	K	Nb	
Concentration (ppm)	0.37	0.15	1029	185	3.7	124	22,101	7	4	33,606	1587	
Standard deviation	0.2	0.02	860	14	3	4	240	0.2	5.4	1168	24	

Elements	Ta	La	Ce	Sr	Nd	Pr	P	Hf	Zr	Sm	Gd	
Concentration (ppm)	81	571	537	2543	67	32	2.1	26	787	4.3	9.6	
Standard deviation	8	7	13	89	5	2	1.4	2	12	2	1.7	

Elements	Dy	Ho	Er	Y	Tm	Yb	Lu	Ga	Ge	Sc	V	Co
Concentration (ppm)	2.3	1.6	0.35	0.52	0.27	1.4	0.37	119	329	34	1830	233
Standard deviation	0.7	0.3	0.29	0.37	0.04	3.6	0.28	5	15	2	63	29

Standard deviations of two analyses (for light elements) and three analyses (for heavy elements) are given.

4.3. Trace element chemistry

Yimengite inclusion 5 is enriched in large ion lithophile elements, e.g., K, Ba, Sr and Rb and contains elevated levels of Nb, Sc, V, Co, Zr and REE (Table 3). The chondrite-normalised REE pattern obtained for the yimengite has a distinctive shape, showing enrichment in LREE and depletion in HREE, similar to other published data for mantle titanates (Fig. 6). A clear difference between the yimengite included in the diamond and other mantle titanates (e.g., Jones and Ekambaram, 1985) is the lower concentration of REE in general in the Sese yimengite. In detail, the REE pattern of the yimengite from the diamond is very similar to those of kimberlite and lamproite rocks themselves (Fig. 6). The identical concentration and distribution of LILE and trace elements in kimberlites, MARIDS and mantle titanates led Jones (1989) to conclude that they have close genetic relationships, which seems to apply to Sese yimengite diamond inclusions as well.

4.4. $^{40}Ar/^{39}Ar$ dating of yimengite

Following irradiation, yimengite grains were dated by the $^{40}Ar/^{39}Ar$ laser probe method. Breaching of the buried inclusions using the UV laser to release gas held at the inclusion–diamond interface resulted in variable yields of radiogenic Ar. Inclusions 2 and 4 yielded between 2.8 and 8.8×10^{-12} c.c. of ^{40}Ar (Table 4). In contrast, inclusion 3 yielded considerably more ^{40}Ar $(71 \times 10^{-12}$ c.c.). Ages calculated from this gas trapped at the inclusion–diamond interface have large errors and are not meaningful until combined with the lattice-held Ar, released via heating with the IR laser. These combined ages are labelled "Total" ages in Table 4. As expected, there was no significant trapped gas present surrounding inclusion 1, which had been exposed by the polishing.

Interpretation of the ^{40}Ar–^{39}Ar age systematics may be complicated by the possibility of artefacts being introduced into the K–Ar system during the preparation of the polished diamond plate. Polishing can cause the plate to become very hot and cracking

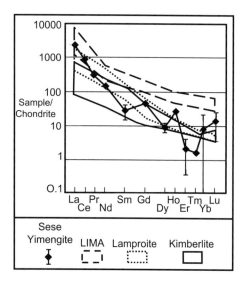

Fig. 6. REE concentrations in yimengite from Sese diamond compared with compositional fields of other mantle titanates (lindsleyite–mathiasite series) (Haggerty, 1987; Jones and Ekambaram, 1985), kimberlites and lamproites (Jaques et al., 1984).

Table 4
^{40}Ar*/^{39}Ar data for yimengite inclusions from Sese diamond

	^{40}Ar	±	^{39}Ar	±	^{38}Ar	±	^{37}Ar	±	^{36}Ar	±	^{40}Ar/^{39}Ar	±	Age (Ma)	±
Yim. 1														
UV total	0.244	0.389	0.022	0.008	0.034	0.020	0.106	0.088	0.038	0.020	− 503.52	− 331.4	no age	
IR total	139.595	0.414	4.034	0.028	0.082	0.028	0.000	0.011	0.150	0.013	23.637	0.9914	472.0	17.5
Total	139.839	0.568	4.056	0.029	0.116	0.034	0.093	0.089	0.187	0.024			472.0	17.5
Gas in UV	0.2%		0.5%											
Yim. 3														
UV total	71.084	0.503	0.119	0.010	− 0.031	0.025	0.033	0.113	− 0.022	0.026	654.248	85.688	4018.6	211.0
IR total	274.659	0.460	5.631	0.026	0.114	0.026	0.000	0.011	0.209	0.012	37.8137	0.6763	705.3	10.8
Total	345.742	0.682	5.749	0.028	0.083	0.036	0.014	0.113	0.187	0.029	50.5473	1.5079	891.8	21.3
Gas in UV	20.6%		2.1%											
Yim. 2														
UV total	2.787	0.316	0.080	0.008	− 0.023	0.016	0.000	0.071	− 0.001	0.016	39.8808	58.568	736.9	888.4
IR total	65.598	0.069	1.832	0.004	0.037	0.004	0.000	0.002	0.068	0.002	24.8508	0.3313	493.2	6.1
Total	68.385	0.323	1.912	0.009	0.014	0.017	0.000	0.071	0.067	0.016	25.4818	2.4764	504.1	42.8
Gas in UV	4.1%		4.2%											
Yim. 4														
UV total	8.781	0.319	0.189	0.008	0.002	0.016	0.000	0.070	0.007	0.016	35.4273	25.652	668.0	404.3
IR total	32.506	0.023	0.866	0.002	0.018	0.002	0.000	0.001	0.032	0.001	26.5592	0.3309	522.6	6.0
Total	41.287	0.319	1.056	0.008	0.019	0.016	0.000	0.070	0.039	0.016	28.1507	4.6091	549.6	77.6
Gas in UV	21.3%		17.9%											

J value = 0.01265 ± 0.00006. Amounts are × 10^{-12} cc STP.

can be induced. This raises the possibility that daughter Ar gas accumulated at the diamond–inclusion interface could have escaped, making the total gas ages of some buried inclusions younger than their real age. In addition, heating and partial release of lattice-held ^{40}Ar during polishing is a likely explanation for the younger than kimberlite age for the exposed inclusion 1. Microcracking around the inclusion cavity could also have taken place during eruption of the diamond to the Earth's surface, although, as stated earlier, we have seen no evidence that these cracks have propagated to the surface of the diamond. An indication that gas has escaped from some of the inclusions is given by the low Ar trapped at the inclusion–diamond interface for two of them. This situation is analogous to that observed by Kelley et al. (1997) in which metamorphic biotites included in garnet had largely lost radiogenic Ar trapped at the biotite–garnet interface following a heating/deformation event.

The alternatives of partial gas loss during eruption of the kimberlite or by heating during diamond polishing have likely influenced three of the four inclusions analysed. The gaseous nature of Ar renders it extremely mobile in situations where little or no other mass exchange occurs, so that syn-eruption gas loss can be possible, with little or no infiltration of external fluid/melt phases into the diamond. We recommend further investigation of the possibilities of Ar loss during polishing of diamond plates before routine use is made of such plates for Ar–Ar dating of inclusions.

Inclusion 3 shows different systematics to the other inclusions analysed. Over 25% of the total radiogenic Ar for yimengite-3 was released when the inclusion interface was breached with the UV laser. The large volume of ^{40}Ar released indicates significant retention of interface Ar compared to the other inclusions. Despite this, it is dangerous to assume that all the Ar has been retained and therefore the total gas age must be viewed as a minimum. The total gas age for inclusion 3 (892 ± 21 Ma) is significantly older than the host kimberlite (538 ± 11 Ma) and is a clear indication that some of the inclusions, and by inference the diamond, formed well before the emplacement age of the Sese kimberlite.

Sulphide inclusions of peridotitic paragenesis from Sese diamonds have given Re–Os model ages as old as 3.4 Ga (D.G. Pearson, unpublished data). However, Spetsius et al. (2002) have shown that young minerals such as zircon can encapsulate ancient mantle sulphides and so in the absence of isochron systematics, Re–Os model ages on inclusions have to be viewed with uncertainty. Hence, these ages do not offer strong constraints on the age of the yimengite-bearing diamond.

5. Discussion: origin of the yimengite

5.1. Constraints from yimengite chemistry

The origin of all previously found yimengites has been attributed to the metasomatic alteration of Upper Mantle chrome spinel. The identification of two oriented lamellae of chromite in yimengite inclusion-5 (Fig. 2c) provides some further evidence about the relationship of the two minerals. Two possibilities may be considered: (I) the spinel represents a first (early) stage of chromite alteration into yimengite; (II) spinel lamellae were exsolved from a solid solution of chromite and yimengite. The first scenario is the currently accepted model for yimengite formation during metasomatic replacement of Cr-spinel. The low chromium–titanium spinel lamellae in yimengite (Table 2) is similar in chemistry to some chrome spinel inclusions in other Sese diamonds (RioTinto, unpublished data), but it is not common for chromite diamond inclusions worldwide (Fipke et al., 1995). Similar low Cr and high Ti spinel was found in association with yimengite from Venezuela, and was interpreted as a transitional type between typical high chromium spinel and yimengite (Nixon and Condliffe, 1989). Therefore, the distinctive chemistry of the Ti-bearing chromite lamellae could also indicate the first stage of chemical change in chromite to yimengite during the mantle metasomatism by a fluid enriched in Ti, K, Ba, Sr and LREE.

To explore the second possibility, the primary bulk analysis of the whole inclusion was calculated from the visually estimated proportion of the exsolved phase relative to the host of 10% chromite to 90% yimengite (Table 2). The composition and stoichiometry obtained for the primary mineral still matches that

of yimengite, with a larger deficiency of cations on the A-site and excess on the B-site:

$$(K_{0.647}Ba_{0.098}Sr_{0.034}Ca_{0.014})_{0.793}$$
$$(Cr_{4.67}Ti_{2.98}Al_{1.328}Si_{0.143}Mg_{1.991}Fe^{2+}_{0.69}$$
$$Fe^{3+}_{0.403})_{12.20}O_{19}.$$

So it is possible to interpret the chromite lamellae in the yimengite as exsolution during decompression. This seems in agreement with the lamellae geometry and the sharp contacts with yimengite.

It can be concluded that Sese yimengite likely formed by metasomatism of the Upper Mantle chromite, and that subsequently chromite lamellae were re-exsolved during decompression. Direct evidence of metasomatic events affecting harzburgite mantle xenoliths similar to those leading to the yimengite formation has been identified solely for the LILE yimengite analogue, hawthorneite (Haggerty et al., 1989). Yimengite has only been found as an alteration product of chromite from kimberlite heavy mineral concentrate. However, the concentrate is considered to consist largely of minerals derived from disintegrated Upper Mantle peridotites together with megacrysts and eclogites. The chemistry of other peridotitic diamond inclusions from Sese kimberlites in general provides evidence of their much depleted dunite–harzburgite paragenesis (RioTinto, unpublished). This restite Upper Mantle assemblage, consisting mainly of olivine and chromite, could have been affected by fluids enriched in incompatible elements such as K, Ba, Sr, Ti and REE to form the yimengite. It is important to note that diamond growth continued during these metasomatic events.

The most obvious example of metasomatic fluids coexisting with diamonds would be the carbonatitic component of melt/fluid found as microinclusions in the coats of coated diamonds (Navon et al., 1988). Such melts are enriched in Mg, Ca, K, Fe, Ti, Ba, and P and are uniform in composition within coated diamonds from many different locations. This homogeneity is believed to reflect an origin from a large convecting mantle reservoir such as the asthenospheric Upper Mantle (Navon et al., 2003). The Sese yimengite diamond inclusion occurrence may be part of a wider metasomatic regional event. Kopylova et al. (1997) has described from River Ranch kimberlite the occurrence of other REE titanites as diamond inclusions (Cr-chevkinite, Sr–K–Cr–loparite).

5.2. Age constraints

Notional time constraints from N aggregation systematics for the Sese diamond give a range of possible formation ages from 530 Ma of the kimberlite emplacement to the Archean. The extreme temperature sensitivity of nitrogen aggregation systematics therefore does not help to resolve the interpretation of the radiometric ages for the yimengite-bearing diamond and cannot offer strong independent constraints.

The safest interpretation of the Ar–Ar age constraints is to regard the 892 Ma age for inclusion 3 as a minimum age for the yimengite, and as such, the diamond-host. Whether this age equates to a realistic age or not, it is clearly older than the host kimberlite and so is indicative of an early metasomatic event that formed diamonds within the lithosphere prior to Sese kimberlite eruption at 538 ± 11 Ma. This situation is similar to the early MARID-producing mantle metasomatic events in the Kaapvaal Craton at 120–150 Ma recorded by zircons at Wesselton and Kampfersdam (Konszett et al., 1995; Hamilton et al., 1998) prior to eruption of these Group I kimberlites at 84 and 87 Ma, respectively. Thus, formation of the Sese yimengite-diamond association may well be part of the spectrum of protokimberlitic metasomatic events affecting the lithospheric mantle, the products of which become sampled by later kimberlite magmatism.

6. Conclusions

Syngenetic inclusions of yimengite identified in situ in peridotitic diamond from the Sese kimberlite demonstrate for the first time the stable genetic association of these minerals. The textural evidence of crystallisation of yimengite throughout the whole period of growth of the diamond-host indicates their coexistence within metasomatized continental Upper Mantle. The yimengite inclusions are characterised by the highest contents of Al, Mg and Sr and lower concentrations of Fe^{3+} compared with all previously published compositions for this mineral. The levels of REE-elements in the yimengite inclusions from the Sese diamond are similar to those in kimberlite and lamproite rocks. The N content of the host diamond varies from 576 at ppm in the core to 146 at. ppm in the rim, with a corresponding slight decrease in the degree of N aggregation from 40% to 30% 1aB. The diamond has unusual, very high hydrogen concentration in the rim zone.

$^{40}Ar/^{39}Ar$ dating of the Sese yimengite inclusions from the single diamond are probably complicated by the high-temperature diamond polishing processes used to prepare the diamond plate in addition to possible syn-eruption gas loss. The best age constraint for one inclusion that retained a significant amount of radiogenic Ar at the inclusion–diamond interface is 892 ± 21 Ma, which could be considered as a minimum age. Although technically feasible, $^{40}Ar/^{39}Ar$ dating of high-K minerals such as yimengite that are included within polished diamond plates is subject to significant problems and uncertainties.

Our study in general agrees with other previous models of yimengite genesis in the mantle whereby yimengite forms within the continental Upper Mantle from a fluid enriched in Ti, K, Ba, Sr and REE during metasomatic replacement of the chrome spinel. The clear association of yimengite and diamond, shown for the first time here, suggests that this fluid may be related to that sampled by fibrous diamonds. Significantly, diamond growth continued during this mantle metasomatic event.

Acknowledgements

Rio Tinto Zimbabwe and Rio Tinto Mining and Exploration are thanked for kind permission to publish the paper. R. Trautmann is thanked for taking FTIR measurements of the diamond and S. Kearns for producing the SEM image of the yimengite. The Guest Editor, T. Stachel, and reviewers M. Hutchinson and P. Roedder, are thanked for their useful suggestions that have helped to improve the paper.

References

Blundy, J.D., Dalton, J.A., 2000. Experimental comparison of trace element partitioning between clinopyroxene and melt in carbonate and silicate systems, and implications for mantle metasomatism. Contrib. Mineral. Petrol. 139, 356–371.

Burgess, R., Turner, G., Harris, J.W., 1992. $^{40}AR-^{39}AR$ laser probe studies of clinopyroxene inclusions in eclogitic diamonds. Geochim. Cosmochim. Acta 56, 389–402.

Chinn, I., Kyser, K., Viljoen, F., 2000. Microdiamonds from the Lake (Akluilak) Dyke, Northwest Territories, Canada. Goldschmidt 2000. J. Conf. Abstr. 5 (2), 307.

Dong, Z., Zhou, J., Lu, Q., Peng, Z., 1983. Yimengite, $K(Cr,Ti,Fe,Mg)_{12}O_{19}$, a new mineral from China. Kexue Tongbao, Bull. Sci. 15, 932–936 (in Chinese).

Evans, T., Harris, J.W., 1989. Nitrogen aggregation, inclusion equilibration temperatures and the age of diamonds. In: Ross, J. (Ed.), Kimberlites and Related Rocks, Vol. 2, Their Mantle/Crust Setting, Diamonds and Diamond Exploration. Proc. Fourth International Kimberlite Conference, Perth, Geol. Soc. Australia Spec. Pub., vol. 14, pp. 1001–1006.

Finger, L.W., 1972. The uncertainty in the calculated ferric iron content of a microprobe. Carnegie Inst. Year Book 71, 600–603.

Fipke, C.E., Gurney, J.J., Moore, R.O., 1995. Diamond exploration techniques emphasising indicator mineral geochemistry and Canadian examples. Bull.-Geol. Surv. Can. 123, 1–86.

Foley, St., Hofer, H., Brey, G., 1994. High pressure synthesis of priderite and members of the lindsleyite–mathiasite and hawthorneite–yimengite series. Contrib. Mineral. Petrol. 117, 164–174.

Frenklach, M., Skokov, S., Weiner, B., 1994. An atomic model for stepped diamond growth. Nature 372, 535–537.

Grey, I.E., Madsen, I.C., Haggerty, S.E., 1987. Structure of a new upper mantle magnetoplumbite-type phase, $Ba(Ti_3Cr_4Fe_4Mg)O_{19}$. Am. Mineral. 72, 633–666.

Haggerty, S.E., 1987. Metasomatic mineral titanates in upper mantle xenoliths. In: Nixon, P.H. (Ed.), Mantle Xenoliths. Wiley, Chichester, pp. 671–690.

Haggerty, S.E., 1991. Oxide mineralogy of the upper mantle. In: Lindsley, D.H. (Ed.), Oxide Minerals: Petrologic and Magnetic Significance. Reviews in Mineralogy, vol. 25, pp. 355–416.

Haggerty, S.E., Grey, I.E., Madsen, I.C., Criddle, A.J., Stanley, C.J., Erlank, A.J., 1989. Hawthorneite, $Ba(Ti_3Cr_4Fe_4Mg)O_{19}$: a new metasomatic magnetoplumbite-type mineral from the upper mantle. The American Mineralogist 74 (5-6), 668–675.

Hamilton, M.A., Pearson, D.G., Stern, R.A., Boyd, F.R., 1998. Constraints on MARID petrogenesis: SHRIMP II U–Pb evidence for pre-eruption metasomatism at Kampfersdam. Ext. Abstr. 7th International Kimberlite Conference, pp. 296–298.

Jaques, A.L., Lewis, J.D., Smith, C.B., Gregory, G.P., Ferguson, J., Chappell, B.W, McCulloch, M.T., 1984. The diamond-bearing ultrapotassic (lamproitic) rocks of the West Kimberley region, Western Australia. In: Kornprobst, J. (Ed.), Kimberlites I: Kimberlites & Related Rocks, Proc. 3rd Int. Kimb. Conf., Developments in Petrology IIA. Elsevier, Amsterdam, pp. 225–254.

Jones, A.P., 1989. Upper mantle enrichment by kimberlitic or carbonatitic magmatism. In: Bell, K. (Ed.), Carbonatites; Genesis and Evolution. Allen & Unwin, Ottawa, p. 448.

Jones, A.P., Ekambaram, V., 1985. New INAA analysis of mantle derived titanate mineral of crichtonite series, with particular reference to the rare earth elements. Am. Mineral. 70, 414–418.

Kelley, S.P., Bartlett, J.M., Harris, N.B.W., 1997. Pre-metamorphic ages from biotite inclusions in garnet. Geochim. Cosmochim. Acta 61, 3873–3878.

Kiviets, G.B., Phillips, D., Shee, S.R., Vercoe, S.C., Barton, E.S., Smith, C.B., Fourie, L.F., 1998. $^{40}Ar/^{39}Ar$ dating of yimengite from Turkey Well kimberlite, Australia: the oldest and the rarest. Ext. Abstr. 7th International Kimberlite Conference, pp. 432–434.

Klemme, S., Blundy, J.D., Wood, B.J., 2002. Experimental constraints on major and trace element partitioning during partial melting of eclogite. Geochim. Cosmochim. Acta 66, 3109–3123.

Konszett, J., Sweeney, R.J., Compston, W., 1995. The correlation of kimberlite activity with mantle metasomatism. Ext. Abstr. 6th International. Kimberlite Conference, Novosibirsk, pp. 285–286.

Kopylova, M.G., Rickard, R.S., Kleyensueber, A., Taylor, W.R., Gurney, J.J., Daniels, L.R., 1997. First occurrence of strontian K–Cr loparite and Cr chevkinite in diamonds. Proc. Sixth Intern. Kimberlite Conference, Vol. 2. Diamonds: Characterization, Genesis and Exploration, Russian Geology and Geophysics 38, pp. 405–420.

Mendelssohn, M.J., Milledge, H.J., 1995. Geologically significant information from routine analysis of the mid-infrared spectra of diamonds. Int. Geol. Rev. 37 (2), 95–110.

Navon, O., Hucheon, I.D., Rossman, G.R., Wassenburg, G.J., 1988. Mantle derived fluids in diamond microinclusions. Nature 335 (N6193), 784–789.

Navon, O., Izraeli, E.S., Klein-BenDavid, O., 2003. Fluid inclusions in diamonds—the carbonatitic connection. Extended Abstract FLA_0107. 8th International Kimberlite Conference, Victoria, Canada.

Nixon, P.H., Condliffe, E., 1989. Yimengite of K–Ti metasomatic origin in kimberlitic rocks from Venezuela. Min. Mag. 53, 305–309.

Peng, Z., Lu, Q., 1985. The crystal structure of yimengite. Sci. Sin., Ser. B, Chem. Biol. Agric. Med. Earth Sci. 28, 882–887.

Reed, S.J.B., 1997. Electron Microprobe Analysis, 2nd ed. Cambridge Univ. Press.

Smith, C.B., Sims, K, Chimuka, L., Duffin, A, Beard, A.D., Townend, R., these proceedings. Kimberlite metasomatism at Murowa and Sese Pipes, Zimbabwe. Proc. 8th International Kimberlite Conference. Lithos.

Sobolev, N.V., Yefimova, E.S., Kaminsky, F.V., Lavrentiev, Y.G., Usova, L.V., 1988. Titanate of complex composition and phlogopite in the diamond stability field. In: Sobolev, N.V. (Ed.), Composition and Processes of Deep Seated Zones of Continental Lithosphere. Nauka, Novosibirsk, pp. 185–186.

Spetsius, Z.V., Belousova, E.A., Griffin, W.L., O'Reilly, S.Y., Pearson, N.J., 2002. Archean sulphide inclusions in Paleozoic zircon megacrysts from the Mir kimberlite, Yakutia: implications for the dating of diamonds. Earth Planet. Sci. Lett. 6175, 1–16.

Taylor, W.R., Bulanova, G.P., Milledge, H.J., 1995. Quantitative nitrogen aggregation study of some Yakutian diamonds: constraints on the growth, thermal and deformation history of peridotitic and eclogitic diamonds. Ext. Abs. 6th Int. Kimb. Conf., Novosibirsk, pp. 608–610.

Available online at www.sciencedirect.com

Lithos 77 (2004) 193–208

www.elsevier.com/locate/lithos

Aspects of diamond mineralisation and distribution at the Helam Mine, South Africa ☆

N. Mc Kenna[a,*], J.J. Gurney[a], J. Klump[a], J.M. Davidson[b]

[a] Department of Geological Sciences, University of Cape Town, Rondebosch 7700, South Africa
[b] Helam Diamond Mine, PO Box 2, Swartruggens 2835, South Africa

Received 27 June 2003; accepted 15 January 2004
Available online 20 May 2004

Abstract

The diamonds from the Swartruggens dyke swarm are mainly tetrahexahedra, with subsidiary octahedral and cuboid crystals. They are predominantly colourless, with subordinate yellows, browns, and greens. The existence of discrete cores and oscillatory growth structures within the diamonds, together with the recognition of harzburgite, lherzolite, at least two eclogitic and a websteritic diamond paragenesis, variable nitrogen contents, and both Type IaAB and Type Ib–IaA diamonds provides evidence for episodic diamond growth in at least six different environments. The predominance of plastic deformation in the diamonds, the state of nitrogen aggregation, and the suite of inclusion minerals recovered are all consistent with a xenocrystic origin for the diamonds, with the Type Ib–IaA diamonds being much younger than the rest. Mantle storage at a time-averaged temperature of ±1100 °C is inferred for the Type IaAB diamonds. The distribution of mantle xenocrysts of garnet and chromite within the high-grade Main kimberlite dyke compared to the low-grade Changehouse kimberlite dyke strongly suggests that the difference in diamond content is due to an increased eclogitic component of diamonds in the Main kimberlite dyke.
© 2004 Elsevier B.V. All rights reserved.

Keywords: Diamond; Xenocryst; Megacryst; Majorite; Type IaAB; Type Ib; Group I eclogite

1. Introduction

The Swartruggens kimberlite dyke swarm exploited under the name of Helam Mine is situated approximately 60 km west of the town of Rustenburg, North–West Province, South Africa (Fig. 1). The dyke swarm, comprising numerous east–west-trend-ing, near-vertical, kimberlite dykes (Harris et al., 1979), intrudes the central region of the Kaapvaal Craton on the southwestern edge of the Bushveld Complex (which outcrops ± 30 km to the east of the dyke system). The western sections of the occurrence have been reported to carry 1.5 ct/ton, with grades as high as 5 ct/ton having been recorded, making this, locally, the highest-grade kimberlite occurrence in South Africa.

As many as six kimberlite dykes have been identified (Main, Changehouse, Muil, John, North, and South dykes), each of which can be distinguished petrographically and have all been genetically linked to a common magma source (Klump, 1995) with the

---

☆ Supplementary data associated with this article can be found in the online version, at doi 10.1016/j.lithos.2004.04.004.

* Corresponding author. De Beers GeoScience Centre, PO Box 82232, Southdale, Johannesburg, South Africa. Tel.: +27-11-374-7868; fax: +27-11-835-1315.

E-mail address: neil.mckenna@debeersgroup.com
(N. Mc Kenna).

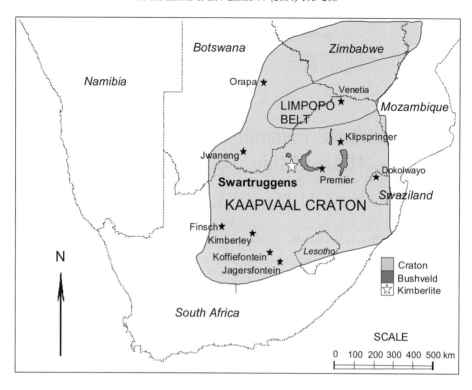

Fig. 1. Locality map illustrating the location of the Swartruggens kimberlite.

possible exception of the barren Muil dyke (Coe et al., 2003). The diamond grades within the different dykes vary considerably, from as high as 5ct/ton in the Main dyke to zero in the Muil dyke. Consequently, it has become essential to mine the dykes selectively. The Helam Diamond Mine currently only exploits the Main and Changehouse dykes for their diamonds, the latter being marginally economic but spatially closely associated with the Main dyke.

Radiometric dating based on Rb–Sr mica isochrons has reported ages of 147 ± 4 Ma (Allsopp and Barrett, 1975) and 156 ± 13 Ma (Smith et al., 1985). Isotopically, the dyke is classified as a group II kimberlite according to the definition of Smith (1983). Consistent with that classification, it is highly micaceous and falls within the chemical compositional field of other group II kimberlites as shown by Smith et al. (1985).

Field relationships suggest that the Main dyke is the oldest phase. It can also be shown (from field relations) that the Changehouse dyke is younger than the Main dyke, but older than the Muil dyke. On this

basis, there appears to be a drop in grade with each successive intrusion.

Mantle xenoliths of peridotite and eclogite commonly associated with kimberlite intrusions are very rare at this locality. None have been described in published literature and none have been found in the current study. Large single crystal garnets (>2 mm in largest dimension) reporting regularly in the heavy mineral concentrates produced in the diamond recovery process were noted in this study. Evidence for the presence of peridotite, eclogite, and websterite in the lithospheric mantle sampled by the Swartruggens dyke swarm is, to all practical purposes, confined to xenocrysts of the minerals derived from disaggregated xenoliths, which as is normal for kimberlites are ubiquitously present in all the Swartruggens dykes, except the Muil.

In this study, a carefully selected suite of diamonds and their associated mineral inclusions have been studied. In addition, suites of xenocrystic garnets and chromites (which may be xenocrysts or phenocrysts) have been analysed with a view of investigating the differences in diamond content between the Main and Changehouse dykes.

2. Samples

Three hundred and thirty-nine diamonds selected from approximately 500 carats of run-of-mine diamond production were described in terms of their colour, morphology, and surface features using a binocular microscope. Cathodoluminescence (CL) studies and Fourier transform infrared spectroscopy (FTIR) were conducted on the smaller whole crystals (<3 mm) and on polished diamond plates cut from the larger (>3 mm) diamonds. A subset of 60 diamonds was selected for their mineral inclusions. The latter were liberated mechanically using a steel diamond cracker, and analysed by electron microprobe.

Kimberlite samples collected from the Main and Changehouse dykes were crushed using a standardized procedure and a heavy mineral concentrate produced, from which a representative suite of garnets and chromite was analysed by electron microprobe.

3. Analytical methods

The diamonds were studied by both CL and FTIR techniques. CL imaging of the whole diamonds and the diamond plates was performed at the University of Cape Town using a Cambridge S200 scanning electron microscope (SEM) at 15 kV and an electron beam current of between 1.5 and 2.0 nA. In addition, colour CL images were captured using an optical microscope attached to a Technosyn luminescence generator with an accelerating voltage of 15 kV and an electron beam current set at 0.8 mA.

Hydrogen and nitrogen concentrations, nitrogen aggregation states, as well as the presence and position of platelets were calculated on all Type IaAB diamonds studied, by the deconvolution of FTIR absorbance spectra using the equations and absorption coefficients outlined in Mendelssohn and Milledge (1995). These parameters were also measured along traverses across the diamond plates. FTIR spectra were measured at the University of Cape Town using a Nicolet Magna-IR 560 spectrometer, with an attached KBr beam splitter and a MCT/A detector that

was cooled with liquid nitrogen. In addition, the instrument was purged with nitrogen during analysis, preventing the build-up of CO_2. Spectra were recorded over the range 4000–650 cm^{-1}, at a resolution of 8 cm^{-1}. Background spectra were recorded after each diamond analysis. These analyses were preformed online using Nicolet OMNIC 5.0 software.

Deconvolution and peak extraction from the spectra of the Type IaAB diamonds were preformed using Bruker OPUS/3D software. Peak information was then transferred electronically into a Quatro-Pro spreadsheet developed by Mendelssohn and Milledge (1995). Deconvolution and quantitative analysis of the spectra from the natural Type Ib diamonds were performed using a deconvolution program at the De Beers GeoScience Centre (Johannesburg, South Africa).

The diamond inclusions and mantle xenocrysts recovered from the kimberlite samples were analysed for their major element compositions using a Cameca

Table 1
Surface feature characteristics of diamonds from Swartruggens ($n = 339$)

Feature		%
Main form	Octahedra	30
	Cubes	1
	Cubo-octahedra	4
	Tetrahexahedra	53
	Fragments	12
Body colour	Colourless	65
	Brown	20
	Green	9
	Canary yellow	5
	Cape yellow	1
Octahedral features	Trigons	16
	Hexagonal pits	1
	Triangular plates	25
	Serrate laminae	20
Cubic features	Tetragonal etch pits	1
Tetrahexahedral features	Elongate hillocks	40
	Microhillocks	4
	Terraces	15
	Corrosion sculpture	7
	Circular microdisks	11
	Lamination lines	42
Unrestricted features	Ruts	14
	Inclusion cavities	7
Crystal regularity	Equidimensional	44
	Elongate	16
	Flattened	16
	Irregular	24

SX50 electron microprobe at the De Beers GeoScience Centre using standard techniques.

4. Results

4.1. Physical characteristics of the diamonds

The physical characteristics of the diamonds have been studied in detail (Table 1). While octahedra, cubo-octahedra, and cuboid diamonds all exist in significant quantities, the majority of the diamonds appear to have been extensively resorbed and commonly occur in the form of tetrahexahedra (Fig. 2).

While etching is neither ubiquitous nor pervasive among the diamonds, the size, depth, and abundance of the various etch features are variable between diamonds, and even between the different surfaces of individual diamonds. The canary yellow diamonds are associated with more intense etching and appear to have been *more* susceptible to the processes of resorption.

Various colour groups of diamonds have been identified including colourless, brown, green, pale (Cape) yellow, and canary yellow (Table 1).

Plastic deformation, manifested in the form of surface lamination lines parallel to the diamond cleavage planes, is observed on approximately 42% of the tetrahexahedral diamonds. It has been shown (Urusovskaya and Orlov, 1964) that brown colouration is caused by plastic deformation, and it follows therefore that features associated with such deformation are more commonly associated with brown diamonds. At Helam, surface lamination lines appear to be ubiqui-

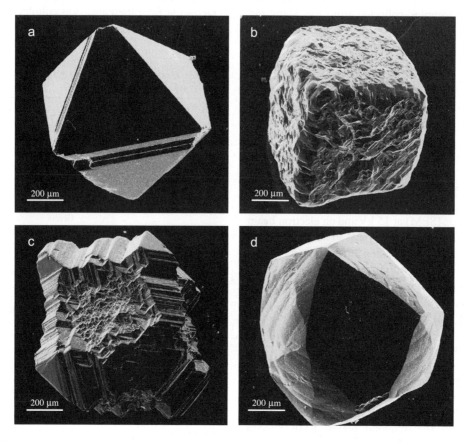

Fig. 2. SEM photomicrographs. (a) Unresorbed, unetched octahedron. (b) Highly etched cuboid diamond. (c) Differentially resorbed cubo-octahedron with sharp-edged octahedral faces and tetragonal etch pits on cubic faces. (d) Tetrahexahedron displaying low-relief features.

tous within *both* the brown and canary yellow categories, whilst they may also be visible on some other resorbed crystal forms.

4.2. Diamond internal structure and growth history: cathodoluminescence analysis

While some of the diamonds studied exhibit simple octahedral zonation, having grown by a spiral mechanism (layer by layer) within relatively stable environments (Sunagawa, 1984), the majority of the diamonds show more complex growth features. A common feature is the existence of diamond 'cores' around which later generations of diamond have nucleated (Fig. 3a–c). Both cubic and octahedral cores have been observed. These cores are overgrown either by octahedral or complexly zoned diamond.

The cubo-octahedral diamonds (and their resorbed equivalents) are characterized by mixed-habit, sector growth features (Fig. 3d) with normal growth on the octahedral surfaces and nonfaceted growth on the cubic surfaces. These features indicate growth under conditions transitional between that ideal for octahedral and cubic growth.

Many of the diamonds exhibit features of plastic deformation, internal resorption, and dark zones (inferred as nitrogen-free, Type II diamond), all of which reflect changes in the crystallization environment. The

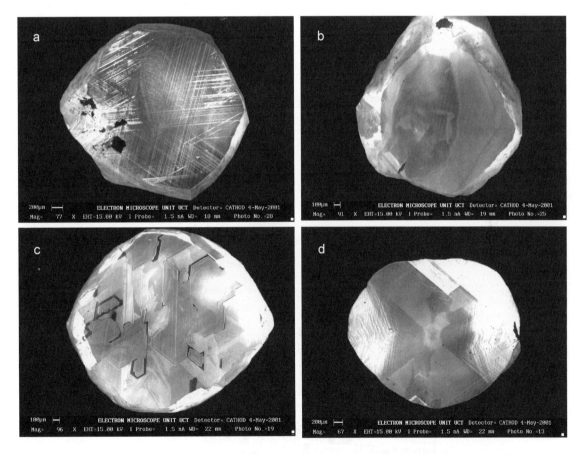

Fig. 3. CL photomicrographs of some diamond plates. (a) Low-N, peridotitic diamond tetrahexahedron with numerous chromite inclusions. Scotch-plaid deformation lines mask much of the growth zonation. Notice diamond core displaying hummocky (cubic) growth zones. (b) Tetrahexahedron characterized by hummocky or cubic zonation at the core and overgrown by octahedral zonation. (c) Tetrahexahedron characterized by numerous octahedrally zoned diamond cores, which have been overgrown by complexly zoned diamond. Notice that the dark layers (inferred as Type II diamond) occur at the transition between the octahedral and complex growth. (d) Tetrahexahedron exhibiting sectoral growth structures.

dark zones are only found in association with octahedral growth and commonly occur at the interface between changes in growth habit and/or layering (Fig. 3c).

4.3. FTIR analysis

The majority (95%) of the diamonds have been classified as Type IaAB diamonds based on their infrared absorption properties. These diamonds are associated with relatively high nitrogen contents (between 200 and 2500 ppm) and relatively mature aggregation states (between 10% and 65% nitrogen occurring as B aggregate). These diamonds display well-constrained time-averaged mantle residence temperatures of approximately 1093 °C (using the calibration of Taylor, et al., 1990), assuming a mantle residence time of 3 Ga (Fig. 4). Although the Swartruggens diamonds have not been dated, this is considered a reasonable assumption of the mantle residence period based on Archean ages reported for both harzburgitic (e.g., Richardson et al., 1984) and eclogitic (e.g., Richardson et al., 2001) diamonds from the Kaapvaal Craton. Incidentally, if one were to assume younger ages of 2 and 1 Ga for mantle residence, the calculated mantle residence temperatures for the Swartruggens diamonds would be remain similar, averaging 1102 and 1119 °C, respectively.

The highest nitrogen concentrations and most mature nitrogen aggregation states are associated with the diamonds exhibiting nonpenetrative green coloration (which is attributed to radiation damage).

In general, the Type IaAB diamonds show a predominant trend of decreasing nitrogen and hydrogen concentrations outward from the crystal centers. Nitrogen content can vary by as much as 500 ppm between the crystal center and rim. When diamond cores exist, however, the core–overgrowth boundary is marked by a significant increase in nitrogen concentration (as much as 400 ppm), which appears then to systematically decrease towards the rim as for diamonds with no observed core. No increase in hydrogen concentration has been observed across core–overgrowth boundaries.

Platelet peaks are visible in the majority of the spectra for the Type IaAB diamonds; however, some of these diamonds (40%) show no evidence of platelet peak development. It has been found that platelet peak position is related to platelet size (e.g., Hanley et al., 1977). Among the Swartruggens diamonds, the maximum intensity of the platelet peaks on the infrared spectra varies between 1361 and 1378 cm^{-1} (Fig. 5). It is evident, therefore, that a range of platelet sizes is present, with a peak position of 1361 cm^{-1} corresponding to relatively large platelets (i.e., >10

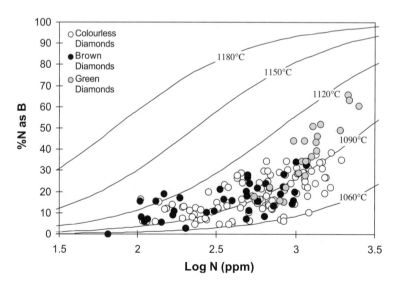

Fig. 4. Nitrogen concentration vs. the percentage of nitrogen occurring as B aggregate (% N as B) for the Type IaAB Swartruggens diamonds. Isotherms have been calculated for a mantle residence time of 3 Ga according to the calibration of Taylor et al. (1990).

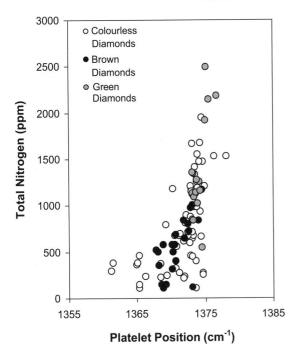

Fig. 5. Plot of nitrogen content vs. platelet peak position (cm^{-1}) for the Type IaAB Swartruggens diamonds.

μm) and a position of 1378 cm^{-1} corresponding to platelets smaller than 20 nm (Mendelssohn and Milledge, 1995). Platelet size development in diamonds is typically correlated with nitrogen content of the dia-

monds (Woods, 1986) and this is also, generally, the case for the Swartruggens diamonds (Fig. 5).

The concentration of hydrogen in the diamonds studied (as determined by the intensity of the peak at 3107 cm^{-1}) is variable among the Swartruggens diamonds. There is a tendency for the hydrogen content to be higher in the diamonds with more advanced aggregation states for each colour group (Fig. 6).

The FTIR studies have shown that there exists a sectoral dependence of nitrogen concentration and aggregation associated with the cubo-octahedral diamonds, and their resorbed equivalents. The cubo-octahedral diamonds display higher nitrogen concentrations and aggregation within the cubic growth sectors than within the octahedral growth sectors of the cubo-octahedral diamonds.

Approximately 5% of the diamond population ('canary yellow' diamonds) is characterized by low nitrogen concentrations (<150 ppm) and aggregation states assigned to the Type Ib–IaA aggregation series. Combined CL and FTIR studies indicate that these diamonds comprise complexly zoned Type Ib and Type IaAB diamonds.

4.4. Diamond inclusion minerals

The study of syngenetic mineral inclusions within the diamonds provides evidence supporting mul-

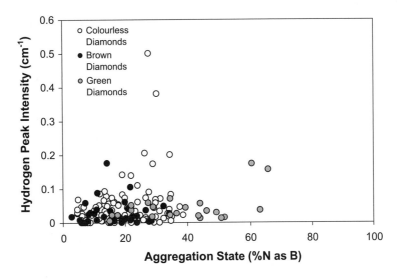

Fig. 6. Plot of hydrogen peak intensity vs. the percentage of nitrogen occurring as B aggregate (% N as B) for the Swartruggens diamonds.

Table 2
Swartruggens diamond inclusion abundances, diamond colour, and morphology

Diamond	Cor	Maj	Cpx	Opx	Coes	Grt	Olv	Chr	Pyr	Mss	Pnt	Pnt-Hz	Hz	Paragenesis	Colour	Morphology
HM006					3									Eclogitic	Y	THH
HM018			1											Eclogitic	Y	THH
HM019				1										Eclogitic	Y	THH
HM034				1										Eclogitic	C	THH
HM059									1					Eclogitic	B	THH
HM112	1													Eclogitic	C	THH
HM114	1			3										Eclogitic	C	THH
HM115				1					1					Eclogitic	C	F
HM119				1										Eclogitic	C	CO
HM124									1					Eclogitic	C	THH
HM126									1					Eclogitic	C	THH
HM211				1										Eclogitic	Y	O
HM236				1										Eclogitic	Y	THH
HM1A2									5	2				Eclogitic	C	THH
HM1A7									1					Eclogitic	C	F
HM1B2				1	1									Eclogitic	C	THH
HM2A4									1					Eclogitic	C	O
HM2B2	16			1										Eclogitic	B	O
HM2B3	1								3	2				Eclogitic	C	THH
HM2B1	1													Eclogitic	C	THH
HM2B11	5													Eclogitic	C	O
HM3A11									2	1				Eclogitic	LY	O
HM3C10									1	2				Eclogitic	B	THH
HM3D6a									1					Eclogitic	C	O
HM007			1											Peridotitic	C	THH
HM008			2				3			1	1		1	Peridotitic	C	O
HM009		2												Peridotitic	C	THH
HM031								4						Peridotitic	C	O
HM032								2						Peridotitic	C	THH
HM055								2						Peridotitic	LY	THH
HM067								1						Peridotitic	C	O
HM086								1						Peridotitic	C	O
HM109								3						Peridotitic	C	THH
HM111													1	Peridotitic	C	THH
HM113							1					2		Peridotitic	C	THH
HM203										2				Peridotitic	Y	THH
HM208							3							Peridotitic	Y	THH
HM209							1							Peridotitic	C	THH
HM210							1							Peridotitic	C	O
HM221										1	1			Peridotitic	C	THH
HM235									1				2	Peridotitic	C	THH
HM240							3							Peridotitic	C	F
HM2B7										2				Peridotitic	C	THH
HM1A9							10							Peridotitic	C	F
HB1A10							1							Peridotitic	C	THH
HM3B4							5		1	1				Peridotitic	C	THH
HM3B7													4	Peridotitic	C	THH
HM3B9								2			2		1	Peridotitic	C	THH
HB3C3								3						Peridotitic	C	THH
HM108			1	1										Websteritic	Y	F
HM1A12					1									Websteritic	C	THH
HM3A5					1									Websteritic	C	O
HM3B11			1		4									Websteritic	C	O

Table 2 (*continued*)

Diamond	Cor	Maj	Cpx	Opx	Coes	Grt	Olv	Chr	Pyr	Mss	Pnt	Pnt-Hz	Hz	Paragenesis	Colour	Morphology
HM033											1			Unknown	Y	O
HM123											1			Unknown	C	THH
HM127											1			Unknown	C	THH
HM217											1			Unknown	B	THH
HM220											1			Unknown	C	THH
HM3B10											1			Unknown	C	O
HM3D1												1		Unknown	B	THH
TOTAL	25	2	5	1	12	10	17	29	20	12	12	3	9	157		

Cor: corundum; Maj: majorite; Cpx: clinopyroxene; Opx: orthopyroxene; Coes: coesite; Olv: olivine; Chr: chromite; Mss: monosulfide solid solution; Pyr: pyrrhotite; Pnt: pentlandite; hz: heazlewoodite; C: colourless; Y: canary yellow; LY: Cape yellow; B: brown; O: octahedron; THH: tetrahexahedron.

tiple diamond source regions. A total of 157 inclusion analyses from 60 diamonds were recovered (Table 2). Of these, 47 inclusions were silicates (Fig. 7), 54 were oxides, and 56 were sulfides (Fig. 8). Tables 3 and 4 record representative analytical data.

Peridotitic and eclogitic diamond mineral inclusions dominate, along with subordinate websteritic and minor calc-silicate diamond mineral inclusions. It is observed that while peridotitic diamonds are dominant among the smaller (<3 mm) diamonds, the reverse appears true for the larger diamonds (>3 mm). The intermediate, websteritic component (which is more closely associated with the eclogitic paragenesis) also

appears more dominant among the larger diamond sizes studied. This is consistent with observations made by Gurney (1989) that there is a general trend for a greater eclogitic component among the large diamond sizes. This observation re-emphasizes the importance of the eclogitic component in establishing grade at this locality.

4.4.1. Peridotitic paragenesis

Peridotitic diamond mineral inclusions include olivine, Mg-chromite, diopside, and high-Ni sulfides, suggesting that the peridotitic diamonds crystallized within harzburgite and lherzolite substrates. The olivine

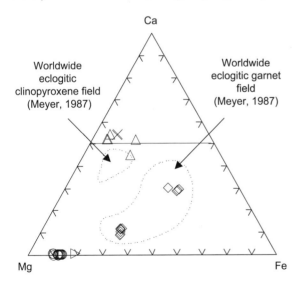

Fig. 7. Ca–Mg–Fe ternary diagram illustrating the compositions of silicate inclusions (circles = olivine, diamonds = garnet, upright triangle = pyroxene, tilted triangle = orthopyroxene, cross = majorite). Inserted fields reflect worldwide compositions of eclogitic garnet and clinopyroxene inclusions (Meyer, 1987).

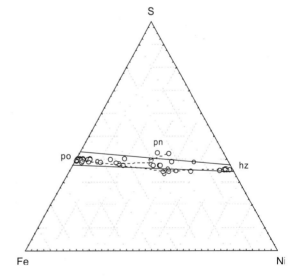

Fig. 8. S–Fe–Ni ternary diagram showing the compositions of the sulfide phases liberated from the diamonds, with respect to monosulfide solid solution stability at 650 °C (solid lines). Dashed lines indicate coexisting sulfide phases within individual diamonds. po = pyrrhotite, pn = pentlandite, hz = heazlewoodite.

Table 3

Chemical analyses for selected silicate and oxide inclusions from some Swartruggens diamonds

	Olv-p	Olv-p	Olv-p	Olv-p	Olv-p	Olv-p	Olv-p	Chr-p	Chr-p	Chr-p
	HM008d	HM008c	HM240a	HM240c	HM240d	HM3B4d	HM3B4f	HM031a	HM031d	HM055a
SiO_2	40.89	41.66	41.47	40.71	41.25	41.2	41.03	0.14	0.12	0.13
TiO_2	N.D.	0.03	N.D.	N.D.	N.D.	N.D.	N.D.	0.76	0.74	0.02
Al_2O_3	N.D.	N.D.	N.D.	0.03	N.D.	N.D.	N.D.	4.69	4.68	4.80
Cr_2O_3	0.04	0.05	0.04	0.04	0.04	0.06	0.05	64.30	64.75	66.91
FeO	7.85	8.03	7.15	6.94	7.15	6.06	6.14	16.09	16.02	14.87
MnO	0.12	0.11	0.10	0.08	0.08	0.08	0.10	0.20	0.21	0.28
MgO	51.24	50.21	51.35	51.72	51.19	53.11	53.36	13.52	13.83	12.93
CaO	0.07	0.08	0.04	0.06	0.06	0.01	0.02	N.D.	N.D.	N.D.
Na_2O	N.D.	N.D.	N.D.	0.05	0.05	0.04	N.D.	N.D.	N.D.	N.D.
K_2O	N.D.	N.D.	N.D.	N.D.	N.D.	N.D.	N.D.	N.D.	N.D.	0.08
NiO	0.32	0.32	0.40	0.40	0.38	0.39	0.39	0.11	N.D.	0.03
Total	100.53	100.49	100.55	100.03	100.20	100.95	101.09	99.81	100.35	100.05

	Chr-p	Chr-p	Chr-p	Chr-p	Cpx-p	Opx-p	Gnt-e	Gnt-e	Gnt-e	Cpx-e
	HB3C3c	HM067c	HM032e	HM109a	HM007a	HM108a	HM1B2a	HM006b	HM006c	HM018a
SiO_2	0.15	0.13	0.13	0.11	54.36	57.60	40.38	38.83	38.34	54.53
TiO_2	0.05	0.82	0.03	0.17	N.D.	N.D.	0.29	0.32	0.32	0.21
Al_2O_3	5.24	4.83	5.89	2.38	2.11	0.06	21.66	22.84	23.17	7.81
Cr_2O_3	66.45	64.19	65.14	69.39	3.02	0.40	0.09	0.07	0.06	0.05
FeO	14.35	16.23	15.23	15.31	2.13	8.04	15.43	16.99	17.08	5.97
MnO	0.25	0.22	0.22	0.26	0.06	0.05	0.32	0.35	0.35	0.10
MgO	13.43	13.47	13.37	12.58	15.50	34.13	10.60	9.62	9.50	11.37
CaO	0.11	N.D.	N.D.	0.07	19.23	0.22	11.38	10.87	11.02	14.23
Na_2O	N.D.	N.D.	N.D.	N.D.	2.74	0.12	0.12	0.11	0.11	5.10
K_2O	N.D.	N.D.	0.10	N.D.	N.D.	N.D.	N.D.	N.D.	N.D.	0.08
NiO	N.D.	0.08	0.03	0.05	0.05	0.03	N.D.	N.D.	N.D.	N.D.
Total	100.03	99.97	100.14	100.32	99.20	100.65	100.27	100.00	99.95	99.45

	Cor-e	Cor-e	SiO_2-e	SiO_2-e	Gnt-w	Cpx-w	Opx-w	Cpx-cs	Majorite	Majorite
	HM2B2k	HM2B2n	HM114c	HM119b	HM3A5b	HM008e	HM108a	HM3B11c	HM009a	HM009b
SiO_2	N.D.	N.D.	99.13	99.24	42.13	54.54	57.60	52.99	47.44	47.98
TiO_2	N.D.	N.D.	N.D.	N.D.	0.23	0.18	N.D.	0.52	0.60	0.60
Al_2O_3	98.90	98.85	0.02	N.D.	22.95	0.26	0.06	5.35	10.19	9.71
Cr_2O_3	N.D.	N.D.	N.D.	N.D.	0.21	0.15	0.40	0.06	2.19	2.20
FeO	N.D.	N.D.	N.D.	0.08	11.05	2.62	8.04	6.82	3.33	3.39
MnO	N.D.	N.D.	0.01	0.01	0.43	0.06	0.05	0.06	0.18	0.20
MgO	N.D.	N.D.	N.D.	N.D.	19.72	17.50	34.13	11.87	13.67	13.70
CaO	N.D.	N.D.	0.06	0.40	3.10	24.16	0.22	20.33	20.18	20.45
Na_2O	N.D.	N.D.	0.12	0.02	0.12	0.33	0.12	2.19	0.99	1.00
K_2O	N.D.	N.D.	0.02	N.D.	N.D.	N.D.	N.D.	N.D.	N.D.	N.D.
NiO	N.D.	N.D.	N.D.	0.03	N.D.	N.D.	N.D.	N.D.	N.D.	N.D.
Total	98.90	98.85	99.36	99.78	99.94	99.80	100.62	100.19	98.77	99.23

Olv: olivine; Cpx: clinopyroxene; Gnt: garnet; Opx: orthopyroxene; Cor: corundum; Chr: chromite; p: peridotitic; e: eclogitic; w: websteritic; cs: calc silicate paragenesis; N.D.: not detected.

inclusions have forsterite 92–94 contents with restricted ranges of Cr_2O_3 (0.02–0.07 wt.%) and NiO (0.32–0.42 wt.%). Chromite inclusions are common, characterized by high Cr_2O_3 (63.7–69.4 wt.%) contents.

A single peridotitic diopsidic clinopyroxene (HM007a) was identified (Table 3), based on its relatively high Cr_2O_3 and MgO content (3.02 and 15.2 wt.%, respectively) and low Na_2O and Al_2O_3

Table 4
Chemical analyses for sulfide inclusion minerals from some Swartruggens diamonds

	mss	mss	pyrr	pyrr	pnt	pnt	hz	hz
	HM3C10a	HB3B4d	HM059a	HM124a	HM3B9d	HM3B10a	HM235a	HM3B7a
S	38.13	36.38	39.24	38.64	36.70	34.18	35.34	35.32
Cr	ND	ND	ND	ND	ND	ND	ND	ND
Mn	ND	ND	ND	ND	ND	ND	ND	ND
Fe	54.75	41.61	58.09	58.06	27.47	27.76	1.48	2.49
Co	0.77	2.55	0.29	0.20	1.12	0.37	0.14	1.01
Ni	5.31	19.25	1.13	2.21	33.91	36.60	62.67	61.03
Cu	0.31	0.07	ND	0.11	0.10	0.05	0.36	0.07
Zn	ND	ND	ND	ND	ND	ND	ND	ND
Total	99.27	99.86	98.75	99.22	99.30	98.96	99.99	99.92

mss = monosulfide solid solution; pyrr = pyrrhotite; pnt = pentlandite; hz = heazlewoodite.

contents (2.74 and 2.11 wt.%, respectively). This suggests a small lherzolitic diamond component.

High Ni-sulfide inclusions are also found in abundance. Inclusions of monosulfide solid solution with Ni contents greater than 16 wt.% have all been assigned to the peridotitic parageneses, after Yefimova et al. (1983).

4.4.2. Eclogitic paragenesis

The eclogitic diamonds are characterized by inclusions of low-Ni sulfides, Group I eclogitic garnet, inferred coesite (SiO_2), and corundum (Table 3). The latter two minerals suggest crystallization within per-aluminous, grospyditic substrates. Two groups of eclogitic garnet have been identified, a single eclogitic inclusion from a colourless diamond (HM1B2) exhibits a Mg # of 57. The second group was identified in a 'canary yellow' diamond (HM006) and is characterized by Mg # between 51 and 54. A single omphacitic clinopyroxene (HM018a) is assigned to the eclogitic paragenesis. It has Na_2O and Al_2O_3 contents of 5.1 and 5.4 wt.%, respectively. This inclusion has 0.05 wt.% Cr_2O_3 and exhibits a K_2O content of 0.08 wt.%. The most notable feature of the eclogitic diamond mineral inclusion suite is the abundance of a SiO_2 phase (inferred to represent coesite) and corundum. Eclogitic, low Ni-sulfide inclusions have Ni contents less than 10 wt.%, as identified by Yefimova et al. (1983).

4.4.3. Websteritic paragenesis

The websteritic diamonds are characterized by inclusions of garnet, clinopyroxene, and orthopyrox-

ene (Table 3). Garnet inclusions display Mg # between 77 and 83, with low Cr_2O_3 (0.2–0.3 wt.%). Two websteritic clinopyroxene inclusions were identified from a single diamond (HM008), characterized by high MgO (approximately 17.5 wt.%), CaO contents of 22.0 and 24.2 wt.%, Cr_2O_3 of 0.08 and 0.15 wt.%, and Na_2O of 0.12 and 0.33 wt. %. A single orthopyroxene inclusion (HM108a) with 0.4 wt.% Cr_2O_3 and Mg # of 88.33 has also been assigned to the websteritic paragenesis.

4.4.4. Calc-silicate paragenesis

An unusual clinopyroxene inclusion (HM3B11c), characterized by high CaO and Al_2O_3 (20.3 and 5.4 wt.%, respectively) and Mg # of 75.63 (Table 3), is similar to Ca-rich pyroxenes from diamonds in southeastern Australia (Sobolev et al., 1984) and to a calc-silicate inclusion recovered from the Premier Mine, South Africa (Tsai et al., 1979). This inclusion has consequently been assigned to the calc-silicate paragenesis.

4.4.5. Deep diamond inclusions

The compositions of two unusual garnet diamond mineral inclusions (Table 3), liberated from a single diamond (HM009), are interpreted to represent majorite. An attempt to verify the majorite structure using X-ray diffraction was unsuccessful due to the small size of both inclusions. This association is therefore made on the basis of the major element compositions reported in Table 3. A similar solid solution observed among diamonds from other worldwide localities (Moore and Gurney, 1985; Harte and Harris, 1994; Kaminsky et al., 2001) is consistent with pressures and temperatures associated with the sublithospheric upper mantle.

4.5. Mantle xenocrysts

Following crushing of the kimberlite samples from both the Main and the Changehouse kimberlite dykes, the heavy minerals were recovered using heavy liquid, and a representative suite of garnet and chromite xenocrysts was selected for analysis by electron microprobe.

4.5.1. Main kimberlite dyke

The sample from the Main dyke was characterized by garnets from harzburgites and lherzolites, a significant chromite population, Group I (as defined by McCandless and Gurney, 1989) eclogitic garnets, and a minor megacryst garnet population (Fig. 9).

4.5.2. Changehouse kimberlite dyke

The sample of xenocrysts from the Changehouse dyke was similar to the Main dyke with respect to the lherzolite garnets and the chromite macrocrysts, but differed in having fewer harzburgitic garnets, markedly fewer Group I eclogitic garnets, and the presence of two compositionally discrete megacryst garnet populations (Fig. 9).

5. Discussion

5.1. Diamond formation

5.1.1. Episodic crystallization

There are correlations between diamond colour and chemical characteristics. The colourless and brown diamonds exhibit nitrogen concentrations in

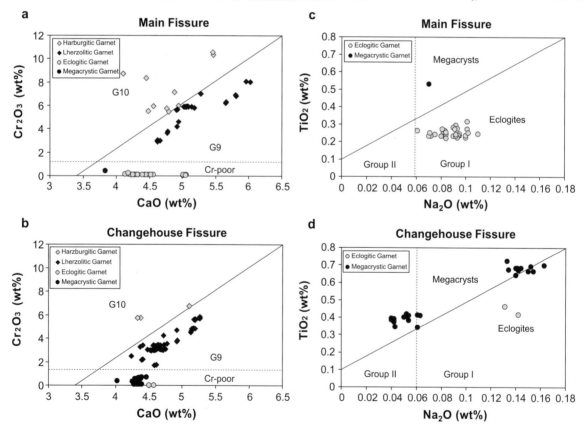

Fig. 9. (a and b) Garnet xenocryst compositions from the Main and Changehouse dykes; solid line (85% line) defines lherzolitic trend at 50 kbar pressure (Gurney and Zweistra, 1985). (c and d) Low-Cr garnet compositions. Solid line separates Cr-poor megacrysts and eclogitic garnets (Kimberlite Research Group database, University of Cape Town).

the range 200–1300 ppm, with nitrogen aggregation states between 10% and 35% N as B aggregate. The green diamonds exhibit higher nitrogen concentrations (up to 2500 ppm) and aggregation states (up to 65% N as B aggregate). The Cape yellow diamonds display nitrogen characteristics that are identical to the colourless and brown diamonds on these criteria, but the colour is believed to be associated with the presence of optical N_3 centers (Orlov, 1977; Collins, 2001).

The colourless, brown, green, and Cape yellow diamonds can all be classified as Type IaAB diamonds, consistent with geologically significant residence times within the mantle for diamonds linked through inclusions to peridotitic, eclogitic, and websteritic parageneses.

Integrated CL and FTIR studies of some of the peridotitic diamond plates (e.g., Fig. 3a) show that the peridotitic diamonds show further evidence for episodic crystallization (diamond cores overgrown by younger Type IaAB diamond) within a single diamond.

The canary yellow diamonds, however, are ubiquitously associated with very low nitrogen concentrations (between 50 and 150 ppm). Electron paramagnetic resonance (EPR) studies have revealed the presence of low concentrations of unaggregated, dispersed nitrogen, confirming that these diamonds belong to the Ib–IaA aggregation series. It is believed that the presence of single, dispersed nitrogen atoms imparts the vivid canary yellow/amber colouration to these diamonds (Sobolev, 1978; Collins, 1982). The relatively immature aggregation states associated with these natural Type Ib diamonds suggest a significantly younger formation age than for the Type IaAB diamonds.

Only eclogitic inclusion minerals have been liberated from the canary yellow, natural Type Ib diamonds. In addition, it has been shown by Logan (1999) that these canary yellow diamonds only display highly fractionated carbon isotopic signatures (between −10.25‰ and −19.48‰), uniquely associated with diamonds of the eclogitic paragenesis (e.g., Sobolev et al., 1979; Harris, 1987). It is reasonable then to ascribe all these Type Ib diamonds to a separate, younger eclogitic paragenesis. These natural Type Ib diamonds, while only comprising a small percentage of the total diamond budget at this locality, are generally exceedingly rare in nature (Evans,

1992), and thus a contribution of 5% to the diamond budget is significant and unusual.

It is clear that several diamond growth events have contributed, through time, to the Swartruggens diamond population. On a small scale, integrated FTIR and CL studies of individual diamond plates demonstrate that hiatus in growth and regrowth is a common phenomenon associated with diamonds, often marked by significant changes in growth morphology and/or the nitrogen concentration and distribution within the diamond.

5.1.2. Environment of crystallization

In the absence of mantle xenoliths for study and detailed geothermobarometric analysis of diamond inclusions, inferences regarding the mantle environment must be made principally from FTIR data analysis. Assuming a mantle residence time of ±3 Ga for all Type IaAB diamonds, the diamonds' record time averaged mantle residence temperatures of approximately 1100 °C. While the thermal environment of crystallization appears to have been consistent over extended periods of geological time, IR analysis of individual diamond plates shows that significant, often systematic chemical variations have occurred during diamond growth, leading to the suspicion that temperature, too, may have varied from time to time.

Octahedral and cubic diamonds can be interpreted to be representative of diamond crystallization within environments of low- and high-speed growth, respectively (Sunagawa, 1984). The large population of cubo-octahedral diamonds represents crystallization within intermediate growth rates and is characterized by growth sectoral dependence of nitrogen. With relatively constant thermal residence temperatures being recorded over extended periods of geological time, it is reasonable to suggest that marked differences in growth forms within individual diamonds may have been caused by significant variations in the level of carbon supersaturation within the mantle over time.

The observed variation in the nature of growth within specific diamonds is attributed to variations in composition of the carbon reservoir from which the diamonds grew, probably caused by episodic fluid ingress rather than temperature variations. Based on the consistent time-averaged mantle residence temperatures derived from the diamond plates, it is proposed that while such fluids may have resulted in localized thermal fluctuations, the perturbations have been short-lived and relatively small as not to have a sustained effect on the N aggregation.

Cathodoluminescence studies of diamond plates have shown that diamond 'cores' are relatively common. Where such diamond cores are observed, the transition between the core and the overgrowth layers is commonly marked by an abrupt *increase* in nitrogen concentration (by as much as 400 ppm), and only then a decrease with successive growth. Perhaps these abrupt changes mark temporary 'open-system' conditions of N- and C-enriched fluid ingress, from which younger diamonds may have crystallized.

5.2. Significance of the majorite component

The observation of probable majorite as an inclusion in diamonds from the Swartruggens kimberlite dyke swarm may provide evidence for a sublithospheric origin for Group II kimberlites.

5.3. Geological history of the diamonds subsequent to crystallization

The individual diamonds are characterized by distinguishable physical and chemical properties, which together reflect differences in their respective geological histories.

Brown colouration and the presence of lamination lines suggest that a large proportion of the diamonds has been affected by deformation. It has been observed that the crystals displaying the most significant deformation features are those associated with relatively low nitrogen contents (<400 ppm) and an absence of platelets. The canary yellow (Type Ib) diamonds, which are all of low nitrogen contents (<150 ppm) and display no platelets, are ubiquitously associated with deformation features. The inference therefore is that the diamonds with low nitrogen concentration and an absence of platelets have been more susceptible to the effects of deformation. This is consistent with observations made by Woods et al. (1990).

While these diamonds reflect ductile deformation subsequent to crystallization, it is not clear if deformation occurred in a single event, or episodically through time. The observed deformation in the youn-

ger canary yellow diamonds certainly suggests a more recent deformation event, possibly associated with kimberlite emplacement.

It is interesting to note that recently developed high-pressure–high-temperature treatment methods of diamonds showing plastic deformation, particularly those with low nitrogen content, can be annealed within a short period of time (~ 10 h) at temperatures of 1900 °C and pressures of 7 GPa (Van Royen and Palyanov, 2002). Such experiments suggest that signs of plastic deformation in diamond may not survive for geologically significant periods of time within the mantle. Nor is it likely to occur after incorporation of the diamond into the kimberlite as grain boundary contact is required to convert the deviatoric stress needed to create the deformation. In which case, by default, it must occur during crack propagation as the kimberlite disaggregates and assimilates, with the diamondiferous lithospheric mantle en route to the surface.

The abundance of tetrahexahedroidal diamonds attests to extensive diamond dissolution. Processes of mantle metasomatism may contribute to this dissolution (Griffin and Ryan, 1995). Elevated TiO_2 contents among many of the garnet xenocrysts (Mc Kenna, 2002) and highly resorbed chromite macrocrysts (Klump, 1995), especially in the Changehouse dyke, are evidence that support such metasomatism.

The large varieties of etch features associated with the diamonds attest to the presence of oxidizing conditions, over a wide range of temperatures as the kimberlite decompress and cool (Robinson, 1979). The canary yellow diamonds commonly exhibit more intense etching than the Type IaAB diamonds, suggesting that their unique chemical characteristics predispose them to the effects of oxidation.

Radiation damage was perhaps the most recent natural process to affect some of the diamonds. This process has resulted in the green surface colouration of some of the diamonds at low temperatures (<600 °C) within the oxidized portions of the kimberlite. FTIR analysis of these green diamonds revealed that they are commonly associated with very high nitrogen contents and aggregation states. This suggests that the nature of the nitrogen content of diamond facilitates the development of the lattice disorders that create the green colour.

5.4. Diamond content of the Main and Changehouse dykes

The abundances and compositions of the component parts of the garnet and chromite populations in the Main and Changehouse kimberlites show significant differences that provide an explanation for the widely disparate diamond content. The high-grade Main dyke has sampled a substantial component of diamondiferous Group I eclogite on the basis of the presence of Group I xenocryst garnets in the kimberlite. The lower-grade Changehouse dyke has, by comparison, much fewer Group I garnets as well as fewer diamonds (see Fig. 9).

Since the Changehouse dyke in places crosscuts the Main dyke and commonly incorporates fragments of the Main, it is clearly younger. The obvious and economically important difference between these two spatially closely associated intrusive events demonstrates that while xenocryst minerals within kimberlite may be very useful in providing information on the mantle rocks sampled by kimberlite, individual intrusions do not accurately sample the cratonic mantle in a manner that reflects either the full range of rocks present or their abundance in the lithospheric mantle.

6. Conclusions

Detailed observations on a small subset of run-of-mine diamonds from the Helam Mine indicate that diamond formation has been episodic, and has occurred in the lithospheric mantle associated with harzburgite $\pm$ garnet $\pm$ chromite, lherzolite, at least two events associated with eclogite, and garnet websterite. Sulfide minerals have been closely associated with both the peridotitic and eclogitic parageneses and are the most common inclusions in the diamonds. A minor diamond component from the deep upper mantle is indicated by a single diamond containing two inclusions with a major element composition most similar to majorite.

Nitrogen concentrations in the diamonds vary from below detection limit to >2500 ppm. Nitrogen aggregation within most of the diamonds is consistent with lengthy storage of the diamonds within the cratonic lithosphere at a time-averaged mantle temperature

close to 1100 °C. Approximately 5% of the diamonds are canary yellow Type Ib diamonds, with much shorter mantle residence time. Octahedral and cuboid forms are present, but the majority of the diamonds are tetrahexahedra showing evidence of plastic deformations. Diamonds with low nitrogen content appear to be more susceptible to plastic deformation and resorbtion, whilst green surface colour is associated with high nitrogen diamonds. The pronounced difference in diamond grade between the Main and Changehouse dykes appears to be due to the presence of a significant eclogitic component of diamonds in the Main dyke, and its near absence in the Changehouse dyke.

Acknowledgements

Jim Davidson and Helam Diamond Mine are thanked for making available the diamond and kimberlite samples. Howard Bell (Zlotowski's Diamond Cutting Works) is thanked for arranging for the cutting and polishing of diamonds into plates. Craig Smith, Fanus Viljoen, and Ingrid Chinn of the De Beers GeoScience Centre are thanked for their technical support. René Dobbe of the De Beers GeoScience Centre is thanked for his assistance during the acquisition of the microprobe data. Thanks, too, to Graeme Hill of DebTech for conducting EPR on some of the diamonds. At the University of Cape Town, Miranda Waldon and Dane Gernicke are thanked for assisting in obtaining scanning electron microscopy and cathodoluminescence images. Eva Anckar of the Kimberlite Research Group is thanked for help in accessing data from the Kimberlite Research Group database. This project was supported by the National Research Foundation (Pretoria) and the De Beers Consolidated Mines.

References

Allsopp, H.L., Barrett, D.R., 1975. Rb–Sr age determinations of South African kimberlites. Phys. Chem. Earth 9, 605–617.

Coe, N., le Roex, A.P., Gurney, J.J., 2003. The petrology and geochemistry of the Swartruggens and Star kimberlite dyke swarms, South Africa. Extended Abstracts. 8th International Kimberlite Conference.

Collins, A.T., 1982. Colour centres in diamond. J. Gemol. 18, 35–37.

Collins, A.T., 2001. The colour of diamonds and how it may be changed. J. Gemol. 27, 341–359.

Evans, T., 1992. Aggregation of nitrogen in diamond. In: Field, J.E. (Ed.), The Properties of Natural and Synthetic Diamond. Academic Press, London, pp. 259–290.

Griffin, W.L., Ryan, C.G., 1995. Trace elements in indicator minerals, area selection, and target evaluation in diamond exploration. J. Geochem. Explor. 53, 311–337.

Gurney, J.J., 1989. Diamonds. In: Ross, J. (Ed.), Kimberlites and Related Rocks: 2. Their Mantle/Crust Setting, Diamonds and Diamond Exploration. Geological Society of Australia Special Publication, vol. 14. Blackwell, Oxford, pp. 935–965.

Gurney, J.J., Zweistra, P., 1985. The interpretation of the major element compositions of mantle minerals in diamond exploration. J. Geochem. Exploration 53, 293–309.

Hanley, P.L., Kiflawi, I., Lang, A.R., 1977. On topographically identifiable sources of cathodoluminescence in natural diamond. Philos. Trans. Roy. Soc., A 284, 329–368.

Harris, J.W., 1987. Recent physical, chemical, and isotopic research of diamond. In: Nixon, P.H. (Ed.), Mantle Xenoliths. Wiley, New York, pp. 477–500.

Harris, J.W., Hawthorne, J.B., Oosterveld, M.M., 1979. Regional and local variations in the characteristics of diamonds from some southern African kimberlites. Proceedings of the 2nd International Kimberlites Conference, 27–41.

Harte, B., Harris, J.W., 1994. Lower mantle mineral associations preserved in diamonds. Mineral. Mag. v58A, 384–385.

Kaminsky, F.V., Zakharchenko, O.D., Davies, R., Griffin, W.L., Khachatryan-Blinova, G.K., Shirayev, A.A., 2001. Superdeep diamonds from the Juina area, Mato Grosso Stake, Brazil. Contrib. Mineral. Petrol. 40, 734–753.

Klump, J., 1995. A pilot study of the Swartruggens Kimberlite dyke swarm. Unpublished Honor's Thesis. University of Cape Town, South Africa.

Logan, F., 1999. A mineralogical and isotope study of macrodiamonds from the Helam Mine, South Africa. Unpublished Honor's Thesis. Queens University, Canada.

McCandless, T.E., Gurney, J.J., 1989. Sodium in garnet and potassium in clinopyroxene: criteria for classifying mantle xenoliths. Kimberlites and Related Rocks, 2. 4th International Kimberlite Conference. Geological Society of Australia Special Publication, vol. 14, Blackwell, Oxford, pp. 827–832.

Mc Kenna, N., 2002. A study of the diamonds, diamond inclusion minerals and other mantle minerals from the Swartruggens Kimberlite, South Africa. Unpublished MSc Thesis. University of Cape Town, South Africa.

Mendelssohn, M.J., Milledge, H.J., 1995. Geologically significant information from routine analysis of the mid-infrared spectra of diamonds. Int. Geol. Rev. 37, 95–110.

Meyer, H.O.A., 1987. Inclusions in diamond. In: Nixon, P.H. (Ed.), Mantle Xenoliths, Wiley, New York, pp. 501–523.

Moore, R.O., Gurney, J.J., 1985. Pyroxene solid-solution in garnets included in diamonds from Monastery Kimberlite, South Africa. Kimberlites and Related Rocks: 2. Their Mantle/Crust Setting, Diamonds and Diamond Exploration. Geo-

logical Society of Australia Special Publication, vol. 14, Blackwell, Oxford pp. 1029–1041.

Orlov, Yu.L., 1977. Mineralogy of the Diamond (translated from the Russian language). Wiley, New York original published in Russian in 1973 by Izdatel'stvo, Nauka, USSR.

Richardson, S.H., Gurney, J.J., Erlank, A.J., Harris, J.W., 1984. Origin of diamonds in old enriched mantle. Nature 310, 198–202.

Richardson, S.H., Shirey, S.B., Harris, J.W., Carlson, R.W., 2001. Archean subduction recorded by Re–Os isotopes in eclogitic sulfide inclusions in Kimberley diamonds. Earth Planet. Sci. Lett. 5925, 1–11.

Robinson, D.N., 1979. Surface textures and other features of diamonds. Unpublished PhD Thesis. University of Cape Town.

Smith, C.B., 1983. Pb, Sr and Nd isotopic evidence for sources of southern African cretaceous kimberlites. Nature 304, 51–54.

Smith, C.B., Allsopp, H.L., Kramers, J.D., Hutchinson, G., Roddick, J.C., 1985. Emplacement ages of Jurassic–Cretaceous South African kimberlites by the Rb–Sr method on phlogopite and whole-rock samples. Trans. Geol. Soc. S. Afr. 88, 249–266.

Sobolev, E.V., 1978. Nitrogen centers and natural diamond crystal growth. In: Sobolev, E.V. (Ed.), Problems of Lithospheric and Upper Mantle Petrology. Nauka Press, Novosibirsk, pp. 245–255.

Sobolev, N.V., Galimov, E.M., Ivanovskaya, I.N., Yefimova, E.S., 1979. The carbon isotope compositions of diamonds containing crystallographic inclusions. Dokl. Akad. Nauk SSSR 249, 1217–1220.

Sobolev, N.V., Yefimova, E.S., Laverent'yev, Yu.G., Sobolev, V.S., 1984. Dominant calc-silicate association of crystalline inclusions in placer diamonds from southeastern Australia. Dokl. Akad. Nauk SSSR 274, 148–153.

Sunagawa, I., 1984. Growth of crystals in nature. In: Sunagawa, I. (Ed.), Growth of Crystals in Nature Materials Science of the Earth's Interior. Terra Scientific, Tokyo, pp. 63–105.

Taylor, W.R., Jaques, A.L., Ridd, M., 1990. Nitrogen-defect aggregation characteristics of some Australasian diamonds: time-temperature constraints on the source regions of pipe and alluvial diamonds. Am. Mineral. 75, 1290–1310.

Tsai, H.M., Meyer, H.O.A., Moreau, J., Milledge, H.J., 1979. Mineral inclusions in diamond: premier, Jaggersfontein and Finsch kimberlites, South Africa, and Williamson Mine, Tanzania. In: Boyd, F.R., Meyer, H.O.A. (Eds.), Kimberlites, Diatremes and Diamonds: Their Geology, Petrology and Chemistry. American Geophysical Union, Washington, DC, pp. 16–26.

Urusovskaya, A.A., Orlov, Yu. L., 1964. Nature of plastic deformation of diamond crystals. Dokl. Akad. Nauk SSSR 154, 112–115.

Van Royen, J., Palyanov, Yu.N., 2002. High-pressure–high-temperature treatment of natural diamonds. J. Phys. Condensed Matter 14, 10953–10956.

Woods, G.S., 1986. Platelets and the infrared absorption of type Ia diamonds. Proc. R. Soc. Lond., A 407, 219–238.

Woods, G.S., Purser, G.C., Mtimkulu, A.S.S., Collins, A.T., 1990. The nitrogen content of Type Ia natural diamonds. J. Phys. Chem. Solids 51, 1191–1197.

Yefimova, E.S., Sobolev, N.V., Pospelova, L.N., 1983. Sulfide inclusions in diamond and specific features of their paragenesis. Zap. Vses. Mineral. Obsestva 112, 300–310 (in Russian).

Available online at www.sciencedirect.com

Lithos 77 (2004) 209–223

www.elsevier.com/locate/lithos

Peridotitic diamonds from Namibia: constraints on the composition and evolution of their mantle source

Jeff W. Harris[a,*], Thomas Stachel[b], Ingrid Léost[c], Gerhard P. Brey[c]

[a] Division of Earth Sciences, University of Glasgow, Glasgow G12 8QQ, UK
[b] University of Alberta, Earth and Atmospheric Sciences, Edmonton, AB, Canada T6G 2E3
[c] Institut für Mineralogie, Universität Frankfurt, Senckenberganlage 28, 60054 Frankfurt, Germany

Received 27 June 2003; accepted 17 February 2004
Available online 12 May 2004

Abstract

About half the diamonds studied from the Cenozoic placer deposits along the Namibian coast belong to the peridotitic suite. The peridotitic mantle source is heterogeneous ranging from lherzolitic to strongly Ca depleted (down to 0.24 wt.% CaO in garnet) and shows large variations in Cr/Al ratio, illustrated by very low to very high Cr_2O_3 contents in garnet (2.6–17.3 wt.%). The Cr-rich end of this range includes exceptionally high Cr_2O_3 contents in Mg-chromite (70.7 wt.%) and clinopyroxene (3.6 wt.%). Garnet-olivine thermometry appears to indicate two groups, one that equilibrated at temperatures between 1200 and 1220°C and a second between 960 and 1100°C. Combined estimates of pressure and temperature based on garnet-orthopyroxene pairs indicate a large variance in geothermal gradients, corresponding to 38–42 mW/m² surface heat flow.

The trace-element composition of peridotitic garnet inclusions (determined by SIMS) also indicates large diversity. Two principal groups, corresponding to different styles of metasomatic source enrichment, are recognized. The first group ranges from extremely $LREE_N$-depleted patterns, through trough-shaped REE_N to sinusoidal patterns with the position of the first peak gradually moving from the $LREE_N$ to the $MREE_N$. This series of REE patterns is interpreted to reflect a range of metasomatic agents with decreasing LREE/HREE. Only in the case of the two garnets with REE_N peaking at Sm–Eu is this process connected with enrichment in Zr, without significant introduction of Y and Ti. The metasomatism responsible is interpreted as reflecting percolation of CHO-fluids through harzburgite under sub-solidus conditions. A second group of garnets shows an increase from $LREE_N$–$MREE_N$ and almost flat (lherzolitic garnet) to moderately declining $MREE_N$–$HREE_N$ at super-chondritic levels. This second style of metasomatism is caused by an agent carrying HFSE and showing only moderate enrichment in LREE over HREE, which points towards silicate melts.
© 2004 Elsevier B.V. All rights reserved.

Keywords: Diamond; Inclusion; Peridotite; Mantle metasomatism; REE; Placer deposit; Namibia

1. Introduction

Through combined studies on the geothermobarometry and the major- and trace-element composition of mantle xenoliths and xenocrysts from deep-seated

* Corresponding author.
E-mail address: jharris@earthsci.gla.ac.uk (J.W. Harris).

magmas, a detailed picture on the state of the subcontinental lithospheric mantle has emerged (see Griffin et al., 1999a for a recent review). Since the original study of Frey and Green (1974), it has been recognized globally that subcontinental lithospheric mantle is characterized by a two-stage evolution, primary depletion followed by metasomatic re-enrichment (for a review, see, for example, Harte and Hawkesworth, 1989). Metasomatic processes in Archean lithospheric mantle are of particular interest because of their relationship to the formation and destruction of diamonds. Detailed studies on trace-element zonation in xenolith garnets (e.g., Griffin et al., 1999b; Burgess and Harte, 2004) have shown that the metasomatic overprint affecting a single source rock can be very complex, documenting multiple episodes of melts/fluids of vastly different composition passing through these rocks. A highly evolved end-member composition of such metasomatic fluids has recently been recognized as fluid inclusions in diamonds (Izraeli et al., 2001; Klein-BenDavid et al., 2004). Silicate inclusions in diamonds may provide important information on the evolutionary path of mantle metasomatism, as they are completely shielded from all events affecting the surrounding rocks subsequent to their encapsulation. By studying a series of inclusions, it may thus be possible to derive steps in the compositional evolution of deep lithospheric rocks. Here, we present major- and trace-element data on peridotitic inclusions in diamonds from Namibia.

1.1. Samples

The inclusion-bearing diamonds were obtained from alluvial mines along the Orange River close to Oranjemund, as well as from all the alluvial productions along the coastal belt from Oranjemund north to Elisabeth Bay, a distance of approximately 240 km. Additionally, similar diamonds were recovered from the Namibian offshore deposits. Out of 106 diamonds from which inclusions were recovered, 49 (46%) contained inclusions of the peridotitic suite (Table 1). The remaining diamonds belong to the eclogitic suite (41%); an "undetermined suite" (12%), which is characterized by lamellar exsolutions of clino- and orthopyroxene with compositional variability and characteristics of both peridotitic and websteritic diamond sources (Leost

Table 1
Inclusion abundance table for peridotitic diamonds from Namibia

	Diamonds
Single inclusions	
Pyrope garnet	9
Orthopyroxene	6
Mg-chromite	6
Olivine	5
Clinopyroxene	2
Inclusions pairs	
Garnet + olivine	10
Garnet + orthopyroxene	4
Orthopyroxene + olivine	2
Clinopyroxene + olivine	1
2 olivines	1
Inclusions triples	
Clinopyroxene + orthopyroxene + olivine	1
Garnet + orthopyroxene + chromite	1
Garnet + orthopyroxene + olivine	1
Total diamonds	49
Total inclusions	73

et al., 2003) and "normal" websteritic inclusions (cf. Gurney et al., 1984a; Deines et al., 1993) at 1%. However, the relative proportions of the different parageneses in our relatively small sample may not be truly representative of the Namibian production, as additional selection criteria, like the preferential collection of garnet-bearing diamonds for REE analysis and geothermobarometry has probably led to significant bias.

1.2. Analytical methods

Inclusions were released by crushing the host diamonds, before embedding in Araldite® and polishing. Major- and minor-element analyses were performed by EPMA (Jeol JXA-8900 RL at Frankfurt University) at 20 kV gun potential and 20 nA beam current using silicate, oxide and metal standards. Count times range between 30 and 90 s to ensure detection limits of 100 ppm or better for all oxides except Na_2O (200 ppm).

Trace-element analyses of 17 garnet inclusions were carried out using the CAMECA IMS 4f ion microprobe with Charles Evans and Associates interface and control system at Edinburgh University. Positive secondary ions were sputtered by bombardment of the sample with an 8-nA primary beam (beam size about 20–25 μm) of negatively charged oxygen ions. The nominal impact energy of the primary beam on the sample surface was 14.5 keV. An energy offset of 75 ± 20 eV was applied to suppress molecular interferences. Count rates for each element and the background were collected for 50 s. Two to three spots were analysed per sample and then averaged. Initially, corrections were made for the molecular interference of $Si + Fe$ on Rb, but as the correction was much bigger than the value of any Rb present, these results are not presented in Table 4. Corrections were also made for the molecular interference of ZrH on Nb, BaO on Eu, and the LREE oxides on the HREE. Ba concentrations in garnet are generally very low (≤ 0.05 ppm, see Table 4) and were determined to check the magnitude of the BaO interference on Eu. Ba analyses are therefore not considered further in this study. Element abundances were normalised to the SRM610 glass standard and concentrations were calculated from intensity ratios against silica. Results were verified by analysis of other standard minerals (Dutsen Dushowa garnet of Irving and Frey, 1978 and Kilbourne Hole clinopyroxene of Irving and Frey, 1984) and by glass standards. Analytical uncertainties are determined by the elemental abundances but also by the ion yields which are different for each element. In general, analytical errors for concentrations in the ppm level are small ($< 1\%$ for concentrations >100 ppm, $< 10\%$ between 1 and 10 ppm) but become increasingly large at sub-ppm levels. For concentrations of less than 0.01 ppm, errors exceed 50%, and therefore, values <0.005 ppm are reported as 0.00 ppm.

2. Mineral inclusion chemistry

2.1. Garnet

Garnet (25 inclusions from 25 diamonds) compositions are dominantly Ca-undersaturated, indicating clinopyroxene-free sources, with only one garnet being lherzolitic (Fig. 1). CaO contents as low as 0.2 wt.% suggest that some garnets represent highly refractory, dunitic lithologies. Such extreme depletion in Ca has previously only been observed in pyrope garnet inclusions from the Kaapvaal block (Finsch, Gurney et al., 1979; Monastery, Moore and Gurney, 1989; De Beers Pool, Phillips and Harris, 1995; Letseng, McDade and Harris, 1999) and the Limpopo belt (River Ranch,

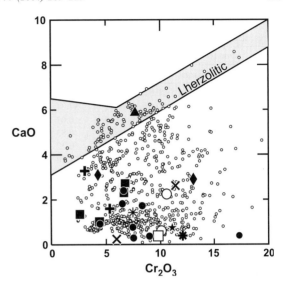

Fig. 1. $CaO–Cr_2O_3$ (wt.%) plot for peridotitic garnets from Namibia and from worldwide sources (small open circles, for references, see Stachel and Harris, 1997a; Stachel et al., 2000). The compositional field for lherzolitic garnets is from Sobolev et al. (1973). The symbols for Namibian garnets correspond to groups with similar REE patterns, as presented in Fig. 7; garnets not analyzed for trace elements are shown as filled black circles.

Kopylova et al., 1997; Venetia, Viljoen et al., 1999) of the Kalahari Craton. The observed large range in chromium contents (Cr_2O_3 of 2.6–17.3 wt.%) of garnets is rare among diamond sources worldwide, and on the Kalahari Craton, such diversity has only been observed at Venetia. A further unusual feature is high TiO_2 (0.42 wt.%) combined with detectable Na_2O (0.04 wt.%) in garnet Nam-46 (Table 2).

2.2. Olivine

Olivine (22 inclusions from 21 diamonds) also is mainly of harzburgitic–dunitic paragenesis (see Table 2). One olivine coexists with clinopyroxene in the same diamond (Nam-51B, Table 2) and two more olivines have similarly high CaO contents (= 0.05 wt.%) and low Mg numbers (91.8 and 92.8) indicative of a possible lherzolitic paragenesis. All other olivines are too Ca poor (CaO ≤ 0.03 wt.%) to have formed in equilibrium with clinopyroxene. This also implies that olivine from diamond Nam-215 (0.02 wt.% CaO, Mg number of 93.5) is not in equilibrium with a clinopyroxene inclusion released from the same sample. Oli-

Table 2
Representative major and minor element composition (EPMA) of peridotitic inclusions from Namibia

Sample inclusion	Nam-16		Nam-18	Nam-24	Nam-29		Nam-31	Nam-33	Nam-46	Nam-50	Nam-51B				Nam-55
	olivine	garnet	chromite	lhz. garnet	opx	garnet	garnet	garnet	olivine	garnet	garnet	olivine	opx	cpx	garnet
P_2O_5	≤0.01	≤0.01	≤0.01	0.02	≤0.01	0.02	0.01	0.04	≤0.01	0.04	0.02	≤0.01	≤0.01	≤0.01	≤0.01
SiO_2	39.96	41.16	0.07	41.12	57.21	42.72	42.17	42.21	40.47	41.51	42.06	40.11	56.67	54.39	42.41
TiO_2	≤0.01	0.07	0.25	0.14	≤0.01	≤0.01	0.01	0.01	≤0.01	0.42	0.02	≤0.01	0.03	0.05	0.02
Al_2O_3	0.03	15.18	1.70	17.68	0.83	22.55	20.71	15.02	0.03	14.45	19.02	0.04	0.73	1.91	19.20
Cr_2O_3	0.07	11.42	70.70	7.71	0.22	2.60	4.26	12.08	0.04	10.58	6.76	0.04	0.19	0.92	5.96
FeO	7.48	6.74	13.00	6.51	3.34	4.93	5.50	4.40	6.16	6.24	5.78	8.44	5.71	2.56	4.70
MnO	0.11	0.37	0.25	0.35	0.06	0.18	0.24	0.22	0.09	0.34	0.27	0.11	0.14	0.11	0.19
NiO	0.39	0.01	0.08	0.01	0.17	0.01	0.01	≤0.01	0.33	0.01	0.02	0.37	0.12	0.06	0.02
MgO	50.79	22.27	14.38	20.13	36.82	25.19	23.22	26.04	51.93	23.43	23.33	49.80	34.37	17.30	26.44
CaO	0.02	2.62	≤0.01	5.86	0.13	1.35	3.10	0.39	0.01	2.24	2.71	0.06	0.58	20.29	0.24
Na_2O	≤0.02	0.04	≤0.02	0.03	0.02	≤0.02	≤0.02	≤0.02	0.02	0.04	≤0.02	0.02	0.10	1.32	≤0.02
K_2O	≤0.01	≤0.01	≤0.01	0.01	≤0.01	≤0.01	≤0.01	≤0.01	≤0.01	≤0.01	≤0.01	≤0.01	≤0.01	0.01	≤0.01
Total	98.85	99.86	100.42	99.56	98.81	99.55	99.24	100.40	99.07	99.31	99.99	98.99	98.64	98.92	99.19

	Nam-64	Nam-65		Nam-71	Nam-92				Nam-94	Nam-101		Nam-104			Nam-108
	garnet	olivine	garnet	garnet	olivine	garnet	opx	olivine	olivine	garnet	olivine	garnet	opx	garnet	garnet
P_2O_5	≤0.01	≤0.01	0.01	≤0.01	≤0.01	0.02	≤0.01	≤0.01	≤0.01	0.01	≤0.01	≤0.01	≤0.01	≤0.01	≤0.01
SiO_2	40.94	41.34	41.47	41.21	40.43	41.73	57.47	40.90	41.03	43.58	40.83	42.33	57.40	42.22	40.88
TiO_2	0.02	≤0.01	0.04	0.01	≤0.01	0.03	≤0.01	≤0.01	≤0.01	0.02	≤0.01	≤0.01	0.01	0.01	0.07
Al_2O_3	11.37	0.03	16.87	16.88	0.02	18.43	0.53	0.02	0.03	21.48	0.03	21.18	1.09	18.92	13.61
Cr_2O_3	17.28	0.04	10.08	9.78	0.03	7.47	0.33	0.07	0.02	3.07	0.01	4.39	0.61	7.53	13.09
FeO	4.68	4.65	4.34	5.11	5.31	5.36	3.48	6.24	6.04	5.39	5.30	4.61	2.81	3.94	5.81
MnO	0.24	0.07	0.20	0.25	0.08	0.23	0.08	0.09	0.08	0.25	0.06	0.19	0.07	0.19	0.32
NiO	0.01	0.32	≤0.01	≤0.01	0.33	0.02	0.19	0.34	0.40	≤0.01	0.38	≤0.01	0.13	0.01	≤0.01
MgO	24.73	53.18	25.61	25.21	52.03	24.48	36.41	51.71	51.57	24.43	52.56	25.41	36.97	26.43	22.20
CaO	0.39	0.01	0.60	0.38	≤0.01	1.43	0.17	0.02	0.03	3.28	0.01	1.02	0.05	0.29	2.90
Na_2O	≤0.02	≤0.02	≤0.02	≤0.02	≤0.02	≤0.02	0.03	≤0.02	≤0.02	≤0.02	≤0.02	≤0.02	≤0.02	≤0.02	≤0.02
K_2O	≤0.01	≤0.01	≤0.01	≤0.01	≤0.01	≤0.01	≤0.01	≤0.01	≤0.01	≤0.01	≤0.01	≤0.01	≤0.01	≤0.01	≤0.01
Total	99.66	99.63	99.23	98.82	98.22	99.20	98.68	99.40	99.20	101.50	99.19	99.12	99.16	99.54	98.89

	Nam-109		Nam-110		Nam-113		Nam-118		Nam-214
	garnet	opx	opx	garnet	olivine	garnet	garnet	olivine	cpx
P_2O_5	≤0.01	≤0.01	≤0.01	≤0.01	≤0.01	≤0.01	0.05	≤0.01	≤0.01
SiO_2	41.78	57.63	57.17	41.85	40.64	42.00	41.28	41.12	54.59
TiO_2	≤0.01	≤0.01	≤0.01	≤0.01	≤0.01	≤0.01	≤0.01	≤0.01	0.14
Al_2O_3	19.17	0.36	0.71	18.76	0.02	20.23	16.07	0.03	2.66
Cr_2O_3	6.64	0.19	0.25	6.65	0.03	5.34	11.10	0.03	3.57
FeO	5.94	3.70	4.13	5.88	6.59	6.35	5.82	5.72	1.89
MnO	0.29	0.09	0.10	0.29	0.09	0.32	0.31	0.07	0.09
NiO	0.01	0.12	0.13	≤0.01	0.36	≤0.01	≤0.01	0.36	0.04
MgO	23.44	36.51	35.63	23.17	51.18	23.76	24.43	52.97	15.05
CaO	2.41	0.23	0.50	2.31	0.01	1.60	0.74	≤0.01	18.03
Na_2O	≤0.02	≤0.02	0.03	≤0.02	≤0.02	≤0.02	≤0.02	≤0.02	3.10
K_2O	≤0.01	≤0.01	≤0.01	≤0.01	≤0.01	≤0.01	≤0.01	≤0.01	0.01
Total	99.67	98.84	98.64	98.91	98.91	99.60	99.80	100.30	99.19

vines have a poorly defined mode in Mg number between 93.5 and 95 (Fig. 2) and thus fall on the Mg-rich side of the worldwide data set (mode at class 92.5– 93.0) but are very similar to the Mg-rich olivines from the Kaapvaal, in particular the forsterite rich olivines from De Beers Pool (Wilding et al., 1994; Phillips and

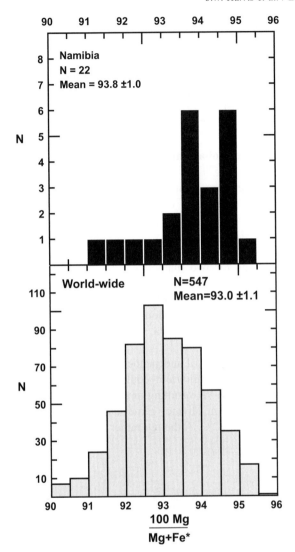

Fig. 2. Forsterite content in olivine inclusions from Namibia and from worldwide sources.

Harris, 1995), Roberts Victor (Gurney et al., 1984b) and Finsch (Gurney et al., 1979).

2.3. Pyroxene

Pyroxene inclusions of peridotitic and websteritic paragenesis are described in detail in Leost et al. (2003). Out of 15 peridotitic orthopyroxenes (from 15 diamonds), one with a low Mg number (91.5) and high Na_2O (0.10 wt.%) is associated with a clinopyroxene. Four more orthopyroxenes with Mg

numbers ≤ 94.3 also have high CaO (≥ 0.45 wt.%, Fig. 3) and in one case (Nam-37) high Na_2O (0.19 wt.%) and thus may be of lherzolitic paragenesis as well. However, one of these inclusions (Nam-110) occurs together with harzburgitic garnet. The remaining, more magnesian orthopyroxenes (Mg number of 94.6–95.9) have low CaO (0.23–0.04 wt.%, Fig. 3) and thus are all of harzburgitic paragenesis.

Only four clinopyroxene inclusions (from four diamonds) were observed which were not associated with an orthopyroxene exsolution ("undetermined paragenesis" of Leost et al., 2003) or with eclogitic or websteritic chemistries. Two of these lherzolitic clinopyroxenes are very high in Cr_2O_3 (3.3 and 3.6 wt.%), compared to both the other Namibian samples (0.34–0.09 wt.%) and worldwide data. Nam-214 contains 3.1 wt.% Na_2O (Table 2), which is not commonly observed for peridotitic clinopyroxenes.

2.4. Chromite

Chromite occurred as single inclusions in six diamonds. A seventh chromite associated with harzburgitic garnet and orthopyroxene yielded only poor totals and is not considered here. The spinels are homoge-

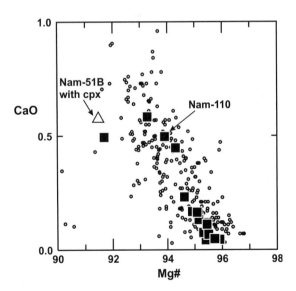

Fig. 3. CaO content (wt.%) vs. Mg number for orthopyroxene inclusions from Namibia (squares and triangle) and from worldwide sources (open circles). The orthopyroxene in diamond Nam-110 occurs together with a garnet inclusion of harzburgitic paragenesis.

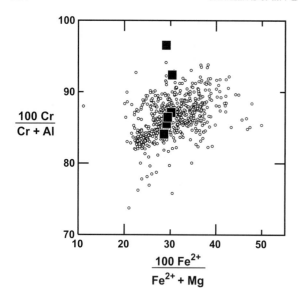

Fig. 4. Cr number vs. FFM ratio for chromite inclusions from Namibia (filled squares) and from worldwide sources (open circles).

neous in their FFM ($100Fe^{2+}/[Fe^{2+}Mg^{2+}]$) ratio (29–30, Fig. 4). Apart from Nam-36 (ferric iron ratio of 12), all have about 20% iron in the ferric state. The positive correlation between Cr number and FFM ratio, which is usually characteristic for

spinels, is not observed. With a Cr_2O_3 content of 70.7 wt.% Mg-chromite, Nam-18 (Table 2) approaches pure end-member composition. If this spinel had formed in a garnet harzburgite source, then it indicates unusually high pressure (or low temperature) of formation (Doroshev et al., 1997).

3. Geothermobarometry

Temperature estimates based on the Mg–Fe exchange between olivine–garnet (O'Neill and Wood, 1979; O'Neill, 1980) and orthopyroxene–garnet (Harley, 1984) were possible for 14 diamonds of harzburgitic paragenesis. Temperature conditions for two diamonds with lherzolitic inclusions were assessed based on the two-pyroxene and the Ca-in-orthopyroxene thermometers of Brey and Köhler (1990). Calculations were carried out for a fixed pressure of 5 GPa, and results are listed in Table 3.

Combinations of independent exchange reactions as a test for equilibrium could only be applied to diamond Nam-92, resulting in a difference in temperatures based on olivine–garnet and orthopyroxene–garnet of 109 °C. Although rather large, this temper-

Table 3

Temperature (°C) and pressure (GPa) estimates for coexisting (nontouching) peridotitic inclusions from Namibia

Sample	Inclusions	T_{BKN}	$T_{Ca-in-opx}$	T_{OW}	T_{Harley}	$T_{Har/BKN}$	$P_{BKN/Har}$
Nam-16	gt,olivine			1079			
Nam-29	gt,opx				1141	1130	4.8
Nam-46	gt,olivine			982			
Nam-65	gt,olivine			1101			
Nam-71	gt,olivine			1063			
Nam-72	gt,olivine			1223			
Nam-92	gt,opx,olivine			1203	1094	1139	5.7
Nam-94	gt,olivine			1209			
Nam-101	gt,olivine			1206			
Nam-104	gt,opx				1212	1232	5.3
Nam-109	gt,opx				1041	1102	6.0
Nam-110	gt,opx				1138	1108	4.5
Nam-113	gt,olivine			1041			
Nam-118	gt,olivine			961			
Nam-51B	cpx,opx,olivine	1092	1090				

T_{BKN} and $T_{Ca-in-opx}$ are the two-pyroxene and the Ca-in-orthopyroxene thermometers of Brey and Köhler (1990), T_{OW} is the garnet–olivine thermometer of O'Neill and Wood (1979) and O'Neill (1980), T_{Harley} is the garnet–orthopyroxene thermometer of Harley (1984). Calculations are based on an assumed pressure of 5 GPa. $T_{Har/BKN}$ and $P_{BKN/Har}$ are simultaneous estimates of temperature and pressure based on a combination of the garnet–orthopyroxene thermometer (Harley, 1984) and barometer (Brey and Köhler, 1990).

ature difference is consistent with the predicted accu-racy of the two thermometers (Brey and Köhler, 1990).

Olivine–garnet thermometry appears to indicate a bimodal inclusion temperature distribution (Fig. 5), with a relatively "hot" set (four diamonds) sitting at 1200–1220°C and a second, "cooler" group (six diamonds) ranging between 960 and 1100°C. The five diamonds with garnet–orthopyroxene pairs fall between 1040 and 1210°C and clinopyroxene–orthopyroxene thermometry for lherzolitic diamond Nam-51B indicates formation at around 1090°C. These temperatures fall within the ranges observed for diamond formation worldwide (Stachel et al., 2002).

For the five diamonds containing garnet and ortho-pyroxene, pressure and temperature may be estimated simultaneously by combining the thermometer of Harley (1984) with the barometer of Brey and Köhler (1990). Apart from one "hot diamond," the other four indicate a narrow temperature range (1100–1140°C) but span a large range in pressure (4.5–6GPa) corresponding to a range in geothermal gradients (Fig. 6). Taking the results at face value, a shallow group (140–165 km) falling at a conductive geotherm corresponding to 42 mW/m² surface heat flow may be distinguished from a second deeper group (180–190

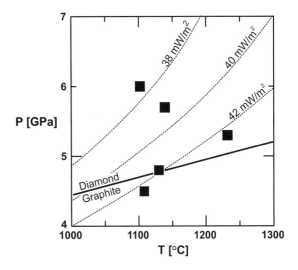

Fig. 6. Pressure and temperature estimates for coexisting inclusions of garnet and orthopyroxene in diamonds from Namibia. Geother-mal gradients are calculated after Pollack and Chapman (1977) and correspond to surface heat flow of 38–42 mW/m². The diamond–graphite transition is from Kennedy and Kennedy (1976).

km) plotting close to a gradient corresponding to 38 mW/m² surface heat flow.

4. Discussion of major-element chemistry

Peridotitic diamonds from Namibia represent com-positionally heterogeneous mantle sources character-ized by high harzburgite/lherzolite and Mg/Fe ratios, and this inclusion suite resembles that from known diamond sources on the Kaapvaal block. Such simi-larities include a high proportion of garnets with less than 2 wt.% CaO and high Mg numbers in olivine. A distinguishing feature to known inclusions in dia-monds from the Kaapvaal is the large range in the Cr content of garnet (Cr_2O_3 of 2.6–17.3 wt.%). The largest variation in garnet composition from the Kaapvaal block is observed for inclusions in the De Beers Pool diamonds (Phillips and Harris, 1995 and unpublished data supplied by the De Beers Geosci-ence Centre), with Cr_2O_3 concentrations between 3.0 and 15.1 wt.%. This compositional array appears very similar to the present data, but the DeBeers Pool range is based on the study of a very large sample (>200 garnet inclusions), and thus, differences in Cr_2O_3 may be significant. Sampling of unusually Cr-rich diamond

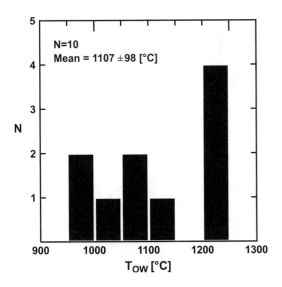

Fig. 5. Garnet–olivine thermometry for Namibian diamonds. Class size is 50 °C.

sources is also indicated by the very high Cr contents of some clinopyroxenes and chromites and a very Cr-rich "lherzolitic" garnet (Nam-217B, 15.6 wt.% Cr_2O_3) described by Leost et al. (2003) as part of the "undetermined" suite.

A comparison of geothermobarometric data for diamonds from Namibia and the Kaapvaal is hampered by statistically insufficient data sets for both localities. Olivine–garnet thermometry for the Kaapvaal is dominated by data from Finsch (Gurney et al., 1979; Deines et al., 1984) which typically yields temperatures below 1100°C, but other mines, such as Roberts Victor (Gurney et al., 1984b), extend the temperature range for the Kaapvaal to >1300°C. The apparent large spread in geothermal gradients observed for Namibia may indicate mixing of more than one source. However, a similarly large range (conductive geotherms corresponding to surface heat flow of 37–41 mW/m²) was documented by Phillips and Harris (1995) for diamonds from the De Beers Pool kimberlites alone. Still, the "high" geothermal gradient (corresponding to ≥ 42 mW/m² surface heat flow) obtained for three out of five garnet–orthopyroxene pairs from Namibia is not commonly observed for the Kaapvaal, where PT data are mainly available from De Beers Pool (Phillips and Harris, 1995), Koffiefontein (Rickard et al., 1989) and Letseng (McDade and Harris, 1999).

5. Trace-element composition of garnet inclusions

REE_N (N = chondrite normalized) pattern of peridotitic garnets are extremely heterogeneous, and based on overall similarities, a large number of different groups may be recognized (Fig. 7A–D).

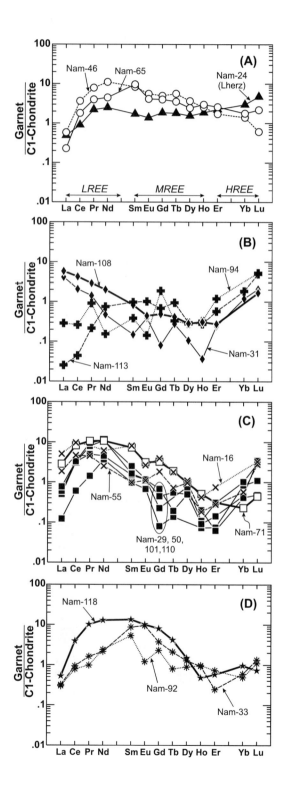

Fig. 7. REE concentrations in peridotitic garnet inclusions from Namibia normalized to the C1-chondrite composition of McDonough and Sun (1995). (A) Shown is the only lherzolitic garnet studied (Nam-24) and two harzburgitic garnets with hump shaped patterns. (B) A sequence from highly LREE-depleted garnets (Nam-94 and 113) to trough-shaped patterns. (C) Garnets with typical sinusoidal REE_N pattern. The first peak is located at Ce_N–Pr_N for garnets Nam-16 and -55 and at Pr_N–Nd_N for Nam-16, -29, -50, -71, -101 and -110. Nam-71, however, is different from the latter group in lacking a steep positive slope within the $HREE_N$. (D) For garnet Nam-118, the first peak is at Nd_N–Sm_N and for Nam-33 and -92, it is located entirely within the $MREE_N$ (Sm_N–Eu_N).

Table 4
Trace-element analyses (SIMS) of peridotitic garnet inclusions from Namibia

Sample inclusions	Nam-24 grt (lherz)	Nam-46 grt, ol	Nam-65 grt, ol	Nam-31 grt	Nam-108 grt	Nam-94 grt, ol	Nam-113 grt, ol	Nam-55 grt	Nam-16 grt, ol	Nam-29 grt, ol	Nam-50 grt	Nam-101 grt, ol	Nam-110 grt, opx	Nam-71 grt, ol	Nam-118 grt, ol	Nam-33 grt	Nam-92 grt, opx, ol
Sr	0.26	4.91	1.09	0.81	15.32	0.72	0.10	1.65	11.00	0.29	1.58	1.23	3.94	14.30	9.57	0.49	0.17
Y	2.84	3.57	3.13	0.21	0.19	0.24	0.39	0.37	0.39	0.11	0.33	0.11	0.23	0.36	0.59	0.69	0.79
Zr	7.23	25.90	20.40	0.30	1.13	0.15	1.37	3.79	2.10	0.03	2.24	0.50	1.99	1.28	6.03	44.00	12.00
Nb	1.00	0.23	0.22	0.09	0.41	0.03	0.13	0.00	1.30	0.27	0.14	0.44	0.06	0.41	0.09	0.57	0.99
Ba	0.01	0.00	0.02	0.01	0.03	0.01	0.01	0.04	0.05	0.01	0.02	0.03	0.01	0.02	0.02	0.02	0.02
La	0.11	0.14	0.05	0.97	1.37	0.07	0.01	0.44	1.20	0.03	0.12	0.14	0.18	0.68	0.13	0.08	0.07
Ce	0.55	2.24	1.11	1.26	2.59	0.16	0.03	2.84	5.90	0.37	1.91	1.97	2.01	5.03	2.41	0.58	0.48
Pr	0.21	0.73	0.37	0.13	0.27	0.08	0.02	0.42	0.77	0.13	0.74	0.46	0.88	0.98	0.94	0.15	0.09
Nd	1.14	5.06	2.05	0.22	0.91	0.07	0.34	1.16	2.80	1.69	4.97	2.08	5.12	4.91	5.88	0.99	1.10
Sm	0.25	1.23	1.43	0.02	0.12	0.06	0.14	0.15	1.20	0.14	0.37	0.22	0.24	1.08	1.98	1.29	0.79
Eu	0.08	0.31	0.23	0.02	0.03	0.01	0.06	0.06	0.15	0.04	0.07	0.07	0.07	0.18	0.57	0.53	0.07
Gd	0.37	0.99	0.80	0.02	0.10	0.37	0.14	0.36	0.78	0.02	0.05	0.14	0.09	0.65	1.59	0.75	0.46
Tb	0.07	0.20	0.13	0.01	0.02	0.01	0.03	0.03	0.06	0.01	0.00	0.01	0.02	0.07	0.15	0.08	0.03
Dy	0.38	0.90	0.60	0.03	0.07	0.07	0.07	0.24	0.27	0.00	0.19	0.12	0.12	0.24	0.37	0.31	0.22
Ho	0.10	0.14	0.16	0.00	0.02	0.02	0.02	0.01	0.02	0.00	0.01	0.01	0.01	0.03	0.03	0.05	0.05
Er	0.36	0.28	0.42	0.04	0.04	0.19	0.09	0.05	0.12	0.01	0.05	0.02	0.01	0.05	0.09	0.04	0.12
Yb	0.48	0.23	0.29	0.19	n.a.	n.a.	0.30	0.09	n.a.	0.14	0.17	0.07	0.07	0.04	0.16	0.10	0.08
Lu	0.12	0.02	0.05	0.05	0.04	0.13	0.12	0.07	0.08	0.03	0.07	0.09	0.01	0.01	0.02	0.03	0.03

Concentrations are given in wt. ppm and are rounded to the second decimal place; concentrations of 0.00 ppm therefore refer to values <0.005 ppm. "n.a." stands for "not analyzed."

The single lherzolitic garnet studied (Nam-24) shows depletion relative to C1-chondrite for La and Ce (Fig. 7A) and flat MREE (MREE = Sm–Ho) and HREE at about two to three times chondritic levels. Two harzburgitic garnets (Nam-46 and -65) are similar to the lherzolitic pattern in their straight MREE$_N$–HREE$_N$, but have higher Ce–Ho and lower HREE (Fig. 7A). Fig. 7B depicts four garnets with unusual REE$_N$ patterns: Nam-31 and -108 are trough shaped and MREE$_N$ depleted, Nam-94 is depleted and has flat (apart from spikes at Pr and Gd) LREE$_N$–MREE$_N$ and Nam-113 shows extreme depletion in La$_N$–Ce$_N$, flat and almost chondritic values from Nd–Tb, a trough at Dy$_N$–Ho$_N$ and then, like the other three samples, a steep positive slope towards Lu$_N$.

The majority of harzburgitic garnets from Namibia have sinusoidal REE$_N$ patterns, as shown in Fig. 7C and D. This type of pattern is characteristic for harzburgitic garnets from both cratonic peridotite

xenoliths and diamonds (Shimizu and Richardson, 1987; Stachel et al., 1998; Shimizu et al., 1999). Subtle differences are recognized within this sinusoidal group: for garnets Nam-16 and -55, the first peak already occurs at Ce$_N$–Pr$_N$, whereas all other garnets in Fig. 7C peak at Pr$_N$–Nd$_N$. From this latter group, garnet Nam-71 is distinct because the second minimum is shifted from around Ho$_N$ to Yb$_N$, with only a small upward kink towards Lu$_N$. For the three garnets in Fig. 7D, the first peak is shifted even further towards the MREE and is very broad for Nam-118. A strong upward kink within the HREE$_N$ is only observed for Nam-33.

Four of the harzburgitic inclusions show a distinct correlation between REE$_N$ patterns and HFSE concentrations (Table 4). Nam-33 and 92, which show weakly sinusoidal REE$_N$ peaking at Sm (Fig. 7D), both have elevated Zr (44 and 12 ppm, respectively) at relatively low Y (< 1 ppm, see Fig. 8), and the hump-shaped REE$_N$ of Nam-46 and -65 are accompanied by elevated contents in both Zr (>20 ppm) and Y (about 3 ppm, Fig. 8). All four garnets have CaO <2.3 wt.% and Cr$_2$O$_3$ >7wt.% (Fig. 1). The other two garnets plotting in the same low Ca–high Cr field in Fig. 1 (Nam-71 and -118) also lack the significant MREE depletion otherwise typical for harzburgitic garnets.

6. Discussion of trace-element chemistry

Sinusoidal REE$_N$ patterns, as depicted in Fig. 7C, are discussed controversially in the current literature. Hoal et al. (1994) proposed a model relating sinusoidal REE$_N$ to incomplete reequilibration of preexisting, depleted garnet with steep positive LREE$_N$–HREE$_N$ with a liquid with high LREE and high LREE/HREE. This model was modified by Shimizu et al. (1997), who suggested that disequilibrium growth would cause precipitation of garnet that mimics the REE of the crystallizing medium. Subsequent reequilibration proceeds from LREE to HREE, with sinusoidal patterns resulting at intermediate stages. This latter model requires rapid crystallization as observed in super-cooled magmas, which indeed causes trace-element distribution coefficients to approach towards a value of 1. However, mantle metasomatism involving strong thermal disequilibrium between a percolating melt/ fluid and the surrounding peridotite is not feasible. In

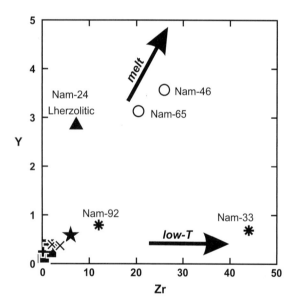

Fig. 8. Y vs. Zr (in ppm) for garnet inclusions from Namibia. The majority of analyses indicate highly depleted concentrations in these elements. Nam-92 and Nam-33 show re-enrichment in Zr without significant addition of Y. This coincides with the "low-temperature" metasomatic trend of Griffin et al. (1999b). Nam-46 and -65 are the two garnets with hump-shaped REE$_N$, together with lherzolitic garnet Nam-24, they fall on a re-enrichment trend with "high" Y/Zr ('melt metasomatism' trend). Note that with the exception of Nam-33, all garnets plot in the "depleted" field of Griffin et al. (1999b), indicating that the observed metasomatism by melts and fluids was relatively mild.

addition, the disequilibrium models of Hoal et al. (1994) and Shimizu et al. (1997) require strongly decreasing diffusion rates from LREE to HREE, implying a connection between ionic radius and diffusion coefficients. Van Orman et al. (2002) have shown experimentally that diffusion coefficients for Ce, Sm. Dy and Yb in pyrope garnet at a pressure of 2.8 GPa and temperatures of 1200–1450 °C are indistinguishable from each other within the measurement uncertainty.

Thus, the REE patterns of garnet inclusions are considered as snapshots reflecting equilibrium with the bulk rock and/or the metasomatic medium (fluid/melt) at the moment of encapsulation in the host diamond. Under this premise, the large diversity in trace-element patterns observed for Namibia allows us to assess the composition and history of peridotitic diamond sources in far more detail than in the cases where only one or two end-member types of REE patterns are observed (e.g., Stachel et al., 1998).

From low MREE–HREE and Y contents, it is apparent that all Namibian garnets, including the one lherzolitic sample, come from depleted source rocks. As noted previously (e.g., Hoal et al., 1994; Stachel and Harris, 1997b; Shimizu, 1999), the original depleted pattern of these source rocks is often well preserved in steep positive slopes within the $HREE_N$. Corresponding whole rock patterns with such steep positive slopes from $LREE_N$ to $HREE_N$ are observed, for example, in highly depleted oceanic residues such as the Voykar Massif, Polar Urals (Sharma and Wasserburg, 1996). Fig. 7B and C shows excellent examples for such steep positive slopes in the $HREE_N$. Super-chondritic Lu concentrations in these garnets must not be interpreted to reflect near-primitive HREE in the diamond source region: HREE in these rocks are quantitatively hosted in garnet which occurs in low modal abundance (average for cratonic harzburgite is as low as 2%, McDonough and Rudnick, 1998), thus a relative "enriching" HREE by more than an order of magnitude over the whole rock value.

On the premise that the source rocks for all the garnets studied here initially were LREE depleted and possessed low $LREE_N/HREE_N$, it follows that the variability in REE_N pattern seen in Fig. 7A–D cannot be satisfactory explained by varying degrees of melt depletion combined with variable modal amounts of

garnet. The extent of these effects can be estimated from the variation in concentration and slope in the $HREE_N$ in Fig. 7B and D and accounts for variability of about an order of magnitude in HREE concentrations. Variations in LREE and MREE exceed this range by far, clearly suggesting that different styles and degrees of metasomatic re-enrichment have to be invoked to explain this observation.

As inclusions in diamonds are shielded from the influence of later metasomatic events, the range of REE pattern in Fig. 7 may be interpreted as a sequence representing various stages of source metasomatism. Based on linear interpolation of the $HREE_N$ slopes in Fig. 7B and D, the pre-metasomatic La content must have been less than 0.001 of the chondritic abundance. Although probably more complex in detail, Fig. 7B may be interpreted to show the rotation of LREE from these ultra-depleted values through intermediate concentrations (Nam-94 and 113) to La and Nd enrichment (relative to primitive garnet compositions). This metasomatic process is not associated with re-enrichment in HREE, Y and Zr, implying the involvement of a highly fractionated fluid/melt. "Chromatographic" processes analogous to laboratory methods (Hofmann, 1972; Navon and Stolper, 1987; Bodinier et al., 1990) provide a feasible model to explain selective enrichment in the most incompatible trace elements (LREE) only. Following such a "chromatographic" fractionation model, the REE_N pattern in Fig. 7C and D represent "snap shots" over a sequence of subsequent metasomatic pulses or along a chromatographic column, with the initial peak at La_N (Fig. 7B) gradually moving to the $MREE_N$ (Sm–Eu, Fig. 7D). Thus, the metasomatic agents affecting the sources of the garnets shown in Fig. 7D are the least fractionated in LREE/HREE which is consistent with the observed overprint of the initial positive slope within the $HREE_N$ and with enrichment in other HFSE, most notably Zr. LREE concentrations in Nam-33, -92 and -118 (Fig. 7D) are close to garnets from primitive mantle bulk rock compositions (PHN 1611 Shimizu, 1975; J-4, data of Jagoutz and Spettel in table 1 of Stachel et al., 1998), HREE are depleted by a factor of about 10. Despite the relative HREE enrichment indicated by the difference in slope (Fig. 7D), the HREE concentrations are lower than for the majority of garnets depicted in Fig. 7B and C which is interpreted as an expression of increasing modal garnet rather than lower HREE in the bulk rock.

The sequence depicted in Fig. 7B–D appears to represent an evolution from a highly fractionated metasomatic agent with high $LREE_N/HREE_N$ and low HFSE to a fluid/melt with decreasing $LREE_N/HREE_N$ and increasing contents in more incompatible HFSE such as Zr. The very steep positive slopes from La_N to Pr_N–Nd_N for garnets Nam-29, -50 and -110 (Fig. 7C) suggest that fractionation of a LREE phyllic phase (e.g., crichtonite group) may have occurred, leading to relative MREE enrichment. The lack of significant enrichment in HREE, Y and Ti even in the most advanced stages of this style of metasomatism seems to rule out known types of silicate melts as metasomatic agents. This is consistent with the observation that equilibration temperatures of silicate inclusions are below the harzburgitic solidus (in the presence of CHO, Wyllie, 1987), precluding percolation of non-hydrous melts as they would freeze on equilibration with the host rock. Griffin et al. (1999b) classified metasomatism characterized by high Zr/Y ratios (such as evidenced by Nam-33 and 92 in Fig. 8) as "low temperature" or "phlogopite-metasomatism". Elevated Sr contents in some of the samples suggest that carbonatites may have been involved to cause this re-enrichment. However, the low Sr contents (≤ 1 ppm) in two of the three most Zr-rich samples (Nam-33 and -92) imply that carbonatites probably are not the dominant metasomatic agent. In the absence of constraints on the fO_2 conditions for both metasomatic agent and host rock, a range of CHO fluids from carbonatite–H_2O to CH_4–H_2O mixes are possible. In addition, "percolative fractional crystallization" (Harte et al., 1993) may lead to Ti and HFSE depleted hydrous silicate liquids suitable to cause the observed style of metasomatism, without ever erupting at the Earth's surface (Burgess and Harte, 2004).

A different style of source metasomatism is documented by Nam-46 and -65 (Fig. 7A): The characteristic evidence for previous depletion, in the form of steep positive slopes for $HREE_N$, is completely erased and instead, a negative slope from $MREE_N$ to $HREE_N$ is observed, corresponding to near-primitive (i.e., similar to garnet J-4, see above) concentrations for La–Eu and depleted HREE. Despite this HREE

depletion, the change in slope documents significant modification of the HREE pattern, in accordance with the observed enrichment in Y, Zr (see Fig. 8) and Ti (the latter only in the case of Nam-46). Such metasomatism connected with HFSE enrichment is usually assigned to silicate melts (e.g., Griffin et al., 1999b). Taking the high Cr content (7.7 wt.% Cr_2O_3) of the single lherzolitic garnet (Nam-24) as an indication of previous source depletion, then metasomatic re-enrichment by silicate melts may be proposed for the source of this diamond as well.

The five harzburgitic garnets showing the strongest effects of metasomatic overprint (Fig. 7A and D) all have high Cr_2O_3 (7.6–12.1 wt.%), which appears to suggest a possible connection between major- and trace-element composition. However, as Fig. 9 shows, this characteristic is not shared by all high Cr garnets, excluding a direct relationship. High Cr garnets reflect source rocks with high Cr/Al, which is considered to be a consequence of melt depletion. Thus, it may be concluded that the most depleted rocks are the most sensitive to metasomatic re-enrichment because of very low garnet contents. However, it is the less fractionated character of the metasomatic fluids/melts (i.e., their lower LREE/HREE and LREE/HFSE) that distinguishes the overprint for these rocks. Therefore, a correlation between the depth of formation and

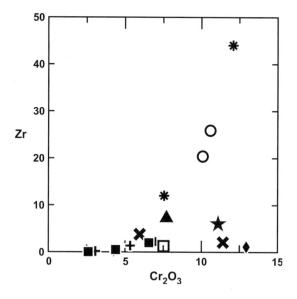

Fig. 9. Zr (ppm) vs. Cr_2O_3 (wt.%) in garnet from Namibia. High Zr is restricted to garnets with high Cr_2O_3.

metasomatic overprint may be invoked, relating the style of metasomatism to the distance from the base of the lithosphere. Experimental (Doroshev et al., 1997) and empirical (Grütter and Sweeney, 2000) studies have shown that the Cr content of garnet in equilibrium with Mg-chromite is positively correlated to pressure, temperature and Ca content. Cr-rich, Ca-poor garnets such as Nam-33, -65, and -118 thus must be derived from depth several tens of kilometers below the intersection of the local geotherm with the graphite–diamond transition. However, because equilibrium with Mg-chromite is the prerequisite for this depth sensitivity, this statement cannot be reversed to deduce a shallow origin for Cr-poor garnets, as these may well be derived from deep sources with low Cr/Al. Considering that Namibian diamonds represent placer deposits such apparent correlations may actually reflect sampling of more than one primary diamond source and thus could also relate to lateral variations in the mantle lithosphere.

7. Conclusions

The mantle source of peridotitic diamonds from Namibia is characterized by extreme depletion in Ca and high Mg numbers and has a very wide range in Cr/Al ratios. The first of these characteristics is a feature of the diamond mines in the Kimberley area. The inclusions also define a thermal regime corresponding to $38-42$ mW/m^2 surface heat flows, the upper limit of which is slightly high for the central Kaapvaal, although information is again largely confined to diamonds from DeBeers Pool. The high frequency (about 40%) of eclogitic diamonds in the Namibian sample may indicate a Proterozoic origin resulting from tectonothermal events affecting the Kaapvaal roots (Shirey et al., 2002), which may in turn suggest at least part of the Namibian diamonds are derived from a peripheral cratonic setting.

Trace-element patterns in pyrope garnet inclusions from Namibia reveal two principal styles and a range in the intensity of source metasomatism. The majority of garnets are affected by a metasomatic agent characterized by low HREE and HFSE (in particular Y and Ti) and very high LREE/HREE. The resulting REE$_N$ pattern can be arranged into a sequence beginning with trough-shaped REE$_N$ and then proceeding to

sinusoidal patterns with the first peak gradually moving from the LREE$_N$ to the MREE$_N$. This sequence corresponds to a continuous decrease in the LREE/HREE ratio of the metasomatic agent. In some cases, the LREE/MREE ratio in the metasomatic medium was probably modified by fractionation of a LREE phyllic phase. It is assumed that the sequence of decreasing LREE/HREE reflects decreasing fractionation of the metasomatic agent, which in a chromatographic model may correspond to a progression in time or to increasing proximity to the source of the metasomatic medium. This style of metasomatism is probably related to CHO fluids percolating through harzburgite under sub-solidus conditions.

A second style of metasomatic overprint is characterized by less fractionated LREE$_N$/HREE$_N$ and consequently leads to significant re-enrichment in HREE and HFSE as well, causing hump-shaped REE$_N$ pattern with a progressive decrease in the negative slope from MREE$_N$ to HREE$_N$. It is assumed that the apparently "primitive" patterns of lherzolitic garnets (positive slope in LREE$_N$, flat, super-chondritic MREEN–HREE$_N$) are the final result of this style of re-enrichment. This kind of metasomatism is commonly associated with silicate melts (e.g., Griffin et al., 1999b) possibly corresponding to the megacryst magma (e.g., Burgess and Harte, 1999).

Acknowledgements

For their outstanding support, we thank the members of the Edinburgh ion probe facility: John Craven, Richard Hinton and Paula McDade. The project was funded through the German Research Foundation (DFG), T.S. acknowledges the Canada Research Chairs program (CRC). Samples and additional support were provided by DeBeers Consolidated Mines. The manuscript benefited from constructive reviews by Galina Bulanova and Oded Navon.

References

Bodinier, J.L., Vasseur, G., Vernières, J., Dupuy, C., Fabriès, J., 1990. Mechanisms of mantle metasomatism: geochemical evidence from the Lherz orogenic peridotite. J. Petrol. 31, 597–628.

Brey, G.P., Köhler, T., 1990. Geothermobarometry in four-phase lherzolites, II: New thermobarometers, and practical assessment of existing thermobarometers. J. Petrol. 31, 1353–1378.

Burgess, S.R., Harte, B., 1999. Tracing lithospheric evolution trough the analysis of heterogeneous G9/G10 garnets in peridotite xenoliths, I: Major element chemistry. In: Gurney, J.J., Gurney, J.L., Pascoe, M.D., Richardson, S.H. (Eds.), The J.B. Dawson Volume, Proceedings of the VIIth International Kimberlite Conference. Red Roof Design, Cape Town, pp. 66–80.

Burgess, S.R., Harte, B., 2004. Tracing lithosphere evolution through the analysis of heterogeneous G9/G10 garnets in peridotite xenoliths, II: REE chemistry. J. Petrol. 45, 609–634.

Deines, P., Gurney, J.J., Harris, J.W., 1984. Associated chemical and carbon isotopic composition variations in diamonds from Finsch and Premier kimberlite, South Africa. Geochim. Cosmochim. Acta 48 (2), 325–342.

Deines, P., Harris, J.W., Gurney, J.J., 1993. Depth-related carbon isotope and nitrogen concentration variability in the mantle below the Orapa kimberlite, Botswana, Africa. Geochim. Cosmochim. Acta 57 (12), 2781–2796.

Doroshev, A.M., Brey, G.P., Girnis, A.V., Turkin, A.I., Kogarko, L.N., 1997. Pyrope–knorringite garnets in the Earth's upper mantle: experiments in the MgO–Al$_2$O$_3$–SiO$_2$–Cr$_2$O$_3$ system. Russian Geol. Geophys. 38, 559–586.

Frey, F.A., Green, D.H., 1974. The mineralogy, geochemistry and origin of lherzolite inclusions in Victorian basanites. Geochim. Cosmochim. Acta 38, 1023–1059.

Griffin, W.L., O'Reilly, S.Y., Ryan, C.G., 1999a. The composition and origin of subcontinental lithospheric mantle. In: Fei, Y., Bertka, C.M., Mysen, B.O. (Eds.), Mantle Petrology: Field Observations and High Pressure Experimentation: A tribute to Francis R. (Joe) Boyd. Special Publication. The Geochemical Society, Houston, TX, pp. 13–45.

Griffin, W.L., Shee, S.R., Ryan, C.G., Win, T.T., Wyatt, B.A., 1999b. Harzburgite to lherzolite and back again: metasomatic processes in ultramafic xenoliths from the Wesselton kimberlite, Kimberley, South Africa. Contrib. Mineral. Petrol. 134 (2–3), 232–250.

Grütter, H.S., Sweeney, R.J., 2000. Tests and constraints on single-grain Cr-pyrope barometer models: some initial results. GAC-MAC Annual Joint Meeting, Calgary. CD. (CD-ROM, GeoCanada, 2000).

Gurney, J.J., Harris, J.W., Rickard, R.S., 1979. Silicate and oxide inclusions in diamonds from the Finsch kimberlite pipe. In: Boyd, F.R., Meyer, H.O.A. (Eds.), Kimberlites, Diatremes and Diamonds. AGU, Washington, DC, pp. 1–15.

Gurney, J.J., Harris, J.W., Rickard, R.S., 1984a. Minerals associated with diamonds from the Roberts Victor Mine. In: Kornprobst, J. (Ed.), Kimberlites II: The Mantle and Crust–Mantle Relationships. Elsevier, Amsterdam, pp. 25–32.

Gurney, J.J., Harris, J.W., Rickard, R.S., 1984b. Silicate and oxide inclusions in diamonds from the Orapa Mine, Botswana. In: Kornprobst, J. (Ed.), Kimberlites II: The Mantle and Crust–Mantle Relationships. Elsevier, Amsterdam, pp. 1–9.

Harley, S.L., 1984. An experimental study of the partitioning of iron and magnesium between garnet and orthopyroxene. Contrib. Mineral. Petrol. 86, 359–373.

Harte, B., Hawkesworth, C.J., 1989. Mantle domains and mantle xenoliths. In: Ross, J., et al., (Eds.), Kimberlites and Related Rocks. GSA Spec. Publ., vol. 14. Blackwell, Carlton, pp. 649–686.

Harte, B., Hunter, R.H., Kinny, P.D., 1993. Melt geometry, movement and crystallization. Relation to Mantle Dykes, Veins and Metasomatism. Philos. Trans. R. Soc. Lond. Ser. A, Math. Phys. Eng. Sci., vol. 342 (1663), pp. 1–21.

Hoal, K.E.O., Hoal, B.G., Erlank, A.J., Shimizu, N., 1994. Metasomatism of the mantle lithosphere recorded by rare-earth elements in garnets. Earth. Planet. Sci. Lett. 126 (4), 303–313.

Hofmann, A.W., 1972. Chromatographic theory of infiltration metasomatism and its application to feldspars. Am. J. Sci. 272, 69–90.

Irving, A.J., Frey, F.A., 1978. Distribution of trace elements between garnet megacrysts and host volcanic liquids of kimberlitic and rhyolitic composition. Geochim. Cosmochim. Acta 42, 771–787.

Irving, A.J., Frey, F.A., 1984. Trace element abundances in megacrysts and their host basalts: constraints on partition coefficients and megacryst genesis. Geochim. Cosmochim. Acta 48, 1201–1221.

Izraeli, E.S., Harris, J.W., Navon, O., 2001. Brine inclusions in diamonds: a new upper mantle fluid. Earth. Planet. Sci. Lett. 187 (3–4), 323–332.

Kennedy, C.S., Kennedy, G.C., 1976. The equilibrium boundary between graphite and diamond. J. Geophys. Res. 81, 2467–2470.

Klein-BenDavid, O., Izraeli, E.S., Hauri, E.H., Navon, O., 2004. Mantle fluid evolution—a tale of one diamond. Lithos 77, 243–253 (this volume).

Kopylova, M.G., Gurney, J.J., Daniels, L.R.M., 1997. Mineral inclusions in diamonds from the River Ranch kimberlite, Zimbabwe. Contrib. Mineral. Petrol. 129 (4), 366–384.

Léost, I., Stachel, T., Brey, G.P., Harris, J.W., Ryabchikov, I.D., 2003. Diamond formation and source carbonation: mineral associations in diamonds from Namibia. Contrib. Mineral. Petrol. 145, 15–24.

McDade, P., Harris, J.W., 1999. Syngenetic inclusion bearing diamonds from Letseng-la-Terai, Lesotho. In: Gurney, J.J., Gurney, J.L., Pascoe, M.D., Richardson, S.H. (Eds.), The P.H. Nixon Volume, Proceedings of the VIIth International Kimberlite Conference. Red Roof Design, Cape Town, pp. 557–565.

McDonough, W.F., Rudnick, R.L., 1998. Mineralogy and composition of the upper mantle. Rev. Mineral. 37, 139–164.

McDonough, W.F., Sun, S.-S., 1995. The composition of the Earth. Chem. Geol. 120, 223–253.

Moore, R.O., Gurney, J.J., 1989. Mineral inclusions in diamond from Monastery kimberlite, South Africa. In: Ross, J., et al., (Eds.), Kimberlites and Related Rocks. GSA Spec. Publ., vol. 14. Blackwell, Carlton, pp. 1029–1041.

Navon, O., Stolper, E., 1987. Geochemical consequences of melt percolation: the upper mantle as a chromatographic column. J. Geol. 95, 285–307.

O'Neill, H.S.C., 1980. An experimental study of the iron–magnesium partitioning between garnet and olivine and its calibration as a geothermometer: corrections. Contrib. Mineral. Petrol. 72, 337.

O'Neill, H.S.C., Wood, B.J., 1979. An experimental study of the iron–magnesium partitioning between garnet and olivine and its calibration as a geothermometer. Contrib. Mineral. Petrol. 70, 59–70.

Phillips, D., Harris, J.W., 1995. Geothermobarometry of diamond inclusions from the De Beers Pool Mines, Kimberley, South Africa. Sixth International Kimberlite Conference, Novosibirsk, Extended, Abstracts, Siberian Branch, Russian Academy of Sciences, Novosibirsk, pp. 441–443.

Pollack, H.N., Chapman, D.S., 1977. On the regional variation of heat flow, geotherms, and lithospheric thickness. Tectonophysics 38, 279–296.

Rickard, R.S., Harris, J.W., Gurney, J.J., Cardoso, P., 1989. Mineral inclusions from Koffiefontein mine. In: Ross, J., et al., (Eds.), Kimberlites and Related Rocks. GSA Spec. Publ., vol. 14. Blackwell, Carlton, pp. 1054–1062.

Sharma, M., Wasserburg, G.J., 1996. The neodymium isotopic compositions and rare earth patterns in highly depleted ultramafic rocks. Geochim. Cosmochim. Acta 60, 4537–4550.

Shimizu, N., 1975. Rare earth elements in garnets and clinopyroxenes from garnet lherzolite nodules in kimberlites. Earth. Planet. Sci. Lett. 25, 26–32.

Shimizu, N., 1999. Young geochemical features in cratonic peridotites from southern Africa and Siberia. In: Fei, Y., Bertka, C.M., Mysen, B.O. (Eds.), Mantle Petrology: Field Observations and High Pressure Experimentation: A tribute to Francis R. (Joe) Boyd. The Geochemical Society, Houston, TX, pp. 47–55.

Shimizu, N., Richardson, S.H., 1987. Trace element abundance patterns of garnet inclusions in peridotite-suite diamonds. Geochim. Cosmochim. Acta 51 (3), 755–758.

Shimizu, N., Sobolev, N.V., Yefimova, E.S., 1997. Chemical heterogeneities of inclusion garnets and juvenile character of peridotitic diamonds from Siberia. Russian Geol. Geophys. 38-2, 356–372.

Shimizu, N., Pokhilenko, N.P., Boyd, F.R., Pearson, D.G., 1999. Trace element characteristics of garnet dunites/harzburgites, host rocks for Siberian peridotitic diamonds. In: Gurney, J.J., Gurney, J.L., Pascoe, M.D., Richardson, S.H. (Eds.), The P.H. Nixon Volume, Proceedings of the VIIth International Kimberlite Conference. Red Roof Design, Cape Town, pp. 773–782.

Shirey, S.B., et al., 2002. Lithospheric seismic structure and petrogenesis of southern African diamonds. Science 297, 1683–1686.

Sobolev, N.V., Lavrent'ev, Y.G., Pokhilenko, N.P., Usova, L.V., 1973. Chrome-rich garnets from the kimberlites of Yakutia and their paragenesis. Contrib. Mineral. Petrol. 40, 39–52.

Stachel, T., Harris, J.W., 1997a. Syngenetic inclusions in diamond from the Birim field (Ghana)—a deep peridotitic profile with a history of depletion and re-enrichment. Contrib. Mineral. Petrol. 127 (4), 336–352.

Stachel, T., Harris, J.W., 1997b. Diamond precipitation and mantle metasomatism—evidence from the trace element chemistry of silicate inclusions in diamonds from Akwatia, Ghana. Contrib. Mineral. Petrol. 129 (2–3), 143–154.

Stachel, T., Viljoen, K.S., Brey, G., Harris, J.W., 1998. Metasomatic processes in lherzolitic and harzburgitic domains of diamondiferous lithospheric mantle: REE in garnets from xenoliths and inclusions in diamonds. Earth. Planet. Sci. Lett. 159 (1–2), 1–12.

Stachel, T., Brey, G.P., Harris, J.W., 2000. Kankan diamonds (Guinea) I: From the lithosphere down to the transition zone. Contrib. Mineral. Petrol. 140, 1–15.

Stachel, T., Harris, J.W., Aulbach, S., Deines, P., 2002. Kankan diamonds (Guinea) III: δ^{13}C and nitrogen characteristics of deep diamonds. Contrib. Mineral. Petrol. 142 (4), 465–475.

Van Orman, J.A., Grove, T.L., Shimizu, N., Layne, G.D., 2002. Rare earth element diffusion in a natural pyrope single crystal at 2.8 GPa. Contrib. Mineral. Petrol. 142 (4), 416–424.

Viljoen, K.S., Phillips, D., Harris, J.W., Robinson, D.N., 1999. Mineral inclusions in diamonds from the Venetia kimberlites, Northern Province, South Africa. In: Gurney, J.J., Gurney, J.L., Pascoe, M.D., Richardson, S.H. (Eds.), The P.H. Nixon Volume, Proceedings of the VIIth International Kimberlite Conference. Red Roof Design, Cape Town, pp. 888–895.

Wilding, M.C., Harte, B., Fallick, A.E., Harris, J.W., 1994. Inclusion chemistry, carbon isotopes and nitrogen distribution in diamonds from the Bultfontein Mine, South Africa. In: Meyer, H.O.A., Leonardos, O.H. (Eds.), Diamonds: Characterization, Genesis and Exploration. CPRM Spec Publ Jan/94, Brasilia, pp. 116–126.

Wyllie, P.J., 1987. Metasomatism and fluid generation in mantle xenoliths. In: Nixon, P.H. (Ed.), Mantle Xenoliths. Wiley, Chichester, pp. 609–621.

Available online at www.sciencedirect.com

ELSEVIER

Lithos 77 (2004) 225–242

www.elsevier.com/locate/lithos

Mineral inclusions in microdiamonds and macrodiamonds from kimberlites of Yakutia: a comparative study

N.V. Sobolev[a,*], A.M. Logvinova[a], D.A. Zedgenizov[a], Y.V. Seryotkin[a],
E.S. Yefimova[a], C. Floss[b], L.A. Taylor[c]

[a] Institute of Mineralogy and Petrography, Russian Academy of Sciences, Siberian Branch 630090 Novosibirsk, Russian Federation
[b] Laboratory for Space Sciences, Washington University, St. Louis, MO 63130, USA
[c] Planetary Geosciences Institute, Department of Earth and Planetary Sciences, The University of Tennessee, Knoxville, TN 37996, USA

Received 27 June 2003; accepted 14 December 2003
Available online 1 June 2004

Abstract

Chemical compositions were determined on mineral inclusions recovered from 290 microdiamonds (< 1 mm) from 8 operating diamond mines in Yakutia. The sampled diamond mines include Mir, Udachnaya, Internatsionalnaya, Aykhal, Sytykanskaya, Yubileynaya, Komsomolskaya and Krasnopresnenskaya. The mineral inclusions include both ultramafic (peridotitic) suite (U-type) and eclogitic suite (E-type) examples. Olivines, chromites, Cr-pyropes, Cr-diopsides and enstatite were studied from U-type diamonds. Mg–Ca–Fe-garnets and omphacitic clinopyroxenes were studied from E-type microdiamonds. Abundances and compositions of these inclusions were compared with published and unpublished data on inclusions available from approximately 2000 macrodiamonds (>1 mm) from the same sources, and worldwide data for olivines and chromites. Although there are general similarities, notable exceptions were detected in about 10% of the inclusions from microdiamonds. For each of the pipes, anomalous compositions occur between the micro- and macrodiamond inclusions, but in different proportions, sometimes as high as 50% of the inclusions. Our study has demonstrated that mineral inclusions in microdiamonds are considerably more variable in their compositions and parageneses compared with inclusions in macrodiamonds.

Significant compositional anomalies in inclusions from microdiamonds include: (1) garnets containing pyroxene solid solution (majoritic component) both in U- and E-type microdiamonds from three pipes: Yubileynaya, Komsomolskaya and Krasnopresnenskaya. The moles of Si (pfu) in these garnets range from 3.07 to 3.13 and as high as 3.29, on the basis of 12 oxygens, along with a notable contents of Na_2O in two eclogitic garnets (0.43 and 0.93 wt.%) and uniquely high Cr_2O_3 and CaO contents in an ultramafic garnet of wehrlitic paragenesis; (2) coexisting wehrlitic garnets in a single microdiamond, one majoritic, the other normal, both with distinct + Eu anomalies, considered as signatures of crustal protoliths for the precursors to these garnets; (3) olivines with relatively low Fo (86–89) and high-NiO contents (0.46–0.64 wt.%), from Yubileynaya and Sytykanskaya microdiamonds; (4) chromites containing high-TiO_2 (up to 4.7 wt.%) and some extremely rich in MgO (Mg# 80). It is concluded that many of these compositional features observed may be related to a deeper origin for the microdiamond source region (>300 km), for at least a 10–30% portion of microdiamonds from each Yakutian pipe.
© 2004 Elsevier B.V. All rights reserved.

Keywords: Diamond; Inclusions; Siberian craton; Yakutia; Eclogite; Peridotite; Kimberlite

* Corresponding author. Fax: +7-3832-332-792.
E-mail address: sobolev@uiggm.nsc.ru (N.V. Sobolev).

0024-4937/$ - see front matter © 2004 Elsevier B.V. All rights reserved.
doi:10.1016/j.lithos.2004.04.001

1. Introduction

Kimberlites contain diamonds with a large range in size, varying from microdiamonds (<1 mm) weighting on average about 1 mg (0.005 carats) up to large diamonds which may exceed several hundred carats. This gives a size range for diamonds in kimberlite spanning four to five orders of magnitude. The question logically follows: do the different sizes of diamonds vary in their mineral inclusions?

The diamond deposits of Yakutia (Russia) located in the northwestern region of the Siberian craton (Fig. 1) attracted the attention of scientists immediately after their discovery in mid the of 1950s (Sobolev, 1959, 1960, 1964; Sobolev and Burov,

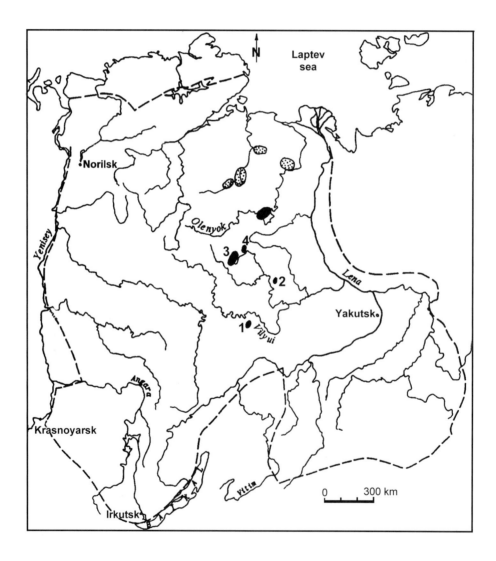

Fig. 1. Location of major kimberlite fields of the Siberian Platform of Paleozoic (solid symbols) and Mezozoic (dotted symbols) ages. Kimberlite fields with operating diamond mines: Mirny (1), Mir and Internatsionalnaya mines; Nakyn (2); Alakit (3), Aykhal, Sytykanskaya, Yubileynaya, Komsomolskaya and Krasnopresnenskaya mines; Daldyn (4), Udachnaya mine. All other fields include mostly barren and low grade kimberlite pipes. The boundaries of the Siberian craton are shown by dotted line. Modified after Sobolev et al. (1995).

1957). The most important diamond mines, Mir and Udachnaya, became the source of many samples of industrial-quality diamonds containing mineral inclusions that have been systematically studied and described, with the earliest results summarized by Sobolev (1974). Even in these early studies, both micro- and macrodiamonds were found within the same eclogite xenolith from the Mir pipe, thereby

demonstrating a probable similarity in the conditions of formation for both sizes of diamonds, at least for this xenolith and some others (e.g., Anand et al., 2004).

Further exploration activities led to the discovery of a number of pipes within the fields 1–4 on the map in Fig. 1, and some of these kimberlites subsequently became operating mines. These include the Internationalnaya, Aykhal, Yubileynaya,

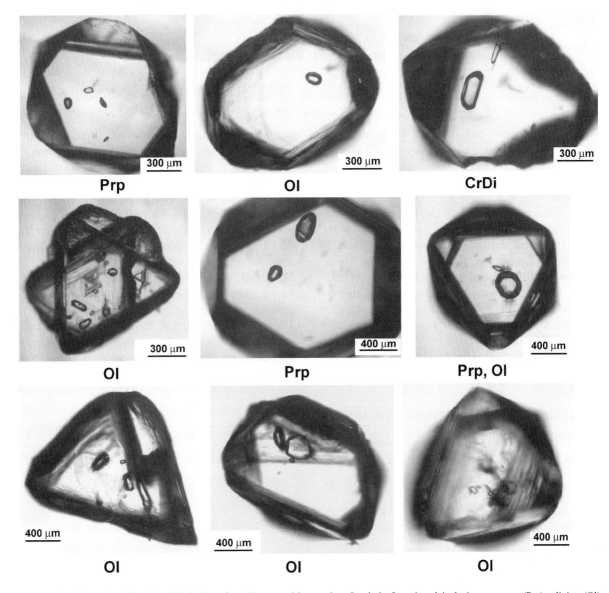

Fig. 2. Microdiamonds with mineral inclusions from Komsomolskaya mine. Symbols for mineral inclusions: pyrope (Prp), olivine (Ol), chromediopside (CrDi).

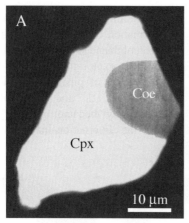 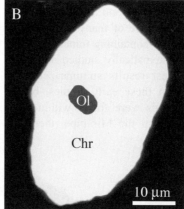

Fig. 3. Backscattered images of some polymineralic inclusions in microdiamonds from Udachnaya mine exposed at the polished surface of microdiamond crystals: omphacite (Cpx)–coesite (Coe)—A; olivine (Ol) inclusion in chromite (Chr)—B.

Sytykanskaya, Komsomolskaya and Krasnopresnen-skaya pipes. As demonstrated by U–Pb zircon (Davis et al., 1980) and perovskite ages (Kinny et al., 1997), all listed pipes are Upper Devonian–Early Carboniferous.

Microdiamonds (<1 mm) exhibit a wide range of physical characteristics that suggests the presence of different microdiamond populations at each location (McCandless et al., 1994; Bulanova, 1995; Pattison and Levinson, 1995; Trautman et al., 1997). In spite of the small dimensions of microdiamonds, their mineral inclusions are of comparable size (50–200 μm) to those from macrodiamonds (>1 mm) (Fig. 2). On occasions touching inclusion pairs were observed (Fig. 3) on the polished surface of a microdiamond sample. Recent inves-

tigations of mineral inclusions from a limited number of Yakutian microdiamonds have brought new results compared with inclusions from macro-diamonds of the same pipes. These include: (a) the most magnesian Group A garnet of eclogitic paragenesis from Mir diamonds, reported to date; (b) anomalously high NiO in low-Fo olivines; (c) the first report of ferropericlase from Udachnaya diamond inclusions; and (d) the first occurrence of Mg-spinel, containing no Cr_2O_3, also from Udach-nayan diamonds (Zedgenizov et al., 1998, 2001; Sobolev et al., 2000; Logvinova et al., 2001). These results stimulated more extensive investigations microdiamonds from other Yakutian operating diamond mines, as presented in this study (Table 1).

Table 1
Mineral inclusions in microdiamonds from Yakutian kimberlites

N	Pipe	n	U-type					E-type				
			Ol	Chr	Prp	En	Cr-Di	Grt	Omph	Coe	Fe-per	Sp
1	Udachnaya	67	41	19	2	–	–	1	3	1	1	2
2	Yubileynaya	79	60	9	6*	–	–	1	1	–	–	–
3	Sytykanskaya	66	52	7	2	–	–	4	2	–	–	–
4	Aykhal	34	20	9	3	–	–	2	–	–	–	–
5	Mir	5	3	1	–	–	–	1	1	–	–	–
6	Internatsionalnaya	5	2	–	2	–	–	1	–	–	–	–
7	Komsomolskya	32	20	2	5	1	2	2*	–	–	–	–
8	Krasnopresnenskaya	2	–	–	1	–	–	1*	–	–	–	–
	Total	290	198	47	21	1	2	13	7	1	1	2

n = number of diamonds studied; * includes one majoritic garnet; Ol, olivine; Chr, chromite; Prp, pyrope; En, enstatite; Cr-Di, Cr-diopside; Grt, Mg-Fe garnet; Omph, omphacite; Coe, coesite; Fe-per, ferropericlase; Sp, Mg-spinel.

About 290 microdiamonds (<1 mm) from operating diamond mines in Yakutia were found to contain mineral inclusions (Table 1), which were subsequently analyzed for major- and minor-element compositions. Most of inclusions are similar in composition when compared with mineral inclusions studied from about 2000 macrodiamonds (>1 mm) from the same pipes. Recent studies of large macrodiamonds containing mineral inclusions has provided a means to visually estimate inclusions abundance and paragenesis. Such studies of inclusions exposed at the surface of some large rough diamonds (10–108 carats) from Yakutian mines have confirmed a general similarity of chemistry for all the mineral inclusions from diamonds over a wide range of sizes (Sobolev et al., 2001; Taylor et al., 2003). Macrodiamond inclusion data are mainly from publications of the senior author, as well as from unpublished data from this same research team. The aim of this contribution is to summarize all available results of mineral inclusions in Yakutian microdiamonds and compare them with inclusions from macrodiamonds.

2. Analytical methods

Micro-Raman spectroscopy was used for nondestructive inclusion identification. Mineral inclusions were liberated from diamonds both by crushing and burning as well as by polishing host diamonds exposing inclusions. As shown earlier by Taylor et al. (1996) in a comparative study of pyroxene inclusions liberated both by diamond burning versus simply crushing, the pyroxene from different fragments of the same diamond showed no differences in major- and trace-element contents—i.e., the burning process did not modify the mineral composition. During the present study, inclusion grains released by burning, as well as polished fragments of diamonds with exposed inclusions, were mounted on epoxy resin and polished for analysis.

X-ray diffraction, single-crystal analysis was performed on one of the garnet inclusions from microdiamond Yum-27, associated with pyrope-uvarovitic garnet and olivine. The crystal-structural study and refinement was performed with a Stoe STADI-4 diffractometer (graphite-monochromated MoKα radiation; scintillation counter) at room temperature. Unit-cell parameters were refined by centering 24 reflections in the 2θ range of $21-28°$, and a total of 1554 diffraction intensities were collected up to $2\theta = 50°$, for the triclinic symmetry. The diffraction-intensity distribution revealed a cubic symmetry with observed systematic extinctions indicative of space group $Ia\bar{3}d$. The crystal structure was solved using SHELXS-86 (Sheldrick, 1986) and refined using SHELXL-93 (Sheldrick, 1993). Experimental details are given in Table 2 and confirmed that the Yum-27 inclusion has a garnet structure.

Major- and minor-element analyses were performed with a CAMEBAX electron microprobe at Novosibirsk and with CAMECA SX-50 electron microprobe at the University of Tennessee. The analyses were performed at 15 kV, with a 30-nA beam current and a 5–10-μm spot size. Counting times varied from 20 s for major elements to 100 s for minor-trace components. All analyses were fully corrected using the Cameca PAP software. It should be noted that a recent comparative study of analytical data on inclusions from large diamonds obtained with both instruments demonstrated a good agreement (2–3% of absolute amount) in the analyses from both Institutions (Sobolev et al., 2001; Taylor et al., 2003).

Trace-element analyses of the diamond inclusions were performed with the modified CAMECA IMS-3f ion microprobe at Washington University. Details of the experimental procedures are described

Table 2

X-ray diffraction single-crystal analysis for Yubileynaya inclusion Yum-27

Instrument	Stoe STADI-4 diffractometer
Crystal size (mm^3)	$0.04 \times 0.03 \times 0.02$ mm^3
2θ range (°)	$8.48-49.72°$
$h_{min,max}$, $k_{min,max}$, $l_{min,max}$	$-13,13$; $0,13$; $-13,13$
Number of I_{hkl} measured	1554
Number of unique F_{hkl}^2	122
Crystal system	Cubic
Space group	$Ia\bar{3}d$
a (Å)	11.775(1) Å
V (Å^3)	1632.6(2) Å^3
Identified mineral	Garnet
a_{theor} (0.57 uvarovite + 0.43 pyrope)	11.78

by Zinner and Crozaz (1986a,b), Alexander (1994), Hsu (1995) and Fahey et al. (1987). Detection limits are variable, depending on the element and phase being analyzed, but may be as low as a few ppb in favorable cases.

3. Mineral inclusions

The mineral inclusions in our study are related to two main types of diamond parageneses: ultramafic (or peridotitic)—U/P-type and eclogitic (E-type), as classified by Sobolev (1974), Meyer (1987) and Gurney (1989) and unanimously accepted in the scientific literature.

3.1. U-type mineral inclusions

3.1.1. Olivine

Olivine is the most abundant mineral inclusion in Yakutian diamonds (Yefimova and Sobolev, 1977). Available olivine compositions from macro-

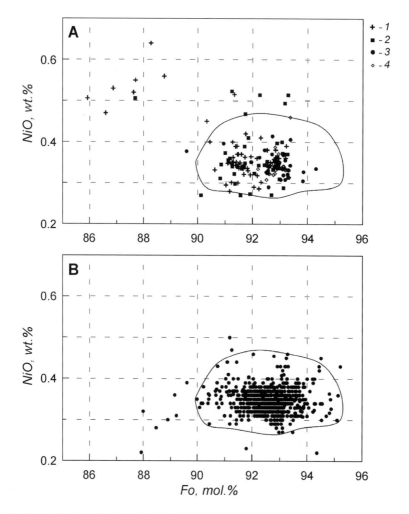

Fig. 4. NiO (wt.%) vs. Fo [100Mg/(Mg+Fe)] in olivine inclusions from microdiamonds of Yubileynaya (1), Sytykanskaya (2), Udachnaya (3) and Aykhal (4) mines (A). Plots of olivine inclusions from macrodiamonds worldwide (about 700 plots) are shown in (B) with a solid line surrounding about 98% of plots. Data sources for B: Daniels and Gurney (1989), Davies et al. (1999), Griffin et al. (1992), Gurney et al. (1979, 1985), Harris et al. (1991), Hervig et al. (1980), Jaques et al. (1998), Kopylova et al. (1997), McDade and Harris (1999), Meyer and Boyd (1972), Otter and Gurney (1989), Sobolev et al. (1993, 1997a,b, 2000), Stachel and Harris (1997), Stachel et al. (2000), Viljoen et al. (1999).

Table 3
Major-element compositions of selected olivine inclusions from microdiamonds

Sample	Yum-10	Yub-317	Yub-13	Yum-27	Yum-162	Yum-165	Yum-170	Yum-174	STI-303	STI-51	UVI-20	STI-52	STI-18	UDV-2	UD-8/01	UD-7	Mrm-2	Mrm-8
SiO_2	40.5	40.3	40.7	41.3	41.1	41.2	41.4	41.2	40.7	41.6	41.1	41.9	41.1	40.7	40.9	41.4	41.6	41.7
Cr_2O_3	0.11	0.09	0.15	0.04	0.06	0.09	0.04	0.03	<0.03	0.13	0.18	0.09	0.09	0.09	0.03	0.04	<0.03	<0.03
FeO	11.6	12.0	9.15	8.32	9.25	8.57	8.41	7.66	11.6	8.77	9.96	7.98	8.45	6.55	6.70	7.09	5.99	5.85
MnO	0.08	0.06	0.11	0.11	0.10	0.16	0.12	0.14	0.08	0.10	0.09	0.13	0.10	0.10	0.11	0.11	0.09	0.07
MgO	46.4	46.3	49.1	50.2	49.1	49.6	49.5	50.5	46.2	49.8	48.1	49.6	49.0	51.8	51.2	50.7	52.20	52.8
CaO	<0.03	<0.03	<0.03	0.08	<0.03	0.06	0.03	0.03	0.03	0.06	<0.03	0.06	0.03	<0.03	<0.03	0.03	<0.03	<0.03
NiO	0.55	0.52	0.33	n.d.	0.40	0.28	0.35	0.36	0.51	0.37	0.38	0.39	0.36	0.41	0.33	0.41	0.35	0.35
Total	99.2	99.3	99.5	100.1	100.03	99.96	99.84	99.91	99.1	100.8	99.8	100.2	99.1	99.7	99.3	99.8	100.2	100.8
Ox	4	4	4	4	4	4	4	4	4	4	4	4	4	4	4	4	4	4
Si	1.009	1.006	1.000	1.004	1.005	1.005	1.009	1.002	1.015	1.007	1.010	1.016	1.010	0.989	0.997	1.005	1.000	0.997
Cr	0.002	0.002	0.003	0.001	0.001	0.002	0.001	0.001	0.000	0.002	0.003	0.002	0.002	0.002	0.000	0.001	0.000	0.000
Fe	0.242	0.250	0.188	0.169	0.189	0.175	0.171	0.156	0.242	0.177	0.205	0.162	0.174	0.133	0.137	0.144	0.120	0.117
Mn	0.002	0.001	0.002	0.002	0.002	0.003	0.002	0.003	0.002	0.002	0.002	0.003	0.002	0.002	0.002	0.002	0.002	0.001
Mg	1.723	1.722	1.798	1.818	1.789	1.803	1.798	1.829	1.716	1.796	1.761	1.792	1.794	1.875	1.859	1.834	1.870	1.881
Ca	0.000	0.000	0.000	0.002	0.001	0.002	0.001	0.001	0.001	0.002	0.000	0.002	0.001	0.000	0.000	0.001	0.000	0.000
Ni	0.011	0.011	0.007	0.000	0.008	0.006	0.007	0.007	0.010	0.007	0.008	0.008	0.007	0.008	0.007	0.008	0.007	0.007
Total	2.989	2.992	2.998	2.997	2.994	2.995	2.990	2.998	2.985	2.993	2.988	2.984	2.989	3.009	3.002	2.995	2.999	3.002
Fo	87.7	87.3	90.5	91.5	90.4	91.2	91.3	92.2	87.7	91.0	89.6	91.7	91.2	93.4	93.2	92.7	94.0	94.1

Data source: Yum-10, Yub-317, Yub-13—from Sobolev et al. (2000); symbols Yum and Yub = Yubileynaya; STI = Sytykanskaya; UDV, UVI and UD = Udachnaya, and Mrm = Mir mines; n.d. = not determined.

diamonds worldwide are summarized by Meyer (1987), Sobolev et al. (2000), references to caption to Fig. 4 and new data in this paper. Information on NiO (wt.%) and Fo = 100 Mg/(Mg + Fe) show that the majority of compositions fall in the range of Fo 92–93 with NiO = 0.30–0.38 wt.% (Fig. 4). However, olivines from Yubileynaya microdiamonds demonstrate surprising exceptions with several samples having compositions in the range Fo − 86 to 89 and NiO = 0.46–0.64 wt.% (Table 3). These extreme compositions represent about 20% of all studied olivines from Yubileynaya microdiamonds (Sobolev et al., 2000). This discovery stimulated the additional study of a number of selected microdiamonds containing olivine inclusions from the Udachnaya (41 samples), Sytykanskaya (52 samples) and Aykhal (20 samples) mines (Table 1). All 41 olivine inclusions from Udachnaya microdiamonds average Fo 92.8, similar to 87 macrodiamond olivines from the same pipe. From averages of Yubileynaya olivines, significant differences are present between macro- (18 inclusions) and microdiamonds (61 inclusions) with Fo 92.8 and 91.7, respectively. A less pronounced but notable difference is also found for Sytykanskaya macro- (91)

and microdiamond (52) olivines with Fo 92.7 and 92.2, respectively.

Five olivines from Sytykanskaya and two from Yubileynaya microdiamonds also plot outside of a 98% field for olivines from diamonds worldwide (Fig. 4). They also demonstrate unusual high-NiO contents, but their Fo contents fall within the range typical for olivine inclusions. These data do not correlate with a worldwide NiO-Fo positive correlation established by Simkin and Smith (1970). However, the occurrence of a positive correlation of low Fo and high NiO is pronounced and is probably related to unusual assemblages of Fe-enriched harzburgites with high-Opx contents (Kelemen et al., 1998; Sobolev et al., 2000).

3.1.2. Chromite

Chromite is a common inclusion in Yakutian diamonds (Yefimova and Sobolev, 1977) and also an important mineral in diamond exploration (Sobolev, 1971, 1974). The proportion of chromite-bearing diamonds from Yakutia is within 45–56% of the total of all inclusion-bearing diamonds. Forty-seven chromite samples from microdiamonds of the Udachnaya, Aykhal, Sytykanskaya, Mir, Komsomol-

Table 4
Selected compositions of chrome spinels from microdiamonds

Sample	AL-1	AL-4	AL-5	AL-10	AL-11	SYT-14	S-2/99	UD-PL/1	UD-PL/2	UD-4/01	UD-8	UV-608	Yum-16
TiO_2	0.31	0.11	0.14	1.41	0.21	4.15	1.73	0.20	0.18	0.51	0.22	0.06	0.14
Al_2O_3	6.13	7.00	6.26	6.64	7.47	9.63	5.57	5.28	5.47	4.93	16.2	7.09	6.36
Cr_2O_3	63.0	62.7	62.8	62.1	61.2	52.4	64.0	66.2	64.0	63.6	52.5	63.2	66.0
FeO	17.0	17.0	17.5	15.1	17.2	17.6	15.5	12.5	16.0	18.0	14.9	15.6	9.42
MnO	0.19	0.17	0.18	0.17	0.19	0.17	0.18	0.15	0.18	0.19	0.15	0.17	0.13
MgO	12.6	12.4	12.2	14.3	12.7	15.0	11.4	14.1	13.1	11.9	15.2	13.5	17.1
NiO	0.10	0.09	0.06	0.13	0.09	0.20	0.15	0.08	0.10	0.11	0.11	0.12	0.08
Total	99.3	99.5	99.1	99.9	99.1	99.2	98.5	98.5	99.0	99.2	99.3	99.7	99.2
Oxygen	4	4	4	4	4	4	4	4	4	4	4	4	4
Ti	0.008	0.003	0.004	0.035	0.005	0.100	0.044	0.005	0.004	0.013	0.005	0.001	0.003
Al	0.239	0.273	0.245	0.255	0.291	0.365	0.222	0.207	0.214	0.195	0.596	0.273	0.241
Cr	1.651	1.637	1.652	1.598	1.597	1.332	1.711	1.737	1.681	1.685	1.296	1.633	1.677
Fe	0.471	0.470	0.487	0.411	0.475	0.473	0.438	0.347	0.444	0.504	0.389	0.426	0.253
Mn	0.005	0.005	0.005	0.005	0.005	0.005	0.005	0.004	0.005	0.005	0.004	0.005	0.004
Mg	0.622	0.610	0.605	0.694	0.625	0.719	0.575	0.697	0.648	0.594	0.707	0.658	0.819
Ni	0.003	0.002	0.002	0.003	0.002	0.005	0.004	0.002	0.003	0.003	0.003	0.003	0.002
Total	2.999	3.000	3.000	3.001	3.000	2.999	2.999	2.999	2.999	2.999	3.000	2.999	2.999
Mg#	56.9	56.5	55.4	62.8	56.8	60.3	56.7	66.8	59.3	54.1	64.5	60.7	76.4
Cr/Cr + Al	87.3	85.7	87.1	86.3	84.6	78.5	88.5	89.4	88.7	89.6	68.5	85.7	87.4

Symbols: AL for Aykhal; SYT and S for Sytykanskaya; UD and UV for Udachnaya; Yum and Yub for Yubileynaya mines.

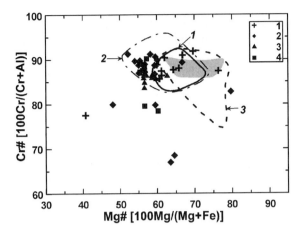

Fig. 5. Cr#—100Cr/(Cr + Al) vs. Mg#—100Mg/(Mg + Fe) in chromite inclusions from Yakutian (fields 1 and 2) and South African (field 3) macrodiamonds (Griffin et al., 1994; Sobolev et al., 1997b) and from microdiamonds of Yubileynaya (1), Udachnaya (2), Aykhal (3) and Sytykanskaya (4) pipes, Yakutia.

In spite of their small size, some microdiamonds contain multiple inclusions of chromites. These features were found in nine diamonds samples from the available collection, and two chromite grains were analyzed from each of these samples. For the most contrasting compositions, data are presented in Table 4. In keeping with earlier studies of chromite inclusions in macrodiamonds (Sobolev and Yefimova, 1998), the majority of samples were found to contain chromite grains with distinct differences in compositions between individual inclusions within the same diamonds, but homogeneous within a single grain. Two trends of inhomogeneties are confirmed: (1) simultaneous differences in Al_2O_3 and Cr_2O_3 and MgO-FeO contents; (2) variations in only the MgO-FeO contents. It should be mentioned that the chemical "pristinity" (i.e., non−open-system behavior) of all these diamond inclusions is unknown.

skaya and Yubileynaya pipes have been analyzed in this study (see Table 1), with the results of selected analyses presented in Table 4 and plotted in Figs. 5 and 6. The expanded compositional field of chromites from Yakutian diamonds (field 2 in Fig. 5) is based upon 700 data points and is considerably broader compared with that defined by Griffin et al. (1994). Only 10% of chromites from Yakutian macrodiamonds containing < 62 wt.% Cr_2O_3 and >0.7 wt.% TiO_2 and plot outside of this field (Sobolev, 1971, 1974; Sobolev et al., 1992, 1997a) .

Our first attempt to study chromite inclusions from microdiamonds has demonstrated extreme variations in their compositions. About 75% of all analyzed chromites plot within an expanded field of chromite compositions from Yakutian diamonds. This field includes analytical data for 34 chromite grains from an Udachnaya macrodiamond single crystal (Ud-34). This field is plotted as the shaded area in Fig. 5 (Sobolev and Yefimova, 1998). More than 25% of chromites from this study are significantly different compared to the major field. Compositions of some chromites are extremely Mg-rich with Mg# approaching 80. Some unusual chrome spinels include those containing high TiO_2.

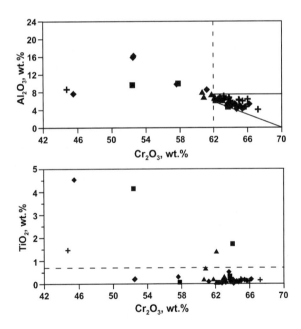

Fig. 6. Al_2O_3 vs. Cr_2O_3 and TiO_2 vs. Cr_2O_3 in chromite inclusions from macrodiamonds worldwide (solid field and dotted boundary) and from microdiamonds of Yakutian diamond mines (see Fig. 5 for symbols). Boundaries: 62 wt.% of Cr_2O_3 and 0.7 wt.% of TiO_2 for typical diamond inclusions and chromite related to diamonds from heavy concentrates of diamondiferous kimberlites are modified after Sobolev (1971) and Sobolev et al. (1975, 1992).

3.1.3. Cr-pyropic garnet

Cr-pyropes are rare as inclusions both in macro- and microdiamonds, compared with olivines and chromites. Only 19 microdiamonds from our available collection contained purple, lilac or dark-green inclusions characteristic of Cr-rich garnets, and confirmed by subsequent EMP analyses (Table 5). Along with a general similarity of Cr-rich garnets composition in both micro- and macrodiamonds, a unique Cr–Ca-rich majoritic garnet was discovered in single microdiamond from the Yubileynaya mine. It coexists with Cr–Ca rich non-majoritic garnet and olivine within the same diamond (Table 6, Fig. 7). Unfortunately, no details about relative position of these inclusions within a diamond crystals were available before burning of crystal. Positive identification of majoritic garnet was preliminary obtained by a single-crystal X-ray diffraction study (Table 2). The Si moles calculated from the microprobe analyses show 3.29 and 3.02 Si in the coexisting majoritic and non-majoritic garnets, respectively (Table 6). The coexisting olivine is Fo 91.5 (see Table 3 for analysis), which is consistent with the relatively low Mg# of the non-majoritic garnet. Furthermore, the Ca–Cr component of the majoritic garnet is unusually high (more than 50%). Fig. 8 shows the REE patterns of these garnets, in addition to another unusual majoritic garnet from Komsomolskaya. The complete trace-element contents of these garnets are given in Table 7. Both REE patterns from the majoritic and normal garnet inclusions from the Yubileynaya, Ym-27, diamonds are very similar and contain a notable anomaly. The presence of a distinct +Eu anomaly in both patterns is unique for Ca–Cr-rich garnets, having never been described previously. Such +Eu anomalies are indicative of the involvement of plagioclase feldspar sometime in the genesis of the garnet, and this is thought to be a signature of a low-T and low-P protolith for the rock from which the garnet was sampled by the diamond. This is evidence for the ancient subduction of oceanic crust beneath the Siberian craton.

Several more Cr-rich pyropes that were recovered from Yakutian microdiamonds are similar in composition to typical Cr-rich harzburgitic pyropes from macrodiamonds. Our extensive data base of Cr-rich pyropes included in Yakutia diamonds consists of about 650 samples, with an overwhelming majority of the samples coming from the Udachnaya and Mir mines, with lesser numbers from the Aykhal, Sytykanskaya and Yubileynaya pipes (Fig. 9).

3.1.4. Cr-diopside

Chrome diopside was recovered and studied from three Komsomolskaya microdiamonds. Two samples are enriched in Cr_2O_3 (6.5 and 6.8 wt.%) and contain up to 19 mol.% kosmochlor. One sample (100/23) is typical chrome diopside (e.g., Meyer, 1987), as shown in Table 8.

3.1.5. Enstatite

Enstatite was found in one microdiamond only from the Komsomolskaya mine (Table 8). This sample contained enstatite as an isolated grain. In general, this enstatite composition is different from typical enstatite inclusions in macrodiamonds (e.g. Sobolev, 1974; Meyer, 1987) in its lack of Al_2O_3 and very low contents of Cr_2O_3.

3.2. E-type mineral inclusions

3.2.1. Garnets

A small number of E-type garnets from microdiamonds fall outside the typical range of Mg–Ca–Fe contents (Fig. 9), containing up to 45% Ca (grossular) component along with high Mg# 67.8% and 30% Ca component (Aykhal sample) along with low Mg# 29.2 (Yubileynaya sample). Elevated (0.09–0.21 wt.%) Na_2O is typical of most of these garnets (Sobolev and Lavrent'ev, 1971). Their analyses are presented in Tables 5 and 6 and the REE pattern in Fig. 8. A series of Mg-rich garnets classified as indicative of Group A eclogites was detected for the first time in Mir diamonds (Logvinova et al., 2001). Two E-type majoritic garnets with the range of Si (pfu) from 3.05 to 3.13 and high Na_2O contents (up to 0.93 wt.%) were discovered in microdiamonds. The majorite garnets are similar to some garnet inclusions from Monastery mine in South Africa (Moore and Gurney, 1985). On a Na_2O (wt.%) versus Mg# diagram (Fig. 10), modified from Stachel (2001), both garnet compositions clearly plot within the field of majoritic garnets. High $P–T$ experimental results on the origin of majoritic garnets (Gasparik,

Table 5
Selected compositions of garnets from Yakutian microdiamonds

Sample	Syt-1	Syt-1	Im-10	Im-21	Yub-212	Yub-322	Yum-150	AL-8	AL-9	Km-42/35	Km-64/23	Km-68/23	Km-69/23	Km-71/49	Im-3	STI-203/98	STI-203/00	AL-1	AL-2	Yum-31	Km-94/49
										[10]	[15]	[10]	[10]	[12]							[8]
SiO_2	40.8	40.9	42.8	42.4	40.9	41.4	41.5	41.8	41.5	41.7 (2)	41.0 (4)	41.9 (3)	41.3 (3)	40.6 (3)	41.1	40.6	40.6	39.8	41.2	38.6	40.2 (3)
TiO_2	0.05	0.03	0.01	0.74	0.13	0.06	0.06	0.08	0.13	0.02 (1)	0.11 (0)	0.19 (1)	<0.02	0.02 (1)	0.78	0.23	0.6	0.03	0.24	0.05	0.55 (3)
Al_2O_3	12.4	13.0	19.5	20.5	13.2	15.6	16.6	17.9	16.5	17.6 (2)	15.4 (1)	17.5 (1)	18.2 (1)	15.7 (1)	21.2	22.0	21.0	21.6	22.1	21.1	22.6 (0)
Cr_2O_3	14.5	13.4	4.95	2.40	12.2	10.1	8.94	7.55	8.8	8.87 (20)	11.2 (0)	7.43 (7)	7.93 (8)	11.2 (1)	0.23	0.22	0.12	0.03	0.09	0.01	0.02 (1)
FeO	6.45	6.29	5.57	6.31	6.60	6.96	6.51	6.06	6.22	7.28 (7)	6.84 (5)	6.61 (8)	6.20 (5)	6.19 (6)	15.9	13.0	15.9	19.9	8.54	23.0	16.5 (1)
MnO	0.37	0.37	0.23	0.25	0.33	0.43	0.35	0.32	0.36	0.44 (1)	0.35 (1)	0.30 (1)	0.33 (1)	0.37 (2)	0.28	0.41	0.36	0.55	0.22	1.04	0.38 (1)
MgO	23.1	23.1	23.5	22.6	20.0	22.2	21.7	22.6	22.7	21.7 (1)	20.5 (2)	20.4 (1)	20.6 (2)	19.9 (1)	16.1	9.33	13.1	9.2	10.1	5.33	13.8 (1)
CaO	0.94	0.99	3.37	4.32	6.05	3.04	4.12	2.70	3.59	3.10 (3)	4.56 (2)	6.05 (2)	5.40 (4)	5.81 (4)	4.33	14.1	7.55	8.6	17.1	10.2	5.85 (3)
Na_2O	0.01	0.07	0.02	0.09	0.02	0.04	0.03	0.08	0.02	<0.02	<0.02	<0.02	0.03 (5)	0.03 (6)	0.18	0.20	0.21	0.03	0.09	0.01	0.17 (12)
Total	98.6	98.2	100.0	99.6	99.4	99.8	99.8	99.1	99.8	100.7	100.0	100.4	100.0	99.8	100.1	100.1	99.4	99.7	99.7	99.3	100.
Oxygen	12	12	12	12	12	12	12	12	12	12	12	12	12	12	12	12	12	12	12	12	12
Si	3.023	3.032	3.031	3.013	3.031	3.011	3.008	3.016	3.000	2.992	2.998	3.019	2.980	2.975	3.017	3.024	3.030	3.024	3.038	3.016	2.969
Ti	0.003	0.002	0.001	0.040	0.007	0.003	0.003	0.004	0.007	0.001	0.006	0.010	–	0.001	0.043	0.013	0.034	0.002	0.013	0.003	0.030
Al	1.083	1.136	1.627	1.717	1.153	1.337	1.418	1.522	1.406	1.486	1.326	1.484	1.550	1.353	1.834	1.931	1.847	1.934	1.921	1.943	1.969
Cr	0.849	0.785	0.277	0.135	0.715	0.581	0.512	0.431	0.503	0.503	0.648	0.423	0.452	0.649	0.013	0.013	0.007	0.002	0.005	0.001	0.002
Fe	0.400	0.390	0.330	0.375	0.409	0.423	0.395	0.366	0.376	0.437	0.418	0.398	0.374	0.379	0.976	0.810	0.992	1.264	0.527	1.503	1.019
Mn	0.023	0.023	0.014	0.015	0.021	0.026	0.021	0.020	0.022	0.027	0.022	0.018	0.020	0.023	0.017	0.026	0.023	0.035	0.014	0.069	0.024
Mg	2.550	2.552	2.479	2.393	2.209	2.406	2.344	2.430	2.445	2.323	2.230	2.192	2.219	2.169	1.761	1.035	1.457	1.042	1.110	0.621	1.521
Ca	0.075	0.079	0.256	0.329	0.480	0.237	0.320	0.209	0.278	0.238	0.357	0.467	0.417	0.456	0.341	1.125	0.604	0.700	1.351	0.854	0.462
Na	0.001	0.010	0.003	0.012	0.003	0.006	0.004	0.011	0.003	0.002	0.002	0.001	0.004	0.004	0.026	0.029	0.030	0.004	0.013	0.002	0.024
Total	8.007	8.009	8.017	8.027	8.028	8.029	8.025	8.008	8.039	8.011	8.007	8.012	8.016	8.009	8.028	8.005	8.024	8.008	7.991	8.010	8.020
Mg#	86.5	86.7	88.3	86.5	84.4	85.0	85.6	86.9	86.7	77.4	74.3	71.6	74.0	85.1	64.3	56.1	59.5	45.2	67.8	29.2	50.7
Cr/Cr+Al	44.0	40.9	14.6	7.28	38.3	30.3	26.5	22.1	26.4	25.3	33.3	22.2	22.6	32.4	0.72	0.67	0.38	0.09	0.27	0.03	0.10

Symbols: Im=Internatsionalnaya; Syt, ST and STI=Syrykanskaya; AL=Aykhal; Km=Komsomolskaya; Yub and Yum=Yubileynaya mines.

Table 6
Chemical composition of majoritic and normal garnets included in microdiamonds from Yakutian kimberlites (Ym-27, Km-88/23, Kr-119/13) and Arkhangelsk kimberlite (Po-99)

Gt type	Yum-27		Km-88/23	Kr-119/13	Po-99
	Majorite	Normal	Majorite	Majorite	Majorite
# analysis	[8]	[10]	[5]	[17]	[17]
P_2O_5	0.05 (1)*	0.05 (1)	0.211 (6)	n.d.	n.d.
SiO_2	42.8 (2)	38.9 (1)	41.3 (1)	40.8 (2)	44.9 (2)
TiO_2	0.33 (1)	0.37 (1)	1.90 (2)	0.41 (1)	0.71 (2)
Al_2O_3	6.79 (3)	10.9 (1)	18.0 (1)	20.9 (1)	16.6 (1)
Cr_2O_3	10.2 (9)	13.4 (1)	0.05 (1)	0.07 (1)	1.23 (4)
FeO	5.67 (6)	9.11 (5)	16.0 (1)	14.1 (2)	8.65 (11)
MnO	0.28 (2)	0.31 (1)	0.29 (2)	0.31 (1)	0.21 (1)
MgO	12.2 (2)	12.9 (1)	8.82 (8)	9.41 (5)	23.5 (5)
CaO	20.8 (1)	12.8 (1)	12.2 (1)	12.9 (1)	3.77 (3)
Na_2O	0.04 (2)	<0.02	0.93 (9)	0.43 (1)	0.25 (1)
Total	99.16	98.74	99.70	99.3	99.8
Oxygen	12	12	12	12	12
P	0.003	0.003	0.014	–	–
Si	3.293	3.022	3.131	3.068	3.200
Ti	0.019	0.022	0.108	0.023	0.038
Al	0.616	0.996	1.606	1.852	1.394
Cr	0.621	0.825	0.003	0.004	0.069
Fe	0.365	0.591	1.010	0.887	0.515
Mn	0.018	0.020	0.018	0.020	0.013
Mg	1.411	1.496	0.995	1.055	2.496
Ca	1.723	1.065	0.988	1.039	0.288
Na	0.005	0.000	0.136	0.063	0.035
Total	8.074	8.041	8.009	8.012	8.048
#Mg	79.3	71.6	49.6	54.3	82.9

*Numbers in ( ) are the one sigma variance in analyses for the least unit cited.

2002) clearly support the unusual deep origins for these garnets (i.e., >300 km).

3.2.2. Omphacite

A limited number of E-type clinopyroxene inclusions from microdiamonds were recovered, but demonstrate some interesting compositional features (Table 8). Sample Ud-2 represents an omphacite intergrowth with coesite, shown in Fig. 3, containing about 31% jadeite and high K_2O. The omphacite from a Yubileynaya microdiamond (Yum-20) is enriched in FeO. A single microdiamond from the Mir pipe (Logvinova et al., 2001) containing hundreds of minute mineral inclusions is characterized by the presence of clinopyroxene grains classified from Group-A eclogites, as classified by Taylor and Neal (1989).

This is the first example of a Group-A pyroxene coexisting with high-Mg Group-A garnet from Mir diamonds, as shown in Fig. 11 (Sobolev et al., 1998).

4. Discussion

Mineral inclusions in 290 microdiamond from 8 Yakutian diamond mines have been characterized and form the basis for comparison with similar minerals from macrodiamonds from these same kimberlite pipes. Compositional data for inclusions of U- and E-type garnet and pyroxene, as well as olivine and chromite, from 98 of these microdiamonds were published earlier (Logvinova et al., 2001; Sobolev et al., 2000; Zedgenizov et al., 1998, 2001). About 70% of the microdiamond inclusions are represented by olivines and 16% Cr-spinels. Both U and E-type garnets represent only 11% of the collection, with the remaining samples represented by pyroxenes. At the present time, the inclusion data base from Yakutian macrodiamonds is almost an order of magnitude larger than that from microdiamonds. In spite of this limited and unequal sampling, some important compositional differences are readily apparent. These differences include the following specific features of inclusions from microdiamonds: (a) Cr–Ca-rich majoritic garnet from the Yubileynaya mine, coexisting in the same diamond with another Ca–Cr-rich non-majoritic garnet, but both with similar REE patterns, including +Eu anomalies; (b) majoritic eclogitic (E-type) garnets with a considerably wider range in compositions compared with inclusions in macrodiamonds (e.g., Meyer, 1987); (c) relatively high-NiO (0.45–0.64 wt.%) and low-Fo (<90) contents in olivines from the Yubileinaya (Sobolev et al., 2000) and Sytykanskaya mines; (d) Mg-spinels containing but traces of Cr (Zedgenizov et al., 1998), as well as high-magnesian (Mg# >75) and Ti-rich (>4 wt.% TiO_2) chromites; (e) ferropericlase inclusions in a microdiamond from the Udachnaya mine (Zedgenizov et al., 2001); (f) enstatite inclusion from Komsomolskaya microdiamond with extremely low Al_2O_3 and Cr_2O_3 contents.

A limited yet highly significant number (3) of majoritic garnets have been found both in U-type

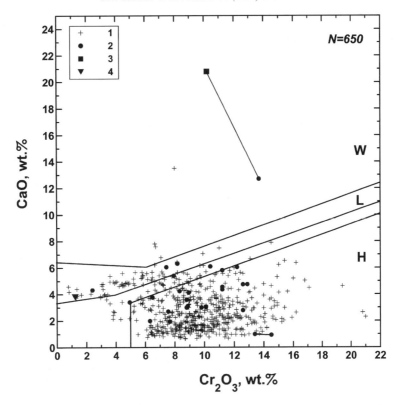

Fig. 7. CaO vs. Cr₂O₃ in Cr-bearing pyropes from macrodiamonds of major Yakutian diamond mines (1) and from microdiamonds of the same mines (2). Solid boundaries for garnet parageneses are from Sobolev (1971, 1974). H—harzburgitic, L—lherzolitic, W—wehrlitic parageneses. Majoritic garnet from microdiamond of Yubileynaya mine (3) associated with a "normal" garnet in the same diamond—two plots connected by solid line. The plot of majoritic garnet from Arkhangelsk microdiamond (4) is shown for comparison. Data source: Griffin et al. (1993), Kovalsky (1979), Sobolev (1974), Sobolev et al. (1997a, 2001), Zedgenizov et al. (1998) and authors database. *N*—number of analyses.

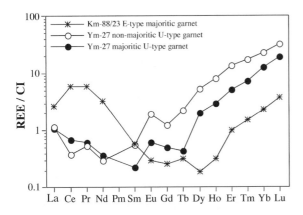

Fig. 8. Normalized REE patterns for two majoritic and one associated non-majoritic garnets (see Table 6 for major and Table 7 for REE and trace element analyses).

and E-type microdiamonds, but not in Yakutian macrodiamonds to date, in spite of there being at least 770 garnet inclusions from Yakutian macrodiamonds. Majoritic garnet of lherzolite-websterite paragenesis has also been documented from an Arkhangelsk microdiamond (Sobolev et al., 1997a,b). These observations, in an addition to other majoritic garnet inclusions by Stachel (2001), lead to the paradigm that *30–40% of kimberlitic pipes worldwide contain these unusual majoritic garnets as inclusions only in microdiamonds*. It should also be note that a significant number of majoritic garnets have been discovered from placer diamonds with unknown primary sources (e.g., Stachel, 2001; Gasparik, 2002).

The discovery of majoritic garnets in microdiamonds from three of the Yakutian diamond

Table 7
Trace-element concentrations of garnet inclusions determined by SIMS

	Km-88/23		Yum-27 non-majoritic		Yum-27 majoritic	
	Conc (ppm)	Conc/CI	Conc (ppm)	Conc/CI	Conc (ppm)	Conc/CI
K	7.2	0.013	2743	4.9	1124	2.0
Sc	149	26	213	37	225	39
V	323	5.7	627	11	474	8.4
Mn	2772	1.5	3190	1.6	2900	1.5
Rb	0.32	0.14	2.1	0.92	1.1	0.49
Sr	11	1.4	2.7	0.34	4.2	0.53
Y	0.95	0.61	15	9.6	4.9	3.2
Zr	0.42	0.11	5.7	1.5	26	6.5
Nb	0.68	2.7	12	49	4.3	18
Ba	0.12	0.053	41	18	4.0	1.7
La	0.62	2.7	0.27	1.2	0.26	1.1
Ce	3.6	6.0	0.23	0.39	0.42	0.70
Pr	0.53	6.0	0.048	0.54	0.054	0.61
Nd	1.5	3.3	0.13	0.30	0.17	0.37
Sm	0.084	0.57	0.083	0.56	0.034	0.23
Eu	0.017	0.30	0.11	1.9	0.035	0.62
Gd	0.052	0.26	0.24	1.2	0.099	0.51
Tb	0.012	0.32	0.080	2.2	0.016	0.43
Dy	0.046	0.19	1.3	5.3	0.50	2.0
Ho	0.018	0.32	0.45	8.0	0.17	3.0
Er	0.16	1.0	2.2	14	0.81	5.1
Tm	0.037	1.5	0.42	17	0.18	7.3
Yb	0.38	2.3	3.7	23	2.1	13
Lu	0.091	3.7	0.79	33	0.47	19

microdiamond from the Yubileynaya pipe (Tables 5 and 6; Fig. 8). Although normal garnets enriched both in Cr and Ca are very rare, only one additional sample has been described from Yakutian diamonds (Sobolev, 1974).

This Yubileynaya majoritic garnet associated in a single diamond with a normal garnet, as isolated grains, however, with uncertain relative positions, is most significant. They both are rich in Cr and Ca related to wehrlitic paragenesis. In spite of their differences in bulk compositions and majoritic component, they have very similar REE patterns, each displaying a distinct + Eu anomaly. In addition, the REE patterns (Fig. 8) do not have any HREE negative slope, characteristic of harzburgitic garnets (Taylor et al., 2003). We suggest that the majoritic garnet was encapsulated by the microdiamond at the depth >300 km, but that the microdiamond crystal continued to grow in a silicate environment of similar chemical composition. At considerably lower pressure, albeit still within the diamond stability field, a normal Cr–Ca garnet was encapsulated, in addition to an olivine grain.

mines is of special interest and importance. Until recently, all majoritic garnets reported, dominantly eclogitic, were from but six kimberlitic pipes worldwide and two alluvial sources, as summarized by Stachel (2001) and Gasparik (2002). Our present study adds three additional pipes this still limited statistic. Along with recent discovery of a number of majoritic garnets in both U- and E-type diamonds from Snap Lake kimberlite, Canada (Pokhilenko et al., 2001, 2004), it is now possible to conclude that majoritic garnets represent virtually all known mineral parageneses of U-type diamonds (harzburgitic, lherzolitic, wehrlitic).

Wehrlitic garnet with significant majorite component is the rarest among all described majoritic garnets. In our study, we have presented the first discovery of such a wehrlitic majoritic garnet, as verified by X-ray diffraction data (see Table 2). This garnet coexists with a normal garnet in a single

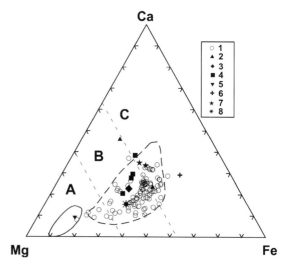

Fig. 9. Ternary diagram for garnets from E-type paragenesis in macrodiamonds from Mir and Udachnaya pipes (1), microdiamonds from Aykhal (2), Udachnaya (3), Sytykanskaya (4), Mir (5), Yubileynaya (6), Komsomolskaya, majoritic (7) and normal (8). Data source: this study, Sobolev et al. (1998) and Logvinova et al. (2001). Group boundaries are from Coleman et al. (1995).

Table 8
Selected compositions of pyroxenes from Yakutian microdiamonds

Sample	Kmsm-6 Cr-Di	Kmsm-7 Cr-Di	Kmsm-21 En	UD-2* Omph	STI-33 Omph	Yum-20 Omph
SiO_2	55.3	56.2	58.5	55.8	55.3	55.6
TiO_2	<0.03	<0.03	<0.03	0.41	0.20	0.21
Al_2O_3	1.69	2.36	0.01	8.51	8.44	8.65
Cr_2O_3	6.50	6.80	0.08	0.04	0.04	0.03
FeO	1.90	1.96	4.62	3.52	3.88	7.26
MnO	0.09	0.09	0.13	0.10	0.05	0.07
MgO	14.9	14.2	36.5	11.4	11.4	8.99
CaO	15.3	14.6	0.27	14.5	16.1	14.5
Na_2O	3.20	3.91	0.1	4.46	4.02	4.56
K_2O	0.76	0.48	n.d.	0.58	0.12	0.66
Total	99.7	100.6	100.2	99.3	99.6	100.5
Oxygen	6	6	6	6	6	6
Si	2.009	2.016	1.996	1.998	1.982	2.001
Ti	0.000	0.000	0.000	0.011	0.005	0.006
Al	0.072	0.100	0.000	0.359	0.357	0.367
Cr	0.187	0.193	0.002	0.001	0.001	0.001
Fe	0.058	0.059	0.132	0.105	0.116	0.218
Mn	0.003	0.003	0.004	0.003	0.002	0.002
Mg	0.807	0.759	1.855	0.608	0.609	0.482
Ca	0.595	0.561	0.010	0.556	0.618	0.559
Na	0.225	0.272	0.007	0.310	0.279	0.318
K	0.035	0.022	—	0.026	0.005	0.030
Total	3.991	3.985	4.018	3.977	3.975	3.984
Mg#	93.3	92.8	93.4	85.2	84.0	68.8
Ca/Ca+Mg	42.5	42.5	0.53	47.8	50.4	53.7

Symbols: UD=Udachnaya; STI=Sytykanskaya; Yum=Yubiley-naya; Kmsm=Komsomolskaya mines; n.d.=not determined.
 * Associated (touching) with coesite (see Fig. 3).

A similar association of majoritic (Si, pfu=3.17) and three normal E-type garnets were described in a diamond from DO-27 pipe, Canada (Davies et al., 1999). No information about the relative position of these garnet grains within the diamond was noted. This example supports our suggestion about a possibility of a single diamond growth within a considerably pressure range but in the environment of a similar chemical composition.

Our lack of knowledge of the relative garnet positions from the Yubileynaya diamond and the same lacking with the study of Davies et al. (1999), as discussed above, exemplifies the importance of *in-situ* examination of inclusions on polished surfaces of diamonds, as stressed by Taylor et al. (2000) and Taylor and Anand (in press). The findings that we have presented in this study for the inclusions in the Ym-27 diamond, although highly significant, would have been more valuable

if the mineral observations had been made while still in the diamond. Then, the CL zoning, the N-aggregation from FTIR, and the chemistry of the diamond (e.g., $\delta^{13}C$, $\delta^{15}N$) could have been factored into the paragenesis of the inclusions and their adjoining host diamond.

5. Summary

We conclude that mineral inclusions in microdiamonds are considerably more variable in their compositions and parageneses compared with inclusions in macrodiamonds. The inhomogeneities between different grains of inclusions within the same microdiamond provide evidence for a complex growth history for at least some microdiamonds.

The percentage of mineral inclusions of unusual compositions in microdiamonds, particularly the olivines from Yubileynaya and Sytykanskaya; chromites from all Yakutian mines, and significantly majoritic garnets, leads us to conclude that many of these compositional features may be related to a deeper

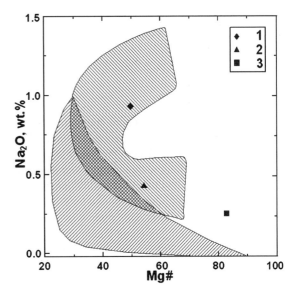

Fig. 10. Na_2O (wt.%) vs. molar pyrope content (Mg#) for "normal" (lower area) and majoritic (upper area) garnets. Note a very little overlap between high-Na normal and low-Na majoritic garnets. Modified from Stachel (2001). Komsomolskaya (1) and Krasnopresnenskaya (2) microdiamonds E-type majoritic garnets; Arkhangelsk microdiamond majoritic garnet (3); after Sobolev et al. (1997a).

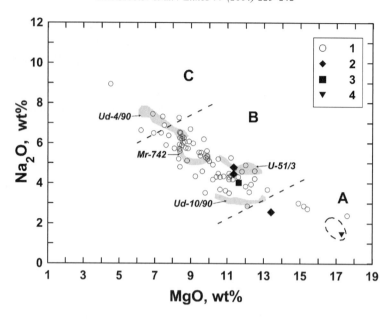

Fig. 11. Na₂O (wt.%) vs. MgO (wt.%) in clinopyroxenes from E-type macrodiamonds of Mir and Udachnaya pipes (1) and from microdiamond of the Mir pipe. For symbols, see Figs. 5 and 9. Group boundaries are from Taylor and Neal (1989). The hatching outlines the compositional domains of multiple pyroxene inclusions in individual macrodiamonds from Mir (Mr) and Udachnaya (Ud, U) pipe diamonds. Data source: Sobolev et al. (1998).

origin for the microdiamond source region (>300 km) for at least a 10–30% portion of microdiamonds from each Yakutian pipe.

Acknowledgements

Rory Moore and Rondy Davies are acknowledged for their constructive comments and fine reviews of an earlier version of this manuscript. We are also grateful to Thomas Stachel for his thoughtful comments and editorial handling of this paper. The SIMS analyses at Washington University were supported by NSF Grant EAR 99-80394 (G. Crozaz). This study was supported by Russian Foundation for Basic Research (Grants # 02-05-64248 and 932003.5) to NVS and by NSF Grant EAR 99-09430 to LAT.

References

Alexander, C.M.O.D., 1994. Trace element distributions with ordinary chondrite chondrules: implications for chondrule formation conditions and precursors. Geochim. Cosmochim. Acta 58, 3451–3467.

Anand, M., Taylor, L.A., Misra, K.C., Carlson, W.D., Sobolev, N.V., 2004. Nature of diamonds in Yakutian eclogites: views from eclogite tomography and mineral inclusions in diamonds, Lithos 77, 333–348 (this volume).

Bulanova, G.P., 1995. The formation of diamonds. J. Geochem. Explor. 53, 1–23.

Coleman, R.G., Lee, D.E., Beatty, L.B., Brannock, B.B., 1995. Eclogites and eclogites: their differences and similarities. Geol. Soc. Amer. Bull. 76, 483–508.

Daniels, L.R., Gurney, J.J., 1989. The chemistry of the garnets, chromites and diamond inclusions from the Dokolwayo kimberlite, Kingdom of Swaziland. GSA Special Publication No. 14. Kimberlites and Related Rocks, vol. 2, Blackwell, Oxford, pp. 1012–1021.

Davies, R.M., Griffin, W.L., Pearson, N.J., Andrew, A.S., Doyle, B.J., O'Reilly, S.Y., 1999. Diamonds from the Deep: Pipe DO-27, Slave Craton, Canada. Proceedings of the VII IKC, vol. 1, Red Roof Design, Cape Town, pp. 148–155.

Davis, G.L., Sobolev, N.V., Kharkiv, A.D., 1980. New data on the age of Yakutian kimberlites obtained by U–Pb method on zircon. Dokl. Akad. Nauk SSSR 254, 53–57.

Fahey, A.J., Zinner, E.K., Crozaz, G., Kornacki, A.S., 1987. Microdistribution of Mg isotopes and REE abundances in a type A calcium–aluminum-rich inclusion from Efremovka. Geochim. Cosmochim. Acta 51, 3215–3229.

Gasparik, T., 2002. Experimental investigation of the origin of majoritic garnet inclusions in diamonds. Phys. Chem. Miner. 29, 170–180.

Griffin, W.L., Gurney, J.J., Ryan, C.G., 1992. Variations in trapping temperatures and trace elements in peridotite-suite inclusions from African diamonds: evidence for two inclusion suites, and implications for lithosphere stratigraphy. Contrib. Mineral. Petrol. 110, 1–15.

Griffin, W.L., Sobolev, N.V., Ryan, C.G., Pokhilenko, N.P., Win, T.T., Yefimova, E.S., 1993. Trace elements in garnets and chromites: diamond formation in the Siberian lithosphere. Lithos 29, 235–256.

Griffin, W.L., Ryan, C.G., Gurney, J.J., Sobolev, N.V., 1994. Chromite macrocrysts in kimberlites and lamproites: geochemistry and origin. In: Meyer, H.O.A., Leonardos, O.H.Kimberlites, Related Rocks and Mantle Xenoliths. Spec. Publ.-CPRM, vol. 1A/93, CPRM, Rio de Janeiro, pp. 366–377.

Gurney, J.J., 1989. Diamonds. In: Ross, J., et al.Kimberlites and Related Rocks. Spec. Publ.-GSA, vol. 2, Blackwell, Oxford, pp. 935–965.

Gurney, J.J., Harris, J.W., Rickard, R.S., 1979. Silicate and oxide inclusions from diamonds of the Finsch kimberlite pipe. In: Boyd, F.R., Meyer, H.O.A.Kimberlites. Diatremes and Diamonds: Their Geology, Petrology and Geochemistry. Amer. Geophys. Union, Washington, pp. 1–15.

Gurney, J.J., Harris, J.W., Rickard, R.S., Moore, R.O., 1985. Inclusions in Premier mine diamonds. Trans. Geol. Soc. S. Afr. 88, 301–310.

Harris, J.W., Duncan, D.J., Zhang, F., Miao, Q., Zhu, Y., 1991. The physical characteristics and syngenetic inclusion geochemistry of diamonds from pipe 50, Liaoning Province, People's Republic of China. Proceedings of the Fifth Intern. Kimberlite Conference. Araxa, Brasil, pp. 106–115.

Hervig, R.L., Smith, J.V., Steele, I.M., Gurney, J.J., Meyer, H.O.A., Harris, J.W., 1980. Diamonds: minor elements in silicate inclusions: pressure–temperature implications. J. Geophys. Res. 85 (B12), 6919–6929.

Hsu, W., 1995 Ion microprobe studies of the petrogenesis of enstatite chondrites and eucrites, PhD thesis, Washington University, St. Louis, MO, p. 380.

Jaques, A.L., Hall, A.E., Sheraton, J., Smith, C.B., Sun, S.-S., Drew, R.M., Fouddis, C., Ellingsen, K., 1998. Composition of crystalline inclusions and C-isotopic composition of Argyle and Ellendale diamonds. In: Ross, J., et al.Kimberlite and Related Rocks. Special Publication-GSA, vol. 2, Blackwell, Oxford, pp. 66–989.

Kelemen, P.B., Hart, S.R., Bernstein, S., 1998. Silica enrichment in the continental upper mantle via melt/rock reaction. Earth Planet. Sci. Lett. 164, 387–406.

Kinny, P.D., Griffin, B.J., Heaman, L.M., Brakhvogel, F.F., Spetsius, Z.V., 1997. SHRIMP U-Pb ages of perovskite from Yakutian kimberlites. Russian Geol. Geophys. 38, 97–105.

Kopylova, M.G., Gurney, J.J., Daniels, L.R.M., 1997. Mineral inclusions in diamonds from the River Ranch kimberlite Zimbabwe. Contrib. Mineral. Petrol. 129, 366–384.

Kovalsky, V.V., Bulanova, G.P., Nikishov, K.N., Botkunov, A.I., Makhotko, V.F., Shestakova, O.E., Gotovtsev, V.V., 1979. Composition of garnets, chromites and rutiles associated with diamonds from Yakutian kimberlite pipes. Dokl. Akad. Nauk SSSR 247, 946–951.

Logvinova, A.M., Zedgenizov, D.A., Sobolev, N.V., 2001. Pyrox-

enite paragenesis of abundant mineral and probable fluid inclusions in microdiamons from the Mir kimberlite pipe, Yakutia. Dokl. Acad. Nauk. 380, 363–367 (in Russian). English Translation: Dokl. Earth Sci. 380, 795–799.

McCandless, T.E., Waldman, M.A., Gurney, J.J., 1994. Microdiamonds from Murfeesboro lamproites, Arkansas: morphology, mineral inclusions and carbon isotope geochemistry. Proc. 5th International Kimberlite Conf. Araxa, Brazil. Spec. Publ.-CPRM, vol. 2, CPRM, Rio de Janeiro, pp. 78–89.

McDade, P., Harris, J.W., 1999. Syngenetic inclusion bearing diamonds from Letseng-la-Terai Lesotho. Proceedings of the VII IKC 2, 557–565.

Meyer, H.O.A., 1987. Inclusions in diamonds. In: Nixon, P.H.Mantle Xenoliths. Wiley, Chichester, England, pp. 501–522.

Meyer, H.O.A., Boyd, F.R., 1972. Composition and origin of crystalline inclusions in natural diamonds. Geochim. Cosmochim. Acta 36 (11), 1255–1273.

Moore, R.O., Gurney, J.J., 1985. Pyroxene solid solution in garnets included in diamond. Nature 318, 583.

Otter, J.J., Gurney, J.J., 1989. Mineral inclusions in diamonds from the Sloan diatremes, Colorado-Wyoming State Line kimberlite district, North America. GSA Special Publication No. 14. Kimberlites and Related Rocks, vol. 2, Blackwell, Oxford, pp. 1042–1053.

Pattison, D.R.M., Levinson, A.A., 1995. Are euhedral microdiamonds formed during ascent and decompression of kimberlite magma? Implications for use of microdiamond in diamond grade estimation. Appl. Geochem. 10, 725–740.

Pokhilenko, N.P., Sobolev, N.V., McDonald, J.A., Hall, A.E., Yefimova, E.S., Zedgenizov, D.A., Logvinova, A.M., Reimers, L.F., 2001. Crystalline inclusions in diamonds from kimberlites of the Snap Lake area (Slave Craton, Canada): new evidences for the anomalous lithospheric structure. Dokl. Earth Sci. 380, 806–811.

Pokhilenko, N.P., Sobolev, N.V., Reutsky, V.N., Hall, A.E., Taylor, L.A., 2004. Crystalline inclusions and C isotope ratios in diamonds from the Snap Lake/King Lake kimberlite dyke system: evidence of ultradeep and enriched lithospheric mantle. Lithos 77, 57–67 (this volume).

Sheldrick, G.M., 1986. SHELXS-86. Program for Solving Crystal Structures. Univ. Göttingen, Germany.

Sheldrick, G.M., 1993. SHELXL-93. Program for Refinement of Crystal Structures. Univ. Göttingen, Germany.

Simkin, T., Smith, J.V., 1970. Minor-element distribution in olivine. J. Geol. 78, 304–325.

Sobolev, V.S.1959. The Diamond Deposits of Yakutia. Gosgeoltekhizdat, Moscow (in Russian).

Sobolev, V.S., 1960. Conditions of formation of diamond deposits. Geol. Geofiz. 1, 3–20 (in Russian with English abstract).

Sobolev, V.S.1964. Petrography and Mineralogy of the Kimberlitic Rocks of Yakutia. Nedra Press, Moscow (in Russian).

Sobolev, N.V., 1971. On mineralogical criteria of a diamond potential of kimberlites. Geol. Geofiz. 12 (3), 70–78 (in Russian with English Abstract).

Sobolev, N.V., 1974. Deep-seated Inclusions in Kimberlites and the Problem of the Composition of the Upper mantle Nauka Publ. House, Novosibirsk (in Russian). English Translation: 1977,

Boyd, F.R. (Ed.), Washington. D.C. American Geological Union.

Sobolev, V.S., Burov, A.P.1957. The Diamonds of Siberia. Geoltekhizdat, Moscow (in Russian).

Sobolev, N.V., Lavrent'ev, Y.G., 1971. Isomorphic sodium admixture in garnets formed at high pressures. Contrib. Mineral. Petrol. 31, 1–12.

Sobolev, N.V., Yefimova, E.S., 1998. Compositional variations of chromite inclusions as an indicators of the zonation of diamond crystals. Dokl. Akad. Nauk 358, 649–652 (in Russian). English Translation: Dokl. Earth Sci. 359, 163–166.

Sobolev, N.V., Pokhilenko, N.P., Lavrent'ev, Y.G., Usova, L.V., 1975. Distinctive featutes of the composition of chrome-spinels in the diamonds and kimberlites of Yakutia. Geol. Geofiz. 16 (11), 7–24. English Translation: Sov. Geol. Geophys. 16 (11), 4–15.

Sobolev, N.V., Pokhilenko, N.P., Grib, V.P., Skripnichenko, V.A., Titova, V.E., 1992. Specific composition and conditions of formation of deep-seated minerals in exploration pipes of the Onega peninsula and kimberlites of Zimnii Coast in the Arkhangelsk province. Geol. Geofiz. 33 (10), 84–92 (in Russian). English Translation: Russ. Geol. Geophys. 33 (10), 71–78.

Sobolev, N.V., Galimov, E.M., Yefimova, E.S., Sobolev, E.V., Usova, L.V., 1993. Crystalline inclusions, isotope composition of carbon, nitrogen centers in diamonds and features of garnet composition from the concentrate of the Madjgawan pipe (India). Geol. Geofiz. 34 (12), 85–91 (in Russian). English Translation: Russ. Geol. Geophys. 34 (12), 77–83.

Sobolev, N.V., Zuev, V.M., Afanasiev, V.P., Pokhilenko, N.P., Zinchuk, N.N.1995. Kimberlites of Yakutia. Field Guide Book, Novosibirsk.

Sobolev, N.V., Kaminsky, F.V., Griffin, W.L., Yefimova, E.S., Win, T.T., Ryan, C.G., Botkunov, A.I., 1997a. Mineral inclusions in diamonds from the Sputnik kimberlite pipe Yakutia. Lithos 39, 135–157.

Sobolev, N.V., Yefimova, E.S., Reimers, L.F., Zakharchenko, O.D., Makhin, A.I., Usova, L.V., 1997b. Mineral inclusions in diamonds of the Arkhangelsk kimberlite province. Russ. Geol. Geophys. 38, 379–393.

Sobolev, N.V., Taylor, L.A., Zuev, V.M., Bezborodov, S.M., Snyder, G.A., Sobolev, V.N., Yefimova, E.S., 1998. The specific features of eclogitic paragenesis of diamonds from Mir and Udachnaya kimberlite pipes (Yakutia). Geol. Geofiz. 39 (in Russian). English Translation: Russ. Geol. Geophys 39, 1653–1663.

Sobolev, N.V., Logvinova, A.M., Zedgenizov, D.A., Yefimova, E.S., Lavrent'ev, Y.G., Usova, L.V., 2000. Anomalously high Ni admixture in olivine inclusions from microdiamonds, the Yubileynaya kimberlite pipe, Yakutia. Dokl. Akad. Nauk. 375, 393–396 (in Russian). English Translation: Dokl. Earth Sci. 375A, 1403–1406.

Sobolev, N.V., Yefimova, E.S., Logvinova, A.M., Sukhodolskaya, O.V., Solodova, Yu.P., 2001. Abundance and composition of mineral inclusions in large diamonds from Yakutia. Dokl.

Akad. Nauk 376, 382–386 (in Russian). English Translation: Dokl. Earth Sci. 376, 34–38.

Stachel, T., 2001. Diamonds from the asthenosphere and the transition zone. Eur. J. Mineral. 13, 883–892.

Stachel, T., Harris, J.W., 1997. Syngenetic inclusions in diamond from the Birim field (Ghana)—a deep peridotitic profile with a history of depletion and re-enrichment. Contrib. Mineral. Petrol. 27, 336–352.

Stachel, T., Brey, G.P., Harris, J.W., 2000. Kankan diamonds (Guinea) I: from lithosphere down to the transition zone. Contrib. Mineral. Petrol. 140, 1–15.

Taylor, L.A., Anand, M., 2004. Diamonds: time capsules from the Yakutian Mantle. Chemie der Erde-Geochem. 64, 1–74.

Taylor, L.A., Neal, C.R., 1989. Eclogites with oceanic crustal and mantle signatures from the Belsbank kimberlite, South Africa: Part 1. Mineralogy, petrography and whole-rock chemistry. J. Geol. 97, 551–567.

Taylor, L.A., Snyder, G.A., Crozaz, G., Sobolev, V.N., Yefimova, E.S., Sobolev, N.V., 1996. Eclogitic inclusions in diamonds: evidence of complex mantle processes over time. Earth Planet. Sci. Lett. 142, 535–551.

Taylor, L.A., Keller, R.A., Snyder, G.A., Wang, W., Carlson, W.D., Hauri, E.H., McCandless, T., Kim, K.R., Sobolev, N.V., Bezborodov, S.M., 2000. Diamonds and their mineral inclusions and what they tell us: a detailed "Pull-Apart" of a diamondiferous eclogite. Int. Geol. Rev. 42, 959–983.

Taylor, L.A., Anand, M., Promprated, P., Floss, C., Sobolev, N.V., 2003. The significance of mineral inclusions in large diamonds from Yakutia Russia. Am. Mineral. 88, 912–920.

Trautman, R.L., Griffin, B.J., Taylor, L.A., Spetsius, Z.V., Smith, C.B., Lee, D.C., 1997. A comparison of the microdiamonds from kimberlite and lamproite of Yakutia and Australia. Russ. Geol. Geophys. 38, 341–355.

Viljoen, K.S., Philips, D., Harris, J.W., Robinson, D.N., 1999. Mineral inclusions in diamonds from the Venetia Kimberlites, Northern Province, South Africa. Proceedings of the VII IKC, vol. 2, Red Roof Design, Cape Town, pp. 888–895.

Yefimova, E.S., Sobolev, N.V., 1977. Abundance of crystalline inclusions in Yakutian diamonds. Dokl. Akad. Nauk SSSR 237, 1475–1478 (in Russian).

Zedgenizov, D.A., Logvinova, A.M., Shatskii, V.S., Sobolev, N.V., 1998. Inclusions in microdiamonds from some kimberlite diatremes of Yakutia. Dokl. Earth Sci. 359, 204–208.

Zedgenizov, D.A., Yefimova, E.S., Logvinova, A.M., Shatsky, V.S., Sobolev, N.V., 2001. Ferropericlase inclusions in a diamond microcrystal from Udachnaya kimberlite pipe Yakutia. Dokl. Earth Sci. 377a, 319–321.

Zinner, E., Crozaz, G., 1986a. A method for the quantitative measurement of rare earth elements in the ion microprobe. Int. J. Mass Spectrom. Ion Process. 69, 17–38.

Zinner, E., Crozaz, G., 1986b. In: Benninghoven A. et al. (EDS.), Secondary Ion Mass Spectrometry (SIMS V). Springer, Berlin, pp. 444–446.

Available online at www.sciencedirect.com

Lithos 77 (2004) 243–253

www.elsevier.com/locate/lithos

Mantle fluid evolution—a tale of one diamond

Ofra Klein-BenDavid[a,*], Elad S. Izraeli[a], Erik Hauri[b], Oded Navon[a]

[a] *Institute of Earth Sciences, The Hebrew University of Jerusalem, Jerusalem 91904, Israel*
[b] *Department of Terrestrial Magnetism, Carnegie Institution of Washington, Washington, DC 20015, USA*

Received 27 June 2003; accepted 12 November 2003
Available online 25 May 2004

Abstract

Microinclusions analyzed in a coated diamond from the Diavik mine in Canada comprise peridotitic minerals and fluids. The fluids span a wide compositional range between a carbonatitic melt and brine. The diamond is concentrically zoned. The brine microinclusions reside in an inner growth zone and their endmember composition is $K_{19}Na_{25}Ca_5Mg_8Fe_3Ba_2Si_4Cl_{32}$ (mol%). The carbonatitic melt is found in an outer layer and its endmember composition is $K_{11}Na_{21}Ca_{11}Mg_{26}Fe_7Ba_2Si_{10}Al_3P_2Cl_5$. The transition in inclusion chemistry is accompanied by a change in the carbon isotopic composition of the diamond from $-8.5‰$ in the inner zone to $-12.1‰$ in the outer zone. We suggest that this transition reflects mixing between already evolved brine and a freshly introduced carbonatitic melt of different isotopic composition.

The compositional range found in diamond ON-DVK-294 is the widest ever recorded in a single diamond. It closes the gap between brine found in cloudy octahedral diamonds from South Africa and carbonatitic melt analyzed in cubic diamonds from Zaire and Botswana. Thus, all microinclusions analyzed to date fall along two arrays connecting the carbonatitic melt composition to either a hydrous-silicic endmember or to a brine endmember. This connection suggests that many diamonds are formed from fluids derived form a mantle source not significantly influenced by local heterogeneities.
© 2004 Elsevier B.V. All rights reserved.

Keywords: Inclusion; Brine; Carbonatitic melt; Peridotitic; Coated diamond; Fibrous diamond

1. Introduction

Diamond inclusions provide a unique opportunity to investigate the mantle rocks where diamonds are formed and the fluids from which they crystallize. The strength of the diamond and its low reactivity in a silicate environment ensure that the trapped substances remain shielded from their changing environment. Mineral inclusions are widespread in diamonds and range from less than a micrometer to millimeter size. They consist of silicates, oxides, sulfides, and rare carbonate and phosphate minerals and provide information on host rock lithology and on the temperature and pressure of diamond crystallization (Meyer, 1987). Most mineral inclusions indicate peridotitic and eclogitic host rocks at temperatures of 900–1300 °C and depth of 150–200 km (Meyer, 1987; Harris, 1992). Based on the internal structure of diamonds (Milledge et al., 1984; Bulanova, 1995), the low density of crystallographic dislocations (Sunagawa, 1984) and the association of diamonds with

* Corresponding author. Fax: +972-2-5662581.
E-mail address: ofrak@vms.huji.ac.il (O. Klein-BenDavid).

cracks within their host xenoliths (Taylor et al., 2000), it is believed that most natural diamonds grow from fluids. Fluid inclusions are the best source of information about the nature and chemical composition of the diamond's growth medium, and shed light on its evolution.

Fluid inclusions are found in fibrous diamonds, in the fibrous coat of coated diamonds and as internal clouds in octahedral diamonds (Navon, 1999). The inclusion-bearing zones are populated by millions of submicrometer inclusions separated from each other and enclosed in the diamond matrix. TEM analyses of individual inclusions identified a multiphase assemblage of apatite, quartz, mica and carbonate surrounded by material of low electron density (Lang and Walmsley, 1983; Guthrie et al., 1991). The common interpretation is that the inclusions trapped a fluid and that the minerals are secondary phases that grew during cooling. Water and carbonates detected by infrared (IR) spectroscopy (Chrenko et al., 1967; Navon et al., 1988) are probably the main volatile components.

Electron probe microanalysis of the bulk composition of microinclusions reveals a wide range of compositions among three endmembers (Schrauder and Navon, 1994; Izraeli et al., 2001): (a) hydrous-silicic melt rich in water, Si, Al, and K; (b) carbonatitic melt rich in carbonate, Mg, Ca, Fe, K and Na and (c) brine-rich in Cl, K and Na.

Intermediate compositions between carbonatitic and hydrous-silicic melts were found in fibrous diamond (Navon et al., 1988; Schrauder and Navon, 1994). Limited mixing between carbonatitic melt and brine were found in cloudy diamonds (Izraeli et al., 2001). No intermediate compositions between brine and hydrous-silicic melt were detected. All compositions are rich in K and many other incompatible elements and are characterized by steep REE patterns similar to those in kimberlites and lamproites (Schrauder et al., 1996; Raga et al., 2003). Halogens and Ar are also enriched and the halogen abundance ratios are similar to those of MORB (Johnson et al., 2000; Burgess et al., 2002).

In spite of the high abundance of sulfide minerals in diamonds, such inclusions were never found in fluid-bearing diamonds, reminding us that diamonds may grow from other media as well, e.g., sulfide melts (Bulanova, 1995; Sobolev et al., 1997; Bulanova et

al., 1998) or reduced carbon-bearing fluids (Tomilenko et al., 1997). It is interesting to note that Bulanova et al. (1998) and Klein-BenDavid et al. (2003) suggested involvement of carbonatitic melt in the formation of sulfide-bearing diamonds.

The carbon isotopic composition of fibrous diamonds from localities worldwide ranges over relatively narrow $\delta^{13}C$ values, between $-5‰$ and $-8‰$. Their nitrogen isotopic composition ranges between $0‰$ and $-10‰$. These values lie close to those of carbon in MORB and kimberlites (Galimov, 1991; Boyd et al., 1992; Cartigny et al., 1998, 2003).

Fluid inclusions are rarely found together with mineral inclusions. Tal'nikova (1995) reported the finding of omphacite along with silicic melt in a Siberian diamond. Izraeli et al. (2001, in press) reported microinclusions bearing peridotitic and eclogitic minerals as well as hydrous minerals residing together with brine inclusions in cloudy diamonds from Koffiefontein. This, together with the evidence for metasomatism in diamondiferous xenoliths suggest that the fluids penetrated the peridotitic or eclogitic host rock and took part in diamond formation in both rock types.

Here we report the composition of 167 microinclusions in one concentrically zoned diamond from the Diavik mine in Lac de Gras, Slave Craton, Canada. Microinclusions in this diamond carry peridotitic minerals and fluids with a broad compositional range. In all previous reports (Navon et al., 1988; Schrauder and Navon, 1994; Izraeli et al., 2001), intradiamond variation was limited and the range was spanned because of the variation between different diamonds. Recently, Shiryaev et al. (2003) reported the finding of a wide compositional range in individual diamond from Brazil. Here we present the first detailed study of a diamond with such a wide compositional range. This, and the clear radial compositional zoning along profiles, allowed us to examine in details the formation of the diamond. The bridging between brine and carbonatitic endmember composition constrains the possible relation among the three endmembers and their evolution.

2. Lac de Gras diamonds

The Diavik mine is composed of four high-grade kimberlites (A154 North, A154 South, A418 and

A21). The kimberlites are located just off East Island beneath the shallow waters of Lac de Gras. The kimberlites are small (<2000 m^2), steep sided, and are hosted in a complex of Archean granitoids and micaceous metasediments of the Slave Craton (Graham et al., 1999). Griffin et al. (1999) suggested that the lithospheric mantle underneath the central Slave Craton consists of an ultradepleted, olivine-rich upper layer (140–150 km) and a less depleted lower layer where eclogites are more abundant (200–220 km).

Over 150 pipes have been identified in the Lac de Gras area, most of those are Cretaceous to Paleocene age (Heaman and Kjarsgaard, 2000; Creaser et al., 2003). A quarter of the kimberlites carry macrodiamonds. Color and morphology was determined for diamonds from the neighboring Ekati kimberlites in the Lac de Gras vicinity. The three most common groups are white octahedra, brown octahedra and fibrous diamonds (Gurney et al., 2003). The inclusion population found in pipes in the Lac de Gras area consists of peridotitic, eclogitic and super deep paragenesis (ferropericlase and MgSi and CaSi perovskite). However the abundance of the different suits varies between pipes. For example, Tappert et al. (2003) and Chinn et al. (1999) found dominance of peridotitic inclusions in diamonds from Panda, Misery, Sable and Jay pipes, while Davies et al. (1999, 2003) recorded high abundance of eclogitic mineral inclusions in DO-27, A154, A418, A 21, Ranch Lake, DO18 and DD17 pipes.

3. Analytical methods

The diamond was polished and cleaned in HF (60%), HNO$_3$ (69%) and ethanol and rinsed in distilled water in order to remove all organic and inorganic surface contamination prior to analysis. It was then carbon-coated and analyzed using a JEOL JXA 8600 electron probe micro analyzer (EPMA). Cathodoluminescence (CL) images of the diamond faces were collected using a Gatan MiniCL attached to the electron probe and were used as maps of the diamond internal structure.

Individual, shallow, subsurface microinclusions were detected using backscattered electron imaging and were accurately related to specific diamond growth layers using the CL imaging. Individual inclusions

were measured using a focused 15 keV, 10 nA electron beam. The inclusions were analyzed for 100 s using a Pioneer-Norvar energy depressive spectrometer (EDS). The beam interaction volume is about 4 μm^3, and is larger then the common inclusion volume (<1 μm^3).

The spectral data were reduced using the PROZA correction procedure supplied by Noran (Bastin and Heijligers, 1991). As the inclusion material comprises only a few percent of the total analyzed volume it was assumed that the difference to 100 wt.% is comprised of pure carbon. The small and variable totals obtained for the inclusions varied between 2 and 26 wt.%. As the total reflects mostly the size and depth of each inclusion, it made sense to renormalize all results to 100% (water and CO$_2$-free basis). The original total is also reported. Izraeli et al. (in press) verified the accuracy of the measurements by analyzing olivine microinclusions of restricted composition. They concluded that the accuracy is better than 15% for the major elements in individual inclusions. Precision is even better, of the order of a few percent.

Carbon isotopic composition was determined along a cross-section through one of the diamond faces (Face I) using a Cameca 6F secondary-ion mass spectrometer (SIMS) with Cs$^+$ source (Hauri et al., 2002). The δ^{13}C values were determined using extreme energy filtering; the total uncertainty (accuracy + precision) of the result is $\pm 0.4‰$. δ^{13}C values are calculated relative to a working standard with composition of $6.51 \pm 0.1‰$ and reported relative to Pee Dee Belemnite (PDB) standard (Hauri et al., 2002).

No infrared spectroscopic data are available for this diamond due to its opaque nature.

4. Description of the diamond

Diamond ON-DVK-294 is 2.6 mm in diameter and weighs 22.8 mg. The diamond is opaque, gray in color and has an external octahedral morphology.

Two parallel faces were polished to form a 1-mm-thick slab (Fig. 1). One of the faces (Face I) intersected the central part of the diamond. It exposed a transparent core (A) surrounded by a white layer containing cavities of approximately 20 μm (B). A gray opaque zone (C) surrounds the hollowed layer. Under a microscope, the gray zone appears similar to the fibrous material of coated diamonds. It is sur-

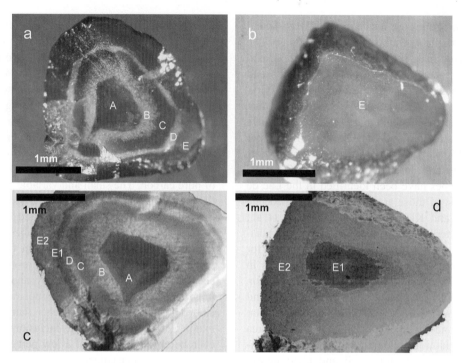

Fig. 1. Diamond ON-DVK-294. (a) A reflected light image of the polished Face I reveals concentric zoning: (A) transparent core, (B) internal cavity-filled layer, (C) internal gray layer, (D) external cavity-filled layer, (E) an external gray layer. (b) A reflected light image of polished face II exposing layer E only. (c) CL image of Face I. All layers visible by reflected light are apparent. The outermost zone (E) is divided into a darker inner zone (E1) and a brighter outer zone (E2), not visible in reflected light. (d) CL image of Face II. This face exposes the inner darker zone (E1) and an outer brighter zone (E2) (colors in the CL images were inverted to allow a better view).

rounded by another cavity-rich layer (D) and a gray fibrous rim (E) (Fig. 1a).

Cathodoluminescence (CL) of the core reveals some brighter areas that may reflect pressure deformation and some resorption of the core margins. The CL image of the outer fibrous layer reveals that it consists of two zones (Fig. 1c): a thin inner darker zone (E1), and an outer brighter zone (E2).

Face II exhibits a uniform gray opaque surface in normal light (Fig. 1b). The CL image of this face reveals an inner darker zone (E1) and an external brighter zone (E2) (Fig 1d). This transition is similar to the one recognized in the outer fibrous layer of Face I.

5. Results

One hundred and sixty-seven microinclusions were analyzed in the fibrous zones exposed on both diamond faces. Most inclusions are indicative of fluids with compositions varying between carbonatitic melt and brine, but 29 microinclusions represent peridotitic minerals. All mineral inclusions were detected on face I. No inclusions were detected in the cavity filled zones (Fig. 1a, zones B and D).

5.1. Mineral microinclusions

All the mineral inclusions detected belong to the peridotitic suite. They are all smaller than 1 μm and their average total oxide content is 12 ± 8 wt.%.

5.1.1. Cr-diopside

Twenty-six inclusions were detected. The composition of one representative individual inclusion along with the average composition is presented in Table 1.

Table 1
Composition of mineral inclusions in diamond ON-DVK-294

Phase (wt.%)	Cr-diopside			Chromite	
	Average (11)[a] Standard deviation				
Inclusion number	182[b]			189	190
SiO$_2$	52.3	51.2	2.79	n.d.	1.52
TiO$_2$	1.34	0.24	0.42	0.25	n.d.
Al$_2$O$_3$	1.98	2.37	0.44	8.28	9.00
FeO	1.55	1.84	1.09	17.1	16.6
MgO	14.3	14.5	1.51	13.1	12.6
CaO	17.2	16.5	2.01	0.27	n.d.
BaO	n.d.	0.11	0.36	n.d.	2.26
Na$_2$O	4.02	5.22	2.22	0.62	0.96
K$_2$O	1.45	1.79	1.05	0.28	0.98
P$_2$O$_5$	0.55	0.12	0.21	n.d.	n.d.
Cl	1.45	1.64	0.94	0.11	n.d.
Cr$_2$O$_3$	4.12	4.44	1.20	60.0	56.1
#Mg	0.94	0.93	0.04		
Analyzed total[c]	12.5	13.4	6.4	26.2	9.9
Formula units	(6 oxygen)			(32 oxygen)	
Si	1.94	1.92	0.05	n.d.	0.40
Ti	0.04	0.01	0.01	0.05	n.d.
Al	0.09	0.11	0.02	2.58	2.82
Fe	0.05	0.06	0.04	3.78	3.67
Mg	0.79	0.81	0.07	5.15	4.99
Ca	0.68	0.67	0.09	0.08	n.d.
Ba	n.d.	0.002	0.01	n.d.	0.24
Na	0.29	0.38	0.16	0.32	0.49
K	0.07	0.09	0.05	0.09	0.33
P	0.02	0.004	0.01	n.d.	n.d.
Cl	0.09	0.10	0.06	0.05	n.d.
Cr	0.12	0.002	0.01	12.54	11.78

n.d.—not detected.

[a] Average composition of 11 inclusions that have low Cl and K concentration (contain only a small amount of brine).

[b] Representative analysis.

[c] The original total contents of oxide and chlorine before renormalization to 100%.

The average Mg# is 0.94 ± 0.04 and the Ca/(Ca + Mg + Fe) ratio is 0.43 ± 0.05. The Cr$_2$O$_3$ content is 4.4 ± 1.2 wt.%. Some of the diopside-bearing inclusions are high in K and Cl (up to 8% Cl), indicating the entrapment of some brine together with the mineral phase in a single inclusion.

5.1.2. Chromite

Two inclusions share a Cr/(Cr + Al) ratio of 0.82 (Table 1).

5.1.3. Olivine

One inclusion has Mg# of 0.93 but (Mg + Fe)/Si ratio of 2.2. High fluorine content indicates contamination during sample preparation. The inclusion probably contained an olivine crystal, but the contamination precludes accurate analysis.

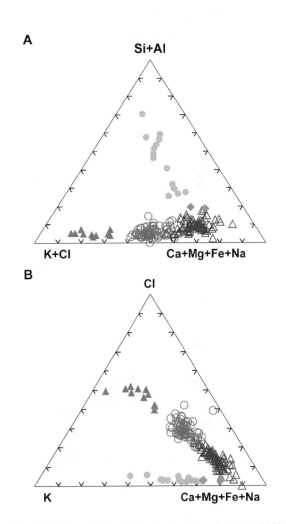

Fig. 2. (A) Composition of the microinclusions in ON-DVK-294 and in diamonds from other localities. Hydrous silicic melt endmember composition falls close to the Si + Al apex. The carbonatitic melt falls close to the Ca + Mg + Fe + Na apex and the K + Cl apex represents the brine. △: carbonatitic melt inclusions in ON-DVK-294; ○: brine-rich inclusions in ON-DVK-294. ▲: Koffiefontein brine (Izraeli et al., 2001); ◆: Koffiefontein melt (Izraeli, unpublished data); ●: Botswanan melt (Schrauder and Navon, 1994). (B) A K, Cl and Ca + Mg + Fe + Na projection of the same data set.

Fig. 3. Map of the analyzed inclusions on Face II. The different symbols represent analyses performed on different days.

5.2. Fluid microinclusions

Microinclusions containing fluids were found on both sides of the diamond. On Face I they were detected in layers C and E (Fig. 1a). Fig. 2 presents the wide range of composition spanned by the inclusions between the brine and carbonatitic melt endmembers.

5.2.1. Spatial relation between the two components

Face II was mapped in detail in order to detect any zoning in the composition of the inclusions. Fig. 3

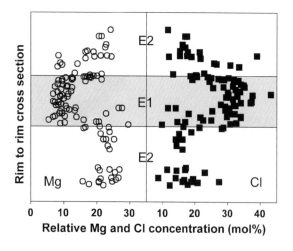

Fig. 4. Mg (○) and Cl (■) variations along a profile across the diamond. The brine resides in the inner zone (high Cl and low Mg concentration) and the carbonatitic melt in the outer zone (low Cl and high Mg concentration).

presents the location of the inclusions relative to the zones revealed by CL. Careful examination of the data showed that inclusions located in the inner growth zone are enriched in the brine component, while the inclusions in the outer zone are enriched in the

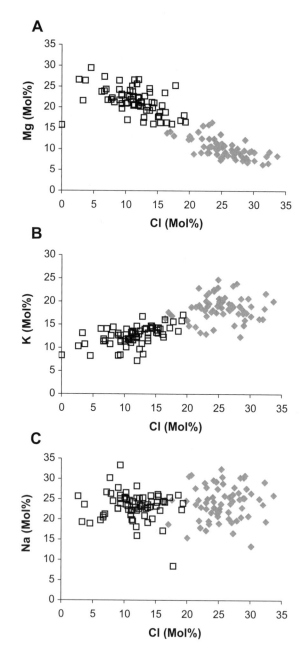

Fig. 5. Variation diagrams of inclusions containing brine and carbonatitic melts. □: Carbonatitic melt, ◆: brine.

carbonatitic melt component. Face II (Fig. 4) clearly presents the compositional difference between the two zones but the transition is not as sharp as exhibited by CL. Some inclusions with intermediate composition were analyzed in the contact area between the two zones. The inclusions detected on Face I of this diamond have similar chemical compositions and obey the same spatial relations. The inclusions in the outermost zone contain carbonatitic melt, while the inner zones contain brine inclusions.

As an example a negative correlation between Mg and Cl is exhibited in Fig. 5. Magnesium decreases from 28 mol% in the carbonatitic composition to almost zero in the brine-rich compositions, while Cl rises from zero to more than 30%. Calcium behaves

similarly (not shown). Potassium is present in both fluid compositions, but reaches its maximum concentration in the brine-rich inclusions. The sodium content is high in all fluid compositions.

From these observations it is clear that compositions vary between two endmembers—*The brine-enriched composition*: The average composition of the fluid in all the brine-bearing inclusions is $K_{18}Na_{24}Ca_6Mg_{10}Fe_4Ba_3Si_5PCl_{26}$ (mol%) and the composition of five inclusions that show the strongest enrichment in Cl is $K_{19}Na_{25}Ca_5Mg_8Fe_3Ba_2Si_4Cl_{32}$ (Table 2). In these brine-rich inclusions, the most abundant cation is sodium, the average Na/Cl ratio is 0.8 ± 0.1. The K/Cl ratio is 0.6 ± 0.1 and the Mg/Cl and Ca/Cl ratios are 0.2 ± 0.04 and 0.14 ± 0.05 re-

Table 2
Composition of fluid inclusions in diamond ON-DVK-294

	Brine								Carbonatitic melt							
	Representative inclusions					End member	Average[a]	Standard deviation	Representative inclusions					End member	Average[a]	Standard deviation
Inclusion number	66	70	71	129	153				87	92	143	145	180			
(wt.%)																
SiO_2	5.8	4.4	5.7	4.0	7.0	6.7	6.4	2.4	12.3	9.8	7.4	9.0	7.3	12.4	9.5	2.9
TiO_2	1.1	n.d.	n.d.	n.d.	n.d.	1.0	0.8	1.5	1.2	n.d.	n.d.	2.8	2.5	0.7	1.0	1.7
Al_2O_3	0.4	1.0	n.d.	n.d.	2.1	–	0.9	1.4	2.5	0.8	0.8	1.3	2.2	2.7	1.5	1.3
FeO	7.6	11.0	10.5	3.8	6.7	5.3	6.3	3.0	8.2	6.9	7.9	10.3	8.0	10.2	7.7	2.1
MgO	9.9	8.9	9.4	6.6	10.9	7.7	9.1	2.4	19.4	18.2	18.6	21.1	19.1	21.3	18.6	2.8
CaO	8.7	8.5	7.7	6.7	9.0	6.4	8.1	2.4	15.2	12.9	16.3	12.9	12.6	12.5	12.8	3.0
BaO	13.1	11.9	12.0	12.0	14.5	8.7	10.2	4.6	4.3	10.3	4.0	5.0	5.5	6.8	7.9	4.7
Na_2O	12.8	15.1	16.7	22.4	13.1	18.9	17.6	4.7	12.0	15.6	15.5	15.2	16.6	13.6	15.7	2.9
K_2O	21.1	19.3	21.0	21.5	19.4	21.5	20.0	4.6	12.2	13.8	15.0	13.3	12.1	11.1	12.8	2.2
P_2O_5	2.5	1.9	1.8	1.1	2.2	0.6	2.0	1.8	3.8	3.1	2.4	2.5	3.1	3.5	3.1	1.8
Cl	19.3	21.0	18.1	28.3	16.7	28.2	21.6	5.1	9.0	9.5	12.6	6.8	8.3	3.5	8.8	3.1
Analysed total (wt.%)	7.6	6.1	6.1	5.7	6.0		3.9	2.0	7.0	5.0	5.1	7.4	5.0		5.0	2.3
(mol%)																
Si	4.5	3.3	4.3	2.7	5.4	4.5	4.6	1.8	9.6	7.6	5.4	7.0	5.6	10.1	7.4	2.3
Ti	0.6	n.d.	n.d.	n.d.	n.d.	0.6	0.4	0.8	0.7	n.d.	n.d.	1.6	1.4	0.4	0.6	1.0
Al	0.4	0.9	n.d.	n.d.	1.9	–	0.8	1.2	2.3	0.8	0.7	1.2	2.0	2.5	1.4	1.2
Fe	4.9	6.9	6.6	2.1	4.3	3.0	3.8	1.9	5.3	4.5	4.9	6.7	5.1	6.9	5.0	1.4
Mg	11.4	9.9	10.4	6.7	12.6	7.8	9.8	2.8	22.5	21.1	20.5	24.3	22.0	25.6	21.5	3.2
Ca	7.1	6.8	6.2	4.8	7.5	4.7	6.3	2.1	12.7	10.7	12.9	10.7	10.4	10.8	10.6	2.6
Ba	3.9	3.5	3.5	3.2	4.4	2.4	2.9	1.4	1.3	3.1	1.2	1.5	1.6	2.2	2.4	1.5
Na	19.0	21.8	24.2	29.2	19.7	24.7	24.4	5.1	18.1	23.5	22.2	22.7	24.7	21.3	23.6	4.4
K	20.7	18.4	20.1	18.5	19.2	18.5	18.4	3.8	12.1	13.6	14.1	13.1	11.9	11.4	12.6	2.1
P	1.6	1.2	1.2	0.6	1.4	0.4	1.2	1.1	2.5	2.0	1.5	1.6	2.0	2.4	2.0	1.2
Cl	25.1	26.6	23.0	32.3	21.9	32.4	26.1	4.7	11.9	12.5	15.7	9.0	10.8	4.8	11.4	4.0

[a] Average composition of 71 brine inclusions and 67 carbonatitic melt inclusions with their standard deviation.

spectively. On the average, the mono and divalent positive ions account for 80 positive charges of which only 32 are compensated by Cl ions, the rest may be compensated by carbonate ions (Izraeli et al., 2001).

The carbonatitic melt enriched composition: The average composition of all the inclusions that carry carbonatitic melt is $K_{13}Na_{24}Ca_{11}Mg_{22}Fe_5Ba_2Si_7AlP_2Cl_{11}$ (mol%) and the composition of the five inclusions that are richest in carbonatitic melt is $K_{11}Na_{21}Ca_{11}Mg_{26}Fe_7Ba_2Si_{10}Al_3P_2Cl_5$ (Table 2). In this carbonatitic melt endmember sodium is also the most abundant cation. However the Cl concentration is much lower compared with the brine, The Na/Cl ratio of the carbonatitic melt is 5.1 ± 2.5, the K/Cl ratio is 2.6 ± 1.0 and the Mg/Cl ratio is 6.0 ± 2.3.

The average total content of the oxides and chloride for all fluid inclusions is 4.4 ± 2.2 wt.% and is lower than that obtained for the mineral inclusions. A possible reason for that difference may be a smaller size of the fluid inclusions. Alternatively, if the size of the different microinclusions in this diamond is similar, the lower totals measured for the nonmineral inclusions may reflect a high concentration of light elements not measured by EPMA (Izraeli et al., 2001). We could not analyze the water and carbonate of the inclusions by InfraRed spectroscopy. However, in analogy with the Koffiefontein and Jwaneng diamonds (Schrauder and Navon, 1994; Izraeli et al., 2001), we suggest that water and carbonate are also present in the inclusions of ON-DVK-294.

5.3. Carbon isotopic composition

Table 3 presents C isotopic composition of the diamond matrix, as measured by SIMS. $\delta^{13}C$ was measured at seven points across Face I. The core (A)

Table 3
Carbon isotopic composition of selected points across Face I of diamond ON-DVK-294

Point location	$\delta^{13}C$ ($\pm 0.4‰$)
Zone A	− 6.0
Zone A	− 5.2
Zone B	− 8.9
Zone C	− 8.6
Zone D	− 8.1
Zone E2	− 12.8
Zone E2	− 11.4

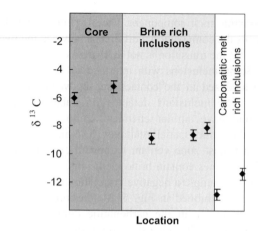

Fig. 6. Variation in the C isotopic composition ($\delta^{13}C$) along a line from the diamond core through its fibrous layers (error bars $\pm 0.4‰$).

yielded values of $- 5.2‰$ and $- 6‰$. The coat is depleted in ^{13}C relative to the core. Three measurements in the inner zone (B to E1), where brine inclusions were found, had similar $\delta^{13}C$ of $- 8.1‰$ to $- 8.9‰$. Additional depletion in ^{13}C was observed in the outer zone (E2), where the carbonatitic inclusions were found. The $\delta^{13}C$ measured in two points was $- 11.4‰$ and $- 12.8‰$ (Fig. 6). All five measurements in the fibrous zones revealed isotopic compositions that are more depleted than the worldwide range of fibrous diamonds (Cartigny et al., 1998, 2003).

6. Discussion

6.1. Comparison to other mantle derived fluids

Carbonatitic melts and brine inclusions were previously reported for diamonds from Jwaneng (Schrauder and Navon, 1994) and Koffiefontein (Izraeli et al., in press). Partial analyses of chlorine-rich inclusions (0.2–2 μm in size) were also reported by Chen et al. (1992) in Chinese diamonds. However, no full analysis was provided. Recently, Kamenetsky et al. (2002) reported Na–K–Cl and C-rich inclusions in olivine phenocrysts from the Udachnaya kimberlite.

Fig. 2 presents the chemical composition of the fluids from Jwaneng and Koffiefontein along with the compositions detected in ON-DVK-294. It is clear that the microinclusions in this diamond bridge the compositional gap between the brine-rich inclusions from

Koffiefontein and the carbonatitic melt endmember of the Jwaneng diamonds.

The composition of brine microinclusions in diamond ON-DVK-294 is broadly similar to the brine detected in the Koffiefontein cloudy diamonds. The Canadian brine inclusions carry higher proportions of the carbonatitic endmember, but their K/Cl ratio (0.7 ± 0.1) is similar to that of Koffiefontein (0.64 ± 0.06). Thus, the fundamental components of the brine endmember are similar in diamonds from different cartons. The fact that the ON-DVK-294 brine was detected in a fibrous coat while the Koffiefontein brine was detected in inner clouds within octahedral diamonds indicates that diamonds of both habits can grow from similar solutions. The main difference between the two brines is the strong enrichment in Na concentration in the Diavik diamond. It is interesting that Izraeli et al. (2001) noted higher concentration of Na in diamonds carrying peridotitic inclusions. The Na/K ratio was 0.4 ± 0.1 in two peridotitic diamonds compared with 0.12 ± 0.06 in the brine of five eclogitic diamonds. ON-DVK-294 also contains peridotitic mineral inclusions, but its enrichment in Na (Na/K $= 1.3$) in the brine-rich inclusions is even higher that in any of the Koffiefontein peridotitic diamonds.

The carbonatitic melt inclusions in ON-DVK-294 (Fig. 2) show similar characteristics to the carbonatitic melt endmember in the Jwaneng cubic diamonds (Schrauder and Navon, 1994) and to the carbonatitic melt detected recently in Koffiefontein (Izraeli et al., unpublished data). All the fluids are enriched in divalent ions such as Mg, Ca and Fe; however, the ON-DVK-294 carbonatitic melt is characterized by higher proportions of Mg. It is also unique in its high Na and notable Ba content.

The similar nature of the Botswanan, Koffiefontein and ON-DVK-294 carbonatitic melts become is impressive considering the fact that the Botswanan melts align along the compositional range between the carbonatitic and the hydrous silicic melt, whereas the Diavik fluids represent mixtures along an array between carbonatitic melt and brine. The fact that both arrays meet at a similar endmember composition suggests that the carbonatitic melt is the link connecting all fluids and that all fluids along both arrays are genetically related.

Johnson et al. (2000) and Burgess et al. (2002) studied the composition of Ar, Cl, Br, I, Ca and K in fibrous diamond from Africa, Russia and Canada. In most diamonds, they found uniform I/Cl and Br/Cl ratios that are similar to depleted astenospheric mantle ratios derived from MORB data. Only in some Canadian diamonds (from the Ekati kimberlites, Lac de Gras area) they detected much higher Br/Cl and I/Cl ratios that deviated significantly from the restricted range of all other diamonds. The analyses of all microinclusions in diamonds from Congo, Botswana and Siberia revealed only carbonatitic to hydrous silicic melts (Navon et al., 1988; Schrauder and Navon, 1994; Klein-BenDavid and Logvinova, unpublished data), but no brine. It is possible that the high I/Cl and Br/Cl ratios in some of the Canadian diamonds are related to the formation of the halogen-rich brine similar to that found in ON-DVK-294 diamond. The high ratios may represent preferred fractionation of Br and I into the brine, whereas Cl, as a major element is enriched to a lesser amount.

6.2. Evolution of the fluid

The similar compositions of carbonatitic melts associated with brine and with hydrous silicic melt suggest that both fluids are the products of a primary carbonatitic melt. If so, fluids should evolve from carbonatitic to brine-rich compositions, but the fluid inclusions in ON-DVK-294 show the reversed transition from brine-enriched fluids in the inner part of the diamond to carbonatitic melt at the diamond rim. However, the inclusions in ON-DVK-294 do not show a continuous, uniform evolution. The transition between the brine-rich central zones (C and E1) and the carbonatitic outer zone (E2) is abrupt with only a few inclusions close to the interface showing intermediate compositions (Fig. 4). This reflects a rapid but continues transition from brine-rich to carbonatitic composition. The observed compositional zoning is accompanied by an even sharper transition in the CL image (Fig. 1c and d) and a significant depletion in ^{13}C (Fig. 6) from the inner fibrous zones (-8.5‰) to the outer zone (-12.1‰).

We suggest that the sharp compositional and isotopic transition reflects mixing between evolved brine and freshly introduced carbonatitic melt infiltrating into the area. Most probably, the change in the isotopic composition of the diamond reflects growth from a more ^{13}C depleted carbonatitic melt. More data are needed before the source of the fluids can be inferred.

7. Conclusions

The compositional range of microinclusions trapped in diamond ON-DVK-294 sheds light on fluid evolution in the diamond-forming environment in the mantle. The compositional gap between brine found in cloudy octahedral diamonds from South African and carbonatitic melt analyzed in cubic diamonds from Zaire and Botswana is now closed. The compositions of all microinclusions analyzed to date fall along two arrays connecting the carbonatitic melt composition to either a hydrous silicic endmember or to a brine endmember (Fig. 2).

The connection between all fluids found in microinclusions in fibrous and cloudy diamonds from various cratons suggest that diamonds are formed from fluids derived form a mantle source not significantly influenced by local heterogeneities.

The composition of microinclusions in diamond ON-DVK-294 changes from brine-rich to carbonatitic. This transition is associated with depletion in the carbon isotopic composition of the diamond. We suggest that this transition reflects mixing between an already evolved brine and a freshly introduced carbonatitic melt of different isotopic composition.

References

Bastin, G., Heijligers, J., 1991. In: Heinrich, K., Newbury, D. (Eds.), Electron Probe Quantitation, Workshop at the National Bureau of Standards, Gaithersburg, Maryland. Plenum, New York, pp. 145–161.

Boyd, S.R., Pillinger, C.T., Milledge, H.J., Seal, M.J., 1992. C-Isotopic and N-Isotopic composition and the infrared-absorption spectra of coated diamonds-evidence for the regional uniformity of CO_2–H_2O rich fluids in lithospheric mantle. Earth and Planetary Science Letters 108 (1–3), 139–150.

Bulanova, G.P., 1995. The formation of diamond. Journal of Geochemical Exploration 53 (1–3), 1–23.

Bulanova, G.P., Griffin, W.L., Ryan, C.G., 1998. Nucleation environment of diamonds from Yakutian kimberlites. Mineralogical Magazine 62 (3), 409–419.

Burgess, R., Layzelle, E., Turner, G., Harris, J.W., 2002. Constraints on the age and halogen composition of mantle fluids in Siberian coated diamonds. Earth and Planetary Science Letters 197 (3–4), 193–203.

Cartigny, P., Harris, J.W., Phillips, D., Girard, M., Javoy, M., 1998. Subduction-related diamonds? The evidence for a mantle-derived origin from coupled δC^{13}–δN^{15} determinations. Chemical Geology 147 (1–2), 147–159.

Cartigny, P., Harris, J.W., Taylor, A., Davies, R., Javoy, M., 2003. On the possibility of a kinetic fractionation of nitrogen stable isotopes during natural diamond growth. Geochimica et Cosmochimica Acta 67 (8), 1571–1576.

Chen, F., Guo, J., Chen, C., Liu, C., 1992. High-K and high-Cl inclusions in diamond and mantle metasomatism. Acta Mineralogica Sinica 12 (3), 193–198.

Chinn, I., Gurney, J., Kyser, K., 1999. Diamonds and mineral inclusions from the NWT, Canada. 7th International Kimberlite Conference. Addendum to Extended Abstracts. Cape Town, South Africa.

Chrenko, R., McDonald, R., Darrow, K., 1967. Infrared spectrum of diamond coat. Nature 214, 474–476.

Creaser, R., Grutter, H., Carlson, J., Crawford, B., 2003. Macrocrystal phlogopite Rb–Sr dates for the Ekati property kimberlites, Slave province, Canada: evidence for multiple intrusive episodes in the Paleocene and Eocene. 8th International Kimberlite Conference, Extended Abstracts, Victoria, Canada.

Davies, R., Griffin, W., Pearson, N., Andrew, A., Doyle, B., O'Reilly, S., 1999. Diamonds from the deep; Pipe DO-27, Slave Craton, Canada. Proceedings of the 7th International Kimberlite Conference. Red Roof Design Cc, Cape T, South Africa, pp. 148–155.

Davies, R., Griffin, W., O'Reilly, S., Doyle, B., 2003. Geochemical characteristics of microdiamonds from kimberlites at Lac De Gras, central Slave craton. 8th International Kimberlite Conference, Extended Abstracts, Victoria, Canada.

Galimov, E.M., 1991. Isotope fractionation related to kimberlite magmatism and diamond formation. Geochimica et Cosmochimica Acta 55 (6), 1697–1708.

Graham, I., Burgess, J., Bryan, D., Ravenscroft, P., Thomas, E., Doyle, B., Hopkins, R., Armstrong, K., 1999. Exploration history and geology of the Diavik Kimberlites, Lac de Gras, Northwest Territories, Canada. Proceedings of the 7th International Kimberlite Conference. Red Roof Design, Cape Town, South Africa, pp. 262–279.

Griffin, W.L., Doyle, B.J., Ryan, C.G., Pearson, N.J., O'Reilly, S.Y., Davies, R., Kivi, K., Van Achterbergh, E., Natapov, L.M., 1999. Layered mantle lithosphere in the Lac de Gras area, Slave Craton: composition, structure and origin. Journal of Petrology 40 (5), 705–727.

Gurney, J., Hildebrand, P., Carlson, J., Dyck, D., Fedortchouk, F., 2003. Diamonds from the Ekati core and buffer zone Properties. 8th International Kimberlite Conference, Extended Abstracts, Victoria, Canada.

Guthrie, G., Veblen, D., Navon, O., Rossman, G., 1991. Submicrometer fluid inclusions in turbid-diamond coats. Earth and Planetary Science Letters 105, 1–12.

Harris, J., 1992. Diamond geology. In: Field, J. (Ed.), The Properties of Natural and Synthetic Diamonds. Academic Press, Oxford, UK, pp. 384–385.

Hauri, E.H., Wang, J., Pearson, D.G., Bulanova, G.P., 2002. Microanalysis of delta C-13, delta N-15, and N abundances in diamonds by secondary ion mass spectrometry. Chemical Geology 185 (1–2), 149–163.

Heaman, L., Kjarsgaard, B., 2000. Timing of eastern North American kimberlite magmatism: continental extension of the Great Meteor Hotspot track? Earth and Planetary Science Letters 178, 253–268.

Izraeli, E.S., Harris, J.W., Navon, O., 2001. Brine inclusions in diamonds: a new upper mantle fluid. Earth and Planetary Science Letters 187 (3–4), 323–332.

Izraeli, E.S., Harris, J.W., Navon, O., 2003. Fluid-and mineral inclusions in cloudy diamonds from Koffiefontein, South Africa. Geochmica et Cosmochemica Acta (in press).

Johnson, L., Burgess, R., Turner, G., Milledge, H., Harris, J., 2000. Noble gas and halogen geochemistry of mantle fluids: comparison of African and Canadian diamonds. Geochimica et Cosmochimica Acta 64 (4), 717–732.

Kamenetsky, M., Sobolev, A., Kamenetsky, V., Maas, R., Danyushevsky, L., Sobolev, N., Pokhilenko, N., 2002. Origin, composition and fractionation of the Udachnaya pipe kimberlitic melts: constraints from melt, fluid and crystal inclusions in olivine. Goldschmidt Conference, Extended Abstracts, Davos, Switzerland.

Klein-BenDavid, O., Logvinova, A.M., Izraeli, E., Sobolev, N.V., Navon, O., 2003. Sulfide melt inclusions in Yubileinayan (Yakutia) diamonds. 8th International Kimberlite Conference, Extended Abstracts, Victoria, Canada.

Lang, A., Walmsley, J., 1983. Apatite inclusion in natural diamond coat. Physics and Chemistry of Minerals 9, 6–8.

Meyer, H.O.A., 1987. Inclusions in diamond. In: Nixon, P.H. (Ed.), Mantle Xenoliths. Wiley, Chichester, UK, pp. 501–522.

Milledge, H., Mendelssohn, M., Woods, P., Seal, M., Pillinger, C., Mattey, D., Carr, L., Wright, I., 1984. Isotopic variations in diamond in relation to cathodoluminescence. Acta Crystallographica. Section A, Foundations of Crystallography 40, 255.

Navon, O., 1999. Formation of diamonds in the Earth's mantle. Proceedings of the 7th International Kimberlite Conference. Red Roof Design, Cape Town, South Africa, pp. 584–604.

Navon, O., Hutcheon, I.D., Rossman, G.R., Wasserburg, G.L., 1988. Mantle-derived fluids in diamond micro-inclusions. Nature 335, 784–789.

Raga, S., Davies, R.M., Griffin, W.L, Jackson, S., O'Reilly, S.Y., 2003. Trace element analysis of diamond by LAM ICPMS: preliminary results. 8th International Kimberlite Conference, Extended Abstracts, Victoria, Canada.

Schrauder, M., Navon, O., 1994. Hydrous and carbonatitic mantle fluids in fibrous diamonds from Jwaneng, Botswana. Geochimica et Cosmochimica Acta 58 (2), 761–771.

Schrauder, M., Koeberl, C., Navon, O., 1996. Trace element analyses of fluid-bearing diamonds from Jwaneng, Botswana. Geochimica et Cosmochimica Acta 60 (23), 4711–4724.

Shiryaev, A., Izraeli, E., Hauri, E.H., Galimov, E.M., Navon, O., 2003. Fluid inclusions in Brazilian coated diamonds. 8th International Kimberlite Conference, Extended Abstracts, Victoria, Canada.

Sobolev, N., Kaminsky, F., Griffin, W., Yefimova, E., Win, T., Ryan, C., Botkunov, A., 1997. Mineral inclusions in diamonds from the Sputnik kimberlite pipe, Yakutia. Lithos 39 (3–4), 135–157.

Sunagawa, I., 1984. Morphology of natural and synthetic diamond crystals. Materials Science of the Earth's Interior, Terra Sci. Publ. Co., 303–330.

Tal'nikova, S.B., 1995. Inclusions in natural diamonds of different habits. 6th International Kimberlite Conference, Extended Abstracts, Novosibirsk, Russia, pp. 603–605.

Tappert, R., Stachel, T., Harris, J., Brey, G., 2003. Mineral inclusions in diamonds from Panda kimberlite, Slave province (Canada). 8th International Kimberlite Conference, Extended Abstracts, Victoria, Canada.

Taylor, L.A., Keller, R.A., Snyder, G.A., Wang, W.Y., Carlson, W.D., Hauri, E.H., McCandless, T., Kim, K.R., Sobolev, N.V., Bezborodov, S.M., 2000. Diamonds and their mineral inclusions, and what they tell us: a detailed "pull-apart" of a diamondiferous eclogite. International Geology Review 42 (11), 959–983.

Tomilenko, A.A., Chepurov, A.I., Palyanov, Y.N., Pokhilenko, L.N., Shebanin, A.P., 1997. Volatile components in the upper mantle (from data on fluid inclusions). Geologiya I Geofizika 38 (1), 276–285.

Available online at www.sciencedirect.com

Lithos 77 (2004) 255–271

www.elsevier.com/locate/lithos

The relationship between the distribution of nitrogen impurity centres in diamond crystals and their internal structure and mechanism of growth

Felix V. Kaminsky*, Galina K. Khachatryan

KM Diamond Exploration Ltd., 2446 Shadbolt Lane, West Vancouver, BC, Canada V7S 3J1

Received 27 June 2003; accepted 30 October 2003
Available online 24 June 2004

Abstract

Fifty diamond crystals of different morphological types (octahedra, dodecahedroids, cubes and single tetrahexahedroid) with differing internal structures were examined using methods of cathodoluminescence (CL), anomalous birefringence and local infrared (IR) analysis. The main objective of the study was to examine the regularities of nitrogen impurity distribution in diamond with differing internal structures. Almost all the analyzed octahedra, as well as dodecahedroids with zonal structures and the blocky dodecahedroids, are characterized either by nearly isothermic growth conditions or by a decrease in formation temperature during the crystallization process. In contrast to zoned octahedra and dodecahedroids, dodecahedroids with zonal–sectorial and sectorial internal structures show a notably different distribution of nitrogen defects, with N_{tot} generally decreasing from crystal cores to marginal areas, and degree of nitrogen aggregation increasing in the same direction. From this, it would follow that in these crystals, the temperature of diamond formation of the outer crystal zones is approximately 40–50 °C higher than that of the inner zones. The same result (15 to 80 °C) was obtained for diamond crystals with cubic habit, which generally show a fibrous internal structure, reflecting normal mechanisms of growth. The anomalous distribution of nitrogen centres in diamond crystals that grew through the normal mechanism, with a high rate of growth and in an oversaturated medium, might point to non-equilibrium relationships between the concentrations of different nitrogen centres. It is likely that in crystals of this type, the rate of growth is higher than the rate of structural nitrogen aggregation. Thus, it appears that in these peculiar crystals of diamond we deal with non-equilibrium concentrations of nitrogen B centres and, consequently, with anomalous, non-actual diamond formation temperatures.

Keywords: Diamond; Internal structure; Nitrogen; Crystallization; Cathodoluminescence; Optical birefringence

1. Introduction

In accordance with the results of our previous studies (Kaminsky and Khachatryan, 2001; Khachatryan and Kaminsky, 2003), an individual primary diamond deposit is characterized by quite distinct

---

* Corresponding author. Tel.: +1-604-925-8755; fax: +1-604-925-8754.

E-mail address: felixvkaminsky@cs.com (F.V. Kaminsky).

0024-4937/$ - see front matter © 2004 Elsevier B.V. All rights reserved.
doi:10.1016/j.lithos.2004.04.035

concentration and distribution parameters of nitrogen and hydrogen defects, in its host diamond crystals. Specifically, it has been established that 'equilibrium' and 'non-equilibrium' diamond crystals differ significantly in the character of their internal distribution of nitrogen centres and the degree of nitrogen aggregation (Khachatryan and Kaminsky, 2003). The heterogeneity of nitrogen distribution in diamond has previously been intimated by a number of researchers (e.g., Klyuev et al., 1969; Sobolev, 1974, 1978; Suzuki and Lang, 1976; Taylor et al., 1995a,b; Davies et al., 1999; Hauri et al., 1999). However, agreement among these scientists regarding the interpretation of this phenomenon has not been met.

To date, the lack of strong data on the heterogeneity of nitrogen distribution in diamond was the result of studies that were focused on just a limited number of diamond crystals. It also reflects the fact that not all known diamond varieties have been investigated. To address this, we have performed a systematic and comprehensive study of the distribution of impurity centres in diamond crystals with differing (as diversified as possible) crystal habits and internal structures from widely different regions. In all, 50 diamonds crystals were examined: 11 octahedra, 32 dodecahedroids, 6 cubes and 1 tetrahexahedroid; crystals with differing internal structures being present in each morphological group.

The main objective of this study is to reveal and examine the regularities of nitrogen impurity distribution in diamond for a representative collection of natural diamond with differing internal structures.

2. Background information

Natural diamond is characterized by a wide variety of internal structures (Seal, 1965; Frank, 1969; Genshaft et al., 1977; Beskrovanov, 1992; Zezin et al., 2001), reflecting the temporal evolution of conditions during the formation of diamond.

As Patel and Patel (1969), Tolansky (1974) and Bokii et al. (1986) have shown, the most common internal structure of diamond crystals is octahedral zoning. This zoning appears as growth layers alternating along the $<111>$ direction, formed as a result of layer-by-layer growth. The 'tangential' mechanism of crystal growth either occurs through 2-D nucleation or

through screw dislocation (Chernov et al., 1980). As was shown by Sunagawa (1984), layer-by-layer octahedral growth of natural diamond crystals proceeds in a slightly oversaturated environment, with a low rate of growth. In general, the layer-by-layer octahedral zoning can be taken as an indicator of equilibrium conditions of diamond formation.

On the basis of the internal structural features in cubic diamond crystals, Moore and Lang (1972) proposed an alternative model, involving a 'normal' mechanism of diamond growth, which, they suggest, occurs by infilling the space by a branching columnar structure (Lang, 1974). Crystals of this type have fibrous internal structures, with fibril axes oriented along the growth direction. According to this interpretation, the concentric zoning, that develops parallel to cubic faces, is related to an occasionally undulating crystallization front of the columnar structure, as opposed to layer-by-layer growth along the $<100>$ direction. Normal (continuous) crystal growth suggests that diamond matter particles adhere to the growing crystal with equal probability at any point of any crystal face (Chernov et al., 1980). Evidence in support of normal growth, where found in diamond, indicates that the crystal formed in a strongly oversaturated medium, with a high rate of growth, i.e., under non-equilibrium conditions (Bokii et al., 1986).

In addition to crystals of diamond with zonal and fibrous internal structures, among natural and synthetic diamond crystals also occur crystals with a sectorial structure (Seal, 1965). Diamond crystals of this type are composed of growth pyramids with differing crystallographic orientations, most commonly $<111>$ and $<100>$ and, more rarely, $<110>$, $<311>$ and some others (Sobolev, 1974; Palyanov et al., 1997).

The cores of diamond crystals with sectorial and zonal–sectorial internal structures appear as wedge-like $<111>$ sectors with rectilinear zoning. Varshavsky (1968) analyzed a representative set of samples by a polarization optical method to show that octahedral growth pyramids have a layered structure, whereas cubic growth pyramids show no lamination. Subsequently, Moore and Lang (1972) suggested that diamond crystals with this internal structure have a combination-type mechanism of growth, with $<111>$ pyramids forming through the tangential growth mechanism, and $<100>$ pyramids growing through the normal mechanism. If so, natural diamond crystals with sectorial and zonal–sectorial structures likely formed under non-equilibrium conditions, as compared to octahedral crystals of diamond with concentric zonal structures.

In general, equilibrium and non-equilibrium crystallization conditions could alternate repeatedly during the diamond formation process. Moreover, diamond growth could sometimes give way to dissolution, where upon it could resume, and so on. Evidence in favor of this appears to be found in diamond crystals with very complex internal structures, reflecting several distinct stages of diamond formation and dissolution. Generally, the structural heterogeneities that reflect the changes in diamond growth conditions correlate with changes in nitrogen and hydrogen impurity concentrations in different zones of the crystals. This being so, diamond crystals with zonal and sectorial structures are usually characterized by a heterogeneous distribution of structural defects (optically active centres) (Seal, 1966; Sobolev, 1978; Bokii et al., 1986; Field, 1992; Milledge et al., 1995; Chen et al., 1999, 2000).

According to the presently accepted 'annealing' model for the formation of nitrogen impurity centres in diamond (Evans, 1992; Taylor et al., 1996), the concentration of different nitrogen defects depends upon the temperature and the duration of post-crystallization annealing ('mantle residence' parameters). At the same time, recent studies (Palyanov et al., 1997; Khachatryan and Kaminsky, 2003) reveal that the relative abundance of A, B, and N defects in diamond depends not only on the temperature and duration of annealing, but also on a number of other factors (i.e., primarily on the rate of crystal growth). According to experimental data on diamond crystals synthesized in a Ni–Fe–C system, by the temperature difference method, at 1500 °C, A centres dominate in crystals grown with low growth rates, and N centres prevail in crystals of diamond produced with a high rate of growth (Palyanov et al., 1997).

3. Experimental procedure

Recent trends in research on structural defects in diamond are toward an increased use of infrared (IR) spectroscopy methods, which provide a means of estimating the concentrations of nitrogen (A and B)

and hydrogen (H) impurity centres and 'platelets' (P), i.e., the optically active defects that are most characteristic of diamond. Consistent with current ideas, the A-nitrogen centre in diamond involves a pair of nitrogen atoms jointly replacing a single carbon atom (Davies, 1976; Sobolev, 1978), whereas the B-defect is an aggregate of nitrogen atoms tetrahedrally arranged around a vacancy (Bursill and Glaisher, 1985). The 'platelets' (P-centres) are linear defects several atoms thick (Woods, 1986), and the H-centres are structural defects involving hydrogen (Sobolev and Lisoivan, 1971; Woods and Collins, 1983).

In order to evaluate the concentrations of A and B nitrogen centres, we use the analytical formulae of Boyd et al. (1994, 1995), according to which the concentration of A + B nitrogen centres in diamond is directly proportional to IR absorption coefficient values for the spectral peak at 1282 cm^{-1}. As in most cases, natural diamond belongs to the combined IaAB type; we decided to follow the method of interpretation of spectral characteristics proposed by Mendelssohn and Milledge (1995). The relative proportions of the 'platelets' (P) and the hydrogen (H) structural impurities in the diamond were estimated in this study in arbitrary units, more precisely, in absorption coefficient values measured, respectively, at 1365 and 3107 cm^{-1}.

IR absorption spectra of the diamond samples in our study were recorded using the Specord M-80 spectrophotometer (Carl Zeiss, Jena), with a beam condenser operating within the spectral range of 4000–400 cm^{-1}. Spectral resolution is 6–10 cm^{-1}. The average relative error of the concentration of nitrogen impurity is ± 2.5%; this produces the temperature estimate error less than 10 °C.

Local IR analyses were made with plates cut off along the (110) direction from diamond crystals with due regard to their internal structural features (zones and growth sectors). The thickness of the plates varies from 0.89 to 1.73 mm. In these tests, the area of the spot diaphragmed from crystal surfaces to record local IR spectra varied from 0.3 to 0.5 mm^2.

In this study, examination of the internal structural features of diamond crystals was essentially performed using the *method of cathodoluminescence* (CL) which provides a resolution of up to 0.2–0.5 μm. CL analyses of diamond were carried out by H.J. Milledge of University College, London, with a Newclide luminoscope, and by Prof. G.V. Saparin, using a

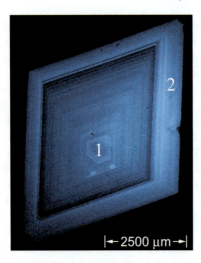

Fig. 1. Octahedron with layer-by-layer 'closed' octahedral zoning: CL pattern of sample #M1 (0.06 ct) from the Mir pipe, Yakutia (type O-1). Plate thickness is 1.45 mm.

Stereoscan MK-I A scanning electron microscope complete with a color CL attachment, in the Physics Department of the Moscow State Lomonossov University. CL images were obtained in a common operating mode ($U = 20$ kV; $I = 1 - 1000$ nA). In these experiments, polished diamond plates, gold-coated by

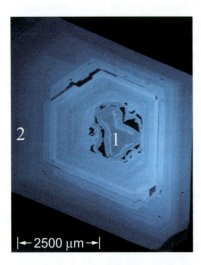

Fig. 2. Octahedron with 'open' layered octahedral zoning and crystal core with agate-like structure: CL pattern of sample #M4 (0.69 ct) from the Mir pipe, Yakutia (type O-3). Rectilinear-step-like octahedral zoning was formed as a result of the incomplete growth of layers around the crystal apices and along crystal edges. Plate thickness is 1.73 mm.

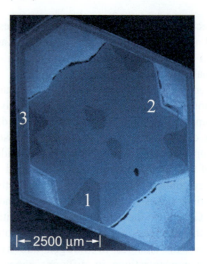

Fig. 3. Octahedron with zonal–sectorial structure: CL pattern of sample #U3 (0.14 ct) from the Udachnaya pipe, Yakutia (type O-4). The <100> growth pyramids originate from the very centre of the crystal (in the early stage of growth they would occupy the whole crystal volume), while the growth of <111> faces starts somewhat later, which is evident in the rim-to-core 'wedging-out' of octahedral faces. Plate thickness is 1.27 mm.

Fig. 4. Dodecahedroid with layer-by-layer 'closed' octahedral zoning: anomalous birefringence pattern of sample #NU1 from Vishera placers, Northern Urals (type D-1). Plate thickness is 1.10 mm.

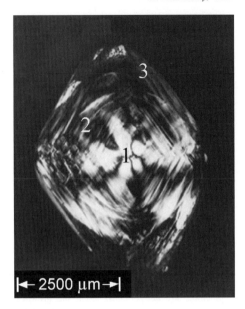

Fig. 5. Dodecahedroid with zonal–sectorial structure: anomalous birefringence pattern of sample #NU12 from Vishera placers, Northern Urals (type D-4). Plate thickness is 0.98 mm.

the method of sputtering, were placed into a cathode camera, and the resulting CL image was viewed on a monitor and captured via a computer.

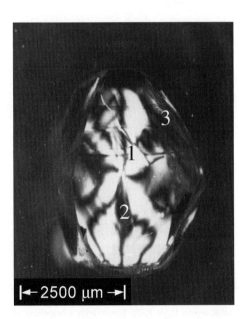

Fig. 6. Dodecahedroid with sectorial structure: anomalous birefringence pattern of sample #NU14 from Vishera placers, Northern Urals (type D-5). Plate thickness is 1.21 mm.

Fig. 7. Dodecahedroid with sectorial structure: anomalous birefringence pattern of sample #NU15 from Vishera placers, Northern Urals (type D-5). Plate thickness is 1.03 mm.

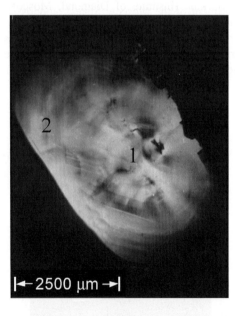

Fig. 8. Dodecahedroid with concentric, agate-like, rounded zoning #NU13 from Vishera placers, Northern Urals (type D-3). Rounded zoning in the core might reflect normal growth mechanisms, probably like the agate-like core of octahedra of type O-3. However, concentric zones in the intermediate portion of the crystal have a combination-type, curvi-linear + layer-by-layer-step-like appearance, which would be difficult to explain by normal growth alone. Plate thickness is 0.89 mm.

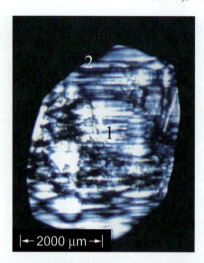

Fig. 9. Dodecahedroid with blocky structure: anomalous birefringence pattern of sample #K3 from pipe Karpinsky-1, Arkhangelsk region (type D-6). Plate thickness is 0.97 mm.

In addition, some diamond samples (primarily, those showing no luminescence) were analyzed by G.A. Gurkina (Institute of Diamond, Moscow) for anomalous birefringence by the *polarization optical method* with a MPS-2 stereomicroscope. Anomalous birefringence of diamond is related to internal stress in diamond crystals, owing to which it quite adequately

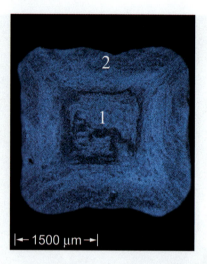

Fig. 11. Cube with concentric zonal structure: CL pattern of sample #U6 (0.12 ct) from the Udachnaya pipe, Yakutia (type K-1). Fibrils have a branching structure oriented along the growth sectors, particularly in the outer zone of the diamond. Plate thickness is 1.28 mm.

reflects the layered octahedral zoning, zonal–sectorial and purely sectorial internal structures of diamond crystals, and structural imperfections due to plastic deformation of diamond (Tolansky, 1966; Seal, 1966; Varshavsky, 1968).

Fig. 10. Cube with concentric zonal structure: anomalous birefringence pattern of sample #U5 from the Udachnaya pipe, Yakutia (type K-1). Well-defined fibrous structure of this crystal is evident in the <111> direction and in the outer zone of the diamond. Plate thickness is 1.33 mm.

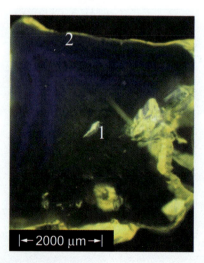

Fig. 12. Cube with undulating boundaries of growth zones: CL pattern of sample #L3 (0.12 ct) from the Lomonosov pipe, Arkhangelsk region (type K-2). Fibrils with a branching structure are developed in the inner zone of the diamond. Undulation is visible in the intermediate and outer zones. Plate thickness is 1.00 mm.

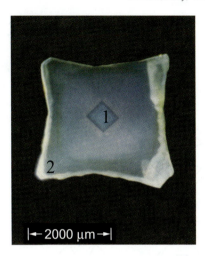

Fig. 13. Cube with octahedral core: CL pattern of sample #K4 (0.12 ct) from pipe Karpinsky-1, Arkhangelsk region (type K-3). Some evidence of normal crystal growth can be observed in the rounded boundary of the intermediate zone. Plate thickness is 0.98 mm.

The CL and anomalous birefringence studies of diamond generally complemented each other. Since these two methods show variable effectiveness examining different crystals of diamond from the analyzed

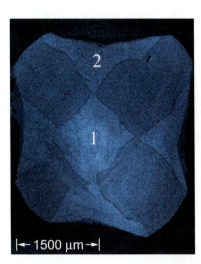

Fig. 14. Cube with sectorial structure and agate-like crystal core: CL pattern of sample #U8 (0.12 ct) from the Udachnaya pipe, Yakutia (type K-4). Pseudo-octahedral 'faces' are layer-by-layer overgrown by (111) faces with a rather high growth rate, as a result of which the sectorial structure with peculiarly undulating (111) surfaces is formed. In this example, we deal with a combined (normal + tangential) mechanism of growth. Plate thickness is 1.14 mm.

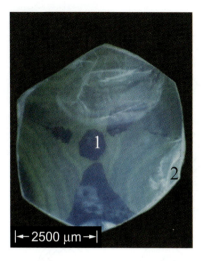

Fig. 15. Tetrahexahedroid with sectorial structure: CL pattern of sample #K5 from pipe Karpinsky-1, Arkhangelsk region (type K-5). Undulation is visible throughout the entire crystal volume. Plate thickness is 0.99 mm.

collection, we herein present both CL and birefringence images for comparison (see Figs. 1–15).

4. Characteristics of the diamond samples

Analyses were made of plates cut from octahedral, rounded-rhombic-dodecahedral and cubic diamond crystals from different deposits of Yakutia (pipes Mir, XXIII Congress of the CPSU, Yubileynaya, Prognoznaya and Udachnaya), Arkhangelsk region (pipes Lomonosov, Karpinsky-1 and Pomorskaya), Sayan region (placers of the Udinsky and Tumanshet–Biryusinsky districts) and Northern Urals (placers of the Vishera region). In all, 50 diamond crystals were analyzed: 11 octahedra, 32 dodecahedroids, 6 cubic crystals and 1 tetrahexaedroid; all of them with various internal structural heterogeneities. A more detailed description of internal structural features of crystals of diamond with differing morphology is given below.

5. Results

5.1. Internal structural features of diamond crystals

Analysis of CL and anomalous birefringence images of diamond crystals, the most typical of which

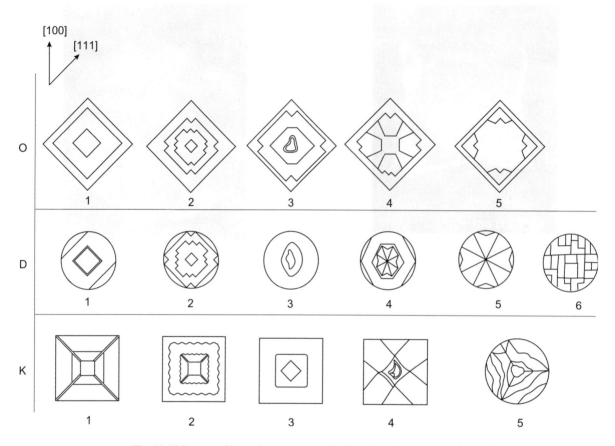

Fig. 16. Main types of internal structure occurring in analyzed crystals of diamond.

Octahedra

O-1   octahedron with layer-by-layer 'closed' octahedral zoning (sample #M1, #M2, #X1, #U1, #A1 and #K1);
O-2   octahedron with 'open' layered octahedral zoning (sample #M3);
O-3   octahedron with 'open' layered octahedral zoning and crystal core with agate-like structure (sample #M4);
O-4   octahedron with zonal-sectorial structure (sample #U2 and #M5);
O-5   octahedron with zonal-sectorial structure and crystal core of 'Maltese-cross' type (sample #U3).

Dodecahedroids

D-1   dodecahedroid with layer-by-layer 'closed' octahedral zoning (sample #Yu1, #NU1, #NU2, #NU3, #NU4, #NU5, #NU6, #U4, #P1, #V3 and #V68);
D-2   dodecahedroid with 'open' layered octahedral zoning (sample #Pr1, #Pr2, #NU7, #NU8, #NU9, #NU10, #NU11, #L1 and #V61);
D-3   dodecahedroid with rounded zoning (sample #L2, #NU13 and #NU19);
D-4   dodecahedroid with zonal-sectorial structure (sample #NU12);
D-5   dodecahedroid with sectorial structure of 'Maltese-cross' type (sample #NU14 and #NU15);
D-6   dodecahedroid with blocky structure (sample #K3, #NU16, #NU17, #NU18 and #S1).

Crystals of cubic habit

K-1   cube with concentric zonal structure (sample #U5 and #U6);
K-2   cube with undulating zone boundaries (sample #L3);
K-3   cube with octahedral core (sample #K4);
K-4   cube with sectorial structure and agate-like crystal core (sample #U8);
K-5   tetrahexahedroid with sectorial structure (sample #K5).

are given in Figs. 1–15, allowed us to identify the main types of internal structure, for the analyzed crystals of diamond which are presented in Fig. 16.

Among *octahedral* crystals of diamond, the most abundant are crystals with layer-by-layer 'closed' octahedral zoning (Fig. 1), which we will refer to here as O-1 type of zoning. Rectilinear-step-like layer-by-layer 'open' octahedral zoning, which will be referred to as O-2 type, likely formed as a result of the incomplete growth of layers around the crystal apices and along crystal edges, resembling the linear decorations in internal octahedral zones in natural diamond, previously reported by Frank et al. (1994). Similar diamond samples with irregularly shaped cores, having an agate-like structure (Fig. 2), were classified as diamond with zoning of O-3 type. This peculiar, agate-like structure reflects the normal mechanism of growth during the early stages of diamond formation.

In addition, there is a group of octahedral crystals of diamond with zonal–sectorial structures having cuboctahedral cores (Fig. 3) of variable shape and size. The zonal–sectorial diamond structure occurs as several distinct varieties, with differing relationships between growth rates of <111> and <100> faces. In cases of predominance of sectorial growth, we have type O-4 crystals. In cases where this proportion is nearly constant, growth pyramids of both octahedra and cube result in a structure known as 'Maltese-cross' (zoning type O-5).

In addition to the five main types of internal structure of octahedral crystals of diamond listed above, there occur other varieties, which are described in the literature. For example, Zezin et al. (2001) describe diamond crystals with polycentric structures, that is thought as evidence of growth from more than one crystallization centre. In some cases, polycentrism is only evident in the core of a crystal, while the bulk of the crystal comprises octahedral faces that have grown layer-by-layer (Genshaft et al., 1977).

Since the scheme and mechanism of crystal growth may change repeatedly in the course of diamond formation, many crystals of diamond show a combination of several distinct types of internal structure, a feature that has also been noted by previous workers (e.g., Genshaft et al., 1977). The structure of crystal cores is particularly diversified (Beskrovanov, 1992). In our study, when describing diamond internal struc-

tures, we decided to ignore crystal cores in those cases where they are very small ($\ll 1\%$ of the crystal volume), in order to avoid unnecessary complication of structural patterns (Fig. 16).

The internal structure of *dodecahedroids* is generally similar to that of octahedral crystals, however, being somewhat more diversified. In particular, the D-1 (Fig. 4) and D-2 dodecahedroid types, with differing patterns of octahedral zoning, can be correlated with the O-1 and O-2 octahedral types, and the D-4 zonal–sectorial crystal type would be correlated to octahedral types O-4 and O-5 (Figs. 3, 5 and 16). At the same time, quite common among dodecahedroids are crystals with sectorial structures developed through almost the entire crystal volume (Figs. 6 and 7), which were classified here as type D-5.

A peculiar feature of dodecahedroids that differentiates them from octahedra is the presence of dodecahedroid crystals with concentric, agate-like, rounded zoning (type D-3, Fig. 8), the nature of which is as yet unclear. On the one-hand, rounded zoning may result from alternating growth and dissolution, as suggested by Varshavsky (1968). On the other, zoning of this type may be related to normal growth mechanisms (Gurkina, 1980).

Five of the analyzed dodecahedroids exhibited a blocky internal structure lacking any zonal–sectorial features (Fig. 9). This type of diamond structure might be related to interaction between dislocations active as foci for electrons and holes (Sumida and Lang, 1981). We classified this type of structure as D-6.

In addition to the six main internal structure types of dodecahedral diamond crystals, some researchers (e.g., Genshaft et al., 1977; Zezin et al., 2001) describe dodecahedroid crystals of diamond with quite homogenous internal structures, showing little or no zoning and rather evenly distributed impurity centres.

Like octahedral crystals of diamond, some dodecahedroids also show various combinations of different internal structure types.

The group of diamond *crystals with cubic habit*, in the analyzed collection, included six cubes and one tetrahexahedroid, all of them formed through the normal mechanism of growth (Figs. 10–15). Types K-1 and K-2 have a concentric zonation following the cubic scheme (Figs. 10–12), with nearly rectilinear zoning in the case of type K-1 and undulating zoning in the case of type K-2. Evidence of normal growth is

Table 1
Distribution of structural defects (optically active centres) in crystals of diamond with different internal structures

Sample no.	Deposit, province	Type of internal structure	Zone	N_A (at.ppm)	N_B (at.ppm)	$N_{tot.}$ (at.ppm)	$\%N_B$	P (cm^{-1})	H (cm^{-1})	Figure
Octahedra										
M1	Mir pipe, Yakutia	O-1	Internal <111>	610	398	1008	39	15.8	0.6	Fig. 1
			External <111>	770	248	1018	24	3.3	0	
M2	Mir pipe, Yakutia	O-1	Internal <111>	459	96	555	18	9.5	0.6	
			External <111>	421	107	528	20	8.4	0.2	
M3	Mir pipe, Yakutia	O-2	Internal <111>	457	351	808	43	18.9	–	
			External <111>	266	124	390	32	2.8	–	
M4	Mir pipe, Yakutia	O-3	Internal <100>?	337	154	491	31	6.6	1.7	Fig. 2
			External <111>	660	167	827	20	2.4	0.1	
M5	Mir pipe, Yakutia	O-4	Internal <111>	–	–	–	–	–	–	
			Internal <100>	622	190	812	23	2.3	2	
			External <111>	536	160	696	23	5.1	0	
X1	XXIII Congress of CPSU, Yakutia	O-1	Internal <111>	26	25	51	50	0	0	
			External <111>	425	23	448	5	1.5	0.2	
U1	Udachnaya pipe, Yakutia	O-1	Internal <111>	90	195	285	67	12.7	0.2	
			External <111>	43	21	64	33	1	0.3	
U2	Udachnaya pipe, Yakutia	O-4	Internal <111>	951	–	–	–	6.6	1	
			Internal <100>	571	281	852	33	7.7	13.7	
			External <111>	595	250	845	30	2	0.3	
U3	Udachnaya pipe, Yakutia	O-5	Internal <111>	767	490	1257	39	13.9	1.1	Fig. 3
			Internal <100>	579	232	811	29	7	5.9	
			External <111>	871	270	1141	24	9.5	0.3	
A1	Aykhal pipe, Yakutia	O-1	Internal <111>	279	103	382	27	9.2	0.1	
			External <111>	260	85	345	25	3.3	0	
K1	Karpinsky-1 pipe, Arkhangelsk	O-1	Internal <111>	19	480	499	96	16.6		
			External <111>	24	107	131	82	4	4.7	
		Mean		421	198	619	32	7.1	1.7	
		σ		273	134	334	20	5.3	3.1	
Dodecahedroids										
U4	Udachnaya pipe, Yakutia	D-1	Internal <111>	136	346	482	72	18.2	0.4	
			External <111>	80	158	238	67	9	1.5	
Yu1	Yubileynaya pipe, Yakutia	D-1	Internal <111>	615	95	710	13	0	0.2	
			External <111>	783	70	853	8	0	0.8	
Pr1	Prognoznaya pipe, Yakutia	D-2	Internal <111>	40	303	343	88	22.5	0.3	
			External <111>	22	35	57	61	0.9	0.4	
Pr2	Prognoznaya pipe, Yakutia	D-2	Internal <111>	135	441	576	77	24.1	0	
			External <111>	75	60	135	44	5.5	0.8	
NU1	Northern Urals placers	D-1	Internal <111>	300	207	507	41	13.4	1.6	Fig. 4
			External <111>	318	231	549	42	14.6	1.3	
NU2	Northern Urals placers	D-1	Internal <111>	756	147	903	16	11	0	
			External <111>	376	14	390	3	5.8	0	
NU3	Northern Urals placers	D-1	Internal <111>	177	36	213	17	2.8	0.2	
			External <111>	447	81	528	15	2.7	0	
NU4	Northern Urals placers	D-1	Internal <111>	289	107	396	27	7	0.7	
			External <111>	835	123	958	13	5.5	0	
NU5	Northern Urals placers	D-1	Internal <111>	173	327	500	65	15.6	1.3	
			External <111>	390	180	570	32	15.7	0.5	
NU6	Northern Urals placers	D-1	Internal <111>	344	77	421	18	3	0.5	
			External <111>	372	59	431	14	2.5	0.3	

Table 1 (*continued*)

Sample no.	Deposit, province	Type of internal structure	Zone	N_A (at.ppm)	N_B (at.ppm)	$N_{tot.}$ (at.ppm)	$\%N_B$	P (cm^{-1})	H (cm^{-1})	Figure
Dodecahedroids										
NU7	Northern Urals placers	D-2	Internal <111>	229	139	368	38	11.7	0.6	
			External <111>	211	129	340	38	10.8	0.5	
NU8	Northern Urals placers	D-2	Internal <111>	300	173	473	37	6	4.1	
			External <111>	–	–	–	–	–	2.7	
NU9	Northern Urals placers	D-2	Internal <111>	169	163	332	49	9	0	
			External <111>	336	140	476	29	11	0	
NU10	Northern Urals placers	D-2	Internal <111>	440	91	531	17	7.1	0.5	
			External <111>	487	52	539	10	5.1	0	
NU11	Northern Urals placers	D-2	Internal <111>	27	62	89	70	2.8	0	
			External <111>	21	25	46	54	1.8	0	
NU12	Northern Urals placers	D-4	Internal <111>	–	–	–	–	–	–	Fig. 5
			Internal <100>	750	319	1069	30	13.4	3.6	
			External <111>	121	61	182	34	2.6	0.8	
NU13	Northern Urals placers	D-3	Internal <*hhl*>?	366	143	509	28	7.5	1.5	Fig. 8
			External <111><*hhl*>	445	188	633	30	7.4	1.6	
NU14	Northern Urals placers	D-5	Internal 1 <111>	849	191	1040	18	11.7	4.7	Fig. 6
			Internal 2 <100>	524	134	658	20	2.6	9.4	
			External <111>	194	66	260	25	2.3	0.9	
NU15	Northern Urals placers	D-5	Internal <111>	1256	325	1581	21	11.4	0.9	Fig. 7
			Internal <100>	1154	396	1550	26	11.7	0.5	
			External <111>	513	373	886	42	4	0.2	
NU16	Northern Urals placers	D-6	Internal ?	58	43	101	43	2.1	0	
			External ?	49	33	82	40	1.9	0	
NU17	Northern Urals placers	D-6	Internal ?	11	33	44	75	0	0	
			External ?	19	50	69	72	0	0.2	
NU18	Northern Urals placers	D-6	Internal ?	352	78	430	18	3.9	0	
			External ?	376	61	437	14	2.4	0.3	
NU19	Northern Urals placers	D-3	Internal <111>	89	71	160	44	0	0	
			External <100>	96	29	125	23	2.7	0	
L1	Lomonosov pipe, Arkhangelsk	D-2	Internal <111>	689	189	878	22	2.2	7.6	
			External <111>	564	157	721	22	1.8	6.5	
L2	Lomonosov pipe, Arkhangelsk	D-3	Internal <*hhl*>	1050	323	1373	24	15.4	15	
			External <*hhl*>	946	276	1222	23	12	10	
K2	Karpinsky-1 pipe, Arkhangelsk	?	Internal ?	919	232	1151	20	10.9	4.9	
			External ?	446	85	531	16	7.1	0	
K3	Karpinsky-1 pipe, Arkhangelsk	D-6	Internal ?	244	49	293	17	4.5	0	Fig. 9
			External ?	181	28	209	13	2.3	0	
P1	Pomprskaya pipe, Arkhangelsk	D-1	Internal <111>	807	684	1491	46	19.5	1	
			External <111>	590	326	916	36	17.1	6.7	
S1	Sayan region placers	D-6?	Internal ?	24	15	39	39	1.3	0.6	
			External ?	42	28	70	39	0	0	
V3	?	D-1	Internal <111>	454	289	743	39	20.9	0.5	
			External <111>	419	294	713	41	14.9	0.9	
V61	?	D-2	Internal <111>	414	46	460	10	0.3	0	
			External <111>	483	73	556	13	0.4	0	
V68	?	D-1	Internal <111>	554	250	804	31	11	0	
			External <111>	541	224	765	29	10.3	0	
			Mean	391	157	548	29	7.5	1.5	
			σ	303	130	384	20	6.3	2.8	

(continued on next page)

Table 1 (*continued*)

Sample no.	Deposit, province	Type of internal structure	Zone	N_A (at.ppm)	N_B (at.ppm)	$N_{tot.}$ (at.ppm)	$\%N_B$	P (cm^{-1})	H (cm^{-1})	Figure
Cubes and tetrahexahedroid										
U5	Udachnaya pipe, Yakutia	K-1	Internal <100>	419	20	439	5	0	2.3	Fig. 10
			External <100>	125	17	142	12	0	1	
U6	Udachnaya pipe, Yakutia	K-1	Internal <100>	726	307	1033	30	1.1	2.2	Fig. 11
			External <100>	502	190	692	27	2.9	4.6	
U7	Udachnaya pipe, Yakutia	?	Internal <100>	1102	0.01	1102	0.01	1.5	3.2	
			External <100>	476	167	643	26	0	2.6	
U8	Udachnaya pipe, Yakutia	K-4	Internal <*hhl*><111>?	566	0.01	566	0.01	2.2	2	Fig. 14
			External <100>	427	314	741	42	2.7	0.01	
L3	Lomonosov pipe, Arkhangelsk	K-2	Internal <100>	997	93	1090	9	1.4	5.6	Fig. 12
			External <100>	665	92	757	12	0	0.01	
K4	Karpinsky-1 pipe, Arkhangelsk	K-3	Internal <111>	1362	77	1439	5	0	0.01	Fig. 13
			External <100>	511	29	540	6	0	0.01	
K5	Karpinsky-1 pipe, Arkhangelsk	K-5	Internal <*hhl*>	929	117	1046	11	0	2.4	Fig. 15
			External <*hhl*>	531	65	596	11	0	3	
		Mean		667	106	773	14	0.8	2	
		σ		314	104	322	12	1.1	1.7	

found in the well-defined fibrous structure of these crystals.

One of the examined cubic crystals of diamond has an octahedral core (Fig. 13). This is a good example of change in zoning type and growth mechanism, and likely reflects some change in crystallization conditions (most probably, in the temperature of the crystallization medium). We assigned this peculiar diamond to an individual structural type K-3.

Diamond sample #U8 has a specific, agate-like structure combined with sectorial internal structures in its outer zone (type K-4, Fig. 14). In the case of cubic crystals, we consider agate-like structure as a variety of undulating structure.

The *tetrahexahedroid* diamond #K5, we suggest, also formed under normal growth mechanisms. It has sectorial internal structure throughout their entire crystal volume (type K-5; Fig. 15). likely having undergone oxidative dissolution during the final stage of crystal formation, evident by the fact that the outer crystal surface truncates the undulating zones (Fig. 15).

5.2. Results of local IR analysis of diamond crystals

The concentrations of nitrogen and hydrogen centres and platelets in the analyzed diamond crystals are given in Table 1. As can be seen from the Table, total nitrogen and hydrogen content in diamond depends upon where it is from (i.e., which deposit it is from). In particular, octahedron #K1 (Type O-1) from pipe Karpinsky-1 in the Arkhangelsk kimberlite province has lower N_{tot} and higher hydrogen concentration values, when compared to O-1 octahedra from the Mir pipe (sample #M1 and #M2). Dodecahedroids #P1 and #L1 (correspondingly, type D-1 and type D-2) from pipes of the Arkhangelsk region have generally higher nitrogen and hydrogen contents compared to dodecahedroids #NU1 to #NU11 of the same structural types, from North Uralian placers (Table 1). These data provide additional support for the view that concentrations of optically active centres in diamond provide a specific means of 'fingerprinting' crystals of diamond from different deposits (Kaminsky et al., 1988; Kaminsky and Khachatryan, 2001).

In addition, as we will demonstrate below, the distribution of structural defects in diamond is also related to the appearance and internal structural peculiarities of diamond crystals.

5.2.1. Octahedra

IR spectral features of the analyzed crystals of diamond with octahedral habit are indicative of sizeable concentrations of nitrogen A and B centres, platelets and hydrogen centres (Table 1). In all octahedral crystals, with the exception of #K1, nitrogen defects are dominated by A centres, with average N_A proportions of about 69% of N_{tot}. The concentration of B centres varies between 23 and 490 at.ppm (Table 1). Crystals of diamond with octahedral zoning (types O-1, O-2 and O-3) have N_A varying between 19 and 770 at.ppm, which is generally lower than in crystals of diamond with zonal–sectorial structures (536–951 at.ppm, types O-4 and O-5).

In some octahedral crystals of diamond (#M1, #M4 and #X1), N_{tot} increases from core to rim, whereas in other octahedral crystals (#M2, #M3, #M5, #U1, #U2, #A1 and #K1) the reverse is true. The degree of nitrogen aggregation (%N_B) in the outer zones of octahedral crystals does not usually exceed that of the crystal cores (Table 1).

Platelets are present in almost all the octahedral crystals of diamond studied, with concentrations varying from 0 to 18.9 cm^{-1} (Table 1) and typically showing a core-to-rim decrease in all zonal crystals with the exception of #M5 and #X1.

Hydrogen impurity concentration varies from 0 to 13.7 cm^{-1}, being generally lower in zoned crystals than in octahedra with zonal–sectorial structures (Table 1).

Optically active centres are heterogeneously distributed in <111> and <100> growth pyramids. In particular, in diamond #U3 N_{tot}, %N_B and platelet concentration values are significantly higher in <111> pyramids than in ⟨100⟩ ones. Along with this, growth pyramids of cubic faces are enriched with hydrogen centres which contrasts with octahedral faces of the same crystal. In diamond #U2, hydrogen impurity concentrations in <100> growth pyramids are approximately an order of magnitude higher than in <111> ones (Table 1).

5.2.2. Dodecahedroids

The distribution of optically active centres in dodecahedroids is quite similar to that observed in the octahedra. In dodecahedroids, N_A varies between 11 and 1256 at.ppm, and N_B between 14 and 684 at.ppm. Dodecahedroids with zonal structures (types D-1 and D-2), much like zoned octahedra, are characterized by generally lower N_A values (21–807 at.ppm) when compared to crystals with sectorial structures ($N_A = 194–1256$ at. ppm).

The concentration of platelets in dodecahedroids is nearly the same as in octahedra, varying between 0 and 22.5 cm^{-1} (Table 1).

Hydrogen impurity content of dodecahedroids varies from 0 to 9.4 cm^{-1}, which is, on average, close to hydrogen concentration in octahedra (Table 1).

Zoned dodecahedroids, much like octahedra, show no regularity in the distribution of structural defects. Some crystals show a core-to-rim increase in nitrogen content, while in other crystals, the concentration of nitrogen defects decreases in the same direction.

In contrast to zoned dodecahedroids, dodecahedroids with zonal–sectorial (#NU12) and sectorial structures (#NU14 and #NU15) do show certain regularity in the distribution of nitrogen impurity centres. In these samples, N_{tot} shows a decrease by a factor of 2 to 5 when moving from core to rim (Table 1). In this respect, dodecahedroids with zonal–sectorial and sectorial structures are similar to crystals with cubic habit (see below).

5.2.3. Crystals of cubic habit

The analyzed cubes and tetrahexahedroid differ significant from octahedra and zoned dodecahedroids in concentration of optically active centres. In cubic diamond, N_A varies from 125 to 1362 at.ppm, which is, on average, higher than in octahedra and dodecahedroids. By contrast, the concentration of nitrogen B centres is quite the reverse: it is less than half that in octahedral crystals and dodecahedroids (Table 1). Consequently, the degree of nitrogen aggregation (%N_B) in cubic crystals is 2–3 times lower than in octahedra and dodecahedroids.

Another characteristic feature of diamond crystals with cubic habit (as compared to other morphological varieties of diamond), which is of diagnostic importance, is their very low platelets content (in sample #U5, #K4 and #K5, i.e., in almost half of the cubic crystals of diamond studied, no platelets were detected). These data corroborate the findings of Orlov et al. (1978) who showed that grey and colorless diamond cubes have very low concentrations of platelets and nitrogen B centres as compared to octahedra and dodecahedroids. The maximum platelet concentration (2.7 cm^{-1}) among the analyzed crystals of cubic habit was found in diamond #U8 which was probably formed by a combined (normal + tangential) mechanism of growth (Table 1, Fig. 14).

Table 2
Temperatures of formation estimated for different zones in diamond crystals. Temperature estimate for 3 Ga (Taylor and Milledge, 1995)

Sample no.	Zones of crystals (see Table 1)							$\Delta T = T_{ext} - T_{int}$
	Internal <111>			Internal <100> or others		External <111><100> and others		
	Main form	Zone no.	T (°C)	Zone no.	T (°C)	Zone no.	T (°C)	
M1	Octahedron	1	1100			2	1085	−15
M2	Octahedron	1	1090			2	1090	0
M3	Octahedron	1	1110			2	1110	0
M4	Octahedron	1	1110			2	1085	−25
M5	Octahedron	1	–	2	1090	3	1095	+5
X1	Octahedron	1	1175			2	1060	−15
U1	Octahedron	1	1160			2	1160	0
U2	Octahedron	1	–	2	1100	3	1060	−40
U3	Octahedron	1	1095	2	1095	3	1060	−35
A1	Octahedron	1	1110			2	1110	0
K1	Octahedron	1	1200			2	1200	0
U4	Dodecahedroid	1	1155			2	1160	+5
Yu1	Dodecahedroid	1	1075			2	1055	−20
Pr1	Dodecahedroid	1	1190			2	1190	0
Pr2	Dodecahedroid	1	1160			2	1160	0
NU1	Dodecahedroid	1	1125			2	1125	0
NU2	Dodecahedroid	1	1075			2	1050	−25
NU3	Dodecahedroid	1	1110			2	1090	−20
NU4	Dodecahedroid	1	1110			2	1070	−40
NU5	Dodecahedroid	1	1145			2	1105	−40
NU6	Dodecahedroid	1	1100			2	1095	−5
NU7	Dodecahedroid	1	1125			2	1125	0
NU8	Dodecahedroid	1	1120			2	–	
NU9	Dodecahedroid	1	1135			2	1105	−30
NU10	Dodecahedroid	1	1090			2	1075	−15
NU11	Dodecahedroid	1	1190			2	1185	−5
NU12	Dodecahedroid	–	–	1	1090	2	1140	–
NU13	Dodecahedroid	1	1105			2	1105	0
NU14	Dodecahedroid	1	1075	2	1090	3	1115	+40
NU15	Dodecahedroid	1	1070	2	1075	3	1110	+40
NU16	Dodecahedroid	1	1160			2	1160	0
NU17	Dodecahedroid	1	1220			2	1210	−10
NU18	Dodecahedroid	1	1090			2	1085	−5
NU19	Dodecahedroid	1	1150			2	1140	−10
L1	Dodecahedroid	1	1085			2	1090	+5
L2	Dodecahedroid	1	1080			2	1080	0
K2	Dodecahedroid	1	1070			2	1085	+15
K3	Dodecahedroid	1	1105			2	1105	0
P1	Dodecahedroid	1	1100			2	1100	0
S1	Dodecahedroid	1	1185			2	1170	−15
V3	Dodecahedroid	1	1110			2	1110	0
V61	Dodecahedroid	1	1075			2	1075	0
V68	Dodecahedroid	1	1100			2	1100	0
U5	Cube			1	1060	2	1110	+50
U6	Cube			1	1085	2	1100	+15
U7	Cube			1	1020	2	1100	+80
U8	Cube			1	1025	2	1110	+85
L3	Cube			1	1055	2	1070	+15
K4	Cube	1	1025			2	1065	+40
K5	Tetrahexahedroid			1	1060	2	1075	+15

In the overwhelming majority of analyzed crystals of diamond with cubic habit, N_{tot} decreases from crystal cores to marginal zones. This being so, $\%N_B$ in the outer zones of these crystals is generally not lower than in crystal cores (Table 1).

6. Discussion

Analysis of the internal structure of diamond crystals (Figs. 1–15) demonstrates that the appearance of a crystal quite commonly provides information pertaining only to the final stages of diamond formation. Thus, for the correct interpretation of data on the distribution of structural impurities in diamond, it is also necessary to take into account the internal structure of crystals.

Proceeding from the assumption that the geological time of formation can be taken as equivalent, for different zones within a common crystal, we have estimated the formation temperatures for all the examined crystals of diamond using the diagram proposed by Taylor and Milledge (1995). The results of this estimation are presented in Table 2.

A peculiar feature of octahedral diamond with zonal structures is the trend of decreasing degree of nitrogen aggregation from the inner crystal zones to the outer ones (sample #M1, #M4 and #X1), which agrees with the data presented by Bokii et al. (1986). This is evidence of a decrease in the temperature of diamond formation by some 15–25 °C, during the final stages of crystallization (Table 2). In zonal–sectorial octahedra (sample #U2 and #U3), formation temperatures of the outer zones appears to be 35–40 °C lower than that of the inner crystal zones. Approximately one half of the analyzed octahedra (#M2, #M3, #U1, #A1 and #K1; Table 2) formed under nearly isothermic conditions ($\Delta T = 0$). Only in the zonal–sectorial diamond #M5 is formation temperature of the outer zone slightly (not more than by 5 °C) higher than that of the inner zone, although this temperature difference is within the accuracy of nitrogen concentration measurements.

To summarize, almost all analyzed octahedra (with the exception of diamond #M5) are characterized either by nearly isothermic growth conditions or by a decrease in formation temperature during the crystallization process. This indicates that the distribution of nitrogen centres in octahedral crystals is quite adequately described in the context of the 'annealing model' (Evans, 1992; Taylor et al., 1996).

Dodecahedroids with zonal structures (types D-1, D-2 and D-3) and the blocky dodecahedroid (type D-6), much like octahedral crystals of diamond, also bear evidence either of isothermic growth (sample #Pr1, #Pr2, #K3, #L2, #P1, #NU7, #NU13, #NU16, #V61, #V68 and #V3) or a decrease in formation temperature from crystal cores to marginal areas (sample #Yu1, #NU2 to #NU6, #NU9 to #NU11, #NU17 to #NU19, and S1) (Table 2). Dodecahedroids #U4 and #L1 (types D-1 and D-2) and the octahedral diamond #M5 show an insignificant (within the accuracy of the experimental method) increase in formation temperatures in the outer crystal zones (Table 2).

In contrast to zoned octahedra and dodecahedroids, dodecahedroids with zonal–sectorial and sectorial internal structures (sample #NU12, #NU14 and #NU15; types D-4 and D-5) show a notably different distribution of nitrogen defects, with N_{tot} generally decreasing from crystal cores to marginal areas, and degree of nitrogen aggregation increasing in the same direction. From this, it would follow that in these crystals, the temperature of diamond formation of the outer crystal zones is approximately 40–50 °C higher than that of the inner zones (Table 2). This is somewhat puzzling, considering that formation temperature of the outermost zone of a diamond crystal is, at the same time, the annealing temperature for all the internal zones of the same crystal. Data on similar anomalous distributions of nitrogen centres in diamond, apparently inconsistent with the 'annealing model', were reported previously for synthetic diamond (Palyanov et al., 1997) and for diamond microcrystals from the Udachnaya pipe (Zedgenizov et al., 1999). For natural diamond macrocrystals, this phenomenon has already been highlighted and discussed in our recent study (Khachatryan and Kaminsky, 2003).

Of prime interest among the examined crystals of diamond is the group of crystals with cubic habit, which generally show a fibrous internal structure, reflecting normal mechanisms of growth. In diamond of cubic habit, like the dodecahedroids with sectorial structures, the outer zones are characterized by higher formation temperatures when compared to crystal cores, with temperature differences varying from 15 to 80 °C. The anomalous distribution of nitrogen centres in diamond crystals that grew through the normal mechanism, with a high rate of growth and in an oversaturated medium, might point to non-equilibrium relationships between the concentrations of different nitrogen centres. It is likely that in crystals of this type, the rate of growth is higher than the rate of structural nitrogen aggregation. Thus, it appears that in these peculiar crystals of diamond, we deal with non-equilibrium concentrations of nitrogen B centres and, consequently, with anomalous diamond formation temperatures.

To summarize, for diamond crystals with a high rate of growth, which is usually reflected in sectorial or fibrous internal structures, calculated formation temperature values do not necessarily correspond to actual temperatures of growth, being essentially dependent on the kinetic parameters of the diamond formation process. In particular, internal structural heterogeneities have certain effects on diamond quality and on the behavior of diamond crystals during processing. In some cases, examination of the internal structure of diamond crystals in advance of processing could help prevent excessive losses in cutting and grinding procedures.

Acknowledgements

We are thankful to G.A. Gurkina, who analyzed some diamond crystals for anomalous birefringence, to H.J. Milledge and G.V. Saparin for performing CL analyses of diamonds, and to Ian Coulson for helping us editing the text of the manuscript. We are grateful to N. Mc Kenna and T. Stachel for the constructive remarks which helped us to improve the manuscript.

References

Beskrovanov, V.V., 1992. Ontogeny of Diamond. Nauka Press, Moscow. In Russian.

Bokii, G.B., Bezrukov, G.N., Klyuev, Yu.A., Naletov, A.M., Nepsha, V.I., 1986. Natural and Synthetic Diamonds. Nauka Press, Moscow. In Russian.

Boyd, S.R., Kiflawi, I., Woods, G.S., 1994. The relationship between infrared absorption and A-defect concentration in diamond. Philos. Mag. 69 (6), 1149–1153.

Boyd, S.R., Kiflawi, I., Woods, G.S., 1995. Infrared absorption by the B nitrogen aggregate in diamond. Philos. Mag. 72 (3), 351–361.

Bursill, L.A., Glaisher, R.W., 1985. Aggregation and dissolution of small and extended defect structures in type Ia diamond. Am. Mineral. 70, 608–618.

Chen, M., Lu, F., Zheng, J., 1999. Cathodoluminescence features of diamond in Fuxian, Liaoning China and their implications. Earth Sci., J. China Univ. Geosc. 24 (2), 179–182.

Chen, M., Lu, F., Di, J., Zheng, J., 2000. CL and FTIR analysis on the diamonds in Wafangdian, Liaoning Province. Chin. Sci. Bull. 45 (21), 1986–1990.

Chernov, A.A., Givargizov, E.I., Bagdasarov, Kh.S., Demyanetz, L.N., Kuznetsov, V.A., Lobachev, A.N., 1980. Modern Crystallography. Crystallogeny, vol. 3. Nauka Publishing House, Moscow. 406 pp. In Russian.

Davies, G., 1976. The A nitrogen aggregate in diamond: its symmetry and possible structure. J. Phys. C9, L537–L542.

Davies, R.M., O'Reilly, S.Y., Griffin, W.L., 1999. Growth structures and nitrogen characteristics of Group B alluvial diamond crystals from Bingara and Wellington, Eastern Australia. In: Gurney, J.J., Gurney, J.L., Pascoe, M.D., Richardson, S.H. Proc. VIIth Internat. Kimberlite Conf., vol. 1. Red Roof Design, Cape Town, pp. 156–163.

Evans, T., 1992. Aggregation of nitrogen in diamond. In: Field, J.E. The Properties of Natural and Synthetic Diamond. Academic Press, London, pp. 259–290.

Field, J.E. 1992. The Properties of Natural and Synthetic Diamond. Academic Press, London. 710 pp.

Frank, F.C., 1969. Diamonds and deep fluids in the upper mantle. In: Runcorn, S.K. The Application of Modern Physics to the Earth and Planetary Interiors. Wiley, London, pp. 247–250.

Frank, F.C., Harris, J.V., Kaneko, K., Lang, A.R., 1994. Linear decorations defining edges of an internal octahedron within a natural diamond: observations and an explanation. J. Cryst. Growth 143 (1–2), 46–57.

Genshaft, Y.S., Yakubova, S.A., Volkova, L.M., 1977. Internal morphology of natural diamonds. In: Genshaft, Yu.S. Investigation of High-Pressure Minerals. Publishing House of the Institute of Physics of the Earth, Moscow, pp. 5–131. In Russian.

Gurkina, G.A., 1980. Research on the internal morphology of spherical diamonds by the methods of X-ray topography and birefringence. In: Orlov, Yu.L., Ivankin, P.F., Kaminskiy, F.V. Combined Studies on Diamonds. TSNIGRI Publishing House, Moscow, pp. 43–51. In Russian.

Hauri, E.H., Pearson, D.G., Bulanova, G.P., Milledge, H.J., 1999. Microscale variations and C and N isotopes within mantle diamond revealed by SIMS. In: Gurney, J.J., Gurney, J.L., Pascoe, M.D., Richardson, S.H. Proc. VIIth Internat. Kimberlite Conf., vol. 1. Red Roof Design, Cape Town, 341–347.

Kaminsky, F.V., Khachatryan, G.K., 2001. Characteristics of nitrogen and other impurity in diamond, as revealed by infrared absorption data. Can. Mineral. 39 (6), 1735–1745.

Kaminsky, F.V., Bartoshinsky, Z.V., Blinova, G.K., Galimov, E.M., Gurkina, G.A., Krasnikov, V.I., Lapushkov, V.M., Sobolev, E.V., Sobolev, N.V., 1988. Principles of the Comprehensive Study of Diamonds in the Course of Prospecting for Primary Diamond Deposits, TSNIGRI Publishing House, Moscow. in Russian.

Khachatryan, G.K., Kaminsky, F.V., 2003. 'Equilibrium' and 'non-equilibrium' diamonds from deposits in the East European platform, as revealed by infrared absorption data. Can. Mineral. 41 (1), 171–184.

Klyuev, Y.A., Rykov, A.N., Khosak, L.A., 1969. Examination of a peculiar case of heterogeneous nitrogen distribution in a diamond crystal. Almazi (Diam.) 5, 5–9 (in Russian).

Lang, A.R., 1974. Space-filling by branching columnar single-crystal growth: an example from crystallization of diamond. J. Cryst. Growth 23, 151–153.

Mendelssohn, M.J., Milledge, H.J., 1995. Geologically significant information from routine analysis of the mid-infrared spectra of diamonds. Int. Geol. Rev. 37, 95–110.

Milledge, J., Bulanova, G.P., Taylor, W.R., Woods, P.A., Turner, P.H., 1995. Internal morphology of Yakutian diamonds—a cathodoluminescence and infrared mapping study. Extended Abstr. Sixth Internat. Kimberlite Conf. Novosibirsk, August 1995, 384–386.

Moore, M., Lang, A.R., 1972. On the internal structure of natural diamonds of cubic habit. Philos. Mag. 26 (6), 1313–1326.

Orlov, Yu.L., Dudenkov, Yu.A., Solodova, Yu.P., 1978. Fibrous growth, IR spectra and carbonate inclusions in cubic diamond crystals. New Data on Minerals in the USSR. Nauka Press, Moscow, pp. 109–112. In Russian.

Palyanov, Y.N., Khokhryakov, A.F., Borzdov, Y.M., Sokol, A.G., Gusev, V.A., Rylov, G.M., Sobolev, N.V., 1997. Growth conditions and real structure of synthetic diamond crystals. Russ. Geol. Geophys. 38 (5), 920–945.

Patel, A.R., Patel, M.M., 1969. Studies on the dodecahedral face of diamond. Am. Mineral. 54, 1324–1329.

Seal, M., 1965. Structures in diamond as revealed by etching. Am. Mineral. 50, 105–131.

Seal, M., 1966. Inclusions, birefringence and structure in natural diamonds. Nature (Lond.) 212 (5070), 1528–1531.

Sobolev, E.V., 1974. Impurity nitrogen in natural diamond crystals. Geology and Prognostication of Diamond Deposits. Ext. Abstr. IIIrd Diamond Conf., Moscow, 52–54. In Russian.

Sobolev, E.V., 1978. Nitrogen centers and crystal growth of natural diamond. In: Sobolev, V.S. Problems of Lithosphere and Upper Mantle Petrology. Nauka Press, Novosibirsk, pp. 245–255. In Russian.

Sobolev, E.V., Lisoivan, V.I., 1971. Impurity centers in diamond. Abstr. VIII Sci. Conf. Novosibirsk, 60–61 in Russian.

Sumida, N., Lang, A.R., 1981. Cathodoluminescence evidence of dislocation interactions in diamond. Philos. Mag. A43 (5), 1277–1287.

Sunagawa, I., 1984. Morphology of natural and synthetic diamond crystals. In: Sunagawa, I. Materials science of the Earth's Interior. Terra Scientific Publishing, Tokyo, pp. 303–330.

Suzuki, S., Lang, A.R., 1976. Internal structures of natural diamond crystals revealing mixed-habit growth. Diamond Res., Suppl. Ind. Diamond Rev., 39–47.

Taylor, W.R., Milledge, H.J., 1995. Nitrogen aggregation character, thermal history and stable isotope composition of some xenolith-derived diamonds from Roberts Victor and Finch. Extended Abstr. Sixth Internat. Kimberlite Conf., Novosibirsk, August 1995, 620–622.

Taylor, W.R., Bulanova, G.P., Milledge, H.J., 1995a. Quantitative nitrogen aggregation study of some Yakutian diamond crystals: constraints on the growth, thermal, and deformation history of peridotitic and eclogitic diamond crystals. Extended Abstr. Sixth Internat. Kimberlite Conf. Novosibirsk, August 1995, 608–610.

Taylor, W.R., Kiviets, G., Gurney, J.J., Milledge, H.L., Woods, P.A., Harte, B., 1995b. Growth history of an eclogitic diamond from the Vaal Valley kimberlite, South Africa—an infrared, cathodoluminescence and carbon isotope study. Extended Abstr. Sixth Internat. Kimberlite Conf., Novosibirsk, August 1995, 617–619.

Taylor, W.R., Canil, D., Milledge, H.J., 1996. Kinetics of Ib to IaA nitrogen aggregation in diamonds. Geochim. Cosmochim. Acta 60 (23), 4725–4733.

Tolansky, S., 1966. Birefringence of diamond. Nature (Lond.) 211 (5045), 158–160.

Tolansky, S., 1974. A comparison of synthetic and natural cuboctahedral diamond crystals. Synthetic Diamond in Industry. Naukova Dumka Press, Kiev, pp. 36–41. In Russian.

Varshavsky, A.V., 1968. Anomalous Birefringence and Internal Morphology of Diamond. Nauka Publishing House, Moscow. In Russian.

Woods, G.S., 1986. 'Platelets' and the infrared absorption of type Ia diamonds. Proc. R. Soc. Lond., A 407, 219–238.

Woods, G.S., Collins, A.T., 1983. Infrared absorption spectra of hydrogen complexes in Type I diamonds. J. Phys. Chem. Solids 44 (5), 471–475.

Zedgenizov, D.A., Rylov, G.M., Shatskiy, V.S., 1999. Internal structure of microdiamonds from Udachnaya kimberlite pipe. Geologiya i Geofizika 40 (1), 113–120 (in Russian).

Zezin, R.B., Smirnova, E.P., Saparin, G.V., Obyden, S.K., 2001. Diagnostics of natural and man-made cut diamonds based on a study of their internal structure. Vestn. Gemmologii 2, 7–16 (in Russian).

Available online at www.sciencedirect.com

Lithos 77 (2004) 273–286

LITHOS

www.elsevier.com/locate/lithos

ELSEVIER

Corundum inclusions in diamonds—discriminatory criteria and a corundum compositional dataset[☆]

Mark T. Hutchison[a,b,*], Peter H. Nixon[c], Simon L. Harley[d]

[a] Lunar and Planetary Laboratory, University of Arizona, Tucson, AZ, USA
[b] Research School of Earth Sciences, Australian National University, Canberra, ACT 0200, Australia
[c] University of Leeds, Leeds, UK
[d] University of Edinburgh, Edinburgh, UK

Received 27 June 2003; accepted 14 December 2003
Available online 1 June 2004

Abstract

Mineral inclusions of corundum are reported from diamonds from alluvial deposits of tributaries of the Rio Aripuanã, Juina, Brazil. We present the first recorded occurrence of sapphire as an inclusion in diamond and expand on the database of ruby and white corundum inclusions. Ruby inclusions are found to occur both as isolated and touching grains with aluminous pyroxene and associated with ferropericlase. Mineral chemistry and phase relations place the origin of such ruby-bearing diamonds within the lower mantle at ~ 770 km. Mineral associations involving other corundum inclusions were not observed; hence, their depth of origin is less certain.

Compositions of corundum samples were characterised by electron and ion microprobe. Given the scarcity of literature data, corundum samples from a variety of other geological settings were also analysed. Samples comprised corundums associated with granitic emplacement, metasomatism, amphibolite-facies and granulite-facies rocks, gem and industrial synthetic origins and carmine-coloured corundums recovered from kimberlite drill cores.

In addition to variable amounts of Cr, Fe, Ti, Mg and Si, measurable quantities of other transition elements and high field strength elements were also detected. Corundums from similar geological settings show very similar compositions and are easily distinguishable from other settings. Irrespective of locality, rubies from Norwegian, Tanzanian and Kenyan amphibolite-facies rocks are compositionally indistinguishable. Additionally, corundums from metasomatised zones associated with contact metamorphism from Arizona and Japan were very similar, particularly characterised by unusually high abundance of mobile Zr and Nb (tens of ppm). All Juina inclusions are particularly distinguishable from other corundums by high concentrations of Ni (18–171 ppm weight), typically at least an order of magnitude enriched over the same corundum varietal types from elsewhere. Furthermore, the sapphire inclusion exhibited much larger ratios of Ga and Ge to HFSE elements compared to otherwise similar samples, and ruby inclusions are distinguished by high Mg/Fe ratios (0.27–1.56 by weight). Compositional differences between inclusions in diamonds and corundums from other settings in

---

☆ Supplementary data associated with this article can be found, in the online version, at doi:10.1016/j.lithos.2004.04.006.
* Corresponding author. Research School of Earth Sciences, Australian National University, Mills Road, Canberra ACT 0200, Australia.
E-mail address: mhutchis@lpl.arizona.edu (M.T. Hutchison).

addition to corundum's physical and chemical durability suggest that with the employment of rapid identification tools such as energy dispersive spectrometry (EDS) and laser-ICPMS, corundum has promise as an indicator of diamond prospectivity.

Keywords: Corundum; Diamond inclusion; Mantle; Trace element; Nickel

1. Introduction

Corundum has been previously reported as occurring as syngenetic inclusions in diamonds from the alluvial deposits of tributaries of the Rio Aripuanã, Juina, Brazil (Watt et al., 1994; Hutchison et al., 2001). Diamonds from this area are of particular interest as they contain the most comprehensive source of deep transition zone and lower mantle material available for study (Hutchison, 1997; Harte et al., 1999; Gasparik and Hutchison, 2000; Hutchison et al., 2001). Here we report the first occurrence of corundum var. sapphire as a syngenetic inclusion in diamond and present data for white corundum inclusions from the same source. We also expand upon the first description of corundum var. ruby occurring as part of a mineral association within diamond (Hutchison et al., 2001) and reflecting a lower mantle origin. By studying the geochemistry of corundum inclusions, we aim to provide further insights into characteristics that discriminate the deep mantle from other more commonly observed sources of corundum.

Corundum is often observed by prospectors in alluvial settings in association with diamond (Mousseau Tremblay, personal communication, 2002). Given the range of common rock types in which corundum can occur, its existence in heavy mineral separates is perhaps not surprising. However, the identification of corundum inclusions within diamonds supports at least an occasional genetic link. As corundum is also resistant to weathering, the propensity exists to use this mineral as an indicator for diamond prospectivity. For corundum to be useful for this purpose, however, means must be established to isolate corundums with a diamond affinity from those from other mantle and crustal sources.

Despite the low abundances of trace elements in corundums, they do contain measurable quantities. Furthermore, because the Al_2O_3 structure contains sites appropriate for hosting high field strength ele-

ments and a number of transition metals that are fractionated in widely differing fashions in different mantle and crustal settings, trace elements could provide a means of identifying corundums grown in different settings. Here we present preliminary discriminatory tools for corundum genesis on the basis of minor and trace element data.

Although quantitative analyses of corundums do appear in the literature, data presented are most commonly routine electron microprobe (EPMA) analyses where trace element data is not available and minor element analyses are subject to large errors. Precision trace element data for corundums occur only sparsely (Schreyer et al., 1981; Kerrich et al., 1987; Limtrakun et al., 2001; Emmet et al., 2003). To supplement available data, we have collected and analysed corundums from a variety of source environments in addition to inclusions in diamonds. Samples associated with granitic emplacement were procured in addition to those from amphibolite-facies and granulite-facies rocks, and various synthetic origins. Finally, carmine-coloured corundum grains from kimberlite drill cores from Saskatchewan, Canada, were included in the sample set. Secondary ion mass spectrometry (SIMS) was used for trace element analyses; however, to obtain absolute concentration data, it was also necessary to fabricate and characterise appropriate standards.

2. Corundum inclusions in diamonds

Corundum-occluding diamonds were collected from a wide area of alluvial gravels from the Juina district of Mato Grosso, Brazil, centred on $-59°05'50''$ West, $-11°23'33''$ South, which corresponds to (271015, 8736200) on Brazilian 1:250 000 Map SC21-Y-C (Juina). The samples came from the Rio Vinte e Um and the Rio Cinta Larga and its tributaries the Rio Mutum, Igarapé Porcão, Rio Juininha and Rio Juina-Mirim. Diamonds from this area are the subject of a number of previous publications (Harris et al., 1997; Harte et al., 1999; Hutchison et al., 1999, 2001) and thesis work of Hutchison (1997). Observations have been repeated on similar samples from the same area more recently by Kaminsky et al. (2001). Kimberlitic sources of these diamonds have not so far been identified; however, anomalous electromagnetic signatures and an abundance of kimberlitic ilmenites and G9 garnets in associated gravels suggests diamonds were recovered close to their source rocks. Furthermore, Diagem International Resources Corp. have recently discovered kimberlite in a different catchment area approximately 8 km away from the present source area. The corundum inclusions were recovered from their host diamonds by fracturing in a purpose-built steel anvil as described below.

Table 1 details the occurrence of corundum inclusions in Juina diamonds and summarises the mineralogy and geochemistry of diamond hosts and associated mineral inclusions. Three inclusions of ruby have been recovered with one 60-µm sample (BZ241C) being a touching grain with a high-Al (10 wt.%) Mg-pyroxene (BZ241B1). This composite inclusion was recovered from within the same diamond as a syngenetic inclusion of ferropericlase [(Mg,Fe)O]. The other two ruby inclusions are a small grain released from the same diamond and the separate ruby sample of Watt et al. (1994). Additionally, a single 300-µm intense blue sapphire inclusion and two white corundums were

recovered from three separate diamonds. A colour photograph of sample BZ241B1,C polished and mounted within epoxy and taken in transmitted light is available from the supplementary data set.

Where analysed, occluding diamonds are Type IIa, which means that they have subdetection-level (~ 15 ppm) concentrations of nitrogen (Kaiser and Bond, 1959) as determined by Fourier transform infrared spectrometry (FTIR). Occluding diamonds also exhibit typical deep-mantle carbon isotopic compositions of $-5‰$ (e.g., Deines and Wickman, 1985) with the exception of the diamond associated with ruby inclusion BZ214A, which has a lighter isotopic composition ($-11.56‰$) more usually associated with eclogitic mantle parageneses.

The occurrence of ruby within the same diamonds as high alumina pyroxene in conjunction with their association with ferropericlase establishes a mineral paragenesis and provides evidence that the ruby inclusions have an origin within the lower mantle. Specifically, phase relations with Mg,Al-pyroxene (originally perovskite structured) place their origin within a depth range of ~ 720–820 km depending on ambient temperature (Hutchison et al., 2001). At shallower depths in the lower mantle, the mineral phase TAPP (tetragonal almandine-pyrope phase; Harris et al., 1997) acts as principal Al-host whereas with greater pressure, all Al is able to be accommodated within the perovskite-structured polymorph of Mg,Al-pyroxene (Irifune et al., 1996).

Table 1
Occurrence and associated mineralogy and geochemistry of corundum inclusions in Juina diamonds

Inclusion	Description	Size (µm)	Associated phases	N (ppm)[a]	$\delta^{13}C$
BZ214A	Bright red	–	None	II	-11.56
BZ241C	Red comp. tip to green bfg. grain BZ241B1	$60 \times 40 \times 30$[b]	BZ241A ferropericlase BZ241B1 and B2 Mg,Al-pyroxene	II	-5.30
BZ241Ac	Red-brown	$30 \times 20 \times 20$	same host as above	II	-5.30
BZ227A	Powder blue vitreous	$300 \times 200 \times 150$	None	II	-4.99
					-4.41
BZ228A	White isotropic	$60 \times 30 \times 40$	None	n.d.	n.d.
BZ229B	Light blue	$120 \times 50 \times 40$	None	n.d.	n.d.

comp.—composite; bfg.—birefringent; n.d.—not determined.
Data were from this study and Hutchison et al. (1999). Carbon isotopic composition and FTIR analytical techniques are described in Hutchison et al. (1999).

[a] Nitrogen was measured by Fourier transform infrared spectrometry (FTIR); II refers to Type IIa diamond, where N concentration is less than 20 ppm.

[b] Ruby portion comprises approximately half of this grain.

3. Additional samples

Carmine-coloured corundums PHN6087/20 were recovered from kimberlite drill cores from Forte à la Corne, Canada. In a different study, diamonds with inclusions of ferropericlase have been recovered from elsewhere in Canada, at Lac de Gras (Davies et al., 1999). Without further inclusions to identify a paragenesis and thus constrain a depth of origin, ferropericlase-bearing diamonds may reflect a deep upper mantle origin (Brey et al., 2003) rather than a lower mantle origin. In any case, however, these diamonds have a likely deep mantle origin and therefore it was considered pertinent to question whether the Forte à la Corne samples also had a similar flavour. These samples were also of interest because of their unusual colouration and compositional zonation. The samples were recovered from heavy mineral separation of in excess of 300 kg of kimberlite borehole cores from the Forte à la Corne kimberlite field. The separation procedure involved jaw and roller crushing, HCl acidification and screening, drying at 105–110 °C, electromagnetic separation and heavy mineral separation using bromoform and clerici solution-thallium formate. Separation and handpicking was overseen by Kevin Leahy (Exploration Consultants Ltd.), Oleg von Knorring (deceased) and P.H. Nixon. One representative sample [PHN6087/20(2) and abbreviated herein to PHN20(2)] was chosen for analysis.

Samples obtained from the University of Arizona Mineral Museum consisted of white corundum from the Rockford Granite, Tallapoosa County, AL (UA3640), which is an s-type felsic plutonic suite (Drummond et al., 1988) and granite intruding, felsite dyke-hosted, light-brown corundum from the Sacatan Mountains, Pinal County, AZ (UA12976 of the type recorded in Larrabee, 1969). Claret-coloured corundums var. ruby associated with clinozoisite from Kenya (UA9997) and from high-grade Proterozoic orthoamphibolites of the Bamble Sector, Froland, Arundel, Norway (UA1855 of the type described by Vissor and Senior, 1990; Nijland et al., 1993) were obtained from the same source.

Claret-coloured corundum associated with clinozoisite from Tanzania (PHNz) was provided from the collection of P.H. Nixon in addition to synthetic corundum (SYNCOR) manufactured commercially

by the reaction of chromite with metallic Al to form Cr-corundum plus Cr metal and catalysed with undisclosed Ca- and K-bearing compounds. A synthetic ruby (BURM), believed to be from Burma and obtained in Kuala Lumpur, was also obtained for study.

High-grade granulite-facies hosted samples from Japan and Antarctica were included from the collection of S.L. Harley. The Japanese sample (HIGO) is from Zone D of the Higo Mesozoic metamorphic belt, which is part of the high T/low P Ryoke Belt in Kyushu. The sample is from a high-temperature reaction zone intimately associated with an ultramafic sliver interpreted as being a high-temperature intrusion (800 °C, 5–6 kbar; Osanai et al., 1996). This metasomatised corundum is associated with pale blue Mg-sapphirine, spinel and altered plagioclase. Sample 91–38 (Harley, 1998) is from a garnet sillimanite phlogopite gneiss where corundums are surrounded by coronas of sillimanite, sapphirine and cordierite and constitutes a metasomatic zone between mafic and pelitic rocks hosted in Archaean (2800 Ma) orthogneisses. The remaining samples are all from the Taynaya supracrustals of the Vestfold Hills, Antarctica (Snape and Harley, 1996); samples (65178, 65305 and SH-9296) all come from what are termed 'Type II' white sugary textured boudin cores containing assemblages of K-feldspar, corundum, sapphirine and sillimanite. The protoliths are thought to be pyrophyllite or muscovite-bearing rocks, probably claystones of hydrothermally altered clay/sericite/chlorite rocks and are hosted in 2520–2500 Ma TTG suite orthogneisses or as rafts in 2500–2486 Ma monzodioritic gneisses. Sample 65305 was obtained from locality B of Harley (1993), sample 65178 from locality A and sample SH-9296 from within Crooked Lake Gneisses approximately 3 km to the northwest.

4. Inclusion release

It is a significantly delicate task to break diamond samples and retrieve included phases relatively intact and there exists a risk of missing useful inclusion fragments in the resulting residue. However, given the advantages that the retention of diamond hosts for separate analysis provide, in addition to the risk of sample alteration by heating, inclusion release by fracturing was preferred over burning. Although previously published work has employed similar techniques for inclusion release and sample preparation, a description of the technique has not been widely available and is hence discussed in the following.

An 80/80 stainless-steel anvil housing was employed with a flat base and two removable glass viewing windows toughened with nail polish on either side of the anvils. The anvils, made from hardened silver steel, lie vertically within the housing with the lower anvil being static and the upper anvil being driven by a screw through the top of the housing. The junction between the upper anvil and screw is occupied by a ball bearing to minimise torque. Diamond samples were placed on the static anvil in such a fashion to exploit diamond cleavage planes and to protect the inclusion. Stones occasionally prove to be very hard to break and fractured explosively. To minimise inclusion fracturing, pressure was increased slowly whilst observing the stone by optical macroscope. Close to final fracture, energy was often released in the form of high-pitched sound waves and an iridescent interference pattern developed.

Broken material was subsequently brushed into a flat-based petri dish. Potential inclusion grains were measured against a graticule and transferred by wooden needle with finger or silica grease to a lightly greased, glass slide. A ring of epoxy was smeared around the inclusion with a small trail being taken over the inclusion to minimise inclusion movement and the formation of air bubbles on addition of more epoxy. A brass cylinder (~ 4-mm diameter) was then coated internally with epoxy and placed over the inclusion. Small amounts of epoxy were then bled down the inside surface of the cylinder and allowed to harden overnight. To bring the inclusion to the surface, the dried sample was carefully ground to 1/4 μm with diamond grit. Denture fixative (Stachel, personal communication, 1997) can be used rather than epoxy, which although more difficult to handle can reduce sample rounding during polishing.

5. Analytical methods

SIMS was chosen as the most suitable technique for characterisation of the sometimes sub-ppm level quantities of trace elements in unknown corundums.

The ~ 20-μm beam size was considered particularly suitable for analysis of inclusions in diamonds, given their rarity and small size.

Analyses were conducted on inclusions in diamonds using the Cameca ims-4f of the University of Edinburgh/NERC using the duo-plasmatron O^- source. An 8nA beam current was used with a contrast diaphragm of 150 μm, field aperture setting of 1 and an energy offset of 80 V. Masses ^{40}Ca, ^{48}Ti, ^{52}Cr, ^{56}Fe, ^{58}Ni, ^{69}Ga, ^{74}Ge, ^{85}Rb, ^{88}Sr, ^{89}Y, ^{90}Zr and ^{208}Pb were measured as single positively charged ions, backgrounds were measured at mass 130.5 and high mass standardisation was carried out on ^{197}Au. Because of the high abundance of mono-isotopic Al, measurement was made at mass 13.5 corresponding to Al^{2+}. A ratio of Al^{2+} to Al^+ of 0.00125 based on Faraday Cup measurement was used in standardisation. Additional corundums, including the standard described in the following, were measured using similar techniques on the Cameca ims-3f of Arizona State University (ASU) and using a 12.5-kV duo-plasmatron source and for masses ^{24}Mg, ^{28}Si, ^{55}Mn, ^{63}Cu, ^{85}Rb, ^{93}Nb, ^{95}Mo, ^{101}Ru and ^{114}Cd in addition to those listed previously. Contrast aperture of 150 μm, field aperture of 1800 μm and energy offset of 75 V were employed.

To fully deconvolute SIMS analysis of unknowns, ion yield values relative to a known element are required. Such values are ideally obtained for standard materials as similar in major element composition and crystal structure to the unknown as possible. For the purposes of this study, a doped Al_2O_3 standard would have been ideal; however, given the absence of historical SIMS analyses of corundums, such standards were not available. Given the difficulty in manufacturing Al_2O_3 glass, eutectic La-aluminate of composition $La_3Al_5O_{12}$ was chosen as a suitable compromise material with a dominance of Al_2O_3 and comparable structure to corundum. The standard material was prepared as a finely ground powder of oxides of the following elements: La, Al, Si, Cd, Pb, Y, Sr, Ca, Ge, Zr, Nb, Ga, Mg, Mo, Ti, Cu, Ru, Cr, Ni, Mn, Rb and Fe with minor elements introduced as 2000 ppm atomic. Glasses were manufactured by levitation laser vitrification where initial sample chips were fused and heated in an insulating environment held in place by tetrahedraly orientated refractory gas jets. Doping elements were chosen so as to avoid molecular and ion interferences on SIMS analyses and X-ray overlapping on EPMA. A number of intense green glasses were formed, of which two glasses (1 and 5) were chosen that showed good visual microscopic homogeneity. These samples were characterised by 25 EPMA spot analyses and two line transects on each grain that showed the samples to be closely homogeneous within analytical error for most elements. Most elements were found in concentrations between 50 and 3000 ppm atomic depending on the degree to which the oxides volatilised or were concentrated during manufacture of the glass. Elements Y, Mo and Ru were found to be below detection limit of EPMA analysis and Rb was not measured due to lack of appropriate standards. To optimise the usefulness of the standard glasses, measurements were also conducted using laser inductively coupled plasma mass spectrometry (laser-ICPMS) at the Australian National University's RSES. Five analyses were conducted on each glass by measuring background first and standardising against SRM 610 glass. ^{27}Al was chosen for internal standardisation. As with EPMA measurements, analyses were found to be closely similar within each grain and similar concentrations and depletions of doped elements were usually obtained. Some discrepancies were observed for Si, Cr, Ge, Y and Pb between laser-ICPMS and EPMA analyses. Careful consideration was therefore given to which data to use to represent the best approximation to the standard glass compositions. Broadly, ICPMS analyses were chosen due to superior detection limits although consideration was also given to possible laser-ICPMS interferences and standard deviations of analyses. Preferred average compositions as elemental wt.% for each standard glass are available in Table 1 of the supplementary data set.

Relative ion yield (RIY) values were subsequently calculated from isotope-corrected SIMS measurements of counts per second conducted on both glasses at ASU under the conditions described above and using Al concentrations determined by EPMA. Typically, RIY is expected to be roughly linearly related to first-order ionisation energy and most elements give a good correlation. The obvious outlier is Pb, and given the very low concentration of this element in the standard glass, a RIY value was

estimated to intercept the relationship with ionisation energy of the other elements. RIY values were similarly calculated for Ru and Rb, which were not measured in the standard glasses. Values for RIY are presented in Table 2 of the supplementary data set. Given the close proximity of composition and RIY values of both glasses, averaged data were used for conversion of isotope-corrected cps data for unknowns. RIY values calculated show a good approximation to those obtained during routine SIMS analyses of glass and crystalline standards.

6. Results

All elements were detected in measurable quantities in both natural and synthetic samples with a few exceptions—notably Ru in BURM and Y, Rb and Ru in the low-K portions of PHN20(2). Averaged concentrations for each sample are presented in Table 2 as ppm weight and arranged broadly according to corundum varietal type. Observations are made in the following, where it should be noted that varietal classifications do not follow gemological criteria but are based principally on sample colouration. It is particularly noticeable from Table 2 that corundums that can be broadly classified by colouration show strong similarities of composition, which are distinctly different from corundums of other varieties. Further and as detailed below, corundums from widely differing geographical settings yet similar geological settings are often indistinguishable from each other. All corundums have significant Fe concentrations (≥ 2000 ppm), intermediary Mg and Si (~ 10–1000 ppm) and are low in Ca (< 10 ppm). However, rubies and sapphires both have intermediary Ti contents but are, respectively, enriched and depleted in Cr. White sapphires contain intermediate concentrations of Cr and are also more enriched in Ti.

6.1. Ruby-type corundums

Concentrations of trace elements in ruby inclusions from Juina diamonds are presented in Fig. 1. Comparison with rubies from other localities (Fig. 2; Table 2) shows that ruby inclusions are particularly identifiable by high Ni concentrations (18–101 ppm)—often two orders of magnitude greater than other rubies (0.32–

Table 2
Compositions of corundums as determined by SIMS expressed as ppm weight

	BZ227A	65178	SH-9296	91-38	BZ228A	BZ229B	UA12976	HIGO	UA3640	65305	BURM	BZ214A	BZ241Ac	BZ241C	UA1855	PHNz	UA9997	SYN-Hi-K	PHN20-Hi-K	SYN-Lo-K	PHN20-Lo-K
n	2	2	2	2	1	1	2	2	3	2	2	1	1	1	2	2	2	1	1	1	1
**	Blue-D	M/M	M/M	M/M	Wh-D	Wh-D	LBr-Ig	M-Ig	Wh-Ig	M/M	Red-Syn	Ruby-D	Ruby-D	Ruby-D	Ct-M/M	Ct-M/M	Ct-M/M	Br.Red-Syn	Carm-Syn?	Br.Red-Syn	Carm-Syn?
Al	522400	525500	523400	526100	528200	524200	521600	526313	522200	526600	526500	476400	524000	473000	522400	523300	522500	430500	466300	522200	511300
Mg	873.4	93.34	82.43	49.26	*201.0*	615.1	158.8	170.9	108.2	137.9	5.46	*2426*	n.d.	*2629*	39.82	20.98	16.48	897.4	707.8	15.40	14.49
Si	*1792*	108.1	85.84	148.1	*bdl*	245.4	318.3	286.2	330.7	168.8	90.02	*2732*	n.d.	*3664*	452.8	136.8	301.4	1711	1717	2857	2584
Ca	2.84	1.33	2.67	3.411	1144	3816	4.78	2.04	6.63	9.74	4.83	13.42	132.1	120.6	3.36	2.85	4.76	346.2	1309	3.102	8.032
Ti	414.7	159.2	134.2	63.00	1181	11290	7556	3199	1114	100.0	26.03	355.4	46.05	328.2	69.12	35.02	15.08	20.97	14.82	3.87	3.064
Cr	9.80	32.32	82.62	87.11	534.9	135.3	710.9	97.02	1106	3.48	2933	49960	5630	92180	5926	5023	5281	28090	12965	6895	5376
Mn	*28.16*	0.12	0.10	0.07	*bdl*	bdl	1.50	13.20	0.60	104.6	0.15	*307.3*	n.d.	*492.2*	0.32	0.09	0.20	23.39	8.00	0.68	1.06
Fe	3128	4090	6527	3245	1729	418.0	4083	2309	6677	2952	5.51	1557	175.6	9497	1405	2239	1878	6.95	136.5	0.40	0.65
Ni	36.79	0.42	0.50	0.41	171.2	40.85	0.71	0.34	0.81	2.75	0.33	43.96	18.16	101.1	1.479	0.33	0.45	1.75	12.09	0.34	1.04
Cu	n.d.	14.84	6.65	8.31	n.d.	n.d.	299.2	129.2	46.28	10.35	7.97	n.d.	n.d.	6.62	6.62	5.99	8.31	4.72	13.18	2.57	8.76
Ga	181.9	36.82	12.30	24.21	1.18	29.44	122.3	38.49	21.17	11.16	1.29	42.12	13.26	62.76	24.13	12.14	9.65	11.31	6.61	1.46	1.37
Ge	198.5	20.48	29.66	13.57	44.96	742.2	110.1	72.23	80.13	67.20	6.49	178.6	3.22	152.3	10.58	11.67	10.05	6.71	1.83	4.09	5.77
Rb	0.03	0.02	0.03	0.03	0.60	2.37	0.03	0.03	0.06	0.12	0.02	0.03	0.07	0.12	0.01	0.01	0.03	1.54	1.25	0.03	bdl
Sr	0.01	0.01	0.02	0.01	2.38	248.8	0.04	0.03	0.30	0.04	0.02	0.04	0.21	0.03	0.02	0.02	0.02	7.99	3.92	0.00	0.01
Y	0.02	0.02	0.01	0.01	0.40	83.28	0.11	0.36	0.01	0.44	0.01	0.00	0.02	0.02	0.00	0.01	0.01	0.08	0.01	bdl	0.01
Zr	0.01	0.01	0.02	0.04	32.37	4403	36.14	1.53	0.02	0.04	0.04	0.06	0.86	0.53	0.01	0.01	0.02	3.36	0.06	0.04	0.02
Nb	0.01	0.03	0.01	0.02	5.57	1.11	22.63	17.92	0.04	0.03	0.05	0.03	0.18	0.18	0.04	0.01	0.03	0.07	0.04	0.01	0.03
Mo	n.d.	0.07	0.12	0.12	n.d.	n.d.	4.81	1.04	0.15	0.12	0.24	n.d.	n.d.	n.d.	0.16	0.27	0.31	0.99	0.41	0.36	0.22
Ru	n.d.	0.05	0.06	0.05	n.d.	n.d.	0.16	0.12	0.05	0.05	bdl	n.d.	n.d.	n.d.	0.17	0.11	0.056	0.029	bdl	0.08	0.07
Cd	n.d.	2.31	3.49	1.74	n.d.	n.d.	11.15	5.62	2.93	2.21	3.25	n.d.	1.73	2.90	2.06	2.01	1.05	1.42	2.59	1.31	2.07
Pb	0.45	0.33	0.36	0.42	30.07	3.76	0.58	0.75	0.42	1.26	0.51	1.15	1.73	2.90	0.46	0.37	0.42	0.16	1.16	0.28	0.545
Cr/Ni	0.27	77.49	165.1	211.2	3.12	3.31	994.7	287.6	1372	1.27		1136	310.05	912.03	4008	15450	11770	Synthetic	Synthetic?	Synthetic	Synthetic?
Al/Ga	2872	14270	42560	21730	446300	17800	4265	13670	24660	47180		11310	39520	7537	21650	43120	54150				

ppm values in italics were determined by EPMA; accuracy of sup-ppm analyses are about ±30% and ppm level analyses range from ±5% to 20%; n.d.—not determined; bdl—below detection limit; SYN—sample SYNCOR; n—no. of analyses; **—sample descriptions; D—inclusion; M/M—metamorphic; Ig—igneous; Syn—synthetic; Wh—white; LBr—light brown; Ct—claret; Br.Red—browny red; Crm—carmine; M-Ig—meta-igneous.

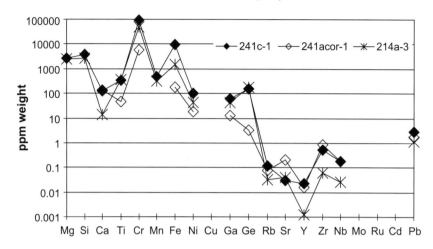

Fig. 1. Concentrations of minor and trace elements in inclusions of ruby from Juina diamonds expressed in ppm weight. Analyses are particularly distinguishable from rubies and corundums from other localities by high Ni concentrations.

1.5 ppm). Kerrich et al. (1987) report high Ni (up to 2000 ppm) in analyses of "corundum"; however, these figures come from neutron activation whole-rock analysis of massive corundums associated with rutile and chlorite. As the same 'corundums' are reported to contain upwards of 2 wt.% TiO_2, it is concluded that these unusual Ni values arise from included phases. Throughout the literature, the highest Ni concentrations reported by precision techniques for natural non-diamond inclusion corundums are from sample 65305 of the current study (2.75 ppm). Juina ruby inclusions are also less evolved having high Mn and ratios of Mg/Fe by weight of 0.27–1.56 compared to 0.008–0.028

for clinozoisite associated rubies and 0.012–0.084 for other natural corundums. It is also notable that notwithstanding the large uncertainties associated with EPMA analyses of minor elements in corundum, analyses that are reported for corundums associated with kimberlites (Exley et al., 1983; Mazzone and Haggerty, 1989) show intermediary ratios of Mg/Fe ranging from 0.07 to 0.24.

Analyses of samples UA9997, UA1855 and PHNz show closely similar compositions as demonstrated in Fig. 2 to the point of being almost indistinguishable within analytical error. As with ruby inclusions, HFSE and LILE concentrations are low ($\sim$ 0.01–0.03 ppm).

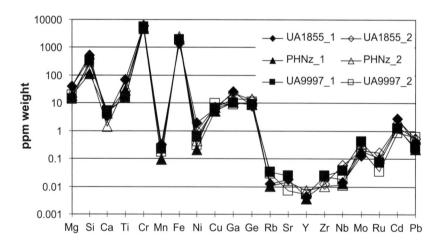

Fig. 2. Concentrations of minor and trace elements in rubies associated with clinozoisite from meta-basic rocks from Norway (UA1855), Kenya (UA9997) and Tanzania (PHNz). The three samples from widely differing geographical origin are strikingly similar in composition.

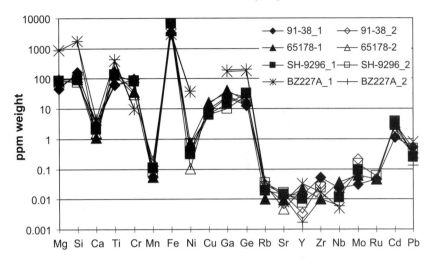

Fig. 3. Concentrations of minor and trace elements in sapphire inclusion BZ227A in comparison with samples 91-38, 65178 and SH-9296 from Antarctic granulites. The inclusion is distinguishable by elevated Ga, Ge and Ni and depleted Cr.

6.2. Sapphire-type corundums

The large, blue sapphire inclusion BZ227A distinguishes itself from other corundums in particular by high Ga and Ge (Fig. 3). It contains ~ 200 ppm weight for both elements compared to tens of ppm for metamorphic origin grains and ~ 100 ppm for igneous sourced samples. Ga concentration data is also available for a small number of gem-quality sapphires from Thailand (Limtrakun et al., 2001) and elsewhere (Emmet et al., 2003). With concentrations ranging from 29.6 to 187.5 ppm weight, these sapphires are intermediate in Ga concentration between the more Ga-rich Juina sapphire inclusions and grains in Antarctic granulites (samples 65178, SH-9296, 91-38 and 65305). Thai sapphires have ratios of Al/Ga ranging from 2820 to 6180 in comparison with 2872 for BZ227A and >14000 for granulite-hosted sapphire-type corundums. Ni contents in BZ227A are about two orders of magnitude higher than otherwise broadly similar samples 65178, SH-9296 and 91-38 from Antarctic granulites whereas Cr contents are lower. Furthermore, BZ227A is very impoverished in LILE and particularly in HFSE compared to otherwise similar igneous and meta-igneous associated corundums UA12976 and HIGO.

65305 and UA3640 have slightly unusual compositions compared to other corundums. They seem to

be most closely allied to sapphire-type compositions except that 65305 has high Y, Ga and Ge and Mn is three orders of magnitude greater (consistently ~ 100 ppm in all three analyses) than comparable grains. UA3640 is unusual as Ti and Cr ppm weight concentrations are almost equal and some analyses show relatively high Sr (~ 0.5 ppm).

6.3. Other natural corundums

White corundum inclusions in Juina diamonds are particularly striking in comparison with all other natural corundums analysed by their relatively high abundances of Ca (>1000 ppm), high field strength elements (HFSE) Y, Zr and Nb and large ion lithophile elements (LILE) Rb and Sr. The closest other sample in terms of HFSE is the felsite-dyke occluding UA12976, which is still significantly depleted in Y. White corundum inclusions are relatively Fe-poor (~ 400–1700 ppm weight) compared to similarly coloured samples that typically contain ~ 3000–7000 ppm weight. In contrast, Mg contents as determined by separate EPMA analysis are high (MgO of ~ 200–600 ppm weight compared to ~ 100–150 ppm weight). Like all the other corundum inclusions in diamonds analysed, Ni contents in white corundums are significantly higher than corundums from other localities. The two inclusions analysed yielded 171.2 and 40.85 ppm weight whereas the highest non-Juina

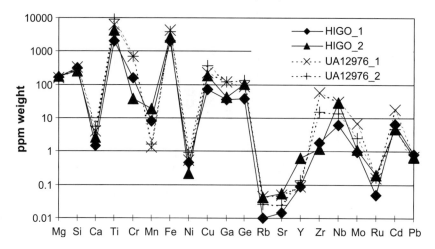

Fig. 4. Concentrations of minor and trace elements in corundums from high-temperature reaction zone associated with ultramafic intrusion from Japan (HIGO) and felsite dyke associated with granite from Arizona (UA12976).

corundum (sample 65305) yielded 2.75 ppm weight. Notably, sample UA3640, which exhibits some characteristics similar to sapphires, shares the characteristic of white corundum inclusions in diamonds in that Ti and Cr have almost equal ppm weight concentrations (∼ 1000 ppm).

HIGO and UA12976 show broadly similar compositions (Fig. 4) and are particularly characterised by low (<0.1 ppm) LILE, intermediary Y (∼ 0.1 ppm) and high HFSE concentrations (∼ 10 ppm).

6.4. Synthetic corundums

SYNCOR and PHN20(2) and other associated grains from PHN6087/20 all show complex intergrowths of regions of varying mean atomic number under back-scattered electron imaging. In contrast, no compositional inhomogeneity was observed for any natural corundums. Analysis by EPMA shows that these zones can broadly be classified into high K and low K. The high-K portions of SYNCOR and

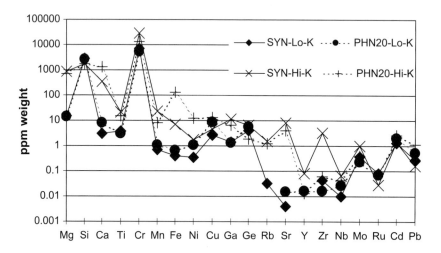

Fig. 5. Concentrations of minor and trace elements in the two principal compositional subdivisions (high and low K) of synthetic corundum (SYNCOR) manufactured by reaction of chromite with metallic Al and PHN20(2) carmine corundum obtained from heavy mineral separate PHN6087/20 from Forte à la Corne kimberlitic drillcore. Both composition of subdivisions and textural characteristics of the two samples are closely similar.

PHN20(2) contain 12.02 and 7.10 wt.% K_2O, respectively, as determined by EPMA analyses and these regions of SYNCOR also contain measurable Na_2O (0.15 wt.%). High-K regions are also distinguishable by relative enrichment in Ca, Fe, Mg and Sr. Low-K regions in SYNCOR and PHN20(2) contain ~ 50 ppm of K_2O. PHN20(2) additionally contains zones containing 25.12 wt.% Cr_2O_3, and 4.72 wt.% CaO, which were too narrow to be measured by SIMS. As can be seen from Table 2 and Fig. 5, equivalent zones in SYNCOR and PHN20(2) are extremely similar in composition with only relatively minor differences between Fe, Ni, Zr and Pb (SYNCOR is relatively deficient in Fe, Ni and Pb and enriched in Zr). Furthermore, synthetic gem ruby BURM, although containing different concentrations of Si, Ti and Fe to low-K zones in SYNCOR and PHN20(2), shows strongly similar HFSE, LILE, Cu, Cd, Pb, Ga and Ge concentrations. It is notable that Ga and Ge in SYNCOR, PHN20(2) and BURM are lower than any natural sample with the marginal exceptions of Ga in BZ228A and Ge in BZ241C.

7. Discussion

The relative abundance of Cr compared to Ti appears to be the most likely influence on the colouration of corundums where the relative quantity of blue-light-absorbing Cr dictates the colouration of rubies and blue-sapphire varieties. Whereas the particularly high abundance of Ti in white corundums contributes to an almost equal absorption across the visible spectrum. More subtle effects are likely due to variable quantities of Fe and trace elements as described by Emmet et al. (2003).

7.1. Origin of corundum inclusions in diamonds

The Ni content of all corundum inclusions typically being orders of magnitude higher than corundums from shallow-sourced settings is regarded as a reflection of the mantle source of the corundum inclusions. Similar arguments are used to support deep mantle sources for ferropericlase and olivine (0.30–0.41 wt.% NiO) within Juina diamonds (Hutchison, 1997) and elsewhere (Brey et al., 2003). With the exception of one very Fe-rich ferropericlase outlier, NiO contents vary from 0.3

to 1.49 wt.% with high Ni corresponding to high Mg content. Furthermore, ferropericlase BZ241A associated with ruby inclusions BZ241C and BZ241Ac contains the highest concentration of Ni observed amongst the Juina suite. The high Mg/Fe ratios observed in Juina rubies are thus consistent with the compositions of Ni and Mg in associated ferropericlase, further supporting a syngenetic origin of the mineral association.

Ruby inclusions

Ga is frequently associated with Al in rocks and minerals and the ratio of Ga to Al is used as an indication of mantle affinity (Schreyer et al., 1981). Schreyer et al. (1981) attribute the very low Ga/Al ratio measured in corundum-fuschite rocks associated with meta-ultramafics from southern Africa to be a reflection of the unusually high-pressure origin of the rocks they describe. Other ruby samples in the literature from likely crustal rocks (Emmet et al., 2003) have higher Ga concentrations and correspondingly higher Ga/Al ratios.

Although phase relation arguments already place the origin of diamond BZ241 within the lower mantle (Hutchison et al., 2001), low Ga/Al ratios in Juina ruby inclusions in comparison with rubies from other localities and ratios of Mg/Fe higher than those recorded from xenoliths in kimberlites (Exley et al., 1983; Mazzone and Haggerty, 1989), which in turn are higher than crustal samples, lend further support to a mantle origin.

Sapphire inclusion

Following the arguments above, the particularly high abundance of Ga in sapphire inclusion BZ227A is also consistent with a mantle origin. With the exception of association with diamond, which places the origin of BZ227A at greater than ~ 80 km, the absence of other associated mineral phases precludes the use of phase relation arguments to shed further light on a depth of origin. However, as phase relation and other arguments for additional Juina corundum inclusions place their origin within the lower mantle, a deep mantle origin for BZ227A would not be considered unlikely. Furthermore, Limtrakun et al. (2001) link the compositions of their gem-sapphire samples to formation in K- and CO_2-rich alkali magmatism. Given that their Ga concentrations are not as high as those of BZ227A, this observation is considered to lay further weight to a deep magmatic origin for the Juina sapphire inclusion. Additionally, inclusion study and measurement of nitrogen characteristics (Hutchison et al., 1999) in run-of mine production of Juina diamonds are compatible with the majority of diamonds being of transition-zone and lower-mantle origin. Lower mantle origin diamonds besides having carbon isotopic compositions of ~ − 5‰ are also almost exclusively Type IIa (Hutchison et al., 1999). As Table 1 shows, similar measurements were obtained for diamond BZ227; hence, again, the evidence is at least consistent with a lower mantle origin for sapphire BZ227A.

Other corundum inclusions

The relatively high Mg contents over Fe in white corundum inclusions in diamonds compared to similar corundums from other sources would point to a more mafic origin. This observation would appear to be at odds, however, with the large abundance of volatile elements. Given the association with diamond, it would appear likely that corundum inclusions of this type would have formed from a partially fluid phase responsible for the syngenetic formation of diamond. This fluid could conceivably act as a carrier for volatile elements of the type observed. However, it is not generally held that LILE are associated with diamond formation from other localities. In addition, LILE concentrations in other Juina inclusions are unremarkable and it is only notable that Zr and Nb in Juina rubies are somewhat elevated above rubies from other localities. In the absence of associated mineral phases by which to estimate pressure and temperature conditions of origin, geochemical indicators are used to support a preferred model that the Juina white corundum-occluding diamonds were formed at shallower depth than Juina ruby inclusions, still within the mantle but involving a more significant crustal component.

7.2. Other natural corundums

The high concentrations of highly mobile HFSE seen in corundums associated with magmatic metasomatism (HIGO and UA12976) is not surprising considering the likelihood of high concentrations of these elements in metasomatising fluids. Metamor-

phism of the HIGO sample may have further contributed to this aspect of its composition. The unusually high Mn concentration of sample 65178 is most likely a reflection of the metamorphic host's claystone protolith.

7.3. Synthetic corundums

The occurrence of the same types of zonation with very strongly similar compositions between synthetic corundum SYNCOR and PHN20(2) is striking given the differing sources for the two samples and especially as sample PHN20(2) was originally considered to be a natural constituent of the manual disintegration of kimberlitic core. Furthermore, all of SYNCOR, PHN20(2), synthetic gem ruby (BURM) and synthetic ruby and sapphire of Emmet et al. (2003) have unusually low concentrations of Fe, Ti and Ga compared to natural rubies, strongly suggesting that PHN20(2) also has an artificial origin. Although mineral separation from core was conducted carefully and corundum was not understood to have been a constituent of the drilling process, neither its use nor its existence as a contaminant can be totally excluded.

8. Conclusions and possible applications

As corundum is shown to occur syngenetically with diamond and its hardness and chemical stability lend it to survival in conditions more extreme than other diamond-associated minerals, potential exists for corundum to be used as a tool for diamond prospectivity. We have demonstrated with a small sample set that minor and trace element compositions of corundum show strong similarities amongst corundums from related geological yet varied geographical settings. At the same time, identifiable differences occur between corundums from differing geological settings and in particular corundum inclusions in diamonds are shown to be distinguishable from other samples. In this case, all inclusion corundums, namely sapphire, ruby and white corundum, are particularly identifiable by high Ni concentrations in excess of 20 ppm; at least an order of magnitude greater than Ni in other samples and large ratios of Mg relative to Fe. Additional precision would be achieved by expanding the present data set into more samples and a larger range of geological settings. At this stage, the data indicates that mantle-sourced corundums are separable from crustal corundums. Corundums are reported from mantle rocks (notably Padovani and Tracey, 1981; Exley et al., 1983; Mazzone and Haggerty, 1989; Rossman and Smyth, 1990) although trace element data have not been obtained. Opportunity exists therefore for an expanded trace element study involving more mantle material where it is hoped that subdivision within the mantle could be achievable.

As corundum is a common accessory phase in rocks from a range of georiclal settings, it is envisaged that corundums with a diamond association may often be significantly outnumbered by those from nondiamond-bearing country rocks. At this stage, application as a prospecting tool would encounter limitations; however, it is likely that high Mg/Fe ratios could be easily detectable by rapid energy dispersive spectrometry (EDS) techniques such as currently being developed by Australia's CSIRO and the Geological Survey of Denmark and Greenland. Furthermore, elevated Ni concentrations may also be detectable by these techniques and certainly suitable samples should be rapidly identifiable by laser-ICPMS and with minimal sample preparation. Once suitable grains are identified, fully quantitative analyses could then be achieved. It is therefore envisaged that corundum has potential to be developed as a tool to compliment conventional indicator mineral techniques in diamond prospectivity.

Acknowledgements

Mimi Hill (University of Liverpool), Peter Gummer and the Rhonda Mining, Shirley Wetmore (University of Arizona), Sopemi and the DTC are thanked for supply of samples. Martin Wilding (Arizona State University) is acknowledged for advice on the formulation of standard La-aluminate glass carried out by Jean Tangeman (Containerless Research Inc.). John Craven (University of Edinburgh), Rick Hervig (Arizona State University), Eric Condliffe (University of Leeds) and Hugh O'Neill and Charlotte Allen (ANU) are recognised for analytical support. The work presented in this paper was supported by N.E.R.C. postgraduate funding under Ben Harte (University of Edinburgh) and Jeff Harris (University of Glasgow), the N.S.F. (EAR 97-06024, 00-01945), an ANU Visiting Fellowship and Diagem International Resources Corp. for conference attendance to M.T.H. This manuscript benefited from the constructive criticism of Julie Hollis (GEUS) and reviews by Steve Haggerty and Craig B. Smith.

References

Brey, G.P., Bulatov, V., Girnis, A., Harris, J.W., Stachel, T., 2003. Ferropericlase—a lower mantle phase in the upper mantle. In: Mitchell, R., et al. (Eds.), 8th International Kimberlite Conference Extended Abstracts. On CD-ROM.

Davies, R., Griffin, W.L., Pearson, N.J., Andrew, A., Doyle, B.J., O'Reilly, S.Y., 1999. Diamonds from the Deep: Pipe DO-27, Slave Craton, Canada. In: Gurney, J.J., et al., (Eds.), Proc. VIIth Internat. Kimberlite Conf., vol. I. Red Roof Design, Cape Town, pp. 148–155.

Deines, P., Wickman, F.E., 1985. The stable carbon isotopes in enstatite chondrites and Cumberland Falls. Geochim. Cosmochim. Acta 49, 89–95.

Drummond, M.S., Wesolowski, D., Allison, D.T., 1988. Generation, diversification and emplacement of the Rockford Granite, Alabama appalachians: mineralogic, pertrologic, isotopic (C & O) and P–T constraints. J. Petrol. 29, 869–897.

Emmet, J.L., Scarratt, K., McClure, S.F., Moses, T., Douthit, T.R., Hughes, R., Novak, S., Shigley, J.E., Wang, W., Bordelon, O., Kane, R.E., 2003. Beryllium diffusion of ruby and sapphire. Gems Gemol. 39, 84–135.

Exley, R.A., Smith, J.V., Dawson, J.B., 1983. Alkremite, garnetite and eclogite xenoliths from Bellsbank and Jagersfontein, South Africa. Am. Mineral. 68, 512–516.

Gasparik, T., Hutchison, M.T., 2000. Experimental evidence for the origin of two kinds of inclusions in diamonds from the deep mantle. Earth Planet. Sci. Lett. 181, 103–114.

Harley, S.L., 1993. Sapphirine granulites from the Vestfold Hills, East Antarctica: geochemical and metamorphic evolution. Antarct. Sci. 5, 389–402.

Harley, S.L., 1998. Ultrahigh temperature granulite metamorphism (1050 °C, 12 kbar) and decompression in garnet (Mg70)-orthopyroxene-sillimanite gneisses from the Rauer Group, East Antarctica. J. Metamorph. Geol. 16, 541–562.

Harris, J.W., Hutchison, M.T., Hursthouse, M., Light, M., Harte, B., 1997. A new tetragonal silicate mineral from the lower mantle. Nature 387, 486–488.

Harte, B., Harris, J.W., Hutchison, M.T., Watt, G.R., Wilding, M.C., 1999. Lower mantle mineral associations in diamonds from São Luiz, Brazil. In: Fei, Y., Bertka, C., Mysen, B.O. (Eds.), Mantle Petrology: Field Observations and High Pressure Experimentation: A Tribute to Francis R. (Joe) Boyd. Geochem. Soc. Spec. Publ., vol. 6. The Geochemical Society, Houston, pp. 125–153.

Hutchison, M.T., 1997. Constitution of the deep transition zone and lower mantle shown by diamonds and their inclusions. PhD Thesis of the University of Edinburgh, p. 660. CD-ROM.

Hutchison, M.T., Cartigny, P., Harris, J.W., 1999. Carbon and nitrogen composition and cathodoluminescence characteristics of transition zone and lower mantle diamonds from São Luiz, Brazil. In: Gurney, J.J., et al. (Eds.), Proceedings of the VIIth International Kimberlite Conference, vol. I. Red Roof Design, Cape Town, pp. 372–382.

Hutchison, M.T., Hursthouse, M.B., Light, M.E., 2001. Mineral inclusions in diamonds: associations and chemical distinctions around the 670 km discontinuity. Contrib. Mineral. Petrol. 142, 119–126.

Irifune, T., Koizumi, T., Ando, J.-I., 1996. An experimental study of the garnet–perovskite transformation in the system $MgSiO_3$–$Mg_3Al_2Si_3O_{12}$. Phys. Earth Planet. Inter. 96, 147–157.

Kaiser, W., Bond, W.L., 1959. Nitrogen, a major impurity in common type I diamond. Phys. Rev. 115, 857–863.

Kaminsky, F.V., Zakharchenko, O.D., Davies, R., Griffin IV, W.L., Khachatryan-Blinova, G.K., Shiryaev, A.A., 2001. Superdeep diamonds from the Juina area, Mato Grosso State, Brazil. Contrib. Mineral Petrol. 140, 353–374.

Kerrich, R., Fyfe, W.S., Barnett, R.L., Blair, B.B., Willmore, L.M., 1987. Corundum, Cr-muscovite rocks at O'Briens, Zimbabwe: the conjunction of hydrothermal desilicification and LIL-element enrichment—geochemical and isotopic evidence. Contrib. Mineral. Petrol. 95, 481–498.

Larrabee, D.M., 1969. Corundum. Mineral and water resources. Bull. Ariz. Bur. Mines 180, 336–337.

Limtrakun, P., Zaw, K., Ryan, C.G., Mernagh, T.P., 2001. Formation of the Denchai gem sapphires, northern Thailand: evidence from mineral chemistry and fluid/melt inclusion characteristics. Mineral. Mag. 65, 725–735.

Mazzone, P., Haggerty, S.E., 1989. Peraluminous xenoliths in kimberlite: metamorphosed restites produced from partial melting of pelites. Geochim. Cosmochim. Acta 53, 1551–1561.

Nijland, T.G., Liauw, F., Visser, D., Maijer, C., Senior, A., 1993. Metamorphic petrology of the Froland corundum-bearing rocks: cooling and uplift history of the Bamble Sector, South Norway. Bull. Norges Geol. Unders. 424, 51–63.

Osanai, Y., Hamamoto, T., Kamei, A., Owada, M., Kagami, H., 1996. High-temperature metamorphism and crustal evolution of the Higo metamorphic terrane, central Kyushu, Japan (in Japanese with English abstr.). Tectonics and Metamorphism. SOUBUN, Japan, pp. 113–123.

Padovani, E.R., Tracey, R., 1981. A pyrope–spinel (alkremite) xenolith from Moses Rock Dike: first known North American occurrence. Am. Mineral. 66, 741–745.

Rossman, G.R., Smyth, J.R., 1990. Hydroxyl contents of accessory minerals in mantle eclogites and related rocks. Am. Mineral. 75, 775–780.

Schreyer, W., Werding, G., Abraham, K., 1981. Corundum–Fuchsite rocks in greenstone belts of southern Africa: petrology, geochemistry, and possible origin. J. Petrol. 22, 191–231.

Snape, I., Harley, S.L., 1996. Magmatic history and the high-grade geological evolution of the Vestfold Hills, East Antarctica. Terra Antarct. 3, 23–38.

Vissor, D., Senior, A., 1990. Aluminous reaction textures in orthoamphibole-bearing rocks: the pressure–temperature evolution of the high-grade Proterozoic of the Bamble sector, south Norway. J. Metamorph. Geol. 8, 231–246.

Watt, G., Harris, J., Harte, B., Boyd, S., 1994. A high-chromium corundum (ruby) inclusion in diamond from the São Luiz alluvial mine, Brazil. Mineral. Mag. 58, 490–492.

Available online at www.sciencedirect.com

Lithos 77 (2004) 287–294

ELSEVIER

LITHOS

www.elsevier.com/locate/lithos

High-pressure experimental growth of diamond using $C–K_2CO_3–KCl$ as an analogue for Cl-bearing carbonate fluid

Emma Tomlinson*, Adrian Jones, Judith Milledge

Department of Earth Sciences, University College London, Gower Street, London WC1E 6BT, UK

Received 27 June 2003; accepted 28 November 2003
Available online 19 May 2004

Abstract

High-pressure, high-temperature diamond growth experiments have been conducted in the system $C–K_2CO_3–KCl$ at 1050–1420 °C, 7.0–7.7 GPa. KCl is of interest because of the strong effect of halogens on the phase relations of carbonate-rich systems [Geophys. Res. Lett. 30 (2003) 1022] and because of the occurrence of KCl coexisting with alkali silicate–carbonate fluids in natural-coated diamond [Geochim. Cosmochim. Acta 64 (2000) 717]. We have used system $C–K_2CO_3–KCl$ as an analogue for these mantle fluids in diamond growth experiments. The presence of KCl reduces the potassium carbonate liquidus to ≤ 1000 °C at 7.7 GPa, allowing it to act as a solvent catalyst for diamond growth at temperatures below the continental geotherm. This is a reduction on the minimum diamond growth temperature reported in the alkali-carbonate–C–O–H system [Lithos 60 (2002) 145]. Diamond growth using carbonate solvent catalysts is characterised by a relatively long induction period. However, the addition of KCl also reduced the period for diamond growth in carbonate to $\ll 5$ min; no such induction period appears to be necessary. It is suggested that KCl destabilises carbonate, allowing greater solubility and diffusion of carbon.
© 2004 Elsevier B.V. All rights reserved.

Keywords: High experiments; Diamond growth; Potassium carbonate; Potassium chloride

1. Introduction

Crystalline (Stachel and Harris, 1997) and fluid (Navon, 1991) inclusions indicate many natural diamonds form at conditions of 5–6 GPa and 900–1400 °C in the mantle. The presence of carbonate-rich, alkali-silicate (Navon et al., 1988; Schrauder and Navon, 1994) and KCl (Burgess et al., 2002; Johnson et al., 2000; Turner et al., 1990) fluids in inclusions

in the fibrous coats of coated octahedral diamond indicates that these fluids are involved in the growth of these diamonds.

During the synthesis of diamond under high pressure–high temperature (HPHT) conditions, solvent catalysts are used to reduce the pressure–temperature conditions required for diamond nucleation and growth. Experiments in the carbonate–carbon system (Akaishi et al., 1990) demonstrated the solvent catalytic properties of carbonates. Since then, a variety of inorganic compounds have been used in experimental investigations of diamond growth and nucleation, e.g. kimberlitic melt (Arima et al., 1993);

---

* Corresponding author.
E-mail address: emma.tomlinson@ucl.ac.uk (E. Tomlinson).

0024-4937/$ - see front matter © 2004 Elsevier B.V. All rights reserved.
doi:10.1016/j.lithos.2004.04.029

halides (Wang and Kanda, 1998); carbonates (Kanda et al., 1990; Litvin et al., 1997, 1998a,b, 1999a; Pal'yanov et al., 1998, 1999a,b; Sato et al., 1999; Sokol et al., 1998, 2000; Taniguchi et al., 1996); and multi-component carbonate–silicate melts (Litvin et al., 1999; Pal'yanov et al., 2002b).

Carbonates have remained the centre of attention, because of the occurrence of carbonates in inclusions in natural diamond coat (Guthrie et al., 1991; Lang and Walmsley, 1983; Walmsley and Lang, 1992). The lower boundary temperature for diamond synthesis is considered to be related to the melt temperature of the solvent catalyst. In the carbonate system, introducing an alkali component can lower the solidus temperature (Pal'yanov et al., 1999a,b, 2002a). Previous studies suggest that diamond growth in the presence of carbonate is characterised by an induction period (before which no diamond growth occurs) that can reach tens of hours (Pal'yanov et al., 2002a). The addition of C–O–H fluid to carbonate substantially decreases the minimum temperature required for diamond synthesis (from ∼ 1700 °C in dolomite–carbon to ∼ 1420 °C in dolomite-fluid–carbon (Sokol et al., 2001) and duration of the induction period (Pal'yanov et al., 2002a; Sokol et al., 2001) required for diamond growth.

Diamond synthesis experiments using KCl–C at 6 GPa, 1620 °C (Wang and Kanda, 1998), and KCl–H_2O–C at 7–8 GPa, 1200–1700 °C (Litvin, 2003) have also been successful. Halogens are of interest because of the profound effect they have on the phase relations of carbonate-rich systems (Williams and Knittle, 2003); and because of the occurrence of KCl in inclusions in coated diamond (Izraeli et al., 2001) along with alkali silicate–carbonate fluids (Navon et al., 1988; Schrauder and Navon, 1994). Chlorine has been shown to be an important component of mantle fluids, by comparisons of experimentally derived Cl partition coefficients for apatite with the Cl-content of mantle-derived apatite (Brenan, 1993). High-pressure experiments (1.5–2.0 GPa) indicate that Cl forms complexes with alkalis (K, Na, Ba, Rb), altering their partition coefficients $D^{fluid/melt}$ and causing them to fractionate into the fluid phase (Ayers and Eggler, 1995). There is a correlation between the concentration of Cl and K in micro-inclusions in natural-coated diamond.

We have used the system C–K_2CO_3–KCl as an analogue for Cl-rich mantle fluids in diamond growth experiments. The purpose of this study is to extend the range of fluids used to experimentally model diamond growth in the carbonate-fluid system. This article describes the preliminary results.

2. Experimental set-up

2.1. Starting materials

Starting materials were high purity graphite powder (UCP1-100: by Ultra Carbon Group, impurity concentration < 5 ppm). For the solvent catalyst, potassium carbonate (K_2CO_3: BDH limited AnalaR, 99.9%) was used with KCl (Fisons Analytical Reagents, 99.8%). Graphite (50 mol%), K_2CO_3 (35 mol%) and KCl (15 mol%) were weighed and mixed by lightly grinding together in an agate mortar. K_2CO_3 is strongly hygroscopic; in order to minimise this effect, the powdered starting materials were dried at 120 °C for 24 h before loading the capsule, and again between loading and crimping, and between crimping and welding of the capsule.

One synthetic cubo-octahedral seed (0.9–1.1 mm) and one natural octahedral seed (0.6–0.9 mm) (Aikhal, Russia) were embedded in the powdered mix and positioned 1/3 and 2/3 along the length of the capsule so that both were the same distance from the hotspot (∼ 5 mm). Both natural and synthetic diamond seeds were used because their differing morphologies provide different potential nucleation surfaces. This was also done so that results could be compared both to the natural system (natural diamond coats grow on octahedral seeds) and published experimental work (mostly using synthetic cubo-octahedral seeds). The size, shape and surface texture of seed crystals were examined by both by optical microscopy and Scanning Electron Microscopy (SEM) prior to experiments: the natural octahedra had smooth surfaces and sharp edges (Fig. 2a); the synthetic seeds also had sharp edges, but the cubic (100) faces were characterised by a dendritic surface texture (Fig. 2b), which may have aided nucleation and growth on these faces. The seeds were then cleaned using isopropanol, acetone and then distilled water.

2.2. High-pressure experiments

Multi-Anvil Press (MAP) experiments were carried out using a Walker module (modified double-stage HP apparatus similar to 6–8 type HP devices). The WC cubic inner anvils were 26- with 8-mm truncated edge length. The pressure-transmitting medium was an $MgO-Cr_2O_3$ octahedron with an edge length of 14- and 5-mm diameter hole, combined with pyrophyllite fin gaskets. The cylindrical graphite furnace with 7.9-mm length and 3.4-mm inner diameter sits within a ZrO thermal insulator. A platinum sample capsule of length 3 mm and inner diameter 1.7 mm was positioned so that its centre was at the hottest region in the furnace. The sample assembly is shown in Fig. 1.

One-hour duration experiments were carried out at 7.0–7.7 GPa and 1050–1420 °C, the heating rate was 52 °C min^{-1}. Pressure calibration curves were constructed for the phase transitions of bismuth (2.55 and 7.7 GPa) at room temperature. No high-temperature pressure calibration has been conducted. Temperatures were estimated from the calibration between applied electrical power and temperature using a $W_{97\%}Re_{3\%}-W_{75\%}Re_{25\%}$ thermocouple. The accuracy of the pressure and temperature determinations was ± 0.5 GPa and ± 100 °C, respectively. The temperature distribution within the capsule was not measured, however, thermal gradients were minimised by the use of a highly conductive capsule, and by the

positioning of the hotspot at the centre of the sample. The thermocouple was aligned axially, so measured temperatures represent the lower bound of the temperature of the sample space. Quenching was achieved by shutting of the electric power supply, while the system was still at experimental pressure.

After experiments, the capsules were sliced open using a razor. The run product was characterised by Electron Probe Micro-Analysis (EPMA), SEM observation, and with a crushed grain mount under oil using a petrological microscope. Seed diamonds were cleaned using HCl and distilled water to remove carbonate and KCl, and characterised by optical microscopy and SEM. Diamond growth was established by observing changes in the surface morphology of the seeds, and was confirmed by an increase in mass at the limit of balance precision. The K_2CO_3- KCl host was not analysed for the presence of spontaneously grown diamond. Melting of carbonate was confirmed both by quench textures in optical mount and by the sinking of diamond seeds.

3. Results

Carbonate crystals in the recovered sample are irregularly shaped with acicular and dendritic forms, while KCl forms small spherules (Fig. 2h). These textures are typical of quench crystals, indicating that the K_2CO_3-KCl mix melted in all experimental runs. Large (20 μm) graphite flakes are also present, and the seed diamond had sunk to the bottom of the capsule in all runs.

At 1050 °C, 7.0 GPa (SK-1), 1 h was sufficient for diamond growth on both the synthetic and natural seeds in the system $50C-35K_2CO_3-15KCl$. Diamond growth was most prominent in the 1-h experiment at 1260 °C, 7.7 GPa (SK-4). During SK-3, the capsule ruptured during heating at 7.7 GPa, 1420 °C. However, the 5 min spent above 1050 °C (the lowest successful growth temperature) was sufficient for growth on both the natural and synthetic seed. Experimental results are presented in Table 1.

Diamond growth took place on the octahedral (111) faces of both the natural and synthetic diamond, and also on the cubic (100) face of the synthetic seed. Two growth morphologies are observed: (1) Development of numerous epitaxial octahedra {111} up to

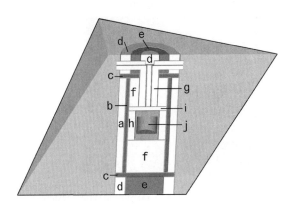

Fig. 1. Cross-section of cell assembly in the 14 mm MgO octahedra. (a) ZrO sleeve, (b) Graphite heater, (c) Graphite discs, (d) ZrO end caps, (e) Cu contact, (f) Al$_2$O$_3$ pressure medium and spacer parts, (g) Thermocouple tube, (h) MgO jacket, (i) MgO Cap, (j) Pt capsule.

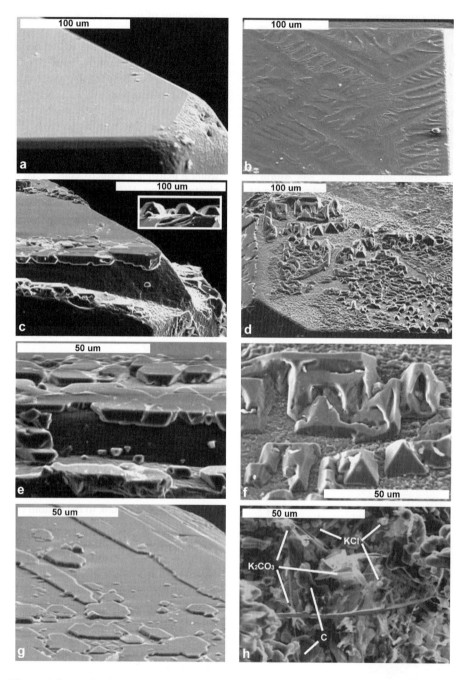

Fig. 2. Scanning Electron Micrographs of diamond growth on seeds in the system $50C-35K_2CO_3-15KCl$. (a) Smooth surface texture of pre-run natural diamond seed face; (b) dendritic texturing on surface of a pre-run synthetic diamond seed (100) face; (c) new diamond growth along the edges of the octahedral face (111) of natural seed in SK-4, inset window (width 50 μm) shows octahedra developed at the apex of the natural octahedral seed; (d) high density growth of epitaxial octahedra on the cubic (100) face of the synthetic seed from SK-4; (e) epitaxial octahedra grown on an octahedral face of the natural seed in SK-4; (f) skeletal morphology of octahedra grown on the synthetic seed in SK-4; (g) layered growth and octahedra on the natural seed from SK-3, layer edges are aligned along the (111) direction; (h) graphite flakes, KCl spherules and dendrites and needles of K_2CO_3 from SK-1.

Table 1
Experimental results for diamond growth on seeds in the system $50C-35K_2CO_3-15KCl$

Run no.	Run conditions				Growth layer thickness (μm)*		Tetrahedral pyramids, surface coverage			
	T (°C)	P (GPa)	t (min)	Seed	(100)	(111)	Morphology	(100)	(111)	Size (μm)
SK-1	1050	7.0	60	Syn	<2	<2	111>100	–	–	–
				Nat	–	0	111>100	–	–	–
SK-3	1420	7.7	<5	Syn	5	5	111>100	5%	<2%	5–10
				Nat	–	3	111>100	–	<2%	5–10
SK-4	1260	7.7	60	Syn	10	10	111>100	60%	45%	10–30
				Nat	–	8	111>100	–	5%	5–20

* ± 30%.

20 μm in size (Fig. 2c–f). On the natural seed, the octahedra have nucleated along the edges of the seed facets (Fig. 2c and d); on the synthetic seed, octahedra appear to be aligned and in places are so densely packed that they form a growth layer parallel to the seed surface (Fig. 2d). The apparent alignment of growth features and presence of areas of high nucleation density for octahedra {111} may be related to the original dendritic surface texture of the synthetic seed, because defects and steps at the surface of the seed face will provide high-energy nucleation sites. Many octahedra {111} on both the natural and synthetic seeds from SK-4 display skeletal forms (Fig. 2f); the development of skeletal morphologies testifies to high crystallisation rates. (2) Layered growth occurred on both seed types in all runs. Layers generally grow inwards from the edge of the seed faces (Fig. 2g); however, they also developed away from the crystal edge (again this may be related to textures on the original seed face). In both cases, the edges of the layers are aligned parallel to the edges of the seed face (Fig. 2g). In SK-3, layered growth is the dominant growth mechanism; this is surprising since peak temperatures in SK-3 were ~ 200 °C higher than in SK-4 (1420 and 1260 °C, respectively), in which skeletal growth is prevalent. This may be because SK-3 spent a greater proportion of its total run time at lower temperatures, and also because conditions in SK-3 were unstable. All the developed growth features show flat surfaces with sharp edges, these features are characteristic of solution growth.

The development of octahedra and growth layers are characteristics similar to those observed using $K_2CO_3-H_2O-CO_2$ at 5.7 GPa, 1150–1420 °C during experiments in excess of 20 h (Pal'yanov et al., 2002a). Skeletal growth forms have been observed in experiments using KCl–C at 1200–1600 °C at 7–8 GPa (Litvin 2003).

4. Discussion and conclusions

The straight edges and flat surfaces of grown diamond indicate a solution process, i.e. growth was by precipitation process from fluid-rich liquid. It is likely that graphite was preserved as a metastable phase in the starting mix, until the conditions of eutectic melting of $KCl-K_2CO_3$ were reached. Graphite is more soluble than diamond because it is thermodynamically metastable at the experimental conditions. The dissolution of graphite leads to formation of carbon saturated melt solutions, from which carbon is precipitated as diamond. In practice, levels of carbon concentration in the sample capsule are also likely to be affected by temperature gradients.

Halides are known to reduce carbonate liquidus temperatures: at 1 bar, 15 mol% KF and KCl lower the K_2CO_3 liquidus temperature from 896, to 840 and 800 °C, respectively (Nyankovskaya, 1952, cited in Levin et al., 1956). Williams and Knittle (2003) used F to reduce the liquidus temperature of carbonate to 700 °C at 1.6 GPa, in experiments investigating the structural complexity of carbonate. These authors observe a shift of the C–O symmetric stretch (from 1050–1075 cm^{-1} in non-fluorinated carbonate glasses to 970 cm^{-1}) in the Raman spectra, which they suggest is most likely to be due to disruption of the carbonate group by F, which allows extensive bridging between C–O bond units and cations (in this case K$^+$) in the melt. It is likely that chlorine has a similar affect on the carbonate liquidus, and is responsible for the reduction of the K_2CO_3 liquidus temperature of potassium carbonate from ~ 1300 to ≤1000 °C at 7.7 GPa. Therefore, the presence of liquid KCl allows K_2CO_3 to act as a solvent catalyst at conditions below the continental geotherm (Fig. 3).

A reduction of the minimum diamond growth temperature is 1050 °C at 7.7 GPa is a reduction of the minimum growth temperature reported in the alkali-carbonate–C–O–H system (Pal'yanov et al., 2002a). A minimal temperature of 1050 °C cannot be considered a minimum temperature, i.e. diamond growth occurred in all experimental runs so we cannot rule out the possibility of growth at even lower temperatures. Furthermore, the experiments in this study were conducted in a narrow pressure range and are limited to durations of 1 h or less, the minimum growth temperature will be reduced at lower pressures. Minimum synthesis temperatures are likely to be controlled by phase relations in the system K_2CO_3–KCl, which are poorly constrained.

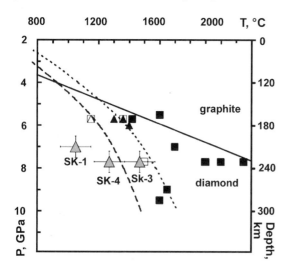

Fig. 3. PT conditions of successful diamond growth experiments (grey triangles). Error bars are 100 °C, 0.5 GPa. Also shown, continental geotherms 35 mW/m² (– –) and 40 mW/m² (- - -), and published conditions of diamond synthesis in carbonate on seeds (triangles) and spontaneous (squares). Solid symbols show synthesis in carbonate only (Akaishi et al., 1990; Pal'yanov et al., 1998, 1999a, 2002a; Sato et al., 1999; Sokol et al., 1998, 2000, 2001; Taniguchi et al., 1996); Open symbols show synthesis using carbonate-fluid (CO_2, H_2O, H_2O–CO_2) (Pal'yanov et al., 2002a; Pal'yanov et al., 1999b; Sokol et al., 2001; Yamaoka et al., 2002) systems.

Previous diamond growth experiments in carbonate–carbon systems below 1500 °C (Pal'yanov et al., 1999a,b, 2002a,b; Sokol et al., 2000, 2001) suggest that diamond growth is characterised by a relatively long induction period before diamond nucleation (<30 h at 5.7 GPa and 1420 °C using K_2CO_3; Pal'yanov et al., 2002a). C–O–H fluids reduce, but do not eliminate, this induction period (<20 h at 5.7 GPa, 1420 °C in the K_2CO_3–$H_2C_2O_4$.$2H_2O$–C system; Pal'yanov et al., 2002a). However, diamond growth in carbonate has been achieved in 20 min at conditions above 1500 °C (Taniguchi et al., 1996). The concept of an induction period is completely at odds with our experimental data; In the K_2CO_3–KCl system, significant growth occurred in 5 min as the temperature was increased from 1050 to 1460 °C in SK-3, indicating a very short induction period. In the 1-h runs, growth rates are sufficiently high as to allow the formation of a skeletal growth of diamond octahedra and layers. Perhaps the high reactivity of KCl in carbonate means that carbonate decomposition is a rapid process and so diamond growth in the carbonate–carbon–alkali chloride system is less limited by kinetic factors than in the carbonate–carbon systems.

There are several possible reasons for this accelerated growth rate: (1) KCl reduces the carbonate melting by destabilising the carbonate unit, as suggested by Williams and Knittle (2003) for CaF. This is supported by the melting of K_2CO_3 melted in all experiments. (2) the presence of KCl may increase the solubility of graphite in carbonate, which is then re-crystallised as diamond. (3) KCl may increase the diffusion rate of carbon, thus improving the supply to the diamond growth face. High growth rates, and therefore high carbon diffusion (supply) rates are a necessary feature of mantle regions in which fibrous diamond is grown. (4) There may be compositional effects at the carbonate–diamond interface, caused by a change in the anion ligands at the surface. Further work is needed to identify which mechanism(s) is responsible for the accelerated diamond growth rates in the system C–K_2CO_3–KCl.

Acknowledgements

Thanks to the Diamond Trading Company (DTC) for sponsoring this research and to Dr. David Dobson for his helpful comments on this manuscript. Thanks also to Dr. George Harlow for the useful criticism in review.

References

Akaishi, M., Kanda, H., Yamaoka, S., 1990. Synthesis of diamond from graphite–carbonate systems under very high temperature and pressure. Journal of Crystal Growth 104, 578–581.

Arima, M., Nakayama, K., Akaishi, M., Yamaoka, S., Kanda, H., 1993. Crystallisation of diamond from a silicate melt of kimberlite composition in high-pressure high-temperature experiments. Geology 21, 968–970.

Ayers, J.C., Eggler, D.H., 1995. Partitioning of elements between silicate melt and H_2O–NaCl fluids at 1.5 and 2.0 GPa pressure—implications for mantle metasomatism. Geochimica et Cosmochimica Acta 59 (20), 4237–4246.

Brenan, J.M., 1993. Partitioning of fluorine and chlorine between apatite and aqueous fluid at high pressure and temperature: implications for the F and Cl content of high P–T fluids. Earth and Planetary Science Letters 117, 251–263.

Burgess, R., Layzelle, E., Turner, G., Harris, J.W., 2002. Constraints on the age and halogen composition of mantle fluids in Siberian coated diamonds. Earth and Planetary Science Letters 197 (3–4), 193–203.

Guthrie, G.D., Veblen, D.R., Navon, O., Rossman, G.R., 1991. Sub-micrometer fluid inclusions in turbid-diamond coats. Earth and Planetary Science Letters 105 (1–3), 1–12.

Izraeli, E.S., Harris, J.W., Navon, O., 2001. Brine inclusions in diamonds: a new upper mantle fluid. Earth and Planetary Science Letters 187 (3–4), 323–332.

Johnson, L.H., Burgess, R., Turner, G., Harris, J.W., 2000. Noble gas and halogen geochemistry of mantle fluids: comparison of African and Canadian diamonds. Geochimica et Cosmochimica Acta 64 (4), 717–732.

Kanda, H., Akaishi, M., Yamaoka, S., 1990. Morphology of synthetic diamonds grown from Na_2CO_3 solvent-catalyst. Journal of Crystal Growth 106, 471–475.

Lang, A.R., Walmsley, J.C., 1983. Apatite inclusions in natural diamond coat. Physics and Chemistry of Minerals 9 (1), 6–8.

Levin, E.M., McMurdie, H.F., Hall, F.P., 1956. Phase Diagrams for Ceramists. The American Ceramic Society, Westerville, OH, USA.

Litvin, Y.A., 2003. Alkaline–chloride components in processes of diamond growth in the mantle and high-pressure experimental conditions. Doklady Earth Sciences 389 (3), 388–391.

Litvin, Y.A., Chudinovskikh, L.T., Zharikov, V.A., 1997. Crystallization of diamond and graphite in the mantle alkaline–carbonate melts in the experiments at pressure 7–11 GPa. Doklady Akademii Nauk 355 (5), 669–672.

Litvin, Y.A., Chudinovskikh, L.T., Zharikov, V.A., 1998a. Crystallization of diamond in the system $Na_2Mg(CO_3)_2$–$K_2Mg(CO_3)_2$–C at 8–10 GPa. Doklady Akademii Nauk 359 (5), 668–670.

Litvin, Y.A., Chudinovskikh, L.T., Zharikov, V.A., 1998b. Seed growth of diamond in the system $Na_2Mg(CO_3)_2$–$K_2Mg(CO_3)_2$–C at 8–10 GPa. Doklady Akademii Nauk 359 (6), 818–820.

Litvin, Y.A., Aldushin, K.A., Zharikov, V.A., 1999. Synthesis of diamond at 8.5–9.5 GPa in the system $K_2Ca(CO_3)(2)$–$Na_2Ca(CO_3)(2)$–C corresponding to the composition of fluid-carbonatitic in inclusions diamond from kimberlites. Doklady Akademii Nauk 367 (4), 529–532.

Navon, O., 1991. High internal-pressures in diamond fluid inclusions determined by infrared-absorption. Nature 353 (6346), 746–748.

Navon, O., Hutcheon, I.D., Rossman, G.R., Wasserburg, G.J., 1988. Mantle-derived fluids in diamond micro-inclusions. Nature 335, 784–789.

Pal'yanov, Y.N., Sokol, A.G., Borzdov, Y.M., Khokhryakov, A.F., Sobolev, N.V., 1998. Crystallization of diamond in the $CaCO_3$–C and $MgCO_3$–C and $CaMg(CO_3)(2)$–C systems. Doklady Akademii Nauk 363 (2), 230–233.

Pal'yanov, Y.N., Sokol, A.G., Borzdov, Y.M., Khokhryakov, A.F., Shatsky, A.F., Sobolev, N.V., 1999a. The diamond growth from Li_2CO_3, Na_2CO_3, K_2CO_3 and Cs_2CO_3 solvent-catalysts at $P=7$ GPa and $T=1700$–1750 degrees C. Diamond and Related Materials 8 (6), 1118–1124.

Pal'yanov, Y.N., Sokol, A.G., Borzdov, Y.M., Khokhryakov, A.F., Sobolev, N.V., 1999b. Diamond formation from mantle carbonate fluids. Nature 400 (6743), 417–418.

Pal'yanov, Y.N., Sokol, A.G., Borzdov, Y.M., Khokhryakov, A.F., 2002a. Fluid-bearing alkaline carbonate melts as the medium for the formation of diamonds in the Earths mantle: an experimental study. Lithos 60 (3–4), 145–159.

Pal'yanov, Y.N., Sokol, A.G., Borzdov, Y.M., Khokhryakov, A.F., Sobolev, N.V., 2002b. Diamond formation through carbonate–silicate interaction. American Mineralogist 87 (7), 1009–1013.

Sato, K., Akaishi, M., Yamaoka, S., 1999. Spontaneous nucleation of diamond in the system $MgCO_3$–$CaCO_3$–C at 7.7 GPa. Diamond and Related Materials 8 (10), 1900–1905.

Schrauder, M., Navon, O., 1994. Hydrous and carbonatitic mantle fluids in fibrous diamonds from Jwaneng, Botswana. Geochimica et Cosmochimica Acta 58 (2), 761–771.

Sokol, A.G., Pal'yanov, Y.N., Borzdov, Y.M., Khokhryakov, A.F., Sobolev, N.V., 1998. Crystallization of diamond from Na_2CO_3 melt. Doklady Akademii Nauk 361 (3), 388–391.

Sokol, A.G., Tomilenko, A.A., Pal'yanov, Y.N., Borzdov, Y.M., Pal'yanova, G.A., Khokhryakov, A.F., 2000. Fluid regime of diamond crystallisation in carbonate–carbon systems. European Journal of Mineralogy 12 (2), 367–375.

Sokol, A.G., Borzdov, Y.M., Pal'yanov, Y.N., Khokhryakov, A.F., Sobolev, N.V., 2001. An experimental demonstration of diamond formation in the dolomite–carbon and dolomite-fluid–carbon systems. European Journal of Mineralogy 13 (5), 893–900.

Stachel, T., Harris, J.W., 1997. Diamond precipitation and mantle metasomatism—evidence from the trace element chemistry of silicate inclusions in diamonds from Akwatia, Ghana. Contributions to Mineralogy and Petrology 129 (2–3), 143–154.

Taniguchi, T., Dobson, D., Jones, A.P., Rabe, R., Milledge, H.J., 1996. Synthesis of cubic diamond in the graphite–magnesium carbonate and graphite–$K_2Mg(CO_3)_2$ systems at high pressure of 9–10 GPa region. Journal of Materials Research 11 (10), 2622–2632.

Turner, G., Burgess, R., Bannon, M., 1990. Volatile-rich mantle fluids inferred from inclusions in diamond and mantle xenoliths. Nature 344 (6267), 653–655.

Walmsley, J.C., Lang, A.R., 1992. On sub-micrometer inclusions in diamond coat-crystallography and composition of ankerites and related rhombohedral carbonates. Mineralogical Magazine 56 (385), 533–543.

Wang, Y., Kanda, H., 1998. Growth of HPHT diamonds in alkali halides: possible effects of oxygen contamination. Diamond and Related Materials 7 (1), 57–63.

Williams, Q., Knittle, E., 2003. Structural complexity in carbonatite liquid at high pressures. Geophysical Research Letters 30 (1), 1–4.

Yamaoka, S., Kumar, M.D.S., Kanda, H., Akaishi, M., 2002. Formation of diamond from $CaCO_3$ in a reduced C–O–H fluid at HP–HT. Diamond and Related Materials 11 (8), 1496–1504.

Available online at www.sciencedirect.com

Lithos 77 (2004) 295–316

www.elsevier.com/locate/lithos

Nature and origin of eclogite xenoliths from kimberlites

D.E. Jacob*

Institut für Geowissenschaften, Universität Mainz, Becherweg 21, D-55099 Mainz, Germany

Received 21 June 2003; accepted 25 January 2004
Available online 25 May 2004

Abstract

Eclogites from the Earth's mantle found in kimberlites provide important information on craton formation and ancient geodynamic processes because such eclogites are mostly Archean in age. They have equilibrated over a range of temperatures and pressures throughout the subcratonic mantle and some are diamond-bearing. Most mantle eclogites are bimineralic (omphacite and garnet) rarely with accessory rutiles. Contrary to their overall mineralogical simplicity, their broadly basaltic-picritic bulk compositions cover a large range and overlap with (but are not identical to) much younger lower grade eclogites from orogenic massifs. The majority of mantle eclogites have trace element geochemical features that require an origin from plagioclase-bearing protoliths and oxygen isotopic characteristics consistent with seawater alteration of oceanic crust. Therefore, most suites of eclogite xenoliths from kimberlites can be satisfactorily explained as samples of subducted oceanic crust. In contrast, eclogite xenoliths from Kuruman, South Africa and Koidu, Sierra Leone stem from protoliths that were picritic cumulates from intermediate pressures (1–2 Ga) and were subsequently transposed to higher pressures within the subcratonic mantle, consistent with craton growth via island arc collisions. None of the eclogite suites can be satisfactorily explained by an origin as high pressure cumulates from primary melts from garnet peridotite.
© 2004 Elsevier B.V. All rights reserved.

Keywords: Eclogites; Xenoliths; Major elements; Trace elements; Isotopes; Mantle

1. Introduction

Eclogitic xenoliths are found in most kimberlites and from all cratonic areas are also abundant at off-craton localities (e.g. Appleyard, 2003; Pearson et al., 1995c; Robey, 1981). Most xenolith populations are dominated by peridotite (Gurney et al., 1991; Sobolev et al., 1977), but eclogite can make up the majority of mantle nodules at some localities, such as Roberts Victor, Bellsbank, Newlands (South Africa), and Zagadochnaya (Siberia). This overabundance may reflect

local enrichment of eclogite within the predominantly peridotitic subcontinental mantle, but is in some cases interpreted to be an effect of differences in xenolith preservation during sampling by the kimberlite. Schulze (1989) showed that despite the overabundance of eclogite nodules at Roberts Victor (estimated to be 80–98%, Hatton, 1978; MacGregor and Carter, 1970), garnet peridotite dominates the kimberlite heavy mineral concentrate. This author estimated the overall abundance of eclogite in the upper 200 km of subcontinental mantle to be < 1 vol.%.

Eclogite is the original host rock from which diamond was first recovered (Bonney, 1899) and diamond with eclogitic affinity dominates the dia-

---

* Tel.: +49-6131-3923170; fax: +49-6131-3923070.
  E-mail address: jacobd@uni-mainz.de (D.E. Jacob).

0024-4937/$ - see front matter © 2004 Elsevier B.V. All rights reserved.
doi:10.1016/j.lithos.2004.03.038

mond population at some localities, for example at Orapa (Botswana; Gurney et al., 1991). Best studied, therefore, are eclogite suites from major diamond mines on the Kaapvaal craton, Siberian platform, Congo and West African cratons. More recently, investigations have begun on those from the Slave and the Karelian cratons.

Off-craton eclogites are distinct from those from on-craton localities in that they are often associated with mafic garnet granulites that are less common (although not absent) in on-craton kimberlites (Griffin et al., 1979). Furthermore, they are finer grained than on-craton eclogites and show lower equilibration pressures (1–2.3 GPa; Robey, 1981). Equilibration temperatures overlap with those for the on-craton eclogites (Pearson et al., 1995c).

Although eclogite is apparently only a minor component of the Earth's mantle, it plays an important role in geodynamic processes, i.e. subduction, for which the gabbro-eclogite transition was believed to be the major driving force (Ringwood and Green, 1966). Since the onset of plate tectonics, this process transferred large amounts of basaltic material into the Earth's mantle, today impressively imaged by seismic tomography (e.g., van der Hilst, 1995) giving rise to tonalitic magmas upon subduction (e.g. Foley et al., 2002) and adding to mantle heterogeneity (e.g. Albarede and van der Hilst, 2002). Subducted oceanic crust in the form of eclogite is discussed by some authors as a component in continental flood basalts (e.g. Cordery et al., 1997; Lassiter and DePaolo, 1997) or as representing a "hidden reservoir" for Nb, Ta, Ti counterbalancing the mass imbalance for these elements in the silicate Earth (Rudnick et al., 2000). Furthermore, as most dated eclogite suites are Archean in age, the study of mantle eclogites allows insight into Archean geodynamic processes (e.g. Foley et al., 2003).

2. Hypotheses for the origin of eclogite xenoliths from kimberlites

Models for the origin of eclogites cover a wide range between two extremes. Those put forward mainly in the 1960s focussed on bimineralic eclogitic xenoliths and interpreted them to be of purely mantle origin, originating as high-pressure cumulates from mantle melts (Caporuscio and Smyth, 1990; Hatton

and Gurney, 1987; O'Hara, 1969; O'Hara et al., 1975; O'Hara and Yoder, 1967). Ringwood and Green (1966) proposed the basalt–eclogite transition as the driving force for subduction and later models consequently interpreted the eclogitic xenoliths as remnants of subducted oceanic crust (e.g. Helmstaedt and Doig, 1975; Jagoutz et al., 1984; MacGregor and Manton, 1986). This latter hypothesis experienced growing attention, refinement and alteration over time leading to a number of variants, e.g. models in which eclogites are residual subducted oceanic crust after tonalitic melt removal at high pressure (e.g. Barth et al., 2001; Hatton and Gurney, 1987; Ireland et al., 1994; Jacob and Foley, 1999). More recently, some eclogite suites were interpreted as subducted cumulates formed at low-pressures in the oceanic crust, such as gabbros and spinel-gabbros, (Barth et al., 2002b; Jacob et al., 2003b) attesting to the petrological diversity of oceanic crust. Finally, Schmickler et al. (in press) proposed sanidine-bearing eclogites from the margin of the Kaapvaal craton (Kuruman province) to be derived from phlogopite–pyroxenite veins in the upper lithospheric mantle. From this plethora of models involving a variety of subduction-related processes it is clear that eclogite in the Earth's mantle is not of uniform origin and it seems unjustified that the so-called "mantle hypothesis" has not received more attention over time. This may, however, be caused by the fact that evidence in favour of a subduction origin is much more evident and unequivocal, being based mainly on a deviation of the $\delta^{18}O$ values from those of the Earth's mantle. Such oxygen isotopic anomalies can only be caused at low pressures and temperatures (Clayton et al., 1975) and are commonly interpreted as a seawater alteration signature of oceanic crust (e.g. Barth et al., 2001; Jacob et al., 1994; Jagoutz, 1984). Similarly striking evidence is not available for the identification of high-pressure cumulates or melts. However, as will be shown here, constraints on the characteristics of such rocks can be derived from forward-modelling and trace element systematics (e.g. Jacob et al., 2003b).

3. Petrography and classification

Eclogites from the Earth's mantle are high-grade metamorphic rocks with large grain sizes and few primary mineral phases. Whereas eclogites from orogenic massifs may contain abundant inclusions (e.g. allanite), garnet and clinopyroxene in eclogite xenoliths from kimberlites are generally of high purity and very rarely contain mineral inclusions that give evidence for their prograde metamorphic history. Exceptions are coesite reported from some eclogites (e.g. Hatton, 1978; Sobolev, 1977; Schulze et al., 2000) and sanidine–coesite inclusions in garnets form Kuruman eclogites, interpreted as phlogopite (or phengite) breakdown products (Schmickler et al., in press). The large majority of mantle eclogites are biminerallic garnet- and clinopyroxene-bearing rocks, commonly with rutile as an accessory. Other primary minor or accessory phases comprise coesite (typically transformed to quartz), ilmenite, sanidine, orthopyroxene, diamond, graphite, kyanite, corundum, apatite, zircon and sulphides. Recently, olivine has been described from eclogites of the Slave craton (Fung, 1998; Kopylova et al., 1999) and spinel-bearing eclogites are known from the Kaalvaallei and Orapa kimberlites (Shee and Gurney, 1979; Viljoen, 1994). Coesite or quartz was only rarely reported from mantle eclogites and this apparent scarcity used to be interpreted as one main difference between mantle- and massif eclogites, where quartz is abundant (e.g. Rudnick, 1995). More recent detailed studies, however, show that its abundance in mantle eclogites may have been underestimated, at least at Roberts Victor, RSA (Schulze et al., 2000). Amphibole, plagioclase and phlogopite in mantle eclogites are of secondary origin.

Exsolution features are common and comprise, for example, garnet exsolving from clinopyroxene (e.g. Udachnaya, Siberian platform, Jerde et al., 1993a), orthopyroxene exsolution from clinopyroxene (e.g. Kuruman, Kaapvaal craton; Schmickler et al., in press), rutile from garnet (common in Roberts Victor group II eclogites), sanidine from clinopyroxene (e.g. Kuruman, Kaapvaal craton; Schmickler et al., in press), and apatite from garnet (e.g. Ekati, Slave craton and Koidu, West African craton; Haggerty et al., 1994; Jacob et al., 2003a).

Classifications of eclogites are based on southern African and Siberian samples and can be applied reasonably well to mantle eclogites worldwide. MacGregor and Carter (1970) introduced a combined textural and geochemical classification scheme for Roberts Victor eclogites dividing them into groups I

and II. Group I eclogites contain subhedral or rounded garnets in a clinopyroxene matrix, sometimes displaying layering, whereas group II rocks have interlocking texture of anhedral garnet and clinopyroxene and are more gneissic in appearance, often very fresh. Group I garnets are higher in Na_2O and MgO than those of group II, group I clinopyroxenes have more K_2O, FeO, Cr_2O_3, CaO and MnO than group II clinopyroxenes. McCandless and Gurney (1989) refined and extended the classification of MacGregor and Carter (1970) and differentiated the often diamondiferous group I eclogites by their higher Na_2O_{gt} (≥ 0.09 wt.%) and K_2O_{cpx} (≥ 0.08 wt.%) from the non-diamondiferous group II eclogites at Roberts Victor. The elevated Na_2O contents of garnets and K_2O contents in clinopyroxenes most likely reflect higher pressures of equilibration for group I eclogites, as results from experimental petrology and diamond inclusion studies attest to the pressure dependent incorporation of these elements (Erlank and Kushiro, 1970; Harlow, 1997; Sobolev and Lavrentev, 1971). An alternative classification scheme was developed by Taylor and Neal (1989) on eclogite xenoliths from Bellsbank, Kaapvaal craton and is inspired by the eclogitic garnet classification by Coleman et al. (1965). This classification differentiates three groups based on MgO and Na_2O contents (diopside and jadeite molecules) in clinopyroxenes: Group A rocks contain diopsidic clinopyroxenes with high MgO and low Na_2O contents, group B eclogites are intermediate and group C rocks contain pronounced jadeite-rich clinopyroxenes. Rocks in group A are ultramafic, predominantly pyroxenites or show transitions towards this rock group; kyanite and corundum eclogites are group C according to this classification. Groups I and II eclogites of MacGregor and Carter (1970) and McCandless and Gurney (1989) are found in group B and C of Taylor and Neal (1989) and are not distinguished by this classification.

Pyroxenites are more common in off craton or craton-rim localities. Very few are described from the well-sampled kimberlite pipes on the cratons (e.g. Eggler et al., 1987) and only at Orapa/Lethlakane is a considerable percentage of pyroxenite documented. Because some eclogite suites show compositional transitions towards pyroxenites (e.g. at Orapa or the Kuruman suite, Kaapvaal craton) it is sometimes difficult to draw the line between eclogite and garnet pyroxenite. A useful distinction is the ratio of Tschermak's to jadeite molecules in clinopyroxene expressed as Al [6]/Al [4] ratio (Aoki and Shiba, 1973). Eclogites have Al [6]/Al [4] ratios ≥ 2, reflecting higher equilibration pressures, but differentiation is difficult at low total Al-contents or when high-pressure pyroxenites are encountered (e.g. Green, 1966). The classification scheme of McCandless and Gurney (1989) helps in the identification of diamond-friendly conditions and is probably the most widely applied.

4. Major element chemistry and equilibration conditions

4.1. Mineral compositions

Garnets in mantle eclogites are Cr-poor pyrope–grossular–almandine mixtures with wide variations in grossular-contents. Eclogites in which the grossular component dominates are kyanite- and/or corundum-bearing, high Al bulk compositions and are termed grospydites (grossular–pyroxene–disthene (=kyanite)) rocks (Sobolev et al., 1977). Most eclogite suites have a small grospydite component. Data published on eclogite xenoliths from the Siberian Zagadochnaya pipe (Fig. 1G) represent an effort to sample explicitly the grospydite component at this locality (Sobolev, 1977: the plotted data are therefore not representative for the Zagadochnaya eclogite population). Available data on the Jagersfontein eclogite suite has a very restricted chemical composition (Fig. 1C). Similarly, no grospydites have been reported from eclogite suites of kimberlites from the Slave craton and Colorado/Wyoming (Fig. 1H). All eclogites here classify as group II, based on their garnet composition. The Orapa and Lethlakane mines (Botswana, Fig. 1F) yield a large amount of pyrope-rich, mostly pyroxenitic garnets that are group II. For most pipes, group I and group II eclogites overlap compositionally. The only pipe without compositional overlap between the two eclogite groups is Kaalvallei (South Africa, Fig. 1D), implying that two compositionally distinct suites of eclogite (of different age?) are stored at different depths in the litho-

sphere at this locality. At Koidu, Sierra Leone (Fig. 1E) two eclogite suites are identified by their bulk MgO-contents (e.g. Hills and Haggerty, 1989) and distinct origins are suggested for the low MgO and high MgO suites (Barth et al., 2001, 2002b). Some garnets of the "high MgO" suite, however, plot to high Fe-contents in Fig. 1E (triangles), illustrating the effects of kimberlite infiltration on the eclogite bulk compositions (see below).

Clinopyroxenes in eclogites are omphacites with jadeite contents typically between 20 and 70 mol%, corresponding to Na_2O-contents between approximately 2 and 8 wt.%. They are distinguished from pyroxenitic clinopyroxenes by their higher jadeite component, as outlined above.

Orthopyroxene-bearing eclogites are not very common, but are more abundant in off-craton and craton-rim kimberlites than at on-craton localities (e.g. Rietfontein, Kuruman: Appleyard, 2003; Schmickler et al., in press). Some opx-eclogites are known from on-craton kimberlites such as Roberts Victor, Kaalvallei and Jagersfontein. Orthopyroxenes from Kuruman and Rietfontein have Mg-numbers (Mg-number = $100 \times Mg/(Mg + Fe)$) between 74 and 89 and contain between 0.58 and 0.95 wt.% Al_2O_3.

Unfortunately, no suitable barometers exist for bimineralic eclogites. Equilibrium pressures can therefore not be determined for the majority of the on-craton suites, although the occurrence of coesite and diamond clearly show that they represent high-pressure assemblages. For orthopyroxene-bearing eclogites pressures can be determined using several geobarometers developed for peridotites. The most widely used and most accurate are those based on the Al exchange between garnet and orthopyroxene (e.g. Brey and Köhler, 1990).

Equilibration temperatures are usually calculated using the $Mg-Fe^{2+}$ exchange between garnet and clinopyroxene at an assumed pressure (often 5 GPa). The most widely used geothermometer is that by Ellis and Green (1979), which has the advantage of being calibrated for the range of Mg-numbers found in eclogites. The Fe^{3+} correction required is commonly bypassed by assuming all Fe as Fe^{2+}, which results in minimum temperatures. The newer refined calibration by Ai (1994) for a $P-T$ range of 1 to 6 GPa and 600 to 1500 °C is believed to be the most accurate for Mg-rich eclogites.

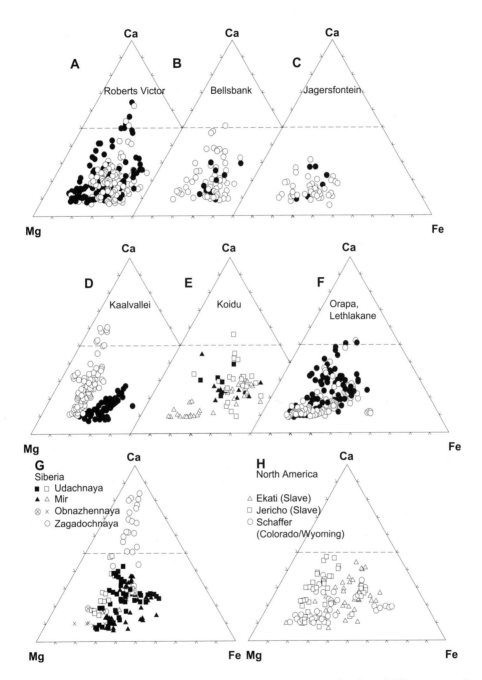

Fig. 1. A–H: Trilateral garnet compositions for bimineralic eclogites from individual kimberlite pipes of different cratons. Open symbols are garnets from group II eclogites; solid symbols are from group I eclogites after McCandless and Gurney (1989). The high-MgO suite of the Koidu pipe (label E) are plotted as triangles, low MgO eclogites are squares. (Data sources: database of the Kimberlite Research Group, University of Cape Town (queries are carried out on request) and Hatton, 1978; Hills and Haggerty, 1989; Jerde et al., 1993a,b; Kopylova et al., 1999; MacGregor and Manton, 1986; McCulloch, 1989; Pearson et al., 1999; Schulze et al., 2000; Shee and Gurney, 1979; Smith et al., 1989; Snyder et al., 1997; Sobolev et al., 1994; Stiefenhofer et al., 1997; Taylor and Neal, 1989; Viljoen, 1994; Viljoen et al., 1996.)

To derive a pressure estimate for bimineralic eclogites, equilibration temperatures are often projected to a paleogeotherm, either deduced from peridotite xenoliths from the same pipe or to a shield geotherm (Pollack and Chapman, 1977). For the Kuruman eclogite suite, however, it has been shown that peridotite xenoliths and orthopyroxene-bearing eclogites describe distinct paleogeotherms. Extrapolation of the temperatures obtained for bimineralic eclogites to the peridotite paleogeotherm would result in erroneously high pressures for the bimineralic eclogites (Fig. 2; Schmickler et al., in press).

Equilibration temperatures (at an assumed pressure of 5 GPa) calculated with major element data plotted in Fig. 1 generally vary between approximately 800 and 1380 °C (Fig. 3). Eclogites from the Siberian pipes yield lower average temperatures than from pipes on other cratons, consistent with a slightly lower geothermal gradient proposed for this region (e.g. Griffin et al., 1999) and those from the Obnazhennaya pipes record the lowest (700 to 960 °C at 5 GPa). Group I eclogites yield overall higher average temperatures than group II eclogites, supporting their higher pressure origin. Canadian eclogites show a very similar range of temperatures (800–1350 °C at 5 GPa).

MacGregor and Manton (1986) suggested that most eclogites are derived from the base of the lithosphere, from depths in excess of 150 km, and Gurney (1990) proposed that a pre-existing peridotitic cratonic keel may have been periodically underplated with diamondiferous eclogites. The presence of diamond in group I eclogites clearly indicates a high pressure origin of >5GPa (Kennedy and Kennedy, 1976) and latest results from seismic tomography showing a cratonic keel of at least 200–300 km below the Kaapvaal craton (Fouch et al., in press) provide the prerequisites for a deep lithospheric origin. In contrast, ranges of equilibration pressures and temperatures in group II eclogites, where diamond as a high pressure indicator mineral is lacking, are much larger and suggest that eclogites reside throughout the depth profile of the continental lithosphere.

4.2. Bulk compositions

Generally of picritic to basaltic character, the original bulk major element composition can be affected by mantle metasomatism and infiltration by kimberlitic melts as well as by reactions between primary phases and the kimberlite. As a cautionary note, it should be pointed out that *all* eclogite xenoliths are infiltrated by kimberlitic material, and bulk analyses of trace elements and radiogenic isotopes never represent the eclogite composition, but are rather mixtures of eclogite and kimberlite (or its lower pressure alteration products). Pre-entrainment metasomatic changes in composition, less common than kimberlite infiltration, have been recognized in individual samples from most pipes, as well as in whole eclogite suites, as for example that from Kimberley, South Africa. Metasomatic effects may be both cryptic and patent (Dawson, 1984), and although trace elements are affected much more than major elements (see below) metasomatic overprints on the major element concentration have been recognized. Ireland et al. (1994) for example, compared Siberian eclogitic diamond inclusions occurring within

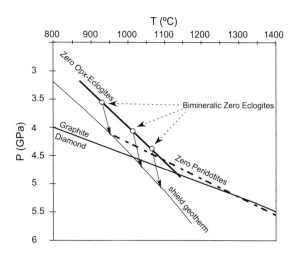

Fig. 2. *P–T* plot for eclogites from the Kuruman kimberlites (Zero pipe, Schmickler et al., in press). Bimineralic eclogites (open circles) are plotted onto the geotherm derived for opx-bearing eclogites from the same locality. Vectors denote loci of the bimineralic samples if their temperatures are extrapolated onto a typical shield geotherm (Pollack and Chapman, 1977). This example shows that extrapolation of temperatures onto a shield geotherm may result in erroneous pressures for bimineralic eclogites. Graphite–diamond boundary and geotherm for the Kuruman peridotites (Shee et al., 1989) are shown for comparison.

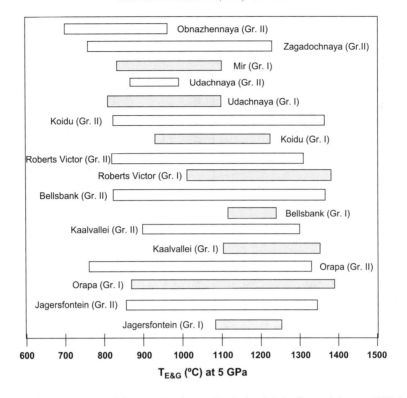

Fig. 3. Temperature ranges (Ellis and Green, 1979) for group I and group II eclogites (McCandless and Gurney, 1989) for individual kimberlite pipes calculated at 5 GPa. In general, group II eclogites cover a wider range of temperatures supporting the view that they are derived from a larger section throughout the lithosphere. Group I eclogites, derived from higher pressures than group II eclogites show generally more restricted and higher temperature ranges. Siberian eclogite suites plot to slightly lower temperatures, consistent with the lower Siberian geotherm (Griffin et al., 1999).

xenoliths with their eclogite host minerals. These authors found lower Mg-numbers in the diamond inclusions than in the eclogite host minerals and concluded that chemical exchange with kimberlitic melts with higher Mg-numbers and rich in incompatible elements changed the composition of those minerals not encapsulated in diamond. A similar observation was made by Barth et al. (2001) for some of the eclogite xenoliths from Koidu, West Africa. However, the major element bulk compositions of most eclogite xenoliths do not show signs of metasomatism, which prevents their reconstruction using mineral and modal analyses that would introduce further uncertainties.

Eclogitic xenoliths are distinct from those exposed in high-pressure orogenic terrains ("massif eclogites") in that the former are on average picritic rather than basaltic (average MgO = 13.8 wt.% as opposed to 9.7 wt.% in the latter) and have lower average bulk

SiO_2, TiO_2, Na_2O (Fig. 4) and Cr_2O_3 contents at comparable CaO contents. The lower SiO_2 content also manifests itself by the general scarcity of quartz in eclogite xenoliths, although quartz is a common mineral in massif eclogites. Both rock groups are explained as subducted oceanic crust (e.g. Stosch and Lugmair, 1990), which in the case of massif eclogites is supported by the fact that many of their bulk compositions are similar to that of MORB. Rudnick (1995) for example, explained the discrepancies between eclogitic xenoliths and MORB/massif eclogites by loss of a tonalitic melt component at high pressures in the case of the xenoliths.

Although some eclogite suites have examples with high MgO values up to 20 wt.% (e.g. Koidu, West Africa: Barth et al., 2002b), none extend to the very high MgO values of komatiites (Fig. 4). Bulk TiO_2 contents range between 0.06 to about 2 wt.%, similar to the range found in komatiites, island arc

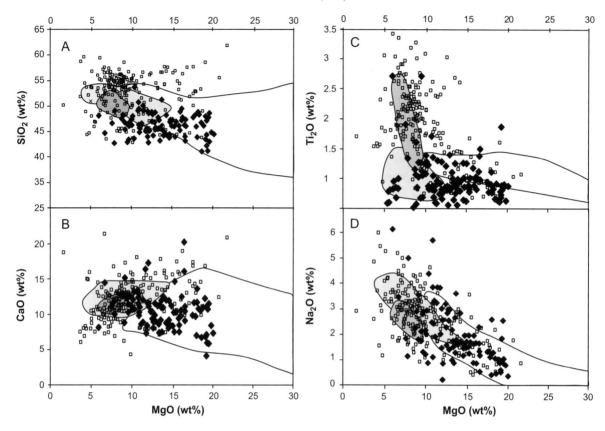

Fig. 4. A–D: Bulk major element plots for mantle eclogites (black diamonds, compiled from the literature) compared with eclogites from orogenic massifs (open squares) and fields for komatiites (white), oceanic gabbros (light grey, Bach et al., 2001) and MORB (dark grey). On average, mantle eclogites are distinct from massif eclogites and have lower SiO_2, TiO_2 and higher MgO. Massif eclogites are more similar to MORB, especially in TiO_2.

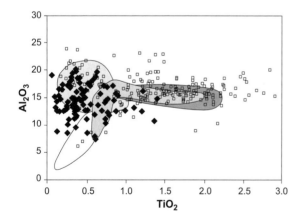

Fig. 5. Bulk Al_2O_3 vs. TiO_2 in weight percent for mantle eclogites and massif eclogites compared with fields for komatiites (white), Archean basalts (grey), oceanic gabbros (light grey) and MORB (dark grey). Symbols and data sources as in Fig. 4.

related volcanics and oceanic gabbros, but are lower than the majority of mid-ocean ridge basaltic glasses (Fig. 5).

5. Trace element compositions

In contrast to major elements, the trace element bulk composition of *every* eclogite xenolith is significantly influenced by kimberlite infiltration. Barth et al. (2001) demonstrated by mass balance that the measured bulk trace element composition of Koidu eclogites was affected by 1% to 5% infiltration of kimberlitic melts (Fig. 6), significantly altering the incompatible trace element budget. Accordingly, trace element bulk compositions must be reconstructed using measured mineral compositions and estimated

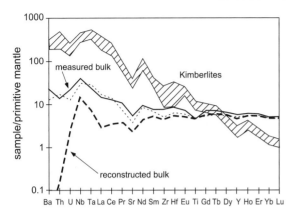

Fig. 6. Example for the effect of kimberlite infiltration upon the bulk trace element budget of mantle eclogites after Barth et al. (2001). The reconstructed trace element bulk composition of the eclogite sample (based on modal analyses and mass balance) is significantly depleted in LREE and LILE compared to the measured eclogite composition. However, addition of 5% of kimberlite to the reconstructed composition can mimic the measured whole rock eclogite.

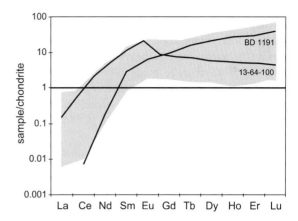

Fig. 7. Rare Earth element patterns of garnets from mantle eclogites worldwide (shaded field compiled from the literature and own unpublished data). The two black lines are examples for extreme LREE depletion found in some eclogitic garnets and patterns with positive Eu anomalies and flat HREE (sample 13-64-100 taken from Jacob et al., 2003b).

modal mineralogy, as described below. During in situ mineral trace element analyses by Laser-Ablation ICP-MS, secondary alteration products may be encountered along even the finest cracks in otherwise pristine primary minerals. This, if unrecognized, can result in artificially high concentrations of elements that are enriched in kimberlite compared to eclogite, as for example Ba, Cs, Nb, La, Ce, U, Th. Most pristine, unaltered eclogitic garnets, for example, contain less than approximately 0.2 ppm Ba, so that this element can be used as an alteration monitor during analysis.

5.1. Rare earth elements

There are two typical types of patterns for rare earth elements in *garnets* from eclogitic xenoliths: Most rare earth element patterns of eclogitic garnets are light rare earth element (LREE) depleted with Ce_N as low as 0.01 (Jacob et al., 2002; Fig. 7). Heavy rare earth elements (HREE) are variably enriched; extreme enrichments of $Lu_N = 70$ are found in Roberts Victor eclogites (Harte and Kirkley, 1997). The other type of REE pattern has nearly flat HREE with $Lu_N < 10$ and positive Eu anomalies (Figs. 7 and 8) and is displayed by roughly half of all published garnet data (49% of the worldwide database). This second type is espe-

cially common amongst coesite-bearing eclogite xenoliths (e.g. Jacob et al., 2003b) as well as amongst those containing kyanite and/or corundum (e.g. Snyder et al., 1997; Jerde et al., 1993a,b; Harte and Kirkley, 1997). These type II patterns are very unlike equilibrium REE patterns for garnet from high-pressure rocks that are expected to have high abundances of HREE due to its mineral-melt partitioning charac-

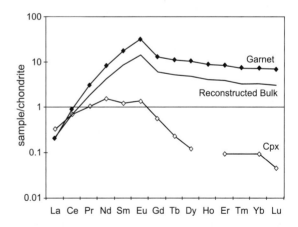

Fig. 8. A typical example for a mantle eclogite displaying positive Eu-anomalies in clinopyroxene, garnet and consequently in the reconstructed whole-rock. Note also the flat HREE pattern of garnet, not consistent with the mineral-melt equilibrium partitioning behaviour of this mineral (data for sample 13-64-100 from Jacob et al., 2003b).

teristics. These anomalies, when traced by the reconstructed bulk composition and associated with bulk positive Sr anomalies (see below), are interpreted as evidence of a prograde metamorphic reaction from plagioclase to garnet and are used in support of the subduction theory that explains the mantle eclogites as subducted oceanic crustal rocks (e.g. Jacob et al., 2003b; Jagoutz et al., 1984).

The majority of *clinopyroxenes* in mantle eclogites display convex-upward REE patterns with moderately enriched LREE concentrations of around $Ce_N = 30$ relative to HREE ($Yb_N = 0.02$, Fig. 9) and may show positive Eu-anomalies when coexisting with garnets showing such anomalies (Fig. 8). Very LREE depleted ($Ce_N = 0.005$) and enriched clinopyroxenes with $Ce_N = 70$ to 100 are only reported from Roberts Victor and from Bellsbank (Jacob et al., 2002, 2003b; Taylor and Neal, 1989). The enrichment is most likely a result of reaction of primary cpx with metasomatic agents prior to entrainment in the kimberlite (Jacob et al., 2003b).

5.2. High field strength elements

Abundances of High Field Strength Elements (HFSE: Nb, Ta, Zr, Hf) in garnets and clinopyroxenes vary and are buffered by rutile, where present. Silicates of rutile-free assemblages show higher Ti and HFSE concentrations than rutile-bearing eclogites.

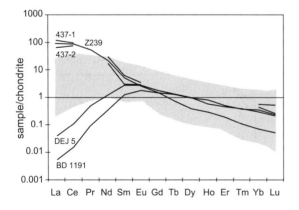

Fig. 9. Rare Earth element patterns of clinopyroxenes from mantle eclogites worldwide (shaded field). The black lines are examples for extreme LREE depletion and enrichment found in some eclogitic clinopyroxenes (Neal et al., 1990; Schmickler et al., in press and worldwide eclogite database).

Rutile, in turn, has very high partition coefficients for Ti and the HFSE and controls the bulk eclogite budget for these elements. Nb/Ta ratios in eclogitic rutiles are highly variable (Rudnick et al., 2000): very high Nb/Ta ratios (≤ 343) and heterogeneous Nb and Ta concentrations (16,000 ppm Nb, 900–1700 ppm Ta) in addition to sometimes skeletal textures of rutiles in Koidu eclogites were used by Barth et al. (2001) to argue for a metasomatic origin of these rutiles. In contrast, very low and even undetectable Nb and Ta concentrations, low Nb/Ta and Zr/Hf ratios (Nb/Ta = 3, Zr/Hf = 16, Jacob et al., 2003b) in some Roberts Victor eclogites show that a partial melting event affected a rutile-free bulk composition, either under amphibolite facies conditions, below the rutile stability field, or at higher pressures, deep in the upper mantle, where all TiO_2 is dissolved in cpx and garnet (Green and Sobolev, 1975; Klemme et al., 2002; Konzett, 1997). The low HFSE rutiles now present in these eclogites exsolved later from garnet upon cooling and decompression (Jacob et al., 2002).

Other minor and trace elements commonly analyzed are Cr, Mn, Co, Ni, Zn, Cu Sc, V, Ga and recently also Li. Nickel in eclogitic garnets ranges between 4 and 300 ppm. The average for non-diamondiferous eclogite parageneses is 43 ppm, whereas garnets from diamondiferous eclogites have an average Ni content of 109 ppm. Principal carrier of Li in eclogites is clinopyroxene, which has about an order of magnitude higher Li contents than coexisting garnet (0.72–11.3 ppm as opposed to 0.052–0.82 ppm) (Woodland et al., 2002). Based on the small amount of data currently available, eclogite xenoliths have lower reconstructed bulk Li contents than eclogites from orogenic massifs.

5.3. Bulk rock trace element characteristics

Bulk rock reconstructions for the HFSE obviously are very sensitive to the modal abundances of rutile. Large xenoliths provide the opportunity to estimate rutile modal abundances via Ti mass balance (e.g. Barth et al., 2001, 2002b), but more often samples are too small for this approach. Modal abundances of garnet and clinopyroxene, in turn, may be estimated with an uncertainty of about 10–20%, depending on the size and homogeneity of the sample. Usually, garnet and clinopyroxene modal abundances range between 70:30 and 40:60. Compositions dominated by garnet are more common than those where clinopyroxene makes up the majority. Jerde et al. (1993b) calculated bulk REE patterns for a suite of modal compositions with garnet/clinopyroxene ratios between 60:40% and 30:70%. They observed that many patterns are relatively insensitive to variance in mode and qualitative information is preserved even when the mode is changed by 30%.

Perhaps the most notable recalculated bulk trace element patterns are those that display positive Eu (Fig. 8) and Sr (not shown) anomalies and flat HREE. Eclogites with this type of pattern are known from many localities worldwide. Otherwise, trace element patterns are very variable with HREE enrichments in the order of $Lu_N = 30$ (e.g. Snyder et al., 1997) and variably depleted or enriched LREE. A sample with extreme HREE enrichment of $Lu_N = 100$ is known from the Koidu kimberlite pipe (Barth et al., 2002b), whereas eclogites from the Kuruman region have flat, chondritic REE patterns (Schmickler et al., in press). Metasomatism, related to the kimberlite but caused by passing melts before entrainment in the kimberlite affects the highly incompatible elements, such as Ba, Sr, Nb, Ta, Zr and the LREE (for a good assessment, see Barth et al., 2001, 2002b), resulting in "wavy" LREE patterns and overall enrichment of the affected trace elements. Ti and the HREE, however, are less influenced by metasomatism and have been found to give reliable information on the original eclogite composition. Compatible elements are also useful to constrain the eclogite precursor rocks. Reconstructed bulk Ni contents are between 60 and 900 ppm (100 to 300 ppm in diamondiferous eclogites), Co ranges between 30 and 100 ppm and Zn from 16 to 100 ppm.

A large number of mantle eclogites show Zr depletions relative to Sm together with depleted LREE in their recalculated bulk compositions. Barth et al. (2001) showed that in the case of Koidu eclogites this is not due to overlooked accessory phases such as zircons, because recalculated and measured bulk compositions both show Zr depletions. These depletions go with the observed general depletion of eclogite xenoliths in SiO_2 (see above) and are a strong indication that many have lost a partial melt. Jacob and Foley (1999) reached a similar conclusion for a suite of diamondiferous eclogites from Udachnaya, which they interpreted as subducted oceanic

crust. While Mg-numbers of the samples are relatively high (up to 82) and could be an indication of olivine-accumulation in a picritic protolith, moderate reconstructed bulk Ni contents (102–237 ppm) do not support olivine-accumulation, that would have led to very high Ni contents. Loss of a silica-rich melt could explain this, as it would cause an increase in Mg-numbers but would give rise to only a slight increase in Ni content of the residue. It appears therefore that many eclogite xenoliths are residues after partial melting. Coesite/quartz-bearing eclogites may be an exception, because large amounts of partial melting would eliminate any free SiO_2. Yaxley and Green (1998) demonstrated experimentally that quartz is stable in an eclogite assemblage only up to 13% partial melting.

Melts produced by partial fusion of eclogites are of tonalitic composition (Rapp and Watson, 1995) and have been discussed as the major contributor to continental crustal growth. The HFSE systematics of continental crust, however, require amphibolite melting rather than eclogite melting (Foley et al., 2002).

6. Radiogenic and stable isotopes

Eclogite xenoliths are isotopically one of the most diverse rock groups on Earth. Their ε_{Nd}-values, for example, cover 646 epsilon units between -38ε (cpx from Mbuji Mayi, El Fadili and Dmaiffe, 1999) and $+684\varepsilon$ (gt from Roberts Victor, Jagoutz et al., 1984). This diversity is testimony to their antiquity as well as a reflection of a number of processes that affected the eclogites over time.

Bulk isotopic compositions for eclogite xenoliths are measured using the Re–Os isotopic system, whereas whole-rock compositions must be reconstructed from mineral analyses for the other radiogenic isotopic systems, introducing an uncertainty connected with the estimation of the mode, as discussed above. Neodymium, Sm, Lu and Hf have relatively similar partitioning behaviour for garnet and clinopyroxene, whereas Sr contents and most often Pb contents differ by up to two orders of magnitude between these two minerals. Strontium contents in garnet rarely exceed 1 ppm and are often less than 0.5 ppm (20 to 800 ppm in cpx). Lead concentrations measured by isotope dilution range

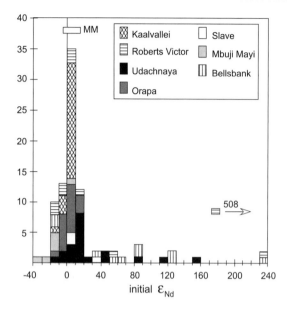

Fig. 10. Histogram of calculated bulk eclogite ε_{Nd} at the time of kimberlite eruption. Mantle eclogites cover an immense range of Nd isotopic compositions between -26 and $+508$ ε_{Nd}. A vertical line denotes the composition of bulk Earth and the box labelled "MM" shows the isotopic composition of the mantle array onto which all major mantle melts fall. (Data sources: El Fadili and Demaiffe, 1999; Jacob and Jagoutz, 1995; Jacob et al., 1994; Jagoutz, 1988; Jagoutz et al., 1984; McCulloch, 1989; Neal et al., 1990; Pearson et al., 1995a; Roden et al., 1999; Smith et al., 1989; Snyder et al., 1997; Viljoen, 1994; Viljoen et al., 1996 and own unpubl. data.)

between 10 and 200 ppb and 170 to 523 ppb in coexisting clinopyroxenes (Jacob and Jagoutz, 1995; Jacob and Foley, 1999). Analysis of Pb isotopic data of garnets from kimberlitic xenoliths (peridotite and eclogite) shows that these very often form a linear array in $^{206}Pb/^{204}Pb-^{207}Pb/^{204}Pb$ space whose apparent age is identical to the kimberlite emplacement age, indicating U/Pb contamination by the kimberlite. It is therefore advisable to use only the Sr and Pb isotopic compositions of clinopyroxenes for comparisons, particularly as the garnet contributes little to the bulk rock composition. All data used here for interpretation and those shown in Fig. 10 are therefore recalculated bulk compositions using published modes or 60% gt: 40% cpx for Nd and Hf, whereas Pb and Sr data are from clinopyroxenes. The range of Nd and Sr isotopic compositions cover a much greater compositional space than all major magmas produced by the Earth's mantle today, occupying all four quadrants of the Nd–Sr diagram. Some extremely radiogenic $^{143}Nd/^{144}Nd$

ratios (Fig. 10) are consistent with strongly LREE depleted patterns and old Nd model ages found in the same rocks. These samples also possess some of the most unradiogenic initial $^{87}Sr/^{86}Sr$ ratios (Fig. 11), at time of kimberlite emplacement (0.70091, Jacob et al., 2003b; Jagoutz et al., 1984) and these characteristics are interpreted as effects of an ancient melting event. Very unradiogenic $^{143}Nd/^{144}Nd$ ratios found in eclogites range down to initial ε_{Nd} values of -26 (El Fadili and Demaiffe, 1999) and indicate long-term light REE enrichment, but are less extreme than those found in peridotitic rocks (Pearson et al., 1995a).

Few hafnium isotopic data exist for eclogite xenoliths. Initial measurements on Roberts Victor eclogites (Jacob et al., 2002) indicate extreme Hf isotopic variability reflecting the image shown by Sm–Nd systematics. Initial ε_{Hf} range between -9.2 to $+166$ at initial ε_{Nd}-values of -22 to 484. Most of the analysed samples plot below the mantle array, which defines the loci of all major magmas from the Earth's mantle.

Only few Pb isotopic studies exist and these concentrate on Roberts Victor and Udachnaya eclogites (Jacob and Jagoutz, 1995; Jacob and Foley, 1999; Kramers, 1977, 1979). Clinopyroxene data plot

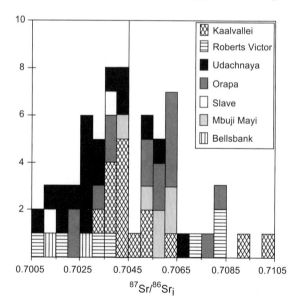

Fig. 11. Histogram of $^{87}Sr/^{86}Sr$-ratios of clinopyroxenes from mantle eclogites recalculated to the kimberlite eruption ages. Vertical bar denotes present bulk Earth composition of 0.7045. Data sources as in Fig. 10.

on both sides of the Geochron, and samples from both localities span a similar range of Pb isotopic ratios (Fig. 12).

Os isotopic compositions are commonly very radiogenic with initial γOs values ranging from close to chrondritic to >6500. They are similar to those of Archean basalts and komatiites (e.g. Barth et al., 2002a; Menzies et al., 1998; Pearson et al., 1995a, Fig. 13) and are indicative of long-term evolution with basaltic–picritic Re/Os ratios.

6.1. Age determinations

Not many eclogite suites are dated, but those that are give Archean ages that most likely reflect emplacement into the lithosphere. Jagoutz et al. (1984) obtained a 2.7 ± 0.1 Ga Sm–Nd isochron age for recalculated bulk eclogites from Roberts Victor. Jacob and Foley (1999) reported a Pb–Pb isochron age for eclogites from the Udachnaya kimberlite, Siberia of 2.57 ± 0.2 Ga. This is within error of the age obtained using Re–Os on bulk eclogites for this suite by Pearson et al. (1995b) of 2.9 ± 0.4 Ga. Osmium isotopic data, although a popular and useful tool in dating eclogites, often show considerable scatter and ages are burdened with relatively large errors. Never-

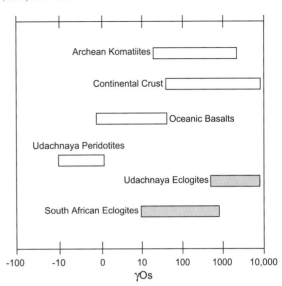

Fig. 13. Osmium isotopic compositions of mantle eclogites (modified after Pearson et al., 1995b).

theless, Re–Os isotopic systematics clearly indicate Archean ages for the Koidu and Newlands eclogites suite (3.44 ± 0.76 Ga: Barth et al., 2002a and 3.6 ± 0.6 Ga: Menzies et al., 1998). Eclogites from the Slave craton (Jericho and Ekati/Lac de Gras pipes)

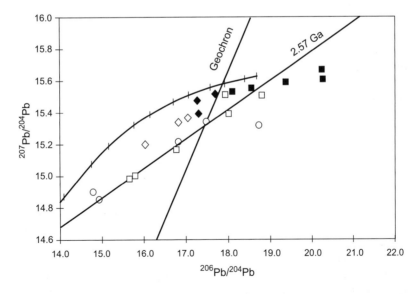

Fig. 12. Lead isotopic compositions of cpx (open symbols) and garnets (solid symbols) from mantle eclogites. Squares are samples from Udachnaya (Siberia, from Jacob and Foley, 1999) with a fitted 2.57 Ga isochron, interpreted as the age of eclogitization. Diamonds and circles are eclogites from Roberts Victor, South Africa. (Jacob and Jagoutz, 1995; Kramers, 1977, 1979). Also shown are the Geochron for 4.55 Ga, and the one-stage Pb model evolution line (Stacey and Kramers, 1975).

yield Proterozoic Nd-model ages (1.0–1.3 Ga, Heaman et al., 2003; Jacob et al., 2003a) that record carbonatite metasomatism within the eclogite facies. Eclogitization itself is currently estimated to be at least 1.79 Ga (Heaman et al., 2003), possibly connected to east-dipping subduction beneath the Slave craton during the 1.88–1.84 Ga Great Bear Magmatic arc event.

Mineral isochrons show a much more complicated picture. Eclogite xenoliths are stored in the subcratonic lithosphere at significantly higher temperatures than the closure temperature for Nd diffusion in pyrope (ca. 800 °C, van Orman et al., 2002) and garnet–clinopyroxene isochrons could therefore be expected to yield kimberlite emplacement ages for the eclogites. However, apparent Sm–Nd internal ages scatter over several orders of magnitude between 4 Ga and ages in the future, but only 7 out of 94 eclogite xenoliths worldwide show Sm–Nd isotopic equilibrium at the time of kimberlite emplacement. The most extreme apparent ages can occur among kyanite/corundum-bearing eclogites from the Kaalvallei and Bellsbank kimberlites (Neal et al., 1990; Viljoen, 1994), but eclogite suites from kimberlites worldwide show ages that deviate more than 100% from the age of the respective kimberlite. One-third of the database yield apparent ages in the future (so-called futurechrons, Jagoutz, 1995), i.e. the regression line between cpx and garnet shows a negative slope.

Some of the mineral ages can be attributed to preserved older equilibration events. Jagoutz (1988), for example showed in a very detailed study on a single eclogite xenolith from Tanzania that its Sm–Nd isotopic ratios evolved in a closed system to vastly different $^{143}Nd/^{144}Nd$ ratios between the mineral phases (>200ε difference), and thus demonstrated that the garnet-cpx age of 1.75 ± 0.014 Ga preserved information on an ancient equilibration event. Many other apparent inter-mineral ages might represent chemical equilibria that are "frozen in" near the closure temperature (Dodson, 1973) or, as has been shown recently, reflect slow cooling of the lithosphere at a temperature range above the closure temperature of the respective isotopic system (Albarede, 2003). Apparent future ages, however, cannot be explained by these hypotheses, but are results of metamorphic reactions involving finite

diffusion couples (Jenkin et al., 1995; Jacob and Jagoutz, in preparation). Similar to processes in high-grade metamorphic rocks (e.g. Mørk and Mearns, 1986), minerals newly formed during reactions such as exsolution or prograde formation of garnet from plagioclase inherit their precursors isotopic and trace element characteristics. As garnet and cpx in bimineralic eclogites have very different and rather complementary equilibrium concentrations of REE, they represent finite diffusion couples for these elements. In this case, the rate of diffusion strongly depends on the concentration of the specific element in the exchange partner and diffusional equilibrium may not be reached over large geological timescales. In kyanite- or corundumbearing eclogites these low-REE phases limit bulk diffusional exchange between cpx and garnet even further, and chemical heterogeneities are effectively preserved leading to extreme aberrant internal ages, as described above.

6.2. Oxygen isotopic compositions

Garlick et al. (1971) were the first to establish the unusually wide range of $\delta^{18}O$-values of mantle eclogites, now known to be characteristic of many eclogite suites worldwide (Fig. 14). Later studies established a minimal pressure-dependence of oxygen isotopic fractionation (Clayton et al., 1975) and a rather constant oxygen isotopic composition of the unchanged Earth's mantle of $5.5 \pm 0.4‰$ (Mattey et al., 1994) and provided a framework for the interpretation of mantle eclogites as subducted seawateraltered oceanic crust. The oxygen isotope argument still plays a key role in the subduction hypothesis and is based on similarities between the ranges of $\delta^{18}O$ values observed in eclogite xenoliths and those of ophiolites (e.g. Gregory and Taylor, 1981) and modern oceanic crust (e.g. Alt et al., 1989). Low $\delta^{18}O$ values can only be caused close to the Earth's surface by hydrothermal alteration at elevated temperatures, such as in the dike sections of modern oceanic crust. Especially the low $\delta^{18}O$ values found in eclogite suites from kimberlites are therefore truly indicative of their derivation from subducted altered oceanic crust.

Eclogites from the Roberts Victor mine show the widest range of $\delta^{18}O$-values from 2.5‰ to 8.0‰

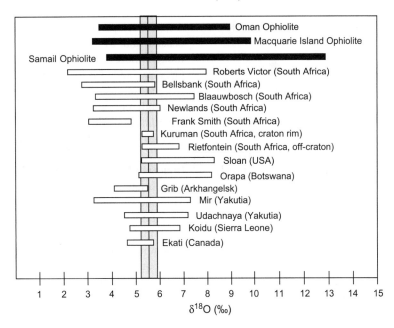

Fig. 14. Oxygen isotopic compositions of mantle eclogite suites worldwide compared to those of ophiolites (black bars). Vertical line denotes $\delta^{18}O$ value of unchanged mantle with a grey error envelope (5.5‰ ± 0.4, Mattey et al., 1994). Note that only the Kuruman suite falls into the mantle field, whereas $\delta^{18}O$-values of all other suites deviate significantly. (Data sources: Oman ophiolite: Gregory and Taylor, 1981, Macquarie Island ophiolite: Cocker et al., 1982; Samail ophiolite: Gregory and Taylor, 1981; Roberts Victor eclogites: MacGregor and Manton, 1986; McDade, 1999, Lowry unpubl. data; Bellsbank: Neal et al., 1990; Blaauwbosch, Newlands, Frank Smith and Sloan: Schulze et al., 2003b; Kuruman: Schmickler et al., in press; Rietfontein: Appleyard, 2003; Orapa: Viljoen et al., 1996; Grib: Malkovets et al., 2003; Mir: Beard et al., 1996, Udachnaya: Jacob et al., 1994; Snyder et al., 1997; Koidu: Barth et al., 2001, 2002b; Ekati: Jacob et al., 2003a; Aulbach, 2003.)

(Fig. 14) and this variation originally measured by conventional fluorination techniques was confirmed by laser fluorination analysis (2.40‰ to 6.98‰, Lowry unpubl. data). Fractionation between clinopyroxene and garnet is approximately 0.3‰, the laser technique giving more systematic fractionation factors than the conventional method.

7. Discussion

The extreme range of possible solid solution in the clinopyroxene and garnet structures in the eclogite metamorphic facies can accommodate a wide variety of bulk compositions (Spear, 1993) and it is therefore not surprising that mantle eclogites represent a rather heterogeneous group of rocks. This fact alone shows that it is not justified to postulate a single origin for all eclogites found in the Earth's mantle. To model their petrogenesis can be rather challenging, as a number of modifying processes need to be taken into account.

Most prominent are the effects of kimberlite infiltration that generally can be bypassed by using ultra clean mineral separates for isotopic analyses and in situ methods for trace element determinations. However, the benefits are limited when pronounced metasomatism leads to growth of clinopyroxene and/or garnet. Further care has to be taken in recalculation of clean bulk compositions as accessory phases, e.g. rutile or quartz can easily be overlooked. Quartz is especially vulnerable as it is not in equilibrium with the quartz-undersaturated kimberlitic melt and, in addition to the fact that it maybe overlooked, reacts out easily. Furthermore, most eclogites from the Earth's mantle show trace element depletions and are thus residues after partial melting. Even coesite/quartz-bearing specimens show evidence for this (Jacob et al., 2003b), although the presence of free SiO_2 limits the amount of partial melt lost to ca. 13% (Yaxley and Green, 1998). All these effects need to be recognized and unravelled in each individual sample, because their time-integrated effects have influence on the radiogenic isotopic ratios and may lead to erroneous interpretation.

8. Hypotheses for the origin of eclogites evaluated

8.1. Recycling of seawater-altered oceanic crust

Many eclogites show evidence for protoliths of low-pressure (i.e. not mantle) origin. A crustal origin is well constrained for those eclogites that contain coesite/quartz or kyanite, because these phases are not in equilibrium with the peridotitic paragenesis in the Earth's mantle. Phase equilibrium considerations show that at pressures within the diamond stability field kyanite reacts with olivine to form pyroxenes and garnet, thus kyanite-bearing eclogites cannot be high-pressure cumulates from melts derived from garnet-peridotite (Jacob et al., 1998). The occurrence of coesite/quartz in eclogites also excludes a direct origin of the eclogites as melts from the Earth's mantle. Partial melts of peridotite are not quartz-normative at any pressure higher than 0.8 GPa, but are olivine + hypersthene-normative at higher pressures (Green and Falloon, 1998). Although free silica may occur in olivine tholeiites or olivine basalt compositions in eclogite facies conditions, these melts could not have originated by melting at pressures greater than 2 GPa (Green and Falloon, 1998).

Geochemical evidence for low-pressure protoliths comes from distinct trace element patterns with positive Eu- and Sr-anomalies and flat HREE in reconstructed bulk compositions (Fig. 8) that can be found not only in coesite/quartz- and kyanite-bearing eclogites, but also in many bimineralic samples. These trace element patterns are very unusual for garnet-bearing high-pressure rocks, but are characteristic of metamorphosed plagioclase-bearing rocks (Mørk and Brunfelt, 1988; Mørk and Mearns, 1986). These characteristics clearly point to a low-pressure origin, but the key arguments for a derivation from oceanic crust are the variation in $\delta^{18}O$ values observed in most suites of eclogites (Fig. 14). Clayton et al. (1975) showed that the effect of pressure on the fractionation of oxygen isotopes is negligible, thus excluding fractionation processes at mantle pressures and temperatures as a reason for the observed variations. However, seawater alteration of oceanic crust at different temperatures, as recorded in modern oceanic basalts and by ophiolites, causes very similar (although more extreme) ranges of $\delta^{18}O$ values (e.g. Gregory and Taylor, 1981). The match with the oxygen isotopic variation in mantle eclogites is close, although not identical, which has led to criticism regarding its proposed analogous origin (e.g. Haggerty, 1999). However, more extreme $\delta^{18}O$ values of +16.1‰ are reported for coesites included in diamond (translating to +14.6‰ for hypothetical garnet in equilibrium with this coesite (Schulze et al., 2003a). This shows that minerals shielded by diamond from equilibrium with surrounding mantle do indeed preserve their original more extreme $\delta^{18}O$ values.

Within ophiolites and modern oceanic crust, the variations in $\delta^{18}O$ values are caused by alteration by circulating seawater at variable temperatures, leading to elevated $\delta^{18}O$ values at low temperatures, close to the seafloor and low $\delta^{18}O$ values at higher temperatures (see for example Alt, 1995). In contrast to meteoric water that can have extreme negative $\delta^{18}O$ values, seawater has a $\delta^{18}O$ value close to 0, and thus shifts the isotopic composition of the altered rocks to significantly less extreme values (Fig. 15). This provides a tool to differentiate between those eclogites whose precursors were altered by meteoric waters (e.g. at Dabie Shan, China, Yui et al., 1997) and those that show a seawater alteration signature. Fig. 15 also illustrates that apparently unchanged $\delta^{18}O$ values in individual samples are not necessarily evidence for an unaltered state of the rocks. The $\delta^{18}O$ trend caused by seawater alteration shows a cross-over with the magmatic fractionation line which is simply a function of the endmember compositions and creates magmatic, apparently unchanged $\delta^{18}O$ values in strongly seawater altered samples (Gregory and Taylor, 1981).

Further supporting evidence for a seawater alteration history of the protoliths of mantle eclogites can be found in trace element and isotopic compositions (e.g. Jacob et al., 1994), as the effects of this process on the geochemical signature of oceanic crust are well constrained. Perhaps most significant are Cs concentrations in eclogitic minerals from Udachnaya and Roberts Victor that overlap with those of MORB (Jacob et al., 1994). Due to the incompatible behaviour of the alkali elements, high-pressure cumulate rocks from melts from garnet peridotite are not expected to contain measurable Cs concentrations,

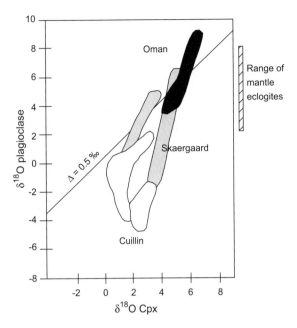

Fig. 15. Oxygen fractionation between plagioclase and clinopyroxene in layered intrusions (Skaergaard and Cuillin) altered by meteoric waters compared to fractionation effects in the seawater-altered Oman ophiolite (redrawn after Gregory and Taylor, 1981). Meteoric waters have more extreme (negative) $\delta^{18}O$-values than seawater and consequently generate more extreme ranges in altered mineral pairs. The black line denotes the magmatic equilibrium fractionation between clinopyroxene and plagioclase. Note that, although altered by seawater with a $\delta^{18}O$ value of 0, some minerals, by coincidence, fall on the magmatic fractionation line.

whereas oceanic basalts contain significant amounts (Hofmann and White, 1982).

An origin of a *suite* of mantle eclogites from seawater-altered oceanic protoliths is well justified in cases where evidence for plagioclase in the protolith ("ghost-plagioclase signatures", such as positive Eu-anomalies) coincides with variations in $\delta^{18}O$ values. However, it is more difficult on the scale of *individual* samples for the following reasons: an apparently pristine $\delta^{18}O$ value similar to that of the unchanged Earth's mantle does not necessarily exclude a seawater alteration history due to the crossover issue outlined above. Furthermore, as the oceanic crust is made up of basalts and ultramafic cumulates in addition to gabbros, a lack of "ghost-plagioclase" signature, indicative for a gabbro precursor rock, does also not necessarily exclude an oceanic origin for the eclogite protoliths.

8.2. Models explaining eclogites as cumulates

Models explaining mantle eclogites as cumulates from primary melts from garnet peridotite were the first to be put forward (e.g. O'Hara, 1969; O'Hara et al., 1975). One of the major problems with this hypothesis, however, is that the accumulation of garnet and clinopyroxene as liquidus minerals from a mantle-derived melt is unlikely; the first phase to accumulate from a primary mantle melt in geologically realistic conditions should be olivine, which is a rare mineral in mantle eclogites.

This is shown in Fig. 16 using the example of high-pressure experimental results on a synthetic olivine tholeiite (Green and Ringwood, 1967). Garnet and clinopyroxene occur together at the liquidus at a pressure of 2.5 GPa for this rock, and close to the liquidus above 1.7 GPa. They are replaced to lower pressures by orthopyroxene alone and then by olivine. Although it appears that it is reasonable that garnet and clinopyroxene should co-crystallize from this

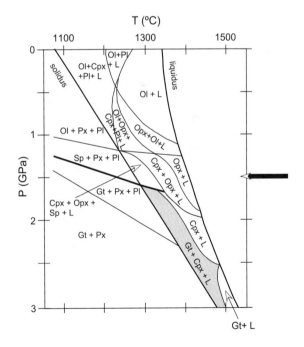

Fig. 16. Phase diagram of olivine tholeiite (modified after Green and Ringwood, 1967). Grey field shows $P–T$ area in which accumulation of cpx and garnet occurs. This is, however, at greater pressures than the origin of the magma (black arrow, Jaques and Green, 1980; Falloon and Green, 1988) and therefore geodynamically impossible. Fat black line is the garnet-in line.

melt, it must be taken into account that the pressure of origin for olivine tholeiites is known to be on the order of 1.5–1.7 GPa from partial melting experiments on peridotite (arrow in Fig. 16; Jaques and Green, 1980; Falloon and Green, 1988), meaning that the garnet and clinopyroxene liquidus fields are exclusively at higher pressures than the origin of the melt. This would require the geodynamically unlikely situation of displacement of a melt to higher pressures in order to generate eclogites by accumulation.

The lack of a multiphase saturation point at the liquidus in Fig. 16 demonstrates that the rock does not represent a primary melt, but that it must have lost olivine, which would be the liquidus phase at pressures below the pressure of origin (Foley, 1989). Although this is only an example it is a general rule that multiphase saturation points, although required for primary melts, are generally not found in experimental studies of basaltic rocks. Furthermore, garnet is mostly restricted to pressures above 1.5 GPa even at temperatures well below the liquidus, so that considerable fractionation must occur to bring it to the liquidus. This means that garnet should be an intercumulus, rather than a cumulus phase in any mantle cumulate derived from a partial melt of peridotite. Garnet may eventually crystallize as a liquidus mineral from a strongly fractionated rest-melt, but in this case the Mg-number of the garnet would not correspond to those seen in the eclogite xenoliths, but instead would be much lower. Phase topologies similar to that shown in Fig. 16 are commonly found in experiments on basaltic rocks, indicating that these conclusions are generally applicable.

The formation of cumulates from primitive melts may be modelled using the pMELTS program (Ghiorso and Hirschmann, 2002) and this has been applied to investigate the possible formation of eclogite suites by Jacob et al. (2003a) and Schmickler et al. (in press). For the Kuruman eclogite suite, it could be shown that the eclogites can be explained by accumulation from a picritic melt, but only by low pressure accumulation of olivine, clinopyroxene and plagioclase followed by subsequent metamorphic transformation to eclogites by an increase in pressure. The pressure of accumulation of around 1 GPa is too high for modern oceanic crust, but may apply for thick ocean crust in late Archean or earliest Proterozoic times (Schmickler et al., in press). Al-

though this may appear to imply a convergence of the originally opposed subduction and mantle cumulate hypotheses, it retains the main feature of the subduction hypothesis that eclogite does not result directly by crystallization of garnet and clinopyroxene from a melt.

In summary, the vast majority of eclogite suites are dominated by samples that correspond to parts of subducted oceanic crust; either volcanics or plagioclase-bearing cumulates. Recent results have revived the idea of an origin by accumulation of liquidus minerals from mantle-derived melts, but at pressures below the eclogite facies followed by eclogite metamorphism.

Acknowledgements

Reviews by Dan Schulze and Chris Hatton are gratefully acknowledged. C. Appleyard, P. McDade, K.S. Viljoen, C.Hatton and D. Lowry generously provided unpublished Theses and data. Financial support from the 8IKC and the DFG made my attendance to the Conference and this publication possible.

References

Ai, Y., 1994. A revision of the garnet–clinopyroxene Fe^{2+}-Mg exchange geothermometer. Contrib. Mineral. Petrol. 115 (4), 467–473.

Albarede, F., 2003. The thermal history of leaky chronometers above their closure temperature. Geophys. Res. Lett. 30 (1), 1015 (doi 10.1029/2002GL016484).

Albarede, F., van der Hilst, R.D., 2002. Zoned mantle convection. Philos. Trans. R. Soc. Lond., A 360 (1800), 2569–2592.

Alt, J.C., 1995. Subseafloor processes in Mid-ocean ridge hydrothermal systems. In: Humphris, S.E., Zierenberg, R.A., Mullineaux, L.S., Thomson, R.E. (Eds.), Seafloor Hydrothermal Systems: Physical, Chemical, Biological and Geological Interactions. AGU, Washington, pp. 85–114.

Alt, J.C., Anderson, T.F., Bonnell, L., Muehlenbachs, K., 1989. Mineralogy, chemistry, and stable isotopic composition of hydrothermally altered sheeted dikes: ODP hole 504B, Leg 111. Proc. ODP, Sci. Results 111, 27–40.

Aoki, K., Shiba, I., 1973. Pyroxenes from lherzolite inclusions of Itinomegata, Japan. Lithos 6, 41–51.

Appleyard, C., 2003. The geochemistry of a suite of eclogite xenoliths from the Rietfontein kimberlite, South Africa. Ext. Abstr. 8th Internat. Kimberlite Conference.

Aulbach, S., 2003. The listhospheric mantle beneath the Slave

Craton and Alberta, Canada, PhD thesis, Macquarie University. 282 pp.

Bach, W., Alt, J.C., Nau, Y., Humphris, S.E., Erzinger, J., Dick, H.J.B., 2001. The geochemical consequences of late-stage low-grade alteration of lower ocean crust at the SW Indian Ridge: results from ODP Hole 735B (Leg 176). Geochim. Cosmochim. Acta 65, 3267–3288.

Barth, M., Rudnick, R.L., Horn, I., McDonough, W.F., Spicuzza, M., Valley, J.W., Haggerty, S.E., 2001. Geochemistry of xenolithic eclogites from West Africa: Part I. A link between low MgO eclogites and Archean crust formation. Geochim. Cosmochim. Acta 65, 1499–1527.

Barth, M., Rudnick, R.L., Carlson, R.W., Horn, I., 2002a. Re–Os and U–Pb geochronological constraints on th eclogite–tonalite connection in the Archean Man Shield, West Africa. Precambrian Res. 118, 267–283.

Barth, M., Rudnick, R.L., Horn, I., McDonough, W.F., Spicuzza, M., Valley, J.W., Haggerty, S.E., 2002b. Geochemistry of xenolithic eclogites from West Africa: Part II. Origins the high MgO eclogites. Geochim. Cosmochim. Acta 66, 4325–4345.

Bonney, T.G., 1899. The parent-rock of the diamond in South Africa. Geol. Mag. 6, 309–321.

Brey, G.P., Köhler, T., 1990. Geothermobarometry in four-phase lherzolites: II. New thermobarometers, and practical assessment of existing thermobarometers. J. Petrol. 31, 1353–1378.

Caporuscio, F.A., Smyth, J.R., 1990. Trace element crystal chemistry of mantle eclogites. Contrib. Mineral. Petrol. 105, 550–561.

Clayton, R.N., Goldsmith, J.R., Karel, V.J., Mayeda, T.K., Newton, R.C., 1975. Limits on the effect of pressure on isotopic fractionation. Geochim. Cosmochim. Acta 39, 1197–1201.

Cocker, J.D., Griffin, B.J., Muehlenbachs, K., 1982. Oxygen and carbon isotope evidence for seawater-hydrothermal alteration of the Macquarie Island ophiolite. Earth Planet. Sci. Lett. 61, 112–122.

Coleman, R.G., Lee, D.E., Beatty, L.B., Brannock, W.W., 1965. Eclogites and eclogites: their differences and similarities. Geol. Soc. Amer. Bull. 76, 483–508.

Cordery, M.J., Davies, G.F., Campbell, I.H., 1997. Genesis of flood basalts from eclogite-bearing mantle plumes. J. Geophys. Res. 102, 20179–20197.

Dawson, J.B., 1984. Contrasting types of upper-mantle metasomatism. In: Kornprobst, J. (Ed.), Kimberlites: II. The Mantle and Crust–mantle Relationships. Elsevier, Amsterdam, pp. 289–294.

Dodson, M.H., 1973. Closure temperature in cooling geochronological and petrological systems. Contrib. Mineral. Petrol. 40, 259–274.

Eggler, D.H., McCallum, M.E., Kirkley, M.B., 1987. Kimberlite-transported nodules from Colorado–Wyoming; a record of enrichment of shallow portions of an infertile lithosphere. In: Morris, E.M., Pasteris, J.D. (Eds.), Mantle Metasomatism and Alkaline Magmatism. Special Paper - Geol. Soc. America, vol. 215, pp. 77–90.

El Fadili, E.S., Demaiffe, D., 1999. Petrology of eclogite and granulite nodules from the Mbuji Mayi kimberlites (Kasai, Congo): significance of kyanite–omphacite intergrowths. In: Gurney, J.J., Gurney, J.L., Pascoe, M., Richardson, S.H. (Eds.), Proc.

7th International Kimberlite Conference. Red Roof Design cc, Cape Town, pp. 205–213.

Ellis, D.J., Green, D.H., 1979. An experimental study of the effect of Ca upon garnet–clinopyroxene Fe–Mg exchange equilibria. Contrib. Mineral. Petrol. 71, 13–22.

Erlank, A.J., Kushiro, I., 1970. Potassium contents of synthetic pyroxenes at high temperatures and pressures. Carnegie Inst. Washington, Yearbook 68, 433–439.

Falloon, T.J., Green, D.H., 1988. Anhydrous partial melting of peridotite from 8 to 35 kb and the petrogenesis of MORB. J. Petrol., 379–414 (Special Lithospheric Issue).

Foley, S., 1989. The genesis of lamproitic magmas in a reduced fluorine-rich mantle. In: Ross, J., et al. (Eds.), Kimberlites and Related Rocks. Special Publication - Geol. Soc. Australia, vol. 14, pp. 616–631. Perth.

Foley, S., Tiepolo, M., Vannucci, R., 2002. Growth of early continental crust controlled by melting of amphibolite in subduction zones. Nature 417 (6891), 837–840.

Foley, S.F., Buhre, S., Jacob, D.E., 2003. Evolution of the Archaean crust by delamination and shallow subduction. Nature 421 (6920), 249–252.

Fouch, M.J., James, D.E., Vandecar, J.C., van der Lee, S. Kaapvaal Seismic Group, 2004. Mantle seismic structure beneath the Kaapvaal and Zimbabwe cratons. S. Afr. J. Geol. 107 (1–2), 35–46.

Fung, A.T., 1998. Petrochemistry of upper mantle eclogites from the Grizzly, Leslie, Pigeon and Sable kimberlites in the Slave Province, Canada. Ext. Abs. 7th. Int. Kimberlite Conf., pp. 230–232.

Garlick, G.D., MacGregor, I.D., Vogel, D.E., 1971. Oxygen isotope ratios in eclogites from kimberlites. Science 172, 1025–1027.

Ghiorso, M.S., Hirschmann, M.M., 2002. pMELTS: a revision of MELTS for improved calculation of phase relations and major element partitioning related to partial melting of the mantle to 3 GPa. Geochem.Geophys. Geosyst. 3, U1–U36 (May 31, 2002).

Green, D.H., 1966. The origin of the 'eclogites' from Salt Lake Crater, Hawaii. Earth Planet. Sci. Lett. 1 (6), 103–190.

Green, D.H., Falloon, T.J., 1998. Pyrolite: a Ringwood concept and its current expression. In: Jackson, I. (Ed.), The Earth's Mantle. Cambridge Univ. Press, Melbourne, pp. 311–380.

Green, D.H., Ringwood, A.E., 1967. The genesis of basaltic magmas. Contrib. Mineral. Petrol. 15, 217–229.

Green, D.H., Sobolev, N.V., 1975. Coexisting garnets and ilmenites synthesized at high pressures from pyrolite and olivine basanite and their significance for kimberlitic assemblages. Contrib. Mineral. Petrol. 50, 217–229.

Gregory, R.T., Taylor, H.P., 1981. An oxygen isotope profile in a section of cretaceous oceanic crust, Samail ophiolite, Oman: evidence for $\delta^{18}O$ buffering of the oceans by deep (>5 km) seawater-hydrothermal circulation at mid-ocean ridges. J. Geophys. Res. 86 (B4), 2737–2755.

Griffin, W.L., Carswell, D.A., Nixon, P.H., 1979. Lower crustal granulites from Lesotho, South Africa. In: Boyd, F.R., Meyer, H.O.A. (Eds.), The Mantle Sample: Inclusions in Kimberlites and other Volcanics. American Geophysical Union, pp. 59–86.

Griffin, W.L., Ryan, C.G., Kaminsky, F.V., O'Reilly, S.Y., Natapov, L.M., Win, T.T., Kinny, P.D., Ilupin, I.P., 1999. The Siberian lithosphere traverse: mantle terranes and the assembly of the Siberian Craton. Tectonophysics 310 (1–4), 1–35.

Gurney, J.J., 1990. The diamondiferous root of our wandering continents. S. Afr. J. Geol. 93, 424–437.

Gurney, J.J., Moore, R.O., Otter, M.L., Kirkley, M.B., Hops, J.J., McCandless, T.E., 1991. Southern African kimberlites and their xenoliths. In: Kampunzu, A.B., Luballa, R.T. (Eds.), Magmatism in Extensional Structural Settings. Springer, pp. 495–536.

Haggerty, S.E., 1999. A diamond trilogy: superplumes, supercontinents, and supernovae. Science 285, 851–860.

Haggerty, S.E., Fung, A.T., Burt, D.M., 1994. Apatite, phosphorus and titanium in eclogitic garnet from the upper mantle. Geophys. Res. Lett. 21, 1699–1702.

Harlow, G.E., 1997. K in clinopyroxene at high pressure and temperature: an experimental study. Am. Mineral. 82, 259–269.

Harte, B., Kirkley, M.B., 1997. Partitioning of trace elements between clinopyroxene and garnet: data from mantle eclogites. Chem. Geol. 136, 1–24.

Hatton, C.J., 1978. The geochemistry and origin of xenoliths from the Roberts Victor mine. PhD thesis, University of Cape Town.

Hatton, C.J., Gurney, J.J., 1987. Roberts Victor eclogites and their relation to the mantle. In: Nixon, P.H. (Ed.), Mantle Xenoliths. Wiley, London, pp. 453–463.

Heaman, L.M., Creaser, R.A., Cookenboo, H.O., Chacko, T., 2003. Multi-stage modification of the mantle lithosphere beneath the Slave Craton: evidence from an unusual suite of zircon-bearing eclogite xenoliths entrained in the Jericho kimberlite, Canada. Ext. Abs. 8th. Int. Kimberlite Conf.

Helmstaedt, H., Doig, R., 1975. Eclogite nodules from kimberlite pipes in the Colorado plateau—samples of subducted Franciscan type oceanic lithosphere. Phys. Chem. Earth 9, 95–111 (First international conference on kimberlites).

Hills, D.V., Haggerty, S.E., 1989. Petrochemistry of eclogites from the Koidu Kimberlite Complex, Sierra Leone. Contrib. Mineral. Petrol. 103, 397–422.

Hofmann, A.W., White, W.M., 1982. Ba, Rb and Cs in the Earth's mantle. Z. Naturforsch. 38a, 256–266.

Ireland, T.R., Rudnick, R.L., Spetsius, Z., 1994. Trace elements in diamond inclusions reveal links to Archean granites. Earth Planet. Sci. Lett. 128, 199–213.

Jacob, D.E., Foley, S.F., 1999. Evidence for Archean ocean crust with low high field strength element signature from diamondiferous eclogite xenoliths. Lithos 48, 317–336.

Jacob, D., Jagoutz, E., 1995. A diamond–graphite bearing eclogitic xenolith from Roberts Victor (South Africa): indications for petrogenesis from Pb-, Nd- and Sr-isotopes. In: Meyer, H.O.A., Leonardos, O.H. (Eds.), Kimberlites, Related Rocks and Mantle Xenoliths. CPRM Spec. Publ., vol. 1/94, pp. 304–317.

Jacob, D.E., Jagoutz, E., in preparation. Chemical Equilibrium between Garnet and Clinopyroxene in Eclogitic Rocks from Kimberlites.

Jacob, D., Jagoutz, E., Lowry, D., Mattey, D., Kudrjavtseva, G., 1994. Diamondiferous eclogites from Siberia: remnants of Archean oceanic crust. Geochim. Cosmochim. Acta 58, 5191–5207.

Jacob, D.E., Jagoutz, E., Lowry, D., Zinngrebe, E., 1998. Comment on "The origin of Yakutian eclogite xenoliths". J. Petrol. 39, 1527–1533.

Jacob, D.E., Bizimis, M., Salters, V.J.M., 2002. Lu–Hf isotopic systematics of subducted ancient oceanic crust: Roberts Victor eclogites. Geochim. Cosmochim. Acta 66 (15A), A360.

Jacob, D.E., Fung, A.T., Jagoutz, E., Pearson, D.G., 2003a. Petrology and geochemistry of eclogite xenoliths from the Ekati kimberlites area. Ext. Abstr. 8th Internat. Kimberlite Conference.

Jacob, D.E., Schmickler, B., Schulze, D.J., 2003b. Trace element geochemistry of coesite-bearing eclogites from the Roberts Victor kimberlite, Kaapvaal craton. Lithos 71, 337–351.

Jagoutz, E., 1988. Nd and Sr systematics in an eclogite xenolith from Tanzania: evidence for frozen mineral equilibria in the continental mantle. Geochim. Cosmochim. Acta 52, 1285–1293.

Jagoutz, E., 1995. Isotopic constraints on garnet equilibration. Terra Abstr. 7, 339.

Jagoutz, E., Dawson, J.B., Hoernes, S., Spettel, B., Wänke, H., 1984. Anorthositic oceanic crust in the Archean Earth. 15th Lunar Planet. Sci. Conf., pp. 395–396. Abs.

Jaques, A.L., Green, D.H., 1980. Anhydrous melting of peridotite at 0–15 Kb pressure and the genesis of tholeiitic basalts. Contrib. Mineral. Petrol 73 (3), 287–310.

Jenkin, G.R.T., Rogers, G., Fallick, A.E., Farrow, C.M., 1995. Rb–Sr closure temperatures in bi-mineralic rocks: a mode effect and test for different diffusion models. Chem. Geol. 122, 227–240.

Jerde, E.A., Taylor, L.A., Crozaz, G., Sobolev, N.V., 1993a. Exsolution of garnet within clinopyroxene of mantle eclogites: major- and trace-element chemistry. Contrib. Mineral. Petrol. 114, 148–159.

Jerde, E.A., Taylor, L.A., Crozaz, G., Sobolev, N.V., Sobolev, V.N., 1993b. Diamondiferous eclogites from Yakutia, Siberia: evidence for a diversity of protoliths. Contrib. Mineral. Petrol. 114, 189–202.

Kennedy, C.S., Kennedy, G.C., 1976. The equilibrium boundary between graphite and diamond. J. Geophys. Res. 81, 2467–2470.

Klemme, S., Blundy, J.D., Wood, B.J., 2002. Experimental constraints on major and tarce element partitioning during partial melting of eclogite. Geochim. Cosmochim. Acta 66, 3109–3123.

Konzett, J., 1997. Phase relations and chemistry of Ti-rich K-richterite-bearing mantle assemblages: an experimental study to 8.0 GPa in a Ti-KNCMASH system. Contrib. Mineral. Petrol. 128, 385–404.

Kopylova, M.G., Russell, J.K., Cookenboo, H.O., 1999. Mapping the lithosphere beneath the North Central Slave Craton, The J B Dawson volume. Proceedings of the VIIth International Kimberlite Conference, Cape Town, vol. 1. Red Roof Design cc, pp. 468–479.

Kramers, J.D., 1977. Lead and strontium isotopes in cretaceous kimberlites and mantle-derived xenoliths from Southern Africa. Earth Planet. Sci. Lett. 34, 419–431.

Kramers, J.D., 1979. Lead, uranium, strontium, potassium and rubidium in inclusion-bearing diamonds and mantle-derived xenoliths from Southern Africa. Earth Planet. Sci. Lett. 42, 58–70.

Lassiter, J.C., DePaolo, D.J., 1997. Plume/lithosphere interaction in the generation of continental and oceanic flood basalts:

chemical and isotopic constraints. In: Mahoney, J.J. (Ed.), Large Igneous Provinces. American Geophysical Union, Monograph 100, pp. 335–355.

MacGregor, I.D., Carter, J.L., 1970. The chemistry of clinopyroxenes and garnets of eclogite and peridotite xenoliths from the Roberts Victor mine, South Africa. Phys. Earth Planet. Inter. 3, 391–397.

MacGregor, I.D., Manton, W.I., 1986. Roberts Victor eclogites: ancient oceanic crust. J. Geophys. Res. 91 (B14), 14063–14079.

Malkovets, V., Taylor, L.A., Griffin, W., O'Reilly, S., Pokhilenko, N., Verichev, E., Golovin, N., Litasov, K., Valley, J., Spicuzza, M., 2003. Eclogites from the Grib Kimberlite Pipe Arkhangelsk, Russia. Ext. Abstr. 8th International Kimberlite Conference.

Mattey, D., Lowry, D., MacPherson, C., 1994. Oxygen isotope composition of mantle peridotite. Earth Planet. Sci. Lett. 128, 231–241.

McCandless, T.E., Gurney, J.J., 1989. Sodium in garnet and potassium in clinopyroxene; criteria for classifying mantle eclogites. In: Ross, J., Jaques, A.L., Ferguson, J., Green, D.H., O'Reilly, S.Y., Danchin, R.V., Janse, A.J.A. (Eds.), Kimberlites and Related Rocks. Special Publication—Geological Society of Australia, vol. 14, pp. 827–832. Perth.

McCulloch, M.T., 1989. Sm–Nd systematics in eclogite and garnet peridotite nodules from kimberlites: implications for the early differentiation of the earth. In: Ross, J., Jaques, A.L., Ferguson, J., Green, D.H., O'Reilly, S.Y., Danchin, R.V., Janse, A.J.A. (Eds.), Kimberlites and Related Rocks. Special Publication—Geological Society of Australia, vol. 14, pp. 864–876. Perth.

McDade, P., 1999. An Experimental Study of the Products of Extreme Metamorphic Processes: Ultrahigh-temperature Granulites and the Roberts Victor Eclogites University of Edinburgh. 390 pp.

Menzies, A.H., Shirey, S.B., Carlson, R.W., Gurney, J.J., 1998. Re–Os isotope systematics of diamond bearing eclogites and peridotites from Newlands kimberlite. Ext. Abstr. 7th Internat. Kimberlite Conference, Cape Town, pp. 579–581.

Mørk, M.B.E., Brunfelt, A.O., 1988. Geochemical comparisons of coronitic olivine gabbro and eclogites: metamorphic effects and the origin of eclogite protoliths (Flemsøy, Sunnmøre, Western Norway). Nor. Geol. Tidsskr. 68, 51–63.

Mørk, M.B.E., Mearns, E.W., 1986. Sm–Nd isotopic systematics of a gabbro-eclogite transition. Lithos 19, 255–267.

Neal, C.R., Taylor, L.A., Davidson, J.P., Holden, P., Halliday, A.N., Nixon, P.H., Paces, J.B., Clayton, R.N., Mayeda, T.K., 1990. Eclogites with oceanic crustal and mantle signatures from the Bellsbank kimberlite, South Africa, Part 2: Sr, Nd, and O isotope geochemistry. Earth Planet. Sci. Lett. 99, 362–379.

O'Hara, M., 1969. The origin of eclogite and ariégite nodules in basalt. Geol. Mag. 106, 322–330.

O'Hara, M.J., Yoder, H.S., 1967. Formation and fractionation of basic magmas at high pressure. Scott. J. Geol. 3, 67–117.

O'Hara, M.J., Saunders, M.J., Mercy, E.L.P., 1975. Garnet–peridotite, primary ultrabasic magma and eclogite; interpretation of upper mantle processes in kimberlite. Phys. Chem. Earth 9, 571–604 (First international conference on kimberlites).

Pearson, D.G., Shirey, S.B., Carlson, R.W., Boyd, F.R., Pokhilenko, N.P., Shimizu, N., 1995a. Re–Os, Sm–Nd, and Rb–Sr isotope

evidence for thick Archean lithospheric mantle beneath the Siberian craton modified by multistage metasomatism. Geochim. Cosmochim. Acta 59, 959–978.

Pearson, D.G., Snyder, G.A., Shirey, S.B., Taylor, L.A., Carlson, R.W., Sobolev, N.V., 1995b. Archaean Re–Os age for Siberian eclogites and constraints on Archaean tectonics. Nature 374, 711–713.

Pearson, N.J., O'Reilly, S.Y., Griffin, W.L., 1995c. The crust–mantle beneath cratons and craton margins: a transect across the southwest margin of the Kaapvaal craton. Lithos 36, 257–287.

Pearson, N.J., Griffin, W.L., Doyle, B.J., O'Reilly, S.Y., van Achterbergh, E., Kivi, K., 1999. Xenoliths from kimberlite pipes of the Lac de Gras area, Slave Craton, Canada. In: Gurney, J.J., Gurney, J.L., Pascoe, M., Richardson, S.H. (Eds.), Proceedings of the VIIth International Kimberlite Conference, Cape Town, pp. 644–658.

Pollack, H., Chapman, D., 1977. On the regional variation of heat flow, geotherms, and lithospheric thicknesses. Tectonophysics 38, 279–296.

Rapp, R.P., Watson, E.B., 1995. Dehydration melting of metabasalt at 8–32 kbar. Implications for continental growth and crust–mantle recycling. J. Petrol. 36, 891–931.

Ringwood, A., Green, D., 1966. An experimental investigation of the gabbro-eclogite transformation and some geophysical implications. Tectonophysics 3, 383–427.

Robey, J.V., 1981. Kimberlites of the Central Cape Province, RSA. PhD thesis, University of Cape Town. 261 pp.

Roden, M.F., Lazko, E.E., Jagoutz, E., 1999. The role of garnet pyroxenites in the Siberian lithosphere: evidence from the Mir kimberlite. In: Gurney, J.J., Gurney, J.L., Pascoe, M., Richardson, S.H. (Eds.), Proc. 7th International Kimberlite Conference. Red Roof Design cc, Cape Town, pp. 714–720.

Rudnick, R.L., 1995. Eclogite xenoliths: samples of Archean ocean floor. Ext. Abstr. 6th Intern. Kimberlite, 473–475.

Rudnick, R.L., Barth, M., Horn, I., McDonough, W.F., 2000. Rutile-bearing refractory eclogites: missing link between continents and depleted mantle. Science 287, 278–281.

Schmickler, B., Jacob, D.E., Foley, S.F., 2004. Eclogite xenoliths from the Kuruman kimberlites, South Africa: geochemical fingerprinting of deep subduction and cumulate processes. Lithos (in press).

Schulze, D.J., 1989. Constraints on the abundance of eclogite in the upper mantle. J. Geophys. Res. 94 (B4), 4205–4212.

Schulze, D.J., Valley, J.W., Spicuzza, M., 2000. Coesite eclogites from the Roberts Victor kimberlite, South Africa. Lithos 54, 23–32.

Schulze, D.J., Harte, B., Valley, J.W., Brenan, J.M., Channer, D.M.D.R., 2003a. Extreme crustal oxygen isotope signatures preserved in coesite in diamond. Nature 423, 68–70.

Schulze, D.J., Valley, J.W., Viljoen, K.S., Spicuzza, M.J., 2003b. Oxygen isotope composition of mantle eclogites. Ext. Abstr. 8th Internat. Kimberlite Conference.

Shee, S.R., Gurney, J.J., 1979. The mineralogy of xenoliths from Orapa, Botswana. In: Boyd, F.R., Meyer, H.O.A. (Eds.), The Mantle Sample: Inclusions in Kimberlites and Related Rocks. AGU, Washington, pp. 37–49.

Shee, S.R., Bristow, J.W., Bell, D.R., Smith, C.B., Allsopp, H.L.,

Lo, C.H., 1989. The petrology of kimberlites, related rocks and associated mantle xenoliths from the Kuruman Province, South Africa. In: Ross, J., Jaques, A.L., Ferguson, J., Green, D.H., O'Reilly, S.Y., Danchin, R.V., Janse, A.J.A. (Eds.), Kimberlites and Related Rocks. Spec. Publ. - Geol. Soc. Australia, vol. 14, pp. 60–82. Perth.

Smith, C.B., et al., 1989. Sr and Nd isotopic systematics of diamond-bearing eclogitic xenoliths and eclogite inclusions in diamond from southern Africa. In: Ross, J., Jaques, A.L., Ferguson, J., Green, D.H., O'Reilly, S.Y., Danchin, R.V., Janse, A.J.A. (Eds.), Kimberlites and Related Rocks. Special Publication - GSA, vol. 14, pp. 853–863. Perth.

Snyder, G.A., Taylor, L.A., Crozaz, G., Halliday, A.N., Beard, B.L., Sobolev, V.N., Sobolev, N.V., 1997. The origins of Yakutian eclogite xenoliths. J. Petrol. 38, 85–113.

Sobolev, N.V., 1977. Deep-seated inclusions in kimberlites and the problem of the composition of the upper mantle. American Geophysical Union, Washington, DC. 279 pp.

Sobolev, N.V., Lavrentev, Y.G., 1971. Isomorphic sodium admixture in garnets formed at high pressure. Contrib. Mineral. Petrol. 31, 1–12.

Sobolev, N.V., Pokhilenko, N.P., Lavrentev, Y.G., Yefimova, E.S., 1977. Deep-seated xenoliths, xenocrysts in kimberlites and crystalline inclusions in diamonds from "Udachnaya" kimberlite pipe, Yakutia. Ext. Abstr. 2nd Internat. Kimberlite Conference: Unpaginated.

Sobolev, V.N., Taylor, L.A., Snyder, G.A., 1994. Diamondiferous eclogites from the Udachnaya kimberlite pipe, Yakutia. Int. Geol. Rev. 36, 42–64.

Spear, F.S., 1993. Metamorphic Phase Equilibria and Pressure-Temperature-Time Paths. Min. Soc. America Monograph, Washington, DC. 799 pp.

Stacey, J.S., Kramers, J.D., 1975. Approximation of terrestrial lead isotope evolution by a two-stage model. Earth Planet. Sci. Lett. 26, 207–221.

Stiefenhofer, J., Viljoen, K.S., Marsh, J.S., 1997. Petrology and geochemistry of peridotite xenoliths from the Letlhakane kimberlites, Botswana. Contrib. Mineral. Petrol. 127 (1–2), 147–158.

Stosch, H.-G., Lugmair, G.W., 1990. Geochemistry and evolution of MORB-type eclogites from the Münchberg massif, southern Germany. Earth Planet. Sci. Lett. 99, 230–249.

Taylor, L.A., Neal, C.R., 1989. Eclogites with oceanic crustal and mantle signatures from the Bellsbank kimberlite, South Africa. Part I: mineralogy, petrography and whole rock chemistry. J. Geol. 97 (5), 551–567.

van der Hilst, R.D., 1995. Complex morphology of subducted lithosphere in the mantle beneath the Tonga trench. Nature 374, 154–157.

van Orman, J.A., Grove, T.L., Shimizu, N., Layne, G.D., 2002. Rare earth element diffusion in a natural pyrope single crystal at 2.8 GPa. Contrib. Mineral. Petrol. 142 (4), 416–424.

Viljoen, K.S., 1994. The petrology and geochemistry of a suite of mantle-derived eclogite xenoliths from the Kaal-vallei kimberlite, South Africa. PhD thesis, Witwatersrand, Johannesburg.

Viljoen, K.S., Smith, C.B., Sharp, Z.D., 1996. Stable and radiogenic isotope study of eclogite xenoliths from the Orapa kimberlite, Botswana. Chem. Geol. 131, 235–255.

Woodland, A.B., Seitz, H.M., Altherr, R., Marschall, H., Olker, B., Ludwig, T., 2002. Li abundances in eclogite minerals; a clue to crustal or mantle origin? Contrib. Mineral. Petrol. 143 (5), 587–601.

Yaxley, G.M., Green, D.H., 1998. Reactions between eclogite and peridotite: mantle refertilisation by subduction of oceanic crust. Schweiz. Mineral. Petrogr. Mitt. 78, 243–255.

Yui, T.F., Rumble, D., Chen, C.H., Lo, C.H., 1997. Stable isotope characteristics of eclogites from the ultra-high-pressure metamorphic terrain, east-central China. Chem. Geol. 137 (1–2), 135–147.

Available online at www.sciencedirect.com

www.elsevier.com/locate/lithos

Lithos 77 (2004) 317–332

A study of eclogitic diamonds and their inclusions from the Finsch kimberlite pipe, South Africa

C.M. Appleyard*, K.S. Viljoen, R. Dobbe

De Beers GeoScience Centre, P.O. Box 82232, Southdale 2135, South Africa

Received 21 June 2003; accepted 12 December 2003
Available online 1 June 2004

Abstract

Previous studies of diamonds from Finsch have shown that eclogitic inclusions are rare at Finsch and that the eclogitic garnet and clinopyroxenes are iron and manganese-rich. In order to expand the current database of information, 93 eclogitic diamonds were selected for this study. Eight diamonds were polished into plates for cathodoluminescence studies and infrared examination of diamond growth and 31 diamonds were cracked to retrieve inclusions. The eclogitic garnets analysed in this study are enriched in Fe and are relatively depleted in Ca and Mg relative to worldwide data. FeO contents for garnet range from 15 to 27 wt.% and MnO contents reach a maximum value of 1.6 wt.%. The eclogitic clinopyroxenes have relatively high FeO contents, up to 14.8 wt.% and K_2O contents are low (<0.4 wt.%). Three non-touching garnet–clinopyroxene mineral pairs produce equilibration temperatures of 1138–1179 °C at an assumed pressure of 50 kb. No Type II diamonds were found during this study, all diamonds are of Type IaAB. Total nitrogen contents of Type IaAB diamonds range from 11 to 1520 ppm, with variable aggregation states (up to 84% nitrogen aggregated as B-defects). Distinct infrared characteristics suggest that the Finsch kimberlite sampled either more than one mantle source region of similar age but differing temperature, or two different populations of diamonds with different ages. The diamonds provide evidence of changing mantle conditions during crystallisation. Continuous diamond growth is illustrated by the presence of regular octahedral growth zones, although in some diamonds cubic growth is noted. One diamond shows evidence of platelet degradation, suggesting exposure to high temperatures and/or shearing stresses.
© 2004 Elsevier B.V. All rights reserved.

Keywords: Proterozoic; Dodecahedra; Deformation; Type IaAB; Platelet degradation

1. Introduction

Minerals occurring as inclusions in diamond will have experienced the same physical conditions as the diamond and, due to the chemically inert nature of

diamond, are protected from further changes within the mantle. Although rare, such inclusions provide vital geological information with regards to the formation of diamond in the mantle as well as mantle dynamics and mantle chemistry within cratonic root zones (Richardson et al., 1984). Two dominant parageneses of diamond inclusions have been identified, the peridotitic and eclogitic suites (Meyer, 1987). Peridotitic inclusion assemblages typically include minerals such as olivine, chromite, orthopyroxene

---

* Corresponding author. Tel.: +27-11-374-7906; fax: +27-11-835-1315.
E-mail address: clare.appleyard@debeersgroup.com (C.M. Appleyard).

and purple (chrome pyrope) garnet whereas eclogitic assemblages are predominantly comprised of orange pyrope-almandine garnet and pale green omphacitic clinopyroxene. A third, less commonly encountered websteritic paragenesis is also recognised, as is the case at Orapa (Gurney et al., 1984).

The Finsch kimberlite is of the Group II variety (Smith, 1983) with an estimated Rb–Sr pipe emplacement age of 118 Ma (Smith et al., 1985). The inclusion suite of small diamonds of approximately 2 mm diameter at Finsch is dominated by inclusions of the peridotitic paragenesis (Gurney et al., 1979; Harris and Gurney, 1979), with eclogitic inclusions accounting for approximately 2% of the inclusion population. Peridotitic diamonds at Finsch crystallised at a depth of 150–200 km (Boyd et al., 1985) and have an Archean model age of 3.3 Ga (Richardson et al., 1984). In contrast, eclogitic diamonds are significantly younger at 1580 ± 50 Ma (Richardson et al., 1990). In addition, Smith et al. (1991) reported Nd model ages for eclogitic garnets in large single diamonds that range from 1443 to 2408 Ma.

During some of the early work on diamonds (Robertson et al., 1934), defects in diamonds were recognised and impurities such as N, H, O and B were analysed. More recently, diamonds have been classified as Type I (nitrogen-bearing) or Type II (nitrogen-free) varieties (Allen and Evans, 1981), on the basis of their infrared spectroscopy. Nitrogen in Type Ib diamonds occurs as isolated substitutional nitrogen atoms (typical of synthetic diamonds). In Type Ia diamonds, the nitrogen is present as clusters of two atoms substituting for a carbon atom (Type IaA) or four or more clusters of nitrogen substituting for carbon in the highly aggregated Type IaB diamonds (Jones et al., 1992). The aggregation of nitrogen within natural diamond is a kinetic phenomenon whereby the degree of aggregation of the nitrogen atoms depends on factors such as the mantle residence time, diamond nitrogen content and mantle residence temperature (Evans and Qi, 1982; Evans and Harris, 1989; Taylor et al., 1996). With increasing time and/or temperature, nitrogen aggregation in diamond proceeds from C-centres (singly substituted nitrogen atoms) in Type Ib diamonds, to A-centres (nitrogen pairs) in Type IaA diamond through to B-centres (tetrahedrally arranged nitrogen atoms) in Type IaB diamond (Kiflawi and Bruley, 2000). Most natural diamonds with aggregat-

ed nitrogen also contain platelets; these are possibly clusters of carbon and/or nitrogen atoms (Allen and Evans, 1981) several atoms thick (Woods, 1986). A study of 50 diamonds by Woods (1986) showed that typically a positive correlation exists between the amount of nitrogen present in a diamond and the area of platelets per unit volume. Such diamonds were termed 'regular' as opposed to 'irregular' diamonds where such a correlation does not exist.

It is possible to estimate diamond mantle residence temperatures and mantle residence times (e.g. Evans and Harris, 1989; Taylor et al., 1990; Richardson and Harris, 1997) based on nitrogen contents and nitrogen aggregation states. From this, one can deduce the thermal evolution of continental lithosphere (Mendelssohn and Milledge, 1995; Taylor et al., 1990).

2. This study

Previous studies of inclusions in diamond from the Finsch Mine concentrated on the peridotitic inclusion suite, with only six eclogitic garnets and four eclogitic clinopyroxenes analysed by Gurney et al. (1979), and with the diamonds subsequently studied for their carbon isotopic composition and diamond nitrogen content by Deines et al. (1989). Smith et al. (1991) published mineral chemistry and isotopic data for an additional five large (2–4 carats) eclogitic diamonds from Finsch.

In view of this comparatively small data set for eclogitic inclusions at Finsch, it was considered desirable to expand the size of the eclogitic inclusion database. Furthermore, the studies of both Gurney et al. (1979) and Smith et al. (1991) reported surprisingly high iron and manganese contents in eclogitic garnet and clinopyroxene inclusions in Finsch diamonds and an expansion of the data set would validate these findings. Infrared studies on eclogitic diamonds from Finsch have also been rare, with only 10 nitrogen contents and aggregation states having been published by Deines et al. (1989). A detailed study of growth zonation in association with nitrogen and cathodoluminescence studies has not, until now, been undertaken on Finsch eclogitic diamonds. Hence, a total of 131 inclusion-bearing diamonds were collected from the Finsch run-of-mine production. These range in size from 0.03 to 0.650 carats. From this collection, a subset

Table 1
Details of mineral inclusions recovered in the present study

Minerals recovered	Number of inclusions
Diamonds cracked	31
Orange garnets	26
Eclogitic clinopyroxene	18
Sulphide	1
Coesite (with included cpx)	1
Garnet–clinopyroxene pairs	3

of 93 diamonds with eclogitic inclusions such as orange garnet and/or pale green clinopyroxene were studied visually and analysed by infrared spectroscopy. Eight of the larger diamonds were polished into plates for cathodoluminescence studies and infrared examination of diamond growth, and a further 31 diamonds were cracked to retrieve 26 orange garnets and 18 pale green clinopyroxenes (Table 1). In addition, a single large sulphide as well as one coesite inclusion (with a tiny clinopyroxene included in the coesite host) was also recovered from the sample set.

3. Methods

Inclusions were released by crushing the diamonds in a stainless steel cracker. These were then mounted in brass stubs using epoxy resin and polished down to a 1-μm finish. Electronprobe microanalyses were performed with a wavelength-dispersive spectrometer-equipped Cameca SX-50 operated at an acceleration potential of 20 kV and a probe current of 40 nA. Counting times were 20 s for elements Mg, Cr, Fe, Mn, Ti, Al, Ca and Si and 40 s for Na, P, K and Ni. K-α lines were used for all elements. Apparent concentrations were corrected for matrix effects with the on-line PAP computer program. MgO (Mg), Cr_2O_3 (Cr), Fe_3O_4 (Fe), $MnTiO_3$ (Mn, Ti), Ni metal (Ni), almandine (Al), orthoclase (K), albite (Na), diopside (Ca, Si) and apatite (P) were used as standards.

Nitrogen contents and aggregation states were determined by transmission infrared spectroscopy using a Nicolet 760 Magna-IR infrared spectrometer linked to a Nicplan infrared microscope. The signal was recorded with a liquid nitrogen cooled MCT-A detector. Spectra were acquired in transmission mode by signal-averaging 50 scans with a resolution of 8 cm^{-1} using a fixed 100-μm aperture. Infrared

detection limits are dependent on the spectral quality with poor spectral quality resulting in higher detection limits. For most of the Finsch diamonds, the detection limits are approximately 50 atomic ppm. However, very high quality spectra were obtained on some of the polished diamond plates and on these samples nitrogen concentrations of < 50 ppm were successfully measured.

Nitrogen contents and the degree of nitrogen aggregation were determined by deconvolution of the infrared spectra in the 1400–1000-cm^{-1} region into components generated by the different centres (Davies, 1981), utilising a least squares approach and a similar methodology to that outlined in Taylor et al. (1990). Spectra were base-lined before processing by taking a straight line between 4000 and 650 cm^{-1}. Nitrogen contents were calculated from the infrared absorption coefficients of Boyd et al. (1994, 1995), i.e. 16.5 and 79.4 atomic ppm ccm for IaA and IaB nitrogen, respectively.

The internal structure of the polished diamond plates was investigated using a Cambridge Instruments CITL CCL 8200MK3 Cold Cathode Luminescence instrument. Accelerating voltages of 15–18 kV of electron energy and an electron beam current of 1.2 mA were used. Specimens were mounted on a thick aluminium block with conductive colloidal graphite. Cathodoluminescence images were recorded using a Leica DC200 digital camera mounted on a Leica Wild M10 stereomicroscope.

4. Results

A summary of the physical and infrared characteristics of the Finsch diamonds is presented in Table 2.

4.1. Diamond characteristics

The majority (51%) of the Finsch eclogitic diamonds examined are colourless, with the remainder of the population consisting of yellow and brown diamonds. Dodecahedra are the dominant diamond morphology, accounting for 87% of the population. Only 8% of the sample suite can be considered main form octahedra. A small proportion of diamonds (5%) is fragments. Octahedral surfaces show common surface features such as trigons, and hillocks are common on

Table 2
Physical and infrared characteristics of the diamonds

Sample	Col	Shp	Incl	Type	NA	NB	NT	%B
F004	Co	Do	Cpx, S	IaAB	509.6	178.7	688.3	26.0
F005	Co	Do	Og	IaAB	591.8	269.8	862.5	31.3
F006	Ye	Do	Og	IaAB	143.0	282.8	425.9	66.4
F007	Ye	Do	Og, S	IaAB	203.9	666.8	870.7	76.6
F008	Br	Do	Cpx, S	IaAB	97.6	7.7	105.3	7.3
F009	Ye	Do	Og	IaAB	207.0	680.4	887.4	76.7
F010	Co	Do	Og, S	IaAB	258.5	9.5	268.0	3.6
F011	Ye	Do	Og	IaAB	186.2	663.9	850.1	78.1
F012	Co	Fr	Og	IaAB	8.9	47.1	56.0	84.1
F013	Br	Do	Og	IaAB	507.8	860.3	1368.2	62.9
F014	Br	Do	Og	<50	–	–	>18	–
F015	Ye	Do	Og	IaAB	193.1	561.0	754.1	74.4
F016	Co	Do	Cpx, S	IaAB	232.2	1141.8	1374.0	83.1
F017	Co	Do	Og	<50	–	–	>46	–
F018	Co	Do	Og	IaAB	652.9	117.2	770.2	15.2
F019	Co	Do	Og, S	IaAB	603.8	124.9	728.7	17.1
F020	Ye	Do	Og	ND	–	–	–	–
F021	Br	Do	Og	IaAB	564.4	497.5	1061.9	46.8
F022	Co	Fr	Og	IaAB	202.6	700.2	902.8	77.6
F023	Co	Do	Og	IaAB	509.0	150.5	659.5	22.8
F024	Co	Do	Og, S	ND	–	–	–	–
F025	Br	Do	Cpx, S	<50	–	–	>15	–
F026	Ye	Do	Cpx, S	IaAB	201.9	752.6	954.5	78.9
F027	Co	Fr	Og	IaAB	134.9	203.9	338.9	60.2
F029	Ye	Do	Og, S	IaAB	162.6	389.5	552.1	70.5
F030	Co	Fr	Og, S	IaAB	513.6	151.5	665.2	22.8
F031	Co	Do	Cpx	IaAB	430.1	11.3	441.4	2.6
F032	Co	Do	Og, S	IaAB	243.8	23.5	267.3	8.8
F033	Co	Fr	Og, S	IaAB	501.2	188.8	689.9	27.4
F035	Br	Do	Cpx	IaAB	365.7	366.0	731.6	50.0
F036	Br	Do	Og, S	IaAB	24.1	30.4	54.4	55.8
F037	Co	Do	Og	IaAB	336.3	105.3	441.7	23.8
F038	Ye	Do	Og, S	ND	–	–	–	–
F040	Br	Do	Og	IaAB	67.8	23.8	91.6	26.0
F041	Br	Do	Og, S	IaAB	201.9	40.5	242.5	16.7
F042	Ye	Do	Og, S	IaAB	255.6	615.5	871.1	70.7
F043	Br	Do	Og	IaAB	271.8	191.7	463.4	41.4
F044	Co	Do	Og, S	ND	–	–	–	–
F045	Br	Do	Og	<50	–	–	>16	–
F046	Ye	Do	Og	IaAB	260.6	1260.1	1520.7	82.9
F047	Br	Do	Og, cpx, S	IaAB	59.4	24.0	83.3	28.8
F048	Ye	Do	Og	IaAB	212.5	496.9	709.4	70.0
F052	Br	Do	Og	IaAB	118.3	71.7	190.0	37.7
F053	Ye	Do	Og, S	IaAB	184.9	650.6	835.5	77.9
F054	Co	Oc	Og	IaAB	350.3	3.4	353.7	1.0
F055	Br	Do	Og, S	ND	–	–	–	–
F056	Co	Do	Og	ND	–	–	–	–
F057	Ye	Do	Og, S	IaAB	233.2	733.8	967.1	75.9
F058	Co	Do	Og, S	IaAB	165.9	383.8	549.7	69.8
F059	Co	Do	Og, cpx	IaAB	348.7	171.7	520.4	33.0
F060	Co	Do	Og	IaAB	274.1	310.3	584.4	53.1
F061	Co	Do	Og	IaAB	60.6	71.9	132.6	54.3
F062	Co	Oc	Og	IaAB	406.9	38.1	445.1	8.6

Table 2 (*continued*)

Sample	Col	Shp	Incl	Type	NA	NB	NT	%B
F063	Br	Do	Cpx	IaAB	692.8	81.9	774.7	10.6
F065	Ye	Do	Og, S	IaAB	252.4	851.7	1104.1	77.1
F066	Co	Do	Og, S	IaAB	35.6	16.2	51.8	31.3
F067	Co	Do	Cpx	IaAB	502.7	55.7	558.4	10.0
F068	Br	Do	Og	ND	–	–	–	–
F071	Co	Do	Og	IaAB	527.5	232.9	760.5	30.6
F072	Co	Do	Og, S	ND	–	–	–	–
F073	Co	Oc	Og, S	IaAB	568.2	62.4	630.6	9.9
F074	Br	Oc	Og, cpx	IaAB	67.8	58.7	126.6	46.4
F075	Co	Do	Og	IaAB	520.4	197.3	717.6	27.5
F076	Co	Do	Og	IaAB	436.2	218.9	655.1	33.4
F077	Co	Oc	Og, S	IaAB	420.0	94.1	514.0	18.3
F078	Br	Do	Og, S	IaAB	98.8	65.0	163.8	39.7
F079	Co	Do	Cpx	IaAB	150.7	248.6	399.3	62.3
F080	Br	Do	Og, co, cpx	IaAB	205.6	82.2	287.8	28.6
F082	Br	Do	Og, S	<50	–	–	>15	–
F083	Br	Do	Og, cpx	IaAB	90.9	139.0	229.8	60.5
F084	Co	Do	Og	IaAB	−198.6	702.7	901.3	78.0
F085	Co	Do	Og	ND	–	–	–	–
F086	Co	Do	Og	IaAB	395.2	108.7	503.9	21.6
F087	Br	Do	Cpx, S	IaAB	149.1	81.1	230.3	35.2
F088	Br	Do	Og	IaAB	55.4	100.8	156.2	64.5
F089	Co	Do	Cpx	IaAB	274.7	464.4	739.2	62.8
F090	Co	Do	Og, cpx, S	IaAB	154.9	376.7	531.5	70.9
F092	Co	Do	Og, cpx, S	IaAB	64.1	1.8	65.9	2.8
F093	Co	Oc	Og	IaAB	474.6	132.8	607.4	21.9
F094	Ye	Do	Og, S	IaAB	190.5	620.6	811.0	76.5
F095	Ye	Do	Og	IaAB	722.9	144.7	867.7	16.7
F096	Ye	Do	Og, cpx, S	IaAB	200.0	677.0	877.0	77.2
F097	Ye	Do	Og, S	ND	–	–	–	–
F098	Co	Do	Og, S	IaAB	626.1	189.4	815.5	23.2
F099	Ye	Do	Cpx, S	IaAB	151.1	339.6	490.7	69.2
F101	Co	Do	Cpx	IaAB	538.7	221.0	759.7	29.1
F102	Co	Do	Og	IaAB	139.6	289.9	429.5	67.5
F122	Co	Do	Cpx	IaAB	–	–	519.9	39.4
F127	Co	Do	Og	IaAB	–	–	951.0	35.9
F128	Co	Do	Og	IaAB	–	–	973.1	35.2
F129	Br	Do	Og	IaAB	–	–	859.6	75.4
F130	Co	Do	Og	IaAB	–	–	755.2	29.7
F131	Co	Do	Og	IaAB	–	–	639.9	21.5

Col = Colour; Shp = Shape; Incl = inclusion. Co = colourless; Do = dodecahedron; Og = orange garnet. Br = Brown; Oc = octahedron; Cpx = pale clinopyroxene. Ye = yellow; Fr = fragment; S = Sulphide. Co = coesite. Type = Diamond type. ND = non-determinate (unable to deconvolute data). <50 = low nitrogen diamonds (<50ppm N). NA = ppm Nitrogen in A-centres. NB = ppm Nitrogen in B-centres. NT = Total nitrogen content in ppm (NA + NB). %B = N aggregation. Please note that for diamonds F122–F131, whole stone IR analyses were not available, so the NT and %B values presented are an average of values obtained during IR traverses.

dodecahedral surfaces. Lamination lines have been noted and are evidence of plastic deformation of the diamonds.

4.2. Cathodoluminescence

Seven eclogitic diamond plates from Finsch have been examined in detail for their growth and infrared characteristics and the cathodoluminescence images are presented in Fig. 1. Most of the polished diamond plates show regular, octahedral growth zones, such as those illustrated by diamond F127 (Fig. 1). Black octahedral growth zones indicate the presence of low luminescence Type II material in the diamonds. Associated with the orange garnets in some diamonds (e.g. F126, F127) are distinct black haloes where the luminescence has been quenched. Milledge et al. (1989) suggested that the presence of aluminium in inclusions such as garnet may getter nitrogen, leading to non-luminescent areas surrounding the inclusion. A core–rim traverse of diamond F127 (Fig. 2) indicates that the lowest total nitrogen contents are found in the zones of Type II material, scattered throughout the diamond. Although no whole stone Type II diamonds were found, zones of Type II material are common in the diamonds. Due to the fact that FTIR analyses are taken over areas of ~ 100 μm and that growth zones are fairly small, analyses represent an average of several growth zones. As a consequence, analysis of a thin Type II zone will not yield nitrogen-free results, but instead the total nitrogen content will be contaminated by the concurrent analysis of the surrounding Type IaAB material. The highest total nitrogen contents are found in areas intermediate to the core and rim, where totals exceeding 1100-ppm nitrogen are noted. These areas exhibit the brightest blue cathodoluminescence colours. Nitrogen aggregation is found to be highest in the core of the diamond, with a uniform decrease in the degree of aggregation towards the rims of the diamond. This suggests that the rim of the diamond is much younger, or formed at much cooler temperatures than the core of the diamond. Possible evidence of plastic deformation, appearing in cathodoluminescence images as closely spaced yellow slip planes, is visible in one diamond, F126 (Fig. 1).

A more complex growth pattern is illustrated by diamond F129 (Fig. 3). Here, a central zone of hummocky cubic growth is noted, with late stage octahedral growth visible on the corners of the diamond. The octahedral growth material is clearly younger and cuts into the original cubic material of the diamond. Traverses across the diamond illustrate that the highest degree of nitrogen aggregation in the diamond is found in the cubic core of the diamond, with a progressive drop-off in aggregation state towards the octahedral rim of the diamond. Total nitrogen contents are found to be highest in the core of the diamond, with lower nitrogen contents in the darker octahedral growth zones where some black Type II diamond is present. Growth patterns for the eclogitic Finsch polished plates are consistent with a growth history of decreasing nitrogen availability during the growth event.

Smooth core to rim decreases of total nitrogen content and aggregation state may be an artefact of the analytical method. If the diamonds have clear core–rim contacts, infrared spectroscopy will measure a gradual zonation across the contacts, even if there is a sharp boundary between the two (Fitzsimmons et al., 1999). Cathodoluminescence images typically show small-scale zonation which are generally too small to be resolved by the infrared analysis technique. Both Fitzsimmons et al. (1999) and Harte et al. (1999) have illustrated that small-scale zonation is associated with changes (often large) in nitrogen content. This small-scale zoning coupled with the infrared analysis technique generates a three-dimensional effect that may be responsible for the apparent gradual decrease in nitrogen content and nitrogen aggregation state between the core and rim.

The Finsch eclogitic diamond plates from this study can all be considered "regular" after Woods (1986), with the exception of F129 which shows both "regular" and "irregular" areas. Examination of the infrared traverses of F129 (Fig. 3), in combination with the cathodoluminescence image and Fig. 9, shows that the central, cubic growth zone area (A'–B') of F129 is platelet degraded (or irregular), whereas the younger octahedral growth zones are regular.

4.3. Mineral chemistry

Garnet and clinopyroxene are the dominant eclogitic inclusions in this suite of Finsch diamonds,

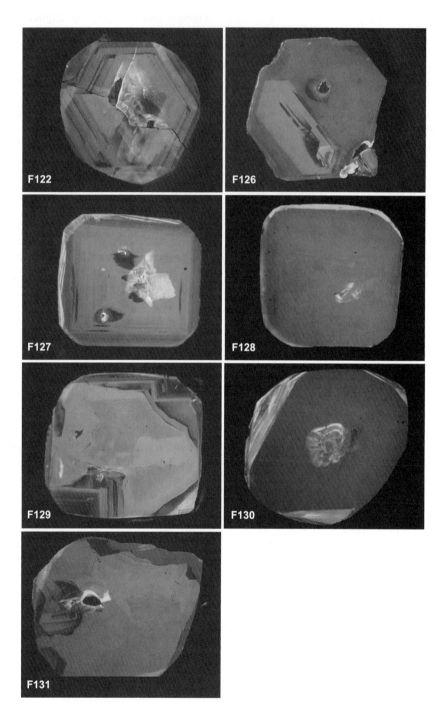

Fig. 1. Cathodoluminescence images of the seven eclogitic diamond plates from Finsch. Most of the diamonds, with the exception of F129 and F131, exhibit regular octahedral growth patterns. Black patches of Type II material are visible in many of the diamonds.

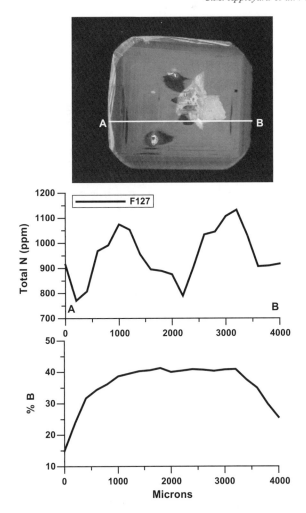

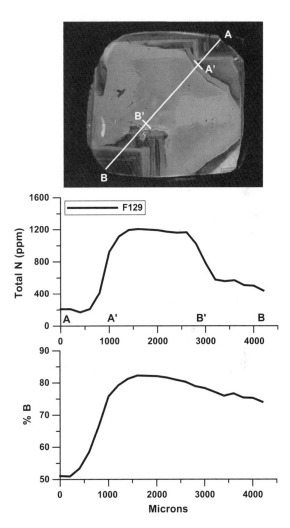

those reported by Smith et al. (1991) (maximum FeO = 25 wt.%) and Gurney et al. (1979) (maximum FeO = 22.6 wt.%). MgO and CaO contents vary between 8–15.6 and 3.1–11.5 wt.%, respectively, and these maximums are also higher compared to the MgO and CaO contents measured by Gurney et al. (1979) and Smith et al. (1991). Al_2O_3 contents have a

Fig. 2. Cathodoluminescene image for Finsch diamond F127, showing regular octahedral growth zones. A sulphide rosette in the plate is visible as a bright white colour on this black and white image. Black haloes are seen surrounding the orange garnet inclusions, where the luminescence has been quenched by the Al content of the inclusion. Plotted below the CL image are the variations in total nitrogen and degree of aggregation for the A–B traverse. A regular decrease in nitrogen aggregation can be seen from the core to the rim of the diamond. A dip in the total nitrogen traverse can be seen in the centre of the diamond where there is dilution by Type II material.

Fig. 3. Finsch diamond plate F129 exhibiting both hummocky (cubic) growth and regular octahedral growth. The core of the diamond is represented by a "maltese cross". Late stage growth on the diamond resulted in octahedral growth layers and two periods of growth are represented here. The traverses show distinctly higher nitrogen contents in the hummocky core of the diamond, and lower nitrogen contents in the octahedral growth zones at the corners. Aggregation is also noticeably higher in the core of the diamond.

however, analyses of sulphide and coesite have also been obtained. Garnet and clinopyroxene analyses are presented in Tables 3 and 4, respectively. Finsch garnets are enriched in Fe and are relatively depleted in Ca and Mg (Fig. 4). FeO contents for garnet range from 15 to 27 wt.% and these are higher values than

Table 3
Major element mineral composition for eclogitic garnets from Finsch diamonds

	F5A	F5B	F5C	F9A	F12A	F23A	F23B	F27A	F29A	F29B	F29C	F33B	F33C	F37A	F37B	F46A	F53A	F59B	F59C	F71B	F77A	F77B	F90A	F92C	F96A	F96B
SiO_2	38.83	37.97	39.17	39.30	40.66	38.85	39.02	39.60	37.63	38.70	38.28	38.01	38.29	38.47	38.63	38.74	39.13	39.34	39.10	38.25	38.80	39.75	38.40	40.01	38.22	38.30
TiO_2	0.34	0.32	0.33	0.31	0.49	0.35	0.37	0.32	0.35	0.38	0.35	0.32	0.29	0.27	0.31	0.13	0.38	0.59	0.61	0.40	0.29	0.31	0.42	0.37	0.34	0.37
Al_2O_3	21.10	21.58	21.32	21.45	22.25	21.22	21.09	21.01	22.16	21.93	22.10	21.98	21.09	22.02	21.99	22.04	21.59	21.47	21.47	21.06	22.64	22.69	20.83	22.88	21.05	20.97
Cr_2O_3	0.04	0.04	0.03	0.08	0.02	0.05	0.05	0.05	0.05	0.05	0.07	0.05	0.05	0.06	0.07	0.13	0.07	0.03	0.05	0.05	0.02	0.02	0.08	0.09	0.08	0.10
FeO	24.08	23.91	23.95	22.35	15.14	24.03	22.31	22.49	22.49	22.53	22.55	25.47	25.49	22.09	21.63	22.67	21.98	20.95	20.72	27.21	17.00	16.84	26.94	16.84	25.62	25.70
MnO	0.92	0.92	0.97	1.09	0.27	0.95	0.70	0.95	0.92	0.95	0.92	1.25	1.04	1.04	1.01	1.61	0.92	0.33	0.34	1.31	0.41	0.38	1.36	0.37	1.43	1.37
MgO	8.97	9.19	9.09	10.69	12.46	8.63	8.14	8.58	9.29	9.32	9.36	9.14	8.74	9.31	9.24	11.89	9.61	8.09	8.10	7.98	9.05	9.23	7.95	15.61	8.19	8.24
CaO	5.25	5.19	5.19	4.45	9.00	5.62	7.49	6.52	6.20	6.12	6.11	4.00	3.99	6.18	6.88	3.16	6.38	9.35	9.43	3.88	11.52	11.44	4.10	4.23	4.53	4.54
Na_2O	0.13	0.14	0.14	0.15	0.20	0.13	0.14	0.17	0.13	0.17	0.15	0.16	0.15	0.11	0.14	0.11	0.17	0.18	0.19	0.14	0.13	0.11	0.16	0.11	0.15	0.15
K_2O	0.00	0.00	0.00	0.00	0.01	0.00	0.00	0.00	0.00	0.00	0.00	0.00	0.00	0.00	0.00	0.00	0.00	0.00	0.00	0.00	0.00	0.00	0.00	0.00	0.00	0.00
NiO	0.02	0.03	0.02	0.01	0.00	0.04	0.05	0.01	0.00	0.01	0.01	0.01	0.01	0.03	0.03	0.02	0.02	0.01	0.00	0.01	0.02	0.01	0.01	0.01	0.01	0.00
P_2O_5	0.05	0.03	0.07	0.02	0.05	0.04	0.05	0.06	0.05	0.09	0.04	0.04	0.05	0.07	0.08	0.02	0.08	0.08	0.05	0.02	0.23	0.10	0.04	0.02	0.03	0.04
Total	99.73	99.32	100.27	99.90	100.55	99.86	99.35	99.76	99.27	100.25	99.94	100.44	99.40	99.66	100.02	100.52	100.32	100.43	100.07	100.31	100.11	100.88	100.29	100.53	99.65	99.78

Table 4
Major element mineral composition for eclogitic clinopyroxenes from the Finsch diamonds

	F17A	F17B	F59A	F63A	F63B	F67A	F67B	F73A	F80A	F80B	F80C	F89A	F90B	F90C	F92D	F92E	F99A
SiO_2	54.61	55.45	53.64	54.93	54.15	54.41	52.93	54.90	53.83	54.05	53.80	54.64	54.11	53.82	55.40	55.12	54.22
TiO_2	0.57	0.56	0.58	0.41	0.44	0.26	0.26	0.26	0.42	0.43	0.39	0.44	0.46	0.42	0.49	0.49	0.31
Al_2O_3	8.58	8.68	10.70	9.65	10.03	5.44	5.47	8.71	6.38	6.22	6.24	11.81	8.17	7.96	8.25	7.89	6.17
Cr_2O_3	0.11	0.12	0.03	0.07	0.06	0.25	0.23	0.14	0.14	0.13	0.14	0.06	0.09	0.07	0.13	0.14	0.29
FeO	5.27	5.2	6.24	7.00	6.79	13.85	14.81	5.56	6.02	6.10	6.23	6.89	11.86	12.25	5.51	5.48	8.16
MnO	0.10	0.10	0.04	0.27	0.27	0.58	0.59	0.09	0.07	0.07	0.07	0.18	0.37	0.38	0.10	0.10	0.31
MgO	11.53	11.64	7.95	9.30	9.53	11.85	10.98	11.86	12.55	12.40	12.49	7.82	8.16	8.27	11.14	11.27	13.05
CaO	12.60	12.54	12.54	11.87	11.69	9.09	9.00	12.54	15.46	15.71	15.76	9.34	10.01	10.04	12.62	13.01	11.71
Na_2O	4.99	5.19	5.59	5.37	5.17	3.85	3.78	4.77	3.34	3.40	3.32	7.01	5.34	5.24	5.36	5.11	3.80
K_2O	0.13	0.14	0.00	0.34	0.42	0.12	0.12	0.13	0.14	0.14	0.15	0.18	0.43	0.42	0.09	0.09	0.33
NiO	0.02	0.02	0.03	0.07	0.07	0.06	0.08	0.02	0.04	0.04	0.03	0.06	0.09	0.07	0.03	0.03	0.10
P_2O_5	0.01	0.02	0.24	0.05	0.04	0.02	0.03	0.03	0.02	0.08	0.05	0.01	0.04	0.03	0.04	0.03	0.03
Total	98.52	99.73	97.59	99.32	98.66	99.77	98.28	99.02	98.41	98.78	98.66	98.42	99.12	98.96	99.16	98.75	98.48

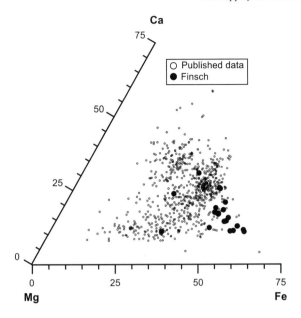

Fig. 4. Ternary Ca–Mg–Fe diagram for the Finsch eclogitic garnet inclusions. Also shown is data for world-wide localities (Sobolev et al., 1976; Gurney et al., 1979; Tsai et al., 1979; Gurney et al., 1985; Moore, 1986; Wilding, 1990; Jaques et al., 1989; Rickard et al., 1989; Kirkley et al., 1991; Smith et al., 1991; Chinn, 1995; Kopylova et al., 1997; Stachel and Harris, 1997; Wang, 1998; McDade and Harris, 1999; Richardson et al., 1999; Viljoen et al., 1999).

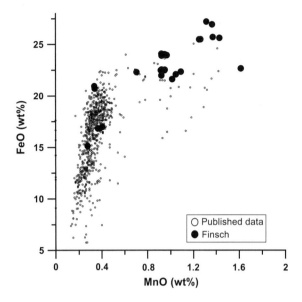

Fig. 5. Distribution of FeO and MnO contents in eclogitic garnet inclusions from world-wide localities and Finsch. The high FeO and MnO contents of the Finsch inclusions are clearly obvious. References as in Fig. 1.

restricted range (as expected) between 20.8 and 22.8 wt.%, and Na_2O contents are low (0.11–0.2 wt.%). These Al_2O_3 and Na_2O contents are similar to those previously reported (Gurney et al., 1979; Smith et al., 1991). However, MnO contents are high (0.7–1.6 wt.%) and these Finsch garnets represent some of the highest MnO contents thus measured in diamond inclusions (Fig. 5). Garnets from the same diamond are generally uniform in composition.

Finsch eclogitic clinopyroxenes also have relatively high FeO contents, from 5.2 to 14.8 wt.% (Fig. 6). CaO contents vary between 9 and 15.7 wt.% (lower than Gurney et al., 1979), whereas MgO contents have a more restricted range from 7.8 to 13 wt.% (higher than Gurney et al., 1979). Na_2O, MgO and Al_2O_3 contents are intermediate compared to worldwide localities. Na_2O contents range from 3.3 to 7 wt.% and Al_2O_3 contents vary from 5.4 to 11.8 wt.%. These maximums are both higher than those recorded by Gurney et al. (1979). K_2O contents in the clinopyroxene inclusions are also relatively low compared to the worldwide database, reaching a maximum of 0.4 wt.%. The

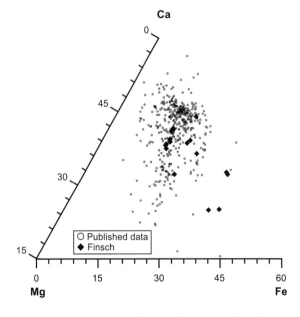

Fig. 6. Ca–Mg–Fe ternary diagram for eclogitic clinopyroxenes from Finsch and world-wide localities. The relatively higher proportions of Fe in the Finsch inclusions are clear. Other localities with high proportions of Fe include Jagersfontein and Monastery. References as in Fig. 1.

clinopyroxene inclusion in the coesite shows similar mineral chemistry to the other inclusions, but has a lower Na_2O content (3.3 wt.%) and a higher CaO content (15.5 wt.%). Clinopyroxenes from the same diamond are mostly uniform in composition.

The single sulphide retrieved (intact) from diamond F92 contains a significant amount of Ni (8.7%) although it co-exists with an orange garnet inclusion of definite eclogitic paragenesis. This sulphide consists predominantly of Fe (49.8 wt.%) and contains minor concentrations of Co (0.5 wt.%) and Cu (0.9 wt.%).

No significant correlations can be found between the nitrogen characteristics of the Finsch diamonds and the inclusion mineral chemistry. However, a broad positive correlation is noted between the total nitrogen content of the host diamond and the MnO contents of garnet inclusions.

4.4. Inclusion thermometry

Three pristine (fresh) non-touching garnet–clinopyroxene pairs have been recovered from this particular Finsch sample suite. These three mineral pairs have been used to calculate temperatures of equilibration based on the Fe–Mg exchange thermometer of Ellis and Green (1979). Temperatures of 1138, 1159 and 1179 °C have been obtained for the Finsch garnet–clinopyroxene pairs at an assumed pressure of 50 kb (Fig. 7 and Table 5).

Equilibration temperatures from co-existing garnet–clinopyroxene pairs from Orapa (Gurney et al., 1984) are very similar to those temperatures obtained from Finsch inclusions (Fig. 7). Premier diamonds (Gurney et al., 1985) exhibit much higher temperatures, between 1200 and 1400 °C. Argyle eclogitic diamonds show the widest range in temperatures, and also the highest temperatures of equilibration, most likely due to its mobile belt setting (Jaques et al., 1989).

4.5. Infrared characteristics

No nitrogen-free Type II diamonds were found during this study, a result which correlates with that of Deines et al. (1989). Five diamonds contained nitrogen contents of less than 50 ppm, approximately the lower limit of detection of the spectrometer. Fifteen diamonds exhibited poor spectra that could

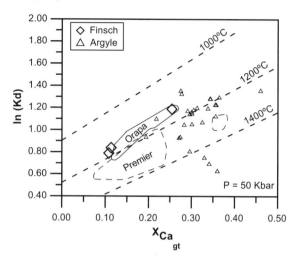

Fig. 7. Ln K_d vs. molar proportion of Ca in garnet for three garnet–clinopyroxene pairs from the Finsch diamonds. These pairs give temperatures ranging from 1138 to 1179 °C. Data from Orapa (Gurney et al., 1984), Argyle (Jaques et al., 1989) and Premier (Gurney et al., 1985) is plotted for comparative purposes. Equilibration temperatures of Finsch diamonds, based on garnet–clinopyroxene thermometry, are very similar to equilibration temperatures obtained for Orapa.

not be deconvoluted, and these are referred to as non-determinate (or ND) diamonds. The nitrogen contents of Type IaAB diamonds range from 11 to 1520 ppm, with an average value of 585 ppm. Nitrogen aggregation states vary from almost pure Type IaA diamond to highly aggregated Type IaAB diamond. The percentage of B-defects in the diamonds ranges from 1% to 84% aggregation.

Intra-grain variations in nitrogen characteristics have been noted in the eclogitic plates, with the largest intra-diamond variations found in sample F129. This is a complex diamond showing two stages of growth (refer to Section 4.2) and total nitrogen contents range from 209 to 1337ppm, whereas nitrogen aggregation states vary from 51% to 82% nitrogen aggregated as B-defects. Fig. 8 illustrates the variation in total nitrogen contents and aggregation states for the eclogitic Finsch diamonds. Isotherms on the diagram represent temperatures for an assumed mantle residence time of 1.5 Ga (Eclogitic diamond age of 1580 Ma; Richardson et al., 1990 minus kimberlite emplacement age of 118Ma; Smith et al., 1985). The diamonds in Fig. 8 can be broadly divided into two categories, those with aggregation states less

Table 5
Equilibration temperatures calculated from garnet–clinopyroxene pairs (Ellis and Green, 1979) at 50 kb

Sample	K_d	Ln K_d	X_{Ca}^{Gt}	T (°C)
F59	3.278	1.187	0.253	1138
F90	2.312	0.838	0.113	1159
F92	2.201	0.789	0.108	1179

than 40% and those with aggregation states greater than 60%. On the right-hand side of the diagram (%B>60%), a trend that broadly approximates an isothermal trend is visible in the vicinity of the 1150 and 1175 °C isotherms. This is best illustrated by the core-to-rim trend of decreasing total nitrogen with decreasing aggregation state in diamond F129. On the left-hand side of the diagram, a cluster of diamonds between the 1150 and 1050 °C isotherms show no isothermal trends at all, whether they be whole stones or infrared traverses.

The maximum intensity of the platelet peaks on the infrared spectra varies between 1359 and 1376 cm^{-1}. The platelet peak positions of 1359 cm^{-1} correspond to large platelets of >10 μm, whereas platelet peak positions of approximately 1374 cm^{-1} correspond to smaller platelets of <20 nm (Mendelssohn and Milledge, 1995). Fig. 9 illustrates the positive correlation of the area of platelet peaks (i.e. the density of platelets in a specific volume of diamond) with the nitrogen aggregation state, for the Finsch eclogitic diamond plates. These diamonds are thus "regular" after the classification of Woods (1986). Only one diamond is considered "irregular", diamond F129, although parts of F129 are also "regular", as discussed in Section 4.2.

Diamond mantle residence temperatures as a function of mantle residence time were calculated using the equation of Evans and Qi (1982). Here, t is the mantle residence time in seconds, C_0 is the total

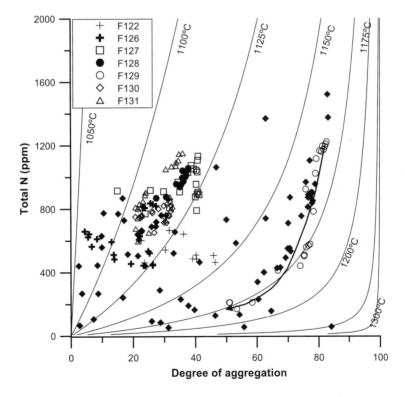

Fig. 8. Total nitrogen contents vs. degree of aggregation for eclogitic diamonds from Finsch. Infrared of whole stones is indicated by solid diamonds; each other individual symbol represents infrared traverses across diamond plates. Isotherms represent temperatures based on an assumed mantle residence time of 1.5 Ga. Highlighted by the arrow is the core–rim variation of total nitrogen and aggregation states (core to rim in direction of arrow) for sample F129, showing an isothermal trend.

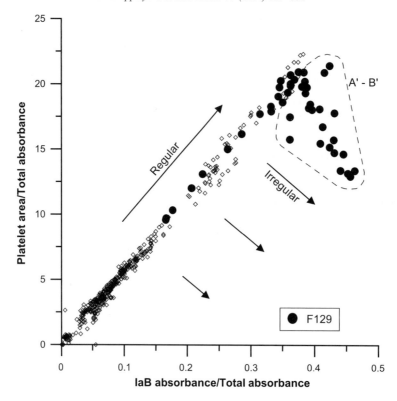

Fig. 9. Plot of the ratio of the platelet peak area to the total absorbance at the 1282 cm^{-1} peak position (i.e. equivalent to the platelet peak strength) vs. the ratio of the absorbance associated with B defects at the 1282 cm^{-1} peak position to the total absorbance at the 1282 cm^{-1} position (i.e. equivalent to increasing nitrogen aggregation state). It can be seen that all Finsch polished plates, with the exception of F129, can be considered regular (Woods, 1986).

nitrogen concentration of the diamond and C is the concentration of nitrogen remaining in A centres in the case of IaA to IaB aggregation:

$$T\ (^\circ C) = \frac{-E}{R}\ \frac{1}{ln\left[\dfrac{\left(\dfrac{C_0}{C}-1\right)}{C_0 tA}\right]} - 273.15$$

Isotherms and mantle residence temperatures were calculated using an activation energy (E) of 674.8 kJ mol^{-1} and an Arrhenius constant (A) of 2.68×10^5 s^{-1} ppm^{-1}. R is the gas constant with a recommended value of 8.31441 J K^{-1} mol^{-1} (Cohen and Taylor, 1973). Using this formula, at an assumed mantle residence time of 1.5 Ga, the Finsch eclogitic diamonds have mantle residence temperatures ranging between 1049 and 1251 °C. These results com-

pare favourably with mantle residence temperatures obtained from inclusion geochemistry (Table 6).

5. Discussion

The mineral chemistry results of this study confirm the earlier observations of Gurney et al. (1979) and Smith et al. (1991), that the eclogitic diamond para-

Table 6
Equilibration temperatures for the Finsch eclogitic diamonds, calculated using Ellis and Green, (1979) geothermometer (EG79) and the equation of Evans and Qi (1982), based on mantle residence time (1.5 Ga)

	EG79 (°C)	1.5 Ga (°C)
F59	1138	1130
F90	1159	1170
F92	1179	1111

genesis at Finsch represent a highly unusual, iron- and manganese-rich population. Such iron-rich compositions are extremely rare and have only been found before at Roberts Victor (Deines et al., 1987). At Roberts Victor, the eclogitic diamonds exhibit a bimodal distribution of $\delta^{13}C$, with values ranging from $-5.5\%o$ to very light isotopic compositions of $-16.3\%o$ (Deines et al., 1987). These very low isotopic compositions could be a result of organic carbon being the source of the Roberts Victor diamonds (Deines et al., 1987). In contrast, diamonds at Finsch with unusually Fe- and Mn-enriched garnet inclusions show a range in carbon isotopic compositions from $-3\%o$ to $-7.8\%o$ (Deines et al., 1984). These Finsch diamonds have values close to the $\delta^{13}C$ value of $5\%o$ which has been identified as the major isotopic signature for the mantle (Deines, 2002). Despite the similarities in inclusion compositions from Roberts Victor and Finsch, the isotopic compositions indicate different sources of carbon. Eclogitic garnet inclusions at Finsch show a positive correlation between MnO contents and Sr isotopic compositions; this prompted Smith et al. (1991) to postulate that these diamonds may have crystallised in a protolith consisting of altered, subducted oceanic crust.

Cathodoluminescence images as well as infrared zonation patterns of the diamonds are consistent with diamond growth under equilibrium conditions for most of their history. The variations in diamond growth zones, appearing as variations in the cathodoluminescence images, are directly associated with variations in the abundance of nitrogen, an observation noted by Harte et al. (1999). As was the case with their study, in this study the brightness of the blue CL is directly proportional to total nitrogen contents. Nitrogen contents are more consistent in cubic growth zones than they are in octahedral growth zones. It has been suggested that changes in nitrogen uptake during diamond growth may have a greater impact on octahedral growth zones, relative to cubic growth zones, as octahedral zones have a greater nitrogen uptake and slower growth rates than cubic zones (Sunagawa, 1984). Thin high nitrogen rims are observed on three of the diamonds (F122, F126, and F127); these rims are typically in a low state of nitrogen aggregation. This is most noticeable in diamond F127 (Fig. 2) where nitrogen contents rise distinctly at the outermost rim. This could result from diamond growth in a

cooler mantle environment relative to the bulk of the diamond, or may reflect a comparatively young diamond crystallisation event.

Evidence of two populations of diamonds is provided in Fig. 10, where it can be seen that there is essentially a bimodal distribution of mantle residence temperatures for the Finsch diamonds. Peak distribution of temperatures occurs at 1125 and 1175 °C. Correlation of this diagram with Fig. 8 illustrates that diamonds with residence temperatures of <1125 °C have nitrogen aggregation states <40%, whereas diamonds with residence temperatures >1150 °C mostly have aggregation states >50%. The two populations of diamonds that are present amongst this eclogitic suite may differ in age and/or mantle equilibration temperature.

This study confirms the results of Deines et al. (1989) in that there are no Type II macro-diamonds found at Finsch. However, many of the cathodoluminescence images (Fig. 1) show zones of low-luminescent (or even black) material, generally believed to be Type II material. Alternating zones of Type II and Type IaAB material in the diamonds imply increases and decreases in nitrogen abundances during growth of the diamonds. It suggests that there were distinct

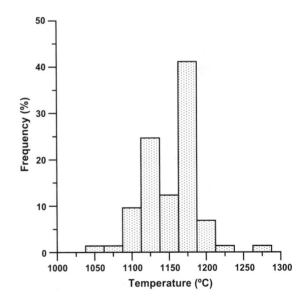

Fig. 10. Histogram of distribution of mantle residence temperatures for the Finsch eclogitic diamonds, based on an assumed mantle residence time of 1.5 Ga.

fluctuations in the availability of nitrogen from the fluid reservoir in which diamonds are believed to have grown (Harte et al., 1999). The fact that none of the diamonds consist completely of Type II material implies that the fluid reservoir was large enough to be unaffected by diamond crystallisation; diamond growth did not entirely drain the source of nitrogen. Instead, growth of the diamonds would have diminished nitrogen contents (leading to zones of Type II material), but replenishment of the reservoir would have occurred with further fluid influx, leading to renewed growth of Type IaAB material.

Mantle equilibration temperatures of 1111–1170 °C have been calculated for these Finsch diamonds (Table 6), at an assumed mantle residence time of 1.5 Ga. However, it should be remembered that during FTIR studies, the measured nitrogen abundances most likely represent the averages of several growth zones with differing total nitrogen contents. As a result of this, the accuracy of calculations of mantle residence times from aggregation states may be affected (Harte et al., 1999). It would thus be more reliable to use equilibration temperatures calculated from mineral inclusion pairs (1138–1179 °C) for accurate estimations of mantle equilibration temperatures. These equilibration temperatures are similar to the range of equilibration temperatures obtained from Finsch peridotitic inclusions, viz. 900–1250 °C (Harris, 1992).

Fig. 9 plots the ratio of platelet peak area to total absorbance ($=\mu_A + \mu_B + \mu_D$; where μ_A, μ_B and μ_D represent contributions to the lattice absorption at 1282 cm^{-1} of each of the A, B and D spectral components, respectively) vs. the ratio of μ_B to total absorbance. In Fig. 9, it is noticeable that all Finsch diamonds, with the exception of F129, can be termed "regular" after Woods (1986). F129 is a brown diamond with a complex, multistage growth history and sector zoning. The brown colour is evidence that the diamond has undergone shear stresses and has been plastically deformed. Close examination of the infrared data shows that the central, cubic growth region of the diamond is platelet degraded, with values plotting away from the linear trend. The octahedral growth zones consist of perfectly "regular" diamond that plots on the linear trend in Fig. 9. Platelet degradation is the collapse of the linear relationship between platelet peak area and aggregation state of nitrogen in a diamond.

It may thus be possible that the central zone of diamond F129 is older than the other Finsch diamonds examined, and may have been exposed to extreme temperatures or shearing stresses, which may lead to platelet degradation (Woods, 1986). Such events must have occurred prior to the start of octahedral diamond growth in this diamond. Thus, there have been at least two episodes of diamond growth in the lithosphere underlying Finsch, with probable heating and/or shearing events leading to platelet degradation in the initial diamond population. The fact that there is only one "irregular" diamond in the sample suite suggests that the processes typically associated with platelet degradation were not particularly active in the region of the mantle from which Finsch diamonds have been derived.

6. Conclusions

The major and minor element chemistry of eclogitic inclusions in diamonds from the Finsch Mine, analysed in the present study, mirror and extend the ranges reported in previous studies of this inclusion suite in that an unusually Fe- and Mn-rich protolith is represented, as was previously suggested and discussed by Smith et al. (1991). Additional studies of these inclusions and their host diamonds incorporating carbon and nitrogen isotopic compositions as well as REE contents and patterns are highly desirable.

The diamonds provide evidence of changing mantle conditions during their crystallisation, as evidenced by the range in aggregation states and total nitrogen contents, with a possibility of at least two populations of diamond being represented by the eclogitic suite. These may represent two diamond crystallisation events of differing age or, alternatively, derive from regions of the mantle of differing depth and hence different ambient temperatures. Additional evidence of a heating event or episode of shearing stress is provided by a platelet degraded, "irregular" diamond encountered.

Acknowledgements

The authors would like to thank the management of De Beers Consolidated Mines for the donation of study material and permission to publish. Verlece Anderson, Gill Parker, Dianne Hart, Edna van Blerk, Ray Ferraris, Ronel Hamman and Wanita Moore are thanked for the collection of inclusion-bearing diamonds. David Fisher of the Diamond Trading Company Research Laboratory in Maidenhead, England kindly provided his spectral deconvolution software.

References

Allen, B.P., Evans, T., 1981. Aggregation of nitrogen in diamond, including platelet formation. Proceedings of the Royal Society of London 375, 93–104.

Boyd, F.R., Gurney, J.J., Richardson, S.H., 1985. Evidence for a 150–200 km thick Archaean lithosphere from diamond inclusion thermobarometry. Nature 315, 387–389.

Boyd, S.R., Kiflawi, I., Woods, G.S., 1994. The relationship between infrared absorption and the A-defect concentration in diamond. Philosophical Magazine. B 69, 1149–1153.

Boyd, S.R., Kiflawi, I., Woods, G.S., 1995. Infrared absorption by the B-nitrogen aggregation in diamond. Philosophical Magazine. B 72, 351–361.

Chinn, I.L., 1995. A study of unusual diamonds from the George Creek K1 kimberlite dyke, Colorado. PhD thesis. University of Cape Town, South Africa.

Cohen, E.R., Taylor, B.N., 1973. The 1973 least-squares adjustment of the fundamental constants. Journal of Physical Chemistry Reference Data 2, 663–734.

Davies, G., 1981. Decomposing the IR absorption spectra of diamonds. Nature 290, 40–41.

Deines, P., 2002. The carbon isotope geochemistry of mantle xenoliths. Earth-Science Reviews 58, 247–278.

Deines, P., Gurney, J.J., Harris, J.W., 1984. Associated chemical and carbon isotopic composition variations in diamonds from Finsch and Premier kimberlite, South Africa. Geochimica et Cosmochimica Acta 48, 325–342.

Deines, P., Harris, J.W., Gurney, J.J., 1987. Carbon isotopic composition, nitrogen content and inclusion composition of diamonds from the Roberts Victor kimberlite, South Africa, evidence for ^{13}C depletion in the mantle. Geochimica et Cosmochimica Acta 51, 1227–1243.

Deines, P., Harris, J.W., Spear, P.M., Gurney, J.J., 1989. Nitrogen and ^{13}C content of Finsch and Premier diamonds and their implications. Geochimica et Cosmochimica Acta 53, 1367–1378.

Ellis, D.J., Green, D.H., 1979. Experimental study of the effect of Ca upon garnet–clinopyroxene Fe–Mg exchange equilibria. Contributions to Mineralogy and Petrology 71, 13–22.

Evans, T., Harris, J.W., 1989. Nitrogen aggregation, inclusion equilibration temperatures and the age of diamonds. In: Ross, J. (Ed.), Kimberlites and Related Rocks Vol. 2: Their Mantle/Crust Setting. Special Publication - Geological Society of Australia, vol. 14, pp. 1002–1006.

Evans, T., Qi, Z., 1982. The kinetics of the aggregation of nitrogen in diamond. Proceedings of the Royal Society of London 381, 159–178.

Fitzsimmons, I.C.W., Harte, B., Chinn, I.L., Gurney, J.J., Taylor, W.R., 1999. Extreme chemical variation in complex diamonds from George Creek, Colorado, a SIMS study of carbon isotope compositions and nitrogen abundance. Mineralogical Magazine 63, 857–878.

Gurney, J.J., Harris, J.W., Rickard, R.S., 1979. Silicate and oxide inclusions in diamonds from the Finsch kimberlite pipe. In: Boyd, F.R., Meyer, H.O.A. (Eds.), Kimberlites, Diatremes and Diamonds, their Geology, Petrology and Geochemistry. American Geophysical Union, Washington, pp. 1–15.

Gurney, J.J., Harris, J.W., Rickard, R.S., 1984. Silicate and oxide inclusions in diamonds from the Orapa Mine, Botswana. In: Kornprobst, J. (Ed.), Kimberlites II: The Mantle and Crust–Mantle Relationships. Developments In Petrology, vol. 11B. Elsevier, Amsterdam. 393 pp.

Gurney, J.J., Harris, J.W., Rickard, R.S., Moore, R.O., 1985. Inclusions in Premier mine diamonds. Transactions of the Geological Society of South Africa 88, 301–310.

Harris, J.W., 1992. Diamond geology. In: Field, J.E. (Ed.), The Properties of Natural and Synthetic Diamond. Academic Press, London, pp. 345–393.

Harris, J.W., Gurney, J.J., 1979. Inclusions in diamonds. In: Field, J.E. (Ed.), The Properties of Diamond. Academic Press, London, pp. 555–591.

Harte, B., Fitzsimmons, I.C.W., Harris, J.W., Otter, M.L., 1999. Carbon isotope ratios and nitrogen abundances in relation to cathodoluminescence characteristics for some diamonds from the Kaapvaal Province, South Africa. Mineralogical Magazine 63, 829–856.

Jaques, A.L., Hall, A.E., Sheraton, J.W., Smith, C.B., Sun, S.-S., Drew, R.M., Foudoulis, C., Ellingsen, K., 1989. Composition of crystalline inclusions and C-isotopic composition of Argyle and Ellendale diamonds. In: Ross, J., Jaques, A.L., Ferguson, J., Green, D.H., O'Reilly, S.Y., Danchin, R.V., Janse, A.J.A. (Eds.), Kimberlites and Related Rocks Vol. 2: Their Mantle/Crust Setting, Diamonds and Diamond exploration. GSA Special Publication 14, 966–989.

Jones, R., Briddon, P.R., Öberg, S., 1992. First-principles theory of nitrogen aggregates in diamond. Philosophical Magazine Letters 66, 67–74.

Kiflawi, I., Bruley, J., 2000. The nitrogen aggregation sequence and the formation of voidites in diamond. Diamond and Related Materials 9, 87–93.

Kirkley, M.B., Gurney, J.J., Rickard, R.S., 1991. Jwaneng framesites, Carbon isotopes and intergrowth compositions. Proceedings of the 5th International Kimberlite Conference Vol. 2. CPRM, Rio de Janeiro, pp. 127–135.

Kopylova, M.G., Gurney, J.J., Daniels, L.R.M., 1997. Mineral inclusions in diamonds from the River Ranch kimberlite, Zimbabwe. Contributions to Mineralogy and Petrology 129, 366–384.

McDade, P., Harris, J.W., 1999. Syngenetic inclusion bearing diamonds from Letseng-la-Terai, Lesotho. In: Gurney, J.J., Gurney, J.L., Pascoe, M.D., Richardson, S.H. (Eds.), Proceedings of the 7th International Kimberlite Conference Volume 2. Red Roof Design, Cape Town, pp. 557–565.

Mendelssohn, M.J., Milledge, H.J., 1995. Geologically significant information from routine analysis of the mid-infrared spectra of diamonds. International Geology Review 37, 95–110.

Meyer, H.O.A., 1987. Inclusions in diamond. In: Nixon, P.H. (Ed.), Mantle Xenoliths. Wiley, London, pp. 501–522.

Milledge, H.J., Mendelssohn, M.J., Boyd, S.R., Pillinger, C.T., van Heerden, L.A., Seal, M., 1989. IR, CL and MS data for Finsch diamonds and an Argyle stone exhibiting giant platelets. Diamond Conference Abstracts, Bristol. Unpublished.

Moore, R.O., 1986. A study of the kimberlites, diamonds and associated rocks and minerals from the Monastery Mine, South Africa. PhD thesis. University of Cape Town, South Africa.

Richardson, S.H., Harris, J.W., 1997. Antiquity of peridotitic diamonds from the Siberian craton. Earth and Planetary Science Letters 151, 271–277.

Richardson, S.H., Gurney, J.J., Erlank, A.J., Harris, J.W., 1984. Origin of diamonds in old enriched mantle. Nature 310, 198–202.

Richardson, S.H., Erlank, A.J., Harris, J.W., Hart, S.R., 1990. Eclogitic diamonds of Proterozoic age from Cretaceous kimberlites. Nature 346, 54–56.

Richardson, S.H., Chinn, I.L., Harris, J.W., 1999. Age and origin of eclogitic diamonds from the Jwaneng kimberlite, Botswana. In: Gurney, J.J., Gurney, J.L., Pascoe, M.D., Richardson, S.H. (Eds.), Proceedings of the 7th International Kimberlite Conference Vol 2. Red Roof Design, Cape Town, pp. 709–713.

Rickard, R.S., Harris, J.W., Gurney, J.J., Cardoso, P., 1989. Mineral inclusions in diamonds from Koffiefontein Mine. In: Ross, J., Jaques, A.L., Ferguson, J., Green, D.H., O'Reilly, S.Y., Danchin, R.V., Janse, A.J.A. (Eds.), Kimberlites and Related Rocks, Vol. 2: Their Mantle/Crust Setting, Diamonds and Diamond Exploration. GSA Special Publication, vol. 14, pp. 1054–1062.

Robertson, R., Fox, G.G., Martin, A.E., 1934. Two types of diamonds. Philosophical Transactions of the Royal Society 232 (719), 463–535.

Smith, C.B., 1983. Pb, Sr and Nd isotopic evidence for sources of southern African Cretaceous kimberlites. Nature 304, 51–54.

Smith, C.B., Allsopp, H.L., Kramers, J.D., Hutchinson, G., Roddick, J.C., 1985. Emplacement ages of Jurassic–Cretaceous South African kimberlites by the Rb–Sr method on phlogopite and whole-rock samples. Transactions of the Geological Society of South Africa 88, 249–266.

Smith, C.B., Gurney, J.J., Harris, J.W., Otter, M.L., Kirkley, M.B.,

Jagoutz, E., 1991. Neodymium and strontium isotope systematics of eclogite and websterite paragenesis inclusions from single diamonds, Finsch and Kimberley Pool RSA. Geochimica et Cosmochimica Acta 55, 2579–2590.

Sobolev, N.V., Lavrentyev, Y.G., Pokhilenko, N.P., Ponomarenko, A.I.G., Sobolev, V.S., 1976. Diamond-bearing grospidite and diamond-bearing disthene eclogites from the kimberlite pipe "Udachnaya", Yakutia. Doklady Akedemii Nauk 226, 927–930.

Stachel, T., Harris, J.W., 1997. Syngenetic inclusions in diamond from the Birim fields (Ghana)—a deep peridotitic profile with a history of depletion and re-enrichment. Contributions to Mineralogy and Petrology 127, 336–352.

Sunagawa, I., 1984. Morphology of natural and synthetic diamond crystals. In: Sunagawa, I. (Ed.), Materials Science of the Earth's interior. Terra Scientific Publishing, Tokyo, pp. 303–330.

Taylor, W.R., Jaques, A.L., Ridd, M., 1990. Nitrogen-defect aggregation characteristics of some Australasian diamonds, Time–temperature constraints on the source regions of the pipe and alluvial diamonds. American Mineralogist 75, 1290–1310.

Taylor, W.R., Canil, D., Milledge, H.J., 1996. Kinetics of Ib to IaA nitrogen aggregation in diamond. Geochimica et Cosmochimica Acta 23, 4725–4733.

Tsai, H.-M., Meyer, H.O.A., Moreau, J., Milledge, H.J., 1979. Mineral inclusions in diamond, Premier, Jagersfontein and Finsch kimberlites, South Africa and Williamson Mine, Tanzania. In: Boyd, F.R., Meyer, H.O.A (Eds.), Kimberlites, Diatremes and Diamonds, their Geology, Petrology and Geochemistry. American Geophysical Union, Washington, DC, pp. 16–26.

Viljoen, K.S., Phillips, D., Harris, J.W., Robinson, D.R., 1999. Mineral inclusions in diamonds from the Venetia kimberlites, Northern Province, South Africa. In: Gurney, J.J., Gurney, J.L., Pascoe, M.D., Richardson, S.H. (Eds.), Proceedings of the 7th International Kimberlite Conference Volume 2. Red Roof Design, Cape Town, pp. 888–895.

Wang, W., 1998. Formation of diamond with mineral inclusions of "mixed" eclogite and peridotite paragenesis. Earth and Planetary Science Letters 160, 831–843.

Wilding, M.C.O., 1990. A study of diamonds with syngenetic inclusions. PhD thesis, University of Edinburgh, Scotland.

Woods, G.S., 1986. Platelets and the infrared absorption of Type Ia diamonds. Proceedings of the Royal Society of London 407, 219–238.

Available online at www.sciencedirect.com

Lithos 77 (2004) 333–348

www.elsevier.com/locate/lithos

ELSEVIER

Nature of diamonds in Yakutian eclogites: views from eclogite tomography and mineral inclusions in diamonds

Mahesh Anand[a,*], Lawrence A. Taylor[a], Kula C. Misra[a],
William D. Carlson[b], Nikolai V. Sobolev[c]

[a] Planetary Geosciences Institute, Department of Geological Sciences, University of Tennessee, 306 GS Building Knoxville,
TN 37996-1440, USA
[b] Department of Geological Sciences, University of Texas, Austin, TX 78712, USA
[c] Institute of Mineralogy and Petrography, Russian Academy of Sciences, Novosibirsk 630090, Russia

Received 21 June 2003; accepted 11 November 2003
Available online 2 July 2004

Abstract

We have performed dissections of two diamondiferous eclogites (UX-1 and U33/1) from the Udachnaya kimberlite, Yakutia in order to understand the nature of diamond formation and the relationship between the diamonds, their mineral inclusions, and host eclogite minerals. Diamonds were carefully recovered from each xenolith, based upon high-resolution X-ray tomography images and three-dimensional models. The nature and physical properties of minerals, in direct contact with diamonds, were investigated at the time of diamond extraction. Polished sections of the eclogites were made, containing the mould areas of the diamonds, to further investigate the chemical compositions of the host minerals and the phases that were in contact with diamonds. Major- and minor-element compositions of silicate and sulfide mineral inclusions in diamonds show variations among each other, and from those in the host eclogites. Oxygen isotope compositions of one garnet and five clinopyroxene inclusions in diamonds from another Udachnaya eclogite (U51) span the entire range recorded for eclogite xenoliths from Udachnaya. In addition, the reported compositions of almost all clinopyroxene inclusions in U51 diamonds exhibit positive Eu anomaly. This feature, together with the oxygen isotopic characteristics, is consistent with the well-established hypothesis of subduction origin for Udachnaya eclogite xenoliths. It is intuitive to expect that all eclogite xenoliths in a particular kimberlite should have common heritage, at least with respect to their included diamonds. However, the variation in the composition of multiple inclusions within diamonds, and among diamonds, from the same eclogite indicates the involvement of complex processes in diamond genesis, at least in the eclogite xenoliths from Yakutia that we have studied.
© 2004 Elsevier B.V. All rights reserved.

Keywords: Eclogite; Xenolith; Diamonds; Diamond mineral inclusions; Udachnaya

1. Introduction

Significant advances in our understanding of mantle processes have resulted from the petrological studies of diamonds and their mineral inclusions. However, the physical and chemical conditions of

* Corresponding author. Present address: Department of Mineralogy, The Natural History Museum, Cromwell Road, London SW7 5BD, UK. Tel.: +44-207-942-6023; fax: +44-207-942-5537.
E-mail address: m.anand@nhm.ac.uk (M. Anand).

0024-4937/$ - see front matter © 2004 Elsevier B.V. All rights reserved.
doi:10.1016/j.lithos.2004.03.026

the growth of diamonds and their relationships to their host rock (mantle xenoliths) and its mineralogy are not fully understood. Almost all diamonds occur in kimberlites as loose crystals, being liberated from their mantle host rocks during the transport to the Earth's surface. In some instances diamonds arrive at the surface in their mantle hosts, such as peridotite and eclogite xenoliths. However, due to its friable nature, peridotites are usually disaggregated during their transport in the kimberlitic magma, and their diamonds are released into the kimberlite. In contrast, due to relatively greater resistance of constituent minerals, eclogite xenoliths, including the diamondiferous ones, commonly remain intact, thereby preserving the textural context in which their diamonds originally formed. Thus, studies of diamondiferous eclogite xenoliths provide a unique opportunity to understand the textural and spatial relationships between diamonds and their hosts.

For several years, our research group has been involved in the study of diamondiferous eclogite xenoliths from Yakutia, Russia. We have made significant progress in our endeavors to understand the origin of diamonds and their mineral inclusions (DIs) in relation to their host rocks, particularly from diamondiferous eclogites (e.g., Keller et al., 1999; Taylor et al., 2000). In this contribution, we present results of detailed dissections of two additional diamondiferous eclogites (UX-1 and U33/1) from the Udachnaya kimberlite. We also present oxygen isotope data on five clinopyroxene and one garnet inclusion in diamonds from U51 eclogite, also from Udachnaya kimberlite. As diamondiferous eclogites are rare, few detailed studies have been performed with an aim to understand the relationship between diamond and the host eclogite. Taylor et al. (2000) presented the first detailed dissection of a Yakutian eclogite (U51), in which they compared the geochemical characteristics of the host minerals to those of DIs. To gain a better understanding of the formation of Yakutian eclogites and their included diamonds, we have integrated into the present study some of the mineral-inclusion data of Taylor et al. (2000).

2. Samples and analytical techniques

The diamondiferous eclogite xenoliths were recovered from the "Udachnaya East" kimberlite pipe in Yakutia. This intrusion is part of the Daldyn–Alakit region of kimberlitic magmatism in the Aldan shield in Siberia and is one of the most studied kimberlites in Russia (e.g., Sobolev, 1977; Jerde et al., 1993; Jacob et al., 1994; Pearson et al., 1995; Snyder et al., 1993, 1997; Sobolev et al., 1998b, 1999). The Udachnaya kimberlite has an age of ~ 389 Ma (Snyder et al., 1993), although the Archean ages (2.9 ± 0.4 Ga, Pearson et al., 1995) of diamondiferous eclogite xenoliths from this intrusion have confirmed the antiquity of these samples.

High-resolution computed X-ray tomography (HRCXT) of two diamondiferous eclogites (UX-1 and U33/1) was performed at the University of Texas, Austin, following the procedure of Carlson and Denison (1992) and Denison and Carlson (1997a). A tungsten X-ray source of 100 kV voltage and 0.24 mA current was used for this purpose. The in-plane resolution is of the order of 100–200 μm. The "slice" thickness was 0.16 mm and the inter-slice distance was 0.12 mm, thereby allowing for some overlap between two consecutive X-ray "slices".

Major- and minor-element mineral compositions were determined using an automated Cameca SX-50 electron microprobe at the University of Tennessee. The standard ZAF (PAP) corrections were applied to the resulting dataset. Analytical conditions for silicates employed an accelerating voltage of 15 keV, a beam current of 20 nA, beam size of 1 μm, and 20 s counting times for all elements, except K in clinopyroxene and Na in garnet (60 s each). Analytical conditions for sulfides employed a beam current of 10 nA with variable beam sizes ranging from 2 to 5 μm depending upon the size of the analyzed grain. Counting time for Ni, S, Fe, and Cu (also Zn and Si in some cases) was 30 s and for Co it was 40 s.

Ion microprobe measurement of $^{18}O/^{16}O$ values of the mineral inclusions in the U51 diamonds were determined with a Cameca 4f at Oak Ridge National Laboratory, using a Cs^+ primary beam. Analytical details are described in Riciputi et al. (1998). REE data from U51 are used in the present work to highlight the nature of Eu anomaly in clinopyroxene inclusions that are probably related to the oxygen isotope characteristics of mineral inclusions from U51 diamonds.

3. Results and discussion

3.1. High-resolution computed X-ray tomography (HRCXT)

The industrial X-ray tomographic analyzer used in this study is conceptually similar to a medical CAT-scan, however, it is capable of significantly higher X-ray intensity and markedly higher spatial resolution, on the order of microns (Carlson and Denison, 1992; Carlson et al., 1995; Denison and Carlson, 1997a,b; Rowe et al., 1997).

The HRCXT technique provides a unique opportunity to investigate the spatial and textural relationships between diamonds and other minerals in the host eclogite by a non-destructive method. Application of computed tomography X-ray scans to diamondiferous eclogites was first demonstrated by Schulze et al. (1996) in which they successfully imaged several diamonds in an eclogite sample from Roberts Victor kimberlite in South Africa. The HRCXT images used in our studies were obtained as a series of 350–400 two-dimensional (2-D) X-ray "slices" each of a diamondiferous eclogite using a micro-focal X-ray source and an image-intensifier

detector system to measure the absorption of X-rays along numerous coplanar paths through the sample. Different minerals have different X-ray attenuation values, which primarily depends on the density and mean atomic number of the object. The higher the density and the mean atomic number, the higher the X-ray attenuation. In addition, the presence of micro- or macro-porosity and micro-crystallinity reduces X-ray attenuation. Thus, in the 2-D image, different minerals appear as different shades of color. We have assigned the brightest color to the phase with highest X-ray attenuation (sulfides) and the darkest color to the phase with lowest attenuation (diamonds). In this manner, it is easy to differentiate between diamonds, sulfides, silicates, and their alteration products in a xenolith (Fig. 1). Zones of secondary mineralization appear as dark thin lines on 2-D X-ray images presumably due to their porous nature, resulting in lower X-ray attenuation.

Several 2-D X-ray "slices" are stacked together using volume-visualization software to produce a three-dimensional (3-D) model of the xenolith. This essentially represents a density map of the sample, and provides sizes, shapes, textures, and locations of individual crystals that have dimensions larger than

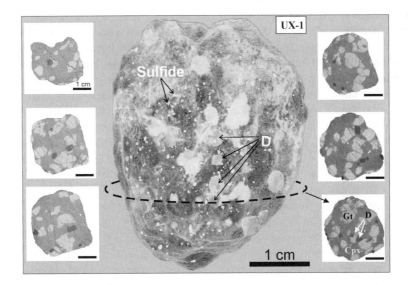

Fig. 1. (Center) A three-dimensional (3-D) model of diamondiferous eclogite xenolith, UX-1, based upon results of high-resolution X-ray tomography. Dark-gray grains are diamonds, and white specs are sulfide minerals distributed throughout the sample. Four diamonds are seen aligned parallel to the central vertical axis. Figures on the left and right of the central image are individual 2-D X-ray "slices" obtained by the HRCXT technique. The dotted girdle around the eclogite indicates the level from which the bottom-right 2-D X-ray "slice" was obtained. In these "slices", diamond appears black, garnet as light-gray, clinopyroxene as dark-gray, and sulfide minerals as white.

the spatial resolution of the scans. Such 3-D models reveal clearly the spatial relationships between diamonds and their surroundings, providing clues to the processes that control diamond crystallization. These relationships are determined by rotating and viewing the model at different perspectives to look for any mineral associations, alignments, or fabric. Fig. 1 is a three-dimensional HRCXT model of UX-1 eclogite xenolith. For clarity, the garnet and clinopyroxene grains are rendered invisible so that locations and sizes of diamonds can be easily seen. Only five diamonds were visible on the surface of the eclogite; an additional 69 macro-diamonds (≥ 1 mm) were located in the interior of this sample, using tomography (i.e., a total of 74 macro-diamonds in the 66 g UX-1 eclogite $\approx$ 144,000 carats/tonne). In general, diamonds are distributed evenly through the xenolith but in some cases, they seem to occur in linear alignment (Fig. 1). The even distribution of diamonds throughout the xenolith suggests that the diamonds could not have formed from an igneous melt (Schulze et al., 1996); instead metasomatic growth appears reasonable. On either side of the central image in Fig. 1, six of the total 360 individual 2-D X-ray "slices" are shown that were originally collected by the X-ray analyzer. The in-plane spatial resolution of this technique greatly aided in the imaging of macro diamonds (≥ 1 mm), and of some micro-diamonds (< 1 mm).

From the 2-D X-ray images, it is apparent that many diamonds are associated with alteration zones with prominent sub-planar fabric of secondary mineralization. These alteration zones appear prominently on 2-D HRCXT images due to their lower densities (Fig. 1). It is interesting to note that diamonds were never observed in direct contact or enclosed within fresh garnet or clinopyroxene. This would seem to indicate that diamonds are likely to have formed later than the primary silicates. Sulfide minerals are distributed evenly throughout the xenolith and are not preferentially associated with the diamonds. The suggestion that diamonds precipitated (crystallized) from an immiscible-sulfide melt (Bulanova, 1995; Spetsius, 1995) would require a close association between diamonds and significant quantities of sulfide minerals. The fact that the majority of DIs are sulfides, yet the diamonds are not preferentially associated with sulfides, would seem to negate the formation of

diamonds from an immiscible-sulfide melt, at least in the eclogites that we have studied.

3.2. Eclogite dissection

Based on the HRCXT imaging, each diamondiferous eclogite was dissected along selected planes to recover diamonds. The nature and physical properties of minerals, in direct contact with diamonds, were especially noted at the time of the diamond extraction. Subsequently, polished sections of the eclogites were prepared, containing the mould areas left by the extracted diamonds, to further investigate the mineralogy and composition of the host minerals that were in contact with the diamonds.

3.2.1. Textures and mineralogy of the host eclogites

Misra et al. (2004) have presented a detailed account of the texture and mineralogy of the primary and secondary mineral phases in the host eclogites, UX-1 and U33/1. For the sake of completeness, we will briefly review the important aspects of host mineral compositions with emphasis on the nature of diamond–silicate contact zone. The eclogites from the present study consist of almandine- and grossular-rich garnet, omphacitic pyroxene, and minor sulfide minerals (Fig. 2A). In terms of major-element compositions of clinopyroxene and garnet, the eclogites studied have been classified as Group B eclogites (Figs. 3 and 4). Sulfide grains occur throughout the xenoliths either as inclusions in clinopyroxene and garnet or interstitially. In both modes of occurrence, pentlandite exsolution occurs in pyrrhotite, mostly as oriented lamellar intergrowths. In some instances, minor chalcopyrite exsolution also occurs.

The primary garnet-omphacite assemblage of the eclogite has experienced multiple metasomatic events. The textural evidence for the metasomatism is seen in the form of a "spongy texture" (Fig. 2B), formed by alteration of the primary clinopyroxene (Cpx 1), kelyphitization of garnets, and the presence of phlogopite. Veins of secondary clinopyroxene crosscut the primary clinopyroxene grains (Fig. 2B). There is a general coarsening of grain size in secondary clinopyroxenes away from the primary clinopyroxene grain boundary, and this feature is commonly pronounced near the diamond–clinopyroxene contact (Fig. 2B). The secondary-mineral assemblage, formed from the

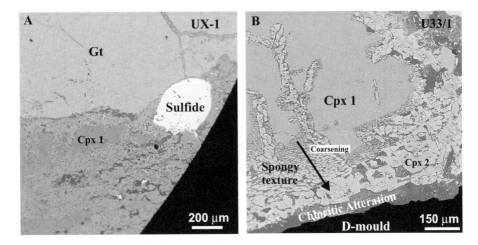

Fig. 2. (A) Back-scattered electron (BSE) image showing a sulfide grain enclosed within the clinopyroxene area in UX-1 eclogite. (B) BSE image showing the contact between the diamond mould and the host minerals in U33/1 eclogite. A zone of chloritic alteration forms a rim around the diamond in the eclogite. The diamond–silicate contact is well-defined, but ragged indicating resorption of diamond, perhaps by invading metasomatic fluids. The primary clinopyroxene (Cpx 1) is altered to Na-depleted, spongy textured clinopyroxene (Cpx 2) that tends to be coarser in grain size away from the unaltered clinopyroxene.

alteration of omphacite generally defines the diamond–silicate contact, which in most cases, is overprinted by chlorite, formed by low-temperature hydrothermal alteration. Secondary clinopyroxenes formed by the alteration of Cpx 1 are termed Cpx 2.

Similarly, clinopyroxenes formed by the alteration of garnets are termed Cpx 3, following the terminology of Misra et al. (2004). The major-element compositions of secondary clinopyroxenes are highly variable and are characterized by Na-depleted, Mg-enriched

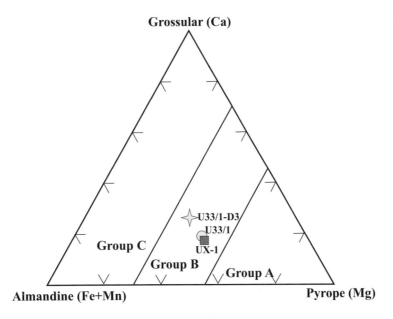

Fig. 3. The composition of U33/1-D3 garnet inclusion in diamond compared with the average compositions of garnets in xenoliths (U33/1 and UX-1 host data are from Misra et al., 2004). The eclogite group classification is adopted from Coleman et al. (1965).

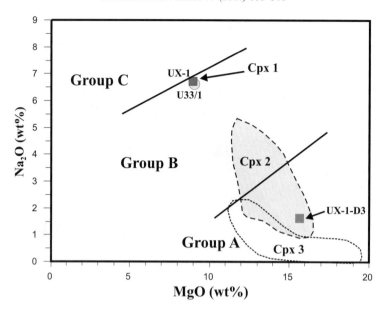

Fig. 4. MgO vs. Na$_2$O plot for clinopyroxene. The primary clinopyroxene (Cpx 1) plots in the Group B eclogite field of Taylor and Neal (1989). Secondary clinopyroxenes (Cpx 2 and Cpx 3) derived from alteration of Cpx 1 and garnet, respectively, are also plotted for comparison (data from Misra et al., 2004). Note that the inclusion composition in the UX-1-D3 diamond appears to represent a Cpx 2-type altered clinopyroxene.

compositions (Fig. 4). Other secondary minerals include spinel, which occurs exclusively in the kelyphitic rim of the garnet, amphibole, and K-rich glass that occurs in association with Cpx 2.

Diamonds are invariably associated with clinopyroxenes, rarely with garnet, being only separated by narrow alteration zones. Fresh clinopyroxene and garnet were never found in direct contact with diamond. It is not clear if this spatial relationship of diamond and clinopyroxene alteration zones has any specific genetic significance. However, it does appear as if the diamond-forming fluid selectively invaded the clinopyroxene grain boundaries. Alternatively, the diamond–clinopyroxene interfaces might have provided an increased permeability con-trast thereby promoting avenues for the invading metasomatic fluids.

3.2.2. Diamonds and their mineral inclusions

The majority of the extracted diamonds occur as perfect octahedra, others as cubo-octahedra, with well-developed crystal faces; some also occur as macles (twins). In one instance, a cluster of diamonds, weighing over 0.75 carat, was recovered, but due to the polycrystalline nature of this cluster, it was not possible to separate individual diamonds. This diamond cluster was dark-gray in color, whereas the color of the single recovered diamonds varied from colorless-to-yellowish white. The size and weight of the diamonds varied from <1 to 4 mm and 0.03 to 1 carat.

Fig. 5. (A) CL image of diamond U33/1-D6 showing multiple growth zones. Notice the resorption, termination, and regrowth of some of the diamond growth zones indicating a complex evolutionary history of this diamond. To the right, the two figures are enlargements of the two squares marked on image on the left. (B) CL image of diamond U33/1-D16 showing multiple growth zones. Notice the actual extent of the diamond marked by the white dashed line. Such dark CL response from outer zones of the diamond is probably due to absence of significant nitrogen aggregation, because these zones formed after the incorporation of the host xenolith into the kimberlite and not long before the kimberlite eruption. (C) BSE and CL image of the enlarged area in B marked by the square. A calcite grain was found enclosed in the outer portion of this diamond, consistent with the latest growth of diamond in a Ca-rich fluid environment. A crystal of chlorite nearby is probably related to a fracture, although no evidence for this was seen in the CL image.

From the UX-1 and U33/1 eclogites, a total of 30 diamonds were examined optically for visible mineral inclusions. In the majority of the cases, sulfide min-

erals are the most common DIs (Anand et al., 2001). These sulfide phases are invariably associated with radiating fractures—apparently not extending up to

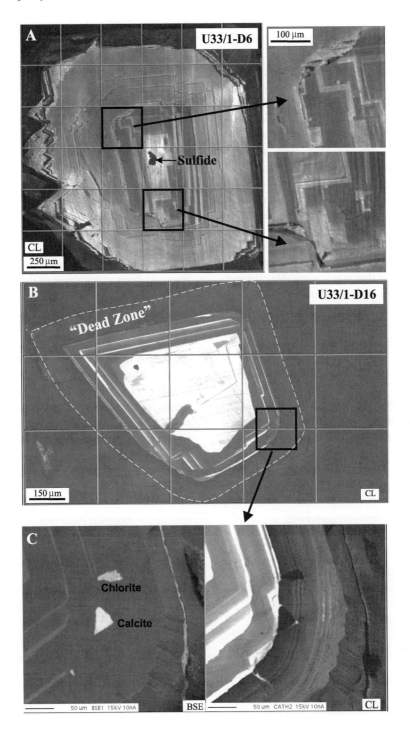

the diamond surface—and occur as thin films coating these fractures. The "smearing" of sulfides is a function of their plasticity at high temperatures (Pearson and Shirey, 1999). The association of sulfide minerals with fractures has been thought to be a response to stress cracking during differential thermal contraction of the inclusion and the host diamond during eruption (Meyer, 1987). In some cases, euhedral to subhedral sulfide inclusions were also noticed, displaying negative diamond crystal forms (Fig. 9A). In such cases, sulfide inclusions were exposed by polishing the host diamonds. The diamonds from UX-1 and U33/1 eclogites are unusually poor in macroscopic (>20 μm) silicate mineral inclusions, a stark contrast to U51 diamonds that were exceptionally rich in silicates (25 clinopyroxene and three garnet inclusions from eight diamonds). Only one garnet inclusion has been identified to date in a U33/1 diamond (U33/1-D3). A clinopyroxene inclusion was originally identified in the diamond UX-1-D3. Subsequent EMP analysis indicated that this is a low-Na clinopyroxene, comparable to many secondary clinopyroxenes in the host eclogite. In addition, this grain is located inside a healed crack extending up to the diamond surface, and thus, it not considered as a pristine mineral inclusion.

Polished surfaces of diamonds were subjected to detailed examination with cathodoluminescence, attached to Cameca SX-50 electron microprobe, to determine their growth patterns. Most of the diamonds have CL zonations recording their commonly torturous, and contorted growth histories. These zones have different CL colors, probably reflecting different nitrogen-aggregation states and nitrogen contents within the diamond structure. The degree of nitrogen aggregation in a diamond is a function of its residence time and temperature in the mantle (Mendelssohn and Milledge, 1995). At longer mantle residence times and at higher mantle temperatures, diamonds develop higher nitrogen-aggregation states and thereby show the brightest CL colors. In addition, the majority of the diamonds show hiatuses in their growth, which is obvious from the CL images, such as those shown in Fig. 5. In the case of U33/1-D6 (Fig. 5A), the CL pattern shows that the growth history of this diamond involved resorption and re-deposition, when the diamond was partially dissolved back into the fluids from which it may have originally grown. Continued new growth of diamond over resorbed zones was commonly accompanied by change in growth mode from cubic to octahedral. The outermost portions of many diamonds (e.g., U33/1-D16) show weak to no CL (Fig. 5B), termed "dead zones". Such dark CL response from outer zones of the diamond is probably due to absence of significant nitrogen aggregation, because these zones formed after the incorporation of the host xenolith into the kimberlite and not long before the kimberlite eruption. This is a signature commonly observed in the late-stage fibrous diamond growth. Based on CL patterns, this diamond appears to have experienced at least three distinct growth events (Fig. 5B). The central portion is the oldest portion of the diamond that grew in the mantle over a considerable period of time. As mentioned above, the outermost zone grew not long before kimberlite eruption, and the intermediate zones probably grew during

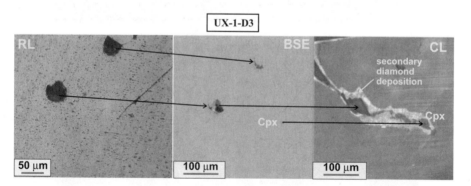

Fig. 6. Reflected-light (RL), BSE, and CL image of diamond UX-1-D3. In RL and BSE, no cracks are visible, whereas in the CL image, a clinopyroxene inclusion is seen clearly enclosed by a large healed fracture, which extends up to the surface of the diamond, confirming the open-system nature of this inclusion.

some time in between. Boundaries between these three zones show the cessation of growth, followed by resorption, with renewed growth. A calcite inclusion was found in the outermost dark CL region of this diamond, perhaps related to the latest growth of diamond in a Ca-rich fluid medium. Each diamond from a single eclogite appears to have specific CL features, depending upon the growth conditions. These observations are intriguing, but consistent with our previous suggestions that diamond growth is seldom simple and probably occurs over a significantly long geological time-period and under constantly changing fluid/melt compositions (Taylor et al., 2000; Anand et al., 2002).

Our CL studies have also demonstrated that some of the diamonds from the studied eclogites have minute, optically invisible cracks (Fig. 6), which extend from the DIs to the surfaces of the diamonds, thereby providing evidence of probable "open system" behavior of at least some of the DIs. For example, in UX-1-D3 diamond, EMP analysis of the clinopyroxene inclusion, located in the healed fracture, confirms the secondary nature of this inclusion (Fig. 4; Table 1). These optically invisible cracks are only discernible in CL images. This observation highlights the usefulness, indeed, the necessity for CL imaging in the study diamonds and diamond inclusions, as it can provide the stratigraphy of the

Table 1

Representative analysis of garnet, clinopyroxene, K-feldspar and sulfide mineral inclusions in diamonds from UX-1 and U-33/1 eclogites

Mineral	U33/1-D3	UX1-D3	UX-1-D1	wt.%	U33-Di19	U33-Di19	U33-Di19	U33-Di19	Ux1-Di5	Ux1-Di5	Ux1-Di12	Ux1-Di12	Ux1-Di13	Ux1-Di13
	Gt	Cpx	K-fel		Sulfide 1	Sulfide 1	Sulfide 2	Sulfide 2	Sulfide	Sulfide	Sulfide	Sulfide	Sulfide	Sulfide
Average	$n=5$	$n=9$	$n=3$	Type	Low-Ni	High-Ni	Low-Ni	High-Ni	Low-Ni	High-Ni	Low-Ni	High-Ni	Low-Ni	High-Ni
SiO_2	39.0	52.1	64.30	S	37.0	36.7	36.6	37.4	36.0	35.1	39.4	38.0	39.2	37.7
Al_2O_3	21.9	3.13	16.48	Fe	48.6	55.2	51.4	52.4	53.7	52.1	56.2	38.0	56.7	48.2
TiO_2	0.64	0.40		Co	0.09	0.10	0.20	0.25	0.45	0.48	0.08	1.23	0.05	0.58
Cr_2O_3	0.06	0.11		Ni	2.62	4.29	3.57	3.90	8.00	9.50	1.96	19.77	1.18	8.25
MgO	9.94	15.7		Cu	11.2	3.75	7.14	4.89	0.31	0.49	0.13	0.01	0.06	0.00
CaO	9.54	19.5	0.06	Zn	n.d.	n.d.	n.d.	n.d.	n.d.	n.d.	0.00	0.00	0.00	0.02
MnO	0.30	0.10		Si	n.d.	n.d.	n.d.	n.d.	n.d.	n.d.	0.32	0.24	n.d.	n.d.
FeO*	17.7	5.71	2.00	Total	99.5	100.0	98.9	98.9	98.5	97.7	98.0	97.2	97.2	94.8
Na_2O	0.23	1.64	0.18											
K_2O	n.d.	0.02	16.14	at%										
P_2O_5	0.05			S	51.39	50.47	50.98	51.79	50.27	49.60	53.84	53.10	54.12	53.69
NiO	0.02			Fe	38.71	43.63	41.13	41.66	43.07	42.34	44.06	30.50	44.91	39.42
Total	99.4	98.4	99.15	Co	0.07	0.08	0.16	0.19	0.34	0.37	0.06	0.93	0.03	0.45
				Ni	1.99	3.22	2.72	2.95	6.10	7.34	1.46	15.09	0.89	6.42
O Basis	12	6	8	Cu	7.85	2.61	5.02	3.42	0.22	0.35	0.09	0.01	0.04	0.00
Si	2.960	1.94	3.03	Zn	–	–	–	–	–	–	0.00	0.00	0.00	0.02
Al	1.960	0.14	0.91	Si	–	–	–	–	–	–	0.50	0.38	–	–
Ti	0.036	0.01		Total	100	100	100	100	100	100	100	100	100	100
Cr	0.003	0.00		S	37.2	36.7	37.0	37.8	36.6	35.9	40.3	39.2	40.4	39.8
Mg	1.123	0.87		Fe*	60.1	59.0	59.2	58.0	54.9	53.9	57.6	39.2	58.4	50.9
Ca	0.775	0.78	0.00	Ni*	2.7	4.4	3.8	4.2	8.6	10.2	2.1	21.6	1.3	9.3
Mn	0.019	0.00												
Fe	1.119	0.18	0.08											
Na	0.034	0.12	0.02											
K		0.00	0.97											
P	0.003													
Ni	0.001													
Total	8.03	4.04	5.01											
Mg#	50.08	83.03												
Or			98.07											

FeO*=all Fe reported as FeO; Mg#=cation $100*Mg/(Mg+Fe)$; Or=$100*K/(K+Na+Ca)$; Fe*=Fe* $(Fe+Cu)$; Ni*=Ni* $(Ni+Co)$.

diamonds, and can help to recognize open nature of the mineral inclusions in the diamonds. Such information can also be crucial to the isotopic studies of mineral inclusions in diamonds, particularly sulfide minerals.

An inclusion of K-feldspar was also found in the diamond UX-1D-1. The EMP analysis of this inclusion yields an almost pure orthoclase end-member (sanidine) composition of the alkali feldspar solid solution series. This is consistent with the reported occurrence of sanidine as inclusions in diamonds

(Meyer, 1987; Meyer and McCallum, 1986) from other locations. Inclusions of sanidine and calcite in the outer regions of some diamonds indicate involvement of K-rich melt/fluid in the formation of diamonds in these eclogites, at least during the latest diamond growth event.

The garnet inclusion in diamond U33/1-D3 exhibits well-developed cubo-octahedral morphology (Fig. 7A) and is orange-red in color, characteristic of almandine-rich eclogitic garnets. The CL image of this diamond also shows complex growth zones,

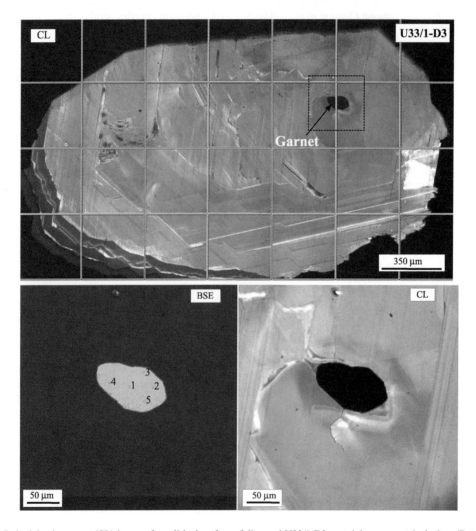

Fig. 7. (Top) Cathodoluminescence (CL) image of a polished surface of diamond U33/1-D3 containing a garnet inclusion. (Bottom) BSE and CL image of the area shown by dashed square in the top figure. Numbers on BSE image indicates the location of spots for EMP analysis. The CL image shows a dark halo around the garnet inclusion. Evidence for some healed fractures are seen in the vicinity of the garnet, but these fractures do not appear to reach the surface of the diamond.

similar to other diamonds from this eclogite. A dark halo around the garnet DI is present in the CL image; although such haloes are commonly observed around DIs, their origin is not known. The CL image of the diamond area containing the garnet inclusion also shows the presence of some healed fractures, but they do not extend to the surface of the diamond (Fig. 7B). This garnet inclusion is Ca-rich and Mg-poor, compared to the garnets in the host eclogite U33/1 (Fig. 3, Table 1). Such compositional differences between silicates inside the diamonds and in the host eclogites have also been reported previously (Taylor et al., 2000). The discordance between the inclusion composition and the host mineral composition is in agreement with the opinion that mineral inclusions in diamonds have remained closed since their encapsulation, whereas the host has undergone multi-stage metasomatic processes. Extreme chemical variability among multiple diamond mineral inclusions has also been recorded previously (Sobolev et al., 1998a; Taylor et al., 2000). Such variations in mineral inclusion compositions have been interpreted as reflecting the encapsulation of DIs at various stages of metasomatic diamond growth.

Clinopyroxene inclusions from U51 diamonds studied by Taylor et al. (2000) were also different in terms of their rare-earth-element (REE) compositions. Especially, almost all of those clinopyroxene inclusions also showed positive Eu anomalies. This compositional diversity is highlighted for three inclusions in Fig. 8A. At relatively low oxygen fugacity, divalent Eu behaves like Sr, being partitioned into plagioclase in preference to other REEs and substituting for Ca in the crystal structure. Rocks derived from the fractionation of a basaltic magma may develop positive or negative Eu anomaly depending upon the presence or absence of plagioclase. Since plagioclase is only stable at lower pressures (10–15 kbar), such Eu anomalies are only seen in basaltic rocks that have fractionated at crustal pressures, such as at mid-ocean ridges. Presence of positive Eu anomaly in eclogitic pyroxene is commonly taken as independent evidence for the involvement of crustal protoliths in their formation via the subduction cycle. The amplitude of Eu anomaly in an inclusion, therefore, may be a function of its Ca content. Fig. 8B is a plot of CaO vs. Eu* for all the clinopyroxene inclusions from U51 diamonds. The dataset does not show any apparent

correlation. In fact, multiple inclusions from the same diamond have widely different Eu* at almost same CaO content. These observations highlight the complexity of eclogite petrogenesis. The nature of Eu anomaly, considered together with oxygen isotopic data on mineral inclusions from U51 diamonds (see below), confirm the subduction hypothesis for the origin of Group B eclogites from Udachnaya kimberlite pipe (e.g., Jacob et al., 1994).

Four diamonds (one from U33/1 and three from UX-1) were polished to expose their sulfide inclusions. One of the diamonds (U33/1-D19) has two sulfide inclusions simultaneously exposed on the same polished surface. In general, sulfide inclusions occur in variety of shapes and sizes, and are composed almost entirely of "pyrrhotite" with variable Ni contents. X-ray elemental maps of these sulfide inclusions have shown that fine-scale intergrowths of Cu- and Ni-bearing phases are quite common. Two examples are shown in Fig. 9. In diamond U33/1-D19, a Cu-rich area exists at the southwest periphery of the grain; in diamond UX-1-D12, a small Cu-rich area occurs at the northeastern edge of the grain. The presence of such small high-Cu areas at the margins of sulfide grains, probably representing chalcopyrite, is a common feature of sulfide inclusions in diamonds (Ruzicka et al., 1999). In diamond UX-1-D12, there are two Ni-rich bands that suggest the presence of pentlandite exsolved from an original mono-sulfide solid solution. Attempts to measure the compositions of the suspected chalcopyrite and pentlandite with electron microprobe were unsuccessful because of the extremely small (<2 μm) width of the exsolved bodies. The composition of "pyrrhotite" in these inclusions is highly variable in terms of its Ni content, and invariably higher than the practically Ni-free pyrrhotite coexisting with pentlandite in sulfide grains of the host eclogites. This is illustrated in Fig. 10, which shows the most Ni-rich and Ni-poor "pyrrhotite" compositions in each of the sulfide inclusions, along with compositions of pyrrhotite and pentlandite in the host eclogites. The DI "pyrrhotite" compositions are within the range for eclogitic sulfides reported by Bulanova et al. (1996), and fall within the Ni-rich portion of the Mss field at 650 °C and 1 atm pressure. It is, however, possible that the relatively high Ni concentrations measured in the DI "pyrrhotites" are due to con-

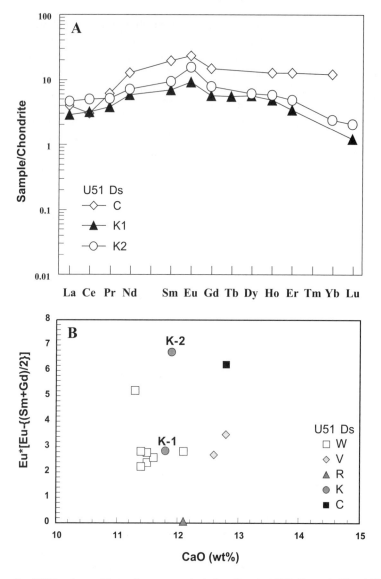

Fig. 8. (A) Chondrite-normalized REE patterns of three clinopyroxene inclusions from two U51 diamonds (C and K) (data from Taylor et al., 2000). All three inclusions exhibit a positive-Eu anomaly. (B) Plot of CaO vs. Eu* for clinopyroxene inclusions in diamonds from U51 eclogites (data are from Taylor et al., 2000). The plot clearly shows that there is no apparent correlation between CaO and Eu anomaly, indicating complex processes involved in the eclogite petrogenesis.

tamination by submicroscopic bodies of exsolved pentlandite. In any case, a comparison of the sulfide compositions indicate that the bulk composition of the melt from which the DI sulfide phases precipitated was different from that responsible for the sulfide phases in the host eclogites.

Fig. 9. (A) BSE and X-ray elemental maps of a sulfide inclusion exposed on a polished surface of diamond U33/1-D19. Note the Cu-rich area on the southwest corner of the inclusion. (B) X-ray elemental maps of a sulfide inclusion in diamond U33/1-D12 showing Cu-rich and Ni-rich patches at the level of the exposed surface in the polished diamond plate. Such Cu- and Ni-rich areas, visible only in X-ray maps, are common in almost all sulfide inclusions that we have studied. Lamellar intergrowths of pyrrhotite and exsolved pentlandite, optically visible in almost all sulfide grains in the host eclogite, were not seen in the sulfide inclusions.

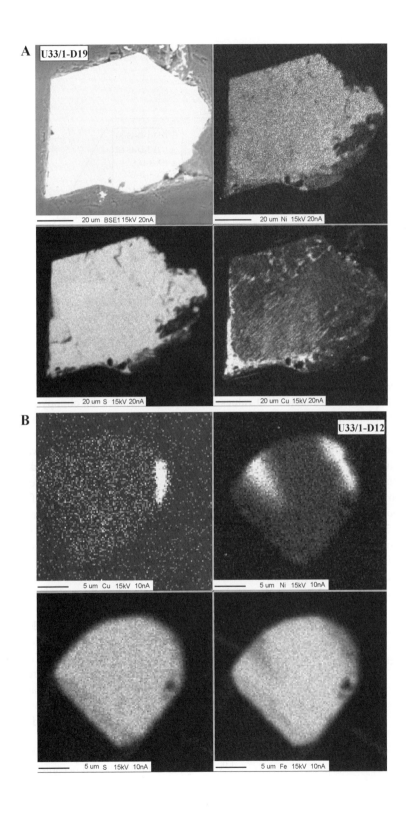

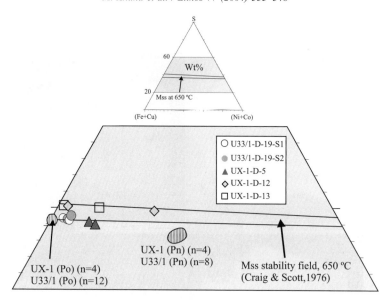

Fig. 10. The highest-Ni and lowest-Ni compositions (in wt.%) of sulfide inclusions in diamonds are plotted in Fe–Ni–S ternary space. Cu is assumed to substitute for Fe and Co for Ni. Almost all sulfide inclusions plot in the mono-sulfide solid solution (Mss) stability field at 650 °C towards the Ni-poor region, consistent with their eclogitic heritage (Bulanova et al., 1996). Compositions of sulfide occurring in the host eclogites are also plotted. These have exsolved pyrrhotite and pentlandite and plot in the respective fields.

3.2.2.1. Oxygen isotopes in diamond mineral inclusions.

There are relatively few available $\delta^{18}O$ values on eclogitic diamond inclusions compared to the data on host eclogite garnet and clinopyroxene. In fact, only two other studies have presented $\delta^{18}O$ data on eclogitic mineral inclusions in diamonds (10 eclogitic garnets in Finch diamonds (Lowry et al., 1999); coesite inclusions in Gauniamo diamonds (Schulze et al., 2003)). In both studies, the ranges in $\delta^{18}O$ values were large (5.7–8‰ and 10.2–16.9‰, respectively), presenting evidence for the subduction related origin of eclogitic mineral inclusions as well.

We have obtained oxygen isotope data on five clinopyroxene and one garnet inclusion in diamonds

Table 2
Oxygen isotopic compositions of Cpx and Gt inclusions in diamonds from U51 eclogite xenolith

Sample No.	$\delta^{18}O$ (‰)	2σ
W-1 cpx	4.4	0.9
W-5 cpx	2.4	1.0
W-14 cpx	2.4	1.0
V-19 cpx	2.9	0.8
L-29 cpx	4.3	0.9
L-27 gt	8.6	1.0

from U51 eclogite (Table 2). The overall variation in the $\delta^{18}O$ values in clinopyroxene inclusions is between 2.4‰ and 4.4‰, significantly lower than the mantle value of 5.5 ± 0.4‰. In contrast, one garnet diamond inclusion has $\delta^{18}O$ value of 8.6, the highest oxygen isotope value reported for an eclogitic garnet inclusion. This is only the second time that such a high $\delta^{18}O$ value has been found in any eclogitic garnet. Deines et al. (1991) determined $\delta^{18}O$ values up to 9.2‰ in eclogitic garnets from Orapa, albeit not as DIs. Such high $\delta^{18}O$ values in the eclogitic garnets are in agreement with the hypothesis that they were ultimately derived from subducted oceanic crust. The $\delta^{18}O$ ratios in clinopyroxene inclusions, however, define the lower range seen in eclogites. It is even more interesting that multiple clinopyroxene inclusions from the same diamonds show different $\delta^{18}O$ values (W DIs in Table 2). These observations highlight the variability in oxygen isotopic compositions of eclogitic DIs, and are consistent with variations in major- and trace-element compositions of multiple DIs in and between these diamonds, as discussed above.

The longstanding debate about the origin of eclogites from subducted oceanic crust is mainly based on anomalously high and low oxygen isotopic values in eclogite xenoliths (Jagoutz et al., 1984). The accepted mantle oxygen isotopic composition is $5.5 \pm 0.4‰$ (Mattey et al., 1994). In the case of Udachnaya eclogites, Jacob et al. (1994) showed that $\delta^{18}O$ ratios in garnets and clinopyroxenes range from 5.19‰ to 7.38‰. Snyder et al. (1997) also reported $\delta^{18}O$ values of 4.9–7‰ on garnets and clinopyroxenes from Udachnaya eclogites. These ranges in oxygen isotopic composition in Udachnaya eclogites are relatively smaller than those seen in the case of South African eclogites (e.g., Roberts Victor, 2–8‰, Garlick et al., 1971; but these were whole-rock values, which are usually unreliable). Nonetheless, these data taken together with oxygen isotope data on eclogitic mineral inclusions, having both lower and higher $\delta^{18}O$ ratios than typical mantle, confirm the origin of many eclogites by subduction of oceanic crust.

4. Concluding remarks

1. High-resolution computed X-ray tomography is a powerful tool for mapping xenoliths as to the relative positions of the minerals, including diamonds. The Udachnaya eclogite xenoliths in the present study are exceptionally rich in diamonds; >70–90% of the total diamonds in these xenoliths occur in the interior of the xenolith and were located and their positions mapped with 3-D images, without destroying the sample.

2. Diamonds are always near omphacite, along zones with a prominent sub-planar fabric of metasomatic alteration assemblages. Sulfide minerals, although abundant, are not preferentially associated with diamonds. Thus, there are insufficient quantities of sulfide minerals to call upon formation of these diamonds from an immiscible sulfide melts, as proposed by some authors in other cases.

3. Diamonds in association with secondary minerals indicate metasomatic formation of diamonds, obviously post-dating the formation of the host eclogite. These observations are also consistent with our other studies that suggest that diamonds are not syngenetic with the garnet and clinopyroxene in the host eclogite.

4. Inclusions in diamonds show variable chemical compositions. Multiple inclusions from the same diamonds have different compositions, suggesting episodic encapsulation of DIs by the diamonds under changing melt/fluid conditions brought about by various metasomatic fluid fronts in the mantle.

5. Oxygen isotopic compositions of garnet and clinopyroxene DIs deviate significantly from the accepted mantle values. The $\delta^{18}O$ value for one of the garnet inclusions is the highest reported to date, for any eclogite garnet DI. This is consistent with the well-accepted hypothesis of origin for Udachnaya eclogites by the subduction of ancient oceanic crust.

Acknowledgements

We are grateful to Allan Patchen for his assistance with Electron Microprobe analyses. Mostafa Fayek is thanked for providing oxygen isotope data on garnet and clinopyroxene inclusions. We thank Dawn Taylor for help with drafting figures. Critical reviews by Oded Navon, Roger Mitchell and an anonymous reviewer helped to improve the manuscript considerably, for which we are grateful. This study was supported by NSF grant EAR-09430 to LAT.

References

Anand, M., Misra, K.C., Taylor, L.A., Sobolev, N.V., 2001. In situ chemical analyses of mineral inclusions in diamonds in kimberlitic eclogites from Yakutia. Eos Trans. AGU 82 (47), F1289. (Fall Meet. Suppl., Abstract V12C-1006).

Anand, M., Taylor, L.A., Carlson, W.D., Taylor, D.-H., Sobolev, N.V., 2002. Stratigraphy of diamonds: complex growth histories highlighted by cathodoluminescence. Eos Trans. AGU 83 (47), F1403. (Fall Meet. Suppl., Abstract V51B-1267).

Bulanova, G.P., 1995. The formation of diamond. J. Geochem. Explor. 53, 1–23.

Bulanova, G.P., Griffin, W.L., Ryan, C.G., 1996. Trace elements in sulfide inclusions from Yakutian diamonds. Contrib. Mineral. Petrol. 124, 111–125.

Carlson, W.D., Denison, C., 1992. Mechanisms of porphyroblast crystallization: results from high-resolution computed X-ray tomography. Science 257, 1236–1239.

Carlson, W.D., Denison, C., Ketcham, R.A., 1995. Controls on the nucleation of porphyroblasts: kinetics from natural textures and numerical models. Geol. J. 30, 207–225.

Coleman, R.G., Lee, D.E., Beatty, L.B., Brannock, W.W., 1965. Eclogites and eclogites: their differences and similarities. Bull. Geol. Soc. Am. 76, 483–508.

Deines, P., Harris, J.W., Robinson, D.N., Gurney, J.J., Shee, S.R., 1991. Carbon and oxygen isotope variations in diamond and graphite eclogites from Orapa, Botswana, and the nitrogen content of their diamonds. Geochim. Cosmochim. Acta 55, 515–524.

Denison, C., Carlson, W.D., 1997a. Three-dimensional quantitative textural analysis of metamorphic rocks using high-resolution computed X-ray tomography: Part I. Methods and techniques. J. Metamorph. Geol. 15, 29–44.

Denison, C., Carlson, W.D., 1997b. Three-dimensional quantitative textural analysis of metamorphic rocks using high-resolution computed X-ray tomography: Part II. Application to natural samples. J. Metamorph. Geol. 15, 45–57.

Garlick, G.D., McGregor, I.D., Vogel, D.E., 1971. Oxygen isotope ratios in eclogites from kimberlites. Science 172, 1025–1027.

Jacob, D., Jagoutz, E., Lowry, D., Mattey, D., Kudrjavtseva, G., 1994. Diamondiferous eclogites from Siberia: remnants of Archean oceanic crust. Geochim. Cosmochim. Acta 58, 5191–5207.

Jagoutz, E., Dawson, J.B., Hoernes, S., Spettel, B., Wanke, H., 1984. Anorthositic oceanic crust in the Archean Earth (abstract). Proc. 15th Lunar Planet Sci. Conf. Lunar and Planetary Science Institute, Houston, pp. 395–396.

Jerde, E.A., Taylor, L.A., Crozar, G., Sobolev, N.V., Sobolev, V.N., 1993. Diamondiferous eclogites from Yakutia, Siberia: evidence for a diversity of protoliths. Contrib. Mineral. Petrol. 114, 189–202.

Keller, R.A., Taylor, L.A., Snyder, G.A., Sobolev, V.N., Carlson, W.D., Bezborodov, S.M., Sobolev, N.V., 1999. Detailed pull-apart of a diamondiferous eclogite xenolith: implications for mantle processes during diamond genesis. Proc. 7th Intl. Kimb. Conf., vol. I. pp. 397–402.

Lowry, D., Mattey, D.P., Harris, J.W., 1999. Oxygen isotope composition of syngenetic inclusions in diamond from the Finch Mine, RSA. Geochim. Cosmochim. Acta 63, 1825–1836.

Mattey, D., Lowry, D., MacPherson, C., 1994. Oxygen isotope composition of mantle peridotite. Earth Planet. Sci. Lett. 128, 231–241.

Mendelssohn, M.J., Milledge, H.J., 1995. Morphological characteristics of diamond populations in relation to temperature-dependent growth and dissolution rates. Int. Geol. Rev. 37, 285–312.

Meyer, H.O.A., 1987. Inclusions in Diamond. Wiley, Chichester. 501–522 pp.

Meyer, H.O.A., McCallum, M.E., 1986. Mineral inclusions in diamonds from the Sloan kimberlites, Colorado. J. Geol. 94, 600–612.

Misra, K.C., Anand, M., Taylor, L.A., Sobolev, N.V., 2004. Multistage metasomatism of diamondiferous eclogite xenoliths from the Udachnaya kimberlite pipe, Yakutia, Siberia. Contrib. Mineral. Petrol. 146, 696–714.

Pearson, D.G., Shirey, S.B. 1999. Isotopic dating of diamonds. Rev. Econ. Geol. 12, 143–171.

Pearson, D.G., Snyder, G.A., Shirey, S.B., Taylor, L.A., Carlson, R.W., Sobolev, N.V. 1995. Archean Re–Os age for Siberian eclogites and constraints on Archean tectonics. Nature 374, 711–713.

Riciputi, L.R., Paterson, B.A., Ripperdan, R.L., 1998. Measurement of light stable isotope ratios by SIMS: matrix effects for oxygen, carbon, and sulfur isotopes in minerals. Int. J. Mass Spectrosc. 178, 81–112.

Rowe, T., Kappelman, J., Carlson, W.D., Ketcham, R.A., Denison, C., 1997. High-resolution computed tomography: A breakthrough technology for earth scientists. Geotimes, 23–27 (Sobolev, V.N., Taylor, L.A., Snyder, G.A., Sobolev, N.V. 1994. Diamondiferous eclogites from the Udachnaya kimberlite pipe, Yakutia. Int. Geol. Rev. 36, 42–64).

Ruzicka, A., Riciputi, L.R., Taylor, L.A., Snyder, G.A., Greenwood, J., Keller, R.A., Bulanova, G.P., Milledge, H.J., 1999. Petrogenesis of mantle-derived sulfide inclusions in Yakutian diamonds: chemical and isotopic disequilibrium during quenching from high temperatures. Proc. 7th Int. Kimb. Conf., vol. 2, pp. 741–749.

Schulze, D.J., Wiese, D., Steude, J., 1996. Abundance and distribution of diamonds in eclogite revealed by volume visualization of CT X-ray scans. J. Geol. 104, 109–113.

Schulze, D.J., Harte, B., Valley, J.W., Brenan, J.M., Channer, D.M.DeR., 2003. Extreme crustal oxygen isotope signatures preserved in coesite in diamond. Nature 423, 68–70.

Snyder, G.A., Jerde, E.A., Taylor, L.A., Halliday, A.N., Sobolev, V.N., Sobolev, N.V., 1993. Nd and Sr isotopes from diamondiferous eclogites, Udachnaya kimberlite pipe, Yakutia, Siberia: evidence of differentiation in the early Earth? Earth Planet. Sci. Lett. 118, 91–100.

Snyder, G.A., Taylor, L.A., Crozaz, G., Halliday, A.N., Beard, B.L., Sobolev, V.N., Sobolev, N.V., 1997. The origins of Yakutian eclogite xenoliths. J. Petrol. 38, 85–122.

Sobolev, N.V., 1977. Deep-Seated Inclusions in Kimberlites and the Problem of the Composition of the Upper Mantle. American Geophysical Union, Washington, DC. 304 pp.

Sobolev, N.V., Snyder, G.A., Taylor, L.A., Keller, R.A., Yefimova, E.S., Sobolev, V.N., Shimizu, N., 1998a. Extreme chemical diversity in the mantle during eclogitic diamond formation: evidence from 35 garnet and 5 pyroxene inclusions in a single diamond. Int. Geol. Rev. 40, 567–578.

Sobolev, N.V., Taylor, L.A., Zuev, V.M., Bezborodov, S.M., Snyder, G.A., Sobolev, V.N., Yefimova, E.S., 1998b. The specific features of eclogitic paragenesis of diamonds from Mir and Udachnaya kimberlite pipes (Yakutia). Russ. Geol. Geophys. 39, 1653–1663.

Sobolev, N.V., Sobolev, V.N., Snyder, G.A., Yefimova, E.S., Taylor, L.A., 1999. Significance of eclogitic and related parageneses of natural diamonds. Int. Geol. Rev. 41, 129–140.

Spetsius, Z.V., 1995. Occurrence of diamond in the mantle: a case study from the Siberian platform. J. Geochem. Explor. 53, 25–39.

Taylor, L.A., Neal, C.R., 1989. Eclogites with oceanic crustal and mantle signatures from the Bellsbank kimberlite, South Africa: part I. Mineralogy, petrography, and whole-rock chemistry. J. Geol. 97, 551–567.

Taylor, L.A., Keller, R.A., Snyder, G.A., Wang, W.Y., Carlson, W.D., Hauri, E.H., McCandless, T., Kim, K.R., Sobolev, N.V., 2000. Diamonds and their mineral inclusions, and what they tell us: a detailed "pull-apart" of a diamondiferous eclogite. Int. Geol. Rev. 42, 959–983.

Available online at www.sciencedirect.com

SCIENCE @ DIRECT®

Lithos 77 (2004) 349–358

www.elsevier.com/locate/lithos

Evidence of subduction and crust–mantle mixing from a single diamond

Daniel J. Schulze[a,*], Ben Harte[b], John W. Valley[c], Dominic M. DeR. Channer[d]

[a] Department of Geology, Erindale College, University of Toronto, Mississauga, Ontario, Canada L5L 1C6
[b] Department of Geology and Geophysics, University of Edinburgh, Edinburgh EH9 3JW, UK
[c] Department of Geology and Geophysics, University of Wisconsin, Madison, WI 53706, USA
[d] Guaniamo Mining Company, Centro Gerencial Mohedano, 9D Urb. La Castellana, Caracas, Venezuela

Received 27 June 2003; accepted 17 February 2004
Available online 18 May 2004

Abstract

Cathodoluminescence (CL) imaging of polished sections of a diamond from the Guaniamo region of Venezuela suggests a history of the diamond involving two periods of growth separated by a period of resorption and possibly brittle deformation. In situ electron probe analysis of multiple eclogitic garnet inclusions reveals a correlation between garnet composition and location in the stone. An early-formed garnet in the diamond core has higher $Ca/(Ca+Mg)$ and lower $Mg/(Mg+Fe)$ values than later garnets associated with the second period of diamond growth. This variation conforms to an extensive trend of variation in the suite of eclogitic garnets extracted from Venezuelan diamonds. The diamond is zoned in carbon isotope composition (in situ secondary ion mass spectrometry, SIMS, data). The core compositions ($\delta^{13}C$ PDB), corresponding to the first stage of growth, average $-17.7\permil$. The second period of growth is apparently in two sub-sets of CL zones with mean values of $-13.0\permil$ and $-7.9\permil$. Nitrogen contents of diamond are low (30–300 atomic ppm) and do not correlate with carbon isotope composition. Oxygen isotope ratios of the garnet inclusions are elevated substantially above those expected for "common mantle"; $\delta^{18}O$ VSMOW of early garnet is approximately $+10.5\permil$ and two late garnets average $+8.8\permil$. The evolutionary trend of magnesium enrichment in garnet is unlikely to represent igneous fractionation. The stable isotope data are consistent with diamond formation in subducted meta-basic rocks that had interacted with sea water at low temperatures at or near the sea floor and contained a substantial biogenic carbon component. During or following subduction, diamonds continued to form in an evolving system that was progressively modified by interaction with mantle material.
© 2004 Elsevier B.V. All rights reserved.

Keywords: Diamond; Garnet; Carbon isotopes; Oxygen isotopes; Subduction

1. Introduction

The study of the minute minerals found as primary (syngenetic) inclusions in diamonds has provided a wealth of information bearing on the origin of diamonds within Earth's interior (e.g., Sobolev, 1976; Meyer, 1987; Gurney, 1989). Due to technical limitations, and expediency, however, most studies to date have involved breaking or burning diamonds to liberate their inclusions. In many cases, multiple inclusions of the same mineral in a single diamond have been found to have virtually the same compo-

---

* Corresponding author. Tel.: +1-905-828-3970; fax: +1-905-828-3717.
E-mail address: dschulze@utm.utoronto.ca (D.J. Schulze).

0024-4937/$ - see front matter © 2004 Elsevier B.V. All rights reserved.
doi:10.1016/j.lithos.2004.04.022

sition (e.g., Prinz et al., 1975). In cases in which multiple inclusions of one mineral have been found to vary in composition within a single diamond, however, information on their position within the diamond is lost if they are extracted by breaking or burning (e.g., Jaques et al., 1989; Sobolev et al., 1998). Furthermore, with the extraction method, the composition of the host diamond, typically analysed as bulk fragments in separate studies, cannot be directly correlated with compositions of specific inclusions.

Recently, investigators have begun to study diamonds and their mineral inclusions in situ, in polished sections or plates of diamonds, using a variety of microbeam methods (e.g., Rudnick et al., 1993; Bulanova, 1995). This approach has provided new insights into diamond formation, as individual mineral compositions can be correlated with specific stages of diamond growth.

We have studied polished sections of inclusion-bearing diamonds of the eclogite suite from the Guaniamo region of Venezuela, on the Guyana Shield (e.g., Meyer and McCallum, 1993; Channer et al., 2001), using electron microprobe, cathodoluminescence (CL) and secondary ion mass spectrometry (SIMS) techniques. In one stone, we have found a correlation between major element composition of garnet inclusions and carbon isotope composition of the diamond host that can be related using CL imaging to position in the stone, and thus with growth history of the diamond. Comparison of these data with the large existing data set for Guaniamo diamond inclusion minerals (Sobolev et al., 1998; Kaminsky et al., 2000) allows us to place constraints on the source material and processes involved in formation of eclogitic diamonds.

2. Diamond #13-127-27

2.1. CL textures

Within a small suite of diamonds from the Guaniamo region of Venezuela, cut and polished to expose mineral inclusions, we identified eight stones with garnet inclusions exposed by polishing. One of these, diamond 13-127-27, cut approximately parallel to (100), has six garnet grains exposed on its polished

surfaces, three on each half of the laser-cut stone. A CL image of one half of this stone is illustrated in Fig. 1, and a portion of the other half is shown in Fig. 2. The major portion of the stone is relatively dark, with only faint zoning visible in CL, and contains a single large garnet (#27-1). A brighter, more complex CL response characterises much of the rim, which is only partially preserved, but also fills a "re-entrant" into the stone (Fig. 1). (The term "re-entrant", as used here, does not refer to a physical hole or opening in the stone, but to the elongate bright CL portion of the diamond projecting into the centre from the rim. Anand et al. (2003) have described a somewhat similar CL feature in a diamond and attributed it to late diamond forming in a healed fracture.) Two smaller garnets (#27-2 and #27-3) occur in the late-formed diamond in the re-entrant. Three other garnets (#27-4 to -6) are exposed on the polished surface of the other half of the stone (Fig. 2), and also occur within the re-entrant portion of the late, bright complex CL zone, although the re-entrant portion in half B does not extend fully to the rim. The faint CL zoning visible in the dark, major portion of the stone appears, in half A, to be abruptly truncated on the upper right margin and overgrown by the light, complex CL diamond rim phase (Fig. 1).

2.2. Electron microprobe data

The garnet inclusions were analysed with a Cameca SX-50 electron microprobe at University of Toronto using standard WDS techniques. Compositions are given in Table 1. Garnet #27-1 is more calcic (molar $Ca/(Ca + Mg)$ (ca) = 0.271) and less magnesian (molar $Mg/(Mg + Fe)$ (mg) = 0.561) than are garnets 2 and 3, which are virtually identical in composition (ca = 0.218, mg = 0.610). Garnets 4–6 are intermediate in composition between these two extremes, though very similar to garnets 2 and 3 (ca = 0.226–0.233, mg = 0.596–0.600). Garnet 1 also has lower contents of TiO_2 (0.60 wt.%) and Na_2O (0.19 wt.%) than do the later-formed garnets 2–6 (0.77–0.85 wt.% TiO_2 and 0.27–0.28 wt.% Na_2O).

In Fig. 3, compositions of the garnets in diamond 13-127-27 are compared with those from other eclogitic diamonds from the Guaniamo region. The suite, as a whole, has a very distinct negative correlation between $Ca/(Ca + Mg)$ and $Mg/(Mg + Fe)$, and the

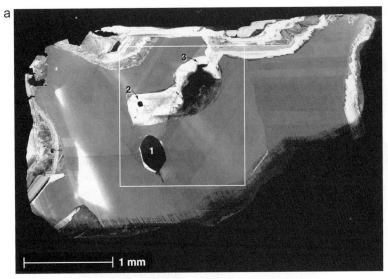

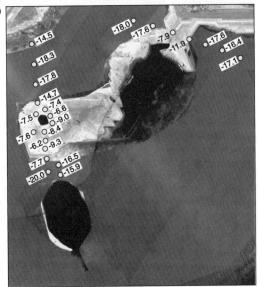

Fig. 1. (a) Cathodoluminescence image of polished section of half A of diamond #13-127-27. Three non-luminescent garnet inclusions are labelled (1, 2, 3). Adjacent to garnet #27-3 is a large, irregular hole with a poor to absent CL response due to the presence of epoxy. The sawtooth pattern and general darkness on the bottom margin of the stone is due to incomplete removal during polishing of amorphous carbon build-up remaining after laser cutting of the stone, which partly obscures the CL response in the region. The area outlined in white is enlarged in (b). (b) Enlargement of region indicated in (a) showing positions of the ion probe points where carbon isotope ratios were measured. Values of $\delta^{13}C$ are indicated for each point.

garnets included in diamond #13-127-27 form part of this trend and typify it.

Fig. 4 illustrates the co-variation of Mg/(Mg + Fe) and Cr content of the Guaniamo diamond inclusion garnets. There is a subtle, though distinct, positive correlation between Mg/(Mg + Fe) and Cr.

2.3. SIMS data

Using secondary ion mass spectrometry (SIMS), we have determined the carbon isotope composition and nitrogen abundance of the diamond in both of the distinct CL zones, and the oxygen isotope composi-

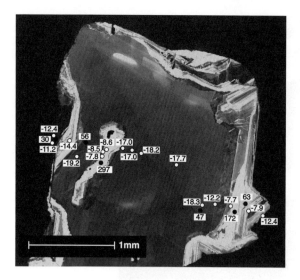

Fig. 2. Cathodoluminescence image of a portion of half B of diamond 13-127-27. Positions of points analysed for carbon isotope values ($\delta^{13}C$) are indicated in light circles and those for nitrogen abundance (ppm, atomic) in dark circles.

tion of three of the garnet inclusions. SIMS data were acquired using a Cameca ims-4f ion microprobe at the University of Edinburgh using a primary beam of $^{133}Cs^+$ defocussed to a spot approximately 30 μm in diameter and with an impact potential of 14,150 eV. An energy offset of 350 V was used for oxygen and carbon analysis. Detailed descriptions of the analytical techniques in use at the Edinburgh Ion Microprobe Facility can be found in the works of Eiler et al. (1997), Valley et al. (1998), Harte et al. (1999) and Fitzsimmons et al. (2000).

The carbon and nitrogen measurements of diamond 13-127-27 were standardized against synthetic diamond "SYNA" ($\delta^{13}C$ PDB $= -23.93‰$, N $= 230.4$ ppm by weight; Harte et al., 1999). In a given analytical session, typical reproducibility on the standards is approximately $\pm 0.7‰$ to $1.0‰$ (1 S.D.) for carbon isotope measurements and $\pm 7\%$ (1 S.D.) for N abundance.

In the absence of standards close in chemical composition to the unknowns (not a difficulty encountered in diamond work, as a diamond standard is used to analyse unknown diamonds), corrections for instrumental mass fractionation due to composition are required (e.g., Eiler et al., 1997). An instrumental mass fractionation correction procedure applicable to determination of the oxygen isotope composition of

eclogitic garnets has been developed at the University of Edinburgh (J. Craven, personal communication, 2003) and applied to the garnets in diamond 27. The garnet oxygen isotope standards used to develop the instrumental mass fractionation correction include the garnet end-members almandine, spessartine, grossular and pyrope and nine additional garnets close in composition to, and surrounding, the range of Guaniamo diamond inclusion garnets. These garnet standards were determined to be homogeneous by the ion probe, and their $\delta^{18}O$ values were calibrated by laser fluorination at the University of Wisconsin. The details of this method will be presented elsewhere (Schulze et al., manuscript in preparation). The instrumental mass fractionation corrections are very sensitive to compositional variation, and reproducibility on the unknowns is on the order of 1‰ to 2‰.

Two ion probe traverses were made in the diamond in the vicinity of garnets 1, 2 and 3. The locations of the analysed points are shown in Fig. 1b.

Table 1
Compositions of garnets included in diamond #13-127-27 (weight percent oxide)

Garnet #	27-1	27-2	27-3	27-4	27-5	27-6
SiO_2	40.09	40.73	40.60	40.44	39.97	40.56
TiO_2	0.60	0.85	0.84	0.81	0.79	0.77
Al_2O_3	22.54	22.09	21.96	21.91	21.89	22.07
Cr_2O_3	0.05	0.06	0.03	0.04	0.06	0.06
FeO^a	17.56	16.26	16.19	16.51	16.21	16.63
MnO	0.35	0.36	0.35	0.33	0.27	0.37
MgO	12.60	14.31	14.23	13.94	13.47	13.74
CaO	6.52	5.54	5.54	5.67	5.66	5.80
Na_2O	0.19	0.28	0.27	0.27	0.28	0.27
Total	100.51	100.48	100.02	99.92	98.61	100.26
Cations normalised to 12 oxygens						
Si	2.971	2.993	2.997	2.994	2.996	2.994
Ti	0.034	0.047	0.047	0.045	0.045	0.043
Al	1.969	1.913	1.911	1.912	1.934	1.920
Cr	0.003	0.003	0.002	0.003	0.004	0.003
Fe	1.088	0.999	1.000	1.022	1.016	1.027
Mn	0.022	0.022	0.022	0.021	0.017	0.023
Mg	1.392	1.567	1.565	1.538	1.505	1.512
Ca	0.518	0.436	0.438	0.449	0.455	0.459
Na	0.027	0.040	0.039	0.039	0.041	0.039
Total	8.023	8.021	8.020	8.023	8.012	8.020
Mg/(Mg + Fe)	0.561	0.611	0.610	0.601	0.597	0.596
Ca/(Ca + Mg)	0.271	0.218	0.219	0.226	0.232	0.233

[a] Total Fe reported as FeO.

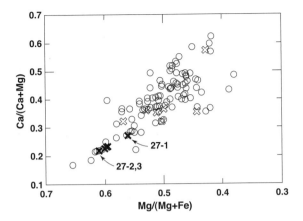

Fig. 3. Compositions of garnets included in Guaniamo diamonds in terms of molar Ca/(Ca+/Mg) and Mg/(Mg + Fe) values. The six garnet inclusions in diamond #13-127-27 analysed in this study are indicated as black " × " symbols, and inclusions 27-1,-2 and -3 (27-2 and 27-3 plots at the same point) are labelled. Inclusions 27-4,-5 and -6 have compositions intermediate between 1 and 2/3. The other new data in this study (open " × " symbols) confirm the early data of Sobolev et al. (1998) and Kaminsky et al. (2000), indicated by open circles.

One traverse (16 points in the left portion of Fig. 1b) extends from the dark, early CL zone beside garnet #27-1, across the re-entrant characterised by bright and complicated CL response and containing garnet #27-2, continues back into the dark CL region, and finishes just inside the bright rim CL zone near the top of the stone. The second traverse (seven spots in the upper right of Fig. 1b) starts and finishes in the dark CL zone and includes data in the bright re-entrant near garnet 3.

The diamond in the dark CL region of diamond 13-127-27 ranges in $\delta^{13}C$ from $-20.0‰$ to $-15.9‰$, with a mean of $-17.6 \pm 1.2‰$ (9 analyses). $\delta^{13}C$ of the diamond in the bright, complex CL re-entrant ranges from $-9.3‰$ to $-6.2‰$, with a mean value of $-7.8 \pm 1.0‰$ (10 analyses). Data for points that overlap the two zones ($-16.5‰$ and $-14.7‰$ from top and bottom of the bright CL zone in the left traverse and $-11.9‰$ from the right traverse) are not included in the mean values summarized above. The single point in the bright CL rim at the top of the left traverse, which is completely within the rim zone, has a $\delta^{13}C$ value of $-14.5‰$.

Carbon isotope data were also acquired for the other half (27B) of the stone (Fig. 2). The $\delta^{13}C$ values of six spots in the dark core range from $-19.2‰$ to

$-17.0‰$ and average $-18.2‰ \pm 0.8$. Three points within the re-entrant are in the range $-8.6‰$ to $-7.8‰$ (average $= -8.3 \pm 0.5‰$). Within the rim zone, some points have carbon isotope values similar to those of the re-entrant (two have values of $\delta^{13}C$ of $-7.7‰$ and $-7.9‰$) and others are similar to the single rim point analysed in the other half of the stone (five points are in the range $\delta^{13}C = -14.4‰$ to $-11.2‰$ and average $-12.5 \pm 1.2‰$).

Considering the entire stone, the three different carbon isotope populations are summarized as follows: Nineteen analyses in the dark region yield an average $\delta^{13}C$ value of $-17.7 \pm 1.0 ‰$; thirteen analyses in the re-entrant together with the two anomalously high rim values ($-7.7‰$ and $-7.9‰$) yield an average $\delta^{13}C$ value of $-7.9 \pm 0.8 ‰$; five analyses of the rim yield an average $\delta^{13}C$ value of $-13.0 \pm 1.4‰$.

Nitrogen abundance was measured at six points in half 13-127-27B (Fig. 2). Two points in the dark core (56 and 47 ppm) average 53 ppm. One point in the re-entrant has 297 ppm N, and three points in the bright rim (30, 63 and 172 ppm) have an average of 88 ppm.

Oxygen isotope data were obtained at five spots on the three garnets in 13-127-27A. Two values of $\delta^{18}O$ in garnet 27-1 in the early dark portion of the stone ($+9.7‰$ and $+11.3‰$) average $+10.5‰$. Three $\delta^{18}O$ values in garnets 27-2 ($+10.6‰$ and $+8.0‰$) and 27-3 ($+7.9‰$) yield an average of $+8.8‰$.

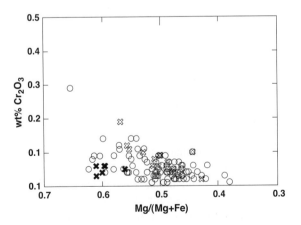

Fig. 4. Compositions of garnets included in Guaniamo diamonds in terms of molar Mg/(Mg + Fe) values and wt.% Cr_2O_3. Symbols as in Fig. 3.

3. Discussion

3.1. Diamond structure

The re-entrant into the interior of diamond 13-127-27 is a rather unusual feature for diamonds studied to date using CL imaging, most of which display reasonably concentric zoning patterns (e.g., Bulanova, 1995; Harte et al., 1999; Hauri et al., 1999). As the diamond within the re-entrant has CL characteristics similar to those of the physical rim on the stone, and is connected to it, we interpret the re-entrant to be a late feature. Truncation of the faint CL zonation visible in the dark portion of the stone suggests that at one time, the stone may have been substantially larger and concentrically zoned. The brittle deformation event that appears to have broken the stone may also have caused incipient {111}cleavage planes now occupied by the diamond in the re-entrant. A diamond resorption event, following diamond breakage, is suggested by the convex inward regions occupied by the late-formed diamond, well-developed along the left and top sides in Fig. 1a. A second generation of diamond growth, post-dating resorption, appears as the bright and complex CL region around much of the present rim of the stone, and occupying the re-entrant.

3.2. Garnet compositions

The positions of the garnets, in the context of the CL variations in diamond 13-127-27, suggest that garnet 27-1 in the dark CL portion formed before garnets 27-2 to 27-6, which occur in the late-formed, bright CL diamond in the re-entrant. The differences in Mg/(Mg + Fe) and Ca/(Ca + Mg) values of the two generations of garnet indicate a changing geochemical environment over the course of diamond formation, evolving from relatively iron-rich and calcic compositions, to a more magnesian and calcium-poor crystallization medium. This trend, which has a relative age significance in diamond 27, is within and parallel to that of the larger suite of garnet inclusions extracted from Guaniamo diamonds (Sobolev et al., 1998; Kaminsky et al., 2000), for which there is no relative age information (Fig. 3). If the spatial–compositional relationships of the garnets in diamond 13-127-27 can be applied to the suite as a whole, it implies that the Guaniamo eclogitic diamonds have formed in an

evolving chemical system in which the mg value of coexisting silicates increased over time. The evolution of the eclogite parent bodies of the diamonds to more magnesian (and less calcic for garnet) compositions was accompanied by the subtle, though distinct, increase in Cr of their garnets (Fig. 4).

A similar compositional change in garnets, and coexisting clinopyroxenes (those from diamond interior enriched in Ca and Fe relative to those from the diamond periphery), was documented in an eclogitic diamond from Yakutia (Bulanova, 1995). The same trend was also noted in clinopyroxenes in two other zoned eclogitic diamonds, though they lacked garnet (Bulanova, 1995), and Taylor et al. (1996) documented similar variations between diamond inclusion minerals and those in the eclogite xenoliths hosting the diamonds in three of the four Yakutian diamond eclogites that they studied. Bulanova (1995) interpreted this spatial–composition variation as due to "differentiation of eclogitic melts". Igneous differentiation during diamond formation in the mantle has also been proposed to account for geochemical trends in suites of eclogitic mineral inclusions extracted from diamonds at Orapa (Gurney et al., 1984), Sloan (Otter and Gurney, 1989) and Argyle (Jaques et al., 1989), for example, but there is no relative age information for the trends of any of these suites other than the Yakutian diamonds. In magmatic silicate systems, however, Mg partitions into the crystallizing silicates in preference to Fe (e.g., Bowen and Schairer, 1935). Precipitation of sub-equal amounts of garnet and clinopyroxene, therefore, would result in an iron-enrichment trend in cumulates and evolving liquids, which is the opposite of the evolutionary trends documented by Bulanova (1995) and in this study. The apparent rise in Cr (a compatible element in garnet) as the system evolves is also inconsistent with an igneous fractionation model.

It is possible that the garnets in the re-entrant formed entirely during the second diamond growth event, and thus are younger than the bulk of the diamond, which contains garnet 27-1. Alternatively, they may have existed prior to the brittle deformation event, their presence perhaps focussing the stresses that caused the {111} cleavage, and their present compositions reflect re-equilibration from a composition like that of garnet 27-1 towards the garnets in equilibrium with the second stage diamond-forming medium. This latter scenario, termed "open system behaviour" by Anand et al. (2003), could also provide an explanation for the apparently young trace element zoning patterns of peridotite-suite diamond inclusion garnets from Yakutia (Shimizu and Sobolev, 1995), although peridotitic diamonds are considered by most workers to have ancient (i.e., Archean) origins (e.g., Richardson et al., 1984). Note that although within our suite of polished diamonds from Guaniamo, we have found only a single example of a diamond with such a complex history of breakage, resorption and renewed diamond growth revealed by CL imaging, Anand et al. (2003) cite many examples of Yakutian diamond inclusion minerals associated with similar "healed fractures" from their CL studies.

3.3. Oxygen isotope composition of garnets

Many mantle eclogite xenoliths have $\delta^{18}O$ values that are significantly outside of the range accepted as normal for mantle materials. In contrast to garnets from most mantle peridotites, which have $\delta^{18}O$ values in the very small range $+5.36 \pm 0.18$ (Lowry et al., 1999), the $\delta^{18}O$ of garnets in mantle eclogites have been shown to be approximately $+2.3‰$ to $+9.2‰$ (e.g., Garlick et al., 1971; Deines et al., 1991; Mattey et al., 1994). By analogy with basaltic rocks from ophiolite sequences (e.g., Gregory and Taylor, 1981), eclogites (or their constituent garnets and clinopyroxenes) with $\delta^{18}O$ values above those of "common mantle" are thought to represent subducted ocean-floor basalts that have been altered by interaction with sea water at temperatures below about 350 °C, whereas eclogites with lower $\delta^{18}O$ values represent basic rocks that have interacted with higher temperature hydrothermal fluids (e.g., MacGregor and Manton, 1989).

In a previous ion probe study of diamond inclusion minerals in Guaniamo eclogitic diamonds, Schulze et al. (2003) found $\delta^{18}O$ values of coesite inclusions to be in the range $+10.2‰$ to $+16.9‰$, higher than documented previously for mantle eclogite minerals or diamond inclusion minerals (Lowry et al., 1999). The oxygen isotope ratios of the garnets in diamond 27 ($\delta^{18}O = +7.9‰$ to $+11.3‰$) overlap the upper end of those from mantle eclogites and the lower end of Guaniamo coesite inclusions. The overlap of the $\delta^{18}O$ values of the Guaniamo garnet inclusions (a common

diamond inclusion mineral) with the oxygen isotope values of the included coesites (a less common diamond inclusion mineral) confirms the earlier findings for the coesites. The isotope data for both minerals are clearly highly anomalous relative to typical upper mantle values and consistent with the subducted basalt hypothesis for certain mantle eclogites and their diamonds.

3.4. Carbon isotope ratios of host diamond

The two main diamond regions in stone 13-127-27 are characterised by distinctly different carbon isotope ratios. Diamond in the early (dark CL) portion of the stone is homogeneous and has a mean $\delta^{13}C$ value near $-17.7‰$. The late, bright CL diamond in the re-entrant has a mean $\delta^{13}C$ value near $-7.9‰$, and within the physical rim of the stone, the diamond seems to have two distinct carbon isotope populations, one similar to that of the re-entrant material, and the other with $\delta^{13}C$ near $-13.0‰$.

An increase in the $\delta^{13}C$ value (i.e., to less negative values) of precipitating diamond over time is the opposite of the carbon isotope evolution trend predicted by the "differentiation of mantle melts" model of Cartigny et al. (2001). Such a change is, however, consistent with the blending of subducted biogenic crustal carbon (suggested by its low $\delta^{13}C$ values—e.g., Sobolev et al., 1979; Kirkley et al., 1991) with "common" mantle carbon, such as that in peridotite-suite diamonds, which typically have $\delta^{13}C$ values near $-5‰$ to $-6‰$ (e.g., Kirkley et al., 1991). This model for carbon isotope evolution in eclogite-suite diamonds is in complete agreement with the geochemical evolution of the garnets in diamond 13-127-27 and the Guaniamo diamond inclusion garnet suite as a whole, in which the progressive increase in Mg/(Mg + Fe) (and Cr) suggests an increasing mantle contribution over time. Taylor et al. (2003) have suggested that similar mixing of subducted oceanic crust with more refractory mantle material was an important process in the evolution of the non-diamondiferous Obnazhennaya (Siberia) eclogite suite.

Although it has been suggested that mantle fractionation models might be able to explain derivation of low $\delta^{13}C$ values from "normal" mantle carbon reservoirs with initial $\delta^{13}C$ near $-5‰$ (e.g., Cartigny et al., 2001), this type of model does not account for the anomalously high oxygen isotope values of coexisting silicates. Oxygen isotope crystal–liquid fractionation is minimal at mantle temperatures, and differences in pressure do not cause a measurable change (Clayton et al., 1975). Igneous processes in the upper mantle thus cannot account for the wide range of $\delta^{18}O$ values of mantle eclogites. As discussed above, most workers interpret the anomalously high and low oxygen isotope ratios of mantle eclogite xenoliths as the result of subduction and prograde metamorphism of oceanic lithosphere that has undergone oxygen isotope exchange with sea water on or near the sea floor (e.g., Jagoutz et al., 1984; MacGregor and Manton, 1989). This explanation clearly applies to the elevated $\delta^{18}O$ values of the garnet inclusions in diamond 13-127-27 ($\delta^{18}O = +7.9\%o$ to $+11.3\%o$) and thus, the simplest explanation for the origin of the low $\delta^{13}C$ diamond host (as low as $-20\%o$ in the core) is that biogenic (i.e., crustal) carbon was present in the package of altered oceanic basaltic crust that was subducted to form the parent eclogite bodies hosting the diamonds (Fig. 5).

3.5. Nitrogen abundance

The overall nitrogen content of diamond 13-127-27 is fairly low (<300 ppm). It is lowest in the dark $\delta^{13}C$-depleted core (47 and 56 ppm) but it is high in the late-formed diamond in the re-entrant (297 ppm) and the rim diamond has N values above and below that of the core (30 to 172 ppm). There is not a regular correlation between nitrogen content and carbon isotope values (Fig. 2). The variability of both carbon isotope ratio and nitrogen content in the second generation of diamond (re-entrant and rim) suggests that both of these parameters were controlled by variations in the fluid composition from which the diamond crystallized. In part, the fluid composition will be a reflection of the extent of mixing of crust and mantle sources.

3.6. Origin and evolution of diamond #13-127-27

The geologic history of this diamond is summarized as follows. Subduction of ocean-floor basalt that had experienced low-temperature alteration by sea water (imposing elevated oxygen isotope ratios on the rock) resulted in prograde metamorphism of the

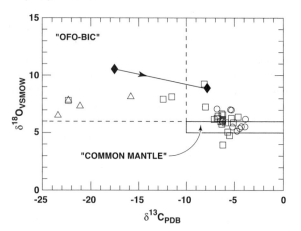

Fig. 5. Correlation between carbon isotope composition of diamond and oxygen isotope composition of silicates (garnet or clinopyroxene) from Guaniamo diamond #13-127-27 and samples of diamond eclogites from the published literature. The solid diamond symbols represent garnet 27-1 and the average value of early (dark CL) diamond, connected to the average value of the late diamond in the re-entrant and its coexisting garnets (garnets 27-2 and 27-3). Open squares represent diamond eclogites from Orapa (Deines et al., 1991), open circles represent diamond eclogites from Udachnaya (Snyder et al., 1995) and open triangles represent coexisting garnet and diamond in polycrystalline diamond aggregates from Venetia (Jacob et al., 2000). Note that all of the samples with $\delta^{13}C$ values below those typical of the "common mantle" (i.e., $\delta^{13}C < -10\%o$) also have $\delta^{18}O$ values higher than "common mantle" (i.e., $\delta^{18}O > +6\%o$), corresponding to values in the "OFO-BIC" (ocean-floor oxygen-biogenic carbon) quadrant.

basalt to the eclogite-facies, with the resultant silicate minerals (e.g., garnet) retaining the elevated $\delta^{18}O$ of the altered oceanic protolith. Growth of a low nitrogen octahedral diamond from a reservoir of subducted biogenic carbon ($\delta^{13}C$ in the range $-16\%o$ to $-20\%o$) within this eclogite body occurred following emplacement of the eclogite into the diamond stability field. A brittle deformation event (such as a deep-seated earthquake) broke the diamond, truncating some of the concentric CL zonation pattern and leaving smaller fragments that contained {111} cleavage fractures parallel to the octahedral outline of the original stone. A diamond-resorbing event occurred, possibly due to an influx of oxidizing fluids, dissolving diamond from portions of the outer edge of the broken stone and from along the cleavage crack. A second period of diamond growth then occurred, overgrowing the rim and filling in the re-entrant, with the new diamond having substantially lower, and more variable, $\delta^{13}C$ values and variable nitrogen contents. The new diamond-precipitating medium, which was also in equilibrium with eclogitic garnet, appears to have been a blend of crustal and mantle components, as it was more magnesian (reflected in the second generation garnet composition) and had higher $\delta^{13}C$ values than the original diamond-precipitating medium.

4. Concluding remarks

It is unclear how applicable the genesis of this stone is to the origin of eclogitic diamonds in general. A similar chemical evolution of included silicates has been documented for diamonds from Yakutia (Bulanova, 1995; Taylor et al., 1996), however, and many other suites of garnets extracted from eclogitic diamonds exhibit similar trends of Mg/(Mg + Fe), Ca/(Ca + Mg) and Cr values (e.g., Jwaneng—Gurney et al., 1995; Richardson et al., 1999 and Sloan—Otter and Gurney, 1989; Orapa—Gurney et al., 1984). The Orapa and Jwaneng diamond inclusion suites contain a "websteritic" paragenesis of diamond that is intermediate in composition (especially in mg and Cr_2O_3 content) between typical eclogitic minerals and those of the peridotite-suite. Mixing of a subducted eclogitic component with the ambient mantle, as we suggest for the Guaniamo suite, and Taylor et al. (2003) suggested for the non-diamondiferous Obnazhennaya eclogites, could explain the genesis of the websteritic diamond suite. Furthermore, the radiogenic isotope composition of eclogitic minerals in Jwaneng diamonds also suggests a mixing of subducted components with mantle lithosphere or asthenosphere (Richardson et al., 1999), and so our conclusions may indeed have a wider applicability to the genesis of eclogitic diamonds.

Acknowledgements

We are particularly indebted to John Craven of the University of Edinburgh Ion Microprobe Facility for his assistance and support in this project. It would not have been possible without his dedication. We also wish to thank Nicola Cayzer for SEM assistance, Robert Cooper for providing the diamonds, Barney Schumacher and co-workers for polishing the diamonds, Claudio Cermignani and Martina Miklos for their help with the electron microprobe, Mike Spicuzza for the laser fluorination work on the garnet oxygen isotope standards, Nick Sobolev and Patrick Anderson for providing unpublished data and Alison Dias and Brandon Beshara for help with the figures. Reviews by Felix Kaminsky and Larry Taylor helped to improve the manuscript. DJS was supported by NSERC and JWV by DOE and NSF.

References

Anand, M., Taylor, L.A., Misra, K.C., Carlson, W.D., Sobolev, N.V., 2003. Diamondiferous eclogite dissections: anomalous diamond genesis? Ext. Abstr. 8th Int. Kimb. Conf., Victoria, BC.

Bowen, N.L., Schairer, J.F., 1935. The system $MgO-FeO-SiO_2$. Am. J. Sci. 26, 151–217.

Bulanova, G.P., 1995. The formation of diamond. J. Geochem. Explor. 53, 1–24.

Cartigny, P., Harris, J.W., Javoy, M., 2001. Diamond genesis, mantle fractionations and mantle nitrogen content: a study of $\delta^{13}C$–N concentrations in diamonds. Earth Planet. Sci. Lett. 185, 85–98.

Channer, D.M.DeR., Egorov, E., Kaminsky, F., 2001. Geology and structure of the Guaniamo diamondiferous kimberlite sheets, south–west Venezuela. Rev. Bras. Geocienc. 31, 615–630.

Clayton, R.N., Goldsmith, J.R., Karel, K.J., Mayeda, T.K., Newton, R.P., 1975. Limits on the effect of pressure in isotopic fractionation. Geochim. Cosmochim. Acta 39, 1197–1201.

Deines, P., Harris, J.W., Robinson, D.N., Gurney, J.J., Shee, S.R., 1991. Carbon and oxygen isotope variations in diamond and graphite eclogites from Orapa, Botswana, and the nitrogen content of their diamonds. Geochim. Cosmochim. Acta 55, 515–524.

Eiler, J.M., Graham, C., Valley, J.W., 1997. SIMS analysis of oxygen isotopes: matrix effects in complex minerals and glasses. Chem. Geol. 138, 221–244.

Fitzsimmons, I.C.W., Harte, B., Clark, R.M., 2000. SIMS stable isotope measurement: counting statistics and analytical precision. Min. Mag. 64, 59–83.

Garlick, G.D., MacGregor, I.D., Vogel, D.E., 1971. Oxygen isotope ratios in eclogites from kimberlites. Science 172, 1025–1027.

Gregory, R.T., Taylor Jr., H.P. 1981. An oxygen isotope profile in a section of Cretaceous oceanic crust, Samail Ophiolite, Oman: evidence for $\delta^{18}O$ buffering of the oceans by deep (5 km) seawater-hydrothermal circulation at mid-ocean ridges. J. Geophys. Res. 86, 2737–2755.

Gurney, J.J., 1989. Diamonds. In: Ross, J. (Ed.), Kimberlites and Related Rocks. Their Mantle/Crust Setting, Diamonds and Diamond Exploration, vol. 2. Blackwell, Carlton, Australia, pp. 935–965.

Gurney, J.J., Harris, J.W., Rickard, R.S., 1984. Silicate and oxide inclusions in diamonds from the Orapa Mine, Botswana. In: Kornprobst, J. (Ed.), Kimberlites II: Their Mantle and Crust–Mantle Relationships. Amsterdam, Elsevier, pp. 3–9.

Gurney, J.J., Harris, J.W., Otter, M.L., Rickard, R.S., 1995. Jwaneng diamond inclusions. Ext. Abstr. 6th Int. Kimb. Conf. Novosibirsk. Siberian Branch, Russian Academy of Sciences, Novosibirsky, pp. 208–210.

Harte, B., Fitzsimmons, I.C.W., Harris, J.W., Otter, M.L., 1999. Carbon isotope ratios and nitrogen abundances in relation to cathodoluminescence characteristics for some diamonds from the Kaapvaal Province. S. Africa Min. Mag. 63, 829–856.

Hauri, E.H., Pearson, D.G., Bulanova, G.P., Milledge, H.J., 1999. Microscale variations in C and N isotopes within mantle diamonds revealed by SIMS. In: Gurney, J.J., Gurney, J.L., Pascoe, S.H., Richardson, S.H. (Eds.), Proc. 7th Int. Kimb. Conf., vol. 1. Red Roof Design, Cape Town, South Africa, pp. 341–347.

Jacob, D.E., Viljoen, K.S., Grassineau, N., Jagoutz, E., 2000. Remobilization in the cratonic lithosphere recorded in polycrystalline diamond. Science 289, 1182–1185.

Jagoutz, E., Dawson, J.B., Hoernes, S., Spettel, B., Wanke, H., 1984. Anorthositic oceanic crust in the Archean. Lunar Planet. Sci. 15, 395–396.

Jaques, A.L., Hall, A.E., Sheraton, J.W., Smith, C.B., Sun, S.-S., Drew, R.M., Foudoulis, C., Ellingsen, K., 1989. Composition of crystalline inclusions and C-isotopic composition of Argyle and Ellendale diamonds. In: Ross, J. (Ed.), Kimberlites and Related Rocks. Their Mantle/Crust Setting, Diamonds and Diamond Exploration, vol. 2. Blackwell, Carlton, Australia, pp. 966–989.

Kaminsky, F.V., Zakharchenko, O.D., Griffin, W.L., Channer, D.M.DeR., Khachatrayan-Blinova, G.K., 2000. Diamond from the Guaniamo area, Venezuela. Can. Mineral. 38, 1347–1370.

Kirkley, M.B., Gurney, J.J., Otter, M.L., Hill, S.J., Daniels, L.R., 1991. The application of C isotope measurements to the identification of the sources of C in diamonds: a review. Appl. Geochem. 6, 477–494.

Lowry, D., Mattey, D.P., Harris, J.W., 1999. Oxygen isotope composition of syngenetic inclusions in diamond from the Finsch Mine, RSA. Geochim. Cosmochim. Acta 63, 1825–1836.

MacGregor, I.D., Manton, W.I., 1989. Roberts Victor eclogites: ancient oceanic crust. J. Geophys. Res. 91, 14063–14079.

Mattey, D.P., Lowry, D., Macpherson, C.G., Chazot, G., 1994. Oxygen isotope composition of mantle minerals by laser fluorination analysis: homogeneity in peridotites, heterogeneity in eclogite. Min. Mag. 58A, 573–574.

Meyer, H.O.A., 1987. Inclusions in diamond. In: Nixon, P.H. (Ed.), Mantle Xenoliths. Wylie, London, pp. 501–522.

Meyer, H.O.A., McCallum, M.E., 1993. Diamonds and their sources in the Venezuelan portion of the Guyana shield. Econ. Geol. 88, 989–998.

Otter, M.L., Gurney, J.J., 1989. Mineral inclusions in diamonds from the Sloan diatremes, Colorado–Wyoming State Line kimberlite district, North America. In: Ross, J. (Ed.), Kimberlites and Related Rocks. Their Mantle/Crust Setting, Diamonds and Diamond Exploration, vol. 2. Blackwell, Carlton, Australia, pp. 1042–1053.

Prinz, M., Manson, D.V., Hlava, P.F., Keil, K., 1975. Inclusions in diamonds. Garnet lherzolite and eclogite assemblages. Phys. Chem. Earth 9, 797–815.

Richardson, S.H., Gurney, J.J., Erlank, A.J., Harris, J.W., 1984. Origin of diamonds in old enriched mantle. Nature 310, 198–202.

Richardson, S.H., Chinn, I.L., Harris, J.W., 1999. Age and origin of diamonds from Jwaneng kimberlite, Botswana. In: Gurney, J.J., Gurney, J.L., Pascoe, M.D., Richardson, S.H. (Eds.), Proc. 7th Int. Kimb. Conf., vol. 2. Red Roof Design, Cape Town, South Africa, pp. 709–713.

Rudnick, R.L., Eldridge, C.S., Bulanova, G.P., 1993. Diamond growth history from in situ measurements of Pb and S isotopic compositions of sulfide inclusions. Geology 21, 13–16.

Schulze, D.J., Harte, B., Valley, J.W., Brenan, J.M., Channer, D.M.DeR., 2003. Extreme crustal oxygen isotope values preserved in coesite in diamond. Nature 423, 68–70.

Shimizu, N., Sobolev, N.V., 1995. Young peridotitic diamonds from the Mir kimberlite pipe. Nature 375, 394–397.

Snyder, G.A., Taylor, L.A., Jerde, E.A., Clayton, R.N., Mayeda, T.K., Deines, P., Rossman, G.R., Sobolev, N.V., 1995. Archean mantle heterogeneity and the origin of diamondiferous eclogites, Siberia: evidence from stable isotopes and hydroxyl in garnet. Am. Mineral. 80, 799–809.

Sobolev, N.V., 1976. Deep-Seated Inclusions in Kimberlites and the Problem of the Composition of the Upper Mantle. American Geophysical Union, Washington, DC.

Sobolev, N.V., Galimov, E.M., Ivanovskaya, N.N., Yefimova, E.S., 1979. Isotopic composition of the carbon from diamonds containing inclusions. Dokl. Akad. Nauk Ukr. SSR 249, 1217–1220 (in Russian).

Sobolev, N.V., Efimova, E.S., Channer, D.M.DeR., Anderson, P.F.N., Barron, K.M., 1998. Unusual upper mantle beneath Guaniamo, Guyana Shield, Venezuela: evidence from diamond inclusions. Geology 26, 971–974.

Taylor, L.A., Snyder, G.A., Crozaz, G., Sobolev, V.N., Yefimova, E.S., Sobolev, N.V., 1996. Eclogitic inclusions in diamonds: evidence of complex processes over time. Earth Planet. Sci. Lett. 142, 535–551.

Taylor, L.A., Snyder, G.A., Keller, R., Remley, D., Anand, M., Wiesli, R., Valley, J., Sobolev, N.V., 2003. Petrogenesis of group A eclogites and websterites: evidence from the Obnazhennaya kimberlite, Yakutia. Contrib. Mineral. Petrol. 145, 424–443.

Valley, J.W., Graham, C.M., Harte, B., Eiler, J.M., Kinny, P.D., 1998. Ion microprobe analysis of oxygen, carbon, and hydrogen isotope ratios. SEG Rev. Econ. Geol. 7, 73–98.

Available online at www.sciencedirect.com

Lithos 77 (2004) 359–373

LITHOS

www.elsevier.com/locate/lithos

ELSEVIER

Constraining diamond metasomatic growth using C- and N-stable isotopes: examples from Namibia

Pierre Cartigny[a,*], Thomas Stachel[b,c], Jeff. W. Harris[d], Marc Javoy[a]

[a]IPGP et Universite Paris 7, Laboratoire de Géochimie des Isotopes Stables, Institut de Physique du Globe, UMR CNRS 7047, 4 Place Jussieu, T54-64 E1, 75251 Paris cedex 05, France
[b]Department of Earth and Atmospheric Science, University of Alberta, Edmonton, AB, Canada T6G 2E3
[c]Institut für Mineralogie, Universität Frankfurt, Senckenberganlage 28, 60054 Frankfurt, Germany
[d]Division of Earth Sciences, University of Glasgow, Gregory Building, Glasgow G12 8QQ, Scotland, UK

Received 21 June 2003; accepted 25 January 2004
Available online 24 May 2004

Abstract

The present paper provides C- and N-stable isotope characteristics, N-contents and N-aggregation states for alluvial diamonds of known paragenesis from placers along the Namibian coast. The sample set includes diamonds with typical peridotitic and eclogitic inclusions and the recently reported "undetermined" suite of Leost et al. [Contrib. Mineral. Petrol. 145 (2003) 15] which resulted from infiltration of high temperature, carbonate-rich melts. $\delta^{13}C$-values range from $-20.3‰$ to $-0.5‰$ ($n=48$) for peridotitic diamonds and from $-38.5‰$ to $-1.6‰$ ($n=45$) for eclogitic diamonds. Diamonds belonging to the "undetermined" suite span a narrower range in $\delta^{13}C$ from $-8.5‰$ to $-2.7‰$ ($n=13$). When compared with previous studies, diamonds from Namibia are characterised by unusually low proportions of N-free (i.e. Type II) peridotitic and eclogitic diamonds (3% and 2%, respectively) and an unprecedented high proportion of N-rich diamonds (15% and 73%, respectively, have N-contents >600 ppm). $\delta^{15}N$-values for diamonds of the peridotitic, eclogitic and "undetermined" suites range from $-10‰$ to $+13‰$ without correlations with either N-content or $\delta^{13}C$. The similarity in N-isotopic composition and the N-rich character of diamonds belonging to the eclogitic, peridotitic and "undetermined" suites is striking and suggests a close genetic relationship. We propose that a large part of the diamonds mined in Namibia formed during metasomatic events of similar style that introduced carbon and nitrogen into a range of different host lithologies.
© 2004 Elsevier B.V. All rights reserved.

Keywords: Diamond; Placer deposit; Namibia; Nitrogen; Stable isotopes; Metasomatism

1. Introduction

Increasing evidence suggests that *some* eclogite nodules are recycled oceanic crust (e.g. Jagoutz et

al., 1984; Jacob et al., 1994; Barth et al., 2000, 2002) of Archean age (see Shirey et al., 2002 for review). Inferring that diamonds in these rocks actually represent carbon subducted together with the oceanic crust critically depends on the question if the carbon was already present during subduction or was brought after incorporation of the eclogite hosts into the cratonic lithosphere. This assumption may not be true,

* Corresponding author. Tel.: +33-1-44-27-60-88; fax: +33-1-44-27-28-30.
E-mail address: cartigny@ipgp.jussieu.fr (P. Cartigny).

0024-4937/$ - see front matter © 2004 Elsevier B.V. All rights reserved.
doi:10.1016/j.lithos.2004.03.024

as mounting evidence documents diamond growth during metasomatic episodes introducing carbon into eclogites *after* their formation and incorporation into the cratonic lithosphere.

Evidence for de-coupling of the formation of diamonds from their eclogite hosts comes from several independent studies including 3D X-Ray tomography of eclogite xenoliths (Schulze et al., 1996; Keller et al., 1999), trace element analysis of inclusions in diamonds (e.g. Taylor et al., 1996), fluid(s) trapped during diamond growth (e.g. Izraeli et al., 2001 and references therein) and the presence of metasomatized inclusions within diamonds (Loest et al., 2003, see below). A number of these studies demonstrated that CO_2-rich (carbonatitic) melts/fluids are involved in the crystallization of diamonds.

Although *some* eclogitic and peridotitic diamonds apparently are related to metasomatism, it is not clear whether *all* diamonds from the mantle share this origin. In addition, the spectrum of possible CHO fluids for diamond precipitation is not limited to (hydrous) carbonatites but may extend also to more reduced methane–water mixtures. The significant metasomatic enrichment evident from the trace element patterns of inclusions in diamonds (e.g. Shimizu and Richardson, 1987; Griffin et al., 1992; for review see Stachel et al., 2004) bringing elements such as Rb, Sr, Sm and Nd may have implications for the interpretation of diamond formation ages. Up to now, mass balances for the metasomatic input into eclogites and/ or peridotites leading to diamond growth remain however to be established.

In this context of metasomatic versus non-metasomatic diamond formation, diamonds from the placer deposits along the Namibia coast may provide important clues. A detailed study of the mineral inclusion content of these diamonds not only led to the identification of typical peridotitic (Harris et al., 2004) and eclogitic (Stachel et al., 2004) inclusions, but also to the recognition of an "undetermined" suite (Loest et al., 2003).

The "undetermined" suite of inclusions (in 13 diamonds) is characterised by (1) unusual textural features such as lamellar exsolution of orthopyroxene from clinopyroxene, (2) in part low Mg-values and high K, Ba, Sr, and (3) the presence of additional unusual inclusions such as $MgCO_3$, $CaCO_3$, a Ti-rich

phase (lindsleyite?), phlogopite and coesite. The chemistry of this inclusion suite points towards diamond formation during infiltration of a CO_2 (carbonate)-rich fluid/melt with an unusually high fluid/rock ratio leading to an olivine-free mineral association. To completely eliminate olivine via carbonation reactions requires that the source for the "undetermined" suite was locally "flooded by the CO_2 (carbonate)-rich fluid/melt" (Loest et al., 2003, p. 23). The paragenesis of the pre-metasomatic host rocks can still be recognized using chemical criteria (Mg-number, Ni and Cr contents of the inclusions) and comprises both former peridotites and eclogites.

This unique combination of diamonds from typical peridotitic and eclogitic sources with the unusual "undetermined" suite offers an opportunity to examine possible links between these different diamond groups. The present work provides the C- and N-stable isotope characteristics, N-contents and N-aggregation levels for diamonds from all three growth environments.

2. Samples and analytical techniques

All 106 diamonds from Namibia analysed by us were previously studied for their syngenetic mineral inclusion content (Loest et al., 2003: "undetermined" suite; Harris et al. (2004: peridotitic suite; Stachel et al., 2004: eclogitic suite). On the basis of the inclusion studies, our sample set contains 49 peridotitic, 43 eclogitic, one websteritic and 13 diamonds from the "undetermined" suite. From the broken fragments left after inclusion release, two diamond chips were chosen, weighing between 0.15 and 2.30 mg. The first diamond chip was combusted for $\delta^{13}C$ analysis. After sample combustion in an O_2 atmosphere (see Boyd et al., 1995a) and quantification (to check for total combustion), $\delta^{13}C$-values were determined from the resulting CO_2 gas using a conventional dual-inlet mass spectrometer. The results are expressed in the conventional delta-notation, where $\delta^{13}C = (^{13}C/^{12}C_{sample}/^{13}C/^{12}C_{PDB} - 1) \times 1000$. The second diamond chip was analysed by micro-Fourier transform infrared (FTIR) spectroscopy to determine both nitrogen content (N_{FTIR}) and nitrogen aggregation state. Because of limited sample material, this second step could only be carried out for 88 of the 106 diamonds. Nitrogen contents and aggregation states were calculated from FTIR spectra applying the absorption coefficients for nitrogen A- and B-centres determined by Boyd et al. (1994a, 1995b). Nitrogen aggregation states are expressed as the relative proportion of the B-defect of the total nitrogen content (%B). Errors in nitrogen content (N_{FTIR}) and aggregation state are estimated to be better than 20% and 5%, respectively.

Fifty-five diamonds from the set of 106 have so far been analysed for $\delta^{15}N$, total nitrogen content (N_{comb}) and $\delta^{13}C$ (as a second analysis) by bulk combustion (BC). These analyses follow the experimental procedure given by Boyd et al. (1995a) with accuracies of 0.5‰, 5% and 0.1‰ (all 2σ) for $\delta^{15}N$, N-contents and $\delta^{13}C$, respectively. The 55 samples analysed using BC include 12 (out of 13) diamonds from the "undetermined" suite, 17 peridotitic and 26 eclogitic diamonds. Peridotitic and eclogitic diamonds were chosen to cover the full range in $\delta^{13}C$ and of N-contents (e.g. Nam-27 with $\delta^{13}C$ as low as −38.5‰ or Nam-43 with a N-contents as high as 1900 ppm) or for containing inclusions displaying negative (Nam-5 and Nam-74) or positive (Nam-59 and Nam-89) Eu-anomalies.

3. Results

Columns 3 to 6 in Tables 1–4 report the weight (mg), $\delta^{13}C$, N_{FTIR} and nitrogen aggregation states (%B) determined on the 106 diamonds of the eclogitic, peridotitic, and "undetermined" suite and those of unknown paragenesis from the Namibian placer deposits. Columns 7 to 10 of Tables 1–3 show the weights, $\delta^{13}C$, $\delta^{15}N$ and N_{Comb} for the 55 diamonds referred to earlier.

As illustrated in Fig. 1, the Namibian diamonds cover an unusually large range of $\delta^{13}C$-values. Eclogitic diamonds vary from −38.5‰ to −1.6‰, thereby extending the known range for diamonds from worldwide sources to lower (isotopically lighter) values. Two of the four eclogitic diamonds containing inclusions with Eu anomalies in their REE_N patterns have mantle-like $\delta^{13}C$-values of −5 ± 1‰ (Nam-59 and 89). Nam-5 and Nam-74, containing a garnet inclusion with a negative Eu-

anomaly have very light carbon isotopic compositions (< −24.7‰). Peridotitic diamonds range from −20.3‰ to −0.5‰ with the one very light value occurring in a zoned sample (Nam-62, ranging from −20.3‰ to −4.8‰; Table 2). For the undetermined suite, a narrower distribution in $\delta^{13}C$ between −8.5‰ and −2.7‰ is observed and the six diamonds of unknown paragenesis (no inclusions recovered) range from −21.1‰ to −2.6‰ (Table 4).

Column 5 in Tables 1–3 lists the nitrogen contents (based on FTIR analyses) in Namibian diamonds which range from 0 to 1860 ppm for the eclogitic suite, 0 to 1090 ppm for the peridotitic suite and 0 to 875 ppm for the "undetermined" suite. Compared to diamonds from worldwide sources (e.g. Harris and Spear, 1986; Deines et al., 1999 and references therein), our Namibian samples are characterised by a low proportion of Type II (i.e. N-free) diamonds in both the peridotitic (3%) and eclogitic (2%) suites and a high proportion of N-rich diamonds (>600 ppm, see Fig. 2). Nitrogen aggregation states (%B, Fig. 3 and Tables 1–4) vary from IaA (poorly aggregated; N-pairs) to IaB (highly aggregated; clusters of four N-atoms surrounding a vacancy) diamonds, with no clear relation to either paragenesis or N-content (Fig. 3).

Of the 55 diamonds in the second analytical set, the 17 peridotitic diamonds (Table 2) have $\delta^{15}N$-values ranging from −9.8‰ up to +12.4‰. No correlations between $\delta^{15}N$ and $\delta^{13}C$ (−6.7‰ to −0.5‰; Fig. 4A) or N-content (40 to 800 ppm, BC; Fig. 4B) are observed. However, it is important to note that the two peridotitic diamonds with the highest N-content (Nam-16 and 36, >700 ppm) have positive $\delta^{15}N$-values (+12.4‰ and +3.7 ‰, respectively). This relationship contrasts with results for the central part of the Kaapvaal (Kimberley Pool) where peridotitic diamonds with the highest N-contents show negative $\delta^{15}N$-values (Cartigny et al., 2004).

The 26 eclogitic diamonds have $\delta^{15}N$-values from −8.5‰ up to +14.8‰. The large spread in nitrogen isotopic composition is not related to variations in $\delta^{13}C$: for a range in carbon isotopic composition of −5.0 ± 1.5‰ (Table 1; Fig. 4A) an associated spread in $\delta^{15}N$ of −8.5‰ to +14.8‰ is observed. No correlation between nitrogen isotopic composition and content is observed but again, high N-contents

Table 1
$\delta^{13}C$, $\delta^{15}N$, N contents (determined by infrared spectroscopy and/or bulk combustion) and percentage of the B species in eclogitic diamonds from Namibia

Sample	Paragenesis	Weight (mg)	$\delta^{13}C$ (‰)	N FTIR (ppm)	%B	Weight (mg)	$\delta^{13}C$ (‰)	$\delta^{15}N$ (‰)	N Comb. (ppm)
Nam-005	E	0.7528	− 24.70	265	54.7	1.8834	− 26.63	+ 11.8	188
Nam-011	E	0.3567	− 3.55	1198	20.5	1.7834	− 4.42	+ 1.2	1254
Nam-013	E	1.3320	− 4.77	428	66.2				
Nam-014	E	0.5609	− 6.50	919	6.7				
Nam-019	E	0.8154	− 2.35	867	22.8				
Nam-020	E	0.4077	− 5.94	1285	21.4	1.3601	− 5.10	− 8.5	1044
Nam-021	E	1.6516	− 5.21	617	35.8				
Nam-022	E	0.8766	− 6.02	1194	21.9	2.1614	− 5.97	+ 1.4	1367
Nam-026	E	0.7040	− 4.64	680	15.3				
Nam-027	E	0.5134	− 38.54	0	–	2.2640	− 38.59	− 3.8	9
		0.5644	− 38.52			2.2336	− 38.65	− 2.2	16
Nam-034	E	2.2037	− 4.76	1483	39.9	1.5379	− 4.91	− 6.7	1361
Nam-035	E	0.4886	− 5.57	764	78.6	1.3446	− 5.37	+ 3.3	745
Nam-038	E	1.0909	− 3.42	1437	29.4	1.7211	− 3.58	+ 7.2	1140
Nam-041	E	0.4500	− 7.94	1003	76.1	0.1326	− 6.89	− 0.2	1270
Nam-042	E	0.3450	− 5.76	1134	18.8				
Nam-043	E	1.0766	− 16.93	1859	70.3	2.6501	− 16.68	+ 7.2	2116
Nam-044	E	1.0806	− 19.97	106	41.8	2.8455	− 21.35	+ 14.8	92
Nam-047	E	1.3689	− 5.00	46	49.6				
Nam-053	E	0.4507	− 7.14	998	77.0				
Nam-056	E	0.7318	− 29.33	35	70.6	1.9358	− 29.26	+ 6.7	55
Nam-059	E	0.7390	− 5.29	753	24.0				
Nam-063	E		− 6.36						
Nam-068	E	2.3939	− 4.11	614	17.6				
Nam-074	E	0.7964	− 25.68	327	90.6	0.9238	− 27.09	+ 3.5	255
Nam-078	E	0.3076	− 5.81	1008	11.6	0.2348	− 5.94	− 4.8	883
Nam-079	E	0.3717	− 8.08	863	30.6				
Nam-080	E	0.6379	− 6.30	1262	27.6				
Nam-080-Bis	E	0.7157	− 1.63	957	68.0				
Nam-081	E	0.3326	− 6.17	1574	34.1	0.2847	− 6.66	− 2.7	1704
Nam-086	E	0.3241	− 6.21	844	15.8	0.3275	− 5.69	− 1.5	797
Nam-089	E	0.7095	− 4.05	812	50.8	0.5137	− 4.23	− 1.4	805
Nam-095	E	0.6319	− 5.96	725	59.4				
Nam-096	E	0.1737	− 3.71	1332	23.1	0.2042	− 3.74	− 7.5	1410
Nam-097	E	0.3464	− 5.84	834	83.9				
Nam-098	E	0.7609	− 5.93	277	40.7	1.2413	− 5.61	+ 3.1	515
Nam-102	E	0.3673	− 28.70	711	90.1	0.5398	− 28.70	+ 0.3	1178
Nam-114	E	0.5386	− 26.93	35	80.9	1.2173	− 26.93	+ 4.0	28
Nam-202	E	1.0498	− 5.76	546	66.0				
Nam-203	E	0.9623	− 3.91	946	21.9	1.3088	− 3.46	− 1.4	1005
Nam-205-Bis	E	0.7389	− 4.18	57	21.4				
Nam-207	E	0.6795	− 6.33	1032	23.7				
Nam-208	E	1.0626	− 5.70	653	75.5	1.6902	− 5.54	− 6.8	637
Nam-212	E (W)	0.3459	− 4.53	128	100.0	1.5296	− 4.66	+ 11.6	34
Nam-216	E	0.5140	− 4.44	1308	28.9	0.6010	− 5.90	− 0.5	1483
Nam-218	E	0.6134	− 4.66	1070	30.7	1.6456	− 5.14	− 1.3	1204

Note that sample numbers with the extension-B refer to separate samples and not just analytical duplicates.

are associated with positive $\delta^{15}N$-values. Eclogitic diamonds with $\delta^{13}C$-values between − 16‰ and − 10‰ also show positive $\delta^{15}N$-values, the lowest $\delta^{13}C$-value (i.e. − 38.5‰, Nam-27, Table 1), however, is associated with a negative $\delta^{15}N$-value of about − 3‰ (Fig. 4A).

Table 2
$\delta^{13}C$, $\delta^{15}N$, N contents (determined by infrared spectroscopy and/or bulk combustion) and percentage of the B species in peridotitic diamonds from Namibia

Sample	Paragenesis	Weight (mg)	$\delta^{13}C$ (‰)	N FTIR (ppm)	%B	Weight (mg)	$\delta^{13}C$ (‰)	$\delta^{15}N$ (‰)	N Comb. (ppm)
Nam-016	P	0.6466	− 5.83	484	97.5	0.9254	− 5.47	+ 12.4	722
Nam-018	P	0.9052	− 3.95	128	3.6				
Nam-024	L	0.9080	− 4.59	1093	49.0				
Nam-028	P	1.9980	− 7.79	144	48.8				
Nam-029	P	0.7811	− 4.17	117	24.0	1.6620	− 4.89	− 1.3	147
Nam-030	P	0.2722	− 7.19						
Nam-031	P	0.7675	− 4.95						
Nam-033	P	0.3952	− 6.26	0	−				
Nam-036	P	0.5450	− 5.02	889	32.3	0.9028	− 5.16	+ 3.7	806
Nam-037	L	0.6634	− 5.05	614	43.5				
Nam-040	P	0.6493	− 3.57	487	39.9	0.4754	− 4.20	− 6.8	434
Nam-046	P	0.7010	− 7.43	176	97.0	2.3291	− 6.71	+ 10.2	215
Nam-050	P	0.8665	− 4.66	163	94.6				
Nam-051-Bis	P	0.7126	− 3.98	308	11.1	2.0913	− 4.26	− 6.2	350
Nam-052	P	0.1205	− 6.57						
Nam-054	P	0.4635	− 5.69			0.7677	− 5.66	− 3.2	52
Nam-055	P	2.1756	− 5.64						
Nam-061	P	1.0213	− 7.24	484	53.1				
Nam-062	P	0.6039	− 20.26	557	73.0	1.3945	− 4.84	− 5.4	536
Nam-064	P	0.2285	− 4.47						
Nam-065	P	0.5407	− 4.99						
Nam-070	P	0.1048	− 7.28						
Nam-071	P	0.6964	− 5.44	165	40.5				
Nam-072	P	0.5058	− 5.61						
Nam-073	P	0.4808	− 5.28	224	31.5	0.5935	− 3.69	− 1.0	200
Nam-075	P	0.5995	− 5.47						
Nam-077	P	0.3331	− 0.53	386	20.4	0.4267	− 0.53	− 9.3	473
Nam-082	P	0.1244	− 7.81						
Nam-087	P	0.2739	− 5.59						
Nam-092	P	0.4250	− 7.77						
Nam-093	P	0.4476	− 5.15	102	79.6				
Nam-094	P	0.5431	− 5.19	174	50.2	0.7637	− 5.35	+ 2.4	278
Nam-101	P	0.4034	− 5.93	279	10.4	1.3116	− 5.98	+ 5.9	232
Nam-103	P	0.9748	− 0.42	343	54.6				
Nam-104	P	0.2997	− 5.59	151	34.8	1.2284	− 5.49	+ 11.0	143
Nam-105	P	0.3842	− 5.92	151	34.8				
Nam-108	P	0.3957	− 1.52	573	72.3				
Nam-109	P	0.3898	− 7.03	51	56.3	1.2008	− 5.97	+ 1.3	40
Nam-110	P	0.7013	− 4.26	13	15.7				
Nam-111	P	0.7777	− 6.00	58	66.9				
Nam-112	P	0.5889	− 5.10	147	98.6				
Nam-113	P	0.6909	− 4.46	363	16.2	1.3656	− 4.57	+ 1.1	351
Nam-115	P	0.5114	− 4.82						
Nam-116	P	1.3464	− 3.64	161	34.6				
Nam-118	P	0.4481	− 5.95	25	9.1				
Nam-210-Bis	P	0.9653	− 4.60			0.6907	− 4.52	− 9.8	475
Nam-214	P	0.4670	− 4.83	669	0.0	0.9254	− 5.04	− 4.9	475
Nam-215	P	0.7363	− 4.66	345	27.8				

Table 3
δ^{13}C, δ^{15}N, N contents (determined by infrared spectroscopy and/or bulk combustion) and percentage of the B species in diamonds from Namibia belonging to the "undetermined" suite of Loest et al. (2003)

Sample	Paragenesis	Weight (mg)	δ^{13}C (‰)	N FTIR (ppm)	%B	Weight (mg)	δ^{13}C (‰)	δ^{15}N (‰)	N Comb. (ppm)
Nam-217	Und. (l)	1.7878	− 6.75	357	99.1	0.6552	− 6.90	+ 1.8	1572
Nam-210	Und. (l)	0.4968	− 5.79	247	15.8	1.1491	− 6.33	+ 3.1	16
Nam-211	Und. (l)	0.4963	− 3.99	383	63.5	1.2575	− 4.55	− 8.3	432
Nam-217-Bis	Und. (l)	0.4056	− 4.89	13	57.0	0.9654	− 5.30	n.d.	15
Nam-083	Und. (l)	0.8318	− 3.76	0					
Nam-100	Und. (l)	0.3095	− 2.69	490	51.6	0.4289	− 2.57	− 6.4	492
Nam-060	Und. (w)	0.4750	− 6.23	383	77.0	0.6563	− 6.00	+ 13.2	417
Nam-088	Und. (w)	0.6483	− 5.04			0.8017	− 5.39		1
Nam-204	Und. (w)	0.7449	− 5.93	810	14.6	1.1783	− 5.60	− 2.8	670
Nam-205	Und. (w)	0.7853	− 8.48			1.6404	− 8.90	n.d.	4
Nam-206	Und. (w)	0.4215	− 7.30	9	0.0	2.3383	− 9.51	+ 4.1	91
Nam-208-Bis	Und. (w)	0.7199	− 4.40	875	33.7	1.4022	− 4.53	− 1.7	807
Nam-209	Und. (w)	0.3804	− 4.43	686	20.7	1.9794	− 3.85	− 2.1	719

δ^{15}N could be determined for only 9 of the 13 samples from the "undetermined" suite (range of − 8.3‰ to + 13.2‰), the remaining four samples were too low in N-content. Again, co-variations with δ^{13}C (from − 9.5‰ to − 2.6‰) or N-contents (from 0 to 1600 ppm) are not observed. A division of the "undetermined" suite according to inclusion chemistry (lherzolitic or websteritic affinity, see Table 3) is not reflected in the isotopic composition and nitrogen concentration of the host diamonds.

A comparison of δ^{13}C-values measured on two different fragments of the same diamonds (Fig. 5A and columns 4 and 8 in Tables 1–3) reveals isotopic homogeneity better than 2‰ for all but one sample (Nam-62, Table 2). The observed internal δ^{13}C-variation of 15‰ in this sample is unusual but not unprecedented (cf. Deines et al., 1991 and Cartigny et al., 2004 for similar zonation in diamonds from

Jagersfontein and Kimberley Pool, respectively). N-contents measured on the same fragment by combustion and by FTIR compare well (Fig. 5B), with only one sample (Nam-217, see Table 3) being strongly heterogeneous.

4. Discussion

4.1. Nitrogen thermometry and inclusions-based geothermobarometry

It is now well accepted that the different nitrogen-bearing defects in diamond are linked by a diffusion process following second-order kinetics (Chrenko et al., 1977; Evans and Qi, 1982), with the percentage of the B-species depending upon the initial nitrogen concentration and the time-integrated thermal history of the diamond. Evans and Harris (1989) noticed that the activation energy for nitrogen in the A-centre to diffuse and form B-defects is very high and hence, the nitrogen aggregation state depends chiefly on nitrogen content and temperature and very little on time. This relationship is illustrated in Fig. 3 by the small difference between two sets of isopleths calculated for mantle residence times of 3 and 1 Ga, respectively. For a typical nitrogen content, this corresponds to a difference of less than 25 °C in residence temperature (see also Evans and Harris, 1989; Taylor et al., 1990; Navon, 1999). Nitrogen aggregation, therefore, is a good geothermometer.

Table 4
δ^{13}C, N contents (determined infrared spectroscopy) and percentage of the B species in diamonds of unknown paragenesis (no inclusions recovered)

Sample	Paragenesis	Weight (mg)	δ^{13}C (‰)	N FTIR (ppm)	%B
Nam-009	Unk.	0.6221	− 2.55	440	8.4
Nam-025	Unk.	1.0767	− 5.35	813	27.4
Nam-057	Unk.	0.4837	− 6.01	547	20.5
Nam-058	Unk.	0.4083	− 5.75	1127	46.8
Nam-117	Unk.	0.9812	− 5.35		
Nam-204-Bis	Unk.	0.7867	− 21.08	228	34.4

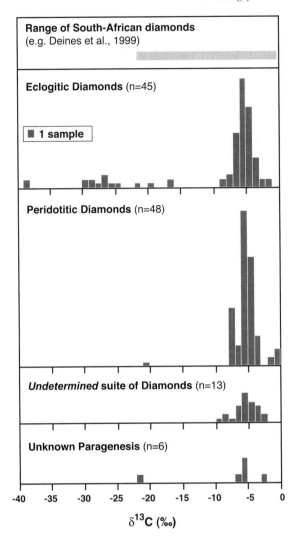

Range of South-African diamonds
(e.g. Deines et al., 1999)

Eclogitic Diamonds (n=45)

■ 1 sample

Peridotitic Diamonds (n=48)

Undetermined **suite of Diamonds** (n=13)

Unknown Paragenesis (n=6)

$\delta^{13}C$ (‰)

Fig. 1. $\delta^{13}C$ distributions of Namibian diamonds based on column 4 in Tables 1–4. Note that eclogitic diamonds extend the known range in $\delta^{13}C$ to a new minimum value of − 38.5‰ and thereby clearly exceed the $\delta^{13}C$ range for known southern Africa diamond sources.

A comparison of temperatures derived from nitrogen aggregation and from exchange equilibria between silicate inclusions can provide further insights into the conditions during diamond formation and during mantle residence. In the case of Namibia, equilibration temperatures between 960 and 1300 °C were obtained for 31 inclusion bearing diamonds (Loest et al., 2003; Harris et al., 2004) of which only 23 were nitrogen-bearing. Table 5 compares the results of inclusion and nitrogen based

temperature estimates, for the latter assuming mantle residence times of 1 (T_{1Ga}) and 3 Ga (T_{3Ga}). Nitrogen aggregation confines temperatures to a narrow interval ($T_{1\ Ga}$: 1106–1272 °C). The extremely poor agreement between the two datasets is also illustrated in Fig. 6. Plastic deformation, heating, and possibly also the presence of other impurities within the diamond structure may enhance nitrogen aggregation (see Taylor et al., 1990; Navon, 1999) thereby causing differences between nitrogen and inclusion thermometry. However, these processes should lead to consistently higher temperatures derived from nitrogen aggregation and not to the scatter observed in Fig. 6. The large temperature range for the inclusions (340 °C) may also result from chemical disequilibrium between some non-touching inclusion pairs but poor agreement with estimates based on nitrogen aggregation is also evident where temperatures were derived from touching inclusions.

Based on inclusion thermometry and the presence of opx exsolutions from originally homogenous high-T clinopyroxenes, Loest et al. (2003) concluded that diamonds of the "undetermined" suite formed at high temperatures of about 1300–1500 °C before cooling to more usual lithospheric temperatures (1100–1300 °C). In view of generally lower equilibration temperatures for peridotitic and eclogitic inclusions (960 to 1300 °C; Harris et al. 2003 and our data), it may be expected that diamonds of the "undetermined" suite display higher nitrogen aggregation states (and consequently higher nitrogen temperatures). However, Fig. 3 shows that nitrogen aggregation and residence temperature for the "undetermined" suite of diamonds is not different from the rest of the dataset.

The only way to reconcile nitrogen aggregation states with the mineral equilibration temperatures is to assume diamond formation during short-lived thermal perturbations. Precipitation of diamonds may have occurred during infiltration of high-temperature melts (probably up to 1500 °C in the case of the "undetermined" suite) into colder subcontinental lithosphere (1100–1300 °C). In this scenario, the non-touching inclusions and the re-combined cpx–opx intergrowths of the "undetermined" suite (Loest et al., 2003) record the temperature of diamond formation, whereas N-aggregation reflects a time averaged thermal histo-

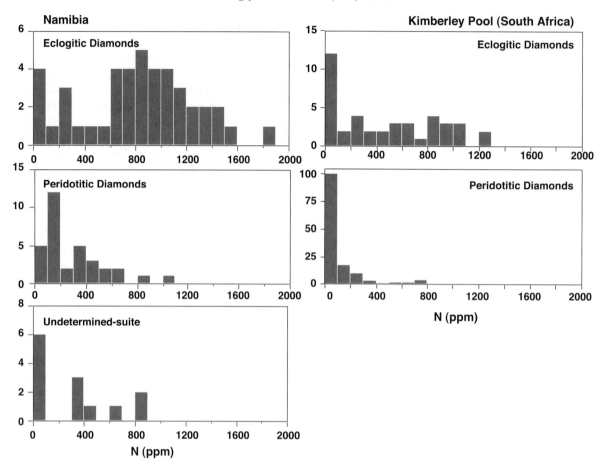

Fig. 2. Histograms comparing the N-contents (ppm, i.e. μg/g) in diamonds from Namibia and from the Kimberley Pool, South Africa (Cartigny et al., 2001a) as a typical example of diamond populations elsewhere.

ry. This implies that cooling and readjustment to a local geotherm occurred fairly rapidly after diamond formation, otherwise the rather narrow range in residence temperatures (1106–1272 °C) for diamonds of the "undetermined", the peridotitic and the eclogitic suite should not exist.

4.2. Are the eclogitic, peridotitic and undetermined diamond suites related?

The headwaters of the combined Orange and Vaal rivers cover a large portion of southern Africa. Assuming that the diamonds from the Namibian placer deposits are indeed derived from the hinterland of South Africa (e.g. Meyer 1991), it seems

unlikely that the suite of diamonds studied here should be derived from a single kimberlitic source. However, our $\delta^{13}C - \delta^{15}N - N$ data appear to suggest the opposite, i.e. that most, if not all, the Namibian diamonds studied were derived from a common source. Our Namibian samples are set apart from diamonds derived from worldwide sources by their overall N-rich character. Previous studies on total nitrogen contents in diamonds ($n \sim 3000$)—mostly derived from the Kaapvaal (e.g. Deines et al., 1987, 1989, 1991) but also from the Sino-Korean Craton (Cartigny et al., 1997), the Slave (unpublished data) and West Africa (Stachel and Harris, 1997)—have shown nitrogen arrays similar to the distribution for Kimberley Pool diamonds (Cartigny

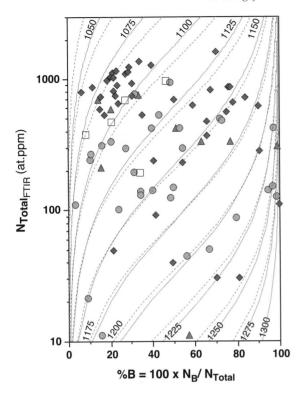

Fig. 3. Relationships of nitrogen aggregation state with total nitrogen content (determined by FTIR). Diamond shaped symbols denote eclogitic samples, circles indicate peridotitic diamonds, triangles represent the "undetermined" suite and open squares are diamonds of unknown paragenesis. Time integrated isopleths (temperature in °C) are constructed for mantle residence times of 3 Ga (solid lines) and 1 Ga (dashed lines).

our samples share a common source and origin. This proposed uniqueness of the primary source for Namibian diamonds is supported by three additional observations:

(a) On a worldwide basis, a relationship of decreasing maximum nitrogen content with decreasing $\delta^{13}C$ has been identified (Stachel and Harris, 1997; Cartigny et. al., 2000a,b) and the corresponding "limit sector" is depicted in Fig. 7. Four eclogitic diamonds from Namibia fall on the "forbidden", nitrogen-rich side of the limit sector. Thus N-rich diamonds from Namibia are not restricted to mantle-like $\delta^{13}C$-values ($\sim -5 \pm 3\permil$) but may also show strongly negative (i.e. $\delta^{13}C < -10\permil$) isotopic compositions.

(b) Worldwide data indicate a decrease in maximum nitrogen content with increasing $\delta^{15}N$ (solid line in Fig. 4B) and again, five diamonds (eclogitic, peridotitic and "undetermined" suite) from Namibia plot outside the worldwide array, being too nitrogen-rich for their high $\delta^{15}N$.

(c) With the exception of eight eclogitic diamonds with low $\delta^{13}C$, the overlap of the eclogitic, peridotitic and "undetermined" suites in $\delta^{13}C-\delta^{15}N-N$ space (Fig 4) is striking. For samples with mantle-like $\delta^{13}C$ (grey horizontal band in Fig. 4B), all three suites of Namibian diamonds extend the previously known $\delta^{15}N$-range towards more positive values.

The observation that Namibian diamonds share certain unique characteristics irrespective of their inclusion paragenesis (peridotitic, eclogitic and "undetermined" suites) suggests a commonality in growth mechanism that is independent of source rock composition. We propose that the common characteristic of overall high nitrogen contents implies a similar metasomatic origin for all Namibian diamonds. Differences in crystallization temperature (e.g. a hot origin for the "undetermined" suite) and in the buffering capacity of the source for CO_2 (facilitating isotopic fractionation for eclogitic sources, Cartigny et al. 1998a,b) introduced some diversity among diamonds of different paragenesis without completely masking the common characteristics of the metasomatic event(s).

et al., 2001a) presented in Fig. 2. Worldwide data also typically indicate a significant proportion of Type II diamonds (20% to 80% in some cases, see Harris and Spear, 1986; Deines et al., 1987, 1989, 1991). In contrast, our Namibian sample set contains relatively few Type II diamonds and reveals an unusually high proportion of eclogitic and peridotitic diamonds (there are no worldwide reference data for the undetermined suite) with high N-contents (Fig. 2). For example, ~ 70% of eclogitic diamonds from Namibia have N-contents higher than 600 ppm, compared to only ~ 40% at Kimberley Pool (Cartigny et al., 2001a). The overall N-rich character of Namibian diamonds is unlikely to result from random mixing of sources with characteristics like Kimberley Pool, but rather suggests that most of

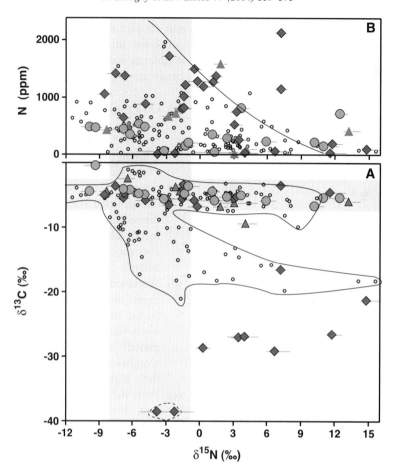

Fig. 4. (A, bottom) δ^{15}N versus N-content for diamonds from Namibia. The decrease in maximum N-content with increasing δ^{15}N identified in previous studies on samples from worldwide sources in indicated by a solid black line. (B, top) δ^{13}C versus δ^{15}N for diamonds from Namibia (same symbols as Fig. 3). Small open circles correspond to published analyses of peridotitic and eclogitic diamonds worldwide (Cartigny et al., 1998a,b; 1999 and references therein), their compositional variation is outlined as a solid black line. The two analyses for Nam-027 are enclosed by a dashed line. The ranges for "mantle-like" δ^{13}C and δ^{15}N are indicated as grey fields in A and B.

4.3. Origin and formation of diamonds from Namibia

A number of isotopic tracers have been employed to establish a link between eclogitic diamonds and subducted oceanic crust (see Navon, 1999 for review), involving, e.g. the oxygen isotopic composition of eclogite xenoliths (Jagoutz et al., 1984; Jacob et al., 1994; Barth et al., 2002), and eclogitic silicate inclusions in diamonds (Lowry et al., 1999; Schulze et al., 2003) or the sulfur isotopic signature of sulfide inclusions in diamonds (Chaussidon et al. 1987; Rudnick et al., 1993; Farquhar et al., 2002). These studies have produced strong evidence that diamond formation in subducted source rocks does indeed

occur. However, the oxygen and sulfur isotopic data cannot be used to constrain the source of the carbon from which the diamonds formed as its origin may well be de-coupled from that of the host rocks.

Consequently, the role of isotopically heterogeneous, subducted carbon in the formation of eclogitic diamonds is still subject of considerable debate. Direct formation of eclogitic diamonds from metasedimentary carbon may explain the low δ^{13}C character observed for many of these diamonds. However, considering that about 80% of subducted carbon is inorganic with δ^{13}C $\approx$ 0‰ (see Cartigny et al., 1998a,b, such a process should lead to a large amount of eclogitic diamonds with positive δ^{13}C-

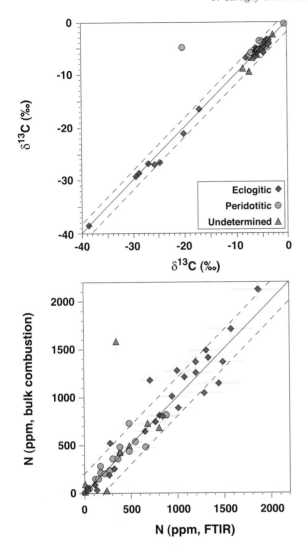

Fig. 5. (A, top) Comparison of $\delta^{13}C$-values measured on two cleavage chips from the same diamond. Dashed lines indicate variations of 2‰. (B, bottom) N-contents as determined by FTIR and by bulk combustion using the same diamond fragment. Dashed lines indicate a difference of 200 ppm. Most samples show heterogeneity smaller than 2‰ and 200 ppm in $\delta^{13}C$ and N-contents, respectively.

values. In the present paper and in numerous previous studies, positive $\delta^{13}C$-values were shown to be very scarce. Positive $\delta^{13}C$-values would also be expected if eclogitic diamonds resulted from mixing of mantle-derived and recycled metasedimentary carbon (e.g. Kirkley et al., 1991; Navon, 1999).

A variation of the above mixing model is the selective combination of recycled carbon significantly depleted in ^{13}C (e.g. < − 25‰, typical for organic matter) and mantle-derived carbon with a $\delta^{13}C$ of around − 5‰ (e.g. Navon, 1999). Such a mixing process could occur during injection of mantle-derived carbon into eclogite containing recycled metasedimentary carbon. However, $\delta^{13}C$–$\delta^{15}N$ relationships displayed by eclogitic diamonds from Jwaneng and Orapa are inconsistent with such a mixing model (Cartigny et al., 1998b, 1999). In particular, a mixing model would require that the recycled end-member is low in nitrogen (Cartigny et al. 2001a), contrasting with the nitrogen-rich character of indisputably subduction-related diamonds such as the metamorphic diamonds from the Kokchetav massif (see Sobolev and Schatsky, 1990; De Corte et al., 1998; Cartigny et al. 2001b). In addition, mixing of mantle-like carbon into a metasedimentary source should lead to strong zonation within eclogitic diamonds from cores with low $\delta^{13}C$ (typical of recycled organic carbon) to rims with $\delta^{13}C$-values around − 5‰ (reflecting the composition of the injected primitive carbon). As shown, e.g. in Fig. 5, heterogeneity of diamonds in $\delta^{13}C$ is usually restricted to a few per mil. More complex models involving the dissolution of preexisting diamonds are not supported by data on fibrous diamonds (see Boyd et al., 1994a,b). Homogenisation after diamond crystallisation through isotopic diffusion (Harte et al., 1999) also is not supported by new experimental data (Koga et al., 2003).

A subduction origin of diamond carbon can be further tested using N-isotopes. Nitrogen is not only the main impurity in diamond (Kaiser and Bond, 1959) but also has the advantage that its isotopic composition ($\delta^{15}N$) varies strongly from mantle material (mostly negative) to metasediments (strictly positive) (see Cartigny et al., 1998a and references therein). In this framework, available data have revealed strong similarity between eclogitic and peridotitic diamonds with both displaying largely negative $\delta^{15}N$-values (Cartigny et al., 1998a,b, 1999). This observation implies a strong contribution of mantle-derived carbon for both diamond suites. Accordingly, differences in the $\delta^{13}C$-distribution of eclogitic and peridotitic diamonds are in general more readily explained by variations in source min-

Table 5

Temperature estimates based on exchange equilibria between co-existing silicate inclusions and nitrogen aggregation characteristics for 23 Type I diamonds from Namibia

Sample	Paragenesis	Temperature	[Method]	N FTIR (ppm)	%B	$T_{1\ Ga}$	$T_{3\ Ga}$
Nam-034	E	1236	[Krogh]	1483	39.9	1125	1099
Nam-038	E	1080	[Krogh]	1437	29.4	1115	1089
Nam-086	E	1165	[Krogh]	844	15.8	1109	1083
Nam-089	E	1207	[Krogh]	812	50.8	1151	1124
Nam-114	E	961	[O'Neill]	35	80.9	1115	1090
Nam-203	E	1193	[Krogh]	946	21.9	1123	1097
Nam-218	E	1304	[Krogh]	1070	30.7	1275	1243
Nam-016	P	1079	[O'Neill]	484	97.5	1123	1097
Nam-029	P	1141	[Harley]	117	24.0	1170	1142
Nam-046	P	982	[O'Neill]	176	97.0	1177	1149
Nam-051-Bis	P	1092	[BKN]	308	11.1	1230	1200
Nam-071	P	1063	[O'Neill]	165	40.5	1214	1185
Nam-094	P	1209	[O'Neill]	174	50.2	1263	1231
Nam-101	P	1206	[O'Neill]	279	10.4	1287	1255
Nam-104	P	1212	[Harley]	151	34.8	1181	1153
Nam-109	P	1041	[Harley]	51	56.3	1190	1161
Nam-110	P	1138	[Harley]	13	15.7	1123	1098
Nam-113	P	1041	[O'Neill]	363	16.2	1129	1103
Nam-210	Underterm (L)	1185	[BKN]	247	15.8	1138	1112
Nam-211	Underterm (L)	1101	[BKN]	383	63.5	1183	1155
Nam-217-Bis	Underterm (L)	1222	[BKN]	13	57.0	1269	1238
Nam-208-Bis	Undeterm. (W)	1269	[BKN]	875	33.7	1132	1105
Nam-209	Undeterm. (W)	1297	[BKN]	686	20.7	1121	1095

$T_{1\ Ga}$ and $T_{3\ Ga}$ are temperatures estimated based on mantle residence times of 1 and 3 Ga, respectively. For the calculations, an activation energy of 7.0 eV and an Arrhenius constant of 2.94×10^5 atomic $ppm^{-1}\ s^{-1}$ for the IaA to IaB transition (Cooper, 1990; Taylor et al., 1990) was used.

eralogy (causing large differences in the buffering capacity for CO_2) than by different carbon sources (Cartigny et al., 1998b).

Interpreting the isotopic characteristics and nitrogen concentrations of Namibian diamonds in the context of this discussion reveals a complex picture that fits neither a pure mantle origin nor a subduction model. About half the diamonds have negative $\delta^{15}N$-values in support of mantle-derived nitrogen and carbon. In such a scenario, lower $\delta^{13}C$ and/or positive $\delta^{15}N$-values could result from stable-isotope fractionation of mantle-derived metasomatic fluids with initial $\delta^{13}C$ and $\delta^{15}N$ of about − 5‰. However, such a fractionation model is not feasible in light of the high nitrogen concentrations of some diamonds with "evolved" compositions. For example, in a fractionation model, the eight eclogitic diamonds with $\delta^{13}C$-values < − 10‰ should define a trend of decreasing nitrogen content with decreasing $\delta^{13}C$ (Cartigny et al., 2001a).

In contrast to other locations (e.g. Orapa and Jwaneng), this is not the case for Namibian diamonds where two eclogitic samples with strongly negative $\delta^{13}C$ have nitrogen contents >1000 ppm. Similar observations apply to the nitrogen isotopic composition, where some peridotitic and eclogitic diamonds from Namibia with $\delta^{13}C$ ≈ − 5‰ display high N-contents and positive $\delta^{15}N$-values. Again this is inconsistent with a fractionation model that predicts that positive $\delta^{15}N$ should be associated with low N-contents.

Thus, the model derived from eclogitic and peridotitic diamonds from Orapa, Jwaneng and Kimberley Pool (Cartigny et al., 2001a and references therein), stating that variations in C- and N-isotopic composition resulted from fractionation in the mantle, cannot be extended to the present dataset. Diamonds from Namibia (eclogitic, peridotitic and "undetermined" suite) with positive $\delta^{15}N$ and simultaneously high N-content may either reflect

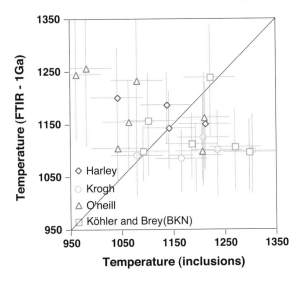

Fig. 6. Comparison of temperatures deduced from nitrogen characteristics assuming a mantle residence time of 1 Ga with results of conventional inclusion thermometry (results of Loest et al., 2003; Harris et al., 2004). The following mineral exchange equilibria were used: Harley = grt–opx; Krogh = grt–cpx; O'Neill = grt–olivine; BKN = cpx–opx.

formation from subduction-related metasomatic fluids/melts, as envisaged for the metamorphic diamonds from the Kokchetav massif (Cartigny et al., 2001a,b) or resulted from complex fractionation (from methane-rich fluids?) or mixing processes which have not been previously recognized.

5. Conclusions

Compared with diamonds from worldwide sources, alluvial diamonds from Namibia are distinctive because of their unusually high N-contents and in part uncommon C and N isotopic compositions. High N contents are not restricted to a single inclusion paragenesis but occur among samples of the eclogitic, peridotitic and "undetermined" suite (the latter related to high-temperature metasomatism). This overall similarity of all parageneses is unlikely to be a coincidence, but suggests a genetic relationship. Consequently, we suggest that most if not all diamonds from Namibia formed during a common metasomatic process that introduced both carbon and nitrogen into the various source rocks.

Negative $\delta^{15}N$-values imply that at least half the diamonds (eclogitic, peridotitic and "undetermined" suite) crystallised from mantle-derived fluids. Whether diamonds with high N-contents together with positive $\delta^{15}N$-values (in some cases associated with $\delta^{13}C$ below $-10‰$) are related to the same fluid source is uncertain. The observation that diamonds with such anomalous compositions are not restricted to the eclogitic suite is not in support of a formation directly from subducted carbon. More likely, this second group documents an additional fluid source which may be a subducting slab.

In any case, the close similarity of diamonds formed in a large variety of source rocks clearly supports precipitation from external carbon sources.

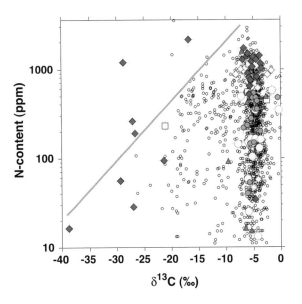

Fig. 7. N-content versus $\delta^{13}C$ for eclogitic (diamonds), peridotitic (circles) and undetermined-paragenesis diamonds (triangles) from Namibia. Filled symbols correspond to $\delta^{13}C$–N data obtained by combustion on a single diamond fragment. Open symbols indicate that the N-content was measured by FTIR using a different fragment than the one analysed for $\delta^{13}C$. Diamond from worldwide sources (open circles) and the proposed limit sector of Cartigny et al. 2001a, solid line) are shown for comparison. From the distribution of the data, it is apparent that diamonds from Namibia are characterised by unusually high N-contents (see also Fig. 2 and Fig. 4A).

Acknowledgements

Funding by the German Research Foundation (DFG) and the French INSUE-CNRS is gratefully acknowledged. DeBeers Consolidated Mines Limited is thanked for kindly providing the specimens used in this study. This manuscript benefited from constructive reviews by Peter Deines and Elad Izraeli. IPGP contribution 1960.

References

Barth, M., Rudnick, R.L., Horn, H., McDonough, W.F., Spicuzza, M.J., Valley, J.W., Haggerty, S.E., 2000. Geochemistry of xenolithic eclogites from West Africa: Part I. A link between low MgO eclogites and archean crust formation. Geochim. Cosmochim. Acta 65, 1499–1527.

Barth, M., Rudnick, R.L., Horn, H., McDonough, W.F., Spicuzza, M.J., Valley, J.W., Haggerty, S.E., 2002. Geochemistry of xenolithic eclogites from West Africa: Part 2. Origins of the high MgO eclogites. Geochim. Cosmochim. Acta 66, 4325–4345.

Boyd, S.R., Kiflawi, I., Woods, G.S., 1994a. The relationship between infrared absorption and the A defect concentration in diamond. Philos. Mag., B 69, 1149–1153.

Boyd, S.R., Pineau, F., Javoy, M., 1994b. Modelling the growth of natural diamonds. Chem. Geol. 116, 29–42.

Boyd, S.R., Réjou-Michel, A., Javoy, M., 1995a. Improved techniques for the extraction, purification and quantification of nanomole quantities of nitrogen gas: the nitrogen content of a diamond. Meas. Sci. Technol. 6, 297–305.

Boyd, S.R., Kiflawi, I., Woods, G.S., 1995b. Infrared absorption by the B nitrogen aggregate in diamond. Philos. Mag., B 72, 351–361.

Cartigny, P., Boyd, S.R., Harris, J.W., Javoy, M., 1997. Nitrogen isotopes in peridotitic diamonds from Fuxian, China: the mantle signature. Terra Nova 9, 175–179.

Cartigny, P., Harris, J.W., Javoy, M., 1998. Formation of eclogitic diamonds at Jwaneng: no room for a recycled component. Science 280, 1421–1423.

Cartigny, P., Harris, J.W., Philips, D., Girard, M., Javoy, M., 1998. Subduction-related diamonds?—The evidence for a mantle derived origin from coupled $\delta^{13}C$–$\delta^{15}N$ determinations. Chem. Geol. 147, 147–159.

Cartigny, P., Harris, J.W., Javoy, M., 1999. Eclogitic, peridotitic and metamorphic diamonds and the problems of carbon recycling: the case of Orapa (Botswana). In: Gurney, J.J., Gurney, J.L., Pascoe, M.D., Richardson, S.H. (Eds.), Proceedings of the 7th International Kimberlite Conference, vol. 1. Red Roof Design, Cape Town, pp. 117–124.

Cartigny, P., Harris, J.W., Javoy, M., 2001a. Diamond genesis, mantle fractionations and mantle nitrogen content: a study of $\delta^{13}C$–N concentrations in diamonds. Earth Planet. Sci. Lett. 185, 85–98.

Cartigny, P., et al., 2001b. The origin and formation of metamorphic microdiamonds from the Kokchetav massif, Kazakhstan: a nitrogen and carbon isotopic study. Chem. Geol. 176, 267–283.

Cartigny, P., Harris, J.W., Girard, M., Javoy, M., 2004. Eclogitic and peridotitic diamonds from Kimberley Pool kimberlites (South Africa): C- and N stable isotope evidence for a main mantle-derived source. Geochim. Cosmochim. Acta.

Chaussidon, M., Albarède, F., Sheppard, S.M.F., 1987. Sulfur isotope heterogeneity in the mantle from ion microprobe measurements of sulphide inclusions in diamonds. Nature 330, 242–244.

Chrenko, R.M., Tuft, R.E., Strong, H.M., 1977. Transformation of the state of nitrogen in diamond. Nature 270, 141–144.

Cooper, G.I., 1990. Infrared microspectroscopy of diamond in relation to mantle processes. PhD thesis, University of London, p. 262.

De Corte, K., Cartigny, P., Shatsky, V.S., Sobolev, N.V., Javoy, M., 1998. First evidence of fluid inclusions in metamorphic microdiamonds from the Kokchetav massif, Northern Kazakhstan. Geochim. Cosmochim. Acta 62, 3765–3773.

Deines, P., Harris, J.W., Gurney, J.J., 1987. Carbon isotopic composition, nitrogen content and inclusion composition of diamonds from the Roberts Victor kimberlite, South Africa: evidence for a ^{13}C depletion in the mantle. Geochim. Cosmochim. Acta 51, 1227–1243.

Deines, P., Harris, J.W., Spear, P.M., Gurney, J.J., 1989. Nitrogen and ^{13}C content of Finsch and Premier diamonds and their implications. Geochim. Cosmochim. Acta 53, 1367–1378.

Deines, P., Harris, J.W., Gurney, J.J., 1991. The carbon isotopic composition and nitrogen content of lithospheric and asthenospheric diamonds from the Jagersfontein and Koffiefontein Kimberlite, South Africa. Geochim. Cosmochim. Acta 55, 2615–2625.

Deines, P., Viljoen, K.S., Harris, J.W., 1999. Implications of the carbon isotope and mineral inclusion record for the formation of diamonds in the mantle underlying a mobile belt; Venetia, South Africa. Geochim. Cosmochim. Acta 65, 813–838.

Evans, T., Harris, J.W., 1989. Nitrogen aggregation, inclusion equilibration temperatures and the age of diamonds. In: Ross, J. (Ed.), Kimberlites and Related Rocks—Their Mantle/Crust Setting, Diamonds and Diamond Exploration. Spec. Publ.-Geol. Soc., vol. 14. Blackwell, Oxford, UK, pp. 1001–1006.

Evans, T., Qi, Z., 1982. The kinetics of the aggregation of nitrogen atoms in diamond. Phil. Trans. R. Soc. Lond., A 381, 159–178.

Farquhar, J., Wing, B.A., McKeegan, K.D., Harris, J.W., Cartigny, P., Thiemens, M.H., 2002. Mass-independent sulfur of inclusions in diamond and sulfur recycling on early Earth. Science 298, 2369–2371.

Griffin, W.L., Gurney, J.J., Ryan, C.G., 1992. Variations in trapping temperatures and tarce elements in peridotite-suite inclusions from African diamonds: evidence for two inclusion suites, and implications for lithosphere stratigraphy. Contrib. Mineral. Petrol. 110, 1–15.

Harris, J.W., Spear, P.M., 1986. Systematic studies of nitrogen in diamonds from known sources. Ext. Abstracts, 4th Int. Kimberlite Conf. Spec. Publ.- Geol. Soc. Aust., vol. 16. Perth, Oxford, UK, pp. 398–400.

Harris, J.W., Stachel, T., Loest, I., Brey, G.P., 2004. Peridotitic diamonds from Namibia: constraints on the composition and evolution of their mantle source 77, 209–223 this issue.

Harte, B., Fitzsimons, I.C.W., Harris, J.W., Otter, M.L., 1999. Carbon isotope ratios and nitrogen abundances in relation to cathodoluminescence characteristics for some diamonds from the Kaapvaal Province, S. Africa. Min. Mag. 63, 829–856.

Izraeli, E.S., Harris, J.W., Navon, O., 2001. Brine inclusions in diamonds: a new upper mantle fluid. Earth Planet. Sci. Lett. 187, 323–332.

Jacob, D., Jagoutz, E., Lowry, D., Mattey, D., Kudrajavtseva, G., 1994. Diamondiferous eclogites from Siberia: remnants of Archean oceanic crust. Geochim. Cosmochim. Acta 58, 5191–5207.

Jagoutz, E., Dawson, J.B., Hoernes, S., Spettel, B., Wänke, H., 1984. Anorthositic oceanic crust in the Archean Earth. 15th Lunar Planet. Sci. Conf., J. of Geophys. Res., Suppl., 395–396.

Kaiser, W., Bond, L., 1959. Nitrogen, a major impurity in common type I diamond. Phys. Rev. 115, 857–863.

Keller, R.A., Taylor, L.A., Snyder, G.A., Sobolev, V.N., Carlson, W.D., Bezborodov, S.M., Sobolev, N.V., 1999. Detailed pull-apart of a diamondiferous eclogite xenolith: implications for mantle processes during diamond genesis. In: Gurney, J.J., Gurney, J.L., Pascoe, M.D., Richardson, S.H. (Eds.), Proceedings of the 7th International Kimberlite Conference, vol. 1. Red Roof Design, Cape Town, pp. 397–402.

Kirkley, M.B., Gurney, J.J., Otter, M.L., Hill, S.J., Daniels, L.R., 1991. The application of C isotope measurement to the identification of the sources of C in diamonds: a review. Appl. Geochem. 6, 447–494.

Koga, K.T., Van Orman, J.A., Walter, M.J., 2003. Diffusive relaxation of carbon and nitrogen isotope heterogeneity in diamond: a new thermochronometer. Phys. Earth Planet. Inter. 139, 35–43.

Loest, I., Stachel, T., Brey, G.P., Harris, J.W., Ryabchikov, I.D., 2003. Diamond formation and source carbonation: mineral associations in diamonds from Namibia. Contrib. Mineral. Petrol. 145, 15–24.

Lowry, D., Mattey, D.P., Harris, J.W., 1999. Oxygen isotope composition of syngenetic inclusions in diamond from the Finsch Mine, RSA. Geochim. Cosmochim. Acta 63, 1825–1836.

Meyer, H.O.A., 1991. Marine diamonds of southern Africa. Diam. Int., 49–58.

Navon, O., 1999. Diamond formation in the Earth's mantle. In: Gurney, J.J., Gurney, J.L., Pascoe, M.D., Richardson, S.H. (Eds.), Proceedings of the 7th International Kimberlite Conference, vol. 2. Red Roof Design, Cape Town, pp. 584–604.

Rudnick, R.L., Eldridge, C.S., Bulanova, G.P., 1993. Diamond growth history from in situ measurement of Pb and S isotopic compositions of sulfide inclusions. Geology 21, 13–16.

Schulze, D.J., Wiese, D., Steude, J., 1996. Abundance and distribution of diamonds in eclogite revealed by volume visualization of CT X-Ray scans. J. Geol. 104, 109–113.

Schulze, D.J., Harte, B., Valley, J.W., Brenan, J.M., Channer, D.M.R., 2003. Extreme crustal oxygen isotope signatures preserved in coesite in diamond. Nature 423, 68–70.

Shimizu, N., Richardson, S.H., 1987. Trace element abundance patterns of garnet inclusions in peridotite-suite diamonds. Geochim. Cosmochim. Acta 51, 755–758.

Shirey, S.B., Harris, J.W., Richardson, S.H., Fouch, M.J., James, D.E., Cartigny, P., Deines, P., Viljoen, K.S., 2002. Diamond Genesis, seismic Structure, and evolution of the Kaapvaal–Zimbabwe Craton. Science 297, 1683–1686.

Sobolev, N.V., Schatsky, V.S., 1990. Diamond inclusions in garnet from metamorphic rocks: a new environment for diamond formation. Nature 343, 742–745.

Stachel, T., Harris, J.W., 1997. Syngenetic inclusions in diamond from the Birim field (Ghana)—a deep peridotitic profile with a history of depletion and re-enrichment. Contrib. Mineral. Petrol. 127, 336–352.

Stachel, T., Aulbach, S., Brey, G.P., Harris, J.W., Loest, I., Tappert, R., Viljoen, K.S., 2004. The Trace Element Composition of Silicate Inclusions in Diamonds: A Review, 77, 1–19 (this issue).

Taylor, W.R., Jaques, A.L., Ridd, M., 1990. Nitrogen-defect aggregation characteristics of some Australian diamonds: time–temperature constraints on the source regions of pipe and alluvial diamonds. Am. Mineral. 75, 1290–1310.

Taylor, L.A., Snyder, G.A., Crozaz, G., Sobolev, V.N., Yemimova, E.S., Sobolev, N.V., 1996. Eclogitic inclusions in diamonds: evidence of complex mantle processes over time. Earth Planet. Sci. Lett. 142, 535–551.

Available online at www.sciencedirect.com

ELSEVIER

Lithos 77 (2004) 375–393

www.elsevier.com/locate/lithos

Mildly incompatible elements in peridotites and the origins of mantle lithosphere[☆]

Dante Canil[*]

School of Earth and Ocean Sciences, University of Victoria, Petch building, Room 280, 3800 Finnerty Rd., Victoria, BC, Canada V8W 3P6

Received 27 June 2003; accepted 4 February 2004
Available online 15 June 2004

Abstract

The abundances of the mildly incompatible elements Al, Cr, V, Sc and Yb in more than 1700 mantle peridotite bulk rock analyses are interpreted in the light of a fractional melting model based on experimentally measured partition coefficients (D) and melting reaction stoichiometries. All peridotites examined, irrespective of sample type (abyssal peridotites, orogenic massifs, ophiolites, on/off craton xenoliths), tectonic environment (divergent/convergent/passive margin, intraplate) or the pressure (P) they last equilibrated at in the mantle (plagioclase-, spinel-, or garnet facies), originated as residues at less than 3 GPa, mainly within the spinel-facies. Mantle rocks currently in the garnet facies likely were originally spinel-facies lithosphere underthrust or subducted to greater depths in convergent margins. This view is inescapable even within the widest range of D values employed in the calculations, and is furthermore strengthened when metasomatic effects on the abundances of the mildly incompatible elements in residues are considered. A pressure of origin of below ~ 3 GPa for most mantle lithosphere creates difficulties for any model ascribing a significant volume of deep, cratonic mantle roots to plume sub-cretion or any other vertical tectonic mechanism.
© 2004 Elsevier B.V. All rights reserved.

Keywords: Mantle; Lithosphere; Peridotite; Melting; Trace element; Geochemistry; Modeling

1. Introduction

The composition of mantle lithosphere dictates its thermal and mechanical properties, stability and lifetime during the geodynamic evolution of our planet (Jordan, 1975; Poudjom Djomani et al., 2001). It would be useful to know or predict if fundamentally different compositions of mantle lithosphere are produced in different settings, and if so, why, and how they control the greater dynamics of continental and oceanic plates of varying age and tectonothermal history.

A wealth of information about the composition of the mantle lithosphere can be obtained from the study of hundreds of volcanic-hosted xenoliths (Griffin et al., 1999a; Jagoutz et al., 1979; Maaloe and Aoki, 1977; McDonough and Sun, 1995; Pearson et al., 2003). Although xenoliths can be placed into a spatial context in the mantle, they are accidental samples when entrained by their host magma, and age information and their original tectonic environment of formation is often obscure. In contrast, samples of the mantle in outcrop as orogenic massifs, ophiolites and dredged from the modern ocean basins provide a large in situ example of lithosphere in contrasting tectonic settings,

---

☆ Supplementary data associated with this article can be found, in the online version, at doi: 10.1016/j.lithos.2004.04.014.

* Tel.: +1-250-472-4180; fax: +1-250-721-6200.
E-mail address: dcanil@uvic.ca (D. Canil).

Table 2

Data sources to world peridotite compilation

Peridotite type	Data source
Samples in 'Outcrop'	
Abyssal	Aumento and Loubat (1971)
	Brandon et al. (2000)
	Casey (1997)
	Coogan et al. (in press)
	Gillis et al. (1993)
	Miyashiro et al. (1969)
	Niu and Hekinian (1997)
	Prinz et al. (1976)
	Snow and Dick (1995)
	Stephens (1997)
Ophiolite	Berry (1981)
	Coogan et al. (in press)
	Gruau et al. (1991)
	Gruau et al. (1998)
	Jaques and Chappell (1980)
	Loney et al. (1971)
	Rampone et al. (1993)
	Rampone et al. (1996)
	Zhou et al. (1996)
Forearc	Parkinson and Pearce (1998)
	Pearce et al. (2000)
Orogenic Massif	Becker (1996)
	Bodinier (1988)
	Bodinier et al. (1988)
	Burnham et al. (1998)
	Canil et al. (2003)
	Chauvel and Jahn (1984)
	Ernst (1978)
	Fabries et al. (1991)
	Frey et al. (1985)
	Gueddari et al. (1996)
	Hartmann and Wedepohl (1993)
	McPherson et al. (1996)
	Scambelluri et al. (2001)
	Shervais and Mukasa (1991)
	Takazawa et al. (2000)
Passive margin	Bonatti et al. (1986)
	Seifert and Brunotte (1996)
	Zhang et al. (2000)
Xenoliths	
Cratonic xenolith	Bernstein et al. (1998)
	Boyd and Mertzman (1987)
	Boyd et al. (1993)
	Boyd et al. (1997)
	Boyd (1999)
	Carlson et al. (1999)
	Jaques et al. (1990)
	Kopylova and Russell (2000)
	Lee and Rudnick (1999)
	Peltonen et al. (1999)
	Rudnick et al. (1993)

Table 2 (*continued*)

Peridotite type	Data source
	Schmidberger and Francis (2001)
	Winterburn et al. (1990)
Xenoliths	
Ocean Island	Ehrenberg (1982)
	Gregoire et al. (2000)
	Hauri et al. (1993)
	Siena et al. (1991)
Cont. Rift	Bedini et al. (1997)
	Ionov et al. (1993)
	Press et al. (1986)
Cont. Intraplate	Aoki (1981)
	Dautria and Girod (1986)
	Dautria et al. (1992)
	Dupuy et al. (1987)
	Embey-Isztin et al. (1989)
	Francis (1987)
	Frey and Green (1974)
	Griffin et al. (1987)
	Hunter and Upton (1987)
	Jagoutz et al. (1979)
	Laurora et al. (2001)
	Lee et al. (2003)
	Lenoir et al. (2000)
	Lenoir et al. (2001)
	Menzies and Hawkesworth (1987)
	Morten (1987)
	Peslier et al. (2002)
	Qi et al. (1995)
	Shi et al. (1998)
	Smith and Levy (1976)
	Smith et al. (1999)
	Song and Frey (1989)
	Stolz and Davies (1991)
	Vaselli et al. (1995)
	Xue et al. (1990)
	Yaxley et al. (1998)
	Zangana et al. (1999)
Cont. Arc	Liang and Elthon (1990)
	Luhr and Aranda-Gomez (1997)
Oceanic Arc	Franz et al. (2002)
	Maury et al. (1992)
	McInnes et al. (2001)

usually of known age, but are less frequently sampled and may also undergo modification when exhumed or exposed in environments with complex geological histories.

This contribution examines a large data compilation of the bulk composition of mantle peridotites in outcrop and as xenoliths from a range of tectonic settings. The database is used in concert with experimentally measured partition coefficients (D) and mantle melting reactions, to distinguish the pressure (P) and oxygen fugacity (fO_2) of melting and potentially the tectonic environment for the formation of lithosphere later sampled accidentally as xenoliths in alkaline magmas.

2. The mantle peridotite data set

More than 1700 peridotite bulk rock analyses were compiled from over 80 data sources in the literature. Similar though less extensive compilations have been published previously (Hart and Zindler, 1986; Maaloe and Aoki, 1977; McDonough and Sun, 1995; Palme and Nickel, 1985), but in this case a full table of analyses (Table 1) is available as a supplementary online electronic data set accessible via the Elsevier/ Lithos website. The main focus here is on the abundances of Al, Cr, Sc, V and Yb because they are mildly incompatible elements less affected by metasomatism, as will be shown below.

All data sources are post-1969. In all cases, element concentrations were determined by X-ray fluorescence (XRF), neutron activation or inductively coupled plasma mass spectrometry (ICPMS). Unfortunately, Sc, V and Yb are not always analysed in the same or all samples in a suite. Interlaboratory consistency or accuracy for these elements is not always demonstrable and no robust assessment of this issue can be presented at this time. Problems arise from this because some peridotites have V or Sc contents near the typical detection limits for these elements using XRF ($\sim$ 10 ppm). A recent assessment by Lee et al. (2003) for V showed that this element is likely known to within 10% relative.

The peridotite samples were divided according to the tectonic environment in which they were sampled either in outcrop or as accidental xenoliths in volcanic rocks (Table 2). High-temperature peridotites with 'sheared' or porphyroclastic textures derived from kimberlites are often strongly overprinted by trace element budgets related to melt-metasomatism and are excluded from this study. Some tectonic environments are not well represented (e.g., forearc exposures, arc xenoliths), mainly due to a lack of studies in the literature. Nonetheless, the division of tectonic environments shows a good correlation with the average

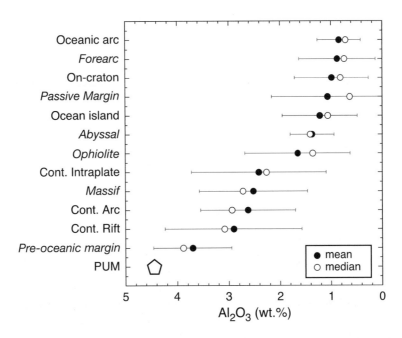

Fig. 1. The median and mean level of depletion (Al$_2$O$_3$ content) of mantle peridotites in the data compilation from this study ($n > 1700$), classified according to the tectonic environment in which they were sampled. Samples from outcrop are listed in italics, all others are xenolith data. Error bars are one standard deviation of the mean. The composition of primitive upper mantle (PUM) used in this and subsequent figures is from McDonough and Sun (1995).

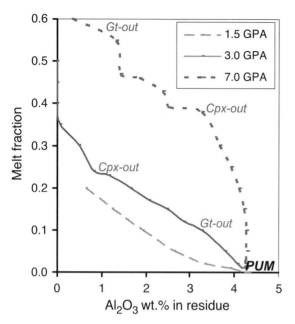

Fig. 2. Al_2O_3 contents in peridotite residues as a function of melt fraction (F) at pressures of 1.5, 3.0 and 7.0 GPa. Trends are calculated using a fractional melting model in which Al is treated as a mildly incompatible element, with primitive upper mantle (PUM) as the source material. Bulk $D_{Al_2O_3}$ residue/liq is parameterized as a function of (F) using chemical data from peridotite melting experiments and weighted according to experimentally determined stoichiometry of melting reactions (see Canil, 2002 for details). The inflections in the melting trends at each pressure correspond to consumption of a phase and corresponding change in the stoichiometry of the melting reaction. Note that although garnet is stable on the solidus at 3.0 GPa, it is consumed at ~ 10% melting, corresponding to ~ 3.5 wt.% Al_2O_3 in the residue.

level of depletion as indexed by the bulk rock Al_2O_3 content (Fig. 1). Al_2O_3 is used as a depletion index because it is immobile during metamorphism and allows for comparison of samples from all states of preservation. Moreover, the behaviour of Al_2O_3 during partial melting is sensitive to P (Herzberg, 1995; Walter, 1998) and simple to model empirically with experimental data on peridotite melting (Fig. 2).

3. Empirical partial melting model

Experimentally measured element partition coefficients (D) in mafic and ultramafic systems (Table 3, Fig. 3) are used in a fractional melting model (Johnson et al., 1990) that weights the D's according to the

stoichiometry of the melting reactions along the peridotite solidus, also determined by experiment. This empirical melting model is described in detail in Canil (2002) and used to derive the composition of residues as functions of P, degree of melting and, in the case of V, oxygen fugacity (fO$_2$). Because experimentally measured D values vary considerably (by factors of 2 to 5), the melting models used to derive residue trends in this study were calculated as end-member cases (i.e., highest and lowest D) for comparison with chemical data for mantle residues. The elements Yb, Cr, Sc and V all partition differently amongst key mantle minerals (garnet, spinel, clino-pyroxene, orthopyroxene, olivine) involved during the partial melting of peridotite (Table 3). Of these elements, V is redox sensitive, and the order of partition differs, with the general relation: $D_{gt/liq} > D_{cpx/liq} \gg D_{sp/liq}$ for Sc and Yb; $D_{sp/liq} > D_{cpx/liq} \gg D_{gt/liq}$ for V; $D_{sp/liq} \gg D_{cpx/liq} > D_{gt/liq}$ for Cr. Thus, correlations amongst Yb, Cr, Sc, V and Al are predicted to be illustrative of the effects of P, fO$_2$ and extent of melting to form a peridotite residue.

4. Effects of metasomatism

Numerous isotopic and trace element studies have revealed that all examples of mantle lithosphere have suffered some degree of chemical modification (metasomatism) that post-dates the original melting process. An assessment of the impact of metasomatic processes on the bulk rock chemistry of samples considered in this study is required to evaluate the degree to which they retain their chemical signature as residues.

Elements that are highly incompatible are most compromised by metasomatism, either over long periods in the lithosphere or immediately prior to or during entrainment in their host magma. The residence of many highly incompatible elements (bulk $D < 0.01$) is dominated by minor phases and grain boundary phenomena, some of which are introduced by entrainment in the host magma (Pearson et al., 2003). For this reason, this study does not concern itself with highly incompatible elements. The main focus here is on Al, Cr, Sc, V and Yb, because these elements generally have a bulk D (residue/liq) between 1 and 0.1 and thus their abundances should be less affected by chemical modifications impregnated

Table 3
Data sources for partition coefficients

Mineral	Bulk comp.	Element	D value	Data source
Garnet	komatiite	Sc	1.1	Yurimoto and Ohtani, 1992
	basalt	Sc	2.62	Hauri et al., 1994
	komatiite	Yb	0.9	Yurimoto and Ohtani, 1992
	basalt	Yb	6.6	Johnson, 1998
	peridotite	V	**	Canil, 2002
Clinopyroxene	picrite	Sc	0.51	Ulmer, 1989
	basalt	Sc	1.31	Hart and Dunn, 1993
	basalt	Yb	0.22	Gaetani and Grove, 1995
	basalt	Yb	0.623	Hauri et al., 1994
	Ab-An-Di	V	**	Canil and Fedortchouk, 2000
	basalt	V	**	Canil and Fedortchouk, 2000
Olivine	basalt	Sc	0.12	Beattie, 1994
	chondrule	Sc	0.47	Kennedy et al., 1992
	basalt	Yb	0.0157	Beattie, 1994
	chondrule	Yb	0.017	Kennedy et al., 1992
	komatiite	V	**	Canil and Fedortchouk, 2001
Low Ca Pyroxene	picrite	Sc	0.33	Ulmer, 1989
	chondrule	Sc	0.48	Kennedy et al., 1992
	chondrule	Yb	0.032	Kennedy et al., 1992
	basalt	Yb	0.08	Schwandt and McKay, 1998
	chondrule	V	**	Canil, 1999
Spinel	Fo-An-Di	Sc	0.36	Horn et al., 1994
	synthetic	Sc	0.0478	Nagasawa et al., 1980
	basalt	Yb	0.01	McKenzie and O'Nions, 1991
	synthetic	Yb	0.0076	Nagasawa et al., 1980
	komatiite	V	**	Canil, 2002

**D values for V vary with fO_2 according to relationships given in the data source. See Fig. 2 for range of D.

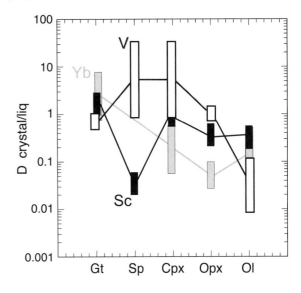

Fig. 3. End-member (highest and lowest) partition coefficients (D) measured experimentally for mantle minerals in mafic and ultramafic systems (Table 3). Note the order of partition for these mildly incompatible elements differs for each phase. The spread in V is due its change with fO_2 (Table 3).

upon mantle peridotites during their residence in the lithosphere, as is demonstrated with detailed studies of the metasomatic process.

A well-constrained study of metasomatism at the outcrop scale by McPherson et al. (1996) examined trace element abundances in peridotite surrounding amphibole-bearing veins in the Lherz orogenic massif. The veins were interpreted to be channels for melt that percolated outward into the host peridotite creating a metasomatic front. Fig. 4 shows no real change can be discerned for whole rock abundances of Sc, V, Al and Cr (not shown) with distance from the veins into the host peridotite. Only the most incompatible Yb has

been disturbed, but only for small mantle volumes near the veins. This example shows that metasomatism may only serve to inflate the concentrations of all incom-

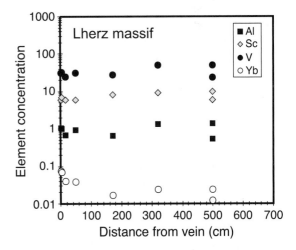

Fig. 4. Chemical data for peridotites sampled in outcrop at intervals away from metasomatic veins in the Lherz orogenic peridotite massif, France (McPherson et al., 1996). Note little change in Al, V and Ti (used as a proxy for Sc) away from the veins, and inflation of Yb relative to the former three elements, but only within a decimeter scale near the veins. Data for Cr are not shown to reduce the scale but show an identical trend to Al.

patible elements relative to the original residue, and will preferentially increase the abundances of Yb relative to Al, V, Sc or Cr.

Grain-scale heterogeneity in trace and major elements suggests that many mantle samples contain too much garnet and pyroxene for their degree of depletion, and have experienced addition of these components to an original residue, in some cases shortly before or during entrainment in their host magma (Griffin et al., 1999a,b; Pearson et al., 2002; Simon et al., 2003). The degree to which clinopyroxene and garnet components in mantle rocks are not primary and have been introduced to a residue during its residence in the lithosphere needs to be evaluated because the budgets of Al, Yb, Sc, Cr and V in bulk rock peridotites are dominated by these phases (e.g., Glaser et al., 1999; Schmidberger and Francis, 2001). The modes of clinopyroxene and garnet in highly depleted rocks are only adequately sampled in large specimens (Boyd and Mertzman, 1987), so only high-quality data from large samples is useful in this analysis. Fig. 5 shows the observed modes of these rocks compared with those expected from a partial melting model. Such a comparison is only semi-quantitative because modes of clinopyroxene and garnet in a melting model are dependent on starting compositions, and are lower than those recorded in natural samples equilibrated at temperatures below the solidus.

Most residues with greater than 2 wt.% Al_2O_3 do not contain inordinate amounts of clinopyroxene, whereas those with less than ~ 1.5 wt.% Al_2O_3 contain an excess of several percent (Fig. 5). A similar trend is observed by Pearson et al. (2002) using a different melting model and Mg# as a depletion index. The origin of the excess clinopyroxene in such rocks has been debated. Such residues may have cooled to form clinopyroxene from higher temperature Ca- and Al-rich pyroxene above the solidus, as originally proposed on textural grounds (Cox et al., 1987) and demonstrated by experiment (Canil, 1991). More quantitative mass balance and consideration of trace element abundances and heterogeneity show that exsolution does not account for sufficient clinopyroxene in these rocks, and that much of this clinopyroxene is instead "added" to a residue (Boyd and Mertzman, 1987; Canil, 1992; Shimizu et al., 1997; Simon et al., 2003). This would explain the scatter of clinopyroxene abundance in residues show-

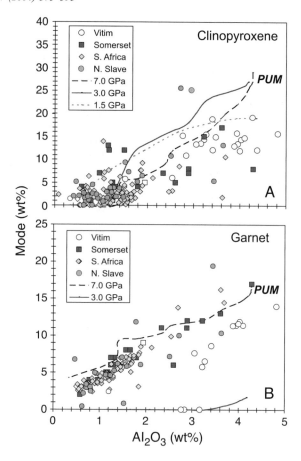

Fig. 5. Modal garnet and clinopyroxene in large (>500 g) samples of cratonic garnet peridotites from Vitim, the northern Slave, Kaapvaal and Somerset Island, Canada plotted against Al_2O_3 (see Table 2 for references). The latter element is used as a depletion index. Partial melting trends assuming a primitive upper mantle starting material (PUM) are based on the same model detailed in Canil (2002) and summarized in Fig. 2. Note the regular change in garnet, but relative scatter in clinopyroxene, with depletion. Peridotites with less than ~ 1.5 wt.% Al_2O_3 appear to contain clinopyroxene and or garnet in excess of that predicted by their degree of depletion by partial melting.

ing a range of depletion (e.g., Al_2O_3 content—Fig. 5). The effect of this added clinopyroxene on trace element abundances is evaluated further below.

The presence of garnet overgrowths in several studies shows that some component of this phase has also been introduced in the mantle lithosphere. No study has yet provided a quantitative estimate of the total mass of garnet added to a mantle rock by metasomatism, relative to the amount originally present. A survey of the size of metasomatic garnet overgrowths described in the literature (Smith et al.,

1991; Smith and Boyd, 1992; Griffin et al., 1999a,b; Simon et al., 2003) suggest that 40% to 70% of the volume of garnet was added in one or more metasomatic processes. This figure is determined by approximating the garnets as spheres and using the grain radii reported in these studies to calculate relative volumes of the original core and the overgrown rims. The volumes of the overgrowths are maxima because the garnets were assumed to have been sectioned in their geometric center, which is not likely the case.

Some authors report garnet in excess of that expected for a given degree of depletion (Pearson et al., 2002), but comparing observed modes to partial melting trends shows that garnet shows a very regular trend with depletion (Fig. 5B), rather than a scatter as observed for clinopyroxene. As in the case of clinopyroxene, the spatial association of garnet and orthopyroxene in some mantle specimens has led to the proposal for an exsolution origin from high Ca–Al orthopyroxene. Experiments on a typical cratonic peridotite residue with only 1.5 wt.% Al_2O_3 show exsolution of 3% to 7% garnet during cooling (Canil, 1991), similar to that actually observed in modes of these rocks (Fig. 5). The mass balance for exsolution of garnet also fits much better than for clinopyroxene (Canil, 1992; Simon et al., 2003).

Estimates of the effect of garnet and clinopyroxene component, if added by metasomatism, on the abundances of mildly incompatible trace elements can be made in some well-characterized, large garnet peridotites from the Nikos kimberlite, Somerset Island, Canada, for which modes, and trace element abundances in garnet, clinopyroxene and bulk rocks are known (Schmidberger and Francis, 2001). In this example, the authors did not analyze for Sc–Ti is substituted as an element with broadly similar behav-

Fig. 6. Abundances of Ti, V, Yb and Cr in cratonic garnet peridotites from Somerset Island, Canada, plotted against Al_2O_3. Shown are the original bulk analyses (measured), and re-calculated whole rock analyses in which clinopyroxene (no cpx), garnet (no gt), and both clinopyroxene and garnet (no cpx + gt), are subtracted from the bulk rock, assuming they were 'introduced' to the residue by metasomatism. The calculations used whole rock, modal and mineral chemical data for large specimens (Schmidberger and Francis, 2001). Note that relative to Al, garnet addition/subtraction does not affect trends for Yb, Sc, Cr and V, whereas clinopyroxene addition/subtraction has a notable affect, but only for V and Ti (the latter element is assumed to behave like Sc). The modeled removal of garnet and clinopyroxene also incorporates the budget for Al.

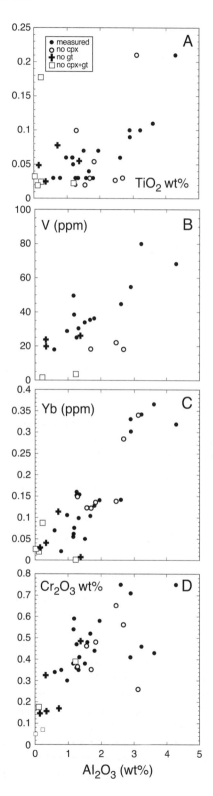

iour in the mantle. Fig. 6 shows a regular trend for mildly incompatible elements for these samples that could be consistent with simple melt depletion. If it is assumed that clinopyroxene was added to these rocks in a metasomatic process, subtraction of this phase would serve to lower the abundances of only V and Sc (Ti in this example), for a given level of depletion (Al), and would produce far more scatter, compared to the regular depletion trend (Fig. 6A,B). In contrast, subtraction of garnet, if considered to be introduced to these samples, moves samples parallel to the depletion array for all elements, towards lower Al (or level of depletion).

Thus, the effects of clinopyroxene and garnet added by a metasomatic process differ on the mildly incompatible element arrays (Fig. 6). Addition of garnet by metasomatism is indistinguishable from the melt depletion trend, whereas clinopyroxene addition will inflate some element abundances at a constant level of depletion. Relative to Al, significant garnet addition to a peridotite is not substantial for its Yb, Sc, Cr and V contents, whereas clinopyroxene has a notable affect on V and Sc (the latter assumed to behave like Ti). This was also an extreme end-member case because 100% of the garnet and clinopyroxene was assumed to be introduced; as shown above, this is certainly an overestimate for garnet. The 'depletion trends' for mildly incompatible element abundances in peridotites from the database compiled in this study can now be viewed in this light.

5. Element correlations in peridotites

5.1. Cr vs. Al

Covariation of Cr with Al_2O_3 shows essentially no change in Cr content with increasing depletion, irrespective of peridotite sample type or facies. Notable scatter in Cr is generally not present in massif peridotites from outcrop, but restricted to the xenolith and abyssal peridotite data sets. This scatter is ascribed to a 'nugget' effect for spinel. Spinel is a modally minor phase (< 2%) in peridotites but one that dominates the bulk rock budget for Cr. Spinel is adequately sampled in the typically larger specimens of massif peridotites taken in outcrops, but poorly sampled in the smaller xenolith and/or abyssal peridotite specimens (Fig. 7).

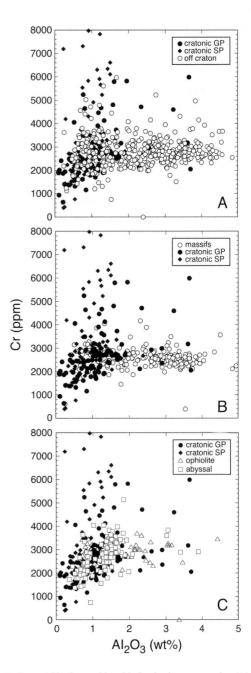

Fig. 7. Cr vs. Al in the world peridotite database comparing (A) on- and off-craton xenoliths (B) cratonic xenoliths and massif peridotites and (C) cratonic xenoliths with abyssal peridotites and ophiolites (Table 2). Note significant scatter in Cr abundances for xenolith specimens, especially at low Al content. SP—spinel peridotite, GP—garnet peridotite.

The latter specimens in the data set are mainly from Ocean Drilling Program shipboard reports where limited sample is permitted for analysis.

Chemical data from peridotite melting experiments show that bulk $D_{Cr/Al}$ between residue/melt decreases substantially with increasing pressure (Fig. 8). In order for Cr to remain constant, and the Cr/Al ratio of a residue to change with depletion, this simple observation from experimental data must require that melting occur mainly at $P < 3.0$ GPa where $D_{Cr/Al}$ between melt and residue along the solidus is large. This observation applies to all peridotite samples irrespective of their current pressure of equilibration (i.e., in the garnet- or spinel-facies). Peridotites from both the cratonic data sets (both spinel and garnet bearing), and abyssal spinel peridotites trend to low Cr at less than 2 wt.% Al_2O_3 (Fig. 7C). This trend might suggest that these peridotites formed at P greater than 4 GPa, where $D_{Cr/Al}$ is lower, leaving less Cr in the residue. An origin for abyssal peridotites at greater than 4 GPa can be ruled out on several grounds. The trend of some peridotites to low Cr/Al is most likely due to exhaustion of spinel at high degrees of depletion. Qualitatively, it suggests the garnet-

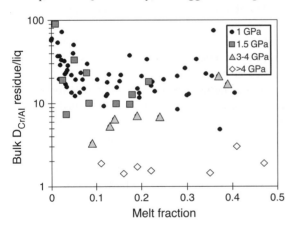

Fig. 8. Compilation of $D_{Cr/Al}$ between bulk residue and melt derived from peridotite melting experiments at various pressures plotted against the degree of partial melting. $D_{Cr/Al}$ is calculated from mass balance of chemical data from the melting experiments (Baker and Stolper, 1994; Falloon et al., 2001; Pickering-Witter and Johnston, 2000; Robinson et al., 1998; Schwab and Johnston, 2001; Walter, 1998). Experiments that did not mass balance for Cr or Al are omitted. Note melting at pressures greater than 3 GPa does not significantly fractionate Cr from Al as is observed in the mantle peridotites (Fig. 7).

bearing cratonic peridotite samples must have formed in the spinel stability field.

5.2. Yb vs. Al

The correlation of Al with Yb in mantle rocks is well known and utilized. The coherency of these two elements in meteorites and their covariation in peridotites has had much utility in estimates of primitive upper mantle (PUM) (Hart and Zindler, 1986; Jagoutz et al., 1979; McDonough and Sun, 1995). These previous studies show a straight line for this array in mantle samples but detailed inspection of samples in this larger database reveals an inflection at ~ 2 wt.% Al_2O_3 (Fig. 9).

The trend of Yb and Al in residues and its inflection are fit by the melting model only at P of 3 GPa and with the highest D_{Yb} values for all mantle minerals stable at the solidus (Fig. 9). Melting at lower pressure and high D_{Yb} predicts too little Yb in the residue, whereas too much Yb is retained for melting at 7 GPa, even using low D_{Yb} values in the calculations. This trend might suggest that all mantle residues are produced in the garnet stability field but this not the case. At 3 GPa, garnet is not stable in the melting interval of peridotite beyond 10% melting (Robinson and Wood, 1998; Walter, 1998) which corresponds to a residue with only ~ 3.5% Al_2O_3 (Fig. 2). Most of the melting interval to produce more depleted residues with less than 3 wt.% Al_2O_3 does not involve garnet, but rather clinopyroxene, which plays a major role until its exhaustion at larger degrees of melting (~ 2 wt.% Al_2O_3—Fig. 2), corresponding almost exactly to the inflection in the Yb–Al array for natural samples (Fig. 9). The match between the inflection derived from the melting model and that in the array of peridotite compositions suggests that the former, although clearly empirical, is a satisfactory description of the residues produced in nature.

Thus, most lithosphere in ophiolites, orogenic massifs and the modern ocean basins (abyssal peridotites) formed at P less than or equal to 3 GPa. Garnet is exhausted with limited melt depletion at this pressure (Fig. 2) and so does not play a major role during melting to form most mantle peridotite residues. Furthermore, the fit of melting model to the residue trend only at high values of

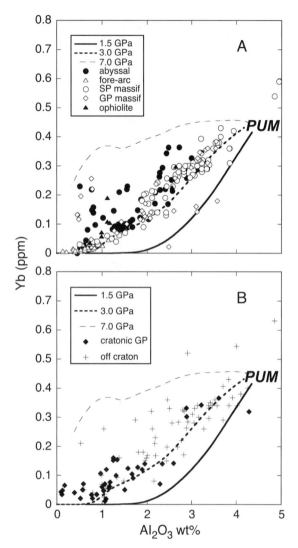

Fig. 9. Covariation of Yb and Al in mantle peridotites. (A) Data for samples from outcrop in 'known' tectonic settings (Table 1). Lines are partial melting trends calculated using a fractional melting model, melting reactions from experiment (see Canil, 2002), and lower limits for D_{Yb} values from Table 2. Higher D_{Yb} values give erratic results. (B) As above but showing cratonic garnet peridotites (GP) and off craton xenolith samples, with same partial melting models as in (A). SP—spinel peridotite, GP—garnet peridotite.

bulk D_{Yb} suggests that Yb and other HREE are quite likely very compatible in near-solidus clinopyroxene, as has been argued on other experimental evidence (Blundy et al., 1998).

The calculated Yb–Al trends can be used to interpret the P of origin of accidental volcanic-

hosted xenoliths. All peridotite xenoliths, whether from on- or off-craton, in the garnet or spinel facies, appear to form as residues at P less than 3 GPa (Fig. 9B). Almost no mantle residues plot near the trend exhibited by the 7 GPa melting model. Most interesting is that at a given level of depletion (i.e., Al content), few garnet-bearing mantle samples show the higher Yb expected for residues produced in the garnet stability field because of the large $D_{gt/liq}$ for Yb (Fig. 3). This feature further amplifies the case that garnet is not present during melting to form the residues represented by almost all of the mantle peridotites in this compilation.

5.3. V vs. Al

The behaviour of V during mantle melting is sensitive to fO_2 and variations of this element in peridotite can be a useful paleoredox indicator (Canil and Fedortchouk, 2000). Covariation between V and Al in mantle peridotites has been reviewed (Canil, 2002). Examination of more data for samples from "known" geological environments in this study adds to that analysis. It must be emphasized, however, that the modeling of peridotite V contents in terms of their 'paleoredox' during formation can still only be illustrative because of the pack of internally consistent analytical data for V (Lee et al., 2003).

Abyssal peridotites have the highest V for a given level of depletion and plot along a melting trend consistent with an fO_2 between NNO-2 and NNO-3 (Fig. 10). This fO_2 is identical to that recorded by mid-ocean ridge basalts (Carmichael, 1991) considered to be the complement of abyssal peridotites (Baker and Beckett, 1999). Many abyssal peridotite samples clearly have V in excess of what could be explained by the melting model. It is uncertain whether such samples have experienced V addition (relative to Al) by hydrothermal alteration or impregnation with melt, or have a mantle source with a V content considerably different from PUM. Some of the samples in the database are shipboard measurements requiring relatively small samples, and there may be a bias due to coarse grain sizes.

Mantle tectonites in ophiolites are also interpreted to have formed beneath spreading centers in oceanic

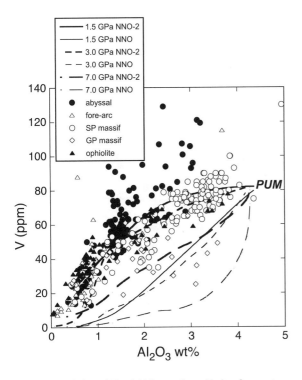

Fig. 10. Covariation of V and Al in mantle peridotites from outcrop in 'known' tectonic settings (Table 2). Comparisons of peridotite compositions with partial melting trends calculated for melting at 1.5, 3.0 and 7.0 GPa and two different log fO_2's relative to the nickel–nickel oxide (NNO) buffer using D_V values from Canil (2002). Note the lack of fit of most residue trends with melting at pressures of 7 GPa, and much better fit between 1.5 and 3 GPa. SP—spinel peridotite, GP—garnet peridotite.

lithosphere, though not necessarily within extensive ocean basins. Ophiolite mantle is slightly less enriched in V when compared to abyssal peridotites, and forms a coherent array along a depletion trend consistent with depletion at a fO_2 similar to that for abyssal peridotites today (Fig. 10). This would suggest that perhaps the scatter to high V contents in many abyssal rocks is a secondary effect.

In contrast, massif peridotites, considered to be sub-continental mantle exhumed in orogenic settings (Den Tex, 1969; Menzies and Dupuy, 1991) form an array with even less V, likely due to melting at higher fO_2 in the continental environment (see also Woodland et al., 1992). There is also no real difference between garnet- or spinel-bearing massif peridotites, though the ultrahigh P garnet peridotites from the Dabie Sulu region form a distinctly low V

array (Fig. 10). This has been attributed to formation at very high fO_2 in a convergent margin environment (Canil, 2002). Note from the analysis above that metasomatism in the continental lithosphere would serve to increase the levels of V in the rock at the expense of Al. This means that trends on this diagram require an even higher fO_2 if the rocks had V disturbed by metasomatism.

A large proportion of cratonic peridotite xenoliths form a trend distinct from many other peridotite types. Many of these rocks have low V for a given Al_2O_3 content (Fig. 11). One interpretation is that most cratonic peridotites were formed by melt extraction at a fO_2 higher than that during formation of other types of continental or oceanic mantle lithosphere, perhaps in a convergent margin setting (Canil, 2002). The bulk D_V for melting in the garnet facies is distinctly lower than that in the spinel facies (Canil and Fedortchouk, 2000) and so the low V in many cratonic peridotites could also be equally attributed to higher P of origin. In the absence of other data, it is difficult to separate the effects of P from fO_2 based on

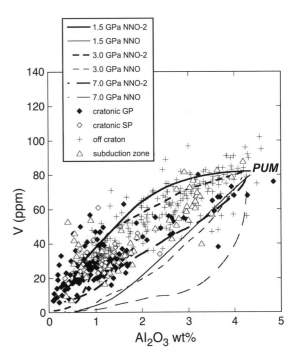

Fig. 11. Covariation of V and Al in mantle peridotites sampled as xenoliths compared with partial melting trends calculated as in Fig. 10. SP—spinel peridotite, GP—garnet peridotite.

V content of cratonic peridotites alone. The covariation of V and Yb, however, can be used to unravel the
effect of P versus that of fO_2 in causing the distinct V
contents of different types of mantle residues, because
Yb is very sensitive to the pressure of melting (Fig. 9).

5.4. V vs. Yb

The Yb–Al covariation in most peridotites is
difficult to explain at any P of melting greater than
3 GPa (Fig. 9). This pressure is a maximum, because
if the rocks were metasomatized, Yb contents are
inflated relative to Al (Fig. 4). Furthermore, the
covariation of V and Yb in residues from known
geological environments (Fig. 12A) also makes clear
that none of the residues from orogenic massifs,
ophiolites or abyssal peridotites is consistent with a
pressure of melting greater than 3 GPa. The distinctly
high V at a given Yb content for abyssal peridotite
remains consistent with a lower fO_2 for formation of
these peridotites compared to those in ophiolites and
massifs.

There is relatively less data for Yb and V in xenoliths, but many cratonic peridotites plot along a distinct array of low V for a given Yb content (Fig. 12B), a
distribution similar to that observed in V–Al space
(Fig. 11). Thus, it appears that the lower V in these
residues is caused by a low bulk D_V during melting at
higher fO_2, and pressures below 3 GPa, rather than by
melting at higher pressures and lower fO_2.

5.5. Sc vs. Al

Covariations of Sc with Al show an opposite trend
than would be predicted by trends with V and Cr. The
abyssal peridotite array does not extrapolate to primitive upper mantle values for Sc, but spreads along a
distinct trend with high Sc for a given Al content (Fig.
13A). The reasons for this are not presently understood—they may be analytical, or may concern the
different behavior of Sc in the mantle than can be
modelled here. Alternatively, it is possible the source
for abyssal peridotite has a Sc content higher than that
in PUM.

The trends for Sc–Al, as well as the inflections at a
given degree of depletion in residues are fit by the
melting trends at pressures between 1 and 3 GPa and
the range of D_{Sc} values encompassed in the experi-

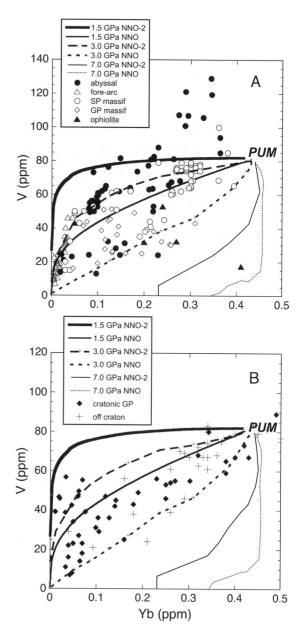

Fig. 12. (A) Covariation of V and Yb in mantle peridotites from
outcrop in 'known' tectonic settings (Table 1) compared with partial
melting trends calculated for melting at 1.5, 3.0 and 7.0 GPa and
two different log fO_2's relative to the nickel–nickel oxide (NNO)
buffer using D_V values from Canil (2002) and D_{Yb} from Table 3.
(B) As above but comparing melting models with xenolith data.
Note the lack of fit of the residue trends with melting at pressures
above 3 GPa. SP—spinel peridotite, GP—garnet peridotite.

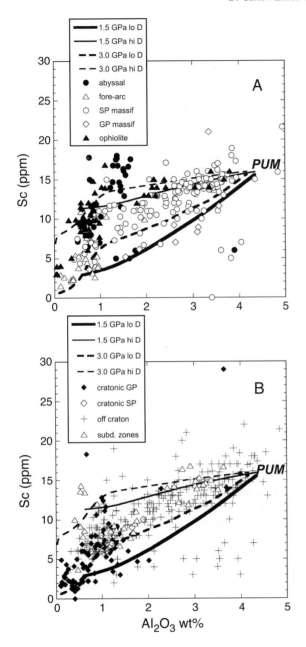

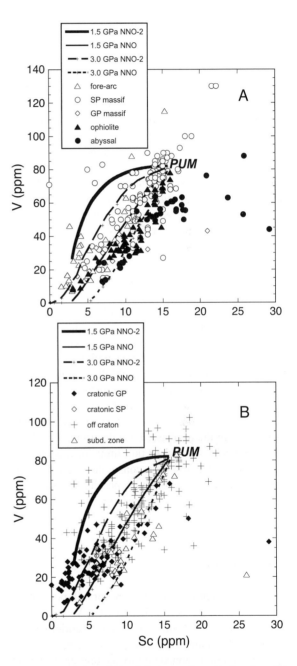

Fig. 13. Covariation of Sc and Al in mantle peridotites from (A) outcrop in 'known' tectonic settings (Table 1), and (B) in xenoliths, compared with partial melting trends calculated for melting at 1.5 and 3.0 GPa using high and low D_{Sc} values from Table 3.

mental measurements. This is especially true for the ophiolite data sets. The xenolith data sets show the same distribution, but there is considerable scatter that may represent an analytical effect (Fig. 13B).

Fig. 14. Covariation of V and Sc in mantle peridotites from (A) outcrop in 'known' tectonic settings (Table 1) and (B) in xenoliths, compared with partial melting trends calculated for melting at 1.5 and 3.0 GPa and two different fO_2's using D_V values from Canil (2002) and low D_{Sc} values from Table 3.

5.6. V vs. Sc

The covariation of these two elements in samples of known geological environment show curious sub-parallel arrays for abyssal, massif and ophiolite-hosted peridotites (Fig. 14A). If it is assumed that Sc behaves similar to Al, then it would appear that abyssal peridotites are low, rather than high in V at a given level of depletion. This is unlikely, because a strong case can be made that V contents are mainly controlled by fO_2 during melting, and that the lower fO_2 of melting at mid-ocean ridges produces the higher V in abyssal peridotites, their complementary residue. The Sc–V trend in abyssal peridotites is perplexing and cannot be explained by any melting model at any reasonable D_{Sc} value (Fig. 14A). One interpretation could be that abyssal peridotites have a mantle source with Sc and V contents that differ from those of PUM. More and better data for the latter samples are required to test this idea.

The xenolith data sets show much scatter, again likely due to analytical effects. A significant number of cratonic peridotites at high levels of depletion have low Sc/V whereas Sc/V increases with level of depletion in abyssal peridotites, ophiolites, and massifs (Fig. 14B). The change in this ratio to low values in the cratonic peridotite data could be due to a detection limit effect in the analyses of these rocks. In most samples from the database, Sc is determined by XRF, which typically has a detection limit for this element near 10 ppm. Because few constraints are offered in publications which report bulk peridotite data, more robust statements cannot be made on the quality of the analytical data for Sc and V.

6. Conclusions

An over-arching observation for all peridotites examined with respect to their Al, Cr, V, Sc and Yb abundances is that almost all mantle peridotites, regardless of sample type (abyssal, orogenic massif, ophiolite, on/off craton xenolith), tectonic environment (divergent margin, convergent margin, passive margin, craton, peri-craton) or the final P of equilibration (plagioclase-, spinel-, or garnet facies) likely originated as lithosphere produced mainly at less than 3 GPa.

Based on whole rock data and residue trends, the original depletion events to form the cratonic mantle lithosphere occurred at lower depths than those at which the mantle was later sampled by kimberlite (> 120 km depth), requiring tectonic transport of lithosphere to these depths beneath the craton, likely by underthrusting, stacking or subduction of originally shallow, spinel-facies residues (Helmstaedt and Schulze, 1989). A similar conclusion has been made on the basis of other evidence (Canil and Wei, 1992; Kelemen et al., 1998; Stachel et al., 1998) and convincing evidence for this process is provided by seismic images of 'frozen subduction' at mantle depths along the western margin of the Archean Slave Province (Bostock, 1998; Cook et al., 1998). In this light, it is difficult to ascribe any large volume of the mantle roots beneath cratons to plume sub-cretion (e.g., Griffin et al., 1999a) or analogous vertical tectonics that arise in plate-less numerical simulations (De Smet et al., 2000). The former models require the generation of residues at pressures far above those required by the melting models using known partitioning of mildly incompatible elements between mantle minerals and melts.

Acknowledgements

This paper is dedicated to F.R. 'Joe' Boyd, a beacon in science. I am grateful to I. Wada, R. Rhodes, J. Thom and P. Leong for entering much of the literature data in this compilation. I also thank C. McCammon, L. Coogan, W. Griffin and H. Grutter for reviews of the paper. This research is supported by a Discovery Grant from NSERC of Canada.

References

Aoki, K., 1981. Major element geochemistry of chromian spinel peridotite xenoliths in the Green Knobs Kimberlite, New Mexico. Sci. Rep. Tohoku Univ., Ser. 3. Mineral. Petrol. Econ. Geol. 1, 127–130.

Aumento, F., Loubat, H., 1971. The Mid-Atlantic ridge near 45 degrees N; XVI, Serpentinized ultramafic intrusions. Can. J. Earth Sci. 8, 631–633.

Baker, M.B., Beckett, J.R., 1999. The origin of abyssal peridotites, a reinterpretation of constraints based on primary bulk compositions. Earth Planet. Sci. Lett. 171, 49–61.

Baker, M.B., Stolper, E.M., 1994. Determining the composition of high-pressure mantle melts using diamond aggregates. Geochim. Cosmochim. Acta 58, 2811–2827.

Beattie, P., 1994. Systematics and energetics of trace element partitioning between olivine and silicate melts: implications for the nature of mineral/melt partitioning. Chem. Geol. 117, 57–71.

Becker, H., 1996. Geochemistry of garnet peridotite massifs from lower Austria and the composition of deep lithosphere beneath a Palaeozoic convergent plate margin. Chem. Geol. 134, 49–65.

Bedini, R.M., Bodinier, J.-L., Dautria, J.-M., Morten, L., 1997. Evolution of LILE-enriched small melt fractions in the lithospheric mantle: a case study from the East African Rift. Earth Planet. Sci. Lett. 153, 67–83.

Bernstein, S., Kelemen, P.B., Brooks, C.K., 1998. Depleted spinel harzburgite xenoliths in Tertiary dykes from East Greenland: Restites from high degree melting. Earth Planet. Sci. Lett. 154, 221–235.

Berry, R.F., 1981. Petrology of the Hili Manu lherzolite, East Timor. J. Geol. Soc. Aust. 28, 453–469.

Blundy, J.D., Robinson, J.A.C., Wood, B.J., 1998. Heavy REE are compatible in clinopyroxene on the spinel lherzolite solidus. Earth Planet. Sci. Lett. 160, 493–504.

Bodinier, J.L., 1988. Geochemistry and petrogenesis of the Lanzo peridotite body, Western Alps. Tectonophysics 149, 67–88.

Bodinier, J.L., Dupuy, C., Dostal, J., 1988. Geochemistry and petrogenesis of Eastern Pyrenean peridotites. Geochim. Cosmochim. Acta 52, 2893–2907.

Bonatti, E., Ottonello, G., Hamlyn, P.R., 1986. Peridotites from the island of Zabargad (St. John), Red Sea: petrology and geochemistry. J. Geophys. Res. 91, 599–631.

Bostock, M.G., 1998. Mantle stratigraphy and evolution of the Slave province. J. Geophys. Res. 103, 21183–21200.

Boyd, F.R., 1999. Spinel-facies peridotites from the Kaapvaal root. In: Gurney, J. (Ed.), Proc. 7th Inter. Kimb. Conf.. Red Roof Designs, Capetown, pp. 40–48.

Boyd, F.R., Mertzman, S.A., 1987. Composition and structure of the Kaapvaal lithosphere, southern Africa. In: Mysen, B.O. (Ed.), Magmatic Processes, Physicochemical Principles. Geochemical Society, Washington, DC, pp. 13–24.

Boyd, F.R., Nixon, P.H., Pearson, D.G., Mertzman, S.A., 1993. Low-calcium garnet harzburgites from southern Africa: their relations to craton structure and diamond crystallization. Contrib. Mineral. Petrol. 113, 352–366.

Boyd, F.R., et al., 1997. Composition of the Siberian cratonic mantle: evidence from Udachnaya peridotite xenoliths. Contrib. Mineral. Petrol. 128, 228–246.

Brandon, A.D., Snow, J.E., Walker, R.J., Morgan, J.W., Mock, T.D., 2000. 190Pt–186Os and 187Re–187Os systematics of abyssal peridotites. Earth Planet. Sci. Lett. 177, 319–335.

Burnham, O., Rogers, N.W., Pearson, D.G., van Calsteren, P.W., Hawkesworth, C.J., 1998. The petrogenesis of the eastern Pyrenean peridotites: An integrated study of their whole rock geochemistry and Re–Os isotope composition. Geochim. Cosmochim. Acta 62, 2293–2310.

Canil, D., 1991. Experimental evidence for the exsolution of cra-

tonic peridotite from high-temperature harzburgite. Earth Planet. Sci. Lett. 106, 64–72.

Canil, D., 1992. Orthopyroxene stability along the peridotite solidus and the origin of cratonic lithosphere beneath southern Africa. Earth Planet. Sci. Lett. 111, 83–95.

Canil, D., 1999. The Ni-in-garnet geothermometer: calibration at natural abundances. Contrib. Mineral. Petrol. 136, 240–246.

Canil, D., 2002. Vanadium in peridotites, mantle redox and tectonic environments: Archean to present. Earth Planet. Sci. Lett. 195, 75–90.

Canil, D., Fedortchouk, Y., 2000. Clinopyroxene–liquid partitioning for vanadium and the oxygen fugacity during formation of cratonic and oceanic mantle lithosphere. J. Geophys. Res. 105, 26003–26016.

Canil, D., Fedortchouk, Y., 2001. Olivine–liquid partitioning of vanadium and other trace elements, with applications to ancient and modern picrites. Can. Mineral. 39, 319–330.

Canil, D., Wei, K., 1992. Constraints on the origin of mantle-derived low Ca garnets. Contrib. Mineral. Petrol. 109, 421–430.

Canil, D., Johnston, S.T., Evers, K., Shellnutt, J.G., Creaser, R.C., 2003. Mantle exhumation in an early Paleozoic passive margin, northern Cordillera, Yukon. J. Geol. 111, 313–327.

Carlson, R., Irving, A.J., Hearn Jr., B.C. 1999. Peridotites of the Wyoming craton. In: Gurney, J. (Ed.), Proc. 7th Inter. Kimb. Conf. Red Roof Designs, Capetown, pp. 440–448.

Carmichael, I.S.E., 1991. The redox states of basic and silicic magmas: a reflection of their source regions? Contrib. Mineral. Petrol. 106, 129–141.

Casey, J.F., 1997. Comparison of major and trace-element geochemistry of abyssal peridotites and mafic plutonics with basalts from the MARK region of the Mid-Atlantic Ridge. In: Karson, J.A., Cannat, M., Miller, D.J., et al. (Eds.), Proc. Sci. Results ODP Leg, vol. 153. Ocean Drilling Program, College station, Texas, pp. 181–241.

Chauvel, C., Jahn, B.M., 1984. Nd–Sr isotope and REE geochemistry of alkali basalts from the Massif Central, France. Geochim. Cosmochim. Acta 48, 93–110.

Coogan, L., et al. Abyssal peridotites and basalts along a flow line from the southwest Indian Ridge. Chem. Geol., in press.

Cook, F.A., van der Velden, A.J., Hall, K.W., Roberts, B.J., 1998. Tectonic delamination and subcrustal imbrication of the Precambrian lithosphere in northwestern Canada mapped by LITHOPROBE. Geology 26, 839–842.

Cox, K.G., Smith, M.R., Beswetherick, S., 1987. Textural studies of garnet lherzolites: evidence of exsolution origin from high-temperature harzburgites. In: Nixon, P.H. (Ed.), Mantle Xenoliths. Wiley, Chichester, pp. 537–550.

Dautria, J.M., Girod, M., 1986. Les enclaves de lherzolite a spinelle et plagioclase du volcan DeDibi (Adamoua, Camerou):des temoins d'un manteau superieur anormal. Bull. Mineral. 109, 275–288.

Dautria, J.M., Dupuy, C., Takherist, D., Dostal, J., 1992. Carbonate metasomatism in the lithospheric mantle: peridotitic xenoliths from a melilitic district of the Sahara basin. Contrib. Mineral. Petrol. 111, 37–52.

Den Tex, E., 1969. Origin of ultramafic rocks, their tectonic setting and history: a contribution to the discussion of the paper "The origin of ultramafic and ultrabasic rocks" by P.J. Wyllie. Tectonophysics 7, 457–488.

De Smet, J., Van den Berg, A.P., Vlaar, N.J., 2000. Early formation and long-term stability of continents resulting from decompression melting in a convecting mantle. Tectonophysics 322, 19–33.

Dupuy, C., Dostal, J., Bodinier, J.L., 1987. Geochemistry of spinel peridotite inclusions in basalts from Sardinia. Mineral. Mag. 51, 561–568.

Ehrenberg, S.N., 1982. Petrogenesis of garnet lherzolite and megacrystalline nodules from the Thumb, Navajo volcanic field. J. Petrol. 23, 505–545.

Embey-Isztin, A., Scharbert, H.G., Dietrich, H., Poultidis, H., 1989. Petrology and geochemistry of peridotite xenoliths in alkali basalts from the Transdanubian volcanic region, west Hungary. J. Petrol. 30, 79–105.

Ernst, W.G., 1978. Petrochemical study of lherzolitic rocks from the Western Alps. J. Petrol. 19, 341–392.

Fabries, J., Lorand, J.P., Bodinier, J.L., Dupuy, C., 1991. Evolution of the upper mantle beneath the Pyrenees: evidence from orogenic spinel lherzolite massifs. J. Petrol., 55–76 (Spec. Lherzolites Issues).

Falloon, T.J., Danyushevsky, L.V., Green, D.H., 2001. Peridotite melting at 1 GPa: reversal experiments on partial melt compositions produced by peridotite–basalt sandwich experiments. J. Petrol. 42, 2363–2390.

Francis, D., 1987. Mantle–melt interaction recorded in spinel lherzolite xenoliths from the Alligator Lake volcanic complex, Yukon, Canada. J. Petrol. 28, 569–597.

Franz, L., Becker, K.-P., Kramer, W., Herzig, P.M., 2002. Metasomatic mantle xenoliths from the Bismarck microplate (Papua New Guinea)—Thermal evolution, geochemistry and extent of slab-induced metasomatism. J. Petrol. 43, 315–343.

Frey, F.A., Green, D.H., 1974. The mineralogy, geochemistry and origin of lherzolite inclusions in Victorian basanites. Geochim. Cosmochim. Acta 38, 1023–1059.

Frey, F.A., Suen, C.J., Stockman, H.W., 1985. The Ronda high temperature peridotite: geochemistry and petrogenesis. Geochim. Cosmochim. Acta 49, 2469–2491.

Gaetani, G.A., Grove, T.L., 1995. Partitioning of rare earth elements between clinopyroxene and silicate melt: crystal chemical controls. Geochim. Cosmochim. Acta 59, 1951–1962.

Gillis, K., Mevel, C., Allan, J., et al., 1993. Initial Reports-Proceedings of Ocean Drilling Program Hess Deep, vol. 147. Ocean Drilling Program, College Station, Texas, pp. 1–29.

Glaser, S., Foley, S.F., Gunther, D., 1999. Trace element compositions of minerals in garnet and spinel peridotite xenoliths from the Vitim volcanic field, Transbaikalia, eastern Siberia. Lithos 48, 263–285.

Gregoire, M., Moine, B.N., O'Reilly, S.Y., Cottin, J.Y., Giret, A., 2000. Trace element residence and partitioning in mantle xenoliths metasomatized by highly alkaline, silicate- and carbonate-rich melts (Kerguelen Islands, Indian Ocean). J. Petrol. 41, 477–509.

Griffin, W.L., Sutherland, F.L., Hollis, J.D., 1987. Geothermal profile and crust–mantle transition beneath east-central Queensland: volcanology, xenolith petrology and seismic data. J. Volcanol. Geotherm. Res. 31, 177–203.

Griffin, W.L., O'Reilly, S.Y., Ryan, C.G., 1999a. The composition and origin of sub-continental lithospheric mantle. In: Fei, Y., Bertka, C.M., Mysen, B.O. (Eds.), Mantle Petrology, Field Observations and High Pressure Experimentation. The Geochemical Society, Washington, DC.

Griffin, W.L., Shee, S.R., Ryan, C.G., Win, T.T., Wyatt, B.A., 1999b. Harzburgite to lherzolite and back again: metasomatic processes in ultramafic xenoliths from the Wesselton kimberlite, Kimberley, South Africa. Contrib. Mineral. Petrol. 134, 232–250.

Gruau, G., Lecuyer, C., Bernard-Griffiths, J., Morin, N., 1991. Origin and petrogenesis of the Trinity Ophiolite Complex (California): new constraints from REE and Nd isotope data. J. Petrol., 229–242 (Spec. Lherz. Issue).

Gruau, G., Bernard-Griffiths, J., Lecuyer, C., 1998. The origin of U-shaped rare earth patterns in ophiolite peridotites: assessing the role of secondary alteration and melt/rock reaction. Geochim. Cosmochim. Acta 62, 3545–3560.

Gueddari, K., Piboule, M., Amossé, J., 1996. Differentiation of platinum-group elements (PGE) and of gold during partial melting of peridotites in the lherzolitic massifs of the Betico-Rifean range (Ronda and Beni Bousera). Chem. Geol. 134, 181–197.

Hart, S.R., Dunn, T., 1993. Experimental cpx/melt partitioning of 24 trace elements. Contrib. Mineral. Petrol. 113, 1–8.

Hart, S.R., Zindler, A., 1986. In search of a bulk earth composition. Chem. Geol. 57, 247–267.

Hartmann, G.H., Wedepohl, K.H., 1993. The composition of peridotite tectonites from the Ivrea Complex, northern Italy: residues of melt extraction. Geochim. Cosmochim. Acta 57, 1761–1782.

Hauri, E., Shimizu, N., Dieu, J.J., Hart, S.R., 1993. Evidence for hotspot-related carbonatite metasomatism in the oceanic upper mantle. Nature 365, 221–227.

Hauri, E.H., Wagner, T., Grove, T.L., 1994. Experimental and natural partitioning of Th, U, Pb and other trace elements between garnet, clinopyroxene and basaltic. Chem. Geol. 117, 149–166.

Helmstaedt, H., Schulze, D.J., 1989. Southern African kimberlites and their mantle sample: implications for the Archaean tectonics and lithosphere evolution. Spec. Publ.-Geol. Soc. Aust. 14, 358–368.

Herzberg, C.H., 1995. Generation of plume magmas through time: an experimental perspective. Chem. Geol. 126, 1–16.

Horn, I., Foley, S.F., Jackson, S.E., Jenner, G.A., 1994. Experimentally determined partitioning of high field strength and selected transition elements between spinel and basaltic melt. Chem. Geol. 117, 193–218.

Hunter, R.H., Upton, B.G.J., 1987. The British Isles—A Paleozoic mantle sample. In: Nixon, P.H.N. (Ed.), Mantle Xenoliths. Wiley, New York, pp. 107–118.

Ionov, D., Ashchepkov, I.V., Stosch, H.-G., Witt-Eickschen, G., Seck, H.A., 1993. Garnet peridotite xenoliths from the Vitim volcanic field, Baikal region: the nature of the garnet–spinel peridotite transition zone in the continental mantle. J. Petrol. 34, 1141–1175.

Jagoutz, E., et al., 1979. The abundances of major, minor and trace elements in the earth's mantle as derived from primitive ultramafic nodules. 10th Lunar and Planetary Science Conference. NASA, Houston, TX, pp. 2031–2050.

Jaques, A.L., Chappell, B.W., 1980. Petrology and trace element geochemistry of the Papuan Ultramafic belt. Contrib. Mineral. Petrol. 75, 55–70.

Jaques, A.L., O'Neill, H.S.C., Smith, C.B., Moon, J., Chappell, B.W., 1990. Diamondiferous peridotite xenoliths from the Argyle (AK1) lamproite pipe, western Australia. Contrib. Mineral. Petrol. 104, 255–276.

Johnson, K.T.M., 1998. Experimental determination of partition coefficients for rare earth and high-field strength elements between clinopyroxene, garnet and basaltic melt at high pressures. Contrib. Mineral. Petrol. 133, 60–68.

Johnson, K.T.M., Dick, H.J.B., Shimizu, N., 1990. Melting in the oceanic upper mantle: an ion microprobe study of diopsides in abyssal peridotites. J. Geophys. Res. 95, 2661–2678.

Jordan, T.H., 1975. The continental tectosphere. Rev. Geophys. Space Phys. 13, 1–12.

Kelemen, P.B., Hart, S.R., Bernstein, S., 1998. Silica enrichment in the continental upper mantle via melt/rock reaction. Earth Planet. Sci. Lett. 164, 387–406.

Kennedy, A.K., Lofgren, G.E., Wasserburg, G.J., 1992. An experimental study of trace element partitioning between olivine, orthopyroxene and melt in chondrules: equilibrium values and kinetic effects. Earth Planet. Sci. Lett. 115, 177–195.

Kopylova, M., Russell, J.K., 2000. Chemical stratification of cratonic lithosphere: constraints from the northern Slave craton, Canada. Earth Planet. Sci. Lett. 181, 71–87.

Laurora, A., et al., 2001. Metasomatism and melting in carbonated peridotite xenoliths from the mantle wedge: the Gobernador Gregores Case (southern Patagonia). J. Petrol. 42, 69–87.

Lee, C.-T., Rudnick, R.L., 1999. Compositionally stratified cratonic lithosphere: petrology and geochemistry of peridotite xenoliths from the Labait volcano, Tanzania. 7th International Kimberlite Conference. Red Roof Designs, Capetown, pp. 503–521.

Lee, C.T., Brandon, A., Norman, M.D., 2003. Vanadium as a proxy for paleo-fO2 during partial melting: prospects, limitations and implications. Geochim. Cosmochim. Acta 67, 3045–3064.

Lenoir, X., Garrido, C.J., Bodinier, J.-L., Dautria, J.-M., 2000. Contrasting lithospheric mantle domains beneath the Massif Central (France) revealed by geochemistry of peridotite xenoliths. Earth Planet. Sci. Lett. 181, 359–375.

Lenoir, X., Garrido, C.J., Bodinier, J.-L., Dautria, J.-M., Gervilla, F., 2001. The recrystallization front of the Ronda Peridotite: evidence for melting and thermal erosion of subcontinental lithospheric mantle beneath the Alboran Basin. J. Petrol. 42, 141–158.

Liang, Y., Elthon, D., 1990. Geochemistry and petrology of spinel lherzolite xenoliths from Xalapasco de La Joya, San Luis Potosi, Mexico, Partial melting and mantle metasomatism. J. Geophys. Res. 95, 15859–15878.

Loney, R.A., Himmelberg, G.R., Coleman, R.G., 1971. Structure and petrology of the Alpine-type peridotite at Burro Mountain, California, USA. J. Petrol. 12, 245–310.

Luhr, J.F., Aranda-Gomez, J.J., 1997. Mexican peridotite xenoliths and tectonic terranes: correlations among vent location, texture, temperature, pressure, and oxygen fugacity. J. Petrol. 38, 1075–1112.

Maaloe, S., Aoki, K., 1977. The major element composition of the upper mantle estimated from the composition of lherzolites. Contrib. Mineral. Petrol. 63, 161–173.

Maury, R., Defant, M., Joron, J.-L., 1992. Metasomatism of the sub-arc mantle inferred from trace elements in Philippine xenoliths. Nature 360, 661–663.

McDonough, W.F., Sun, S.S., 1995. The composition of the earth. Chem. Geol. 120, 223–253.

McInnes, B.I.A., Gregoire, M., Binns, R.A., Herzig, P.M., Hannington, M.D., 2001. Hydrous metasomatism of oceanic sub-arc mantle, Lihir, Papua New Guinea: petrology and geochemistry of fluid-metasomatised mantle wedge xenoliths. Earth Planet. Sci. Lett. 188, 169–183.

McKenzie, D., O'Nions, R.K., 1991. Partial melt distributions from inversion of rare earth element concentrations. J. Petrol. 32, 1021–1092.

McPherson, E., Thirlwall, M.F., Parkinson, I.J., Menzies, M.A., Bodinier, J.L., Woodland, A., Bussod, G., 1996. Geochemistry of metasomatism adjacent to amphibole-bearing veins in the Lherz peridotite massif. Chem. Geol. 134, 135–157.

Menzies, M.A., Hawkesworth, C.J., 1987. Upper mantle processes and composition. In: Nixon, P.H.N. (Ed.), Mantle Xenoliths. Wiley, New York, pp. 725–738.

Menzies, M., Dupuy, C., 1991. Orogenic massifs: protolith, process and provenance. J. Petrol., 1–16 (Spec. Lherzolites Issues).

Miyashiro, A., Shido, F., Ewing, M., 1969. Composition and origin of serpentinites from the mid-Atlantic Ridge near 24° and 30° north latitude. Contrib. Mineral. Petrol. 23, 117–127.

Morten, L., 1987. Italy: a review of xenolithic occurrences and their comparison with Alpine peridotites. In: Nixon, P.H. (Ed.), Mantle Xenoliths. Wiley, New York, pp. 135–148.

Nagasawa, H., Schreiber, H.D., Morris, R.V., 1980. Experimental mineral/liquid partition coefficients of the rare Earth elements, Sc and Sr for perovskite, spinel and melilite. Earth Planet. Sci. Lett. 46, 431–437.

Niu, Y., Hekinian, R., 1997. Basaltic liquids and harzburgitic residues in the Garrett Transform: a case study at fast spreading ridges. Earth Planet. Sci. Lett. 146, 243–258.

Palme, H., Nickel, K., 1985. Ca/Al ratio and composition of the Earth's mantle. Geochim. Cosmochim. Acta 49, 2123–2132.

Parkinson, I.J., Pearce, J.A., 1998. Peridotites from the Izu-Bonin-Mariana Forearc (ODP Leg 125): evidence for mantle melting and melt–mantle interaction in a supra-subduction zone setting. J. Petrol. 39, 1577–1618.

Pearce, J.A., Barker, P.F., Edwards, S.J., Parkinson, I.J., Leat, P.T., 2000. Geochemistry and tectonic significance of peridotites from the South Sandwich arc-basin system, South Atlantic. Contrib. Mineral. Petrol. 139, 36–53.

Pearson, D.G., Irvine, G.J., Carlson, R.W., Kopylova, M.G., Ionov, D.A., 2002. The development of lithospheric keels beneath the earliest continents: time constraints using PGE and Re–Os isotope systematics. In: Fowler, C.M.R., Ebinger, C.J., Hawkesworth, C.J. (Eds.), The Early Earth Physical, Chemical and

Biological Development. Spec. Publ.-Geol. Soc. London, vol. 199, pp. 65–90.

Pearson, D.G., Canil, D., Shirey, S.B., 2003. Mantle samples included in volcanic rocks: xenoliths and diamonds. In: Carlson, R.W. (Ed.), Treatise in Geochemistry, vol. 2. Elsevier, Amsterdam, pp. 171–276.

Peltonen, P., Huhma, H., Tyni, M., Shimizu, N., 1999. Garnet peridotite xenoliths from kimberlites of Finland: nature of the continental mantle at an Archean craton–Proterozoic mobile belt transition. In: Gurney, J. (Ed.), Proc. 7th Int. Kimb. Conf. Red Roof Designs, Capetown, pp. 664–670.

Peslier, A.H., Francis, D., Ludden, J., 2002. The lithospheric mantle beneath continental margins: melting and melt–rock reaction in Canadian Cordillera Xenoliths. J. Petrol. 43, 2013–2047.

Pickering-Witter, J., Johnston, A.D., 2000. The effects of variable bulk composition on the melting systematics of fertile peridotitic assemblages. Contrib. Mineral. Petrol. 140, 190–211.

Poudjom Djomani, Y.H., O'Reilly, S.Y., Griffin, W.L., Morgan, P., 2001. The density structure of subcontinental lithosphere through time. Earth Planet. Sci. Lett. 184, 605–621.

Press, S., Witt, G., Seck, H.A., Eonov, D., Kovalenko, V.I., 1986. Spinel peridotite xenoliths from the Tariat Depression, Mongolia: II. Geochemistry and Nd and Sr isotopic composition and their implications for the evolution of the subcontinental lithosphere. Geochim. Cosmochim. Acta 50, 2587–2599.

Prinz, M., et al., 1976. Ultramafic and mafic dredge samples from the equatorial Mid-Atlantic Ridge and fracture zones. J. Geophys. Res. 81, 4087–4103.

Qi, Q., Taylor, L.A., Zhou, X., 1995. Petrology and geochemistry of mantle peridotite xenoliths from SE China. J. Petrol. 36, 55–79.

Rampone, E., Piccardo, G.B., Vannucci, R., Bottazzi, P., Ottoline, L., 1993. Subsolidus reactions monitored by trace element partitioning: the spinel- to plagioclase-facies transition in mantle peridotes. Contrib. Mineral. Petrol. 115, 1–17.

Rampone, E., et al., 1996. Trace element and isotope geochemistry of depleted peridotites from an N-MORB type ophiolite (internal Liguride, N. Italy). Contrib. Mineral. Petrol. 123, 61–76.

Robinson, J.A.C., Wood, B.J., 1998. The depth of the spinel to garnet transition at the peridotite solidus. Earth Planet. Sci. Lett. 164, 277–284.

Robinson, J.A.C., Wood, B.J., Blundy, J.D., 1998. The beginning of melting of fertile and depleted peridotite at 1.5 GPa. Earth Planet. Sci. Lett. 155, 97–111.

Rudnick, R.J., McDonough, W.F., Chappell, B.W., 1993. Carbonatite metasomatism in the northern Tanzanian mantle: petrographic and geochemical characteristics. Earth Planet. Sci. Lett. 114, 463–475.

Scambelluri, M., Rampone, E., Piccardo, G., 2001. Fluid and element cycling in subducted serpentinite: a trace-element study of the Erro–Tobbio High-Pressure Ultramafites (Western Alps, NW Italy). J. Petrol. 42, 55–67.

Schmidberger, S.S., Francis, D., 2001. Constraints on the trace element composition of the Archean mantle root beneath Somerset Island, Arctic Canada. J. Petrol. 42, 1095–1117.

Schwab, B.E., Johnston, A.D., 2001. Melting systematics of modally variable, compositionally intermediate peridotites and effects of mineral fertility. J. Petrol. 42, 1789–1811.

Schwandt, C., McKay, G.A., 1998. Rare earth element partition coefficients from enstatite/melt synthesis experiments. Geochim. Cosmochim. Acta 62, 2845–2848.

Seifert, K., Brunotte, D., 1996. Geochemistry of serpentinized mantle peridotite from Site 897 in the Iberia Abyssal Plain. Proc. Ocean Drill. Program Sci. Result 149, 143.

Shervais, J.W., Mukasa, S.B., 1991. The Balmuccia orogenic lherzolite massif, Italy. Journal of Petrology, 155–174 (Special issue).

Shi, L., Francis, D., Ludden, J., Frederiksen, A., Bostock, M., 1998. Xenolith evidence for lithospheric melting above anomalously hot mantle under the northern Canadian Cordillera. Contrib. Mineral. Petrol. 131, 39–53.

Shimizu, N., Sobolev, N.V., Yefimova, E.S., 1997. Chemical heterogeneities of inclusion garnets and juvenile character of peridotitic diamonds from Siberia. Russ. Geol. Geophys. 38, 356–372.

Siena, F., Beccaluva, L., Coltorti, M., Marchesi, S., Morra, V., 1991. Ridge to Hot-spot evolution of the Atlantic lithospheric mantle: evidence from Lanzarote peridotite xenoliths (Canary Islands). J. Petrol., 269–289 (Special Lherzolites Issue).

Simon, N.S.C., Irvine, G.J., Davies, G.R., Pearson, D.G., Carlson, R.W., 2003. The origin of garnet and clinopyroxene in "depleted" Kaapvaal peridotites. Lithos 71, 289–322.

Smith, D., Boyd, F.R., 1992. Compositional zonation in garnets of peridotite xenoliths. Contrib. Mineral. Petrol. 112, 134–147.

Smith, D., Levy, S., 1976. Petrology of Green Knobs diatreme, New Mexico, and implications for the mantle below the Colorado Plateau. Earth Planet. Sci. Lett. 19, 107–125.

Smith, D., Griffin, W.L., Ryan, C.G., Sie, S.H., 1991. Trace-element zonation in garnets from The Thumb: heating and melt infiltration below the Colorado Plateau. Contrib. Mineral. Petrol. 107, 60–79.

Smith, D., Riter, J.C.A., Mertzman, S.A., 1999. Water rock interactions, orthopyroxene growth and Si-enrichment in the mantle: evidence in xenoliths from the Colorado Plateau, southwestern United States. Earth Planet. Sci. Lett. 167, 347–356.

Snow, J.E., Dick, H.J.B., 1995. Pervasive magnesium loss by marine weathering of peridotite. Geochim. Cosmochim. Acta 59, 4219–4235.

Song, Y., Frey, F.A., 1989. Geochemistry of peridotite xenoliths in basalt from Hannuoba, eastern China: Implications for subcontinental mantle heterogeneity. Geochim. Cosmochim. Acta 53, 97–113.

Stachel, T., Viljoen, K.S., Brey, G., Harris, J.W., 1998. Metasomatic processes in lherzolitic and harzburgitic domains of diamondiferous lithospheric mantle: REE in garnets from xenoliths and inclusions in diamonds. Earth Planet. Sci. Lett. 159, 1–12.

Stephens, C.J., 1997. Heterogeneity of oceanic peridotite from the western canyon wall at MARK: results from Site 920. In: Karson, J.A., Cannat, M., Miller, D.J., Elthon, D. (Eds.), Proc. Ocean Drill. Program Sci. Results Leg, vol. 153. Ocean Drilling Program, College station, Texas, pp. 233–285.

Stolz, A.J., Davies, G.R., 1991. Chemical and isotopic evidence from spinel lherzolite xenoliths for episodic metasomatism of the upper mantle beneath southeastern Australia. J. Petrol., 303–330 (Special Volume, Special Lherzolites Issue).

Takazawa, E., Frey, F.A., Shimizu, N., Obata, M., 2000. Whole rock compositional variations in an upper mantle peridotite (Horoman, Hokkaido, Japan): are they consistent with a partial melting process? Geochim. Cosmochim. Acta 64 (4), 695–716.

Ulmer, P., 1989. Partitioning of high field strength elements among olivine, pyroxenes, garnet and calc-alkaline picrobasalt: experimental results and application. Year b.-Carnegie Inst. Wash. 88, 42–47.

Vaselli, O., et al., 1995. Ultramafic xenoliths in plio-Pleistocene alkali basalts from the eastern Transylvanian basin: depleted mantle enriched by vein metasomatism. J. Petrol. 36, 25–53.

Walter, M.J., 1998. Melting of garnet peridotite and the origin of komatiite and depleted lithosphere. J. Petrol. 39, 29–60.

Winterburn, P.A., Harte, B., Gurney, J.J., 1990. Peridotite xenoliths from the Jagersfontein kimberlite pipe: I. Primary and primary metasomatic mineralogy. Geochim. Cosmochim. Acta 54, 329–341.

Woodland, A.B., Kornprobst, J., Wood, B.J., 1992. Oxygen thermobarometry of orogenic lherzolite massifs. J. Petrol. 33, 203–230.

Xue, X., Baadsgaard, H., Irving, A.J., Scarfe, C.M., 1990. Geochemical and isotopic characteristics of lithospheric mantle beneath West Kettle River, British Columbia: evidence from ultramafic xenoliths. J. Geophys. Res. 95, 15879–15891.

Yaxley, G.M., Green, D.H., Kamenetsky, V., 1998. Carbonatite metasomatism in the southeastern Australian lithosphere. J. Petrol. 39, 1917–1930.

Yurimoto, H., Ohtani, E., 1992. Element partitioning between majorite and liquid: a secondary ion mass spectrometric study. Geophys. Res. Lett. 19, 17–20.

Zangana, N.A., Downes, H., Thirlwall, M.F., Marriner, G.F., Bea, F., 1999. Geochemical variation in peridotite xenoliths and their constituent clinopyroxenes from Ray Pic (French Massif Central): implications for the composition of the shallow lithospheric mantle. Chem. Geol. 153, 11–35.

Zhang, R.Y., Liou, J.G., Yang, J.S., Yui, T.-F., 2000. Petrochemical constraints for dual origin of garnet peridotites from the Dabie-Sulu UHP terrane, eastern-central China. J. Metamorph. Geol. 18, 149–166.

Zhou, M.F., Robinson, P.T., Malpas, J., Li, Z., 1996. Podiform chromitites in the Luobusa ophiolite, southern Tibet: implications for melt–rock interaction and chromite segregation in the upper mantle. J. Petrol. 37, 3–21.

Available online at www.sciencedirect.com

Lithos 77 (2004) 395–412

www.elsevier.com/locate/lithos

Peridotitic mantle xenoliths from kimberlites on the Ekati Diamond Mine property, N.W.T., Canada: major element compositions and implications for the lithosphere beneath the central Slave craton[☆]

Andrew Menzies[a,*], Kalle Westerlund[b], Herman Grütter[c], John Gurney[a], Jon Carlson[d], Agnes Fung[e], Tom Nowicki[c]

[a] Mineral Services South Africa, P.O. Box 38668, Pinelands 7430, South Africa
[b] Department of Geology, University of Cape Town, Rondebosch 7700, South Africa
[c] Mineral Services Canada, 205-930 Harbourside Drive, North Vancouver, B.C., Canada
[d] BHP Billiton Diamonds, #8-2604 Enterprise Way, Kelowna, B.C., Canada V1X 7Y5
[e] CF Minerals, 1677 Powick Road, Kelowna, B.C., Canada V1X 4L1

Received 27 June 2003; accepted 17 February 2004
Available online 15 July 2004

Abstract

The composition, structure and thermal state of the lithosphere beneath the Slave craton have been studied by analysing over 300 peridotitic mantle xenoliths or multiphase xenocrysts entrained within kimberlites in the Lac de Gras area. These xenoliths are derived from seven kimberlites located on the Ekati Diamond Mine™ property and define a detailed stratigraphic profile through the central Slave lithosphere from less than 120 km down to ~ 200 km. Two dominant peridotite types are present, namely garnet-bearing harzburgite and lherzolite with rare occurrences of chromite-facies peridotite, websterite and wehrlite. The pressures and temperatures (P–T's) defined by the entire data-set range from 28 to 62 kbar and 650 to 1250 °C, respectively, and approximately intersect the diamond stability field at 900 °C and 42 kbar. There is no apparent change in the geotherm with depth that is discernable beyond the resolution of the various thermobarometers. The peridotites can be divided into two compositional zones—a shallow layer dominated by garnet harzburgite that straddles the diamond–graphite boundary and a deeper layer that is strongly dominated by garnet lherzolite. Compositionally, the harzburgites (and to a lesser extent, the shallow lherzolites) are ultra-depleted relative to the more fertile deeper layer, irrespective of whether they reside within the graphite or diamond stability field. This ultra-depleted layer beneath Ekati continues to ~ 150 km.
© 2004 Elsevier B.V. All rights reserved.

Keywords: Slave; Peridotite; Harzburgite; Lithosphere; Geotherm; Diamond–graphite boundary

1. Introduction

☆ Supplementary data associated with this article can be found, in the online version, at doi 10.1016/j.lithos.2004.04.013.

* Corresponding author. Tel.: +27-21-5313162; fax: +27-21-5314209.
E-mail address: Andrew.menzies@minserv.co.za (A. Menzies).

The composition, structure and thermal state of the lithosphere beneath the Slave craton have been studied by analysing over 300 peridotitic mantle xenoliths or multiphase xenocrysts entrained within kimberlites in the Lac de Gras area. Mantle samples brought to the

Earth's surface (most commonly in kimberlites) provide the only direct information of the underlying lithosphere. Such samples have been extensively documented from the Kaapvaal and Siberian cratons. In contrast, relatively little has been reported in the scientific literature from the Slave craton. The Canadian diamond rush of the 1990s has resulted in the discovery of numerous kimberlite fields on the Slave craton and the subsequent availability of underlying lithospheric mantle xenoliths for scientific study. Recent studies of samples derived from the lithosphere beneath the central Slave craton (for example Griffin et al., 1999a,b; Pearson et al., 1999) have noted compositional and thermal features distinctly different to previous observations based on samples from the Kaapvaal and Siberian lithospheres. This paper reports in detail the major element composition of 325 peridotitic xenoliths and xenocrysts entrained in seven kim-

berlites located on the Ekati Diamond Mine™ property on the central Slave craton.

The xenoliths and xenocrysts analysed in this study were obtained from the Arnie, Grizzly, Leslie, Mark, Panda, Pigeon, and Sable kimberlites, located on the Ekati Diamond Mine™ property (henceforth termed Ekati), Northwest Territories, Canada (Fig. 1). These kimberlites occur within ~ 30 km of each other at surface, range between 47.5 and 53 Ma in emplacement age (for those dated; Davis and Kjarsgaard, 1997; Creaser et al., this volume), and are all diamondiferous to various degrees. The Panda kimberlite contains economically viable concentrations of diamonds and has been mined since Ekati opened in 1998 (Carlson et al., 1999; Dyck et al., this volume). An overview of Ekati and its history is given in Carlson et al., (1999) whilst details of the various kimberlites,

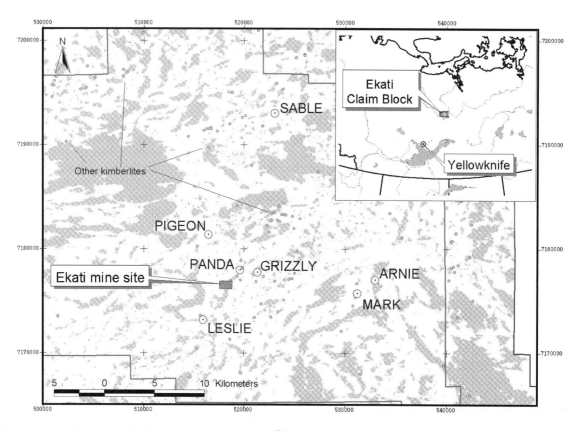

Fig. 1. Location of the seven kimberlites on the Ekati Diamond Mine™ property Northwest Territories, Canada from which all the xenoliths and xenocrysts analysed in this study were derived. Light grey shaded areas represent lakes. Co-ordinates in UTM.

local geology and geophysics, diamonds, and economic resources models are presented at this conference (for example, Nowicki et al., this volume; Lockhart et al., 2004; Gurney et al., this volume; Dyck et al., 2004).

2. Sample description and petrography

This study focused on xenoliths of "peridotitic" affinity derived from kimberlites located on the Ekati property (Fig. 1). The term peridotite is used here to encompass any xenolith or multiphase xenocryst de-

rived from lherzolite, harzburgite, dunite, wehrlite, pyroxenite, or websterite. No eclogites were analysed in this study, but are relatively common at Ekati (Fung, 1998; Jacob et al., 2003). The xenolith samples were obtained from the coarse concentrate tailings produced during the processing of bulk kimberlite samples. They range up to ~ 2 cm in maximum dimension (with many less than 1 cm) and are irregularly shaped, commonly with "fresh" breakage surfaces, most likely the result of crushing in the processing plant. Larger xenoliths were observed in various drill cores, however, these tended to be significantly more altered and friable. There was a

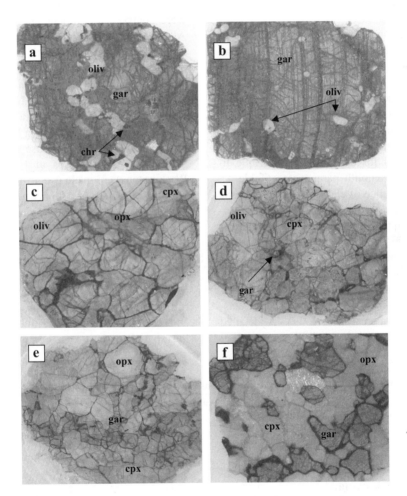

Fig. 2. Photomicrographs of the various types of xenoliths from Panda. In order (from right-to-left) are xenoliths PH-15, PH-17, PL-3, PL-2, PW-3 and PW-1. Field of view for each image is ~ 2 cm. Note the variation in modal mineralogy between the samples. Figures (a) and (b) represent garnet macrocrysts with inclusions of other mineral phases whilst (c) through (f) are multiphase xenoliths where garnet is generally subordinate. Gar = garnet, Cpx = clinopyroxene, Oliv = olivine, Opx = orthopyroxene, Chr = chromite.

sampling bias towards multiphase xenoliths with fresh minerals that contained garnet.

The 325 xenoliths examined and analysed in this study can be divided into two groups based on hand specimen appearance—namely, garnet xenocrysts with inclusions of other mineral phases and multiphase xenoliths where garnet is subordinate or even absent (Fig. 2). Unless otherwise stated, the term "xenolith" is used below to describe both groups. In addition, the seven kimberlites from which these xenoliths are derived are collectively termed the Ekati kimberlites in this paper, even though over 150 kimberlites have been discovered on the Ekati Diamond Mine™ property to date (Nowicki et al., 2004).

No detailed petrography or point counting was undertaken as the majority of samples were too small and the respective grain size too large to modally represent the source mantle rocks from which they were derived. Consequently, the presence or absence of a mineral cannot reliably be used to distinguish between various parageneses and thus mineral compositions have been used, in conjunction with known compositional ranges for different upper mantle lithologies, to assign xenolith parageneses (Table 1). For example, the absence of clinopyroxene cannot be used to distinguish between lherzolite and harzburgite—of 213 xenoliths mineralogically classified as harzburgite, only 77 contain sub-calcic garnet indicative of equilibration in a clinopyroxene-free assemblage. None of the xenoliths in this study display features commonly associated with "sheared" peridotites and all the peridotites may be broadly classified as "coarse" (after Harte, 1977). A detailed petrographic description of xenoliths entrained in

kimberlites from Ekati can by found in Doyle, (2002).

3. Analytical methods

The xenoliths were mounted in epoxy resin discs as either intact xenoliths or individual grains (small chips of the various mineral phases). Major and minor element concentrations were determined using electron microprobes (EMP) housed at CF Minerals and at the Carnegie Institute of Washington. The vast majority of analyses were made at CF Minerals, whilst a select subset (primarily the samples from Leslie) were analysed at Carnegie Institute of Washington. All analyses at CF Minerals were performed at an accelerating voltage of 15 kV and a beam current of 30 nA, with a beam diameter of ~ 1 μm. Count times were 10 s on peak positions and 5 s on background positions with the exception of Na-in-garnet and K-in-clinopyroxene, where peak times of 30 s were used. Data was reduced on-line using PAP matrix correction procedures based on in-house natural and synthetic standards. All analyses at Carnegie Institute of Washington were made using similar setup conditions with the exception of a 40 nA beam current and using ZAF matrix correction procedures. Prior to analysis at CF Minerals, elemental scans were undertaken for Mg and Al in all xenolith probe mounts. These scans allowed the various mineral phases of olivine, orthopyroxene, garnet, clinopyroxene and chromite to be identified and precise positions (i.e. core and rim) to be selected for analysis. In addition, the scans and core-rim analyses provided an indication of the spatial variation in composition and allowed

Table 1

Number of peridotitic xenoliths analysed in this study from the various kimberlites located on the Ekati Diamond Mine™

Location	Lherzolite (mineralogical)	Harzburgite (mineralogical)	Mineralogical, Other	Lherzolite (Gar Min Chem)	Harzburgite (Gar Min Chem)	Other (Gar Min Chem)
Pigeon	20	7	0	27	0	0
Mark	0	21	0	15	6	0
Arnie	10	33	1	40	2	1
Panda	43	69	2	71	39	2
Sable	23	10	1	31	2	1
Leslie	9	65	0	43	28	0
Grizzly	3	8	0	1	6	0

Classifications are based on mineralogy (left columns) and garnet mineral chemistry (right columns). Note that 11 of the 325 xenoliths analysed in this study did not contain garnet.

checking for both inter- and intra-grain homogeneity that is critical for the correct application of geothermobarometers. For those samples mounted as grain fragments, equilibrium conditions are assumed as all inter- and intra-grain spatial resolution is lost. In most cases, multiple analyses were made of the various mineral phases present in each xenolith. The data were rigorously checked and any results displaying heterogeneity and disequilibrium have been omitted from further discussion in this paper. All results reported in this paper represent averages for that particular mineral phase.

4. Mineral compositions

Examples of major-element mineral compositions are given in Table 2 whilst a more extensive listing is provided in Supplementary Electronic Appendix I. The data can be obtained in an electronic format from the primary author upon request. A detailed discussion of the key compositional characteristics of each mineral phase is provided below.

4.1. Garnet

Garnet is present in 312 of the xenoliths including samples derived from all seven localities. The garnets are chrome pyropes and display a wide range of Cr_2O_3 and CaO contents (Fig. 3a). They include numerous harzburgitic "G10" garnets (after Gurney, 1984), in particular high-Cr sub-calcic grains that overlap Ekati diamond inclusion compositions (Chinn et al., 1998; Stachel et al., 2003). The balance is predominantly lherzolitic "G9" garnets (after Gurney, 1984) with possible minor examples of websteritic garnet also present. The Cr_2O_3 concentrations range from a minimum of 1 wt.% up to a maximum of 14.5 wt.% whilst CaO contents range down to ~ 1 wt.%. There are distinct variations between localities evident with the greatest compositional range in the large data sets for Panda and Leslie (Fig. 3a). The G10 garnets can be divided into two groups (high-Cr and low-Cr) that display a good separation by the "graphite–diamond constraint" of Grütter and Sweeney (2000). The graphite–diamond constraint represents a multiply constrained equilibrium feature of garnet coexisting with chromite in natural peridotites, including whether

diamond or graphite is stable (assuming the xenoliths reside on a "normal" cratonic geotherm). Noticeably, the vast majority of garnets defining the high-Cr G10 group are derived from the Panda kimberlite.

The high Mg-numbers are indicative of the highly depleted nature of the garnets, with the harzburgitic G10 garnets relatively more depleted than the lherzolitic G9 garnets. The G10 and G9 garnets also display significantly different compositional ranges in TiO_2 and MnO concentrations (Fig. 3b). Specifically, the harzburgitic G10 garnets are depleted in TiO_2 with values ranging from below detection limits up to a maximum of ~ 0.05 wt.%, whilst MnO concentrations are all above 0.37 wt.%. In contrast, the lherzolitic G9 garnets display a wide range of TiO_2 values that extend up to a maximum of 0.85 wt.%, with a notable proportion exceeding 0.4 wt.%, a cutoff value often used as a generalised indicator of metasomatic activity in the mantle (Harte and Gurney, 1975; Gurney et al., 1975; Harte, 1983; Matthews et al., 1992). In addition, the G9 garnets also display a relatively wider range of MnO concentrations. Significantly only two of the high-Ti G9 garnets yield a MnO concentration above 0.47 wt.%.

Garnet compositions in xenoliths that contain primary clinopyroxene all plot on the lherzolitic trend and display the full range in Cr concentrations observed at Ekati, whilst those that contain primary chromite are predominantly high-Cr highly sub-calcic G10 garnet that plot above the GDC line (figures not shown).

4.2. Olivine

Olivine is present in 214 xenoliths including samples derived from all seven localities. The olivines are classified as forsterites with Fo contents ranging from ~ 86 up to a maximum of ~ 94, the vast majority clustering between 90 and 93 (Fig. 4a). The concentration of CaO ranges from below detection limits up to 0.05 wt.% (Fig. 4b) and displays a broad linear negative correlation with Fo content.

The harzburgitic olivines predominantly have a Fo content greater than 92 and a CaO concentration less than 0.02 wt.%. In contrast, whilst some lherzolitic olivines overlap these compositions, the vast majority extend to lower Fo contents and higher CaO concentrations, respectively. This explains the variations

Table 2
Representative mineral composition analyses from selected xenoliths

Sample ID	Location	Mineral ID	SiO$_2$	TiO$_2$	Al$_2$O$_3$	Cr$_2$O$_3$	FeOt	MnO	MgO	CaO	Na$_2$O	K$_2$O	NiO	Total	Mg#
JJG 5050-105	Panda	Opx	58.64	0.01	0.47	0.27	4.17	0.10	35.66	0.20	0.03	0.00	0.10	99.63	93.9
JJG 5050-105	Panda	Oliv	41.51	0.00	0.01	0.02	7.04	0.09	50.24	0.01	0.01	0.01	0.38	99.33	92.7
JJG 5050-105	Panda	Gar	41.45	0.01	17.62	8.84	7.10	0.47	20.77	3.70	0.01	n.a.	n.a.	99.96	84.0
JJG 5050-105	Panda	Chr	0.03	0.01	7.04	62.58	14.87	0.34	12.50	0.00	n.a.	n.a.	0.10	97.46	60.1
JJG 5050-106	Panda	Opx	58.46	0.01	0.45	0.29	4.29	0.10	35.52	0.22	0.01	0.00	0.09	99.44	93.7
JJG 5050-106	Panda	Oliv	41.19	0.01	0.02	0.02	7.20	0.10	50.06	0.01	0.01	0.00	0.36	98.97	92.5
JJG 5050-106	Panda	Gar	41.53	0.01	17.73	8.43	7.05	0.42	20.80	3.73	0.01	n.a.	n.a.	99.70	84.0
JJG 5050-106	Panda	Chr	0.05	0.02	6.72	62.38	16.55	0.27	12.23	0.01	n.a.	n.a.	0.07	98.29	56.9
JJG 5050-108	Panda	Opx	58.89	0.01	0.50	0.31	4.28	0.09	35.36	0.14	0.06	0.00	0.07	99.71	93.6
JJG 5050-108	Panda	Oliv	41.43	0.01	0.00	0.02	7.09	0.09	50.02	0.01	0.00	0.00	0.34	99.01	92.6
JJG 5050-108	Panda	Gar	41.83	0.01	20.25	5.49	7.94	0.48	20.69	3.37	0.02	n.a.	n.a.	100.08	82.3
JJG 5050-116	Panda	Opx	58.20	0.01	0.49	0.36	4.18	0.11	35.11	0.28	0.10	0.00	0.10	98.93	93.7
JJG 5050-116	Panda	Oliv	41.28	0.00	0.01	0.05	6.96	0.11	49.93	0.01	0.00	0.00	0.36	98.73	92.8
JJG 5050-116	Panda	Gar	41.00	0.01	16.78	9.49	6.79	0.43	20.02	4.64	0.02	n.a.	n.a.	99.17	84.1
JJG 5050-116	Panda	Chr	0.04	0.47	9.03	57.22	19.00	0.38	12.39	0.01	n.a.	n.a.	0.10	98.63	53.8
JJG 5050-120	Panda	Opx	58.59	0.00	0.46	0.28	4.19	0.11	35.61	0.20	0.03	0.00	0.10	99.56	93.8
JJG 5050-120	Panda	Oliv	41.21	0.01	0.01	0.02	6.99	0.11	50.07	0.01	0.01	0.00	0.36	98.80	92.7
JJG 5050-120	Panda	Gar	41.64	0.01	18.68	7.42	7.09	0.48	20.78	3.69	0.01	n.a.	n.a.	99.81	83.9
JJG 5050-120	Panda	Cpx	53.67	0.04	2.70	1.15	4.75	0.57	20.35	15.27	0.47	0.01	n.a.	98.98	88.4
JJG 5050-19	Panda	Opx	58.56	0.01	0.52	0.34	4.18	0.13	35.61	0.12	0.02	0.01	0.11	99.62	93.8
JJG 5050-19	Panda	Gar	41.66	0.00	17.43	9.25	6.82	0.42	22.20	1.83	0.01	n.a.	n.a.	99.62	85.3
JJG 5050-27	Panda	Oliv	41.18	0.03	0.01	0.02	7.81	0.09	49.18	0.02	0.02	0.01	0.37	98.74	91.8
JJG 5050-27	Panda	Gar	42.43	0.38	22.17	2.37	7.25	0.34	20.91	4.10	0.07	n.a.	n.a.	100.00	83.7
JJG 5050-27	Panda	Cpx	55.00	0.29	2.82	1.50	2.62	0.09	15.98	18.08	2.48	0.03	n.a.	98.88	91.6
JJG 5050-29	Panda	Opx	57.86	0.03	0.52	0.34	4.46	0.11	34.77	0.48	0.15	0.02	0.12	98.87	93.3
JJG 5050-29	Panda	Oliv	41.50	0.00	0.00	0.02	7.22	0.10	49.91	0.03	0.01	0.04	0.35	99.18	92.5
JJG 5050-29	Panda	Gar	41.39	0.32	17.31	8.53	6.68	0.39	19.13	5.95	0.04	n.a.	n.a.	99.76	83.6
JJG 5050-29	Panda	Cpx	54.76	0.11	1.90	2.97	2.16	0.07	15.75	18.71	2.44	0.04	n.a.	98.96	92.8
JJG 5050-49	Panda	Opx	58.17	0.04	0.25	0.15	7.19	0.16	33.66	0.35	0.02	0.01	0.12	100.11	89.3
JJG 5050-49	Panda	Oliv	40.74	0.02	0.00	0.02	12.08	0.10	46.25	0.02	0.00	0.00	0.38	99.61	87.2
JJG 5050-49	Panda	Gar	40.82	0.17	17.86	7.21	10.84	0.57	15.29	7.37	0.01	n.a.	n.a.	100.12	71.6
JJG 5050-49	Panda	Cpx	55.52	0.03	0.45	0.75	2.55	0.09	17.36	22.47	0.50	0.02	n.a.	99.75	92.4
JJG 5050-62	Panda	Opx	58.69	0.00	0.49	0.20	4.41	0.11	35.62	0.16	0.06	0.00	0.08	99.81	93.5
JJG 5050-62	Panda	Oliv	41.30	0.00	0.00	0.01	7.04	0.08	50.19	0.01	0.00	0.00	0.37	99.01	92.7
JJG 5050-62	Panda	Gar	41.71	0.00	20.29	5.60	8.12	0.54	20.68	3.45	0.01	n.a.	n.a.	100.39	82.0
JJG 5050-63	Panda	Opx	58.00	0.08	0.75	0.42	5.23	0.16	34.03	0.83	0.15	0.00	0.11	99.76	92.1
JJG 5050-63	Panda	Oliv	40.98	0.03	0.02	0.05	9.62	0.11	48.36	0.03	0.02	0.01	0.29	99.51	90.0
JJG 5050-63	Panda	Gar	41.43	0.54	19.60	4.57	8.27	0.39	19.02	6.01	0.05	n.a.	n.a.	99.88	80.4
JJG 5050-63	Panda	Cpx	54.98	0.17	1.55	1.01	2.89	0.11	17.13	20.11	1.30	0.05	n.a.	99.29	91.4
JJG 5050-69	Panda	Opx	57.53	0.02	0.31	0.19	7.87	0.16	32.66	0.45	0.02	0.00	0.07	99.27	88.1
JJG 5050-69	Panda	Oliv	40.01	0.00	0.01	0.02	13.43	0.16	44.74	0.03	0.01	0.00	0.29	98.70	85.6
JJG 5050-69	Panda	Gar	39.19	0.17	13.85	10.97	11.52	0.60	13.02	9.75	0.00	n.a.	n.a.	99.06	66.8
JJG 5050-69	Panda	Cpx	54.68	0.01	0.36	0.53	3.11	0.11	17.26	22.49	0.35	0.02	n.a.	98.93	90.8
JJG 5050-82	Panda	Opx	58.55	0.01	0.45	0.28	4.22	0.13	35.62	0.15	0.02	0.00	0.09	99.51	93.8
JJG 5050-82	Panda	Oliv	41.37	0.01	0.01	0.03	7.02	0.10	50.18	0.01	0.00	0.01	0.36	99.08	92.7
JJG 5050-82	Panda	Gar	41.76	0.01	18.56	7.75	7.14	0.47	21.16	3.09	0.01	n.a.	n.a.	99.95	84.1
JJG 5050-82	Panda	Chr	0.03	0.01	7.35	62.23	16.57	0.31	12.21	0.00	n.a.	n.a.	0.06	98.76	56.8
JJG 5050-87	Panda	Opx	58.37	0.01	0.48	0.28	4.26	0.11	35.36	0.16	0.05	0.00	0.08	99.16	93.7
JJG 5050-87	Panda	Oliv	41.45	0.00	0.00	0.02	7.06	0.11	50.22	0.00	0.00	0.00	0.32	99.18	92.7
JJG 5050-87	Panda	Gar	41.80	0.00	20.39	5.25	7.93	0.55	21.46	2.16	0.02	n.a.	n.a.	99.57	82.6
JJG 5050-90	Panda	Opx	58.19	0.00	0.50	0.34	4.38	0.12	35.08	0.23	0.11	0.00	0.10	99.04	93.5
JJG 5050-90	Panda	Oliv	41.19	0.01	0.01	0.02	7.13	0.10	49.85	0.01	0.00	0.01	0.38	98.71	92.6
JJG 5050-90	Panda	Gar	41.25	0.03	19.14	6.66	7.54	0.47	19.72	4.84	0.04	n.a.	n.a.	99.68	82.6
JJG 5050-90	Panda	Chr	0.02	0.01	7.46	59.57	19.10	0.30	11.69	0.02	n.a.	n.a.	0.03	98.19	52.2

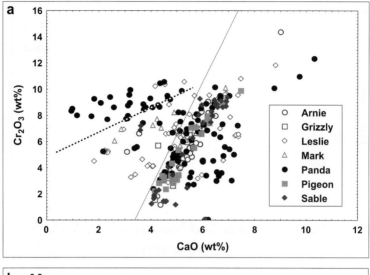

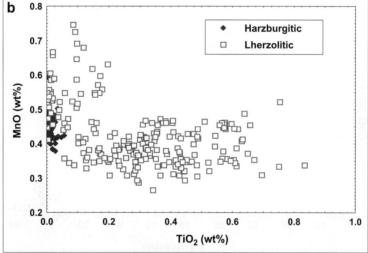

Fig. 3. Garnet: (a) Garnet Cr_2O_3 and CaO compositions separated by locality. The solid line of Gurney (1984) separates the grains into the G10 (sub-calcic) and G9 (calcic) fields, whilst the dashed line represents the graphite–diamond constraint of Grütter and Sweeney (2000) and separates the high-Cr and low-Cr G10 garnets. (b) Garnet TiO_2 and MnO compositions separated by garnet paragenesis.

observed between individual localities—for example, in this study, the xenoliths at Arnie are predominantly lherzolitic and, thus, compared to xenoliths from Panda (which contains numerous harzburgites), they show a lower average Fo content and higher average Ca concentration (figures not shown).

4.3. Orthopyroxene

Orthopyroxene is present in 122 xenoliths including samples derived from all seven localities. The orthopyroxenes are defined as enstatites with Mg-numbers ranging from ~ 86 up to a maximum of ~ 94, the vast majority clustering between 93 and 94 (Fig. 5a). The orthopyroxenes display very low Al_2O_3 concentrations (relative to known worldwide mantle derived compositions) and yield a broad positive correlation with CaO concentrations (Fig. 5b). In general, harzburgitic orthopyroxenes are low in Mn, Ca, Na, Ti and higher in Cr compared to those derived from lherzolite, although there is some overlap and the distinction is not definitive. As with olivines, this

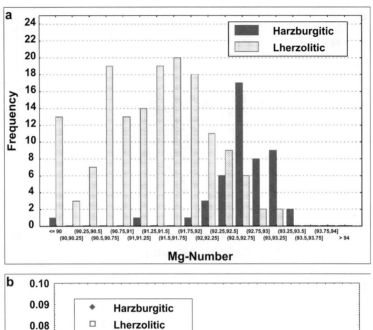

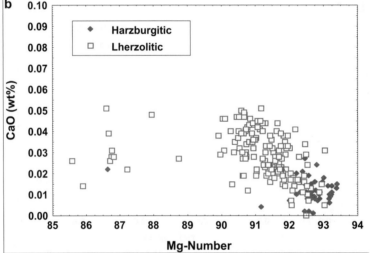

Fig. 4. Olivine: (a) histogram of the Forsteritic content separated into harzburgitic and lherzolitic parageneses. (b) CaO concentration vs. Fo content.

results in noticeable compositional differences between individual localities (figures not shown).

4.4. Clinopyroxene

Clinopyroxene is present in 107 xenoliths including samples derived from six localities—no samples from Mark contained clinopyroxene, even though many of the garnets are of lherzolitic affinity. The majority of clinopyroxenes analysed are classified as chromian diopsides. There is a general overlap in clinopyroxene compositions from the various localities with the Mg-numbers clustering predominantly between 91 and 93, whilst the Ca-numbers display a more diverse range of between 42 and 50. The majority of clinopyroxenes have Cr_2O_3 and Al_2O_3 concentrations that plot in the field associated with garnet-bearing lherzolites (Fig. 6a), with many straddling the Al–Mg boundary that distinguishes low-Al peridotites (Fig. 6b). A small subset with very low Cr and Al concentrations are likely not in equilibrium with garnet.

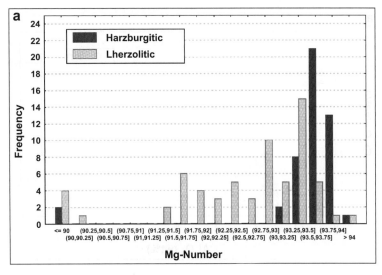

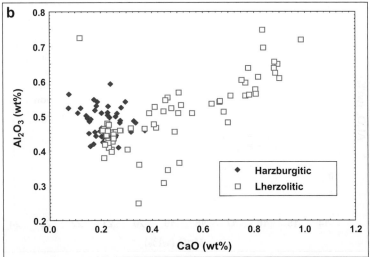

Fig. 5. Orthopyroxene: (a) histogram of Mg-number separated into harzburgitic and lherzolitic parageneses. (b) Al_2O_3 concentration vs. CaO.

4.5. Chromite

Chromite is present in 26 xenoliths with samples derived from only two localities (Leslie and Panda)—noticeably the least abundant of the five major mineral phases commonly observed in mantle xenoliths. The chromites display a limited compositional range relative to those known for mantle-derived chromites, with Cr_2O_3 and MgO concentrations ranging from 58 to 65 wt.% and 11 to 12.5 wt.%, respectively. The chromites are highly depleted, with Cr-numbers greater than 79 and TiO_2 concentrations

commonly less than 0.15 wt.%. The chromites all contain low levels of SiO_2 (less than 0.12 wt.%) and MnO contents range from 0.27 to 0.41 wt.%. The samples display a relatively narrow range of stoichiometrically calculated ferric iron ratios (primarily between 82% and 88% Fe^{2+}).

5. Geothermobarometry

Over the last two decades, numerous geothermometers and geobarometers applicable to xenoliths de-

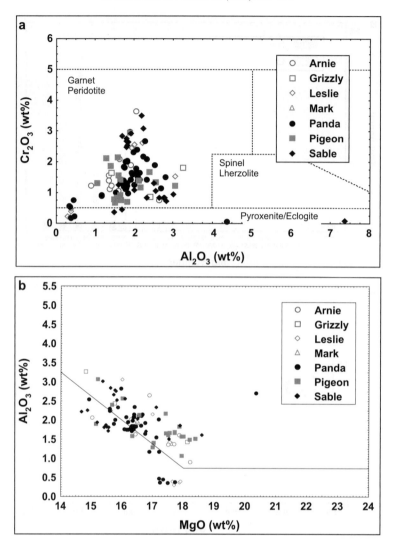

Fig. 6. Clinopyroxene: (a) Cr_2O_3 and Al_2O_3 compositions separated by locality. (b) Al_2O_3 and MgO compositions separated by locality. Fields are derived from Ramsay (1992) and Nimis (1998).

rived from the upper mantle have been developed based on experimental studies, empirical observations and thermodynamic considerations. Detailed assessments and reviews of widely used geothermobarometers are beyond the scope of this paper and the reader is referred to the following overviews, e.g. Carswell and Gibb (1980, 1987), Finnerty and Boyd (1984, 1987), Nickel and Green (1985), Finnerty (1989), Brey and Köhler (1990), Smith (1999), Pearson et al. (1999), and Grütter and Moore (this volume). These detailed investigations yield varying interpretations as to which geothermobarometers provide the best preci-

sion and/or accuracy, particularly over the large compositional and pressure–temperature (P–T) ranges observed in cratonic upper mantles. The systems considered of primary interest and discussed in this paper are the single grain clinopyroxene geothermobarometer of Nimis and Taylor (2000) (T-NT00 and P-NT00), the Al-in-orthopyroxene barometer and two-pyroxene thermometer calibration of Brey and Köhler (1990) and Brey et al. (1990) (P-BKN90, T-BKN90), and the olivine–garnet thermometer of O'Neill and Wood (1979) (T-ONW79) due to its applicability to both lherzolitic and harzburgitic parageneses. The latter thermometer has been combined with the Al-in-ortho-pyroxene barometers of Brey and Köhler (1990) and Nickel and Green (1985) (P-NG85). A selection of the various $P-T$ arrays are presented graphically in Fig. 7.

The xenoliths yield a wide range of pressures (P) and temperatures (T), indicating that these seven kimberlites sampled and entrained material from a large section of the lithosphere beneath Ekati. Depending on the thermobarometer combination applied, $P-T$'s may range from ~ 28 to 62 kbar and 650 to 1250 °C, respectively (Fig. 7). At any one individual kimberlite, the $P-T$ range of the majority of xenoliths is generally narrower, reflecting sampling of specific levels within the mantle by particular kimberlites. On average, the $P-T$ values for Pigeon and Arnie are higher than those of Panda and Leslie that, in turn, are greater than those obtained for Sable

and Grizzly. It should be noted, however, that there is sufficient overlap between the various localities to provide a continuous $P-T$ profile from ~ 28 kbar and 650 °C up to 60 kbar and 1200 °C beneath Ekati. For any given geothermobarometer combination, the resultant $P-T$ data display some scatter, but they generally indicate a coherent trend with no obvious differences between the various kimberlites. Noticeably the various $P-T$ arrays commonly yield an increase in scatter with depth. In the case of T-ONW79 vs. P-NG85 (Fig. 7c), it could be argued that the xenoliths display an inflected geotherm similar to that observed for the Lesotho kimberlites (Boyd and Nixon, 1975). However, none of these deeper xenoliths display the high-T deformed (shared) textures commonly associated with samples lying on an inflected geotherm. The resolution of the various

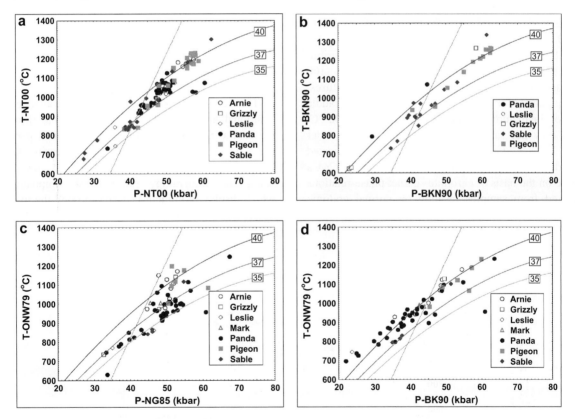

Fig. 7. Pressure–temperature plots and paleogeotherms using selected geothermobarometry combinations: (a) thermometer and barometer of Nimis and Taylor (2000), (b) thermometer and barometer of Brey and Köhler (1990), (c) thermometer of O'Neill and Wood (1979) against barometer of Nickel and Green (1985), and (d) thermometer of O'Neill and Wood (1979) against barometer of Brey and Köhler (1990). Solid lines represent conductive model geotherms of Pollack and Chapman (1977) whilst the dashed black line represents the diamond–graphite boundary of Kennedy and Kennedy (1976).

geothermobarometers (commonly on the order of ± 30 °C and ± 2 kbar at best) is not sufficient to determine the significance of this increased scatter— in particular, it is not possible to conclude whether the deeper xenoliths reside on a higher geotherm than the shallower xenoliths. On average, the various P–T arrays (including those not shown) intersect the diamond–graphite boundary (Kennedy and Kennedy, 1976) at approximately 42 kbar and 900 °C.

A comparison between T-ONW79 and T-BKN90, both calculated in conjunction with P-BKN90, for lherzolites, indicates that at any given pressure T-BKN90 is systematically higher than T-ONW79 on the order of ~ 100 °C (Fig. 8). Geothermal arrays defined with the T-BKN90 thermometer are accordingly systematically displaced to higher P–T values than when T-ONW79 is used (e.g. Fig. 7).

At an assumed pressure of 40 kbar, T-ONW79 yields a temperature range for harzburgites of 700 to 1000 °C and for lherzolites of 650 to 1200 °C (Fig. 9). The lherzolites show a skewed temperature distribution, with the dominant mode between 950 and 1100 °C and a significant tail to lower temperatures. Temperatures for garnet harzburgite xenoliths straddle that of the diamond–graphite boundary at ~ 900 °C, with only one sample yielding a temperature greater than 1000 °C. Grütter and Sweeney (2000) projected the diamond–graphite boundary into garnet Cr-Ca space on the basis of chromite–garnet phase-relations derived from samples residing on normal cratonic

geotherms and labelled it as the graphite–diamond constraint. Equilibration temperatures (T-ONW79) for the Ekati xenolith data set are, in general, consistent with their graphite–diamond constraint—that is, harzburgitic garnets residing in the graphite stability field (T < 900 °C) have compositions lower in Cr than the graphite–diamond constraint, whilst those with higher Cr contents have T > 900 °C and reside within the diamond stability field (Fig. 10). Accordingly, at Ekati it would be expected that graphite could be associated with harzburgite that is characterized by low-Cr G10 garnets plotting below the graphite–diamond constraint. In contrast, diamond would be associated with harzburgite characterized by high-Cr G10 garnets plotting above the graphite–diamond constraint. It should be noted that the speciation of different carbon polymorphs does not imply a different petrogenetic origin for the harzburgite in which the diamond or graphite occurs.

6. Discussion

The xenoliths from Ekati define a detailed compositional and thermal stratigraphic profile through the central Slave lithosphere. Two dominant peridotite types are present, namely garnet-bearing harzburgite and lherzolite (paragenetically classified using both mineralogy and mineral composition), which display prominent layering through the mantle

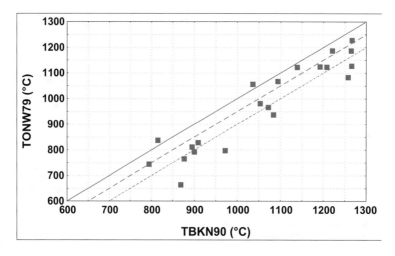

Fig. 8. Comparison of temperatures determined using the thermometers of T-ONW79 and T-BKN90 both calculated in conjunction with P-BKN90.

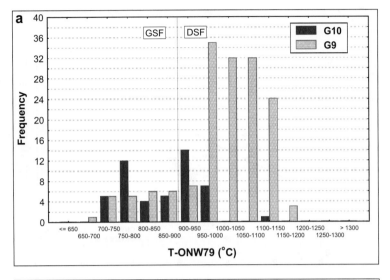

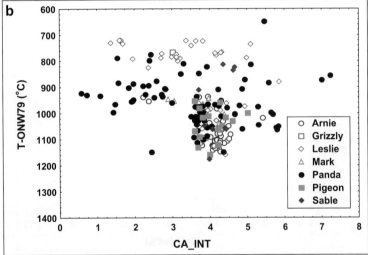

Fig. 9. (a) Histogram of T-ONW79 (at an assumed pressure of 40 kbar) separating G10 and G9 garnets. (b) T-ONW79 (at an assumed pressure of 40 kbar) vs. Ca_Int. The Ca_Int variable is a projection of the Cr-Ca content of a garnet to the Cr axis (Grütter et al., 2004). Accordingly, G10 garnets yield Ca_Int values less than 3.375, whilst G9 garnets yield greater values.

section sampled at Ekati. Chromite-facies peridotites and websterites and wehrlites appear to be rare. No single kimberlite has sampled over the entire calculated *P–T* range but, taken together, they represent a reasonably continuous stratigraphic profile from between ~ 120- and 200-km depth. It cannot be concluded that the maximum calculated depth is the extent of undeformed coarse peridotite beneath Ekati since no high-temperature sheared peridotite xenoliths were recovered in this study. Nor can a

chemical lithosphere–asthenosphere boundary be inferred. However, studies of mantle xenoliths from the DIAVIK kimberlites immediately to the south and south-east of the Ekati property yielded a few such xenoliths, all recording *P–T*'s of greater than 55 kbar and 1200 °C (Pearson et al., 1999). Furthermore, on the basis of trace element compositions of concentrate garnets from the same location combined with other samples from across the Slave craton, Griffin et al. (1999a,b) determined the chemical lithosphere–

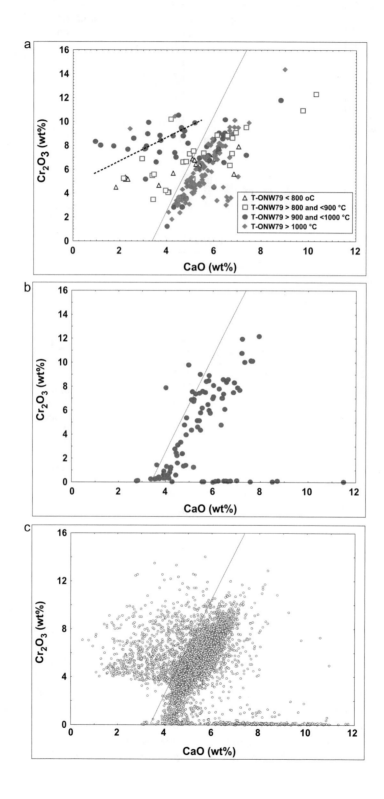

asthenosphere boundary at ~ 1250 °C for this region.

The harzburgites beneath Ekati (and to a lesser extent, the shallow lherzolites) are ultra-depleted, irrespective of whether they are derived from within the graphite or diamond stability field. All the garnets yield TiO_2 concentrations below 0.05 wt.%, olivines with Fo contents above 93, and orthopyroxenes with Mg-numbers above 92. Furthermore, preliminary trace element analyses by LA-ICPMS indicate that these harzburgitic G10 garnets, as well as shallow lherzolitic G9 garnets, record extremely depleted concentrations of incompatible elements (such as Zr, Ti and Ga) as well as elements such as Y and HREE, commonly regarded as compatible in garnet. Specifically, the harzburgite ($n = 29$) and shallow lherzolite ($n = 8$) garnets yield median values of Ti = 48 ppm, Y = 0.7 ppm, Ga = 1.5 ppm, and Zr = 6 ppm, and Ti = 59 ppm, Y = 0.8 ppm, Ga = 2.0 ppm, and Zr = 9 ppm, respectively. In addition, the harzburgitic garnets display the Sr and LREE enrichment commonly associated with ancient metasomatism (Shimizu and Richardson, 1987). It is noted that any metasomatism commonly associated with an increase in Ti concentration has not infiltrated into the harzburgitic garnets (see Fig. 3b). In contrast, the few deeper lherzolitic G9 garnets ($n = 4$) all yield significantly more fertile compositions. Consequently, the peridotites can be divided into two compositional zones—a shallow ultra-depleted layer dominated by garnet harzburgite that straddles the diamond–graphite boundary and a deeper layer that is strongly dominated by garnet lherzolite. This is consistent with the results of other studies of samples from the central Slave lithosphere (Pearson et al., 1999; Griffin et al., 1999a,b) where it was determined that the shallower ultra-depleted layer is comprised of up to 60% harzburgite, whilst the deeper fertile layer comprises less than 20%.

The majority of harzburgite xenoliths (in this study) almost entirely reside in the ultra-depleted layer at levels shallower than 150 km and straddle the diamond–graphite stability field—only one sample

yielded a temperature indicative of derivation from within the deeper fertile layer. On the basis of trace element results, Griffin et al. (1999a,b) place a "sharp" compositional boundary between the shallower ultra-depleted layer and deeper fertile layer at just over 900 °C (see Fig. 7 in Griffin et al., 1999a). In contrast, the results of this study indicate that the ultra-depleted layer beneath Ekati continues to ~ 1000 °C, as defined by the upper temperature limit of the ultra-depleted harzburgite. This difference could be due to real depth (thermal) changes over a localized area of the ultra-depleted layer, or more likely, a lack of high-Cr highly sub-calcic harzburgitic garnets available for analysis in the studies at DIAVIK—either due to the kimberlites themselves not sampling this layer or the lack of adequate samples available for scientific study. For example, in the study of Pearson et al. (1999), no xenoliths with medium or highly sub-calcic garnet compositions were analysed (Fig. 10). Indeed, of the seven kimberlites in this study, only Panda has entrained a significant number of harzburgites with such garnet compositions. It should be noted that there are in likelihood additional deeper, garnet-bearing harzburgite populations that are not part of the ultra-depleted layer and are not represented in this work due to their paucity (e.g. Fig. 10). Diamond inclusions from Sable (Chinn et al., 1998) yield garnets with high-Cr sub-calcic garnet compositions with calculated temperatures using T–Ni (Ryan et al., 1996) that are consistent with an origin in the deeper fertile layer and not the shallower ultra-depleted layer. High-Cr sub-calcic garnets similar in composition to those found included in diamond constitute a small portion (~ 1%) of those present in an extremely large garnet data set for the central Slave craton (Fig. 10; Armstrong, 2001), but do not occur in any of the xenolith samples obtained from the central Slave craton. Accordingly, the data set in this study does not preclude the possibility of other peridotitic rock types being present in the mantle beneath Ekati and sampled by the kimberlite but not represented in this study either because there are too few or because they were

Fig. 10. (a) Garnet Cr_2O_3 and CaO compositions separated by T-ONW79 (at assumed pressure of 40 Kbar). The solid line of Gurney (1984) separates the grains into the G10 (sub-calcic) and G9 (calcic) fields, whilst the dashed line represents the graphite–diamond constraint of Grütter and Sweeney (2000). (b) Garnet Cr_2O_3 and CaO compositions from xenoliths analysed by Pearson et al. (1999). (c) Garnet Cr_2O_3 and CaO compositions derived from diamond exploration samples collected in the central Slave craton. For clarity, the compositions shown are restricted to a randomly selected subset of 4872 analyses in the much larger data set of Armstrong (2001).

preferentially excluded (e.g. the samples were altered and/or disaggregated).

On the basis of their geothermobarometry results, Pearson et al. (1999) proposed that this portion of the Slave craton lithosphere displays a change in the gradient of the paleogeotherm at ~ 900–1000 °C, and that this coincides with the boundary between two compositionally distinct layers (Griffin et al., 1999a,b). Pearson et al. (1999) suggested two possibilities to explain the apparent difference in geotherms for low-T xenoliths and high-T xenoliths. It could either be a transient feature related to heating of the xenoliths at the time of kimberlite eruption or it may reflect two compositionally and thermally distinct layers in the upper mantle. Such an interpretation is contrary to the results of this study, that, significantly, incorporate xenoliths derived from around the graphite–diamond boundary, and do not display any clear geothermal distinction between the shallower and deeper xenoliths, at least not beyond the expected resolution of the various geothermobarometer combinations. Thus, if any differences are present, they must be minor.

Garnet-based lithospheric mapping revealed that the compositionally distinct ultra-depleted layer is confined to a northeast trending zone within the central Slave craton (Grütter et al., 1999) and is situated at < 150-km depth, partly within the graphite stability field (this study; Griffin et al., 1999a). The ultra-depleted layer, which contains significant proportions of harzburgite, is largely absent from lithospheric sections to the immediate north and south of the central Slave lithosphere (e.g. Kopylova and Russell, 2000; Carbno and Canil, 2002), implying that three compositionally distinct lithospheric sections exist over a horizontal distance of less than 300 km (Grütter et al., 1999; Davis et al., 2003). In addition, the ultra-depleted layer occurs beneath both the western and eastern crustal terranes, as defined by Pb and Nd isotopic constraints (Thorpe et al., 1992; Davis and Hegner, 1992), suggesting that crust and mantle may have been decoupled at that time (Grütter et al., 1999; Davis et al., 2003). Furthermore, the ultra-depleted layer is noted to be broadly spatially correlated in three dimensions with a regional conductive zone (Jones et al., 2001; Grütter et al., 1999).

The overall composition of the central Slave lithosphere broadly appears to display key differences to that of other well-documented Archaean craton lithospheres, e.g. Kaapvaal and Siberia. An unusual feature is the presence of an ultra-depleted layer that contains significant amounts of harzburgite that resides partly within the graphite stability field. Whilst graphite-bearing harzburgites have been recorded from the Kaapvaal and Siberian lithospheres (see Grütter and Sweeney, 2000 and references therein for a summary), it appears that such a layer is relatively rare and certainly not as prominent as that beneath the central Slave craton.

7. Conclusions

1. The xenoliths from Ekati define a detailed compositional and thermal stratigraphic profile through the central Slave lithosphere from less than 120 up to ~ 200 km.
2. The xenoliths yield a wide range of pressures (P) and temperatures (T), indicating that these seven kimberlites sampled and entrained material from a large section of the lithosphere beneath Ekati ranging from ~ 28 to 62 kbar and 650 to 1250 °C. On average, the various P–T arrays intersect the diamond–graphite boundary at approximately 42 kbar and 900 °C. There is no apparent change in the geotherm with depth that is discernable beyond the resolution of the various thermobarometers.
3. Two dominant peridotite types are present, namely garnet-bearing harzburgite and lherzolite. They define two compositional layers—a shallow ultra-depleted layer dominated by harzburgite and a deeper more fertile layer that is very strongly dominated by lherzolite. The harzburgites almost entirely reside below 150 km in the ultra-depleted layer, which straddles the diamond–graphite boundary.
4. The ultra-depleted layer beneath Ekati continues to ~ 1000 °C, as defined by the upper limit of the ultra-depleted harzburgite.

References

Armstrong, J.P., 2001. Kimberlite Indicator Mineral Chemistry Database (KIMC): a preliminary digital compilation of Kimberlite Indicator Mineral Chemistry (KIMC) extracted from publically available assessment filings; Slave Craton and environs, Northwest Territories and Nunavut, Canada. DIAND NWT Geology Division, DIAND EGS Open Report 2001-02 (CD-ROM).

Boyd, F.R., Nixon, P.H., 1975. Origins of the ultramafic nodules from some kimberlites of northern Lesotho and the Monastery Mine, South Africa. Phys. Chem. Earth 9, 431–454.

Brey, G.P., Köhler, T., 1990. Geothermobarometry in four-phase lherzolites: II. New thermometers and practical assessment of existing thermobarometers. J. Petrol. 31, 1353–1378.

Brey, G.P., Köhler, T., Nickel, K.G., 1990. Geothermobarometry of four-phase lherzolites: I. Experimental results from 10 to 60 kb. J. Petrol. 6, 1313–1352.

Carbno, G.B., Canil, D., 2002. Mantle structure beneath the SW Slave craton, Canada: constraints from garnet geochemistry in the Drybones Bay kimberlite. J. Petrol. 43, 129–142.

Carlson, R.W., Pearson, D.G., Boyd, F.R., Shirey, S.B., Irvine, G., Menzies, A.H., Gurney, J.J., 1999. Re-Os systematics of lithospheric peridotities: implications for lithosphere formation and preservation: Proc. 7th Int Kimberlite Conf, Cape Town, vol. 1, 99–108.

Carswell, D.A., Gibb, F.G.F., 1980. Geothermometry of garnet lherzolite nodules with special reference to those from the kimberlites of Northern Lesotho. Contrib. Mineral. Petrol. 74, 403–416.

Carswell, D.A., Gibb, F.G.F., 1987. Garnet lherzolite xenoliths in the kimberlites of northern Lesotho: revised $P-T$ equilibration conditions and upper mantle palaeogeotherm. Contrib. Mineral. Petrol. 97, 473–487.

Chinn, I.L., Gurney, J.J., Kyser, K.T., 1998. Diamonds and mineral inclusions from the NWT, Canada. Addendum to Ext. Abs. 7th Int. Kimb. Conf., Cape Town. Unpaginated.

Creaser, R.A., Grutter, H.S., Carlson, J., Crawford, B., this volume. Macrocrystal phlogopite Rb–Sr dates for the Ekati property kimberlites, Slave Province, Canada: Evidence for multiple intrusive episodes in the Paleocene and Eocene

Davis, W.J., Hegner, E., 1992. Neodymium isotopic evidence for the tectonic assembly of late Archean crust in the slave province, northwest Canada. Contrib. Mineral. Petrol. 111, 493–504.

Davis, W.J., Kjarsgaard, B.A., 1997. A Rb–Sr isochron age for a kimberlite from the recently discovered Lac de Gras field, Slave province, northwest Canada. J. Geol. 105, 503–509.

Davis, W., Jones, A.G., Bleeker, W., Grütter, H.S., 2003. Lithosphere development in the Slave Craton: a linked crustal and mantle perspective. Lithos 71 (2–4), 575–589.

Doyle, P., 2002. A petrographic and geochemical study of selected periodititic and pyroxenitic xenoliths from three kimberlite localities in the Lac de Gras region, Northwest territories, Canada. Unpublished MSc Thesis, University of Capetown, unpaginated

Dyck, D., Oshust, P., Carlson, J., Nowicki, T.E., Mullins, M., this volume. Effective resource estimates for primary deposits from the Ekati diamond mine, Canada

Finnerty, A.A., 1989. Xenolith-derived mantle geotherms: whither the inflection? Contributions to Mineralogy and Petrology 102, 367–375.

Finnerty, A.A., Boyd, F.R., 1984. Evaluation of thermobarometers for garnet peridotites. Geochim. Cosmochim. Acta 48, 15–27.

Finnerty, A.A., Boyd, F.R., 1987. Thermobarometry for garnet peridotites: basis for the determination of thermal and compositional structure of the upper mantle. In: Nixon, P.H. (Ed.), Mantle Xenoliths. Wiley, New York, pp. 381–402.

Fung, A.T., 1998. Petrochemistry of upper mantle eclogites from the Grizzly, Leslie, Pigeon and Sable kimberlites in the Slave Province, Canada. Ext. Abs. 7th Int. Kimb., 230–232.

Griffin, W.L., Doyle, B.J., Ryan, C.G., Pearson, N.J., O'Reilly, S.Y., Davies, R., Kivi, K., van Achterbergh, E., Natapov, L.M., 1999a. Layered mantle lithosphere in the Lac de Gras area, Slave craton: composition, structure and origin. J. Petrol. 40, 705–727.

Griffin, W.L., Doyle, B.J., Ryan, C.G., Pearson, N.J., O'Reilly, S.Y., Natapov, L.M., Kivi, K., Kretschmar, U., Ward, J., 1999b. Lithosphere structure and mantle terranes: Slave Craton, Canada. In: Gurney, J.J., Gurney, J.L., Pascoe, M.D., Richardson, S.H. (Eds.), Proc. 7th Int. Kimb. Conf. J. B. Dawson Volume, vol. 1. Red Roof Design, Cape Town, pp. 299–306.

Grütter, H.S., Moore, R.O., 2003. Pyroxene geotherms revisited—an empirical approach based on Canadian xenoliths. Ext. Abstr. 8th Int. Kimb. Conf., Victoria, (on CD), File FLA_272.

Grütter, H.S., Sweeney, R.J., 2000. Tests and constraints on single-grain Cr-pyrope barometer models: some initial results. Ext. Abstr. GAC/MAC Annual Joint Meeting, Calgary. CD-ROM, GeoCanada 2000.

Grütter, H.S., Apter, D.B., Kong, J., 1999. Crust–mantle coupling: evidence from mantle-derived xenocrystic garnets. In: Gurney, J.J., Gurney, J.L., Pascoe, M.D., Richardson, S.H. (Eds.), Proc. 7th Int. Kimb. Conf. J.B. Dawson Volume, Red Roof Design, Cape, pp. 307–313.

Grütter, H.S., Gurney, J.J., Menzies, A.H., Winter, F., 2004. An updated classification scheme for mantle-derived garnet, for use by diamond explorers. 77, 841–857.

Gurney, J.J., 1984. A correlation between garnets and diamonds in kimberlites. In: Glover, J.E., Harris, P.G. (Eds.), Kimberlite Occurrence and Origin: A Basis for Conceptual Models in Exploration. Geology Department and University Extension, vol. 8. University of Western Australia, Perth, pp. 143–166.

Gurney, J.J., Harte, B., Cox, K.G., 1975. Mantle xenoliths in the Matsoku kimberlite pipe. Phys. Chem. Earth 9, 507–529.

Gurney, J.J., Hildebrand, P., Carlson, J.C., Dyck, D., Fedortchouk, Y., 2004. The morphological characteristics of diamonds from the Ekati property, Northwest Territories, Canada. Lithos 77, 21–38.

Harte, B., 1977. Rock nomenclature with particular relation to deformation and recrystallisation textures in olivine-bearing xenoliths. J. Geol. 85, 279–288.

Harte, B., 1983. Mantle peridotites and processes—the kimberlite sample. In: Hawkesworth, C.J., Norry, M.J. (Eds.), Continental Basalts and Mantle Xenoliths: Shivaa, Nantwick, pp. 46–91.

Harte, B., Gurney, J.J., 1975. Ore mineral and phlogopite mineralisation within ultramafic nodules from the Matsoku kimberlite pipe Lesotho. Carnegie Inst. Washington Yearbook 74, 528–536.

Jacob, D.E., Fung, A., Jagoutz, E., Pearson, G., 2003. Petrology and geochemistry of eclogite xenoliths from the Ekati kimberlites area. Ext. Abs. 8th Int. Kimb. Conf., Victoria (on CD). File FLA_0239.

Jones, A.G., Ferguson, I.J., Chave, A.D., Evans, R.L., McNeice, G.W., 2001. Electric lithosphere of the Slave craton. Geology 29, 423–426.

Kennedy, C.S., Kennedy, G.C., 1976. The equilibrium boundary between graphite and diamond. J. Geophys. Res. 81, 2467–2470.

Kopylova, M.G., Russell, J.K., 2000. Chemical stratification of cratonic lithosphere: Constraints from the Northern Slave craton, Canada. Earth Planet. Sci. Lett. 181, 71–87.

Lockhart, G, Grütter, H.S., Carlson, J.A., 2004. Temporal, geomagnetic and related attributes of kimberlite magmatism at Ekati, Northwest Territories, Canada. Lithos 77, 665–682 (this volume).

Matthews, M., Harte, B., Prior, D., 1992. Mantle garnets: a cracking yarn. Geochim. Cosmochim. Acta 56, 2633–2642.

Nickel, K.G., Green, D.H., 1985. Empirical geothermobarometry for garnet peridotites and implications for the nature of the lithosphere, kimberlites and diamonds. Earth Planet. Sci. Lett. 73, 158–168.

Nimis, P., 1998. Evaluation of diamond potential from the composition of peridotitic chromian diopside. Eur. J. Mineral. 10, 505–519.

Nimis, P., Taylor, W.R., 2000. Single clinopyroxene thermobarometry for garnet peridotites: Part 1. Calibration and testing of a Cr-in-Cpx barometer and an enstatite-in-Cpx thermometer. Contrib. Mineral. Petrol. 139, 541–554.

Nowicki, T.E., Crawford, B., Dyck, D., Carlson, J., McElroy, R., Helmstaedt, H., Oshust, O., this volume. The geology of kimberlite pipes of the Ekati property, Northwest Territories, Canada.

O'Neill, H.St.C., Wood, B.J., 1979. An experimental study of Fe–Mg partitioning between garnet and olivine, and its calibration as a geothermometer. Contrib. Mineral. Petrol. 70, 59–70.

Pearson, N.J., Griffin, W.L., Doyle, B.J., O'Reilly, S.Y., Van Achterbergh, E., Kivi, K., 1999. Xenoliths from kimberlite pipes of the Lac de Gras area, Slave Craton, Canada. In: Gurney, J.J., Gurney, J.L., Pascoe, M.D., Richardson, S.H. (Eds.), Proc. 7th Int. Kimb. Conf. P.H. Nixon Volume Red Roof Design, Cape Town, pp. 644–658.

Pollack, H.N., Chapman, D.S., 1977. On the regional variation of heat flow, geotherms, and lithospheric thickness. Tectonophysics 38, 279–296.

Ramsay, R.R., 1992. Geochemistry of Diamond Indicator Minerals. PhD thesis, University of Western Australia, Perth.

Ryan, C.G., Griffin, W.L., Pearson, N.J., 1996. Garnet geotherms: pressure–temperature data from Cr-pyrope garnet xenocrysts in volcanic rocks. J. Geophys. Res. B3, 5611–5625.

Shimizu, N., Richardson, S.H., 1987. Trace element abundance patterns of garnet inclusions in peridotitic suite diamonds. Geochim. Cosmochim. Acta 51, 755–758.

Smith, D., 1999. Temperatures and pressures of mineral equilibration in peridotite xenoliths: review, discussion, and implications. In: Fei, Y., Bertka, C.M., Mysen, B.O. (Eds.), Mantle Petrology: Field Observations and High Pressure Experimentation: A tribute to Francis R. (Joe) Boyd. Geochemical Society Special Publication, vol. 6. pp. 171–188.

Stachel, T., Harris, J.W., Tappert, R., Brey, G.P., 2003. Peridotitic diamonds from the Slave and the Kaapvaal cratons—similarities and differences based on a preliminary data set. Lithos 71 (2–4), 489–503.

Thorpe, R.I., Cumming, G.L., Mortensen, J.K., 1992. A significant Pb isotope boundary in the Slave province and its probable relation to ancient basement in the western Slave province. Geol. Surv. Can. Open File 3079, 46.

Available online at www.sciencedirect.com

Lithos 77 (2004) 413–451

www.elsevier.com/locate/lithos

Genesis and evolution of the lithospheric mantle beneath the Buffalo Head Terrane, Alberta (Canada)☆

S. Aulbach[a,*], W.L. Griffin[a,b], S.Y. O'Reilly[a], Tom E. McCandless[c]

[a] GEMOC ARC National Key Centre, Macquarie University, Australia
[b] CSIRO Exploration and Mining, North Ryde, Australia
[c] Ashton Mining Canada Inc., North Vancouver, Canada

Received 27 June 2003; accepted 17 February 2004
Available online 10 June 2004

Abstract

Mantle xenoliths and xenocrysts were retrieved from three of the 88–86 Ma Buffalo Hills kimberlites (K6, K11, K14) for a reconnaissance study of the subcontinental lithospheric mantle (SCLM) beneath the Buffalo Head Terrane (Alberta, Canada). The xenoliths include spinel lherzolites, one garnet spinel lherzolite, garnet harzburgites, one sheared garnet lherzolite and pyroxenites. Pyroxenitic and wehrlitic garnet xenocrysts are derived primarily from the shallow mantle and lherzolitic garnet xenocrysts from the deep mantle. Harzburgite with Ca-saturated garnets is concentrated in a layer between ~ 135–165 km depth. Garnet xenocrysts define a model conductive paleogeotherm corresponding to a heat flow of 38–39 mW/m². The sheared garnet lherzolite lies on an inflection of this geotherm and may constrain the depth of the lithosphere–asthenosphere boundary (LAB) beneath this region to ca 180 km depth.

A loss of >20% partial melt is recorded by spinel lherzolites and up to 60% by the garnet harzburgites, which may be related to lithosphere formation. The mantle was subsequently modified during at least two metasomatic events. An older metasomatic event is evident in incompatible-element enrichments in homogeneous equilibrated garnet and clinopyroxene. Silicate melt metasomatism predominated in the deep lithosphere and led to enrichments in the HFSE with minor enrichments in LREE. Metasomatism by small-volume volatile-rich melts, such as carbonatite, appears to have been more important in the shallow lithosphere and led to enrichments in LREE with minor enrichments in HFSE. An intermediate metasomatic style, possibly a signature of volatile-rich silicate melts, is also recognised. These metasomatic styles may be related through modification of a single melt during progressive interaction with the mantle. This metasomatism is suggested to have occurred during Paleoproterozoic rifting of the Buffalo Head Terrane from the neighbouring Rae Province and may be responsible for the evolution of some samples toward unradiogenic Nd and Hf isotopic compositions.

Disturbed Re–Os isotope systematics, evident in implausible model ages, were obtained in situ for sulfides in several spinel lherzolites and suggest that many sulfides are secondary (metasomatic) or mixtures of primary and secondary sulfides. Sulfide in one peridotite has unradiogenic $^{187}Os/^{188}Os$ and gives a model age of 1.89 ± 0.38 Ga. This age coincides with the inferred emplacement of mafic sheets in the crust and suggests that the melts parental to the intrusions interacted with the lithospheric mantle.

A younger metasomatic event is indicated by the occurrence of sulfide-rich melt patches, unequilibrated mineral compositions and overgrowths on spinel that are Ti-, Cr- and Fe-rich but Zn-poor. Subsequent cooling is recorded by fine exsolution lamellae in the pyroxenes and by arrested mineral reactions.

☆ Supplementary data associated with this article can be found, in the online version, at doi: 10.1016/j.lithos.2004.04.020.
* Corresponding author. Fax: +61-2-9850-8943.
E-mail address: saulbach@els.mq.edu.au (S. Aulbach).

If the lithosphere beneath the Buffalo Head Terrane was formed in the Archaean, any unambiguous signatures of this ancient origin may have been obliterated during these multiple events.

Keywords: Mantle xenoliths; Mantle xenocrysts; Kimberlites; Mantle trace elements; Os and Hf isotopes; Mantle metasomatism; Canada lithosphere evolution

1. Introduction

In areas with thick sedimentary cover and recent glacial deposits, such as the Buffalo Head Terrane (hereafter, BHT) in northern Alberta, information on the nature of the crust and mantle is often restricted to remote sensing and drill cores. Using these methods, a new kimberlite province, erupted at 88–86 Ma (Buffalo Hills kimberlites), was identified in the centre of the BHT (Carlson et al., 1999). This discovery has provided an opportunity to directly study the lithosphere beneath the BHT by means of mantle samples transported to the surface. Xenoliths and xenocrysts were retrieved from three kimberlites (K6, K11, K14) for a reconnaissance geochemical study (major-element, trace-element and isotopic) in order to constrain the paleogeotherm and the stratigraphy of the subcontinental lithospheric mantle (SCLM) beneath this region, and to assess its origin and evolution. In a parallel study, xenocryst minerals from 29 of the 37 known kimberlite bodies from Buffalo Hills have been investigated at the province scale (Hood and McCandless, 2003). Preliminary results indicate that there are regional variations in distribution and composition of the xenocrysts. Furthermore, kimberlites in the southern part of the province have higher abundances of metasomatic minerals and are dominated by a lower-Cr assemblage, which may be directly related to their lower diamond grade (Hood and McCandless, 2003).

Geophysics and crustal geology of the area provide important constraints on any interpretation of data from the SCLM, as major tectonothermal events in the crust are expressions of mantle processes. A prolonged episode of rifting, subduction beneath the BHT, emplacement of mafic sheets and kimberlite magmatism are some of the events recognised in the crust that may entail reworking of the underlying SCLM. Ross and Eaton (2002) suggest that the BHT represents a collage of Paleoprotero-

zoic crust formed on an Archaean microcontinent that has been extensively modified during Proterozoic magmatism. This study is designed to investigate whether and how these events have affected the SCLM beneath the BHT, and whether any Archaean signatures have survived the postulated Proterozoic reworking. Furthermore, comparisons will be made with peridotite suites from the Slave craton to the north, in order to establish whether any relationship exists between these mantle sections, given their present-day proximity.

2. Geology

The Paleoproterozoic BHT is part of the western Canadian Shield in northern Alberta. Because of its Phanerozoic rock cover, the boundaries of the BHT and other covered domains have been defined based on aeromagnetic data, using the overall signature, internal fabric and contrasts with adjacent domains (Ross, 2000), and geochronology on drill core recovered during petroleum exploration (Thériault and Ross, 1991). The BHT is truncated to the south by the Snowbird Tectonic Zone, whereas to the north, it is attenuated with the Great Slave Lake Shear Zone (Fig. 1). In the east, the BHT is separated from the Rae Province by the younger (1.987 to 1.900 Ga) Taltson magmatic zone. In the west, it is separated from the Nova Terrane by the magmatic rocks of the Ksituan High (Ross et al., 1991; Ross and Eaton, 1997; Carlson et al., 1999). The BHT is transected in the south by the Peace River Arch, a lineament where elevated heat flow and anomalous crustal uplift have been observed and which has been active since the Proterozoic (Carlson et al., 1999).

The tectonic setting of the BHT is not well constrained because its lithosphere has been overprinted by younger events (Ross, 2002). Archaean inheritance ages were obtained for Proterozoic basement rocks of

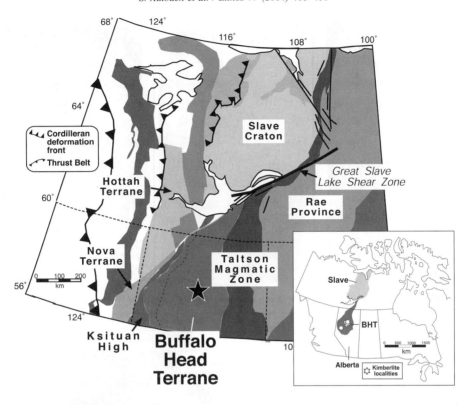

Fig. 1. Domain map showing the BHT and neighbouring structural elements (after Hoffman, 1989). Star shows sample locality. Inset shows map of western Canada with outlines of BHT and Slave craton.

the BHT by Villeneuve et al. (quoted in Ross and Eaton, 1997), suggesting an older component is present. The similarity of basement ages (2.4–2.1 Ga) and Nd isotope systematics for the BHT and the adjacent Rae Province suggests that they formed a single crustal domain by 2.4 Ga (McNicoll et al., 2000). The occurrence of 2.34 Ga mafic to ultramafic rocks in the contact area between the BHT and the Rae Province may indicate that these two entities were separating by that time (Ross et al., 1991; Bostock and van Breemen, 1994). Rifting was followed by subduction at the eastern and western terrane margin, with the subduction polarity facing away from the BHT. This resulted in coeval arc magmatism that produced the 1980–1920 Ma Taltson magmatic zone and the 1986–1900 Ma Ksituan High (Ross et al., 1991; Bostock and van Breemen, 1994).

Postcollisional underplating of mafic rocks in the core of the BHT may be manifested by the 1.89 to 1.76 Ga sheet-like mafic sills of the Winagami reflection sequence, a package of reflectors in the upper

crust of the Peace River Arch region, (Ross and Eaton, 1997, and references therein). The high-grade metamorphic Buffalo Head Terrane was eventually covered by up to 1600 m of Devonian and Cretaceous sediments (Carlson et al., 1999). Between 88 and 86 Ma, the centre of the BHT was intruded by the kimberlites of the Buffalo Hills kimberlite province (Carlson et al., 1999; Ross et al., 2000).

3. Petrography

Thirty-one xenoliths were retrieved from three different kimberlites (K6, K11, K14) for this study. Two major xenolith types are present: peridotites and pyroxenites. Our study includes 22 peridotites and nine pyroxenites. Fig. 2 shows examples of the rock types sampled by the kimberlites. Petrographic details are given in Table 1.

Peridotites encompass four mineralogical varieties: 15 spinel peridotites, one garnet spinel lherzolite

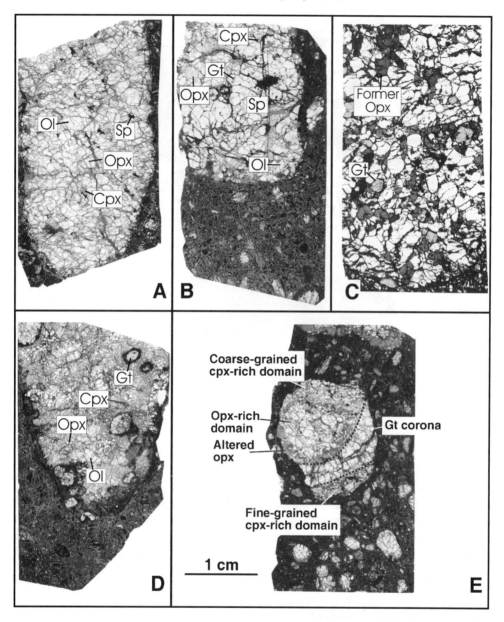

Fig. 2. Xenoliths from the Buffalo Hills kimberlites in thin section (plane-polarised light). Some minerals are indicated (ol: olivine, sp: spinel, opx: orthopyroxene, cpx: clinopyroxene, gt: garnet). (A) Spinel lherzolite; (B) garnet spinel lherzolite showing spinel blebs in large garnet clusters; (C) garnet harzburgite with altered opx; (D) sheared garnet lherzolite; (E) corona-structured garnet pyroxenite; garnet-free pyroxenite is visible at top of section. The scale is the same for all sections as indicated in (E).

(K14-3a), three garnet peridotites (with altered pyroxene; all from kimberlite K11) and one sheared garnet lherzolite (K14-1). Most spinel lherzolites are medium-grained rocks with equant to tabular, foliated microstructure (Fig. 2A). Spinel peridotites have ol-

ivine modes ranging from 56% to 87%, orthopyroxene modes from 10% to 38% and clinopyroxene modes from 1% to 10%. The garnet spinel lherzolite has a medium-grained equant microstructure with large garnet clusters that contain numerous inclusions

Table 1
Petrographic details

Sample #	Rock type	Size [cm]	Microstructure	Assemblage	Modes
K6-1	sp lherz	5	medium-eq/tab fol	ol,opx,cpx,sp,sf	87,11,1,1,<1
K6-2a	sp lherz	2.5	medium-eq	ol,opx,cpx,sp,sf	84,10,5,1,<1
K6-2b	sp lherz	5	medium-eq/tab fol	ol,opx,cpx,sp	na
K6-2b2	gt pyrox	0.3	fine-eq	gt,opx	na
K6-3	sp lherz	5	coarse-eq/tab fol	ol,opx,cpx,sp	75,23,1,1
K6-4	sp lherz	5	medium-eq/tab fol	ol,opx,cpx,sp	70,26,4,<1
K6-5a	sp-lherz	2.5	medium-eq/tab fol	ol,opx,cpx,sp	na
K6-5b	pyrox	2	fine-mosaic	opx,cpx,sf	89,10,1
K11-3	gt perid	5	coarse-tab fol	ol,alt opx,gt,sp	85,14,1,<1
K11-5	gt perid	5	coarse-eq/tab fol	ol,alt opx,gt	78,19,3
K11-6	gt perid	3	coarse-tab	ol,alt opx,gt	80,19,1
K14-1	gt lherz	4	porph-tab	ol,opx,cpx,gt	78,11,2,9
K14-2a	gt pyrox	2	mosaic-porph	gt,cpx,opx,rut,sf	8,35,57,<1,<1
K14-2ba	sp lherz	4	medium-tab fol	ol,opx,cpx,sp,sf	81,15,2,2,<1
K14-2bb	gt pyrox	2	mosaic-porph	gt,cpx,opx,sf	4,53,43,<1
K14-2c	sp lherz	2	coarse-tab	ol,opx,cpx,sp	59,38,2,1
K14-3a	gt sp lherz	2.5	medium-eq/tab fol	ol,opx,cpx,gt,sp,ilm,sf	43,11,2,42,2,<1
K14-3b2	perid	0.4	?porph	ol,cpx,sf	na
K14-3c	gt pyrox	2	medium-eq	gt,cpx,alt opx,rut,sf	4,49,47,<1,<1
K14-4a	sp lherz	5	coarse-tab grano	ol,opx,cpx,sp	72,16,10,2
K14-4b	gt pyrox	1.5	medium-eq	gt,cpx	na
K14-5a	sp harz	1	medium-eq/tab grano	ol,opx,sp	56,44,<1
K14-5b	sp lherz	2.5	coarse-eq/tab fol	ol,opx,cpx,sp,sf	75,18,5,2,<1
K14-5c	gt pyrox	2	corona	gt,cpx,opx,sf	na
K14-5c2	pyrox	2	mosaic-porph	cpx,opx	na
K14-5d	sp lherz	1.5	medium-eq/tab grano	ol,opx,cpx,sp,sf	67,28,2,3,<1
K14-5f	gt pyrox	2.5	mosaic-porph	gt,cpx,opx,sf	7,21,72,<1
K14-5g	harz	1	na	ol,opx	85,15
K14-6	sp lherz	5	medium-eq/tab grano	ol,opx,cpx,sp,sf	80,14,5,1,<1
K14-7a	sp lherz	5	coarse-eq/tab fol	ol,opx,cpx,sp,sf	82,11,7,<1,<1
K14-8	sp lherz	5	coarse-eq/tab grano	ol,opx,cpx,sp	77,21,1,1

Petrographic data: sp: spinel; gt: garnet; lherz: lherzolite; harz: harzburgite; perid: peridotite; pyrox: pyroxenite; eq: equant; tab: tabular; fol: foliated; porph: porphyroclastic; grano: granoblastic; ol: olivine; opx: orthopyroxene; cpx: clinopyroxene; sf: sulfide; ilm: ilmenite; rut: rutile; alt: altered; na: not available.

of spinel and are arranged around large blebs of spinel (Fig. 2B). Its modal proportions of 43% olivine, 11% orthopyroxene, 2% clinopyroxene, 42% garnet and 2% spinel may not be representative of the actual modes, considering the sample size (2.5 cm) and the size of the garnet clusters (up to 1 cm). The three garnet peridotites are inferred to have been pyroxene-bearing before alteration (Fig. 2C). They have medium- to coarse-grained equant to tabular microstructure, with olivine modes ranging from 78% to 85%, garnet modes 1% to 3% and modes of former pyroxene of 14–19%. The sheared garnet lherzolite has a mosaic–porphyroclastic foliated microstructure (Fig. 2D). Its mineral modes are 78% olivine, 11% orthopyroxene, 2% clinopyroxene and 9% garnet. Some

xenoliths from the Buffalo Hills kimberlites are quite small, and this contributes to uncertainty in the point-counted mode estimates. Jerde et al. (1993) considered the error in estimating modes to be around 10% for medium- to coarse-grained rocks.

Eight of the peridotites contain accessory sulfides, most of which have polyhedral or spherical shapes. Interstitial sulfides are slightly more frequent than enclosed ones, which are hosted by olivine, orthopyroxene and clinopyroxene. Spinel lherzolite K14-5b contains an oval melt patch ca. 0.5 mm in diameter, along with several smaller melt patches. Half of the large melt patch is made up of a cluster of polyhedral sulfide grains, one of which is intergrown with skeletal spinel.

Xenoliths of the pyroxenite suite include mostly medium-grained equant or mosaic porphyroclastic (as an analogy to peridotites) varieties. Orthopyroxene modes range from 43% to 89%, clinopyroxene modes from 0% to 53% and garnet modes from 0% to 8%. Six of the pyroxenites contain accessory sulfide and two contain accessory rutile. One sample has a striking corona structure that implies former grain sizes >1 cm (Fig. 2E). It consists of five distinct domains: (1) a broad garnet corona; (2) a domain of large orthopyroxene grains with garnet necklaces, rimmed by (3) a ring of almost completely altered orthopyroxene; (4) a domain of large clinopyroxene grains and garnet strings adjacent to (1) and (3); (5) a domain of small clinopyroxene and garnet grains adjacent to (1). Garnet necklaces around orthopyroxene, suggestive of exsolution processes, occur in several other garnet pyroxenites.

4. Analytical methods

4.1. Electron probe microanalysis (EPMA)

Major-element contents were collected using the CAMECA Camebax SX50 electron microprobe in the GEMOC National Key Centre, Department of Earth and Planetary Sciences, Macquarie University. Silicates were analysed with a 15 kV acceleration voltage and counting times of 10 s on the peak, 5 s on each side of the background. In general, totals between 99% and 101% oxides were accepted; in some cases, they lie between 98.5 and 101.5. Sulfide analyses were performed with an accelerating voltage of 20 kV and a beam current of 20 nA. Counting times were 20 s on the peak and 10 s on each side of the background. The PAP matrix correction procedure (Pouchou and Pichoir, 1984) was applied to the raw data.

4.2. Laser ablation ICPMS

Laser ablation analyses were performed with a custom-built laser-ablation system (designed by S.E. Jackson) or a Merchantek LUV 266 Nd:YAG UV laser system, both linked to an Agilent 7500 ICPMS. The methods followed those described by Norman et al. (1996) and Xu et al. (1999). Raw data were processed on-line using the GLITTER-software (van Achterberg et al., 1999; see www.es.mq.edu.au/GEMOC).

4.3. Solution Sr–Nd–Hf analyses of garnet and clinopyroxene

Hand-picked clinopyroxene and garnet separates were washed in 6 N HCl, and rinsed in Milli-Q water prior to dissolution in concentrated HF and HNO_3. Sample digestion and chemical separation of the elements of interest follow conventional procedures described in the literature (Patchett and Tatsumoto, 1980; Blichert-Toft et al., 1997; Blichert-Toft, 2001). Sr, Nd and Hf aliquots were analysed on a Nu plasma multicollector ICP–MS at Macquarie University, with blanks generally below 1000, 80 and 60 pg, respectively. Repeated measurements of standard materials during the data acquisition (February to July 2003) yielded the following values: $^{87}Sr/^{86}Sr$ of 0.71024 ± 0.00010 (2 S.D.; $n = 51$) for the SRM-987 standard; $^{87}Sr/^{86}Sr$ of 0.70352 ± 0.00004 ($n = 15$) for BHVO-1; $^{143}Nd/^{144}Nd$ of 0.511138 ± 0.000038 ($n = 42$) for the JMC-321 standard, $^{143}Nd/^{144}Nd$ of 0.513003 ± 0.000043 ($n = 10$) for BHVO-1; $^{176}Hf/^{177}Hf$ of 0.282163 ± 0.000002 ($n = 20$) for the JMC-475 standard and $^{176}Hf/^{177}Hf$ of 0.283128 ± 0.000090 for BHVO-1 ($n = 8$).

4.4. In situ Re–Os isotope analyses of sulfides

Re and Os isotopic ratios were collected with a New Wave Research LUV266 laser microprobe attached to a Nu plasma multicollector ICP–MS at Macquarie University. Typical laser operating conditions are 4–5 Hz frequency, a beam energy of 2–5 mJ/pulse and spot sizes of 60–80 μm. The set-up used for collection of ion beams consisted of eight Faraday cups, using Os as internal isotope standard for mass fractionation correction, or of a combination of eight Faraday cups plus two ion counters (for masses 185 and 187). Drift on the ion counters was corrected for by bracketing sample analyses with analyses of standard PGE-A, a synthetic PGE-doped NiS bead with known Os isotopic composition. The calibration procedures and the corrections applied, as well as reproducibility of standard material are described in detail in Pearson et al. (2002).

5. Major elements

Representative major-element compositions of minerals in xenoliths are shown in Tables 2–9 (online Tables 2–9). Many orthopyroxenes and clinopyroxenes have exsolved complementary pyroxene and/or spinel; some garnets have exsolved rutile. If not indicated otherwise, the compositions discussed here are obtained from exsolution-free areas of the grains; analyses obtained with defo-

cused electron beam (i.e., where exsolved phases are also sampled) are given in Tables 4 and 5.

5.1. Garnet

In the Cr_2O_3 against CaO plot (Sobolev et al., 1973; Gurney, 1984), most garnets (xenocrystic and in xenoliths) plot in the lherzolite field, but several groups may be distinguished (Fig. 3). One group has correlated CaO and Cr_2O_3 and lies entirely in

Table 2
Representative major-element composition of garnet in xenoliths

Sample	K6-2b2	K11-5	K11-5	K11-6	K14-1	K14-2a	K14-3a
n	43	5	21	8	56	14	23
Comment		?primary	?metasomatised	core			
Type	gt pyrox	alt gt harz	alt gt harz	alt gt harz	gt lherz	gt pyrox	gt lherz
SiO_2	40.93	41.29	41.17	42.67	42.01	41.45	41.83
TiO_2	<0.06	0.79	0.11	0.27	0.79	<0.06	<0.06
Al_2O_3	23.01	17.98	17.69	18.91	19.35	23.00	21.70
Cr_2O_3	0.50	6.96	7.59	5.89	3.73	0.35	2.01
FeO^{total}	14.49	7.21	7.21	6.95	8.19	13.44	9.90
MnO	0.47	0.21	0.24	0.21	0.21	0.46	0.51
MgO	15.79	21.72	19.40	20.88	20.81	16.43	18.09
CaO	5.27	4.10	6.50	5.18	4.81	5.16	5.87
NiO	<0.08	<0.08	<0.08	<0.08	<0.08	<0.08	<0.08
Na_2O	<0.04	<0.04	<0.04	<0.04	0.09	<0.04	<0.04
K_2O	<0.04	<0.04	<0.04	<0.04	<0.04	<0.04	<0.04
Total	100.5	100.3	100.0	101.0	100.0	100.4	100.0
Mg#	66.0	84.3	82.7	84.3	81.9	68.6	76.5
Cr#	1.4	20.6	22.3	17.3	11.5	1.0	5.9

Sample	K14-5c (2) max	K14-5c (2) avg	K14-5c (4) min	K14-5c (4) avg	K14-5c (5) max	K14-5c (5) max	K14-5c (5) avg
n		11		9			5
Comment	domain 2	domain 2	domain 4	domain 4	domain 5	domain 5	domain 5
Type	gt pyrox	gt pyrox	gt pyrox	gt pyrox	gt pyrox	gt pyrox	gt pyrox
SiO_2	41.47	41.32	41.32	41.66	42.45	42.22	42.34
TiO_2	<0.06	<0.06	<0.06	<0.06	0.07	<0.06	<0.06
Al_2O_3	23.28	23.08	22.85	23.02	23.43	23.10	23.24
Cr_2O_3	0.23	0.17	<0.09	0.11	0.15	0.10	0.13
FeO^{total}	16.31	15.89	14.62	15.10	13.84	13.41	13.60
MnO	0.36	0.31	0.19	0.22	0.25	0.23	0.24
MgO	15.43	15.07	15.19	15.36	17.23	17.00	17.09
CaO	5.05	4.97	5.25	5.40	4.97	4.69	4.83
NiO	0.10	<0.08	<0.08	<0.08	0.08	<0.08	<0.08
Na_2O	0.05	<0.04	<0.04	<0.04	<0.04	<0.04	<0.04
K_2O	<0.04	<0.04	<0.04	<0.04	<0.04	<0.04	<0.04
Total	101.3	100.9	100.2	100.9	102.2	101.1	101.5
Mg#	63.6	62.8	63.7	64.5	69.4	68.8	69.1
Cr#	0.6	0.5	0.2	0.3	0.4	0.3	0.4

Representative major-element composition of garnet in xenoliths (wt.%); *n*: number of analyses; Mg# = molar [100 Mg/(Mg + Fe)]; Cr# = molar [100 Cr/(Cr + A1)]; max, min, avg are maximum, minimum and average values for compositionally zoned minerals; domains denote distinct areas in corona-structured sample; xenocr: xenocryst; for other abbreviations, see Table 1.

Table 3
Average major-element composition of garnet xenocrysts

Sample	Avg fertile lherz	Avg depl lherz	Avg sili metas lherz	Avg phlog metas lherz	Avg fertile harz	Avg depl harz	Avg fertile gt sp lherz	Avg depl gt sp lherz	Avg sili metas gt sp lherz	Avg fertile wehr	Avg depl wehr
n	4	40	29	19	2	7	10	4	3	9	4
SiO_2	41.40	41.08	41.70	41.07	39.90	40.09	41.46	41.53	41.53	41.30	40.38
TiO_2	0.22	0.09	0.44	0.09	0.08	0.14	0.06	0.04	0.05	0.08	0.04
Al_2O_3	19.55	19.12	20.40	19.06	11.65	12.46	22.22	22.10	22.67	22.16	16.98
Cr_2O_3	5.27	6.12	3.85	6.39	15.25	14.06	2.36	2.60	2.09	2.35	8.50
FeO^{total}	8.39	7.63	8.06	7.69	6.79	7.00	9.02	8.75	9.68	9.56	8.39
MnO	0.43	0.43	0.36	0.43	0.38	0.38	0.51	0.55	0.51	0.52	0.51
MgO	19.35	19.00	20.11	19.02	18.60	18.41	18.65	18.90	18.33	18.14	16.93
CaO	5.63	6.19	5.02	6.07	6.91	6.87	5.70	5.64	5.57	6.04	7.96
Na_2O	0.02	0.03	0.07	0.04	0.04	0.03	0.03	0.02	0.02	0.04	0.06
Total	100.3	99.7	100.0	99.9	99.6	99.4	100.0	100.1	100.5	100.2	99.7
Mg#	80.4	81.6	81.6	81.5	83.0	82.4	78.7	79.4	77.1	77.2	78.3
Cr#	15.4	17.7	11.3	18.4	46.8	43.2	6.7	7.3	5.8	6.7	25.2

Average major element composition of different classes of garnet xenocrysts (analyses of Griffin et al., 2004) in wt.%; the basis for the classification is the Cr–Ca relationship (lherzolite, harzburgite, garnet–spinel lherzolite, wehrlite) combined with Zr–Y relationships (used to distinguish fertile, depleted, silicate-metasomatised and "phlogopite"-metasomatised sources; Griffin et al., 1999a); *n*: number of grains averaged; avg: average; depl: depleted; sili metas: silicate melt metasomatised; phlog met: "phlogopite"-metasomatised; wehr: wehrlite; for other abbreviations see Table 2.

the lherzolite field, with Cr_2O_3 contents up to ~ 10 wt.% (group 1). A second group (group 2) scatters toward higher CaO contents at a given Cr_2O_3 content. This is similar to what is observed for spinel-bearing garnet peridotites from the north-central Slave craton, which were interpreted to be less strongly enriched in Cr_2O_3 at a given CaO content due to coexisting high-Cr spinel (Kopylova et al., 2000). The group of garnets with the highest Cr_2O_3 contents (14–18 wt.%) plots parallel to the lherzolite field but is displaced toward lower CaO contents (group 3).

One group of garnets, which includes those from garnet spinel lherzolite where spinel and garnet share high-energy amoeboid grain boundaries, is restricted to low Cr_2O_3 contents (~ 1.5–3 wt.%; group 4). This group may represent peridotites from the garnet spinel transition zone. The higher abundance of spinel and the presence of Cr-rich orthopyroxene in these samples, compared to chromite-bearing garnet peridotites, may limit the uptake of Cr_2O_3 in coexisting garnet. Garnets in pyroxenite xenoliths from Buffalo Hills have < 1 wt.% Cr_2O_3 and show a steeper correlation between Cr_2O_3 and CaO contents than those in peridotites.

Garnets in peridotite xenoliths have TiO_2 contents up to 0.79 wt.%, Cr_2O_3 from 2.00 to 6.96 wt.% and

CaO from 4.61 to 6.50 wt.%. The highest TiO_2 contents are observed in the sheared lherzolite K14-1. Mg-numbers [Mg#: $100Mg/(Mg + Fe)$] range from 71.9 to 84.3. Garnet in garnet spinel lherzolite K14-3a has the lowest Mg-number and Cr-number [Cr#: $100Cr/(Cr + Al)$]. Of the 26 spot analyses sampling five garnets in garnet harzburgite K11-5, 21 gave a Ca-saturated lherzolitic composition (CaO = 6.5 ± 0.19 wt.%). Five analyses in two of the garnets gave a Ca-undersaturated composition, with distinctly and relatively uniform lower CaO (4.1 ± 0.34 wt.%) and higher TiO_2 content.

Garnets from pyroxenite xenoliths have lower TiO_2 contents (< 0.06–0.08) and Cr_2O_3 contents (0.09–1.00 wt.%) than peridotitic garnets, and CaO varies from 4.69 to 5.90 wt.%. Mg-numbers range from 62.2 to 70.3 and do not overlap those of peridotitic garnets. The pyroxenite with corona structure has garnets with unequilibrated major-element composition, depending on their position relative to the broad garnet corona. Average, minimum and maximum values are given in Table 2. Garnet arranged in necklaces around orthopyroxene in the orthopyroxene-rich domain (domain 2) and garnet of the corona have average CaO of 5.40 and 4.83 wt.%, respectively, and Mg-numbers of 64.5 and 69.1, respectively.

Table 4
Representative major-element composition of clinopyroxene

Sample	K6-2b	K6-2b	K6-2b	K6-3	K6-4	K6-5b	K14-1	K14-2a	K14-2a
n	20	10	20	14	15	20	22	18	6
Comment	core	rim	20 mic		core			core	rim
Type	sp lherz	sp lherz	sp lherz	sp lherz	sp lherz	pyrox	gt lherz	gt pyrox	gt pyrox
SiO_2	53.55	53.78	53.31	55.07	54.95	52.89	55.01	53.80	54.14
TiO_2	<0.06	<0.06	<0.06	<0.06	<0.06	0.42	0.21	0.33	0.28
Al_2O_3	2.55	2.08	2.54	2.28	2.76	5.68	1.79	4.88	4.25
Cr_2O_3	0.90	0.75	0.90	0.81	0.98	0.91	1.01	0.33	0.36
FeO^{total}	1.74	1.70	1.82	1.58	1.71	3.43	3.80	2.75	2.50
MnO	<0.09	<0.09	<0.09	<0.09	<0.09	<0.09	0.10	<0.09	<0.09
MgO	16.72	17.02	17.16	16.63	16.52	14.03	18.62	14.82	15.07
CaO	23.73	23.91	23.28	24.15	23.60	22.11	17.46	22.09	22.52
NiO	<0.08	<0.08	<0.08	<0.08	<0.08	<0.08	<0.08	<0.08	<0.08
Na_2O	0.56	0.53	0.57	0.47	0.63	1.40	1.61	1.45	1.39
K_2O	<0.04	<0.04	<0.04	<0.04	<0.04	<0.04	<0.04	<0.04	<0.04
Total	99.9	100.0	99.8	101.1	101.3	101.0	99.7	100.5	100.6
Mg#	94.5	94.7	94.4	94.9	94.5	87.9	89.7	90.6	91.5
Cr#	19.1	19.5	19.2	19.3	19.2	9.7	28.2	4.4	5.5

Sample	K14-2c	K14-2c	K14-2c	K14-3a	K14-3b2	K14-5b	K14-5c (4) max	K14-5c (4) min	K14-5c (4) avg
n	12	6	15	17	12	33			10
Comment	core	rim	20 mic				domain 4	domain 4	domain 4
Type	sp lherz	sp lherz	sp lherz	gt lherz	perid	sp lherz	gt pyrox	gt pyrox	gt pyrox
SiO_2	53.72	54.21	53.64	54.64	54.59	53.62	55.27	52.87	54.32
TiO_2	<0.06	<0.06	<0.06	0.11	0.19	0.07	1.83	0.15	0.51
Al_2O_3	2.34	1.84	2.31	1.66	0.75	1.75	5.03	2.71	4.23
Cr_2O_3	0.90	0.62	0.81	0.70	1.14	0.63	0.17	<0.09	0.09
FeO^{total}	1.82	1.74	1.89	1.75	2.75	1.44	3.63	2.38	3.04
MnO	<0.09	<0.09	<0.09	<0.09	<0.09	<0.09	<0.09	<0.09	<0.09
MgO	16.80	17.35	17.31	17.30	17.62	16.94	16.60	14.27	14.97
CaO	23.82	23.88	23.27	23.62	21.95	24.28	24.06	22.03	22.50
NiO	<0.08	<0.08	<0.08	<0.08	<0.08	<0.08	<0.08	<0.08	<0.08
Na_2O	0.52	0.48	0.49	0.65	0.77	0.61	1.80	0.33	1.40
K_2O	<0.04	<0.04	<0.04	<0.04	<0.04	<0.04	<0.04	<0.04	<0.04
Total	100.1	100.2	99.9	100.5	99.9	99.5	101.4	100.8	101.1
Mg#	94.3	94.7	94.2	94.6	92.0	95.5	91.9	87.7	89.8
Cr#	20.4	18.3	19.0	21.4	50.4	19.3	2.7	0.6	1.5

Sample	K14-5c (5) max	K14-5c (5) min	K14-5c (5) avg	K14-6	K14-6	K14-6	K14-7a	K14-7a	K14-7a
n			12	25	3	20	17	10	30
Comment	domain 5	domain 5	domain 5		incl	20 mic	core	rim	20 mic
Type	gt pyrox	gt pyrox	gt pyrox	sp lherz	sp lherz	sp lherz	sp lherz	sp lherz	sp lherz
SiO_2	55.06	52.90	54.45	52.94	53.49	53.42	53.63	53.81	53.34
TiO_2	0.57	0.34	0.47	<0.06	0.06	<0.06	0.17	0.15	0.19
Al_2O_3	6.02	5.25	5.61	2.63	2.01	2.56	2.78	2.11	3.10
Cr_2O_3	0.14	<0.09	<0.09	0.91	0.74	0.89	1.08	0.87	1.17
FeO^{total}	5.17	2.51	3.19	1.78	1.47	1.89	1.44	1.91	1.56
MnO	<0.09	<0.09	<0.09	<0.09	<0.09	<0.09	<0.09	<0.09	<0.09
MgO	16.51	13.97	14.48	16.52	16.71	16.95	16.25	17.28	16.57
CaO	21.50	18.06	20.78	23.79	24.36	23.38	23.59	23.00	22.96
NiO	<0.08	<0.08	<0.08	<0.08	<0.08	<0.08	<0.08	<0.08	<0.08

(continued on next page)

Table 4 (*continued*)

Sample	K14-5c (5) max	K14-5c (5) min	K14-5c (5) avg	K14-6	K14-6	K14-6	K14-7a	K14-7a	K14-7a
n			12	25	3	20	17	10	30
Comment	domain 5	domain 5	domain 5		incl	20 mic	core	rim	20 mic
Type	gt pyrox	gt pyrox	gt pyrox	sp lherz	sp lherz	sp lherz	sp lherz	sp lherz	sp lherz
Na_2O	2.25	1.70	2.13	0.55	0.54	0.55	1.04	0.80	1.08
K_2O	0.14	<0.04	<0.04	<0.04	<0.04	<0.04	<0.04	<0.04	<0.04
Total	101.8	100.6	101.3	99.3	99.5	99.8	100.1	100.0	100.1
Mg#	91.1	83.7	89.1	94.3	95.3	94.1	95.3	83.8	95.0
Cr#	1.7	0.3	1.0	18.7	19.9	18.8	20.9	19.2	20.1

Representative major-element composition of clinopyroxene in wt.%; 20 mic denotes cpx with exsolved opx analysed with 20 μm electron beam; incl: inclusion; for other abbreviations and explanations, see Table 2.

5.2. Clinopyroxene

Clinopyroxenes in spinel lherzolites have Al_2O_3 from 1.75 to 3.34 wt.%, Cr_2O_3 from 0.62 to 1.08 wt.% and Na_2O from 0.44 to 1.05 wt.% (Fig. 4). Mg-numbers range from 91.3 to 95.5. Clinopyroxene in the sheared lherzolite has a distinctly lower Mg-number and higher FeO, MgO, Na_2O and TiO_2 contents than other peridotites, while also having the lowest CaO of all samples.

Compared to peridotites, clinopyroxenes in pyroxenites have higher Al_2O_3 (2.71–6.30 wt.%) and Na_2O (0.33–2.25 wt.%) and lower Cr_2O_3 contents (<0.09–0.91 wt.%). Mg-numbers are lower and more variable than for peridotitic garnets (82.6–92.1). Garnet-free pyroxenites have the highest Al_2O_3 and Cr_2O_3 contents. Like garnet, clinopyroxenes in the corona-structured pyroxenite have differing compositions, depending on their position in the sample. Clinopyroxenes within the corona (coarse-grained domain) has Al_2O_3 and CaO contents of 4.72 and 22.08 wt.%, respectively, compared to 5.61 and 20.78 wt.%, respectively, for clinopyroxene outside the corona (fine-grained domain). Cr_2O_3 contents in clinopyroxenes of the corona-structured pyroxenite are the lowest of all samples analysed in this study, on average below the detection limit (0.09 wt.%), whereas Na_2O contents are highest for the corona-structured pyroxenite, with 1.40 wt.% for clinopyroxene within and 2.13 wt.% for clinopyroxene outside the garnet corona.

Some clinopyroxenes have rim compositions that differ from those in the cores. Al_2O_3, Cr_2O_3 and Na_2O contents of clinopyroxene rims are higher than in the cores in most cases. Clinopyroxene occurs as inclusions in olivine, orthopyroxene and garnet. Included grains have lower Al_2O_3 and Cr_2O_3 contents and higher CaO contents than non-included grains in the same sample, while their TiO_2 and Na_2O contents and Mg-numbers show no appreciable difference. Many clinopyroxenes contain exsolution lamellae of both orthopyroxene and spinel. Compositions prior to the exsolution were estimated by averaging 20–30 analyses obtained with a defocused microbeam, sampling an area ca 20 μm wide. Pre-exsolution clinopyroxenes tend to have higher Al_2O_3 and lower CaO contents than postexsolution clinopyroxenes, while TiO_2 and Na_2O are indistinguishable.

5.3. Orthopyroxene

Orthopyroxenes in spinel lherzolites have Al_2O_3 contents ranging from 2.20 to 3.07 wt.%, Cr_2O_3 from 0.33 to 0.55 wt.% and CaO from 0.23 to 1.10 wt.%. Mg-numbers lie between 91.3 and 92.0. CaO contents are generally lower than in garnet peridotites. This is consistent with derivation from the spinel stability field, and hence lower temperatures (Brey and Köhler, 1990). Orthopyroxene from the sheared garnet lherzolite has the lowest Mg-number and the lowest Al_2O_3, Cr_2O_3 and MgO contents of all peridotites. It has high CaO (1.0 wt.%) and Na_2O content (0.25 wt.%). Orthopyroxene from the garnet spinel peridotite has Al_2O_3, Cr_2O_3 and CaO contents intermediate between the sheared garnet lherzolite and spinel lherzolites. Orthopyroxene in garnet pyroxenite generally has higher Al_2O_3

Table 5
Representative major-element composition of orthopyroxene

Sample	K6-2b	K6-2b	K6-3	K6-3	K6-4	K6-5b	K14-1	K14-2a
n	18	16	4	8	5	17	26	13
Comment		20 mic		20 mic				
Type	sp lherz	sp lherz	sp lherz	sp lherz	sp lherz	pyrox	gt lherz	gt pyrox
SiO_2	56.15	56.12	57.52	56.32	57.71	55.02	57.67	56.75
TiO_2	<0.06	<0.06	<0.06	<0.06	<0.06	<0.06	0.12	<0.06
Al_2O_3	2.66	2.68	2.46	2.66	2.46	4.60	0.73	2.47
Cr_2O_3	0.55	0.54	0.46	0.52	0.46	0.51	0.19	0.15
FeO^{total}	5.54	5.43	5.68	5.61	5.54	10.21	5.98	8.08
MnO	0.11	0.14	0.13	0.15	0.12	0.22	0.12	0.11
MgO	34.53	34.62	34.58	34.41	34.48	30.28	34.15	33.09
CaO	0.57	0.61	0.31	0.51	0.35	0.25	1.00	0.20
NiO	0.08	0.09	<0.08	0.09	0.10	<0.08	0.13	<0.08
Na_2O	<0.04	<0.04	<0.04	<0.04	<0.04	<0.04	0.25	<0.04
K_2O	<0.04	<0.04	<0.04	<0.04	<0.04	<0.04	<0.04	<0.04
Total	100.3	100.3	101.2	100.3	101.2	101.2	100.4	100.9
Mg#	91.7	91.9	91.6	91.6	91.7	84.1	91.1	88.0
Cr#	12.2	12.0	11.0	11.6	11.1	7.0	14.8	3.8

Sample	K14-2c	K14-2c	K14-2c	K14-3a	K14-5b	K14-5b	K14-5c (2) max	K14-5c (2) min
n	22	12	2	33	21	30		
Comment	core	rim	20 mic			20 mic	domain 2	domain 2
Type	sp lherz	sp lherz	sp lherz	gt lherz	sp lherz	sp lherz	gt pyrox	gt pyrox
SiO_2	56.48	56.75	56.48	57.62	56.06	56.13	57.05	55.79
TiO_2	<0.06	<0.06	<0.06	<0.06	<0.06	<0.06	0.08	<0.06
Al_2O_3	2.53	2.31	2.48	1.40	2.20	2.69	3.12	1.70
Cr_2O_3	0.54	0.47	0.51	0.31	0.33	0.50	0.14	<0.09
FeO^{total}	5.52	5.59	5.40	5.75	5.68	5.68	9.95	9.03
MnO	0.13	0.12	0.12	0.11	0.13	0.13	0.14	<0.09
MgO	34.52	34.80	34.77	35.22	34.85	35.00	33.16	31.28
CaO	0.62	0.42	0.56	0.18	0.30	0.39	0.58	0.14
NiO	<0.08	<0.08	0.08	0.08	<0.08	<0.08	0.09	<0.08
Na_2O	<0.04	<0.04	<0.04	<0.04	<0.04	<0.04	0.09	<0.04
K_2O	<0.04	<0.04	<0.04	<0.04	<0.04	<0.04	<0.04	<0.04
Total	100.4	100.5	100.4	100.7	99.7	100.6	101.5	100.4
Mg#	91.8	91.7	92.0	91.6	91.6	91.7	86.5	85.1
Cr#	12.5	11.9	12.2	11.5	9.2	10.9	3.7	0.0

Sample	K14-5c (2) avg	K14-5c2	K14-6	K14-6	K14-7a	K14-7a
n	32	22	31	23	34	29
Comment	domain 2			20 mic		20 mic
Type	gt pyrox	pyrox	sp lherz	sp lherz	sp lherz	sp lherz
SiO_2	56.45	53.59	55.93	56.32	56.59	56.45
TiO_2	<0.06	0.06	<0.06	<0.06	0.06	<0.06
Al_2O_3	2.32	5.24	2.43	2.66	2.26	2.32
Cr_2O_3	<0.09	0.47	0.42	0.50	0.47	0.53
FeO^{total}	9.49	9.57	5.63	5.62	5.47	5.48
MnO	<0.09	0.19	0.14	0.15	0.13	0.12
MgO	32.30	30.90	34.31	34.28	35.25	35.29
CaO	0.21	0.23	0.42	0.52	0.32	0.40
NiO	<0.08	<0.08	0.08	0.09	<0.08	0.08
Na_2O	<0.04	<0.04	<0.04	<0.04	<0.04	<0.04

(continued on next page)

Table 5 (*continued*)

Sample	K14-5c (2) avg	K14-5c2	K14-6	K14-6	K14-7a	K14-7a
n	32	22	31	23	34	29
Comment	domain 2			20 mic		20 mic
Type	gt pyrox	pyrox	sp lherz	sp lherz	sp lherz	sp lherz
K_2O	<0.04	<0.04	<0.04	<0.04	<0.04	<0.04
Total	101.0	100.3	99.4	100.2	100.6	100.8
Mg#	85.9	85.2	91.6	91.6	92.0	92.0
Cr#	1.7	5.6	10.3	11.2	12.1	13.3

Representative major-element composition of orthopyroxene in wt.%; for abbreviations, see Tables 2 and 4.

(2.31–3.63 wt.%) and lower Cr_2O_3 (<0.09–0.53 wt.%), CaO contents (0.15–0.58 wt.%) and Mg-number (85.1–88.5) than peridotitic orthopyroxene.

Orthopyroxene in garnet-free pyroxenite ranges to even higher Al_2O_3, FeO and lower MgO contents and has the lowest Mg-number of all samples (84.1).

Table 6
Representative major-element composition of olivine

Sample	K6-2b	K6-2b	K6-3	K6-4	K6-4	K11-5	K11-6
n	18	6	21	14	2	21	3
Comment	core	rim			incl in opx		
Type	sp lherz	sp lherz	sp lherz	sp lherz	sp lherz	alt gt harz	alt gt harz
SiO_2	40.94	40.60	41.98	41.91	42.10	41.04	41.23
TiO_2	<0.06	<0.06	<0.06	<0.06	<0.06	<0.06	<0.06
Al_2O_3	<0.03	<0.03	<0.03	<0.03	<0.03	<0.03	<0.03
Cr_2O_3	<0.09	<0.09	<0.08	<0.05	<0.04	<0.09	<0.09
FeO^{total}	8.53	8.26	8.44	8.41	8.21	8.29	7.80
MnO	0.12	0.12	0.13	0.13	0.10	0.12	0.11
MgO	50.60	50.01	50.16	50.38	50.42	50.92	50.64
CaO	<0.05	<0.05	<0.05	<0.05	<0.05	<0.05	<0.05
NiO	0.40	0.39	0.39	0.39	0.49	0.36	0.41
Na_2O	<0.04	<0.04	<0.04	<0.04	<0.04	<0.04	<0.04
K_2O	<0.04	<0.04	<0.04	<0.04	<0.04	<0.04	<0.04
Total	100.7	99.5	101.2	101.3	101.4	100.8	100.3
Mg#	91.4	91.5	91.4	91.4	91.6	91.6	92.1

Sample	K14-1	K14-2c	K14-3a	K14-3b2	K14-5b	K14-6	K14-7a
n	44	24	25	17	44	27	26
Comment							
Type	gt lherz	sp lherz	gt sp lherz	perid	sp lherz	sp lherz	sp lherz
SiO_2	40.86	41.09	40.89	40.97	40.38	40.29	41.20
TiO_2	<0.06	<0.06	<0.06	<0.06	<0.06	<0.06	<0.06
Al_2O_3	<0.03	<0.03	<0.03	<0.03	<0.03	<0.03	<0.03
Cr_2O_3	<0.09	<0.09	<0.09	<0.09	<0.09	<0.09	<0.09
FeO^{total}	9.91	8.50	8.65	9.18	8.38	8.56	8.12
MnO	0.11	0.12	0.11	0.11	0.11	0.12	0.09
MgO	49.09	50.45	50.56	49.63	50.63	50.08	51.21
CaO	0.05	<0.05	<0.05	<0.05	<0.05	<0.05	<0.05
NiO	0.39	0.39	0.37	0.38	0.37	0.43	0.37
Na_2O	<0.04	<0.04	<0.04	<0.04	<0.04	<0.04	<0.04
K_2O	<0.04	<0.04	<0.04	<0.04	<0.04	<0.04	<0.04
Total	100.5	100.6	100.6	100.4	99.9	99.5	101.1
Mg#	89.8	91.4	91.2	90.6	91.5	91.2	91.8

Representative major-element composition of olivine in wt.%; for abbreviations, see Tables 2 and 4.

Table 7
Representative major-element composition of spinel

Sample	K6-2b	K6-2b	K6-2b	K6-2b	K6-3	K6-3	K6-4
n	11	6	6	6	21	9	16
Comment	core	rim	skeletal	overgr	core	rim	core
Type	sp lherz	sp lherz	sp lherz	sp lherz	sp lherz	sp lherz	sp lherz
SiO_2	0.10	0.16	0.38	0.57	<0.07	<0.07	<0.07
TiO_2	<0.06	<0.06	2.65	0.09	<0.06	<0.06	<0.06
Al_2O_3	35.78	35.62	11.87	28.68	36.81	36.91	37.58
Cr_2O_3	31.97	31.33	42.99	37.99	31.30	30.97	30.84
FeO	12.17	11.21	14.06	10.51	13.29	12.57	11.92
Fe_2O_3	2.74	2.99	11.88	4.87	1.74	2.10	2.15
MnO	0.23	0.19	0.34	0.19	0.22	0.21	0.19
MgO	16.23	16.64	14.35	17.24	15.50	15.95	16.58
CaO	<0.05	0.06	0.09	<0.05	<0.05	<0.05	<0.05
NiO	0.10	0.12	0.09	0.06	0.12	0.12	0.13
ZnO	0.29	0.35	0.13	0.26	0.26	0.26	0.27
V_2O_5	0.17	0.16	0.06	0.21	0.15	0.16	0.15
Total	99.8	98.9	98.9	100.7	99.5	99.3	99.9
Mg#	66.4	68.1	50.8	67.4	65.0	66.3	68.1
Cr#	37.5	37.1	70.3	47.2	36.3	36.0	35.5
Fe^{3+}#	16.9	19.4	42.7	29.1	10.5	13.1	14.0

Sample	K6-4	K6-4	K14-3a	K14-3a	K14-3a	K14-3a	K14-5b
n	6	7	4	4	9	4	7
Comment	rim	overgr	contact w/ilm	contact w/sulf	incl type I	incl type II	core
Type	sp lherz	sp lherz	gt sp lherz	gt sp lherz	gt sp lherz	gt sp lherz	sp lherz
SiO_2	0.10	0.15	0.12	0.10	0.07	0.08	<0.07
TiO_2	<0.06	0.06	1.33	0.27	0.29	0.16	<0.06
Al_2O_3	35.73	33.02	26.99	27.24	30.56	39.22	39.26
Cr_2O_3	32.11	34.69	36.16	37.59	33.97	26.65	28.34
FeO	10.13	9.62	17.80	14.78	15.42	12.89	11.79
Fe_2O_3	3.54	4.31	4.04	5.62	5.10	4.35	3.70
MnO	0.20	0.14	0.27	0.45	0.22	0.21	0.20
MgO	17.61	17.83	12.47	13.80	13.79	16.32	17.14
CaO	<0.05	<0.05	<0.05	<0.05	<0.05	<0.05	<0.05
NiO	0.17	0.18	0.14	0.38	0.13	0.21	0.14
ZnO	0.28	0.23	0.24	0.14	0.34	0.39	0.29
V_2O_5	0.13	0.16	na	na	na	na	na
Total	100.1	100.4	99.6	100.4	99.9	100.5	100.9
Mg#	70.2	70.2	50.9	55.4	55.1	63.4	66.9
Cr#	37.7	41.4	47.3	48.1	42.7	31.3	32.6
Fe^{3+}#	23.9	28.7	17.0	25.5	22.9	23.3	22.1

Sample	K14-5b	K14-5b	K14-5b	K14-6	K14-6	K14-6	K14-7a
n	3	11	8	20	12	10	5
Comment	rim	incl in opx	overgr	core	rim	overgr	incl in ol
Type	sp lherz	sp lherz	sp lherz	sp lherz	sp lherz	sp lherz	sp lherz
SiO_2	<0.07	<0.07	0.11	0.09	0.08	0.15	<0.07
TiO_2	<0.06	0.07	0.19	0.06	0.07	0.10	0.17
Al_2O_3	39.60	38.82	33.91	38.57	38.08	32.52	32.34
Cr_2O_3	27.80	28.22	33.61	30.22	30.53	34.39	36.01
FeO	10.40	11.06	8.76	12.51	11.87	11.39	9.42
Fe_2O_3	4.05	4.12	5.49	1.81	2.41	4.77	4.03
MnO	0.15	0.14	0.18	0.19	0.20	0.18	0.14

(continued on next page)

Table 7 (*continued*)

Sample	K14-5b	K14-5b	K14-5b	K14-6	K14-6	K14-6	K14-7a
n	3	11	8	20	12	10	5
Comment	rim	incl in opx	overgr	core	rim	overgr	incl in ol
Type	sp lherz	sp lherz	sp lherz	sp lherz	sp lherz	sp lherz	sp lherz
MgO	18.03	17.55	18.86	16.57	16.96	16.83	18.05
CaO	<0.05	<0.05	<0.05	<0.05	<0.05	<0.05	<0.05
NiO	0.12	0.14	0.20	0.12	0.13	0.14	0.15
ZnO	0.24	0.21	0.20	0.29	0.28	0.25	0.25
V_2O_5	na	na	na	na	na	na	na
Total	100.5	100.4	101.5	100.4	100.6	100.7	100.6
Mg#	69.6	67.9	71.0	67.6	68.3	65.7	71.1
Cr#	32.0	32.8	40.0	34.5	35.0	41.5	42.8
Fe^{3+}#	26.0	25.2	36.1	11.5	15.4	27.3	27.8

Representative major-element composition of spinel in wt.%; Fe^{3+}# = molar $(100\ Fe^{3+}/(Fe^{3+}+Fe^{2+}))$ after Droop (1987); overgr: overgrowth; w/ilm: with ilmenite; w/sulf: with sulfide; na: not analysed or not available; for other abbreviations, see Tables 2 and 4.

Like clinopyroxene, many orthopyroxenes contain exsolution lamellae of the complementary pyroxene and of spinel. Orthopyroxenes after exsolution have lower Al_2O_3 and CaO contents, while Mg-numbers are mostly comparable.

5.4. Olivine

Mg-numbers of olivine in spinel lherzolites show little variation (91.0 to 91.8). The sheared garnet lherzolite has the lowest Mg-number (89.8) and garnet spinel lherzolite has a low Mg-number (91.2). Garnet peridotites with altered pyroxene, which are assumed to be garnet harzburgites, trend towards the highest Mg-numbers (91.6 to 92.9) and have NiO contents ranging from 0.36 to 0.41 wt.%. There are no covariations between Mg-number and NiO or MnO content.

5.5. Spinel

Spinel in many xenoliths consists of primary spinel cores and secondary overgrowths (see Fig. 14B).

Table 8
Major-element composition of rutile and ilmenite

Sample	K14-2a rutile	K14-5c rutile	K14-5c rutile	K14-2a ilmenite	K14-3a ilmenite	K14-5c ilmenite
n	8	8	4	3	4	1
Comment			incl in cpx	rim on rut		rim on rut
Type	gt pyrox	gt pyrox	gt pyrox	gt pyrox	gt sp lherz	gt pyrox
SiO_2	0.10	<0.07	0.08	0.13	0.50	0.05
TiO_2	97.50	98.58	98.29	54.94	53.77	56.87
Al_2O_3	0.16	0.25	0.21	0.30	0.83	0.60
Cr_2O_3	0.55	0.13	0.13	0.65	2.15	0.22
FeO	0.34	0.59	1.07	32.87	27.81	31.36
Fe_2O_3	na	na	na	na	1.17	
MnO	<0.09	<0.09	<0.09	0.75	0.50	0.58
MgO	<0.06	<0.06	<0.06	9.12	11.32	9.25
CaO	0.07	0.07	<0.05	0.21	0.19	0.10
NiO	<0.08	<0.08	<0.08	<0.08	0.16	<0.08
ZnO	<0.04	<0.04	<0.04	<0.04	<0.04	0.10
V_2O_5	0.31	0.65	0.44	0.23	<0.04	0.53
Total	99.1	100.4	100.3	99.0	98.4	99.7
Mg#	na	na	na	33.1	57.9	65.5
Cr#	71.9	23.2	29.5	58.9	63.7	19.9

Major-element composition of rutile and ilmenite in wt.%; na: not analysed or not available; for other abbreviations and explanations, see Tables 2 and 4.

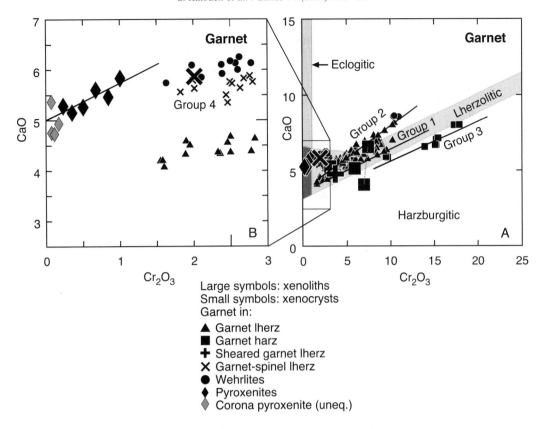

Fig. 3. Cr_2O_3 against CaO (wt.%) in garnet. Xenocrysts from Griffin et al. (2004), fields of Sobolev et al. (1973) and Gurney (1984). Stippled line connects two compositionally distinct garnet populations in sample K11-5. Diagram on the left represents the enlarged area of the rectangle shown in the diagram on the right. Groups 1 to 4 are discussed in the text.

Core–rim zonations are commonly observed, as are skeletal growth textures. Spinel is assumed to be primary in origin if it has TiO_2 contents from <0.06 to 0.11 wt.%, ZnO from 0.25 to 0.34 wt.%, Mg-number from 63.1 to 71.3, Cr-number from 29.0 to 40.6 and Fe^{3+}-number [Fe^{3+}#: $100Fe^{3+}/(Fe^{3+}+Fe^{2+})$] from 9.2 to 28.4. By contrast, secondary overgrowths on spinel trend towards substantially higher TiO_2 contents (0.06–3.43 wt.%), Cr-numbers (36.6–79.6) and Fe^{3+}-numbers (22.7–54.1) and lower ZnO contents (0.06–0.34 wt.%) and Mg-numbers (49.6–72.4). Zn contents in mantle spinel are inversely correlated with temperature (Griffin et al., 1994), and the lower ZnO contents recorded for many of the secondary overgrowths could indicate that they formed in a hotter environment than primary spinel cores. Rim compositions of spinels vary according to the composition of the phase with which they are in contact. For

example, spinel next to ilmenite contains significantly more TiO_2, whereas spinel adjacent to sulfide contains more NiO than spinel adjacent to silicate phases (sample K14-3a).

5.6. Rutile and ilmenite

Rutile occurs in two garnet pyroxenites, one of which has the corona structure described above. Oxides of Al, Cr, Fe and V are present at levels between 0.13 and 1.07 wt.%. Rutile shows minor ilmenite exsolution. One small ilmenite grain occurs in the garnet spinel peridotite. It has an Mg-number of 57.9 and Cr-number of 63.7.

5.7. Sulfide

Bulk sulfides were reconstructed from low-temperature assemblages using mineral compositions

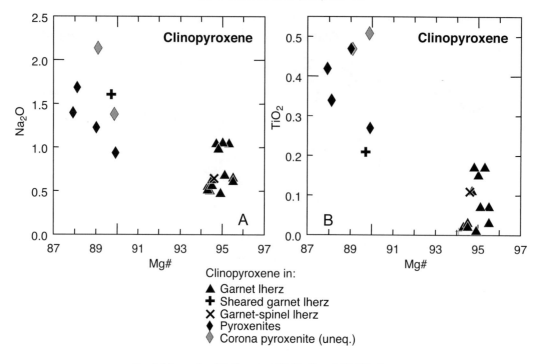

Fig. 4. Mg-number (Mg#) against (A) Na_2O and (B) TiO_2 in cpx.

and modal abundances. Their grain sizes, shapes and low-temperature mineral assemblages are given in Table 9. Sulfides in spinel lherzolites and in pyroxenites are mostly single-phase pentlandites and reconstructed sulfur-poor (?altered) monosulfide solid solution (mss). Two pyroxenites contain chalcosite (Cu_2S). Ni- and Co-rich sulfide occurs in one spinel lherzolite and in a pyroxenite, where it coexists with mss and pentlandite. Sulfide is unusually abundant in the garnet spinel lherzolite (K14-3a), suggesting a secondary origin during influx of sulfur. Spinel lherzolite K14-5b contains a melt patch, half of which is occupied by an aggregate of polygonal pentlandites. A spherical pendlandite grain occurs in another melt patch in the same sample. Many sulfides contain high amounts of Si (up to 4.2 wt.%), which are possibly due to silicate melt inclusions. Most sulfides have high O contents (up to 16.1 wt.%) regardless of their Si contents, and this is ascribed to oxidation and alteration of the sulfides, although discrete secondary minerals, such as hematite, are not observed.

5.8. Melt

The major-element composition of the melt patch in sample K14-5b is difficult to estimate because it contains large oxide and sulfide phenocrysts, which appear to have crystallised from this melt. Phenocryst-free areas contain 43.1 wt.% SiO_2, 0.81 wt.% Al_2O_3, 4.6 wt.% FeO, 38.9 wt.% MgO, 0.11 wt.% CaO and 0.13 wt.% NiO; TiO_2, Na_2O and K_2O are below detection limits (0.07, 0.04 and 0.04 wt.%, respectively). The sum from the microprobe analysis is only 88 wt.%, and trace-element analyses have not revealed any elements at concentrations higher than in the thousands of ppm. This suggests that light elements, such as carbon and/or hydrogen, may be present.

6. Trace elements

6.1. Garnet

Representative trace-element data for xenolithic garnets and clinopyroxenes, and for the melt patch

Table 9
Representative characteristics and major-element composition of sulfide

Sample	K6-1	K6-5b-2	K6-5b-3	K14-2a	K14-2ba	K14-3a	K14-5b	K14-5d	K14-6	K14-7a
n										
Context	i	e opx	i	e cpx	e ol	i+e opx+e ol	i melt	i	i	e opx
Grain size [μm]	125	120	100	50	70	div	div	95	455	155
Shape	irregular	elongate	elongate–irregular	irregular	polyhedral	polyhedral/spherical	polyhedral/spherical	polyhedral	polyhedral	spherical
Mineral/assemblage	pn	mss	gs–hz	cs	mss	pn	pn	pn	pn	pn
Lithotype	sp lherz	gt pyrox	gt pyrox	gt pyrox	sp lherz	grt lherz	sp lherz	sp lherz	sp lherz	sp lherz
Cu	0.5	2.9	3.1	58.3	13.4	3.2	2.6	0.9	2.4	1.4
Fe	27.9	46.0	4.0	3.6	26.4	33.5	30.6	26.5	31.3	32.0
Ni	37.4	13.1	44.3	1.2	23.3	22.3	23.9	24.0	29.2	25.7
Co	0.7	0.7	5.4	0.1	0.5	0.5	0.6	0.5	0.6	0.6
S	32.0	33.5	21.4	15.9	27.8	27.2	27.1	24.7	29.2	28.7
Si	0.2	<0.07	na	na	1.2	1.2	2.1	4.2	1.0	1.5
K	na	0.0	0.0	<0.04	<0.04	na	<0.04	<0.04	<0.04	<0.04
O	1.0	2.0	11.3	13.8	5.3	8.6	8.7	16.1	4.6	7.9
Total	99.8	98.3	89.6	92.9	98.0	96.5	95.7	97.0	98.3	97.8
Me/S	1.16	1.06	1.46	2.03	1.27	1.23	1.19	1.18	1.22	1.16
Ni/(Ni+Fe)	0.56	0.21	0.91	0.24	0.46	0.38	0.43	0.46	0.47	0.43

Representative characteristics and major-element composition of sulfide in wt.%; Me/S = atomic (Cu + Fe + Ni + Co)/S; Ni/(Ni + Fe) from at.%; e: enclosed; i: interstitial; div: diverse; pn: pentlandite; gs: godlevskite; hz: heazlewoodite; cs: chalcosite; for other abbreviations, see Tables 2 and 4.

in sample K14-5b are given in Table 10 (full data set, also for xenocrysts, available online). Garnet from sheared lherzolite K14-1 has a chondrite-normalised (indicated by the subscript $_N$) REE pattern with a steep positive slope in the LREE and MREE, and unfractionated (flat) HREE (Fig. 5A). The HFSE abundances are high relative to most other garnets. Garnet from the garnet spinel lherzolite K14-3a has a pattern similar to garnet from the sheared lherzolite, with the exception of a steep negative slope between La_N and Pr_N, and higher Nb and markedly lower Ti abundance. Garnet in sample K11-6 is zoned with regard to trace elements and shows an order of magnitude difference in La_N, and smaller differences in Ce_N and Pr_N; the MREE and HREE pattern is flat at about 10 × chondrite (Fig. 5B). The two other garnet harzburgites have pronounced sinusoidal REE patterns, with the hinge at Eu_N ~ 50 for one sample and at Nd_N ~ 7 for the other sample. Garnet trace-element data are available for one pyroxenite. Four of five analyses of garnet in this sample show a very steep, smooth slope between Ce_N and Lu_N and one analysis shows markedly higher La and Ce abundances and La/Ce > 1 (out-

side the analytical error). $LILE_N$ and $HFSE_N$ increase almost smoothly with increasing compatibility. Apparent negative Ga anomalies are probably due to a choice of compatibility order and do not have any geological significance.

6.2. Clinopyroxene

Clinopyroxenes in the sheared lherzolite and in the garnet spinel lherzolite have similar REE patterns, starting at La_N = 10, peaking at slightly higher Pr_N and descending smoothly towards Lu_N (Fig. 6B). All but one clinopyroxene in spinel lherzolites have $LREE_N$ of ~ 10, a negative slope in the $MREE_N$ and flat or smoothly decreasing $HREE_N$ (Fig. 6D). One clinopyroxene has a sinusoidal REE pattern. The LREE enrichment is not accompanied by LILE or HFSE enrichment. HFSE and LILE abundances in clinopyroxenes in spinel lherzolites are similar to those in the garnet lherzolites. Clinopyroxenes in several spinel lherzolites also have low HREE abundances similar to those in garnet lherzolites. This may indicate equilibration with garnet, although garnet was not identified petrographically (Kempton et al., 1999). Clinopyroxene in a garnet-

Table 10
Representative trace-element composition of garnet, cpx and melt patch

Sample Mineral Lithotype n	K6-5b cpx pyrox 5	S.D./1σ	K11-5 grt gt harz 5	S.D./1σ	K11-6 (1) grt gt harz 1	1σ	K11-6 (2) grt gt harz 1	1σ	K14-1 grt gt lherz 5	S.D./1σ	K14-1 cpx gt lherz 5
Li	4.5	0.3	0.08	0.02	0.08	0.03	0.10	0.02	0.20	0.02	1.0
Be	0.10	0.03	<0.29		<0.05		<0.034		<0.035		0.19
B	13	2	70	10	17	1	19	1	1.3	0.2	1.5
Al	30000	1000	105000	5000	na		102924	4000	111000	4000	10500
P	40	7	120	30	120	20	118	20	130	30	32
Ca	158000	5000	46000	1000	na		37022	1000	34000	1000	125000
Sc	104	4	165	6	122	4	124	4	94	3	12.3
Ti	2400	100	400	400	2800	100	2868	100	5100	200	1280
V	350	10	440	30	300	10	278	9	270	10	200
Co	24	3	43	6	36	2	36	2	45	2	31
Ni	150	10	60	10	48	3	57	3	101	5	500
Ga	5.1	0.3	6.1	0.9	5.9	0.3	5.3	0.3	12.1	0.5	5.7
Rb	0.5	0.1	<0.2		<0.160		0.37	0.06	<0.067		0.9
Sr	83	4	<1.8		2.1	0.3	0.9	0.2	0.9	0.2	152
Y	16.0	0.7	0.8	0.3	14.5	0.5	13.0	0.5	17.4	0.6	2.4
Zr	22.3	0.8	16	10	86	3	70	3	58	2	6
Nb	0.5	0.3	1.09	0.08	0.93	0.05	0.67	0.03	0.59	0.05	0.5
Cs	<0.064		<0.10		<0.10		0.20	0.04	<0.028		<0.019
Ba	0.9	0.2	<0.15		0.47	0.07	0.09	0.03	<0.096		5
La	3.2	0.2	0.11	0.02	0.62	0.03	0.044	0.005	0.05	0.01	2.6
Ce	10.6	0.5	1.2	0.2	1.20	0.05	0.44	0.02	0.46	0.02	9
Pr	1.50	0.06	0.41	0.04	0.17	0.01	0.152	0.008	0.15	0.01	1.4
Nd	6.0	0.2	3.1	0.3	1.56	0.10	1.46	0.07	1.43	0.07	6.5
Sm	1.35	0.07	0.9	0.1	1.07	0.08	1.04	0.06	1.01	0.06	1.42
Eu	0.45	0.02	0.24	0.03	0.48	0.03	0.52	0.02	0.50	0.02	0.43
Gd	1.77	0.09	0.4	0.2	2.0	0.1	1.85	0.09	1.92	0.09	1.09
Dy	2.6	0.1	0.28	0.08	2.5	0.1	2.34	0.09	2.9	0.1	0.64
Ho	0.61	0.03	0.05	0.01	0.58	0.03	0.47	0.02	0.67	0.03	0.100
Er	1.9	0.1	0.12	0.06	1.53	0.09	1.29	0.06	1.97	0.09	0.21
Yb	1.9	0.1	0.4	0.1	1.8	0.1	1.21	0.06	2.0	0.1	0.12
Lu	0.26	0.02	0.08	0.02	0.27	0.02	0.22	0.01	0.32	0.02	0.013
Hf	0.93	0.07	0.71	0.08	2.4	0.1	1.38	0.07	1.45	0.09	0.35
Ta	0.017	0.005	0.08	0.02	0.039	0.009	0.036	0.005	0.036	0.005	0.02
Pb	4.1	0.5	0.19	0.04	<0.125		0.13	0.05	<0.205		0.3
Th	0.16	0.04	0.08	0.03	0.041	0.009	0.012	0.003	0.015	0.003	0.06
U	0.03	0.01	0.09	0.04	0.017	0.007	0.018	0.003	0.025	0.005	0.01

Representative trace-element abundances of garnet and clinopyroxene in xenoliths in ppm; S.D.: standard deviation (in italics); 1σ: one sigma error (reflecting the within-run precision); where multiple analyses were performed, standard deviation is presented (in italics) if it was greater than 1σ; for other abbreviations, see Tables 2 and 4.

bearing pyroxenite has La_N to Gd_N of ~ 3–5, which decreases smoothly toward $Lu_N = 0.5$. Clinopyroxene in a garnet-free pyroxenite has an almost flat REE pattern with all abundances at 8–20 × chondritic. HREE and most LILE and HFSE abundances are higher in clinopyroxenes from the garnet-free pyroxenite.

7. Geothermobarometry

Different geothermobarometers were applied to mantle xenoliths from Buffalo Hills kimberlites, depending on metamorphic facies and assemblage (see Table 11). For exsolution-free spinel lherzolites, temperature estimates range from 823 and 1152 °C

(mean 946 °C), using $T_{Ca\text{-in-orthopyroxene}}$ of Brey and Köhler (1990) and between 787 and 847 °C (mean 826 °C; at a nominal pressure of 30 kbar), using T_{WES91} of Witt-Eickschen and Seck (1991). Temperatures and pressures were solved simultaneously for the garnet spinel lherzolite and the sheared garnet lherzolite, using $T_{Ca\text{-in-orthopyroxene}}$ and P_{NG} of Nickel and Green (1985). The garnet spinel lherzolite gives a $T_{Ca\text{-in-orthopyroxene}}$ of 851 °C and P_{NG} of 27 kbar, while the sheared garnet lherzolite gives $T_{Ca\text{-in-orthopyroxene}}$ of 1270 °C and P_{NG} of 58 kbar. Temperature estimates for the three inferred garnet harzburgites vary between 890 and 1126 °C (at a nominal pressure of 50 kbar, using T_{OW79}). Temperatures and pressures were solved simultaneously for three garnet pyroxenites: temperatures vary between 729

Table 10 (continued)

Sample	S.D./1σ	K14-3a	S.D./1σ	K14-3a	S.D./1σ	K14-5b	S.D./1σ	K14-5b	S.D./1σ	K14-6	S.D./1σ	K14-7	S.D./1σ
Mineral		grt		cpx		5		melt		cpx		cpx	
Lithotype		gt sp lherz		gt sp lherz		sp lherz				sp lherz		sp lherz	
n		7		5		5		5		5		6	
Li	0.2	0.26	0.05	0.8	0.2	0.8	0.4	2.2	0.9	1.3	0.2	1.0	0.2
Be	0.02	<0.043		0.06	0.02	0.04	0.02	0.13	0.02	0.04	0.01	0.08	0.02
B	1.0	1.0	0.4	1.5	0.7	1.3	0.9	90	30	23	2	25	2
Al	400	131000	7000	14000	3000	17000	1000	3500	100	15600	700	17400	900
P	7	44	9	30	20	21	4	30	20	18	3	20	4
Ca	4000	42000	1000	169000	5000	169000	5000	2100	700	170000	5000	169000	5000
Sc	0.4	140	10	30	1	79	8	3.7	0.2	51	2	41	1
Ti	60	900	200	800	100	430	10	170	10	300	10	1020	50
V	10	160	30	180	30	200	10	27	5	200	10	240	10
Co	3	37	4	23	2	25	3	300	300	19	3	15	2
Ni	40	7	1	260	30	280	40	<10000		310	30	230	20
Ga	0.5	3.7	0.4	4.6	0.5	1.9	0.2	1.6	0.2	1.5	0.1	2.6	0.2
Rb	0.6	<0.07		0.6	0.3	0.2	0.1	11	2	0.4	0.4	2	1
Sr	6	0.4	0.2	176	30	135	4	210	50	77	2	63	2
Y	0.1	24	2	0.9	0.1	2.4	0.1	0.23	0.02	1.60	0.06	1.7	0.1
Zr	2	36	10	5.2	0.6	9.2	0.5	7	2	8.0	0.3	12.3	0.4
Nb	0.5	2.2	1.0	1.9	0.9	0.12	0.03	9	3	0.39	0.06	0.4	0.1
Cs	<0.031			0.04	0.01	<0.04		0.22	0.05	0.13	0.02	<0.08	
Ba	6	0.2	0.1	40	20	6	2	28	7	1.5	0.7	7	4
La	0.5	0.3	0.5	3	1	2.16	0.07	2	1	2.6	0.1	2.0	0.1
Ce	1	0.4	0.4	8	2	5.9	0.2	4	3	7.7	0.5	7.1	0.3
Pr	0.1	0.04	0.01	1.14	0.18	0.74	0.05	0.3	0.2	1.13	0.04	1.14	0.06
Nd	0.4	0.39	0.09	4.87	0.69	3.2	0.1	0.8	0.5	4.8	0.2	5.1	0.3
Sm	0.07	0.5	0.1	0.89	0.12	0.66	0.05	0.10	0.05	0.84	0.04	1.07	0.06
Eu	0.03	0.30	0.02	0.2	0.0	0.20	0.01	0.03	0.01	0.22	0.01	0.28	0.01
Gd	0.06	1.59	0.09	0.6	0.1	0.59	0.06	0.06	0.02	0.57	0.03	0.82	0.05
Dy	0.04	3.6	0.3	0.29	0.03	0.48	0.04	0.03	0.01	0.36	0.02	0.47	0.07
Ho	0.006	1.0	0.1	0.036	0.008	0.09	0.01	0.008	0.003	0.065	0.004	0.07	0.01
Er	0.01	3.2	0.6	0.06	0.01	0.25	0.02	0.02	0.01	0.15	0.02	0.14	0.03
Yb	0.02	3.9	0.9	<0.02		0.27	0.06	0.02	0.01	0.13	0.02	0.07	0.01
Lu	0.004	0.7	0.2	0.003	0.001	0.048	0.005	0.004	0.002	0.020	0.003	0.008	0.002
Hf	0.03	0.8	0.2	0.33	0.03	0.35	0.03	0.15	0.03	0.29	0.02	0.69	0.05
Ta	0.02	0.019	0.007	0.03	0.02	0.019	0.005	0.28	0.09	0.049	0.007	0.051	0.006
Pb	0.2	<0.23		1.3	0.2	0.3	0.1	0.7	0.6	0.5	0.2	0.7	0.4
Th	0.05	0.03	0.02	0.2	0.2	0.06	0.02	0.12	0.09	0.10	0.03	0.08	0.03
U	0.01	0.05	0.05	0.03	0.02	0.016	0.003	0.4	0.1	0.026	0.005	0.02	0.02

and 793 °C (using $T_{\text{Ca-in-orthopyroxene}}$) and pressures vary between 12 and 13 kbar (using P_{NG}). The temperature and pressure estimate for one clinopyroxene-free garnet pyroxenite is 628 °C and 9 kbar, respectively (when T_{Harley} of Harley, 1984, is solved simultaneously with P_{NG}). The temperature estimate for two garnet-free pyroxenites is 742 and 750 °C (using $T_{\text{Ca-in-orthopyroxene}}$ at a nominal pressure of 15 kbar) and for two orthopyroxene-free garnet pyroxenites, it is 713 and 906 °C (using T_{Krogh} of Krogh, 1988).

Ni-in-garnet thermometry was applied to garnet xenocrysts (Griffin et al., 2004) classified as garnet from garnet spinel lherzolite, wehrlite, lherzolite and Ca-harzburgite based on their Cr–Ca relationships. Garnets from garnet spinel lherzolites give temperatures from 660 to 690 °C (mean 670 °C), wehrlite garnets from 710 to 1040 °C (mean 790 °C), lherzolite garnets from 610 to 1340 °C (mean 940 °C) and Ca-harzburgite garnets from 710 to 1200 °C (mean 1060 °C). Garnets from garnet spinel lherzolites give minimum pressures ($P_{\text{Cr-in-gt}}$; Ryan et al., 1996) from 19 to 21 kbar (mean 20 kbar, $n=3$), wehrlite garnets from 11 to 46 kbar (mean 29 kbar, $n=13$), lherzolite garnets from 24 to 55 kbar (mean 49 kbar, $n=127$) and Ca-harzburgite garnets from 24 to 55 kbar ($n=9$).

Application of $P_{\text{Cr-in-grt}}$ presupposes equilibrium with chromite and application of this method in the absence of chromite results in a minimum pressure; with a large enough database, the envelope of maximum pressures at a given temperature defines the

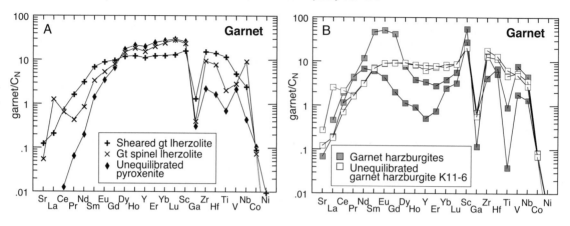

Fig. 5. Extended trace-element patterns in garnet. C_N denotes normalisation to chondrite of McDonough and Sun (1995).

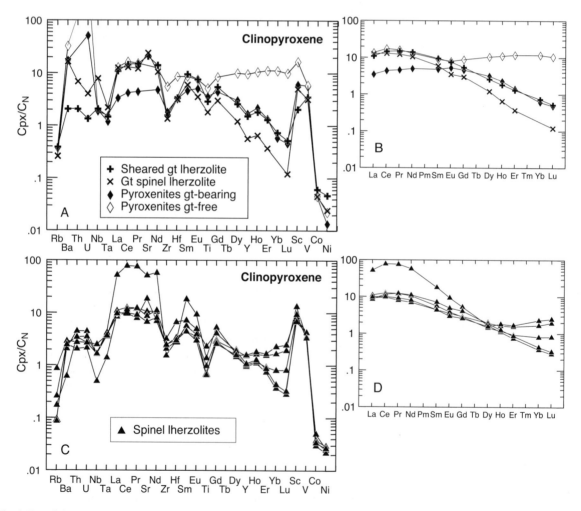

Fig. 6. Extended trace-element patterns (A and C) and REE patterns (B and D) in cpx. C_N denotes normalisation to chondrite of McDonough and Sun (1995).

geotherm (Ryan et al., 1996). For the mantle beneath the BHT, a geotherm of 38–39 mW/m^2 is obtained (Fig. 7). The sheared lherzolite gives a very high temperature and the highest pressure of all samples, consistent with its microstructure and enrichment in basaltic components, which is similar to high-temperature sheared peridotites interpreted to have been infiltrated by asthenospheric melts shortly before entrainment in the kimberlite (Smith and Boyd, 1987; Griffin et al., 1987). Together with several harzburgitic garnet xenocrysts, it may define a kink in the geotherm. The garnet spinel lherzolite and pyroxenites plot off the geotherm, possibly due to major-element disequilibrium caused by sluggish reaction rates in the cool shallow lithosphere. Another group of xenocrysts gives low temperatures for their pressures (grey field in Fig. 7). The cause for this is not known.

8. Whole-rock reconstruction

Whole rocks were reconstructed by combining modal abundance data (obtained by counting on average >1000 points) and mineral analyses (data available online). Garnet peridotites with assumed altered orthopyroxene were reconstructed using the relationship between olivine MgO and FeO^t contents and orthopyroxene MgO, FeO and SiO$_2$ contents in 55 garnet peridotites from Lac de Gras (Pearson et al., 1999 and unpublished data) and average contents of other major elements in orthopyroxene (orthopyroxene MgO = 0.66*ol MgO + 2.12, r^2 = 0.84; orthopyroxene FeO = 0.511*ol FeO = 0.73, r^2 = 0.89; orthopyroxene SiO$_2$ = 0.495*ol MgO + 32.6, r^2 = 0.31). Reconstructed whole rocks will be used to estimate the degree of partial melting they have experienced. Uncertainties in mineral modes from medium- to coarse-grained samples are estimated to be about 10% (Jerde et al., 1993). Because the grain sizes of most xenolith samples from the Buffalo Hills kimberlites are small compared to the size of the thin sections, the actual error may be closer to 5%.

9. Sr–Nd–Hf isotope data

Sr, Nd and Hf isotope data were obtained for garnet in two garnet harzburgites, garnet and clinopyroxene in the sheared garnet lherzolite, garnet in the garnet spinel peridotite and clinopyroxene in one garnet pyroxenite and in two spinel peridotites (Table 12). The terms unradiogenic and radiogenic refer to isotopic compositions relative to a chondritic reservoir (CHUR) at the time of kimberlite eruption (ca. 85 Ma ago). ε_{Hf} and ε_{Nd} indicate the parts per 10,000 deviation of the isotopic composition of the sample from that of a model chondritic source. Model ages are calculated relative to a model depleted mantle source (T_{DM}).

Mineral separation of harzburgite K11-3 revealed the presence of two garnet populations, of which one is purple and transparent, with smooth shiny surfaces and one is red and opaque, with a dull surface. Their Nd and Hf isotope ratios are almost indistinguishable but different from all other separates measured and confirm that these two separates were derived from a single sample and are not due to contamination. Garnets in both garnet harzburgites (K11-3 and K11-5) have similar ^{143}Nd/^{144}Nd around the chondritic value at the time of kimberlite eruption (ε_{Nd} = −1.2 to 0.0) and radiogenic ^{176}Hf/^{177}Hf (ε_{Hf} = 16.3 to 37.0). Sm–Nd T_{DM} are negative for K11-3 and at least Proterozoic for K11-5; Lu–Hf T_{DM} are Proterozoic for K11-3 and implausible for K11-5.

The garnet in K14-1 has relatively low ε_{Hf} (10.1) and ε_{Nd} (4.2), whereas the clinopyroxene has identical Nd (within the uncertainty) but radiogenic Hf (ε_{Hf} = 63.4). Garnet and clinopyroxene yield an internal Sm–Nd isochron age of 86 ± 17 Ma, similar to the kimberlite eruption age. However, the Lu–Hf isochron age is negative, indicating isotopic disequilibrium. Sr in the clinopyroxene is unradiogenic (^{87}Sr/^{86}Sr$_{measured}$ = 0.703546). Garnet from garnet spinel lherzolite K14-3 has radiogenic Hf and Nd (ε_{Hf}>80, ε_{Nd}>50); not enough clinopyroxene could be separated for isotope analysis.

Clinopyroxenes in two spinel lherzolites have unradiogenic Nd and Sr (ε_{Nd} = −16.7 and −8.9, respectively, ^{87}Sr/^{86}Sr$_{measured}$ = 0.703997 and 0.703727, respectively) and yield Early Proterozoic T_{DM}. Clinopyroxenes in K14-6 have radiogenic Hf and a young model age (ε_{Hf} = 18.3; 27 Ma). Clinopyroxenes in K14-7 have unradiogenic Hf (ε_{Hf} = −11.1; T_{DM} = 1.2 Ga). Garnet in the pyroxenite has radiogenic Nd and unradiogenic Hf (ε_{Nd} = 10.3;

Table 11
Temperature–pressure estimates

Sample	Lithotype	Paragenesis used	T_{BKN}	$T_{Ca-in-opx}$	T_{EG}	T_{Krogh}	T_{OW79}	T_{Harley}	T_{WES91}	$T_{Ni-in-grt}$
K6-1	sp lherz	ol, opx, cpx	810	1001					833	
K6-2a	sp lherz	ol, opx, cpx core	699	1012					844	
K6-2a	sp lherz	ol, opx, cpx rim	590	1012					844	
K6-2b	sp lherz	ol + cpx core, opx	718	991					847	
K6-2b	sp lherz	ol + cpx rim, opx	686	991					847	
K6-2b2	gt pyrox	gt, opx core						657		
K6-2b2	gt pyrox	gt, opx rim						646		
K6-3	sp lherz	ol, opx, cpx	744	866					818	
K6-4	sp lherz	ol, opx, cpx core	818	888					825	
K6-4	sp lherz	ol, opx, cpx rim	944	888					825	
K6-5a	sp lherz	ol, opx, cpx	712	1152					845	
K6-5b	pyrox	opx, cpx	742	750					919	
K11-3	gt harz	ol, gt					890			755
K11-5	gt harz	ol, Ca-poor gt					1126			1093
K11-6	gt harz	ol rim, gt core					1096			1080
K11-6	gt harz	ol rim, gt rim					1060			1080
K14-1	gt lherz	ol, opx, cpx, gt	1288	1233	1267	1236	1194	1142	787	1303
K14-2a	gt pyrox	gt, opx, cpx core	757	736	792	705		705	754	616
K14-2a	gt pyrox	gt, opx, cpx rim	682	736	758	669		705	754	
K14-2ba	sp lherz	ol, opx, cpx	590	938					811	
K14-2bb	gt pyrox	gt opx, cpx core	575	753	753	670		678	797	
K14-2bb	gt pyrox	gt opx, cpx rim	535	753	735	650		678	797	
K14-2c	sp lherz	ol, opx + cpx core	725	1011					838	
K14-2c	sp lherz	ol, opx + cpx rim	743	925					815	
K14-3a	gt lherz	ol, opx, cpx, gt	765	817	822	749	742	842	747	
K14-3c	gt pyrox	gt cpx			791	713				
K14-4a	sp lherz	ol + opx, cpx core	524	926					823	
K14-4a	sp lherz	ol + opx, cpx rim	892	926					823	
K14-4b	gt pyrox	Ca-poor cpx, gt			969	906				
K14-5a	sp lherz	ol, opx		947					839	
K14-5b	sp lherz	ol, opx, cpx	355	862					787	
K14-5c	gt pyrox	gt, cpx domain 5			848	758				
K14-5c	gt pyrox	gt, cpx domain 4			767	684				
K14-5c	gt pyrox	gt, opx domain 2		742				674	750	
K14-5c2	pyrox	opx cpx	716	742					929	
K14-5d	sp lherz	ol, opx, cpx	273	823					833	
K14-5f	gt pyrox	gt, opx + cpx core	539	801	853	785		716	842	
K14-5f	gt pyrox	gt, opx + cpx rim	588	765	789	715		681	815	
K14-6	sp lherz	ol, opx, cpx	638	928					812	
K14-7a	sp lherz	ol, opx, cpx core	435	872					795	
K14-7a	sp lherz	ol, opx, cpx rim	801	872					795	
K14-8	sp lherz	ol, opx + cpx core	491	976					840	
K14-8	sp lherz	ol, opx + cpx rim	790	915					822	

Sample			$T_{Ca-in-opx}$ at P_{NG}	P_{NG} at $T_{Ca-in-opx}$	T_{Harley} at P_{NG}	P_{NG} at T_{Harley}	T_{BKN} at P_{NG}	P_{NG} at T_{BKN}
K6-2b2	gt pyrox	gt, opx core			628	9		
K14-1	gt lherz	ol, opx, cpx, gt	1270	58	1167	54	1311	60
K14-2a	gt pyrox	gt, opx, cpx core	729	13	685	11	673	10
K14-2bb	gt pyrox	gt opx, cpx core	746	13	649	9	552	5
K14-3a	gt sp lherz	ol, opx, cpx, gt	851	27	746	22	737	21
K14-5f	gt pyrox	gt, opx + cpx core	793	12	679	8	522	2

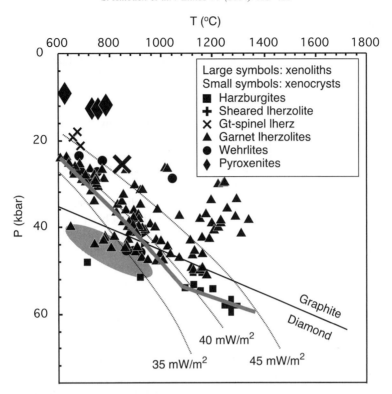

Fig. 7. Simultaneously solved pressure and temperature for garnet xenocrysts, garnet lherzolites and garnet pyroxenites, as discussed in the text. The highest pressures at a given temperature for garnet geothermobarometry define the local geotherm (thick grey line). Sheared lherzolite K14-1 forms part of an inflected array. Graphite–diamond phase boundary of Kennedy and Kennedy (1976); model geotherms after Pollack and Chapman (1977).

$\varepsilon_{Hf} = -2.3$) and yields a Proterozoic model age for both isotope systems.

10. Re–Os isotope systematics of sulfides

Re–Os isotope data were obtained for one sulfide enclosed in orthopyroxene and five interstitial sulfides in six peridotites. $^{187}Re/^{188}Os$ ratios range from 0.0938 ± 0.0024 to 10.71 ± 0.38. $^{187}Os/^{188}Os$ ratios range from 0.1172 ± 0.0026 to 2.4000 ± 0.0051, corresponding to positive, i.e., suprachondritic, γ_{Os} for four of the samples and negative (subchondritic)

γ_{Os} for two of the samples (Table 13; γ_{Os} is percent deviation from $^{187}Os/^{188}Os$ of a model chondritic source). These latter give Re-depletion ages (T_{RD} as defined by Walker et al., 1989) of 0.40 ± 0.18 and 1.46 ± 0.38 Ga. Sulfide in the melt patch in sample K14-5b is one of the samples with suprachondritic $^{187}Os/^{188}Os$ and gives a model age of 1.73 ± 0.13 Ga. This sample has $^{187}Re/^{188}Os > 10$, requiring large corrections for the overlap of ^{187}Re on ^{187}Os and its $^{187}Os/^{188}Os$ probably has a larger uncertainty than indicated by the internal error. In addition, a second analysis of sulfide in this sample has revealed significant heterogeneity of $^{187}Os/^{188}Os$. The sample giving

---

Note to Table 11:

Temperature estimates at nominal pressures of 50 kbar (harzburgites), 30 kbar (spinel lherzolites) and 15 kbar (pyroxenites), using geothermobarometers of Ellis and Green (1979, T_{EG}), O'Neill and Wood (1979, T_{OW79}), Harley (1984, T_{Harley}), Nickel and Green (1985, P_{NG}), Krogh (1988, T_{Krogh}), Brey and Köhler (1990, T_{BKN}, $T_{Ca-in-opx}$), Witt-Eickschen and Seck (1991, T_{WES91}); pressures and temperatures were solved simultaneously in rocks with garnet and opx ± cpx; for other abbreviations and explanations, see Tables 2 and 4.

Table 12
Sr, Nd and Hf isotope ratios and model ages for garnet and cpx (solution analyses)

Sample	Type	$^{87}Sr/^{86}Sr$	2 S.E.	$^{147}Sm/^{144}Nd^a$	$^{143}Nd/^{144}Nd$	2 S.E.	$\varepsilon_{Nd(T)}^b$	T_{DM}^c	$^{176}Lu/^{177}Hf^a$	$^{176}Hf/^{177}Hf$	2 S.E.	$\varepsilon_{Hf(T)}^b$	T_{DM}^c
K11-3 purple gt	gt harz	na	na	0.796	0.512780	0.000010	−1.2	−193	0.036	0.283239	0.000011	16.3	1677
K11-3 red gt	gt harz	na	na	0.796	0.512844	0.000026	0.0	−176	0.036	0.283268	0.000024	17.4	981
K11-5 gt	gt harz	na	na	0.180	0.512476	0.000010	−0.5	3178	0.016	0.283766	0.000019	36.1	−1192
K11-5 gt repeat	gt harz	na	na	0.180	0.512495	0.000014	−0.1	3094	0.016	0.283791	0.000011	37.0	−1252
K14-1 cp	sheared gt lherz	0.703546	0.000024	0.137	0.512698	0.000007	4.3	901	0.005	0.284519	0.000132	63.4	−2066
K14-1 gt	sheared gt lherz	na	na	0.443	0.512862	0.000012	4.2	−305	0.031	0.283055	0.000011	10.1	1681
K14-1 gt repeat	sheared gt lherz	na	na	0.443	0.512877	0.000020	4.5	−295	0.031	0.283034	0.000004	9.4	1824
K14-3 gt	gt–sp lherz	na	na	0.838	0.515460	0.000013	50.6	469	0.119	0.285434	0.000009	89.3	1310
K14-3 gt repeat	gt–sp lherz	na	na	0.838	0.515443	0.000010	50.3	465	0.119	0.285272	0.000010	83.6	1203
K14-5f cp	gt pyrox	0.703721	0.000012	0.201	0.513038	0.000008	10.3	1785	0.005	0.282661	0.000015	−2.3	936
K14-6 cp	sp lherz	0.703997	0.000079	0.110	0.511606	0.000011	−16.7	2246	0.010	0.283251	0.000079	18.3	27
K14-7 cp	sp lherz	0.703727	0.000022	0.131	0.511982	0.000008	−8.9	2097	0.002	0.282407	0.000031	−11.1	1213

Sr–Nd–Hf isotope data for solutions of clinopyroxene and garnet separates.

[a] Calculated from LA–ICPMS trace-element analyses and isotopic abundances of Sm and Nd and arbitrarily quoted to three significant figures.

[b] Parts per 10,000 deviation from chondrite (values of Wasserburg et al., 1981; Blichert-Toft and Albarède, 1997, respectively) at the time of kimberlite eruption ca. 85 Ma ago.

[c] Depleted mantle source (DM) of DePaolo (1981) and Griffin et al. (2000), respectively.

a 1.46 Ga T_{RD} has a model age of 1.89 ± 0.38 Ga. Future and implausibly old model ages suggest disturbance of the Re–Os budget by secondary Re-addition and/or ^{187}Os-addition.

11. Discussion

11.1. Lithosphere structure and lithosphere–asthenosphere boundary (LAB)

A preliminary stratigraphy of the SCLM beneath the BHT has been derived by comparing the relative abundances of peridotite classes (based on major- and trace-element abundances in garnet xenocrysts) and Mg-numbers in olivine in different depth or temperature intervals, respectively (Griffin et al., 2004; Fig. 8A–C). Xenolith types (this study) provide a cross-check (Fig. 8D). Fertile garnet lherzolites are concentrated at shallow depths (< 140 km) and prevail at depths < 110 km; depleted peridotites are concentrated between ca. 120 and 160 km and prevail between ca. 130 and 150 km; melt-metasomatised lherzolites, similar to sheared lherzolites (Griffin et al., 2003), are concentrated between ca. 140 and 180 km and prevail at depths >170 km; a relatively steady proportion of depleted metasomatised peridotites occurs between ca. 110 and 180 km depth (Fig. 8A). Mg-numbers in olivine, derived by inversion of the garnet–olivine thermometer of O'Neill and Wood (1979) (Gaul et al., 2000), are relatively steady between 90 and 140 km depth at 92.0–92.2% and decrease toward 90.3% at ~ 180 km depth (Fig. 8B). This is consistent with the increasing proportion of melt-metasomatised peridotites in this depth interval. Data for the xenoliths agree well with the xenocryst-derived stratigraphy, as pyroxenite and garnet spinel lherzolite are restricted to the shallow mantle and harzburgite with Ca-saturated garnet is concentrated in a layer between ~ 140 and 160 km (~ 1000–1200 °C; Fig. 8C,D; Table 14).

Temperatures for spinel lherzolites were calculated using the Ca-in-orthopyroxene thermometer of Brey and Köhler (1990) and the formulation of Witt-Eickschen and Seck (1991), both of which tend to overestimate for temperatures < 900 °C (Smith, 1999). This may explain their apparent persistence to temperatures above the spinel–garnet transition, which is

Table 13
Re–Os isotope composition of sulfide (in situ analyses)

Sample	K6-1-2	K14-3a	K14-5b-1	K14-5b-2	K14-5d-1	K14-6	K14-7a-1
$^{187}Re/^{188}Os$	0.540	0.513	10.71	9.777	0.0938	1.220	0.5471
2 S.E.	0.024	0.026	0.38		0.0024	0.032	0.0048
$^{187}Os/^{188}Os$	0.1679	0.1809	0.4298	1.016	0.1172	2.4000	0.1244
2 S.E.	0.0012	0.0027	0.0160	0.018	0.0026	0.0051	0.0012
γOs	32.19	42.4	240	700	− 7.7	1789	− 2.12
2 S.E.	0.94	2.1	10	10	2.1	4	0.96
T_{RD} [Ma]	na	na	na	na	1455	na	402
+/−					381		182
T_{CHUR} [Ma]	15492	23549	1737	5437	1893	79707	− 1114
+/−	2808	4832	132	371	378	7811	114

Re–Os isotope data from in situ analysis of sulfides; γOs is the percent deviation from chondritic $^{187}Os/^{188}Os$; T_{RD} is rhenium depletion age (Walker et al., 1989), T_{CHUR} is derived by referral to chondrite Os isotopic evolution (chondrite of Walker et al., 1994).

constrained by garnet spinel lherzolite. The relatively Cr-rich nature of the spinel peridotites may also act to delay the stabilisation of garnet (Webb and Wood, 1986). The absence of nonsheared garnet lherzolites in the xenolith suite contrasts with the abundance of lherzolitic garnet xenocrysts and may indicate that this rock type was preferentially comminuted during entrainment in the kimberlite. The possibility that the sheared lherzolite defines an inflection of the 38–39 mW/m² geotherm, combined with its microstructure

and enrichment in basaltic components, suggests that it was derived from close to the lithosphere–asthenosphere boundary (Smith and Boyd, 1987). It constrains the depth of the SCLM beneath the BHT to somewhere in the vicinity of 180 km.

11.2. Melting

The degree of depletion can be assessed by comparison of whole rocks, reconstructed from min-

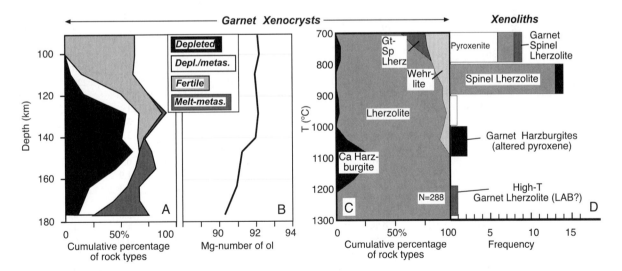

Fig. 8. Lithosphere structure beneath the BHT from (A) garnet classes against depth (derived by multivariate analyses of trace-element abundances in garnet xenocrysts (Griffin et al., 2002) and Ni-in-garnet thermometry (Ryan et al., 1996)); (B) Mg-number in olivine against depth (converted from major-element composition of garnet xenocrysts using the formulation of Gaul et al., 2000); (C) garnet classes against temperature (using Cr_2O_3–CaO relationships of garnet xenocrysts); (D) xenolith lithologies plotted against temperatures (different geothermometers were used, according to assemblage and metamorphic facies of the sample, as outlined in the text). Sections (A) to (C) from Griffin et al. (2004).

eral compositions weighted by modes, with experimental melting residues (Fig. 9A,B). Keeping in mind that the accuracy of such estimates is limited by the uncertainties in the modal analyses, most samples from the BHT have FeO–SiO_2 and FeO–MgO relationships consistent with a loss of between 20–40% partial melt at pressures between 1 and 4 GPa, assuming a fertile peridotite starting composition (Walter, 1999). Two of the garnet harzburgites appear to have been depleted more strongly, at pressures up to 7 GPa. The sheared garnet lherzolite and garnet spinel lherzolite are too FeO-rich to represent samples depleted by an extraction of melt and plot outside the scale chosen in Fig. 9. Three samples lie on a trend to high SiO_2, indicating high orthopyroxene/olivine similar to peridotites from Kaapvaal and to a lesser extent Siberia, although displaced toward higher SiO_2 for a given FeO content (Fig. 9C).

11.3. Old metasomatism

Cryptic metasomatism is evidenced by incompatible element enrichment in garnet and clinopyroxene in otherwise depleted samples from Buffalo Hills, both in the shallow mantle (spinel stability field) and in the deep mantle (garnet stability field). Most of these samples have well-equilibrated microstructures, and homogeneous garnet and clinopyroxene compositions with regard to major and trace elements, suggesting that the enrichment predates entrainment in the host kimberlite.

Zr–Y relationships can be used to distinguish garnets from different sources: fertile, depleted and melt-metasomatised. In addition, sources that were metasomatised by a silicate melt can be distinguished from sources that were metasomatised by a distinct metasomatic agent associated with phlogopite-bearing peridotite ("phlogopite" metasomatism; Griffin

Table 14
Average trace-element composition of garnet xenocrysts

Sample	Fertile lherz	Depl lherz	Sili metas lherz	Phlog metas lherz	Fertile harz	Depl harz	Fertile gt sp lherz	Depl gt sp lherz	Sili metas gt sp lh	Fertile wehr	Depl wehr
n	4	40	29	19	2	7	10	4	3	9	4
Ni	18	39	65	34	24	68	15	16	12	14	36
Co	34	36	44	37	42	38	38	37	39	35	36
V	150	290	250	240	200	320	130	100	130	160	410
Sc	170	160	120	180	230	210	200	160	160	220	180
Ti	680	430	2600	420	500	850	370	35	440	430	340
Y	33	3.2	17	5.2	24	3.4	22	5.20	21	30	2.5
Ce	0.59	1.3	1.8	0.85	na	1.7	1.2	0.45	3.0	1.2	2.8
Nd	1.9	3.5	2.4	3.0	na	4.6	1.6	2.3	1.8	1.8	3.7
Sm	1.3	1.5	1.4	2.2	na	1.4	1.1	1.75	1.4	1.3	1.4
Eu		0.45	0.64	0.90	0.87	0.49	0.51	0.64	0.77	0.77	na
Gd	2.0	1.07	2.51	2.5	1.9	1.1	1.9	1.6	1.9	1.9	1.6
Dy	5.2	0.88	3.1	1.61	4.5	1.5	3.5	0.98	3.3	4.2	na
Ho	1.1	0.18	0.73	0.32	1.4	0.33	0.82	0.18	0.92	1.1	na
Er	3.8	0.64	2.0	0.65	3.7	0.91	2.4	0.65	2.6	4.0	1.1
Yb	4.0	0.90	2.2	0.79	2.9	0.89	3.0	0.84	3.2	4.8	1.4
Nb	0.33	1.1	0.68	0.69	na	2.1	0.41	0.37	0.90	0.68	0.85
Zr	15	14	54	57	21	13	15	16	34	20	12
Hf	0.55	0.63	1.2	0.92	na	0.80	0.54	0.34	0.83	0.84	na
Sr	2.0	2.3	1.4	0.57	na	3.1	0.74	0.26	6.4	1.5	1.6
Ga	5.2	5.8	10	4.4	11	7.0	2.9	2.2	5.3	3.7	6.8
T_{Ni}	774	957	1121	919	827	1140	737	747	696	716	933

Average trace-element abundances of garnet xenocrysts (analyses of Griffin et al., 2004) in ppm; the basis for the classification is the Cr–Ca relationship combined with trace-element relationships explained in the Discussion; abundances are arbitrarily given to two significant figures; Ni-in-garnet thermometer ($T_{Ni-in-grt}$) of Ryan et al. (1996); for abbreviations, see Tables 2 and 3.

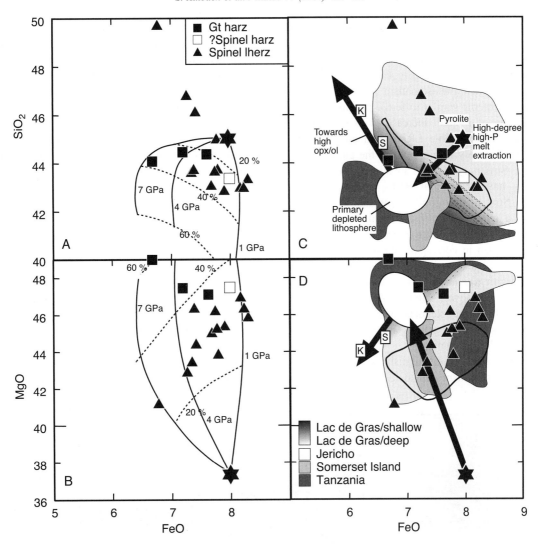

Fig. 9. Whole-rock FeO against (A) SiO₂ and (B) MgO of peridotites from the Buffalo Hills kimberlites (reconstructed from mineral data and modal estimates). (A) and (B) compared to experimental residues of isobaric batch melting at different pressures (solid lines, pressures in GPa) of pyrolite (star) for different degrees of partial melting (hatchured lines, percentages; from Walter, 1999); (C, D) partial melting trend, primary depleted lithosphere, trend towards high opx/ol ratios and average compositions of peridotites from the Kaapvaal craton (K) and Siberia (S) (from Walter, 1999); also shown are fields for peridotites from Lac de Gras (central Slave craton: Pearson et al., 1999; unpublished data), Jericho (north-central Slave craton: Kopylova et al., 1999), Somerset Island (Schmidberger and Francis, 1999) and Tanzania (Lee and Rudnick, 1999).

et al., 1999c; Fig. 10A). Garnet from the high-temperature sheared lherzolite, from the garnet spinel lherzolite and from the garnet harzburgite that has a nonsinusoidal REE pattern (see Fig. 5B) plot in the silicate melt metasomatism field. Garnets in the two garnet harzburgites with sinusoidal REE patterns plot in the depleted field and between the melt and "phlogopite" metasomatism field, respectively. Most

garnet xenocrysts plot in the depleted or silicate melt-metasomatised field; only a few garnet xenocrysts have Zr–Y relationships indicative of fertile peridotite.

Xenocrysts were grouped according to their lithologies (CaO–Cr₂O₃ relationships) and Zr–Y relationships shown in Fig. 10A, and average compositions were calculated. A comparison of their averaged chondrite-normalised REE patterns (Fig. 11) shows that garnets in fertile lherzolites and garnet spinel lherzolites have smoothly increasing REE_N abundances from Ce_N to Lu_N. In both lithologies, silicate melt metasomatism produces garnets with REE_N patterns similar to fertile peridotites, with the exception of a small increase ($2-3 \times$ chondrite) in the most incompatible LREE, and an additional positive Eu anomaly in garnet from the garnet spinel lherzolite. Strikingly, garnets in all depleted peridotite types have sinusoidal REE_N patterns with a hinge at Sm_N. "Phlogopite" metasomatism also produced a sinusoidal REE_N pattern, with a hinge at Eu_N. $HREE_N$ in depleted types are three to five times lower than in corresponding undepleted types, and this is evident also in their low Sc/Y. REE data for harzburgitic xenocrysts are incomplete, and their patterns are not shown. Some of the averaged REE patterns of the garnet xenocrysts show positive Eu anomalies. By contrast, garnets in analysed xenoliths do not show Eu anomalies.

In order to further distinguish between different styles of metasomatism and to identify the cause of sinusoidal REE patterns in otherwise depleted peridotites, the Zr/Sm–Ce/Yb relationships in garnets were investigated. Interaction with silicate melt tends to increase HFSE, which are more incompatible and therefore enriched in melts, relative to MREE, which are less incompatible (i.e., bulk distribution coefficients controlled by garnet/melt and clinopyroxene/melt are higher for Sm and Eu than for Zr and Ti; e.g., Hauri et al., 1994). By contrast, increased LREE/HREE at restricted HFSE/MREE ratios are diagnostic of interaction with carbonatite melts, and these relationships may be used to distinguish carbonatite from silicate melt metasomatism (Rudnick et al., 1993; Klemme et al., 1995; Ionov, 1998; Coltorti et al., 1999, 2000).

In a plot of Zr/Sm_N against Ce/Yb_N in garnet, many lherzolitic garnets that are identified as silicate melt-metasomatised in the Zr–Y diagram show evidence for enrichment in Zr/Sm without or with a small concomitant enrichment in Ce/Yb (Fig. 12A). It is striking that most fertile garnet xenocrysts have low Zr/Sm and Ce/Yb ratios, whereas virtually all depleted garnets are enriched in Ce/Yb at low Zr/Sm

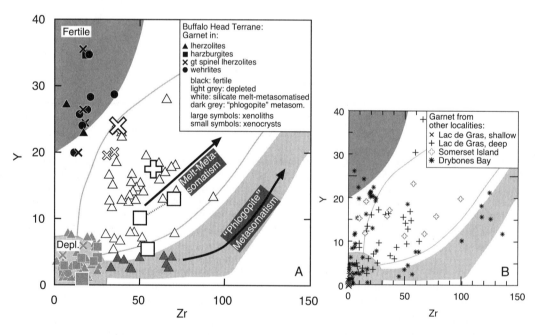

Fig. 10. (A) Zr against Y (ppm) in garnet from xenoliths and xenocrystic garnet. Fields and metasomatic trends of Griffin et al. (1999a). (B) Garnets from the central and marginal Slave craton, and from Somerset Island, are shown for comparison (Lac de Gras, central Slave craton: Pearson et al., 1999; unpublished data; Drybones Bay, Slave craton margin: Carbno and Canil, 2002; Griffin et al., 2004; Somerset Island: Schmidberger and Francis, 1999).

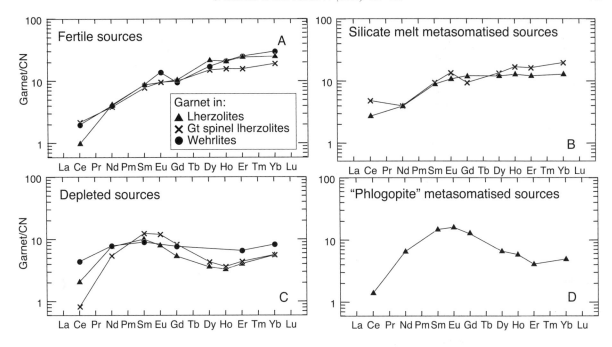

Fig. 11. REE patterns of average garnet xenocryst types (classed according to CaO–Cr₂O₃ and Zr–Y relationships) from (A) fertile sources, (B) silicate melt metasomatised sources, (C) depleted sources and (D) "phlogopite" metasomatised sources. Normalised to chondrite of McDonough and Sun (1995).

(Fig. 12B). This is suggested to be a signature of interaction with carbonatite or similar low-volume volatile-rich melt. The sinusoidal REE patterns that are observed in the depleted garnets are ascribed to the steep negative REE pattern of carbonatite, combined with reduced LREE accommodation in garnet due to the increasing size misfit (Griffin et al., 1999b; see also review by Stachel et al., 1998). Samples with positive Eu anomalies do not plot in any particular region of Zr/Sm–Ce/Yb or Zr–Y space. Average "phlogopite" metasomatised lherzolitic garnets are enriched in both Ce/Yb and Zr/Sm. This may indicate that they sequentially interacted with silicate melt and carbonatite (Powell et al., in press), or it may reflect the higher solubility of LREE relative to HFSE in volatile-rich silicate melts such as those that precipitate phlogopite.

The shallow lithosphere (spinel stability field) is represented by garnet spinel peridotites and spinel peridotites, where clinopyroxene is the main host for incompatible elements. A plot of Ti/Eu against La/Yb in clinopyroxene has been used to discriminate between carbonatite and silicate metasomatism (Coltorti

et al., 1999) and shows that clinopyroxenes in two of the spinel lherzolites lie between the trends for carbonatite and silicate metasomatism, whereas clinopyroxenes in two other spinel lherzolites have high La/Yb relative to Ti/Eu and may have interacted with carbonatitic melt (Fig. 13). Pure silicate melt metasomatism appears not to have been as important as in the deep lithosphere (garnet stability field), but more data points from the shallow lithosphere are needed to confirm this. A comparison between garnet in Fig. 12A and coexisting clinopyroxene in Fig. 13 is hampered by the absence of either garnet or clinopyroxene in most of the xenolith samples. Furthermore, the scale used by Coltorti et al. (1999) for Phanerozoic and Proterozoic samples may not be appropriate for ancient metasomatised mantle as one clinopyroxene plots far outside the diagram.

Fig. 8A shows that silicate melt metasomatism is the dominant metasomatic style in the deep lithosphere. Depleted and metasomatised rocks, which correspond to volatile-rich silicate melt metasomatised and carbonatite metasomatised sources, occur throughout the lithosphere column but represent the

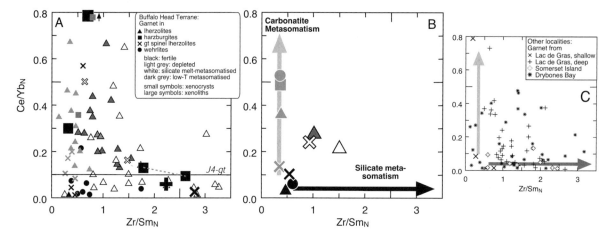

Fig. 12. Zr/Sm$_N$–Ce/Yb$_N$ relationships (A) in garnet xenocrysts (data of Griffin et al., 2004) and xenoliths; primitive J4-garnet of Jagoutz and Spettel (in Stachel et al., 1998); chondrite of McDonough and Sun (1995), (B) averaged for different garnet classes (see text for details) with trends for silicate and carbonatite metasomatism and (C) in garnets from the central and marginal Slave craton (references as in Fig. 10).

dominant metasomatic style in the shallower lithosphere. This distribution of temperature versus metasomatic style may indicate an evolution of the mantle beneath the BHT similar to that beneath East Africa. During extension in the East African Rift, asthenosphere-derived basaltic melts infiltrated and modified the lower lithosphere; reaction with the surrounding peridotite led to a decrease in the melt volume and evolution of small LILE/LREE/volatile-rich and Nb–Ta-depleted melt fractions that are highly mobile and can penetrate upward in the mantle (Bedini et al., 1997). Signatures of silicate melt metasomatism also occur at shallower depth (e.g., in garnet spinel lherzolites) but, from our limited data set, appear to be volumetrically not as important, perhaps because they are restricted to aureoles along magma conduits (Menzies et al., 1987).

Unradiogenic Nd in the spinel lherzolites indicates ancient enrichment in Nd relative to Sm, which translates into T_{DM} of 2.1 to 2.2 Ga. One spinel lherzolite also has unradiogenic Hf, which would be consistent with enrichment in Hf relative to Lu at least in the Mesoproterozoic. The garnet harzburgites (K11-3 and K11-5) have ε_{Nd} lower than a model depleted mantle indicative of Nd-addition but high ε_{Hf} expected of melt residues. This suggests that Nd, but not Hf, was added at some stage during their evolution and would be typical of carbonatite metasomatism, consistent with the sinusoidal REE patterns of garnets in K11-3 and K11-5. While K11-3 has high Sm/Nd relative to

its Nd isotopic composition, yielding a negative Sm–Nd model age, K11-5 has Sm–Nd isotope systematics that translate into a minimum T_{DM} of 2.1 Ga. By contrast, the Lu/Hf of K11-5 is very low to have produced the amount of radiogenic Hf, indicative of Hf addition at some stage during its evolution. K11-3 gives a Mesoproterozoic Lu–Hf model age. In summary, the isotopic evidence suggests an episode of metasomatism beneath the BHT some time around the Meso- or Paleoproterozoic.

11.4. Young (precursory) heating and metasomatism

Several features in peridotites from Buffalo Hills point to disequilibrium or require high surface energies that are not stable under upper mantle conditions, and therefore appear to be relatively recent. These include compositional disequilibrium, melt patches, secondary overgrowths, exsolution and arrested mineral reactions.

The presence of unequilibrated LREE-enriched garnets (see Fig. 5B) suggests that there was not enough time for diffusive equilibrium to be attained after the enrichment. Ca-undersaturated domains in garnet in sample K11-5 may represent relics of the original depleted composition, whereas high-Ca domains are due to interaction with a metasomatic component.

The major-element composition of the melt patch in sample K14-5b (Fig. 14A) is dissimilar to kim-

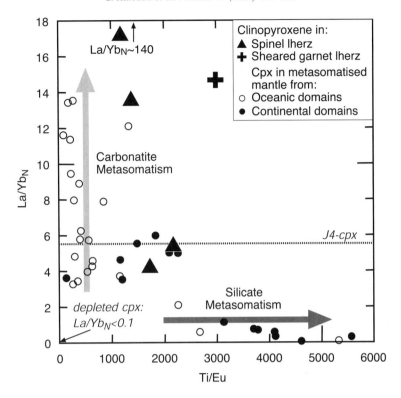

Fig. 13. Ti/Eu against La/Yb$_N$ in cpx from Buffalo Hills. Shown for comparison are clinopyroxenes in carbonatite and silicate melt metasomatised samples from oceanic and continental mantle (from Coltorti et al., 1999). Depleted cpx from Norman (1998), primitive J4 cpx of Jagoutz and Spettel (in Stachel et al., 1998); chondrite of McDonough and Sun (1995).

berlites and intracratonic basalts, and to silicate melt inclusions and melt patches in mantle xenoliths reported in the literature (e.g., Dawson, 1980; Hauri et al., 1993; Schiano and Clocchiatti, 1994; Wiechert et al., 1997; Furman and Graham, 1999). Formation by decompression-induced melting of hydrous phases, such as amphibole (e.g., Chazot et al., 1996), is also precluded by the mismatch in major element content, such as the lack of alkali elements in the melt. By contrast, the high-MgO content, the low contents of other major elements except SiO_2 (FeO below 1 wt.%) and low total oxides obtained by microprobe analysis of the melt patch (see Section 5.8) is similar to volatile-rich ultramafic liquids that occur in spinel lherzolites from Spitsbergen and are interpreted to have exsolved from a primitive volatile-rich magma due to liquid immiscibility (Amundsen, 1987). The ultramafic melt appears to have precipitated spinel and pentlandite, which occur within the melt patches (see Fig. 14A). Precipitation

of many sulfides from a melt may be indicated by their high Si contents (Table 9), ascribed to inclusion of silicate melt during sulfide growth, and by high Re/Os and Os isotopic ratios (Table 13). In addition, most sulfides in xenoliths from the Buffalo Hills kimberlites are pentlandites, which have been interpreted as a metasomatic sulfide (Alard et al., 2000). A secondary origin would be consistent with implausible Re–Os model ages obtained for many sulfides in this study, including one sulfide in the melt patch (sample K14-5b; Table 12).

Most spinels in peridotites from Buffalo Hills have distinct TiO_2-, Cr_2O_3- and Fe^{3+}-rich overgrowths (Fig. 14B), which may indicate interaction with an oxidized silicate melt shortly before entrainment in the kimberlite. Skeletal crystals with compositions similar to the overgrowths are observed in two samples and can be explained by rapid crystallisation from a melt. These phenomena may be related to magmas parental to the megacryst

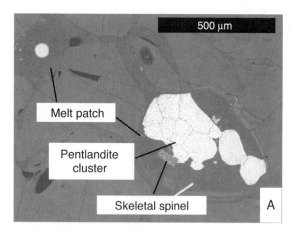

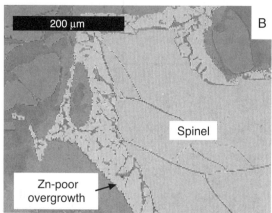

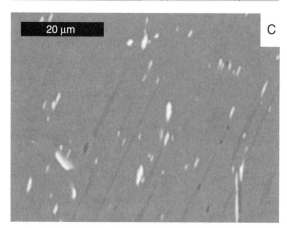

Fig. 14. Backscattered electron images of (A) melt patches containing pentlandite and skeletal spinel, (B) spinel overgrown by Fe^{3+}-, Cr- and Ti-rich, and Zn-poor spinel and (C) cpx with exsolved opx and spinel.

suite which are often temporally related to kimberlite magmatism; their compositions change after fractionation of megacrysts at the base of the lithosphere and assimilation of peridotitic wall-rock during ascent, becoming more Fe- and Ti- rich (e.g., Griffin et al., 2000). Overprinting by similar Ti-rich oxidizing magmas that acquired high Cr contents through interaction with the lithospheric mantle may explain secondary overgrowths on preexisting spinels.

The similarity of the composition of secondary overgrowths to skeletal spinel in the melt patch in sample K14-5b suggests that, like the precipitation of some pentlandites, these are manifestations of the same event. The infiltration of melts was accompanied by heating of the lithosphere, consistent with the lower Zn contents observed in spinel overgrowths and skeletal spinel, as Zn abundance in spinel decreases with increasing temperature (Griffin et al., 1994).

The solubility of the enstatite component in clinopyroxene and of the diopside component in orthopyroxene is temperature dependent (Brey and Köhler, 1990). Therefore, the ubiquitous exsolution of complementary pyroxene from clinopyroxenes and orthopyroxenes in spinel lherzolites (Fig. 14C) records cooling, subsequent to the heating and melt intrusion described above. Garnet and spinel in a garnet spinel lherzolite xenolith share amoeboid and vermicular grain boundaries (see Fig. 2B) that represent high energy states that cannot survive for extended periods of time in the upper mantle (e.g., Field and Haggerty, 1994). This indicates that there was not enough time for these grains to texturally re-equilibrate before entrainment in the kimberlite.

Many of the samples that show evidence for heating and young metasomatism are stored in the shallow mantle at relatively low temperatures, which could have prevented equilibration for substantial amounts of time. Therefore, disequilibrium features do not constrain the timing of the disturbance to immediately preceding entrainment in the kimberlite. They do, however, establish a sequence of events where these phenomena postdate the older modification of the mantle that is manifest in equilibrated assemblages with homogeneous compositions.

11.5. Pyroxenite genesis

Several pyroxenites show evidence for cooling and re-equilibration from high temperatures. One sample with corona and necklace structures is interpreted to have preserved at least three stages of re-equilibration. The garnet corona and orthopyroxene- and clinopyroxene-rich domains may have originated during an initial stage of exsolution from high-temperature Al-rich and Ca-poor clinopyroxene during cooling, as suggested for garnet pyroxenites from eastern Australia (Griffin et al., 1984; O'Reilly et al., 1988; Pearson et al., 1991). In a second phase, both the orthopyroxene and clinopyroxene domains exsolved garnet, now arranged as necklaces. In a third phase, large grains of orthopyroxene and clinopyroxene exsolved spinel and pyroxenes, and garnet exsolved rutile. The garnet corona protected the assemblages in each of the domains and slowed diffusive exchange and equilibration (e.g., Pearson et al., 1991). For example, garnet within the orthopyroxene domain has higher FeO and lower MgO than garnet in other domains. Clinopyroxene on either side of the garnet corona has distinct Al_2O_3, CaO and Na_2O contents. Garnet necklaces around orthopyroxene, without marked corona structures, are observed in other pyroxenites and suggest a more advanced state of re-equilibration subsequent to cooling-related exsolution.

Garnet and clinopyroxene from Buffalo Hills pyroxenites and those from eastern Australia have compositions that partially overlap and show similar trends (Fig. 15). Garnet pyroxenites from eastern Australia are suggested to have formed during underplating of mafic magmas into the lower crust and upper mantle (Wass and Hollis, 1983). Pyroxenites from Buffalo Hills may have a similar origin. Cooling to the ambient geotherm led to progressive exsolution and reaction processes that were arrested at low temperatures.

11.6. Comparison with Slave craton mantle sections

Mantle sections from the Slave craton include the north-central Jericho kimberlites (Kopylova et al., 1999; Griffin et al., 1999c), central Lac de Gras kimberlites (Pearson et al., 1999; Griffin et al., 1999c; 2004), the Torrie kimberlite which lies between the two (Griffin et al., 1999c; MacKenzie and Canil, 1999; Orr and Luth, 2000) and the Drybones Bay kimberlite at the craton margin (Carbno and Canil, 2002; Griffin et al., 2004). The mantle beneath Jericho and Torrie may have been reworked in connection with formation of the Kilohigok basin (Griffin et al., 1999c) or the 1.27 Ga Mackenzie plume event (Irvine et al., 1999). The mantle beneath Drybones Bay may have been heated and strongly metasomatised (Carbno and Canil, 2002) and may have been reworked or entirely replaced during the Proterozoic in the course of subduction events, or of movement along the Great Slave Lake shear zone (Griffin et al., 1999c).

With regard to major-element composition, reconstructed peridotites from Buffalo Hills have a range in FeO, MgO and SiO_2 similar to those in the deeper layer of the Archaean central Slave craton (Fig. 9). It is conceivable that samples that have not been silicate melt-metasomatised have preserved major-element relationships that were established during the melting related to lithosphere formation. If so, this would suggest that the mantle beneath the BHT and the central Slave craton formed under similar conditions. The lower lithosphere of the central Slave craton is suggested to have formed by subcretion of a plume that originated in the lower mantle (Griffin et al., 1999c). Inclusions in diamond from the Buffalo Hills kimberlites encompass minerals from the transition zone and the lower mantle as well as mixed parageneses, and the host diamonds are suggested to have grown in a plume (Davies et al., 2003), strengthening the plume connection.

For peridotites from both BHT and central Slave craton, there is a negative correlation of SiO_2 and FeO that is displaced toward higher SiO_2 relative to peridotites from the Kaapvaal and Siberian cratons, which are characterised by high orthopyroxene/olivine ratios. This suggests that, whatever the mechanism by which ancient lithosphere is enriched in orthopyroxene relative to olivine (see review by Walter, 1998), it differed for the mantle beneath the BHT and central Slave craton on the one hand and the Kaapvaal and Siberian cratons on the other.

Garnets from the Slave craton margin setting (Drybones Bay) have a range of Zr–Y relationships similar to garnets from the BHT (Fig. 10B), although

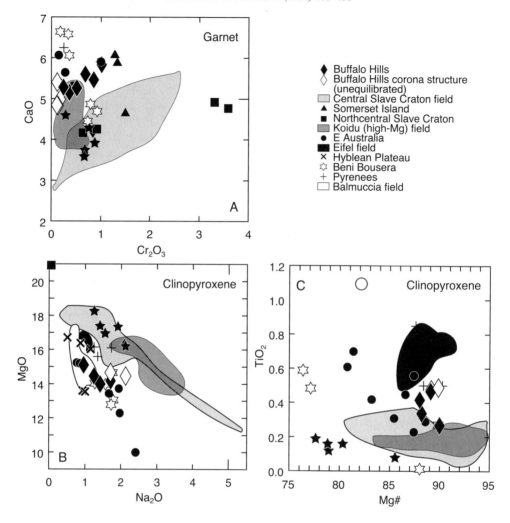

Fig. 15. (A) Cr$_2$O$_3$ against CaO in garnet, (B) Na$_2$O against MgO in cpx and (C) Mg# against TiO$_2$ in cpx in pyroxenites from Buffalo Hills. Shown for comparison are pyroxenite xenoliths and pyroxenites in peridotite massifs (Beni Bousera, Morocco: Kornprobst, 1969; Pyrenées: Bodinier et al., 1987; Balmuccia: Rivalenti et al., 1995; Eastern Australia: O'Reilly and Griffin, 1995; Hyblean Plateau, Sicily: Nimis and Vannucci, 1995; Eifel volcanic fields, Germany: Witt-Eickschen et al., 1998; other references as in Fig. 10).

carbonatite metasomatism appears to have been much more important for the mantle beneath the BHT (Fig. 12C). This similarity could stem from the fact that the mantle across the BHT, because it is a narrow continental block, has been affected by processes at its margins, such as rifting, similar to the mantle beneath Drybones Bay. In contrast, trace-element relationships differ for the central Slave craton in that garnets from fertile and "phlogopite" metasomatised sources are absent, and this may reflect its position in the craton interior, away from processes occurring at the craton margins.

12. Summary and conclusions

Several features of peridotites from the Buffalo Hills kimberlites hint at the presence of Archaean mantle beneath the BHT. Sinusoidal trace element patterns resemble those of inclusions in diamond occurring in Archaean cratonic areas of the world, and diamonds have been recovered from host kimberlites. The occurrence of relict Ca-undersaturated Cr$_2$O$_3$-rich garnet in one of the garnet harzburgites may be an Archaean inheritance because true Ca-undersaturated garnet is virtually exclusive to Archaean SCLM (Griffin et al., 1998). However, there is no unequivocal isotopic evidence for Archaean mantle.

The SCLM beneath the BHT experienced at least two episodes of modification. The oldest recognised metasomatic event is suggested to be responsible for enriched trace-element abundances in compositionally equilibrated xenoliths. Three metasomatic styles are recognised: silicate melt metasomatism, carbonatite metasomatism and an intermediate style possibly involving volatile-rich silicate melt. Silicate melt metasomatism prevailed in the deep lithosphere and led to enrichment of HFSE with minor enrichment in LREE. Carbonatite metasomatism and volatile-rich melt metasomatism led to enrichment of LREE without, or with only a small, concomitant enrichment in HFSE, respectively. These metasomatic styles are more common in the shallower lithosphere. This distribution of metasomatic styles in the lithosphere column is interpreted in terms of a model, whereby rift-related asthenosphere-derived silicate melts extensively interact with the mantle close to the lithosphere–asthenosphere boundary, which leads to formation of more mobile residual volatile-rich melts that are able to penetrate shallower lithosphere levels (Bedini et al., 1997). Rifting-related metasomatism of the SCLM beneath the BHT may have occurred between 2.4–2.1 Ga ago, when the BHT and adjacent Rae Province were separated by oceanic crust (Bostock and van Breemen, 1994; McNicoll et al., 2000). Clinopyroxenes with 2.10–2.25 Nd model ages in two spinel lherzolites may be related to this event.

Mesoproterozoic Re–Os T_{CHUR} model ages were obtained for sulfides in two spinel peridotites. The 1.74 Ga model age obtained for sulfide in sample K14-5b is considered fortuitous because this sulfide has highly radiogenic Os and very high Re/Os ratio requiring large corrections for the overlap of ^{187}Re on ^{187}Os. In addition, a second analysis of sulfides in this sample revealed that the Re–Os isotope systematics and resultant model ages are variable. By contrast, sulfide in K14-5d is the only sample of all samples measured that has a low Re/Os and unradiogenic Os isotopic composition. The minimum age (Re-depletion age, T_{RD}; Walker et al., 1989) is 1.46 Ga; the model age of 1.84 Ga is considered to reflect the true age of this sulfide. This age coincides with the inferred emplacement age of the mafic sheets of the

Winagami Reflection Sequence in the crust (WRS; Ross and Eaton, 1997) and suggests that the sulfide precipitated from melts related to this igneous event. Emplacement of mafic dykes and sills may be related to impingement of a plume at the base of the lithosphere (e.g., 1.27 Ga Mackenzie plume and dike swarm; LeCheminant and Heaman, 1989). The presence of magmatic mantle sulfide having a similar age to the igneous sheets may indicate that the plume source was located beneath the BHT, and that the lithospheric mantle interacted with the plume-derived melts at about 1.9 to 1.8 Ga. Garnet in the pyroxenite K14-5f gives an Nd model age of 1.8 Ga. This may indicate that pyroxenites also formed from the asthenosphere-derived melts parental to the mafic sheets of the WRS but were stored at temperatures too low to allow textural re-equilibration.

A younger metasomatic event occurred some time before kimberlite magmatism at 88–86 Ma. Heating and interaction with oxidizing silicate melts are indicated by heterogeneous mineral compositions, overgrowths on spinel that are Ti- and Fe^{3+}-rich but Zn-poor and the occurrence of melt patches. Fine-scale exsolutions of orthopyroxene and clinopyroxene in the complementary pyroxene are observed in many samples and testify to subsequent cooling. Some of these magmas may have pooled at shallow depth where they exsolved ultramafic volatile-rich liquids similar to melt patches found in one spinel lherzolite. Megacryst magmas are inferred to be temporally, although not necessarily genetically, related to kimberlite magmatism (e.g., Griffin et al., 2000). Migration of such melts may have heated and interacted with the mantle prior to eruption of the kimberlites. The isotope data set shows that some samples have unsupported radiogenic Hf or Nd or high parent/daughter ratios relative to their isotopic composition and this may also relate to relatively recent decoupling of parent and daughter elements during metasomatism.

Our data show that samples from the SCLM beneath the BHT are witnesses to multiple metasomatic episodes, and it is plausible that these processes are linked to events expressed at crustal levels. The emplacement of plume-derived diamonds (Davies et al., 2003) may have occurred during rifting at 2.4–2.3 Ga, assuming this event was caused by a plume. Alternatively, formation of these diamonds may be related to intrusion of the mafic sills of the Winagami Reflection Sequence at ~ 1.8 Ga, which may be plume-related similar to the prominent Mackenzie dike swarm in the Slave craton. The variety of events documented within the BHT and along its margins may well have obliterated any unambiguous Archaean signature that was originally present.

Acknowledgements

We are grateful to S. Elhlou, C. Lawson, N. Pearson, A. Sharma and P. Wieland for help with the analytical facilities. Detailed and constructive reviews by P. Kempton and G. Pearson, as well as editorial comments from L. Heaman, are much appreciated. This work was funded by a Macquarie University International Postgraduate Award and Postgraduate Research Fund (S.A.), by an ARC SPIRT grant sponsored by Kennecott Canada and by an ARC Large Grant to W.L.G. and S.Y.O'R. This is publication no. 346 from the ARC National Key Centre for Geochemical Evolution and Metallogeny of Continents (www.es.mq.edu.au/GEMOC/).

References

Alard, O., Griffin, W.L., Lorand, J.P., Jackson, S.E., O'Reilly, S.Y., 2000. Non-chondritic distribution of the highly siderophile elements in mantle sulfides. Nature 407, 891–894.

Amundsen, H.E.F., 1987. Evidence for liquid immiscibility in the upper mantle. Nature 327, 692–695.

Bedini, R.M., Bodinier, J.-L., Dautria, J.-M., Morten, L., 1997. Evolution of LILE-enriched small melt fractions in the lithospheric mantle: a case study from the East African Rift. Earth Planet. Sci. Lett. 153, 67–83.

Blichert-Toft, J., 2001. On the Lu-Hf Isotope Geochemistry of Silicate Rocks. J. Geost. Geoanal. 25, 41–56.

Blichert-Toft, J., Albarède, F., 1997. The Lu–Hf isotope geochemistry of chondrites and the evolution of the mantle–crust system. Earth Planet. Sci. Lett. 148, 243–258.

Blichert-Toft, J., Chauvel, C., Albarède, F., 1997. Separation of Hf and Lu for high precision isotope analysis of rock samples by magnetic sector-multiple collector ICPMS. Contrib. Mineral. Petrol. 127, 248–260.

Bodinier, J.J., Guiraud, M., Fabries, J., Dostal, J., Dupuy, C., 1987. Petrogenesis of layered pyroxenites from the Lherz, Freychinède and Prades ultramafic bodies (Ariège, French Pyrenées). Geochim. Cosmochim. Acta 51, 279–290.

Bostock, H., van Breemen, O., 1994. Ages of detrital and metamor-phic zircons and monazites from a pre-Taltson magmatic zone basin at the western margin of Rae Province. Can. J. Earth Sci. 31, 1353–1364.

Brey, G.P., Köhler, T., 1990. Geothermobarometry in four-phase lherzolites: II. New thermobarometers, and practical assessment of existing thermobarometers. J. Petrol. 31, 1353–1378.

Carbno, G.B., Canil, D., 2002. Mantle structure beneath the SW Slave craton, Canada: constraints from garnet geochemistry in the Drybones Bay kimberlite. J. Petrol. 43, 129–142.

Carlson, S.M., Hillier, W.D., Hood, C.T., Pryde, R.P., Skelton, D.N., 1999. The Buffalo Hills kimberlites: a newly-discovered diamondiferous kimberlite province in north-central Alberta, Canada. In: Gurney, J.J., Gurney, J.L., Pascoe, M.D., Richardson, S.H. (Eds.), Proc. 7th Intl. Kimb. Conf. Red Roof Design cc, Cape Town, pp. 109–116.

Chazot, G., Menzies, M.A., Harte, B., 1996. Determination of partition coefficients between apatite, clinopyroxene, amphibole, and melt in natural lherzolites from Yemen: implications for wet melting in the lithospheric mantle. Geochim. Cosmochim. Acta 60, 423–437.

Coltorti, M., Bonadiman, C., Hinton, R.W., Siena, F., Upton, B.G.J., 1999. Carbonatite metasomatism of the oceanic upper mantle: evidence from clinopyroxenes and glasses in ultramafic xenoliths of Grande Comore, Indian Ocean. J. Petrol. 40 (1), 133–165.

Coltorti, M., Beccaluva, L., Bonadiman, C., Salvini, L., Siena, F., 2000. Glasses in mantle xenoliths as geochemical indicators of metasomatic agents. Earth Planet. Sci. Lett. 183, 303–320.

Davies, R.M., Griffin, W.L., O'Reilly, S.Y., McCandless, T.E., 2003. Inclusions in diamonds from the K10 and K14 kimberlites, Buffalo Hills, Alberta, Canada: diamond growth in a plume? Ext. Abstr., 8th Intl. Kimb. Conf., Victoria BC.

Dawson, J., 1980. Kimberlites and Their Xenoliths. Springer, Berlin 252 pp.

DePaolo, D.J., 1981. Neodymium isotopes in the Colorado Front Range and crust–mantle evolution in the Proterozoic. Nature 291, 193–196.

Droop, G., 1987. A general equation for estimating Fe^{3+} concentrations in ferromagnesian silicates and oxides from microprobe analyses, using stoichiometric criteria. Geol. Mag. 51, 431–435.

Ellis, D., Green, D., 1979. An experimental study of the effect of Ca upon garnet–clinopyroxene Fe–Mg exchange equilibria. Contrib. Mineral. Petrol. 71, 13–22.

Field, S., Haggerty, S., 1994. Symplectites in upper mantle peridotites: development and implications for the growth of subsolidus garnet, pyroxene and spinel. Contrib. Mineral. Petrol. 118, 138–156.

Furman, T., Graham, D., 1999. Erosion of lithospheric mantle beneath the East African Rift system: geochemical evidence from the Kivu volcanic province. Lithos 48, 237–262.

Gaul, O.F., Griffin, W.L., O'Reilly, S.Y., Pearson, N.J., 2000. Mapping olivine composition in the lithospheric mantle. Earth Planet. Sci. Lett. 182, 223–235.

Griffin, W.L., Wass, S.Y., Hollis, J.D., 1984. Ultramafic xenoliths from Bullenmerri and Gnotuk Maars, Victoria, Australia: petrology of a sub-continental crust–mantle transition. J. Petrol. 25, 53–87.

Griffin, W.L., Smith, D., Boyd, F.R., 1987. Trace-element zoning in garnets from sheared mantle xenoliths. Geochim. Cosmochim. Acta 53, 561–567.

Griffin, W.L., Ryan, C.G., Gurney, J.J., Sobolev, N.V., Win, T.T., 1994. Chromite macrocrysts in kimberlites and lamproites: geochemistry and origin. In: Meyer, H., Leonardo, O. (Eds.), Kimberlites, Related Rocks and Mantle Xenoliths. CPRM Spec. Publ., Brasilia, pp. 366–377. Jan/94.

Griffin, W., O'Reilly, S., Ryan, C., Gaul, O., Ionov, D., 1998. Secular variation in the composition of subcontinental lithospheric mantle. In: Braun, J., Dooley, J., Goleby, B., van der Hilst, R., Klootwijk, C. (Eds.), Structure and Evolution of the Australian Continent. Geodynamics Series. Amer. Geophys. Union, Washington, DC, pp. 1–26.

Griffin, W.L., Shee, S.R., Ryan, C.G., Win, T.T., Wyatt, B.A., 1999a. Harzburgite to lherzolite and back again: metasomatic processes in ultramafic xenoliths from the Wesselton kimberlite, Kimberley, South Africa. Contrib. Mineral. Petrol. 134 (2–3), 232–250.

Griffin, W.L., Fisher, N.I., Friedman, J.H., Ryan, C.G., O'Reilly, S.Y., 1999b. Cr-pyrope garnets in the lithospheric mantle. I: compositional systematics and relations to tectonic setting. J. Petrol. 40, 679–705.

Griffin, W.L., Doyle, B.J., Ryan, C.G., Pearson, N.J., O'Reilly, S.Y., Davies, R.M., Kivi, K., Van Achterbergh, E., Natapov, L.M., 1999c. Layered mantle lithosphere in the Lac de Gras Area, Slave craton: composition, structure and origin. J. Petrol. 40, 705–727.

Griffin, W.L., Pearson, N.J., Belousova, E., Jackson, S.E., van Achterbergh, E., O'Reilly, S.Y., Shee, S.R., 2000. The Hf isotope composition of cratonic mantle: LAM–MC–ICPMS analysis of zircon megacrysts in kimberlites. Geochim. Cosmochim. Acta 64, 133–147.

Griffin, W.L., Fisher, N.I., Friedman, J.H., O'Reilly, S.Y., Ryan, C.G., 2002. Cr-pyrope garnets in the lithospheric mantle: II. Compositional populations and their distribution in time and space. Geochem. Geophys. Geosyst. 3, 1073.

Griffin, W.L., O'Reilly, S.Y., Abe, N., Aulbach, S., Davies, R.M., Pearson, N.J., Doyle, B.J., Kivi, K., 2003. The origin and evolution of Archean lithospheric mantle. Precambrian Res. 127, 19–41.

Griffin, W.L., O'Reilly, S.Y., Doyle, B.J., Kivi, K., 2004. Lithospheric Mapping Beneath the North American Plate. Lithos 77, 873–922 (this volume).

Gurney, J., 1984. A correlation between garnets and diamonds in kimberlites. Publ. Geol. Dept. & Univ. Extension, vol. 8. Univ. West Aust, Perth, pp. 143–166.

Harley, S., 1984. An experimental study of the partitioning of iron and magnesium between garnet and orthopyroxene. Contrib. Mineral. Petrol. 86, 359–373.

Hauri, E.H., Shimizu, N., Dieu, J.J., Hart, S.R., 1993. Evidence for hotspot-related carbonatite metasomatism in the oceanic upper mantle. Nature 365, 221–227.

Hauri, E.H., Wagner, T.P., Grove, T.L., 1994. Experimental and natural partitioning of Th, U, Pb and other trace elements between garnet, clinopyroxene and basaltic melts. Chem. Geol. 117, 149–166.

Hoffman, P.F., 1989. Precambrian geology and tectonic history of North America. In: Bally, A., Palmer, A. (Eds.), The Geology of North America—An Overview. Geol. Soc. Amer., 447–512.

Hood, C.T., McCandless, T.E., 2003. Systematic variations in xenocryst mineral composition at the province scale, Buffalo Hills kimberlites, Alberta, Canada. Ext. Abstr., 8th Intl. Kimb. Conf., Victoria BC.

Ionov, D., 1998. Trace element composition of mantle-derived carbonate and coexisting phases in peridotite xenoliths from alkali basalts. J. Petrol. 39, 1931–1941.

Irvine, G.J., Kopylova, M.G., Carlson, R.W., Pearson, D.G., Shirey, S.B., Kjarsgaard, B.A., 1999. Age of the lithospheric mantle beneath and around the Slave craton: a Re–Os isotope study of peridotite xenoliths from the Jericho and Somerset Island kimberlites. Abstr. 9th Goldschmidt Conf. LPI, Boston, MA, pp. 134–135.

Jerde, E.A., Taylor, L.A., Crozaz, G., Sobolev, N.V., Sobolev, V.N., 1993. Diamondiferous eclogites from Yakutia, Siberia—evidence for a diversity of protoliths. Contrib. Mineral. Petrol. 114, 189–202.

Kempton, P.D., Hawkesworth, C.J., Lopez-Escobar, L., Pearson, D.G., Ware, A.J., 1999. Spinel ± garnet lherzolite xenoliths from Pali Aike. Part 2: trace element and isotopic evidence bearing on the evolution of lithospheric mantle beneath southern Patagonia. In: Gurney, J.J., Gurney, J.L., Pascoe, M.D., Richardson, S.H. (Eds.), 7th Intl. Kimberlite Conf. Red Roof Design cc, Cape Town, pp. 415–428.

Kennedy, C., Kennedy, G., 1976. The equilibrium boundary between graphite and diamond. J. Geophys. Res. 81, 2467–2470.

Klemme, S., van der Laan, S.R., Foley, S.F., Guenther, D., 1995. Experimentally determined trace and minor element partitioning between clinopyroxene and carbonatite melt under upper mantle conditions. Earth Planet. Sci. Lett. 133, 439–448.

Kopylova, M.G., Russell, J.K., Cookenboo, H., 1999. Petrology of peridotite and pyroxenite xenoliths from the Jericho kimberlite: implications for the thermal state of the mantle beneath the Slave craton, Northern Canada. J. Petrol. 40 (1), 79–104.

Kopylova, M.G., Russell, J.K., Stanley, C., Cookenboo, H., 2000. Garnet from Cr and Ca-saturated mantle: implications for diamond exploration. J. Geochem. Explor. 68, 183–199.

Kornprobst, J., 1969. Le massif ultrabasique des Beni Bouchera (Rif Interne, Maroc): Etude des peridotites de haute temperature et de haute pression, et des pyroxénolites, a grenat ou sans grenat. Contrib. Mineral. Petrol. 23, 283–322.

Krogh, E., 1988. The garnet–clinopyroxene iron–magnesium geothermometer—a reinterpretation of existing experimental data. Contrib. Mineral. Petrol. 99, 44–48.

LeCheminant, A.N., Heaman, L.M., 1989. Mackenzie igneous events, Canada: middle Proterozoic hotspot magmatism associated with ocean opening. Earth Planet. Sci. Lett. 96, 38–48.

Lee, C.-T., Rudnick, R.L., 1999. Compositionally stratified cratonic lithosphere: petrology and geochemistry of peridotite xenoliths from the Labait tuff cone, Tanzania. In: Gurney, J.J., Gurney, J.L., Pascoe, M.D., Richardson, S.H. (Eds.), Proc. 7th Int. Kimb. Conf. Red Roof Design cc, Cape Town, pp. 503–521.

MacKenzie, J.M., Canil, D., 1999. Composition and thermal evolution of cratonic mantle beneath the central Archean

Slave Province, NWT, Canada. Contrib. Mineral. Petrol. 134, 313–324.

McDonough, W.F., Sun, S.-S., 1995. The composition of the Earth. Chem. Geol. 120, 223–253.

McNicoll, V.J., Thériault, R.J., McDonough, M.R., 2000. Taltson basement gneissic rocks: U–Pb and Nd isotopic constraints on the basement to the Paleoproterozoic Taltson magmatic zone, northeastern Alberta. Can. J. Earth Sci. 37, 1575–1596.

Menzies, M.A., Rogers, N., Tindle, A., Hawkesworth, C.J., 1987. Metasomatic and enrichment processes in lithospheric peridotites, an effect of asthenosphere–lithosphere interaction. In: Menzies, M.A., Hawkesworth, C.J. (Eds.), Mantle Metasomatism. Academic Press, London, pp. 313–361.

Nickel, K.G., Green, D.H., 1985. Empirical geothermobarometry for garnet peridotites and implications for the nature of the lithosphere, kimberlites and diamonds. Earth Planet. Sci. Lett. 73 (1), 158–170.

Nimis, P., Vannucci, R., 1995. An ion microprobe study of clinopyroxenes in websteritic and megacrystic xenoliths from Hyblean Plateau (Se Sicily, Italy)—constraints On HFSE/REE/Sr fractionation at mantle depth. Chem. Geol. 124 (3–4), 185–197.

Norman, M.D., 1998. Melting and metasomatism in the continental lithosphere: laser ablation ICPMS analysis of minerals in spinel lherzolites from Eastern Australia. Contrib. Mineral. Petrol. 130, 240–255.

Norman, M.D., Pearson, N.J., Sharma, A.L., Griffin, W.L., 1996. Quantitative analysis of trace elements in geological materials by laser ablation ICPMS: instrumental operating conditions and calibration values of NIST glasses. Geostand. Newsl. 20 (2), 247–261.

O'Neill, H., Wood, B., 1979. An experimental study of the iron–magnesium partitioning between garnet and olivine and its calibration as a geothermometer. Contrib. Mineral. Petrol. 70, 59–70.

O'Reilly, S.Y., Griffin, W.L., 1995. Trace-element partitioning between garnet and clinopyroxene in mantle-derived pyroxenites and eclogites—P–T–X controls. Chem. Geol. 121 (1–4), 105–130.

O'Reilly, S.Y., Griffin, W.L., Stabel, A., 1988. Evolution of Phanerozoic Eastern Australian lithosphere: isotope evidence for magmatic and tectonic underplating. J. Petrol., Spec, 89–108.

Orr, P., Luth, R.W., 2000. Petrology and oxygen-isotope geochemistry of the Yamba Lake kimberlite rocks, NWT. Can. J. Earth Sci. 37, 1053–1071.

Patchett, P.J., Tatsumoto, M., 1980. Hafnium isotope variations in oceanic basalts. Geophys. Res. Lett. 7, 1077–1080.

Pearson, N.J., O'Reilly, S.Y., Griffin, W.L., 1991. The granulite to eclogite transition beneath the eastern margin of the Australian craton. Eur. J. Mineral. 3 (2), 293–322.

Pearson, N.J., Griffin, W.L., Doyle, B.J., O'Reilly, S.Y., van Achterbergh, E., Kivi, K., 1999. Xenoliths from kimberlite pipes of the Lac de Gras area, Slave craton, Canada. In: Gurney, J.J., Gurney, M.D., Pascoe, M.D., Richardson, S.H. (Eds.), 7th Intl. Kimberlite Conf. Red Roof Design cc, Cape Town, pp. 644–658.

Pearson, N.J., Alard, O., Griffin, W.L., Jackson, S.E., O'Reilly, S.Y., 2002. In situ measurement of Re–Os isotopes in mantle sulfides by laser ablation multicollector-

inductively coupled plasma mass spectrometry: analytical methods and preliminary results. Geochim. Cosmochim. Acta 66, 1037–1050.

Pollack, H.N., Chapman, D.S., 1977. On the regional variation of heat flow, geotherms, and lithospheric thickness. Tectonophys 38, 279–296.

Pouchou, J.L., Pichoir, F., 1984. A new model for quantitative X-ray microanalysis of homogeneous samples. Rech. Aérosp. 5, 13–38.

Powell, W., Zhang, M., O'Reilly, S.Y., Tiepolo, M., 2004. Mantle amphibole trace-element and isotopic signatures trace multiple metasomatic episodes in lithospheric mantle, western Victoria, Australia. Lithos (in press).

Rivalenti, G., Mazzucchelli, M., Vannucci, R., Hofmann, A.W., Ottolini, L., Bottazzi, P., Obermiller, W., 1995. The relationship between websterite and peridotite in the Balmuccia peridotite massif (NW Italy) as revealed by trace element variations in clinopyroxene. Contrib. Mineral. Petrol. 121, 275–288.

Ross, G.M., 2000. Introduction to special issue of Canadian Journal of Earth Sciences: the Alberta basement transect of lithoprobe. Can. J. Earth Sci. 37, 1447–1452.

Ross, G.M., 2002. Evolution of Precambrian continental lithosphere in western Canada: results from lithoprobe studies in Alberta and beyond. Can. J. Earth Sci. 39, 413–437.

Ross, G.M., Eaton, D.W., 1997. Winagami reflection sequence: seismic evidence for post-collisional magmatism in the Proterozoic of western Canada. Geology 25, 199–202.

Ross, G.M., Eaton, D.W., 2002. Proterozoic tectonic accretion and growth of western Laurentia: results from lithoprobe studies in northern Alberta. Can. J. Earth Sci. 39, 313–329.

Ross, G.M., Parrish, R.R., Villeneuve, M.E., Bowring, S.A., 1991. Geophysics and geochronology of the crystalline basement of the Alberta basin, western Canada. Can. J. Earth Sci. 28, 512–522.

Ross, G.M., Eaton, D.W., Boerner, D.W., Miles, W., 2000. Tectonic entrapment and its role in the evolution of the continental lithosphere: an example from the Precambrian of western Canada. Tectonics 19, 116–134.

Rudnick, R.L., McDonough, W.F., Chappell, B.W., 1993. Carbonatite metasomatism in the northern Tanzanian mantle: petrographic and geochemical characteristics. Earth Planet. Sci. Lett. 114, 463–475.

Ryan, C.G., Griffin, W.L., Pearson, N.J., 1996. Garnet geotherms—pressure–temperature data from Cr-pyrope garnet xenocrysts in volcanic rocks. J. Geophys. Res. 101 (B3), 5611–5625.

Schiano, P., Clocchiatti, R., 1994. Worldwide occurrence of silica-rich melts in sub-continental and sub-oceanic mantle minerals. Nature 368, 621–624.

Schmidberger, S., Francis, D., 1999. Nature of the mantle roots beneath the North American craton: mantle xenolith evidence from Somerset Island kimberlites. Lithos 48, 195–216.

Smith, D., 1999. Temperatures and pressures of mineral equilibration in peridotite xenoliths: review, discussion, and implications. In: Fei, Y., Bertka, C.M., Mysen, B.O. (Eds.), Mantle Petrology: Field Observations and High Pressure Experimentation: A Tribute to Francis R. (Joe) Boyd. Spec. Publ. - Geochem. Soc., vol. 6, pp. 171–188.

Smith, D., Boyd, F.R., 1987. Compositional heterogeneities in a high-temperature lherzolite nodule and implications for mantle processes. In: Nixon, P.H. (Ed.), Mantle Xenoliths. Wiley, New York, pp. 551–561.

Sobolev, N., Lavrent'yev, Y., Pokhilenko, N., Usova, L., 1973. Chrome-rich garnets from the kimberlites of Yakutia and their paragenesis. Contrib. Mineral. Petrol. 40, 39–52.

Stachel, T., Viljoen, K.S., Brey, G., Harris, J.W., 1998. Metasomatic processes in lherzolitic and harzburgitic domains of diamondiferous lithospheric mantle: REE in garnets from xenoliths and inclusions in diamonds. Earth Planet. Sci. Lett. 159 (1–2), 1–12.

Thériault, R.J., Ross, G.M., 1991. Nd isotopic evidence for crustal recycling in the ca. 2.0 Ga subsurface of western Canada. Can. J. Earth Sci. 28, 1149–1157.

van Achterberg, E., Ryan, C.G., Griffin, W.L., 1999. Glitter: on line intensity reduction for the laser ablation inductively coupled plasma spectrometry. Abstr. 9th Goldschmidt Conf. LPI, Boston, MA, pp. 905.

Walker, R.J., Carlson, R.W., Shirey, S.B., Boyd, F.R., 1989. Os, Sr, Nd, and Pb isotope systematics of Southern African peridotite xenoliths: implications for the chemical evolution of subcontinental mantle. Geochim. Cosmochim. Acta 53, 1583–1595.

Walker, R.J., Morgan, J.W., Horan, M.F., Czamanske, G.K., Krogstad, E.J., et al., 1994. Re–Os isotopic evidence for an enriched-mantle source for the Noril'sk-type, ore-bearing intrusions, Siberia. Geochim. Cosmochim. Acta 58, 4179–4197.

Walter, M.J., 1998. Melting of garnet peridotite and the origin of komatiite and depleted lithosphere. J. Petrol. 39 (1), 29–60.

Walter, M.J., 1999. Melting residues of fertile peridotite and the origin of cratonic lithosphere. In: Fei, Y., Bertka, C.M., Mysen, B.O. (Eds.), Mantle Petrology: Field Observation and High Pressure Experimentation. Spec. Publ. - Geochem. Soc., pp. 225–239.

Wass, S.Y., Hollis, J.D., 1983. Crustal growth in south-eastern Australia—evidence from lower crustal eclogitic and granulitic xenoliths. J. Metamorph. Geol. 1, 24–45.

Wasserburg, G.J., Jacobsen, S.B., DePaolo, D.J., McCulloch, M.T., Wen, T., 1981. Precise determination of Sm/Nd ratios, Sm and Nd isotopic abundances in standard solutions. Geochim. Cosmochim. Acta 45, 2311–2323.

Webb, S., Wood, B., 1986. Spinel–pyroxene–garnet relationships and their dependence on Cr/Al ratio. Contrib. Mineral. Petrol. 92, 471–480.

Wiechert, U., Ionov, D.A., Wedepohl, K.H., 1997. Spinel peridotite xenoliths from the Atsagin–Dush volcano, Dariganga lava plateau, Mongolia: a record of partial melting and cryptic metasomatism in the upper mantle. Contrib. Mineral. Petrol. 126, 345–364.

Witt-Eickschen, G., Seck, H., 1991. Solubility of Ca and Al in orthopyroxene from spinel peridotite: an improved version of an empirical geothermometer. Contrib. Mineral. Petrol. 106, 431439.

Witt-Eickschen, G., Kaminsky, W., Kramm, U., Harte, B., 1998. The nature of young vein metasomatism in the lithosphere of the west Eifel (Germany): geochemical and isotopic constraints from composite mantle xenoliths from the Meerfelder Maar. J. Petrol. 39 (1), 155–185.

Xu, X., O'Reilly, S.Y., Griffin, W.L., Zhou, X., 1999. Genesis of young lithospheric mantle in southeastern China: an LAM–ICPMS trace element study. J. Petrol. 41, 111–148.

Available online at www.sciencedirect.com

Lithos 77 (2004) 453–472

www.elsevier.com/locate/lithos

Timing of Precambrian melt depletion and Phanerozoic refertilization events in the lithospheric mantle of the Wyoming Craton and adjacent Central Plains Orogen

Richard W. Carlson[a,*], Anthony J. Irving[b], Daniel J. Schulze[c], B. Carter Hearn Jr.[d]

[a] Department of Terrestrial Magnetism, Carnegie Institution of Washington, 5241 Broad Branch Road, NW, Washington, DC 20015, USA
[b] Department of Earth and Space Sciences, University of Washington, Seattle, WA 98195, USA
[c] University of Toronto at Mississauga, Mississauga, Toronto, Canada L5L 1C6
[d] U.S. Geological Survey, Reston, VA 22092, USA

Received 27 June 2003; accepted 17 February 2004
Available online 28 May 2004

Abstract

Garnet peridotite xenoliths from the Sloan kimberlite (Colorado) are variably depleted in their major magmaphile (Ca, Al) element compositions with whole rock Re-depletion model ages generally consistent with this depletion occurring in the mid-Proterozoic. Unlike many lithospheric peridotites, the Sloan samples are also depleted in incompatible trace elements, as shown by the composition of separated garnet and clinopyroxene. Most of the Sloan peridotites have intermineral Sm–Nd and Lu–Hf isotope systematics consistent with this depletion occurring in the mid-Proterozoic, though the precise age of this event is poorly defined. Thus, when sampled by the Devonian Sloan kimberlite, the compositional characteristics of the lithospheric mantle in this area primarily reflected the initial melt extraction event that presumably is associated with crust formation in the Proterozoic—a relatively simple history that may also explain the cold geotherm measured for the Sloan xenoliths.

The Williams and Homestead kimberlites erupted through the Wyoming Craton in the Eocene, near the end of the Laramide Orogeny, the major tectonomagmatic event responsible for the formation of the Rocky Mountains in the late Cretaceous–early Tertiary. Rhenium-depletion model ages for the Homestead peridotites are mostly Archean, consistent with their origin in the Archean lithospheric mantle of the Wyoming Craton. Both the Williams and Homestead peridotites, however, clearly show the consequences of metasomatism by incompatible-element-rich melts. Intermineral isotope systematics in both the Homestead and Williams peridotites are highly disturbed with the Sr and Nd isotopic compositions of the minerals being dominated by the metasomatic component. Some Homestead samples preserve an incompatible element depleted signature in their radiogenic Hf isotopic compositions. Sm–Nd tie lines for garnet and clinopyroxene separates from most Homestead samples provide Mesozoic or younger "ages" suggesting that the metasomatism occurred during the Laramide. Highly variable Rb–Sr and Lu–Hf mineral "ages" for these same samples suggest that the Homestead peridotites did not achieve intermineral equilibrium during this metasomatism. This indicates that the metasomatic overprint likely was

* Corresponding author. Tel.: +1-202-478-8474; fax: +1-202-478-8821.
E-mail addresses: carlson@dtm.ciw.edu (R.W. Carlson), irving@ess.washington.edu (A.J. Irving), dschulze@credit.erin.utoronto.ca (D.J. Schulze), chearn@usgs.gov (B.C. Hearn).

introduced shortly before kimberlite eruption through interaction of the peridotites with the host kimberlite, or petrogenetically similar magmas, in the Wyoming Craton lithosphere.

Keywords: Xenoliths; Lithosphere; Major elements; Sr–Nd–Hf–Os–Pb isotopes; Laramide orogeny

1. Introduction

Mantle xenoliths provide an important glimpse of the characteristics of the continental lithospheric mantle. The picture they provide, however, is complicated by the fact that many xenoliths carry an integrated record of the number and variety of events that work to chemically modify the mantle. The predominant

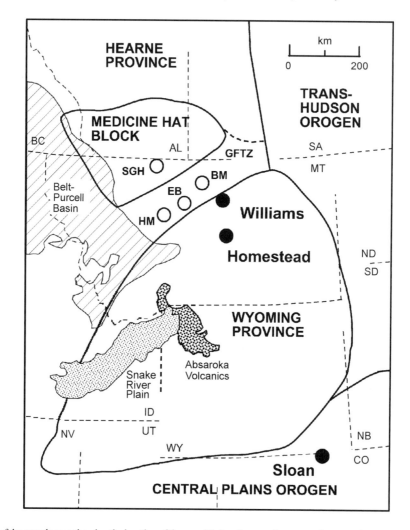

Fig. 1. Tectonic map of the sample area showing the location of the xenolith-bearing eruptive centers in comparison to basement geologic terranes that include the Archean Hearne, Medicine Hat, and Wyoming Cratons and the Proterozoic Trans-Hudson and Central Plains Orogens. The Great Falls Tectonic Zone (GFTZ) is an area of poorly exposed basement interpreted as the suture zone between the Medicine Hat and Wyoming Cratons. Phanerozoic xenolith-bearing eruptive centers in this area include Sweet Grass Hills (SGH), the Highwood Mountains (HM), Eagle Buttes (EB), and the Bearpaw Mountains (BP). The three xenolith-bearing kimberlites that are the subject of this study are shown by the large black circles.

chemical signature of most peridotite xenoliths from the continental lithospheric mantle is one of melt depletion (Boyd and Mertzman, 1987) indicating that most of the continental mantle is the residual complement to the igneous rocks that now make up the overlying crust. This melt depletion gives continental mantle peridotite a compositional buoyancy (Boyd and McCallister, 1976) that most likely plays a key role in stabilizing continental lithosphere at Earth's surface (Jordan, 1978).

Once buoyantly stabilized in the continental lithospheric mantle, these depleted residues participate in the continuing chemical differentiation of the Earth by interacting with passing melts and fluids. The Wyoming Craton and surrounding Proterozoic Cratons of the northwestern United States provide a particularly good setting to examine the evolution of subcontinental lithospheric mantle subjected to a variety of tectonic and magmatic events. Xenoliths from the Devonian Sloan kimberlite (Fig. 1), which erupted through mid-Proterozoic crust along the Colorado–Wyoming border, provide a geotherm typical of old, cold, continental lithosphere (Eggler et al., 1988). In contrast, present day heat flow in this area is high (Decker et al., 1984), reflecting the major late Cretaceous–early Tertiary tectonothermal event in this area, the Laramide Orogeny. Near the end of the Laramide event, the mantle of this area was sampled again by a variety of xenolith-bearing diatremes in central and northern Montana (Hearn, 1989; Carlson and Irving, 1994).

We report here an attempt to deconvolve the events that occurred in the lithospheric mantle of the Precambrian western United States through a multipronged approach:

- Examination of xenoliths contained in kimberlites that erupted within the Archean Wyoming Craton, from its tectonized northern boundary, and from south of the Craton in order to compare mantle samples from regions that may have experienced distinct tectonothermal histories.
- Comparison of xenolith characteristics between kimberlites erupted well before, and those erupted near the end of the Laramide Orogeny to examine the extent to which this large-scale event overprinted mantle characteristics throughout this area.
- Use of a number of radiometric systems that have variable susceptibility to the effects of metasoma-

tism in an attempt to distinguish chemical and mineralogical modifications caused by interaction with the host magmas or their precursors from more ancient metasomatic events that may have occurred in the mantle.

To accomplish this task, we have obtained new chemical and Sr, Nd, Hf, Os and Pb isotopic data for garnet and spinel harzburgite, lherzolite and websterite xenoliths from the Homestead and Sloan kimberlites and new Hf isotopic data for previously studied (Carlson et al., 1999a) garnet peridotite xenoliths from the Williams kimberlite.

2. Sampling the tectonic evolution of the Wyoming Craton

The xenolith localities selected for study sample mantle from beneath areas that have distinct crustal tectonomagmatic histories (Fig. 1). From north to south, the Eocene Highwood Mountains and Eagle Butte olivine minette magmas (Carlson and Irving, 1994) erupted through the Great Falls Tectonic Zone (GFTZ), a NE–SW trending belt that was tectonically, but not magmatically, active in the Proterozoic (O'Neill and Lopez, 1985). The 48 Ma Williams kimberlite is part of the Missouri Breaks diatremes that were erupted near the border of the GFTZ and the Archean Wyoming Province (Hearn, 1989). The recently discovered (Ellsworth, 2000) Homestead kimberlite erupted through the central Wyoming Craton in an area that has experienced minimal post-Archean tectonism, at least up until the Laramide. Though undated, the Homestead kimberlite is presumed to be similar in age to other circa 50 Ma mafic-alkalic intrusions in the Grassrange province (Irving et al., 2003; Hearn, 2004). The 380 Ma Sloan kimberlite (Smith et al., 1979) erupted along the Colorado–Wyoming border well south of the Wyoming Craton in the 1.7–1.8 Ga juvenile arc crust of the Central Plains Orogen (Dutch and Neilsen, 1990).

Recent geophysical studies of this area (Gorman et al., 2002; Karlstrom et al., 2002) reveal deep crustal and upper mantle structures that aid in interpreting the mechanism of tectonic assembly. Though originally interpreted as an intracratonic tectonic zone (O'Neill and Lopez, 1985), recent results from the Lithoprobe project (Gorman et al., 2002) suggest that the Medicine

Hat Block and the Wyoming Province are separate Archean terranes, and that the GFTZ represents the suture zone between these separate blocks. The timing of the suturing of the Wyoming-Medicine Hat-Hearne blocks is uncertain. Gorman et al. (2002) suggest that the suturing occurred in the late Archean through northward-dipping subduction, but the very strong signature of 1.8 Ga metasomatic overprinting of the shallow mantle sampled by the Eagle Buttes and High-wood Mountains xenoliths (Carlson and Irving, 1994; Rudnick et al., 1999) suggests either mid-Proterozoic reactivation of this boundary, or perhaps that the assembly did not occur until the mid-Proterozoic. The importance of Proterozoic events in this area is further highlighted by the presence of a 20–30 km thick high velocity lower crustal layer under the northern Wyoming Province-GFTZ interpreted to be magmatically underplated material (Gorman et al., 2002). The age of the underplating is indicated by a number of radiometric ages for lower crustal xenoliths from this area. A garnet granulite xenolith from the Sweet Grass Hills in northern Montana provides an Sm–Nd isochron age of 1696 ± 6 Ma with initial $\varepsilon_{Nd} = +0.3$ (Irving et al., 1997). Lower crustal xenoliths from the Sloan kimberlite give a wide range of U–Pb zircon ages from Devonian through early Archean, but the dominant population falls in the age range of 1640–1750 Ma (Karlstrom et al., 2002), similar to the Sweet Grass Hills result.

3. Xenolith suites

Petrographic features of the analyzed xenolith samples are given in Appendix A. The Williams kimberlite xenoliths provide the northernmost sample set that extends well into the garnet stability field (Hearn and Boyd, 1975). Hearn and McGee (1984) showed that Williams peridotite xenoliths define a slightly warm (44 mW/m²) conductive geotherm with a marked step to much higher temperatures at relatively shallow depth (< 150 km).

The Homestead xenolith suite consists of harzburgites (containing garnet, spinel or both), garnet lherzolites and sparse chromite dunites (Irving et al., 2003; Hearn, 2004). A few samples are low-Ca harzburgites containing garnets within the G10 field (Gurney, 1984). Evidence of younger events is seen in composite xenoliths that contain garnet, garnet–

spinel, or spinel pyroxenite veins or selvedges, many of which also contain phlogopite. Phlogopite also occurs as disseminated grains, as rims on garnet, and in clinopyroxene–spinel clots. Peridotite xenoliths from Homestead originate from depths of at least 150 km (Irving et al., 2003; Hearn, 2004), but are offset to much higher temperatures than typical cratonic mantle, consistent with a much hotter Eocene geotherm approaching 60 mW/m² (Irving et al., 2003).

The Sloan kimberlite xenolith suite contains a variety of crustal and mantle xenoliths that have received considerable study (Smith et al., 1979; Eggler et al., 1987, 1988). Perhaps the most interesting aspect of the Sloan peridotite xenolith suite is that they define a 40 mW/m² conductive geotherm to depths of order 200 km (Eggler et al., 1987), which is more typical of Archean than mid-Proterozoic sections of lithospheric mantle. The geotherm shown by Sloan xenoliths provides a picture of mantle temperatures 380 My ago when they were transported to the surface by the Sloan kimberlite. Present day surface heat flow (Decker et al., 1980, 1984) suggests that mantle temperatures beneath Sloan currently exceed 1500 °C at depths as shallow as 100 km (Eggler et al., 1988) indicating that the cool geotherm recorded by the xenoliths has been severely disturbed by Laramide tectonism and magmatism.

Seismic imaging (Gorman et al., 2002; Karlstrom et al., 2002) shows that crustal and upper mantle structures dip to the north at the Cheyenne belt, the boundary between the Wyoming and the Central Plains Orogen, suggesting northward convergence and underthrusting of the southern block during Proterozoic assembly (Karlstrom et al., 2002) and the likelihood that only Proterozoic crust and upper mantle underlie Sloan. In contrast, studies of the mantle xenoliths from Sloan distinguish an upper depleted layer from a lower layer displaying the metasomatic enrichment in incompatible elements typical of Archean lithospheric mantle. This feature of the xenoliths led Eggler et al. (1988) to conclude that the Wyoming Province underthrust the Proterozoic lithosphere of the Central Plains Orogen.

The Sloan samples analyzed here were collected from an access road paved with disaggregated material from the Sloan 2 pipe. They are dominantly garnet harzburgites, most containing no modal clinopyroxene at all and garnet lherzolites, with minor spinel–garnet lherzolites, spinel harzburgites and garnet websterites.

4. Analytical procedures

Whole rock powders were prepared from representative interior portions of each xenolith by grinding in a Spex alumina mill. Major element compositions and the trace element abundances reported in Table 1 were determined by XRF spectrometry using procedures described by Boyd and Mertzman (1987). Mineral compositions were determined by electron microprobe at the University of Washington, and equilibration temperatures and pressures (Table 1) were calculated by the methods of Brey et al. (1990) utilizing the iterative PTEXEL program. Splits of most of the samples were selected for gentle crushing and mineral separation by handpicking only. Before

analysis, all mineral separates were washed in an ultrasonic bath for 15 min in approximately 5 ml of 4 N HCl followed by an additional 5 min with an added 0.5 ml of concentrated HF. The minerals were then washed repeatedly in distilled water before the addition of isotopically enriched spikes and dissolution acids. The samples were not weighed after leaching in order to avoid sample loss caused by static electricity in the beakers, so the concentrations reported here are calculated using the pre-leaching weights. Before dissolution, spike solutions containing [87]Rb, [84]Sr, [150]Nd, [149]Sm, [176]Lu, [180]Hf and [205]Pb were added to the samples to determine sample concentrations of these elements by isotope dilution. All samples were dissolved in a 2 to 1 mixture of concentrated HF and HNO_3. Garnet separates were dissolved in sealed Parr digestion vessels in an oven for 24 h at 170 °C. All other samples were dissolved in capped Teflon beakers on hot plates for times ranging from 1 to 3 days.

Table 1
Major element composition and mineral thermobarometry of xenoliths from the Homestead and Sloan kimberlites

Sample name	Homestead									Sloan							
	HK 1-13H	HK 1-13W	HK 1-19	HK 1-21	HK 1-24	HK 1-25	HK 1-26	HK 4-1	HOO 10-22	13-16-3	13-19-2	13-19-6	13-19-11	13-19-15	13-19-20	13-19-22	13-19-27
Rock Type	Sp Hz	Sp Web	Ga Hz	Dunite	Ga Lhz	Ga Lhz	Ga Lhz	Ga Lhz	Ga Hz	Ga Hz	Ga Web	Sp Hz	Ga Hz	Ga Lhz	Sp-Ga Lhz	Ga Lhz	Ga Lhz
T (°C)		1172	1136		1211	1213	1090	1127	1124	1255	711	758	1249	757	815	1328	777
P (GPa)			3.42		2.45	3.79	4.21	3.74	3.41	5.13			5.05	2.49	2.19	6.01	2.27
SiO_2	41.61	48.62	44.77	39.49	41.82	41.52	43.01	41.24	41.13	39.49	48.14	39.05	41.37	41.39	41.38	40.30	40.17
TiO_2	0.00	0.02	0.01	0.01	0.03	0.02	0.02	0.04	0.01	0.01	0.25	0.05	0.12	0.10	0.07	0.09	0.14
Al_2O_3	0.35	1.12	1.33	0.15	1.98	1.84	1.29	2.70	1.26	0.33	3.19	1.14	0.77	3.96	2.71	0.95	3.30
Fe_2O_3	3.30	2.75	3.06	4.23	3.95	3.13	3.30	2.75	2.53	3.36	3.46	4.34	2.76	3.43	3.41	4.15	4.32
FeO	3.54	2.60	3.60	3.46	3.71	4.32	3.44	3.99	3.39	3.48	3.39	4.01	4.59	4.78	4.77	3.99	3.86
MnO	0.11	0.13	0.12	0.10	0.13	0.13	0.12	0.13	0.10	0.14	0.16	0.14	0.12	0.18	0.13	0.14	0.19
MgO	43.28	35.56	39.52	42.92	37.72	40.00	40.16	40.30	42.48	43.00	20.80	41.55	42.40	35.70	37.78	40.76	35.78
CaO	0.36	2.48	0.99	0.38	2.06	1.60	0.87	1.88	0.26	0.51	15.47	0.73	0.69	3.34	2.64	1.10	2.94
Na_2O	0.14	0.28	0.21	0.13	0.27	0.19	0.15	0.19	0.13	0.18	0.52	0.16	0.18	0.30	0.30	0.18	0.27
K_2O	0.00	0.02	0.06	0.03	0.04	0.00	0.01	0.07	0.01	0.08	0.44	0.02	0.15	0.13	0.00	0.14	0.24
P_2O_5	0.01	0.01	0.02	0.05	0.02	0.02	0.02	0.02	0.02	0.03	0.04	0.03	0.05	0.02	0.02	0.06	0.05
LOI	6.96	5.93	6.05	8.41	7.90	7.00	7.33	6.44	8.03	9.11	3.97	8.52	6.47	6.45	6.48	7.87	8.48
Total	99.66	99.52	99.74	99.36	99.63	99.77	99.72	99.75	99.35	99.72	99.83	99.74	99.67	99.78	99.69	99.73	99.74
Rb	<2	3	2	<2	<2	5	<2	4	<2	2	20	<2	9	<2	5	6	17
Sr	42	80	99	50	75	55	42	86	177	98	156	12	75	8	52	93	84
Zr	19	20	16	15	16	15	16	20	14	13	21	16	19	16	14	17	19
V	55	80	69	50	78	72	62	74	52	57	198	68	65	100	93	71	107
Ni	2640	1730	2510	3380	2510	2640	2570	2065	2415	2785	960	2925	2755	2040	2560	2710	2520
Cr	2280	5210	2955	2085	2625	2760	2390	2820	2770	2080	3470	2400	4000	2555	2270	2125	2440
Co	106	72	87	118	105	90	88	93	99	119	55	109	107	93	98	117	102

Pb was separated first from the samples using the procedure described in Walker et al. (1989). The non-Pb fraction eluted in HBr was dried, converted to chlorides with 4 N HCl, dried and then redissolved in a mixture of 1 N HCl–0.1 N HF. This mixture was loaded onto a 20 cm long, 6 mm ID column filled with AG50W-X8 cation exchange resin. The Hf fraction was eluted with 1 N HCl–0.1 N HF, followed by Rb and Sr in 2.5 N HCl, and an REE fraction in 4 N HCl. Hf was then purified using the first column procedure described by Blichert-Toft et al. (1997) followed by Ti removal on the cation column loading the sample in 2.5 N HCl with a trace of H_2O_2, eluting Ti with 2.5 N HCl, and then Hf with 2.5 N HCl–0.3 N HF. Later samples involved a single Hf clean up column developed by Mark Schmitz, and modeled after the procedure described by Muenker et al. (2001) that uses a 3.8 cm long, 6 mm ID column filled with EICHROM© Ln-Spec resin. In this procedure, the Hf cut from the cation column is loaded in 3 M HCl, washed in sequence with 3 M HCl, 6 M HCl, H_2O, 0.09 M citric acid + 0.45 M

HNO_3 + 1 wt.% H_2O_2, 0.09 M citric acid + 0.45 M HNO_3, 6 M HCl + 0.06 M HF, with Hf then eluted with 6 M HCl + 0.4 M HF. The REE cut from the cation column was separated into heavy REE, Sm and Nd fractions using a cation column with methylactic acid as eluant (Walker et al., 1989). Lu was then separated from the HREE fraction using a 10 cm long by 4 mm ID column filled with Ln-Spec resin. For this column, the sample was loaded in 0.05 N HCl, washed with 2.5 N HCl and the Lu eluted in 6 N HCl. Total procedural blanks measured during these analyses were Rb = 95 pg, Sr = 300 pg, Nd = 60 pg, Sm = 13 pg, Lu = 7 pg, Hf = 80 pg, and Pb = 33 pg.

Re–Os measurements were obtained on 1 g of whole rock powder using procedures described by Carlson et al. (1999b). Re (1 pg) and Os (2 pg) blank corrections were applied to all the data reported in Table 2. Some of the spike was lost during Carius tube loading for sample 13-16-3. Since a mixed Re–Os spike was used, this results in inaccurate concentration data, but does not affect the Re/Os ratio or Os isotopic composition reported in Table 2.

Table 2
Re–Os data for Homestead and Sloan whole rock samples

Sample	Re (ppb)	Os (ppb)	$^{187}Re/^{188}Os$	$^{187}Os/^{188}Os$	Error	γ_{Os} (I)	T_{RD} (Ga)	T_{MA} (Ga)
Homestead								
HK1-13H	0.0903	3.878	0.1120	0.11037	0.00014	−14.7	2.61	3.47
HK1-13W	0.2682	3.061	0.4214	0.11291	0.00021	−12.9	2.30	47.37
HK1-19	0.0486	4.934	0.0473	0.11022	0.00026	−14.8	2.62	2.93
HK1-21	0.1879	0.696	1.298	0.11193	0.00017	−14.2	2.53	−1.24
HK1-24	0.0174	0.845	0.0991	0.11820	0.00046	−8.6	1.56	2.00
HK1-25	0.0295	0.805	0.1761					
HK1-25/2	0.0299	0.595	0.2418	0.10925	0.00028	−15.6	2.77	6.00
HK1-26	0.0637	4.353	0.0704	0.11173	0.00018	−13.6	2.42	2.87
HK4-1	0.0109	0.934	0.0562	0.11000	0.00019	−14.9	2.65	3.03
HK-3a[a]	0.1152	0.700	0.7983	0.17764	0.00102	36.9		7.46
Sloan								
13-16-3			0.0884	0.11617	0.00011	−8.8	1.90	2.28
13-19-2	0.1681	0.261	6.073	7.379	0.029	5688		49.63
13-19-6	0.1051	2.277	0.2219	0.11285	0.00014	−12.1	2.45	4.54
13-19-11	0.0815	4.572	0.0859	0.12472	0.00085	−2.1	0.74	0.83
13-19-15	0.2964	2.462	0.5805	0.12664	0.00018	−3.1	0.91	−1.24
13-19-20		3.211		0.12441	0.00010			
13-19-22	0.4309	9.842	0.2108	0.11939	0.00013	−6.9	1.57	2.67
13-19-27	0.5497	2.672	0.992	0.12758	0.00037	−4.4	1.14	−0.22
13-19K[a]	2.7276	0.759	17.33	0.13196	0.00034	−82.7		

[a] Host kimberlites. Os fractionation corrected to $^{192}Os/^{188}Os$ = 3.082614 and corrected for oxygen isotopic composition using the values reported by Nier (1950).

Sr, Re, Os and Pb isotopic compositions were determined by thermal ionization mass spectrometry. Os isotopic compositions were measured with OsO_3^- ions, and Re as ReO_4^- ions, with procedures similar to that described by Carlson et al. (1999b). Precision of concentration determinations are estimated at better than 0.5% for Sr, Re, Os and Pb based on repeat measurements of standard solutions.

Rb, Sm and Lu concentrations and Nd and Hf isotopic compositions were determined by MC-ICPMS using the DTM VG-P54. Nd was run with a three step multidynamic routine, monitoring ^{140}Ce, ^{141}Pr and ^{147}Sm, with all ratios except ^{150}Nd/^{144}Nd

being calculated in dynamic mode. Hf was run statically monitoring ^{173}Yb, ^{175}Lu, ^{181}Ta and ^{182}W for potential interferences. Rb and Lu concentrations were determined by comparison of the measured isotopic compositions of the samples with the average of standards run repeatedly during an analytical session. The accuracy of the Yb correction to Lu was checked each analytical session with runs of Lu standard to which various amounts of Yb had been added. Estimated concentration precision for Rb and Lu is better than 1%. Repeat measurements of a Sm concentration standard show a concentration reproducibility near 0.1%. Rb, Sr and Pb

Table 3
Rb, Sr and Pb concentration and isotopic composition of analyzed samples

Homestead	Type	Rb (ppm)	Sr (ppm)	^{87}Rb/^{86}Sr	^{87}Sr/^{86}Sr	Pb (ppm)	^{206}Pb/^{204}Pb	^{207}Pb/^{204}Pb	^{208}Pb/^{204}Pb
HK1-13H	WR		11.81		0.706931	0.0441	18.137	15.522	37.98
HK1-13W	Cpx	0.0286	236.2	0.000350	0.705456	0.285	18.024	15.504	38.02
HK1-19	Cpx	0.377	392.3	0.00278	0.704910	0.394	19.146	15.614	39.77
HK1-19	Ga	0.0416	1.098	0.109	0.706354	0.00531	29.060	16.430	38.99
HK1-21	WR	1.614				0.419	18.086	15.530	38.29
HK1-24	Cpx	0.131	164.4	0.00230	0.705466	0.222	18.214	15.918	38.90
HK1-24	Ga	0.0551	1.442	0.111	0.705927	0.0288	19.137	16.016	39.02
HK1-25	Cpx	0.0162	262.0	0.000179	0.705094	0.313	17.977	15.504	37.98
HK1-25	Ga	0.0361	2.858	0.0365	0.705200	0.00215			
HK1-26	Cpx	0.0248	432.9	0.000166	0.705248	0.341	19.620	15.631	40.58
HK1-26	Ga	0.0063	0.658	0.0275	0.705440				
HK4-1	Cpx	0.156	373.0	0.00121	0.704824	0.584	18.435	15.596	38.38
HK4-1	Ga	0.0542	2.185	0.0717	0.705343	0.00174			
HK-3[a]	WR	25.3	887	0.0825	0.705774		18.134	15.571	38.50
Sloan									
13-16-3	Ga	1.46	3.263	0.494	0.708584				
13-19-2	WR	23.40	128.4	0.527	0.708010	9.64	20.563	15.991	39.72
13-19-6	Cpx	0.452	53.23	0.0245	0.702477	0.188	16.910	15.375	36.77
13-19-11	WR	10.7				0.322	20.936	15.815	40.07
13-19-15	WR	4.12				0.288	20.315	15.768	39.95
13-19-20	Cpx	1.405	25.18	0.161	0.705556	0.039	19.180	15.380	38.84
13-19-22	WR	8.20				1.17	20.652	15.790	39.46
13-19-27	Cpx	0.460	72.80	0.0183	0.702400	0.149	16.686	15.332	36.55
13-19-27	Ga	0.0825	0.8871	0.269	0.708954				
13-19-27	WR				0.709230				
13-19K[a]	WR	116	609	0.0784	0.707696		22.923	15.943	43.74
Williams									
H69-15F	Cpx	0.0469	82.76	0.00164	0.703203				
H81-21	Cpx	0.126	153.4	0.00238	0.704290				
WP 28-3-1	Cpx	0.116	108.8	0.00308	0.704594				

[a] Host Kimberlite. Sr fractionation corrected to ^{86}Sr/^{88}Sr = 0.1194 and reported relative to ^{87}Sr/^{86}Sr = 0.710250 for the NIST987 Sr standard. The average value for this standard at DTM is 0.710260 ± 0.000013 (2σ-mean for 12 runs). Internal run errors for all samples reported above are below the external reproducibility of the standard. Pb is corrected to the isotopic composition reported by Todt et al. (1996) for the NIST981 standard with uncertainties of 0.14% (^{206}Pb/^{204}Pb), 0.17% (^{207}Pb/^{204}Pb), and 0.25% (^{208}Pb/^{204}Pb).

Table 4

Sm, Nd, Lu, and Hf concentrations and isotopic compositions for Homestead, Sloan and Williams xenoliths and host kimberlites

Homestead	Type	Sm (ppm)	Nd (ppm)	$\frac{^{147}Sm}{^{144}Nd}$	$\frac{^{143}Nd}{^{144}Nd}$	ε_{Nd} (i)	T_{DM} (Ga)	Lu (ppm)	Hf (ppm)	$\frac{^{176}Lu}{^{177}Hf}$	$\frac{^{176}Hf}{^{177}Hf}$	Error	ε_{Hf} (I)	T_{DM} (Ga)
HK1-13H	WR	0.0296	0.1357	0.1318	0.512487	−2.5	1.23		0.0216					
HK1-13W	Cpx	1.868	7.880	0.1433	0.512451	−3.3	1.51	0.0084	0.5791	0.00205	0.282894	0.000008	5.3	0.59
HK1-19	WR		0.8818						0.0642		0.282545	0.000036		
HK1-19	Cpx	4.909	30.71	0.0967	0.512279	−6.3	1.13	0.0503	0.8537	0.00837	0.282491	0.000005	−9.1	1.40
HK1-19	Ga	3.138	5.496	0.3451	0.512309	−7.3	−0.98	0.0148	0.3804	0.00552	0.282353	0.000011	−13.9	1.50
HK1-21	WR								0.0494		0.283891	0.000079		
HK1-24	WR	0.2933	1.192	0.1487	0.512398	−4.3	1.76	0.0229	0.0956	0.03394				
HK1-24	Cpx	1.606	7.693	0.1262	0.512409	−4.0	1.29	0.0181	0.3770	0.00681	0.283119	0.000015	13.1	0.30
HK1-24	Ga	0.6975	1.611	0.2617	0.512600	−1.1	−1.75	0.318	0.2829	0.1597	0.285436	0.000031	90.0	0.94
HK1-25	WR	0.1657	0.7125	0.1406	0.512488	−2.5	1.38	0.0208	0.0551	0.05349				
HK1-25	Cpx	1.301	7.584	0.1037	0.512477	−2.5	0.93	0.0072	0.2193	0.00463	0.284022	0.000008	45.2	−1.14
HK1-25	Ga	0.984	2.384	0.2495	0.512602	−1.0	−2.34	0.514	0.2733	0.2667	0.287386	0.000007	155.5	0.95
HK1-26	WR								0.0575		0.282746	0.000064		
HK1-26	Cpx	3.514	9.001	0.1587	0.512494	−2.5	1.82	0.0046	0.7668	0.00085	0.282359	0.000004	−13.5	1.31
HK1-26	Ga	2.770	3.637	0.4605	0.512770	0.9	−0.23	0.0664	0.5899	0.01598	0.282202	0.000006	−19.6	2.51
HK4-1	WR	0.2254	0.7275	0.1873				0.0315	0.0663	0.06757	0.284332	0.000060	54.0	1.90
HK4-1	Cpx	1.557	9.360	0.1005	0.512202	−7.9	1.28	0.0066	0.2042	0.00456	0.283836	0.000015	38.6	−0.84
HK4-1	Ga	1.773	2.966	0.3613	0.512317	−7.3	−0.86	0.398	0.3992	0.1415	0.284503	0.000007	57.6	0.63
H00-10-22	WR	0.0613	0.2689	0.1380	0.512393	−4.4	1.52	0.154	0.0286	0.07610				
H00-10-22	Ga	0.6215	0.8839	0.4250	0.512442	−5.2	−0.51	0.069	0.3729	0.02620	0.282986	0.000045	7.8	1.26
HK-3[a]	WR	8.58	50.45	0.1028	0.512324	−5.4	1.14	0.16	3.92	0.00578	0.282461	0.000003	−10.0	1.30
Sloan														
13-16-3	Ga	0.4408	1.819	0.1465	0.512587	1.5	1.28	0.1347	0.2692	0.0710	0.283828	0.000033	27.9	0.87
13-16-3	WR	0.0877	0.7916	0.0670	0.512476	3.2	0.70		0.0224		0.283346	0.000088		
13-19-2	WR	0.973	4.742	0.1241	0.512548	1.8	1.02	0.0419	0.3180	0.0187	0.283173	0.000020	17.8	0.34
13-19-6	Cpx	1.369	3.194	0.2591	0.513558	15.0	1.37	0.2032	1.248	0.0231	0.282767	0.000013	2.4	1.78
13-19-11	WR	0.2673	1.588	0.1017	0.512567	3.3	0.79	0.0071	0.1779	0.0057	0.282469	0.000081	−3.8	1.32
13-19-15	WR	0.3036	1.157	0.1586	0.513077	10.5	0.20	0.089	0.1556	0.0812				
13-19-20	Cpx	0.921	1.489	0.3740	0.514887	35.3	1.65	0.1978	0.3957	0.0710	0.284144	0.000005	39.1	1.39
13-19-22	WR	0.2290	1.556	0.0890				0.0116	0.1521	0.0108	0.282970	0.000052	12.6	0.63
13-19-27	WR	0.3534	1.910	0.1119	0.512717	5.7	0.65	0.0336	0.1877	0.0254	0.283665	0.000033	33.6	−1.47
13-19-27	Cpx	1.457	4.522	0.1948	0.513132	9.8	0.14		1.111		0.282639	0.000008		
13-19-27	Ga	0.4975	0.2375	1.2661	0.519489	81.8	0.92	0.605	0.3239	0.2649	0.286015	0.000011	56.4	0.64
13-19K[a]	WR	10.14	74.05	0.1369	0.512428	−3.6	1.43	0.13	3.53	0.0052	0.282621	0.000003	−4.3	1.03
Williams														
H68-16b	WR							0.0062	0.1149	0.00759	0.283363	0.000040	21.7	−0.11
H68-16B	Ga	0.2203	1.554	0.08566	0.511739	−16.8	1.68	0.0763	0.0489	0.2208	0.290995	0.000035	284.9	2.22
H69-15F	Cpx	1.039	4.133	0.1519	0.512750	2.5	0.99	0.0052	0.1833	0.00401	0.285952	0.000027	113.4	−4.24
H69-15F	Ga	1.742	2.280	0.4617	0.512849	2.5	−0.18	0.2177	1.422	0.0217	0.282927	0.000034	5.8	1.15
H81-21	Cpx	0.5268	4.946	0.06438	0.512281	−6.1	0.89	0.002	0.0381	0.00727	0.283545	0.000076	28.2	−0.42
H81-21	Ga	1.448	4.813	0.1818	0.512209	−8.2	4.47	0.1337	0.6721	0.0282	0.283292	0.000045	18.4	0.04
WP 28-3-1	Cpx	0.7506	4.055	0.1119	0.512636	0.5	0.77	0.004	0.1648	0.00344	0.282916	0.000028	6.0	0.58
W4-002[a]	WR							0.25	3.65	0.0097	0.282731	0.000023	−0.7	1.03

[a] Host kimberlite. Nd fractionation corrected to $^{146}Nd/^{144}Nd = 0.7219$ and reported relative to $^{143}Nd/^{144}Nd = 0.511860$ for the La Jolla Nd standard. Average value obtained for this standard is $^{143}Nd/^{144}Nd = 0.511831 \pm 0.000009$ (2σ mean of 26 measurements). Internal precisions for all Nd runs reported here are less than the external error of the standard. Hf is fractionation corrected to $^{179}Hf/^{177}Hf = 0.7325$ and reported relatively to $^{176}Hf/^{177}Hf = 0.282160$ for the JMC475 Hf standard. At DTM, repeat measurements of this standard give $^{176}Hf/^{177}Hf = 0.282166 \pm 0.000006$ (2σ mean of 19 runs). Sm is fractionation corrected to $^{147}Sm/^{152}Sm = 0.56081$.

results are reported in Table 3, Nd and Hf results in Table 4.

All line fitting done in this paper was carried out with the Isoplot program (Ludwig, 1991) with errors reported at the 95% confidence level. Decay constants used are $^{87}Rb = 1.42 \times 10^{-11}$ year^{-1}, $^{235}U = 9.8485 \times 10^{-10}$ year^{-1}, $^{238}U = 1.551 \times 10^{-10}$ year^{-1} (Steiger and Jager, 1977), $^{147}Sm = 6.54 \times 10^{-12}$ year^{-1} (Lugmair and Marti, 1978), $^{176}Lu = 1.865 \times 10^{-11}$ year^{-1} (Scherer et al., 2001), and $^{187}Re = 1.666 \times 10^{-11}$ year^{-1} (Smoliar et al., 1996). Bulk earth parameters for the calculation of Rb–Sr model ages and ε_{Nd}, ε_{Hf} and γ_{Os} are $^{87}Rb/^{86}Sr = 0.09$, $^{87}Sr/^{86}Sr = 0.705$, $^{147}Sm/^{144}Nd = 0.1966$, $^{143}Nd/^{144}Nd = 0.512636$, $^{176}Lu/^{177}Hf = 0.0332$, $^{176}Hf/^{177}Hf = 0.282772$ (Vervoort and Blichert-Toft, 1999), $^{187}Re/^{188}Os = 0.4353$, $^{187}Os/^{188}Os = 0.1296$ (Meisel et al., 2001). Depleted mantle parameters used in the calculation of Nd and Hf model ages are $^{147}Sm/^{144}Nd = 0.2152$, $^{143}Nd/^{144}Nd = 0.5132$, $^{176}Lu/^{177}Hf = 0.03915$, $^{176}Hf/^{177}Hf = 0.2833$ (Vervoort and Blichert-Toft, 1999).

5. Sample chemical characteristics

Whole rock major element analyses (Table 1, Fig. 2) show that both the Homestead and Sloan peridotites analyzed here display a range of compositions, from fairly fertile (e.g., high Al, Ca, low Mg) to the very depleted compositions typical of cratonic mantle. Several of the Sloan samples have Al and Ca concentrations approaching those estimated for undepleted mantle (McDonough and Sun, 1995; Fig. 2). The Williams peridotites included in this study (Carlson et al., 1999a) cluster towards the more depleted, high MgO, end of the compositional arrays shown in Fig. 2. The Sloan and Homestead samples tend towards low SiO_2 contents as expected for residues of melt extraction (Johnson et al., 1990), whereas the Williams samples cluster at higher SiO_2 (Fig. 2), similar to peridotites from both the Kaapvaal (Boyd and Mertzman, 1987) and Siberian (Boyd et al., 1997) cratons. This trend towards high SiO_2, and hence high modal orthopyroxene, has most recently been interpreted as an indication of interaction of peridotite with through-going melts (Kelemen et al., 1998; Simon et al., 2003).

Perhaps the most striking compositional contrast between samples from the different localities is that both the Homestead and Williams samples show strong incompatible element enrichment whereas the Sloan samples do not. This is seen most clearly by the primitive mantle normalized incompatible element patterns determined for minerals separated from these samples (Fig. 3). Clinopyroxenes from both the Wil-

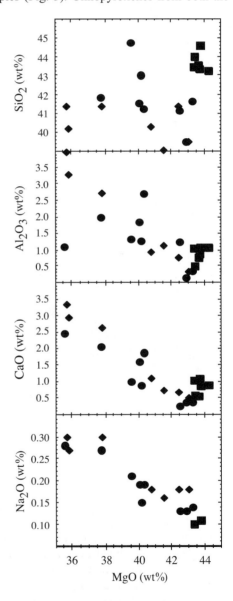

Fig. 2. Major element Mg variation diagrams for Sloan (diamonds), Homestead (circles) and Williams (squares) analyzed here and by Carlson et al. (1999a).

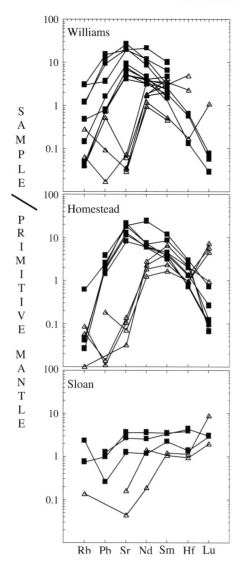

Fig. 3. Primitive mantle (McDonough and Sun, 1995) normalized trace element abundances from diopside (squares) and garnet (open triangles) mineral separates.

liams and Homestead samples show strongly humped-shaped patterns with dramatic enrichments in LREE and Sr. In contrast, the Sloan pyroxenes have smooth patterns of decreasing abundance with increasing incompatibility, as would be expected for simple single-stage residues of melt extraction. In other words, the Sloan samples analyzed here do not show evidence for metasomatic enrichment of their incompatible element abundances whereas both the

Williams and Homestead samples do. Eggler et al. (1988) report trace element data for pyroxenes from Sloan xenoliths that are similar to those shown in Fig. 3, but they also measured samples from deeper in the stratigraphic column that have the strong LREE enrichment typical of our Williams and Homestead samples. Samples of such LREE enriched Sloan peridotites were not available for this study. Without isotopic analysis of these LREE enriched samples, it is not clear whether they represent old metasomatized mantle or instead are xenoliths that experienced substantial interaction with their host kimberlite.

6. Timing of mantle chemical modification

The range of major and trace element compositions observed in these mantle samples, and in particular, the contrasting major magmaphile element depletion yet incompatible trace element enrichment of the Homestead and Williams samples are indicative of multiple stages of chemical modification of the mantle beneath this area. The isotopic data obtained for these samples can be used to document when these events occurred at the various localities and provide insights into the processes responsible and their connection with the tectonic history of this area. As is common in cratonic mantle peridotites, however, the isotopic systematics shows little or no correlation with major element indices of enrichment/depletion. Explanations for this decoupling are: (1) the elements involved in the radiometric isotope systems are not sensitive to the same chemical modification processes as are the major elements, or (2) the chemical modification occurred close enough in time to xenolith capture that the xenoliths did not have time to evolve isotopic compositions consistent with their new chemical characteristics. As will be seen in the following discussion, both of these explanations may apply to the samples studied here.

6.1. Depletion

Removal of partial melt is the most obvious explanation for the magmaphile major element depletion that is characteristic of most of the peridotites analyzed here. This event is perhaps best dated with the Re–Os system, since the Re–Os system has been shown to be relatively insensitive to mantle metasomatic processes, at least compared to the Sr, Nd and Pb isotopic systems (Carlson et al., 1999b; Walker et al., 1989). Indeed, all of the peridotites have $^{187}Os/^{188}Os$ substantially less than that of fertile mantle, indicating that the Re–Os system in these samples does primarily record the depletion event since Re is incompatible, while Os is compatible during peridotite melting. As a result of their limited range in $^{187}Os/^{188}Os$, the Homestead samples have a fairly narrow range in Re-depletion model ages (T_{RD}—Walker et al., 1989) with seven out of eight samples having a T_{RD} between 2.30 and 2.77 Ga (mean = 2.61 Ga). The fact that all but one Homestead sample has a T_{RD} either in the Archean or earliest Proterozoic indicates that the sub-Homestead mantle most likely acquired its melt depleted characteristics in the Archean. The range of T_{RD} ages displayed by the Homestead samples is slightly older than the ages obtained previously for low-T peridotites from Williams, where the range in T_{RD} was 1.77 to 2.55 Ga (Carlson et al., 1999a) when recalculated relative to the bulk-mantle Re–Os parameters reported by Meisel et al. (2001). High-temperature samples from Williams give much younger T_{RD} model ages (0.83 to 0.94 Ga, Carlson et al., 1999a).

The Os isotopic composition of the Homestead samples does not correlate with indices of fertility such as MgO, Al_2O_3 (Fig. 4) or Re/Os, which is a sign

that post-depletion metasomatism has affected these samples. The effect of metasomatism on the Re–Os system is shown clearly in the three Homestead samples that have Re–Os mantle model ages (T_{MA}—Walker et al., 1989) either negative or older than the age of the Earth (Table 2). The Homestead websterite xenolith has Os concentration and isotopic composition quite similar to the peridotites. The slightly higher Re content, and hence Re/Os of this sample is in contrast to its relatively unradiogenic Os isotopic composition, which indicates relatively recent decoupling of Re and Os concentrations from the isotopic systematics recorded by this sample. The same conclusion applies to the Homestead dunite (HK1-21) that has a significantly superchondritic Re/Os, yet subchondritic Os isotopic composition similar to the other Homestead samples. Given the quite high Re/Os ratio of the dunite, the $^{187}Os/^{188}Os$ of this sample will increase by 0.0022 every 100 My. Working backwards in time, this shift in Os isotopic composition corresponds to an increase in Re-depletion model age of approximately 290 My per 100 My of evolution with this high Re/Os. Given that the measured T_{RD} of the Homestead dunite is 2.53 Ga, the high Re/Os of this sample thus could not have existed for more than a few hundred million years, at most, prior to the capture of this sample by the kimberlite. Given the similarity in Os isotopic compositions for the dunite and most of the Homestead peridotites, the Re addition to the dunite may well have occurred very close in time to the kimberlite eruption event, if not directly by interaction with the host kimberlite. With the exception of the three samples with anomalous T_{MA} ages, the remaining five samples have T_{MA} ages ranging from 2.00 to 3.47 Ga. The fact that the Re/Os ratios of these samples may have been modified by recent Re addition, either by metasomatism in the mantle or through interaction with the host kimberlite, makes it impossible to define a precise age for the depletion of this mantle section from the T_{MA} ages. The mean T_{MA} of these five samples is 2.86 Ga (or 3.07 Ga without the apparently younger sample HK1-24), and given the lower bound mean T_{RD} age of 2.61 Ga, a depletion age in the late Archean seems likely.

Sloan xenoliths provide a wide range of Re-depletion model ages (0.74 to 2.45 Ga, mean = 1.45). Four of the Sloan xenoliths show a rough correlation between

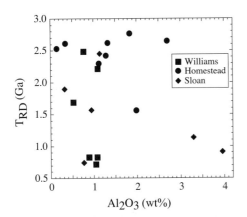

Fig. 4. Re depletion model age versus Al_2O_3 concentration in the analyzed xenoliths.

Os isotopic composition and indicators of melt depletion like MgO and Al_2O_3 (Fig. 4). Using the "alumichron" approach of Reisberg and Lorand (1995), this correlation extrapolates to an Os isotopic composition at zero Al_2O_3 of approximately 0.1154, corresponding to a T_{RD} age of 1.93 Ga. The generally younger Re depletion model ages for the Sloan samples compared to the Homestead peridotites is consistent with the possibility that Sloan is underlain by mantle of similar age to the circa 1.8 Ga crust in this area, but the scatter in Sloan Re–Os model ages does not strongly constrain the depletion age of this mantle section. The single spinel peridotite 13-19-6 is the only Sloan sample with a T_{RD} age approaching the late Archean.

Of some interest in the Sloan results is the observation that several of the samples have clinopyroxenes with low $^{87}Sr/^{86}Sr$ and high $^{143}Nd/^{144}Nd$ and $^{176}Hf/^{177}Hf$ (Tables 3 and 4) as would be expected for samples that have experienced incompatible element depletion. Sample 13-19-20 is particularly notable in having among the highest $^{143}Nd/^{144}Nd$ and $^{176}Hf/^{177}Hf$ yet reported for a mantle clinopyroxene. Surprisingly, 13-19-20 has moderately high CaO and Al_2O_3 contents, near bulk-earth $^{87}Sr/^{86}Sr$ and moderately radiogenic Pb, none of which would be expected given the strongly depleted signal conveyed by the Nd and Hf isotopic composition of this sample. Given the magnitude of deviation of the Nd and Hf isotopic composition of this sample from that of "normal" mantle as sampled by oceanic basalts, the Nd and Hf depleted mantle model ages of 1.65 and 1.39 Ga, respectively, for 13-19-20 clinopyroxene may provide an indication of the timing of the depletion experienced by this sample.

Compared to 13-19-20, 13-19-6 has lower CaO and Al_2O_3 and the very low $^{87}Sr/^{86}Sr$, $^{206}Pb/^{204}Pb$ and $^{187}Os/^{188}Os$ expected for an old residue of partial melt extraction, but its $^{176}Hf/^{177}Hf$ is well below that expected based on its Nd isotopic composition. Since this sample has no garnet, the relatively unradiogenic Hf may be a characteristic of the bulk sample. Nd and Hf depleted mantle model ages for 13-19-6 are 1.37 and 1.78 Ga, respectively, again suggestive that the major melt depletion event affecting the sub-Sloan mantle occurred in the mid-Proterozoic. Additional support for mid-Proterozoic events in the development of the depleted character of the Sloan samples are ages of 1550 ± 360 Ma (MSWD = 17), 1521 ± 2500 Ma

(MSWD = 408—age changes to 1490 ± 19 Ma including only 13-19-20 and 13-19-27) and 1521 ± 17 Ma for best fit lines to the Sloan clinopyroxene data for the Rb–Sr, Sm–Nd and Lu–Hf systems, respectively. In spite of a moderate range in $^{206}Pb/^{204}Pb$ between, for example, the clinopyroxenes from samples 13-19-20 and 13-19-6, these samples have the same $^{207}Pb/^{204}Pb$ within error and thus define a near zero age Pb–Pb isochron. Thus, a number of model ages for the Sloan samples are consistent with formation of this mantle section by melt extraction occurring in the time interval 1.4 to 1.8 Ga.

6.2. The complications of internal isochron systematics

Whole-rock or mineral model ages, in the best case, provide the times of major chemical differentiation events, but their chronological accuracy depends on how well the evolution of the samples match the simple one-stage differentiation assumed in the calculation of model ages. Isochron systematics between minerals in a rock provide both the time of last equilibration of the minerals and the isotopic composition of the sample at that time independent of any assumptions about the chemical evolution of a "model" reservoir. As will be seen, however, mineral systematics are susceptible to disturbance from a variety of causes including intermineral diffusion at the high storage temperatures typical of many xenoliths and the growth of new minerals during metasomatism that do not reach isotopic equilibrium with the whole rock.

The only Sloan sample for which two mineral and a whole rock analysis are available (garnet lherzolite 13-19-27) provides an example of the complex history experienced by some xenoliths. 13-19-27 has nearly the highest Al_2O_3 and CaO, and lowest MgO contents of the Sloan samples and on a major element basis is close in composition to undepleted mantle. The whole rock analysis of this sample shows strong incompatible element enrichment (i.e. subchondritic Sm/Nd and Lu/Hf) and radiogenic Sr, yet the clinopyroxene in this sample has near chondritic Sm/Nd, radiogenic Nd and Hf, and unradiogenic Sr and Pb, all consistent with this being a depleted sample. Given that both clinopyroxene and garnet in this sample have higher Sm/Nd than the whole rock analysis, the whole rock clearly contains an extraneous component enriched in

LREE and with relatively low $^{143}Nd/^{144}Nd$, and the likely candidate is the host kimberlite. Only 1–2% by weight addition of the host kimberlite to 13-19-27 will cause its Nd isotopic composition the change from that measured for the clinopyroxene to that measured for the whole rock. This sample also has a Re/Os ratio approximately a factor of 2 higher than chondritic, in spite of its subchondritic initial (at kimberlite eruption age) Os isotopic composition, again indicative of recent introduction of a high Re/Os material such as the host kimberlite. The degree to which the kimberlite has interacted with the minerals in the xenoliths thus is of some concern.

The two-point Rb–Sr tie line between 13-19-27 garnet and pyroxene corresponds to an age of 1817 ± 21 Ma (Table 5), consistent with a mid-Proterozoic formation of this sample. However, the garnet in this sample has an Sr isotopic composition similar to that of the host kimberlite at the kimberlite eruption age (at 380 Ma, garnet $^{87}Sr/^{86}Sr = 0.70750$, kimberlite $^{87}Sr/^{86}Sr = 0.70727$). This could be coincidence, but given that the Sr concentration of the host kimberlite is almost 700 times that of the garnet, it could also suggest some exchange of Sr between garnet and kimberlite. If the latter, then the age defined by the pyroxene–garnet tie line need have no chronological significance.

In the Sm–Nd system of this sample, both the clinopyroxene and garnet have Sm/Nd ratios and Nd isotopic ratios higher than that of the kimberlite. The tie line connecting the Sm–Nd data for these minerals corresponds to an age of 905 ± 4 Ma. The whole rock datum for this sample lies near the garnet–clinopyroxene tie line, as does the host kimberlite. Consequently the Sm–Nd "isochron" defined by the 13-19-

27 minerals and whole rock may simply represent a mixing line between the peridotite and the host kimberlite, a possibility supported by the fact that the Rb–Sr and Sm–Nd ages do not agree. The sense of the ages returned from these two systems, with Rb–Sr giving the oldest age and Sm–Nd the youngest, is opposite that expected if these systems are recording the closure times of a rock undergoing slow cooling since Rb–Sr has the lower closure temperature. This suggests that the internal isochron data are not providing chronologically significant information, but instead are tracking incomplete chemical exchange between the minerals and infiltrating melts (Gunther and Jagoutz, 1994; Carlson et al., 1999a). The fact that the ages provided by Rb–Sr and Sm–Nd do not agree is a sign that this exchange occurred close enough in time to xenolith capture that equilibrium between the minerals was not achieved.

6.3. The nature of interaction between peridotite minerals and metasomatic agent

In spite of the evidence described above for possible interaction between xenoliths and host kimberlite, the Sloan peridotites still contain many chemical and isotopic features consistent with their being the relatively little metasomatized residues of melt extraction occurring at some time in the mid-Proterozoic. In contrast, the Sr and Nd isotopic systematics of the Homestead and Williams peridotites appear to have been extensively modified by interaction with an LIL-rich fluid/melt. This is shown both in the humped-shaped incompatible element patterns of their garnets and clinopyroxenes (Fig. 3) and in their unradiogenic Nd and moderately radiogenic Sr isotopic compositions. All the Homestead clinopyroxenes have Nd isotopic compositions within ± 3 epsilon units, and $^{87}Sr/^{86}Sr$ no more than 0.0009 lower than that of the host kimberlite. This could be interpreted as indicating that peridotites with the compositional characteristics of the Homestead samples are the source of the Homestead kimberlite. As will be discussed, however, several lines of evidence suggest that the Nd and Sr characteristics of the Homestead peridotites reflect extensive interaction with the host kimberlite or similar genetically related magmas.

The complicated effect of this interaction is illustrated by the scattered internal isochron systematics

Table 5
Garnet–clinopyroxene tie line ages in Ma

Sample	Rb–Sr	Sm–Nd	Lu–Hf	Pb–Pb
HK1-19	947 ± 21	18 ± 13	2531 ± 210	1253 ± 82
HK1-24	297 ± 19	215 ± 24	806 ± 4	1734 ± 630
HK1-25	205 ± 57	131 ± 22	684 ± 3	
HK1-26	493 ± 75	140 ± 11	-559 ± 42	
HK4-1		67 ± 12	261 ± 5	
13-19-27	1817 ± 21	905 ± 4		
H68-16B		-2463 ± 100		
H81-21		-100 ± 28	-627 ± 98	
H69-15F		49 ± 11	-9667 ± 190	

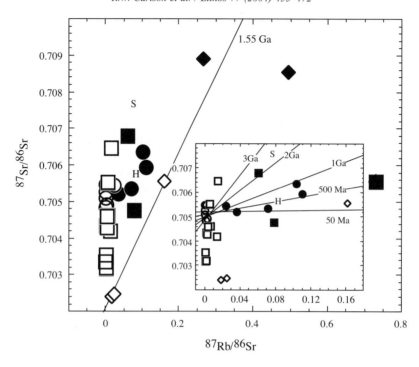

Fig. 5. Rb–Sr isochron diagram for mineral separates from the Williams, Homestead, and Sloan xenoliths. Homestead samples are shown by the circles, open for clinopyroxene, filled for garnets. Williams samples are shown by squares, open for clinopyroxene, filled for garnets. Sloan diopside data are shown by the open diamonds with garnets shown by the filled diamonds. Isotopic compositions for the Sloan (S) and Homestead (H) kimberlites are shown by the corresponding letters. The 1.55 Ga line shown in the large figure is the best-fit line to the five Sloan diopside analyses. Inset shows the same data with expanded scale. Lines provide the slopes expected for arrays of the ages indicated.

shown in Figs. 5–7 that result in the variable garnet–clinopyroxene tie line "ages" listed in Table 5. The Rb–Sr isochron diagram of Fig. 5 shows the Homestead and Williams clinopyroxenes to define a nearly vertical array at very low Rb/Sr. The Homestead clinopyroxenes cluster at $^{87}Sr/^{86}Sr$ similar to the host kimberlite whereas the Williams clinopyroxenes describe a wide range in Sr isotopic compositions ranging from quite low values to values similar to that of the Williams kimberlite. $^{87}Sr/^{86}Sr$ values for garnets scatter away from the clinopyroxene array on slopes ranging from similar to the age of kimberlite eruption to as much as 950 Ma (Table 5). On the Sm–Nd isochron diagram of Fig. 6, the Sm–Nd data for both the Williams and Homestead whole rocks and clinopyroxenes cluster close to the Sm–Nd characteristics of the host kimberlites with garnet data trending off on shallow slopes approaching those expected for ages close to that of the kimberlite eruption. Two point clinopyroxene–garnet Sm–Nd tie lines corre-

spond to ages ranging from 18 to 215 Ma for Homestead and −2463 to 49 Ma for Williams (Carlson et al., 1999a). In these characteristics, the constituent minerals of both the Homestead and Williams peridotites indicate that they underwent extensive chemical and isotopic exchange with metasomatic agents similar in composition to the host kimberlite. That complete equilibrium was not achieved during this interaction is shown by the fact that the garnet–clinopyroxene tie lines for the different radiometric systems in the same sample do not provide ages that agree, and in the case of two of the Williams samples the Sm–Nd tie lines provide negative ages (Carlson et al., 1999a).

Both Williams and Homestead clinopyroxenes show a wide range in Hf isotopic compositions that are not supported by corresponding variation in Lu/Hf (Fig. 7). Initial Hf isotopic compositions of the Homestead clinopyroxenes range from near the value for the host kimberlites to very radiogenic values that would

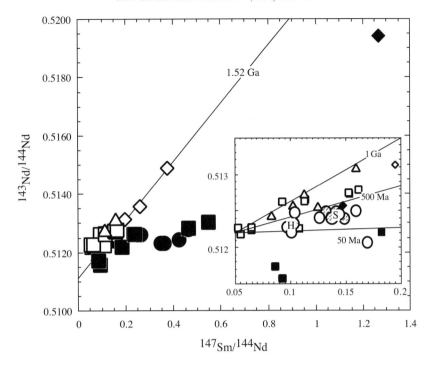

Fig. 6. Sm–Nd isochron diagram for mineral separates and whole rocks from the analyzed xenoliths. Symbols are the same as defined in Fig. 5 with the addition of hatched circles (Homestead) and open triangles (Sloan) to show the location of whole rock data. The 1.52 Ga line shown in the large figure is the best-fit line to the Sloan clinopyroxenes.

be expected for old LIL-depleted residual mantle. Assuming, just for example, that the Homestead mantle was depleted at 2.7 Ga, in order to reach the highest ^{176}Hf/^{177}Hf measured for a Homestead clinopyroxene (ε_{Hf}=+45) would require an ^{176}Lu/^{177}Hf=0.058, which is within the range of the Lu/Hf ratios measured for the depleted Sloan whole rocks. This suggests that the Lu–Hf system in some Homestead peridotites indeed carries memory of the depletion event recorded by the major element composition and Re–Os isotopic system of these samples.

In contrast, Homestead samples HK1-19 and HK1-26, both of which have moderately low CaO and Al_2O_3 contents, appear to have completely exchanged their Sr, Nd, Hf and Pb with the host kimberlite. Given the isotopic overlap of the clinopyroxenes in these two samples with the host kimberlite, the clinopyroxenes may have crystallized from kimberlite melt introduced into the xenolith shortly before eruption. This is supported by evidence for textural disequilibrium within many of the

Homestead peridotites (Irving et al., 2003), including the heterogeneous distribution and irregular shapes of large clinopyroxene grains. The exchange is not restricted to the clinopyroxenes, however, as garnets in both HK1-19 and HK1-26 have low Lu/Hf, in the case of HK-19, lower than in coexisting clinopyroxene, a further indicator of extensive interaction with an LIL-rich melt or fluid. As a result, the garnet–clinopyroxene Lu–Hf tie lines for these two samples provide the extreme ages among the Homestead data set, and in the case of HK1-26 the age is negative, a sure sign that intermineral isotopic equilibria was not achieved during this chemical modification event.

In contrast to these two samples, clinopyroxenes from Homestead samples HK1-24, HK1-25 and HK4-1 have quite radiogenic Hf isotopic compositions, overlapping the values measured for the Sloan samples. Radiogenic Hf is a signature of incompatible element depletion, and is quite distinct from the unradiogenic Hf of the host kimberlite. Garnet–clinopyroxene tie lines for the Lu–Hf data

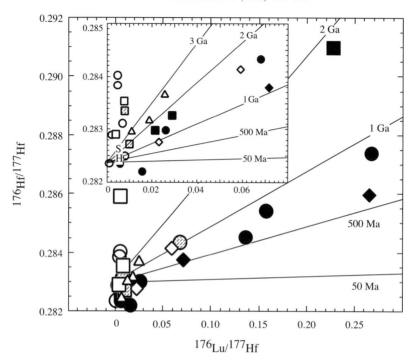

Fig. 7. Lu–Hf data for analyzed xenoliths. Symbols as defined in Figs. 5 and 6 with the addition of hatched squares to show the location of whole rock data for xenoliths from the Williams kimberlite.

for these samples correspond to ages of 806 ± 4, 684 ± 3, and 261 ± 5 Ma, respectively, all of which substantially predate the presumed circa 50 Ma eruption age of the host kimberlite. The $^{207}Pb/$ $^{204}Pb-^{206}Pb/^{204}Pb$ tie line for garnet and clinopyroxene from HK1-24 corresponds to an age of 1734 ± 630 Ma. Whether or not any of these ages have chronological significance is unclear, particularly in light of the fact that these same three samples have significantly younger Sm–Nd garnet–clinopyroxene tie line ages of 215 ± 24, 131 ± 22, 67 ± 12 Ma, respectively, and all have negative initial ε_{Nd} at these ages, inconsistent with the depletion implied by their positive ε_{Hf}. Since the garnet–clinopyroxene tie line Lu–Hf and Sm–Nd "ages" are based on mineral systematics, and the sequence of the ages for both systems is the same, the ages might be interpreted as reflecting various closure times for the Lu–Hf and Sm–Nd systems as these samples cooled while residing in the lithosphere (Bedini et al., 2002). The sequence of ages, from oldest for sample HK1-24 (equilibra-

tion temperature $= 1211$ °C) to youngest for HK4-1 (equilibration temperature $= 1127$ °C) is anticorrelated with the temperatures determined by mineral thermometry for these samples, casting doubt on the cooling age interpretation of the Lu–Hf mineral ages.

Our interpretation is that the Homestead samples provide a snapshot of disequilibrium conditions caused by the arrested interaction of previously depleted peridotites with LIL-rich magmas that had isotopic compositions similar to the Homestead kimberlite. The magnitude of disequilibrium observed in the isotope systematics suggest that this interaction occurred close enough in time to kimberlite eruption that the xenoliths did not have time to reach chemical and/or isotopic equilibrium with these magmas. This interpretation may explain the scatter in geothermobarometry results for the Homestead samples (Irving et al., 2003, Hearn, 2004) and the fact that they are offset to much higher temperatures at a given depth compared to most cratonic geotherms.

7. Conclusions

The results presented here for mantle lithosphere samples from the Wyoming Craton and surrounding areas reflect the complicated multistage chemical history experienced by these samples. The Sloan samples appear to have experienced melt extraction in the mid-Proterozoic. It is tempting to associate this event with the circa 1.8 Ga formation age of the overlying crustal section, but the chronological constraints, at best, can only be said to be consistent with this possibility. Several of the Sloan samples appear to have escaped significant post-depletion re-enrichment of incompatible trace elements, suggesting a lack of regional metasomatism of the mantle in this area between its Proterozoic formation and the Devonian eruption of the Sloan kimberlite. This lack of metasomatic activity may explain the relatively cold geotherm recorded by the Sloan peridotites (Eggler et al., 1988).

The Homestead peridotites show a small range of Re depletion model ages consistent with the main melt depletion event in the mantle of this region occurring in the late Archean. Further to the north, at Williams, the signature of this Archean event is less persistent and limited in depth extent (Carlson et al., 1999a) suggesting that the mantle lithosphere near the northern boundary of the Wyoming Craton has been more severely affected by post-Archean events occurring on the borders of the Craton. Both the Williams and Homestead samples show clear evidence of extensive interaction with incompatible-element-rich melts. The timing of this interaction is not well resolved, but likely occurred within a few hundred million years of the eruption of the Homestead and Williams kimberlites. Given the unradiogenic Nd isotopic compositions of the required metasomatic agent, the melt metasomatism may reflect internal melting and melt migration of an old metasomatized mantle, of which we have yet to obtain any sample from the western US. Alternatively, the evolved isotopic compositions of the Homestead and Williams peridotites, and the Montana Eocene alkalic magmatic rocks in general (Dudas et al., 1987; Fraser et al., 1986; O'Brien et al., 1995), may reflect the introduction of isotopically evolved fluids/melts from the subducting Farallon plate into the depleted mantle lithosphere of the Wyoming Craton during the Laramide Orogeny.

Acknowledgements

This work was supported by NSF Grant EAR-0106475. Development of the Lu–Hf isotopic method at DTM was greatly aided by the assistance of Mark Schmitz and Mary Horan. The smooth operation of the various mass spectrometers employed in this work was made possible through the efforts of Tim Mock. Their role in this work is much appreciated. We thank Stan Mertzman and Scott Kuehner for their skilled production of the XRF major element and electron-microprobe mineral chemical data discussed here. Comments from two anonymous reviewers and the editor helped clarify some of the complicated arguments presented herein.

Appendix A. Petrography of analyzed xenoliths from the Homestead, Sloan and Williams kimberlite pipes

A.1. Homestead kimberlite

This xenolith suite contains a variety of lithologies, including garnet lherzolites, garnet harzburgites, spinel harzburgites, dunites, websterite and clinopyroxenite. All the olivine-bearing lithologies exhibit minor secondary alteration (probably in part related to infiltration by the host kimberlite) in the form of veinlets of serpentine, chlorite, sulfide and other minerals. Garnet grains range widely in size (even in a single xenolith) and exhibit variable replacement by kelyphite, and in one sample total replacement by intergrown chlorite + phlogopite + chromite. In several of the garnet lherzolites, the garnet grains are mantled completely by green, Cr-rich clinopyroxene. The websterite occurs as a 6-mm-wide vein cutting a spinel harzburgite xenolith, and one sample (HK1-18, not analyzed here) contains a narrow (1-mm-wide) veinlet of clinopyroxenite. Several peridotite xenoliths contain sparse patches of phlogopite and apatite, which probably represent additions from the host kimberlite. Large, angular grains of Cr-rich clinopyroxene in some samples of garnet lherzolite, harzburgite and dunite appear to be texturally unequilibrated.

A.1.1. Garnet lherzolites HK-1-24, HK-1-25, HK-1-26 and HK-4-1

The garnet lherzolites have protogranular (rather than sheared) fabrics, but exhibit varying degrees of textural equilibration. Some appear well-equilibrated, with evenly distributed grains of orthopyroxene, garnet and clinopyroxene of uniform size, in addition to the predominant olivine. However, in several samples (e.g., HK1-24) the garnets have a very uneven distribution and quite variable grain size. In such samples, the clinopyroxene grains are large and angular with re-entrant grain boundaries, and may represent porphyroclasts that have not recrystallized completely following a mantle deformation event. Primary chromite is absent from the garnet lherzolites, except for rare grains in HK1-26.

A.1.2. Garnet harzburgite HK-1-19

This sample has a sheared fabric, and contains large, ellipsoidal orthopyroxene porphyroclasts, sparse garnets, few clinopyroxene grains and primary chromite grains.

A.1.3. Garnet harzburgite H00-10-22

This sample is composed of large olivines (1–10 mm, 89 modal%), orthopyroxenes (1–3 mm, 10 modal%), garnets (1–3 mm, 0.5 modal%), and primary spinels (0.1–0.3 mm, 0.1 modal%). Clinopyroxene and phlogopite are absent. Coarse tabular texture is shown by parallel elongate orthopyroxenes and chains of orthopyroxenes. Garnets (of G10 composition) contain sparse, small (0.03–0.1 mm) grains and blades of brown to black spinel, show inner rims of fine-grained kelyphite, and partial outer rims of orthopyroxene grains (0.2–0.5 mm) containing irregular brown spinel inclusions. Primary brown to black spinels occur in olivine and orthopyroxene. Olivines and orthopyroxenes locally show sparse single or widely spaced kink bands, but many appear to be strain-free.

A.1.4. Spinel harzburgite HK-1-13H

This sample is a protogranular aggregate of olivine, orthopyroxene and sparse chromite, with no clinopyroxene.

A.1.5. Dunite HK-1-21

This sample is composed of a mosaic of relatively large recrystallized Mg-rich olivine ($Fo_{92.5}$) grains with sparse chromite and minor interstitial phlogopitic mica.

A.1.6. Websterite HK-1-13W

Occurring as a narrow vein cutting harzburgite, this lithology is composed of well-equilibrated orthopyroxene with subordinate clinopyroxene, and may represent an igneous precipitate from a mantle-derived magma.

A.2. Sloan kimberlite

A.2.1. Garnet websterite 13-19-2

Coarse, polygonal aggregate of clinopyroxene, orthopyroxene, garnet and chromite. Exsolved blades of chromite are abundant in all the silicate minerals. Minor kelyphite occurs around garnet and along grain boundaries.

A.2.2. Spinel harzburgite 13-19-6

Protogranular aggregate of partly serpentinized olivine, orthopyroxene, minor irregularly shaped clinopyroxene and Cr–Al–spinel.

A.2.3. Spinel–garnet lherzolite 13-19-20

Protogranular aggregate of olivine (slightly serpentinized), orthopyroxene (with minor clinopyroxene exsolution lamellae), minor clinopyroxene and chromite. Sparse garnet occurs as rims along some olivine–orthopyroxene grain contacts, and may be a reaction product from isobaric cooling.

A.2.4. Garnet lherzolites 13-19-15, 13-19-22 and 13-19-27

These samples have protogranular textures, and consist mainly of olivine and orthopyroxene with isolated garnet and clinopyroxene grains. Garnet and clinopyroxene are spatially associated in 13-19-15. Garnets have narrow kelyphite rims, and olivine is variably veined by serpentine (especially in 13-19-27, where orthopyroxene also has been partly altered to serpentine). Exsolved chromite blades are present in clinopyroxene grains in 13-19-27.

A.2.5. Garnet harzburgite 13-19-11

Large grains of garnet (partly altered to kelyphite with minor phlogopite) and orthopyroxene occur with serpentinized olivine (some of which has been recrystallized to finer, polygonal grain aggregates). Clinopyroxene is absent.

A.2.6. Garnet harzburgite 13-16-3

Small, isolated grains of garnet (with kelyphite rims) and orthopyroxene occur with moderately serpentinized olivine. Clinopyroxene is absent.

A.3. Williams kimberlite

The xenoliths from the Williams pipes are generally much fresher than those from the Homestead and Sloan pipes, and exhibit only minor serpentinization of olivine. The samples analyzed here (and previously by Carlson et al., 1999a) are all garnet lherzolites with dominantly protogranular textures. H68-16B consists of coarse olivine with isolated garnet, clinopyroxene, orthopyroxene and minor chromite; some phlogopite is present along grain boundaries. H68-15F contains smaller, polygonal, recrystallized olivine grains, with larger orthopyroxenes, isolated garnets (with kelyphite rims) and sparse, irregular clinopyroxenes. H81-21 and WP28-3-1 contain isolated grains of garnet (with kelyphite rims) and clinopyroxene (some irregular and elongated in WP28-3-1) with orthopyroxene and olivine (partly recrystallized to polygonal grain aggregates in H81-21).

References

Bedini, R.M., Blichert-Toft, J., Boyet, M., Albarede, F., 2002. Lu–Hf isotope geochemistry of garnet–peridotite xenoliths from the Kaapvaal craton and the thermal regime of the lithosphere. Geochim. Cosmochim. Acta 66, A61.

Blichert-Toft, J., Chauvel, C., Albarede, F., 1997. Separation of Hf and Lu for high-precision isotope analysis of rock samples by magnetic sector-multiple collector ICP-MS. Contrib. Mineral. Petrol. 127, 248–260.

Boyd, F.R., McCallister, R.H., 1976. Densities of fertile and sterile garnet peridotites. Geophys. Res. Lett. 3, 509–512.

Boyd, F.R., Mertzman, S.A., 1987. Composition and structure of the Kaapvaal lithosphere, Southern Africa. In: Mysen, B.O. (Ed.), Magmatic Processes: Physicochemical Principles. The Geochemical Society, University Park, pp. 3–12.

Boyd, F.R., Pokhilenko, N.P., Pearson, D.G., Mertzman, S.A., Sobolev, N.V., Finger, L.W., 1997. Composition of the Siberian cratonic mantle: evidence from Udachnaya peridotite xenoliths. Contrib. Mineral. Petrol. 128, 228–246.

Brey, G.P., Koehler, T., Nickel, K.G., 1990. Geothermobarometry in four-phase lherzolites: I. Experimental results from 10 to 60 kb. J. Petrol. 31, 1313–1352.

Carlson, R.W., Irving, A.J., 1994. Depletion and enrichment history of subcontinental lithospheric mantle: an Os, Sr, Nd and Pb isotopic study of ultramafic xenoliths from the northwestern Wyoming Craton. Earth Planet. Sci. Lett. 126, 457–472.

Carlson, R.W., Irving, A.J., Hearn Jr., B.C., 1999a. Chemical and isotopic systematics of peridotite xenoliths from the Williams kimberlite, Montana: clues to processes of lithosphere formation, modification and destruction. In: Gurney, J.J., Gurney, J.L., Pascoe, M.D., Richardson, S.H. (Eds.), Proc. 7th Int. Kimberlite Conf. Red Roof Design, Cape Town, pp. 90–98.

Carlson, R.W., Pearson, D.G., Boyd, F.R., Shirey, S.B., Irvine, G., Menzies, A.H., Gurney, J.J., 1999b. Re–Os systematics of lithospheric peridotites: implications for lithosphere formation and preservation. In: Gurney, J.J., Gurney, J.L., Pascoe, M.D., Richardson, S.H. (Eds.), Proc. 7th Int. Kimberlite Conf. Red Roof Design, Cape Town, pp. 99–108.

Decker, E.R., Baker, K.R., Bucher, G.J., Heasler, H.P., 1980. Preliminary heat flow and radioactivity studies in Wyoming. J. Geophys. Res. 85, 311–321.

Decker, E.R., Bucher, G.J., Buelow, K.L., Heasler, H.P., 1984. Preliminary interpretation of heat-flow and radioactivity in the Rio Grande rift zone in central and northern Colorado. In: Baldridge, W.S., Dickerson, P.W., Riecker, R.E., Zidek, J. (Eds.), Rio Grande Rift, northern New Mexico. Guidebook—New Mexico Geological Society, vol. 35, pp. 45–50.

Dudas, F.O., Carlson, R.W., Eggler, D.H., 1987. Regional middle Proterozoic enrichment of the subcontinental mantle source of igneous rocks from central Montana. Geology 15, 22–25.

Dutch, S.I., Neilsen, P.A., 1990. The Archean Wyoming Province and its relations with adjacent Proterozoic provinces. In: Lewry, J.F., Stauffer, M.R. (Eds.), The Early Proterozoic Trans-Hudson Orogen of North America. Spec. Pap. - Geol. Soc. Can., vol. 37, pp. 287–300.

Eggler, D.H., McCallum, M.E., Kirkley, M.B., 1987. Kimberlite-transported nodules from Colorado–Wyoming: a record of enrichment of shallow portions of an infertile lithosphere. In: Morris, E.M., Pasteris, J.D. (Eds.), Mantle Metasomatism and Alkaline Magmatism. Spec. Pap. - Geol. Soc. Am., pp. 77–90.

Eggler, D.H., Meen, J.K., Welt, F., Dudas, F.O., Furlong, K.P., McCallum, M.E., Carlson, R.W., 1988. Tectonomagmatism of the Wyoming Province. In: Drexler, J., Larson, E.E. (Eds.), Colorado Volcanism. Colorado School of Mines Quarterly, Boulder, pp. 25–40.

Ellsworth, P.C., 2000. Homestead kimberlite: new discovery in central Montana. Guidebook, 25th Annual Field Conference, Tobacco Root Geological Society, pp. 14–20.

Fraser, K.J., Hawkesworth, C.J., Erlank, A.J., Mitchell, R.H., Scott-Smith, B.H., 1986. Sr, Nd, and Pb isotope and minor element geochemistry of lamproites and kimberlites. Earth Planet. Sci. Lett. 76, 57–70.

Gorman, A.R., Clowes, R.M., Ellis, R.M., Henstock, T.J., et al., 2002. Deep probe: imaging the roots of western North America. Can. J. Earth Sci. 39, 375–398.

Gunther, M., Jagoutz, E., 1994. Isotopic disequilibria (Sm/Nd, Rb/Sr) between minerals of coarse grained, low temperature garnet peridotites from Kimberly floors, southern Africa. In: Meyer, H.O.A., Leonardos, O.H. (Eds.), Fifth Int. Kimberlite Conf. CPRM, Araxa, pp. 354–365.

Gurney, J.J., 1984. A correlation between garnets and diamonds. In: Glover, J.E., Harris, P.G. (Eds.), Kimberlite Occurrence and Origins: A Basis for Conceptual Models in Exploration. Univ. Western Australia, Perth, pp. 143–166.

Hearn Jr., B.C., 1989. Alkalic ultramafic magmas in north-central Montana, USA: genetic connections of alnoite, kimberlite, and carbonatite. In: Jaques, A.L., Ferguson, J., Green, D.H. (Eds.), Kimberlites and Related Rocks—Volume 1. Spec. Publ. - Geol. Soc. Aust., vol. 14. Blackwell, Oxford, pp. 109–119.

Hearn, B.C., 2004. Upper-mantle xenoliths in the Homestead kimberlite, central Montana, USA: depleted and re-enriched Wyoming Craton samples. Proc. 8th Int. Kimb. Conf., Lithos.

Hearn Jr., B.C., Boyd, F.R., 1975. Garnet peridotite xenoliths in a Montana, U.S.A., kimberlite. Phys. Chem. Earth 9, 247–256.

Hearn Jr., B.C., McGee, E.S., 1984. Garnet peridotites from Williams kimberlites, north-central Montana, USA. In: Kornprobst, J. (Ed.), Kimberlites II: The Mantle and Crust–Mantle Relationships. Elsevier, Amsterdam, pp. 57–70.

Irving, A.J., Kuehner, S., Carlson, R.W., 1997. 1.70 Ga Sm–Nd age for a garnet granulite xenolith from a minette sill, Sweetgrass Hills, Northern Montana supports Proterozoic collision of Hearne and Wyoming cratons. EOS, Trans.-Am. Geophys. Union 78, 786.

Irving, A.J., Kuehner, S., Ellsworth, P., 2003. Petrology and thermobarometry of mantle xenoliths from the Eocene Homestead kimberlite pipe, central Montana, USA. Extended Abstracts, 8th Int. Kimb. Conf.

Johnson, K.T.M., Dick, H.J.B., Shimizu, N., 1990. Melting in the oceanic upper mantle: an ion microprobe study of diopsides in abyssal peridotites. J. Geophys. Res. 95, 2661–2678.

Jordan, T.H., 1978. Composition and development of the continental tectosphere. Nature 274, 544–548.

Karlstrom, K.E., Bowring, S.A., Chamberlain, K.R., Dueker, K.G., et al., 2002. Structure and evolution of the lithosphere beneath the Rocky Mountains: initial results from the CD-ROM experiment. GSA Today 12, 4–10.

Kelemen, P.B., Hart, S.R., Bernstein, S., 1998. Silica enrichment in the continental upper mantle via melt/rock reaction. Earth Planet. Sci. Lett. 164, 387–406.

Ludwig, K.R., 1991. ISOPLOT: a plotting and regression program for radiogenic-isotope data. Open-File Rep. - U.S. Geol. Surv. 91-445 (39 pp.).

Lugmair, G.W., Marti, K., 1978. Lunar initial 143Nd/144Nd: differential evolution of the lunar crust and mantle. Earth Planet. Sci. Lett. 39, 349–357.

McDonough, W.F., Sun, S.-S., 1995. The composition of the Earth. Chem. Geol. 120, 223–253.

Meisel, T., Walker, R.J., Irving, A.J., Lorand, J.P., 2001. Osmium isotopic compositions of mantle xenoliths: a global perspective. Geochim. Cosmochim. Acta 65, 1311–1323.

Muenker, C., Weyer, S., Scherer, E., Mezger, K., 2001. Separation of high field strength elements (Nb, Ta, Zr, Hf) and Lu from rock samples for MC-ICPMS measurements. Geochem. Geophys. Geosystems. doi:2001GC000183.

Nier, A.O., 1950. A redetermination of the relative abundances of the isotopes of carbon, nitrogen, oxygen, argon, and potassium. Phys. Rev. 77, 789–793.

O'Brien, H.E., Irving, A.J., McCallum, I.S., Thirwall, M.F., 1995. Strontium, neodymium and lead isotopic evidence for interaction of post-subduction asthenospheric potassic mafic magmas of the Highwood Mountains, Montana, USA, with ancient Wyoming craton lithospheric mantle. Geochim. Cosmochim. Acta 59, 4539–4562.

O'Neill, J.M., Lopez, D.A., 1985. Character and regional significance of Great Falls tectonic zone, east-central Idaho and west-central Montana. Am. Assoc. Pet. Geol. Bull. 69, 437–447.

Reisberg, L.C., Lorand, J.-P., 1995. Longevity of sub-continental mantle lithosphere from osmium isotope systematics in orogenic peridotite massifs. Nature 376, 159–162.

Rudnick, R.L., Ireland, T.R., Gehrels, G., Irving, A.J., Chesley, J.T., Hanchar, J.M., 1999. Dating mantle metasomatism: U–Pb geochronology of zircons in cratonic mantle xenoliths from Montana and Tanzania. In: Gurney, J.J., Gurney, J.L., Pascoe, M.D., Richardson, S.H. (Eds.), Proc. 7th Int. Kimberlite Conf. Red Roof Design, Cape Town, pp. 728–735.

Scherer, E., Muenker, C., Mezger, K., 2001. Calibration of the lutetium–hafnium clock. Science 293, 683–687.

Simon, N.S.C., Carlson, R.W., Davies, G.R., Nowell, G.M., Pearson, D.G., 2003. Os–Sr–Nd–Hf isotope evidence for the ancient depletion and subsequent multi-stage enrichment history of the Kaapvaal cratonic lithosphere. Extended Abstracts, 8th Int. Kimb. Conf.. FLA-0117.

Smith, C.B., McCallum, M.E., Coopersmith, H.G., Eggler, D.H., 1979. Petrochemistry and structure of kimberlites in the Front Range and Laramie Range, Colorado–Wyoming. In: Boyd, F.R., Meyer, H.O.A. (Eds.), Kimberlites, Diatremes, and Diamonds; Their Geology, Petrology, and Geochemistry. Amer. Geophys. Union, Washington, pp. 178–189.

Smoliar, M.I., Walker, R.J., Morgan, J.W., 1996. Re–Os ages of group IIA, IIIA, IVA, and IVB iron meteorites. Science 271, 1099–1102.

Steiger, R.H., Jager, E., 1977. Subcommission on geochronology: convention on the use of decay constants in geo- and cosmochronology. Earth Planet. Sci. Lett. 36, 359–362.

Todt, W., Cliff, R.A., Hanser, A., Hofmann, A.W., 1996. Evaluation of a 202Pb–205Pb double spike for high precision lead isotope analysis. In: Hart, S.R., Basu, A. (Eds.), Earth Processes: Reading the Isotope Code. Amer. Geophys. Union, Washington, pp. 429–437.

Vervoort, J.D., Blichert-Toft, J., 1999. Evolution of the depleted mantle: Hf isotope evidence from juvenile rocks through time. Geochim. Cosmochim. Acta 63, 533–556.

Walker, R.J., Carlson, R.W., Shirey, S.B., Boyd, F.R., 1989. Os, Sr, Nd, and Pb isotope systematics of southern African peridotite xenoliths: implications for the chemical evolution of subcontinental mantle. Geochim. Cosmochim. Acta 53, 1583–1595.

Available online at www.sciencedirect.com

Lithos 77 (2004) 473–491

LITHOS

www.elsevier.com/locate/lithos

The Homestead kimberlite, central Montana, USA: mineralogy, xenocrysts, and upper-mantle xenoliths ☆

B. Carter Hearn Jr. *

U.S. Geological Survey, 954 National Center, Reston, VA 20192, USA

Received 27 June 2003; accepted 25 February 2004
Available online 28 May 2004

Abstract

The Homestead kimberlite was emplaced in lower Cretaceous marine shale and siltstone in the Grassrange area of central Montana. The Grassrange area includes aillikite, alnoite, carbonatite, kimberlite, and monchiquite and is situated within the Archean Wyoming craton. The kimberlite contains 25–30 modal% olivine as xenocrysts and phenocrysts in a matrix of phlogopite, monticellite, diopside, serpentine, chlorite, hydrous Ca–Al–Na silicates, perovskite, and spinel. The rock is kimberlite based on mineralogy, the presence of atoll-textured groundmass spinels, and kimberlitic core-rim zoning of groundmass spinels and groundmass phlogopites.

Garnet xenocrysts are mainly Cr-pyropes, of which 2–12% are G10 compositions, crustal almandines are rare and eclogitic garnets are absent. Spinel xenocrysts have MgO and Cr_2O_3 contents ranging into the diamond inclusion field. Mg-ilmenite xenocrysts contain 7–11 wt.% MgO and 0.8–1.9 wt.% Cr_2O_3, with (Fe^{+3}/Fe^{tot}) from 0.17–0.31. Olivine is the only obvious megacryst mineral present. One microdiamond was recovered from caustic fusion of a 45-kg sample.

Upper-mantle xenoliths up to 70 cm size are abundant and are some of the largest known garnet peridotite xenoliths in North America. The xenolith suite is dominated by dunites, and harzburgites containing garnet and/or spinel. Granulites are rare and eclogites are absent. Among 153 xenoliths, 7% are lherzolites, 61% are harzburgites, 31% are dunites, and 1% are orthopyroxenites. Three of 30 peridotite xenoliths that were analysed are low-Ca garnet–spinel harzburgites containing G10 garnets. Xenolith textures are mainly coarse granular, and only 5% are porphyroclastic.

Xenolith modal mineralogy and mineral compositions indicate ancient major-element depletion as observed in other Wyoming craton xenolith assemblages, followed by younger enrichment events evidenced by tectonized or undeformed veins of orthopyroxenite, clinopyroxenite, websterite, and the presence of phlogopite-bearing veins and disseminated phlogopite. Phlogopite-bearing veins may represent kimberlite-related addition and/or earlier K-metasomatism.

Xenolith thermobarometry using published two-pyroxene and Al-in-opx methods suggest that garnet–spinel peridotites are derived from 1180 to 1390 °C and 3.6 to 4.7 GPa, close to the diamond–graphite boundary and above a 38 mW/m² shield geotherm. Low-Ca garnet–spinel harzburgites with G10 garnets fall in about the same T and P range. Most spinel peridotites with assumed 2.0 GPa pressure are in the same T range, possibly indicating heating of the shallow mantle. Four of 79 Cr

☆ Supplementary data associated with this article can be found, in the online version, at doi: 10.1016/j.lithos.2004.04.030.

* Tel.: +1-703-648-6768; fax: +1-703-648-6383.
E-mail address: chearn@usgs.gov (B. Carter Hearn).

0024-4937/$ - see front matter. Published by Elsevier B.V.
doi:10.1016/j.lithos.2004.04.030

diopside xenocrysts have $P–T$ estimates in the diamond stability field using published single-pyroxene $P–T$ calculation methods.

Published by Elsevier B.V.

Keywords: Kimberlite; Xenolith; Xenocryst; Peridotite; Upper mantle; Thermobarometry; Montana

1. Introduction and regional setting

The north-central and central Montana igneous province contains unusual igneous rocks that have drawn the attention of petrologists for more than 100 years. Within this region, the larger areas of igneous activity are the Highwood, Bearpaw, Little Rocky, Judith, Moccasin, Little Belt, Castle, and Crazy Mountains, Adel Mountain Volcanics, and Sweet Grass Hills, that are cored by nested intrusions and some eruptive deposits (Bearpaw, Highwood, Adel, Hearn et al., 1989; Irving and Hearn, 2003, and references therein) (Fig. 1A). Ages for these mountain uplifts are Late Cretaceous (Adel Mountain Volcanics, Harlan et al., in

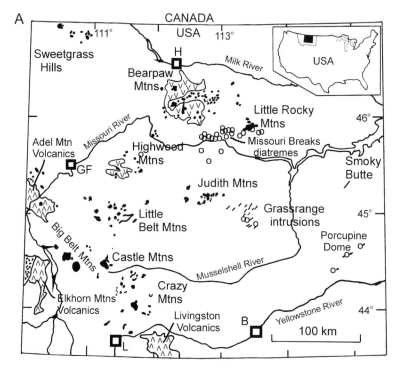

Fig. 1. (A) Mesozoic and Cenozoic igneous occurrences in central Montana. Inset shows map area in Montana outline in USA. Alkalic rocks: solid black, intrusive; V pattern, extrusive. Melnoitic, lamproitic, and kimberlitic rocks: black lines, intrusive; circles, diatremes. Calc-alkalic rocks, Late Cretaceous: random dash pattern, intrusive (Boulder Batholith); inverted V pattern, extrusive. B, Billings; GF, Great Falls; H, Havre; L, Livingston. (B) Igneous occurrences in the Grassrange area, Montana. Locations from Johnson and Smith (1964), Porter and Wilde (1999a,b), Doden and Gold (1993), Nelson (1993), P.C. Ellsworth (unpublished mapping), and B.C. Hearn Jr. (unpublished mapping). Size of some bodies is slightly exaggerated for clarity. Abbreviations: BB1,2 = Button Butte 1,2; BBNE = Button Butte northeast; BC = Buffalo Creek pipes 1,2,3; BCW = Buffalo Creek west pipe; BT = Blacktail Creek dike; C = Crescent breccia; CC1,2,3 = Chippewa Creek dikes 1,2,3; CL = County Line west and east pipes; EW = Eklund west diatreme; EC1 = Elk Creek breccia; EC2 = Elk Creek dike; HC = Hal's Claim breccia; LR = Lund Ranch dike; RCB = Rooster Comb Butte dike; SB = Schultz breccia; SBU = Schultz Butte breccia; SD1,2 = Schultz dikes 1,2; TNE = Teigen northeast dike; TNW = Teigen northwest dikes and sills; TFD = Tennessee Flat dike; TR1,2,3,4 = Timber Ridge dikes 1,2,3,4; TV = TV Hill breccias; WHL = War Horse Lake 1–5 diatremes; YWR = Yellow Water Reservoir dike. (C) Geologic map of the Homestead kimberlite.

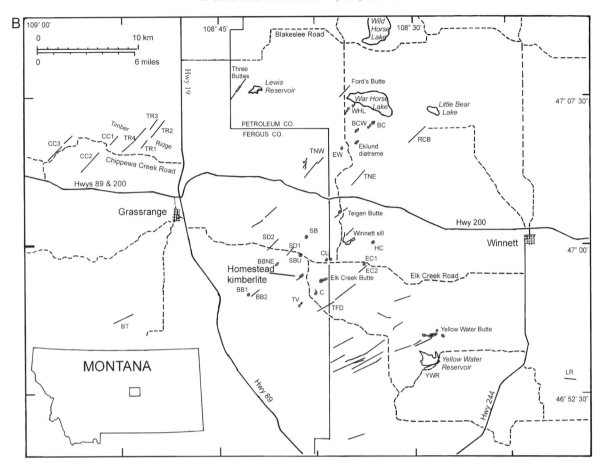

Fig. 1 (*continued*).

press), Late Cretaceous to Paleocene (Little Rocky, Judith, and Moccasin Mountains, Marvin et al., 1980), and middle Eocene (Highwood and Bearpaw Mountains, Sweet Grass Hills, Marvin et al., 1980; Little Belt Mountains, Marvin et al., 1973; Harlan, 1996; Crazy Mountains, Dudas, 1991; du Bray and Harlan, 1996; Castle Mountains, Chadwick, 1980; Irving and Hearn, 2003). Some of the most unusual rocks occur in four separate clusters (Fig. 1A) located in the plains and badlands between the mountain uplifts, such as the Missouri Breaks area (diatremes, dikes, and plugs, Hearn, 1968, 1979; Hearn et al., 1989), Grassrange–Winnett area (diatremes and dikes, Johnson and Smith, 1964; Doden and Gold, 1993), Porcupine Dome area (diatremes and dikes, Doden, 1996; Hearn, 1999), and Smoky Butte (dike and diatremes, Mitchell et al., 1987; Hearn et al., 1989; Irving and Hearn, 2003). The

youngest is the Smoky Butte lamproite cluster, emplaced at 27 Ma (Marvin et al., 1980; O'Brien et al., 1995). Igneous rocks in the other three clusters range in age from 46 to 51 Ma, and rock types vary from alkalic ultramafic (alnoite and monticellite–peridotite; both generally referred to as alnoitic, polzenitic, or melnoitic), to carbonate-rich, olivine–phlogopite-bearing varieties (aillikites), to carbonatites (Doden and Gold, 1993; Doden, 1996; Hearn, 1999; Irving and Hearn, 2003 and references therein). Many of these rock types are difficult to classify accurately in the field because they are altered, or if unaltered, have the similar macroscopic appearance of abundant olivine with less abundant phlogopite in a black, dark gray, or greenish gray groundmass. Even in apparently fresh rocks, microscopic examination confirms that some or many groundmass phases have been altered. Thus

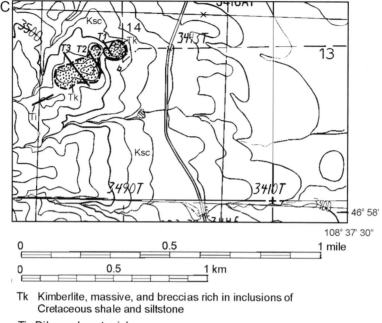

Fig. 1 (*continued*).

Tk Kimberlite, massive, and breccias rich in inclusions of
      Cretaceous shale and siltstone

Ti Dike, carbonate-rich

Ksc Skull Creek Member of Thermopolis Formation
      (Lower Cretaceous)

T1, T2, T3 Trenches T1, T2, T3

characterization usually requires optical petrography, electron microprobe analyses, and whole-rock major element analyses.

Basement rocks underlying the central Montana igneous province are part of the Archean Wyoming craton, of which the nearest exposures are in the Little Rocky and Little Belt Mountains uplifts (Fig. 1A). Parts of the Wyoming craton were affected by igneous activity and metamorphism at 1.7–1.8 Ga (Mueller et al., 2002; Holm and Schneider, 2002). The western part of the central Montana igneous province is underlain by the northeast-trending Great Falls Tectonic Zone. The term Great Falls Tectonic Zone was originally based on aligned Cretaceous to Eocene plutons in southwest Montana and extrapolation through the Eocene Highwood and Bearpaw Mountains igneous foci (O'Neill and Lopez, 1985). The present definition identifies a zone of basement geophysical anomalies interpreted as a Proterozoic collisional suture between the Archean Medicine Hat block to the northwest and the Wyoming craton to the southeast (Gorman et al., 2002), also interpreted as a zone of Proterozoic metamorphism separating the northern extension of the Wyoming craton from the main part to the southeast (Buhlmann et al., 2002). Mueller et al. (2002) and Holm and Schneider (2002) extend the Great Falls Tectonic Zone as a 150 km wide feature across Montana and into south-central Saskatchewan, although the actual boundaries of the zone are still unknown. Studies of xenoliths from other localities within this zone may provide a better definition of the boundaries. The Sweet Grass Hills are the only Montana igneous cluster located directly above the Medicine Hat block.

2. Grassrange area

In the Grassrange area, approximately 30 igneous bodies are known, based on mapping by Johnson and Smith (1964), Nelson (1992), Porter and Wilde

(1999a,b), Doden and Gold (1993), Ellsworth (2000), and Irving and Hearn (2003) (Fig. 1B). The igneous bodies are mainly dikes, plugs, or diatremes, and sills are rare. Most of the diatremes are breccia pipes that lack the bedded pyroclastic deposits and descended slices of wall-rock formations that are characteristic of the more evolved Missouri Breaks diatremes (Hearn, 1968; McGee and Hearn, 1989; Hearn et al., 1989; Irving and Hearn, 2003). Grassrange igneous rocks are dominantly alnoite, aillikite, and carbonatite. Two are classified as kimberlite (Homestead and Three Buttes), and monchiquite has been recognized in one location (Ford's Butte or War Horse Lake NW). The Grassrange area igneous rocks occur in a 53 × 48 km area that is slightly elongate in the northeast–southwest direction. The southwesternmost occurrence, the Round Butte pipe, is 25 km southwest from the nearest known occurrence. If igneous rocks were confirmed at an isolated occurrence of baked Fox Hills Formation (Upper Cretaceous) the NE–SW dimension of the area of igneous occurrences would extend to 87 km. Most of the individual arrays of close-spaced dike segments, breccia pipes and plugs trend in northeastly directions, with a tendency to fan from about N 65–70 E for southern arrays, to N 45 E for northern arrays (Johnson and Smith, 1964). In addition, clusters of dike-diatreme arrays have northeasterly trends. The distribution of igneous rocks in the Grassrange area represents a broad area of magma generation and ascent from the upper mantle. Clusters of arrays may reflect a NE–SW orientation of the stress field in the basement rocks of the upper crust, and similar alignment within individual arrays is a reflection of the stress field in the supracrustal sequence in the middle Eocene.

3. Homestead kimberlite

The Homestead kimberlite (Fig. 1C) was discovered by consulting geologist Peter Ellsworth in 1999 (Ellsworth, 2000). Three trenches (T1, T2, T3) exposed variable proportions of massive, variably altered hypabyssal kimberlite, tuffisitic kimberlite breccias, altered kimberlite breccias, and brecciated shale with minor admixture of kimberlite fragments. The kimberlite underlies two low hills, and may be either a single pipe 450 × 150 m (Ellsworth, 2000), or a double pipe with an intervening zone of brecciated shale (Fig. 1C). The wall rock consists of black shale and thin beds of rusty-weathering siltstone and sandstone that belong to the Lower Cretaceous Skull Creek Shale Member of the Thermopolis Formation.

The eastern pipe, 130 × 90 m, underlies the East Hill and the saddle. The East Hill contains orange altered kimberlite breccia with xenoliths of baked sedimentary rocks and altered peridotites with olivine and orthopyroxene replaced by quartz and carbonates. The saddle area exposes gray kimberlite breccia, and locally has massive, dark gray fresh hypabyssal kimberlite which contains partially serpentinized peridotite xenoliths.

The western pipe, 300 × 150 m, underlying the western hill, contains gray to orange breccias of kimberlite fragments and abundant sedimentary rock fragments, tuffisitic kimberlite breccia (in part containing sparse juvenile lapilli in the form of pelletal autoliths), soft gray, green, brown, or orange altered kimberlite, and massive, dark gray fresh kimberlite. Locally, the kimberlite contains sparse phlogopite-carbonate vugs or phlogopite-carbonate veins. Some breccia fragments are silicified. Peridotite xenoliths are abundant on the east and west sides of the western hill. On the northeast, a 3 × 3 m area of light gray baked siliceous shale containing fish-scale imprints is probably from the Mowry Shale (Upper Cretaceous), and has descended about 150 m. This is the only large subsided or downfaulted xenolith recognized so far in the pipe. Southwest of the kimberlite, an orange altered carbonate-rich dike trending S 60°W, away from the pipe, contains apparent pelletal lapilli with pseudomorphed olivine cores in a carbonate matrix. The pseudomorphed olivine cores are rimmed by altered silicates containing tangentially oriented lath-shaped pseudomorphs with tapered ends. The tapered ends suggest that these are from phlogopite rather than melilite.

The Homestead kimberlite has not been dated. It is probably middle Eocene, similar to the nearby Elk Creek Butte (50.34 Ma) and Yellow Water Butte (51.41 Ma, [40/39]Ar phlogopite, Mitchell and Bergman, 1991), and the Winnett sill (50.2 Ma, conventional K/ Ar age, Marvin et al., 1980).

3.1. Kimberlite

The rock (Fig. 2A) is a phlogopite–monticellite–diopside–calcite–serpentine kimberlite. It is termed

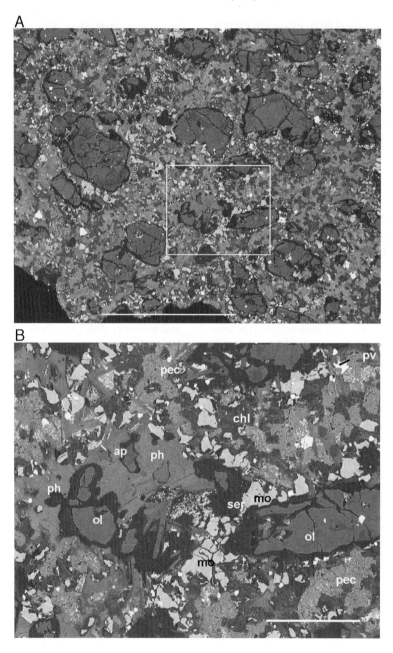

Fig. 2. (A) Back-scattered electron image of kimberlite sample H009A from Trench 2. Rectangle is area of (B). Scale bar 1 mm. (B) Back-scattered electron image of kimberlite matrix. Olivine, ol; phlogopite, ph; monticellite, mo; apatite, ap; chlorite, chl; serpentine, ser; pectolite?, pec; perovskite, pv. Scale bar 200 μm.

kimberlite based on the groundmass mineralogy, presence of sparse atoll-textured groundmass spinels, kimberlitic Trend 2 core-rim zoning in groundmass spinels, and kimberlitic core-rim zoning in ground-

mass phlogopites. The kimberlite contains 25 to 30 modal% olivine, from 0.1 to 5 mm in size. The matrix contains phlogopite, monticellite, sparse diopside, serpentine, chlorite, calcite, perovskite, spinel, pectolite, and hydrous Na-bearing Ca and Ca–Al silicates (Fig. 2B). Many of the larger olivines (0.5 to 5 mm) have rounded shapes and are xenocrystic. Smaller olivines, less than 0.25 mm, have irregular to euhedral shapes, and their borders tend to contain FeTi oxide grains. Some smaller olivines may have xenocrystic cores with magmatic rims developed in the kimberlite.

Olivines have a wide range of Fo contents, from 82 to 96.5. Some systematic variations are related to core-rim zoning and size (Fig. 3). Most large and medium-size olivines (0.25 to 5 mm) have cores of Fo 90.5 to 96.5, NiO 0.36 to 0.48 wt.%, with rims of Fo 87 to 90.5 and NiO 0.07 to 0.15 wt.%. Fo values greater than 93 are unusual for peridotitic olivines, but some Homestead xenoliths contain olivines with Fo values as high as 93.8. High Fo cores may be xenocrysts from disaggregated peridotite xenoliths, or could be in part magmatic (similar to magmatic olivines with Fo contents as high as 94.2 in the Zortman and Ricker Butte alnoites (Irving and Hearn, 2003). Small and very small olivines (<0.25 mm) tend to cluster in an intermediate range of Fo 90.3 to 90.8, with intermediate NiO contents of 0.20 to 0.35 wt.%. It appears that 0.25–5 mm xenocrystic olivines developed more iron-rich rims, and subsequently, olivines of<0.25 mm size crystallized in a higher Mg environment, resulting in intermediate Fo contents.

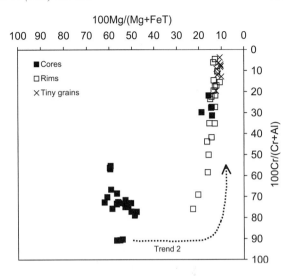

Fig. 4. Zoning of groundmass spinels; kimberlite trend 2 from Mitchell (1986, 1995).

Olivines tend to show an antithetic relationship of Ni and Ca, as expected for xenocrystic vs. magmatic olivine grains or zones. As Fo content decreases, Ni decreases and Ca increases. However, variation of CaO content of olivines is inconsistent. Although many large and medium size olivines tend to show low CaO in cores (0.02–0.20 wt.%), and higher CaO in rims (0.40–1.05 wt.%), some have about 1 wt.% CaO in cores and 0.2–0.3 wt.% in rims. For the small olivines <0.25 mm, about two-thirds have CaO content from 0.9 to 1.1 wt.%, and the rest have CaO contents of 0.10–0.35 wt.%. Variable olivine compositions may be related to mixing of separate magma batches, or to increasingly local approaches to equilibrium as the melt crystallized.

The larger spinel grains (0.04 to 0.1 mm) in the groundmass are zoned from cores of Cr–Fe–Al–Mg spinel to rims of Mg–Ti magnetite. Some spinels have poorly developed atoll texture. Variations of Cr/(Cr + Al), Mg/(Mg + Fe), and Ti content (Fig. 4) show core-rim trends of decreasing Mg and Cr contents, similar to Trend 2 of kimberlites (Mitchell, 1986, 1995), which is also closely similar to the orangeite trend. MnO contents are 0.2–0.5 wt.%, and Nb_2O_5 contents range from 0.00 to 0.12 wt.%.

Monticellite occurs as fresh to partially altered, anhedral to irregular 0.01–0.1 mm grains scattered throughout the matrix and also as multi-crystalline

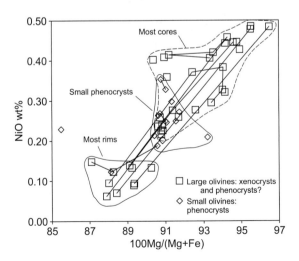

Fig. 3. NiO–Fo compositions of olivine phenocrysts and xenocrysts.

areas on borders of some forsteritic olivines. The X_{Ca} contents (100Ca/(Ca + Mg)) of monticellite, 47.9–49.8, imply crystallization temperatures of 800 to 900 °C, based on the experimental data of Davidson and Mukhopadhyay (1984), if equilibrium with small olivines can be assumed. Diopside grains are sparse, small (0.02–0.05 mm), and irregular in the matrix.

Phlogopites in the matrix have low Al_2O_3 (8.1–10.1 wt.%) and TiO_2 (0.9–1.6 wt.%) contents. Limited core-rim zoning in most grains is kimberlitic, showing a slight decrease in TiO_2 with increasing Al_2O_3; a few show opposite trends of decreasing Al_2O_3 from cores to rims. BaO increases from cores (0.39–0.93 wt.%) to rims (0.65–1.15 wt.%). Fluorine ranges from 1.13 to 1.52 wt.% F, with no core-rim variation; Cl is below detection. FeO and Mg# show no consistent core-rim trends.

Phlogopites in a 2 cm calcite–phlogopite vug show a more pronounced kimberlitic core-rim trend, of strongly increasing Al_2O_3 (cores 6.9–11.3 wt.%, rims 19.5–22.2 wt.%) and decreasing TiO_2 (cores 0.48–1.14 wt.%, rims 0.13–0.43 wt.%), and increasing kinoshitalite component. BaO is strongly enriched in rims, to as much as 15 wt.%, compared to core compositions of less than 1 wt.%. Magnetite, apatite, barite, and fluorite also occur in the vug. The latter phase assemblage suggests carbonatitic affinity.

A chemical analysis of fresh hypabyssal kimberlite (Fig. 2A and B) is given in Table 1. In comparison to Williams kimberlite chemical data (Hearn et al. 1989), Homestead kimberlite contains comparable amounts of SiO_2, Al_2O_3, FeO total, and CaO; Na_2O is considerably higher (0.1–0.2 wt.% in Williams), and K_2O, TiO_2 and P_2O_5 are lower (average 3.2, 2.0, and 1.1 wt.%, respectively, in Williams). In comparison to the compositional ranges in African and Russian kimberlites, the Homestead values of K_2O and P_2O_5 are within those ranges, Na_2O is higher, and TiO_2 is lower.

The chrondite-normalized rare-earth element (REE) pattern of the Homestead kimberlite is steeply light REE enriched and nearly linear (Fig. 5A). Homestead is less enriched in light REE than Williams and Three Buttes kimberlites, Elk Creek Butte aillikite, and Yellow Water Butte alnoite/aillikite (Fig. 5A), but is more enriched than the Winnett sill and other Montana alnoites.

A plot of Zr vs. Nb can discriminate among kimberlites, orangeites, lamproites, and minettes

Table 1

Chemical composition, hypabyssal kimberlite sample H009A, Trench 2, Homestead kimberlite, MT

	wt.%		ppm		ppm
SiO_2	36.89	Ni[a]	858	La[a]	62.7
TiO_2	0.48	Cr[a]	1120	Ce[a]	120
Al_2O_3	4.53	Co	62	Pr	14.5
Fe_2O_3	3.26	Cu	61	Nd	55.3
FeO	5.59	Zn[a]	66	Sm	8.56
MnO	0.153	Pb	6	Eu	2.21
MgO	29.27	V	66	Tb	0.71
CaO	9.62	Ga	7	Dy	3.37
Na_2O	1.52	Ge	0.9	Ho	0.58
K_2O	1.18	Bi	<0.1	Er	1.46
P_2O_5	0.61	Sb	<0.2	Tm	0.187
H_2O^+	4.53	As	<5	Yb	1.07
H_2O^-	0.34	Sc	21	Lu	0.136
CO_2	1.25	Be	6		
Cl	0.02	Tl	0.22		
F	0.52	Cs	0.8		
Total	99.77	Ba	820		
− Cl,F	−0.22	Rb	30		
Total	99.55	Sr	984		
LOI 925 °C	6.61	Zr	129		
		Hf	3.3		
		Nb	26.3		
		Ta	1.76		
		Th	6.32		
		U	1.26		
		Y	14.8		

Analyses by Activation Laboratories Major and trace elements by induction-coupled plasma mass spectrometry, except as noted below. Other methods: $H_2O^{+/-}$, gravimetric; CO_2, infrared; Cl, instrumental neutron activation analysis (INAA); F, ion selective electrode.

[a] Elements by INAA.

(Mitchell, 1995; Mitchell and Bergman, 1991). Fig. 5B shows the Zr and Nb contents of various central Montana igneous rocks, relative to the nearly separate fields of kimberlites, orangeites, lamproites, and minettes. The Zr–Nb content of Homestead kimberlite is within the kimberlite and orangeite fields. However, melnoites and aillikites of the Missouri Breaks and Grassrange areas are within or close to the same fields, so the diagram does not discriminate between these two types, or discriminate the two types from kimberlites, orangeites, lamproites, and minettes.

3.2. Xenocrysts

In prospecting for kimberlites, garnet color has commonly been used to select Cr-rich pyrope garnets

A

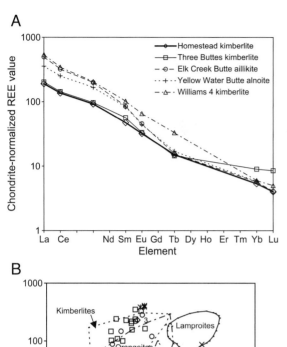

B

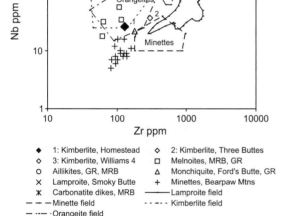

Fig. 5. (A) A selection of chondrite-normalized rare-earth elements for Homestead kimberlite, Three Buttes kimberlite, Williams 4 kimberlite, Elk Creek Butte aillikite, and Yellow Water Butte alnoite. (B) Zr–Nb in central Montana igneous rocks. Lamproite and minette fields from Mitchell and Bergman (1991); kimberlite and orangeite fields from Mitchell (1995). MRB, Missouri River Breaks; GR, Grassrange.

that are pertinent to the presence of upper mantle material and evaluation of diamond prospectivity. Other workers have described Cr pyropes as purple, lilac, or lavender; fewer are termed red or deep red. Size is a factor in apparent color, as smaller fragments

of purple Cr pyrope can appear pink. For the Homestead kimberlite, heavy mineral concentrates were derived by panning approximately 3 kg of surface and near-surface material of less than 2 mm size at five sites: the east hill, adjacent to trenches T1, T2, and T3, and in colluvium 150 m south–southeast of the east hill. Garnets were grouped by color: purple, pink, and orange. Garnet xenocrysts in the Homestead kimberlite are mainly purple G9 Cr-pyropes, with compositions ranging between 2.3 and 6.2 wt.% Cr_2O_3 (Fig. 6A–E). They show a lherzolitic trend that is parallel to and about 0.5 to 1.5 wt.% CaO above the 85% line of Gurney (1984). Compositions of purple garnets from the East Hill are clustered in a narrower range of 3 to 4 wt.% Cr_2O_3 (Fig. 6C). Purple G10 low-calcium pyropes are present in each concentrate, accounting for 2% to 12% of the purple garnet populations. Subordinate pink and orange garnets, with <0.2 wt.% Cr_2O_3, are mostly almandines, derived from crustal rocks (Fig. 6A). Eclogitic garnets are absent (using the criterion of Na_2O greater than 0.06 wt.%; Gurney and Moore, 1993; Schulze, 2003). Some pink garnets and a few orange garnets that have low TiO_2 contents of 0.1 to 0.4 wt.% and Cr_2O_3 contents of 1.4 to 3.2 wt.% may be from garnet pyroxenites or from low-Cr garnet lherzolites. Pyrope grains that are larger than about 0.5 mm and have higher Cr_2O_3 contents showed a metameric color change from purple in incandescent light, to gray, blue-gray, or bluish in daylight-type fluorescent light (Springfield and Mansker, 1985), which is a useful qualitative aid for picking out garnets with higher Cr contents.

Cr-spinel xenocrysts have MgO and Cr_2O_3 contents (Fig. 7A–D) that are clustered in the ranges 11 to 16 wt.% MgO and 48 to 59 wt.% Cr_2O_3, with a few containing 60 to 65 wt.% Cr_2O_3. About 10% are in the diamond inclusion field of Smith et al. (1994). Magnetic and non-magnetic spinels (separated by hand magnet) show no systematic differences in Cr_2O_3 or MgO content, suggesting that the magnetic character is related to the presence of thin rims of Ti magnetite.

Mg-ilmenite xenocrysts are uncommon, with most found in a pan concentrate from the T1 trench area. Scarcity of ilmenite xenocrysts may be related to absence of evolved garnet pyroxenites at depth. Ilmenite xenocrysts are kimberlitic (Fig. 8A). MgO contents range from 7 to 11 wt.%, whereas Cr_2O_3

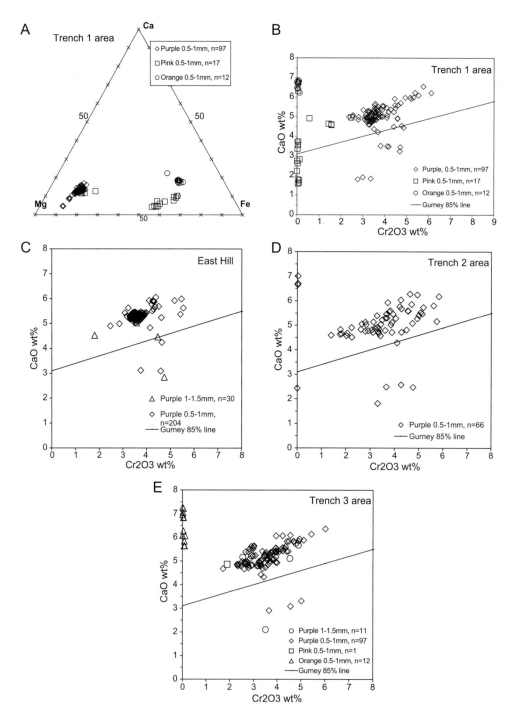

Fig. 6. Garnet xenocrysts from pan concentrates. (A) Ca–Mg–Fe compositions of garnets 0.5–1 mm size from T1 trench area. (B) CaO–Cr₂O₃ compositions of garnets 0.5–1 mm size from T1 trench area. Angled line is the boundary between G9 lherzolitic garnets and G10 harzburgitic garnets, after Gurney (1984). (C) CaO–Cr₂O₃ compositions of garnets of 0.5–1 and 1–1.5 mm size from the East Hill. (D) CaO–Cr₂O₃ compositions of garnets 0.5–1 mm size from T2 trench area. (E) CaO–Cr₂O₃ compositions of garnets 0.5–1 mm size from T3 trench area.

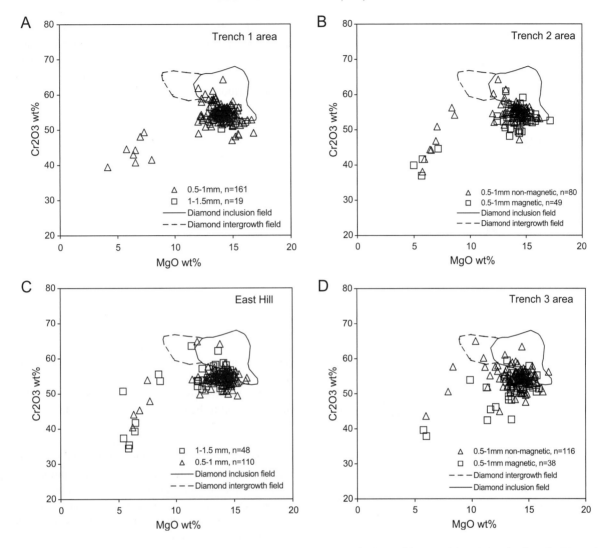

Fig. 7. Spinel xenocrysts from pan concentrates. (A) Cr_2O_3–MgO compositions of spinels of 0.5–1 and 1–1.5 mm size from the T1 trench area. Diamond inclusion and diamond intergrowth fields are from Smith et al. (1994). (B) Cr_2O_3–MgO compositions of magnetic and non-magnetic spinels 0.5–1 mm size from the T2 trench area. (C) Cr_2O_3–MgO compositions of spinels of 0.5–1 and 1–1.5 mm size from the East Hill area. (D) Cr_2O_3–MgO compositions of magnetic and non-magnetic spinels of 0.5–1 and 1–1.5 mm size from the T3 trench area.

contents are 0.8–1.9 wt.% (Fig. 8B). Fe^3/Fe^{tot} values are 0.17 to 0.31, suggesting a low oxidation range thought by some workers to be favorable for diamond preservation in kimberlites (Gurney and Moore, 1993). Others have shown that the relative oxidation value may not be related to diamond preservation (Schulze et al., 1995).

Clinopyroxene xenocrysts are Cr-diopsides, with a wide range in Al_2O_3, Cr_2O_3, and Na_2O (0.46–7.69, 1.03–3.86, and 1.31–3.75 wt.%, respectively,

for Cr-diopsides from the T1 trench area). Cr-diopsides from other parts of the Homestead kimberlite contain as much as 4.2 wt.% Cr_2O_3. In the MgO vs. Al_2O_3 plot (Nimis and Taylor, 2000), 96% to 98% of the xenocrysts are in the field of garnet-bearing peridotites.

The presence of G10 garnets and high Cr–Mg spinels suggests that the kimberlite could contain diamonds. One microdiamond was recovered by caustic fusion of a 45-kg sample of kimberlite (Ells-

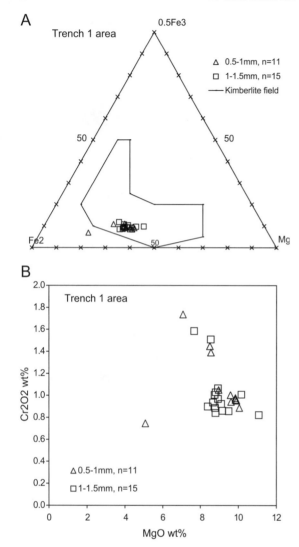

Fig. 8. Ilmenite xenocrysts from pan concentrate, T1 trench area. (A) Ferric Fe–ferrous Fe–Mg compositions of ilmenites of 0.5–1 and 1–1.5 mm size with the kimberlitic field of Mitchell (1986). (B) Cr_2O_3–MgO compositions of ilmenites.

worth, 2000). The processing of multi-ton samples necessary for full evaluation of diamond potential has not been done for the Homestead kimberlite.

3.3. Xenoliths

Xenoliths identified in the Homestead kimberlite are Paleozoic limestone and dolomite, sparse crustal granulites, amphibolites, metagabbros, abundant upper-mantle peridotites and sparse upper-mantle pyrox-

enites. Locally, peridotite xenoliths are remarkably abundant and up to 0.7 m in maximum dimension, forming closely packed arrays that resemble boulder conglomerates. These are among the largest known garnet peridotites known in North America. No eclogites are present, signifying that eclogites were not sampled by the kimberlite. Megacrysts (size >1 cm) of garnet (gar), clinopyroxene (cpx), orthopyroxene (opx), and ilmenite, common in many kimberlites, are absent, probably because coarse-grained garnet pyroxenite bodies are absent from the subjacent upper mantle. Megacrysts of olivine may be present, but most large olivines have higher Fo contents than typical megacrysts and are likely to have been derived from peridotites.

The peridotite xenoliths show slight to extensive serpentinization. Xenoliths in the East Hill area show extensive replacement of olivine and orthopyroxene by quartz and calcite. The xenolith suite is dominated by dunites and harzburgites, containing garnet (gar), garnet and spinel (sp), or sp only. Many contain gar and sp in apparent equilibrium, as separate grains or as sp included in gar. Composite xenoliths show spatial relationships among peridotites, pyroxenites, and phlogopite-bearing veins. Pyroxenites, found mainly as veins or selvages in peridotite xenoliths, are sp websterite, gar–sp websterite, gar websterite, and gar orthopyroxenite; some pyroxenites also contain phlogopite. Many peridotites contain phlogopite as disseminated grains, as rims around garnets, as part of cpx-sp clots, or as crosscutting veins with or without cpx or opx. Phlogopite-rich zones form one or more borders of some peridotite xenoliths, indicating that the peridotite probably has split along phlogopite-bearing veins.

Proportions of xenolith lithologies are variable in different parts of the kimberlite. In a suite of 153 xenoliths (a somewhat biased selection, in part based on the presence of unusual textures, macroscopic gar, cpx, veins, or composite nature) (Table 2), 7.2% are lherzolite, 61.4% are harzburgite, 30.0% are dunite, and 0.7% are orthopyroxenite (using < 5% cpx content to define harzburgites, and < 5% of either pyroxene to define dunites). Three of 30 peridotite xenoliths selected for detailed study are low-Ca gar–sp harzburgites. A more representative estimate, for a 1.5 × 3 m area of exposed kimberlite near Trench T1, is given in Table 3. Of 107 xenoliths greater than 5 cm size,

Table 2
Lithology of 153 collected upper-mantle xenoliths

Lithology	Garnet	Gar + sp	Spinel	Without gar, sp	Total	Percent	Phlogopite present	Composite
Lherzolite	4	5	2	0	11	7.2	0	2
Harzburgite	39	33	22	0	94	61.4	22	13
Wehrlite	0	0	1	0	1	0.7	1	0
Dunite	3	9	27	7	46	30.0	13	6
Orthopyroxenite	0	1	0	0	1	0.7	0	0
Total	46	48	52	7	153		36	21
Percent	30.0	31.4	34.0	4.6			23.5	13.7

Composite xenoliths contain veins or bands of clinopyroxenite, orthopyroxenite, or websterite.

21.5% are harzburgite, 77.9% are dunite, and 0.6% are crustal amphibolite. About 4% of the peridotites contain macroscopic phlogopite.

Both xenolith counts indicate that the suite is dominantly depleted in major elements, similar to other xenolith suites from the Wyoming craton, such as Williams kimberlites (Hearn and McGee, 1984), Macdougal Springs diatreme (McGee and Hearn, 1989), Froze-to-Death Butte and Ingomar south dike (Hearn, 1999), Highwood Mountains (O'Brien et al., 1995), and Bearpaw Mountains (Hearn and McGee, 1987). Mineral compositions also indicate that harzburgites and pyroxene-bearing dunites are more depleted in comparison to lherzolites. Clinopyroxenes in lherzolites tend to have higher Al_2O_3 and TiO_2, and lower Cr_2O_3. A complete data set of mineral compositions for 30 xenoliths is given in Supplemental Data Tables 1 and 2.

Textures are mainly coarse granular or coarse tabular (Table 4), and only 5% are porphyroclastic, in contrast to xenoliths in the Williams kimberlites, of which about 50% are porphyroclastic (Hearn and McGee, 1984). Xenoliths with coarse tabular texture commonly have sub-parallel small opx grains among

Table 3
Count of upper-mantle xenoliths larger than 5 cm, in 1.5 × 3 m area of exposed kimberlite on west side of trench T1

Lithology	Spinel or probable spinel	Garnet, gar–sp	Total	Percent	Phlogopite present
Lherzolite	0	0	0	0	0
Harzburgite	33	2	35	32.7	6
Dunite	70	2	72	67.3	1
Total	103	4	107	100	7
Percent	96.3	3.7			6.5

large olivine grains, suggesting coarsening or annealing from a former porphyroclastic texture. Younger, post-depletion enrichment is evidenced by tectonized or undeformed veins of orthopyroxenite, clinopyroxenite, and websterite with gar and/or sp. Veins with abundant Cr-diopsides may represent tectonized earlier clinopyroxenite or websterite veins. Later undeformed Cr-diopside clinopyroxenite veins cut across some peridotites. Phlogopite-bearing veins may represent earlier K-metasomatism and/or late kimberlite-related addition.

Garnets in some peridotites show complex textures of small spinel inclusions within garnet, sometimes with additional small spinel inclusions in parts of opx grains adjacent to garnet. These textures could have resulted from incomplete re-equilibration near the garnet–spinel stability boundary, due to change in T or P or both.

Clinopyroxenes in peridotite xenoliths show various textural settings: irregular to sub-equant grains among opx and olivine grains, inclusions in large garnets, and multiple grains clustered around garnets or spinels (Table 4). Similar textural settings for clinopyroxenes are seen in peridotite xenoliths from Williams kimberlites. Large angular clinopyroxene grains in Homestead xenoliths have been interpreted by Irving et al. (2003) as an indication of lack of textural equilibrium. Isotopic data suggest a crystallization age of a few hundred million years or less for some clinopyroxenes (Carlson et al., 2004). One composite xenolith (H01.4J) appears to show conversion of garnet–spinel–phlogopite harzburgite to clinopyroxene-bearing, orthopyroxene-free spinel dunite. Along a rather sharp lithologic boundary, large orthopyroxenes in the harzburgite have been converted to aggregates of smaller olivines and clinopyroxenes.

Table 4
Lithology of peridotite xenoliths and calculated temperatures and pressures

Sample no.	Lithology[a]	Texture[b]	Pyroxenes[a]	#av	T BKN[c] (°C)	P BKN[c] (GPa)
H00.10.02	Gar–sp harzburgite + phl orthopyroxenite vein	CG	cpx in and adj to gar, adj to opx	4	1248	3.83
H00.10.02	Orthopyroxenite vein in gar–sp harzburgite	CG	opx in vein, cpx in matrix	1	1242	3.88
H00.10.14	Gar–sp harzburgite	CT	cpx adj to opx	3	1269	3.87
H00.10.20	Gar–sp harzburgite + dunite	FMP	cpx adj to gar	1	1384	4.47
H00.10.22	Gar–sp harzburgite, low Ca	CT	no cpx; calc P for est T 1200	1	*1200*	3.80
H00.10.23	Gar–sp harzburgite	CT	cpx in and adj to gar	1	1186	3.76
H00.10.24	Gar–sp harzburgite, low Ca	CT	no cpx; calc P for est T 1200	1	*1200*	3.74
H00.10.26	Gar–sp harzburgite + clinopyroxenite vein	CG	cpx in, adj to, and away from gar	3	1303	3.81
H00.10.26	Clinopyroxenite vein in gar–sp harzburgite	CG	cpx in vein, opx adj to vein	1	1302	3.71
H00.10.44	Gar–sp harzburgite	CG	cpx in and adj to gar	2	1203	4.79
H00.10.50	Gar–sp harzburgite	CG	cpx away from gar	2	1193	4.04
H00.11.03	Gar–sp harzburgite, low Ca	CG	cpx away from gar	3	1267	4.01
H00.11.11	Gar–sp harzburgite	CG	cpx adj to opx	4	1242	3.57
H00.11.19	Gar–sp harzburgite	CG	cpx in and away from gar, on sp	4	1336	3.89
H00.11.35	Gar–sp harzburgite	CG	cpx, opx in gar; cpx adj to opx	3	1243	3.92
H00.11.42	Gar–sp harzburgite	CG	cpx adj to gar	2	1184	4.40
H00.12.02	Gar–sp harzburgite	vCG	cpx adj to opx	1	1221	3.87
H00.12.09	Gar–sp harzburgite	CG	cpx adj to opx	4	1228	3.92
H01.04J	Gar–sp-phl harz + sp dunite	CG	cpx adj opx in gar–sp-phl harz	1	1225	3.85
H01.04J	Gar–sp-phl harz + sp dunite	CG	cpx adj and away from sp in dunite	3	1179	*3.85*
H00.10.51	Gar lherzolite	P	cpx adj to opx	2	1249	3.89
H00.11.16	Gar–sp lherzolite w/zoned gar	CG	cpx adj to opx; cpx adj to gar	3	1250	4.03
H01.03C	Gar–sp harzburgite + phl vein	CG	cpx adj to opx and near opx	3	1290	40.9
H01.04C	Gar-sulfide lherzolite	CG	cpx adj to opx	2	1264	4.02
H00.11.04	Gar–sp-ol-cpx orthopyroxenite	CG	in cpx-rich band and adj to sp	3	1225	4.56
H00.11.08	Gar–sp dunite	CG	w/opx; cpx in and adj to gar	4	1174	4.65
H00.12.05	Gar–sp dunite + clinopyroxenite veins	P	cpx adj to and away from gar	1	1288	3.98
H00.12.05	Clinopyroxenite vein in gar–sp dunite	CG	cpx adj to and away from sp in vein	2	1291	3.99
H00.10.01	Sp dunite + sp clinopyroxenite vein	CG	cpx, opx in vein	1	1109	*2.0*
H00.10.28	Sp dunite	vCG	w/opx; cpx adj to sp	3	1290	*2.0*
H00.12.06	Sp-phl dunite + sp-ol clinopyroxenite veins	CG	cpx in vein; T FB86[c]	2	1234	*2.0*
H00.12.07	Sp dunite + phl clinopyroxenite vein	CG	cpx adj to sp in sp dunite	2	1273	*2.0*
H00.12.07	Phl clinopyroxenite vein in sp dunite	CG	opx in vein, cpx in dunite	1	1256	*2.0*
H01.04O	Dunite + clinopyroxenite vein	vCG	cpx in vein; T FB86[c]	1	1203	*2.0*
H00.10.32	Sp-phl harzburgite	CG	cpx away from sp	2	919	*2.0*

[a] gar, garnet; sp, spinel; cpx, clinopyroxene; opx, orthopyroxene; ol, olivine; phl, phlogopite; adj, adjacent.

[b] CG, coarse granular; vCG, very coarse granular; P, porphyroclastic; FMP, foliated mosaic porphyroclastic.

[c] Temperature–pressure methods: BKN, Brey et al., 1990; FB86, Finnerty and Boyd, 1987; numbers in italics are estimated T or P.

Xenolith olivines, orthopyroxenes, and spinels in general are unzoned. Large garnets show up to 1 wt.% variation in Cr_2O_3 content, which generally is not related to core-rim position. Some clinopyroxenes show limited variation within single grains.

4. Thermobarometry

Xenolith temperatures and pressures were calculated with the DOS program "PT", based on methods in Brey et al. (1990) (Table 4). Garnet-bearing peridotites have a range of 1180–1390 °C and 3.6–4.7 GPa, using Brey et al. (1990) two-pyroxene and Al-in-opx methods ('BKN'). These points lie in P–T space in the graphite stability field near to the diamond–graphite boundary, and above a 38 mW/m^2 geotherm (Fig. 9A). The three low-Ca garnet harzburgites that contain G10 garnets fall in a similar T range, using BKN methods for one xenolith with sparse clinopyroxene, and calculating P for Al-in-opx for assumed T of 1200 °C for the two xenoliths that lack clinopyr-

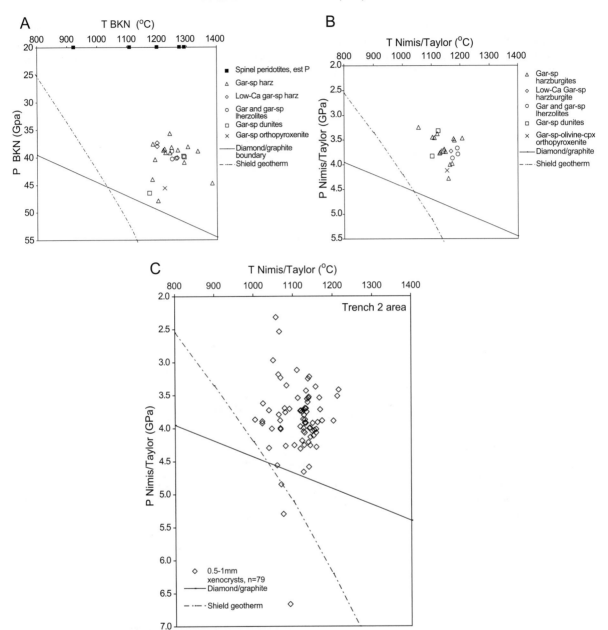

Fig. 9. (A) Calculated equilibrium temperatures and pressures of Homestead peridotite xenoliths, calculated by BKN methods (two-pyroxene *T*, Al-in-opx *P*). For spinel peridotites, *P* is assumed to be 2.0 GPa. For two low-Ca harzburgites lacking cpx, *P* was calculated for estimated *T* of 1200 °C. (B) Calculated equilibrium temperatures and pressures of clinopyroxenes in Homestead peridotite xenoliths, calculated by NT2000 method. (C) Calculated equilibrium temperatures and pressures of 79 clinopyroxene xenocrysts 0.5–1 mm size from the T2 trench area, by NT2000 method.

oxene. Calculated olivine–spinel Fe/Mg temperatures, using the Li et al. (1995) calibration, are consistently much lower (similar to the olivine–spinel *T*'s for peridotite xenoliths from Williams kimberlites), indicative of re-equilibration, and thus are not useful for thermometry.

For most spinel peridotites, TBKN at an assumed P of 2.0 GPa is in the same range as for garnet peridotites, possibly indicating heating of the shallow mantle. Pyroxenite veins have TBKN estimates that are similar to those of the host peridotite. The calculated T and P values by the Nimis and Taylor (2000) method ('NT2000') for clinopyroxenes in garnet peridotite xenoliths (Fig. 9B) cluster about 100 °C and 0.3 GPa lower than the BKN values, but are still above the geotherm. Fig. 10 shows calculated temper-

atures and pressures for other Montana xenolith suites, in comparison with Homestead kimberlite xenoliths.

The calculated NT2000 temperatures and pressures for 79 Cr diopside xenocrysts from the T2 trench area are mostly in the ranges of 1030–1160 °C and 3.5–4.3 GPa, also above the geotherm (Fig. 9C). Four of 79 xenocrysts are in the diamond stability field. Cr-diopsides in other pan concentrates show similar ranges of calculated temperatures and pressures using NT2000.

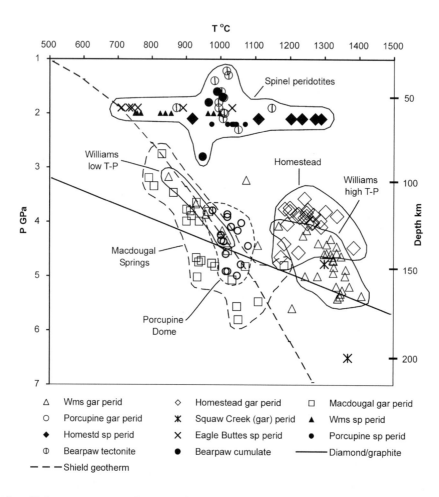

Fig. 10. Calculated equilibrium temperatures and pressures for Montana peridotite and pyroxenite xenoliths, and composite megacrysts. Williams, Homestead, Squaw Creek, and Macdougal Springs gar and gar–sp peridotites with surviving opx: T and P are by BKN methods. Macdougal Springs, Porcupine Dome, and Squaw Creek gar and gar–sp peridotites without surviving opx: T and P are by NT2000 method. Spinel peridotites: T is by BKN or by Eggler et al. (1987) (Ca in opx method), at assumed P values of 1.9 GPa for Eagle Buttes, 2.0 GPa for Williams, 2.1 GPa for Homestead, and 2.2 GPa for Porcupine Dome (P values were chosen arbitrarily to separate suites from different localities in the figure). Bearpaw Mountains spinel peridotites (tectonite or cumulate textures): T is by BKN and P is by Kohler and Brey (1990) (Ca in olivine method).

Ellsworth (2000) reports that the nickel-in-garnet geothermometer shows that some Cr pyrope xenocrysts are in the diamond window. However, similar analyses by Griffin et al. (this volume) show no garnets in the diamond stability field.

5. Conclusions

1. The Homestead igneous occurrence is a phlogopite – monticellite – diopside – calcite – serpentine kimberlite, based on (a) groundmass mineralogy, (b) groundmass spinels with core-rim zoning similar to kimberlite trend 2, and (c) groundmass phlogopites with kimberlitic core-rim $TiO_2 - Al_2O_3$ zoning trends. Late-magmatic phlogopites in a vug show a stronger kimberlitic trend, with high content of kinoshitalite component in the rims, with as much as 15 wt.% BaO.

2. Most xenocrysts are derived from disaggregated peridotites or pyroxenites. Garnet xenocrysts are dominantly lherzolitic G9 type, but 2% to 12% are harzburgitic low-Ca G10 type. Garnets from crustal rocks are minor. Most spinel xenocrysts are Cr-rich with about 10% in the diamond inclusion field of Smith et al. (1994). Ilmenite xenocrysts are kimberlitic with low contents of ferric iron. Clinopyroxene xenocrysts are mainly from garnet peridotites, and a small proportion has calculated temperatures and pressures in the diamond stability field. The presence of G10 garnets, high Cr–high Mg spinels, and low-oxidation ilmenites is favorable for the presence and preservation of diamond.

3. The kimberlite has been remarkably efficient at transporting upper mantle material to the surface, resulting in a world-class mantle xenolith locality. Spinel and garnet peridotite xenoliths up to 0.7 m in size are among the largest known in North America. Textures are mostly coarse granular or coarse tabular; few xenoliths show porphyroclastic texture. Peridotites are dominantly major-element depleted as shown by the high proportion of dunites and harburgites in comparison to lherzolites. Pyroxenite veins and bands in peridotites indicate later enrichment by melt addition. Clinopyroxene clusters around garnets and around spinels indicate late addition or mobilization of clinopyroxene. Potassic enrichment is shown by the presence of disseminated phlogopite and phlogopite-bearing veins. These peridotites show that the Wyoming craton has undergone major-element depletion and later enrichment. Similar earlier depletion and later enrichment characteristics are seen in xenolith suites from Williams, Macdougal Springs, Porcupine Dome, and Highwood Mountains.

4. Calculated temperatures and pressures of equilibration for garnet and garnet–spinel peridotite mineral assemblages cluster in the graphite field, above a 38 mW/m^2 shield geotherm, suggesting heating prior to entrainment. Calculated T and P for spinel peridotites are mainly in the same temperature range for an assumed pressure of 2.0 GPa, suggesting heating of the shallow part of the upper mantle. The xenolith TP values are not favorable for the co-existence of diamond.

Acknowledgements

I am grateful to Pete Ellsworth for introducing me to the puzzles of the Homestead kimberlite. I especially thank Russ and Betty Gjerde for access to their ranch for geological fieldwork and research. In addition, many other ranchers in the area have provided access to fascinating localities, and I appreciate their continuing patience with the geologists who arrive to visit sites of petrological interest in the central Montana area. H.E. Belkin has provided considerable advice and solutions relating to electron microprobe analysis problems. Temperatures and pressures were calculated by use of the DOS program "PT" of G.P. Brey and T. Kohler. Reviews by H. Coopersmith, N. Simon, C. Hood, T. McCandless, and the editor made substantial improvements in readability and organization of the paper.

References

Brey, G.P., Kohler, T., Nickel, K.G., 1990. Geothermobarometry in four-phase lherzolites: Part II. New thermobarometers, and practical assessment of existing thermobarometers. J. Petrol. 31, 1353–1378.

Buhlmann, A.L., Cavell, P., Burwash, R.A., Creaser, R.A., Luth,

R.W., 2002. Minette bodies and cognate mica-clinopyroxenite xenoliths from the Milk River area, southern Alberta: records of a complex history of the northernmost part of the Archean Wyoming craton. Can. J. Earth Sci. 37, 1629–1650.

Carlson, R.W., Irving, A.J., Schulze, D.J., Hearn Jr., B.C., 2004. Timing of lithospheric mantle modification beneath the Wyoming craton. Lithos (this volume).

Chadwick, R.A., 1980. Radiometric ages of some Eocene volcanic rocks, southwestern Montana. Isochron-West 27, 11.

Davidson, P.M., Mukhopadhyay, D.K., 1984. Ca–Fe–Mg olivines: phase relations and a solution model. Contrib. Mineral. Petrol. 86, 256–263.

Doden, A.G., 1996. Ultramafic lamprophyres from Porcupine dome, east-central Montana, and their potential for diamond. In: Jones, R.W., Harris, R.E. (Eds.), Proceedings of the 32nd Annual Forum on the Geology of Industrial Minerals, Wyoming State Geological Survey. Public Information Circular, vol. 38, pp. 241–256.

Doden, A.G., Gold, D.P., 1993. Diatreme-dike associations of central Montana. In: Hunter, L.D.V. (Ed.), Energy and Mineral Resources of Central Montana, 1993 Field Conference Guidebook. Montana Geological Society, Billings, Montana, pp. 215–226.

du Bray, E.A., Harlan, S.S., 1996. The Eocene Big Timber stock, south-central Montana: development of extensive compositional variation in an arc-related intrusion by side-wall crystallization and cumulate glomerocryst remixing. Geol. Soc. Amer. Bull. 108, 1404–1424.

Dudas, F.O., 1991. Geochemical features of igneous rocks from the Crazy Mountains, Montana, and tectonic models for the Montana igneous province. J. Geophys. Res. 96, 13261–13277.

Ellsworth, P.C., 2000. Homestead kimberlite: new discovery in central Montana. Guidebook, 25th Annual Field Conference. Tobacco Root Geological Society, Missoula, Montana, pp. 14–20.

Finnerty, A.A., Boyd, F.R., 1987. Thermobarometry for garnet peridotites: basis for the determination of thermal and compositional structure of the upper mantle. In: Nixon, P.H. (Ed.), Mantle Xenoliths. John Wiley and Sons, Chichester, pp. 381–402.

Gorman, A.R., Clowes, R.M., Ellis, R.M., Henstock, T.J., Spence, G.D., Keller, G.R., Levander, A.R., Snelson, C.M., Burianyk, M.J.A., Kanasewich, E.R., Asudeh, I., Hajnal, Z., Miller, K.C., 2002. Deep probe: imaging the roots of western North America. Can. J. Earth Sci. 39, 375–398.

Gurney, J.J., 1984. A correlation between garnets and diamonds in kimberlites. Glover, J.E., Harris, P.G. (Eds.), Kimberlite Occurrence and Origin: A Basis for Conceptual Models in Exploration, vol. 8. The University of Western Australia Publ., Perth, pp. 143–166.

Gurney, J.J., Moore, R.O., 1993. Geochemical correlations between kimberlitic indicator minerals and diamonds. Diamonds: Exploration, Sampling and Evaluation, Short Course Proceedings. Prospectors and Developers Association of Canada, Toronto, pp. 147–171.

Harlan, S.S., 1996. Timing of emplacement of the sapphire-bearing Yogo dike, Little Belt Mountains, Montana. Econ. Geol. 91, 1159–1162.

Harlan, S.S., Snee, L.W., Mehnert, H.H., Schmidt, R.G., Sheriff, S.D., Irving, A.J., in press. $^{40}Ar/^{39}Ar$ and K–Ar geochronology of the Late Cretaceous Adel Mountain Volcanics and spatially associated igneous rocks, northwestern Montana. U.S. Geol. Surv. Bull.

Hearn Jr., B.C., 1968. Diatremes with kimberlitic affinities in north-central Montana. Science 159, 622–625.

Hearn Jr., B.C., 1979. Preliminary map of diatremes and alkalic ultramafic intrusions, Missouri River Breaks and vicinity, north-central Montana. U. S. Geol. Surv. Open-File Report, 79–1128.

Hearn Jr., B.C., 1999. Peridotite xenoliths from Porcupine Dome, Montana, USA: depleted subcontinental lithosphere samples in an olivine–phlogopite–carbonate magma. In: Gurney, J.J., Gurney, J.L., Pascoe, M.D., Richardson, S.H. (Eds.), Proc. 7th Internat. Kimberlite Conf., Dawson Volume. Red Roof Design, Cape Town, pp. 353–360.

Hearn Jr., B.C., McGee, E.S., 1984. Garnet peridotites from Williams kimberlites, north-central Montana, USA. In: Kornprobst, J. (Ed.), Kimberlites II: The Mantle and Crust–Mantle Relationships. Elsevier, Amsterdam, pp. 57–70.

Hearn Jr., B.C., McGee, E.S., 1987. Crust and upper mantle beneath the northern plains: evidence from xenoliths. U.S. Geol. Surv. Circ. 956, 32–34.

Hearn Jr., B.C., Dudas, F.O., Eggler, D.H., Hyndman, D.W., O'Brien, H.E., McCallum, I.S., Irving, A.J., Berg, R.B., 1989. Montana high-potassium igneous province. 28th International Geological Congress, Field Trip Guidebook T346. Am. Geophys. Union., Washington D.C., 86 pp.

Holm, D., Schneider, D., 2002. $^{40}Ar/^{39}Ar$ evidence for ca. 1800 Ma tectonothermal activity along the Great Falls tectonic zone, central Montana. Can. J. Earth Sci. 39, 1719–1728.

Irving, A.J., Hearn Jr., B.C. 2003. Alkalic rocks of Montana: kimberlites, lamproites, and related magmatic rocks. Montana Field Trip Guidebook, 16–21 June 2003, 8th International Kimberlite Conference. Geological Survey of Canada Bookstore, Ottawa. 44 pp.

Irving, A.J., Kuehner, S.M., Ellsworth, P.C., 2003. Petrology and thermobarometry of mantle xenoliths from the Eocene Homestead kimberlite pipe, central Montana, USA. Program with Abstracts, 8th International Kimberlite Conference, Victoria, BC, Canada, p. 53.

Johnson Jr., W.D., Smith, H.R. 1964. Geology of the Winnett–Mosby area, Petroleum, Garfield, Rosebud, and Fergus Counties, Montana. U.S. Geol. Surv. Bull. 1149 (91 pp.).

Kohler, T., Brey, G.P., 1990. Calcium exchange between olivine and clinopyroxene calibrated as a geothermobarometer for natural peridotites from 2 to 60 kb with applications. Geochim. Cosmochim. Acta 54, 2375–2388.

Li, J., Kornprobst, J., Vielzeuf, D., 1995. An improved experimental calibration of the olivine–spinel geothermometer. Chin. J. Geochem. 14, 68–77.

Marvin, R.F., Witkind, I.F., Keefer, W.R., Mehnert, H.H., 1973. Radiometric ages of intrusive rocks in the Little Belt Mountains, Montana. Geol. Soc. Amer. Bull. 84, 1977–1986.

Marvin, R.F., Hearn Jr., B.C., Mehnert, H.H., Naeser, C.W., Zartman, R.E., Lindsey, D.A., 1980. Late Cretaceous–Paleocene–

Eocene igneous activity in north-central Montana. Isochron-West 29, 5–25.

McGee, E.S., Hearn Jr., B.C., 1989. Primary and secondary mineralogy of carbonated peridotites from the MacDougal Springs diatreme. In: O'Reilly, S.Y., et al. (Ed.), Kimberlites and Related Rocks, Volume 2. Spec. Publ.-Geol. Soc. Aust., vol. 14, pp. 725–734.

Mitchell, R.H., 1986. Kimberlites: Mineralogy, Geochemistry, and Petrology. Plenum, New York.

Mitchell, R.H., 1995. Kimberlites, Orangeites, and Related Rocks. Plenum, New York.

Mitchell, R.H., Bergman, S.C., 1991. Petrology of Lamproites. Plenum, New York.

Mitchell, R.H., Platt, R.G., Downey, M., 1987. Petrology of lamproites from Smoky Butte, Montana. J. Petrol. 28, 6455–6677.

Mueller, P.A., Heatherington, A.L., Kelly, D.M., Wooden, J.L., Mogk, D.W., 2002. Paleoproterozoic crust within the Great Falls tectonic zone: implications for the assembly of southern Laurentia. Geology 30, 127–130.

Nelson, W.J., 1993. Structural geology of the Cat Creek anticline and related features, central Montana. Mont. Bur. Mines Geol. Memoir 64 (44 pp).

Nimis, P., Taylor, W.R., 2000. Single clinopyroxene thermobarometry for garnet peridotites: Part 1. Calibration and testing of a Cr-in-Cpx barometer and an enstatite-in-Cpx thermometer. Contrib. Mineral. Petrol. 139, 541–554.

O'Brien, H.E., Irving, A.J., McCallum, I.S., Thirlwall, M.F., 1995. Sr, Nd and Pb isotopic evidence for interaction of post-subduction asthenospheric potassic mafic magmas of the Highwood Mountains, Montana with ancient Wyoming craton lithospheric mantle. Geochim. Cosmochim. Acta 59, 4539–4556.

O'Neill, J.M., Lopez, D.A., 1985. Character and regional significance of Great Falls Tectonic Zone, east-central Idaho and west-central Montana. Am. Assoc. Pet. Geol. Bull. 69, 437–447.

Porter, K.W., Wilde, E.M., 1999a. Geologic map of the Musselshell 30′ × 60′ quadrangle, central Montana. Open File Rep.-Mont. Bur. Mines Geol., 386.

Porter, K.W., Wilde, E.M., 1999b. Geologic map of the Winnett 30′ × 60′ quadrangle, central Montana. Open File Rep.-Mont. Bur. Mines Geol., 307.

Schulze, D.J., 2003. A classification scheme for mantle-derived garnets in kimberlite: a tool for investigating the mantle and exploring for diamonds. Lithos 71, 195–213.

Schulze, D.J., Anderson, P.F.N., Hearn Jr., B.C., Hetman, C.M., 1995. Origin and significance of ilmenite megacrysts and macrocrysts from kimberlite. Int. Geol. Rev. 37, 780–812.

Smith, C.B., Lucas, H., Hall, A.E., Ramsay, R.R., 1994. Diamond prospectivity and indicator mineral chemistry: a Western Australian perspective. In: Meyer, H.O.A., Leonardos, O.H. (Eds.), Diamonds: Characterization, Genesis and Exploration. Proc. 5th Int. Kimb. Conf., vol. 2. CPRM, Brasilia, pp. 312–318.

Springfield, J.T., Mansker, W.L., 1985. Factors affecting garnet metamerism and applications in kimberlite evaluation/exploration. Abst. Progr.-Geol. Soc. Amer. 17, 193.

Available online at www.sciencedirect.com

Lithos 77 (2004) 493–510

www.elsevier.com/locate/lithos

Petrological constraints on seismic properties of the Slave upper mantle (Northern Canada) ☆

M.G. Kopylova[a,*], J. Lo[a], N.I. Christensen[b]

[a] Geological Sciences Division, Earth and Ocean Sciences, The University of British Columbia, 6339 Stores Road, Vancouver, British Columbia, Canada, V6T 1R9
[b] Department of Geology and Geophysics, University of Wisconsin-Madison, Madison, WI, USA

Received 27 June 2003; accepted 21 January 2004
Available online 28 May 2004

Abstract

Modes and compositions of minerals in Slave mantle xenoliths, together with their pressures and temperatures of equilibrium were used to derive model depth profiles of P- and S-wave velocities (Vp, Vs) for composites equivalent to peridotite, pyroxenite and eclogite. The rocks were modeled as isotropic aggregates with uniform distribution of crystal orientations, based on single-crystal elastic moduli and volume fractions of constituent minerals. Calculated seismic wave velocities are adjusted for in situ pressure and temperature conditions using (1) experimental P- and T- derivatives for bulk rocks' Vp and Vs, and (2) calculated P- and T- derivatives for bulk rocks' elastic moduli and densities. The peridotite seismic profiles match well with the globally averaged *IASP91* model and with seismic tomography results for the Slave mantle. In peridotite, an observed increase of seismic wave velocities with depth is controlled by lower degrees of chemical depletion in the deeper upper mantle. In eclogite, seismic velocities increase more rapidly with depth than in peridotite. This follows from contrasting first-order pressure derivatives of bulk isotropic moduli for eclogite and peridotite, and from the lower compressibility of eclogite at high pressures. Our calculations suggest that depletion in cratonic mantle has a distinct seismic signature compared to non-cratonic mantle. Depleted mantle on cratons should have slower Vp, faster Vs and should show lower Poisson's ratios due to an orthopyroxene enrichment. For the modelled Slave craton xenoliths, the predicted effect on seismic wave velocities would be up to 0.05 km/s.
© 2004 Elsevier B.V. All rights reserved.

Keywords: Seismic velocity; Slave mantle; Eclogite; Peridotite; Chemical depletion; Density

1. Introduction

More than 20 years ago, the pioneering work of Jordan (1979) established links between density,

elastic properties and the bulk composition of peridotitic mantle. Since then we have begun to understand better the compositional contrast between Archean cratonic mantle and younger non-cratonic mantle (Boyd, 1999; Kelemen et al., 1998). This paper aims at analysing compositional effects on seismic properties of cratonic mantle.

The analysis is carried out for mantle rocks of the Slave craton found as kimberlite-derived xenoliths.

☆ Supplementary data associated with this article can be found, in the online version, at doi: 10.1016/j.lithos.2004.03.012.
* Corresponding author. Tel.: +1-604-822-0865.
E-mail address: mkopylov@eos.ubc.ca (M.G. Kopylova).

These mantle rocks are characterized mineralogically and geochemically, and comprehensive data on their modal mineralogy, mineral chemistry and pressures and temperatures of equilibria are available. For this study, these bulk characteristics are combined with the latest experimental data on elastic properties of end-member components of mineral solid solutions (i.e. reviews by Vacher et al., 1998; Hofmeister and Mao, 2003; Ji et al., 2003). Sufficient experimental data now exist for estimating the bulk elastic properties of a mantle rock with good accuracy if the mineralogy of the rock is known (Gregoire et al., 2001; Hofmeister and Mao, 2003).

The Slave mantle is an excellent natural laboratory for analysing compositional effects on seismic properties of cratonic mantle as it provides samples of variously depleted peridotite, pyroxenite and compositionally diverse eclogite. The modelling used in this study determines the contrast in seismic properties in different mantle lithologies and the range of seismic velocities expected for a naturally heterogeneous mantle section. We find that eclogite stands out when compared with other mantle rocks because of its faster increase of compressional and sheared wave velocities (Vp and Vs) with depth. Our calculations also suggest a distinct seismic signature indicating depletion in the cratonic mantle.

2. Samples and analytical techniques

This study is based on mantle xenoliths derived from the Jericho and 5034 kimberlites on the Slave craton (northern Canada). The Slave craton, stabilized at 2.6 Ga, represents a small Archean nucleus to the larger Proterozoic North American craton. The kimberlites are located in the northern and southeastern parts of the Slave Craton (Fig. 1). The Jericho pipe is dated as Middle Jurassic (172 ± 2 Ma by Rb-Sr and U-Pb geochronology) and the 5034 kimberlite of the Gahcho Kue cluster as Middle Cambrian (539 ± 2 Ma by the Rb-Sr method on phlogopite) (Heaman et al., 2003). We studied all mantle lithologies typical of the cratonic mantle, i.e., peridotite, pyroxenite and eclogite. The mantle peridotites and pyroxenite that are the basis for this study were previously described in several papers (Table 1) which reported the petrology and bulk and mineral compositions.

Eclogitic xenoliths make up ~ 25% of all mantle xenoliths in the Jericho kimberlite, and ~ 18% in the 5034 kimberlite. This work presents new data on mineral compositions, pressures and temperatures of equilibrium, and modal mineralogy for 13 new samples of the Jericho eclogite whose elastic properties are modelled (Supplementary Electronic Tables 1 and 2).

Mineral modes were estimated using image analysis techniques applied to scanned images of 2 cm × 4 cm thin sections of eclogite. Digital images were acquired using a Polaroid Sprintscan® 35 on thin sections that were cut thicker than normal (>30 microns) to lend stronger body colour to garnet. The images were captured with a polarized light source, enhanced, and analysed using free image analysis software NIH Image 1.62 (http://rsb.info.nih.gov/nih-image/). It was possible to determine modal abundances of secondary clinopyroxene and garnet because of their distinctly finer grain size which give them darker colours. The precision on the modal estimates by the image analysis method is estimated to be 2.5% (Kopylova and Russell, 2000).

Minerals in the eclogites were analysed using an automated CAMECA SX-50 microprobe (Department of Earth and Ocean Sciences, University of British Columbia, Canada) at an accelerating voltage of 15 kV and with a 20-mA beam current. On-peak counting times were 10 s for major elements, and 60 s for K in clinopyroxene and Na in garnet. Primary phases in a sample were analysed as 8–15 points in cores and rims of 4–5 grains. Analyses with poor stoichiometry and totals were excluded and mineral compositions were averaged over 3–12 analyses for homogeneous phases or presented as individual analyses for inhomogeneous minerals (Supplementary Electronic Table 2).

In eclogite of the 5034 kimberlite, all primary minerals, with the exceptions of garnet (20–30%) and rutile (2%), are completely altered to serpentine, chlorite and phlogopite. Limited observations on mineralogical and textural characteristics of these rocks suggest that the 5034 eclogite resemble that of Jericho. Severe weathering of the 5034 eclogite prevents detailed petrographic and petrophysical work, and only Jericho eclogite was used as the basis for this study.

Eclogite xenoliths from Jericho are fresh and are composed of primary pyrope, omphacite and rutile, with occasional zircon, olivine, orthopyroxene, kya-

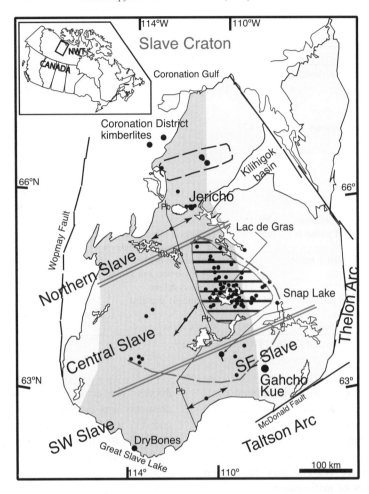

Fig. 1. Schematic map of the Slave craton (Northwest Territories, Canada—see inset), showing the location of kimberlite pipes (black dots). Double lines designate the boundaries between northern, central and southern lithospheric domains as distinguished by distinct compositions of garnet in kimberlite concentrates (Grütter et al., 1999). The SW and SE Slave terranes may be separated by the Pb isotopic boundary of Thorpe et al. (1992) (thin solid line). The darker area is the postulated surface and subsurface extent of the Central Slave Basement Complex (protocraton of Ketchum and Bleeker, 2001). Orientation of the S-wave polarization (bars with arrows) indicates the directions of anisotropy of the mantle (Bank et al., 2000). Also shown are the minimum extents of a shallow ultra-depleted layer (horizontal lined pattern) and of the deeper Archean lherzolitic layer (Griffin et al., 1999b) of the Central Slave (dashed outlines).

nite, apatite and ilmenite. Omphacitic clinopyroxene commonly comprises 60–75% of the eclogite (Supplementary Electronic Table 2). Detailed mineralogical work on these specimens revealed the presence of two late mineral assemblages. The first, which we consider mantle metasomatic and pre-kimberlitic, relates to partial recrystallization of garnet and clinopyroxene and their replacement by phlogopite and amphibole. Spongy rims of recrystallized, secondary clinopyroxene enriched in Ti, Ca and Mg, and depleted in the jadeitic component (Supplementary Electronic Table 1), may overgrow primary omphacite and replace up to half of it. Pyrope may also be overgrown by rims of late, Mg ± Ti-rich and Ca-poor garnet or, more commonly, by amphibole. The second mineral assemblage comprises epidote, chlorite and serpentine that mark shallow retrograde alteration of the eclogite. The Jericho eclogite is subdivided into two groups based on the presence of massive or foliated fabric; the foliated texture is partly controlled by preferential replacement of garnet and clinopyroxene by secondary volatile-rich phases along specific planes. The

Table 1
Types and characteristics of Slave mantle xenoliths

Location	Rock type	Depth, km[a]	Comments	Reference
Jericho (N Slave)	Low-*T* spinel peridotite	35–100	Chemically depleted	mineral analyses are from McCammon and Kopylova (accepted pending revisions); modes and bulk compositions are from Kopylova and Russell (2000)
Jericho (N Slave)	Low-*T* spinel-garnet peridotite	80–170		Petrology and mineral analyses are from Kopylova et al. (1999a); modes and bulk compositions are from Kopylova and Russell (2000)
Jericho (N Slave)	Low-*T* garnet peridotite	120–185		
Jericho (N Slave)	High-*T* garnet peridotite	165–194	Deformed, not equilibrated on a steady-state geotherm	
Jericho (N Slave)	Low-*T* and high-*T* fertile garnet peridotite		Enriched in modal clinopyroxene and garnet	
Jericho (N Slave)	Pyroxenite	200–215	Megacrystalline	
5034 (SE Slave)	Low-*T* spinel peridotite	35–100	Chemically depleted	Kopylova and Caro (2004)
5034 (SE Slave)	Low-*T* garnet peridotite	215–260	Coarse or deformed	

[a] Estimated according to the Brey and Köhler (1990) thermobarometry.

massive and foliated eclogites differ in mineral composition (Kopylova et al., 1999b), bulk composition and origin (Kopylova, 2003). The protolith for the foliated eclogite may have been low-pressure mafic rocks that formed partly as plagioclase cumulates. The protoliths for the massive eclogite may have been deeper high-P cumulates of mafic magmas.

3. Experimental methods

Laboratory determination of acoustic velocity and density was performed at the Rock Physics Lab at Purdue University on mini-cores of approximately 2 cm in diameter and more than 3 cm in length. Where possible, three cores were cut for a sample; A-core was perpendicular to foliation, B-core was parallel to lineation in the foliation plane, and C-core was perpendicular to lineation in the foliation plane. For smaller samples only A- and B-cores were cut. Compressional and shear wave velocities polarized at A and B directions were measured on each core samples at hydrostatic confining pressures up to 1000 MPa using the pulse transmission technique (Christensen, 1965) and 1-MHz trans-

ducers. The error in the laboratory velocity measurements was evaluated to be less than 0.5% for Vp and 1% for Vs (Christensen and Shaw, 1970). The bulk density of each core was calculated from its mass and dimensions.

4. Measured elastic properties

Acoustic velocities and density were measured in four eclogite xenoliths chosen to represent the foliated and massive types. Measured Vp and Vs at different confining pressures (Supplementary Electronic Table 3) are plotted on Fig. 2. An extrapolation of the linear fits of velocity–pressure curves at $P > 600$ MPa, where microcracks are closed, yields Vp and Vs at room temperature and pressure (Fig. 2). Vp in Jericho eclogite varies from 6.25 to 7.9 km/s, and Vs from 3.5 to 4.4 km/s. The velocities are strongly controlled by modal mineralogy. In fresh eclogite rocks 26-6 and F6NEcl Vp = 7.25–7.9 and Vs = 3.9–4.4 km/s, whereas in retrograde eclogite, where primary omphacite is almost totally replaced by chlorite + epidote ± phlogopite and amphibole, the velocities are much lower. Massive eclogite 26-6

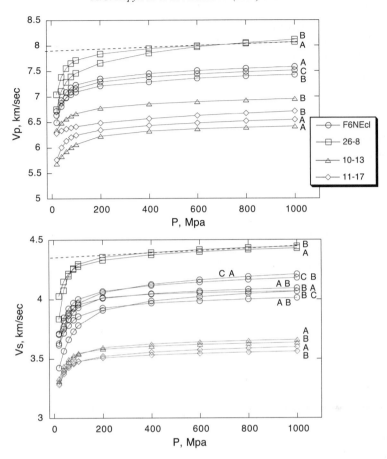

Fig. 2. Laboratory-measured velocity–pressure curves for the Jericho eclogite. Letter after sample number indicates direction of the wave propagation. For Vs measurements, the first letter designates the propagation direction, second letter—the vibration direction. Dashed lines extrapolate seismic velocities for atmospheric pressures.

shows Vp and Vs values that exceed those for all samples of foliated eclogite.

Vp anisotropy varies in Jericho eclogite from 0% to 8.1%. Massive fresh eclogite 26-6 shows no anisotropy, fresh anisotropic sample F6NEcl has 2% Vp anisotropy, and a maximum anisotropy of 8% is recorded in sample 10–13, where it is density-related. Large variations in density from 2.971 in the A direction to 3.108 in the B direction reflect uneven development of the secondary chlorite–epidote–phlogopite aggregate (65% of the rock).

Vs anisotropy (1.3–3.4%) is detected only in sample F6NEcl. In all other samples it does not exceed 1%. Shear wave splitting was measured only in sample F6Necl. It is relatively large (2%) only in the A direction and is practically nonexistent in the B and C directions. It reflects strong polarizing properties of the foliation plane where waves parallel and normal to the lineation propagate with different velocities.

5. Calculated elastic properties

5.1. Elastic properties at the surface

Seismic velocities (Vp, Vs) of the mantle rocks were estimated from high-precision single crystal elastic moduli and volume fractions of constituent minerals using appropriate mixture rules. They describe variations of effective elastic moduli of polymineralic composites as a function of their end-member elastic moduli and volume fractions. It has been shown that the mean Vp of a polymineralic rock is exclusively controlled by the volume fractions of its constituent minerals, while grain shape, crystallographic preferred orientations, anisotropy and other perturbations have minimal effects (Ji et al., 2003). Therefore, we computed average velocities for mantle rocks as for isotropic aggregates with uniform distribution of crystal orientations. Several mixture rules were proposed for such calculations: the Reuss average, the Voigt average, their arithmetic (known as the Hill) average or geometrical mean (reviewed in Ji et al., 2003), and the average of the Hashin–Shtrikman bounds (Bina and Helffrich, 1992). Since the Hill average remains the most widely used mixture rule for predictions of seismic velocities in isotropic polycrystalline mixtures (e.g. Long and Christensen, 2000; Gregoire et al., 2001), and gives results similar to the more complex iterative Hashin–Shtrikman method (Bina and Helffrich, 1992), it was employed for the computations.

These methods require knowledge of the volume percentage of mineral end-members in a rock. Here, volume percentages were calculated based on mineral modes and volume fractions of end-member components which, in turn, were estimated based on compositions of mineral solid solutions. We expressed the mineral solid solutions as ideal mixtures of appropriate components with known elastic properties. Olivine, orthopyroxene and spinel were recalculated as mixtures of Fe and Mg end-members, whereas clinopyroxene and garnet were modelled more elaborately (Table 2). Cr_2O_3 in garnet, which may range up to 20 wt.% in cratonic peridotite, was accounted for using an uvarovite end-member. Although the andradite component in mantle garnets is not very significant, we assigned all Fe^{3+} estimated stoichiometrically to andradite. Clinopyroxene in studied rocks contains significant amounts of Na, Al and Cr. Since elastic moduli data are not available for Cr end-members of clinopyroxene, and the elastic

Table 2
Densities and elastic moduli for minerals and mineral end-members

Component	Composition	Adiabatic bulk modulus K_S, GPa	Shear modulus G, GPa	Density, g/cm^3[a]	Source
Forsterite	Mg_2SiO_4	128	81	3.222	Duffy and Ahrens (1995), Li et al. (1996), Zha et al. (1996, 1997)
Fayalite	Fe_2SiO_4	128	50	4.404	Duffy and Ahrens (1995), Li et al. (1996), Zha et al. (1996, 1997)
Enstatite	$Mg_2Si_2O_6$	104	74.9	3.215	Flesch et al. (1998), Vacher et al. (1998)
Ferrosilite	$Fe_2Si_2O_6$	124	54	4.014	Angel and Hugh-Jones (1994), Chai et al. (1997)
Jadeite	$NaAlSi_2O_6$	126	84	3.320	Zhao et al. (1997), Kandelin and Weidner (1998)
Diopside	$CaMgSi_2O_6$	105	67	3.277	Zhang et al. (1997), Sumino and Anderson (1984)
Hedenbergite	$CaFeSi_2O_6$	118	61	3.657	Zhang et al. (1997), Sumino and Anderson (1984)
Pyrope	$Mg_3Al_2(SiO_4)_3$	173	92	3.600	Vacher et al. (1998)
Almandine	$Fe_3Al_2(SiO_4)_3$	180	99	4.328	Chen et al. (1996)
Grossular	$Ca_3Al_2(SiO_4)_3$	168	107	3.597	Isaak et al. (1992)
Andradite	$Ca_3Fe_2(SiO_4)_3$	157	90	3.836	Bass, 1986
Uvarovite	$Ca_3Cr_2(SiO_4)_3$	162	92	3.85	Bass, 1986
Spinel	$MgAl_2O_4$	197.9	108.5	3.582	Yoneda (1990), Chang and Barsch (1973)
Hercinite	$FeAl_2O_4$	210.3	84	4.258	Wang and Simmons, 1972
Phlogopite[b]		50	25.2	2.820	Yang and Prewitt, 2000
Hornblende		87.1	43.2	3.120	Hearmon, 1984
Ilmenite		212.3	132.3	3.795	Weidner and Ito, 1985
Apatite		84.3	60.7	3.200	Hearmon, 1984
Rutile		211.5	113.1	4.240	Isaak et al., 1998
Chlorite		81.0	43.1[c]	2.800	Collins and Catlow (1992); Welch and Crichton (2002)

[a] Densities are after Duffy and Anderson (1989) except when indicated.

[b] Density is from Christensen (1989), and G is calculated from Vs of Christensen (1989).

[c] Data for mica.

moduli of Ca-tchermakite (Ca-Ts) were shown to be reasonably well approximated by those of jadeite (Gregoire et al., 2001), the "jadeite" abundances in Table 3 refer to combined volume percentages of cosmochlore ($NaCrSi_2O_6$), Ca-tchermakite ($CaAlAlSiO_6$), and jadeite ($NaAlSi_2O_6$). Fortunately, in eclogites, where modes of clinopyroxene are high, molar fractions of Ca-Ts and cosmochlore in omphacites are negligible, and the absence of experimentally determined elastic parameters for them does not affect the calculations. We used the most recent experimentally determined densities and elastic properties available for the mineral end-members (Table 2).

The resulting volume proportions of fixed composition minerals and end-member components for individual xenoliths of peridotite and pyroxenite are given in Supplementary Electronic Table 4. Various types of peridotites were also characterized by mean mineral modes (Kopylova and Russell, 2000; Kopylova and Caro, 2004). Calculated abundances of end-member components for them and for individual eclogite xenoliths are listed in Table 3.

Seismic velocities modelled at the surface can be checked against those measured experimentally. An excellent match between the two is observed in sample 26-6 (7.92 vs. 7.82 km/s and 4.56 vs. 4.36 km/s; Table 3). In all other samples, measured velocities are lower than those calculated, even when secondary chlorite is compensated for. The difference is 0.2 km/s in a sample with 25% chlorite pseudomorphs, and 1.1 km/s in severely altered samples 11–17 and 10–13 with 65% chlorite. A factor that contributes to this discrepancy may be a higher porosity of altered retrograde eclogite, which makes the pores remain open at higher pressures and leads to underestimation of measured ultrasonic velocities in xenoliths (Soedjatmiko and Christensen, 2000).

5.2. Elastic properties at depth

Knowledge of accurate ambient pressures and temperatures for mantle lithologies is central to modelling mantle seismic profiles because seismic properties are strongly dependent on $P-T$ variations. Pressures and temperatures were estimated by two methods, both of which satisfy available petrological constraints for the Jericho and 5034 xenolith suites, i.e. (1) the two-pyroxene geothermometer of Brey and

Köhler (1990) (BKN) and the Al-in-Opx geobarometer of Brey and Köhler (1990) (BK); and (2) the geothermometer of Finnerty and Boyd (1987) (FB) and the geobarometer of MacGregor (1974) (MC) (Table 3). Pseudo-univariant pressure–temperature lines for eclogite samples have been calculated by garnet-clinopyroxene geothermometry using both the Ellis and Green (1970) (EG) and the Ai (1994) formulations with all Fe calculated as Fe^{2+}. The rationale for this choice was discussed in detail in Kopylova et al. (1999b). The Ai thermometry was used only in conjunction with the FB-MC peridotitic geotherm, whereas the EG thermometry was used only in conjunction with the BKN-BK peridotitic geotherm (Table 3). These combinations of eclogitic and peridotitic thermobarometry were chosen to satisfy a petrological constraint based on the diamondiferous character of some of the Jericho eclogites (Kopylova et al., 1999b).

Two methods were used to estimate seismic wave velocity at depth. First, calculated seismic wave velocities at the surface were adjusted for pressures and temperatures corresponding to mineral equilibration at depth. Temperature and pressure derivatives employed in these computations are listed in Table 4. Pressure derivatives for eclogite were calculated from high-pressure measurements on samples JDF6Necl and 26-6, which are the least altered samples and are practically chlorite-free. The temperature derivatives for eclogites and all derivatives for peridotite and pyroxenite were calculated based on high-frequency ultrasonic studies reported elsewhere (Table 4). We considered only measurements for unaltered samples that were close in mineralogy to the samples studied here. Table 4 also shows that the generalized values of seismic wave velocity derivatives for all mantle and lower crust lithologies, and for pyrolite, are similar to those used in this study. The pyrolite derivatives were estimated based on the Mie–Gruneisen's equation of state and single crystal elastic data at temperatures of 500, 1000 and 1500 °C (Bina and Helffrich, 1992).

The second method directly computes Vp and Vs of mineral end-members at high P and T by accounting for the pressure and temperature dependence of their elastic moduli and densities. We used the method of Fei (1995) for a subset of representative samples of eclogite and peridotite. The solution requires tabulated

Table 3
Calculated abundances (vol.%) of end-member components and seismic speeds at surface and at depth for averaged types of peridotite and for individual samples of eclogite

Rock type	Jericho peridotites and pyroxenite						5034 peridotites	
	Spl peridotite	Spl-Gar peridotite	Low-T Gar peridotite	Fertile Gar peridotite	High-T Gar peridotite	Pyroxenite	Spl peridotite	Low-T Gar peridotite
Average of	12	7	6	5	9	2	5	24
Forsterite	67.7	66.6	67.9	56.5	71.5	20.4	74.9	72.1
Fayalite	4.0	4.2	4.9	3.4	5.5	1.8	4.1	4.6
Enstatite	24.3	20.2	16.3	15.8	14.3	26.4	18.0	14.0
Ferrosilite	1.5	1.2	1.1	0.9	1.0	2.4	1.2	0.8
Jadeite	0.1	0.3	0.6	1.6	0.3	3.4	0.0	0.3
Diopside	1.3	2.9	3.9	8.7	2.1	25.1	0.6	2.2
Heldenbergite	0.1	0.2	0.3	0.6	0.2	2.8	0.0	0.2
Almandine	0.0	0.6	0.8	1.6	0.7	3.1	0.0	0.8
Pyrope	0.0	2.7	3.7	9.2	3.7	12.5	0.0	4.0
Grossular	0.0	0.1	0.0	0.0	0.0	0.4	0.0	0.0
Andradite	0.0	0.0	0.0	0.0	0.0	0.0	0.0	0.1
Uvarovote	0.0	0.5	0.6	1.7	0.7	1.7	0.0	1.0
$MgAl_2O_4$	0.6	0.2	0.0	0.0	0.0	0.0	0.8	0.0
$FeAl_2O_4$	0.4	0.4	0.0	0.0	0.0	0.0	0.4	0.0
Total	100.0	100.0	100.0	100.0	100.0	100.0	100.0	100.0
Rock Mg-number	92.9	92.0	91.8	89.0	90.6		92.6	91.2
Density	3.287	3.308	3.318	3.340	3.319	3.380	3.286	3.314
Vp Hill, km/s	8.27	8.29	8.29	8.31	8.31	8.11	8.32	8.34
Vs Hill, km/s	4.85	4.85	4.84	4.84	4.84	4.73	4.87	4.86
Lower P–T limit								
T FB, deg C	450	650	760	1020	1120	1115	450	970
P MC, GPa	1.09	2.42	3.40	5.17	5.77	5.65	1.10	5.20
Vp at depth, km/s	8.14	8.18	8.18	8.27	8.28	7.68	8.19	8.33
Vs at depth, km/s	4.77	4.78	4.80	4.83	4.83	4.73	4.79	4.86
T BKN, deg C	450	640	840	1070	1280	1200	610	1130
P BK, GPa	1.1	2.5	3.6	4.9	5.0	6.0	1.1	6.5
Vp at depth, km/s	8.14	8.19	8.19	8.21	8.10	7.63	8.09	8.37
Vs at depth, km/s	4.77	4.79	4.78	4.79	4.72	4.73	4.74	4.89
Higher P–T limit								
T FB, deg C	780	860	1090	1190	1250	1135	680	1190
P MC, GPa	3.30	4.00	5.70	5.85	6.35	5.71	3.00	5.80
Vp at depth, km/s	8.18	8.22	8.18	8.24	8.26	7.66	8.25	8.26
Vs at depth, km/s	4.80	4.81	4.83	4.81	4.82	4.73	4.83	4.83
T BKN, deg C	750	1000	1100	1120	1310	1250	830	1290

Partial continuation at top of table (BK geotherm rows):

P BK, GPa	3.00	4.90	5.50	5.80	5.80	6.50	3.00	7.90
Vp at depth, km/s	8.16	8.23	8.24	8.28	8.17	7.63	8.16	8.43
Vs at depth, km/s	4.79	4.82	4.81	4.83	4.77	4.74	4.78	4.93

Jericho eclogite — first 13 columns (6-11 … JDF6NEcl) are Primary paragenesis; last 5 columns (55-4 … 47-8) are Mantle metasomatic paragenesis.

Sample	6-11	20-7	16-4	55-4	JDF6N-Ecl13	52-5	47-2	42-3	47-8	26-6	11-17	10-13	JDF6NEcl	55-4	52-5	47-2	42-3	47-8
"Jadeite"	9.4	13.5	8.2	18.1	17.9	22.1	20.1	17.9	26.3	7.4	36.4	38.6	28.2	16.2	21.7	18.9	14.1	13.9
Diopside	52.9	30.6	52.8	24.7	35.3	37.0	40.5	33.4	24.0	53.2	22.8	25.2	33.2	26.1	36.3	40.5	38.7	36.0
Hedenbergite	12.9	5.8	7.3	4.7	8.0	5.0	8.5	9.8	10.2	6.3	8.9	9.8	7.1	5.0	4.9	8.6	8.4	12.0
Almandine	7.3	15.4	8.0	15.3	15.1	8.0	11.8	14.0	15.6	5.1	12.2	10.4	9.9	17.3	7.8	12.7	13.9	13.3
Pyrope	6.9	18.0	19.0	21.8	16.2	19.7	12.0	11.2	8.4	19.8	6.7	5.7	9.2	20.6	18.4	10.8	9.2	5.9
Grossular	2.2	8.7	3.8	15.1	6.5	4.7	6.6	5.4	8.3	2.6	9.5	7.9	7.9	14.2	4.1	5.9	4.5	5.9
Andradite	0.3	0.0	0.6	0.0	0.4	0.3	0.3	0.3	0.7	0.3	0.0	0.0	0.0	0.0	0.3	0.3	0.3	0.5
Uvarovite	0.0	0.0	0.0	0.0	0.0	0.0	0.0	0.0	0.0	0.3	0.0	0.0	0.0					
Orthopyroxene	0.3	0.0	0.0	0.0	0.0	0.0	0.0	0.0	0.0	0.0	0.5	0.0	0.0					
Apatite	0.0	0.0	0.0	0.3	0.0	0.0	0.2	0.0	0.5	0.0	3.0	0.0	0.0	0.3	0.0	0.2	0.0	0.0
Rutile	2.1	0.0	0.0	0.3	0.2	0.0	0.0	5.1	0.0	0.0	0.0	2.5	0.0	0.3	0.0	0.0	5.1	0.5
Ilmenite	0.0	0.0	0.4	0.0	0.0	0.0	0.0	0.0	0.0	0.4	0.0	0.0	0.0	0.0	3.2	1.1	2.9	5.9
Phlogopite	5.7	8.0	0.0	0.0	0.4	3.0	0.0	2.9	6.0	4.7	0.0	0.0	0.0	0.0	3.1	1.0	2.9	5.9
Chlorite													4.5					
Total	100.0	100.0	100.0	100.0	100.0	100.0	100.0	100.0	100.0	100.0	100.0	100.0	100.0	100.0	100.0	100.0	100.0	100.0
Density, g/cm³	3.431	3.516	3.471	3.585	3.549	3.453	3.505	3.560	3.526	3.413	3.536	3.508	3.455	3.599	3.439	3.503	3.538	3.476
Vp Hill, km/s	7.75	7.97	8.10	8.44	8.20	8.13	8.16	8.09	8.00	7.92	8.29	8.25	8.03	8.40	8.09	8.09	7.98	7.82
Vs Hill, km/s	4.46	4.57	4.68	4.87	4.73	4.69	4.72	4.65	4.61	4.56	4.82	4.80	4.65	4.84	4.69	4.68	4.58	4.52
Vp measured, km/s										7.82								
Vs measured, km/s										4.36								
T Ai-FB, deg C	713	902	841	1200	861	1013	840	861	1012	n/d[a]	1140	1125	983					
P Ai-FB, GPa	2.93	4.38	4.00	6.20	4.05	5.21	3.90	4.04	5.17	n/d	5.95	5.90	5.00					
Vp at depth, km/s	8.03	8.45	8.53	9.14	8.63	8.71	8.57	8.52	8.58		8.97	8.92	8.59					
Vs at depth, km/s	4.53	4.70	4.80	5.07	4.85	4.87	4.83	4.77	4.78		5.02	5.00	4.82					
T EG-BK, deg C	803	911	928	1290	886	1003	871	867	921	n/d	1012	1002	946					
P EG-BK, GPa	3.70	4.13	4.27	5.50	3.97	4.80	3.86	3.82	4.20	n/d	4.86	4.77	4.38					
Vp at depth, km/s	8.07	8.40	8.55	8.98	8.61	8.64	8.55	8.48	8.44		8.81	8.76	8.49					
Vs at depth, km/s	4.54	4.68	4.80	5.00	4.84	4.84	4.82	4.75	4.72		4.96	4.94	4.78					
Mean Vp anisotropy										0.20%	2.36%	8.10%	2.03%					
Mean Vs anisotropy										0.36%	0.97%	0.36%	1.3–3.38%					
Vs splitting[b]													0.5–2.04%					

[a] Not determined as mineral compositions are outside the range of calibration for the used thermometers.

[b] SKS splitting is calculated as an average for measurements at 600, 800 and 1000 MPa as $200\% * (V_A - V_B)/(V_A + V_B)$.

Table 4
Pressure and temperature derivatives of seismic velocities for mantle rocks

Rock type	Description and source	dVp/dP, km/s $\times$ Mpa $\times 10^{-5}$	dVp/dT, km/s $\times$ deg $\times 10^{-5}$	dVs/dP, km/s $\times$ MPa $\times 10^{-5}$	dVs/dT, km/s $\times$ deg $\times 10^{-5}$
Peridotite	Harzburgite with <16% serpentine (Long and Christensen, 2000)	10.7		6.4	
	Peridotite (Christensen, 1989)		−60		−35
	Computation for pyrolite with 15% garnet (Bina and Helffrich, 1992)		−44 at 500–1000 °C		−34
			−56 at 1000–1500 °C		
Pyroxenite	Pyroxenite (Christensen, 1989)		−100		−35
Eclogite	Measurements for 26-6A, F6NEcl	18.1		7.9	
	Chlorite-free eclogite (Kern et al., 1999)		−36.1		−24.1
Mantle and lower crust rocks	Jackson et al., 1990; Gregoire et al., 2001	10	−50		

Temperature derivatives are calculated at $P = 300–600$ MPa for $T = 600–1000$ °C (100–300 °C for pyroxenite).
Pressure derivatives are calculated for 25 °C, $P = 600–1000$ MPa to eliminate the effects of microfracture closing (Long and Christensen, 2000; Kern et al., 1999).

Table 5
P and *T* derivatives of elastic moduli, thermal expansion coefficients and Andersen–Gruneisen parameters calculated for selected Slave peridotite and eclogite

	Spl peridotite		Eclogite	Source of elastic data for mineral end-members
	11–18	44–12	6–11	
Thermal expansion coefficient, αo (10^{-6})[a]	27.8	27.2	29.4	Fei (1995)
Andersen–Gruneisen parameter, δ_T	4.1	4.2	6.1	Bina and Helffrich (1992), Fei (1995)
First-order *P* derivative K_S'	4.0	4.3	5.8	Bina and Helffrich (1992), Vacher et al. (1998), Hofmeister and Mao (2003)
First-order *P* derivative G'	1.1	1.1	1.6	Bina and Helffrich (1992), Vacher et al. (1998), Hofmeister and Mao (2003)
First-order *T* derivative K_S/dT, GPa/deg	−0.016	−0.015	−0.015	Bina and Helffrich (1992), Vacher et al. (1998)
First-order *T* derivative G/dT, GPa/deg	−0.014	−0.013	−0.010	Bina and Helffrich (1992), Vacher et al. (1998)
At surface				
Density, g/cm^3	3.284	3.291	3.431	
K_S, GPa	125.6	122.1	110.0	
G, GPa	77.9	77.2	65.2	
Vp, km/s	8.36	8.27	7.58	
Vs, km/s	4.87	4.84	4.36	
At depth				
T, °C	715	647	713	
P, GPa	3.00	2.50	2.93	
Density, g/cm^3	3.296	3.301	3.439	
K_S, GPa	127.0	123.6	116.8	
G, GPa	68.5	69.0	62.9	
Vp, km/s	8.138	8.083	7.637	
Vs, km/s	4.558	4.572	4.275	
Vp calculated using seismic speed derivatives, km/s	8.27	8.16	7.86	
Vs calculated using seismic speed derivatives, km/s	4.82	4.78	4.43	

[a] αo independent of temperature as tabulated in Fei (1995).

data on first-order individual end-member derivatives of isothermal and adiabatic bulk moduli K_T and K_S, shear modulus G, as well as their thermal expansion coefficients and Andersen–Gruneisen parameters (Table 5). These data are now published for almost all end-member components, including non-quadrilateral pyroxenes and garnets, and cover real mineralogies of rocks with sufficient accuracy. Seismic velocities at depth computed in this manner are listed in Table 5.

6. Results

Calculated seismic velocities for the Slave peridotites and eclogite are plotted against depth on Fig. 3. The figure shows individual samples using two sets of pressures and temperatures that are equally

good at describing their equilibrium conditions of origin. These two P–T solutions define the similar xenolith-derived geotherm, but may significantly shift depth position of individual samples within the geotherm. The scatter on the resulting plot thus reflects (1) uncertainty in the estimated P–T conditions of rocks, and (2) varying mineral proportions in rocks found at a given depth. The former uncertainty translates to an average velocity variation of 0.02 km/s, although in rare samples the difference can be as high as 0.15 km/s in Vp and 0.1 km/s in Vs. The calculated spread of Vp at a given depth that relates to varying mantle mineralogy reaches 0.3 km/s in peridotite and 0.4 km/s in eclogite. This variation reflecting mantle heterogeneity exceeds errors inherent to methodology of Vp–Vs calculations. An example of the latter is the difference of 0.05 km/s

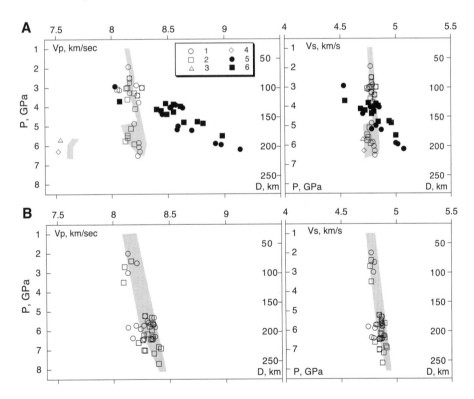

Fig. 3. Variation of seismic velocities with depth in the Jericho mantle (A) and the 5034 mantle (B). Symbols for seismic velocities estimated for individual xenoliths are: 1 and 2—peridotites at P–T conditions calculated according to the FB-MC method and the BKN-BK method; 3 and 4—pyroxenites at P–T conditions calculated according to the FB-MC method and the BKN-BK method; 5 and 6—eclogites estimated according to the Ai-FB method and the EG-BK method. Estimates for eclogites and pyroxenites are not available for the 5034 kimberlite. Grey fields are V–depth profiles computed for average pyroxenite and average peridotites by type at P–T conditions estimated according to the FB-MC method and the BK-BKN method (Table 3). There is a good agreement between seismic velocity estimates for individual and averaged peridotites, but the pyroxenite sample is not representative of a larger group of averaged pyroxenites.

(<1 rel.%) and 0.15 km/s (<3 rel.%) between the Reuss velocities and the Hill velocities for peridotite and eclogite (Supplementary Electronic Table 4).

Fig. 3 also illustrates seismic wave velocities in various peridotite types with averaged mineralogy. These velocities are computed for lower and upper limits of pressure and temperature determined for the rock types of Table 3, e.g. for spinel peridotite immediately below the Moho, and to $P = 33$ kb, respectively.

The profiles show that Vp in the Jericho peridotite increases slightly from 8.2 km/s below the crust to 8.3 km/s at 210 km. A similar gentle increase with depth from 8.1–8.2 km/s below the Moho to 8.45 km/s at 260-km depth is observed for the 5034 peridotites. The shear wave velocity in peridotite increases with depth even less, from 4.8 to 4.85 km/s. Because recalculation of surface seismic wave velocities to ambient P–T conditions at depth decreases them (Table 3), commonly observed increasing velocity profiles in the upper mantle (Kennett and Engdahl, 1991) must be controlled by consistent changes in the mantle mineralogy with depth. Pyroxenites have significantly lower Vp's (7.5–7.6 km/s) and Vs's similar to those of peridotite. The seismic velocity profiles for eclogite feature a steep increase with depth, from 8.0 km/s at 100 km to 9.2 km/s at 200 km, "cutting through" the peridotite profile. The Vs profile for eclogite also cross-cuts the analogous peridotite profile.

The contrasting velocity–depth gradients of peridotite and eclogite are expected based on experimental pressure and temperature derivatives of their elastic constants (Table 4). In peridotite, an increase in seismic wave velocities due to higher pressure ($\Delta Vp^P \sim 0.35$ km/s) is offset by a decrease due to higher temperature ($\Delta Vp^T \sim -0.38$ km/s). This leads to practically constant modelled velocities independent of depth, or to a slight decrease in Vp with depth along higher geotherms (Jackson et al., 1990, O'Reilly et al., 1990; Weiss et al., 1999). In eclogite, seismic waves propagate more quickly with increasing pressure ($\Delta Vp^P = 0.5$ km/s) but the temperature effect is less pronounced ($\Delta Vp^T = -0.2$ km/s). The disparity in the derivatives for eclogite and peridotite arises from contrasting first-order pressure derivatives of bulk isotropic moduli. They are much lower in peridotite than in eclogite (Table 5) because the olivine modulus increases with pressure at a lower rate than that of pyroxenes and

garnet (compare their Ks' in Table 1 of Hofmeister and Mao, 2003). Another effect that contributes to the contrasting behavior of eclogite and peridotite at depth is the lower compressibility of eclogite at high pressures. Lower Andersen–Gruneisen parameters for eclogite reflect its smaller increase in density at depth ($\Delta \rho \sim 0.008$ in eclogite vs. 0.012 g/cm^3 in peridotite, Table 5).

The modelled profiles represent the expected wave velocities in mantle lithologies at depth at times preceding kimberlite eruption, i.e. in the Jurassic for Jericho and in the Cambrian for the 5034. As such, they reflect Jurassic and Cambrian steady-state geotherms. However, the deeper part (>160 km) of the Jericho profiles may not be representative of the steady-state mantle. Temperatures in this part of the mantle were reported to be 100–200 °C higher due to transient thermal perturbations and interactions with asthenospheric fluids (Kopylova et al., 1999a). Thus, at Jericho, the modelled velocities at depths greater than 160–190 km may be slightly lower than those typical of the ambient mantle, recording the unusual state of the upper mantle that precedes generation of kimberlitic magma.

7. Discussion

7.1. Model profiles in comparison with seismic surveys of the Slave craton

The predominantly peridotitic mantle of the Slave craton is modelled to have Vp increasing from 8.2 to 8.4 km/s and Vs from 4.8 to 4.9 km/s at depths of 35–260 km. These velocities can be compared to corresponding values derived from seismic studies of the Slave mantle. Two sets of data based on travel time Ps tomography are available for the Slave craton. Ramesh et al. (2002) report that the Canadian region has a normal upper mantle, in accordance with the *IASP91* model of Kennett and Engdahl (1991), and is underlain by a uniform normal mantle transition zone. The study of Bank et al. (2000) found that the Slave mantle is slightly faster than the global average of the *IASP91* model. An analysis of receiver functions with respect to the peak defining the discontinuity at 410 km can estimate the magnitude of the higher velocities postulated for the Slave mantle. Assuming that the shift of the peak to earlier times relative to *IASP91* was produced in the upper 410 km of the mantle, S-wave velocities should be higher by 0.05 km/s (Bostock, personal communication). These are the only velocity estimates available for the Slavic mantle. Here, the quilt-like structure of the mantle consists of laterally and vertically distinct domains (Grütter et al., 1999; Jones et al., 2001). This prevents any extrapolations of absolute seismic velocites estimated for the SW Slave in refraction studies (Fernandez Viejo and Clowes, 2003) to other parts of the Slave craton.

The range of seismic velocities permitted by mineralogy of Slave mantle rocks would expand if anisotropy is accounted for. It is generally thought to relate to lattice preferred orientations (LPO) of olivine and pyroxenes which can be measured microstructurally and experimentally (i.e. Soedjatmiko and Christensen, 2000; Ben-Ismail et al., 2001). The contribution of LPO to anisotropy of cratonic peridotite was calculated to be in the range of 2–8% for P-waves and 1–6% for S-waves for ~ 50 Kaapvaal xenoliths irrespective of their coarse or sheared textures (Ben-Ismail et al., 2001). A combined study of laboratory measurements and numerical calculations on a smaller set of cratonic

xenoliths yielded similar values, 4.4–5.4% for Vp and 3.4–4.4% for Vs (Long and Christensen, 2000). In the absence of our own LPO data for Slave peridotite samples, we accept these estimates as an approximation of their average anisotropy. The anisotropy differentially expands the permissible limits of seismic velocities in the Jericho peridotitic mantle to higher values (Fig. 4). The regular *IASP91* mantle, as suggested for the Slave craton by Ramesh et al. (2002), and a 0.05 km/s faster mantle (Bostock, personal communication) fit equally well within the range of modelled peridotitic Vp's.

7.2. Effect of chemical depletion on elastic properties

Our calculations permit a quantitative assessment of bulk compositional effects on seismic velocities of cratonic mantle. The first and foremost of them is the effect of chemical depletion. It is well known based on model calculations that fertile mantle is slower (Jordan, 1979). Seismic velocities in the primitive mantle are up to 0.07 km/s slower than those in the more depleted subcontinental lithosphere (8.07 vs. 8.00 km/s for Vp and 4.68 vs. 4.65 km/s for Vs at 30 km; Weiss et al., 1999). Our data suggest that this

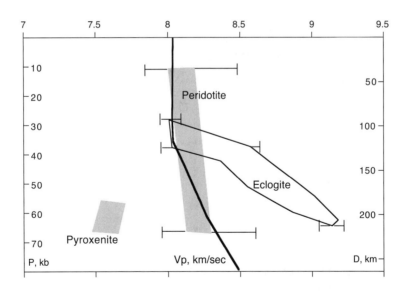

Fig. 4. Calculated Vp–depth profiles for Jericho mantle lithologies in comparison with a globally averaged *IASP91* model (Kennett and Engdahl, 1991) profile (bold). Ranges of velocities in isotropic peridotite and pyroxenite (grey) and eclogite (open field) are from Fig. 3. Bars designate the extent of seismic anisotropy expected in these rocks and equal 2% in eclogite (based on experimental values for homogeneous and unaltered samples) and 5% in peridotite. The bars are asymmetric (2 times closes to the minimum than maximum anisotropic Vp) about the range of isotropic velocity in peridotite.

pattern is true only of circumcratonic and oceanic mantle, i.e. of the mantle stabilized in the Phanerozoic and found as abyssal peridotites, peridotite massifs and mantle xenoliths in Phanerozoic settings. The composition of cratonic peridotite is anomalous in many aspects. Cratonic harzburgites and lherzolites show inverse correlations of olivine mode with depletion expressed as Mg-number, whereas olivine mode increases with the peridotite Mg-number in off-craton samples (Fig. 9 in Griffin et al., 1999a,b). Many xenoliths from Archean cratons are characterized by extreme depletion, abundant orthopyroxene, and low olivine/orthopyroxene ratios (Boyd, 1989; Kelemen et al., 1998; Griffin et al., 1999a,b).

In the Slave mantle, P-waves travel slower and S-waves faster in more depleted peridotites, thus defining a visible decrease in the Poisson's ratio

(Fig. 5B–D). We ascribe this to a correlation between depletion and orthopyroxene modes (Fig. 5A). This correlation is apparent in data averaged by peridotite types, but is masked by sample heterogeneity if sought in individual specimens of peridotite.

Peridotites below other cratons do not show the visible correlation between orthopyroxene modes and rock Mg-numbers, but an intrinsic link between the two is suggested by common orthopyroxene enrichment in high Mg-number peridotites (Kelemen et al., 1998). The correlation may be implied by several other chemical characteristics of cratonic peridotites, i.e. a negative correlation between Mg-number and olivine mode (Griffin et al., 1999a,b), and between bulk SiO_2 and FeO (Boyd, 1989, 1999; Kopylova and Russell, 2000). Since orthopyroxene is the second

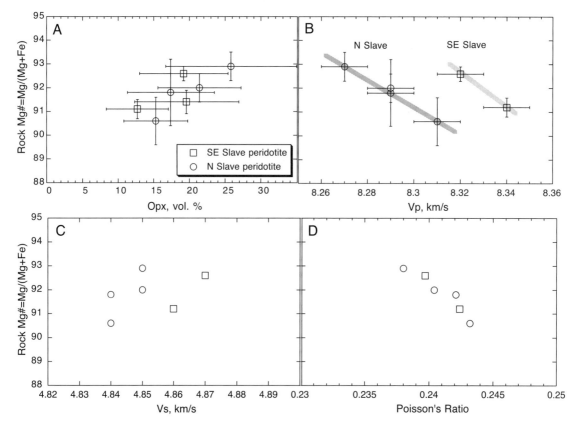

Fig. 5. Variation of geochemical and elastic parameters of the Slave peridotitic mantle with depletion, expressed as rock Mg-number. The data are averaged by rock type and can be found in Table 3 of this work and Table 3 of Kopylova and Caro (2004). Error bars plotted for Opx modes and Mg-numbers represent 2σ standard deviations of the data sets. Mg-numbers are plotted against orthopyroxene mode (A), Vp (B), Vs (C), and Poisson's ratio (D). The latter is defined as $\sigma=(3K_S - 2G)/(6K_S + 2G)$.

major phase of cratonic peridotite, and modes of clinopyroxene and garnet rarely exceed 5%, the cratonic olivine-poor peridotites with high Mg-numbers should be richer in orthopyroxene. Because bulk FeO and MgO are inversely related, and SiO_2 is controlled mainly by the orthopyroxene/olivine ratio, there should be a correlation between Mg-number of cratonic peridotite and its orthopyroxene mode. These predicted correlations can be hard to isolate if mineral modes and mineral compositions are recalculated based on bulk chemical compositions of peridotites (e.g. Kelemen et al., 1998), which are rarely fresh. Thorough petrographic studies of cratonic peridotite and direct measurements of mineral modes and compositions in thin sections are needed to map the orthopyroxene–Mg# correlation.

The above data can be summarized to suggest a distinct seismic signature for depletion in cratonic mantle. The depleted mantle below cratons should be slower in Vp, faster in Vs and should show lower Poisson's ratios than the less depleted mantle. The calculated effect should depend on the overall compositional range of peridotite and may differ from craton to craton. On the Slave craton, the restricted range of depleted bulk compositions suggests the effect on Vp or Vs would be up to 0.05 km/s.

Our modelling of seismic velocities forecasts different responses of Vp and Vs to depletion of cratonic mantle. Similar conclusion was reached in a recent study on modelled elastic properties of mantle peridotite (Lee, 2003). Lee found that the trend of correlated increase of Vs with increasing Mg-number is tight and well defined ($R^2 = 0.71$). In contrast, the covariation of Vp and Mg-number cannot be characterised by an apparent trend for the entire data set. A positive correlation of Vp and Mg-number is evident only for spinel peridotite that plot on a melt depletion trend ("oceanic array" of Boyd, 1989). Garnet peridotites, many of which are from cratonic localities and plot off the melt depletion trend, form a large scattered field on the Vp-Mg-number diagram with barely recognizable negative correlation (Fig. 11 in Lee, 2003).

The only reason model seismic wave velocities increase over the range of 30–250-km depth in the Slave peridotite is the consistently lower degree of chemical depletion in the deeper mantle. A mantle with a uniform composition would show decreasing Vp and Vs with depth according to our models. A consistent decrease in degree of depletion with depth is recorded in garnet concentrate data (Gaul et al., 2000; Griffin et al., 2003) for many Archean, Proterozoic and some Phanerozoic subcontinental mantle columns. Commonly observed increasing Vp, Vs depth profiles in the mantle as tabulated in *IASP91* suggest that the progressively lower depletion of the peridotite with depth should be a widespread phenomenon.

Another important elastic parameter of the mantle is density. Calculated densities of various peridotite types increase consistently from shallow spinel peridotite to deep garnet-bearing peridotite types (Table 4). Density increases from 3.286–3.287 g/cm^3 in depleted spinel peridotite to 3.319 and 3.340 g/cm^3 in the high-T and fertile peridotites. This density profile is gravitationally stable and is consistent with the long-stabilized mantle system. Densities of 3.28 g/cm^3 in the Slave depleted spinel peridotite are lower than the average (3.31 ± 0.016 g/cm^3, Poudjom Djomani et al., 2001) for the Archean cratonic mantle.

7.3. Effect of metasomatism on elastic properties of eclogite

Mantle metasomatism partially melts Jericho eclogite and leads to partial replacement of the primary omphacite–garnet assemblage. The secondary assemblage thus produced comprises fine-grained diopside, pyrope, phlogopite and amphibole. We calculated model seismic velocities for the metasomatized eclogite (Table 3) and found them to be invariably lower than those in eclogite not affected by metasomatism. Recrystallization of 14% primary clinopyroxene and 31% garnet into corresponding secondary phases lowers Vp from 8.13 to 8.09 km/s and leaves Vs unchanged (Sample 52-5). Introduction of 6% amphibole and recrystallization of 36% clinopyroxene, as in sample 47-8, lowers Vp more substantially, from 8.00 to 7.82 km/s, and Vs from 4.61 to 4.52 km/s. We conclude that metasomatic recrystallization of eclogite in the Slave mantle decreases P-wave velocities in the order of 0.05–0.1 km/s.

8. Conclusions

1. Seismic profiles calculated for the Slave mantle peridotites based on their mineralogy match well

with the globally averaged *IASP91* model. The calculated spread of Vp related to variations in mantle mineralogy at a given depth reaches 0.3 km/s in peridotite and 0.4 km/s in eclogite.

2. Seismic velocities in eclogite increase faster with depth than those in peridotite. A faster mantle below 90-km depth on the Slave craton can be explained by an unusually high proportion of eclogite.

3. Depletion in the cratonic mantle has a distinct seismic signature compared to the non-cratonic mantle stabilized in the Phanerozoic. The depleted mantle on cratons should have slower Vp, faster Vs and should show lower Poisson's ratios due to an orthopyroxene enrichment. On the Slave craton, the predicted effect on seismic wave velocities would be up to 0.05 km/s.

4. The consistently lower degree of chemical depletion in the deeper Slave mantle is the sole reason for the increase in the modelled seismic wave velocities at 30–250—km depth.

Acknowledgements

Funding for this research is derived from the Natural Sciences and Engineering Research Council (NSERC) of Canada. We are grateful to M Bostock and C. Bank for useful discussions on the topic of the manuscript, and to D. Mainprice, A. Jones and H. Grütter for reviews and editorial suggestions.

References

Ai, Y., 1994. A revision of the garnet-clinopyroxene Fe^{2+}–Mg exchange geothermometer. Contrib. Mineral. Petrol. 115, 467–473.

Angel, R.J., Hugh-Jones, D.A., 1994. Equations of state and thermodynamic properties of enstatite pyroxenes. J. Geophys. Res. 99, 19777–19783.

Bank, C.G., Bostock, M.G., Ellis, R.M., Cassidy, J.F., 2000. A reconnaissance teleseismic study of the upper mantle and transition zone beneath the Archean Slave craton in NW Canada. Tectonophysics 319, 151–166.

Bass, J.D., 1986. Elasticity of uvarovite and andradite garnets. J. Geophys. Res 91, 7505–7515.

Ben-Ismail, G., Barruol, D., Mainprice, A., 2001. The Kaapvaal seismic anisotropy: petrophysical analyses of upper mantle kimberlite nodules. Geophys. Res. Lett. 28 (13), 2497–2500.

Bina, C.R., Helffrich, G.R., 1992. Calculations of elastic properties from thermodynamic equation of state principles. Annu. Rev. Earth Planet Sci. 20, 527–552.

Boyd, F.R., 1989. Compositional distinction between oceanic and cratonic lithosphere. Earth Planet. Sci. Lett. 96, 15–26.

Boyd, F.R., 1999. The origin of cratonic peridotites: a major-element approach. In: Snyder, G.A., Neal, C.R., Ernst, W.G. (Eds.), Planetary Petrology and Geochemistry; the Lawrence A. Taylor 60th Birthday Volume. Bellwether Publishing, Columbia, pp. 5–14.

Brey, G.P., Köhler, T., 1990. Geothermobarometry in four-phase lherzolites: II. New thermobarometers, and practical assessment of existing thermobarometers. J. Petrol. 31, 1353–1378.

Chai, M., Brown, M., Slutsky, L.J., 1997. The elastic constants of an aluminous orthopyroxene to 12.5 GPa. J. Geophys. Res. 102, 14779–14785.

Chang, Z.P., Barsch, G.R., 1973. Pressure dependence of single-crystal elastic constants and anharmonic properties of spinel. J. Geophys. Res. 78, 2418–2433.

Chen, G.L., Spetzler, H.A., Getting, I.C., Yoneda, A., 1996. Selected elastic moduli and their temperature derivatives for olivine and garnet with different Mg/(Mg + Fe) contents: results from GHz ultrasonic interferometry. Geophys. Res. Lett. 23, 5–8.

Christensen, N.I., 1965. Compressional wave velocities in metamorphic rocks at pressures to 10 kb. J. Geophys. Res. 70, 6147–6164.

Christensen, N.I., 1989. Seismic velocities. In: Carmichael, R.S. (Ed.), CRC Handbook of Physical Properties of Rocks. CRC Press, Boca Raton, Florida, pp. 206–531.

Christensen, N.I., Shaw, G.H., 1970. Elastisity of mafic rocks from the mid-Atlantic ridge. Geophys. J. R. Astron. Soc. 20, 271–284.

Collins, M.D., Catlow, C.R.A., 1992. Computer simulation of structures and cohesive properties of micas. Am. Mineral. 77, 1172–1181.

Duffy, T.S., Ahrens, T.J., 1995. Compressional sound velocity, equation of state and constitutive response of shock-compressed magnesium oxide. J. Geophys. Res. 100, 529–542.

Duffy, T.S., Anderson, D.L., 1989. Seismic velocities in mantle minerals and the mineralogy of the upper mantle. J. Geophys. Res. 94, 1895–1912.

Ellis, D.J., Green, D.H., 1970. An experimental study of the effect of Ca upon garnet–clinopyroxene Fe–Mg exchange equilibria. Contrib. Mineral. Petrol. 71 (1), 13–22.

Fei, Y., 1995. Thermal expansion. In: Ahrens, T.J. (Ed.), Mineral Physics and Crystallography: A Handbook of Physical Constants. AGU, Washington, pp. 29–44.

Fernandez Viejo, G., Clowes, R.M., 2003. Lithospheric structure beneath the Archean Slave Province and Proterozoic Wopmay orogen, northwestern Canada, from a LITHOPROBE refraction/wide-angle reflection study. Geophys. J. Int. 153, 1–19.

Finnerty, A.A., Boyd, J.J., 1987. Thermobarometry for garnet peridotites: basis for the determination of thermal and compositional structure of the upper mantle. In: Nixon, P.H. (Ed.), Mantle Xenoliths. Wiley, New York, pp. 381–402.

Flesch, L.M., Li, B., Liebermann, R.C., 1998. Sound velocities of polycrystalline $MgSiO_3$–orthopyroxene to 10 GPa at room temperatures. Am. Mineral. 83, 444–450.

Gaul, O.F., Griffin, W.L., O'Reilly, S.Y., Pearson, N.J., 2000. Mapping olivine composition in the lithospheric mantle. Earth Planet. Sci. Lett. 182, 223–235.

Gregoire, M., Jackson, I., O'Reilly, S.Y., Cottin, J.Y., 2001. The lithospheric mantle beneath the Kerguelen Islands (Indian Ocean): petrological and petrophysical characteristics of mantle mafic rock types and correlation with seismic profiles. Contrib. Mineral. Petrol. 142, 244–259.

Griffin, W.L., O'Reilly, S.Y., Ryan, C.G., 1999a. The composition and origin of sub-continental lithospheric mantle. In: Fei, Y., Bertka, C.M., Mysen, B.O. (Eds.), Mantle Petrology: Field Observations and High-Pressure Experimentation: A Tribute to F.R. Boyd. The Geochemical Society Special Publication, No., 6. The Geochemical Society, Houston, pp. 13–45.

Griffin, W.L., Doyle, B.J., Ryan, C.G., Pearson, N.J., O'Reilly, S.Y., Natapov, L., Kivi, K., Kretschmar, U., Ward, J., 1999b. Lithosphere structure and mantle terranes: Slave craton, Canada. In: Gurney, J.J., Gurney, J.L., Pascoe, M., Richardson, S.R. (Eds.), Proc. 7th Int. Kimberlite Conference. Red Roof Designs, Cape Town, pp. 299–306.

Griffin, W.L., O'Reilly, S.Y., Doyle, B.J., Kivi, K., Coopersmith, H.G., 2003. Lithospheric mapping benarth the North American plate. Extended Abstract, 8th Intern Kimb Conf., June 2003, Victoria, Canada.

Grütter, H.S., Apter, D.B., Kong, J., 1999. Crust–mantle coupling: evidence from mantle-derived xenocrystic garnets. In: Gurney, J.J., Gurney, J.L., Pascoe, M., Richardson, S.R. (Eds.), Proc. 7th Int. Kimberlite Conference. Red Roof Designs, Cape Town, pp. 307–312.

Heaman, L.M., Kjarsgaard, B., Creaser, R.A., 2003. The timing of kimberlite magmatism and implications for diamond exploration; a global perspective. Lithos 71, 153–184.

Hearmon, R.F.S., 1984. The elastic constants of crystals and other anisotropic materials. In: Hellwage, K.H., Hellwage, A.M. (Eds.), Landolt-Bornstein Tables, III/18. Springer, Berlin, pp. 1–54.

Hofmeister, A.M., Mao, H.K., 2003. Pressure derivatives of shear and bulk moduli from the thermal Gruneisen parameter and volume–pressure data. Geochim. Cosmochim. Acta 67 (7), 1207–1227.

Isaak, D.G., Anderson, O.L., Oda, H., 1992. High-temperature thermal-expansion and elasticity of calcium-rich garnets. Phys. Chem. Miner. 19, 106–120.

Isaak, D.G., Carnes, J.D., Anderson, O.L., Cynn, H., Hake, E., 1998. Elastisity of TiO2 rutile to 1800 K. Phys. Chem. Miner. 26, 31–43.

Jackson, I., Rudnick, R.L., O'Reilly, S.Y., Bezant, C., 1990. Measured and calculated elastic wave velocities for xenoliths from the lower crust and upper mantle. Tectonophysics 173, 207–210.

Ji, S., Wang, Q., Xia, B., 2003. P-wave velocities of polymineralic rocks: comparison of theory and experiment and test of elastic mixture rules. Tectonophysics 166, 165–185.

Jones, A., Ferguson, I., Chave, A., Evans, R., McNeice, G., 2001. Electric lithosphere of the Slave craton. Geology 29, 423–426.

Jordan, T.H., 1979. Mineralogies, densities and seismic velocities of garnet lherzolites and their geophysical implications. In: Boyd, F.R., Meyer, H.O.A. (Eds.), The Mantle Sample: Inclusions in Kimberlites and Other Volcanics. AGU, Washington, pp. 1–15.

Kandelin, J., Weidner, D.J., 1998. The single crystal elastic properties of jadeite. Phys. Earth Planet. Inter. 50, 251–260.

Kelemen, P.B., Hart, S.R., Bernstein, S., 1998. Silica enrichment in the continental upper mantle via melt/rock reaction. Earth Planet. Sci. Lett. 164, 387–406.

Kennett, B.L.N., Engdahl, E.R., 1991. Travel times for global earthquake location and phase identification. Geophys. J. Int. 105, 429–465.

Kern, H., Gao, S., Jin, Z., Popp, T., Jin, S., 1999. Petrophysical studies of rocks from the Dabie ultrahigh-pressure (UHP) metamorphic belt, Central China: implications for the composition and delamination of the lower crust. Tectonophysics 301, 191–215.

Ketchum, J.W.F., Bleeker, W., 2001. 4.03–2.85 Ga growth and modification of the Slave protocraton, NW Canada. Abstract, The Slave–Kaapvaal Workshop, Sept. 2001, Merrickville.

Kopylova, M.G., 2003. Two distinct origins of the Northern Slave eclogites. Ext. Abstract of the 8th Intern. Kimb. Conf., Victoria, June.

Kopylova, M.G., Caro, G., 2004. Mantle xenoliths from the Southeastern Slave craton: The evidence for a thick cold stratified lithosphere. J Petrol 45 (5), 1045–1067.

Kopylova, M.G., Russell, J.K., 2000. Chemical stratification of cratonic lithosphere: constraints from the Northern Slave Craton, Canada. Earth Planet. Sci. Lett. 181, 71–87.

Kopylova, M.G., Russell, J.K., Cookenboo, H., 1999a. Petrology of peridotite and pyroxenite xenoliths from the Jericho kimberlite: implications for the thermal state of the mantle beneath the Slave Craton, Northern Canada. J. Petrol. 40 (1), 79–104.

Kopylova, M.G., Russell, J.K., Cookenboo, H., 1999b. Mapping the lithosphere beneath the North Central Slave Craton. In: Gurney, J.J., Gurney, J.L., Pascoe, M.D., Richardson, S.H. (Eds.), Proceedings of the VIIth International Kimberlite Conference, vol. 1 Red Roof Design, Cape Town, pp. 468–479.

Lee, C.A., 2003. Compositional variation of density and seismic velocities in natural peridotites at STP conditions: implications for seismic imaging of compositional heterogeneities in the upper mantle. J. Geophys. Res. 108 (B9), 2441–2462.

Li, B., Gwanmesia, G.D., Liebermann, R.C., 1996. Sound velocities of olivine and beta polymorphs of Mg_2SiO_4 at Earth's transition zone pressures. Geophys. Res. Lett. 23, 2259–2262.

Long, C., Christensen, N.I., 2000. Seismic anisotropy of South African upper mantle xenoliths. Earth Planet. Sci. Lett. 179, 551–565.

MacGregor, I.D., 1974. The System $MgO-Al_2O_3-SiO_2$: solubility of Al_2O_3 in enstatite for spinel and garnet peridotite compositions. Am. Mineral. 59, 110–119.

McCammon, C., Kopylova, M.G., accepted pending revisions. A redox profile of the Slave mantle and oxygen fugacity control in the cratonic mantle. Contrib. Miner. Petrol.

O'Neill, H.S.C., Wood, B.J., 1979. An experimental study of Fe–Mg partitioning between garnet and olivine and its calibration as a geothermometer. Contrib. Mineral. Petrol. 70, 59–70 (corrected in Contrib Mineral Petrol, 1980, 72, 337).

O'Reilly, S.Y., Jackson, I., Bezant, C., 1990. Equilibration temperatures and elastic wave velocities for upper mantle rocks from Eastern Australia: implications for the interpretation of seismological models. Tectonophysics 185, 67–82.

Poudjom Djomani, Y.H., O'Reilly, S.Y., Griffin, W.L., Morgan, P., 2001. The density structure of subcontinental lithospher through time. Earth Planet. Sci. Lett. 184, 605–621.

Ramesh, D.S., Kind, R., Yuan, X., 2002. Receiver function analysis of the North American crust and upper mantle. Geophys. J. Int. 150, 91–108.

Soedjatmiko, B., Christensen, N.I., 2000. Seismic anisotropy under estended crust: evidence from upper mantle xenoliths, Cima Volcanic Field, California. Tectonophysics 321, 279–296.

Sumino, Y., Anderson, O.L., 1984. Elastic constants of minerals. In: Carmichael, R.S. (Ed.), Handbook of Physical Properties of Rocks. CRC Press, Boca Raton, pp. 39–137.

Thorpe, R.I., Cumming, G.L., Mortensen, J.K., 1992. A major Pb-isotope boundary in the Slave Province and its probable relation to ancient basement of the western Slave. Canada-NWT MDA summary volume. Geol. Surv. Can., Open File 2484, 179–184.

Vacher, P.A., Mocquet, A., Sotin, C., 1998. Computation of seismic profiles from mineral physics: the importance of the non-olivine components for explaining the 660 km depth discontinuity. Phys. Earth Planet. Inter. 106, 275–298.

Wang, H., Simmons, G., 1972. Elastisity of some mantle crystal structures: 1. Pleonaste and hercynite spinel. J. Geophys. Res. 77, 4379–4392.

Weidner, D.J., Ito, E., 1985. Elasticity of $MgSiO_3$ in the ilmenite phase. Phys. Earth Planet. Inter. 40, 65–70.

Weiss, T., Siegesmund, S., Bohlen, T., 1999. Seismic. structural and petrological models of the subcrustal lithosphere in Southern Germany: a quantitative re-evaluation. Pure Appl. Geophys. 156, 53–81.

Welch, M.D., Crichton, W.A., 2002. Compressibility of clinochlore to 8 GPa at 298 K and a comparison with micas. Eur. J. Mineral. 14, 561–565.

Yang, H.M., Prewitt, C.T., 2000. Chain and layer silicates at high temperatures and pressures. In: Hazen, R.M., Downs, R.T. (Eds.), High-Temperatures and High-Pressure Crystal ChemistryReviews in Mineralogy and Geochemistry, vol. 41. Mineralogical Society of America and Geochemical Society of America, Washington, pp. 211–255.

Yoneda, A., 1990. Pressure derivatives of elastic constants of single crystal MgO and $MgAl_2O_4$. J. Phys. Earth 38, 19–55.

Zha, C.S., Duffy, T.S., Downs, R.T., Mao, H.K., Hemley, R.J., 1996. Sound velocity and elasticity of single-crystal forsterite to 16 GPa. J. Geophys. Res. 101 (B8), 17535–17545.

Zha, C.S., Duffy, T.S., Mao, H.K., Downs, R.T., Hemley, R.J., Weidner, D.J., 1997. Single-crystal elasticity of beta-Mg_2SiO_4 to the pressure of the 410 km seismic discontinuity in the Earth's mantle. Earth Planet. Sci. Lett. 147, 9–15.

Zhang, L., Ahsbahs, H., Hafner, S.S., Kutoglu, A., 1997. Single-crystal compression and crystal structure of clinopyroxene up to 10 GPa. Am. Mineral. 82, 245–258.

Zhao, Y.S., VonDreele, R.B., Shankland, T.J., Weidner, D.J., Zhang, J.Z., Wang, Y.B., Gasparik, T., 1997. Thermoelastic equation of state of jadeite $NaAlSi_2O_6$: an energy-dispersive reitveld refinement study of low symmetry and multiple phases diffraction. Geophys. Res. Lett. 24, 5–8.

Available online at www.sciencedirect.com

Lithos 77 (2004) 511–523

ELSEVIER

www.elsevier.com/locate/lithos

Megacrysts from the Grib kimberlite pipe (Arkhangelsk Province, Russia)☆

S.I. Kostrovitsky[a,*], V.G. Malkovets[b,c], E.M. Verichev[d], V.K. Garanin[e], L.V. Suvorova[a]

[a] Institute of Geochemistry, P.O. Box 4019, Irkutsk 664033, Russia
[b] GEMOC, Department of Earth and Planetary Sciences, Macquarie University, Sydney, Australia
[c] Institute of Mineralogy and Petrography SB RAS, Novosibirsk, Russia
[d] ArkhangelskGeolRazvedka, Archangel, Russia
[e] Moscow State University, Moscow, Russia

Received 27 June 2003; accepted 26 January 2004
Available online 17 June 2004

Abstract

The megacryst suite of the Grib kimberlite pipe (Arkhangelsk province, Russia) comprises garnet, clinopyroxene, magnesian ilmenite, phlogopite and garnet-clinopyroxene intergrowths. Crystalline inclusions, mainly of clinopyroxene and picroilmenite, occur in garnet megacrysts. Ilmenite is characterized by a wide range in the contents of MgO (10.6–15.5 wt.%) and Cr_2O_3 (0.7–8.3 wt.%). Megacryst garnets show wide variations in Cr_2O_3 (1.3–9.6 wt.%) and CaO (3.6–11.0 wt.%) but relatively constant MgO (15.4–22.3 wt.%) and FeO (5.2–9.9 wt.%). The pyroxenes also show wide variations in such oxides as Cr_2O_3, Al_2O_3 and Na_2O (0.56–2.95; 0.86–3.25; 1.3–3.0 wt.%, respectively). The high magnesium and chromium content of all these minerals puts them together in one paragenetic group. This conclusion was confirmed by studies of the crystalline inclusions in megacrysts, which demonstrate similar variations in composition. Low concentration of hematite in ilmenite suggests reducing conditions during crystallization. P–T estimates based on the clinopyroxene geothermobarometer (Contrib. Mineral. Petrol. 139 (2000) 541) show wide variations (624–1208 °C and 28.8–68.0 kbars), corresponding to a 40–45 mW/m^2 conductive geotherm. The majority of Gar-Cpx intergrowths differ from the corresponding monomineralic megacrysts in having higher Mg contents and relatively low TiO_2. The minerals from the megacryst association, as a rule, differ from the minerals of mantle xenoliths, but garnets in ilmenite-bearing peridotite xenoliths are compositionally similar to garnet megacrysts. The common features of trace element composition of megacryst minerals and kimberlite (they are poor in Zr group elements) suggest a genetic relationship. The origin of the megacrysts is proposed to be genetically connected with kimberlite magma-chamber evolution on the one hand and with associated mantle metasomatism on the other. We suggest that, depending on the primary melt composition, different paragenetic associations of macro/megacrysts can be crystallized in kimberlites. They include: (1) Fe–Ti (Mir, Udachnaya pipes); (2) high-Mg, Cr (Zagadochna, Kusova pipes); (3) high-Mg, Cr, Ti (Grib pipe).
© 2004 Elsevier B.V. All rights reserved.

Keywords: Kimberlite; Megacrysts; High-chromium association; Genesis

☆ Supplementary data associated with this article can be found, in the online version, at doi:10.1016/j.lithos.2004.03.014.

* Corresponding author.
E-mail address: serkost@igc.irk.ru (S.I. Kostrovitsky).

1. Introduction

The Grib kimberlite pipe is located in the Kepino-Pachuga field of the Arkhangelsk kimberlite province

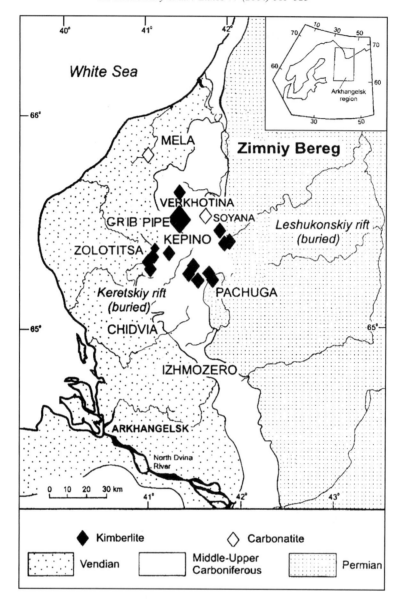

Fig. 1. Simplified geological sketch map of the Arkhangelsk Alkaline Igneous Province (Makhotkin et al., 2000).

(Fig. 1). The pipe, discovered in 1996, differs from certain other Arkhangelsk province kimberlites discovered earlier (e.g. Lomonosovo, Pionerskaya) in that the latter deposits contain abundant chrome spinel and lack typical megacryst phases like garnet and magnesian ilmenite. In contrast, the kimberlites from the high-grade Grib pipe contain abundant magnesian ilmenite, garnet, clinopyroxene, phlogopite and garnet-clinopyroxene intergrowths, whereas chrome spinel is a minor phase. Many of these phases occur as grains that are over 1 cm across and belong to the megacryst association.

Crystalline inclusions, mainly of clinopyroxene and picroilmenite occur in garnet megacrysts from the Grib kimberlite. Mineral compositions were analyzed by electron microprobe "JEOL" JXA-50A at Moscow State University, and JCXA-733 at the Institute of Geochemistry, Irkutsk. Trace

Table 2
Mineral analyses of megacryst garnets and ilmenite inclusions from Grib pipe

| | 1 | 2 | 3 | 4 | 5 | 6 | 7 | 8 | 9 | 10 | 11 | 12 | 13 |
| | gr1-1 | gr1-1-1 | gr1-1-2 | gr1-6 | gr1-6-1 | gr1-6-2 | gr1-8 | gr1-8-1 | gr1-8-2 | gr1-8-3 | gr1-9 | gr1-9-1 | gr1-9-2 |
	Gar	Ilm	Ilm	Gar	Ilm	Ilm	Gar	Ilm	Ilm	Ilm	Gar	Ilm	Ilm
SiO$_2$	40.91	0.35	0.29	41.19	0.14	0.09	42.07	0.18	0.32	0.39	42.14	0.14	0.07
TiO$_2$	0.91	54.61	56.13	1.13	54.85	54.22	1.12	56.23	55.1	53.9	0.76	55.64	56.25
Al$_2$O$_3$	19.32	0.61	1.15	18.92	0.93	0.68	17.92	0.57	0.82	0.99	20.19	0.65	0.91
Cr$_2$O$_3$	2.89	2.81	3.2	3.26	3.46	4.37	3.93	3.01	3.65	3.63	2.76	2.03	2.41
Fe$_2$O$_3$		2.75	0		0	1.13		0	0	2.27		0.64	0
FeO	9.49	25.6	27.36	9.31	28.65	27.51	9.37	27.69	28.43	25.56	9.7	27.2	27.8
MnO	0.33	0.25	0.19	0.14	0.15	0.28	0.5	0.25	0.27	0.28	0.31	0.48	0.22
MgO	19.89	13.21	11.78	20.01	11.55	11.82	19.14	12	11.18	12.87	18.54	12.83	11.83
CaO	4.6	0.1	0.16	4.8	0.08	0	5.38	0.06	0.06	0.12	4.79	0	0
Total	98.34	100.29	100.26	98.76	99.81	99.99	99.43	99.99	99.83	99.78	99.19	99.61	99.49
mg#	78.9			79.3			78.4				77.3		

elements were analyzed by Laser-Ablation Inductively Coupled Plasma Mass Spectrometry (LAM-ICPMS; Hewlett Packard 7500) at GEMOC, Macquarie University, Australia. Methods and operating conditions have been described by Norman et al. (1996). Representative compositional data for megacryst-suite and related minerals are listed in Tables 1, 4, 6, 8 and 9. The compositions of ilmenite and clinopyroxene intergrowths with garnet are listed in Tables 2 and 5, respectively. Tables 1, 4–6 and

8 are presented as supplementary digital data sets on line.

2. Megacryst occurrence and compositions

2.1. Magnesian ilmenite

Mg-rich ilmenite is a dominant mineral in the kimberlite heavy fraction. It occurs as rounded,

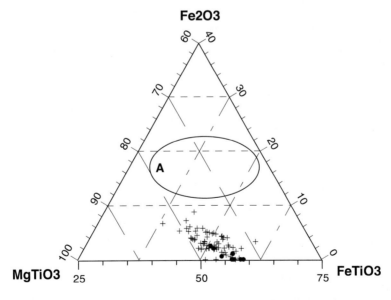

Fig. 2. Plot of magnesian ilmenite compositions: plus-megacrysts from the Grib pipe, black circle-inclusions in garnets from the Grib pipe, oval A-field of compositions of picroilmenite from Yakutian province.

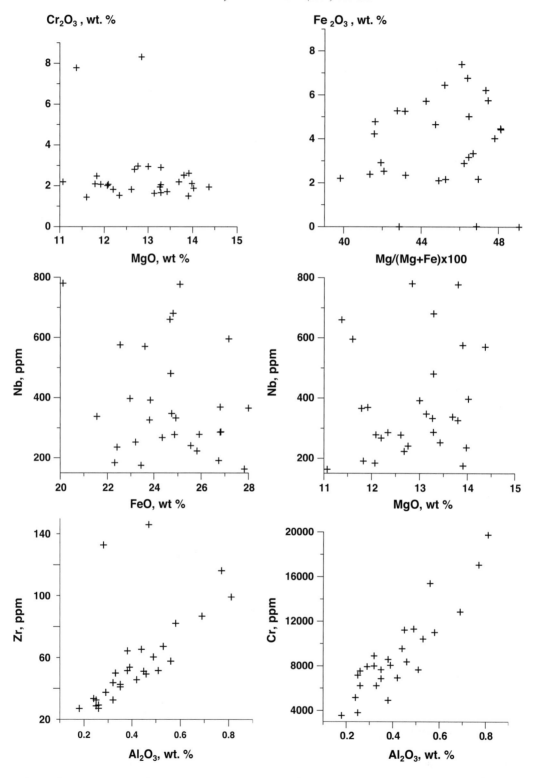

Fig. 3. Binary plots of correlation between major, minor oxides and rare elements in the ilmenites from the Grib pipe.

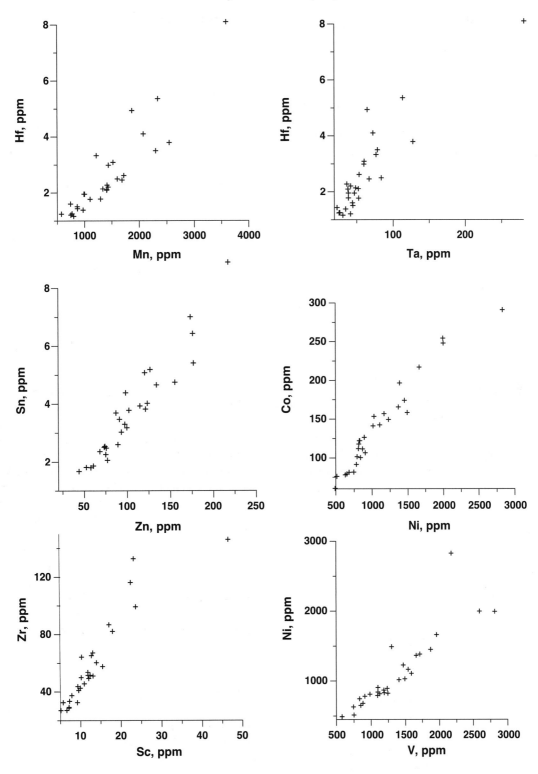

Fig. 3 (*continued*).

oval-flattened and fragmental-angular grains up to 15 mm across and as microcrystalline inclusions in garnet and phlogopite megacrysts. We have studied the composition of megacryst ilmenite grains varying in size from 5 to 15 mm and of ilmenite inclusions measuring 300×20 to 400×250 μm that occur in other phases. In addition to rounded and rounded-oval inclusions some of them are sliced and angular with resorbed margins. Tables 1 and 2 give representative picroilmenite compositions.

Mg-ilmenite megacrysts from Grib pipe are marked by wide variations in Fe and Mg concentrations, low iron oxidation state (Fig. 2) and high chromium contents. Average Cr_2O_3 and Fe_2O_3 contents are 2.4 and 3.6 wt.%, respectively. Megacryst ilmenite with such compositions has not been found in kimberlites of the Yakutian province. For example, in ilmenites from the Daldyn field (Kostrovitsky et al., 2003) the average Cr_2O_3 concentrations can reach as high as 1.5–1.8 wt.% (Osennya pipe), while minimum Fe_2O_3 contents are 10 wt.% (Leningradska pipe). The composition of ilmenite inclusions in garnet from Grib is similar to that of ilmenite macrocrysts, but is richer in Cr_2O_3 and poorer in Fe_2O_3 (Table 2). The composition of ilmenite inclusions from the same garnet can vary significantly (Table 2).

The megacryst ilmenites from the Grib kimberlite pipe are marked by a wide variation of trace element compositions and low Zr, Nb and Hf contents (Table 1). Most of the analyzed trace elements (V, Ga, Cu, Sn, Co,

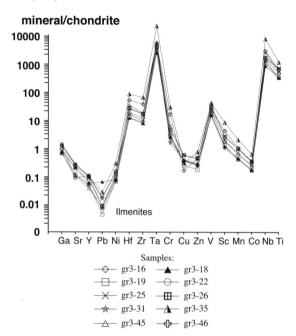

Fig. 4. Chondrite-normalized (McDonough and Sun, 1995) spidergram of trace elements in Grib ilmenites.

Y, Mn, Ni, Nb, Zr, Hf, Ta) are well correlated with one another (Figs. 3 and 4) and with minor oxides (Al_2O_3, Cr_2O_3, MnO). However, the correlation between major oxides and the minor oxides and trace elements is poor. Table 3 gives the trace element composition of picroilmenites from the Yakutian province and South African pipes for comparison. Chondrite-normalized spidergrams (Fig. 4) show groups of rare earth elements that behave similarly in ilmenite: (1) Ta, Nb (three to four orders higher than chondrite); (2) Hf, Zr, V (one to two orders higher than chondrite); (3) Ga, Cr, Sc, Mn (most similar to chondrite); (4) Cu, Zn, Co, Sr (two to eight times lower than chondrite); (5) Y, Ni, Pb (one to two orders lower than chondrite). The succession of groups reflects a real degree of element compatibility during the crystallization of magnesian ilmenite.

2.2. Garnet

Garnet occurs as individual megacrysts, which are typically rounded, in some cases with relics of crystallographic shape. Megacrysts are as a rule not over 10 mm in size. In addition, intergrowths of garnet and chrome-diopside are frequently found. Clinopyroxene and garnet commonly occur in equal proportions in

Table 3
Variations, average content of trace elements (in ppm) in picroilmenite megacrysts from different locations. Data for South-African province from Griffin et al. (1997), for Yakutian province from Genshaft et al. (1983). Number of analyses is given in brackets. The numerator means variation range, the denominator means average content

	Grib pipe (29)	South-African Province (63)	Yakutian province (572)
Ni	(489-2023)/1108	(111-1350)/726	
Cu	(18-92)/51	(19-39)/20	
Zn	(43-176)/105	(75-130)/110	
Ga	(4-18)/9	(11-22)/12	
Zr	(27-132)/62	(415-1020)/592	
Nb	(163-1790)/415	(294-3220)/925	(577-1223)/916
Co	(60-291)/138		(175-207)/192
Sc	(5-46)/13		(18-31)/24
Hf	(1-5)/2		(8-26)/17
Ta	(23-281)/62	(47-695)/159	(65-211)/148

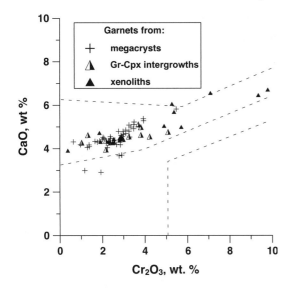

Fig. 5. Plot of Cr_2O_3 vs. CaO for the garnets from the Grib pipe. Fields shown for garnets are taken from Sobolev et al. (1974). Data for xenoliths are from Sablukova et al. (2003) and Malkovets et al. (2003).

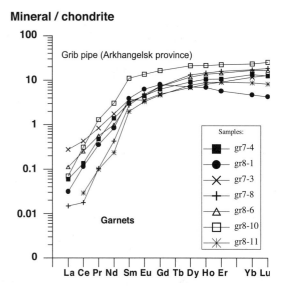

Fig. 7. Chondrite-normalized (McDonough and Sun, 1995) REE patterns for garnet megacrysts of the Grib pipe.

these intergrowths. These inclusions can occur both in the center and on the margins of megacrysts. In some cases, clinopyroxene is found on margins of the garnet

megacrysts, forming separate grains with uneven borders or as a discontinuous rim. Intergrowths with a predominant clinopyroxene, and garnet as fine inclusions, are rare.

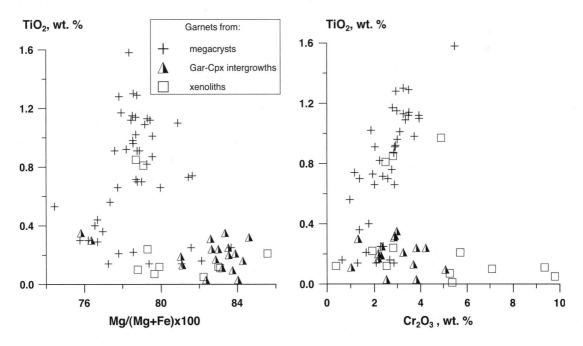

Fig. 6. Compositions of garnets from megacrysts, intergrowths and xenoliths from the Grib pipe. Data for xenoliths are from Sablukova et al. (2003). Mg/(Mg + Fe) is given as a molar ratio.

Table 7
The average content of Zr, Nb, Ta and Hf (in ppm) in garnet megacrysts and kimberlites from Grib pipe (data from Krotkov et al., 2001) and Yakutian province (data from Ilupin et al., 1978)

	Garnet megacrysts		Kimberlite	
	Grib pipe (n = 16)	Yakutian province (not published data) (n = 14)	Grib pipe (Krotkov et al., 2001) (n = 1)	Yakutian province (Ilupin et al., 1978) (n = 18)
Zr	28.1	79.4	59	133
Nb	0.12	0.67	38	121.9
Hf	0.47	2.3	1.1	2.7
Ta	<0.01		3.6	5.8

Garnets occurring as individual megacrysts (Table 4) and in Gar-Cpx intergrowths (Table 5) have high TiO_2 contents, and as such belong to groups 1, 2 and 9 of Dawson (1980). The majority of studied grains represent high-titanium pyropes of megacryst groups 1 and 2, but those from Grib are richer in Cr_2O_3. Based only on their CaO and Cr_2O_3 contents (Fig. 5) the megacrysts and Gar-Cpx intergrowths belong to the lherzolite paragenesis of Sobolev (1977). Despite wide variations in Cr_2O_3 (1.3–9.6 wt.%) and CaO (3.6–11.0 wt.%), the megacryst garnets are characterized by relatively constant

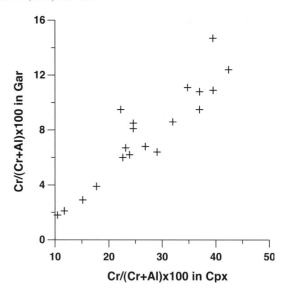

Fig. 9. High correlation of compositions of garnet and clinopyroxene from intergrowths of the Grib pipe.

MgO (15.4–22.3 wt.%) and FeO (5.2–9.9 wt.%). Variations in garnet composition are mainly due to variations in the uvarovite/pyrope ratio. Garnet and chrome-diopside in intergrowths demonstrate wide

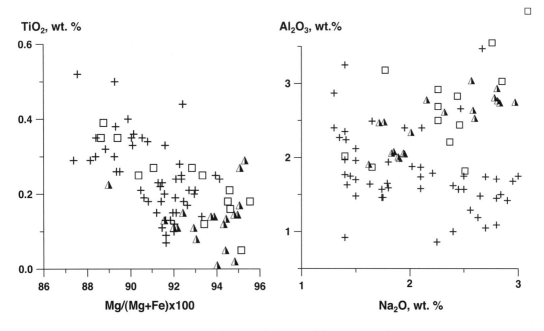

Fig. 8. Compositions of clinopyroxenes from megacrysts, intergrowths and xenoliths from the Grib pipe, with symbols as for Fig. 6. Data for xenoliths are from Sablukova et al. (2003) and Malkovets et al. (2003). Mg/(Mg + Fe) is given as a molar ratio.

Mineral /chondrite

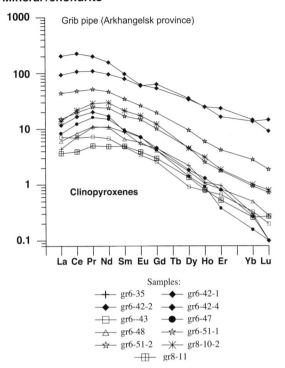

Fig. 10. Chondrite-normalized (McDonough and Sun, 1995) REE patterns for clinopyroxene megacrysts from the Grib pipe.

variations in composition. Compared to megacrysts, garnets from the Gar-Cpx intergrowths are characterized by relatively low TiO_2 contents and higher MgO (Fig. 6). Most of the investigated garnet megacrysts are homogenous, but a few show a weak zoning with a slight increase of Cr_2O_3 and decrease in TiO_2 towards their rims.

Among the garnets analyzed for trace element compositions (Table 6) only two grains (Gr 7-4 and Gr 8-1) could be reliably referred to as megacrysts. The distribution patterns of rare earth elements in garnets are similar for megacrysts and intergrowths (Fig. 7). The only difference is Zr and Hf enrichment in the megacryst garnets, correlated with relatively high titanium content. As in the ilmenite, the garnet megacrysts are depleted in the rare elements of the HFSE group (Zr, Nb, Hf, Ta). Table 7 gives average contents of those elements for garnet megacrysts and hosting kimberlite from the Grib pipe and the Yakutian province.

2.3. Clinopyroxene

Clinopyroxenes occur as discrete megacrysts which are rounded, elongated in some cases with a subidiomorphic crystallographic shape, and over 10 mm in size. By the classification of Dawson (1980) the pyroxenes of megacrysts and intergrowths with garnet from the Grib pipe are chrome-diopside and ureyitic diopside. Megacryst pyroxenes (Table 8) are marked by wide variations in the contents of Cr_2O_3, Al_2O_3 and Na_2O (0.56–2.95; 0.86–3.25; 1.3–3.0 wt.%, respectively) and in Mg/(Mg + Fe) and Ca/(Ca + Mg) ratios, which range from 87.4–94.3 and 43.1–50.7, respectively.

Pyroxenes from Gar-Cpx intergrowths, like the garnets, differ from the megacrysts by relatively low TiO_2 and slightly increased Mg/(Mg + Fe) (Fig. 8). Garnet and clinopyroxene from intergrowths show a clear correlation in terms of Cr/(Cr + Al) ratio (Fig. 9). Rare pyroxene megacrysts are inhomogeneous in composition. Margins of grains are relatively rich in Cr_2O_3

Table 9
Representative compositions of phlogopite megacrysts from Grib pipe

	Gr10-15	Gr10-15-3	Gr10-16	Gr10-16-3	Gr10-14	Gr10-1	Gr10-4	Gr10-8	Gr10-6	Gr9-20	Gr9-26
SiO_2	41.98	40.75	41.88	40.54	42.70	42.02	41.51	42.67	41.54	40.14	40.93
TiO_2	0.60	0.57	0.59	1.18	0.75	0.65	0.64	0.72	0.58	1.01	0.51
Al_2O_3	11.50	10.83	11.37	11.60	11.48	12.01	11.59	11.96	11.41	10.76	10.87
Cr_2O_3	0.31	0.40	0.64	0.62	0.72	0.84	0.67	0.68	0.69	0.55	0.56
FeO	3.70	4.86	3.58	4.22	3.58	3.57	3.78	3.63	4.42	3.59	3.31
MnO	0.05	0.00	0.09	0.06	0.15	0.00	0.03	0.10	0.07	0.06	0.05
MgO	25.00	27.60	25.71	24.64	24.61	24.88	24.97	25.27	26.15	23.77	25.59
K_2O	11.43	8.72	11.14	10.68	11.58	11.37	11.73	11.75	10.38	11.17	10.53
Total	94.57	93.73	95.00	93.54	95.57	95.34	94.92	96.78	95.24	91.04	92.35
mg#	92.33	91.01	92.74	91.23	92.46	92.55	92.16	92.54	91.38	92.19	93.22

and Al_2O_3 and poor in MgO. The REE contents of the clinopyroxenes are high and widely variable (Fig. 10).

2.4. Phlogopite

Phlogopite is found as large lamellar crystals with rounded margins and range in length from 2 to 15 mm. Their compositions (Table 9) are characterized by high MgO (23–30.2, average 25.4 wt.%), moderate TiO_2 (0.32–1.18, average 0.67 wt.%) and moderate Cr_2O_3 (0.22–0.85, average 0.6 wt.%) contents, values that are typical of the primary phlogopite from mantle peridotite xenoliths (Dawson and Smith, 1975).

3. Discussion

The high-chromium megacryst association (garnet, clinopyroxene, orthopyroxene, olivine) was for the first time described from kimberlites of the North-American province (Eggler et al., 1979). However, this association does not include picroilmenite. Kimberlites from the Grib pipe contain abundant megacrysts of picroilmenite, garnet, clinopyroxene and phlogopite. High MgO and TiO_2 and moderate to high Cr_2O_3 contents are common to all megacryst phases from Grib. The genetic similarity of the minerals and a common paragenesis are confirmed by studies of co-existing phases found as crystalline inclusions in the megacrysts. Magnesian ilmenite macrocrysts and crystalline ilmenite inclusions in garnet megacrysts are characterized by high chromium contents and similar MgO concentrations. This is the strongest argument to include picroilmenite in the same paragenetic association as garnet and chrome-diopside. A relatively high TiO_2 content is one of the essential features used to distinguish phases as megacrysts, irrespective of whether they belong to the low-chromium or high-chromium megacryst association.

Pressure–temperature ($P–T$) parameters of crystallization were calculated for clinopyroxene megacrysts, Gar-Cpx intergrowths and ultramafic xenoliths (data of Sablukova et al., 2003) using the clinopyroxene geothermobarometer of Nimis and Taylor (2000). The results show wide and similar $P–T$ variations common to all parageneses (Fig. 11). Crystallization temperatures range from 624 to 1208 °C for clinopyroxene megacrysts, 730 to 1077 °C for intergrowths, and 733–

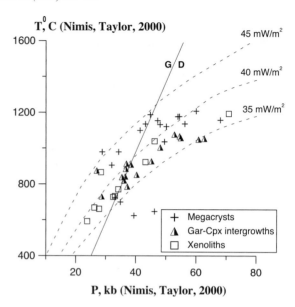

Fig. 11. Equilibrium $P–T$ estimates for the Grib megacrysts, intergrowths and xenoliths according to Nimis and Taylor (2000). Data for xenoliths are from Sablukova et al. (2003) and Malkovets et al. (2003).

1194 °C for xenoliths. Pressure estimates range from 28.8–68 kbars for clinopyroxene megacrysts, 27–62.7 kbars for intergrowths and 28.4–70.8 kbars for xenoliths. Intergrowths and xenoliths give $P–T$ estimates that correspond to a 35–40 mW/m^2 conductive geotherm, whereas estimates for megacrysts correspond to a slightly higher geotherm (40–45 mW/m^2).

It should be noted that the majority of studied minerals from Gar-Cpx intergrowths have higher MgO and lower TiO_2 than the corresponding monomineralic megacrysts (Figs. 6 and 8). During a decrease of temperature the recrystallization of Gar-Cpx intergrowths will lead to an increase in the MgO and Cr_2O_3 contents of clinopyroxenes and a decrease of TiO_2 abundance in garnets. Despite the large size of the garnet and clinopyroxene grains in the intergrowths, the latter cannot be referred to the megacryst association. The garnet and clinopyroxene from intergrowths are similar in composition to those from ultramafic xenoliths (lherzolites and pyroxenites) from the Grib pipe (Sablukova et al. 2003; Malkovets et al., 2003). On the other hand, individual megacrysts of garnet and clinopyroxene from the Grib pipe generally differ in their composition from minerals in xenoliths (Figs. 6 and 8). Only two garnets from ilmenite-bearing ultramafic xenoliths demonstrate a compositional affinity with garnet megacrysts. The clinopyroxene megacrysts differ from xenolith clinopyroxenes in their low Al_2O_3 content, and the two clinopyroxene populations form two distinct trends in the Na_2O vs. Al_2O_3 plot (see Fig. 8).

REE distribution in garnets and clinopyroxenes (Figs. 7 and 10) corresponds to the model of their joint crystallization and is typical of megacryst association minerals. A wide range in the composition of rare earth elements is common to all phases, in particular clinopyroxene.

The minerals of the megacryst association are similar to the association of minerals from the micaceous kimberlite pipes in Yakutia (e.g. Zagadochna, Kusova and Bukovinska). On $Ca/(Ca+Mg)$ vs. $Cr/(Cr+Al)$ and $Ca/(Ca+Mg)$ vs. $Mg/(Mg+Fe)$ plots, the chrome-diopsides from Grib pipe and those from the Yakutian mica-bearing kimberlites form a common cluster dissimilar to clinopyroxenes occurring in ultramafic xenoliths from the Udachnaya pipe (Fig. 12). According to Kostrovitsky and De Bruin (1998), the macrocryst mineral association and the host micaceous kimberlite are genetically connected. On the other hand, we believe that metasomatic processes played a significant role in the formation of the megacryst mineral association. The upper mantle metasomatism may have been spatially and temporally related to the formation of a kimberlitic source that is represented by the mineral compositions of the Grib megacryst assemblage. The formation of the megacrysts could have been related to the kimberlites as evidenced by low HFSE in both kimberlites and megacrysts and/or to mantle metasomatism as evidenced by similar mineral compositions of megacrysts and metasomatized ilmenite-bearing mantle xenoliths. We suggest that Grib megacrysts crystallized from a kimberlitic melt at a protomagmatic stage.

In conclusion we would like to note that Group-1 type kimberlites have elevated bulk FeO and TiO_2 contents (for example pipes Mir, Udachnaya and Yubileynaya, Yakutian province) and their heavy mineral fraction contains picroilmenite and orange-red garnet megacrysts, which belong to the low-chromium association. High-MgO kimberlites are less abundant and their heavy fraction is lacking in low-Cr megacrysts, but rich in high-Mg garnet and chrome spinels (for example Aikhal, Nachalnaya and Svet-

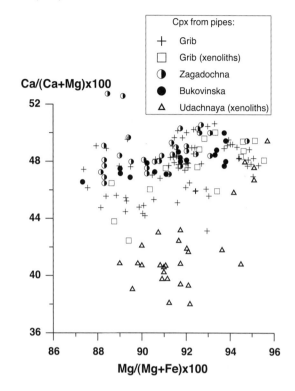

Fig. 12. Plot of $Ca/(Ca+Mg) \times 100$ vs. $Mg/(Mg+Fe) \times 100$ for clinopyroxenes from different locations (see legend). Data for xenoliths from Grib pipe are from Sablukova et al. (2003) and Malkovets et al. (2003), for xenoliths from Udachnaya pipe from Sobolev (1977) and for macrocrysts from Zagadochna and Bukovinska by the authors (unpublished).

laya pipes from Alakit field, Yakutian province). Micaceous high-Mg kimberlites are even less abundant (a few pipes from Bukovinska and Zagadochna clusters, Daldyn field, Yakutian province), and they do not contain low-Cr megacrysts. However, the high-Cr, high-Mg macrocryst mineral association (garnet, chrome-diopside, phlogopite) is widespread in these kimberlites (Kostrovitsky and De Bruin, 1998). The low-Cr and high-Cr associations of megacrysts have been described from the kimberlites of the North-American Province (Eggler et al., 1979). Their host kimberlites do not contain mantle xenoliths, which could be the source of the corresponding minerals. This is consistent with the assumption that the origin of megacrysts can be associated with the protocrystallization of different paragenetic mineral associations depending on the primary kimberlite melt composition.

4. Conclusions

The kimberlite from the Grib pipe contains a high-chromium megacryst association, marked by the occurrence of magnesian high-Cr picroilmenite. Low concentrations of the hematite molecule in ilmenite indicate reducing conditions during the crystallization of this assemblage. P–T estimates for clinopyroxene megacrysts based on the clinopyroxene geothermobarometer (Nimis and Taylor, 2000) show wide variations (624–1208 °C and 28.8–68 kbars), which correspond to a 40–45 mW/m^2 conductive geotherm. P–T estimates for garnet-clinopyroxene intergrowths and possibly related xenoliths show a similar wide range but correspond to a lower conductive geotherm (35–40 mW/m^2). The concentration of rare elements in megacrysts varies over a wide range, and strong correlations are found between most elements. A depletion in the HFSE group is common to megacrysts and the host kimberlite. The origin of megacrysts therefore is thought to be genetically connected with the evolution of the kimberlite magmatic chamber, and with associated mantle metasomatism.

We propose that, depending on the primary melt composition, chemically different paragenetic associations of megacrysts can be crystallized in kimberlites. They include: (1) low-Cr megacrysts in high Fe–Ti kimberlites (Mir, Udachnaya pipes), (2) high-Mg phlogopite + garnet + chrome-diopside associations in high-Mg, Cr kimberlites (Zagadochna, Kusova pipes) and (3) high-Cr megacrysts in high-Mg, Cr, Ti kimberlites (Grib pipe). Compositional differences in terms of megacryst major elements are mainly due to the heterogeneous lithosphere sources of the melts. The trace element composition of megacrysts, in particular for the incompatible elements, indicates close affinity with an asthenospheric source.

Acknowledgements

The authors are grateful to Prof. Barry Dawson and guest editor Dr. Herman Grutter for detailed reviews and tedious editing of the manuscript and for showing maximal patience and respect to the authors. We thank Prof. Bill Griffin for editing the final English text. This research was funded by Russian Foundation for Basic Research, No. 02-05-64793 to S.I. Kostrovitsky. Vlad Malkovets was supported by a Macquarie University Research Fellowship and by a Postdoctoral Grant from the Ministry of Education of the Russian Federation, NO. PD02-1.5-434.

References

Dawson, J.B., 1980. Kimberlites and Their Xenoliths. Springer-Verlag, Berlin.

Dawson, J.B., Smith, J.V., 1975. Chemistry and origin of phlogopite megacrysts in kimberlite. Nature 253, 336–338.

Eggler, D.H., McCallum, M.E., Smith, C.B., 1979. Megacryst assemblages in kimberlites from northern Colorado and southern Wyoming: petrology, geothermometry, barometry and areal distribution. In: Boyd, F.R., Meyer, H.O.A. (Eds.), The Mantle Sample: Inclusions in Kimberlites and Other Volcanics. Am. Geophys. Union, Washington, DC, pp. 213–226.

Genshaft, Yu.S., Ilupin, I.P., Kuligin, V.M., Vitozhents, G.Ch., 1983. Typomorphism of ilmenites from deep magmatic rocks. Composition and Properties of Deep Rocks of Earth Crust and Upper Mantle of Platform. Nauka, Moscow, pp. 95–190. in Russian.

Griffin, W.L., Moore, R.O., Ryan, C.G., Gurney, J.J., Win, T.T., 1997. Geochemistry of magnesian ilmenite megacrysts from Southern African kimberlites. Russian Geology and Geophysics 38, 398–419.

Kostrovitsky, S.I., De Bruin, D., 1998. Ultramafic association of minerals (garnet-ureyite diopside-chromspinelid) in micaceous kimberlites of Yakutian province. Ext. Abs. 7th Int. Kimb. Conf. Geological Survey of South Africa, Cape Town, pp. 463–465.

Kostrovitsky, S.I., Alymova, N.V., Ivanov, A.S., Serov, V.P., 2003. Structure of the Daldyn field (Yakutian Province) based on the study of picroilmenite composition. Ext. Abs. 8th Int. Kimb. Conf., vol. 207. DIAVIK (Diamond Mines Inc.), Victoria. CD-ROM.

Krotkov, V.V., Kudryavtseva, G.P., Bogatikov, O.A., Valuev, E.P., Verzhak, V.V., Garanin, V.K., 2001. New Technologies of Diamond Deposits Exploration GEOS, Moscow. 310 pp.

Makhotkin, I.L., Gibson, S.A., Thompson, R.N., Zhuravlev, D.Z., Zherdev, P.U., 2000. Late Devonian diamondiferous kimberlite and alkaline picrite (proto-kimberlite?) magmatism in the Arkhangelsk region, NW Russia. Journal of Petrology 41, 201–227.

Malkovets, V., Taylor, L., Griffin, W., O'Reilly, S., Pearson, N., Pokhilenko, N., Verichev, E., Litasov, K., 2003. Cratonic conditions beneath Arkhangelsk, Russia: garnet peridotites from the Grib kimberlite. Ext. Abs. 8th. Kimb. Conf., Victoria, vol. 220. Springer-Verlag, Berlin. CD-ROM.

McDonough, W.F., Sun, S.S., 1995. The composition of the Earth. Chemical Geology 120, 223–253.

Nimis, P., Taylor, W.R., 2000. Single clinopyroxene thermobrometry for garnet peridotites: Part 1. Calibration and testing of a Cr-in Cpx barometer and an enstatite-in-Cpx thermometer. Contributions to Mineralogy and Petrology 139, 541–554.

Norman, M.D., Pearson, N.J., Sharma, A., Griffin, W.L., 1996. Quantitative analysis of trace elements in geological materials by laser ablation ICPMS: instrumental operating conditions and calibration values of NIST glasses. Geostandards Newsletter 20, 247–261.

Sablukova, L.I., Sablukov, S.M., Verichev, E.M., Golovin, N.N., 2003. Petrography and mineral chemistry of mantle xenoliths from the Grib pipe, Zimny Bereg area, Russia. Plumes and Problems of Deep Sources of Alkaline Magmatism. Proceedings of Intern. Seminar. Institute of Geochemistry, SB RAS, Habarovsk, pp. 65–95.

Sobolev, N.V., 1977. Deep-seated Inclusions in Kimberlites and the Problem of the Composition of the Upper Mantle. American Geophys. Union, Washington, DC.

Available online at www.sciencedirect.com

Lithos 77 (2004) 525–538

ELSEVIER

LITHOS

www.elsevier.com/locate/lithos

Petrology of highly aluminous xenoliths from kimberlites of Yakutia

Z.V. Spetsius

Institute of Diamond Industry, ALROSA Co. Ltd., Mirny, Russia

Received 21 June 2003; accepted 16 December 2003
Available online 18 May 2004

Abstract

Highly aluminous xenoliths include kyanite-, corundum- and coesite-bearing eclogites, grospydites and alkremites. These xenoliths are present in different kimberlites of Yakutia but have most often been found in Udachnaya and other pipes of the central Daldyn–Alakitsky region. Kimberlites of this field also contain eclogite-like xenoliths with kyanite and corundum that originate in the lower crust or the lower crust–upper mantle transition zone. Petrographic study shows that two rock groups of different structure and chemistry can be distinguished among kyanite eclogites: fine- to medium-grained with mosaic structure and coarse-grained with cataclastic structure. Eclogites with mosaic structure are characterized by the occurrences of symplectite intergrowths of garnet with kyanite, clinopyroxene and coesite; only in this group do grospydites occur. In cataclastic eclogites, coarse-grained coesite occurs, corresponding in size to other rock-forming minerals. Highly aluminous xenoliths differ from bimineralic eclogites in their high content of Al_2O_3 and total alkali content. Coesite-bearing varieties are characterized by low MgO content and higher Na/K and Fe^{2+}/Fe^{3+} ratios, as well as high contents of Na_2O. Geochemical peculiarities of kyanite eclogites and other rocks are exhibited by a sloping chondrite-normalized distribution of rare earth elements (REE) in garnets and low Y/Zr ratio, in contrast to bimineralic rocks. Coesite is found in more than 20 kyanite eclogites and grospydites from Udachnaya. Grospydites with coesite from Zagadochnaya pipe are described. Three varieties of coesite in these rocks are distinguished: (a) subhedral grains with size of 1.0–3.0 mm; (b) inclusions in the rock-forming minerals; (c) sub-graphic intergrowths with garnet. The presence and preservation of coesite in eclogites indicate both high pressure of formation (more than 30 kbar) and set a number of constraints on the timing of xenolith cooling during entrainment and transport to the surface. Different ways of formation of the highly aluminous eclogites are discussed. Petrographic observations and geochemistry suggest that some highly aluminous rocks have formed as a result of crystallization of anorthosite rocks in abyssal conditions. $\delta^{18}O$-estimations and other petrologic evidence point out the possible origin of some of these xenoliths as the result of subduction of oceanic crust. Diamondiferous samples have been found in all varieties except alkremites. Usually these eclogites contain cubic or coated diamonds. However, two sample corundum-bearing eclogites with diamonds from the Udachnaya pipe contain octahedra that show evidence of resorption.
© 2004 Elsevier B.V. All rights reserved.

Keywords: Xenolith; Eclogite; Lithosphere; Diamond; Kyanite; Coesite; Siberia

1. Introduction

Highly aluminous xenoliths include kyanite-, corundum- and coesite-bearing eclogites, grospydites and alkremites. These xenoliths are present in many

E-mail address: Spetsius@yna.alrosa-mir.ru (Z.V. Spetsius).

different Yakutian kimberlites but are most common in Udachnaya and other pipes of the central Daldyn–Alakitsky region (Bobrievich et al., 1959; Ponomarenko et al., 1976; Spetsius and Serenko, 1990). Granulites or eclogite-like xenoliths with kyanite and corundum belonging to the lower crust or transitional lower crust–upper mantle zone are also present in kimberlites of this field. All varieties of these highly aluminous xenoliths except alkremites and eclogite-like rocks include diamondiferous examples. Evidence discussed below indicates that these xenoliths represent a specific group of mantle rocks that could not be formed by the differentiation of primitive mantle but represent the remnants of subducted crust. The presence of coesite reflects a number of con-

straints bearing on conditions of the xenoliths origin and kimberlite formation.

The Siberian craton occupies about 4×10^9 km^2, mostly buried beneath Riphean–Phanerozoic sedimentary cover 1–8 km thick, averaging about 4 km. The main structural blocks and tectonic zones are shown in Fig. 1. According to the terrane concept, the craton's structure results from the collision and amalgamation (accretion) of heterochronous microcontinents, which were transformed into terranes or tectonic blocks of varying genesis (Rosen et al., 2002). Their bounding shear zones show traces of tectonic compression and overthrust in zones of collision. The accretion of terranes seemed to have occurred in several stages, and generally larger units,

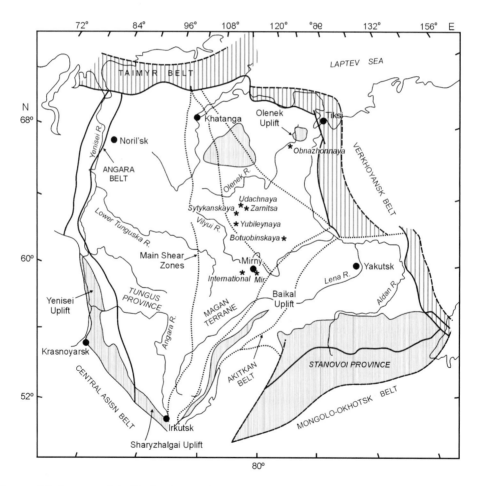

Fig. 1. Sketch map of the basement of the Siberian Craton (after Rosen et al., 2002 with addition), showing main structural elements and localities of xenoliths from kimberlite pipes (star ornament) mentioned in the text.

superterranes or tectonic provinces had appeared before they consolidated as the craton. Fig. 1 shows the main structural domains of the craton as well as the location of the main well-known pipes referred to in this paper. Kimberlite diatremes occur from the Vilui River in the south to the lower reaches of the Olenek and Kotui Rivers in the north, over an area of more than 1100 km in longitude and 800 km in latitude. Situated in the northeastern part of the Siberian craton, the Yakutian kimberlite province occupies mostly the territory of the Anabar super-terraine, including the Magan and Daldyn granulite–gneiss terranes and the Markha granite–greenstone terrane. A more detailed description is given in Rosen et al. (2002).

2. Samples and analytical techniques

More than 200 samples of mantle xenoliths from the kimberlite pipes situated in different parts of the Yakutian kimberlite province were studied. Modal analyses have been performed for the major part of xenoliths. For most samples major rock chemistry was determined. Major-element analyses were performed for the rock-forming and minor minerals. Trace element compositions were obtained for some minerals. All the samples were classified into different varieties of eclogites according to their petrographic and chemical features.

Major element compositions of silicate and oxide minerals in the xenoliths were determined with a Superprobe JXA-8800R electron microprobe at the ALROSA (Mirny, Russia) and partly using a CAMECA SX-50 electron microprobe at the Institute of Geology (Yakutsk). Part of the rock-forming garnets and clinopyroxenes and also different secondary phases of eclogites were investigated by ESM with EDS at the University of Western Australia (Perth). Analytical conditions included an accelerating voltage of 15 keV, a beam current of 20 nA, beam size of 5 μm, and 20 s counting time for all elements. All analyses underwent a full ZAF correction.

The trace elements (TRE) have been measured in rock-forming and some secondary minerals of eclogites by laser ablation ICP-MS (LAM) at the RSES, Australian National University, Canberra, using NIST 610 glass as external standard and Ca as internal standard; pit diameters were 40–50 mm.

3. Results

Highly aluminous xenoliths are predominantly presented by eclogites typically containing kyanite (up to 30%) as an additional phase to the high-Ca garnets and high-jadeite clinopyroxenes (up to 11% Na_2O). Wide variations of garnet (20–80%) and clinopyroxene (20–60%) are typical of these rocks. Coesite and corundum are common, with rare rutile, sulfides, and ilmenite. The petrographic peculiarities of kyanite eclogites showed two rock groups of different texture which coincide with differences in chemical composition: fine- to medium-grained with mosaic structure and coarse-grained rocks with cataclastic or more seldom granoblastic structure. Eclogites with mosaic structure are characterized by medium-grained (0.5–2.5 mm size) constitution, banding, and occurrences of symplectic intergrowths of garnet with kyanite, clinopyroxene and coesite. In a number of samples, there is later development of kyanite. Grospydites were found only in this group. Coarse-grained kyanite eclogites are characterized by (1–4 mm size) constitution, strong garnet kelyphitization and euhedral kyanite, which are often replaced by mullite and corundum. In this group of kyanite eclogites, coarse-grained coesite with the same size as the other rock-forming minerals occurs.

Highly aluminous xenoliths differ from bimineralic eclogites in their high content of Al_2O_3 and total alkali content (Table 1). Coesite-bearing varieties are characterized by low MgO content and higher Na/K and Fe^{2+}/Fe^{3+} ratios, as well as high contents of Na_2O. Geochemical peculiarities of kyanite eclogites and other rocks exhibit themselves in a sloping chondrite-normalized distribution of rare earth element in garnets, in contrast to bimineralic rocks, as well as in low Y/Zr ratio. Preliminary data for the trace elements that were taken by dissolution on garnet separated from alkremites and analysed by quadrupole ICP-MS at Durham University suggest an extremely high $^{176}Lu/^{177}Hf$ ratio (>20) at least for one sample from the Udachnaya pipe (Nowell et al., 2003).

Table 1

Average major compositions of highly aluminous xenoliths from kimberlites of Yakutia (analyses recalculated on 100% dry weight basis)

	SiO_2	TiO_2	Al_2O_3	Fe_2O_3	FeO	MnO	MgO	CaO	Na_2O	K_2O	P_2O_5	#
1	46.36	0.34	24.08	1.32	4.19	0.08	7.13	13.85	1.89	0.72	0.04	8
2	44.34	0.37	17.27	3.6	5.73	0.17	14.64	11.77	1.14	0.91	0.06	33
3	44.99	0.28	25.16	2.96	3.97	0.11	8.09	11.36	1.61	1.4	0.07	43
4	46.31	0.3	25.22	2.68	4.53	0.1	6.42	11.22	2.23	0.93	0.07	10
5	30.02	0.16	32.86	5.49	2.84	0.22	25.02	2.88	0.07	0.4	0.04	12
6	44.85	–	28.18	1.71	2.84	–	7.72	12.45	1.57	0.68	–	11

Analyses: 1, eclogite-like xenoliths from the Udachnaya pipe; 2, bimineralic magnesian (group A) eclogites from the Udachnaya pipe; 3, kyanite eclogites from the Udachnaya pipe; 4, coesite eclogites from the Udachnaya pipe; 5, alkremites from the Udachnaya pipe; 6, grospydites from the Zagadochnaya pipe (after Sobolev, 1977). # = number of analyzed samples.

3.1. Mineralogy of highly aluminous xenoliths

The mineralogy of the eclogite-like rocks with kyanite is not so variable as eclogites. The main minerals are garnet, clinopyroxene, and plagioclase with addition of secondary formed clinopyroxene and kyanite in intergrowths. Representative compositions of all minerals are given in Table 2. In these xenoliths the modal kyanite content varies considerably (1% to 15%). Kyanite occurs as spectacular intergrowths with clinopyroxene. This texture consists of fine subhedral laths (0.1 × 1.2 mm) of both minerals (Fig. 2). These intergrowths are usually associated with garnet or situated between garnet and plagioclase and sometimes they have a star-like appearance. Similar textures have been described by Fadili and Demaiffe (1999) in xenoliths from the Mbuji Mayi kimberlites. Rare samples of these rocks with dispersed corundum

Table 2

Compositions of minerals in kyanite-bearing eclogite-like xenolith from the Udachnaya (sample U-2295)

	Gt (1)	Pl (2)	Cpx (3)	Ky (4)	Cpx (5)	Gt (6)
SiO_2	41.35	52.31	50.29	36.51	52.6	41.39
TiO_2	0.1	0.03	0.56	0.05	0.1	0.03
Al_2O_3	22.75	29.82	10.26	61.31	5.13	22.75
Cr_2O_3	0.14	0.05	0.34	0.22	0.41	0.3
FeO	12.19	0.03	2.25	0.39	2.26	12.13
MnO	0.14	0.02	0.03	0.07	0.13	0.14
MgO	14.05	0.14	12.26	0.16	14.61	11.71
CaO	9.1	12.75	22.25	0.05	22.6	10.75
Na_2O	0.08	4.69	1.66	0.04	1.84	0.32
K_2O	0.01	0.23	0.02	0.01	0.03	0.02
Total	99.91	99.84	99.92	98.81	99.71	99.54

Numbers in parenthesis correspond to the points of analyses on Fig. 2.

grains (0.2–1.5 mm in size) are present among xenoliths of the Udachnaya pipe. It should be pointed out that samples exist that are transitional from the kyanite-bearing eclogite-like rocks with single plagioclase grains to the kyanite eclogites (Spetsius and Serenko, 1990).

3.2. Eclogite xenoliths

Clinopyroxene in kyanite eclogites is omphacite characterized by high Mg# and high-Al content (Table 3). There is a clear correlation of jadeite content in omphacites of bimineral and kyanite eclogites with their CaO content; this relationship holds for eclogite clinopyroxenes from all the pipes (Spetsius and Serenko, 1990). The comparison of eclogite clinopyroxenes from pipes Mir, Obnazhonnaya and Udachnaya shows that the clinopyroxenes of the last have a more variable composition due to the absence of eclogites of high-Al composition in the first two pipes. Omphacites in eclogites from pipe Udachnaya are enriched in lithophile trace elements, perhaps resulting from the action of metasomatic fluids (Spetsius, 1995).

Garnet from highly aluminous xenoliths has a wide variation in Fe, Mg and Ca content (Table 4). Garnets of these xenoliths differ by higher Ca# from the garnets from bimineral eclogites and usually they have high Mg#. They contain over 40 mol% of pyrope, 10–20 mol% of almandine and variable number (20–60 mol%) of grossular components. Garnets relatively high in Cr_2O_3 (containing up to 1.0 wt.%) are found in some kyanite eclogites. A wide range of content of Ca, Mg, Fe, Cr and Ti is typical of garnets from xenoliths of the eclogite suite from individual pipes (Spetsius and Serenko, 1990). The

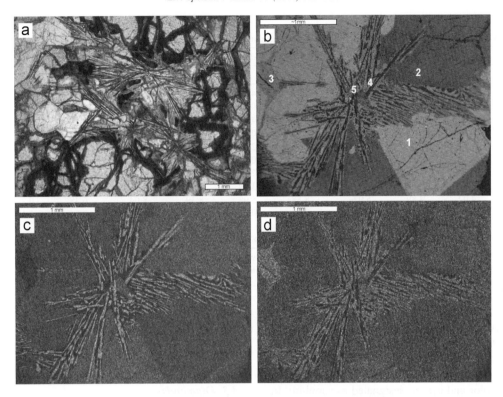

Fig. 2. Kyanite-bearing eclogite-like xenoliths from the kimberlites of Yakutia. (a) Plane-polarized light (intergrowths of Cpx and Ky between grains of Gt and Plag are obvious, sample Zr-18 from Zarnitsa pipe); (b–d) scanning electron microscope images of eclogite-like xenoliths from the kimberlite pipe Udachnaya: (a) backscattered image of Ky-Cpx intergrowth (sample U-2295); (c,d) images in Al K_α and Ca K_α. Points of microprobe analyses of minerals are shown (see data in Table 2).

garnets from kyanite eclogites of Udachnaya vary widely in Ca/(Ca + Mg) from 22% to 76%. The most calcic samples (>50% Ca#) are thus grospydites. There are smaller variations in Ca# of eclogite garnets from the Mir pipe where kyanite eclogites and grospydites are absent. It should be pointed that garnets from kyanite eclogites with coesite have wide variations in Ca/(Ca + Mg) (Fig. 3) and in some cases have occupied the field of C-group eclogites and B-group as well.

Kyanite differs by its euhedral and "fresh" outlook from the other rock-forming minerals, except in rare cases, when it is replaced by corundum and mullite. Two types of kyanite are recognisable: (1) crystals of tabular shape, of macroscopic blue colour, sometimes with a greenish tint (chrome containing); (2) needle-shaped colourless kyanite, typical for eclogite-like rocks sometimes occurring in kyanite eclogites where it is the latest. In some samples,

there are symplectite intergrowths of kyanite with garnet.

Coesite is found in more than 20 samples of kyanite eclogites and grospydites from the Udachnaya pipe where it was first discovered in mantle xenoliths (Ponomarenko et al., 1977). The abundance of coesite ranges from trace quantities (single grains and inclusions) to approximately 10% by volume. Grospydites with coesite also occur in the Zagadochnaya pipe. Three morphological varieties of coesite are recognized: (a) subhedral coesite grains with a size of 1.0–3.0 mm (Fig. 4a–c); (b) coesite inclusions in the rock-forming eclogite minerals; (c) subgraphic intergrowths with garnet (Fig. 4d). The presence and preservation of coesite in eclogites indicate high pressure of formation (more than 30 kbar) and set a number of limitations on the timing of cooling of xenoliths during their capture and transportation to surface by kimberlitic magma.

Table 3
Representative compositions of omphacites from highly aluminous xenoliths from the Udachnaya pipe

Sample	U-9	U-154	U-157	U-947	U-163	U-188	U-2290	U-2310	Ud-45	Ud-155
Analyses	1	2	3	4	5	6	7	8	9	10
SiO_2	55.3	54.62	54.78	56.21	54.99	55.67	56.46	54.49	55.87	59.58
TiO_2	0.23	0.16	0.17	0.3	0.24	0.24	0.22	0.28	0.22	0.25
Al_2O_3	14.83	13.42	12.27	16.52	14.47	16.12	16.88	1652	9.05	18.16
Cr_2O_3	0.05	0	0.13	0.06	0.02	0.04	0.05	0.03	0.07	0.04
FeO	2.82	1.74	1.99	1.98	2.57	3.07	1.77	1.95	2.56	1.04
MnO	<0.03	<0.03	0.03	<0.03	<0.03	<0.03	0.01	0	0.03	<0.03
MgO	7.05	8.4	9.64	5.8	7.1	5.67	6.56	6.04	10.96	4.44
CaO	13.12	15.62	16.44	10,06	13.38	9.9	9,98	12.88	15.99	9.21
Na_2O	4.02	4.89	5.07	8.88	5.98	8.39	8.04	6.48	0.2	6.66
K_2O	0.07	0.02	0.04	0.22	0.02	0.03	0.03	0.04	5.19	0.12
Total	97.52	98.87	100.56	100.03	98.79	99.14	100	98.7	100.14	99.49

Analyses: 1 and 2, kyanite eclogites; 3, grospydite; 4, kyanite eclogite with sanidine; 5–7, coesite eclogites; 8, coesite-bearing grospydite; 9–10, diamondiferous eclogites (9, corundum-bearing eclogite, 10, kyanite eclogite with corundum).

Corundum is present in the form of small needles 0.01–0.2 mm long. Usually corundum replaces kyanite (Fig. 5a) and more rarely forms separate lamellar crystals up to 2 mm in kyanite eclogites. It has a dark-blue color and can be designated as sapphire. In some eclogites, the corundum is ruby-red in color and forms elongated crystals up to 1–3 mm. Such corundum is present in a diamondiferous sample that contains more than 90% of garnet in the rock (Fig. 5b). It is more likely that corundum has a secondary origin in this case also. In their compositions these samples are close to the xenoliths described as corganites by Mazzone and Haggerty (1989).

3.3. Diamonds

Diamondiferous samples have been found in all varieties of these xenoliths except alkremites. Usually kyanite eclogites from the Udachnaya pipe contain cubic or coated diamonds, which are small in size

Table 4
Representative compositions of garnets of highly aluminous xenoliths from the Udachnaya pipe

Sample	U-9	U-154	U-157	U-947	U-163	U-2290	U-2310	Ud-114	Ud-155	U-306	U-294
Analyses	1	2	3	4	5	6	7	8	9	10	11
SiO_2	39.32	39.65	40.3	40.21	39.26	39.68	38.99	38.93	40.56	47.72	42.92
TiO_2	0.14	0.13	0.09	0.01	0.2	0.07	0.31	0.19	0.32	0.02	0.05
Al_2O_3	22.48	22.42	22.6	22.65	22.34	22.74	21.98	22.72	22.92	24.08	22.45
Cr_2O_3	0.01	<0.03	0.17	0.06	0.01	<0.03	0.01	0.01	0.06	0.1	0.5
FeO	16.39	8.43	8.86	12.35	12.2	15.58	8.41	7.04	10.77	5.96	5.52
MnO		0.04	0.14	0	0.17	0.26	0.02	0.04	0.11	0.34	0.1
MgO	9.36	6.35	6.9	8.28	5.37	8.87	3.61	10.83	8.97	24.55	14.55
CaO	11.86	22.6	21.19	16.56	20.42	12.01	25.67	19.67	17.35	1.42	14.12
Na_2O	<0.03	<0.03	0.05	0.11	<0.03	0.19	<0.03	0.24	0.03	n.d	n.d
Total	99.77	99.62	100.31	100.24	99.97	99.41	99.01	99.67	101.09	100.19	100.29

Analyses: 1 and 2, kyanite eclogites; 3, grospydite; 4, kyanite eclogite with sanidine; 5–7, coesite eclogites; 8, corundum-bearing diamondiferous garnetite; 9, kyanite eclogite with corundum; 10 and 11, alkremites (after Ponomarenko and Leskova, 1980). n.d. = not detected element.

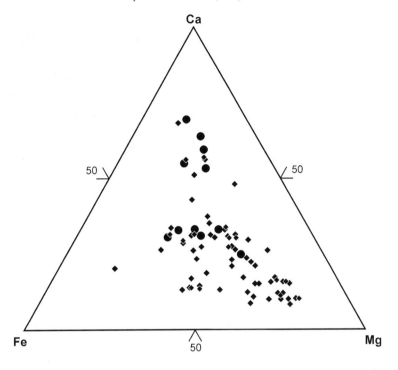

Fig. 3. Garnets compositions of eclogite xenoliths from the Udachnaya pipe (filled circles—coesite eclogites).

(about 1 mm) in most cases (Spetsius, 1995). There exists some evidence for their secondary metasomatic origin due to introduction of fluids enriched in H₂O and other volatile components (Spetsius, 1999). However, a new discovery of two samples of corundum-bearing eclogites with diamonds from the Udachnaya pipe is not consistent with this origin; the xenoliths contain large diamonds (size 4–5 mm) that are present as planar octahedral crystals and show smooth faces (Fig. 5c,d). The diamond crystals are colorless. A black, small mineral inclusion is seen in the center of one of the crystals and is probably sulfide. In addition, the diamonds show obvious features of dissolution and resorption in one sample (Fig. 3c). Most probably its unresorbed parts were inside the xenolith while the resorbed faces projected from it. The surface textures of the diamond crystals imply that the present resorbed surfaces seen on diamonds most probably resulted from their dissolution in the kimberlite melt, since the parts of the diamonds that resided inside the xenolith remained unresorbed. The petrography, mineral composition and geochemistry of these two samples of corundum-bearing eclogites show close

affinities with garnetites from this pipe. Thin section study clearly showed secondary formation of corundum in these samples and so a secondary late origin of diamonds cannot be excluded as well.

3.4. Trace element distribution in eclogite minerals from the Udachnaya pipe

Trace element data for mantle xenoliths has important implications in many aspects: (a) the estimation of distribution in minerals and correct definition of partitioning of trace elements between minerals of mantle rocks in relationship with the PT conditions of their formation, (b) the deciphering of the complex history and evolution of mantle eclogites, which is a subject of much discussion (Ireland et al., 1994; Snyder et al., 1997), (c) the elucidation of possible distinctions in the behavior of trace elements in different mantle processes.

A suite of about 20 xenoliths from the Udachnaya kimberlite pipe have been studied for the trace element composition of their minerals. Only six of the samples were simple bimineral or kyanite eclo-

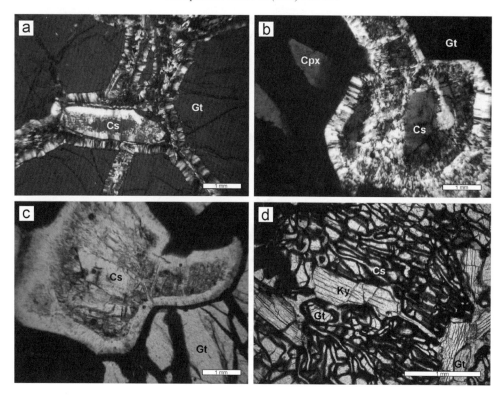

Fig. 4. Coesite-bearing eclogite xenoliths from the Udachnaya kimberlite pipe. (a,b) Crossed polarizers and (c,d) plane-polarized light. Legend for these and next photomicrographs is as follows: Gt = garnet, Cpx = clinopyroxene, Cs = coesite, Ky = kyanite, Cd = corundum. (a) View of coesite relict with pronounced palisade texture of surrounding secondary quartz between grains of garnet (sample U-2290). (b,c) Relicts of coesite surrounded by rims of palisade quartz with addition of fine grained quartz (sample U-256). (d) Intergrowth of coesite with garnet (sample U-168).

gites without diamonds. Diamondiferous xenoliths include not only bimineral eclogites, but one sample of garnet clinopyroxenite and two xenoliths of garnetites (with content of clinopyroxene less than 1%). Abundance of trace elements has been studied in coexisting garnets and clinopyroxenes in nearly all samples, and in secondary clinopyroxenes in some xenoliths. The special checking of core and rim parts of garnet grains shows that they are homogeneous in major and trace element composition. A little zonation of garnet is found only in two samples where rims are slightly enriched in Nd, Sm, Eu, Dy and Ho. It is necessary to stress that during the analysis of primary garnet and clinopyroxene in intensively metasomatized and partially melted xenoliths only the fresh relicts of these minerals were chosen.

Salient features of the results of LAM analyses can be summarized as follows:

– Chondrite-normalized REE pattern of garnets usually shows a convex shapes and varies from slightly to strongly enriched in LREE (Fig. 6), whereas clinopyroxenes have characteristically high LREE abundance and show broad variations in MREE (Fig. 7).
– On the basis of the REE distribution three different types of garnets are distinguished in the studied eclogite xenoliths: (1) a "normal" group having upwardly convex pattern with initial steep progressive increase in LREE followed by a slower rate of increase in HREE, (2) a "HREE depleted" group in which the HREE showed no marked increase from Dy to Yb, and (3) an "Eu-anomalous group" in which there is a small positive Eu anomaly and a generally flat HREE pattern (Spetsius et al., 1998).
– The trace element content of most of the group 1 garnets are similar, excluding garnetite sample

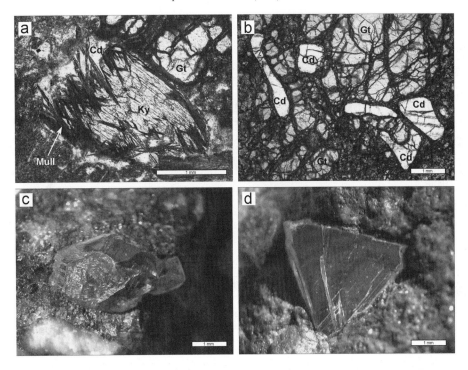

Fig. 5. Diamondiferous corundum-bearing eclogite xenoliths from the Udachnaya kimberlite pipe. (a,b) Plane-polarized light and (c,d) microphotographs. (a) Needles of secondary corundum (sapphire variety) in association with mullite replacing kyanite (sample U-820). (b) Corundum of ruby variety in diamondiferous garnetite (sample Ud-114). (c) Intergrowth of two octahedral diamonds in corundum-bearing eclogite. Resorption texture on the bigger crystal is obvious (sample Ud-155). (d) Plane-faced octahedron in corundum-bearing garnetite (sample Ud-114).

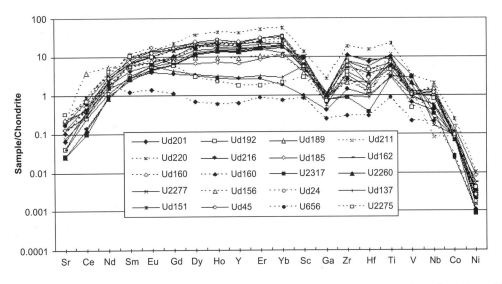

Fig. 6. Chondrite-normalized REE diagram for garnets of eclogites from the Udachnaya kimberlite pipe (Spetsius et al., 1998; unpublished data, ICP-MS, National University of Australia). Data normalized against chondrite values of McDonough and Sun (1995).

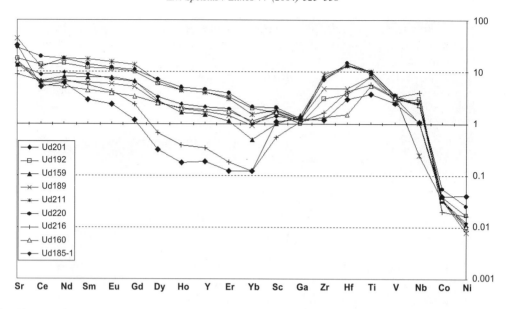

Fig. 7. Chondrite-normalized REE diagram for clinopyroxenes of eclogites from the Udachnaya kimberlite pipe (Spetsius et al., 1998; unpublished data, ICP-MS, National University of Australia). Data normalized against chondrite values of McDonough and Sun (1995).

(Ud-220) where this mineral is very rich in Sc, Ga, Yt, Nb, Zr, Ce, Gd, Dy, Ho, Er, Yb, Lu and Hf (Fig. 6).

- Wide variations in REE and also in Sr are observed for clinopyroxenes (Fig. 7); the most suitable explanation for this enrichment is partial melting connected with metasomatism. Clinopyroxene Ud-2260 has especially high Y, Sm, Dy, Ho, Er and Yb.
- The highly aluminous xenoliths have slightly positive Eu-anomalies and low HFSE abundances that give some evidence for crustal protoliths, in accordance with the findings of Jacob and Foley (1999).
- Primary clinopyroxenes of most samples are enriched in LREE suggesting widespread cryptic metasomatism in many eclogite xenoliths from Udachnaya, this confirms the results of petrographic observation (Spetsius, 1995).

It should be pointed out that the substitution of Sr, Ba and such incompatible elements as V, Zr, Ni, and others into garnet and clinopyroxene is not controlled essentially by T and P or bulk composition of minerals (Spetsius et al., 1998). These observations indicate that part of the trace elements has independent behav-

ior and support the suggestion (Ireland et al., 1994) that enrichment or depletion of eclogite xenolith rock-forming minerals from Udachnaya pipe in some trace elements is due to complicated metasomatic and partial melting events. The REE abundance in coexisting garnets and clinopyroxenes and Cpx/Gt partition coefficients for some trace elements in eclogite xenoliths from Udachnaya (Spetsius et al., 1998) are similar to those found by O'Reilly and Griffin (1995) for eclogites from South Africa, but these occurrences differ in Sr, Zr and Ni suggesting some distinct differences in the evolution of the mantle eclogites of the Siberian and South African cratons.

4. Discussion

Highly aluminous xenoliths are common amongst the eclogite xenolith suite from many pipes of the Yakutian province, especially in kimberlites of the Daldyn–Alakitsky region. By their petro- and geochemistry these rocks belong to the group C eclogites of Coleman et al. (1965). Aside from the presence of coesite or corundum, there are no apparent differences in texture of xenoliths and major chemistry of garnets and clinopyroxenes. In addition, most xenoliths of this suite contain kyanite. Coesite is present in silica oversaturated- and corundum in silica undersaturated bulk-composition rocks. The presence of coesite implies that the eclogites from kimberlites must have been subducted to a depth of greater than 100 km, at pressures in excess of 3 GPa (Spetsius and Serenko, 1990 and references therein). Highly aluminous xenoliths including kyanite-, corundum- and coesite-bearing eclogites, grospydites and alkremites constitute a minor portion of the mantle-derived xenoliths and are predominantly of the eclogite suite. The mineralogy of these rocks and isotopic compositions of minerals suggest that they could represent subducted oceanic lithosphere (Ireland et al., 1994).

It should be pointed out that only rare estimations for oxygen isotope composition of garnet from these rocks have been made but values of the $\delta^{18}O$ are in the range 4.5–7.0‰ (Spetsius and Taylor, 2003). These data are in accordance with the oxygen isotope estimation for garnets of coesite eclogites from Roberts Victor (Schulze et al., 2000) and provide strong evidence for their possible origin as the products of oceanic crust subduction.

It is possible to use well-documented representatives of mantle (xenoliths and diamonds) to estimate the distribution of subducted crustal remnants in the upper mantle underlying the Siberian platform. The populations of mantle xenoliths from the main well-investigated kimberlite pipes of the Yakutian province are well known, and there are published data and estimations of the distribution of light C isotope diamonds in populations from different kimberlite pipes (e.g., Bulanova et al., 2002; Galimov, 1991). The carbon isotopic ratios of diamonds from the Yakutian kimberlites have shown that many diamonds have $\delta^{13}C$ distinct from the typical mantle values (e.g., Bulanova et al., 1999; Galimov, 1991). The reasonable explanation for the high $\delta^{13}C$ is a contribution from subducted oceanic crust. These results show that in many pipes kyanite eclogites or isotopically light diamonds are present, or both. This is confirmed first of all for the kimberlites of central Daldyn–Alakitsky region. The presence and preservation of coesite in eclogites and diamonds (Sobolev et al., 1998) give strong proof of the involvement of subducted oceanic crust in the formation of the sub-continental lithospheric mantle (SCLM).

The study of xenoliths in kimberlite from pipes of the Yakutian province indicates an obvious difference in their distribution in separate pipes and fields (Spetsius, 1995). Comparison of eclogite suite xenoliths on the profile from the south to the north of the province shows that there is a lateral mantle heterogeneity over a distance of about 1000 km. It is expressed by the presence of a suite of the highly aluminous rocks in kimberlite pipes of the Daldyn–Alakitsky field in the central part of the province. Such xenoliths were found in Udachnaya, Zagadochnaya, Zarnitsa and other pipes. It should be noted that these mantle xenoliths are found and coupled with some crustal granulite xenoliths such as eclogite-like rocks sometimes containing kyanite.

Eclogitic mantle xenoliths occur in all kimberlite provinces worldwide and various origins have been postulated for them. Two contrasting petrogeneses are in favour: either mantle eclogites represent (1) high-pressure magmatic cumulates which occur as magma chambers or dykes within the Upper Mantle (Dawson and Carswell, 1990; Smyth et al., 1989; Snyder et al., 1993) or (2) recycled and metamorphosed Archaean oceanic crust (Ireland et al., 1994; Jacob et al., 1994; Jacob and Foley, 1999; Barth et al., 2001, 2002). There exists mineralogical and isotopic evidence that mantle eclogites may have multiple origins and both types may occur even within one kimberlite pipe (e.g., Taylor and Neal, 1989; Snyder et al., 1997). The petrology of these rocks has been properly characterised but the origin of eclogite xenoliths is still a matter of debate. At the same time it should be pointed out that eclogites and especially xenoliths with such rare and unique minerals as coesite and corundum may provide valuable information about asthenosphere–lithosphere interaction and the conditions of the SCLM formation and diamond origin.

It is possible that some eclogites are the products of subducted oceanic crust. This is confirmed for the coesite-bearing and diamondiferous eclogites from kimberlites of Yakutia, as well as from other kimberlite provinces (e.g., Jacob et al., 1994; Barth et al., 2002; Spetsius and Taylor, 2003). According to the distribution of different types of eclogites, and the presence of isotopically light as well as cubic crystals of diamonds in kimberlites of the Yakutian province, it is possible to estimate the amount of crustal sources added to the mantle under Siberian platform. In the central part of the Siberian platform the remnants of subducted crust could be about 5–10% of the total upper mantle. If we assume that all isotopic light diamonds were crystallized in remnants of subducted crust we could use this for the estimation of the contribution of subducted oceanic crust in the formation of the cratonic roots under the Siberian platform. According to these data about 10% of the diamond population from kimberlites of Yakutia originated in subducted rocks.

There exist two possible ways for highly aluminous eclogites to form: (i) as a result of transformation of initial rocks of gabbro-anorthosite composition through the intermediate stage of eclogite-like rocks during the process of subduction or delamination; (ii) as a result of fractional crystallization of ultra-mafic melts under mantle conditions. The first alternative is suggested by petrographic, petrochemical and isotopic investigations that show a genetic relationship of all the series of these rocks (Spetsius and Serenko, 1990). The second alternative is proved by the presence of subsolidus changes in eclogites and by continuous sets of highly aluminous xenoliths from kyanite eclogites up to alkremites and by linear differentiation trends of their compositions (Exley et al., 1983; Spetsius, 1995). The possibility of forming some eclogites via subduction and subsequent metamorphism of oceanic crust, as is proved by simplified isotopic composition of diamonds in different eclogites, is not excluded. Such diversity of origin and also the subsequent evolution of eclogites during the process of partial melting and mantle metasomatism have been considered in a number of papers (Spetsius and Serenko, 1990; Snyder et al., 1993, 1997; Ireland et al., 1994; Pearson et al., 1995; Spetsius and Taylor, 2002 and references therein), and, in turn, determine the specific nature of these rocks in separate pipes.

Based on highly enriched Sr isotope composition ($^{87}Sr/^{86}Sr>0.8$), a crustal origin was proposed for alkremites and it was suggested that these rocks could represent the restites from subducted, melted pelitic sediments (Mazzone and Haggerty, 1989). Preliminary data on trace element content in garnet and high Lu/Hf ratio do not support this model (Nowell et al., 2003). It is more likely that these rocks were formed as a result of fractionation by melting of lithosphere as

suggested first by Ponomarenko et al. (1977). In my opinion, at least some alkremites are formed as a result of fractional crystallization of ultramafic melts enriched in Al and Mg and probably these unique rocks could represent a residual melt after separation firstly of some kyanite eclogites.

5. Conclusions

Highly aluminous mantle xenoliths found in kimberlite, including kyanite and coesite eclogites and alkremites, comprise a specific rock group. The wide abundance of these types of xenoliths in the central region of the Yakutian Kimberlite province suggests a lateral petrographic heterogeneity of the SCLM under Siberian craton. The confinement of cubic diamonds to highly aluminous rocks is also matched by the lateral heterogeneity in the morphology of the diamond populations from the south to the north of the province. Two possible ways of formation of highly aluminous rocks could be proposed.

The correlation between the oxygen isotopes of eclogite minerals and carbon isotopes of diamonds gives strong support for subduction of oceanic crust having a role in the formation of the mantle root beneath the Yakutian Kimberlite Province within the Siberian platform, especially its central part. The presence of Ky- and Cs-bearing eclogites in kimberlite as well as light isotopic diamonds and cubic crystals in the diamond populations could be used for the estimation of the intensity of the subduction process at the time of the SCLM formation beneath any given kimberlite pipe or field.

Acknowledgements

Wayne Taylor is thanked for the collaborative work on doing trace elements by ICPMS and Brendon Griffin for his assistance during the author's visit to Australia and support with the electron microprobe analyses of primary and secondary phases in eclogites. I am grateful to Alexander Ivanov who kindly helped with the electron microprobe analyses and SEM images. Thanks are given to Graham Pearson for cooperative work on trace elements in alkremites and discussions on eclogite petrology that helped in the understanding of their nature. This paper has benefited from constructive remarks by Chris Smith, and I appreciate deeply his efforts in helping me to accurately present data and interpretation. The author is grateful to the reviewers, J. B. Dawson and N.P. Pokhilenko, who made invaluable comments and suggestions that greatly improved the clarity of the paper.

References

Barth, M.G., Rudnick, R.L., Horn, I., McDonough, W.F., Spicuzza, M.J., Valley, J.W., Haggerty, S.E., 2001. Geochemistry of xenolithic eclogites from West Africa: Part I. A link between low MgO eclogites and Archaean crust formation. Geochim. Cosmochim. Acta 65, 1499–1527.

Barth, M.G., Rudnick, R.L., Horn, I., McDonough, W.F., Spicuzza, M.J., Valley, J.W., Haggerty, S.E., 2002. Geochemistry of xenolithic eclogites from West Africa: Part 2. Origin of the high MgO eclogites. Geochim. Cosmochim. Acta 66, 4325–4345.

Bobrievich, A.P., Bondarenko, M.N., Gnevushev, M.A., Krasov, A.M., Smirnov, G.I., Yurkevich, R.K., 1959. The Diamond Deposits of Yakutia. Gosgeoltekhizdat, Moscow (in Russian).

Bulanova, G.P., Griffin, W.L., Kaminsky, F.V., Davies, R., Spetsius, Z.V., Ryan, C.G., Andrew, A., Zahkarchenco, O.D., 1999. Diamonds from Zarnitsa and Dalnaya kimberlites (Yakutia), their nature and lithospheric mantle source. Proceedings VIIth Intern. Kimberlite Conf., Cape Town, South Africa, vol. 1, pp. 49–56.

Bulanova, G.P., Pearson, D.G., Hauri, E.H., Griffin, B.J., 2002. Carbon and nitrogen isotope systematics within a sector-growth diamond from the Mir kimberlite, Yakutia. Chem. Geol. 188, 105–123.

Coleman, R.G., Lee, D.E., Beatty, L.B., Brannock, W.W., 1965. Eclogites and eclogites: their differences and similarities. Geol. Soc. Am. Bull. 76, 483–508.

Dawson, J.B., Carswell, D.A., 1990. High temperature and ultra-high pressure eclogites. In: Carswell, D.A. (Ed.), Eclogite–Facies Rock. Chapman Hall, New York, pp. 316–319.

Exley, R.A., Smith, J.F., Dawson, J.B., 1983. Alkremite, garnetite and eclogite xenoliths from Bellsbank and Jagersfontein, South Africa. Amer. Mineral. 68, 512–516.

Fadili, S.El., Demaiffe, D., 1999. Petrology of eclogite and granulite nodules from the Mbuji Mayi Kimberlites (Kasai, Congo): significance of kyanite–omphacite intergrowths. Proceedings VIIth Intern. Kimberlite Conf., Cape Town, South Africa, vol. 1, pp. 205–213.

Galimov, E.M., 1991. Isotope fractionation related to kimberlite magmatism and diamond formation. Geochim. Cosmochim. Acta 55, 1697–1708.

Ireland, T.R., Rudnick, R.L., Spetsius, Z.V., 1994. Trace elements in diamond inclusions from eclogites reveal link to Archean granites. Earth Planet. Sci. Lett. 128, 199–213.

Jacob, D.E., Foley, S.F., 1999. Evidence for Archean ocean crust

with low high field strength element signature from diamondiferous eclogite xenoliths. Lithos 48, 317–336.

Jacob, D., Jagoutz, E., Lowry, D., Mattey, D., Kudrjavtseva, G., 1994. Diamondiferous eclogites from Siberia: remnants of Archean oceanic crust. Geochim. Cosmochim. Acta 58, 5191–5207.

Mazzone, P., Haggerty, S.E., 1989. Peraluminous xenoliths in kimberlite: metamorphosed restites produced by partial melting of pelites. Geochim. Cosmochim. Acta 53 (7), 151–156.

McDonough, W.F., Sun, S.S., 1995. The composition of the Earth. Chem. Geol. 120, 223–253.

Nowell, G.M., Pearson, D.G., Jacob, D.J., Spetsius, Z.V., Nixon, P.H., Haggerty, S.E., 2003. The origin of alkremites and related rocks: a trace element, Lu–Hf, Rb–Sr and Sm–Nd isotope study. Ext. Abstr. 8th Intern. Kimberlite Conf., Victoria, Canada.

O'Reilly, S.Y., Griffin, W.L., 1995. Trace-element partitioning between garnet and clinopyroxene in mantle-derived pyroxenites and eclogites: P-T-X controls. Chem. Geol. 121, 105–130.

Pearson, D.G., Shirey, S.B., Carlson, R.W., Boyd, F.R., Pokhilenko, N.P., Shimizu, N., 1995. Re–Os, Sm–Nd, and Rb–Sr isotope evidence for thick Archaean lithospheric mantle beneath the Siberian craton modified by multistage metasomatism. Geochim. Cosmochim. Acta 59, 959–977.

Ponomarenko, A.I., Leskova, N.V., 1980. Peculiarities of chemical composition of minerals of alkremites from kimberlite pipe "Udachnaya". Dokl. Akad. Nauk SSSR. 252 (3), 707–711 (in Russian).

Ponomarenko, A.I., Sobolev, N.V., Pokhilenko, N.P., 1976. Diamondiferous grospydite and diamondiferous disthene eclogites from kimberlite pipe Udachnaya, Yakutia. Dokl. Akad. Nauk SSSR. 226 (4), 927–930 (in Russian).

Ponomarenko, A.I., Spetsius, Z.V., Lybushkin, V.A., 1977. Kyanite eclogite with coesite. Dokl. Akad. Nauk SSSR 236 (1), 215–219 (in Russian).

Rosen, O.M., Serenko, V.P., Spetsius, Z.V., Manakov, A.V., Zinchuk, N.N., 2002. Yakutian kimberlite province: position in the structure of the Siberian craton and composition of the upper and lower crust. Russ. Geol. Geophys. 43, 3–26 (in Russian).

Schulze, D.J., Valley, J.W., Spicuzza, K.J., 2000. Coesite eclogites from the Roberts Victor kimberlite, South Africa. Lithos 54, 23–32.

Smyth, J.R., Caporuscio, F.A., McCormick, T.C., 1989. Mantle eclogites: evidence of igneous fractionation in the mantle. Earth Planet. Sci. Lett. 93, 133–141.

Snyder, G.A., Jerde, E.A., Taylor, L.A., Halliday, A.N., Sobolev, V.N., Sobolev, N.V., 1993. Nd and Sr isotopes from diamondiferous eclogites, Udachnaya Kimberlite Pipe, Yakutia, Siberia: evidence of differentiation in early Earth? Earth Planet Sci. Lett. 118, 91–100.

Snyder, G.A., Taylor, L.A., Crozaz, G., Halliday, A.N., Beard, B.L., Sobolev, V.N., Sobolev, N.V., 1997. The origins of Yakutian eclogite xenoliths. J. Petrol. 38, 85–113.

Sobolev, N.V., 1977. Deep-Seated Inclusions in Kimberlites and the Problem of the Composition of the Upper Mantle American Geophysical Union, Washington, DC.

Sobolev, N.V., Taylor, L.A., Zuev, V.M., Bezborodov, S.M., Snyder, A., Sobolev, V.N., Yefimova, E.S., 1998. The specific features of eclogitic paragenesis of diamonds from Mir and

Udachnaya kimberlite pipes (Yakutia). Russ. Geol. Geophys. 39, 1653–1663.

Spetsius, Z.V., 1995. Occurrence of diamond in the mantle: a case study from the Siberian Platform. J. Geochem. Explor. 53, 25–39.

Spetsius, Z.V., 1999. Two generation of diamonds in the eclogite xenoliths. Proceedings VIIth Intern. Kimberlite Conf., Cape Town, South Africa, vol. 2, pp. 823–828.

Spetsius, Z.V., Serenko, V.P., 1990. Composition of Continental Upper Mantle and Lower Crust Beneath the Siberian Platform Nauka, Moscow (in Russian).

Spetsius, Z.V., Taylor, L.A., 2002. Partial melting in mantle eclogite xenoliths: connection with diamond paragenesis. Int. Geol. Rev. 44, 973–987.

Spetsius, Z.V., Taylor, L.A., 2003. Kimberlite xenoliths as evidence for subducted oceanic crust in the formation of the Siberian craton. Proceedings Intern. Conf. Plumes and Problems of Deep Sources of Alkaline Magmatism. Publ. Univ. House, Irkutsk, pp. 5–19.

Spetsius, Z.V., Taylor, W.R., Griffin, B.J., 1998. Major and trace-element partitioning between mineral phases in diamondiferous and non-diamondiferous eclogites from the Udachnaya kimberlite pipe, Yakutia. Extended Abstracts of 7th International Kimberlite Conference, Cape Town, South Africa., pp. 856–858.

Taylor, L.A., Neal, C.R., 1989. Eclogites with oceanic crustal and mantle signatures from the Bellsbank kimberlite, South Africa: Part 1. Mineralogy, petrography, and whole rock chemistry. J. Geol. 97, 551–567.

Available online at www.sciencedirect.com

Lithos 77 (2004) 539–552

www.elsevier.com/locate/lithos

Petrology and geochemistry of a diamondiferous lherzolite from the Premier diamond mine, South Africa

K.S. (Fanus) Viljoen[a,*], René Dobbe[a], Braam Smit[a],
Emilie Thomassot[b], Pierre Cartigny[b]

[a] GeoScience Centre, De Beers Consolidated Mines Limited, P.O. Box 82232, Southdale 2135, South Africa
[b] Laboratoire de Géochimie des Isotopes Stables, UMR CNRS 7047, Institut de Physique du Globe and Université de Paris 7,
2 Place Jussieu, T54-64 E1, 75251 Paris, Cedex 05, France

Received 27 June 2003; accepted 12 January 2004
Available online 19 May 2004

Abstract

This paper reports on the petrology and geochemistry of a diamondiferous peridotite xenolith from the Premier diamond mine in South Africa.

The xenolith is altered with pervasive serpentinisation of olivine and orthopyroxene. Garnets are in an advanced state of kelyphitisation but partly fresh. Electron microprobe analyses of the garnets are consistent with a lherzolitic paragenesis (8.5 wt.% Cr_2O_3 and 6.6 wt.% CaO). The garnets show limited variation in trace element composition, with generally low concentrations of most trace elements, e.g. Y (<11 ppm), Zr (<18 ppm) and Sr (<0.5 ppm). Garnet rare earth element concentrations, when normalised against the C1 chondrite of McDonough and Sun (Chem. Geol. 120 (1995) 223), are characterised by a rare earth element pattern similar to garnet from fertile lherzolite.

All diamonds recovered are colourless. Most crystals are sharp-edged octahedra, some with minor development of the dodecahedral form. A number of crystals are twinned octahedral macles, while aggregates of two or more octahedra are also common. Mineral inclusions are rare. Where present they are predominantly small black rosettes believed to consist of sulfide. In one instance a polymineralic (presumably lherzolitic) assemblage of reddish garnet, green clinopyroxene and a colourless mineral is recognised.

Infrared analysis of the xenolith diamonds show nitrogen contents generally lower than 500 ppm and variable nitrogen aggregation state, from 20% to 80% of the 'B' form. When plotted on a nitrogen aggregation diagram a well defined trend of increasing nitrogen aggregation state with increasing nitrogen content is observed. Carbon isotopic compositions range from -3.6 ‰ to -1.3‰ These are broadly correlated with diamond nitrogen content as determined by infrared spectroscopy, with the most negative C-isotopic compositions correlating with the lowest nitrogen contents.

Xenolith mantle equilibration temperatures, calculated from nitrogen aggregation systematics as well as the Ni in garnet thermometer are on the order of 1100 to 1200 °C.

* Corresponding author. Tel.: +27-11-3747873; fax: +27-11-8351315.
E-mail address: fanus.viljoen@debeersgroup.com (K.S. Viljoen).

It is concluded that the xenolith is a fertile lherzolite, and that the lherzolitic character may have resulted from the total metasomatic overprinting of pre-existing harzburgite. Metasomatism occurred prior to, or accompanied, diamond growth.

Keywords: Peridotite; Lherzolite; Diamond; Kimberlite; Lithosphere; Infrared

1. Introduction

This paper reports on the petrology and geochemistry of a diamondiferous peridotite from the Premier diamond mine. Such xenoliths are surprisingly rare on the Kaapvaal Craton, with four specimens having been described from Finsch (Shee et al., 1982; Viljoen et al., 1992), eight from Roberts Victor (Viljoen et al., 1994), and one from Mothae in Lesotho (Dawson and Smith, 1975). A total of 18 diamond-bearing peridotitic garnet macrocrysts have also been recognised at the Newlands kimberlite (Daniels et al., 1995; Menzies et al., 1999).

The X-ray diamond recovery circuit (the Sortex Plant) at the Premier diamond mine in South Africa typically generates about three specimens of diamond partially enclosed in kimberlite each day. These are sent to the sorthouse at the mine where a small press is used to liberate the diamonds from the kimberlite matrix. Staff at the sorthouse recognised the importance of the specimen when crushing revealed numerous diamonds. A total of 59 diamonds, ranging in size from less than 1 mm to approximately 2 mm, along with 50-cm-sized fragments of the complete xenolith (many of which still contain diamonds in situ) were provided by the Premier sorthouse.

2. The mantle sample at Premier

The Premier kimberlite is located in the Central Terrain of the Archaean Kaapvaal Craton (De Wit et al., 1992), with a preferred emplacement age of 1180 Ma (Allsopp et al., 1989; Richardson et al., 1993). Underground mining operations reveal that the kimberlite, which is cut by a 1100 Ma gabbroic sill, penetrates a norite phase of the Bushveld Complex. The Bushveld Complex is among the largest layered intrusions known, comprising a sequence of basic rocks 7–9 km thick, with an areal extent of 66,000 km² (the Rustenburg Layered Suite) and subsequent granites. The Premier diamond mine is renowned as the source of many of the world's largest gem diamonds, e.g. the Cullinan and Centenary diamonds. The locality has been the subject of xenolith studies (e.g. Danchin, 1979; Boyd and Mertzman, 1987; Boyd, 1989; Boyd et al., 1993) as well as minerals occurring

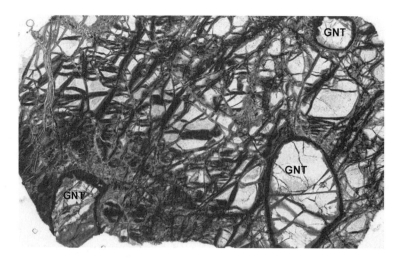

Fig. 1. Photomicrograph of Premier diamond-bearing peridotite DO40. Rounded garnets (GNT) are set in a coarse matrix of serpentine pseudomorphs after possible olivine and orthopyroxene. Field of view = 1.5 cm.

as inclusions in diamond (Gurney et al., 1985). The Premier garnet lherzolites can be clearly subdivided into coarse and deformed varieties on the basis of differences in mineral composition and texture. Deformed lherzolites originate from greater depths than those with coarse textures which are in turn more depleted in Fe, Al, Ca and Ti relative to the deformed rocks. Constituent minerals of the deformed xenoliths are extremely titaniferous and clinopyroxenes contain significant potassium. Garnet harzburgites are also progressively deformed and less depleted with respect to Fe/Mg with increasing depth of origin (Danchin, 1979). Inclusions in diamonds from Premier have major element compositions covering the worldwide range for such inclusions. About 40% of the inclusions are peridotitic and 60% eclogitic (Gurney et al., 1985). The peridotitic inclusion suite of olivines, orthopyroxenes and garnets are bimodal in composition, reflecting a more magnesian, clinopyroxene-free harzburgitic assemblage and a more calcic, clinopyroxene-bearing lherzolitic assemblage (Gurney et al., 1985; Griffin et al., 1992). A difference in carbon isotopic composition between harzburgitic and lherzolitic diamonds is observed (Deines et al., 1984; 1989).

Harzburgitic garnet inclusions in diamonds from Cretaceous kimberlites at Kimberley and Finsch, and their counterparts found as isolated macrocrysts in the kimberlite host rocks, have neodymium and strontium isotope signatures consistent with an Archaean age (>3000 Ma) and a metasomatised mantle source (Richardson et al., 1984). At Premier, the harzburgitic garnets occurring as inclusions in diamond have raised Nd and Sr concentrations, which belie their residual major element compositions. Furthermore, their Sm/Nd and $^{143}Nd/^{144}Nd$ ratios are lower, and their $^{87}Sr/^{86}Sr$ ratios higher, than bulk Earth values for 1180 Ma ago. These characteristics are similar to those for harzburgitic garnet inclusions in diamonds from Kimberley and Finsch. Although this suggests a similar Archaean age and origin for the Premier harzburgitic diamonds, their garnet Sm/Nd and $^{143}Nd/^{144}Nd$ ratios are not sufficiently different from each other to yield an isochron age, or from bulk Earth values to yield a meaningful model age (Richardson et al., 1993).

Lherzolitic garnet and clinopyroxene inclusions in diamonds from Premier yield a preferred Sm–Nd isochron age of 1930 Ma, which is ~ 100 Ma less than that of the adjacent Bushveld Complex, suggesting a link analogous to that between harzburgitic diamond formation and komatiitic magmatism in the Archaean (Richardson et al., 1993; Shirey et al., 2002). The crystallisation of eclogitic diamonds at Premier is essentially contemporaneous with the age of pipe emplacement (Richardson, 1986).

Table 1
Chemical composition (as well as averages and 1σ standard deviation) and garnet–olivine nickel equilibration temperature (°C) of garnets from Premier diamond-bearing peridotite DO40

Garnet	3	4	5	Average	Sdev
wt.%				n = 11	n = 11
SiO₂	40.73	40.62	40.80	40.57	0.18
TiO₂	0.23	0.25	0.23	0.24	0.01
Al₂O₃	17.25	16.96	17.67	17.12	0.20
Cr₂O₃	8.23	8.57	7.92	8.50	0.23
FeO	6.53	6.60	6.42	6.54	0.05
MnO	0.33	0.33	0.33	0.32	0.02
MgO	20.32	20.28	20.53	20.33	0.09
CaO	6.54	6.69	6.41	6.61	0.07
Na₂O	0.01	0.02	0.02	0.02	0.01
Total	100.17	100.32	100.33		
ppm				n = 3	n = 3
Ni	63.2	63.0	61.9	62.7	0.7
Ga	6.98	6.95	6.73	6.9	0.14
Y	10.58	10.86	8.64	10.0	1.21
Zr	18.01	17.93	14.74	16.9	1.86
La	0.084	0.055	0.104	0.081	0.025
Ce	0.711	0.680	0.662	0.684	0.025
Pr	0.213	0.221	0.175	0.203	0.024
Nd	1.530	1.565	1.269	1.455	0.162
Sm	0.643	0.531	0.482	0.552	0.083
Eu	0.200	0.297	0.191	0.230	0.059
Gd	1.001	0.986	0.812	0.933	0.105
Tb	0.201	0.226	0.174	0.201	0.026
Dy	1.557	1.664	1.517	1.579	0.076
Ho	0.409	0.378	0.319	0.369	0.046
Er	1.226	1.271	1.053	1.183	0.115
Yb	1.841	1.936	1.675	1.817	0.132
Lu	0.318	0.328	0.276	0.307	0.028
Hf	0.691	0.615	0.496	0.601	0.099
Nb	1.259	1.119	1.013	1.130	0.123
Pb	0.294	0.208	0.294	0.265	0.050
Sr	0.485	0.477	0.448	0.470	0.020
Ba	<0.076	<0.067	<0.042	–	–
Th	0.036	0.036	0.039	0.037	0.002
TNi (°C)					
Ryan et al. (1996)	1111	1110	1104	1108	4
Canil (1999)	1107	1106	1102	1105	2

3. Methods

Garnets were manually extracted from the xenolith, mounted in epoxy resin and the surface polished in preparation for analysis by electron microprobe and laser ablation inductively coupled plasma mass spectrometer (LA-ICP-MS).

Major element chemical analysis of garnet was performed at the De Beers GeoScience Centre in Johannesburg with a Cameca SX-50 wavelength-dispersive spectrometer operated at an acceleration potential of 20 kV and at a probe current of 40 nA. Counting times were 20 s for Al, Mg, Cr, Ti, Ca, Fe and 30 s for K, Na, Ni. K-α lines were used for all elements. Apparent concentrations were corrected for matrix

effects with the online PAP procedure. MgO (Mg), Cr_2O_3 (Cr), Fe_3O_4 (Fe), $MnTiO_3$ (Mn, Ti), almandine garnet (Si, Al), diopside (Ca), albite (Na), sanidine (K) and Ni metal (Ni) were used as standards.

Concentrations of selected trace elements in garnet were determined by LA-ICP-MS analysis (Norman, 1998) at the De Beers GeoScience Centre in Johannesburg, utilising a frequency-quadrupled Cetac LSX200 Nd:YAG UV laser beam of 266 nm wavelength at a 4 Hz repetition rate, and operating at maximum power. The ablated material was carried to a Perkin Elmer Sciex ELAN 6000 inductively coupled plasma mass spectrometer in a high purity Ar–He mixture gas stream. A typical analysis consists of 100 replicates with each replicate representing three sweeps of the

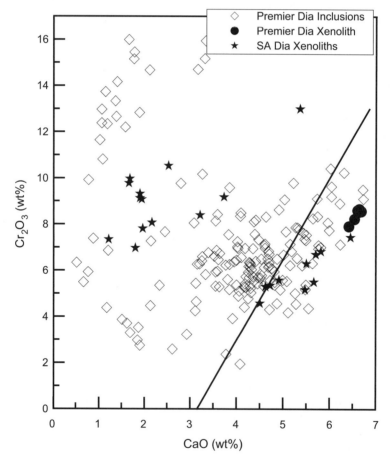

Fig. 2. Distribution of CaO and Cr_2O_3 in peridotitic garnets occurring as inclusions in diamonds from Premier (Gurney et al., 1985; Richardson et al., 1993) as well as the diamond-bearing peridotite from Premier. Also shown are garnets from other diamond-bearing peridotites and diamond-bearing macrocrysts in Southern Africa (Shee et al., 1982; Viljoen et al., 1992, 1994; Daniels et al., 1995; Menzies et al., 1999). The Premier xenolith garnets plot within the lherzolite field as defined by the 85% line of Gurney (1984).

selected mass range, a data collection period of 60–120 s during ablation, and an additional 40 s for background counting. All analytical data were examined in real-time to preclude processing of data for mineral inclusions or alteration products. Typical pit diameters were on the order of 50 μm. The NIST 610 and 612 glass standards were used to calibrate relative element sensitivities for the analyses, using the data of Norman et al. (1996). Each analysis was normalised using Ca values determined by electron microprobe.

The aggregation of nitrogen in diamond is a kinetic phenomenon in which the degree of aggregation depends on the mantle residence temperature, diamond nitrogen content and the mantle residence time (Evans and Qi, 1982; Evans and Harris, 1989; Evans, 1992). Experimental studies have established that, with increasing temperature and/or time, the aggregation proceeds from initially incorporated single nitrogen atoms (C-centres) in type Ib diamond, to nitrogen pairs (A-centres) in type IaA diamond and, finally to nitrogen in tetrahedral arrangement (B-centres) in type IaB diamond (Kiflawi and Bruley, 2000). Nitrogen contents and aggregation states were determined at the GeoScience Centre by transmission infrared spectroscopy using a Nicolet 760 Magna-IR infrared spectrometer linked to a Nicplan infrared microscope (Viljoen, 2002). Nitrogen contents were calculated from the infrared absorption coefficients specified by Boyd et al. (1994, 1995a) for nitrogen in the A and B forms at 1282 cm^{-1}, i.e. 16.5 and 79.4 atomic ppm cm, respectively. Isotherms and mantle residence temperatures were calculated using an activation energy of 674.8 kJ mol^{-1} ($-E/R = -81164$; Cooper, 1990) and an Arrhenius constant (A) of 2.683373×10^5 s^{-1} ppm^{-1}.

Diamond $\delta^{13}C$ values were measured with a conventional dual-inlet mass spectrometer after sample combustion in an oxygen atmosphere (Boyd et al., 1995b) and quantification to check for total combustion. Values are expressed in the conventional delta-notation where $\delta^{13}C = (^{13}C$ sample$/^{13}C$ PDB $-1) \times 1000$.

4. Xenolith petrography and diamond content

The xenolith is altered with pervasive serpentinisation of constituent minerals (Fig. 1). Garnets are generally partially kelyphitised and range in size up to 2 mm. These occur interstitially with serpentine pseudo-

morphs after possible olivine and/or orthopyroxene. The pervasive alteration complicates petrographic analysis and it is not possible to accurately determine volume percentages of these two minerals, assuming that both were present. Other minerals typical of peridotites such as clinopyroxene and accessory spinel are not seen. The xenolith texture is coarse (i.e. not sheared).

A total of 59 individual diamonds with a total estimated weight of 2.371 carats were observed in the 85.4 g of xenolith fragments provided for study. The xenolith is hence estimated to contain a minimum of 27,700 carat/metric ton diamond. This is significantly higher than the previously estimated average grade of diamond-bearing peridotite xenoliths (125 carats per metric ton with a maximum of 651 carats per metric ton), and comparable to that of diamond-bearing eclo-

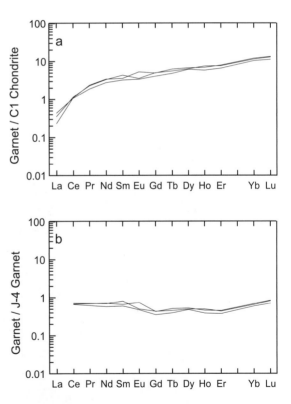

Fig. 3. (a) REE patterns for three garnets from Premier xenolith DO40 normalised to the C1 carbonaceous chondrite of McDonough and Sun (1995). (b) REE patterns for three garnets from Premier xenolith DO40 normalised to garnet from a well equilibrated garnet lherzolite xenolith with a primitive mantle composition (J-4 from Jagersfontein, analysis of Jagoutz and Spettel, see Table 1 in Stachel et al., 1998).

gites with a typical average value of 14,000 carats per metric ton and a maximum value of 90,000 carats per metric ton (Viljoen et al., 1992; Schulze, 1992; Peltonen et al., 2002).

5. Mineral major and trace element compositions

The cores of 11 individual garnets show extremely limited compositional variability between grains (Table 1). They contain, on average, 0.24 wt.% TiO_2, 8.5 wt.% Cr_2O_3 and 6.6 wt.% CaO. The Cr and Ca values are consistent with a lherzolitic paragenesis (Fig. 2). Garnets occurring as inclusions in diamonds from the Premier mine (Gurney et al., 1985; Richardson et al., 1993) as well as garnets from diamond-bearing peridotites and diamond-bearing macrocrysts from the Kaapvaal craton are characterised by a range in Ca compositions, with both Ca-saturated lherzolitic as well as low-Ca harzburgitic varieties represented (Shee

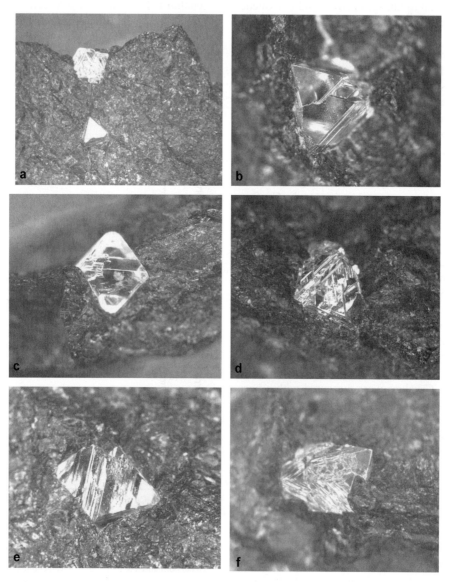

Fig. 4. Diamonds in xenolith DO40. Field of view = 6 mm.

et al., 1982; Viljoen et al., 1992, 1994; Daniels et al., 1995; Menzies et al., 1999). Compared to these compositions the garnets analysed in the present study are calcic and almost at the maximum CaO-envelope defined by garnet inclusions in diamonds from Premier (Fig. 2).

The three garnets from DO40 analysed for their trace element contents show limited variation in trace element composition (Table 1), with generally low concentrations of most trace elements, e.g. Y (< 11 ppm), Zr (< 18 ppm) and Sr (< 0.5 ppm). Ni content in the garnets analysed is on the order of 63 ppm (Table 1). This corresponds to an average garnet–olivine equilibration temperature of 1108 °C using the Ni in garnet thermometer of Griffin et al. (1989) (after the reformulation of Ryan et al., 1996) and a temperature of 1105 °C using the formulation of Canil (1999).

Rare earth element patterns for the garnets from Premier xenolith DO40 show an increase in the light REE from La at 0.2 to approximately 10 × chondritic values for middle REE and heavy REE (Fig. 3a). In terms of REE compositions, the garnets are chemically similar to garnets from high temperature, sheared garnet lherzolite xenoliths PHN1611 and J4 (Shimizu, 1975; Wolf-Boenisch, 1994) with a xenolith bulk chemistry similar to that of primitive mantle (Stachel et al., 1998; Fig. 3b). Such garnets closely match the REE pattern of hypothetical garnet compositions calculated for primitive mantle compositions using known element partition coefficients and mass balance considerations (Stachel et al., 1998). Garnets with similar primitive REE profiles are now known to occur as inclusions in diamonds (e.g. Akwatia, Birim field, Ghana; Stachel and Harris, 1997), in diamond-bearing xenoliths (e.g. Roberts Victor; Stachel et al., 1998) and in heavy mineral concentrates derived from kimberlite (e.g. Hoal et al., 1994).

Fig. 5. Diamonds recovered from xenolith DO40.

Table 2
Sample, carbon isotope and infrared data for diamonds from Premier xenolith DO40

Diamond	Morphology	$\delta^{13}C$	Nitrogen content (ppm)	%B	Platelet peak position (cm^{-1})	Platelet peak area (cm^{-1})	Hydrogen peak area (cm^{-1})
1	octa		65	29.5	1361.52	4.16	4.6
2	octa aggregate		60	13.1	1361.52	8.07	9
4	octa macle		418	81.3	1368.27	248.03	13.6
5	octa aggregate		178	48.8	1361.52	51.34	4.9
6	octa macle		67	47.3	1361.52	4.13	3
7	octa macle		201	71.4	1366.34	91.33	10.1
8	octa aggregate		225	68.4	1365.38	98.78	9.2
9	octa aggregate		296	80.3	1365.38	168.58	3.6
10	octa aggregate		1226	84.9	1368.27	701.42	19.8
12	octa		59	28.3	1361.52	9.08	3.4
13	octa macle		229	72.0	1365.38	113.28	10.9
14	octa		51	29.7	1361.52	9.36	2.8
15	2 octa intergrowth		209	58.1	1364.41	67.54	9.7
17	pseudohemimorphic		47	13.3	1361.52	6.36	4
18	octa aggregate	− 1.84	168	54.7	1361.52	30.71	9.1
19	octa		235	64.7	1364.41	102.34	8.9
21	octa		35	13.0	1361.52	2.90	3.4
22	pseudohemimorphic		47	31.7	1361.52	5.19	2.7
23	octa aggregate		405	85.1	1366.34	257.35	25.1
24	octa aggregate		26	21.5	–	–	5.2
25	2 octa intergrowth		64	37.8	1361.52	6.11	4.7
26	octa		91	34.3	1361.52	15.25	4.3
27	2 octa intergrowth		84	37.6	1363.45	12.47	6.3
28	octa aggregate		35	10.0	1361.52	7.05	3.7
30	fragment		152	46.0	1362.48	34.94	8.2
31	octa aggregate		234	57.0	1363.45	91.92	8.2
32	2 octa intergrowth		79	29.9	1361.52	11.38	4.4
33	octa aggregate		141	49.5	1362.48	34.11	12
34	octa	− 2.07	80	34.1	1361.52	12.71	5.5
35	octa		164	49.5	1361.52	46.38	12.1
36	octa macle		59	46.1	1361.52	4.27	5.6
37	octa aggregate	− 1.77	225	56.8	1362.48	59.77	17
38	octa	− 2.56	103	53.0	1361.52	10.32	7.1
40	fragment	− 1.45	259	58.5	1364.41	101.80	15.1
41	octa		105	56.3	1366.34	37.16	7.4
42	octa	− 3.31	65	57.6	1361.52	5.23	4.9
43	2 octa intergrowth		66	39.3	1362.48	7.28	9.3
44	octa		100	57.8	1363.45	33.08	4.8
45	octa macle		69	50.6	1361.52	5.73	9.8
46	fragment		76	57.6	1361.52	9.46	2.9
47	fragment		170	50.2	1362.48	29.91	14.2
48	octa		70	39.1	1362.48	13.58	6.4
49	octa		132	54.8	1362.48	40.47	14.1
50	octa macle	− 1.68	207	52.3	1361.52	63.40	17.1
51	fragment		82	36.5	1361.52	14.25	14.9
52	fragment		136	39.5	1361.52	32.91	7
53	octa aggregate		282	79.6	1369.23	158.10	21.3
54	octa aggregate	− 3.62	99	57.7	1364.41	33.58	9.4
55	pseudohemimorphic	− 3.61	56	42.6	1362.48	4.68	8.6
56	octa		76	42.1	1362.48	10.88	9.2
57	octa aggregate	− 1.3	266	60.9	1364.41	93.46	19.9
58	aggregate		53	63.4	1360.56	4.28	1.9
59	octa		83	39.0	1361.52	13.95	6.2

6. Diamonds

All diamonds recovered are colourless. Most crystals are sharp-edged octahedra (Figs. 4 and 5), some with minor development of the dodecahedral form. The resulting pseudohemimorphism (Robinson, 1979) is considered to represent original octahedra that protruded from the protective xenolith and which were partially resorbed where in contact with the kimberlite magma (Robinson et al., 1998; see Fig. 5b). A number of crystals are twinned octahedral macles, while aggregates of two or more octahedra are also common (Fig. 5c and d). Surface textures on the diamonds (Robinson et al., 1989) are limited to negatively oriented trigons on octahedral surfaces, minor ribbing on flat-faced dodecahedral surfaces (Fig. 5d) and hillock development on rounded dodecahedral surfaces associated with pseudohemimorphic forms (Fig. 5b; Robinson et al., 1998). Graphite coats as well as knob-like asperities which are typically observed on diamonds from xenoliths (Robinson, 1978; Robinson et al., 1984) are conspicuously absent.

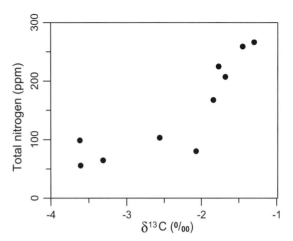

Fig. 7. Carbon isotopic composition plotted as a function of diamond nitrogen content (determined by infrared spectroscopy) for diamonds from xenolith DO40.

Mineral inclusions in the diamonds are rare. Where present they are predominantly small black rosettes which may be sulfide. In one instance a polymineralic (presumably lherzolitic) assemblage of reddish garnet, green clinopyroxene and a colourless mineral is recognised, while another diamond exhibits a small colourless inclusion. Breakout of the inclusions has not been attempted on account of their small size (~ 50 μm), their particular location within the host diamonds, and the considered risk of considerable fragmentation of the specimens.

Infrared analysis of the xenolith diamonds show nitrogen contents generally lower than 500 ppm and variable nitrogen aggregation state, from 10% to 80% of the 'B' form (Table 2). Many of the diamond spectra show a peak at 3107 cm^{-1}, consistent with the presence of structurally bonded hydrogen in the crystal lattice (Woods and Collins, 1983; Table 2). When plotted on a nitrogen aggregation diagram a well-defined trend of increasing nitrogen aggregation state with increasing nitrogen content is observed (Fig. 6). This equates to a calculated mantle residence temperature on the order of 1200 °C at an assumed mantle residence time of 750 My (i.e. the inferred mantle residence time of lherzolitic diamonds at Premier if a diamond crystallisation age of 1930 Ma (Richardson et al., 1993) is accepted along with a kimberlite emplacement age of 1180 Ma (Allsopp et al., 1989)).

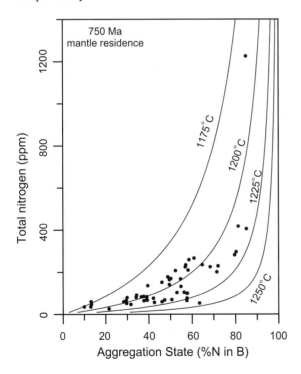

Fig. 6. Nitrogen content and aggregation state for diamonds from the Premier xenolith. Isotherms calculated for a mantle residence time of 750 Ma.

Carbon isotopic compositions range from − 3.6‰ to − 1.3‰ (Table 2). These are broadly correlated with diamond nitrogen content as determined by infrared spectroscopy, with the most negative C-isotopic compositions correlating with the lowest nitrogen contents (Fig. 7).

7. Discussion

The presence of a lherzolitic inclusion suite in a diamond from Premier xenolith DO40, coupled with the Ca-saturated (lherzolitic) major element composition of the garnets from the host xenolith and their primitive REE profiles, is consistent with diamond crystallisation in relatively undepleted lherzolitic mantle. The temperature distribution of peridotitic garnets occurring in heavy mineral concentrates from kimberlites suggests that depleted harzburgites and relatively undepleted lherzolites are intermixed over the depth range 150–180 km beneath the Kaapvaal craton (Griffin et al., 1992). This mixture may reflect local variations in the degree of partial melting during an ancient depletion event, with diamonds crystallising in both lithologies (Gurney et al., 1985; Griffin et al., 1992; Richardson et al., 1993), presumably as a result of metasomatic volatile transfer (Navon, 1999).

Intermixed harzburgite and lherzolite in the mantle may also be the result of a later metasomatic event subsequent to initial ultradepletion, effectively converting harzburgite to lherzolite (Griffin et al., 1992, 1999; Carbno and Canil, 2002; Grégoire et al., 2003). The high chrome contents (e.g. >4 wt.% Cr_2O_3) of many mantle derived peridotitic garnets are compatible with a model of protolith formation as a residue of partial melting in the spinel peridotite stability field to produce very high bulk Cr/Al ratios in the protolith (Kesson and Ringwood, 1989; Canil and Wei, 1992; Stachel et al., 1998). Cr-rich garnets would grow from such compositions upon subduction into the garnet stability field. On this basis, Stachel et al. (1998) argues that high-Cr, mantle-derived lherzolitic garnets with primitive REE patterns must derive from a mantle source which has experienced a later re-enrichment at pressures in the stability field of garnet. This last event re-introduced previously depleted major elements such as Ca, Fe and Si and causes REE enrichment such that all evidence of earlier

depletion is overprinted, with resulting REE patterns typical of an undepleted source. Griffin et al. (1999) presents evidence for such a process in the form of rimwards increases in Ca, Zr, Y, Ti and the heavy REE, and decreased Cr and Mg in garnets from harzburgite xenoliths at the Wesselton diamond mine. REE patterns in particular change from sinuous in the cores to that typical of magmatic garnets, or garnets showing a low degree of depletion (i.e. primitive garnet REE patterns). This they interpret as the result of a metasomatic process that has converted harzburgite to lherzolite at temperatures near 1000 °C, and consider it to be distinct from the high-temperature (1200–1400 °C) metasomatism associated with the infiltration of asthenosphere-derived melts into mantle wall rocks (Griffin et al., 1999 and references therein). Sinuous REE patterns in the cores of some of the garnets analysed by Griffin et al. (1999) are regarded as 'primary' features reflecting an ancient metasomatic event superimposed on a depleted protolith, as is seen in some peridotitic garnets occurring as inclusions in diamond (e.g. Shimizu et al., 1989; Stachel and Harris, 1997; Stachel et al., 1998, 1999). On the basis of garnet Cr-content, trace element chemistry and REE profiles, and as per the views of Stachel et al. (1998) and Griffin et al. (1999), it is proposed that the lherzolitic character of Premier diamond-bearing xenolith DO40 results from the total metasomatic overprinting of pre-existing harzburgite. The presence of a lherzolitic diamond inclusion paragenesis at Premier (Gurney et al., 1985; Richardson et al., 1993) is indicative of metasomatic enrichment of existing depleted harzburgite prior to, or during, diamond growth.

The sharp-edged, octahedral morphology of the diamonds examined is consistent with crystallisation in situ within the high pressure mantle environment, with limited resorption by the carrier kimberlite. Unresorbed, octahedral diamonds up to 2 mm in diameter have previously been recovered from peridotitic (McCallum and Eggler, 1976) and eclogitic (McCandless and Collins, 1989) xenoliths in the Colorado-Wyoming Sate Line district, while unresorbed, octahedral diamonds up to 1.8 mm in diameter have also been reported in peridotite xenoliths from other localities, e.g. the Finsch and Roberts Victor kimberlites in South Africa (Shee et al., 1982; Viljoen et al., 1994).

Xenolith diamond mantle residence temperatures calculated for a mantle residence time of 750 Ma (see Section 6) range from 1171 to 1241 °C with an average of 1203 °C for the 52 diamonds analysed. This is significantly higher than the 1105 °C temperature calculated with the Ni in garnet thermometer of Canil (1999) for Ni exchange between garnet and olivine. However, the nitrogen aggregation state of a diamond is time-averaged for the period between diamond crystallisation and kimberlite entrainment. Furthermore, rate constants for the aggregation of nitrogen in diamond are extremely difficult to determine due to sluggish reaction rates (Cooper, 1990; Taylor et al., 1990, 1996; Mendelssohn and Milledge, 1995), and it is therefore concluded that the temperatures derived from the nitrogen aggregation data are essentially within error of the 1105 °C temperature calculated with the Ni in garnet thermometer of Canil (1999), and which presumably represent the ambient mantle temperature at the time of sampling of the xenolith by the carrier kimberlite.

The carbon isotopic compositions of 36 inclusion-bearing peridotitic diamonds from Premier range from − 1.88 ‰ to − 12.21 ‰ (Deines et al., 1984, 1989). Nitrogen contents of these diamonds range from below the detection limit of approximately 20 ppm nitrogen, to 1086 ppm, with most diamonds containing < 500 ppm nitrogen. The carbon isotopic compositions and nitrogen contents of the diamonds analysed from xenolith DO40 in the present study is therefore within the compositional range previously defined for diamonds from Premier. The 2.6 ‰ range in carbon isotopic compositions for diamonds from xenolith DO40 is surprising as the limited domain of a mantle xenolith is expected to have no compositional variation in carbon isotopic composition, and perhaps indicate that the formation of the high pressure mineralogy of the xenolith and the contained diamonds are two separate events. Large internal compositional variations in the carbon isotopic composition of diamonds are commonly observed (e.g. Harte et al., 1999; Hauri et al., 2002), and are probably a consequence of the crystallisation of diamond from fractionating carbon- and nitrogen-bearing fluids which have penetrated a pre-existing mantle substrate (Navon, 1999). It is therefore considered unlikely that correlations between carbon isotopic composition and associated peridotitic inclusion mineralogy will be meaningful. It is note-

worthy that Deines et al. (2001) found no correlation between peridotitic inclusion mineral chemistry and diamond carbon isotopic composition at the Venetia diamond mine.

The estimated diamond grade of Premier xenolith DO40 of 27,700 carats/metric ton is equivalent to a diamond (carbon) concentration of 5540 ppm. This carbon content is within the range reported by Deines (2002) for mantle xenoliths, from below 1 ppm to close to 10,000 ppm. However, it is noteworthy that 95% of the samples considered in the review by Deines (2002) contain less than 500 ppm C.

8. Conclusions

The conclusions of this study are summarised as follows:

1. Premier diamond-bearing xenolith DO40 is a lherzolite.
2. It is proposed that the lherzolitic character of Premier diamond-bearing xenolith DO40 may result from the total metasomatic overprinting of pre-existing harzburgite. Metasomatism occurred prior to, or accompanied, diamond growth.
3. The octahedral morphology of the diamonds is typical of xenolith-derived diamonds, and indicative of no, or very limited, resorption by the host kimberlite.
4. Carbon isotopic compositions and nitrogen contents are within the range reported previously for inclusion-bearing peridotitic diamonds from Premier.
5. The range in carbon isotopic compositions and nitrogen contents is probably a consequence of diamond crystallisation from fractionating carbon- and nitrogen-bearing fluids which have penetrated pre-existing mantle peridotite.

Acknowledgements

The authors would like to thank the management of De Beers Consolidated Mines Limited and the Premier Diamond Mine for the donation of study material and permission to publish. David Fisher of the Diamond Trading Company Research Laboratory in Maidenhead, England kindly provided his spectral deconvolution software. Maggie Mahlase of the Premier Sorthouse deserves special mention for discovering the specimen. Andrew Menzies, Kalle Westerlund and Herman Grütter are acknowledged for their efforts in improving the manuscript.

References

Allsopp, H.L., Bristow, J.W., Smith, C.B., Brown, R., Gleadow, A.J.W., Kramers, J.D., Garvie, O.G., 1989. A summary of radiometric dating methods applicable to kimberlites and related rocks. Kimberlites and Related Rocks, vol. 1, Their composition, occurrence, origin and emplacement. Spec. Publ.-Geol. Soc. Aust., vol. 14, pp. 343–357.

Boyd, F.R., 1989. Compositional distinction between oceanic and cratonic lithosphere. Earth Planet. Sci. Lett. 96, 15–26.

Boyd, F.R., Mertzman, S.A., 1987. Composition and structure of the Kaapvaal lithosphere, southern Africa. In: Mysen, B.O. (Ed.), Magmatic Processes: Physicochemical Principles. Spec. Publ.-Geochem. Soc., vol. 1, pp. 13–24.

Boyd, F.R., Pearson, D.G., Nixon, P.H., Mertzman, S.A., 1993. Low-calcium garnet harzburgites from southern Africa: their relations to cratonic structure and diamond crystallisation. Contrib. Mineral. Petrol. 113, 352–366.

Boyd, S.R., Kiflawi, I., Woods, G.S., 1994. The relationship between infrared absorption and the A defect concentration in diamond. Philos. Mag., B 69, 1149–1153.

Boyd, S.R., Kiflawi, I., Woods, G.S., 1995a. Infrared absorption by the B nitrogen aggregation in diamond. Philos. Mag., B 72, 351–361.

Boyd, S.R., Réjou-Michel, A., Javoy, M., 1995b. Improved techniques for the extraction, purification and quantification of nanomole quantities of nitrogen gas: the nitrogen content of a diamond. Meas. Sci. Technol. 6, 297–305.

Canil, D., Wei, K.J., 1992. Constraints on the origin of mantle-derived low Ca garnets. Contrib. Mineral. Petrol. 109, 421–430.

Canil, D., 1999. The Ni-in-garnet geothermometer calibration at natural abundances. Contrib. Mineral. Petrol. 136, 240–246.

Carbno, G.B., Canil, D., 2002. Mantle structure beneath the SW Slave craton, Canada: constraints from garnet geochemistry in the Drybones Bay kimberlite. J. Petrol. 43, 129–142.

Cooper, G.I., 1990. Infrared microspectroscopy of diamond in relation to mantle processes. PhD Thesis, Univ. College, London.

Danchin, R.V., 1979. Mineral and bulk chemistry of garnet lherzolite and garnet harzburgite xenoliths from the Premier mine, South Africa. In: Boyd, F.R., Meyer, H.O.A. (Eds.), The Mantle Sample: Inclusions in Kimberlites and other Volcanics. Proceedings of the 2nd International Kimberlite Conference, vol. 2. American Geophysical Union, Washington, pp. 104–126.

Daniels, L.R.M., Richardson, S.R., Menzies, A.H., De Bruin, D., Gurney, J.J., 1995. Diamondiferous garnet macrocrysts in the Newlands kimberlite, South Africa-Rosetta stones from the Kaapvaal craton. Extended Abstracts 6th International Kimberlite Conference.

Dawson, J.B., Smith, J.V., 1975. Occurrence of diamond in a mica-garnet lherzolite xenolith from kimberlite. Nature 254, 580–581.

De Wit, M.J., Roering, C., Hart, R.J., Armstrong, A., de Ronde, C.E.J., Green, R.W.E., Tredoux, M., Pederby, E., Hart, R.A., 1992. Formation of an Archaean continent. Nature 357, 553–562.

Deines, P., 2002. The carbon isotopic composition of mantle xenoliths. Earth-Sci. Rev. 58, 247–278.

Deines, P., Gurney, J.J., Harris, J.W., 1984. Associated chemical and carbon isotopic composition variations in diamonds from Finsch and Premier kimberlite, South Africa. Geochim. Cosmochim. Acta 48, 325–342.

Deines, P., Harris, J.W., Spear, P.M., Gurney, J.J., 1989. Nitrogen and ^{13}C content of Finsch and Premier diamonds and their implications. Geochim. Cosmochim. Acta 53, 1367–1378.

Deines, P., Viljoen, F., Harris, J.W., 2001. Implications of the carbon isotope and mineral inclusions record for the formation of diamonds in the mantle underlying a mobile belt: Venetia, South Africa. Geochim. Cosmochim. Acta 65, 813–838.

Evans, T., 1992. Aggregation of nitrogen in diamond. In: Field, J.E. (Ed.), The Properties of Natural and Synthetic Diamond. Academic Press, London, pp. 259–290.

Evans, T., Harris, J.W., 1989. Nitrogen aggregation, inclusion equilibration temperatures and the age of diamonds. In: Ross, J. (Ed.), Kimberlites and Related Rocks, vol. 2: Their Mantle/Crust Setting. Spec. Publ.-Geol. Soc. Aust., vol. 14, pp. 1002–1006.

Evans, T., Qi, Z., 1982. The kinetics of the aggregation of nitrogen in diamond. Proc. R. Soc. Lond., A 381, 159–178.

Grégoire, M., Bell, D.R., Le Roux, A.P., 2003. Garnet lherzolites from the Kaapvaal craton (South Africa): trace element evidence for a metasomatic history. J. Petrol. 44, 629–657.

Griffin, W.L., Cousens, D.R., Ryan, C.G., Sie, S.H., Suter, G.F., 1989. Ni in chrome pyrope garnets: a new geothermometer. Contrib. Mineral. Petrol. 103, 199–202.

Griffin, W.L., Gurney, J.J., Ryan, C.G., 1992. Variations in trapping temperatures and trace elements in peridotite-suite inclusions from African diamonds: evidence for two inclusion suites, and implications for lithosphere stratigraphy. Contrib. Mineral. Petrol. 110, 1–15.

Griffin, W.L., Shee, S.R., Ryan, C.G., Win, T.T., Wyatt, B.A., 1999. Harzburgite to lherzolite and back again: metasomatic processes in ultramafic xenoliths from the Wesselton kimberlite, Kimberley, South Africa. Contrib. Mineral. Petrol. 134, 232–250.

Gurney, J.J., 1984. A correlation between garnets and diamonds in kimberlites. In: Glover, J.E., Harris, P.G. (Eds.), Kimberlite Occurrence and Origin: A Basis for Conceptual Models in Exploration. Publ.-Geol. Dep. Ext. Serv., Univ. West. Aust., vol. 8, pp. 143–166.

Gurney, J.J., Harris, J.W., Rickard, R.S., Moore, R.O., 1985. Inclusions in Premier Mine diamonds. Trans. Geol. Soc. S. Afr. 88, 301–310.

Harte, B., Fitzsimmons, I.C.W., Harris, J.W., Otter, M.L., 1999. Carbon isotope ratios and nitrogen abundances in relation to cathodoluminescence characteristics for some diamonds from the Kaapvaal Province, South Africa. Mineral. Mag. 63 (6), 829–856.

Hauri, E.H., Wang, J., Pearson, D.G., Bulanova, G.P., 2002. Microanalysis of $\delta^{13}C$, $\delta^{15}N$, and N abundances in diamonds by secondary ion mass spectrometry. Chem. Geol. 185, 149–163.

Hoal, K.E.O., Hoal, B.G., Erlank, A.J., Shimizu, N., 1994. Metasomatism of the mantle lithosphere recorded by rare earth elements in garnets. Earth Planet. Sci. Lett. 126, 303–313.

Kesson, S.E., Ringwood, A.E., 1989. Slab-mantle interactions: 2. The formation of diamonds. Chem. Geol. 78, 97–118.

Kiflawi, I., Bruley, J., 2000. The nitrogen aggregation sequence and the formation of voidites in diamond. Diam. Relat. Mater. 9, 87–93.

McCallum, M.E., Eggler, D.H., 1976. Diamonds in an upper mantle peridotite nodule from kimberlite in southern Wyoming. Science 192, 253–256.

McCandless, T.E., Collins, D.S., 1989. A diamond-graphite eclogite from the Sloan 2 kimberlite, Colorado, USA. In: Ross, J. (Ed.), Kimberlites and Related Rocks, vol. 2. Spec. Publ.-Geol. Soc. Aust., vol. 14, pp. 1063–1069.

McDonough, W.F., Sun, S.-S., 1995. The composition of the earth. Chem. Geol. 120, 223–253.

Mendelssohn, M.J., Milledge, H.J., 1995. Geologically significant information from routine analysis of the mid-infrared spectra of diamonds. Int. Geol. Rev. 37, 95–110.

Menzies, A.H., Carlson, R.W., Shirey, S.B., Gurney, J.J., 1999. Re–Os systematics of Newlands peridotite xenoliths: implications for diamond and lithosphere formation. In: Gurney, J.J., Gurney, J.L., Pascoe, M.D., Richardson, S.R. (Eds.), Proceedings of the 7th International Kimberlite Conference Red Roof Design cc, vol. 2. National Book Printers, Goodwood, South Africa, pp. 566–573.

Navon, O., 1999. Diamond formation in the earth's mantle. In: Gurney, J.J., Gurney, J.L., Pascoe, M.D., Richardson, S.R. (Eds.), Proceedings of the 7th International Kimberlite Conference Red Roof Design cc, vol. 2. National Book Printers, Goodwood, South Africa, pp. 584–604.

Norman, M.D., 1998. Melting and metasomatism in continental lithosphere: laser ablation ICPMS analysis of minerals in spinel lherzolites from eastern Australia. Contrib. Mineral. Petrol. 130, 240–255.

Norman, M.D., Pearson, N.J., Sharma, A., Griffin, W.L., 1996. Quantitative analysis of trace elements in geological materials by laser ablation ICPMS: instrumental operating conditions and calibration of NIST glasses. Geostand. Newsl. 20, 247–261.

Peltonen, P., Kinnunen, K.A., Huhma, H., 2002. Petrology of two diamondiferous eclogite xenoliths from the Lahtojoki kimberlite pipe, eastern Finland. Lithos 63, 151–164.

Richardson, S.H., 1986. Latter-day origin of diamonds of eclogitic paragenesis. Nature 332, 623–626.

Richardson, S.H., Gurney, J.J., Erlank, A.J., Harris, J.W., 1984. Origin of diamonds in old enriched mantle. Nature 310, 198–202.

Richardson, S.H., Harris, J.W., Gurney, J.J., 1993. Three generations of diamonds from old continental mantle. Nature 366, 256–259.

Robinson, D.N., 1978. Diamond and graphite in eclogite xenoliths from kimberlite. In: Boyd, F.R., Meyer, H.O.A. (Eds.), The Mantle Sample: Inclusions in Kimberlites and Other Volcanics. Proceedings of the 2nd International Kimberlite Conference, vol. 1. American Geophysical Union, Washington, pp. 104–126.

Robinson, D.N., 1979. Surface textures and other features of diamonds. Unpublished PhD thesis, University of Cape Town.

Robinson, D.N., Gurney, J.J., Shee, S.R., 1984. Diamond eclogite and graphite eclogite xenoliths from Orapa, Botswana. In: Kornprobst, J. (Ed.), Kimberlites: II. The mantle and crust–mantle relationships. Proceedings of the 3rd International Kimberlite Conference, Elsevier, Amsterdam, pp. 11–24.

Robinson, D.N., Scott, J.A., Van Niekerk, A., Anderson, V.G., 1989. The sequence of events reflected in the diamonds of some southern African kimberlites. Kimberlites and Related Rocks: Vol. 2: Their Mantle/Crust Setting, Diamonds and Diamond Exploration. Spec. Publ.-Geol. Soc. Aust., vol. 14, pp. 990–1000.

Robinson, D.N., Ferraris, R., Anderson, V.G., Parker, G.M., Van Blerk, E., Hart, D., 1998. Colour, morphological and surface textural characteristics of the diamonds in the Venetia kimberlites. Abstract volume 7th International Kimberlite Conference, Cape Town, pp. 737–739.

Ryan, C.G., Griffin, W.L., Pearson, N.J., 1996. Garnet geotherms: pressure–temperature data from Cr-pyrope garnet xenocrysts in volcanic rocks. J. Geophys. Res. 101, 5611–5625.

Schulze, D.J., 1992. Diamond eclogite from Sloan Ranch, Colorado, and its bearing on the diamond grade of the Sloan kimberlite. Econ. Geol. 87, 2175–2179.

Shee, S.R., Gurney, J.J., Robinson, D.N., 1982. Two diamond-bearing peridotite xenoliths from the Finsch kimberlite, South Africa. Contrib. Mineral. Petrol. 81, 79–87.

Shimizu, N., 1975. Rare earth elements in garnets and clinopyroxenes from garnet lherzolite nodules in kimberlites. Earth Planet. Sci. Lett. 25, 26–32.

Shimizu, N., Gurney, J.J., Moore, R., 1989. Trace element geochemistry of garnet inclusions in diamonds from the Finsch and Koffiefontein kimberlite pipes. 28th Int. Geol. Congr., workshop on diamonds, Ext. The Geophysical Laboratory, Canaegie Institution of Washington, pp. 100–101.

Shirey, S.B., Harris, J.W., Richardson, S.H., Fouch, M.J., James, D.E., Cartigny, P., Deines, P., Viljoen, F., 2002. Diamond genesis, seismic structure, and evolution of the Kaapvaal-Zimbabwe craton. Science 297, 1683–1686.

Stachel, T., Harris, J.W., 1997. Diamond precipitation and mantle metasomatism-evidence from the trace element chemistry of silicate inclusions in diamonds from Akwatia, Ghana. Contrib. Mineral. Petrol. 129, 143–154.

Stachel, T., Viljoen, K.S., Brey, G., Harris, J.W., 1998. Metasomatic processes in lherzolitic and harzburgitic domains of diamondiferous lithospheric mantle: REE in garnets from xenoliths and inclusions in diamonds. Earth Planet. Sci. Lett. 159, 1–12.

Stachel, T., Harris, J.W., Brey, G.P., 1999. REE patterns of peridotitic and eclogitic inclusions in diamonds from Mwadui (Tanzania). In: Gurney, J.J., Gurney, J.L., Pascoe, M.D., Richardson, S.R. (Eds.), Proceedings of the 7th International Kimberlite Conference, Red Roof Design cc, vol. 2. National Book Printers, Goodwood, South Africa, pp. 829–835.

Taylor, W.R., Jaques, A.L., Ridd, M., 1990. Nitrogen-defect aggregation characteristics of some Australasian diamonds: time–temperature constraints on the source regions of pipe and alluvial diamonds. Am. Mineral. 75, 1290–1310.

Taylor, W.R., Canil, D., Milledge, H.J., 1996. Kinetics of Ib to IaA nitrogen aggregation in diamond. Geochim. Cosmochim. Acta 60, 4725–4733.

Viljoen, K.S., 2002. An infrared investigation of inclusion-bearing diamonds from the Venetia kimberlite, Northern Province, South Africa: implications for diamonds from craton-margin settings. Contrib. Mineral. Petrol. 144, 98–108.

Viljoen, K.S., Swash, P.M., Otter, M.L., Schulze, D.J., Lawless, P.J., 1992. Diamondiferous garnet harzburgites from the Finsch kimberlite, Northern Cape, South Africa. Contrib. Mineral. Petrol. 110, 133–138.

Viljoen, K.S., Robinson, D.N., Swash, P.M., Griffin, W.L., Otter, M.L., Ryan, C.G., Win, T.T., 1994. Diamond- and graphite-bearing peridotite xenoliths from the Roberts Victor kimberlite, South Africa. In: Meyer, H.O.A., Leonardos, O.H. (Eds.), Kimberlites, Related Rocks and Mantle Xenoliths, vol. 1. Proceedings of the Fifth International Kimberlite Conference (1991). Companhia de Pesquisa de Recursos Minerais-Special Publication, pp. 285–303. 1/A Jan/94, Brasilia.

Wolf-Boenisch, B., 1994. Aufbau der analytik zur messung der Osmium-isotopie sowie die strontium-, neodym-und osmium-systematik eines hoch-temperatur-granat-lherzoliths (J4) von Jagersfontein (Südafrika). Dipl. Thesis, Mainz.

Woods, G.S., Collins, A.T., 1983. Infrared absorption spectra of hydrogen complexes in Type I diamonds. J. Phys. Chem. Solids 44, 471–475.

Available online at www.sciencedirect.com

Lithos 77 (2004) 553–569

www.elsevier.com/locate/lithos

A fertile harzburgite–garnet lherzolite transition: possible inferences for the roles of strain and metasomatism in upper mantle peridotites

J.B. Dawson*

Department of Geology and Geophysics, University of Edinburgh, West Mains Road, Edinburgh EH9 3JW, UK

Received 27 June 2003; accepted 3 January 2004
Available online 28 July 2004

Abstract

Porphyroclastic enstatite in a garnet lherzolite xenolith from the Monastery Mine kimberlite, South Africa, has exsolved pyrope garnet, Cr-diopside and Al-chromite, and the specimen is interpreted as representing a transition from fertile harzburgite, (containing high Ca-Al-Cr enstatite) to granular garnet lherzolite. Although the exsolved phases occur in morphologically different forms (fine and coarse lamellae; equant, ripened grains), indicating textural disequilibrium, the exsolved grains are very constant in composition, indicating chemical equilibrium. Theoretically, the exsolution could have been due to a fall in temperature, but the close association of exsolution and deformation of the host enstatite suggests that exsolution was also aided by straining of the enstatite lattice. The phase compositions can be broadly matched with those in other mantle peridotites, except that all phases are characterised by a virtual absence of Ti. In the garnet and diopside Ti, Co, Zr and most of the REE are lower than in published analyses of garnet and diopside in both granular and sheared garnet lherzolites from Southern African kimberlites, and diopside/garnet partitioning for Sr and the REE is higher. Comparison with the trace element chemistry of an enstatite from a fertile harzburgite indicates that, except for Nb, the trace element content and distribution found in the Monastery phases could arise by isochemical exsolution from such an enstatite. On the assumption that (a) the Monastery specimen represents a transition from harzburgite to garnet lherzolite, and (b) many garnet lherzolites are of exsolution origin (as suggested by their modal compositions), the inference is that most garnet lherzolites, and not just the sheared variety, have been subject to varying degrees of Ti, Zr, Sr and REE metasomatism.
© 2004 Elsevier B.V. All rights reserved.

Keywords: Harzburgite; Lherzolite; Deformation; Metasomatism; Southern Africa

1. Introduction

It is now generally accepted that garnet lherzolite is a major rock type in the Earth's upper mantle and the potential parent of a variety of basic/ ultrabasic magmas. In most earlier models of upper mantle magmatism, garnet lherzolite is the starting material, and the relationship between the various types of upper mantle peridotite was most simply expressed by starting with a lherzolite containing appreciable amounts of garnet and diopside (plus minor amounts of phlogopite and magnesite) which, after partial melting and extraction of basic or

* Fax: +44-131-66-83-184.
  E-mail address: jbdawson@glg.ed.ac.uk (J.B. Dawson).

0024-4937/$ - see front matter © 2004 Elsevier B.V. All rights reserved.
doi:10.1016/j.lithos.2004.03.016

ultrabasic liquids, gives rise to a series of residual rocks ranging from lherzolites containing small amounts of garnet and diopside (depleted lherzolite) through garnet harzburgite to highly refractory harzburgite and dunite (e.g. Yoder, 1976).

More recent work has shown that some of these relationships are more complex. It is apparent that "garnet lherzolite" is a term that has been applied rather loosely to what is, in fact, a heterogeneous group of rocks containing varying modal contents of the four major phases forsterite, enstatite, Cr-diopside and pyrope garnet; for example, in 97 rocks described as "garnet lherzolite" in the literature, the volume of garnet varies from 1% to 15% (mostly < 8%) and that of clinopyroxene from 1% to 21% (mainly < 2%) (Dawson et al., 1980). In addition, the composition of the pyroxenes and garnets in these 97 rocks is variable and there are also chemical, grain size and textural variations (e.g. Gurney et al., 1975). These differences suggest that all "garnet peridotites" may not have a simple, common origin.

When considering if garnet lherzolite really is pristine, unmodified mantle, account must be taken of the alternative proposal that primitive mantle had been completely freed of garnet and clinopyroxene by partial melting to produce harzburgite, and subsequently generated minor amounts of garnet and clinopyroxene by exsolution from orthopyroxene on cooling (O'Hara et al., 1975). This possibility, based upon the experimentally determined high mutual solubility between the three phases at high temperature, has been reinforced by further experimental work that shows the exsolution history is dependent upon the bulk-rock composition of the harzburgite protolith (Canil, 1991).

The postulated exsolution origin for at least some garnet lherzolites has been confirmed by direct observation on lherzolite specimens from South African kimberlites in which porphyroclasts of enstatite contain exsolved garnet, diopside and spinel (Dawson et al., 1980), and the fact that integrated analyses of the enstatite plus exsolved garnet and diopside match the compositions of high-Ca-Al enstatites in "fertile" harzburgites. In these enstatites there is significant garnet and diopside (up to 6 and 4 wt.% respectively) compared with much smaller amounts in the enstatites in refractory "barren" harzburgites (Hervig et al., 1980). Further, the amounts of garnet and diopside

in solid solution in fertile harzburgite enstatites are very similar to the amounts in many garnet "lherzolites" (see above), suggesting that most garnet lherzolites may be of exsolution origin. Additional indirect evidence is provided by textural studies on garnet lherzolites in which, although garnet and diopside exsolution lamellae no longer exist in the orthopyroxene, an exsolution origin can be inferred from the consistent, close spatial association of enstatite, garnet and clinopyroxene (Cox et al., 1987; Saltzer et al., 2001).

Although a case for an exsolution origin for some lherzolites can be made on the major and minor element chemistry of the phases, the trace element composition of the fertile harzburgite → garnet lherzolite transition is as yet not investigated. Published trace element analyses of upper mantle enstatite (Irving and Frey, 1984; Stosch, 1982) show very low concentrations of the REE and transition elements and, theoretically, garnet and diopside exsolving from enstatites should also contain low amounts. Paradoxically, garnets and diopsides in both granular and sheared garnet lherzolites from Southern African kimberlites contain appreciable concentrations of the REE, Sr, Ti and Zr (Shimizu, 1975; Shimizu and Allègre, 1978; Hoal et al., 1994; Stachel et al., 1998; Grégoire et al., 2003).

The present paper is particularly concerned with the trace element chemistry of garnet and diopside formed by exsolution from orthopyroxene in a lherzolite xenolith from the kimberlite of the Monastery Mine, South Africa. Reconnaissance major element analyses of the main phases were given in Dawson et al. (1980), but that study is extended here to investigate the major, minor and trace element chemistry of the enstatite and of texturally different exsolved phases, and comparison with the chemistry of enstatite in a fertile harzburgite—a potential parental rock in the exsolution hypothesis.

2. Sample descriptions

Exsolved lherzolite BD1366 from Monastery is an ovoid peridotite block that was approximately 11 cm (maximum dimension) before sectioning. It is inequigranular with large (up to 15 mm) orthopyroxene grains set in a finer matrix mainly of olivine with rarer grains of red garnet and green diopside. Even in hand-specimen, deformation of the orthopyroxene grains can be deduced from uneven reflections from their undulose crystal surfaces.

In thin section, the olivines are up to 2 mm and equant, with straight grain boundaries and well-developed triple-junctions. The enstatite porphyroclasts are cut by numerous brittle micro-fractures (kink bands); some are normal to the enstatite cleavage whereas others are at a high angle and, between the high-angle fractures, the enstatite subgrains have been rotated giving rise to stepped or saw-tooth grain boundaries (Fig. 1). Within the enstatite crystals, garnet, diopside and chromite grains of differing morphology have exsolved in close association with both the kink bands and the cleavage. The kink bands are decorated by discontinuous or linked chains of equant grains of garnet and diopside (150 to 200 μm), with garnet more abundant than diopside (Figs 2 and 3). Coarse abundant rod-shaped lamellae or elongate pods of garnet and rarer diopside (50 to 100 μm wide) parallel the cleavage, and occasionally coalesce to form very coarse (500 μm wide) cleavage-parallel aggregates (Fig. 4). Finer lamellae (up to 20 μm wide) of garnet

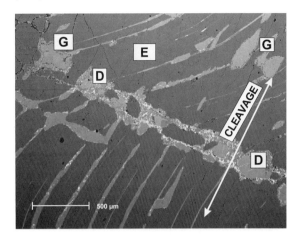

Fig. 2. Back-scattered electron (BSE) image. Most of the image is of dark enstatite (E). A kink band, running NW–SE across the image and normal to the cleavage (fine dark lines), contains exsolved garnet (G) and diopside (D) (the BSE coefficients of garnet and diopside, expressed as "brightness", are very similar). Garnet is also present as bright pods and lamellae. Note that the garnet lamellae are approximately parallel to the cleavage south of the kink band but, to the north, cut across the cleavage.

and chromite form "swarms" that usually parallel the cleavage, but other curved "swarms" cut across any visible cleavage (Figs. 2 and 3); the fine garnet lamellae, which often contain minute bright (in

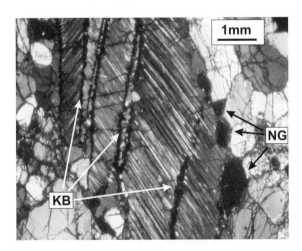

Fig. 1. Photomicrograph (crossed polarisers) of an enstatite porphyroclast cut by several kink bands (KB) that are emphasised by isotropic (black) garnet and that are at ~ 45° to the cleavage; the cleavage is picked out by black isotropic garnet lamellae, and also by bright bands that are due to cleavage-parallel distortion of the enstatite lattice. Note the rotation of some enstatite subgrains between the kink bands. The porphyroclast margin is decorated by necklace garnet (NG). The other grains are olivine.

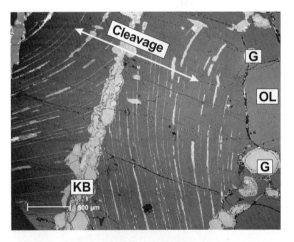

Fig. 3. BSE image of a kink band (KB) containing mainly exsolved garnet cutting across an enstatite porphyroclast, normal to the cleavage (fine dark lines). Fine exsolution lamellae of garnet cut across the cleavage in different directions on either side of the kink band. The margin of the enstatite grain is decorated with necklace garnet (G). Olivine (OL) is the other phase.

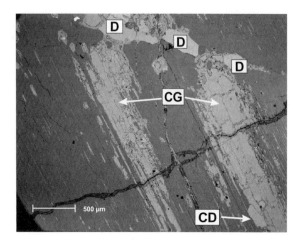

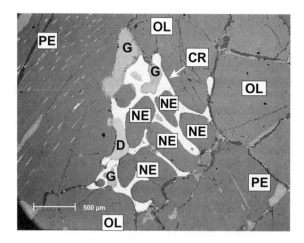

Fig. 4. BSE image of very coarse lamellae of garnet (CG) and diopside (CD), together with finer lamellae, exsolved from a porphyroclastic enstatite. Compare the width of these lamellae with the finer ones shown in Figs. 2 and 3. A kink band picked out by exsolved diopside (D) runs across the north part of the image.

Fig. 6. BSE image of necklace garnet (G), diopside (D) and enstatite neoblasts (NE) embedded in bright intergranular chromite (CR) at the interface of two porphyroclastic enstatite grains (PE). Note the absence of exsolution lamellae in the margins of the porphyroclastic enstatites, and the presence of brighter fine lamellae of chromite in the larger (left-hand) porphyroclast. Other grains are olivine (OL).

B.S.E. imaging) grains, usually coalesce with larger kink-band grains where swarms converge with the kink bands. Although lamellae are typically abundant in the enstatite interiors, the enstatite margins tend to be devoid of lamellae (Figs. 5 and 6). Equant, coarser (up to 700 μm) grains of abundant garnet and rarer

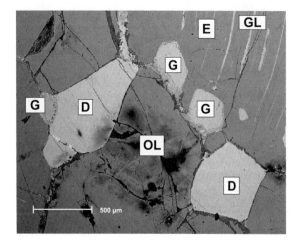

Fig. 5. BSE image of the margin of an enstatite porphyroclast (E), with necklace diopside (D). Note coarse grains of garnet (G) in the margin of the enstatite, contrasting with fine garnet lamellae (GL) in the enstatite interior.

diopside form "necklaces" around the edge of enstatite porphyroclasts, and some coarser garnet is embedded in the margins of the enstatite grains (Fig. 5). The bimodality of garnet and diopside grain size has its parallel in regionally metamorphosed rocks where porphyroblast growth is at the expense of smaller grains (Ostwald ripening). The garnet and diopside necklace grains may be either isolated (Fig. 5) or clustered with lamella-free polygonal enstatite neoblasts (Fig. 7); occasionally the clusters are surrounded by intergranular chromite, in which case the silicate grains are rounded (Fig. 6). Most chromite outside the porphyroclasts occurs in these intergranular areas but also, more rarely, as isolated necklace grains that, at around 50–100 μm, are smaller than the necklace garnets and diopsides.

Overall, garnet, both as lamellae or necklace grains, is a more abundant phase than diopside, and both are much more abundant than chromite. However, due to the extreme inequigranularity of the rock and the difficulty in distinguishing between olivine and enstatite neoblasts, an accurate modal analysis has not been possible. An approximate mode, made on one sawn face of the xenolith, is: porphyroclastic enstatite (including exsolved kink-band and lamellar garnet, diopside and chromite) ~35 vol.%; remainder

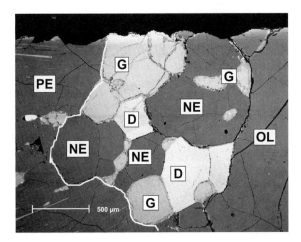

Fig. 7. BSE image of a cluster of necklace garnets (G) and diopsides (D), and enstatite neoblasts (NE) at the margin of an enstatite porphyroclast (PE). The margin of the porphyroclast has been inked in for clarity.

(mainly olivine, but including necklace garnet, diopside and chromite and neoblast ensatite) ~ 65 vol.%. Very thin (10 to 20 μm) reaction rinds sometimes occur between enstatite and garnet; but the phases are too small to be identified.

Harzburgites are fertile harzburgite BD2004 from the Bultfontein kimberlite, and barren harzburgite BD1919 from Letseng le Terae, Lesotho. The modes for the two harzburgites, respectively, are: olivine 65 and 80 vol.%; enstatite 34.5 and 17 vol.%; spinel 0.5 and 3 vol.%. Both are coarse-grained (5 to 15 mm) with most grains meeting at triple junctions. The olivines are generally surrounded and veined by serpentine and are smaller than the enstatite crystals. Fresh enstatite grains are glassy green, but altered grains have a bronzy lustre. Small (up to 0.5 mm) Cr-Al spinel grains occur in both specimens.

3. Analytical methods

Major and minor element analyses of the minerals were made by W.D.S. techniques on a Camebax Microbeam electron microprobe at the University of Edinburgh. Details of the methodology and standards are given in Dawson and Hill (1998). Calculation of ferric iron was made following the method of Droop (1981). Analyses for the trace elements were made in situ on polished, gold-coated thin sections by secondary ion mass spectrometry using a Cameca imf-4f ion microprobe in the Department of Geology and Geophysics at the University of Edinburgh. Details of the methods of analysis are given in Dawson (2002).

4. Mineral chemistry

4.1. BD 1366 major and minor element chemistry

4.1.1. Olivine

Analyses of the olivine show no detectable compositional variation between grain cores and rims, or between different grains. The olivine is forsterite that contains the very low concentrations of CaO typical of mantle olivines (Table 1). Its high $Mg/(Mg + Fe)$ ratio of 0.932 groups it more closely with the olivines in harzburgites than with those in garnet lherzolites.

4.1.2. Orthopyroxene

Analyses were made in different areas in the porphyroclasts (close to, and away from, areas of densely packed exsolution lamellae; recrystallised, lamella-free grains in kink bands; the lamella-free grain-margins) and of lamella-free neoblasts in the necklace zone. A striking feature is its remarkably

Table 1
Olivines in BD1366 and other mantle peridotites

	Average in BD1377	GGLK[a]	SGLK[a]	FH[a]	BH[a]
n	7				
SiO_2	40.44	41.2	41.0	41.6	41.7
TiO_2	0	0.01	0.02	n.a.	n.a.
Al_2O_3	0.02	0.01	0.05	0.005	0.004
Cr_2O_3	0.02	0.02	0.05	0.004	0.002
FeO	6.71	8.10	10.0	6.92	6.36
MnO	0.09	0.10	0.12	0.09	0.08
NiO	0.43	0.39	0.35	0.37	0.39
MgO	52.20	50.4	48.8	51.6	51.4
CaO	0.01	0.02	0.09	0.009	0.007
Total	99.91	100.25	100.48	100.598	99.943
$Mg/(Mg + Fe)$	0.932	0.918	0.898	0.930	0.936

[a] Abbreviations: GGLK granular, cold garnet lherzolite in kimberlite; SGLK sheared (hot) garnet lherzolite in kimberlite (Hervig et al., 1986); FH fertile harzburgite, BH barren harzburgite (Hervig et al., 1980).

Table 2
Orthopyroxene in BD1366 and other mantle peridotites

	Average opx interior away from visible lamellae	Average opx in lamella-rich areas	Average kink-band opx	Average lamella-free opx margin	Average necklace zone neoblast	Average opx neoblast in intergranular chromite	Average overall	CGLK[a]	HGLK[a]	FH[a]	BH[a]
n	18	4	3	5	3	6	39	42	18	11	10
SiO_2	56.91	57.20	57.09	57.04	57.43	56.98	57.02	57.8	56.7	56.5	58.6
TiO_2	0.00	0.00	0.00	0.00	0.00	0.00	0.00	0.05	0.17	n.a.	n.a.
Al_2O_3	0.87	0.89	0.89	0.86	0.83	0.86	0.87	0.84	1.25	3.05	1.10
Cr_2O_3	0.26	0.26	0.26	0.25	0.24	0.32	0.27	0.32	0.29	0.85	0.35
Fe_2O_3	2.10	1.71	1.26	1.85	1.45	3.16	2.08				
FeO	2.32	2.70	3.09	2.52	2.93	1.38	2.35				
MnO	0.10	0.11	0.12	0.10	0.11	0.09	0.10	0.11	0.13	0.11	0.10
NiO	0.10	0.10	0.10	0.09	0.09	0.10	0.10	0.09	0.09	0.07	0.07
MgO	36.47	36.47	36.22	36.48	36.46	37.07	36.55	35.7	33.9	34.8	36.0
CaO	0.27	0.23	0.23	0.23	0.22	0.23	0.25	0.43	1.35	0.91	0.24
Na_2O	0.04	0.04	0.02	0.04	0.06	0.04	0.04	0.09	0.31	0.05	0.02
Total	99.44	99.70	99.29	99.46	99.83	100.22	99.61	100.30	100.19	100.72	100.61
Total FeO	4.22	4.24	4.22	4.18	4.24	4.22	4.22	4.90	6.00	4.38	4.13
Mg/(Mg+ total Fe)							0.939	0.929	0.910	0.916	0.940

[a] Abbreviations as in Table 1.

uniform composition; although analyses were made at various points relative to other phases, the compositions are virtually identical within the bounds of analytical precision. Analyses are given in Table 2. The phase is now, after exsolution of the other phases, a magnesian enstatite which has a higher $Mg/(Mg+Fe^{2+})$ ratio (~ 0.94) than enstatites in other mantle peridotites with the exception of those in barren harzburgites ($Mg/[Mg+Fe]$ 0.94—Hervig et al., 1980). Its low CaO content is also most similar to

Table 3
Garnet in BD1366

	Average kink band	Average necklace	Average surrounded by chromite	Average fine lamella parallel to cleavage	Average fine lamella cutting across cleavage	Average coarse lamellae	Overall average	Standard deviation
n	13	10	6	6	9	6	50	
SiO_2	41.63	41.79	41.80	41.81	41.80	41.77	41.72	0.23
TiO_2	0.00	0.00	0.00	0.00	0.00	0.00	0.00	0.00
Al_2O_3	22.10	22.10	22.11	22.11	22.11	22.11	22.00	0.34
Cr_2O_3	2.50	2.58	2.57	2.57	2.57	2.56	2.55	0.27
Fe_2O_3	1.73	1.45	1.44	1.41	1.42	1.49	1.75	0.51
FeO	5.65	5.94	5.95	5.97	5.96	5.89	5.64	0.47
MnO	0.41	0.42	0.42	0.42	0.42	0.42	0.41	0.16
NiO	0.02	0.01	0.01	0.01	0.01	0.01	0.02	0.01
MgO	20.88	20.85	20.85	20.85	20.85	20.86	20.96	0.21
CaO	4.99	4.96	4.96	4.95	4.96	4.97	4.98	0.09
Na_2O	0.02	0.02	0.02	0.02	0.02	0.02	0.01	0.01
Total	99.93	100.12	100.12	100.12	100.12	100.08	100.05	0.53
Total FeO	7.20	7.24	7.24	7.24	7.24	7.23	7.21	0.07

Table 4
Clinopyroxenes in BD1366

	Average kink band	Average coarse lamella	Average necklace	Average surrounded by intergranular chromite	Overall average	Standard deviation
n	14	3	14	6	37	
SiO_2	54.00	53.96	53.89	53.96	53.95	0.25
TiO_2	0.00	0.00	0.00	0.00	0.00	0.00
Al_2O_3	2.01	2.03	1.97	1.93	1.98	0.04
Cr_2O_3	1.53	1.49	1.48	1.48	1.50	0.09
Fe_2O_3	1.37	1.62	1.44	1.59	1.45	0.27
FeO	0.21	0.00	0.11	0.00	0.12	0.25
MnO	0.06	0.05	0.05	0.05	0.05	0.01
NiO	0.05	0.07	0.05	0.07	0.06	0.01
MgO	16.83	16.99	17.05	17.31	17.01	0.28
CaO	21.84	22.18	21.83	21.80	21.86	0.20
Na_2O	1.46	1.44	1.46	1.45	1.46	0.04
Total	99.35	99.84	99.32	99.64	99.43	0.43
Total FeO	1.44	1.46	1.40	1.43	1.43	0.03

those in barren harzburgites, though its Al and Na contents resemble those in enstatites in granular, or cold, garnet lherzolites, i.e. those equilibrated at temperatures < 1130 °C (Hervig et al., 1986). Together with total Fe, these minor elements are appreciably higher in the enstatites in sheared (or hot) garnet lherzolites which have been subject to varying degrees of metasomatism (Table 2).

4.1.3. Garnet

Analyses (Table 3) were made on all the morphologically different types of garnet, though small grain size has limited the analysis of the fine lamellae. A striking feature of the analyses is their uniformity irrespective of the grain morphology and their location relative to the host enstatite or other phases. The garnet composition lies within the range for chrome pyrope (group 9) as defined by Dawson and Stephens (1975) on the basis of cluster analysis. This is the commonest type of garnet in cold granular garnet lherzolites.

4.1.4. Clinopyroxene

The clinopyroxene, like the garnet, is of uniform composition regardless of its morphology and relationship of the analysed individual grains to the parental enstatite (Table 4). Compositionally, it lies within the Cr, Fe, Mg, Ca and Na ranges for chrome-diopside as defined by Stephens and Dawson (1977). This is the commonest group of clino-

pyroxene in both granular and sheared garnet lherzolites and variants such as garnet-olivine pyroxenite and garnet pyroxenite.

4.1.5. Spinel

Like the garnet and chrome-diopside, the spinel is of uniform composition (Table 5), whether it occurs as lamellae within the enstatite or as intergranular to other necklace phases. The spinel is an aluminous chromite whose $Cr/(Cr+Al)$ value of 0.69 and $Mg/(Mg+Fe)$ ratio of 0.64 are best matched by spinels in garnet-chromite lherzolites

Table 5
Al-chromite in BD1366

	Average lamellar	Average intergranular	Average necklace	Overall average	Standard deviation
n	3	6	3	12	
SiO_2	0.06	0.04	0.05	0.05	0.01
TiO_2	0.00	0.00	0.00	0.00	0.00
Al_2O_3	16.24	16.02	16.29	16.15	0.15
Cr_2O_3	53.14	53.35	53.28	53.28	0.18
Fe_2O_3	3.57	4.03	3.05	3.64	0.45
FeO	13.81	13.38	14.35	13.76	0.45
MnO	0.25	0.26	0.26	0.26	0.01
NiO	0.10	0.09	0.13	0.10	0.02
MgO	13.44	13.85	13.07	13.52	0.35
CaO	0.01	0.00	0.00	0.00	0.01
Na_2O	0.03	0.00	0.02	0.01	0.01
Total	100.65	101.01	100.50	100.77	0.27
Total FeO	17.03	17.00	17.09	17.03	0.09

Table 6
Trace elements in phases in BD1366 and in harzburgite enstatites (in ppm)

Textural type	Sc	Ti	V[a]	Cr[b]	Mn[b]	Co	Y	Zr	Sr	Nb	Ba	La	Ce	Pr	Nd	Sm	Eu	Gd	Dy	Er	Yb
In BD1366																					
Orthopyroxenes																					
1-1 kink band	4.98	6.66	29.9	1711	697	52.2	0.005	0.066	0.143	0.173	0.241	0.020	0.046	0.009	0.058	0.012	bd	bd	0.014	bd	0.025
2-1 kink band	4.92	7.06	29.4	1779	774	54.0	0.013	0.092	0.141	0.145	0.027	0.028	0.069	0.019	0.269	0.076	0.044	0.153	0.039	0.009	0.135
1-3- Clear margin	5.04	7.08	29.3	1847	774	54.0	0.009	0.054	0.172	0.203	0.011	0.014	0.046	0.007	0.039	bd	0.002	0.016	0.032	0.015	bd
1-4 Clear margin	5.39	7.05	29.0	1847	620	54.1	0.004	0.202	0.165	0.139	0.083	0.025	0.041	0.010	0.089	0.036	0.006	0.007	bd	0.068	bd
1-4 neoblast1	5.24	6.71	29.7	1779	774	56.5	0.010	0.049	0.122	0.143	0.008	0.020	0.062	0.022	0.132	0.089	bd	bd	0.018	0.022	0.112
1-4 neoblast 2	5.41	7.33	29.5	1779	697	55.1	0.004	0.063	0.104	0.121	0.021	0.018	0.078	0.012	0.312	bd	0.01	0.023	bd	0.044	0.11
2-4 neoblast 1	5.88	6.98	31.3	1711	929	55.3	0.013	0.036	0.149	0.197	0.015	0.016	0.037	0.004	0.053	bd	0.002	bd	bd	bd	0.054
2-4 neoblast 2	5.65	7.43	30.5	1711	929	54.2	0.007	0.055	0.163	0.102	0.028	0.017	0.051	0.020	0.033	0.083	0.023	bd	0.005	0.006	0.010
2-5 grain interior	4.94	6.67	30.3	1711	774	54.4	0.015	0.069	0.175	0.324	0.040	0.041	0.084	0.025	0.149	0.091	0.025	0.096	0.008	0.082	0.009
Average	5.27	6.99	29.9	1764	774	54.4	0.009	0.076	0.148	0.172	0.053	0.022	0.057	0.014	0.126	0.065	0.016	0.059	0.019	0.035	0.065
orthopyroxene																					
Clinopyroxenes																					
1-1 kink band	20.1	27.1	244.1	10606	387	17.4	0.095	4.69	389.7	0.590	0.975	26.5	50.5	4.52	14.9	1.15	0.276	0.757	0.288	0.107	bd
2-1 kink band	19.4	27.1	257.2	11701	697	18.7	0.079	4.73	258.9	0.680	0.087	21.8	48.8	4.07	13.6	1.22	0.311	0.867	0.450	0.198	0.082
2-2 coarse lamella	18.7	28.1	269.6	10264	387	18.8	0.095	5.17	286.4	0.695	0.109	23.3	48.4	4.52	14.4	1.25	0.299	0.932	0.373	0.109	0.152
1-3 necklace(core)	19.9	27.8	265.1	10127	310	18.5	0.104	4.69	257.5	0.686	0.099	23.0	45.6	4.14	12.3	0.99	0.253	0.758	0.25	0.129	bd
1-3 necklace (rim)	21.5	29.0	265.5	10127	310	17.9	0.079	4.86	303.9	0.678	1.210	24.6	48.9	4.47	14.9	1.26	0.226	0.654	0.444	0.138	bd
1-3 another necklace	20.9	27.8	254.5	10401	465	18.7	0.109	4.77	327.3	0.715	0.084	25.2	50.0	4.51	14.7	1.35	0.312	0.745	0.103	0.099	0.187
1-4 necklace (anal.1)	19.8	28.8	249.0	9648	387	17.9	0.109	4.38	469.1	0.599	0.339	27.4	53.3	4.94	15.8	1.04	0.332	1.06	0.224	0.057	bd
1-4 necklace (anal.2)	20.4	29.1	246.0	9648	387	19.5	0.107	4.66	465.6	0.660	0.281	27.0	52.7	4.69	16.8	1.45	0.309	1.30	0.271	0.024	bd
2-4 necklace	20.8	28.5	250.2	11016	310	19.4	0.104	4.93	377.9	0.653	3.810	25.4	49.4	4.52	16.0	1.40	0.301	1.589	0.270	0.247	bd
Average diopside	20.2	28.1	255.8	10393	404	18.5	0.098	4.76	348.5	0.662	0.777	24.9	49.7	4.49	14.9	1.23	0.291	0.962	0.297	0.123	0.140

Garnets

Sample																					
2-1 kink band	114.1	12.1	95.6	16422	3330	65.7	1.31	2.57	0.078	0.197	0.023	0.053	0.274	0.082	0.430	0.251	0.094	0.293	0.370	0.236	0.582
2-3 kink band 1	113.0	13.6	93.0	19501	3098	65.4	1.08	1.91	0.053	0.238	0.039	0.041	0.206	0.053	0.546	0.141	0.073	0.295	0.226	0.182	0.357
2-3 kink band garnet 2	108.1	12.6	86.9	20528	3253	64.9	0.98	1.45	0.055	0.231	0.031	0.039	0.184	0.045	0.314	0.318	0.075	0.212	0.323	0.337	0.382
2-3 fine lamella	107.7	11.5	89.9	14575	3098	63.9	1.64	4.40	0.144	0.161	0.220	0.052	0.315	0.121	1.150	0.844	0.192	0.593	0.628	0.339	0.657
1-1 Coarse lamella(pod-shaped)	104.6	12.3	88.7	15738	3253	65.3	1.58	4.71	0.109	0.166	0.002	0.032	0.286	0.085	0.741	0.353	0.164	0.159	0.384	0.212	0.570
1-1 Coarse lamella(pod-shaped)	101.5	10.5	85.5	15738	3253	64.2	1.59	5.19	0.109	0.174	0.034	0.263	0.312	0.084	0.793	0.338	0.156	0.531	0.476	0.229	0.397
2-5 Coarse lamella (rod)	116.3	12.2	100.9	18338	3330	65.2	1.34	2.71	0.098	0.218	0.023	0.059	0.247	0.077	0.517	0.447	0.111	0.379	0.254	0.091	0.795
1-3 In opx margin	113.2	12.9	94.5	19159	3098	65.7	1.20	2.35	0.052	0.189	0.008	0.03	0.225	0.053	0.523	0.138	0.078	0.299	0.239	0.192	0.373
1-3 In opx margin	113.3	12.7	95.8	16422	3098	65.3	1.23	2.37	0.101	0.160	0.008	0.041	0.225	0.063	0.534	0.144	0.089	0.317	0.175	0.216	0.524
1-4 necklace	111.3	11.9	100.3	17927	3098	64.8	1.31	3.28	0.145	0.174	0.099	0.078	0.297	0.093	0.514	0.229	0.073	0.480	0.162	0.276	0.802
1-4 necklace	109.6	12.1	96.8	17927	3098	64.5	1.31	3.28	0.090	0.156	0.027	0.043	0.301	0.907	0.703	0.392	0.154	0.471	0.268	0.123	0.213
2-4 necklace	112.0	12.9	101.9	19159	3407	67.0	1.18	2.57	0.067	0.177	0.019	0.059	0.022	0.062	0.413	0.109	0.090	0.229	0.172	0.14	0.421
2-3 necklace	114.0	12.9	101.4	19159	3330	68.2	0.94	1.00	0.083	0.200	0.027	0.044	0.174	0.051	0.353	0.239	0.071	0.195	0.236	0.270	0.717
2-2 Very coarse lamella	92.1	14.4	63.6	16969	3253	52.1	1.38	4.24	0.361	0.484	0.153	0.335	0.463	0.234	2.03	1.41	0.022	1.45	1.59	0.916	1.580
Average (excluding v.coarse lamella)	110.7	12.3	94.7	17738	3211	65.4	1.28	2.91	0.091	0.188	0.043	0.064	0.236	0.137	0.579	0.303	0.109	0.343	0.301	0.219	0.522
Average (including v.coarse lamella)	109.3	12.5	92.3	17683	3214	64.4	1.29	3.00	0.110	0.209	0.051	0.083	0.252	0.144	0.683	0.382	0.103	0.422	0.393	0.269	0.598
Detection limits	0.7	0.5	10.0				1.0	0.003	0.005	0.009	0.010	0.009	0.010	0.008	0.070	0.060	0.016	0.050	0.035	0.024	0.080
In harzburgites																					
Enstatite in fertile harzburgite 2004								0.377	0.795	0.33	0.018	0.07	0.076	0.093	0.63	0.077	0.091	0.076	0.174	0.029	0.133
Enstatite in barren harzburgite 1919								0.012	0.113	0.102	0.01	0.033	bd	0.008	bd	bd	bd	bd	bd	bd	0.019

Point annotation: first digit = slide number, second digit = area. Thus, 2-3 means BD1366, slide 2, area 3.

[a] Values have been reduced for MgAl molecular interference by: Opx 9.8 ppm, cpx 7.3 ppm, garnet 48 ppm.

[b] Recalculatd from electron microprobe data; bd = below detection.

rather than in refractory chromite harzburgites (Smith and Dawson, 1975).

One particular common chemical feature of all the phases in BD1366 is the absence of TiO_2. This contrasts with the phases, particularly garnet and clinopyroxene, in other mantle lherzolites that generally contain small but detectable amounts.

4.2. Conditions of formation

The chemical homogeneity of the phases suggests they are equilibrated and thus suitable for estimating their formation conditions. Calculations using the enstatite-in-diopside thermometer, and the Al-in-enstatite barometer of Brey and Köhler (1990), calculating all the iron as Fe^{2+}, give 785 °C and 28.9 kb, approximating to a depth of 90–95 km. The temperature is towards the lower end of the temperature range for mantle peridotites (e.g. 705–>1300 °C, Shimizu et al., 1997), whilst the depth is close to the garnet in boundary between Kaapvaal spinel- and garnet-lherzolites estimated at around 100 km (Boyd et al., 1999).

4.3. Trace element chemistry

The trace element contents of the texturally different grains of enstatite, diopside and garnet, determined by ion-microprobe analysis, are given in Table 6, and shown graphically in Fig. 8. Note, however, for Table 6 and in Figs. 8–10 the values for Cr and Mn are converted from electron-probe analyses. Overall, the trace element concentrations are low. Compared with chondrites, all three phases have very low amounts of Ti, Co and Ba; Ti, which is often in minor element concentrations in mantle garnets and diopsides (see Grégoire et al., 2003), is present in diopside at <30 ppm and in garnet at ~12 ppm. Sc, V, Cr and Nb in all phases equal or are slightly higher than chondrite values, as is Zr in garnet and diopside. Garnet contains more Sc, Cr and Mn, Co and Y than diopside and enstatite, whereas diopside contains the most Ti, V, Sr and Zr. Enstatite holds more Mn and Co than diopside. For the REE, the LREE are strongly concentrated in the diopside with La 105 × chondrite and a steep pattern, with La/Yb_{CN} 362. The patterns for both garnet and enstatite are

relatively flat; chondrite-normalised values for garnet are between 0.35 (La) and 3.7 (Yb), whereas those for enstatite are all <1.

Compared with data on garnet and diopside in granular and sheared garnet lherzolites from Southern African kimberlites (Shimizu, 1975; Shimizu and Allègre, 1978), BD1366 garnet and diopside contain low V and Zr and very low Ti; the diopside, however, contains more Sr than the lherzolite clinopyroxenes (Fig. 9). For the REE, BD1366 garnet has a pattern more similar to that in sheared lherzolite garnets (differing from the sinusoidal pattern for those in granular lherzolites), but the chondrite-normalised values are around one order of magnitude less; the pattern is at variance with the observation of Shimizu et al. (1997) that "sinusoidal REE patterns occur ubiquitously in garnets in low-temperature peridotites". The BD1366 diopside pattern is more similar to the pattern in granular lherzolite clinopyroxenes, but is steeper because of its higher amounts of LREE but lower MREE and HREE.

Further differences are apparent between BD1366 and the Shimizu peridotites when the partitioning of elements between diopside and garnet are compared. For the transition elements, the concentration of Ti, V, Cr and Zr in clinopyroxene relative to garnet is higher, whereas Mn and Co are present in relatively low concentrations in diopside (Fig. 10). For the REE, although the overall pattern for cpx/gt partitioning for the MREE and HREE is similar to that in the peridotites, in BD1366 the REE, especially the LREE, are more strongly concentrated in the clinopyroxene (Fig. 10). This particular feature, the relatively high amounts of the LREE in the diopside, is in broad agreement with the findings of Harte et al. (1996) that the relative concentrations of the LREE increase with decreasing temperature of equilibration.

Compared with a more recent data set on both granular and deformed garnet lherzolites from the Kaapvaal Craton kimberlites (Grégoire et al., 2003), BD1366 garnet is low in Ti, Zr and the REE (as with the Shimizu specimens) but also high in Co; BD1366 diopside contains low Ti, Y and Zr but high Sr and REE. The enstatites in Grégoire et al.'s specimens contain less Sc and Co than BD1366 enstatite, but more Ti, Y, Zr and the REE.

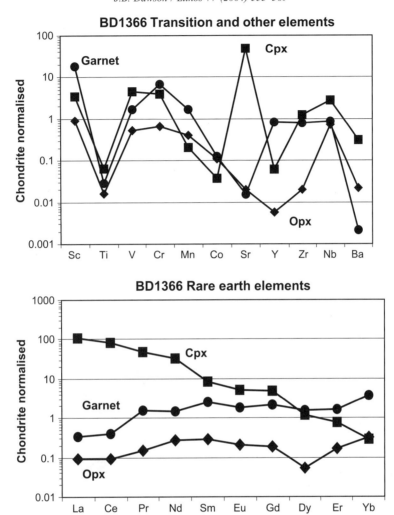

Fig. 8. Chondrite-normalised plots of the concentrations of transition and rare earth elements in the phases in BD1366. In this, and Figs 9 and 11, data are normalised against CI values of McDonough and Sun (1995, Table 2).

Another relevant difference from the Southern African peridotites has been reported for a suite of xenoliths from the Udachnaya kimberlite in which, although the cpx/garnet partitioning in high-T peridotites is similar to that in the Southern African peridotites, in contrast, the partitioning in relatively low-T rocks is much more random, suggesting trace element disequilibrium, possibly due to LREE metasomatism (Shimizu et al., 1997). Although the uniformity of composition of the phases in BD1366 indicates equilibrium, the cpx/gt partitioning pattern in BD1366, and also the absolute concentrations of the REE and Ti, is most similar to that in one of these unequilibrated peridotites (UV25/91) which formed at a somewhat lower temperature (705 °C).

4.4. Composition of harzburgite enstatites

It can be inferred that, prior to exsolution of the garnet, Cr-diopside and chromite, the parental enstatite in BD1366 must have contained considerably more Ca, Al and Cr than at present, and that it would have been most closely comparable to enstatites in

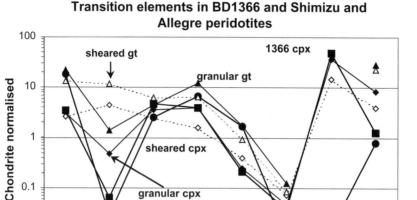

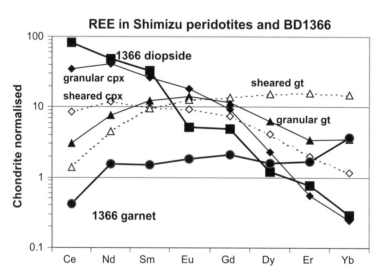

Fig. 9. Chondrite-normalised plots of transition and rare earth elements in average BD1366 garnet (gt) and clinopyroxene (cpx), and in average garnets and clinopyroxenes in 10 granular and 4 sheared Southern African garnet lherzolites (Shimizu and Allègre, 1978). The differences apparent between the BD1366 phases and those in the Shimizu and Allègre peridotites are even greater for Shimizu and Allègre metasomatised granular peridotites (not plotted). In the rare earth element plot, the average garnets and clinopyroxenes are from three granular and three sheared Southern African garnet lherzolites (Shimizu, 1975).

fertile harzburgite (Hervig et al., 1980). For comparison with the BD1366 data, new major, minor and trace element analyses have been made on enstatites from one fertile (FH) and one barren harzburgite (BH) (Tables 6 and 7). Major elements that highlight the differences between the FH and BH orthopyroxenes are Al, Cr and Ca. Accepting that these represent garnet and diopside in solid solution in the enstatite,

and on the broad assumption that on exsolution Al and Cr would be sited in garnet, and all Ca in diopside, the analyses shown in Table 7 represent 16.9 wt.% potential garnet and 3.4 wt.% potential diopside in the FH enstatite. Using the mode for typical fertile harzburgite (olivine 68%, enstatite 32%) given by Hervig et al. (1980), this converts to 5.4 wt.% garnet and 1.7 wt.% diopside in a rock that

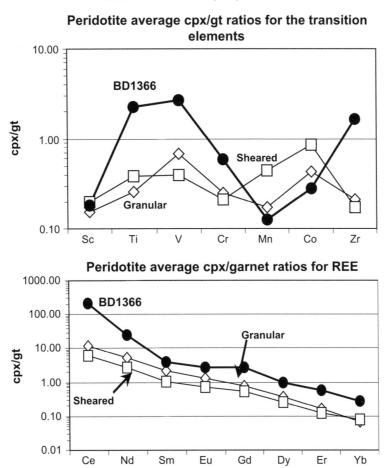

Fig. 10. Clinopyroxene/garnet transition element and rare earth element partitioning in BD1366 and in average granular and sheared Southern African garnet lherzolites (Shimizu, 1975; Shimizu and Allègre, 1978).

could be potentially formed by exsolution from a fertile harzburgite protolith.

The trace element data given in Table 6 show the extremely low trace-element concentrations in these enstatites. The enstatite in BH1919 contains very low amounts of all the analysed trace elements, for many elements the amounts being below detection. The REE pattern for FH2004 is slightly humped with the highest chondrite-normalised amounts in the MREE; any phases exsolving from this enstatite would clearly have been able to inherit only very low amounts of trace elements.

Comparative trace element data for other mantle orthopyroxenes are not abundant. Data on enstatites in Southern African garnet lherzolites (Grégoire et al.,

2003) have been referred to above. Enstatites in anhydrous spinel peridotites from the Eifel area, Germany, contain 0.08–0.18 wt.% TiO_2, 14–30 ppb Eu and 37–74 ppb Sm (McDonough et al., 1992). Chondrite-normalised REE patterns for the same enstatites together with others from USA and Mongolia (Stosch, 1982) are unlike that in the fertile harzburgite, in that most show low LREE (~ 0.1 to 0.5 × chondrite) and then rise linearly to around 1 to 3 × chondrite for the HREE. Other data, on an enstatite from a Roberts Victor garnet lherzolite, are believed to be compromised by host-kimberlite contamination (Philpotts et al., 1972), and orthopyroxene megacrysts in a range of basaltic rocks from Australia, New Zealand and USA (Irving and Frey, 1984) are not strictly comparable as

Table 7
Compositions of orthopyroxenes in fertile (FH) and barren (BH) harzburgites

Sample	FH	BH
	BD2004	BD1919
SiO_2	55.4	56.9
TiO_2	0.01	0.01
Cr_2O_3	0.90	0.52
Al_2O_3	3.34	1.84
FeO	4.40	3.97
MnO	0.13	0.10
MgO	34.8	35.9
CaO	0.75	0.35
Na_2O	0.03	0.04
Sum	99.76	99.63
Wt.% Garnet[a]	16.9	9.4
Wt.% Cpx[b]	3.4	1.6
Rock Garnet[c]	5.4	
Rock Cpx[c]	1.1	

[a] Assumption: in garnet $Al_2O_3 + Cr_2O_3 = 25$ wt.%.
[b] Assumption: in cpx $CaO = 22$ wt.%.
[c] Weight percent of rock, assuming 68% olivine, 32% opx.

they are more iron rich ((7.2 to 12.8 wt.% total FeO vs. 4.4% for the FH enstatites).

5. Comparison between trace element concentrations in BD1366 and fertile harzburgite enstatite

Chondrite-normalised trace element data for the phases in BD1366 and the fertile harzburgite enstatite are shown on Fig. 11. The data points for most

elements in FH BD2004 enstatite lie intermediate between those for BD1366 enstatite and the other phases, with BD1366 diopside showing the greatest departure from FH BD2004 enstatite; for several of the REE, FH BD2004 data points coincide with the values for BD1366 garnet. Assuming the parental enstatite in BD1366 had a trace element budget similar to that of FH BD2004, the relative abundances of the elements in the BD1366 phases could indicate depletion of most trace elements in the now-depleted enstatite with the greatest partitioning of most elements into exsolved diopside; particularly for the LREE; exsolved garnet would appear to be relatively neutral in this trace element redistribution. The high concentrations of the LREE in diopside are not improbable, considering the low amounts of exsolved diopside compared with the relatively high volumes of both the parental enstatite and the exsolved garnet (see specimen description). However, although a case might be made for an isochemical redistribution of most of the trace elements (though strongly dependent upon the relative abundance of receptor phases formed during the exsolution process), the low amounts of Nb in FH BD2004 enstatite cannot provide a budget for the higher concentrations found in all the phases in BD1366.

6. Discussion and conclusions

1. The major and minor element chemistry of the phases in BD1366 shows that the olivine most

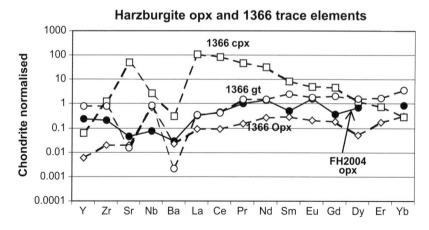

Fig. 11. Chondrite-normalised plot of trace elements in the phases in BD1366 and in the orthopyroxene in fertile harzburgite BD2004.

closely matches the chemistry of harzburgite olivines and, prior to expulsion of Ca, Al and Cr from its lattice in the form of the exsolved garnet, diopside and chromite, the enstatite would have been most similar to the high-Ca-Al enstatites found in fertile harzburgites. Further, the amount of TiO$_2$ at only the ppm level in the phases in BD1366 is matched by absence of TiO$_2$ in enstatites from fertile harzburgites (Hervig et al., 1980). With the exception of the extremely low TiO$_2$, the present enstatite, garnet, diopside and Al-chromite all can be matched most closely with analogous phases in granular, cold garnet lherzolites. Thus, on textural grounds, and on the major and minor element chemistry of the phases, there are grounds for interpreting BD1366 as representing a transition from fertile harzburgite to cold, granular garnet lherzolite.

2. A feature of the phases in BD1366 is their remarkably uniform composition. Compositional gradients between the exsolved phases and the host enstatite have not been found, regardless of whether the exsolved phases are fine lamellae in the host enstatite, or coarser grains at enstatite margins. This suggests that the phases are in chemical equilibrium. However, the different morphologies of the phases (fine lamellae, coarse lamellae, equant necklace grains) indicate that the xenolith had not achieved textural equilibrium as might have occurred if the rock had undergone a prolonged period of annealing in the high-thermal regime of the upper mantle. The inference is that the exsolution (and associated deformation) might have taken place shortly before entrainment and eruption.

3. The close association between deformation features in the host enstatite and exsolution strongly suggests that strain, by setting up lattice dislocations in the enstatite and thereby enhancing element diffusion, was an essential part of the exsolution process. Further, as the result of strain-induced recrystallisation, with the exception of the enstatite porphyroclaststs, the resultant garnet lherzolite is of finer grain size than the coarser-grained harzburgite protolith.

4. The chemical equilibrium found for BD1366 contrasts with that in a texturally similar eclogite from the Roberts Victor kimberlite, in which lamellar and granular garnet have exsolved from a formerly high-Al clinopyroxene, with considerable compositional variation both within single clinopyroxene crystals and also between the morphologically different garnets. Changing Fe/Mg partitioning between clinopyroxene and successively exsolved garnet indicates a large fall in temperature (Harte and Gurney, 1975) but, despite this, gradients for Al in the pyroxene testify to incomplete exsolution (Sautter and Harte, 1988). One main difference is that the eclogite does not show the same effects of strain as BD1366. This reinforces the case for the possible importance of strain-induced lattice deformation in enhancing element diffusion.

5. Theoretically, after their formation, fertile harzburgite enstatites should exsolve garnet and diopside when the temperature falls to that of the ambient upper mantle (Canil, 1991). The failure of some fertile harzburgites to invert to garnet lherzolite suggests that there must be some reason(s) additional to temperature decline to assist their inversion to garnet lherzolite. The link between exsolution and deformation seen in BD1366 may indicate that initial exsolution also depends upon strain; the visible effects of this may subsequently be obscured by the more complete recrystallisation typical of granular garnet lherzolites. Conversely, the very survival of the metastable fertile harzburgites may depend upon their siting in strain-free areas of the upper mantle.

6. Assuming that the initial trace element concentration in the enstatite in BD1366 was similar to that in fertile harzburgite BD2004, the contents of most trace elements in the residual host enstatite and exsolved garnet and diopside are compatible with iso-chemical exsolution. The exception is Nb, for which the enstatite in FH BD2004 cannot provide the amounts found in the phases in BD1366.

7. Compared with the garnets and diopsides in granular garnet lherzolites, those in BD1366 contain lower Ti, Co, Zr and, for garnet, most of the REE. Compared with the phases in sheared garnet lherzolites, the low amounts of Ti and Zr are even more pronounced. Further, the garnet in BD1366 lacks the sinusoidal REE pattern that is inferred to result from metasomatism (Shimizu and Richardson, 1987; Hoal et al., 1994; Stachel et al., 1998; Van Achterbergh et al., 2001).

8. The modal compositions (5–10% garnet, 2–5% diopside) of many garnet lherzolites suggests that they may be of exsolution origin (Dawson et al., 1980). BD 1366 is a convincing example of exsolution, but the data presented here are on a single sample and much more research is needed on similar samples from more localities before extrapolating the results to a general commentary on mantle processes. Nonetheless, this is a tantalising pointer in that, if it can be shown that the enstatites in more parental fertile harzburgites have trace element budgets similar to that in FH BD2004, the trace element compositions of the garnets and diopsides from granular garnet lherzolites (e.g. Shimizu, 1975; Shimizu et al., 1997) could not have been inherited from parental enstatite during isochemical exsolution. It would follow that metasomatic addition of Ti, Zr, the LREE and, in the case of diopsides, Sr would need to have taken place. Thus, in this respect, granular lherzolites would be similar to sheared garnet lherzolites which display evidence of metasomatism—major element as well as trace element. The fact that *all* garnet lherzolites xenoliths have been subject to metasomatic influences has indeed been proposed by Grégoire et al. (2003), though from different evidence.

9. BD1366 with its large enstatite porphyroclasts is envisaged as being texturally intermediate between a coarse-grained harzburgite precursor and a granular garnet lherzolite. There are two possibilities for the timing of the further strain that would be necessary to produce the more complete recrystallisation typical of the granular garnet lherzolites. Either it could be a continuation of the strain event linked with the exsolution or, alternatively, it might have been caused by a later, unconnected event. The timing of the inferred metasomatism vis a vis the subsequent deformation and recrystallisation is a worthy subject for further study.

Acknowledgements

Peter Hill, John Craven, Richard Hinton, Simone Kasemann and Nicola Cayzer provided sterling assistance with the analytical work and SEM imaging, and Tom Stachel made the PT calculations. Reviewers Nikolai Sobolev and Michael Roden, and guest editor Herman Grütter made constructive comments on the original version of this paper. I am grateful to them all. The electron- and ion-microprobe facilities at the University of Edinburgh are supported by NERC.

References

Boyd, F.R., Pearson, D.G., Mertzman, S.A., 1999. Spinel-facies peridotites from the Kaapvaal root. In: Gurney, J.J., Gurney, J.L., Pascoe, M.D., Richardson, S.H. (Eds.), Proc. 7th Internat. Kimberlite Conf. Red Roof Design, Cape Town, vol. 1, pp. 40–48.

Brey, G.P., Köhler, T., 1990. Geothermobarometry in four-phase lherzolites: new thermometers and a practical assessment of existing thermobarometers. J. Petrol. 31, 1353–1378.

Canil, D., 1991. Experimental evidence for the exsolution of cratonic peridotite from high-temperature harzburgite. Earth Planet. Sci. Lett. 106, 64–72.

Cox, K.G., Smith, M.R., Beswetherick, S., 1987. Textural studies of garnet lherzolites: evidence of exsolution from high-temperature harzburgites. In: Nixon, P.H. (Ed.), Mantle Xenoliths. Wiley, Chichester, pp. 537–550.

Dawson, J.B., 2002. Metasomatism and melting in upper mantle peridotite xenoliths from the Lashaine volcano, Northern Tanzania. J. Petrol. 43, 1749–1777.

Dawson, J.B., Hill, P.G., 1998. Mineral chemistry of a peralkaline combeite-lamprophyllite nephelinite from Oldoinyo Lengai Tanzania. Min. Mag. 62, 179–196.

Dawson, J.B., Stephens, W.E., 1975. Statistical classification of garnets from kimberlite and associated xenoliths. J. Geol. 83, 589–607.

Dawson, J.B., Smith, J.V., Hervig, R.L., 1980. Heterogeneity in upper-mantle lherzolites and harzburgites. Philos. Trans. R. Soc. Lond., A 297, 323–331.

Droop, G.T.R., 1981. A general equation for estimating Fe^{3+} concentrations in ferromagnesioan silicates and oxides from microprobe analyses, using stoichiometric critera. Mineral. Mag. 51, 431–435.

Grégoire, M., Bell, D.R., Le Roux, A.P., 2003. Garnet lherzolites from the Kaapvaal Craton (South Africa): trace element evidence for a metasomatic history. J. Petrol. 44, 629–657.

Gurney, J.J., Harte, B., Cox, K.G., 1975. Mantle xenoliths in the Matsoku kimberlite pipe. Phys. Chem. Earth 9, 507–523.

Harte, B., Gurney, J.J., 1975. Evolution of clinopyroxene and garnet in an eclogite nodule from the Roberts Victor kimberlite pipe South Africa. Phys. Chem. Earth 9, 367–387.

Harte, B., Fitzsimons, I.C., Kinny, P.D., 1996. Clinopyroxene-garnet trace element partition coefficients for mantle peridotite and melt assemblages. J. Conf. Abstr. 1, 235.

Hervig, R.L., Smith, J.V., Steele, I.M., Dawson, J.B., 1980. Fertile and barren Al-Cr spinel harzburgites from the upper mantle: ion and electron probe analyses of trace elements in olivine and orthopyroxene: relation to lherzolites. Earth Planet. Sci. Lett. 50, 41–58.

Hervig, R.L., Smith, J.V., Dawson, J.B., 1986. Lherzolite xenoliths in kimberlites and basalts: petrogenetic and crystallochemical significance of some minor and trace elements in olivine, pyroxenes, garnet and spinel. Proc. R. Soc. Edinb., Earth Sci. 77, 181–201.

Hoal, K.E.O, Hoal, B.G., Erlank, A.J., Shimizu, N., 1994. Metasomatism of the mantle lithosphere recorded by rare earth elements in garnet. Earth Planet. Sci. Lett. 126, 303–313.

Irving, A.J., Frey, F.A., 1984. Trace element abundances in megacrysts and their host basalts: constraints on partition coefficients and megacryst genesis. Geochim. Cosmochim. Acta 48, 1201–1221.

McDonough, W.F., Sun, S.-S., 1995. The composition of the Earth. Chem. Geol. 120, 223–253.

McDonough, W.F., Stosch, H.-G., Ware, N.G., 1992. Distribution of titanium and the rare earth elements between peridotitic minerals. Contrib. Mineral. Petrol. 110, 321–328.

O'Hara, M.J., Saunders, M.J., Mercy, E.P.L., 1975. Garnet peridotite, primary ultrabasic magma and eclogite; interpretation of upper mantle processes in kimberlite. Phys. Chem. Earth 9, 571–604.

Philpotts, J.A., Schnetzler, C.C., Thomas, H.H., 1972. Petrogenetic implications of some new geochemical data on eclogitic and ultrabasic inclusions. Geochim. Cosmochim. Acta 36, 1131–1166.

Saltzer, R.L., Chatterjee, N., Grove, T.L., 2001. The spatial relationship of garnets and pyroxenes in mantle peridotites; pressure-temperature history of peridotites from the Kaapvaal craton. J. Petrol. 42, 2215–2229.

Sautter, V., Harte, B., 1988. Diffusion gradients in an eclogite xenolith from the Roberts Victor kimberlite pipe; (1) mechanism and evolution of garnet exsolution in Al_2O_3-rich clinopyroxene. J. Petrol. 29, 1325–1358.

Shimizu, N., 1975. Rare earth elements in garnets and clinopyroxenes from garnet lherzolite nodules in kimberlites. Earth Planet. Sci. Lett. 25, 26–32.

Shimizu, N., Allègre, C., 1978. Geochemistry of transition elements in garnet lherzolite nodules in kimberlites. Contrib. Mineral. Petrol. 67, 41–50.

Shimizu, N., Richardson, S.H., 1987. Trace element abundance patterns of garnet inclusions in peridotite-suite diamonds. Geochim. Cosmochim. Acta 51, 755–758.

Shimizu, N., Pokhilenko, N.P., Boyd, F.R., Pearson, D.G., 1997. Geochemical characteristics of mantle xenoliths from the Udachnaya kimberlite pipe. Russ. Geol. Geophys. 38, 205–217.

Smith, J.V., Dawson, J.B., 1975. Chemistry of Ti-poor spinels, ilmenites and rutiles from peridotite and eclogite xenoliths. Phys. Chem. Earth 9, 309–322.

Stachel, T., Viljoen, K.S., Brey, G., Harris, J.W., 1998. Metasomatic processes in lherzolitic and harzburgitic domains of diamondiferous lithosphereic mantle; REE in garnets from xenoliths and inclusions in diamonds. Earth Planet. Sci. Lett. 159, 1–12.

Stephens, W.E., Dawson, J.B., 1977. Statistical comparison between pyroxenes from kimberlites and their associated xenoliths. J. Geol. 85, 433–449.

Stosch, H.-G., 1982. Rare earth element partitioning between minerals from anhydrous spinel peridotite xenoliths. Geochim. Cosmochim. Acta 46, 793–811.

Van Achterbergh, E., Griffin, W.L., Stiefenhofer, J., 2001. Metasomatism in mantle xenoliths from the Letlhakane kimberlites: estimation of element fluxes. Contrib. Mineral. Petrol. 141, 397–414.

Yoder, H.S., 1976. Generation of Basaltic Magma. National Academy of Sciences, Washington, DC.

Available online at www.sciencedirect.com

LITHOS

ELSEVIER

Lithos 77 (2004) 571

www.elsevier.com/locate/lithos

Editorial

In memoriam Francis R. "Joe" Boyd

The following paper was presented by Joe Boyd at the Eighth International Kimberlite Conference; sadly, this was to be his last presentation in this forum. Although it is impossible to list all of Joe's achievements here, we nevertheless wish to pay tribute to his outstanding contributions to mantle petrology over the past 40 years and his association with International Kimberlite Conferences.

In the mid 1960s, whilst working at the Geophysical Laboratory of the Carnegie Institute of Washington, Joe first became interested in the mantle-derived xenoliths carried by kimberlites. On the basis of experimental studies on the diopside–enstatite solvus, coupled with data on the solubility of alumina in garnet, he was the first to recognize that the composition of pyroxenes and garnets in these rocks could be used to calculate their equilibrium temperatures and pressures. These experimental results were initially applied to xenoliths and megacrysts from Lesotho collected by Peter Nixon.

This work, presented at the First International Kimberlite Conference in Cape Town and at the 1973 meeting of the Geochemical Society in Washington, led to the development of the new field of "paleogeothermobarometry", together with the introduction of the now common-place terms, "kinked geotherm" and "sheared nodule", and the concepts of "fertile" and "barren" mantle.

These ideas resulted in a flood of controversy, in particular, as mantle petrologists hotly debated the origin and significance of "kinked geotherms". The controversies engendered by Joe's ideas stimulated further experimental and geochemical work on the upper mantle and proved fruitful in the development of the careers of many scientists currently working in such diverse fields as mantle geophysics and diamond exploration—which are based directly on his work. Importantly, the work of scientists studying the isotopic and trace element geochemistry of the mantle has been greatly enhanced by Joe's generosity in making available to the community his substantial collection of xenoliths. Over the past 30 years, his original contributions have been further refined by a subsequent generation of mantle petrologists—but never discredited.

Joe's association with International Kimberlite Conferences began when he and the late Henry Meyer suggested that a conference be held in southern Africa to bring together scientists interested in kimberlites, diamonds, and the upper mantle. This, the First International Kimberlite Conference, was organized by John Gurney in cooperation with Barry Hawthorne of De Beers. Joe, working with Peter Nixon, provided an important contribution to the seminal Lesotho Kimberlites volume, published in conjunction with this conference, by providing data on the composition of several hundred minerals in Lesotho xenoliths.

The outstanding success of this conference resulted in Joe and Henry being asked to organize the Second International Kimberlite Conference in Santa Fe, New Mexico. This too was memorable, as geoscientists presented the results of work undertaken on material collected on field trips associated with the South African conference, thus setting the tone for all subsequent kimberlite conferences. Joe published the results of his later work in the proceedings of every conference while remaining an active member of the International Kimberlite Conferences Advisory Committee.

Joe will be recognized by geoscience historians as one of the great contributors to the development of our knowledge of the evolution of the upper mantle. We thank him and deeply regret his passing; he will always be remembered fondly as an approachable generous person and outstanding geoscientist.

8IKC Editorial Committee

doi:10.1016/j.lithos.2004.06.001

Available online at www.sciencedirect.com

Lithos 77 (2004) 573–592

www.elsevier.com/locate/lithos

Garnet lherzolites from Louwrensia, Namibia: bulk composition and *P*/*T* relations ☆

F.R. Boyd[a], D.G. Pearson[b,*], K.O. Hoal[c], B.G. Hoal[d], P.H. Nixon[e], M.J. Kingston[f], S.A. Mertzman[g]

[a] *Geophysical Laboratory, 5251 Broad Branch Rd. N.W., Washington, DC 20015, USA*
[b] *Department of Earth Sciences, Durham University, South Road, Durham DH1 3LE, UK*
[c] *Hazen Research Inc., 4601 Indiana St., Golden, CO 80403, USA*
[d] *Society of Economic Geologists, 7811 Shaffer Parkway, Littleton, CO 80127, USA*
[e] *School of Earth Sciences, University of Leeds, LS2 9JT, UK*
[f] *U.S. Geological Survey, MS 954, Reston, VA 20192, USA*
[g] *Department of Geology, Franklin and Marshall College, P.O. Box 3003, Lancaster PA, 17604-3003, USA*

Received 27 June 2003; accepted 23 January 2004
Available online 10 June 2004

Abstract

Bulk, mineral and trace element analyses of garnet lherzolite xenoliths from the Louwrensia kimberlite pipe, south-central Namibia, together with previously published Re–Os isotopic data [Chem. Geol. (2004)], form the most extensive set of chemical data for off-craton suites from southern Africa. The Louwrensia suite is similar to those from the Kaapvaal craton in that it includes both predominantly coarse-grained, equant-textured peridotites characterised by equilibration temperatures < 1100 °C and variably deformed peridotites, frequently sheared, with higher equilibration temperatures >1200 °C. Re-depletion ages range back to 2.1 Gy, concordant with the age of the crustal basement and about 1 Gy younger than the older peridotites of the adjacent Kaapvaal craton root. The coarse, low-temperature Louwrensia peridotites have an average *Mg* number for olivine of 91.6 in comparison to 92.6 for low-temperature peridotites from the craton. Orthopyroxene content averages 24 wt.% with a range of 11–40 wt.% for Louwrensia low-temperature peridotites, in comparison to a mean of 31.5 wt.% and a range of 11–44 wt.% for low-temperature peridotites from the Kaapvaal craton. Other major, minor and trace element concentrations in minerals forming Louwrensia lherzolites are more similar to values in corresponding Kaapvaal peridotite minerals than to those in lithospheric peridotites of Phanerozoic age as represented by off-craton basalt-hosted xenoliths and orogenic peridotites. Proportions of clinopyroxene and garnet in both the Louwrensia and Kaapvaal lherzolites overlap in the range up to 10 wt.% forming a trend extending towards pyrolite composition. Disequilibrium element partitioning between clinopyroxene and garnet for some incompatible trace elements is evidence that some of the trend is caused by enrichment following depletion. The disequilibrium is interpreted to have been caused by relatively recent growth of diopside, as previously suggested for cratonic peridotites. Attempts to constrain the depth of melting required to produce the Louwrensia peridotites suggests formation at pressures < 3 GPa. Estimates of temperature and depth of equilibration for low-temperature Louwrensia lherzolites yield a trend that is in approximate agreement with the average geotherm for the Kaapvaal craton and

---

☆ Supplementary data associated with this article can be found, in the online version, at doi: 10.1016/j.lithos.2004.03.010.
* Corresponding author. Fax: +44-191-374-2510.
E-mail address: d.g.pearson@durham.ac.uk (D.G. Pearson).

may indicate that the lithospheric mantle beneath this region was at one time as thick as that beneath the craton (>200 km). Temperature–depth plots for the high-temperature Louwrensia rocks, however, form pronounced, apparent higher-temperature thermal anomalies at depths of 140 km and above. These anomalies are believed to reflect regional igneous activity, perhaps associated with thermal erosion of an originally thicker lithosphere, a short time prior to eruption.

Keywords: Xenolith; Mantle; Namibia; Lithosphere; Geotherms; Peridotite; Kaapvaal

1. Introduction

Studies of abundant kimberlite xenoliths and concentrates from the Archean Kaapvaal craton in southern Africa have provided a relatively detailed picture of the petrology and chemical structure of the cratonic lithospheric mantle to a depth of about 200 km.

Much less is known about the lithospheric mantle beneath the Proterozoic circum-cratonic crustal terranes. Petrographic and electron probe investigations have been carried out for mantle materials from East Griqualand kimberlites, southeast of the craton (Boyd and Nixon, 1979, 1980), from the Karoo kimberlites south of the craton (Robey, 1981), and from the

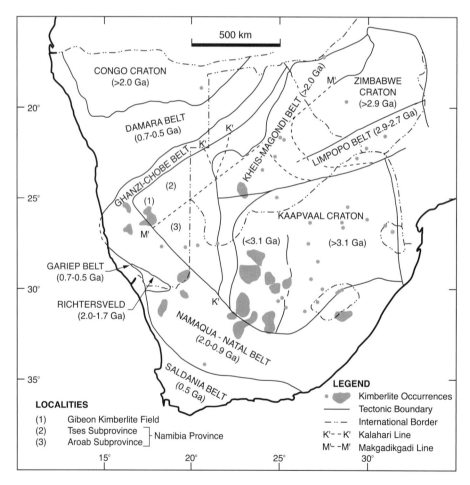

Fig. 1. Location map of the Gibeon Kimberlite Province, Namibia. The Louwrensia kimberlite pipe lies in the Gibeon Kimberlite Field (labeled 1). Modified after Hoal et al. (1995).

Gibeon kimberlites in Namibia ~ 400 km west of the Kalahari Line, the western boundary of the exposed Kaapvaal craton (e.g., Mitchell, 1984; Franz et al., 1996a,b). Comparison of the compositions of the lithospheres beneath the craton and surrounding terranes, however, requires bulk as well as mineral analyses. Off-craton xenoliths from lithosphere surrounding the Kaapvaal craton studied in earlier investigations were for the most part too small or too altered to permit high quality bulk analyses (Franz et al., 1996a,b). Nevertheless, garnet lherzolite xenoliths weighing up to several kilograms have been found in

small numbers at Louwrensia, near the center of the Gibeon kimberlite cluster, south central Namibia. This investigation is based on bulk and mineral analyses of 26 of these large lherzolites, supplemented by electron probe study of eight additional specimens that are too small for bulk analysis, together with ion-probe trace element analyses of garnet-diopside pairs from three samples. The whole rock analyses of the large Louwrensia xenoliths presented here form the most extensive set of such data for any off-craton locality in southern Africa. Platinum group element and Os isotope data for 11 of

Table 1
Louwrensia xenolith petrography for samples with bulk analyses

New number	Old number	Donor	Texture	Assemblage	Wt, g
FRB 1180	L70	IDM	Coarse	SPN-GAR harzburgite	500
FRB 1181	L69	IDM	Coarse	GAR lherzolite	500
FRB 1183	L61A	IDM	Coarse	GAR lherzolite	500
FRB 1625	L41	IDM	Coarse	GAR lherzolite	160
FRB 1626	L37	IDM	Deformed	GAR lherzolite	250
FRB 1650	L93-001	GSN	Coarse	SPN-GAR lherzolite. Primary PHLOG and chromite	525
FRB 1651	L93-002	GSN	Coarse	GAR lherzolite	430
FRB 1652	L93-201	GSN	Coarse	GAR lherzolite	445
FRB 1680	175/36/K14/18	AARL	Coarse	GAR lherzolite	400
FRB 1681	175/36/K14/23	AARL	Deformed	GAR lherzolite	330
FRB 1682	175/36/K14/22	AARL	Coarse	SPN-GAR lherzolite. Lindsleyite in a GAR	525
FRB 1683	175/36/K14/24	AARL	Deformed	SPN-GAR lherzolite. Chromite present	445
FRB 1684	175/36/k14/17	AARL	Coarse	GAR lherzolite	435
FRB 1685	175/36/k14/15	AARL	Coarse	GAR lherzolite. Foliation of GAR and OPX.	375
FRB 1686	175/36/k14/21	AARL	Coarse	GAR lherzolite	430
FRB 1687	175/36/k14/20	AARL	Coarse	GAR lherzolite	335
JJG 2513	JJG 2513	JJG	Coarse	GAR lherzolite	500
JJG 2514	JJG 2514	JJG	Coarse	SPN. Harzburgite. Primary chromite	500
JJG 2517	JJG 2517	JJG	Coarse	GAR lherzolite	500
PHN 5304	PHN 5304	PHN	Deformed	GAR lherzolite	500
PHN 5315	PHN 5315	PHN	Coarse	GAR lherzolite	500
PHN 5316	PHN 5316	PHN	Coarse	GAR lherzolite	360
PHN 5364	PHN 5364	PHN	Coarse	GAR lherzolite Some deformation	500
PHN5365	PHN5365	PHN	Coarse	GAR lherzolite	500
PHN6199	A599	TNC	Coarse	SPN-GAR lherzolite Primary chromite	500
E-11	E-11	PHN	Coarse	GAR lherzolite	500

Also given is sample origin/re-numbering for specimens obtained from other collections. Codes for sample donors; IDM = I.D. MacGregor, RHM = R.H. Mitchell, GSN = Geological Survey Namibia, AARL = Anglo American Research Labs, JJG = J.J. Gurney, TNC = T.N. Clifford. GAR = garnet, SPN = spinel, OPX = orthopyroxene. PHLOG = phlogopite. Wt, g = weight of specimen from which whole rock powder was prepared for bulk analysis.

these samples, with additional specimens studied by Jones (1984), are described by Pearson et al. (2004). Whole rock Re–Os isotope analyses of 21 Louwrensia peridotites yield maximum Re depletion ages of 2.1 Gy (Pearson et al., 1994; Hoal et al., 1995; Pearson et al., in press). Analysis of sulfides from the same xenoliths gives the same early Proterozoic age for the oldest Re depletion ages (Griffin, personal communication, 2003). There is no evidence of an Archean age in any of the >60 peridotite xenoliths now analysed from off-craton areas surrounding the Kaapvaal craton (Pearson et al., 2002). Hence, the lithosphere beneath this region appears to be formed in the Proterozoic, at a broadly similar time to the overlying crustal basement (Hoal et al., 1995) and is at least 1 Gy younger than the adjacent Kaapvaal craton (Pearson et al., 1995; Pearson et al., in press).

The Louwrensia kimberlite is located about 20 km south of the town of Gibeon in a cluster of over 70 non-diamondiferous pipes (Fig. 1). This cluster of kimberlites is referred to as the Gibeon Kimberlite Province (e.g., Janse, 1975; Franz et al., 1996a). The kimberlites were emplaced into the Namibia Tectonic Province, a poorly exposed region west of the Kalahari Line that regionally includes ca. 1700- to 2000-My-old basement rocks (Hoal et al., 1995). Their eruption ages, near 70 My (Davies et al., 2001), are distinctly younger than the ~ 90-My-old Group 1 kimberlites on the craton. Contacts of the Louwrensia kimberlite against ca. 300-My-old Dwyka tillite and sedimentary rocks are poorly exposed but the pipe appears to be oval, approximately 60×100 m in size. The kimberlite and surrounding area are level except for a large prospecting pit. Wind erosion and surface wash left an original abundance of xenoliths on the surface but these have now been largely removed. The erosion surface surrounding the diatreme may have been relatively long-lived, reflecting the absence of sedimentation between deposition of Karoo (Carboniferous–Cretaceous) and Kalahari (post-emplacement Tertiary–Quaternary) sediments. Some calcrete is evident at the site.

There are pronounced similarities in lithology, but important differences in abundance, of mantle xenoliths between the Louwrensia kimberlite and xenolith suites from the Kaapvaal craton. Predominantly coarse-grained peridotites are common at Louwrensia and a number of deformed (porphyroclastic) perido-

tites have also been found. Clinopyroxene-free harzburgites are rare and only one low-Ca garnet harzburgite has been described from a Namibian kimberlite (Franz et al., 1996a). Spinel-facies peridotites are also rare and this paucity may reflect the presence of a thicker crust with a correspondingly thinner zone in the mantle where spinel-facies peridotites are stable. There are no eclogites among our suite of over 400 Louwrensia xenoliths. These distinctions were also found in peridotites from other kimberlites in the Gibeon area by Franz et al. (1996a,b).

A number of the larger peridotites that we have analyzed were collected in previous investigations and generously provided by colleagues. The origins and original specimen numbers of these xenoliths are identified in Table 1. Most of the original numbers were changed to FRB or PHN numbers for convenience in storing the data.

2. Petrography

Coarse-grained peridotites from Louwrensia have equant or equigranular, undeformed textures, with only 3 out of 22 having significant development of neoblasts. Neoblast development is estimated to be 10% or less. Coarse-grained olivine commonly ranges up to 1 cm, more in a few specimens. Pyroxenes are finer-grained than olivines and some orthopyroxenes are mantled with secondary clinopyroxene, exhibiting textures similar to those described for Udachnaya peridotites (Boyd et al., 1997). The habit of garnet is irregular, ranging from grains over 1 cm to 1.5 mm rounded crystals. The occurrence of kelyphite reaction rims around garnets is variable. Some garnets have no kelyphite whereas some have rims as thick as 0.5 mm. Garnets in some specimens are irregularly bordered by phlogopite. In one low-temperature peridotite, the garnet is replaced by round clots of polycrystalline mica that are mantled by clinopyroxene. Primary phlogopite is scarce at Louwrensia and in other peridotite xenoliths from kimberlites of the Gibeon Province (Franz et al., 1996a,b). The abundance and size of primary clinopyroxene grains varies widely but they are most commonly sparse and dispersed in grains a millimeter or less in diameter. Some are included in garnet. Spinel (chromite) is commonly dispersed in

Table 2
Bulk analyses of Louwrensia lherzolite xenoliths

(A) Low-temperature lherzolites

	FRB 1180	FRB 1181	FRB 1183	FRB 1625	FRB 1650	FRB 1651	FRB 1652	FRB 1680	FRB 1682	FRB 1684	FRB 1685
SiO_2	42.22	40.34	40.01	40.41	41.36	40.15	40.81	41.6	41.58	40.60	41.79
TiO_2	0.02	0.02	0.01	0.03	0.05	0.03	0.02	0.04	0.10	0.10	0.02
Al_2O_3	0.48	0.81	0.58	0.57	0.57	0.63	0.63	1.33	0.43	1.09	1.32
Fe_2O_3	3.33	4.66	4.65	4.10	3.60	4.54	4.11	4.55	5.47	4.37	4.35
FeO	3.78	2.94	3.65	3.53	3.92	3.10	3.88	2.90	2.95	3.37	3.57
MnO	0.11	0.10	0.09	0.11	0.11	0.11	0.11	0.10	0.10	0.11	0.10
MgO	41.60	41.27	42.30	43.05	43.31	42.25	42.80	39.57	40.10	40.08	39.47
CaO	0.37	0.54	0.38	0.25	0.53	0.40	0.47	0.84	0.35	1.18	1.04
Na_2O	0.07	0.05	0.07	0.00	0.01	0.00	0.01	0.03	0.02	0.04	0.03
K_2O	0.02	0.00	0.01	0.00	0.02	0.00	0.01	0.05	0.01	0.09	0.00
P_2O_5	0.00	0.00	0.00	0.02	0.02	0.02	0.02	0.03	0.02	0.03	0.02
LOI	7.66	9.20	8.34	8.00	6.69	8.65	7.33	8.77	8.80	8.75	8.11
Total	99.66	99.9	100.1	100.1	100.2	99.9	100.2	99.8	99.9	99.8	99.8
Rb	2	2	2	<1	1	1	<1	1	0.3	3	0.5
Sr	2	17	9	6	13	14	10	25	11	26	14
Y	<0.3	1.5	0.4	2.2	2.2	1.9	2.3	1.1	0.8	2	1
Zr	9	25	6	10	12	11	11	17	15	21	11
Nb								4.1	1.6	5.2	2.5
V	23	26	14	14	22	30	31	34	24	35	41
Ni	2360	2446	2744	2293	2353	2240	2460	2129	2519	2244	2398
Cr	2338	1872	1109	1920	2203	1904	1493	3085	1475	1846	2167
Ga				1.2	1.3	0.7	1.2	1	0.8	0.8	0.7
Cu				6	6	8	5	4.3	7.1	5.6	4.6
Zn				44	45	43	46	44	52	48	48
Ba	25	27	32	16	31	19	45	74	53	73	56
Co	111	102	119	104	104	105	114	105	116	104	111
Sc	6	5.2	2.8	5	6.4	5.6	6.5	12	3.7	7.5	9.9

	PHN 1686	PHN 1687	JJG 2513	JJG 2514	JJG 2517	PHN 5315	PHN 5316	PHN 5364	PHN 5365	PHN 6199	E11
SiO_2	41.26	42.34	43.57	41.48	41.64	41.1	43.86	42.42	42.07	42.16	43.02
TiO_2	0.11	0.03	0.01	0.02	0.01	0.10	0.03	0.05	0.03	0.04	0.03
Al_2O_3	0.51	1.04	1.87	0.25	1.43	1.90	2.50	0.61	1.40	1.07	1.61
Fe_2O_3	6.08	4.07	2.80	3.42	2.11	3.56	2.78	3.00	3.58	3.27	3.27
FeO	1.51	3.41	4.25	3.98	5.32	3.81	4.65	4.43	3.89	4.21	4.25
MnO	0.10	0.10	0.12	0.09	0.11	0.12	0.12	0.10	0.13	0.11	0.10
MgO	38.3	40.82	40.47	43.15	43.23	40.42	39.41	42.17	41.18	41.9	40.44
CaO	0.38	0.85	1.47	0.69	1.29	1.38	2.08	0.95	0.74	0.93	0.89
Na_2O	0.03	0.02	0.09	0.07	0.08	0.08	0.1	0.08	0.08	0.04	0.08
K_2O	0.05	0.02	0.01	0.03	0.01	0.04	0.01	0.01	0.04	0.05	0.02
P_2O_5	0.02	0.03	0.00	0.00	0.00	0.00	0.02	0.00	0.00	0.03	0.02
LOI	11.60	7.04	5.06	6.79	4.42	7.20	4.65	6.23	6.72	6.31	6.39
Total	99.9	99.8	99.7	99.9	99.7	99.7	100.2	100.1	99.7	100.1	100.1
Rb	2	0.3	2	1	3	3	1	3	3	2	1.5
Sr	20	20	11	10	12	32	18	22	16	22	14
Y	1.7	0.7	<0.3	<0.3	<0.3	2.2	2.5	0.3	1.8	2.6	2
Zr	16	13	10	15	11	7	9	10	8	12	11
Nb	3.2	4				4.3	2.5	3.7		4.4	2.9
V	26	37	42	26	38	43	55	21	27	26	40
Ni	2393	2272	2215	2459	2298	2175	2045	2297	2224	2243	2248

(continued on next page)

Table 2 (*continued*)

(A) Low-temperature lherzolites

	PHN 1686	PHN 1687	JJG 2513	JJG 2514	JJG 2517	PHN 5315	PHN 5316	PHN 5364	PHN 5365	PHN 6199	E11
Cr	1477	2465	2342	1250	2146	2668	3118	1334	2691	2219	3223
Ga	0.5	0.6				1.6	1.8	1.4		0.9	1.4
Cu	4.6	2.5				12	5	8		2	3
Zn	45	44				46	46	52		44	44
Ba	82	75	60	26	33	67	106	71	54	48	83
Co	106	103	102	113	109	103	103	111	102	93	107
Sc	4.3	9	10	1	9	11	13	2.4	10	8	

(B) High-temperature lherzolites

	FRB 1626	FRB 1681	FRB 1683	PHN 5304
SiO_2	39.71	40.80	40.77	40.24
TiO_2	0.12	0.10	0.13	0.13
Al_2O_3	0.87	0.89	0.54	1.29
Fe_2O_3	5.68	4.39	4.46	5.04
FeO	4.63	3.67	3.78	4.95
MnO	0.12	0.11	0.11	0.12
MgO	40.25	41.11	42.47	40.12
CaO	0.83	0.82	0.53	0.97
Na_2O	0.05	0.03	0.04	0.08
K_2O	0.01	0.01	0.00	0.01
P_2O_5	0.01	0.01	0.01	0.00
LOI	8.03	8.15	7.14	6.89
Total	100.3	100.1	100.0	99.8
Rb	1	0.4	0.4	3
Sr	8	6.4	5.1	2
Y	2.6	0.5	1	1.6
Zr	11	11	10	5
Nb	1.9	1.6	2.1	
V	43	44	32	43
Ni	2162	2269	2292	2206
Cr	1510	2245	2558	1760
Ga	2.1	1.6	0.9	
Cu	13	14	12	
Zn	69	48	49	
Ba	45	12	10	13
Co	103	110	113	117
Sc	5.7	7.3	6.6	7.4

Analyses were performed by X.R.F. at Franklin and Marshall College. Oxides in wt.%. Elements in ppm.

1–3-mm rounded grains that appear primary, but in several specimens it is intimately intergrown with garnet. In one coarse chromite-garnet lherzolite (FRB 1682), a grain of the LIMA mineral lindsleyite was found included in a garnet.

The deformed peridotites that we have analyzed are all lherzolites and are variably sheared (porphyroclastic) with neoblast proportions ranging from 10% to 90%. Neoblast grain size is predominantly in the range 0.1–0.3 mm, but in two specimens neoblasts range up to 1 mm. None are fluidized (porphyroclastic/mosaic). Garnets in the coarse peridotites are commonly rounded with diameters less than 5 mm and with thin to very thin kelyphite rims. Irregularly shaped garnet, however, is intergrown with rutile in PHN 5358. Primary spinel is absent in high-temperature peridotites from the Kaapvaal craton but has been found in one specimen from Louwrensia (FRB

1683; Table 1) and three from East Griqualand (Boyd and Nixon, unpublished).

3. Mineral compositions

Electron probe analyses for minerals in the large specimens of Louwrensia lherzolite were performed at the Geophysical Laboratory, Carnegie Institution, Washington with a JEOL JXA-8900L microprobe. A full set of analyses are available at the on-line Lithos database (http://www.XXXX). Homogeneity was checked by analyzing cores and rims of six grains of each mineral in each section, where possible. Analyses for a few of the small high-temperature peridotites were obtained some years ago with a MAC 400 instrument. The two data sets are concordant.

The olivines within coarse-grained peridotites have an average Mg number of 91.6 with a limited range of 91.1–92.2. They are consistently more Fe-rich than Kaapvaal lherzolite olivines which have an average Mg number of 92.6. Orthopyroxenes in the coarse-grained peridotites are systematically zoned to higher Al_2O_3 in the margins (Table 2). Franz et al. (1996b) have noted similar zoning, accompanied by increasing Ca and Cr towards the rims in peridotite xenoliths from other Gibeon kimberlites such as Hanaus and Anis Kubub. The magnitude of the zoning is variable but commonly about 10 relative percent and indicates significant recent disturbance to equilibrium. As pointed out by Franz et al. (1996a), increasing Ca and Al contents towards the rims of orthopyroxenes indicate a strong mantle-heating event, possibly related to thermal disturbances associated with kimberlite magmatism.

The compositions of garnets and diopsides in the coarse-grained lherzolites are similar to those in Kaapvaal cratonic peridotite counterparts. Garnets that coexist with chromite have Cr_2O_3 in the range 4–7 wt.% whereas those in chromite-free rocks predominantly range from 3 to 4 wt.% Cr_2O_3. Garnets from Louwrensia coarse-grained lherzolites are poor in TiO_2 with most containing less than 0.1 wt.% . The diopsides are calcic with average Na_2O in the range 1–2 wt.%, somewhat below values of 1.5 to >2 wt.% found in diopsides of suspected metasomatic origin by Van Achterberg et al. (2001) and Gregoire et al. (2003). Atomic proportions of Na are less than the sums of Al + Cr, most likely reflecting a Tschermaks

component. There is a wide range of composition for phlogopite but mica mantles on garnet tend to be richer in Fe and Ti (Hoal et al., 1995). Chromites have average Cr_2O_3 near 51 wt.%.

Garnets within the deformed lherzolites are markedly richer in TiO_2 and have lower Mg numbers than those in the coarse peridotites. These relations mirror compositional relations for high-temperature (commonly deformed) and low-temperature (coarse) peridotites from the Kaapvaal craton. The average Mg number for olivines from deformed Louwrensia peridotites is 90.6 with a range of 88.5–92.2, in comparison with an average of 91.1 for deformed, high-temperature peridotites from the southern Kaapvaal craton. Average TiO_2 in garnets from Louwrensia

Table 3

Mg numbers of olivines together with modal abundance (wt.%) for Louwrensia lherzolites calculated from the bulk and mineral analyses

Specimen	Mg no.	OLV	OPX	GAR	CPX	SPN	Total
Coarse, low-temperature							
FRB 1180	91.7	68.63	29.84	0.69	0.66	0.21	100.03
FRB 1181	91.8	73.56	21.23	3.40	1.52		99.71
FRB 1183	91.6	76.78	18.82	2.26	0.90	0.01	98.77
FRB 1625	91.8	77.41	19.43	1.93	0.36	0.25	99.38
FRB 1650	91.8	76.65	19.59	1.80	1.49	0.15	99.68
FRB 1651	91.9	77.20	18.97	2.19	1.09	0.21	99.66
FRB 1652	91.6	76.90	19.08	2.22	1.47	0.10	99.77
FRB 1680	91.7	63.02	28.29	5.72	2.22	0.12	99.37
FRB 1682	91.3	64.68	33.25	0.31	0.82	0.05	99.11
FRB 1684	91.6	68.41	22.00	4.13	4.78		99.32
FRB 1685	91.3	63.29	27.70	5.15	3.48		99.62
FRB 1686	92.2	57.40	39.47	0.45	0.79	0.01	98.12
FRB 1687	91.7	65.07	28.04	3.91	2.61		99.63
JJG 2513	91.3	62.66	25.46	7.54	4.76		100.42
JJG 2514	91.5	75.91	20.84		2.92	0.15	99.82
JJG 2517	91.2	78.74	11.34	6.01	4.60		100.69
PHN 5315	91.6	69.80	17.19	8.34	4.83		100.16
PHN 5316	91.1	59.55	23.07	10.21	7.41		100.24
PHN 5364	91.2	70.77	24.51		4.57		99.69
PHN 5365	91.6	65.90	25.94	5.10	1.94	0.36	99.24
PHN 6199	91.6	69.85	22.85	3.78	3.26	0.16	99.90
E-11	91.9	60.40	29.97	6.92	1.91		99.20
Average	91.6	69.22	23.94	4.12	2.65	0.15	100.08
Sheared, high-temperature							
FRB 1626	88.5	77.53	16.31	2.77	2.69		99.30
FRB 1681	91.2	71.32	23.56	2.48	2.11		99.47
FRB 1683	90.9	77.61	20.41	0.83	1.11	0.13	100.09
PHN 5304	88.8	72.93	18.57	4.79	2.81		99.10

OLV = olivine, OPX = orthopyroxene, GAR = garnet, CPX = clinopyroxene, SPN = spinel.

deformed rocks is 0.55 wt.%. In contrast to the coarse peridotites, orthopyroxenes in the deformed Louwrensia peridotites are not zoned in Al. Some of the garnets, however, exhibit zoning in Ti, Cr, and Fe (Table 3) as well as differences between grains. The diopsides contain somewhat higher Ca, and hence have lower equilibration temperatures, than has been found for some occurrences on the craton (e.g., Nixon and Boyd, 1973).

4. Bulk compositions

Bulk analyses for 26 of the large Louwrensia peridotites (Table 2) were performed at Franklin and Marshall College using procedures described in Boyd et al. (1993, 1997). Compositional variations in peridotite xenoliths are most easily evaluated and compared by use of modes (Table 3), calculated from the bulk and probe analyses (see Boyd and Mertzman, 1987), as well as by using major element oxide variation.

Olivine contents in coarse-grained Louwrensia lherzolites vary from 57 to 79 wt.%, similar to, but

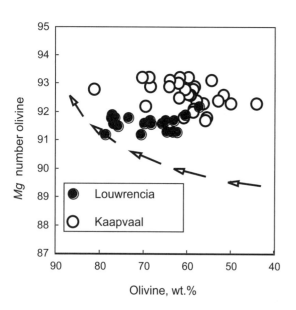

Fig. 2. Modal olivine (wt.%), calculated from bulk and electron probe compositions, plotted against the *Mg* number of olivine. The arrows describe an oceanic melt depletion trend (Boyd, 1989). The Kaapvaal data are from Boyd and Mertzman (1987), Boyd et al. (1993) and Boyd (unpublished).

not as large as the range of 44–81 wt.% in Kaapvaal cratonic lherzolites (Fig. 2). There is comparable variation in orthopyroxene contents for Louwrensia (11 to 40 wt.%) and for the coarse-grained peridotites from the Kaapvaal craton (11 to 44 wt.%). In contrast to the orthopyroxene-rich nature of the coarse peridotites from Louwrensia, Franz et al. (1996a) suggested that the coarse peridotites from the Gibeon Townsland 1 pipe were orthopyroxene-poor compared to Kaapvaal craton coarse peridotites. They also suggested that the deformed peridotites from this pipe were orthopyroxene-rich, more comparable to the Kaapvaal coarse peridotites. We note however, that the samples used in the study of Franz et al. (1996a) were only 2 to 10 cm in size and that this, combined with the modal point-counting method they employed, may have biased their results (see Boyd and Mertzman, 1987 for a detailed discussion of these effects). Until similar, large analytical specimens are available from Gibeon Townsland, we take the modal mineralogy results obtained at Louwrensia to be more representative of the lithospheric mantle beneath this region.

The wide ranges of modal olivine and orthopyroxene at restricted *Mg* number are incompatible with an origin as unmodified residues. Whatever process of segregation (Boyd et al., 1997) or metasomatism (Kelemen et al., 1998) has caused the development of these relations in the Kaapvaal root may also have operated in some degree in the formation of the Namibian lithosphere.

Both coarse and deformed Louwrensia lherzolites are richer in bulk Fe than respective Kaapvaal craton lherzolites (e.g., Fig. 3). The comparison is best made with analyses that have been normalized to exclude variable loss-on-ignition. The normalized values of total Fe as FeO (wt.%) are 7.77 (coarse) and 9.35 (deformed) for Louwrensia and 6.30 and 7.88 for Kaapvaal peridotites, respectively. The Louwrensia coarse peridotites have lower FeO at higher MgO than xenoliths from Vitim or orogenic peridotites such as the Beni Bousera peridotite (Fig. 3; Pearson et al., in press). Fe has been introduced in many peridotite xenoliths included in kimberlites during the final stages of eruption (e.g., Boyd et al., 1997). The amount of Fe introduced can be estimated from the difference between FeO in a bulk xenolith analysis and whole-rock FeO calculated from the probe anal-

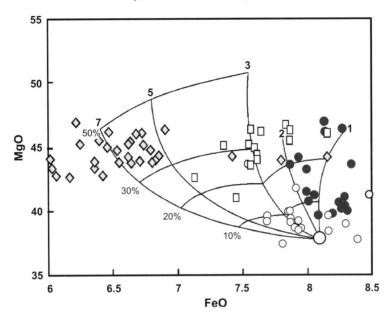

Fig. 3. Bulk rock MgO–FeO relations (anhydrous basis) for Kaapvaal craton low-temperature peridotites (diamonds; data from Boyd and Mertzman, 1987; Boyd et al., 1993; Boyd, unpublished; Pearson et al., 2004), Louwrensia (squares), Vitim (open circles) and Beni Bousera (solid circles) peridotites. Data for Vitim and Beni Bousera samples from Pearson et al. (2004). The melting grid outlines the expected composition of melt residues and is contoured for percentage melt extracted (in 10% intervals) at different melting pressures, in GPa (from 1 to 7 GPa; Walter, 1998). The effect of small amounts of introduced Fe in kimberlite-derived peridotites (e.g., Boyd et al., 1997) results in a small decrease in the apparent pressure of melting and degree of melting.

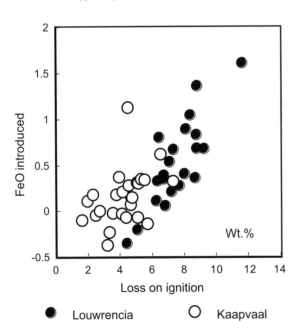

Fig. 4. "FeO introduced" against loss on ignition. "FeO introduced" is the difference between total Fe as FeO in a bulk analysis and FeO calculated from the electron probe analyses and mode.

yses and mode calculated excluding FeO. This difference is commonly positive and ranges to over 1 absolute wt.%. A plot of "introduced FeO" against loss-on-ignition (Fig. 4) has a pronounced positive correlation very similar to that found for Udachnaya peridotites (Boyd et al., 1997). The correlation is stronger for both Louwrensia and Udachnaya peridotites than for those from the Kaapvaal because the former have higher ranges of loss-on-ignition. These correlations could be taken as evidence that the introduction of at least some of the Fe is associated with serpentinisation of the peridotite or perhaps some associated or later form of hydrous alteration. Higher FeO in Louwrensia peridotites relative to Kaapvaal peridotites, however, is not entirely secondary, because the *Mg* number of Louwrensia olivines is homogeneous and systematically lower than that for cratonic peridotites. This difference is clearly not the product of a late-stage introduction and it is likely that the higher FeO of the Proterozoic Louwrensia peridotites compared to the Archean Kaapvaal peridotites is a feature of their igneous petrogenesis.

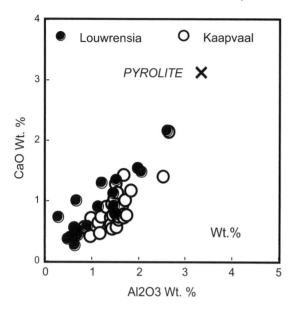

Fig. 5. A comparison of CaO versus Al_2O_3 in low-temperature, coarse peridotites from Louwrensia (solid points) and the Kaapvaal craton (open circles). Sources of Kaapvaal data are those listed for Fig. 2.

Proportions of clinopyroxene and garnet in Louwrensia and Kaapvaal suites overlap in the range up to 10 wt.%. On average, the Louwrensia lherzolites are a

little richer in CaO but there are a few with less combined CaO and Al_2O_3 than even the Kaapvaal low-Ca harzburgites. Peridotites in both suites are strongly depleted in CaO and Al_2O_3 relative to pyrolite (Fig. 5), but their overlapping trends could reflect enrichment events following initial melt depletion. In general, cratonic peridotites have lower CaO/Al_2O_3 ratios than chondrite and also off-craton samples (see compilation plot in Pearson et al., 2003). CaO/Al_2O_3 ratios in Louwrensia bulk peridotites are similar to cratonic peridotites. The cause of this is not understood. In contrast to the fairly invariant trend of bulk Cr versus Al_2O_3 in cratonic and non-cratonic peridotites, the relatively small Louwrensia data set shows a general decrease in Cr with decreasing Al_2O_3 (Fig. 6).

5. Trace element compositions of diopside and garnet

Diopside and garnet from three coarse-textured Louwrensia peridotites (JJG2513, PHN6199 and E-11) have been analysed using an ion-microprobe to determine trace element concentrations (Table 4).

Table 4
Trace element data (ppm) for diopside and garnet in Louwrensia peridotites

	PHN 6199	PHN 6199	E-11	E-11	JJG 2513	JJG 2513
	CPX	GAR	CPX	GAR	CPX	GAR
La	9.94	0.043	13.9	0.054	6.55	0.018
Ce	37.6	0.388	42.5	0.576	23.6	0.137
Nd	34.2	3.88	32.7	0.984	15.5	0.175
Sm	5.74	2.29	5.37	0.316	2.96	0.076
Eu	1.52	0.828	1.24	0.119	0.556	0.041
Gd	2.73	2.04	2.42	n.d.	1.10	0.200
Dy	1.25	1.05	1.17	0.68	0.59	0.345
Ho	0.168	0.145	n.d.	n.d.	0.073	0.117
Er	0.277	0.443	0.423	0.59	0.136	0.788
Yb	0.224	0.773	0.242	0.678	0.100	1.48
Ti	135	39.7	196	194	105	80.9
Sr	311	0.191	273	0.129	183	0.06
Y	1.53	3.71	1.82	3.09	1.33	3.23
Zr	55.2	24.3	n.d.	4.12	39.1	1.27
Nb	1.70	0.374	0.943	0.802	0.649	0.205
Ba	2.23	0.018	0.26	0.041	0.065	0.014
Yb*	0.037		0.052		0.116	

CPX = clinopyroxene, GAR = garnet. n.d. = not determined. Yb* = bulk rock Yb calculated using modes in Table 3.

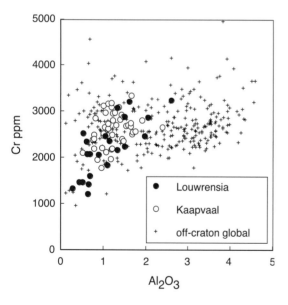

Fig. 6. Whole rock Cr versus Al_2O_3 in low-temperature Louwrensia peridotites compared to Kaapvaal craton low-temperature peridotites and a global compilation of off-craton peridotite xenoliths (Canil, this volume; Pearson et al., 2003).

Data for JJG2513 and PHN6199 have been presented by Pearson and Nowell (2002). These data are in addition to those obtained by Jones (1984) for bulk separates. Diopside and garnet from the coarse Louwrensia xenoliths show similar trace element characteristics to those in Kaapvaal coarse lherzolites (Shimizu, 1975; Harte et al., 1993; Pearson and Nowell, 2002; Gregoire et al., 2003; Simon et al., 2003a,b; this report, Fig. 7). Diopsides are consistently LREE-enriched, with La and Ce concentrations over $10 \times$ chondritic. Levels of LREE enrichment and LREE/HREE ratios lie between those of clinopyroxenes from the MARID and those from phlogopite–ilmenite–clinopyroxene (PIC) xenoliths from the Kimberley area analysed by Gregoire et al. (2002).

Garnets from the Louwrensia lherzolites show highly variable REE patterns, as found for concentrate garnets analysed by Hoal et al. (1994). All garnets are

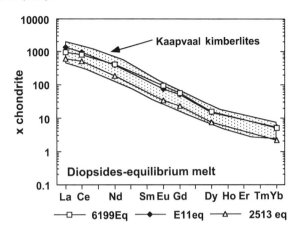

Fig. 8. Chondrite-normalised rare earth element profiles of calculated equilibrium liquids to Louwrensia diopsides compared to Kaapvaal and Namibian kimberlites (shaded field; data from Pearson, unpublished).

LREE depleted to varying degrees (Fig. 4) but HREE/MREE varies considerably from >1 for JJG2513 to $\ll 1$ for PHN6199. PHN6199, together with two garnets reported by Jones (1984) and several garnets from concentrate reported by Hoal et al. (1994), shows "sigmoidal" REE patterns of the type commonly observed in sub-calcic garnets (e.g., Shimizu et al., 1999). The Louwrensia garnets, however, are all Ca-saturated. Hoal et al. (1994) showed that more fertile garnets (as a function of Fe) showed normal, LREE-depleted garnet REE patterns.

Calculated REE patterns of melt compositions in equilibrium with the diopsides analysed from our Louwrensia peridotites are very LREE enriched and broadly kimberlitic in composition (Fig. 8). This was also found by Hoal et al. (1994) for concentrate garnets. The level of trace element equilibrium between garnet and diopside can be assessed by reference to partition coefficient ratios, e.g., Zack et al. (1997). This approach has been used by Simon et al. (2003a) to show that incompatible trace elements in some garnets coexisting with diopside in lherzolite peridotites from N. Lesotho have not equilibrated. The same result is obtained irrespective of the choice of partition coefficient data sets. Partitioning of Sr between garnet and clinopyroxene in the Louwrensia samples is close to equilibrium and is consistent with the high-diffusion coefficients for divalent cations (Simon et al., 2003a). In contrast, partitioning of

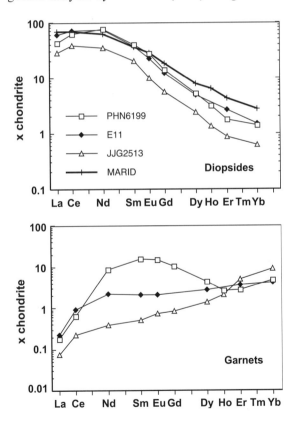

Fig. 7. Chondrite-normalised rare earth element profiles of diopside and garnet in Louwrensia peridotites analysed by ion probe. For analytical details, see Pearson and Nowell (2002). Normalising values from McDonough and Sun (1995).

some LREE, most HREE and particularly HFSE such as Hf, diverge significantly from equilibrium values. This indicates a lack of equilibrium for these elements between garnets and diopsides as is further illustrated by the sinuous garnet REE profile for PHN6199 (Fig. 7).

6. Temperature/depth estimates

There are significant inhomogeneities for some major, minor and trace elements in minerals of the Louwrensia peridotites, documented here and previously by Franz et al. (1996a,b). However, we believe that these have not substantially affected the estimated temperatures and depths of origin of the coarse-textured rocks. Failures of equilibration, or the inadequacies in thermobarometer formulations may, however, have affected P/T estimates made for the deformed suite, as discussed hereafter. Systematically higher Al in the rims of orthopyroxenes from the coarse peridotites affects pressure estimates for individual specimens by an amount that is commonly less than 5%. The aggregate effect on a geotherm plot is small, although a rim plot is slightly shallower than a core plot (Fig. 9). Significant disequilibrium exists for some trace elements between garnet and diopside in the Louwrensia lherzolites but divalent cations such as strontium appear to have equilibrated. The much more rapidly diffusing divalent species such as Mg and Fe are likely to have equilibrated also. Since temperature estimates are mainly dependant on divalent major element exchange, we therefore suggest that the calculated temperatures are likely to closely represent equilibria. For example, this appears to be the case for Kaapvaal peridotites that are out of isotopic disequilibrium for Nd, but which give reasonable temperature estimates using Fe–Mg exchange thermometry (Simon et al., 2003a,b).

The compositions of either cores or rims of crystals in the Louwrensia lherzolites can be used for thermobarometry. Smith (1999) suggests that in general, rims compositions are the most likely to have approached inter-mineral equilibrium. While this may be the case for some xenolith suites, there are two reasons why we prefer to use core compositions in the case of the Louwrensia peridotites. Firstly, to maintain compatibility with the previous

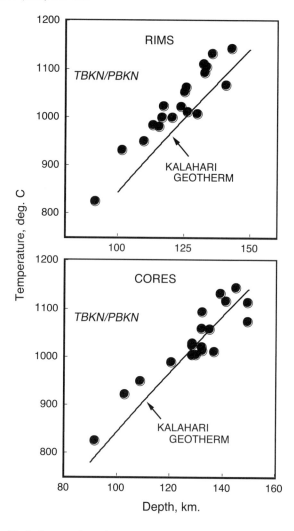

Fig. 9. A comparison of temperature–depth estimates calculated for cores and rims of pyroxene and garnet grains in coarse, low-temperature Louwrensia lherzolites. The TBKN/PBKN thermobarometer of Brey and Kohler (1990) was used. The Kalahari geotherm is an average for the Kaapvaal craton (Rudnick and Nyblade, 1999).

P/T estimates of Mitchell (1984) for Louwrensia peridotites and of Franz et al. (1996a,b) for other Gibeon peridotites. Secondly, as noted by Franz et al. (1996a,b) diopsides in these peridotites have rims that appear heavily reacted with a metasomatic fluid and contain many micro-inclusions, making accurate probe analyses almost impossible in some circumstances. This observation and the resulting likely lack of equilibrium of such rims, together with Al-zoning in orthopyroxene leads us to believe

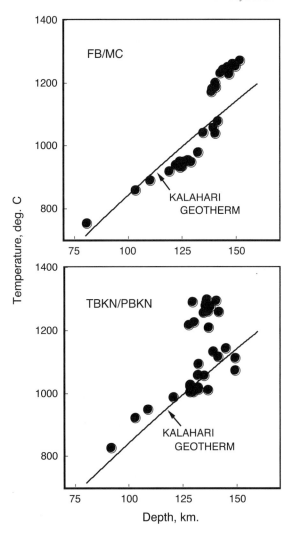

Fig. 10. A comparison of temperature–depth estimates calculated with the FB/MC thermobarometer (Finnerty and Boyd, 1987) and the TBKN/PBKN thermobarometer of Brey and Kohler (1990) for both high- and low-temperature Louwrensia lherzolites Core compositions of pyroxenes and garnets were used. The Kalahari geotherm is an average for the Kaapvaal craton (Rudnick and Nyblade, 1999).

that the core compositions best reflect equilibrium conditions at depth, prior to disturbance by any heating event associated with host magmatism. As such, they will be used in the subsequent discussion. Nevertheless, the conclusions of this study are not significantly affected by choice of either core or rim compositions for thermobarometry and plots (Figs. 9 and 10). Calculations of results obtained

using both sets of data are provided in an on-line database.

The Kalahari geotherm (Rudnick and Nyblade, 1999) is an average best-fit mantle geotherm for the Kaapvaal craton and is used as a reference curve for the temperature/depth plots in Figs. 9–11 rather than the traditional 40 mW/m^2 geotherm of Pollack and Chapman (1977). The latter curve is calculated with an assumed heat production below 120 km that is unreasonably high for low-temperature peridotites in a craton root (Rudnick and Nyblade, 1999). As a result, the Pollack and Chapman geotherm has too great a curvature below that depth. The Kalahari geotherm is an average cratonic geotherm based on both high- and low-temperature peridotite data for xenoliths from the Lesotho, Kimberley and Letlhakane kimberlites, calculated with the TBKN/PBKN thermobarometer of Brey and Kohler (1990).

The Louwrensia peridotite suite can be classified into two groups on the basis of their equilibration temperatures, in the same way that this can be done for peridotites hosted in Kaapvaal craton kimberlites.

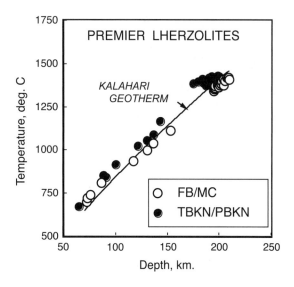

Fig. 11. Temperature–depth estimates for both high- and low-temperature garnet lherzolites from the Premier kimberlite, located in the central part of the Kaapvaal craton. The FB/MC thermobarometer is from Finnerty and Boyd (1987). The TBKN/PBKN thermobarometer is from Brey and Kohler (1990). The Kalahari geotherm is an average for the Kaapvaal craton (Rudnick and Nyblade, 1999).

The coarse peridotites form a low-temperature group, with equilibration temperatures < 1100 °C. The deformed peridotites form a high-temperature group with equilibration temperatures generally above 1200 °C (Figs. 8 and 9; on-line *P/T* data base). Estimates of equilibrium conditions for the low-temperature Louwrensia lherzolites form well-defined geotherms in the depth range of 75–150 km. Plots using either the TBKN/PBKN thermobarometer of Brey and Kohler (1990) or the FB/MC model system approach advocated by Finnerty and Boyd (1987) agree relatively well (Fig. 10). This is also true for plots made with the O'Neill/Wood thermometer (O'Neill and Wood, 1979) or with TA97/NGmod thermobarometer combination (Taylor, 1998; not plotted). Moreover, all such temperature–depth plots for the Louwrensia low-temperature peridotites are remarkably close to the average Kalahari cratonic geotherm calculated by Rudnick and Nyblade (1999).

The high-temperature Louwrensia peridotites plot at systematically higher temperatures, in clusters of varying depth and shape depending on which thermobarometer is used (Fig. 10). The slope of the high-temperature anomaly is relatively gentle in a plot made with FB/MC, but with TBKN/PBKN points for the high-temperature rocks overlap the depth interval of the low-temperature suite (Fig. 10). The thermal anomalies exhibited in Louwrensia plots are in some cases more extreme if thermobarometers other than FB/MC or TBKN/PBKN are used. Similar anomalies for Louwrensia high-temperature lherzolites have previously been noted in plots made by MacGregor (1975), Mitchell (1984) and Franz et al. (1996a,b).

Aspects of the origin of high-temperature peridotites are not well understood as yet, but it can be reasonably argued that they have been metasomatised and erupted from near the base of the lithosphere (e.g., Griffin et al., 2003). By comparison, peridotites from the cratonic Premier kimberlite (Fig. 11) do not show a thermal anomaly or inflection, despite having sampled a wide range of variably deformed and metasomatised mantle (700–1400 °C, 75–200 km) in the vicinity of the huge Bushveld Igneous Complex emplaced some 800 My earlier. This observation, together with the elemental zoning noted above and by Franz et al. (1996a,b) leads us to conclude that the anomalies recorded in high-temperature Louwrensia peridotites may reflect a heat-ing event affecting the lithosphere prior to entrainment of the peridotites by the kimberlite.

7. Discussion

7.1. Significance of the bulk compositional variations

The high bulk rock MgO, high olivine *Mg* numbers, and low bulk rock Al_2O_3 and CaO of the Louwrensia peridotites relative to fertile mantle clearly indicate an origin as residues of partial melting. Although their mean orthopyroxene content (24 wt.%) could be generated by high-pressure partial melting at *Mg* numbers >91 (e.g., Walter, 1998), a number of the peridotites have orthopyroxene contents greatly exceeding this value (up to 40 wt.%; Table 3). Our current understanding of the phase equilibria of mantle melting precludes such high orthopyroxene contents from being the product of partial melting at any pressure and an origin via either crystal segregation (Boyd et al., 1993; Boyd et al., 1997) or melt-solid reaction (e.g., Kelemen et al., 1998) is more likely. Further indications that the Louwrensia peridotites cannot be explained by a simple single-stage melting model come from the wide variation of bulk Al_2O_3 and CaO contents at near constant *Mg* number, together with the ubiquitous LREE-enriched nature of the diopside and complex garnet REE patterns (Figs. 5 and 7). As with many well-documented examples of mantle metasomatism of on-craton xenoliths, some/much of the garnet and diopside may have been introduced after initial depletion (e.g., Shimizu, 1999; Pearson et al., 2001; Gregoire et al., 2003, Simon et al., 2003a,b). Diopside–phlogopite rims on some garnets further attest to different stages of mineral replacement over time.

Despite these complications, it is important to attempt to constrain further the melting environment of these rocks and what they may tell us about the processes of lithosphere generation through time. The FeO–MgO relationships shown in Fig. 3 indicate a distinct difference between Archean cratonic peridotites and the Proterozoic Namibian peridotites. It is possible that secondary introduction of Fe into the bulk rock may cause some of this difference. However, there is also a systematic difference in olivine

Mg number (Hoal et al., 1995; Pearson, 1999) that cannot reflect secondary processes. The Namibian peridotite whole rock analyses lie on experimentally determined melting curves for relatively low pressure melting (3 GPa and less; Fig. 3), at higher degrees of melting than most peridotites from off-craton areas such as Vitim or orogenic massifs such as Beni Bousera (Fig. 3). This relatively low pressure of melt extraction implied by the whole rock data is sensitive to Fe addition, but agrees with the recent conclusions of Canil (2004, this volume), that most lithospheric peridotites formed in low-pressure melting environments where garnet was not stable on the solidus.

$Cr-Al_2O_3$ relationships can also be used to constrain the likely pressure of melt extraction. Canil (in press) interprets the relatively invariant Cr concentration at varying Al_2O_3 of most peridotite xenolith suites world-wide as reflecting melt extraction below 3 GPa, where $D_{Cr/Al}$ between melt and residue at solidus temperatures is high. The majority (75%) of the Louwrensia peridotites have Cr contents >2000 ppm, in keeping with those of cratonic and non-cratonic peridotites (Fig. 6). Six samples, with the lowest Al_2O_3 (< 1 wt.%) have Cr contents substantially below this level. The sparse data appear to define a continuous trend that could be interpreted as melt extraction at pressures above 4 GPa where $D_{Cr/Al}$ is low, leaving less Cr in the residue. Alternatively, low Cr/Al could be due to the exhaustion of spinel at high degrees of melting at $P < 3$ GPa and this could explain the sub-vertical nature of the trend on Fig. 6 below 1% Al_2O_3. One problem with this interpretation is that one of the low Cr samples (JJG2514) is a chromite–harzburgite. Cr–Al relationships in the present database are thus equivocal concerning the pressure of melting.

An additional approach, employed by Kelemen et al. (1998) and Canil (in press) is to look at either Ca–Yb or Al–Yb relationships, on the basis that residual garnet buffers Yb to high levels until phase exhaustion. Of the Louwrensia sample suite, the bulk rock Yb contents can be determined for only three samples in this study and seven from Jones (1984). This can be done by mass balance using the modal abundances in Table 3 together with trace element determinations of the phases (Table 4), assuming negligible Yb in OPX (those analysed were below the 5 ppb detection limit). With this relatively small database ($n = 10$) that spans

a representative range of CaO (0.69 to 2.4 wt.%) and Al_2O_3 (1.0 to 2.6 wt.%) contents, the relationship between these two elements and Yb clearly indicates that the Louwrensia peridotites were unlikely to have experienced melt extraction at pressures higher than 3 GPa. This is in accord with the results obtained for cratonic and off-craton peridotite suites (Kelemen et al., 1998; Canil, in press). The present mineralogies of the Louwrensia and other garnet-bearing peridotites is then a function of their subsequent metamorphic evolution. Future detailed investigation of the affects of metasomatism on these relations must be carried out to confirm these findings.

The variation in average olivine *Mg* number and bulk rock *Mg* number is likely to reflect a primary igneous process. The mean *Mg* number of the Louwrensia peridotites or their olivines is higher than that for massif-type and abyssal peridotites but lower than for cratonic peridotites (Pearson et al., 1994; Hoal et al., 1995; Pearson, 1999), i.e., there is a variation of *Mg* number with age of melting. If we accept the reasoning of Canil (in press), that all these peridotites are the residues of melting at or below 3 GPa then pressure effects on Fe partitioning will not be the main cause of the variation in *Mg* number and the dominant cause is varying degrees of melt extraction. Hence, it is possible that the variation in melt depletion as a function of time simply records the cooling of the Earth's mantle through time and that melt residues with *Mg* numbers equal to the Louwrensia peridotites or higher, will not be the normal modern-day product of low-pressure melting of the mantle (Pearson et al., 1994; Pearson et al., 2003).

7.2. Trace element variations, metasomatism and evidence for new mineral growth

CaO/Al_2O_3 ratios in Louwrensia bulk peridotites are similar to cratonic peridotites and distinct from off-craton peridotites erupted by alkali basalts (Pearson et al., 2003). Although the reasons for this are not understood, the similarity of Ca/Al relations between the Louwrensia and Kaapvaal peridotites suggests that they have been affected by similar processes, either relating to melt-depletion, or to metasomatism. Strong evidence that the Louwrensia and Kaapvaal lithospheric mantle have been subject to similar metasomatic processes comes from the observation of a

LIMA mineral in one of the Louwrensia peridotites (Table 1) and the very similar REE patterns of the garnets and diopsides from both regions (Jones, 1984; Hoal et al., 1994; Fig. 7).

Calculated melts in equilibrium with diopsides from the Louwrensia peridotites are highly LREE-enriched and comparable in composition to kimberlitic and other alkalic melt compositions, as has been noted for cratonic lherzolites (Shimizu, 1999; Simon et al., 2003a,b). Both trace element and isotopic systematics in the diopsides from on- and off-craton low-T peridotites indicate that much or all of the diopside in low-T peridotites could have formed from a metasomatic melt with an alkaline affinity such as kimberlite (Van Achterberg et al., 2001; Simon et al., 2003a,b). Gregoire et al. (2003) find independent support for this idea from the similarity of the trace element characteristics of Kimberley garnet lherzolites to diopsides within the metasomatic MARID and PIC suites. The metasomatic rocks are thought to be the crystallized products of kimberlitic melts (e.g., Gregoire et al., 2002). The partitioning of some trace elements between diopside and garnet indicates that the two minerals had not fully equilibrated with each other at the time of sampling by the kimberlite, at least for trace elements with $+4$ or $+5$ valence. The diffusion coefficients for divalent cation species such as Mg and Fe are likely to be orders of magnitude faster than the high-valence species and therefore more likely to represent equilibrium partitioning relationships between the two minerals. The lack of equilibration of highly charged and highly incompatible cations between garnet and diopside together with Nd isotopic disequilibrium for some Louwrensia coarse peridotites (e.g., JJG2513; Pearson and Nowell, unpublished data) and low-T Kaapvaal peridotites in general (Simon et al., 2003b; Pearson et al., 2003) support a relatively recent introduction of the diopside.

An important feature that must be explained by a model invoking recent introduction of at least diopside into the Namibia lherzolites is how there can be a difference in major element compositions of the diopsides in lherzolites and the diopsides in MARID and PIC lithologies, if the latter are also introduced by a similar melt. It may be that the difference has arisen because the major element composition of introduced diopside (or garnet) in lherzolites is buffered by the depleted Mg-rich nature of the host peridotite. In vein-like metasomes such as MARID rocks, the melt/rock ratio is considerably higher than for the percolative flow (e.g., Harte et al., 1993) that is believed to have crystallized diopside in lherzolites. Hence, major element compositions of MARID diopsides are more reflective of their parental melt than lherzolitic diopsides. In contrast, the trace element signature of diopside crystallising in lherzolites is not substantially affected by the depleted host rock and hence better reflects the nature of the parental metasomatic melt. Alternatively, it may be that multiple episodes of kimberlitic or alkalic melt infiltration in the mantle have resulted in subtle mineral compositional differences. An example may be the igneous xenoliths from Premier (Hoal, 2003) that differ from the host kimberlite but were clearly present in the vicinity of the conduit in order to have been sampled.

Highly fractionated, sigmoidal REE patterns in some Louwrensia garnets are similar to those observed in low-Ca garnets from cratonic peridotites. These have been interpreted as Archean features (Richardson, 1990), but their occurrence in Louwrensia lherzolites is evidence that they may be at least as young as Proterozoic. Burgess and Harte (in press) suggest that the sigmoidal REE patterns can be explained as having been caused by interaction with melt that has fractionated the Cr-poor megacryst assemblage. Hoal et al. (1994) suggested interaction with melts of kimberlitic affinity. Studies on other Kaapvaal peridotites support this idea (Simon et al., 2001, 2003a) and there is a possibility that the imposed REE patterns, in some instances, might be of a recent origin (Shimizu, 1999; Shimizu et al., 1999). This type of metasomatic process could be accompanied by introduction of additional garnet. Further combined trace element and isotope work on the Louwrensia rocks is required to address this possibility.

7.3. Thermal structure of the lithosphere and its evolution

It is possible to compare aspects of the thermal structure and evolution of the Archean and Proterozoic lithospheres beneath southern Africa by comparison of thermobarometry results from peridotite xenoliths. As noted previously (Finnerty and Boyd, 1987; Pearson et al., 1994; Bell et al., 2003), low-temperature peridotites from both Archean and Proterozoic lithospheric terranes record conductive geotherms that are characteristic of cratonic heat flow (circa 40–45 mW m^{-2}; Rudnick and Nyblade, 1999), indicative of cooling of the shallow lithosphere since the last major thermal disturbance. In contrast, some peridotites erupted by kimberlites to the South of the Kaapvaal craton, in the Karoo, record anomalous and disturbed P/T arrays and shallow lithospheric depth (Finnerty and Boyd, 1987; Bell et al., 2003) and this seems consistent with the most recent seismic imaging of lithospheric structure (e.g., James et al., 2001).

At Louwrensia, at the time of kimberlite eruption, the thickness of lithospheric mantle estimated from the maximum depth of derivation of coarse, low-temperature peridotites is approximately 140–150 km. On the Kaapvaal craton prior to 90 My, the thickness was in the range 175–200 km. Both lithospheres, however, had thermal gradients in the Mesozoic that were close to the average Kalahari geotherm (Figs. 8 and 9). Under steady-state conditions, a thinner conductive lithosphere would be expected to have a steeper T/P slope. The fact that the low-temperature Louwrensia and cratonic gradients are nearly the same implies that the Louwrensia lithosphere was approximately the same thickness as that of the craton a relatively short time before kimberlite eruption. Bell et al. (2003) have reached a similar conclusion. These observations are interpreted to mean that the Namibian lithosphere was thermally eroded and thinned prior to kimberlite eruption. The time interval must have been sufficiently brief in order that the shallower portion of the geotherm be unaffected. This thermal disturbance is likely reflected in the zonation of minerals in coarse peridotite (increasing Al in orthopyroxene rims) and elevated temperatures of the deformed, "high-temperature" Louwrensia peridotites. We interpret the thermobarometry results from Louwrensia, together with recent fission-track studies of the uplift history of this area (Brown et al., 1998), as indicating that the Namibian lithosphere was thermally eroded.

Other processes have been invoked to explain the complex signatures recorded in Namibian peridotite xenoliths. Tectonic transport of Kaapvaal lithospheric mantle slices during plume emplacement (Franz et al., 1996a) would have to involve extensive modification of the mantle to lower Mg numbers and does not explain the observation that the peridotite ages are all distinct from the Kaapvaal lithosphere. The consistent observation of post-Archean Re-depletion ages in every xenolith suite outside the accepted craton margin (Pearson et al., 2002) also refutes this suggestion.

Local magmatism prior to kimberlite eruption could account for small-scale disturbances, particularly as recorded by the deformed peridotites at Louwrensia. Such small-scale igneous events can be accommodated in models for the generation of deformed high-temperature peridotites such as those put forward by Gurney and Harte (1980). In contrast, Bell et al. (2003) propose that the entire region of southern Africa experienced a protracted heating event in the Mesozoic that advected heat into both the lithosphere and crust. This heating event is proposed to have been affecting the Namibian lithosphere shortly before kimberlite eruption at 70 Ma and may have been related to plume/supercontinent breakup. No thermal erosion is envisaged by these authors.

The widespread nature of alkaline igneous activity in southern Namibia, where outcrops of Group I kimberlites, melnoites and parakimberlites occur over an area of 180,000 km^2, strongly supports the presence of a regional igneous event that could thermally modify a thick, refractory lithosphere and culminate in the emplacement of the kimberlites pipes. Whether this activity lead to a reduction in lithospheric thickness by thermal erosion must await confirmation from other studies.

8. Conclusions

Re–Os isotope systematics indicate that the Louwrensia peridotites formed by melt depletion in the early Proterozoic (~ 2.1 Gy ago; Pearson et al., 1994; Hoal et al., 1995; Pearson et al., in press), consistent with their emplacement 400 km outside the accepted craton boundary as denoted by crustal rocks and geophysical signatures (Hoal et al., 1995). The textures, mineral and bulk compositions of the Louwrensia garnet lherzolite xenoliths are far more similar to peridotites from the Kaapvaal craton than to samples of Phanerozoic lithosphere sampled as xenoliths in basalt, or as abyssal and alpine peridotites. Low-temperature peridotites from both Namibia and the Kaapvaal craton have wide ranges of modal olivine combined with restricted ranges in *Mg* number of olivine. The principal difference between these suites is that average olivine *Mg* number for Louwrensia is 91.6 in comparison to 92.6 for the Kaapvaal. This difference is more likely to be primary than to be the product of metasomatism because of the relative constancy of the Louwrensia *Mg* number. The lower *Mg* number for the Louwrensia peridotites may reflect a lesser degree of melt depletion during the Proterozoic compared to the higher degrees of melting for melt depletion in the Archean (Pearson et al., 1994; Pearson et al., 2003). Both Kaapvaal and Louwrensia peridotite suites are depleted in Ca and Al relative to pyrolite but there is virtually complete overlap between the suites. The overlap in Ca and Al contrasts with the difference in *Mg* number between the suites and provides support for the idea that Al and Ca contents of these rocks have increased from very depleted levels following the introduction of diopside and possibly garnet. Most other elements appear also to have been affected by secondary processes.

The wide ranges of orthopyroxene contents are identical for Louwrensia and Kaapvaal craton peridotites. The origin of orthopyroxene-rich peridotites remains a matter of conjecture, but there is little reason to suppose that the formation of Namibian lithosphere differed considerably from that of the Kaapvaal root, despite the difference in age. If a secondary process of melt-interaction caused the orthopyroxene enrichment in the Kaapvaal lithospheric mantle then the same process seems to have affected the Proterozoic surrounding lithosphere to the west. Attempts to constrain the depth of melting of the Louwrensia peridotites using FeO–MgO and Al–Cr relations are equivocal. If unaffected by significant post-melting disturbance, Yb–Ca or Yb–Al relations for samples spanning a range of apparent depletion, argue for a relatively low-pressure (<3 GPa) melting environment.

REE contents and fractionations within diopsides from Louwrensia peridotites are comparable to those observed in Kaapvaal low-*T* lherzolites plus MARID and PIC xenoliths. Melts in equilibrium with the diopsides have similar, extreme LREE enrichment similar to those of kimberlitic magmas. The data indicate introduction of diopside to the peridotite assemblage, preceding or coincident with the magmatic events that produced the host kimberlite. Partitioning of a number of trace elements between garnet and diopside is not at equilibrium levels, suggesting a lack of complete equilibrium between these phases. These data and the highly fractionated REE patterns of some garnets may indicate recent introduction of some of the garnet also. These features of the trace element geochemistry of the Louwrensia peridotites are similar to those identified in Kaapvaal craton peridotites and also suggest that similar metasomatic processes have acted to modify the mineralogy and trace element budget of both suites of rocks.

Plots of estimated equilibration temperatures and depths for the low-temperature Louwrensia lherzolites made with a number of thermobarometers are concordant with each other and with an average geotherm for peridotites from the Kaapvaal craton. Plots for the high-temperature lherzolites, however, exhibit a marked thermal anomaly whose *P/T* position differs with use of different thermobarometers. Although it is unlikely that the thermometers accurately reflect the temperature experienced by these xenoliths, mineral zoning studies made here and in other peridotites from within the Gibeon Province, show clear evidence of a recent thermal disturbance. We interpret the agreement between the geotherm derived from low-temperature Louwrensia peridotites and an average Kaapvaal peridotite geotherm as evidence that the Louwrensia lithosphere was approximately as thick as that of the craton prior to thermal disturbance and that some thickness has been thermally eroded.

Acknowledgements

Particular thanks are owed to colleagues who provided large specimens from their collections for the bulk analyses we have obtained: I.D. MacGregor, R.H. Mitchell, J.J. Gurney and the Anglo American Research Laboratories. Douglas Smith updated his thermobarometry program TP02v2 at our request. We thank T. McCandless and D. Bell for reviews. The junior authors are greatly indebted to R.W. Carlson, M.

Morel, B. Mysen, B. Kjarsgaard and H. Grutter for timely assistance in revising the manuscript.

References

Bell, D.R., Schmidt, M.D., Janney, P.E., 2003. Mesozoic thermal evolution of the southern African mantle lithosphere. Lithos 71, 273–287.

Boyd, F.R., 1989. Compositional distinction between oceanic and cratonic lithosphere. Earth Planet. Sci. Lett. 96, 15–26.

Boyd, F.R., Mertzman, S.A., 1987. Composition and structure of the Kaapvaal lithosphere, southern Africa. In: Mysen, B.O. (Ed.), Magmatic Processes: Physiochemical Principles. Geochem. Soc. Spec. Publ., vol. 1, pp. 13–24.

Boyd, F.R., Nixon, P.H., 1979. Garnet lherzolite xenoliths from the kimberlites of East Griqualand, South Africa. Carnegie Inst. Wash. Yearbk. 78, 488–492.

Boyd, F.R., Nixon, P.H., 1980. Discrete nodules from the kimberlites of East Griqualand, South Africa. Carnegie Inst. Wash. Yearbk. 79, 296–302.

Boyd, F.R., Pearson, D.G., Nixon, P.H., Mertzman, S.A., 1993. Low-calcium garnet harzburgites from southern Africa: their relation to craton structure and diamond crystallization. Contrib. Mineral. Petrol. 113, 352–366.

Boyd, F.R., Pokhilenko, N.P., Pearson, D.G., Mertzman, S.A., Sobolev, N.V., Finger, L.W., 1997. Composition of the Siberian cratonic mantle: evidence from Udachnaya peridotite xenoliths. Contrib. Mineral. Petrol. 128, 228–246.

Brey, G.P., Kohler, T., 1990. Geothermobarometry in four phase lherzolites: II. New thermobarometers and practical assessment of existing thermobarometers. J. Petrol. 31, 1353–1378.

Brown, R.W., Gallagher, K., Griffin, W.L., Ryan, C.G., deWit, M.C.J., Belton, D.X., Harman, R., 1998. Kimberlites, accelerated erosion and evolution of the lithospheric mantle beneath the Kaapvaal craton during the Mid-Cretaceous. Extended Abstracts, 7th International Kimberlite Conference, Red Roof Designs, Cape, 105–107.

Burgess, S.R., Harte, B., in press. Tracing lithosphere evolution through the analysis of heterogeneous G9/G10 garnets in peridotite xenoliths: II. Trace element chemistry. Jour. Petrol.

Canil, D., 2004. Mildly incompatible Elements in Peridotites and the Origin of Mantle Lithosphere. Lithos, these proceedings doi:10.1016/j.lithos.2004.04.014.

Davies, G.R., Spriggs, A.J., Nixon, P.H., 2001. A non cognate origin for the Gibeon kimberlite megacryst suite, Namibia: implications for the origin of Namibian kimberlites. J. Petrol. 142, 159–172.

Finnerty, A.A., Boyd, F.R., 1987. Thermobarometry for garnet peridotites: basis for the determination of thermal and compositional structure of the upper mantle. In: Nixon, P.H. (Ed.), Mantle Xenoliths. Wiley, New York, pp. 381–402.

Franz, L., Brey, G.P., Okrusch, M., 1996a. Steady state geotherm, thermal disturbances, and tectonic development of the lower lithosphere underneath the Gibeon kimberlite province, Namibia. Contrib. Mineral. Petrol. 126, 181–198.

Franz, L., Brey, G.P., Okrusch, M., 1996b. Reequilibration of ultramafic xenoliths from Namibia by metasomatic processes at the mantle boundary. J. Geol. 104, 599–615.

Gregoire, M., Bell, D.R., le Roux, A.P., 2002. Trace element geochemistry of glimmerite and MARID mantle xenoliths: their classification and relationship to phlogopite-bearing peridotites and to kimberlites revisited. Contrib. Mineral. Petrol. 142, 603–625.

Gregoire, M., Bell, D.R., le Roux, A.P., 2003. Garnet lherzolites from the Kaapvaal craton (South Africa): Trace element evidence for a metasomatic history. J. Petrol. 44, 629–657.

Griffin, W.L., O'Reilly, S.Y., Natapov, L.M., Ryan, C.G., 2003. The evolution of lithospheric mantle beneath the Kalahari craton and its margins. Lithos 71, 215–241.

Gurney, J.J., Harte, B., 1980. Chemical variations in upper mantle nodules from southern African kimberliktes. Philos. Trans. R. Soc. Lond., A 297, 273–293.

Harte, B., Hunter, R.H., Kinny, P.D., 1993. Melt geometry, movement and crystallisation, in relation to dykes, veins and metasomatism. Philos. Trans. R. Soc. Lond., A 342, 1–21.

Hoal, K.O., 2003. Samples of Proterozoic iron-enriched mantle from the Premier kimberlite. Lithos 72, 259–272.

Hoal, K.E.O., Hoal, B.G., Erlank, A.J., Shimizu, N., 1994. Metasomatism of the mantle lithosphere recorded by rare earth elements in garnets. Earth Planet. Sci. Lett. 126, 303–314.

Hoal, B.G., Hoal, K.E.O., Boyd, F.R., Pearson, D.G., 1995. Age constraints on crustal and mantle lithosphere beneath the Gibeon kimberlite field, Namibia. S. Afr. J. Geol. 98, 112–118.

James, D.E., Fouch, M.J., VanDecar, J.C., van der Lee, S., 2001. Tectospheric structure beneath Southern Africa. Geophys. Res. Lett. 28, 2485–2488.

Janse, A.J.A., 1975. Kimberlite and related rocks from the Nama Plateau of South-West Africa. In: Ahrens, L.H., Dawson, J.B., Duncan, A.P., Erlank, A.J. (Eds.), Physics and Chemistry of the Earth, vol. 9. Pergamon Press, Amsterdam, pp. 81–94.

Jones, R.A., 1984. Geochemical and isotopic studies of some kimberlites and included ultrabasic xenoliths from southern Africa. PhD thesis, University of Leeds, UK.

Kelemen, P.B., Hart, S.R., Bernstein, S., 1998. Silica enrichment in the continental upper mantle via melt/rock reaction. Earth Planet. Sci. Lett. 164, 387–406.

MacGregor, I.D., 1975. Petrologic and thermal structure of the upper mantle beneath South Africa in the Cretaceous. In: Ahrens, L.H., Dawson, J.B., Duncan, A.P., Erlank, A.J. (Eds.), Physics and Chemistry of the Earth, vol. 9. Pergamon Press, Amsterdam, pp. 455–466.

McDonough, W.F., Sun, S.S., 1995. The composition of the Earth. Chem. Geol. 120, 223–253.

Mitchell, R.H., 1984. Garnet lherzolites from the Hanaus-1 and Louwrensia kimberlites of Namibia. Contrib. Mineral. Petrol. 86, 178–188.

Nixon, P.H., Boyd, F.R., 1973. Petrogenesis of the granular and Sheared ultrabasic nodule suite in kimberlites. In: Nixon, P.H. (Ed.), Lesotho Kimberlites. Lesotho Natl. Devel., Maseru Lesotho, pp. 48–56.

O'Neill, H.St.C., Wood, B.J., 1979. An experimental study of Fe–

Mg partitioning between garnet and olivine and its calibration as a geothermometer. Contrib. Mineral. Petrol. 70, 59–70.

Pearson, D.G., 1999. Evolution of cratonic lithospheric mantle: an isotopic perspective. In: Fei, Y., Bertka, C., Mysen, B.O. (Eds.), Mantle Petrology: Field Observations and High-pressure Experimentation: A Tribute to Francis R. (Joe) Boyd. Geochem. Soc. Spec. Pub., vol. 6. The Geochemical Society, San Antonio, pp. 57–78.

Pearson, D.G., Nowell, G.M., 2002. The continental lithospheric mantle: characteristics and significance as a mantle reservoir. Philos. Trans. R. Soc. Lond., A 360, 1–28.

Pearson, D.G., Boyd, F.R., Hoal, K.E.O., Hoal, B.G., Nixon, P.H., Rogers, N.W., 1994. A Re–Os isotopic and petrological study of Namibian peridotites: contrasting petrogenesis and composition of on- and off-craton lithospheric mantle. Mineral. Mag. 58A, 703–704.

Pearson, D.G., Carlson, R.W., Shirey, S.B., Boyd, F.R., Nixon, P.H., 1995. The stabilization of Archean lithospheric mantle: A Re–Os isotope study of peridotite xenoliths from the Kaapvaal craton. Earth Planet. Sci. Lett. 134, 341–357.

Pearson, D.G., Boyd, F.R., Simon, N.S.C., 2001. Modal mineralogy and geochemistry of Kaapvaal peridotites: the origin of garnet and diopside and implications for craton stability. Extended Abstracts, The Slave-Kaapvaal Workshop. Merrickville, Ontario, Canada.

Pearson, D.G., Irvine, G.J., Carlson, R.W., Kopylova, M.G., Ionov, D.A., 2002. The development of lithospheric mantle keels beneath the earliest continents: time constraints using PGE and Re–Os isotope systematics. In: Fowler, M., Ebinger, D.J., Hawkesworth, C.J. (Eds.), The Early Earth: Physical, Chemical and Biological Development. Geol. Soc. Spec. Pub., vol. 199. The Geological Society, Bath, pp. 65–90.

Pearson, D.G., Canil, D., Shirey, S.B., 2003. Mantle samples included in volcanic rocks: xenoliths and diamonds. In: Holland, H.D., Turekian, K.K. (Eds.), Treatise on Geochemistry, vol. 2. Elsevier, Amsterdam, pp. 171–275. Chap. 5.

Pearson, D.G., Irvine, G.J., Ionov, D.A., Boyd, F.R., Dreibus, P.H., 2004. Re–Os isotope systematics and Platinum Group Element fractionation during mantle melt extraction: A study of peridotite xenoliths from N. Lesotho and S. Namibian kimberlites, the Vitim volcanic field and massif peridotites from Beni Bousera. Chem. Geol. (in press).

Pollack, H.N., Chapman, D.S., 1977. On the regional variation of heat flow, geotherms and lithosphere thickness. Tectonophysics 38, 279–296.

Richardson, S.H., 1990. Age and early evolution of the continental mantle. In: Menzies, M.A. (Ed.), Continental Mantle. Clarendon Press, Oxford, pp. 55–65.

Robey, J.V.A., 1981. Kimberlites of the Central Cape Province, R.S.A. Unpublished. PhD thesis, University of Cape Town, South Africa.

Rudnick, R.L., Nyblade, A.A., 1999. The thickness and heat production of Archean lithosphere: constraints from xenolith ther-

mobarometry and surface heat flow. In: Fei, Y., Bertka, C.M., Mysen, B.O. (Eds.), Mantle petrology: field observations and high-pressure experimentation. The Geochemical Society Special Publications, vol. 6. University of Pennsylvania, pp. 3–12.

Shimizu, N., 1975. Rare earth elements in garnets and clinopyroxenes from garnet lherzolite nodules in kimberlites. Earth Planet. Sci. Lett. 25, 26–32.

Shimizu, N., 1999. Young geochemical features in cratonic peridotites from southern Africa and Siberia. In: Fei, Y., Bertka, C.M., Mysen, B.O. (Eds.), Mantle Petrology: Field Observations and High-Pressure Experimentation. The Geochemical Society Special Publication, vol. 6. University of Pennsylvania, pp. 47–55.

Shimizu, N., Pokhilenko, N.P., Boyd, F.R., Pearson, D.G., 1999. Trace element characteristics of garnet dunites/harzburgites, host rocks for Siberian peridotitic diamonds. In: Gurney, J.J., Gurney, J.L., Pascoe, M.D., Richardson, S.H. (Eds.), Proc. 7th International Kimberlite Conference, Cape Town. Red Roof Designs, Cape Town, pp. 773–832.

Simon, N.S.C., Pearson, D.G., Carlson, R.W., Davies, G.R., 2001. Origin of garnet and clinopyroxene in Kaapvaal low-T peridotite xenoliths: implications from secondary ionisation mass spectrometry (SIMS) data. Extended Abstracts. The Slave-Kaapvaal Workshop, Merrickville, Ontario, Canada.

Simon, N.S.C., Irvine, G.J., Davies, G.R., Pearson, D.G., Carlson, R.W., 2003a. The origin of garnet and clinopyroxene in "depleted" Kaapvaal peridotites. Lithos 71, 289–322.

Simon, N.S.C., Carlson, R.W., Davies, G.R., Nowell, G.M., Pearson, D.G., 2003b. Os–Sr–Nd–Hf Isotope Evidence for the Ancient Depletion and Subsequent Multi-Stage Enrichment History of the Kaapvaal Craton, Ext. Abstracts. 8th International Kimberlite Conference, Victoria..

Smith, D.C., 1999. Temperatures and pressures of mineral equilibration in peridotite xenoliths; review, discussion and implications. In: Fei, Y., Bertka, C., Mysen, B.O. (Eds.), Mantle Petrology: Field Observations and High-pressure Experimentation: A Tribute to Francis R. (Joe) Boyd. Geochem. Soc. Spec. Pub., vol. 6. The Geochemical Society, San Antonio, pp. 171–188.

Taylor, W.R., 1998. An experimental test of geothermometer and geobarometer formulations for upper mantle peridotites with application to the thermobarometry of fertile lherzolite and garnet websterite. Neues Jahrb. Mineral. Abh. 172, 381–408.

Van Achterberg, E., Griffin, W.L., Steinhofer, J., 2001. Metasomatism in mantle xenoliths from the Letlhakane kimberlites: estimation of element fluxes. Contrib. Mineral. Petrol. 141, 397–414.

Walter, M.J., 1998. Melting of garnet peridotite and the origin of komatiite and depleted lithosphere. J. Petrol. 39, 29–60.

Zack, T., Foley, S.F., Jenner, G.A., 1997. A consistent partition coefficient set for clinopyroxene, amphibole and garnet from laser ablation microprobe analysis of garnet pyroxenites from Kakanui, New Zealand. Neues Jahrb. Mineral. Abh. 172, 23–41.

Available online at www.sciencedirect.com

Lithos 77 (2004) 593–608

ELSEVIER

www.elsevier.com/locate/lithos

Layered mantle at the Karelian Craton margin: *P–T* of mantle xenocrysts and xenoliths from the Kaavi–Kuopio kimberlites, Finland

M.L. Lehtonen*, H.E. O'Brien, P. Peltonen, B.S. Johanson, L.K. Pakkanen

Geological Survey of Finland, P.O. Box 96, FIN-02151 Espoo, Finland

Received 27 June 2003; accepted 5 November 2003
Available online 18 May 2004

Abstract

Peridotitic clinopyroxene (cpx) and pyrope garnet xenocrysts from four kimberlite pipes in the Kaavi–Kuopio area of Eastern Finland have been studied using major and trace element geochemistry to obtain information on the vertical compositional variability of the underlying mantle. The xenocryst data, when combined with the petrological constraints provided by peridotite xenoliths, yield a relatively complete section through the lithospheric mantle. Single-grain cpx thermobarometry fits with a 36-mW/m^2 geotherm calculated using heat flow constraints and xenolith modes and geophysical properties. Ni thermometry on pyrope xenocrysts gives 700–1350 °C and, based on the cpx xenocryst/xenolith geotherm, indicates a wide sampling interval, ca. 80–230 km. Plotting pyrope major and trace element compositions as a function of temperature shows there are three distinct layers in the local lithospheric mantle:

(1) A low-temperature (<850 °C) harzburgite layer distinguished by Ca-rich but Ti-, Y- and Zr-depleted pyropes. The xenoliths originating from this layer are all fine-grained garnet-spinel harzburgites with secondary cpx.
(2) A variably depleted lherzolitic, harzburgitic and wehrlitic horizon from 950 to 1150 °C or 130 to 180 km.
(3) A deep layer from 180 to 240 km composed largely of fertile material.

The peridotitic diamond window at Kaavi–Kuopio stretches from the top of the diamond stability field at 140 km to the base of the harzburgite-bearing mantle at about 180 km, implying a roughly 40-km-wide prospective zone.
© 2004 Elsevier B.V. All rights reserved.

Keywords: Kimberlite; Xenocryst; Xenolith; Lithosphere; Thermobarometry; Karelian Craton

1. Introduction

The Kaavi–Kuopio Kimberlite Province situated at the edge of the Karelian Craton comprises two distinct kimberlite clusters containing at least 19 kimberlites with mineralogy typical of Group I kimberlite (Tyni, 1997; O'Brien and Tyni, 1999). The pipes have intruded Archean (3.5–2.6 Ga) basement gneisses and allochthonous Proterozoic (1.9–1.8 Ga) metasediments (Kontinen et al., 1992; Fig. 1). Several methods have been used to date the

* Corresponding author. Fax: +358-20-550-12.
E-mail address: marja.lehtonen@gsf.fi (M.L. Lehtonen).

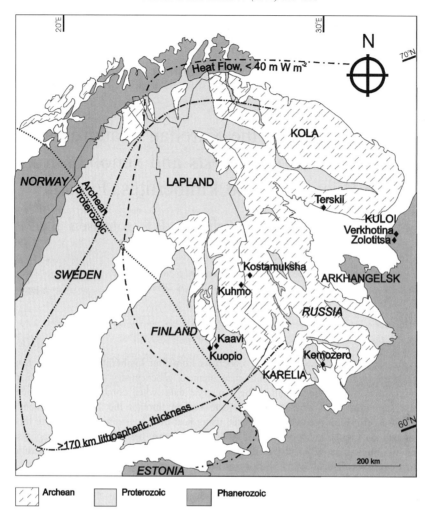

Fig. 1. Map illustrating the diamond prospective area of Northern Europe characterized by (a) low heat flow (simplified from Kukkonen and Jõeleht, 1996) and (b) lithosphere thicker than 170 km based on seismic data (Calcagnile, 1982). Generalized geology modified after Gaál and Gorbatschev (1987). The Archean/Proterozoic boundary marks the subsurface extent of the Archean craton. The black diamonds represent diamond-bearing kimberlites and lamproites.

kimberlite magmatism (Tyni, 1997; Peltonen et al., 1999; Peltonen and Mänttäri, 2001), but the U–Pb ion probe ages of 589–626 Ma from perovskites (O'Brien et al., in press) are considered to be the most reliable. Mantle xenoliths from the Kaavi–Kuopio pipes have been studied by Peltonen et al. (1999), and a geotherm of 36 mW/m^2 has been calculated using heat flow constraints and xenolith modes and geophysical properties (Kukkonen and Peltonen, 1999). The xenolith suite provides evidence of a compositionally stratified lithospheric mantle adjacent to the ancient suture zone between

the Archean Karelian Craton and the Proterozoic Svecofennian mobile belt. A shallow zone (<900 °C) of garnet-spinel harzburgites is underlain by a zone of garnet facies peridotites (180–240 km) including lherzolitic, harzburgitic, olivine websteritic and wehrlitic varieties (Peltonen et al., 1999). Mantle eclogite xenoliths are also present, some of them being highly diamondiferous (Peltonen et al., 2002). The upper mantle beneath the Karelian Craton is interpreted to have a mixed origin, consisting of remnants of thinned Archean subcontinental lithospheric mantle, possibly underlain by early Proterozoic lithosphere and/or younger material underplated around 1.8 Ga (Peltonen et al., 1999). The small number of xenoliths available gives a limited sampling of the lithosphere relative to the xenocryst record. The aim of this study was therefore to obtain additional information on the vertical compositional variability of the lithosphere by carrying out a systematic study of peridotitic clinopyroxene and pyrope garnet xenocrysts from four Kaavi–Kuopio kimberlites.

2. Samples

Samples were selected from four kimberlites: the 2-ha Lahtojoki (Pipe no. 7 with 26 cpht of + 0.8 mm diamonds), the 2-ha Kylmälahti (no. 17, marginally diamondiferous), the 700×30 m Kärenpää (no. 5, diamond grade unknown) and the 300×50 m Niilonsuo (no. 2, microdiamond-rich). Xenocryst grains (0.25–2.0 mm) were liberated by lightly crushing the kimberlite material—except for the hard magmatic Niilonsuo kimberlite which was fragmented electrodynamically at Forschungszentrum Karlsruhe—followed by heavy medium separation and hand picking. Hundreds of garnet and clinopyroxene (cpx) xenocrysts as well as 20 garnet-cpx aggregates were recovered from the concentrates. Other mineral phases appropriate for $P–T$ calculations (i.e., orthopyroxene) were not found to coexist with garnet or cpx. Garnets in this study are unsorted and represent the entire range of garnet populations within the kimberlites. However, for the cpx grains, an attempt was made to choose only peridotitic varieties, based on their bright green color.

3. Analytical techniques

Mineral compositions were determined by a Cameca Camebax SX50 electron microprobe (EMP) at the Geological Survey of Finland. For garnet analyses, an acceleration voltage of 25 kV, probe current of 48 nA and beam diameter of 1 μm were applied. The parameters for cpx analyses were 15 kV, 30 nA and 5 μm, respectively. Selected garnet xenocrysts were analyzed for trace elements by LA-ICP-MS at the University of Cape Town using the

methodology and equipment described in Grégoire et al. (2003). Detection limits are in the range of 10–20 ppb for Zr and Y and 2 ppm for Ti and Ni. Trace Ni, Mn and Ti data by EMP were obtained on pyrope grains employing 500-nA probe current, 600-s counting times on peak plus background positions and were reduced by the CSIRO TRACE program for the SX50 (Robinson and Graham, 1992). Cross-checking of the two trace methods shows that Ni, Ti and Mn analyses in pyrope by EMP can achieve similar precision to those of LA-ICP-MS down to the level of ca. 10 ppm.

The garnets were classified into harzburgitic, lherzolitic and nonperidotitic varieties according to Gurney (1984); the compositional field for wehrlitic garnet was separated from the lherzolite field using the division of Sobolev et al. (1973). Equilibration pressures and temperatures of the peridotitic cpx xenocrysts were calculated using the single-grain cpx thermobarometer of Nimis and Taylor (2000, NT hereafter). The cpx xenocrysts were screened according to the compositional and crystal structural criteria of Nimis and Taylor (2000) in order to calculate pressures and temperatures only for those grains that had probably coexisted with orthopyroxene and garnet. Xenolith $P–T$ data were calculated using NT and the method of Brey et al. (1990, BKN hereafter). For garnet xenocrysts, the Ni thermometer (Griffin et al., 1989a) was applied using both the calibration of Ryan et al. (1996) and that of Canil (1999).

4. Results

4.1. Garnet major element geochemistry

Representative EMP analyses of garnet xenocrysts are presented in Table 1 and representative garnet analyses from the xenoliths were taken from Peltonen et al. (1999). CaO and Cr_2O_3 contents of pyropes from peridotite xenoliths and pyrope xenocrysts (Fig. 2) show that lherzolitic garnets (G9 according to Dawson and Stephens, 1975) predominate. In addition, orange Ti-rich pyropes of megacryst composition (G1/G2) are common among the xenocrysts. Subcalcic harzburgitic garnets (G10) represent about 6% of the analyzed grains. All of the G10 garnet analyses came from xenocrysts, suggesting that their host

Table 1
Electron microprobe and LA-ICP-MS analyses of garnet xenocrysts from Finnish kimberlites

Sample ID	4682.02	10657.02	4398.96	4403.96	#43	15947.01	15946.01	18279.01	15995.01	#106
Classification	CCGE	CCGE	CCGE	HRZ	HRZ	HRZ	HRZ	HRZ	WHR	WHR
SiO_2 (wt.%)	39.61	40.17	40.83	42.19	41.00	41.05	41.34	41.53	41.03	40.04
TiO_2 (wt.%)	0.05	0.00	0.00	0.03	0.28	0.11	0.31	0.01	0.58	0.53
Al_2O_3 (wt.%)	20.21	17.68	17.00	19.35	17.24	16.81	18.08	16.04	19.20	17.31
Cr_2O_3 (wt.%)	5.01	7.55	8.62	6.05	8.08	8.49	6.59	10.24	4.49	6.89
FeO (wt.%)	8.74	8.25	7.74	6.44	6.22	6.31	6.49	6.27	9.93	8.98
MnO (wt.%)	0.48	0.53	0.56	0.36	0.32	0.32	0.29	0.37	0.50	0.53
MgO (wt.%)	18.35	17.16	17.60	23.19	20.81	21.26	21.04	22.92	17.29	17.58
NiO (wt.%)	0.00	0.00	0.00	0.01	0.01	0.01	0.01	0.01	0.01	0.01
CaO (wt.%)	6.38	7.44	7.67	3.06	5.18	4.24	4.83	2.62	7.28	6.85
Na_2O (wt.%)	0.00	0.00	n.a.	n.a.	n.a.	n.a.	n.a.	n.a.	n.a.	n.a.
K_2O (wt.%)	0.00	0.00	n.a.	n.a.	0.00	0.00	0.00	0.00	0.00	0.00
Total	98.92	98.83	100.08	100.69	99.16	98.68	99.00	100.08	100.33	98.76
Mg/(Mg + Fe)	0.789	0.787	0.802	0.865	0.856	0.857	0.853	0.867	0.756	0.777
Ni (ppm)	12	20	23	50	62	65	65	66	44	46
Y (ppm)	2.8	0.2	0.5	1.1	3	3	3	1.3	25	23
Zr (ppm)	3.8	0.8	2.4	3.9	16	5.7	21	1.2	64	64
T_{Ni} (°C) RGP	687	795	826	1034	1105	1122	1124	1129	993	1006
T_{Ni} (°C) Canil	833	916	938	1082	1127	1137	1139	1142	1055	1064

Sample ID	#94	15,924.01	#9	4674.02	#11	10,635.02	13,358.95	4694.02	18,078.01	4611.02
Classification	WHR	WHR	LHZ	LHZ	LHZ	LHZ	LHZ	LHZ	LHZ	LHZ
SiO_2 (wt.%)	39.99	40.14	40.18	39.58	40.36	40.68	41.13	40.19	41.16	40.37
TiO_2 (wt.%)	0.91	0.88	0.04	0.15	0.61	0.54	0.55	0.35	0.34	0.86
Al_2O_3 (wt.%)	15.99	17.62	17.83	20.11	16.74	19.37	19.75	20.05	17.46	18.67
Cr_2O_3 (wt.%)	7.71	5.70	7.39	4.55	7.86	4.08	4.35	4.88	7.57	5.49
FeO (wt.%)	9.27	9.35	7.85	8.56	8.19	8.52	7.05	6.57	7.03	7.29
MnO (wt.%)	0.42	0.42	0.54	0.42	0.48	0.40	0.41	0.32	0.41	0.38
MgO (wt.%)	17.64	17.84	19.07	19.91	19.04	19.39	20.80	21.72	20.26	20.46
NiO (wt.%)	0.01	0.01	0.00	0.01	0.00	0.01	0.01	0.01	0.01	0.00
CaO (wt.%)	6.87	6.37	5.54	5.20	5.97	5.55	4.83	4.70	5.56	5.32
Na_2O (wt.%)	n.a.	n.a.	n.a.	0.05	n.a.	0.09	n.a.	0.04	n.a.	0.11
K_2O (wt.%)	0.00	0.00	0.00	0.00	0.00	0.00	n.a.	0.00	0.00	0.00
Total	98.82	98.38	98.44	98.59	99.27	98.69	99.06	98.93	99.82	98.99
Mg/(Mg + Fe)	0.772	0.773	0.812	0.806	0.806	0.802	0.840	0.855	0.837	0.833
Ni (ppm)	56	59	21	34	40	46	51	53	55	63
Y (ppm)	17	17	17	13	14	16	11	1.4	1.7	14
Zr (ppm)	50	46	45	46	43	42	28	39	38	45
T_{Ni} (°C) RGP	1071	1084	803	918	964	1004	1040	1050	1061	1109
T_{Ni} (°C) Canil	1106	1114	922	1004	1035	1063	1086	1092	1099	1129

Sample ID	18,065.01	8101.96	10,634.02	18,155.01	18,147.01	18,077.01	4672.02	8114.96	4655.02	18,193.01
Classification	LHZ	LHZ	LHZ	LHZ	LHZ	LHZ	TiP	TiP	TiP	TiP
SiO_2 (wt.%)	40.86	40.65	40.33	41.55	41.44	41.43	40.23	41.30	40.50	41.29
TiO_2 (wt.%)	0.29	0.24	0.89	0.56	0.54	0.56	0.77	0.64	0.64	0.43
Al_2O_3 (wt.%)	16.97	14.50	17.28	18.95	19.45	17.25	20.23	20.74	21.04	22.39
Cr_2O_3 (wt.%)	8.52	10.87	6.28	5.05	4.10	6.96	2.91	1.48	2.16	0.34
FeO (wt.%)	6.55	6.62	8.45	7.06	7.58	6.73	8.50	9.07	8.03	8.89
MnO (wt.%)	0.34	0.39	0.44	0.30	0.28	0.28	0.34	0.37	0.28	0.29
MgO (wt.%)	20.21	19.34	19.02	21.00	21.20	20.91	20.67	20.76	21.31	20.01
NiO (wt.%)	0.00	0.01	0.01	0.01	0.01	0.01	0.01	0.02	0.01	0.01
CaO (wt.%)	5.72	6.70	6.04	5.25	4.80	5.57	5.00	4.23	4.76	5.43

Table 1 (*continued*)

Sample ID	18,065.01	8101.96	10,634.02	18,155.01	18,147.01	18,077.01	4672.02	8114.96	4655.02	18,193.01
Classification	LHZ	LHZ	LHZ	LHZ	LHZ	LHZ	TiP	TiP	TiP	TiP
Na_2O (wt.%)	n.a.	n.a.	0.08	n.a.	n.a.	n.a.	0.06	n.a.	0.04	n.a.
K_2O (wt.%)	0.00	n.a.	0.00	0.00	0.00	0.00	0.00	n.a.	0.00	0.01
Total	99.49	99.48	98.86	99.76	99.42	99.75	98.79	98.79	98.88	99.18
Mg/(Mg + Fe)	0.846	0.839	0.800	0.841	0.833	0.847	0.812	0.803	0.825	0.800
Ni (ppm)	64	71	71	79	84	118	63	71	80	92
Y (ppm)	1.1	20	15	13	14	10	15	16	14	11
Zr (ppm)	25	41	44	43	35	41	43	39	37	23
T_{Ni} (°C) RGP	1117	1154	1155	1196	1222	1383	1108	1155	1201	1264
T_{Ni} (°C) Canil	1135	1157	1158	1182	1198	1288	1129	1157	1186	1222

RGP = Ryan et al. (1996), Canil = Canil (1999); n.a. = not analyzed.

peridotites were more susceptible to fragmentation than other xenolith varieties. Moreover, a well-developed pyrope trend exists in the wehrlitic field similar to that described by Kopylova et al. (1999, 2000) and Carbno and Canil (2002) from the Slave Craton. The trend, named CCGE by Kopylova et al. (2000) after

"chromite–clinopyroxene–garnet equilibrium," is characterized by Ca and Cr saturation but shows less enrichment in Cr than the usual lherzolitic trend, and this is taken to indicate that the garnets equilibrated with coexisting cpx and chromite. The Kaavi–Kuopio xenolith record supports this conclusion since the

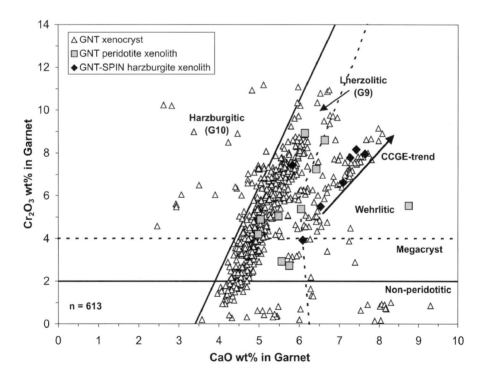

Fig. 2. Cr_2O_3 vs. CaO of Kaavi–Kuopio kimberlite-derived garnets. The xenocryst data are representative of the entire garnet population of the pipes, whereas the xenolith data represent more limited sampling. The harzburgite, lherzolite and nonperidotite fields are redrawn after Gurney (1984) and the wehrlite field is separated according to Sobolev et al. (1973). Arrow marks the "CCGE" garnet trend, i.e., chromite–clinopyroxene–garnet equilibrium, recognized in spinel-garnet peridotites from the Jericho kimberlite (Kopylova et al., 2000).

fine-grained garnet-spinel harzburgite xenoliths with minor amounts ($<5\%$) of secondary cpx (Peltonen et al., 1999) contain garnets that plot along the same CCGE trend.

4.2. Clinopyroxene major element geochemistry

Clinopyroxenes in the garnet-cpx aggregates are chrome diopsides that contain $0.50-4.67$ wt.% Cr_2O_3 and $1.21-4.43$ wt.% Na_2O. Their $Mg/(Mg+Fe)$ ratios range from 0.90 to 0.94. The corresponding values for the entire cpx xenocryst population that qualified for $P-T$ calculations are $0.50-3.21$ wt.% Cr_2O_3, $0.77-$

3.03 wt.% Na_2O and $0.88-0.96$ $Mg/(Mg+Fe)$. Representative EMP analyses of cpx xenocrysts are presented in Table 2. Fig. 3, where cpx analyses from the xenoliths (Peltonen et al., 1999) and xenocrysts (this study) are plotted in a Cr_2O_3 vs. $Mg/(Mg+Fe)$ diagram, shows that the two populations coincide rather well.

The CCGE trend seen in garnet is reflected by cpx geochemistry as described by Kopylova et al. (1999). The cpx grains from garnet-spinel peridotite xenoliths and most of the low-T (<900 °C, NT) cpx xenocrysts show an unusual negative correlation in Mg number and Cr (Fig. 3) due to partitioning of Cr and Mg into

Table 2
Electron microprobe analyses of clinopyroxene xenocrysts from Finnish kimberlites

Sample ID	13,751.01	17,915.01	17,887.01	13,743.01	17,895.01[a]	17,857.01	17,896.01	17,905.01	13,801.01	13,773.01
SiO_2 (wt.%)	54.80	54.61	54.70	55.32	54.98	54.80	54.92	55.34	55.14	54.98
TiO_2 (wt.%)	0.00	0.00	0.29	0.30	0.26	0.22	0.32	0.14	0.33	0.31
Al_2O_3 (wt.%)	1.64	1.34	1.78	2.10	1.54	1.67	1.80	2.50	1.75	1.78
Cr_2O_3 (wt.%)	1.10	0.94	0.87	1.11	1.37	1.39	2.01	1.85	1.92	1.65
FeO (wt.%)	1.36	1.63	2.70	3.22	2.17	2.66	2.30	2.39	2.95	2.55
MnO (wt.%)	0.00	0.01	0.14	0.01	0.11	0.07	0.14	0.00	0.00	0.02
MgO (wt.%)	16.89	17.38	16.29	15.97	16.89	16.48	16.41	16.44	16.93	17.15
NiO (wt.%)	0.08	0.00	0.07	0.03	0.02	0.04	0.09	0.03	0.03	0.02
CaO (wt.%)	22.43	22.55	21.39	20.71	21.11	20.63	20.10	18.96	19.44	19.36
Na_2O (wt.%)	1.12	0.87	1.37	1.82	1.45	1.46	1.75	2.23	1.79	1.75
K_2O (wt.%)	0.04	0.02	0.03	0.01	0.01	0.02	0.01	0.05	0.01	0.03
Total	99.46	99.35	99.63	100.63	99.95	99.44	99.87	99.93	100.28	99.62
$Mg/(Mg+Fe)$	0.957	0.950	0.915	0.898	0.933	0.917	0.927	0.925	0.911	0.923
T (°C) NT	779	855	881	893	937	975	984	1017	1066	1078
P (kbar) NT	34	37	40	43	45	45	44	45	49	51

Sample ID	13,707.01	#3_104[a]	13,706.01	17,835.01	17,850.01	17,832.01	13,767.01	13,850.01[a]	17,848.01	13,824.01
SiO_2 (wt.%)	55.24	54.43	55.34	55.04	55.22	55.10	55.47	55.45	55.77	55.25
TiO_2 (wt.%)	0.37	0.30	0.37	0.31	0.24	0.29	0.14	0.20	0.23	0.16
Al_2O_3 (wt.%)	1.89	1.71	1.85	1.62	1.48	1.58	1.72	1.73	1.63	2.73
Cr_2O_3 (wt.%)	1.58	1.34	1.41	2.07	2.25	1.76	1.09	0.94	2.62	0.67
FeO (wt.%)	3.00	3.02	3.15	2.27	2.20	2.33	3.30	3.46	3.37	4.53
MnO (wt.%)	0.03	0.03	0.01	0.04	0.08	0.10	0.06	0.05	0.22	0.10
MgO (wt.%)	16.86	17.00	17.30	17.51	17.53	18.19	18.93	19.15	19.07	19.71
NiO (wt.%)	0.05	0.07	0.05	0.02	0.11	0.11	0.06	0.06	0.07	0.07
CaO (wt.%)	18.92	19.13	19.02	18.61	18.52	18.54	17.77	17.49	15.86	14.04
Na_2O (wt.%)	1.97	1.64	1.81	1.83	1.82	1.62	1.33	1.42	1.98	1.89
K_2O (wt.%)	0.03	0.01	0.03	0.04	0.04	0.01	0.03	0.02	0.02	0.01
Total	99.94	98.70	100.44	99.41	99.48	99.63	99.90	100.07	100.89	99.15
$Mg/(Mg+Fe)$	0.909	0.909	0.907	0.932	0.934	0.933	0.911	0.908	0.910	0.886
T (°C) NT	1091	1102	1117	1140	1157	1190	1279	1292	1309	1364
P (kbar) NT	57	54	56	54	55	54	56	61	57	58

NT = Nimis and Taylor (2000).
[a] Intergrown with garnet.

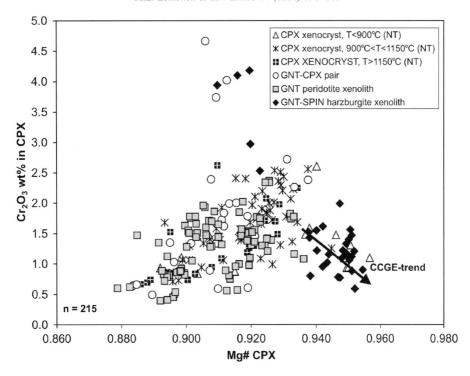

Fig. 3. Cr_2O_3 vs. Mg number for Kaavi–Kuopio clinopyroxenes. The temperatures for cpx xenocrysts are calculated after Nimis and Taylor (2000, NT). Cpx grains from garnet-spinel harzburgites and the low-temperature cpx xenocrysts ($T < 900\ °C$), in particular, follow the CCGE trend described by Kopylova et al. (1999).

silicate phases controlled by the presence of spinel with a constant Cr/Mg ratio.

4.3. Clinopyroxene thermobarometry

Single-grain cpx thermobarometry fits reasonably well with a geotherm of 36 mW/m^2 calculated using heat flow constraints and Kaavi–Kuopio xenolith modes and geophysical properties (Kukkonen and Peltonen, 1999). Fig. 4 shows the correspondence between the xenolith and xenocryst data. Although there is a shift to lower pressures and slightly lower temperatures in NT results relative to BKN for data from the same xenoliths, these shifts are along the modeled heat flow curves and therefore do not affect the fit of the data to the preferred Kukkonen–Peltonen geotherm (Fig. 4, KP). The BKN P–T values for xenoliths on this geotherm indicate a wide sampling range, approximately 90–230 km. The cpx xenocrysts with compositions along the CCGE trend (Fig. 3) have NT temperatures and

pressures in the range of 800–900 °C and 37–41 kbar, respectively, indicating derivation from the same depth range as the garnet-spinel peridotites, i.e., shallower than 130 km according to the local geotherm.

4.4. Garnet thermometry and trace element geochemistry

A subset of the pyrope xenocrysts shown in Fig. 2 was analyzed for trace elements. Representative analyses are presented in Table 1. The T_{Ni} histograms (Fig. 5) show bimodal distributions using both the calibration of Ryan et al. (1996, used hereafter in T_{Ni} values) and that of Canil (1999). The two peaks in the sampling include a strong low-temperature peak at 700–850 °C consisting dominantly of the CCGE pyropes described previously, and a stronger sampling peak at 1000–1150 °C, which includes all but one of the G10 pyropes analyzed for Ni and a second population of more

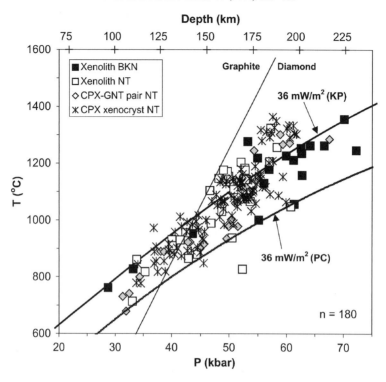

Fig. 4. *P–T* calculated for xenoliths, cpx-garnet pairs and cpx xenocrysts using the thermobarometers of Brey et al. (1990, BKN) and Nimis and Taylor (2000, NT). Reference model geotherm for 36 mW/m² (PC) from Pollack and Chapman (1977) and 36 mW/m² (KP) from Kukkonen and Peltonen (1999), the latter calculated for the Karelian craton at 600 Ma. Depth in kilometers is converted from pressure according to Kukkonen et al. (2003). The diamond–graphite transition is redrawn after Kennedy and Kennedy (1976).

typical wehrlitic grains (i.e., with compositions that do not plot along the CCGE trend). In addition to being the highest temperature at which the latter two pyrope types occur, 1150 °C also represents a break in the stratigraphic section of the lithosphere based on the Zr and Y contents of garnets, separating mantle that contains strongly depleted pyropes from more enriched mantle (Fig. 6). The intermediate temperature (1000–1150 °C) wehrlitic population is enriched in Y and Zr relative to all other pyrope varieties (Fig. 6) and to the ultra-depleted low-temperature CCGE garnets in particular. Fig. 6 also shows megacryst composition Ti-rich pyropes (TiPs), which have an enigmatic origin, but are nevertheless included here for illustrative purposes. We defend the reasoning for plotting at least the most Mg-rich TiPs on these diagrams as they reach the same Mg and Cr contents as lherzolitic pyropes, implying equilibration with olivine similar

to that in the lherzolites. The great majority of these Mg-rich TiP analyses plot at temperatures above the 1150 °C edge.

The two breaks in the mantle stratigraphy underlying Kaavi–Kuopio also show up well using TiO_2 contents of the pyropes, as seen in Fig. 7, where the T_{Ni} temperatures have been extrapolated to the local geotherm (Kukkonen and Peltonen, 1999) to give pressures and depths for each grain. Overall, the garnet xenocrysts cover a sampling interval from 80 to 230 km, without any major discontinuities. The main horizon of sampling at 1000–1150 °C corresponds to the mantle section from 140 to 180 km, from which only few xenoliths have been recovered. The lower temperature boundary at about 850 °C corresponds to the break in TiO_2 at approximately 110 km depth above which only extremely low TiO_2 content pyropes exist, most of these being CCGE in composition. The second boundary roughly at 1150

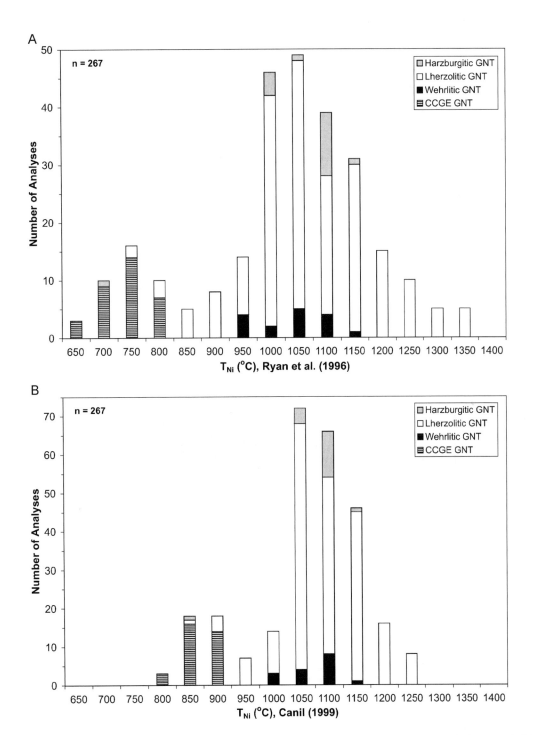

Fig. 5. Distribution of T_{Ni} for trace element analyzed Kaavi–Kuopio pyrope xenocrysts comparing the calibrations of (A) Ryan et al. (1996) and (B) Canil (1999).

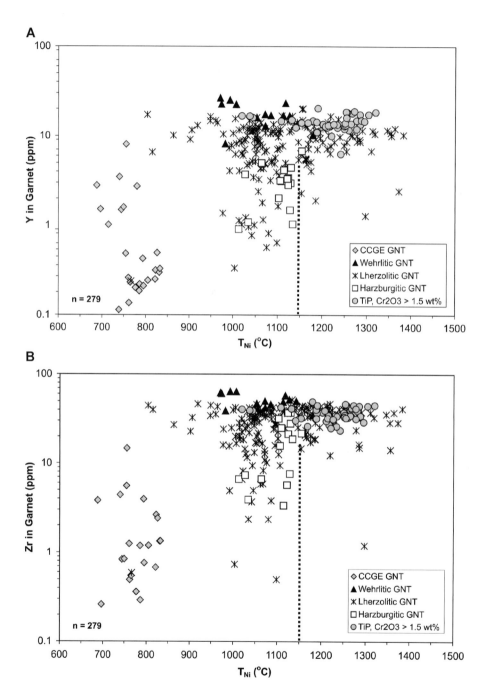

Fig. 6. (A) Y and (B) Zr contents of pyropes vs. T_{Ni} (Ryan et al., 1996) both show an edge at about 1150 °C. The edge does not mark the lower boundary of the lithosphere, as all xenoliths from >1150 °C are coarse granular peridotites typical of subcontinental lithospheric mantle.

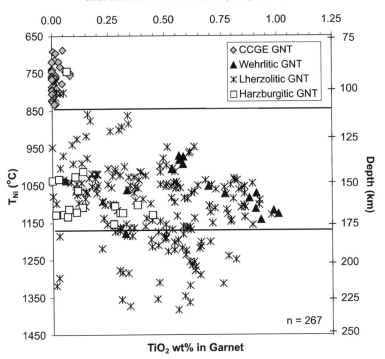

Fig. 7. TiO$_2$ vs. T_{Ni} (Ryan et al., 1996) and depth for the Kaavi–Kuopio pyropes showing three distinct layers of the underlying mantle. Depths are calculated by extrapolating the T_{Ni} temperatures of pyropes to the local geotherm of Kukkonen and Peltonen (1999).

°C or ca. 180 km is the limit below which only a very few TiO$_2$-depleted pyropes occur.

5. Discussion

5.1. The CCGE trend in garnet

The compositional trend of Ca- and Cr-saturated garnets (CCGE) seen in the Kaavi–Kuopio garnet-spinel peridotites and low-T garnet xenocrysts is only rarely observed in kimberlites or associated xenoliths. Kopylova et al. (1999, 2000) and Carbno and Canil (2002) described it from garnet-spinel peridotite xenoliths in the Jericho and Drybones Bay kimberlites in the Slave Craton. Similar garnets are also found in minor proportion in the Ranch Lake kimberlite concentrate (Cookenboo, 1996). The few other localities with this type of trend include the chromite-bearing garnet lherzolites from the Thumb primitive minette in the Colorado Plateau (Ehrenberg, 1982; Smith et al., 1991), which show the trend strongly, and a set of rare peridotites from the Lesotho xenolith suite (Smith and Boyd, 1992).

Smith and Boyd (1992) noted that the CCGE trend parallels the compositional trend in garnet produced by experiments on natural (spinel-bearing) lherzolite (Brey et al., 1990). The CCGE garnets from Kaavi–Kuopio are plotted in Fig. 8, along with the compositional fields for garnets from Jericho (Kopylova et al., 1999, 2000), Thaba Putsoa, Lesotho (Smith and Boyd, 1992) and the lherzolite experiments (Brey et al., 1990). The Kaavi–Kuopio CCGE trend parallels that of the other two kimberlites, but at lower temperatures and pressures, and cuts slightly across the isotherms derived from the experimental data. The temperatures and pressures estimated from Fig. 8 for Kaavi–Kuopio CCGE-bearing xenoliths are reasonably consistent with the P–T of the cpx from these xenoliths and the T_{Ni} values of the garnets (Fig. 5). Taken together, this indicates derivation from temperatures and pressures below 900 °C and 40 kbar, i.e., shallower than 130 km according to the local geotherm.

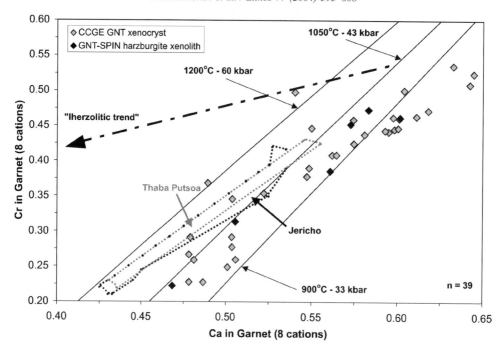

Fig. 8. Ca–Cr diagram for garnet showing the Kaavi–Kuopio CCGE xenocrysts and garnets from garnet-spinel harzburgite xenoliths. For comparison, the spinel-garnet equilibration trends from Jericho (Kopylova et al., 1999), Thaba Putsoa, Lesotho (Smith and Boyd, 1992) and from experiments on natural lherzolite compositions at 900–1200 °C and 33–60 kbar (Brey et al., 1990). Arrow denotes the common lherzolitic trend.

Kopylova et al. (2000) concluded that Ca-saturated cpx-bearing garnet-spinel cratonic peridotites are restricted to mantle segments of special character since Cr-rich rocks are normally poor in Ca and vice versa. Garnet and spinel-bearing mantle, which is harzburgitic under most Archean cratons (Griffin et al., 1999a), tends to be too depleted to provide sufficient cpx for development of the CCGE trend. Moreover, the rarity of the trend may derive from the disequilibrium between phases in cratonic spinel-garnet peridotites (Kopylova et al., 2000; Smith and Boyd, 1992). The mantle beneath Jericho (Kopylova et al., 2000) and the Colorado Plateau (Ehrenberg, 1982) are special because their spinel-garnet mantle is lherzolitic or cpx-bearing harzburgitic. These locations are also characterized by the complete absence of subcalcic harzburgitic garnet (G10) and the presence of lherzolitic Ca-saturated garnet at all mantle depths (Kopylova et al., 2000; Griffin and Ryan, 1995). The Kaavi–Kuopio mantle segment is similar to that of Jericho in the respect that the CCGE trend derives from garnets in garnet-spinel harzburgites with minor

amounts (<5%) of cpx. However, despite this similarity in host rock for the CCGE garnets, the Kaavi–Kuopio mantle also differs significantly from that of Jericho (and the Colorado Plateau) since it contains G10s in the deeper layers of the lithosphere, and moreover, one G10 grain falls within the temperature range of the CCGE garnets (Fig. 7). Overall, the CCGE trend in Kaavi–Kuopio exists only in a narrow, upper layer of the lithosphere (<120 km), as its lower parts are of a completely different nature.

The ultradepletion of CCGE garnets with respect to Ti, Y and Zr, combined with their contradictory saturation in Ca, indicate that their host peridotite was not produced by a single partial melting event of the primitive mantle. The CCGE garnets may have been derived from a mantle source that has been ultradepleted by melt extraction and later refertilized in Ca by some chemical enrichment process such as carbonatite metasomatism (Kopylova et al., 2000; Carbno and Canil, 2002). The addition of Ca led to crystallization of secondary cpx (as seen in Kaavi–Kuopio garnet-spinel facies xenoliths) and the transformation of harzburgitic garnets to wehrlitic. The transformation from harzburgite to lherzolite has been described from mantle garnets from suites in Kimberley, South Africa (Schulze, 1995; Griffin et al., 1999b). In contrast to the peridotitic minerals in the mantle beneath Jericho (Kopylova et al., 1999) and the Colorado Plateau (Smith and Ehrenberg, 1984; Smith et al., 1991), the Kaavi–Kuopio grains do not display extensive zoning that would indicate such secondary refertilization. This suggests that there was sufficient time for equilibration with the metasomatizing agents. This conclusion is also supported by the garnet-cpx Sm–Nd isochrons of garnet-spinel facies xenoliths that yield ages of 600 Ma (Peltonen et al., 1999), the same age as the host kimberlites, suggesting continuous re-equilibration of garnets at mantle temperatures up to the time of emplacement.

Based on the whole rock REE abundances, regional geology and model presented for the local geodynamic evolution, Peltonen et al. (1999) concluded that the low-T garnet-spinel peridotites could represent harzburgitic residues, i.e., remnants of the reworked Archean lithosphere, metasomatized shortly before or during the invasion of a kimberlite-derived melt or fluid completely obscuring their ancient isotopic signature. Carbno and Canil (2002), however, suggested that the abundance of ultradepleted (Zr, Y, Ti) but Ca-enriched garnets in Drybones Bay kimberlite may be explained by carbonatite metasomatism, where cpx forms from opx as a result of the interaction of carbonate melt with harzburgite (Yaxley et al., 1998).

5.2. The 180-km boundary

The petrographic evidence for modal metasomatism seen in the Kaavi–Kuopio garnet facies xenoliths (Peltonen et al., 1999) is in agreement with the garnet xenocrysts. The Y–Zr diagram (Fig. 9) shows that the xenocryst record is comprised of depleted and melt metasomatized grains. The wehrlitic pyropes, in particular, preserve evidence of melt metasomatism, whereas the lherzolitic grains show this chemical signature to a lesser extent. Most of the harzburgitic

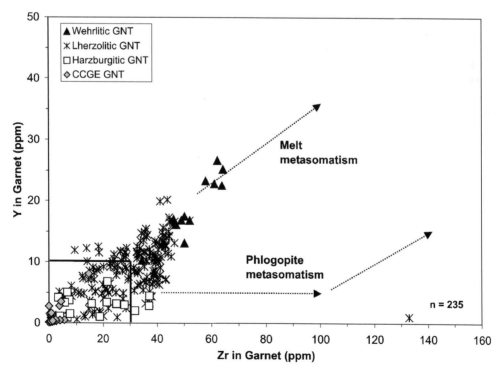

Fig. 9. Plot of Y and Zr for the Kaavi–Kuopio pyrope xenocrysts. Melt and phlogopite metasomatism trends after Griffin et al. (1999b). Solid black line defines the depleted field.

and all of the CCGE xenocrysts plot in the "depleted" field.

As illustrated in Figs. 6 and 7, there is a distinct upper temperature limit for strongly depleted garnets defined by low Ti, Zr and Y contents at 1150 °C (T_{Ni}), corresponding to a depth of 180 km on the local geotherm. According to Griffin and Ryan (1995), this kind of garnet composition "edge" marks the position of the inferred base of the lithosphere. Another interpretation, however, is supported by the existence of a few depleted garnet xenocrysts of T_{Ni}—temperatures higher than 1150 °C in the Kaavi–Kuopio suite (Figs. 6 and 7). These are significant because they may represent remnants of a previously existing depleted layer or deeper extension of the 130–180 km layer that was refertilized by melt metasomatism. Such metasomatism could have caused the destruction of all previously existing harzburgitic material in this layer and forced the crystallization of abundant Ti-rich pyropes.

The T_{Ni} of the compositional edge in garnet also corresponds to the "kink" or "step" seen in xenolith geotherms from various kimberlite fields (Griffin and Ryan, 1995) reflecting a change from an essentially conductive geotherm to a steeper temperature gradient (Finnerty and Boyd, 1987; Griffin and Ryan, 1995). Xenoliths along the "kink" or above the high-T "step" of the geotherm commonly have foliated microstructures and show geochemical evidence of infiltration by asthenosphere-derived melts (Smith et al., 1991; Griffin et al., 1989b). However,

none of the Kaavi–Kuopio xenoliths, from a stratigraphic record that goes down to 240 km, shows any of the textural or geochemical features of sheared high-T peridotites (Kukkonen and Peltonen, 1999). Thus, the 1150 °C/180-km edge determined by the garnet compositional break does not mark the base of the lithospheric mantle at Kaavi–Kuopio, but instead represents a sublithospheric mantle compositional discontinuity.

6. Conclusions

Lithosphere mantle xenoliths in Group I kimberlites from the Kaavi–Kuopio area of Eastern Finland provide evidence of a stratified mantle underlying the SW margin of the Archean Karelian Craton. This craton edge has experienced a complicated history including rifting around 1.95 Ga and continental collision around 1.88 Ga (Nironen, 1997; Peltonen et al., 1998), and consequently, the mantle stratigraphy is complex. With the present resolution, at least three lithospheric mantle layers can be distinguished (Fig. 10): (1) an upper layer, shallower than 130 km, composed of secondary cpx-bearing CCGE pyrope-spinel harzburgites; (2) a middle garnet peridotite layer, from 130 to 180 km, ranging from lherzolite to harzburgite to wehrlite; and (3) a lower, more fertile layer, from 180 to 240 km, from where the bulk of the garnet facies peridotite and eclogite xenoliths were derived.

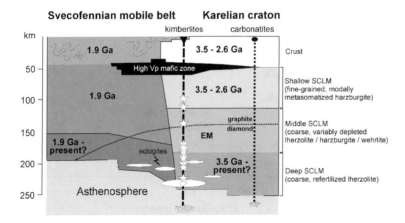

Fig. 10. Schematic cross section through the rifted and redocked edge of the Karelian craton showing the three distinct petrologic layers in the mantle inferred from pyrope and xenolith compositions. Stars mark the depth of origin of the xenoliths based on BKN thermobarometry.

The ages of these distinct mantle layers are still to be resolved, but we believe at least the middle, and likely the upper layer, to be Archean, with a similar age to that of the overlying crust. The lowermost layer may represent Archean mantle that has been considerably refertilized by melt metasomatism. This is supported by the few low-Ti pyropes thought to be from this depth and representing remnant depleted material. Alternatively, it is possible that this section of the mantle has been tectonically emplaced during the collision event at 1.88 Ga. In terms of diamond prospectivity, the graphite–diamond stability curve (Kennedy and Kennedy, 1976) transects the middle mantle layer and implies a diamond window in the Kaavi–Kuopio mantle between 140 and 180 km, defined by the lower temperature stability limit of diamond and the deepest extent to which G10 garnets occur, a roughly 40-km-thick prospective zone. Additionally, a well-sampled component of diamondiferous eclogites adds considerably to the diamond potential of this mantle.

Acknowledgements

This study was funded by the Geological Survey of Finland and the Academy of Finland. Professor John Gurney and Dr. Andreas Späth from the University of Cape Town are acknowledged for the access to the LA-ICP-MS instrument. We are also grateful to the head of the GTK Research Laboratory, Dr. Jukka Marmo, as well as the laboratory personnel for their encouragement and skillful assistance. The support of Professor Ilmari Haapala from the University of Helsinki is appreciated.

References

Brey, G.P., Köhler, T., Nickel, K.G., 1990. Geothermobarometery in four-phase lherzolites: I. Experimental results from 10 to 60 kb. J. Petrol. 31, 1313–1352.

Calcagnile, G., 1982. The lithosphere-asthenosphere system in Fennoscandia. Tectonophysics 90, 19–35.

Canil, D., 1999. The Ni-in-garnet geothermometer: calibration at natural abundances. Contrib. Mineral. Petrol. 136, 240–246.

Carbno, G.B., Canil, D., 2002. Mantle structure beneath the SW Slave Craton, Canada: constraints from Garnet Geochemistry in the Drybones Bay kimberlite. J. Petrol. 43 (1), 129–142.

Cookenboo, H., 1996. Ranch Lake kimberlite in the Central Slave Craton: the mantle sample. The Gangue, vol. 52, Geol. Assoc. of Canada, Miner. Deposits Div., Victoria, Canada, pp. 12–13.

Dawson, J.B., Stephens, W.E., 1975. Statistical analysis of garnets from kimberlites and associated xenoliths. J. Geol. 83, 589–607.

Ehrenberg, S.N., 1982. Petrogenesis of garnet lherzolite and megacrystalline nodules from the Thumb, Navajo volcanic field. J. Petrol. 23 (4), 507–547.

Finnerty, A.A., Boyd, F.R., 1987. Thermobarometry for garnet peridotites: basis for the determination of thermal and compositional structure of the upper mantle. In: Nixon, P.H. (Ed.), Mantle Xenoliths. Wiley, Chichester, pp. 381–412.

Gaál, G., Gorbatschev, R., 1987. An outline of the Precambrian evolution of the Baltic Shield. Precambrian Res. 35, 15–52.

Grégoire, M., Bell, D.R., le Roex, A.P., 2003. Garnet lherzolites from the Kaapvaal craton (South Africa): trace element evidence for a metasomatic history. J. Petrol. 44 (4), 629–657.

Griffin, W.L., Ryan, C.G., 1995. Trace elements in indicator minerals: area selection and target evaluation in diamond exploration. In: Griffin, W.L. (Ed.), Diamond Exploration: Into the 21st Century. J. Geochem. Explor., vol. 53, pp. 311–337.

Griffin, W.L., Cousens, D.R., Ryan, C.G., Sie, S.H., Suter, G.F., 1989a. Ni in chrome pyrope garnets: a new geothermometer. Contrib. Mineral. Petrol. 103, 199–202.

Griffin, W.L., Smith, D., Boyd, F.R., Cousens, D.R., Ryan, C.G., Sie, S.H., Suter, G.F., 1989b. Trace element zoning in garnets from sheared mantle xenoliths. Geochim. Cosmochim. Acta 53, 561–567.

Griffin, W.L., Fisher, N.I., Friedman, J., Ryan, C.G., O'Reilly, S.Y., 1999a. Cr-pyrope garnets in the lithospheric mantle: I. Compositional systematics and relations to tectonic setting. J. Petrol. 40 (5), 679–704.

Griffin, W.L., Shee, S.R., Ryan, C.G., Win, T.T., Wyatt, B.A., 1999b. Harzburgite to lherzolite and back again: metasomatic processes in ultramafic xenoliths from the Wesselton kimberlite, Kimberley, South Africa. Contrib. Mineral. Petrol. 134, 232–250.

Gurney, J.J., 1984. A correlation between garnets and diamonds. In: Glover, J.E., Harris, P.G. (Eds.), Kimberlite Occurrence and Origin: A Basis for Conceptual Models in Exploration, vol. 8, University of Western Australia, Perth, Australia, pp. 143–166.

Kennedy, C.S., Kennedy, G.C., 1976. The equilibrium boundary between graphite and diamond. J. Geophys. Res. 81, 2467–2470.

Kontinen, A., Paavola, J., Lukkarinen, H., 1992. K–Ar ages of hornblende and biotite from Late Archean rocks of eastern Finland; interpretation and discussion of tectonic implications. Geol. Surv. Finland Bull. 365 (31 pp.).

Kopylova, M.G., Russell, J.K., Cookenboo, H., 1999. Petrology of peridotite and pyroxenite xenoliths from the Jericho kimberlite: implications for the thermal state of the mantle beneath the Slave Craton, northern Canada. J. Petrol. 40 (1), 79–104.

Kopylova, M.G., Russell, J.K., Stanley, C., Cookenboo, H., 2000. Garnet from Cr- and Ca-saturated mantle: implications for diamond exploration. J. Geochem. Explor. 68, 183–199.

Kukkonen, I.T., Jõeleht, A., 1996. Geothermal modelling of the lithosphere in the central Baltic Shield and its southern slope. Tectonophysics 255, 24–45.

Kukkonen, I.T., Peltonen, P., 1999. Xenolith controlled geotherm for the central Fennoscandian Shield: implications for lithosphere–asthenosphere relations. Tectonophysics 304 (4), 301–315.

Kukkonen, I.T., Kinnunen, K.A., Peltonen, P., 2003. Mantle xenoliths and thick lithosphere in the Fennoscandian shield. Phys. Chem. Earth 28 (9–11), 349–360.

Nimis, P., Taylor, W.R., 2000. Single clinopyroxene thermobarometry for garnet peridotites: Part I. Calibration and testing of a Cr-in-cpx barometer and an enstatite-in-Cpx thermometer. Contrib. Mineral. Petrol. 139, 541–554.

Nironen, M., 1997. The Svecofennian orogen: a tectonic model. Precambrian Res. 86, 21–44.

O'Brien, H.E., Tyni, M., 1999. Mineralogy and geochemistry of kimberlites and related rocks from Finland. In: Gurney, J.J., Gurney, J.L., Pascoe, M.D., Richardson, S.H. (Eds.), Proceedings of the 7th International Kimberlite Conference. Red Roof Design cc, Cape Town, South Africa, pp. 625–636.

O'Brien, H.E., Peltonen, P, Vartiainen, H., in press. Kimberlites, carbonatites and alkaline rocks. In: Lehtinen, M., Nurmi, P.A., Rämö, O.T. (Eds.), Precambrian Geology of Finland—Key to the Evolution of the Fennoscandian Shield. Elsevier Science, Amsterdam.

Peltonen, P., Mänttäri, I., 2001. An ion microprobe U–Th–Pb study of zircon xenocrysts from the Lahtojoki kimberlite pipe, eastern Finland. Bull. Geol. Soc. Finl. 73, 47–58.

Peltonen, P., Kontinen, A., Huhma, H., 1998. Petrogenesis of the mantle sequence of the Jormua Ophiolite (Finland): melt migration in the upper mantle during Palaeoproterozoic continental break-up. J. Petrol. 39, 297–329.

Peltonen, P., Huhma, H., Tyni, M., Shimizu, N., 1999. Garnet peridotite xenoliths from kimberlites of Finland: nature of the continental mantle at an Archaean craton—Proterozoic mobile belt transition. In: Gurney, J.J., Gurney, J.L., Pascoe, M.D., Richardson, S.H. (Eds.), Proceedings of the 7th International Kimberlite Conference. Red Roof Design cc, Cape Town, South Africa, pp. 664–675.

Peltonen, P., Kinnunen, K.A., Huhma, H., 2002. Petrology of two diamondiferous eclogite xenoliths from the Lahtojoki kimberlite pipe, eastern Finland. Lithos 63 (3–4), 151–164.

Pollack, H.N., Chapman, D.S., 1977. On the regional variations of heat flow, geotherms and lithosphere thickness. Tectonophysics 38, 279.

Robinson, B.W., Graham, J., 1992. Advances in electron microprobe trace element analysis. J. Comput.-Assist. Microsc. 4, 263–265.

Ryan, C.G., Griffin, W.L., Pearson, N.J., 1996. Garnet geotherms: pressure–temperature data from Cr-pyrope garnet xenocrysts in volcanic rocks. J. Geophys. Res. 101, 5611–5625.

Schulze, D.J., 1995. Low-Ca garnet harzburgites from Kimberley, South Africa: abundance and bearing on the structure and evolution of the lithosphere. J. Geophys. Res. 100 (12), 513–526.

Smith, D., Boyd, F.R., 1992. Compositional zonation in garnets in peridotite xenoliths. Contrib. Mineral. Petrol. 112, 134–147.

Smith, D., Ehrenberg, S.N., 1984. Zoned minerals in garnet peridotite nodules from the Colorado plateau: implications from mantle metasomatism and kinetics. Contrib. Mineral. Petrol. 86, 274–285.

Smith, D., Griffin, W.L., Ryan, C.G., Sie, S.H., 1991. Trace-element zonation in garnets from the Thumb: heating and melt infiltration below the Colorado Plateau. Contrib. Mineral. Petrol. 107, 60–79.

Sobolev, N.V., Lavrentiev, Yu.G., Pokhilenko, N.P., Usova, N.P., 1973. Chrome-rich garnets from the kimberlites of Yakutia and their paragenesis. Contrib. Mineral. Petrol. 40, 39–52.

Tyni, M., 1997. Diamond prospecting in Finland—a review. In: Papunen, H. (Ed.), Mineral Deposits: Research and Exploration, Where do They Meet? Proceedings of the 4th SGA Meeting. A.A. Balkema, Rotterdam, pp. 789–791.

Yaxley, G.M., Green, D.H., Kamenetsky, V., 1998. Carbonatite metasomatism in the southeastern Australian lithosphere. J. Petrol. 39, 1917–1930.

Available online at www.sciencedirect.com

SCIENCE DIRECT®

Lithos 77 (2004) 609–637

ELSEVIER

LITHOS

www.elsevier.com/locate/lithos

Petrology and geochemistry of spinel peridotite xenoliths from Hannuoba and Qixia, North China craton

Roberta L. Rudnick[a,*], Shan Gao[b,c], Wen-li Ling[b],
Yong-shen Liu[b], William F. McDonough[a]

[a] Geochemistry Laboratory, Department of Geology, University of Maryland, College Park, MD 20742, USA
[b] Faculty of Earth Sciences, China University of Geosciences, Wuhan 430074, China
[c] Key Laboratory of Continental Dynamics, Department of Geology, Northwest University, Xi'an 710069, China

Received 27 June 2003; accepted 17 February 2004
Available online 25 May 2004

Abstract

We report mineralogical and chemical compositions of spinel peridotite xenoliths from two Tertiary alkali basalt localities on the Archean North China craton (Hannuoba, located in the central orogenic block, and Qixia, in the eastern block). The two peridotite suites have major element compositions that are indistinguishable from each other and reflect variable degrees (0–25%) of melt extraction from a primitive mantle source. Their compositions are markedly different from typical cratonic lithosphere, consistent with previous suggestions for removal of the Archean mantle lithosphere beneath this craton. Our previously published Os isotopic results for these samples [Earth Planet. Sci. Lett. 198 (2002) 307] show that lithosphere replacement occurred in the Paleoproterozoic beneath Hannuoba, but in the Phanerozoic beneath Qixia. Thus, we see no evidence for a compositional distinction between Proterozoic and Phanerozoic continental lithospheric mantle. The Hannuoba xenoliths equilibrated over a more extensive temperature (hence depth) interval than the Qixia xenoliths. Neither suite shows a correlation between equilibration temperature and major element composition, indicating that the lithosphere is not chemically stratified in either area. Trace element and Sr and Nd isotopic compositions of the Hannuoba xenoliths reflect recent metasomatic overprinting that is not related to the Tertiary magmatism in this area.
© 2004 Elsevier B.V. All rights reserved.

Keywords: Peridotite xenolith; Archean craton; North China craton; Major and trace element geochemistry; Sr and Nd isotopes; Thermometry

1. Introduction

Archean cratons are underlain by mantle lithosphere that is thick, cold and refractory (Jordan, 1975, 1988; van der Hilst and McDonough, 1999). Such lithosphere has a high viscosity because it is cold and nearly anhydrous, and thus contributes significantly to craton stability (Pollack, 1986; Hirth et al., 2000). However, not all regions of Archean-aged crust are underlain by such refractory mantle lithosphere, and these regions are characterized by a more protracted history of tectonism and magmatism than their cratonic counterparts. There are at least two possible reasons for the absence of thick mantle keels beneath Archean-aged crust: (1) they may have never formed, or (2) they may have formed but were subsequently removed.

* Corresponding author. Tel.: +1-301-405-1311.
E-mail address: rudnick@geol.umd.edu (R.L. Rudnick).

0024-4937/$ - see front matter © 2004 Elsevier B.V. All rights reserved.
doi:10.1016/j.lithos.2004.03.033

An example of the first possibility is the Mojave terrain in SW U.S. Here, the crust has Paleoproterozoic to late Archean Nd model ages (Bennett and DePaolo, 1987; Raymo and Calzia, 1998) but middle Proterozoic crystallization ages (Wooden and Miller, 1990). Mojavia is underlain by late Archean lithospheric mantle that is considerably more fertile and dense than typical cratonic mantle (Lee et al., 2001). Hence, this mantle lithosphere did not grow to the same thickness as that beneath Archean cratons. Lee et al. (2001) proposed that the thinner lithosphere beneath Mojavia failed to shield this small fragment of Archean lithosphere from tectonic reworking. This study demonstrated that thick lithospheric keels do not always form beneath Archean crust.

An example of the second possibility is the North China craton, where multiple lines of evidence (surface geology, xenolith studies, seismic and heat flow data) show that this craton formed with a thick lithospheric keel in the Archean that was subsequently removed (e.g., Menzies et al., 1993; Griffin et al., 1998). The timing and

mechanisms of lithospheric mantle removal beneath the North China craton are yet to be fully understood.

The present paper reports the petrography, mineral chemistry, thermometry and major element compositions of spinel peridotite xenoliths from the Hannuoba and Qixia localities for which Os data have previously been reported (Gao et al., 2002). In addition, for the Hannuoba peridotites, we present trace element and Sr and Nd isotope geochemistry. We show that the Sr and Nd isotopes reflect recent metasomatic overprinting not related to the Tertiary hosts and that both Hannuoba and Qixia peridotites are indistinguishable in terms of their bulk compositions, despite the fact that their formation ages differ by nearly 2 billion years.

2. Geologic setting

The North China craton is divided into three regions based on geology, tectonic evolution and $P–T–t$ paths

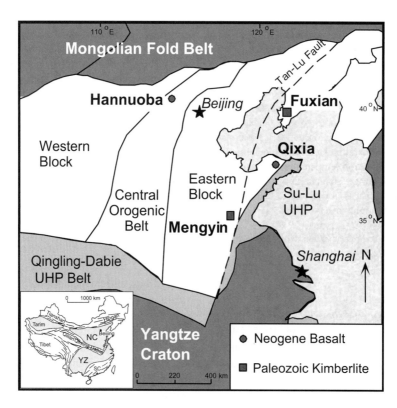

Fig. 1. Map of North China craton (white) showing xenolith localities mentioned in the text. Tectonic subdivisions are based on Zhao et al. (2000, 2001). Inset shows location of the North China craton (NC) relative to other cratonic blocks (e.g., YZ—Yangtze craton) and intervening fold belts.

of metamorphic rocks (Fig. 1; Kusky et al., 2001; Zhao et al., 2000, 2001). The western block forms a stable platform composed of late Archean to Paleoproterozoic metasedimentary belts that unconformably overly Archean basement (Wu et al., 1998; Li et al., 2000; Zhao et al., 2000). The latter consists of granulite facies tonalite–trondhjemite–granodiorite (TTG) gneiss and charnockite (3300 Ma; Kröner et al., 1987) with minor mafic granulite and amphibolite. The Central Orogenic Belt, or Trans-North China orogen, separates the western and eastern blocks of the craton. This belt is composed of late Archean amphibolites and granulites and 2500 Ma granite–greenstone terrains (Kröner et al., 1988; Kern et al., 1996; Zhao et al., 2000, 2001). These are overlain by Paleoproterozoic bimodal volcanic rocks in the southern part of the orogen and thick carbonate and terrigenous sedimentary rocks intercalated with basaltic flows in the central part of the orogen; these deposits may reflect a continental rift

setting during the Paleoproterozoic. The basement of the eastern block of the North China craton is composed of early to late Archean orthogneisses intruded by 2500 Ma syntectonic granitoids (Zhao et al., 2000, 2001). A variety of geochronological and $P–T–t$ evidence, cited in Zhao et al. (2000, 2001), document a major collisional event between the eastern and western blocks of the North China craton between 1800 and 2000 Ma. This event formed the Central Orogenic Belt and represents the final amalgamation of the North China craton.

Unlike other Archean cratons, the North China craton experienced widespread tectonothermal reactivation during the late Mesozoic and Cenozoic. This is documented by the emplacement of voluminous Mesozoic granitic and volcanic rocks (Qiu et al., 2002; Yang et al., 2003), which were followed by extensive Tertiary basaltic volcanism. The latter carries a variety of mantle and crustal xenoliths (Cao and Zhu, 1987),

Table 1
Major and trace element analysis of basalt standards at Northwest University in Xi'an

	BCR-2						GSR-3					
	Meas.	1σ	n	RSD (%)	Rec.	RE (%)	Meas.	1σ	n	RSD (%)	Rec.	RE (%)
SiO_2 (wt.%)	54.0	0.2	4	0.45	54.1	0.16	44.6	0.02	3	0.05	44.64	−0.10
TiO_2	2.26	0.02	4	0.91	2.26	0.11	2.40	0.03	3	1.20	2.37	1.13
Al_2O_3	13.4	0.08	4	0.57	13.5	1.00	13.86	0.10	3	0.74	13.83	0.24
$Fe_2O_3{}^a$	13.9	0.04	4	0.29	13.8	−0.36	13.29	0.08	3	0.63	13.40	−0.85
MnO	0.183	0.005	4	2.74	0.19	3.95	0.16	0.01	3	3.53	0.17	−3.92
MgO	3.69	0.01	4	0.27	3.59	−2.65	7.79	0.04	3	0.49	7.77	0.21
CaO	7.17	0.02	4	0.24	7.12	−0.67	8.81	0.03	3	0.30	8.81	0
Na_2O	3.10	0.10	4	3.34	3.16	1.98	3.47	0.09	3	2.49	3.38	2.56
K_2O	1.80	0.01	4	0.28	1.79	−0.70	2.31	0.02	3	0.66	2.32	−0.29
P_2O_5	0.35	0	4	0	0.35	0	0.94	0.01	3	0.61	0.95	−0.70
V (ppm)	445	19	3	4.18	416	−7.05	200	1	3	0.58	167	19.96
Cr	60	1	3	1.94	18	−231	125	7	3	5.31	134	−6.47
Co	50	2	3	3.08	37	−34	54	2	3	2.81	46.5	16.85
Ni	27	1	3	4.33			137	1	3	0.73	140	−2.14
Zn	128	17	3	13.5	127	−1.05	151	2	3	1.01	150	0.44
Ga	23	1	3	4.35	23	0	24	1	3	4.17	24.8	−3.23
Rb	48.7	0.6	3	1.19	48	−1.39	38.7	0.6	3	1.49	37	4.50
Sr	330	2	3	0.61	346	4.62	1104	12	3	1.08	1100	0.33
Y	32	1	3	3.65	37	14.4	23	3	3	13.48	22	3.03
Zr	179	2	3	1.16	188	4.61	290	1	3	0.50	277	4.69

BCR-2G and GSR-3 are USGS and Chinese National standards, respectively.

n: number of analyses; Meas.: measured value; Rec.: recommended value; RSD: relative standard deviation; RE: relative error between measured and recommended values.

Recommended values are from http://minerals.cr.usgs.gov/geo_chem_stand/ and Govindaraju (1994).

[a] All Fe reported as Fe_2O_3.

some of which are the focus of this paper. The region also experienced the development of extensive sedimentary basins (most of the eastern portion of the craton is covered by Quaternary sediments) and presently has high heat flow (60 mW/m^2; Hu et al., 2000) compared to other Archean cratons (Nyblade et al., 1990). The changes in tectonic and magmatic activity are also reflected in a change in mantle xenolith compositions (Menzies et al., 1993; Griffin et al., 1998); xenoliths carried in Ordovician kimberlites are deep-seated garnet-facies peridotites that are highly refractory, like cratonic xenoliths elsewhere. These xenoliths record the presence of a cold and thick lithospheric keel, consistent with the occurrence of diamonds in the kimberlites. In contrast, xenoliths carried in the Tertiary alkali basalts derive from shal-

Table 2
Analyses of USGS basalt (BHVO-1) and andesite (AGV-1) standards by ICP–MS at the Xi'an laboratory

	Isotope	Blank	BHVO-1 ($n=2$)					AGV-1 ($n=2$)				
			Meas.	1σ	RSD (%)	Rec.	RE (%)	Meas.	1σ	RSD (%)	Rec.	RE (%)
Li	6	99	5.08	0.17	3.4	4.6	10.4	10.5	0.34	3.2	12.0	− 12.4
Be	9	4	0.99	0.01	0.8	1.1	− 10.4	2.21	0.06	2.7	2.1	5.2
Sc	45	32	31.8	0.07	0.2	31.8	− 0.2	12	0.06	0.5	12.2	1.1
V	51	903	314	0.17	0.1	317	− 0.9	122	0.38	0.3	121	0.8
Cr	52	604	285	5	1.8	289	− 1.2	12	0.13	1.1	10.1	19.3
Co	59	55	45	0.29	0.6	45	0.2	15	0.08	0.5	15.3	− 0.6
Ni	60	146	121	3.35	2.8	121	− 0.1	15	1.68	11.1	16	− 5.7
Cu	65	131	138	0.65	0.5	136	1.8	56	0.48	0.8	60	− 6.6
Zn	66	260	110	0.68	0.6	105	4.4	80	0.25	0.3	88	− 8.8
Ga	71	30	21	0.06	0.3	21	0.5	20.3	0.15	0.7	20	1.4
Ge	74	11	1.65	0.03	1.6	1.64	0.4	1.28	0.04	3.3	1.25	2.6
Rb	85	219	9.6	0.13	1.4	11	− 12.6	66	0.18	0.3	67.3	− 2.3
Sr	88	458	399	2.06	0.5	403	− 0.9	662	1.33	0.2	662	0.001
Y	89	30	27.3	0.05	0.2	27.6	− 1.2	21	0.09	0.4	20	5.3
Zr	90	263	173	0.67	0.4	179	− 3.6	233	0.87	0.4	227	2.5
Nb	93	32	19.3	0.07	0.4	19	1.4	15	0.1	0.4	15	− 1.3
Cs	133	10	0.11	0.01	6.0	0.13	− 12.9	1.34	0.01	0.4	1.28	5.0
Ba	135	559	138	0.32	0.2	139	− 0.8	1234	8	0.6	1226	0.6
La	139	102	15.6	0.05	0.3	15.8	− 1.5	38.4	0.11	0.3	38	1.1
Ce	140	135	38.3	0.01	0.0	39	− 1.7	68.4	0.10	0.1	67	2.1
Pr	141	25	5.44	0.02	0.4	5.7	− 4.6	8.4	0.03	0.4	7.6	10.9
Nd	146	107	25.6	0.10	0.4	25.2	1.6	32.8	0.07	0.2	33	− 0.6
Sm	147	27	6.24	0.03	0.5	6.2	0.7	5.9	0.04	0.7	5.9	− 0.8
Eu	151	3	2.01	0.01	0.4	2.06	− 2.2	1.69	0.02	1.0	1.64	2.8
Gd	157	25	6.17	0.06	1.0	6.4	− 3.7	5.40	0.05	0.8	5	8.0
Tb	159	3	0.96	0.003	0.4	0.96	− 0.2	0.70	0.0001	0.0	0.7	0.1
Dy	161	5	5.20	0.01	0.1	5.2	0.0	3.63	0.01	0.3	3.6	0.7
Ho	165	3	0.98	0.003	0.3	0.99	− 1.1	0.68	0.01	1.4	0.67	1.1
Er	166	5	2.36	0.01	0.5	2.4	− 1.6	1.75	0.03	1.8	1.7	3.1
Tm	169	3	0.32	0.002	0.6	0.33	− 2.7	0.25	0.0004	0.2	0.34	− 26
Yb	172	6	2.03	0.02	1.0	2.02	0.6	1.70	0.02	1.4	1.72	− 1.0
Lu	175	5	0.30	0.0001	0.0	0.29	4.1	0.26	0.01	2.5	0.27	− 2.1
Hf	178	10	4.41	0.03	0.6	4.38	0.6	5.10	0.03	0.6	5.1	− 0.02
Ta	181	3	1.23	0.005	0.4	1.23	0.1	0.90	0.01	0.7	0.90	− 0.4
Pb	208	123	2.33	0.11	4.7	2.6	− 10.5	36.3	0.7	1.9	36	0.8
Th	232	7	1.25	0.02	1.2	1.08	16.0	6.39	0.05	0.8	6.5	− 1.6
U	238	4	0.42	0.01	1.4	0.42	0.8	1.88	0.02	1.3	1.92	− 2.2

n: number of analyses; Meas.: measured value; Rec.: recommended value; RSD: relative standard deviation; RE: relative error between measured and recommended values.
Recommended values are from http://minerals.cr.usgs.gov/geo_chem_stand/ and Govindaraju (1994).

Table 3
Major and trace element data for Hannuoba and Qixia spinel lherzolite xenoliths

Sample	SiO$_2$	TiO$_2$	Al$_2$O$_3$	Cr$_2$O$_3$	FeO	MnO	NiO	MgO	CaO	Na$_2$O	K$_2$O	P$_2$O$_5$	Total	Mg#
Hannuoba peridotites														
DMP-04	44.40	0.06	2.29		7.33	0.13	0.31	42.05	1.94	0.28	0.015	0.010	98.79	91.1
DMP-05	44.15	0.13	2.83	0.36	7.53	0.13	0.31	41.55	2.18	0.26	0.020	0.010	99.46	90.8
DMP-19	44.83	0.03	1.91	0.34	7.07	0.12	0.31	40.74	1.80	0.19	0.025	0.010	97.35	91.1
DMP-23a	44.22	0.10	2.32	0.42	7.77	0.13	0.31	41.25	1.64	0.24	0.100	0.030	98.53	90.4
DMP-25	44.39	0.08	1.61	0.36	7.11	0.12	0.31	43.88	1.00	0.30	0.135	0.030	99.32	91.7
DMP-41	44.75	0.06	2.76	0.40	7.74	0.13	0.26	40.15	2.12	0.27	0.004	0.011	98.67	90.2
DMP-51	44.83	0.05	1.96	0.38	7.41	0.12	0.29	41.97	1.89	0.24	0.006	0.003	99.14	91.0
DMP-56	44.79	0.13	3.49	0.35	7.97	0.14	0.26	38.15	3.21	0.36	0.007	0.006	98.85	89.5
DMP-56-rep.	44.59	0.16	3.73	0.34	7.84	0.13	0.27	38.47	3.03	0.28	0.020	0.010	98.87	89.7
DMP-57	44.34	0.06	1.96	0.37	7.41	0.12	0.32	42.47	1.56	0.16	0.020	0.010	98.79	91.1
DMP-58	44.87	0.08	3.16	0.34	7.92	0.13	0.27	38.82	2.76	0.34	0.012	0.005	98.72	89.7
DMP-59	43.99	0.06	2.58	0.40	7.99	0.13	0.29	40.38	2.43	0.27	0.007	0.019	98.54	90.0
DMP-60	46.34	0.11	3.67	0.38	7.48	0.13	0.25	36.68	3.47	0.38	0.006	0.009	98.90	89.7
DMP-60	45.92	0.14	3.90	0.39	7.41	0.13	0.28	37.20	3.24	0.35	0.020	0.015	98.99	89.9
DMP-67c	44.00	0.20	3.78	0.42	8.39	0.14	0.26	37.81	2.88	0.30	0.260	0.020	98.46	88.9
KD-03	44.45	0.11	3.08	0.37	7.70	0.13	0.29	41.05	2.31	0.28	0.020	0.010	99.80	90.5
KD-04	44.23	0.14	3.29	0.34	8.01	0.14	0.27	40.29	2.47	0.32	0.023	0.010	99.53	90.0
Qixia peridotites														
Q1	44.15	0.09	2.56		8.21	0.14		39.94	2.79	0.30	0.064	0.064	98.31	89.7
Q4	44.48	0.10	3.52		7.83	0.13		38.54	3.23	0.32	0.111	0.047	98.31	89.8
Q5	43.94	0.04	1.61		7.83	0.13		43.63	1.08	0.11	0.011	0.020	98.38	90.9
Q6	43.27	0.01	1.02		7.87	0.13		45.15	0.54	0.11	0.010	0.007	98.12	91.1
Q8	44.30	0.01	1.44		7.64	0.13		42.23	1.95	0.06	0.057	0.129	97.94	90.8
Q17	44.18	0.13	3.40	0.35	8.35	0.14	0.24	38.36	2.98	0.33	0.095	0.034	98.23	89.1
QX-07	42.72	0.11	3.41	0.42	7.96	0.13	0.29	40.07	2.45	0.28	0.130	0.070	97.62	90.0
QX-09	41.43	0.02	4.01	0.44	7.56	0.13	0.32	43.40	1.11	0.09	0.040	0.010	98.12	91.1
QX-11	43.59	0.10	2.99	0.34	8.12	0.13	0.29	40.80	2.52	0.30	0.080	0.020	98.94	90.0
QX-13	43.83	0.11	3.09	0.42	7.95	0.13	0.29	39.99	2.53	0.23	0.040	0.060	98.25	90.0
QX-14	45.22	0.02	2.13	0.46	7.23	0.13	0.29	41.44	1.59	0.18	0.120	0.010	98.36	91.1

Sample	C	S	S dup	Li	Be	Sc	Ti	V (XRF)	V	Cr (XRF)	Cr	Co	Ni (XRF)	Ni	Cu	Zn (XRF)	Zn
Hannuoba peridotites																	
DMP-04	272	73		1.64	0.018	12.7		70	51	2739	2477	106	2429	2322	17	57	37
DMP-05	243	5	203					63		2466			2407			64	
DMP-19	1729	91	46	2.34	0.059	10.1		25	39	2322	2238	108	2413	2454	10	37	33
DMP-23A				4.22	0.16	10.6		55	45	2879	2814	103	2432	2432	9	85	56
DMP-25	1202	20	25	2.41	0.105	10.0		41	38	2466	2318	108	2462	2377	10	53	35
DMP-41	283	110		1.55	0.028	12.9	357	52	59	2751	2654	105	2071	2224	15	53	40
DMP-51	264	130		1.85	0.030	11.9	238	43	49	2620	2619	107	2257	2403	14.7	50	36
DMP-56	280	260		2.23	0.043	16.2	715	67	74	2372	2318	103	2029	2006	27.3	57	42
DMP-56-rep.								88		2342			2145			56	
DMP-57	248	70	53					56		2509			2509			49	
DMP-58	184	230		2.13	0.027	14.9	417	57	65	2343	2224	103	2112	2190	20.5	53	38
DMP-59	294	200		2.09	0.049	12.8	357	50	55	2727	2628	110	2307	2366	20.8	56	42
DMP-60	245	320		2.20	0.021	17.2	596	68	75	2579	2530	97	1949	1967	22.4	51	39
DMP-60								85		2690			2209			69	
DMP-67C	183	23	35	4.67	0.16	16.0		83	73	2880	2634	99	2046	2120	12.5	76	54
KD-03	136	17	4					65		2526			2271				53
KD-04								80		2325			2144				59

(continued on next page)

Table 3 (continued)

Sample	C	S	S dup	Li	Be	Sc	Ti	V	V (XRF)	Cr	Cr (XRF)	Co	Ni	Ni (XRF)	Cu	Zn	Zn (XRF)
Qixia peridotites																	
Q1	472	72															
Q4	439	66															
Q5	316	26															
Q6																	
Q8	744	65															
Q17	198	38					780		63		2394			1912			54
QX-07	208	18							68		2853			2254			61
QX-09									27		2991			2529			54
QX-11	199	9							59		2300			2303			56
QX-13									82		2843			2249			59
QX-14									62		3143			2256			51

Sample	Ga (XRF)	Ga	Ge	Rb (XRF)	Rb	Sr (XRF)	Sr	Y (XRF)	Y	Zr (XRF)	Zr	Nb	Cs	Ba	La	Ce	Pr
Hannuoba peridotites																	
DMP-04		2.1	0.9		0.28		4.38		1.46		2.16		0.18	0.008	0.77	0.141	0.35	0.050
DMP-05																	
DMP-19		2.97	0.91		0.81		17.9		0.73		3.43		0.73	0.011	1.40	0.23	0.51	0.068
DMP-23A		2.63	0.92		2.04		43.2		1.99		9.45		2.79	0.042	16.01	2.04	5.08	0.73
DMP-25		1.35	0.92		2.07		17.7		0.80		6.24		2.08	0.005	13.82	2.27	3.56	0.42
DMP-41	8	8.44	0.9	b.d.	0.15	6	6.27	1.8	1.93	3	2.84	0.26	0.003	0.364	0.77	1.40	0.16
DMP-51	2	1.91	0.94	0.20	0.73	4	4.23	1.3	1.3	2	2.27	0.18	0.026	1.20	0.21	0.40	0.075
DMP-56	4	3.73	1.03	0.30	0.35	11	9.83	3.5	3.7	7	6.72	0.49	0.009	0.78	0.17	0.59	0.116
DMP-56-rep.																	
DMP-57																	
DMP-58	3	3.14	0.98	0.50	0.46	5	4.57	2.6	2.7	2	2.57	0.35	0.013	0.545	0.088	0.19	0.037
DMP-59	2	2.39	0.96	0.40	0.44	9	8.31	1.8	1.9	3	3.20	0.53	0.020	3.25	1.32	2.02	0.22
DMP-60	3	3.13	1.08	0.20	0.26	16	15	3.3	3.7	5	5.14	0.22	0.005	5.26	3.06	1.09	0.087
DMP-60																	
DMP-67C		4.62	1.05		4.19		29.7		4.23		15.3	1.23	0.009	16.43	1.24	3.38	0.52
KD-03																	
KD-04																	
Qixia peridotites																	
Q1																	
Q4																	
Q5																	
Q6																	
Q8																	
Q17	4			1.9		45		3.2		8							
QX-07																	
QX-09																	
QX-11																	
QX-13																	
QX-14																	

Sample	Nd	Sm	Eu	Gd	Tb	Dy	Ho	Er	Tm	Yb	Lu	Hf	Ta	Pb	Th	U
Hannuoba peridotites																
DMP-04	0.26	0.10	0.036	0.13	0.030	0.20	0.050	0.14	0.025	0.18	0.031	0.101	0.012	0.116	b.d.	0.005
DMP-05																
DMP-19	0.31	0.074	0.031	0.09	0.016	0.11	0.025	0.08	0.013	0.09	0.017	0.087	0.029	0.318	0.013	0.062
DMP-23A	3.59	0.82	0.26	0.71	0.10	0.48	0.080	0.18	0.023	0.15	0.024	0.26	0.14	0.296	0.149	0.045

Table 3 (continued)

DMP-25	1.71	0.34	0.086	0.29	0.040	0.17	0.028	0.062	0.009	0.060	0.010	0.22	0.11	0.190	0.278	0.058
DMP-41	0.68	0.17	0.062	0.21	0.045	0.30	0.070	0.20	0.032	0.24	0.040	0.15	0.013	0.145	0.015	0.031
DMP-51	0.38	0.13	0.040	0.15	0.031	0.20	0.050	0.14	0.028	0.17	0.034	0.19	0.011	0.172	0.018	0.014
DMP-56	0.74	0.29	0.11	0.38	0.09	0.58	0.14	0.37	0.061	0.41	0.069	0.28	0.009	0.144	b.d.	0.015
DMP-56-rep.																
DMP-57																
DMP-58	0.25	0.15	0.061	0.22	0.057	0.40	0.10	0.29	0.049	0.34	0.055	0.13	0.011	0.133	b.d.	0.029
DMP-59	0.88	0.20	0.062	0.23	0.045	0.29	0.072	0.21	0.034	0.24	0.041	0.14	0.022	0.167	0.072	0.023
DMP-60	0.50	0.24	0.10	0.36	0.082	0.57	0.14	0.40	0.062	0.45	0.075	0.22	0.008	0.220	0.642	0.062
DMP-60																
DMP-67C	2.64	0.74	0.24	0.75	0.14	0.82	0.16	0.44	0.066	0.43	0.071	0.50	0.081	0.514	0.115	0.041
KD-03																
KD-04																
Qixia peridotites																
Q1																
Q4																
Q5																
Q6																
Q8																
Q17																
QX-07																
QX-09																
QX-11																
QX-13																
QX-14																

Major elements as wt.% oxides. FeO as total Fe, originally measured as Fe_2O_3 and converted to FeO. Trace elements in micrograms per gram. Sample numbers listed in italics show samples for which XRF data are from the Xi'an lab, all others are from the University of Massachusetts.

lower depths, are relatively hot and have less refractory compositions. Collectively, these observations indicate the loss of ~ 80–140 km of Archean lithosphere from beneath the eastern portion of the North China craton.

Our previously published Os results (Gao et al., 2002) show evidence for two episodes of replacement of Archean lithospheric mantle. One episode occurred during the Paleoproterozoic beneath the Trans-North China orogen, as the Hannuoba peridotites yield a Paleoproterozoic (1.9 Ga) Re–Os age that overlaps the period of cratonization documented by Zhao et al. (2000, 2001). In contrast, and as predicted from earlier studies (Menzies et al., 1993; Griffin et al., 1998), thick Archean lithosphere persisted under the eastern block of the North China craton through the Ordovician. Garnet peridotite xenoliths carried in the Fuxian kimberlite pipe record Archean Re depletion ages (Gao et al., 2002). The single sample from Mengyin yielded a Mesoproterozoic Re depletion age, but the relatively high Re/Os of this sample suggests disturbance to the Re–Os system. This ancient lithosphere was replaced

after the Ordovician by mantle lithosphere that has Os isotopic characterisitics indistinguishable from modern convecting upper mantle, as observed in xenoliths from Qixia (Gao et al., 2002).

3. Samples and previous work

The samples investigated here are spinel-facies peridotites from the Hannuoba and Qixia xenolith localities (Fig. 1). In this section, we review the xenolith associations at each locality and some results from previous investigations. A more complete discussion of previous results for the peridotite xenoliths, in the context of our new data, is provided in Sections 6 and 7.

3.1. Hannuoba

The Hannuoba basalts (10–22 Ma; Zhu, 1998) carry a remarkable variety of deep-seated xenoliths from both the lower crust and upper mantle. These include mafic to felsic granulites (Gao et al., 2000; Chen et al., 2001; Liu et al., 2001; Zhou et al., 2002; Wilde et al., 2003), spinel- and garnet-bearing pyroxenites (Song and Frey, 1989; Tatsumoto et al., 1992; Chen et al., 2001; Xu, 2002), abundant spinel lherzolites and harzburgites (Song and Frey, 1989; Tatsumoto et al., 1992; Fan et al., 2000; Chen et al., 2001) and rare spinel–garnet lherzolites (Fan and Hooper, 1989; Chen et al., 2001).

Although mafic granulites dominate the granulite xenolith populations, intermediate and felsic granulites are common and metapelite xenoliths are also present. The mafic and intermediate granulites are cumulates interpreted to have formed by magmatic underplating and subsequent fractional crystallization at the base of the crust (Chen et al., 2001; Liu et al., 2001; Zhou et al., 2002), possibly during the Mesozoic, based on zircon U–Pb ages (Fan et al., 1998; Liu et al., 2001). However, more recently, Wilde et al. (2003) suggested that the Phanerozoic zircons in the granulite xenoliths are metamorphic in origin and the granulite protoliths are Precambrian. Liu et al. (2004) found oscillatory-zoned igneous zircon in an olivine pyroxenite to be Mesozoic,

suggesting that the melts that gave rise to the pyroxenites provided the heat to metamorphose the lower crust. The estimated lower crust composition, based on xenolith and geophysical studies, is intermediate (Gao et al., 2000; Liu et al., 2001).

Garnet pyroxenites yield a narrow $P–T$ range (1.6–1.9 GPa, 990–1030 °C) that reflects an elevated geotherm (Chen et al., 2001). However, the narrow pressure window makes its shape difficult to define. The Hannuoba pyroxenites have been variably interpreted as metamorphic segregations (Chen et al., 2001) or young (late Mesozoic) cumulates from basaltic magmas (Xu, 2002). The distinctive trace element patterns of some Al-pyroxenites (Eu anomalies, high field-strength element depletions), coupled with their very wide range of Nd and Sr isotopic compositions, are interpreted to reflect mixing between mantle and crustal sources to generate the basalts from which the pyroxenites precipitated (Xu, 2002).

The samples investigated here are all spinel-facies peridotites from the Damaping (DMP) locality in the Hannuoba basalt field. They range in size from 10 to 35 cm and in composition from lherzolite to harzburgite

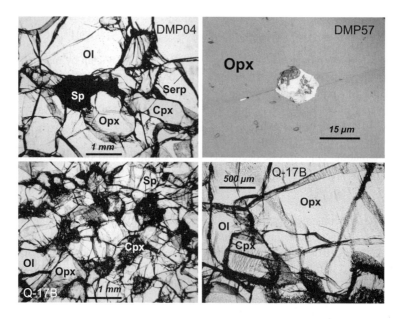

Fig. 2. Photomicrographs of representative textures of Hannuoba and Qixia spinel-facies peridotites. Upper panels: Hannuoba lherzolite DMP-04 (left) shows typical granuloblastic texture of Hannuoba xenoliths and serpentinite along grain boundary (plane polarized light). Primary sulfide (monosulfide solid solution) within orthopyroxene in lherzolite DMP-57 (right), reflected light. Lower panels (plane polarized light): Qixia lherzolite QX-17B (left) shows breakdown texture of clinopyroxene, whereas right panel shows a large orthopyroxene porphyroblast adjacent to clinopyroxene and olivine. Both pyroxenes contain exsolution lamellae.

(i.e., <5% clinopyroxene). Although composite xenoliths are relatively common at Hannuoba, none of the samples investigated here contain any pyroxenite veins. Os isotope results for these samples demonstrate that they formed ~ 1.9 Ga ago as residues of partial melting (Gao et al., 2002). Trace element and Sr, Nd and Pb isotopic compositions of clinopyroxene separates reflect later metasomatic overprinting (Song and Frey, 1989; Tatsumoto et al., 1992; see Discussion).

3.2. Qixia

In contrast to Hannuoba, mantle xenoliths from the 9–18 Ma Qixia olivine nephelinites have re-

ceived less attention (Fan and Hooper, 1989; Zheng et al., 1998), perhaps because of their relatively small size <6 cm and comparative scarcity. In addition to spinel-facies peridotites, the Qixia nephelinite also carries clinopyroxenites (Zheng et al., 1998) and olivine websterites (Fan et al., 2000), the latter of which can be considerably Fe-rich (Fo_{83}) and have evolved Sr and Nd isotopic compositions ($^{143}Nd/^{144}Nd = 0.51141$, $^{87}Sr/^{86}Sr = 0.70956$; Fan et al., 2000). Trace element patterns of clinopyroxenes have been used to infer the origin of the spinel peridotites as residues of up to 20% fractional melting, followed by metasomatic enrichment of the highly incompatible elements, perhaps by carbo-

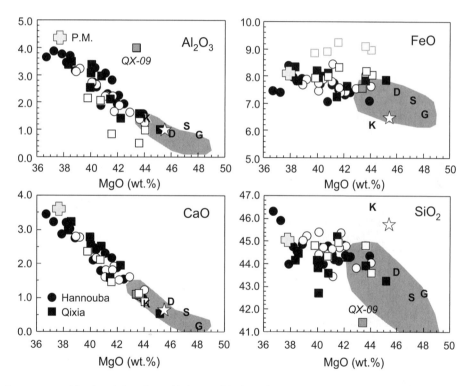

Fig. 3. Major element compositional variations for peridotite xenoliths investigated here. Circles are data for Hannuoba spinel peridotites: ○ represent previously published data (Song and Frey, 1989; Chen et al., 2001); ● are data from this study. ■ are data from Qixia xenoliths (this study) and □ are previously published data (Zheng et al., 1998). The iron data published by Zheng et al. is reported as FeO. However, if true, these samples would plot significantly above all others on the MgO vs. FeO plot (shown as faint gray squares). We assume that the data are in fact Fe_2O_3 and have plotted the data accordingly. One Qixia xenolith from this study (QX-09, gray symbol) shows an anomalously high Al_2O_3 content and low SiO_2 content, reflecting an overabundance of spinel in the mode. Cross denotes primitive mantle (P.M.) from McDonough and Sun (1995). Gray field represents cratonic peridotite xenoliths from the Tanzanian craton (Lee and Rudnick, 1999). These do not show the extensive metasomatic overprints that are more typical of cratonic peridotites from the Kaapvaal craton (including orthopyroxene enrichment; Boyd, 1989). Averages for other cratons given by letters: K: Kaapvaal (Boyd, 1989); D: Daldyn, Siberia; S: Slave (both from Griffin et al., 1999); G: east Greenland (from Bernstein et al., 1998); ☆ represents average "Archon" from Griffin et al. (1999).

natites (Zheng et al., 1998). These results are discussed further in Discussion in the context of the new data presented here.

The Qixia xenoliths investigated here are spinel lherzolites, harzburgites and a unique dunite. All xenoliths are found within the host lava, most of which are only relatively thin veneers of peridotite on an exposed weathering surface. The size of the xenoliths investigated here is small, ranging from 2 to 5 cm. The dunite (QX-18), which consists entirely of mosaic textured olivine, has an extremely forsteritic olivine composition (Fo$_{98}$, Table 4). The origin of this unusual sample is unclear and it is not discussed further in this paper. Re–Os investigations of the spinel-facies peridotites show them to have $^{187}Os/^{188}Os$ that is indistinguishable from modern convecting mantle (Gao et al., 2002), consistent with

recent (Mesozoic or younger) formation of this section of lithospheric mantle.

4. Analytical methods

The xenoliths were sawn from their lava hosts and the cut surfaces were abraded with quartz in a sand blaster to remove any possible contamination from the saw blade. The samples were then disaggregated between thick plastic sheets with a rock hammer and reduced to powder using first an alumina disk mill followed by an alumina ring mill. A portion of the crushed fraction was sieved and clinopyroxene separates were handpicked under a binocular microscope to a purity of >98%. They were cleaned in an ultrasonic bath in distilled water before isotopic analysis.

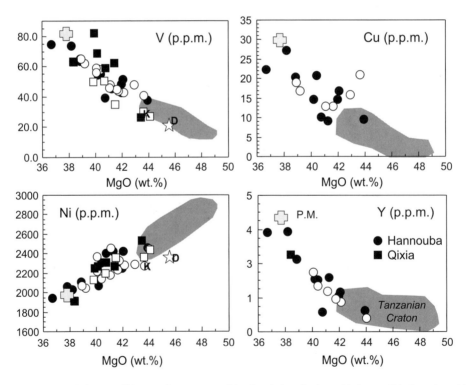

Fig. 4. Compatible and moderately incompatible trace element compositional variations for the peridotite xenoliths investigated here. Circles are data for Hannuoba spinel peridotites: ○ represent previously published data (Song and Frey, 1989; Chen et al., 2001); ● are data from this study. ■ are data from Qixia xenoliths (this study) and □ are previously published data (Zheng et al., 1998). Cross denotes primitive mantle (P.M.) from McDonough and Sun (1995). Gray field represents cratonic peridotite xenoliths from the Tanzanian craton (Lee and Rudnick, 1999). Averages for other cratons given by letters: K: Kaapvaal; D: Daldyn (Griffin et al., 1999); and ☆ represents average "Archon" from Griffin et al. (1999).

Mineral analyses were carried out on polished thick sections both at Harvard University, using a Camebax MBX electron microprobe (EMP), and at the University of Maryland on a JEOL 8900 Electron Probe Microanalyzer. Analyses on the Camebax were performed in wavelength dispersive mode with 15 keV accelerating voltage and a 15 nA beam current. Samples analysed at the University of Maryland Laboratory for Microscopy and Microanalysis utilized the following operating conditions: 15 keV accelerating voltage, 20 nA cup current and a 5–10 μm beam. Natural standards were used for the analysis of olivine (Fe, Mg, Ni and Si—San Carlos olivine; Al and Ca—Kakanui hornblende; Mn—Rockport fayalite), pyroxene (Mn and Fe—Rockport fayalite; Mg and Ni—San Carlos olivine; Cr—Johnstown hypersthene; Ca—Mammoth Lakes wollasonite; Al—augite; Na and Ti—Kakanui hornblende) and chromite (Cr, Mn and Si—Bushveld chromite; Mg, Al and Fe—spinel). Raw intensities were corrected using the Bence–Albee (olivine) and CIT–ZAF (chromite and pyroxene) algorithms.

Major element compositions of whole rocks were determined by XRF on fused glass disks at the University of Massachusetts at Amherst (see Rhodes, 1996, for analytical details) and Northwest University in Xi'an, China. Samples analysed in the latter laboratory are designated with italicized labels in Table 3. Selected trace elements (V, Cr, Ni, Zn, Rb, Sr, Y and Zr) were also determined on pressed powder pellets by XRF in both laboratories. Accuracy and precision of the XRF data from the Xi'an lab can be evaluated from results obtained for USGS standard BCR-2 and Chinese National standard GSR-3 (Table 1). Precision (RSD) is better than 6% for the major and trace elements. Accuracy, as indicated by relative difference (RE) between measured and recommended values, is better than 4% for major elements and 14% for most of the trace elements. The only exceptions are Cr and Co in BCR-2, whose measured values are significantly higher than the recommended values for this standard, by a factor of 230% and 34%, respectively. Both elements are in low concentration in BCR-2; the correspondence between measured and recommended values is better for GSR-3, which has higher concen-

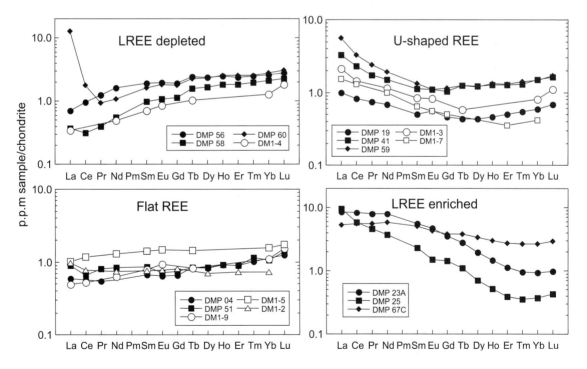

Fig. 5. Chondrite-normalized REE patterns for Hannuoba whole rock peridotites. Filled symbols are data from this work and open symbols are data from Song and Frey (1989). Chondrite values from McDonough and Sun (1995).

trations of both elements. Moreover, the close correspondence between peridotite Cr values obtained in
both XRF laboratories and between the XRF and
ICP–MS data (Table 3 and discussion below), leads
us to conclude that the Xi'an XRF Cr values for the
peridotites are accurate to within 10%.

Whole rock trace element compositions were determined by ICP–MS (Elan 6100 DRC) after acid
digestion of samples in Teflon bombs at Northwest
University, Xi'an, China. Table 2 shows results for
two USGS standards (BHVO-1 and AGV-1) analysed
in this laboratory during the course of these analyses.
Based on these analyses, precision is generally better
than 5% for most elements and accuracy is better than
10%, with many elements agreeing to within 2% of
the reference values. Exceptions are Rb, Cs and Th in
BHVO-1 (which differ by up to 16%) and Cr and Tm
in AGV-1, which differ by up to 26% from the
reference values. The Tm value we obtained for
AGV-1 (0.25 ppm) is significantly lower than the

recommended value (0.34 ppm); the latter is probably
in error, based on the positive Tm anomaly created by
this value in the chondrite-normalized rare earth
element (REE) pattern for AGV-1. The higher Cr
values we obtained for AGV-1 may reflect polyatomic
interferences (e.g., $^{36}Ar^{16}O$ on ^{52}Cr). However, a
similar discrepancy in Cr data is not observed for
BHVO-1, which has a higher Cr content. With the
exception of Zn, elemental concentrations determined
by both XRF and ICP–MS (V, Cr, Ni, Ga, Rb, Sr, Y
and Zr) agree to within 10% (Table 3).

Nd and Sr isotopic compositions were determined
using a multicollector Finnigan MAT-261 mass spectrometer operated in static multicollector mode at the
Isotope Laboratory of the China University of Geosciences, Wuhan. Two aliquots of sample powder
(200 mesh), ~100 mg each, were weighed. To one,
a known quantity of mixed ^{84}Sr, ^{85}Rb, ^{149}Sm and
^{145}Nd spike solution was added. Samples were
digested in Teflon bombs with a mixture of concen-

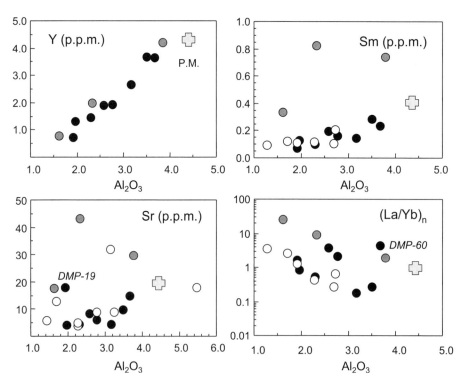

Fig. 6. Moderately incompatible to highly incompatible trace element compositional variations for the Hannuoba xenoliths. ● from this work, ○
from Song and Frey (1989). Gray circles represent the three strongly LREE-enriched peridotites (DMP-23A, DMP-25 and DMP-67C; Fig. 5).
Cross denotes primitive mantle (P.M.) from McDonough and Sun (1995). DMP-60, labeled on the $(La/Yb)_n$ plot (lower right panel), shows a
strong increase in La and Ce (see Fig. 5).

trated HF, HNO_3 and $HClO_4$. The sealed bombs were kept in an oven at 185 °C for 48 h. The decomposed samples were then dried on a hot plate and converted into chlorates by adding more concentrated $HClO_4$. This was followed by dry down and addition of concentrated HCl to form chlorides, followed by a final evaporation. The dried salts were dissolved again in 500 µl of dilute HCl and then loaded onto columns of AG50W-X8 resin for separation and purification of Rb, Sr and REE, with the REE cut finally loaded on to HDEHP columns for separation of Nd and Sm by HCl eluants. The measured $^{143}Nd/^{144}Nd$ and $^{87}Sr/^{86}Sr$ ratios were normalized to $^{146}Nd/^{144}Nd = 0.721900$ and $^{86}Sr/^{88}Sr = 0.11940$, respectively. External reproducibility of the isotope measurements can be judged from repeat analyses of international standards. The average $^{143}Nd/^{144}Nd$ ratio of the La Jolla standard

measured during the sample runs is 0.511862 ± 5 (2σ, $n = 15$). Analysis of BCR-2 gave $^{143}Nd/^{144}Nd = 0.512635 \pm 4$, $^{147}Sm/^{144}Nd = 0.1369$, $Nd = 29.10$ ppm and $Sm = 6.591$ ppm, which fall within uncertainty of the recommended values ($^{143}Nd/^{144}Nd = 0.512647 \pm 22$, 2σ; White and Patchett, 1984; $Nd = 28.8$ ppm; $Sm = 6.59$ ppm; Govindaraju, 1994) for BCR-1. Six analyses of the NBS-987 standard gave $^{87}Sr/^{86}Sr = 0.710236 \pm 16$ (2σ, $n = 15$).

5. Petrography

5.1. Hannuoba

The Hannuoba spinel-facies peridotites range from coarse- to medium-grained and have granulo-

Table 4
Average EMP analyses of olivines from spinel peridotites

	Qixia													
	Sp lherzolites						Dunite							
	Q1	Q4	Q5	Q6	Q8	Q17	QX-02	QX-07	QX-10	QX-11	QX-13	QX-14	QX-17	QX-18
n	5	4	5	6	8	5	5	8	5	8	8	5	5	8
SiO_2	40.93	40.41	40.76	40.25	40.81	40.33	40.58	41.11	40.60	40.64	41.17	40.28	40.71	41.97
FeO	9.45	9.80	8.82	7.77	8.57	10.23	9.23	9.59	9.75	9.53	9.56	9.25	9.24	1.72
MnO	0.13	0.14	0.12	0.12	0.10	0.15	0.13	0.13	0.14	0.11	0.14	0.14	0.12	0.07
MgO	49.49	48.87	50.03	49.28	48.88	48.92	49.51	49.42	49.24	48.03	49.63	48.98	48.99	54.96
NiO	0.39	0.38	0.41	0.33	0.36	0.37	0.36	0.36	0.35	0.35	0.36	0.36	0.38	0.02
CaO	0.02	0.04	0.03	0.02	0.05	0.02	0.03	0.01	0.03	0.02	0.01	0.02	0.02	0.01
Total	100.41	99.64	100.18	97.76	98.77	100.03	99.85	0.00	100.11	98.68	100.88	99.03	99.45	98.74
Fo	90.3	89.9	91.0	91.9	91.0	89.5	90.5	90.2	90.0	90.0	90.3	90.4	90.4	98.3

	Hannuoba										
	Sp Lherzolites										
	DMP04	DMP19	DMP23a	DMP25	DMP41	DMP51	DMP56	DMP58	DMP59	DMP60	DMP67c
n	5	6	6	5	5	4	5	5	5	5	6
SiO_2	41.27	41.07	40.85	41.22	40.36	40.49	40.18	40.49	40.60	40.15	40.89
FeO	8.66	8.38	8.53	8.15	9.34	8.62	9.82	9.42	9.25	9.60	10.10
MnO	0.12	0.11	0.12	0.10	0.14	0.12	0.15	0.13	0.14	0.14	0.14
MgO	49.48	49.49	49.01	49.83	49.50	49.21	48.83	48.72	49.01	48.78	48.16
NiO	0.36	0.37	0.36	0.38	0.39	0.39	0.37	0.37	0.38	0.36	0.35
CaO	0.03	0.04	0.04	0.06	0.04	0.07	0.06	0.06	0.05	0.05	0.05
Total	99.91	99.46	98.90	99.75	99.77	98.91	99.39	99.20	99.42	99.07	99.68
Fo	91.1	91.3	91.1	91.6	90.4	91.1	89.9	90.2	90.4	90.1	89.5

n: number of individual spot analyses.
Fo: forsterite component $= 100 \times$ molar $Mg/(Mg + Fe)$.

blastic textures (Fig. 2). Most samples are massive, but a few show foliation defined by aligned spinel grains. In general, both clinopyroxene and orthopyroxenes are homogenous, showing no exsolution lamellae. All xenoliths show alteration along grain boundaries and fractures (typically pale brown serpentine), but the degree of alteration is highly variable from one sample to the next (Fig. 2). A characteristic of the Hannuoba xenoliths is the relative abundance of fresh sulfides, which are found both as inclusions within silicate phases (Fig. 2) or decorating healed fractures.

5.2. Qixia

Qixia spinel-facies peridotites are typically medium-grained with granuloblastic textures. A few samples are coarse-grained (Q8, QX-09) and/or contain deformed olivine porphyroclasts. Several (QX-11, Q17B) show a pronounced foliation, but most are massive. All show pockets of partially crystallized melt, presumably derived from the host basalt, as the melt pocket abundance increases towards the xenolith margin. Orthopyroxene and clinopyroxene typically contain exsolution lamellae.

6. Results

6.1. Whole rock major and trace element data

Major and trace element analyses of whole rock samples are reported in Table 3 and plotted in Figs. 3–6. Both Hannuoba and Qixia peridotites show a con-

Table 5
Average EMP analyses of orthopyroxene from spinel peridotites

	Qixia											
	Q1	Q4	Q5	Q8	Q17	QX-02	QX-07	QX-10	QX-11	QX-13	QX-14	QX-17
n	4	5	4	7	5	8	8	5	8	8	7	3
SiO_2	56.00	55.71	56.14	56.09	55.32	55.09	55.62	55.39	55.39	55.92	54.83	54.72
TiO_2	0.07	0.09	0.09	0.01	0.12	0.08	0.08	0.07	0.05	0.08	0.08	0.11
Al_2O_3	3.66	3.81	2.96	2.45	4.06	3.71	4.13	3.57	3.57	3.78	3.46	4.10
Cr_2O_3	0.35	0.19	0.41	0.46	0.25	0.29	0.27	0.27	0.28	0.33	0.23	0.33
FeO	6.32	6.20	5.90	5.54	6.73	6.21	6.28	6.25	5.98	6.30	6.20	6.10
MnO	0.14	0.15	0.12	0.13	0.15	0.15	0.15	0.16	0.14	0.16	0.14	0.13
NiO				0.09			0.09		0.07	0.08		
MgO	34.03	32.26	34.55	33.35	33.19	33.94	33.28	33.86	33.35	33.48	33.60	33.44
CaO	0.47	0.48	0.47	0.55	0.44	0.44	0.39	0.42	0.41	0.39	0.43	0.40
Na_2O	0.04	0.05	0.03	0.02	0.05	0.03	0.05	0.06	0.03	0.04	0.04	0.05
Total	101.08	98.93	100.67	98.69	100.29	99.92	100.35	100.05	99.27	100.57	99.01	99.39

	Hannuoba										
	DMP-04	DMP-19	DMP-23a	DMP-25	DMP-41	DMP-51	DMP-56	DMP-58	DMP-59	DMP-60	DMP-67c
n	6	6	9	6	8	7	7	5	9	9	6
SiO_2	55.64	55.60	55.63	56.52	55.45	55.47	54.97	55.25	55.29	54.67	54.94
TiO_2	0.06	0.03	0.05	0.04	0.07	0.08	0.13	0.09	0.09	0.09	0.13
Al_2O_3	4.21	3.69	3.84	2.85	3.77	3.67	4.61	4.44	3.95	4.25	5.17
Cr_2O_3	0.46	0.55	0.50	0.52	0.35	0.46	0.31	0.33	0.41	0.31	0.35
FeO	5.49	5.15	5.31	5.02	5.90	5.39	6.18	5.94	5.79	6.01	6.26
MnO	0.12	0.11	0.12	0.11	0.14	0.13	0.12	0.14	0.13	0.13	0.14
NiO	0.09	0.07	0.08	0.10							0.11
MgO	33.07	33.39	32.35	33.91	33.87	33.65	32.70	33.08	33.14	33.42	31.96
CaO	0.64	0.59	0.62	0.61	0.51	0.69	0.70	0.68	0.59	0.56	0.69
Na_2O	0.08	0.07	0.14	0.03	0.08	0.09	0.11	0.12	0.08	0.09	0.14
Total	99.86	99.24	98.63	99.71	100.14	99.64	99.84	100.08	99.45	99.53	99.89

n: number of individual spot analyses.

siderable spread in major element compositions, ranging from fertile compositions approaching primitive mantle to refractory harzburgites with up to 45% MgO. These refractory compositions slightly overlap the compositional field of cratonic peridotites, as exemplified by samples from the Tanzanian craton (Rudnick et al., 1994; Lee and Rudnick, 1999) and are distinct from cratonic peridotite averages (Boyd, 1989; Griffin et al., 1999). The Tanzanian xenoliths were chosen as representatives of cratonic lithosphere as they show good correlations on MgO vs. major oxide diagrams (Fig. 3) and do not show the same extent of metasomatic overprinting that some other well-studied cratonic xenoliths exhibit (e.g., Kaapvaal craton samples). In particular, the SiO_2-enrichment that is prevalent in the

Kaapvaal low-temperature peridotites (Boyd, 1989) is rare in the Tanzanian samples (Rudnick et al., 1994; Lee and Rudnick, 1999).

The Qixia samples show more scatter on Al_2O_3 and CaO vs. MgO plots than the Hannuoba xenoliths. This is probably due to their very small sample size and consequently biased mineralogical sampling, and probably also to the presence of the pockets of host basalts described above. For example, sample QX-09 has the highest Al_2O_3 content of the suite and also one of the highest MgO contents, causing it to fall off the negative correlation between Al_2O_3 and MgO (Fig. 3). However, this sample is anomalous in no other way and plots within the data array on the CaO, FeO and SiO_2 vs. MgO diagrams and has mineral compositions

Table 6
Average EMP analyses of clinopyroxenes from spinel peridotites

	Qixia												
	Q1	Q4	Q5	Q6	Q8	Q17	QX-02	QX-07	QX-10	QX-11	QX-13	QX-14	QX-17
n	5	5	5	7	6	5	7	7	7	7	8	7	5
SiO_2	52.57	52.68	52.53	54.44	53.72	52.08	52.42	52.51	52.36	52.20	52.68	51.32	51.35
TiO_2	0.31	0.50	0.36	0.03	0.03	0.62	0.42	0.50	0.39	0.43	0.46	0.43	0.55
Al_2O_3	5.46	6.87	4.58	1.90	2.62	7.16	5.84	7.14	6.46	6.20	6.38	6.14	6.45
Cr_2O_3	0.77	0.72	1.21	0.77	0.84	0.70	0.73	0.80	0.76	0.77	0.91	0.69	0.71
FeO	2.45	2.43	2.15	1.92	2.14	2.62	2.27	2.32	2.23	2.39	2.43	2.40	2.35
MnO	0.08	0.08	0.07	0.07	0.08	0.09	0.08	0.07	0.09	0.07	0.08	0.07	0.09
MgO	15.31	14.57	15.66	17.26	17.11	14.21	15.16	14.32	14.83	14.66	14.76	14.86	14.56
CaO	21.49	20.74	21.67	22.95	22.69	20.58	21.49	20.49	21.02	21.26	21.12	21.31	20.61
Na_2O	1.49	1.94	1.45	0.47	0.31	2.15	1.55	1.87	1.54	1.53	1.60	1.43	1.94
Total	99.93	100.51	99.69	99.80	99.54	100.21	99.96	100.03	99.67	99.50	100.42	98.66	98.61
Mg#	91.8	91.5	92.9	94.1	93.4	90.6	92.3	91.7	92.2	91.6	91.6	91.7	91.7

	Hannuoba										
	DMP-04	DMP-19	DMP-23a	DMP-25	DMP-41	DMP-51	DMP-56	DMP-58	DMP-59	DMP-60	DMP-67c
n	7	6	5	7	9	5	8	9	10	10	7
SiO_2	53.00	52.88	52.73	53.41	52.56	52.81	52.23	52.54	52.51	52.08	51.31
TiO_2	0.26	0.14	0.13	0.13	0.35	0.27	0.60	0.38	0.39	0.45	0.92
Al_2O_3	5.66	5.02	3.25	3.49	6.13	4.96	6.62	6.32	5.65	6.49	4.56
Cr_2O_3	0.92	1.28	1.24	1.09	0.99	0.96	0.66	0.73	0.91	0.75	0.85
FeO	2.35	1.96	2.45	2.03	2.26	2.33	2.79	2.66	2.33	2.33	3.43
MnO	0.08	0.08	0.09	0.08	0.08	0.07	0.08	0.09	0.08	0.09	0.11
MgO	15.71	15.67	16.66	16.75	15.30	16.10	15.23	15.51	15.44	15.23	16.19
CaO	20.26	20.46	21.33	21.25	20.58	20.35	19.61	19.79	20.62	20.06	21.10
Na_2O	1.35	1.31	0.51	0.79	1.77	1.35	1.73	1.72	1.57	1.71	0.46
Total	99.58	98.82	98.39	99.00	100.03	99.21	99.54	99.74	99.50	99.19	98.93
Mg#	92.3	93.4	92.4	93.6	92.3	92.5	90.7	91.2	92.2	92.1	89.4

n: number of individual spot analyses.
Mg# = 100 × molar Mg/(Mg + Fe).

similar to other samples (Tables 4–7; Fig. 7). The high Al_2O_3 content may reflect an oversampling of spinel in the whole rock powder. Similar sampling biases may explain the two samples from the literature that fall below the Al_2O_3–MgO array (Fig. 3)—in this case, an undersampling of spinel. For these reasons, the mineral chemical data for Qixia xenoliths are viewed as a more reliable indication of rock composition than the whole rock data.

Compatible (Ni) and moderately incompatible (V, Cu, Y) trace elements show positive and negative correlations, respectively, when plotted against MgO

(Fig. 4). In contrast, highly incompatible trace elements (La, Ce, Nb, Ta, Ba, Sr, Th, U) show no correlation with MgO and are only poorly correlated with each other (not shown). As for the major elements, the compatible and moderately incompatible trace element abundances of these Chinese samples are distinct from those of cratonic peridotites (Fig. 4).

Rare earth element (REE) patterns of the Hannuoba spinel-facies peridotites are plotted in Fig. 5. The patterns range dramatically from light REE (LREE) depleted to LREE-enriched, with a number of samples having flat and U-shaped REE patterns. A comparison

Table 7
Average EMP analyses of spinel from spinel peridotites

	Qixia												
	Q1	Q4	Q5	Q6	Q8	Q17	QX-02	QX-07	QX-10	QX-11	QX-13	QX-14	QX-17
n	5	5	5	9	6	5	5	4	5	6	6	5	5
TiO_2	0.05	0.05	0.12			0.06	0.05		0.05			0.04	0.06
V_2O_5	0.09	0.07	0.15			0.05	0.07		0.07			0.08	0.05
Al_2O_3	56.44	60.07	45.79	29.01	35.94	60.72	58.83	59.22	59.05	59.22	56.20	58.63	59.43
Cr_2O_3	11.81	8.54	22.50	39.67	32.48	8.01	10.38	9.45	10.00	9.84	12.23	8.84	9.10
FeO	11.64	10.41	12.94	16.58	14.36	11.26	10.38	10.62	10.76	10.45	11.58	10.78	10.49
MnO	0.10	0.09	0.12	0.19	0.13	0.11	0.10	0.07	0.08	0.08	0.10	0.09	0.09
MgO	19.99	20.52	18.55	15.31	17.63	20.50	20.54	21.37	20.24	21.46	20.82	20.45	20.43
NiO	0.36	0.40	0.28			0.40	0.32		0.36			0.37	0.35
ZnO	0.09	0.08	0.13			0.08	0.10		0.10			0.09	0.09
Total	100.56	100.23	100.57	100.75	100.53	101.19	100.77	100.72	100.70	101.05	100.92	99.38	100.10
Cr#	12.3	8.7	24.8	47.9	37.8	8.1	10.6	9.7	10.2	10.0	12.7	9.2	9.3
Mg#	75.4	77.8	71.9	62.2	68.6	76.5	77.9	78.2	77.0	78.5	76.2	77.2	77.6

	Hannuoba										
	DMP-04	DMP-19	DMP-23a	DMP-25	DMP-41	DMP-51	DMP-56	DMP-58	DMP-59	DMP-60	DMP-67c
n	6	5	5	5	5	6	5	5	5	5	6
TiO_2					0.09	0.14	0.16	0.11	0.11	0.10	
V_2O_5					0.07	0.11	0.06	0.08	0.08	0.07	
Al_2O_3	52.87	46.35	43.89	37.39	57.09	49.32	58.66	57.36	54.85	59.11	57.34
Cr_2O_3	15.90	22.39	24.33	31.33	12.49	18.65	8.76	10.39	13.33	9.00	9.43
FeO	10.34	10.21	11.94	11.72	10.12	10.65	10.23	10.19	10.15	9.60	12.84
MnO	0.10	0.09	0.11	0.13	0.09	0.10	0.09	0.09	0.10	0.09	0.09
MgO	21.20	20.14	19.39	18.85	20.50	19.68	21.05	20.94	20.36	20.84	20.63
NiO					0.34	0.30	0.40	0.34	0.33	0.40	
ZnO					0.06	0.06	0.05	0.06	0.06	0.07	
Total	100.41	99.18	99.66	99.41	100.85	99.02	99.46	99.56	99.36	99.26	100.33
Cr#	16.8	24.5	27.1	36.0	12.8	20.2	9.1	10.8	14.0	9.3	9.9
Mg#	78.5	77.9	74.3	74.1	78.3	76.7	78.6	78.6	78.1	79.5	74.1

n: number of individual spot analyses.
Cr# = $100 \times$ molar Cr/(Cr + Al).
Mg# = $100 \times$ molar Mg/(Mg + Fe).

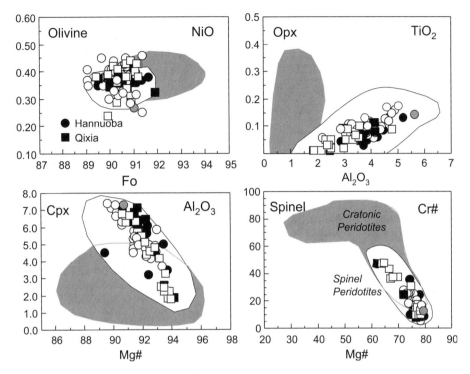

Fig. 7. Mineral compositional plots for Hannuoba and Qixia spinel peridotites and a single garnet lherzolite from Hannuoba (HT-28 from Fan and Hooper, 1989). Circles are data for Hannuoba spinel peridotites: ○ are previously published data (Fan and Hooper, 1989; Song and Frey, 1989; Chen et al., 2001); ● are data from this study; the gray circle is garnet peridotite HT-28. ■ are data from Qixia xenoliths (this study) and □ are previously published data (Fan and Hooper, 1989; Zheng et al., 1998). OPX: orthopyroxene; CPX: clinopyroxene. Open field is for worldwide spinel peridotites and gray field encompasses minerals from cratonic peridotites (including Kaapvaal, east Greenland, Siberia and Tanzania).

between clinopyroxene REE patterns and whole rock patterns by Song and Frey (1989) showed that whole rocks consistently have higher La/Sm than the clinopyroxene, indicating that some LREE reside elsewhere in the peridotite (e.g., on grain boundaries or in accessory phases). Furthermore, mass balance calculations, described below, show that two of the samples investigated here (DMP-41 and DMP-59) contain a significant amount of LREE in phases other than clinopyxoxene. Both these samples have U-shaped REE patterns and relatively high concentrations. The origin of the LREE mass balance discrepancies is considered further in the Discussion. These differences notwithstanding, Song and Frey (1989) generally found that clinopyroxene and whole rock REE patterns are quite similar, especially for the middle to heavy REE (HREE), implying that the whole rock REE pattern faithfully records the pre-entrainment REE pattern of the peridotite. This conclusion is supported by LA–ICP–MS clinopyroxene analyses (Gao et al.,

unpublished data), which generally mimic the whole rock REE pattern, including the unusual LREE enrichment seen in sample DMP-60 (Fig. 5). Clinopyroxenes from this sample and DMP-59 show zoned La and Ce concentrations, with marked La and Ce enrichment on the mineral rims (Pr and heavier REE do not vary from rim to core). Other incompatible trace elements (e.g., Rb, Sr, Ba) also showed enrichments on the rims. No other clinopyroxenes were observed to exhibit REE zoning in this study, but Song and Frey (1989) describe one of their samples as also having compositionally heterogeneous clinopyroxene (DM1–3). This sample also has a U-shaped REE pattern.

Excellent correlations are observed for Y vs. Al_2O_3 (Fig. 6), which are similar to those observed globally in anhydrous peridotites (McDonough and Frey, 1989). The HREE show similar correlations (not shown). The middle REE (Sm to Gd) also show positive correlations when plotted against Al_2O_3, except that samples with the greatest LREE enrichments (the LREE enriched

Table 8
Modal mineralogy[a] of Hannuoba peridotites

	Ol	Opx	Cpx	Sp	Total	SSQ[b]
DMP-04	67.5	21.9	8.9	1.6	99.9	0.0
DMP-19	61.7	29.8	7.8	0.8	100.0	0.1
DMP-23a	64.5	27.1	6.3	2.0	100.0	0.5
DMP-25	70.3	24.2	3.8	1.5	99.8	0.2
DMP-41	57.5	30.9	9.6	1.9	99.8	0.1
DMP-51	64.4	27.0	7.6	0.9	100.0	0.1
DMP-56	54.9	27.7	15.3	2.1	100.0	0.1
DMP-58	56.7	28.4	13.0	2.0	100.0	0.2
DMP-59	63.5	23.6	11.0	1.9	100.0	0.2
DMP-60	44.7	37.4	16.1	1.8	100.0	0.2
DMP-67c	55.7	27.8	12.9	3.2	99.6	0.2

Ol: olivine; Opx: orthopyroxene; Cpx: clinopyroxene; Sp: spinel.

[a] Calculated using MINSQ (Hermann and Berry, 2002). See text for details.

[b] Sum of squared residuals.

samples in Fig. 5) fall off the trend defined by the other samples. In contrast, there is no correlation between LREE concentrations and Al_2O_3 (not shown), or between the most highly incompatible trace elements and any other elements or equilibration temperature. There is a weak, negative correlation between the degree of LREE enrichment [as reflected by $(La/Yb)_n$] and Al_2O_3, with the data for our samples showing considerably more scatter than those of Song and Frey (1989), and the most LREE-enriched samples falling off the trend established by the others.

6.2. Mineral chemistry, modes and thermometry

Mineral chemical data are reported in Tables 4–7 and plotted in Fig. 7. Minerals in these xenoliths are homogenous (based on several core and rim analyses in each sample), and thus, average compositions are reported. Mineral compositions of Qixia and Hannuoba peridotites overlap significantly, although it appears that Qixia samples range to slightly more refractory compositions than the Hannuoba samples, based on the Cr# of spinels (Fig. 7). The minerals in the Chinese xenoliths are identical to those of off-craton spinel peridotite xenoliths, worldwide, but are clearly distinct from those in cratonic peridotites (Fig. 7). Although the general absence of garnet in the Chinese samples will result in higher Al_2O_3 contents in pyroxenes and spinels (hence, lower Cr# in the latter) compared with garnet-bearing cratonic peridotites, it is interesting to note that the single

garnet-bearing, refractory (Fo_{91}) spinel lherzolite from Hannuoba (sample HT-28 of Fan and Hooper, 1989) contains pyroxenes with high Al_2O_3 and TiO_2 contents and spinels with low Cr# and falls within the spinel-facies peridotite data arrays in Fig. 7. Therefore, these mineral compositional differences between cratonic peridotites and the North China craton samples primarily reflect differences in bulk rock compositions, rather than differences in facies (garnet vs. spinel) or temperature, and high-

Table 9
Equilibration temperatures (°C) of spinel lherzolite and pyroxenite xenoliths, calculated from average mineral compositions

Sample	T Ca in Opx	T_{BKN}	Wells	N&T
Qixia				
Q1	880	850	860	785
Q4	890	880	875	790
Q5	880	815	840	750
Q6				920
Q8	920	950	970	955
Q17	870	820	825	720
QX-02	870	835	850	765
QX-07	850	925	895	840
QX-09	915	950	895	865
QX-10	860	910	890	840
QX-11	860	865	865	790
QX-13	847	902	888	830
QX-14	870	840	850	770
QX-17	855	810	825	720
Hannouba				
DMP-04	950	1050	1005	990
DMP-19	930	1020	985	965
DMP-23a	945	1035	1035	990
DMP-25	940	1030	1010	990
DMP-41	900	935	910	990
DMP-51	965	1030	990	990
DMP-56	970	1040	985	980
DMP-56px	960	1020	965	980
DMP-58	960	1030	980	985
DMP-59	930	1000	960	990
DMP-60	920	990	945	990
DMP-67c	965	1055	1040	960
JSB-01px	960	1015	965	980

Temperatures are rounded to nearest 5 °C.

T Ca in Opx: calcium in opx thermometer of Brey and Köhler (1990).

T_{BKN}: opx–cpx thermometer of Brey and Köhler (1990).

Wells: two-pyroxene thermometer of Wells (1977).

N&T: cpx thermometer of Nimis and Taylor (2000).

Pressure of 1.5 GPa assumed throughout.

"px": pyroxenite.

light further the noncratonic character of the Hannuoba and Qixia peridotites.

Modal mineralogy, calculated from least squares mixing of mineral proportions to match whole rock compositions, is presented in Table 8 for the Hannuoba samples. The calculations were performed using the MINSQ program of Hermann and Berry (2002). All whole rock analyses used in the calculation were summed to 100% anhydrous and all Fe is reported as FeO. Hermann and Berry suggested that a value of the sum of squared residuals (SSQ) of less than 0.5 is an acceptable outcome in most cases. All of the modes shown in Table 8 conform to this criterion. Similar calculations were not performed for the Qixia xenoliths because the overall small sample sizes makes the validity of any modes calculated from whole rock data uncertain.

Equilibration temperatures calculated from selected cation exchange thermometers are given in Table 9 and are plotted in Figs. 8 and 9. Several features are

apparent from these diagrams. Firstly, although there is a positive correlation between temperatures calculated from different thermometers, the degree of correlation between different thermometers is variable. For example, Ca-in-orthopyroxene temperatures (Brey and Köhler, 1990) show considerable scatter when plotted against two-pyroxene temperatures for both the Hannuoba and Qixia data sets (Fig. 8, upper panels). In contrast, the Wells (1977) and Brey and Köhler (1990) calibrations of the two-pyroxene thermometer show a tighter correlation (Fig. 8, lower panels), with Hannuoba data falling within a fairly narrow array and Qixia data showing more scatter. Similar correlations are observed between temperatures calculated using the Nimis and Taylor (2000) clinopyroxene thermometer and the two-pyroxene temperatures (not shown).

Secondly, there appears to be a systematic offset in temperatures calculated using data from different laboratories for all but the lower left panel in Fig. 8. The

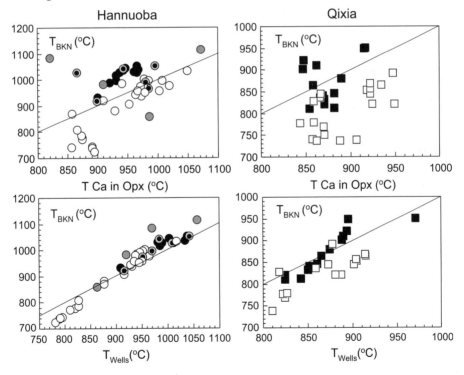

Fig. 8. Comparison of equilibration temperatures calculated from Ca-in-orthopyroxene (Brey and Köhler, 1990) and two-pyroxene thermometers given by Wells (1977) and Brey and Köhler (1990) (BKN) for Hannuoba (circles) and Qixia (squares) peridotite xenoliths. Filled symbols are temperatures calculated from data in Table 9 (this study); open symbols are calculated from the published data of Chen et al. (2001) for Hannuoba, and Zheng et al. (1998) for Qixia. Gray circles are calculated from data of Fan and Hooper (1989) and gray circles with black centers are calculated from data of Song and Frey (1989).

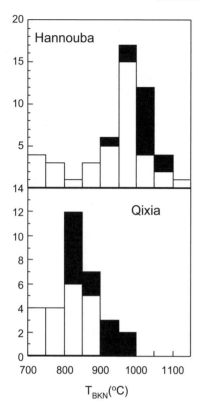

Fig. 9. Two pyroxene equilibration temperature histogram for Hannuoba (upper) and Qixia (lower) spinel peridotite xenoliths. Temperatures calculated using the Brey and Köhler (1990) calibration. Filled symbols are temperatures calculated from data in Table 9 (this study); open symbols are calculated from the published data of Song and Frey (1989), Fan and Hooper (1989) and Chen et al. (2001) for Hannuoba, and Zheng et al. (1998) for Qixia.

open symbols in Fig. 8 were calculated from microprobe data gathered at Macquarie University (Zheng et al., 1998; Chen et al., 2001), and these temperatures are systematically offset from those calculated from the data presented here, which were gathered at Harvard University and the University of Maryland. The published data of Song and Frey (1989) scatter between the two data sets in the upper panels but follow the good correlation in the lower panels, whereas results from the published data of Fan and Hooper (1989—gray circles) scatter the most of any data set. It is likely that the offset in temperatures reflects EMP calibration differences for Ca in orthopyroxene between different labs, highlighting the sensitivity of this thermometer to small differences in calibration. Because of this scatter,

we will use only two pyroxene temperatures in further discussion of equilibration temperatures.

Finally, there appears to be a real temperature difference between the Hannuoba and Qixia xenoliths. The Hannuoba data set shows a large range in equilibration temperatures, from 725 to 1100 °C, whereas the Qixia temperatures are generally < 900 °C (Fig. 9). In addition, there is no correlation between equilibration temperature and major element, trace element or isotopic composition for either xenolith suite, indicating that the lithospheric mantle is not compositionally stratified beneath these two areas of the North China craton.

6.3. Sr and Nd isotopes

The Sr and Nd isotopic compositions of both whole rocks and clinopyroxene separates from the Hannuoba xenoliths are given in Table 10 and plotted in Figs. 10–12 along with previously published data for both Hannuoba (Song and Frey, 1989; Fan et al., 2000) and Qixia (Fan et al., 2000). Two of the clinopyroxene separates have been run in duplicate. For one of these (DMP-58 cpx), the $^{143}Nd/^{144}Nd$ reproduces within uncertainty, but not the $^{87}Sr/^{86}Sr$, which is quite different between the two duplicates. For the other (DMP-60 cpx), $^{87}Sr/^{86}Sr$ is reproduced within uncertainty but $^{143}Nd/^{144}Nd$ is not. This irreproducibility of Nd in clinopyroxene from DMP-60 is consistent with the zoning in LREE, as revealed by LA–ICP–MS analyses (Gao et al., unpublished data) and the unusual whole rock REE pattern (Fig. 5), and suggests that the zoning may be accompanied by isotopic variations (cf. Schmidberger et al., 2003, who found Sr isotope heterogeneities within single clinopyroxenes from peridotites from Somerset Island). The only other sample that shows similar zoning of LREE in clinopyroxene is DMP-59, which is also characterized by extreme enrichments of LREE on the rims.

The whole rock Nd and Sr isotopic compositions generally do not match those of the clinopyroxenes (Fig. 10). In most cases, the whole rocks have more radiogenic $^{87}Sr/^{86}Sr$ and/or less radiogenic $^{143}Nd/^{144}Nd$ compared to the clinopyroxene. However, one sample (DMP-19) has clinopyroxene with *less* radiogenic Nd than the whole rock, and another sample (DMP-41) has clinopyroxene with *more* radiogenic Sr than the whole rock. Table 11 shows mass balances for

Table 10
Sr and Nd isotopic data for whole rock and Cr-diopside separates from Hannuoba peridotites

		Nd (ppm)	Sm (ppm)	$^{147}Sm/^{144}Nd$	$^{143}Nd/^{144}Nd$	$\pm 2\sigma$	ε_{Nd}	Sr (ppm)	Rb (ppm)	$^{87}Rb/^{86}Sr$	$^{87}Sr/^{86}Sr$	$\pm 2\sigma$
DMP-04	Whole rock	0.22	0.08	0.2183	0.513185	40	10.7	4.51	0.26	0.1685	0.704442	61
DMP-04-cpx	Cr diopside	2.82	0.99	0.2122	0.513143	8	9.9	39.6	0.58	0.0426	0.703634	20
DMP-05-cpx	Cr diopside	4.95	1.83	0.2230	0.513165	8	10.3	66.8	0.17	0.0074	0.703026	18
DMP-19	Whole rock	0.25	0.06	0.1554	0.513068	25	8.4	18.8	0.74	0.1138	0.704369	24
DMP-19-cpx	Cr diopside	3.28	0.97	0.1781	0.512832	8	3.8	76.4	0.24	0.0090	0.704353	21
DMP-23a	Whole rock	3.24	0.75	0.1392	0.513033	16	7.7	47.6	1.97	0.1195	0.703709	17
DMP-25	Whole rock	1.78	0.33	0.1139	0.512921	8	5.5	20.4	2.09	0.2965	0.704585	19
DMP-41	Whole rock	0.62	0.19	0.1885	0.513025	13	7.5	6.63	0.14	0.0632	0.703396	40
DMP-41-cpx	Cr diopside	3.10	1.16	0.2269	0.513359	9	14.1	57.8	0.21	0.0105	0.703696	22
DMP-51	Whole rock	0.2	0.08	0.2387	0.513280	22	12.5	3.64	0.32	0.2558	0.703889	55
DMP-56	Whole rock	0.65	0.26	0.2409	0.513294	10	12.8	10.8	0.34	0.0920	0.703337	75
DMP-56-cpx	Cr diopside	4.32	1.74	0.2428	0.513241	7	11.8	65.1	0.30	0.0132	0.702571	19
DMP-58	Whole rock	0.23	0.13	0.3395	0.513149	25	10.0	5.05	0.49	0.2789	0.704485	60
DMP-58-cpx	Cr diopside	1.45	0.99	0.4132	0.513645	10	19.6	27.3	0.16	0.0166	0.703246	29
DMP-58-cpx duplicate	Cr diopside	1.45	1.00	0.4161	0.513651	10	19.8	27.3	0.16	0.0173	0.703759	38
DMP-59	Whole rock	0.8	0.17	0.1269	0.513127	11	9.5	8.96	0.43	0.1395	0.704515	32
DMP-59-cpx	Cr diopside	2.96	1.00	0.2047	0.513327	8	13.4	69.4	0.51	0.0214	0.703717	18
DMP-60	Whole rock	0.46	0.21	0.2759	0.513154	78	10.1	16.3	0.29	0.0515	0.703232	18
DMP-60-cpx	Cr diopside	2.74	1.21	0.2669	0.513831	16	23.3	154	0.29	0.0055	0.703252	18
DMP-60-cpx duplicate	Cr diopside	3.00	1.35	0.2719	0.513780	15	22.3	155	0.31	0.0058	0.703153	29
DMP-67c	Whole rock	2.38	0.64	0.1642	0.512963	9	6.3	32.4	4.11	0.3670	0.705446	17

Nd and Sr based on the measured clinopyroxene and whole rock concentrations and the modal clinopyroxene calculations from Table 8. Because the latter are estimated to have uncertainties on the order of $\pm 10\%$, only mass balance discrepancies greater than 10% are considered significant. As can be seen from the table, all but one sample show significantly more Sr in the whole rock than can be accounted for by the clinopyroxene alone, and several samples also show significantly more Nd in the whole rock than can be accounted for by the clinopyroxene. Indeed, one sample (DMP-59) has one and one-half times more Nd in the whole rock than is present in the clinopyroxene. This sample has an elevated, U-shaped REE pattern and has clinopyroxene that is strongly zoned in LREE. The significance of these differences between clinopyroxene and whole rock Sr and Nd compositions are returned to in the Discussion.

Our Hannuoba clinopyroxene data range to higher $^{143}Nd/^{144}Nd$ than previously published data, and generally have higher $^{87}Sr/^{86}Sr$ for a given $^{143}Nd/^{144}Nd$ (Fig. 11). The latter may reflect a labile Sr component in the unleached clinopyroxenes analysed here. The clinopyroxene data form a positive trend on a Sm–Nd isochron diagram (Fig. 12) but do not show any correlation on a Rb–Sr isochron diagram (not shown). The clinopyroxene Nd and Sr data (both published and our new results) correlate with major element compositions (as noted by Song and Frey), with more fertile compositions having higher $^{143}Nd/^{144}Nd$ and lower $^{87}Sr/^{86}Sr$ (Fig. 13). As for the Sr–Nd isotope plot (Fig. 11), our $^{87}Sr/^{86}Sr$ data are offset to higher values than published data (Fig. 13).

7. Discussion

7.1. Origin of the peridotites and major element systematics

The major element, compatible and moderately incompatible trace element and mineral chemical data for the Hannuoba and Qixia spinel peridotite xenoliths reflect their origin as residues from variable degrees of partial melting of a primitive mantle composition (Figs. 3, 4, 7). Based on major element systematics

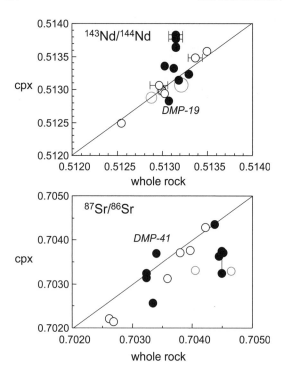

Fig. 10. Comparison of Nd and Sr isotopic compositions of clinopyroxenes and whole rocks for Hannuoba peridotites. Analytical uncertainties (2 σ) are generally equal to or smaller than the symbol size, except where shown. Filled circles are data from Table 10; open black circles are data from Song and Frey (1989) and open gray circles are data from Tatsumoto et al. (1992). Diagonal line marks condition of whole rock = clinopyroxene. Only a few whole rocks have isotopic compositions within uncertainty of the clinopyroxene values. Most whole rocks have elevated ^{87}Sr/^{86}Sr and lower ^{143}Nd/^{144}Nd than the clinopyroxene separates.

(Fig. 3; Table 3) and published experimental melting studies, these peridotites may reflect between 0% to 25% removal of a batch melt from a primitive mantle composition at pressures between 1 and 3 GPa (Walter, 1999). The lack of correlation between equilibration temperature and major element composition in either suite shows that the lithosphere is not chemically stratified beneath the North China craton. Thus, variations in degree of partial melting are not easily related to a simple lithospheric column generated during a single upwelling event, where shallower residues are predicted to be more refractory than deeper residues.

These same geochemical parameters also demonstrate that both lithospheric sections overlap substantially in terms of their bulk composition, although the

partial melting events in which they formed were widely separated in time (by >1300 Ma; Gao et al., 2002). There is thus no indication in these data that Phanerozoic lithosphere is any less refractory than Proterozoic lithosphere (Griffin et al., 1999; O'Reilly et al., 2001). Indeed, on average, the Mesozoic lithosphere studied may be slightly more refractory than Proterozoic lithosphere, based on spinel compositions that range to slightly higher Cr# in the Qixia samples (Fig. 7).

Another important observation from these data is that both lithospheric sections are compositionally distinct from the highly refractory mantle lithosphere that characterizes Archean cratons (Figs. 3, 4, 7). This is consistent with previous observations from the North China craton (Menzies et al., 1993; Griffin et al., 1998; Zheng et al., 1998; Fan et al., 2000) and, coupled with Os isotope results (Gao et al., 2002), documents replacement of cratonic lithosphere that occurred in two separate episodes: (1) the Proterozoic, beneath the Central Orogenic Belt and (2) the Mesozoic, beneath the Eastern Block (Fig. 1).

7.2. Sr and Nd variations between clinopyroxene and whole rocks

Discrepancies in Nd and Sr mass balance and differences in Nd and especially Sr isotopic compositions between whole rocks and clinopyroxenes in the Hannuoba peridotites may reflect one or both of the following: (1) addition of Sr and/or Nd on grain boundaries during metasomatism, host basalt infiltration or posteruption alteration; (2) the presence of Sr- or Nd-bearing accessory phases that may or may not be in isotopic equilibrium with the clinopyroxene.

Whole rocks with higher ^{87}Sr/^{86}Sr and/or lower ^{143}Nd/^{144}Nd than their paired clinopyroxene may be explained by a grain boundary phase (glass or alteration) that is isotopically more evolved than the clinopyroxene (e.g., Zindler and Jagoutz, 1988). Such an explanation can account for most of the whole rock–clinopyroxene Sr isotope differences observed here. In all but one sample (DMP-56), the whole rock Sr abundance is significantly (i.e., >10%) greater than that accounted for by the clinopyroxene. However, not all samples with elevated Sr contents have radiogenic whole rock compositions. In fact, the samples with the largest mass balance discrepancies for Sr (DMP-19

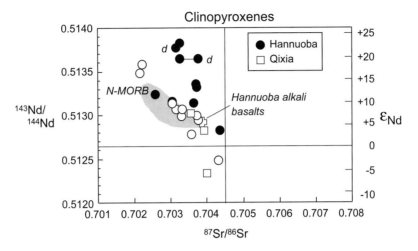

Fig. 11. Nd and Sr isotopic compositions of clinopyroxenes from Hannuoba and Qixia spinel peridotites. Duplicate analyses are connected by tielines or labeled with "d". Filled circles are data from this study. Open circles are previously published data from Song and Frey (1989), Tatsumoto et al. (1992) and Fan et al. (2000). Qixia data (squares) are from Fan et al. (2000). Open cross represents range in isotopic composition of Hannuoba alkali basalts (from Song et al., 1990). Field of N-MORB is from PetDB (Lehnert et al., 2000) for Atlantic and Pacific MORB.

and DMP-60, for which 70–85% of the whole rock Sr is *not* in clinopyroxene; Table 11) do not exhibit differences in the Sr isotopic compositions between clinopyroxene and host rock. This suggests one of two

possibilities. Firstly, the "missing" Sr may occur in a Sr-rich accessory phase that is in isotopic equilibrium with the clinopyroxene. However, we are at a loss to identify this phase from either thin section or bulk rock

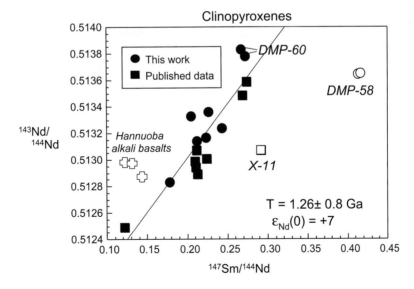

Fig. 12. Sm–Nd isochron plot for Hannuoba clinopyroxene separates from this study (circles) and previously published data (squares) from Song and Frey (1989) and Tatsumoto et al. (1992). Open crosses represent compositions of Damaping basalts that are the hosts to the xenoliths ($^{143}Nd/^{144}Nd$ data from Song et al., 1990; $^{147}Sm/^{144}Nd$ calculated from Sm and Nd data in Zhi et al., 1990). Linear regression yields "errorchron" of ca. 1.3 Ga with a high initial ε_{Nd} of +7. Samples plotted as labeled open symbols were not included in the regression.

Table 11
Mass balance of Sr and Nd in Hannuoba peridotites

Sample	Cpx mode (%)	Sr wr meas. (ppm)	Sr wr calc. (ppm)	Difference (meas. − calc.)	Percentage of difference (%)	Nd wr meas. (ppm)	Nd wr calc. (ppm)	Difference (meas. − calc.)	Percentage of difference (%)
DMP-04	8.9	4.51	3.53	1.0	22	0.22	0.25	− 0.03	− 12
DMP-19	7.8	18.8	5.96	12.8	68	0.25	0.26	− 0.01	− 2
DMP-41	9.6	6.63	5.55	1.1	16	0.62	0.30	0.32	108
DMP-56	15.3	10.83	9.96	0.9	8	0.65	0.66	− 0.01	− 2
DMP-58	13.0	5.05	3.54	1.5	30	0.23	0.19	0.04	22
DMP-59	11.0	8.96	7.63	1.3	15	0.80	0.33	0.47	146
DMP-60	16.1	16.3	2.62	13.7	84	0.46	0.44	0.02	4

chemistry. For example, neither of these samples have particularly high P_2O_5, which suggests that apatite is not the culprit. Alternatively, the Sr may occur in Sr-rich grain boundary melts that have equilibrated with the clinopyroxene in some, but not all, xenoliths prior to their entrainment in the host basalt. One of the samples with "missing" Sr (DMP-60) also has clinopyroxenes that are zoned with respect to Sr and the LREE, and shows the greatest discrepancy in Nd isotopic composition between clinopyroxene and whole rock (the clinopyroxene is 10 ε units lower than the whole rock). Infusion of a LILE-enriched melt from an isotopically evolved source, coupled with faster Sr diffusion and isotopic equilibration compared to LREE (Sneeringer et al., 1984; Van Orman et al., 2001) may therefore explain these observations. However, this scenario does not explain why Nd mass balances in this sample. That is, all of the Nd appears to be accounted for by the zoned clinopyroxene.

A radiogenic, Sr-rich grain boundary phase is consistent with the results of Song and Frey (1989) who found leached clinopyroxenes to yield generally lower $^{87}Sr/^{86}Sr$ than the whole rocks and the leachates to have higher $^{87}Sr/^{86}Sr$ than the whole rocks. However, one of our samples (DMP-41) has a whole rock composition that is less radiogenic than the clinopyroxene, which accounts for only 84% of the total Sr. This result suggests either the presence of an additional Sr-bearing phase that is less radiogenic than the clinopyroxene (0.7037), or the presence of isotopically zoned clinopyroxene, which may have formed during recent metasomatic overprinting (e.g., Schmidberger et al., 2003).

The correspondence between the Nd isotopic composition of clinopyroxenes and whole rocks is generally much better than for Sr (Fig. 10) and where discrepancies do exist, the clinopyroxene is generally more radiogenic than the whole rock, consistent with

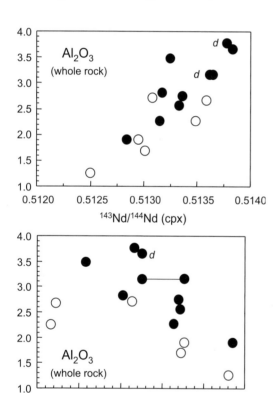

Fig. 13. Sr and Nd isotopes measured in clinopyroxene vs. whole rock Al_2O_3 concentration. Duplicate analyses are connected by tie-lines or labeled with "d". The most fertile and least metasomatized peridotites have the highest $^{143}Nd/^{144}Nd$ and lowest $^{87}Sr/^{86}Sr$. ● are data from Table 10; ○ are data from Song and Frey (1989).

some unradiogenic Nd (along with radiogenic Sr?) being introduced during alteration or recent metasomatism. The unradiogenic Nd isotopic composition of clinopyoxene leachates measured by Song and Frey (1989) supports this interpretation. However, one of our samples (DMP-19) contains clinopyroxene that is *less* radiogenic in Nd than the whole rock (Fig. 10). This sample does not exhibit REE zoning in the clinopyroxene and mass balance using the clinopyroxene mode (Table 11), which shows that the clinopyroxene accounts for all of whole rock Nd (Nd mass balances to within 2%). This mass balance makes it difficult to argue for an additional REE-bearing phase being present in the rock. The reason for the discrepancy between the Nd isotopic composition of the clinopyroxene and whole rock is not known, but it may also indicate isotopically heterogeneous clinopyroxene. Peridotites with U-shaped REE patterns or anomalously-enriched La and Ce (DMP-60) show the greatest discrepancy in Nd mass balance and/or greatest discrepancy in Nd isotopic composition between clinopyoxene and whole rock. For the other LREE-enriched peridotites (Fig. 5, lower right panel), we have no clinopyroxene data. The correspondence between LREE-enrichment and Nd isotopic heterogeneity is consistent with the lack of correlation between LREE concentration and other parameters of melt depletion (e.g., Al_2O_3) and suggests that LREE and Sr have been introduced to the peridotite after partial melting.

7.3. The timing of metasomatic overprinting

The Hannuoba peridotites experienced overprinting of the highly incompatible trace elements following melt depletion, with the most refractory samples showing the strongest overprint, as reflected in the LREE/HREE ratios (Fig. 6). Such systematics were noted long ago for peridotites from SE Australia (Frey and Green, 1974) and appear to be the norm worldwide, but are not fully understood (e.g., McDonough and Frey, 1989 and references therein). What constraints, if any, can be placed on the timing of this metasomatic overprinting?

The LREE zoning observed in two of the Hannuoba clinopyroxenes and the discrepancy between Nd isotopes in clinopyroxene vs. whole rocks reflects a lack of equilibrium in the LREE that will dissipate over time

due to diffusion. However, diffusivity of La in clinopyroxene at the temperatures of Hannuoba xenolith equilibration is quite slow: $\sim 10^{-24}$ m^2 s^{-1} at 1000 °C, extrapolated from data of Van Orman et al. (2001). This diffusion coefficient predicts that such LREE zoning may persist on a several millimeter scale for more than seven billion years at these temperatures. In contrast, Sr diffusivity is considerably faster. Using the data of Sneeringer et al. (1984) for Sr diffusion in clinopyroxene (10^{-19} to 10^{-21} m^2 s^{-1}, extrapolated to 1000 °C) we calculate that the observed zoning will be obliterated within 30 Myr. These results suggest that the addition of LILE-rich material occurred during the Tertiary beneath Hannuoba.

Further constraints on the timing of metasomatism may be had from the correlation between Sr and Nd isotope composition and whole rock composition. These correlations (Fig. 13) can be interpreted in two ways: (1) ancient melt depletion (at 1.9 Ga from Os) followed shortly thereafter (i.e., in the Proterozoic) by variable LILE metasomatism, or (2) ancient melt depletion followed by recent metasomatism by a fluid or melt having an evolved isotopic signature. The first explanation is consistent with the positive correlation on the Nd isochron plot (Fig. 11), which yields a Proterozoic "errorchron". In this scenario, the LREE depleted samples grow progressively more radiogenic relative to LREE enriched samples. In contrast, there is no correlation between Rb/Sr and $^{87}Sr/^{86}Sr$, therefore, the variation in $^{87}Sr/^{86}Sr$ is not due to ^{87}Sr ingrowth since the Proterozoic, and we therefore view the first alternative as unlikely.

In the second hypothesis, the metasomatic overprinting occurred recently (probably Mesozoic or younger) because the variable Sm/Nd in the peridotites would cause the trend in Fig. 13 to disappear over time due to radioactive in-growth (a similar correlation between major element composition, and Sr and Nd isotopic data with similar chronological implications was observed in lower crustal xenoliths from north Queensland, Australia; Rudnick et al., 1986). A rough positive correlation between 1/Nd and $^{143}Nd/^{144}Nd$ supports this mixing hypothesis. However, a similar (negative) correlation does not exist for 1/Sr vs. $^{87}Sr/^{86}Sr$, suggesting that additional processes may have affected the Sr isotopic composition. The scatter on the Sr–Nd isotope plot (Fig. 11) also suggests that Sr has been affected by additional processes. If the mixing hypothesis is correct, the metasomatic agents had a maximum ^{143}Nd/^{144}Nd of 0.51250 ($\varepsilon_{Nd} = -3$) and a minimum ^{87}Sr/^{86}Sr of 0.7045 (based on the most evolved isotopic compositions observed in the Hannuoba clinopyroxenes). Such isotopic compositions are more evolved than any of the Hannuoba basalts (Song et al., 1990; Tatsumoto et al., 1992), and therefore, the metasomatism likely predates this episode of magmatism.

8. Conclusions

The data presented here, coupled with the previously published Os isotopic results for these same samples provide insights into the nature of the mantle lithosphere underlying the North China craton. The main conclusions are:

(1) The spinel peridoitite xenolith suites from Hannuoba and Qixia are compositionally indistinguishable from each other and yet markedly different from cratonic mantle lithosphere found beneath other Archean cratons.

(2) The lithospheric mantle beneath Hannuoba and Qixia formed by the loss of zero to ~ 25% partial melt from a primitive mantle source.

(3) Hannuoba peridotites are melt residues formed during the Paleoproterozoic (~ 1.9 Ga) and Qixia peridotites are melt residues formed during the Phanerozoic (probably Mesozoic; Gao et al., 2002). Thus, there is no indication that Proterozoic lithosphere is chemically distinct from Phanerozoic lithosphere (cf., Griffin et al., 1999; O'Reilly et al., 2001).

(4) Qixia peridotite xenoliths equilibrated over a narrower temperature interval than the Hannuoba peridotites, and the latter suite persists into the garnet stability field. This is consistent with derivation of the highest temperature Hannuoba xenoliths from greater depths.

(5) Neither suite shows a correlation between equilibration temperature and major element composition. Thus, there is no evidence for chemical stratification of the lithosphere as might be expected if the samples formed from a simple single upwelling column of mantle.

(6) Sr and Nd isotopic compositions of clinopyroxenes separated from Hannuoba peridotites correlate with bulk rock composition: the more refractory the sample, the more evolved the isotopic composition (Song and Frey, 1989). This, coupled with a lack of correlation between Rb/Sr and ^{87}Sr/^{86}Sr, suggests that the peridotites are the product of recent mixing (Mesozoic or younger) between a Proterozoic lithospheric mantle with high ε_{Nd} (up to +23) and low ^{87}Sr/^{86}Sr and a metasomatic component with low ε_{Nd} (< −2) and higher ^{87}Sr/^{86}Sr. The latter is significantly more evolved than the Hannuoba host basalts. Sr enrichment on clinopyroxene rims also reflects a very recent (Cenozoic) introduction of incompatible trace element-rich melts.

Acknowledgements

We thank Mike Rhodes for providing some of the XRF whole rock analyses. David Lange and Phil Piccoli provided guidance in electron mircroprobe analyses. Trisha Fiore helped with electron microprobe analyses at Harvard and Xiao-min Liu with XRF and ICP–MS analyses in Xi'an. We thank Stephanie Schmidberger and an anonymous reviewer for very helpful comments. The EPMA used in this study at the University of Maryland was purchased with grants from Department of Defense-Army/ARO (DAAG 559710383) and NSF (EAR-9810244), and UMD funding. This work was also supported by N.S.F. grant EAR99031591 and Nature Science Foundation of China grants 40133020 and 40373013 as well as Chinese Ministry of Science and Technology (grant G1999043202).

References

Bennett, V.C., DePaolo, D.J., 1987. Proterozoic crustal history of the western United States as determined by neodymium isotopic mapping. Geological Society of America Bulletin 99, 674–685.

Bernstein, S., Kelemen, P.B., Brooks, C.K., 1998. Depleted spinel harzburgite xenoliths in Tertiary dykes from east Greenland: restites from high degree melting. Earth and Planetary Science Letters 154, 221–235.

Boyd, F.R., 1989. Compositional distinction between oceanic and cratonic lithosphere. Earth and Planetary Science Letters 96, 15–26.

Brey, G.P., Köhler, T., 1990. Geothermobarometry in four-phase lherzolites: II. New thermobarometers, and practical assessment of existing thermobarometers. Journal of Petrology 31, 1353–1378.

Cao, R.L., Zhu, S.-H., 1987. Mantle xenoliths and alkali-rich host rocks in eastern China. In: Nixon, P.H. (Ed.), Mantle Xenoliths. Wiley, Chichester, pp. 167–180.

Chen, S., O'Reilly, S.Y., Zhou, X., Griffin, W.L., Zhang, G., Sun, M., Feng, J., Zhang, M., 2001. Thermal and petrological structure of the lithosphere beneath Hannuoba, Sino–Korean craton, China: evidence from xenoliths. Lithos 56, 267–301.

Fan, Q.C., Hooper, P.R., 1989. The mineral chemistry of ultramafic xenoliths of Eastern China—implications for upper mantle composition and the paleogeotherms. Journal of Petrology 30, 1117–1158.

Fan, Q.C., Liu, R.X., Li, H.M., Li, N., Sui, J.L., Lin, Z.R., 1998. Zircon geochronology and rare earth element geochemistry of granulite xenoliths from Hannuoba. Chinese Science Bulletin 43, 133–137 (in Chinese).

Fan, W.M., Zhang, H.F., Baker, J., Jarvis, K.E., Mason, P.R.D., Menzies, M.A., 2000. On and off the North China craton: where is the Archaean keel? Journal of Petrology 41, 933–950.

Frey, F.A., Green, D.H., 1974. The mineralogy, geochemistry and origin of lherzolite inclusions in Victorian basanites. Geochimica et Cosmochimica Acta 38, 1023–1059.

Gao, S., Kern, H., Liu, Y.S., Jin, S.Y., Popp, T., Jin, Z.M., Feng, J.L., Sun, M., Zhao, Z.B., 2000. Measured and calculated seismic velocities and densities for granulites from xenolith occurrences and adjacent exposed lower crustal sections: a comparative study from the North China craton. Journal of Geophysical Research, [Solid Earth] 105 (B8), 18965–18976.

Gao, S., Rudnick, R.L., Carlson, R.W., McDonough, W.F., Liu, Y., 2002. Re–Os evidence for replacement of ancient mantle lithosphere beneath the North China craton. Earth and Planetary Science Letters 198, 307–322.

Govindaraju, K., 1994. 1994 compilation of working values and sample description for 383 geostandards. Geostandards Newsletter 18, 1–158.

Griffin, W.L., Zhang, A.D., O'Reilly, S.Y., Ryan, C.G., 1998. Phanerozoic evolution of the lithosphere beneath the Sino–Korean craton. In: Flower, M., Chung, S.-L., Lo, C.-H., Lee, T.-Y. (Eds.), Mantle Dynamics and Plate Interactions in East Asia. American Geophysical Union, Washington, DC, pp. 107–126.

Griffin, W.L., O'Reilly, S.Y., Ryan, C.G., 1999. The composition and origin of sub-continental lithospheric mantle. In: Fei, Y., Bertka, M., Mysen, B.O. (Eds.), Mantle Petrology: Field Observations and High-Pressure Experimentation. A Tribute to France R. (Joe) Boyd. Spec. Publ.-Geochemical Society, pp. 13–46. Houston, TX.

Hermann, W., Berry, R.F., 2002. MINSQ—a least squares spreadsheet method for calculating mineral proportions from whole rock major element analyses. Geochemistry: Exploration, Environment, Analysis 2, 361–368.

Hirth, G., Evans, R.L., Chave, A.D., 2000. Comparison of continental and oceanic mantle electrical conductivity: is the Archean lithosphere dry? Geochemistry, Geophysics, Geosystems 1 2000GC0000-48.

Hu, S., He, L., Wang, J., 2000. Heat flow in the continental area of China: a new data set. Earth and Planetary Science Letters 179, 407–419.

Jordan, T.H., 1975. The continental tectosphere. Reviews of Geophysics and Space Physics 13, 1–12.

Jordan, T.H., 1988. Structure and formation of the continental lithosphere. In: Menzies, M.A., Cox, K. (Eds.), Oceanic and Continental Lithosphere: Similarities and DifferencesJ. Petrol., pp. 11–37. Special Lithosphere Issue.

Kern, H., Gao, S., Liu, Q.-S., 1996. Seismic properties and densities of middle and lower crustal rocks exposed along the North China geoscience transect. Earth and Planetary Science Letters 139, 439–455.

Kröner, A., Compston, W., Zhang, G., Guo, A., Cui, W., 1987. Single grain zircon ages for Archean rocks from Henan, Hebei and Inner Mongolia, China and tectonic implications. Int. Symp. Tectonic Evolution and Dynamics of Continental Lithosphere. Abst.

Kröner, A., Comproston, W., Zhang, G.W., Guo, A.L., Todt, W., 1988. Ages and tectonic setting of Late Archean greenstone–gneiss terrain in Henan Province, China as revealed by single-grain zircon dating. Geology 16, 211–215.

Kusky, T.M., Li, J.-H., Tucker, R.D., 2001. The Archean Dongwanzi Ophiolite Complex, North China craton: 2.505-billion-year-old oceanic crust and mantle. Science 292, 1142–1145.

Lee, C.-T., Rudnick, R.L., 1999. Compositionally stratified cratonic lithosphere: petrology and geochemistry of peridotite xenoliths from the Labait volcano, Tanzania. In: Gurney, J.J., Gurney, J.L., Pascoe, M.D., Richardson, S.R. (Eds.), The P.H. Nixon Volume. Proceedings VIIth International Kimberlite Conference. Red Roof Design, Cape Town, pp. 503–521.

Lee, C.-T., Yin, Q., Rudnick, R.L., Jacobsen, S.B., 2001. Preservation of ancient and fertile lithospheric mantle beneath the southwestern United States. Nature 411, 69–73.

Lehnert, K., Su, Y., Langmuir, C.H., Sarbas, B., Nohl, U., 2000. A global geochemical database structure for rocks. Geochemistry, Geophysics, Geosystems-G 3 1 (Paper number 1999GC000026).

Li, J.H., Qian, X.L., Huang, X.N., Liu, S.W., 2000. Tectonic framework of the North China craton and its cratonization in the early Precambrian. Acta Petrolei Sinica 16, 1–10 (in Chinese).

Liu, Y.S., Gao, S., Jin, S.Y., Hu, S.H., Sun, M., Zhao, Z.B., Feng, J.L., 2001. Geochemistry of lower crustal xenoliths from Neogene Hannuoba Basalt, North China craton: implications for petrogenesis and lower crustal composition. Geochimica et Cosmochimica Acta 65, 2589–2604.

Liu, Y.S., Yuan, H.L., Gao, S., Hu, Z.C., Wang, X.C., Liu, X.M., Ling, W.-L., 2004. Zircon U–Pb geochronology of Hannuoba olivine pyroxenites: linking the 157–97 Ma basaltic underplating and granulite-facies metamorphism. Chinese Science Bulletin (in press).

McDonough, W.F., Frey, F.A., 1989. Rare earth elements in upper mantle rocks. In: Lipin, B.R., McKay, G.A. (Eds.), Geochemistry and Mineralogy of Rare Earth Elements. Reviews in Mineralogy. Mineralogical Society of America, Washington, DC, pp. 99–145.

McDonough, W.F., Sun, S.-S., 1995. Composition of the earth. Chemical Geology 120, 223–253.

Menzies, A., Fan, W.-M., Zhang, M., 1993. Paleozoic and Cenozoic lithoprobes and the loss of >120 km of Archean lithosphere, Sino–Korean craton, China. In: Prichard, H.M., Alabaster, H.M., Harris, T., Neary, C.R. (Eds.), Magmatic Processes and Plate Tectonics. Geological Soc., London, pp. 71–81.

Nimis, P., Taylor, W.R., 2000. Single clinopyroxene thermobarometry for garnet peridotites: Part I. Calibration and testing of a Cr-in-cpx barometer and an enstatite-in-cpx thermometer. Contributions to Mineralogy and Petrology 139, 541–554.

Nyblade, A.A., Pollack, H.N., Jones, D.L., Podmore, F., Mushayandebvu, M., 1990. Terrestrial heat flow in East and Southern Africa. Journal of Geophysical Research 95, 17371–17384.

O'Reilly, S.Y., Griffin, W.L., Poudjom Djomani, Y.H., Morgan, P., 2001. Are lithospheres forever? Tracking changes in subcontinental lithospheric mantle through time. GSA Today, 4–9 (April issue).

Pollack, H.N., 1986. Cratonization and thermal evolution of the mantle. Earth and Planetary Science Letters 80, 175–182.

Qiu, Y., Groves, D.I., McNaughton, N.J., Wang, L.-G., Zhou, T., 2002. Nature, age, and tectonic setting of granitoid-hosted, orogenic gold deposits of the Jiaodong Peninsula, eastern North China craton. China Mineral Deposita 37, 283–305.

Raymo, O.T., Calzia, J.P., 1998. Nd isotopic composition of cratonic rocks in the south Death Valley region: evidence for a substantial Archean source component in Mojavia. Geology 26, 891–894.

Rhodes, J.M., 1996. Geochemical stratigraphy of lava flows sampled by the Hawaiian scientific drilling project. Journal of Geophysical Research 101, 11726–11746.

Rudnick, R.L., McDonough, W.F., McCulloch, M.T., Taylor, S.R., 1986. Lower crustal xenoliths from Queensland, Australia: evidence for deep crustal assimilation and fractionation of continental basalts. Geochimica et Cosmochimica Acta 50, 1099–1115.

Rudnick, R.L., McDonough, W.F., Orpin, A., 1994. Northern Tanzanian peridotite xenoliths: a comparison with Kaapvaal peridotites and inferences on metasomatic interactions. In: Meyer, H.O.A., Leonardos, O. (Eds.), Kimberlites, Related Rocks and Mantle Xenoliths. Proceedings Fifth Int. Kimb. Conf., vol. 1. CPRM, Brasilia, pp. 336–353.

Schmidberger, S.S., Simonetti, A., Francis, D., 2003. Small-scale Sr isotope investigation of clinopyroxenes from peridotite xenoliths by laser ablation MC–ICP–MS—implications for mantle metasomatism. Chemical Geology 199, 317–329.

Sneeringer, M., Hart, S.R., Shimizu, N., 1984. Strontium and samarium diffusion in diopside. Geochimica et Cosmochimica Acta 48, 1589–1608.

Song, Y., Frey, F.A., 1989. Geochemistry of peridotite xenoliths in basalt from Hannuoba, Eastern China: implications for subcontinental mantle heterogeneity. Geochimica et Cosmochimica Acta 53, 97–113.

Song, Y., Frey, F.A., Zhi, X.C., 1990. Isotopic characteristics of Hannuoba basalts, Eastern China—implications for their petrogenesis and the composition of subcontinental mantle. Chemical Geology 88, 35–52.

Tatsumoto, M., Basu, A.R., Wankang, H., Junwen, W., Guanghong,

X., 1992. Sr, Nd and Pb isotopes of ultramafic xenoliths in volcanic rocks of Eastern China: enriched components EMI and EMII in subcontinental lithosphere. Earth and Planetary Science Letters 113, 107–128.

van der Hilst, R.D., McDonough, W.F. (Eds.), 1999. Composition, Deep Structure and Evolution of Continents. Developments in Geotectonics, vol. 24. Elsevier, Amsterdam. 342 pp.

Van Orman, J.A., Grove, T.L., Shimizu, N., 2001. Rare earth element diffusion in diopside: influence of temperature, pressure, and ionic radius, and an elastic model for diffusion in silicates. Contributions to Mineralogy and Petrology 141, 687–703.

Walter, M.J., 1999. Melting residues of fertile peridotite and the origin of cratonic lithosphere. In: Fei, Y., Bertka, M., Mysen, B.O. (Eds.), Mantle Petrology: Field Observations and High-Pressure Experimentation. A Tribute to France R. (Joe) Boyd. Spec. Publ.-Geochemical Society, pp. 225–240. Houston, TX.

Wells, P.R.A., 1977. Pyroxene thermometry in simple and complex systems. Contributions to Mineralogy and Petrology 62, 129–139.

White, W.M., Patchett, P.J., 1984. Hf–Nd–Sr isotopes and incompatible element abundances in island arcs: implications for magma origins and crust–mantle evolution. Earth and Planetary Science Letters 67, 167–185.

Wilde, S.A., Zhou, X.H., Nemchin, A.A., Sun, M., 2003. Mesozoic crust–mantle interaction beneath the North China craton: a consequence of the dispersal of Gondwanaland and accretion of Asia. Geology 31, 817–820.

Wooden, J.L., Miller, D.M., 1990. Chronologic and isotopic framework for early Proterozoic crustal evolution in the eastern Mojave Desert region, SE California. Journal of Geophysical Research, [Solid Earth and Planets] 95, 20133–20146.

Wu, J.S., Geng, Y.S., Shen, Q.H., Wan, Y.S., Liu, D.Y., Song, B., 1998. Archean Geology and Tectonic Evolution of the North China Craton (in Chinese). Geological Publishing House, Beijing. 212 pp.

Xu, Y.G., 2002. Evidence for crustal components in the mantle and constraints on crustal recycling mechanisms: pyroxenite xenoliths from Hannuoba, North China. Chemical Geology 182, 301–322.

Yang, J.H., Wu, F.Y., Wilde, S.A., 2003. A review of the geodynamic setting of large-scale Late Mesozoic gold mineralization in the North China craton: an association with lithospheric thinning. Ore Geology Reviews 23, 125–152.

Zhao, G.C., Cawood, P.A., Wilde, S.A., Sun, M., 2000. Metamorphism of basement rocks in the Central Zone of the North China craton: implications for Paleoproterozoic tectonic evolution. Precambrian Research 103, 55–88.

Zhao, G.C., Wilde, S.A., Cawood, P.A., Sun, M., 2001. Archean blocks and their boundaries in the North China craton: lithological, geochemical, structural and P–T path constraints and tectonic evolution. Precambrian Research 107, 45–73.

Zheng, Z., O'Reilly, S.Y., Griffin, W.L., Lu, F., Zhang, M., 1998. Nature and evolution of Cenozoic lithospheric mantle beneath Shandong Peninsula, Sino–Korean craton, eastern China. International Geology Review 40, 471–499.

Zhi, X.C., Song, Y., Frey, F.A., Feng, J.L., Zhai, M.Z., 1990. Geochemistry of Hannuoba basalts, eastern China—constraints on the origin of continental alkalic and tholeiitic basalt. Chemical Geology 88, 1–33.

Zhou, X.H., Sun, M., Zhang, G.H., Chen, S.H., 2002. Continental crust and lithospheric mantle interaction beneath North China: isotopic evidence from granulite xenoliths in Hannuoba, Sino–Korean craton. Lithos 62, 111–124.

Zhu, B.-Q., 1998. Theory and Applications of Isotope Systematics in Geosciences: Evolution of Continental Crust and Mantle in China (in Chinese). Science Press, Bejing.

Zindler, A., Jagoutz, E., 1988. Mantle cryptology. Geochimica et Cosmochimica Acta 52, 319–333.

Available online at www.sciencedirect.com

Lithos 77 (2004) 639–646

www.elsevier.com/locate/lithos

The influence of Cr on the garnet–spinel transition in the Earth's mantle: experiments in the system $MgO–Cr_2O_3–SiO_2$ and thermodynamic modelling

Stephan Klemme[*]

Mineralogisches Institut, Universität Heidelberg, Im Neuenheimer Feld 236, 69120 Heidelberg, Germany

Received 27 June 2003; accepted 3 January 2004
Available online 18 May 2004

Abstract

The position of the transition from spinel peridotite to garnet peridotite in a simplified chemical composition has been determined experimentally at high pressures and high temperatures. The univariant reaction $MgCr_2O_4 + 2Mg_2Si_2O_6 = Mg_3Cr_2Si_3O_{12} + Mg_2SiO_4$, has a negative slope in $P–T$ space between 1200 °C and 1600 °C. The experimental results, combined with assessed thermodynamic data for $MgCr_2O_4$, $MgSiO_3$ and Mg_2SiO_4 give the entropy and enthalpy of formation of knorringite garnet ($Mg_3Cr_2Si_3O_{12}$). Thermodynamic calculations in simplified chemical compositions indicate that Cr shifts the garnet-in reaction to much higher pressures than previously anticipated. Moreover, in Cr-bearing systems a pressure–temperature field exists where garnet and spinel coexist. The width of this divariant field strongly depends on the $Cr/(Cr + Al)$ of the system.
© 2004 Elsevier B.V. All rights reserved.

Keywords: Garnet lherzolite; Spinel lherzolite; Experimental petrology; Chromium; Garnet spinel transition; $MgO–Cr_2O_3–SiO_2$; $MgO–Al_2O_3–Cr_2O_3–SiO_2$

1. Introduction

From geophysics and experimental petrology it is well known that the Earth's mantle is stratified. The Earth's uppermost mantle consists of only four main minerals such as olivine, clinopyroxene, orthopyroxene and spinel. At higher pressures, however, the spinel bearing assemblage converts into a garnet bearing mineral assemblage. Among others, the transition from spinel lherzolite to garnet lherzolite is one of the major phase boundaries in the Earth's upper mantle (Asimov et al., 1995; Green and Ringwood,

1967; Hales, 1969; Klemme and O'Neill, 2000a,b; O'Neill, 1981). This transition is of particular petrogenetic relevance as the interpretation of magmatic processes at mid-ocean ridges, for example, requires a sound knowledge of the position of the transition from garnet to spinel lherzolite. For example, geochemists have argued for beginning of melting just within the stability field of garnet lherzolite (e.g. LaTourette et al., 1993; Salters and Hart, 1989), although other interpretation of so-called 'garnet-signatures' have been put forward (Allègre et al., 1984; Asimov et al., 1995; Hirschmann and Stolper, 1996; Klemme and O'Neill, 2000a,b; Shen and Forsyth, 1995; Wood, 1979). To resolve these matters, the garnet–spinel transition needs to be better con-

---

* Tel.: +49-6221-546039; fax: +49-6221-544805.
E-mail address: sklemme@min.uni-heidelberg.de (S. Klemme).

strained as a function of pressure, temperature and, most importantly, composition.

The position of the spinel–garnet transition is well understood in simple chemical systems (e.g. CaO–MgO–Al$_2$O$_3$–SiO$_2$) and in complex, fertile compositions (Gasparik and Newton, 1984; Green and Ringwood, 1967; Hales, 1969; Klemme and O'Neill, 2000a,b; MacGregor, 1965; O'Hara et al., 1971; O'Neill, 1981; Robinson and Wood, 1998). However, a strong influence of minor components such as chromium on the garnet–spinel transition has long been suggested (Brey et al., 1999; MacGregor, 1970; Nickel, 1986; O'Neill, 1981; Webb and Wood, 1986; Wood, 1978) but there is scant experimental data in Cr-rich compositions (Brey et al., 1999; Doroshev et al., 1997; Girnis et al., 2003; Nickel, 1986; Webb and Wood, 1986). This may be of relevance to the deeper continental mantle which is clearly depleted in Al and more or less constant in Cr when compared to normal fertile mantle (e.g. Liang and Elthon, 1990). This trend is believed to be caused by melting events (e.g. Falloon and Green, 1988; Walter et al., 2002) and it seems that the mantle shows increasing Cr/Cr + Al with depth as extremely Cr-rich garnets are commonly found in diamond inclusions (e.g. Stachel and Harris, 1997; Stachel et al., 1998, 2000).

Moreover, rigorous thermodynamic modelling of spinel–garnet reactions in the upper mantle over a range of temperatures, pressures and compositions requires reliable thermodynamic data for Cr-bearing minerals such as Cr-spinels (Klemme and O'Neill, 1997; Klemme et al., 2000; Klemme and van Miltenburg, 2002), Cr-bearing pyroxenes (Klemme and O'Neill, 2000a,b) and Cr-bearing garnets, the latter of which are, unfortunately, rather unconstrained (Doroshev et al., 1997; Irifune et al., 1982; Turkin et al., 1983). The present study tries to fill this gap.

The simplest reaction describing the transition from spinel-bearing peridotite to garnet peridotite may be written as follows

MgAl$_2$O$_4$ + 2 Mg$_2$Si$_2$O$_6$
spinel   orthopyroxene

= Mg$_3$Al$_2$Si$_3$O$_{12}$ + Mg$_2$SiO$_4$ (1)
pyrope garnet   olivine

This reaction has been studied previously in the systems MgO–Al$_2$O$_3$–SiO$_2$ and CaO–MgO–

Al$_2$O$_3$–SiO$_2$ which are excellent proxies for fertile mantle compositions, i.e. undepleted and poor in Cr (e.g. Danckwerth and Newton, 1978; Gasparik and Newton, 1984; Klemme and O'Neill, 2000a,b; MacGregor, 1965; O'Neill, 1981). Although there was some recent debate about the transition at higher temperature (Klemme and O'Neill, 2000a,b; Longhi, 2002; Walter et al., 2002) there is general agreement, however, that in both these simple systems the garnet–spinel transition is univariant and has a positive slope in pressure–temperature space (Fig. 1). From previous experimental results (Doroshev et al., 1997; Nickel, 1986) and early thermodynamic calculations (O'Neill, 1981; Wood, 1978) it is well known that addition of Cr to the system dramatically increases the stability of the spinel phase assemblage relative to the garnet assemblage (or the plagioclase stability field at lower pressures). But thermodynamic modelling in realistic mantle compositions was unreliable (e.g. Asimov et al., 1995) as thermodynamic data for Cr-spinels were subsequently shown to be in error (Klemme and O'Neill, 1997; Klemme et al., 2000) and reliable thermodynamic data for Cr-bearing garnets were unavailable.

1.1. The stability of knorringite garnet

Three previous studies investigated experimentally the following knorringite forming reaction (Iri-

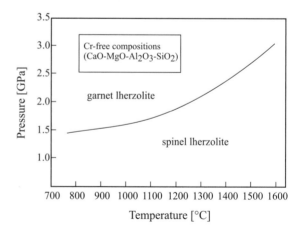

Fig. 1. The transition from garnet lherzolite to spinel lherzolite in Cr-free (fertile) mantle. The curve shown is based on experimental data in the system CaO–MgO–Al$_2$O$_3$–SiO$_2$ as given in Klemme and O'Neill (2000a,b).

fune et al., 1982; Ringwood, 1977; Turkin et al., 1983).

$$Cr_2O_3 + 3\ MgSiO_3 = Mg_3Cr_2Si_3O_{12} \qquad (2)$$
eskolaite   enstatite   knorringite

All these previous experimental studies on the stability of knorringite disagree considerably with each other (Fig. 2). This may be due to experimental uncertainties such as pressure calibration or temperature measurements. Moreover, it is well known that garnets tend to nucleate very sluggishly in high pressure experiments. The fact that Irifune et al. (1982) describe large garnets (mm size) in their experiments may indicate a rather large overstepping of the true position of the reaction. Summarising, the stability of knorringite garnet and, consequently, the thermodynamic properties of knorringite garnet were rather unconstrained.

1.2. Experimental strategy

To both investigate the influence of chromium on the garnet–spinel transition in a Cr-rich bulk composition and to derive thermodynamic properties of

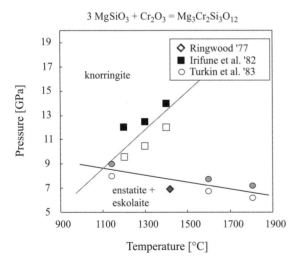

$$3\ MgSiO_3 + Cr_2O_3 = Mg_3Cr_2Si_3O_{12}$$

Fig. 2. Previous experimental results on the stability of knorringite ($Mg_3Cr_2Si_3O_{12}$) garnet (Irifune et al., 1982; Ringwood, 1977; Turkin et al., 1983). Filled symbols represent garnet stability. Note considerable differences between the individual studies that are probably due to experimental problems (see text for further information).

knorringite garnet, the analogue reaction to (1) was studied experimentally:

$$MgCr_2O_4 + 2\ Mg_2Si_2O_6$$
spinel   orthopyroxene

$$= Mg_3Cr_2Si_3O_{12} \quad + Mg_2SiO_4 \qquad (3)$$
knorringite garnet   olivine

This reaction, which has hitherto not been investigated experimentally, describes the maximum stability of spinel in the Earth's mantle. Extraction of thermodynamic data for knorringite garnet is facilitated because reaction (3) is univariant and thermodynamic properties of $MgCr_2O_4$, $Mg_2Si_2O_6$ and Mg_2SiO_4 are well understood (Klemme, 1998; Klemme and O'Neill, 2000a,b) and solid solutions are not significant.

2. Experimental and analytical techniques

Reversal high-pressure high-temperature experiments on reaction (3) were performed between 4.5 and 16 GPa and at temperatures between 1200 and 1600 °C in a multi-anvil apparatus at the Bayerisches Geoinstitut, Germany. Analysis of run products with X-ray diffraction and electron microprobe indicated which phase assemblage grew and which was consumed. The starting material consisted of all four phases partaking in the reaction. Enstatite ($MgSiO_3$), magnesiochromite ($MgCr_2O_4$) and forsterite (Mg_2SiO_4) were synthesised previously and details were described elsewhere (Klemme, 1998; Klemme and O'Neill, 1997; Klemme and O'Neill, 2000a,b). Knorringite ($Mg_3Cr_2Si_3O_{12}$) starting material was synthesised from stoichiometric mixtures of MgO, Cr_2O_3 and SiO_2 in a multi-anvil press in two runs at 16 GPa and 1600 °C for 4 h. X-ray diffraction indicated knorringite only with only very minor peaks of unreacted eskolaite. The experiments were performed in 14/8 and 18/11 (octahedral edge length/ truncation edge length) octahedral sample assemblies made from Cr-doped MgO. Starting material were tightly contained in capsules made of thin Re foil. $LaCrO_3$ heaters surrounded by zirconia sleeves were employed with $W_3Re/W_{25}Re$ thermocouples. Experimental run conditions and run times are given in Table 1.

Electron microprobe analyses were done at Heidelberg University using a five spectrometer Cameca SX 51 electron microprobe. Several mineral standards were used (garnets, spinels, orthopyroxene and synthetic eskolaite). Microprobe analyses (Table 2) of the phases indicate essentially pure $MgCr_2O_4$, whereas enstatites contained about 1 wt.% Cr_2O_3. The Cr content of enstatite does not vary significantly with pressure or temperature. The olivines contain very little chromium (< 0.3% (wt.) Cr_2O_3). This may be taken as evidence that almost all chromium is present in the trivalent state as significant amounts of Cr^{2+} in the charge would stabilize the Cr_2SiO_4 component in olivines (Li et al., 1995; Schreiber and Haskin, 1976;

Table 1

Experimental run conditions and results

Run #	T (°C)	P (GPa)	Duration (h)	Result
S2685	1600	16	4	Kn (very minor eskolaite)
S2686	1600	16	4	Kn (very minor eskolaite)
S2688	1600	11	4	Kn + For (minor Sp + En)
S2689	1400	11	5	Kn + For (minor Sp + En)
S2690	1200	11	12	Kn + For (minor Sp + En)
S2691	1600	9	4	Kn + For (minor Sp + En)
S2692	1400	9	10	Kn + For (minor Sp + En)
S2693	1600	7	5 h 20 min	Kn + For (minor Sp + En)
S2806	1200	6.4	12	Sp + En (minor Kn + For)
S2807	1600	4.5	1 h 10 min	Sp + En (minor Kn + For)
S2808	1600	5.5	1 h 40 min	Sp + En (minor Kn + For)
S2809	1200	6	11	Sp + En (minor Kn + For)
S2812	1600	6.7	1 h 35 min	Kn + For (minor Sp + En)
S2815	1400	8	6	Kn + For (minor Sp + En)
S2904	1200	7.75	10	Sp + En (minor Kn + For)
S2905	1200	9	12	Kn + For (minor Sp + En)
S2906	1500	6.7	5 h 20 min	Kn + For (minor Sp + En)
S2907	1400	6.4	5 h 20 min	Sp + En (minor Kn + For)

Individual experimental run conditions, experimental results and starting materials. T is temperature in °C, P is calculated run pressure using a number of different calibration runs. Kn = knorringite ($Mg_3Cr_2Si_3O_{12}$), For = forsterite (Mg_2SiO_4), Sp = magnesiochromite ($MgCr_2O_4$), En = enstatite ($MgSiO_3$). All experiments were run with starting material GS2 which contained all four phases that were synthesised prior to the commencement of the study (see text for details). S2685 and S2686 were knorringite synthesis runs and contained $MgCr_2O_4$, MgO and SiO_2 as starting material. Estimated pressure uncertainties are in the order of 0.2 GPa and temperature uncertainties are estimated to be ± 15 °C. Note that for runs S2906 and S2907 the thermocouples were dysfunctional, therefore the estimated temperature uncertainties are probably in the order of ± 50 °C.

Table 2

Mineral	Kn	For	Sp	En
SiO_2	39.7 (2)	42.4 (3)	0.1 (1)	59.3 (4)
MgO	26.9 (3)	56.9 (4)	21.2 (2)	40.4 (2)
Cr_2O_3	33.6 (2)	0.15 (15)	78.8 (3)	1.1 (3)
Total	100.2 (3)	99.5 (4)	100.1 (2)	100.7 (3)
Si	2.990	0.998	0.003	1.975
Mg	3.020	1.998	1.009	2.006
Cr	2.000	0.004	1.989	0.029
Total	8.010	3.000	3.001	4.010

Representative electron microprobe analyses of experimental run products. Mineral compositions are virtually identical in all runs. Note that most minerals are very close to their ideal composition. Numbers show element abundances in oxide components (wt.%) as well as recalculated on the basis of 4, 6 and 12 oxygens, respectively. Numbers in parenthesis indicate the analytical uncertainties. Kn = knorringite ($Mg_3Cr_2Si_3O_{12}$), For = forsterite (Mg_2SiO_4), Sp = magnesiochromite ($MgCr_2O_4$), En = enstatite ($MgSiO_3$).

Seifert and Ringwood, 1988). The knorringite garnets were stoichiometric and showed no majorite component. It should be noted that almost all experiments contain coexisting enstatite and knorringite garnets. It is rather surprising that no majorite component was found even at pressures as high as 16 GPa, as significant majorite substitution was reported in pyrope garnets at much lower pressures (Ito and Taka-

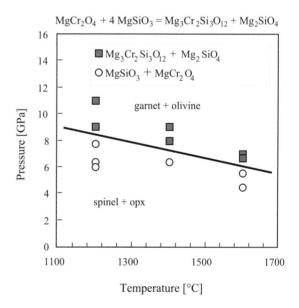

Fig. 3. Experimental results on the garnet–spinel transition in the system $MgO–Cr_2O_3–SiO_2$. Filled squares represent garnet stability, whereas circles represent spinel stability.

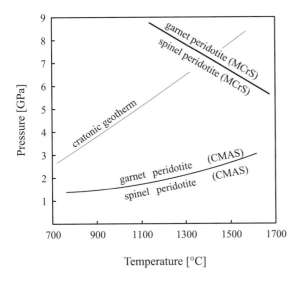

Fig. 4. The garnet–spinel transition in both Cr-free (CMAS: CaO–MgO–Al$_2$O$_3$–SiO$_2$) and Al-free (MCrS: MgO–Cr$_2$O$_3$–SiO$_2$) mantle. A cratonic geotherm (40mW/m^2) intersects the garnet–spinel transition in Al-free mantle at around 7 GPa and 1400 °C.

hashi, 1987; Liu, 1977; Ringwood, 1967; Ringwood and Major, 1971). The present experiments seem to indicate that knorringite is far less susceptible to the majorite substitution than pyrope garnets which would be an interesting subject for further research.

3. Experimental results and discussion

The experimental results on the garnet–spinel transition in the system MgO–Cr$_2$O$_3$–SiO$_2$ (reaction (2)) are given in Table 1 and depicted in Fig. 3. The experiments that closely bracket the reaction are highlighted in the diagram. The garnet–spinel transi-

tion in Al-free compositions (i.e. MgO–Cr$_2$O$_3$–SiO$_2$) exhibits a negative slope in pressure–temperature space which is in stark contrast to the analogue reaction in a Cr-free or fertile compositions (Fig. 4). Note that Doroshev et al. (1997) and Girnis and Brey (1999) proposed a roughly similar position of reaction (3) based on thermodynamic extrapolations. Thermodynamic evaluation of the experimental data yields $S°_{298} = 377$ J mol^{-1} K^{-1} and $\Delta_f°H = -5542$ kJ mol^{-1} for knorringite garnet (Mg$_3$Cr$_2$Si$_3$O$_{12}$), given the thermodynamic data for MgCr$_2$O$_4$, MgSiO$_3$ and Mg$_2$SiO$_4$ (Table 3) and neglecting the effects of possible solid solutions (Klemme and O'Neill, 1997, 2000a,b).

The present experiments define the maximum stability of spinel in the Earth's upper mantle. It may be concluded that no Cr-rich spinel, commonly found in diamond inclusions and in mantle xenoliths has originated at pressures higher than 7.5 GPa assuming a normal cratonic mantle geotherm (Griffin et al., 1999; Kopylova et al., 1999; Russell and Kopylova, 1999). This implies that Cr-rich spinels, commonly transported to the Earth's surface in mantle xenoliths in kimberlites and frequently reported as diamond inclusions, cannot have originated from depths greater than about 240 km.

Moreover, new thermodynamic data for knorringite (this study), Cr-bearing spinels (Klemme et al., 2000) and Cr-bearing pyroxenes (Klemme and O'Neill, 2000a,b) enable thermodynamic calculations to investigate the influence of Cr on phase equilibria with mantle minerals (see Table 3). The thermodynamics of Cr–Al spinel solid solutions, and Cr–Al pyroxene solid solutions are reasonably well understood (Klemme and O'Neill, 2000a,b; Oka et al., 1984),

Table 3
Thermodynamic data

	$\Delta_f H°$ [J mol^{-1}]	Source	$S°$ [J mol^{-1} K^{-1}]	Source	$V°_{298}$ [J bar^{-1} mol^{-1}]	Source
MgCr$_2$O$_4$	−1762000	K00	119.6	K00	4.356	R79
Mg$_2$SiO$_4$	−2171870	HP90	94.01	HP90	4.366	HP90
Mg$_3$Cr$_2$Si$_3$O$_{12}$	−5542310	this study	376.7	this study	11.738	IRI82
Mg$_3$Al$_2$Si$_3$O$_{12}$	−6282900	K98	266.4	K98	11.318	HP90
Mg$_2$Si$_2$O$_6$	−3089400	HP90	132.5	HP90	6.262	HP90
MgAl$_2$SiO$_6$	−3201000	K98	115.7	K98	5.892	K98
MgCr$_2$SiO$_6$	−2637800	K98	169.3	K98	6.116	K98

References: K00 = (Klemme et al., 2000), HP90 = (Holland and Powell, 1990), IRI82 = (Irifune et al., 1982), R79 = (Robie et al., 1979). K98 = (Klemme, 1998), Solid solutions used for modelling: MgAl$_2$O$_4$–MgCr$_2$O$_4$ (Oka et al., 1984), Mg$_2$Si$_2$O$_6$–MgCr$_2$SiO$_6$ (Klemme, 1998; Klemme and O'Neill, 2000a,b), Mg$_3$Al$_2$Si$_3$O$_{12}$–Mg$_3$Cr$_2$Si$_3$O$_{12}$ (ideal).

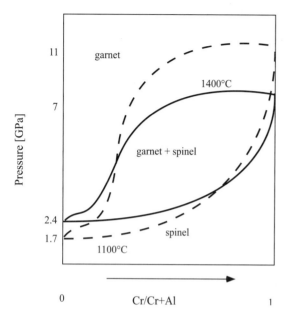

Fig. 5. The effect of Cr on the stability of garnet and spinel, based on thermodynamic models in the system MgO–Al$_2$O$_3$–Cr$_2$O$_3$–SiO$_2$. The garnet–spinel transition is univariant in compositions with Cr/(Cr+Al)=0 and Cr/(Cr+Al)=1. At intermediate compositions a divariant field exists where garnet and spinel coexist. The width of this garnet+spinel stability field is strongly dependent on Cr/(Cr+Al). In relatively fertile compositions (Cr/(Cr+Al) ≈ 0.1) the garnet+spinel stability field is relatively narrow. If the bulk composition has Cr/(Cr+Al) ≥ 0.2 the garnet+spinel stability field expands quite dramatically and spinel is stable to pressures of 7 GPa and higher. The solid lines show coexisting garnet and spinel compositions at 1400 °C, whereas the dashed lines depict their compositions at 1100 °C.

thermodynamics of the pyrope-knorringite solid solution, however, remain somewhat uncertain. As data is lacking, ideal solution was assumed, based on near-ideal chromium-aluminium mixing along the join Ca$_3$Al$_2$Si$_3$O$_{12}$–Ca$_3$Cr$_2$Si$_3$O$_{12}$ (Wood and Kleppa, 1984). Using free energy minimisation techniques (Klemme, 1998), results from thermodynamic modelling indicate that the divariant field where garnet and spinel coexist, is strongly dependent on bulk composition (Fig. 5). Previous studies reported a strong influence of Cr on the spinel stability relative to garnet stability (Brey et al., 1999; Doroshev et al., 1997; Girnis et al., 2003; O'Neill, 1981; Webb and Wood, 1986). However, the present results from thermodynamic modelling (Fig. 5) extend to much more Cr-rich compositions. In fertile compositions (0 < Cr/

(Cr+Al) < 0.2), the divariant garnet+spinel stability field is only fairly narrow, in agreement with experiments in simple and complex fertile bulk compositions (e.g. Green and Ringwood, 1967; Nickel, 1986). The effect of Cr on spinel stability is only small in such relatively Cr-poor compositions. In compositions with higher Cr/(Cr+Al), however, the garnet+spinel stability field expands extraordinarily when the bulk composition exceeds a Cr/(Cr+Al) ratio of about 0.2 (Fig. 5).

The experiments and the results from thermodynamic modelling show that the influence of Cr on the garnet–spinel transition is extraordinary. Therefore, phase equilibria calculations in realistic mantle compositions must account for the influence of Cr (Asimov et al., 1995). Although the present thermodynamic calculations were performed in a simplified chemical composition (MgO–Al$_2$O$_3$–Cr$_2$O$_3$–SiO$_2$) and other components such as Ca or Fe may influence calculated phase equilibria, it is believed that it is Cr that exerts the major control on phase relations and that the proposed trends should also hold for more complex compositions.

Acknowledgements

Whilst at Bristol, the author acknowledges funding by the European Union as a Marie-Curie Individual Fellow. Additional funding for this project was provided by the European Union 'Access to Large Scale Facilities program' at Bayerisches Geoinstitut, Bayreuth (EU IHP-Access to research Infrastructures Programme. Contract No HPRI-1999-CT-00004 to D.C. Rubie). I am indebted to Dan Frost for his help with the multi-anvil experiments at Bayreuth. Discussions on this subject with Drs H. StC. O'Neill, D.H. Green, R. Altherr among many others at Bristol, Bayreuth and Heidelberg, were of great value.

References

Allègre, C.J., Hamelin, B., Dupré, B., 1984. Statistical analysis of isotopic ratios in MORB: the mantle blob cluster model and the convective regime of the mantle. Earth Planet. Sci. Lett. 71, 71–84.

Asimov, P.D., Hirschmann, M.M., Ghiorso, M.S., O'Hara, M.J., Stolper, E.M., 1995. The effect of pressure induced solid–solid phase transitions on decompression melting of the mantle. Geochim. Cosmochim. Acta 59, 4489–4506.

Brey, G.P., Doroshev, A.M., Girnis, A.V., Turkin, A.I., 1999. Garnet–spinel–olivine–orthopyroxene equilibria in the FeO–MgO–Al₂O₃–SiO₂–Cr₂O₃ system: I. Composition and molar volumes of minerals. Eur. J. Mineral. 11, 599–617.

Danckwerth, P.A., Newton, R.C., 1978. Experimental determination of the spinel peridotite to garnet peridotite reaction in the system MgO–Al₂O₃–SiO₂ in the range 900–1100 °C and Al₂O₃ isopleths of enstatite in the spinel field. Contrib. Mineral. Petrol. 66, 189–201.

Doroshev, A.M., Brey, G.P., Girnis, A.V., Turkin, A.I., Kogarko, L.N., 1997. Pyrope-knorringite garnets in the Earth's Mantle: experiments in the MgO–Al₂O₃–SiO₂–Cr₂O₃ system. Russ. Geol. Geophys. 38, 559–586.

Falloon, T.J., Green, D.H., 1988. Anhydrous partial melting of peridotite from 8 to 35 kb and the petrogenesis of MORB. J. Petrol., 379–414 (Special Lithosphere Issue).

Gasparik, T., Newton, R.C., 1984. The reversed alumina contents of orthopyroxene in equilibrium with spinel and forsterite in the system MgO–Al₂O₃–SiO₂. Contrib. Mineral. Petrol. 85, 186–196.

Girnis, A.V., Brey, G.P., 1999. Garnet–spinel-olivine-orthopyroxene equilibria in the FeO–MgO–Al₂O₃–SiO₂–Cr₂O₃ system: II. Thermodynamic analysis. Eur. J. Mineral. 11, 619–636.

Girnis, A.V., Brey, G.P., Doroshev, A.M., Turkin, A.I., Simon, N., 2003. The system MgO–Al₂O₃–SiO₂–Cr₂O₃ revisited: reanalysis of Doroshev et al.'s (1977) experiments and new experiments. Eur. J. Mineral. 15, 953–964.

Green, D.H., Ringwood, A.E., 1967. The stability field of aluminous pyroxene peridotite and garnet peridotite and their relevance in upper mantle structure. Earth Planet. Sci. Lett. 3, 151–160.

Griffin, W.L., et al., 1999. The Siberian SCLM Traverse: composition, structure and thermal state of the lithospheric mantle beneath the Siberian kimberlite province. Tectonophysics 310, 1–35.

Hales, A.L., 1969. A seismic discontinuity in the lithosphere. Earth Planet. Sci. Lett. 7, 44–46.

Hirschmann, M.M., Stolper, E.M., 1996. A possible role for garnet pyroxenite in the origin of the "garnet signature" in MORB. Contrib. Mineral. Petrol. 124, 185–208.

Holland, T.J.B., Powell, R., 1990. An enlarged and updated internally consistent thermodynamic dataset with uncertainties and correlations: the system K₂O–Na₂O–CaO–MgO–MnO–FeO–Fe₂O₃–Al₂O₃–TiO₂–SiO₂–C–H–O₂. J. Metamorph. Geol. 8, 89–124.

Irifune, T., Ohtani, E., Kumazawa, M., 1982. Stability field of knorringite Mg₃Cr₂Si₃O₁₂ at high pressure and its implication to the occurrence of Cr-rich pyrope in the upper mantle. Phys. Earth Inter. 27, 263–272.

Ito, E., Takahashi, E., 1987. Ultrahigh-pressure phase transformations and the constitution of the deep mantle. In: Manghnani, M.H., Syono, Y. (Eds.), High Pressure Research In Mineral Physics. Terra Scientific Publishing, Tokyo, pp. 221–230.

Klemme, S., 1998. Experimental and Thermodynamic Studies of Upper Mantle Phase Relations. PhD thesis, Australian National University, Canberra.

Klemme, S., O'Neill, H.S.C., 1997. The reaction MgCr₂O₄ + SiO₂ = Cr₂O₃ + MgSiO₃ and the free energy of formation of magnesiochromite (MgCr₂O₄). Contrib. Mineral. Petrol. 130, 59–65.

Klemme, S., O'Neill, H.S.C., 2000a. The effect of Cr on the solubility of Al in orthopyroxene: experiments and thermodynamic modelling. Contrib. Mineral. Petrol. 140, 84–98.

Klemme, S., O'Neill, H.S.C., 2000b. The near-solidus transition from garnet lherzolite to spinel lherzolite. Contrib. Mineral. Petrol. 138, 237–248.

Klemme, S., van Miltenburg, J.C., 2002. Thermodynamic properties of nickel chromite (NiCr₂O₄) based on adiabatic calorimetry at low temperatures. Phys. Chem. Miner. 29, 663–667.

Klemme, S., O'Neill, H.S.C., Schnelle, W., Gmelin, E., 2000. The heat capacity of MgCr₂O₄, FeCr₂O₄ and Cr₂O₃ at low temperatures and derived thermodynamic properties. Am. Mineral. 85, 1686–1693.

Kopylova, M.G., Russell, J.K., Cookenboo, H., 1999. Petrology of peridotite and pyroxenite xenoliths from the Jericho kimberlite: implications for the thermal state of the mantle beneath the Slave craton, Northern Canada. J. Petrol. 40, 79–104.

LaTourette, T.K., Kennedy, A.K., Wasserburg, G.J., 1993. Thorium-uranium fractionation by garnet: evidence for a deep source and rapid rise of oceanic basalts. Science 261, 739–742.

Li, J.-P., O'Neill, H.S.C., Seifert, F., 1995. Subsolidus phase relations in the system MgO–SiO₂–Cr–O in equilibrium with metallic Cr, and their significance for the petrochemistry of chromium. J. Petrol. 36, 107–132.

Liang, Y., Elthon, D., 1990. Evidence from chromium abundances in mantle rocks for extraction of picrite and komatiite melts. Nature 343, 551–553.

Liu, L., 1977. The system enstatite-pyrope at high pressures and temperatures and the mineralogy of the Earth's mantle. Earth Planet. Sci. Lett. 36, 237–245.

Longhi, J., 2002. Some phase equilibrium systematics of lherzolite melting: I. Geochem. Geophys. Geosys. 3 (3), 1020.

MacGregor, I.D., 1965. Stability fields of spinel and garnet peridotites in the synthetic system MgO–CaO–Al₂O₃–SiO₂. Year b.-Carnegie Inst. Wash. 64, 126–134.

MacGregor, I.D., 1970. The effect of CaO, Cr₂O₃, Fe₂O₃ and Al₂O₃ on the stability of spinel and garnet peridotites. Phys. Earth. Inter. 3, 372–377.

Nickel, K.G., 1986. Phase equilibria in the system SiO₂–MgO–Al₂O₃–CaO–Cr₂O₃ (SMACCR) and their bearing on spinel/garnet lherzolite relationships. N. Jahrb. Mineral. Abh. 155, 259–287.

O'Hara, M.J., Richardson, S.W., Wilson, G., 1971. Garnet-peridotite stability and occurrence in crust and mantle. Contrib. Mineral. Petrol. 32, 48–68.

O'Neill, H.S.C., 1981. The transition between spinel lherzolite and garnet lherzolite, and its use as a geobarometer. Contrib. Mineral. Petrol. 77, 185–194.

Oka, Y., Steinke, P., Chatterjee, N.D., 1984. Thermodynamic mixing properties of Mg(Al,Cr)2O4 spinel crystalline solutions at high pressures and temperatures. Contrib. Mineral. Petrol. 87, 196–204.

Ringwood, A.E., 1967. The pyroxene garnet transformation in the Earth's mantle. Earth Planet. Sci. Lett. 2, 255–263.

Ringwood, A.E., 1977. Synthesis of pyrope-knorringite solid-solution series. Earth Planet. Sci. Lett. 36, 443–448.

Ringwood, A.E., Major, A., 1971. Synthesis of majorite and other high-pressure garnets and perovskites. Earth Planet. Sci. Lett. 12, 411–441.

Robie, R.A., Hemingway, B.S., Fisher, J.R., 1979. Thermodynamic properties of minerals and related substances at 298.15 K and 1 bar (1e5 Pa) pressures and at higher temperatures. U.S. Geol. Surv. Bull. 1452, 1–456.

Robinson, J.A.C., Wood, B.J., 1998. The depth of the garnet/spinel transition in fractionally melting peridotite. Earth Planet. Sci. Lett. 164, 277–284.

Russell, J.K., Kopylova, M.G., 1999. A steady state conductive geotherm for the north central Slave, Canada: inversion of petrological data from the Jericho Kimberlite pipe. J. Geophys. Res. 104, 7089–7101.

Salters, V.J.M., Hart, S.R., 1989. The hafnium paradox and the role of garnet in the source of mid-ocean-ridge basalts. Nature 342, 420–422.

Schreiber, H.D., Haskin, L.A., 1976. Chromium in basalts: experimental determination of redox states and partitioning among silicate phases, 7th Lunar Science Conference. Geochim. Cosmochim. Acta, Suppl., pp. 1221–1259.

Seifert, S., Ringwood, A.E., 1988. The lunar geochemistry of chromium and vanadium. Earth Moon, Planets 40, 45–70.

Shen, Y., Forsyth, D.W., 1995. Geochemical constraints on initial and final depths of melting beneath mid-ocean ridges. J. Geophys. Res. 100, 2211–2237.

Stachel, T., Harris, J.W., 1997. Syngenetic inclusions in diamond from the Birim field (Ghana)—a deep peridotitic profile with a history of depletion and re-enrichment. Contrib. Mineral. Petrol. 127, 336–352.

Stachel, T., Viljoen, K.S., Brey, G.P., Harris, J.W., 1998. Metasomatic processes in herzolitic and harzburgitic domains of diamondiferous lithospheric mantle—REE in garnets from xenoliths and inclusions in diamonds. Earth Planet. Sci. Lett. 159, 1–12.

Stachel, T., Harris, J.W., Brey, G.P., Joswig, W., 2000. Kankan diamonds (Guinea): II. Lower mantle inclusion parageneses. Contrib. Mineral. Petrol. 140, 16–27.

Turkin, A.I., Doroshev, A.M., Yu, I., 1983. Study of the phase composition of garnet-bearing associations of the system $MgO–Al_2O_3–SiO_2–Cr_2O_3$ system at high temperatures and pressures [in Russian]. Silicate Systems Under High Pressure, p. 5. Novosibirsk.

Walter, M., et al., 2002. Spinel–garnet lherzolite transition in the system $CaO–MgO–Al_2O_3–SiO_2$ revisited: an in situ X-ray study. Geochim. Cosmochim. Acta 66, 2109–2121.

Webb, S.A.C., Wood, B.J., 1986. Spinel-pyroxene-garnet relationships and their dependence on Cr/Al ratio. Contrib. Mineral. Petrol. 92, 471–480.

Wood, B.J., 1978. The influence of Cr_2O_3 on the relationships between spinel- and garnet-peridotites. In: MacKenzie, W.S. (Ed.), Progress in Experimental Petrology. Natural Enviroment Research Council, Manchester, pp. 78–80.

Wood, D.A., 1979. A variably veined suboceanic upper mantle-genetic significance for mid-ocean ridge basalts from geochemical evidence. Geology 7, 499–503.

Wood, B.J., Kleppa, O.J., 1984. Chromium and aluminum mixing in garnet: a thermochemical study. Geochim. Cosmochim. Acta 48, 1373–1375.

Available online at www.sciencedirect.com

Lithos 77 (2004) 647–653

ELSEVIER

LITHOS

www.elsevier.com/locate/lithos

Status report on stability of K-rich phases at mantle conditions

George E. Harlow*, Rondi Davies

Department of Earth and Planetary Sciences, American Museum of Natural History, New York, NY 10024-5192, USA

Received 27 June 2003; accepted 17 February 2004
Available online 18 May 2004

Abstract

Experimental research on K-rich phases and observations from diamond inclusions, UHP metamorphic rocks, and xenoliths provide insights about the hosts for potassium at mantle conditions. K-rich clinopyroxene ($Kcpx–KM^{3+}Si_2O_6$) can be an important component in clinopyroxenes at $P>4$ GPa, dependent upon coexisting K-bearing phases (solid or liquid) but not, apparently, upon temperature. Maximum Kcpx content can reach ∼ 25 mol%, with 17 mol% the highest reported in nature. Partitioning $^{(K)}D_{(cpx/liquid)}$ above 7 GPa = 0.1–0.2 require ultrapotassic liquids to form highly potassic cpx or critical solid reactions, e.g., between Kspar and Di. Phlogopite can be stable to about 8 GPa at 1250 °C where either amphibole or liquid forms. When fluorine is present, it generally increases in Phl upon increasing P (and probably T) to about 6 GPa, but reactions forming amphibole and/or $KMgF_3$ limit F content between 6 and 8 GPa. The perovskite $KMgF_3$ is stable up to 10 GPa and 1400 °C as subsolidus breakdown products of phlogopite upon increasing P. $^{(M4)}$K-substituted potassic richterite (ideally $K(KCa)Mg_5Si_8O_{22}(OH,F)_2$) is produced in K-rich peridotites above 6 GPa and in Di + Phl from 6 to ∼ 13 GPa. K content of amphibole is positively correlated with P; Al and F content decrease with P. In the system 1Kspar + 1H₂O K-cymrite (hydrous hexasanidine–$KAlSi_3O_8 \cdot nH_2O$–Kcym) is stable from 2.5 GPa at 400 to 1200 °C and 9 GPa; Kcym can be a supersolidus phase. Formation of Kcym is sensitive to water content, not forming within experiments with H₂O < Kspar, and melt + aluminous phases forming with H₂O>Kspar. Phase X, a potassium di-magnesium acid disilicate (($(K_{1-x-n})_2(Mg_{1-n}M_n^{3+})_2Si_2O_7H_{2x}$)), forms in mafic compositions at $T = 1150–1400$ °C and $P = 9–17$ GPa and is a potential host for K and H₂O at mantle conditions with a low-T geotherm or in subducting slabs. The composition of phase-X is not fixed but actually represents a solid solution in the stoichiometries $\square_2Mg_2Si_2O_7H_2–(K\square)Mg_2Si_2O_7H–K_2Mg_2Si_2O_7$ ($\square$ = vacancy), apparently stable only near the central composition. K-hollandite, $KAlSi_3O_8$, is possibly the most important K-rich phase at very high pressure, as it appears to be stable to conditions near the core–mantle boundary, 95 GPa and 2300 °C. Other K-rich phases are considered.
© 2004 Elsevier B.V. All rights reserved.

Keywords: Potassium; Mantle; K-rich-phases; High-pressure

1. Introduction

Potassium is an important element in Earth, because as a large-ion lithophile element, it is an "incompatible" element that has been strongly parti- tioned into the crust from the mantle by differentia- tion, it is important to many important mineral families (feldspars, micas, amphiboles) and, in the form of ^{40}K, it contributes to Earth's heat budget. How potassium can be retained in the mantle, whether via recycling of the crust or by inherent stabilities that prevented thorough differentiation into the crust, can only be addressed by understanding mineral reservoirs

* Corresponding author. Fax: +1-212-769-5339.
E-mail address: gharlow@amnh.org (G.E. Harlow).

0024-4937/$ - see front matter © 2004 Elsevier B.V. All rights reserved.
doi:10.1016/j.lithos.2004.04.010

at mantle conditions. Much experimental research has been carried out on K-rich phases at high pressure in recent years. Combined with observations from diamond inclusions, UHP metamorphic rocks, and xenoliths and enclaves from extrusives, these provide an expanding understanding of the hosts for potassium at mantle conditions. An update of the extant data is presented here.

2. K-rich phases

2.1. Sanidine

In a "dry" environment this feldspar is stable to 6.5 GPa at 1000 °C (see Table 1), breaking down to $K_2Si_4O_9 + Al_2SiO_5 + SiO_2$ (Yagi et al., 1994; Urakawa et al., 1994) and has been found as inclusions in diamonds and kyanite eclogites (always in eclogite association). Experiments in alkali basalt by Tsuruta and Takahashi (1998) found sanidine as a minor subsolidus phase up to 6 GPa. Abundant fluid inclusions (Smyth and Hatton, 1977) or strong H_2O IR-signature in sanidine, in certain cases, makes K-cymrite a logical precursor phase (see below).

2.2. K-rich clinopyroxene

K-rich clinopyroxene ($Kcpx–KM^{3+}Si_2O_6$): Experiments show that Kcpx can become an important component in clinopyroxenes at pressures above 4–5 GPa, depending upon coexisting K-bearing phases (or the lack thereof) but not, apparently, upon temperature (Luth, 1997; Harlow, 1997; Schmidt and Poli, 1998; Tsuruta and Takahashi, 1998; Perchuk et al., 2002). Maximum Kcpx content can reach ~ 25 mol% with 17 mol% the highest reported from a natural (UHP) sample (see Perchuk et al., 2002), other K-rich clinopyroxene being found as inclusions in diamonds (e.g., Harlow and Veblen, 1991; see Perchuk et al., 2002). The partition coefficient for calcic clinopyroxene (cpx) and melt, $^KD_{cpx/liquid}$, above 7 GPa $\cong 0.1–0.2$ and requires either ultrapotassic liquids (inferred to require carbonatitic melt or a fluid rather than a silicate melt; e.g., Konzett and Fei, 2000; Perchuk et al., 2002) to form highly potassic cpx (>1.5 wt.% K_2O) or a "critical" cpx solid-solution reaction with a liquid (Safonov et al., 2003). In studying melt parti-

tioning for diopside–jadeite solid solutions, Chamorro et al. (2002) have confirmed $^KD_{cpx/liquid}$ values ≥ 0.2 for diopsidic cpx and lower values, ~ 0.01, for jadeitic cpx. As cpx can be a liquidus-to-subsolidus product, the only critical requirements in the formation of K-rich cpx is calcic cpx stability and K content in the phase assemblage, however, coexisting more K-rich solids (e.g., sanidine, K-rich amphibole) can reduce the K-content of the cpx coexisting with a melt relative to that in a cpx–melt pair where no K-rich solids are present (e.g., Konzett and Fei, 2000; Mitchell, 1995). Reporting on experiments with a K-spiked, water-saturated MORB composition (0.49 wt.% K_2O in their KMB-7), Schmidt and Poli (1998) found that above 4 GPa K-content of cpx increases upon increasing T up to the solidus but then decreases dramatically at supersolidus conditions. Some, thus, interpret high Kcpx contents (>1 wt.% K_2O) in natural cpx to be the result of growth in the absence of another coexisting K-rich phase and melt or, alternatively, growth from a very K-rich fluid/melt that, in all cases, was followed by encapsulation–preservation to avoid reequilibration/unmixing that would reduce the Kcpx content. A review of experimental results is provided by Perchuk et al. (2002). Oxidized iron-rich assemblages (e.g., some lamproites/lamprophyres/etc.; Plá Cid et al., 2003; Mitchell, 1995; Mellini and Cundari, 1989) may enable a ferric Kcpx at somewhat lower pressures than 5 GPa and is being investigated by the authors.

2.3. Phlogopite

In peridotites, phlogopite is stable to >6 GPa at 1100 °C (Konzett et al., 1997) and to between 8 and 12 GPa at 1250–1350 °C in diopside (Di) plus phlogopite (Phl) assemblages (depending on bulk composition, see Harlow, 2002) above which P and T either amphibole or liquid is more stable. When fluorine is present, it generally increases in Phl upon increasing P (and probably T) to about 6 GPa, but reactions to form amphibole and/or $KMgF_3$ limit F content between 6 and 8 GPa (Harlow, 2002). In peralkaline KNCMASH, Phl persists to 10 GPa at ~ 1300 °C (see Konzett and Fei, 2000). In more pelitic compositions, Phl breaks down at low P and T (~ 3 GPa at 1000 °C) to phengite ± sanidine ± fluid (see Massone, 1999).

Table 1
Potential K-rich mantle minerals

Mineral	Composition	Stability	Citations
Sanidine	$KAlSi_3O_8$	to ~ 6 GPa, 1200 °C	Yagi et al. (1994), Urakawa et al. (1994)
Phengitic muscovite	$KAl_{2-x}Mg_xAl_{1-x}Si_{3+x}O_{10}(OH)_2$	to 9.5–10, 750–1050 °C	Schmidt (1996), Schmidt and Poli (1998)
		to 8–11 GPa, 750–1050 °C	Domanik and Holloway (2000)
Phlogopite	$KMg_3AlSi_3O_{10}(OH)_2$	to ~ 9 GPa, 1400 °C	Luth (1997)
	$KMg_3AlSi_3O_{10}(F)_2$	to ~ 10 GPa, 1400 °C	Harlow (2002)
KK-richterite	$KKCaMg_5Si_8O_{22}(OH)_2$	from 6 GPa at 1000 °C to >15 GPa at 1400 °C from 8 GPa at 1100 °C to 14 GPa at 1400 °C	Inoue et al. (1998), Luth (1997), Sudo and Tatsumi (1990), Harlow (2002) Konzett and Fei (2000)
	$K(Ca,K,Na)_2Mg_5(Si,Al)_8O_{22}(F,OH)_2$	Konzett et al. (1997) to 13 GPa at 1400 °C	Konzett and Ulmer (1999) Konzett and Fei (2000)
Phase X	$(K\square)Mg_2Si_2O_7H$	9–23 GPa at 1150–1700 °C	Luth (1997), Harlow (1997), Inoue et al. (1998), Konzett and Ulmer (1999), Konzett and Fei (2000)
(21)-MHP	$KNa_2Ca_2Mg_6AlSi_{12}O_{34}(OH)_2$	Konzett and Fei (2000) 5–18 GPa at 1100–1600 °C	Konzett and Japel (2003)
K-clinopyroxene	$K(Al,Cr)Si_2O_6 - Cpx_{ss}$	5–15 GPa at 1000–1500 °C Kcpx1 to Kcpx20	Harlow (1997), Wang and Takahashi (1999), Safonov et al. (2003)
Si-wadeite (or K-wadeite)	$K_2Si_4O_9$	from 6 GPa at 1000 °C to ~ 16 GPa at 1400 °C to 12 GPa at 1500 °C	Yagi et al. (1994), Kanzaki et al. (1998) Harlow (1997)
Wadeite	$K_2ZrSi_3O_9$	from 1 atm at 800 to >3 GPa at 1200 °C	Arima and Edgar (1980) Orlando et al. (2000)
K-cymrite (or sanidine hydrate)	$KAlSi_3O_8 \bullet nH_2O$ ($n \leq 1$)	from 2.5 GPa at 400 °C to 9 GPa at 1200 °C	Massone (1995), Fasshauer et al. (1997), Thompson et al. (1998) This work
$KMgF_3$-perovskite $KMgF_3$		to $\geq$ 8 GPa at 1400 °C	Harlow (2002), Thibault (1993)
Al-rich phase	$[K,Na]_{0.9}[Mg,Fe]_2[Mg,Fe,Al,Si]_6O_{12}$	$\geq$ 24 GPa at 1700–1800 °C	Gasparik and Litvin (2002)
K-phase 1	$(K,Na)_2FeMg_4Si_5O_{16}$ or $(K,Na)_2Mg_4Si_4O_{13}$	15 GPa at 1400 °C to 20 GPa at <1900 °C	Wang and Takahashi (2000), Gasparik and Litvin (2002)
K-hollandite	$KAlSi_3O_8$	from ~ 9 GPa at >1000 °C to >25 GPa at >1600 °C to 95 GPa at 2300 °C	Yagi et al. (1994), Urakawa et al. (1994), Wang and Takahashi (1999) Tutti et al. (2001)
K-Ti silicates	$K_2TiSi_3O_9$	from 1 to 3 GPa at 1100 °C	Gulliver et al. (1998)
	$K_4Ti_2Si_7O_{20}$ to $K_4TiSi_8O_{20}$	from ~ 4 to >6 GPa at 1100–1400 °C	Mitchell (1995)
Priderite	$(K,Ba,\square)_2(Ti,Fe^{3+},Fe^{2+})_8O_{16}$	1200–1500 °C, 3.5–5 GPa	Foley et al. (1994)
Mathiasite	$K(Ti_{13}Cr_4FeZrMg_2)O_{38}$	1300 °C, 5 GPa	Foley et al. (1994)
LIMA	$(Ba_{0.5}K_{0.5})(Ti_{13}Cr_{3.5}FeZrMg_{2.5})O_{38}$	1200–1300 °C, 3.5–5 GPa	Foley et al. (1994)
Yimengite	$K(Ti_3Cr_5Fe_2Mg_2)O_{19}$	1200 °C, 5 GPa	Foley et al. (1994)
HAWYIM	$(Ba_{0.5}K_{0.5})(Ti_3Cr_{4.5}Fe_2Mg_{2.5})O_{19}$	1150–1350 °C, 4.3–5 GPa	Foley et al. (1994)

2.4. Phengitic muscovite

Phengite stability has been examined in "wet" graywacke and K-rich basalt compositions (Schmidt, 1996; Schmidt and Poli, 1998), a bulk composition approximating something between mica and a pelitic sediment (Domanik and Holloway, 1996), and calcareous metapelite (Domanik and Holloway, 2000) over the general range of 7–11 GPa and 750 to 1100 °C (see Table 1). Upon increasing pressure phengite yields to K-hollandite stability between 8 and 11 GPa and 750–900 °C (the reaction has a slightly negative P/T slope), with the transition pressure depending perhaps on bulk composition, such as alumina content. Melt breakdown of the mica occurs above 900 °C at low bulk Al content and above 1075–1150 °C at 7–8 GPa and high Al content. Phengite is the most significant host for potassium in the subducting oceanic crust (sediments + basaltic components) and, in that role, also carries water to pressures from 2 to perhaps 10 GPa, as pointed out by Schmidt and Poli (1998).

2.5. Amphibole–^{M4}K-substituted potassic richterite

This amphibole, sometimes called KK-richterite (ideally K[KCa]Mg$_5$Si$_8$O$_{22}$[OH,F]$_2$), must be differentiated from K-richterite (ideally K[NaCa]Mg$_5$Si$_8$O$_{22}$[OH,F]$_2$) that is a moderate to high-pressure ($\sim$ 2 to < 6 GPa) amphibole found in xenoliths, such as from the Kimberley (RSA) pipes (e.g., Erlank et al., 1987), and in lamproites (e.g., Mitchell and Bergman, 1991; Mitchell, 1995). The substitution of K into the M4 site, analogous to K in M2 in cpx, is the defining high-pressure feature of this K-rich amphibole (Yang et al., 1999). It has been produced from of Di + Phl and K-rich peridotitic compositions at P>6 to 15 GPa (Trønnes, 1990; Sudo and Tatsumi, 1990; Luth, 1997; Konzett et al., 1997; Inoue et al., 1998; Konzett and Ulmer, 1999; Konzett and Fei, 2000; Harlow, 2002; see Table 1), above which it breaks down to phase X, Si-wadeite, and/or liquid. Various experimental studies have shown that K content is positively correlated with P; Al and F content decrease with P; and F content is positively correlated with T but lowered by coexisting KMgF$_3$ (Harlow, 2002). In less K-rich bulk compositions, such as natural KLB-1, amphibole breaks down at 12–13 GPa and 1200 °C (see Konzett and Fei, 2000). Amphibole is a major potential reservoir in the upper mantle at depths exceeding $\sim$ 200 km and may be important in fertilized mantle wedge during subduction (e.g., Schmidt and Poli, 1998).

2.6. Phase X

Phase X, a potassium di-magnesium acid disilicate ([K$_{1-x-n}$]$_2$[Mg$_{1-n}$M$_n^{3+}$]$_2$Si$_2$O$_7$H$_{2x}$), was discovered

in synthesis products in various studies at T = 1150–1400 °C and P = 9–17 GPa (e.g., Luth, 1997; Harlow, 1997; Inoue et al., 1998; Konzett and Ulmer, 1999). It is the "cold" breakdown product of amphibole: < 1200 °C at $P \sim$ 16 GPa (Inoue et al., 1998). Its maximum stability in KNCMASH and KLB-1 peridotite reaches 20–23 GPa at 1500–1700 °C where it breaks down to K-hollandite; the stability limit is reduced by a few GPa in subalkaline KNCMASH (Konzett and Fei, 2000). Thus, phase X is a potential host for K and H$_2$O in the mantle to the bottom of the transition zone. The composition of phase-X is not fixed but actually represents a solid solution in the stoichiometries $\square_2$Mg$_2$Si$_2$O$_7$H$_2$—(K$\square$)Mg$_2$Si$_2$O$_7$H—K$_2$Mg$_2$Si$_2$O$_7$, where the center part of the solid solution series appears to be the most stable portion at the conditions examined so far. There is also solid solution with a Na-version of phase X, but experiments show that K is preferred over Na at higher pressures (Konzett and Fei, 2000).

2.7. K-hollandite

KAlSi$_3$O$_8$, in which Al and Si are both 6-coordinated forming a channel for the K site, is stable from $\sim$ 8 GPa through conditions of the lower mantle (95 GPa to 2300 °C in DAC experiments) (e.g., Yagi et al., 1994; Urakawa et al., 1994; Konzett and Fei, 2000; Tutti et al., 2001). This phase appears to be the main K-rich solid-phase reservoir for potassium through the transition zone to the lower mantle when considering silicate compositional systems, as most K–Al silicates react to yield K-hollandite upon increasing pressure (e.g., Faust and Knittle, 1994; Irifune et al., 1994; Schmidt, 1996). One inclusion in diamond (Kankan, Guinea) has been reported where K-hollandite might have been the precursor phase (Stachel et al., 2000).

2.8. KMgF$_3$ perovskite

Pure KMgF$_3$ was studied by Gulliver et al. (1998) and shown to be stable to 2.6 GPa and 1400 °C. Thibault (1993) and Edgar and Pizzolato (1995) found it as a breakdown of F-phlogopite up to 8 GPa. It (with F-bearing clinohumite and chondrodite) is stable up to at least 10 GPa and 1400 °C as subsolidus breakdown products of F-bearing phlogopite (+ Di) upon increasing P (Harlow, 2002), thus some phlogopite (or melts) may be a reaction product of this phase upon uplift.

2.9. K-cymrite

K-cymrite (or hydrous hexasanidine–$KAlSi_3O_8 \cdot nH_2O$, $n \leq 1$) has been shown to be stable above 2.5 GPa at 400 to 1000 °C and $\sim$ 4 GPa (Massone, 1995; Fasshauer et al., 1997; Thompson et al., 1998) in the pure system $KAlSi_3O_8 + H_2O$. Producing K-cymrite (Kcym) at high pressure appears to be highly dependent upon H_2O content of the system. At low water contents ($H_2O < KAlSi_3O_8$) experiments yield the anhydrous assemblages plus vapor or melt without ever forming a hydrous crystalline phase, whereas high water content ($H_2O > KAlSi_3O_8$) produces extensive melting with coexisting aluminous crystalline phases such as muscovite, kyanite, and corundum. At conditions of $H_2O \approx KAlSi_3O_8$, Kcym is stable to at least 9 GPa at 1200 °C and 8 GPa at 1250 °C, suggesting a negative P/T slope for breakdown to K-hollandite + stishovite + fluid. Kcym can be a supersolidus phase, which may be very significant to its survival at mantle conditions. Study of K-cymrite stability conditions continues by the authors. In experiments with a mixture of Di + sanidine + Phl, Kcym was found at 6–8 GPa and 1200 °C coexisting with Cpx + Phl $\pm$ kyanite $\pm$ enstatite.

Whereas no preserved examples of K-cymrite have been identified, the existence of sanidine with an abundance of fluid inclusions from a sanidine grospydite (Smyth and Hatton, 1977) and sanidine + quartz + biotite inclusions coexisting with apparent graphite pseudomorphs of diamond in garnets from high-pressure enclaves of the Erzgebirge (Massone and Nasdala, 2003) appear to be the retrograde assemblages after K-cymrite. K-cymrite appears to be an important potential reservoir for both potassium and water at UHP and upper mantle conditions for eclogites and continental crust components.

2.10. Wadeite and Si-wadeite

Wadeite, $K_2ZrSi_3O_9$, a common minor constituent in lamproites, and some other highly potassic rocks and carbonatites, is stable from 1 atm to at least 3 GPa at 800–1200 °C (Arima and Edgar, 1980; Orlando et al., 2000). It forms a complete solid solution with $K_2TiSi_3O_9$ in binary experiments, but in the presence of melt or phlogopite or Ti content decreases, indicating the more refractory nature of wadeite. Si-wadeite ($K_2Si_4O_9$) is a common product in K-rich experiments with stability from 6 to >12 GPa, typically limited by reactions forming sanidine, K-cymrite, or K-hollandite. Pure $K_2Si_4O_9$ melts congruently to at least 12 GPa (Kanzaki et al., 1998). It was found as a breakdown product of K-rich amphibole at $T > 1200$ °C and P from 14 to 16 GPa (Inoue et al., 1998). There are no citations of Si-wadeite in natural samples.

2.11. K–Ti-silicates

$K_2TiSi_3O_9$, a wadeite analog (see above), is stable to at least 3 GPa in the pure K–Ti-silicate system, but in nature has only been found in one lamproite (Gulliver et al., 1998). Perhaps related to $BaTi_9O_{20}$ (hollandite-like structure), a phase with apparent formula $K[Si,Ti]_9O_{20}$ was produced from sanidine–phlogopite lamproite in experiments at >4.5 GPa and $\sim$ 1400–1500 °C (Mitchell, 1995). These K–Ti silicates bear a passing resemblance to members of the crichtonite group, particularly mathiasite, $(K,Ca,Sr)(Ti,Cr,Fe,Mg)_{21}O_{38}$, which are interpreted as forming by kimberlite-related metasomatism in the upper mantle (Haggerty et al., 1983).

2.12. K–Ba–Ti–Fe-oxides

Experiments have been carried out on priderite $[(K,Ba,\square)_2(Ti,Fe^{3+})_8O_{16}$—with a hollandite-like structure] and the lindsleyite–mathiasite (LIMA—with a rhombohedral layer structure) and hawthorneite–yimengite (HAWYIM—with a magnetoplumbite structure) series, a group of K–Ba–Ti–Cr–Zr oxides (see Table 1), considered important in the origin of deep alkaline magmas such as lamproites and kimberlites. These experiments have shown that priderite and LIMA phases are stable to pressures between 3.5 and 5 GPa and HAWYIM phases between 4.3 and 5 GPa at 1200 to 1500 °C (Foley et al., 1994), which do not represent their maximum stabilities as the experiments only reached 5 GPa.

2.13. (21)-MHP

The mixed-chain (21)-hydrous clinopyribole, nominally $KNa_2Ca_2Mg_6AlSi_{12}O_{34}(OH)_2$—a combination of K-richterite plus 2 omphacite formulae, was found in experiments on a KNCMASH bulk composition by Konzett and Fei (2000). In recent work with compositions approximating its own composition, (21)-MHP has been shown to be stable from 5–18 GPa and 1100–1600 °C (Konzett and Japel, 2003) and, potassium has been shown to be necessary to produce this clinopyribole.

2.14. K-phase-1

$(K,Na)_2Mg_4Si_4O_{13}$ and related K-rich phases have been reported at high P in experiments by Wang and Takahashi (2000) and Gasparik and Litvin (2002) (see Table 1), but the low-melting temperatures (subgeotherm) make them unlikely candidates as a sink or sampleable phase in the mantle.

2.15. Al-rich phase

A phase with a composition like $[K,Na]_{0.9}[Mg, Fe]_2[Mg,Fe,Al,Si]_6O_{12}$ with up to ~ 7 wt.% K_2O has been reported in experiments (Gasparik et al., 2000; Gasparik and Litvin, 2002) at 24 GPa from the breakdown of garnet to perovskite in the presence of potassium and invoked as a partial explanation for a cpx-corundum pair from a São Luiz, Brazil diamond (Hutchison, 1997).

Acknowledgements

We gratefully acknowledge the National Science Foundation (EAR-9314819 & EAR-9903203) for support of this research and thank Thomas Stachel, Sergei Matveev, and an anonymous reviewer for their helpful criticisms.

References

Arima, M., Edgar, A.D., 1980. Stability of wadeite $(Zr_2K_4Si_6O_{18})$ under mantle conditions: petrological implications. Contrib. Mineral. Petrol. 72, 191–195.

Chamorro, E.M., Brooker, R.A., Wartho, J.-A., Wood, B.J., Kelly, S.P., Blundy, J.D., 2002. Ar and K partitioning between clinopyroxene and silicate melt to 8 GPa. Geochim. Cosmochim. Acta 66, 507–519.

Domanik, K.J., Holloway, J.R., 1996. The stability of phengitic muscovite and associated phases from 5.5 to 11 GPa: implications for deeply subducted sediments. Geochim. Cosmochim. Acta 60, 4133–4150.

Domanik, K.J., Holloway, J.R., 2000. Experimental synthesis and phase relations of phengitic muscovite from 6.5 to 11 GPa in a calcareous metapelite from the Dabie Mountains, China. Lithos 52, 51–77.

Edgar, A.D., Pizzolato, L.A., 1995. An experimental study of partitioning of fluorine between K-richterite, apatite, phlogopite, and melt at 20 kbar. Contrib. Mineral. Petrol. 121, 247–257.

Erlank, A.J., Waters, F.G., Hawkesworth, C.J., Haggerty, S.E., Allsopp, H.L., Rickard, R.S., Menzies, M., 1987. Evidence for mantle metasomatism in peridotite nodules from the Kimberley pipes, South Africa. In: Menzies, M.A., Hawkesworth, C.J. (Eds.), Mantle Metasomatism. Academic Press, London, pp. 221–311.

Fasshauer, D.W., Chatterjee, N.D., Marler, B., 1997. Synthesis, structure, thermodynamic properties and stability of K-cymrite, $K[AlSi_3O_8] \times H_2O$. Phys. Chem. Miner. 24, 455–462.

Faust, J., Knittle, E., 1994. The equation of state, amorphization, and high-pressure phase diagram of muscovite. J. Geophys. Res. 99, 19785–19792.

Foley, S., Höfer, H., Brey, G., 1994. High-pressure synthesis of priderite and members of the lindsleyite–mathiasite and hawthorneite–yimengite series. Contrib. Mineral. Petrol. 117, 164–174.

Gasparik, T., Litvin, Y.A., 2002. Experimental investigation of the effect of metasomatism by carbonatic melt on the composition and structure of the deep mantle. Lithos 60, 129–143.

Gasparik, T., Tripathi, A., Parise, J.B., 2000. Structure of a new Al-rich phase, $[K,Na]_{0.9}[Mg,Fe]_2[Mg,Fe,Al,Si]_6O_{12}$, synthesized at 24 GPa. Am. Mineral. 85, 613–618.

Gulliver, C.E., Edgar, A.D., Mitchell, R.H., 1998. Stability and composition of K–Ti silicates, K–Ba phosphate and K–Mg fluoride at 0.85–2.6 GPa; implications for the genesis of potassic alkaline magmas. Can. Mineral. 36, 1339–1346.

Haggerty, S.E., Smyth, J.R., Erlank, A.J., Rickard, R.S., Danchin, R.V., 1983. Lindsleyite (Ba) and mathiasite (K): two new chromium–titanates in the crichtonite series from the upper mantle. Am. Mineral. 68, 494–505.

Harlow, G.E., 1997. K in clinopyroxene at high pressure and temperature: an experimental study. Am. Mineral. 82, 259–269.

Harlow, G.E., 2002. Diopside + F-rich phlogopite at high P and T: systematics and the stability of $KMgF_3$, clinohumite and chondrodite. Geol. Mater. Res. v4n3, 1–28.

Harlow, G.E., Veblen, D.R., 1991. Potassium in clinopyroxene inclusions from diamonds. Science 251, 652–655.

Hutchison, M.T., 1997. The constitution of the deep transition zone and lower mantle shown by diamonds and their inclusions. PhD thesis, Univ. Edinburgh, Edinburgh.

Inoue, T., Irifune, T., Yurimoto, T., Yurimoto, H., Miyagi, I., 1998. Decomposition of K-amphibole at high pressures and implications for subduction zone volcanism. Phys. Earth Planet. Inter. 107, 221–231.

Irifune, T., Ringwood, A.E., Hibberson, W.O., 1994. Subduction of continental crust and terrigenous and pelagic sediments: an experimental study. Earth Planet. Sci. Lett. 77, 351–368.

Kanzaki, M., Xue, X., Stebbins, J.F., 1998. Phase relations in $Na_2O–SiO_2$ and $K_2Si_4O_9$ systems up to 14 GPa and ^{29}Si NMR study of the new high-pressure phases; implications to the structure of high-pressure silicate glasses. Phys. Earth Planet. Inter. 107, 9–21.

Konzett, J., Fei, Y., 2000. Transport and storage of potassium in the Earth's upper mantle and transition zone: an experimental study to 23 GPa in simplified and natural bulk compositions. J. Petrol. 41, 583–603.

Konzett, J., Japel, S.L., 2003. High PT phase relations and stability of an ordered (21)-hydrous clinopyribole in the system $K_2O–Na_2O–CaO–MgO–Al_2O_3–SiO_2–H_2O$: an experimental study to 18 GPa. Am. Mineral. 88, 1073–1083.

Konzett, J., Ulmer, P., 1999. The stability of hydrous potassic phases in lherzolitic mantle—an experimental study to 9.5 GPa in simplified and natural bulk compositions. J. Petrol. 40, 629–652.

Konzett, J., Sweeney, R.J., Thompson, A.B., Ulmer, P., 1997. Potassium amphibole stability in the upper mantle: an experimental study in a peralkaline KNCMASH system to 8.5 GPa. J. Petrol. 38, 537–568.

Luth, R.W., 1997. Experimental study of the system phlogopite–diopside from 3.5 to 17 GPa. Am. Mineral. 82, 1198–1209.

Massone, H.-J., 1995. Experimental and petrogenetic study of UHPM. In: Coleman, R., Wang, X. (Eds.), Ultrahigh Pressure Metamorphism. Cambridge Univ. Press, New York, pp. 33–95.

Massone, H.-J., 1999. Experimental aspects of UHP metamorphism in pelitic systems. Int. Geol. Rev. 41, 623–638.

Massone, H.-J., Nasdala, L., 2003. Characterization of an early metamorphic stage through inclusions in zircon of a diamondiferous quartzofelspathic rock from the Erzgebirge, Germany. Am. Mineral. 88, 883–889.

Mellini, M., Cundari, A., 1989. On the reported presence of potassium in clinopyroxene from potassium-rich lavas: a transmission electron microscope study. Mineral. Mag. 53, 311–314.

Mitchell, R.H., 1995. Melting experiments on a sanidine phlogopite lamproite at 4–7 GPa and their bearing on the sources of lamproite magmas. J. Petrol. 36, 1455–1474.

Mitchell, R.H., Bergman, S.C., 1991. Petrology of Lamproites. Plenum, New York.

Orlando, A., Thibault, Y., Edgar, A.D., 2000. Experimental study of the $K_2ZrSi_3O_9$ (wadeite)–$K_2TiSi_3O_9$ and $K_2(Zr,Ti)Si_3O_9$–phlogopite systems at 2–3 GPa. Contrib. Mineral. Petrol. 139, 136–145.

Perchuk, L.L., Safonov, O.G., Yapaskurt, V.O., Barton, J.M., 2002. Crystal-melt equilibria involving potassium-bearing clinopyroxene as indicator of mantle-derived ultrahigh-potassic liquids: an analytical review. Lithos 60, 89–111.

Plá Cid, J., Nardi, L.V.S., Stabel, L.Z., Conceição, R.V., Balzaretti,

N.M., 2003. High-pressure minerals in mafic microgranular enclaves: evidences for com-mingling between lamprophyric and syenitic magmas at mantle conditions. Contrib. Mineral. Petrol. 145, 444–459.

Safonov, O.G., Litvin, Y.A., Perchuk, L.L., Bindi, L., Menchetti, S., 2003. Phase relations of potassium-bearing clinopyroxene in the system $CaMgSi_2O_6–KAlSi_2O_6$ at 7 GPa. Contrib. Mineral. Petrol. 146, 120–133.

Schmidt, M.W., 1996. Experimental constraints on recycling of potassium from subducted oceanic crust. Science 272, 1927–1930.

Schmidt, M.W., Poli, S., 1998. Experimentally based water budgets for dehydrating slabs and consequences for arc magma generation. Earth Planet. Sci. Lett. 163, 361–379.

Smyth, J.R., Hatton, C.J., 1977. A coesite–sanidine grospydite from the Roberts Victor kimberlite. Earth Planet. Sci. Lett. 34, 284–290.

Stachel, T., Harris, J.W., Brey, G.P., Joswig, W., 2000. Kankan diamonds (Guinea) II: lower mantle inclusion parageneses. Contrib. Mineral. Petrol. 140, 16–27.

Sudo, A., Tatsumi, Y., 1990. Phlogopite and K-amphibole in the upper mantle: implication for magma genesis in subduction zones. Geophys. Res. Lett. 17, 29–32.

Thibault, Y., 1993. The role of pressure and fluorine content in the distribution of K, Na, and Al in the upper mantle. EOS Trans. 74, 321.

Thompson, P., Parsons, I., Graham, C., Jackson, B., 1998. The breakdown of potassium feldspar at high water pressures. Contrib. Mineral. Petrol. 130, 176–186.

Trønnes, R.G., 1990. Low-Al, high-K amphiboles in subducted lithosphere from 200–400 km depth: experimental evidence. EOS Trans. 71, 1587.

Tsuruta, K., Takahashi, E., 1998. Melting study of an alkali basalt JB-1 up to 12.5 GPa: behavior of potassium in the deep mantle. Phys. Earth Planet. Inter. 107, 119–130.

Tutti, F., Dubrovinsky, L.S., Saxena, S.K., Carlson, S., 2001. Stability of $KAlSi_3O_8$ hollandite-type structure in the Earth's lower mantle conditions. Geophys. Res. Lett. 28, 2735–2738.

Urakawa, S., Kondo, T., Igawa, N., Shimomura, O., Ohno, N., 1994. Synchrotron radiation study on the high-pressure and high-temperature phase relations of $KAlSi_3O_8$. Phys. Chem. Miner. 21, 387–391.

Wang, W., Takahashi, E., 1999. Subsolidus and melting experiments of a K-rich basaltic composition to 27 GPa: implication for the behavior of potassium in the mantle. Am. Mineral. 84, 357–361.

Wang, W., Takahashi, E., 2000. Subsolidus and melting experiments of a K-doped peridotite KLB-1 to 27 GPa: its geophysical and geochemical implications. J. Geophys. Res. 105, 2855–2868.

Yagi, A., Suzuki, T., Akaogi, M., 1994. High pressure transition in the system $KAlSi_3O_8–NaAlSi_3O_8$. Phys. Chem. Miner. 21, 12–17.

Yang, H., Konzett, J., Prewitt, C.T., Fei, Y., 1999. Single-crystal structure refinement of synthetic ^{M4}K-substituted potassic richterite, $K(KCa)Mg_5Si_8O_{22}(OH)_2$. Am. Mineral. 84, 681–684.

Available online at www.sciencedirect.com

Lithos 77 (2004) 655–663

www.elsevier.com/locate/lithos

Ferropericlase—a lower mantle phase in the upper mantle

Gerhard P. Brey[a,*], Vadim Bulatov[b], Andrei Girnis[c],
Jeff W. Harris[a,d], Thomas Stachel[a,e]

[a] *Johann Wolgang Goethe-Universitat Institut für Mineralogie, Fachbereich Geowissenschaften, Universität Frankfurt,
Senckenberganlage 28, D-60054 Frankfurt, Germany,*
[b] *Vernadsky Institute, Moscow, Russia*
[c] *IGEM, Moscow, Russia*
[d] *University of Glasgow, UK*
[e] *University of Alberta, Edmonton, Canada*

Received 27 June 2003; accepted 22 January 2004
Available online 28 May 2004

Abstract

Experiments on compositions along the join $MgO - NaA^{3+}Si_2O_6$ ($A = Al$, Cr, Fe^{3+}) show that sodium can be incorporated into ferropericlase at upper mantle pressures in amounts commonly found in natural diamond inclusions. These results, combined with the observed mineral parageneses of several diamond inclusion suites, establish firmly that ferropericlase exists in the upper mantle in regions with low silica activity. Such regions may be carbonated dunite or stalled and degassed carbonatitic melts. Ferropericlase as an inclusion in diamond on its own is not indicative of a lower mantle origin or of a deep mantle plume. Coexisting phases have to be taken into consideration to decide on the depth of origin. The composition of olivine will indicate an origin from the upper mantle or border of the transition zone to the lower mantle and whether it coexisted with ferropericlase in the upper mantle or as ringwoodite. The narrow and flat three phase loop at the border transition zone—lower mantle together with hybrid peridotite plus eclogite/sediments provides an explanation for the varying and Fe-rich nature of the diamond inclusion suite from Sao Luiz, Brazil.
© 2004 Elsevier B.V. All rights reserved.

Keywords: Ferropericlase; Olivine; Diamond; Upper mantle; Lower mantle; Inclusions

1. Introduction

The Earth's mantle may be subdivided into a number of layers, that correspond to phase transitions of its constituent minerals. The 400-km discontinuity marks the top of the transition zone at the onset of the transformation of olivine into a spinel-like β-phase named wadsleyite (Ringwood and Major, 1966; Akaogi et al., 1989). The 650-km discontinuity, leading into the lower mantle, is attributed to the decomposition of Mg_2SiO_4 spinel (γ-phase) or ringwoodite to Mg–Si perovskite plus periclase (Liu, 1974, 1975). Ferropericlase is therefore a major constituent of the Earth's lower mantle. Minor amounts also occur in the Earth's crust, where it forms in high-temperature contact metamorphic processes. When ferropericlase was found as inclusion in diamonds from Orroroo and together with enstatite in diamonds from Koffiefontein, it was considered as an indicator of a lower mantle origin of the host diamond (Scott Smith et al., 1984). On this basis Wilding (1990), Harte and Harris (1994) and Hutchison

* Corresponding author. Fax: +49-69-798-28066.
E-mail address: brey@em.uni-frankfurt.de (G.P. Brey).

(1997) identified a number of lower mantle parageneses in diamonds from Sao Luiz (Brazil). A critical voice came when Gurney (1989) and later Stachel et al. (1998) pointed out, that some of these "lower mantle" parageneses could be explained in terms of disequilibrium between different inclusion generations within individual diamonds.

Stachel et al. (2000) reported further evidence that ferropericlase may form as inclusion in diamonds at upper mantle conditions, provided that the Si activity was low enough. In a sample collection from the Kankan district, Guinea, they found a diamond (KK84) with ferropericlase and olivine both as separate monomineralic inclusions and as a contacting pair. The olivine crystal structure was confirmed for both olivines by in situ single crystal X-ray measurements (see Stachel et al., 2000). They show sharp reflexes characteristic for untwinned crystals. Twinning might be expected if the olivines were the retrograde products of ringwoodite through an ex

tremely rapid uprise of the diamond from 670 km. The single olivine inclusion is identical in composition to that in contact with ferropericlase and vice versa (Fig. 1). This demonstrates their entrapment in a single stage of diamond growth as a ferropericlase–olivine pair and excludes the possibility of the formation as an original ferropericlase–ringwoodite pair in the narrow divariant field at the 670-km discontinuity (where ferropericlase, ringwoodite and Mg–perovskite coexist). In the latter case, ringwoodite would have to transform to olivine which would necessitate Fe- and Mg-exchange of the touching pair. Thus, this diamond has grown in the upper mantle in a region with a particularly low Si activity. Disequilibrium between different inclusion generations is highly unlikely because the chemical similarity includes all measured trace elements (Stachel et al., 2000). Yet, the ferropericlases have elevated concentrations of Na and Cr, which was considered to be indicative of a lower mantle origin (Kesson and Gerald, 1992). The solubility of Na in periclase has been determined experimentally so far only at high pressures corresponding to the transition zone (Gasparik and Litvin, 1997; Gasparik, 2000, 2002) but not at upper mantle pressures. This led us to study the solubility of Na in periclase at upper mantle pressures in the presence of the trivalent cations Al, Fe^{3+} and Cr.

2. Experiments

2.1. Starting materials and experimental procedures

Starting compositions along the joins MgO–$NaAlSi_2O_6$, –$NaCrSi_2O_6$, –$NaFe^{3+}Si_2O_6$ were selected such that MgO was stable together with olivine, spinel (chromite, magnetite) and a Na-bearing phase, in our case melt (Fig. 2). In order to generate large amounts of periclase, the bulk MgO:-SiO_2 was chosen to be 2.5 in experiments at 1300–1400 °C and 4.0 at 1600 °C. The mixtures were prepared from crystalline jadeite and fired oxide mixtures of $NaCrSi_2O_6$ and $NaFeSi_2O_6$ composition added in appropriate amounts to crystalline MgO. The two latter mixtures were fired at ~900 °C in air, which provided a highly oxidized state. The starting materials were welded shut into Pt capsules. The

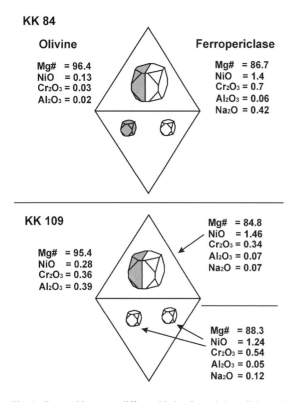

Fig. 1. Compositions two different kinds of coexisting olivine and ferropericlase inclusions in diamond.

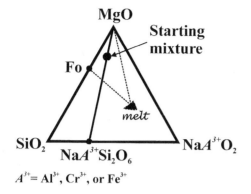

Fig. 2. Composition of the starting mixes in mol% and of the melts produced in the sandwich experiments plotted in a diagram MgO–SiO_2–$NaA^{3+}O_2$.

experiments were carried out under dry conditions in a belt apparatus at 30 and 50 kbar and 1300, 1400 and 1600 °C. One experiment at 30 kbar in the Cr-bearing system was carried out in a Pt double capsule with hematite + H_2O in the outer capsule to ensure highly oxidizing conditions (denoted in Fig. 3 with fO_2). There is no difference in Cr_2O_3 content to the equivalent run without an outer buffer capsule. In order to determine the melt composition more accurately in the Al-bearing system, a sandwich technique was applied. A crude estimate of melt composition was made from earlier runs and a fired oxide mix of that composition sandwiched between two layers of a jadeite–MgO mixture in a Pt capsule.

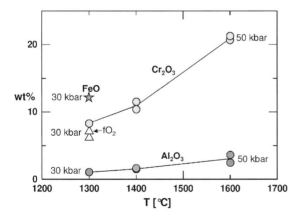

Fig. 3. FeO, Al_2O_3 and Cr_2O_3 contents in periclase vs. experimental temperature at 30 and 50 kbar. "fO_2" indicates a run carried out in a double capsule with hematite as a buffer.

2.2. Analytical methods

All runs were analysed by EPMA with a JEOL 8900 Superprobe with 15 kV acceleration voltage and 20 nA. For the analysis of solid phases the beam was focussed to 250 nm. It was widened to 30 μm for the quenched melt analysis.

2.3. Experimental results

Experimental run conditions, results and the composition of the synthesized periclases are given in Table 1. The structural formula of periclases calculated with different oxidation states of Fe and Cr are given in Table 2. In all but one run, the experimental products were always periclase, olivine, spinel (chromite, magnetite) and interstitial, quench-modified liquid. A hitherto unknown Na- and Cr-rich phase grew in the subsolidus experiment run at 50 kbar and 1300 °C in the Cr-bearing system. Significant amounts of

Table 1
Run conditions and experimental run products

P	T	Run no.	Phases	Remarks	Computed (wt.%) of periclase		
			MgO–$NaAlSi_2O_6$		Al_2O_3	MgO	Na_2O
30	1300	295/2	P, ol, sp, L		0.98	96.92	0.59
30	1300	299	P, ol, sp, L	held at 1450	0.87	96.87	0.38
50	1300	1214/2	P, ol, sp, L		1.06	96.05	0.50
50	1400	1209/1	P, ol, sp, L		1.49	94.90	0.73
50	1600	1218/1	P, ol, sp, L	MgO-richer	3.64	94.86	0.78
50	1600	1221/1	P, ol, sp, L	MgO-richer	2.56	94.53	0.67
30	1300	300	P, ol, sp, L	Sandwich	0.84	98.39	0.33
50	1400	1216	P, ol, sp, L	Sandwich	1.63	97.32	0.64
			MgO–$NaCrSi_2O_6$		Cr_2O_3	MgO	Na_2O
30	1300	295/1	P, ol, chr, L		6.18	89.30	0.78
30	1300	296	P, ol, chr, L	high fo_2	7.10	89.60	0.65
50	1300	1214/1	P, ?, L		8.32	87.34	1.80
50	1400	1209/2	P, ol, chr, L		11.49	84.78	1.58
50	1400	1215/2	P, ol, chr, L	synthesis	10.38	89.15	0.57
50	1600	1218/2	P, ol, chr, L	MgO-richer	20.66	77.17	0.77
50	1600	1221/2	P, ol, chr, L	MgO-richer	21.19	76.99	1.21
			MgO–$NaFe^{3+}Si_2O_6$		FeO	MgO	Na_2O
30	1300	292	P, ol, mt, L		11.88	87.75	0.27

Table 2
Structural formula of periclase based on one oxygen and calculated for Al^{3+}, Cr^{3+}, Cr^{2+}, Fe^{3+} and Fe^{2+}

Run no.	Al^{3+}		Mg		Na		Sum
295/2	0.008		0.984		0.008		1.000
299	0.007		0.989		0.005		0.999
1214/2	0.009		0.984		0.007		0.999
1209/1	0.121		0.977		0.010		0.999
1218/1	0.029		0.952		0.010		0.991
1221/1	0.021		0.965		0.009		0.994
300	0.007		0.988		0.004		0.999
1216	0.013		0.976		0.084		0.998

Run no.	Cr^{3+}	Mg	Na	Sum	Cr^{2+}	Mg	Na	Sum
295/1	0.035	0.943	0.011	0.988	0.035	0.959	0.011	1.001
296	0.039	0.937	0.009	0.985	0.040	0.955	0.009	1.005
1214/1	0.046	0.918	0.025	0.989	0.048	0.940	0.025	1.013
1209/2	0.064	0.893	0.022	0.979	0.066	0.922	0.022	1.011
1215/2	0.056	0.912	0.008	0.976	0.058	0.938	0.008	1.004
1218/2	0.117	0.820	0.011	0.947	0.124	0.871	0.011	1.006
1221/2	0.119	0.814	0.017	0.949	0.126	0.865	0.018	1.009

Run no.	Fe^{3+}	Mg	Na	Sum	Fe^{2+}	Mg	Na	Sum
292	0.068	0.896	0.004	0.968	0.071	0.928	0.003	1.002

Table 3
Composition (wt.%) of melts from sandwich experiments

Run no.	SiO_2	Al_2O_3	MgO	Na_2O
300	38.4 ± 7	14.1 ± 6	15.0 ± 1.8	29.7 ± 1.8
1216	42.0 ± 6	11.4 ± 3	18.1 ± 7	30.1 ± 1.0

Na_2O were present in the synthetic periclases at all run conditions (Table 1 and Fig. 4). Their compositions duplicate the range of values observed in natural ferropericlases.

The sodium content of periclase is lowest in the one Fe-bearing run, is higher at the same conditions when Al is present, and is the highest in the Cr-bearing system. In the Al-bearing system Na solubility in periclase seemingly increases with temperature. It decreases in the Cr system with temperature yet is favoured by pressure. These results cannot be applied directly to natural ferropericlases because no other Na-bearing phase is present and the melt coexisting in the experimental charges presumably varies in amount and composition. The melts from the two sandwich runs are very rich in sodium and are highly silica undersaturated (Table 3). Partition coefficients $D_{Na}^{Fe-Per/melt}$ are 0.011 at 30 kbar, 1300 °C and 0.021 at 50 kbar, 1400 °C.

Al and Cr in periclase show a positive dependency on pressure and very pronounced dependence on temperature, with Cr_2O_3 reaching more than 20 wt.% at 50 kbar and 1600 °C (Fig. 3). These elements (especially Cr) are incorporated into periclase in amounts far in excess of what is needed for charge balancing Na (Fig. 5). Al is present at 1300 and 1400 °C in just the correct amounts, but at 1600 °C it is double the amount of sodium and the ferropericlases must be nonstoichiometric. This is much more so the case for Cr which, however, may also partly be incorporated in the divalent state. Fe is also far above the amount necessary for charge balance,

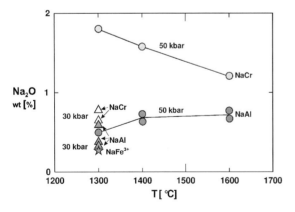

Fig. 4. Na_2O in periclase vs. experimental temperature at 30 and 50 kbar. Sodium solubility lowest in the Fe-bearing system, increases with Al and is highest in the Cr-bearing system.

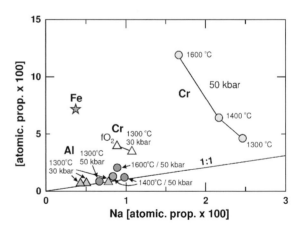

Fig. 5. Atomic proportions of Fe, Al and Cr vs. the atomic proportion of Na in periclase at the various experimental conditions. The 1:1 line is the reference line for charge balance.

but much of it will be Fe^{2+}. The effects of non-stoichiometry and varying oxidation states may be judged from the structural formula calculations given in Table 2.

The new and unknown crystalline phase produced at 50 kbar and 1300 °C in the Cr-bearing system forms large poikilitic grains with abundant periclase and chromite inclusions. It is pleochroic in blue and green shades. It is extremely low in silica and Cr-rich. An idealized formula would be $Na_2Mg_3Si_2O_8$ with significant $2Cr->MgSi$ and $NaCr->2Mg$ isomorphism to yield $(Na,Mg)_2$ $(Mg,Cr)_3(Si,Cr)_2O_8$.

3. Ferropericlase–olivine relations in diamonds

3.1.1. Ferropericlase–olivine pairs

Ferropericlases occurring together with olivine in a single diamond are reported from Guinea, Brazil and Canada. They are compared with the worldwide data set of inclusions in Fig. 6 and individually discussed in the following.

The olivines (structure confirmed by in situ X-ray diffraction) in diamond KK84 are distinct from other olivine inclusions in diamonds worldwide (Fig. 6, see also Fig. 1) in their extremely high Mg-numbers (96.5) and very low NiO (0.13 wt.%) and low Cr_2O_3 (0.03 wt.%) contents, an expression of the bulk composition of the source rock and of equilibrium with ferropericlase at upper mantle pressures. The ferropericlases have high Mg-numbers and Ni, Cr and Na and fall within the range of the worldwide database; that is, they do not occupy a separate compositional field.

A second diamond (KK109) from Guinea (Fig. 6, see also Fig. 1) has a ferropericlase/olivine contacting pair and two separate ferropericlases with a common composition which is different to that of the ferropericlase in contact with olivine (Stachel et al., 2000). Because of micron to submicron spinel exsolutions and generally higher Cr and Al contents in the olivine than found in the worldwide database, a case could be made that this diamond inclusion had grown near the lower mantle boundary, that the touching pair was originally ferropericlase + Mg-perovskite and that it reacted on uprise to form ferropericlase + ringwoodite

(a spinel structure, which, in comparison to the olivine structure, takes in trivalent cations and whose partioning behaviour with ferropericlase is such that it takes more Mg, Ni and Co compared to olivine). The final adjustment, but not to completion, to coexisting ferropericlase + olivine occurred in the upper mantle before the kimberlite eruption. The chemical composition of the ferropericlases falls within the spread of the worldwide database.

Diamond KK 44 from Guinea seems unusual since it contained the three major lower mantle phases and, in addition, a touching pair of former Mg-perovskite (now orthopyroxene) and olivine with very low nickel (Fig. 6). This pair must have originally formed as an intergrowth of small amounts of ferropericlase and abundant perovskite which kept the bulk Ni abundance of the inclusion at a very low level. Thus, after upward transport and on final adjustment to ambient conditions in the upper mantle, the olivine was left with only little nickel.

Two examples of joint occurrences of olivine and ferropericlase in a single diamond are described from the Juina district in Brazil (Kaminsky et al., 2000; Hutchison, 1997; Hutchison et al., 2000). In both cases, the ferropericlases are indistinguishable from the world database (except for the relatively Fe-rich nature of the whole brazilian suite), while the "olivines" show very high Ni, Co and Cr (Fig. 6). In both cases, the authors used this as an indication for the origin of these diamonds from the boundary to the lower mantle. A further olivine with similar high Ni, Co and Cr coexists with $CaSiO_3$, fortifying the conclusions about a very high-pressure origin of this suite.

The diamond suite from Panda, Canada (unpublished) also yields olivine–ferropericlase pairs (PA 39 and PA 54), but with much lower Ni and also Cr contents in the olivines relative to the worldwide database, but similar to olivine–ferropericlase pairs thought to be of upper mantle origin. Diamond DO 27-300 (Davies et al., pers. comm. and submitted to Lithos) has olivine and $MgSiO_3$ as separate inclusions with the olivine being far too low in Ni and Cr (Fig. 6) to have originated as ringwoodite. It may, however, be a former coupled pair of Mg-perovskite in excess and ferropericlase which reacted to olivine and some leftover $MgSiO_3$ not observed during the study.

From the above discussion, it appears that the composition of the ferropericlases does not allow

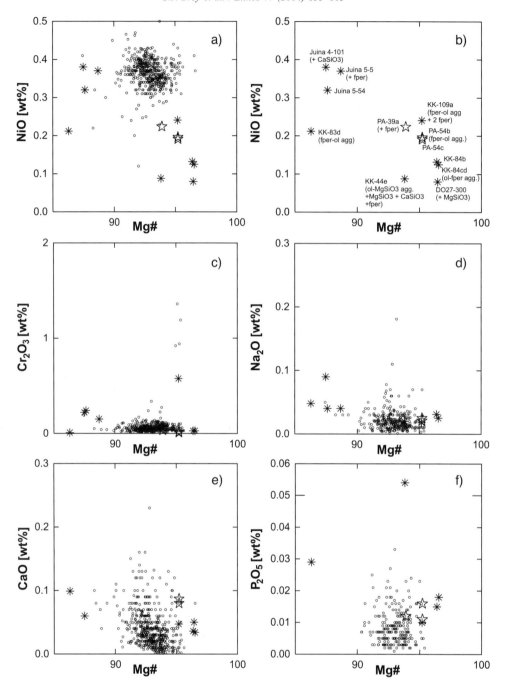

Fig. 6. Composition of olivine inclusions from the worldwide database. Olivines coexisting with ferropericlase (stars and double crosses) plot away from the field of peridotitic olivines of upper mantle origin (small circles), both to low and high Ni and Cr contents. A number of olivines from the worldwide database plot in the vicinity of these uncommon types. They indicate a similar origin as those described in this paper but detailed descriptions are lacking in the literature. (a) All available data are compiled. (b) Only olivines with coexisting ferropericlase are shown and their sample numbers are given. (c) Shows the Cr content which is analoguous to Ni low in olivines considered to have an origin in the upper mantle and high in those considered to have originated at or within the perovskite–ferropericlase–ringwoodite loop. Sodium (d), calcium (e) and phosphorous (f) do not differ significantly from the worldwide database.

distinction between an origin in the upper mantle, the transition zone or the lower mantle, but that olivines could carry this information.

3.2. Single ferropericlases, single olivines

Single ferropericlases and their chemical compositions are not indicative of a lower or an upper mantle origin. Because there is no structural change occurring in this pressure range, its uptake of suitable elements does not change drastically with pressure. Olivine, however, undergoes two phase transitions with concomitant changes in partition coefficients of divalent cations and the enhanced intake of trivalent cations. Olivines with low Ni, Co and Cr contents at high Mg-values are indicative of an upper mantle origin, and equlibrium with ferropericlase, whereas olivines with high Cr (and Al) and Ni and Co at "normal" mantle values indicate an origin in the transition zone or the boundary to the lower mantle.

4. Discussion and conclusions

We have shown that sodium can be incorporated into ferropericlase at upper mantle pressures in amounts commonly found in natural diamond inclusions. Our experiments and the diamond inclusions establish firmly that ferropericlase exists in the upper mantle in regions with low silica activity. Such regions could be a magnesite bearing dunite in which ferropericlase may form by reduction of magnesite during diamond growth by the following reactions:

$$MgCO_3 + CH_4 = MgO + 2C + 2H_2O$$

concomitant with the Fe–Mg exchange reaction

$$Mg_2SiO_4 + FeO = Fe_2SiO_4 + MgO$$

The difficulty is how to generate a magnesite-bearing dunite free of orthopyroxene in the first place. This can only be via the introduction of a carbonatite melt into dunite, since carbonation reactions in peridotite generally increase the activity of silica by converting olivine to orthopyroxene. Another scenario to be imagined is that ferropericlase crystallizes from a melt containing 8–10 wt.% Na_2O (this number is

deduced from our sandwich experiments). Carbonated peridotite may generate Na-rich carbonatite melts near the mantle solidus (Wallace and Green, 1988) in equilibrium with orthopyroxene and olivine. CO_2-degassing of the carbonatite melt at decreasing pressures may produce a silicate melt with very low Si content, which crystallizes olivine and periclase. Experimental evidence for this is presented by Brey and Ryabchikov (1994) who found a vastly expanded field of crystallisation of ferropericlase in carbonated, highly Si-undersaturated magmas with decreasing pressure. This and the above model are similar except that instead of a solid carbonate we have carbonate liquid, which can easily move and produce mantle lithologies with very low Si/Mg ratios.

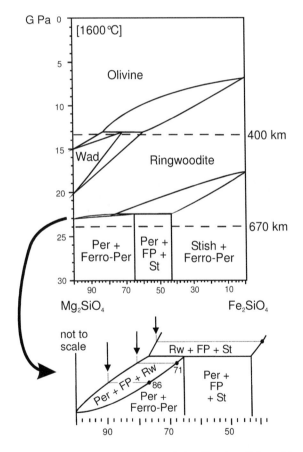

Fig. 7. Largely schematic phase relationships for olivine with increasing pressure. The numbers written on the loop for coexisting ringwoodite, ferropericlase and perovskite correspond to the Mg-value of ferropericlase.

In summary, ferropericlase as an inclusion in diamond on its own is not indicative of a lower mantle origin, let alone of a mantle plume. The composition of coexisting phases has to be taken into account to decide on an upper mantle, transition zone or lower mantle origin.

From the presently known diamond inclusion inventory we have clear indications that diamonds come from the upper mantle, the transition zone, from the boundary to the lower mantle and from its uppermost part. Ferropericlase with very variable Fe/Mg-ratios in diamond inclusions from Juina, Brazil has been inferred to originate in the D" layer (Harte et al., 1999; Kaminsky et al., 2000). We argue here that they also stem from the border zone of the lower mantle from the loop with coexisting Mg-perovskite, ferropericlase and ringwoodite (Fig. 7). Their protoliths were hybrid rocks of peridotite plus subducted eclogite or sediments to increase bulk Fe/Mg. The partitioning behaviour of Fe and Mg (e.g., Frost and Langenhorst, 2002) between coexisting Mg-perovskite and ferropericlase, between ringwoodite and ferropericlase (Frost et al., 2001) and the very narrow and flat loop readily explain the observed variation in Fe/Mg ratios in both phases. From their work, a ringwoodite with an Mg-value of 90 would coexist with ferropericlase with Mg# $\approx$ 86 and Mg-perovskite with Mg# $\approx$ 95, ringwoodite with Mg# $\approx$ 80 with ferropericlase$_{71}$ and perovskite$_{85}$ and ringwoodite with Mg# $\approx$ 75 with a ferropericlase with Mg# $\approx$ 40 (schematically shown in Fig. 7) which coincides with the range observed in Sao Luis.

Acknowledgements

We would like to thank D.H. Green for helpful discussions and Kai Klama, Marina Lazarov and Iris Sonntag for help with the final preparation of the manuscript. We thank the reviewers Kathy McCammon, Tibor Gasparik and Reidar Tronnes for helpful and interesting suggestions and Herman Grütter for careful editorial handling.

References

Akaogi, M., Ito, E., Navrotsky, A., 1989. Olivine-modified spinel–spinel transition in the system Mg$_2$SiO$_4$–Fe$_2$SiO$_4$: calorimetric measurements, thermochemical calculation, and geophysical application. J. Geophys. Res. 94, 15671–15685.

Brey, G.P., Ryabchikov, I.D., 1994. Carbon dioxide in strongly undersaturated melts and origin of kimberlitic magmas. Neues Jahrb. Mineral., Monatsh. H. 10, 449–463.

Frost, D.J., Langenhorst, F., van Aken, P.A., 2001. Fe-Mg partitioning between ringwoodite and magnesiowüstite and the effect of pressure, temperature and oxygen fugacity. Phys. Chem. Miner. 28, 455–470.

Frost, D.J., Langenhorst, F., 2002. The effect of Al$_2$O$_3$ on Fe–Mg partitioning between magnesiowüstite and magnesium silicate perovskite. Earth Planet. Sci. Lett. 199, 227–241.

Gasparik, T., 2000. Evidence for the transition zone origin of some [Mg,Fe]O inclusions in diamonds. Earth Planet. Sci. Lett. 183 (1–2), 1–5.

Gasparik, T., 2002. Experimental investigation of the origin of majoritic garnet inclusions in diamonds. Phys. Chem. Miner. 29 (3), 170–180.

Gasparik, T., Litvin, Y.A., 1997. Stability of Na$_2$Mg$_2$Si$_2$O$_7$ and melting relations on the forsterite–jadeite join at pressures up to 22 GPa. Eur. J. Mineral. 9 (2), 311–326.

Gurney, J.J., 1989. Diamonds. In: Ross, J., et al., (Eds.), Kimberlites and Related Rocks. Spec. Publ.-Geol. Soc. Aust. 14, vol. 2. Blackwell, Carlton, pp. 935–965.

Harte, B., Harris, J.W., 1994. Lower mantle mineral associations preserved in diamonds. Min. Mag. 58, 384–385.

Harte, B., Harris, J.W., Hutchison, M.T., Watt, G.R., Wilding, M.C., 1999. Lower mantle mineral associations in diamonds from Sao Luiz, Brazil. In: Fei, Y., Bertka, C., Mysen, B.O. (Eds.), Mantle Petrology: Field Observations and High Pressure Experimentation: A Tribute to Francis R. (Joe) Boyd. Publication-The Geochemical Society, Houston, vol. 6, pp. 125–153.

Hutchison, M.T., 1997. Constitution of the deep transition zone and lower mantle shown by diamonds and their inclusions. PhD thesis. University of Edinburgh, UK, vol. 1, 340 pp. vol 2 (Tables and Appendices), 306 pp.

Hutchison, M.T., Hursthouse, M.B., Light, M.E., 2000. Mineral inclusions in diamonds: associations and chemical distinctions around the 670-km discontinuity. Contrib. Mineral. Petrol. 142, 119–126.

Kaminsky, R.V., Zakharchenko, O.D., Davies, R., Griffin, W.L., Khachatryan-Blinova, G.K., Shiryaev, A.A., 2000. Superdeep diamonds from the Juina area, Mato Grosso State, Brazil. Contrib. Mineral. Petrol. 140, 734–743.

Kesson, S.E., Gerald, J.D.F., 1992. Partitioning of MgO, FeO, NiO, MnO and Cr$_2$O$_3$ between magnesian silicate perovskite and magnesiowustite—implications for the origin of inclusions in diamond and the composition of the lower mantle. Earth Planet. Sci. Lett. 111, 229–240.

Liu, L.-G., 1974. Silica perovskite from phase transformations of pyrope–garnet at high pressure and temperature. Geophys. Res. Lett. 1, 277–280.

Liu, L.-G., 1975. Post-oxide phases of forsterite and enstatite. Geophys. Res. Lett. 2, 417–419.

Ringwood, A.E., Major, A., 1966. Synthesis of Mg$_2$SiO$_4$–Fe$_2$SiO$_4$ spinel solid solutions. Earth Planet. Sci. Lett. 1, 241–245.

Scott Smith, B.H., Danchin, R.V., Harris, J.W., Stracke, K.J., 1984.

Kimberlites near Orroroo, South Australia. In: Kornprobst, J. (Ed.), Kimberlites I: Kimberlites and Related Rocks. Elsevier, Amsterdam, pp. 121–142.

Stachel, T., Harris, J.W., Brey, G.P., 1998. Rare and unusual mineral inclusions in diamonds from Mwadui, Tanzania. Contrib. Mineral. Petrol. 132, 34–47.

Stachel, T., Harris, J.W., Brey, G.P., Joswig, W., 2000. Kankan diamonds (Guinea): II. Lower mantle inclusion paragenesis. Contrib. Mineral. Petrol. 140, 34–47.

Wallace, M.E., Green, D.H., 1988. Experimental determination of primary carbonatite magma composition. Nature 335, 343–346.

Wilding, M.C., 1990. A study of diamonds with syngenetic inclusions. Unpubl PhD thesis. University of Edinburgh, UK. 281 pp.

Available online at www.sciencedirect.com

Lithos 77 (2004) 665–682

LITHOS

www.elsevier.com/locate/lithos

ELSEVIER

Temporal, geomagnetic and related attributes of kimberlite magmatism at Ekati, Northwest Territories, Canada

Grant Lockhart[a,*], Herman Grütter[b], Jon Carlson[a]

[a] BHP-Billiton Diamonds Inc., #8-2604 Enterprise Way, Kelowna, B.C., Canada V1X 7Y5
[b] Mineral Services Canada Inc., Vancouver, B.C., Canada

Received 27 June 2003; accepted 17 February 2004
Available online 1 June 2004

Abstract

This paper outlines the development of a multi-disciplinary strategy to focus exploration for economic kimberlites on the Ekati property. High-resolution aeromagnetic data provide an over-arching spatial and magnetostratigraphic framework for exploration and kimberlite discovery at Ekati, and hence also for this investigation. The temporal, geomagnetic, spatial and related attributes of kimberlites with variable diamond content have been constrained by judiciously augmenting the information gathered during routine exploration with detailed, laboratory-based or field-based investigations. The natural remanent magnetisation of 36 Ekati kimberlites has been correlated with their age as determined by isotopic dating techniques, and placed in the context of a well-constrained geomagnetic polarity timescale. Kimberlite magmatism occurred over the period 75 to 45 Ma, in at least five temporally discrete intrusive episodes. Based on current evidence, the older kimberlites (75 to 59 Ma) have low diamond contents and are distributed throughout the property. Younger kimberlites (56 to 45 Ma) have moderate to high diamond contents and occur in three distinct intrusive corridors with NNE to NE orientations. Economic kimberlite pipes erupted at 55.4 ± 0.4 Ma along the A154-Lynx intrusive corridor, which is 7 km wide and oriented at $015°$, and at 53.2 ± 0.3 Ma along the Panda intrusive corridor, which is 1 km wide and oriented at $038°$. The intrusion ages straddle a paleopole reversal at Chron C24n, consistent with the observation that the older economic kimberlites present as aeromagnetic "low" anomalies while the younger economic pipes are characterised as aeromagnetic "highs". The aeromagnetic responses for these kimberlites are generally muted because they contain volcaniclastic rock types with low magnetic susceptibility. Kimberlites throughout the Ekati property carry a primary natural magnetic remanence (NRM) vector in Ti-bearing groundmass magnetite, and it dominates over vectors related to induced magnetisation. Magnetostratigraphic correlation of Ekati kimberlites may therefore present a powerful adjunct to existing exploration techniques, mainly because the diamond content of Ekati kimberlites apparently is related more to the age of eruption than to any other parameter investigated in this work.
© 2004 Elsevier B.V. All rights reserved.

Keywords: Ekati; Kimberlite magmatism; Geomagnetism; Magnetic anomaly; Exploration

---

* Corresponding author. Fax: +1-250-860-7242.
E-mail address: grant.d.lockhart@bhpbilliton.com
(G. Lockhart).

1. Introduction

The Lac de Gras kimberlite province occurs in a northwest trending zone, with dimensions of 100 by

0024-4937/$ - see front matter © 2004 Elsevier B.V. All rights reserved.
doi:10.1016/j.lithos.2004.03.029

200 km, in the central portion of the Archean Slave craton, Canada (Fig. 1). More than 270 confirmed kimberlites have been discovered in this important diamondiferous kimberlite province as a result of intense exploration following the initial discovery of the Point Lake kimberlite in September 1991. The discovery of potentially economic kimberlites was foreshadowed by the presence of high counts of mantle-derived heavy mineral indicators in till samples near Exeter Lake. Modern geophysical techniques rapidly gained wide-spread use as the preferred

method of locating prospective kimberlites under shallow lakes and a veneer of glacial sediment (Fipke et al., 1995). By 1993, the acquisition of high-resolution airborne magnetic and resistivity surveys over areas of 100 to 1000 km^2 became standard practice in diamond exploration in the Slave craton. Four such surveys, covering 4000 km^2, were flown over the Ekati property between 1992 and 2000 at a cost of US$4.8 million. Subsequent drilling of the geophysical anomalies identified 152 kimberlites within the Ekati property, most of which occur as

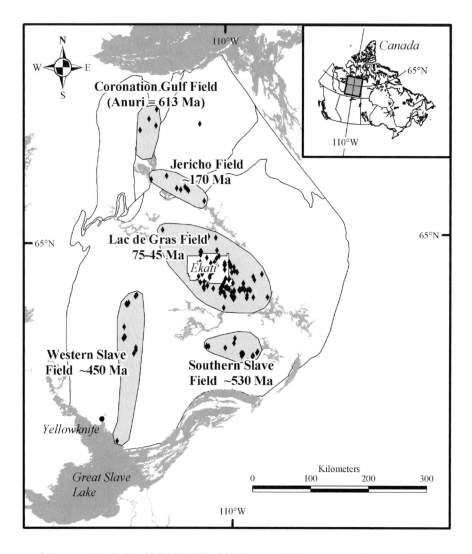

Fig. 1. Location map of the greater Lac de Gras kimberlite field within the central Slave craton, Northwest Territories and Nunavut, Canada.

kimberlite pipes with surface areas of less than 5 ha. Nine of these kimberlites are currently classified as economic resources (Dyck et al., this volume). The work reported herein outlines the data sets collected and general approach used by BHP Billiton to correlate the temporal, geomagnetic and spatial attributes of highly diamondiferous kimberlites with other kimberlites at Ekati, in order to refine exploration efforts at the scale of the property.

Establishing the relative ages of Ekati kimberlites was recognised as a key aspect of the investigation, and a comprehensive radiometric dating program was in progress by 1998 (e.g. Armstrong and Moore, 1998). Fresh macrocrystic phlogopite and/or groundmass perovskite was systematically collected from kimberlite core and rock chips generated by exploration drilling programmes. Radiometric dating by Rb–Sr or U–Pb methods was performed at commercial facilities (e.g. GeoSpec Consultants, 1996, 1999). This work, described in detail by Creaser et al. (this volume), resolved five temporally discrete episodes of magmatism amongst the kimberlites present at Ekati. Magmatism occurred from about 75 to 45 Ma, during which 36 known polarity reversals of the Earth's magnetic field took place (Cande and Kent, 1995). The natural remanent magnetism of kimberlites would record such rapidly variant paleopolarity vectors, implying that magnetostratigraphic principles could be used to additionally characterise and isolate kimberlites with possibly higher economic potential. We therefore compare our radiometrically determined ages with a well-defined global geomagnetic timescale and investigate the correlation between the paleopole predicted for a given kimberlite and the magnetic response observed in the field. Additional characterisation of economically important kimberlites is achieved by cross-correlating other attributes like structural setting and kimberlite type (volcaniclastic/magmatic). Our results to date indicate that highly diamondiferous kimberlites in the Ekati property occur within spatially constrained intrusive corridors that were magmatically active only during specific intrusive episodes. The location and orientation of the corridors of interest appear to be largely unrelated to basement geology and show evidence of rotation with time that speculatively could be linked to far-field tectonic stresses.

2. Dataset and methods

Table 1 presents a compilation of the temporal, geomagnetic and other attributes of interest for the 36 dated Ekati kimberlites that form the basis of this study. Corresponding data for the economic A154S, A154N, A418 and A21 kimberlites in the adjacent Diavik property are also included (from Graham et al., 1999). The spatial distribution of kimberlite localities listed in Table 1 is shown in Fig. 2. The nature and origin of the primary information given in Table 1 is summarised here by heading, as follows:

- Isotopic Age, Error and Age Type: This represents the currently available, radiometrically determined age of sample(s) from individual kimberlites. Errors are listed at two standard deviations (2σ). Ages determined by the Rb–Sr technique on macrocrystal phlogopite include predominately single-point model ages, a two-point model age (Arnie), multi-point Rb–Sr regressions (Leslie and Lynx) and several multi-point Rb–Sr isochron ages, two of which are newly published (Panda and Cobra South). The Rb–Sr ages, their error estimates and the internal consistency of these isotopic data for phlogopite macrocrysts are discussed in detail by Creaser et al. (this volume).

- Age Array: This denotes the formal grouping of kimberlites of similar age based only on their radiometric attributes. The age arrays represent temporally discrete episodes of kimberlite magmatism at Ekati (see Creaser et al., this volume) and have been labelled MAA, PAA, 154AA and CAA after the Rb–Sr isochron-dates obtained for the Mark (~ 48 Ma), Panda (~ 53 Ma), A154 (~ 55 Ma) and Cobra South (~ 59 Ma) kimberlite pipes. The MSK age array informally denotes "mid-sixties kimberlites", for which U–Pb model ages on perovskite range from 75 to 64 Ma, with comparatively large errors. The MSK age group probably defines one or more further episode(s) of kimberlite magmatism at Ekati. U–Pb dating results are given in Table 2 and discussed further below.

- Mag Anomaly Polarity (Predicted): This is the predicted polarity of the primary natural remanent

Table 1
Summary of ages, geomagnetic attributes, principal lithology and microdiamond abundance in isotopically dated Lac de Gras kimberlites

Kimberlite	Isotopic age (Ma)	Error (Ma, 2σ)	Age type and reference	Age array[a]	Magnetic anomaly polarity (predicted)[b]	Magnetic anomaly polarity (observed)[c]	Magnetic anomaly intensity (nT)[d]	Magnetic susceptibility (× 10^-6 SI units)	Main rock type[e]	Microdiamond abundance[f]
Aaron	45.2	1.3	1-pt Rb–Sr model (Ref. 1)	UC	Reverse	Reverse	−74	1020	MK	high
Brent	47.1	0.9	1-pt Rb–Sr model (Ref. 1)	MAA	Normal	Reverse	−503	6800	MK	moderate
Arnie	47.5	1.6	2-pt Rb–Sr model (Ref. 2)	MAA	Normal	Reverse	−1530	5100	MK	very high
Mark	47.5	0.5	4-pt Rb–Sr isochron (Ref. 3)	MAA	Normal	Reverse	−1200	4600	MK	very high
Giraffe	47.8	1.4	1-pt Rb–Sr model (Ref. 1)	MAA	Normal	Normal	34	510	VK	low
Gazelle	47.9	2.1	1-pt Rb–Sr model (Ref. 1)	MAA	Normal	Indistinct	0	2100	VK	moderate
Hawk	48.0	1.3	1-pt Rb–Sr model (Ref. 1)	MAA	Reverse	Reverse	−170	13,600	MK	low
Grizzly	50.8	4.8	1-pt Rb–Sr model (Ref. 1)	UC	Normal	Reverse	−1920	9500	MK	high
Falcon	51.5	1.7	1-pt Rb–Sr model (Ref. 1)	UC	Normal	Reverse	−10	1100	VK	very high
Point Lake	51.5	0.8	1-pt Rb–Sr model (Ref. 1)	UC	Normal	Reverse	−7	480	VK	very high
Leslie	52.1	1.0	3-pt Rb–Sr regression (Ref. 2)	PAA	Normal	Normal (Ref. 5)	7585	11,000	MK	very high
Zach	52.8	0.8	1-pt Rb–Sr model (Ref. 1)	PAA	Normal	Indistinct	0	1170	VK	moderate
Panda	53.3	0.6	7-pt Rb–Sr isochron (Ref. 1)	PAA	Normal	Normal (Ref. 5)	115	633	VK	very high
Beartooth	53.1	1.0	1-pt Rb–Sr model (Ref. 1)	PAA	Normal	Normal	24	1240	VK	high
Koala	53.3	0.9	1-pt Rb–Sr model (Ref. 1)	PAA	Normal	Normal (Ref. 5)	659	941	VK	very high
Koala North	53.3	0.9	1-pt Rb–Sr model (Ref. 1)	PAA	Normal	Normal (Ref. 5)	7	1560	VK	moderate
Bison	54.7	0.9	1-pt Rb–Sr model (Ref. 1)	154AA	Reverse	Indistinct	0	830	VK	low
Cardinal	54.8	1.9	1-pt Rb–Sr model (Ref. 1)	154AA	Reverse	Reverse	−21	800	VK	very high
Lynx	55.8	4.8	5-pt Rb–Sr regression (Ref. 1)	154AA	Reverse	Reverse (Ref. 5)	0	794	VK	high
Piranha	55.8	1.6	1-pt Rb–Sr model (Ref. 1)	154AA	Reverse	Reverse (Ref. 5)	0	468	VK	very high
A154 South	55.5	0.7	Rb–Sr isochron (Ref. 4)	154AA	Reverse	Indistinct (Ref. 4)	–	–	VK	very high

Name	Age	±	Dating method	Array[a]	Predicted polarity[b]	Observed polarity[c]	Amplitude[d]	Recovery	Class[e]	Category[f]
A418	55.2	0.3	Rb–Sr isochron (Ref. 4)	154AA	Reverse	Indistinct (Ref. 4)	—	—	VK	very high
A21	55.7	2.1	Rb–Sr isochron (Ref. 4)	154AA	Reverse	Reverse (Ref. 4)	—	—	VK	very high
A154 North	56.0	0.7	Rb–Sr isochron (Ref. 4)	154AA	Reverse	Reverse (Ref. 4)	—	—	VK	very high
Panther	58.3	5.1	1-pt Rb–Sr model (Ref. 1)	CAA	Reverse	Reverse	−10	1020	VK	very low
Crab	58.4	0.9	1-pt Rb–Sr model (Ref. 1)	CAA	Reverse	Reverse	−60	1230	VK	low
Shark	58.4	1.7	1-pt Rb–Sr model (Ref. 1)	CAA	Reverse	Reverse	−24	1380	VK	high
Rooster	58.7	2.3	1-pt Rb–Sr model (Ref. 1)	CAA	Reverse	Reverse	−53	1780	VK	low
Antelope	59.5	2.4	1-pt Rb–Sr model (Ref. 1)	CAA	Reverse	Reverse	−50	14,000	MK	low
Cobra South	59.7	0.4	8-pt Rb–Sr isochron (Ref. 1)	CAA	Reverse	Reverse	−36	3800	VK	low
Rattler	59.7	1.5	1-pt Rb–Sr model (Ref. 1)	CAA	Reverse	Reverse	−4	1380	VK	low
Glory	61.3	3.4	1-pt Rb–Sr model (Ref. 1)	CAA	Reverse	Reverse	−53	1050	VK	high
Rufus	61.1	2.1	1-pt Rb–Sr model (Ref. 1)	CAA	Normal	Normal	57	1520	VK	very low
King	63.9	3.4	1-pt U–Pb model (Table 2)	MSK	Reverse	Reverse	−87	6200	VK	very low
Springbok	63.9	2.6	1-pt U–Pb model (Table 2)	MSK	Reverse	Reverse	−120	34,000	MK	low
Husky	64.1	4.0	1-pt U–Pb model (Table 2)	MSK	Normal	Reverse	−53	4500	MK	very low
Roger	67.6	9.1	2-pt U–Pb model mean (Table 2)	MSK	Normal	Reverse	−16625	45,000	MK	barren
Caribou West	68.2	3.8	1-pt U–Pb model (Table 2)	MSK	Normal	Indistinct	0	2100	VK	very low
Jaeger	69.1	6.4	1-pt U–Pb model (Table 2)	MSK	Reverse	Reverse (Ref. 6)	−1437	364	VK	low
Kudu	74.7	6.8	1-pt U–Pb model (Table 2)	MSK	Normal	Reverse	−80	15,382	MK	very low

References: Ref. 1 = Creaser et al., this volume, Ref. 2 = Armstrong and Moore, 1998, Ref. 3 = Davis and Kjarsgaard, 1997, Ref. 4 = Graham et al., 1999, Ref. 5 = Enkin, 2003, Ref. 6 = Wynne, 1995.

[a] UC = Unclassified, MAA = Mark Age Array, PAA = Panda Age Array, A154AA = A154 Age Array, CAA = Cobra Age Array, MSK = Mid-Sixties Kimberlites.

[b] Magnetic polarity predicted based on isotopic age. Normal = Magnetic High, Reverse = Magnetic Low. See text.

[c] Total-field magnetic polarity as observed in airborne or ground magnetic surveys. Polarities measured in oriented core are underlined. Normal = Magnetic High, Reverse = Magnetic Low. See text.

[d] Amplitude (peak less background) of magnetic anomaly associated with kimberlite. Negative values denote magnetic lows. See text.

[e] VK = Volcaniclastic kimberlite; MK = Magmatic kimberlite.

[f] Categories reflect microdiamond recoveries per 100 kg. Barren = 0 stones recovered, very low = 1–5, low = 6–24, moderate = 25–49, high = 50–99, very high = 100 plus.

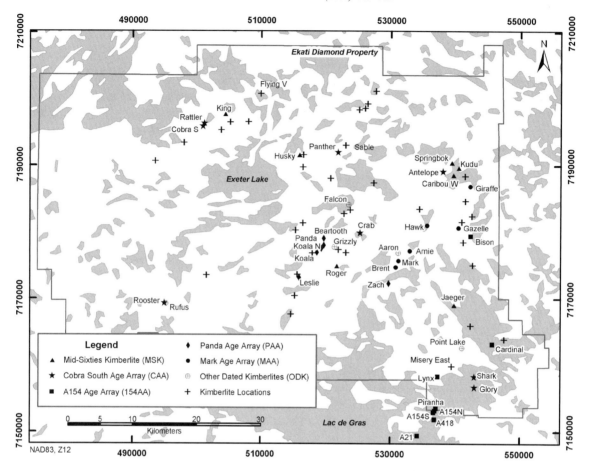

Fig. 2. Location of dated kimberlites considered in this work and extent of the Ekati diamond property. Different symbols denote kimberlites broken down by age arrays, as discussed in the text. The datum and projection is NAD83, UTM zone 12N.

magnetism (NRM) for a kimberlite, based on the intersection of its radiometric age with the global geomagnetic paleopole reversal timescale of Cande and Kent (1995).

- Mag Anomaly Polarity (Observed): This represents the characteristic total-field magnetic response of a kimberlite, based on its appearance as an anomalous magnetic "low" or "high" in high-resolution aeromagnetic or ground-magnetic data. Magnetic anomaly "lows" should correspond with periods of paleopole reversal and "highs" with normal paleopole stance if the total-field magnetic response of a kimberlite is dominated by NRM. Kimberlites with an observed indistinct aeromagnetic anomaly are indicated as such in Table 1. Certain NRM polarities have been measured in core and are

underlined in Table 1. The corresponding data are listed in Table 3 and discussed further below.

- Mag Anomaly Intensity: This is the maximum intensity (in nT) of the aeromagnetic anomaly (peak less background) associated with a kimberlite. This value is measured by a sensor at a nominal altitude of 20 m above ground level. The value is listed as positive for relative magnetic "highs" and negative for relative magnetic "lows".

- Mag Susc: Apparent magnetic susceptibility measurements (10^{-6} SI units) were made on kimberlite drill core, at 2.0 m intervals, using a KT-6 Scintex hand-held instrument and applying a calibration factor for core size. The value listed is a formal average for core recovered during exploration drilling.

Table 2
U–Pb data for groundmass perovskite and ^{238}U/^{206}Pb model ages

Sample	Kimberlite	U (ppm)	Pb (ppm)	Model age (Ma)	Error (Ma, 2σ)
99-K	King	66	22	63.9	3.4
12-02	Springbok	106	29	63.9	2.6
10-20	Husky	91	31	64.1	4.0
09-17A	Roger	44	48	64.1	9.0
09-17B	Roger	44	49	71.0	9.2
11-01	Caribou West	70	29	68.2	3.8
04-10	Jaeger	89	71	69.1	6.4
13-06	Kudu	79	28	74.7	6.8

- Microdiamond Abundance: This represents the microdiamond content of kimberlite samples that were subjected to caustic fusion. Sample weights ranged from 40 to 325 kg, with an average weight of 100 kg. Microdiamond abundance is classified into six categories: barren (0 stone recoveries), very low (1–5 stones per 100 kg), low (6–24 stones per 100 kg), moderate (25–49 stones per 100 kg), high (50–99 stones per 100 kg) and very high (100+ stones per 100 kg).

U–Pb data for groundmass perovskite and corresponding ^{206}Pb/^{238}U model ages for the six Ekati kimberlites in the MSK age group are given in Table 2. The U–Pb age constraints listed represent single-point ^{206}Pb/^{238}U model-ages, corrected for common

Pb content using the Pb isotope evolution curve of Stacey and Kramers (1975). The 2σ errors range from ± 3 Ma to ± 9 Ma and are large compared to U–Pb age constraints typically obtained using kimberlitic perovskite. The poor precision is attributed to the low relative abundance and fine grain size of perovskite in the samples submitted for dating, and a comparatively low content of radiogenic lead as a function of their young age (Heaman, pers. comm., 2003). All attempts by BHP Billiton to recover macrocrystal phlogophite in these kimberlites for Rb–Sr dating were unsuccessful.

In order to model the total-field magnetic response of typical Ekati kimberlites, the natural remanent magnetisation (NRM) contained in a selection of kimberlites was measured by thermal and alternating field demagnetisation techniques under laboratory conditions (Wynne, 1995; Enkin, 2003). Table 3 contains summary results for seven dated and two undated kimberlites for which oriented, vertical, or near-vertical core samples were available. Table heading details are as follows:

- Sample and Kimberlite: These denote the drill hole and corresponding kimberlite from which demagnetisation samples were obtained.
- Mag Anomaly Intensity: This is the amplitude (in nT) of the aeromagnetic anomaly (peak less

Table 3
Demagnetisation data for selected Ekati kimberlites

Sample	Kimberlite	Magnetic anomaly intensity (nT)[a]	Magnetic susceptibility (× 10^{-6} SI units)	NRM (mA/m)	Kn	NRM polarity[b]	n	Ds (°)	Is (°)	ks	a95s (°)
LDC-06	Leslie	7585	11,000	8030	18.43	Normal (Ref. 1)	5	–	–	–	–
PGT-22	Panda	115	478	16	0.84	Normal (Ref. 1)	10	337.2	66.1	–	5
PDC-09	Panda	115	787	216	3.87	Normal (Ref. 1)	8	326.5	76.0	136.0	4.8
KDC-41	Koala	659	941	154	4.13	Normal (Ref. 1)	3	–	74.8	18.1	i+11.3–9.1
KNGT-10	Koala North	7	1560	60	0.94	Normal (Ref. 1)	10	157.8	80.8	15.6	12.6
SGT-15	Sable[c]	− 22	412	25	1.53	Reverse (Ref. 1)	6	121.2	− 59.1	69.7	8.1
LXDC-02	Lynx	0	823	88	2.67	Reverse (Ref. 1)	5	–	− 75.2	34.3	i+6.2–10.7
A841	Piranha	0	468	19	1.03	Reverse (Ref. 1)	9	–	− 68.3	86.2	i+4.0–5.1
MGT-27	Misery East[c]	− 364	18,400	5840	7.97	Reverse (Ref. 1)	4	169.3	− 77.8	1766.8	2.2
93-10	Jaeger	− 1437	364	104	6.52	Reverse (Ref. 2)	27	–	–	–	–

[a] Relative amplitude (peak less background) of magnetic anomaly associated with kimberlite. Negative values denote magnetic lows.

[b] NRM polarities measured in core. Normal=Magnetic high, Reverse=Magnetic low. References: Ref. 1=Enkin, 2003 and Ref. 2=Wynne, 1995.

[c] Not successfully dated.

background) as measured by a sensor at a nominal altitude of 20 m above ground level.

- Mag Susc, NRM, and Kn: These are the mean magnetic susceptibility (in 10^{-6} SI units), NRM intensity (in mA/m), and the Koeninsberger ratio (NRM to the induced magnetisation) for each kimberlite.

- NRM Polarity, Ds, Is, k, a95s: The polarity is normal (+ve) when the magnetisation has the same polarity as the present geomagnetic field and reversed (−ve) when opposite. Ds (stratigraphic declination) and Is (stratigraphic inclination) are component directions of the NRM. The precision parameter is denoted as k and a95s is the half-radius of the 95% confidence cone.

3. Analysis of relationships

3.1. Orientation and spatial distribution of kimberlite intrusions

The intrusion of kimberlite dykes and pipes is often related at a local scale (i.e. 10s to 100s of metres) with structural discontinuities in country rocks, and on a property scale (i.e. kilometres) with linear intrusive corridors (e.g. Dawson, 1984; Mitchell, 1986, pp. 105–135). This appears to be confirmed at Ekati. At the local scale, elongation of dykes or pipe contacts can be seen in NW and E directions (e.g. Flying V, Fig. 3a), in a NW direction (e.g. Jaeger, Fig. 3b) and in NE and SE directions (e.g. Panda, see McElroy et al., 2003). The elongation is related to a variety of struc-

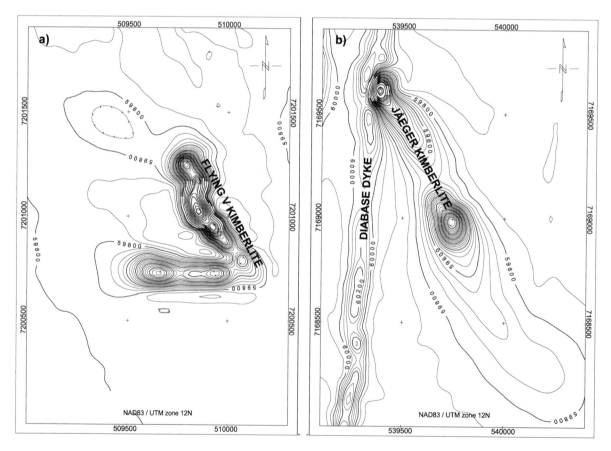

Fig. 3. (a, b) Total field aeromagnetic data over Flying V (a) and Jaeger kimberlites (b). Both examples demonstrate how local geology can influence the shape of a kimberlite. Flying V "limbs" follow strongly foliated metasediments in NW and E directions. Jaeger is elongated in a NW direction, following a granite (east) and metasediments (west) contact and terminates against a dyke contact. The contour interval is 50 nT in both panels. The nominal flight path is E–W, with a 75 m line spacing and a 20 m sensor height.

tural controls in country rocks, including faults, prominent fractures, diabase dyke contacts, strongly foliated regions in country rock metasediments, the contacts of country rock metasediments with granitoids, and to joint patterns in unfoliated, massive granitoid country rocks. These localised structural controls on kimberlite intrusion may in turn be related to linear intrusive corridors that are useful to target exploration at the scale of a property. Because kimberlite dykes are comparatively rare at Ekati, the identification of economically significant intrusive corridors involves grouping together spatially separated kimberlite pipes using radiometric and magnetostratigraphic data.

Five temporally discrete episodes of kimberlite magmatism are recognised within the 36 radiometrically dated kimberlites that occur on the Ekati property (Table 1; Creaser et al., this volume). The spatial distribution of dated kimberlites in each magmatic episode is displayed in Fig. 4. U–Pb model ages (with large errors) indicate that MSK group magmatism commenced at ~ 75 Ma and continued to ~ 64 Ma, taking the form of dominantly hypabyssal kimberlite pipes and dykes (Table 1). The available data indicate that MSK kimberlites are distributed across the Ekati property and do not occur in a narrow intrusive corridor (Fig. 4a). Dominantly volcaniclastic kimberlite pipes, belonging to the well-constrained ~ 59 Ma CAA, also occur throughout the Ekati property and also do not provide a clear spatial constraint on magmatic activity (Fig. 4b).

Kimberlite magmatism of the well-dated ~ 55 Ma 154AA is spatially constrained to the SE portion of the Ekati property and the adjacent Diavik property, and occurs within a corridor some 7 km wide oriented at 015° (Fig. 4c). We label this feature as the "154 intrusive corridor" (154IC) and note that it contains several economically important pipes. The economically important Koala and Panda pipes occur in a distinct and spatially separate intrusive corridor near the centre of the Ekati property (Fig. 4d). The Panda intrusive corridor (PIC) is defined exclusively by ~ 53 Ma PAA magmatism and comprises five kimberlites located within a 1 km wide corridor that trends at 038°. The moderately diamondiferous Zach kimberlite, which occurs to the east of the currently defined PIC, is of identical age (Table 1), and could be related to the PIC by a parallel, en-echelon or conjugate intrusive corridor (Fig. 4d). Magmatism of the

~ 48 Ma MAA appears to occur exclusively in the east-central portion of the Ekati property (Fig. 4e). The Mark intrusive corridor (MIC) is oriented at 043° and is approximately 4 km wide.

The available data indicate that spatial relationships between separate kimberlite intrusions in the Ekati property depend strongly on the age of kimberlite magmatism. Kimberlite intrusions belonging to both the MSK and the CAA age groups are widely distributed throughout the property and show little or no alignment that could be used to target areas for further exploration (Fig. 4a and b). In contrast, kimberlites of the 154AA, PAA and MAA age groups occur in distinct linear intrusive corridors. The location and orientation of the defined intrusive corridors show discernable changes over a geologically short time period (55 to 48 Ma, see Fig. 4f), implying that the intrusive corridors are unlikely to have a coherent or predictable relationship to static fracture or lineament patterns in the surficial country rocks at Ekati. This inference is consistent with the results of several field and geophysical investigations, in which kimberlite pipes forming linear arrays could not readily be connected with one another along structural features present in the basement rocks at Ekati (Wright, 1999 and authors' unpublished data). These observations suggest that fundamental structural controls on intrusive corridors may reside in the lower crust or upper mantle (see also Anderson, 1979). Trends of the 154IC, PIC and MIC show a clear clockwise rotation with time (Fig. 4f), signifying that time-variant plate-scale tectonic stresses likely contribute to the formation, orientation and location of intrusive corridors. The trends defined by the 154IC, PIC and MIC are useful guides to target further exploration at Ekati.

3.2. Observed versus predicted magnetic polarity

Table 1 shows that 18 of the 21 kimberlites with reversed-predicted magnetic polarity appear as aeromagnetic "lows", and 7 of the 19 kimberlites with normal-predicted magnetic polarity appear as aeromagnetic "highs". A higher success rate for the latter could be achieved if higher-precision U–Pb dates were available for perovskite-bearing kimberlites of the MSK age group (Tables 1 and 2). The predicted polarities are remarkably accurate for those kimberlites dated with high precision using the Rb–

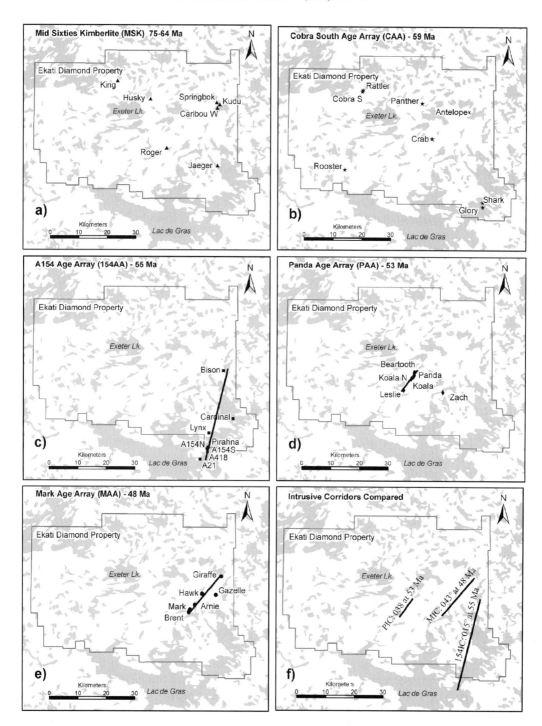

Fig. 4. (a–f) Location map of the Ekati and Diavik kimberlites distinguished by age array in order to demonstrate their spatial relationships. Older kimberlites (>59 Ma) appear to be distributed throughout the Ekati property (a,b), whereas younger kimberlites (<55 Ma) form intrusive corridors along 015°, 038° and 043° trends (c,d,e,). The 154AA group (~ 55 Ma) and the PAA group (~ 53 Ma) include economically important diamond-bearing pipes.

Sr technique on macrocrystal phlogopite. In general, this indicates that the remanent component of kimberlite magnetisation is relatively large (in comparison to the geomagnetically induced component), that it has retained a stable, primary vector since emplacement, and that total-field aeromagnetic data can be used to confidently predict the NRM polarity of Ekati kimberlites.

Samples from selected Ekati kimberlites were submitted for demagnetisation studies in order to verify certain NRM polarities predicted from total-field aeromagnetic data, to determine the NRM polarity of kimberlites with a neutral aeromagnetic response, and to possibly identify the magnetic carrier.

Although demagnetisation was taken only as far as the confident determination of paleomagnetic polarity, the results point to a single NRM carrier. Thermal demagnetisation of core samples from the Leslie and Misery East show near-complete unblocking of NRM at about 500 °C (Fig. 5). The observed unblocking temperatures, although a poor estimate of the Curie temperature, are well under the 580 °C Curie temperature of pure magnetite. Since Curie temperature decreases with increasing titanium content in the magnetite–ulvospinel solid solution, magnetite composition can be estimated for a specific Curie temperature (Merrill et al., 1996). An approximate 500 °C Curie temperature implies a $magnetite_{90}ulvospinel_{10}$ to magnet-

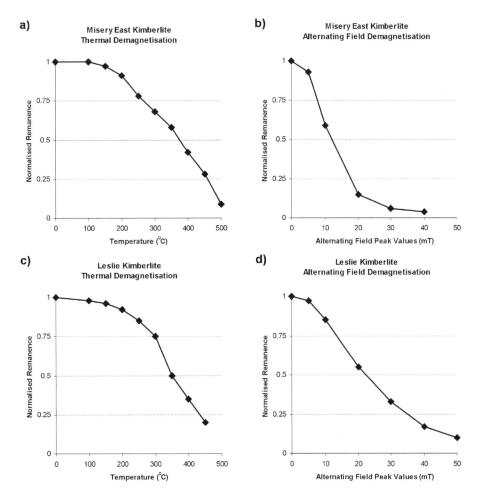

Fig. 5. Incomplete thermal and alternating field demagnetisation plots for Ekati's Misery East (not dated) and Leslie (52.5 ± 1.0 Ma) hypabyssal kimberlites (from Enkin, 2003). Unblocking temperatures of around 500 °C suggest a single-domain NRM carrier is Ti-magnetite, which has compositions of $magnetite_{90}ulvospinel_{10}$ to $magnetite_{80}ulvospinel_{20}$. Demagnetisation was carried only as far as to determine NRM polarity.

ite$_{80}$ulvospinel$_{20}$ composition for the NRM carrier (Enkin, 2003), which is well within the compositional range observed for groundmass Ti-magnetite that is commonly observed in kimberlites (Mitchell, 1986).

A secondary NRM residing in a fine-grained magnetite carrier may be formed by secondary, post-emplacement alteration of kimberlite. The contribution of such a remanent magnetic vector to the Ekati samples is considered subordinate to the initial thermal magnetisation set at emplacement, because the paleomagnetic pole measured in core samples occurs within error of the North American Apparent Polar Wander Path (APWP) over the time interval from 60 to 10 Ma, and furthermore, has its best fit to the APWP at 50 Ma (Fig. 6, Enkin, 2003). These observations indicate that

the NRM found in the Ekati kimberlites represents a primary magnetic vector that has not been measurably reset or degraded by post-intrusion processes.

3.3. Magnetic versus radiometric timescale

The geomagnetic timescale of Cande and Kent (1995) incorporates 36 paleopole reversals during the 30 million years corresponding to kimberlite emplacement at Ekati. The radiometric ages and observed magnetic polarities of Ekati kimberlites are compared with the geomagnetic paleopolarity timescale in Fig. 7. The diagram highlights the frequency of paleopole reversals at irregular intervals that are substantially shorter than the temporal resolution

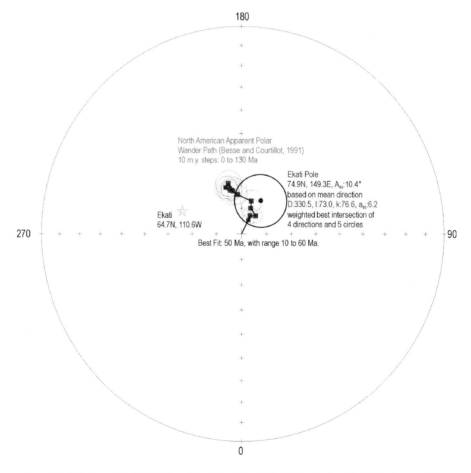

Fig. 6. Paleomagnetic age determination of the Ekati paleopole position. The dot is the Ekati paleopole, with a heavy circle representing its error. The squares are the North American Apparent Polar Wander Path in 10 Ma steps from 0 to 130 Ma, each with grey error circles. The best-fit age is 50 Ma, with a range of 60 to 10 Ma (from Enkin, 2003).

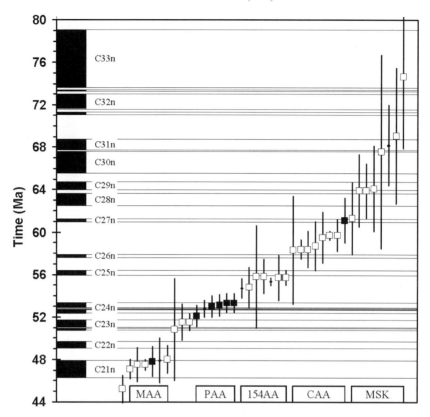

Fig. 7. Age and magnetic character of 40 Lac de Gras kimberlites compared with the geomagnetic polarity timescale of Cande and Kent (1995, Y-axis). Kimberlite magmatism occurred in at least five temporally discrete intrusive episodes labeled MAA to MSK (see text). Black and white bars respectively denote periods of normal or reverse paleomagnetic polarity, also identified by their Chron names (C21n to C33n). Box-whisker plots denote radiometrically determined kimberlite ages with 2σ errors. Filled boxes denote kimberlites with observed normal NRM polarity, open boxes those with observed reverse NRM polarity, and small solid dots those with an indistinct polarity based on aeromagnetic data. See text for discussion of relationships.

obtained by radiometric dating of the kimberlites. The magnetic and radiometric timescales neverthe-less show very satisfactory correspondence of ne-gative kimberlite anomalies with reversal periods between Chrons C24n to C29n, spanning some 10 million years between 53.35 and 63.98 Ma. Kimber-lite intrusion during periods of paleopole reversal in this interval is accurately punctuated by the corres-pondence of Chron C27n (60.92 to 61.28 Ma) with a 61.1 ± 2.1 Ma model-age for the positive anomaly of the Rufus kimberlite (Fig. 7). Weak to strong posi-tive aeromagnetic anomalies are characteristic of 5 out of 6 kimberlites defining the Panda age array (Table 1). A NRM normal polarity, *measured* in 36 samples from four of these pipes, is consistent with the observed positive total-field aeromagnetic re-

sponse of the PAA kimberlites, and indicates that the PAA kimberlites erupted during period(s) of normal geomagnetic polarity (see Table 3). Radio-metric ages for individual PAA kimberlites are con-strained to within ± 1.0 Ma and show outstanding agreement with an intricate paleopole reversal pattern dominated by normal polarities at Chron C24n (52.36 to 53.35 Ma, Fig. 7).

There is excellent consistency between the mag-netic and radiometric timescales for kimberlites in the 154AA group. The Bison, Cardinal, Lynx, Piranha, A418, A21, A154N and A154S pipes have radiomet-ric ages that place them formally within error of Chrons C24n to C26n. Their isotopic correspondence with the well-constrained 55.7 ± 0.7 Ma isochron date for the 154AA group indicates that they very likely intruded during the single geomagnetic reversal period between Chrons C24n and C25n (Fig. 7). This is supported by *measured* reversely polarised remanent vectors in 14 samples from two kimberlites (Piranha and Lynx) and is consistent with neutral or very weak negative aeromagnetic anomalies (see Table 3).

An apparent complication in magnetic versus temporal attributes occurs at ~ 47 Ma (Fig. 7). Although the reasonably well-constrained radiometric ages for Brent, Arnie and Mark overlap predominantly with a period of normal geomagnetic polarity at Chron C21n, the pipes show strong negative aeromagnetic anomalies (Table 1). The source of this apparent discrepancy is unclear, although paleomagnetic characterisation of these kimberlites should resolve this issue.

The observations presented here for kimberlites of the PAA and 154AA clearly demonstrate the variable aeromagnetic anomaly response(s) that may be obtained from economic kimberlite pipes that occur in tightly constrained magmatic corridors (Fig. 4). An integrated approach involving detailed laboratory-based geochronology and geomagnetic studies is required to correctly group kimberlites with similar, desirable attributes and differentiate them from other kimberlites (e.g. Fig. 7).

3.4. Intensity of aeromagnetic anomalies

The total magnetic intensity (TMI) over the Ekati kimberlites, as measured by an airborne magnetometer, varies considerably. The Leslie kimberlite at +7585 nT and the Roger kimberlite at − 16,625 nT represent extreme peak values of the 150 known Ekati kimberlites (Table 1). Most of the kimberlites in this study have amplitudes that are in the range of ±100 nT. In terms of aeromagnetic polarity, 27 out of 40 kimberlites are reversely magnetised, 6 out of 40 are neutral, while 7 out of 40 have normal polarity (Table 1).

The intensity of net magnetisation of Lac de Gras kimberlites is dependent on the vector addition of NRM and induced magnetisation. Since, other than changing polarity, there has been little change in the inclination of the geomagnetic field in the Lac de Gras area (84° at present) over the past 70 Ma (Wynne, 1995), the result of the vector addition is that kimberlites appear as "monopolar" aeromagnetic anomalies.

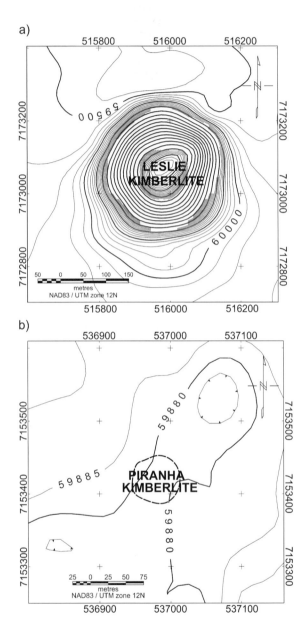

Fig. 8. (a–b) Total field aeromagnetic response over the Leslie (a) and Piranha kimberlites (b). Leslie has a strong normal NRM vector, an exceptionally high Koeninsberger Ratio of 18.43 (Table 3) and the resulting total field peak aeromagnetic value is +7585 nT. The contour interval is 100 nT. In contrast, Piranha has a weak reversed NRM vector, a Koenisberger Ratio of 1.03 (Table 3) and is "invisible" in the aeromagnetic data. The contour interval for the Piranha panel is 5 nT. The nominal flight path is E–W, with a 75 m line spacing and a 20 m sensor height.

Where the NRM dominates the induced magnetisation and the polarities are both normal, the net result is a strong and normal magnetisation. The Leslie kimberlite exemplifies the case where the NRM is normal and exceptionally strong (Koeninsberger Ratio of 18.43, Table 3 and Fig. 8a) and the resulting total field aeromagnetic anomaly is +7585 nT. When the NRM is strong and has a reversed polarity, the net result is a strong negative magnetic anomaly, which is exemplified by the Jaeger kimberlite (Koeninsberger Ratio of 6.52, Table 3 and Fig. 3b). The resulting aeromagnetic peak intensity of the latter is −1437 nT. An unusual case occurs when the magnitude of the NRM is approximately equal to the induced magnetisation. In this case, the resulting magnetisation is almost zero and the kimberlite may be "invisible" in the aeromagnetic data, as appears to be the case for Lynx and Piranha (Table 3 and Fig. 8b). The variety in aeromagnetic signature is essentially a product of frequent geomagnetic pole reversals occurring within the 30-million-year time-span during which kimberlites intruded episodically at Ekati.

The magnetic susceptibilities of Ekati kimberlites appear to be closely related to the two dominant kimberlite rock types present (Table 1). Fig. 9 shows that a magnetic susceptibility of 4000×10^{-6} SI units readily distinguishes volcaniclastic kimberlite (VK) from magmatic kimberlite (MK). The average magnetic susceptibilities of VK and MK are 1100×10^{-6} SI units and $11,000 \times 10^{-6}$ SI units, respectively (i.e. a full order of magnitude difference). This is likely the result of varying magnetite content and is based on bulk density differences between the two kimberlite lithologies. On this basis, hypabyssal kimberlites typically appear in aeromagnetic data as stronger anomalies than kimberlite pipes filled with volcaniclastic material.

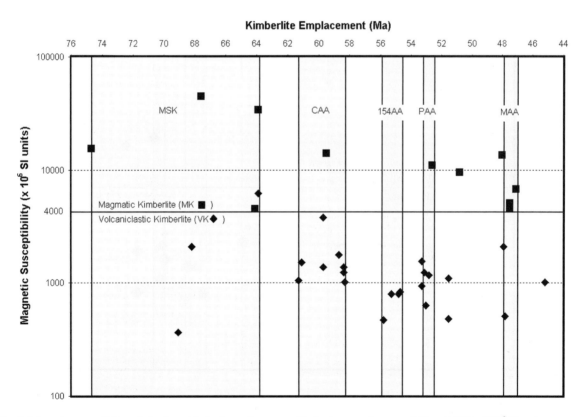

Fig. 9. Magnetic susceptibility of dominant kimberlite rock types at Ekati. A magnetic susceptibility of 4000×10^{-6} SI units separates magmatic kimberlite (MK) from volcaniclastic kimberlite (VK), which can be explained by the relative abundance of magnetite in each rock type.

3.5. Intrusive episodes and diamond content

The diamond content of kimberlites at Ekati is represented by microdiamond abundance categories in Table 1 and displayed as a function of kimberlite emplacement age and magnetic susceptibility in Fig. 10. High to very high microdiamond abundance is related to episodes of kimberlite magmatism dated at 56 to 45 Ma, while very low to moderate microdiamond contents are observed in kimberlites dated in the 75 to 59 Ma age range (Fig. 10). Known economic phases of kimberlite occur exclusively in the ~ 55 Ma 154AA group (Lynx, A154N, A154S, A418 and A21) and the ~ 53 Ma PAA group (Panda, Koala, Koala North and Beartooth). The economic phases generally occur as volcaniclastic kimberlite and are characterised by relatively low magnetic susceptibilities. Kimberlites, such as Crab and Roger from older age groups, are located near the Panda cluster (Fig. 4) and have very low microdiamond contents (Table 1). This temporal and spatial relationship indicates that kimberlites within a particular intrusive episode have a common ability to sample, transport and preserve diamondiferous mantle, and that these aspects of kimberlite magmatism can vary dramatically between intrusive episodes, even when they are closely related spatially.

4. Summary and conclusions

Kimberlite magmatism at Ekati occurred over a 30 million year period (75 to 45 Ma) in at least five temporally discrete episodes. Our data show that the younger (56 to 45 Ma) 154AA, PAA and MAA group kimberlites have significantly higher diamond

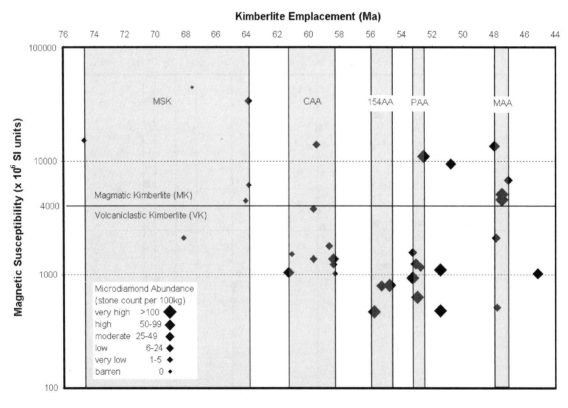

Fig. 10. Magnetic susceptibility and Ekati microdiamond abundance categories plotted against emplacement age. Younger (<56 Ma) kimberlites at Ekati have a relatively high microdiamond abundance (>100 stones per 100 kg) compared to older (>59 Ma) kimberlites. Economically important diamond-bearing kimberlites appear to be restricted to the 154AA (~ 55 Ma) and the PAA (~ 53 Ma) groups. Economic kimberlites are volcaniclastic and have very low magnetic susceptibilities, generally manifesting as neutral or weak magnetic "highs" or "lows".

contents than the older (75 to 64 Ma) perovskite-bearing MSK group kimberlites and the intermediate age (59 Ma) phlogopite-bearing CAA kimberlite group. Known economic kimberlites erupted at 55.4 ± 0.4 Ma (154AA group) and at 53.2 ± 0.3 Ma (PAA group), each along a linear intrusive corridor, respectively trending $015°$ (154IC) and $038°$ (PIC). A third discrete intrusive corridor trends $043°$ (MIC) and contains the 47.8 ± 0.3 Ma MAA kimberlites. The economically important 154AA kimberlites (Lynx, A154N, A154S, A418 and A21) and PAA kimberlites (Panda, Koala, Koala North and Beartooth) are dominated by volcaniclastic rock types that manifest in aeromagnetic surveys as neutral to weak "low" (154AA) or weak magnetic "high" (PAA) anomalies. Their magnetic anomaly expression reflects a relatively low magnetite content and the paleopole polarity carried by natural remanent magnetisation.

Based on the results of this study, a refined property-scale exploration model for Ekati should consider the following:

- Kimberlite magmatism has occurred episodically, along narrow intrusive corridors in NNE to NE directions. The location and orientation of the intrusive corridors appears to be poorly correlated, if at all, with the structural geology of Archean basement rocks at Ekati.
- Economic kimberlites are generally dominated by volcaniclastic phases that tend to produce neutral to weak aeromagnetic anomalies.
- Kimberlites belonging to a coeval intrusive episode will have a consistent polarity in aeromagnetic data, which is that of natural remanent magnetisation (NRM) because this tends to dominate over induced magnetisation. This feature presents the opportunity to accurately correlate coeval kimberlites at Ekati by magnetostratigraphic means, thereby aiding in the discrimination of economically significant intrusive episodes.

Notwithstanding the above, the possibility of finding a high-grade kimberlite within an age group generally characterised or dominated by low-grade kimberlites cannot be excluded. Diamond grade and quality are typically related to their mantle source-rocks and the magmatic processes that preserve or dilute them en route to surface, rather than to purely temporal considerations (e.g. Helmstaedt, 1993).

5. Uncited reference

Besse and Courtillot, 1991

Acknowledgements

The authors gratefully acknowledge the encouragement and continued financial support of this investigation by BHP Billiton Diamonds, and are thankful for having been granted permission to publish the results at the 8IKC. Rob Creaser and Larry Heaman provided expert advice and superb geochronological data, commercially available through GeoSpec Consultants. Fugro Airborne Surveys provided the high-resolution geophysical data that underpins successful exploration drilling campaigns at Ekati. We thank Mineral Services Canada, Randy Enkin, Karen Wright, and Herb Helmstaedt for their input as consultants and Barbara Crawford, Sara Harrison, Melissa Kirkely, Susannah Price, and Pauline Orr for diligently searching core samples for macrocrystic phlogopite. Vadim Kravchinsky and Ian Basson provided constructive reviews of the manuscript.

References

Anderson, O.L., 1979. The role of fracture dynamics in kimberlite pipe formation. In: Boyd, F.R., Meyer, H.O.A.Kimberlites, Diatremes and Diamonds: Their Geology, Petrology, and Geochemistry. Proc. Second Int. Kimberlite Conf.. Am. Geophys. Union, Washington, D.C., pp. 344–353.

Armstrong, R.A., Moore, R.O., 1998. Rb–Sr ages on kimberlite from Lac de Gras area, Northwest Territories, Canada. South African Journal of Geology 101 (2), 155–158.

Besse, J., Courtillot, V., 1991. Revised and synthetic apparent polar wander paths of the African, Eurasian, North American and Indian Plates, and True Polar Wander since 100 Ma. Journal of Geophysical Research 96, 4029–4050.

Cande, S.C., Kent, D.V., 1995. Revised calibration of the geomagnetic polarity timescale for the Late Cretaceous and Cenozoic. Journal of Geophysical Research 100, 6093–6095.

Creaser, R.A., Grutter, H., Carlson, J.A., Crawford, B.B., this volume. Macrocrystal phlogopite Rb–Sr dates for the Ekati property kimberlites, Slave Province, Canada: evidence for multiple intrusive episodes in the Paleocene and Eocene.

Davis, W.J., Kjarsgaard, B.A., 1997. A Rb–Sr isochron age for a kimberlite from the recently discovered Las de Gras Field, Slave Province, Northwest Territories, Canada. Journal of Geology 105, 503–509.

Dawson, J.B., 1984. Ascent and emplacement of kimberlite magma. In: Glover, J.E., Harris, P.G.Kimberlite Occurrence and Origin: A Basis for Conceptual Models in Exploration. Publication-Geology Department and University Extension, University of Western Australia, vol. 8, pp. 113–124.

Dyck, D.R., Oshust, P.A., Carlson, J.A., Nowicki, T.E., Mullins, M.P, this volume. Effective Resource Estimates for Primary Diamond Deposits-Ekati Diamond Mine™, Canada.

Enkin, R., 2003. Paleomagnetic analysis of selected Ekati kimberlites. Unpublished report for BHP Billiton Diamonds by Paleomagnetism, Geological Survey of Canada, January 2003.

Fipke, C.E., Dummett, H.T., Moore, R.O., Carlson, J.A., Ashley, R.M., Gurney, J.J., Kirkley, M.B., 1995. History of the discovery of diamondiferous kimberlites in the Northwest Territories, Canada. Extended Abstracts, Sixth International Kimberlite Conference, Novosibirsk, United Institue of Geophysics and Mineralogy, pp. 158–160.

GeoSpec Consultants, 1996. U–Pb and Rb–Sr isotopic analyses for BHP Minerals. Canada. Unpublished report dated December 1996.

GeoSpec Consultants, 1999. Rb–Sr and U–Pb isotopic analyses for BHP Minerals. Canada. Unpublished report dated August 1999.

Graham, I., Burgess, D., Bryan, D., Ravencroft, P.J., Thomas, E., Doyle, B.J., Hopkins, R., Armstrong, K.A., 1999. Exploration history and geology of the Diavik Kimberlites, Lac de Gras, Northwest Territories, Canada. In: Gurney, J.J., Gurney, J.L.,

Pascoe, M.D., Richardson, S.H.J.B. Dawson volume, Proceedings of the VIIth International Kimberlite Conference. Red Roof Design, Cape Town, pp. 262–279.

Helmstaedt, H.H., 1993. Natural diamond occurrences and tectonic setting of "primary" diamond deposits. Diamonds: Exploration, Sampling and Evaluation. Proc. PDAC Short Course, March 27, 1993, Toronto, Prospectors and Developers Assoc., Canada, pp. 3–72.

McElroy, R., Nowicki, T., Dyck, D., Carlson, J., Todd, J., Roebuck, S., Crawford, B., Harrison, S., 2003. The geology of the Panda kimberlite, Ekati Diamond Mine™, Canada. Extended Abstracts, Eighth International Kimberlite Conference, Victoria.

Merrill, R.T., McElhinny, M.W., McFadden, P.L., 1996. The Magnetic Field of the Earth: Paleomagnetism, the Core, and the Deep Mantle. International Geophysics Series, vol. 63. Academic Press, San Diego, CA, pp. 83–87.

Mitchell, R.H., 1986. Kimberlites: Mineralogy, Geochemistry, and Petrology. Plenum, New York, pp. 219–239.

Stacey, J.S., Kramers, J.D., 1975. Approximation of terrestrial lead isotopic evolution by a two-stage model. Earth and Planetary Science Letters 26, 207–221.

Wright, K.J., 1999. Possible structural controls of kimberlite in the Lac de Gras Region, central Slave Province, Northwest Territories, Canada. MSc Thesis. Queen's University, Kingston, Ontario.

Wynne, P.J., 1995. Preliminary results from the paleomagnetic study of kimberlites from the NWT Diamond Project. Unpublished report for BHP Diamonds by Paleomagnetism, Geological Survey of Canada, June 1995.

Available online at www.sciencedirect.com

SCIENCE DIRECT®

Lithos 77 (2004) 683–693

LITHOS

www.elsevier.com/locate/lithos

Spatial distribution of kimberlite in the Slave craton, Canada: a geometrical approach

M.P. Stubley*

Stubley Geoscience, 158 Toki Road, Cochrane, AB, Canada T4C 2A2

Received 25 June 2003; accepted 27 November 2003
Available online 18 May 2004

Abstract

Exploration within the Slave craton has revealed clusters of kimberlite intrusions, commonly with internally consistent geochemical and temporal characteristics. Translation diagrams ("Fry analysis") allow an unbiased geometrical examination of the distance and direction between each kimberlite occurrence and all others in the database. Recurrent patterns are visually accentuated due to the square function in data density. Circular histograms quantify the azimuthal density of kimberlite at various distances. For this study, the database comprises the geographic position of 212 kimberlite occurrences of which 70% are from the Lac de Gras field (LDG). Analyses are presented separately for the LDG data and for all non-LDG data in order to test for regional variations and to avoid overwhelming the craton-scale studies by the high density of LDG data.

Empirical grouping of kimberlite locations results in delineation of five elliptical clusters that encompass all but four kimberlite occurrences. Clusters within the western part of the craton are elongate to the north–northeast and align within a narrow zone ("Western Corridor"). Elsewhere, the clusters are elongate to the northwest or west–northwest and appear to be arranged en echelon within a poorly defined north–northwest trending zone ("Central Corridor"). Geometrical spatial analyses of kimberlite locations highlight the craton-scale pattern of emplacement within the two main corridors. At regional and local scales, individual intrusions are preferentially located towards the west–northwest (ca. 280°) and north–northeast (ca. 015°) of other intrusions, and these orientations are interpreted to reflect upper mantle trends in magma generation. At local scales (10–25 km), kimberlite of the central and southern craton tends to be located to the northeast (ca. 045°), and possibly weakly to the east–northeast (ca. 070°), of other intrusions, and these orientations correspond to major crustal fractures systems. It is proposed that kimberlite emplacement is controlled primarily by the interaction of elongate 280° and 015° source regions with near-surface deviations influenced by crustal fracture systems.

The 015° trend evident at craton, regional, and local scales is parallel to a swarm of alkaline diabase dykes that are concentrated in a ca. 30-km-wide corridor passing through Lac de Gras. A profound spatial association between significantly diamondiferous kimberlite and the margins of the dyke corridor suggests the corridor is the surface expression of a mantle-depth structure. It remains unclear whether the proposed mantle structure coincides with a diamond-rich zone near the base of the lithosphere, or delineates pathways favorable for diamond preservation during

* Tel.: +1-403-851-1311; fax: +1-403-851-1312.
E-mail address: stubley@pathcom.ca (M.P. Stubley).

emplacement. The linear array of kimberlite within the western craton forms a parallel corridor that may be an analogous mantle structure, but which to date has failed to yield economic diamond concentrations.
© 2004 Elsevier B.V. All rights reserved.

Keywords: Diamonds; Mantle structure; Diabase; Lac de Gras

---

1. Introduction

The recognition of features that control the generation and emplacement of kimberlite is fundamental to diamond exploration methodologies, yet an understanding of these features is in its infancy. Exploration within the Slave craton of northwestern Canada has revealed a clustered pattern of kimberlite occurrences, but local-scale patterns within and between the clusters remain unresolved. Several explanations for the location of kimberlite have been proposed, including cryptic zonation within the upper mantle and influence by crustal fracture systems. These proposals are, however, fraught with uncertainty due to a variety of factors including availability and consistency of data, scale of observation, and personal bias.

Vearncombe and Vearncombe (2002) present a novel, unbiased, and purely geometrical approach to investigate tectonic controls on kimberlite location within southern Africa. A similar analysis has been conducted for the Slave craton and forms the basis for this paper. The results of the spatial analysis and of empirical observations of cluster distribution are compared to crustal geological features to search for correlations that could aid future diamond exploration.

2. Kimberlite clusters of the Slave craton

Archean rocks of the Slave craton are exposed over approximately 229,000 km^2 and dip below Proterozoic and Paleozoic cover to the northwest and north, respectively (Fig. 1). More than 310 kimberlite occurrences are known from the Slave craton of the mainland Northwest Territories and Nunavut (Kjarsgaard and Levinson, 2002), and geographic positions of approximately 227 of these are reported in the public domain (e.g. Armstrong and Chatman, 2001; Jones, 2002). The distribution of kimberlite intruding the Archean Slave craton and bounding Proterozoic orogens is illustrated in Fig. 2. Minimum-enclosing

ellipses delimit, empirically, five broad kimberlite clusters or "fields". All but four of the 227 reported occurrences are accommodated in this simplistic subdivision. The Central Slave cluster is further subdivided to delimit the dense distribution of the Lac de Gras field.

The Coronation and Southwest Slave clusters are elongate to the north–northeast and occupy a narrow (<45 km wide) co-linear zone, herein termed the Western Corridor (Fig. 2). The three kimberlite occurrences at Drybones Bay (South Slave; Fig. 2), which may constitute another mini-cluster, also occupy this corridor. In contrast, the other clusters are elongate to the northwest or west–northwest and appear to be arranged en echelon within a poorly defined north–northwest trending zone (Central Corridor of Fig. 2).

The above-defined clusters are classified solely on spatial proximity and without regard for features of the crustal geology or kimberlite characteristics. Nevertheless, publicly available data reveals distinctive features of kimberlite and the underlying mantle of each cluster, and supports a geological basis for these fundamental subdivisions. For example, kimberlite of the Central Slave cluster is dominated by small steep-sided pipes lacking diatreme-facies infilling that were intruded from depths of up to ca. 220 km during the Cretaceous and Eocene (Field and Scott Smith, 1999; Griffin et al., 1999). In contrast, the Southeast Slave cluster contains abundant diatreme facies (Field and Scott Smith, 1999) and was emplaced during the Paleozoic through a significantly thicker (230 to 300 km) lithosphere (Pokhilenko et al., 1998; Kopylova and Caro, 2001), with approximately one-third of the kimberlite occurrences at surface represented by dyke morphologies. Data from the other clusters is less comprehensive but supports distinctions in age of emplacement (Jurassic in the North Contwoyto cluster; Paleozoic for the Southwest Slave and Drybones occurrences; and Neoproterozoic to Cambrian for the Coronation cluster) and inferred lithospheric thickness. Grütter et al. (1999) recognized a chemical

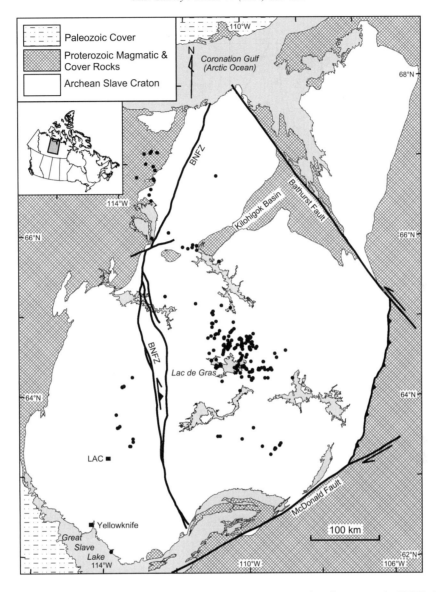

Fig. 1. Kimberlite occurrences (dots; $n = 227$) in the Slave craton and the overlying Proterozoic sedimentary rocks. BNFZ = Beniah–Napaktulik fault zone. LAC = Leith alkaline complex.

similarity between garnet populations originating within the Central and Southwest Slave clusters that contrasts with those from both the Southeast Slave and North Contwoyto fields. Interestingly, the pan-Slave Beniah–Napaktulik fault zone (Stubley, 2003) separates kimberlite of the Western and Central corridors (Fig. 1). The westernmost kimberlite of the North Contwoyto cluster (Nanurjuk; Fig. 2) is the only exception to this last observation, and

suggests that a geological-driven delineation of the clusters might preferentially assign Nanurjuk to the Coronation field.

3. Spatial analyses

Centre-to-centre (point-to-point) techniques of spatial analysis may be used to emphasize inherent

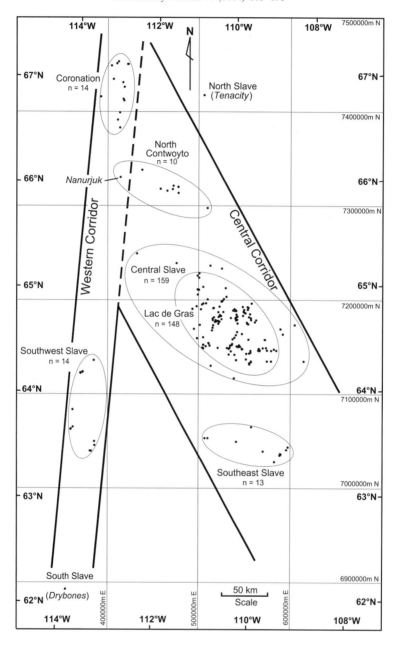

Fig. 2. Distribution of reported kimberlite within the Slave craton, as used for the template of the spatial analyses reported herein. Ellipses delimit clusters of kimberlite as discussed in text. All kimberlite locations are projected to Universal Transverse Mercator (UTM) Zone 12 using North American Datum 1983 (NAD83) and are shown relative to the orthogonal UTM grid.

patterns in point data. Fry (1979) presented a simple, yet elegant, graphical variant of this technique to evaluate strain in deformed rocks. The so-called "Fry analysis" displays the spatial relationship between each point and all others in the database, without the use of numerical calculations or statistics. Each point is, in turn, placed at a common origin and all other points are plotted in their respective new positions resulting in a composite plot with n^2 points (including n points on the origin). Recurrent patterns are visually accentuated due to the square function in data density. In particular, enhanced point density illustrates a more common spacing and directional relationship between pairs of points. Conversely, regions of reduced point density mark less common distances and orientations between any two points.

The present analysis uses the geographic position of all distinct and reported kimberlite occurrences in the Slave craton ($n = 212$) as the base template (Fig. 2). Multiple occurrences of contiguous kimberlite dykes (e.g. Snap and King lakes) and close-spaced pipes that may be contiguous at depth (e.g. Anuri pipes; CL-25 and CL-174) are treated as single entities. Publicly available locations are taken primarily from Armstrong and Chatman (2001), Jones (2002), and Kjarsgaard et al. (2002), and are believed to be positionally accurate to within 500 m for more than 95% of the occurrences. The computer-assisted copying and overlaying of the entire template that

places each kimberlite in turn on the origin results in a symmetrical Fry Plot with nearly 45,000 points. Circular histograms are constructed to display data density within 5° azimuthal sectors for a variety of distances from the origin. Peaks in the histograms effectively quantify the most common directions between any kimberlite, represented by the origin, and all others in the database. It should be noted that quoted orientations are relative to the Universal Transverse Mercator grid and vary up to 2.7° from cardinal directions. Similar errors may be introduced by positional inaccuracies in the database when considering distances of less than 10 km from the origin, but are deemed insignificant at regional scales.

The point-to-point graphical analysis has been conducted for the entire Slave database. The dense distribution within the Lac de Gras field (70% of the data) tends to overwhelm and mask the relationships in other clusters. In an attempt to test for

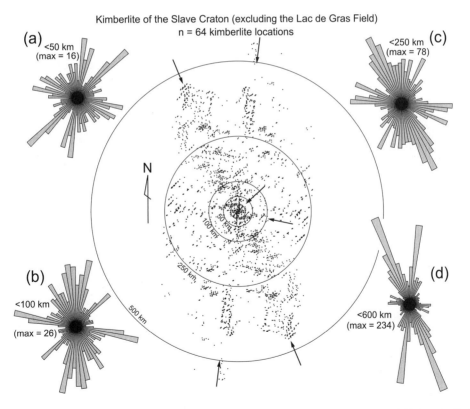

Fig. 3. "Fry Plot" for kimberlite of the Slave craton, excluding data and translation effects of the Lac de Gras field ($n = 64$). Circles superposed on the data cluster represent distances of 25, 50, 100, 250, and 500 km from the origin. Arrows highlight principal trends. Circular histograms display data density within 5° azimuthal sectors (maximum radii of histograms are indicated).

local and regional variations, the analysis has also been conducted separately on kimberlite locations within, and external, to the Lac de Gras field.

Fig. 3 presents the results of analysis of the data external to the Lac de Gras field. The Fry Plot shows the tendency for kimberlite to be located to the north–northwest or north–northeast (or their complementary directions) of other occurrences, reflecting the broad distribution within the two first-order corridors. Fig. 3d illustrates these craton-scale relationships, and its smaller peaks reflect the west–northwest alignment of data clusters superposed on the Central Corridor. At regional scales of observation (e.g. <100 km radius; Fig. 3a and b), the prominent orientations (west–northwest and north–northeast) reflect the distribution within the elongate clusters. A prominent northeast–southwest alignment is constrained within 25 km of the origin and likely reflects the linear distribution of kimberlite in the Kennady Lake area of the Southeast Slave cluster (see Fig. 2).

Similar, although less distinct, regional and local relationships are evident from Fry analysis of the Lac de Gras kimberlite field (Fig. 4). A tendency for kimberlite to be located northeast–southwest of other occurrences is evident only at distances of less than 10 km (Fig. 4a). More-prominent preferential alignments to the north–northeast and west–northwest (or their complementary directions) persist to distances of about 50 km (Fig. 4c). At greater distances, the general northwest elongation of the field dominates the distribution pattern.

4. Discussion

4.1. Kimberlite emplacement models

Numerous studies have suggested a correlation between crustal fracture systems and kimberlite emplacement. Most authors favor a model wherein

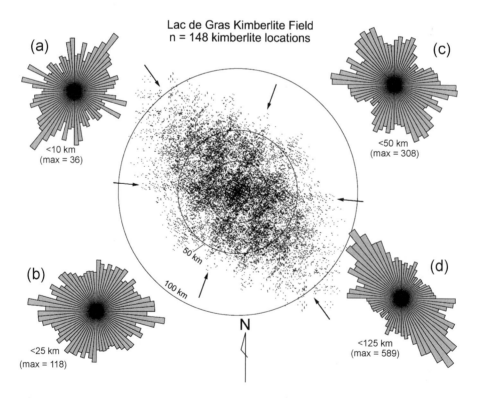

Fig. 4. "Fry Plot" for kimberlite of the Lac de Gras field (n = 148). Circles superposed on the data cluster represent distances of 10, 25, 50, and 100 km from the origin. Arrows highlight regional-scale trends; note the prominent northeast-oriented trend within 10 km of the origin. Circular histograms display data density within 5° azimuthal sectors (maximum radii of histograms are indicated).

ascending magma follows a path of least resistance and exploits suitably oriented pre-existing anisotropies (e.g. White et al., 1995 and references therein). Vearncombe and Vearncombe (2002) present an opposing view in which kimberlite is emplaced in corridors of competent crust between major fracture zones. For the most part, neither view addresses structural control on kimberlite generation. However, White et al. (1995) acknowledge that crustal fracture systems may overlie structures that penetrate the upper mantle and possibly extend to the base of the lithosphere. To date, no steeply oriented structures have been recognized in the upper mantle beneath the Slave craton.

Deviations from vertical ascent of 3° over 200 km from a point source could produce kimberlite dispersal over ca. 20 km diameter at surface. Although reasonable estimates of angular deviation are uncertain, the size of the kimberlite clusters of the Slave craton (>100 km maximum dimension) likely precludes variable ascent paths as the primary cause of kimberlite dispersion. Instead, elongate, and possibly quasi-linear, source regions for kimberlite magma are suggested. It is probable that kimberlite emplacement is controlled by the interaction of elongate source regions and near-surface fracture systems.

4.2. Principal kimberlite distribution and fracture orientations

Prominent systems of northeast (ca. 045°) and east–northeast (ca. 070°) trending faults and fractures characterize the south and central Slave craton. Diabase dykes of the ca. 2.23 Ga "Malley" and ca. 2.21 Ga "MacKay" swarms accentuate these fracture systems, respectively (Wilkinson et al., 2001). Kimberlite is commonly found within or adjacent to these fractures (Wright, 1999; Cookenboo, 1999), and this empirical spatial correlation has aided successful exploration. The Fry analyses support the association between kimberlite location and 045° fractures at distances of up to 10 km within the Lac de Gras field and up to 25 km in the Southeast Slave field. Possible associations between kimberlite location and the 070° fracture system are weak or absent in all Fry analyses at scales exceeding 5–10 km. It is postulated that these two fracture systems exerted only near-surface (crustal) influence on kimberlite emplacement, and do not reflect upper mantle structures or source-region geometry.

The Fry analyses reveal a strong tendency for kimberlite to be located to the west–northwest (ca. 280°) of other occurrences within all clusters of the Central Corridor. At a regional scale, this pattern corresponds approximately to the general elongation of each of the eastern clusters (Fig. 2). Kilometre-scale emplacement patterns are exemplified by a swath of pipes (including Diavik's "T-" series) in the southern Lac de Gras field (Fig. 2). West–northwest oriented features of the crustal geology are common in the area extending north–northeast from the North Contwoyto cluster (Fig. 2), but are sparse and apparently insignificant in all other areas of the Central Corridor. This general paucity of corresponding crustal features may suggest lower- or sub-crustal control on the preferential kimberlite localization.

A preference for kimberlite to be emplaced north–northeast (ca. 015°) of other occurrences is evident in all analyses and at all scales. The Western Corridor and its contained clusters reflect this distribution at craton, regional, and local scales (Fig. 2). Within the Lac de Gras field of the Central Corridor, two prominent zones of kimberlite follow this trend (Fig. 2) and incorporate all pipes slated for exploitation at both the Ekati™ and Diavik mine sites. North–northeast striking faults are uncommon in the central Slave craton. However, at least one swarm of diabase dykes follows this trend, and its possible association with kimberlite is discussed further below.

4.3. Kimberlite, diamonds, and "Lac de Gras" diabase association

Within the Central Corridor, a prominent system of diabase dykes ("Lac de Gras" swarm) parallels the north–northeast kimberlite emplacement trend. Dating of two dykes of the system has returned consistent ages of ca. 2023–2030 Ma (LeCheminant, 1994; Kjarsgaard et al., 2002) and all published geochemical data reveal a consistent and distinct alkaline character (Wilkinson et al., 2001). However, more than one episode of parallel dyke emplacement is suggested by along-strike diabase transecting mid-Proterozoic gabbro sills at Great Slave Lake to the south (Hoffman, 1988) and, from aeromagnetic signatures, transection of ca. 1.9 Ga sedimentary rocks of the Kilohigok Basin (Fig. 1) to the north. Within the Lac de Gras area, dykes of the "Lac de Gras" swarm typically attain widths of 30–50 m; approximately 50 km farther north, a 120-m-wide dyke occupies a fault with a significant component of vertical displacement.

A GIS-based weights-of-evidence analysis by Wilkinson et al. (2001) noted a moderate to strong spatial association between kimberlite pipes of the Lac de Gras field and diabase dykes of the "Lac de Gras"

swarm despite a paucity of coincident intrusions. A compilation of north–northeast striking dykes of the Central Corridor, based on geological mapping and interpretation of aeromagnetic data, indicates the dykes are concentrated within a ca. 30-km-wide corridor that passes through Lac de Gras. All Slave kimberlite pipes within current mine plans are located along the margins of the dyke corridor, as is the highly diamondiferous Snap Lake deposit (Fig. 5). This strong spatial association between kimberlite ore and

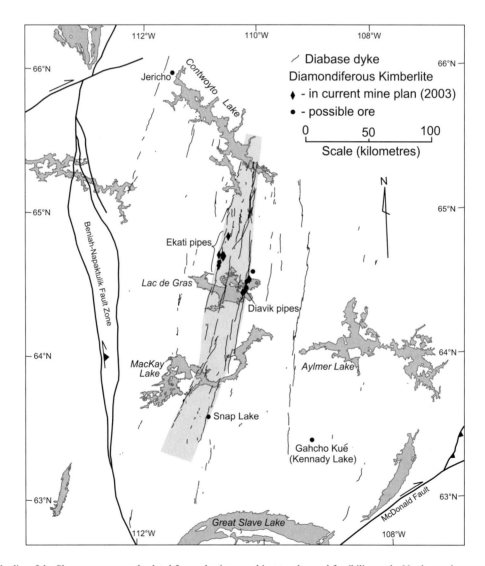

Fig. 5. Kimberlite of the Slave craton currently slated for production or subject to advanced feasibility study. North–northeast striking diabase dykes, dominated by the 2.0 Ga alkaline "Lac de Gras" swarm, are illustrated for the area south of Contwoyto Lake and east of the Beniah–Napaktulik fault zone, and are concentrated in a ca. 30-km-wide corridor (shaded).

the "Lac de Gras" diabase corridor is currently being evaluated. Kimberlite ore has not been identified more than ca. 4200 m from the margins of the dyke corridor, although not all kimberlite within the corridor is significantly diamondiferous. Preliminary interpretations suggest the dyke corridor may overlie a diamond-enriched zone at, or near, the base of the lithosphere. Alternatively, the diamond enrichment may indicate "diamond-friendly" transport and preservation processes were facilitated along the corridor margins. The apparent restriction of ore to narrow linear zones suggests minimal lateral deviation in kimberlite ascent paths, although evaluation of deviations parallel to the zones is unconstrained. Interestingly though, kimberlite within the dyke corridor fails, in general, to exploit the dykes during ascent.

4.4. Steep structures of the upper mantle

Diabase of the "Lac de Gras" swarm has a distinctly alkaline geochemical signature (Wilkinson et al., 2001) and projects northward to the coeval (ca. 2023 Ma) alkalic Booth River intrusive suite (LeCheminant, 1994; Roscoe et al., 1987). Intrusion of the Snap Lake kimberlite at ca. 523–535 Ma (Agashev et al., 2001) and of many of the pipes of the Lac de Gras field between 47 and 84 Ma (Carlson et al., 1999) indicates repeated alkaline magmatism exploited the north–northeast trending dyke corridor. The commonly cited association between alkaline intrusions and deep-seated structures (e.g. White et al., 1995) supports the assertion of a north–northeast trending feature underlying the central Slave craton; the coincidence of diamond-enrichment with the proposed structure may suggest it extends to depths of diamond stability within the upper mantle. If valid, this model suggests that the crust and lithospheric mantle below the central craton have been coupled for the last 2.0 Ga.

The north–northeast trending Western Corridor (Fig. 2) is parallel to, and of similar width to, the proposed upper mantle structure below the central Slave craton. Although this corridor is defined solely from alignment of kimberlite intrusions, one could speculate it also represents a deep-seated structure. The Leith Alkaline Complex (Fig. 1: Stubley and Cairns, 1998; Armstrong, 2001), one of only two known Archean carbonatite-bearing intrusions in the

Slave craton, is hosted within the narrow Western Corridor and adds credence to this speculation. In addition, a series of north–northeast striking Archean lamprophyre dykes at Yellowknife (Armstrong, 2001) effectively coincides with the western margin of the proposed Western Corridor.

5. Conclusions

Kimberlite intrusions through the Slave craton are grouped in elongate clusters, commonly with internally consistent geochemical and temporal characteristics. The size of the clusters is thought to reflect elongate source regions for kimberlite magma. Co-linear alignment of north–northeast trending clusters in the western craton contrast with an apparent en echelon arrangement of northwest-aligned clusters in the central craton. Two corridors defined by the contrasting cluster alignments fail to correspond to structural features recognized in the crustal geology.

Geometrical spatial analyses of kimberlite locations highlight the craton-scale pattern of emplacement within the two main corridors. At regional and local scales, individual intrusions are preferentially located towards the west–northwest (ca. 280°) and north–northeast (ca. 015°) of other intrusions, and these orientations are interpreted to reflect upper mantle trends in magma generation. At local scales (10–25 km), kimberlite of the central and southern craton tends to be located to the northeast (ca. 045°), and possibly weakly to the east–northeast (ca. 070°), of other intrusions, and these orientations correspond to major crustal fracture systems. It is proposed that kimberlite emplacement is controlled primarily by the interaction of elongate 280° and 015° source regions with near-surface deviations influenced by crustal fracture systems.

The 015° trend evident at craton, regional, and local scales is parallel to a swarm of alkaline diabase dykes that are concentrated in a ca. 30-km-wide corridor passing through Lac de Gras. A moderate to strong spatial relationship between the diabase dykes and kimberlite location is evident in the Lac de Gras area, as is also shown by other studies, despite a paucity of coincident intrusions. However, when considering only significantly diamondiferous kimberlite, the association with the diabase dyke corridor is remarkably strong. All kimberlite ore, and much of the probable ore, within the Slave craton is located within a few kilometres of the margins of the corridor. The alkaline character of the dyke corridor, and the association with diamond-enhanced kimberlite, suggest the corridor is the surface expression of a mantle-depth structure. It remains unclear whether the proposed mantle structure coincides with a diamond-rich zone near the base of the lithosphere, or delineates pathways favorable for diamond preservation during emplacement. The linear array of kimberlite within the western craton forms a parallel corridor that may be an analogous mantle structure, but which to date has failed to yield economic diamond concentrations.

Acknowledgements

This paper results from parts of a long-term project investigating the crustal architecture of the Slave craton supported by Diamondex Resources, and its predecessor Winspear Resources. The management of these companies is gratefully acknowledged for the opportunity to present these data and interpretations. J.A. McDonald, K.-J. Wright, and J.S. Gebert provided helpful suggestions on earlier versions of this manuscript. Comments by B.H. Scott Smith (editor), S. Verma, and an anonymous reviewer led to improved clarity.

References

Agashev, A.M., Pokhilenko, N.P., McDonald, J.A., Takazawa, E., Vavilov, M.A., Sobolev, N.V., Watanabe, T., 2001. A unique kimberlite–carbonatite primary association in the Snap Lake dyke system, Slave Craton: evidence from geochemical and isotopic studies. Proceedings of the Slave-Kaapvaal Workshop, Merrickville, Canada, Sept. 2001.

Armstrong, J.P., 2001. Alkaline magmatic events—Leith lake carbonatites and Yellowknife lamprophyres: evidence for Archean mantle metasomatism, Southern Slave Craton. Proceedings of the Slave-Kaapvaal Workshop, Merrickville, Canada, Sept. 2001.

Armstrong, J.P., Chatman, J., 2001. Kimberlite Indicator and Diamond Database (KIDD): Update: A compilation of publicly available till sample locations and kimberlite indicator mineral picking results, Slave Craton and environs, Northwest Territories and Nunavut, Canada. DIAND-NWT Geology Division, Open EGS Report 2001-01.

Carlson, J.A., Kirkley, M.B., Thomas, E.M., Hillier, W.D., 1999. Recent Canadian kimberlite discoveries. In: Gurney, J.J., Gurney, J.L., Pascoe, M.D., Richardson, S.H. (Eds.), Proceedings of the 7th International Kimberlite Conference, vol. 1. Red Roof Design, Cape Town, pp. 81–89.

Cookenboo, H.O., 1999. History and process of emplacement of the Jericho (JD-1) kimberlite pipe, northern Canada. In: Gurney, J.J., Gurney, J.L., Pascoe, M.D., Richardson, S.H. (Eds.), Proceedings of the 7th International Kimberlite Conference, vol. 1. Red Roof Design, Cape Town, pp. 125–133.

Field, M., Scott Smith, B.H., 1999. Contrasting geology and near-surface emplacement of kimberlite pipes in southern Africa and Canada. In: Gurney, J.J., Gurney, J.L., Pascoe, S.H., Richardson, S.H. (Eds.), Proceedings of the 7th International Kimberlite Conference, vol. 1. Red Roof Design, Cape Town, pp. 214–237.

Fry, N., 1979. Random point distributions and strain measurements in rocks. Tectonophysics 60, 89–105.

Griffin, W.L., Doyle, B.J., Ryan, C.G., Pearson, N.J., O'Reilly, S.Y., Natapov, L., Kivi, K., Kretschmar, U., Ward, J., 1999. Lithosphere structure and mantle terranes: Slave craton, Canada. In: Gurney, J.J., Gurney, J.L., Pascoe, M.D., Richardson, S.H. (Eds.), Proceedings of the 7th International Kimberlite Conference, vol. 1. Red Roof Design, Cape Town, pp. 299–306.

Grütter, H.S., Apter, D.B., Kong, J., 1999. Crust–mantle coupling: evidence from mantle-derived xenocrystic garnets. In: Gurney, J.J., Gurney, J.L., Pascoe, M.D., Richardson, S.H. (Eds.), Proceedings of the 7th International Kimberlite Conference, vol. 1. Red Roof Design, Cape Town, pp. 307–313.

Hoffman, P.F., 1988. Geology and tectonics, East Arm of Great Slave Lake, Northwest Territories. Geological Survey of Canada, Map 1628A, scales 1:125,000 and 1:250,000.

Jones, G.E., 2002. Lac de Gras Exploration Area. Mineral Information Maps. Calgary, Canada.

Kjarsgaard, B.A., Levinson, A.A., 2002. Diamonds in Canada. Gems and Gemology 38, 208–238.

Kjarsgaard, B.A., Wilkinson, L., Armstrong, J.A., 2002. Geology, Lac de Gras kimberlite field, central Slave province, Northwest Territories-Nunavut. Geological Survey of Canada, Open File 3238, scale 1:250,000.

Kopylova, M.G., Caro, G., 2001. Lithospheric terranes of the Slave craton: contrasting North and South. Proceedings of the Slave-Kaapvaal Workshop, Merrickville, Canada, Sept. 2001.

LeCheminant, A.N., 1994. Proterozoic diabase dyke swarms, Lac de Gras and Aylmer Lake areas, District of Mackenzie. Northwest Territories. Geological Survey of Canada, Open File 2975.

Pokhilenko, N.P., McDonald, J.A., Melnyk, W., Hall, A.E., Shimizu, N., Vavilov, M.A., Afanasiev, V.P., Reimers, L.F., Irvin, L.N., Pokhilenko, L.N., Vasilenko, V.B., Kuligin, S.S., Sobolev, N.V., 1998. Kimberlites of Camsell Lake field and some features of construction, and composition of lithosphere roots of southeastern part of Slave Craton, Canada. Extended Abstracts, 7th International Kimberlite Conference, Cape Town, pp. 699–700.

Roscoe, S.M., Henderson, M.N., Hunt, P.A., van Breemen, O., 1987. U–Pb zircon age of an alkaline granite body in the Booth River Intrusive Suite, N.W.T.. Radiogenic Age and Isotopic Studies: Report 1. Paper - Geological Survey of Canada, vol. 87-2, pp. 95–100.

Stubley, M.P., 2003. Interpretive compilation of the bedrock geology of the Slave craton. Extended Abstracts, 8th International Kimberlite Conference, Victoria.

Stubley, M.P., Cairns, S.R., 1998. Geology of the Fishing Lake area, southern Slave Province (Parts of NTS 85 O/1 and 8 and 85 P/5). NWT Geology Division-DIAND, Yellowknife, EGS 1998-5.

Vearncombe, S., Vearncombe, J.R., 2002. Tectonic controls on kimberlite location, southern Africa. Journal of Structural Geology 24, 1619–1625.

White, S.H., de Boorder, H., Smith, C.B., 1995. Structural controls of kimberlite and lamproite emplacement. Journal of Geochemical Exploration 53, 245–264.

Wilkinson, L., Kjarsgaard, B.A., LeCheminant, A.N., Harris, J., 2001. Diabase dyke swarms in the Lac de Gras area, Northwest Territories, and their significance to kimberlite exploration: initial results. Current Research - Geological Survey of Canada 2001-C8. 17 p.

Wright, K.-J., 1999. Possible structural controls of kimberlites in the Lac de Gras region, central Slave Province, Northwest Territories, Canada. MSc thesis, Queen's University, Kingston, Canada. 103 pp.

Available online at www.sciencedirect.com

Lithos 77 (2004) 695–704

LITHOS

www.elsevier.com/locate/lithos

ELSEVIER

Kimberlite AT-56: a mantle sample from the north central Superior craton, Canada

K.A. Armstrong[a],*, T.E. Nowicki[b], G.H. Read[c]

[a] Navigator Exploration Corp., 1300-409 Granville St., Vancouver, BC, Canada V6C 1T2
[b] Mineral Services Canada Inc., 205-930 Harbourside Dr., North Vancouver, Canada V7P 3S7
[c] Canabrava Diamond Corporation, 1650-701 West Georgia St., Vancouver, Canada V7Y 1C6

Received 27 June 2003; accepted 19 December 2003
Available online 19 May 2004

Abstract

Kimberlite AT-56, discovered in February 2001, represents the most recent addition to the Attawapiskat kimberlite cluster, located in the James Bay Lowlands of Ontario, Canada. AT-56 is a small kimberlite body with a surface diameter of approximately 40 m and a steep southeastern plunge. It consists of a medium to coarse-grained matrix supported kimberlite with abundant olivine, clinopyroxene, garnet, ilmenite and mica macrocrysts in a green-black to orange-black matrix. The kimberlite is classified as a hypabyssal facies sparsely macrocrystic calcite kimberlite. Heavy mineral concentrates from two representative samples of AT-56 have been analyzed to characterize the mantle sampled by the kimberlite. Both samples yielded large heavy mineral concentrates comprised of roughly equal proportions of Mg-ilmenite, Cr-diopside, high-Cr garnet and low-Cr garnet. Mg-chromite is also present in quantities an order of magnitude less than the other constituents.

The high-Cr peridotitic garnet macrocrysts are only slightly more abundant than the low-Cr varieties, the population being dominated by G9 (lherzolitic) types with only a few (less than 10%) weakly sub-calcic G10 (probable harzburgitic) garnets present. Ni thermometry results for a representative selection of G9 and G10 garnets indicate that the majority equilibrated at temperatures ranging from 1000 to 1250 °C. A significant proportion of the low-Cr garnet population derived from AT-56 is characterized by relatively low-Ti (0.2 to 0.4 wt.% TiO_2) and elevated Na (0.07 to 0.13 wt.% Na_2O) contents characteristic of Group 1, diamond inclusion type eclogite garnets. These sodic garnets have elevated Cr_2O_3 contents (typically 1 to 2 wt.% Cr_2O_3), suggesting they may be websteritic in origin rather than eclogitic. Comparison of AT-56 garnet compositions with published data available for other Attawapiskat kimberlites suggests websteritic mantle has also been sampled by kimberlite bodies elsewhere in the Attawapiskat cluster and it may be an important diamond reservoir in this area.
© 2004 Elsevier B.V. All rights reserved.

Keywords: Attawapiskat; Websteritic mantle; Eclogite; Garnet; Ni thermometry

---

* Corresponding author. Tel.: +1-604-668-8355; fax: +1-604-668-8366.
E-mail address: nvr_karmstrong@telus.net (K.A. Armstrong).

0024-4937/$ - see front matter © 2004 Elsevier B.V. All rights reserved.
doi:10.1016/j.lithos.2004.03.011

1. Introduction and setting

The Attawapiskat kimberlite cluster is located approximately 110 km west of the community of Attawapiskat in the James Bay Lowlands of Ontario, Canada (Fig. 1). The kimberlite bodies intrude the flat-lying shallow marine rocks of the Palaeozoic Moose River basin. These rocks are predominantly limestones and dolomites and reach thicknesses of approximately 300 m in the Attawapiskat area, unconformably overlying the eastern extension of the Sachigo Subprovince of the Archean Superior craton. U–Pb perovskite ages for three of the kimberlites in the area range from approxi-

mately 175 to 180 Ma (Heaman and Kjarsgaard, 2000).

AT-56 is the most recent kimberlite discovered in the area, bringing the total number of kimberlites in the cluster to 19 (Fig. 1). Evaluation of AT-56 has been conducted under a joint venture agreement between Navigator Exploration and Canabrava Diamond. It is located within 5 km of three other kimberlite bodies, including the Victor kimberlite, which is currently the subject of a mine development feasibility study. This paper presents the results of an investigation of the mantle sampled by AT-56, providing some insight into the lithosphere beneath this portion of the Superior craton. Comparisons are also

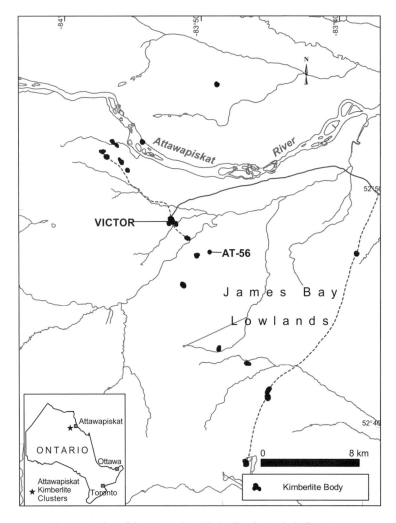

Fig. 1. Location of the Attawapiskat kimberlite cluster, including AT-56.

made with the mantle sampled by other kimberlites in the cluster (Kong et al., 1999; Sage, 2000; Scully, 2000; Fowler et al., 2001).

2. Kimberlite AT-56

AT-56 is a small kimberlite with a surface diameter of approximately 40 m and a steep south-eastern plunge of about 70°. The body lies beneath approximately 3 m of overburden, consisting of mixed black organic material (peat) and glacial till. Overburden thickness may increase towards the edges of the kimberlite where it is in contact with adjacent limestone country rocks. AT-56 is a medium to coarse grained, matrix to locally near-clast supported kimberlite with abundant olivine, garnet, ilmenite, clinopyroxene and mica macrocrysts in a green-black to orange-black matrix. Average macrocryst size varies from < 0.5 to 1.5 mm. Mantle derived xenoliths are very rare, and where identified tend to be in the form of mineral pairs (i.e. olivine plus garnet, olivine plus clinopyroxene). Uncommon country rock xenoliths are predominantly tan to beige coloured limestone fragments with lesser basement (gneissic) fragments. The dark matrix is dominated by carbonate; however, it becomes lighter in colour in areas of increased serpentinisation, most commonly near the kimberlite margins. Spherical globular segregations, typical of hypabyssal kimberlite, are locally identifiable, particularly where increased serpentinisation has resulted in a lighter coloured matrix. Thin, late stage calcite veinlets are common, averaging less than 2 mm but ranging locally up to 10 mm in thickness. The veinlets increase in concentration towards the kimberlite/country rock contact and cut country rock xenoliths. Measured orientations of these veinlets in drill core indicate that they parallel the plunge of the kimberlite. They are interpreted to result from late-stage, post-emplacement degassing of the cooling kimberlite.

In thin section, AT-56 is classified as a hypabyssal facies, sparsely macrocrystic, calcite kimberlite (terminology after Field and Scott-Smith, 1998). Macrocrysts and abundant olivine phenocrysts are set in a fine-grained calcite dominated matrix. Olivine macrocrysts average 1.5 mm in size, are subrounded and

tend to be partially altered to serpentine and secondary opaque mineral phases. Phlogopite laths (up to 1 mm) are present and display pervasive chloritization. Subhedral to anhedral altered olivine phenocrysts range in size from 0.2 to 0.7 mm and are evenly distributed throughout the groundmass. The matrix consists of elongate calcite laths, possible relict melilite (mostly pervasively altered to carbonate) and relatively abundant, small (< 0.3 mm), subhedral opaques set in a base made up predominantly of fine-grained carbonate.

3. Mantle xenocryst compositions

Two representative samples of AT-56 have been analyzed to characterize the mantle sampled by the kimberlite. An initial 600 g sample (A1020) was crushed and delivered to I&M Morrison Geological of Delta, British Columbia for visual extraction of indicator minerals. Selected mineral grains from the sample were then sent to RL Barnett Geological of Lambeth, Ontario for determination of mineral compositions using a JEOL Superprobe 733 equipped with five wavelength spectrometers and a Tracer Northern energy dispersive detector and operating system. All analyses were performed at 15 kV accelerating voltage and 10 nA sample current. Counting times were 20 s for all routine elements. Sodium in garnet was determined with a counting time of 50 s with the background function enabled.

A second sample (2701) weighing approximately 4.8 kg was submitted to Mineral Services Canada for a more representative investigation of the indicator mineral suite of AT-56. Electron microprobe analyses were carried out at the South African Council for Geoscience using a JEOL 733 electron microprobe. An accelerating voltage of 15 kV and beam current of 40 nA were used, with counting times of 10 s for all elements other than Na and K which were analyzed for 60 s.

Both samples of AT-56 yielded large heavy mineral concentrates, with the concentrate of sample 2701 representing 26.9% of the processed sample weight (0.3 to 2.36 mm size fraction; S.G. ≥ 2.9). Ilmenite, clinopyroxene, high-Cr garnet and low-Cr garnet were all present in similar high abundance. Chromite was

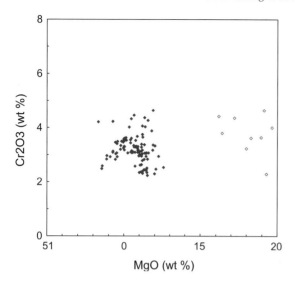

Fig. 2. Composition of ilmenite xenocrysts recovered from AT-56.

also present, although its abundance was an order of magnitude lower than that of the other minerals.

3.1. Ilmenite

Ilmenite represents approximately 25% of the indicator mineral suite (i.e. garnet, Cr-diopside, chromite and ilmenite) and displays a typical kimberlitic

trend with MgO contents ranging between 8 and 12 wt.% and elevated Cr_2O_3 contents of between 2 and 4 wt.%. A number of ilmenite grains are zoned, with outer rims elevated in MgO relative to their cores (Fig. 2).

3.2. Clinopyroxene

Clinopyroxene represents just under 20% of the indicator mineral content of the sample. Compositionally the clinopyroxenes are Cr-diopsides, typically having CaO contents ranging from 14 to 23 wt.% and Cr_2O_3 contents of 0.5 to 3 wt.%. However, a distinct subpopulation is characterized by elevated chromium (2.5 to 4.0 wt.% Cr_2O_3) at calcium contents of 15 to 18 wt.% (Fig. 3) and sodium contents of 3.2 to 5.0 wt.% Na_2O (not illustrated).

3.3. Chromite

The few chromite macrocrysts analyzed display Cr_2O_3 contents of 31 to 57 wt.%, and TiO_2 contents of either <0.7 wt.% (four grains) or between 2 and 6 wt.% (six grains). The four low TiO_2 grains are interpreted as xenocrysts whereas the high TiO_2 chromites are considered to be probable phenocrysts,

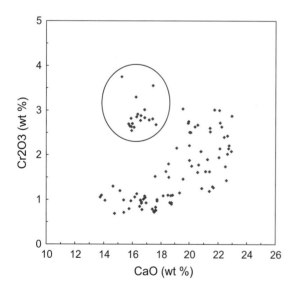

Fig. 3. Composition of Cr-diopside xenocrysts recovered from AT-56. The oval field encloses a high-Cr subpopulation that also shows high Na_2O contents (see text).

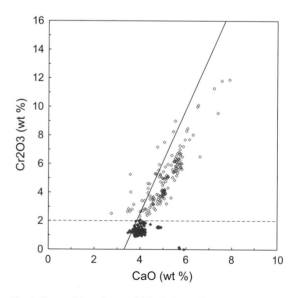

Fig. 4. Composition of garnet high-Cr (open diamonds) and low-Cr (filled diamonds) garnet xenocrysts recovered from AT-56.

Table 1
Representative compositions of AT-56 garnet xenocrysts

Sample	Grain	Type[a]	SiO$_2$	TiO$_2$	Al$_2$O$_3$	Cr$_2$O$_3$	FeO	MgO	MnO	CaO	Na$_2$O	Total	Mg#[b]
2701	12	G9	40.84	0.12	17.32	8.03	7.83	18.43	0.33	6.30	nd	99.20	80.76
2701	16	G10	41.32	0.33	18.51	5.94	8.38	19.92	0.34	4.83	nd	99.57	80.91
2701	46	G10	42.02	0.35	19.71	4.60	7.70	20.32	0.28	4.41	nd	99.39	82.47
2701	15	G9	41.17	0.18	20.69	3.38	7.60	20.30	0.33	4.90	nd	98.55	82.65
A1020	71	G9	40.99	0.33	17.28	8.29	7.07	19.46	0.38	6.01	nd	99.81	83.07
2701	67	MEG	41.00	0.57	20.80	1.54	10.51	19.34	0.36	4.91	0.070	99.10	76.64
2701	79	WEB	41.46	0.30	21.33	1.19	9.43	20.82	0.41	3.66	0.130	98.73	79.74
2701	88	WEB	41.46	0.38	21.74	0.96	9.67	20.49	0.45	3.76	0.110	99.02	79.07
A1020	15	WEB	41.9	0.31	22.85	1.36	9.69	19.78	0.42	4.19	0.08	100.58	78.45
A1020	40	WEB	42.37	0.3	22.86	1.06	9.64	19.45	0.38	3.75	0.09	99.9	78.25

[a] MEG = megacryst; WEB = websteritic; G9 and G10 after classification of Gurney (1984).

[b] Mg# = 100(Mg/(Mg + Fe)).

showing decreasing MgO contents with decreasing Cr$_2$O$_3$.

3.4. Garnet

The garnet population of AT-56, comprising approximately 50% of the indicator mineral suite, consists of high-Cr and low-Cr varieties, with the former being only slightly more abundant than the latter. The high-Cr peridotitic garnet population is

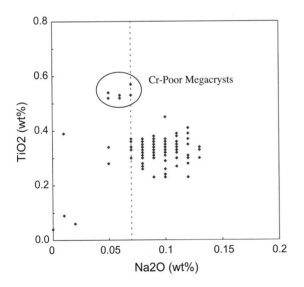

Fig. 5. Composition of low-Cr (<2 wt.% Cr$_2$O$_3$) garnet xenocrysts recovered from AT-56. Circled cluster of grains with >0.5 wt.% TiO$_2$ are considered megacrysts. Xenocrysts with elevated Na$_2$O (>0.07 wt.%; vertical dashed line) are interpreted herein as being derived from websteritic mantle.

made up predominantly of G9 (lherzolitic) types with only a few (approximately 13%) weakly sub-calcic G10 (probable harzburgitic) garnets present (classification of Gurney, 1984). Chrome contents range up to 12 wt.% Cr$_2$O$_3$, with most grains ranging from 4 to 8 wt.% Cr$_2$O$_3$ (Fig. 4). The lherzolitic trend defined by AT-56 G9 garnets is displaced to slightly higher CaO values with respect to the '85% line' used to define G9 and G10 garnet populations (Gurney, 1984).

The low-Cr garnet population, characterized by Cr$_2$O$_3$ contents of between 1 and 2 wt.% and MgO contents of 19–20 wt.% (Table 1), is an important component of the indicator mineral population of AT-56. A small group of these grains displays elevated TiO$_2$ contents (>0.4 wt.%) and likely represents Cr-poor megacrysts (Fig. 5). However, a significant proportion of the low-Cr garnet population is characterized by relatively low TiO$_2$ (0.2 to 0.4 wt.%) and elevated Na$_2$O (0.07 to 0.13 wt.%) contents typical of Group 1, diamond inclusion type eclogite garnets.

4. Ni thermometry

A representative population of 24 high-Cr (>2 wt.% Cr$_2$O$_3$) garnets (20 G9 grains and 4 G10 grains) was submitted for trace element analysis by laser ablation, inductively coupled plasma mass spectrometry (LA-ICP-MS), at the University of Cape Town, South Africa. The analyses were undertaken using a Perkin Elmer/Sciex Elan 6000

Table 2
Trace element compositions of peridotitic garnet xenocrysts from AT-56

Grain	Paragenesis	Sc	Ti	Mn	Ni	Ga	Rb	Sr	Y	Zr	La	Ce	Pr	Nd	Sm	Eu	Gd	Dy	Er	Yb	Lu	Hf	Canil94	Ryan96
48	G9	224	132	4209	21	2	0	0	22	55	0	3	0	1	2	1	4	4	2	1	0	0	923.04	797.56
38	G9	132	1610	3538	44	12	1	1	14	27	1	1	0	1	1	1	1	2	2	3	0	1	1040.37	992.89
3	G9	123	802	4086	49	9	1	1	14	37	0	1	0	2	1	1	1	2	2	3	1	1	1056.34	1021.83
42	G9	123	2485	3147	50	16	0	0	15	32	0	0	0	1	1	0	1	2	2	2	0	1	1062.06	1032.33
60	G9	115	1278	3036	51	9	0	0	13	18	0	0	0	0	0	0	1	2	2	3	0	1	1063.52	1035.03
19	G9	130	490	3544	51	8	1	0	5	18	0	0	0	2	1	0	1	0	1	2	0	0	1065.58	1038.85
30	G9	116	970	3142	52	8	1	0	14	47	0	0	0	1	2	1	2	3	1	2	0	1	1068.77	1044.77
46	G10	102	1180	2610	54	8	0	0	11	24	0	0	0	1	0	0	1	2	1	2	0	0	1075.6	1057.55
40	G9	110	1156	3148	57	8	0	0	12	20	0	0	0	0	0	0	1	2	1	2	0	0	1085.09	1075.47
12	G9	163	644	3210	59	3	1	1	1	14	0	0	0	2	1	0	0	0	0	1	0	0	1090.94	1086.65
29	G10	124	1765	3427	59	9	1	0	16	28	0	0	0	1	1	1	2	3	2	3	0	1	1091.66	1088.04
23	G9	131	1176	2909	66	6	1	0	6	21	0	0	0	1	1	0	1	1	1	1	0	0	1112.52	1128.65
59	G9	119	1102	2598	67	9	0	0	13	19	0	0	0	1	0	0	0	2	2	3	0	1	1113.55	1130.67
16	G10	151	1300	3126	74	6	0	0	7	26	1	0	0	1	1	0	1	1	1	1	0	0	1133.47	1170.63
1	G9	145	1587	4143	74	9	0	1	3	24	0	1	0	3	2	1	1	1	0	1	0	1	1133.62	1170.93
28	G10	151	1883	3236	75	9	1	1	12	34	1	0	0	1	1	1	2	2	2	1	0	1	1135.7	1175.16
2	G9	142	1108	2601	76	8	0	1	9	31	0	0	0	1	1	0	1	2	1	2	0	1	1137.99	1179.84
35	G9	122	343	3239	81	9	1	0	6	3	1	0	0	1	0	0	0	1	0	2	0	0	1151.23	1207.18
15	G9	107	721	3211	83	7	1	0	6	13	1	0	0	1	2	0	1	1	1	2	1	1	1156.13	1217.43
45	G9	86	1701	2217	87	12	0	0	18	31	0	0	0	1	0	0	2	3	2	3	0	1	1166.84	1240.08
11	G9	157	932	2997	88	10	1	0	9	17	0	0	0	1	1	0	1	1	1	2	0	0	1168.84	1244.35
22	G9	150	1291	2524	90	9	1	0	11	31	0	0	0	2	2	1	2	2	1	2	0	1	1172.16	1251.46
31	G9	150	1826	2842	102	9	0	1	13	41	0	0	0	2	1	1	2	2	1	2	0	1	1197.86	1307.67
33	G9	161	2171	3059	102	12	2	1	15	51	0	1	0	2	2	1	3	3	2	2	0	1	1198.94	1310.1

Canil94—temperature in °C, determined using the Ni thermometer of Canil (1994).
Ryan96—temperature in °C, determined using the Ni thermometer of Ryan et al. (1996).

ICP-MS with sample ablation using a Cetac LSX-200 laser ablation module. Temperatures of equilibrium were calculated for all 24 garnet grains, using the Ni thermometry calibrations of Ryan et al. (1996) and Canil (1994) (Table 2). The resultant Ni temperatures indicate equilibration at temperatures ranging from 1000 to 1250 °C (Fig. 6). On an assumed geotherm of 40 mW/m^2, the majority of peridotitic garnets from AT-56 would be derived from within the diamond stability field (>1000 °C).

A 40 mW/m^2 geotherm for the Attawapiskat area is consistent with thermobarometry results obtained for eight garnet- and orthopyroxene-bearing xenoliths from four nearby Attawapiskat kimberlites (Scully, 2000). Kong et al. (1999) reported that garnet Ni thermometry results for xenocrysts recovered from other Attawapiskat kimberlites fell in the range of 750 to 1200 °C and defined a poorly constrained, somewhat cooler geotherm of 37 mW/m^2.

5. Comparison to other Attawapiskat kimberlites

AT-56 appears to share many characteristics with other kimberlites of the Attawapiskat cluster (Kong et al., 1999). Broad similarities include its macrocrystic nature, the presence of segregationary textures, abundant groundmass carbonate and a general lack of mantle derived xenoliths (Scully, 2000). Kong et al. (1999) considered the Attawapiskat kimberlites unusual in terms of their high abundances of groundmass carbonate. In addition to groundmass carbonate, AT-56 contains abundant calcite veinlets that are interpreted as being related to late, post emplacement degassing of the cooling kimberlite.

Mineral compositions for recovered ilmenite, clinopyroxene, chromite, and garnet grains from AT-56 are similar to those reported from xenoliths and mineral concentrates from the other kimberlites of the Attawapiskat cluster (Sage, 2000; Scully, 2000; Kong et al., 1999; Hetman, 1996). For example, heavy mineral concentrates of the Victor kimberlite

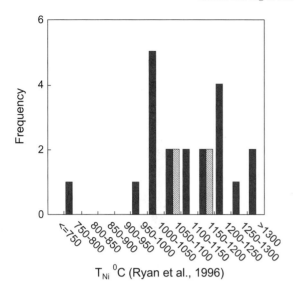

Fig. 6. T_{Ni} histogram showing the estimated temperature distribution for a representative sample of peridotitic garnets from AT-56. Solid and striped bars represent temperature estimates for G9 and G10 garnets, respectively.

include a population of high-Cr garnets that form a lherzolitic (G9) trend with Cr_2O_3 contents ranging from 2 to 9 wt.% (Fig. 7). Significantly sub-calcic G10 garnets are apparently absent in the Victor garnet population and appear to be rare in all of the Attawapiskat kimberlites (Sage, 2000).

6. Nature and origin of low-Cr garnets in Attawapiskat kimberlites

As indicated above, AT-56 hosts a significant population of low-Cr garnet characterized by Ti and Na contents that are similar to garnets found in Group 1 type, diamond eclogite xenoliths and as eclogitic inclusions in diamonds (McCandless and Gurney, 1989). However, the Cr_2O_3 and MgO contents of these grains are elevated relative to those typical of an eclogitic paragenesis. Compositions of the low-Cr garnets are therefore suggestive of derivation from a more mafic source, likely garnet websterite.

Comparison of AT-56 garnet compositions with data available for other Attawapiskat kimberlites suggests that similar Cr-poor, sodic garnet is an important component elsewhere in the cluster. In particular, low-Cr garnets within the Victor concentrates are compositionally very similar to those from AT-56, with elevated Cr_2O_3 (1–2 wt.%) compared to typical eclogite and clustering around 0.38 wt.% TiO_2 at greater than 0.07 wt.% Na_2O (Fig. 8). This suggests that the Victor kimberlite also sampled a significant component of possible websteritic mantle.

Use of 2 wt.% Cr_2O_3 as a cut off for distinguishing high-Cr peridotitic garnets from low-Cr (eclogitic,

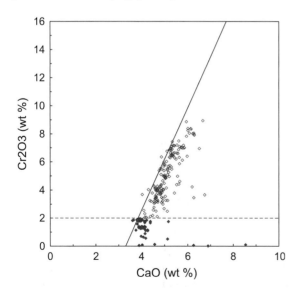

Fig. 7. Composition of garnet xenocrysts recovered from the Victor kimberlite. Data from Sage (2000). Symbols as for Fig. 4.

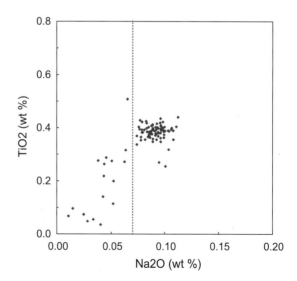

Fig. 8. Composition of low-Cr (< 2 wt.% Cr_2O_3) garnet xenocrysts recovered from Victor. Xenocrysts with elevated Na contents (>0.07 wt.% Na_2O; vertical dashed line) are interpreted as being derived from websteritic mantle. Data from Sage (2000).

websteritic, megacrysts) garnets is somewhat arbitrary and open to interpretation (e.g. Aulbach et al., 2002). Based on similar CaO, MgO and FeO contents, the low-Cr garnet populations of AT-56 and Victor may

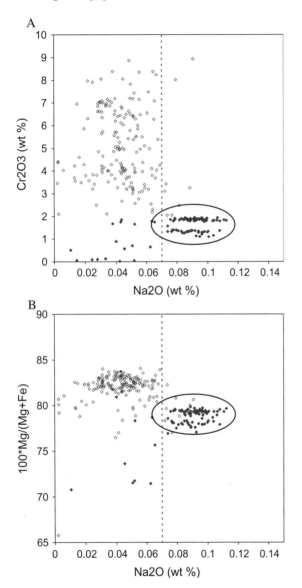

Fig. 9. Composition of garnet xenocrysts recovered from Victor. Symbols as for Fig. 4. Two websteritic garnet populations (circled, defined by horizontal dashed line) with elevated Na_2O (>0.07 wt.%; vertical dashed line) at low Cr_2O_3 (1–2 wt.%; A) and 100(Mg/(Mg+Fe)) (~ 77–80; B) are clearly distinct from peridotitic garnets with low Na (<0.07 wt.% Na_2O), elevated 100(Mg/(Mg+Fe)) (~ 81 to 84) and variable Cr_2O_3. Data from Sage (2000).

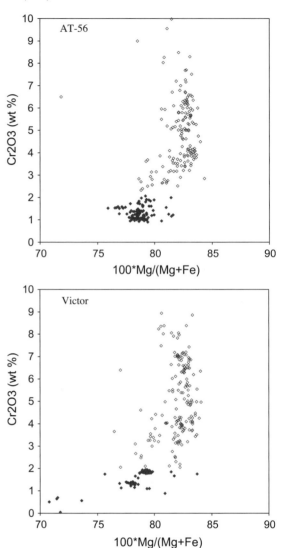

Fig. 10. Composition of garnet xenocrysts from Victor and AT-56. Symbols as for Fig. 4. Low-Cr websteritic garnets from both kimberlites are circled. Websteritic garnets from Victor are distinguishable as tight clusters of points at 1.9 and 1.3 wt.% Cr_2O_3. Low-Cr websteritic garnets from AT-56 show a more broad clustering of data; however, at least two subpopulations can be distinguished around 1.1 and 1.5 wt.% Cr_2O_3 at 100(Mg/(Mg + Fe)) values of 77–79 and 76–77, respectively. Victor data from Sage (2000).

be superficially interpreted as a low-Cr extension of the peridotitic suite (e.g. Fig. 7). Classification of these low-Cr garnets as being of websteritic paragenesis is largely based on their elevated Na contents. Analytical data from the present study do not include the Na_2O contents of the high-Cr garnet population from AT-56. However, as part of a detailed investigation of the mineralogy of the Attawapiskat cluster kimberlites, Sage (2000) reported the sodium contents for 271 high- and low-Cr garnets recovered from Victor. These data clearly show that the Victor low-Cr garnet population is distinct from the high-Cr, peridotitic garnets based on their elevated Na_2O contents and slightly lower Mg# values (Mg# = 100(Mg/(Mg + Fe)) (Fig. 9). While there is some overlap in Mg# and Cr_2O_3 values, the low-Cr, websteritic garnets can be clearly distinguished by their elevated Na contents. In fact, based on Mg# and Cr_2O_3 content, the Victor websteritic garnets appear to form two distinct subpopulations with variable but similar Na_2O (Fig. 9). The compositions of garnets from AT-56 and Victor are broadly comparable on a plot of Cr_2O_3 versus Mg# (Fig. 10). The two subpopulations of websteritic garnets from Victor are distinguishable as tight clusters of points at 1.9 and 1.3 wt.% Cr_2O_3, respectively. AT-56 low-Cr websteritic garnets also appear to have at least two subpopulations, clustered around 1.1 and 1.5 wt.% Cr_2O_3 at Mg#'s of 77–79 and 76–77, respectively.

7. Upper mantle composition of the north central Superior craton

Analysis of the abundance and composition of mantle-derived xenocrysts from AT-56 indicates that the kimberlite preserved an extensive sample of the lithosphere underlying the Superior craton at the time of eruption. The high abundance of Cr-diopside and G9 garnet in AT-56 indicates a strong lherzolitic component to the mantle. The absence of significantly sub-calcic G10 garnets indicates that the peridotitic mantle material that was sampled did not include significant quantities of depleted harzburgite.

A significant population of low-Cr garnet, characterized by high sodium and magnesium contents and 1–2 wt.% Cr_2O_3, is present within AT-56 and some of the other Attawapiskat cluster kimberlites. This population is interpreted to indicate the presence of a substantial websteritic component to the mantle beneath the north central Superior craton. Evidence for distinct subpopulations of websteritic garnet within the AT-56 and Victor kimberlites further suggests that this

websteritic component may be derived from more than one compositionally distinct source region. Although there is no direct method to determine the depth of origin of these low-Cr garnets, it may be reasonable to assume they were derived from within the depth range indicated by the peridotitic garnet, and therefore potentially from within the diamond stability field.

Sodic websteritic garnets identified as diamond inclusions and in diamond-bearing websterite xenoliths have been reported from several African kimberlites including Venetia (Aulbach et al., 2002; Viljoen et al., 1999), Orapa (Deines et al., 1993), Monastery (Moore and Gurney, 1989), and Letseng-la-Terai (McDade and Harris, 1999). Compositions of websteritic garnet inclusions in diamond from Venetia are somewhat more variable than, but overlap with, the low-Cr garnets from AT-56 and Victor (Fig. 11). Sodium contents of these low-Cr websteritic garnets are generally ≥ 0.07 wt.% Na_2O, an empirically determined threshold that has been used to determine a diamond association in garnets from low-Mg to moderate-Mg bulk composition eclogites (Gurney, 1984; Gurney and Zweistra, 1995). Recent studies have highlighted the fact that, whereas the Na content of low-Cr garnets generally increases with depth, this relationship is strongly dependent on bulk composi-

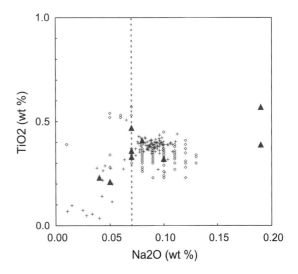

Fig. 11. Comparison of low-Cr websteritic garnet xenocrysts from AT-56 (open diamonds) and Victor (crosses) with websteritic garnet inclusions in diamond from Venetia (filled triangles). Vertical dashed line represents 0.07 wt.% Na_2O. Diamond inclusion compositions from Aulbach et al. (2002).

tion (Grütter and Quadling, 1999). Of particular relevance is that there is a general decrease in Na content with increased MgO concentration and, for high-Mg bulk compositions (reflected in Mg-rich garnet compositions), a sodium threshold *lower* than 0.07 wt.% Na_2O may be more appropriate for determining a potential diamond association (Cookenboo et al., 1998; Grütter and Quadling, 1999). It is possible, therefore, that the low-Cr, high-Mg websteritic garnet population recovered from AT-56 (and Victor) is derived from diamond-bearing mantle. Given the paucity of harzburgitic garnet in AT-56 and other kimberlites of the Attawapiskat cluster (indicating low peridotitic diamond potential), garnet websterite may be an important, if not dominant, source of diamonds in these bodies.

Acknowledgements

This work was supported by Navigator Exploration and Canabrava Diamond. This manuscript has benefited from the careful reviews of M. Muggeridge, H. Grütter and an anonymous reviewer. Rory Moore, Eira Thomas, Robin Hopkins and Bill Mosher are thanked for useful discussions relating to AT-56 and its mantle sample.

References

Aulbach, S., Stachel, T., Viljoen, K.S., Brey, G.P., Harris, J.W., 2002. Eclogitic and websteritic diamond sources beneath the Limpopo Belt—is slab-melting the link? Contrib. Mineral Petrol. 143, 56–70.

Canil, D., 1994. An experimental calibration of the "Ni-in-garnet" geothermometer with applications. Contrib. Mineral Petrol. 117, 410–420.

Cookenboo, H.O., Kopylova, M.G., Daoud, D.K., 1998. A chemically and texturally distinct layer of diamondiferous eclogite beneath the central Slave craton, northern Canada. Extended Abstracts of the 7th International Kimberlite Conference, Cape Town, 1998, pp. 164–166.

Deines, P., Harris, J.W., Gurney, J.J., 1993. Depth-related carbon isotope and nitrogen concentration variability in the mantle below the Orapa kimberlite, Botswana, Africa. Geochim. Cosmochim. Acta 57, 2781–2796.

Field, M., Scott-Smith, B.H., 1998. Textural and genetic classification schemes for kimberlites: a new perspective. Extended Abstracts of the 7th International Kimberlite Conference, Cape Town, 1998, pp. 214–216.

Fowler, J.A., Grütter, H.S., Kong, J.M., Wood, B.D., 2001. Diamond exploration in northern Ontario with reference to the Victor kimberlite, near Attawapiskat. Explor. Min. Geol. 10, 67–75.

Grütter, H.S., Quadling, K.E., 1999. Can sodium in garnet be used to monitor eclogitic diamond potential? In: Gurney, J.J., Gurney, J.L., Pascoe, M.D., Richardson, S.H.J.B. Dawson volume, Proc. 7th Int. Kimb. Conf., Red Roof Design, Cape Town, pp. 314–320.

Gurney, J.J., 1984. A correlation between garnets and diamonds in kimberlites. In: Glover, J.E., Harris, P.G. Kimberlite Occurrence and Origin: A Basis for Conceptual Models in Exploration, vol. 8. University of Western Australia Publ., pp. 143–166.

Gurney, J.J., Zweistra, P., 1995. The interpretation of major element compositions of mantle minerals in diamond exploration. J. Geochem. Explor. 53, 293–309.

Heaman, L.M., Kjarsgaard, B.A., 2000. Timing of eastern North American kimberlite magmatism: continental extension of the Great Meteor hotspot track? Earth Planet. Sci. Lett. 178, 253–268.

Hetman, C.M., 1996. The ilmenite association of the Attawapiskat kimberlite cluster, Ontario, Canada. MSc thesis. University of Toronto, Canada. 126 pp.

Kong, J.M., Boucher, D.R., Scott-Smith, B.H., 1999. Exploration and geology of the Attawapiskat kimberlites, James Bay lowlands, northern Ontario, Canada. In: Gurney, J.J., Gurney, M.D., Pascoe, M.D., Richardson, S.H.J.B. Dawson volume, Proc. 7th Int. Kimb. Conf., Red Roof Design, Cape Town, pp. 452–468.

McCandless, T.E., Gurney, J.J., 1989. Sodium in garnet and potassium in clinopyroxene: criteria for classifying mantle eclogites. In: Ross, J. Kimberlites and Related Rocks, Volume 2, Their Mantle/crust Setting, Diamonds, and Diamond Exploration. Spec. Publ. - Geol. Soc. Aust., vol. 14, pp. 827–832.

McDade, P., Harris, J.W., 1999. Syngenetic inclusion bearing diamonds from Letseng-la-Terai, Lesotho. In: Gurney, J.J., Gurney, J.L., Pascoe, M.D., Richardson, S.H.P.H. Nixon volume, Proc. 7th Int. Kimb. Conf., Red Roof Design, Cape Town, pp. 557–565.

Moore, R.O., Gurney, J.J., 1989. Mineral inclusions in diamond from the Monastery kimberlite, South Africa. In: Ross, J. Kimberlites and Related Rocks, Volume 2, Their Mantle/crust Setting, Diamonds, and Diamond Exploration. Spec. Publ. - Geol. Soc. Aust., vol. 14, pp. 1029–1041.

Ryan, C.G., Griffin, W.L., Pearson, N.J., 1996. Garnet geotherms: pressure–temperature data from Cr-pyrope garnet xenocrysts in volcanic rocks. J. Geophys. Res. 101, 5611–5625.

Sage, R.P., 2000. Kimberlites of the Attawapiskat area, James Bay Lowlands, northern Ontario. Ontario Geological Survey, Open File Report 6019. 341 pp.

Scully, K.R., 2000. Mantle xenoliths from the Attawapiskat kimberlite field, James Bay Lowlands, Ontario. MSc Thesis. University of Toronto, Canada. 64 pp.

Viljoen, K.S., Phillips, D., Harris, J.W., Robinson, D.N., 1999. Mineral inclusions in diamonds from the Venetia kimberlites, northern Province, South Africa. In: Gurney, J.J., Gurney, J.L., Pascoe, S.H., Richardson, S.H.P.H. Nixon volume, Proc. 7th Int. Kimb. Conf., Red Roof Design, Cape Town, pp. 888–895.

Available online at www.sciencedirect.com

Lithos 77 (2004) 705–731

www.elsevier.com/locate/lithos

Indicator mineralogy of kimberlite boulders from eskers in the Kirkland Lake and Lake Timiskaming areas, Ontario, Canada

Ingrid M. Kjarsgaard[a],[*], M. Beth McClenaghan[b], Bruce A. Kjarsgaard[b], Larry M. Heaman[c]

[a] 15 Scotia Place, Ottawa, ON, Canada K1S 0W2
[b] Geological Survey of Canada, 601 Booth Street, Ottawa, ON, Canada K1A 0E8
[c] Department of Earth and Atmospheric Sciences, University of Alberta, Edmonton, AB, Canada T6G 2E3

Received 27 June 2003; accepted 23 January 2004
Available online 1 June 2004

Abstract

Sixteen kimberlite boulders were collected from three sites on the Munro and Misema River Eskers in the Kirkland Lake kimberlite field and one site on the Sharp Lake esker in the Lake Timiskaming kimberlite field. The boulders were processed for heavy-mineral concentrates from which grains of Mg-ilmenite, chromite, garnet, clinopyroxene and olivine were picked, counted and analyzed by electron microprobe. Based on relative abundances and composition of these mineral phases, the boulders could be assigned to six mineralogically different groups, five for the Kirkland Lake area and one for the Lake Timiskaming area. Their indicator mineral composition and abundances are compared to existing data for known kimberlites in both the Kirkland Lake and Lake Timiskaming areas. Six boulders from the Munro Esker form a compositionally homogeneous group (I) in which the Mg-ilmenite population is very similar to that of the A1 kimberlite, located 7–12 km N (up-ice), directly adjacent to the Munro esker in the Kirkland Lake kimberlite field. U–Pb perovskite ages of three of the group I boulders overlap with that of the A1 kimberlite. Three other boulders recovered from the same localities in the Munro Esker also show some broad similarities in Mg-ilmenite composition and age to the A1 kimberlite. However, they are sufficiently different in mineral abundances and composition from each other and from the A1 kimberlite to assign them to different groups (II–IV). Their sources could be different phases of the same kimberlite or—more likely—three different, hitherto unknown kimberlites up-ice of the sample localities along the Munro Esker in the Kirkland Lake kimberlite field. A single boulder from the Misema River esker, Kirkland Lake, has mineral compositions that do not match any of the known kimberlites from the Kirkland Lake field. This suggests another unknown kimberlite exists in the area up-ice of the Larder Lake pit along the Misema River esker. Six boulders from the Sharp Lake esker, within the Lake Timiskaming field, form a homogeneous group with distinct mineral compositions unmatched by any of the known kimberlites in the Lake Timiskaming field. U–Pb perovskite age determinations on two of these boulders support this notion. These boulders are likely derived from an unknown kimberlite source up-ice from the Seed kimberlite, 4 km NW of the Sharp Lake pit, since indicator minerals with identical compositions to those of the Sharp Lake boulders have been found in till samples collected down-ice from Seed. Based on abundance and composition of indicator minerals, most importantly Mg-ilmenite, and supported by U–Pb age

---

* Corresponding author.
E-mail address: ikjarsgaard@sympatico.ca (I.M. Kjarsgaard).

dating of perovskite, we conclude that the sources of 10 of the 16 boulders must be several hitherto unknown kimberlite bodies in the Kirkland Lake and Lake Timiskaming kimberlite fields.

Keywords: Kimberlite indicator minerals; Ilmenite; U–Pb perovskite age; Glaciofluvial; Esker

1. Introduction

Northeastern Ontario is covered by up to 75 m of glacial sediments, deposited by the Laurentide Ice Sheet which retreated from the region approximately 8000 years ago. Over the past 30 years, numerous kimberlite boulders and cobbles have been found in gravel pits in glaciofluvial sediments (eskers) that cross the two major kimberlite fields (Fig. 1) in the region (e.g., Baker, 1982; Brummer et al., 1992a,b; McClenaghan, 1993). Kimberlite boulders are easily identified by their distinct light green or orange colour. They are commonly found in fresh faces of active gravel pits in matrix supported, poorly sorted pebble to small boulder gravel.

The study of kimberlite boulders from eskers in glaciated regions provides an additional exploration tool for identifying the presence of kimberlites. This method has potential application in areas where certain kimberlites are difficult to locate by geophysical techniques or to ascertain if unknown kimberlites occur in the region. A total of 16 kimberlite boulders and cobbles collected from eskers in the Kirkland Lake and Lake Timiskaming kimberlite fields have been examined by the Geological Survey of Canada (GSC) to determine their kimberlite indicator mineralogy and to attempt to identify their sources. Herein, kimberlite boulders and cobbles are referred to collectively as kimberlite boulders. Boulders with indicator mineral compositions different from known kimberlites in the region indicate the presence of undiscovered kimberlite(s) and are therefore of possible interest in diamond exploration.

1.1. Glacial and glaciofluvial transport

Boulders were transported as competent fragments of kimberlite, deposited in glaciofluvial sediments and subsequently subjected to physical and chemical weathering since deglaciation (8000–

10,000 years ago). Because the boulders weathered rapidly, they readily disintegrate, and it is rare to find them in gravel pits in which the pit faces are not fresh.

Glacial striation orientations and esker transport directions must be considered when attempting to determine the source of kimberlite boulders. In the vicinity of the Kirkland Lake kimberlite field, striations indicate ice movement towards the west to southwest during the main phase of glaciation, then south, and finally southeast during deglaciation (McClenaghan et al., 1995; Veillette and McClenaghan, 1996). Glaciofluvial sediments were deposited in the south-trending Munro and Misema River Eskers (Fig. 2; Baker and Storrison, 1979; Baker et al., 1982).

Striated bedrock displays evidence of three major ice flow phases across the Lake Timiskaming kimberlite field (Veillette, 1996; McClenaghan and Veillette, 2001). The oldest flow, associated with the main phase of the Laurentide ice sheet, was towards the southwest. Glaciofluvial sediments were deposited in the south and southeast-trending Montreal River outwash system (Veillette, 1986, 1996) and in generally south-trending eskers such as the Sharp Lake Esker (Fig. 3).

2. Methods

2.1. Sample collection

Nine of the kimberlite boulders were collected from two active gravel pits in the Munro Esker which transects the Kirkland Lake kimberlite field; five from one pit on the north side of Highway 66 and four from a second pit 6 km to the north on the east side of Victoria Lake (see Table 1 and Fig. 2). Another boulder (7846) was collected from an inactive gravel pit in the Misema River esker (Killarney Lake pit), on

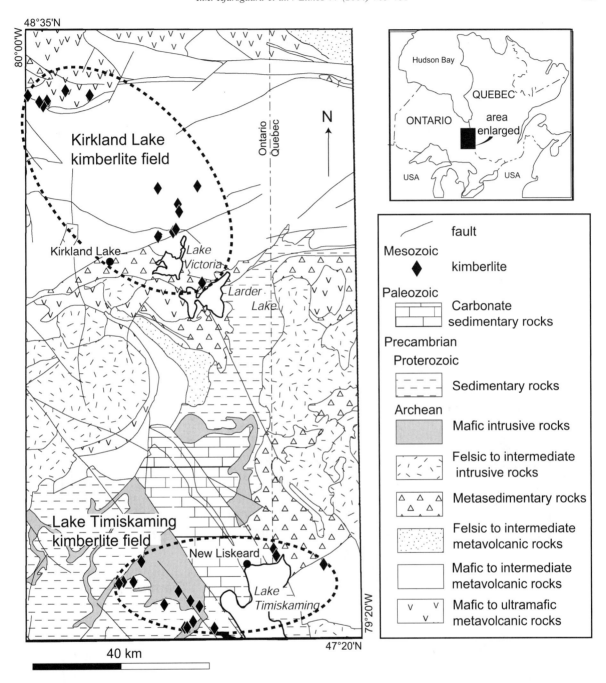

Fig. 1. Location of the Kirkland Lake and Lake Timiskaming kimberlite fields in northeastern Ontario and western Quebec. Bedrock geology from Ontario Geological Survey (1991).

the south side of Highway 66 just east of Larder Lake (Fig. 2). The six Lake Timiskaming boulders (samples 7839–7845 and 7848) were collected from an active gravel pit in the Sharp Lake esker east of Highway 11 at the north end of Sharp Lake, 10 km south of New Liskeard (Fig. 3).

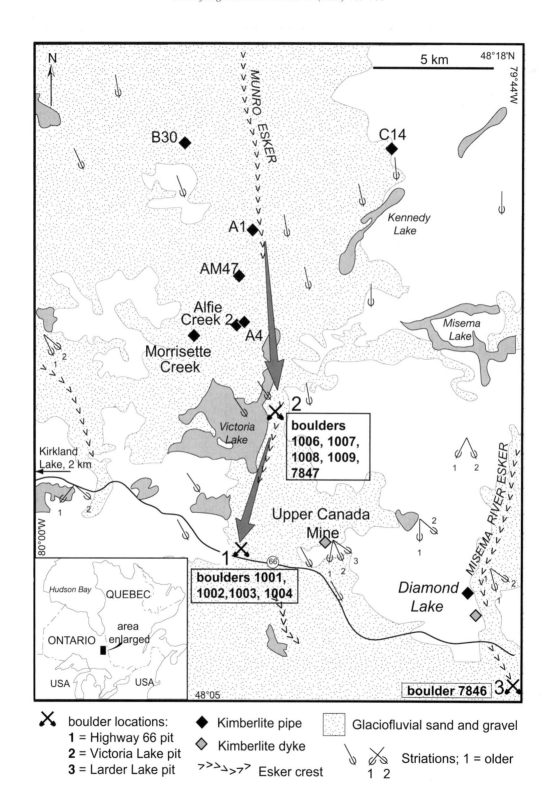

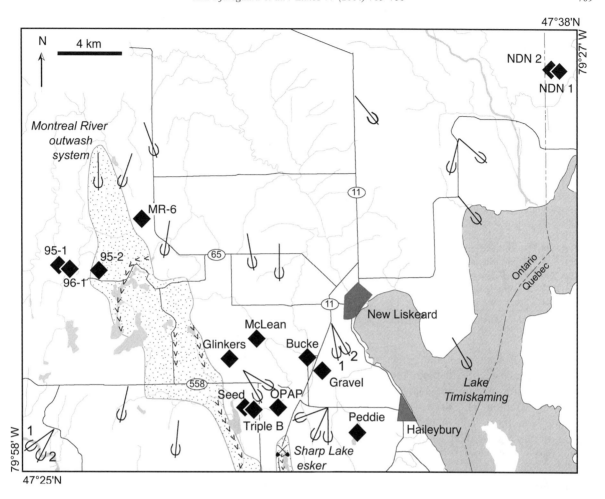

Fig. 3. The Lake Timiskaming kimberlite field with selected striation data, areas covered by glaciofluvial sand and gravel deposits, locations of known kimberlites and the Sharp Lake esker gravel pit, where six boulders were found. Legend as in Fig. 2, geology from Veillette (1986).

2.2. Sample preparation

Kimberlite boulders 1001–1009 were processed using methods similar to those reported in McClenaghan et al. (1999a). Samples 7839–7848 were processed using heavy-liquid separation only (see McClenaghan et al., 2002 for further details). Two samples (7840 and 7841) in the second batch were subjected to mechanical crushing prior to sample processing; the other 14 boulders were sufficiently weathered that they disintegrated after being soaked in a calgon/water solution for 24–48 h. Three size fractions (0.25–0.5, 0.5–1.0 and 1.0–2.0 mm) of non-ferromagnetic heavy-mineral concentrate were prepared from each boulder for indicator mineral picking.

Potential kimberlite indicator minerals were selected on the basis of visual properties (e.g., Mug-

Fig. 2. Areas covered by glaciofluvial sand and gravel deposits and location of gravel pits in which kimberlite boulders have been found along the Munro Esker (center) and the Misema River esker (lower right) within the southern part of the Kirkland Lake kimberlite field. Known kimberlites indicated by diamonds. Large arrows indicate transport direction of boulders within the Munro Esker. Geology from Baker and Storrison (1979) and Baker et al. (1982).

Table 1
Location and characteristics of kimberlite boulders included in this study

Sample number	Location	Esker	UTM-E[a]	UTM-N[a]	Colour	Weathering	Crushed	Weight (kg)	U–Pb age (Ma)
93MPB1001	Hwy 66 pit, KL	Munro	584250	5332100	dark green	moderate	no	2.50	157.9 ± 4.2
93MPB1002	Hwy 66 pit, KL	Munro	584250	5332100	orange	high	no	1.95	
93MPB1003	Hwy 66 pit, KL	Munro	584250	5332100	pale green	high	no	0.83	159.6 ± 4.8
93MPB1004	Hwy 66 pit, KL	Munro	584250	5332100	orange	high	no	0.50	157.8 ± 4.8
93MPB1006	Victoria Lake pit, KL	Munro	584450	5337800	light green/ orange	high	no	0.41	
93MPB1007	Victoria Lake pit, KL	Munro	584450	5337800	olive green	high	no	0.55	157.3 ± 2.9
93MPB1008	Victoria Lake pit, KL	Munro	584450	5337800	dark green	high	no	0.27	158.4 ± 4.4
93MPB1009	Victoria Lake pit, KL	Munro	584450	5337800	dark green	high	no	0.28	
97MPB7839	Sharp Lake pit, LT	Sharp Lake	595000	5252000	orange	high	no	0.46	145.0 ± 1.5
97MPB7840	Sharp Lake pit, LT	Sharp Lake	595000	5252000	light green	high	yes	0.65	144.7 ± 1.0
97MPB7841	Sharp Lake pit, LT	Sharp Lake	595000	5252000	dark green	moderate	yes	0.90	
97MPB7844	Sharp Lake pit, LT	Sharp Lake	595000	5252000	light green	high	no	0.16	
97MPB7845	Sharp Lake pit, LT	Sharp Lake	595000	5252000	orange	high	no	0.15	
97MPB7846	Larder Lake pit, KL	Misema River	594000	5327800	light green	high	no	1.43	
97MPB7847	Victoria Lake pit, KL	Munro	584450	5337800			no	0.17	
97MPB7848	Sharp Lake pit, LT	Sharp Lake	595000	5252000	light green	high	no	0.15	

[a] NAD27. Abbreviations: KL: Kirkland Lake kimberlite field; LT: Lake Timiskaming kimberlite field (see Fig. 1).

geridge, 1995), such as colour, grain morphology and/or the presence of adhering kimberlite matrix material. Electron microprobe analyses were carried out at the Geological Survey of Canada using operating conditions and mineral sorting routines similar to those described by McClenaghan et al. (1999a,b,c).

2.3. U–Pb age determinations

Splits of kimberlite boulders 1001, 1003, 1004, 1007 and 1008 from the Munro Esker and 7839 and 7840 from the Sharp Lake esker were selected (based on the abundance of perovskite in concentrates) for U–Pb age determination at the Radiogenic Isotope Laboratory, Department of Earth and Atmospheric Sciences, University of Alberta. Analytical details and processing techniques are provided in Heaman and Kjarsgaard (2000). All age uncertainties reported in the text and Table 1 are quoted at 2σ.

3. Boulder mineralogy

3.1. Kimberlite indicator mineral abundance

Because of the large number (several thousands) of indicator mineral grains in each sample, only a subset of grains were analyzed by electron microprobe. The total number of grains of each mineral in a sample was normalized to a sample weight of 1.0 kg (Table 2) in order to compare results for boulders of varying weight.

From Table 2, considerable differences in the indicator mineral populations of the boulders are apparent. Based on these differences alone, six different groups were established which were confirmed by mineral compositions and U–Pb ages (see below). Samples 1002, 1003, 1004, 1006, 1008 and 1009 (group I) are dominated by abundant olivine and oxides (with chromite>Mg-ilmenite in all samples, except 1009) and lesser garnet and clinopyroxene. Sample 1001 (group II) is similar to group I

Notes to Table 2:

[a] Includes pyrope–almandine with MgO >5 wt.% and FeO_{tot} <22 wt.%.

[b] Perovskite was not picked systematically but found by accident in the oxide fraction. The numbers are not necessarily representative of the actual amount occurring in the sample.

Table 2
Number of kimberlite indicator minerals in individual kimberlite boulders normalized to 1.0 kg sample weight

Sample number	Group	(a) Mg-ilmenite				(b) Chromite			
		0.25–0.50	0.50–1.00	1.0–2.0 mm	Total	0.25–0.50	0.50–1.00	1.0–2.0 mm	Total
93MPB1001	II	885	237	0	1122	0	0	0	0
93MPB1002	I	3519	615	99	4233	5058	312	7	5377
93MPB1003	I	573	288	45	906	1080	394	30	1504
93MPB1004	I	543	383	56	982	6247	905	14	7166
93MPB1006	I	1472	538	61	2071	5964	717	7	6688
93MPB1007	III	598	117	51	766	425	124	9	558
93MPB1008	I	405	162	4	571	1095	174	7	1276
93MPB1009	I	2050	538	18	2606	1414	124	0	1538
97MPB7839	VI	2139	543	0	2682	78	3	0	81
97MPB7840	VI	2531	810	0	3341	35	11	0	46
97MPB7841	VI	1728	595	0	2323	60	12	0	72
97MPB7844	VI	933	195	18	1146	0	0	0	0
97MPB7845	VI	2965	760	130	3855	7	0	0	7
97MPB7846	V	159	33	1	193	5	1	0	6
97MPB7847	IV	5529	432	0	5961	2652	259	0	2911
97MPB7848	VI	2155	525	79	2759	13	7	0	20
		(c) Pyrope garnet[a]				(d) Diopside			
93MPB1001	II	72	129	92	293	9	10	0	19
93MPB1002	I	606	89	118	813	1083	188	2	1273
93MPB1003	I	389	110	96	595	52	12	11	75
93MPB1004	I	516	238	280	1034	884	254	42	1180
93MPB1006	I	499	237	242	978	138	56	65	259
93MPB1007	III	289	166	614	1069	42	18	0	60
93MPB1008	I	397	161	112	670	34	7	90	131
93MPB1009	I	162	63	423	648	0	7	4	11
97MPB7839	VI	243	91	347	681	0	0	0	0
97MPB7840	VI	674	260	264	1198	5	0	0	5
97MPB7841	VI	371	134	100	605	2	0	0	2
97MPB7844	VI	329	37	122	488	0	0	0	0
97MPB7845	VI	445	75	274	794	27	0	0	27
97MPB7846	V	4	3	14	21	3	0	0	3
97MPB7847	IV	3590	587	978	5155	322	46	0	368
97MPB7848	VI	205	410	265	880	13	0	0	13
		(e) Olivine				(f) Perovskite[b]			
93MPB1001	II	>10000	>10000	160	>10000	46	28	0	74
93MPB1002	I	>10000	>10000	400	>10000	60	60	0	119
93MPB1003	I	63	2545	2452	5060	0	7	0	7
93MPB1004	I	>10000	>10000	1580	>10000	0	17	0	17
93MPB1006	I	4262	15763	10925	>10000	46	46	0	91
93MPB1007	III	94	43	4	141	95	38	0	133
93MPB1008	I	899	2929	6367	>10000	0	4	0	4
93MPB1009	I	1394	10352	7292	>10000	0	7	0	7
97MPB7839	VI	0	0	0	0	353	84	0	437
97MPB7840	VI	0	0	0	0	4	25	4	32
97MPB7841	VI	0	0	0	0	79	48	0	127
97MPB7844	VI	0	0	0	0	1353	123	0	123
97MPB7845	VI	0	0	0	0	178	222	0	222
97MPB7846	V	104	74	17	195	2	2	0	3
97MPB7847	IV	0	23	6	29	480	0	0	480
97MPB7848	VI	0	0	0	0	264	168	0	432

samples in that it contains abundant olivine and Mg-ilmenite, but it lacks chromite and contains only minor garnet and trace Cr-diopside. Coarse-grained perovskite is also present. Sample 1007 (group III) contains little olivine but considerable amounts of garnet and oxides (including perovskite). Sample 7847 (group IV) contains abundant Mg-ilmenite, garnet and chromite, but no olivine. Sample 7846 (group V) contains predominantly Mg-ilmenite, with some olivine and very little garnet, chromite and clinopyroxene. Samples 7839–7845 and 7848 (group VI) do not contain any olivine and are dominated by Mg-ilmenite, with minor garnet and very little chromite and clinopyroxene.

3.2. Mineral compositions

The compositional ranges for each indicator mineral in individual boulders are summarized in Table 3 and below. The complete data set for all indicator mineral grains from boulders examined in this study is presented in McClenaghan et al. (2002).

3.2.1. Mg-ilmenite

Mg-ilmenite compositions in samples of group I do not differ significantly from each other and have a very limited compositional range, between 10 and 13

wt.% MgO and < 1.2 wt.% Cr_2O_3 (Fig. 4, Table 3). Cr_2O_3 content increases with increasing MgO. Mg-ilmenite in group II sample 1001 (Fig. 5a) contains between 10 and 14 wt.% MgO and < 0.6 wt.% Cr_2O_3, with a few outliers at lower MgO and higher Cr_2O_3 (up to 1.7 wt.%). Mg-ilmenite in group III sample 1007 (Fig. 5b) occupies a similar compositional field as that of sample 1001, although with no low-MgO, high-Cr_2O_3 outliers and one high Cr-ilmenite with 3 wt.% Cr_2O_3 and 13 wt.% MgO. Mg-ilmenite in group IV sample 7847 (Fig. 5c) has a slightly wider variation in both MgO and Cr_2O_3 content as compared to the boulder samples described above (Fig. 4 and 5a, b). Mg-ilmenite from group V sample 7846 (Fig. 5d) is distinctly more MgO-rich (11–15 wt.%) with a few outliers with lower (< 11 wt.%) and higher (16–17 wt.%) MgO levels. Mg-ilmenite from the Sharp Lake esker boulders (group VI, Fig. 6) are characterized by a large variation in MgO content (from 6 to 14 wt.%), at comparatively low Cr_2O_3 (< 0.8 wt.%), with some additional MgO-rich grains scattering towards higher Cr_2O_3 (2–5 wt.% Cr_2O_3; Fig. 6). Mg-ilmenite from these samples also displays a distinct break in MgO content between 8 and 9 wt.%. Furthermore, samples with >9 wt.% MgO typically also exhibit lower Cr_2O_3 content than those with < 8 wt.% MgO.

Table 3
Summary of indicator mineral compositional ranges for kimberlite boulders

Sample Number	Group	(a) Mg-ilmenite		(b) Chromite	(c) Pyrope	(d) Diopside	(e) Olivine	
		MgO (wt.%)	Cr_2O_3 (wt.%)	Cr_2O_3 (wt.%)	Cr_2O_3 max. (wt.%)	Cr_2O_3 (wt.%)	Fo	NiO (wt.%)
93MPB1001	II	8.59–13.73	0.08–1.66	–	10.51	0.77–2.78	84.9–93.7	0.22–0.49
93MPB1002	I	9.93–13.29	0.04–1.05	7.32–62.60	7.51	0.36–2.59	86.1–94.8	0.14–0.49
93MPB1003	I	8.96–14.05	0.00–0.94	7.41–62.82	8.33	0.15–2.52	80.5–95.0	0.03–0.50
93MPB1004	I	10.84–12.43	0.05–1.21	6.51–58.77	6.88	0.02–2.59	79.9–94.4	0.02–0.45
93MPB1006	I	10.34–12.67	0.02–1.02	6.35–59.62	9.25	0.17–2.23	80.5–95.1	0.02–0.50
93MPB1007	III	10.77–13.87	0.00–3.05	22.75–58.57	7.41	0.30–2.51	85.0–93.4	0.11–0.46
93MPB1008	I	10.53–13.15	0.03–0.84	14.63–60.79	5.77	0.64–2.16	84.4–94.6	0.14–0.54
93MPB1009	I	10.01–12.67	0.02–1.15	15.20–59.62	5.31	–	86.6–95.3	0.20–0.53
97MPB7839	VI	5.89–14.42	0.00–5.64	44.60–54.38	7.89	–	–	–
97MPB7840	VI	5.73–14.38	0.00–2.66	46.75–56.53	11.30	2.09–2.41	–	–
97MPB7841	VI	6.02–13.76	0.00–3.02	34.03–54.82	9.83	1.56–2.29	–	–
97MPB7844	VI	5.95–14.00	0.00–2.24	–	6.45	–	–	–
97MPB7845	VI	5.97–13.98	0.00–4.18	56.14	8.04	2.65–2.78	–	–
97MPB7846	V	9.50–16.94	0.04–4.66	48.95–58.53	1.68	0.69–1.37	85.8–92.6	0.15–0.47
97MPB7847	IV	9.06–14.25	0.00–1.78	10.57–60.13	6.32	0.40–2.81	88.7–89.3	0.32–0.42
97MPB7848	VI	5.78–14.64	0.07–3.92	36.51–54.23	4.89	1.16–1.87	–	–

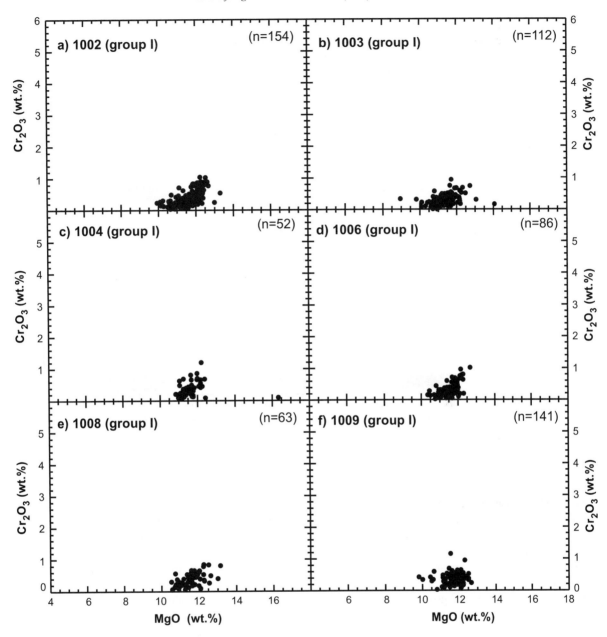

Fig. 4. Cr_2O_3 vs. MgO content of Mg-ilmenite from group I kimberlite boulders 1002 (a), 1003 (b), 1004 (c), 1006 (d), 1008 (e) and 1009 (f), found in gravel pits on the Munro Esker, Kirkland Lake kimberlite field.

3.2.2. Cr-spinel

Cr-spinel in samples from groups I (Fig. 7), III and IV (Fig. 8a and b), which are all from the Munro Esker, are characterized by a strong trend from FeMg-chromite towards spinel compositions ($MgAl_2O_4$), i.e., decreasing Cr_2O_3 with increasing

MgO (xenocryst trend, or MAC-AMC trend of Mitchell, 1986). The scatter towards Cr- and MgO-poor and Fe and Ti enrichment is typical of the Ti-magnetite trend (i.e., kimberlite groundmass spinels). These compositional trends are also displayed by Cr-spinel in known Kirkland Lake kimberlites (Sage,

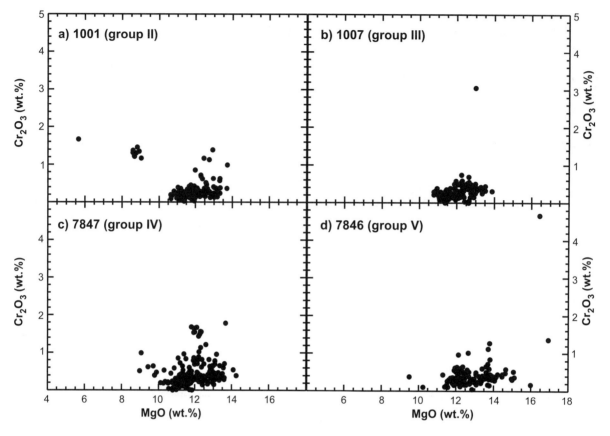

Fig. 5. Cr_2O_3 vs. MgO content of Mg-ilmenite from boulder groups II–V: 1001 (a), 1007 (b) and 7847 (c) from the Munro Esker; and 7846 (d) from the Misema River Esker, Kirkland Lake kimberlite field.

1996; McClenaghan et al., 1996, 1998, 1999a,b). Samples 1002 and 1003 each contain one chromite that plots into the chromite diamond intergrowth field (Fig. 7a and b). No obvious differences in chromite composition are apparent between samples of groups I and IV (Table 3). Insufficient chromite grains were recovered from group V sample 7846 (Misema River Esker; Fig 8c) and the Sharp Lake esker boulders (group VI, Fig. 8d) to determine compositional variations. However, in the Sharp Lake esker samples, there is an increased scatter towards Fe- and Ti-rich compositions rather than Fe-chromite to MgAl-spinel solid solutions. This pattern is consistent with available analyses of Cr-spinel from Lake Timiskaming kimberlites, which show a less pronounced MAC-AMC trend (Sage, 1996, 2000; McClenaghan et al., 1999c, 2003 and unpublished data).

3.2.3. Garnet

Five different compositional groups of garnet occur in the kimberlite boulders: (1) peridotitic Cr-pyrope garnet; (2) websteritic low-Cr-pyrope garnet; (3) titanian pyrope garnet (i.e., megacryst garnet); (4) eclogitic/pyroxenitic Mg-almandine garnet; and (5) crustal almandine garnet (Figs. 9–12). Not all of the five groups are present in each boulder. The range of Cr_2O_3 content is quite variable (Table 3) and reaches values of >10 wt.% Cr_2O_3 in garnet from boulders 7840 (11.3 wt.%) and 1001 (10.5 wt.%).

Group I boulders from the Munro Esker contain abundant orange garnets that are absent in the other samples studied. These orange garnets range in composition from Mg-almandine of potentially eclogitic origin (<22 wt.% FeO_{tot}, between 5 and 15 wt.% MgO, with variable but high CaO up to 10

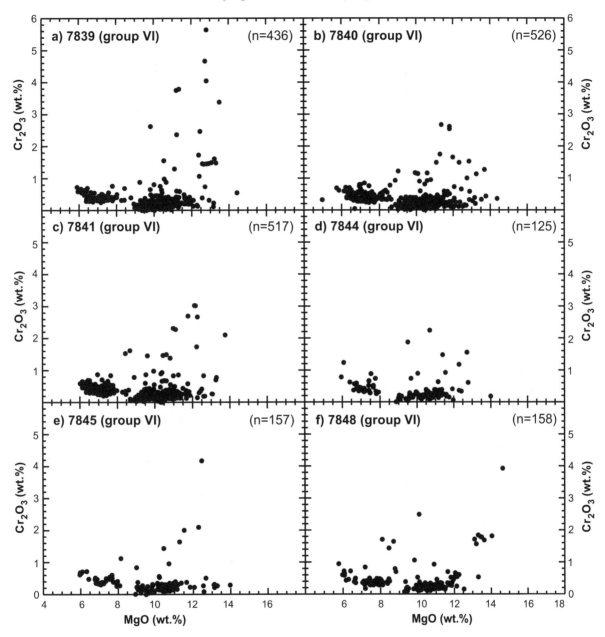

Fig. 6. Cr_2O_3 vs. MgO content of Mg-ilmenite from group VI kimberlite boulder samples 7839 (a), 7840 (b), 7841 (c), 7844 (d), 7845 (e) and 7848 (f), from the Sharp Lake gravel pit in the Lake Timiskaming kimberlite field.

wt.%) to more FeO-rich almandine and a few spessartine (the latter both of crustal origin). For the Mg-almandine garnets, Na_2O determinations were not precise enough to discriminate between potentially diamond-associated eclogitic garnets (high Na_2O) and non-diamondiferous eclogitic garnets (low

Na_2O). Group I samples contain a population of low Ca, Cr pyrope garnets that straddle the 2-wt.% Cr_2O_3 line (Fig. 9). These garnets plot below the 85% of Gurney (1984) but do not contain sufficient Cr_2O_3 to be considered subcalcic harzburgitic (or G10) garnets which would indicate diamond potential.

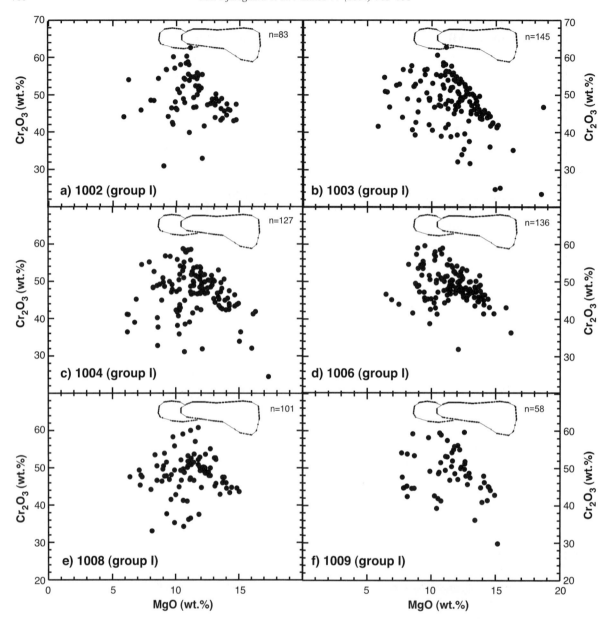

Fig. 7. Cr_2O_3 vs. MgO content of chromite from group I kimberlite boulders 1002 (a), 1003 (b), 1004 (c), 1006 (d), 1008 (e) and 1009 (f). Diamond inclusion and intergrowth fields from Fipke et al. (1995).

Groups II, III and IV samples (Fig. 10a–c) are different from group I samples in that they lack eclogitic Mg-almandine garnet and the low-Ca/low-Cr pyrope garnet populations. They contain, however, abundant megacryst garnets that form two different compositional fields which converge at Cr_2O_3 < 0.25 wt.% and CaO = 3.8 wt.%. One group has variable

CaO between 4.0 and 5.5 wt.% at more or less constant and very low Cr_2O_3. The other group of megacryst garnets shows decreasing Cr_2O_3 with decreasing CaO, extending the lherzolite trend below the 2-wt.% Cr_2O_3 line (Fig. 11a–c). These low-Cr megacrysts plot into the same area on the CaO vs. Cr_2O_3 discriminant plot as the low-Cr (websteritic) garnets,

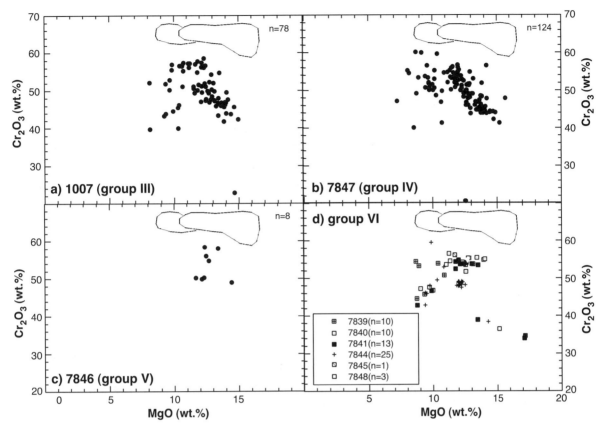

Fig. 8. Cr_2O_3 vs. MgO content of chromite from groups III to VI kimberlite boulders: 1007 (a), 7847 (b), 7846 (c) and all Sharp Lake boulders samples shown in (d). No chromite was recovered from group II boulder (1001). Fields as in Fig. 7.

but have distinctly higher TiO_2 values. Group V sample 7846 contains only a few, almost identical, low-Cr-pyrope garnet, which probably represent fragments from a single grain (Fig. 10d). Group I samples 1001, 1003 and 1006 each contain one subcalcic G10 garnet that not only plots below the 85% line, but also in the subcalcic harzburgite field of Sobolev et al. (1973; Figs. 9b, d and 10a).

Group VI samples (Fig. 11) are characterized by tight compositional fields of megacryst garnets (thought to represent a few large, crushed grains) with ~ 0.5 wt.% TiO_2, a few Cr-pyrope garnets that plot along the lherzolite trend, as well as a small population of websteritic low-Cr, low-Ti pyrope garnets. No Mg-almandine garnets were found in any kimberlite boulder samples from the Sharp Lake Esker. Only a few peridotitic garnets from samples 7839, 7840 and 7841 plot (just) below the 85% line (Fig. 11a–c).

3.2.4. Diopside

Group I boulders contain very magnesian (peridotitic) Cr-diopside, with Mg numbers 94–96 and Cr_2O_3 ranging from 0.5 to 2.5 wt.% (Fig. 12a). In addition, these samples also contain a small population of less magnesian Cr-diopside grains which have approximately 1 wt.% Cr_2O_3 with Mg numbers 84–88 (Fig. 12a). Cr-diopside with Mg numbers 84–88 are commonly found in till samples from the Kirkland Lake and Lake Timiskaming areas (McClenaghan et al., 1998, 1999a,b,c, 2003). These Cr-diopside grains do not contain sufficient MgO to originate from mantle-derived xenoliths and too much Cr_2O_3 to be megacrysts. They are likely derived from regional ultramafic rocks (e.g., Dundonald sill; Naldrett and Mason, 1968) that might have been incorporated in the kimberlites as crustal xenoliths. Group III sample 1007 contains Cr-diopside

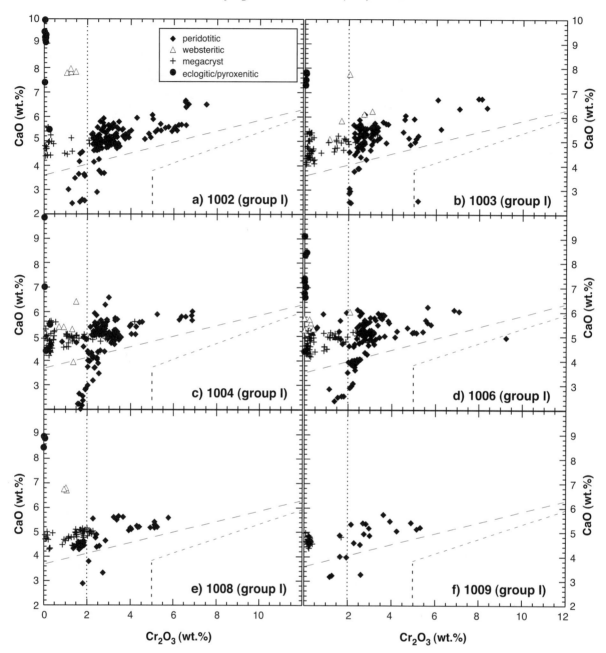

Fig. 9. CaO vs. Cr_2O_3 content of garnet from group I kimberlite boulder samples 1002 (a), 1003 (b) , 1004 (c), 1006 (d), 1008 (e) and 1009 (f). The vertical line at 2 wt.% Cr_2O_3 separates Cr-rich peridotitic garnet from Cr-poor megacryst and eclogitic or pyroxenitic garnet. The diagonal dashed line is the 85% line defined by Gurney (1984), and the dashed line below the 85% line is the field of subcalcic garnet of Sobolev (1977). For legend, see (a).

grains similar in composition to those from mantle peridotite xenoliths and regional ultramafic rocks, as found in group I samples. In addition, sample 1007

contains Cr-diopside with high Cr_2O_3 (2.0–2.5 wt.%) and moderate Mg numbers (88–90) and augitic clinopyroxene with high Mg numbers (91–93) and

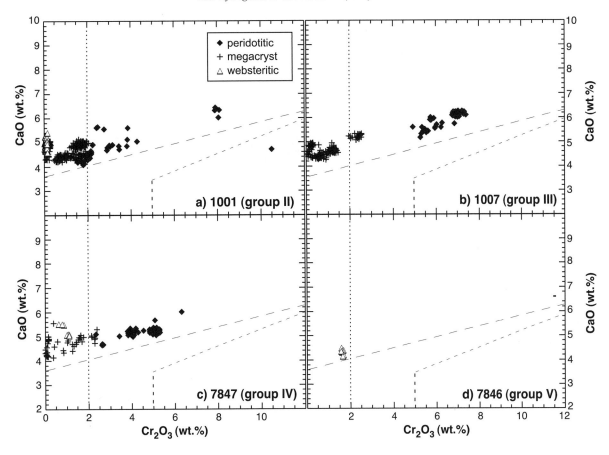

Fig. 10. CaO vs. Cr_2O_3 of garnet from kimberlite boulder samples 1001 (a), 1007 (b), 7847 (c) and 7846 (d), all from the Kirkland Lake area. Fields and symbols as in Fig. 9.

very low Cr_2O_3 (below 0.2 wt.%). This latter group is likely derived from mantle-derived pyroxenite xenoliths, or basaltic xenoliths. Cr-diopside in group II sample 1001 (Fig. 12b) is characterized by Mg numbers 90–94 with variable Cr_2O_3 content of 0.7–3.0 wt.%. Group IV sample 7847 contains three diopside populations: Mg numbers 90–94 and Cr_2O_3 ~0.5 wt.%; Mg numbers 90–94 and Cr_2O_3 between 2.0 and 3.0 wt.%; and Mg number 86 with Cr_2O_3 ~1 wt.% (Fig. 13b). The few grains of Cr-diopside found in group VI boulders (Fig. 12c) have Mg numbers 91–95 and Cr_2O_3 levels of 1.0–3.0 wt.%.

3.2.5. Olivine

The compositional variation of olivine is similar in all samples, characterized by a dense cluster of Mg-rich (Fo 90–94), Ni-rich (>0.30 wt.% NiO) olivine of peridotitic origin (Fig. 13). Grains with forsterite composition <Fo 90 show a trend of decreasing NiO with decreasing Fo content and are predominantly observed in the two smallest (0.25–0.5 and 0.5–1.0 mm) size fractions of the heavy-mineral concentrates (Fig. 13). These olivine are likely kimberlite phenocrysts, which are generally smaller than peridotite-derived (macrocrystic) olivine (Mitchell, 1986) and are therefore more abundant in the smaller size fractions. A few olivine grains in samples 1003 (not plotted) and 1006 (Fig. 13b) have Fo contents (<82) and NiO levels (<0.06 wt.%) which are too low for typical kimberlite phenocrysts. In samples 1003 and 1006, green pargasitic amphibole was also observed. This amphibole is locally common in mafic to ultramafic rocks, and therefore, the low Ni, low Fo olivine in combination with amphibole suggest a

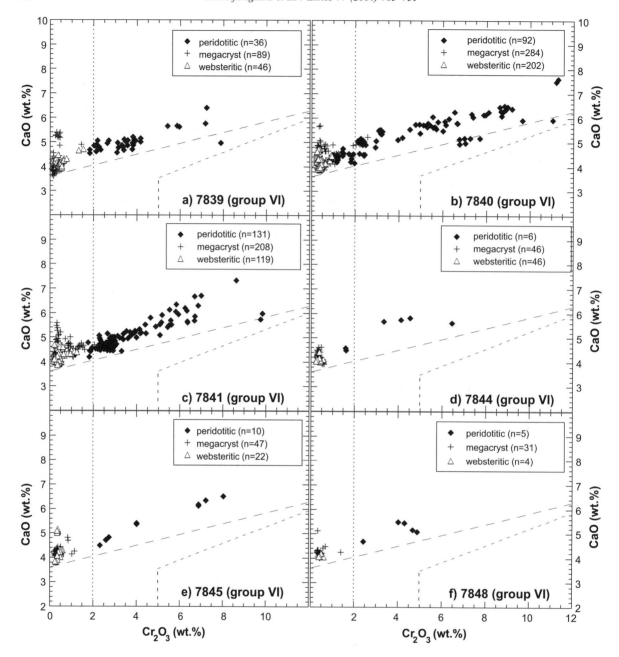

Fig. 11. CaO vs. Cr_2O_3 of garnet from group VI kimberlite boulder samples from the Sharp Lake esker in the Lake Timiskaming area. Fields and symbols as in Fig. 9.

derivation from crustal mafic or ultramafic rocks. Sample 1001 (Fig. 13c) contains a cluster of peridotitic olivine (Fo 91–93), with a separate cluster of kimberlite phenocrystal olivine (Fo 85–90). Sample 1007 (Fig. 13d) has populations of kimberlite phe-

nocrystal olivine and peridotitic olivine. Sample 7847 contains only a few olivine grains (Fig. 13e). Sample 7846 (Fig. 13f) has a kimberlite phenocrystal olivine population (Fo 85–90) and a peridotitic olivine population (Fo 92–94).

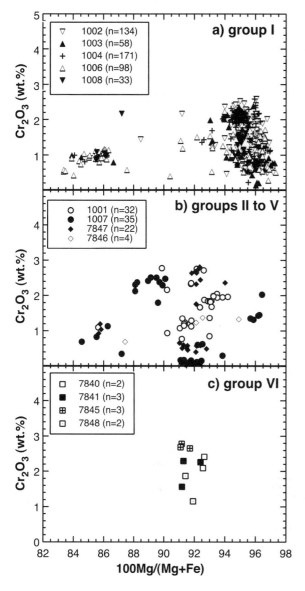

Fig. 12. Mg number $[100 \times Mg/(Mg + Fe)]$ vs. Cr_2O_3 for clinopyroxene from group I boulder samples (a), groups II–V boulder samples 1001, 1007, 7847 and 7846 (b) and group VI boulder samples 7840, 7841, 7845 and 7848 from the Sharp Lake pit, Lake Timiskaming area (d). No Cr-diopside was recovered from samples 7839 and 7844.

3.2.6. Perovskite

A few perovskite grains, although not initially targeted in the indicator mineral picking, were found in the 0.25–0.5 and 0.5–1.0 mm fraction of almost every boulder (except 7847; Table 2).

They are unusually coarse grained for groundmass perovskite in kimberlite (typically < 0.2 mm in size; Mitchell, 1986). Coarse perovskite is particularly numerous in the group VI boulders. The perovskite grains contain between 0.5 and 3.5 wt.% FeO_{tot} and between 0.02 and 2.03 wt.% Nb_2O_5 (Fig. 14), which is characteristic of kimberlitic perovskite. The composition of perovskite from the Kirkland Lake samples (Fig. 14a) is generally less FeO and Nb_2O_5 rich than that of perovskite from the Sharp Lake samples (Fig. 14b).

3.3. U–Pb age determinations

U–Pb ages for selected boulder samples are given in Table 1 and are compared to Kirkland Lake and Lake Timiskaming kimberlite ages in Fig. 15. The average age for three group I boulders is 158.3 ± 4.8 Ma, which falls in the upper range of the ages determined for Kirkland Lake kimberlites and is identical (within error) to that of the A1 kimberlite (158.9 ± 3.7 Ma; Heaman and Kjarsgaard, 2000). However, the uncertainty associated with these ages covers nearly the entire range of all Kirkland Lake kimberlites dated thus far (Fig. 15). Samples 1001 (group II) and 1007 (group III) are essentially contemporaneous with each other and slightly younger in age (but overlapping within error) than group I samples (Fig. 15).

Perovskite separated from the Sharp Lake esker boulder samples 7838 and 7840 (group VI) yielded younger ages of 145.0 ± 1.5 and 144.7 ± 1.0 Ma, respectively. These ages are within the range obtained for kimberlites in the Lake Timiskaming field (Fig. 15; Heaman and Kjarsgaard, 2000) but do not match any specific kimberlite in that field. Group VI boulder ages overlap within error the age for the McLean kimberlite (141.9 ± 2.8 Ma). The McLean kimberlite, however, has vastly different indicator mineral abundances and compositions, being rich in olivine and very poor in all other indicator minerals, particularly Mg-ilmenite (McClenaghan et al., 2003), contrary to the characteristics of the Sharp Lake esker boulders.

4. Discussion

Kimberlite is a very heterogeneous rock. Indicator minerals in kimberlite are derived from (a)

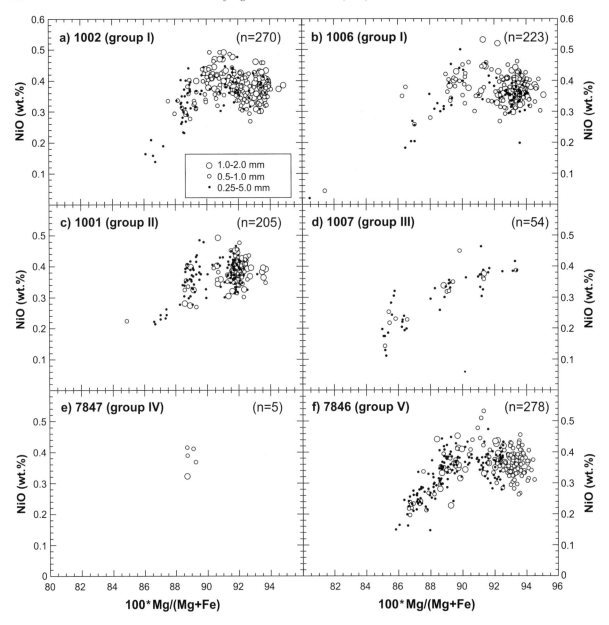

Fig. 13. Forsterite content [100 × Mg/(Mg + Fe)] vs. NiO for olivine in three size fractions of heavy minerals from boulder samples 1002 (a), 1006 (b), 1001 (c), 1007 (d), 7847 (e) and sample 7846 (d) from the Misema River Esker. No fresh olivine was recovered from group VI boulders.

disaggregated mantle xenoliths (e.g., Cr-pyrope, chromite, olivine and Cr-diopside); (b) megacryst suite minerals (titanian pyrope, Mg-ilmenite and Cr-poor diopside); and (c) phenocryst minerals (olivine, spinel). In addition, crustal xenocrysts derived from regional mafic or ultramafic rocks may contribute

additional olivine, chromite and Cr-diopside. The relative abundances of all these minerals varies between kimberlites in a field or cluster, and different populations of indicator minerals with different compositions can be distinctive for individual kimberlites. This observation allows the indicator

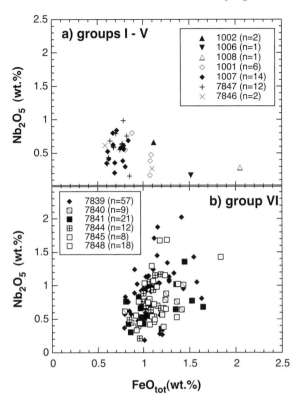

Fig. 14. FeO_{tot} vs. Nb_2O_5 content of perovskite grains from kimberlite boulder samples from (a) the Kirkland Lake area and from (b) the Sharp Lake esker, Lake Timiskaming area.

mineral populations from boulder samples to be compared and contrasted with known kimberlites within a field in order to trace the provenance of the boulders. Previous studies have suggested that megacryst Mg-ilmenite compositions in particular are distinctly different for individual kimberlite bodies (e.g., Mitchell, 1986; Schulze et al., 1995, Sage, 1996, 2000). This difference provides a valuable tool for tracing indicator minerals from glacial sediments and boulders to their potential host kimberlites (provided there are sufficient data, i.e., >50 grains to characterize the compositional variation within a given mineral species).

Based on mineral chemistry and modal abundance, as well as U–Pb perovskite age determinations, the 16 boulders examined in this study are assigned to six compositionally distinct groups and compared to existing data for kimberlites from the Kirkland Lake and Lake Timiskaming kimberlite fields.

4.1. Group I boulders—Munro Esker, Victoria Lake and Highway 66 gravel pits

Kimberlite boulders 1002, 1003, 1004, 1006, 1008 and 1009 are from two gravel pits along the Munro Esker, within the Kirkland Lake kimberlite

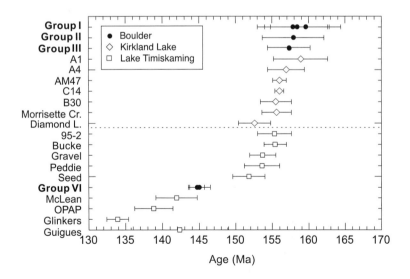

Fig. 15. U–Pb perovskite ages of dated boulder samples (this study, Table 1) compared to known ages of Kirkland Lake and Lake Timiskaming kimberlites dated by the same technique (Heaman and Kjarsgaard, 2000). Error bars indicate 2σ errors.

field (Fig. 3). They form a homogeneous group characterized by abundant olivine and chromite, high Mg-ilmenite counts and minor proportions of garnet and Cr-diopside (Table 2). The similarities between these six boulders are so close that they are thought to represent fragments of one phase of the same kimberlite that was eroded and transported by glacial processes. The mineral chemical characteristics of this group are (1) Mg-ilmenite with a narrow range of MgO (between 10 and 13 wt.%) and <1.2 wt.% Cr_2O_3; (2) chromite that exhibits a MAC-AMC compositional trend; (3) numerous crustal almandine garnet and also eclogitic Mg-almandine garnet that occur in addition to megacryst Ti–Cr pyrope, lherzolitic Cr-pyrope and low Ca, Cr pyrope; (4) Cr-diopside that forms two populations, one very Mg- and Cr-rich and of peridotitic origin, the other with lower Mg number and probably of crustal xenocryst origin; (5) abundant olivine of peridotitic and/or macrocryst origin, with possible crustal xenocrystic olivine in samples 1003 and 1006.

Three boulder samples of this group have been dated by the U–Pb perovskite technique and give an average age of 158.3 ± 4.8 Ma, which most closely resembles the age of the A1 kimberlite (158.9 ± 3.7 Ma; Heaman and Kjarsgaard, 2000; Fig. 15). The A1 kimberlite is also the only candidate among the kimberlites in the Kirkland Lake area (for which indicator mineral data are available, see Figs. 16 and 17) which has Mg-ilmenite compositions almost identical to that of the group I boulders. However, the A1 kimberlite has a characteristic outlier population of high-Cr/low-Mg-ilmenite, which is not observed in the boulders (Fig. 16a and b). Furthermore, the garnet population of the A1 kimberlite lacks the crustal and eclogitic almandine garnets and the low-Ca/low-Cr pyropes seen in the group I boulders (Fig. 16e and f). This discrepancy could be due to the fact that Sage (1996) used drill core material for his study of the A1 kimberlite, whereas the boulders—if they were derived from the A1 kimberlite—must have been eroded from the surface (and possibly a different facies with a different mantle xenolith population) of the kimberlite. Sage (1996) noted common serpentinized olivine megacrysts with some fresh relicts, Mg-ilmenite and perovskite in the thin sections of the A1 kimberlite

samples he studied which matches the characteristics of the group I boulders.

The only other kimberlite in the Kirkland Lake field that contains numerous almandine garnet similar to the group I boulders is the Morrisette Creek kimberlite (Sage, 1996). This kimberlite also contains abundant chromite and pargasitic amphibole (Sage, 1996). Pargasitic amphibole was also recovered from two of the group I boulders. However, the Morrisette Creek kimberlite lacks the low-Ca/low-Cr pyrope garnets found in the group I boulders (Fig. 16e and g) and is exceptionally poor in Mg-ilmenite (Fig. 16c). Only five of the nine available Mg-ilmenite analyses from Morrisette Creek overlap with Mg-ilmenite compositions from the group I boulders (Fig. 16c). The only other kimberlite in the Kirkland Lake field that contains low-Ca/low-Cr pyrope similar to those in the group I boulders is the A4 kimberlite (Fig. 16h), which, however, has a very different proportion of indicator minerals dominated by garnet with comparatively little Mg-ilmenite of very different compositions (Fig. 16d; McClenaghan et al., 1999b) than that of the group I boulders (Fig. 16a).

Based on the similarity in Mg-ilmenite composition, mineral content and age, the A1 kimberlite is the most likely source for the group I boulders, even though the garnet populations are different. The latter is likely due to variations in (mantle) xenolith content in the different portions of the kimberlite sampled by drill core (examined by Sage, 1996) and by the glacier (this study). There is also the possibility that the group I boulders are from an unknown kimberlite in the Kirkland Lake field that has a similar age and indicator mineralogy as A1 and is located within the source area of the Munro Esker.

The A1 kimberlite is also geographically the most likely kimberlite source of boulders found in the Munro Esker since it directly underlies the esker approximately 7 km up-ice (north) of the Victoria Lake gravel pit (Fig. 2). Kimberlite boulders from the A1 kimberlite were likely eroded and incorporated into the Munro Esker glacial drainage system (Fig. 2) and deposited in poorly sorted gravel now exposed at the Victoria Lake pit (samples 1006–1009) and the Highway 66 pit (samples 1002–1004).

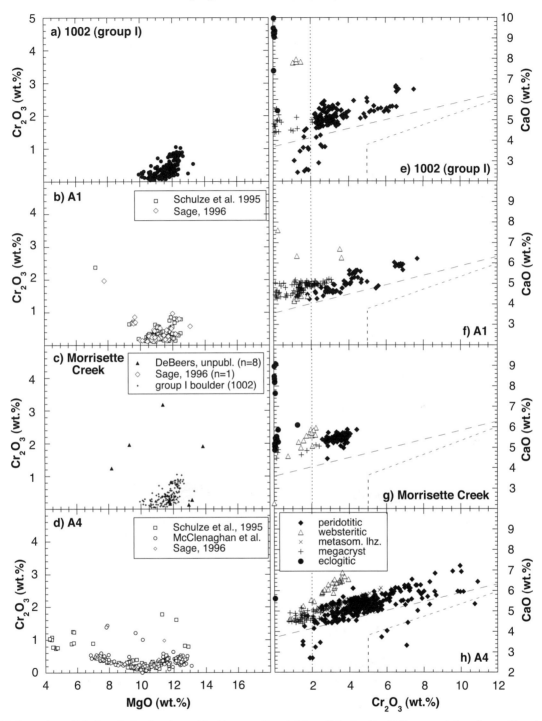

Fig. 16. Comparison of Mg-ilmenite (a–d) and garnet (e–h) compositions in Munro Esker boulder 1002, representative of group I (a and e), the A1 kimberlite (b and f), the Morrisette Creek kimberlite (c and g) and the A4 kimberlite (d and h) from the Kirkland Lake kimberlite field. Garnet and Mg-ilmenite data for the A1 and Morrisette Creek kimberlites are from Sage (1996), with additional Mg-ilmenite data for A1 from Schulze et al., 1995 and for Morrisette Creek provided by DeBeers Canada Exploration. Data for A4 are from McClenaghan et al. (1999a). Legend for (e–h) in (h).

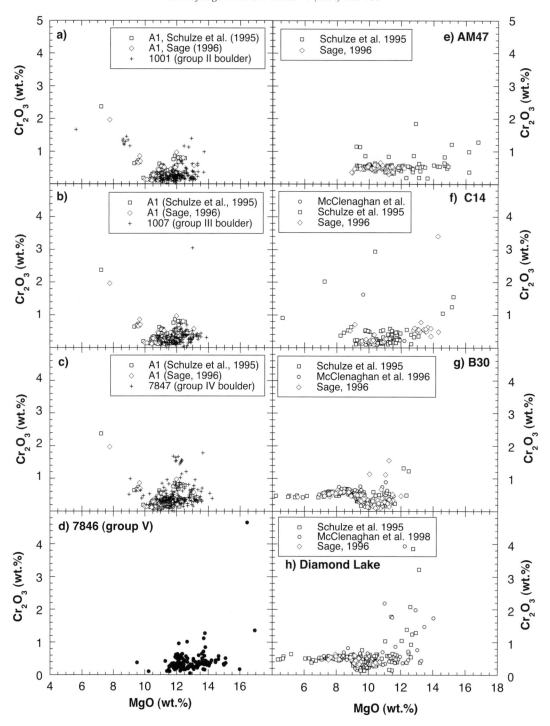

Fig. 17. Mg-ilmenite compositions in Munro Esker boulder samples 1001 (a), 1007 (b) and 7847 (c) compared to those of the A1 kimberlite, and of Misema Esker boulder sample 7846 (d) compared to Mg-ilmenite from known kimberlites of the Kirkland Lake field: AM47 (e), C14 (f) B30 (g) and Diamond Lake (h). Data sources: Schulze et al. (1995) (open square); Sage (1996) (open diamond); and McClenaghan et al. (1996) (solid circles) (B30), McClenaghan et al. (1998) (Diamond Lake) and McClenaghan et al. (1999b) (C14). Data for A4 are shown in Fig. 16.

4.2. Group II (sample 1001)—Munro Esker, Highway 66 pit

The most striking feature of boulder 1001 is its complete lack of chromite, which together with low abundances of garnet and Cr-diopside (Table 2), points to an absence of (chromite-bearing) mantle xenoliths in this particular boulder. The Mg-ilmenite composition of sample 1001 is similar to that of the A1 kimberlite, albeit slightly more magnesian and therefore not as good a match as those from group I (Fig. 17a). Garnet compositions from boulder 1001 (Fig. 10a) resemble those of megacryst and peridotitic garnets in the A1 kimberlite (Fig. 16f) but comprise some low-Cr websteritic garnet which is not found in the A1 kimberlite (Sage, 1996). Cr-diopside in the group II boulder is compositionally different from that of group I boulders (Fig. 12a and b) and resembles Cr-diopside from A1 mantle xenoliths studied by Vicker (1997). A U–Pb age determined for sample 1001 (157.9 ± 4.2 Ma) is slightly younger but well within error of that of the A1 kimberlite (158.9 ± 3.7 Ma). It is suggested that boulder 1001 was derived either from a different (xenolith-poor) phase of the A1 kimberlite or from an unknown kimberlite within the drainage area of the Munro Esker with features very similar to those of the A1 kimberlite.

4.3. Group III (sample 1007)—Munro Esker, Victoria Lake pit

Boulder 1007 contains very little olivine compared to other samples from the Munro Esker. The few recovered olivine grains consist of equal amounts of kimberlitic phenocryst olivine and peridotitic olivine (Fig. 13d) suggesting that most peridotitic olivine succumbed to alteration. Mg-ilmenite compositions are again broadly similar to those of the A1 kimberlite, but offset to slightly higher MgO and lacking the MgO-poor/Cr-rich outlier compositions (Fig. 17b). Garnet compositions consist of two or three different megacryst populations and a group of lherzolitic garnets with >5 wt.% Cr_2O_3. Websteritic and low-Cr lherzolitic garnets found in the A1 kimberlite are absent in the group III boulder. Cr-diopside compositions show little overlap with those of group I or II boulders (Fig. 12a and b). The age of

boulder sample 1007 (157.3 ± 2.9 Ma) is slightly younger than that of the group I and II boulders and lies between that of the A1 and the A4 kimberlites (158.9 ± 3.7 and 156.9 ± 2.5 Ma, respectively (Heaman and Kjarsgaard, 2000); Fig. 15). Although a few similarities (in Mg-ilmenite and garnet composition and age) link this boulder to the A1 kimberlite, the similarities are less pronounced than in group I or II boulders, which leads us to invoke another undiscovered kimberlite in the Kirkland Lake area located N to NW of the Victoria Lake pit as its source.

4.4. Group IV (sample 7847)—Munro Esker, Victoria Lake pit

Sample 7847 contains abundant indicator minerals, particularly oxides but very little olivine (Table 2). Mg-ilmenite compositions cover the compositional field of A1 Mg-ilmenite but extend beyond it to higher MgO (Fig. 17c). Garnet compositions are similar to those of the group III boulder with a few additional websteritic garnets (Fig. 10b and c). Cr-diopside compositions overlap with those of boulder 1001 (group II) (Fig. 12). No age has been determined for this boulder. Similar to group II and III boulders, this boulder might have originated from another phase of the A1 kimberlite or from yet another unknown kimberlite, located N-NNW of the Victoria Lake pit, in the Kirkland Lake field.

4.5. Group V (sample 7846)—Misema River Esker, Larder Lake pit

Sample 7846 is characterized by very low concentrations of garnet, chromite and diopside. However, it contains abundant olivine and Mg-ilmenite, which tends to be very MgO rich (up to 17 wt.% MgO; Fig. 17e). These characteristics are unlike those of the nearby Diamond Lake kimberlite (Fig. 17f; McClenaghan et al., 1998), as well as any other kimberlite in the Kirkland Lake area whose Mg-ilmenite compositions are shown in Figs. 16b–d and 17f–h (Sage, 1996, McClenaghan et al., 1996, 1999a, b). Therefore, the Misema River esker boulder must be derived from an unknown kimberlite in the Kirkland Lake field, located upstream and up-ice of the Larder Lake pit.

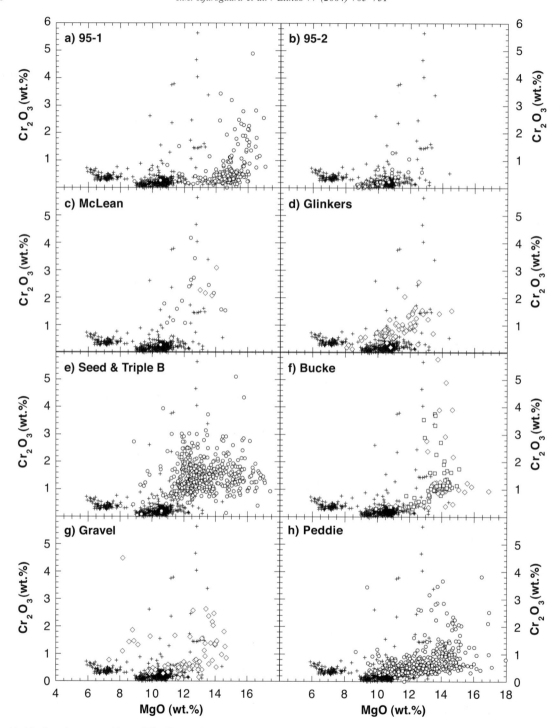

Fig. 18. Mg-ilmenite compositions of the Sharp Lake boulder 7839 (crosses), representative of group VI, compared to Mg-ilmenite data of known kimberlites west of Lake Timiskaming. Data sources and symbols: Schulze et al. (1995) (open squares); Sage (1996, 2000) (open diamonds); McClenaghan et al. (1999c, in press, unpublished) (solid circles).

4.6. Group VI—Sharp Lake Esker gravel pit, Lake Timiskaming area

Boulder samples 7839, 7840, 7841, 7844, 7845 and 7848 recovered from the Sharp Lake pit have almost identical indicator mineral compositions. They are characterized by megacryst garnet with ~ 0.5 wt.% TiO_2, low concentrations of chromite and clinopyroxene and a lack of fresh olivine. The samples also contain coarse perovskite. Radiometric age dating of Sharp Lake pit boulder samples 7839 and 7840 produced almost identical ages of 145.0 ± 1.5 and 144.7 ± 1.0 Ma, respectively, which are well within the range of known ages for kimberlites from the Lake Timiskaming field (Fig. 15), but do not exactly match any of the dated kimberlites in the area (Heaman and Kjarsgaard, 2000). Based on their identical mineral compositions (and ages), the group VI boulders are interpreted to be derived from the same kimberlite source.

Mg-ilmenite compositions in group VI boulders are characterized by a wide variation of MgO with a distinct break in the pattern between 8 and 9 wt.% MgO, which is unlike any published Mg-ilmenite data for Lake Timiskaming area kimberlites (Fig. 18). A possible exception is the 95-2 kimberlite in the NW part of the Lake Timiskaming kimberlite field (Fig. 3) which overlaps with the more MgO-rich Mg-ilmenite of the group VI boulders (Fig. 18b), but does not contain any Mg-ilmenite with <8.5 wt.% MgO or >1.2 wt.% Cr_2O_3 (McClenaghan et al., unpublished data). However, the 95-2 kimberlite contains very little Mg-ilmenite but abundant garnet and chromite and is much older (Fig. 15). It is therefore not considered a likely source for the group VI boulders. Although indicator mineral data are not (yet) available for the MR-6 and 96-1 kimberlites (Fig. 3), MR-6 can probably be ruled out as a source for the Sharp Lake boulders because Mg-ilmenite in till samples collected down-ice (S) of this kimberlite has a markedly different composition than that in the group VI boulders (McClenaghan et al., unpublished data). However, abundant Mg-ilmenite with compositions identical to those of the group VI boulders has recently been discovered in till collected in the vicinity of the Seed kimberlite (McClenaghan et al., in press), which is situated 4 km NW of the Sharp Lake pit. Therefore, an unknown kimberlite located up-ice (north to northwest) of and relatively close to the Seed kimberlite is likely the source of the group VI boulders.

5. Conclusions

The mineralogical and compositional criteria used in this study to group the boulders and identify their source indicate that Mg-ilmenite is the most important "fingerprint" mineral, because the compositional variations of this high-pressure, cognate phase is slightly but distinctly different from kimberlite to kimberlite. The second most important "fingerprint" mineral is garnet, due to a wide variation in chemical composition and abundances (i.e., megacryst, peridotite, pyroxenite, eclogite and crustal garnet types). Similar to garnet, (Cr-)diopside also has a wide range of compositions and lithological sources, but this mineral is commonly much less abundant in kimberlite and therefore less useful for discrimination purposes. Compositional variation in chromite and olivine provides useful information as to the presence of mantle peridotite or mafic to ultramafic crustal xenoliths. In addition to mineral composition, the relative abundance of all kimberlite indicator minerals in boulder samples is a useful tool for discriminating between potential kimberlite sources. U–Pb perovskite age determinations on the boulder samples supported the findings based on indicator mineral abundances and composition. In areas in which kimberlite boulders are found at an early stage of prospecting, we suggest that U–Pb age dating of perovskite can provide useful baseline information for the diamond exploration.

Six kimberlite boulders from two localities on the Munro Esker near Kirkland Lake form a homogeneous group (I) with regards to their indicator mineral abundances and compositions and are considered to be from the same kimberlite source. The group I boulders show the closest similarities to the A1 kimberlite (Sage, 1996) in age, Mg-ilmenite compositions and the occurrence of fresh olivine and perovskite. However, the A1 kimberlite studied by Sage (1996) lacks the abundant almandine garnets and low-Cr/low-Ca pyropes found in the boulders. Garnets of these types were found in the Morrisette Creek and A4 kimberlites, respectively, but both of these kimberlites do not have Mg-ilmenite compositions or ages that match the group I boulders. Therefore, we consider the A1 to be the most likely source of the group I boulders. The A1 kimberlite directly underlies the Munro Esker, in which the boulders were transported south to where they have been deposited in poorly sorted gravel units now exposed at the Victoria Lake and the Highway 66 pits.

Boulders 1001 (group II), 1007 (group III) and 7847 (group IV) from the Munro Esker have slightly differing Mg-ilmenite compositions, each broadly similar to that of the A1 kimberlite. U–Pb ages for samples 1001 and 1007 overlap within error with each other and with that of the A1 kimberlite. However, the three boulders are distinctly different in indicator mineral content and compositions and were therefore assigned to different groups. The differences in mineral contents can be explained by variations in the abundances of mantle xenoliths (chromite, garnet and Cr-diopside) and megacrysts (olivine, Mg-ilmenite, garnet), but the differences in mineral compositions (particularly Mg-ilmenite) suggest different (although possibly related) sources. The three boulders might represent different phases of the same kimberlite or—more likely—two or three unknown kimberlite sources up-ice from the sample localities and in the vicinity of the Munro Esker in the Kirkland Lake kimberlite field.

Boulder sample 7846, from the Misema River esker, does not show any similarities in indicator mineral abundance or composition with the nearby Diamond Lake kimberlite or any other kimberlite of the Kirkland Lake field. Another undiscovered kimberlite in the area of the Kirkland Lake kimberlite field up-ice of the Larder lake pit and in the vicinity of the Misema River esker must be considered as its source.

Kimberlite boulders from the Sharp Lake pit at the south end of the Lake Timiskaming field (group VI) share similar mineral compositions and abundances (and in two cases also age) and are likely derived from the same kimberlite source. These boulders, however, do not match any of the known kimberlites in the Lake Timiskaming field in mineral composition, relative abundances and age. Mg-ilmenite of identical composition as that of the group VI boulders has been found in abundance in till samples adjacent to and S of the Seed kimberlite, 4 km NW of the Sharp Lake pit. This suggests the unknown kimberlite source of the group VI boulders is likely located up-ice of the Seed kimberlite.

Only 6 of the 16 boulders presented here can be matched to a known kimberlite source (A1), whereas the remaining 10 boulders are likely derived from 3 to 4 unknown kimberlites in the Kirkland Lake field and 1 unknown kimberlite in the Lake Timiskaming field. Results of till sampling in the areas of the MR-6 and 95-2 kimberlites (McClenaghan et al., in preparation) and around Seed (McClenaghan et al., in press) might aid in identifying the location of the unknown Lake Timiskaming kimberlite.

Acknowledgements

This study was funded by the Geological Survey of Canada under the Minerals Program of the Canada-Ontario Northern Ontario Development Agreement (NODA) 1991–1995 and the Targeted Geoscience Initiative (TGI-1) 2000–2003. The 1980s discoveries of kimberlite boulders in eskers near Kirkland Lake by C.L. Baker and G. Grabowski inspired the authors' interest in kimberlite boulders. Kimberlite boulders from the Munro Esker were collected by M.B. McClenaghan and R.N.W. DiLabio. Kimberlite boulders from the Sharp Lake pit were collected by Mr. H. Walton. Excellent service was provided by Overburden Drilling Management, I.&M. Morrison Geological Services and Lakefield Research. Additional Mg-ilmenite data were kindly provided by D. Boucher of De Beers Canada Exploration for the Morrisette Creek kimberlite and by P. Sobie of MPH Consulting for the Sudbury Contact Mines 95-2 kimberlite. The manuscript benefited from suggestions by R. Knight, T. Morris, an anonymous reviewer and guest editor H. Grütter.

This is GSC Contribution No. 2003218.

References

Baker, C.L., 1982. Report on the sedimentology and provenance of sediments in eskers in the Kirkland Lake Area, and on finding of kimberlite float in Gauthier Township. In: Wood, J., White, O.L., Barlow, R.B., Colvine, A.C. (Eds.), Summary of Field Work. Ontario Geological Survey Miscellaneous Paper, vol. 106, pp. 125–127.

Baker, C.L., Storrison, D.J., 1979. Quaternary geology of the Larder Lake area, District of Timiskaming. Ontario Geological Survey, Preliminary Map, p. 2290, scale 1:50 000.

Baker, C.L., Steele, K.G., Seaman, A.A., 1982. Quaternary geology of the Magusi River area, Cochrane and Timiskaming Districts. Ontario Geological Survey, Preliminary Map, p. 2483, scale 1:50 000.

Brummer, J.J., MacFayden, D.A., Pegg, C.C., 1992a. Discovery of kimberlites in the Kirkland Lake area, Northeastern Ontario: Part I. Early surveys and surficial geology. Exploration and Mining Geology 1, 339–350.

Brummer, J.J., MacFayden, D.A., Pegg, C.C., 1992b. Discovery of kimberlites in the Kirkland Lake area, Northeastern Ontario: Part II. Kimberlite discoveries, sampling, diamond content, ages and emplacement. Exploration and Mining Geology 1, 351–370.

Fipke, C.E., Gurney, J.J., Moore, R.O., 1995. Diamond exploration techniques emphasizing indicator mineral geochemistry and Canadian examples. Geological Survey of Canada Bulletin 423 (86 pp.).

Gurney, J.J., 1984. A correlation between garnets and diamonds in kimberlites. In: Kimberlite Occurrence and Origin: A Basis for Conceptual Models in Exploration. Geology Department and University Extension, University of Western Australia, Publication No. 8, pp. 143–166.

Heaman, L.M., Kjarsgaard, B.A., 2000. Timing of eastern North American kimberlite magmatism: continental extension of the Great Meteor hotspot track. Earth and Planetary Science Letters 178, 253–268.

McClenaghan, M.B., 1993. Location of known kimberlite bedrock, float and indicator minerals in drift in the Kirkland Lake area. Geological Survey of Canada, Open File Map 2636, scale 1:100 000.

McClenaghan, M.B., Veillette, J.J., 2001. Surficial geology: ice flow indicators for the New Liskeard—Temagami area. Geological Survey of Canada, Open File 3385, 1 sheet.

McClenaghan, M.B., Veillette, J.J., DiLabio, R.N.W., 1995. Ice flow patterns in the Timmins and Kirkland Lake area, northeastern Ontario. Geological Survey of Canada Map, Open File 3014, scale 1:200 000.

McClenaghan, M.B., Kjarsgaard, I.M., Schulze, D.J., Stirling, J.A., Pringle, G., Berger, B.R., 1996. Mineralogy and Geochemistry of the B30 Kimberlite and Overlying Glacial Sediments, Kirkland Lake, Ontario. Geological Survey of Canada, Open File 3295 (245 pp.).

McClenaghan, M.B., Kjarsgaard, I.M., Schulze, D.J., Stirling, J.A., Pringle, G., Berger, B.R., 1998. Mineralogy and Geochemistry of the Diamond Lake Kimberlite and Associated Esker Sediments, Kirkland Lake, Ontario. Geological Survey of Canada, Open File 3576 (200 pp.).

McClenaghan, M.B., Kjarsgaard, I.M., Stirling, J.A.R., Pringle, G., Kjarsgaard, B.A., Berger, B., 1999a. Mineralogy and Geochemistry of the C14 Kimberlite and Associated Glacial Sediments, Kirkland Lake, Ontario. Geological Survey of Canada, Open File 3719 (147 pp.).

McClenaghan, M.B., Kjarsgaard, I.M., Kjarsgaard, B.A., Stirling, J.A.R., Pringle, G., Berger, B., 1999b. Mineralogy and Geochemistry of the A4 Kimberlite and Associated Glacial Sediments, Kirkland Lake, Ontario. Geological Survey of Canada, Open File 3769 (162 pp.).

McClenaghan, M.B., Kjarsgaard, B.A., Kjarsgaard, I.M., Paulen, R.C., 1999c. Mineralogy and Geochemistry of the Peddie Kimberlite and Associated Glacial Sediments, Lake Timiskaming, Ontario. Geological Survey of Canada, Open File 3775 (190 pp.).

McClenaghan, M.B., Kjarsgaard, I.M., Kjarsgaard, B.A., Heaman, L.M., 2002. Mineralogy of kimberlite boulders from eskers in the Lake Timiskaming and Kirkland Lake areas, northeastern Ontario. Geological Survey of Canada, Open File 4361, CD-ROM.

McClenaghan, M.B., Kjarsgaard, I.M., Kjarsgaard, B.A., 2003. Mineralogy and geochemistry of the McLean kimberlite and associated glacial sediments, Lake Timiskaming, Ontario. Geological Survey of Canada, Open File 1762, CD-ROM.

McClenaghan, M.B., Kjarsgaard, I.M., Kjarsgaard, B.A., in press. Mineralogy and geochemistry of the Seed and Triple B kimberlites and associated glacial sediments, Lake Timiskaming, Ontario. Geological Survey of Canada, Open File XXXX.

Mitchell, R.H., 1986. Kimberlites: Mineralogy, Geochemistry, and Petrology. Plenum, New York (442 pp.).

Muggeridge, M.T., 1995. Pathfinder sampling techniques for locating primary sources of diamond; recovery of indicator minerals, diamonds and geochemical signatures. Journal of Geochemical Exploration 53, 183–204.

Naldrett, A.J., Mason, G.D., 1968. Contrasting Archean ultramafic igneous bodies in Dundonald and Clergue Townships, Ontario. Canadian Journal of Earth Sciences 5, 111–143.

Ontario Geological Survey, 1991. Bedrock geology of Ontario, east-central sheet. Ontario Geological Survey, Map 2543, scale 1:1 000 000.

Sage, R.P., 1996. Kimberlites of the Lake Timiskaming Structural Zone. Ontario Geological Survey, Open File Report 5937 (435 pp.).

Sage, R.P., 2000. Kimberlites of the Lake Timiskaming Structural Zone; Supplement. Ontario Geological Survey, Open File Report 6018 (123 pp.).

Schulze, D.J., Anderson, P.F.N., Hearn, B.C., Hetman, C.M., 1995. Origin and significance of ilmenite megacrysts and macrocrysts from kimberlite. International Geology Review 37, 780–812.

Sobolev, N.V., 1977. Deep Seated Inclusions in Kimberlites and the Problem of the Composition of the Upper Mantle. American Geophysical Union, Washington (279 pp.).

Sobolev, N.V., Lavrent'ev, Y.G., Pospelova, L.N, Sobolev, E.V., 1973. Chrome-rich garnets from the kimberlites of Yakutia and their parageneses. Contributions to Mineralogy and Petrology 40, 39–52.

Veillette, J.J., 1986. Surficial geology, Haileybury, Ontario-Quebec. Geological Survey of Canada, Map 1642A, scale 1:100 000.

Veillette, J.J., 1996. Géomorphologie et Géologie du Quaternaire du Témiscamingue, Québec et Ontario. Geological Survey of Canada Bulletin 476 (269 pp.).

Veillette, J.J., McClenaghan, M.B., 1996. Sequence of ice flow in the Abitibi-Timiskaming region: implications for mineral exploration and dispersal of carbonate rocks from the Hudson Bay Basin, Quebec and Ontario. Geological Survey of Canada, Open File 3033, scale 1:500 000.

Vicker, P.A., 1997. Garnet peridotite xenoliths from kimberlite near Kirkland Lake, Canada. Unpubl. M.Sc. thesis, University of Toronto (125 pp.).

Available online at www.sciencedirect.com

Lithos 77 (2004) 733–747

www.elsevier.com/locate/lithos

Systematic variations in xenocryst mineral composition at the province scale, Buffalo Hills kimberlites, Alberta, Canada

Chris T.S. Hood*, Tom E. McCandless

Ashton Mining of Canada Inc., Unit 116-980 W. 1st Street, North Vancouver, BC, Canada

Received 27 June 2003; accepted 3 January 2004
Available online 28 May 2004

Abstract

The Buffalo Hills kimberlites define a province of kimberlite magmatism occurring within and adjacent to Proterozoic crystalline basement termed the Buffalo Head Terrane in north-central Alberta, Canada. The kimberlites are distinguished by a diverse xenocryst suite and most contain some quantity of diamond. The xenocryst assemblage in the province is atypical for diamondiferous kimberlite, including an overall paucity of mantle indicator minerals and the near-absence of compositionally subcalcic peridotitic garnet (G10). The most diamond-rich bodies are distinguished by the presence of slightly subcalcic, chromium-rich garnet and the general absence of picroilmenite, with the majority forming a small cluster in the northwestern part of the province. Barren and near-barren pipes tend to occur to the south, with increasing proximity to the basement structure known as the Peace River Arch. Niobian picroilmenite, compositionally restricted low-to moderate-Cr peridotitic garnet, and megacrystal titanian pyrope occur in kimberlites closest to the arch. Major element data for clinopyroxene and trace element data for garnet from diamond-rich and diamond-poor kimberlites suggests that metasomatism of lithospheric peridotite within the diamond stability field may have caused destruction of diamond, and diamond source rocks proximal to the arch were the most affected.
© 2004 Elsevier B.V. All rights reserved.

Keywords: Mineral chemistry; Proterozoic mantle; Pyrope; Chromian spinel

1. Introduction

The Buffalo Hills kimberlite province is located approximately 350 km north of the city of Edmonton in north-central Alberta. A joint venture consisting of Ashton Mining of Canada, EnCana Corporation (formerly Alberta Energy Company) and Pure Gold Minerals discovered the first kimberlites in the region

* Corresponding author. Tel.: +1-604-983-7753; fax: +1-604-987-7107.

E-mail address: chris.hood@ashton.ca (C.T.S. Hood).

in early 1997 and exploration efforts since that time have led to the identification of a new diamondiferous kimberlite province in Canada (Carlson et al., 1999; Skelton et al., 2003). Thirty-eight bodies have been identified as of May 2003. The area of kimberlite magmatism extends over 6000 km^2, with most bodies concentrated in the core area of the province (Fig. 1).

The kimberlites of the Buffalo Hills are emplaced through Devonian and Cretaceous sedimentary sequences mantling a regionally significant, geophysically defined block of crystalline basement known as

0024-4937/$ - see front matter © 2004 Elsevier B.V. All rights reserved.
doi:10.1016/j.lithos.2004.03.015

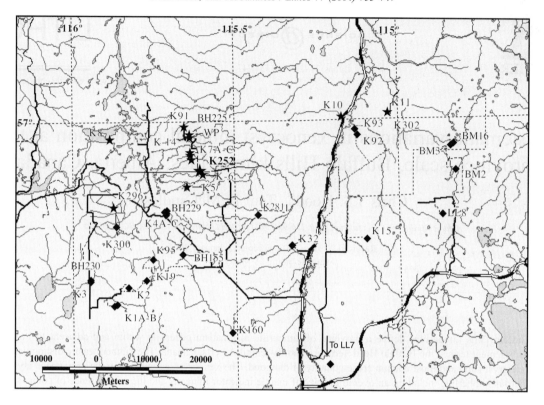

Fig. 1. Location map for the Buffalo Hills kimberlites. Stars represent northern group kimberlites; diamonds represent southern group kimberlites. The town of Red Earth Creek is situated just off the south-central boundary of the map.

the Buffalo Head Terrane. Samples of basement from oil company boreholes have generated Paleoproterozoic ages for the terrane (Ross et al., 1991), although Sm–Nd model ages for some samples have indicated a possible Archaean component (Villeneuve et al., 1993). The terrane is also considered a part of the tectonometamorphic Red Earth granulite domain of Burwash et al. (2000), with basement rocks interpreted to have experienced granulite facies metamorphism. A broad, northeast-trending upward known as the Peace River Arch bisects the southern part of the terrane and is associated with a long history of epei-orogeny.

Most of the kimberlites range from 1 to 40 ha in surface area and are presently established as crater facies to depths of 200 m (Boyer et al., 2003), with two groupings based on geographic position and diamond content. The northern group includes bodies containing higher diamond contents, typically with a sub-population of larger (+1 mm) stones.

The southern/eastern group is dominated by barren to weakly diamondiferous kimberlites, usually with microdiamond results suggesting poor commercial diamond potential. The K252 kimberlite has the highest diamond content (55 carats per hundred tonnes, cpht) and is situated within the northern group.

Twenty-six of thirty-eight bodies are known to contain diamond as of mid-2003. Most of the kimberlites also contain a diverse xenocryst mineral assemblage, as discrete grains or as part of a rarer mantle nodule suite. The mantle nodule suite is well-represented in only a few bodies and is part of a separate study that suggests a Paleoproterozoic or reworked Archaean history (Aulbach et al., 2003). In this study, the xenocryst suites of representative kimberlites are characterized with respect to relative abundance, mineral compositions and parental mantle rock types, diamond cogenesis, and lithosphere evolution.

2. Methodology

In comparing xenocryst minerals from different kimberlites, it is important that grains are selected such that the mantle sampled by the kimberlite is objectively represented. Grain selection for the Buffalo Hills kimberlites was accomplished by visual examination of non-magnetic mineral concentrates exceeding a 2.85 specific gravity in the 0.4–2.0 mm size fractions. Characterization of the overall mineralogical constituents was first completed and garnets were then removed as the first 100 encountered, or all grains were removed from the sample if less than 100 were present. Chromite was similarly removed on a "first 50" basis, while all picroilmenite grains were removed when present. Approximately 3500 grains were derived for the study, covering most of the Buffalo Hills kimberlites.

Subsequent to visual characterization of colour and morphology, grains were cast in resin and polished. Major and minor elements were analyzed by the IXION Research Group (Montreal, PQ), using a Jeol JXA-8900L microprobe, located at McGill University. Trace element abundances in a subsample of 254 garnets were determined by laser ablation ICP-MS using a Perkin Elmer/ Sciex Elan 6000 ICP-MS with a Cetac LSX-200 laser ablation module housed at the Department of Geological Sciences, University of Cape Town. Trace element abundances were internally standardized against SiO_2 and CaO (wt.%) concentrations of each garnet, previously determined through microprobe analysis. Relative variations are less than 10%, which is acceptable for trace elements discussed in this study.

3. The xenocryst mineral suite

3.1. Relative abundance and distribution

Most of the Buffalo Hills kimberlites contain some amount of the standard kimberlite xenocryst suite, which are in most cases visually distinctive. This includes peridotitic pyrope, eclogitic pyrope-almandine, titanian pyrope of the low-Cr megacryst suite, chromite/Cr-spinel, Cr-diopside/augite and forsteritic olivine. Magnesian ilmenite, kimberlitic zircon, enstatite, Cr-corundum, Mg–Al spinel, uvarovite and ede-

nitic hornblende are also occasionally present. Olivine is by far the most abundant mineral in most concentrates, but is excluded from further consideration in this study. Xenocryst occurrence varies widely between bodies and appears uncorrelated with geographical location, pipe morphology or diamond content. Table 1 summarizes xenocryst abundance in typical bodies from the two defined groups and for the K252 kimberlite (included with the northern group).

The peridotitic suite of garnets is variously represented in concentrates from the Buffalo Hills province, ranging from forming an important indicator in pipes such as K14, K15 and K4B to only trace amounts in other bodies (K7C, K8, LL7). The suite is sourced from lherzolitic, wehrlitic, websteritic, pyroxenitic, dunitic and harzburgitic protoliths, with intact nodules of the parent rocks occasionally present in drill core samples (Aulbach et al., 2003). Grains occasionally contain small ($\ll 1$ mm) euhedral inclusions of chromian spinel, suggesting an equilibrium relationship between the two phases. Compositions for individual garnet populations are dominated by species of lherzolitic

Table 1
Xenocryst mineralogy of representative kimberlites

Kimberlite	Grade (cpht)	P-pyrope, kg^a	Low-Cr pyrope, kg^a	Chromite, kg^a	Mg Ilmenite, kg^a	Cpx, kg^a
Northern group						
K252	55.0	0.3	0.1	1.0	0	0.3
K14	11.8	63.7	21.2	148.7	0	26.8
K6	7.2	1.7	~ 0	13.1	0	1.8
K11	4.4	170.4	8.9	1049.0	4.5	5.4
K10	1–5	333.2	7.8	238.2	3.9	3.1
K5	0.36	0.6	0.1	22.6	0	0
K8	<1	0	0	0.1	0	0
K7C	0	0	0	1.7	0	4.0
K7A	0	0	3.3	3.3	0	6.7
Southern group						
K2	<1	31.8	2.5	43.6	2.5	8.0
K1B	<1	3.2	0.4	0.4	0.2	9.8
K95	~ 0	75.9	6.5	12.0	0.7	2.1
K32	~ 0	0.5	1.1	5.4	0.5	2.2
K92	~ 0	6.6	1.3	60.4	1.3	0
K4B	~ 0	665.5	14.6	1167.5	1.4	1.8
K15	0	32.6	0.6	38.8	1.7	0.6
LL7	0	0.2	0	2.0	0	0
K3	0	0.1	0	2.0	0.1	0.2

[a] Based on standard 0.4–2 mm, non-magnetic concentrate. Grain abundances are per kilogram of kimberlite.

derivation, paralleling the rough trends defined by Sobolev et al. (1973) and Gurney (1984) as the line of calcium saturation.

Eclogitic garnets form a relatively minor component of most of the Buffalo Hills bodies, normally less than a few percent of the total garnet content. In some examples, such as the K160 and K7A pipes, orange (i.e. low-Cr) garnets form a significant portion of the total garnet content, and in the latter comprise the only easily recognizable mantle xenocryst phase. Orange, low-Cr garnets tend to be more abundant in the southern group kimberlites as a whole, particularly in the more indicator-rich examples. There is, however, a possible chemical distinction between the low-Cr suite from the northern and southern groups that is discussed in Section 3.2.

Chromite (and chromian spinel) is the most abundant and widely distributed xenocryst mineral in the Buffalo Hills province, occurring in at least trace amounts in all of the bodies for which indicator concentrates have been generated. Chromite tends to be the dominant indicator in the more xenocryst-poor kimberlites, including the LL7 and K6 kimberlites (Table 1). In most bodies, the mineral may form as much as 80% of the (non-olivine) xenocryst assemblage and is comprised of a range of morphologies showing varying degrees of resorption. No correlation of composition with morphology has been attempted, but grain morphology appears to be roughly consistent throughout the province. Individual grain sizes may be as much as several millimeters. The widespread occurrence of chromite makes it an ideal candidate to monitor compositional trends within the Buffalo Hills kimberlites.

Picroilmenite is a minor constituent of the Buffalo Hills kimberlites, with only seven bodies containing greater than trace amounts (K1A, K1B, K2, K3, K10, K11, and K15). The phase is interpreted to be xenocrystic or megacrystic in nature, with most grains occurring within the coarser fractions. Several other kimberlites, entirely within the southern group, contain small amounts of the mineral (Table 1), but limited recovery of grains restricts its application in evaluating diamond preservation potential (e.g. Haggerty, 1975; Gurney and Zweistra, 1995) to the few bodies where it is present in sufficient quantities.

Chromian clinopyroxene (diopside to augite) has been identified in most kimberlites from the province. The mineral is assumed to derive dominantly from disaggregated lherzolites and is most abundant in kimberlites from the southern cluster, where pyroxene alteration appears to be less severe. Orthopyroxene is much less abundant and is also believed to source from peridotitic and pyroxenitic mantle, based on comparable minerals in mantle xenoliths (Aulbach et al., 2003).

Minor components of the xenocryst population include edenitic amphibole, which has been identified in minor amounts in some barren and near-barren bodies of the southern group (Table 1), as well as trace amounts in a few kimberlites in the northern group. The formation of edenitic amphibole has been linked to hydrous metasomatism of lherzolitic mantle (Field et al., 1989). Kimberlitic zircon is relatively rare in the Buffalo Hills and is most frequently encountered in the southern group kimberlites, where it occurs as large, generally colourless and highly resorbed grains. The mineral is believed to be related to the Cr-poor megacryst suite and is thought to be the result of fractionation in an incompatible element-enriched melt (LeCheminant, 1998). Corundum and Mg–Al spinel have also been identified in a number of kimberlites in the Buffalo Hills, including the diamondiferous K252 kimberlite. Corundum and spinel are present in the essentially barren K7A and 7B kimberlites, implying a distribution that is controlled by lateral rather than vertical variations in the source region. Chromian corundum is present throughout the Buffalo Hills kimberlites, although the highest concentrations are found in the northern part of the province. The presence of the mineral in both diamondiferous (K6) and essentially barren (K7A) pipes is indicative of a source that is unrelated to diamond-bearing lithosphere. Xenocrystic Mg–Al spinel is an important phase in the K6 kimberlite complex in the northern cluster and is commonly associated with chromian corundum Trace amounts of corundum, Mg–Al spinel, and white, Cr-free pyrope have been identified within concentrates from the southern group; these grains may be similar to the aluminous mantle assemblages described by Mazzone and Haggerty (1989) within the Jagersfontein kimberlite.

In summary, the xenocryst mineral suite is dominated by species characteristic of peridotitic mantle, with some amount of eclogitic and pyroxenitic mantle also represented. The presence of both orthopyroxene

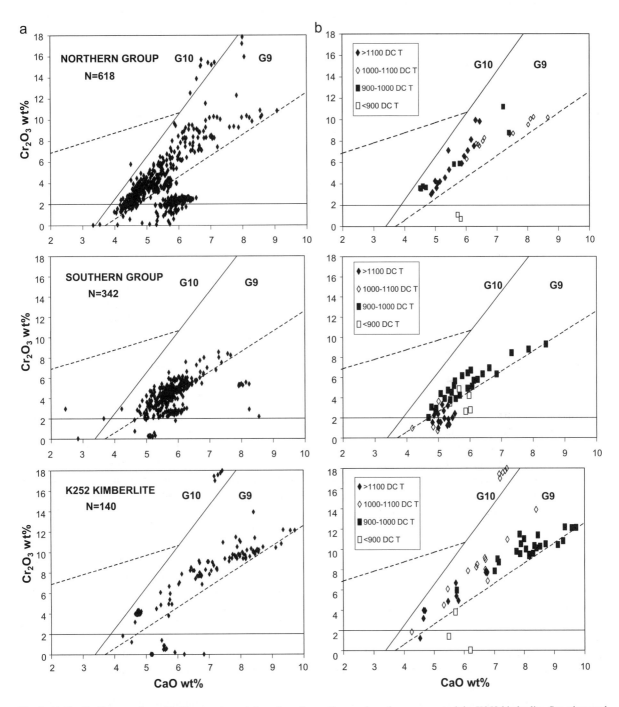

Fig. 2. (a) Cr–Ca diagrams for peridotitic garnet populations from the northern and southern groups, and the K252 kimberlite. Superimposed fields are G9/G10 (Gurney, 1984), with dashed lines for spinel–garnet equilibrium (Kopylova et al., 2000) and graphite–diamond (Grutter and Sweeney, 2000). (b) Temperature distribution for representative northern group, southern group and K252 garnets in Cr–Ca space. Ni-in-garnet temperature calculated according to method of Canil (1994).

and clinopyroxene is additionally suggestive of a possible lherzolitic character to the peridotite. These xenocryst data are matched by xenolith and diamond studies that also demonstrate a dominantly lherzolitic mantle in the area (Aulbach et al., 2003; Davies et al., 2003). Mineral abundance does not appear to parallel diamond content, as some diamond-rich kimberlites contain relatively low concentrations of peridotitic xenocrysts, while a few diamond-poor kimberlites contain large numbers of xenocrysts. The presence (or absence) of some species, such as magnesian ilmenite, may correlate with diamond content.

3.2. Mineral compositions in the northern and southern groups

3.2.1. Peridotitic garnets

In addition to relative abundance, the northern and southern groups can also be defined on the basis of contrasting xenocryst mineral compositions. Peridotitic garnets from the Buffalo Hills kimberlites are plotted in $CaO–Cr_2O_3$ space, using the traditional division of Gurney (1984) for "G10" subcalcic pyrope for reference (Fig. 2). A few pyrope lie within the G10 field, but the grains are not strongly subcalcic and form a very small percentage of the total population, suggesting that subcalcic harzburgite is very rare beneath the Buffalo Hills. Garnets from the northern pipes display lherzolitic trends with a closer proximity to the 85th percentile line of Gurney (1984) (Fig. 2). Also present in these diamondiferous kimberlites are sparsely distributed greenish, knorringitic grains containing up to 20 wt.% Cr_2O_3. Data for the northern group resolves into a number of distinct clusters and trends (Fig. 2), implying a diversity of bulk composition and possibly depth of the source lithosphere. The extent of this diversity is more pronounced than is evident in the southern group kimberlites.

A somewhat diverse source is also implied by pyrope from the bodies in the K11 area of the northern group, where the lherzolitic trends have a broader range of Ca-values and, in the macrodiamond-bearing K10 and K11 pipes, relatively chromium-rich and calcium-poor ranges (Fig. 3). The K10 kimberlite (not included in Fig. 2) in particular is distinctive in that four recovered grains are classifiable as "G10's", with two of these showing probable affinity to a depleted, harzburgitic parent. The breadth of the

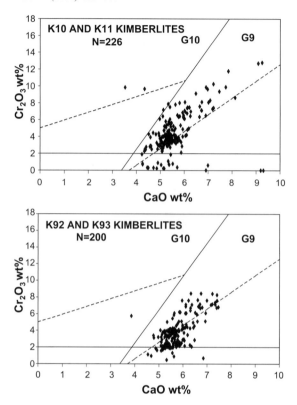

Fig. 3. Cr–Ca diagrams for garnet populations from the K10 and K11 kimberlites (northern group) and K92 and K93 kimberlites (southern group). Superimposed fields are G9/G10 (Gurney, 1984), with dashed lines for spinel–garnet equilibrium (Kopylova et al., 2000) and graphite–diamond (Grutter and Sweeney, 2000).

garnet compositions in Cr–Ca space may also define a contribution from a wehrlitic or websteritic source, or possibly a complex history for the mantle lithosphere from which they derive.

As with the northern group, garnet populations from the southern group are dominated by lherzolitic (calcic) compositions. In the southern bodies from the core area (K4A and B, K229, K2, and K95), the lherzolitic garnets are even more calcic and show an obvious lack of high-chromium compositions, although a few low chromium, subcalcic grains do occur within the G10 field (Fig. 2). Although the subcalcic grains are few in number, they may represent the contribution of a shallow (as implied by the low Cr content), depleted source within a dominantly lherzolitic population. A particular note may be made of the more calcium-rich lherzolitic pyrope of the barren K93 and weakly diamondiferous K92 kimberlites. In the same vicinity, the more diamondiferous K10 and K11 kimberlites define lherzolitic trends with lower average CaO values (Fig. 3).

The range of CaO–Cr$_2$O$_3$ analyses associated with the southeastern, "peripheral" bodies (K15, BM2, and LL8) is distinguished by densely clustered, low chromium garnets that may source from wehrlitic or websteritic lithosphere. The apparent shallow, fertile source material is expected to have a very low prospectivity for diamonds, confirmed by the essentially barren nature of these kimberlites. In most bodies from the southern group, garnet compositional trends follow the spinel–garnet equilibrium trend ("CCGE") defined for xenoliths from the Jericho kimberlite (Kopylova et al., 2000).

3.2.2. Low-Cr garnets

When plotted in Na$_2$O–TiO$_2$ space, low-Cr garnets from several bodies in the northern cluster reveal a number of unusual trends. The K10 and K11 kimberlites include distinct clusters of garnets with relatively high sodium contents, whereas garnets from the K5 and K14 kimberlites are dominated by very low sodium and titanium (Fig. 4). Application of the McCandless and Gurney (1989) model, using a cutoff of 0.09 wt.% Na$_2$O, would thus conclude that eclogitic diamond potential within the K10 and K11 kimberlites is relatively high, whereas the K14 area is dominated by peridotitic or perhaps "exotic" diamond protoliths. Davies et al. (2003) conversely recognized lherzolitic inclusions in one large diamond found in K10, with a

number of diamonds containing eclogitic inclusions from K14. At present, the diamond data are too sparse to assess this apparent discrepancy.

Garnets that can visually be defined as low-Cr are also present in many pipes in the southern group, becoming an important component in a few bodies such as K2 and K95. Many of the grains contain elevated MgO and TiO$_2$ contents, suggesting that they are representative of the low-Cr megacryst suite rather than of true eclogitic mantle (Fig. 4). A subsidiary population of eclogitic grains is also present, with a small percentage containing elevated Na$_2$O (to ~ 0.12 wt.%). The diamond-poor nature of the southern group kimberlites suggests that the high-Na component of the eclogitic population is probably not associated with diamondiferous mantle, or that it is a minor contributor to the diamond population. Alternatively, the Na discriminant may not be useful in assessing the eclogitic diamond potential of this particular province (e.g. Grutter and Quadling, 1999).

3.2.3. Chromite/chromian spinel

Chromite Cr$_2$O$_3$ versus MgO plots have been generated for several pipes from the northern and southern groups (Fig. 5). A compositional field of "spinel inclusions in diamond" has been outlined (after Gurney and Zweistra, 1995) in order to apply a measure of diamond prospectivity for the spinel source region. Gurney (2000) and Griffin et al. (1994) note that compositional trends proposed to be associated with early stage fractional crystallization generally follow a positive slope, whereas xenocryst trends are normally negatively sloping in MgO–Cr$_2$O$_3$ space. Although the terminology assigned to these trends is questionable (McCandless and Dummett, 2003), the terms are used in the ensuing discussion.

As with the garnet populations, chromites from the northern and southern groups display distinct geochemical trends. In the northern group, low titanium chromite follow a prominent xenocryst trend, with the apex at high Cr$_2$O$_3$ content and in the diamond inclusion field. A possible fractionation trend is also expressed. Some bodies (ie. K14, K6, and K91) appear to contain multiple populations within the xenocryst trend, implying a diversity of mantle sources. In the diamondiferous K10 and K11 pipes, however, the fractionation trend is dominant, with few grains falling within the inclusion field. The distinctly indicator-and

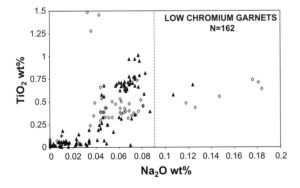

Fig. 4. Ti–Na diagram for low chromium garnet-bearing kimberlites from the Buffalo Hills. Open diamonds are northern group, filled triangles are southern group, filled diamonds are for K252. Field for diamond association compositions set at >0.09 wt.% Na$_2$O, as per McCandless and Gurney (1989).

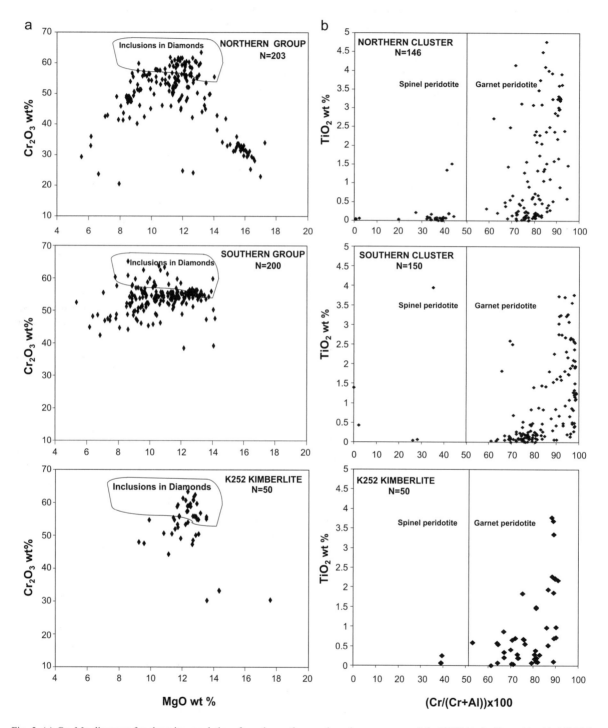

Fig. 5. (a) Cr–Mg diagrams for chromite populations from the northern and southern groups, and the K252 kimberlite, with added field for inclusions in diamond (after Gurney and Zweistra, 1995). (b) Ti–Cr# diagrams of chromite and chromian spinel grains from the northern and southern groups, and the K252 kimberlite.

diamond-poor K7A pipe has a prominent low chromium spinel population as the only feature of note. These grains probably represent xenocrysts derived from shallow-depth spinel lherzolites and are unusual in the Buffalo Hills indicator population.

The nearly barren kimberlites of the southern group include a few with chromites of the fractionation trend within the diamond inclusion field, but with an overall diffuse distribution that may imply a more complex geochemical history for the spinel protoliths (Fig. 5). These grains may represent a mix of xenocryst and fractionation trend chromites, as is apparent for TiO_2 contents within K2 chromites. Unlike the northern group, the fractionated phenocryst association is still the dominant paragenesis. The dominance of this trend in diamond-poor bodies is indicative of a negative association with diamond-bearing mantle in the Buffalo Hills kimberlites.

Many bodies of the southern group (eg. LL7, LL8, BM2, K15, and K32) also include grains within the diamond inclusion field, but with most analyses consisting almost entirely of tight clusters of grains associated with the fractionation trend, in agreement with the indicator-poor nature of these bodies and their associated low diamond content.

TiO_2 versus $Cr/(Cr + Al)$ discriminant plots have been generated for chromite from several kimberlites in the northern and southern groups, after Ramsay (1992) (Fig. 5). The diagrams also define a narrow "diamond association field", as well as provide a rough division between chromian spinels derived from spinel or garnet peridotites. All northern pipes consist dominantly of high chromium populations with grains falling within the diamond-associated field. Interestingly, the diamondiferous K5 and K6 pipes have a lower percentage of grains within the prospective field, as well as lower overall $Cr/(Cr + Al)$ ratios. These distinct populations within the spinel peridotite field are an additional feature of the diamondiferous bodies and hint at a complex mantle lithosphere beneath the K5 and K6 kimberlites, or a complex sampling process during kimberlite ascent.

Plots of TiO_2 versus $Cr/(Cr + Al)$ from the southern group show a general similarity with results for the northern group, including numerous grains within the field for spinel inclusions in diamonds (Ramsay, 1992) (Fig. 5). Cr-number appears to be slightly higher for the southern group, with very few grains within the spinel peridotite field. The higher average Cr-number values, along with single defined paragenesis, may be produced through derivation from garnet peridotites dominated by low chromian garnet. As with the garnet, the existence of the positive trend may thus signal a shallower, less complex mantle source.

3.2.4. Picroilmenite

Picroilmenite has not been identified in the majority of pipes defined within the northern group, although the moderately diamondiferous K10 and K11 kimberlites both contain minor concentrations. Picroilmenite geochemical trends in these two pipes define a number of features (Fig. 6). Of particular interest is the elevated Nb_2O_5 in picroilmenite from the K10 kimberlite, with individual values up to almost 3 wt.%. This enrichment trend has been suggested to reflect increasing degrees of fractional crystallization in the source magma (e.g. Griffin et al., 1991). The lowest niobium contents are found in picroilmenite from the moderately diamondiferous K11 pipe. Most K10 and K11 ilmenites show a negative correlation between Cr and Nb, implying that fractional crystallization provides the most important control on ilmenite chemistry. The K10 pipe occupies an intermediate position in both diamond content and niobium values, suggesting a possible negative correlation between niobium and diamond content.

In the southern group, significant amounts of picroilmenite have been recovered from four pipes (K1A, K2, K3, and K15). Ilmenite from these bodies consistently contains elevated Nb_2O_5, with individual values up to almost 4 wt.%. The absence of a strong negative correlation with Cr content in the southern cluster bodies argues for an influence on ilmenite composition other than fractional crystallization (Fig. 6).

Evaluation of the mantle oxidation state and thus "diamond preservation potential" through plotting of Cr_2O_3 versus MgO (Haggerty, 1975) reveals a similar differentiation between the K11 and K2/K1A bodies (Fig. 6). Ilmenite data from both K11 and K10 cluster on the "reduced" limb of the parabola, indicating a relatively high degree of diamond preservation; K10 also contains some points plotting towards successively more oxidized compositions and therefore lower degrees of inferred diamond preservation. In the weakly diamondiferous K1A and K2 pipes, ilmenite compositions do not fall on a simple oxidation/reduction parabola, instead forming loose linear trends with

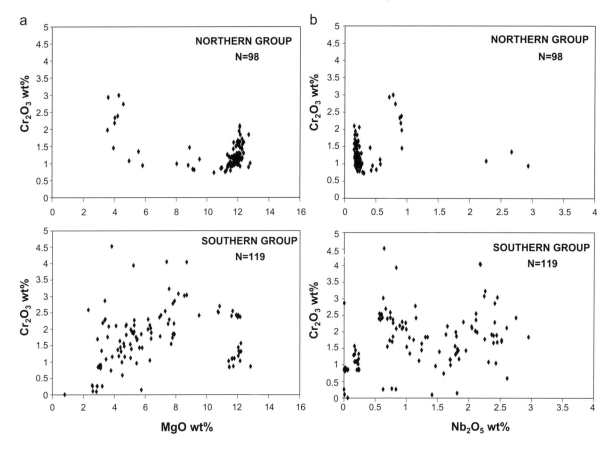

Fig. 6. (a) Cr–Mg plots for picroilmenite from the northern and southern group concentrates. Picroilmenite from the northern group is limited to two bodies (K10 and K11). (b) Cr–Nb plots for picroilmenite in northern and southern group concentrates.

positive slopes (Fig. 6). The association of these unusual trends with elevated niobium contents may be reflective of an additional metasomatic influence that disturbed host oxidation state and provided supplementary element contributions that affected overall Cr_2O_3 content. The overall implication for diamond potential is probably negative due to the implications of isobaric thermochemical modification of the source mantle lithosphere.

3.3. K252: a sample of diamond-bearing Proterozoic mantle?

The K252 kimberlite is geographically situated within the area encompassed by the northern group kimberlites and has the highest diamond content of the Buffalo Hills kimberlites: 55 carats per hundred tonnes from 22.8 tonnes of material. The xenocryst mineral

assemblage present within K252 is consistent with other kimberlites nearby, but some chemical characteristics appear to be enhanced, implying a possible relation with diamondiferous mantle lithosphere. As with the other kimberlites, however, the mineral compositions are considered atypical for Archaean mantle.

3.3.1. Peridotitic garnets

In comparison with other kimberlites of the northern group, garnets from the K252 kimberlite are distinguished by a pronounced trend of elevated Ca and Cr, with some grains showing significant knorringite components (Fig. 2). As with the northern group kimberlites, a small proportion of grains with greenish tints contain Cr_2O_3 up to wt.%. The abundance of high-chromium garnet in the most diamondiferous kimberlite implies a relationship between this garnet population and diamond. The complete garnet population projects into the compositional region associated with wehrlites (Sobolev et al., 1973), but its proximity to the spinel–garnet equilibrium line of Kopylova et al. (2000) coupled with the implied high-chromium nature of the peridotite source may be associated with deeper mantle.

3.3.2. Low-Cr garnets

A small population of eclogitic garnets was identified in K252, with grains tending to contain very low TiO_2 and Na_2O (Fig. 4). No low-Cr megacryst-suite garnets were identified in concentrates.

3.3.3. Chromite/chromian spinel

Chromite and chromian spinel from the K252 kimberlite display a strong clustering at higher chromium (45–63 wt.%) contents, with approximately one-third of the analyzed grains falling within the "spinel inclusions in diamond" field of Gurney and Zweistra (1995) (Fig. 5). The population does not strongly follow either the xenocryst or phenocryst compositional trend, although a few grains approximate the xenocryst trend which is exhibited by the northern cluster kimberlites. Low chromium, aluminous spinels from K252 plot near the lower terminus of the xenocryst trend, suggesting a possible genetic relationship with the higher chromium population. The dominance of both high-chromium spinel and garnet in the K252 concentrate may reflect a possible co-genetic relationship, the result of either bulk-compositional differences or higher pressure regime.

Higher average Al contents are also implied from Cr numbers for the spinel population (Fig. 5), suggesting derivation from garnet peridotites containing garnet with elevated Cr. As with other kimberlites in the northern cluster, a subordinate population of spinel peridotite compositions is present. Overall, chromite compositions from K252 typify a "diamond favourable" mantle in the traditional interpretation of xenocryst mineral chemistry (Fipke et al., 1995).

4. Trace element analysis and geothermobarometry

Representative subsamples of garnet, chromite, and pyroxene populations from selected northern and southern kimberlites were subjected to a more rigorous

analytical program for determination of trace element contents, including analysis of Ni in peridotitic garnet and Zn in chromian spinel for geothermometry (Ryan et al., 1996; Canil, 1999). Garnet populations from the highly diamondiferous K252 (northern group) and weakly diamondiferous K160 (southern group) kimberlites were examined in particular, for comparison of the two defined groupings. A comparison of the Ni-in-garnet temperatures to those obtained by application of the Mn-in-garnet geothermometer (Grutter et al., 1999) is illustrated in Fig. 7. The relatively good correlation allowed for an evaluation of Mn-temperature for garnets from other kimberlites for which Ni-data were not obtained.

Calculated Ni temperatures for the K6 (northern group), K160 (southern group) and K252 (northern group) kimberlites have been plotted in $CaO–Cr_2O_3$ space, allowing for closer examination of compositional changes associated with changing temperature of equilibration (Fig. 2). Garnet populations from the K252 kimberlite, and to a lesser extent the K6 kimberlite, include grains with relatively high Cr_2O_3 contents that also have calculated temperatures exceeding 1000 °C, the approximate position of the diamond stability field on a 40 mW/m^2 geotherm. These results compare well with inclusions from diamonds, in which lherzolitic garnet–clinopyroxene pairs give equilibration temperatures of 1100 to 1200 + 50 °C on a 40 mW/m^2 geotherm (Davies et al., 2003). In contrast, the

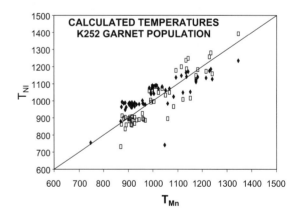

Fig. 7. Comparison of single garnet thermometric techniques for K252 garnets. Ni-in-garnet temperatures are calculated via the methods of Ryan et al. (1996) (open squares) and Canil (1994) (filled diamonds). Mn-in-garnet temperatures calculated according to method of Grutter et al. (1999).

K160 kimberlite consists primarily of temperatures below 1000 °C, with the higher temperature grains having much lower Cr_2O_3 contents. The coincident occurrence of lower Cr_2O_3 and higher temperature (e.g. >1000 °C) in the southern group kimberlite may indicate the presence of a thinner lithosphere, or more extensive thermal alteration at depths coincident with the diamond stability field. This trend has been observed in other kimberlites from the southern group.

The garnet populations also show distinct differences in their trace element composition. The K252 kimberlite consistently displays lower levels of the "metasomatic indicator" elements (Zr, Y) relative to temperature (Fig. 8). In contrast, the K160 kimberlite contains a well-defined population of higher Zr, Y and TiO_2 garnets over the calculated temperature spectrum, suggesting thermal and chemical modification at the base of the lithosphere (e.g. Griffin et al., 1999a,b). The metasomatized garnet population within a diamond-poor kimberlite may imply a thinner (shallower) lith-

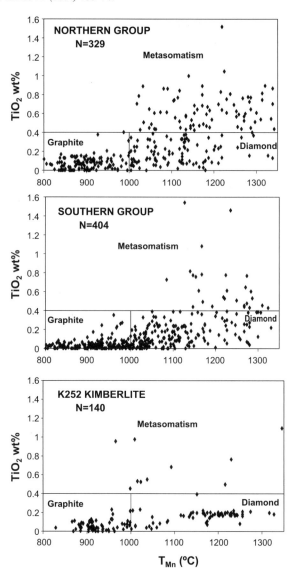

Fig. 9. T_{Mn}–Ti diagrams for peridotitic garnet populations from the northern and southern groups, and for peridotitic garnets from the K252 kimberlite. Fields defined according to a 40 mW/m^2 geotherm.

osphere associated with the K160 garnets, or a diamond-destructive event that has altered the bulk of the lithosphere within the diamond stability field.

Garnet TiO_2 contents from kimberlites in the Buffalo Hills exhibit little apparent contrast between northern and southern group bodies when plotted against Mn-temperatures (Fig. 9). In the K252 pipe, however, garnet TiO_2 contents define a relatively

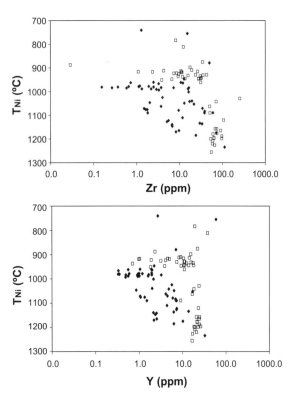

Fig. 8. Comparative trace element chemistry for peridotitic garnets from K252 (filled diamonds) and K160 (hollow squares). T_{Ni} is calculated using the method of Canil (1999).

narrow, even representation up to the upper limit of the Mn-thermometry method, including pronounced sampling from within the diamond stability field at $T_{Mn} > 1000$ °C (Fig. 9). In other northern and southern group bodies, a large number of grains fall within the high-Ti field of metasomatism regardless of grouping. The implication is that the best potential diamond source has the highest number of low-Ti garnets with temperatures above ~ 1000 °C.

Although unaltered clinopyroxene grains are rare, single grain thermobarometry has also been applied to clinopyroxene grains from northern and southern kimberlites, using the method defined by Nimis and Taylor (2000). Results do not define a particular geotherm (e.g. Pollack and Chapman, 1977) and difficulties with the application limit use of the technique. Clinopyroxene has been identified in some Buffalo Hills diamonds (Davies et al., 2003).

5. Discussion and conclusions

A traditional assessment of xenocryst mineral compositions for the Buffalo Hills kimberlite province would conclude that it has poor economic diamond potential. All of the chemical plots, with the possible exception of those for chromite, imply that diamond from conventional parent rocks such as harzburgite and eclogite would be lacking. To some degree the assessment is correct, as most of the kimberlites for which tonnage-sized samples have been treated indicate diamond contents of less than 20 carats per hundred tonnes (Skelton et al., 2003). However, the interpretation would also extend to the K252 kimberlite, which based on its pyrope chemistry should have a diamond content comparable to or less than 10 carats per hundred tonnes (Fipke et al., 1995). With a grade of 55 carats per hundred tonnes and diamonds approaching one carat in size, K252 is distinguished by a singular lack of typical G10 pyrope, sodic eclogitic garnet, and exceptional numbers of favourable chromite. More elaborate assessments such as P–T constraints for garnets and clinopyroxenes also produce results that would indicate low diamond potential. An argument could be made that the xenocryst mineral suite does not represent the parent rocks from which diamond was derived. This may be true for clinopyroxene xenocrysts, if comparable to

clinopyroxenes from xenoliths (Aulbach et al., this volume), but not for clinopyroxene inclusions in diamond (Davies et al., this volume). For the pyrope garnets, the Mn- and Ni-temperatures demonstrate that some could have derived from within the diamond stability field for a typical cratonic geotherm. Even though the pyrope are lherzolitic in composition, that they could have derived from the diamond stability field is supported in part by the fact that the only pyrope inclusion found in a Buffalo Hills diamond to date is lherzolitic in composition (Davies et al., this volume).

Although the xenocryst mineral chemistry for the Buffalo Hills kimberlites in many ways contradicts conventional thought regarding diamond co-genesis, this study indicates that some regional variations can be related to diamond content, and to geographic position. In the latter context, this has implications with respect to lithosphere evolution beneath the Buffalo Head Terrane. The mantle xenocryst compositions support a variably metasomatised lithosphere, also supported by xenolith studies (Aulbach et al., 2003) and by inclusions of disparate origin in single diamonds (Davies et al., 2003). The most significant feature in the Buffalo Head Terrane is the absence of typical Archaean material, either as xenocrysts, xenoliths, or inclusions in diamond. This combined evidence suggests either that Archaean mantle is missing beneath this terrane, or that chemical evidence of its existence has been completely obliterated from the geologic and geochemical record.

Closer examination of garnet, chromite, and picroilmenite data from various bodies from the Buffalo Hills indicates that a number of subtle trends are present within the more diamondiferous pipes. High chromium populations and slightly lower average calcium lherzolitic trends distinguish the peridotitic garnet populations from all of the more diamond-rich bodies, with an eclogitic component implied by data from the K10 and K11 kimberlites. Chromite data highlight the importance of identifying the trend of interest, particularly when considering its dominance in the diamond-bearing pipes of the northern group and K252. An additional significance may be attached to the presence of a complex mantle lithosphere, as implied by both garnet and chromite populations from bodies within the northern group. Picroilmenite is of more limited application due to its restricted occurrence, but a rough association may be made between low niobium contents and higher diamond content in the pipes where it is present.

The presence of the high-chromium, slightly sub-calcic garnets within the northern kimberlites may be a high-pressure analogue of peridotitic mantle. Co-existence with chromian spinel implies that garnet chromium content can be qualitatively linked with depth (Doroshev et al., 1997; Grutter et al., 1999), suggesting that the grains may represent the deeper, potentially diamond-favourable mantle lithosphere.

Two external influences on the Buffalo Head mantle lithosphere may be reflected in the xenocryst data. The first control is defined by the history of the terrane itself, with a crystalline block of probable Paleoproterozoic age forming the core area and represented by lherzolitic, wehrlitic and websteritic garnet compositions. Development of compressional regimes during formation of the Ksituan and Taltson-Thelon magmatic arcs (Ross et al., 1991) allowed for accretion of eclogitic material to the base of the lithosphere, with associated volatile release affecting the margins of the block. Further melt-generated, incompatible element-rich metasomatism associated with inflation of the Peace River Arch and related structures also appears to have affected the vicinity of the arch, producing a diamond-destructive event in lithosphere sampled by the southern kimberlites. The association of higher temperature, Cr-rich pyrope (and, to a lesser extent, xenocrystic Mg–Al spinel) with the more diamondiferous northern cluster bodies may be related to the first event, defining a thicker, more complex section of lithosphere favourable for the formation of diamond. The absence of highly depleted mantle signatures throughout the Buffalo Hills xenocryst suite, and in the least metasomatized northern group kimberlites argues for an overall Proterozoic age for the mantle lithosphere, suggesting that similar, relatively stable crystalline blocks can potentially host diamond-bearing kimberlite.

Acknowledgements

The Buffalo Hills kimberlites are the product of a joint venture between Ashton Mining of Canada, EnCana Corporation, and Pure Gold Minerals. The authors thank these companies for their support of research into the xenocryst mineralogy of the Buffalo Hills kimberlite province.

Other individuals who have contributed to the development of this paper include Brooke Clements, Alberta Project Manager Dave Skelton, Cristiana Mircea, Emma Gofton, and Volodomyr Zhuk for feedback with respect to xenocryst mineralogy and diamond distribution within the Buffalo Hills. The authors are also indebted to Liane Boyer for insights into kimberlite geology in the region. Comments from Drs. John Gurney, Nick Pokhilenko and Herman Grutter allowed for considerable improvement to the manuscript. Finally, we thank Robert Pryde, whose insight and initiative made all of this work possible.

References

Aulbach, S., Griffin, W.L., O'Reilly, S.Y., McCandless, T.E., 2003. The lithospheric mantle beneath the Buffalo Head Terrane, Alberta, Canada: xenoliths from the Buffalo Hills kimberlites. 8th International Kimberlite Conference Extended Abstract.

Boyer, L., Hood, C., McCandless, T., Skelton, D., Tosdal, R., 2003. Volcanology of the Buffalo Hills kimberlites, Alberta, Canada. 8th International Kimberlite Conference Extended Abstract.

Burwash, R.A., Krupicka, J., Wijbrans, J.R., 2000. Metamorphic evolution of the Precambrian basement of Alberta. Can. Mineral. 38, 423–434.

Canil, D., 1994. An experimental calibration of the 'Nickel in Garnet' geothermometer with applications. Contrib. Mineral. Petrol. 117, 410–420.

Canil, D., 1999. The Ni-in-garnet geothermometer: calibration at natural abundances. Contrib. Mineral. Petrol. 136, 240–246.

Carlson, S.M., Hillier, W.D., Hood, C.T., Pryde, R.P., Skelton, D.P., 1999. The Buffalo Hills kimberlites: a newly discovered kimberlite province in north-central Alberta, Canada. In: Gurney, J.J., Gurney, J.L., Pascoe, M.D., Richardson, S.H. (Eds.), Proceedings of the VIIth International Kimberlite Conference, Red Roof Design, Cape Town, pp. 109–116.

Davies, R.M., Griffin, W.L., O'Reilly, S.Y., McCandless, T.E., 2003. Inclusions in diamonds from the Buffalo Hills, Alberta, Canada: diamond growth in a plume? 8th International Kimberlite Conference Extended Abstract.

Doroshev, A.M., Brey, G.P., Girnis, A.V., Turkin, A.I., Kogarko, L.N., 1997. Pyrope-Knorringite garnets in the Earth's mantle: experiments in the $MgO-Al_2O_3-SiO_2-Cr_2O_3$ system. Russ. Geol. Geophys. 38, 559–586.

Field, S.W., Haggerty, S.E., Erlank, A.J., 1989. Subcontinental metasomatism in the region of Jagersfontein, South Africa. In: Ross, J. (Ed.), Kimberlites and Related Rocks Volume 2: Their Mantle/Crust Setting, Diamonds, and Diamond Exploration. Special Publication-Geological Society of Australia, vol. 14, pp. 771–783.

Fipke, C.E., Gurney, J.J., Moore, R.O., 1995. Diamond exploration techniques emphasizing indicator mineral geochemistry and Canadian examples. Bull.-Geol. Surv. Can. 423, 86.

Griffin, W.L., Ryan, C.G., Schulze, D.J., 1991. Ilmenite and silicate megacrysts from Hamilton branch: trace element geochemistry and fractional crystallization. Extended Abstracts. 5th International Kimberlite Conference. Campanhia de Pesquisa de Recursos Minerais, Brasilia, pp. 148–150.

Griffin, W.L., Ryan, C.G., Gurney, J.J., Sobolev, N.V., Win, T.T., 1994. Chromite macrocrysts in kimberlites and lamproites: geochemistry and origin. In: Meyer, H.O.A., Leonardos, O.H. (Eds.), Kimberlites, Related Rocks and Mantle Xenoliths. CPRM Spec. Publ., vol. 1A/93, Campanhia de Pesquisa de Recursos Minerais, Brasilia, pp. 366–377.

Griffin, W.L., Doyle, B.J., Ryan, C.G., Pearson, N.J., O'Reilly, S.Y., Natapov, L., Kivi, K., Kretschmar, U., Ward, J., 1999a. Lithosphere structure and mantle terranes: slave Craton, Canada. In: Gurney, J.J., Gurney, J.L., Pascoe, M.D., Richardson, S.H. (Eds.), Proceedings of the VIIth International Kimberlite Conference, Red Roof Design, Cape Town, pp. 299–306.

Griffin, W.L., Fisher, N.I., Friedman, J., Ryan, C.G., O'Reilly, S.J., 1999b. Cr-pyrope garnets in the lithospheric mantle: I. Compositional systematics and relations to tectonic setting. J. Pet. 40, 679–704.

Grutter, H.S., Quadling, K.E., 1999. Can sodium in garnet be used to monitor eclogitic diamond potential? In: Gurney, J.J., Gurney, J.L., Pascoe, M.D., Richardson, S.H. (Eds.), Proceedings of the VIIth International Kimberlite Conference, Cape, Red Roof Design, Town, pp. 314–320.

Grutter, H.S., Sweeney, R.J., 2000. Tests and constraints on single-grain Cr-pyrope barometer models: some initial results. GEOCANADA 2000 Extended Abstract, University of Calgary. May.

Grutter, H.S., Apter, D.B., Kong, J., 1999. Crust-mantle coupling: evidence from mantle-derived xenocrystic garnets. In: Gurney, J.J., Gurney, J.L., Pascoe, M.D., Richardson, S.H. (Eds.), Proceedings of the VIIth International Kimberlite Conference, Red Roof Design, Cape Town, pp. 307–313.

Gurney, J.J., 1984. A correlation between garnets and diamonds in kimberlite. In: Harris, P.G., Glover, J.E. (Eds.), Kimberlite occurrence and origin: a basis for conceptual models in exploration. Publication-Geology Department and University Extensions, University of Western Australia, vol. 8, pp. 143–146.

Gurney, J.J., 2000. Diamond indicator mineral interpretations: a discussion of some recent developments. GEOCANADA 2000 Extended Abstract, University of Calgary, May.

Gurney, J.J., Zweistra, P., 1995. The interpretation of the major element compositions of mantle minerals in diamond exploration. J. Geochem. Explor. 53, 293–310.

Haggerty, S.E., 1975. The chemistry and genesis of opaque minerals in kimberlites. Phys. Chem. Earth 9, 295–308.

Kopylova, M.G., Russell, J.K., Stanley, C., Cookenboo, H., 2000. Garnet from Cr-and Ca-saturated mantle: implications for diamond exploration. J. Geochem. Explor. 68, 183–199.

LeCheminant, A.N., 1998. Paragenesis of zircon in Buffalo Hills kimberlites, north-central Alberta. AMCI Internal Report.

Mazzone, P., Haggerty, S.E., 1989. Corganites and corgaspinites: two new types of aluminous assemblages from the Jagersfontein kimberlite pipe. In: Ross, J. (Ed.), Kimberlites and Related Rocks Volume 2: Their Mantle/Crust Setting, Diamonds, and Diamond Exploration. Special Publication-Geological Society of Australia, vol. 14 (640), pp. 795–808.

McCandless, T.E., Dummett, H.T., 2003. Some aspects of chromian spinel (chromite) in relation to diamond exploration (abstract). Geological Association of Canada-Mineralogical Association of Canada, Program with Abstracts 1 page CD-ROM.

McCandless, T.E., Gurney, J.J., 1989. Sodium in garnet and potassium in clinopyroxene: criteria for classifying mantle eclogites. In: Ross, J. (Ed.), Kimberlites and Related Rocks Volume 2: Their Mantle/Crust Setting, Diamonds, and Diamond ExplorationSpecial Publication-Geological Society of Australia, vol. 14, pp. 827–832.

Nimis, P., Taylor, W.R., 2000. Single clinopyroxene thermobarometry for garnet peridotites. Part 1. Calibration and testing of a Cr-in-cpx barometer and an enstatite-in-cpx thermometer. Contrib. Mineral. Petrol. 139, 541–554.

Pollack, H.N., Chapman, D.S., 1977. On the regional variation of heat flow, geotherms and lithosphere thickness. Tectonophysics 38, 279–296.

Ramsay, R.R., 1992. Geochemistry of diamond indicator minerals. PhD dissertation. Key Centre for Strategic Mineral Resources, University of Western Australia, September, 1992.

Ross, G.M., Parrish, R.R., Villeneuve, M.E., Bowring, S.A., 1991. Geophysics and geochronology of the crystalline basement of the Alberta Basin, western Canada. Can. J. Earth Sci. 28, 512–522.

Ryan, C.G., Griffin, W.L., Pearson, N.J., 1996. Garnet geotherms: a technique for derivation of P–T data from Cr-pyrope garnets. J. Geophys. Res. 101, 5611–5625.

Skelton, D.N., Clements, B., McCandless, T.E., Hood, C., Aulbach, S., Davies, R., Boyer, L.P., 2003. The Buffalo Head Hills kimberlite province. In: Kjarsgaard, B.A. (Ed.), Kimberlites of Northern Alberta and Slave Province. 8th International Kimberlite Conference Field Trip guidebook.

Sobolev, N.V., Lavrentiev, Yu.G., Pokhilenko, N.P., Usova, L.V., 1973. Chrome-rich garnets from the kimberlites of Yakutia and their paragenesis. Contrib. Mineral. Petrol. 40, 39–52.

Villeneuve, M.E., Ross, G.M., Theriault, R.J., Miles, W., Parrish, R.R., Broome, J., 1993. Tectonic subdivisions and U–Pb geochronology of the crystalline basement of the Alberta Basin, western Canada. Bull.-Geol. Surv. Can., 447.

Available online at www.sciencedirect.com

Lithos 77 (2004) 749–764

LITHOS

www.elsevier.com/locate/lithos

Concepts for diamond exploration in "on/off craton" areas—British Columbia, Canada

George J. Simandl*

British Columbia Ministry of Energy and Mines and University of Victoria, Victoria, BC, Canada

Received 25 June 2003; accepted 6 January 2004
Available online 28 May 2004

Abstract

The tectonic setting of British Columbia (BC) differs from classic diamond-bearing intracratonic regions such as the Northwest Territories and South Africa. Nevertheless, several diamond occurrences have been reported in BC. It is also known that parts of the province are underlain by Proterozoic and possibly Archean basement. Because the continents of today are composites of fragments of ancient continents, it is possible that some of the regions underlain by old crystalline basement in eastern British Columbia were associated with a deep crustal keel. The keel may have predated the break-up of the early Neoproterozoic supercontinent called Rodinia and was preserved possibly until the Triassic. Some of these old continental fragments may have been displaced relative to their position of origin and dissociated from their keel, or the keel may have since been destroyed. Such fragments represent favourable exploration grounds in terms of the "Diamondiferous Mantle Root" model (DMR model) if they were intersected by kimberlites or lamproites prior to displacement or destruction of their underlying deep keel. Therefore, extrapolation of fragments of the diamond-bearing Precambrian basement from the Northwest Territories or Alberta to BC provides a sufficient reason for initiating reconnaissance indicator mineral surveys. The "Eclogite Subduction Zone" model (ES model) predicts formation of diamonds at lower pressure (i.e., depth) than required by the DMR model in convergent tectonic settings. Although not proven, this model is supported by thermal modeling of cold subduction zones and recent discoveries of diamonds in areas characterized by convergent tectonic settings. If the ES model is correct, then the parts of BC with a geological history similar to today's "cold" subduction zones, such as Honshu (Japan), or to continental collision zones, such as Kokchetav massif (Kazakhstan) and the Dabie–Sulu Terrane (east central China), may be diamondiferous. The terranes where geological evidences suggest an ultrahigh pressure (UHP) metamorphic event followed by rapid tectonic exhumation (which could have prevented complete resorption of diamonds on their journey to the surface) are worth investigating. If UHP rocks were intercepted at depth by syn- or post-subduction diamond elevators, such as kimberlites, lamproites, lamprophyres, nephelinites or other alkali volcanic rocks of deep-seated origin, the diamond potential of the area would be even higher.
© 2004 Elsevier B.V. All rights reserved.

Keywords: Diamond; Subduction zone; Craton; Rodinia; British Columbia; Exploration; Tectonic elements

1. Introduction

Classical diamond-producing areas, such as the Northwest Territories, are located within old, stable

* Fax: +1-250-952-0381.
E-mail address: george.simandl@gems2.gov.bc.ca
(G.J. Simandl).

0024-4937/$ - see front matter © 2004 Elsevier B.V. All rights reserved.
doi:10.1016/j.lithos.2004.03.018

cratons and conform to the so called "Clifford's Rule" as considered by Janse (1994a). Such areas host most of the large primary diamond orebodies (Helmstaedt, 1993). A major exception is the Argyle Mine (Australia), which is located within the Proterozoic Hall's Creek mobile belt, adjacent to the Kimberley block (O'Neill et al., 2003). If Clifford's Rule is applied on a global scale, British Columbia (BC) does not rate as a "highly prospective" area for diamond exploration; however, eastern portions of BC are underlain by Proterozoic or Archean basement (Hoffman, 1988, Ross et al., 1991, 1995; Villeneuve et al., 1993). The same basement hosts a number of diamond-bearing pipes in neighbouring Alberta (Eccles, 2002; Ettlinger, 1998; Carlson et al., 1999); therefore, there is a possibility that similar pipes will be discovered in eastern BC.

In the past, most diamond exploration was confined to the belt-shaped Alkaline Province, which follows the Omineca–Foreland belt boundary and is characterized by a variety of alkaline rocks (Fig. 1). All diamond occurrences were discovered using tra-ditional prospecting and indicator minerals based on the "Diamondiferous Mantle Root" (DMR) concept.

1.1. Diamondiferous mantle root model

The diamondiferous mantle root model (DMR Model) is well described by Haggerty (1986), Boyd and Gurney (1986), Mitchell (1991), Kirkley et al. (1991), Helmstaedt (1993) and Helmstaedt and Gurney (1995). The model (Fig. 2) explains the origin of the world's main diamond deposits which occur within ancient stable cratons as predicted by Clifford's Rule. Classical examples of diamond-producing areas fitting this tectonic setting are the Slave (Canada), Kaapvaal (Southern Africa) and Siberian cratons. The model (Fig. 2) consists of a depleted mantle root and accreted mobile belt(s). The stable craton and underlying lithospheric keel may be more than 200 km thick. The diamond–graphite stability boundary is convex upward and diamonds that originate within the keel (the diamond stability field) are expected to be predominantly peridotitic, although eclogitic dia-

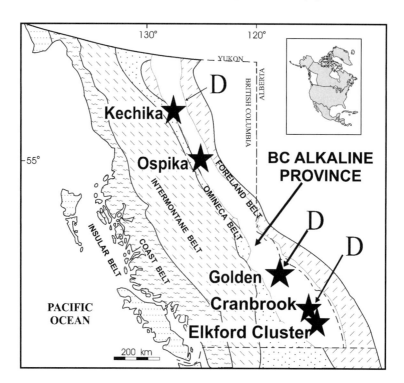

Fig. 1. British Columbia's Alkaline Province (shown in white) and the locations of major ultrabasic diatreme-bearing areas indicated by stars. D—identifies reported diamond/microdiamond localities. G8, G9 or G10 garnets have not been found outside of the Elkford Cluster.

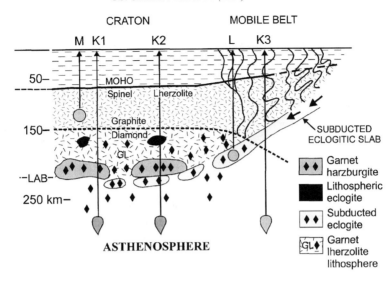

Fig. 2. Diamondiferous Mantle Root Model showing accretion of a mobile belt against Archean craton (after Mitchell, 1991). L—lamproite; M—melilitic magma; K—kimberlite; LAB—lithosphere–asthenosphere boundary. ♦—presence of diamond in host rock. K1 will carry peridotitic garnets, K2 and L will carry peridotitic and eclogitic garnets and K3 will be barren.

monds may also be present. Mobile belts are associated with a subduction zone (Fig. 2), and diamonds in this setting will be mostly eclogitic. Kimberlites or lamproites that intersect both the deep subducting slab and the keel below the diamond–graphite stability boundary may contain both eclogitic and peridotitic diamonds. Kimberlite and lamproite magmas act as elevators bringing diamonds to the surface. These magmas travel quickly, minimizing resorption of diamonds as they ascend to the surface (when outside of the diamond stability field).

1.2. Eclogite subduction zone model

The eclogite subduction zone model (ES model) provides an alternative for the selection of exploration targets in BC and an explanation for the origin of known BC diamond occurrences. The "Eclogite Subduction Zone" model, as proposed by Barron et al. (1994) and Barrows et al. (1996), corresponds to a typical convergent plate boundary (Fig. 3), where the key elements are the oceanic plate and underlying asthenosphere, crustal subduction characterized by dehydration as well as blueschist and eclogite facies metamorphism, oceanic trench (which may show obducted blueschist or eclogite facies metamorphic rocks), cold forearc, volcanic arc, hot backarc and

stable continent or craton. No thickening of the crust and continental lithosphere with increasing distance from volcanic arc and subcontinental mantle wedge is evident until the craton is reached (Hyndman and Lewis, 1999). Continental collision could be part of the subduction scenario but may not be essential. The main attraction of the ES model relative to the DMR model is created by the positive slope of the diamond–graphite stability line on the $P–T$ diagram (Fig. 4). In general, it is accepted that the temperatures are higher in oceanic settings. Under DMR model, primary diamond deposits are not expected to be associated with oceanic plates, which are young (few are older than 180 Ma), thin (Fig. 3) and hot relative to cratonic settings. Heat flow in the Archean shield areas is estimated at approximately 40 mW/m^2 and average heat flows for continental and oceanic crusts, which are highly variable, are estimated at 56.5 and 78.2 mW/m^2, respectively (Sclater et al., 1980). Important deviations to this generalization occur in convergent tectonic settings, where temperatures within the subducting slab may be anomalously low and the diamond stability field may be reached at depths of less than 100 km. This is substantiated by modern studies describing pressure–temperature–time ($P–T–t$) paths of subducting slabs and thermal models of several of "cold" subduction zones (e.g., van

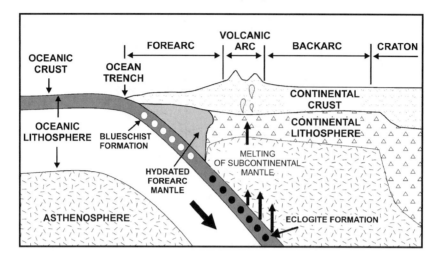

Fig. 3. Eclogite subduction zone model showing the typical convergent tectonic setting 1 (not to scale), including the locations of blueschist and eclogite formation. Microcontinent subduction or continental collision may be part of the scenario.

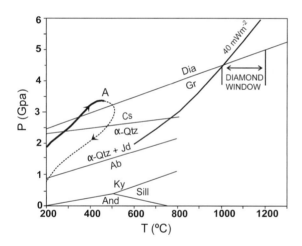

Fig. 4. Contrasting pressure–temperature conditions required for the formation of diamonds under the ES model (e.g., continental collisional setting) and DMR model. (A) depicts the metamorphic path of a central point within a subducting slab as described by Roselle and Engi (2002). This simulation is in line with conditions encountered in known UHP terranes of eastern China, western Norway and the western Alps. The prograde portion of path (A) is shown by a thick line and the retrograde portion of the path is shown by a dashed line. Dia/Gr—boundary between stability fields of diamond and graphite (Kennedy and Kennedy, 1976). "Diamond Window", as defined by Griffin and Ryan (1995), corresponds to approximate temperature conditions in cratonic areas required to form diamond under the DMR model. ES diamonds could therefore form at much shallower depths than DMR diamonds. Metamorphic grid modified from Roselle and Engi (2002). Dia—diamond; Gr—graphite; Cs—coesite; Ab—albite; Ky—kyanite; Sill—sillimanite; And—andalusite; α-Qtz—Alpha quartz; Jd—jadeite.

Keken et al., 2002). These studies indicate temperatures as low as 500–600 °C at 100 km depth within the slab, as described by Ponco and Peacock (1995). Slightly higher temperatures are suggested for ultra-high pressure metamorphic settings by Roselle and Engi (2002) and van Keken et al. (2002). For comparison, in the continental setting within the mantle root area (in the DMR model), diamonds are expected to form mainly at temperatures ranging from 1000 to 1200 °C and pressures around or in excess of 45 kilobars (Fig. 4). In the setting described by Roselle and Engi (2002), the subducting slab may reach the diamond stability field at temperatures below 500 °C and pressures below 35 kilobars.

The contrast between pressure–temperature conditions needed to reach the diamond stability field in the cold subduction zone setting (under the ES model) and in the intracratonic setting (under the DMR model) can be demonstrated using existing high-resolution models of the Honshu subduction zone in Japan (van Keken et al., 2002) and superposing on it the diamond window and continental geotherm from Griffin and Ryan (1995), as shown in Fig. 5. Curves A and B in this figure summarize $P–T$ conditions in different parts of the subducting slab. Because the Honshu subduction zone is characterized by rapid convergence, there is a strong predicted temperature gradient through the subducting crust. The top of the subducting crust (curve B) experiences fast tempera-

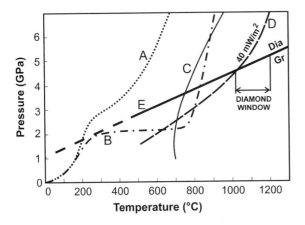

Fig. 5. Contrasting pressure temperature conditions required for formation of diamonds under ES model and DMR model. Pressure–temperature paths of subducting oceanic crust (an example from Honshu subduction zone, Japan). Curves account for stress- and temperature-dependent viscosity. (A) Cold curve represents conditions 7 km below the surface of the subducting slab (van Keken et al., 2002). (B) Hotter curve represents conditions at the boundary between subducting crust and overlying wedge (van Keken et al., 2002). (C) Wet solidus. (D) Xenolith-derived geotherm for northern Lesotho (40 mW/m^2) and associated "diamond window" (Griffin and Ryan, 1995). (E) Graphite–diamond stability boundary (Kennedy and Kennedy, 1976).

ture increases due to contact with the hot overlying asthenosphere wedge. Conversely, the base of the crust remains relatively cold (curve A). Deeper portions of the subducting crust reach the diamond–graphite stability boundary without melting; however, there could be melting along the contact between the subducting crust and overlying wedge. Fig. 5 supports the ES model and indicates that under favourable conditions, portions of the subducting slab may reach the diamond–graphite boundary at pressures under 30 kilobars. Depending on the kinetics of the graphite–diamond reaction, oxygen fugacity, and number of subduction-related parameters, diamonds may form at relatively low pressures (<100 km depth) within subducting oceanic crust.

The discovery of microdiamonds in the ultrahigh pressure (UHP) metamorphic rocks and ophiolites (Katayama et al., 2001; Bai et al., 1993; Xu et al., 1992; Liou et al., 2002; Massonne, 2001) demonstrates that the diamondiferous mantle root or meteorite impacts are not required for diamond formation in natural settings and that diamonds can form within convergent plate settings. The validity of the ES

model in diamond exploration remains to be proven in the field, but the model is worthwhile testing.

2. British Columbia's tectonic and geological setting

The tectonic setting of British Columbia consists of the North American craton and the Canadian Cordillera. The latter has been divided into five orogen-parallel morphogeological belts (Gabrielse et al., 1991). They are, from east to west, the Foreland, Omineca, Intermontane, Coast and Insular belts (Fig. 1).

The Canadian Cordillera has been located on an ocean–continent boundary since the supercontinent called Rodinia started to rift apart between 530 and 570 Ma ago (Price and Monger, 2000; Colpron et al., 2002). Fig. 6 shows the relative position between the West coast of North America and East coast of Australia and Antarctica as proposed by Hoffman (1991) and Young (1992). Similar interpretations are also advanced by Li et al. (1995), Rainbird et al. (1996) and Dutch (2002). This paleogeographic relationship between continents can be further exploited

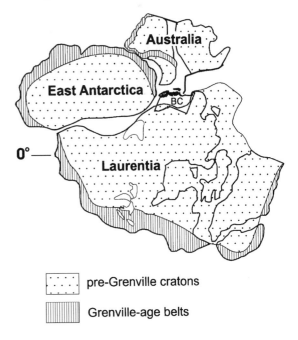

Fig. 6. Relative position of Laurentia, Australia and East Antarctica in the time interval of 1050 to 700 Ma (modified from Hoffman, 1991; Li et al., 1995); 0°—equator.

to draw possible parallels between unexplained diamond occurrences in North America and Eastern Australia and to explain an excellent diamond exploration potential of Antarctica, which was highlighted by Levinson et al. (1992). An alternative interpretation of Rodinia's configuration which links Siberia and Laurentia proposed by Sears and Price (2003) provides the same or an even higher level of exploration potential for eastern BC.

On a more detailed scale, the tectonic setting of BC can be described in terms of the North American craton, adjacent pericratonic and displaced terranes, and accreted superterranes (Fig. 7). The subduction along the western margin of the North American craton started at least 390 Ma ago (Price and Monger, 2000). The Cache Creek, Wrangellia and Alexander (Fig. 7) are interpreted as far-traveled terranes (Price and Monger, 2000; Monger and Nokleberg, 1996).

The Quesnellia, Stikinia and Slide Mountain terranes are considered as ancient portions of the North American plate, but are displaced relative to their original position. Pericratonic terranes formed near the craton; however, some of them were also displaced. Substantial information is available about the current setting on the convergent margin of western North America (Clowes et al., 1995; Cox et al., 1989; Hyndman, 1995, Hyndman et al., 2003; Mazzotti et al., 2002; Wang, 2000; Wang et al., 2001; Dragert et al., 2001; Oleskevich et al., 1999). However, no hard data exists on the past variations in the geometry (dip of subducting slab) and physical properties (subduction speed of the oceanic slab, its thickness, age and temperature) during that time.

All modern studies agree that episodic arc magmatism associated with the subduction zone did persist from the Mid-Devonian to present (Price and Monger,

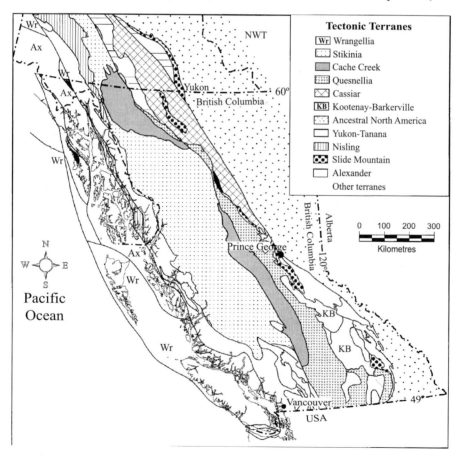

Fig. 7. Tectonic terrane map of British Columbia (modified from Wheeler et al., 1991).

2000). However, the interpreted location of volcanic arcs and the direction of the movement of the subducting plate relative to the current convergent plate boundary varies at any given point in time (Johnston, 2001).

In summary, most of the near-surface geology west of the Foreland belt (including large portions of the Omineca Belt) may be described as pericratonic or accreted terranes (Fig. 7 of Price and Monger, 2000). The Lithoprobe project and modern geophysical studies suggest that, at depth, a substantial portion of the province may be underlain by subcontinental lithosphere (Clowes et al., 1995). However, Os isotopic systems in mantle xenoliths suggest that the mantle that underlies the Canadian Cordillera is probably not a simple extension of the North American cratonic lithosphere (Peslier et al., 2000b). Furthermore, according to Hyndman and Lewis (1999), there is a thermal transition between Cordillera (80–100 mW/m^2) and Craton (40–60 mW/m^2). This transition is located approximately 100 km west of the limit of Cordilleran deformation front and is less than 200 km wide (Hyndman and Lewis, 1999). The thermal transition coincides roughly with the Rocky Mountain trench and BC's Alkaline Province.

3. British Columbia's Alkaline Province

British Columbia's Alkaline Province (Fig. 1) is characterized by a variety of alkaline rocks including carbonatites, nepheline syenites, kimberlites and other alkaline rocks (Pell, 1994). Four major ultrabasic pipe/dyke areas were investigated for their diamond potential by the industry (Fig. 1). They are described here as the Ospika River, Kechika, Golden and Cranbrook areas and were discussed by a number of authors including Ijewliw (1991), Pell (1994), Pell and Ijewliw (2003), McCallum (1994) and Helmstaedt et al. (1988). The term Elkford Cluster, as used in this paper, comprises the Cross, Bonus and Ram group of pipes. This cluster contains BC's only known pipes containing mantle-derived pyrope garnets indicating derivation from depths in excess of 50 km.

Pipes and dykes within the Alkaline Province include calc–alkaline and alkaline lamprophyres, possibly limburgites, olivine melilitites, lamproites, kimberlites and rocks of basaltic affinity. A variety of basaltic and ultramafic dykes and lamprophyre rocks within the province have been also reported outside of the above highlighted areas. Many of known dykes and pipes contain Cr-diopside, spinels, ilmenites and other indicator minerals. However, to date, no sites outside of the Elkford cluster have been found to contain garnets that can be classified as G8, G9 or G10. The Cross pipe (Smith et al., 1988) and adjacent Bonus pipes are confirmed as true kimberlites while Ram 5 and 6.5 pipes may be kimberlites, lamproites or lamprophyres.

The ages of some of the BC's pipes were determined by radiometric dating. However, in most cases, the ages are constrained only by overlying unconformities, cross cutting and stratigraphic relationships, or the presence/absence of metamorphic and tectonic overprint. The Ospika pipe was dated at 334 ± 7 and 351 ± 3 Ma by Rb/Sr and at 323 ± 10 Ma using K/Ar methods (Pell, 1994). Pipes of the Golden cluster cut Upper Cambrian rocks and are believed to be Late Silurian to Early Devonian in age. The Rb/Sr date on mica from HP pipe is 391 ± 5 Ma and the K/Ar method gives 391 ± 12Ma (Pell, 1994). The Cross Kimberlite's Rb/Sr dates on phlogopite are 240–250 Ma (Grieve, 1982; Smith et al., 1988). Parrish and Reichenbach (1991) studied a number of pipes including Cross, Blackfoot and Joff in the Cranbrook area as well as Jack, Mark, Mike and HP pipes, which are part of the Golden Cluster. Using U–Pb method, zircon dates vary from 2.7 Ga to 440 Ma. These age variations do not represent ages of pipe emplacements, as the zircon xenocrysts were probably derived from a combination of sources, including basement gneisses and local intrusive rocks. The oldest zircons may have been liberated by weathering from the western Canadian Shield, shed westward, incorporated into pericratonic terranes and later extracted by pipes. It is also possible that these old dates indicate the age of the lower crust, which was penetrated by the pipes. A number of the pipes terminate against Mid- or Late Devonian unconformities. If diamond-bearing pipes were eroded, there is a possibility of diamond placers.

4. British Columbia's diamond occurrences

Diamonds were reported from the Jack (Lens Mountain) and Mark (Valenciennes River) pipes (Northcote, 1983a,b) within the Golden area (Fig.

1). A single microdiamond was reported from a breccia subcrop located near the RAR-5 pipe on the Xeno property located in the Kechika area (Roberts, 2002). This property was extensively explored for Yttrium and rare earth elements (Pell, 1994; Pell et al., 1989). Diamonds were reported in the Cranbrook cluster, from the Bonus and Ram 5 pipes (Anonymous, 1994; Anderson, 1999; Allan, 1999, 2002). However, no diamonds were recovered from core drilled by Skeena Resources Limited at the Ram 6 pipe (Allan, 2002). Indicator mineral data in selected deposits in British Columbia is covered in several of the assessment reports of the BC Ministry of Energy and Mines and other industry reports, such as Stapleton (1997), McCallum (2000), and to some extent, Fipke et al. (1995) and Hall (1991). Unfortunately, the overall quality of the microprobe analyses is inconsistent and the discrimination diagrams in some reports are not supported by tabulated data.

Currently, there is little information regarding the morphology and composition of BC diamonds and their inclusions. In the future, it would be worthwhile to confirm reported diamond occurrences. Recovered diamonds should be physically, chemically and isotopically characterized and their inclusions as well as related indicator minerals systematically studied and documented. The benefits of such costly and labour-intensive studies are commonly questioned. However, in many cases, such studies represent a huge leap forward in our understanding of nontraditional diamond occurrences. An excellent example of such research is given by De Corte et al. (1999) regarding the Kokchetav Massif (Kazakhstan). In BC's context, such studies may point to the DMR model, the Argyle (mobile-belt) setting or the ES model. Regardless of the outcome, the detailed studies of diamonds would contribute to the better understanding of the province's diamond potential and provide better focus diamond exploration in BC.

5. High-pressure metamorphic rocks

High-pressure (blueschist and eclogite facies) rocks are known along the margin of Ancestral North America from Alaska to Mexico (Fig. 8A). Isotopic ages of these rocks range from 37 to 447 Ma (Erdmer et al., 1998). Within BC alone, there are at least 7 such localities. Those shown in Fig. 8A are Dease Lake (172 Ma), Pinchi Lake (211–223 Ma) and Bridge River (230 Ma). These localities correspond to numbers 13, 14 and 15, respectively. It is not clear if these radiometric dates represent peak metamorphic conditions, near-peak retrograde metamorphic conditions, later cooling or resetting. French Range and Jennings River localities are also important, but no dates are available for these. High-pressure rocks in BC are not known to carry diamonds; however, diamonds are known in ultrahigh pressure metamorphic rocks at several localities around the world (Bai et al., 1993; De Corte et al., 1999; Liou et al., 2002).

The high-pressure ultramafic rocks in BC can be also used to locate ancient subduction zone trenches or continental collision zones. The position of trenches in relation to associated volcanic arcs could in turn be used to determine the dip direction of the corresponding subduction zone and to select exploration areas in terms of the ES model (i.e., down dip from the ancient subduction zone trenches). According to Hausel (1994) and Janse (1994b), some of the unexplained alluvial diamond occurrences reported in the literature may have been derived from ultrahigh pressure rocks (Fig. 8B) or other unconventional sources.

6. Alkali basalts and related rocks

Cenozoic alkali basalts, related basanitoids, nephelinites and ankaramite are known in BC and some of them contain mantle xenoliths (Fuji and Scarfe, 1982; Mitchell, 1987; Peslier, 1998; Peslier et al., 2000a,b; Carignan et al., 1996; Edwards and Russell, 2000; Abraham et al., 2000; Shi et al., 1998): mainly, spinel lherzolite and much less abundant olivine websterite, websterite, clinopyroxenite and wehrlite. These rocks are located mostly west of the Alkaline Province, are Tertiary or younger in age and originated at greater depths than typical basalts and andesites that are common in subduction zone-related volcanic arcs. Osmium isotopes were used by Peslier et al. (2000b) to characterize the lherzolites and harzburgite xenoliths from these volcanic rocks as derived in an off-craton setting. The $^{187}Os/^{188}Os$ and Lu data has been used to infer a model age of 1.12 $\pm$ 0.26 Ga (Mesoproterozoic) for the stabilization of the mantle lithosphere beneath the Canadian Cordillera (Peslier,

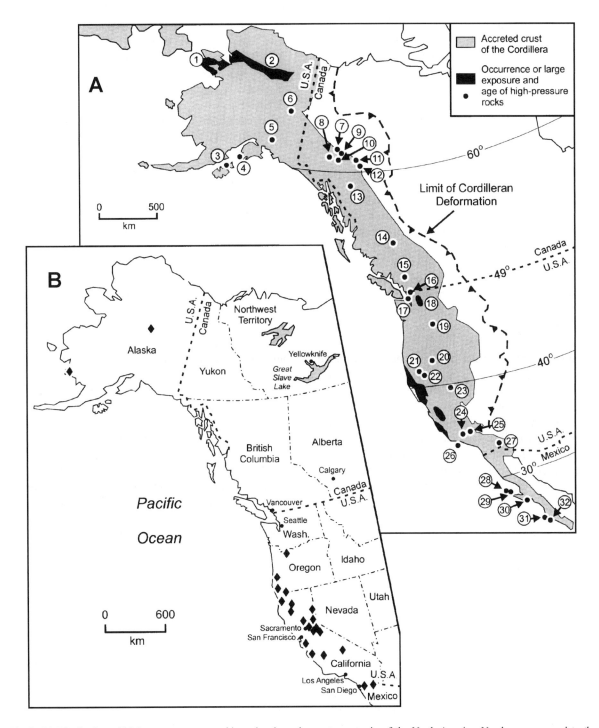

Fig. 8. (A) Distribution of high-pressure metamorphic rocks along the western margin of the North America. Numbers correspond to the locations of high-pressure rocks described by Erdmer et al. (1998). (B) Distribution of alluvial diamonds in western North America (Hausel, 1994).

1998). The age difference between upper crust terranes and formation of underlying mantle suggest that
they are not genetically related and that crustal terranes were probably thrust on top of mantle lithosphere. The $^{87}Sr/^{86}Sr$ and $^{143}Nd/^{144}Nd$ and Rb
concentrations in lavas are specific to individual
tectonic belts (Intermontane, Omineca and Coast)
and such distinctive signatures appear to be inherited
from lithospheric mantle (Abraham et al., 2000).
Available minimum depth estimates for the origin of
these rocks vary from 30 to 55 km based on nature of
mantle xenoliths that they contain (Fuji and Scarfe,
1982; Canil and Scarfe, 1989). This is consistent with
the lack of peridotite garnet within the mantle xenoliths; however, some of the xenoliths may have
originated at greater depth. In addition to pressure
and temperature, other factors (such as the chemistry
of the protolith) affect the stability field of garnet.
Mineral assemblages predicted within the subducting
slab along a given pressure–temperature path differ
depending on the nature of subducted material.
Assemblages containing chlorite, lawsonite, zoisite,
amphibole, chloritoid and clinopyroxene are expected
if subducted material is a water-saturated basalt.
Chlorite, clinopyroxene, orthopyroxene, garnet, amphibole, olivine are expected if subducted material is a
water-bearing lherzolite (e.g., van Keken et al., 2002).
Therefore, depending on their depth of origin and
tectonic setting and age, some of the alkali volcanic
rocks are potential diamond transporters in the context
of the ES model.

7. Discussion

This section discusses the relevance of the DMR
and ES models to diamond exploration in BC with
brief comments on the diamond placer potential of
the province.

7.1. Applicability of DMR to British Columbia

The Precambrian basement rocks of similar age
and nature as those that underlie the diamondiferous
kimberlite occurrences east of the Alberta–BC border
and within the Northwest Territories described by
Carlson et al. (1999) do extend into eastern BC
(Fig. 9). For example, the Archean Nova Terrane

may be a fragment of the Slave Craton. Intersections
of the NW–SE trending controls on BC geology
(including the Alkaline Province itself) with the major
NE–SW lineament sets oriented parallel to, or coinciding with a tectonic feature commonly referred to as
the Peace River Arch should be examined in more
detail (Fig. 10), particularly those associated with
known diamond occurrences (such as the Mountain
L.—Buffalo Head—Birch Mountain).

There are also other reasons for suggesting that a
"deep keel" may have underlain parts of BC and
Alberta's basement in the past. For example, before
the breakup of Rodinia (Fig. 6), the west coast of
North America was probably joined with, or adjacent
to, Eastern Australia and Antarctica (Li et al., 1995;
Dutch, 2002), and was underlain by a mantle root at
this time. This may explain the presence of placer
diamond deposits in Eastern Australia, the fact that
most of the BC occurrences do not fit the traditional
DMR setting, and why portions of Antarctica appear
to have excellent diamond exploration potential. More
recently, it was proposed that Laurentia (including
Eastern BC) was linked with Siberia (Sears and Price,
2003). Regardless of the configuration, the remnants
of the keel in southeastern BC probably persisted in a
modified form at least until the Triassic. This is
supported by the presence of garnet bearing-mantle
xenoliths within the Cross Kimberlite, which was
dated at 240–250 Ma (Grieve, 1982; Smith et al.,
1988). An analogous example of the setting described
above may be eastern China's Sino–Korean Craton,
where a diamond-bearing area may have been underlain by the Archean lithospheric keel, which was since
destroyed (Xu, 2001).

In summary, independently of the paleotectonic
continent matching hypothesis, the DMR model can
be applied to eastern parts of BC, where Archean or
Proterozoic basement (Fig. 9) can be extrapolated
from Alberta and Northwest Territories by a combination of geology and geophysics.

7.2. Applicability of the ES model to British Columbia

The discovery of the lamproite-hosted Argyle Mine
in northwest Australia opened new diamond exploration areas, and for several years attracted attention to
"Proterozoic Mobile Belts" and other "off craton"
prospects that are still considered by number of junior

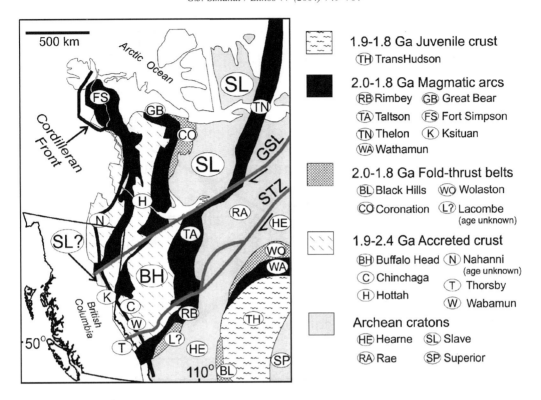

Fig. 9. Generalized tectonic map showing projections of Proterozoic basement rocks in Alberta and British Columbia. Note also that a portion of the Slave Craton (Archean) may have been displaced along the Great Slave Lake shear zone all the way to the British Columbia Cordilleran Front. GSL—Great Slave Shear Zone. STZ—Snowbird Tectonic Zone. Modified from Gehrels and Ross (1998).

exploration companies. Ultrabasic pipes within BC's Alkaline Province are traditionally interpreted as rift-related (Pell, 1994). However, some of the pipes that are younger than 390 Ma postdate the start of the subduction. These younger pipes may have intercepted a UHP portion of subducting slab during its ascent and incorporated diamonds with eclogitic origins. Some of the zircons that were recovered from BC's pipes have postsubduction ages; others have Proterozoic and Archean ages. The latter two may have been derived from an old lower crust intersected by the pipes, or alternately, they may have weathered out of the Precambrian Shield and became incorporated into younger pericratonic terranes and subsequently extracted by pipes.

Diamond discoveries in nontraditional settings, such as Alaska, Yukon, Australia, (Janse, 1994b; Hausel, 1994, 1998; Liedtke, 2002; Dow, 2003) unexplained alluvial diamond occurrences in the Western USA (Fig. 8B) and elsewhere throughout the world, indicate that if these reports are correct, the future research and exploration should not be focused exclusively to intracratonic settings. New models, including the ES model, should not be discarded without careful consideration. If the ES model is valid, BC has a good potential to host diamond-bearing rocks west of the Rocky Mountain Trench. The tasks undertaken during the selection of exploration target areas would be to examine the geological history, determine geographic location of past "cold" subduction zones as defined by Peacock and Wang (1999), their ages and identify alkali rocks that could have intercepted the subducting slab and carried diamonds to the surface. Alternately, slivers of the subducting slab containing diamonds could have been exhumed from the diamond stability field by tectonic activity (obduction). All known diamond-bearing occurrences that have been tectonically exhumed contain only small diamonds that cannot be recovered economically. Such diamonds occur main-

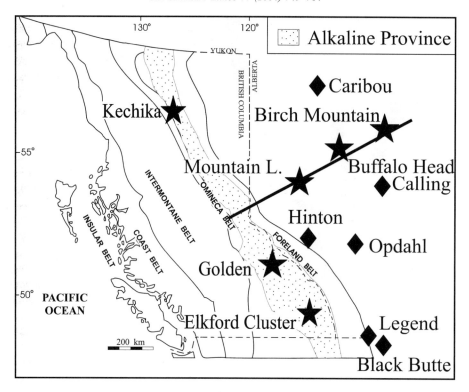

Fig. 10. Compilation of BC and Alberta diamond occurrences. Primary host rock diamond occurrences are identified by ★. Locations of alluvial occurrences are indicated by ◆ (Dufresne et al., 1994; Ettlinger, 1998).

ly as inclusions within other minerals such as garnet, zircon and clinopyroxene, or as mineral aggregates replacing garnet (Sobolev and Shatsky, 1990) and zoisite, kyanite and biotite (De Corte et al., 1999). The host minerals are commonly believed to act as armor and protect diamonds during their tectonic ascent. Diamonds formed in a subduction environment are expected to be associated with a different indicator mineral suite than diamonds from intracratonic settings. Thus, the standard indicator mineral exploration methodology would have to be modified to fit the ES model.

7.3. Placer potential of British Columbia

BC is well known for its placer gold deposits. Diamonds were reported in the drainage of the Tulamene ultramafic complex near Princeton (Camsell, 1911); however, the presence of diamonds was not independently confirmed (Lang, 1970). In summary, no diamonds were recovered during placer gold/platinum mining in the province. This may be because there are no diamonds present, or it may be related to recovery techniques. The equipment used in placer gold extraction is not ideally suited for diamond exploration, which involves specialized gigs, heavy-media separation grease tables and X-ray sorters.

The diamond-placer potential of British Columbia has not been seriously evaluated, but several regions may be worth examining. The number of pipes in southeastern BC (some of them believed to be diamondiferous) appears to terminate against local or regional unconformities (Pell, 1994) and current erosional surfaces. Indicator minerals in till and in glaciolacustrine sediments in northeastern BC may have also originated within the Slave Craton (Northwest Territories) or Buffalo Hills (Alberta), or they may have originated from local undiscovered primary deposits. For example, the Fort St. John area (near British Columbia/Alberta border) may have some potential. The Peace River Arch structure is considered as a favourable zone for kimberlite emplacement, where there are several diamond-bearing pipes with an economic potential east of BC/Alberta border (Eccles, 2002; Carlson et al., 1999). The last ice to have overlain the Fort St. John area was probably of Wisconsinian age, which created a series of large ice-dammed lakes (Mathews, 1980). Today, riverbanks and gravel pits within the area display well-developed cross-beddings and paleochannels enriched in heavy minerals. Fine visible gold, abundant zircon and garnet were observed during the field preconcentration of heavy mineral samples, although the analyses of diamond indicator minerals are not available at the time of writing.

Elsewhere in the province, diamond placer deposits could conceivably be derived from unconventional diamond occurrences. Diamonds were reported in alpine and ophiolitic peridotites (Helmstaedt, 1993; Hausel, 1994), komatiitic rocks (Capdevila and Arndt, 1999), lamprophyres, other alkali rocks and basalts (Janse, 1994b; Hausel, 1994, 1998). Such rocks are relatively common in BC, although they are not known to be diamondiferous. A map of BC's ultramafic rocks (Voormeij and Simandl, 2004) in conjunction with current and paleodrainage maps, locations of high pressure rocks and known pipe-bearing areas could be useful in selecting areas for grassroots exploration programs aimed at unconventional primary and placer diamond deposits.

8. Summary

It is unlikely that BC will match the diamond potential of the Northwest Territories. However, parts of the province along the BC/Alberta border (east of the Rocky Mountain Trench) are underlain by Proterozoic and Archean basement and have acceptable exploration potential in terms of the DMR model. Regions where diamond-related lineaments and unexplained indicator mineral anomalies extend from Alberta to the BC border merit particular attention. It is also conceivable that before the break-up of Rodinia, British Columbia may have been underlain by a deep keel, which was destroyed during the subsequent rifting and subduction activity. This is independent of the exact configuration of Rodinia, and the hypotheses of both Hoffman (1988) and

Sears and Price (2003) are compatible with our interpretation.

The ES model is supported by recent $P-T-t$ modeling studies of a variety of "cold" subduction zones and by documented diamond occurrences in convergent tectonic settings such as Kokchetav massif (Kazakhstan). If gem-quality diamonds can form within the subducting slab and be preserved during a rapid tectonic exhumation process, or brought to the surface by volcanic rocks, then BC has much larger diamond exploration potential than we realize.

Acknowledgements

The authors gratefully acknowledge Brian Grant of British Columbia Ministry of Energy and Mines, and Danae Voormeij and Nicole Robinson from the University of Victoria who edited an earlier version of this document. The document benefited from constructive comments of Dr. Suzanne Paradis and Dr. Roy D. Hyndman of Geological Survey of Canada, Victoria; Dr. Jennifer Pell of Dunsmuir Ventures, Vancouver, Dr. Barbara H. Scott Smith of Scott-Smith Petrology, Vancouver, and an anonymous reviewer.

References

Abraham, A.C., Francis, D., Polvé, M., 2000. Recent alkaline basalts as probes of the lithospheric mantle roots of the northern Canadian Cordillera. Chemical Geology 175, 361–386.

Allan, R.J., 1999. Micro-diamond results from Ice Claim Project, B.C. Skeena Resources, News release—August 20, 1999.

Allan, R.J., 2002. Diamond exploration corporate up-date. Skeena Resources, News release—March 22, 2002.

Anderson, D., 1999. Assessment Report on Rock Sampling and Micro-Diamond Testing–Ice Property, Elkford Area, BC, Canada. Assessment Report 26 030. British Columbia Ministry of Energy and Mines, Victoria.

Anonymous, 1994. BC Diamonds discovered—Consolidated Ramrod Gold. George Cross News Letter, No.225.

Bai, W., Robinson, P.T., Zhou, M., 1993. Diamond-bearing peridotites from Tibetan ophiolites: implication for subduction-related origin of diamonds. In: Dunne, K.P.E., Grant, B. (Eds.), Mid-Continental Diamonds. GAC-MAC Symposium Volume. Geological Association of Canada, Edmonton, Alberta, pp. 77–82.

Barron, L.M., Lishmund, S.R., Oakes, G.M., Barron, B.J., 1994. Subduction diamonds in New South Wales: implications for exploration in Eastern Australia. Quarterly Notes-Geological Survey of New South Wales 94, 1–23.

Barrows, L.M., Lishmunwg, R., Oakes, M., Barron, B.J., Sutherland, F.L., et al., 1996. Subduction model for the origin of some diamonds in the Phanerozoic of eastern New South Wales. Australian Journal of Earth Sciences 43, 257–267.

Boyd, F.R., Gurney, J.J., 1986. Diamonds and the African lithosphere. Science 232, 472–477.

Camsell, C., 1911. A new diamond locality in the Tulameen district, British Columbia. Economic Geology 6, 604–611.

Canil, D., Scarfe, C., 1989. Origin of phlogopite in mantle xenoliths from Costal Lake, Wells Gray Park, British Columbia. Journal of Petrology 30, 1159–1179.

Capdevila, R., Arndt, N., 1999. Diamonds in volcanoclastic komatiite from French Guiana. Nature 399, 456–458.

Carignan, J., Ludden, J., Francis, D., 1996. On the recent enrichment of subcontinental lithosphere: a detailed U–Pb study of spinel lherzolite xenoliths, Yukon, Canada. Geochimica et Cosmochimica Acta 60, 4241–4252.

Carlson, S.M., Hillier, W.D., Hood, C.T., Pryde, R.P., Skelton, D.N., 1999. The Buffalo Hills kimberlites: a newly-discovered diamondiferous kimberlite province in north-central Alberta, Canada. In: Gurney, J.J., Gurney, J.L., Pascoe, M.D., Richardson, S.H. (Eds.), Proceedings of the VIIth International Kimberlite Conference, vol. 1. Red Roof Design, Capetown, pp. 109–116.

Clowes, R.M., Zelt, C.A., Amor, J.R., Ellis, R.M., 1995. Lithospheric structure in the southern Canadian Cordillera from a network of seismic refraction lines. Canadian Journal of Earth Sciences 32, 1485–1513.

Colpron, M., Logan, J.M., Mortensen, J.K., 2002. U–Pb zircon age constrain for late Neoproterozoic rifting and initiation of the lower Paleozoic passive margin of western Laurentia. Canadian Journal of Earth Sciences 39, 133–143.

Cox, A., Debiche, M.G., Enggebretson, D.C., 1989. Terrane trajectories and plate interactions along continental margin in the North Pacific Basin. In: Ben-Avraham, Z. (Ed.), The Evolution of Pacific Ocean Margin. Oxford Press, New York, pp. 20–35.

De Corte, K., Cartigny, P., Shatsky, V.S., De Paepe, P., Sobolev, N.V., Javoy, M., 1999. Characteristics of microdiamonds from UHPM rocks of Kokchetav Massif, Kazakhstan. In: Gurney, J.J., Gurney, J.L., Pascoe, M.D., Richardson, S.H. (Eds.), Proceedings of the VIIth International Kimberlite Conference, vol. 1. Red Roof Design, Capetown, pp. 174–182.

Dow, R.B., 2003. Patrician Finds More Diamonds, Sapphires and Kimberlitic Indicators in Yukon. Patrician Diamonds. Press release November 21.

Dragert, H., Wang, K., James, T.S., 2001. A silent slip event on the deeper Cascadia subduction interface. Science 292, 1525–1528.

Dufresne, M.B., Olson, R.A., Schmitt, D.R., McKinstry, B., Eccles, D.R., Fenton, M.M., Pawlowics, J.G., Edwards, W.A.D., Richardson, R.J.H., 1994. The diamond potential of Alberta: a regional synthesis of the structural and stratigraphic setting, and other preliminary indication of diamond potential. Open File Report 94 -10, Canada–Alberta MDA Project M93-04-037. Alberta Geological Survey, Edmonton.

Dutch, S., 2002. History of global plate motions; www.uwgb.edu/dutchs/platetec/plhist94.htm.

Eccles, D.R., 2002. Enzyme leach-based soil geochemistry of the Mountain Lake diatreme, Alberta. In: Dunlop, S., Simandl, G.J. (Eds.), Industrial Minerals in Canada. Canadian Institute of Mining and Metallurgy and Petroleum, Special, vol. 53, pp. 355–360, Montréal, QC.

Edwards, B.R., Russell, J.K., 2000. The distribution, nature and origin of Neogene–Quaternary magmatism in the Northern Cordilleran Volcanic Province, northern Canadian Cordillera. Geological Society of America Bulletin 112, 1280–1295.

Erdmer, P., Ghent, E.D., Archibald, D.A., Stout, M.Z., 1998. Paleozoic and Mesozoic high pressure metamorphism at the margin of ancestral North America in Central Yukon. Geological Society of America Bulletin 110, 615–629.

Ettlinger, A.D., 1998. Diamonds in Alberta. Yorkton Securities Mining Research, Vancouver.

Fipke, C.E., Gurney, J.J., Moore, K.O., 1995. Diamond exploration techniques emphasizing indicator mineral geochemistry and Canadian examples. Geological Survey of Canada Bulletin, 423.

Fuji, T., Scarfe, C.M., 1982. Petrology of ultramafic nodules from West Kettle River, near Kelowna, southern British Columbia. Contributions to Mineralogy and Petrology 80, 297–306.

Gabrielse, H., Monger, J.W.R., Wheeler, J.O., Yorath, C.J., 1991. Part A. Morpho–geological belts, tectonic assemblages and terranesGabrielse, H., Yorath, C.J. (Eds.), Geology of the Cordilleran Orogen in Canada, vol. 4. Geological Survey of Canada, Ottawa, ON, pp. 15–28 (Chapter 2).

Gehrels, G.E., Ross, G.M., 1998. Detrital zircon geochronology of Neoproterozoic to Permian miogeoclinal strata in British Columbia and Alberta. Canadian Journal of Earth Sciences 35, 1380–1402.

Grieve, D.A., 1982. Diatreme breccias in the southern Rocky Mountains. Geological Fieldwork 1980, Paper 1981-1. British Columbia Ministry of Energy, Mines and Petroleum Resources, pp. 96–103.

Griffin, W.L., Ryan, C.G., 1995. Trace elements in indicator minerals: area selection and target evaluation in diamond exploration. Journal of Geochemical Exploration 53, 311–317.

Haggerty, S.E., 1986. Diamond genesis in a multiply constrained model. Nature 320, 34–38.

Hall, D.C., 1991. A petrological investigation of the Cross Kimberlite occurrence, Southeastern British Columbia, Canada. PhD thesis. Queens University, Kingston, Ontario.

Hausel, W.D., 1994. Pacific Coast diamonds—an unconventional source terrane. Mineral Report MR94-8. Wyoming State Geological Survey, Laramie, Wyoming.

Hausel, W.D., 1998. Diamonds and mantle source rocks in the Wyoming Craton, with a discussion of other US occurrences. Report of Investigations, vol. 53. Wyoming State Geological Survey.

Helmstaedt, H.H., 1993. Primary diamond deposits—what controls their size, grade, and location. In: Whiting, B.H., Mason, R., Hodgson, C.J. (Eds.), Giant Ore Deposits, Special Publication, vol. 2. Society of Economic Geologists, Littleton, CO, pp. 13–80.

Helmstaedt, H.H., Gurney, J.J., 1995. Geotectonic controls of primary diamond deposits: implications for area selection. Journal of Geochemical Exploration 53, 125–144.

Helmstaedt, H.H., Mott, J.A., Hall, D.C., Schulze, D.J., Dixon,

J.M., 1988. Stratigraphic and structural setting of intrusive breccia diatremes in the White River–Bull River area, southeastern British Columbia. Geological Fieldwork 1987, Paper 1988-1. British Columbia Ministry of Energy, Mines and Petroleum Resources, pp. 363–368.

Hoffman, P.F., 1988. United plates of America: the birth of a craton. Annual Review of Earth and Planetary Sciences 16, 543–603.

Hoffman, P.F, 1991. Did the breakout of Laurentia turn Gondwanaland inside–out? Science 252, 1409–1412.

Hyndman, R.D., 1995. Review: the thermal regime along the southern Canadian Cordillera lithoprobe corridor. Canadian Journal of Earth Sciences 32, 1611–1617.

Hyndman, R.D., Lewis, T.J., 1999. Geophysical concequences of the Cordillera–Craton thermal transition in southwestern Canada. Tectonophysics 306, 397–422.

Hyndman, R.D., Mazzotti, S., Weichert, D.H., Rogers, G.C., 2003. Frequency of large crustal earthquakes in Puget Sound—S. Georgia Strait predicted from geodetic and geological deformation rates. Journal of Geophysical Research 108, 2033, pp. 12-1–12-12.

Ijewliw, O.J., 1991. Petrology of the Golden cluster lamprophyres, southeastern British Columbia, Canada. MSc thesis. Queen's University, Kingston, Ontario.

Janse, A.J.A., 1994a. Is Clifford's Rule still valid? In: Meyer, H.O.A., Leonardos, O.H. (Eds.), Kimberlites, Related Rocks and Mantle Xenoliths. Proceedings of the 5th International Kimberlite Conference, Araxa, Special Publication 1/A Jan 94, vol. 2. Companhia de Pesquisa de Recursos Minerais—CPRM, Brasilia, pp. 215–235.

Janse, A.J.A., 1994b. Review of supposedly non-kimberlitic and non-lamproitic diamond host rocks. In: Meyer, H.O.A., Leonardos, O.H. (Eds.), Kimberlites, Related Rocks and Mantle Xenoliths. Proceedings of the 5th International Kimberlite Conference, Araxa, Special Publication 1/A Jan 94, vol. 2. Companhia de Pesquisa de Recursos Minerais—CPRM, Brasilia, pp. 144–159.

Johnston, S.T., 2001. The great Alaskan terrane wreck: reconciliation of paleomagnetic and geological data in the northern Cordillera. Earth and Planetary Science Letters 123, 259–272.

Katayama, I., Maruyama, S., Parkinson, C.D., Terada, K., Sano, Y., 2001. Ion microprobe U–Pb zircon geochronology of peak and retrograde stages of ultrahigh-pressure metamorphic rocks from the Kokchetav Massif, Northern Kazakhstan. Earth and Planetary Science Letters 188, 185–196.

Kennedy, C.S., Kennedy, G.C., 1976. The equilibrium boundary between graphite and diamond. Journal of Geophysical Research 81, 2467–2470.

Kirkley, M.B., Gurney, J.J., Levinson, A.A., 1991. Age, origin and emplacement of diamonds: scientific advances in the last decade. Gems and Gemology, Spring, 2–25.

Lang, A.H., 1970. Prospecting in Canada. Economic Geology Report, vol. 7. Geological Survey of Canada, Ottawa, ON.

Levinson, A.A., Gurney, J.J., Kirkley, M.B., 1992. Diamond sources and production: past, present and future. Gems and Gemology 26, 234–254.

Li, Z.-X., Zhang, L., Powell, C.McA., 1995. South China in Rodi-

nia: part of the missing link between Australia–East Antarctica and Laurentia? Geology 23, 407–410.

Liedtke, G.J., Diamonds discovered on Shulin Lake Proerty, Alaska. Press Release—July 30.2002. Golcanda Resources.

Liou, J.G., Zhang, R.-Y., Katayama, I., Maruyama, S., Ernst, G.W., 2002. Petrotectonic characterization of the Kokchetav Massif and the Dabie–Sulu Terrane—ultrahigh-P metamorphism in the so-called P–T Forbidden Zone. Western Pacific Earth Sciences 2, 119–148.

McCallum, M.E., 1994. Lamproitic (?) diatremes in the Golden area of the Rocky Mountain fold and thrust belt, British Columbia, Canada. In: Meyer, H.O.A., Leonardos, O.H. (Eds.), Kimberlites, Related Rocks and Mantle Xenoliths. Proceedings of the 5th International Kimberlite Conference, Araxa, Special Publication 1/A Jan 94, vol. 1. Companhia de Pesquisa de Recursos Minerais—CPRM, Brasilia, pp. 195–209.

McCallum, M.E., 2000. Chemistry of definite and probable/possible kimberlite indicator minerals from heavy mineral concentrate recovered from samples collected in 1999 for Skeena Resources from Ice prospect, Southeastern British Columbia. Unpublished Report. Skeena Resources.

Massonne, H.-J., 2001. A new occurrence of microdiamonds in quartzofeldspathic rocks of the Saxonian Erzgebirge, Germany, and their metamorphic evolution. In: Gurney, J.J., Gurney, J.L., Pascoe, M.D., Richardson, S.H. (Eds.), Proceedings of the VIIth International Kimberlite Conference, vol. 2. Red Roof Design, Goodwood, SA, pp. 532–539.

Mathews, W.H., 1980. Retreat of the last ice sheets in northeastern British Columbia and adjacent Alberta. Bulletin, vol. 331. Geological Survey of Canada, Ottawa, ON.

Mazzotti, S., Dragert, H., Hyndman, R.D., Miller, M.M., Henton, J.A., 2002. GPS deformation in a region of high crustal seismicity: N. Cascadia forearc. Earth and Planetary Science Letters 198, 41–48.

Mitchell, R.H., 1987. Mantle-derived xenoliths in Canada. In: Nixon, P.H. (Ed.), Mantle Xenoliths. Wiley, New York, pp. 33–40.

Mitchell, R.H., 1991. Kimberlites and lamproites: primary sources of diamonds. Geoscience Canada 18, 1–16.

Monger, J.W.H., Nokleberg, W.J., 1996. Evolution of the northern North American Cordillera: generation, fragmentation, displacement and accretion of successive North American plate–margin arcs. In: Coyner, A.R., Fahey, P.L. (Eds.), Geology and Ore Deposits of the American Cordillera, Geological Society of Nevada Symposium Proceedings, Reno/Sparks, Nevada, pp. 1133–1152.

Northcote, K.E., 1983a. Report on Mark property, Pangman Peek (82N/15W). Assessment Report 13596. British Columbia Ministry of Energy Mines and Petroleum Resources.

Northcote, K.E., 1983b. Report on Jack Claims Lens Mountain (82N/14E). Assessment Report 13597. British Columbia Ministry of Energy, Mines and Petroleum Resources.

Oleskevich, D., Hyndman, R.D., Wang, K., 1999. The up and downdip limits to great subduction earthquakes: thermal and structural models of Cascadia, S.W. Japan, Alaska and Chile. Journal of Geophysical Research 104, 965–14991.

O'Neill, C., Moresi, L., Lenardic, A., Cooper, C.M., 2003. Inferences on Australia's heat flow and thermal structure from mantle convection modeling results. In: Hillis, R.R., Müller, R.D. (Eds.), Special Publication. Evolution and Dynamics of the Australian Plate, vol. 22. Geological Society of Australia, Sydney, Australia, pp. 163–178.

Parrish, R.R., Reichenbach, J., 1991. Age of xenocrystic zircons from diatremes of western Canada. Canadian Journal of Earth Sciences 28, 1159–1238.

Peacock, S.M., Wang, K., 1999. Seismic consequences of warm versus cool subduction zone metamorphism: examples from northeast and southwest Japan. Science 286, 937–939.

Pell, J., 1994. Carbonatites, Nepheline Syenites, Kimberlites and Related Rocks in British Columbia. Bulletin, vol. 88. British Columbia Ministry of Energy and Mines, Victoria, BC.

Pell, J.A., Ijewliw, O.J., 2003. Kimberlites, melnoites and look-alikes in British Columbia, Canada. Extended Abstract FLA 200. CD-ROM, 8th International Kimberlite Conference, Victoria, BC, Canada.

Pell, J., Cullbert, R.R., Fox, M., 1989. The Kechika yttrium and rare earth prospect. Geological Fieldwork 1988, Paper 89-1. British Columbia Ministry of Energy and Mines, pp. 417–421.

Peslier, H.A., 1998. Pétrologie et Géochimie Isotopique de Xénolites Mantelliques de la Cordillère Canadienne. PhD thesis. Université de Montréal, Montréal, Canada.

Peslier, A., Reisberg, L., Ludden, J., Francis, D., 2000a. Re–Os constraints on harzburgite and lherzolite formation in the lithospheric mantle: a study of northern Canadian Cordilleran xenoliths. Geochimica et Cosmochimica Acta 64, 3061–3071.

Peslier, A.H., Reisberg, L., Ludden, J., Francis, D., 2000b. Os isotopic systematics in mantle xenoliths; age constraints on the Canadian Cordilleran lithosphere. Chemical Geology 166, 85–101.

Ponco, S.C., Peacock, S.M., 1995. Thermal modeling of the southern Alaska subduction zone: insight into the petrology of the subducting slab and overlying mantle wedge. Journal of Geophysical Research 100, 22117–22118.

Price, R.A., Monger, J.W.H., 2000. A transect of the southern Canadian Cordillera from Calgary to Vancouver. A Fieldtrip Guidebook. Geological Association of Canada—Cordilleran Section, Vancouver.

Rainbird, R.H., Jefferson, C.W., Young, G.M., 1996. The early Neoproterozoic sedimentary succession B of north-western Laurentia: correlations and paleogeographic significance. Geological Society of America Bulletin 108, 454–470.

Roberts, W.J., 2002. Diamond discovery at Xeno. News Release— Wednesday, March 13, 2002, Pacific Ridge Exploration.

Roselle, G.T., Engi, M., 2002. Ultra high pressure (UPH) terrains: lessons from thermal modeling. American Journal of Science 302, 410–441.

Ross, G.M., Parish, R.R., Villeneuve, M.E., Bowring, S.A., 1991. Geophysics and geochronology of the crystalline basement of the Alberta Basin, western Canada. Canadian Journal of Earth Sciences 28, 512–522.

Ross, G.M., Milkerreit, B., Eaton, D., White, D., Kanasewich, E.R.,

Buryanyk, M.J.A., 1995. Paleoproterozoic collisional orogen beneath western Canada sedimentary basin imaged by lithoprobe crustal seismic reflection data. Geology 23, 195–199.

Sears, J.W., Price, R.A., 2003. Tightening the Siberian connection to western Laurentia. GSA Bulletin 115, 943–953.

Sclater, J.G., Jaupart, C., Galson, D., 1980. The heatflow through oceanic and continental crust and the heat loss of the earth. Reviews of Geophysics and Space Physics 18, 269–311.

Shi, L., Francis, D., Ludden, J., Frederiksen, A., Bostock, M., 1998. Xenolith evidence of lithospheric melting above anomalously hot mantle under the northern Canadian Cordillera. Contributions to Mineralogy and Petrology 131, 39–53.

Smith, C.B., Colgan, E.A., Hawthorne, J.B., Hutchinson, G., 1988. Emplacement age of the Cross Kimberlite, southeastern British Columbia, by Rb–Sr phlogopite method. Canadian Journal of Earth Sciences 25, 790–792.

Sobolev, N.V., Shatsky, V.S., 1990. Diamond inclusions in garnets from metamorphic rocks: a new environment for diamond formation. Nature 343, 742–746.

Stapleton, B.A., 1997. Moose creek and neighbouring claims; Peace Diamond Project, Assessment Report 25 081. British Columbia Ministry of Energy and Mines.

van Keken, P.E., Kiefer, B., Peacock, S.M., 2002. High resolution models for subduction zones: implications for mineral dehydration reactions and the transport of water into the deep mantle. G^3 Geochechemistry Geophysics Geosystems. AGU and Geochemical Society 3, 1–20.

Villeneuve, M.E., Ross, G.M., Thérialt, R.J., Miles, W., Parrish, R.R., Broome, J., 1993. Tectonic subdivision and U–Pb geochronology of the crystalline basement of the Alberta basin, western Canada. Geological Survey of Canada Bulletin, 447.

Voormeij, D.A., Simandl, G.J., 2004. A Map of Ultramafic Rocks in British Columbia: Applications in CO_2 Sequestration and Mineral Exploration, GEOFILE 2004-1. British Columbia Ministry of Energy and Mines, Victoria.

Wang, K., 2000. Stress–strain paradox, plate coupling, and forearc seismicity at the Cascadia and Nankai subduction zones. Tectonophysics 319, 321–338.

Wang, K., He, J., Dragert, H., James, T., 2001. Three-dimensional viscoelastic interseismic deformation model for the Cascadia subduction zone. Earth Planets Space 53, 295–306.

Wheeler, J.O., Brookfield, A.J., Gabrielse, H., Monger, J.W.H., Tipper, H.W., Woodsworth, G.J., 1991. Terrane Map of Canadian Cordillera. Geological Survey of Canada, Map 1713A, scale 1:2,000,000.

Xu, Y.-G., 2001. Thermo–tectonic destruction of the Archean lithospheric keel beneath the Sino–Korean craton in China: evidence, timing and mechanism. Physics and Chemistry of the Earth: Part A. Solid Earth and Geodesy 26, 747–757.

Xu, S., Okay, A.I., Ji, S., Sengor, A.M.C., Su, W., Liu, Y., Jiang, L., 1992. Diamond from Dabie Shan metamorphic rocks and its implication for tectonic setting. Science 256, 80–82.

Young, G.M., 1992. Late Proterozoic stratigraphy and the Canada–Australian connection. Geology 20, 215–218.

Available online at www.sciencedirect.com

Lithos 77 (2004) 765–782

www.elsevier.com/locate/lithos

Area selection for diamond exploration using deep-probing electromagnetic surveying

Alan G. Jones*, James A. Craven

Geological Survey of Canada, 615 Booth Street, Ottawa, ON, Canada K1A 0E9

Received 25 June 2003; accepted 4 January 2004
Available online 7 July 2004

Abstract

Previously proposed methods of area selection for diamond-prospective regions have predominantly relied on till geochemistry, airborne geophysics, and/or an appraisal of tectonic setting. Herein we suggest that a novel, deep-probing geophysical technique—electromagnetic studies using the natural-source magnetotelluric (MT) method—can contribute to such an activity. Essentially, diamondiferous regions must have (1) old lithosphere, (2) thick lithosphere, and (3) lithosphere that contains high concentrations of carbon. Deep-probing MT studies are able to address all three of these. The second and the third of these can be accomplished independently using MT, but for the first the geometries produced from modelling the MT observations must be interpreted with appropriate interaction with geologists, geochemists and other geophysicists. Examples are given from the Slave and Superior cratons in North America, with a brief mention of an area of the Rae craton, and general speculations about possible diamondiferous regions.
© 2004 Published by Elsevier B.V.

Keywords: Magnetotellurics; Geophysics; Slave craton; Superior craton; Rae craton

1. Introduction

As diamond exploration activities proceed into frontier areas there is a critical need for an effective area selection methodology. Previously proposed methods of area selection for diamond-prospective regions have predominantly relied on till geochemistry (e.g., Griffin and Ryan, 1995; Gurney and Zweistra, 1995; Jennings, 1995), airborne geophysics (e.g.,

Macnae, 1995), and/or an appraisal of tectonic setting (e.g., Helmstaedt and Gurney, 1995). As discussed by Morgan (1995), regional scale geophysical methods can aid in area selection for diamondiferous provinces, and, of the available geophysical methods, magnetotellurics and teleseismics are the two methods able to "look" into the mantle and compliment each other well. They are, in fact, the only geophysical methods with true depth resolving capability of material property variations—the other methods use inference rather than direct detection. To date, teleseismics has been used predominantly to derive the geometries of sub-cratonic Archean lithospheric mantles. Herein we propose that deep-probing electromag-

* Corresponding author. Now at: Dublin Institute for Advanced Studies, 5 Merrion Square, Dublin 2, Ireland. Fax: +1-613-943-9285.
E-mail addresses: ajones@nrcan.gc.ca, ajones@cp.dias.ie (A.G. Jones).

0024-4937/$ - see front matter © 2004 Published by Elsevier B.V.
doi:10.1016/j.lithos.2004.03.057

netic studies, using the natural-source magnetotelluric (MT) technique (Jones, 1998, 1999), offer an attractive additional, and on occasion alternate, cost-effective means for rapid area selection of diamond-prospective regions.

Essentially, diamondiferous regions must have (1) old lithosphere, (2) thick lithosphere, and, what is not as appreciably discussed in the literature, (3) lithosphere that contains high concentrations of carbon. All three of these are important: lithosphere that is "young", generally taken as Proterozoic and younger, does not appear to host diamonds. Kimberlites intruded through lithosphere that is thinner than the diamond–graphite stability field will not be diamondiferous. Lithosphere that is poor in carbon content will most likely be poor in diamonds, possibly both in diamond quantity and in diamond grade.

As we will present and demonstrate in this paper, deep-probing MT studies are able to address all three of these. The second and the third can be undertaken independently by using MT alone, but for the first—old lithosphere—the geometries produced from modelling the MT observations must be interpreted with appropriate interaction with geologists, geochemists and other geophysicists, especially seismologists. We will show the results from two cratons, the Slave craton and the western part of the Superior craton, mention briefly an area of the Rae craton, and deduce areas prospective for diamond exploration based on our results.

2. The magnetotelluric technique

The magnetotelluric (MT) method is a natural-source, electromagnetic (EM) geophysical surveying technique for obtaining information about the variations in electrical conductivity in all three dimensions, both laterally and vertically. One records the time-varying EM field components, the two horizontal electric field components (Ex and Ey) and all three magnetic field components (Hx, Hy and Hz), typically along profiles at a number of locations on the surface of the Earth. From these time series one computes response functions (the so-called MT impedance tensors), that relate the electric and magnetic fields to each other, and these responses are then analyzed and interpreted for information about Earth structure. Older reviews of the method are given in Vozoff

(1972, 1986, 1991), and Jones (1998, 1999) discusses modern MT methods and practices, especially the imaging of the continental lithospheric and asthenospheric mantle. More specialised, technical reviews are published in issues of Kluwer's journal *Surveys in Geophysics* devoted to the review papers presented at the biennial series of EM Induction Workshops (see Surveys in Geophysics volume 13, pages 305–505, 1992; volume 17, 361–556, 1996; volume 18, 441–510, 1997; volume 20, 197–375, 1999; and volume 23, 99–273, 2002).

The basic physical phenomenon that governs all EM methods is the skin-depth effect by which penetration into a medium (in this case the Earth) is a function of the resistivity of the medium and the period of the incident EM wave. The skin depth, defined mathematically in a uniform medium as the depth at which the amplitude of the wave has been attenuated by $1/e$ (approx. 37%) of its surface value, is given by

$$\delta \approx 500\sqrt{\rho T} \ (\mathrm{m})$$

where δ is the skin depth (in m), ρ is the resistivity of the medium (in Ω m), and T is the period (in s). Thus, the shortest periods observed in MT (0.000025 to 0.001 s, i.e., 40 to 1 kHz in frequency) give information about the near surface (top tens to hundreds of metres), whereas the longest periods (10,000 s and on occasion beyond) give information about deep structures (typically to the base of the continental lithosphere and into the asthenosphere). This is shown schematically in Fig. 1 with an *apparent resistivity* curve, deduced from the scaled squared magnitude of the electric field divided by the perpendicular magnetic field, for a simple two-layered Earth. The short periods (high frequencies) do not penetrate deeply, sensing only the top layer and thus give the resistivity of the top layer, ρ_1. The long period (low frequencies) EM waves pass through the top layer without being affected by it, and give the resistivity of the bottom layer, ρ_2. The period where the curve crosses ρ_1 gives the thickness of the top layer, h. In addition to deriving a magnitude relationship between the electric and magnetic fields, then squaring and scaling it to yield an apparent resistivity at each period, in MT one also calculates the phase difference between the two fields. This phase difference is $45°$ for a uniform

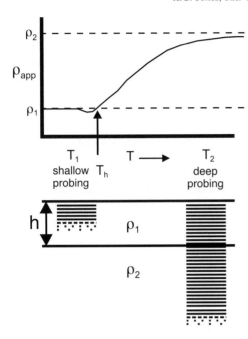

Fig. 1. Schematic apparent resistivity curve for a two-layered Earth with a less resistive layer (ρ_1) over a more resistive one (ρ_2). The apparent resistivity curve departs from the resistivity of the top layer at a period that gives the thickness of the top layer.

Earth, is less than 45° when moving to a more resistive layer with depth, and greater than 45° when moving to a conductive layer with depth.

One of MT's advantages is that exactly the same method, and almost the same instrumentation, can be used for investigations of geothermal energy resources at 100 m (e.g., Meju, 2002), mineral resources at 500–1000 m (e.g., Meju, 2002; Jones and Garcia, 2003), crustal structures to 40+ km (e.g., Jones, 1998), and lithospheric and deeper mantle to >200 km (e.g., Schultz et al., 1993; Jones et al., 2001, 2003), albeit with decreasing resolution with depth. Different magnetic sensors need to be used for the different applications, as not one single sensor exists that covers the whole period range required to image from the surface down to the base of the lithosphere.

At the low frequencies used in MT the propagating EM waves can be treated mathematically as diffusive in nature, rather than as wave propagation, and consequently MT has lower spatial resolution than most seismic methods. The MT response to a structure is the integration of the conductivity from the surface down to that structure, akin to seismic surface wave

studies. However, in contrast to the small percentages of seismic impedance variations, electrical resistivity varies by over many orders of magnitude, from 0.001 Ω m and lower for graphite to 100,000 Ω m and higher for competent, unfractured silicate rock (see, e.g., Beblo, 1982; Haak, 1982), thus enabling high sensitivity to anomalous structures containing an interconnected, conducting minor phase. In addition, and in comparison to potential field methods, the MT method is not inherently non-unique; a uniqueness theorem for one-dimensional (1-D) earths, i.e., where resistivity varies with depth alone, exists (Bailey, 1970).

One-dimensional models can be fitted to the data, by objective inversion approaches, which represent minimum structure models of one sense or another. The models have either a minimum number of layers (e.g., Fischer and Le Quang, 1981) or minimum change (gradient) between many multiple layers in an over-parameterized problem (Constable et al., 1987; Smith and Booker, 1988). These two represent end-member acceptable models, and the "truth", the actual resistivity–depth distribution, lies in between. Remembering the uniqueness theorem, non-uniqueness in MT model parameter resolution is due to data inaccuracy (bias errors), imprecision (statistical errors) and inadequacy (insufficient bandwidth and insufficient estimates per decade), but not due to inherent non-uniqueness.

As well as 1-D layer-cake models of the Earth, MT data can be modelled using two-dimensional (2-D) and three-dimensional (3-D) methods, which include both forward trial-and-error approaches (e.g., Wannamaker et al., 1984, 1985) and formal inversion (e.g., deGroot-Hedlin and Constable, 1990; Smith and Booker, 1991; Siripunvaraporn and Egbert, 2000; Rodi and Mackie, 2001). Two-dimensional models are appropriate for MT data that exhibit two-dimensionality in their responses, whereas 3-D models are appropriate for array data or MT data that cannot be validly modelled using 2-D. Dimensionality validity tests are undertaken within a statistical framework that considers the effects of local, electric field distortion and also noise in the data, and yields the regional 2-D (Groom and Bailey, 1989; Groom et al., 1993; McNeice and Jones, 2001) or 3-D (Garcia and Jones, 2001) responses.

Finally, akin to seismic anisotropy revealed through teleseismic shear wave SKS studies (see Snyder et al., 2004), various workers present MT anisotropy determined qualitatively from the MT strike information (e.g., Mareschal et al., 1995; Simpson, 2001; Bahr and Simpson, 2002). However, given the vast range of electrical conductivity, one must be careful that the EM information from different directions is coming from the same depths. We present here MT anisotropy information from a large portion of the western part of the Superior craton that is consistent with Archean tectonic processes.

3. Old lithosphere

That diamonds occur primarily in Archean regions is the well-known *Clifford's Rule* (Clifford, 1966). However, a dilemma facing diamond exploration is the development of a reasonably useful predictive model for the tectonic development of Mesoarchean lithosphere. Whereas the plate tectonic paradigm can be used as a well-established predictive exercise in mineral exploration, plate tectonics, in the modern sense, only became the dominant tectonic process by Paleoproterozoic times, although some are appealing to it to explain Neoarchean events (e.g., Davis et al., 2003; White et al., 2003). In Paleoarchean times it has been suggested that plate tectonics, sensu stricto, was not the dominant tectonic process (articulated most recently by Bleeker, 2002), and the transition between the dominance of plume or sagduction (e.g., Davies, 1992; Hamilton, 1998) tectonics over plate tectonics must have occurred as the mantle became more ordered and less chaotic, likely during the Mesoarchean. The conundrum is that many diamonds are reported to be dated as Mesoarchean in age, so their search is compounded by this difficulty in defining an appropriate predictive model.

Geometries of structures imaged by EM studies within the sub-continental lithospheric mantle aid in the development of a model for the tectonic history of the craton, which relates to its age. This has been demonstrated for the Slave craton, the western part of the Superior craton, and part of the Rae craton, and is discussed in greater detail below for each of these.

4. Thick lithosphere

Diamonds only exist stably in thick, cold lithosphere at depths where the continental lithosphere–

asthenosphere transition, taken typically as an isotherm of about 1300 °C denoting a change from a conductive geotherm to convective geotherm (e.g., Jaupart and Mareschal, 1999), lies below the graphite–diamond stability field. In the absence of any interconnected conducting material, the sub-continental lithospheric mantle (SCLM) is electrically highly resistive. Laboratory studies of the resistivity variation with temperature of mantle materials, olivine, orthopyroxene and clinopyroxene, show that resistivity values of many hundreds to tens of thousands of ohm-metres are to be expected (Xu et al., 2000), decreasing with increasing temperature. This variation is shown in Fig. 2 for those three minerals, calculated from the formulae in Xu et al. (2000) using the values shown in the figure for each mineral in the solid-state Arrhenius equation (see Xu et al., 2000, for explanation). Using appropriate mixing relationships, it is possible to determine the resistivity of the mantle for any given mineral modal composition (Ledo and Jones, 2003), and Ledo and Jones (2003) show an application for a region of the Yukon known to have harzburgitic lithospheric mantle. Using minimum resistivity estimates for the upper mantle of 5000 Ω m and the observed harzburgitic mineralogy, Ledo and Jones (2003) deduce a maximum possible temperature

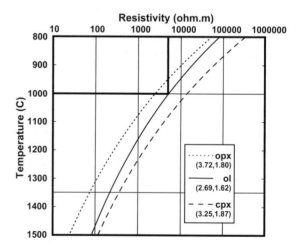

Fig. 2. Resistivity–temperature curves for olivine, Opx and Cpx mineralogies based on laboratory measurements in Xu et al. (2000). The numbers given for each mineralogy are the pre-exponent term ($\log_{10}(\sigma_0)$, where σ_0 is in S/m) and activation enthalpies (exponent term, ΔH, in eV) respectively in the Arrhenius equation for conductivity given by $\sigma = \sigma_0 \exp(-\Delta H/kT)$, where k is Boltzmann's constant and T is the temperature in Kelvin.

of 1000 °C, and a less-well defined minimum temperature of 850 °C, for the Intermontane Belt.

MT can determine the thickness of the lithosphere, i.e., the depth to the lithosphere–asthenosphere boundary (LAB), due to the two orders of magnitude increase in electrical conductivity at the onset of partial melt. Sensitivity to the onset of partial melt is shown in Fig. 3 derived from the melting experimental data in Schilling et al. (1997) and Partzsch et al. (2000). The figure shows the results of heating experiments on a sample of pyroxene granulite—not a mantle rock but illustrative of the effect. At low temperatures (<1050 °C), the material conducts electric currents by the flow of electrons, and this can be described by the solid-state Arrhenius equation. At high temperatures (>1100 °C) ionic conduction becomes dominant and can be described by an appropriate mixing law.

Between these two, in the transition zone where partial melt is low (<2%), both mechanisms are important and resistivity decreases by over 1.5 orders of magnitudes within 50 °C. Quenching studies show that the melt becomes interconnected, which is critical for reducing electrical resistivity, at very low partial melt fractions. Minarik and Watson (1995) and Drury and Fitz Gerald (1996) both demonstrate that interconnectivity can be achieved with partials melts below 0.1%.

It should be appreciated that these laboratory experiments are very difficult to perform; the results

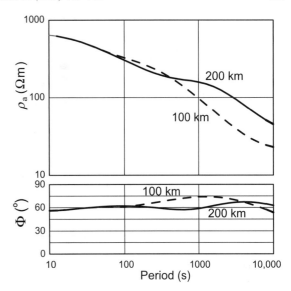

Fig. 4. Apparent resistivity and phase responses for a standard continental lithospheric mantle (see Fig. 5) with the base of the lithosphere at 200 km (full lines) and 100 km (dashed lines).

of Schilling et al. (1997) and Partzsch et al. (2000) required holding the temperature constant for 200 h to obtain a stable result at each data point. Accordingly, there are no studies of appropriate mantle rocks being taken from solid to partially molten.

Such high sensitivity to the onset of partial melt, with resistivity decreasing by over 1.5 orders of magnitude over a temperature range of less than 50 °C, means that precise MT data have the highest potential precision of any geophysical method to the depth to the LAB. This boundary is generally taken to have a temperature in the region of 1250–1350 °C (1280 °C, McKenzie and Bickle, 1988; 1250–1350 °C, Ryan et al., 1996), so the resistivity is expected to decrease from some hundreds of Ω m to about 10 Ω m over some 25 km in depth (which is equivalent to a temperature increase of around 50 °C).

Fig. 4 shows the effect on the apparent resistivity and phase curves when the depth to the LAB is changed from 200 to 100 km. Note in particular the phase responses in the 100–1000 s band of periods. The shallower LAB has higher phases than the deeper one.

As an example of LAB determination using MT, Fig. 5 shows the data, and corresponding 1-D layered Earth model, for a site near Kiruna in northern Sweden (Jones, 1982). The LAB is estimated to be

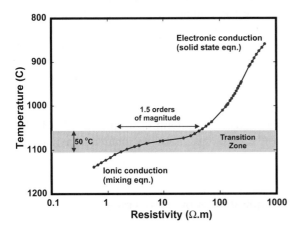

Fig. 3. Variation of resistivity of pyroxene granulite with increasing temperature as the rock is taken through partial melt. Based on the data of Schilling et al. (1997) and Partzsch et al. (2000).

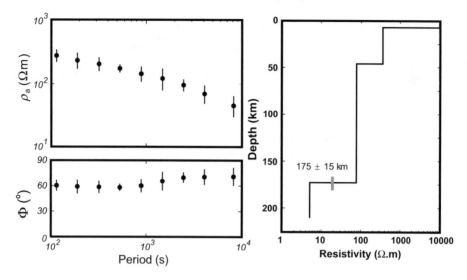

Fig. 5. EM responses from Kiruna, northern Sweden, expressed as MT parameters, plus the best-fitting 1-D model.

at a depth of 175 ± 15 km, which correlates excellently with lithospheric thickness estimates derived from seismic surface waves studies (Calcagnile, 1982, 1991; Calcagnile and Panza, 1987).

In 1977, Alekseyev et al. (1977) produced maps of the Former Soviet Union showing properties of the asthenosphere deduced from seismological and electromagnetic experiments. The maps are reproduced in Fig. 6, and demonstrate that electrical and seismological properties spatially correlate.

Accordingly, the MT method can detect whether the LAB exceeds the graphite–diamond (G–D) stability field, or not.

5. Presence of carbon

There are many candidates that can be introduced into mantle rocks for reducing electrical resistivity, and these are discussed below. For both the Slave and Superior conductors (discussed below), we exclude all of them with the exception of interconnected carbon on grain boundaries or in conducting graphite form. Carbon is highly unusual in that it can exhibit extremely low resistivity in one form, and extremely high resistivity in another. When in graphite form the carbon atoms have one S orbit electron and two P orbit electrons, and the atoms form themselves into a lattice with triangular planar symmetry with a low

energy barrier between the valence band and the conduction band. Besides the other petrophysical properties that this structure possesses, graphite has conductivity (inverse of resistivity) of 10^4 S/m (10^{-4} Ω m resistivity) across the layers, and 10^{10} S/m (10^{-10} Ω m resistivity) along the layers. In stark contrast, when carbon is in diamond form it has one S orbit electron and three P orbit electrons, and the atoms are arranged in a lattice with tetrahedral symmetry with a very high energy barrier between the valence and conduction bands. Consequently, the conductivity of diamond is 10^{-12} S/m (10^{12} Ω m resistivity).

In the sub-continental lithospheric mantle, above the graphite–diamond (G–D) transition, an interconnected graphite phase decreases electrical resistivity by two or more orders of magnitude over that predicted from laboratory studies and petrophysical modelling for an olivine or pyroxene mineralogy dominant upper mantle. Below the G–D transition, carbon exists in the form of diamond, and is highly resistive, and thus invisible, unfortunately, to EM observations.

6. Application to the Slave craton

Deep-probing magnetotelluric studies have been conducted on the Slave craton since 1996 as part of a number of programs (Jones et al., 2001, 2003). The

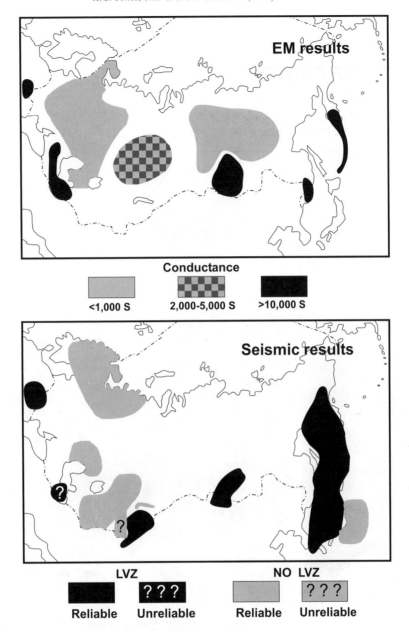

Fig. 6. Comparison of EM and seismological observations of parameters of the asthenosphere in the Former Soviet Union. Regions of high conductance for the asthenosphere correlate spatially with regions of observed seismic low velocity zones. Similarly, regions of low conductance correlate with regions where the LVZ is absent. Re-drawn from Alekseyev et al. (1977).

locations where MT data have been acquired are shown in Fig. 7 with reference to the important kimberlites, the isotope boundaries of Thorpe et al. (1992) and Davis and Hegner (1992), and the G-10 garnet lines of Grütter et al. (1999). MT coverage is greatest in the southwest part of the craton, and lowest in the northern third with only the four lake-bottom sites providing information.

Averaging the responses from all sites together, and scaling to account for local distortion effects,

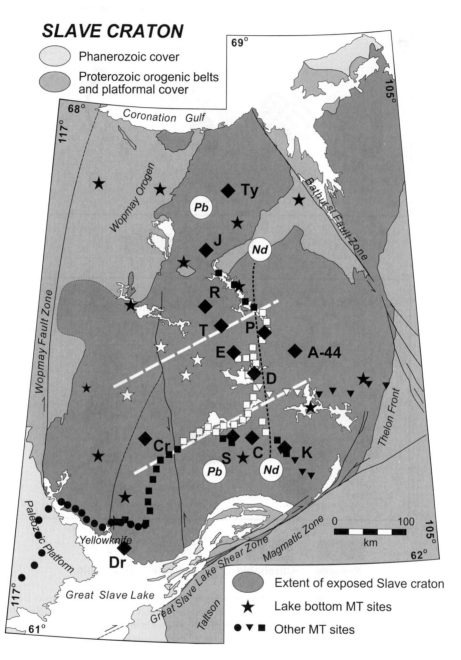

Fig. 7. Locations of magnetotelluric sites on the Slave craton, northwestern Canada. Dots: 1996 survey. Squares: 1998, 1999 and 2000 winter road surveys. Inverted triangles: 2001 Targeted Geoscience Initiative survey. Stars: 1999–2000 and 2000–2001 lake bottom locations. Open symbols show sites interpreted to be on top of the Central Slave Mantle Conductor. White dashed lines: G10 boundaries of Grütter et al. (1999). Pb and Nd isotope lines taken from Thorpe et al. (1992) and Davis and Hegner (1992). Diamonds: Important kimberlites: Ty—Tenacity; J—Jericho; R—Ranch; T—; P—Point Lake; E—Ekati; D—Diavik; A—44; K—Kennady; C—Camsell; S—Snap; Cr—Cross Lake; Dr—Drybones.

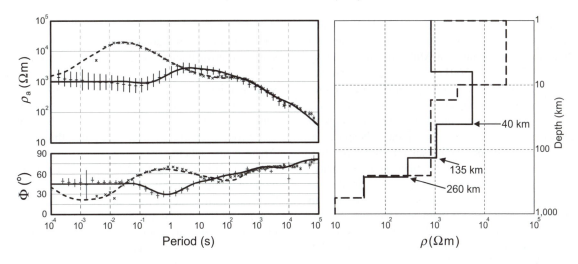

Fig. 8. Solid lines: Averaged Slave craton MT responses and 1-D best-fitting minimum layers model. Dashed lines: MT data from the central part of the Superior craton (Schultz et al., 1993) and best-fitting 1-D model.

yields the overall Slave MT apparent resistivity and phase responses shown in Fig. 8 (more details are given in Jones et al., 2003). The 1-D model that fits these data with the minimum number of layers is also shown in Fig. 8, and comprises a two-layered crust over a two-layered mantle lithosphere of thick-

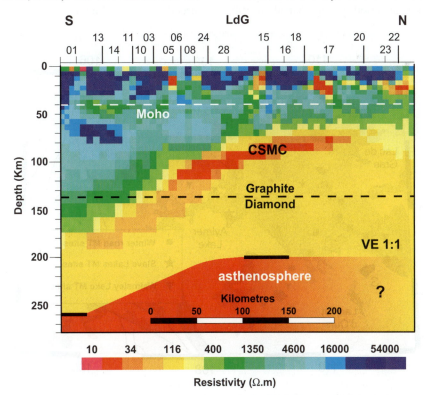

Fig. 9. Best-fitting 2-D N–S model of the winter road data from the Slave craton. LdG = Lac de Gras. CSMC = Central Slave Mantle Conductor. The depths to the base of the lithosphere are given by 1-D inversions. Taken from Jones et al. (2003).

ness 260 km. The base of the crust at 40 km was a fixed parameter in the inversion, given the sharp and distinct resistivity change at Moho depths noted by Jones and Ferguson (2001). The intra-mantle lithosphere discontinuity at 135 km approximately correlates with the graphite–diamond stability field, but it is weakly resolved and the resistivity change is in the wrong direction. The base of the lithosphere at 260 km is one of the well-resolved model parameters, but nevertheless has a rather large statistical error of ± 50 km given the large error estimates for the averaged data. A thickness of 260 km for an average dominated by southern Slave sites correlates well with the thickness of 260 km derived petrologically by Kopylova et al. (2001) from xenolith material recovered from the Kennady kimberlite (K in Fig. 7).

Comparing the averaged Slave MT response with the most accurate and precise MT response determined for any craton, namely the central part of the Superior craton (Schultz et al., 1993), shows an astounding and remarkable similarity at periods greater than about 300

s. Both the apparent resistivity curves and the phase curves from the two cratons are within each other's statistical error bounds. The 1-D model derived from this part of the Superior craton also exhibits a lithospheric thickness of 260 km, but with a far higher precision (± 20 km). That these two both display the same electrical lithospheric thickness attests to physical process limits on the maximum thickness of Archean lithosphere.

A 2-D North–South model, derived from the data along the Lupin mine winter road, is shown in Fig. 9. The prominent feature of the model is the detection of a zone of low resistivity beneath the Lac de Gras kimberlite region at depths of some 80–120 km. This Central Slave Mantle Conductor (CSMC) lies within the graphite stability field. 2-D models were also constructed along the other profiles (see Jones et al., 2003), and the stations that lie directly above the CSMC are shown in Fig. 7 in open symbols, and a plan of the CSMC is shown in Fig. 10. Note in Fig. 10 that the eastern limit of the CSMC along the profile directly east of Lac de Gras correlates spatially

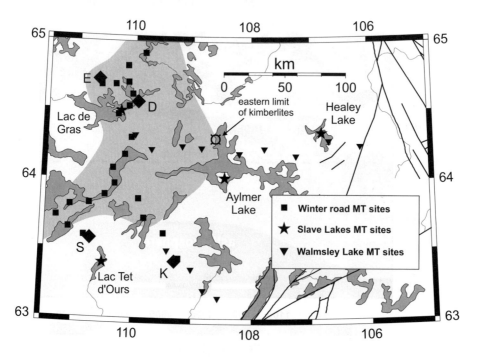

Fig. 10. Eastern boundary of the Central Slave Mantle Conductor compared to the known eastern limit of Eocene kimberlites. MT sites from the different MT surveys are depicted with different symbols. The stars represent the locations of Ekati (E), Diavik (D), Snap Lake (S) and Kennady Lake (K).

precisely with the eastern limit of Eocene kimberlites in the northern arm of Aylmer Lake.

A 3-D model of the data has been constructed, and a depth slice at 111 km is shown in Fig. 11. The CSMC can be seen to be NE–SW trending. Slices at deeper depths (not shown) imply a NW dip to the anomaly.

MT sites on the CSMC are shown in Fig. 7 as open symbols, and the spatial correlation with Grütter et al.'s (1999) G10 garnet boundaries is clearly evident. Note that the CSMC does not respect the N–S trending isotope boundaries, but, as with Grütter's lines, crosses them at a sharp angle. Also, the conductor correlates spatially and in depth extent with an ultra-depleted mantle region mapped by Griffin et al. (1999).

Based on xenolith studies from northern, central and southern Slave (Kopylova et al., 1997, 2001; Pearson et al., 1999), the expected temperatures at 80–100 km depths are 700–750 °C, which, by reference to Fig. 2, suggests an ambient resistivity of the order of >30,000 Ω m is to be expected, as is seen in the southern part of the Slave craton (Fig. 9). In contrast, the CSMC has a resistivity <15 Ω m (Jones et al., 2003). The modelled 3-D geometry of this body, with a NE–SW strike and a NW dip, was one of the key geometric factors used by Davis et al. (2003) in their argument for proposing that the SCLM beneath

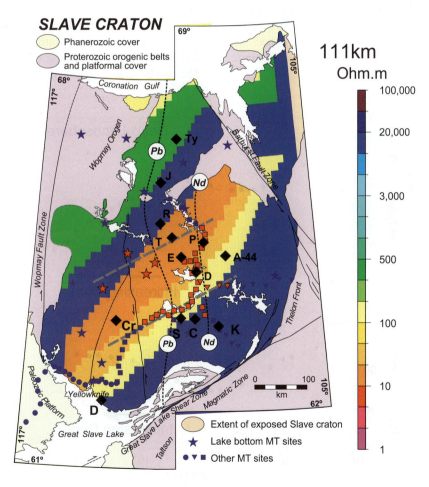

Fig. 11. Depth slice at 111 km of the 3-D resistivity model of the Slave craton. The colours on the Archean parts of the craton are the model resistivities at 111 km depth, with highly resistive (50,000–100,000 Ω m) in dark brown to highly conductive (1 Ω m) in red. The MT site locations are shown with different symbols for the different surveys, and the model is best constrained in regions of high site density.

the Slave today is not the original SCLM formed during Mesoarchean time. Davis et al. (2003) postulate that the Slave's SCLM is the consequence of subcretion of exotic lithosphere during an orogenesis at 2630–2590 Ma, postdating final Slave crustal amalgation at 2690 Ma.

7. Application to the Superior craton

In the western part of the Superior craton, Craven et al. (2001) have discovered distinctive electrical anisotropy over a wide region. An example is shown in Fig. 12 of the apparent resistivity and phase curves from two orthogonal directions from a representative site in the southern part of the region. Note that the short period responses are the same, testifying to the 1-D nature of the crust. With increasing period, the two responses diverge, indicative of either intrinsic anisotropy or structural anisotropy.

The strengths of electrical anisotropy, given by the maximum phase difference in the two orthogonal directions, and the directions of maximum conductivity are shown in Fig. 13. Three zones are apparent, and representative MT phases from each zone are shown in Fig. 14. The northern zone (yellow arrows and sites) exhibits NW–SE trending anisotropy that includes the lower crust and lithospheric mantle. The middle zone (green arrows and sites) exhibits very low phase differences, indicative of 1-D structure. The southern zone (blue arrows and sites) exhibits E–W trending anisotropy of the lithospheric mantle only.

Models of the data show an isotropic conducting body beneath the central zone within the SCLM that displays the same properties as the body in the central Slave craton. This conductor lies within the North Caribou Terrane (NCT), a Mesoarchean terrane within the mélange of predominantly Neoarchean terranes that form the western region of the Superior craton. The two bounding regions that display high anisotro-

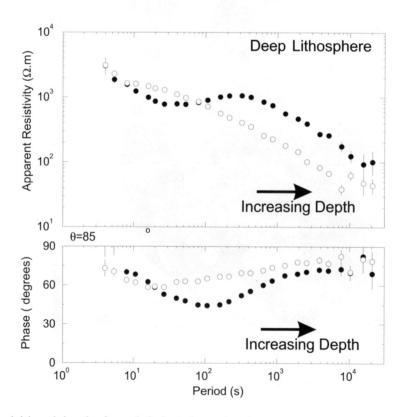

Fig. 12. MT apparent resistivity and phase data from a site in the southern region of the western Superior craton. The full symbols are those in a direction of N85E, and the open symbols perpendicular to that, i.e., N05W.

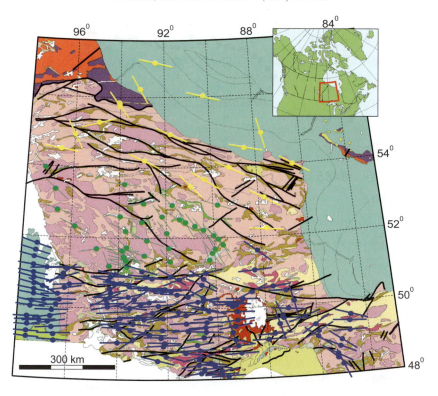

Fig. 13. Electrical anisotropy for western Ontario and eastern Manitoba, Canada, in the western Superior craton. The data fall into three groups: a northern group (yellow) with a strike of NW–SE. A central group (green) virtually 1-D. A southern group (blue) with a strike of E–W. Also shown on the figure are major faults in the region.

py have electrical strike directions parallel to the major syn- and post-Kenoran zones of transpression on either side of the NCT.

8. Application to the Rae craton

A low resolution MT experiment across Baker Lake, Nunavut, northern Canada, crossed the Rae–Hearne boundary designated by the Snowbird Tectonic Zone (Jones et al., 2002). The data yielded crustal and mantle information that addressed the Neoarchean geometry of Rae–Hearne collision. Of particular note is that despite regionally extensive and pervasive Proterozoic metasomatism (Cousens et al., 2001) the Rae craton lithospheric mantle is highly resistive and can be explained by the dry mineralogies represented in Fig. 2. This observation is counter to the suggestion of Boerner et al. (1999) that metasomatism will induce

enhanced conductivity. Thus, although the lithosphere in this location is undoubtedly old and thick, it does not contain any measurable amount of interconnected graphite. By induction, this would suggest that there are unlikely to be significant diamonds within kimberlites that came through this region.

9. Cause of conductvity enhancement and redox conditions

Given our understanding of the conductivity of olivine from laboratory studies (Xu et al., 2000), some features in our models are too conductive, by 2–3 orders of magnitude, for an olivine-dominant mantle without the addition of an interconnected minor conducting phase. Such differences are often attributed to interconnected water or free H+, partial melt phases, grain boundary carbon, carbon in the form of graphite

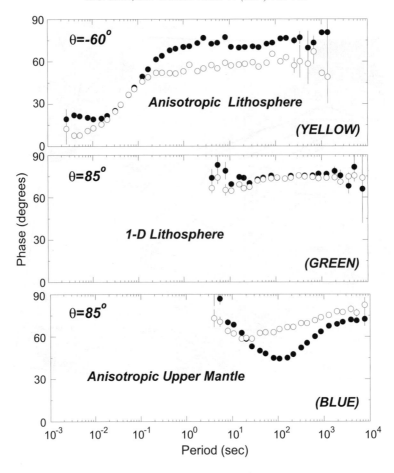

Fig. 14. MT phase data from three representative sites in the three regions shown in Fig. 13.

or sulphides within the upper mantle (for discussion see Constable and Heinson, 1993).

The observation of >0.03 S/m material within continental upper mantle requires olivine water contents of the order 1000 H/10^6 Si (Hirth et al., 2000). Such an amount of water may be tenable at sub-Moho depths if the olivine a-axis is strongly aligned within a region. The alignment of the a-axis would imply a strongly anisotropic conductor with the anisotropic axes aligned with the teleseismic fast direction. These correlations cannot be made easily for either the central Slave where the CSMC is isotropic, nor for the Superior where the electrical anisotropy is oblique to the observed teleseismic SKS anisotropy fast directions (Kay et al., 1999).

Partial melting enhances conductivity due to the high mobility of charge carriers within a melt fraction;

however, it is unlikely such shallow zones of partial melt exist as heat flow values at the surface of the Slave craton (Lewis and Hyndman, 1998) and Superior Province (Jaupart and Mareschal, 1999) are low, and petrological estimates of geothermal gradient also infer thick cold lithosphere as recently as the Eocene (Russell and Kopylova, 1999; MacKenzie and Canil, 1999; Russell et al., 2001).

Sulphides are a recent suggestion as a cause for upper mantle conductivity (Ducea and Park, 2000): sulphides have electrical conductivities in the range 10^2–10^6 S/m. Interstitial sulphides are observed in quantities as high as 330 ppm in xenoliths from the Kaapvaal craton (Alard et al., 2000), but they would have to be almost perfectly interconnected to result in a conductivity of the order of 0.05–0.1 S/m, and be interconnected over the large areas we observe in our two study areas. In addition, the CSMC correlates laterally and in depth with an ultradepleted region mapped by Griffin et al. (1999).

Grain boundary carbon may also contribute to elevated conductivity in the SCLM (Duba and Shankland, 1982) given an appropriate network of faults/joints along which the charge carriers may diffuse. Carbon in graphite form behaves electrically as a metal and therefore has very high electrical conductivity, typically 10^5 S/m (Duba and Shankland, 1982). Graphite is an accessory phase in many xenoliths observed worldwide (Pearson et al., 1994). We suggest a combination of graphite and possibly grain boundary carbon are the most attractive explanations for the source of the conductivity enhancement within the upper mantle.

Craven et al. (2001), following Foley (1988), suggest a redox melting model for emplacing graphite in the lithosphere as a consequence of plume and plate tectonic processes during Meso- and Neoarchean times. Given that Fe^{3+} is generally incompatible, then geochemical depletion and 'redox melting' are coupled processes, and MT may therefore identify regions where ancient C-H fluid-present partial melting occurred under reducing or depleted conditions. In the highly anisotropic regions of the western Superior Province, the graphite residual from syn- to post-tectonic partial melting may have been deposited along the (weak) fault zones, and thus explains the parallelism between the MT anisotropy directions and major zones of transpression. Within the Slave craton, the direct spatial correlation of the Central Slave Mantle Conductor with the ultradepleted zone of Griffin et al. (1999) suggests 'redox' processes.

10. Conclusions

Area selection for potentially diamondiferous provinces in frontier regions is problematic given the lack of a predictive model. Accordingly, one needs to acquire data indicative of such provinces, and herein we proposed that deep probing MT surveys can address the three fundamental questions that need to be answered. Is the region old? Is the lithosphere thick? Does the lithosphere contain observable amounts of carbon? MT is a powerful technique, especially when coupled with appropriate teleseismic studies. MT surveys have

the advantage that the technique is sensitive to electronic conduction in carbon (when in graphite form), and that the data acquisition generally requires only 1 month per site, rather than the 1 to 2 years that is required for teleseismic surveys.

Examples have been shown of the electrical geometries of the Slave craton and the western part of the Superior craton. In contrast, Rae craton mantle lithosphere around Baker Lake does not contain any observable conductivity enhancements, thus excluding interconnected graphite or grain boundary carbon. Taking the points presented in this paper, we suggest that

1. the central part of the Slave craton west of the Pb isotope boundary is likely to be as productive as the Lac de Gras Corridor of Hope,
2. the North Caribou Terrane of western Superior Province is also likely to be as productive, and
3. the Rae craton around Baker Lake will not yield high quantities of diamonds.

Of course, (1) and (2) are predicated on kimberlites successfully bringing the diamonds to the surface.

Acknowledgements

The Slave craton and Western Superior province work was made possible through the efforts of many people. We gratefully acknowledge the logistical and financial support of Lithoprobe, Geological Survey of Canada, National Science Foundation's Continental Dynamics program, Indian and Northern Affairs Canada, De Beers Canada Exploration, Kennecott Canada Exploration, Diavik Diamond Mines, BHP Billiton Diamonds, Falconbridge, Royal Oak Mines and Miramar Mining. Reviews of an earlier draft of this manuscript by B. Corner and G.H. Read are gratefully acknowledged. Geological Survey of Canada publication 2004044. Lithoprobe publication 1368. Dublin Institute for Advanced Studies publication GP165.

References

Alard, O., Griffin, W.L., Lorand, J.P., Jackson, S.E., O'Reilly, S.Y., 2000. Non-chondritic distribution of the highly siderophile elements in mantle sulphides. Nature 407, 891–894.

Alekseyev, A.S., Vanyan, L.L., Berdichevsky, M.N., Nikolayev, A.V., Okulessky, B.A., Ryaboy, V.Z., 1977. Map of asthenospheric zones of the Soviet Union. Dokl. Akad. Nauk 234, 22–24.

Bahr, K., Simpson, F., 2002. Electrical anisotropy below slow- and fast-moving plates: paleoflow in the upper mantle? Science 295, 1270–1272.

Bailey, R.C., 1970. Inversion of the geomagnetic induction problem. Proc. R. Soc. Lond., Ser. A 315, 185–194.

Beblo, M., 1982. Electrical conductivity of minerals and rocks at ordinary temperatures and pressures. Numerical Data and Functional Relationships in Science and Technology. Group V: Geophysics and Space Research, vol. 1. Physical Properties of Rocks, Springer, Berlin, pp. 239–253. Chapter. 5.1.

Bleeker, W., 2002. Archean tectonics: a review, with illustrations from the Slave craton. In: Fowler, C.M.R., Ebinger, C.J., Hawkesworth, C.J. (Eds.), The Early Earth, Physical, Chemical and Biological Development. Special Publication - Geological Society of London, vol. 199, pp. 151–181.

Boerner, D.E., Kurtz, R.D., Craven, J.A., Ross, G.M., Jones, F.W., Davis, W.J., 1999. Electrical conductivity in the Precambrian lithosphere of Western Canada. Science 283, 668–670.

Calcagnile, G., 1982. The lithosphere–asthenosphere system in Europe. Tectonophysics 90, 19–35.

Calcagnile, G., 1991. Deep structure of Fennoscandia from fundamental and higher mode dispersion of Rayleigh waves. Tectonophysics 195, 139–149.

Calcagnile, G., Panza, G.F., 1987. Properties of the lithosphere–asthenosphere system in Europe with a view towards Earth conductivity. Pure Appl. Geophys. 125, 241–254.

Clifford, T.N., 1966. Tectono-magmatic-units, metallogenic provinces of Africa. Earth Planet. Sci. Lett. 1, 421–434.

Constable, S., Heinson, G., 1993. In defence of a resistive oceanic upper mantle: reply to comment by Tarits Chave and Schultz. Geophys. J. Int. 114, 717–723.

Constable, S.C., Parker, R.L., Constable, C.G., 1987. Occam's inversion: a practical algorithm for generating smooth models from electromagnetic sounding data. Geophysics 52, 289–300.

Cousens, B.L., Aspler, L.B., Charenzelli, J.R., Donaldson, J.A., Sendeman, H., Peterson, A.D., LeCheminant, A.N., 2001. Enriched Archean lithospheric mantle beneath Western churchill province tapped during paleoproterozoic orogenesis. Geology 29, 827–830.

Craven, J.C., Kurtz, R.D., Boerner, D.E., Skulski, T., Spratt, J., Ferguson, I.J., Wu, X., Bailey, R.C., 2001. Conductivity of western Superior Province upper mantle in northwestern Ontario. Curr. Res. - Geol. Surv. Can. (2001-E6, 6 pp.).

Davies, G.F., 1992. On the emergence of plate tectonics. Geology 20, 963–966.

Davis, W.J., Hegner, E., 1992. Neodymium isotopic evidence for the accretionary development of the Late Archean Slave Province. Contrib. Mineral. Petrol. 111, 493–504.

Davis, W.J., Jones, A.G., Bleeker, W., Grütter, H., 2003. Lithospheric development in the Slave Craton: a linked crustal and mantle perspective. Lithos 71, 575–589.

deGroot-Hedlin, C., Constable, S., 1990. Occam's inversion to generate smooth two-dimensional models from magnetotelluric data. Geophysics 55, 1613–1624.

Drury, M.R., Fitz Gerald, J.D., 1996. Grain boundary melt films in an experimentally deformed olivine-orthopyroxene rock; implications for melt distribution in upper mantle rocks. Geophys. Res. Lett. 23, 701–704.

Duba, A.L., Shankland, T.J., 1982. Free carbon and electrical conductivity in the Earth's mantle. Geophys. Res. Lett. 9, 1271–1274.

Ducea, M.N., Park, S.K., 2000. Enhanced mantle conductivity from sulfide minerals, southern Sierra Nevada, California. Geophys. Res. Lett. 27, 2405–2408.

Fischer, G., Le Quang, B.V., 1981. Topography and minimization of the standard deviation in one-dimensional magnetotelluric modelling. Geophys. J. R. Astron. Soc. 67, 279–292.

Foley, S.F., 1988. The genesis of continental basic alkaline magmas—an interpretation in terms of redox melting. J. Petrol. Spec. Lithos. 139–161.

Garcia, X., Jones, A.G., 2001. Decomposition of three-dimensional magnetotelluric data. In: Zhdanov, M.S., Wannamaker, P.E. (Eds.), Three-Dimensional Electromagnetics. Methods in Geochemistry and Geophysics, vol. 35. Elsevier, Amsterdam, pp. 235–250. ISBN 0 444 50429 X.

Griffin, W.L., Ryan, C.G., 1995. Trace elements in indicator minerals: area selection and target evaluation in diamond exploration. In: Griffin, L. (Ed.), Diamond Exploration into the 21st Century. J. Geochem. Explor., vol. 53, pp. 311–337.

Griffin, W.L., Doyle, B.J., Ryan, C.G., Pearson, N.J., O'Reilly, S., Davies, R., Kivi, K., van Achterbergh, E., Natapov, L.M., 1999. Layered mantle lithosphere in the Lac de Gras area, Slave craton: composition, structure and origin. J. Petrol. 40, 705–727.

Groom, R.W., Bailey, R.C., 1989. Decomposition of magnetotelluric impedance tensors in the presence of local three-dimensional galvanic distortion. J. Geophys. Res. 94, 1913–1925.

Groom, R.W., Kurtz, R.D., Jones, A.G., Boerner, D.E., 1993. A quantitative methodology for determining the dimensionality of conductive structure from magnetotelluric data. Geophys. J. Int. 115, 1095–1118.

Grütter, H.S., Apter, D.B., Kong, J., 1999. Crust–mantle coupling; evidence from mantle-derived xenocrystic garnets. In: Dawson, J.B. (Ed.), Proc. 7th International Kimberlite Conference, Conference Proceedings University of Cape Town, South Africa, vol. 1, pp. 307–313.

Gurney, J.J., Zweistra, P., 1995. The interpretation of the major element compositions of mantle minerals in diamond exploration. In: Griffin, W.L. (Ed.), Diamond exploration into the 21st century. J. Geochem. Explor., vol. 53, pp. 293–309.

Haak, V., 1982. Electrical conductivity of minerals and rocks at high temperatures and pressures. Numerical Data and Functional Relationships in Science and Technology. Group V: Geophysics and Space Research, vol. 1. Physical Properties of Rocks, Springer, Berlin, pp. 291–307. Chapter 5.5.

Hamilton, W.B., 1998. Archean magmatism and deformation were not products of plate tectonics. Precambrian Res. 91, 143–179.

Helmstaedt, H.H., Gurney, J.J., 1995. Geotectonic controls of primary diamond deposits; implications for area selection. In: Griffin, W.L. (Ed.), Diamond Exploration into the 21st century. J. Geochem. Explor., vol. 53, pp. 125–144.

Hirth, G., Evans, R.L., Chave, A.D., 2000. Comparison of continental and oceanic mantle electrical conductivity: is the Archean lithosphere dry? Geochem. Geophys. Geosyst. 1 (Article, 2000GC000048).

Jaupart, C., Mareschal, J.C., 1999. The thermal structure and thickness of continental roots. In: van-der-Hilst, R.D., McDonough, W.F. (Eds.), Composition, Deep Structure and Evolution of Continents. Lithos, vol. 48, pp. 93–114.

Jennings, C.M.H., 1995. The exploration context for diamonds. In: Griffin, W.L. (Ed.), Diamond Exploration into the 21st Century. J. Geochem. Explor., vol. 53, pp. 113–124.

Jones, A.G., 1982. On the electrical crust–mantle structure in Fennoscandia: no Moho and the asthenosphere revealed? Geophys. J. R. Astron. Soc. 68, 371–388.

Jones, A.G., 1998. Waves of the future: superior inferences from collocated seismic and electromagnetic experiments. Tectonophysics 286, 273–298.

Jones, A.G., 1999. Imaging the continental upper mantle using electromagnetic methods. Lithos 48, 57–80.

Jones, A.G., Ferguson, I.J., 2001. The electric moho. Nature 409, 331–333.

Jones, A.G., Garcia, X., 2003. The Okak Bay MT dataset case study: a lesson in dimensionality and scale. Geophysics 68, 70–91.

Jones, A.G., Ferguson, I.J., Chave, A.D., Evans, R., McNeice, G.W., 2001. The electric lithosphere of the Slave craton. Geology 29, 423–426.

Jones, A.G., Snyder, D., Hanmer, S., Asudeh, I., White, D., Eaton, G., Clarke, G., 2002. Magnetotelluric and teleseismic study across the Snowbird Tectonic Zone, Canadian shield: a neoarchean mantle suture? Geophys. Res. Lett. 29 (17), 10-1–10-4 (doi: 10.1029/2002GL015359).

Jones, A.G., Lezaeta, P., Ferguson, I.J., Chave, A.D., Evans, R.L., Garcia, X., Spratt, J., 2003. The electrical structure of the Slave craton. Lithos 71, 505–527.

Kay, I., Sol, S., Kendall, J.-M., Thomson, C., White, D., Asudeh, I., Roberts, B., Francis, D., 1999. Shear wave splitting observations in the Archean Craton of Western Superior. Geophys. Res. Lett. 26, 2669–2672.

Kopylova, M.G., Russel, J.K., Cookenboo, H., 1997. Petrology of peridotite and pyroxenite xenoliths from Jericho Kimberlite; implications for the thermal state of the mantle beneath the Slave Craton, northern Canada. J. Petrol. 40, 79–104.

Kopylova, M.G., Caro, G., Russell, J.K., 2001. Kimberlite and xenoliths from the Southern Slave—a terrain with the highest diamond potential in the Slave craton. In: Cook, F., Erdmer, P. (comp), Slave-Northern Cordillera Lithospheric Evolution (SNORCLE) Transect and Cordilleran Tectonics Workshop Meeting (February 22–25), Pacific Geoscience Centre Sidney, BC Lithoprobe Report No. 79, 18–24.

Ledo, J., Jones, A.G., 2003. Temperature bounds of the Intermontane Belt mantle (northern Canadian Cordillera) using its mineral composition and electrical conductivity. Presented at the European Geophysical Society–European Union of Geosciences–American Geophysical Union joint meeting, Nice, France, April, 6–11.

Lewis, T.J., Hyndman, R., 1998. Heat flow transitions along the SNORCLE Transect; asthenospheric flow. In: Slave-NORthern Cordillera Lithospheric Evolution (SNORCLE) and Cordilleran tectonics workshop. Compiled by: F. Cook and P. Erdmer-Philippe. Lithoprobe Report 64, 92.

MacKenzie, J.M., Canil, D., 1999. Composition and thermal evolution of cratonic mantle beneath the central Archean Slave Province NWT, Canada. Contrib. Mineral. Petrol. 134, 313–324.

Macnae, J.C., 1995. Applications of geophysics for the detection and exploration of kimberlites and lamproites. In: Griffin, W.L. (Ed.), Diamond Exploration into the 21st Century. J. Geochem. Explor., vol. 53, pp. 213–243.

Mareschal, M., Kellett, R.L., Kurtz, R.D., Ludden, J.N., Bailey, R.C., 1995. Archean cratonic roots, mantle shear zones and deep electrical anisotropy. Nature 373, 134–137.

McKenzie, D.P., Bickle, M.J., 1988. Volume and composition of melt generated by extension of the lithosphere. J. Petrol. 29, 625–679.

McNeice, G., Jones, A.G., 2001. Multisite, multifrequency tensor decomposition of magnetotelluric data. Geophysics 66, 158–173.

Meju, M., 2002. Geoelectromagnetic exploration for natural resources: models, case studies and challenges. Surv. Geophys. 23, 133–206.

Minarik, W.G., Watson, E.B., 1995. Interconnectivity of carbonate melt at low melt fraction. Earth Planet. Sci. Lett. 133, 423–437.

Morgan, P., 1995. Diamond exploration from the bottom up: regional geophysical signatures of lithosphere conditions favorable for diamond exploration. In: Griffin, W.L. (Ed.), Diamond Exploration into the 21st Century. J. Geochem. Explor., vol. 53, pp. 145–165.

Partzsch, G.M., Schilling, F.R., Arndt, J., 2000. The influence of partial melting on the electrical behavior of crustal rocks: laboratory examinations, model calculations and geological interpretations. Tectonophysics 317, 189–203.

Pearson, D.G., Boyd, F.R., Haggerty, S.E., Pasteris, J.D., Field, S.W., Nixon, P.H., Pokhilenko, N.P., 1994. The characterisation and origin of graphite in cratonic lithospheric mantle: a petrological carbon isotope and Raman spectroscopic study. Contrib. Mineral. Petrol. 115, 449–466.

Pearson, N.J., Griffin, W.L., Doyle, B.J., O'Reilly, S.Y., Van Achterbergh, E., Kivi, K., 1999. Xenoliths from kimberlite pipes of the Lac de Gras area, Slave craton, Canada. In: Gurney, J.J., et al., (Ed.), Proceedings of the 7th International Kimberlite Conference, P.H. Nixon Volume 2: Red Roof Design, Cape Town, pp. 644–658.

Rodi, W., Mackie, R.L., 2001. Nonlinear conjugate gradients algorithm for 2-D magnetotelluric inversion. Geophysics 66, 174–187.

Russell, J.K., Kopylova, M.G., 1999. A steady-state conductive geotherm for the north-central Slave: inversion of petrological data from the Jericho kimberlite pipe. J. Geophys. Res. 104, 7089–7101.

Russell, J.K., Dipple, G.M., Kopylova, M.G., 2001. Heat production and heat flow in the mantle lithosphere to the Slave craton, Canada. Phys. Earth Planet. Inter. 123, 27–44.

Ryan, R.G., Grffin, W.L., Pearson, N.J., 1996. Garnet Geotherms: a technique for derivation of $P–T$ data from Cr-pyrope garnets. J. Geophys. Res. 101, 5611–5625.

Schilling, F.R., Partzsch, G.M., Brasse, H., Schwarz, G., 1997.

Partial melting below the magmatic arc in the central Andes deduced from geoelectromagnetic field experiments and laboratory data. Phys. Earth Planet. Inter. 103, 17–31.

Schultz, A., Kurtz, R.D., Chave, A.D., Jones, A.G., 1993. Conductivity discontinuities in the upper mantle beneath a stable craton. Geophys. Res. Lett. 20, 2941–2944.

Simpson, F., 2001. Resistance to mantle flow inferred from the electromagnetic strike of the Australian upper mantle. Nature 412, 632–634.

Siripunvaraporn, W., Egbert, G., 2000. An efficient data-subspace inversion method for 2-D magnetotelluric data. Geophysics 65, 791–803.

Smith, J.T., Booker, J.R., 1988. Magnetotelluric inversion for minimum structure. Geophysics 53, 1565–1576.

Smith, J.T., Booker, J.R., 1991. Rapid inversion of two and three-dimensional magnetotelluric data. J. Geophys. Res. 96, 3905–3922.

Snyder, D.B., Bostock, M.G., Lockhart, G.D., 2004. Mapping the mantle lithosphere for diamond potential using teleseismic methods. Lithos. 77, 859–872 doi:10.1016/j.lithos.2004.03.049.

Thorpe, R.I., Cumming, G.L., Mortensen, J.K., 1992. A significant Pb isotope boundary in the Slave Province and its probable relation to ancient basement in the western Slave Province. Project Summaries, Canada-Northwest Territories Mineral Development Subsidiary Agreement. Open File Rep. - Geol. Surv. Can., vol. 2484, pp. 279–284.

Vozoff, K., 1972. The magnetotelluric method in the exploration of sedimentary basins. Geophysics 37, 98–141.

Vozoff, K. (Ed.), 1986. Magnetotelluric Methods. Soc. Explor. Geophys. Reprint Ser. No. 5: Tulsa, OK, ISBN 0-931830-36-2.

Vozoff, K., 1991. The magnetotelluric method. Electromagnetic Methods in Applied Geophysics—Applications. Soc. Explor. Geophys., pp. 641–712. Tulsa, OK, Chapter 8.

Wannamaker, P.E., Hohmann, G.W., Ward, S.H., 1984. Magnetotelluric responses of three-dimensional bodies in layered earths. Geophysics 49, 1517–1533.

Wannamaker, P.E., Stodt, J.A., Rijo, L., 1985. PW2D: Finite Element Program for Solution of Magnetotelluric Responses of Two-Dimensional Earth Resistivity Structure. Earth Sci. Lab. Univ. Utah Res. Inst, Salt Lake City.

White, D.J., Musacchio, G., Helmstaedt, H.H., Harrap, R.M., Thurston, P.C., van der Velden, A., Hall, K., 2003. Images of a lower-crustal oceanic slab: direct evidence for tectonic accretion in the Archean western Superior province. Geology 31, 997–1000.

Xu, Y., Shankland, T.J., Poe, B.T., 2000. Laboratory-based electrical conductivity of the Earth's mantle. J. Geophys. Res. 105, 27865–27875.

Available online at www.sciencedirect.com

LITHOS

Lithos 77 (2004) 783–802

www.elsevier.com/locate/lithos

ELSEVIER

Patterns and controls on the distribution of diamondiferous intrusions in Australia

A.L. Jaques*, P.R. Milligan

Geoscience Australia, GPO Box 378, Canberra, ACT 2601, Australia

Received 25 June 2003; accepted 28 November 2003
Available online 28 May 2004

Abstract

The distribution of kimberlite, lamproite and related alkaline volcanism in Australia can be broadly related to the structure of the Australian continent and lithosphere. Diamondiferous kimberlites and lamproites, with the apparent exception of the weakly diamondiferous Orrorro kimberlites in the Adelaide Fold Belt, lie within the large Precambrian shield where seismic tomographic models and heat flow data indicate the presence of relatively cold, high seismic wave speed lithosphere (tectosphere) typically some 200 km thick or more beneath the Archaean cratons and up to 300 km in parts of central Australia. Many of the diamondiferous intrusions appear to lie at the margins rather than in the centre of the lithosphere domains. The highest concentration of diamondiferous intrusions (kimberlites and lamproites) is on and around the Kimberley Craton where seismic data indicate crustal thicknesses of 35–40 km and a lithosphere up to 275 km thick that is distinct from Proterozoic northern Australia.

Many, but clearly not all, of the intrusions show evidence of regional and local structural controls. Some are spatially associated with known crustal structures, especially regional faults. Others are aligned, either singly or in clusters, along or near discontinuities and/or gradients evident in regional scale potential field data, especially the total horizontal gradients of gravity data continued upward tens to hundreds of kilometres. Many of these features are not evident in the original datasets as their signatures are masked by shorter wavelength (near surface) anomalies. In some cases, the kimberlites and associated rocks lie within crustal blocks and domains defined by discontinuities in the potential field data rather than at domain boundaries.

Our overview suggests that analysis of potential field data, especially horizontal gradients in upwardly continued potential field data, at all scales can assist definition of crustal and, potentially, lithospheric structures that may influence the distribution of diamond pipes. However, more definitive mapping of Australia's diamond prospective regions requires the integration of data on crustal structures, especially trans-lithospheric faults, and geodynamic settings with high resolution tomographic models and other geophysical, petrologic, and isotopic information on the nature of the lithosphere beneath the Australian continent.
© 2004 Elsevier B.V. All rights reserved.

Keywords: Diamond exploration; Geophysics; Kimberlite; Lamproite; Gravity; Aeromagnetics; Crustal structure; Potential field; Lithosphere; Australia

* Corresponding author. Tel.: +61-2-62499745; fax: +61-2-62499983.

E-mail addresses: lynton.jaques@ga.gov.au (A.L. Jaques), peter.milligan@ga.gov.au (P.R. Milligan).

0024-4937/$ - see front matter © 2004 Elsevier B.V. All rights reserved.
doi:10.1016/j.lithos.2004.03.042

1. Introduction

The distribution of diamondiferous kimberlites is commonly zoned with respect to craton structure, with diamondiferous kimberlites restricted to ancient cratons underlain by well-preserved lithospheric roots and non-diamondiferous kimberlites and related alkaline ultrabasic rocks typically found off-craton and associated with thinner lithosphere (e.g., Clifford, 1966; Janse, 1994; Helmstaedt and Gurney, 1995). At the regional scale kimberlites, lamproites and related alkaline rocks occur in individual provinces and fields that may be associated with particular structural and tectonic features, notably deep fractures and shear zones, and exhibit petrologic zoning within structurally defined corridors (e.g. Kaminsky et al., 1995; White et al., 1995; Vearncombe and Vearncombe, 2002). Intra-plate continental alkaline rocks are commonly related to continental rifting, but a wide range of geotectonic models have been proposed for kimberlites, including rifting, mantle plumes and subduction (see Helmstaedt and Gurney, 1995, for summary).

Here, we review the distribution of diamondiferous kimberlite, lamproite and related intrusions in relation to features evident in the Australian crust and mantle. Our aim is to develop improved targeting criteria for area selection in diamond exploration by better constraining the relationship between lithospheric—especially crustal—features and structures, and the locations of diamond pipes, enabling better definition of prospective domains and corridors. We have focussed particularly on seismic and potential field (gravity and magnetic) data because these provide information on the structure and physical properties at depth in the Earth.

2. Distribution of kimberlites, lamproites and related rocks

There are more than 450 occurrences of kimberlite, lamproite and related alkaline rocks known in Australia, most found in the course of diamond exploration over the past 25 years (Atkinson et al., 1990; Jaques, 1998). The distribution of these intrusions is shown in relation to the crustal mega-elements mapped from potential field data (Shaw et al., 1996; Wellman, 1998) in Fig. 1 and to the major geological regions in Fig. 2. A more detailed map at 1:5 million scale is freely available on-line from the Geoscience Australia website (Jaques, 2002). Kimberlites occur both within and on the margins of the major Australian Precambrian cratons (Yilgarn, Pilbara, Gawler cratons) and within

the large north Australia crustal mega-element (also termed North Australia craton) that includes the Kimberley Craton. Lamproite intrusions are restricted to the Halls Creek and King Leopold Orogens and the Fitzroy Trough at the SE and SW margins, respectively, of the Kimberley Craton (Fig. 2). Lamprophyres and carbonatites appear to be distributed both within, but particularly at the margins of, cratons (Fig. 2; see Jaques, 2002). In addition to the above intrusions which range in age from 2020 to 20 Ma, there are a number of Mesozoic and Cainozoic alkali basalt breccia pipes and lamprophyre dykes that carry high pressure megacrysts and mantle-derived xenocrysts and peridotite xenoliths (including a number carrying garnet) within the Tasman Orogenic System of Eastern Australia (Figs. 1 and 2; Stracke et al., 1979; O'Reilly et al., 1989; Johnson et al., 1989; Sutherland, 1996). Leucitite volcanic vents and flows of Miocene age are distributed in several fields in the Lachlan Orogen and its northerly extension under sedimentary cover (Fig. 2; Johnson et al., 1989).

All confirmed diamondiferous intrusions lie within the Precambrian provinces and are typically either of kimberlite, lamproite or, rarely, ultramafic lamprophyre composition (Atkinson et al., 1990; Jaques, 1994; Graham et al., 1999a). Diamonds have been reported from a number of basaltic breccia pipes in the Tasman orogenic system (summarised by Sutherland, 1996) where alluvial diamonds are widely distributed but, to our knowledge, none of these have been confirmed by subsequent testing of large samples. Petrologic studies (especially pressure–temperature estimates) of xenoliths from the basalts indicate that none have come from deeper than 70 km and the prevailing geotherm within the lithosphere at the time of basalt eruption was high and strongly curved (advective): this makes it unlikely that diamond was stable in the lithosphere at that time (O'Reilly and Griffin, 1985; O'Reilly et al., 1989).

Commercial diamond deposits are associated with lamproite intrusions at Argyle (1180 Ma, Boxer and Jaques, 1990) and Ellendale (20 Ma, Hughes and Smith, 1990) and kimberlites in the Merlin field dated at 380 Ma (Hell et al., 2003). The Argyle pipe is the world's largest single producer of primary diamonds and more than 500 million cts of diamonds have been produced from the Argyle mine since mining commenced in 1986. Trial mining at Merlin was con-

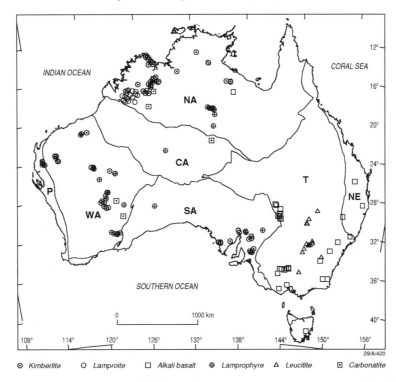

Fig. 1. Map showing distribution of kimberlites, lamproites, lamprophyres, carbonatites, alkali basalts carrying garnet megacrysts and/or garnet peridotite, and leucitites in relation to Australian crustal mega-elements (Wellman, 1998). Crustal mega-elements are: CA—central Australia; NA—north Australia; NE—New England; P—Pinjarra (orogen); SA—south Australia; T—Tasman (orogenic system); WA—western Australia.

ducted in the period 1998–2003 and mining of Ellendale Pipe 9 commenced in 2002 (Kimberley Diamond Company NL, 2003).

3. Data and methods

Good quality coverage of the Australian continent with magnetic and gravity data, as well as seismic data and a digital elevation model, is available as part of Geoscience Australia's National Geoscience Data Sets (Geoscience Australia, 2003). About half (54%) of the continent is covered by modern (post-1985) aeromagnetic data flown at 500 m line spacing and 80 m terrain clearance or better and the remainder by regional data flown at 150 m terrain clearance and flight line spacing at least 1500 m. Continental gravity coverage is mostly at 11 km station spacing with 35% of the continent covered with station spacing of 4–7 km and 21%

with station spacing less than 4 km (Richardson et al., 2002). Grids with cell sizes of 400 m (aeromagnetic data) and 800 m (gravity data) are available from the Geoscience Australia website. Heat flow data are available but the quantity and distribution are poor. The three-dimensional structure beneath the Australian continent has been mapped by recording natural earthquake events using portable seismic recorders deployed in stages across the continent as part of the Skippy Project (Van der Hilst et al., 1998); subsequent experiments, such as Wacraton, have improved the data over the western third of the continent (Kennett, 2003). These data allow construction of the three-dimensional variation in seismic wave speed below 60 km depth, with a horizontal resolution of around 200 km, and results so far have enabled resolution of structures to 400 km depth. The data are presented as variations in seismic velocities from a global continental reference, with higher velocities indicating denser, cooler and probably older rocks, and lower velocities representing less-dense, hotter and probably younger material.

The continental potential field datasets have been used to define the major geophysical domains and structures of the Australian continent (e.g. Wellman,

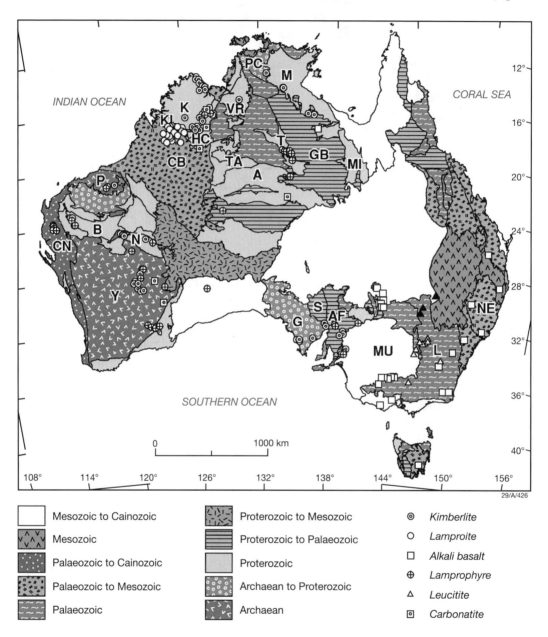

Fig. 2. Map showing distribution of kimberlites, lamproites, lamprophyres, carbonatites, alkali basalts carrying garnet megacrysts and/or garnet peridotite, and leucitites in relation to the major Australian geological provinces. Province names referred to in text are: A—Arunta Block; AF—Adelaide Fold Belt; B—Bangemall Basin; CB—Canning Basin; CN—Carnarvon Basin; G—Gawler Craton; GB—Georgina Basin; HC—Halls Creek Orogen; K—Kimberley Basin; KL—King Leopold Orogen; L—Lachlan Orogen; M—McArthur Basin; MI—Mount Isa Block; MU—Murray Basin; N—Nabberu Basin; NE—New England Orogen; P—Pilbara Craton; PC—Pine Creek Geosyncline; S—Stuart Shelf; T—Tennant Creek Block; TA—Tanami Block; VR—Victoria River Basin; Y—Yilgarn Craton.

1978, 1988, 1998) and analysis of these, employing contour and image maps and modelling, resulted in the Australian Crustal Elements Map (Shaw et al., 1996; Wellman, 1998).

Archibald et al. (1999) and Holden et al. (2000) presented an alternative method of analysing potential field data using a "multiscale" approach based in part on edge-detection methods, such as those developed and used by Blakely and Simpson (1986) and Grauch et al. (2001). Assuming that the positions of maxima in the horizontal gradient of potential field data represent the edges of source bodies, such maxima can be detected and mapped as points, thus providing the interpreter with an unbiased estimate of their positions compared with visual inspection. When extended to many different levels of upward continuation of the data, sets of points are generated that can be displayed in three dimensions by using the upward continuation level as the *z*-dimension. Another method to separate the responses of sources

from various depths is separation filtering of gravity and magnetic data either by the 'matched filter' method using the log-energy spectrum (Spector and Grant, 1970) or other filtering methods (Cowan and Cowan, 1993). An underlying assumption in both gradient mapping and separation filtering is that lower levels of upward continuation map near-surface sources (shorter wavelengths) whereas higher levels of continuation map deeper sources (longer wavelengths), viz. multiple scales of information are mapped. While this assumption is generally true, the non-uniqueness of potential-field solutions should always be borne in mind, and additional constraining information used wherever possible (Archibald et al., 1999; Hornby et al., 1999).

The total horizontal derivative of gravity upward continued to 6 km for the Australian continent are given in Fig. 3 and can be compared with the mega-element boundaries (Fig. 1). The points calculated for the maxima of the horizontal gradient are better

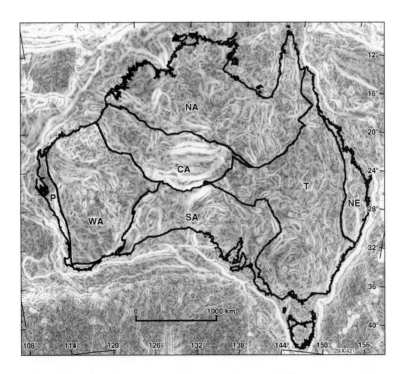

Fig. 3. Greyscale image of the total horizontal derivative (thd) or gradient of the Australian Bouguer gravity, upward continued to 6000 m, and overlain with the mega-element boundaries (Wellman, 1998). Continuation to this level removes the short-wavelength effects of near-surface sources and noise. Maxima in the thd of the upward-continued gravity are closely associated with boundaries in density variations within the Earth. The different domains show distinct trend and/or textural characteristics in the thd image.

visualised and interpreted by converting them into polylines, or "strings", and this has been done for areas of interest in this study. The points are attributed by the strength of the horizontal gradient from which they are derived, and colouring by this attribute may also be diagnostic in terms of source information. If the points or strings are viewed in three-dimensions, the attitude of the "surfaces" they form in the vertical sense may also provide information regarding the attitude of the physical property contrast from which they are derived (Archibald et al., 1999). Hobbs et al. (2000) provide examples of applying multi-scale edge analysis to Australian gravity data, and Milligan et al. (2003) extended the method to include directional analysis of linear gradients. Hildenbrand et al. (2001) used potential field data and a similar approach to map deep-seated structures and regional controls on mineralisation in the western USA.

4. Continental scale features and structures

4.1. Evidence for mantle roots

The seismic tomographic data obtained in the SKIPPY and subsequent experiments show that the Precambrian shield of Australia is underlain by anomalous high-velocity material that extends to 200 km and locally deeper to as much as 300 km. This contrasts with the Phanerozoic crust of eastern Australia that is underlain by only a thin zone of high wave speeds in the lithosphere at about 100 km, below which there is a zone of lowered wave speeds (Van der Hilst et al., 1998; Debayle and Kennett, 2000; Kennett, 2003). The 3D shear wave models obtained by waveform inversion show smaller-scale heterogeneities within the lithosphere beneath the Precambrian shield. More recent data have significantly improved the models over the western third of the continent to show heterogeneities within the Yilgarn Craton and confirmed that the Archaean Yilgarn and Pilbara cratons have strong, fast shear wave anomalies extending to more than 200 km (B. Kennett and S. Fishwick, personal communication, 2003). The Proterozoic basement of central Australia (Arunta region) is underlain by mantle with high wave speeds extending to more than 250 km. No Archaean age rocks are exposed on the Kimberley Craton but

isotopic studies (Os, Sr, Nd, Pb) and U–Pb zircon dating indicate the presence of Archaean basement, and Re–Os isotopic data for peridotite xenocrysts from the Seppelt kimberlite and the Argyle lamproite suggest the stabilization of continental lithosphere beneath the Kimberley Craton in the Archaean (Graham et al., 1999b). Lithosphere under the Kimberley Craton is characterized by fast wave speeds to more than 250 km and a distinct character compared with its surrounds within the dominantly Proterozoic North Australian mega-element (Debayle and Kennett, 2000; Kennett, 2003). However, the relationship between lithospheric and crustal characteristics is only loosely defined at present. Simons et al. (1999) found a broad relationship between lithospheric wave speed and increasing crustal age but noted significant variations in deep continental structure within regions of similar crustal age and tectonic setting.

The diamondiferous kimberlites and lamproites all lie within regions underlain by lithosphere greater than 125 km and mostly greater than 200 km thick: these lithospheric keels potentially lie in the diamond stability field. Fig. 4 shows the distribution of diamondiferous kimberlites, lamproites and related rocks in relation to the three-dimensional shear-wave model at 125, 175 and 225 km (B. Kennett and S. Fishwick, personal communication, 2003).

The seismic tomographic model suggests that the lithosphere under the Kimberley Craton, host to the largest number of diamondiferous intrusions, is distinct from surrounding lithosphere of northern Australia and underlain by a lithospheric root that is more than 250 km thick under the central Kimberley and thins to the east and south (Van der Hilst et al., 1998; Kennett, in press). The tomographic model is consistent with petrologic data from studies of peridotite xenoliths and inclusions in diamonds from Argyle (Jaques et al., 1990) and Ellendale (Jaques et al., 1994), that indicate derivation of diamond from depleted lithospheric mantle at depths of ~ 180 and ~ 150 km, respectively. Garnet concentrates from kimberlites in the North Kimberley suggest that lithosphere in that region may be ~ 160 km thick, with diamonds derived from mantle depths of 130–160 km (O'Reilly et al., 1997; Wyatt et al., 1999).

The Merlin and Roper field kimberlites (Blackjack, Packsaddle and Abner Range) lie at the NE margin of the lithospheric root evident in the tomographic mod-

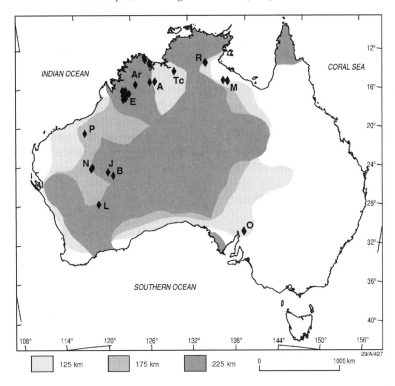

Fig. 4. Distribution of diamondiferous kimberlites and lamproites in relation to simplified outlines of anomalously fast shear wave velocities based on map views of the three-dimensional shear wave gradient model at 125, 175 and 225 km (Van der Hilst et al., 1998; Kennett, 2003; B. Kennett and S. Fishwick, written communication, 2003). The Kimberley Craton is underlain by a deep zone of anomalously high velocities representing denser, colder lithosphere. Diamondiferous intrusions are: A—Argyle (AK1) lamproite pipe; Ar—Aries kimberlite pipe; B—Buljah lamprophyres; E—Ellendale lamproite pipes; J—Jewill kimberlites; L—Leonora kimberlites; M—Merlin kimberlite pipes; N—Nabberu kimberlites; O—Orroroo kimberlites; P—Brockman kimberlite dyke.

els (Fig. 4). The lithosphere beneath the Timber Creek, Pilbara and Yilgarn diamondiferous kimberlites appears to be around 175 km thick or more whereas the lithosphere beneath the Orroroo kimberlites is distinctly thinner.

The lithosphere thicknesses indicated by the seismic tomographic models for the Archaean cratons of Australia are broadly consistent with the model Precambrian lithospheric thicknesses estimated by Artemieva and Mooney (2001) from heat flow data and thermal conductivity models. However, the anomalously thick lithosphere beneath central Australia is not predicted by the heat flow and thermal conductivity models. The thermal models suggest lithosphere thicknesses of 175–225 km beneath the Yilgarn and Pilbara Cratons and 150–175 km under the Halls Creek Orogen-Musgrave region, and less than 150 km elsewhere (Artemieva and Mooney, 2001).

In summary, although the lateral resolution of current seismic tomographic models is not adequate to precisely define prospective lithospheric domains within the Australian continent, current models provide good evidence for the existence of lithosphere extending well into the diamond stability field beneath the known diamondiferous intrusions of northern Australia, especially the Kimberley Craton, as well as the Yilgarn and Pilbara cratons and large areas of central Australia.

4.2. Crustal thickness

Collins et al. (2003) assembled available seismic data to map the depth of the crust–mantle boundary (Moho) beneath continental Australia and showed that the thickness varies considerably from 24 to 56 km with an average of 38 km. Broadly, the crust is thinnest beneath the Tasman Orogenic System and thickest under the McArthur Basin to Arunta Block region and the Mt. Isa Block in northern Australia. The patterns of crustal thickness do not match the crustal mega-elements mapped by Shaw et al. (1996) using continent-scale gravity and magnetic data sets (Collins et al., 2003). The unusually thick crust (and associated seismic wave speed anomaly) under Central Australia may be the result of tectonic thickening associated with the Alice Springs orogeny (450–300 Ma) that included the development of deep crustal-penetrating shear zones that resulted in substantial offsets to the Moho.

The distribution of kimberlites, lamproites, and lamprophyres shows little apparent correlation with crustal thickness other than that alkali basalts (with the exception of Coanjula) and leucitites are restricted to the generally thinner crustal regions of eastern Australia (Fig. 1). With the exception of the Brockman kimberlite in the Pilbara where the crust is only 30 km thick (Reading and Kennett, 2003), the diamondiferous intrusions all lie within regions with crustal thicknesses of 35 km or more. However, only the Merlin kimberlite pipes in the McArthur Basin lie within regions of anomalously thick crust. Crust under the Kimberley Craton is ∼ 35 km thick, but thickens towards the SE to ∼ 40 km under the Argyle pipe. Apart from the Palaeoproterozoic alkaline intrusions at Coanjula (Lee et al., 1994) and lamprophyres at Tennant Creek (Duggan and Jaques, 1996) no other alkaline ultramafic intrusions are known to be associated with the thick crust in the Mt. Isa or the southern Georgina Basin-Arunta regions, although this area is currently being explored for diamonds.

In summary, nearly all the known diamondiferous kimberlites, lamproites and related intrusions lie within regions with crustal thicknesses of 35 km or more. Variation in crustal thickness alone—at least as presently understood—does not appear to provide a basis for defining those regions within the Australian Precambrian shield that are more prospective for diamonds. The coincidence of anomalously thick lithosphere (>200 km) and thickened crust (>45 km) in large areas of central northern Australia may suggest conditions favourable for the preservation of diamonds in mantle roots but the thickening of crust and lithosphere under the Arunta region could reflect relatively young tectonic events.

4.3. Heat flow

Cull (1991), from a compilation of heat flow data, recognised three basic heat flow provinces in the Australian continent, corresponding roughly to an Archaean western domain, a central Proterozoic domain, and a Palaeozoic eastern domain. A compilation of estimated crustal temperatures at 5 km from bottom-hole temperatures in ∼ 3500 boreholes (Somerville et al., 1994) suggests a more complex heat flow distribution. In this data set the Yilgarn and Pilbara cratons, the Halls Creek Orogen – Tanami – Arunta region, the southern McArthur Basin, and the western Gawler Craton and Adelaide Fold Belt are all characterised by low temperatures but large regions of much higher apparent heat flow are found in other regions underlain by dominantly Proterozoic basement. The diamondiferous kimberlites of the North Kimberley Province, as well as the Timber Creek and Roper kimberlites, all lie in regions where the currently available heat flow data suggest anomalously high temperatures.

5. Craton scale features

Potential field data, especially the total horizontal gradients of upward continued gravity data (Archibald et al., 1999; Milligan et al., 2003), show discontinuities and structures within the larger crustal elements defined by Shaw et al. (1996) on the basis of continental-scale gravity (Fig. 3) and magnetic data. These are interpreted to represent large-scale structures deep in the crust. At the coarsest scale, i.e. total horizontal gradients of gravity and magnetic upward continued to more than 100 km, some of the gradients coincide with or approximate terrane boundaries and/or crustal subdivisions within the mega-elements (Fig. 3). Many of the features evident in the upward continued gravity data, however, are not shown on published geological maps and some are discordant to mapped faults suggesting that they may reflect more fundamental changes in density deeper in the crust.

A number of kimberlite and lamproite fields, and individual intrusions within them, have been previously shown to align along known faults and other structures, suggesting that crustal structures influenced emplacement (Jaques et al., 1986; White et al., 1995; Shee et al., 1999). According to Kaminsky et al. (1995), kimberlites globally typically occur in zones of 'high permeability' characterised by "a system of long-lived deep-seated major faults, controlling the intrusion of mantle (mafic and ultramafic) magmatism in cratons". Several of the Australian diamondiferous intrusion fields appear to be spatially associated with features evident in the horizontal gradients of gravity data upward continued from 16 to 100 km (e.g. Fig. 5) and, to a much lesser extent, reduced-to-pole (RTP) magnetic data. These are examined in more detail below.

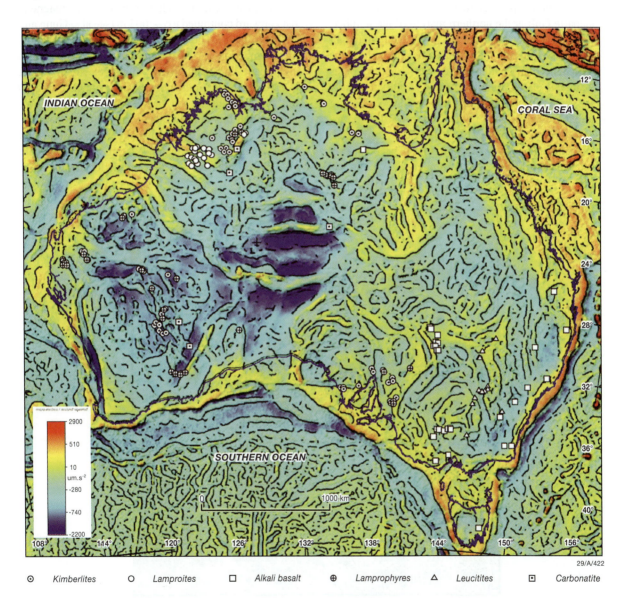

Fig. 5. Bouguer gravity anomaly map of Australia (Murray et al., 1997) with locations of horizontal gradient maxima of the gridded data upward continued to 20 km plotted as dots. Locations of kimberlites, lamproites, and related rocks as in Fig. 1. The gravity image is derived from the digital gravity grid of the Australian region (Murray, 1997, 2001). This grid combines accurate onshore gravity measurements with levelled free air marine gravity measurements supplemented by satellite data in areas where there are no marine data. The horizontal gradients define domains and structures within the Bouguer gravity anomaly data.

5.1. North Kimberley province

The North Kimberley kimberlite province comprises more than 30 kimberlite pipes and dykes of Neoproterozoic age (~ 800 Ma) clustered in several fields along a broad NW-trending gravity high that is aligned roughly parallel to the coast in the Joseph Bonaparte Gulf at the northern margin of the Kimberley Craton (Fig. 5; Jaques et al., 1986). More than half of the kimberlites are diamondiferous and include the significantly diamondiferous Seppelt 01 and 02 pipes (Wyatt et al., 1999; Reddicliffe et al., 2003). The bulk of the kimberlites lie along a pronounced NW-trending gradient evident in the gravity data upwardly continued to 16 km or more (Fig. 5). Locally, however, emplacement appears to have been influenced by a set of dominant NE and less prominent NW- and N-

trending structures evident in RTP magnetic images and in upwardly continued magnetic data (Gunn and Meixner, 1998; Fig. 6). For example, the Ashmore-Bulgurri-Dioro, Seppelt, and Berkeley kimberlite clusters all show a strong NE alignment (see also Reddicliffe et al., 2003) coincident with NE-trending structures evident in both the gravity and magnetic data upward continued to ~ 60 km (see also Gunn and Meixner, 1998). The Delancourt kimberlites, however, are aligned on a NNW-trending fault evident as a significant lineament in the digital elevation model (DEM) and linear gradient in the magnetic data (Fig. 6) but not evident in the gravity anomaly map. We suggest that this fault may be a near-surface feature.

Both the gravity and, especially, the magnetic data show the marked conjugate NW–NE and subordinate N-trending structures in the Kimberley Craton. The

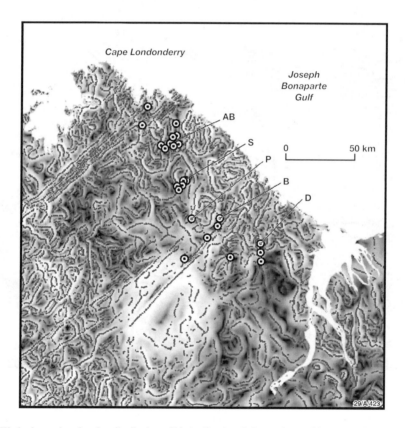

Fig. 6. Map of North Kimberley region showing distribution of kimberlites in relation to the total horizontal derivative (upward continued to 1600 m) of total magnetic intensity (RTP) as a greyscale image overlain with points calculated at the maxima of the horizontal gradients. Individual kimberlites and kimberlite clusters named in the text are: AB—Ashmore-Bulgurri-Dioro; B—Berkeley; D—Delancourt; P—Pteropus; S—Seppelt.

conjugate NW–NE series of magnetic anomalies in the North Kimberley are infilled by dolerite dykes and have been interpreted as having formed by differential basement extension during Devonian-Carboniferous rifting in the central Joseph Bonaparte Gulf (Gunn and Meixner, 1998). The kimberlites are arrayed along on a broad NW-trending gravity high flanking the rift that may reflect thinned crust but their locations appear to be controlled by structures parallel to NE-trending transform faults that accommodated differential extension in the basin (Meixner and Gunn, 1997). The apparent spatial association of these structures with the 800 Ma kimberlites suggests that at least some of the structures, especially those defined in the gravity data, may be older, perhaps associated with the breakup of Rhodinia and later reactivated during basin formation.

5.2. East Kimberley province

The East Kimberley Province includes the 1180 Ma Argyle (AK1) lamproite pipe, a suite of more than 22 kimberlites of late Proterozoic age (~ 800 Ma), and a number of lamprophyre dykes of similar or slightly older age (Jaques et al., 1986). Apart from the Maude Creek kimberlite that has a grade of ~ 1 ct/ 100 t, the kimberlites and lamprophyres are barren, whereas the Argyle pipe and the nearby associated Lissadell Road and Seagull dykes are diamondiferous. The world-class Argyle pipe has unusually high diamond grades (Boxer and Jaques, 1990).

The Argyle lamproite pipe, together with nearby Seagulls and Lissadell Road lamproite dykes and the Bow Hill ultramafic lamprophyre, lies within the Halls Creek Orogen at the SE margin of the Kimberley Craton (Fig. 2). The kimberlites of the East Kimberley Province, in contrast, lie west of the Greenvale Fault—the western margin of the Halls Creek Orogen—within the Kimberley Basin (Fig. 7). The Halls Creek Orogen itself is defined by a broad magnetic low and a marked discontinuous gravity high, and is bordered by zones of geophysical overprinting (Wellman, 1998; Shaw et al., 2000). The strength and consistency of orientation of the feature defined by the horizontal gradients associated with the Halls Creek Orogen and the persistence of this feature in gravity data upward continued to more than 100 km suggest that the Orogen extends to significant depths.

This is reinforced by the abrupt change in upper mantle structure at ~ 200 km observed beneath the Halls Creek Orogen (Van der Hilst et al., 1998; Kennett, 2003) which may represent a Palaeoproterozoic plate boundary (Shaw et al., 2000). Formation of the Halls Creek Orogen is believed to have resulted from collision between the Kimberley and North Australia cratons as a result of W-dipping subduction at around 1.91–1.85 Ga (Myers et al., 1996; Betts et al., 2002). A feature of the Argyle diamonds is the very high proportion of eclogitic paragenesis stones with a wide range of ^{13}C-depleted isotopic compositions that may represent recycled crustal material (Jaques et al., 1989).

The Argyle (AK1) pipe, together with the associated lamproites and Bow Hill Dykes, lies at the northern end of the NE-trending major Bouguer gravity high that marks the boundary between two of the geophysical sub-domains identified within the Halls Creek Orogen (Shaw et al., 1996, 2000). The intrusions lying immediately west of the Halls Creek Orogen, such as the Maude Creek, Blackfellow Creek and Devils Elbow kimberlite dykes, are aligned along NE-trending structures evident in both the upwardly continued gravity (Fig. 7) and magnetic data. Some, but not all, of these intrusions are spatially associated with mapped faults shown on the WA 1:500,000 scale and other published Government maps (Fig. 7). Further west in the Phillips Range the Aries kimberlite (Towie et al., 1994) lies near a prominent NNE-trending discontinuity in the upwardly continued gravity data (Fig. 7).

5.3. West Kimberley province

The West Kimberley lamproite province (Fitzroy Lamproites) comprises more than 150 lamproite pipes, plugs, sills and dykes distributed in three major fields at the northern margin of the Canning Basin—Ellendale, Calwynyardah and Noonkanbah—and several outlying clusters at the southern edge of the King Leopold Orogen (Jaques et al., 1986). Most of the diamondiferous bodies lie within the Ellendale Field where approximately two thirds of the olivine lamproite pipes and plugs are diamondiferous with the two highest grade pipes—Ellendale 4 and 9—having average grades of about 14 and 5 ct/100 t, respectively (Hughes and Smith, 1990). Higher grades have been

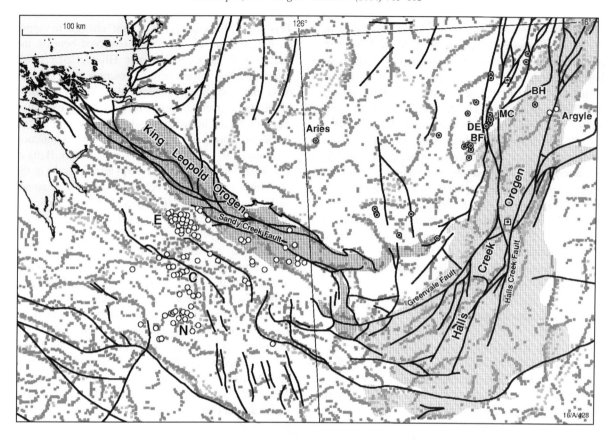

Fig. 7. Map showing the distribution of kimberlites, lamproites, carbonatites and lamprophyres in relation to King Leopold and Halls Creek Orogens, major faults, and the maxima of the total horizontal gradient of upward continued gravity data. The dark points are the maxima of the total horizontal gradient of gravity data continued upward to 4 km and the lighter points are for data continued upward to higher levels (11–64 km). Labelled intrusions are: BH—Bow Hill dykes, DE—Devil's Elbow, MC—Maude Creek. Letters E, C and N refer to the Ellendale, Calwynyardah and Noonkanbah lamproite fields. The separation of maxima of gradients calculated at low and high levels of upward continuation is attributable to dispersion either dip of the density contrast and/or dispersion with upward continuation. Different structures identified at different levels of upward continuation may indicate the presence of both shallow and deep sources in the crust.

reported from near surface and are currently being mined (Kimberley Diamond Company NL, 2003).

The distribution of the lamproites, especially in the Ellendale field, appears to be influenced by basement structures at the northern margin of the Fitzroy Trough evident in the upwardly continued gravity and RTP magnetic data. The gravity data continued upward to low levels show fine-scale structures trending north–south as well as the dominant NW-trending gradient (marked Bouguer gravity high) associated with the King Leopold Orogen (Fig. 7). At higher levels of upward continuation the major NW trend is disrupted by an E to ENE-trending gradient (Fig. 7) that

Archibald et al. (1999) equated with NE-trending transfer faults that segmented the main NW–SE trend of the Fitzroy Trough rift (Drummond et al., 1988). White et al. (1995) suggested that the transfer faults controlled the rough N–S alignment of the three major lamproite fields.

A number of lamproites in the three main fields form clusters aligned along the NE-trending features evident in the upwardly continued gravity data (Fig. 7). However, the majority of lamproite clusters, and the long axes of individual vents and dykes, in the Ellendale Field appear to have been controlled—at least in the near-surface environment—by approximately E-trending structures (Jaques et al., 1986). These structures, which probably represent shallow faults and fractures as they are much less evident in gravity data upward continued above 10 km, are clearly shown in high resolution aeromagnetic data as linear magnetic anomalies trending 105° filled by fissure intrusions of lamproite (Kimberley Diamond Company NL, 2001; D. Jones, personal communication, 2003). The fissure systems comprise four parallel structures some 400–1000 m apart that can be traced over 3.5 km with the fissure lamproites typically 40–60 m wide (Kimberley Diamond Company NL, 2001).

White et al. (1995) related the distribution of lamproites in the West Kimberley to the major mantle-penetrating shear system, evident in deep seismic reflection data (Drummond et al., 1988), that forms the contact between the King Leopold Orogen and the Lennard Shelf at the northern margin of the Fitzroy Trough. This Proterozoic compressional structure was reactivated as a detachment zone during late Palaeozoic basin extension (Braun and Shaw, 1998).

5.4. Merlin-roper province

At least two ages of diamondiferous kimberlites are known from the North Australia mega-element outside the Kimberley Craton (Figs. 1 and 4). The Mesozoic (inferred age) Roper and Stow intrusions in the McArthur Basin region (Berryman et al., 1999) roughly coincide with a broad NW-trending discontinuity evident in gravity datasets, upwardly continued to 100 km and above, that extends from the Mount Isa to the Pine Creek Inliers (Figs 2 and 3). Hobbs et al. (2000) termed this major structure the 'Barramundi worm' and suggested that it represented a boundary between two lithospheric blocks of different ages and densities. At lower levels of upward continuation shallower features dominate in the gravity field, and the major broad discontinuity is much less evident and replaced by a meshwork of relatively small NW, NE, and E-trending discontinuities (Figs. 3 and 8).

The Merlin kimberlite field comprises 14 small (all < 5 ha, most < 0.5 ha) kimberlite pipes of Devonian age (380 Ma) that lie in several discrete clusters at the eastern margin of the Batten Trough in the southern McArthur Basin (Lee et al., 1998; Reddicliffe, 1999, Hell et al., 2003). The pipes appear to coincide with a prominent NE-trending gravity discontinuity (Fig. 8)

and are located some 6 km E of the NNW-trending Emu Fault that forms the eastern boundary of the Batten Trough, a 70-km wide zone of extensive faulting trending NNW. The pipes lie at the intersection of the NE-trending gravity gradient with a subordinate N-trending gravity gradient and an inferred extension of the NW-trending Calvert Fault (Fig. 8). At the local scale the kimberlites are controlled by NNE (015°)-trending fractures (Reddicliffe, 1999). The NW, NE and E-trending structures are believed to reflect structures imposed during the formation and uplift of the Palaeoproterozoic McArthur Basin with features such as the Calvert Fault formed relatively late (post-Roper Group) in the tectonic history (Leaman, 1998).

The Timber Creek kimberlite field consists of 5 kimberlites—a small pipe TC-01 and 4 dykes—that intrude Paleo-mesoproterozoic sediments of the Victoria River Basin (Berryman et al., 1999). U–Pb

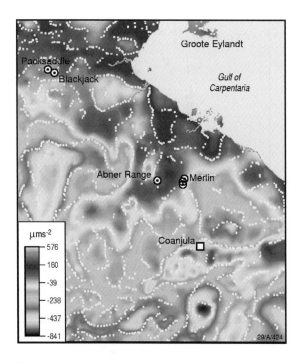

Fig. 8. Gravity image map of the McArthur Basin, Northern Territory showing the distribution of Merlin, Abner Range and Roper field kimberlites (circled dots) and Coanjula alkaline intrusions (squares) in relation to ground level Bouguer gravity anomaly overlain with points calculated at the maxima of the horizontal gradient for an upward continued level of 5600 m.

dating of zircon megacrysts from the Timber Creek pipe indicates a likely Jurassic (179 Ma) eruption age (Belousova et al., 2001). The kimberlites are aligned E–W near, although not apparently on, horizontal gradients in the upward-continued gravity field.

5.5. Gawler Craton

Kimberlites occur in several discrete provinces and fields within the Gawler Craton (Fig. 9) with the main clusters being at Elliston (Mount Hope) and Cleve on the Eyre Peninsula (Wyatt et al., 1994), Port Augusta in the Stuart Shelf, and Orroroo (Scott-Smith et al., 1984) and Terowie in the Adelaide Fold Belt at the eastern margin of the craton (Stracke et al., 1979; see Jaques, 2002). Available isotopic dating indicates all are of Jurassic age ~ 170–180 Ma (Stracke et al., 1979; Scott-Smith et al., 1984; Wyatt et al., 1994) but it has been suggested that there may be two ages of kimberlite–diamondiferous Triassic–Jurassic kimberlites and barren Jurassic kimberlites (see O'Reilly et al., 1997). The Orroroo kimberlites are the only

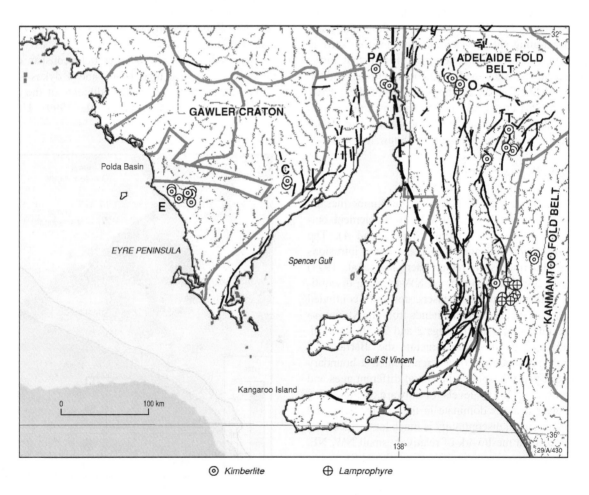

Fig. 9. Distribution of kimberlites and lamprophyres in the Gawler Craton and surrounding Adelaide and Kanmantoo Fold Belts in relation to faults, crustal element boundaries (in grey, Shaw et al., 1996), and points calculated at the maxima of the horizontal gradient for upward continued gravity data. Darker points are data upward continued to 800 m, grey points are maxima for data upward continued to 4 km. The dark dashed line indicates the inferred position of the eastern edge of the Gawler Craton. Main kimberlite fields: C—Cleve, E— Elliston (Mt Hope), O—Orroroo (Euralia), PA—Port Augusta, and T—Terowie (from Stracke et al., 1979; Wyatt et al., 1994; Jaques, 2002).

known diamondiferous kimberlites in the southern mega-element. They are unusual both in terms of their location within a Neoproterozoic–early Palaeozoic fold belt and in the presence of enstatite and magnesio-wustite inclusions in diamond, indicating an anomalously deep (i.e. lower mantle) origin (Scott-Smith et al., 1984).

The Adelaide Fold Belt kimberlites and associated lamprophyres are located on NW to NE-trending structures that parallel the fold grain in the belt and are evident as distinct linear anomalies in magnetic data, and much less obvious in upward continued gravity data. Locally, the Orroroo kimberlites lie at the intersection of NNW-trending and NW-trending structures evident in high resolution TMI data and on the extension of mapped NW-trending regional faults. Stracke et al. (1979) and White et al. (1995) noted that the kimberlites in the Adelaide Fold Belt are aligned with a major transform fault from the Antarctic Ridge (Fig. 5) and proposed that the location of the kimberlites was controlled by a postulated landward extension of the transform.

Two fields of barren kimberlites lie at margins of the E–W trending Polda Basin, an intra-cratonic Paleozoic – Cainozoic graben at the western margin of the Gawler Craton: the Cleve kimberlites lie at the eastern end and the Mt. Hope (Elliston) kimberlites cluster at the southern margin of the graben (Figs. 2 and 9). The Polda Basin is marked by distinct gravity and magnetic lows and its location is believed to have been controlled by tectonic rejuvenation of a pre-existing fracture system. In detail, the Cleve kimberlites, comprising several small dykes and associated blows that intrude Proterozoic schists (Wyatt et al., 1994), lie on NNE-trending structures evident in TMI data. The northernmost of the Mt Hope kimberlites lie close to a major anomaly evident in gravity data upward-continued to around 50 km. At the local scale the kimberlites are not obviously associated with particular structures evident in the potential field data, at least at the resolution used in this study.

5.6. Yilgarn Craton

Four fields of Precambrian age kimberlites and lamprophyres lie on the northern Yilgarn Craton and its northern margin (Fig. 10; Shee et al., 1999; Graham et al., 1999a). The Nabberu field comprises a suite of some 12 kimberlite pipes and dykes, four of which carry diamond at sub-economic grade, and associated ultramafics emplaced at about 1900–1700 Ma (Shee et al., 1999). The Jewill field includes four small (< 2 ha) marginally diamondiferous kimberlite pipes and the Buljah field, some 70 km to the SE, comprises four ultramafic lamprophyre (melnoite) pipes and sills: all have been dated at ∼ 1300 Ma (Graham et al., 1999a). Further south in the Eastern Goldfields Province of the Yilgarn Craton there are at least 14 kimberlites and 13 lamprophyres in addition to the carbonatite intrusions at Mount Weld and Pontin Creek (Cundelee). Four of the kimberlites are believed to be marginally diamondiferous. Graham et al. (2003) showed that the alkaline ultramafic intrusions as well as the carbonatites of the Eastern Goldfields Province have a similar age of emplacement (∼ 2020 Ma) and suggested a common enriched mantle source.

Shee et al. (1999) reported that the Nabberu kimberlites and lamprophyres exhibited strong structural control with E–W, N–S and NE–SW structures dominant. The bodies are both extensively sheared and highly altered. The Buljah and Jewill intrusions lie on NW-trending gradients in the Bouguer gravity map that align with the dominant regional trend of greenstones and shear zones of Eastern Goldfields Province of the Yilgarn Craton. The Nabberu intrusions lie at or near the intersection of the prominent regional NW-trending gravity gradient of the Yilgarn Craton with strong NE–SW and ENE- to E-trending gravity gradients that mark the southern boundary of the Capricorn Orogen (Fig. 10). The Leonora-Menzies kimberlites and lamprophyres are distributed in several clusters that mostly lie on either major NW-trending regional gravity gradients that commonly mark major shear systems or on ENE-trending gradients evident in both magnetic and gravity data that mark Proterozoic mafic dyke swarms.

The Jurassic Wandagee field lying in the Phanerozoic Carnarvon Basin at the western edge of the Yilgarn Craton (Fig. 1) consists of some 16 pipes and 6 sills of lamprophyre (mostly picritic monchiquites, Jaques et al., 1986). The intrusions lie in a 50-km long belt on a northerly-trending basement horst that is outlined by gradients in the upward-continued gravity data. Early sampling returned four small diamonds from four separate pipes (Jaques et al., 1986) but these have not been confirmed. Some 10 barren kimberlites

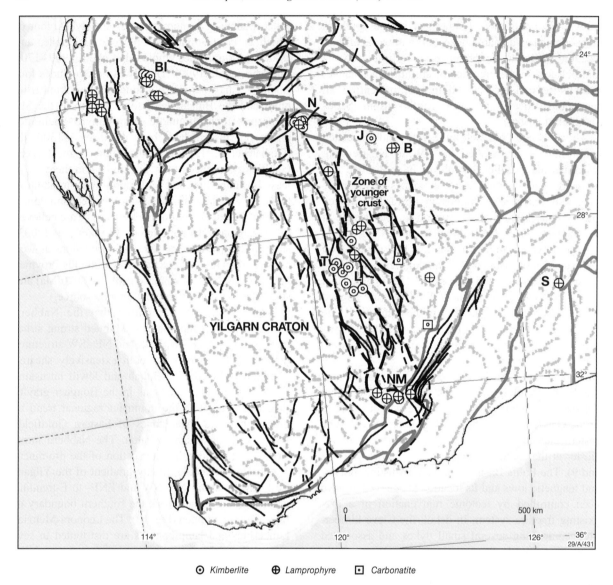

⊙ *Kimberlite*   ⊕ *Lamprophyre*   ☐ *Carbonatite*

Fig. 10. Distribution of kimberlites and lamprophyres in the northern Yilgarn Craton region, Western Australia in relation to faults, crustal element boundaries (in grey, Shaw et al., 1996), and points calculated at the maxima of the horizontal gradient for gravity data upward continued to 10 km. Main kimberlite-lamprophyre fields are: B—Buljah, BI—Barlee-Ilirrie, J—Jewill, L—Leonora, N—Nabberu, NM—Norseman, S—Shell Lakes, T—Teutonic Bore, W—Wandagee. Heavy dashed lines are crustal domains identified on the basis of model Nd isotopic compositions of Yilgarn granites (Cassidy, 2003).

and a lamprophyre have also been found by company exploration in the Barlee Range area at the western margin of the Bangemall Basin where they straddle the boundary between two major crustal elements evident in gravity and magnetic data (Fig. 10).

5.7. Pilbara Craton

The diamondiferous (sub-economic) 1900 Ma Brockman kimberlite in the Pilbara Craton occurs as a 20-km long dyke some 1–12 m wide with a marked photogeological and magnetic expression (Wyatt et al., 2003). Originally mapped as a fault, the dyke parallels significant regional NE-trending structures and gradients evident in both the gravity and magnetic data. Wyatt et al. (2003) suggested that the dyke may have exploited a pre-existing en echelon fault system formed during D_2 (Archaean) deformation.

6. Conclusions

This overview suggests that many of the kimberlite, lamproite and lamprophyre intrusions in the Australian continent are associated with crustal structures—especially major fault systems—and that many of these can be identified in potential field data, at local, regional and even continental scales. Frequently these structures are not marked on geological maps and are either not evident or only weakly visible in primary magnetic and gravity datasets where they are masked by shorter wavelengths (near surface anomalies) but are revealed in the total horizontal gradients of the upward continued data. The origin of these features is uncertain and remains an area of investigation: some of the density discontinuities evident in gravity data upward continued for many tens to hundreds of kilometres may mark different lithospheric blocks. Analysis of horizontal gradients in upwardly continued potential field data at all scales assists definition of crustal and potentially lithospheric structures, and hence, we suggest, is a potentially useful additional approach to defining prospective domains and corridors in diamond exploration. Although crustal structure, in particular major faults and faults systems, can be an important factor in localising the emplacement of kimberlites, lamproite and related alkaline rocks in the crust, the nature of the underlying lithosphere is the key determinant as to whether the intrusions are potentially diamondiferous (e.g. Helmstaedt and Gurney, 1995; Morgan, 1995). The present state of knowledge of the lithosphere under the Australian continent is sufficient to say that large areas of the Precambrian shield have lithospheric roots that are potentially in the diamond stability field. However, it is clear from pressure–temperature estimates and isotopic studies of mantle xenoliths and xenocrysts that some regions of this lithosphere have undergone thermal erosion that has likely destroyed the diamond potential (O'Reilly et al., 1997; Graham et al., 1999a). A more complete understanding of the distribution of potentially diamondiferous lithosphere under the Australian continent requires the integration of higher resolution seismic tomography with petrologic and isotopic studies of lithospheric evolution and with more rigorous study of crustal structure (including potential field modelling) and a better understanding of geodynamics.

Acknowledgements

We gratefully acknowledge Brian Kennett and Stewart Fishwick of the Australian National University for providing us with the revised (work in progress) seismic tomographic models of the Australian continent. We also thank Joe Mifsud for drafting the figures and Clive Collins, Subhash Jaireth, Tony Meixner and Alan Whitaker for valuable comments on the draft manuscript. Helpful reviews by 8IKC conference reviewers Bram Janse and Kevin Kivi improved the presentation of the paper. Publication is with permission of the Chief Executive Officer, Geoscience Australia.

References

Archibald, N., Gow, P., Boschetti, F., 1999. Multiscale edge analysis of potential field data. Exploration Geophysics 30, 38–44.

Artemieva, I.M., Mooney, W.D., 2001. Thermal thickness and evolution of Precambrian lithosphere; a global study. Journal of Geophysical Research 106 (8), 16387–16414.

Atkinson, W.J., Smith, C.B., Danchin, R.V., Janse, A.J.A., 1990. Diamond deposits of Australia. In: Hughes, F.E. (Ed.), Geology of the Mineral Deposits of Australia and Papua New Guinea. Australasian Institute of Mining and Metallurgy, Melbourne, pp. 69–76.

Belousova, E.A., Griffin, W.L., Shee, S.R., Jackson, S.E., O'Reilly, S.Y., 2001. Two age populations of zircons from the Timber Creek kimberlites, Northern Territory as determined by laser-ablation ICP-MS analysis. Australian Journal of Earth Sciences 48, 757–765.

Berryman, A.K., Steifenhoffer, J., Shee, S.R., Wyatt, B.A., Beousova, E.A., 1999. The discovery and geology of the Timber Creek kimberlites, Northern Territory, Australia. In: Gurney, J.J., Gurney, J.L., Pasco, M.D., Richardson, S.H. (Eds.), 7th International Kimberlite Conference, Cape Town Proceedings, vol. 1. Red Roof Design, Cape Town, pp. 30–39.

Betts, P.G., Giles, D., Lister, G.S., Frick, L.R., 2002. Evolution of
the Australian lithosphere. Australian Journal of Earth Sciences
49, 661–695.

Blakely, R.J., Simpson, R.W., 1986. Locating edges of source bodies from magnetic and gravity anomalies. Geophysics 51,
1394–1396.

Boxer, G.L., Jaques, A.L., 1990. The Argyle (AK1) diamond deposit.
In: Hughes, F.E. (Ed.), Geology of the Mineral Deposits of Aus-
tralia and Papua New Guinea. Monograph, vol. 14. Australasian
Institute of Mining and Metallurgy, Melbourne, pp. 697–706.

Braun, J., Shaw, R.D., 1998. Extension in the Fitzroy Trough,
Western Australia: an example of reactivation tectonics. In:
Braun, J., Dooley, J., Goleby, B., van der Hilst, R., Klootwijk,
C. (Eds.), Structure and Evolution of the Australian Continent.
AGU Geodynamics Series American Geophysical Union, Wash-
ington, DC, pp. 157–174.

Cassidy, K., 2003. Temporal framework of the Yilgarn Craton:
implications for mineral systems. Geoscience Australia Mineral
Exploration Seminar, Presentation.

Clifford, T.N., 1966. Tectono-metallogenic units and metallogenic
provinces of Africa. Earth and Planetary Science Letters 1,
421–434.

Collins, C.D.N., Drummond, B.J., Nicoll, M.G., 2003. Crustal
thickness patterns in the Australian continent. Geological Society of Australia, Special Publication 22, 115–122.

Cowan, D.R., Cowan, S., 1993. Separation filtering applied to
aeromagnetic data. Exploration Geophysics 24, 429–436.

Cull, J.P., 1991. Heat flow and regional geophysics in Australia. In:
Cermak, V., Rybach, L. (Eds.), Terrestrial Heat Flow and Lithospheric Structure. Springer-Verlag, Berlin, pp. 486–500.

Debayle, E., Kennett, B.L.N., 2000. The Australian continental upper mantle: structure and deformation inferred from surface
waves. Journal of Geophysical Research 105, B11, 25, 423–450.

Drummond, B.J., Etheridge, M.A., Davies, P.J., Middleton, M.F.,
1988. Half-graben model for the structural evolution of the
Fitzroy Trough, Canning Basin, and implications for resource
exploration. The APEA Journal 28 (1), 76–86.

Duggan, M.B., Jaques, A.L., 1996. Mineralogy and geochemistry
of Proterozoic shoshonitic lamprophyres from the Tennant
Creek Inlier, Northern Territory. Australian Journal of Earth
Sciences 43, 266–278.

Geoscience Australia, 2003. Online National Geoscience Datasets.
http://www.ga.gov.au/rural/projects/20011023_32.jsp.

Graham, S., Lambert, D.D., Shee, S.R., Smith, C.B., Hamilton, R.,
1999a. Re–Os and Sm–Nd Isotopic Constraints on the Sources
of Kimberlites and Melnoites, Earaheedy Basin, Western Australia. In: Gurney, J.J., Gurney, J.L., Pascoe, M.D., Richardson,
S.H. (Eds.), 7th International Kimberlite Conference, Cape
Town, Proceedings. Red Roof Design, Cape Town, pp. 280–290.

Graham, S., Lambert, D.D., Shee, S.R., Smith, C.B., Reeves, S.,
1999b. Re–Os isotopic evidence for Archaean lithospheric
mantle beneath the Kimberley Block, Western Australia. Geology 27, 431–434.

Graham, S., Lambert, D., Shee, S., 2003. Geochemical and isotopic
evidence of a kimberlite–melnoite–carbonatite genetic link.
Extended Abstracts, 8th International Kimberlite Conference,
Victoria, BC, Canada.

Grauch, V.J.S., Hudson, M.R., Minor, S.A., 2001. Aeromagnetic
expression of faults that offset basin fill, Albuquerque basin,
New Mexico. Geophysics 66, 707–720.

Gunn, P.J., Meixner, A.J., 1998. The nature of the basement to the
Kimberley Block, Northwestern Australia. Exploration Geophysics 29, 506–511.

Hell, A., Ramsay, R., Rheinberger, G., Pooley, S., 2003. The geology, age, near surface features, and mineralogy of the Merlin
kimberlites, Northern Territory, Australia. Extended Abstracts,
8th International Kimberlite Conference, Victoria, BC, Canada.

Helmstaedt, H., Gurney, J.J., 1995. Geotectonic controls primary
diamond deposits: implications for area selection. Journal of
Geochemical Exploration 53, 125–144.

Hildenbrand, T.G., Berger, B., Jachens, R.C., Ludington, S., 2001.
Utility of magnetic and gravity data in evaluating regional con-
trols on mineralization: examples from the Western United
States. Society of Economic Geology Reviews 14, 75–109.

Hobbs, B.F., Ord, A., Archibald, N.J., Walshe, J.L., Zhang, Y.,
Brown, M., Zhao, C., 2000. Geodynamic modeling as an exploration tool. After 2000: The Future of Mining. Australasian
Institute of Mining and Metallurgy Publication Series 2/2000,
pp. 34–49.

Holden, D.J., Archibald, N.J., Boschetti, F., Jessell, M.W., 2000.
Inferring geological structures using wavelet-based multiscale
edge analysis and forward models. Exploration Geophysics
31, 617–621.

Hornby, P., Boschetti, F., Horowitz, F.G., 1999. Analysis of potential field data in the wavelet domain. Geophysical Journal International 137, 175–196.

Hughes, F.E., Smith, C.B., 1990. Ellendale diamond deposits. In:
Hughes, F.E. (Ed.), Geology of the mineral deposits of Australia
and Papua New Guinea. Monograph, vol. 14. Australasian Institute of Mining and Metallurgy, Melbourne, pp. 1115–1122.

Janse, J.A., 1994. Is Clifford's Rule still valid? Affirmative examples from around the World. In: Meyer, H.O.A., Leonardos,
O.H. (Eds.), Proceedings of the Fifth International Kimberlite
Conference, Araxa, Brazil 1991, Volume 2. Diamonds: Characterisation, Genesis and Exploration. CPRM Special Publication,
vol. 1B. Companhia de Pesquisa de Recursos Minerais, Brasilia,
pp. 215–235.

Jaques, A.L., 1994. Diamonds in Australia. In: Solomon, M.,
Groves, D.I. (Eds.), The Geology and Origin of Australian Mineral Deposits. Clarendon Press, Oxford, pp. 787–820.

Jaques, A.L., 1998. Kimberlite and lamproite diamond pipes.
AGSO Journal of Australian Geology and Geophysics 17 (4),
153–162.

Jaques, A.L., 2002. Australian diamond deposits, kimberlites, and
related rocks. 1:5 million map. Geoscience Australia. www.ga.
gov.au/pdf/RR0114.pdf.

Jaques, A.L., Lewis, J.D., Smith, C.B., 1986. The Kimberlites and
Lamproites of Western Australia. Geological Survey of Western
Australia Bulletin, 132 (268 pp.).

Jaques, A.L., Hall, A.E., Sheraton, J.D., Smith, C.B., Sun, S.-S.,
Drew, R.M., Foudoulis, C., Ellingsen, K., 1989. Composition of
crystalline inclusions and C-isotopic composition of Argyle and
Ellendale diamonds. In: Ross, J., et al. (Ed.), Kimberlites and
Related Rocks-Volume 2: Their Mantle/Crust Setting, Diamonds

and Diamond Exploration. Geological Society of Australia Special Publication, vol. 14, pp. 966–989.

Jaques, A.L., O'Neill, H.St.C., Smith, C.B., Moon, J., 1990. Diamond-bearing peridotite xenoliths from the Argyle (AK1) pipe. Contributions to Mineralogy and Petrology 104, 255–276.

Jaques, A.L., Hall, A.E., Sheraton, J., Smith, C.B., Roksandic, Z., 1994. Peridotitic planar octahedral diamonds from the Ellendale lamproite pipes, Western Australia. In: Meyer, H.O.A., Leonardos, O.H. (Eds.), Proceedings of the Fifth International Kimberlite Conference, Araxa, Brazil 1991, Volume 2. Diamonds: Characterisation, Genesis and Exploration. CPRM Special Publication Companhia de Pesquisa de Recursos Minerais, Brasilia, pp. 69–77.

Johnson, R.W., Knutson, J., Taylor, S.R., 1989. Intraplate Volcanism in Eastern Australia and New Zealand. Cambridge Univ. Press, Cambridge, United Kingdom.

Kaminsky, F.V., Feldman, A.A., Varlamov, V.A., Boyko, A.N., Olofinsky, L.N., Shofman, I.L., Vaganov, V.I., 1995. Prognostication of primary diamond deposits. Journal of Geochemical Exploration 53, 167–182.

Kennett, B.L.N., 2003. Seismic structure in the mantle beneath Australia. Geological Society of Australia, Special Publication 22, 7–23.

Kimberley Diamond Company NL, 2001. Annual report to shareholders.

Kimberley Diamond Company NL, 2003. Annual report to shareholders.

Leaman, D.E., 1998. Structure, contents and setting of Pb–Zn mineralisation in the McArthur Basin, northern Australia. Australian Journal of Earth Sciences 45, 3–20.

Lee, D.C., Boyd, S.R., Griffin, B.J., Griffin, W.L., Reddicliffe, T.H., 1994. Coanjula diamonds, Northern Territory, Australia. In: Meyer, H.O.A., Leonardos, O.H. (Eds.), Proceedings of the Fifth International Kimberlite Conference, Araxa-Brazil, June 1991, Diamonds: Characterisation, Genesis and Exploration. CPRM Special Publication, vol. 1B, pp. 51–68.

Lee, D.C., Reddicliffe, T.H., Scott Smith, B.H., Taylor, W.R., Ward, L.M., 1998. Merlin diamondiferous kimberlite pipes. In: Berkman, D.A., MacKenzie, D.H. (Eds.), Geology of Australian and Papua New Guinean Mineral Deposits, Monograph, vol. 22. Australasian Institute of Mining and Metallurgy, Melbourne, pp. 461–464.

Meixner, A.J., Gunn, P.J., 1997. Three-dimensional kinematic modelling of the magnetic field of the southern Joseph Bonaparte Gulf. Exploration Geophysics 28, 260–264.

Milligan, P.R., Lyons, P., Direen, N.G., 2003. Spatial and directional analysis of potential field gradients—new methods to help solve and display three-dimensional crustal architecture. Extended Abstracts, Australian Society of Exploration Geophysicists 16th Geophysical Conference and Exhibition, Adelaide.

Morgan, P., 1995. Diamond exploration from the bottom up: regional geophysical signatures of lithosphere conditions favourable for diamond exploration. Journal of Geochemical Exploration 53, 145–165.

Murray, A.S., 1997. The Australian National Gravity Database. AGSO Journal of Australian Geology and Geophysics 17, 145–155.

Murray, A.S., 2001. Digital gravity grid of the Australian region, CD-ROM. AGSO-Geoscience Australia, Canberra.

Murray, A.S., Morse, M., Milligan, P.R., Mackey, T., 1997. Gravity Anomaly Map of the Australian Region, scale 1:5,000,000 (2nd edition), Australian Geological Survey Organisation, Canberra.

Myers, J.S., Shaw, R.D., Tyler, I.M., 1996. Tectonic evolution of Proterozoic Australia. Tectonics 15, 1431–1446.

O'Reilly, S.Y., Griffin, W.L., 1985. A xenolith-derived geotherm for southeastern Australia and its geophysical implications. Tectonophysics 111, 41–63.

O'Reilly, S.Y., Nicholls, I.A., Griffin, W.L., 1989. Xenoliths and megacrysts from Eastern Australia. In: Johnson, R.W., Knutson, J., Taylor, S.R. (Eds.), Intraplate Volcanism in Eastern Australia and New Zealand. Cambridge Univ. Press, Cambridge, pp. 249–287.

O'Reilly, S.Y., Griffin, W.L., Gaul, O., 1997. Paleo-geothermal gradients in Australia: key to 4-D lithosphere mapping. AGSO Journal of Australian Geology and Geophysics 17, 63–72.

Reading, A.M., Kennett, B.L.N., 2003. Lithospheric structure of the Pilbara Craton, Capricorn Orogen and northern Yilgarn Craton, Western Australia, from teleseismic receiver functions. Australian Journal of Earth Sciences 50, 439–445.

Reddicliffe, T.H., 1999. Merlin Kimberlite Field Batten Province, Northern Territory. MSc. by coursework. Research Report, Department of Geology and Geophysics, University of Western Australia.

Reddicliffe, T.H., Jakimowicz, J., Hell, A.J., Robins, J.A., 2003. The geology, mineralogy, and near surface characteristics of the Ashmore and Seppelt kimberlite clusters, North Kimberley, Australia. Extended Abstracts, 8th International Kimberlite Conference, Victoria, BC, Canada.

Richardson, L.M., Wynne, P, Hone, I., 2002. Geophysical data sets over continental Australia. Preview October 2002, 48–54.

Scott-Smith, B.H., Danchin, R.V., Harris, J.W.K., Stracke, K.J., 1984. Kimberlites near Orroroo, South Australia. In: Kornprobst, J. (Ed.), Kimberlites I: Kimberlites and Related Rocks. Elsevier, Amsterdam, pp. 121–142.

Shaw, R.D., Wellman, P., Gunn, P., Whitaker, A.J., Tarlowski, C., Morse, M., 1996. Guide to Using the Australia Crustal Elements Map. Australian Geological Survey Organisation Record 1996/30, Canberra.

Shaw, R.D., Meixner, T.J., Murray, A.S., 2000. Regional geophysical setting and tectonic implications of the mafic-ultramafic intrusions. In: Hoatson, D.M., Blake, D.H. (Eds.), Geology and Economic Potential of the Palaeoproterozoic Layered Mafic-ultramafic Intrusions in the East Kimberley, Western Australia. AGSO Bulletin, vol. 246, pp 63–98.

Shee, S.R., Vercoe, S.C., Wyatt, B.A., Hwang, P.H., Campbell, A.N., Colgan, E.A., 1999. Discovery and geology of the Nabberu Kimberlite Province, Western Australia. In: Gurney, J.J., Gurney, J.L., Pascoe, M.D., Richardson, S.H. (Eds.), 7th International Kimberlite Conference, Cape Town, Proceedings. Red Roof Design, Cape Town, pp. 912–922.

Simons, F.J., Zielhuis, A., van der Hilst, R.D., 1999. The deep structure of the Australian continent from surface wave tomography. Lithos 48, 17–43.

Somerville, M., Wyborn, D., Chopra, P.N., Rahman, S.S., Estrella,

D., van der Meulen, T., 1994. Hot Dry Rocks Feasibility Study, Australian Energy Research and Development Corporation Report 94/243, pp. 133.

Spector, A., Grant, F.S., 1970. Statistical methods for interpreting aeromagnetic data. Geophysics 35, 293–302.

Stracke, K.J., Ferguson, J., Black, L.P., 1979. Structural setting of kimberlites in southeastern Australia. In: Boyd, F.R., Meyer, H.O.A. (Eds.), Kimberlites, Diatremes, and Diamonds: their Geology, Petrology and Geochemistry. American Geophysical Union, Washington, pp. 71–91.

Sutherland, F.L., 1996. Alkaline rocks and gemstones, Australia: a review and synthesis. Australian Journal of Earth Sciences 43, 323–343.

Towie, N.J., Bush, M.D., Manning, E.R., Marx, M.R., Ramsay, R.R., 1994. The Aries diamondiferous kimberlite pipe, central Kimberley Block, Western Australia: exploration, setting and evaluation. In: Meyer, H.O.A., Leonardos, O.H. (Eds.), Proceedings of the Fifth International Kimberlite Conference, Araxa, Brazil 1991, Volume 2. Diamonds: Characterisation, Genesis And Exploration. CPRM Special Publication, vol. 1B. Companhia de Pesquisa de Recursos Minerais, Brasilia, pp. 319–328.

Van der Hilst, R.D., Kennett, B.L.N., Shibatani, T., 1998. Upper mantle structure beneath Australia from portable array deployment. In: Dooley, J., Goleby, J., van der Hilst, B., Klootwijk, R. (Eds.), Structure and Evolution of the Australian continent. AGU Geodynamics Series. American Geophysical Union, Washington, DC, pp. 39–57.

Vearncombe, S., Vearncombe, J.R., 2002. Tectonic controls on kimberlite location, southern Africa. Journal of Structural Geology 24, 1619–1625.

Wellman, P., 1978. Gravity evidence for abrupt changes in mean crustal density at the junction of Australian crustal blocks. BMR Journal of Australian Geology and Geophysics 3, 153–162.

Wellman, P., 1988. Development of the Australian Proterozoic crust as inferred from gravity and magnetic anomalies. Precambrian Research 40/41, 89–100.

Wellman, P., 1998. Mapping of geophysical domains in the Australian Continental crust using gravity and magnetic anomalies. In: Braun, J., Dooley, J., Goleby, B., van der Hilst, R., Klootwijk, C. (Eds.), Structure and Evolution of the Australian ContinentAGU Geodynamics Series American Geophysical Union, Washington, DC, pp. 59–71.

White, S.H., de Boorder, H., Smith, C.B., 1995. Structural controls of kimberlite and lamproite emplacement. Journal of Geochemical Exploration 53, 245–264.

Wyatt, B.A., Shee, S.R., Griffin, W.L., Zweistra, P., Robinson, H.R., 1994. The petrology of the Cleve Kimberlite. Eyre Peninsula, South Australia. In: Meyer, H.O.A., Leonardos, O.H. (Eds.), Proceedings of the Fifth International Kimberlite Conference, Araxa, Brazil 1991, Volume 1. Kimberlites, related rocks and mantle xenoliths. CPRM Special Publication, vol. 1A. Companhia de Pesquisa de Recursos Minerais, Brasilia, pp. 62–79.

Wyatt, B.A., Sumpton, J.D.H., Stiefenhofer, J., Shee, S.R., Smith, T.W., 1999. Kimberlites in the Forrest River Area, Kimberley Region, Western Australia. In: Gurney, J.J., Gurney, J.L., Pascoe, M.D., Richardson, S.H. (Eds.), 7th International Kimberlite Conference, Cape Town, Proceedings. Red Roof Design, Cape Town, pp. 912–922.

Wyatt, B.A., Mitchell, M., Shee, S.R., Griffin, W.L., Tomlinson, N., White, B., 2003. The Brockman Creek Kimberlite, East Pilbara, Australia. Extended Abstracts, 8th International Kimberlite Conference, Victoria, BC, Canada.

Available online at www.sciencedirect.com

Lithos 77 (2004) 803–818

www.elsevier.com/locate/lithos

Stratigraphic relations, kimberlite emplacement and lithospheric thermal evolution, Quiricó Basin, Minas Gerais State, Brazil

George Read[a,*], Herman Grutter[b,1], Stewart Winter[c,2], Nigel Luckman[d,3], Frank Gaunt[e,4], Fernando Thomsen[f,5]

[a] Shore Gold Inc., Canada
[b] Mineral Services Canada Inc., Canada
[c] Winterbourne Explorations Inc., Canada
[d] Superior Diamonds Inc., Canada
[e] Parimá Mineração S.A., Brazil
[f] Resende and Thomsen Pesquisas Geologicas, Brazil

Received 27 June 2003; accepted 19 January 2004
Available online 25 May 2004

Abstract

The Quiricó Basin covers an area of 10,000 km^2 and is situated to the west of the conventionally defined southwestern margin of the Archean São Francisco craton in Minas Gerais State, Brazil. The sedimentary infill of the Quiricó Basin consists of lightly metamorphosed shallow marine clastic bedrock sediments of the Bambuí Group ($\sim 650 \pm 15$ Ma), unconformably overlain by Early Cretaceous terrigenous lacustrine (Quiricó Formation), alluvial fan (Abaeté Formation) and fluvial/aeolian (Três Barras Formation) deposits of the Areado Group. Rare kimberlites and ubiquitous kamafugites of the Alto Paranaíba Igneous Province (APIP) erupted through the recently deposited sediments of the Quiricó Basin in the time period 95–61 Ma. The 120-m-thick Mata da Corda Group overlies the Late Cretaceous Areado Group over an area of 8000 km^2 and is composed largely of extrusive kamafugite and related volcanosedimentary material. Unusually large diamonds with proximal surface features and population characteristics are well known to occur in rivers and streams that drain the stratigraphic succession in the Quiricó Basin, prompting the search for their presumably local primary source(s) and a possibly associated Archean basement or cratonic root. Conceptual exploration models for this setting may in part be based on the diamondiferous 120 Ma Canastra and 95 Ma Três Ranchos kimberlites, but require reconciliation with the observed abundance of 85–61 Ma old diamond-free kamafugites. Field relations and carefully controlled stratigraphic samples show that a distinctive mantle-derived indicator mineral suite occurs in the Maxixe Member, a volcaniclastic breccia unit that occurs at the base of the Mata da Corda Group. A detailed thermobarometric comparison of mantle-derived xenocrystic clinopyroxene compositions from this member with clinopyroxene populations derived from kimberlites and kamafugites situated in the Quiricó Basin shows a distinct and

* Corresponding author. Fax: +1-306-664-7181.
E-mail addresses: gread@shoregold.com (G. Read), herman.grutter@mineralservices.com (H. Grutter), swinter@vianet.on.ca (S. Winter), nluckman@superiordiamonds.ca (N. Luckman), frankgaunt@yahoo.com (F. Gaunt), fthomsen@bol.com.br (F. Thomsen).
[1] Fax: +1-604-602-9855.
[2] Fax: +1-705-524-6368.
[3] Fax: +1-604-688-5175.
[4] Fax: +55-61-345-2901.
[5] Fax: +55-61-335293.

abrupt change in geothermal conditions at the onset of kamafugite magmatism. Kimberlites entrained fragments of garnet peridotite along a cool, craton-like geotherm which intersected the diamond stability field in the temperature range 880–1000 °C. Clinopyroxene xenocrysts in early-stage kamafugites (Maxixe Member) record heating by ~ 240 °C, placing much of the deep lithosphere inside the graphite stability field. Clinopyroxenes contained in samples from common APIP kamafugites indicate the loss of the entire garnet-facies lithospheric section, estimated at 75 km thick at the time. This critical event fundamentally changed the lithospheric architecture and geothermal regime of the Quiricó Basin from cratonic and "diamond-friendly" to noncratonic and "diamond-unfriendly" over a short time interval. The lithospheric evolution constrained here by judicious investigation of geological relationships during diamond exploration by Canabrava Diamond confirms the potential of the Quiricó Basin to host diamondiferous kimberlites that have tapped cool cratonic lithosphere at times prior to eruption of Mata da Corda kamafugites.

Keywords: Areado; Clinopyroxene; Diamond; Kamafugite; Mata da Corda; Thermobarometry

1. Introduction

The Quiricó Basin occurs between the towns of Carmo do Paranaíba in the west, Tiros in the east and extends north of João Pinheiro in Minas Gerais State, Brazil as shown in Fig. 1. The Quiricó Basin is restricted to the southern part of the San Franciscana Basin, which has previously been defined by Campos and Dardenne (1997a,b). Cretaceous infilling of the Quiricó basin occurred at a crucial time in the thermal evolution of the southwestern part of the São Fran-

cisco Craton and the development of the Alto Para-naíba Igneous Province (APIP). Alkaline volcanoes that form part of the APIP erupted through the recently deposited sediments of the Quiricó Basin. Field relations, petrography and geochronology show that the initial alkaline volcanism was kimberlitic, while the final event of the APIP was kamafugitic (Araujo et al., 2001). Kamafugitic eruptive centres are ubiquitous in the Quiricó Basin and are far more common than true kimberlites. The kamafugites and related alkaline volcanics form the Mata da Corda

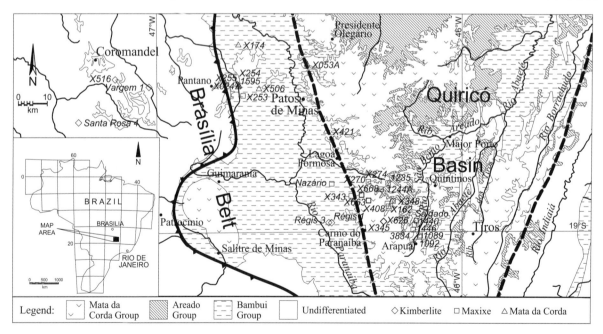

Fig. 1. Map of the Quiricó Basin and the adjacent Brasilia Belt, illustrating stratigraphic groups, locations of alkaline eruptives, major drainages and cities.

Group, which is the world's largest known occurrence of extrusive kamafugite.

This region has long been known for its high-value diamonds mined from alluvial deposits associated with the Abaeté, São Bento, Areado and Tiros rivers. There are strong suggestions that the alluvial diamonds, found in the drainage systems of the Quiricó Basin, are locally derived from kimberlite source rocks, but such sources have long eluded exploration geologists. Conceptual exploration models can be based in part on the age and stratigraphic relations of the diamondiferous Três Ranchos and Canastra kimberlites, which occur over 100 km to the west of Archean rocks that define the western margin of the São Francisco craton. This setting and the relatively close proximity, in space and time, of diamondiferous kimberlites and barren kamafugites suggest a complex evolution of possibly non-Archean lithosphere below the Quiricó Basin during the Late Cretaceous.

The kimberlites and kamafugites of the APIP have transported to the surface xenocrysts, which are inferred to be representative of the mantle regions sampled by these magmas. Some 10 years of diamond exploration in this region by Canabrava Diamond has resulted in a large database of indicator mineral analyses. Clinopyroxene, specifically chrome diopside, is a peridotitic silicate that is common to both kimberlites and kamafugites of this region. Xenocrystic clinopyroxene compositions provide insight into the relative abundance of spinel-facies or garnet-facies peridotite entrained by kimberlite and kamafugite magmas from the upper mantle. Application of the Nimis and Taylor (2000) thermobarometer to garnet-facies Cr-diopsides from kimberlites or kamafugites enables us to constrain the thermal and temporal evolution of lithospheric mantle during the Late Cretaceous infilling and development of the Quiricó Basin.

2. Stratigraphic relations

The Quiricó Basin is situated somewhat west of the conventionally defined southwestern margin of the Archean São Francisco craton. Radiometric U–Pb dating of granitoid intrusions and greenstone belt volcanism indicates that the cratonic nucleus stabilized at ~2720 Ma (Machado et al., 1992). The conventional definition of the São Francisco craton in the literature suggests that the southwestern margin of the craton lies to the east of the cities of São Gotardo and Bambuí in central Minas Gerais (Fig. 63a of Dardenne and Schobbenhaus, 2001). This definition is consistent with the occurrence of the ubiquitous kamafugitic eruptive centres of the Mata da Corda Group. However, the spatial distribution of alluvial diamonds and the occurrence of diamondiferous kimberlite at Três Ranchos-04 in the west and Canastra-01 in the south indicate that, at the time of kimberlite emplacement in the Early Cretaceous, thick, diamond-bearing lithosphere was far more extensive than the conventional (Archean) definitions of a craton (e.g. Janse, 1994). This paper aims to explain the occurrence of diamondiferous kimberlites within a complex stratigraphy in the apparently off-craton setting of the rapidly evolving Quiricó Basin during the Cretaceous. Fig. 2 illustrates the temporal relationships of the lithostratigraphic units discussed in the paper.

The shallow marine clastic sediments of the Bambuí Group (~650±15 Ma) constitute the basement rocks underlying the Quiricó Basin. During the Late Proterozoic Brasiliano Orogeny (700–450 Ma), the rocks of the Bambuí Group were thrust eastwards over the São Francisco cratonic basement. This was caused by the collision of the São Francisco and Amazonia cratonic nuclei during the amalgamation of the Gondwana supercontinent (Gibson et al., 1995). These Proterozoic rocks are presently preserved as a succession of thrust slices known as the Brasília Belt. The Ibiá Formation rocks form the base of the Bambuí Group and consist of strongly foliated green phyllites with quartz and carbonate stringers. The remainder of the Bambuí Group is composed of slates and arkosic metasediments (Byron, 1999). The weathered slates of the Bambui are easily recognizable in the field by their characteristic pale pink colour and cleavage.

The Paraná Basin, bordering the Brasília Belt to the west, is covered by tholeiitic lavas erupted from 127 to 137 Ma (Turner et al., 1994). The Paraná basalts are closely correlated with the basalts, rhyolites and intrusive complexes of the Etendeka Province of northwestern Namibia. This correlation strongly suggests that the opening of the south Atlantic had not commenced by 127 Ma, with the Quiricó Basin hence being located in a stable intracontinental setting. The

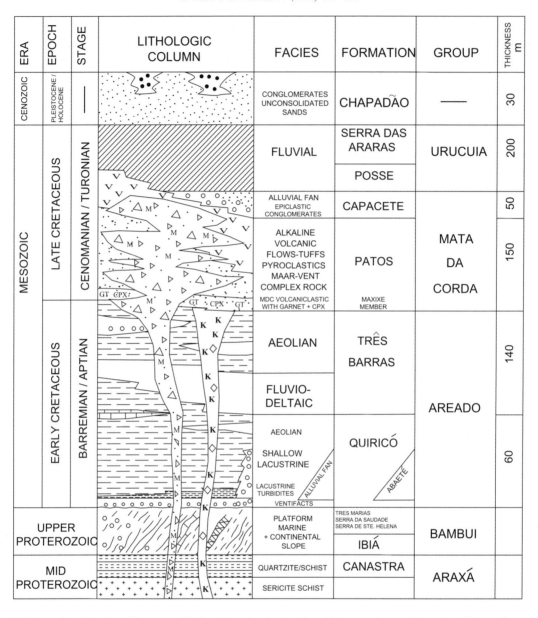

Fig. 2. This lithostratigraphic column illustrates and defines the rock units from the mid-Proterozoic crystalline rocks of the Araxá Group to the Late Cretaceous sediments of the Urucuia Group. The Araxá and Urucuia do not occur within the Quiricó Basin but are found across the Brasilia Belt to the west (Araxá) and east and west of the São Francisco River north of Pirapora (Urucuia). Note the relative stratigraphic positions of Early Cretaceous kimberlite (K) and Late Cretaceous Mata da Corda kamafugite (M), including the Maxixe Member. Modified after Seer et al. (1989).

Early Cretaceous stability of the Quiricó Basin has implications for the emplacement of diamondiferous kimberlites during that period. The Early Cretaceous Paraná and Etendeka volcanism is unique and distinct from the Jurassic (183 Ma) Karoo Igneous Province (Duncan et al., 1997) that was active further east in Gondwana.

The Quiricó Basin is situated to the east of the Paraná Basin across the Brasília Belt. Brazilian stratigraphers define these Cretaceous volcanosedimentary rocks to the east of the Brasília Belt as the San Franciscana Basin (Campos and Dardenne, 1997a,b). Their definition of the San Franciscana Basin includes all Phanerozoic cover rocks of the São Francisco craton. However, the area of the Quiricó defined in this study is more restricted than former definitions of the San Franciscana Basin. In particular, the Quiricó Basin as defined in this study does not include the fluvial sediments (conglomerates and arenites) of the Late Cretaceous Urucuia Group that accounts for most of the outcrop of the San Franciscana Basin north of 17°S. The onset of Cretaceous sedimentation in the Quiricó Basin is defined by a ventifact marker horizon that has developed at the base of the Areado Group on the Early Cretaceous weathering surface of the Proterozoic Bambuí basement rocks. Sgarbi (2000) notes that the conglomerates of the Abaeté Formation include transported ventifacts. Most of the authors of this paper have observed quartzite and banded ironstone ventifacts in situ on a Proterozoic Bambuí weathering surface near Tiros (east of the Rio Abaeté), which is unconformably covered by aeolianites of the Early Cretaceous Três Barras Formation. These relationships strongly suggest that winds at the onset of the Cretaceous faceted the ventifacts and subsequently deposited the aeolianites of the lower part of the Quiricó Basin.

The Areado Group includes three formations: the Quiricó Formation (lacustrine; siltstones and argillites), the Abaeté Formation (alluvial fan; conglomerates) and the Três Barras Formation (fluvial and aeolian; arenites) (Fig. 2). Ostracodes from the lacustrine Quiricó Formation are assigned an Aptian–Albian age of 110–115 Ma (Arai et al., 1995). Detailed mapping suggests that there are no Quiricó sediments east of the Rio Abaeté, with the dominant Areado unit in this region being the aeolian quartz arenite, which was deposited directly on the lag/ventifact surface on the Bambuí. This suggests that this area was elevated at the time of the Quiricó sedimentation. The transition from a lacustrine arenite at the top of the Quiricó in the west to an aeolian arenite in the east suggests that the two are time equivalent. The lacustrine arenite may represent sand blown into the lake from the adjacent highland areas (Winter, 2000). While the rocks of the Quiricó, Abaeté and Três Barras are stratigraphically classified

as individual formations, it is possible they represent various, more or less contemporaneous, sedimentary facies operating in adjacent parts of the depositional basin.

The Late Cretaceous volcanic rocks of the Mata da Corda Group overlie the Early Cretaceous rocks of the Areado Group. Outcrop geology consists of intrusions, flows and pyroclastic and epiclastic deposits of ultramafic/mafic composition. Exposures are intensely weathered and fresh outcrops are rare. The rocks of the Mata da Corda Group include the world's largest known occurrence of extrusive kamafugite. Kamafugites are silica undersaturated kalsilite-bearing lavas, which have olivine, clinopyroxene, phlogopite, melilite and kalsilite (silica depleted feldspar) as the major phases. Leucite and augite appear with increasing silica activity (Mitchell and Bergman, 1991). Sahama (1974) defined "kamafugite" as a contraction of the names katungite–mafurite–ugandite. Araujo et al. (2001) examined 52 kamafugite occurrences from the APIP, which include mafurite (with kalsilite) and ugandite (with leucite) end members. Kamafugites differ from lamproites with respect to the degree of silica saturation as expressed petrographically by the presence of kalsilite and melilite and the absence of sanidine (Mitchell and Bergman, 1991).

The Mata da Corda volcanics presently cover an area of some 4500 km^2 between the cities of Bambui (south), Pirapora (north) and Coromandel (west). The Mata da Corda Group consists of two formations: the Patos Formation includes the primary alkaline volcanic rocks, while the younger Capacete Formation includes distal epiclastic sediments with an important aeolian sand contribution (Campos and Dardenne, 1997a,b). In areas where volcanic feeders have intruded the Areado Group Três Barras Formation arenites, the sandstones are disrupted, brecciated, fluidized and silicified (Byron, 1999). The fluidization of the Areado sandstones by the Mata da Corda volcanics may indicate that the Areado sediments were only semiconsolidated at the time of Mata da Corda volcanism. Byron (1999) has documented the volcanology and lithogeochemistry of the Mata da Corda volcanics in detail. Literature ages for Mata da Corda volcanics are in the range of 84–61 Ma (Bizzi et al., 1994; Gibson et al., 1995; Sgarbi et al., 2000b). The features of the APIP are best described by

Leonardos and Meyer (1991), Gibson et al. (1995) and Carlson et al. (1996). The geochemical evolution of the APIP in the Cretaceous is not well understood in spite of these review papers.

Sampling completed during diamond exploration programs has shown that the Mata da Corda kamafugites carry a suite of heavy minerals that can easily be confused with indicator minerals from kimberlites. The kamafugites produce abundant ilmenite, chromite and clinopyroxene, while garnet is essentially absent. In the late 1990s, geologist Fernando Thomsen, employed by Canabrava Diamond, located a basal Mata da Corda unit, in contact with Areado sandstones, that carried abundant peridotitic garnet and clinopyroxene. The garnet and clinopyroxene grains were initially observed in heavy mineral lag deposits in local streams. Sampling and processing of nearby rock outcrops soon showed that the garnet and clinopyroxene xenocrysts were weathering out of a breccia unit occurring at the base of the Mata da Corda volcanic sequence. Thomsen mapped the first outcrop of this garnet-bearing, basal Mata da Corda unit on the north slope of Serra do Maxixe (UTM coordinates: E360402, N7910657 Zone 23 South, Corrego Alegre Datum), some 11 km north of the city of Carmo do Paranaíba. Soon after the discovery of this unit, it was found that a number of the peridotitic garnet xenocrysts had chrome-rich harzburgitic (subcalcic) compositions, indicating that their host rock may have sampled the lithosphere at temperatures and pressures favourable for the crystallization of diamond. The presence of the harzburgitic garnets encouraged further mapping and sampling, which located similar garnet-bearing breccia units, always proximal to the contact with the Areado, over an area of some 600 km^2. These garnet-bearing units have been named the Maxixe Member after the type section on Serra do Maxixe. The Maxixe Member rocks are typically volcaniclastic kamafugite breccias that uniquely contain harzburgitic and lherzolitic garnet and lherzolitic clinopyroxene. The juxtaposition of the Maxixe Member rocks on the Areado sandstones confirms that they represent the initial pulse of Mata da Corda kamafugite volcanism. They are mineralogically consistent and stratigraphically conformable with the overlying Patos Formation kamafugites of the Mata da Corda Group. The Maxixe Member is distinct in composition, mineralogy and chronostratigraphic setting from

the Early Cretaceous kimberlites. Maxixe Member outcrops were mapped by Canabrava Diamond geologists in the vicinity of the cities of Patos de Minas, Carmo do Paranaíba, Quintinos and Tiros. However, examination of data given in Ramsay and Tompkins (1994) suggests that the Boa Esperança and Cana Verde bodies represent subvolcanic equivalents of the Maxixe Member further south, beyond the southern limit of mapped Mata da Corda.

The final formation represented on the lithostratigraphic column is the Chapadão Formation, which is composed of Quaternary unconsolidated sands and some conglomerates. Within the Quiricó Basin, these Quaternary sediments have developed in association with the larger modern-day drainages such as the Rios Abaeté, São Bento, Areado, Paranaíba, Santo Antônio do Bonito and Santo Inácio.

3. Setting and age of alkaline magma types

The recovery of many high-value diamonds from placer deposits in this region motivated geologists to locate their primary source rocks. Sopemi (De Beers) discovered the first kimberlites at Vargem on the Santo Inácio River in 1968. Over the next 30 and more years, exploration programs applied the classical techniques of stream sediment sampling for indicator minerals combined with airborne and ground geophysical surveys. This exploration work resulted in the discovery of many alkaline magmatic rocks, which included abundant kamafugites and rare kimberlites. Discrimination of kimberlites from kamafugites using the data obtained by classical exploration techniques has proved extremely difficult because their geophysical signatures overlap and because the mantle-derived indicator minerals ilmenite, spinel and clinopyroxene occur in abundance in kimberlites and in kamafugites. Discrimination of these alkaline rock types is, however, possible based on their field relationships, xenolith content and overall stratigraphic setting. We outline these geological and available radiometric age relationships below to establish a stratigraphically controlled temporal framework for the alkaline magmatic rocks that occur within the Quiricó Basin (Fig. 2).

A large number of Mata da Corda magmatic and volcaniclastic extrusive rocks and related maar-type kamafugitic volcanic vents have been discovered within the Quiricó Basin (Fig. 1). Accurate age dates for these rocks are difficult to obtain due to their generally altered nature. Gibson et al. (1995) provide K/Ar dates for fresh phlogopite from a kamafugite intrusion and a lava outcrop that coincides within error at 84–83 Ma. The dates punctuate the 90–80 Ma period during which alkaline magmas are thought to have intruded throughout the APIP (see Table 1 and overview of Gibson et al., 1995). Radiometric ages obtained by U–Pb dating of perovskite from Mata da Corda kamafugites range from 81 to 61 Ma (Sgarbi et al., 2000b). The available data indicate that major kamafugitic magmatism occurred at ~84 Ma and lasted for a period of about 10 or 20 Ma. While the ratio of kamafugites to kimberlites discovered is high,

Table 1
Famous diamonds recovered from rivers draining the Quiricó Basin

Name	Carats	Year	Drainage
Abaeté (carbonado)	827.5	1935	Abaeté
Presidente Vargas	726.6	1938	Santo Antônio do Bonito
Darcy Vargas	460	1939	Santo Antônio do Bonito
Charneca I	428	1940	Santo Inácio
Presidente Dutra	407.68	1949	Dourados
Diário de Minas	375.1	1941	Santo Antônio do Bonito
Vitória I	375	1945	Abaeté
Tiros I	354	1940	Abaeté
Bonito I	346	1948	Santo Antônio do Bonito
Vitória II	328	1943	Abaeté
Patos	324	1937	São Bento
Abaeté	238	1926	Abaeté
Coromandel III	228	1936	Santo Inácio
Regente de Portugal	215	1732	Abaeté
Tiros II	198	1935	Abaeté
Tiros III	182	1935	Abaeté
Estrela de Minas	179.38	1910	Dourados
Brasília	176	1944	Preto
Tiros 4	173	–	Abaeté
Minas Gerais	172.5	1937	Santo Antônio do Bonito
Princesa do Carmo de Paranaíba	165	1986	São Bento
Nova Estrela do Sul	140	1937	Abaeté
Charneca III	132	1972	Santo Antônio do Bonito
Vargem I	110	1940	Santo Inácio
Abadia dos Dourados	104	–	Dourados
Rosa de Abaeté	80.3	1935	Abaeté

a number of true kimberlites have been discovered such as the Três Ranchos, Santa Rosa, Santa Clara, Vargem, Douradinho, Regis and X270. All known kimberlites within the APIP crosscut host rocks of the Araxá Group (Santa Rosa 4), the Bambui Group (Regis 1) or Areado Group (X270). None of the known kimberlites have been found to intrude the Mata da Corda Group, nor to contain xenoliths of the Maxixe Member or xenoliths of other kamafugitic material. These important, though often overlooked, stratigraphic relationships establish that intrusion of all known kimberlites immediately predates volcanic extrusion of the Maxixe Member. Topographic, stratigraphic and field relationships constrain the source of alluvial diamonds to the same stratigraphic horizon (i.e. post-Areado, pre-Maxixe, see Fig. 2), indicating that only the kimberlite vents would constitute viable local primary source rocks for some of the alluvial diamonds (e.g. Kaminsky et al., 2001).

The X270 kimberlite pipe typifies these stratigraphic relations. The pipe occurs some 40 km south of the city of Patos de Minas and has a surface area of at least 10 ha. X270 was initially discovered using airborne and ground geophysical surveys as it is capped by up to 12 m of laterite and exhibits no local indicator mineral anomaly. By contrast, drill core samples have yielded abundant garnet, clinopyroxene, ilmenite and chromite as indicator minerals, as well as one 10-point diamond (0.10 carat). Petrographic study classifies X270 as a crater facies, primary volcaniclastic kimberlite breccia with Group-1 type kimberlite groundmass mineralogy (Grütter, 2002). Accurate geochronology using U–Pb isotopes on perovskite defined an eruption age of 89.5±3.4 Ma (Heaman, 2002), consistent with X270 being hosted in Proterozoic Bambuí basement. However, the kimberlite petrography reveals many arenite clasts and quartz and feldspar sand grains, which confirm the presence of overlying semiconsolidated Areado Group sediments at the time of eruption. Large Areado clasts are also observed in drill core from depth. The isotopic age determination and the presence of Areado sandstone clasts confirm the post-Areado eruption age for the X270 kimberlite, while the absence of kamafugite clasts suggests a pre-Maxixe age.

Kimberlites and diamonds are found on both sides of the Brasília Belt. The diamondiferous Três Ranchos-04 kimberlite in the west near Catalão has been dated at 95 Ma (Bizzi, 1995). The significantly diamondiferous Canastra-01 kimberlite is located in southwestern Minas Gerais, west of the Brasília Belt, and has been dated at 120 Ma (Doyle and De Beers, personal communication, 2002). The Vargem and Santa Rosa kimberlite clusters are also west of the Brasília Belt and produce high Cr_2O_3 harzburgitic and lherzolitic garnets suggesting that the diamond stability field lithosphere has been sampled by these kimberlites. This study presents clinopyroxene data for two kimberlites east of the Brasília Belt: X270 and Regis. The distribution of these kimberlites suggests that the mobilization along the Brasília Belt did not affect the adjacent lithosphere. However, an explanation is required for the geographic juxtaposition of the petrologically different magmas of the ∼ 120–89 Ma kimberlites, the ∼ 85 Ma Maxixe Member and the ∼ 84–61 Ma kamafugites.

4. Distribution of alluvial diamonds

Diamonds were first discovered in Brazil in 1729 by alluvial gold miners working placer deposits at Diamantina in central Minas Gerais. As settlers and agricultural development moved westward, so did the *garimpeiros* (prospectors and diggers) and diamond discoveries were made in the rivers of western Minas Gerais. Before the end of the 1700s, garimpeiros were actively mining diamonds from gravels associated with a number of rivers in the region of the Quiricó Basin. Brazil soon became the world's principal diamond producer and held this position until 1870 when diamonds were discovered near Kimberley in South Africa. By 1880, the South African mines were in production and Brazil's proportion of world production declined to 3%. Present production from Minas Gerais amounts to some 220,000 carats per annum with 98% of the diamonds mined from alluvial deposits and less than 2% from the Proterozoic metaconglomerates in the vicinity of Diamantina (Karfunkel et al., 1994).

Accurate diamond production statistics are difficult to obtain for this part of Minas Gerais or for most alluvial diamond mining districts in Brazil. Virtually all diamonds are mined by *garimpeiros* who operate alluvial diggings which range from one-man pick and shovel operations to larger scale diggings using trucks, earth-moving equipment and a processing plant, usually consisting of jigs. In spite of these somewhat primitive mining operations, a number of significant diamond discoveries have been made and continue to be made every year. Garimpeiros are usually sponsored by a landowner, diamond buyer or wealthy benefactor. The profits from the sale of diamonds are split between the garimpeiro, landowner and sponsor according to prearranged proportions. Many diamond transactions are completed in cash, without documentation and outside the tax regime, hence, the difficulty compiling past diamond production records for specific drainages and the monitoring of annual production statistics. Many diamond records are based on the "oral tradition".

The western margin of the Quiricó Basin essentially forms the drainage divide between the west and south flowing Rio Paranaíba and the east and north flowing Rio São Francisco. The principal diamond bearing rivers of the São Francisco drainage are the Abaeté, São Bento, Areado, Tiros, Borrachudo, Indaiá and da Prata. Significant diamonds have been recovered from all these drainages, in particular the Rio Abaeté. The Santo Antônio do Bonito, Santo Antônio das Minas Vermelhas, Santo Inácio, Douradinho and Dourados drain westwards from the western margin of the Quiricó Basin into the Paranaíba. The Paranaíba has its source in the southern part of the Quiricó Basin and initially flows north along the western side of the basin before it turns west across the Brasília Belt (see Fig. 1).

The largest gem diamond found in Brazil and the world's seventh largest is the 727-carat Presidente Vargas, which was recovered from diggings on the Santo Antônio do Bonito in 1938. The westward draining Santo Antônio do Bonito and Santo Inácio and the northward draining Abaeté have a remarkable history of yielding large high-quality stones. Some significant pink diamonds have been recovered from the Abaeté—notably the 79-carat pink of May 1999. Table 1 lists some of the famous Brazilian diamonds recovered from the vicinity of the Quiricó Basin. It is the authors' opinion that many large diamonds have been recovered from these drainages and are not part of the historic record.

The origin of these diamonds is both enigmatic and controversial. Tompkins and Gonzaga (1989) and Gonzaga et al. (1994) argue in favour of a glacial origin from undiscovered kimberlites in the central portion of the São Francisco craton. They state that the alkaline magmas of the APIP have been emplaced in a Proterozoic mobile belt that is unfavourable for the generation of magmas that have sampled within the diamond stability field. Kaminsky et al. (2001) examined over 1000 diamonds from the São Francisco and Paranaíba drainages and argue in favour of possible local sources. It is the authors' opinion that the size and abundance of diamonds in local drainages, combined with diamond population differences between drainages and the presence of the Early Cretaceous diamondiferous Canastra and Três Ranchos kimberlites, support derivation of high-value alluvial diamonds from local kimberlites. A model is required to reconcile the cool cratonic geotherm determined by Carvalho and Leonardos (1997) for the Três Ranchos 4 kimberlite with the voluminous eruption of barren kamafugites in an apparently off-craton setting. This is pursued below.

5. Clinopyroxene thermobarometry

5.1. Rationale, methods and database

Independent from using the occurrence, distribution and physical attributes of alluvial diamonds in the Quiricó Basin, the diamond exploration model for the area may be tested and validated by reconciling the mineralization model with the pressure–temperature conditions recorded by mantle-derived indicator minerals. With this in mind, a total of 64 field samples of magmatic rock types (kimberlite and kamafugite) were systematically collected at stratigraphically selected control points. Ilmenite, chromite, garnet and clinopyroxene were extracted from the samples and commercially analysed by electron microprobe in a statistically representative manner. Constraints on the thermal evolution of the lithosphere underlying the Quiricó Basin were obtained by applying the single-grain thermobarometric technique of Nimis and Taylor (2000) to the compositions of the recovered clinopyroxene grains. The quality of each clinopyroxene analysis was checked to comply within 2% relative of a 6-oxygen and 4-cation formula, and nominal ferric iron contents were calculated using the algorithm of Droop (1987). The compositional screening techniques of Ramsay (1992), Ramsay and Tompkins (1994), Nimis (1998) and Nimis and Taylor (2000) were employed and partially modified in order to classify each clinopyroxene grain into one of four paragenetic categories, as follows:

- Crustal clinopyroxene (CRUS): These grains have diopsidic, salitic or augitic compositions similar to phenocrystic or groundmass clinopyroxene in basaltic rocks. They are most common in samples of the Mata da Corda Group where they are correlated petrographically with the occurrence of basaltic microxenoliths.
- Phenocrystic clinopyroxene (PHEN): Occasional grains of Al-poor diopside with notable Fe^{3+} contents have compositions similar to groundmass or phenocrystic clinopyroxenes analysed in alkaline magma types like kimberlites, lamproites and kamafugites (Sgarbi et al., 2000a). These grains are interpreted to represent a phenocrystic or coarse groundmass component of the samples.
- Spinel-facies clinopyroxene (CPXSP): Mg-rich chromian diopsides with variable, but elevated Al contents are common in the samples and represent mantle-derived clinopyroxene xenocrysts. Their derivation from disaggregated spinel-facies lherzolite is signaled by high Tschermacks contents (Al+Cr−Na−K cations>0.05 on a 6-oxygen formula).
- Garnet-facies clinopyroxene (CPXGT): Mg-rich chromian diopsides with moderate Al contents, low Tschermacks contents (Al+Cr−Na−K cations<0.05 on a 6-oxygen formula) and near-zero Fe^{3+} contents are very common in many of the samples. They are interpreted as mantle xenocrysts derived from disaggregated garnet-facies lherzolite xenoliths.

Table 2 summarises the abundance of the above clinopyroxene types in the samples collected, broken down by their stratigraphic setting. Also shown are the numbers of garnet-facies clinopyroxene grains for which reliable pressure estimates cannot be made (GTNOP) because they fail one of four separate compositional criteria that delimit the calibrated range of the Cr-in-Cpx barometer (see Nimis and Taylor, 2000). Reliable mantle temperatures were, however, calculated for these grains by iterative projection on an independently established mantle geotherm (see below). Mantle temperatures were also calculated for all the spinel-facies clinopyroxene grains by substi-

Table 2
Abundance of clinopyroxene grains summarised by compositional
category (see text) and stratigraphic setting of samples

Setting	Kimberlites	Maxixe	Mata da Corda	Total
Samples	n=13	n=38	n=13	n=64
Localities	n=7	n=22	n=10	n=39
CRUS	6	28	159	193
PHEN	62	363	139	564
CPXSP	321	376	503	1200
CPXGT	584	1001	5	1590
GTNOP	59	49	0	108
Total	973	1724	853	3547

tuting an assumed pressure of 26 kbar into Eq. (17) of Nimis and Taylor (2000). The temperature estimates hence obtained are considered realistic because the thermometer used is quite insensitive to pressure; we found an average pressure dependence of only 1.9 °C/kbar for our spinel-facies clinopyroxene compositions. The temperatures calculated in this investigation would represent minima in the unlikely event that a clinopyroxene grain occurred in an orthopyroxene-free assemblage (i.e. a wehrlitic assemblage). Nimis and Taylor (2000) and Nimis (2002) have assessed the precision of the single-grain clinopyroxene thermobarometer at ±3.0 kbar and ±50 °C for the type of application presented here.

5.2. Thermobarometry results

Pressures and temperatures (henceforth PNT00 and TNT00, respectively) calculated for mantle-derived clinopyroxene xenocrysts derived from samples of kimberlite are illustrated in Fig. 3. Spinel-facies clinopyroxenes are typically subordinate to garnet-facies clinopyroxenes in the kimberlites, but account for ~60% of the concentrate clinopyroxenes in two of the seven kimberlite localities represented. They occur predominantly in the temperature interval 625–875 °C, overlapping the lower end of the 700–1000 °C temperature range in which garnet-facies clinopyroxenes predominate (Fig. 3).

Calculated pressures for garnet-facies clinopyroxenes typically range across an ~12-kbar interval at any given temperature and impart a broad P–T scatter to the geothermal array evident in Fig. 3. The scatter is characteristic of PNT00–TNT00 results for clinopyroxene xenocryst populations from each of the kimberlite samples investigated in this work and is similar

to that obtained for clinopyroxenes included in diamond and for concentrate clinopyroxenes from Kimberley, South Africa (Figs. 6 and 8 in Nimis and Taylor, 2000). We hence regard the P–T scatter to be intrinsic to the single-clinopyroxene thermobarometric technique and therefore justify fitting a single empirical curve to all the garnet-facies P–T data that fall within a 600–1000 °C temperature range. The curve is expressed in simplified form by T=815.2× ln[P/14.11] with fit parameters r^2=0.63, σT±72 °C and σP±3.2 kbar. Much of the fitted T and P variance can be ascribed directly to the temperature dependence of the PNT00 barometer, which, for our garnet-facies clinopyroxene compositions, averages 3.5 kbar/100 °C. A reconciliation of this internally correlated P–T effect shows that thermal aspects of the fitted curve are actually resolved to a precision of ±20°C, substantially better than baric aspects. We interpret the curve to represent the lithospheric geotherm at the time of kimberlite eruption. The geotherm closely approximates a 37-mW/m^2 heat flow model for steady-state conductive lithosphere (Pollack and Chapman, 1977), and we note its similarity to the cratonic geotherm defined by clinopyroxene compositions in xenoliths from kimberlites in the Slave craton (Grütter and Moore, 2003). The geotherm intersects the diamond stability field at T>880 °C and is interpreted to remain within lithospheric peridotite up to temperatures of about 1000 °C (Fig. 3). These thermobarometric results are consistent with microdiamond recoveries in some of the Quiricó Basin kimberlites and impart a regional dimension to a similarly cold, cratonic geotherm recorded by orthopyroxenes with 0.3–0.5 wt.% Al$_2$O$_3$ in coarse-textured garnet lherzolite microxenoliths from the diamondiferous Três Ranchos-4 kimberlite (Carvalho and Leonardos, 1997).

Clinopyroxene xenocrysts recovered from samples of the Maxixe Member are derived from spinel-facies and garnet-facies peridotite (Fig. 4). The spinel-facies grains constitute less than one-fifth of the clinopyroxene concentrate in most of the samples, but attain an abundance of 60% or more in 2 of the 22 localities represented. They are derived predominantly from the temperature range 625–850°C where a small proportion of garnet-facies clinopyroxene grains also occur (Fig. 4). Garnet-facies clinopyroxenes make up 72% of all the xenocrystic clinopyroxenes recovered from the

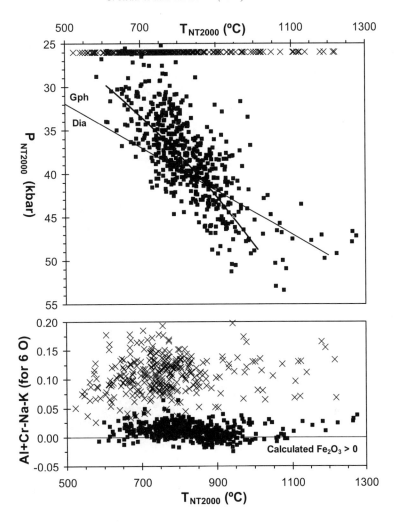

Fig. 3. *P–T* conditions and compositions of clinopyroxenes in samples from kimberlites. Garnet-facies clinopyroxene *P–T* data (solid squares) are fitted empirically with a conductive geotherm (solid curve) that intersects the graphite/diamond stability field of Kennedy and Kennedy (1976) at 880 °C. Temperatures for spinel-facies clinopyroxenes (crosses) were calculated at an assumed pressure of 26 kbar. See text for discussion.

Maxixe Member (Table 1), consistent with the presence of common lherzolitic Cr-pyrope garnet xenocrysts in the same samples. These clinopyroxenes are derived predominantly from relatively high mantle temperatures, in the range 950–1175 °C, and occur within a geothermal array displaced to temperatures some 150–250 °C higher at any given pressure than that observed for clinopyroxenes derived from samples of kimberlite (Fig. 4). Curves fitted to the geothermal array over the temperature interval 850–1250 °C do not yield robust regressions, most likely indicating that

a variety of geothermal states are recorded by samples from the Maxixe Member. Selective modeling of the high-temperature *P–T* data suggest, however, that the geothermal conditions probably intersect the diamond stability field at temperatures exceeding 1240–1280 °C. As this falls somewhat above the temperatures recorded by most of the recovered clinopyroxene grains (Fig. 4), it is inferred that the magmatic sources for the Maxixe Member volcanic breccias probably do not contain primary diamonds. No microdiamonds have as yet been recovered from any of our samples

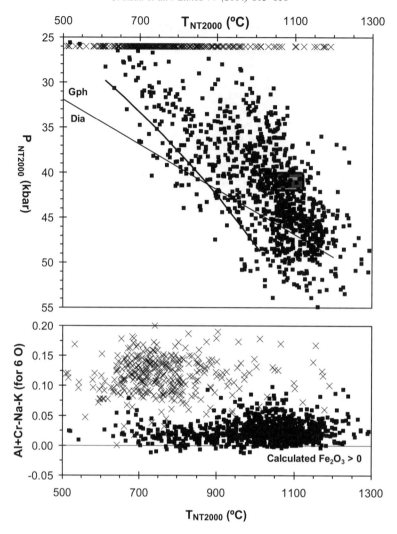

Fig. 4. $P–T$ conditions and compositions of clinopyroxenes in samples from the Maxixe Member. Symbols and empirical geotherm as in Fig. 3. The shaded box indicates calculated $P–T$ conditions for orthopyroxene grains from the Boa Esperança pipe (Ramsay and Tompkins, 1994). See text for discussion.

of the Maxixe Member, despite the presence of G10-type diamond-associated garnets in the samples, and we know of no other magmatic samples of this Member in which diamonds have been recognized. Ramsay and Tompkins (1994) reported the presence of peridotitic orthopyroxene grains with ~1.0 wt.% Al_2O_3 in concentrate from the Boa Esperança pipe and calculated that they equilibrated in garnet lherzolite at 40–42 kbar and 1050–1120 °C (their Table 5). These $P–T$ conditions fall within the geothermal regime experienced by clinopyroxene grains in the

Maxixe Member (Fig. 4) and are also consistent with the reported absence of diamond from the Boa Esperança pipe.

Mantle-derived clinopyroxene xenocrysts in samples of the Mata da Corda Group kamafugites are strongly dominated by spinel-facies grains derived from temperatures in the range 650–850 °C (Fig. 5). The samples are effectively devoid of garnet-facies clinopyroxene grains (Table 1), with the few such grains present recording $P–T$ conditions similar to that calculated for clinopyroxene in the Maxixe Mem-

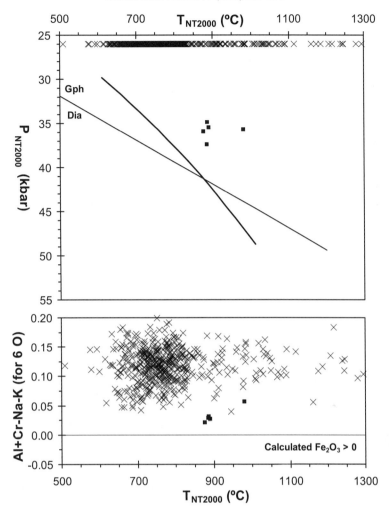

Fig. 5. *P–T* conditions and compositions of clinopyroxenes in samples from the Mata da Corda Group. Symbols and empirical geotherm as in Fig. 3. See text for discussion.

ber. Since nearly one in every five clinopyroxenes derived from the Mata da Corda Group samples can be attributed to contamination by upper crustal basaltic sources (Table 1), and our Mata da Corda samples only very rarely contain lherzolitic Cr-pyrope garnets, it is suspected that some, if not all, of the garnet-facies clinopyroxenes may be derived from fragments of the Maxixe Member that were incorporated as xenoliths in kamafugites of the Mata da Corda Group. The *P–T* conditions indicated by garnet-facies clinopyroxenes from the Mata da Corda Group must therefore be considered as circumspect until such time as they can be confirmed by independent work on bona fide mantle xenoliths that show primary relationships with their kamafugite hosts. Our current data suggest that the Mata da Corda Group kamafugites may have sampled only spinel lherzolite and not garnet lherzolite from the lithosphere underlying the Quiricó Basin or that garnet lherzolite may have been largely absent from the lithospheric section at that time.

6. Lithospheric thermal evolution

The thermal evolution of mantle lithosphere underlying the Quiricó Basin shows a progressive history that can be constrained temporally using the stratigraphic setting and ages of specific alkaline magma types. Kimberlites erupted in the time period 95–89 Ma and entrained spinel lherzolite and garnet lherzolite along a cold geotherm typical of conductive cratonic settings. The lithospheric mantle root extended into the diamond stability field at 880–1000 °C and contains evidence for limited thermal disturbance at temperatures in excess of 1300 °C (Fig. 6).

Magmatic deposits of the ~85-Ma-old Maxixe Member contain a record of substantive thermal disturbance(s), best visualized as a +240 °C shift in the mode of temperatures experienced by deep-seated

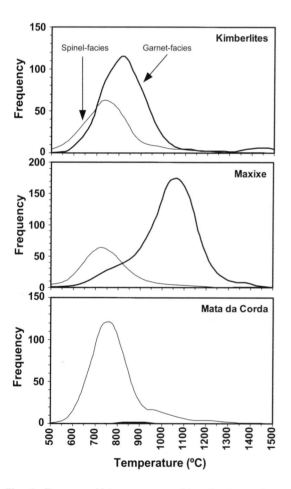

Fig. 6. Frequency histograms summarising the temperatures experienced by spinel-facies or garnet-facies mantle clinopyroxenes, in three stratigraphically constrained time periods during the evolution of the Quiricó Basin (top to bottom). See text for discussion.

garnet lherzolite (Figs. 4 and 6). The temperature increase placed much of the lithospheric root within the graphite stability field, even though the available thermobarometric data possibly indicate a thickening of the lithosphere at this time (compare Figs. 3 and 4). Temperatures recorded by spinel lherzolites are essentially unchanged, however (Fig. 5), and therefore do not support propagation of the deep-seated thermal anomaly to depths shallower than about 30 kbar.

Spinel lherzolite with an unchanged thermal profile is by far the dominant mantle lithology encountered in ~84–61 old kamafugites of the Mata da Corda Group (Fig. 6). The little evidence that does exist for the presence of lithospheric garnet lherzolite at this time could be explained as resulting from contamination. Thus the apparent disappearance of garnet lherzolite from the mantle section at this time leads us to speculate that a major portion of the deep lithosphere beneath the Quiricó Basin may have detached and foundered into the asthenosphere. Stratigraphic and thermobarometric relations pin this postulated event to a very short time period just after eruption of the Maxixe Member volcanics at ~85 Ma, and hence effectively coincident with eruption of the first major kamafugites of the Mata da Corda Group. The postulated lithospheric detachment would expose shallow, previously protected, metasomatised mantle to suprasolidus temperatures and hence provide the impetus for fundamentally switching the style of alkaline magmatism in the Quiricó Basin from low-volume deep-seated kimberlitic or lamproitic melts to extrusion of an enormous volume of kamafugitic melts derived from shallower portions of the lithosphere. The evolutionary model demands rapid and significant shifts to occur in the petrologic and isotopic character of these alkaline magmas, in agreement with a recent investigation of an extensive suite of fresh kimberlite and kamafugite samples from the region (see Araujo et al., 2001).

The lithospheric thermal evolution depicted in Fig. 6 could also be considered as compatible with heating related to lithospheric extension and incipient rifting during opening of the South Atlantic ocean or may have resulted from the impact of a mantle plume (e.g. Gibson et al., 1995). However, neither of those evolutionary scenarios implicitly provide for the apparent disappearance of an ~75-km-thick section of garnet-facies cratonic lithosphere directly after deposition of the Maxixe Member, and coincident with major, voluminous kamafugite magmatism of the Mata da Corda Group. Further detailed work on stratigraphic relations within the kamafugites of the Mata da Corda Group, and their contained mantle xenolith suite, may provide the physical evidence required to validate the lithospheric detachment model postulated here.

7. Conclusions

Cretaceous alkaline volcanics of divergent compositions, erupting through the Areado Group sediments of the Quiricó Basin from 95 to 61 Ma, transport to surface xenocryst mineral suites representative of the underlying lithospheric mantle. The recovery of high-value alluvial diamonds from the drainage system in and around the Quiricó Basin has encouraged more than 30 years of diamond exploration. This exploration, focused on location of primary diamond bearing host rocks, has generated a large database of compositions of the xenocryst mineral suites of the alkaline volcanics. Simultaneously, rigorous petrographic studies of the alkaline volcanic eruptives, located using airborne geophysics and heavy mineral sampling, have shown that the area contains rare kimberlites and ubiquitous kamafugites. Stratigraphic relations and geochronological studies of the compositional spectrum of volcanics show that the alkaline volcanism commenced with kimberlites at ~95 Ma and culminated with kamafugites during ~84–61 Ma. During this time, the xenocryst mineral suite sampled by the volcanics evolved from garnet to spinel facies lithosphere. Garnet-facies and spinel-facies peridotite xenocrysts of harzburgitic and lherzolitic compositions are well represented in the kimberlites and the initial kamafugitic eruptives. However, the later, voluminous, kamafugites contain only spinel-facies peridotitic lithospheric material. The juxtaposition of high-value alluvial diamonds, kimberlites and kamafugites in an apparently off-craton setting posed a seemingly irresolvable problem for diamond exploration. Recent developments in clinopyroxene classification and thermobarometry have enabled the authors to show that these kimberlites erupted in the Early Cretaceous through a cool cratonic lithosphere, suitable for the preservation of diamond, and that a major thermal event occurred in the deep lithospheric mantle beneath the Quiricó Basin during eruption of the Maxixe Member volcanic breccias. The available data suggest that a 75-km-thick section of garnet-facies lithosphere vanished at the onset of voluminous Mata da Corda kamafugite magmatism, possibly by foundering into the asthenosphere. Further work is required to establish the relationship of this dramatic lithospheric incident to the impact of a mantle plume and to lithospheric extension related to the breakup of Gondwana and the opening of the South Atlantic.

Acknowledgements

The authors acknowledge Canabrava Diamond for permission to publish the data and financial support in generating these ideas. George Read, Herman Grütter and Frank Gaunt thank De Beers for introducing them to the diamond producing regions around Coromandel. Charles Skinner and Luiz Bizzi are acknowledged for many stimulating discussions on the origins of the diamonds in the Quiricó Basin.

References

Arai, M., Dino, R., Milhomen, P. da S., Sgarbi, G.N.C., 1995. Micropaleontologia da Formação Areado, cretaceo da bacia sanfranciscan: estudo de ostracodes e palinologia. XIV Congr. Bras. Paleont., Uberaba, 2–3.

Araujo, A.L.N., Carlson, R.W., Gaspar, J.C., Bizzi, L.A., 2001. Petrology of kamafugites and kimberlites from the Alto Paranaíba Alkaline Province, Minas Gerais, Brazil. Contrib. Mineral. Petrol. 142, 163–177.

Bizzi, L.A., 1995. Mesozoic Alkaline Volcanism and Mantle Evolution of the Southwest São Francisco Craton, Brazil. PhD Thesis. University of Cape Town, Cape Town, South Africa.

Bizzi, L.A., Smith, C.B., Meyer, H.O.A., Armstrong, R., 1994. Mesozoic kimberlites and related rocks in southwestern São Francisco craton, Brazil: a case for local mantle reservoirs and their interaction. In: Meyer, H.O.A., Leonardos, O.H. (Eds.), Proc. 5th Int. Kimb. Conf., Araxá. Brasilia, D.F.CPRM, Spec. Publ. 2/91, CPRM, Brasilia, DF Brazil, pp. 156–177.

Byron, M.J., 1999. Physical volcanology and lithogeochemistry of the Mata da Corda Formation, Minas Gerais, Brazil. PhD Thesis. Carleton University, Ottawa, Canada.

Campos, J.E.G., Dardenne, M.A., 1997a. Estratigrafia e sedimentação da bacia Sanfraciscana: uma revisão. Rev. Bras. Geociênc. 27 (3), 269–282.

Campos, J.E.G., Dardenne, M.A., 1997b. Origem e evolução tectônica da bacia Sanfraciscana: uma revisão. Rev. Bras. Geociênc. 27 (3), 283–294.

Carlson, R.W., Esperança, S., Svisero, D.P., 1996. Chemical and Os isotopic study of cretaceous potassic rocks from Southern Brazil. Contrib. Mineral. Petrol. 125, 393–405.

Carvalho, J.B., Leonardos, O.H., 1997. Garnet peridotites from the Três Ranchos 4 kimberlitic pipe, Alto Paranaiba Igneous Province, Brazil: geothermobarometric constraints. Proc. 6th Int. Kimb. Conf. Russ. Geol. Geophys., vol. 38, pp. 168–181.

Dardenne, M.A., Schobbenhaus, C., 2001. Metalogênese do Brasil. Universidade de Brasília, Brasília.

Droop, G.T.R., 1987. A general equation for estimating Fe^{3+} concentrations in ferromagnesian silicates and oxides from microprobe analyses using stoichiometric criteria. Min. Mag. 51, 431–435.

Duncan, R.A., Hooper, P.R., Rhacek, J., Marsh, J.S., Duncan, A.R., 1997. The timing and duration of the Karoo igneous event, southern Gondwana. J. Geophys. Res. 102, 18127–18138.

Gibson, S.A., Thompson, R.N., Leonardos, O.H., Dickin, A.P., Mithchell, J.G., 1995. The late cretaceous impact of the Trinidade Mantle Plume: evidence from the large-volume, mafic, potassic magmatism in SE Brazil. J. Petrol. 36, 189–229.

Gonzaga, G.M., Teixeira, N.A., Gaspar, J.C., 1994. The origins of diamonds in western Minas Gerais, Brazil. Miner. Depos. 29, 414–421.

Grütter, H.S., 2002. The petrography and classification of drill core samples from anomaly X270. Internal company report prepared by Mineral Services Canada.

Grütter, H., Moore, R., 2003. Pyroxene geotherms revisited—an empirical approach based on Canadian xenoliths. Ext. Abs. 8th Int. Kimb. Conf., p. 272. CD-ROM.

Heaman, L.M., 2002. Report on U–Pb Perovskite Dating of Sample C02-0690 (X270) for Canabrava Internal company report prepared by Geospec Consultants.

Janse, A.J.A., 1994. Is Clifford's Rule still valid? Affirmative examples from around the World. In: Meyer, H.O.A., Leonardos, O.H. (Eds.), Proc. 5th Int. Kimb. Conf. 2, Kimberlites, Related Rocks and Mantle Xenoliths. CPRM Spec. Publ., CPRM, Brasilia, DF Brazil, pp. 215–235.

Kaminsky, F.V., Zakharchenko, O.D., Khachatryan, G.K., Shiryaev, A.A., 2001. Diamonds from the Coromandel area, Minas Gerais, Brazil. Rev. Bras. Geociênc. 31, 583–596.

Karfunkel, J., Chaves, M.L.S.C., Svisero, D.P., Meyer, H.O.A., 1994. Diamonds from Minas Gerais, Brazil: an update on sources, origin and production. Int. Geol. Rev. 36, 1019–1032.

Kennedy, C.S., Kennedy, G.C., 1976. The equilibrium boundary between graphite and diamond. J. Geophys. Res. 81, 2467–2470.

Leonardos, O.H., Meyer, H.O.A., 1991. Outline of the geology of Western Minas Gerais. Field Guide Book 5th International Kimberlite Conference, Araxá, Brazil. CPRM, Brasila, DF Brazil, pp. 17–24.

Machado, N., Noce, C.M., Ladeira, E.A., Belo de Oliveira, O., 1992. U–Pb geochronology of Archean magmatism and Proterozoic metamorphism in the Quadrilatero Ferrifero, southern São Francisco craton, Brazil. Geol. Soc. Amer. Bull. 104, 1221–1227.

Mitchell, R.H., Bergman, S.C., 1991. Petrology of Lamproites. Plenum, New York.

Nimis, P., 1998. Evaluation of diamond potential from the composition of peridotitic chromian diopside. Eur. J. Mineral. 10, 505–519.

Nimis, P., 2002. The pressures and temperatures of formation of diamond based on thermobarometry of chromian diopside inclusions. Can. Mineral. 40, 871–884.

Nimis, P., Taylor, W.R., 2000. Single clinopyroxene thermobarometry for garnet peridotites: Part I. Calibration and testing of a Cr-in-Cpx barometer and an enstatite-in-Cpx thermometer. Contrib. Mineral. Petrol. 139, 541–554.

Pollack, H.N., Chapman, D.S., 1977. On the regional variation of heat flow, geotherms and lithosphere thickness. Tectonophysics 38, 279–296.

Ramsay, R.R., 1992. Geochemistry of diamond Indicator Minerals. PhD Thesis. University of Western Australia, Perth, Australia.

Ramsay, R.R., Tompkins, L.A., 1994. The geology, heavy mineral concentrate mineralogy, and diamond prospectivity of the Boa Esperança and Cana Verde pipes, Corrego D'anta, Minas Gerais, Brazil. In: Meyer, H.O.A., Leonardos, O.H. (Eds.), Proc. 5th Int. Kimb. Conf. 2, Kimberlites, Related Rocks and Mantle Xenoliths, CPRM, Spec. Publ., CPRM, Brasilia, DF Brazil, pp. 329–345.

Sahama, T.G., 1974. Potassium-rich alkaline rocks. In: Sorenson, H. (Ed.), The Alkaline Rocks. Wiley, New York, pp. 96–109.

Seer, H.J., Moraes, L.C., Fogaca, A.C.C., 1989. Roteiro geológico para região de Lagoa Formosa-Chumbo-Carmo do Paranaíba, MG. Sociedade Brasileira de Geológia. Núcl. Minas Gerais, Bol. 9, 58.

Sgarbi, G.N.C., 2000. The Cretaceous Sanfranciscan Basin, Eastern Plateau of Brazil. Rev. Bras. Geociênc. 30 (3), 450–452.

Sgarbi, P.B.A., Gaspar, J.C., Valença, J.G., 2000a. Clinopyroxene from Brazilian kamafugites. Lithos 53, 101–116.

Sgarbi, P.B.A., Heaman, L.M., Gaspar, J.C., 2000b. U–Pb perovskite ages for Brazilian kamafugites. Abstract Submitted to the 31st Int. Geol. Congress, Rio de Janeiro, Brazil, August, 6–17.

Tompkins, L.A., Gonzaga, G.M., 1989. Diamonds in Brazil and a proposed model for the origin and distribution of diamonds in the Coromandel region, Minas Gerais, Brazil. Econ. Geol. 84, 591–602.

Turner, S., Regelous, M., Kelley, S., Hawkesworth, C., Mantovani, M., 1994. Magmatism and continental break-up in the South Atlantic: high precision $^{40}Ar–^{39}Ar$ geochronology. Earth Planet. Sci. Lett. 121, 333–348.

Winter, L.D.S., 2000. Stratigraphy and Maxixe units. Nova Era Mineração. Internal Company Report.

Available online at www.sciencedirect.com

Lithos 77 (2004) 819–840

www.elsevier.com/locate/lithos

Compositional classification of "kimberlitic" and "non-kimberlitic" ilmenite

Bruce A. Wyatt[a,*], Mike Baumgartner[b], Eva Anckar[c], Herman Grutter[d]

[a] De Beers Canada Exploration Inc., 1 William Morgan Drive, Toronto, Ontario, Canada M4H 1N6
[b] Mineral Services South Africa, Cape Town 7430, South Africa
[c] University of Cape Town Kimberlite Research Group, Cape Town 7700 South Africa
[d] Mineral Services Canada, North Vancouver, British Columbia, Canada V7P 3S7

Received 27 June 2003; accepted 14 December 2003

Abstract

Ilmenite is one of the common kimberlitic indicator minerals recovered during diamond exploration, and its distinction from non-kimberlitic rock types is important. This is particularly true for regions where these minerals are present in relatively low abundance, and they are the dominant kimberlitic indicator mineral recovered. Difficulty in visually differentiating kimberlitic from non-kimberlitic ilmenite in exploration concentrates is also an issue, and distinguishing kimberlitic ilmenite from those derive from other similar rocks, such as ultramafic lamprophyres, is practically impossible. Ilmenite is also the indicator mineral whose compositional variety has the most potential to resolve provenance issues related to mineral dispersions with contributions from multiple kimberlite sources.

Various published data sets from selected kimberlitic (including kimberlites, lamproites, and various ultramafic lamprophyres) and non-kimberlitic rock types have been compiled and evaluated in terms of their major element compositions. Compositional fields and bounding reference lines for ilmenites derived from kimberlites (sensu stricto), ultramafic lamprophyres, and other non-kimberlitic rock types have been defined primarily on $MgO–TiO_2$ graphs as well as $MgO–Cr_2O_3$ relationships.
© 2004 Elsevier B.V. All rights reserved.

Keywords: Picroilmenite; Geikielite; Hematite; Kimberlite; Exploration; Classification

1. Introduction

Ilmenite, together with pyrope garnet and chromite, is one of the dominant indicator minerals found in kimberlite. The ilmenite present in kimberlites derives from a number of sources (see, e.g., Mitchell, 1973, 1977, 1986; Haggerty, 1975, 1976, 1991), but most commonly as discrete ilmenite xenoliths belonging to the megacryst suite of minerals (Schulze, 1987; Schulze et al., 1995), ilmenites of metasomatic origin (Wyatt and Lawless, 1984; Harte, 1987; Haggerty, 1989; Dawson et al., 2001; Moore and Lock, 2001), ilmenites intergrown with megacrysts, MARID (Dawson and Smith, 1977), and Granny Smith silicates (Boyd et al., 1984), and groundmass ilmenite and phenocrysts in the host magma (Tompkins and Haggerty, 1985; Mitchell, 1986; Moore, 1987). Less com-

* Corresponding author. Tel.: +1-416-423-5811x225; fax: +1-416-423-9944.
  E-mail address: bruce.wyatt@ca.debeersgroup.com (B.A. Wyatt).

0024-4937/$ - see front matter © 2004 Elsevier B.V. All rights reserved.
doi:10.1016/j.lithos.2004.04.025

mon sources of ilmenite in kimberlite include primary ilmenites from disaggregated peridotite (both cold-coarse and hot-deformed varieties) and eclogite xenoliths, and rare ilmenites included in diamond (Meyer and Svisero, 1975; Sobolev and Yefimova, 2000). Ilmenites are also present in a wide variety of non-kimberlitic igneous (gabbros, norites, granites, and anorthosites) and metamorphic (orthogneisses) rocks that may occur in areas hosting kimberlite intrusions. In Southern Africa, ilmenites are present in the volcanic successions of Karoo lavas (Bristow, 1980), and in other gabbroic and picritic intrusions such as those found in the Mount Ayliff Intrusion of the Insizwa Complex in Transkei (Cawthorn et al., 1988).

The distinction of ilmenites derived from kimberlitic versus non-kimberlitic rocks is important in the context of diamond exploration in regions in which ilmenites are present in relatively low abundance but where they are the dominant kimberlitic indicator mineral recovered. Ilmenite is also the indicator mineral whose compositional variety could be used to greatest effect in provenance studies related to mineral dispersions with contributions from one or more kimberlite sources. This study focuses on a simple and practical scheme for separating kimberlitic from non-kimberlitic ilmenite on the basis of major element compositions, but does not address the detailed analysis of crystal–chemical issues and phase relationships that are the subject of comprehensive studies such as Haggerty (1976, 1991), Haggerty and Tompkins (1984), Tompkins and Haggerty (1985), and references referred to therein.

Because ilmenite is a key kimberlitic indicator mineral, its correct identification is critical, yet the visual distinction of kimberlitic from non-kimberlitic ilmenite during the extraction of these grains from exploration sample concentrates is imperfect. In addition, ilmenite populations derived from kimberlites or lamprophyres share similar mantle-derived petrogenetic origins but have different significance in the context of diamond exploration. Visually differentiating kimberlite derived ilmenite from grains derived from similar ultramafic rocks, such as ultramafic lamprophyres, is practically impossible, even for highly trained mineral sorters. The latter problem is a direct result of a compositional overlap in the range 4 to 6 wt.% MgO, as well as the visual similarity of ilmenites having 4 to 18 wt.% MgO

which represents the compositional range of ilmenite from kimberlites and related rocks. This study highlights the need to determine the compositions and paragenesis of ilmenites extracted from exploration sample concentrates by analytical means, particularly during early phase reconnaissance.

2. Methods and data sources

The compositions of ilmenite derived from potentially diamondiferous sources (kimberlites and lamproites) and other non-kimberlitic sources (e.g., ultramafic lamprophyres, basalt, and gabbro) have been compiled from selected published and internal Mineral Services and De Beers data sets. Compositional fields for ilmenites derived from kimberlites (sensu stricto), and other non-kimberlitic rock types have been defined on selected bivariate graphs and form the basis of a robust and simple classification scheme. Ilmenite $MgO–TiO_2$ diagrams (Sobolev, 1977) are particularly useful to discriminate kimberlitic from non-kimberlitic ilmenite compositions. Equations for kimberlitic and non-kimberlitic reference lines that are given in the text below are for convenience and can be used to filter ilmenite analyses for classification purposes using a simple spreadsheet.

Sobolev (1977, Fig. 42) utilised a $MgO–TiO_2$ plot overlayed with Fe_2O_3 contours to display various kimberlitic ilmenite populations, which was also used to make some inferences regarding the oxidising environment. In this study, we have inserted reference lines of constant hematite content in scatter plots of TiO_2 versus MgO. The Fe_2O_3 contents of the ilmenites were calculated stoichiometrically using Finger (1972), and the isopleths were established empirically using contours of the data on the TiO_2 versus MgO plots. These stoichiometric Fe_2O_3 reference contours are based on the kimberlite data set discussed below, and are only an indication of the hematite content. They can be used to assess data quality. For example, data points should not plot above the 0% Fe_2O_3 contour (implying negative Fe_2O_3 content) and such analyses should be scrutinised for quality. In many cases inconsistencies, especially for non-kimberlitic ilmenite, may be due to the microprobe beam inadvertently impinging on submicroscopic rutile inclusions resulting in abnormally high TiO_2 content (and often high totals). In some cases, such analyses may in fact derive from other high-TiO_2 minerals such as the pseudobrookite series or Nb–Ta-rich rutiles. While the contours illustrate that the Fe_2O_3 content of ilmenites increases with decreasing TiO_2, extreme caution should be used in attempting to infer the fO_2 environment associated with the ilmenites. The reader is referred to Haggerty (1976, 1991), Haggerty and Tompkins (1984), Tompkins and Haggerty (1985) and references therein for a full discussion of ilmenite, Fe_2O_3 and fO_2 systematics.

Bivariate graphs of MgO versus Cr_2O_3 (after Haggerty, 1975, 1976, 1991) are also presented for selected data sets. Such graphs, which are commonly used by the diamond exploration fraternity, illustrate certain compositional criteria that also aid in the distinction of kimberlitic from non-kimberlitic ilmenites.

Table 1
Paragenesis and data sources used in this study

Paragenesis[a]	Region	Localities (no. grains)[b]	Data source[c]
Kimberlites Off-craton	Southern Africa	Abiquaputs[o](5), Amalia[o](1), Andries[o](46), Berseba[o](11), Brandvlei[o](20), Deutsche Erde II[o](3), East Griqualand[o](6), Gibeon[o](51), Hebron[o](37), Lichtenfels[o](6), Nouzees[o](78), Pofadder[o](1), Uintjiesberg[o](33), Witputs[o](94)	UCT–KRG
Kimberlites On-craton	Southern Africa	Balmoral[l](50), Borrelskop[l](37), Bultfontein*(46), De Beers*(2), Dutoitspan*(2), Frank Smith*(62), Franspoort[l](835), Goedehoop[l](47), Good Hope[l](49), Jagersfontein*(1), Kimberley*(53), Koffiefontein*(6), Last Hope[l](50), Monastery*(159), Montrose[l](6), Palmietfontein[l](21), Premier*(329), Riverton[l](4), Schuller[l](222), Smithdale[l](7), Victoria[l](50), Washington[l](52), Wesselton*(3)	UCT–KRG
Kimberlites	West Africa	Sierra Leone-Koidu*(52), Liberia*(33)	UMASS
Kimberlites	North America	Attawapiskat*(33), Dry Bones Bay[l](148), Iron Mountain[l](54), Kelsey Lake[l](2), Kirkland Lake[l](558), Lake Ellen[l](4), Lake Temiskaming[l](135), Mt. Horeb Church[l](2), Stockdale[l](42), Williams[l](93)	Schulze et al. (1995)
Kimberlites	Australia	Cleve[b](1343), Skerring[b](261)	DBGSC
Kimberlites	Siberia	Udachnaya*(975), Mir*(1654)	DBGSC
Kimberlites	Southern Africa, N. America, Russia	Various—see text	UMASS
KRR–Melnoite	Southern Africa	Entilombo (1417)	DBGSC
KRR–Melnoite	Malaita	Malaita (901)	DBGSC
KRR–Melnoite	Australia	Lake Bullenmerrie (92)	DBGC
KRR–Melnoite	Canada	Selco Alnoites (95)	Sage (2000)
KRR	Namibia	Okenyenya (45)–Ultramafic Lamprophyre	Baumgartner (1994)
NK–Gabbro	Namibia	Okenyenya (55)	Le Roux (UCT) (pers. comm.)
NK–Dolerite	South Africa	Insizwa Gabbro (58), Insizwa Picrite (163)	Cawthorne et al. (1988)
NK–Karoo Basalt	Botswana	Bobonong (1564)	DBGSC
Prospecting	Canada	Slave Province (4743)	DIAND NWT
Prospecting	Southern Africa	Soil sampling programme: Southern Africa (142)	Mineral Services

[a] KRR: Kimberlite-related rock. NK: Non-kimberlitic.

[b] Locality or area from which the ilmenites derive. Numbers in ( ) refer to no. of grains. NOTE: Diamond content of primary kimberlites: o—barren; l—at best low (subeconomic); *—near economic or economic. All other rock types are barren.

[c] UCT–KRG: University of Cape Town—Kimberlite Research Group. UMASS: University of Massachusetts 1987 Ilmenite Database given to UCT by Steve Haggerty in the late 1980s. DBGSC: De Beers GeoScience Centre. DIAND NWT: Department of Indian and Northern Affairs, North West Territories, Canada—Geology Division, Kimberlite Indicator Mineral Chemistry Database, Slave Province. Schulze et al. (1995)—data available from web site: http://www.geology.utoronto.ca/faculty/schulze/ilmenite.html.

A wide variety of data sets covering Southern Africa, West Africa, North America, Russia, and Australasia were used in this study (Table 1), and relevant averages are summarised in Table 2 along with the range in major element compositions of each data set. Only data flagged as off-craton and on-craton

Table 2
Mineral composition statistics for some of the key data sets utilised in this study

Statistic	Off-craton group I kimberlites (RSA and Nambia) $n=392$					On-craton group I kimberlites (RSA only) $n=2338$				
	Min	Median	Max	Mean	S.D.	Min	Median	Max	Mean	S.D.
SiO$_2$	0.00	0.01	0.21	0.01	0.01	0.00	0.00	1.87	0.01	0.06
TiO$_2$	41.67	52.42	56.00	51.45	3.04	36.44	52.80	57.92	52.33	3.09
Al$_2$O$_3$	0.00	0.21	1.23	0.26	0.21	0.00	0.32	4.70	0.36	0.27
Cr$_2$O$_3$	0.00	1.17	6.20	1.25	0.87	0.00	0.82	4.68	0.98	0.55
FeOt	26.93	33.53	49.65	35.24	5.09	19.64	34.11	56.20	34.67	5.40
MnO	0.01	0.28	1.27	0.27	0.10	0.00	0.30	1.94	0.30	0.08
MgO	4.72	11.07	15.55	10.66	2.34	3.37	11.16	19.94	10.84	2.54
CaO	0.00	0.00	0.00	0.00	0.00	0.00	0.08	0.63	0.05	0.05

Statistic	North American kimberlites (USA and Canada) $n=1071$					Siberian kimberlites (De Beers) $n=2629$				
	Min	Median	Max	Mean	S.D.	Min	Median	Max	Mean	S.D.
SiO$_2$	0.00	0.02	1.18	0.02	0.05	0.00	0.00	1.58	0.02	0.06
TiO$_2$	31.52	49.88	55.35	48.75	3.81	29.69	47.81	57.08	46.67	4.19
Al$_2$O$_3$	0.00	0.24	2.34	0.27	0.22	0.00	0.58	1.30	0.58	0.12
Cr$_2$O$_3$	0.08	0.54	5.74	0.85	0.85	0.05	0.42	7.03	0.91	1.07
FeOt	24.07	37.16	57.25	38.12	5.56	28.07	39.98	57.55	41.29	4.64
MnO	0.12	0.30	0.68	0.31	0.07	0.02	0.20	4.67	0.22	0.14
MgO	3.25	9.73	17.13	9.45	2.49	1.12	8.99	14.65	8.69	1.75
CaO	0.00	0.02	1.00	0.04	0.07	0.00	0.02	0.85	0.03	0.04
NiO	0.00	0.06	0.59	0.07	0.04					

Statistic	Australian kimberlites (De Beers) $n=1913$				
	Min	Median	Max	Mean	S.D.
SiO$_2$	0.00	0.00	0.41	0.00	0.03
TiO$_2$	41.83	50.07	56.89	49.79	2.45
Al$_2$O$_3$	0.00	0.26	1.36	0.34	0.22
Cr$_2$O$_3$	0.00	0.20	6.67	0.30	0.40
FeOt	28.57	38.12	48.47	38.42	3.78
MnO	0.00	0.30	0.99	0.31	0.08
MgO	5.14	10.24	14.50	9.94	1.90
CaO	0.00	0.08	0.94	0.08	0.05
NiO					

Statistic	Malaita alnoite (De Beers) $n=901$					Australian melnoite (De Beers) $n=92$				
	Min	Median	Max	Mean	S.D.	Min	Median	Max	Mean	S.D.
SiO$_2$	0.00	0.00	0.18	0.00	0.01	0.00	0.00	0.07	0.01	0.02
TiO$_2$	45.52	50.17	55.25	50.04	1.68	34.05	44.46	52.29	45.03	3.22
Al$_2$O$_3$	0.10	0.61	2.27	0.62	0.23	0.00	0.16	0.70	0.18	0.15
Cr$_2$O$_3$	0.00	0.05	2.38	0.11	0.26	0.00	0.01	0.07	0.02	0.02
FeOt	28.42	41.09	48.42	41.41	3.17	37.84	48.49	58.59	48.04	4.23
MnO	0.14	0.25	0.66	0.26	0.07	0.00	0.26	0.39	0.27	0.06
MgO	3.12	7.05	13.06	7.00	1.86	1.58	3.25	8.34	3.63	1.62
CaO	0.00	0.06	0.29	0.08	0.05	0.00	0.01	0.06	0.02	0.01
NiO										

Table 2 (*continued*)

Statistic	Entilombo melnoite (De Beers) $n = 1417$					Selco alnoites (Sage, 2000) $n = 96$				
	Min	Median	Max	Mean	S.D.	Min	Median	Max	Mean	S.D.
SiO_2	0.00	0.00	0.44	0.00	0.02	0.00	0.01	0.10	0.02	0.01
TiO_2	44.48	47.81	57.56	47.77	1.50	41.51	51.31	53.96	50.19	2.56
Al_2O_3	0.04	0.47	0.83	0.47	0.07	0.00	0.33	1.20	0.34	0.17
Cr_2O_3	0.00	0.09	1.44	0.11	0.10	0.00	0.10	2.94	0.20	0.36
FeOt	28.11	43.61	49.49	43.59	2.27	27.48	38.39	49.87	39.77	3.82
MnO	0.25	0.35	0.64	0.35	0.03	0.08	0.27	1.91	0.30	0.19
MgO	0.40	6.55	14.69	6.62	1.03	3.14	8.30	14.67	7.65	1.79
CaO	0.01	0.04	0.13	0.04	0.01	0.00	0.02	0.10	0.02	0.02
NiO						0.00	0.05	0.24	0.06	0.04

Statistic	Non-kimberlitic ilmenite (Insiswa data) $n = 277$					Non-kimberlitic ilmenite (Bobonong Karoo data) $n = 1564$				
	Min	Median	Max	Mean	S.D.	Min	Median	Max	Mean	S.D.
SiO_2	0.00	0.00	0.00	0.00	0.00	0.00	0.00	0.83	0.01	0.04
TiO_2	49.52	53.36	57.43	53.48	2.17	43.79	49.95	54.96	49.77	1.29
Al_2O_3	0.00	0.00	0.00	0.00	0.00	0.00	0.29	5.32	0.30	0.19
Cr_2O_3	0.04	0.48	1.08	0.46	0.26	0.00	0.17	1.16	0.22	0.19
FeOt	32.39	38.65	48.11	40.11	4.79	35.11	45.49	51.08	45.21	1.69
MnO	0.40	0.51	1.35	0.58	0.23	0.09	0.40	2.91	0.42	0.18
MgO	0.28	6.20	10.22	5.06	3.07	0.13	3.31	10.11	3.37	1.10
CaO						0.00	0.02	0.30	0.02	0.02

was extracted from the University of Cape Town (UCT) Kimberlite Research Group (KRG) database (Table 1), and only from kimberlite localities allocated to that specific country and marked as Group I kimberlites. Because the KRG database is constantly being updated and the various locality related information upgraded on an ongoing basis, the South Africa and Namibia kimberlite data do not represent the full set of data available for localities in these two countries. Additional regional analyses in the University of Massachusetts (UMASS) ilmenite database from various localities (Table 1) were utilised as an independent data set to test the consistency of the data used to establish ilmenite reference lines, but as they generally embrace the same localities referred to above, they are not included in the data statistics in Table 2. The non-kimberlitic Insizwa and Karoo ilmenite data sets were obtained from tables in published references (Cawthorn et al., 1988), and through the extraction of element contents directly from incorporated bivariate plots.

It is assumed for this study that the majority of the compiled compositions represent analyses of the cores of ilmenite grains. However, core versus rim compositions are distinguished by Schulze et al. (1995) for several of the North America kimberlite localities. The compositional trends observed in these data and implications for the classification scheme are discussed further in the latter part of this manuscript.

In a sense, we have not taken our own advice, and the mineral composition data utilised in this study have not been screened to exclude poor quality data based on the calculated Fe_2O_3 contours. However, this has been done intentionally to highlight the fact that, in reality, such analyses do occur during the routine analysis of exploration samples, and such data should be viewed as suspicious by the recipient.

3. Results

3.1. Kimberlitic ilmenites

Mineral compositions for ilmenites derived from Southern African kimberlite concentrates (Table 1) were separated into on-craton and off-craton localities. Fig. 1A and B are bivariate $MgO-TiO_2$ plots for the off-craton and on-craton Group I kimberlites, respectively. Fig. 2A and B show the same data plotted in $MgO-Cr_2O_3$ space, which was first used by Haggerty

(1975, 1976) to identify a parabolic arc typical of many kimberlitic ilmenite populations (see also Fig. 13 of Haggerty, 1991). The compositional ranges of the on-and off-craton ilmenites are given in Table 2.

An arc encompassing approximately 90% of the data has been estimated on the MgO–TiO_2 plots. The area to the MgO–rich side of the arc is defined as the "Kimberlitic" ilmenite field. This kimberlitic ilmenite

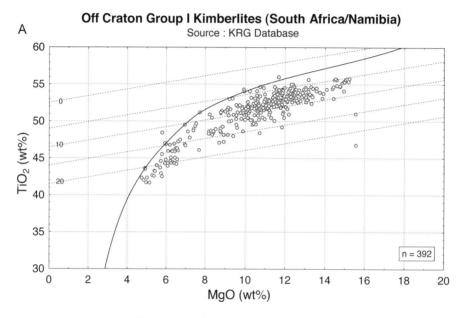

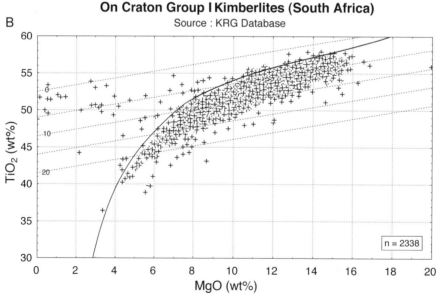

Fig. 1. Plot of MgO versus TiO_2 for (A) off-craton group I kimberlites from South Africa and Namibia, and (B) on-craton group I kimberlites from South Africa. The black line represents the bounding reference line of the kimberlitic ilmenite field. Percentage Fe_2O_3 was calculated using simple ilmenite stoichiometry (Finger, 1972) for individual analyses, and lines of equal Fe_2O_3 were contoured from the data set (dashed light-grey). These lines are also displayed on subsequent MgO–TiO_2 plots for reference (see text).

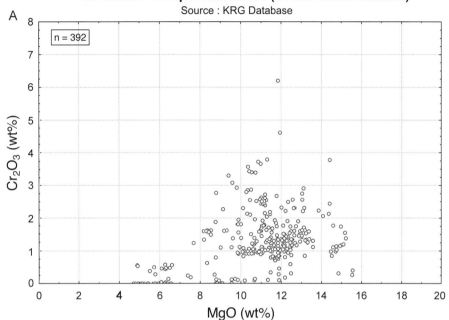

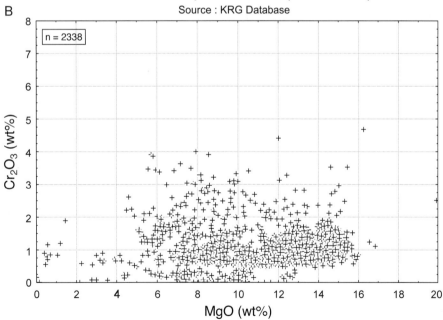

Fig. 2. Plot of MgO versus Cr_2O_3 for (A) off-craton group I kimberlites from South Africa and Namibia, and (B) on-craton group I kimberlites from South Africa.

reference line is well defined by the ilmenite compositions from both the off-craton and on-craton localities at MgO contents between 4 and 15 wt.% (Fig. 1). Below 8 wt.% MgO, the kimberlite compositional arc is defined by the following quadratic equation:

$$y = -51.9078 + 52.8316x - 11.5519x^2 + 1.2003x^3$$
$$- 0.0475x^4.$$

Above 8 wt.% MgO, the arc is defined by the following cubic equation:

$$y = 28.5188 + 4.7521x - 0.287x^2 + 0.0067x^3.$$

It is stressed that these equations are a convenient way to define a bounding or limiting reference curve applicable to kimberlites on a world wide basis. This bounding curve can be used as an aid in assessing the kimberlitic characteristics of individual ilmenite grains found in prospecting grains. However, as discussed further below, individual kimberlite localities comprising a population of ilmenites may define different but approximately parallel arcs. Grade categories for the kimberlites are given in Table 1. All the

off-craton kimberlites are barren or extremely low grade and most of the on-craton bodies have at least some diamonds, and based on these data, there is no obvious systematic relationship between grade and the MgO–TiO$_2$ relationships.

In order to assess the broader applicability of the kimberlitic ilmenite reference arc defined by Southern African sources, the MgO–TiO$_2$ relationships of ilmenite in mineral concentrates from selected North American, Siberian, Australian, and West Africa kimberlites (Table 1) are shown in Figs. 3–6, respectively. The compositional statistics are given in Table 2.

For the North America data, all but one of the 1071 available analyses fall to the MgO-rich side of the defined arc. The North America kimberlitic ilmenite compositions extend to lower MgO content than those from Southern Africa, and hence aid in defining the kimberlitic ilmenite field boundary at very low MgO contents (Fig. 3).

The Siberian data represented on Fig. 4 includes analyses from the Udachnaya and Mir kimberlites, and the vast majority of the data points plot well to the right of the kimberlitic ilmenite reference line. The Siberian data presented here could be used to define its own well-constrained line, but this would not be universally

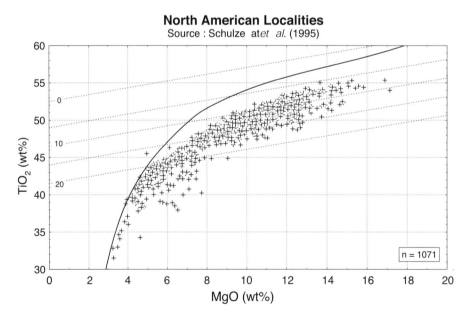

Fig. 3. Plot of MgO versus TiO$_2$ for ilmenite from North America kimberlites (data of Schulze et al., 1995). Symbology as for Fig. 1.

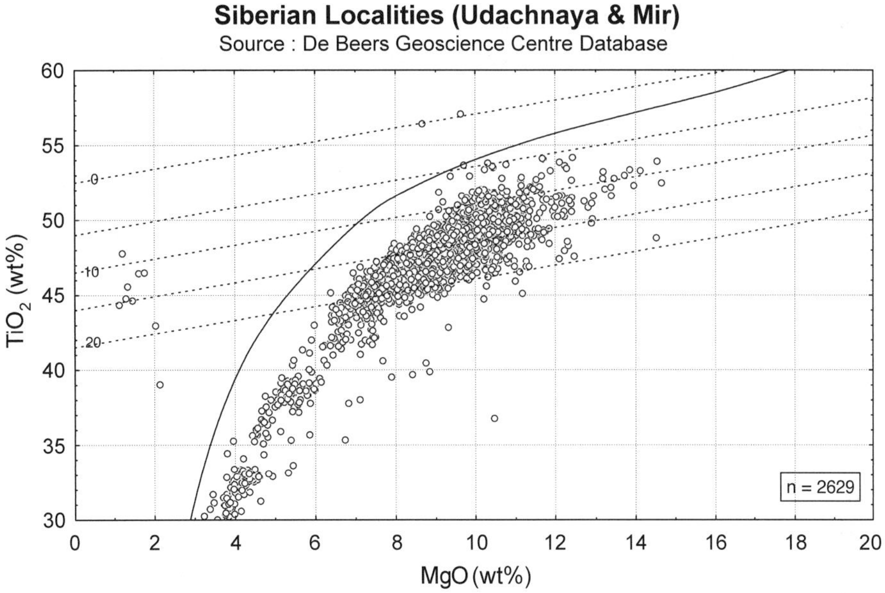

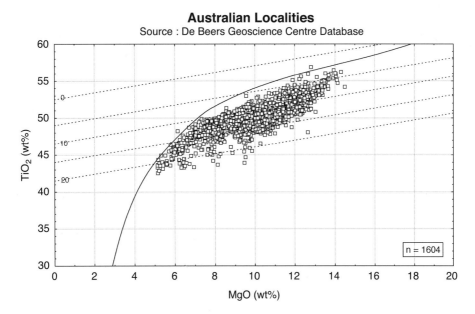

Fig. 4. Plot of MgO versus TiO_2 for ilmenite from Siberia kimberlites. Symbology as for Fig. 1.

applicable because much of the Southern African data would fall to the low-Mg side of such a line.

The Australian data includes ilmenite analyses from the Cleve-1 and Skerring kimberlites located in South and North Australia, respectively (Fig. 5). As with the North America localities, over 90% of the data points plot to the MgO-rich side of the kimberlitic ilmenite reference line. These data actually show a very similar compositional trend to the kimberlites from Southern Africa.

Fig. 5. Plot of MgO versus TiO_2 for ilmenite from Australia kimberlites. Symbology as for Fig. 1.

The West African ilmenites (Fig. 6) are divided into various ilmenite types according to the divisions listed in the UMASS ilmenite database. It is evident in Fig. 6 that a significant proportion of the ilmenites fall to the low-Mg side of kimberlitic reference line, and while these mostly relate to groundmass ilmenites, there are a number of discrete ilmenite nodules and bimineralic associations that also fall in this area. Most of the discrete nodules in the non-kimberlitic field are from the Liberian data, while most of the groundmass ilmenites in the non-kimberlitic field are from Sierra Leone. These ilmenites all contain elevated MnO contents, with the proportion of MnO to MgO increasing to lower MgO contents. A high proportion of these have slightly elevated or elevated Cr_2O_3 (more than approximately 0.25 and 0.5 wt.%, respectively) which approach typical kimberlitic values. Elevated MnO in ilmenites from Koidu, Sierra Leone (Haggerty and Tompkins, 1984; Tompkins and Haggerty, 1985) and MnO enrichment trends evident in some Monastery ilmenites (Haggerty et al., 1979) were attributed to late-stage carbonate and $CO-CO_2$ reactions in the kimberlite. In a practical sense, most groundmass

ilmenites would not report to heavy mineral concentrates in exploration samples, most being less than approximately 0.3 mm in diameter. The relatively few data compiled for the group I kimberlites that fall to the low-MgO side of the kimberlite reference line (Fig. 1b) could be similar late-stage ilmenites, or possibly spurious non-kimberlitic ilmenites incorporated into the kimberlite from disaggregated country rock xenoliths.

It is noted that carbonatites can contain high-MnO ilmenites, often also associated with high-MgO contents (see, e.g., Haggerty, 1976, table Hg-16(3) and Hg-20(9); Haggerty, 1991). Such ilmenites, however, seldom contain high-Cr_2O_3 contents (rarely more than approximately 0.3 wt.%).

Ilmenites from various Southern African, North American, and Russia localities extracted from the UMASS database (Table 1), which were not used in defining the kimberlitic reference line discussed above, are shown in Fig. 7. These data represent an independent confirmation of the applicability of the reference line. While a few of the data clearly fall on the non-kimberlitic side of the reference line, the majority are within the kimberlitic field.

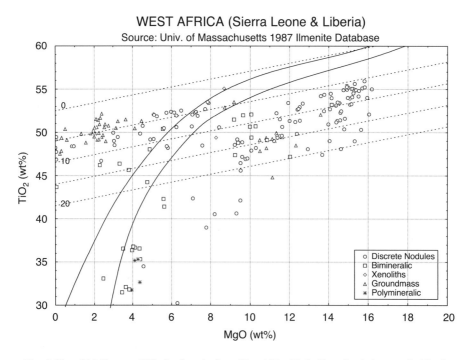

Fig. 6. Plot of MgO versus TiO_2 for ilmenite from West Africa kimberlites. Symbology as for Fig. 1.

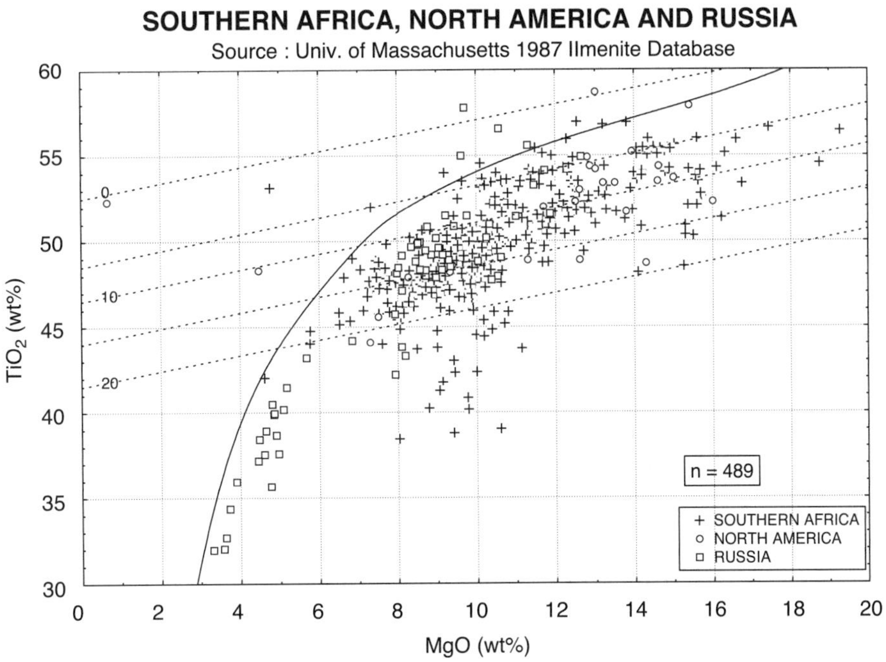

SOUTHERN AFRICA, NORTH AMERICA AND RUSSIA
Source : Univ. of Massachusetts 1987 Ilmenite Database

Fig. 7. Plot of MgO versus TiO$_2$ for ilmenite from Southern Africa, North America, and Russia. Symbology as for Fig. 1.

In summary, we have chosen the arc defined by the Southern African sources as the kimberlitic ilmenite reference line as it would correctly classify kimberlitic ilmenites from a variety of kimberlite sources.

For comparative purposes, the ilmenite data from North America, Siberia, and Australia are also shown in MgO–Cr$_2$O$_3$ compositional space in Figs. 8–10. These data reinforce the notion that a high proportion of kimberlitic ilmenites have elevated Cr$_2$O$_3$, some of which also display a parabolic MgO–Cr$_2$O$_3$ relationship (Haggerty, 1975, 1976, 1991). However, it is noted that while this 'parabolic' relationship is relatively common, it is by no means universal. Often only the right hand limb is present, and both base and position of the limb vary in MgO–Cr$_2$O$_3$ space (see, e.g., Smith, 1977; Eggler et al., 1979; Apter et al., 1984; Schulze, 1984; Moore, 1987; Wyatt et al., 1994; Orr, 1998; Graham et al., 1999).

3.2. Non-kimberlitic ilmenites

A variety of sources were used to define a compositional reference line for non-kimberlitic ilmenites. These included abundant ilmenite compo-sitions from gabbros and picrites that form part of the Mount Ayliff Intrusion (Insizwa Complex), ilmenites from Karoo Basalts in the Bobonong area of Botswana, and groundmass ilmenites in gabbroic phases of the Okenyenya Igneous Complex in Namibia (Table 1). The major element MgO versus TiO$_2$ compositions for the Insizwa and Okenyenya non-kimberlitic ilmenites are presented in Fig. 11, and the Bobonong ilmenites is given in Fig. 12. The non-kimberlitic ilmenites (Table 1) have lower MgO contents at equivalent TiO$_2$ values than ilmenites derived from kimberlites (Figs. 1, Figs. 1, Figs. 3 Figs. 4 Figs. 5). Note that this non-kimberlitic reference line is a measure of the likely maximum MgO limit, at a given TiO$_2$ value, for non-kimberlitic ilmenites, irrespective of the fact that the trend within individual data sets is often subparallel to the Fe$_2$O$_3$ contours and oblique to the reference line. Thus, most non-kimberlitic ilmenites will plot to the left of the non-kimberlitic arc, and kimberlitic ilmenites to the right of the kimberlitic arc. The area between these arcs, therefore, represents an area of uncertainty that will require additional information, such as Cr$_2$O$_3$ and/or MnO

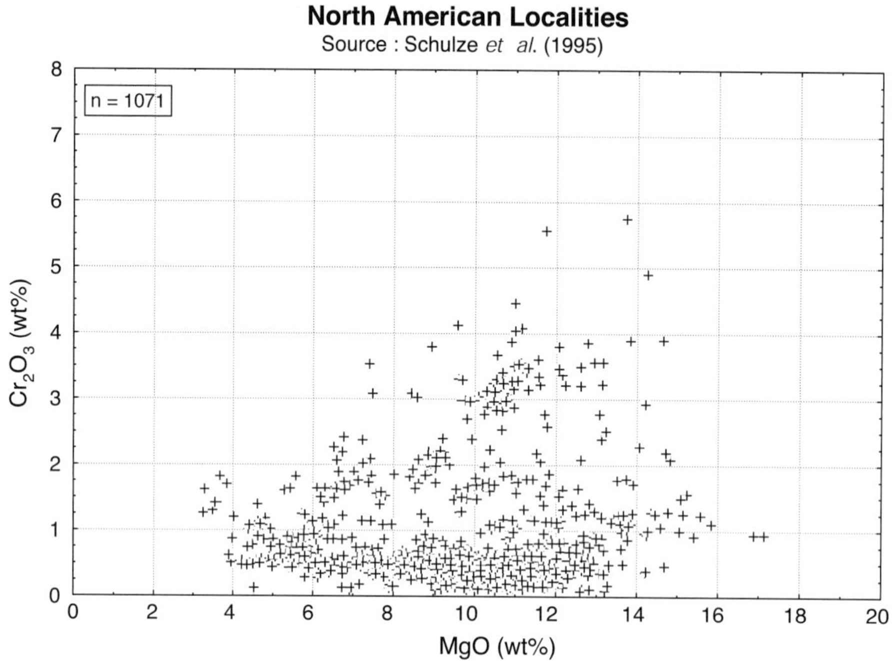

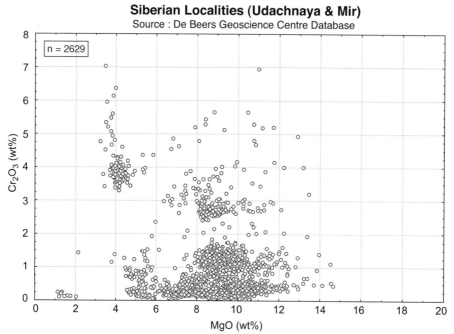

Fig. 8. Plot of MgO versus Cr_2O_3 for ilmenite from North American kimberlites.

Fig. 9. Plot of MgO versus Cr_2O_3 for ilmenite from Siberian kimberlites.

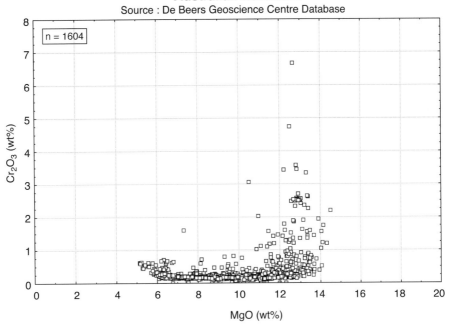

Fig. 10. Plot of MgO versus Cr_2O_3 for ilmenite from Australian kimberlites.

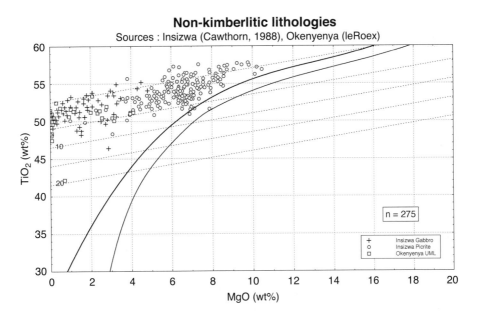

Fig. 11. Plot of MgO versus TiO_2 for for non-kimberlitic rocks from Insiwa and Okenyenya (data from Cawthorn et al., 1988, and le Roex, pers. comm.). Percentage Fe_2O_3 contours (from Fig. 1) are shown as the dashed light-grey lines. The black line at lower MgO represents the chosen bounding reference line of the non-kimberlitic ilmenite field. The black line at higher MgO represents the bounding reference line of kimberlitic ilmenite compositions defined localities in South Africa and Namibia (Fig. 1).

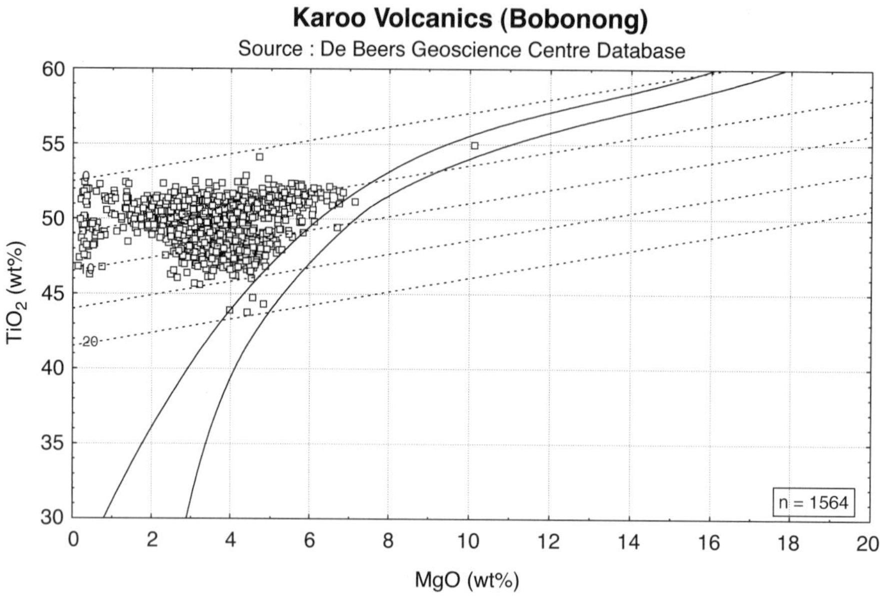

Karoo Volcanics (Bobonong)
Source : De Beers Geoscience Centre Database

Fig. 12. Plot of MgO versus TiO_2 for non-kimberlitic Karoo volcanics from the Bobonong Area in Botswana. Symbology as for Fig. 11.

contents, before a paragenesis can be attributed. Non-kimberlitic ilmenites also usually have less than 1.0 wt.%. Cr_2O_3 and, with few exceptions, have less than 0.5 w% Cr_2O_3. This is well illustrated in the major element statistics presented for the Insizwa and Karoo Data in Table 2. Note that several data points, especially those from Insizwa, fall just above the 0 wt.% Fe_2O_3 line in the MgO– TiO_2 plot, and these data should be viewed as suspicious (perhaps due to rutile or pseudobrookite intergrowths in the ilmenite?). The data set as a whole is nevertheless very useful in defining a nonkimberlitic reference line.

The maximum MgO contents of non-kimberlitic ilmenites were used to estimate a compositional field for these ilmenite varieties. The cubic equation defining the non-kimberlitic bounding reference line is as follows (Figs. 11 and 12):

$$y = 25.4062 + 6.1433x - 0.4187x^2 + 0.0106x^3$$

3.3. Ilmenites from other kimberlite-related rock types (melnoites)

In the previous sections, we have attempted to establish the compositional fields for kimberlitic and non-kimberlitic ilmenites in terms of TiO_2 and MgO. The next step was to evaluate where ilmenites from kimberlite related rock types fall into the classification scheme. These include ultramafic lamprophyres (e.g., alnoites, mellilitites, etc.) and alkali basalts, and are referred to by the term "melnoite" for the purpose of this review (Table 1). Melnoites are known to host phenocrystic and groundmass ilmenites, as well as megacrystic and xenocrystic ilmenite derived from mafic lower crustal or upper mantle lithologies.

Fig. 13 shows the MgO and TiO_2 compositional range of ilmenites present in the Malaita alnoites, as well as ilmenite megacrysts found in the Okenyenya ultramafic lamprophyre (UML) breccia. The ilmenites from Malaita show a trend of slightly decreasing TiO_2 contents with decreasing MgO content, and these ilmenite compositions transect the kimberlitic and non-kimberlitic reference lines. The Okenyenya ilmenite megacrysts plot just to the left of the non-kimberlitic reference line at low MgO content. As with the Malaita data, ilmenites from the Entilombo Melnoite, from the Kwazulu Natal province of South Africa, show a linear trend of decreasing TiO_2 with decreasing MgO contents (Fig. 14). Although the data transect the kimberlitic ilmenite reference line, only 3 of the 1417 analyses plot on the low-MgO side of the non-kimberlitic line.

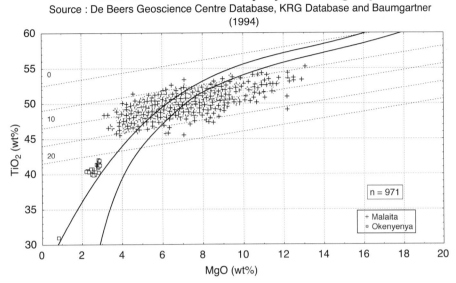

Fig. 13. Plot of MgO versus TiO$_2$ for ilmenites from the Malaita Alnoite and Okenyenya ultramafic lamprophyre. Symbology as for Fig. 11.

Fig. 15 is a plot of MgO versus TiO$_2$ for ilmenites from the Selco alkaline intrusions, Canada, which are described in detail by Janse et al. (1986), and classified petrogenetically as alnoites. These ilmenite data also transect the kimberlitic and non-kimberlitic boundaries. Ilmenites from the Lake Bullenmerri basanite intrusion, Australia, a 'kimberlite related rock' which also comprises abundant upper mantle garnets, are presented in Fig. 16. The majority of the data fall to the low-Mg side of the non-kimberlitic reference line with only two data points within the kimberlitic ilmenite field.

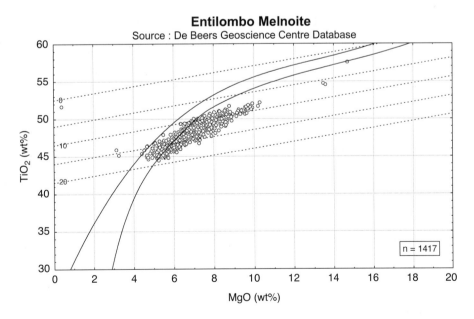

Fig. 14. Plot of MgO versus TiO$_2$ for ilmenites from the Entilombo Melnoite, Kwazulu Natal province, South Africa. Symbology as for Fig. 11.

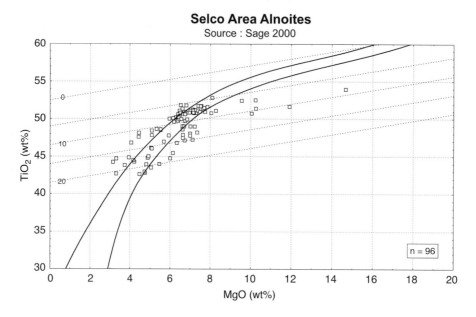

Fig. 15. Plot of MgO versus TiO$_2$ for the North America, Selco Alnoite. Symbology as for Fig. 11.

The observation that these kimberlite 'related rock' ilmenite compositions straddle the kimberlite reference line is consistent with the fact that is no sharp petrological boundary line between them and genuine kimberlites. While it may be difficult to differentiate single isolated 'related rock' ilmenites found in exploration samples from kimberlitic ilmenites on the basis of TiO$_2$–MgO relationships alone, the former are normally low in Cr$_2$O$_3$ (less than approximately 0.3 wt.% Cr$_2$O$_3$; see Table 2). If a population of ilmenites is present, a related rock paragenesis is also suggested if they define a trend at a shallow angle to

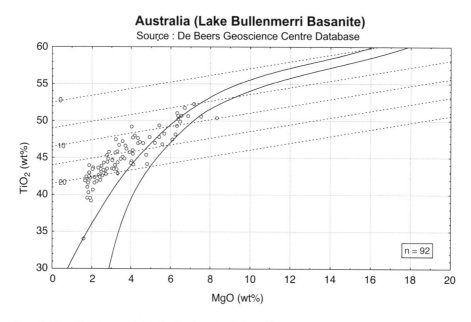

Fig. 16. Plot of MgO versus TiO$_2$ for the Australia, Lake Bullenmerri Basanite. Symbology as for Fig. 11.

the kimberlite reference line (and in some cases, subparallel to the Fe_2O_3 contours).

3.4. Ilmenites from exploration programs

Some exploration data sets are used to illustrate the application of the classification scheme to distinguish kimberlitic from non-kimberlitic ilmenites. Fig. 17 shows the MgO versus TiO_2 contents for a population of ilmenites visually identified as potentially kimberlitic by the Mineral Sorters at Mineral Services diamond laboratory. The majority of the ilmenites are classified as non-kimberlitic in this plot. Importantly, however, four of the 142 grains are in fact classified as kimberlitic. These samples would therefore warrant additional follow-up because these are highly likely to derive from a kimberlite.

The MgO and TiO_2 compositions of the ilmenites in the KIMC exploration database for the Slave craton in Canada (Table 1) are shown in Fig. 18. Clearly several of these analyses plot above the 0 wt.% Fe_2O_3 reference line and are poor analyses, or perhaps could be other high-TiO_2 bearing minerals that have been incorrectly designated as ilmenites in the database. This plot further highlights the apparent difficulties in visually distinguishing kimberlitic from non-kimberlitic ilmenites. The majority of the ilmenites shown in this plot are

clearly kimberlitic, but the data set also contains a large population of non-kimberlitic grains. Any follow-up work conducted over these recoveries prior to mineral analysis may have resulted in misdirected exploration. For comparative purposes, the KIMC ilmenites are also shown in MgO–Cr_2O_3 compositional space in Fig. 19. The majority of these data display a parabolic kimberlite relationship noted by Haggerty (1975, 1976, 1991), but the lower-Mg, low-Cr data clearly fall into a separate population.

3.5. Zoning of ilmenite: implications for the classification scheme

Schulze et al. (1995) demonstrated for several ilmenite populations from North America kimberlites that core and rim compositions might show significant chemical variation. Indeed it appears as though compositional zoning in ilmenite megacrysts and macrocrysts may be fairly common worldwide. Schulze et al. (1995) and O'Brien and Tyni (1999) note that, at most localities where chemical zonation is present, this is represented by increased MgO and/or Cr_2O_3 contents in the outer 100 to 500 μm of the grain. This compositionally distinct rim is different to the perovskite mantle commonly seen on kimberlitic ilmenites derived directly from kimberlite concentrate or recov-

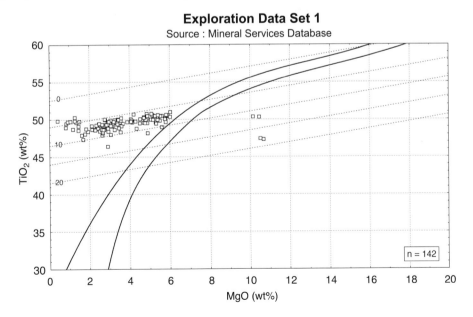

Fig. 17. Plot of MgO versus TiO_2 for an exploration data set. Symbology as for Fig. 11.

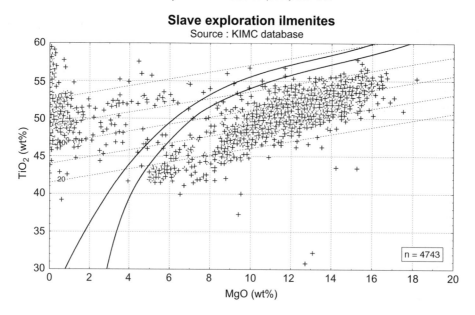

Fig. 18. Plot of MgO versus TiO$_2$ for ilmenites from the Canadian Slave exploration KIMC database. Symbology as for Fig. 11.

ered close to the kimberlite source rock. Schulze et al. (1995) furthermore concur with the conclusions of previous investigations into ilmenite zoning that the rims of elevated MgO and Cr$_2$O$_3$ are a result of late stage magmatic overgrowth of new ilmenite on pre-existing cores or partial re-equilibration of the ilmenite rims with the host magma. O'Brien and Tyni (1999) suggest that magma mixing may also play an

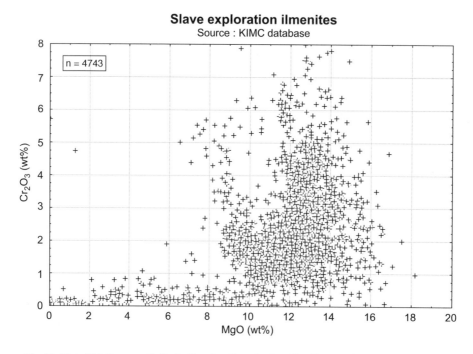

Fig. 19. Plot of MgO versus Cr$_2$O$_3$ for ilmenites from the Canadian Slave exploration KIMC data set.

important role in generating high-MgO ilmenite rims. Certain localities studied by Schulze et al. (1995), and work conducted by O'Brien and Tyni (1999), however, show examples of ilmenite rims with elevated MgO contents and no corresponding Cr_2O_3 increase. The inferred models for these ilmenites are, either that the magma transporting the grains to surface is a more primitive MgO rich variety that had not yet precipitated significant Cr-rich mineral phases, or that magnesite which would have accompanied the ilmenite macrocrysts to surface would have decomposed in the magma resulting in a sudden increase in MgO content of the magma (Schulze et al., 1995).

In addition to the primary chemical zonation seen in kimberlitic ilmenites, alteration of kimberlitic ilmenites in the secondary environment may also be a factor in certain climatic environments. Such secondary alteration is however markedly different from that described above, and is manifested in a distinct increase in MnO and lower MgO contents (e.g., Wyatt, 1979; Agata, 1998; Jiang et al., 1996). Ilmenites showing these features are present in the Premier kimberlite, and were reported at the Second Cambridge Kimberlite Symposium in 1979 (Wyatt, 1979). The data presented in this study, and reproduced in Table 3, have been plotted in $MgO-TiO_2$ space in Fig. 20. Most of the core and intermediate zone analyses fall within the kimberlite field, while the rim analyses fall well within the non-kimberlitic field. Wyatt (1979) interprets the MnO enrichment at Premier as either a late-stage nonmagmatic phenomenon postdating kimberlite consolidation, or being due to low-temperature metasomatism of the grains by circulating Mn-bearing groundwater in a reducing environment below the water table. A characteristic of the individual zoned ilmenites is that they retain their characteristic 'kimberlitic' high-Cr_2O_3 contents ($+0.6$ wt.%) from core to rim. This example illustrates that, when attempting to classify ilmenites extracted from exploration samples, and these turn out to be enriched in MnO (at the possible expense of MgO), and consequently fall on the low-MgO side of the kimberlite reference line (Fig. 1), it would be prudent to scrutinise for high-Cr_2O_3 (more than approximately 0.5 wt.%). Such grains could represent altered kimberlitic ilmenites, or may be kimberlitic but affected by late stage carbonate reactions as described by Haggerty et al. (1979), Haggerty and Tompkins (1984) and Tompkins and Haggerty (1985).

Several of the Cr-rich kimberlitic ilmenites from the Mwenezi kimberlite in Zimbabwe studied by Williams and Robey (1999) showed high MnO contents ranging from 4.5 to 5.4 wt.%, and near zero MgO contents. The authors suggest this may be due to MnO introduction/replacement of MgO in the ilmenites, presumably similar to the Premier case presented above. Indeed, many of these ilmenites have typical kimberlitic high-Cr_2O_3 (more 0.5 wt.%) contents. These Mwenezi ilmenites would plot within the non-kimberlitic field on a bivariate MgO versus TiO_2 plot. Our experience has shown that, in certain cases, the MgO content decreases to the extent that visually kimberlitic ilmenites plot to the low-MgO side of kimberlitic compositional field. It is important to recognise such grains in exploration data sets and to classify their paragenesis correctly.

Table 3
Composition of zoned ilmenites from the Premier Mine, South Africa (from Wyatt, 1979)

	BHS 204			BHS 215			BHS 207			BHS 209		
	Margin	Int	Core	Margin	Int	Core	Margin	Int	Core	Margin	Int	Core
TiO_2	50.65	50.62	51.22	50.04	53.49	54.12	50.73	50.85	54.07	46.13	49.48	51.70
Al_2O_3	0.52	0.34	0.29	0.51	0.48	0.49	0.45	0.24	0.26	0.78	0.99	0.94
Cr_2O_3	1.25	1.21	1.22	1.19	1.23	1.27	0.69	0.63	0.65	5.70	6.35	6.49
FeOt	41.74	41.65	41.00	38.09	32.17	29.90	42.21	42.55	32.43	41.68	32.02	25.62
MnO	5.38	4.44	3.11	9.28	1.36	0.26	3.93	1.97	0.29	2.60	0.74	0.39
MgO	0.08	0.49	2.43	0.18	9.88	12.64	0.65	2.33	11.25	1.18	9.48	13.90
CaO	0.04	0.02	0.01	0.09	0.03	0.03	0.18	0.07	0.03	0.11	0.08	0.09
Total	99.66	98.77	99.28	99.38	98.64	98.71	98.84	98.64	98.98	98.18	99.14	99.13
FeO	39.90	40.12	38.56	35.16	29.64	25.83	40.25	39.49	28.23	36.71	26.74	21.20
Fe_2O_3	2.04	1.70	2.71	3.26	3.44	4.52	2.18	3.40	4.66	5.52	5.87	4.91
Total	99.86	98.94	99.55	99.71	98.99	99.16	99.06	98.98	99.45	98.73	99.73	99.62

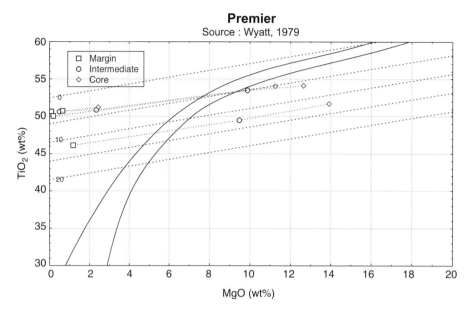

Fig. 20. Plot of MgO versus TiO$_2$ for zoned ilmenites from the Premier kimberlite. Percentage Fe$_2$O$_3$ lines are shown as the dashed light-grey lines. The black lines represent the non-kimberlitic and kimberlitic ilmenite reference lines, respectively.

In contrast to kimberlitic ilmenites, non-kimberlitic ilmenites seldom have zoned rims with increased MgO and/or Cr$_2$O$_3$ contents. A selection of the non-kimberlitic ilmenites from the Mineral Services exploration data set were analysed on their rims at distances of 10 and 20 μm from the grain margin, respectively, and the entire set of ilmenites was also examined using secondary electron backscatter imaging techniques on the in-house Mineral Services Scanning Electron microscope. Both the analytical data and the observations under backscatter confirmed the absence of ilmenite rims of differing composition to their cores.

The analysis of the rims of ilmenites that fall at or close to the compositional boundaries may aid in the interpretation of the paragenesis of these grains. The absence of a compositionally distinct rim does not necessarily mean that the grain is non-kimberlitic, but the presence of a rim with increased MgO or Cr$_2$O$_3$ contents may be significant.

4. Conclusions

The results of this study define a simple and practical classification scheme that can be used to compositionally discriminate ilmenites derived from

kimberlitic sources from those occurring in other sources. The key major elements used in this distinction are MgO and TiO$_2$. In addition, the Cr$_2$O$_3$ content of the ilmenites also needs to be considered because non-kimberlitic ilmenites typically contain very low to zero Cr$_2$O$_3$ contents (seldom above 0.5 wt.%). This is particularly relevant for assessing MnO-enriched ilmenites (more than approximately 1.0 wt.% MnO) that may be zoned or altered ilmenites.

The fact that both non-kimberlitic and kimberlitic ilmenites are recovered in exploration programs in several regions worldwide illustrates the need to be able to discriminate these effectively. The correct identification of the ilmenite source lithology, especially in areas where ilmenite is the key pathfinder mineral, will result in direct cost saving to exploration programs as false anomalies will be easily identified. In addition to this, the ilmenite classification scheme will also aid in finding kimberlitic rocks in areas that contain high background abundances of non-kimberlitic ilmenite.

Acknowledgements

Many of the original concepts presented in this paper were developed within the De Beers organisation in the mid-1970s, the results of which were presented at several internal conferences, meetings, and training courses over approximately the last 25 years. This work benefited from discussions and input from many De Beers employees, but in particular, Dr. R.V. Danchin who first introduced the possibilities of using MgO–TiO$_2$ to one of authors (BAW) and De Beers. De Beers are gratefully thanked for allowing us to present this paper and for permission to use some of their data sets. The following other people and institutions are thanked for their contributions to the data sets that made this study possible. John Armstrong of the Diand–CS Lord Northern Geoscience Centre for providing the KIMC database, the Kimberlite Research Group at UCT for providing records from the KRG Database, Steve Haggerty and the University of Massachusetts (UMASS) for providing their 1987 Ilmenite Database to the KRG, Professor Anton le Roex for supplying the balance of the Okenyenya ilmenite data, Dan Shulze for supplying the data from the North American localities, Grant Cawthorn for providing the Insizwa data, and Mineral Services for supplying one of the exploration data sets. Linda Tompkins and Steve Haggerty are thanked for constructive reviews that improved the paper.

References

Agata, T., 1998. Geochemistry of ilmenite from the Asama ultramafic layered igneous complex, Mikabu greenstone belt, Sambagawa metamorphic terrane, central Japan. Geochem. J. 32, 231–241.

Apter, D.B., Harper, F.J., Wyatt, B.A., Smith, B.H.S., 1984. The geology of the Mayeng sill complex, South Africa. In: Kornprobst, J. (Ed.), Kimberlites I: Kimberlites and Related Rocks. Proc. 3rd Int. Kimb. Conf. Elsevier, Amsterdam, pp. 43–57.

Baumgartner, M.C., 1994. The xenoliths of the Okenyenya Volcanic breccia. Unpub. MSc thesis, Univ. Cape Town, South Africa.

Boyd, F.R., Dawson, J.B., Smith, J.V., 1984. Granny Smith diopside megacrysts from the kimberlites of the Kimberley area and Jagersfontein, South Africa. Geochim. Cosmochim. Acta 48, 381–384.

Bristow, J.W., 1980. The geochronology and geochemistry of Karoo Volcanics in the Lebombo and adjacent areas. Unpub. PhD Thesis, Univ. Cape Town, South Africa.

Cawthorn, R.G., Maske, S., De Wet, M., Groves, D.I., Cassidy, K.F., 1988. Contrasting magma types in the Mount Ayliff Intrusion (Insizwa Complex), Transkei: evidence from ilmenite compositions. Can. Mineral. 26, 145–160.

Dawson, J.B., Smith, J.V., 1977. The MARID (mica–amphibole–rutile–ilmenite–diopside) suit of xenoliths in kimberlite. Geochim. Cosmochim. Acta 41, 309–323.

Dawson, J.B., Hill, P.G., Kinny, P.D., 2001. Mineral chemistry of a zircon-bearing, composite, veined and metasomatised upper-mantle peridotite xenolith from kimberlite. Contrib. Mineral. Petrol. 140, 720–733.

Eggler, D.H., McCallum, M.E., Smith, C.B., 1979. Megacryst assemblages in kimberlite from northern Colorado and southern Wyoming: petrology, geothermomentry–barometry, and areal distribution. In: Boyd, F.R., Meyer, H.O.A. (Eds.), The Mantle Sample: Inclusions in Kimberlites and Other Volcanics. Proc. 2nd Int. Kimb. Conf., vol. 2. Amer. Geophys. Union, Washington, pp. 213–226.

Finger, L.W., 1972. The uncertainty in the calculated ferric iron content of electron microprobe analysis. Year B.-Carnegie Inst. 71, 600–603.

Graham, I., Burgess, J.L., Bryan, D., Ravenscroft, P.J., Thomas, E., Doyle, B.J., Hopkins, R., Armstrong, K.A., 1999. Exploration history and geology of the Diavik kimberlites, Lac de Gras, Northwest Territories, Canada. In: Gurney, J.J., Gurney, J.L., Pascoe, M.D., Richardson, S.H. (Eds.), Proc. VIIth Int. Kimb. Conf., vol. I. Red Roof Design, Cape Town, pp. 262–279.

Haggerty, S.E., 1975. The chemistry and genesis of opaque minerals in kimberlite. In: Ahrens, L.H., Dawson, J.B., Duncan, A.R., Erlank, A.J. (Eds.), Proc. 1st Int. Kimb. Conf., Physics Chem. Earth, vol. 9, pp. 195–307.

Haggerty, S.E., 1976. Opaque mineral oxides in terrestrial igneous rocks. In: Rumble, D. (Ed.), Oxide Minerals. Mineral. Soc. Am. Short Course Notes, vol. 3, pp. Hg 101–Hg 300.

Haggerty, S.E., 1989. Upper mantle opaque stratigraphy and the genesis of metasomites and alkali-rich melts. Kimberlites and Related Rocks, Proc. 4th Int. Kimberlite Conf., Vol. 2. Special Publication-GSA, vol. 14, pp. 687–699.

Haggerty, S.E., 1991. Oxide mineralogy of the upper mantle. Oxide Minerals. Mineralogical Society of America Reviews in Mineralogy, vol. 25, pp. 355–416.

Haggerty, S.E, Tompkins, L.A., 1984. Subsolidus reactions in kimberlitic ilmenites. In: Kornprobst, J. (Ed.), Kimberlites I: Kimberlites and Related Rocks. Proc. 3rd Int. Kimb. Conf., vol. 1. Elsevier, Amsterdam, pp. 335–357.

Haggerty, S.E., Hardie, R.B., McMahon, B.M., 1979. The mineral chemistry of ilmenite nodule associations from the Monastery Daitreme. In: Boyd, F.R., Meyer, H.O.A. (Eds.), The Mantle Sample: Inclusions in Kimberlites and Other Volcanics. Proc. 2nd Int. Kimb. Conf., vol. 2. Amer. Geophys. Union, Washington, pp. 249–256.

Harte, B., 1987. Metasomatic events recorded in mantle xenoliths: an overview. In: Nixon, P.H. (Ed.), Mantle Xenoliths. Wiley, Chichester, pp. 625–640.

Janse, A.J.E., Downie, I.F., Reed, L.E., Sinclair, I.G., 1986. Alkaline intrusions in the Hudson Bay lowlands, Canada: exploration methods, petrology and geochemistry. Kimberlites and Related Rocks, Proc. 4th Int. Kimb. Conf.,Vol. 2. Special Publication-GSA, vol. 14, pp. 1192–1203.

Jiang, S.-Y., Palmer, M.R., Slack, J.F., 1996. Mn-rich ilmenite from the Sullivan Pb–Zn–Ag deposit, British Columbia. Can. Mineral. 34, 29–36.

Meyer, H.O.A., Svisero, D.P., 1975. Mineral inclusions in Brazilian diamonds. In: Ahrens, L.H., Dawson, J.B., Duncan, A.R. Erlank, A.J. (Eds.), Proc. 1st Int. Kimb. Conf. Phys. Chem. Earth, vol. 9, pp. 785–795.

Mitchell, R.H., 1973. Magnesian ilmenite and its role in kimberlite petrogenesis. J. Geol. 81, 301–311.

Mitchell, R.H., 1977. Geochemistry of magnesian ilmenites from kimberlites in South Africa and Lesotho. Lithos 10, 29–37.

Mitchell, R.H., 1986. Kimberlites: Mineralogy, Geochemistry and Petrology. Plenum, New York. 442 pp.

Moore, A.E., 1987. A model for the origin of ilmenite in kimberlite and diamond: implications for the genesis of the discrete nodule (megacryst) suite. Contrib. Mineral. Petrol. 95, 245–253.

Moore, A.E., Lock, N.P., 2001. The origin of mantle-derived megacrysts and sheared peridotites—evidence from kimberlites in the northern Lesotho—Orange Free State (South Africa) and Botswana pipe clusters. S. Afr. J. Geol. 104, 23–38.

O'Brien, H.E., Tyni, M., 1999. Mineralogy and geochemistry of kimberlites and related rocks from Finland. In: Gurney, J.J., Gurney, J.L., Pascoe, M.D., Richardson, S.H. (Eds.), Proc. VIIth Int. Kimb. Conf., vol. II. Red Roof Design, Cape Town, pp. 625–636.

Orr, P., 1998. Geochemistry and petrology of the Yamba Lake kimberlites, Central Slave Province, Northwest Territories. Unpub. MsC thesis, University of Alberta, Edmonton, Alberta. 162 pp.

Sage, R.P., 2000. MRD 60-Kimberlite Heavy Mineral Indicator Data, Attawapiskat Area, James Bay Lowlands, Northern Ontario. Data contained in Appendix A of Open File Report 6019. Ontario Geological Survey, Ontario, Canada.

Schulze, D.J., 1984. Cr-poor megacrysts from the Hamilton Branch kimberlite, Elliot County, Kentucky. In: Kornprobst, J. (Ed.), Kimberlites II: The Mantle and Crust–Mantle Relationships. Proc. 3rd Int. Kimb. Conf. Elsevier, Amsterdam, pp. 97–108.

Schulze, D.J., 1987. Megacrysts from alkalic volcanic rocks. In: Nixon, P.H. (Ed.), Mantle Xenoliths. Wiley, Chichester, pp. 434–451.

Schulze, D.J., Anderson, P.F.N., Hearn, B.C., Hetman, C.M., 1995. Origin and significance of ilmenite megacrysts and macrocrysts from kimberlite. Int. Geol. Rev. 37, 780–812.

Smith, C.B., 1977. Kimberlite and mantle derived xenoliths at Iron Mountain, Wyoming. Unpub. MsC thesis, Colorado State University, Fort Collins, Colorado. 218 pp.

Sobolev, N.V., 1977. Deep-Seated Inclusions in Kimberlites and the Problem of the Composition of the Mantle. Amer. Geophys. Union, Washington, DC. 279 pp.

Sobolev, N.V., Yefimova, E.S., 2000. Composition and petrogenesis of Ti-oxides associated with diamonds. Int. Geol. Rev. 42, 758–767.

Tompkins, L.A., Haggerty, S.E., 1985. Groundmass oxide minerals in the Koidu kimberlite dykes, Sierra Leone, West Africa. Contrib. Mineral. Petrol. 91, 245–263.

Williams, C.M., Robey, J.V.A., 1999. Petrography and mineral chemistry of the Mwenezi-01 kimberlite, Zimbabwe. In: Gurney, J.J., Gurney, J.L., Pascoe, M.D., Richardson, S.H. (Eds.), Proc. VIIth Int. Kimb. Conf., vol. 2. Red Roof Design, Cape Town, pp. 896–903.

Wyatt, B.A., 1979. Manganoan ilmenite from the Premier kimberlite. Proc. 2nd Kimb. Symposium, Cambridge.

Wyatt, B.A., Lawless, P.J., 1984. Ilmenite in polymict xenoliths from the Bultfontein and De Beers Mines, South Africa. In: Kornprobst, J. (Ed.), Kimberlites II: Their Mantle and Crust/Mantle Relationships. Proc. 3rd Int. Kimb. Conf. Elsevier, Amsterdam, pp. 43–56.

Wyatt, B.A, Shee, S.R.S., Griffin, W.L., Zweistra, P., Robinson, H.R., 1994. The petrology of the Cleve kimberlite, Eyre Peninsula, South Australia. In: Meyer, H.O.A., Leonardos, O.H. (Eds.), Kimberlites, Related Rocks and Xenoliths. Proc. 5th Int. Kimb. Conf., Rio de Janeiro. Spec. Publ.-CPRM, vol. 1/A (Jan/94), pp. 62–79.

Available online at www.sciencedirect.com

Lithos 77 (2004) 841–857

www.elsevier.com/locate/lithos

An updated classification scheme for mantle-derived garnet, for use by diamond explorers[☆]

Herman S. Grütter[a,b,*], John J. Gurney[c], Andrew H. Menzies[c], Ferdi Winter[a]

[a] De Beers GeoScience Centre, PO Box 82232, Southdale, Gauteng 2135, South Africa
[b] Mineral Services Canada, 205-930 Harbourside Drive, North Vancouver, BC, Canada V7P 3S7
[c] Mineral Services South Africa, PO Box 38668, Pinelands 7430, South Africa

Received 27 June 2003; accepted 14 December 2003
Available online 17 June 2004

Abstract

Mantle-derived garnets recovered in diamond exploration programs show compositional variations in Cr, Ca, Mg, Fe and Ti that reflect the chemical, physical and lithological environments in which they occur, occasionally together with diamond. The association of diamond with mantle garnet has progressed through a number of geochemical advances, most notably those of Dawson and Stephens (1975) and Gurney (1984), which are integrated in this work with less well known petrological advances made primarily in xenolith and experimental petrology. A simple, robust garnet classification scheme is formulated which accommodates empirical garnet–diamond relationships for peridotitic (G10, G9, G12), megacrystic (G1), Ti-metasomatised (G11), pyroxenitic (G4, G5) and eclogitic (G3) lithologies in eight distinct garnet classes. The calcium-saturation characteristics of harzburgitic (G10), lherzolitic (G9) and wehrlitic (G12) garnets are described by a Ca-intercept projection that also shows promise as a relative barometer for garnet lherzolite (Grütter and Winter, 1997). Thermobarometric aspects of garnet–diamond associations are highlighted in the scheme through the use of the minor elements Mn and Na, though analysis by anything other than an electron microprobe is not required for classification. A "D" suffix is added to the G10, G4, G5 or G3 categories to indicate a strong compositional and pressure–temperature association with diamond. The scheme remains open to improvement, particularly with regard to delineation of pyroxenitic (or websteritic) diamond associations and to advances in Ca-in-garnet and Na-in-garnet thermobarometry.
© 2004 Elsevier B.V. All rights reserved.

Keywords: Diamond exploration; Pyrope; Ca-intercept; Peridotite; Megacryst; Eclogite; Pyroxenite; G10; G9; G1; G3

1. Introduction

Exploration for potentially diamond-bearing intrusives worldwide usually involves a multi-disciplinary approach. One powerful approach is to trace mantle-derived garnet and related indicator minerals to their source. The pathfinder minerals are routinely analysed for major and minor elements by electron microprobe, in part to identify and differentiate them from visually

---

[☆] Supplementary data associated with this article can be found, in the online version, at doi: 10.1016/j.lithos.2004.04.012.

* Corresponding author. Mineral Services Canada, 205-930 Harbourside Drive, North Vancouver, BC, Canada V7P 3S7. Fax: +1-604-980-6751.

E-mail address: herman.grutter@mineralservices.com (H.S. Grütter).

similar minerals of non-mantle origin, but also to enable qualitative predictions to be made regarding the diamond potential of their source rock (e.g. Gurney et al., 1993). Garnet and chromite are the heavy minerals of choice in this application because they are very common amongst the mineral inclusions found in diamond and they usually survive dispersion and alteration at the Earth's surface substantially better than do mantle-derived olivine or pyroxene. Relative to common mantle-derived garnets, the peridotitic and eclogitic varieties found included in diamond have reasonably distinct compositions (e.g. Gurney, 1984) and simplified compositional screens based on bivariate scatterplots are commonly used to classify and prioritise mantle-derived garnets recovered during exploration programmes (e.g. Lee, 1993; Fipke et al., 1995). This methodology developed from the prior use of simple scatterplots to illustrate geochemical relationships of eclogitic and peridotitic garnet to diamond (Sobolev and Lavrent'ev, 1971; Gurney and Switzer, 1973) and to succinctly characterise mantle lithologies in terms of garnet composition (e.g. Sobolev et al., 1973a,b; Switzer, 1975; Schulze, 1995).

Multivariate statistical analysis involving five or more compositional attributes, particularly cluster analysis and derivatives thereof, has also been used to relate garnet xenocryst compositions to their parental mantle lithology (e.g. Dawson and Stephens, 1975; Danchin and Wyatt, 1979), to compare populations of grains within and between different kimberlite intrusions (Jago and Mitchell, 1989), and to characterise the compositional attributes of entire sections of mantle lithosphere (Griffin et al., 2002). Although not specifically concerned with the relationship of garnet to diamond, the diamond exploration industry has, through time and by common use, borrowed the cluster-based nomenclature of Dawson and Stephens (1975) to describe simplified garnet compositional categories. For peridotitic garnets the term "G10" is thus considered short-hand for "subcalcic", "Ca-undersaturated" or "harzburgitic", while "G9" represents "Ca-saturated" or "lherzolitic". The work reported on here continues this convenient practise whilst updating and formalising many of the simple classification thresholds. Additional constraints, based primarily on garnet phase-relations, the results of experimental petrology inves-

tigations and empirical observations on garnet concentrates are added to arrive at a simple, robust garnet classification scheme that is specifically tailored to the requirements of the diamond explorationist. The main features of the updated scheme are (i) reliance only on compositional data obtained by electron microprobe analysis, (ii) backward compatibility with previous work, concepts and nomenclature, (iii) internal consistency with known diamond associations, and (iv) ease and transparency of implementation.

2. Data and methods

2.1. Data sources

The compositions of garnet and associated minerals that occur in some 4500 mantle-derived or granulite-grade lower crustal xenoliths and microxenoliths and as inclusions in about 600 diamonds were compiled from a large number of published sources and selected unpublished theses (see Appendix A of this work, Appendix 1 of Grütter and Moore, 2003 and Appendix 1 of Grütter and Quadling, 1999). Additional data for minerals in xenoliths are derived from the compilations of Schulze (1995, 1996, 1997, 2003). The compositions of garnet inclusions in diamonds are from essentially the same sources as listed by Stachel et al. (2000, and references therein). Garnet compositions used in the multivariate statistical studies of Dawson and Stephens (1975), Danchin and Wyatt (1979) and Jago and Mitchell (1989) were kindly provided on request. Data for minerals occurring in concentrate derived from kimberlites and related rocks are those held by the Kimberlite Research Group at the University of Cape Town and De Beers Consolidated Mines. The compiled data set encompasses a variety of analytical conditions, standardizations and matrix correction procedures for electron microprobes, and in this regard matches the commercial data typically being used in modern-day exploration applications. The available analyses were checked to comply within $\pm 3\%$ of ideal stoichiometry and were screened to eliminate rare non-mantle (i.e. crustal) compositions. Majoritic garnet analyses (>3.1 Si cations per 12 oxygens) were identified for a limited number of garnets occurring as inclusions in diamond; they are not represented elsewhere in the compiled data sets.

2.2. Diamond-facies nomenclature, geotherms and pressure–temperature data

Modern-day single-grain thermometry techniques are in principle capable of assigning individual peridotitic garnet grains recovered during diamond exploration to the stability field of graphite or diamond (e.g. Griffin and Ryan, 1993; Grütter et al., 1999), but doing so requires knowledge of the local geotherm. This is commonly an unknown in exploration programmes and we therefore make the simplifying assumption that a typical cratonic geotherm pertains to the area being explored. The updated classification scheme is thus calibrated by design for geotherms intersecting the graphite/diamond transition at temperatures in the range 920 to 1000 °C, this being equivalent to the 38 to 40 mW/m^2 model conductive geotherms of Pollack and Chapman (1977). Individual grains that have a strong compositional association with diamond and that also fall within the diamond stability field under these conditions are considered to be of "diamond-facies" and are labelled with the suffix "D" in the classification scheme. The general validity of this approach can be tested by comparing garnet compositions in concentrates from cratonic sources that are known to be diamondiferous or barren. The actual geotherm could also be verified by employing single-grain thermobarometers (Ryan et al., 1996; Nimis and Taylor, 2000; Grütter and Moore, 2003).

Current xenolith thermobarometry techniques are capable of discriminating pressure–temperature (P–T) conditions for naturally occurring mantle xenoliths to within approximately 3.0 kbar (Pearson et al., 1994; Table 2 of Taylor, 1998). Allowing for this error margin on either side of the diamond stability curve (P [kbar] = 19.4 + 0.025*T [°C], Kennedy and Kennedy, 1976) permits assignment of mantle xenoliths to the stability fields of graphite (P < 16.4 + 0.025*T) or diamond (P > 22.4 + 0.025*T) with relatively high confidence. Such graphite- or diamond-stable assignments for mantle xenoliths are used below to validate temperature and pressure effects on garnet MnO content and Ca-intercept variations. The requisite pressures and temperatures were calculated for garnet-bearing xenoliths in our compiled data base using the orthopyroxene–garnet barometer of Nickel and Green (1985) in combination with a clinopyroxene–solvus thermometer (Nimis and Taylor, 2000) for lherzolites and pyroxenites, or the olivine–garnet thermometer (O'Neill and Wood, 1979) for garnet harzburgites.

3. Petrological constraints

3.1. Garnet–chromite–carbon relations in peridotite

Compositional and phase-relations between Cr-pyrope garnet, chromite and carbon in model peridotite systems are illustrated for typically cratonic geothermal conditions in Fig. 1. The Cr/(Cr + Al) ratios of Ca-free harzburgitic garnet, Ca-saturated lherzolitic garnet and coexisting chromite increase with pressure and respectively attain values of J, K and L at the graphite/diamond transition. For the fixed geothermal conditions illustrated, this implies that any Cr-pyrope garnet with Cr/(Cr + Al) greater than line segment J–K must derive from inside the diamond stability field. The line segment J–K has been constrained to the simple

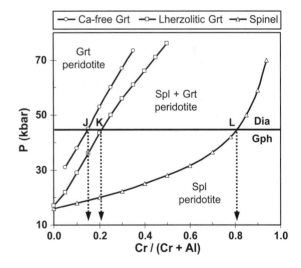

Fig. 1. Partly schematic pressure–composition section illustrating the influence of Cr/(Cr + Al) ratio on spinel–garnet (Spl–Grt) assemblage relations in peridotite (modified after Webb and Wood, 1986). The graphite/diamond (Gph/Dia) transition occurs at fixed pressure for a fixed geotherm and hence specifies garnet and coexisting chromite compositions J, K and L in peridotite assemblages. Line segment J–K represents a compositional tie line which corresponds to the graphite–diamond constraint illustrated in Fig. 3.

relationship $Cr_2O_3 = 5.0 + 0.94*CaO$ (in wt.%) for low-Ca Cr-pyrope garnet compositions by inspection of experimental (Malinovsky and Doroshev, 1977) and natural assemblage relations in chromite-saturated, carbon-bearing garnet harzburgite xenoliths derived from cratonic kimberlites (Pokhilenko et al., 1991, 1993; Grütter, 1994; Grütter and Sweeney, 2000; Menzies, 2001, see Fig. 3). The relationship is designated as the graphite–diamond constraint (the GDC) and we accordingly use $Cr_2O_3 \geq 5.0 + 0.94*CaO$ (wt.%) in the updated classification scheme as a petrologically defined limit that aids in differentiating diamond-facies low-Ca garnet compositions ("D" suffix).

The relations shown in Fig. 1 illustrate that pyrope garnets with $Cr/(Cr + Al)$ less than line segment J–K may also occur inside the diamond stability field, but only when they occur in chromite-free peridotite assemblages. Thus about 46% of the 348 peridotitic Cr-pyrope inclusions in diamond in our database have Cr_2O_3 contents lower than the GDC (i.e. $1.0 < Cr_2O_3 < 5.0 + 0.94*CaO$, see also Fig. 5 of Gurney et al., 1993). An additional constraint is required to correctly identify these Cr-pyrope garnets as being of diamond-facies, and we use their MnO content for this purpose.

3.2. MnO in pyrope garnet

Detailed empirical work on upper mantle xenoliths has documented an inverse relationship between the MnO content of peridotitic garnet and xenolith equilibration temperature (Delaney et al., 1979; Fig. 13 of MacGregor, 1979; Smith et al., 1991). Our thermobarometric categorization of 751 garnet harzburgite and lherzolite xenoliths derived from cratonic, diamond-bearing kimberlites shows a value of $MnO < 0.36$ wt.% (± 0.1 at 1σ) to be characteristic of peridotitic garnets that occur inside the diamond stability field (Fig. 2). The threshold has a large error ($\sim 30\%$) and should be used with care, but since it correctly classifies 268 of 336 peridotitic garnets included in diamond (i.e. 80% of those analysed for MnO), it may be employed to good effect in the updated scheme to differentiate diamond-facies low-Ca garnet compositions (Fig. 2).

It is instructive to note that 52 of the 68 incorrectly classified Cr-pyrope inclusions in diamonds have relatively high Cr_2O_3 contents (8 to 15 wt.%

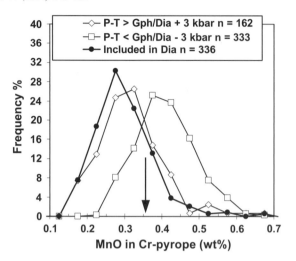

Fig. 2. Frequency histogram of MnO content in garnet for diamond- or graphite-facies peridotite xenoliths from cratonic kimberlites. The diamond- or graphite-facies designation is based on conventional thermobarometry, subject to an error of ± 3 kbar (see Section 2). Garnet MnO contents overlap within the range 0.25 to 0.45 wt.% MnO, but we adopt a graphite/diamond division at 0.36 wt.% MnO (arrow) for the purposes of the classification scheme. This choice correctly classifies as diamond-facies some 268 of 336 peridotitic garnets included in diamond.

Cr_2O_3), and it is possible that their MnO contents have been artificially inflated by as much as 0.20 wt.% during electron microprobe analysis as a result of a Cr-$K\beta$ peak (at 5.947 keV) overlapping that of the Mn-$K\alpha$ peak (at 5.985 keV). This analytical artifact is particularly noticeable for high-Cr_2O_3 garnet compositions reported as inclusions in Chinese diamonds (Wang et al., 2000), but it appears to also be present in a few other data sets investigated in the course of this work. The peak overlap is normally readily resolved during wavelength-dispersive analysis of Mn at low concentrations by selection of a LiF rather than a PET crystal.

3.3. Garnet Ca-intercept variations

The close association of diamond with Ca-undersaturated garnet compositions in preference to Ca-saturated compositions (Gurney and Switzer, 1973; Gurney, 1984) suggests that the G10/G9 divide (Fig. 3) may be empirically correlated with pressure–temperature conditions falling just inside the diamond stability field. Pressure values of 45 to 50 kbar are considered appropriate for cratonic geotherms by

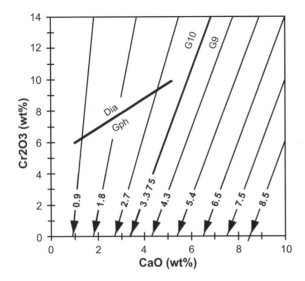

Fig. 3. Conventional garnet Cr_2O_3 vs. CaO diagram showing the 85% line of Gurney (1984) (commonly known as the G10/G9 divide) and the graphite–diamond constraint of Grütter and Sweeney (2000). The downward-pointing arrows with CA_INT values illustrate the geometric effect of the calcium-intercept projection discussed in this work (modified after Grütter and Winter, 1997).

Gurney et al. (1993, p. 2430) and Gurney and Zweistra (1995, p. 297). Prior observations on garnet concentrates from off-craton kimberlites illustrated that, at fixed Cr_2O_3 content, the CaO content of lherzolitic garnet populations increases by ~ 12% relative to cratonic conditions (Boyd and Gurney, 1982), the implication being that the CaO variation may be an effect of decreased pressure. Garnet CaO variations of similar magnitude are well known from high-pressure experimental results in model and natural lherzolitic or pyroxenitic assemblages, where they are found to be non-linear with respect to pressure, dependent on temperature and the Na_2O content of coexisting clinopyroxene, and most useful for relative thermobarometry at pressures inside the graphite stability field (Kushiro et al., 1967; Boyd, 1970; Akella, 1976; Brey et al., 1986, 1990; Brenker and Brey, 1997). These circumstances indicate that the G10/G9 divide may be utilized as an appropriately situated static reference from which to measure the Ca-saturation characteristics of harzburgitic as well as lherzolitic garnets. Given the G10/G9 divide occurs at CaO = 3.375 + 0.25*Cr_2O_3 (in wt.%), a calcium-intercept (CA_INT)

projection to 0 wt.% Cr_2O_3, adapted from that given in Grütter and Winter (1997), is formulated as (in wt.%):

IF CaO ≤ 3.375 + 0.25*Cr_2O_3

THEN CA_INT = 13.5*CaO/(Cr_2O_3 + 13.5)
ELSE CA_INT = CaO − 0.25*Cr_2O_3.

Fig. 3 illustrates that the projection contains geometric elements of the J-score developed by Gurney for harzburgitic garnets (as quoted in Lee, 1993), and also of the "reduced CaO values" calculated and used by Hatton and Gurney (1987, their Fig. 237c) to represent Ca-content relationships amongst eclogitic, websteritic and lherzolitic garnets. The current formulation follows that of Grütter and Winter (1997) in that it sweeps smoothly from a CA_INT value of 0.0 across the G10 field to 3.375 at the G10/G9 divide and continuously extends to substantially higher values (Fig. 3).

Typical CA_INT values calculated for Cr-pyrope garnets from cratonic, peri-cratonic and off-craton settings are shown in histogram format in Fig. 4A to 4C. A general decrease of CA_INT values for lherzolitic garnet populations is observed for these three settings, in agreement with the pressure effects noted by previous workers and summarised above. A notional divide at CA_INT ~ 4.3 partially separates diamond-bearing kimberlite settings (central Slave, central Kaapvaal and Somerset Island) from settings that are barren (Gibeon, Karoo, East Griqualand, The Thumb), but there is substantial overlap of CA_INT values. It follows that CA_INT values for individual lherzolitic garnets cannot be assigned unambiguously to any one of the three cratonic settings considered, nor, by inference, to diamond-stable conditions (Fig. 4A to C). Further work is necessary to improve the accuracy of the Ca-in-garnet thermobarometer and to incorporate additional compositional factors not accounted for by the simple Ca-intercept projection employed here (see also Brenker and Brey, 1997). However, the observed overlap and consequent lack of barometric constraint does not affect the utility of Ca-intercept values in classifying garnet compositions. For instance, it is apparent from Fig. 4 that CA_INT values for lherzolitic garnets range upward from 3.375 and show a natural upper limit at a value of ~ 5.4. We hence regard the latter value as the maximum for lherzolitic garnets derived from the

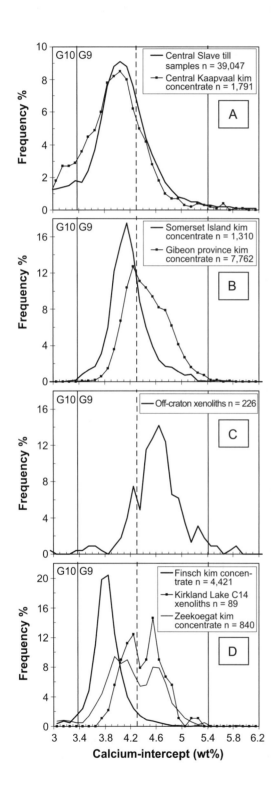

wide variety of settings that are likely to occur in diamond exploration applications.

Bulk-tested kimberlites in the Kirkland Lake and Somerset Island provinces are known to have very low diamond contents (0 to 0.02 ct/ton, Kjarsgaard and Levinson, 2002) while kimberlites in the Gibeon province are not known to contain diamond. The combination of this information with the results illustrated in Fig. 4 suggests that *average* CA_INT values of less than 4.3 for large populations of lherzolitic garnets may be useful to discriminate diamond-stable from graphite-stable conditions in cratonic upper mantle lithospheres. This application of Ca-intercept thermobarometry to populations of lherzolitic garnets should be approached with care because population variances are comparatively high (typically ± 0.3 at 1σ). Since prominent CA_INT modes for lherzolitic garnet populations from individual kimberlites may fall either side of the 4.3 threshold, particularly for small data sets (Fig. 4D), it is considered advisable to closely inspect the data set and to calculate average CA_INT values only for data sets that contain 500 or more lherzolitic garnets.

4. Garnet classifications

4.1. Harzburgitic (G10)

Gurney (1984) correlated 85% of peridotitic garnet inclusions in diamonds from global sources with the Ca-poor, Cr-rich harzburgitic pyrope cluster 10 of Dawson and Stephens (1975) and in doing so "brand-

Fig. 4. (A–D) Histograms of CA_INT values for peridotitic garnets with $Cr_2O_3>1.0$ wt.% from cratonic settings (A), peri-cratonic settings (B), off-craton settings (C), and from individual kimberlites (D). See text for discussion of thresholds indicated at CA_INT values of 3.375, 4.3 and 5.4. Central Slave data from Armstrong (2001), central Kaapvaal from the UCT KRG database, Somerset Island from Jago and Mitchell (1989), Gibeon province, Finsch and Zeekoegat from the De Beers database. Off-craton xenolith data are from The Thumb (Ehrenberg, 1978), Karoo kimberlites (Robey, 1981; Nowicki, 1990), East Griqualand kimberlites (Boyd and Nixon, 1979) and alkali basalt vents at Vitim, Eastern Siberia (Ionov et al., 1993; Glaser et al., 1999). The Zeekoegat kimberlite occurs off-craton in the East Griqualand province, southeast of the Kaapvaal craton. Garnet compositions for most of the southern African localities mentioned here were originally described in Boyd and Gurney (1982), which also includes a locality map.

ed" the G10 garnet standard with which the diamond potential of exploration projects is often judged (Fig. 3). The association made with diamond in this case is primarily geochemical and statistical in nature. The presence of graphite in garnet harzburgite xenoliths (Nixon et al., 1987; Viljoen et al., 1994) implies an association of both polymorphs of carbon with low-Ca G10 garnet compositions. The Cr-saturation characteristics of G10 garnets (Figs. 1 and 3) and/or their MnO content (Fig. 2) may be used to specifically highlight diamond-facies G10 garnets. G10 garnets in our classification scheme are thus compositionally characterised by:

Cr_2O_3 [wt.%]: ≥ 1.0 to < 22.0
CA_INT [wt.%]: 0 to < 3.375
MGNUM: ≥ 0.75 to < 0.95

where MGNUM=$(MgO/40.3)/(MgO/40.3 + FeOt/71.85)$ [oxides in wt.%]. G10D diamond-facies garnets additionally have (in wt.%):

$Cr_2O_3 \geq 5.0 + 0.94*CaO$, or
$Cr_2O_3 < 5.0 + 0.94*CaO$ *and* MnO < 0.36.

4.2. Lherzolitic (G9)

Cr-pyrope garnets derived from lherzolites are by far the most abundant garnet type recovered in diamond exploration applications. Their statistical association with diamond is weak (15% of peridotitic inclusions in diamond, Gurney, 1984), particularly given their high relative abundance as xenocrysts in diamondiferous kimberlites (e.g. Gurney and Switzer, 1973, also Fig. 4). Dawson and Stephens (1975) classified lherzolitic garnets in cluster 9, and in our scheme G9 garnets have the following compositions:

Cr_2O_3 [wt.%]: ≥ 1.0 to < 20.0
CA_INT [wt.%]: ≥ 3.375 to < 5.4
MGNUM: ≥ 0.70 to < 0.90

4.3. Wehrlitic (G12)

Mantle-derived garnets with high CaO and Cr_2O_3 content are only very rarely described as inclusions in diamonds (Sobolev et al., 1970), but are known to occur in xenolith and microxenolith fragments where their green or grey–green colour is often distinctive. Wehrlitic garnets classify predominantly in cluster groups 7 and 12 of Dawson and Stephens (1975), with minor occurrences in their cluster group 11. Here we follow the data compilations of Sobolev et al. (1973a,b) and Schulze (1993, 2003) to establish compositional limits for an updated and simplified wehrlitic garnet category labelled G12, as follows:

Cr_2O_3 [wt.%]: ≥ 1.0 to < 20.0
CA_INT [wt.%]: > 5.4
CaO [wt.%]: < 28.0
MgO [wt.%]: > 5.0

It is noted that crustal uvarovitic garnets commonly have extremely high CaO contents, generally well over 28 wt.% CaO (Schulze, 1993).

4.4. Low-Cr megacrysts (G1)

Garnets belonging to the low-Cr suite of megacryst minerals may occur in high relative abundance in certain kimberlites (e.g. Monastery), but Schulze (1987) notes their occurrence also in other mantle-derived magma types like alnoites (e.g. Solomon Islands) and in certain alkali basalts (e.g. Vitim, see Litasov, 2000). Megacrystic garnets are typically coarse-grained (2–10 cm) and significantly fractured, and due to their relatively Fe-rich and Ti-rich nature generally endure chemical weathering better than several other mantle-derived garnet types. Disaggregation of megacrysts in the secondary environment can hence release disproportionately abundant pathfinder minerals derived from kimberlite or kimberlite-like intrusives. Their isolation as a specific compositional group is therefore important from an exploration perspective, even though megacrystic garnets have no established association with diamond. Garnet megacrysts correspond very closely to cluster group 1 of Dawson and Stephens (1975) and have Cr, Ca and Mg-number characteristics overlapping those of garnets in pyroxenite and websterite xenoliths. However, the latter are generally less titaniferous at any given Mg-number (Fig. 5). Based predominantly on the compilations of megacryst data in Jakob (1977), Bell and Rossman (1992)

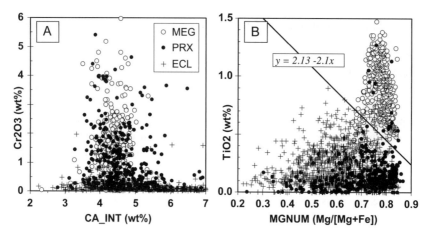

Fig. 5. (A, B) Diagrammatic summary of Cr, Ca, Ti, Mg and Fe compositional characteristics of garnets with moderate to low Cr_2O_3 content derived predominantly from kimberlite sources, but also including occurrences recorded in related rocks and alkali basalts. At any given Mg-number, megacrystic garnets (MEG, $n = 515$) generally have higher Ti content than garnets in eclogite xenoliths (ECL, $n = 1113$) and pyroxenite xenoliths (PRX, $n = 391$). The pyroxenite category includes 139 data for garnet websterite xenoliths.

and Schulze (1997, 2003), we characterise megacryst compositions as follows (see Fig. 5):

Cr_2O_3 [wt.%]: 0 to < 4.0
CA_INT [wt.%]: ≥ 3.375 to < 6.0
MGNUM: ≥ 0.65 to < 0.85
TiO_2 [wt.%]: ≥ 2.13 − 2.1*MGNUM
TiO_2 [wt.%]: < 4.0

Many investigators have shown that Ca–Fe–Ti melt-metasomatism drives garnet compositions in high-temperature mantle peridotites toward those of megacrystic garnets (e.g. Burgess and Harte, 1999 and references therein). A compositional overlap between megacrystic garnets and certain lower-Cr_2O_3, high-TiO_2 peridotitic garnets may thus occur, but we make no discrimination in our classification scheme because neither diamond nor graphite is associated with melt-metasomatic assemblages in peridotites. McCammon et al. (2001) have indicated that the absence of carbon may be a consequence of progressive oxidation during melt-metasomatism.

4.5. High-TiO$_2$ peridotitic (G11)

Classification runs conducted by Dawson and Stephens (1975, pp. 601–602) on high-TiO_2 garnets derived from "sheared" peridotites returned instances in their cluster groups 1, 2, 9, 10 and mostly 11,

depending slightly on the classification method used. We have adopted their group 11 as representative of high-TiO_2 peridotitic garnet compositions and describe them as follows.

Unlike G1 (i.e. G1 grains should be identified first and excluded as possible G11's):
Cr_2O_3 [wt.%]: ≥ 1.0 to < 20.0
CA_INT [wt.%]: ≥ 3.0
CaO [wt.%]: < 28.0
MGNUM: ≥ 0.65 to < 0.90
TiO_2 [wt.%]: ≥ 2.13 − 2.1*MGNUM
TiO_2 [wt.%]: < 4.

4.6. Pyroxenitic, websteritic and eclogitic ("G4" and "G5")

Dawson and Stephens (1975) did not classify garnets that occur in pyroxenite (and websterite) mantle xenoliths into a specific group, but included them within their G9 (dominantly lherzolitic) and G3 (dominantly eclogitic) categories. Pyroxenitic garnets were also left undifferentiated in a recent garnet classification scheme (Schulze, 2003). These moderate- to low-Cr garnets are important to diamond explorers due to a distinct association with diamond (e.g. Gurney et al., 1984; Aulbach et al., 2002), and as possible indicators of lithosphere destruction (see Pokhilenko et al., 1999). Pyroxenitic/websteritic garnets are easily differentiated from megacrysts by Mg-number and TiO_2 content (Fig. 5), but, like previous investigators, we also found significant compositional overlaps to occur with low-Cr peridotitic and eclogitic garnets. A suitable compromise was found by allowing two categories of pyroxenitic garnets.

Pyroxenitic garnets similar to, but richer in Fe than moderate- to low-Cr G9 garnets are designated "G5" in the updated scheme, although the original term referred to an Fe-rich eclogitic category. The updated G5 garnet category is defined by:

TiO_2 [wt.%]: $<2.13 - 2.1*MGNUM$
Cr_2O_3 [wt.%]: ≥ 1 to <4.0
CA_INT [wt.%]: ≥ 3.375 to <5.4
MGNUM: ≥ 0.3 to <0.7.

Pyroxenitic garnets lower in Cr than G9 garnets, but with compositions overlapping low-Ca eclogitic garnets are designated as group "G4". It is recognized that this group contains eclogitic, pyroxenitic and websteritic garnets and that the adopted nomenclature departs somewhat from the titaniferous ferroan eclogitic garnet category originally envisioned by Dawson and Stephens (1975). The updated G4 category is defined by:

TiO_2 [wt.%]: $<2.13 - 2.1*MGNUM$
Cr_2O_3 [wt.%]: <1.0
CaO [wt.%]: ≥ 2.0 to <6.0
MGNUM: ≥ 0.3 to <0.90.

4.7. Eclogitic (G3)

Small diamond-bearing eclogite xenoliths are known to have in-situ grades equivalent to 650 to 20,000 ct/ton (see Helmstaedt, 1993), and for this reason alone eclogitic garnets represent extremely important pathfinder minerals for diamond explorers. Eclogitic garnets are aluminous and show large variations in FeO, MgO and CaO, to the extent that Dawson and Stephens (1975) required five separate cluster groups to describe their compositional variation (their groups 3, 4, 5, 6 and 8). A compilation of garnet compositions in carbon-free and carbonaceous eclogites shows that carbon is not preferentially associated with eclogitic garnets of particular Fe–Mg–Ca compositions (Fig. 6). This implies that subdivision of eclogitic garnets on the basis of variable FeO, MgO or

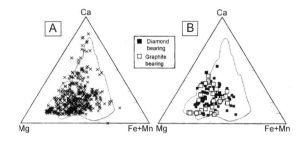

Fig. 6. (A, B) Mg–Ca–(Fe + Mn) diagrams illustrating the compositional range of garnet in (A) carbon-free eclogite xenoliths from kimberlites ($n = 687$), and (B) carbonaceous eclogite xenoliths from kimberlites ($n = 100$). Compositions of worldwide eclogitic garnets outlined in stipple (after Haggerty, 1995). Diagram is from Grütter and Quadling (1999).

CaO content, as in the garnet classification scheme of Schulze (2003), provides little advantage for the diamond explorer and we accordingly choose to define our eclogitic garnet category across a range of compositions, as outlined below. It is noted that these compositional limits also encompass the ranges observed for garnets in alkremite and certain lower crustal garnet granulite xenoliths. Our eclogitic G3 category has:

Cr_2O_3 [wt.%]: 0 to <1.0
CaO [wt.%]: ≥ 6 to <32.0
MGNUM: ≥ 0.17 to <0.86
TiO_2 [wt.%]: $<2.13 - 2.1*MGNUM$
TiO_2 [wt.%]: <2.0

4.8. Na_2O in G3, G4 and G5 garnets

Eclogitic garnet inclusions in diamond are known to commonly have $Na_2O > 0.07$ wt.% (Sobolev and Lavrent'ev, 1971; McCandless and Gurney, 1989), though this threshold provides incomplete discrimination from garnet compositions in graphite-bearing eclogite xenoliths (Grütter and Quadling, 1999). Further investigation of the phase-relations of carbon, garnet and sodic pyroxene may yield a basis for accurately constraining diamond-facies eclogitic, websteritic and pyroxenitic garnet compositions, thereby permitting the suffix "D" to be added to either of these garnet categories with high confidence. In the interim the $Na_2O > 0.07$ wt.% threshold noted in Gurney (1984) and documented further in Gurney et al. (1993, their Fig. 8) could be applied for this purpose, leading to garnet categories G3D, G4D and G5D within the framework of the current classification scheme. The G3D and G4D categories would constitute a replacement for the term "Group 1 eclogite" which originally referred to a coarse-grained eclogite texture (MacGregor and Carter, 1970), but now also has compositional connotations (McCandless and Gurney, 1989).

4.9. Unclassified (G0)

Kimberlites and related mantle-derived magmas occasionally contain xenocrystic garnets derived from uncommon, unusual or "polymict" mantle lithologies. The scheme proposed here makes no specific provision to classify such grains, instead leaving them to collect in an unclassified category labelled G0 by default. Manual inspection of this group may reveal their affinity to the groups defined above.

5. Summary and implementation

The current classification scheme is formulated to be as simple as possible, whilst also trying to address the multivariate nature of the classification problem and the diversity of chemical, physical and lithological environments in which mantle garnets and diamonds occur. The compositional fields for garnet categories outlined in this work are illustrated in terms of Cr_2O_3 and CaO contents in Fig. 7. Compositional overlaps have been resolved in order to keep the scheme robust, implying that certain simplifying choices have been made which reflect the needs of diamond exploration-ists, rather than those of mantle researchers (for which see Schulze, 2003). Thus harzburgitic (G10), lherzo-litic (G9) and wehrlitic (G12) garnet compositions are separated in the scheme by recognizing natural bounds in Ca-intercept values (e.g. Figs. 4 and 7), the latter being a continuous geometric function anchored to the well-known G10/G9 divide of Gurney (1984, see Fig. 3). Megacrystic (G1) and high-TiO_2 peridotitic (G11) garnets occur on the Ti-rich and Mg-rich side of all other garnet compositions (Fig. 5), and their classification presents the only computational directive of the scheme: G1 and G11 categories have to be assigned prior to any other. At low Cr_2O_3 content a compromise is made by excluding pyroxenitic (G5) garnets from

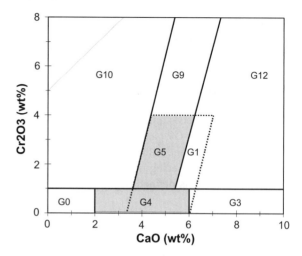

Fig. 7. G-number nomenclature of the classification scheme as viewed in a conventional Cr_2O_3 vs. CaO diagram. Megacryst group G1 (stippled parallelogram) does not actually overlap groups G3, G4, G5, G9 or G12 since it occurs at higher TiO_2 content (see Fig. 5B). Pyroxenitic categories G5 and G4 are indicated by fill pattern. Group G5 garnets are separated from G9 garnets by a Mg-number < 0.7 threshold. Note unclassified category G0 at low CaO and Cr_2O_3 content.

overlapping with lherzolitic (G9) garnets at Mg-num-bers > 0.7 and by defining a very low-Cr "pyroxenitic" (G4) group that also includes all low-CaO eclogitic garnet compositions. A further group of common eclogitic garnets (G3) extends across a range of Mg–Fe compositions to much higher CaO content (Fig. 7). An unclassified category (G0) completes the scheme.

It is recommended that implementation of the scheme sequentially tests an unknown grain for compositional compliance in the order G1–G11–G10–G9–G12–G5–G4–G3–G0, slightly different from the order in which they are defined above. A strong geochemical and petrologic association with diamond is indicated by adding a "D" suffix, currently only applicable to G10, G5, G4 and G3 garnet compositions.

The scheme has been applied to the garnet compositions used in the multivariate studies of Dawson and Stephens (1975), Danchin and Wyatt (1979) and to the databases compiled for this investigation, with results summarised in Tables 1–4. Known petrogenetic and lithological associations are given in the left-hand column of each table and garnet classifications according to the current scheme are aligned in similar petrogenetic associations along the upper row. Moderate to high degrees of correlation are evident as

Table 1
Dawson and Stephens (1975) database

N=398		G10	G9	G11	G1	G5	G4	G3	G12	G0
	n=	83	77	20	45	2	37	86	26	22
DINCL	53	64	–	–	4	–	–	9	17	–	6
HZB	2	–	100	–	–	–	–	–	–	–
LHZ	34	–	71	15	6	–	3	3	3	–
PRX	13	–	39	–	15	–	39	8	–	–
ECL	99	–	1	–	–	–	20	73	–	6
WEH	2	–	–	–	–	–	–	–	100	–
KCONC	180	26	19	7	22	1	3	2	13	7
UNK	15	13	67	–	7	7	–	–	–	7

Classification of garnet compositions by the current scheme. Percentages listed are rounded and calculated as a proportion of the known garnet category given in the left-hand column. Category abbreviations are ALK = Alkremite; DINCL = Inclusion in diamond; DIXEN = Diamondiferous xenolith; ECL = Eclogite; GRAN = Lower crustal granulite; HZB = Harzburgite; KCONC = Concentrate from kimberlite; LHZ = Lherzolite; LHZ_DEF = Deformed lherzolite; MEG = Megacryst; MICXEN = Microxenolith; OTH = Other; PER = Unspecified peridotite; PRX = Pyroxenite; UNK = Unknown; WEB = Websterite; WEH = Wehrlite. The garnet compositions of Dawson and Stephens (1975) and cross-tabulated raw counts for Tables 1–4 are available as digital supplementary data in the online version.

high-valued vectors running from top left to bottom right in each of Tables 1–4, implying that upper mantle garnet compositions are usefully separated and categorized by the current classification scheme. Table 3 shows good correlations to exist for harzburgitic, lherzolitic, unspecified peridotitic, megacrystic

Table 2
Danchin and Wyatt (1979) database

N=1777		G10	G9	G11	G1	G5	G4	G3	G12	G0
	n=	257	455	137	237	8	289	281	86	27
DINCL	191	57	4	3	2	–	9	24	–	2
DIXEN	30	80	7	13	–	–	–	–	–	–
HZB	57	39	30	32	–	–	–	–	–	–
LHZ	168	2	81	13	1	1	1	–	1	–
LHZ_DEF	101	2	33	35	31	–	–	–	–	–
PER	27	15	67	4	11	–	–	–	4	–
MEG	39	–	3	–	67	–	18	–	10	3
PRX	80	18	30	14	13	–	21	4	1	–
WEB	33	–	46	–	3	3	46	–	–	3
ECL	332	–	3	0	3	–	33	59	0	1
ALK	18	–	–	–	–	–	22	72	–	6
WEH	3	–	–	33	–	–	–	–	67	–
KCONC	698	11	27	5	22	1	17	3	11	2

Layout and abbreviations as in Table 1.

Table 3
Xenolith database compiled for this investigation

N=4532		G10	G9	G11	G1	G5	G4	G3	G12	G0
	n=	350	1495	380	601	20	698	872	83	33
HZB	284	97	–	3	–	–	–	–	–	0
LHZ	1378	2	79	13	3	–	2	–	2	0
LHZ_DEF	316	–	33	42	25	–	–	–	1	–
PER	240	14	61	16	3	–	5	–	2	0
MEG	515	–	0	3	88	0	6	1	0	1
PRX	252	0	24	–	2	7	55	8	2	1
WEB	139	–	31	1	2	1	62	–	2	1
ECL	1113	–	1	–	1	–	32	64	1	1
ALK	37	–	8	–	–	–	8	60	3	22
WEH	37	–	8	14	–	–	–	–	78	–
GRAN	153	–	–	–	–	–	26	72	–	2
MICXEN	68	32	57	2	4	–	–	–	4	–

Layout and abbreviations as in Table 1.

and wehrlitic garnets, and similar correlations are also evident in Tables 1, 2 and 4. The re-defined and newly introduced low-Cr "pyroxenitic/websteritic" G4 category shows acceptably low overlap with the low-Cr G3 "eclogitic" category, but separation of G5 Cr-bearing pyroxenitic and websteritic garnets from G9 lherzolitic garnets remains a challenge. Garnets in lower crustal granulite and in alkremite xenoliths classify predominantly as eclogitic, as expected. The current scheme has a low overall incidence of unclassified (G0) garnets, a noteworthy feature given the significantly expanded database compiled for this investigation (Tables 3 and 4).

Updated statistics for garnets included in diamond are given in Table 4. The data set is dominated by peridotitic inclusions and the ratio G10/(G9 + G10) is 82%, still essentially the same as that calculated by Gurney (1984), even though the currently applied definition for peridotite has a lower Cr_2O_3 threshold

Table 4
Inclusions in diamonds compiled for this investigation

N=637		G10	G9	G11	G1	G5	G4	G3	G12	G0
	n=	271	60	16	12	1	63	205	1	8
D suffix		n=494	n=255	–	–	–	n=1	n=50	n=188	–	–
PER	348	78	16	5	1	–	–	–	–	1
WEB	13	–	31	–	15	8	15	–	8	23
ECL	273	–	–	–	3	–	22	75	–	–
OTH	3	–	–	–	–	–	33	–	–	67

Layout and abbreviations as in Table 1.

(1.0 instead of 2.0 wt.% Cr_2O_3). The familiar eclogitic (G3) association with diamond is clearly evident and it is noted that the newly defined G4 category contains more diamond-inclusions than the G9 category, even though the latter garnets are much more abundant in the upper mantle (Table 3). Following the methodology and thresholds presented in the updated scheme, a total of 494 of the 637 garnets in our diamond-inclusion data set are assigned the diamond-facies "D" suffix (i.e. 78%). In particular, the "D" suffix is found to be applicable in 255 of 271 G10 (94%), 1 of 1 G5, 50 of 63 G4 (79%) and 188 of 205 G3 (92%) compositions (Table 4). These statistics should inspire confidence in the use of the G10D, G5D, G4D and G3D categories by diamond explorers.

6. Conclusion

The classification scheme outlined above utilises a few relatively simple criteria to categorise the compositions of garnet grains that may be associated with diamond-bearing intrusives. The scheme is reliant only on the major and minor element compositional data that industry-standard electron microprobe analyses can provide and has superior accuracy compared to historical or contemporary classification schemes because it specifically incorporates both geochemical and petrological constraints that appear to determine the occurrence of peridotitic, eclogitic and "websteritic" diamonds in the lithospheric upper mantle. Improvements in Ca-in-garnet and Na-in-garnet thermobarometry are required to further improve the accuracy of the current scheme.

Acknowledgements

The authors acknowledge the support and encouragement of Mineral Services and the De Beers Group of Companies, particularly during the latter stages of this project. HSG publishes with permission of De Beers Consolidated Mines and acknowledges influential discussions with Gerhard Brey, Dave Apter, Bruce Wyatt and Peter Williamson. We gratefully acknowledge the xenolith and diamond-inclusion mineral composition data published or compiled and made available by Barry Dawson, Bruce Wyatt, Bruce Jago, Dan Schulze, the Kimberlite Research Group at the University of Cape Town and the authors listed in the three cited data appendices (see Section 2.1. Data Sources). This work would have been impossible without their collective effort through many years. Journal reviews by Bruce Jago and Gerhard Brey improved the overall clarity of the manuscript.

Appendix A. Data sources

Benoit and Mercier (1986)
Bloomer and Nixon (1973)
Boyd et al. (2004)
Boyd and Danchin (1980)
Boyd and Nixon (1978)
Boyd and Nixon (1979)
Boyd et al. (1993)
Boyd et al. (1997)
Burgess and Harte (1999)
Carswell et al. (1979)
Cox et al. (1973)
Danchin and Boyd (1976)
Daniels et al. (1995)
Dawson et al. (1978)
Dawson et al. (1980)
Delaney et al. (1979)
Delaney et al. (1980)
Eggler et al. (1987)
Ehrenberg (1978)
Ehrenberg (1982)
Exley et al. (1982)
Field and Haggerty (1994)
Field et al. (1989)
Franz et al. (1996)
Franz et al. (1997)
Griffin et al. (1989)
Griffin et al. (1993)
Hall (1991)
Hervig et al. (1986)
Ionov et al. (1993)
Kopylova et al. (2000)
Logvinova and Sobolev (1995)
Luth et al. (1990)
MacGregor (1979)
McCallum and Eggler (1976)
McGee and Hearne (1989)
Menzies (2001)

Mitchell (1984)
Mofokeng (1998)
Nixon (1987)
Nixon and Boyd (1973)
Nowicki (1990)
Pearson et al. (1990)
Pearson et al. (1994)
Pearson et al. (1999)
Pokhilenko et al. (1977)
Pokhilenko et al. (1991)
Pokhilenko et al. (1993)
Reid et al. (1975)
Robey (1981)
Rudnick et al. (1994)
Schulze (1995)
Schulze (1996)
Schulze et al. (1997)
Schulze et al. (2000)
Shee (1994)
Shee et al. (1989)
Simon et al. (in press)
Skinner (1989)
Smith (1999)
Smith and Boyd (1992)
Smith and Wilson (1985)
Smith et al. (1991)
Sobolev et al. (1973a,b)
Sobolev et al. (1984)
Sobolev et al. (1997)
Stachel et al. (2000)
Stiefenhofer et al. (1999)
van der Westhuizen (1992)
Viljoen et al. (1994)

References

Akella, J., 1976. Garnet pyroxene equilibria in the system CaSiO₃–MgSiO₃–Al₂O₃ and in a natural mineral mixture. Am. Mineral. 61, 589–598.

Armstrong, J.P., 2001. Kimberlite Indicator Mineral Chemistry Database (KIMC): A preliminary digital compilation of Kimberlite Indicator Mineral Chemistry extracted from publically available assessment filings; Slave Craton and environs, Northwest Territories and Nunavut, Canada. DIAND NWT Geology Division, DIAND EGS Open Report 2001-02 (CD-ROM).

Aulbach, S., Stachel, T., Viljoen, K.S., Brey, G.P., Harris, J.W., 2002. Eclogitic and websteritic diamond sources beneath the Limpopo Belt—is slab-melting the link? Contrib. Mineral. Petrol. 143, 56–70.

Bell, D.R., Rossman, G.R., 1992. The distribution of hydroxyl in garnets from the subcontinental mantle of southern Africa. Contrib. Mineral. Petrol. 111, 161–178.

Benoit, V., Mercier, J.-C.C., 1986. Les enclaves ultramafiques du volcanisme alcalin Tertiaire du centre du plateau du Colorado: implications tectoniques. Bull. Soc. Geol. Fr. 6, 1015–1023.

Bloomer, A.G., Nixon, P.H., 1973. The geology of Letseng-la-terae kimberlite pipes. In: Nixon, P.H. (Ed.), Lesotho Kimberlites. Lesotho National Development, Maseru, pp. 20–36.

Boyd, F.R., 1970. Garnet peridotites and the system CaSiO₃–MgSiO₃–Al₂O₃. Spec. Pap.-Miner. Soc. Am. 3, 63–75.

Boyd, F.R., Pearson, D.G., Hoal, K.O., Hoal, B.G., Nixon, P.H., Kingston, M.J., Mertman, S.A., 2004. Garnet lherzolites from Louwrensia, Namibia: Bulk compositions and P/T relations. Lithos 77, 573–592 (this volume).

Boyd, F.R., Danchin, R.V., 1980. Lherzolites, eclogites and megacrysts from some kimberlites of Angola. Am. J. Sci. 280-A, 528–549.

Boyd, F.R., Gurney, J.J., 1982. Low-calcium garnets: keys to craton structure and diamond crystallization. Year B.-Carnegie Inst. Wash. 81, 261–267.

Boyd, F.R., Nixon, P.H., 1978. Ultramafic nodules from the Kimberley pipes, South Africa. Geochim. Cosmochim. Acta 42, 1367–1382.

Boyd, F.R., Nixon, P.H., 1979. Garnet lherzolite xenoliths from the kimberlites of East Griqualand, South Africa. Year B.-Carnegie Inst. Wash. 78, 488–492.

Boyd, F.R., Pearson, D.G., Nixon, P.H., Mertzman, S.A., 1993. Low-calcium garnet harzburgites from Southern Africa: their relations to craton structure and diamond crystallization. Contrib. Mineral. Petrol. 113, 352–366.

Boyd, F.R., Pokhilenko, N.P., Pearson, D.G., Mertzman, S.A., Sobolev, N.V., Finger, L.W., 1997. Composition of the Siberian cratonic mantle: evidence from Udachnaya peridotite xenoliths. Contrib. Mineral. Petrol. 128, 228–246.

Brenker, F.E., Brey, G.P., 1997. Reconstruction of the exhumation path of the Alpe Arami garnet–peridotite body from depths exceeding 160 km. J. Metamorph. Geol. 15, 581–592.

Brey, G.P., Nickel, K.G., Kogarko, L., 1986. Garnet–pyroxene equilibria in the system CaO–MgO–Al₂O₃–SiO₂ (CMAS): prospects for simplified ("T-independent") lherzolite barometry and an eclogite-barometer. Contrib. Mineral. Petrol. 92, 448–455.

Brey, G., Köhler, T., Nickel, K.G., 1990. Geothermobarometry in four-phase lherzolites: I. Experimental results from 10 to 60 kb. J. Petrol. 31, 1313–1352.

Burgess, S.R., Harte, B., 1999. Tracing lithosphere evolution through the analysis of heterogeneous G9/G10 garnets in peridotite xenoliths: I. Major Element Chemistry. In: Gurney, J.J., Gurney, J.L., Pascoe, M.D., Richardson, S.H. (Eds.), J.B. Dawson Volume Proc. 7th Int. Kimb. Conf. Red Roof Design, Cape Town, pp. 66–80.

Carswell, D.A., Clarke, D.B., Mitchell, R.H., 1979. The petrology and geochemistry of ultramafic nodules from Pipe 200, Northern Lesotho. In: Boyd, F.R., Meyer, H.O.A. (Eds.), The Mantle Sample: Inclusions in Kimberlites and Other Volcanics. Proc.

2nd Int. Kimb. Conf., vol. 2. Amer. Geophys. Union, Washington, DC, pp. 127–144.

Cox, K.G., Gurney, J.J., Harte, B., 1973. Xenoliths from the Matsuko pipe. In: Nixon, P.H. (Ed.), Lesotho Kimberlites. Lesotho Nat. Develop, Maseru, pp. 76–100.

Danchin, R.V., Boyd, F.R., 1976. Ultramafic nodules from the Premier kimberlite pipe, South Africa. Year B.-Carnegie Inst. Wash. 75, 531–538.

Danchin, R.V., Wyatt, B.A., 1979. Statistical cluster analysis of garnets from kimberlites and their xenoliths. Ext. Abstr. 2nd Kimb. Symp., Cambridge, pp. 22–27.

Daniels, L.R.M., Richardson, S.H., Menzies, A.H., de Bruin, D., Gurney, J.J., 1995. Diamondiferous garnet macrocrysts in the Newlands kimberlite, South Africa—Rosetta stones from the Kaapvaal craton keel. Extended Abs. 6th Ink. Kimb. Conf., Novosibirsk, pp. 121–123.

Dawson, J.B., Stephens, W.E., 1975. Statistical classification of garnets from kimberlite and associated xenoliths. J. Geol. 83, 589–607.

Dawson, J.B., Smith, J.V., Delaney, J.S., 1978. Multiple spinel–garnet peridotite transitions in upper mantle: evidence from a harzburgite xenolith. Nature 273, 741–743.

Dawson, J.B., Smith, J.V., Hervig, R.L., 1980. Heterogeneity in upper-mantle lherzolites and harzburgites. Philos. Trans. R. Soc. Lond., A 297, 323–331.

Delaney, J.S., Smith, J.V., Dawson, J.B., Nixon, P.H., 1979. Manganese thermometer for mantle peridotites. Contrib. Mineral. Petrol. 71, 157–169.

Delaney, J.S., Smith, J.V., Carswell, D.A., Dawson, J.B., 1980. Chemistry of micas from kimberlites and xenoliths: II. Primary- and secondary-textured micas from peridotite xenoliths. Geochim. Cosmochim. Acta 44, 857–872.

Eggler, D.H., McCallum, M.E., Kirkley, M.B., 1987. Kimberlite-transported nodules from Colorado–Wyoming: a record of enrichment of shallow portions of an infertile lithosphere. Spec. Pap.-Geol. Soc. Am. 215, 77–89.

Ehrenberg, S.N., 1978. Petrology of potassic volcanic rocks and ultramafic xenoliths from the Navajo volcanic field, New Mexico and Arizona, Unpubl. PhD thesis. Univ. California, Los Angeles. 259 pp.

Ehrenberg, S.N., 1982. Petrogenesis of garnet lherzolite and megacrystalline nodules from The Thumb, Navajo volcanic field. J. Petrol. 23, 507–547.

Exley, R.A., Smith, J.V., Hervig, R.L., 1982. Cr-rich spinel and garnet in two peridotite xenoliths from the Frank Smith mine South Africa: significance of Al and Cr distribution between spinel and garnet. Min. Mag. 45, 129–134.

Field, S.W., Haggerty, S.E., 1994. Symplectites in upper mantle peridotites: development and implications for the growth of subsolidus garnet, pyroxene and spinel. Contrib. Mineral. Petrol. 118, 138–156.

Field, S.W., Haggerty, S.E., Erlank, A.J., 1989. Subcontinental metasomatism in the region of Jagersfontein, South Africa. Kimberlites and related rocks: Their mantle/crust setting, diamonds and diamond exploration. Proc. 4th Int. Kimb. Conf. Vol. 2. Spec. Publ.-Geol. Soc. Aust., vol. 14, pp. 771–783.

Fipke, C.E., Gurney, J.J., Moore, R.O., 1995. Diamond explora-

tion techniques emphasising indicator mineral geochemistry and Canadian examples. Bull.-Geol. Surv. Can., vol. 423. 86 pp.

Franz, L., Brey, G.P., Okrusch, M., 1996. Steady state geotherm, thermal disturbances and tectonic development of the lower lithosphere underneath the Gibeon kimberlite province, Namibia. Contrib. Mineral. Petrol. 126, 181–198.

Franz, L., Brey, G.P., Okrusch, M., 1997. Metasomatic reequilibration of mantle xenoliths from the Gibeon kimberlite province (Namibia). Russ. Geol. Geophys. 38, 261–276.

Glaser, S.M., Foley, S.F., Gunther, D., 1999. Trace element compositions of minerals in garnet and spinel peridotite xenoliths from the Vitim volcanic field, Transbaikalia, eastern Siberia. Lithos 48, 263–285.

Griffin, W.L., Ryan, C.G., 1993. Trace elements in garnets and chromites: evaluation of diamond exploration targets. In: Diamonds: Exploration, Sampling and Evaluation. Short Course Proceedings, Prospectors and Developers Assoc. Canada, Toronto, pp. 185–212.

Griffin, W.L., Cousens, D.R., Ryan, C.G., Sie, S.H., Suter, G.F., 1989. Ni in chrome pyrope garnets: a new geothermometer. Contrib. Mineral. Petrol. 103, 199–202.

Griffin, W.L., Sobolev, N.V., Ryan, C.G., Pokhilenko, N.P., Win, T.T., Yefimova, E.S., 1993. Trace elements in garnets and chromites: diamond formation in the Siberian lithosphere. Lithos 29, 235–256.

Griffin, W.L., Fisher, N.I., Friedman, J.H., O'Reilly, S.Y., Ryan, C.G., 2002. Cr-pyrope garnets in the lithospheric mantle: 2. Compositional populations and their distribution in time and space. Geochem. Geophys. Geosyst. 3, 1073, doi:10.1029/2002GC000298.

Grütter, H.S., 1994. Spinel–garnet–carbon phase relations in coarse Kaapvaal-type peridotites and implications for garnet–orthopyroxene equilibration. Ext. Abs. Int. Symp. Phys. Chem. Upper Mantle, Sao Paulo, pp. 5–7.

Grütter, H.S., Moore, R., 2003. Pyroxene geotherms revisited—an empirical approach based on Canadian xenoliths. Ext. Abstr. 8th Int. Kimb. Conf., Victoria, 272 (CD-ROM).

Grütter, H.S., Quadling, K.E., 1999. Can sodium in garnet be used to monitor eclogitic diamond potential? In: Gurney, J.J., Gurney, M.D., Pascoe, M.D., Richardson, S.H. (Eds.), J.B. Dawson Volume Proc. 7th Int. Kimb. Conf. Red Roof Design, Cape Town, pp. 314–320.

Grütter, H.S., Sweeney, R.J., 2000. Tests and constraints on single-grain Cr-pyrope barometer models: some initial results. Ext. Abstr. GAC/MAC Annual Joint Meeting, Calgary (CD-ROM, GeoCanada 2000).

Grütter, H.S., Winter, F., 1997. Pressure, temperature and composition effects on the position of the garnet lherzolite trend. Anglo American Research Laboratories Report KR97/0504. 35 pp.

Grütter, H.S., Apter, D.B., Kong, J., 1999. Crust–mantle coupling: evidence from mantle-derived xenocrystic garnets. In: Gurney, J.J., Gurney, J.L., Pascoe, M.D., Richardson, S.H. (Eds.), J.B. Dawson Volume Proc. 7th Int. Kimb. Conf. Red Roof Design, Cape Town, pp. 307–313.

Gurney, J.J., 1984. A correlation between garnets and diamonds. In:

Glover, J.E., Harris, P.G. (Eds.), Kimberlite occurrence and origins: a Basis for Conceptual Models in Exploration. Geology Department and University Extension, University of Western Australia, Publication 8 143–166.

Gurney, J.J., Switzer, G.S., 1973. The discovery of garnets closely related to diamonds in the Finsch pipe, South Africa. Contrib. Mineral. Petrol. 39, 103–116.

Gurney, J.J., Zweistra, P., 1995. The interpretation of the major element compositions of mantle minerals in diamond exploration. J. Geochem. Explor. 53, 293–309.

Gurney, J.J., Harris, J.W., Rickard, R.S., 1984. Silicate and oxide inclusions in diamonds from the Orapa mine, Botswana. In: Kornprobst, J. (Ed.), Kimberlites II: The mantle and crust–mantle relationships. Proc. 3rd Int. Kimb. Conf., vol. 2. Elsevier, Amsterdam, pp. 3–9.

Gurney, J.J., Helmstaedt, H., Moore, R.O., 1993. A review of the use and application of mantle mineral geochemistry in diamond exploration. Pure Appl. Chem. 65, 2423–2442.

Haggerty, S.E., 1995. Upper mantle mineralogy. J. Geodyn. 20, 331–364.

Hall, D.C., 1991. A petrological investigation of the Cross kimberlite occurrence, Southeastern British Columbia, Canada. Unpubl. PhD thesis. Queen's University, Kingston, Ontario, Canada, 533 pp.

Hatton, C.J., Gurney, J.J., 1987. Roberts Victor eclogites and their relation to the mantle. In: Nixon, P.H. (Ed.), Mantle Xenoliths. Wiley, London, pp. 453–463.

Helmstaedt, H.H., 1993. Natural diamond occurrences and tectonic setting of "primary" diamond deposits. In: Diamonds: Exploration, Sampling and Evaluation. Short Course Proceedings, Prospectors and Developers Assoc. Canada, Toronto, pp. 3–72.

Hervig, R.L., Smith, J.V., Dawson, J.B., 1986. Lherzolite xenoliths in kimberlites and basalts: petrogenetic and crystallographic significance of some minor and trace elements in olivine, pyroxenes, garnet and spinel. Trans. R. Soc. Edinb. Earth Sci. 77, 181–201.

Ionov, D.A., Ashchepkov, I.V., Stosch, H.-G., Witt-Eickschen, G., Seck, H.A., 1993. Garnet peridotite xenoliths from the Vitim volcanic field, Baikal region: the nature of the garnet–spinel peridotite transition zone in the continental mantle. J. Petrol. 34, 1141–1175.

Jago, B.C., Mitchell, R.H., 1989. A new garnet classification technique: divisive cluster analysis applied to garnet populations from Somerset Island kimberlites. In: Ross, J. (Ed.), Kimberlites and Related Rocks, Vol. 1: Their Composition, Origin and Emplacement. Spec. Pub.-Geol. Soc. Aust., vol. 14, pp. 297–310.

Jakob, W.R.O., 1977. Geochemical aspects of the megacryst suite from the Monastery kimberlite pipe. Unpubl. Report, Atomic Energy Board, Pelindaba, 81 pp.

Kennedy, C.S., Kennedy, G.C., 1976. The equilibrium boundary between graphite and diamond. J. Geophys. Res. 81, 2467–2470.

Kjarsgaard, B.A., Levinson, A.A., 2002. Diamonds in Canada. Gems. Gemol. 38, 208–238.

Kopylova, M.G., Russell, J.K., Stanley, C., Cookenboo, H., 2000. Garnet from Cr- and Ca-saturated mantle: implications for diamond exploration. J. Geochem. Explor. 68, 183–199.

Kushiro, I., Syono, Y., Akimoto, S., 1967. Effect of pressure on garnet–pyroxene equilibrium in the system $MgSiO_3$–$CaSiO_3$–Al_2O_3. Earth Planet. Sci. Lett. 2, 460–464.

Lee, J.E., 1993. Indicator mineral techniques in a diamond exploration program at Kokong, Botswana. In: Diamonds: Exploration, Sampling and Evaluation. Short Course Proceedings, Prospectors and Developers Assoc. Canada, Toronto, pp. 213–235.

Litasov, K.D., 2000. Petrology of a megacryst assemblage in Pliocene–Pleistocene basanites from the Vitim Plateau (Transbaikalie). Volc. Seis. 22, 35–53.

Logvinova, A.M., Sobolev, N.V., 1995. Morphology and composition of mineral inclusions in chromite macrocrysts from kimberlites and lamproites. Ext. Abstr. 6th Int. Kimb. Conf., Navosibirsk, pp. 331–332.

Luth, R.W., Virgo, D., Boyd, F.R., Wood, B.J., 1990. Ferric iron in mantle-derived garnets: implications for thermobarometry and for the oxidation state of the mantle. Contrib. Mineral. Petrol. 104, 56–72.

MacGregor, I.D., 1979. Mafic and ultramafic xenoliths from the Kao kimberlite pipe. In: Boyd, F.R., Meyer, H.O.A. (Eds.), The Mantle Sample: inclusions in Kimberlites and Other Volcanics. Proc. 2nd Int. Kimb. Conf. Am. Geophys. Union, Washington, DC, pp. 156–172.

MacGregor, I.D., Carter, J.L., 1970. The chemistry of clinopyroxenes and garnets of eclogite and peridotite xenoliths from the Roberts Victor mine, South Africa. Phys. Earth Planet. Inter. 3, 391–397.

Malinovsky, I.Y., Doroshev, A.M., 1977. Evaluation of P–T conditions of diamond formation with reference to chrome-bearing garnet stability. Ext. Abs. Second Internat. Kimb. Conf., October 3–7, Santa Fe, New Mexico. Unpaginated.

McCallum, M.E., Eggler, D.H., 1976. Diamonds in an upper mantle peridotite nodule from kimberlite in Southern Wyoming. Science 192, 253–256.

McCammon, C.A., Griffin, W.L., Shee, S.R., O'Neill, H.S.C., 2001. Oxidation during metasomatism in ultramafic xenoliths from the Wesselton kimberlite, South Africa: implications for the survival of diamond. Contrib. Mineral. Petrol. 141, 287–296.

McCandless, T.E., Gurney, J.J., 1989. Sodium in garnet and potassium in clinopyroxene: criteria for classifying mantle eclogites. In: Ross, J. (Ed.), Kimberlites and Related Rocks. Geological Society of Australia Special Publication, vol. 14/2, pp. 827–832.

McGee, E.S., Hearne, B.C., 1989. Primary and secondary mineralogy of carbonated peridotites from the Macdougal Springs diatreme. Spec. Publ. - Geol. Soc. Aust. 14, 725–734.

Menzies, A.H., 2001. A detailed investigation into diamond-bearing xenoliths from Newlands kimberlite, South Africa. Unpubl. PhD thesis, Univ. of Cape Town.

Mitchell, R.H., 1984. Garnet lherzolites from the Hanaus-I and Louwrensia kimberlites of Namibia. Contrib. Mineral. Petrol. 86, 178–188.

Mofokeng, S.W., 1998. A comparison of the nickel and the conventional geothermometers with respect to the Jagersfontein and the Matsoku kimberlite peridotite xenoliths. Unpubl. MSc thesis, Univ. Cape Town, 124 pp.

Nickel, K.G., Green, D.H., 1985. Empirical geothermobarometry for garnet peridotites and implications for the nature of the lithosphere, kimberlites and diamonds. Earth Planet. Sci. Lett. 73, 158–170.

Nimis, P., Taylor, W.R., 2000. Single clinopyroxene thermobarometry for garnet peridotites: Part 1. Calibration and testing of a Cr-in-Cpx barometer and an enstatite-in-Cpx thermometer. Contrib. Mineral. Petrol. 139, 541–554.

Nixon, P.H., 1987. Kimberlitic xenoliths and their cratonic setting. In: Nixon, P.H. (Ed.), Mantle Xenoliths. Wiley, London, pp. 215–239.

Nixon, P.H., Boyd, F.R., 1973. Petrogenesis of the granular and sheared ultrabasic nodules suite in kimberlites. In: Nixon, P.H. (Ed.), Lesotho Kimberlites. Lesotho National Development, Maseru, pp. 48–56.

Nixon, P.H., van Calsteren, P.W.C., Boyd, F.R., Hawkesworth, C.J., 1987. Harzburgites with garnets of diamond facies from southern African kimberlites. In: Nixon, P.H. (Ed.), Mantle Xenoliths. Wiley, London, pp. 523–533.

Nowicki, T.E., 1990. Mantle xenoliths from the Abrahamskraal kimberlite: a craton-margin geotherm. Unpubl. MSc Thesis, Rhodes Univ., Grahamstown, 113 pp.

O'Neill, H.S.C., Wood, B.J., 1979. An experimental study of Fe–Mg partitioning between garnet and olivine, and its calibration as a geothermometer. Contrib. Mineral. Petrol. 70, 59–70.

Pearson, D.G., Boyd, F.R., Nixon, P.H., 1990. Graphite-bearing mantle xenoliths from the Kaapvaal craton: implications for graphite and diamond genesis. Ann. Rep. Dir. Geophys. Lab. Carn. Inst. Wash., pp. 11–19.

Pearson, D.G., Boyd, F.R., Haggerty, S.E., Pasteris, J.D., Field, S.W., Nixon, P.H., Pokhilenko, N.P., 1994. The characterisation and origin of graphite in cratonic lithospheric mantle: a petrological carbon isotope and Raman spectroscopic study. Contrib. Mineral. Petrol. 115, 449–466.

Pearson, N.J., Griffin, W.L., Doyle, B.J., O'Reilly, S.Y., Van Achterbergh, E., Kivi, K., 1999. Xenoliths from kimberlite pipes of the Lac de Gras area, Slave Craton, Canada. In: Gurney, J.J., Gurney, J.L., Pascoe, M.D., Richardson, S.H. (Eds.), J.B. Dawson Volume Proc. 7th Int. Kimb. Conf. Red Roof Design, Cape Town, pp. 644–658.

Pokhilenko, N.P., Sobolev, N.V., Lavrent'ev, Y.-G., 1977. Xenoliths of diamondiferous ultramafic rocks from Yakutian kimberlites. Extended Abstracts IInd Int. Kimberlite Conf., Santa Fe. Unpaginated.

Pokhilenko, N.P., Pearson, D.G., Boyd, F.R., Sobolev, N.V., 1991. Megacrystalline dunites and peridotites: hosts for Siberian diamonds. Ann. Rep. Dir. Geophys. Lab. Carn. Inst. Wash., pp. 11–18.

Pokhilenko, N.P, Sobolev, N.V., Boyd, F.R., Pearson, D.G., Shimuzu, N., 1993. Megacrystalline pyrope peridotites in the lithosphere of the Siberian platform: mineralogy, geochemical peculiarities and the problem of their origin. Russ. J. Geol. Geophys. 34, 56–67.

Pokhilenko, N.P., Sobolev, N.V., Kuligin, S.S., Shimuzu, N., 1999. Peculiarities of distribution of pyroxenite paragenesis garnets in Yakutian kimberlites and some aspects of the evolution of the Siberian craton lithospheric mantle. In: Gurney, J.J., Gurney, J.L., Pascoe, M.D., Richardson, S.H. (Eds.), P.H. Nixon Volume Proc. 7th Int. Kimb. Conf. Red Roof Design, Cape Town, pp. 689–698.

Pollack, H.N., Chapman, D.S., 1977. On the regional variation of heat flow, geotherms, and lithospheric thickness. Tectonophysics 38, 279–296.

Reid, A.M., Donaldson, C.H., Brown, R.W., Ridley, W.I., Dawson, J.B., 1975. Mineral chemistry of peridotite xenoliths from Lashaine volcano, Tanzania. Phys. Chem. Earth 9, 525–543.

Robey, J.V.R., 1981. Kimberlites of the Central Cape province, R.S.A. Unpubl. PhD thesis. University of Cape Town, 261 pp.

Rudnick, R.L., McDonough, W.F., Orpin, A., 1994. Northern Tanzanian peridotite xenoliths: a comparison with Kaapvaal peridotites and inferences on metasomatic interactions. In: Meyer, H.O.A., Leonardos, O.H. (Eds.), Kimberlites, Related Rocks and Mantle Xenoliths, Proc. 5th Int. Kimb. Conf.,Spec. Publ.-CPRM, vol. 1A/94, pp. 336–353.

Ryan, C.G., Griffin, W.L., Pearson, N.J., 1996. Garnet geotherms: pressure–temperature data from Cr-pyrope garnet xenocrysts in volcanic rocks. J. Geophys. Res. 101, 5611–5625.

Schulze, D.J., 1987. Megacrysts from alkaline volcanic rocks. In: Nixon, P.H. (Ed.), Mantle Xenoliths. Wiley, London, pp. 433–451.

Schulze, D.J., 1993. Green garnets from South African kimberlites and their relationship to wehrlites and crustal uvarovites. Spec. Publ. - Geol. Soc. Aust. 14, 820–826.

Schulze, D.J., 1995. Low-Ca garnet harzburgites from Kimberley, South Africa: abundance and bearing on the structure and evolution of the lithosphere. J. Geophys. Res. 100, 12513–12526.

Schulze, D.J., 1996. Chromite macrocrysts from southern African kimberlites: mantle xenolith sources and post-diamond re-equilibration. In: Kogbe, C.A. (Ed.), Diamond Deposits in Africa. Spec. Issue-Africa Geosci. Rev., vol 3, pp. 203–216.

Schulze, D.J., 1997. The significance of eclogite and Cr-poor megacryst garnets in diamond exploration. Explor. Min. Geol. 6, 349–366.

Schulze, D.J., 2003. A classification scheme for mantle-derived garnet in kimberlite: a tool for investigating the mantle and exploring for diamonds. Lithos 71, 195–213.

Schulze, D.J., Valley, J.W., Viljoen, K.S., Stiefenhofer, J., Spicuzza, M., 1997. Carbon isotope composition of graphite in mantle eclogites. J. Geol. 105, 379–386.

Schulze, D.J., Valley, J.W., Spicuzza, K.J., 2000. Coesite eclogites from the Roberts Victor kimberlite, South Africa. Lithos 54, 23–32.

Shee, S.R., 1994. Unpublished Compositional Data for Grain Mounts of Peridotite Xenoliths from Wesselton Mine.

Shee, S.R., Bristow, J.W., Bell, D.R., Smith, C.B., Allsopp, H.L., Shee, P.B., 1989. The petrology of kimberlites, related rocks and associated mantle xenoliths from the Kuruman province, South Africa. In: Ross, N. (Ed.), Kimberlites and Related Rocks. Geol. Soc. Aust. Spec. Publ., pp. 60–82.

Simon, N.S.C., Irvine, G.J., Davies, G.R., Pearson, D.G., Carlson, R.W., 2003. The origin of garnet and clinopyroxene in "depleted" Kaapvaal peridotite. Lithos 71, 289–322.

Skinner, C.P., 1989. The petrology of peridotite xenoliths from the Finsch kimberlite, South Africa. S. Afr. J. Geol. 92, 197–206.

Smith, D., 1999. Temperatures and pressures of mineral equilibration in peridotite xenoliths: Review, discussion, and implications. In: Fei, Y., Bertka, C.M., Mysen, B.O. (Eds.), Mantle Petrology: Field Observations and high pressure experimentation: a tribute to Francis R. (Joe) Boyd. Spec. Publ.-Geochem. Soc., vol. 6, pp. 171–188.

Smith, D., Boyd, F.R., 1992. Compositional zonation in garnets in peridotite xenoliths. Contrib. Mineral. Petrol. 112, 134–147.

Smith, D., Wilson, C.R., 1985. Garnet–olivine equilibration during cooling in the mantle. Am. Mineral. 70, 30–39.

Smith, D., Griffin, W.L., Ryan, C.G., Sie, S.H., 1991. Trace-element zonation of garnets in The Thumb: heating and melt-infiltration below the Colorado plateau. Contrib. Mineral. Petrol. 107, 60–79.

Sobolev, N.V., Lavrent'ev, Y.G., 1971. Isomorphic sodium admixture in garnets formed at high pressures. Contrib. Mineral. Petrol. 31, 1–12.

Sobolev, N.V., Bartoshinsky, Z.V., Yefimova, E.S., Lavrentyev, Y.G., Pospelova, L.N., 1970. Olivine–garnet–chrome diopside assemblage from Yakutian diamond. Dokl. Akad. Nauk SSSR 192, 134–137 (in English).

Sobolev, N.V., Kharkiv, A.D., Lavrent'ev, Y.G., Pospelova, L.N., 1973a. Chromite–pyroxene–garnet intergrowths from the Mir kimberlite pipe. Geol. Geofiz. 12, 15–20 (in Russian).

Sobolev, N.V., Lavrent'yev, Y.G., Pokhilenko, N.P., Usova, L.V., 1973b. Chrome-rich garnets from the kimberlites of Yakutia and their paragenesis. Contrib. Mineral. Petrol. 40, 39–52.

Sobolev, N.V., Pokhilenko, N.P., Yefimova, E.S., 1984. Diamond-bearing peridotite xenoliths in kimberlites and the problem of the origin of diamonds. Geol. Geofiz. 25, 63–80.

Sobolev, V.N., Taylor, L.A., Snyder, G.A., Sobolev, N.V., Pokhilenko, N.P., Kharkiv, A.D., 1997. A unique metasomatised xenolith from the Mir kimberlite, Siberian platform. Russ. Geol. Geophys. 38, 218–228.

Stachel, T., Brey, G.P., Harris, J.W., 2000. Kankan diamonds (Guinea) I: from the lithosphere down to the transition zone. Contrib. Mineral. Petrol. 140, 1–15.

Stiefenhofer, J., Viljoen, K.S., Tainton, K.M., Dobbe, R., Hannweg, G.W., 1999. The petrology of a mantle xenolith suite from Venetia, South Africa. In: Gurney, J.J., Gurney, J.L., Pascoe, M.D., Richardson, S.H. (Eds.), P.H. Nixon Volume Proc. 7th Int. Kimb. Conf. Red Roof Design, Cape Town, pp. 836–845.

Switzer, G.S., 1975. Composition of garnet xenocrysts from three kimberlite pipes in Arizona and New Mexico. Smithson. Contrib. Earth Sci. 9, 1–21.

Taylor, W.R., 1998. An experimental test of some geothermometer and geobarometer formulations for upper mantle peridotites with application to the thermobarometry of fertile lherzolite and garnet websterite. Neues Jahrb. Mineral. Abh. 172, 381–408.

van der Westhuizen, A., 1992. The Bellsbank kimberlites, with special reference to a suite of purple garnet megacrysts from the Bobbejaan mine. Unpubl. MSc thesis, University of the Orange Free State. South Africa, 88 pp.

Viljoen, K.S., Robinson, D.N., Swash, P.M., Griffin, W.L., Otter, M.L., Ryan, C.G., Win, T.T., 1994. Diamond- and graphite-bearing peridotite xenoliths from the Roberts Victor kimberlite, South Africa. In: Meyer, H.O.A., Leonardos, O.H. (Eds.), Kimberlites, Related Rocks and Mantle Xenoliths, Proc. 5th Int. Kimb. Conf. Spec. Publ.-CPRM, vol. 1A/94, pp. 285–303.

Wang, W., Sueno, S., Takahashi, E., Yurimoto, H., Gasparik, T., 2000. Enrichment processes at the base of the Archean lithospheric mantle: observations from trace element characteristics of pyropic garnet inclusions in diamonds. Contrib. Mineral. Petrol. 139, 720–733.

Webb, S.A.C., Wood, B.J., 1986. Spinel–pyroxene–garnet relationships and their dependence on Cr/Al ratio. Contrib. Mineral. Petrol. 92, 471–480.

Available online at www.sciencedirect.com

Lithos 77 (2004) 859–872

ELSEVIER

www.elsevier.com/locate/lithos

Mapping the mantle lithosphere for diamond potential using teleseismic methods [☆]

D.B. Snyder[a,*], S. Rondenay[b], M.G. Bostock[c], G.D. Lockhart[d]

[a] Geological Survey of Canada, 615 Booth Street, Ottawa, ON, Canada K1A 0E9
[b] Department of Earth and Planetary Sciences, MIT, Cambridge, MA, USA
[c] Department of Earth and Ocean Sciences, University of British Columbia, Vancouver, BC, Canada V6T1Z1
[d] BHP-Billiton Diamonds Inc., Kelowna, BC, Canada V1X 4L1

Received 27 June 2003; accepted 9 January 2004
Available online 20 July 2004

Abstract

Recent developments in seismic, magnetotelluric and geochemical analytical techniques have significantly increased our capacity to explore the mantle lithosphere to depths of several hundred kilometres, to map its structures, and through geological interpretations, to assess its potential as a diamond reservoir. Several independent teleseismic techniques provide a synergistic approach in which one technique compensates for inadequacies in another. Shear wave anisotropy and discontinuity studies using single seismic stations define vertical mantle stratigraphic columns. For example, beneath the central Slave craton seismic discontinuities at depths of 38, 110, 140 and 190 km appear to bound two distinct anisotropic layers. Tomographic (3-D) inversions of seismic wave travel-times and 2-D inversions of surface or scattered waves use arrays of stations and provide lateral coverage. In combination, and by correlation with electrical conductivity and xenolith petrology studies, these techniques provide maps of key physical properties within parts of the cratons known to host diamonds. Beneath the Slave craton, the discontinuity at 38 km is the base of the crust; the boundaries at 110 and 140 km appear to bound a layer of depleted harzburgite that is interpreted to contain graphite. To date, only some of these techniques have been applied to the Slave and Kaapvaal cratons so that the origin and geological history of the currently mapped mantle structures are not, as yet, generally agreed.
© 2004 Elsevier B.V. All rights reserved.

Keywords: Slave craton; Diamond exploration; Seismic imaging; Lithospheric discontinuity; Craton stabilization

1. Introduction

Diamond deposits are typically identified in four stages: (1) regional targeting, during which a region's potential is assessed, often by mapping of Archean age basement, grid sampling for indicator minerals or global seismological mapping of continental mantle 'keels'; (2) kimberlite detection, during which glacial till mapping and high-resolution aeromagnetic surveys locate individual kimberlite deposits; (3) deposit delineation, in which drill hole core sampling determines a specific deposit's volume and lithology; and (4) evaluation, in which bulk sampling establishes a

☆ Supplementary data associated with this article can be found, in the online version, at doi: 10.1016/j.lithos.2004.03.049.

* Corresponding author. Tel.: +1-613-992-9240; fax: +1-613-943-9285.

E-mail address: dsnyder@nrcan.gc.ca (D.B. Snyder).

deposit's worth and its feasibility to be mined. The diamond exploration industry needs discriminating tools to reduce risks at all of these four stages.

Very approximate statistics from the past few decades indicate that globally, for every 1000 candidate magnetic anomalies identified, 100 are kimberlites, 10 contain gem-quality diamonds and 1 is economic to mine in remote areas. The central part of the Slave craton appears to be more prospective than the remainder, and seismic imaging is helpful in discerning which parts are more likely to host economic deposits. More specifically, at the Ekati Diamond Mine™, 150 kimberlites have been discovered at an 80% success rate and approximately 1 pipe in 15 may be economic. In other parts of the Slave, the Dry Bones Bay area or the Coronation Gulf area, for example, it appears improbable that a commercial property will be found. Seismic techniques can provide three-dimensional (3-D) maps of key physical properties in the mantle to 700 km depth to help accomplish stage one (Nolet et al., 1994; Bostock, 1999). These results can then be used to interpret mantle structure in conjunction with conductivity maps derived from magnetotelluric soundings and the 'ground truth' of actual rock types provided by rare xenolith samples from kimberlites. If such multi-disciplinary upper mantle studies were performed on other cratonic terrains, it may be possible to prioritize where exploration should be focused.

At present, our efforts at refining suitable seismological techniques are concentrated in the central Slave craton of the NW Territories, Canada (Fig. 1), because of the strong geological, geophysical and logistical base currently available. Once a velocity and physical property model is established in the Slave craton with sufficient detail with which to help assess diamond occurrences, this innovative, first-order adaptation of seismic exploration tools can be applied throughout Canada and globally. Here, we describe our progress to date in adapting a number of established and newly developed seismological analysis tools to this special exploration application.

2. Equipment and classification of teleseismic methods

Seismic methods applied to the delineation of Earth structure fall into two broad categories: (1) 3-D map-

ping of seismic wave transmission via travel-times and (2) studies of forward- or back-scattered wave fields. Logistically, these methods divide into different groupings: those using single, independent stations recording many distant (teleseismic) earthquakes and those using an array of stations that all record the same seismic wave source, be it teleseismic or man-made. Each method provides a different feature or characteristic of the Earth model, for example: bulk compressional wave (P-wave) or shear wave (S-wave) velocities, depths and geometries of discontinuities, direction of fabric or anisotropy. In remote, harsh environments such as the central Slave craton, it is prudent to begin with single station studies, assess the results and then develop arrays of stations in key areas. Two methods that utilize a single station are currently being applied in the Slave and their results are assessed herein. A low-resolution, regional-scale array study has also begun.

Seismograms from four years of recording global earthquakes using a few broadband seismometers located near the Ekati Diamond Mine™ can be analyzed by the independent SKS-anisotropy and receiver-function techniques to reveal information about layered structure within the upper mantle of the central Slave craton. Within the past two years, 22 POLARIS satellite-telemetry stations (www.polarisnet.ca) have supplemented recent fieldwork that used a few stations with interchangeable hard-disc recorders (Snyder et al., 2002) and a more regional survey done in 1996–1998 (Bank et al., 2000) (Fig. 1; online Appendix A).

Each POLARIS telemetry station costs about C$60,000 to purchase, transport and install at a remote location in Canada's North. A typical station consists of a GURALP 3ESP sensor with a nominal bandwidth of 0.033 to 50 Hz, a Nanometrics Libra digitizer, satellite uplink electronics, and 16 12-V batteries charged by a 12-panel solar power subsystem. The earlier stations used GURALP 40 T seismometers (0.05 to 50 Hz bandwidth), Nanometrics Orion portable recorders, and similar but smaller power subsystems (Snyder et al., 2002).

3. Discontinuity detection

The first method applied to data from independent, single stations is the so-called receiver function method in which source-deconvolved P-to-S converted

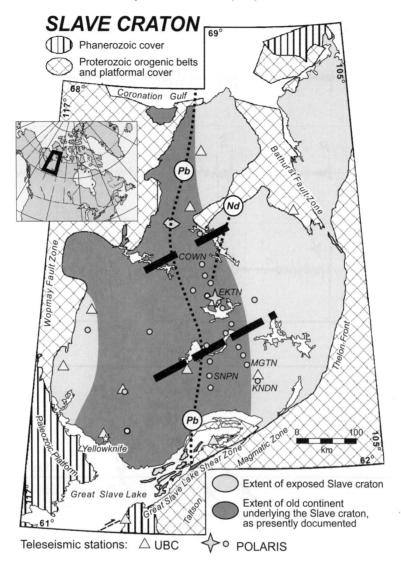

Fig. 1. Location map of the POLARIS and earlier teleseismic arrays in the central Slave craton, Northwest Territories (NWT), Canada. The heavy dashed lines indicate zoning within the mantle as described by Grütter et al. (1999); crustal isotopic zonation is described by Davis et al. (1996).

waves arriving from multiple teleseismic earthquakes reveal discontinuities in seismic wave velocities or density below each seismic station (Fig. 2) (Bostock, 1998, 1999). These discontinuities represent local increases in the rate at which physical properties vary with increasing depth. Changes in velocity, density or anisotropy within the mantle over a depth range that is a fraction of the seismic wavelength scatter upcoming seismic waves and partially convert P-waves into S-waves to reveal the discontinuities. Typically, the sharp (to seismic waves with wavelengths of a few kilometers) increase in both velocity and density with depth at the Moho (roughly the base of the crust) is the most prominent and laterally extensive of these discontinuities (Figs. 3 and 4). Within the central Slave craton, Moho depth determined by this method varies between 36 and 42 km (Fig. 3). At the Ekati Diamond Mine™ station (EKTN), the Moho shows

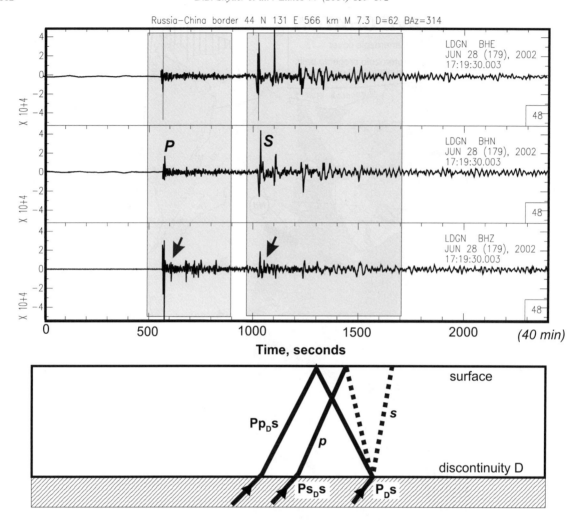

Fig. 2. (Top) Example of an earthquake recorded by the POLARIS NWT array. Here, an earthquake 566 km below the Russia–China border as recorded 62° (7000 km) away at the Lac de Gras station (LDGN). Windows indicate impulsive P-wave and S-wave arrivals and later arrivals from conversions at discontinuities (arrows). (Bottom) Schematic of possible conversions at one discontinuity that cause later arrivals.

little variation with changing direction of arrival (back azimuth) of the seismic wave (Fig. 4a); here, that statement applies within an annular ring (a 'doughnut') between radii of 6 and 12 km. At 130 km depth, a similar ring of illumination has radii of 23 and 48 km because each station senses physical properties within a cone beneath it.

Reverberations or 'multiples' of energy between the surface and strong discontinuities such as the Moho are relatively well understood and can be easily recognized by their polarity and move-out response (changes in travel time with increased offset between

earthquake source and receiver, see Bostock, 1998). These reverberations can often have correlation amplitudes similar to the primary converted S-wave (Figs. 3 and 4a). Within the Slave craton, the Moho discontinuity generally shows no response on the transverse components and therefore no Moho 'multiples' should appear on these components. Our analysis therefore focuses on prominent and consistent discontinuities with polarities different than predicted Moho 'multiples' that manifest themselves on either the transverse or radial components (e.g. 107 km discontinuity beneath EKTN, Figs. 3 and 4).

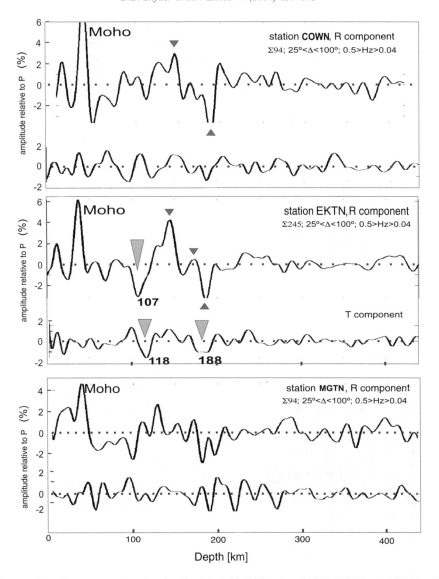

Fig. 3. Samples of summed impulse response or 'receiver functions' for POLARIS stations COWN, EKTN and MGTN (see Fig. 1). Each station has a pair of traces mapped to depth: the top is for radial (R) wave motion, the bottom for transverse (T) motion. Large triangles and associated numbers indicate depths to major discontinuities; small triangles mark reverberations (multiples) from the Moho discontinuity. Note that only the EKTN station has a strong signal on both functions at about 110 km. The receiver functions shown for these three stations used 94, 245 and 94 earthquake sources, respectively.

Other prominent discontinuities at 110–120 and 140–150 km depths observed on the transverse components at multiple stations indicate that a layer of low velocity or distinct anisotropy exists between these depths (Fig. 3). The coherency of pulses at about 13 s on the radial component indicates a strong decrease in velocity at 107 km depth north and west of Ekati, but

no such feature to the south (Fig. 4a). Similarly, the response on the transverse component for station EKTN indicates a change in anisotropic fabric at 118 km depth, a different one at 144 km, and another at 188 km. A flip in polarity at a back azimuth of about 280° occurs in the 118 km discontinuity and marks an axis of symmetry of anisotropy, here probably the dip direction of steep layering or planar fabric.

In the SW Slave craton north of Yellowknife similar discontinuities were observed on the transverse component at nominal depths of 75, 135 and 180–220 km (Fig. 4c, Bostock, 1998). The shallowest of these is the most clearly defined and comprises a sharp-topped layer 10 km thick that exhibits 5% S-wave anisotropy with a polarity flip at about 260°

back azimuth. A number of weak discontinuities at 8, 12 and 20 s can also be recognized beneath the central Kaapvaal craton using the same analytical technique and data from a local, temporary array near Kimberley (Fig. 5; James et al., 2001; but see Gao et al., 2002).

4. Layered anisotropy

A second method, applied to data from independent stations, estimates regional-scale fabric or layering within the mantle using differential travel times of S-waves caused by seismic anisotropy. The anisotropy typically arises from preferred fabric orientations; cracks are an obvious example in near-surface environments. Alignment of minerals such as olivine over large volumes and macroscopic layering of peridotite and eclogite are other possible causes of mantle anisotropy. Seismic waves that originate as S-waves, but travel through the Earth's molten core as P-waves before converting back to S-waves, so-called SKS and SKKS phases, are particularly suitable for anisotropy studies of the mantle beneath the receiver (Silver, 1996; Savage, 1999 are useful reviews). Within the mantle, olivine is the primary anisotropic mineral and fast polarization aligns parallel to the mineral a-axes. Peridotitic xenolith samples from the Kaapvaal exhibit up to 8% anisotropy for P-waves and 6% for S-waves (Ben-Ismail et al., 2001). In general, olivine a-axes concentrate within the foliation plane and parallel to lineations, whatever the cause of the foliation. It thus appears that S-wave energy propagating and vibrating within the plane of a vertical fabric travels faster than does energy vibrating perpendicular to the fabric; the

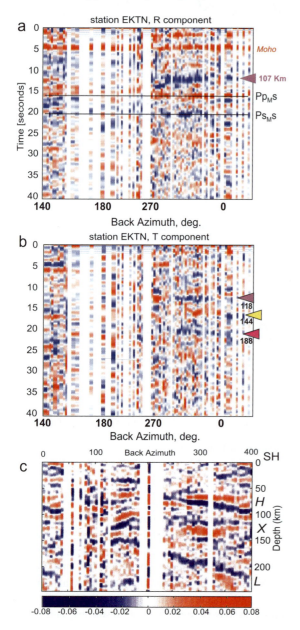

Fig. 4. Radial (a) and transverse (b) impulse responses as a function of back azimuth (incomplete suite of 1° bins) at the Ekati Diamond Mine™ station (EKTN on Fig. 1). Amplitudes are normalized to that of the incident P-wave, red is positive polarity, blue negative. Coverage is incomplete after 3 years of data collection; a back azimuth of 0 is North. The flat, consistent Moho response results in several multiples (Pp_Ms and Ps_Ms, see Fig. 2) at greater delay times. Only phases marked by arrowheads are considered probable mantle discontinuities at the depths indicated. Note the change of polarity at about 280° on the transverse component at these depths; a feature expected from anisotropy. (c) Transverse (SH) impulse response mapped to depth as a function of back azimuth (5° bins) at the Yellowknife array (modified from Bostock, 1998). The absence of significant Moho response on this component affords a window into the mantle where the H, X and L discontinuities are clearly visible.

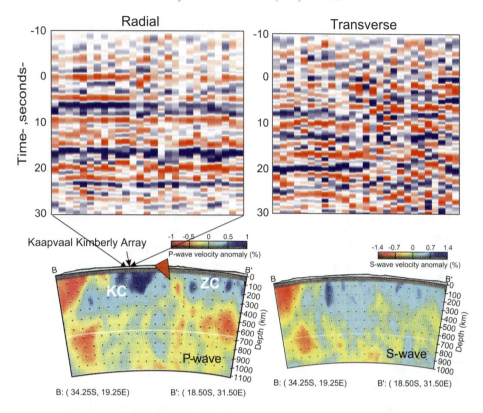

Fig. 5. (Top) Radial and transverse impulse responses as a function of back azimuth at the Kimberley broadband array (James et al., 2001). This is an east–west cross section through the array with each station projected perpendicularly onto the section. (Bottom) Tomographic cross sections for P-wave velocities (left) and S-wave velocities (right) of the Kaapvaal (KC) and Zimbabwe (ZC) cratons using data and methods as in Fig. 9; image from James et al. (2001).

difference in time of arrival is called the delay time, dt. The analysis also reveals the direction of the fast propagation axis, Φ (Fig. 6; online Appendix A). Typically, months of data are analyzed in order to produce an average anisotropy measurement for an assumed single layer of uniform anisotropy beneath each station, but observations made over two or more years often provide much additional and valuable information. Only two teleseismic stations in the NWT (Yellowknife and EKTN) have to date recorded a sufficient number of earthquakes appropriate for SKS analysis to reveal reliable trends.

The variations in the arrival of SKS phases with earthquake back azimuth (direction at which seismic wave arrives) reveal patterns (Fig. 6) that can be modeled by multiple or dipping layers of anisotropy (Vinnik et al., 1992; Rümpker and Silver, 1998; Levin and Park, 1998; Savage, 1999). The observed azi-

muthal variation at station EKTN within the central Slave cannot represent a single uniform layer because systematic variation with back azimuth exceeds uncertainties associated with the measurements. The observed variations can be most simply modeled by assuming two distinct horizontal layers within the lithosphere, each with a horizontal axis of hexagonal anisotropy symmetry. The back azimuths at which minimum delay times and very large delay times occur are particularly diagnostic in such two-layer models. The minimum marks the fast axis of anisotropy in the dominant layer, the associated delay time is that of the secondary anisotropy layer. The large delays mark interference between the two anisotropic layers where individual analyses produce nulls, unstable or large uncertainties. The analysis for EKTN reveals a shallow layer with about a third of the total observed anisotropy and a fast axis oriented at 006°; a

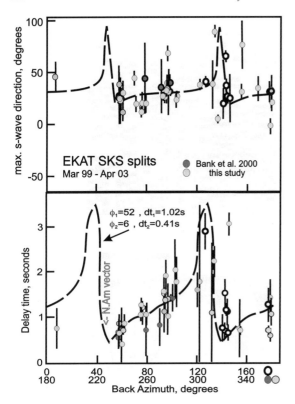

Fig. 6. Plots of the apparent fast polarization direction (top) and delay or splitting time (bottom) as a function of back azimuth for station EKAT (now POLARIS station EKTN) assuming a frequency of 0.08 Hz. Splitting parameters repeat every 180° so that the observations from both hemispheres are superimposed; open circles are observations from 0° to 180°, solid circles from 181° to 360°. The observations have standard uncertainties (e.g. Silver, 1996) shown by vertical bars. Grey dots are measurements at stations EKAT, EKTN, and NORMS. The wide, dashed line shows the predicted variation of these parameters for a two-layer model (Silver and Savage, 1994) with the parameters indicated in the lower graph. The predictive model repeats every 90° so that the observational gap from 0° to 60° is not debilitating.

deeper layer with the remainder of the anisotropy has a fast direction of 052° that coincides with present-day North American plate motion (Bank et al., 2000).

5. Layer interpretation

Together, the two independent, single-station techniques reveal two layers of anisotropy, three mantle discontinuities, and the Moho beneath the Ekati Diamond Mine™ site. The lower anisotropy layer prob-

ably lies deeper than 150 km in the mantle because the anisotropy aligns with both North American plate motion and the ~ N50°E strike of deep mantle structures identified previously by mantle geochemical analyses. The near-coincidence of the slow direction (276°) in the upper anisotropy layer and the polarity flip observed in the 118-km discontinuity suggest a common cause: that the discontinuity marks the top or bottom of the upper anisotropic layer.

The upper layer and 118-km discontinuity coincide with a prominent regional conductor identified by magnetotelluric studies (Jones et al., 2001), as well as with an ultra-depleted harzburgite layer identified from studies of garnets extracted from xenoliths in kimberlite core (Fig. 7) (Griffin et al., 1999; Kopylova and Russell, 2000). Ultra-depleted harzburgite is nearly pure olivine in composition and has potential for high anisotropy if individual olivine crystals are aligned (e.g. Silver, 1996). It appears probable that beneath station EKTN the upper anisotropic layer lies between 118 and 140 km, is anomalously conductive, and is composed of ultra-depleted harzburgite. The 190-km discontinuity may similarly mark the top of the deeper anisotropy layer. However, anisotropy within the crust, above 110 km, and at 140–190 km depths cannot be excluded.

The 110- and 140-km seismic discontinuities observed beneath the central Slave are not typical of cratons globally (Fig. 8). The 'Hales' discontinuity (Hales, 1969) is commonly associated with the spinel to garnet phase change found at 60–80 km depths and the 'Lehman' discontinuity (Lehman, 1955) appears beneath continents at ~ 220 km depth (Bostock, 1999, and references therein). The latter has been hypothesized as the base of an anisotropic layer below which aligned textures in peridotite are annealed and thus made more random and isotropic; it may represent the base of the lithosphere (Jordan, 1988). Thybo and Perchuc (1997) proposed that a layer of lower velocity and anisotropy due to partial melt at 100–150 km depths underlies a generally stratified uppermost mantle.

Because of the poor correlation among global seismic discontinuities, all three discontinuities observed beneath the Yellowknife area were instead interpreted as underthrust or underplated blocks of Archean crust or mantle, and linked to mantle reflectors observed on the LITHOPROBE SNORCLE line

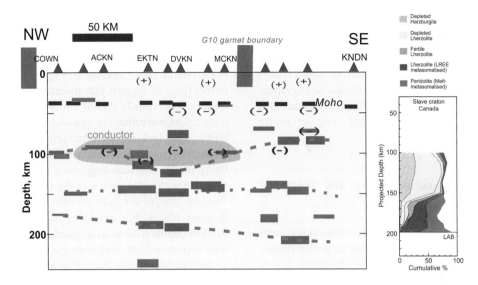

Fig. 7. Discontinuities beneath the central Slave craton. Seismic stations labeled are located at: Kennady Lake (KNDN), NE McKay Lake (MCKN), the Diavik mine airstrip (DVKN), the Ekati mine airstrip (EKTN), Achilles Lake (ACKN) and SW of Contwoyto Lake (COWN). Pluses and minuses represent discontinuities with changes in velocity identified from radial components. Bars represent discontinuities identified beneath individual stations on the transverse component. Smaller bars indicate less reliable estimates. Petrologic column from Griffin et al. (1999); conductor from Jones et al. (2001).

1 seismic reflection profile (Bostock, 1998). A pair of Proterozoic (1.89 – 1.84 Ga) subduction zones is the current preferred interpretation of these mantle reflectors (Cook et al., 1999). It is interesting to speculate that one or both of these Proterozoic underthrust or

subducted layers may continue NW of Yellowknife and into the central Slave near station EKTN. The surface expression of these convergence structures is the Wopmay orogen along the western margin of the Slave craton (Fig. 1). In order to correlate the mantle

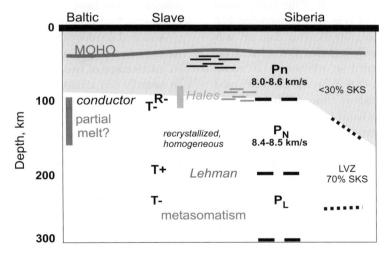

Fig. 8. Cartoon summary of upper mantle discontinuities observed globally beneath cratons and major continental blocks. Siberia structures are based on results from 'peaceful' nuclear explosion surveys (Pavlenkova, 1997; Oreshin et al., 2002). Baltic structures were described by Thybo and Perchuc (1997) and Kovtun and Porokhava (1980). Slave structure as described here and by Bostock (1998, 1999). Groups of short horizontal segments represents 1–2-km-thick layering inferred from scattering of seismic waves to the surface (Tittgemeyer et al., 1996).

discontinuities beneath Yellowknife and EKTN, one must invoke a complex subduction zone geometry with steeply (30°) and flat-dipping segments as is currently observed beneath the Andes of South America (Isacks, 1988). Alternatively, crustal stratigraphic studies suggest that the central Slave may be underlain by an older (2.6 Ga) convergent zone or suture (Bleeker et al., 1999).

6. Tomography, surface wave and scattering studies using arrays

Three other seismic analysis techniques with promising application to the mantle structure studies described here are currently in use elsewhere worldwide, but have only produced preliminary results for the Slave craton due to the lack of a suitable array of stations. One widely used tomography method inverts observed travel times of seismic P-wave arrivals using a large number of rays connecting earthquake–station pairs in order to estimate bulk, average velocities in a 3-D mantle volume (Van Decar, 1991). Because most of these seismic waves travel nearly vertically through the mantle, lateral resolution is generally good whereas vertical resolution is poor.

Within the Slave craton, this method was originally used in a reconnaissance analysis of the entire craton using 14 seismic stations (Fig. 1) that recorded 226 earthquakes (Bank et al., 2000). This data set was enhanced using the newly installed POLARIS array stations that had recorded an additional 120 earthquakes, thus adding 1149 new rays to the 1575 used previously. The new analysis removed all travel-time residuals < 0.2 s and included 10 independent trials at reducing travel-time residuals, in which each trial involved 10,000 iterations. The resulting P-wave tomographic slowness (1/velocity) model has more detail in the central Slave area (Fig. 9) than did the reconnaissance survey, but retains all the major regional trends and features such as slower velocities beneath the Proterozoic terranes to the west (Bank et al., 2000). Depth slices at 170 and 230 km depths show a major transition parallel to and nearly coincident with the northern garnet geochemical boundary as defined by Grütter et al. (1999). POLARIS stations south of their southern garnet geochemical boundary

are underlain by slower velocity mantle than those to the north.

The strong lateral velocity variations shown on a north–south cross section through the tomography model (Fig. 9) suggest that the mantle discontinuities discussed previously do not represent laterally continuous layers, but rather the tops and bottoms of irregular blobs. The lack of well-defined layering is partly attributable to the poor depth resolution in the tomography model, but also suggests a relatively fine-scale structure within the mantle of the central Slave craton. The two analytical results do show consistent features in many places. For example, discontinuities at 80–110 km depths with an associated decrease in velocity with depth correlate well with the velocity model.

Similar tomographic analysis applied to the Kaapvaal and Zimbabwe cratons in southern Africa showed a high-velocity tectospheric keel to 250 km depth and, between the cratons, lower velocities within a zone dipping northeast beneath the Bushveld province (Fig. 5) (James et al., 2001). Velocity variation within the uppermost 200–300 km of the cratons is less pronounced here than in the central Slave craton, but diamond parageneses do generally correlate with bulk mantle velocity variations beneath southern Africa (Shirey et al., 2002). Diamonds from kimberlites erupted through seismically fast lithosphere are dominantly mid-Archean and peridotitic whereas those from slower lithosphere are mostly Proterozoic and eclogitic.

A second seismic method attempts the inversion of surface waveforms (Forsyth and Li, in press). It relies on interference of two incoming waves and depends on laterally and azimuthally varying phase velocities measured between two or more stations. The method recovers 2-D phase velocity maps for a series of narrow frequency bands that match specific depth ranges; longer period waves sample at greater depth. This method can therefore help constrain the location of anisotropy in the mantle, and estimate lithospheric thickness. As in tomography, resolution depends on the density of obliquely intersecting rays over the study area and the trade-off between model roughness and assumed anisotropy. Based on previous results elsewhere (Weeraratne et al., 2003), surface-wave inversion of the Slave array data should resolve anomalies with lateral wavelengths >100–150 km. Results will also be utilized in conjunction with SKS-

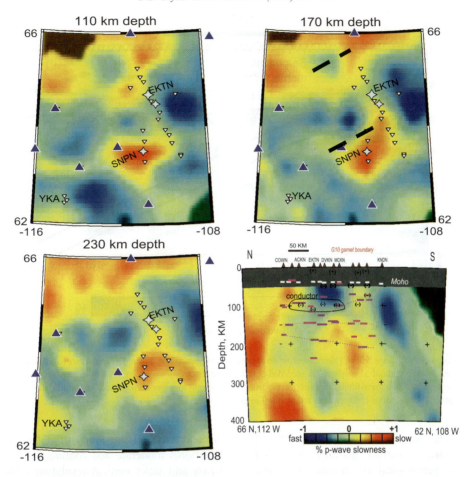

Fig. 9. Tomography-derived 3-D velocity models of the Slave craton; shown are depth slices at 110, 170 and 230 km and a north–south cross section. Modeling used the method and stations (black triangles) used by Bank et al. (2000) supplemented by POLARIS stations (white stars and inverted triangles). Black areas occur where fewer than four rays crossed the model cell. Black dash lines show geochemical boundaries (Grütter et al., 1999). See Fig. 7 caption for annotations on the cross section.

splitting analysis to constrain better the depth of anisotropic layers within the mantle.

A third analysis technique using seismic array data has not been attempted in the central Slave, but is planned once the array has sufficient length and density. Results from its recent application to the Cascadia subduction zone (Rondenay et al., 2001; Bostock et al., 2002) are shown (Fig. 10) here to demonstrate the potential for resolving similar, distinct mantle layers beneath cratons. Multiparameter two-dimensional inversion of scattered teleseismic waves repositions seismic energy arriving after the main P-wave to the depth where it was converted or scattered at major structures or seismic discontinuities

(Bostock et al., 2001). The clear imaging of two different Moho discontinuities as well as subducting oceanic crust beneath Cascadia approaches that achieved with deep seismic reflection profiling techniques and indicates that similar results should be possible where strong mantle discontinuities are already identified, such as in the central Slave craton.

7. Emerging technology drives new models of mantle structure

As was the case in southern Africa, the seismic results aid us in constructing a three-dimensional

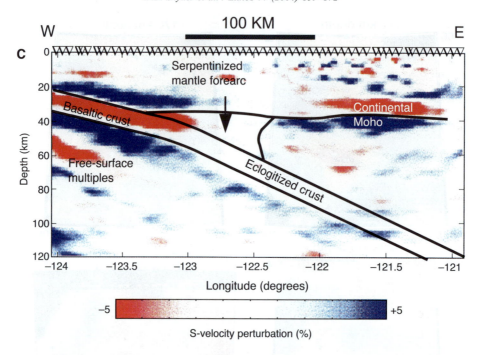

Fig. 10. S-wave velocity perturbations below a dense seismic array across the Cascades of central Oregon, USA (from Rondenay et al., 2001). Scattered waves in the P-wave coda of 31 earthquakes recorded at teleseismic distances were recovered from simultaneous inversion of signal recorded at 69 sites. This is a bandpass-filtered version of the true perturbations to a one-dimensional, smoothly varying reference model; discontinuities are present where steep changes (e.g. red to blue) in polarity occur.

model of the Slave craton from the surface to 700 km depths that is tied to surface geology and other geophysical and petrological constraints on mantle composition. It appears significant that major seismic discontinuities have associated changes in anisotropy more often than with increases or decreases in velocity or density with depth. The change in anisotropy at 118 km depth correlates with high conductivity that has been related to increased graphite content or connectivity (Jones et al., 2001). The change in anisotropy at 140 km depth correlates with an increase of metasomatism in mantle xenoliths (Fig. 7). It is not clear at present if these discontinuities represent the (dis)appearance of seismic anisotropy or just systematic re-orientations of anisotropic mantle structures. It is the combined use of seismic methods as described here that promises new understanding from powerful constraints on mantle structure, constraints not used in previous interpretations of mantle structure to our knowledge (Oreshin et al., 2002).

The preliminary tomography velocity models provide important three-dimensional information to the discontinuity and anisotropy studies in that this method affords moderate and continuous horizontal resolution and lacks vertical resolution which the other methods provide. Early results suggest that the mantle geochemical trends implied from xenolith studies are not laterally continuous or relevant over large areas. For example, regions with slower velocities on the 170 and 230 km depth slices (Fig. 9) indicate that the southeastern-most source region for high-Cr G10 garnets is limited in extent and covers a triangular area of about 5000 km^2 that includes the Diavik, Snap Lake and Kennady Lake sites.

More precise definition of layers with properties such as increased carbon content or metasomatism has direct relevance to mapping potential source regions and long-term mantle reservoirs of diamonds. Once a mantle 'stratigraphy' is established and understood in terms of its seismic and other characteristics, its lateral extent can be mapped and related to broad changes in velocity. If the mantle layers derive from large-scale tectonic events, the resulting improved understanding of a region's geological history from surface mapping and related geochemical and geochronological studies will also guide exploration models and strategies.

Continuing recording of earthquakes at the current 22 POLARIS stations and the infilling of the current array with additional stations will enable techniques such as travel-time tomography and inversions of scattered wavefields to provide more continuous 2-D and 3-D images of the lithosphere beneath the central Slave craton. Commonly observed discontinuities at 410- and 670-km depths appear sensitive to the thermal state of the mantle and may thus provide clues to the more recent kimberlite eruption processes. These various new results from the central Slave craton can also be compared and contrasted with similar studies and their recently published results from the Kaapvaal, Siberia and western Australia (Kennett et al., 1994; Debayle and Kennett, 2000; James et al., 2001; Oreshin et al., 2002; Simon et al., 2002). Meanwhile, we plan to install additional teleseismic stations during the next few years within other major cratons of North America.

Acknowledgements

The work presented here benefited from help from numerous sources, among them: members of the POLARIS consortium, especially Isa Asudeh, Calvin Andrews, Mike Patten and Gerrit Jansen van Beek for their efforts in installing field stations; Carolyn Relf, Hendrik Falck and John Armstrong of the CS Lord Geoscience Centre (Yellowknife, NWT); the staff of the Yellowknife Seismic Observatory; Wouter Bleeker, Bill Davis, Herman Grütter, and Alan Jones for many interesting discussions. This work was funded under Letters of Agreement between the Geological Survey of Canada, BHP-Billiton Diamonds; DeBeers Canada; and Kennecott Canada Exploration. POLARIS equipment was purchased through a Canadian Foundation for Innovation grant to Carleton University. Geological Survey of Canada contribution 2003060.

References

Bank, C.-G., Bostock, M.G., Ellis, R.M., Cassidy, J.F., 2000. A reconnaissance teleseismic study of the upper mantle and transition zone beneath the Archean Slave craton in NW Canada. Tectonophysics 319, 151–166.

Ben-Ismail, G., Barroul, R., Mainprice, D., 2001. The Kaapvaal seismic anisotropy: petrophysical analysis of upper mantle kimberlite nodules. Geophysical Research Letters 28, 2497–2500.

Bleeker, W.J., Ketchum, W.F., Jackson, V.A., Villeneuve, M., 1999. The central slave basement complex. Part I: its structural topology and authochthonous cover. Canadian Journal of Earth Sciences 36, 1083–1109.

Bostock, M.G., 1998. Seismic stratigraphy and evolution of the Slave province. Journal of Geophysical Research 103, 21183–21200.

Bostock, M.G., 1999. Seismic imaging of lithospheric discontinuities and continental evolution. Lithos 48, 1–16.

Bostock, M.G., Rondenay, S., Shragge, J., 2001. Multiparameter two-dimensional inversion of scattered teleseismic body waves. 1: theory for oblique incidence. Journal of Geophysical Research 106, 30771–30794.

Bostock, M.G., Hyndman, R.D., Rondenay, S., Peacock, S.M., 2002. An inverted continental Moho and serpentinization of the forearc mantle. Nature 417, 536–538.

Cook, F.A., van der Velden, A., Hall, K.W., Roberts, B.J., 1999. Frozen subduction in Canada's Northwest Territories: LITHOPROBE deep lithospheric reflection profiling of the western Canadian Shield. Tectonics 18, 1–24.

Davis, W.J., Gariepy, C., van Breeman, O., 1996. Pb isotopic composition of Late Archean granites and the extent of recycling early Archean crust in the Slave Province, northwest Canada. Chemical Geology 130, 255–269.

Debayle, E., Kennett, B.L.N., 2000. Anisotropy in the Australian upper mantle from Love and Rayleigh waveform inversion. Earth and Planetary Science Letters 184, 339–351.

Forsyth, D.W., Li, A., 2003. Array-analysis of two-dimensional variations in surface wave velocity and azimuthal anisotropy in the presence of multipathing interference. In: Levander, A., Nolet, G. Seismic Data Analysis and Imaging with Global and Local Arrays. American Geophysical Union Geophysical Monograph. in press.

Gao, S.S., Silver, P.G., Liu, K.H., the Kaapvaal Seismic Group, 2002. Mantle discontinuities beneath Southern Africa. Geophysical Research Letters 29, 10 (DOI 10.1029/2001GL013834).

Griffin, W.L., Doyle, B.J., Ryan, C.G., Pearson, N.J., O'Reilly, S.Y., Davies, R.M., Kivi, K., van Achterbergh, E., Natapov, L.M., 1999. Layered mantle lithosphere in the Lac de Gras area, Slave Craton: composition, structure, and origin. Journal of Petrology 40, 705–727.

Grütter, H.S., Apter, D.B., Kong, J., 1999. Crust–mantle coupling: evidence from mantle-derived xenocrystic garnets. In: Gurney, J.J., Richardson, S.R. (Eds.). Proc. 7th Kimberlite Conference, Red Roof Design, Cape Town, 307–312.

Hales, A.L., 1969. A seismic discontinuity in the lithosphere. Earth and Planetary Science Letters 7, 44–46.

Isacks, B.L., 1988. Uplift of the central Andean plateau and bending of the Bolivian orocline. Journal of Geophysical Research 93, 3211–3231.

James, D.E., Fouch, M.J., VanDecar, J.C., van der Lee, S., Kaapvaal Seismic Group, 2001. Tectosphere structure beneath southern Africa. Geophysical Research Letters 28, 2485–2488.

Jones, A.G., Ferguson, I.J., Chave, A.D., Evans, R.L., McNeice, G.W., 2001. Electric lithosphere of the Slave craton. Geology 29, 423–426.

Jordan, T.H., 1988. Structure and formation of the continental tectosphere. Journal of Petrology, 11–37 (Special Lithosphere Issue).

Kennett, B.L.N., Gudmunsson, O., Tong, C., 1994. The upper mantle S and P velocity structure beneath northern Australia from broadband observations. Physics of the Earth and Planetary Interiors 86, 85–98.

Kopylova, M.G., Russell, J.K., 2000. Chemical stratification of cratonic lithosphere: constraints from the northern Slave craton, Canada. Earth and Planetary Science Letters 181, 71–87.

Kovtun, A.A., Porokhava, L.N., 1980. Deep conductivity distribution of the Russian platform from the results of combined magnetotelluric and global magnetovariational data interpretation. Journal and Geomagnetism and Geoelectricity 32 (Suppl. 1), 105–133.

Lehman, I., 1955. The times of P and S in northeastern America. Annales Geofisica 8, 351–370.

Levin, V., Park, J., 1998. P–SH conversions in layered media with hexagonal symmetric anisotropy: a cookbook. Pure and Applied Geophysics 151, 669–697.

Nolet, G., Grand, S.P., Kennett, B.L.N., 1994. Seismic heterogeneity in the upper mantle. Journal of Geophysical Research 99, 23753–23766.

Oreshin, S., Vinnik, L., Makeyeva, L., Kosarev, G., Kind, R., Wentzel, F., 2002. Combined analysis of SKS splitting and regional P traveltimes in Siberia. Geophysical Journal International 151, 393–402.

Pavlenkova, N.I., 1997. General features of the upper mantle structure from seismic data. In: Fuchs, K. (Ed.). Upper Mantle Heterogeneities from Active and Passive Seismology. Kluwer Academic Publishing, Netherlands, pp. 225–236.

Rondenay, S., Bostock, M.G., Shragge, J., 2001. Multiparameter two-dimensional inversion of scattered teleseimsic body waves. 3: application to the Cascadia 1993 data set. Journal of Geophysical Research 106, 30795–30807.

Rümpker, G., Silver, P.G., 1998. Apparent shear-wave splitting parameters in the presence of vertically varying anisotropy. Geophysical Journal International 135, 790–800.

Savage, M.K., 1999. Seismic anisotropy and mantle deformation: what have we learned from shear wave splitting. Review of Geophysics 37, 65–106.

Shirey, S.B., Harris, J.W., Richardson, S.H., Fouch, M.J., James, D.E., Cartigny, P., Deines, P., Viljoen, F., 2002. Diamond genesis, seismic structure, and evolution of the Kaapvaal–Zimbabwe craton. Science 297, 1683–1686.

Silver, P.G., 1996. Seismic anisotropy beneath the continents: probing the depths of geology. Annual Review of Earth and Planetary Sciences 24, 385–432.

Silver, P.G., Savage, M.K., 1994. The interpretation of shear wave splitting parameters in the presence of two anisotropic layers. Geophysical Journal International 119, 949–963.

Simon, R.E., Wright, C., Kgaswane, E.M., Kwadiba, M.T.O., 2002. The P wavespeed structure below and around the Kaapvaal craton to depths of 800 km, from traveltimes and waveforms of local and regional earthquakes and mining-induced tremors. Geophysical Journal International 151, 132–145.

Snyder, D.B., Asudeh, I., Darbyshire, F., Drysdale, J., 2002. Field-based feasibility study of teleseismic surveys at high northern latitudes, Northwest Territories and Nunavut. Current Research - Geological Survey of Canada 2002-C3 (10 pp.).

Thybo, H., Perchuc, E., 1997. The seismic 8° discontinuity and partial melting in continental mantle. Science 275, 1626–1629.

Tittgemeyer, M., Wenzel, F., Fuchs, K., Ryberg, T., 1996. Wave propagation in a multiple-scattering upper mantle—observations and modeling. Geophysical Journal International 127, 492–502.

Van Decar, J.C., 1991. Upper-mantle structure of the Cascadia subduction zone from non-linear teleseismic travel-time inversion. PhD thesis, University of Washington, Seattle, WA.

Vinnik, L.P., Makayeva, L.I., Milev, A., Usenko, A.Y., 1992. Global patterns of azimuthal anisotropy and deformations in the continental mantle. Geophysical Journal International 111, 433–447.

Weeraratne, D.S., Forsyth, D.W., Fischer, K.M., Nyblade, A.A., 2003. Rayleigh wave tomography evidence for an upper mantle plume beneath the Tanzanian craton. Journal of Geophysical Research 108 (B9), 2427.

Available online at www.sciencedirect.com

Lithos 77 (2004) 873–922

www.elsevier.com/locate/lithos

Lithosphere mapping beneath the North American plate[☆]

W.L. Griffin[a,b,*], Suzanne Y. O'Reilly[a], B.J. Doyle[c], N.J. Pearson[a],
H. Coopersmith[d], K. Kivi[e], V. Malkovets[a], N. Pokhilenko[f]

[a] Department of Earth and Planetary Sciences, GEMOC ARC Key Centre, Macquarie University, Sydney, NSW 2109, Australia
[b] CSIRO Exploration and Mining, North Ryde, NSW 2113, Australia
[c] Kennecott Canada Exploration Inc., 200 Granville Street, Vancouver, BC, Canada V6C 1S4
[d] Great Western Diamond Co., PO Box 1916, Fort Collins, CO 80522, USA
[e] Kennecott Canada Exploration Inc., Thunder Bay, Ontario, Canada P7B 2Y1
[f] United Institute for Geophysics and Mineralogy, Russian Academy of Science, Novosibirsk, Russia

Received 27 June 2003; accepted 17 February 2004
Available online 2 June 2004

Abstract

Major- and trace-element analyses of garnets from heavy-mineral concentrates have been used to derive the compositional and thermal structure of the subcontinental lithospheric mantle (SCLM) beneath 16 areas within the core of the ancient Laurentian continent and 11 areas in the craton margin and fringing mobile belts. Results are presented as stratigraphic sections showing variations in the relative proportions of different rock types and metasomatic styles, and the mean Fo content of olivine, with depth. Detailed comparisons with data from mantle xenoliths demonstrate the reliability of the sections.

In the Slave Province, the SCLM in most areas shows a two-layer structure with a boundary at 140–160 km depth. The upper layer shows pronounced lateral variations, whereas the lower layer, after accounting for different degrees of melt-related metasomatism, shows marked uniformity. The lower layer is interpreted as a subcreted plume head, added at ca. 3.2 Ga; this boundary between the layers rises to <100 km depth toward the northern and southern edges of the craton. Strongly layered SCLM suggests that plume subcretion may also have played a role in the construction of the lithosphere beneath Michigan and Saskatchewan.

Outside the Slave Province, most North American Archon SCLM sections are less depleted than similar sections in southern Africa and Siberia; this may reflect extensive metasomatic modification. In E. Canada, the degree of modification increases toward the craton margin, and the SCLM beneath the Kapuskasing Structural Zone is typical of that beneath Proterozoic to Phanerozoic mobile belts.

SCLM sections from several Proterozoic areas around the margin of the Laurentian continental core (W. Greenland, Colorado–Wyoming district, Arkansas) show discontinuities and gaps that are interpreted as the effects of lithosphere stacking during collisional orogeny. Some areas affected by Proterozoic orogenesis (Wyoming Craton, Alberta, W. Greenland) appear to retain buoyant, modified Archean SCLM. Possible juvenile Proterozoic SCLM beneath the Colorado Plateau is significantly less refractory. The SCLM beneath the Kansas kimberlite field is highly melt-metasomatised, reflecting its proximity to the Mid-Continent Rift System.

[☆] Supplementary data associated with this article can be found, in the online version, at doi: 10.1016/j.lithos.2004.03.034.

* Corresponding author. Department of Earth and Planetary Sciences, GEMOC, Macquarie University, Sydney, NSW 2109, Australia. Fax: +61-2-9850-8943.

  E-mail address: wgriffin@els.mq.edu.au (W.L. Griffin).

  URL: http://www.es.mq.edu.au/GEMOC/.

A traverse across the continent shows that the upper part of the cratonic SCLM is highly magnesian; the decrease in mg# with depth is interpreted as the cumulative effect of metasomatic modification through time. The relatively small variations in seismic velocity within the continental core largely reflect the thickness of this depleted layer. The larger drop in seismic velocity in the surrounding Proton and Tecton belts reflects the closely coupled changes in SCLM composition and geotherm.

Keywords: SCLM; North America; Lithosphere

1. Introduction

All continental crust is underlain by a complementary shell of subcontinental lithospheric mantle (SCLM), and the composition of this SCLM is broadly related to its tectonothermal age (Boyd, 1989, 1997; Griffin et al., 1998, 1999b; O'Reilly et al., 2001). Seismic tomography shows that the cratonic parts of continents have thick SCLM, and its high Vp and Vs show that it is both cool and depleted compared to the SCLM under younger mobile belts. Its depleted nature makes this type of SCLM not only refractory but buoyant; unlike oceanic or Phanerozoic SCLM, it cannot be

delaminated by gravitational forces alone, and hence is difficult to destroy (Poudjom Djomani et al., 2001; O'Reilly et al., 2001). However, it can be modified by thermal events and the passage of fluids and melts, typically in conjunction with tectonic activity in overlying crust (Griffin et al., 2003a,b). The processes that have formed and modified the SCLM are recorded in xenoliths, and in xenocrysts derived from mantle wall rocks, brought up in volcanic rocks. These samples offer the chance to study the history of the SCLM and its relation to the overlying crust through time.

In this paper we report data on >5900 garnets from >85 kimberlites and related rocks in 27 areas across the

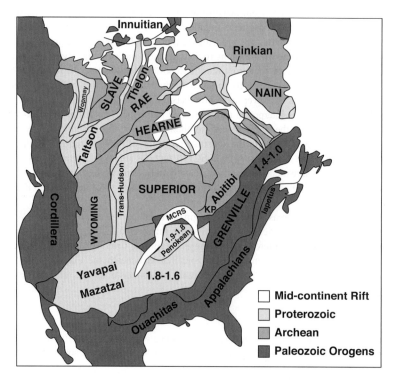

Fig. 1. Tectonic subdivisions of North America, after W.R. Church (http://instruct.uwo.ca/earth-sci/).

Table 1

Localities and number of garnets analysed

Locality	Pipe	No. of grains
Alberta	Three pipes	149
Arkansas	Prairie Creek	65
	Twin Knobs	43
Attawapiskat	Charlie	57
	Delta	59
	Tango	46
	Whiskey	79
Cobalt	Opap	49
	Bucke	43
	Peddie	67
	Nedelec	42
Colorado Plateau	Buell Park	83
	Garnet ridge	131
	Green Knobs	41
	Moses Rock	39
	The Thumb	11
Elliot County	Ison Creek	66
	Hamilton Branch	20
Grass Range	four pipes	236
James Bay lowland	Kyle Lake	59
	Pipe U	50
Kansas	Fancy Creek	52
	Lone Tree	61
	Leonardville	39
	Stockdale	40
	Winkler	33
Kirkland Lake	C-14	75
	A4	69
	Tandem	74
	B30	96
	Diamond Lake	53
Michigan	Eight pipes	380
Sarfartoq	>10 boulders	340
Saskatchewan	Fort a la Corne (n=7)	180
	Sturgeon Lake	10
	Candle Lake	97
Sextant Rapids		55
Slave Province	Anuri	149
	Doyle Lake	50
	Drybones	53
	Jericho	110
	Lac de Gras	785
	Snap Lake	266
	Tenacity	162
	W. Slave	175
Somerset Island	Batty	35
State Line	Chicken Park	50
	George Creek	63
	Iron Mountain	37
	Kelsey Lake 2	56
	Kelsey Lake 1	81
	Sloan 1	57
	Sloan 2	73

Table 1 (continued)

Locality	Pipe	No. of grains
	Sloan 5	52
Sukkertoppen	>5 dikes	252
Tenoma		73
Williams		58
Wisconsin	Six Pack	180
Total		5906

All garnets separated from rock samples except for Jericho and Tenacity (see text).

North American craton and its fringing mobile belts (Fig. 1, Table 1), and use these data to trace the lateral, vertical and temporal variation in the compositional and thermal structure of the continental root.

2. North America—geological setting

We use the tectonothermal-age terminology of Janse (1994) as modified by Griffin et al. (1998): *Archons* are areas where the last major tectonothermal event to affect the upper crust is >2.5 Ga old; *Protons* experienced their last tectonothermal event between 2.5 and 1.0 Ga, and *Tectons* are younger than 1.0 Ga. Archons, Protons and Tectons are typically underlain by different types of SCLM, reflecting a general secular evolution toward less depleted SCLM through time (see Griffin et al., 1998, 1999b for review).

The northern part of the continent (the Canadian Shield) consists of several large blocks of Archean crust, stitched together by PaleoProterozoic mobile belts (Fig. 1). However, some of the Archean blocks, such as the Hearn Province, have been extensively modified during this process of assembly, and thus should be regarded as Protons. Conversely, many Protons contain reworked Archean crust, and might be expected to be underlain by SCLM generated in Archean time. Most studies consider the Yavapai–Mazatzal terranes in the SW part of the continent to consist of juvenile Proterozoic crust, which might in turn be underlain by juvenile Proton SCLM.

The cratonic core built up of Archons and Protons is surrounded by several elongate Tectons, the Grenville (MesoProterozoic), Appalachian (Paleozoic) and Laramide mobile belts; the tectonic fronts of these toward the cratonic core are outlined in Fig. 2. The extensional terrain of the Basin Range Province and

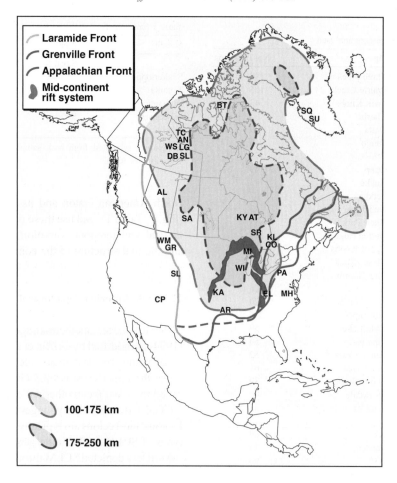

Fig. 2. Sample localities, and outline of the seismic "root" at 100–175 and 175–250 km.

the Cordillera are not considered here, but the Colorado Plateau, a strongly uplifted region of Proterozoic crust on the boundary of the Basin Range Province, is included.

The Proton area of the craton is bisected by the Mid-Continent Rift System, which reflects extension of the Laurentian continent 1.3–1.1 Ga ago (Allen et al., 1995). The main rift system describes a 2200-km concave arc with its apex under Lake Superior. It is geophysically defined by narrow high-amplitude gravity and magnetic anomalies, which reflect the alignment of deep narrow troughs filled by flood basalts and minor rhyolites, related intrusions and sediments; post-extension thermal relaxation produced wider basins filled with fluvial sediments (Cannon et al., 2001). Subparallel troughs, regarded

as parts of the same extensional system, extend through central Michigan and SE into Kentucky (Drahovzal et al., 1992), where the system is overridden by the Grenville Front.

3. North America—geophysical setting

Seismic tomography (Grand, 1994; Ritsema and van Heijst, 2000; Goes and van der Lee, 2002; van der Lee, 2001, 2002) has defined a "root" of high-velocity mantle beneath central North America, extending to depths of at least 250 km; the high velocities (both Vs and Vp) suggest a combination of relatively low temperatures and highly depleted mantle compositions (Griffin et al., 1999b; van der

Lee, 2001). This root is thickest in the northern part of the continent, south and west of Hudson Bay (Fig. 2). Its eastern and southern boundaries broadly parallel the Grenville and Appalachian Fronts, but only locally coincide with them, because these Fronts represent the outer edge of relatively thin-skinned thrusting, rather than deep lithospheric boundaries. Detailed studies across the eastern half of the continent from Missouri to Massachusetts (van der Lee, 2002) show that the edge of the root is defined by a discontinuity in lithosphere thickness, from 180–200 km to ca. 80 km under NE USA. Similarly, the western margin of the root as drawn here parallels, but does not coincide with, the Laramide Front. No root is apparent beneath the Archean portion of southern Greenland, but this probably reflects the thermal effects of the Tertiary opening of the Davis Strait.

Thybo et al. (2000) have defined a transition between a central "cold" part of the continent and an outer "hot" part, using an analysis of delay and scattering in explosion seismic sections. The transition zone between the two regimes corresponds to a narrow, steeply dipping zone with a high concentration of intraplate earthquakes. It closely follows the outline of the root at 100–175 km as shown in Fig. 2, and confirms that the thermal/compositional boundary defined by the seismic tomography studies also reflects a fundamental difference in lithospheric strength.

A map of the crustal magnetisation across North America has been derived by Purucker et al. (2002) using satellite magnetic data, corrected for temperature effects. High magnetic thickness partly reflects lower-crustal temperatures and hence mantle heat flow (Wasilewski and Mayhew, 1982). The "magnetic craton boundary" of Purucker et al. (2002) coincides well with the outline of the cratonic root along its eastern and southern edges, and extends further westward toward the Laramide Front, possibly reflecting the presence of Archean and Proterozoic crust under parts of the Rocky Mountains.

Artemieva and Mooney (2001) have derived lithosphere thicknesses using the downward extrapolation of heat flow data. Their map of North America places the greatest thicknesses (200 km) just south of Hudson Bay, and the lowest thicknesses (<100 km) beneath the Basin Range Province, but otherwise shows little correlation with the root as defined by seismic tomography. The lack of agreement reflects both the quality and distribution of heat flow data, and the many assumptions involved in the extrapolation of surface heat flow data to subcrustal depths (e.g. O'Reilly et al., 1997).

4. Methods

4.1. Analytical methods

The samples used in this study (Fig. 2, Table 1) are peridotitic garnet xenocrysts from kimberlites and other volcanic rocks. The analysed grains are a representative selection of the types present, as reflected in colour variations. Major elements have been analysed by electron microprobe (EMP), and the EMP data have been used to further select representative populations for trace element analysis by proton microprobe (Ni, Zn, Ga, Sr, Y, Zr; before 1995) or laser-microprobe ICPMS (>25 elements; after 1995). Only the Ca–Cr relationships (Fig. 3) and the Ni and Y contents (Fig. 4) of the garnets are presented here, for reasons of space. Detailed discussions of the garnet database, including analytical techniques, data quality and detection limits, are given by Griffin et al. (1999a, 2002b). Analytical data on garnet xenocrysts used here are given in Appendix A (Supplementary Data).

4.2. Thermometry and barometry

The key technique used in constructing the mantle sections presented here is the determination of the equilibration temperature of each garnet grain, using the Ni thermometer as calibrated by Ryan et al. (1996). The use of an alternative calibration by Canil (1994, 1999) (see discussion by Griffin and Ryan, 1996) would simply compress the top and bottom of these sections (in a geologically unrealistic way), without changing the compositional or tectonic relationships discussed below. To estimate the depth from which a grain has been derived, its Ni temperature is referred to a local paleogeotherm. These paleogeotherms can be derived from geothermobarometric analyses of xenolith suites, where available, or through calculation of pressure (P_{Cr}) for each garnet grain (Ryan et al., 1996; Fig. 5). Only garnets coexisting with chromite will give meaningful pressure estimates; others give min-

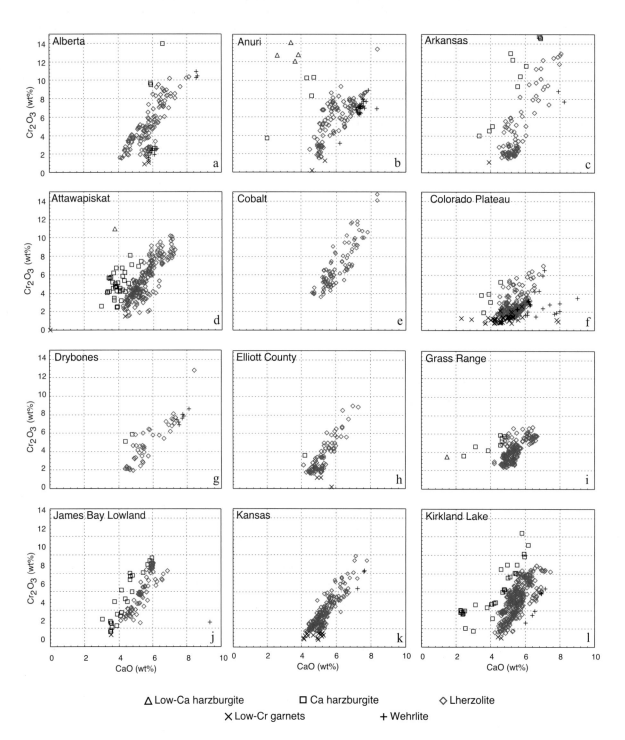

Fig. 3. CaO–Cr$_2$O$_3$ plots for garnets from each locality.

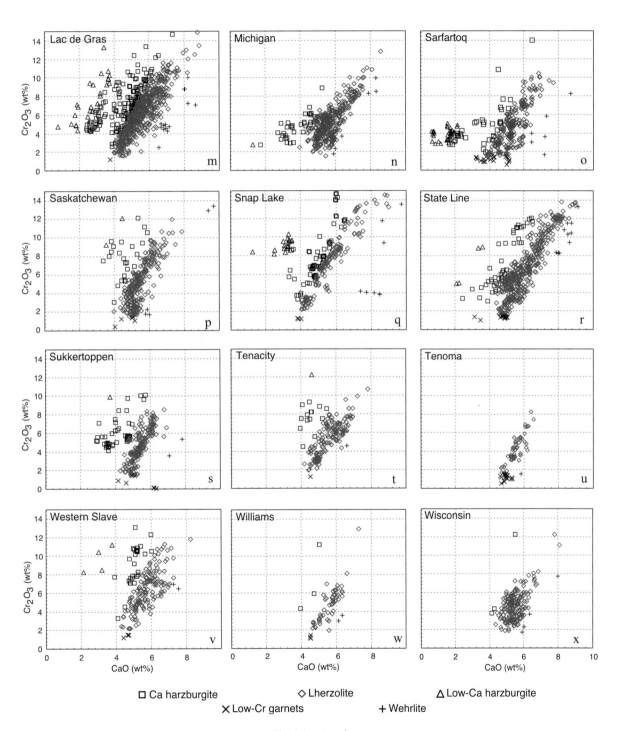

Fig. 3 (*continued*).

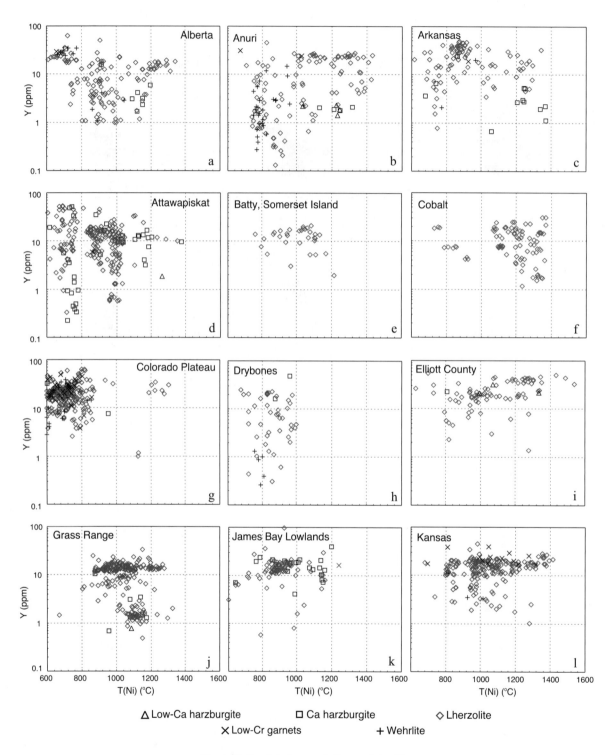

Fig. 4. Y–T_{Ni} plots for garnets from each locality.

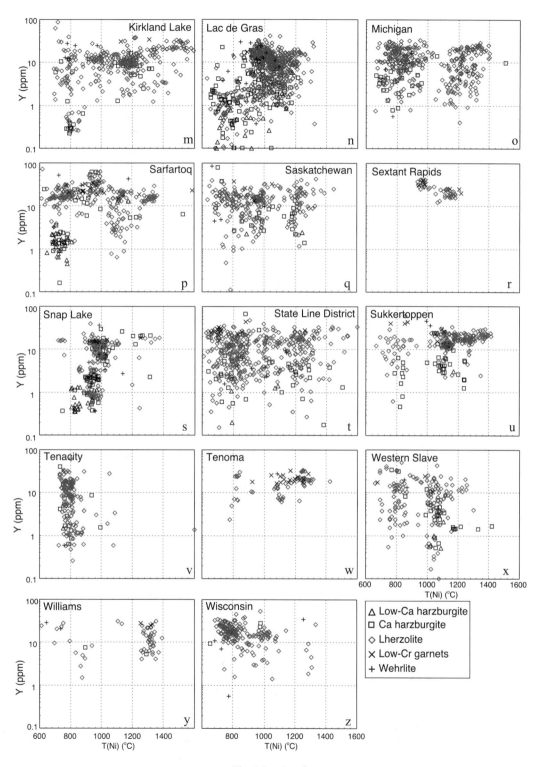

Fig. 4 (*continued*).

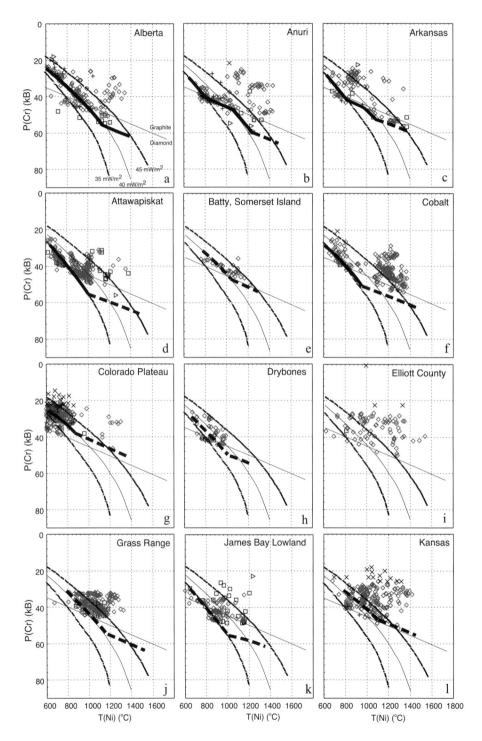

Fig. 5. Garnet geotherm plots for each locality. The inferred geotherm is shown by the thick dashed line. Also shown are model conductive geotherms (for reference) and the diamond–graphite transition.

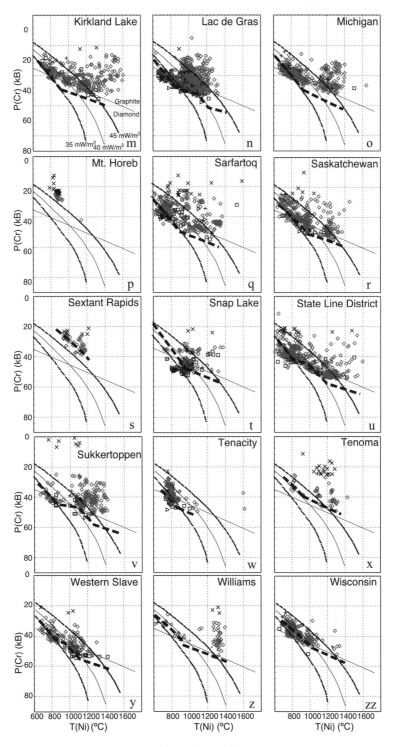

Fig. 5 (*continued*).

Table 2
Summary statistics for garnets from different localities

Locality	N		Sc	V	Co	Ni	Ga	Sr	Y	Zr	Nb	La	Ce	Pr	Nd	Sm	Eu	Gd	Dy	Ho	Er	Yb	Lu	Hf	SiO_2	TiO_2	Al_2O_3	Cr_2O_3	FeO	MnO	MgO	CaO	Na_2O
Alberta	152	average	166	245	38.6	39.7	6.38	1.60	18.0	29.3	0.82	0.69	1.35	0.53	2.83	1.61	0.62	2.41	3.81	0.95	2.76	2.86	0.65	0.94	41.2	0.17	19.6	5.35	8.10	0.43	19.1	5.87	0.04
		S.D.	53.6	106	5.96	25.0	4.28	3.37	69.9	25.5	0.78	0.78	1.56	0.65	2.45	1.75	0.33	6.05	12.4	3.08	8.68	8.45	1.98	0.46	0.60	0.20	2.78	3.41	0.87	0.09	0.99	0.90	0.03
		median	162	228	38.0	32.6	5.35	0.70	8.99	21.5	0.63	0.38	0.83	0.33	1.98	1.33	0.53	1.58	2.56	0.63	1.74	1.83	0.25	0.81	41.3	0.08	20.4	4.54	7.80	0.43	19.1	5.79	0.04
Arkansas	108	average				41.9	6.78	1.32	20.1	27.5															41.2	0.25	19.6	5.25	7.61	0.38	19.7	5.63	0.47
		S.D.				28.7	2.82	1.37	13.6	30.6															0.67	0.23	3.17	3.91	0.85	0.10	0.75	0.96	0.14
		median				30.0	6.37	1.17	18.2	13.0															41.4	0.15	20.9	3.36	7.68	0.39	19.7	5.40	0.48
Attawapiskat	241	average	151	221	38.6	34.0	5.40	0.53	12.7	33.5	0.42	0.26	0.06	0.58	0.23	2.07	1.35	0.57	2.43	0.49	1.34	1.43	0.25	0.58	41.4	0.19	19.6	5.30	7.92	0.35	20.0	5.23	0.01
		S.D.	42.9	87.9	3.54	18.8	2.90	0.68	11.3	25.3	0.26	0.39	0.06	0.64	0.24	1.81	0.90	0.39	2.05	0.43	1.23	1.25	0.19	0.45	0.58	0.26	1.74	2.05	0.73	0.10	1.95	0.97	0.01
		median	140	218	38.6	32.3	5.68	0.33	10.6	29.1	0.39	0.09	0.04	0.39	0.17	1.64	1.14	0.48	1.96	0.41	1.03	1.09	0.20	0.52	41.4	0.16	19.9	4.89	7.79	0.34	20.1	5.18	0.01
Cobalt	201	average	132	313	44.2	66.5	8.52	0.82	11.6	43.5	0.54	9.07	0.50	0.10	1.49	1.06	0.47	1.77	0.45	0.41	1.46	1.57	0.24	1.42	41.6	0.34	19.5	5.32	7.13	0.30	19.8	5.61	0.04
		S.D.	37.1	64.7	4.52	30.6	2.89	0.82	6.99	42.8	0.32	1.99	0.45	0.04	0.90	0.49	0.23	0.72	0.25		0.79	0.79	0.14	1.47	0.52	0.27	1.76	2.43	0.79	0.09	0.79	0.84	0.03
		median	125	308	43.5	76.4	7.85	0.47	11.3	37.3	0.46	8.43	0.34	0.11	1.26	0.97	0.43	1.68	0.41	0.25	1.34	1.45	0.24	1.16	41.7	0.25	19.6	4.92	6.84	0.29	20.0	5.46	0.04
Colorado Plateau	305	average				14.8	6.42	0.89	20.8	19.1															41.7	0.11	21.8	2.40	8.91	0.40	19.0	5.33	0.03
		S.D.				14.0	3.15	0.88	11.5	25.8																0.10	0.92	1.15	1.39	0.08	1.42	0.78	0.02
		median				12.3	5.58	0.67	19.9	10.2															41.7	0.09	22.0	2.10	8.82	0.40	19.0	5.27	0.03
Elliot County	86	average	105	269	49.2	60.4	11.8	1.36	25.2	70.4	0.53	0.14	0.49	0.13	1.15	1.06	0.52	2.19	4.34	1.04	3.19	3.41	0.59	1.70	41.6	0.49	20.5	3.39	8.37	0.32	19.9	5.23	0.02
		S.D.	22.9	73.5	8.46	32.4	4.17	1.66	17.2	56.4	0.29	0.17	0.36	0.07	0.63	0.80	0.40	1.17	2.98	0.74	2.33	2.70	0.54	1.52	0.41	0.40	1.28	1.85	0.97	0.08	0.84	0.65	0.03
		median	104	283	48.3	52.5	11.5	0.73	20.8	46.6	0.47	0.09	0.42	0.11	1.04	0.86	0.42	1.89	3.67	0.89	2.40	2.61	0.41	1.02	41.6	0.31	20.7	2.58	8.53	0.32	20.0	5.14	0.02
Grass Range	236	average	135	182	40.3	54.2	4.76	2.01	10.3	27.7	0.14	0.06	0.47	0.25	2.23	1.49	0.55	1.75	1.90	0.39	1.15	1.37	0.24	0.42	41.6	0.18	21.3	4.19	6.64	0.32	20.5	5.31	
		S.D.	33.3	50.9	5.74	16.5	3.44	11.6	6.24	17.2	0.08	0.10	0.35	0.41	1.29	0.85	0.34	1.11	1.42	0.24	0.68	0.72	0.11	0.20	0.33	0.10	0.81	1.16	0.59	0.07	0.65	0.68	
		median	131	168	40.5	53.8	4.02	0.48	12.4	26.4	0.12	0.04	0.39	0.19	1.98	1.39	0.50	1.70	1.99	0.45	1.33	1.52	0.25	0.40	41.7	0.09	21.3	4.04	6.70	0.35	20.4	5.25	
James Bay lowland (Kyle Lake)	109	average				35.5	7.66	0.78	14.6	50.3															41.7	0.15	19.3	5.78	7.69	0.36	19.9	5.17	0.03
		S.D.				17.2	2.70	0.55	10.2	35.8																							
		median				31.3	7.88	0.78	13.2	38.5																							
Kansas	225	average	109	245	42.1	55.3	8.87	0.31	15.2	29.6	0.45	0.04	0.29	0.09	0.87	0.68	0.34	1.44	2.43	0.59	1.82	2.10	0.34	0.71	41.6	0.32	20.5	3.65	7.90	0.34	19.9	5.35	0.04
		S.D.	29.2	64.8	3.18	25.6	2.35	0.38	6.99	22.3	0.35	0.07	0.29	0.07	0.57	0.35	0.17	0.66	1.14	0.28	0.83	0.84	0.12	0.63	0.42	0.26	1.36	1.87	0.77	0.10	0.82	0.72	0.02
		median	102	239	42.2	49.9	8.99	0.54	10.0	23.6	0.36	0.03	0.23	0.08	0.76	0.65	0.34	1.49	2.57	0.63	1.88	2.15	0.35	0.46	41.7	0.24	20.9	3.22	7.95	0.33	20.3	5.21	0.03
Kirkland Lake	367	average	129	284	46.7	66.2	8.07	0.68	11.7	37.7	0.50	0.11	0.73	0.22	1.81	0.93	0.31	1.15	2.14	0.48	1.43	1.52	0.26	0.94	41.6	0.32	19.5	5.29	7.38	0.36	20.5	5.42	0.03
		S.D.	40.3	71.8	6.55	39.1	3.91	0.42	9.32	30.3	0.26	0.23	1.41	0.29	1.70	0.98	0.40	1.15	1.52	0.35	1.07	1.04	0.16	0.73	0.49	0.17	1.60	1.88	0.73	0.09	0.74	0.89	0.02
		median	123	293	44.8	66.2	7.89	1.16	10.4	29.7	0.45	0.05	0.40	0.16	1.52	0.98	0.40	1.43	1.82	0.41	1.27	1.35	0.23	0.71	41.7	0.32	19.4	5.26	7.62	0.35	19.9	5.24	0.03
Michigan	385	average				49.3	7.54	4.47	11.4	44.0															41.8	0.26	19.5	5.24	7.15	0.34	19.6	5.66	0.19
		S.D.				33.7	3.12	34.4	7.83	35.6																0.83	0.22	1.68	1.85	0.90	0.10	1.38	1.29	0.20
		median				32.5	7.40	1.61	10.0	33.9															41.6	0.17	19.8	5.04	7.01	0.33	19.7	5.51	0.05
Sarfartoq	341	average				40.9	9.82	1.96	16.0	48.6															40.9	0.29	19.8	4.44	8.15	0.34	20.5	4.67	0.04
		S.D.				30.8	4.48	2.50	12.3	57.4															1.09	0.25	1.87	2.21	1.43	0.11	1.60	1.51	0.02
		median				29.6	10.1	1.16	15.4	26.8															41.2	0.20	20.4	4.10	7.76	0.34	20.3	5.05	0.04
Saskatchewan	287	average				44.2	8.07	1.06	15.4	41.5															41.2	0.25	20.4	5.42	7.51	0.37	19.7	5.45	0.02
		S.D.				27.3	3.40	1.20	14.3	40.8															0.27	2.13	2.58	0.96	0.37	0.08	0.94	0.78	
		median				38.4	7.67	0.74	13.1	26.9															0.13	19.6	5.13	7.34	0.36	19.7	5.30	0.04
Sextant Rapids	55	average	126	153	46.8	55.6	5.68	0.16	25.9	32.5	0.18	0.07	0.13	0.06	0.59	0.21	0.11	1.86	3.93	0.93	3.09	3.30	0.54	0.60	41.9	0.16	22.0	2.38	7.70	0.31	19.4	5.51	0.02
		S.D.	26.2	33.1	2.96	14.6	1.48	0.18	8.27	23.2	0.06	0.08	0.08	0.02	0.19	0.11	0.11	0.50	1.21	0.32	0.96	0.96	0.16	0.52	0.19	0.08	0.62	0.81	0.60	0.07	0.50	0.43	0.01
		median	124	149	45.8	57.1	5.67	0.11	24.6	30.8	0.17	0.04	0.11	0.06	0.63	0.72	0.38	1.87	3.78	0.98	2.95	3.30	0.51	0.51	41.9	0.13	22.3	2.13	7.74	0.31	19.3	5.52	0.02

Slave Craton

Lac de Gras

Upper	152	average	163	269	38.6	19.5	4.54	1.86	2.43	9.48	0.63		2.52			3.72	1.06	0.43	0.96		0.45	41.4		0.05	19.1		6.82	7.81	0.46	19.4	5.07 0.04
		S.D.	56.3	73.3	4.42	6.25	3.03	1.95	4.28	16.3	0.55		2.87			3.80	0.94	0.35	1.00		0.40	0.63		0.10	1.80		2.08	0.92	0.10	1.88	1.83 0.06
		median	147	267	38.8	18.1	3.71	1.19	1.12	3.12	0.50		1.40			2.48	0.77	0.34	0.58		0.29	41.4		0.02	19.4		6.57	7.70	0.02	19.1	5.29 0.02
Lower	633	average	129	303	42.5	52.8	8.37	1.00	10.4	33.4	0.75		0.94			2.28	1.18	0.50	0.61		0.91	41.5		0.26	19.1		6.33	7.22	0.39	19.7	5.38 0.06
		S.D.	29.0	71.4	5.47	13.9	2.68	1.33	6.81	21.2	0.60		1.31			2.10	0.70	0.29	0.81		1.93	0.56		0.16	2.03		2.56	0.61	0.85	1.24	0.95 0.07
		median	125	298	42.3	49.7	8.39	0.67	10.2	31.3	0.61		0.59			1.73	1.01	0.44	1.49		0.75	41.5		0.24	19.2		6.18	7.17	0.35	19.9	5.30 0.04

Anuri

Upper	72	average	196	354	45.0	22.1	4.84	0.66	4.47	9.95	0.59		1.27	0.34		2.41	1.10	0.45	1.21		0.33	40.9		0.11	19.4		6.79	7.99	0.51	17.4	7.12 0.05
		S.D.	49.9	107	5.54	5.87	2.92	0.83	7.44	19.0	0.79		4.04	0.60		3.08	1.75	0.79	2.05		0.51	0.49		0.16	1.22		1.52	0.80	0.09	1.69	2.21 0.01
		median	190	362	45.9	20.9	4.26	0.42	1.58	51.0	0.42		0.62	0.20		1.16	0.38	0.14	0.49		0.18	40.9		0.04	19.3		7.07	7.93	0.52	17.4	7.03 0.05
Lower	77	average	135	363	48.4	79.5	10.6	1.09	13.3	51.0	0.58		1.15	0.35		2.60	1.19	0.51	1.79		1.37	41.5		0.56	19.2		5.67	8.07	0.35	19.4	5.44 0.05
		S.D.	27.4	61.9	4.56	32.6	3.65	0.80	9.48	28.8	0.43		1.53	0.42		2.47	0.65	0.24	0.88		0.76	0.83		0.35	2.20		3.42	1.38	0.08	1.42	1.23 0.02
		median	129	352	48.2	79.7	10.1	0.94	11.7	60.3	0.46		0.58	0.25		1.98	1.19	0.51	2.00		1.51	41.4		0.68	20.2		5.28	7.94	0.34	19.5	5.21 0.05
Tenacity	163	average	141	262	39.9	23.3	5.63	0.73	10.4	29.1	0.36		2.24	0.46		2.72	1.37	0.49	1.72		0.81	41.3		0.14	20.2		5.47	7.88	0.46	19.2	5.48
		S.D.	53.7	73.5	3.53	18.1	2.85	2.91	9.32	34.3	0.81		15.7	10.8		6.12	2.27	0.40	1.58		0.79	0.37		0.11	1.54		2.00	0.73	0.08	0.75	0.77
		median	124	255	39.7	20.2	5.51	0.21	8.54	17.1	0.23		0.44	0.13		1.74	0.95	0.38	1.26		0.55	41.3		0.11	20.1		5.67	7.81	0.45	19.2	5.42

Western Slave

Upper	63	average	137	249	39.6	20.4	5.28	0.51	12.2	28.4	0.29		0.82			2.75	1.55	0.55	2.12		0.59	41.3		0.10	19.3		5.35	8.44	0.41	19.0	5.76 0.03
		S.D.	34.0	59.4	3.49	4.77	3.15	0.85	11.3	13.5	0.31		1.19			2.20	0.86	0.29	1.17		0.36	0.44		0.07	1.21		1.59	1.25	0.07	0.98	0.75 0.02
		median	131	241	39.0	20.0	3.66	0.21	10.3	26.5	0.20		0.39			1.93	1.36	0.51	1.89		0.50	41.2		0.09	19.2		5.58	8.11	0.41	19.1	5.64 0.03
Lower	112	average	138	315	40.9	57.6	6.41	1.00	8.30	36.7	0.80		1.51			3.45	1.39	0.49	1.69		0.96	41.4		0.25	17.5		7.58	6.93	0.25	20.2	5.50 0.04
		S.D.	32.5	61.3	4.07	15.0	3.18	0.95	7.82	28.2	0.70		3.80			2.52	0.84	0.31	1.13		0.64	0.49		0.21	1.89		2.50	0.87	0.07	0.95	0.88 0.02
		median	134	319	40.5	55.0	5.69	0.68	5.86	33.6	0.60		0.81			2.55	1.18	0.45	1.49		0.84	41.3		0.19	17.6		7.67	6.69	0.25	20.2	5.39 0.04
Somerset Island	32	average				47.4	7.81	1.08	12.5	31.8												42.3		0.20	19.7		4.92	6.23	0.32	20.3	5.41
		S.D.				16.4	2.94	0.64	5.07	15.2											0.24			0.12	0.92		1.07	0.34	0.05	0.36	0.32
		median				50.4	7.82	0.96	13.4	26.8											42.3			0.18	19.6		4.96	6.17	0.31	20.3	5.46

State Line

Upper	237	average				19.5	6.29	0.99	11.0	28.0											41.4		0.09	20.9		4.83	7.94	0.42	19.4	5.41 0.04
		S.D.				6.36	2.75	2.47	9.52	36.3											0.49		0.16	11.9		1.97	0.90	0.07	0.86	0.89 0.02
		median				19.8	5.87	0.50	8.38	17.4											41.5		0.07	20.2		4.95	7.75	0.43	19.3	5.34 0.04
Lower	231	average				75.1	10.2	1.57	12.4	59.9									40.8	8.50		7.09	0.35		19.0	6.57 0.04
		S.D.				28.4	4.02	2.97	13.0	45.3									0.79	3.13		0.93	0.07		1.11	1.23 0.02
		median				76.3	9.82	1.13	10.1	51.2									40.8	9.10		6.90	0.34		19.0	6.54 0.04
Sukkertoppen	259	average				69.3	11.6	1.55	15.5	64.3											40.5		0.53	19.1		4.85	7.64	0.27	20.6	5.18 0.06
		S.D.				28.2	3.32	2.02	8.20	41.4											0.96		0.31	1.33		1.89	1.53	0.08	1.29	0.77 0.03
		median				68.4	12.4	0.92	16.2	68.6											40.8		0.61	19.2		4.93	7.25	0.26	20.7	5.25 0.05
Tenoma	73	average	122	218	43.2	67.7	7.32	0.68	18.2	64.3	0.55		0.09	0.59		0.19	1.67	1.09	0.50		1.50	41.6		0.39	21.4		3.50	7.71	0.34	19.8	5.43
		S.D.	32.5	44.6	6.43	26.2	1.89	0.61	7.84	31.1	0.29		0.13	0.32		0.09	0.75	0.47	0.20		0.87	0.43		0.21	1.34		1.87	0.84	0.09	0.46	0.48
		median	117	217	41.2	70.7	7.40	0.59	18.6	69.2	0.52		0.06	0.52		0.17	1.58	1.07	0.53		1.55	41.6		0.48	21.6		3.41	7.73	0.33	20.0	5.42
Wisconsin	180	average				29.8	7.84	1.20	17.1	35.4											20.2		4.54	8.11		0.41	19.4		5.50 0.03
		S.D.				18.6	2.32	0.91	8.50	23.2											1.37		1.69	0.96		0.09	0.68		0.59 0.02
		median				22.1	7.66	1.03	16.0	31.5											20.4		4.34	7.93		0.42	19.4		5.43 0.03

imum values. The garnet-based geotherm (Fig. 5) is defined by the highest pressure estimates at each temperature (allowing for ± 50 °C uncertainties), and the geotherm is considered to remain near a conductive model up to the temperature estimated for the base of the depleted lithosphere (see below).

At higher T, where chromite is less likely to be stable, few garnets will give the maximum P_{Cr}, and geotherm is only constrained to a minimum slope. In this case, we have drawn the "geotherm" parallel to the diamond–graphite stability curve. There is no theoretical reason for choosing this slope; it is done by analogy with the "kinked limb" seen in many xenolith-based geotherms (Finnerty and Boyd, 1987). This procedure may underestimate the depths of the hottest garnets, compared to a xenolith-based geotherm, but this uncertainty does not affect the use made of the data in this paper.

4.3. Geochemical information from garnets

Estimates of the temperature corresponding to the base of the depleted lithosphere can be derived from plots of the Y content of garnets (a measure of depletion; Griffin and Ryan, 1995; Griffin et al., 1999a) against T_{Ni} (Fig. 4). The median Y content of Cr-pyrope garnets from Archean and Proterozoic SCLM is 10–15 ppm (Table 2), and values less than this can be regarded as evidence of derivation from strongly depleted peridotites, interpreted here as lithospheric material. These plots typically show a relatively sharp high-temperature limit to the distribution of Y-depleted garnets (e.g. Fig. 4a), which can be regarded as representing the temperature at the base of the depleted lithosphere.

The inter-element correlations in a large database of mantle-derived Cr-pyrope garnets have been described and interpreted in terms of processes (depletion, metasomatism) by Griffin et al. (1999a). This database also was used by Griffin et al. (2002b) to evaluate approaches to the definition of populations using multivariate statistics. The Cluster Analysis by Recursive Partitioning (CARP) technique recognised 15 distinctive populations, which show significant variations in relative abundance and depth distribution in the SCLM across different tectonic settings. By applying the same techniques to garnets from ca. 200 well-described xenoliths

from kimberlites and other volcanic rocks, these populations have been correlated in detail with specific rock types, affected by specific processes (Griffin et al., 2002b).

Based on these correlations, the CARP classes can be grouped into five major categories. *Depleted harzburgites* as defined here contain subcalcic garnets (CaO <4%) depleted in Y, Ga, Zr, Ti and HREE; *depleted lherzolites* have garnets with Ca–Cr relationships indicating equilibration with clinopyroxene (Griffin et al., 1999a), but depleted in HREE, HFSE and Ga. (Note that this thermodynamic definition of "lherzolite" will include "harzburgites" with <5% clinopyroxene). The garnets of *depleted/metasomatised* lherzolites are depleted in Y and HREE, but enriched in Zr and LREE, suggesting that they experienced depletion and subsequent re-fertilisation; xenoliths of this type commonly contain phlogopite±amphibole. The garnets of *fertile lherzolites* have high contents of HREE and near-median contents of HFSE; they retain no evidence of a depletion event. The garnets of *melt-metasomatised peridotites* show a characteristic enrichment in Zr, Ti, Y and Ga ($\pm$Fe), and correspond to the sheared and enriched lherzolite xenoliths found in many kimberlites. In Figs. 6–8, we have plotted the relative abundances of these major categories against depth. The data have been averaged in 100 °C windows, overlapped by 50 °C to smooth local variations. These sections illustrate the distribution of important rock types with depth at each locality.

Given the major-element composition and T_{Ni} of a Cr-pyrope garnet, and an estimate of its depth of origin, it is possible to calculate the mg# (100 Mg/ (Mg+Fe), or %Fo) of the coexisting olivine (Gaul et al., 2000). This is an important parameter in determining the physical properties (including the seismic response) of ultramafic rocks. We have calculated this composition for each garnet grain, and present the data (Figs. 6–8) in terms of the mean olivine composition at each depth, averaged over windows ranging from 50 to 150 °C wide, depending on data density.

Median values of some important compositional parameters are given for the garnets of each locality in Table 2, together with equivalent data for Archon, Proton and Tecton garnets worldwide (Griffin et al., 2002b).

5. Results: the cratonic core

5.1. Slave Craton (Fig. 6)

The Slave Craton in NW Canada is a fragment of a larger Archon, surrounded on three sides by Proterozoic mobile belts. The western part of the craton contains the oldest known rocks, the Acasta gneisses (4.04 Ga) surrounded by belts of younger crust, making up the Central Slave Basement Complex (CSBC, 3.7–3.0 Ga; Bleeker et al., 1999). The basement of the eastern half of the craton appears to be a significantly younger (2.8–2.7 Ga) terrane, and both parts are overlain by a turbidite sequence which has been dated at 2680 Ma at locations across the entire craton (Bleeker, 2001). The nature of the boundary between the younger and older halves of the craton is unclear, but Pb-isotope data on conformable base-metal occurrences, and Pb and Nd-isotope data on late granites (Davis and Hegner, 1992; Davis et al., 1996; Davis, pers. comm.) suggest that the CSBC extends under the Eastern Slave Arc Terrane for at least 100 km east of the surface expression of the terrane boundary (Fig. 6). The boundary may represent a ca. 2.7 Ga suture, with the eastern part of the craton thrust over the western, or the eastern Slave may be underlain by the extended and modified basement of the CSBC. An analysis of lithospheric strength by Poudjom Djomani et al. (abstract, this conference) suggests the presence of a major lithospheric boundary, corresponding roughly to the Nd-isotope line.

Griffin et al. (1999d,e) mapped a pronounced layered structure in the SCLM beneath the Lac de Gras area in the middle of the craton, and Davies et al. (1999, 2004a) have shown that the diamond populations, derived largely or entirely from the lower layer of the SCLM, contain a high proportion of inclusions of the ultradeep paragenesis derived from the lower mantle. These data were used to infer that the lower layer of the SCLM may represent an accreted plume head. In-situ Re–Os analysis of sulfide inclusions in mantle olivines gives model and isochron ages in excess of 3.1 Ga, and these sulfides also have unusual Co- and W-rich compositions that are consistent with derivation from the lower mantle (Aulbach et al., 2004b). Graham et al. (unpubl. data) have obtained a whole-rock Re–Os isochron age of 3.1 ± 0.2 Ga for nine eclogite from the Lac de Gras kimberlites.

In this study we present a new analysis of the concentrate data discussed by Griffin et al. (1999d,e) together with new data from both the northern and the southern parts of the craton.

5.2. Lac de Gras area

Dozens of kimberlites are known in the central part of the craton around Lac de Gras; data presented here are derived from the Diavik pipes on and south of Lac de Gras described by Griffin et al. (1999d,e; $n=14$), and two pipes (Point Lake, Mark) from the north side of the lake. These kimberlites are Eocene in age (50–55 Ma; Heaman et al., 1997; Creaser et al., 2004).

The garnet data (Fig. 3m) show a thick lherzolite trend from 1% to 15% Cr_2O_3, and a large number of mildly subcalcic garnets spanning the same Cr range; strongly subcalcic garnets range from 4% to 13.5% Cr_2O_3; wehrlitic garnets are rare. The $Y–T$ plot (Fig. 4n) shows a pronounced layering; garnets with $T \leq 900$ °C are nearly all extremely depleted, with $Y < 2$ ppm; garnets with $T \geq 900$ °C rarely contain <1 ppm Y, and the median value is >10 ppm. Similar patterns are seen in Zr, Ga and Ti (Griffin et al., 1999d,e); low-T garnets rarely contain >5 ppm Zr, whereas higher-T ones rarely have <5 ppm Zr, and may contain up to 100 ppm. Despite relatively high Zr contents, there are few garnets with $TiO_2 > 0.6\%$. A significant proportion of the garnets, especially the subcalcic types, have sinuous REE patterns; most of these have $T \leq 1000$ °C, and "normal" non-sinuous patterns are rare in the upper layer ($T < 900$ °C). Subcalcic garnets are concentrated in, but not restricted to, the upper ultradepleted layer. Garnets with $T > 1200$ °C are rare, but tend to have the highest Ti contents.

The garnet data from the upper layer lie below or near a 35 mW/m^2 conductive geotherm (Fig. 5n), while those in the lower layer are consistent with a 40 mW/m^2 conductive model; the Y edge at 1200 °C (Fig. 4n) thus corresponds to a lithosphere thickness of 190–200 km. Pearson et al. (1999) showed that $P–T$ estimates for xenoliths from this area, including those described by MacKenzie and Canil (1999) show the same stepped geotherm, regardless of the geothermometer/geobarometer combinations used. The xenolith data therefore validate the garnet geotherm, although it is not clear how the stepped geotherm

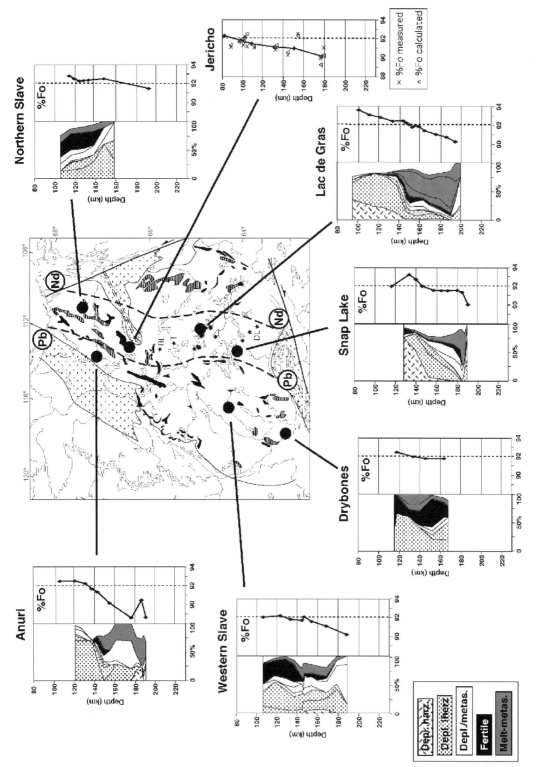

Fig. 6. CARP sections and mean %Fo in olivine vs. depth, for Slave Craton localities. Pb and Nd isotope lines from Davis and Hegner (1992), Davis et al. (1996) and Davis, pers. comm. DL, Doyle Lake, RL, Ranch Lake, T, Torrie.

could be maintained over long time periods (see Griffin et al., 1999e for discussion).

The mean calculated olivine composition of the upper layer (Fig. 6) drops from $Fo_{93.5}$ at shallow depth to $Fo_{92.5}$ at the layer boundary. It drops sharply to $Fo_{91.8}$ from 140–150 km, then declines gradually to $Fo_{90.7}$ at the base of the layer. The deepest olivine is less magnesian ($Fo_{90.2}$) reflecting a higher degree of melt-related metasomatism at depths of ca. 190 km. These estimates are consistent with xenolith data from both the upper and lower layers (Gaul et al., 2000; Pearson et al., 1999). The rapid drop in Mg# near the layer boundary correlates with a pronounced metasomatic signature (high Sr, LREE) in clinopyroxene from the top of the lower layer (data not shown).

The upper boundary of the garnet data in the CARP section (Fig. 6) corresponds to the garnet–spinel peridotite transition near 100 km, which is sharp and relatively deep in these depleted rocks (Griffin et al., 1999e). The extreme depletion of the upper layer is reflected in a high proportion of depleted harzburgites (decreasing downward) and depleted lherzolites. The lower layer is a mixture of depleted harzburgites and lherzolites, depleted/metasomatised lherzolites, and melt-metasomatised lherzolites; the latter type increases in abundance downward, and becomes dominant by 190 km depth.

Snyder et al. (2004), using broadband teleseismic data, has identified a marked seismic discontinuity at 100 km, which corresponds to the spinel–garnet transition mapped here, and to the top of a highly conductive layer in the SCLM (Jones et al., 2001). The teleseismic data also image the sharp 150 km discontinuity between the upper and lower layers of the CARP section, and another discontinuity at 190–200 km, which coincides with the electrically determined lithosphere–asthenosphere boundary and the base of the depleted lithosphere as defined by the garnet data. The seismic and magnetic data thus provide independent validation of the garnet geothermometry/barometry approach used here.

5.3. Snap Lake

The Snap Lake kimberlite is a large sill in the southern part of the craton, with an age of 535 Ma (Pokhilenko et al., 2003).

The garnet data (Fig. 3q) define a very long and narrow lherzolite trend, from 1% to 15% Cr_2O_3, paralleled by mildly subcalcic garnets. Strongly subcalcic garnets are less common than in Lac de Gras, and define a small cluster with 8–10.5% Cr_2O_3; a few wehrlitic garnets also are present. The Y–T plot (Fig. 4s) indicates the presence of layering like that seen at Lac de Gras, but less distinct. Most garnets (nearly all subcalcic) with $T < 900$ °C have Y < 2 ppm; few garnets with $T > 1000$ °C have < 5 ppm Y, and nearly all with $T > 1100$ °C have > 10 ppm Y. However, between 900 and 1000 °C the entire range of Y contents is present. Nearly all garnets with $Cr_2O_3 > 10\%$ are in this T range, while higher-T garnets have distinctly lower Cr contents and higher Zr and Ti contents. Many of the subcalcic garnets and some lherzolitic ones have sinuous REE patterns, and these all have $T \leq 1000$ °C.

The small T range gives poor definition of the geotherm; most data are consistent with a 35 mW/m^2 geotherm up to 1000 °C (Fig. 5t), and the high proportion of melt-related metasomatism above this temperature suggests that the geotherm kinks at ca. 1000 °C. The mean calculated olivine composition is Fo_{92-93} down to 150 km; then $Fo_{91.6}$ to 185 km, and drops rapidly with depth below 185 km.

The CARP section (Fig. 6) can only be constructed from 130 to 190 km. The proportion of subcalcic harzburgites is highest at the top of the section, and decreases rapidly to < 10% at 150 km. The lower part of the section shows a decrease in depleted rock types and an increase in depleted/metasomatised lherzolites and melt-related metasomatism with depth. In general, the layering shown by this section is similar to that seen in the Lac de Gras area, but the transition between the layers is 20–25 km thick, while under Lac de Gras it is 5–10 km. A similar diffuse transition between the upper and lower layers was observed in the Ranch Lake kimberlite north of Lac de Gras (Griffin et al., 1999e). The smaller proportion of melt-related metasomatism in the lower layer may be related to greater age of the Snap Lake intrusion.

5.4. Drybones

The Drybones kimberlite (442–485 Ma; Carbno and Canil, 2002; Heaman et al., 2003) intrudes plutons of the 2620 Ma Defeat Suite on the SW edge of the craton, near the Great Slave Lake Fault Zone. The limited garnet data (Fig. 3g) show a complex lherzolite trend. The main group (2–8.5% Cr_2O_3) trends into the field of wehrlitic garnets, suggesting derivation from spinel–garnet peridotites; a shorter trend at lower Ca/Cr (4–6.5% Cr_2O_3) is derived from more typical garnet lherzolites, and one grain with 12.8% Cr_2O_3 belongs to this group. The dataset includes two weakly subcalcic garnets. The Y–T plot (Fig. 4h) shows no garnets with $T > 1000$ °C; many are strongly Y-depleted; there is no indication of a high-T Y edge, and no indication of layering like that seen at Lac de Gras. However, some tendency to layering is observed in the distribution of Zr contents: garnets with $T < 900$ °C have Zr contents of 3–25 ppm, whereas those with $T > 900$ °C mostly contain 10–200 ppm Zr. None has $TiO_2 > 0.3\%$. Roughly 1/3 of the higher-Cr garnets show HREE depletion and sinuous REE patterns with $Nd/Y)_N > 1$. Maximum Cr contents rise sharply with increasing T, and the maximum Cr content is found in garnets with T near 1000 °C. These data are very similar to those presented by Carbno and Canil (2002), but our dataset contains fewer garnets from spinel-free lherzolites, and a lower proportion of garnets with sinuous REE patterns.

The range of T is too small to define a geotherm but appears to lie near a 35–37 mW/m^2 conductive model (Fig. 5h); this is consistent with the steep increase in maximum Cr contents with T (Griffin and Ryan, 1995). The mean calculated olivine composition is $Fo_{92-92.5}$ above 140 km, and $Fo_{91.8}$ below that. The CARP section also suggests a compositional stratification: lherzolites in the upper part (115–140 km) are dominantly depleted, whereas those from 140–160 km depths are depleted/metasomatised and fertile types. This supports the Zr–T plot, and reflects the abundance of high-Cr garnets at shallow depth.

5.5. Western Slave

The Cross Lake kimberlites (Cross, Orion and Ursa, and Aquilia further north) have been dated to 450 Ma (Heaman et al., 2003). They intrude the Central Slave Basement Complex near the Sleepy Dragon supracrustal complex (Fig. 6).

The garnet data (Fig. 3v) define a long and complex lherzolite trend from 1% to 12% Cr_2O_3, with a minor spinel–garnet lherzolite trend at higher Ca/Cr. There are many mildly to strongly subcalcic garnets, most 7–13% with Cr_2O_3. The Y–T plot (Fig. 4x) shows two groups, 700–900 and 1000–1200 °C; subcalcic garnets and garnets from strongly depleted lherzolites (Y < 1 ppm) mostly occur in the higher-T group. There is a Y edge ca. 1100 °C, but several subcalcic garnets have higher T. Zr shows a similar distribution: garnets in the upper layer mostly contain 10–40 ppm Zr, those in the lower layer 2–100 ppm. In the lower layer, high Zr correlates with high Ti, indicating melt-related metasomatism. Nearly half of the garnets have sinuous REE patterns; a higher proportion, including nearly all of subcalcic garnets, occurs in the lower layer.

The garnets in the upper layer record a 35 mW/m^2 geotherm (Fig. 5y), whereas those in the lower layer are consistent with a 38–40 mW/m^2 conductive model, as at Lac de Gras. The mean olivine composition calculated for the upper layer is $Fo_{92.5}$, dropping to Fo_{92} at 145 km. In the lower layer this reverses to $Fo_{92.5}$, then drops steadily to $Fo_{<91}$ at the deepest levels. While the upper layer is less magnesian than that at Lac de Gras, the lower layer shows a similar distribution of Mg# to the lower layer at Lac de Gras.

The upper part of the CARP section, down to 145 km, shows no subcalcic harzburgites, an abundance of depleted lherzolites, and ca. 30% fertile lherzolites (Fig. 6). The deeper part, from 150 to 170 km, contains subcalcic harzburgites, depleted lherzolites, depleted/metasomatised lherzolites, and minor melt-related metasomatism. There is no strong increase in melt-related metasomatism at the base of the sampled section, and the depleted lithosphere may extend below 190 km.

5.6. Jericho

The Jurassic (172 ± 2 Ma, U–Pb and Rb–Sr; Heaman et al., 1997) Jericho kimberlite occurs at the northern end of Contwyto Lake, ca. 200 km north of Lac de Gras. It is a complex of at least two pipes and several dikes (Cookenboo, 1998). This is probably the best-studied pipe in the Slave Province; extensive descriptions of xenoliths and some concentrate garnet data are given by Kopylova et al. (1999). This offers a rare opportunity to compare garnet-concentrate data with a solid xenolith base. The samples used here include 140 garnets from the till train immediately down-ice from the kimberlite, these show patterns identical to 30 garnets from the pipe itself (Kopylova et al., 1999).

The complex intrudes Archean rocks on the north side of the Proterozoic Kilohigok Basin. It lies east of the Pb-isotope line but west of the Nd-isotope line (W. Davis, pers. comm. 2003), and thus may have penetrated rocks of the Central Slave Basement Complex at depth. Irvine et al. (2003) report Re–Os analyses of a large suite of Jericho xenoliths, with T_{RD} ages from <1 to >3 Ga. The whole-rock Re–Os analyses reflect multiple metasomatic episodes recorded in the xenoliths and garnet data (Alard et al., 2002; see below), and none of the model ages is likely to reflect any specific mantle event. However, the oldest ones, taken as minimum ages, indicate stabilisation of the SCLM before 3 Ga, as for the mantle beneath the Lac de Gras area.

The Ca–Cr data (not shown) show an extended lherzolite trend from 1.5% to 11.5% Cr_2O_3, and a distinct trend of higher Ca/Cr extending to 8% Cr_2O_3, shown by Kopylova et al. (1999) to represent garnets from spinel+garnet lherzolites. The Y–T plot shows a distinct layering; garnets with $T < 900$ °C have Y contents down to 1 ppm, but few >20 ppm; garnets with $T > 900$ °C rarely contain <10 ppm Y, and many have >20 ppm. A cluster of garnets with $T > 1200$ °C includes some very depleted ones. Zr contents define a similar layering, with minimum values at $T < 900$ °C of 2–6 ppm, whereas the minimum values in garnets with $T > 900$ °C is 20 ppm. High-T garnets are all Zr-rich (mean 55 ppm) and Ti-rich. Because garnets with $T > 900$ are nearly all Y-enriched, there is no clear Y edge.

These data give a geotherm rising more steeply than the conductive models, from near the 35 mW/m^2 model at low T to near the 40 mW/m^2 curve at 1100 °C. Kopylova et al. (1999) and Russell and Kopylova (1999) show that xenolith P–T data (using the BK or FB-M74 thermobarometers) scatter along this trend. Data for high-T xenoliths continue along this trend, indicating that the deepest samples may be derived from depths near 200 km. The mean calculated olivine composition is Fo$_{92.2}$ at the top of the section (Fig. 6), and decreases steadily with depth to

reach Fo$_{91}$ at 160 km, and Fo$_{90.1}$ at the deepest levels sampled. Fig. 6 shows calculated Fo contents compared with those measured in xenoliths by Kopylova et al. (1999). The mean difference between calculated and observed values is 0.2% Fo; this includes three samples with >1% deviation (positive), which are interpreted as showing disequilibrium between garnet and olivine, due to metasomatism. Two of these anomalous xenoliths are high-T sheared lherzolites displaying microstructural disequilibrium (Kopylova et al., 1999).

There are too few data to construct a CARP section; the upper part to 130 km depth is dominated by fertile lherzolites, with some depleted and depleted/metasomatised lherzolites; below this the section is dominated by melt-related metasomatism. Although the Re–Os data cited above indicate an Archean protolith, there is little sign of this heritage in mineralogy or mineral compositions. The data suggest that the section was originally similar to the Lac de Gras SCLM, with a more depleted upper layer and less depleted lower layer, but has been strongly affected by metasomatism. This metasomatism must have led to the elimination of harzburgites (cf. Griffin et al., 1999c), and an overall rise in LILE contents, but traces of layering remain, recorded in the Y and Zr contents of the garnets. The metasomatism is interpreted as Proterozoic, and may be related to the extension that produced the Kilohigok Basin. The mineralogy and mineral compositions of the Jericho SCLM are similar to those of mantle beneath the Yamba Lake kimberlites north of Lac de Gras, which also have been interpreted as reflecting metasomatic modification of older SCLM, possibly along structural trends parallel to the Kilohigok Basin (Griffin et al., 1999e; Orr and Luth, 2000).

5.7. Anuri

The Anuri kimberlite lies ca. 100 km north of Jericho, and intrudes rocks of the Central Slave Basement Complex on the Eeast side of the Pb-isotope line (Fig. 6).

The garnet data (Fig. 3b) show a bifurcated lherzolite trend similar to the Jericho data, plus a group of relatively high-Cr (6–8% Cr_2O_3) wehrlite garnets. A small population of mildly to strongly subcalcic garnets extends to 14% Cr_2O_3. The Y–T plot (Fig.

4b) shows that the upper part of the section (to 950 °C) is more depleted and includes the wehrlitic garnets; the subcalcic garnets are all in the lower part of the section. Some Y-depleted garnets have T as high as 1400 °C, but most garnets with $T>1200$ °C contain 20–30 ppm Y. Low-T (<950) garnets are all Cr-saturated, as seen at Lac de Gras. Maximum Cr contents increase with T up to 1250 °C, and higher-T garnets are lower in Cr. Zr also shows layering; many garnets with $T<950$ °C have <1 ppm Zr, while few have >20 ppm; at $T>950$ °C, most garnets have 50–100 ppm Zr. Many of these also have high TiO_2 (0.6–1.5%), indicating melt-related metasomatism. Nearly half of the garnets have sinuous REE patterns; most of these have $T<1100$ °C.

The garnet geotherm (Fig. 5b) rises from near the 35 mW/m^2 curve to near the 38 mW/m^2 one between 750 and 1100 °C; several high-T subcalcic grains constrain the geotherm to $\geq$40 mW/m^2 at 1250 °C. The mean calculated olivine composition is Fo$_{92-92.5}$ down to ca. 130 km, then decreases to Fo$_{91.3}$ at the layer boundary (ca. 145 km). In the lower layer, there is a steady decrease in Fo content with depth, to ca. Fo$_{89}$ (an asthenospheric value) at the base of the section (190 km).

Despite the lack of subcalcic harzburgites in the upper layer, the CARP section strongly resembles the one from Lac de Gras; the upper layer is dominated by depleted lherzolites, and shows a sharp lower boundary at ca. 145 km. The lower layer has ca. 20% subcalcic harzburgites and depleted lherzolites, but is dominated by depleted/metasomatised garnets and those showing melt-related metasomatism.

5.8. Tenacity

The Tenacity kimberlite lies 100 km ENE of Anuri; it intrudes rocks of the Eastern Slave Arc Terrane, east of the Pb isotope line and west of the Nd-isotope line (Fig. 6). The data used here come from the prediscovery till train ($n=62$; Griffin et al., 1999d) and the kimberlite itself ($n=101$); the two datasets are essentially identical.

The garnet data (Fig. 3t) define a long complex lherzolite trend from 1% to 10.7% Cr_2O_3; there is a moderate number of mildly to strongly subcalcic garnets, including one with 12% Cr_2O_3. The Y–T plot (Fig. 4v) shows that the kimberlite mainly sam-

pled a narrow T interval (720–880 °C), with a few grains giving temperatures up to 1100 °C. The garnets range from very depleted (<0.5 ppm Y) to enriched (>40 ppm Y). Similarly, Zr ranges from 0.5 to >100 ppm, but few grains contain >0.3% TiO_2. About 1/4 of the garnets, including most of the subcalcic ones, have sinuous REE patterns.

The data are too concentrated to define a geotherm, but all cluster around a 33–35 mW/m^2 conductive model at 800 °C (Fig. 5w). This is consistent with the high diamond content of the pipe. Some of the higher-T data constrain the geotherm to $\geq$40 mW/m^2 at 1100 °C, and imply that the SCLM is >150 km thick. The mean calculated olivine composition (Fig. 6) is Fo$_{92.5-93}$ over the short section; the deepest samples give values down to Fo$_{92}$. The short CARP section (Fig. 6) shows a clear increase in the proportion of depleted harzburgites and lherzolites, and a decrease in fertile lherzolites, with depth.

5.9. Arctic Canada

5.9.1. Somerset Island (BT)

Nine kimberlite pipes and many dikes occur on Somerset Island, NW of Baffin Island and NE of the Slave Province (Fig. 2; Mitchell, 1978; Schmidberger and Francis, 1999); U–Pb dating of perovskites indicates a Cretaceous age (Heaman, 1989; Smith et al., 1989). The kimberlites intrude Proterozoic sedimentary sequences overlying PaleoProterozoic crust (2.2–2.5 Ga). Irvine et al. (2003) report Re–Os T_{RD} ages on xenoliths ranging from 1.3 to 2.8 Ga, with a peak at 2–2.75 Ga. The younger ages probably represent mixed-sulfide ages rather than specific events (Alard et al., 2002; Griffin et al., 2002a, 2004); the older ones indicate Archean stabilisation of the SCLM, probably >3 Ga ago.

Limited garnet data from the Batty kimberlite (not shown) define a narrow lherzolite trend from 2.9% to 8.4% Cr_2O_3; no subcalcic garnets were found. A Y–T plot (Fig. 4e) shows a temperature range from 760 to 1220 °C, and only weak depletion; 1/3 of the garnets have <10 ppm Y. None has >0.6% TiO_2, indicating only weak melt-related metasomatism. Similar garnets have been described in xenoliths and concentrates by Mitchell (1978), Kjarsgaard and Peterson (1992) and Schmidberger and Francis (1999). The mean calculated olivine composition is Fo$_{93}$; spinel- and garnet-bearing lherzolites described by Mitchell (1978) and Schmidberger and Francis (1999) have olivine Fo_{91-93}, with a mean of $Fo_{92.3}$ in the Nikos kimberlite.

The garnet data are consistent with a non-conductive geotherm rising from ca. 40 mW/m^2 at low T to ca. 45 mW/m^2 at 1000 °C (Fig. 5e). Xenolith data (Mitchell, 1978; Schmidberger and Francis, 1999) suggest a geotherm near the 45 mW/m^2 conductive model at least in the T range 800–1000 °C. The data suggest that the base of the depleted lithosphere lies near 140 km.

There are too few data to construct a CARP section. Lower-T garnets are depleted or depleted/metasomatised types, but above ca. 1000 °C, melt-metasomatised classes dominate.

Except for the high Mg# of the olivine, both the concentrate data and the xenoliths indicate a section that is much less depleted than typical Archean sections (as suggested by the Re–Os data of Irvine et al., 2003). The section may have been strongly modified, but this would have to have involved little lowering of the Mg#. Trace-element and Sr–Hf–Nd–Sr isotopic data (Schmidberger and Francis, 2001; Schmidberger et al., 2001, 2002) show that the upper and lower parts of the section sampled by the Nikos kimberlite are isotopically distinct, and the lower layer is probably younger, or modified at a later time.

5.10. Eastern North America (Fig. 7)

5.10.1. Attawapiskat (AT)

The Attawapiskat field contains >20 kimberlites, which intrude Paleozoic sediments east of James Bay in N. Ontario. The data used here come from the Charlie, Delta, Tango and Whiskey bodies. The Mac-Fayden, Charlie and Bravo pipes have yielded perovskite U–Pb ages of 175–180 Ma (Heaman and Kjarsgaard, 2000). The basement is not exposed, but is inferred to be Archean rocks of either the N. Caribou or Oxford-Stull (Sachigo) terranes (Williams et al., 1992).

The garnet data (Fig. 3d) define a long lherzolite trend from 1.5% to 10.5% Cr_2O_3; a short trend with higher Ca/Cr at low Ca indicates the presence of spinel–garnet peridotites near the top of the section; there are also some mildly subcalcic garnets. The Y–T plot (Fig. 4d) shows a mixture of very fertile and

very depleted garnets at shallow depths, and pronounced Y edge at ca. 1050 °C; most of the higher-T garnets are subcalcic. Zr contents of the garnets are actually higher within the depleted SCLM, whereas high-T garnets contain <40 ppm Zr. The maximum TiO_2 contents are 0.6%, indicating a low level of melt-related metasomatism. Ca 15% of the garnets have sinuous REE patterns, including the low-T subcalcic garnets, whereas the higher-T subcalcic garnets have upward-convex patterns with high HREE contents.

The garnet data suggest a geotherm (Fig. 5d) steeper than the conductive models, rising from near 35 mW/m^2 at 700 °C to ca. 38 mW/m^2 at 1050 °C. There are too few data above 1050 °C to define the lithosphere thickness, but it probably exceeds 165 km. The mean calculated olivine composition (Fig. 7) is $Fo_{92.5-93.2}$ down to 120 km, then drops steadily to reach $Fo_{91.5}$ at 160 km, and $Fo_{90.5}$ at 175 km as the proportion of melt-related metasomatism increases.

The CARP section (Fig. 7) shows a high proportion of fertile lherzolites at the top (<110 km), and the proportions of depleted and depleted/metasomatised lherzolites increase downward. Below 160 km the data are sparse, but indicate that moderate melt-related metasomatism is present. The overall section is similar in many respects to that sampled by the Group 1 kimberlites of the Kaapvaal Craton (Griffin et al., 2002a, 2003a).

5.11. Kyle Lake (KY), James Bay Lowland

Several pipes have been found beneath Paleozoic sediments in the Kyle Lake cluster, SW of the Attawapiskat field (Janse et al., 1995). They have been dated to 1100±40 Ma (Sage, 2000), and their tectonic setting is similar to that of the younger Attawapiskat field.

The limited garnet data (Fig. 3j) define a long narrow lherzolite trend from 3% to 9% Cr_2O_3, paralleled by weakly subcalcic garnets. The Y–T plot (Fig. 4k) shows that few of these garnets are Y-depleted, and those are confined mostly to <1050 °C; as at Attawapiskat most of the high-T garnets are subcalcic and have lower Zr than the shallower ones.

The data are consistent with a 35 mW/m^2 conductive geotherm up to ca. 1050 °C (Fig. 5k). The mean calculated olivine composition is constant at $Fo_{93-93.5}$

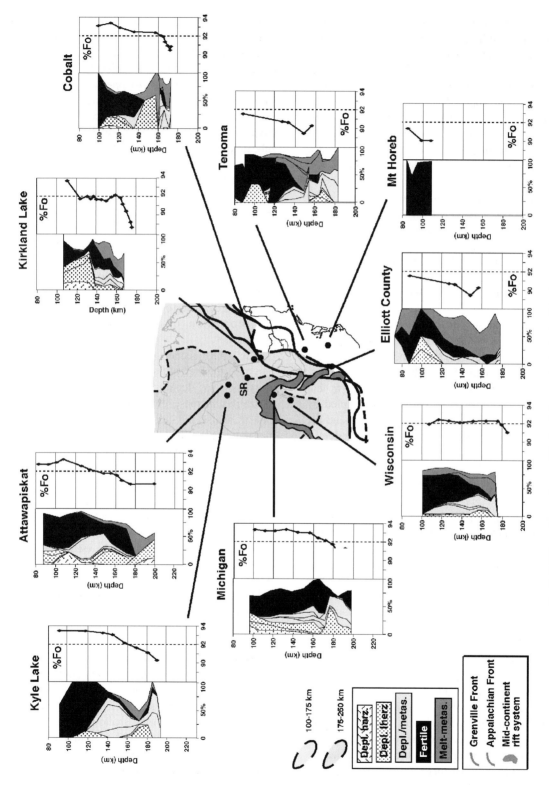

Fig. 7. CARP sections and mean %Fo in olivine vs. depth, for localities in eastern North America.

down to 150 km, then decreases steadily with depth, to reach asthenospheric values ($Fo_{90.3}$) at 190 km. Overall, this is more magnesian than the Attawapiskat section.

There are too few data to provide a good CARP section, but the available data (Fig. 7) show a similar pattern to Attawapiskat, with fertile lherzolites down to 120–130 km, then an increasing proportion of depleted and depleted/metasomatised lherzolites with depth. The highest proportion of melt-related metasomatism is near 160 km, suggesting that this represents the base of the depleted lithosphere. Aside from its overall high Mg#, this section is most similar to several known Proterozoic sections (Griffin et al., 2002b), and it may reflect metasomatic modification of Archean SCLM.

5.12. Sextant Rapids (Coral Rapids) (CR)

The Sextant Rapids locality (also known as Coral Rapids) is a melilitite lying south of the tip of James Bay, between the Attawapiskat and Kirkland Lake kimberlite fields. A 152 Ma perovskite U–Pb age (Heaman and Kjarsgaard, 2000) relates it to the intrusion of the Kirkland Lake kimberlites. The melilitite intrudes the Kapuskasing Structural Zone, a NE-trending belt ca. 70 km wide of uplifted high-grade lower crustal rocks that interrupts the general E–W structural trends of the Archean terranes that make up Superior Province. This uplift has not been dated directly but is inferred on field evidence to relate to 1.8–1.9 Ga compressional events, and was accommodated by thickening of the crust (Percival and West, 1994).

The garnet data (not shown) define a short lherzolite trend from 1.4% to 4.3% Cr_2O_3. The Y–T plot (Fig. 4r) shows that none of these garnets is depleted, but two groups can be recognised: those with $T<1000$ °C have a mean Y content of 30 ppm; those with $T=1100–1200$ °C have slightly lower contents (mean Y=20 ppm). Zr shows the same pattern; the lower-T group have mean Zr=35 ppm, whereas the higher-T ones contain 9–35 ppm. TiO_2 contents are <0.3%; all of these garnets are typical of garnets from fertile lherzolites in Tecton mantle (Griffin et al., 2002b).

The data show a high geotherm (Fig. 5s), near a 50 mW/m^2 conductive model, consistent with the high heat flow over the Kapuskasing zone. The base of the lithosphere cannot be defined, but the deepest garnets give a minimum thickness of 120 km. The mean calculated olivine composition is $Fo_{90.3}$, and shows no significant variation with depth; these values are similar to those of Tecton peridotites. The CARP analysis shows nearly all garnets as derived from fertile lherzolites (Classes L9, L10a, L10b), as is typical of Tecton SCLM (Griffin et al., 2002b).

5.13. Kirkland Lake (KL)

The Kirkland Lake kimberlite field south of Lake Abitibi in southern Ontario consists of two clusters (at least 20 kimberlites) spread over a distance of 70 km NW–SE. Many of the pipes have been dated (U–Pb perovskite) by Heaman and Kjarsgaard (2000). The Tandem body is the oldest at 165 Ma, and the others cluster between 152 and 157 Ma. The data here extend across the field, from Tandem in the NW end, through A4, B30 and C14 to Diamond Lake in the SE.

The garnet data (Fig. 3l) show a thick lherzolite trend with two distinct en echelon components, from 1% to 7% Cr_2O_3 and 3.5–8.5% Cr_2O_3. Mildly sub-calcic garnets range from 2% to 12.5% Cr_2O_3. The Y–T plot (Fig. 4m) shows very strong depletion in garnets with T up to 900 °C, then moderately depleted material continues up to 1400 °C. However, both mean and maximum Zr contents increase with T, and most garnets with $T>1250$ are high-Zr, high-Ti types (TiO_2 to 1.5%) with a strong signature of melt-related metasomatism. Only about 10% of the garnets have sinuous REE patterns, and nearly all of these have $T<900$ °C.

The low-T garnets (to 900 °C) define a 37 mW/m^2 geotherm (Fig. 5m), whereas the higher-T ones lie along a 40 mW/m^2 conductive model; the discontinuity corresponds to the two en-echelon segments of the Ca–Cr plot. We interpret the higher-T geotherm as continuing to the Y edge at 1200 °C; garnets with higher T are largely melt-metasomatised. The Y edge suggests the base of the depleted lithosphere lies at ca. 160 km depth. The mean olivine composition in the upper part (<120 km) is very magnesian ($Fo_{93.7}$), but in the main part of the section it lies between $Fo_{91.5–92}$, and below 160 km it decreases rapidly to $Fo_{<89}$, corresponding to asthenospheric values.

The upper part of the CARP section, down to to 135 km, has a high proportion of depleted and depleted/metasomatised garnets; it shows a sharp change at ca. 140 km, and the lower part of the section has higher proportions of depleted/metasomatised and fertile lherzolites; melt-related metasomatism increases downward. The upper part of the section is typically Archean, despite a relatively low Mg#, but the lower part appears to be very strongly modified, and more similar to Proterozoic sections.

5.14. Cobalt Area (CO)

At least 10 kimberlites occur in a cluster around the town of Cobalt in southern Ontario, and several more occur east of Lake Timiskaming in Quebec. The data used here are from the Opap, Bucke and Peddie kimberlites in the Cobalt cluster, and the Nedelec pipe on Lake Timiskaming. Their ages range from 134 to 154 Ma (perovskite U–Pb; Heaman and Kjarsgaard, 2000). They intrude the basement of the Abitibi Province, near its southern edge where it is affected by 1.8 Ga rejuvenation and the intrusion of the 2.2 Ga Nipissing diabase sills, and ca. 50 km from the Grenville Front, representing the outer edge of basement disturbance by large-scale overthrusting ca. 1050–1020 Ma ago.

The garnet data (Fig. 3e) define a long narrow lherzolite trend from 4% to 14.5% Cr_2O_3, with three low-Cr weakly subcalcic garnets. The Y–T plot (Fig. 4f) shows two groups, one with T=700–900 °C, the other with T mostly >1100 °C; about half of these are Y-depleted. Most garnets show a strong Zr–Ti correlation up to 1% TiO_2, reflecting melt-related metasomatism. A small proportion of the garnets has weakly sinuous REE patterns.

The low-T garnets lie on a 35 mW/m^2 conductive model geotherm (Fig. 5f); the high-T group is largely melt-metasomatised, but Y-depleted material continues to higher T, and this suggests that the base of the depleted lithosphere lies at ca. 160 km. The shallow part of the section has a mean olivine composition of Fo_{93}; the deeper part (below 130 km) has $Fo_{92.5}$ at the top, but the mean composition decreases rapidly to $Fo_{90.5}$ between 160 and 175 km depth.

The CARP section (Fig. 7) is dominated by relatively fertile lherzolites to depths of ca. 140 km; the lower part of the section (140–160 km) is markedly more depleted. Overall, this section is much less Archean in character than Kirkland Lake. The abundance of depleted lherzolite at the bottom and fertile material higher up is similar to many Proton sections that represent reworked Archon SCLM (Griffin et al., 2002b). It lies on the "thinned" edge of the cratonic root (Fig. 2), and this is consistent with its less depleted nature.

5.15. Michigan (MI)

McGee and Hearn (1984) described two kimberlites near Lake Ellen on the Upper Peninsula of Michigan; subsequently ca. 30 others have been discovered by Crystal Exploration, Exmin, Amselco and Ashton Mining. The bodies have not been dated radiometrically, but fossils in sedimentary xenoliths indicate a post-Middle Ordovician (<460 Ma) age. The area lies on the S edge of the Superior Province, adjacent to the 1.8 Ga Penokean mobile belt, and the Archean basement is intruded by Penokean granites. It also lies on the edge of the Paleozoic Michigan basin, and within the arc of the Mid-continent Rift System, but 100–150 km from the axis of rift, represented by Lake Superior syncline (Allen et al., 1995).

The garnet data show a long complex lherzolite trend to 13% Cr_2O_3, and a significant number of mildly subcalcic garnets, all with <7% Cr_2O_3 (Fig. 3n). The Y–T plot (Fig. 4o) shows a gap in T from 950 to 1100 °C; all subcalcic garnets are in the lower-T group, but there are many Y-depleted garnets in both groups. Zr contents reach 150 ppm in the lower-T group and 120 ppm in the higher-T group, but high Zr and Ti are only correlated in the higher-T group, where TiO_2 contents reach 0.5%. The geotherm shows a discontinuity at 950 °C (Fig. 5o); the lower-T garnets define a non-conductive geotherm rising from 35 to 37 mW/m^2 with depth, while the higher-T group is consistent with a 45 mW/m^2 geotherm to ca. 1300 °C. The mean olivine is $Fo_{93–93.5}$ in the upper layer; it is $Fo_{92.3}$ at the top of the deeper layer, and decreases steadily with depth to Fo_{91}.

There are few data on peridotite xenoliths from these kimberlites. Eggler et al. (1987) show four peridotites with mean olivine of Fo_{90}; these have temperatures of 900–1100 °C, and thus correspond to the unsampled gap in the garnet data. Temperatures for many eclogites, pyroxenites and megacrysts range from 850 to 1100 °C with most values between 900 and 1050 °C (McGee and Hearn, 1984). This suggests that the gap in the garnet data represents a zone of mafic rocks and minor Fe-rich peridotites.

In the CARP section (Fig. 7), the upper layer (100–160 km) contains up to 30% depleted lherzolites and harzburgites at the top; these depleted rocks decrease in abundance with depth, balanced by an increase in the depleted/metasomatised types. There also is a large proportion of fertile lherzolites, and the proportion of melt-related metasomatism increases downward. The data from 160 to 175 km depth are not reliable, because this interval is represented by few garnets. The lower layer (175–200 km) contains a high proportion of depleted and depleted/metasomatised lherzolites, and some melt-related metasomatism.

The upper part of the section is similar to many modified Archean sections, and fertilisation and melt-metasomatism may be related to the Penokean event. The lower layer, lacking harzburgites, is similar to some Proterozoic SCLM.

5.16. Wisconsin (WI)

This locality consists of a single ultramafic lamprophyre, known as the Six Pack body, under glacial cover on the outskirts of Milwaukee. Its age is unknown. It intrudes the crust of the 1.8–1.9 Ga Penokean province.

The garnet data (Fig. 3x) show a long lherzolite trend extending to 14.8% Cr_2O_3; some mildly subcalcic garnets span most of the Cr range. The $Y-T$ plot (Fig. 4z) shows that most of the shallow garnets are not depleted (Y>10 ppm), and there is no clear Y edge. Rare garnets have up to 0.7% TiO_2, but melt-related metasomatism not prominent.

The data follow a 37 mW/m^2 conductive geotherm (Fig. 5zz) to at least 1000 °C, giving a minimum SCLM thickness of about 175 km. The calculated olivine composition shows little variation; the mean is $Fo_{92-92.5}$ from 110 to 175 km depth, then drops rapidly to Fo_{91}. This is significantly less magnesian than the Michigan SCLM at equivalent depths.

The CARP section (Fig. 7) shows that the SCLM is only moderately depleted; depleted and depleted/metasomatised garnets increase in abundance with depth, while the proportion of fertile lherzolites expands upward. Melt-related metasomatism is observed through the whole section, and increases downward, especially below 175 km. There is no layering analogous to that seen in the Michigan section.

5.17. Western North America (Fig. 8)

5.17.1. Saskatchewan (SA)

More than 70 kimberlites are known in the Fort à la Corne area of central Saskatchewan; most are of the crater facies and include pyroclastic deposits (Scott-Smith et al., 1995; Nixon and Leahy, 1997; Leckie et al., 1997). The Candle Lake kimberlite lies to the north, and the Sturgeon Lake body ca. 50 km to the west. The kimberlites are Early Cretaceous in age (99 Ma and 101 Ma U–Pb ages on perovskite; Leckie et al., 1997 and 94–96 Rb–Sr ages; Lehnert-Thiel et al., 1992). The data used here are from 7 kimberlites in the Fort a la Corne area, and from Candle Lake; a few data from Sturgeon Lake are similar (Table 1).

The kimberlites intrude Cretaceous sediments overlying the Glennie Domain in the middle of the Trans-Hudson Orogen. The basement rocks are ca. 1.9–1.8 Ga volcanics and gneisses; windows in the exposed part of the Glennie Domain show Archean rocks, and Nd model ages indicate Archean protoliths (2.6–3.0 Ga; Collerson et al., 1989, 1990). These probably represent an Archean microcontinent, over-ridden by Proterozoic rocks during the Trans-Hudson collisions.

Seismic tomography (Bank et al., 1998) shows a series of high- and low-velocity anomalies at 100 km depth, and the kimberlites lie on a high-velocity ridge between two lows; the low-V zones have been interpreted as the traces of plumes related to the Cretaceous kimberlite emplacement (Bank et al., 1998). Deeper tomography (Grand, 1994; van der Lee, 2001) shows the entire area as part of a high-velocity root that is continuous from the Hearne Province, across the Trans-Hudson orogen, into the Superior Province.

Leahy and Taylor (1997) described "Sloan-type" diamonds with degraded platelets, and a bimodal T distribution is suggested by nitrogen aggregation data. They suggested that these features reflect high-T reworking of Archean SCLM. About 1/3 of the diamonds were Type IIA, which is comparable to the proportion observed in the Slave Province kimberlites, where a large plume component has been documented (Davies et al., 1999, 2004a).

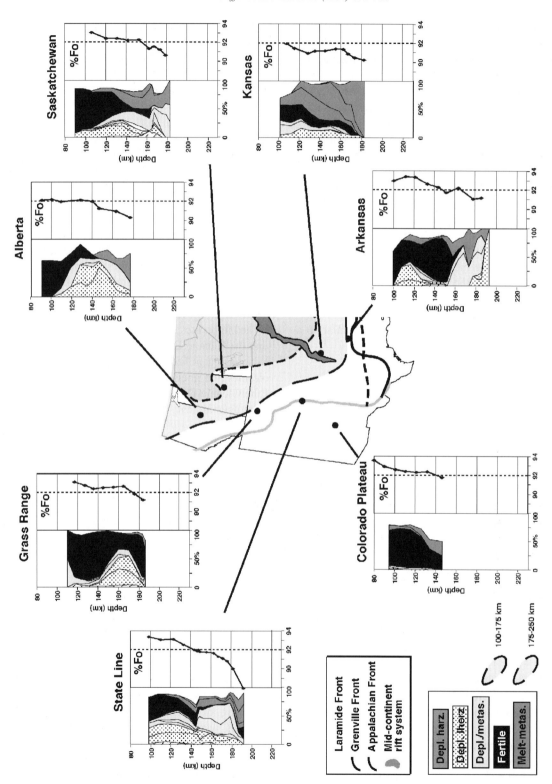

Fig. 8. CARP sections and mean %Fo in olivine vs. depth, for localities in western North America.

The garnet data define a long narrow lherzolite trend from 2% to 12% Cr_2O_3 (Fig. 3p). Two wehrlitic garnets continue this trend to 13.5% Cr_2O_3. A group of garnets with higher Ca/Cr from 1% to 5% Cr_2O_3 probably is derived from spinel–garnet lherzolites. There is a significant number of mildly to strongly subcalcic garnets with Cr_2O_3 3–12%. The $Y–T$ plot (Fig. 4q) is bimodal (650–1050 and 1150–1350 °C), with few data in the gap. Both groups contain Y-depleted garnets. There is a large range in Zr and Y contents and Zr/Y, suggesting a range of metasomatic styles. Zr contents are generally 0–30 ppm up to 900 °C, then jump to 10–140 ppm. High-Zr garnets generally also have high Ti contents, up to 1.25% TiO_2. Ca 30% of our data were used by Leahy and Taylor (1997) to show a bimodal T distribution (850–1000, 1200–1250 °C); the addition of more data has narrowed the gap but left the bimodal pattern.

The low-T garnets follow a 35 mW/m^2 conductive model geotherm to 1050 °C, whereas the higher-T group defines a 43–44 mW/m^2 geotherm to at least 1300 °C (Fig. 5r), suggesting that the base of the depleted lithosphere lies at 175–180 km. The upper layer, down to 150–160 km, has an average olivine composition of $Fo_{92–92.5}$. The lower layer is distinctly less magnesian ($Fo_{91–91.3}$) at the top, decreasing to ca. $Fo_{90.5}$ in the deepest levels.

In the CARP section the upper layer down to 150–160 km shows relatively fertile lherzolites (reflecting phlogopite-related metasomatism) at the top, a higher proportion of depleted and depleted/metasomatised lherzolites toward the base, and melt-related metasomatism increasing with depth. This pattern is very similar to that shown by several Archean sections modified in Proterozoic extension–compression regimes (Griffin et al., 2002b, 2003a), and is consistent with the Archean character of the Glennie Domain. The lower layer shows a lower proportion of "fertile" lherzolites and a higher proportion of depleted/metasomatised lherzolites; it is generally similar to the lower layer of the Slave Craton SCLM.

6. Results: craton margins and Mobile Belts

6.1. Greenland (Fig. 9)

6.1.1. Sarfartoq (SQ)

Numerous kimberlite dikes occur in the mountainous area on the S. side of Kangerdluggsuaq (Søndre

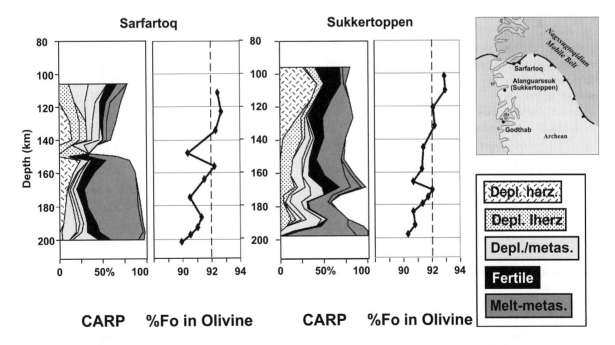

Fig. 9. CARP sections and mean %Fo in olivine vs. depth, for localities in western Greenland.

Strømfjord). The samples studied here are from kimberlite boulders collected on flood plains, but the distribution of blocks and dikes suggests they were derived from within 5–10 km of their locations (L.M. Larsen pers. comm.; Garrit, 2002). Garnet data are taken from Griffin et al. (1995). K–Ar ages for the kimberlites range from 589 to 656 Ma (Larsen et al., 1983; Scott-Smith, 1987). The area lies on the edge of the ancient Archean core of Greenland, with crustal ages ranging from 3.8 to 2.6 Ga. The Sarfartoq localities lie on the margin of the Nagssugtoqidian Mobile Belt, which consists mainly of Archean rocks heavily reworked during the 1.8–1.95 Ga orogeny with S-dipping subduction (Van Gool et al., 1999); crustal thicknesses are 46–50 km (Dahl-Jensen et al., 2003). The samples therefore represent a section through the mantle beneath a complex tectonic front.

The garnet data (Fig. 3o) show a long complex lherzolite trend to >10% Cr_2O_3, a few wehrlitic garnets and a moderate number of mildly to strongly subcalcic garnets; the latter have 3–5% Cr_2O_3. The Y–T plot (Fig. 4p) shows a strongly layered SCLM. The T range 600–800 °C is bimodal with fertile lherzolites and strongly depleted harzburgites. The T range 800–1050 °C contains fertile lherzolites and subcalcic harzburgites, 1050–1200 °C fertile to depleted lherzolites and fertile harzburgites; and samples with T>1200 °C are from fertile lherzolites. Garnets with T<800 °C are very Zr-depleted (mostly <20 ppm); those in the 800–1200 °C range contain 10–150 ppm Zr and have Ti contents up to 1%.

Most of the data are consistent with a 35–40 mW/m^2 conductive geotherm up to the 1200 °C Y edge (Fig. 5q); some low-T garnets would suggest an unusually low geotherm, and may have equilibrated with a low-Ni olivine. The mean calculated Fo contents for the top of the section are relatively depleted (Fo_{93}) and decrease steadily downward to $Fo_{90.7}$ at 170 km, where there is a reversal to Fo_{92}, followed by a decline with depth back to $Fo_{90.7}$.

$P–T$ estimates for eight garnet peridotite xenoliths (Griffin et al., 1995; Garrit, 2002) scatter between the model 35 and 40 mW/m^2 model geotherms from 890 to 1150 °C, and support the continuation of the geotherm to 60 kb (200 km). One xenolith reported by Larsen and Ronsbo (1993) gives a minimum P of 50 kb at 1110 °C, consistent with the garnet geotherm.

The upper part of the CARP section (to 170 km) is dominated by depleted harzburgites; the fertile lherzolites are mostly melt-metasomatised varieties, and proportion of them increases rapidly downward. From 170 to 190 km there is a layer with a high proportion of depleted/metasomatised garnets, and less melt-related metasomatism. Many garnets in this layer do not classify in the CARP classes used here, indicating complex metasomatic processes. At depths >190 km, the SCLM is very strongly melt-metasomatised.

6.2. Sukkertoppen (SU)

Many kimberlite dikes are known around Alanguarssuk, inland from the settlement of Sukkertoppen (Maniitssoq), ca. 100 km SSW of Sarfartoq. Their age is unknown, but they may be either ca. 600 Ma old, like the Sarfartoq kimberlites, or related to the nearby Qaqarssuk carbonatite and lamprophyres, dated at 169–176 Ma (Scott-Smith, 1987). This area lies further inside the Archean craton than Sarfartoq, but is still <150 km from Nagssugtoqidian Front.

The garnets define a long narrow lherzolite trend to 8% Cr_2O_3, with abundant mildly subcalcic harzburgites from 4% to 10% Cr_2O_3 (Fig. 3s); this is a distinctly different distribution from that seen in Sarfartoq. The Y–T plot (Fig. 4u) shows three distinct groups: garnets with T=700–900 °C and 1000–1250 °C are mixtures of fertile to depleted lherzolites and harzburgites; those with T>1250 °C are only fertile lherzolites. Zr contents are <50 ppm in the low-T group, 5–140 ppm in the 1000–1250 °C range, and 50–100 ppm in those with T>1250 °C. Ti and Zr are well-correlated, and most lherzolite garnets have 0.6–1.2% TiO_2 and >10 ppm Ga, indicative of melt-related metasomatism.

The low-T group of garnets lie along a 35 mW/m^2 conductive geotherm (Fig. 5v), while the high-T group defines a 40–42 mW/m^2 geotherm; the gap in the record from 900 to 1000 °C corresponds to the step. The Y edge at 1250 °C suggests an SCLM thickness of ca. 190 km. The curve of the mean calculated olivine composition shows two reversals, defining two layers with $Fo_{92–92.5}$ in the upper part, each decreasing to ca. $Fo_{90.5}$ with depth. The deepest layer has an average $Fo_{91.3}$ at its top, decreasing to less than Fo_{90} with depth.

This layering also is marked in the CARP section (Fig. 9). The upper layer (down to 140 km) is a mixture of depleted, depleted/metasomatised, and fertile lherzolites; ca. 30% of the garnets do not classify, suggesting complex metasomatic signatures. The lower layer (150–190 km) is more depleted at its top, and the proportion of depleted/metasomatised material increases downward, but this layer is dominated by melt-related metasomatism (>60%).

6.3. Eastern North America (Fig. 7)

6.3.1. Tenoma (PA)

The Tenoma kimberlite is a dike up to 15 m wide, observed in coal mines near Dixonville, PA. It is dated to 89 ± 5 Ma (phlogopite+WR, Alibert and Albarede, 1988) and is thus younger than the Masontown kimberlite to the SW (185 ± 10 Ma by K–Ar, 149 ± 5 Ma by Rb–Sr on phlogopite+WR; Alibert and Albarede, 1988). The locality lies in the Appalachian Plateau, east of the Grenville Front. The Tenoma, Masontown and Elliott Co. kimberlites (and the Ithaca and Syracuse kimberlites in NY) all lie on a major zone of structure-parallel faults (down-to-E) extending from the Rome Trough (Parrish and Lavin, 1982), interpreted by Phipps (1988) as reflecting a zone of deep rifting that may be related to early opening of the Atlantic Ocean.

The sample contains only lherzolitic garnets, with Cr_2O_3 ranging from <1% to 8% (Fig. 3u). All have high HREE and convex-downward REE patterns. Garnets with $T_{Ni} < 900$ °C are relatively depleted in Zr and/or Y (Fig. 4w), but those with $T_{Ni} > 1000$ °C are mostly Zr–Ti enriched (up to 0.8% TiO_2), which is interpreted as a signature of mel-related metasomatism. The garnet geotherm (Fig. 5x) lies near a 40 mW/m^2 conductive model up to 900 °C. The "Y edge" is poorly defined because of the generally fertile (high-Y) nature of the garnets, but there are few Y-depleted garnets above ca. 1050 °C, corresponding to a depth of 140 km. The mean composition of the olivine is ca. $Fo_{91.5}$ at the top of the section, and drops to Fo_{90} towards the base, with increasing melt-related metasomatism (Fig. 7).

These data are similar to those for the Masontown kimberlite to the SW; Hunter and Taylor (1984) report Cr_2O_3 contents up to 6%, and TiO_2 contents up to 0.5% in the garnets of three high-T lherzolites. Their

$P–T$ estimates for xenoliths, with olivine of $Fo_{90–93}$, define a geotherm parallel to the diamond/graphite line (Fig. 5x) from 1050 °C/40 kb to 1350 °C/50 kb, equivalent to the kinked limb of the garnet geotherm for Tenoma. Despite the small size of the sample, the CARP data (Fig. 7) show that the whole section is relatively fertile, and the lower part contains a higher proportion of depleted/metasomatised and melt-metasomatised material.

6.4. Elliott County (EL)

The Elliott County area is ca. 130 km east of the Grenville Front in NE Kentucky, and includes the Ison Creek and Hamilton Branch kimberlites and their satellites. The data presented here are mainly from the Ison Creek body; with a few garnets from Hamilton Branch; other data from Hamilton Branch are given by Schulze (1984). Ison Creek is dated to 89 ± 2 Ma (Rb–Sr on phlogopite+WR; Alibert and Albarede, 1988).

The area lies on the East Continental Gravity High, an extension of the Mid-continent Rift System, subparallel to Michigan segment of the MCRS. This structure includes the Fort wayne Rift of McPhee (1983), and extends (as geophysical anomalies) east of the Grenville Front to Elliott County. Drilling has revealed fluvial sediments and bimodal volcanics to depths of more than 7000 m (Drahovzal et al., 1992).

The sample contains lherzolitic garnets, with 1–9% Cr_2O_3 (Fig. 3h), and one mildly subcalcic one with <4% Cr_2O_3. Most garnets have concave-down REE patterns, but a few are flatter and mildly sinuous (MREE>HREE). Most high-T garnets ($T_{Ni} > 1100$ °C) are highly enriched in Y (to 50 ppm; Fig. 4i), Zr (to 140 ppm) and Ti (up to 1.6% TiO_2), giving a clear signature of melt-related metasomatism. The data suggest a geotherm near the 40 mW/m^2 conductive model up to at least 1100 °C (ca 140 km; Fig. 5i). The calculated olivine is relatively magnesian at the top of the section ($Fo_{92–93}$) dropping sharply to ca. Fo_{90} at 140 km, where a reversal occurs, followed by another steep drop to below Fo_{89} at 170–180 km depth.

The geotherm is similar to that derived for xenoliths from the Hamilton Branch body (Finnerty and Boyd, 1987; Eggler et al., 1987), with one cluster of samples around the 40 mW/m^2 model geotherm at 950–1050 °C, and another around 1300 °C and 55

Kb. The shallower group of xenoliths has $Fo_{90-92.5}$ olivine (Eggler et al., 1987), similar to the range calculated for Ison Creek.

The upper part of the CARP section (Fig. 7) shows a strong concentration of depleted and depleted/metasomatised garnets. The proportion of melt-related metasomatism increases rapidly with depth, reaches >50% by 130 km, and makes up nearly all of section by 150 km. These data suggest a thickness for the depleted SCLM of ≤130 km.

6.5. Mount Horeb (MH)

Mount Horeb is a pipe-like body 55 km N of Roanoke in western Virginia (Sears and Bilbert, 1973; Meyer, 1976); it is stratigraphically dated to post-Middle Ordovician. It intrudes the folded rocks of the Appalachian belt, at the inner edge of the Blue Ridge Province. It thus has sampled the SCLM beneath the Grenville Province, where it is overridden by Appalachian mobile belt.

The sample contains only lherzolitic garnets; one contains 5.3% Cr_2O_3, but the remainder contain 1.1–2.1% (data not shown). None has <15 ppm Y, or Ti>0.23%; all have high HREE and very low LREE, consistent with derivation from fertile lherzolites. They are similar in these respects to garnets from beneath many Tectons (Griffin et al., 2002b). All but one of the garnets define a narrow T range (820–910 °C), near a 47 mW/m^2 conductive model geotherm (Fig. 5p). The base of the SCLM is not defined by these data, but probably is not much deeper than the 110 km represented by these garnets. The mean olivine is $Fo_{90.2}$, consistent with derivation of the garnets from fertile lherzolites.

The CARP section (Fig. 7) shows only fertile lherzolites, as is typical of Tecton sections. It is significantly less depleted than the Elliott County and Tenoma sections, and this difference may be related to the greater distance of the Mount Horeb locality from the Tecton front (Fig. 2). Kay et al. (1983) showed that kimberlites at Ithaca (NY) contain garnets with Cr_2O_3 mostly 1.5–2.2%, and olivine with Fo_{89-91}. This locality lies within the Grenville Province, but beyond the Appalachian Front, and the similarity to the Mount Horeb data suggests that both have sampled SCLM produced during the Grenville orogeny.

6.6. Western North America (Fig. 8)

6.6.1. Kansas (KA)

Thirteen known kimberlites lie along the Abilene Anticline in Riley County, KS (Mansker et al., 1987), and include crater-facies, diatreme facies and hypabyssal bodies, of which five are sampled here (Table 1). Radiometric ages are Cretaceous (Rb–Sr 95 Ma, Brookins (1970); Rb–Sr phlogopite+WR 104±4 Ma, Alibert and Albarede (1988)). The area lies on the edge of the Mid-Continent Rift System, where it cuts the Proterozoic basement of the 1.3–1.5 Ga Granite–Rhyolite Province.

The garnets define a long lherzolite trend to extending to 10% Cr_2O_3 (Fig. 3k). Most garnets with normal REE patterns, but a small proportion have flatter and mildly sinuous patterns. Y-depleted garnets are common up to 1100–1150 °C (Fig. 4l), but those above 1150 °C are all Zr–Ti enriched, with up to 1% TiO_2, giving a strong signature of melt-related metasomatism. The geotherm is poorly defined, but is interpreted as following a 40 MW/m^2 conductive model up to 1100 °C (Fig. 5l). The mean olivine is up to Fo_{92} at shallow levels, but most of the section averages Fo_{91}; the mean drops rapidly to Fo_{90} below 160 km.

The CARP section (Fig. 8) shows depleted lherzolites present at low abundances (≤10%) through whole section, but fertile lherzolites are more common. Melt-related metasomatism is registered through whole section, which is unusual. Melt-metasomatised lherzolites make up >50% of the section at 130 km depth, increasing to 80% at 170. The Y edge at 1100 °C corresponds to a lithosphere thickness of ca. 160 km, but this probably is a maximum estimate, since most of the section below 130 km is so strongly affected by melt-related metasomatism. The high geotherm, fertile composition and thin lithosphere suggest that the Proton SCLM has been strongly affected by asthenospheric upwelling related to the Mid-Continent Rift System.

6.7. Arkansas (AR)

At least six bodies of lamproite occur near Murfreesboro, AR; a K–Ar age of 106±3 Ma (Gogineni et al., 1978) is consistent with their mid-Cretaceous stratigraphic age. The intrusions lie on the extension of the Oklahoma Aulocogen at the edge of Mississippi

Embayment, and near the commonly accepted edge of the craton, where Grenville- and Appalachian-age rocks overlie the 1.3–1.5 Ga Granite–Rhyolite Province basement. Data used here are from the Prairie Creek and Twin Knobs bodies (Table 1; Griffin et al., 1994).

The garnet data define a very long lherzolite trend up to 15% Cr_2O_3 (Fig. 3c). Some of the highest-Cr garnets, and a few lower-Cr ones, are mildly subcalcic relative to the lherzolite trend. Maximum Cr content increases monotonically with T, so that the high-Cr subcalcic garnets have the highest T. The $Y–T$ plot (Fig. 4c) shows an alternation of depleted and fertile zones, and thus no clear Y edge. The geotherm is segmented, following a 35 mW/m^2 conductive model in the shallow part, and a 40 mW/m^2 model in the deeper parts (Fig. 5c). The calculated olivine compositions show a similar layering (Fig. 8); they are very depleted (Fo$_{>93}$ mean) down to ca. 130 km, then drop to Fo$_{91.7}$ by 150 km and to Fo$_{91}$ by 180 km, with a short reversal to more than Fo$_{92}$ near 160 km. These segments correspond to those seen on the $Y–T$ plot; 130 km represents the base of the Y-depleted layer, 150 km the base of the intermediate fertile layer, and 170 km the base of the deeper depleted layer.

Dunn et al. (2000) describe xenolith garnets with up to 8.4% Cr_2O_3; this reflects the lower maximum temperature (960 °C) recorded by their xenoliths. The relatively shallow peridotite xenoliths (<100 km) studied by Dunn et al. (2000) have olivine with Fo$_{90-92.6}$ (mean Fo$_{91.2}$), and the deepest samples have Fo$_{92.4}$. Our data suggest a mean value of Fo$_{92.5}$ at 100 km, and the trend suggests shallower samples would be more Fe-rich, consistent with the xenolith data.

The CARP section (Fig. 8) emphasises the strong layering noted above: the upper part of the section down to 155 km is relatively depleted at the top, with an increasing proportion of fertile lherzolites downward. There is a sharp break at ca. 155 km, and the lower part of the section is dominated by depleted/metasomatised lherzolites to ca. 170 km, and depleted lherzolites to harzburgites appear at the greatest depths.

6.8. Colorado Plateau (CP)

The Colorado Plateau is underlain by Proterozoic crust 37–48 km thick (Parsons et al., 1996); a study of crustal xenoliths (Selverstone et al., 1999) suggests that the boundary between the Yavapai (1.8–2.0 Ga) and Mazatzal (1.6–1.8 Ga) terranes runs NE–SW through the area. Mattie et al. (1997) and Condie and Selverstone (1999) regard the lower crust as mafic, but argue that no mafic underplating has occurred since the Proterozoic assembly of the terranes. This view is supported by Sm–Nd data on mineral pairs in lower crustal xenoliths, which record cooling ages of ca. 1.35 Ma, suggesting that no significant heating of the lower crust has occurred since then (Wendlandt et al., 1996). U–Pb ages and Hf-isotope data on zircons from eclogite and garnetite xenoliths indicate protolith ages near 1.8 Ga (Smith and Griffin, 2003).

A cap of marine sediments indicates that the Plateau has been uplifted ca. 2 km since Cretaceous time. Sahagian et al. (2002) argue that uplift was slow from 25 to 5 Ma, and rapid since then; Pedersen et al. (2002) suggest little or no uplift since early Tertiary time (post-Laramide). Gravity data show that the Plateau is in isostatic equilibrium, so the uplift must be compensated by mass reduction at depth. Since the crustal thickness is not great enough to support uplift across most of the region, this compensation requires an anomalously light mantle (Parsons and McCarthy, 1995).

Shallow subduction of the Farallon plate took place under the region in Cenozoic time (Dickinson and Snyder, 1978; Severinghaus and Atwater, 1990). Detachment of this plate could allow upwelling of asthenosphere and rapid heating of lithosphere at ca. 25 Ma, providing both surface uplift at 26–18 Ma (Beghoul and Barazangi, 1989; Riter and Smith, 1996) and the heat source to drive kimberlitic and mafic magmatism. The samples studied here come from five Miocene–Oligocene (20–30 Ma) intrusives in the Four Corners area near middle of the Plateau (Table 1).

The garnets define a short lumpy lherzolite trend from 1% to 7% Cr_2O_3, with most grains containing 1–4%, and some mildly subcalcic garnets are present (Fig. 3f). A pronounced subtrend extending into the wehrlite field (high Ca/Cr) may represent grains from spinel–garnet peridotites. The $Y–T$ plot (Fig. 4g) shows that nearly all garnets are derived from the shallow, low-T (<900 °C) part of the mantle, and that most are from fertile lherzolites, while the proportion of Y-depleted grains increases with depth. Zr and Ti contents are generally low, and only a small cluster of grains with $T>1100$ is enriched in Zr, Ti (0.75% TiO_2) and Y, to give a melt-metasomatised signature. The geotherm is poorly defined (Fig. 5g), but appears to be low, and probably not on a conductive model above 900 °C and 140 km. The mean calculated olivine compositions are magnesian ($Fo_{93.5}$) at the top of the section, and decrease smoothly to $Fo_{92.3}$. The deepest samples are mostly ca. Fo_{90-91}.

$P-T$ estimates for xenoliths from The Thumb (Ehrenberg, 1979; Eggler et al., 1987) lie on a linear trend from 1000 °C/39 kb to 1300 °C/48 kb, equivalent to the kinked limb of the garnet geotherm (Fig. 5g). Ehrenberg (1979, 1982) and Smith et al. (1991) interpreted strong zoning of garnets in high-T xenoliths as reflecting melt-related metasomatism and a sharp rise in T at 130–150 km shortly prior to eruption. Ehrenberg (1979) gave a mean olivine composition of $Fo_{92.2}$ in xenoliths from the Navajo kimberlites, with some values up to $Fo_{94.4}$, and Smith (2000) gives a mean $Fo_{90.5}$ for high-T xenoliths from The Thumb; these values are consistent with our calculated values.

The CARP section (Fig. 8) shows a dominance of fertile peridotites, and a small increase in melt-related metasomatism with depth. An unusually high proportion of unclassified garnets suggests complex metasomatic processes that are not captured by the CARP classes used here.

The garnet data suggest that if the Farallon slab passed beneath this area, it lay deeper than ca. 140 km. Beghoul and Barazangi (1989) measured a Vp of ca. 8.15 km/s at the Moho beneath the Plateau, equivalent to values beneath central N. America. This indicates the presence of a cool depleted root. Smith (2000) estimated a mean $Fo_{89.5}$ for the uppermost mantle beneath the Basin and Range Province, which is ca. 200 °C hotter on average than the Colorado Plateau mantle. A difference of 1% Fo is equivalent to ca. 90 °C in terms of density, which implies that the Colorado Plateau root would need olivine of at least Fo_{92} to provide buoyancy. This is consistent with our estimate of $Fo_{92.3-93.5}$ for the upper SCLM of the Colorado Plateau.

6.9. State Line district (SL)

Over 100 kimberlites are known in several groups scattered across the Colorado–Wyoming border in the Front Ranges of the Rocky Mountains. The data here are from eight kimberlites covering much of the State Line district and the Iron Mountain area (Table 1). The kimberlites intrude a 1.7–2.0 Ga mobile belt off the southern edge of the Wyoming Craton. Stratigraphic control indicates a Devonian age for many of the bodies, which is supported by 377 Ma fission track ages (Naeser and McCallum, 1977) and by Rb–Sr ages (Smith, 1979). Heaman et al. (2003) have dated the Chicken Park kimberlites at 614 Ma and the Iron Mountain bodies at 408 Ma. The area lies on the hot–cold transition of Thybo et al. (2000) and inside the magnetic craton of Purucker et al. (2002), but outside the mantle root visible in seismic tomography (Fig. 2); this may reflect real modification of the root after intrusion of the kimberlites, or simply the thermal effects of the Laramide orogeny.

The garnets define a long and complex lherzolite trend with two en-echelon segments, one from 1% to 8% Cr_2O_3, the other from 4% to 14% (Fig. 3s). A significant number of mildly to strongly subcalcic garnets spans the range from 3% to 12% Cr_2O_3. The maximum Cr content of the garnets increases with T, and the garnets with highest Cr contents have $T=1150-1250$ °C. The Y–T plot (Fig. 4t) shows a T range of 600–1300 °C, with scattered grains to 1600 °C. Y-depleted garnets are present over the whole T range; there is a poorly defined Y edge at 1300 °C. Maximum Zr contents are high (to 150 ppm) over most of the T range, but high Zr is only well-correlated with high Ti (to 1.3% TiO_2) above ca. 1100 °C.

The garnet geotherm (Fig. 5u) lies along a 35 mW/m^2 conductive model below 900 °C, then approximates a 39–40 mW/m^2 conductive model from ca. 900 to 1100 °C; the Y edge at 1300 °C suggests that this geotherm can be extended at least that far. The two sections of the geotherm correspond to the two en-echelon lherzolite trends shown in Fig. 3s. Calculated olivine compositions average Fo_{93} down to 120 km, then decrease to ca. Fo_{92} at 140–150 km, corresponding to the break in the geotherm and garnet Ca/Cr. The lower layer has $Fo_{91.5-92}$ at the top, decreasing to Fo_{90} by 170 km, after which there is a rapid drop with depth to Fo_{88}.

The available xenolith data are heavily biased toward the Sloan 2 and Nix localities. Eggler et al. (1987) show that "infertile" peridotites with $T=$

1000–1300 °C lie along a 40 mW/m^2 conductive model geotherm with no kink, which is consistent with the garnet geotherm. Low-T (650–800 °C) "enriched" peridotites lie above this geotherm. Shallow fertile xenoliths have olivine of Fo$_{90-92.5}$ (mean ca. Fo$_{91}$), while "infertile" ones have Fo$_{92-93}$ up to $T \approx 950$ °C at 140 km (Eggler et al., 1987, 1988). Our data would suggest that the "infertile" type is volumetrically more important at these levels. The Fo values of the infertile xenoliths decrease with depth, with scatter down to Fo$_{90}$. These data are generally consistent with our section, especially the sharp drop in mean Fo at 140 km.

The CARP section (Fig. 8) shows a distinct layering with a break at 140–150 km, corresponding to the discontinuity in the geotherm, and to the break between the lower-T group of garnets with higher Ca/Cr and the deeper higher-Cr group with lower Ca/Cr. The upper layer contains some more depleted rocks, but also a higher proportion of fertile material; the deeper layer contains less depleted material but a high proportion of depleted/metasomatised garnets. This is consistent with the observation of Eggler et al. (1987, 1988) that REE patterns indicate a high degree of cryptic metasomatism in the deeper layer, in rocks with depleted major element compositions. The lower layer also contains a higher proportion of melt-metasomatised lherzolites. However, there is no strong increase in melt-related metasomatism with depth until ca. 190 km; this is consistent with the general rarity of sheared high-T peridotites in the Sloan and Nix pipes (Eggler et al., 1987).

6.10. Williams Ranch diatremes (WM)

Four ultramafic diatremes occur on the Williams Ranch in the Missouri Breaks area of northern Montana; they are dated by K–Ar to 47–52 Ma (Eocene; Hearn and McGee, 1984). The area lies on the edge of the Archean Wyoming Craton, defined by the Great Falls Tectonic Zone, which separates the Wyoming Craton from the tectonically distinct Medicine Hat block to the north. The GFTZ is interpreted as an Archean suture, formerly the site of N-dipping subduction without associated magmatism (Gorman et al., 2002). The basement includes 2.5–2.8 Ga rocks, overprinted by tectonic events at 1.5–1.8 Ga (Hearn et al., 1989).

Carlson et al. (1998) found that three low-T peridotites give Re–Os T_{RD} model ages of 1.7–2.5 Ga, while three deeper ones give T_{RD} ages of ca. 800 Ma, suggesting replacement or reworking of the deep SCLM; they argued that the cratonic root is truncated at 150 km depth. Similar T_{RD} have been obtained for spinel lherzolites from Highwood Mountains and Eagle Buttes minettes (Carlson and Irving, 1994).

Only a small set of garnets has been analysed. Lherzolitic garnets have Cr$_2$O$_3$ contents up to 13%, but most have <7% ; several subcalcic garnets are present (Fig. 3w). The Y–T plot (Fig. 4y) shows two groups, one with $T<900$ °C, the other mostly with $T=1250-1350$ °C; mildly Y-depleted garnets are found in both groups. The high-T group is richer in Zr and Ti (up to 100 ppm Zr, 1% TiO$_2$), suggesting melt-related metasomatism. The low-T data are consistent with a 38–40 mW/m^2 conductive geotherm (Fig. 5z); the high-T ones give minimum P of 50 kb. This pattern suggests a kink in the geotherm at ca. 950 °C and 130 km. The calculated olivine compositions average Fo$_{92.5}$ down to 120 km, while the deeper group is less magnesian (mean Fo$_{90.7}$).

$P-T$ estimates for peridotite xenoliths with $T<1000$ °C lie along a 40 mW/m^2 conductive geotherm, distinct from a cluster at 1300–1400 °C and 50–60 kb (Eggler et al., 1987). Five garnet peridotites described by Carlson et al. (1998) show the same distribution, in good agreement with our data. Eggler et al. (1987) give olivine compositions in five low-T xenoliths from Fo$_{91.8-92.5}$, while high-T xenoliths have Fo$_{89.8-92.2}$. No CARP section is presented, because the data are too few. The shallow garnets are derived from either fertile or melt-metasomatised lherzolites, while the deeper ones mostly represent melt-metasomatised lherzolites.

Hearn (1995) and Hearn et al. (1989) argue that the abundance of megacrysts and high-T sheared peridotites in the Williams diatremes reflects melt infiltration, which is consistent with our data. Carlson et al. (1998) suggest that the high-T peridotites were recently underplated, but the Re–Os data do not require this, because a "younging" of T_{RD} model ages typically accompanies such melt infiltration, due to the addition of asthenosphere-derived sulfides (Griffin et al., 2002a, 2004). We interpret the limited data as reflecting thinning of the SCLM due to upwelling of the asthenosphere along the GFTZ suture, accompanied by melt-related metasomatism of the lower lithosphere.

6.11. Grass Range (GR)

Two possible kimberlites and more than a dozen carbonated or altered breccias are known in the Grass Range area of Montana, ca. 100 km south of the Williams diatremes. The nearby Winnett Sill is dated at 50 Ma (Hearn et al., 1989), similar to the age of the Williams diatremes, and the kimberlites may be of the same age. The area lies within the Wyoming Craton proper. Crustal ages are 3.8–2.6 Ga, but the craton may have a Proterozoic lower crust; the crust is up to 60 km thick, and the lower half has a high seismic velocity and probably is mafic (Gorman et al., 2002). The craton was strongly modified during the Laramide orogeny. Reconnaissance in-situ Re–Os analysis of sulfides in two garnet lherzolite xenoliths shows three groups of T_{RD} model ages at 0.4–0.5, 1.7–2.0 and 2.37–2.95 Ma (as well as negative model ages in high-Re sulfides), consistent with the Proterozoic and Phanerozoic modification of an Archean root (Griffin et al., unpubl. data).

The available garnets show a short lherzolite trend from 2% to 6.5% Cr_2O_3, consisting of two distinct clusters, and a few mildly to strongly subcalcic garnets w/<6% Cr_2O_3 (Fig. 3i). The Y–T plot (Fig. 4j) shows two distinct levels of Y; the low-Y garnets with T=1000–1200 °C correspond to the higher-Cr group in Fig. 3e. There is a distinct Y edge at ca. 1150 °C. Most of the garnets have normal downward-concave REE patterns, but those in the Y-depleted group have sinuous REE patterns with MREE>HREE. The high-T garnets are depleted in Zr and Ti, and have low Nd/Y and Sc/Y.

The data are consistent with a geotherm near a 42 mW/m^2 conductive model (Fig. 5j); if this is extended to the Y edge at 1150 °C, it implies a lithosphere ≥170 km thick. The mean calculated olivine compositions are relatively magnesian at the top of the section (Fo$_{93.2}$), decreasing to ca. Fo$_{92.5}$ around 130 km, where the proportion of fertile types is greatest, and then increasing with depth. The mean olivine composition drops quickly to Fo$_{91–92}$ below 170 km.

The CARP section is strongly layered, with a relatively fertile upper part underlain by a much more depleted lower part. There is little evidence for typical melt-related metasomatism. This pattern, and the Fo values, are typical of many Archean SCLM sections modified by later metasomatism (Griffin et al., 2002b, 2003a). The differences between the Williams and Grass Range SCLM sections can be explained by their positions relative to the craton margin represented by the GFTZ, and suggest the rise of the asthenosphere along the old suture represented by the GFTZ.

6.12. Alberta (AL)

At least 15 crater-facies kimberlites, several known to be diamondiferous, occur in the Buffalo Hills of N-central Alberta, between the Peace River and Loon-Wabasca River drainages. U–Pb perovskite ages date the kimberlites to 83–93 Ma (Carlson et al., 1999). They intrude the Buffalo Head Terrane, consisting of pre-collisional Proterozoic (2.4–2.0 Ga) crust, which Sm–Nd data and zircon inheritance indicate was built on Archean crust (Ross, 2002). This terrane was accreted to the Hearn Province at 2.0–1.9 Ga, and was flanked by outward-dipping subduction zones; the setting implies that the underlying SCLM could be Archean, modified but not destroyed by Proterozoic tectonic activity. Diamonds from the Buffalo Hills kimberlites contain inclusions from the lower mantle and transition zone, as well as peridotitic and eclogitic SCLM, suggesting that plume activity was involved in the generation of the SCLM (Davies et al., 2004b).

The garnet data define a long narrow lherzolite trend extending from 1.5% to 10.5% Cr_2O_3, and continued to 18% Cr_2O_3 by a group of mildly subcalcic garnets (Fig. 3a). A compact group of lower-Cr lherzolite–wehrlite garnets with 1–3.5% Cr_2O_3 have higher Ca/Cr than the main trend, and probably are derived from spinel–garnet lherzolites. These data are equivalent to those described by Carlson et al. (1999). The Y–T plot (Fig. 4a) shows that the lowest-T garnets are derived from fertile spinel–garnet lherzolites. Many garnets with T between 700 and 1150 °C are Y-depleted, and the subcalcic garnets are concentrated at the upper end of this T range; the Y edge lies at ca. 1150 °C. At higher T (>1200 °C) many grains show moderate melt-related metasomatism, with TiO_2 contents to 0.9%. About 1/4 of the garnets (concentrated in the 850–1100 °C range) have sinuous REE patterns, whereas lower-T and higher-T ones have upward-convex patterns and little HREE depletion.

Except for one anomalous group with low T at 40–45 kb, the data are consistent with a 40 mW/m^2 conductive geotherm out to 1200 °C (Fig. 5a). This suggests that the base of the depleted lithosphere is at least 160 km deep. Aulbach et al. (2004a) describe abundant spinel peridotites, and several garnet peridotites, most of which show significant disequilibrium. One sheared garnet lherzolite with Ti-rich garnet lies above the estimated geotherm at 1275 °C and 58 kb, consistent with the Y-edge at 1150 °C (55 kb). The mean calculated olivine composition is Fo$_{92}$ down to ca. 140 km, then decreases steadily to reach Fo$_{90.3}$ at 175 km. Olivine in 14 spinel peridotite xenoliths has mean Fo$_{91.4}$; olivine in five garnet peridotites with $T<1100$ °C is Fo$_{91–93}$; one high-T garnet peridotite has Fo$_{89.8}$ Aulbach et al. (2004a).

The CARP section is dominated by very fertile lherzolites (especially the spinel–garnet lherzolites) down to 115 km. From 115 to 150 km the section consists mainly of depleted and depleted/metasomatised lherzolites; fertile and melt-metasomatised lherzolites dominate below 160 km. The relatively high degree of depletion in the middle of the section is consistent with an Archean SCLM that has been strongly modified in Proterozoic time. The fertility of the upper part of the section may reflect the intrusion of mafic melts at these levels. The xenolith suite Aulbach et al. (2004a) contains abundant garnet pyroxenites with low temperatures and disequilibrium microstructures, which may represent these mafic magmas.

7. Discussion and synthesis

7.1. Characteristics of Archean SCLM: comparisons with Africa, Siberia

Comparison of CARP data for the different sections can be difficult, because different sections have experienced different degrees of melt-related metasomatism, in connection with kimberlite intrusion and other tectonothermal events (Griffin et al., 2002b, 2003a). Removal of the melt-metasomatised classes and renormalisation to 100% allows the relative proportions of the other classes to be compared in cumulate-percentage plots, like those used in sediment classification (Fig. 10).

Fig. 10A shows fields for different types of SCLM, drawn largely on the basis of data from southern Africa (Griffin et al., 2003a) and Siberia (Griffin et al., 1999f). "Ultradepleted" SCLM has been sampled by kimberlites in the Limpopo Belt, and similar material makes up the upper layer of the central parts of the Slave Province (Fig. 11a). The "Archon" field in Fig. 10A represents relatively unmodified SCLM, such as that sampled by the Group 2 kimberlites of the Kaapvaal craton, and the kimberlites of the Daldyn-Alakit and Upper Muna fields in Siberia. The "Metasomatised Archon" field covers SCLM sections sampled by the Group 1 kimberlites of the Kaapvaal Craton, several kimberlite fields in Botswana, and the Malo-Botuobiya, Kharamai and Nakyn fields in Siberia. The characteristics of these sections have been ascribed to metasomatic modification of originally more depleted sections (Griffin et al., 1999c, 2004). The "Proton" field represents the SCLM sampled in the Namaqua–Natal belt of southern Africa, and the off-craton fields of the Birekte Province in Siberia. "Tecton" mantle is found beneath the Appalachians and Phanerozoic mobile belts in eastern Australia, eastern China and Mongolia (Griffin et al., 2002b). The "Ultradepleted" and "Archon" types of SCLM have concave-downward patterns, reflecting high proportions of depleted harzburgites and lherzolites and low proportions of fertile lherzolites. Proton SCLM is dominated by fertile and metasomatised lherzolites, and has a concave-upward pattern, while "Metasomatised Archon" patterns are generally intermediate.

Most of the Archean SCLM sections sampled in this study, outside the Slave Province (see below), are of the "Metasomatised Archon" type (Fig. 10A), suggesting that they have experienced considerable modification. The lower part of the Michigan section is unusual in having a very high proportion of the depleted/metasomatised garnet classes (especially L19), but few of the "fertile" classes. There is a broad similarity between the other curves and that for the inferred plume-related lower layer of the Slave Province, but the latter contains lower proportions of fertile classes and a higher proportion of the depleted/ metasomatised classes, giving it a downward-concave pattern.

The sections from Cobalt, Wisconsin and Sextant Rapids (Fig. 10C) show progressively more concave-

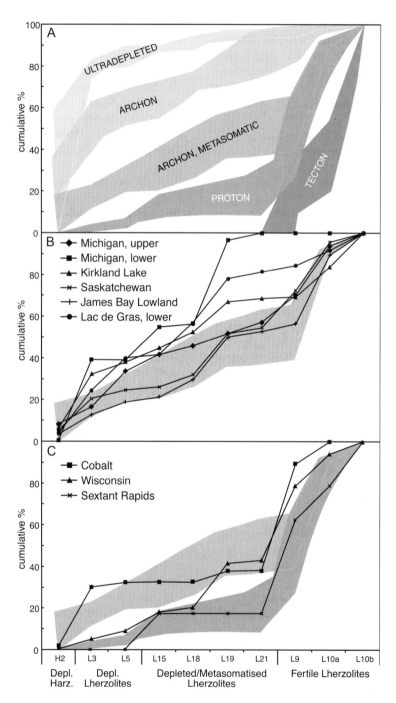

Fig. 10. Cumulative percentage plots of CARP classes, normalised after removal of the classes reflecting melt-related metasomatism. (A) Fields for SCLM types, based on data from southern Africa and Siberia (see text for explanation), (B) N. American Archons, (C) modified Archons.

upward patterns. In the case of the Cobalt and Sextant Rapids sections, this may reflect progressive metasomatism of more typical Archean sections. The Cobalt area lies near the edge of the Archean continent, and may have been modified during Proterozoic magmatic activity along the continental margin. Sextant Rapids lies within the Kapuskasing Structural Zone, which cuts across the grain of the Superior Province. Assuming that the SCLM beneath the KSZ was originally formed in the Archean, the extremely fertile nature of the present SCLM suggests that the Archean SCLM has been either replaced, or severely modified, during or after the formation of the uplift. This in turn suggests that the KSZ experienced significant extension prior to the final compression that uplifted the lower crust.

7.2. Layered SCLM and lithosphere growth: plume subcretion

Nearly all of the SCLM sections described here show stratigraphic variations in depletion and fertility, related to different degrees and styles of metasomatism. But several also show two or more distinct, sharply defined layers, separated in some sections by unsampled gaps. This type of sharply defined layering is not common in cratonic sections worldwide (Griffin et al., 2002b, 2003a). It suggests an episodic construction of the SCLM, which might be caused by two distinct mechanisms: the addition from below of plume-related material ("plume subcretion"), and tectonic stacking related to subduction or continental collision (Snyder, 2002).

The best case for plume subcretion has been made previously in the central Slave Province (Fig. 6), where the presence of abundant lower-mantle inclusions in diamonds, and of highly unusual sulfide compositions (included in both diamonds and olivine macrocrysts) give independent evidence for an Archean plume (Davies et al., 1999, 2004a; Aulbach et al., 2004b). Most of the localities sampled in the Slave Province show layering, suggesting that plume activity like that documented under the Lac de Gras area has been widespread.

When the CARP data for the Slave Province SCLM sections are plotted as in Fig. 10, the upper layers show a wide variety of patterns (Fig. 11a). The Lac de Gras, Anuri and Snap Lake sections are highly depleted, but differ widely in the ratio of harzburgite to lherzolite; the other localities are less strongly depleted. The lower layers, in contrast, show a high degree of uniformity in their degree and style of depletion, and differ mainly in the relative importance of different styles of metasomatism (L18, L19, L21; Fig. 11B). The provinciality of the upper layers, contrasted with the regional similarity of the lower layers, suggests that all of the lower layers represent the plume head material inferred to exist beneath the Lac de Gras area. The Tenacity and Doyle Lake localities (Griffin et al., 1999d) each show only a single layer at relatively shallow depth, and their sections are similar to the inferred plume material of the lower layer beneath Lac de Gras. Both localities lie near the outer edges of the craton, suggesting that the pre-plume SCLM was too thin to be garnet-bearing. A similar conclusion was reached for the Camsell Lake area in the southern part of the craton (Griffin et al., 1999d), using major-element data for garnet concentrates from Pokhilenko et al. (1998).

All of these localities lie on or west of the major lithospheric boundary implied by the elastic-thickness analysis of the craton (Poudjom Djomani et al., abstracts, this conference). This distribution, and the probable pre-3.0 Ga timing of plume emplacement, suggest that the inferred plume head is part of the SCLM beneath the Central Slave Basement Complex, emplaced before the amalgamation of the eastern and western halves of the craton at ca. 2.8 Ga.

The Michigan and Saskatchewan SCLM sections also show distinct layering, each with an unsampled gap, but in these cases the lower layer appears to be significantly thinner than the upper one, and underlies a pre-existing section 165–175 km thick. Each of these localities lies on the margin of a major Paleozoic basin (the Michigan and Williston Basins, respectively), and in each case the kimberlite intrusion coincides broadly with an episode of subsidence within the basin (Kaminski and Jaupart, 2000). These authors have modelled the formation of these basins in terms of mantle plumes intruding the lower lithosphere and replacing part of the older buoyant SCLM with more fertile material. In this model, subsidence follows as the plume cools, causing a net decrease in the buoyancy of the SCLM section. Bank et al. (1998) have proposed that a plume was involved in the emplacement of the Saskatchewan kimberlites and the modi-

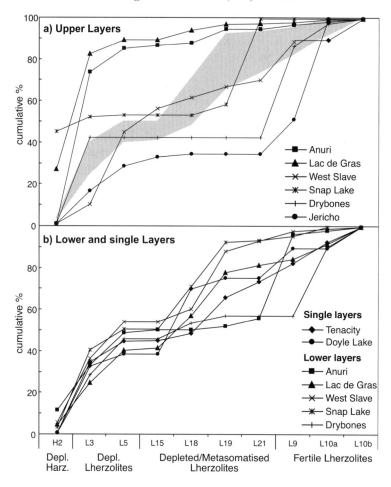

Fig. 11. Cumulative percentage plots of CARP classes for the Slave Craton, normalised after removal of the classes reflecting melt-related metasomatism. (a) Upper layers, with shaded field for bulk of lower layers; (b) lower layers.

fication of the SCLM; it should be noted that the plume mechanism proposed here would have to precede the emplacement of the kimberlites, which have sampled the inferred plume-modified lithosphere.

In both sections, the layer boundary is marked by a sampling gap. In the Michigan case, the gap contains a few peridotites that are much more Fe-rich than those above or below, and the gap also corresponds to the main concentration of eclogites, pyroxenites and megacrysts in the section (McGee and Hearn, 1984). This would be consistent with the concentration of melts near the top of a plume, but could also be explained as the top of a slab subducted under the Archean lithosphere in Proterozoic time. Similarly, the layering of the Saskatchewan section

could reflect its position within the compressional Trans-Hudson orogen.

7.3. Layered SCLM and lithosphere growth: tectonic stacking

The best example of possible tectonic stacking of the SCLM is provided by the two sections from Greenland (Fig. 9), which show strong layering and, in the Sukkertoppen section, a significant gap in the section. Sarfartoq lies near the surface expression of the Nagssuqtoquidian Front, a major suture between the Archean cratonic core and a belt of Proterozoic and strongly reworked Archean crust. The younger rocks are inferred to have been thrust southward

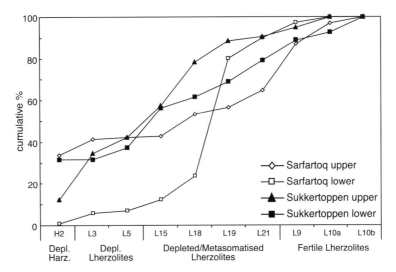

Fig. 12. Cumulative percentage plots of CARP data, normalised after removal of the classes reflecting melt-related metasomatism. Greenland localities, illustrating the similar composition of the upper layer at Sarfartoq and the lower layer at Sukkertoppen.

beneath the older ones. Sukkertoppen lies ca. 150 km to the SW, and ca. 100 km from the Front.

Fig. 12 shows the normalised CARP curves for the upper and lower layers of the Sarfartoq and Sukkertoppen sections. The upper layer at Sarfartoq is typical of many Archean sections, whereas the lower layer is strongly metasomatised, and similar to some Proterozoic sections. The lower layer at Sukkertoppen is essentially identical to the upper layer at Sarfartoq, whereas the upper part of Sukkertoppen contains higher proportions of several depleted/metasomatised classes. The similarity of the upper Sarfartoq and the lower Sukkertoppen sections suggests that the former may have been thrust southward under the latter, at the same time that more "Proterozoic" SCLM from the Nagssuqtoquidian Belt was being thrust under the Sarfartoq area to form its lower layer. The unsampled gaps in these sections may reflect an abundance of mafic, rather than peridotitic, material at the top of the underthrusted sections.

The State Line SCLM section (Fig. 13) may provide another example of tectonic stacking. The upper layer of our section is relatively depleted, and resembles several SCLM sections from the Kaapvaal Craton and Siberia (see below). The lower layer is significantly more metasomatised, with low proportions of depleted lherzolites, and high proportions of the depleted/metasomatised classes L19 and L21; it

resembles several Proterozoic SCLM sections. Eggler et al. (1988) have described fertile spinel lherzolites and pyroxenites from the uppermost part of the State Line section (<80 km depth), and suggested that the boundary between these high-level fertile rocks and the deeper more depleted ones represents thrusting of Proterozoic SCLM above older SCLM on the margin of the Wyoming Craton. The data presented here suggest that the Archean-type SCLM also is bounded by less depleted Proterozoic SCLM, or a mixture of Proterozoic and Archean SCLM, below ca. 145 km. Snyder (2002), drawing on seismic reflection data from several localities, has suggested that this situation, with wedges of older SCLM within younger, more fertile SCLM, is common along cratonic margins and represents an important mechanism of lithospheric growth.

7.4. New proton SCLM, or reworking of archon SCLM?

The observed secular evolution of SCLM composition (Griffin et al., 1998, 1999b) may be interpreted as reflecting changes through Earth's history in the mechanisms that generate the SCLM, or progressive modification of ancient lithosphere. The buoyant nature of depleted Archean SCLM strongly suggests that it will persist and could gradually be modified by

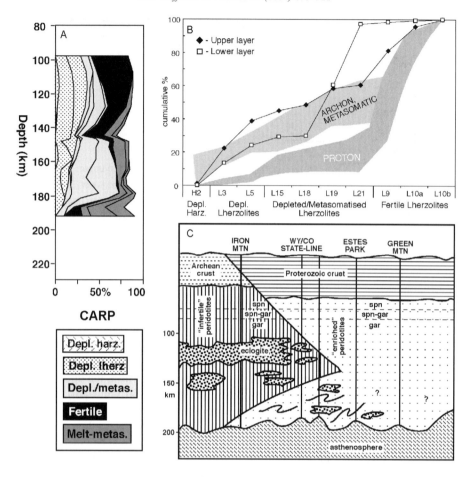

Fig. 13. State Line area. (A) CARP section; (B) Cumulative percentage plots of CARP data for the upper and lower layers of the State Line section, with fields for "Metasomatised Archon" and "Proton" SCLM from Fig. 10; (C) cartoon (after Eggler et al., 1988) illustrating a wedge of Archean mantle separating Proterozoic upper SCLM from a mixed Archean–Proterozoic lower SCLM, produced by Proterozoic collisional tectonics.

metasomatic processes, becoming more fertile with time (Poudjom Djomani et al., 2001; O'Reilly et al., 2001). The presentation shown in Fig. 10A can be used to assess this question for the sections studied here.

Fig. 14A shows that the Alberta SCLM probably represents Archean SCLM; this is consistent with the inferred Archean nature of the basement to the Buffalo Head Terrane. The Alberta section also strongly resembles the lower layer of the SCLM beneath the Slave Province, which we consider to represent subcreted plume material (Griffin et al., 1999e). This is intriguing in light of the evidence from diamond inclusions that the Alberta SCLM

contains a plume-related component (Davies et al., 2004b).

The upper layer (100–145 km) of the State Line section also has a distinctly Archean signature, and despite a higher level of fertility, it bears a strong resemblance to the Alberta SCLM and the lower layer of the Slave Province. The SCLM beneath Kansas, normalised to remove the dominant melt-related component, also bears a striking resemblance to the SCLM beneath parts of the Kaapvaal craton, including the marginal areas in Botswana that have been shown to consist of highly metasomatised Archean SCLM (Griffin et al., 2003a). The Grass Range section contains a lower proportion of depleted mate-

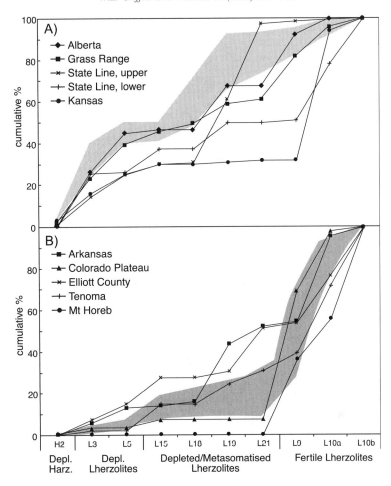

Fig. 14. Cumulative percentage plots of CARP data, normalised after removal of the classes reflecting melt-related metasomatism. (A) Protons with possible Archean SCLM, and shaded field for Archon SCLM from Fig. 10A; (B) Proton and Tecton SCLM, with corresponding fields from Fig. 10A.

rial, and a high proportion of fertile lherzolites, and also resembles SCLM sections from areas such as Botswana. As discussed above, Re–Os data indicate that these samples represent Archean SCLM, but with a strong overprint in Proterozoic and Phanerozoic time.

The Colorado Plateau section (Fig. 14B) underlies juvenile Proterozoic crust, and may represent primary Proterozoic SCLM; it is similar to sections from the Namaqua–Natal belt of southern Africa, and the off-craton areas of Siberia. It contains a higher proportion of fertile lherzolites, but is more depleted than the Tecton section represented by the Mt Horeb sample. The Tenoma section may also represent Proterozoic

SCLM; it contains less depleted material than the Elliott County SCLM, which lies on the edge of the seismic root and may contain remnants of older material. The Arkansas section may be primary Proterozoic SCLM, but the CARP section (Fig. 8) is complex and suggests that tectonic stacking may have juxtaposed SCLM sections of different ages.

7.5. Continental structure and seismic tomography

The temperature at 150 km depth (Fig. 15) is relatively well-constrained for most localities by the geotherms shown in Fig. 5. It shows relatively little variation (most values 950–1025 °C) across most of

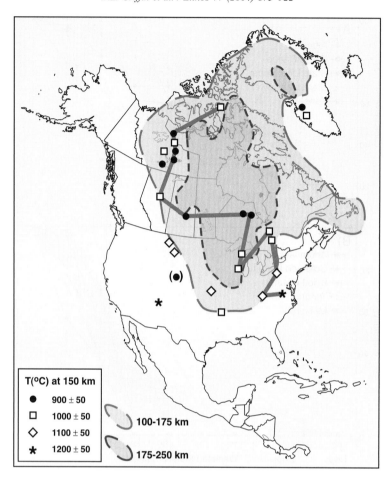

Fig. 15. Map of temperature at 150 km depth, relative to outline of the seismic "root". Thick line shows the traverse illustrated in Fig. 17.

the Laurentian core of the continent, despite a range in surface heat flow between 10–20 and 50–60 mW/m^2 (Artemieva and Mooney, 2001). This uniformity illustrates that the variations in surface heat flow are largely controlled by local variations in upper-crustal heat production, and are not reflected in large T variations at mantle depths. Downward projection of surface heat flow data therefore is likely to give misleading results for both mantle temperatures and the thickness of the thermal lithosphere. This is why the thickness of the thermal lithosphere as derived by Artemieva and Mooney (2001) is generally thinner, but sometimes (Hudson Bay–Kirkland Lake) thicker than those derived here. The largest difference is seen in the Colorado Plateau, where Artemieva and Mooney (2001) derive temperatures ≥1200 °C at 100 km

depth, while the xenolith data discussed above would put such temperatures at 140–160 km depth.

The temperatures at 150 km depth are higher in the mobile belts, where the SCLM is thinner and less depleted, leading to lower seismic velocities and an apparent thinning of the seismically defined continental root. The highest values are found in the outer part of the Appalachians, and beneath the Colorado Plateau. Notable deviations from this pattern are the State Line and Greenland samples, which record low values despite their positions off the seismic root. These localities illustrate the need to consider the fourth dimension (time): in each case the kimberlites are significantly older than the thermal event that has lowered the seismic velocity of the SCLM.

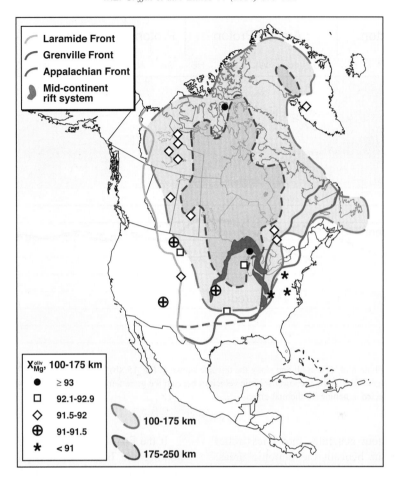

Fig. 16. Map of mean %Fo (X_{Mg}) in olivine for the 100–150 km depth slice, relative to outline of the seismic "root".

Fig. 16 shows the mean Fo content of peridotitic olivine between 100 and 175 km depth. There are few samples in the high-velocity seismic "core" of the continent, but mean values greater than Fo_{93} in the Michigan and Somerset Island sections, and Fo_{92-93} in the Hudson Bay Lowland, suggest that the SCLM beneath the seismic "core" probably has mean Mg#≥93, and that this strong depletion contributes significantly to the high seismic velocities. Most samples from the fringing band of "thinner" lithosphere have mean $Fo_{91.5-92}$, which is significantly lower than most Archon SCLM worldwide (Gaul et al., 2000). These low values contribute to the higher seismic velocities at depth in this zone. The least magnesian olivine compositions are found in the Appalachians, beneath the Colorado Plateau and

northern Montana, and in Kansas. The Fe-rich nature of the Kansas SCLM reflects a very high degree of melt-related metasomatism, almost certainly related to the nearby Mid-Continent Rift System.

The lateral and vertical variation of Mg# in the SCLM along a traverse across the continent is shown in Fig. 17a. Only depths below 70–100 km can be shown here, because few garnets occur at shallower depths on this traverse. The Fo contours emphasise that the upper part of the SCLM is highly magnesian in much of the continental core, including Saskatchewan; the overall composition is less magnesian beneath Alberta and parts of the Slave Province, and lowest beneath the Tectons. The thickest depleted SCLM, with olivine of $Fo_{92->93}$, is found beneath parts of the Superior Province, and beneath Somerset

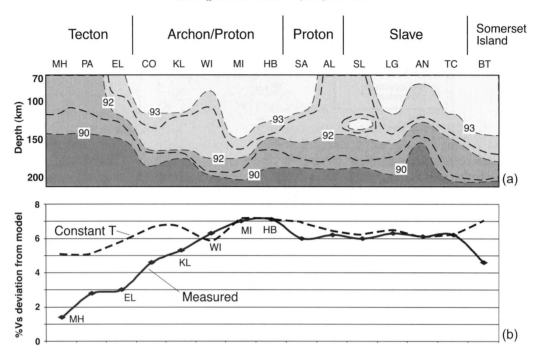

Fig. 17. (a) Contoured distribution of %Fo in olivine along the traverse shown in Fig. 15. (b) Vs anomaly (100–175 km depth slice; S. Grand, pers. comm.) along the traverse, illustrating the lower seismic velocities beneath the more fertile parts of the traverse. Solid line, observed data; dashed line, velocities expected assuming no thermal effects.

Island. The Fo_{92} contour generally shows less relief than the Fo_{93} contour beneath the cratonic areas, which may reflect metasomatic modification of the lower lithosphere over time. It also suggests that much of the continental root outlined in Fig. 16 consists of variably modified Archean SCLM.

There is a strong correlation between the overall Fo content and distribution along the traverse, and the seismic velocity in the 100–175 km depth range (Fig. 17B). The compositional variations alone can only account for ca. 40% of the total observed variation in Vs, and the correlations shown here emphasise the covariation of the compositional and thermal effects that are expressed in seismic tomography: younger SCLM is both warmer and less depleted, and these two effects reinforce one another. The Somerset Island section shows a lower Vs than the Superior Province (MI, HB), despite a similar Fo content. The Somerset Island SCLM, as noted above, consists dominantly of relatively fertile lherzolites, despite its high Mg#, and the lherzolites have lower Vs; the lithosphere here also may have experienced a recent rise in T.

If the Fo_{90} contour in Fig. 17a is taken as approximating the base of the depleted lithosphere, this ranges in depth from ca. 140 km beneath the Appalachians, to 190–210 km beneath most of the continental core. Along this traverse, variations in seismic velocity appear to reflect primarily variations in the degree of depletion, rather than the thickness, of the lithosphere. Detailed analysis of these relationships between composition, density, temperature and seismic velocity are in progress, and will contribute to the interpretation of the seismic tomography.

8. Conclusions

1. Detailed comparisons with data from mantle xenoliths demonstrate that reliable sections showing lithosphere thickness, thermal state, composition and structure can be constructed using major- and trace-element data on garnet concentrates from volcanic rocks. The properties of these sections can be related to geophysical data, including seismic

tomography, in ways that make it possible to map the lateral extent of different mantle types under the continents.

2. The SCLM beneath most parts of the Slave Province shows pronounced layering, with a boundary at 140–160 km. The upper layer shows marked lateral heterogeneity in composition, whereas the lower layer is remarkably uniform. It is suggested that the lower layer represents subcreted plume material, added at ca. 3.2 Ga beneath a variety of pre-existing thinner lithosphere domains. This suggestion is supported by the presence of abundant diamonds from the lower mantle and the dating of unusual sulfide inclusions in olivines and diamonds. This lower layer rises to <100 km depth toward the northern and southern edges of the craton.

3. Strongly layered SCLM suggests that plume subcretion may also have played a role in the construction of the lithosphere beneath Michigan and Saskatchewan, coinciding in each case with the subsidence of major Phanerozoic basins. Gaps in the peridotite sampling of these sections may reflect the accumulation of mafic material at the top of the plume.

4. Outside the Slave Province, most SCLM sections beneath North American Archons are less depleted and contain more fertile lherzolites than many sections in southern Africa and Siberia; this may reflect extensive metasomatic modification. The degree of modification increases toward the Archean craton margin, as shown by comparison of the Attawapiskat, Kirkland Lake and Cobalt sections.

5. The SCLM beneath the Kapuskasing Structural Zone is typical of that beneath Proterozoic to Phanerozoic mobile belts, and implies a replacement, or strong modification, of the Archean SCLM during the Proterozoic development of the KSZ.

6. SCLM sections from several Proterozoic areas around the margin of the Laurentian continental core (W. Greenland, Colorado–Wyoming district, Arkansas) show marked discontinuities that are interpreted as the effects of lithosphere stacking during collisional orogeny. Sampling gaps in these sections appear to reflect accumulations of mafic rocks.

7. Some areas affected by Proterozoic orogeny (Wyoming Craton, Alberta, W. Greenland) appear to retain Archean SCLM, consistent with the inferred buoyant and refractory nature of ancient SCLM (Poudjom Djomani et al., 2001; O'Reilly et al., 2001). Possible juvenile Proterozoic SCLM beneath the Colorado Plateau is significantly less refractory than that beneath Proterozoic terrains that contain reworked Archean crust. The least depleted SCLM is found beneath the Appalachian fold belt.

8. The highly melt-metasomatised SCLM beneath the Kansas kimberlite field probably reflects its proximity to the Mid-Continent Rift System. The Michigan locality, which lies ca. 100 km from the axis of the MCRS, does not show significant modification of this type.

9. A traverse across the continent shows that the upper part of the cratonic SCLM is highly magnesian, and the decrease in mg# with depth is interpreted as the cumulative effect of metasomatic modification through time. The relatively small variations in seismic velocity at 100–175 km depth within the continental core largely reflect the thickness of this depleted layer. The larger drop in seismic velocity in the surrounding Proton and Tecton belts is related to the coupled changes in SCLM composition and temperature.

Acknowledgements

This work would have been impossible without the generous provision of samples and data by many institutions and individuals over many years. Samples from the Kyle Lake, Attawapiskat, Kirkland Lake and Cobalt areas were supplied by Bram Janse, Bruce Kjarsgaard and the Royal Ontario Museum. The Drybones sample was provided by Ulrich Kretschmar, and the Point Lake and Mark samples by Chuck Fipke. Peter Nixon supplied many of the samples from Saskatchewan; others were provided by Ulrich Kretschmar. The Western Slave samples were provided by Ashton Mining of Canada, and the samples from Michigan and Wisconsin by Crystal Exploration, through Shawn Carlson and Bill Jarvis. Samples from the Navajo diatreems, Ison Creek and Williams were supplied by the National Museum of Natural History (Smithsonian Institution), through Sorena Sorensen and Carter Hearn. We also thank Doug Smith for samples from, and information on, the Colorado Plateau. The Alberta samples were provided by

Ashton Mining of Canada. Mike Waldman provided the samples from Arkansas, and Andy McThenia those from Mount Horeb. Some samples from Elliott County were provided by Ashton Mining. Seismic data used in the construction of Figs. 2 and 17 were provided by Steve Grand.

We thank Chris Ryan for years of proton microprobe development, discussions and collaboration, and Bruce Wyatt and Simon Shee for many valuable discussions. We are grateful to Tin Tin Win for analytical assistance on the proton probe, and Carol Lawson, Oliver Gaul, Oleg Belousov, Ashwini Sharma and Suzie Elhlou for help with EMP and LAM-ICPMS analysis. Oleg Belousov constructed the GeoSpeed software that allowed these data to be processed on a human time scale, and Oliver Gaul and Sally Hodgekiss turned them into figures. The final MS was improved through constructive reviews by Richard Walker and Larry Heaman. This is contribution number 348 from the GEMOC ARC National Key Centre (www.es.mq.edu.au/GEMOC/).

References

Alard, O., Griffin, W.L., Pearson, N.J., Lorand, J.-P., O'Reilly, S.Y., 2002. New insights into the Re–Os systematics of subcontinental lithospheric mantle from in-situ analysis of sulfides. Earth and Planetary Science Letters 203, 651–663.

Alibert, C., Albarede, F., 1988. Relationships between mineralogical, chemical, and isotopic properties of some North American kimberlites. Journal of Geophysical Research 93, 7643–7671.

Allen, C.J., Braile, L.W., Hinze, W.J., Mariano, J., 1995. The Midcontinent Rift System, USA: a major Proterozoic continental rift. In: Olsen, K.H. (Ed.), Continental Rifts: Evolution, Structure, Tectonics. Elsevier, Amsterdam, pp. 375–408.

Artemieva, I.M., Mooney, W.D., 2001. Thermal thickness and evolution of Precambrian lithosphere: a global study. Journal of Geophysical Research 106, 16387–16414.

Aulbach, S., Griffin, W.L., O'Reilly, S.Y., McCandless, T.E., 2004a. Genesis and evolution of the lithospheric mantle beneath the Buffalo Head Terrane, Alberta (Canada). Lithos 77, 413–451, this issue; doi:10.1016/j.lithos.2004.04.020.

Aulbach, S., Griffin, W.L., Pearson, N.J., O'Reilly, S.Y., Kivi, K., Doyle, B.J., 2004b. Mantle formation and evolution, Slave Craton: constraints from HSE abundances and Re–Os systematics of sulfide inclusions in mantle xenocrysts. Chemical Geology (in press).

Bank, C.-G., Bostock, M.G., Ellis, R.M., Hajnal, Z., VanDecar, J.C., 1998. Lithospheric mantle structure beneath the Trans-Hudson Orogen and the origin of diamondiferous kimberlite. Journal of Geophysical Research 103, 10103–10114.

Beghoul, N., Barazangi, M., 1989. Mapping high Pn velocity beneath the Colorado Plateau constrains uplift models. Journal of Geophysical Research 94, 7083–7104.

Bleeker, W., 2001. The ca 2680 Ma Raquette Lake formation and correlative units across the Slave Province, Northwest Territories: evidence for a craton-scale overlap sequence. Geological Survey of Canada. Current Research 2001-C7 (11 pp.).

Bleeker, W., Ketchum, J.W.F., Jackson, V.A., Villeneuve, M.E., 1999. The central slave basement complex: Part I. Its structural topology and autochthonous cover. Canadian Journal of Earth Sciences 36, 1083–1109.

Boyd, F.R., 1989. Composition and distinction between oceanic and cratonic lithosphere. Earth and Planetary Science Letters 96, 15–26.

Boyd, F.R., 1997. Origin of peridotite xenoliths: major and trace element considerations. In: Ranalli, G., Ricci Lucchi, F., Ricci, C.A., Trommsdorff, T. (Eds.), High Pressure and High Temperature Research on Lithosphere and Mantle Materials. University of Siena. pp. 89-106.

Brookins, D.G., 1970. The kimberlites of Riley County, Kansas. Bulletin-Kansas Geological Survey 200, 1–32.

Canil, D., 1994. An experimental calibration of the "Nickel in Garnet" geothermometer with applications. Contributions to Mineralogy and Petrology 117, 410–420.

Canil, D., 1999. The Ni-in-garnet geothermometer: calibration at natural abundances. Contributions to Mineralogy and Petrology 136, 240–246.

Cannon, W.F., Daniels, D.L., Nicholson, S.W., Phillips, J., Woodniff, L.G., Chandler, V.W., Morey, G.B., Boerboom, T., Wirth, M.G., Mudrey Jr., M.G., 2001. New map reveals origin and geology of North American Mid-continent rift. EOS Transactions American Geophysical Union 82, 100–101.

Carbno, G.B., Canil, D., 2002. Mantle structure beneath the SW Slave Craton, Canada: constraints from garnet geochemistry in the Drybones Bay kimberlite. Journal of Petrology 43, 129–142.

Carlson, R.W., Irving, A.J., 1994. Depletion and enrichment history of subcontinental lithospheric mantle: an Os, Sr, Nd and Pb isotopic study of ultramafic xenoliths from the northwestern Wyoming Craton. Earth and Planetary Science Letters 126, 457–472.

Carlson, R.W., Irving, A.J., Hearn, B.C., 1998. Chemical and isotopic systematics of peridotite xenoliths from the Williams Kimberlite, Monttana: clues to processes of lithosphere formation, modification and destruction. Proc. 7th Int. Kimberlite Conference. Red Roof Design, Cape Town, 90–98.

Carlson, J.A., Kirkley, M.B., Thomas, E.M., Hillier, W.D., 1999. Recent Canadian kimberlite discoveries. Proc. 7th Int. Kimberlite Conference. Red Roof Design, Cape Town, 81–89.

Creaser, R.A., Grutter, H., Carlson, J., Crawford, B., 2004. Macrocrystal phlogopite Rb-Sr dates for the Ekati property kimberlites, Slave Province, Canada: evidence for multiple intrusive episodes in the Paleocene and Eocene. Lithos, this issue; doi:10.1016/j.lithos.2004.03.039.

Condie, K.C., Selverstone, J., 1999. The crust of the Colorado Plateau: new views of an old arc. Journal of Geology 107, 387–397.

Cookenboo, H.O., 1998. Emplacement history of the Jericho kimberlite pipe, northern Canada. Ext. Abst. 7th Int. Kimberlite Conf., Cape, 161–163.

Dahl-Jensen, T., Larsen, T.B., Woelbern, I., Bach, T., Hanka, W., Kind, R., Gregersen, S., Mosegaard, K., Voss, P., Gudmundsson, O., 2003. Depth to Moho in Greenland: receiver-function analysis suggests two Proterozoic blocks in Greenland. Earth and Planetary Science Letters 205, 379–393.

Davies, R., Griffin, W.L., Pearson, N.J., Andrew, A., Doyle, B.J., O'Reilly, S.Y., 1999. Diamonds from the Deep: pipe DO-27, Slave Craton, Canada. Proc. 7th Int. Kimberlite Conf., Red Roof Design, Cape Town, 148–155.

Davies, R.M., Griffin, W.L., O'Reilly, S.Y., Doyle, B.J., 2004a. Mineral inclusions and geochemical characteristics of microdiamonds from the D027, A154, A21, A418, D018, DD17 and Ranch Lake kimberlites at Lac de Gras, Slave Craton, Canada. Lithos 77, 39–55, this issue; doi:10.1016/j.lithos.2004.04.016.

Davies, R.M., Griffin, W.L., O'Reilly, S.Y., 2004b. Inclusions in diamonds from the K14 and K10 kimberlites, Buffalo Hills, Alberta, Canada: diamond growth in a plume?. Lithos 77, 99–111, this issue; doi:10.1016/j.Lithos.2004.04.008.

Davis, W.J., Hegner, E., 1992. Neodymium isotopic evidence for the tectonic assembly of Late Archean crust in the Slave province, northwest Canada. Contributions to Mineralogy and Petrology 111, 493–504.

Davis, W.J., Gariepy, C., van Breemen, O., 1996. Pb isotopic composition of late Archaean granites and the extent of recycling early Archaean crust in the Slave Province, northwest Canada. Chemical Geology 130, 255–269.

Dickinson, W.R., Snyder, W.S., 1978. Plate tectonics of the Laramide Orogeny. Memoir-Geological Society of America 151, 355–366.

Drahovzal, J.A., Harris, D.C., Wickstrom, L.H., Walker, D., Baranoski, M.T., Keith, B., Furer, L.C., 1992. The east continent rift basin: a new discovery. Information Circular-Ohio Geological Survey 57 (25 pp.).

Dunn, D., Smith, D., McDowell, F.W., Bergman, S.C., 2000. Mantle and crustal xenoliths from the Prairie Creek lamproite province, Arkansas. Abstract with Programs-Geological Society of America A-386.

Eggler, D.H., Dudas, F.O., Hearn, B.C., McCallum, M.E., McGee, E.S., Meyer, H.O.A., Schulze, D.J., 1987. Lithosphere of the continental United States: xenoliths in kimberlites and other alkaline magmas. In: Nixon, P.H. (Ed.), Mantle Xenoliths. Wiley, New York, pp. 41–58.

Eggler, D.H., et al., 1988. Tectonomagmatism of the Wyoming Province. Colorado School of Mines Quarterly 83, 25–40.

Ehrenberg, S.N., 1979. Garnetiferous ultramafic inclusions in minette from the Navajo volcanic field. In: Boyd, F.R., Meyer, H.O.A. (Eds.), The Mantle Sample: Inclusions in Kimberlites and Other Volcanics. Amer. Geophys. Union, Washington, pp. 330–344.

Ehrenberg, S.N., 1982. Petrogenesis of garnet lherzolite and megacrystalline nodules from The Thumb, Navajo volcanic field. Journal of Petrology 23, 507–547.

Finnerty, A.A., Boyd, F.R., 1987. Thermobarometry for garnet peridotite xenoliths: a basis for mantle stratigraphy. In: Nixon, P.H. (Ed.), Mantle Xenoliths. Wiley, New York, pp. 381–402.

Garrit, D., 2002. The nature of the Archaean and Proterozoic lithospheric mantle and lower crust in West Greenland illustrated by the geochemistry and petrography of xenoliths from kimberlites. Unpublished Ph.D. Thesis, University of Copenhagen, Denmark. 289 pp.

Gaul, O.F., Griffin, W.L., O'Reilly, S.Y., Pearson, N.J., 2000. Mapping olivine composition in the lithospheric mantle. Earth and Planetary Science Letters 182, 223–235.

Goes, S., van der Lee, S., 2002. Thermal structure of the North American uppermost mantle inferred from seismic tomography. Journal of Geophysical Research 107 (10/1029/2000JB000049).

Gogineni, S.V., Melton, C.E., Giardini, A.A., 1978. Some petrological aspects of the Prairie Creek diamond-bearing kimberlite diatreme Arkansas. Contributions to Mineralogy and Petrology 66, 251–261.

Gorman, A.R., Clowes, R.M., Ellis, R.M., Henstock, T.J., Spence, G.D., Keller, G.R., Levander, A., Snelson, C.M., Burianyk, M.J.A., Kanasewich, E.R., Asudeh, I., Hajnal, Z., Miller, K.C., 2002. Deep Probe: imaging the roots of western North America. Canadian Journal of Earth Sciences 39, 375–398.

Grand, S.P., 1994. Mantle shear structure beneath the Americas and surrounding oceans. Journal of Geophysical Research 99, 11591–11621.

Griffin, W.L., Ryan, C.G., 1995. Trace elements in indicator minerals: area selection and target evaluation in diamond exploration. Journal of Geochemical Exploration 53, 311–337.

Griffin, W.L., Ryan, C.G., 1996. "An experimental calibration of the "Nickel in Garnet" geothermometer, with applications" by D. Canil: Discussion. Contributions to Mineralogy and Petrology 124, 216–218.

Griffin, W.L., O'Reilly, S.Y., Ryan, C.G., Waldman, M.A., 1994. Indicator minerals from Prairie Creek and Twin Knobs lamproites: relation to diamond grade. In: Meyer, H.O.A., Leonardos, O.H. (Eds.), Diamonds: Characterization, Genesis and Exploration. Spec. Publ., vol. 1B/93. CPRM, pp. 302–311.

Griffin, W.L., Garrit, D., Win, T.T., Ryan, C.G., O'Reilly, S.Y., 1995, Trace elements in diamond indicator minerals from west Greenland kimberlitic rocks. CSIRO Exploration and Mining Report 198R (22 pages+Figures).

Griffin, W.L., O'Reilly, S.Y., Ryan, C.G., Gaul, O., Ionov, D., 1998. Secular variation in the composition of subcontinental lithospheric mantle. In: Braun, J., Dooley, J.C., Goleby, B.R., van der Hilst, R.D., Klootwijk, C.T. (Eds.), Structure and Evolution of the Australian Continent. Geodynamics, vol. 26. Amer. Geopyhys. Union, Washington, DC, pp. 1–26.

Griffin, W.L., Fisher, N.I., Friedman, J.H., Ryan, C.G., O'Reilly, S.Y., 1999a. Cr-pyrope garnets in the lithospheric mantle: I. Compositional systematics and relations to tectonic setting. Journal of Petrology 40, 679–704.

Griffin, W.L., O'Reilly, S.Y., Ryan, C.G., 1999b. The composition and origin of subcontinental lithospheric mantle. In: Fei, C.M., Bertka, C.M., Mysen, B.O. (Eds.), Mantle Petrology: Field Observations and High-Pressure Experimentation: A Tribute to Francis R. (Joe) Boyd. Special Publication-Geochemical Society, vol. 6. The Geochemical Society, Houston, pp. 13–45.

Griffin, W.L., Shee, S.R., Ryan, C.G., Win, T.T., Wyatt, B.A., 1999c. Harzburgite to lherzolite and back again: metasomatic processes in ultramafic xenoliths from the Wesselton kimberlite, Kimberley, South Africa. Contributions to Mineralogy and Petrology 134, 232–250.

Griffin, W.L., Doyle, B.J., Ryan, C.G., Pearson, N.J., O'Reilly, S.Y., Natapov, L., Kivi, K., Kretschmar, U., Ward, J., 1999d. Lithosphere Structure and Mantle Terranes: Slave Craton, Canada. Proc. 7th Int. Kimberlite Conf., Red Roof Design, Cape Town, 299–306.

Griffin,W.L.,Doyle,B.J.,Ryan,C.G.,Pearson,N.J.,O'Reilly,S.Y.,R.M., R.M., Davies, R.M., Kivi, K., van Achterbergh, E., Natapov, L.M., 1999e. Layered Mantle Lithosphere in the Lac de Gras Area, Slave Craton: composition, structure and origin. Journal of Petrology 40, 705–727.

Griffin, W.L., Ryan, C.G., Kaminsky, F.V., O'Reilly, S.Y., Natapov, L.M., Win, T.T., Kinny, P.D., Ilupin, I.P., 1999f. The siberian lithosphere traverse: mantle terranes and the assembly of the Siberian Craton. Tectonophysics 310, 1–35.

Griffin, W.L., Spetsius, Z.V., Pearson, N.J., O'Reilly, S.Y., 2002a. In-situ Re–Os analysis of sulfide inclusions in kimberlitic olivine: new constraints on depletion events in the Siberian lithospheric mantle. Geochemistry, Geophysics, Geosystems 3 (1069).

Griffin, W.L., Fisher, N.I., Friedman, J.H., O'Reilly, S.Y., Ryan, C.G., 2002b. Cr-pyrope garnets in the lithospheric mantle: II. Compositional populations and their distribution in time and space. Geochemistry, Geophysics and Geosystems 3 (1073).

Griffin, W.L., O'Reilly, S.Y., Natapov, L.M., Ryan, C.G., 2003a. The evolution of lithospheric mantle beneath the Kalahari Craton and its margins. Lithos 71, 215–242.

Griffin, W.L., O'Reilly, S.Y., Abe, N., Aulbach, S., Davies, R.M., Pearson, N.J., Doyle, B.J., Kivi, K., 2003b. The origin and evolution of Archean lithospheric mantle. Precambrian Research 127, 19–41.

Griffin, W.L., Graham, S., O'Reilly, S.Y., Pearson, N.J., 2004. Lithosphere evolution beneath the Kaapvaal Craton: Re–Os systematics of sulfides in mantle-derived peridotites. Chemical Geology (in press).

Heaman, L.M., 1989. The nature of the subcontinental mantle from Sr–Nd–Pb isotopic studies on kimberlite perovskite. Earth and Planetary Science Letters 92, 323–334.

Heaman, L.M., Kjarsgaard, B.A., 2000. Timing of eastern North American kimberlite magmatism: continental extension of the Great Meteor hotspot track? Earth and Planetary Science Letters 178, 256–268.

Heaman, L.M., Kjarsgaard, B.A., Creaser, R.A., Cookenboo, H.O., Kretschmar, U., 1997. Multiple episodes of kimberlite magmatism in the Slave Province, North America. LITHOPROBE Report 56, 14–17.

Heaman, L.M., Kjarsgaard, B.A., Creaser, R.A., 2003. The timing of kimberlite magmatism in North America: implications for global kimberlite genesis and diamond exploration. Lithos 71, 153–184.

Hearn, B.C., 1995. Composite megacrysts and megacryst aggregates from the Williams kimberlites, Montana, USA: multiple products of mantle melts. In: Meyer, H.O.A., Leonardos, O.H. (Eds.), Kimberlites, Related Rocks and Mantle Xenoliths. Spec. Publ., vol. 1A. CPRM, pp. 388–404.

Hearn, B.C., McGee, E.S., 1984. Garnet peridotites from Williams kimberlites, north-central Montana. In: Kornprobst, J. (Ed.), Kimberlites: II. The Mantle and Crust–Mantle Relationships. Elsevier, Amsterdam, pp. 57–70.

Hearn, B.C., Dudas, F.O., Eggler, D.H., Hyndman, D.W., O'Brien, H.E., McCallum, I.S., Irving, A.J., Berg, R.B., 1989. Montana high-potassium igneous province. Excursion Guidebook 28th Int. Geological Congress. 7 pp.

Hunter, R.H., Taylor, L.A., 1984. Magma-mixing in the low velocity zone: kimberlitic megacrysts from Fayette Count, Pennsylvania. American Mineralogist 69, 16–29.

Irvine, G.J., Pearson, D.G., Kjarsgaard, B.A., Carlson, R.W., Kopylova, M.G., Dreibus, G., 2003. A Re–Os and PGE study of kimberlite-derived peridotite xenoliths from Somerset Island and a comparison to the Slave and Kaapvaal cratons. Lithos 71, 461–488.

Janse, A.J.A., 1994. Is Clifford's Rule still valid? Affirmative examples from around the world. In: Meyer, H.O.A., Leonardos, O.H. (Eds.), Diamonds, Characterization, Genesis and Exploration. Spec. Publication, vol. 1B. CPRM, Brasilia, pp. 215–235.

Janse, A.J.A., Novak, N.A., MacFayden, D.A., 1995. Discovery of a new type of highly diamondiferous kimberlitic rock in the James Bay Lowlands, northern Ontario, Canada. Ext. Abst. 6th Int. Kimberlite Conf., 260–263.

Jones, A.G., Ferguson, I.J., Chave, A.D., Evans, R.L., McNeice, G.W., 2001. Electric lithosphere of the Slave craton. Geology 29, 423–426.

Kaminski, E., Jaupart, C., 2000. Lithosphere structure beneath the Phanerozoic intracratonic basins of North America. Earth and Planetary Science Letters 178, 139–149.

Kay, S.M., Snedden, W.R., Foster, B.P., Kay, R.W., 1983. Upper mantle and crustal fragments in the Ithaca kimberlites. Journal of Geology 91, 277–290.

Kjarsgaard, B.A., Peterson, T.D., 1992. Kimberlite-derived ultramafic xenoliths from the diamond stability field; a new Cretaceous geotherm for Somerset Island, Northwest Territories. Geological Survey of Canada. Current Research 1992-B, 1–6.

Kopylova, M.G., Russell, J.K., Cookenboo, H., 1999. Petrology of peridotite and pyroxenite xenoliths from the Jericho kimberlite: implications for the thermal state of the mantle beneath the slave craton, northern Canada. Journal of Petrology 40, 79–104.

Larsen, L.M., Ronsbo, J., 1993. Conditions of origin of kimberlites in West Greenland: new evidence from the Sarfartoq and Sukkertoppen regions. Rapport-Grønlands Geologiske Undersøgelse 159, 115–120.

Larsen, L.M., Res, D.C., Secher, K., 1983. The age of carbonatites, kimberlites and lamprophyres from southern West Greenland: recurrent alkaline magmatism during 2500 million years. Lithos 16, 215–221.

Leahy, K., Taylor, W.R., 1997. The influence of the Glennie Domain deep structure on the diamonds in Saskatchewan kimberlites. Russian Geology and Geophysics 38, 481–491.

Leckie, D.A., et al., 1997. Emplacement and reworking of Cretaceous, diamond bearing, crater facies kimberlite of central Saskatchewan, Canada. Geological Society of America Bulletin 109, 1000–1020.

Lehnert-Thiel, K., Loewer, R., Orr, R.G., Robertshaw, P., 1992. Diamond-bearing kimberlites in Saskatchewan, Canada; The Fort à la Corne case history. Exploration Mining Geology 1, 391.

MacKenzie, J.M., Canil, D., 1999. Composition and thermal evolution of cratonic mantle beneath the central Archean Slave Province, NWT, Canada. Contributions to Mineralogy and Petrology 134, 313–324.

Mansker, W.L., Richards, B.D., Cole, G.P., 1987. A note on newly discovered kimberlites in Riley County, Kansas. Special Paper-Geological Society of America 215, 197–204.

Mattie, P.D., Condie, K.C., Selverstone, J., Kyle, P.R., 1997. Origin of the continental crust in the Colorado Plateau: geochemical evidence from mafic xenoliths from the Navajo Volcanic Field, southwestern USA. Geochimica et Cosmochimica Acta 61, 2007–2021.

McGee, E.S., Hearn, B.C., 1984. The Lake Ellen kimberlite, Michigan, USA. In: Kornprobst, J. (Ed.), Kimberlites: I. Kimberlites and Related Rocks. Elsevier, Amsterdam, pp. 143–154.

McPhee, J.P., 1983. Regional gravity analysis of the Anna, Ohio seismogenic region. Unpubl. MS thesis, Purdue University, West Lafayette, IN, 100 pp.

Meyer, H.O.A., 1976. Kimberlites of the continental United States. Journal of Geology 84, 377–403.

Mitchell, R.H., 1978. Garnet lherzolites from Somerset Island, Canada and aspects of the nature of perturbed geotherms. Contributions to Mineralogy and Petrology 67, 341–347.

Naeser, C.W., McCallum M.E., 1997 Fission track dating of kimberlitic zircons. Ext. Abst. 2nd Int. Kimberlite Conf., Santa Fe (unpaged).

Nixon, P.H., Leahy, K., 1997. Diamond-bearing volcaniclastic kimberlites in Cretaceous marine sediments, Saskatchewan, Canada. Russian Geology and Geophysics 38, 17–23.

O'Reilly, S.Y., Griffin, W.L., Gaul, O., 1997. Paleogeotherms in Australia: basis for 4-D lithosphere mapping. Australian Geological Survey Organisation Journal 17, 63–72.

O'Reilly, S.Y., Griffin, W.L., Poudjom Djomani, Y., Morgan, P., 2001. Are lithospheres forever? Tracking changes in subcontinental lithspheric mantle through time. GSA Today 11, 4–9.

Orr, P., Luth, R.W., 2000. Petrology and oxygen-isotope geochemistry of the Yamba Lake kimberlite rocks, N. W. T. Canadian Journal of Earth Sciences 37, 1053–1071.

Parrish, J.B., Lavin, P.M., 1982. Tectonic model for kimberlite emplacement in the Appalachian Plateau of Pennsylvania. Geology 10, 344–347.

Parsons, T., McCarthy, J., 1995. The active southwest margin of the Colorado Plateau: uplift of mantle origin. Geological Society of America Bulletin 107, 139–147.

Parsons, T., McCarthy, J., Kohler, W.M., Ammon, C.J., Benz, H.M., Hole, J.A., Criley, E.E., 1996. Crustal structure of the Colorado Plateau, Arizona: application of new long-offset seismic data analysis techniques. Journal of Geophysical Research 101, 11173–11194.

Pearson, N.J., Griffin, W.L., Doyle, B.J., O'Reilly, S.Y., van Achterbergh, E., Kivi, K., 1999. Xenoliths from kimberlite pipes of the Lac de Gras area, Slave Craton, Canada. Proc. 7th Int. Kimberlite Conf., Red Roof Design, Cape Town, 644–658.

Pedersen, J.L., Mackley, R.D., Eddleman, J.L., 2002. Colorado Plateau uplift and erosion evaluated using GIS. GSA Today 12 (8), 4–7.

Percival, J.A., West, G.F., 1994. The kapuskasing uplift: a geological and geophysical synthesis. Canadian Journal of Earth Sciences 31, 1256–1286.

Phipps, S.P., 1988. Deep rifts as sources for alkaline intraplate magmatism in eastern North America. Nature 334, 27–31.

Pokhilenko, N.P., Mcdonald, J.A., Melnyk, W., Hall, A.E., Shimizu, N., Vavilov, M.A., Afanasiev, V.P., Reimers, L.F., Irvin, J., Pokhilenko, L.N., Vasilenko, V.B., Kuligin, S.S., Soblev, N.V., 1998. Kimberlites of Camsell Lake field and some features of the construction and compsotion of lithospheric roots of the southeastern part of the Slave Craton, Canada. Ext. Abst. 7th Int. Kimberlite Conference, Cape Town, 699–701.

Pokhilenko, N., McDonald, J., Sobolev, N., Reutsky, V., Hall, A., Logvinova, A., Reimers, L., 2003. Crystalline inclusions and C isotope ratios in diamonds from the Snap Lake/King Lake kimberlite dyke system: evidence of ultradeep and enriched lithospheric mantle. Ext. Abst. 8th Int. Kimberlite Conf.

Poudjom Djomani, Y.H., O'Reilly, S.Y., Griffin, W.L., Morgan, P., 2001. The density structure of subcontinental lithosphere: constraints on delamination models. Earth and Planetary Science Letters 184, 605–621.

Purucker, M., Langlais, B., Olsen, N., Hulot, G., Mandea, M., 2002. The southern edge of cratonic North America: evidence from new satellite magnetometer observations. Geophysical Research Letters 29 (DOI 10.1029/2001GL013645).

Riter, J.C.A., Smith, D., 1996. Xenolith constraints on the thermal history of the mantle below the Colorado Plateau. Geology 24, 267–270.

Ritsema, J., van Heijst, H.-J., 2000. Seismic imaging of structural heterogeneity in Earth's mantle: evidence for large-scale mantle flow. Science Progress 83, 243–259.

Ross, G.M., 2002. Evolution of Precambrian continental lithosphere in Western Canada: results from Lithoprobe studies in Alberta and beyond. Canadian Journal of Earth Sciences 39, 413–437.

Russell, J.K., Kopylova, M.G., 1999. A steady state conductive geotherm for the north central Slave, Canada: inversion of petrological data from the Jericho Kimberlite pipe. Journal of Geophysical Research 104, 7089–7101.

Ryan, C.G., Griffin, W.L., Pearson, N.J., 1996. Garnet Geotherms: a technique for derivation of P–T data from Cr-pyrope garnets. Journal of Geophysical Research 101, 5611–5625.

Sage, R.P., 2000. Kimberlites of the Attawapiskat area, James Bay Lowlands, northern Ontario. Open File Report-Ontario Geological Survey, vol. 6019. 341 pp.

Sahagian, D., Proussevitch, A., Carlson, W., 2002. Timing of Colorado Plateau uplift: initial constraints from vesicular basalt-derived paleoelevations. Geology 30, 807–810.

Schmidberger, S.S., Francis, D., 1999. Nature of the mantle roots beneath the North American craton: mantle xenolith evidence from Somerset Island kimberlites. Lithos 48, 195–216.

Schmidberger, S.S., Francis, D., 2001. Constraints on the trace element composition of the Archean mantle root beneath Somerset Island, Arctic Canada. Journal of Petrology 42, 1095–1117.

Schmidberger, S.S., Simonetti, A., Francis, D., 2001. Sr–Nd–Pb isotope systematics of mantle xenoliths from Somerset Island kimberlites: evidence for lithosphere stratification beneath Arctic Canada. Geochimica et Cosmochimica Acta 65, 4243–4255.

Schmidberger, S.S., Simonetti, A., Francis, D., 2002. Probing Archean lithosphere using the Lu–Hf isotope systematics of peridotite xenoliths from Somerset Island kimberlites, Canada. Earth and Planetary Science Letters 197, 245–259.

Schulze, D.J., 1984. Cr-poor megacrysts from the Hamilton Branch kimberlite, Elliott County, Kentucky. In: Kornprobst, J. (Ed.), Kimberlites: II. The Mantle and Crust–Mantle Relationships, pp. 97–108. Elsevier.

Scott-Smith, B.H., Orr, R.G., Robertshaw, P., Avery, R.W., 1995. Geology of the Fort a la Corne kimberlites. Ext. Abst. 6th Int. Kimberlite Conf., Novosibirsk.

Scott-Smith, B.H., 1987. Greenland. In: Nixon, P.H. (Ed.), Mantle Xenoliths. Wiley, New York, pp. 23-32.

Sears, C.E., Bilbert, M.C., 1973. Petrography of the Mt. Horeb, Virginia kimberlite. Abstracts with Programs-Geological Society of America 5, 434.

Selverstone, J., Pun, A., Condie, K.C., 1999. Xenolithic evidence for Proterozoic crustal evolution beneath the Colarado Plateau. Geological Society of America Bulletin 111, 590–606.

Severinghaus, J., Atwater, T., 1990. Cenozoic geometry and thermal state of the subducting slabs beneath western North America. In: Wernicke, B. (Ed.), Basin and Range Extension. Memoir-Geological Society of America, vol. 176, pp. 1–22.

Smith, C.B., 1979. Rb–Sr mica ages of various kimberlites. Abstracts Kimberlite Symposium II, Cambridge, 61–66.

Smith, D., 2000. Insights into the evolution of the uppermost continental mantle from xenolith localities on and near the Colorado Plateau and regional comparisons. Journal of Geophysical Research 105, 16769–16781.

Smith, D., Griffin, W.L., 2003. Navajo garnetites and rock-water interaction in the mantle. EOS Transactions AGU 84 (46) Fall Meeting Supplement, Abstract V22B-0391.

Smith, C.B., Allsopp, H.L., Garvie, O.G., Kramers, J.D., Jackson,

P.F.S., Clement, C.R., 1989. Note on the U–Pb perovskite method for dating kimberlites: examples from the Wesselton and DeBeers mines, South Africa, and Somerset Island, Canada. Chemical Geology 79, 137–145.

Smith, D., Griffin, W.L., Ryan, C.G., Cousens, D.R., Sie, S.H., Suter, G.F., 1991. Trace-element zoning of garnets from The Thumb. A guide to mantle processes. Contributions to Mineralogy and Petrology 107, 60–79.

Snyder, D.B., 2002. Lithospheric growth at margins of cratons. Tectonophysics 355, 7–22.

Snyder, D.B., Rondenay, S., Bostock, M.G., Lockhart, G.D., 2004. Mapping the mantle lithosphere for diamond potential using teleseismic methods. Lithos 77, 859–872, this issue; doi:10.1016/j.lithos.2004.049.

Thybo, H., Perchuc, E., Zhou, S., 2000. Intraplate earthquakes and a seismically defined lateral transition in the upper mantle. Geophysical Research Letters 27, 3953–3956.

van der Lee, S., 2001. Deep below North America. Science 294, 1297–1298.

van der Lee, S., 2002. High resolution estimates of lithospheric thickness from Missouri to Massachusetts, USA. Earth and Planetary Science Letters 203, 15–23.

Van Gool, J.A.M., Kroegs, am.L.M., Marker, M., Nichols, G.T., 1999. Thrust stacking in the inner Nordre Strømfjord are, West Greenland: significance for the tectonic evolution of the paleoproterozoic Nagssugtoqidian orogen. Precambrian Research 93, 71–86.

Wasilewski, P., Mayhew, M.A., 1982. Crustal xenolith magnetic properties and long wavelength anomaly source requirements. Geophysical Research Letters 9, 329–332.

Wendlandt, E., DePaolo, D.J., Baldridge, W.S., 1996. Thermal history of Colorado Plateau lithosphere from Sm–Nd mineral geochronology of xenoliths. Geological Society of America Bulletin 108, 757–767.

Williams, H.R., Stott, G.M., Thurston, P.C., Sutcliffe, R.H., Bennett, G., Easton, R.M., Armstrong, D.K., 1992. Tectonic evolution of Ontario: summary and synthesis. Geology of Ontario. Ontario Geological Survey Special Volume, vol. 4, pp. 1255–1332. Part 2.

Available online at www.sciencedirect.com

LITHOS

Lithos 77 (2004) 923–944

www.elsevier.com/locate/lithos

Integrated models of diamond formation and craton evolution

Steven B. Shirey[a,*], Stephen H. Richardson[b,1], Jeffrey W. Harris[c,2]

[a]*Department of Terrestrial Magnetism, Carnegie Institution of Washington, 5241 Broad Branch Road, NW, Washington, DC 20015, USA*
[b]*Department of Geological Sciences, University of Cape Town, Rondebosch 7701, South Africa*
[c]*Division of Earth Sciences, University of Glasgow, Glasgow G12 8QQ, UK*

Received 27 June 2003; accepted 14 December 2003

Abstract

Two decades of diamond research in southern Africa allow the age, average N content and carbon composition of diamonds, and the dominant paragenesis of their syngenetic silicate and sulfide inclusions to be integrated on a cratonwide scale with a model of craton formation. Individual eclogitic sulfide inclusions in diamonds from the Kimberley area kimberlites, Koffiefontein, Orapa and Jwaneng have Re–Os isotopic ages that range from circa 2.9 Ga to the mid-Proterozoic and display little correspondence with the prominent variations in the P-wave velocity ($\pm 1\%$) that the mantle lithosphere shows at depths within the diamond stability field (150–225 km). Silicate inclusions in diamonds and their host diamond compositions for the above kimberlites, Finsch, Jagersfontein, Roberts Victor, Premier, Venetia, and Letlhakane show a regional relationship to the seismic velocity of the lithosphere. Mantle lithosphere with slower P-wave velocity relative to the craton average correlates with a greater proportion of eclogitic vs. peridotitic silicate inclusions in diamond, a greater incidence of younger Sm–Nd ages of silicate inclusions, a greater proportion of diamonds with lighter C isotopic composition, and a lower percentage of low-N diamonds. The oldest formation ages of diamonds support a model whereby mantle that became part of the continental keel of cratonic nuclei first was created by middle Archean (3.2–3.3 Ga or older) mantle depletion events with high degrees of melting and early harzburgite formation. The predominance of eclogitic sulfide inclusions in the 2.9 Ga age population links late Archean (2.9 Ga) subduction–accretion events to craton stabilization. These events resulted in a widely distributed, late Archean generation of eclogitic diamonds in an amalgamated craton. Subsequent Proterozoic tectonic and magmatic events altered the composition of the continental lithosphere and added new lherzolitic and eclogitic diamonds to the already extensive Archean diamond suite. Similar age/paragenesis systematics are seen for the more limited data sets from the Slave and Siberian cratons.
© 2004 Elsevier B.V. All rights reserved.

Keywords: Diamond; Eclogite; Peridotite; Inclusion; P-wave; Craton; Lithosphere

1. Introduction

* Corresponding author. Fax: +1-202-478-8821.
E-mail addresses: shirey@dtm.ciw.edu (S.B. Shirey), shr@geology.uct.ac.za (S.H. Richardson), jwh@earthsci.gla.ac.uk (J.W. Harris).
[1] Fax: +27-21-650-3783.
[2] Fax: +44-141-330-4817.

A worldwide association between ancient cratons and diamond occurrences has long been known (e.g., Clifford, 1966; Gurney and Switzer, 1973; Boyd and Gurney, 1986; Janse, 1992). This association, plus the intersection of mantle xenolith-derived geotherms

0024-4937/$ - see front matter © 2004 Elsevier B.V. All rights reserved.
doi:10.1016/j.lithos.2004.04.018

with the diamond stability field and the Archean ages of inclusions in diamonds led to the suggestion that diamonds reside in Archean lithospheric mantle keels beneath cratons (e.g., Richardson et al., 1984; Boyd et al., 1985; Boyd and Gurney, 1986; Haggerty, 1986). Within the southern African cratonic keel (the conjoined Kaapvaal–Zimbabwe cratons and the Limpopo mobile belt), Archean mantle peridotite and eclogite host multiple generations of diamonds that are both Archean and Proterozoic in age (Kramers, 1979; Richardson et al., 1984, 1993, 2001; Navon, 1999; Pearson and Shirey, 1999; Shirey et al., 2002) as well as less common occurrences of younger, mostly cubic and fibrous diamonds (Boyd et al., 1987; Akagi and Masuda, 1988; Navon et al., 1988; Schrauder and Navon, 1994; Pearson et al., 1998; Izraeli et al., 2001). Xenoliths of peridotite and eclogite and xenocrysts of diamond have been brought to the surface from depths as great as 150–180 km in kimberlites whose ages are typically young (65–150 Ma) but can be significantly older (240–1600 Ma; Davis et al., 1976; Davis, 1977; Smith et al., 1985).

A major goal of studies of Archean lithosphere is to relate lithospheric structure to the age and chemical variation of peridotite, eclogite, and diamonds and the geological processes that created and assembled the cratons in the hope of arriving at a comprehensive model for diamond formation. This goal is now attainable for a number of reasons: (1) more than two decades of research on the composition and age of diamonds from southern Africa's major diamond deposits (comprising some 4000 individual diamond specimens in more than 30 research papers) have been carried out, (2) new advances have been made over the last decade in applying the Re–Os isotopic system to age determinations of key samples erupted in kimberlites (peridotites, eclogites, and sulfide inclusions in diamond), (3) high-precision U–Pb geochronology and thermochronology of the upper and lower crust are now available, and (4) the Kaapvaal Lithosphere Project has produced a new seismic picture of the lithosphere. In this synthesis, we focus on how the composition of southern Africa's diamonds and the various episodes of diamond growth fit with ideas of cratonic lithosphere creation, stabilization and modification.

2. Methods and materials studied

2.1. Seismic imaging

Seismic imaging of the lithospheric mantle beneath the Kaapvaal and Zimbabwe cratons and the Limpopo mobile belt known as the Southern Africa Seismic Experiment (SASE; James et al., 2001; James and Fouch, 2002) carried out during the multidisciplinary, multinational Kaapvaal Lithosphere Project (Carlson et al., 1996, 2000) has produced a picture of the lithospheric mantle at depths within the diamond stability field. To produce quantitative P-wave anomaly data for each diamond source area in the lithosphere, a 50 km radius cylinder of mantle has been averaged. This cylinder is centered below each diamond mine and extends from 150 to 225 km depth. Seismic data (Table 1) is presented as the P-wave velocity anomaly in % deviation from an average cratonic reference model (James et al., 2001; James and Fouch, 2002).

2.2. Inclusion mineralogy and sample distribution

The compositions of silicate inclusions in diamond can be grouped according to their mineralogical similarity to eclogite or peridotite xenoliths in kimberlite. Diamonds are classified as peridotitic when they contain Cr-pyrope, diopside, enstatite, chromite, or olivine (Meyer and Boyd, 1972; Gurney and Switzer, 1973; Harris and Gurney, 1979) or Fe-sulfide with pentlandite component making up >22 wt.% Ni in the bulk sulfide (Yefimova et al., 1983). Peridotitic diamonds can be further subdivided into harzburgitic or lherzolitic by the degree of depletion indicated by the Cr_2O_3 and CaO content of their included garnet, the absence or presence of diopside, and when applicable, the composition of included olivine or orthopyroxene (e.g., Sobolev et al., 1973; Harris and Gurney, 1979). Diamonds are classified as eclogitic when they contain pyrope–almandine garnet, omphacite, coesite, kyanite or Fe sulfide with pentlandite component making up less than 8 wt.% of the bulk sulfide. Based on the strong correlation between Ni and Os content in sulfide inclusions, Pearson has suggested Os concentration is a good peridotitic vs. eclogitic discriminant (Pearson et al., 1998; Pearson and Shirey, 1999). Thus, South African sulfide inclusions with bulk Ni content >16 wt.% have 2–30 ppm Os and can be classified as peridotitic

Table 1
Seismic velocity of the lithospheric mantle, locality paragenesis, inclusion Re–Os or Sm–Nd age (Ma) and paragenesis of inclusions for southern African diamonds

Kimberlite	Seismic velocity	Locality paragenesis	Re–Os age (Ma)	Re–Os suite paragenesis	Sm–Nd age (Ma)	Sm–Nd suite paragenesis
Jwaneng	− 0.006	E, P	1500–2900	E	1540	E
Letlhakane	− 0.008	E, P				
Orapa	− 0.010	E, P, (W)	1000–2900	E	990	E
Premier	− 0.209	E, P			1150–1930	E, (P)
Venetia	0.194	P, (E, W)				
De Beers Pool	0.245	P, (E)	2900	E	3200	P
Finsch	0.084	P, (E)			1580–3200	P, (E)
Roberts Victor	0.211	P, (E)				
Jagersfontein	0.357	E, P				
Koffiefontein	0.327	P, E	990–2900	E		

Average seismic velocity for P-waves in % deviation from a cratonic reference model for a 50 km radius cylinder of mantle extending from 150 to 225 km depth (James et al., 2001; James and Fouch, 2002). Locality parageneses [peridotitic (P), eclogitic (E), websteritic (W)] are listed in relative order of abundance. Subordinate parageneses are in parentheses. These data are from the work of Deines and Harris and Cartigny. About 100 individual sulfide inclusions have been analyzed for Re–Os ages; about 3000 silicate inclusions (mostly composites but a few individual grains) have been analyzed for Sm–Nd ages. The ages given in the table have been generalized to reflect the major episodes of diamond formation as discussed in the text. Sources of data are from the literature as follows: De Beers Pool (Richardson et al., 1984, 2001; Cartigny, 1998; Cartigny et al., 1998); Finsch (Deines et al., 1984, 1989; Richardson et al., 1984, 1990; Smith et al., 1991); Jagersfontein (Deines et al., 1991a), Jwaneng (Cartigny et al., 1998; Richardson et al., 1999, 2004); Koffiefontein (Deines et al., 1991a; Pearson et al., 1998); Orapa (Richardson et al., 1990; Deines et al., 1991b, 1993; Shirey et al., unpublished data); Premier (Milledge et al., 1983; Deines et al., 1984, 1989; Richardson, 1986; Richardson et al., 1993); Roberts Victor (Deines et al., 1987); Venetia (Deines et al., 2001; Viljoen, 2002).

whereas sulfide inclusions with bulk Ni < 9 wt.% have 200–300 ppb Os and can be classified as eclogitic.

Inclusion suites display significant variability from kimberlite to kimberlite which can lead to sample selection bias in the data set. Also, age determinations from multiple isotopic systems generally have not been made on inclusions in diamonds from the same kimberlite (e.g., Table 1) usually because silicate and sulfide inclusions occur with different frequency in separate diamonds. In fact, eclogitic sulfides and eclogitic silicates occur commonly in only three of the major mines studied so far: Jwaneng, Orapa, and Premier. An ideal situation would be to have silicate and sulfide inclusions in the same diamond, but such diamonds with inclusions large enough for analysis are exceedingly rare in most mines. In addition, larger stones (3–4 mm in diameter) are routinely sought to get sulfides large enough for single grain Re and Os isotopic analysis. The relative abundance of eclogitic diamonds increases with increasing diamond size for both sulfide and silicate inclusions in some of the southern African mines (Gurney, 1989; Sobolev et al., 2001; Viljoen, unpublished data). Thus, in seeking larger stones there can be a greater chance that they will be eclogitic. This may provide an explanation for why,

with the exception of one peridotitic diamond from Koffiefontein, all sulfide inclusions analyzed for Re–Os to date from southern Africa have been eclogitic. All available Kaapvaal inclusion ages obtained prior to 1999 have been summarized recently in Pearson and Shirey (1999) which also contains detailed discussion of inclusion syngenesis and paragenesis.

2.3. Age-dating methods

The dating of diamonds relies on the successful application of radioisotope methods to their mineral inclusions and the indication that the inclusions are syngenetic with respect to diamond crystallization. The question of syngenesis hast been hotly debated over the years (c.f. Pidgeon, 1989; Richardson, 1989; Shimizu and Sobolev, 1995; Navon, 1999; Spetsius et al., 2002) and has recently been reviewed (Pearson and Shirey, 1999; Richardson et al., 2004). The diamond ages used in this paper are based on the conclusion that the dated inclusions are syngenetic. Over the past two decades, ages of diamonds from the major mines have been derived from Sm–Nd and Rb–Sr isotopic studies of silicate inclusions chiefly by Richardson et al. (1984, 1986, 1990, 1993, 1999). Due to the low concentra-

tions of Sm, Nd, Rb, and Sr in garnet and clinopyroxene and the small size of such inclusions, these studies
added together many inclusions of similar mineral
composition to produce isochrons that represent many

tens to hundreds of diamonds from any one kimberlite.
Early attempts to date sulfide inclusions relied on
occasional large grains and the U–Pb isotope system
(e.g., Kramers, 1979). But within the last 5 years,

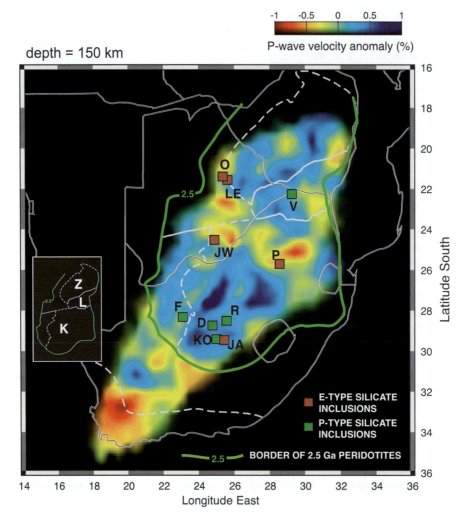

Fig. 1. Tomographic image of the lithospheric mantle derived from seismic P-wave data at a depth of 150 km (James et al., 2001; James and
Fouch, 2002). The color scheme depicts % deviation from an average cratonic lithosphere velocity model. Areal coverage spans the lithospheric
mantle of the Kaapvaal (K) and Zimbabwe (Z) cratons and the Limpopo mobile belt (L; see inset, left). Bold green line indicates the outermost
boundary of the Archean cratons as defined by differences in crustal lithologies, a change in aeromagnetic fabric (Ayres et al., 1998), and a break
between Archean and Proterozoic Re–Os ages on peridotite xenoliths (Carlson et al., 1999; Janney et al., 1999; Irvine et al., 2001). The
locations of diamond mines are shown by colored squares. Red squares are localities whose silicate inclusion in diamond suites are
predominately eclogitic [Jagersfontein (JA), Jwaneng (JW), Letlhakane (LE), Orapa (O), Premier (P)] and green squares are localities whose
silicate inclusion suites are predominately peridotitic [Kimberley area mines of Bultfontein, De Beers, Dutoitspan, and Wesselton termed De
Beers Pool (D), Finsch (F), Koffiefontein (KO), Roberts Victor (R), and Venetia (V)]. Mines located above red-yellowish areas referred to in text
as having diamonds derived from seismically slower mantle; mines located above greenish-blue areas referred to in text as having diamonds
derived from seismically faster mantle. Reprinted with permission from Shirey et al. (2002). Diamond genesis, seismic structure, and evolution
of the Kaapvaal–Zimbabwe craton. (Science, 297: 1683–1686). Copyright 2002 American Association for the Advancement of Science.

improvements in Re–Os analytical sensitivity have made the dating of individual sulfide inclusions in diamonds routine (Pearson et al., 1998; Pearson and Shirey, 1999; Richardson et al., 2001). Diamond ages used in this study come from this more recent literature and ongoing work (Orapa; Shirey et al., unpublished data; Jwaneng; Richardson et al., 2004). Ar–Ar laser probe age measurements can be obtained on single eclogitic and peridotitic clinopyroxene inclusions (e.g., Burgess et al., 1989, 2004; Phillips et al., 1989). These results are not incorporated in this synthesis however, because complexities due to diffusional Ar loss, excess radiogenic Ar, or radiogenic Ar inhomogeneities (Burgess et al., 1992, 1998) have led in some cases to anomalously young ages.

2.4. Direct determinations on diamonds

The carbon isotopic compositions ($\delta^{13}C$) of diamonds from the major mines in southern Africa have been determined over the last two decades by total combustion of inclusion-bearing diamond chips and analysis of the resultant CO_2 by Deines (1980), Deines et al. (1984, 1987, 1989, 1991a,b, 1993, 1997, 2001), Deines and Harris (1995) and Cartigny (1998) and Cartigny et al. (1998, 1999). Infrared spectroscopic (IR) studies of the abundance and crystal-chemical aggregation of trace nitrogen in the diamond structure were carried out on diamond chips from the same samples that were used for carbon isotopic study and Cartigny also has added N isotopes to the C isotopic studies. Based on the type of C isotopic variability of growth horizons seen in ion microprobe analyses of diamond plates (Hauri et al., 1999, 2002), bulk analyses of chips would potentially mix together any C isotopic variability associated with diamond growth zonation. Data from these studies have been summarized previously.

3. Results and analyses

3.1. Seismic structure

Fig. 1, showing the tomographic inversion for P-wave data, presents a picture of the current lithospheric seismic velocity structure at a depth of 150 km. The Kaapvaal–Zimbabwe craton is marked by relatively

high P-wave velocity lithospheric mantle that occurs in two prominent but irregularly shaped lobes separated by a broad west–northwest trending band of relatively lower velocity mantle. The seismic array coverage crosses the margin of the craton only in the southwest (Fig. 1). There, the craton boundary as mapped at the surface is marked by a change in the regional magnetic fabric (Ayres et al., 1998), an abrupt change in Re–Os model age of mantle xenoliths (Janney et al., 1999; Carlson et al., 2000; Irvine et al., 2001), and a sharp decrease of about 1% in seismic velocity. Very low mantle velocities are evident off-craton in the far southwest underneath the western Cape Foldbelt. Kimberlites distributed across southern Africa, therefore, have diamonds that derive from mantle with differences in seismic velocity (Table 1, Fig. 1). Jwaneng, Letlhakane, Orapa, and Premier diamonds were hosted in slower lithospheric mantle with relative P-wave velocity perturbations that vary from − 0.209% to − 0.006% whereas diamonds in the Roberts Victor, Jagersfontein, Finsch, Venetia, Koffiefontein, and the Kimberley area kimberlites (Bultfontein, De Beers, Dutoitspan, Wesselton; known collectively as the 'De Beers Pool') were hosted in seismically faster mantle

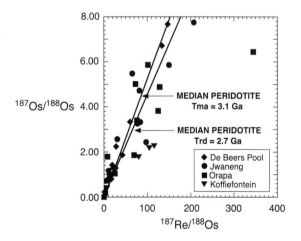

Fig. 2. Re–Os isotopic array for individual sulfide inclusions in single diamonds compared to typical Re–Os model ages on peridotites from the Kaapvaal–Zimbabwe craton. Figure modified after Shirey et al. (2001). Data sources are as follows: De Beers Pool (Richardson et al., 2001), Jwaneng (Richardson et al., 2004), Koffiefontein (Pearson et al., 1998), and Orapa (Shirey et al., unpublished). Peridotite Re–Os model ages from Irvine et al. (2001) and Carlson et al. (1999).

with relative P-wave velocity perturbations that vary from $+0.084\%$ to $+0.357\%$.

3.2. Diamond ages from inclusions

Data on single eclogitic sulfides from the De Beers Pool, Jwaneng, Koffiefontein, and Orapa form a $^{187}Re/^{188}Os$ vs. $^{187}Os/^{188}Os$ data array (Fig. 2) that clearly includes diamonds of different ages. The data are dominated by the greater number of inclusions analyzed from the De Beers Pool, Jwaneng and Orapa. Those from the De Beers Pool show a 2.9 Ga isochron (Richardson et al., 2001) while the other diamond suites have populations that fall around this age but with more scatter (Fig. 2; Shirey et al., 2001). For Koffiefontein, the first sulfide inclusion suite studied with the Re–Os system, only two inclusions out of the six studied correspond to a circa 2.9 Ga age and the majority give a Proterozoic (1 Ga) age (Pearson et al.,

1998). Jwaneng and Orapa each have a number of their analyzed sulfide inclusions which plot at younger ages that match (respectively) the ages obtained from Sm–Nd studies of silicate inclusions. The DeBeers Pool sulfide suite only has one such grain. Thus, in all localities, there appear to be at least two generations of sulfide inclusions: an Archean, circa 2.9 Ga age suite and a Proterozoic suite of an age specific to each locality. The circa 2.9 Ga age of the older suite is not resolvable from the Re–Os model ages obtained on mantle peridotites whose median can be either 2.7 Ga if using a 'time of $\underline{Re\ depletion}$' (Trd) approach or 3.1 Ga if using a 'time of $\underline{mantle\ reservoir\ separation}$' (Tma) approach (Fig. 2; Carlson et al., 1999, 2000; Irvine et al., 2001).

Sm–Nd isochron and mantle model ages for silicate inclusion suites from the Kaapvaal–Zimbabwe craton are summarized in Fig. 3. Peridotitic (harzburgitic) garnets from Finsch and De Beers Pool (Richardson

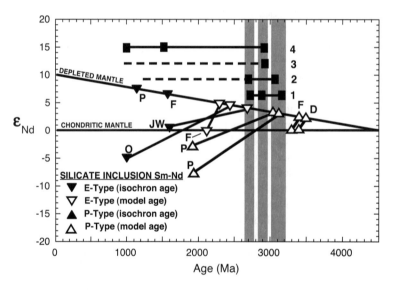

Fig. 3. Sm–Nd isochron and model ages of silicate inclusions in diamond (sources in Table 1 caption) compared to the ages of major crust forming events in the Kaapvaal craton depicted as thick, grey vertical bands (1) (Schmitz, 2002; Schmitz et al., 2004), peridotite xenolith Re–Os model ages (2) (sources in Fig. 2 caption), diamondiferous eclogite xenoliths (3) (Shirey et al., 2001; Menzies et al., 2003), and sulfide inclusions in diamonds (4) (sources in Fig. 2 caption). Horizontal lines that extend to the Proterozoic are dashed to indicate that younger ages on eclogites and peridotites exist but evidence for specific ages are unclear. The two peridotite ages correspond to the dominant Trd (younger) and Tma (older) ages. Growth curves from the isochron age (crystallization) of silicate inclusions to their reservoir separation from depleted mantle are estimated by assuming that the measured Sm/Nd of the clinopyroxene associated with lherzolitic garnet at Premier and the clinopyroxene associated with eclogitic garnet at Finsch, Jwaneng and Orapa represent a maximum (and perhaps typical) Sm/Nd of the protolith. These assumptions can be supported because most of the light REE in an eclogite will reside in clinopyroxene (e.g., Taylor et al., 1996) and lherzolitic garnet and associated clinopyroxene show regular REE patterns (e.g., nonsinusiodal; Stachel and Harris, 1997; Stachel et al., 1998). Note that depleted mantle growth is shown here schematically only as a straight line. Other, more geologically based models for early Archean depleted mantle evolution will lead to slight differences in depleted mantle model ages from those depicted here.

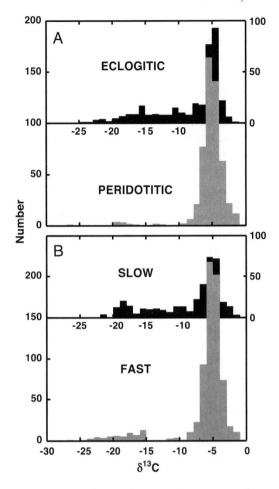

isochron age (1.9 Ga, Richardson et al., 1993) that, with reasonable assumptions of protolith Sm/Nd (see Fig. 3, caption) still would lead to a mid-Archean (3.1 Ga) depleted mantle model age. All other southern African silicate inclusion suites dated with the Sm−Nd system (Richardson, 1986; Richardson et al., 1990, 1999; Smith et al., 1991) are eclogitic and yield Proterozoic isochron or model ages.

Fig. 4. (A) Comparison of the carbon isotopic composition of individual diamond analyses grouped according to eclogitic or peridotitic paragenesis. Note that both eclogitic and peridotitic histograms include diamonds from all nine localities. (B) Comparison of the carbon isotopic composition of southern African diamonds, grouped according to their derivation from a locality in seismically slower (Jwaneng, Letlhakane, Orapa, and Premier) or seismically faster lithospheric mantle (Venetia, De Beers Pool, Finsch, Roberts Victor, Jagersfontein and Koffiefontein). The similarities in the histograms in panels A and B occur because a greater proportion of diamonds of eclogitic paragenesis and isotopically light carbon derive from localities occurring in seismically slower lithospheric mantle. See text and Table 1 for sources of data. Reprinted with permission from Shirey et al. (2002). Diamond genesis, seismic structure, and evolution of the Kaapvaal–Zimbabwe craton. (Science, 297: 1683–1686). Copyright 2002 American Association for the Advancement of Science.

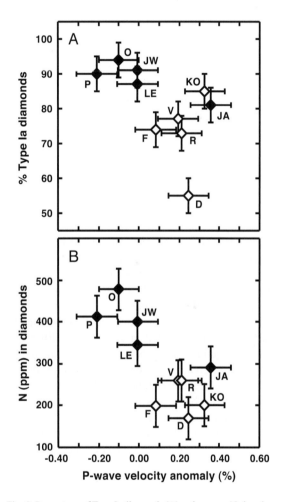

Fig. 5. Percentage of Type Ia diamonds (A) and average N abundance in diamonds (B) vs. P-wave velocity anomaly. Error bars have been set at the ± 5% level for percentage of Type Ia diamonds, ± 50 ppm for average N abundance, and ± 0.1% for P-wave velocity anomaly. Type Ia diamonds are the total of IaA, IaA/B and IaB diamonds (Table 1). Same lettering scheme as in Fig. 1. Peridotitic =◇, eclogitic = ◆. Reprinted with permission from Shirey et al. (2002). Diamond genesis, seismic structure, and evolution of the Kaapvaal–Zimbabwe craton. (Science, 297: 1683–1686). Copyright 2002 American Association for the Advancement of Science.

et al., 1984) have the oldest model ages yet recorded for inclusions in diamond from southern Africa. Premier peridotitic (lherzolitic) garnets have a much younger

3.3. Diamond isotopic compositions

Diamonds selected for C isotopic and N studies perhaps provide a larger diamond sample than those selected for age studies because they include both the peridotitic and eclogitic parageneses that are typically present at each locality. It has long been known (Sobolev et al., 1979) that diamonds of both peridotitic and eclogitic paragenesis show a prevalent mantle-like carbon isotopic composition ($\delta^{13}C = -3\%o$ to $-7\%o$) with an isotopically light subpopulation ($\delta^{13}C = -10\%o$ to $-34\%o$), dominated by diamonds of eclogitic paragenesis (e.g., Gurney, 1989; Galimov, 1991; Kirkley et al., 1991). Compiled data for the Kaapvaal–Zimbabwe cratons (Table 1 and refs. therein, Fig. 4; Shirey et al., 2002, 2003) adhere to this pattern.

Deines et al. (1989, 1993, 1997) previously noted the significantly higher N content of eclogitic diamonds from Finsch, Jwaneng, Orapa, and Premier compared to their peridotitic counterparts from the same kimberlite. Fig. 5 shows that the paragenesis vs. diamond N content relationship extends to all 10 kimberlites with well-studied diamond populations. The kimberlites whose diamonds are mostly eclogitic have diamonds with an average N content above 290 ppm and a percentage of Type Ia diamonds greater than 80% compared to the kimberlites whose diamonds are mostly peridotitic which are lower in both respects. This N content difference has not resulted in a resolvable difference in the aggregation state of N in the Type Ia diamonds (Shirey et al., 2002, 2003) perhaps because of differences in mantle residence times of eclogitic vs. peridotitic diamonds or because N spectra come from multiple growth zones in rough diamonds or chips. The large range in C isotopic composition and N abundance unfortunately cannot be directly correlated with diamond ages because these diamonds are undated.

4. Discussion and conclusion

4.1. Patterns of cratonic keel structure, diamond age, and paragenesis

New age information from the Re–Os system on mantle lithologies and diamonds allows the relationship among diamond ages, paragenesis, and cratonic seismic structure to be explored. The early work of Holmes and Paneth (1936) demonstrated that eclogite xenoliths could be much older than the kimberlites that carried them. Kramers (1979), using common Pb in sulfide inclusions, and Richardson et al. (1984), using Sm–Nd on harzburgitic garnet inclusions, showed that diamonds are early Proterozoic to mid-Archean. These results indicated that at least parts of the craton were underlain by a mantle keel that could be as old as the oldest crust, a result now generally confirmed by Re–Os model ages on whole rock peridotite xenoliths (e.g., Walker et al., 1989; Carlson et al., 1999, 2000) and in situ Re–Os model ages on sulfides hosted in primary silicates (Griffin et al., 2003a,b). Additional work of Richardson et al. (1986, 1990, 1993, 1999) on diverse eclogitic and lherzolitic silicate inclusions in diamonds produced Proterozoic ages suggesting a progression to more fertile silicate inclusion compositions with time. However, recent Re–Os data show that all four of the eclogitic sulfide inclusion suites studied include late Archean ages that overlap the average late Archean Re–Os mantle model ages on cratonic peridotite (Carlson et al., 1999; Irvine et al., 2001; Shirey et al., 2001). Furthermore, the peridotitic silicate inclusion model ages in the 3.2–3.3 Ga range now appear to overlap some of the oldest crustal U–Pb ages obtained in the western Kaapvaal craton (Moser et al., 2000; Schmitz, 2002; Schmitz et al., 2004) and predate the ages of many other materials in the lithosphere (Fig. 3) as suggested by the Re–Os model ages of whole-rock peridotite xenoliths (Carlson et al., 1999; Irvine et al., 2001) and their sulfides (Griffin et al., 2003a,b), sulfide inclusions in diamond (see Fig. 2), diamondiferous eclogite xenoliths (Shirey et al., 2001; Menzies et al., 2003), and the U–Pb ages of crustal metamorphism and plutonism (Schmitz, 2002; Schmitz et al., 2004).

The distribution of diamond ages and paragenesis with respect to seismic structure reveals that peridotitic silicate (harzburgitic) inclusions in diamonds from the De Beers Pool and Finsch which give the oldest mid-Archean model ages (Richardson et al., 1984), were derived from seismically faster lithospheric mantle. The other peridotitic silicate (lherzolitic) inclusion suite dated (Premier; Table 1) is Proterozoic (Richardson et al., 1993) and was derived from seismically slower mantle. All the other silicate inclusion suites dated so far are eclogitic, Proterozoic (Richardson, 1986; Richardson et al., 1990, 1993, 1999; Smith et al., 1991) and were derived from seismically slower mantle. In comparison, sulfide inclusions from the four localities studied so far are eclogitic, have an Archean age population (Pearson et al., 1998; Richardson et al., 2001; Shirey et al., 2001) and were derived from both fast and slow mantle. Of these four localities, De Beers Pool, whose diamonds were derived from seismically faster mantle, has the lowest number of Proterozoic inclusions.

The distribution of C isotope composition for eclogitic diamonds (Fig. 4A) appears nearly identical to that for diamonds from the seismically slow lithospheric mantle (Fig. 4B) but this is chiefly controlled by paragenesis. There are additional complexities, however. The eclogitic diamonds with $\delta^{13}C$ values less than − 9‰ (e.g., those that comprise the isotopically light tail of the $\delta^{13}C$ distribution) are dominated by the large number of specimens from Jwaneng, and Orapa (Shirey et al., 2003). Furthermore, Premier, Jagersfontein, and Roberts Victor directly contradict the isotopically light eclogitic diamond to seismically slow lithosphere correlation; Premier lies above seismically slow mantle but has no isotopically light diamonds in its eclogitic population whereas Jagersfontein and Roberts Victor lie above seismically fast mantle but have their eclogitic populations dominated by diamonds with isotopically light carbon (Shirey et al., 2002, 2003). Therefore, while paragenesis is the controlling factor in the differences in C isotopic composition of diamonds from seismically fast vs. slow mantle, the cratonwide distribution of eclogitic diamonds with isotopically light C is not straightforward, perhaps because of the irregular distribution of petrologically diverse eclogite types.

Nitrogen abundance too, shows some correspondence with lithospheric seismic structure that is linked to paragenesis (Fig. 5A and B; Shirey et al., 2002, 2003). Diamonds from Jwaneng, Letlhakane, Orapa, and Premier which are dominantly eclogitic, have a higher percentage of Ia types, an average N content above 290 ppm, and were derived from seismically slower mantle (Fig. 5A and B). Diamonds from Koffiefontein, Finsch, Roberts Victor, Venetia and the De Beers Pool which are dominantly peridotitic, have a lower percentage of Ia types, an average N content below 290 ppm, and were derived from seismically faster lithospheric mantle. The aggregation state of N in

the Ia diamonds (not shown) displays no clear systematic variation with lithospheric seismic velocity although De Beers Pool, Finsch and Roberts Victor diamonds from seismically fast mantle do display the lowest percentages of aggregation to B-centers (Shirey et al., 2002, 2003).

The best explanation for the complexities in the range and distribution of diamond ages and the distribution of diamond inclusion types lies in the uneven make-up of the overall diamond sample, the variability of eclogitic xenolith suites and the nonoverlapping nature of the extant diamond studies. These complexities are in addition to the episodic nature by which the cratonic keel was generated, stabilized and modified. Table 2 shows the emerging picture from the diamond data of a mantle keel that has retained distinct populations of diamonds that were formed during each of the major tectonothermal episodes to have affected the cratonic keel (e.g., Shirey et al., 2002).

4.2. Age of and compositional constraints on lithospheric seismic velocity structure

Several lines of evidence suggest that the current lithospheric seismic structure of the craton is a mid-Proterozoic overprint to this predominantly 2.9–3.3 Ga keel. Proterozoic eclogitic (silicate) inclusion-bearing diamonds from seismically slow lithosphere (Figs. 1 and 3; Table 1) occur in the same kimberlites that contain Archean eclogitic sulfide inclusions (e.g., Orapa and Jwaneng) or peridotites with Archean Re–Os model ages (e.g., Letlhakane, Irvine et al., 2001). The Premier kimberlite, penetrating seismically slow mantle, contains peridotite xenoliths with Archean Re–Os model ages but also the clearest example of peridotite overprinted in the Proterozoic (Carlson et al., 1999; Irvine et al., 2001). Premier peridotitic silicate inclusions (lherzolitic garnet and clinopyroxene) that were equilibrated at the 1.9 Ga isochron age representing the time of encapsulating diamond growth (Richardson et al., 1993), have Sm–Nd mantle model ages that are Archean (Fig. 3).

Modification of the cratonic keel, which was originally thought to affect chiefly the Premier locality on the basis of Re–Os studies of lithospheric peridotite (Carlson et al., 1999; Irvine et al., 2001), is apparently more widespread, extending across the northern Kaapvaal craton and to the

Table 2
Summary of time periods, inclusions studied and their ages and parageneses, geologic settings proposed for craton formation, and petrologic observations that can be explained by this craton formation model

Time (Ga)	Age, parageneses, locality	Geologic event	Geologic settings proposed	Observations explained by model
3.2–3.7	3.2 Ga; peridotitic (hz) silicates; D, F	Creation of earliest cratonic nuclei	Midocean ridges; Oceanic subduction zones	Silicic, differentiated crust, 3.5–3.7 Ga age hz gt inclusions with depletion and enrichment Formation of SCLM at low pressure Wet, pyroxene spinifex komatiites
2.9–3.2	2.9 Ga; eclogitic sulfides; D, JW, K, O	Craton assembly into present form	Microcontinent collision; Closing of Archean ocean basins; Marginal subduction	Thermochronology of crustal rocks Geologic pattern of terrane accretion Differentiated crustal rocks (e.g., TTG suites) Widely distributed diamonds with eclogitic sulfides Re–Os ages of peridotites and eclogites Silica (opx) enrichment of the lithosphere
<2.1	1.9 Ga; peridotitic (lhz) silicates; P 1–2 Ga; eclogitic silicates; F, JW, O, P 1–2 Ga; eclogitic sulfides; JW, K, O	Modification of the craton	Subcratonic magmatism; Marginal subduction	Proterozoic diamond ages Silicate inclusions from both DM and SCLM Seismic structure (tomography) of lithosphere Correspondence of tomography and inclusions Reworked peridotites (Premier, N Lesotho) Bushveld–Molopo Farms Complexes

Locality letter descriptions follow Fig. 1 (see caption). Acronyms used in observations are as follows: harzburgite (HZ), subcontinental lithospheric mantle (SCLM), tonalite–trondhjemite–granodiorite (TTG), orthopyroxene (opx), depleted mantle (DM), lherzolite (Ihz).

south of the Zimbabwe craton–Limpopo Belt some hundreds of kilometers from the craton edge. Correlation of the seismically slow regions of the northern Kaapvaal Craton that trend ESE–WNW south of the Limpopo belt (Fig. 1) with surface outcrop of the 2.05 Ga Bushveld Complex in South Africa and Molopo Farms Complex in Botswana suggests that the modification may be closely related to Bushveld–Molopo magmatism. For the seismically slow mantle that trends N–S on the west side of the Zimbabwe craton (Fig. 1), regional metamorphism that created the Magondi–Okwa terranes (Carney et al., 1994) is likely to be the surface manifestation of the tectonism that modified the craton on this margin.

Recent thermal structure of the Kaapvaal lithospheric mantle from surface heat flow (Jones, 1988) and cratonic geotherms from xenolith geothermobarometry (Danchin, 1979) show that even in the seismically slow areas, such as near Premier, a normal cratonic geotherm has existed since at least the Premier eruption age of 1.2 Ga. This is evidence that the lithosphere is seismically slower now because it is compositionally different, not hotter.

The seismically slow region of the lithosphere is likely higher in basaltic components (Fe, Ca, clinopyroxene) and metasomatizing veins that hydrate and alter the vein wall of the host peridotite. These are the main petrological differences that would be expected to account for the 1% difference in P-wave velocity seen in Fig. 1.

Diamond suites from seismically slower lithosphere have a greater percentage of eclogitic and lherzolitic inclusions which matches regions of the lithosphere that would have a higher proportion of basaltic components. Also, some diamond suites from these regions with eclogitic inclusions have the isotopically light carbon, the highest percentage of Type Ia stones and a higher average N content (>290 ppm). It is not clear whether higher average N incorporation during growth of these diamonds was due to a diamond-forming fluid with higher N content or due to faster growth. The lack of an obvious difference in N aggregation characteristics between diamonds from seismically slow and fast mantle (Shirey et al., 2002) indicates that diamonds from both types of mantle were stored in the lithosphere at high enough temperatures (e.g.,

1150 ± 50 °C) for long enough to allow substantial aggregation of N to B centers (Evans and Harris, 1989; Navon, 1999). In this case, the low percentage of IaB diamonds in the De Beers Pool, Finsch, and Roberts Victor diamond populations would be related more to the lower average N content of these diamond populations and their perhaps slower growth than to temperature (Shirey et al., 2002, 2003). Simple heating of the lithosphere in the

seismically slow regions, as might be suggested by the resetting of U–Pb ages in low-closure-temperature minerals such as rutile found in lower crustal granulites elsewhere (Schmitz, 2002), is not recorded in the N aggregation data. This could be because any thermal pulse was too short-lived (Danchin, 1979) for substantial N aggregation to occur in a short period (Richardson and Harris, 1997; Navon, 1999). The current N aggregation data also are not

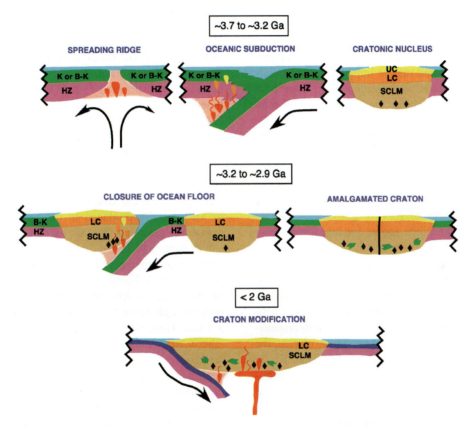

Fig. 6. Proposed model for the creation, stabilization and modification the Archean Kaapvaal–Zimbabwe craton by largely subduction-related processes. In the 3.7–3.2 Ga period, oceanic spreading ridges formed on an Earth hotter than today producing a thicker komatiitic to basaltic–komatiitic oceanic crust and a thicker layer of depleted harzburgite. This early crust was recycled by some form of oceanic subduction which allowed the access of water to the melting of the mantle and began the process of crustal differentiation. The products of this period were differentiated cratonic nuclei with abundant sialic component, a distinct upper and lower crust and an attendant cratonic keel. In the period from 3.2 to 2.9 Ga, amalgamation of cratonic nuclei constructed a composite craton with eastern and western domains sutured along the N–S Colesburg Lineament (vertical black line middle right) and other major features similar to the craton as preserved today. Because the closure of ocean basins floored by high-temperature Archean oceanic lithosphere with attendant subduction was the major way to bring cratonic nuclei together, the process led to abundant C, O, H, S fluids for diamond formation and the emplacement of peridotitic and eclogitic protoliths into the lithospheric mantle while preserving age/terrane differences. In the period after 2 Ga, the composite craton was subject chiefly to modification from the margin by subduction and from below by magmatism and metasomatism. This produced multiple generations of new diamonds with different source characteristics and a correspondence between the diamonds and the seismic structure of the lithospheric mantle. Komatiite (K), basaltic komatiite (BK), harzburgite (HZ), upper crust (UC), lower crust (LC), subcontinental lithospheric mantle (SCLM).

detailed enough to resolve any systematic temperature or depth differences between eclogitic and peridotitic diamonds.

4.3. Model for diamond growth during creation, stabilization, and modification of the craton

A summary of geologic events for the Kaapvaal–Zimbabwe craton and the diamond ages and localities we associate with these events is presented in Table 2 and Fig. 6. Any comprehensive model for diamond formation and craton evolution must explain the ages of diamonds, the distribution of ages relative to paragenesis and craton geology, and the concurrent production of crustal and mantle lithologies. Constraints on lithosphere creation/stabilization come from those diamonds that are Archean: the four suites with eclogitic sulfides that have been dated using the Re–Os system at around 2.9 Ga (Pearson and Shirey, 1999; Richardson et al., 2001, Richardson, Shirey et al., unpublished data) and two suites with peridotitic silicates that have been dated using the Sm–Nd and Rb–Sr system at 3.2–3.3 Ga (Richardson et al., 1984).

Until the advent of Re–Os analyses on sulfide inclusions, the prevalent model for lithosphere creation and stabilization involved depletion caused by degrees of melting high enough to create komatiite (e.g., Richardson et al., 1984; Walker et al., 1989; Boyd et al., 1999). Although this model failed to account for the high orthopyroxene (i.e., silica content) of the lithospheric mantle, it could simultaneously account for the high Mg# of peridotitic olivine, the bouyancy (and hence preservation) of the lithospheric mantle, its low heat production, its highly unradiogenic Os isotopic composition, and the harzburgitic composition of the 3.2–3.3 Ga garnet inclusions. None of the Archean sulfide inclusions yet analyzed for Re–Os are peridotitic, they are all eclogitic, perhaps due to sampling bias as discussed above. As seen in Figs. 2 and 3, typical sulfide ages do not cluster around the oldest peridotitic silicate age of 3.2–3.3 Ga (Richardson et al., 1984) or the dominant crust-forming age of 3.1 Ga (Moser et al., 2000; Schmitz, 2002; Schmitz et al., 2004) but scatter around a younger, 2.9 Ga age. Nowhere is this age distinction clearer than with the De Beers Pool samples from the Kimberley area kimberlites. Here, 3.2–3.3 Ga peridotitic, depleted harzburgitic garnets (Sm–

Nd, Rb–Sr model ages; Richardson et al., 1984) coexist in the same lithospheric mantle section with 2.9 Ga eclogitic sulfides (Re–Os isochron age; Richardson et al., 2001) that have an elevated, enriched initial Os isotopic composition.

Regarding the reliability of these ages, the point has often been made (e.g., Navon, 1999) that Sm–Nd model ages alone do not preclude the formation of the harzburgitic diamonds at some time later than 3.2 Ga which might allow for coeval crystallization of peridotitic and eclogitic diamonds at 2.9 Ga. This is indeed the case since the Sm–Nd system dates the metasomatism producing the low Sm/Nd in the protolith (3.4–3.5 Ga based on chondritic or depleted mantle precursors respectively; Fig. 3) rather than diamond crystallization (3.2–3.3 Ga based on combined Sm–Nd and Rb–Sr model age arguments). However, comparison of the radiogenic Sr isotope signatures of inclusion (encapsulated) and macrocryst (unencapsulated) garnets from disaggregated harzburgitic diamond host rocks indicates that a 2.9 Ga or younger age for the diamonds is unlikely. The key to this argument is that coupled low Sm/Nd and high Rb/Sr are typically produced in the same metasomatic event. Yet these harzburgitic garnets have negligible Rb contents and hence very low Rb/Sr ratios. Both inclusion and macrocryst garnets have low Sm/Nd *but* very low Rb/Sr because garnet includes Sm, Nd, and Sr, but excludes Rb. However, the host rock Rb/Sr remains high beyond diamond crystallization at 3.2–3.3 Ga. Thus, the inclusion and macrocryst garnets inherit their respective Sr isotope signatures by (re)-crystallization at 3.2–3.3 Ga and, in the latter case, continuing diffusive exchange with the radiogenic Sr produced in high Rb/Sr harzburgitic diamond host rocks during lithospheric mantle storage prior to sampling by kimberlite. In particular, Kimberley macrocryst garnets have highly radiogenic present-day Sr isotope compositions ($^{87}Sr/^{86}Sr > 0.755$) as compared to that measured in the corresponding inclusion garnets ($^{87}Sr/^{86}Sr = 0.706$) which were isolated from the high Rb/Sr host after diamond crystallization. If the inclusion garnets were only encapsulated in diamond at 2.9 Ga or later, they would be expected to inherit a much more radiogenic Sr isotope composition ($^{87}Sr/^{86}Sr > 0.710$) from the high Rb/Sr host than that actually observed (0.706; see Fig. 4 in Richardson et al., 1984).

All indications are that diamond formation in Archean cratonic mantle is episodic and such episodicity may apply to the formation and stabilization of the cratonic mantle itself. The occurrence of 3.2–3.3 Ga diamonds with depleted harzburgitic silicate inclusions and 2.9 Ga diamonds with enriched eclogitic sulfide inclusions in the same Kimberley kimberlites indicates that creation and assembly of the craton was a multistage process (Shirey et al., 2002). The Re–Os, Sm–Nd, and Rb–Sr model age relationships indicate a time gap of 300 ± 200 million years between the two Archean diamond formation events.

4.3.1. Craton nuclei creation

Cratonic nuclei initially were created by mantle melting processes that produced severe depletion in the cratonic mantle keel as evidenced by their highly forsteritic olivine (Boyd et al., 1999), high Cr/low Ca garnet (Gurney and Switzer, 1973), and low time averaged Re/Os (Walker et al., 1989). Komatiitic volcanism involving a high degree of mantle melting usually has been implicated in producing this depletion (e.g., Richardson et al., 1984; Walker et al., 1989; Wilson et al., 2003). Classically, komatiite production was thought to involve high degrees of melting above a mantle plume (e.g., Arndt et al., 1997) and in this tectonic setting the cratonic keel could have been produced by vertical underplating of deep mantle (e.g., Haggerty, 1986; Griffin et al., 1999, 2003b), perhaps from transition zone depths. However, more recent experimental and trace element work suggests that such depletion, while still involving high extents of melting, could have occurred at shallower depths (Ringwood, 1977; Canil and Wei, 1992; Stachel et al., 1998; Walter, 1998) perhaps in a hot and wet subduction setting (Parman et al., 2001; Wilson et al., 2003).

A subduction setting for creation of the early cratonic nuclei is advocated (Fig. 6) although its form (if it existed) in the mid-Archean may have been different because plate tectonics then may have been significantly different (e.g., de Wit, 1998). Some form of subduction, nonetheless, is favored over a plume setting for other important reasons (e.g., Table 2). Chief among them is that a subduction setting can produce the ancient rock record of differentiated silicic crust along with the necessary characteristics of the mantle keel. Evolved calc-alkaline plutonic or volcanic rocks such as dacites or members of the tonalite–trondhjemite–granite suite are found as key components of both the eastern (Hunter, 1974; Kroner et al., 1989) and western (Drennan et al., 1990; Anhaeusser and Walraven, 1999; Schmitz, 2002; Schmitz et al., 2004) domains of the Kaapvaal craton and indeed the crust of all of the earliest cratonic nuclei on Earth (Martin, 1993, 1994; Windley, 1995). Furthermore, they occur in most identified oceanic or continental convergent margin settings from the Archean to the present (Barker, 1979; Gill, 1981). The presence of a highly depleted, seismically fast continental keel under the Superior Province (Grand, 1987; van der Lee and Nolet, 1997), a craton formed by arc accretion (Card, 1990) and comprised of more than 50% differentiated crust, demonstrates that Archean subduction is effective at making depleted mantle keels. Indeed, in originally characterizing the properties and evolution of the continental tectosphere, Jordan (1978, 1981, 1988) advocated a subduction model for making the mantle keel that involves depletion in the mantle wedge followed by consolidation and thickening during subsequent collisional orogenesis. In addition, a subduction model for early cratonic nuclei provides a mechanism for producing the light REE enrichment juxtaposed onto the oldest harzburgitic garnet inclusions (Richardson et al., 1984). Mantle plumes could succeed as a way to make cratonic nuclei because they make sufficiently buoyant or depleted mantle residua and can even produce some limited amount of rhyolite if hydrothermally altered basalt is buried to wet-melting depths (e.g., Iceland). However, they fail because they do not provide an efficient mechanism to make an extensive (more than 50% of a crustal section) differentiated silicic crust of the type seen in continents.

Crustal differentiation, the production of sial, and the first phase of crustal preservation may have resulted in only partial preservation of the earliest depleted mantle in the lithosphere. This is hypothesized because there is no seismological evidence for extensive ultramafic residues in the lower crust (James and Fouch, 2002) as required by early crustal differentiation (Martin, 1993). The removal of ultramafic residues, if it occurred by delamination, must have occurred at depths shallower than the lithospheric keel and could not have occurred with a stable keel in place. Thus, the restricted distribution of 3.2–3.3 Ga harzburgitic silicate inclusions (documented so far only from the Kimberley–Finsch area) may be explained partially as a preservation feature. It must be noted that the spatial distribution of extensively studied diamond-bearing kimberlites in southern Africa limit the diamond-based constraints on the above model for episodic creation and stabilization of the Kaapvaal–Zimbabwe craton chiefly to its western part. However, mid-Archean Re–Os model and isochron ages on whole-rock peridotites (Pearson, D.G. et al., 2002; Carlson and Moore, 2004), sulfides hosted in silicate mineral concentrates (Griffin et al., 2003a) and komatiites (Wilson et al., 2003) which have a slightly wider geographic distribution than the diamonds presently studied are proposed similarly to reflect the existence of an older depleted cratonic nucleus in the eastern part of the Kaapvaal craton.

4.3.2. Craton amalgamation and stabilization

In the second phase of the envisioned process (Fig. 6b), early cratonic nuclei were accreted along with Archean oceanic lithosphere to create a larger, composite craton similar to that which is preserved today. In this phase, subduction, which included depleted oceanic harzburgite and perhaps the roots of Archean oceanic plateaus, built the rest of the lithospheric mantle (Shirey et al., 2002). The process envisioned is a geologically more complex version of the 'stacked slab' model proposed by Helmstaedt and Schulze (1989). This would involve processes more akin to typical modern subduction (Richardson et al., 2001) where fluid fluxing, melting, and severe depletion of the deeper mantle wedge (Ringwood, 1977; Kesson and Ringwood, 1989; Wilson et al., 2003) could accompany metasomatism and Si enrichment of the shallower mantle already in place as continental lithosphere. Both processes are as much a part of stable lithosphere creation as are the shallower processes of intracrustal melting, metamorphism and compressional tectonics (Schmitz and Bowring, 2001; Schmitz, 2002; Schmitz et al., 2004). A second phase of lithosphere creation (Table 2) could account for the widespread distribution of circa 2.9 Ga age Archean eclogitic sulfide inclusions in diamonds (Table 1; Shirey et al., 2001), their enriched initial Os isotopic composition (Richardson et al., 2001), the presence of enriched and depleted Archean inclusion chemistry in

diamonds from the same kimberlite, the younger age of these eclogitic diamonds compared to peridotitic (harzburgitic) diamonds, the typical Re–Os model age for mantle peridotite (Carlson et al., 1999; Irvine et al., 2001), and the occurrence of 2.9 Ga diamondiferous eclogite xenoliths with basaltic komatiitic compositions (Shirey et al., 2001; Menzies et al., 2003). This two-step lithosphere creation model fits the detailed geochronological record for the lower and upper crust of the western Kaapvaal craton in which the earliest crustal components are formed from 3.20–3.26 Ga and the eastern and western domains (now separated by the N–S Colesberg Lineament) of the craton are sutured together by subduction convergence at 2.88–2.94 Ga (Schmitz, 2002; Schmitz et al., 2004).

Subduction affecting the western margin of the Kaapvaal–Zimbabwe craton during craton amalgamation is supported by stable isotopic data for Orapa and Jwaneng diamonds. Orapa eclogitic diamonds have light C and heavy N isotopic compositions (Cartigny et al., 1999) and sulfide inclusions with mass-independent sulfur isotopic fractionations (Farquhar et al., 2002) that are consistent with incorporation of C, N, and S from surficial sedimentary endmember reservoirs (Navon, 1999; Farquhar et al., 2002). Also, some Archean Jwaneng megacrystic zircon have light O isotope compositions suggesting that the zircon host lithologies were once hydrothermally altered as oceanic crust (Valley et al., 1998). Subduction is likely not to be the only source of C and N isotopic variability, however, because of the difficulties of finding subducted materials with appropriately high C/N ratios (e.g., Cartigny et al., 1998, 1999). Furthermore, intramantle processes (e.g., Cartigny et al., 2001) that changed the C and N isotopic composition of fluids during diamond growth, are required by the isotopic composition differences observed in the growth horizons of some diamonds (Hauri et al., 1999, 2002). The lack of complete C and N isotopic data sets from all but a few of southern Africa's diamond suites means that it is not possible currently to discuss a cratonwide picture of intramantle isotopic fractionation of C and N in diamonds.

4.3.3. Craton modification

The near one-to-one correspondence of Proterozoic diamond suites having a majority of eclogitic silicate inclusions with seismically slow mantle suggests that craton modification and Proterozoic diamond formation were closely linked (Shirey et al., 2002). Proterozoic craton modification that reduced the seismic velocity of the lithosphere did not apparently result in the growth of the lithospheric mantle. This is indicated by a lack of dominant Proterozoic Re–Os model ages on peridotites from the seismically slower portions of the lithosphere. In the Proterozoic silicate inclusion suite, eclogitic inclusions predominate over peridotitic. Thus, it is likely that the modification process was more closely linked to eclogitic and lherzolitic lithologies, perhaps associated with the mafic to ultramafic Bushveld–Molopo Farms magmatism under the center of the northern Kaapvaal craton or in conjunction with some form of subduction along the western Kaapvaal–Zimbabwe craton margin (Fig. 6). Deep crustal evidence for this thermal event is found in upper U–Pb concordia intercepts of near Bushveld age in garnet–biotite granulite xenoliths from the Orapa kimberlite (Schmitz and Bowring, 2003). Both sublithospheric magmatism and western craton margin subduction were tectonothermal events that altered the composition of the lithosphere and added new diamonds to an already extensive Archean diamond population resident in the lithosphere.

The diamond data do not carry any sign of the young (circa 90 Ma) replacement of the base of the lithosphere below 160–180 km by asthenospheric or metasomatic components as proposed by Griffin et al. (1999, 2002a) from garnet concentrate geothermobarometry and trace element compositions. The chief reason for this is that the nearly all diamonds, being Archean or Proterozoic are too old to carry information on this young event. Furthermore, if the replacement scenario is correct, then most diamonds must originate from depths in the lithosphere shallower than 160–180 km since there is no systematic difference between the diamond content of Group 1 vs. Group 2 kimberlites.

4.4. Secular trends, new south African studies and other cratons

4.4.1. Secular variation of Ni and Re/Os in sulfide inclusions

In the model for craton evolution outlined above, there is a major difference in the composition of

mantle melts associated with Archean processes of craton nuclei creation and craton amalgamation vs. the Proterozoic processes of cratonic lithosphere modification. The Archean processes call for higher-temperature, komatiite to basaltic komatiite magma production to establish the extreme lithosphere depletion and create the oceanic lithosphere that was consumed during craton amalgamation (Fig. 6). Proterozoic processes call for lower temperature, basaltic magmatism to modify the cratonic lithosphere in the Proterozoic. Thus, the diamond-forming fluids/melts would have a greater chance to be in equilibrium with some ultramafic lithologies in the Archean and mafic lithologies in the Proterozoic. Indeed, acknowledging that silicate inclusion suites plotted in Fig. 1 are not dated and that the dated silicate inclusion suites plotted in Fig. 3 have been subjected to the selection bias introduced by the requirement for datable inclusion suites, the much greater proportion of peridotitic silicate inclusions in the Archean and eclogitic silicate inclusions in the Proterozoic (Fig. 3) fits the pattern expected from the model. Eclogitic sulfide inclusions allow an additional test of these ideas because both Archean and Proterozoic inclusions occur in the same study population at each kimberlite (e.g., De Beers Pool, Jwaneng, Orapa, Koffiefontein), each inclusion can be dated separately, and the Re/Os and bulk Ni contents should vary with ultramafic versus basaltic lithologies.

Although there is overlap between the Archean and Proterozoic eclogitic sulfide inclusion populations, Fig. 7A and B confirms a systematic difference between the two age groups. Archean inclusions cover the whole range seen in Os, Ni, and Re content whereas Proterozoic inclusions are largely confined to the lower Os and Ni contents. This follows the observed pattern of Precambrian magmatism on Earth. Komatiites and basalts are formed in the Archean but by the mid- to late-Proterozoic basaltic magmatism dominates.

4.4.2. Predictions testable with new studies

The correlation of existing diamond age, composition and inclusion paragenesis data with lithospheric seismic structure leads to a set of predictions. These can be applied to future studies of new diamond suites at little-studied kimberlites in areas of the craton, such as the eastern or northern Kaapvaal, that have received

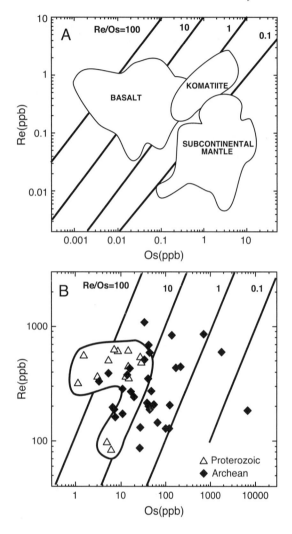

Fig. 7. Re vs. Os for mafic to ultramafic rocks (A) and dated sulfide inclusions (B) grouped according to their Archean or Proterozoic age. Note the lack of low Re/Os sulfide inclusions of Proterozoic age. Panel A is modified from Shirey and Walker (1998) using data from the literature. Data for panel B are from Koffiefontein (Pearson et al., 1998), De Beers Pool (Richardson et al., 2001), Jwaneng (Richardson et al., 2004) and Orapa (Shirey et al., unpublished data).

little previous attention (Table 3). For example, if the western and eastern domains of the Kaapvaal were created separately, the eastern domain, containing older crustal components than the western domain, should have its own clustering of diamond ages that are systematically older than those in the western domain. If diamonds with depleted peridotitic (harzburgitic) inclusions represent portions of the mantle

keel that are relict components after protocraton creation, then newly analyzed suites of peridotitic (harzburgitic) inclusions should be 3.2 Ga old or greater and be rare outside of seismically fast lithosphere. If the major process of craton stabilization was the coalescence of earlier protocratons with accompanying consumption of oceanic lithosphere at about

Table 3
Predictions for future study from new suites of diamonds

Data and/or observation	Implication	Test and/or prediction
Eastern cratonic domain has older crust and lithosphere dating back as far as 3.5 Ga	Eastern cratonic domain and western cratonic domain were created separately	Eastern domain should have a separate clustering of diamond ages/parageneses Ages should be systematically older in eastern domain vs. western domain
Where dated, peridotitic inclusions with depleted compositions are the oldest and are not widespread	Peridotitic inclusions occur in relict mantle keels left from protocraton creation	New suites of peridotitic, depleted inclusions will be rare out side of fast lithosphere, should be 3.2 Ga or older
Eclogitic sulfides inclusions occur in both seismically slow and fast lithosphere and are widely distributed	Assembly by cratonic nuclei accretion made eclogitic diamonds Possible emplacement of eclogite into lithosphere along with subduction	New suites of old (ca. 2.9 Ga) eclogitic diamonds should involve subducted components
Silicate inclusion paragenesis and diamond composition (C and N) correspond with lithospheric type Nd isotopic composition of silicate inclusions show asthenospheric and lithospheric sources	Magmatism and metasomatism around 2 Ga changed the lithospheric composition	Diamonds in the seismically slow regions will have greatest diversity of composition

2.9–3.0 Ga, then additional suites of eclogitic diamonds should be widely distributed in all domains of the craton. Diamond suites as yet comprising small numbers of specimens (e.g., Dokolwayo, Koffiefontein, and Klipspringer) may be part of this late Archean diamond-forming episode or may, with the determination of more precise isochron ages, show fine-scale discrimination that can relate diamond formation to slightly younger geological events such as Ventersdorp magmatism (e.g., Westerlund et al., 2004). If metasomatism of the craton accompanying its modification in the Proterozoic is the cause of the lithospheric seismic structure, then Proterozoic diamonds from the seismically slower part of the lithosphere would typically be expected to have the greatest range in age and composition.

Tests of these ideas will come with future studies, including geographically more complete, overlapping data sets from individual mines, studies of sulfide inclusion composition (Fe, Ni, and Cu content) and C, N, and S isotopic and IR studies on individual diamonds that have been dated with the Re–Os system. These need to be pursued in order to establish the relative importance of intramantle processes and regional variability on the isotopic composition of diamonds and the ultimate source of diamond-forming fluids.

4.4.3. Global patterns in age and paragenesis

There is an interesting parallel to the age distribution of diamonds seen in the Kaapvaal–Zimbabwe craton and other cratons such as the Siberian, Slave and Australian. Presently, other cratons do not have a comparable number of studied diamond or xenolith suites. However, it is significant that those localities where diamond suites have been dated fit the overall pattern shown by the Kaapvaal–Zimbabwe craton. Udachnaya, which is known for its abundant eclogite xenoliths and diamonds with sulfide inclusions, presents the clearest example of this parallel. Peridotitic sulfides from Udachnaya diamonds have been dated with the Re–Os system at 3.2 Ga (Pearson et al., 1999a,b) and sulfide inclusions in olivine in Udachnaya peridotite xenoliths have yielded 3.2 Ga in situ laser ablation ICPMS Re–Os ages (Griffin et al., 2002b; Pearson N.J. et al., 2002). Although no 2.9 Ga diamonds at Udachnaya have been found yet or dated, 2.9 Ga Re–Os whole rock ages have been

Table 4
Summary of diamond ages from four different cratons

Diamond paragenesis, age	Kaapvaal	Siberian	Slave	Australian
3.2 Ga, peridotitic (hz) silicate or sulfide inclusions	D, F	Udachnaya	Panda	
2.9 Ga, eclogitic sulfide inclusions or diamondiferous eclogites	D, JW, KO, N, O, R	Udachnaya		
1.9 Ga, peridotitic (lhz) silicate	F, JW, KO, O, P	Udachnaya		Argyle
1–2 Ga, eclogitic silicates				
1–2 Ga, eclogitic sulfides				

Age references from caption to Table 1 plus additional references for Argyle (Richardson, 1986), Newlands (N; Menzies et al., 2003) Panda (Westerlund et al., 2003b), Premier (Kramers, 1979), Roberts Victor (Shirey et al., 1998, 2001), Udachnaya (Pearson et al., 1995, 1999a,b; Richardson and Harris, 1997). Locality letter designations follow Fig. 1 (see caption).

obtained on diamondiferous Udachnaya eclogite xenoliths (Pearson et al., 1995). If one assumes that the 2.9 Ga diamondiferous eclogites have 2.9 Ga eclogitic diamonds, then this is similar to the distribution of Kaapvaal–Zimbabwe craton Archean diamond ages (Table 4). That is, the Udachnaya peridotitic sulfides match the De Beers Pool and Finsch peridotitic silicates in 3.2 Ga age and depleted harzburgitic paragenesis and the Udachnaya eclogites, presumably dating sulfides in eclogite xenoliths, match eclogitic sulfides from De Beers Pool, Jwaneng, Koffiefontein and Orapa in 2.9 Ga age and paragenesis. Udachnaya even has a Proterozoic, lherzolitic silicate inclusion suite that has negative initial epsilon Nd values yielding an Archean model age (Richardson and Harris, 1997) that is strikingly similar to the lherzolitic silicate inclusion suite from Premier (Richardson et al., 1993). Note, however, that so far only the Kaapvaal cratonic lithosphere records an extended history of Proterozoic diamond formation (Table 4). Both the Slave and Australian cratons have only one diamond suite that has been studied geochronologically but they also roughly fit the Kaapval–Zimbabwe craton age distribution. Peridotitic sulfides in diamonds from the Panda pipe give a 3.2–3.4 Ga isochron age (Westerlund et al., 2003b) and eclogitic silicate inclusions from Argyle give a mid-Proterozoic age (Richardson, 1986). These similarities in age and paragenesis of diamond suites among cratons that are now widely separated may indicate that diamond formation on the Archean Earth was controlled by global scale events in mantle/crustal evolution and/or that the Kaapvaal/ Zimbabwe, Siberian, Slave, and Australian cratons were situated close together in an Archean supercontinent (e.g., Zegers et al., 1998) that may have become disassembled by the mid-Proterozoic.

Acknowledgements

Discussions with D. Bell, F. R. Boyd, K. Burke, R.W. Carlson, P. Cartigny, I. Chinn, P. Deines, M. Fouch, H. Grutter, E. H. Hauri, D. James, O. Navon, D.G. Pearson, M. D. Schmitz, K.S. Viljoen, and K. Westerlund during the preparation of earlier versions of these ideas are greatly appreciated. The authors and all researchers working on inclusions in diamond are grateful to V. Anderson, E. van Blerk, R. Ferraris, R. Hamman, W. Moore, A. Ntidisang, G. Parker and others at Harry Oppenheimer House, Kimberley for their skill in selecting specimens and to the De Beers Diamond Trading Company for making them available. M. Horan and T. Mock are thanked for their help in the DTM chemistry and mass spectrometry labs (respectively). This work was supported chiefly by the NSF EAR Continental Dynamics Grant 9526840.

References

Akagi, T., Masuda, A., 1988. Isotopic and elemental evidence for a relationship between kimberlite and Zaire cubic diamonds. Nature 336, 665–667.

Anhaeusser, C.R., Walraven, F., 1999. Episodic granitoid emplacement in the western Kaapvaal Craton: evidence from the Archean Kraaipan granite–greenstone terrane, South Africa. Journal of African Earth Sciences 28, 289–309.

Arndt, N.T., Kerr, A.C., Tarney, J., 1997. Dynamic melting in plume heads; the formation of Gorgona komatiites and basalts. Earth and Planetary Science Letters 146, 289–301.

Ayres, N.P., Hatton, C.J., Quadling, K.E., Smith, C.B., 1998. An Update on the Distribution in Time and Space of Southern African Kimberlites, 1:5,000,000 Scale Maps. De Beers Geo-Science Centre, Southdale (South Africa).

Barker, F. (Ed.), 1979. Trondhjemites, Dacites and Related Rocks. Elsevier, New York. 659 pp.

Boyd, F.R., Gurney, J.J., 1986. Diamonds and the African lithosphere. Science 232, 472–477.

Boyd, F.R., Gurney, J.J., Richardson, S.H., 1985. Evidence for a 150–200-km thick Archaean lithosphere from diamond inclusion thermobarometry. Nature 315, 387–389.

Boyd, S.R., Mattey, D.P., Pillinger, C.T., Milledge, H.J., Mendelssohn, M., Seal, M., 1987. Multiple growth events during diamond genesis: an integrated study of carbon and nitrogen isotopes and nitrogen aggregation state in coated stones. Earth and Planetary Science Letters 86, 341–353.

Boyd, F.R., Pearson, D.G., Mertzman, S.A., 1999. Spinel-facies peridotites from the Kaapvaal root. In: Gurney, J.J., Gurney, J.L., Pascoe, M.D., Richardson, S.H. (Eds.), The J.B. Dawson Volume—Proceedings of the Seventh International Kimberlite Conference, Cape Town. Red Roof Design, Cape Town, pp. 40–48.

Burgess, R., Turner, G., Laurenzi, M., Harris, J.W., 1989. ^{40}Ar–^{39}Ar laser probe dating of individual clinopyroxene inclusions in Premier eclogitic diamonds. Earth and Planetary Science Letters 94, 22–28.

Burgess, R., Turner, G., Harris, J.W., 1992. ^{40}Ar–^{39}Ar laser probe studies of clinopyroxene inclusions in eclogitic diamonds. Geochimica et Cosmochimica Acta 56, 389–402.

Burgess, R., Phillips, D., Harris, J.W., Robinson, D.N., 1998. Antarctic diamonds in South-eastern Australia? Hints From ^{40}Ar/^{39}Ar Laser Probe Dating of Clinopyroxene Inclusions from Copeton Diamonds. Extended Abstracts of the 7th International Kimberlite Conference, Cape Town, pp. 119–121.

Burgess, R., Kiviets, G., Harris, J.W., 2004. Different age populations of eclogitic diamonds in the Venetia kimberlite: evidence from Ar–Ar dating of syngenetic clinopyroxene inclusions. Lithos.

Canil, D., Wei, K., 1992. Constraints on the origin of mantle-derived low-Ca garnets. Contributions to Mineralogy and Petrology 109, 421–430.

Card, K.D., 1990. A review of the Superior Province of the Canadian Shield, a product of Archean accretion. Precambrian Research 48, 99–156.

Carlson, R.W., Moore, R.O., 2004. Age of the Eastern Kaapvaal mantle: Re–Os isotope data for peridotite xenoliths from the Monastery kimberlite. South African Journal of Geology, 107.

Carlson, R.W., Grove, T.L., de Wit, M.J., Gurney, J.J., 1996. Program to study crust and mantle of the Archean craton in southern Africa. Eos, Transactions-American Geophysical Union 77, 273–277.

Carlson, R.W., Pearson, D.G., Boyd, F.R., Shirey, S.B., Irvine, G., Menzies, A.H., Gurney, J.J., 1999. Re–Os systematics of lithospheric peridotites: implications for lithosphere formation and preservation. In: Gurney, J.J., Gurney, J.L., Pascoe, M.D., Richardson, S.H. (Eds.), The J.B. Dawson Volume—Proceedings of the Seventh International Kimberlite Conference, Cape Town. Red Roof Design, Cape Town, pp. 99–108.

Carlson, R.W., Boyd, F.R., Shirey, S.B., Janney, P.E., Grove, T.L., Bowring, S.A., Schmitz, M.D., Dann, J.C., Bell, D.R., Gurney,

J.J., Richardson, S.H., Tredoux, M., Menzies, A.H., Pearson, D.G., Hart, R.J., Wilson, A.H., Moser, D., 2000. Continental growth, preservation, and modification in southern Africa. GSA Today 10, 1–7.

Carney, J.N., Aldiss, J.T., Lock, N.P., 1994. The Geology of Botswana. Geological Survey Bulletin, vol. 37. Geological Survey Department, Botswana. 113 pp.

Cartigny, P., 1997. Carbon isotopes in diamond. PhD thesis, Concentration, composition isotopique et origine de l'azote dans le manteau terrestre. PhD Thesis (unpublished) Universite de Paris VII.

Cartigny, P., Harris, J.W., Javoy, M., 1998. Eclogitic diamond formation at Jwaneng; no room for a recycled component. Science 280, 1421–1424.

Cartigny, P., Harris, J.W., Javoy, M., 1999. Eclogitic, peridotitic and metamorphic diamonds and the problems of carbon recycling—the case of Orapa (Botswana). In: Gurney, J.J., Gurney, J.L., Pascoe, M.D., Richardson, S.H. (Eds.), The J.B. Dawson Volume—Proceedings of the Seventh International Kimberlite Conference, Cape Town. Red Roof Design, Cape Town, pp. 117–124.

Cartigny, P., Harris, J.W., Javoy, M., 2001. Diamond genesis, mantle fractionations and mantle nitrogen content: a study of d^{13}C–N concentrations in diamonds. Earth and Planetary Science Letters 185, 85–98.

Clifford, T.N., 1966. Tectono-metallogenic units and metallogenic provinces of Africa. Earth and Planetary Science Letters, 421–434.

Danchin, R.V., 1979. Mineral and bulk chemistry of garnet lherzolite and garnet harzburgite xenoliths from the Premier Mine, South Africa. In: Boyd, F.R., Meyer, H.O.A. (Eds.), The Mantle Sample: Inclusions in Kimberlites and Other Volcanics. Proceedings of the Second International Kimberlite Conference, Santa Fe. American Geophysical Union, Washington, pp. 104–126.

Davis, G.L., 1977. The ages and uranium contents of zircons from kimberlites and associated rocks. Year Book-Carnegie Institution of Washington 76, 631–635.

Davis, G.L., Krogh, T.E., Erlank, A.J., 1976. The ages of zircons from kimberlites from South Africa. Year Book-Carnegie Institution of Washington 75, 821–824.

Deines, P., 1980. The carbon isotopic composition of diamonds; relationship to diamond shape, color, occurrence and vapor composition. Geochimica et Cosmochimica Acta 44, 943–962.

Deines, P., Harris, J.W., 1995. Sulfide inclusion chemistry and carbon isotopes of African diamonds. Geochimica et Cosmochimica Acta 59, 3173–3188.

Deines, P., Gurney, J.J., Harris, J.W., 1984. Associated chemical and carbon isotopic composition variations in diamonds from Finsch and Premier kimberlite, South Africa. Geochimica et Cosmochimica Acta 48, 325–342.

Deines, P., Harris, J.W., Gurney, J.J., 1987. Carbon isotopic composition, nitrogen content and inclusion composition of diamonds from the Roberts Victor Kimberlite, South Africa: evidence for d^{13}C depletion in the mantle. Geochimica et Cosmochimica Acta 51, 1227–1243.

Deines, P., Harris, J.W., Spear, P.M., Gurney, J.J., 1989. Nitrogen and ^{13}C content of Finsch and Premier diamonds and their implications. Geochimica et Cosmochimica Acta 53, 1367–1378.

Deines, P., Harris, J.W., Gurney, J.J., 1991a. The carbon isotopic composition and nitrogen content of lithospheric and asthenospheric diamonds from the Jagersfontein and Koffiefontein Kimberlite, South Africa. Geochimica et Cosmochimica Acta 55, 2615–2625.

Deines, P., Harris, J.W., Robinson, D.N., Gurney, J.J., Shee, S.R., 1991b. Carbon and oxygen isotope variations in diamond and graphite eclogites from Orapa, Botswana, and the nitrogen content of their diamonds. Geochimica et Cosmochimica Acta 55, 515–524.

Deines, P., Harris, J.W., Gurney, J.J., 1993. Depth-related carbon isotope and nitrogen concentration variability in the mantle below the Orapa Kimberlite, Botswana, Africa. Geochimica et Cosmochimica Acta 57, 2781–2796.

Deines, P., Harris, J.W., Gurney, J.J., 1997. Carbon isotope ratios, nitrogen content and aggregation state, and inclusion chemistry of diamonds from Jwaneng, Botswana. Geochimica et Cosmochimica Acta 61, 3993–4005.

Deines, P., Viljoen, K.S., Harris, J.W., 2001. Implication of the carbon isotope and mineral inclusion record for the formation of diamonds in the mantle underlying a mobile belt, Venetia, South Africa. Geochimica et Cosmochimica Acta 65, 813–838.

de Wit, M.J., 1998. On Archean granites, greenstones, cratons and tectonics; does the evidence demand a verdict? Precambrian Research 91, 181–226.

Drennan, G.R., Robb, L.J., Meyer, F.M., Armstrong, R.A., de Bruiyn, H., 1990. The nature of the Archean basement in the hinterland of the Witwatersrand Basin: II. A crustal profile west of the Welkom Goldfield and comparisons with the Vredefort crustal profile. South African Journal of Geology 93, 41–53.

Evans, T., Harris, J.W., 1989. Nitrogen aggregation, inclusion equilibration temperatures and the age of diamonds. In: Ross, J., Jaques, A., Ferguson, J., Green, D., O'Reilly, S., Danchin, R., Janse, A. (Eds.), Kimberlites and Related Rocks—Proceedings of the Fourth International Kimberlite Conference, Perth. Blackwell, Melbourne, pp. 1001–1006.

Farquhar, J., Wing, B.A., McKeegan, K.D., Harris, J.W., Cartigny, P., Thiemens, M., 2002. Mass-independent sulfur in inclusions in diamond and sulfur recycling on early Earth. Science 298, 2369–2372.

Galimov, E.M., 1991. Isotope fractionation related to kimberlite magmatism and diamond formation. Geochimica et Cosmochimica Acta 55, 1697–1708.

Gill, J.B., 1981. Orogenic Andesites and Plate Tectonics. Springer-Verlag, New York. 390 pp.

Grand, S.P., 1987. Tomographic inversion for shear velocity beneath the North American plate. Journal of Geophysical Research 92, 14065–14090.

Griffin, W.L., O'Reilly, S.Y., Ryan, C.G., 1999. The composition and origin of sub-continental lithospheric mantle. In: Fei, Y., Bertka, C.M., Mysen, B.O. (Eds.), Mantle Petrology: Field Observations and High-Pressure Experimentation: A Tribute to Francis (Joe) Boyd. Special Publication-Geochemical Society, vol. 6, pp. 13–44.

Griffin, W.L., Fisher, N.I., Friedman, J.H., O'Reilly, S.Y., Ryan, C.G., 2002a. Cr-pyrope garnets in the lithospheric mantle: 2. Compositional populations and their distribution in time and space. Geochemistry, Geophysics, Geosystems 3, 1073 (doi: 1010.1029/2002GC000298).

Griffin, W.L., Spetsius, Z.V., Pearson, N.J., O'Reilly, S.Y., 2002. In-situ Re-Os analysis of sulfide inclusions in kimberlitic olivine: New constraints on depletion events in the Siberian lithospheric mantle. Geochemistry, Geophysics, Geosystems 3, 1069 (doi:10.1029/2001/GC000287).

Griffin, W.L., Graham, S., O'Reilly, S.Y., Pearson, N.J., 2003a. Lithosphere evolution beneath the Kaapvaal Craton: Re–Os systematics of sulfides in mantle-derived peridotites. In: Carlson, R.W., Grutter, H., Jones, A.G. (Eds.), Slave–Kaapvaal Workshop, Merrickville, Ontario.

Griffin, W.L., O'Reilly, S.Y., Abe, N., Aulbach, S., Davies, R.M., Pearson, N.J., Doyle, B.J., Kivi, K., 2003b. The origin and evolution of Archean lithospheric mantle. Precambrian Research 127, 19–41.

Gurney, J.J., 1989. Diamonds. In: Ross, J., Jaques, A., Ferguson, D., Green, D., O'Reilly, S., Danchin, R., Janse, A. (Eds.), Kimberlites and Related Rocks—Proceedings of the Fourth International Kimberlite Conference, Perth. Blackwell, Melbourne, pp. 935–965.

Gurney, J.J., Switzer, G.S., 1973. The discovery of garnets closely related to diamonds in the Finsch Pipe, South Africa. Contributions to Mineralogy and Petrology 39, 103–116.

Haggerty, S.E., 1986. Diamond genesis in a multiply-constrained model. Nature 320, 34–38.

Harris, J.W., Gurney, J.J., 1979. Inclusions in diamond. In: Field, J.E. (Ed.), The Properties of Diamond. Academic Press, New York, pp. 555–591.

Hauri, E.H., Pearson, D.G., Bulanova, G.P., Milledge, H.J., 1999. Microscale variations in C and N isotopes within mantle diamonds revealed by SIMS. In: Gurney, J.J., Gurney, J.L., Pascoe, M.D., Richardson, S.H. (Eds.), The J.B. Dawson Volume—Proceedings of the Seventh International Kimberlite Conference, Cape Town. Red Roof Design, Cape Town, pp. 341–347.

Hauri, E.H., Wang, J., Pearson, D.G., Bulanova, G.P., 2002. Microanalysis of $d^{13}C$, $d^{15}N$, and N abundances in diamonds by secondary ion mass spectrometry. Chemical Geology 185, 149–163.

Helmstaedt, H., Schulze, D.J., 1989. Southern African kimberlites and their mantle sample: implications for Archean tectonics and lithosphere evolution. In: Ross, J., Jaques, A., Ferguson, J., Green, D., O'Reilly, S., Danchin, R., Janse, A. (Eds.), Kimberlites and Related Rocks—Proceedings of the Fourth International Kimberlite Conference, Perth. Blackwell, Melbourne, pp. 358–368.

Holmes, A., Paneth, F.A., 1936. Helium-ratios of rocks and minerals from the diamond pipes of South Africa. Proceedings of the Royal Society of London. Series A, Mathematical and Physical Sciences 154, 385–413.

Hunter, D.R., 1974. Crustal development in the Kaapvaal Craton: I. The Archean. Precambrian Research 1, 259–294.

Irvine, G.J., Pearson, D.G., Carlson, R.W., 2001. Lithospheric mantle evolution of the Kaapvaal craton: a Re–Os isotope study of peridotite xenoliths from Lesotho kimberlites. Geophysical Research Letters 28, 2505–2508.

Izraeli, E.S., Harris, J.W., Navon, O., 2001. Brine inclusions in diamonds; a new upper mantle fluid. Earth and Planetary Science Letters 187, 323–332.

James, D.E., Fouch, M.J., 2002. Formation and evolution of Archean cratons: insights from Southern Africa. In: Ebinger, C., Fowler, M., Hawkesworth, C.J. (Eds.), The Early Earth: Physical, Chemical and Biological Development. Geological Society of London, London, pp. 1–26.

James, D.E., Fouch, M.J., VanDecar, J.C., van der Lee, S., Group, K.S., 2001. Tectospheric structure beneath southern Africa. Geophysical Research Letters 28, 2485–2488.

Janney, P.E., Carlson, R.W., Shirey, S.B., Bell, D.R., le Roex, A.P., 1999. Temperature, pressure, and rhenium–osmium age systematics of off-craton peridotite xenoliths from the Namaqua–Natal Belt, western Southern Africa. Ninth Annual V.M. Goldschmidt Conference Cambridge, MA, August 22–27. Lunar and Planetary Institute, Houston, p. 139. Contribution No. 971.

Janse, A.J.A., 1992. New ideas in subdividing cratonic areas. Russian Geology and Geophysics 33, 9–25.

Jones, M.Q.W., 1988. Heat flow in the Witwatersrand Basin and environs and its significance for the south African shield geotherm and lithosphere thickness. Journal of Geophysical Research 93, 3243–3260.

Jordan, T.H., 1978. Composition and development of the continental tectosphere. Nature 274, 544–548.

Jordan, T.H., 1981. Continents as a chemical boundary layer. Philosophical Transactions of the Royal Society of London A 301, 359–373.

Jordan, T.H., 1988. Structure and formation of the continental tectosphere. Journal of Petrology, Special Lithosphere Issue 11–37.

Kesson, S.E., Ringwood, A.E., 1989. Slab-mantle interactions: 1. Sheared and refertilized garnet peridotite xenoliths; samples of Wadati–Benioff zones? Chemical Geology 78, 83–96.

Kirkley, M.B., Gurney, J.J., Levinson, A.A., 1991. Age, origin, and emplacement of diamonds; scientific advances in the last decade. Gems and Gemology 27, 2–25.

Kramers, J.D., 1979. Lead, uranium, strontium, potassium and rubidium in inclusion-bearing diamonds and mantle-derived xenoliths from southern Africa. Earth and Planetary Science Letters 42, 58–70.

Kroner, A., Compston, W., Williams, I.S., 1989. Growth of early Archean crust in the Ancient Gneiss Complex of Swaziland as revealed by single zircon dating. Tectonophysics 161, 271–298.

Martin, H., 1993. The mechanisms of petrogenesis of the Archean continental crust—comparison with modern processes. Lithos 30, 373–388.

Martin, H., 1994. The Archean grey gneisses and the genesis of continental crust. In: Condie, K.C. (Ed.), Developments in Precambrian Geology: Archean Crustal Evolution. Elsevier, New York, pp. 205–259.

Menzies, A.H., Carlson, R.W., Shirey, S.B., Gurney, J.J., 2003. Re–Os systematics of diamond-bearing eclogites from the Newlands kimberlite. Lithos 71, 323–336.

Meyer, H.O.A., Boyd, F.R., 1972. Composition and origin of crystalline inclusions in natural diamonds. Geochimica et Cosmochimica Acta 36, 1255–1273.

Milledge, H.J., Mendelssohn, M.J., Seal, M., Rouse, J.E., Swart, P.K., Pillinger, P.T., 1983. Carbon isotopic variation in spectral type II diamonds. Nature 303, 791–792.

Moser, D.E., Flowers, R., Hart, R.J., 2000. Birth of the Kaapvaal lithosphere at 3.08 Ga: implications for the growth of ancient continents. Science 291, 465–468.

Navon, O., 1999. Diamond formation in the Earth's mantle. In: Gurney, J.J., Gurney, J.L., Pascoe, M.D., Richardson, S.H. (Eds.), The P.H. Nixon Volume—Proceedings of the Seventh International Kimberlite Conference, Cape Town. Red Roof Design, Cape Town, pp. 584–604.

Navon, O., Hutcheon, I.D., Rossman, G.R., Wasserburg, G.J., 1988. Mantle-derived fluids in diamond micro-inclusions. Nature 335, 784–789.

Parman, S.W., Dann, J.C., Grove, T.L., 2001. The production of Barberton komatiites in an Archean subduction zone. Geophysical Research Letters 28, 2513–2516.

Pearson, D.G., Shirey, S.B., 1999. Isotopic dating of diamonds. In: Lambert, D.D., Ruiz, J. (Eds.), Application of Radiogenic Isotopes to Ore Deposit Research and Exploration. Reviews in Economic Geology, vol. 12. Society of Economic Geologists, Littleton CO, Denver, pp. 143–172.

Pearson, D.G., Snyder, G.A., Shirey, S.B., Taylor, L.A., Carlson, R.W., Sobolev, N.V., 1995. Archaean Re–Os age for Siberian eclogites and constraints on Archaean tectonics. Nature 374, 711–713.

Pearson, D.G., Shirey, S.B., Harris, J.W., Carlson, R.W., 1998. Sulfide inclusions in diamonds from the Koffiefontein kimberlite, S. Africa: constraints on diamond ages and mantle Re–Os systematics. Earth and Planetary Science Letters 160, 311–326.

Pearson, D.G., Shirey, S.B., Bulanova, G.P., Carlson, R.W., Milledge, H.J., 1999a. Re–Os isotope measurements of single sulfide inclusions in a Siberian diamond and its nitrogen aggregation systematics. Geochimica et Cosmochimica Acta 63, 703–711.

Pearson, D.G., Shirey, S.B., Bulanova, G.P., Carlson, R.W., Milledge, H.J., 1999b. Dating and paragenetic distinction of diamonds using the Re–Os isotope system: application to some Siberian diamonds. In: Gurney, J.J., Gurney, J.L., Pascoe, M.D., Richardson, S. (Eds.), The P. H. Nixon Volume. Proceedings of the International Kimberlite Conference, vol. 7. Red Roof Design, Cape Town, pp. 637–643.

Pearson, D.G., Irvine, G.J., Carlson, R.W., Kopylova, M.G., Ionov, D.A., 2002. The development of lithospheric mantle keels beneath the earliest continents: time constraints using PGE and Re–Os isotope systematics. In: Ebinger, C., Fowler, M., Hawkesworth, C.J. (Eds.), The Early Earth: Physical, Chemical and Biological Development. Geological Society of London, London, pp. 665–690.

Pearson, N.J., Alard, O., Griffin, W.L., Jackson, S.E., O'Reilly, S.Y., 2002. In situ measurement of Re–Os isotopes in mantle sulfides by laser ablation multicollector-inductively coupled plasma mass spectrometry: analytical methods and preliminary results. Geochimica et Cosmochimica Acta 66, 1037–1050.

Phillips, D., Onstott, T.C., Harris, J.W., 1989. $^{40}Ar/^{39}Ar$ laser-probe dating of diamond inclusions from Premier kimberlite. Nature 340, 460–462.

Pidgeon, R.T., 1989. Archean diamond xenocrysts in kimberlites—how definitive is the evidence? Journal of the Geological Society of Australia, Special Publication 14, 1007–1011.

Richardson, S.H., 1986. Latter-day origin of diamonds of eclogitic paragenesis. Nature 322, 623–626.

Richardson, S.H., 1989. As definitive as ever: a reply to 'Archean diamond xenocrysts in kimberlite—How definitive in the evidence? by R.T. Pidgeon'. Journal of the Geological Society of Australia, Special Publication 14, 1070–1072.

Richardson, S.H., Harris, J.W., 1997. Antiquity of peridotitic diamonds from the Siberian craton. Earth and Planetary Science Letters 151, 271–277.

Richardson, S.H., Gurney, J.J., Erlank, A.J., Harris, J.W., 1984. Origin of diamonds in old enriched mantle. Nature 310, 198–202.

Richardson, S.H., Erlank, A.J., Harris, J.W., Hart, S.R., 1990. Eclogitic diamonds of Proterozoic age from Cretaceous kimberlites. Nature 346, 54–56.

Richardson, S.H., Harris, J.W., Gurney, J.J., 1993. Three generations of diamonds from old continental mantle. Nature 366, 256–258.

Richardson, S.H., Chinn, I.L., Harris, J.W., 1999. Age and origin of eclogitic diamonds from the Jwaneng kimberlite, Botswana. In: Gurney, J.J., Gurney, J.L., Pascoe, M.D., Richardson, S.H. (Eds.), The P.H. Nixon Volume—Proceedings of the Seventh International Kimberlite Conference, Cape Town. Red Roof Design, Cape Town, pp. 734–736.

Richardson, S.H., Shirey, S.B., Harris, J.W., Carlson, R.W., 2001. Archean subduction recorded by Re–Os isotopes in eclogitic sulfide inclusions in Kimberley diamonds. Earth and Planetary Science Letters 191, 257–266.

Richardson, S.H., Shirey, S.B., Harris, J.W., 2004. Episodic diamond genesis at Jwaneng, Botswana, and implications for Kaapvaal craton evolution. Lithos.

Ringwood, A.E., 1977. Petrogenesis in island arc systems. In: Talwani, M., Pitman, W.C. (Eds.), Island Arcs, Deep sea Trenches and Back-arc Basins. American Geophysical Union, Washington, DC, pp. 311–324.

Schmitz, M.D., 2002. Geology and thermochronology of the lower crust of southern Africa. PhD thesis, (unpublished) Massachusetts Institute of Technology, Cambridge, M.A. 269 pp.

Schmitz, M.D., Bowring, S.A., 2001. Crust formation and stabilization of the western Kaapvaal Craton: evidence from U–Pb geochronology of basement blocks and deep crustal xenoliths from the Kimberley region, South Africa [abs.]. Eos, Transactions-American Geophysical Union 82 (Abs. S61A-11.).

Schmitz, M.D., Bowring, S.A., 2003. Constraints on the thermal evolution of continental lithosphere from U–Pb accessory mineral thermochronometry of lower crustal xenoliths, southern Africa. Contribution to Mineralogy and Petrology 144, 592–618.

Schmitz, M.D., Bowring, S.A., de Wit, M.J., Gartz, V., 2004. Subduction and terrane collision stabilized the western Kaapvaal craton tectosphere 2.9 billion years ago. Earth and Planetary Science Letters (in preparation).

Schrauder, M., Navon, O., 1994. Hydrous and carbonatitic mantle fluids in fibrous diamonds from Jwaneng, Botswana. Geochimica et Cosmochimica Acta 58, 761–771.

Shimizu, N., Sobolev, N., 1995. Young peridotitic diamonds from the Mir kimberlite pipe. Nature 375, 394–397.

Shirey, S.B., Walker, R.J., 1998. The Re–Os isotope system in cosmochemistry and high-temperature geochemistry. Annual Review of Earth and Planetary Sciences 26, 423–500.

Shirey, S.B., Carlson, R.W., Gurney, J.J., van Heerden, L., 1998. Re–Os isotopic systematics of eclogites from Roberts Victor: implications for diamond growth and Archean tectonic processes. Extended Abstracts, 7th International Kimberlite Conference, Cape Town, pp. 808–810.

Shirey, S.B., Carlson, R.W., Richardson, S.H., Menzies, A.H., Gurney, J.J., Pearson, D.G., Harris, J.W., Wiechert, U., 2001. Archean emplacement of eclogitic components into the lithospheric mantle during formation of the Kaapvaal Craton. Geophysical Research Letters 28, 2509–2512.

Shirey, S.B., Harris, J.W., Richardson, S.H., Fouch, M.J., James, D.E., Cartigny, P., Deines, P., Viljoen, K.S., 2002. Diamond genesis, seismic structure, and evolution of the Kaapvaal–Zimbabwe craton. Science 297, 1683–1686.

Shirey, S.B., Harris, J.W., Richardson, S.H., Fouch, M.J., James, D.E., Cartigny, P., Deines, P., Viljoen, K.S., 2003. Regional patterns in the paragenesis and age of inclusions in diamond, diamond composition and the lithospheric seismic structure of southern Africa. Lithos 71, 243–258.

Smith, C.B., Allsopp, H.L., Kramers, J.D., Hutchinson, G., Roddick, J.C., 1985. Emplacement ages of Jurassic–Cretaceous South African kimberlites by the Rb–Sr method on phlogopite and whole-rock samples. Transactions of the Geological Society of South Africa 88, 249–266.

Smith, C.B., Gurney, J.J., Harris, J.W., Robinson, D.N., Kirkley, M.B., Jagoutz, E., 1991. Neodymium and strontium isotope systematics of eclogite and websterite paragenesis inclusions from single diamonds. Geochimica et Cosmochimica Acta 55, 2579–2590.

Sobolev, N.V., Laurent'ev Yu, G., Pokhilenko, N.P., 1973. Chrome-rich garnets from the kimberlites of Yakutia and their paragenesis. Contributions to Mineralogy and Petrology 40, 39–52.

Sobolev, N.V., Galimov, E.M., Ivanovskaya, I.N., Yefimova, E.S., 1979. The carbon isotope composition of diamonds containing crystalline inclusions. Doklady Akademii Nauk Ukrainskoi USSR 189, 133–136.

Sobolev, N.V., Yefimova, E.S., Loginova, A.M., Sukhodol'skaya, O.V., Solodova, Y.P., 2001. Abundance and composition of mineral inclusions in large diamonds from Yakutia. Doklady Earth Sciences 376, 34–38.

Spetsius, Z.V., Belousova, E.A., Griffin, W.L., O'Reilly, S.Y., Pearson, N.J., 2002. Archean sulfide inclusions in Paleozoic zircon megacrysts from the Mir Kimberlite, Yakutia: implications for the dating of diamonds. Earth and Planetary Science Letters 199, 111–126.

Stachel, T., Harris, J.W., 1997. Diamond precipitation and mantle metasomatism; evidence from the trace element chemistry of silicate inclusions in diamonds from Akwatia, Ghana. Contributions to Mineralogy and Petrology 129, 143–154.

Stachel, T., Viljoen, K.S., Brey, G., Harris, J.W., 1998. Metasomatic processes in lherzolitic and harzburgitic domains of diamondiferous lithospheric mantle: REE in garnets from xenoliths and inclusions in diamonds. Earth and Planetary Science Letters 159, 1–12.

Taylor, L.A., Snyder, G.A., Crozaz, G., Sobolev, V.N., Yefimova, E.S., Sobolev, N.V., 1996. Eclogitic inclusions in diamonds; evidence of complex mantle processes over time. Earth and Planetary Science Letters 142, 535–551.

Valley, J.W., Kinny, P.D., Schulze, D.J., Spicuzza, M.J., 1998. Zircon megacrysts from kimberlite: oxygen isotope variability among mantle melts. Contributions to Mineralogy and Petrology 133, 1–11.

van der Lee, S., Nolet, G., 1997. Upper mantle S-velocity structure of North America. Journal of Geophysical Research 102, 22815–22838.

Viljoen, K.S., 2002. An infrared investigation of inclusion-bearing diamonds from the Venetia kimberlite, Northern Province, South Africa: implications for diamonds from craton-margin settings. Contributions to Mineralogy and Petrology 144, 98–108.

Walker, R.J., Carlson, R.W., Shirey, S.B., Boyd, F.R., 1989. Os, Sr, Nd, and Pb isotope systematics of Southern African peridotite xenoliths; implications for the chemical evolution of subcontinental mantle. Geochimica et Cosmochimica Acta 53, 1583–1595.

Walter, M.J., 1998. Melting of garnet peridotite and the origin of komatiite and depleted lithosphere. Journal of Petrology 39, 29–60.

Westerlund, K.J., Shirey, S.B., Richardson, S.H., Gurney, J.J., Harris, J.W., 2003. Re–Os isotope systematics of peridotitic diamond inclusion sulfides from the Panda kimberlite, Slave craton. 8th International Kimberlite Conference, Victoria, BC, abstract, 134.

Westerlund, K.J., Gurney, J.J., Carlson, R.W., Shirey, S.B., Hauri, E.H., Richardson Stephen, H., 2004. A metasomatic origin for late Archean eclogitic diamonds: implications from internal morphology of Klipspringer diamonds and Re–Os and S isotope characteristics of their sulfide inclusions. South African Journal of Geology, 107.

Wilson, A.H., Shirey, S.B., Carlson, R.W., 2003. Archean ultra-depleted komatiites formed by hydrous melting of cratonic mantle. Nature 423, 858–861.

Windley, B.F., 1995. The Evolving Continents, 3rd ed. Wiley, New York. 544 pp.

Yefimova, E.S., Sobolev, N.V., Pospelova, L.N., 1983. Sulfide inclusions in diamonds and specific features of their paragenesis. Zapiski Vsesoyuznogo Mineralogicheskogo Obshchestva 112, 300–310.

Zegers, T.E., de Wit, M.J., Dann, J.C., White, S.H., 1998. Vaalbara, Earth's oldest assembled continent? A combined structural, geochronological, and paleomagnetic test. Terra Nova 10, 250–259.

Available online at www.sciencedirect.com

ELSEVIER

Lithos 77 (2004) 945–949

www.elsevier.com/locate/lithos

Author index to volume 77

doi:10.1016/S0024-4937(04)00241-5